ENCYCLOPEDIA OF PLANT PHYSIOLOGY

EDITED BY

W. RUHLAND

COEDITORS

E. ASHBY · J. BONNER · M. GEIGER-HUBER · W. O. JAMES
A. LANG · D. MÜLLER · M. G. STÅLFELT

VOLUME II

GENERAL PHYSIOLOGY OF THE PLANT CELL

CONTRIBUTORS

G. F. BAHR · H. J. BOGEN · L. BRAUNER · E. BÜNNING · T. CASPERSSON
R. COLLANDER · H. B. CURRIER · H. DRAWERT · F. DUSPIVA · F. EBERHARDT
E. EPSTEIN · H. FISCHER · R. J. HELDER · H. KERN · G. KLEIN · P. J. KRAMER
J. LEVITT · B. S. MEYER · A. MILLERD · K. PAECH · R. N. ROBERTSON
W. SEIFRIZ · W. SIMONIS · D. C. SPANNER · E. STADELMANN · M. G. STÅLFELT · O. STOCKER · C. R. STOCKING · K. UMRATH · V. WARTIOVAARA

SUBEDITORS

H. J. BOGEN AND H. ULLRICH

WITH 204 FIGURES

SPRINGER-VERLAG
BERLIN · GÖTTINGEN · HEIDELBERG
1956

HANDBUCH DER PFLANZENPHYSIOLOGIE

HERAUSGEGEBEN VON

W. RUHLAND

IN GEMEINSCHAFT MIT

E. ASHBY · J. BONNER · M. GEIGER-HUBER · W. O. JAMES
A. LANG · D. MÜLLER · M. G. STÅLFELT

BAND II

ALLGEMEINE PHYSIOLOGIE DER PFLANZENZELLE

BEARBEITET VON

G. F. BAHR · H. J. BOGEN · L. BRAUNER · E. BÜNNING · T. CASPERSSON
R. COLLANDER · H. B. CURRIER · H. DRAWERT · F. DUSPIVA · F. EBERHARDT
E. EPSTEIN · H. FISCHER · R. J. HELDER · H. KERN · G. KLEIN · P. J. KRAMER
J. LEVITT · B. S. MEYER · A. MILLERD · K. PAECH · R. N. ROBERTSON
W. SEIFRIZ · W. SIMONIS · D. C. SPANNER · E. STADELMANN · M. G. STÅL-
FELT · O. STOCKER · C. R. STOCKING · K. UMRATH · V. WARTIOVAARA

REDIGIERT VON

H. J. BOGEN UND H. ULLRICH

MIT 204 ABBILDUNGEN

SPRINGER-VERLAG
BERLIN · GÖTTINGEN · HEIDELBERG
1956

Softcover reprint of the hardcover 1st edition 1956

ISBN-13: 978-3-642-94677-6 e-ISBN-13: 978-3-642-94676-9
DOI: 10.1007/ 978-3-642-94676-9

Inhaltsverzeichnis. — Contents.

IV. Der Stoffaustausch der Zelle (und der Gewebe).

Seite

V. Die allgemeinen Bedingungen des Zellstoffwechsels.

VI. Beziehungen zwischen den physiologischen Bedingungen und der Zellaktivität.

VII. Aktivitätswechsel.

Seite

VIII. Altern und Zelltod.

Mitarbeiter von Band II. — Contributors to volume II.

Dr. med. GÜNTHER F. BAHR, Medizinisches Nobelinstitut für Zellforschung und Genetik, Karolinska Institutet, Stockholm (Schweden).

Dr. HANS J. BOGEN, Professor der Botanik, Botanisches Institut der Technischen Hochschule, Braunschweig.

Dr. LEO BRAUNER, o. Professor der Botanik, Botanisches Institut der Universität München, München 19, Menzingerstraße 67.

Professor Dr. ERWIN BÜNNING, Tübingen, Botanisches Institut der Universität.

Professor Dr. T. CASPERSSON, Medizinisches Nobelinstitut für Zellforschung und Genetik, Karolinska Institutet, Stockholm (Schweden).

Professor Dr. phil. RUNAR COLLANDER, Botanisches Institut der Universität, Helsingfors (Finnland).

Dr. H. B. CURRIER, Associate Professor of Botany, Department of Botany, University of California, Davis, California (USA).

Professor Dr. HORST DRAWERT, Pflanzenphysiologisches Institut der Freien Universität Berlin, Berlin-Dahlem, Königin-Luise-Straße 1—3.

Professor Dr. FRANZ DUSPIVA, Pathologisches Institut der Albert-Ludwigs-Universität, Chem. Abteilung, Freiburg i. Br.

Dr. FRANK EBERHARDT, Botanisches Institut der Universität, Tübingen.

Dr. EMANUEL EPSTEIN, Plant Physiologist, US Plant Industry Station, Beltsville, Maryland (USA).

Dozent Dr. HERMANN FISCHER, Botanisches Institut der Universität, Bonn, Meckenheimer Allee 170.

Dr. R. J. HELDER, Botanisch Laboratorium, Grote Rozenstraat 31, Groningen (Niederlande/Netherlands/Pays Bas).

Dozent Dr. HEINZ KERN, Institut für spezielle Botanik der Eidg. Technischen Hochschule, Universitätsstr. 2, Zürich 6 (Schweiz).

Docent Dr. med. GEORG KLEIN, Medizinisches Nobelinstitut für Zellforschung und Genetik, Karolinska Institutet, Stockholm 60 (Schweden).

Dr. PAUL J. KRAMER, James B. Duke Professor of Botany, Department of Botany, Duke University, Durham, North Carolina (USA).

Dr. J. LEVITT, Professor of Botany, University of Missouri, Columbia, Missouri (USA).

Dr. BERNARD S. MEYER, Professor and Chairman, Dept. Botany and Plant Pathology, Ohio State University (Columbus) and Ohio Agricultural Experiment Station (Wooster) USA.

Dr. ADÈLE MILLERD, Department of Biochemistry, University of Sidney, Sidney (Australia).

Dr. R. N. ROBERTSON, Division of Food Preservation and Transport, Commonwealth Scientific and Industrial Research Organization, Homebush (Australia).

Professor Dr. WILHELM SIMONIS, Botanisches Institut der Tierärztl. Hochschule, Hannover.

Dr. D. C. Spanner, Botany Dept., Imperial College of Science and Technology, London (England).

Dr. phil. Eduard Stadelmann, Assistent, Botanisches Institut der Universität Freiburg (Schweiz).

Dr. M. G. Stålfelt, o. Professor der Pflanzenphysiologie der Universität Stockholm, Kungstensgatan 45, Stockholm Va (Schweden).

Dr. Otto Stocker, Professor der Botanik, Botanisches Institut der Technischen Hochschule, Darmstadt.

Dr. C. Ralph Stocking, Associate Professor of Botany, Department of Botany, University of California, Davis, California (USA).

Professor Dr. Karl Umrath, Graz, Hochsteingasse 59 (Österreich).

Dozent Dr. phil. Veijo Wartiovaara, Botanisches Institut der Universität, Helsinki (Finnland).

Einführung und Übersicht.

Von

H. J. Bogen.

Während der 1955 erschienene erste Band des Handbuches die Konstitution der Pflanzenzelle, insbesondere also die morphologisch faßbaren Zellstrukturen behandelte, ist der vorliegende zweite Band der allgemeinen Physiologie der Pflanzenzelle gewidmet, d.h. der Prozeßanalyse. Sie wird den übrigen Bänden vorangestellt aus der Einsicht, daß alle physiologischen Vorgänge sich in der Zelle abspielen und somit Physiologie und Biologie im wesentlichen Zellphysiologie und Zellbiologie sind.

Bereits die Aufsätze des ersten Bandes zeigten, daß im Bereich der Zelle eine Trennung zwischen Struktur- und Prozeßanalyse nicht mit der Schärfe durchgeführt werden kann, die aus methodischen wie didaktischen Gründen zu wünschen wäre: Die Zellstrukturen können nicht ohne die Vorgänge gedacht werden, die an ihnen ablaufen, sei es, daß sie zur Bildung oder Aufrechterhaltung dieser Struktur führen, oder daß sie an den äußerlich sichtbaren Phänomenen des Lebens, wie z.B. der Zellteilung, beteiligt sind. Ebenso deutlich, vielleicht sogar noch schärfer erkennbar, wird in diesem Bande, daß die physiologischen Vorgänge nicht abgelöst von den Zellstrukturen betrachtet werden dürfen, sondern daß die „Ordnung in der Zeit" ihr Korrelat in der „Ordnung im Raum" hat.

Eine vollständige Physiologie der Pflanzenzelle müßte die Physiologie aller oder doch der wichtigsten Phänomene des Lebens enthalten, also Fortpflanzung und Vermehrung, Vererbung, Entwicklung, Alter und Tod, Stoffwechsel, Reizbarkeit usw. Das ist jedoch beim heutigen Stande der Forschung noch nicht möglich. Wohl sind die Kenntnisse über die Physiologie aller dieser Phänomene schon so umfangreich, daß sie die folgenden Bände füllen, aber hinsichtlich ihrer Einordnung und Verankerung *im Gefüge der Zelle* sind wir noch vielfach auf Vermutungen angewiesen. Das ist in der Hauptsache in den zellphysiologischen Methoden begründet, die bis vor kurzem vorwiegend auf die Erfassung von Stoffen, Mengen und Umsätzen in der Zelle ausgerichtet waren. So ist es nicht zu verwundern, daß, nach einem Kapitel über die Kinetik von Zellvorgängen (II), der Hauptteil des Bandes dem *Stoffwechsel der Zelle* gewidmet ist: den osmotischen Eigenschaften der Zelle (III), dem Stoffaustausch (IV) und den allgemeinen Bedingungen des Zellstoffwechsels (V). Hierin werden die Aufnahme- und Austauschmechanismen der verschiedenen Stoffgruppen, deren Abhängigkeit von äußeren und inneren Faktoren und die Grundlagen des Stoffumsatzes besprochen, wobei (V) auch versucht wird, das Zusammenwirken der Stoffwechselvorgänge und der individuellen Zellbestandteile wenigstens in einigen Zügen zu erfassen.

Abschnitt VI behandelt dann Beziehungen zwischen den physiologischen Bedingungen und der Zellaktivität, wobei auch Resistenz, Reizbarkeit, Narkotisierung und Vergiftung zu Worte kommen. Abschnitt VII und VIII schließlich sind dem Aktivitätswechsel, dem Altern und dem Zelltode gewidmet.

Die Verfasser der einzelnen Aufsätze waren bemüht, jeweils den neuesten Stand der Forschung darzustellen und, soweit es überhaupt zu vertreten war,

Ausblicke auf die künftige Entwicklung zu geben. So unvollkommen unsere Erkenntnisse auch sind, so ermöglichen sie doch die Einsicht, daß in der Zelle eine ungeheuere Vielfalt von Vorgängen und Strukturen gegeben ist, und daß deren Zusammenwirken nicht zufällig, sondern hochgeordnet ist — der heute vielbenutzte Ausdruck „Organisation der Zelle" kennzeichnet freilich weniger die Summe unseres Wissens als vielmehr unsere Hilflosigkeit, diese Vielfalt zu überschauen oder gar zu erfassen.

Um den Umfang der einzelnen Artikel nicht ungebührlich anschwellen zu lassen, haben die Verfasser in der Regel darauf verzichten müssen, die historische Entwicklung der Einzelprobleme aufzuzeigen; sie haben stattdessen auf Spezialwerke verwiesen. Durch dieses Verfahren war es möglich, die moderne Literatur soweit aufzuführen, daß der Benutzer des Handbuches sich über die einschlägigen Fragen hinreichend informieren kann; sein Nachteil ist, daß viele Probleme und manche Antworten als modern erscheinen, obgleich sie so alt wie die experimentelle Zellforschung selbst sind. Sicherlich hat die Zellphysiologie in den letzten 10 Jahren einen großartigen Aufschwung genommen — nicht zuletzt dank dem Einsatz umfangreicher Mittel insbesondere in den angelsächsischen und nordischen Ländern — und sie konnte nicht nur viele neue Probleme aufwerfen, sondern auch auf viele Fragen anscheinend endgültige Antworten geben. Es soll aber nicht vergessen werden, daß alles dies nicht ohne die Vorarbeiten der Klassiker der Zellphysiologie, vor allem von Wilhelm Pfeffer, möglich war, daß manches Neue nur alte Aussagen präzisiert, und daß mehrere Jahrzehnte lang die Last der Forschung nahezu ausschließlich den europäischen Ländern auferlegt war. Das gilt besonders für das wichtige Thema der Permeabilitätsforschung, die mit Namen wie Pfeffer und de Vries, aber auch mit Fitting, Ruhland und seinen Mitarbeitern, der finnischen Schule unter Collanders Leitung und anderen verbunden bleiben wird. Es sei dem Verfasser dieser Einführung gestattet, an dieser Stelle einen generellen Hinweis auf Verdienste zu geben, die heute schon so selbstverständlich sind, daß sie im einzelnen nicht mehr angeführt zu werden brauchen.

Kinetics of cellular processes.

By

William Seifriz.

With 1 figure.

Kinetics is the science of energy transfer. It must, therefore, involve all physical chemistry, thermodynamics, electrokinetics, and physiology. Whatever form the activities of a cell may take, whether respiration, osmosis, the transmission of a stimulus, or the streaming of protoplasm, if work is done, this is *kinetics*, or *dynamics*, which is that branch of *mechanics* involving motion. *Energetics* is a newer term for kinetics; it has the advantage of laying emphasis on energy which is the essential quality of kinetics.

It is customary to express all forms of energy transformations in terms of specific units. Thus, combustion or oxidation is measured in calories. When one gram-molecular weight of glucose is oxidized to carbon dioxide and water, the energy given off is the equivalent of 676,616 calories. Other forms of energy transfer are expressed in other units; for example, oxidation-reduction equilibria in electric potentials, changes in surface energy in ergs, and surface tension in dynes. In each case there is a change in energy level, and the unit by which we express it is determined by the instrument we use to measure it.

Surface energy.

The energy available at surfaces is tremendous and gives rise to a number of familiar activities such as surface tension, capillary action, and adsorption. Broadly viewed, all chemical reactions are surface phenomena. Catalytic activity is often purely a matter of relative surface. Spongy platinum is an example of this.

A chapter on surface energy usually quickly degenerates into a discussion of surface tension. This is true because the latter is more easily measured and expressed. But surface energy and surface tension are not synonymous; tension is but one manifestation of the total energy (BULL 1943, GORTNER 1949).

The kinetic energy of molecules is evident in both liquids and gases, but the liquid, because it has surface, acquires an additional form of energy. The conditions under which the surface molecules exist are such as to impart to them a form of energy which differs from the inner kinetic energy, though it involves the latter.

The surface energy of a system is made up of two qualities, intensity and capacity. Intensity may be expressed in terms of tension; capacity is determined by the extent of the area. Surface energy may, therefore, be represented by the equation: $S_e = \gamma s$, in which S_e is surface energy, γ surface tension (in dynes) and s surface area (BULL 1943).

Although the total surface energy of a pure liquid or true solution is appreciable when measured with a tensiometer, it is small when compared with the inner kinetic energy. It is only in colloidal systems, where solids are finely dispersed in an aqueous medium, that the surface energy can be extremely great because the total surface is great. Every colloidal particle is surrounded by a liquid film;

the surface energy of the system is the product of the tension of this film and the total surface.

Some idea of the extent to which the surface of a given volume of matter is augmented by an increase in the number of particles, is to be had from the statement that a cube 1 cm. on an edge, with a total surface of 6 sq. cm., acquires, when divided into 10^{15} cubes, each 0.1 μ on an edge, an area of 60 sq. m. The relative increase in surface energy which 1 c. cm. of water acquires when subdivided is shown in the following table.

Surface energy, as so far considered, is measured in ergs. Should one seek the total energy at the surface of colloidal particles one should have to take more into consideration, the electric potential, adsorptive forces, and hydration.

Surface energy manifests itself in numerous forms, some, such as surface tension, adhesion, and adsorption are quite familiar, others are less familiar though often of great biological significance. One of the latter is wettability.

Table 1.

Edge of cube	Number of cubes	Total surface	Surface energy in ergs
1 cm.	1	6 Sq. cm.	438
0.1 μ	10^{15}	60 Sq. m.	43,800,000
0.01 μ	10^{18}	600 Sq. m.	438,000,000
1 mμ	10^{21}	6000 Sq. m.	4,380,000,000

When a liquid wets a surface there is a decrease in free surface energy. The study of wettability is a large one involving numerous factors such as contact angles (BARTELL and SLOANE 1929) and adhesion tension (FREUNDLICH 1926, HARKINS and LIVINGSTON 1942). Wettability studies have found numerous applications to biological systems, *e.g.* blood (MUDD 1924, NUGENT 1932) and latex (MOYER 1935).

Surface tension is the most readily recognized form of surface energy and is so well understood that the physics of it need not be gone into here. The biological applications of it should, however, be given some thought.

There has been an excessive use of surface tension in explanation of cell processes. A chemist will insist that surface forces are most certainly responsible for many vital phenomena for it is at surfaces that reactions take place; therefore, rather than deprecate surface energy as an explanation of cell behavior, one would do better to encourage students to view the kinetics of cellular processes in the light of surface forces. There is much truth in this advice, but there is likewise much truth in the view which was first stated, for if we consider one process after another which biologists have interpreted in terms of surface tension, we shall see that only occasionally could the hypothesis be true and rarely is there any actual evidence to support it.

Let us start with one incontrovertible fact: a spherical cell is round because of surface tension forces. If we go beyond this simple case we shall find that most correlations are mere assumptions; thus, it would seem likely that when there is a reduction in surface tension over any given area on the surface of an *Amoeba* or slime mold, inner pressure would force some of the protoplasm outward forming a protrusion or pseudopodium. BERTHOLD advanced such an idea and based a theory of amoeboid movement on it, but JENNINGS (1904) later came to the conclusion that the locomotion of *Amoeba* is demonstrably not due to a local decrease in surface tension on the side toward which the animal is moving.

Surface tension has been presumed to be the force by which an Amoeba holds, severs, and ingests its food. *Paramecium*, on which *Amoeba* feeds, maintains a permanent shape, and must, therefore, possess a more rigid pellicle than *Amoeba*. It would be physically impossible for *Amoeba* with its jelly arms to pinch a firm *Paramecium* in two. MAST found that to cut a *Paramecium* in half with a fine glass fiber requires a pressure of approximately 9 mg., which would mean a reduction in surface tension of at least 1,000 dynes per centimeter. The interfacial tension of protoplasm, according to estimations by CZAPEK (1911), is less than one. The pseudopods of an *Amoeba* could not possibly, therefore, have the same cutting capacity as the glass fiber.

It is very probable that when a "bud" appears on yeast, the tension over a small area of the cell surface has diminished. The bud rounds off into a more or less spherical form. But the parent yeast cell is not round, only the incipient cell formed from it is a spherical fluid drop. The ellipsoidal form of an old yeast cell is maintained because the wall is rigid, and the lack in symmetry was acquired while the wall was young and plastic; though surface tension forces could then operate, the cell, never the less, assumed, as it grew, not a spherical but an ellipsoidal form. Protoplasm is at certain times amenable to surface tension laws but at other times it is not at all amenable to these laws.

Many cellular processes have been said to be due to changes in surface tension; among them are fission, mitosis, muscular action, protoplasmic streaming, nerve impulses, anesthesia, and even memory, with little evidence to support any of the hypotheses. Surface forces are very active in living systems; they are a real source of energy for cellular reactions, but though the cell uses such forces to advantage, all evidence indicates that changes in surface tension accompany rather than give rise to any but a few simple processes. A living cell is as likely to oppose surface tension as to obey its laws.

Adsorption.

Surface condensation, or adsorption, is another form of surface activity. Few phenomena have been the subject of so much unnecessary controversy. The argument has centered primarily around the question of the stoichiometric *vs.* the non-stoichiometric nature of adsorption, and of the number of kinds of adsorption bonds. No such difficulties arise in the GIBBS interpretation of adsorption at liquid surfaces. Here the problem is clearly stated and capable of precise mathematical expression. According to GIBBS (1875, 1878) a dissolved substance is positively adsorbed if it lowers surface tension and negatively adsorbed if it raises it. This is true because free surface energy tends toward a minimum, and any substance that contributes toward the lowering of this energy remains in the surface. If this statement is made a general one by saying that *all energy strives toward a minimum,* we have the second law of thermodynamics upon which GIBBS' principle rests. The GIBBS adsorption equation is:

$$\frac{d\sigma}{dC} = \frac{a}{C}\frac{dP}{dC}$$

where a is the concentration of the solute at the surface in excess of that in solution, C the bulk concentration of the solute, $d\sigma/dC$ the rate of change in surface tension with increasing concentration, and dP/dC the rate of change in osmotic pressure of the solute with increasing concentration.

If, as GIBBS' law implies, substances which lower surface tension are more concentrated at the surface of a solution, whereas substances which raise surface tension are less concentrated there—soaps and fats are in the former category and salts in the latter—then, the surface of the ocean should be less salty than the interior. BULL (1943) calls attention to an ingenious experiment by MCBAIN and HUMPHREY (1932) giving practical proof of the validity of GIBBS' adsorption equation, which GIBBS, like NEWTON, never bothered to establish. A microtome blade was moved across a solution 0.1 mm below the surface thus literally slicing off the top layer. Analysis showed the surface film to be of greater or less concentration than the inner fluid in accordance with GIBBS' adsorption equation.

Adsorption on solid surfaces presents a somewhat different problem. We still have concentration at an interface but the substances come from without, not from within the adsorbent.

As originally viewed by FREUNDLICH (1932), adsorption on solid surfaces is non-stoichiometric. To characterize a substance by its adsorption constant is tacit admission that the reaction is non-stoichiometric. VAN BEMMELEN pointed out that the adsorption of some substances is not stoichiometric but follows the empirical formula:

$$a = \alpha C^{1/n}$$

where a is the amount adsorbed per gram of adsorbent (a/m may be substituted for a where m is the weight of the adsorbent in grams), C is the concentration of the solution in equilibrium, and α and n are constants (α is the adsorption constant). This formula was used by FREUNDLICH (1932) who applied it extensively, showing that it holds in a great number of systems; it is, therefore, known by his name.

The importance of adsorption in a study of cell kinetics lies partly in its non-stoichiometric character. We are dealing with physical attraction, a field of influence. The adsorbed substance may pile up on the surface to whatever depth the field is effective. As no new compound is formed the union is reversible. Thus, anesthesia may be explained in terms of adsorption because the prerequisite of anesthesia, reversibility, is satisfied.

It must be granted that an adsorption bond may be electronic in which case chemical union takes place and a new compound results. Typically, however, adsorption is electrostatic, involving, for example, VAN DER WAALS' intermolecular forces. Attractive forces of this kind are non-stoichiometric.

Fields result when surfaces are electrically charged; they thereby attract and hold substances of opposite electric charge. Surface charges which set up stray fields of electrostatic force, play a role not only in adsorption but in a variety of vital phenomena such as nerve conductance and protoplasmic flow.

The adsorption of substances on the surface of cells and the particles in cells must inevitably play a dominant role in the life of a cell. The transfer of oxygen, of nutritive substances, the occurrence of anesthesia and poisoning may all be adsorption phenomena. FREUNDLICH (1932, 1926) was of the opinion that poisoning followed his adsorption isotherm. He had in mind the taking up of veratrine by the marine snail *Aplysia*.

WARBURG (1921) advanced an adsorption theory of narcosis. The theory had been previously stated by TRAUBE and CZAPEK. WARBURG searched for an explanation of some experiments by FREUNDLICH on the adsorption of oxalic acid by blood charcoal and found that oxidation of the oxalic acid into carbon dioxide and water takes place. He then proved that this inanimate oxidation process can be retarded by narcotics, just as can animate oxidation, *i.e.*, respiration. Furthermore, the retardation of inanimate oxidation by narcotics rises with the adsorption constants of the narcotics used, methyl, ethyl, propyl, and phenyl urethane. Apparently, the oxidation of oxalic acid into carbon dioxide and water in the presence of charcoal takes place on the surface of the charcoal, from which it may be assumed that the oxidation of sugars into carbon dioxide and water in cells, *i.e.* respiration, takes place on the surface of infinitely small protoplasmic particles. Any diminution of these surfaces will mean a decrease in oxidation.

Narcotics are readily adsorbed by charcoal and by blood particles, they thus decrease the surface available for oxidation. Narcosis and anesthesia are thus regarded as a retardation of oxidation due to decrease in the free adsorptive surface of blood particles through adsorption of the narcotics. The hypothesis is a feasible one with several facts to support it; thus, the members of the homologous series of alcohols are adsorbed in increasing amounts as we ascend the series, and the narcotic effects also increase in this wise.

The evidence supporting WARBURG's hypothesis is offset by the fact that methanes and hydrocarbons show little or no surface activity, yet all are good anesthetic agents. Surface activity and adsorption rest in large measure on the polarity of the molecules involved. Hydrocarbons are nonpolar and therefore not adsorbed, but they are strong narcotics. Furthermore, many substances, such as sugars, are readily adsorbed and have no narcotic effect whatever. And, finally, the cyanides, as narcotics, are ten thousand times more effective than they ought to be if they must cover a sufficient surface. They may still function by inhibiting oxidation but in some manner other than adsorption, possibly by uniting with the catalyst iron.

Table 2.

Alcohols	Concentrations required to cause anesthesia of tadpoles, M.
Methyl . . .	0.60
Ethyl	0.30
Propyl . . .	0.10
N-butyl . . .	0.04

Surface forces, and the phenomena attendant thereto, lead to the orientation of surface molecules and the formation of films which are of great biological significance. The spreading of oils and proteins as mono-, pauci-, and multimolecular layers is the basis of studies on cell membranes and selective permeability (LANGMUIR 1917, HARKINS 1917, ADAM 1930, DANIELLI 1934).

Diffusion pressure.

The kinetic energy of molecules within the bulk of a liquid manifests itself in various ways. One of these is diffusion, which results in a *diffusion pressure*, responsible for a number of biological phenomena, notably osmosis, imbibition, turgor, swelling pressure and edema.

Osmosis and imbibition, formerly so sharply distinguished by the biologist, are now regarded as identical phenomena, at least by the physical chemist. The mechanism of the swelling of jellies is an old controversy. Hypotheses such as the filling of capillaries are obviously in error, for the capillary forces of minisci pull in, they do not push out; a gel could not, therefore, expand while filling capillaries. Systems such as baked clay, silica, and pumice take up water by filling capillaries, but they do not swell.

The most satisfactory interpretation of imbibition came from PROCTER and WILSON who said that imbibition and osmosis are identical in so far as the forces involved are concerned; LOEB (1922) added, "the reason that osmotic pressure, the viscosity of protein solutions, and the swelling of protein gels are all influenced in a similar way by electrolytes is that all three properties are in the last analysis functions of one and the same property, namely, osmotic pressure."

The critics of this view, notably FREUNDLICH, stated that it is fundamentally false to regard osmotic pressure as identical to imbibition pressure. There is heat of imbibition, but no heat of osmosis.—This may be merely a matter of degree.—More significant is the fact that osmotic pressure is proportional to concentration, whereas imbibition pressure is proportional to a high power of concentration.

Botanists, concerned as they are with the swelling pressure of cellulose, starch, and agar on the one hand, and the turgor of the cell vacuole, on the other, regard the two as distinct because the respective systems in which they take place are mechanically, *i.e.*, structurally, different.

It may be that imbibition cannot be disposed of as simply as did LOEB. BULL (1943) is of the opinion that the initial isoelectric swelling is simply hydration of the polar groups of the protein plus the osmotic pressure due to the protein alone. The additional swelling resulting from a change of p_H is due to a DONNAN equilibrium established between the interior of the gel and the acid or base on the outside. This leads to electrolyte accumulation by the gel with an increase is osmotic pressure. Water then flows into the gel until the total swelling pressure is equal to the elastic strength of the gel. BULL (1943) says: let us look at the swelling problem by considering the relation between vapor-pressure and osmotic pressure. On the basis of the maximum work principle we obtain:

$$P_h \bar{V}_1 = R\,T \ln \frac{P_0}{P}$$

where P_h is the hydrostatic pressure of a solution with vapor pressure P in equilibrium with the pure solvent whose vapor pressure is P_0. $\bar{V}_1$ is the partial molar volume of the solvent. The swelling pressure of a gel is equal to the pressure calculated by means of the foregoing equation (minus the slight pressure exerted by the elastic property of the gel).

The energy available to the plant in the form of either osmosis or imbibition is tremendous. Seeds will take up water from a saturated lithium chloride solution (osmotic pressure, 1,000 atmospheres), although their salt content is sufficient to account for only a few atmospheres of osmotic pressure. *Cacti* with osmotic pressures not in excess of 6 or 7 atmospheres take up water from very dry soil. NEWTON (1930) put *cactus* stems in a desiccator with sulphuric acid for six months. In this atmosphere of nearly zero humidity, the *cacti* lost less than 10 per cent of their water. It is assumed that the water is held by imbibition forces.

FREY-WYSSLING (1948) in discussing the growth of plant cell walls, presents the energetics of cell elongation as an osmotic phenomenon. Much of growth by elongation, of movement as in the diurnal closing of leaves and flowers, and of the translocation of water and salts is accomplished by diffusion pressure, in one form or another.

Osmosis and imbibition may involve identical forces yet one can pit osmotic forces against imbibition ones. This certainly appears to be the case in the following experiment. If a series of cubes of a gel, say, 25 per cent gelatin, are placed in sugar solutions of different strengths, water passes from the gel to the solution, or in the opposite direction, according to the concentration of the surrounding sugar. One cube in the series may maintain its original size; the others become either larger because of the greater imbibition pressure of the gel, or smaller because of the greater osmotic pressure of the surrounding solutions.

The imbibition *vs.* osmosis controversy when applied to protoplasm brings up an old discussion on the miscibility of protoplasm in water. We may simply be involving ourselves again in a futile polemic, but there is a fundamental question involved in the distinction between soaking up water and going into solution; it is a question of structure. Protoplasm takes up water with avidity; whether it does so osmotically, as would a sugar solution in a sac, or by imbibition, as does gelatin, it does not, in either case, go into solution. Living protoplasm cannot dissolve in water, it holds together. However, osmotic laws hold when protoplasm takes up water, as is shown to be true by LUCKÉ (1932).

Potentials.

Any means for establishing a difference in ionic concentration will result in setting up a difference in potential. Solutions of unlike concentration will, when joined by an electric conductor, yield a *concentration potential.* The name by which we call a potential is often simply a matter of how the difference in ionic concentration is established, or how we measure it. One of the most obvious of these is the difference in duffision rates of ions (LONGSWORTH 1935). Once liberated, ions set up a *diffusion front,* with the faster ions to the fore, H^+ being the fastest.

Some diffusion rates are:

Table 3.

Ion	Absolute velocity cm./sec. × 10^{-4} at 18° C	Ion	Absolute velocity cm./sec. × 10^{-4} at 18° C
H^+	32.50	OH^-.	17.80
K^+	6.78	Cl^-	6.78
Na^+	4.51	NO_3^-.	6.40

The dissociation theory and the concept of H ions were barely well established when criticisms arose. One of the earliest of these held that there is no reason to assume that one hydrogen becomes dislodged from the water molecule any more readily than the other. This criticism is not justified for after one hydrogen ion is separated from any ionizable compound, the others cannot leave half so readily. For example, the dissociation constant of H_3PO_4 is 1.1×10^{-2}, but that of H_2PO_4 is 10^{-7}, and of HPO_4 it is 10^{-13}.

We still speak of the hydrogen ion and write it H^+, but the true situation appears to be:

$$H—O—H + H^+ \rightarrow [H—\overset{\overset{\displaystyle H}{\vdots}}{O}—H]^+.$$

The radicle in brackets is the *hydronium* ion, H_3O^+. It is more convenient to write H^+, which we do, but assume that it is a symbol for H_3O^+.

The hydrogen ion moves at 0.00325 cm. a second, and the nitrate ion at 0.00064 cm. a second. The fact that the faster ions are soon in advance of the others results in an unequal distribution of electric charges. Potentials due to differences in the velocity of ions are called *diffusion potentials.* They are not great and are quickly obliterated through equalization in concentration when the slower ions catch up. Furthermore, one ionic species cannot wander far from the other without setting up a potential difference of great magnitude. The electric force thus produced tends to slow up the faster ions and hasten somewhat the slower ions.

Potentials arise whenever a difference in energy level is established, and this is done when one electrolytic solution is in contact with another of different concentration or of suitably different chemical composition. Such potentials go by various names among which *phase boundary* potential and *liquid junction potential* are two. A *biphasic cell* of this type is represented by the following chain:

$$Ag : AgCl \underset{e_1}{:} NaCl^1 \underset{E}{:} NaCl^2 \underset{e_2}{:} AgCl : Ag$$

$NaCl^1$ and $NaCl^2$ are of different concentrations. Mixing of the two solutions can be prevented by a constant state of flow, or when one of the two miscible solutions is held in a jelly.

In such a cell it is the potential E at the phase boundary of the two salt solutions, $NaCl^1$ and $NaCl^2$, which is presumably measured. Actually, however, what we measure is the sum of three phase boundary potentials (MACINNES 1939), E between the two salt solutions, and e_1 and e_2 between the metal electrodes and their salt solutions; the total emf, E_t, is, therefore, $e_1 + E + e_2$.

Using a cell of this type, and maintaining a sharp phase boundary through use of a flowing junction, MACINNES and YEH (1921) obtained the following potentials between two *kinds* of solutions, at 0.1 M concentrations.

HCl: LiCl = 34.86 millivolts
HCl: NaCl = 33.09 millivolts
HCl: NH_4Cl = 28.40 millivolts
HCl: KCl = 26.78 millivolts
KCl: LiCl = 8.79 millivolts
KCl: NaCl = 6.42 millivolts
KCl: NH_4Cl = 2.16 millivolts

BEUTNER (1927) believes that phase boundaries are the physical basis of all bioelectric potentials. He established such a potential between cresol in contact with aqueous sodium chloride on the one side and aqueous sodium oleate on the other side. An ionic gradient arises in the central solution because the organic salt is more soluble in cresol than is the inorganic salt.

−	aqueous NaCl	{ Na ions in low concentration	Cresol	Na ions in high concentration }	aqueous Na oleate	+

Voltages as high as 0.1 are obtained with such systems.

Entropy.

There is a form of energy, or measure of an energy state, with which the biologist cannot yet deal, at least he cannot apply it, even though it has been cited as characterizing the living state; it is *entropy*. Some of the terms used to define entropy may be of help in understanding it: thus, entropy is said to be a measure of molecular disorder, a measure of the degree of randomness in a system provided the total energy remains constant; a function of the energy of a system, a measure of the probability of a given state in terms of the number of atomic combinations or molecular arrangements which will procedure that state.

Evidence of the reality of entropy lies in the third law of thermodynamics: every substance has a finite positive entropy, but at the absolute zero of temperature the entropy may become zero, and does so become in the case of a perfect crystalline substance.

When entropy is stated in terms of probability, *e.g.* that all the gas molecules in a given space will do a certain thing, it then becomes a measure of the amount and complexity of molecular motion, the more complex the randomness of the molecular motion, the greater the entropy. It is this concept which leads to the definition of entropy as a measure of disorder.

LATIMER (1953), in giving an excellent account of entropy, has added one explanatory note which averts a common misconception. Because diamonds burn as readily as lamp black, it has been said that entropy is not a measure of the degree of order of a system. This is false reasoning because the degree of order of a system does not affect its reactability with other systems. The mere knowledge of the entropy of a reaction does not permit one to determine the driving force of a chemical reaction. For that reason the free energy, ΔF, of the reaction, which is the driving force, is much more useful to the chemist.

Because

$$\Delta F = \Delta H - T\Delta S,$$

the free energy is dependent upon the heat, ΔH, and the entropy, ΔS. Many spontaneous chemical reactions occur with a decrease in the entropies of the substances involved. In such a case the entropy of the surrounding (thermostat) must increase. Entropy applies only to thermal energies whereas the over all free energy involves the bonding of the atoms. If there is no change in bonding for a process, such as the expansion of a gas, then the entropy change is the determining factor.

Free energy and entropy changes are the same in living systems as in non-living systems. A decrease in entropy is accomplished by the aid of external work, the food we eat or the sun-light absorbed by plants. Living organisms are remarkable in being able to utilize energy in complex processes so as to continually maintain and produce order against the tendency, characteristic of inanimate nature, to destroy order, but this is not entropy.

In short, all of life's processes are understandable, except consciousness and life itself and these are not explained by entropy.

Entropy plays as great a role in life as it does any where else, but the form in which it occurs in living matter does not distinguish life, as it has been said to do. LATIMER (1953) holds such speculations to be without foundation.

Electrokinetics.

The coining of the word electrokinetics is attributed to FREUNDLICH, who included under it the four familiar colloidal phenomena which involve the movement of particles and liquids through capillaries. But as kinetics implies any form of molecular activity, then electrokinetics must include all molecular and colloidal processes which are electrical in nature. Those among them which occur in the living world comprise electrophysiology.

The experiments of GALVANI on "animal electricity," and VOLTA's interpretation of them, were the beginning of research in electrophysiology. The date was 1786. GALVANI's experiments initiated a discussion on the significance of electrical forces in the living world which is still with us. He was under the impression that he had discovered the presence of electricity in muscle. He observed that a muscle twitches when touched with a Zn-Cu couple, and rightly concluded that the twitching of the muscle was caused by electricity; but he was wrong in assuming that the cause of the muscular movement was an inner source of electricity. VOLTA correctly interpreted GALVANI's experiment; he realized that the electricity was not of a vital nature, but was generated by the two electrodes immersed in the electrolytic solution which is the natural body fluid. The salts and acids in the tissue formed, with the metal electrodes, a simple Galvanic or Voltaic cell.

But VOLTA, too, made his error in denying, at least tacitly so, that living tissues *per se* can give rise to electric currents. This view prevailed in science for more than half a century until DUBOIS-REYMOND (1848) proved that an electric potential in tissues can be measured with a galvanometer. To account for the potential which he had measured, DUBOIS-REYMOND postulated bipolar molecules specifically oriented and functioning as molecular magnets. The suggestion was a fertile one, and though we still do not know the answer, magnetism is probably not the correct one.

These were the years of FARADAY's great work on magnetism and it was inevitable that magnetism should enter into theories of vital behavior. In fact, such theories are still advanced, but there is little evidence to support them. Faraday himself applied a powerful magnetic field to the human body without noticeable effect. The striking resemblance of the dividing cell in mid-mitosis to a magnetic field has led many an investigator to postulate such a field as the physical basis of cell division and the cause of the migration of chromosomes. One cytologist interpreted the distribution of chromosomes in terms of the arrangement of electromagnets and proved his point by floating numerous small magnets attached to corks on water; the resemblance to chromosome pattern was sufficient to convince him of the soundness of the analogy. Attempts to disturb the mitotic figure by strong magnetic forces have, however, proved futile.

Static electricity was said to have an effect on living organisms, but experiments in electroculture did not prove it.

Electroculture is the growing of plants under the influence of static electricity. Plants out of doors with their roots in the earth and their tops in the air are presumably traversed by minute currents of electricity, because the air is electrically charged with respect to the earth. It was, therefore, reasonable to suppose that an increase or decrease in potential between earth and atmosphere would affect the growth of plants. English investigators obtained evidence demonstrating, so they thought, that increased plant growth follows electric excitation supplied by a charged network suspended above the plants. American investigators working under carefully controlled conditions had only negative results.

Astrophysicists say that the potential difference between sky and earth is very feeble, except at the moment of thunder storms and during snow flurries. Plants are more sensitive than physical instruments. Our radiation environment may be effective.

Electrophysiology is without doubt the proper scientific approach to many fundamental problems, for life processes are largely electrochemical (Beutner 1944). One can hardly today question the belief that electric potentials are present in protoplasm and that such potentials are available to the cell as sources of energy. Nor are we any longer bothered by methods of measuring vital potentials. The experiments of Lund (1931), Michaelis (1925), Fujita (1925), and Burr (1944), the tremendous amount of work done by the neurologists, Lillie (1918, 1920), Adrian (1932), Berger (1936), Gibbs (1936), Jasper (1937), Katz (1939), and the work of Kamiya (1950) and Tasaki (1950) on potentials associated with protoplasmic streaming in slime molds, are all unquestionable evidence of electrical forces at work in living tissues.

Membrane potentials.

The origin of the electrical forces in animate nature is not regarded as an open question by some investigators but due, in all cases, to an unequal distribution of ions. If this unequal distribution is accomplished by a membrane, then the potential receives its name from the membrane. No expression is more familiar in physiology than *membrane potential*. It has been assumed that the membrane is the seat of the potential or materially influences it.

Membrane potentials when discussed along classical lines, remain in very good standing for they were first suggested by Wilhelm Ostwald (1890) and widely accepted among biologists including so able a worker as Michaelis (1925, 1927). Recent opposition to membrane potentials is centered almost entirely in the work of a few investigators, notably Beutner (1944), Barnes (1937), and Amberson (1936).

Apparently, all investigators are agreed that certain membranes in both the living and the non-living worlds permit some ions to pass through and others not, or permit certain ions to pass through more readily than others. Such selective permeability was said by WILHELM OSTWALD (1890) to be the seat of electric forces in the living cell, in muscle, and in nerve. The theory thus places the seat of vital electromotive forces on the surface of, or within, membranes. As WILLARD GIBBS (1875, 1878) had interpreted OSTWALD's theory mathematically, and DONNAN applied it, the theory of membrane potentials gained a strong foothold. GIBBS reference was, however, to surfaces, and not necessarily to membranes, and though DONNAN mentioned membranes he recognized that they were not necessary.

HÖBER (1924), like MICHAELIS and his collaborators (1927), made much use of membrane potentials in interpreting certain vital phenomena. It was held that the living membrane is selectively "permeable for cations because it possesses properties similar to those of dried collodion membrane." The collodion membrane becomes positively charged and anion permeable, when treated with basic dyestuffs, alkaloid bases, or phenylendiamine. Because the same is true of the red blood corpuscle, "the conviction that we have a close analogy to the functioning of the living cell," was strengthened. This work precludes the acceptance of free diffusion in the interpretation of bio-electric potentials.

The fact that the sign of the membrane, *i.e.*, its polarity, can be reversed from the negative, when the amphoteric membrane is in an alkaline medium, to the positive, when the membrane is in an acid medium, is the most convincing evidence the supporters of the membrane hypothesis have in their favor.

At the time OSTWALD advanced his hypothesis, both NERNST and HABER opposed it, maintaining that the potential was not on the membrane but at a phase boundary, the membrane merely serving to maintain the boundary. No one questioned the fact that the selectively permeable qualities of the membrane brought about an unequal distribution of ions, which, in turn, was responsible for a difference in ionic concentration on the two sides of the membrane, and, therefore, for a concentration or phase boundary potential. MICHAELIS (1925) and BEUTNER (1920) had found the apple skin and celloidin membrane to be selectively permeable, and therefore, in a sense, the origin, or the seat, of the potential.

All opinions can be reduced to two major and several minor categories: bioelectric potentials arise 1) from inequalities in electrolytic concentration established by, a) differences in mobilities, b) selective permeability, or c) differential solubility of ions in the constituents of a membrane: or 2) the electromotive force is due to an initial charge on the membrane itself. It is the latter view which OSTWALD, MICHAELIS, and HÖBER held; the first hypothesis based on diffusion phenomena was held by NERNST, HÖBER, BEUTNER, and AMBERSON (1936).

One may, without confusion, use the term "membrane potential" to indicate any electromotive force which depends upon the structure or composition of a membrane. It appears that membranes do impose constraints upon ions of one electric sign, hindering or preventing them from passing through, whereas the companion ion, of opposite sign, can pass through with relative ease. But then, in all cases, it is, in the end, a difference in ionic concentration which gives rise to the electromotive force.

BEUTNER (1912) regards the "vital battery system" as comparable to a galvanic cell in which a phase boundary takes the place of metal electrodes.

Little attention has been given to the kind of ions which produce potentials in tissues. BARNES (1937) lays emphasis on sodium, maintaining that it is the only ion which will bring a potential back to a heart which has been perfused with glucose. Sodium, he says, will also maintain the potential of skin. He believes that neither potassium nor calcium can replace sodium in the matter of holding up a biopotential.

The average potential value obtained in bioelectric work is the same as that generally found on colloidal surfaces, namely, 40 to 50 millivolts, with a theoretical maximum of 55 millivolts for colloidal particles. Phase boundary and "membrane" potentials, however, often exceed this value, reaching 1 volt.

Neither membrane nor phase boundary is the only source of electric forces in living tissues. Chemical reactions involving the transference of electrons between an oxidant and a reductant, are also a source of electric potentials. The potentials set up by oxidation-reduction equilibria are readily measured. One of them in common practice in the laboratory, is used for measuring acidity, by the ionization of quinhydrone, which results in equimolecular concentrations of the reductant hydroquinone and the oxidant quinone. That which is potentiometrically measured is an electromotive force which has a definite relation to the free energy produced in a chemical reaction, in this case brought about by a reversible oxidation-reduction equilibrium. The rate of the reaction, by good fortune, is influenced by the hydrogen-ion concentration of the solution. An interpretation of the complete reaction is:

$$\underset{\text{Quinone}}{C_6H_4O_2} + \underset{\text{electrons}}{2e^-} \rightleftharpoons \underset{\text{anion of hydroquinone}}{C_6H_4O_2^{2-}} + \underset{\text{hydrogen}}{2H^+} \rightleftharpoons \underset{\text{hydroquinone}}{C_6H_6O_2}$$

Oxidation-reduction reactions are of great variety; a mixture of ferrous and ferric salts will yield them. The method most used for determining oxidation-reduction potentials in protoplasm has been the injection of dyes. A dye, such as methylene blue, will be oxidized or reduced in proportion to the relative activities of oxidation and reduction in the cell. BROOKS (1928) found the cell sap of the alga *Valonia* to have a potential of 0.12 to 0.15 volt; and the protoplasm, a potential of 0.21 to 0.48 volt.

Though the presence of oxidation-reduction potentials in living systems can not be doubted, the many conditions which can determine and influence them, the discrepancies in measuring them, and the great heterogeneity of protoplasm has led some investigators to regard interpretation practically out of the question. COHEN goes further and suggests the term "reducing intensity" instead of "oxidation-reduction potential," to meet the criticism that there is no electric potential as such in tissues and that thermodynamic equilibrium does not occur. COHEN's skepticism seems no longer justified when confronted by the more recent work done on brain potentials. However, BARNES (1937) among others discards oxidation-reduction potentials as too feeble to serve any useful purpose in life and also too feeble to be detected by available conductors. Other postulates exist, BURR (1944) has suggested a *Maxwellian* or *electromagnetic field.*

If we accept the presence of electric forces in living matter, and this acceptance seems inescapable, then we are confronted with the problem of function.

The function of bioelectric currents.

Any statement on the function of bioelectric currents is likely to be an assumption. But often there is much justification for the assumption. Correlations with temperature, development, age, metabolism, and even psychological activities, force one not only to recognize the reality of vital electric potentials, but also their possible association with outer environmental conditions and inner physiological activities. Among correlations which indicate that vital potentials are real and intimately associated with physiological activities are the following: oxygen causes an increase in the potential of frog's skin; exposure to a high frequency field stimulates the developing chicken egg and produces a noticeable increase in electric potential; irradiation of the frog's skin with X-rays increases potential differences; ultraviolet light raises the surface potential on chicken eggs; potential differences fall during asphyxia and drop to zero at death.

By studding a plant with platinum pins, which served as electrodes, LUND (1931) found it possible to map the electric plan of a tree. The tree top is positive to the trunk base. Potential differences vary from 30 to 200 mv. LUND believes that the external polarity is evidence of a complex but definite pattern of internal polarities which serve to correlate the activities of one cell with another. It appears to be generally true that physiologically active regions are electronegative to inactive ones; thus, the animal pole of the chicken egg is negative to the vegetal pole. It has also been said that ovulation can be detected by a change in polarity of the sides of the body; that before fertilization the frog's egg is positive at the animal pole and negative at the vegetal, and after fertilization polarity reverses; that glandular changes are associated with tissue potentials, adrenalin secretion causes negative deflection. As connective tissue when wounded is positive to the blood stream, it has been suggested that white cells, which are negative, are attracted by the positively charged wound.

Much of the foregoing is speculative and some of it undoubtedly untrue, but the reality of electric potentials in protoplasm is supported by too many facts to permit doubt as to its existence. It is the meaning of a change in potential with a change in environment, development, or metabolism which we cannot yet state.

Were there doubt, the organism as a whole would dispel it when a ray-fish blows out a 6 volt light bulb, and an electric eel fells a horse?

Body potentials have been extensively measured in animals. Particularly thorough have been the determinations made by BURR (1944). BURR and HOVLAND (1944, 1949) have demonstrated that there is an electric potential gradient between the cephalic and the caudal ends of the embryo, and a gradient between the living organism and the physiological salt solution in which the organism is suspended. Experiments show the living embryo to be negative to its surrounding medium. The potential gradually falls off as the distance between outer electrode and embryo increases. A voltage can be detected at a distance of one-half a millimeter, thus supporting the presence of an electric field, the assumption of which is one of the most controversial subjects in electrophysiology. Of this, I shall say more later.

Of all correlations which link electric potential with physiological processes, that between brain and potential is the most dramatic. Though hypothetical, work has progressed so far that one can hardly question the basic truth of it. It is the meaning and the extent of the correlation which must still be left an open problem. Brain waves, as the plotted curves and their corresponding vital potential rhythms are called, are said to characterize the individual, and to serve in the diagnosis of mental conditions, particularly the pathological state of the brain.

The greater amount of the measured electrical brain potential comes from the cerebral cortex (BEUTNER 1944) though deeper brain currents are known to exist. The frequency is 8 to 40 seconds and the normal voltage 5–75 microvolts.

The 10 per second waves are called alpha, the slower, delta, and the fast ones, beta; they are the same waves moving at different speeds.

A one year old child has waves of less than 4 per second. Foetal brain waves have been recorded through the mother's abdominal wall and are similar to the central waves of the neonate. Electrical maturity is attained at the age of 12 in the visual or occipital regions. There is a slight acceleration at the age of 18 in other areas. Brain waves of very old patients are perfectly normal. The occipital waves of women are said to be faster than those of men, indicating that women are more keenly aware of their immediate surroundings. With this last statement we enter the realm of extreme speculation. Having entered, we might as well explore further. It is possible, that brain waves will be of value in diagnosing encephalitis, epileptics, malingering, faked blindness, faked amnesia, etc., etc. Here the psychiatrist is likely to object, so also the mathematical biophysicist.

There has arisen in the field of cellular studies, a branch of science termed mathematical biophysics (Rashevsky 1944). It is an attempt to label investigations and the conclusions drawn therefrom, somewhat after the manner of Poincaré (1918) who distinguished between science and hypothesis. Most biologists are aware of this distinction. Words of caution and a new method of approach are welcome, but one should not fear hypotheses. Data without theory are sterile. Rashevsky (1944) calls attention to biological speculations on the mechanism of cell division. He shows, by mathematical analysis, that certain theories are not valid. Thus, the hypothesis that "electrical forces of repulsion between particles carrying charges of the same sign may be correct qualitively but it will not stand quantitative test." He suggests, in place of the foregoing hypothesis, the "particularly plausible" one that cell division is "connected somehow with general metabolism."—In the first place, the electric charge hypotheses of many biologic phenomena are at best only tentative, and often wholly unwarranted. But cells certainly possess a charge, which functions biologically, a fact one can hardly escape, but no biologist seriously regards cell division as due to this charge. The metabolism hypothesis is as highly speculative as the electric charge, but has the advantage of being beyond either mathematical or biological proof.

The mathematical, like the biochemical, approach to a problem is as much a part of the science of life processes as is the purely biological approach, *and just as speculative.*

Electric potentials must be involved in metabolic activity because in many forms of such activity a transfer or setting free of electrons occurs. There is the further fact that the potential is often dependent upon oxygen supply.

If final proof of electrical forces in animate nature is still needed, the ray-fish and electric eel supply it.

The electric eel discharges some 550 volts with a load of forty watts—more than enough to stun a man or a horse. Each discharge lasts only two one-thousandths of a second, and of these the eel can send out four hundred or more per second. Measured by volume, nearly half of the fish is electric tissue. It is made of units which are separated by thin walls of electrically resistant tissue; the units act very much like the "cells" in a storage battery. They are the producers of electricity, each one creating about one tenth of a volt. It is by joining these tiny batteries together in series, that the eel builds up its powerful discharge.

Faraday had ascertained that the current from an electric eel flows from head to tail, externally. Coates (1945) "electrolyzed" a tankful of eels by dropping an electrode at each end of their tank and passing a current through the water. All the eels then gathered at the anode. This behavior is sometimes shown by *protozoa.*

Energy emanating from living material.

One can hardly accept any of the truth on the preceding pages without realizing that protoplasm, as a source of energy, must also radiate energy, as indeed it does, and in a variety of forms, heat in warm blooded animals, light in the fire-fly, and electricity in the ray fish. The implication that there may be other forms of energy emanating from living matter which we have not yet measured, brought forth an unaccountable emotional reaction against the idea. The experimental evidence is vague and indirect, but when other forms of energy coming from living matter and capable of physical measurement, are discussed, there will be no need for denial or surprise. Surely, we must recognize protoplasm as in some way more remarkable or at least more intricate, than most non-living matter with which we are familiar.

Plant protoplasm vs. human protoplasm. I wish to close this chapter with a comment which is of some importance to workers on protoplasm. I might pose the question: How far may elementary protoplasm be regarded as nerve, muscle, gland, or any other tissue of plant or animal? Some are opposed to any reference to a nervous system or nerve reaction in an organism which has no nervous system that is comparable to that in man. Those having an opposing view hold that, as protoplasm is a living system it is, *ipso facto*, a nervous system, because it is capable of nervous or irritable response. On the basis of the first of these two views, one finds it difficult to interpret the situation which exists in a protozoan such as *Euplotes* where within a one-celled organism there is a neuromotor apparatus consisting of a neuromotor and fibers which lead to and control the activities of the posterior flagella.

Any droplet of normal protoplasm is a primitive muscle because it contracts—a view further supported by the fact that myxomycete protoplasm apparently contains myosin (LOEWY 1952). In the same way can primitive protoplasm be regarded as a nervous system, for which one may, if one wishes, substitute the expression irritable system. The controversy goes further, it is doubted that an elementary living substance can give rise to an action potential. But action potentials have been found in several forms of plants, in *Mimosa* (BURR 1943) and *Nitella* (BROOKS and GELFAN 1928). TASAKI and KAMIYA (1950) reported an electrical response in the slime mold, *Physarum*. They showed the virtual equivalent of electrical responses to mechanical and electrical stimuli. BURR (1944) interprets an electrical response to a mechanical stimulus as a true action potential, equivalent to the action potential of nerve fibers, except for the parameters of time and magnitude.

Any bit of protoplasm, when properly stimulated, produces a reaction which can be measured in electrical terms. This is true of the slime mold, of plants, of mammalian cells wherever they have been studied, as well of the nervous system. They are all made of protoplasm and all possess the basic properties of protoplasm.

Reference should be made to a form of energy which may prove to be the basis of nerve conduction, muscular activity, and protoplasmic movement, thus correlating three of the most fundamental and least understood of vital processes.

Muscle, nerve, and streaming protoplasm all possess a measurable potential difference between any two points on a fiber or strand of the living material (KAMIYA 1942). Muscle and an elementary form of protoplasm, such as a myxomycete, show, in addition, a visible train of waves or rhythmic pulsations.

The movement of particles within protoplasm, when these are obviously not carried by the protoplasmic stream, has long been a problem in biology, often solved by the assumption that the particles, *e.g.* mitochondria, are capable of autonomous movement; but this conclusion is not justified. The variety of movement of cytoplasmic inclusions is rather great; there are the worm-like progress of mitochondria, the rapid here and there movement of fat droplets the amplitude of which exceeds that of Brownian activity, the spinning of the nucleus prior to mitosis, and the remarkable oscillation of the centrosome area,

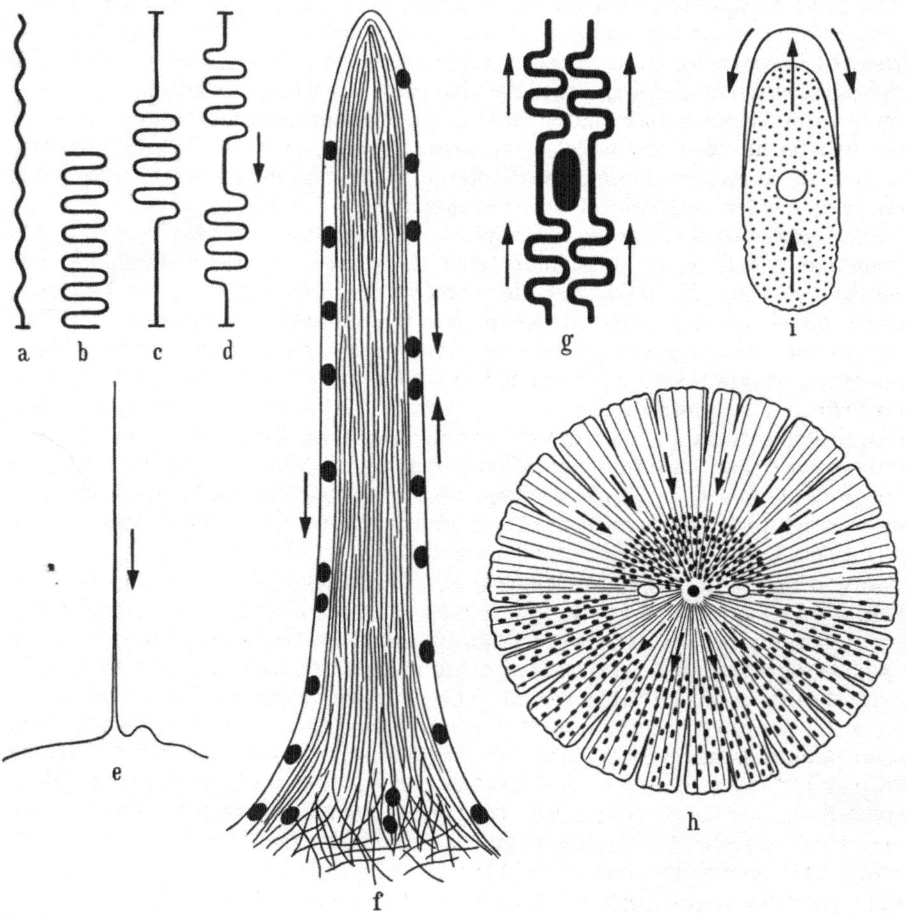

Fig. 1 a—i. An interpretation of protoplasmic streaming in terms of protein contractility: a—d, contracting and expanding protein molecules; a, maximum contraction; b, waves of contraction traversing molecular fibers; e, contraction of a myopodium of *Acanthocystis*; f, axopodium of a foraminiferan showing fibrillar "stereoplasma" in the axis, and "rheoplasma" on the surface with granules moving in the latter; g, transport of a granule by the contractile waves of two adjoining molecular threads; h, melanophore of a fish showing a centrosome (center), two nuclei (right and left of center), and the centripetal flow of granules (above) and the centrifugal flow of granules (below); i, movement of an *Amoeba* by contraction of the surface layer at the rear end. (After W. J. SCHMIDT.)

during cell division. The variety in forms of protoplasmic flow is also rather great, especially difficult of interpretation are the opposing currents of flow within the same cell or plasmodium. SCHMIDT (1928), FREY-WYSSLING (1949), LOEWY (1949), GOLDACRE (1950) *et. al.* interpreted protoplasmic streaming in terms of the contraction and expansion of protein threads for which there is no evidence. Seeking a more convincing hypothesis of protoplasmic movement which would satisfy all conditions and which would serve as well to explain nerve conduction and muscular activity, I concluded that ionic fields, such as those which surround colloidal particles, would, if disturbed by depolarization, provide the necessary stimulus (SEIFRIZ 1953a, b).

For a sound physical-chemical description of these fields, I turned to the work of DEBYE (1929). Molecules similar to H_2O, NH_3, and HCl are surrounded by static electric fields which convert them into dynamic systems, into electrically charged elementary particles. A second molecule coming within close range of such a dipole will acquire a field, or electric moment. Being thus polarized, it will interact with the first molecule. The work of DEBYE and others, notably LONDON has led to the now widely accepted fact that polar molecules are enveloped by electric forces which may be derived from pure coulombic interactions between charges. These are essentially short-range forces which do not extend beyond three or four molecular diameters. Long range forces will accomplish more.

The total force with which a particle, consisting of many molecules, will act on a second particle may be ascertained by integrating over the first sphere and arriving at the reaction between this sphere and a molecule in the second sphere. In a similar manner may one integrate the molecules in the second sphere and thus arrive at the final total reaction. Such a procedure leads to the prediction that in appropriate circumstances short-range molecular forces may build up to appreciably strong long-range forces. OVERBEEK (1948) has recently made direct measurements of the molecular attraction between parallel glass plates at distances of the order of the wave-length of visible light (5000 A).

If we turn to the still larger particles of lyophobic colloidal suspensions, then we have additional forces, namely, the familiar electric charge on the surface, due to preferential adsorption or to dissociation at the surface of the particle. The range of the coulombic interaction of these colloidal charges is determined by the depth of the ionic cloud surrounding the particle. Through such a living kinetic system impulses of vital nature pass and give rise to nerve conductance, muscular contraction, and protoplasmic movement.

Fields.

The introduction of the concept of fields into biology is so new that little can be said of it, other than it will be well not to ignore it. There is little doubt but that we shall have to deal with fields in biology in the near future. There is already a beginning. The physicist speaks of fields and measures them with confidence, but when the biologist attempts to apply them he finds himself the subject of criticism from both the biologists and the physicists, but for different reasons. The biologist objects to the application of an electrical environment because he has never thought in these terms, and the physicist objects because the experiments do not meet his standard in precision, a standard the biologist can rarely meet.

Some work has been done by BURR and MAURO (1949) on electrostatic fields of the sciatic nerve. These investigators were able to pick up a potential from the environment of a nerve. With one electrode on the nerve and the other at a distance from it, potentials from 100 to 400 μv were obtained. When both electrodes touched the nerve a potential of 550 μv was obtained. When one electrode was 1 mm. from the nerve, the potential was 345 μv. This value gradually dropped until at 6 mm. the potential was 200 μv., and at 12 mm. it was 150 μv. The fact that a potential is obtained at so great a distance causes many physiologists to be skeptical of the whole experiment, and to credit biological fields to pure speculation.

It is true that nowhere are artefacts more abundant and misleading than in experiments on the electrokinetics of nerve tissues. But this difficulty does not prove that protoplasm is devoid of an electrical environment which it has itself established. That nerve impulses are electrical in nature is generally accepted.

The presence of an electrostatic field should not, therefore, be difficult to comprehend. The problem thus becomes not so much the reality of a field, but its nature, its magnitude, and the question, have we measured it?

I have (1953) given an interpretation of the forces responsible for protoplasmic streaming, and had recourse to fields such as a physical chemist postulates (DEBYE 1929). We have been so conditioned by years of biochemical kinetics, based on molecular reactions, that physical interpretations are frowned upon. We search for more enzymes, more chemical pathways, and forget entirely that the physical forces between and surrounding particles play a tremendous role in cellular kinetics.

There is one further point to remember, namely, that the electrical properties of one kind of tissue, *e.g.*, nerve, are not characteristic of it alone. The action potentials so typical of the nerve fiber are found in muscle, and in plants such as *Nitella* and *Mimosa* (MOLISCH 1929, DIANNELIDIS and UMRATH 1953), and will in all probability be found in a strand of primitive protoplasm such as that of a myxomycete, in spite of reports to the contrary (TAUC 1953). In fact, although the evidence is not conclusive, we already have published accounts of action potentials from myxomycetes (TASAKI and KAMIYA 1950).

Literature.

ADAM, N. K.: Physics and chemistry of surfaces. Oxford 1930. — ADRIAN, E. D.: Physical mechanism of nervous action. Philadelphia 1952. — AMBERSON, W. R.: On the mechanism of the production of electromotive forces in living tissues. Cold Spring Harbor Symp. Quant. Biol. 4, 53 (1936).

BARNES, T. C.: Textbook of general physiology. Philadelphia 1937. — BARTELL, F. E., and C. K. SLOANE: An interferometric investigation of adsorption. J. Amer. Chem. Soc. **51**, 1637 (1929). — BEMMELEN, J. M. VAN: Die verschiedenen Arten der Verwitterung der Silikatgesteine in der Erdrinde. Z. anorg. u. allg. Chem. **66**, 322 (1910). — BERGER, H.: Über das Elektrencephalogramm des Menschen. Arch. f. Psychiatr. **104**, 678 (1936). — BEUTNER, R.: Die physikalische Natur bioelektrischer Potentialnervendifferenzen. Biochem. Z. **47**, 73 (1912). — Bioelectricity, p. 35 in Medical Physics, ed. O. GLASSER. Chicago 1944. — BROOKS, S. C., and S. GELFAN: Bioelectric potentials in *Nitella*. Amer. J. Physiol. **85**, 356 (1928). — BULL, H. B.: Physical biochemistry. New York 1943. — BURR, H. S.: An electrometric study of *Mimosa*. Yale J. Biol. a. Med. **15**, 823 (1943). — Potential gradients in living systems, p. 1121 in Medical Physics, ed. O. GLASSER. Chicago 1944. — BURR, H. S., and A. MAURO: Electrostatic fields of the sciatic nerve in the frog. Yale J. Biol. a. Med. **21**, 455 (1949).

COATES, C. W., and R. T. COX: A comparison of length and voltage in the electric eel, *Electrophorus electricus*. Zoologica, New York Zool. Soc. **30**, 39 (1945). — COATES, C. W., R. T. COX and L. P. GRANATH: The electric discharge of the electric eel. Zoologica, New York Zool. Soc. **22**, 1 (1937). — COLLANDER, R.: Versuche zum Nachweis elektrolytischer Vorgänge bei der Plasmolyse. Pflügers Arch. **185**, 224 (1920). — COX, R. T., C. W. COATES and M. V. BROWN: Electrical characteristics of electric tissue. Ann. New York Acad. Sci. **47**, 187 (1946). — CZAPEK, F.: Über eine Methode zur direkten Bestimmung der Oberflächenspannung der Plasmahaut von Pflanzenzellen. Jena 1911.

DANIELLI, J. F., and N. K. ADAM: Surface films of ergosterol. Biochemic. J. **28**, 1583 (1934). — DEBYE, P.: Polar molecules. New York 1929. — DIANNELIDIS, T., and K. UMRATH: Über das elektrische Potential bei den Myxomyceten. Protoplasma (Wien) **42**, 312 (1953). — DUBOIS-REYMOND, E.: Untersuchungen über tierische Elektrizität. Berlin 1948. — DUJARDIN, F.: Recherches sur les organismes inférieurs. Ann. des Sci. natur., Sér. II, zool. **4**, 364 (1835). — DUTROCHET, H.: Sur la circulation des fluides chez le *Chara fragilis*. Ann. des Sci. natur., Sér. II, **9** (1838).

FREUNDLICH, H.: Adsorption, in Colloid chemistry, ed. J. ALEXANDER, Vol. 1, p. 796. 1926. — Kapillarchemie. Leipzig 1932. — FREY-WYSSLING, A.: The growth in surface of the plant cell. Growth Symp. **12**, 151 (1948). — Research (Lond.) **2**, 300 (1949). — FUJITA, A.: Untersuchungen über elektrische Erscheinungen und Ionendurchlässigkeit von Membranen. Biochem. Z. **158**, 11 (1925).

GIBBS, F. A., W. G. LENNOX and E. L. GIBBS: The electroencephalogram in diagnosis and in localization of epileptic seizures. Arch. of Neur. **36**, 1225 (1936). — GIBBS, W.: Collected works. New York 1928; also: Trans. Conn. Acad. **3**, 108 (1875); **6**, 343 (1878). — GOLD-

ACRE, R. J., and I. J. LORCH: Folding and unfolding of protein molecules in relation to cytoplasmic streaming. Nature (Lond.) **166**, 497 (1950). — GORTNER, R. A., et al.: Introduction to biochemistry. New York 1949.

HARKINS, W. D., E. C. DAVIES and G. L. CLARK: The orientation of molecules in surfaces. J. Amer. Chem. Soc. **39**, 354, 541 (1917). — HARKINS, W. D., and H. K. LIVINGSTON: Energy relations of the surfaces of solids. J. Chem. Phys. **10**, 342 (1942). — HÖBER, R.: Zur Theorie der bioelektrischen Ketten. Z. physik. Chem. **110**, 142 (1924).

ITERSON, G. VAN, and B. J. D. MEEUSE: The shape of cells in homogeneous plant tissues. Proc. Roy. Acad. Amsterdam **44**, 770 (1941).

JASPER, H. H., and H. L. ANDREWS: Electroencephalography. III. Normal differentiation between occipital and pre-central regions in man. Arch. of Neur. **39**, 96 (1937). — JENNINGS, H. S.: The behavior of lower organisms. Carnegie Instn. Publ. **16**, 129 (1904).

KAMIYA, N.: Physical aspects of protoplasmic streaming. The structure of protoplasm, ed. WILLIAM SEIFRIZ. Ames, Iowa 1942. — KAMIYA, N., and S. ABE: Bioelectric phenomena in the myxomycete plasmodium. J. Colloid. Sci. **5**, 149 (1950). — KATZ, B.: Electric excitation of nerve. London 1939.

LANGMUIR, I.: The constitution of solids and liquids. J. Amer. Chem. Soc. **39**, 1848 (1917). — LATIMER, W. M.: Prediction and speculation in chemistry (entropy). Chem. Eng. News **31**, 3366 (1953). — LILLIE, R. S.: Transmission of activation in passive metals. Science (Lancaster, Pa.) **48**, 51 (1918). — The recovery of transmissivity in passive iron wires. J. Gen. Physiol. **3**, 107 (1920). — LOEB, J.: Proteins and the theory of colloidal behavior. New York 1922. — LOEB, J., u. R. BEUTNER: Über die Potentialdifferenzen. Biochem. Z. **41**, 1 (1913). — LOEWY, A. G.: A theory of protoplasmic streaming. Proc. Amer. Phil. Soc. **33**, 326 (1949). — LONDON, F.: The general theory of molecular forces. London 1937. — LONGSWORTH, L. G.: Transference numbers of aqueous solutions. J. Amer. Chem. Soc. **54**, 2741 (1932); **57**, 1185 (1935). — LORAND, L.: Fibrin clots. Nature (Lond.) **166**, 694 (1950). — LUCKÉ, B., and M. MCCUTCHEON: The living cell as an osmotic system. Physiologic. Rev. **12**, 68 (1932). — LUND, E. J.: Electric correlations between living cells. Plant Physiol. **6**, 631 (1931). — Bioelectric fields and growth. Austin, Texas 1947.

MACINNES, D. A.: The principles of electro-chemistry. New York 1939. — MACINNES, D. A., and Y. L. YEH: The potentials at the junction of monovalent chloride solutions. J. Amer. Chem. Soc. **43**, 2563 (1921). — MAST, S. O.: Structure, movement, locomotion, and stimula in amoeba. J. of Morph. **41**, 347 (1926). — MCBAIN, J. W.: Colloid science. Boston (Mass.) 1950. — MCBAIN, J. W., and G. W. HUMPHREY: The microtome method of determining the absolute amount of adsorption. J. Physic. Chem. **36**, 300 (1932). — MEYER, K. H.: The state of aggregation of rubber. Chem. Rev. **25**, 137 (1939). — Protein and protoplasmic structure. The structure of protoplasm, ed. W. SEIFRIZ. Ames, Iowa 1942. — MICHAELIS, L.: The theory of permeability. J. Gen. Physiol. **8**, 33 (1925). — MICHAELIS, L., et al.: J. Gen. Physiol. **1927**. — MOLISCH, H.: The reflex arc in Mimosa. Nature (Lond.) **123**, 562 (1929). — MOYER, L.: On the surface composition of latex particles. Amer. J. Bot. **22**, 609 (1935). — MUDD, S.: A kinetic mechanism in interfaces. J. of Exper. Med. **40**, 633 (1924).

NÄGELI, C.: Sitzgsber. kgl. bayer. Akad. Wiss. München **2**, 114 (1864). — Die Micellartheorie, OSTWALD's Klassiker der exakten Wissenschaften, Nr 227 (ed. A. FREY). Leipzig 1928. — NERNST, W.: Göttinger Nachr., Math.-physik. Kl. **104** (1899). — Zur Theorie des elektrischen Reizes. Arch. ges. Physiol. **122**, 275 (1903). — NEWTON, R., and W. MARTIN: Physico-chemical studies on the nature of drought resistance in crop plants. Canad. J. Res. **3**, 336 (1930). — NUGENT, R. L.: The application of the interfacial technique in the study of protein films. J. Physic. Chem. **36**, 449 (1932).

POINCARÉ, H.: Science et hypothese. Paris 1918. — PROCTER, H. R., and J. A. WILSON: The acid-gelatin equilibrium. J. Chem. Soc. (Lond.) **1916**, 307.

RASHEVSKY, N.: Mathematical biophysics, p. 706 of Medical Physics, ed. O. GLASSER. Chicago 1944.

SCHMIDT, W. J.: Die Ergebnisse der NAEGELIschen Micellarlehre. Naturwiss. **16**, 900 (1928). — SEIFRIZ, W.: A theory of protoplasmic streaming. Science (Lancaster, Pa.) **86**, 397 (1937). — Swelling and shrinking of protoplasm. Trans. Faraday Soc. B **42**, 259 (1946). — (a) Mechanism of protoplasmic movement. Nature (Lond.) **171**, 1136 (1953). — (b) Le mécanisme des mouvements protoplasmiques. Rev. Cytol. et Biol. végét. **14**, 31 (1953).

TASAKI, I., and N. KAMIYA: Electrical response of a slime mold. Protoplasma (Berl.) **39**, 333 (1950). — TAUC, L.: Quelques observations de bioélectricité cellulaire. J. of Physiol. **45**, 232 (1953). — TRAUBE, J.: Theorie der Narkose. Pflügers Arch. **153**, 276 (1913).

VERWEY, E. J. W., and J. TH. G. OVERBEEK: Theory of the stability of lyophobic colloids. 1948.

WARBURG, O.: Physikalische Chemie der Zellatmung. Biochem. Z. **119**, 134 (1921). — WYK, A. J. A. VAN DER, and K. H. MEYER: Molecular processes during deformation of rubber-like elastic bodies. J. Polymer Sci. **2**, 583 (1947).

Hydration and cell physiology.

By

C. R. Stocking.

I. Water content of cells and tissues.

Because water is commonly the most abundant compound found in living organisms, its necessity for life reactions is self evident, but its significance in these reactions should not be ignored. The mere presence of large amounts of water in an active cell confers on its protoplasm certain physical properties. Thus variations in the water content of protoplasm are reflected in changes in the physical environment in which metabolic reactions occur. For example a decrease in hydration results in an increase in protoplasmic viscosity, an increase in osmotic pressure, and slower metabolism, until if dehydration proceeds too far irreversible changes result and death ensues. Active protoplasm seldom has a moisture content less than about 90 per cent.

The dipole character of the water molecule renders it peculiarly susceptible to orientation in an electrostatic field whether the field is around an ion, another dipole molecule, or at colloidal surfaces such as those found in cell walls or on protein micelles. These interrelations between water molecules and ions, lipoids, proteins, carbohydrates, and various other substances in the cell[1] are of major physiological significance. In the following discussion attention will be paid particularly to the distribution of water among the different parts of the cell and to the effects of variation in water content upon cellular processes.

The influences of environment, age, and type of plant in determining the water content of the various tissues and organs are discussed by Kramer[2] and Meyer[3]. Such wide differences in moisture occur among the tissues of a single plant and among similar tissues of different plants that significant generalizations are difficult. However, a few examples will serve to point out the range in moisture content of plant materials. The apical regions of the plant, where cell division and elongation are proceeding rapidly, characteristically are highly hydrated. Here cell wall material and storage products are at a minimum in proportion to the protoplasmic content of the dividing cells. In the region of cell elongation rapid vacuolation will be proceeding. This is true whether the tissue is stem or root, e.g. apical portions of barley roots were found to be 93 per cent water (Kramer and Wiebe 1952); growing corn stem tips likewise have 92 to 93 per cent moisture (Miller 1938). Similarly, active cambium is known to be highly hydrated.

Plant organs composed chiefly of vacuolated, thin walled parenchymatous tissue such as specialized leaves and certain fruits also have a high moisture content, e.g. inner lettuce leaves, 94.8 per cent; cabbage leaves, 92.4 per cent; tomato fruit, 94.1 per cent; and watermelon fruit, 92.1 per cent water (Chatfield and Adams 1940).

In contrast, if cell differentiation has proceeded with the formation of thick walls or if metabolism has resulted in the deposition of large quantities of storage

---

[1] See this volume, p. 584ff.: Levitt, J., Hydration.

[2] See vol. I, p. 194: Kramer, P. J., Water content and water turnover in plant cells.

[3] See vol. III, p. 596: Meyer, B. S., The hydrodynamic system.

products such as starch, then the amount of water held by the tissue, per unit of dry matter, will be low. For example, wood may have only from 40 to 60 per cent (GIBBS 1935, HUCKENPAHLER 1936, MCDERMOTT 1941) and potato tubers about 78 per cent moisture (CHATFIELD and ADAMS 1940). Even with this low average moisture content, it must be remembered that the protoplasm of the living cells in these tissues would still be about 90 per cent water, if the cells are active.

On the other hand, dormant tissues particularly those characteristics of seeds may have very low moisture contents ranging from 5 to 15 per cent. Here the cell protoplasts themselves are dehydrated, consequently metabolic activity is at a minimum, and only when the cells absorb more water will metabolism increase and growth proceed.

Diurnal variations in the moisture content of the various parts of the plant are a natural result of adjustment to periodic environmental changes. Such variations have been reported by numerous workers particularly for leaves (LLOYD 1913, HAWKINS 1927, STANESCU 1936). Recently a study has been made by WILSON et al. (1953) of the diurnal fluctuations in the moisture content of the roots, stems, and leaves of sunflower (*Helianthus annuus* L.) and *Amoranthus spinosus* L. In 7 week old plants, the leaves reached a minimum moisture content during the afternoon in response to a period of rapid transpiration and a maximum between midnight and 4 a.m. The moisture content of the roots and stems paralleled that for the leaves except that maxima were reached between 6 and 10 a.m. The magnitude of such fluctuations obviously will depend upon the particular environment as well as upon the kind of plant and tissue considered.

Seasonal variations in moisture content are also common in perennial plants.

II. Intracellular distribution of water.

Life is a dynamic condition. Not only are major changes in the environment reflected in measurable shifts in metabolism but on a molecular level equilibria are rarely attained. Continuously and unremittingly within the living cell molecules and ions are diffusing from one location to another. At one point solutes may be increasing through hydrolytic, synthetic, or absorptive processes while at another the activity of water may be increasing through its production in oxidation reactions, through the disappearance of solutes or as a result of other processes. Thus, continuing metabolism results in a constant change in the diffusion pressure of water within the cell. Water diffuses away from regions of increasing diffusion pressure toward regions where water molecules are less active. The basic tendency toward equilibrium of the diffusion pressure of water among the various parts of the cell though often approached from a macroscopic point of view is never reached at the molecular level. Often the osmotic concentration of the intracellular environment will influence the rate and course of a particular cellular reaction or process. One must therefore know the extent of hydration and the factors affecting the hydration of the cell wall, cytoplasm, nucleus, plastids, mitochondria, vacuole, and other cellular structures if he desires to understand the intra and intercellular movement of water and its role in cellular physiology.

A. Cell wall.

The cell wall, separating two adjacent cells, consists of the middle lamella on either side of which lie the thin primary walls of the contiguous cells. These walls are highly hydrated and are capable of growth and of reversible changes in surface area during turgor expansion of the cell. After growth in size is completed, the protoplasts of the cells may deposit on the inner surface of the primary wall a secondary wall. In certain cells this becomes greatly thickened and often is composed of several layers[1] (BAILEY 1938). The water retaining capacity of these walls is a function of their chemical and physical nature that changes during growth and maturity of the cell.

The middle lamella consists almost entirely of pectic substances and appears to be composed chiefly of calcium and magnesium polygalacturonides (BONNER 1950). These materials function as a cementing layer between the primary walls of adjacent cells. They are extremely hydrophilic colloids and because of their numerous free polar groups are capable of holding water through strong adsorptive forces. Similar pectic materials covering the outer wall surface of root hairs play a major role in the absorption of water and minerals[2] (FREY-WYSSLING and MÜHLETHALER 1949).

The bulk of evidence indicates that both the primary and secondary walls have a basic framework of interwoven cellulose fibrils making up a continuous system. Nevertheless as FREY-WYSSLING (1950) points out, the cellulose fibrils make up only a small part of the hydrated primary wall. It has been estimated that about 14 per cent of the wall volume of oat coleoptile parenchyma cells is taken up by cellulose (FREY-WYSSLING 1936). In cambium this value is perhaps 8 per cent (PRESTON and WARDROP 1949). Much of the interfibrilar space in the primary wall is filled with pectic materials, hemicellulose, and water (Table 1).

Table 1. *Dry matter composition of primary cell walls.* (PRESTON and WARDROP 1949.)

Specimen	Cellulose %	Pectin %	Hemicellulose %	Protein %
Conifer cambium (ALLSOPP and MISRA 1940)	25.1	16.6	ca. 6	20,8
Avena coleoptile (THIMANN and BONNER 1933)	42	8	38	12

According to KERR (1951) the elastic extensibility of the primary wall is a property of hydrophilic materials that can change in volume and in surface with changes in the water relations of the protoplast. KERR suggests that in the primary wall protopectin, anhydrogalacturonic acid, could well be responsible for such properties. Changes in hydration of the pectic gel would occur during stretching. Thus he visualizes the possibility of protopectin being the continuous phase with cellulose dispersed in it. EKDAHL (1953) also emphasized the role of pectic compounds in the elongation of root hairs. According to his concept, an increase in the degree of esterification of the anhydrogalacturonic acid residues (of the pectic material in the root hair, particularly at a distance from the tip) would reduce the hydrophilic nature of the wall. This would be a contributing factor in the decreased growth of the more basal parts of the wall. In contrast, CORMACK (1949) stresses the role of calcium in producing a tip to base gradient of calcification of the pectic layers of the root hair. In any event, cell wall growth is intimately associated with the high degree of hydrophily of the growing area.

Although the cellulose texture of the primary wall is not firm and inextensible but rather a network and of very loose structure, the weight of evidence, in contrast to KERR's ideas, points to a continuity of the cellulose fibrils (PRESTON, WARDROP and NICOLAI 1948, PRESTON and WARDROP 1949, FREY-WYSSLING

[1] See vol. I, p. 722: PRESTON, R. D., Microscopic structures of plant cell walls.
[2] See vol. III, p. 173: STOCKING, C. R., Histology and development of the root.

and MÜHLETHALER 1950, HOUWINK and ROELOFSEN 1954). The monographs by FREY-WYSSLING (1953) and PRESTON (1953) present detailed treatments of the subject. These recent advances in our knowledge of the submicroscopic structure of the wall help us in understanding the mechanism of hydration in this part of the cell.

A summary picture would be that the cellulose framework of the primary and of the secondary walls as well consists of long chains of cellulose associated into micelles which are arranged in both crystalline and amorphous regions. In the amorphous regions chains do not lie sufficiently close or parallel to result in strong bondings between them. It is in these amorphous regions that the bulk of the intramicellar water is found and where swelling and shrinking will occur with fluctuations in hydration. The micelles are further associated into fibrils that form a loose (in young primary walls) or a more rigid network (in secondary walls) with anastomosing interfibrilar microcapillaries.

The forces holding the glucose residues together in the chains are strong primary valence linkages between carbon and oxygen. Hydrogen bonds and weaker attraction of the OH dipoles and the permanent electric moment of the C—O—C groups bind the chains together in the micelles (MARK 1944). In the crystalline regions water cannot enter to cause a separation of the chains but the hydrophilic —OH groups that may be exposed on the surfaces of the micelles or in the amorphous regions where the chains are not closely associated may be expected to bind water (HERMANS and WEIDINGER 1946).

Only a small fraction of the water in the wall is adsorbed on the surface of the micelles. This is evident from the results of STAMM's (1936, 1944) studies on the adsorption of water vapor by cellulose. STAMM estimated that the internal surface of one gram of cellulose is between one and ten million square centimeters. Nevertheless, his studies indicated that only about one of every four —OH groups is effective in binding water. Presumably the rest are blocked off in the inter-associations of the cellulose chains or in the attraction with other intermicellar substances such as pectins, lignins, hemicelluloses, and gums. According to STAMM's calculations only about 8 per cent of the water held by the wet cellulose was surface adsorbed water. The bulk of the water found with cell walls must then be loosely held in the microcapillaries among the micelles and fibrils or associated with the highly hydrophilic hemicelluloses and pectic materials in the wall.

The strength with which water is held in the microcapillaries of the wall is a function of capillary size. Water is held relatively firmly in the smaller micro-capillaries that may range down to 1 mμ in diameter. On the other hand, the larger capillaries present enormous channels, in terms of the dimensions of water molecules, up to possibly 100 mμ in diameter, through which molecules and ions may move readily. When the wall is highly hydrated, these capillaries if not filled with non-cellulosic wall material contain loosely held water.

In the original state in which the cellulose is lain down in the wall, many of the polar groups would be surrounded by atmospheres of associated water molecules. During the drying and shrinking of the walls that occur in sorption studies such as those by STAMM, certain of these groups may become satisfied by other cellulose chains so that a hysteresis effect is observed on rewetting. This would be evident in a reduced water uptake by the wall.

The growing cell wall of a living cell is mostly water. Walls of cambial cells may shrink to $^1/_3$ of their original thickness upon drying (PRESTON and WARDROP 1949). Even the more mature cells of the phloem in potato stolons have been observed to decrease to 50 per cent of their size upon drying (Crafts 1931). The hydration of the wall of the living cell is affected by the nature of the infiltrating substances as well as by the presence of solutes in the wall, the state of hydration of the protoplasm that is in intimate contact with the wall, and the nature of the extracellular environment.

Analyses of actively growing cell walls yield high protein values (Table 1). Electron photomicrographs show that although there may be no definite connection between the bulk of the cytoplasm and the wall, in areas where growth is proceeding rapidly, definite protoplasmic connections are present. It appears that these local centers of cellulose formation and of merging of the protoplasm with the wall change with time (Frey-Wyssling 1950, Preston 1952). A particularly high degree of hydration would be expected in these local areas.

Infiltrating substances in the wall are not always hydrophilic. Certain cell walls are impregnated with fatty and waxy materials that reduce the permeability of the wall to water or render it impermeable. In addition to the more obvious cutinized epidermal cells of the aerial plant parts and to the suberized cork cells, an important condition exists in the walls of leaf mesophyll. Bangham and Lewis (1937) observed that the mesophyll walls in contact with the intercellular spaces in the leaves of *Ficus elastica* were not wet by water. On the other hand, non polar materials such as benzene, paraffin, and oils would spread on the surface of the walls. Since then other investigators have demonstrated the hydrophobic nature of the intercellular surfaces of mesophyll cells of other leaves (Frey-Wyssling and Häusermann 1941, Häusermann 1944, Lewis 1945, 1948). It is evident that an accumulation of hydrophobic material even in the outer intermicellar capillaries of the wall would greatly reduce the imbibitional properties of this portion of the wall and might be of primary significance in the reaction of the cell to the drying effects of the intercellular air.

B. Protoplasm.

It has already been stated that one of the most characteristic features of active protoplasm is its high water content that may range from 85 to 95 per cent of the fresh weight. Protein makes up from 7 to 10 per cent of the protoplasm while fatty substances, other organic molecules, and inorganic ions contribute from 1 to 2 per cent each. If we think of protoplasm in terms of the molecular environment in which metabolic reactions occur, we find on the average for every gigantic protein molecule almost 20,000 small water molecules and from 10 to 100 of the relatively small molecules or ions of each of the fatty, other organic, and inorganic materials (Sponsler and Bath 1942).

The properties of protoplasm are dependent on or related to the amount of water held and the type of forces holding this water. Protoplasm may exist in a fluid state or as a gel. Some of its properties are: contractility, elasticity, cohesiveness, rigidity, and tensile strength. Theories of protoplasmic structure must explain these facts, but as Bensley (1942) has pointed out non-living systems also may possess these properties. "Protoplasm on the contrary respires, excretes, performs complicated chemical operations, uses or liberates energy and reproduces its own substance in kind. This metabolism is mediated by a multitude of intracellular enzymes and carriers and can be imitated in part *in vitro* but in the cell is characterized by speed, orderliness and rhythm not to be found in a random mixture of chemical substances in solution" (Bensley 1942, p. 389—390).

Obviously a knowledge of the arrangement of the various molecules into the submicroscopic structure that is characteristic of protoplasm[1] is essential to an understanding of the role of hydration in cell physiology. For reviews of protoplasmic structure see SEIFRIZ (1942), VIRGIN (1953), and FREY-WYSSLING (1953).

1. Hyaloplasm. Proteins are hydrophilic colloids and the principal material in the protoplasm that is able to coordinate and bind water molecules. The affinity of proteins for water rests chiefly in the hydrophilic groups of the amino acid side chains located along the polypeptide backbone of the protein. The side chains may contain only non-polar hydrocarbon units which lack affinity for water or they may have radicles containing oxygen or nitrogen. These radicles, —OH, —NH_2, —COOH, and =NH are hydrophilic. For instance the —OH group on an organic molecule possesses three residual valences capable of coordinating three water molecules in the surrounding medium through hydrogen bonding (PAULING 1939). Similarly the —NH_2 group is capable of associating with three water molecules while the —COOH group may coordinate four or five. Although the =O and =NH groups of the backbones of the peptide chains can act as hydration centers, in the natural proteins they may be partially or entirely blocked because of steric hindrance. In their natural state the protein chains are probably folded or coiled into various shapes and sizes.

Streaming may be observed not only through the bulk of the hyaloplasm but near the surfaces of nuclei, plastids, and other cellular structures. This flow must involve a continual breaking and forming of many bonds. The activation energy for such streaming presumably comes from oxidative reactions.

Although theories have been advanced that water is not present as the dispersing medium (LEPESKIN 1950), experimental evidence, including microinjection experiments (CHAMBERS 1943), indicates that water forms a continuous phase in the hyaloplasm. At the same time, a loose molecular network is visualized consisting of polypeptide chains linked together by junctions of various nature: van der Waals forces, hydrogen bonds, salt and ester bridges, ether and peptide bridges etc. (FREY-WYSSLING 1953).

Not only is water held in the protoplasm by hydrophilic groups but the meshes produced by the polypeptide chains are of considerable size and in the hydrated protoplasm are capable of holding considerable water in a relatively free state. FREY-WYSSLING believes that during the slow decrease in water content accompanying the induction of dormancy, there is a gradual approach of the polypeptide chains. This would result in a masking of hydrophilic areas without changing the essential configuration of molecular structure. Other hydrophilic groups might become screened off by phosphatides and similar materials. He suggests that in this fashion the structural organization of the protoplasm that is necessary for life is maintained in dormant seeds and spores.

Water held by certain of the hydrophilic groups would be more firmly bound than that held by others. Since the water thus bound in the structure of the protoplasm is held by a variety of groups with different binding strengths, it is not surprising that various efforts to determine the amount of bound water in a cell yield different results[2] (GORTNER and GORTNER 1949, BULL 1951).

Although the principal hydrophilic colloids in protoplasm are proteins, protoplasm contains numerous other compounds with polar properties and whose presence may greatly alter protoplasmic hydration. The various ions within the cell not only are active osmotically but themselves are hydrated to varying

[1] See vol. I, p. 301: SEIFRIZ, W., Microscopic and submicroscopic strücture of cytoplasm.

[2] See vol. I, p. 233: KRAMER, P. J., Bound water.

degrees depending upon their ionic radii and the strength of their charge. If present in the protoplasm, the ions will be attracted to charged areas on the proteins and will affect the hydration of these spots. If the ion is small and carries a large amount of water, it may actually induce an increase in hydration of the protein as the ion moves in between adjacent molecules. On the other hand if the ion is large and consequently has a small degree of hydration or if it has a strong charge, it may cause a reduction in the hydration of the protein by decreasing the hydrophilic nature of the group to which it is attracted.

Lipid material, particularly phospholipids are especially important at protoplasmic surfaces where they tend to accumulate. A detailed consideration of the structure of cytoplasmic membranes is given elsewhere[1].

2. Protoplasmic membranes. The plasmalemma or outer cytoplasmic membrane represents the first major obstacle to the free diffusion of solutes into the cell. The characteristic properties of this surface depend upon the presence here of both hydrophilic and hydrophobic molecules. For example the antagonistic effects of potassium and calcium ions upon the penetration of polar substances into cells can be related to their effects on membrane hydration. Potassium ion with its small charge and large hydration shell tends to increase permeability; similarly it generally increases swelling and the hydration of proteins. On the other hand, the divalent calcium ion decreases permeability and likewise usually decreases swelling and hydration of pıoteins.

It should not be assumed that the plasma membranes of different species of plants or even that the membranes of different tissues of the same species are the same. CHOLNOKY (1952) concluded that the plasmalemma of *Senecio cruentus* contained proportionally more lipid than did that of *Calceolaria*. Sodium carbonate causes normal plasmolysis in desmids but does not cause it in angiosperms (HÖFLER 1952). From these observations HÖFLER concluded that major differences exist in the nature of the plasmalemma in different species. HÖFLER even questioned whether the lipid content of the plasmalemma is any greater than that of the rest of the cytoplasm. Nevertheless it is the selective nature of the protoplasmic interfaces that enables the cell to act as an osmometer.

3. Plastids, nuclei, mitochondria. Turning now form the generalized picture of the hyaloplasm and the protoplasmic membranes to a consideration of plastids, mitochondria, and nuclei, we find that each of these protoplasmic bodies is separated from the remainder of the cytoplasm by interfacial membranes (WEIER 1938, FARRANT et al. 1953, PALADE 1952, 1953, LEYON 1954). Consequently each is capable of acting as a micro-osmometer swelling or contracting if bathed by hypotonic or hypertonic solutions, and within the living cell each must be in dynamic equilibrium with its hyaloplasmic environment.

Much of our knowledge of the chemical nature of these cytoplasmic bodies has come from the numerous experiments in which they have been isolated from other cellular constituents. Generally these isolations involve disintegration of the cells, differential centrifugation, and collection of the cellular particulates. In any technique for the isolation of cellular particulates it is necessary to take into consideration the effect of hydration on their morphological characteristics. For example, because of the nature of the plastid membrane, if isolation is attempted by rupturing the cell in a hypotonic solution, water will diffuse into the plastid, causing it to swell. If distilled water is used, granulation followed by plasmoptysis of the plastids of certain species often results, e.g., *Trifolium*,

[1] See vol. I, p. 340: SEIFRIZ, W., The physical chemistry of cytoplasm.

spinach, oats (NEISH 1939, GALSTON 1943, ARONOFF 1946), while plastids of other species may resist plasmoptysis, e.g., tobacco (ARONOFF 1946). Hypertonic isolating media would be expected to cause a corresponding shrinkage and general loss in hydration of the plastids.

Investigations on mitochondria have been much more extensive for animal than for plant cells. Numerous investigations have shown that mitochondria also tend to swell and disintegrate in hypotonic solutions (HOERR 1943, LATIES 1953).

Plastids, nuclei, and perhaps mitochondria of at least some cells (red algal plastids see MCCLENDON and BLINKS 1952, nuclei of amphibian oocytes see CALLAN 1949) are permeable to such low molecular weight molecules as sucrose that are often used to maintain the osmotic environment of the isolated cellular components. In isolation experiments where isotonicity of the bathing medium is desired, it may become necessary to use molecules of high molecular weights such as polyethylene glycol, albumin, or dextrin (MCCLENDON 1954).

Chloroplasts contain proportionally less water per unit dry weight than does the surrounding cytoplasm. This is a result of the concentration of lipophilic materials within the plastid. Chloroplasts contain on a dry weight basis, between 20 to 40 per cent lipids, and 35 to 55 per cent proteins (RABINOWITCH 1945, GRANICK 1949). Although no accurate data are available, it has been estimated that approximately 50 per cent of the active chloroplast is water (MENKE 1938, RABINOWITCH 1945). The hydrophilic parts of the chlorophyll molecules as well as those lipids such as phospholipids are probably associated with plastid protein thus further reducing the capacity for hydration.

Under certain conditions chloroplasts are seen to contain wafer-like grana located throughout an optically uniform matrix, the stroma[1]. Frequently grana are hard to distinguish within plastids inside of the cell but appear during plastid isolation. This effect has been explained by GRANICK (1949) as follows. In uninjured chloroplasts the stroma has normally a high density and high refractive index. During isolation the matrix takes on water, the stroma swells and vacuolates spreading apart the grana. The refractive index of the stroma is lowered and the grana appear. The bulk of the pigment and lipid material appears to be located in the grana. Swelling of grana in water has not been demonstrated. It must also be remembered that during crushing and maceration of cells some proteins may undergo irreversible gelation (HOERR 1943).

Accurate data on the water content of plant mitochondria are likewise not available but BENSLEY (1942) reports that in spite of a relatively high lipid content (34 per cent on a dry weight basis) isolated liver mitochondria contained 82.5 per cent water. As BENSLEY points out, this figure is undoubtedly too high because of dilatancy during isolation.

C. Vacuole.

Of the three cellular phases, cell wall, protoplasm, and vacuole, the vacuole has the highest water content being composed of as much as 98 per cent water. Vacular sap contains a miscellaneous assortment of sugars, organic acids and salts, inorganic salts, pigments, etc.[2]. Both hydrophilic and hydrophobic colloidal substances may be found in the vacuole. The hydrophilic colloids most frequently found are proteins, tannins, gums, and mucilages. However, certain vacuoles may contain large quantities of pectin while in others colloidal substances of a fatty nature such as lipids, phosphatides, and phytosterols may be found (ZIRKLE 1937, GUILLIERMOND 1941).

[1] See vol. I, p. 507: GRANICK, S., Plastid structure etc.
[2] See vol. I, p. 614: PISEK, A., Chemie des Zellsaftes.

Meristematic as well as mature cells are generally vacuolated although at least in the primary meristems the vacuoles are smaller and more numerous than in the more mature cells. The shape and size of vacuoles seem to be conditioned by the activity of the cytoplasm. When the cytoplasm is inactive, they become spherical but during cyclosis they may be drawn out into rod shapes (ZIRKLE 1937).

Vacuoles are known to arise from pre-existing vacuoles although their *de novo* origin is also claimed (GUILLIERMOND 1941). In view of the structure of cytoplasm, it seems logical that vacuoles would arise through the process of de-mixing (FREY-WYSSLING 1953). Within the hyaloplasm there must be a certain dynamic equilibrium between the protein chains with their liophilic and hydrophilic groups on the one hand and the free lipids, water, and ions in the aqueous medium on the other hand. In such a system the constantly changing molecular environment may at times lead to the concentration of one of these components. If such a condition occurs the equilibrium would be destroyed. Similar molecules would come together. A phase boundary would develop between them and the rest of the cytoplasm. Such a phenomenon is known as de-mixing and according to GUILLIERMOND (1941) is responsible for the formation of vacuoles. In this case, hydrophilic colloids in the cytoplasm could orient the water and bring about de-mixing.

FREY-WYSSLING (1952) regards the vacuolar contents as material eliminated from the cytoplasm. Vacuoles would then be excretion areas in which all kinds of materials incompatable with the cytoplasm are stored. According to this concept the osmotic function of the vacuolar contents would be of secondary importance.

The vacuolar sap of most cells has been considered essentially a true solution. Many investigators have demonstrated the ability of cells to accumulate solutes and concentrate them in their vacuoles[1].

In dealing with cell water relations, one must take into consideration the complex nature of the vacuolar contents and the existence of phenomena associated with the colloidal state. In addition the possibility of water moving into the vacuole under electro-osmotic or metabolic forces as suggested by a number of authors can not be neglected (BONNER et al. 1953, BOGEN 1953)[2].

D. Water balance within cells.

Because the overall tendency is toward the establishment of a uniform diffusion pressure of water among the cell parts and the environment, it is possible and convenient to consider the mechanism of water balance in terms of diffusion pressure. If a difference in diffusion pressure is developed within a system, water will move from regions of higher to regions of lower diffusion pressure.

It was emphasized at the beginning of this section that the diffusion pressure of water within each morphological part of the cell is constantly changing as a result of cellular activity (cellular oxidations, absorption or loss of solutes and water, synthesis of cellular constituents). Likewise the diffusion pressure of water in the environment surrounding the cell is seldom constant. Thus, with reference to its water content, each part of the cell is in a dynamic state with water moving into it or out of it according to the continual shifts in the diffusion pressure of water.

If there is an alteration in the diffusion pressure of water in the cellular environment, whether the environment is a contiguous cell, intercellular air

[1] See this volume, p. 449: ROBERTSON, A. N., The mechanism of absorption.
[2] See this volume, p. 316: KRAMER, P. J., The uptake of water by plant cells.

space, or soil solution, this change will be reflected in a shift in the water balance of the cell. Movement of water into or out of the cell wall will be accompanied by corresponding shifts in the water content of the protoplasm and the vacuole. Each part becomes adjusted to the new situation with the tendency toward equilibration of the diffusion pressure throughout the cell. The relative water content of each part, however, may not be the same as it was before the shift was initiated. For example, if a cell were in dynamic equilibrium with its environment, the addition of acid would lower the diffusion pressure of water in the environment. The cell wall would then lose water. This in turn would set up a diffusion gradient so that the protoplasma and vacuole likewise would lose water. Actually, however, a rapid and disproportionally large decrease in vacuolar size may occur, *e.g.* vacuolar contraction in *Cerinthe.* KENDA and WEBER (1952) suggest that in this instance the internal shift in water balance among the cellular phases is a result of the effect of the absorbed acid upon the degree of hydration of pectic substances (pectic acid) in the cell sap.

Under certain conditions when a cell is bathed in a solution of a salt of an alkali metal or in a neutral red solution or when the cell is stimulated by other means, a transferance of water from the vacuole to the cytoplasm results. There is an increase in the proportion of cytoplasm to vacuole and the phenomena of cap plasmolysis[1] or vacuolar contraction are observed. Cap plasmolysis results when there is an accumulation in the protoplasm of ions that induce the protoplasm to swell. Such an accumulation may result if the plasmalemma is more permeable to ions in the bathing solution than is the vacuolar membrane. BOGEN (1951) found that cap plasmolysis would also occur when cells were treated with neutral red solutions made hypertonic with sucrose. He suggested as a working hypothesis that cap plasmolysis, vacuolar contraction and irritation plasmolysis are special cases of the same basic process.

Another instance of such a shift in water balance within the cell has been observed in the case of *Spirogyra* in which the chloroplasts readily contract and give up water to the rest of the cell. This may be induced experimentally through the action of certain salts or it may occur normally (OSTERHOUT 1945).

The actual distribution of water among the cell wall, protoplasm, and vacuole varies markedly with the kind and age of the cell. Although no accurate method has been devised by which the relative amounts of water in the various parts of the cell can be determined, some estimates have been made. The thin parietal layer of protoplasm of highly vacuolated storage parenchyma may occupy only 10 to 30 per cent of the total cell volume (CRAFTS, CURRIER, and STOCKING 1949). The cortical cells of apple twigs contain very bulky protoplasm that fills about half the cross-sectional area of each cell (SCARTH and LEVITT 1937). BUTLER (1953) estimated that 20 to 25 per cent of the water in young wheat roots was in the walls, 5 per cent in the cytoplasm, and the remaining 70 to 75 per cent in the vacuole.

III. Cell hydration and cell reactions.

A condition of water deficit in the plant, whether brought about by low soil moisture, high solute concentration, or desiccating atmospheric conditions greatly modifies transpiration, photosynthesis, assimilation, respiration, and reproduction. In order to understand these effects a more thorough study is necessary of the physical and biochemical changes taking place in the individual cells during a change in hydration.

[1] See this volume, p. 71: STADELMANN, E., Plasmolyse und Deplasmolyse.

A. Growth.

The gross effects of continued water deficit upon the growth of plants are well known. Interpretation of these results in terms of cellular response to specific levels of dehydration is complicated by the internal redistribution of water that occurs within a plant as it dries out[1]. Thus, older highly vacuolated cells often serve as reservoirs from which water is translocated into growing cells. If the volume of the mature cells exceeds that of the enlarging cells, the net result of a water deficit frequently is a decrease in the size of the organ even though young cells are enlarging and cell division is occurring.

Many examples are presented in the literature of the ability of young cells and meristematic tissue to maintain a high degree of hydration and to continue their cell enlargement even when older cells are shrinking. For example, young cotton bolls continue enlarging during periods in which the parent plant is severely wilted. Their ability to gain water from the mature tissues appears to be due to imbibitional rather than osmotic factors (ANDERSON and KERR 1943). Likewise the growth of young sunflower and tomato stems may continue while the older portions are shrinking (MACDOUGAL 1920, WILSON 1948).

One aspect of the problem of the effect of dehydration on cell growth was studied by BURSTRÖM (1953) in his observations of the growth of wheat roots. In this experiment the roots of wheat plants growing in culture solution were surrounded by mannitol solutions of varying concentrations. In hypotonic solutions roots grew to a final length that was independent of the external osmotic value. The hypotonic external solution merely reduced the elastic extension of the wall. This according to BURSTRÖM means that cell wall growth will proceed in apparently the same manner whether walls are expanded elastically or almost completely slackened.

In hypertonic solutions on the other hand, water absorption is a limiting factor for growth. At incipient plasmolysis the protoplast will be separating from the cell wall and cells cease growing. If active cell wall formation involves an association of the protoplast and the wall (page 26), this result is to be expected.

Important as such investigations are, they are not designed to yield information on the role of protoplasmic hydration in cell enlargement. In a moderately wilted plant, the protoplasts of the cell normally are not separated from the walls. Little information is available to indicate at what point in cellular hydration cell growth of higher plants ceases.

However, experiments have been carried out on the growth of fungi on substrates in equilibrium with air of varying moisture contents. SNOW (1948) found that each species of mold has a given range of humidity at which spores can germinate when placed on nutrient gelatin at 25° C. Young germ tubes require a higher humidity for their subsequent active healthy growth than is essential for the early stages of germination from the dormant spore. The reasons for the difference in ability to grow at different degrees of hydration of the cells is not fully understood. Certainly those cells that are able to grow in an atmosphere of 85 per cent relative humidity, equivalent to a DPD of about 214 atmospheres, must be capable of developing tremendous imbibitional and osmotic forces.

WALTER (1931, 1949) classified the microorganisms into three groups depending upon their ability to grow in air at different humidities. 1. Xerophytic species are those in which growth intensity is not markedly influenced by an external humidity from 100 to 95 per cent and growth declines and ceases between

[1] See vol. III, p. 596: MEYER, B. S., The hydrodynamic system.

90 and 85 per cent humidity. Examples are *Aspergillus* and *Penicillium*. 2. Mesophytic species grow at humidities above 97 to 98 per cent and their growth declines and is inhibited between 90 to 95 per cent. Examples are *Rhizopus*, *Phycomyces*, and yeasts. 3. For hydrophytic species the minimum humidity for growth is 95 per cent or higher and growth is affected at humidities below 100 per cent. Examples are *Merulius*, *Chaetocladium*, and numerous bacteria.

In addition to cell enlargement, growth involves cell divisions which also are influenced by protoplasmic hydration. Cell divisions during the reproductive phases of growth may be affected also. Investigations on meiosis in *Oenothera* showed that there was an inverse relationship between the osmotic pressure of the pressed sap of the inflorescences (topped shoots were allowed to dry) and the average frequency of chiasma formation (Kisch 1937, Oehlkers 1937, Huber 1937).

B. Resistance to heat, cold, and desiccation.

Evidence indicates that fundamentally similar processes occur when protoplasm develops a resistance to the destructive effects of heat, cold, or desiccation. Although it is generally agreed that during this "hardening" process changes in protoplasmic hydration occur, agreement does not exist on the mechanism or even on the direction of these changes (Levitt 1951). On the one hand, Stocker (1948) postulated that drought resistance is associated with a decrease in hydration of the protoplasm and a tightening of the binding of protoplasmic fibrils. Bogen (1948) stressed the importance of the specific position and order of hydration centers and areas of polarity on protein molecules in maintaining the structural integrity of the living protoplasm. He suggested that heat resistance is associated with a strengthening of protoplasmic structure though not necessarily with a decrease in total hydration. On the other hand, Scarth's, and Kessler and Ruhland's (see Levitt 1951) theory of frost resistance is based on an increase in protoplasmic hydrophily that is reflected in an increase in permeability to polar substances and a decrease in protoplasmic viscosity.

Detailed treatments of the problems of heat, cold, and drought resistance are found in later sections in this volume.

It has been postulated (see Samish 1954) that the induction of a rest period in buds is associated with an increase in the gel nature of the protoplasm, an increase in the proportion of bound water and a decrease in biochemical reactions.

C. Enzyme activity.

The effect of desiccation on biochemical reactions in the cell has been stressed by Maximov (1941). Enzymatic activity is greatly affected by the degree of cellular hydration. In general hydrolytic activity is high and synthetic processes decrease during periods of water deficit in the plant, however it is difficult to tie these general results to specific intracellular causes.

Catalase, reductase, and the hydrolytic power of phosphatase have all been reported to be high during water deficit in the plant (Golovina 1939, Sisakian and Kobiakova 1940) while the net rate of protein formation from amino acids decreases (Petrie and Wood 1938). Numerous investigations have been carried out on the acceleration of starch dissolution as the water content of leaves decreases (Vassiliev and Vassiliev 1936, Golovina 1939, Spoehr and Milner 1939, Wadleigh, Gouch, and Davies 1943). The exact explanation for this accelerated hydrolysis of starch is not clear but apparently it is a result of increased enzyme activity rather than a p_H shift (SPOEHR and MILNER 1939).

Many of the observations of the effects of dehydration on the rate of photosynthesis have been reviewed by RABINOWITCH (1945). Although water deficit in a higher plant often results in stomatal closure and thus indirectly influences photosynthesis, the rate may be affected by protoplasmic hydration in the photosynthesizing cells. The results of inducing dehydration by use of hypertonic solutions are not always the same as those caused by actual drying (CHRELASHVILI 1941). Investigations with hypertonic solutions, particularly with algae or aquatic plants such as *Elodea*, have shown that the colloidal state of the protoplasm greatly affects photosynthesis, perhaps more so than it does respiration (WALTER 1929, RABINOWITCH 1945).

The osmotic retardation of the rate of photosynthesis was observed by GREENFIELD (1942) only when *Chlorella* plants were in strong light. This would indicate that the hypertonic sucrose solution in which the alga was placed induced a shift in enzymatic activity involving the "dark reactions" of photosynthesis.

Stomata free land plants such as mosses, ferns, and lichens are able to become adapted to the effects of dehydration. For example, RIED (1953) found that lichens growing in a relatively dry habitat showed little change in respiration and photosynthesis upon dehydration to 1 per cent of their normal water content. On the other hand lichens grown near water showed a large increase in respiration over photosynthesis when reduced to 6 per cent of their normal water content while they never recovered if dehydrated to a lower moisture level.

D. Mitochondrial action.

Results of experiments with isolated plastids and mitochondria while not directly applicable to the natural intercellular conditions affecting these protoplasmic bodies, give us some insight into possible mechanisms whereby a changing water content and consequently a changing osmotic environment affects cellular reactions. It has already been pointed out (page 28) that plastids and mitochondria act as osmotic systems. Fundamental alterations in their biochemical properties often follow on osmotically induced physical changes in their structure. Numerous observations particularly on animal mitochondria (GREEN 1951) but with plant mitochondria as well (LATIES 1953) indicate that their remarkable biochemical organization rests upon a precise physical integrity.

It is well known that certain enzyme functions are lost when cellular particles are exposed to a hypotonic environment during isolation procedures. This may result from a spatial disorientation of the components of the mitochondrial complex (GREEN 1951, LATIES 1953) or, in isolated mitochondrial preparations, the tonicity of the environment may also affect the activity of cofactors and of degradative enzymes, LATIES has shown that both the oxidative and phosphorylative capacities of the isolated mitochondria are adversely affected by low tonicity of the surrounding medium.

Fewer studies have been made on the osmotic relationships of mitochondria within intact tissues. Mitochondria in animal tissue slices (liver) swell and vesiculate rapidly when the tissues are placed in water (OPIE 1948, ZOLLINGER 1948). BUVAT (1953) observed major structural changes in the mitochondria (chondriosomes) of *Cichorium intybus* root cells when the cells were immersed in water. The granules of filamentous mitochondria appeared to come together during the immersion and joined at their extremities. These changes are reversible. The cells become adapted to immersion in pure water. Active streaming of the protoplasm which had ceased is resumed and the mitochondria may become vesicular but slowly return to normal. From these observations it appears that immersion of the parenchyma cells of *Cichorium intybus* in water causes a reduction in protoplasmic streaming, and an increased hydration of the mitochondria so that they join together in branching structures. This particular sensitivity to water is attributed by BUVAT to the presence in the mitochondria of phospholipids that possess hydrophilic as well as hydrophobic properties.

LATIES (1954) stresses the importance of the effect of the environment upon the structure and biochemical characteristics of mitochondria *in situ*. In physiological experiments in which the test material is washed in water for long periods of time before the experimental treatment such effects would be particularly significant. As an example, LATIES points to the washing of tissue disks as a preliminary to studies of salt accumulation and the possibility that such washings may affect respiration and salt accumulation through hydration effects on the mitochondrial system in the cells.

Literature.

ABELE, K.: Über die Volumenabnahme des Zellkernes in der Plasmolyse und über das Zustandekommen der Kernplasmarelation. Protoplasma (Wien) **40**, 324—327 (1951). — ACKLEY, W. B.: Seasonal and diurnal changes in the water contents and water deficits of Bartlett Pear Leaves. Plant Physiol. **29**, 445—447 (1954). — ALLSOPP, A., and P. MISRA: The constitution of the cambium, the new wood and the mature sapwood of the common ash, the common elm and the Scotch pine. Biochemic. J. **34**, 1078—1084 (1940). — ANDERSON, D. B., and T. KERR: A note on the growth and behavior of cotton bolls. Plant Physiol. **18**, 261—269 (1943). — ARONOFF, S.: Photochemical reduction by chloroplast grana. Plant Physiol. **21**, 393—409 (1946).

BAILY, I. W.: Cell wall structure of higher plants. Industr. Engin. Chem. **30**, 40—47 (1938). — BANGHAM, D. H., and F. J. LEWIS: Wettability of the cellulose walls of mesophyll in the leaf. Nature (Lond.) **139**, 1107—1108 (1937). — BENSLEY, R. R.: Chemical structure of cytoplasm. Science (Lancaster, Pa.) **96**, 389—393 (1942). — BOGEN, H. J.: Über Kappenplasmolyse und Vakuolenkontraktion. I. Die Wirkung von LiCl und Neutralrot und ihre Abhängigkeit von der Konzentration und dem osmotischen Wert in der Außenlösung. Planta (Berl.) **39**, 1—35 (1951). — Untersuchungen über Hitzetod und Hitzeresistenz pflanzlicher Protoplaste. Planta (Berl.) **36**, 298—340 (1948). — Beiträge zur Physiologie der nichtosmotischen Wasseraufnahme. Planta (Berl.) **42**, 104—155 (1953). — BONNER, J.: Plant Biochemistry. New York: Academic Press Inc. 1950. — BONNER, J., R. S. BANDURSKI and A. MILLERD: Linkage of respiration to auxin induced water uptake. Physiol. Plantarum (Copenh.) **6**, 511—522 (1953). — BULL, H. B.: Physical biochemistry. New York: John Wiley a. Sons, Inc. 1951. — BURSTRÖM, H.: Studies on growth and metabolism of roots. IX. Cell elongation and water absorption. Physiol. Plantarum (Copenh.) **6**, 262—276 (1953). — BUTLER, G. W.: Ion uptake by young wheat plants. I. Time course of absorption of potassium and chloride ions. Physiol. Plantarum (Copenh.) **6**, 594—616 (1953). — BUVAT, R.: Morphological changes in chondriosomes. Endeavour **12**, 33—37 (1953).

CALLAN, H. G.: A general account of experimental work on amphibian oocyte nuclei. Symposia Soc. Exper. Biol. **6**, 242—255 (1952). — CHAMBERS, R.: Electrolytic solutions compatible with the maintenance of protoplasmic structures. Biol. Symp. **10**, 91—109 (1943). — CHATFIELD, C., and G. ADAMS: Proximate composition of American food materials. U. S. Dept. Agric. Circ. **549**, 1—91 (1940). — CHOLNOKY, B. J.: Beobachtungen über die Wirkung der Kalilauge auf das Protoplasma. Protoplasma (Wien) **41**, 57—68 (1952). — CHRELASHVILI, M. N.: The influence of water content and carbohydrate accumulation on the energy of photosynthesis and respiration. Trudy Bot. Inst. Acad. Sci. USSR., Ser. IV. (Exper. Bot.) **5**, 101—137, engl. sum. (1941). — CORMACK, R. G. H.: The development of root hairs in angiosperms. Bot. Review **15**, 583—612 (1949). — CRAFTS, A. S.: Movement of organic materials in plants. Plant Physiol. **6**, 1—42 (1931). — CRAFTS, A. S., H. B. CURRIER and C. R. STOCKING: Water in the physiology of the plant. Waltham, Mass.: Chronica Botanica Co. 1949.

EKDAHL, I.: Studies on the growth and the osmotic conditions of root hairs. Symbolae bot. Upsaliensis **11** (6), 1—83 (1953).

FARRANT, J. L., R. N. ROBERTSON and M. J. WILKINS: The mitochondrial membrane. Nature (Lond.) **171**, 401—402 (1953). — FREY-WYSSLING, A.: Der Aufbau der pflanzlichen

Zellwände. Protoplasma (Berl.) **25**, 261—300 (1936). — Physiology of cell wall growth. Ann. Rev. Plant Physiol. **1**, 169—182 (1950). — Submicroscopic morphology of protoplasm. New York: Elsevier Publ. Co. 1953. — FREY-WYSSLING, A., u. E. HÄUSERMANN: Über die Auskleidung der Mesophyllinterzellularen. Ber. schweiz. bot. Ges. **51**, 430 (1941). — FREY-WYSSLING, A., u. K. MÜHLETHALER: Über den Feinbau der Zellwand von Wurzelhaaren. Mikroskopie (Wien) **4**, 257—266 (1949). — Bau und Funktion der Wurzelhaare. Schweiz. landwirt. Mh. **28**, 212—219 (1950).

GALSTON, A. W.: The isolation, agglutination and nitrogen analysis of intact oat chloroplasts. Amer. J. Bot. **30**, 331—334 (1943). — GIBBS, R. D.: Studies of wood. II. The water content of certain Canadian trees, and changes in the water-gas system during seasoning and flotation. Canad. J. Res. **12**, 727—760 (1935). — GOLOVINA, A. S.: The effect of drought on the biochemical processes of plants. Voronegh. Gosundarst. Univ. Nanch. Roboty Studentov. **1939**, 51—58. — GORTNER, R. A., and W. A. GORTNER: Outlines of biochemistry. New York: John Wiley a. Sons, Inc. 1949. — GRANICK, S.: The chloroplasts: their structure, composition, and development. In: FRANCK and LOOMIS (Editors) Photosynthesis in Plants. Ames, Iowa: Iowa State College Press 1949. — GREEN, D. E.: The cyclophorase complex of enzymes. Biol. Rev. **26**, 410—455 (1951). — GREENFIELD, S. S.: Inhibitory effects of inorganic compounds on photosynthesis in *Chlorella*. Amer. J. Bot. **29**, 121—131 (1942). — GUILLIERMOND, A.: The cytoplasm of the plant cell. Waltham, Mass.: Chronica Botanica Co. 1941.

HÄUSERMANN, E.: The amount of wetting of mesophyllic intercellular spaces. Ber. schweiz. bot. Ges. **54**, 544—589 (1944). — HAWKINS, R. S.: Variations of water and dry matter in the leaves of Pima and Acala cotton. Arizona Stat. Tech. Bull. **17**, 417—444 (1927).— HERMANS, P. H., and A. WEIDINGER: The hydrates of cellulose. J. Colloid Sci. **1**, 185—193 (1946). — HÖFLER, K.: Zur Kenntnis der Plasmahautschichten. Ber. dtsch. bot. Ges. **65**, 391—399 (1952). — HOERR, N. L.: Methods of isolation of morphological constituents of the liver cell. Biol. Symp. **10**, 185—232 (1943). — HOUWINK, A. L., and P. A. ROELOFSEN: Fibrillar architecture of growing plant cell walls. Acta bot. neerl. **3**, 385—395 (1954). — HUBER, B.: Wasserumsatz und Stoffbewegungen. Fortschr. Bot. **7**, 197—207 (1937). — HUCKENPAHLER, B. J.: Amount and distribution of moisture in a living shortleaf pine. J. Forestry **34**, 399—401 (1936).

KENDA, G., u. F. WEBER: Rasche Vakuolen-Kontraction in *Cerinthe*-Blütenzelle. Protoplasma (Wien) **41**, 458—466 (1952). — KERR, T.: Growth and structure of the primary wall. In: F. SKOOG (Editor), Plant Growth Substances. Madison, Wisconsin: Univ. of Wisconsin Press 1951. — KISCH, R.: Die Bedeutung der Wasserversorgung für den Ablauf der Meiosis. Jb. wiss. Bot. **85**, 450—484 (1937). — KRAMER, P. J., and H. H. WIEBE: Longitudinal gradients of P^{32} absorption in roots. Plant Physiol. **27**, 661—674 (1952).

LATIES, G. G.: The physical environment and oxidative and phosphorylative capacities of the higher plant mitochondria. Plant Physiol. **28**, 557—575 (1953). — The osmotic inactivation *in situ* of plant mitochondrial enzymes. J. of Exper. Bot. **5**, 49—70 (1954). — LEPESCHKIN, W. W.: Über die Struktur und den molekularen Bau der lebenden Materie. Protoplasma (Wien) **39**, 222—243 (1950). — LEVITT, J.: Frost, drought, and heat resistance. Ann. Rev. Plant Physiol. **2**, 245—268 (1951). — LEWIS, F. J.: Physical conditions of the surface of the mesophyll cell walls of the leaf. Nature (Lond.) **156**, 407—490 (1945). — Water movements in leaves. Discuss. Faraday Soc. **3**, 159—162 (1948). — LEYON, H.: The structure of chloroplasts. IV. The development and structure of the *Aspidistra* chloroplast. Exper. Cell Res. **7**, 265—273 (1954). — LLOYD, F. E.: Leaf-water, and stomatal movement in *Gossypium* and a method of direct visual observation of stomata *in situ*. Bull. Torrey Bot. Club **40**, 1—14 (1913).

MACDOUGAL, D. T.: Hydration and growth. Carnegie Instn. Washington Publ. No 297 **1920**. — MARK, H.: Cellulose: physical evidence regarding its constitution. In: L. E. WISE, Wood Chemistry, p. 103—136. New York: Reinhold 1944. — MAXIMOV, N. A.: The influence of drought on the physiological processes in plants, p. 299—309. Collection of papers on plant physiology in memory of K. A. TIMIRYAZEV. Acad. Sci. USSR, Inst. Pl. Physiol. in the name of K. A. TIMIRYAZEV 1941. — MCCLENDON, J. H.: The physical environment of chloroplasts as related to their morphology and activity in vitro. Plant Physiol. **29**, 448—457 (1954). — MCCLENDON, J. H., and L. R. BLINKS: Use of high molecular weight solutes in the study of isolated intracellular structures. Nature (Lond.) **170**, 557 (1952). — MCDERMOTT, J. J.: The effect of the method of cutting on the moisture content of samples from tree branches. Amer. J. Bot. **28**, 506—508 (1941). — MENKE, W.: Untersuchungen über das Protoplasma grüner Pflanzenzellen der Chloroplasten aus Spinatblättern. Hoppe-Seylers Z. **263**, 100—103 (1940). — MILLER, E. C.: Plant physiology. New York: McGraw-Hill Co. 1938.

NEISH, A. C.: Studies on chloroplasts. I. Separation of chloroplasts, a study of factors affecting their flocculation and the calculation of the chloroplast content of leaf tissue from chemical analysis. Biochemic. J. **33**, 293—299 (1939).

OEHLKERS, F.: Neue Versuche über zytologisch-genetische Probleme. Biol. Zbl. **57**, 126—149 (1937). — OPIE, E. L.: An osmotic system within the cytoplasm of cells. J. of Exper. Med. **87**, 425—444 (1948). — OSTERHOUT, W. J. V.: Water relations in the cell. J. Gen. Physiol. **29**, 73—78 (1945).

PALADE, G. E.: The fine structure of mitochondria. Anat. Rec. **114**, 427—452 (1952). — An electron microscope study of the mitochondrial structure. J. Histochem. a. Cytochem. **1**, 188—211 (1953). — PAULING, L.: The nature of the chemical bond. Ithaca, N. Y.: Cornell Univ. Press 1939. — PETRIE, A. H. K., and J. G. WOOD: Studies on the nitrogen metabolism of plants. I. The relation between the content of proteins, amino-acids, and water in the leaves. Ann. of Bot., N. S. **2**, 33—60 (1938). — PRESTON, R. D.: Biological units of cellulose structure. Symposia Soc. Exper. Biol. **6**, 348—357 (1952). — The molecular architecture of plant cell walls. London: Chapman & Hall 1953. — PRESTON, R. D., and A. B. WARDROP: The submicroscopic organization of the walls of conifer cambium. Biochim. et Biophysica Acta **3**, 549—559 (1949). — PRESTON, R. D., A. B. WARDROP and E. NICOLAI: Fine structure of cell walls in fresh plant tissues. Nature (Lond.) **162**, 957 (1948).

RABINOWITCH, E. I.: Photosynthesis, Vol. I. New York: Interscience Publ. Inc. 1945. — RIED, A.: Photosynthese und Atmung bei xerostabilen und xerolabilen Krustenflechten in der Nachwirkung vorausgegangener Entquellungen. Planta (Berl.) **41**, 436—438 (1953).

SAMISH, R. M.: Dormancy in woody plants. Ann. Rev. Plant Physiol. **5**, 183—204 (1954). — SCARTH, G. W., and J. LEVITT: The frost-hardening mechanism of plant cells. Plant Physiol. **12**, 51-78 (1937). — SEIFRIZ, W.: The structure of protoplasm. Ames, Iowa: State College Press 1942. — SISSAKIAN, N. M., i A. KOBIAKOVA: The behavior of enzymes as an index of drought resistance in crop plants. IV. The effect of wilting upon the trend of esterification and hydrolysis of phosphoric esters in plants. Biochimija **5**, 225—233 (1940). — SNOW, D.: The germination of mold spores at controlled humidities. Ann. Appl. Biol. **36**, 1—13 (1948). — SPOEHR, H. A., and H. W. MILNER: Starch solution and amylolytic activity in leaves. Proc. Amer. Phil. Soc. **81**, 37—78 (1939). — SPONSLER, O. L., and J. D. BATH: Molecular structure in protoplasm. In: W. SEIFRIZ (Editor), The structure of protoplasm. Ames, Iowa: Iowa State College Press 1942. — STAMM, A. J.: Colloid chemistry of cellulosic materials. U. S. Dept. Agric. Misc. Publ. **240**, 1—90 (1936). — Surface properties of cellulosic materials. In: L. E. WISE (Editor), Wood Chemistry, p. 449—550. New York: Reinhold & Co. 1944. — STANESCU, P. P.: Daily variations in products of photosynthesis, water content, and acidity of leaves toward end of vegetation period. Amer. J. Bot. **23**, 374—379 (1936). — STOCKER, O.: Beiträge zu einer Theorie der Dürreresistenz. Planta (Berl.) **35**, 445—465 (1948).

THIMANN, K. V., and J. BONNER: The mechanism of the action of growth substances of plant. Proc. Roy. Soc. Lond., Ser. B **113**, 126—149 (1933).

VASSILIEV, I. M., and M. G. VASSILIEV: Changes in carbohydrate content of wheat plants during the process of hardening for drought resistance. Plant Physiol. **11**, 115—125 (1936). — VIRGIN, H. I.: Physical properties of protoplasm. Ann. Rev. Plant Physiol. **4**, 363—382 (1953).

WADLEIGH, C. H., H. G. GAUCH and V. DAVIES: The trend of starch reserves in bean plants before and after irrigation of a saline soil. Proc. Amer. Soc. Horticult. Sci. **43**, 201—209 (1943). — WALTER, H.: Plasmaquellung und Assimilation. Protoplasma (Berl.) **6**, 113—156 (1929). — Die Hydratur der Pflanze. Jena: Gustav Fischer 1931. — Grundlagen des Pflanzenlebens, 3. Aufl., Bd. I. Die Hydratur und ihre Bedeutung. Stuttgart: Eugen Ulmer 1949. — WEIER, T. E.: The structure of the chloroplast. Bot. Review **4**, 497—530 (1938). — WILSON, C. C.: Diurnal fluctuations in growth in length of tomato stem. Plant Physiol. **23**, 156—157 (1948). — WILSON, C. C., W. R. BOGGESS and P. J. KRAMER: Diurnal fluctuations in the moisture content of some herbaceous plants. Amer. J. Bot. **40**, 97—100 (1953).

ZIRKLE, C.: The plant vacuole. Bot. Review **3**, 1—30 (1937). — ZOLLINGER, H. V.: Cytologic studies with the phase microscope. II. The mitochondria and other cytoplasmic constituents under various experimental conditions. Amer. J. Path. **24**, 569—589 (1948).

Wall and turgor pressure and tension. Diffusion pressure deficit or suction force.

By

B. S. Meyer.

With 4 figures.

I. Wall and turgor pressure and tension.

The cell-to-cell movement of water in plants, insofar as it is motivated by osmotic and imbibitional mechanisms, can be adequately interpreted only in terms of the three physically distinct quantities of osmotic pressure (OP), turgor pressure (TP), and diffusion pressure deficit (DPD). No one of these three quantities alone adequately characterizes the water relations of a plant cell insofar as they can be interpreted on the basis of a diffusional mechanism. The magnitude of all three of them must be known for the complete evaluation of the water relations of a cell.

1. Turgor pressure.

When a vacuolate plant cell, which is in a flaccid condition, is immersed in water or a hypotonic solution, water moves into the cell by osmosis. As a result of the passage of water into the cell a pressure develops which prevails throughout the cell sap and is exerted against the protoplasm and, in turn, against the cell wall. This actual pressure which develops within such a cell is called *turgor pressure*. This term can be applied with equal appropriateness to the actual pressures developing in purely physical osmotic systems such, for example, as a sucrose solution enclosed within a collodion or cellophane membrane and immersed in water. As will become clear in the following discussion, TPs in osmotic systems can sometimes be negative in value. The term turgor pressure can also be applied to pressures developing as a result of imbibition occuring in closed systems, a phenomenon which occurs in some kinds of plant cells.

The term *osmotic pressure* is preferably not used to designate the actual pressures which develop in osmotic systems, as this only leads to confusion; this term is best employed as an index to certain properties of a solution (MEYER 1945). For example, an aqueous solution of sucrose standing in a beaker may be characterized as having an OP of 15 atm. at 25° C. Obviously this solution is not exerting an actual pressure in the osmotic sense. This index, however, characterizes the solution in two ways.

In the first place this index is an evaluation of the *potential* maximum TP which will develop in a solution if it is permitted to come to equilibrium with pure water in a perfect osmotic system at 25° C. The necessary "ideal conditions" which must prevail if maximum TP is to be attained are that the solution must be confined within a membrane permeable only to the solvent, that the membrane be immersed in pure solvent and that development of a pressure equilibrium must be attained without any appreciable dilution of the solution.

In the second place the OP of an aqueous solution is an index which indicates quantitatively the amount by which the diffusion pressure of the water in the solution is less than that of pure water insofar as this is a result of the presence of solutes. In other words the OP of a solution is an index of the magnitude of the *diffusion pressure deficit* (see later) of the water in the solution insofar as this results from the presence of solutes.

A given solution has an unique OP at a given temperature, but its TP may vary considerably depending upon the circumstances under which it is placed. In many plant cells the TP may be increasing while the OP is diminishing.

No satisfactory methods exist for measuring directly the TP of most kinds of plant cells. When the OP and DPD of a cell are known its TP can be calculated from the equation interrelating these three quantities (see later).

A proposal by BURSTRÖM (1948) to employ the term TP in a radically different sense from the long established one has been drastically criticized by HUBER (1948), BROYER (1950), KRAMER and CURRIER (1950), THODAY (1952), and HYGEN and KJENNERUD (1952). Although his redefinition of TP cannot be accepted, BURSTRÖM'S proposal is of significance in certain other connections which will be discussed later.

2. Wall pressure.

When a TP prevails within a cell it is also exerted against the interior of the walls of the cell. This results in the development of a *wall pressure* which is commonly pictured as being exerted against the cell contents. When the TP is negative, that is, when the cell sap is under tension (see later), the WP is likewise negative, the wall being subjected to a centripetal pull because of its adhesion to the protoplasmic layer and cell sap mass within the cell.

When, as is often the situation in plant tissue, a given cell is subjected to compression by surrounding cells, the pressure thus exerted increases the WP of the cell, and hence also its turgor pressure. Such a cell thereby acquires a lower DPD (see later) than it would possess were it isolated from the tissue in which it is located.

On the basis of fundamental physical principles it is generally considered that the WP and TP of a cell are equal in magnitude and oppositely directed. This situation is not difficult to visualize for a cell in an equilibrium condition. For dynamic states in which water is being gained or lost by a cell, however, the validity of this assumption has been questioned. The basis for this doubt lies in the commonly made assumption that TP is the "driving force" causing the cell to distend when water is being absorbed. This viewpoint has led to much confusion and controversy, because it is difficult to visualize how cell distention can occur under the impact of TP when the WP is considered to be an opposing force which is equal in magnitude.

A number of investigators, including BURSTRÖM (1948), BROYER (1950), ALGEUS (1951), and HYGEN and KJENNERUD (1952) have argued that, in the dynamic state, the TP must at least slightly exceed the WP if cell distention is to occur. Others, including THODAY (1950), SPANNER (1952), WEATHERLEY (1952), and HAINES (1953), subscribe in one form or another to the view that TP and WP must always be equal in the dynamic as well as the static condition.

These contentions appear to be unnecessary as they are apparently based on a misconception. The author is inclined to the view, essentially that of THODAY (1952) and HAINES (1953) that the driving force of cell expansion is to be looked for in the DPD gradient (see later) between the cell and the system from which water is moving, be the latter a solution, a soil, or another cell. This appears also to have been the concept of BURSTRÖM (1948) who unfortunately confused the issue by using the term TP, well established in another sense, as a designation for the DPD gradient between a cell and its environment.

Both TP and cell distention are resultants of the entry of water rather than the first being the cause of the second. Wall pressure is also a resultant of the entry of water and from one point of view may be regarded as just another expression of the turgor pressure. Significant in this connection is the fact that the DPD gradient is greatest at the start of cell distention and has fallen to a zero value when distention ceases.

A consideration of limiting cases may help to clarify this viewpoint. If a cell wall is totally inelastic, influx of water, under such circumstances infinitesmal in amount, results in the development of TP but not in cell enlargement. If on the other hand, we could visualize a cell with an indefinitely extensible wall, an admittedly hypothetical situation, influx of water would result in cell enlargement but no turgor pressure.

3. Tension.

If water is enclosed in a cylinder fitted with a leakproof piston it can be subjected to a pressure by means of the piston. If, instead of a pressure, a pull is exerted on the piston, and if the interior walls of the cylinder and piston are perfectly wetted, the water passes into a state of tension which has the physical dimensions of a negative pressure.

Development of a tension or negative pressure within an enclosed mass of water is possible only because of the mutual attraction ("cohesion") of water molecules and because of the attraction ("adhesion") between the water molecules and the wall of the confining vessel. It is the former of these two physical properties which prevents rupture of the water mass when it is subjected to a tension; it is the latter of these two physical properties which prevents the tearing away of water from the encompassing wall when it is subjected to a tension.

The magnitude of the cohesion of water has been calculated from other physical data in several independent ways; most such calculations yield theoretical values in excess of 10,000 atm. (*cf.* Crafts *et al.* 1949, p. 155). On the basis of a somewhat different analysis of the structure of liquids than the classical one as made by Fürth (1941) it can be calculated that the maximum tension which water will sustain is about 4000 atmospheres.

All experimentally determined values of the cohesion of water are much lower than those considered possible on any theoretical grounds. Relatively low values are obtained by actual measurement principally because the conditions prevailing under experimental conditions do not closely approach the ideal conditions for which computations of the cohesion of water are made.

Dixon (1914), using a capillary tube method first employed by Berthelot in 1850 obtained values for the tensile strength (cohesion) of water ranging up to 158 atm. and values for sap centrifuged from the branches of holly (*Ilex aquifolium* L.) ranging up to 207 atmospheres. In more recent years the magnitude of the tensile strength of water has been reinvestigated by a number of workers (Vincent and Simmonds 1943, Temperely and Chambers 1946, Temperely 1946, 1947, and Scott *et al.* 1948). The consensus of these workers is that the average experimentally determined value for the tensile strength of water as measured by this technique is about 30 atm. which is considerably less than previously accepted values. The higher values obtained by Dixon were attributed by more recent investigators to the false assumption that the pressure in the Berthelot tube at the sealing temperative is zero, whereas it is more nearly of the order of 100 atmospheres. This magnifies the calculated results by a factor of five to seven.

Turning our attention from purely physical systems to situations which prevail in plants it should first be noted that the cell sap in at least some kinds of plant cells can pass into a state of tension. This happens when the enclosed cell sap shrinks sufficiently in volume that the encompassing protoplasm and walls are pulled inwards as a result of adhesion between the walls and sap. The resulting counter pull exerted by the walls on the sap throws it into a state of tension. Visible evidence that the sap in plant cells may pass into a state of tension has been described by THODAY (1921), ENGMANN (1934) and others. These investigators have observed an inward folding or crinkling of the cell walls of the leaves of some species during permanent wilting, presumably as a result of the centripetal pull to which they are subjected by the shrinking volume of cell sap.

Only limited quantitative data are available on the magnitude of the tensions which can be generated inside of plant cells. Measurements made by a vapor pressure equilibrium method indicate that tensions of 300–350 atm. may be engendered in the water in the cells of the annulus of fern sporangia (RENNER

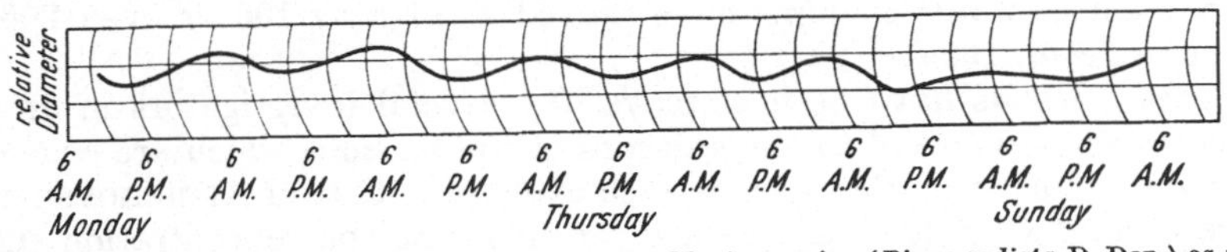

Fig. 1. Daily variations in the diameter of the trunk of a Monterey pine (*Pinus radiata* D. Don.) as measured with a dendrograph. (MACDOUGAL 1936.)

1915, URSPRUNG 1915). The results of CHU (1936) seemingly indicate that tensions of a very considerable magnitude develop in the cell sap of leaf cells both broad-and needle-leaved trees, especially under conditions of severe internal water deficiency.

The turgor pressure of all cells in which the water is under tension is negative, a fact which has important implications for the diffusion pressure deficit (see later) of such cells.

The best known example of the ocurrencce of tension in plants, however, is in the sap of the xylem ducts. Development of tension in the xylem sap is an integral feature of the well-known cohesion theory of the mechanism of the ascent of sap in plants (ASKENASY 1897, RENNER 1911, 1915, DIXON 1914).

Visible evidence that the sap in xylem vessels is often in a state of tension has been obtained by direct microscopic observation (HOLLE 1915, BODE 1923, CRAFTS 1939). The latter investigator, for example, found that the bark could be removed from the stem of *Ribes inerme* RYDB. plants in a state of permanent wilting without rupturing the water filaments in the outer xylem vessels. Tapping the exposed vessels caused immediate rupture of the water columns which was accompanied by a prompt change in the appearance of the tissue from a translucent water-filled to a glistening gas-filled condition.

Indirect evidence which is interpreted as indicating that the sap in the xylem ducts of woody stems is often in a state of tension has been obtained by means of the instrument known as a *dendrograph*. This device is a self-recording instrument with which variations in the diameter of tree trunks or other bulky plant organs can be measured. It is so constructed that its sensitivity to diameter changes is very great and that its recordings are not affected by direct temperature effects upon the instrument. Dendrographs have been used principally to measure periodic variations in the diameter growth of trees. Even in trees in which diameter growth has ceased, or virtually so, small diurnal changes in trunk diameter are of regular occurrence (Fig. 1). The trunk of the tree for which a dendrograph record is shown in this figure attained its minimum diameter during the afternoon hours, at a period when the water columns are undoubtedly under their maximum tension. While under tension the water columns become taut and decrease in diameter. Because of adhesion between water and the walls of the ducts a slight contraction occurs in their diameter. Such daily variations in the diameter of a tree trunk result from alternate contraction of the vessels or tracheids when the water in them is under tension, followed by their dilation when the tension is slackened. This interpretation of the cause of results such as those just described is supported by the direct observaons of BODE (1923) that a decrease of as much as several per cent may occur in the diameter of individual conducting elements when the water in them is in a state of tension.

While there is no doubt that tensions develop in the xylem sap of plants, no direct method of measuring their magnitude has ever been devised. It is commonly estimated that tensions frequently attain magnitudes of a few atmospheres in the xylem sap of herbaceous plants and a few tens of atmospheres in the xylem sap of woody plants. For certain extreme conditions estimates of the tension in water columns have ranged as high as 100 or even 200 atm. (MACDOUGAL *et al.* 1929).

Indirect methods have been employed by several investigators in an effort to arrive at an estimate of the magnitudes of the tensions which are engendered in the water columns of plants. One such attempt is that of ARCICHOVSKIJ *et al.* (1931) who used a "Schlieren" method for observing the concentration currents set up when the bared wood surface of a twig is brought into contact with sugar solutions of different concentrations. Water moves into or out of the cells depending upon whether the immersion solution has a lower or higher diffusion pressure deficit (see later) than the cells. The direction of water movement is ascertained by observing the pattern of the density currents set up in the solution by means of a refractometer. The solution in which there is neither gain nor loss of water is considered to have an osmotic pressure equal to the diffusion pressure deficit (in this case, essentially the tension) of the xylem tissue. Among other plants studied by this method was the central asiatic desert shrub *Arthophytum haloxylon* LITV., in which diffusion pressure deficits (tensions) as high as 142.9 atm. were recorded. A number of objections can be raised against this method, however, so there is considerable doubt as to how much reliance can be put on the results which have been obtained with it.

A somewhat similar, but probably more reliable—because only tissues which are essentially intact are tested—method was used by STOCKING (1945) on squash (*Cucurbita pepo* L.) plants. Sucrose solutions of known osmotic pressures were injected with a fine-pointed hollow needle into the hollow petioles of squash plants and the daily changes in solution concentration followed by withdrawing with the needle one or two drops of solution at intervals and measuring its concentration with a refractometer. The results indicate that the physical status of the water in the xylem of squash ranges from a positive pressure of about 1 atm. (root pressure) during the night to tensions of about 4 atm. on a warm summer day and about 9 atm. during wilting. This relatively low range of tensions is probably to be expected in an herbaceous plant such as squash.

II. Diffusion pressure deficit or suction force.

1. Diffusion pressure.

A concept basic to an understanding of all diffusional phenomena (of which osmosis and imbibition may be regarded as special cases) is that of *diffusion pressure*. In a gas diffusion pressure can be regarded as synonymous with partial pressure and diffusion occurs from regions of greater to regions of lesser diffusion pressure. Liquids can also be considered to possess a diffusion pressure (HALDANE 1918). The existence of such a physical property in a liquid becomes evident only under certain conditions, as, for example, when a solvent and a solution are separated by a differentially permeable membrane. Solutes may also be considered to possess a diffusion pressure. Diffusion pressure is the cause of diffusion, not its result. Just as gases have a pressure whether or not any diffusion is occurring, so also do liquids and solutes have a diffusion pressure whether or not any diffusion is occuring. Diffusion pressure is therefore that physical property of a substance which is responsible for its diffusion whenever other prevailing conditions permit the occurrence of this process. Diffusion of a substance occurs only if a difference in its diffusion pressures exists between two contiguous regions under such circumstances that molecules can move from one region to the other. Discussion of the diffusion pressure of liquids will be restricted to water and aqueous solutions, although the principles considered apply equally well to other liquids.

If pressure be imposed on water, as by a piston in a closed system, or by the confining effect of the walls of a closed osmometer or of a plant cell when the volume of the cell sap is expanding as a result of osmosis, the diffusion pressure of the confined water increases by the amount of the imposed pressure. If the imposed pressure is 13 atm., for example, the diffusion pressure of the water is increased by 13 atmospheres. Correspondingly, if water is subjected to a "negative pressure" (tension), its diffusion pressure decreases by the amount of the negative pressure. The effect of the imposed pressure on the diffusion pressure of the liquid is quantitatively the same whether the liquid is pure, or whether solutes are present. In a solution the diffusion pressure of any solute present is also changed by the amount of the imposed pressure.

If water is confined under such conditions that its volume cannot increase and its temperature is then raised, this is equivalent to putting it under pressure and its diffusion pressure increases. If free expansion of the water is possible, that is, if externally imposed pressure is kept constant instead of volume, increase in temperature has relatively little effect on the diffusion pressure of water. Effects of temperature upon the diffusion pressure and osmotic movement of water are discussed by CURTIS (1937) and URSPRUNG (1939). In general it appears that the effects of temperature on diffusion pressure and osmotic movement of water in plants are small, even when the temperature gradient is steep.

When a substance is dissolved in water the diffusion pressure of the water decreases as compared with that of pure water at the same pressure and temperature. In other words, dilution of the solvent with solute ions or molecules results in a reduction in the diffusion pressure of the solvent. Furthermore, the greater the proportion of solute particles present, the greater the reduction in the diffusion pressure of the solvent. In a solution, therefore, the diffusion pressure of the solvent may be influenced by the three factors of 1. temperature, 2. pressure, and 3. ratio of solute particles to solvent molecules. Ions, molecules and colloidal micelles all are equally effective in influencing the diffusion pressure of water, an important modifying influence in the magnitude of their effects, however, being the degree of hydration of the dissolved or dispersed particles.

In an imbibant such as a small chunk of agar or a seed with coats permeable to water the diffusion pressure of the water is influenced by still another factor—the attraction between the molecules of the water and the molecules of the imbibant. In general the drier the imbibant, the less the diffusion pressure of the water in it. In a "dry" imbibant the diffusion pressure of the water is virtually zero; in a saturated imbibant it is equal to that of pure water. At intermediate values, the greater the water content, the less the diffusion pressure deficit of the water in the imbibant, but the relation is not a linear one.

In any diffusional, osmotic, or imbibitional system the *direction* of diffusion is from the region or regions of greater diffusion pressure of the diffusing substance to the region or regions of lesser diffusion pressure. The *rate* of diffusion, on the other hand, is influenced by a number of factors including the steepness of the diffusion pressure gradient (see below), temperature, the volume and mass of the individual diffusing particles, and the properties of the medium through which diffusion takes place.

Diffusion of any substance, be it a gas, liquid, or solute, always occurs along a diffusion pressure gradient. The "steeper" this gradient, in general, the more rapidly diffusion takes place. The greater the difference in diffusion pressures of the diffusing substance at the two ends of the gradient, the steeper the gradient. Also, for a linear gradient, the shorter the length of a gradient, the steeper. The steepness of a linear gradient can be expressed quantitatively as the decrease in pressure per unit distance, for example, as diminution in partial pressure in atmospheres per centimeter. Many diffusion pressure gradients, however, are non-linear, *i.e.* the diminution in diffusion pressure per unit length is not constant for the entire gradient. For example, if a tube through which gas is diffusing has a larger diameter at one end than at the other, the gradient is steeper toward the end of the smaller bore. The diffusion pressure gradients of carbon dioxide, water-vapor, and oxygen through the stomates are non-linear.

2. Diffusion pressure deficit.

Interpretation of all diffusional phenomena, including osmosis and imbibition, on the basis of differences in diffusion pressure would be a logical and desirable procedure. This concept can be used in the direct sense in the analysis of diffusion phenomena in gases. It has not been found possible, however, to make exact measurements of the magnitudes of the diffusion pressures of liquids or solutes on any absolute scale. It is therefore unfeasible to use this quantity as such in the quantitative interpretation of diffusion phenomena in liquids and solutes. It is often possible, however, to measure quantitatively the amount by which the diffusion pressure of the water in a given system is less than that of pure water at the same temperature and under atmospheric pressure. This quantity, called the *diffusion pressure deficit* (MEYER 1938), can be used in the interpretation of osmotic and imbibitional phenomena in which water molecules participate, even if the exact magnitude of the diffusion pressure of water is unknown. In this indirect fashion, therefore, it is possible to use the concept of diffusion pressure in the analysis of diffusion phenomena in water and other liquids.

It has been pointed out previously that the diffusion pressure of water may be influenced by the temperature, but that under conditions prevailing in plants this effect is seldom of appreciable magnitude. The same statement may be made regarding the DPD of water. In the further discussion an isothermal system will be assumed. In such systems the diffusion pressure and hence the DPD of water are influenced only by the concentration of solutes present and by the pressure (positive or negative) prevailing in the water or solution.

As previously pointed out the OP of a solution, in the sense in which the term is used in this discussion, is a quantitative index of the DPD of the water in the solution insofar as this results from the presence of solutes.

Imposition of a positive pressure on water or a solution increases the diffusion pressure of the water and hence decreases its DPD by the amount of the imposed pressure. Subjection of water or a solution to a "negative pressure" (tension) decreases the diffusion pressure of the water and hence increases its DPD by the amount of the tension.

The fundamental relations among the important osmotic quantities are expressed by the equations:

$$\mathrm{DPD} = \mathrm{OP} - \mathrm{TP}$$

$$\mathrm{WP} = \mathrm{TP}$$

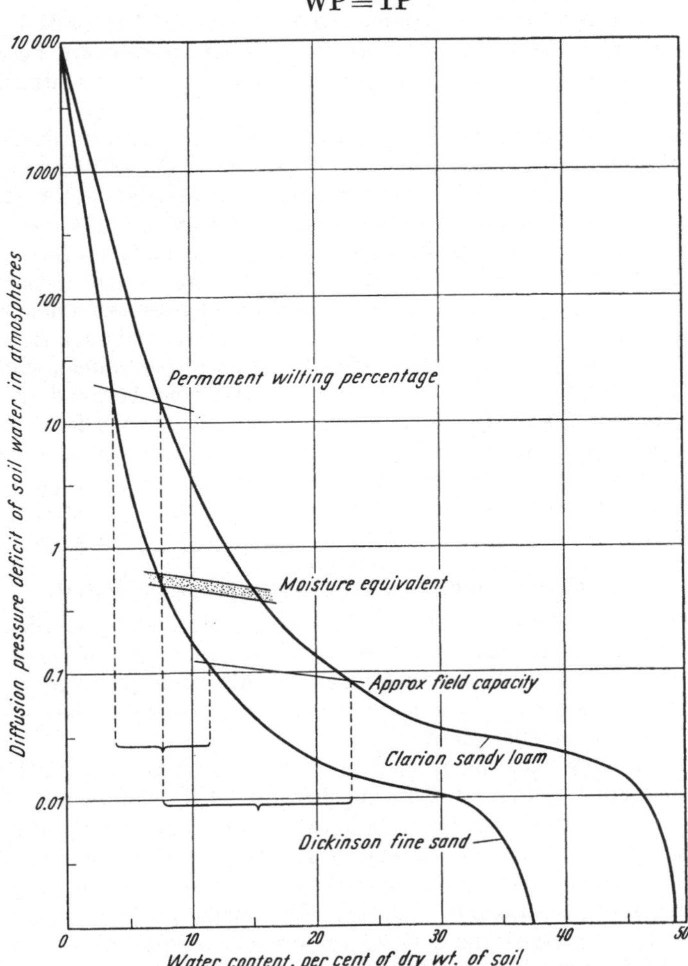

Fig. 2. Relation between the diffusion pressure deficit of the soil water and the soil water content in two soils over the entire soil water range. The horizontal brackets indicate the range over which water is readily available to plants. (RUSSELL 1939.)

For a solution freely exposed to the atmosphere it is obvious that DPD = OP.

For a closed osmometer or vacuolate plant cell with a solution or cell sap OP of 15 atm, and a TP of 7 atm., the DPD would be 8 atm. If a tension of 5 atm. prevailed within the solution or cell sap instead of a TP the DPD would be 20 atm., since the fundamental equation for such a situation is $\mathrm{DPD} = \mathrm{OP} - (-\mathrm{TP})$. In a completely flaccid plant cell or osmometer with a zero TP, DPD = OP; in a completely turgid osmotic system TP = OP, and the DPD is zero (Fig. 3).

In an imbibitional system the imbibition pressure of the imbibant is the analogue of the osmotic pressure in an osmotic system. Hence, in such a system:

$$DPD = IP - TP.$$

The IP is the index of the reduction in diffusion pressure of water in an imbibant insofar as this results from attractions between the molecules of the imbibant and water.

For a freely swelling imbibant immersed in water DPD initially equals IP, since there is no TP. The more nearly saturated such an imbibant becomes the smaller its IP, and hence its DPD. The osmotic analogue of this situation is an unconfined solution.

If an imbibant with an initial IP of 80 atm., is confined in a water-permeable, inelastic membrane and immersed in pure water, the DPD of the water in the imbibant when an equilibrium has been reached will be zero, not because of a shift in IP, but because at equilibrium the TP in such a system is 80 atmospheres.

The DPD of soil water has an important bearing on many problems in plant physiology. In general, the DPD of the water in a soil is low in a relatively moist soil (usually only a few tenths of an atmosphere in a soil at its field capacity, RICHARDS and WEAVER 1944) and increases progressively, although not in linear relation, with decrease in soil content (Fig. 2). At the permanent wilting percentage the DPD of the soil water is often in the neighborhood of 15 atm. (RICHARDS and WEAVER 1943). At high soil water contents the DPD of the soil water results principally from the solutes present; at low soil water contents it results principally from absorptive forces between the soil particles and the water. Several other factors influence the DPD of the soil water in a minor way as mentioned later.

3. Alternative concepts for diffusion pressure deficit.

The concepts to be discussed under this heading are, strictly speaking, alternatives for the concept of diffusion pressure rather than for diffusion pressure deficit. However, since diffusion pressure and diffusion pressure deficit are simple two different aspects of the same fundamental concept, they can appropriately be discussed at this point.

The term "activity" has been used by BROOKS and BROOKS (1941) with a significance analogous to that of diffusion pressure. Diffusion of any substance is considered to occur from regions of its greater to regions of its lesser activity. This term can be used without ambiguity for qualitative designations of the thermodynamic state of water or other substances, but it is not very suitable for quantitative interpretations of diffusion phenomena as they occur in plants.

Quantitatively expressed the *activity* (or *activity coefficient*) of a substance is the ratio of its fugacity in a given state to its fugacity in some standard state which is usually given a value of unity (LEWIS and RANDALL 1923). Fugacity is defined as the ideal or corrected vapor pressure of a substance *i.e.* its vapor pressure if its vapor behaved as an ideal vapor. In water at ordinary temperatures its vapor pressure is essentially identical in value with its fugacity. Hence the activity coefficient of the water in a solution is equal to the ratio of the vapor pressure of the water in the solution to the vapor pressure of pure water at the same temperature. For a given solution the activity coefficient is virtually constant over a wide range of temperatures.

Although the use of activity coefficients in the interpretation of cell-to-cell movements of water in plants is theoretically sound the concept suffers from several serious limitations. In systems which are not isothermal use of this unit may easily lead to misconceptions. Furthermore the working range of the activity scale usually encountered in plants is only a very small part of the entire scale. A difference of 10 atm. in osmotic pressure would usually represent a very steep gradient in a plant, yet the difference in activity coefficient between zero and 10 atm. is only 0.008 at 30° C. While the direction of the movement of water can be interpreted on the basis of differences in activity coefficients, the dimensions of this term are not such as to permit the evaluation of turgor phenomena in plant cells. The importance of being able to do this is discussed below.

The concept of "specific free energy", particularly as applied to analyses of soil moisture problems, has been developed especially by EDLEFSEN (1941) and EDLEFSEN and ANDERSON (1943). The specific free energy of the water at any point in a given system is considered to be the algebraic sum of the energies (see below for terminology) resulting from 1. the hydrostatic pressure the water is under, 2. the presence of dissolved substances, 3. the operation of a force field such as a gravitational or adsorptive field, and 4. the effect of a surface or interfacial tension. Movement of water occurs, whenever possible, in such a direction that its specific free energy is reduced. These investigators have suggested that it might prove desirable to extend this concept to the analysis of soil-plant water relations, to the problem of translocation of water through the plant, and to transpiration problems.

The quantity, "specific free energy" is neither conceptually nor dimensionally the same as diffusion pressure. In the CGS system specific free energy, as the term is used by EDLEFSEN, has the physical dimensions of ergs per gram, while diffusion pressure (or diffusion pressure deficit) has the dimensions of dynes per square centimeter. The two units are thus incommensurable.

While the quantity specific free energy appears to be well suited for thermodynamic interpretations of soil moisture relations, the desirability of extending this concept to plant water relations is dubious. The plant physiologist is often called upon to interpret not only problems involving movement of water in plants, but also, as in certain phases of growth, in the movement of plant organs, and in the behavior of guard cells, with the magnitude of turgor pressures developing in plant cells. Often both the direction of the movement of water and the magnitude of the turgor pressures in certain cells must be evaluated in order to obtain insight into the dynamics of the given physiological situation. Hence pressure units are better suited to the vocabulary of the plant physiologist than energy units, because both water movements and turgor phenomena as they occur in plants can be simultaneously described and interpreted in such units.

It would seem at least as appropriate to extend the diffusion pressure concept and terminology from the plant into the soil, as has been done by some workers, as to extend the concept and terminology of specific free energy from the soil into the plant.

A somewhat similar approach to that described above has been made by BROYER (1947) who has made a comprehensive analysis of the water relations of plant cells in terms of "(osmotic) specific free energies". Specific free energy is expressed as ergs per unit volume, which is dimensionally the same as a pressure; in fact osmotic units are used in the specific examples cited by this author. BROYER's fundamental equation for the movement of water is:

$$\mathrm{NIF} = \sum \mathrm{IF} - \sum \mathrm{EF}$$

in which NIF equals "net inflex specific free energy, $\sum$ IF equals sum of influx specific free energies, and $\sum$ EF, sum of efflux specific free energies". The analogy of this equation with the equation DPD = OP − TP should be apparent.

This section can be very reasonably terminated with a brief discussion of the advantages of the term "diffusion pressure deficit". This is a term which can logically be derived from an analysis of diffusion phenomena on the basis of diffusion pressures. It is a term which has the correct physical dimensions. It can be applied without qualification to solutions, imbibants, soils, and plant cells thus serving as a unifying concept which holds for all phases of the diffusional movement of water in relation to plants. Furthermore the phrase implies that water molecules are the active agents and includes no words with undesirable connotations such as "suction" (see next section).

One objection to diffusion pressure deficit—that it is an unwieldy term—can be met by using the convenient abbreviation DPD. The term has also been criticized on the grounds that it is a negative term, whereas this is one of the advantages of the phrase. It emphasizes the fact that the diffusion pressure of water in plants is usually less than that of pure water at the same temperature and under atmospheric pressure. By implication this indicates that the diffusional movement of water into and through plant tissues is at least largely a result of the "relative negativity" of the diffusion pressure of water in plant cells. Occurrence of negative diffusion pressure deficits in plants is rare; the only common example is when the sap in xylem cells or vessels is subjected to a "root pressure".

4. Diffusion pressure deficit ("suction force") of plant cells.

Dynamic considerations.—In this section a more detailed analysis will be made of the water relations of plant cells, especially in relation to the quantity designated as the diffusion pressure deficit (MEYER 1938). That some quantity equivalent to this term is necessary in the analysis of cell water relations was recognized even by some of the earliest workers in this field, such as DE VRIES (1884) and RENNER (1911, 1915).

The first consistent terminology and logical system of evaluating cell water relations which gained wide recognition was that proposed by URSPRUNG and BLUM (1916). The fundamental equation presented by these authors was:

$$S_z = S_i - W$$

in which S_z represents the suction force ("Saugkraft") of the cell, S_i the suction force of the cell contents and W the wall pressure of the cell. A very similar equation had previously been proposed by RENNER (1915). This equation and the concepts which it represents gained wide publicity through the extensive investigations of URSPRUNG and BLUM and others. For extensive bibliographies of this work papers by URSPRUNG and BLUM (1918), MOLZ (1926), BECK (1928), and URSPRUNG (1935, 1938) should be consulted.

Although the term suction force has been widely used, especially by European investigators, for the quantity termed the diffusion pressure deficit in this discussion, there are several rather serious criticisms which can be made of this term. The most fundamental is that its physical dimensions are incorrect. The physical dimensions of this quantity are such that it should be expressed in pressure units (force per unit area) and it should be named accordingly. The implications of the word "suction" are also undesirable; it is inconsistent with the accepted concept that the water molecules themselves are the active agents in any diffusional phenomenon in which they participate (SHULL 1927).

The development of this entire concept has been hampered by the diversity of the terms which have been used, and by confusions in their application. During the several decades following URSPRUNG and BLUM's pioneer work a number of other terms were proposed or used which had a meaning equivalent to that of diffusion pressure deficit or suction force. These include water absorbing power (THODAY 1918), suction pressure (STILES 1922), suction tension (BECK 1928), turgor deficit (CURTIS and SCOFIELD 1933), effective osmotic pressure (SHULL 1930), net osmotic pressure (SHULL 1939), and osmotic pressure (BROOKS 1940). All of these terms are subject to limitations or objections which have been discussed and analyzed by MEYER (1945). Certain more recent proposals regarding terminology are considered later in the discussion.

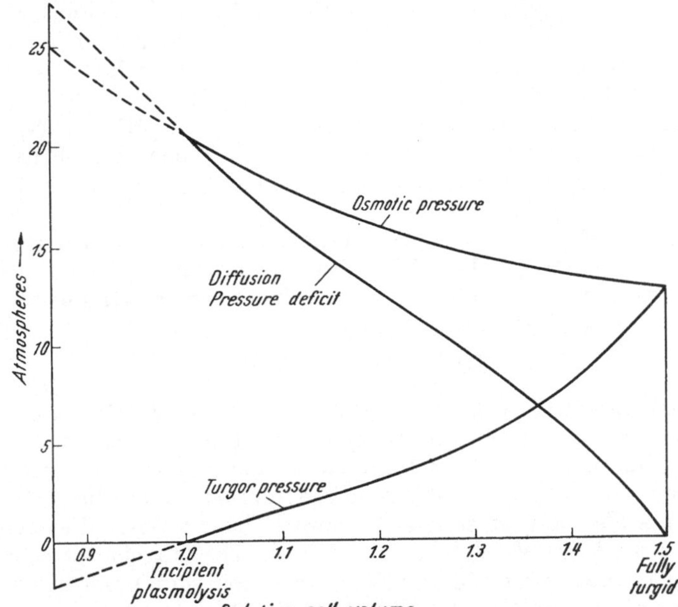

Fig. 3. Inter-relationships among the osmotic pressures, turgor pressures, diffusion pressure deficits, and volumes of a plant cell. (Based on data of HÖFLER 1920.)

The basic equations inter-relating the osmotic quantities of vacuolate plant cells are, as previously noted:

$$\mathrm{DPD} = \mathrm{OP} - \mathrm{TP}$$

$$\mathrm{TP} = \mathrm{WP}.$$

A complete consideration of the osmotic quantities of plant cells requires an evaluation of the influence of volume changes in the cells upon such quantities. Many kinds of plant cells have relatively elastic walls and in passing from the flaccid to the turgid condition undergo a considerable increase in volume. This is true, for example, of most parenchymatous cells, some of which increase 25 per cent or more in volume in shifting from a condition of zero to one of maximum turgidity. In many living cells the walls are considerably less elastic than this and in some, such as many of the cells in the tissues of xerophytes, they are virtually inelastic.

The inter-relationships among the osmotic pressures, turgor pressures, and diffusion pressure deficits of a plant cell are depicted in Fig. 3. In this diagram the influence of volume changes in the cell upon these three physical quantities is also taken into account. When the cell is completely flaccid (relative volume = 1.0) its DPD is equal to its OP at incipient plasmolysis ($O_i = 20$ atm.) while its TP is zero. As the volume of the cell increases as a result of an influx of water, the OP decreases because of a dilution of the cell sap until it reaches its value for the turgid (water-saturated) cell ($O_s = 15$ atm.).

The curve for the change in TP as shown in this figure is drawn somewhat arbitrarily, but apparently does follow such a course in at least some kinds of plant cells. HAINES (1950), has calculated on theoretical grounds that the relation between TP and cell extension is not linear (which has often been assumed) but hyperbolic. However, since the assumptions on which this calculation was based probably do not hold for all kinds of cells, and since the possible influence of surrounding cells on TP (see later) was disregarded, it seems unlikely that this relationship holds for all plant cells. Curves for the turgor pressure-cell volume relationship and discussions there of are to be found in contributions by HÖFLER (1920), ERNEST (1934), TAMIYA (1938), BROYER (1947), and CRAFTS *et al.* (1949).

The DPD, being equal to the difference between the OP und TP, decreases progressively as the turgidity of the cell increases. When the cell has attained a condition of maximum turgidity (relative volume = 1.5) its TP is equal to its OP and its DPD is zero.

As indicated by the dotted extension of the line for TP to the left, this quantity sometimes has a negative value. This situation prevails only when the water in the cell has passed into a state of tension. Under such conditions the equation for the DPD of a cell becomes:

$$\mathrm{DPD} = \mathrm{OP} - (-\mathrm{TP}).$$

In other words, the DPD of a cell in which the TP is negative is greater than its OP by the amount of the prevailing tension (negative pressure).

In some plant tissues many of the cells are under a pressure imposed upon them by surrounding cells. In addition to the internally prevailing TP the water in such a cell is also subject to this pressure. This pressure of external origin also influences the DPD of the water in the cell. The diffusion pressure of such a cell is less than that of the cell considered as an individual unit by the amount of this additional pressure.

Although the terminology and concepts discussed in the preceding paragraphs of this section have gained wide acceptance, modifications in both continue to be suggested. Some of these involve only the terms and symbols which are used (THODAY 1950, LEVITT 1951a, 1951b, SPANNER 1952, WALTER 1952, WEATHERLEY 1952) and will not be discussed in detail since fundamental concepts are not affected. Other suggestions which have been made do involve fundamental concepts. BURSTRÖM's (1948) proposed modification of the term turgor pressure falls into this category and is discussed in a previous section. The proposals of LEVITT (1951b) to introduce a term called the "osmotic potential difference" and of HYGEN and KJENNERUD (1952) to introduce a term called the "osmotic absorption potential" should also be mentioned. Actually both of these terms are designations for the difference in DPDs between a cell and its environmental medium, as is also the term "turgor pressure" as used unorthodoxly by BURSTRÖM (1948). In the opinion of this writer such terms constitute an unnecessary complication and an undesirable proliferation of nomenclature. The relatively simple, straightforward term DPD adequately characterizes the dynamic status of the water in a cell at all times, whether at equilibrium or not, and whether isolated or in contact with a solution or with other cells. Whether water will move into or out of a given cell, assuming water-permeable membranes, depends upon the DPD of that cell relative to the DPD of contiguous cells or of the medium with which the cell is in contact. It does not appear necessary to have a special term for such a difference in DPD values.

The preceding discussion refers primarily to vacuolate plant cells. In some kinds of cells an imbibitional rather than an osmotic mechanism accounts for the movement of water into a cell or from one cell to another. In such a system imbibition pressure (IP) should be substituted for osmotic pressure (OP) in the usual equation for diffusion pressure deficit. Passage of water into dry seeds from a moist or aqueous medium is the most familiar example of movement of water by an imbibitional mechanism. There is some evidence that water also moves into ripening ovules within a young fruit by imbibition (KERR and ANDERSON 1944) and many other examples of movement of water by an imbibitional mechanism in plants can be cited.

In recent years a number of investigators have claimed that metabolic mechanisms, in addition to basically diffusional ones, operate in the movement of water from cell to cell in plants, or from an outside environment into plant cells. It is an inherent assumption that the energy of respiration is utilized in the operation of any such mechanism. This subject is reviewed in considerable detail elsewhere in this volume and will receive only brief attention here.

Evidence for the existence of metabolic mechanisms of water movement has been adduced from various lines of experimentation. No unequivocal evidence of a direct relation between water movement and respiration has been discovered although correlations between rates of water movement or absorption and rates of respiration have been found to exist in some plant tissues. The relationship between these two processes may, however, be very indirect. That complex inter-relationships exist between water movement and other processes occurring in plant cells, and especially in meristematic cells, has long been recognized. STEWARD, STOUT and PRESTON (1940) have stressed the close inter-relationships of aerobic respiration, assimilation of salts, protein synthesis, and absorption of water. Actually even this is a vastly over-simplified picture of the metabolic complex operating in cells which is inter-related with the movement of water into and out of plant cells. Passage of water into cells, or from cell to cell may be linked to such processes as increased solute uptake, increased fabrication of protoplasm, changes in protoplasmic permeability or increased extensibility of the cell wall, all processes which appear to be dependent in some degree upon a respiratory source of energy.

While it is premature to deny categorically the existence of mechanisms of water movement which are directly activated by respiratory energy, it seems almost certain that the magnitude of any such mechanisms which may exist must be relatively small compared to the basically diffusional mechanisms of osmosis and imbibition. Any such mechanism which may be superimposed on the diffusional mechanisms would result in the addition of another component to the DPD of the cell.

It is not always possible to distinguish clearly among the various mechanisms of the movement of water in plant cells. Even when the possibility of a metabolic mechanism of the movement of water is disregarded a distinction between osmotic and imbibitional mechanisms cannot always be sharply made. In some kinds of cells the mass of protoplasm may be relatively large or the cell may be packed with colloidal materials, yet small vacuoles may be present. The relative roles of osmosis and imbibition in the water relations of such cells are difficult to assess, although it seems likely that the osmotic component will be determinative whenever vacuoles of any appreciable magnitude are present. In recognition of this fact the symbol OP is often taken to represent all of the forces which lead to absorption of water by a cell. This is also the sense in which the symbol S_i was used by URSPRUNG.

In accordance with accepted thermodynamic principles, movement of water from cell to cell in plants occurs along gradients of diffusion pressure deficits. If two adjacent cells differ in DPD, water moves from the cell of lower to the cell of higher diffusion pressure deficit. This may occur even if the OP of the cell into which the water moves is greater than that of the cell from which the water moves.

Longer gradients appear to exist in some plant tissues, in which DPDs increase progressively from cell to cell over a series of cells. What appears to be an example of such a gradient is described by KRAMER (1932) in the petiole of the tropical papaw (*Carica papaya* L.). It seems probable that such gradients also exist across the cortex of a root, along which water moves in passing from peripheral cells into the stele. Similar gradients probably also exist in the mesophyll of leaves, in floral parts, in fleshy parts of fruits and in other plant tissues. Movement of water from cell to cell across any tissue implies the existence of a DPD gradient across that tissue.

Magnitudes and variations of diffusion pressure deficits in plant cells.—The range of magnitudes of the DPDs in most plant cells corresponds in general with the range of magnitudes of the OPs of plant cells, since, barring the possibility of the development of tensions, the OP at cell saturation sets the upper limit to the magnitude of the DPD which can be engendered in any plant cell. In general the OPs of the cells of most mesic species of plants lie within a range of 5–40 atm., although lower values occur in some cells, and higher values, often much higher, prevail in the cells of many halophytes and xerophytes (*cf.* this Handbook, vol. III. In some kinds of cells, however, DPD values apparently may exceed OPs as a result of the development of tensions. The results of CHU (1936) appear to indicate that this situation may develop in the leaves of many trees under conditions of severe internal water deficiency.

Many data on specific values of the DPD ("suction force") of cells from many kinds of plant tissues are to be found in many of the papers cited under "Literature."

More or less regular diurnal variations occur in the DPDs of the cells of leaves and of other plant organs. In general, the diurnal variations in the DPDs of leaf cells are greater than the diurnal variations in their osmotic pressures. The reason for this should be obvious from an examination of Fig. 3; the possible range of DPDs in a plant cell is much greater than the possible range of OPs even if tensions do not develop. In cells in which tensions can be generated a still greater range of DPDs relative to the potential range of OPS is possible.

An extensive series of measurements on the daily variations in the DPD (suction force) of the leaf cells of lilac (*Syringa oblata* LINDL.) was made by LI (1929). In general the results obtained (Fig. 4) indicate that the DPDs of the leaf cells of this species increase from about 10–11 atm. in the early morning hours to as much as 17–18 atm. during the mid-afternoon hours of some days, thereafter declining to approximately the early morning values by midnight or not long thereafter. The magnitude of the diurnal fluctuation in DPDs varies with the environmental conditions, being greatest, as would be expected, on bright, clear days.

HERRICK (1933) made parallel measurements of the daily variations in DPDs ("suction tensions") and OPs of the leaf cells of a species of ragweed (*Ambrosia trifida* L.). In general, on hot, clear summer days the DPD of the leaf cells af this species increased from about 7 or 8 atm. in the early morning hours to about 17 atm. at 3 P.M., declining thereafter. Corresponding values for OPs of leaf cells were about 12–13 atm. in the early morning hours, and about 17 atm. in midafternoon, after which a decline set in. During the midafternoon hours the DPDs of the leaf cells were approximately equal to their OPs, and the leaves were in a state of incipient or temporary wilting.

The increase in DPDs of leaf cells which commonly takes place during the forepart of the daylight period results from the simultaneous occurrence of increasing OPs and decreasing TPs within the cells. Increase in OP results partly from an increase in the solute content of the cell sap as an outcome of photosynthesis and partly from a decrease in the water content of the cells. The decrease in TPs results from the shrinkage in volume of the water in the cells, which occurs as a result of the excess of transpiration over water absorption.

The most marked diurnal variations in the DPDs of plant cells, and especially of leaf cells, occur on clear warm days when an adequate soil water supply is

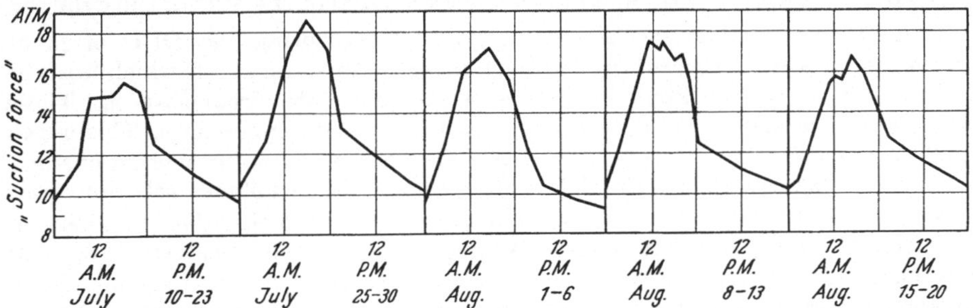

Fig. 4. Daily variations in the "suction force" of the leaves of lilac (*Syringa oblata* LINDL.). Each curve represents the average of the results for several days. (LI 1929.)

available, in other words, under conditions favorable to high rates of transpiration. On cool, cloudy days, or rainy days, DPDs are relatively low, and do not fluctuate very much diurnally. Under drought conditions, on the other hand, DPDs are relatively high, and remain so with relatively little change from one hour of the day to the next.

Measurement of the diffusion pressure deficits of plant cells.—Most methods of measuring the DPD of plant cells are based on the principle that if a cell, or group of cells, is immersed in a solution with an OP equal to the DPD of the cell or cells, a dynamic equilibrium is immediately established and therefore no change takes place in the volume or weight of the cell(s) because no net movement of water occurs. In solutions with a higher OP than the DPD of the cell or cells they will decrease in volume, in solutions of lower OP the cells will increase in volume. The earliest attempts to measure the DPD of plant cells were made upon individual cells (URSPRUNG and BLUM 1916). Without attempting to detail the exact procedure, the principle followed, in brief, was that of finding the OP of the solution in which no change in the volume of the cell occurred. This method is a tedious one, is subject to many sources of error (ERNEST 1931, OPPENHEIMER 1930, 1932, 1936), and has been almost totally superseded by other more convenient methods.

A procedure which has been widely used in measuring the DPD of plant cells is the "simplified method" of URSPRUNG (1923). In this method, as in most other methods currently used for measuring DPD, only the average value for the DPD of a large number of cells is obtained. Tissue strips are cut from such organs as leaves and petals, a common length being about 20 mm.; and width about 2 millimeters. Sharp, double-bladed knives are generally used for cutting uniform strips. The length of each strip is measured immediately under a microscope while immersed in parafin oil. Several strips are than immersed in each of a graded series of sucrose solutions until an equilibrium has been attained in each solution. The OP of the solution in which no change in length occurs is considered to be equal to the average DPD of the cells in the tissue. This has probably been the most widely used method of measuring the DPD of plant cells.

For measurements of the average diffusion pressure deficits of bulky tissues such as potato tubers, masses of tissue such as cylinders cut with a cork borer of suitable dimensions are often used. The equilibrium point can be determined by measuring changes in either the weight or the volume of the cylinder, both methods yielding similar results with potato tuber tissue (MEYER and WALLACE 1941). The solution in which the cylinder neither gains nor loses weight (or volume) is considered to have an OP equal to the average DPD of the cells in the cylinder.

A number of rather specialized methods of measuring the DPD of plant tissues have been described by various investigators, most of which can be applied only to certain kinds of tissues or under certain circumstances. These methods will be mentioned only briefly, with appropriate references. URSPRUNG and BLUM (1927, 1930) have described two methods based on measuring changes in thickness instead of length of leaves, applicable especially to sclerophyllous leaves. ARCICHOVSKIJ *et al.* (1931) described several distinctive methods of measuring the DPD of plant tissues. Among these was the "Schlieren" method previously mentioned in connection with measurements of the tensions which can develop in woody stems. The basic principle of this method consists in finding the point of equality in refractive index between a plant sap and a reference solution. The method used by STOCKING (1945), also mentioned previously, for measuring the DPD of tissues and by inference the tensions in the xylem, is based on the principle of following changes in the refractive index of a sucrose solution which is in equilibrium with the DPD of leaf petiole tissues. Several investigators (URSPRUNG and BLUM 1930, ARCICHOVSKIJ *et al.* 1931, CHU 1936) have employed a vapor pressure method of measuring the DPD of various plant tissues.

Literature.

ALGEUS, S.: Views on turgor pressure and wall pressure. Physiol. Plantarum **4**, 535–541 (1951). — ARCICHOVSKIJ, V., *et al.*: Untersuchungen über die Saugkraft der Pflanzen, I—V. Planta (Berl.) **14**, 517–565 (1931). — ASKENASY, E.: Beiträge zur Erklärung des Saftsteigens. Verh. naturhist.-med. Ver. Heidelberg **5**, 429–448 (1897).

BECK, W. A.: Osmotic pressure, osmotic value, and suction tension. Plant Physiol. **3**, 413–440 (1928). — BODE, H. R.: Beiträge zur Dynamik der Wasserbewegung in den Gefäßpflanzen. Jb. wiss. Bot. **62**, 92–127 (1923). — BROOKS, S. C.: The standardization of osmotic pressure as a term. Science (Lancaster, Pa.) **92**, 428–429 (1940). — BROOKS, S. C., and MATILDA M. BROOKS: The permeability of living cells. Berlin: Gebrüder Bornträger 1941. — BROYER, T. C.: The movement of materials into plants. Part I. Osmosis and the movement of water into plants. Bot. Review **13**, 1–58 (1947). — On the theoretical interpretation of turgor pressure. Plant Physiol. **25**, 135–139 (1950). — BURSTRÖM, H.: A theoretical interpretation of the turgor pressure. Physiol. Plantarum **1**, 57–64 (1948).

CHU, CHIEN-REN: Der Einfluß des Wassergehaltes der Blätter der Waldbäume auf ihre Lebensfähigkeit usw. Flora (Jena) **130**, 384–437 (1936). — CRAFTS, A. S.: Solute transport in plants. Science (Lancaster, Pa.) **90**, 337–338 (1939). — CRAFTS, A. S., H. B. CURRIER and C. R. STOCKING: Water in the physiology of plants. Waltham, Massachusetts: The Chronica Botanica Co. 1949. — CURTIS, O. F.: Vapor pressure gradients, water distribution in fruits and so-called infra-red injury. Amer. J. Bot. **24**, 705–710 (1937). — CURTIS, O. F., and H. T. SCOFIELD: A comparison of the osmotic concentrations of supplying and receiving tissues, and its bearing on the MÜNCH hypothesis of the translocation mechanism. Amer. J. Bot. **20**, 502–512 (1933).

DIXON, H. H.: Transpiration and the ascent of sap in plants. London: Macmillan & Co. 1914.

EDLEFSEN, N. E.: Some thermodynamic aspects of the use of soil moisture by plants. Trans. Amer. Geophysic. Union, part III **22**, 917–926 (1941). — EDLEFSEN, N. E., and A. B. C. ANDERSON: Thermodynamics of soil moisture. Hilgardia **15**, 31–298 (1943). — ENGMANN, K. F.: Studien über die Leistungsfähigkeit der Wassergewebe sukkulenter Pflanzen. Beih. bot. Zbl. A **52**, 381–414 (1934). — ERNEST, E. C. M.: Suction-pressure gradients and the measurement of suction pressure. Ann. of Bot. **45**, 717–731 (1931). — The water relations of the plant cell. J. Linnean Soc. Lond. Bot. **49**, 495–502 (1934).

FÜRTH, R.: On the theory of the liquid state. I. The statistical treatment of the thermodynamics of liquids by the theory of holes. Proc. Cambridge Philos. Soc. **37**, 252–275 (1941).

HAINES, F. M.: The relation between cell dimensions, osmotic pressure and turgor pressure. Ann. of Bot. **14**, 385–394 (1950). — An analysis of turgor and turgor pressure. Ann. of Bot. **17**, 629–640 (1953). — HALDANE, J. S.: The extension of the gas laws to liquids and solids. Biochemic. J. **12**, 464–498 (1918). — HERRICK, E. M.: Seasonal and diurnal variations in the osmotic values and suction tension values in the aerial portions of *Ambrosia trifida*. Amer. J. Bot. **20**, 18–34 (1933). — HÖFLER, K.: Ein Schema für die osmotische Leistung der Pflanzenzelle. Ber. dtsch. bot. Ges. **38**, 288–298 (1920). — HOLLE, H.: Untersuchungen über Welken, Vertrocknen und Wiederstraffwerden. Flora (Jena) **108**, 73–126 (1915). — HUBER, B.: Wasserumsatz und Stoffbewegungen. Fortschr. Bot. **12**, 185–215 (1948). — HYGEN, G., and JULIE KJENNERUD: Osmotic relations during cell expansion. Physiol. Plantarum **5**, 171–182 (1952).

KERR, T., and D. B. ANDERSON: Osmotic quantities in growing cotton bolls. Plant Physiol. **19**, 338–349 (1944). — KRAMER, P. J.: The absorption of water by root systems of plants. Amer. J. Bot. **19**, 148–164 (1932). — KRAMER, P. J., and H. B. CURRIER: Water relations of plant cells and tissues. Annual Rev. Plant Physiol. **1**, 265–284 (1950).

LEWIS, G. N., and M. RANDALL: Thermodynamics. New York: McGraw-Hill Book Co. 1923. — LEVITT, J.: Toward a clearer concept of osmotic quantities in plant cells. Science (Lancaster, Pa.) **113**, 228–231 (1951a). — The osmotic equivalent and osmotic potential difference of plant cells. Physiol. Plantarum **4**, 446–448 (1951b). — LI, TSI-TUNG: Effect of climatic factors on suction force. Quart. Rev. Biol. **4**, 401–414 (1929).

MACDOUGAL, D. T.: Studies in tree growth by the dendrographic method. Carnegie Instn. Wash. Publ. **1936**, No 462. — MACDOUGAL, D. T., J. B. OVERTON and G. M. SMITH: The hydrostatic-pneumatic system of certain trees; movements of liquids and gases. Carnegie Instn. Wash. Publ. **1929**, No 397. — MEYER, B. S.: The water relations of plant cells. Bot. Review **4**, 531–547 (1938). — A critical evaluation of the terminology of diffusion phenomena. Plant Physiol. **20**, 142–164 (1945). — MEYER, B. S., and A. M. WALLACE: A comparison of two methods of determining the diffusion pressure deficit of potato tuber tissue. Amer. J. Bot. **28**, 838–843 (1941). — MOLZ, F. J.: A study of suction force by the simplified method. I, II. Amer. J. Bot. **13**, 433–501 (1926).

OPPENHEIMER, H. R.: Kritische Betrachtungen zu den Saugkraftmessungen von URSPRUNG und BLUM. Ber. dtsch. bot. Ges. **48**, 130–140 (1930). — Untersuchungen zur Kritik der Saugkraftmessungen. Planta (Berl.) **18**, 525–549 (1932). — Remarks on two recent contributions concerning methods used in plant physiology. Palestine J. Bot. a. Hort. Sci. **1**, 84–93 (1936).

RENNER, O.: Experimentelle Beiträge zur Kenntnis der Wasserbewegung. Flora (Jena) **103**, 171–247 (1911). — Theoretisches und Experimentelles zur Kohäsionstheorie der Wasserbewegung. Jb. wiss. Bot. **56**, 617–667 (1915). — RICHARDS, L. A., and L. R. WEAVER: Fifteen-atmosphere percentage as related to the permanent wilting percentage. Soil Sci. **56**, 331–339 (1943). — Moisture retention by some irrigated soils as related to soil-moisture tension. J. Agricult. Res. **69**, 215–235 (1944). — RUSSELL, M. B.: Soil moisture sorption curves for four Iowa soils. Proc. Soil Sci. Soc. Amer. **69**, 51–54 (1939).

SCOTT, A. F., D. P. SHOEMAKER, K. N. TANNER and J. G. WENDEL: Study of the BERTHELOT method for determining the tensile strength of a liquid. J. Chem. Phys. **16**, 495–502 (1948). — SHULL, C. A.: Suction force of plant cells. Bot. Gaz. **83**, 213–214 (1927). — Absorption of water by plants and the forces involved. J. Amer. Soc. Agron. **22**, 459–471 (1930)— Atmospheric humidity and temperature in relation to the water system of plants and soils. Plant Physiol. **14**, 401–422 (1939). — SPANNER, D. C.: The dynamics of cell expansion by turgor. Ann. of Bot. **16**, 133–136 (1952). — STEWARD, F. C., P. R. STOUT and C. PRESTON: The balance sheet of metabolites for potato discs showing the effect of salts and dissolved oxygen on metabolism at 23° C. Plant Physiol. **15**, 409–447 (1940). — STILES, W.: The suction pressure of the plant cell. Biochemic. J. **16**, 727–728 (1922). — STOCKING, C. R.: The calculation of tensions in *Cucurbita pepo*. Amer. J. Bot. **32**, 126–134 (1945).

TAMIYA, H.: Zur Theorie der Turgordehnung und über den funktionellen Zusammenhang zwischen den einzelnen osmotischen Zustandsgrößen. Cytologia 8, 542–562 (1938). — TEMPERELY, H. N. V.: The behaviour of water under hydrostatic tension. II. Proc. Phys. Soc. London 58, 436–443 (1946). — The behaviour of water under hydrostatic tension. III. Proc. Phys. Soc. London 59, 199–208 (1947). — TEMPERELY, H. N. V., and L. G. CHAMBERS: The behaviour of water under hydrostatic tension. I. Proc. Phys. Soc. London 58, 420–436 (1946). — THODAY, D.: On turgescence and the absorption of water by the cells of plants. New Phytologist 17, 108–113 (1918). — On the behaviour during drought of leaves of two cape species of *Passerina*, with some notes on their anatomy. Ann. of Bot. 35, 585–601 (1921). — On the water relations of plant cells. Ann. of Bot. 14, 1–6 (1950). — Turgor pressure and wall pressure. Ann. of Bot. 16, 129–131 (1952).

URSPRUNG, A.: Über die Kohäsion des Wassers im Farnannulus. Ber. dtsch. bot. Ges. 33, 153–162 (1915). — Zur Kenntnis der Saugkaft. VII. Eine neue vereinfachte Methode zur Messung der Saugkraft. Ber. dtsch. bot. Ges. 41, 338–343 (1923). — Osmotic quantities of plant cells in given phases. Plant Physiol. 10, 115–133 (1935). — Die Messung der osmotischen Zustandgrößen pflanzlicher Zellen und Gewebe. In Handbuch der biologischen Arbeitsmethoden von E. ABDERHALDEN, Abt. XI, Teil 4, H. 7, S. 1109–1572. 1938. — Über den Einfluß von Temperaturdifferenzen auf die Osmose. Protoplasma (Berl.) 33, 200–210 (1939). — URSPRUNG, A., u. G. BLUM: Zur Methode der Saugkraftmessung. Ber. dtsch. bot. Ges. 34, 525–539 (1916). — Besprechung unserer bisherigen Saugkraftmessungen. Ber. dtsch. bot. Ges. 36, 599–618 (1918). — Eine Methode zur Messung der Saugkraft von Hartlaub. Jb. wiss. Bot. 67, 334–348 (1927). — Zwei neue Saugkraft-Meßmethoden. Jb. wiss. Bot. 72, 254–334 (1930).

VINCENT, R. S., and G. H. SIMMONDS: Examination of the BERTHELOT method of measuring tension in liquids. Proc. Phys. Soc. London 55, 376–382 (1943). — VRIES, H. DE: Eine Methode zur Analyse der Turgorkraft. Jb. wiss. Bot. 14, 427–601 (1884).

WALTER, H.: Kritisches zur Darstellung der osmotischen Zustandsgrößen in den verschiedenen Lehrbüchern der Botanik. Planta (Berl.) 40, 550–554 (1952). — WEATHERLEY, P. E.: Some theoretical considerations of cell water relations. Ann. of Bot. 16, 137–142 (1952).

Osmotic pressure or osmotic value.

By

C. R. Stocking.

With 2 figures.

I. Introduction.

a) Definitions.

The presence of solutes in solutions or in living cells lowers the activity and the diffusion pressure (MEYER 1945) of the water molecules present. Unless this effect on the diffusion pressure of water is altered in the cell by some other factor such as an increased hydrostatic pressure or turgor pressure, the total concentration of dissolved particles present (molecules, ions, and even colloidal particles) is an index of the water absorbing power of the cell. Usually, however, the presence of solutes is only one of several factors affecting the diffusion pressure of water in a cell. The ability of a cell to absorb water from an adjacent cell or surrounding solution depends upon the magnitudes of all forces affecting the diffusion pressure of water both within the cell and in the adjacent medium[1].

The solute effect can be measured and may be expressed in atmospheres of osmotic pressure (synonyms: osmotic potential, osmotic value). Osmotic pressure will be used here as equal to the theoretical maximum hydrostatic pressure that a solution might develop when transferred to an ideal osmometer (MEYER and ANDERSON 1952). This is also numerically equal to the amount that the diffusion pressure of water is lowered by the solutes present in the solution (CRAFTS, CURRIER, and STOCKING 1949). The osmotic pressure of a cell means the osmotic pressure of its vacuolar solution (HYGEN and KJENNERUD 1952).

b) Early studies on osmosis.

An understanding of the movement of water into living cells rests on a knowledge of the behavior of solutions and semipermeable membranes. As early as 1748, Abbé NOLLET found that when alcohol and water were separated by an animal bladder membrane, the water passed through into the alcohol but the alcohol was unable to pass into the water. In spite of the physiological importance of this phenomenon of *osmosis*, *i.e.* the diffusion of a solvent through a membrane, it was not until almost 100 years later that quantitative measurements of osmosis were seriously attempted (DUTROCHET 1826—1832; and VIERORDT 1848; see FINDLAY 1919). Even these experiments are only of historical interest. Critical quantitative work on osmosis had to await the development of a suitable membrane, a membrane possessing both semipermeable properties and sufficient rigidity to withstand the high pressures which develop as a result of osmosis.

After studying the properties of a number of artificial membranes, MORITZ TRAUB (1867) found that a precipitation membrane of copper ferrocyanide, $Cu_2Fe(CN)_6$, allowed water molecules to pass but prevented the passage of solute molecules and ions. The excellent semipermeable nature of this membrane,

[1] See this volume, p. 38, MEYER, B. S.: Wall and turgor pressure etc.

which has since been used extensively in osmotic studies, was confirmed by MORSE (1914). Thus the first obstacle, that of the development of an effective semi-permeable membrane, was overcome. Still to be solved was the problem of making this membrane strong. PFEFFER (1877) overcame this difficulty simply by causing the precipitation membrane of copper ferrocyanide to be formed in the capillaries of the walls of a porous earthenware pot. Now, a rigid semipermeable membrane was available for use in the study of the osmotic properties of solutions.

PFEFFER next proceeded to determine the osmotic pressure of a number of solutions of different concentrations and at different temperatures. The results of his experiments served for many years as the only quantitative measurements of osmotic pressure and are of historical interest because they were used by VAN'T HOFF in 1885 to formulate his theory of solutions (FINDLAY 1919). With the development of newer techniques, PFEFFER'S results have been superceded by more accurate data, Table 1 (BERKELEY and HARTLEY 1916, MORSE 1914, FRAZER and MYRICK 1916).

Table 1. *Osmotic pressure of sucrose solutions at different temperatures.* (MORSE 1914.)

Concentration in grams per 1000 grams of water	Mean osmotic pressure in atmospheres at								
	0°	5°	10°	15°	20°	25°	30°	40°	50°
0.1	2.462	2.452	2.498	2.540	2.590	2.634	2.474	2.5597	2.6353
0.2	4.722	4.818	4.893	4.985	5.064	5.148	5.0438	5.163	5.2784
0.3	7.085	7.198	7.334	7.476	7.605	7.729	7.6468	7.844	7.9739
0.4	9.442	9.608	9.790	9.949	10.137	10.298	10.295	10.5987	10.724
0.5	11.895	12.100	12.297	12.549	12.748	12.943	12.9779	13.355	13.5037
0.6	14.381	14.605	14.855	15.144	15.388	15.625	15.7132	16.146	16.319
0.7	16.886	17.206	17.503	17.815	18.128	18.435	18.4992	18.932	19.2019
0.8	19.476	19.822	20.161	20.535	20.905	21.254	21.375	21.8055	22.116
0.9	22.118	22.478	22.884	23.305	23.717	24.126	24.2263	24.735	25.1228
1.0	24.825	25.280	25.693	26.189	26.638	27.053	27.2234	27.701	28.213

II. Osmotic pressure and the gas laws.

a) VAN'T HOFF'S equation for dilute solutions.

From a physiological point of view it was useless to speculate about forces causing intercellular water movement and movement of water into and loss from plants until the physical basis causing water movement across membranes between solutions was understood and capable of measurement. Nevertheless, the observations on the effects of salt solutions upon the movement of water into cells (DE VRIES' classical experiments on plasmolysis, 1882) served as a stimulus for a closer investigation of the properties of solutions. It was through the intervention of DE VRIES that VAN'T HOFF (1887, 1888) became acquainted with PFEFFER'S work and realized that here was a method for studying some of the properties of dilute solutions.

PFEFFER'S results had indicated that osmotic pressure is directly proportional to the concentration of the solution under constant temperature. This is analogous to BOYLE'S law for gases. Further observation by PFEFFER showed a direct proportionality between osmotic pressure and absolute temperature, analogous to GAY LUSSAC'S law for gases.

On the basis of the above analogy between the behavior of gases and of solutions and as a result of his thermodynamic analyses, VAN'T HOFF formulated a theory of solutions relating the osmotic pressure with concentration and temperature of the solution. The equation is

$$P_0 V = nRT \qquad (1)$$

where P_0 = osmotic pressure of the solution, V = volume of the solution, n = number of moles of solute, R = the molar gas constant, and T = absolute temperature.

Though it applies only to very dilute solutions VAN'T HOFF's law lead to the formulation of the quantitative relationships between concentration and the colligative properties of solutions. Various modifications of VAN'T HOFF's equation have been proposed to make the equation agree more closely with the actual behavior of solutions and to account for the departure from idealistic behavior caused by such factors as the volume occupied by the solute molecules and the interactions among solute molecules and solvent molecules (FINDLAY 1919, CRAFTS, CURRIER, and STOCKING 1949). Certain of these equations have been so developed that good agreement is obtained between the calculated and the actual osmotic pressures. Nevertheless, the physiologist needs a simple and accurate method of determining the osmotic pressure of solutions or cell sap. Such a method is based upon the relationship between osmotic pressure and the freezing point of a solution.

b) Osmotic pressure, vapor pressure, and freezing point depression.

To one familiar with modern views of molecular kinetics, it now is clear that the diffusion pressure of solvent molecules in solution is related to its vapor pressure. Changes in concentration should, theoretically, alter the pressures in proportion to the mole fraction of solute in the solution (LEWIS 1908).

The thermodynamic relationship between the vapor pressure of the solvent and the osmotic pressure of a solution may be represented by the equation

$$P_0 \bar{V} = RT \ln \frac{f^0}{f} \qquad (2a)$$

in which f^0 is the fugacity of the pure solvent and f is its fugacity from the solution (HILDEBRAND 1955). Fugacity, or escaping tendency is essentially equal to the pressure the vapor would have if it were an ideal gas.

Assuming that the vapor pressure of a solvent obeys the gas laws, the relation between the osmotic pressure of a solution and the vapor pressure is as follows:

$$P_0 \bar{V} = RT \ln \frac{p_0}{p} \qquad (2)$$

where P_0 = osmotic pressure of the solution, $\bar{V}$ = the partial molecular volume of the solvent under standard pressure, p = vapor pressure of the solvent above the solution, and p_0 = vapor pressure of the pure solvent.

The vapor pressure of the solvent vapor above a solution is a function of the escaping tendency of the solvent molecules. Such intermolecular forces as association and hydration that determine the coordination of solvent and solute molecules and hence cause deviations from ideal behavior are integrated in the vapor pressure value. Hence, the vapor pressure reflects the true activity of the solvent in solution. Because it is the difference between this activity and the activity of pure solvent molecules that determinations the true osmotic pressure of the solution (LEWIS 1908), reliable values for the osmotic pressures can be determined from vapor pressure measurements (equation 2).

Of more immediate use to physiologists is the relationship between the freezing point of a solution and its osmotic pressure. It is a relatively simple matter to determine the freezing point of a solution whereas vapor pressure determinations require more elaborate apparatus and are more difficult.

The relationship between freezing point depression and osmotic pressure is given by the formula

$$P_0 = R\,T\,\frac{\Delta}{k} \tag{3}$$

where Δ = the freezing point depression in °C., and $k = 1.858°$ C. for aqueous solutions.

This equation may be written (HITCHCOCK 1945)

$$P_0\,(\text{atm.}) = 0.04416\,T\,\Delta \tag{4}$$

$$P_0\,(\text{atm. } 0°\,\text{C.}) = 12.06\,\Delta, \text{ or} \tag{5}$$

more exactly (LEWIS 1908)

$$P_0 = 12.06\,\Delta - 0.021\,\Delta^2. \tag{6}$$

Temperature corrections for *ideal solutions* may be made by use of the relationship

$$P_t = P_0\left(1 + \frac{t}{273}\right) \tag{7}$$

where P_t is the osmotic pressure at t degrees centigrade.

The cryoscopic method that is widely used by ecologists and plant physiologists, for the indirect determination of the osmotic pressure of plant saps, is based upon this relationship.

III. Determination of the osmotic pressure of plant saps.

The two general methods used in the determination of the osmotic pressure of cells are the well known plasmolytic and cryoscopic methods. In addition, vapor pressure methods have been developed by HALKET (1913), BARGER (1924), and URSPRUNG and BLUM (1930) for the determination of the osmotic pressure of small quantities of sap. Detailed discussions covering these methods and their limitations are presented in a number of papers (BRAUNER 1932, STRUGGER 1935, KÜSTER 1935, URSPRUNG 1938, HÖBER 1945 particularly on plasmolysis and DIXON and ATKINS 1913, WALTER 1931a, HARRIS 1934, STEINER 1939 on cryoscopy).

For a critical consideration of these methods of determining the osmotic pressure of cells, the reader is referred to CRAFTS *et al.* (1949).

Osmotic pressure of a drop or two of sap may be determined by the BALDES and JOHNSON thermo-electric method as modified by VAN ANDEL (1952). This method depends upon the determination of the temperature drop as evaporation takes place from the sap sample and from a reference solution into a chamber saturated with the vapor of a third aqueous solution. The rate of evaporation or condensation and the change in temperature is proportional to the difference between the vapor pressures of the two solutions and that of the third solution. The vapor pressure of the sap may be calculated from the temperature changes. Since vapor pressure varies inversely with the osmotic pressure it is possible to calculate the osmotic pressure of the sap.

a) The plasmolytic method.

In 1884 DE VRIES introduced his plasmolytic method for measuring the osmotic pressure of cells. This technique, which is most successful with cells having colored sap, generally involves the use of thin sections of tissue that are placed in graded sugar solutions, the concentrations of which vary by steps of 0.02 to 0.05 molar. After osmotic equilibrium has been reached between each section and its surrounding solution (30 minutes or longer), the sections are examined microscopically for evidence of plasmolysis. The concentrations of the bathing solutions are so chosen that plasmolysis will occur in the more concentrated solutions and will not occur in the more dilute solutions. Through careful observation of the sections, one can then determine the external concentration that would effect plasmolysis of 50% of the cells. This concentration is taken as isotonic with the cell sap at limiting plasmolysis. At limiting plasmolysis it is assumed that the turgor pressure of the cell is zero and because at equilibrium there is no *net* movement of water between the cell and the solution, the osmotic pressure of the cell is equal to the known osmotic pressure of the external solution.

Difficulties arise in the evaluation of the results of plasmolytic measurements of osmotic pressure because of the changes that occur within the tissue during plasmolysis. Water is lost from the protoplasm and the vacuole to the external solution as the cells lose their normal turgor. This volume change may be of considerable magnitude particularly in succulent and thin walled cells. An inverse relationship exists between volume and concentration. If the volume change is known, appropriate corrections can be made so that the osmotic pressure of the cell in a turgid state can be calculated. The relationship between osmotic pressure and cell volume is as follows:

$$P_n = \frac{P_i V_i}{V_n} \tag{8}$$

where P_n = osmotic pressure of the cell with volume V_n, P_i = osmotic pressure of the cell at incipient plasmolysis and is equal to the osmotic pressure of the plasmolysing solution, and V_i = volume of the cell at incipient plasmolysis.

In addition to the difficulties involved in critically measuring volume changes and in determining incipient plasmolysis, the plasmolytic method has other limitations which, particularly in some tissues, seriously limit the usefulness of this method. Among these limitations (see CRAFTS *et al.* 1949) are changes in permeability of the cell, injury and sap release during sectioning, plastic wall changes, and particularly adhesion of protoplast to the wall (PRINGSHEIM 1931, BUHMANN 1935, HEYN 1940, GASSER 1942, BRAUNER and BRAUNER 1943).

b) The cryoscopic method.

The cryoscopic method has been widely used particularly by ecologists. It involves the determination of the freezing point of expressed sap and a calculation of the osmotic pressure of the sap from the direct relationship between freezing point depression and osmotic pressure (equation 5, 6, and 7). Its advantages over the plasmolytic technique include: 1. the accuracy of the freezing point determination, 2. errors due to changes in permeability and shock are avoided, and 3. measurements are at normal cell volume and are averages for the whole tissue or organ.

Essential to this method is the preparation of plant sap in as nearly unaltered state as possible. Rapid killing of the tissue usually by freezing or hot air treatment is recommended (DIXON and ATKINS 1913, GORTNER and HARRIS 1914, WALTER 1931, STEINER 1939). Sap is expressed from the killed tissue by means of an hydraulic press. For a review of the methods of sap extraction see BROYER (1939). The chief errors and limitations of the cryoscopic method are the inherent difficulties connected with the expression of sap and the changes in the sap resulting from enzymatic action, release of water or adsorption of solutes by the solid cellular materials.

IV. Osmotic pressures of plant cells.

The osmotic pressure of plant sap may be regarded as an expression of environmental interaction with the plant (STODDARD 1935). An enormous number of measurements has been made to obtain quantitative data relative to the role of osmotic pressure in plant functions and plant distribution. It should be remembered, however, that an osmotic pressure value is not a measure of the water absorbing power of a cell at any one instant (except under conditions of zero turgor). The determining factor is of course diffusion pressure deficit[1]. Nevertheless, the maximum diffusion pressure deficit that a cell can develop, without going into a state of tension or increasing its total solute concentration, is equal to the osmotic pressure at incipient plasmolysis.

Although the plasmolytic method has been widely used, the simplicity of cryoscopy and its easy adaptability for field use has made this method a favorite of ecologists. It has already been pointed out that neither of these techniques is without objections. The plasmolytic method measures osmotic pressure at limiting plasmolysis but can be used on single living cells. The cryoscopic method determines the value for sap expressed from bulk tissues at their normal water content, but it is useless in the determination of the osmotic pressures of adjacent cells within a tissue.

a) Cell sap composition and osmotic pressure.

In order to understand the relationships among environmental factors, osmotic pressure, and plant growth, it is necessary to know the extent to which each of the major vacuolar solutes whether molecules, ions, or even colloidal particles contributes to the total osmotic pressure of the sap. Further, it must be established how each of these is influenced by alterations in metabolism induced by environmental changes. For instance, although the problems of cold and frost resistance[2] cannot be explained entirely on an osmotic basis, in many instances increased hardiness is accompanied by changes in osmotic pressure and this change is most frequently a result of sugar increase in the sap. Any plant treatment that results in an increase in the osmotic pressure of a solution is likely to increase frost resistance. The higher concentration of solutes may be effective 1. in lowering the temperature at which sap freezes (MAGISTAD and TRUOG 1925), 2. in decreasing the tendency for internal formation of ice, and 3. in decreasing the total amount of water withdrawn from cells during intercellular ice formation. This latter effect would reduce injuries caused mechanically by such ice (CURTIS and CLARK 1950).

Plants with high osmotic pressures are likewise often more resistant to desiccation injury than similar plants with lower osmotic pressures[3] (ILJIN 1930, WALTER 1931 b). An understanding of these and other similar problems would be advanced if we had more information on the effects of environmental factors on the composition of the cell solutes in different species of plants.

In spite of the vast number of determinations on the total osmotic pressure of plant cells and expressed saps since DE VRIES' (1884) early work[4], relatively few analyses have been attempted on the individual materials contributing to this total solute effect (ILJIN 1932, PITTIUS 1934, FUCHS 1935, KNODEL 1938, STEINER 1939, TAKADA and NAGAI 1953). DE VRIES developed a method of plant

[1] See this volume, p. 38: MEYER, B. S., Wall and turgor pressure etc.
[2] See this volume, p. 632: LEVITT, J., Temperature.
[3] See this volume, p. 639: STOCKER, O., Wassermangel und Zellaktivität.
[4] See vol. I, p. 614: PISEK, A., Chemie des Zellsaftes.

sap analysis and his determinations indicated that sugar, organic acid, and organic and inorganic salts contributed the major fraction of the osmotic pressure.

KNODEL (1938) showed that from 87 to 96 per cent of the osmotic pressure (determined cryoscopically) of a number of plant saps could be accounted for by a summation of the osmotic effects of the 1. total sugars, 2. total salts, and 3. total organic acids determined by sap analysis (Table 2). At the time KNODEL

Table 2. *The components of the osmotic pressure of different plant saps. All values are in atmospheres.* (KNODEL 1938.)

Plant	Time of observation	Total sugar	Monosaccharides	Sucrose	Total salt	Organic salt	Chloride	Ca as $CaCl_2$	NH_4 as NH_4Cl	Organic acids	Total osmotic pressure of sap (cryoscopic)	Sum of components	Per cent of cryoscopic value
Rye	July 1	4.31	2.13	2.18	6.71	2.10	2.46	—	0.60	2.24	15.9	14.0	88
Corn	Aug. 24	3.10	2.16	0.94	3.72	—	—	—	0.22	1.34	10.4	8.2	79
Turnip leaves	Aug. 20	1.60	0.86	0.74	6.70	2.10	—	1.17	0.32	0.23	9.7	8.5	88
Turnip leaves	Sept. 4	1.98	1.98	0.00	7.11	—	—	0.94	0.00	0.33	10.2	9.4	92
Sugar beet. .	Aug. 20	8.69	8.01	0.68	3.69	2.85	0.55	0.12	0.22	0.96	13.8	13.3	96
Elder	July 15	3.90	3.90	0.00	8.42	—	—	—	—	1.80	17.8	14.1	79
Fir	Mar. 15	9.95	9.40	0.55	4.66	2.07	1.01	1.70	0.54	2.65	18.6	17.3	93
Box	Mar. 15	8.58	5.27	3.31	9.85	4.37	2.38	1.90	0.65	1.34	22.7	19.8	87

analyzed his plants, sugars comprised from 19 to 65 per cent of the total osmotic pressure of the saps while from 25 to as much as 75 per cent was contributed by the salts. Chloride was a major sap component and organic salts made up from $^1/_3$ to $^3/_4$ of the total salt effect. A high chloride content is characteristic of certain halophytes and may contribute nearly 80 per cent of the total osmotic value of their saps. The literature in this field has been reviewed by STEINER (1939). Among the conifers, *Metasequoia* is characterized by having a considerable part of the osmotic pressure of the sap contributed by the chloride fraction (KURIMOTO, TAKADA, and NAGAI 1954).

Such observations serve as a confirmation of DE VRIES' earlier work but can only at best give us a general picture ot the situation within the cells for environmental and metabolic effects would be constantly changing the ratios of the various solutes.

It is now well known that species differ markedly in their ability to absorb and accumulate ions. These differences will be reflected in the sap composition and in the contribution that any particular ion plays in the total osmotic pressure of the plant. Such species differences are particularly characteristic of plants adapted to grow on salt or alkaline areas in comparison to species that are not so adapted. The characteristic zonation of vegatetion of the strand as well as that around the periphery of inland salt basins is a result of different degrees of tolerance to salt and to the poor aeration or soil structure which accompanies increasing moisture in these areas. Extensive observations on halophytes have been made (see reviews by WALTER 1931b, HARRIS 1934, STEINER 1939, and MAGISTAD 1945).

An excellent approach to the problem of the role of osmotic pressure and absorption of ions in ecology was made by COLLANDER (1941). Twenty species representing different ecological types were grown in nutrient solution and their ability to select cations from the solution was studied. The greatest differences in the abilities of species to absorb ions were found for sodium and manganese (Table 3). Certain species (irrespective of the year of cultivation) were found

Table 3. *Selective absorption of ions by different plant species. Analysis of entire plants.* (After Collander 1941.)

Species	Per cent of total plant cations made up by				Total cation concentration in mequiv/kg. dry wt.
	Na	K	Mg	Ca	
Avena	3.7	73	14	8	2040
Fagopyrum	0.9	39	27	33	3230
Helianthus	2.3	54	17	27	3020
Solanum	4.1	44	25	27	4290
Vicia	10.6	44	19	26	2470
Zea	2.9	70	16	11	2420
Halophytes					
Atriplex hortense	19.7	39	31	10	4790
Atriplex litorale	10.7	56	23	10	4340
Plantago lanceolata	12.2	45	18	24	3690
Plantago maritima	28.5	39	11	21	4370
Nutrient solution	25	25	25	25	—

to be consistently higher in certain ions and other species were relatively richer in another cation. All halophytes (*Salicornia herbacea, Plantago maritima, Atriplex litorale, A. hortense*) were found to be very high in sodium, with the exception of *Aster trifolium* that contained only moderate amounts. Although these analyses are of entire plants rather than saps, the high sodium absorption by the halophytes would result in a high sodium content in the saps. Many workers have recorded this ability of halophytes to accumulate sodium (see Steiner 1939, Sakazaki *et al.* 1954). Nonhalophytic plants were found by Takada (1954) to have a Na:K ratio of only about one-tenth to one four-hundredth of that of euhalophytes growing under apparently the same conditions of soil and climates.

b) Range of osmotic pressures in plants.

Species variations.

Osmotic pressures of plant tissues may range from as low as one atmosphere or less to an extreme of over 200 atmospheres in the leaves of certain halophytes; *i.e. Atriplex confertifolia* growing in an alkaline area had an osmotic pressure of 202.5 atmospheres (Harris 1934). In general, xerophytic species (salt marsh and desert plants) have saps with high osmotic pressures while most mesophytes exhibit concentrations within a range of 5 to 30 atmospheres and aquatic species have lower values (Maximov 1929). Most agricultural plants usually have an osmotic pressure ranging from 10 to 20 atmospheres (Iljin 1927, Eaton 1942, Magistad 1945).

The osmotic pressure of cell sap is generally highest in trees and shrubs and lowest in annuals (Fitting 1911, Harris and Lawrence 1916a, Harris *et al.* 1921, Korstian 1924). As a result of extensive determinations, Harris (1934) found that woody plants growing in the rain forest of Jamaica had an average osmotic pressure of 11.4 while herbaceous species in the same area averaged 8.8. In the mesophytic regions of Long Island the woody species averaged 14.4 and the herbaceous ones 10.4 atmospheres osmotic pressure. In contrast to this, desert plants in general have high osmotic pressures. In Arizona deserts woody plants were found to have an average value of 25 atmospheres and herbaceous plants averaged 15.2 atmospheres. It must be emphasized that such generalizations represent only trends and that many exceptions are to be expected because of the numerous factors affecting sap concentration.

Host and parasite relations.

Parasites often show an interesting osmotic relationship to their hosts. HARRIS and LAWRENCE (1916b) found that the average osmotic pressure of mistletoes growing on 20 different hosts was 14.43 atmospheres while that of the hosts was 13.59 atmospheres. Similarly the ability of a fungus to grow in an environment of high osmotic pressure is advantageous. Parasitic fungi characteristically have a higher osmotic pressure than the sap of the plants that they parasitize (THATCHER 1939, 1942). Table 4 shows a comparison of the osmotic pressures of some fungi and of their hosts.

Table 4. *Osmotic pressure of parasite and host.* (THATCHER 1942.)

Parasite		Host	
Fungus	Average O. P. atmospheres	Host plant (healthy tissues)	Average O. P. atmospheres
Uromyces fabae		*Pisum Sativum*	
Germ tubes.	44.25	Leaf.	9.15
Haustoria.	21.90	Petiole.	10.10
U. coryophyllinus		*Dianthus*	
Haustoria.	18.6	Leaf base	11.2
Puccinia graminis		*Mindum wheat*	
Haustoria (race 21)	18.9	Leaf.	9.4
Erysiphe Plygoni		*Brassica*	
Hyphae	18.0	Leaf.	10.6
Phytophthora infestans		*Solanum*	
Hyphae (areal)	17.4	Tuber	10.6
Hyphae (intercellular). . .	15.5	Petiole.	8.9
Sporangia	18.1		
Botrytis cinerea		*Apium graveolens*	
Hyphae	29.8	Petiole.	8.3
Sclerotinia Sclerotiorum		*Apium graveolens*	
Hyphae	23.5	Petiole.	13.4 (9.4—17.4)
Phoma lingam		*Brassica*	
Hyphae	41.3	Root	11.3

Variations within the individual plant.

A large sudden increase in the cell sap concentration is an indication that the water balance is disturbed. WALTER (1949) has compared the fluctuations in the osmotic pressure of a plant to the fever curve of an animal, each major increase in osmotic pressure indicating a danger to the plant.

The osmotic pressure of an individual plant will vary diurnally and seasonally or irregularly as a result of environmental changes. Variations in the environment may affect salt absorption and accumulations, water absorption and loss, and the rates of production, utilization, and condensation of soluble organic compounds in any one of a number of ways. Periodic changes in the environment are reflected in the diurnal and seasonal fluctuations in the osmotic pressure in any one plant.

In addition, within a single plant marked differences exist among the osmotic pressures of the different cells, tissues, and organs. Even within an apparently homogeneous tissue, adjacent cells frequently do not have identical cell sap concentrations. Although the cells are connected by plasmodesmata, each cell maintains a structural and functional individuality that is reflected in one way in the osmotic pressure of its contents. Plasmolytic studies clearly indicate this; not all cells within a tissue reach incipient plasmolysis to the same time. In general larger cells are plasmolyzed by somewhat weaker concentrations than are smaller cells within a single tissue.

Generally leaf sap is more concentrated than the sap of roots although this situation may be reversed, *e.g.* during the period of maturation beets have a higher osmotic pressure in their roots than in their leaves. Frequently tissues closer to the supply of water have low concentrations, while those at greater distances have higher concentrations (CURTIS and CLARK 1950). An example of this is found in the common observation that the osmotic pressure of leaves often increases from the lower to the higher levels of insertion on the stem. Among the many investigations on this point are those of DIXON and ATKINS (1916), HARRIS *et al.* (1917), KORSTIAN (1924) and HERRICK (1933).

The latter investigator observed that when the competition for water was not severe, osmotic pressures did not reach a maximum in the top leaf or apical

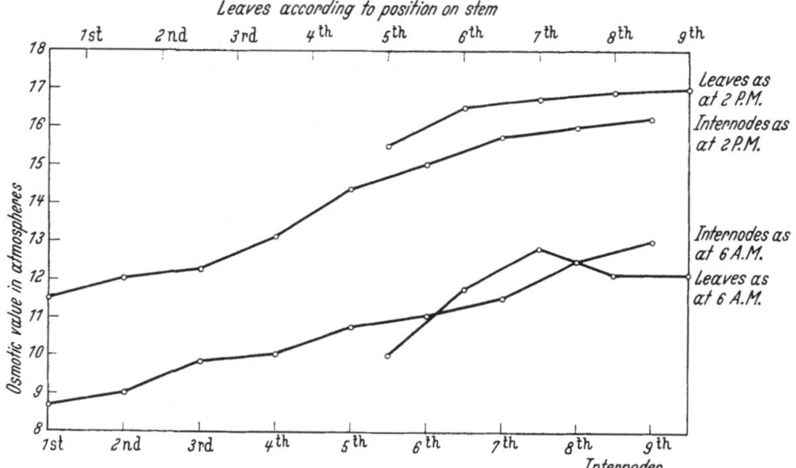

Fig. 1. A comparison of the relative osmotic pressure in the aerial parts of *Ambrosia trifida* under conditions of severe stress at 2:00 P.M. and not severe stress at 6:00 A.M. Parts numbered according to sequence of development, 1st leaf above cotyledons, etc. (From HERRICK 1933.)

portion of the stem, nor was there a consistent gradient in osmotic pressures throughout the plant *(Ambrosia trifida)*. However increasing internal competition for water, whether due to seasonal or diurnal causes, resulted in an increase in the magnitude of osmotic pressures throughout the plant and the establishment of consistently increasing gradients in the osmotic pressure from lower to higher portions of the plant (Fig. 1).

Within an organ gradients often exist among the various tissues and cells. Thus we find an increasing osmotic pressure from root epidermis to stele (URSPRUNG and BLUM 1916). Xylem sap is characteristically dilute frequently ranging between 0.5 to about 2.0 atmospheres osmotic pressure[1] (MAXIMOV 1929, VAN OVERBEEK 1942, STOCKING 1945). On the other hand the sieve tube content is generally concentrated. CRAFTS (1932) observed an average value of 7.0 atmospheres osmotic pressure for the phloem exudate of *Cucurbita pepo*.

URSPRUNG and BLUM (1916) and BUHMANN (1935), using the plasmolytic technique reported a general pattern of osmotic pressure among leaf tissues: palisade parenchyma > spongy parenchyma > upper epidermis > lower epidermis.

These variations in osmotic pressure among the various cells, tissues, and organs as well as variations with the age of the plant part must be taken into consideration when one attempts to interpret the significance of cryoscopic determinations on plant saps derived from whole leaves or other organs.

[1] See Vol. III, p. 489: STOCKING, C. R.: Guttation and bleeding.

Diurnal and seasonal cycles.

Since sugars and other organic solutes often constitute a significant fraction of the total osmotic pressure of plant sap (Table 2), it is to be expected that increasing photosynthesis during the early part of the day will be reflected in an increase in osmotic pressure particularly of the leaves. Coupled with this effect there is a decrease in the water content of the tissues during the period of a most rapid transpiration[1]. Minimum values for sap concentration are usually found in the early morning hours before sunrise and maximum values are reached in the afternoon (Fig. 1) (HERRICK 1933, STODDART 1935, MARSH 1940).

A similar influence of photosynthesis on sap concentration is found between shaded and exposed leaves. Shaded leaves have a lower osmotic pressure than those exposed to sunlight (DIXON and ATKINS 1912, LUTMAN 1919, and HALMA and HAAS 1928).

Seasonal periodicity in osmotic pressures is characteristic of perennial plants. In some, certain conifers for example (Fig. 2), the osmotic pressure may be

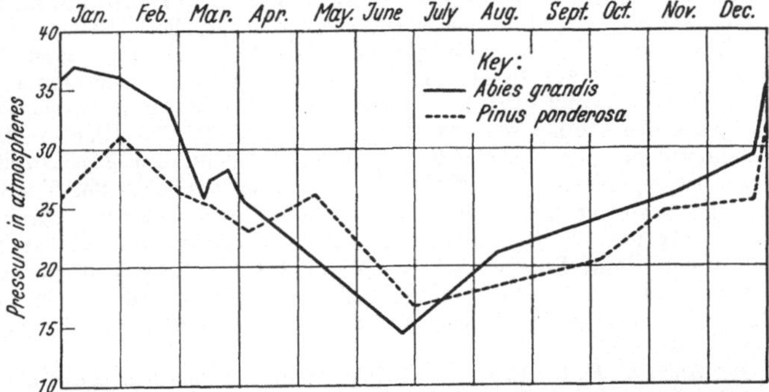

Fig. 2. Seasonal variations in the osmotic pressure of the leaves of *Abies grandis* and *Pinus ponderosa*. (GAIL 1926.)

high during the winter months (LEWIS and TUTTLE 1920, GAIL 1926, MEYER 1928, WALTER 1931b, ULMER 1937, STEINER 1939, PISEK 1950, KURIMOTO *et al.* 1954). This increase in osmotic pressure is often coupled with an increase in the sugar content of the sap brought about by starch hydrolysis induced by low temperatures. In some instances salts may also contribute to this effect. KURIMOTO *et al.* (1954) found significant diurnal and seasonal fluctuations in the osmotic pressure of *Metasequoia* leaf sap. A gradual accumulation of chloride took place accompanying the aging of the leaves and the seasonal and diurnal changes in osmotic pressure were affected by the high potassium and chloride content, as well as by the sugar.

Effects of the external medium.

Only a brief reference to the extensive work on the ecological significance of osmotic pressure in plants and to the characteristics of salt tolerant plants will be made here[2]. The detailed work of WALTER (1931b), HARRIS (1934) and STEINER (1939) should be consulted for a more complete treatment.

It has long been known that under conditions of high osmotic pressure in the external solution some plants become adapted to their environment through an increase in the osmotic pressure of their sap (DRABBLE and DRABBLE 1907, FITTING 1911, MCCOOL and WELDON 1928, HERRICK 1933, STODDART 1935,

[1] See this volume, p. 22: STOCKING, C. R., Hydration and cell physiology.
[2] See Vol. III, p. 902: ADRIANI, M. J., Halophyten.

KNODEL 1939, WALL and HARTMAN 1942, MAGISTAD 1945). Although increasing osmotic pressures are found in plants growing in media of increasing concentration, the osmotic pressure of the sap alone is not a safe criterion of the environment as there may not be a direct proportionality between them (MEYER 1927). The osmotic pressure of the soil solution for productive soils seldom exceeds one to two atmospheres at the permanent wilting percentage. Poor crops are found on soils with an osmotic pressure of ten atmospheres and none at all on soils of 47 atmospheres osmotic pressure (MAGISTAD and REITEMEIER 1943).

Literature.

ANDEL, O. M. VAN: Determinations of the osmotic value of exudation sap by means of the thermo-electric method of BALDES and JOHNSON. Proc. Kon. Ned. Akad. v. Wetensch. **55**, 40–48 (1952).

BARGER, G.: Eine mikroskopische Methode zur Bestimmung des Molekulargewichtes. In ABDERHALDEN's Handbuch der biologischen Arbeitsmethoden, Abt. III, Teil a 1, H. 4, S. 729. 1924. — BRAUNER, L.: Das kleine pflanzenphysiologische Praktikum. II. Die physikalische Chemie der Pflanzenzelle. Jena: Gustav Fischer 1932. — BRAUNER, L., and M. BRAUNER: The relations between water intake and oxybiosis in living plant-tissues. II. The tensility of the cell wall. Rev. Fac. Sci. Univ. Istanbul, Sér. B **8**, 30–75 (1943). — BERKELEY, Earl of, and E. G. J. HARTLEY: Further determinations of direct osmotic pressures. Proc. Roy. Soc. Lond., Ser. A **92**, 477–492 (1916). — BROYER, T. C.: Methods of tissue preparation for analysis in physiological studies with plants. Bot. Rev. **5**, 531–545 (1939). — BUHMANN, A.: Kritische Untersuchungen über vergleichende plasmolytische und kryoskopische Bestimmungen des osmotischen Wertes bei Pflanzen. Protoplasma (Berl.) **23**, 579–612 (1935).

COLLANDER, R.: Selective absorption of cations by higher plants. Plant Physiol. **16**, 691–720 (1941). — CRAFTS, A. S.: Phloem anatomy, exudation, and transport of organic nutrients in cucurbits. Plant Physiol. **7**, 183–225 (1932). — CRAFTS, A. S., H. B. CURRIER and C. R. STOCKING: Water in the physiology of plants. Waltham, Mass.: Chronica Botanica Co. 1949. — CURTIS, O., and D. G. CLARK: An introduction to plant physiology. New York: McGraw-Hill Co. 1950.

DIXON, H. H., and W. R. G. ATKINS: Variations in the osmotic pressure of the leaves of *Hedera helix*. Notes Bot. School Trinity Col. Dublin **2**, 103–110 (1912). — Osmotic pressures in plants. I. Methods of extracting sap from plant organs. Sci. Proc. Roy. Dublin Soc. **13**, 422–433 (1913). — Osmotic pressures in plants. VI. On the composition of the sap of the conducting tracts of trees at different levels and at different seasons of the year. Sci. Proc. Roy. Dublin Soc. **15**, 51–62 (1916). — DRABBLE, E., and H. DRABBLE: The relation between the osmotic strength of cell sap in plants and their physical environment. Biochemic. J. **2**, 117–132 (1907). — DUTROCHET, R. J. H.: Nouvelles observations sur l'endosmose, et sur la cause de ce double phénomène. Ann. de Chim. (Phys.) **35**, 393–400 (1827).

EATON, F. M.: Toxicity and accumulation of chloride and sulfate salts in plants. J. Agricult. Res. **64**, 357–399 (1942).

FINDLAY, A.: Osmotic pressure. London: Longmans, Green & Co. 1919. — FITTING, H.: Die Wasserversorgung und die osmotischen Druckverhältnisse der Wüstenpflanzen. Z. Bot. **3**, 209–275 (1911). — FRAZER, J. C. W., and R. T. MYRICK: The osmotic pressure of sucrose solutions at 30°. J. Amer. Chem. Soc. **38**, 1907–1922 (1916). — FUCHS, W. H.: Der Anteil des Zuckers am osmotischen Wert bei Weizen. Planta (Berl.) **23**, 340–348 (1935).

GAIL, F. W.: Osmotic pressure of cell sap and its possible relation to winter killing and leaf fall. Bot. Gaz. 81, 434–445 (1926). — GASSER, R.: Zur Kenntnis der Änderung der Saugkraft bei Grenzplasmolyse durch Wasserunter- und -überbilanz. Ber. schweiz. bot. Ges. **52**, 47–110 (1942). — GORTNER, R. A., and J. A. HARRIS: Notes on the technique of the determination of the depression of the freezing point of vegetable saps. Plant World **17**, 49–53 (1914).

HALKET, A. C.: On various methods for determining osmotic pressures. New Phytologist **12**, 164–176 (1913). — HALMA, F. F., and A. R. HAAS: Effect of sunlight on sap concentration of citrus leaves. Bot. Gaz. **86**, 102–107 (1928). — HARRIS, J. A.: The physico-chemical properties of plant saps in relation to phytogeography. Minneapolis, Minnesota: Univ. Minnesota Press 1934. — HARRIS, J. A., R. A. GORTNER and J. V. LAWRENCE: The relationship between the osmotic concentration of leaf sap and height of leaf insertion in trees. Torreya **44**, 267–286 (1917). — HARRIS, J. A., and J. V. LAWRENCE: The cryoscopic constants of expressed vegetable saps as related to local environmental conditions in the Arizona deserts. Physiol. Res. **2**, 1–49 (1916a). — On the osmotic pressure of the tissue fluids of Jamaican

Loranthaceae parasitic on various hosts. Amer. J. Bot. **3**, 438–455 (1916b). — HARRIS, J. A., J. V. LAWRENCE and R. A. GORTNER: The osmotic concentration and electrical conductivity of the tissue fluids of ligneous and herbaceous plants. J. Physic. Chem. **25**, 122–146 (1921). — HERRICK, E. M.: Seasonal and diurnal variations in the osmotic values and suction tension values in the aerial portions of *Ambrosia trifida*. Amer. J. Bot. **20**, 18–34 (1933). — HEYN, A. N. J.: The physiology of cell elongation. Bot. Rev. **6**, 515–574 (1940). — HILDEBRAND, J. H.: Osmotic pressure. Science (Lancaster, Pa.) **121**, 116–119 (1955). — HITCHCOCK, D. E.: Selected principles of physical chemistry. In: Physical chemistry of cells and tissues. R. HÖBER, editor. Philadelphia, Pa.: The Blakiston Co. 1945. — HÖBER, R.: Physical chemistry of cells and tissues. Philadelphia, Pa.: The Blakiston Co. 1945. — HYGEN, G., and J. KJENNERUD: Osmotic relations during cell expansion. Physiol. Plant. **5**, 171–182 (1952).

ILJIN, V.: Über die Austrocknungsfähigkeit des lebenden Protoplasmas der vegetativen Pflanzenzellen. Jb. wiss. Bot. **66**, 947–964 (1927). — ILJIN, W. S.: Die Ursachen der Resistenz von Pflanzenzellen gegen Austrocknen. Protoplasma (Berl.) **10**, 379–414 (1930). — Zusammensetzung der Salze in der Pflanze auf verschiedenen Standorten. Kalkpflanzen. Beih. bot. Zbl. **50**, 95–137 (1932).

KNODEL, H.: Eine Methodik zur Bestimmung der stofflichen Grundlagen des osmotischen Wertes von Pflanzensäften. Planta (Berl.) **28**, 704–715 (1938). — Über die Abhängigkeit des osmotischen Wertes von der Saugkraft des Bodens. Jb. wiss. Bot. **87**, 557–564 (1939). — KORSTIAN, C. F.: Density of the cell sap in relation to environmental conditions in the Wasatch Mountains of Utah. J. Agricult. Res. **28**, 845–909 (1924). — KÜSTER, E.: Die Pflanzenzelle. Jena: Gustav Fischer 1935. — KURIMOTO, K., H. TAKADA and S. NAGAI: Physiology of *Metasequoia glyptostroboides* and related species of conifers. I. Osmotic value and salt composition of leaf saps. J. Inst. Polytech. **5**, D 55–65 (1954).

LEWIS, F. J., and G. M. TUTTLE: Osmotic properties of some plant cells at low temperatures. Ann. Bot. Lond. **34**, 405–416 (1920). — LEWIS, G. N.: The osmotic pressure of concentrated solutions, and the laws of the perfect solution. J. Amer. Chem. Soc. **30**, 668–683 (1908). — LUTMAN, B. F.: Osmotic pressures in the potato plant at various stages of growth. Amer. J. Bot. **6**, 181–202 (1919).

MAGISTAD, O. C.: Plant growth on saline and alkali soils. Bot. Rev. **11**, 181–230 (1945). — MAGISTAD, O. C., and R. F. REITEMEIER: Soil solution concentrations at the wilting point and their correlation with plant growth. Soil Sci. **55**, 351–360 (1943). — MAGISTAD, O. C., and E. TRUOG: The influence of fertilizers in protecting corn against freezing. J. Amer. Soc. Agron. **17**, 517–526 (1925). — MARSH, F. L.: Water content and osmotic pressure of certain prairie plants in relation to environment. Nebraska Univ. Stud. **40**, No 3, 3–44 (1940). — MAXIMOV, N. A.: The plant in relation to water, R. H. YAPP, Editor. London: Allen & Univin 1929. — McCOOL, M. M., and M. D. WELDON: The effect of soil type and fertilizer on the composition of expressed sap of plants. J. Amer. Soc. Agron. **20**, 778–793 (1928). — MEYER, B. S.: Studies on the physical properties of leaves and leaf saps. Ohio J. Sci. **27**, 263–288 (1927). — Seasonal variations in the physical and chemical properties of the leaves of the pitch pine, with especial reference to cold resistance. Amer. J. Bot. **15**, 449–472 (1928). — A critical evaluation of the terminology of diffusion phenomena. Plant Physiol. **20**, 142–164 (1945). — MEYER, B. S., and D. B. ANDERSON: Plant physiology. New York: D. van Nostrand Co. 1952. — MORSE, H. N.: The osmotic pressure of aqueous solutions. Carnegie Inst. Wash. Publ. **198**, 1–222 (1914).

NOLLET, M. l'Abbé: Recherches sur les causes du Bouillonnement des liquides. Acad. Roy. Sci. Mém. **1748**, 57–104.

OVERBEEK, J. VAN: Water uptake by excised root systems of the tomato due to nonosmotic forces. Amer. J. Bot. **29**, 677–683 (1942).

PFEFFER, W. F. P.: Osmotische Untersuchungen. Leipzig: W. Engelmann 1877. — PISEK, A.: Frosthärte und Zusammensetzung des Zellsaftes bei *Rhododendron ferrugineum*, *Pinus cembra* und *Picea excelsa*. Protoplasma (Berl.) **39**, 129–146 (1950). — PITTIUS, G.: Über die stofflichen Grundlagen des osmotischen Druckes bei *Hedera helix* und *Ilex aquifolium*. Bot. Archiv **37**, 43–46 (1934). — PRINGSHEIM, E. G.: Untersuchungen über Turgordehnung und Membranbeschaffenheit. Jb. wiss. Bot. **74**, 749–796 (1931).

SAKAZAKI, N., Y. IHARA, Y. TACHIBANA, S. NAGAI and H. TAKADA: Physiology of *Metasequoia glyptostroboides* and related species of conifers. II. Comparative studies of salt tolerance. J. Inst. Polytech. Osaka City Univ. D **5**, 67–78 (1954). — STEINER, M.: Zum Chemismus der osmotischen Jahresschwankungen einiger immergrüner Holzgewächse. Jb. wiss. Bot. **78**, 564–622 (1933). — Die Zusammensetzung des Zellsaftes bei höheren Pflanzen in ihrer ökologischen Bedeutung. Erg. Biol. **17**, 151–254 (1939). — STOCKING, C. R.: The calculation of tensions in *Cucurbita pepo*. Amer. J. Bot. **32**, 126–134 (1945). — STODDART, L. A.: Osmotic pressure and water content of prairie plants. Plant Physiol. **10**, 661–680

(1935). — STRUGGER, S.: Praktikum der Zell- und Gewebephysiologie der Pflanze. Berlin: Gebrüder Bornträger 1935.

TAKADA, H.: Ion accumulation and osmotic value of plants, with special reference to strand plants. J. Inst. Polytech. **5**, 81–96 (1954). — TAKADA, H., and S. NAGAI: Notes on the rich chloride content and the osmotic pressure of *Metasequoia glyptostroboides*. Proc. Jap. Acad. **29**, 274–278 (1953). — THATCHER, F. S.: Osmotic and permeability relations in the nutrition of fungus parasites. Amer. J. Bot. **26**, 449–458 (1939). — Further studies of the osmotic and permeability relations in parasitism. Canad. J. Res. **20**, 283–311 (1942). — TRAUB, M.: Experimente zur Theorie der Zellbildung und Endosmose. Arch. Anat. usw. **1867**, 87–165.

ULMER, W.: Über den Jahresgang der Frosthärte einiger immergrüner Arten der alpinen Stufe, sowie der Zirbe und Fichte. Jb. wiss. Bot. **84**, 553–592 (1937). — URSPRUNG, A.: Die Messung der osmotischen Zustandsgrößen pflanzlicher Zellen und Gewebe. In ABDERHALDEN's Handbuch der biologischen Arbeitsmethoden, Abt. XI, Teil 4, H. 7, S. 1109–1572. 1938. — URSPRUNG, A., u. G. BLUM: Über die Verteilung des osmotischen Wertes in der Pflanze. Ber. dtsch. bot. Ges. **34**, 88–104 (1916). — Zwei neue Saugkraft-Meßmethoden. Jb. wiss. Bot. **72**, 254–334 (1930).

VRIES, H. DE: Eine Methode zur Analyse der Turgorkraft. Jb. wiss. Bot. **14**, 427–601 (1884). — VAN'T HOFF, J. H.: Die Rolle des osmotischen Druckes in der Analogie zwischen Lösungen und Gasen. Z. physik. Chem. **1**, 481–508 (1887). — The function of osmotic pressure in the analogy between solutions and gases. Phil. Mag., Ser. V, **26**, 81–105 (1888).

WALL, R. F., and E. L. HARTMAN: Sand culture studies of the effects of various concentrations of added salts upon the composition of tomato plants. Proc. Amer. Soc. Horticult. Sci. **40**, 460–466 (1942). — WALTER, H.: Die kryoskopische Bestimmung des osmotischen Wertes bei Pflanzen. In ABDERHALDEN's Handbuch der biologischen Arbeitsmethoden, Bd. XI/4, S. 353–371. 1931 a. — Die Hydratur der Pflanze. Jena: Gustav Fischer 1931 b. — Grundlagen des Pflanzenlebens, 3. Aufl., Bd. I. — Die Hydratur und ihre Bedeutung. Stuttgart: Eugen Ulmer 1949.

Plasmolyse und Deplasmolyse.

Von

Eduard Stadelmann.

Mit 8 Abbildungen.

I. Einleitung.

Beim Einlegen pflanzlicher Objekte in hypertonische Lösungen beobachtete NÄGELI 1849 und 1850 (1855, p. 21) erstmalig ein „Zusammenziehen des Primordialschlauches, indem er sich von der Membran lostrennt" (p. 23[1]). Ebenso beschreibt BRAUN (1852, p. 225) bei *Chara* und PRINGSHEIM (1854, p. 12f.) auch bei anderen Pflanzen ein Zurückziehen des Plasmas von der Zellwand in Zucker- und NaCl-Lösungen. Dieses Phänomen wurde dann von DE VRIES (1877, p. 10/370) als *Plasmolyse* bezeichnet, mit welchem Ausdruck seither die *Ablösung des lebenden Protoplasten*[2] *von der Zellwand unter dem Einfluß wasserentziehender Mittel* definiert ist. Hierbei handelt es sich meist um vacuolisierte Zellen (vor allem mit großer Zentralvacuole), bei welchen die Volumverringerung des Protoplasten durch *osmotischen Wasserentzug aus der Vacuole* hervorgerufen wird („*Reaktionsplasmolyse*", SCHÜTT 1895, p. 110). Der zwischen Membran und abgehobenem Plasmaschlauch sich bildende Raum ist dann im Gleichgewichtszustand von Außenlösung erfüllt (*Vorraum*, HÖFLER 1930a, p. 345). Kontraktionen des Inhaltes vacuolenfreier oder plasmareicher Zellen, wie sie in solchen Lösungen auftreten (vgl. PRÁT 1923a; KUCHAR 1950a), gehen hingegen hauptsächlich auf Entquellungen des Plasmas zurück.

Andere Ablösungsvorgänge des Protoplasten (Lit.: PRÁT 1934, p. 1f.) können beobachtet werden z. B. bei der spontanen Abrundung von Eizellen usw. (KÜSTER 1951, p. 29; „*Entwicklungsplasmolyse*", SCHÜTT 1895, p. 109), bei nicht-osmotischem Wasserentzug und Reizplasmolysen (S. 83_1)[3] sowie beim Absterben der Zelle (S. 73). Eine vorübergehende Abhebung zeigt sich manchmal auch bei quellungsfähigen Membranen (S. 74).

Neben der hauptsächlichsten Aufgabe des Plasmolyseversuches, Schlüsse auf die *chemische und physikalische Beschaffenheit des Protoplasten und seiner Elemente* zu ermöglichen, wozu auch die Ermittlung des *osmotischen Wertes* und der *Plasmapermeabilität* zählt, sowie Änderungen dieser Qualitäten nach irgendwelchen Alterationen zu erfassen, dient er noch allgemein als eine einfache Indicatormethode zur *Erkennung des Lebenszustandes* der Zelle, wobei allerdings mit der Plasmolyse verwechselbare Kontraktionserscheinungen an totem Material sorgsam ausgeschlossen werden müssen (KÜSTER 1924, p. 976; SCHNEIDER 1924, p. 38f.).

Der rückläufige Vorgang, die Ausdehnung des kontrahierten Protoplasten unter Wasseraufnahme in die Vacuole, wird *Deplasmolyse* genannt (S. 83_7).

[1] Unterbleibt die Angabe von Name bzw. Jahreszahl, so ist stets der unmittelbar vorher angeführte Autor bzw. die zuletzt zitierte Publikation gemeint.

[2] Als Protoplast (HANSTEIN 1880, p. 9) sei wie üblich das lebende Plasma samt seinen lebenden und leblosen Einschlüssen bezeichnet.

[3] Auf den jeweils in Klammern angegebenen Seiten ist weiteres über den betreffenden Gegenstand angeführt, wobei an der bezugnehmenden Stelle die Seitenzahl des Hinweises wiederholt ist. Bei mehreren Verweisen auf die gleiche Seite wird zusätzlich durch eine tiefgestellte kleine Ziffer angegeben, im wievielten Absatz sich die gesuchte Stelle befindet.

Wenngleich die Plasmolyse den Charakter eines ausgesprochen experimentellen Eingriffes trägt, kann sie bei Algen der Gezeitenzone vorübergehend auch im natürlichen Milieu [HÖFLER 1931b, p. 67; BIEBL 1938b, p. 518; 1939, p. (81)] und bei pathogenen *Bakterien* während ihrer Entwicklung im erkrankten Organismus auftreten (FISCHER 1891, p. 62).

Objekte verschiedenster systematischer Stellung wurden auf ihr Plasmolyseverhalten geprüft. Bei den *Bakterien* (vgl. FISCHER 1891; 1895; RAICHEL 1928; Lit.: RAFFELT 1944, p. 1, 23) zeigen *Beggiatoa* (vgl. RUHLAND und HOFFMANN 1926, p. 8) sowie einige andere keine Plasmolyse. Auch *Blaualgen* lassen sich zum Teil nur schwer plasmolysieren (vgl. BRAND 1903, p. 303; PRÁT 1923a). *Chlorophyceen* (vgl. FRITSCH und HAINES 1923; HÖFLER 1932b; *Vaucheria*, vgl. TIROLD 1932), *Conjugaten* (besonders *Spirogyra*) und *Diatomeen* (vgl. KARSTEN 1899, p. 152f.; CHOLNOKY 1932) wurden häufig untersucht. Die meisten *Rotalgen* sind für ihre Plasmolyseempfindlichkeit bekannt (HÖFLER 1930b), während über *Braunalgen* [vgl. HÖFLER 1932c, p. (56); CHALAUPKA 1939] und *Pilze* (vgl. REINHARDT 1899, p. 443f.; PANTANELLI 1904, p. 304; ÚLEHLA 1939; KUCHAR 1950a) relativ wenige Beobachtungen vorliegen. Neben *Moosen* (vgl. KRESSIN 1935; PECKSIEDER 1947; WILL-RICHTER 1949) und *Farnen* (vgl. HELLWEGER 1935) wurden zahlreiche *Angiospermen*, besonders die in der Zellphysiologie üblichen Objekte (*Tradescantia, Elodea, Allium*), studiert.

II. Zellmembran und Adhäsion.

Die Zellwand wirkt beim Plasmolysevorgang in dreifacher Weise:

a) Als *Bereich mit spezifischen Diffusionseigenschaften*, welchen die Außenlösung passieren muß, um an die Plasmaoberfläche zu gelangen.

b) Als mehr oder weniger dehnbare und feste *Umgrenzung des Zellumens*, von der sich der Plasmamantel ganz oder teilweise ablöst, und welche die Gestalt des plasmolysierten Protoplasten und deren Änderung mitbestimmen kann (S. 80_1, 100).

c) Als *Träger von Plasmaresten* (wandständiges Plasma, S. 74) oder Teilen des Protoplasmamantels (S. 76; Plasmoschise, S. 92; Ausbleiben jeglicher Ablösung, S. 82_2), die trotz der Volumverringerung des Protoplasten dauernd oder vorübergehend mit der Membran verbunden bleiben.

Schließlich kann die Zellwand durch die Einwirkung des Plasmolytikums

d) Veränderungen erfahren, die den Plasmolysevorgang zu beeinflussen vermögen.

Die Membrandurchlässigkeit für die Außenlösung hängt unter anderem von der Teilchengröße des gelösten Stoffes und der Weite der Membranporen ab und ist für die üblichen Plasmolytica normalerweise so groß, daß bei ausreichender Umspülung der Zelle mit dem Plasmolytikum die Abhebung nicht verzögert wird (S. 80_4, 82_3, 104_4). Bei Behinderung des direkten Zutrittes wegen lokaler cuticularer Auflagerungen werden die darunterliegenden Teile der Membran meist von der Seite her durch das submikroskopische Capillarensystem der Wand genügend schnell mit dem Plasmolytikum versorgt.

Für die Oberflächenbenetzung sind zwischen den plasmolytisch abgehobenen und den der Zellwand anliegenden Protoplastenteilen wohl keine Unterschiede zu erwarten, da die Wand ein wasserdurchtränktes Medium darstellt.

Nur bei sehr hoher Wasserpermeabilität des Protoplasten oder geringer Durchlässigkeit der Membran für die gelösten Stoffe (Membranpermeabilität s. Abschn. IV/E) wird die Plasmolyse verzögert (HUBER und HÖFLER 1930, p. 437) oder setzt allein an den leichter durchlässigen Membranseiten ein (S. 104_2). Ist der Zutritt zur Plasmaoberfläche völlig verhindert, unterbleibt eine Abhebung des Protoplasten, und der osmotische Wasserentzug bewirkt eine *Eindellung der Zelle* (S. 82_2). Im Inneren von Zellkomplexen ist die Zulieferung des Plasmolytikums durch die ungünstigen Diffusionsverhältnisse oft erschwert, was zu einer Verspätung des Plasmolyseeintritts führen kann (S. 104_3).

Aus den geometrischen Verhältnissen zwischen Porenweite und Teilchengröße erklären sich auch die bei raschem Wechsel isotonischer Plasmolyticis unterschiedlicher Teilchendurchmesser plötzlich auftretenden und vorübergehenden *Volumänderungen* (*Plasmolytikumwechsel-Effekt*, BENNET-CLARK und BEXON 1946; CASARI 1953) und *Verschiebungen des plasmolysierten Protoplasten* (S. 81).

Die Adhäsion des Protoplasten an der Zellwand beruht wohl nicht allein auf Kohäsion mit den die Wand durchsetzenden Plasmodesmen (die sogar in Epidermen vorkommen mögen; vgl. LAMBERTZ 1954), sondern es ist eine direkte Bindung der Plasmaschicht an der Wand mit einer innigen Verwachsung beider anzunehmen (HECHT 1912, p. 187). Zur Adhäsion wird auch eine wahrscheinlich aus Proteinen bestehende Wandkomponente beitragen, von welcher sich der Plasmaschlauch bei Plasmolyse abheben kann (*Vicia Faba*, CHAYEN 1952, p. 298). Nackte Protoplasten vermögen an zellfremden Substraten zu haften (vgl. PFEIFFER 1934b, c).

Die Adhäsion, die bei *Rotalgen* vielfach besonders groß ist [vgl. HÖFLER 1932c, p. (57)], hat für die Plasmolyseform (S. 75, 80_1, 81_2), aus der sie beurteilt werden kann (S. 80_3), mitunter auch für die plasmolytische Vacuolenkontraktion (S. 91) und Teilprotoplastenbildung (S. 95) gewisse Bedeutung und hängt oftmals vom Alter der Zelle ab (S. 76_7), kann innerhalb derselben oft lokale Unterschiede in ihrer Stärke zeigen (z.B. bei einseitiger Plasmolyse, S. 79), sich im Laufe der Plasmolyse ändern und mit der Viscosität des Plasmas zusammenhängen (vgl. SCARTH 1926, p. 207; S. 93_3). So ist sie bei der Ablösung oft recht stark (HÖFLER 1918a, p. 119, 137, 148), während sie nach längerer Plasmolyse oder bei Deplasmolyse (S. 84) offenbar abnimmt, was zu Änderungen der Plasmolyseform und gelegentlich zu Protoplastenverschiebungen führt (S. 81_7).

Bei der hier üblicherweise unter Adhäsion verstandenen Größe wird es sich meist um die Resultierende aus Kohäsionskräften des Plasmas und dem eigentlichen Haftvermögen des Protoplasten an der Wand handeln (vgl. 1930b, p. 584).

Zusätzliche Einwirkungen, wie von Diäthylharnstoff, Methylurethan, Al-, La-, Li- und Th-Salzen, behindern die plasmolytische Ablösung des Protoplasten, wobei Al-Behandlung weiter noch die Plasmaviscosität ändern soll (vgl. FLURI 1909, p. 91f.; WEBER 1924b, p. 630; 1930d, p. 601; BONTE 1935, p. 223f.; S. 82_4, 93_2). Ebenso vermögen UV-Bestrahlung (TAKAMINE 1940b, p. 583) und wohl auch tiefe Temperaturen (vgl. WEBER 1932c, p. 520) die Adhäsion zu erhöhen, wogegen eine Vorwässerung der Schnitte oft zu einer Adhäsionsverringerung führt (HÖFLER 1918a, p. 141; 1934a, p. 224; S. 93_3).

Die Ablösung des Protoplasten von der Zellwand kann durch verschiedene Faktoren ausgelöst werden (S. 71). Bei Plasmolyse erfolgt sie, wenn eine Zugspannung am Plasmamantel in Richtung zum Zellinneren resultiert (vgl. ÚLEHLA 1926, p. 634). Sie beginnt in ihrer typischen Form nach völliger Entspannung der Membran (S. 74) mit einer Dickenzunahme des Plasmamantels durch Dehnung, worauf sich in diesem Bläschen bilden, die immer größer werden, bis er schließlich *zerreißt* (HECHT 1912, p. 155f.). Dabei bleiben an der Membran nur geringe Reste *(wandständiges Plasma)* oder dickere Plasmaanteile haften (bei *Plasmoschise*, S. 92), was von der Lage der Trennungsfläche abhängt. Diese wird wohl durch das Verhältnis der Adhäsion an der Membran zur Kohäsion innerhalb des Plasmas gegeben, zeigt meist lokale Unterschiede in ihrer Lage und kann stellenweise aussetzen, was für die Ausbildung der Plasmolyseorte (S. 76_5) und der Plasmafäden (S. 94) bestimmend ist (vgl. WEBER 1929d, p. 600).

Die Abhebung bleibt manchmal ganz oder teilweise zeitlich oder räumlich aus. Eine *zeitliche Verzögerung* der Abhebung durch die Adhäsion (Vergrößerung der Grenzplasmolysezeit, S. 80_1) ist dann häufig so stark, daß zur Ablösung des Protoplasten höhere Konzentrationen erforderlich sind als die plasmolytische Grenzkonzentration, wobei der „osmotische Überdruck" zur Überwindung der Adhäsion dient und bisweilen sehr hohe Werte annehmen muß (vgl. HECHT 1912, p. 187; HÖFLER 1918a, p. 119, 148; WEBER 1929d, p. 598; BUHMANN 1935, p. 596). Mitunter bewirkt auch plötzlicher Wasserzusatz die Beseitigung dieses Plasmolyseverzuges in der scheinbar hypotonischen Außenlösung (BONTE 1935, p. 221f.). Schließlich kann die Abhebung auch in jeder Konzentration unterbleiben (S. 82_4). Andererseits erfolgt nach Wässerung (S. 93_3) meist glatte Trennung und rasche Rundung (S. 81_4) des Protoplasten (vgl. hingegen WEIS 1926, p. 154; KAMIYA 1939, p. 390). *Räumlich* ist die

Loslösung behindert, wenn eine schwächere Abhebung auftritt als der Außenkonzentration entsprechen würde (besonders bei Grenzplasmolyse; vgl. HÖFLER 1918a, p. 149; S. 75). In den vom Protoplasten freigegebenen Wandteilen sind *Plasmodesmen* nicht mehr nachzuweisen und fehlen auch nach längerer Zeit vollständiger Deplasmolyse (vgl. STRASBURGER 1901, p. 571; LAMBERTZ 1954, p. 177).

Das wandständige Plasma (S. 72) zeigt sich als ein feinverzweigtes *Netzwerk* unterschiedlicher Dicke mit oft mehr oder weniger deutlichen kleinen Plasmaansammlungen (*Hydrodictyon*: KLEBS 1891, p. 807; *Allium:* HECHT 1912, p. 171f.), an welchen die Plasmafäden ansetzen (S. 94), doch gelingt sein Nachweis nicht bei allen Objekten (p. 176). Es zerfällt nach einiger Zeit oder bei Zusatz von Wasser sofort unter vorübergehender Bildung perlschnurartiger Strukturen zu Plasmatröpfchen (p. 174), die dann die Innenseite der Membran bedecken (KÜSTER 1926, p. 88).

Eine Aufquellung der Zellwand bei Plasmolyse wird entweder allein durch die unmittelbare Einwirkung des Plasmolytikums bzw. einer bestimmten Vorbehandlung ausgelöst (vgl. WEBER 1932e, p. 288), oder es besteht an sich schon ein latentes Quellungsbestreben besonders der inneren Membranschichten (z. B. bei gewissen *Rotalgen*), das bei der Kontraktion des Protoplasten zu einer Volumzunahme der Zellwand führt (Lit.: URSPRUNG 1939, p. 1194; vgl. FÖRSTER 1933, p. 484f.; SCHOCH-BODMER 1936, p. 357).

Die Membranquellungen vermögen aber nicht nur eine Ablösung des Protoplasten zu unterbinden (S. 82), sondern bewirken mitunter sogar die vorübergehende Loslösung (ÚLEHLA 1926, p. 633; S. 71), wie diese auch in ausgetrockneten oder gefrorenen Zellen nach Wasserzugabe einsetzt, wenn sich die Wand rascher streckt als der Inhalt folgen kann („*Pseudoplasmolyse*", ILJIN 1933, p. 421; 1934, p. 108; vgl. KÜSTER 1929b, p. 38).

Eine nicht genügend feste Membran zeigt bisweilen an den abgehobenen Stellen Falten und Einsenkungen (vgl. NÄGELI 1855, p. 22; KOLKWITZ 1896, p. 223f.; SWELLENGREBEL 1905, p. 376). Plasmolyse und Deplasmolyse vermögen ferner zur *Abhebung der Cuticula* (vgl. CHOLNOKY 1953, p. 538f.) sowie zu lokalen *blasenartigen Erweiterungen* der Membran zu führen (*Spirillum:* RAICHEL 1928, p. 338f.; S. 84).

III. Der Ablauf der Plasmolyse.

A. Die osmotischen Verhältnisse beim Plasmolysevorgang.

In ihrer typischen Ausbildung ist die Plasmolyse als osmotisches Phänomen ein Ausgleichsvorgang, wobei die Gleichheit der Saugkräfte von Vacuole und Außenlösung, die durch den je nach Bedingungen mehr oder weniger ideal semipermeablen Plasmamantel getrennt sind, angestrebt wird. Die plasmolytische Kontraktion ist für das gleiche Objekt um so stärker, je höher die Konzentration des Plasmolytikums liegt. Die Raschheit der Kontraktion (bzw. der *Plasmolyseeintritt)* wird durch den Zufluß der Außenlösung, die Adhäsion, die Wasserpermeabilität und das osmotische Gefälle zwischen Zellsaft und Außenlösung bestimmt (vgl. HÖFLER 1930a, p. 300f.; HUBER und HÖFLER 1930, p. 352f.; S. 80) und kann zu einer quantitativen Angabe über die Größe der Wasserpermeation führen (s. Abschn. IV Bb). Von den dabei auftretenden Änderungen des Plasmavolumens oder seines Quellungsgrades kann man, sofern es sich um plasmaarme Zellen handelt und besondere Alterationen fehlen, meist absehen (vgl. URSPRUNG 1939, p. 1197f.; S. 92).

Bevor sich der Protoplast von der Membran trennt (S. 73), entspannt sich zunächst die Zelle, wobe sie je nach Membrandehnbarkeit ihr Volumen bis zu

30% verkleinern kann (vgl. KÜSTER 1951, p. 546). Die schwächste, gerade noch deutlich erkennbare Abhebung des Protoplasten von der Wand wird als *Grenzplasmolyse* bezeichnet; die hierzu nötige *Grenzkonzentration* ist bei günstigen Objekten („Indicatorpflanzen", vgl. DE VRIES 1884, p. 444/153; S. 74) nur um einen verschwindend kleinen Betrag gegenüber der dann vorhandenen Zellsaftkonzentration erhöht, so daß sie stets als dieser osmotisch gleich wirksam (*isotonisch*, p. 430/140; *isosmotisch*, TAMMANN 1888, p. 300) anzusehen ist (S. 79_2). Ihre Maßzahl (*osmotischer Wert der entspannten Zelle*; vgl. HÖFLER 1918a, p. 99, 101; URSPRUNG und BLUM 1920) dient zur Ermittlung des *osmotischen Wertes der wassergesättigten Zelle* sowie der *Saugkraft des Zellinhaltes*, einer charakteristischen *osmotischen Zustandsgröße* (vgl. URSPRUNG 1926; 1930; 1939, p. 1111; s. Abschn. III C). Besonders interessiert weiter, welche Konzentrationen verschiedenartiger Substanzen als Außenlösung bei Plasmolyse die gleiche osmotische Wirksamkeit zeigen. Hierbei wird die molare Konzentration der betreffenden Lösung mit jener einer isotonischen KNO_3- oder Rohrzuckerlösung verglichen (DE VRIES 1884, p. 427/138f.; FITTING 1919) und als Quotient der *isotonische* oder *osmotische Koeffizient* ermittelt (p. 2). Seine Größe entspricht bei den Salzen etwa dem aus der elektrolytischen Dissoziation zu errechnenden Wert (vgl. DE VRIES 1889, p. 109/542), doch ergaben genauere Messungen auch innerhalb der Anelektrolyte oft große und nicht ohne weiteres erklärbare Abweichungen, die auch mit einer Permeabilität des Protoplasten für die betreffende Substanz nicht eindeutig zusammenzuhängen scheinen (FITTING 1919, p. 1f., 80).

B. Plasmolysestadien.

Der Protoplast wird sich solange kontrahieren, bis er seine der betreffenden Außenkonzentration entsprechende *endgültige Größe* erreicht hat, bei welcher ihm die Außenlösung kein Wasser mehr entzieht. Er besitzt dann meist noch nicht seine endgültige Form, die im Zustand der *Endplasmolyse oder perfekten Plasmolyse* erreicht wird, in welchem der Protoplast im osmotischen Gleichgewicht mit der Außenlösung, bei *minimaler Oberfläche* seine Gestalt und Größe nicht mehr verändert (vgl. HÖFLER 1918a, p. 137f.), sofern Konzentrationsänderungen, Endosmose, Exosmose (S. 103), Osmoregulation (S. 84) oder andere sekundäre Einflüsse fehlen. Die vor diesem Endzustand liegenden Plasmolysestadien werden als *imperfekte oder eintretende Plasmolyse* charakterisiert. Unter günstigen Umständen (geringe Adhäsion und Viscosität des Plasmas; S. 73, 93) besitzt der plasmolysierende Protoplast seine endgültige Gestalt schon vor dem Erreichen der endgültigen Oberfläche.

In zylindrischen (und zylindrisch-prismatischen) Zellen nimmt der Protoplast bei Endplasmolyse die Gestalt eines Zylinders (Prismas) mit zwei an dessen Stirnflächen aufsitzenden halbkugeligen oder kugelsegmentförmigen Menisken an (p. 102f.; S. 79_1). In diesem Falle läßt sich sein Volumen berechnen (LEPESCHKIN 1908, p. 209; HÖFLER l. c.) und ein zahlenmäßiger Ausdruck für das Volumverhältnis von plasmolysiertem Protoplasten und dem Innenvolumen der entspannten Zelle gewinnen (*Plasmolysegrad* $G = \mathrm{Vol}_{\mathrm{Protopl.}}/\mathrm{Vol}_{\mathrm{Zelle}}$, HÖFLER 1917, p. 712; $1 - G = D$, die *Verkleinerung des Protoplasten*, HUBER und HÖFLER 1930, p. 369), auf welcher Maßzahl die *plasmometrischen Methoden* zur Berechnung des osmotischen Wertes und der Permeabilität basieren (HÖFLER 1918a, b, 1934a u. a.). Beim *kritischen Plasmolysegrad* (RUHLAND und Mitarbeiter 1938, p. 663) besitzt dabei der Protoplast jenes Volumen, bei welchem die Oberfläche des Zylindermantels allein gleich groß ist jener der Summe der beiden Menisken. Der Plasmolysegrad ändert sich umgekehrt proportional mit der osmotischen

Wirksamkeit (bei Anelektrolyten: Konzentration) der Außenlösung (vgl. HÖFLER 1918a, p. 110). Abweichungen hiervon werden als durch das Protoplasmavolumen (*Protoplasmakorrektur*, p. 113f.; Lit.: BOCHSLER 1948, p. 99) bzw. eine nichtlösende Komponente des Vacuolenvolumens (*nicht-lösender Raum*, S. 101) verursacht angesehen oder auf andere Mechanismen des Wassertransportes (nichtosmotische Wasseraufnahme, BOGEN und PRELL 1953; S. 115) zurückgeführt.

Als *Grad der Plasmolyse* (DE VRIES 1884, p. 469/175f.) wird gelegentlich auch die *Plasmolysehäufigkeit*, d. h. der prozentuale Anteil plasmolysierter Zellen (FITTING 1915, p. 6; KACZMAREK 1928, p. 213; MARKLUND 1936, p. 13; ELO 1937, p. 8; S. 104_5), das Verhältnis der Zellänge im Moment des Einlegens in das Plasmolytikum zur Protoplastenlänge im betrachteten Zeitpunkt (bei *Hefe:* TAKADA 1953, p. 18) oder dasjenige zweier Protoplastenvolumina (BOCHSLER 1948, p. 84) bezeichnet.

C. Plasmolyseort, -form und -zeit.

Die Gestalt des plasmolysierenden Protoplasten, ihre zeitliche Änderung und Unterschiede bei den einzelnen Zellen nach gleicher Behandlung werden unter anderem zur Feststellung „physiologischer Ungleichheit bei morphologischer Gleichheit der Zelle" (*Protoplasmatische Pflanzenanatomie*, vgl. WEBER 1925c; 1930a) sowie in der *vergleichenden Protoplasmatik* (HÖFLER 1932c) ausgewertet. In der *Plasmolyseform-Methode* (vgl. WEBER 1924a; CHOLODNY 1924) und der *Plasmolysezeit-Methode* (WEBER 1929a) dienen solche Beobachtungen zu Schlüssen auf Eigenschaften und Zustand des Protoplasten und seiner Bestandteile.

Plasmolyseort.

Als positiven bzw. negativen Plasmolyseort bezeichnet man jene Stellen, an welchen sich der Protoplast von der Membran abhebt bzw. an ihr haften bleibt (1929d, p. 583; vgl. RUGE 1940, p. 318). Ihre Lage, Zahl und Ausdehnung ändern sich meist im Laufe der Plasmolyse, wobei entweder die anfangs auftretenden positiven Plasmolyseorte bloß *an Ausdehnung zunehmen* (vgl. BAUER 1938, p. 270) oder *neue Plasmolyseorte* entstehen (WEBER 1929d, p. 584). Je nachdem, ob die Abhebung an den Zellecken oder an geraden Wandteilen erfolgt, geht die Vergrößerung des Plasmolyseortes mit einer Verringerung oder mit einer Zunahme der Protoplastenoberfläche einher (vgl. KÜSTER 1929b, p. 5).

Die Lage der Plasmolyseorte dürfte vielfach als „in der Organisation der Zelle begründet" (1951, p. 22; S. 72), durch lokale Unterschiede des Haftvermögens (S. 73_5) sowie der chemischen oder physikalischen Beschaffenheit des Protoplasmas bedingt sein, wie dies auch aus der weitgehenden Symmetrie von Plasmolyseort und -form bei Schwesterzellen (HÖFLER 1940a, p. 99) hervorgeht, doch können auch andere Faktoren mitwirken (S. 79, 104_1).

Beim Plasmolysebeginn liegen in isodiametrischen Zellen die positiven Plasmolyseorte häufig an den Ecken, bei langgestreckten Zellen an den Enden, manchmal an der Längswand (vgl. WEBER 1929d, p. 584) und in der Zellmitte (vgl. CHOLNOKY 1928, p. 453; 1931a, p. 335), und können mit dem Alter der Zelle wechseln (WEBER 1931, p. 139; S. 104_5).

Häufig auftretende negative Plasmolyseorte sind oft junge Zellwände (vgl. CHOLNOKY 1928, p. 497; 1930, p. 280; 1931a, p. 327; 1952c, p. 227; S. 80) oder solche, die gerade im Zustand des Längen- oder Dickenwachstums sind (vgl. REINHARDT 1899, p. 435, 458; CHOLNOKY 1931a, p. 332; 1931b, p. 512f.; RUGE 1940, p. 318; S. 73_2), ferner bei *Pennaten* Bezirke an der Raphe oder allgemein an den Valven (vgl. CHOLNOKY 1930, p. 289; 1928, p. 453f.), in Endodermiszellen die CASPARYschen Streifen, was vielleicht mit deren Verkorkung zusammenhängt und zur Bandplasmolyse führt (vgl. BEHRISCH 1926; BRYANT 1934, p. 231; WEBER 1929d, p. 594; S. 78), in den Schließzellen die Bauchseite (WEBER 1925d, p. 689), in jüngeren Siebröhren die Siebplatte (ESCHRICH 1953, p. 46f.; S. 82), bei *Spirillum* die Zellenden (RAICHEL 1928, p. 335) und im Falle spontaner Systrophe die Fußpunkte der Systrophebrücken (GERM 1950, p. 489; S. 98).

Plasmolyseform.

Der Protoplast kann je nach seinen Qualitäten und den Außenbedingungen (S. 79) mit den verschiedenartigsten Formen kontrahieren, die bei festerer Konsistenz der Vacuole auch durch diese beeinflußt werden (S. 101). Die schon vielfach untersuchten Plasmolyseformen (vgl. KÜSTER 1929b, p. 13f.) können

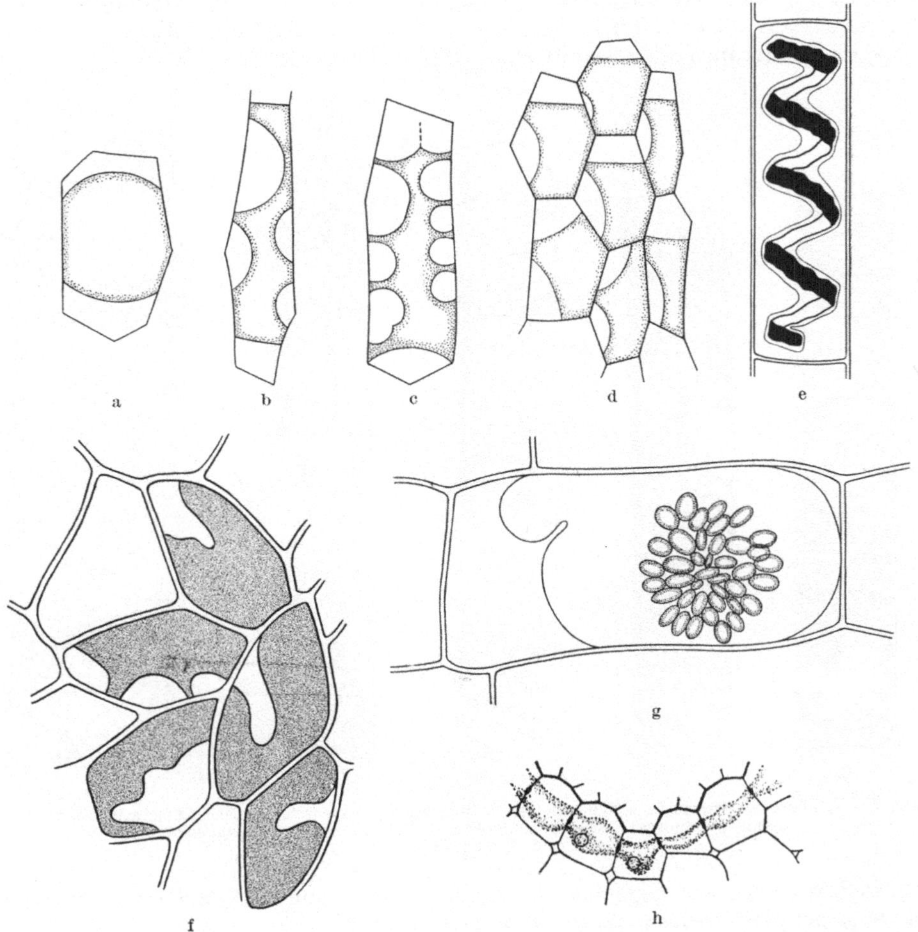

Abb. 1a—h. Plasmolyseformen I. a Konvexplasmolyse; b Konkavplasmolyse; c Konkav- und Krampfplasmolyse (a—c nach KÜSTER 1929b, p. 16); d einseitige Plasmolyse in der Epidermis von *Rhoeo discolor* (nach KÜSTER 1929b, p. 23); e Schraubenplasmolyse bei *Spirogyra* (nach WEBER 1925a, p. 219); f Plasmolyse mit sackartigen Einbuchtungen bei *Plagiochila asplenioides* (grau: Protoplast mit Vacuole; nach WILL-RICHTER 1949, p. 517); g Kerbplasmolyse bei einer Parenchymzelle von *Tradescantia guianensis* (nach HÖFLER 1918a, Tafel II); h Bandplasmolyse in der Endodermis von *Tradescantia virginica* (nach GRAVIS aus WEBER 1929d, p. 593.).

nicht nur je nach der Zellart variieren, sondern auch bei gleichen Zellen je nach deren Alter (S. 104) und Behandlung große Unterschiede zeigen und sich im Laufe der Plasmolyse ändern.

Allgemein unterscheidet man, je nachdem, ob die abgehobenen Protoplastenpartien der Membran eine Hohlfläche oder eine erhabene Fläche zuwenden, zwischen *konkaven* und *konvexen Plasmolyseformen*. Eine stark konkave Kontraktion wird als *Krampfplasmolyse* bezeichnet (WEBER 1921, p. 177; s. Abb. 1a bis c). Behält der Protoplast bei leichtflüssigem Zellsaft die Umrißformen der Zelle in verkleinertem Maßstab noch bei, spricht man von *eckiger Plasmolyse* (hingegen PRUD'HOMME 1935, p. 482), doch werden auch weitergehende Einteilungen vorgeschlagen (SCHAEFER 1955, p. 426).

Durch Ausmessen der Länge der Konturen der von der Zellwand abgelösten Protoplastenteile und derjenigen der ihnen entsprechenden Teile der Zellmembran im optischen Querschnitt läßt sich ein Quotient bilden, welcher ein Maß für die auftretende Oberflächenzunahme oder -abnahme darstellt (CHOLODNY und SANKEWITSCH 1934, p. 69f.). Der Quotient aus der für die vorliegende Plasmolyseform näherungsweise ermittelten Oberflächengröße und der bei gleichem

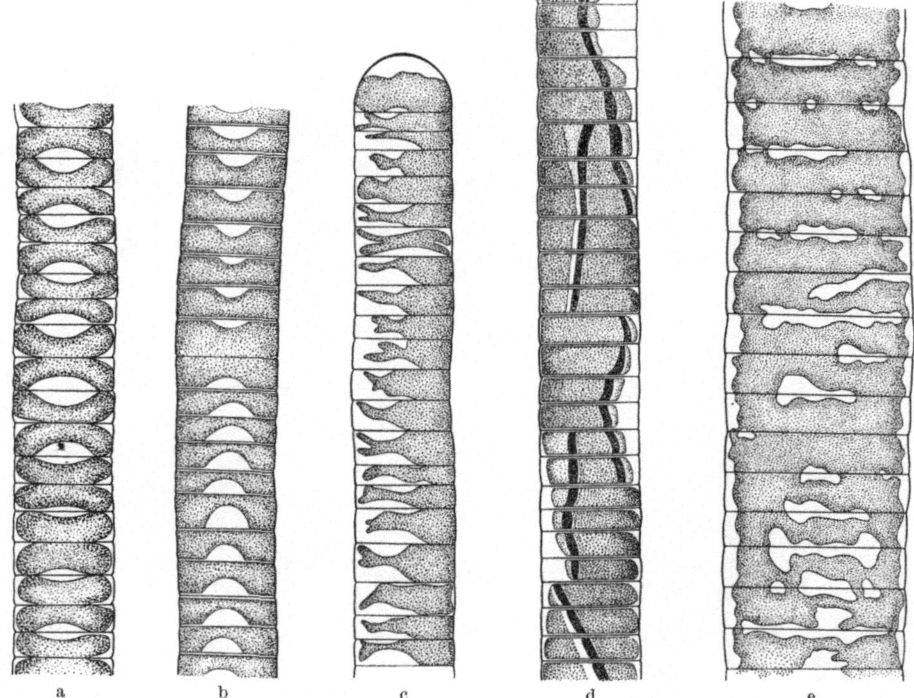

Abb. 2a—e. Plasmolyseformen II. a X-Plasmolyse; b Brückenplasmolyse (a und b nach KUCHAR 1950b, p. 215); c Palisadenplasmolyse; d Spaltenplasmolyse; e Netzplasmolyse bei *Oscillatoria tenuis* (c—e nach KUCHAR 1950b, p. 217.).

Protoplastenvolumen möglichen Minimaloberfläche kann als *Oberflächenverhältnis* (S. 80_4) definiert werden (SCHAEFER 1955, p. 427).

Bei nicht zu hoher Adhäsion und sofern Plasma und Zellsaft eine flüssige Konsistenz besitzen, gehen konkave Formen im Laufe der Plasmolyse stets in konvexe über (vgl. KÜSTER 1942a, p. 135f.), nie aber umgekehrt, was durch die Oberflächenspannung von Plasmamantel und wohl auch Vacuole bewirkt wird (vgl. CHOLNOKY 1952c, p. 227). Hingegen bleiben Krampfplasmolysen bei den meisten Rotalgen dauernd erhalten (HÖFLER 1930b, p. 572).

Besonders auffallende Plasmolyseformen (s. Abb. 1—3) sind die *Schraubenplasmolyse* bei *Spirogyra* (SCARTH 1924, p. 103; WEBER 1925a; S. 80_5), die *Faltenplasmolyse* (vgl. JOST 1929, p. 21; HÖFLER 1932b, p. 200f.), die *Kerbplasmolyse* (S. 95), die *Brückenplasmolyse* (KUCHAR 1950b, p. 218), die *Bandplasmolyse* (SCHNEE 1936; S. 76) sowie in Zellreihen die *Palisaden-*, *Spalten-*, *Netz-* und *X-Plasmolyse* (KUCHAR 1950b, p. 214f.; S. 80_2, 103). Bei *Cladophora* entstehen in den langgestreckten Zellen mitunter rotationssymmetrische *wellenförmige Anschwellungen* (HÖFLER 1932b, p. 195). *Einseitige Plasmolyse* liegt vor, wenn sich der

Protoplast vornehmlich nur an einer Seite der Zelle abhebt (URSPRUNG und BLUM 1926, p. 6; KÜSTER 1929b, p. 24; WEBER 1929c, p. 585; HÖFLER 1930b, p. 582; S. 73, 81). Erfolgt dies zuerst oder überhaupt nur an einer einzigen eng begrenzten Stelle, so bilden sich napfförmige oder sichelförmige Plasmolyseformen mit ausgehöhlter Vacuole (*Napfplasmolyse*, BORRISS 1938, p. 787; KAMIYA 1939, p. 393). An zylindrischen Zellen entstehen oft geometrisch besonders einfache Plasmolyseformen (S. 75_3). Schließlich kann die Loslösung des Protoplasten noch mit einer großen Zahl kleiner hemisphärischer oder anders geformter Abhebungen erfolgen (vgl. JOST 1929, p. 7; HÖFLER 1930b, p. 578).

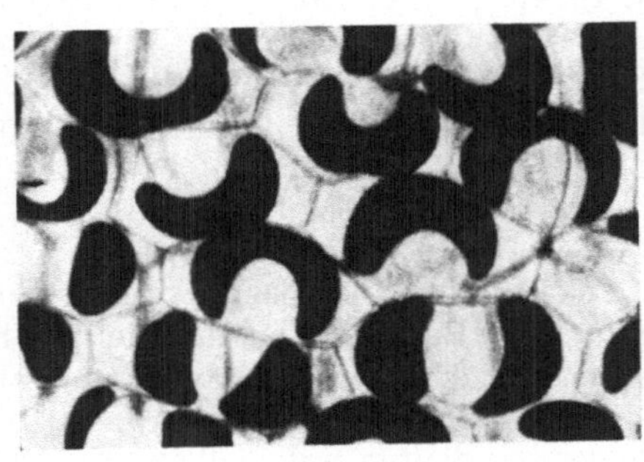

a

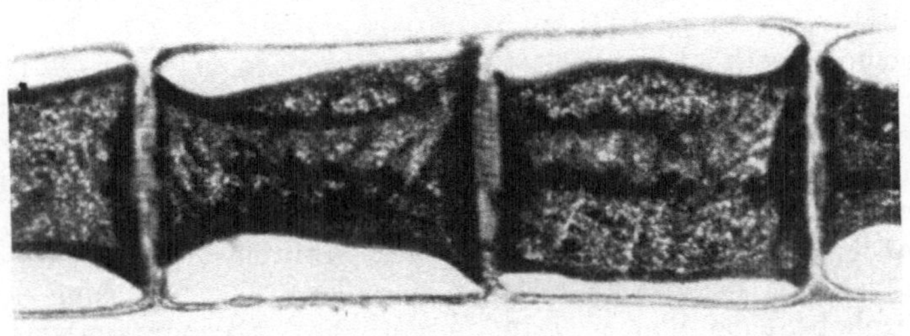

b

Abb. 3a u. b. Plasmolyseformen III. a Sichelförmige Plasmolyse in der Epidermis von *Allium Cepa* (nach KAMIYA 1939, p. 392); b Faltenplasmolyse bei *Chaetomorpha linum* (nach HÖFLER 1932b, p. 200).

Die Zeiten des Plasmolyseablaufes.

Bei der zeitlichen Beurteilung der Plasmolyse wird besonders die Zeitspanne, in allen Fällen vom Einlegen des Objektes in das Plasmolytikum gerechnet, bis zum Auftreten der Grenzplasmolyse (*Grenzplasmolysezeit*, WEBER 1929a, p. 623; DERRY 1930, p. 4; vgl. *Abhebungszeit*, BORRISS 1938, p. 813; S. 75_1), bis zum osmotischen Gleichgewicht (*Plasmolysezeit*, HÖFLER 1932b, p. 199; S. 103), bis zu konvexen Plasmolyseformen der einzelnen Zelle (*Rundungszeit*, HÖFLER 1932b, p. 199, von WEBER l. c. als Plasmolysezeit eingeführt) oder der Hälfte aller Zellen des betreffenden Gewebekomplexes (*Halbwertszeit*, SCHMIDT und Mitarbeiter 1940, p. 571) und bis zur sphärischen Rundung (*Zeit der sphärischen Rundung*, EIBL 1939b, p. 534; Rundungszeit, KREBS 1951, p. 606) beachtet.

Faktoren und Korrelationen für Plasmolyseort, -form und -zeit.

Unterschiede dieser Größen (S. 76, 77), wie sie selbst bei benachbarten und sonst gleichartigen Zellen (z. B. Schwesterzellen, KÜSTER-WINKELMANN 1938, p. 255) auftreten, können betreffen:

Qualitäten des Cytoplasmas und der Zellwand. Während bei plasmareichen Zellen allgemein eine schwerere Kontraktion zu erwarten ist (vgl. URSPRUNG 1939, p. 1202), spielen bei den nur mit einem dünnen Plasmamantel ausgestatteten Zellen die chemischen und physikalischen Eigenschaften des Cytoplasmas eine große Rolle, besonders Viscosität (S. 93_3) und Adhäsion (S. 72_4, 73_2, 73_6, 104), die auch vom *physiologischen Zustand* der Zelle abhängen können (z. B. Öffnungszustand der Stomatazellen, WEBER 1925d). Weitere Veränderungen treten bisweilen im *Stadium der Kopulation* oder *der Zellteilung* auf (z. B. bei *Spirogyra crassa* nur in den kopulierenden Zellen, bei *Spirogyra varians* allgemein im Zustand der Kopulation, 1924a, p. 264; 1924c, p. 148, 154; vgl. CHOLNOKY 1930, p. 290f.).

Cytogenetische Faktoren können an Unterschieden im Plasmolyseort (z. B. *Gynura*-Haare, KÜSTER 1951, p. 24; Drüsenhaare, GICKLHORN 1931; *Diatomeen*, KÜSTER-WINKELMANN 1938, p. 252; KAMIYA 1938, p. 340) oder in der Plasmolyseform (z. B. Bandplasmolyse, teilweise auch Palisaden-, Spalten-, Netz- und X-Plasmolyse, S. 78_4) beteiligt sein. Auch der Wachstumszustand der Membran (S. 76) vermag für die Lokalisierung der Plasmolyseorte und für die Plasmolysezeiten von Bedeutung zu sein. Bei *Pteris*-Prothallien und bestimmten Zellarten des Keimblattes von *Soja hispida* erfolgt Abhebung stets an den plastidenfreien Stellen (HELLWEGER 1935, p. 235; REUTER 1948, p. 392f.).

Konvexe *Plasmolyseformen* deuten auf niedrige Viscosität (Lit.: SCHAEFER 1955, p. 423) besonders der äußeren Plasmaschichten (vgl. KESSLER und RUHLAND 1938, p. 194; RUGE 1940, p. 329; S. 93_3) und geringes Haftvermögen (vgl. WEBER 1924a, p. 261; CHOLODNY 1924, p. 27; HÖFLER 1930b, p. 583; S. 73_2), während eckige Plasmolyse erhöhte Viscosität des *Endoplasmas* anzeigen soll (vgl. WEBER 1925a, p. 223). Neben Viscosität und Haftvermögen wirken noch Oberflächenkräfte vor allem an den abgehobenen Plasmateilen und gegebenenfalls noch andere Faktoren an der Ausbildung der Plasmolyseform mit (vgl. CHOLODNY und SANKEWITSCH 1934, p. 71; BORRISS 1938, p. 812f.; RUGE 1940, p. 329f.; SCHAEFER 1955, p. 423f.). Durch das Verhältnis des Protoplasmawiderstandes bei der Loslösung von der Wand zu jenem bei der bloßen Vergrößerung der Buchten an den abgehobenen Stellen wird im wesentlichen zwischen konvexer und konkaver Plasmolyse entschieden (HÖFLER 1932b, p. 194).

Die Grenzplasmolysezeit ist unter anderem durch das osmotische Gefälle, die Membranpermeabilität (DERRY 1930, p. 4; S. 72_7), die Größe der elastischen Membrandehnung (RUGE 1940, p. 326) und die Adhäsion, die Rundungszeit vor allem durch Adhäsion und Plasmaviscosität bestimmt (vgl. HÖFLER 1930b, p. 583; RUGE 1940, p. 327f.; TAKAMINE 1940a, p. 310; SCHAEFER 1955, p. 423f.; S. 74). Ferner kann auch das Alter der Zelle (REINHARDT 1899, p. 450), Unterschiede zwischen di- und tetraploiden Exemplaren (BOGEN 1949, p. 99) sowie bei endosmierenden Plasmolyticis ihre Eindringgeschwindigkeit eine Rolle spielen (S. 85). Für Protoplasten, die allseitig von der Wand abgelöst sind, läßt sich aus dem Oberflächenverhältnis (S. 78_2) und der Plasmolysezeit ein „*Abrundungskoeffizient*" berechnen, der als „relatives Maß für das Zusammenwirken von Plasmaspannung, Viscosität und Plasmadicke" anzusehen ist (SCHAEFER 1955, p. 428).

Elemente höherer mechanischer Festigkeit im Protoplasten. Chromatophoren (vgl. CHOLNOKY 1928, p. 459; 1931a, p. 322, 328; EIBL 1939b, p. 534; Schraubenplasmolyse, S. 78_4) beeinflussen bisweilen durch ihre Gestalt und den Grad ihrer mechanischen Festigkeit ebenso wie gelegentlich in der Zelle liegende Farbstoffkristalle (vgl. KÜSTER 1929a, p. 193f.; BECKEROWA 1935, p. 386) die Plasmolyseform.

Qualitäten der Vacuole. Die festere Konsistenz eines kolloidreichen Zellsaftes (S. 100) kann eine den Zellumrissen ähnliche Plasmolyseform bewirken (vgl. KENDA und WEBER 1952, p. 461). Beim Rundungsbestreben des Protoplasten scheinen Oberflächenkräfte des Tonoplasten in Frage zu kommen (TIROLD 1932, p. 356, 359) und der osmotische Wert des Zellsaftes beeinflußt mitunter die Plasmolysezeit (SCHAEFER 1955, p. 424; S. 104_4).

Außenfaktoren. Alle Maßnahmen, die eine Änderung der Adhäsion (S. 73_2) oder Viscosität (S. 93) herbeiführen, vermögen auf Plasmolyseort, -form und -zeit zu wirken. Diese können auch von jahreszeitlichen Verschiedenheiten (vgl. KÜSTER 1942a, p. 135), Schnelligkeit, Größe und Richtung des Wasserentzuges sowie Konzentration und Art des Plasmolytikums abhängen (vgl. DERRY 1930, p. 7; SCHAEFER 1955, p. 423).

Oft hebt sich der Protoplast zuerst oder überhaupt nur (z. B. bei einseitiger Plasmolyse, S. 79) an jenen Stellen ab, wo die Zellen mit dem Plasmolytikum *am frühesten in Berührung kommen* (vgl. KÜSTER 1929b, p. 23f.; COLLANDER 1934, p. 226; KRESSIN 1935, p. 30; RUGE 1940, p. 319; S. 104_2). Eine Abhängigkeit von der *Konzentration des Plasmolytikums* zeigt z. B. die Rundungszeit (HUBER und HÖFLER 1930, p. 357; DERRY 1930, p. 6f.; PRUD'HOMME 1935, p. 485; TAKAMINE 1940a, p. 309), der Plasmolyseort (vgl. CHOLNOKY 1928, p. 462, 470f.; WEBER 1929d, p. 593) oder die Plasmolyseform (vgl. PRUD'HOMME 1935, p. 481). Letztere vermag sich noch mit der *Richtung* zu ändern, aus welcher das Plasmolytikum zugegeben wird (EIBL 1939b, p. 532f.).

Plasmolyseort (vgl. DIANNELIDIS 1951, p. 31) und Plasmolyseform (vgl. CHOLODNY und SANKEWITSCH 1934; TAKAMINE 1940b, p. 318f.; CHOLNOKY 1950b, 1952c, p. 227; KUCHAR 1950b) sowie Rundungszeit können auch mit der *Art des Plasmolytikums*, bei *Vorbehandlung* (vgl. BANK 1932; TAKAMINE 1940a, p. 317; S. 73_6) oder bei *elektrischer Reizung* wechseln (DRAWERT 1937, p. 414).

Mit dem jeweiligen p_H-*Wert* variiert bisweilen der Plasmolyseeintritt (Grenzplasmolysezeit, KACZMAREK 1928, p. 263), die Rundungszeit (vgl. PRÁT 1926, p. 250; DERRY 1930, p. 17f.) und die Plasmolyseform (TAKAMINE 1940a, p. 313). Letztere beide sind auch noch *temperaturabhängig* und die Rundungszeit erreicht bei einer bestimmten Temperatur ein Minimum (vgl. KOLKWITZ 1896, p. 224; DERRY 1930, p. 14; PRUD'HOMME 1935, p. 488f.; hingegen ROMIJN 1931, p. 295). Plasmolyseform und Rundungszeit zeigen mitunter auch Zusammenhänge mit der *Beleuchtung, Temperaturbehandlung* oder Art der umgebenden *Atmosphäre*, der das Objekt zuvor ausgesetzt war (WEBER 1929c, p. 257; HELLWEGER 1935, p. 232; TAKAMINE 1940a, p. 312f.; SCHAEFER 1955, p. 433).

Korrelationen von Plasmolyseort, -form und -zeit bestehen manchmal mit der Lage der Zelle im Gewebeverband (S. 103) und mit allgemeinen Zelleigenschaften. Dies kann z. B. für den Grad der Zelldifferenzierung (vgl. RUGE 1940, p. 320; hingegen DIANNELIDIS 1951, p. 33), die Größe der Frostresistenz (KESSLER und RUHLAND 1938, p. 170), die Periode des Streckungswachstums (STRUGGER 1934, p. 427; hingegen BORRISS 1938; S. 104_6) oder die Wuchsstoffleitungsrichtung (RUGE 1937, p. 16) der Fall sein. Ebenso ließ sich hierfür auch ein jahreszeitlicher Rhythmus feststellen (PIRSON und GÖLLNER 1953, p. 493f.).

D. Verschiebungen des plasmolysierten Protoplasten.

Sie erfolgen manchmal spontan, besonders nach längerer Plasmolyse (vgl. RUGE 1940, p. 322), innerhalb des geschlossenen Zellumens, setzen geringe Adhäsion (S. 73_2) voraus und dürften durch (bei zylindrischen Zellen allerdings nur geringfügige) Unterschiede im Zellquerschnitt und die Oberflächenspannung des Protoplasten (HÖFLER 1918c, p. 433), in bestimmten Fällen auch durch vorübergehende lokale Konzentrationsunterschiede verursacht sein (vgl. COLLANDER 1934, p. 228f.). Letztere bewirken (z. B. bei Plasmolytikumwechsel, S. 72) selbst bei einseitig geöffneten Zellen solche Verschiebungen, die bei Deplasmolyse (S. 85) den Austritt des Protoplasten aus dem offenen Lumen herbeizuführen vermögen (vgl. KÜSTER 1910b, p. 353), was zur Fusion von Protoplasten benutzt werden kann (S. 87).

Die Schnelligkeit der Verschiebung ist von der Größe der verschiebenden Kraft, der Adhäsion und Viscosität des Plasmas (S. 93) abhängig, wobei es bei rascher Verschiebung mitunter zu HECHTschen Fäden, Plasmaansammlungen und konkaven Kuppen (CASARI 1953, p. 436) kommt, wie letztere auch bei vorsichtiger Verschiebung von in Mikropipetten aufgenommenen Protoplasten auftreten, wobei das Plasma an den Ablösungsstellen nachgezogen wird (HOFMEISTER 1954b). Die spontane Verschiebung in zylindrischen Zellen erfolgt nur langsam und unter Wahrung der Rundung beider Protoplastenkuppen. Bei Zentrifugierung plasmolysierter Zellen zeigt sich für leichte Verschiebbarkeit ein Konzentrationsoptimum (JUNGERS 1934, p. 358).

IV. Das Ausbleiben jeglicher Plasmolyse.

Unplasmolysierbarkeit, wie sie an verschiedenen Objekten beobachtet wird (vgl. KÜSTER 1929b, p. 36f.), kann resultieren, wenn Adhäsion und Kohäsion des Protoplasten größer sind als die wasserentziehende Kraft, der Protoplast impermeabel ist für Wasser, für die gelösten Stoffe eine außerordentlich hohe Permeabilität besitzt (vgl. WEBER 1929d, p. 597), wenn schließlich die *Zellwand* für den gelösten Stoff undurchlässig ist (vgl. Abschn. IV E; S. 72_5, 72_9) und ihm keinen Zutritt zum Protoplasten gewährt oder bei Plasmolyse *aufquillt* (S. 74). Zu geringe mechanische Festigkeit der Membran führt dann zu deren Einknickung, Eindellung oder Schrumpfelung (*osmotischer Kollaps*, KÜSTER 1929b, p. 37; *Cytorrhyse*, vgl. STRUGGER 1949, p. 92). Vorübergehende Eindellungen können auch auftreten, wenn der Wasserentzug durch stark konzentrierte Lösungen sehr plötzlich einsetzt (KÜSTER 1951, p. 25), die Konzentration zu rasch gesteigert wird (KUCHAR 1950a, p. 60f.) oder die Membran für das Plasmolytikum nur schwer durchlässig ist (FÖRSTER 1933, p. 477; ZEHETNER 1934, p. 509f.).

Das durch eine *Membran-Semipermeabilität* verursachte Ausbleiben der Plasmolyse ist abhängig vom Teilchendurchmesser der gelösten Substanz (S. 72_7); so sind z. B. die Wände bestimmter *Moos*zellen für Rohrzucker und sogar gewisse Salzionen oft nur schwer durchlässig (KRESSIN 1935, p. 26; BRILLIANT 1927, p. 159), was auch von deren Konzentration abhängen kann (p. 157). Auch für die Wandzellen von *Utricularia*-Blasen läßt sich eine selektive Membranpermeabilität, ja für die ganze Pflanze eine semipermeable Hülle annehmen (CZAJA 1922; PRÁT 1923b, p. 227).

Die *Adhäsion* (S. 73_6) allein kann bei dünnen Membranen schon bewirken, daß es eher zur Deformation als zur Abhebung kommt (vgl. WALTER 1923, p. 146) und vermag weiter durch experimentelle Einwirkungen zusätzlich erhöht zu werden (S. 73_4). *Beggiatoa* zeigt schon in schwach hypertonischen Medien Längskontraktion und Einkerbung der Längswände einzelner Zellen (RUHLAND und HOFFMANN 1926, p. 8), für *Euglena* macht die besondere Beschaffenheit der Hülle eine Abhebung unmöglich (Lit.: HILMBAUER 1954). Sie unterbleibt ferner bei der Rotalge *Heterosiphonia* [HÖFLER 1932c, p. (57)] und ist für *Spirotaenia*, wohl wegen hoher Adhäsion, nur schwer zu erreichen (KREBS 1953, p. 107).

Sehr *hohe Permeabilität* (S. 85) soll beim Urmeristem von Wurzeln das Ausbleiben der Plasmolyse in gewissen Außenlösungen verursachen (vgl. REINHARDT 1899, p. 433; TRÖNDLE 1922, p. 50; URSPRUNG 1939, p. 1202), wie dies auch für ausgewachsene Siebröhren diskutiert wird (vgl. CRAFTS 1939, p. 175; SCHUMACHER 1939; S. 76, 115).

Welcher Art die bei alten Zellen mitunter beobachtete Unplasmolysierbarkeit ist, scheint noch nicht geklärt (WEBER 1929d, p. 599; S. 104).

V. Protoplastenkontraktionen ohne hypertonisches Milieu.

Diese können bei äußerem Wasserentzug, spontan oder nach bestimmten Einwirkungen mitunter reversibel auftreten (S. 71_2). Dabei vermag der Plasmamantel aufzuquellen (Vacuolenkontraktion, S. 88_9) und so gelegentlich die Abhebung wieder rückgängig zu machen.

Protoplastenkontraktionen durch *Wasserentzug aus der Vacuole* zeigen sich bei Austrocknung von Zellen (vgl. NÄGELI 1855, p. 1; ILJIN 1933; ETZ 1939, p. 506), ebenso in gefrorenen Geweben (vgl. ILJIN 1934) und bei intracellulärer Eisbildung zwischen Membran und Protoplasten (vgl. CHAMBERS und HALE 1932, p. 345, 349f.; S. 115).

Eine spontane oder nach bestimmten Alterationen einsetzende Kontraktion (*Reizplasmolyse*, SCHÜTT 1895, p. 110; Lit.: PRÁT 1934, p. 1) kann vielleicht unmittelbar durch aktiven Wasseraustritt oder durch eine nach Verlust der Semipermeabilität (JANSE 1887b, p. 393), d. h. also nach Permeabilitätserhöhung (OSTERHOUT 1913, p. 450; BÜNNING 1934, p. 451; S. 103_5) einsetzende Exosmose (S. 85_5) verursacht sein. Bei entsprechend geringer Adhäsion des Protoplasten würden dann dessen Oberflächenkräfte die Kontraktion bewirken (JANSE l. c.). Reizplasmolyse, Vacuolenkontraktion (S. 88_9), Kappenplasmolyse (S. 89) und Synaerese (S. 100) lassen sich unter dem gemeinsamen Gesichtspunkt des Zusammenwirkens von Erhöhung oder Erniedrigung des Hydratationsgrades von Cytoplasma und Vacuole betrachten (BOGEN 1951, p. 4) und zeigen untereinander sowie zur Plasmoschise (S. 92) zahlreiche Übergänge.

Reizplasmolytische Kontraktionen können verschiedentlich *spontan* beobachtet werden (vgl. JANSE 1887b, p. 392), ebenso bei gewissen Objekten nach *mechanischen Einwirkungen* z. B. der Präparation (KÜSTER 1929b, p. 27; 1933a, p. 444f.; vgl. GEITLER 1938, p. 428; KÜSTER-WINKELMANN 1938, p. 254; reversible schwache Abhebung, vgl. GERM 1950, p. 490), durch Klopfen (vgl. KÜSTER 1929c, p. 166; BAUER 1938, p. 285), Deckglasdruck (vgl. SCHÜTT 1895, p. 110; KARSTEN 1899, p. 154; BENECKE 1900, p. 554; BÜNNING 1934, p. 450f.; PRÁT 1934, p. 7f.), lokale Quetschungen (vgl. NÄGELI 1855, p. 12f.; HOFMEISTER 1867, p. 10) sowie durch *Ultraschall* (KÜSTER 1952, p. 87; vgl. YAMAHA und UEDA 1939, p. 527) oder nach *Verwundung* (Lit.: PRÁT 1934, p. 5; vgl. p. 13; KÜSTER 1929b, p. 28; CHATTAWAY 1929, p. 360). Auch starke *Temperaturerniedrigung* (GREELEY 1902, p. 127; vgl. ILJIN zit.: PRÁT 1934, p. 7) oder das *Auftauen* von Zellen mit erst anfänglicher Eisbildung (BUGAEVSKY 1939, p. 133) vermag reizplasmolytisch zu wirken.

Hohe *Beleuchtungsstärke* nach Dunkelheit verursacht manchmal eine bei Wiederverdunkelung reversible, fast momentane Reizplasmolyse (BENECKE 1900, p. 555), ebenso auch mitunter Bogenlampenlicht und starke Sonnenbestrahlung (vgl. BAUER 1938, p. 285f.). Bestimmte hypotonische oder noch isotonische *Milieus* lösen bisweilen Protoplastenkontraktionen aus; so z. B. destilliertes Wasser (OSTERHOUT 1913, p. 446f. „*false plasmolysis*“, die meist irreversibel ist), hypotonische Lösungen von Plasmolyticis mit oder ohne zusätzliche andere Substanzen (vgl. DE VRIES 1885, p. 570/418f.; KARSTEN 1899, p. 152; OSTERHOUT 1908, p. 54; KEMMER 1928, p. 20; CHATTAWAY 1929, p. 363f.; HÖFLER 1918a, p. 157: „*Scheinplasmolyse*“ von am Schnittrand gelegenen Zellen, die auch in hypertonischem Milieu wesentlich stärker plasmolysieren und wobei wohl *traumatische Einflüsse* mitwirken; vgl. BÜNNING 1934, p. 451; S. 104), von Vitalfarbstoffen, die auch Vacuolenkontraktion bewirken (vgl. PRÁT 1934, p. 10; BECKEROWA 1935, p. 388; DIANNELIDIS 1951, p. 40; S. 91) oder Gewebesäfte (KÜSTER 1929c, p. 165; WEBER 1930b). Auch Äther- oder Ammoniakdämpfe können zu Reizplasmolyse führen (vgl. PRÁT 1934, p. 10).

Manchmal setzen Kontraktionen des plasmolysierten Protoplasten während der Wiederausdehnung in diosmierenden Außenlösungen ein, wobei wohl auch an Exosmose zelleigener Substanzen zu denken wäre (STADELMANN 1952, p. 401; PRELL 1953, p. 484; S. 103_5).

VI. Deplasmolyse.

Deplasmolyse (vgl. UNGER 1855, p. 262; HECHT 1912, p. 179; S. 71_4) kann nach Plasmolyse einsetzen bei entsprechender *Konzentrationserniedrigung der Außenlösung*, *Permeation* osmotisch wirksamer Substanzen in die Vacuole (S. 85_5), regulatorischer Erhöhung des osmotischen Wertes des Zellinhaltes durch Stoffwechselvorgänge (*Turgor-* oder *Osmoregulation* durch *Anatonose*, d. h. Erzeugung osmotisch wirksamer Substanz, vgl. Swellengrebel 1905, p. 482f.; Lit.: Ursprung 1939, p. 1193; S. 75, 101) sowie bei Änderungen der *nicht-osmotisch gebundenen Wassermenge* in der Vacuole (vgl. Bogen und Prell 1953; Bogen 1953, p. 145; hingegen Levitt 1953). Nach Vacuolenkontraktion (S. 90) kann Deplasmolyse spontan oder nach bestimmten Einwirkungen erfolgen; gelegentlich wird sie als Reizreaktion bei geringen Erschütterungen (Bogen 1953, p. 142), Druck auf das Deckglas (Bünning 1934, p. 451) und beim Erfrieren plasmolysierter Zellen vor dem Einsetzen der Eisbildung (Bugaevsky 1939, p. 133) beobachtet.

Die Deplasmolyse stellt keineswegs den genau rückläufigen Plasmolysevorgang dar; so werden konkave Protoplasten am Beginn der Ausdehnung zuerst stets konvex, bisweilen sogar unter vorübergehender stärkerer Abhebung (vgl. Prud'homme 1935, p. 490; Küster 1942a, p. 135f.), offenbar durch Viscositätsänderungen (de Haan 1933, p. 311) oder Verringerung der Adhäsion (S. 73). Der Protoplast dehnt sich dann gleichmäßig oder ruckweise aus, ausnahmsweise mit amöboiden Bewegungen (Küster 1942a, p. 144f.) und kann auch quellen (Kappenplasmolyse, S. 89). Am Ende der Rückdehnung bleiben spitze Zellecken oft noch lange frei (Hofmeister 1948b, p. 86), sonst aber wird bei ungeschädigten Zellen vielfach wieder das Bild vor Plasmolysebeginn erreicht, oder die Zelle stirbt unmittelbar nach der Rückdehnung ab (vgl. Albach 1931, p. 258; S. 102_5). Manchmal spontan (vgl. Krebs 1953, p. 107) oder sonst bei Turgorverlust der Nachbarzellen wird die ursprüngliche Zellgröße überschritten, wobei die Zelle mitunter schließlich platzt (vgl. Kuchar 1950b, p. 219) oder Auswüchse bildet (S. 96_7, 102_8). In manchen Fällen erfolgt bei Deplasmolyse Abhebung der Cuticula und blasige Membranerweiterung (S. 74).

Besondere Wirkungen der Plasmolyse können während der Deplasmolyse sich noch weiterentwickeln (z. B. Systrophe nach starkem Plasmolysereiz, S. 99_1), erhalten bleiben (z. B. Teilungen des Protoplasten, vgl. Küster 1909, p. 591; Plasmaverlagerungen, Seemann 1952, p. 118; Veränderungen der Plasmastruktur, vgl. Cholnoky 1950b, p. 75) oder wieder ganz oder teilweise zurückgehen (z. B. Systrophe, S. 98; Verschwinden von Vacuolen in plasmareichen Zellen, S. 99_5; Formänderungen der Plastiden, S. 97).

In einem durch zahlreiche Plasmalamellen unterteilten Zellsaftraum entspricht die Wasseraufnahme der einzelnen Teilvacuolen nicht immer genau ihrem plasmolytischen Wasserverlust (vgl. Höfler 1932b, p. 205). Blaue Blütenblattzellen von *Senecio cruentus* zeigen bei Deplasmolyse bisweilen Farbumschlag nach rot (Cholnoky 1950b, p. 75).

Nackte Protoplasten vermögen sich bis über ihr ursprüngliches Volumen auszudehnen (Törnävä 1939, p. 338f.) und auch Tonoplasten (S. 96_3) sind zur Deplasmolyse fähig.

Die Zeitspanne von der maximalen Kontraktion bis zur vollständigen Deplasmolyse (in Schnitten bis zur Deplasmolyse fast aller oder aller Zellen als *Deplasmolysezeit* meßbar, vgl. de Haan 1933, p. 242; Prud'homme 1935, p. 490; hingegen: Moder 1932, p. 8) kann temperatur- und p_H-abhängig sein und bei einem bestimmten p_H-Wert ein Minimum erreichen (Prud'homme l. c. f.; Kaczmarek 1928, p. 263).

Ist Deplasmolyse die Folge einer Permeation gelöster Stoffe aus der Außenlösung in die Vacuole (*Endosmose*, S. 85), was mitunter auch auf eine Schädigung des Protoplasten zurückgeht (S. 103), so läßt sich aus ihrer Geschwindigkeit auf die Protoplastenpermeabilität schließen (s. Abschn. IV B b).

Deplasmolyse, die vielleicht durch ihre auflockernde Wirkung auf die Protoplastenoberfläche (S. 86_6) oft schwer schädigt (S. 102, 103_1), wird durch Membranbildungen oder Verfestigungen der Plasmaoberfläche behindert, wobei dann der Protoplast vorzeitig platzt (HECHT 1912, p. 180; S. 86_6, 96_4), bleibt manchmal in hypotonischem Milieu oder Wasser völlig oder teilweise aus (vgl. TIROLD 1932, p. 372; DIANNELIDIS 1951, p. 36), führt gelegentlich zu Plasmoschise (S. 92) und setzt, wenn sie spontan erfolgt, bisweilen je nach Alter der Zellen verschieden rasch ein (vgl. BANCHER 1949, p. 90; S. 104_5).

Vorwässerung (KACZMAREK 1928, p. 264f.; S. 93), Ca-Zusatz bei K-Salzen (p. 279) und Schnittrandnähe (S. 104_4) können die Deplasmolyse hemmen; durch Neutralsalze wird sie beschleunigt und erfolgt allgemein bei den eine Quellung bewirkenden Lösungen rascher, wird aber andererseits durch Kappenplasmolyse verzögert (DE HAAN 1933, p. 253, 307; 1935; S. 89_7), wobei die äußere Plasmahaut offenbar wegen Erniedrigung der Wasserpermeabilität der Tonoplastenhaut dieser oft vorauseilt (S. 88_7).

In experimenteller Hinsicht ist die Deplasmolyse für die Fusion von Teilprotoplasten oder Protoplasten (S. 87_3) sowie zu deren Befreiung aus geöffneten Zellen (S. 81) und zur Kontrolle der Vitalität kontrahierter Protoplasten von Bedeutung (vgl. SCHNEIDER 1924, p. 35; URSPRUNG 1939, p. 1124; S. 101).

VII. Verhalten der Plasmaschichten bei Plasmolyse.

Schon PFEFFER (1887, p. 121f.; 1886/8, p. 315f.; vgl. PLOWE 1931a, p. 197) leitete aus dem osmotischen Verhalten des Protoplasten die Existenz besonderer, normalerweise mikroskopisch nicht nachweisbarer Plasmagrenzschichten ab (p. 217). Sie schließen das Protoplasma zur Vacuole hin als *Tonoplastenhaut*[1] und nach außen, d. h. gegen die Zellwand zu, als *Plasmalemma* (vgl. p. 202) ab und gehen bei intaktem und nicht alteriertem Protoplast wohl kontinuierlich in die zentrale Plasmapartie (*Binnenplasma oder Mesoplasma*, p. 202; PFEFFER 1877, p. 124) über.

Vom Binnenplasma und auch untereinander sind die Grenzschichten in der stofflichen Zusammensetzung (vgl. WEIS 1926, p. 163) und in ihren Eigenschaften unterschieden. So scheint in den Grenzschichten, und zwar besonders in der Tonoplastenhaut, der Hauptteil des *Permeationswiderstandes* des Plasmamantels für die osmotisch wirksamen Verbindungen lokalisiert (vgl. HÖFLER 1931a; hingegen CHOLNOKY 1953, p. 521f.), der für die von außen in die Vacuole eintretenden Substanzen (*Endosmose*, S. 80, 82, 83_7, 84) sowie für den Austritt dieser und auch zelleigener Stoffe *(Exosmose)* wirksam ist und bei Plasmolyse wie Deplasmolyse bisweilen Änderungen erfährt (S. 83_3, 89_3, 103_5).

Für eine stoffliche Verschiedenheit beider Grenzschichten spricht das differente Verhalten der Ansatzpunkte von den aus ihren Oberflächen gezogenen Fäden (HÖFLER 1951, p. 427; S. 88_2, 94). In kolloidchemischer Hinsicht soll sich das Binnenplasma dem Solzustand nähern (GAIDUKOV 1906, p. 588; KESSLER und RUHLAND 1938, p. 194), während den Grenzschichten ein mehr gelartiger Charakter zuzuschreiben ist.

Ihre Dicke wird die einer einzigen Moleküllage wohl übersteigen (vgl. KLERCKER 1892, p. 473; CHOLNOKY 1952a, p. 371), worauf auch die Fähigkeit zur Dickenzunahme der Tonoplastenhaut hindeutet (S. 88_6) und an ihren Außenseiten lassen sich vielleicht noch besondere Filme einer oder weniger Moleküllagen annehmen (vgl. KÜSTER 1910c, p. 689; MOTHES 1934, p. 508; wegen Grenzschichtdicke vgl. WEIS 1926, p. 146; PLOWE 1931b, p. 237; HÖBER 1947, p. 307; TÖRNÄVÄ 1939, p. 340). In vollentwickelten Zellen scheint die Substanzmenge der Grenzschichten bei Kontraktion und Ausdehnung konstant zu bleiben (p. 339f.).

---

[1] Auch *Tonoplastenmembran, Vacuolenhülle* (vgl. KÜSTER 1929b), *Vacuolenwand* (wobei an eine Herkunft aus der Vacuole gedacht war, vgl. WEBER 1932b; HÖFLER 1932a; BECKER und BECKEROWA 1934, p. 378f.), oder *Tonoplast* (DE VRIES 1885, p. 469/325) genannt. Mit letzterem Ausdruck wird ebenfalls, wie dies hier *ausschließlich* geschehen soll, die innere Plasmagrenzschicht samt der von ihr umschlossenen Vacuole bezeichnet (vgl. HÖFLER 1932a, p. 465; S. 87_3, 96_2).

A. Das Plasmalemma.

Es bildet die geschlossene äußere Grenzschicht des Protoplasten, die schon in unplasmolysierten Zellen angenommen wird, wenngleich sie sich normalerweise weder dort noch bei Plasmolyse ohne zusätzlichen experimentellen Eingriff zu erkennen gibt (vgl. Hofmeister 1954a, p. 279), ist *flüssig* (vgl. Plowe 1931a, p. 218), zeigt eine gewisse *Klebrigkeit*, die sich beim *Fadenziehen* manifestiert (S. 94_4), und besteht wohl aus *Lipoiden* (vgl. Hansteen-Cranner 1922, p. 134f.; Weis 1926, p. 161). Daneben wird auch eine *Kohlenhydrat-* und eine *Eiweißkomponente* angenommen (vgl. Frey-Wyssling 1955, p. 80; Szücs 1910, p. 770; Weis 1926, p. 179; Bogen 1938, p. 576; Cholnoky 1952a, p. 371), welche die hohe *Elastizität* des Plasmalemma erklären würden (Frey-Wyssling 1953, p. 200; Plowe 1931a, p. 208f.).

Direkt sichtbar wird die Substanz des Plasmalemma vielleicht in jenen dünnen Strängen, welche die einzelnen Tröpfchen in den aus Plasmafäden (S. 94_3) vorübergehend entstehenden Perlschnüren verbinden (p. 203, 216).

Das Plasmalemma übt eine *Schutzfunktion* (vgl. p. 217) für den Protoplasten aus, wobei sich je nach Objekt große Unterschiede in der Resistenz gegenüber KOH (Cholnoky 1952b, p. 62) oder Na_2CO_3 (Höfler 1951) zeigen, die auf stoffliche Verschiedenheiten hinweisen. Die wohl mit der Schutzwirkung zusammenhängende *Permeabilität* des Plasmalemma kann durch bestimmte Außenlösungen (vgl. Strugger 1931, p. 460; 1932, p. 24), den Plasmolysevorgang selbst (p. 30) und durch den osmotischen Druck des Außenmediums beeinflußt werden (vgl. Scarth 1926), wie sie sich offenbar auch bei der plasmolytischen Vacuolenkontraktion erhöht (S. 89, 92).

Bei der Kontraktion des Protoplasten wird durch das Abreißen vom wandständigen Plasma das Plasmalemma wohl zumindest stellenweise zerstört, so daß es hernach neugebildet oder wenigstens wieder instand gesetzt werden muß (vgl. Weber 1932d, p. 522; dagegen Plowe 1931a, p. 218). Auch an neu entstehenden Plasmaoberflächen (vgl. Pardatscher 1953, p. 32; Cholnoky 1953, p. 523, 535) wird Neubildung des Plasmalemma *(sekundäres Plasmalemma)* angenommen. Eine Restitution scheint das Vorhandensein von Ca-Ionen vorauszusetzen, da nach kurzer hypotonischer K-Oxalatvorbehandlung bei Plasmolyse Tonoplastenbildung einzusetzen pflegt, wenn das Calcium zuvor oder während der Plasmolyse nicht ersetzt wird (Weber 1932d, p. 524; 1934; vgl. „surface-precipitation-reaction", Heilbrunn 1927, p. 246f.). Nur gewisse *Desmidiaceen* vermögen auch ohne Ca normal zu plasmolysieren (Höfler 1951, p. 443).

In *Nitella*-Zellen bildet sich nach dem Einstechen von Elektroden in das Plasma um diese eine Membran, welche von Umrath (1932) als neugebildetes Plasmalemma angesehen wird.

Die Plasmaoberfläche kann durch Netzmittelwirkungen und Deplasmolyse aufgelockert werden (Hofmeister 1954a, p. 315; S. 85_1). Sie verändert sich besonders nach langdauernder Plasmolyse und bei bestimmten Außenbedingungen, indem sie sich dann verfestigt (vgl. Cholnoky 1950b, p. 84; Lorey 1929, p. 196; S. 93). So entstehen z. B. in Rohrzucker, Ca-Nitrat usw. starre oder zähe Oberflächenhäutchen (Küster 1910c, p. 692; „*Haptogenmembranen*", Küster 1909, p. 589; vgl. Kamiya 1939, p. 384). Bei rascher Deplasmolyse (S. 85_1) sprengt sie der Protoplast auf und quillt (manchmal als Tonoplast, S. 96) bruchsackartig daraus hervor, während dies bei langsamer Deplasmolyse offenbar wegen Wiederauflösung des Häutchens unterbleibt (Küster 1910c, p. 692f.). Umgekehrt bewirkt stärkere Plasmolyse, daß solche Plasmaoberflächen knittrig werden (p. 716).

Als Übergang dieser mikroskopisch unsichtbaren (p. 696) Plasmaverhärtungen zu den Regenerationsmembranen (S. 96_4) sind Bildungen in den Blattrandzellen von *Elodea* zu bewerten, wo durch Deplasmolyse bei Berührung zweier Teilprotoplasten eine plasmatische Scheinwand entsteht. Jene lösen sich manchmal von einer solchen Wand los, vermögen aber auch um ihre ganze Oberfläche geschlossene feste Hüllen zu entwickeln (1935a).

Die Fusion von Protoplasten oder Teilprotoplasten (vgl. 1939a; S. 96; von Tonoplasten: S. 96_3) hängt von der Beschaffenheit der Plasmaoberfläche ab, und ihr Ausbleiben deutet oft auf deren Veränderung hin. Es kann sich bei den Fusionen um Teilprotoplasten derselben Zelle oder nackte Protoplasten handeln, die zum Ausschlüpfen aus angeschnittenen Zellen veranlaßt wurden (S. 81). Teilprotoplasten derselben Zelle, Protoplasten verschiedener Zellen oder Gewebe artgleicher Pflanzen oder verschiedener Arten lassen sich fusionieren (*autoplastische, hom(ö)oplastische, heteroplastische Fusionen,* vgl. KÜSTER 1939a, p. 1; HOFMEISTER 1954a, p. 282).

Die Fusion wird durch Berühren oder Aneinanderpressen mittels mechanischer Manipulationen oder durch Deplasmolyse (KÜSTER 1909, p. 591; 1910b, p. 353; 1939a, p. 18; vgl. KUCHAR 1950a, p. 60; S. 85_3) eingeleitet; bestimmte Außenbedingungen (vgl. MICHEL 1937, p. 235; KÜSTER 1910c, p. 707) wirken förderlich, doch bleiben Fusionen auch dann noch „eine an sehr enge und kaum zu definierende Bedingungen geknüpfte Ausnahme", wobei die Polypeptidketten des Plasmalemma als fusionshemmend gelten (HOFMEISTER 1954a, p. 280f., 323). Die Fusionsbereitschaft ist zeitlich begrenzt (vgl. KÜSTER 1910c, p. 704; MISSBACH 1928, p. 334f.).

Als Qualitäten der Fusionen können *Verklebungen* (die Partner sind leicht wieder und unter einem Winkel unter 180° zu trennen, vgl. KÜSTER 1928, p. 230), *Verschmelzungen* (der Trennungswinkel ist etwa 180°, die Vacuolen bleiben noch getrennt) und *Vereinigungen* (auch die Vacuolen verschmelzen, was manchmal zum Tode führt, MICHEL 1937, p. 251) unterschieden werden (p. 244f.; vgl. HOFMEISTER 1954a, p. 282). Häufigkeit und Lebensdauer der Endprodukte nehmen in gleicher Reihenfolge ab.

Verschmelzungen von Paraffinöltröpfchen und Protoplasten (CHAMBERS und HÖFLER 1931, p. 346) werden durch bestimmte Vorbehandlungen gefördert (vgl. HOFMEISTER 1954a, p. 314f.).

B. Die Tonoplastenhaut.

Sie ist als die innere Plasmagrenzschicht (S. 85_8) am lebenden Protoplasten meist nicht direkt zu erkennen (vgl. PLOWE 1931a, p. 217; CHOLNOKY 1950b, p. 73) und tritt nach gewissen letalen Schädigungen des übrigen Plasmas als dünnes Häutchen hervor [vgl. DE VRIES 1885, p. 467/324; JANSE 1887b, p. 417; BOKORNY 1893; HÖFLER 1931a, p. (86); CHAMBERS und HALE 1932, p. 345; MEINDL 1934, p. 385]. Bei ihrer Sichtbarmachung sind freilich Veränderungen gegenüber den ursprünglichen Verhältnissen zu erwarten [vgl. HÖFLER 1931a, p. (91); 1952, p. 186; BANK 1935, p. 239]. Die isolierte Tonoplastenhaut umschließt dann die kontrahierte Vacuole und verdankt ihre oft erhebliche Festigkeit wohl der Kohäsion und den Oberflächenspannungsverhältnissen an ihren beiden Grenzflächen (MOTHES 1934, p. 498). Sie behält ihre Semipermeabilität und Spannung noch lange bei und wird in diesem Sinne als „lebend" bezeichnet (vgl. DE VRIES 1885, p. 540/391; KEMMER 1928, p. 18f.; HÖFLER 1932a, p. 468f.).

Die Tonoplastenhaut zeigt ferner große Dehnbarkeit (CHAMBERS und HÖFLER 1931, p. 343) und eine gegenüber dem Plasmalemma allerdings geringere Elastizität (PLOWE 1931a, p. 218), Durchlässigkeit für Farbionen (HÖFLER 1952, p. 186), ist hyalin-durchscheinend, farblos, leicht klebrig (PLOWE 1931a, p. 212), flüssig (CHAMBERS und HÖFLER 1931, p. 346), mit Öltropfen (l. c.), aber nicht mit Wasser mischbar [HÖFLER 1931a, p. (88)] und läßt sich anfärben (LEDERER 1934, p. 413; 1935, p. 215f.; WULFF 1934, p. 76; BECKEROWA 1935, p. 390), wobei ihr auch vielfach eine dünne körnige Plasmaschicht gleicher Färbungseigenschaften anhaftet (*Tonoplastenscheide*, LEDERER 1934, 1935 l. c.; S. 90).

Mit Mikronadeln sind nur kurze Fäden auszuziehen (vgl. PLOWE 1931a, p. 212), deren Ansatzpunkt bei seitlicher Nadelbewegung längs der Oberfläche stets derart wandert, daß die Fäden auf dieser normal stehen (CHAMBERS und HÖFLER 1931, p. 344; S. 85_6). Ähnliche deformierbare Fäden können auch die Tonoplastenhaut von Teilprotoplasten verbinden, wobei sie sich deutlich von umgebendem Plasma unterscheiden (KÜSTER 1926, p. 86f.).

Durch elektrischen Strom läßt sich eine möglicherweise aus der Tonoplastenhaut entstehende Niederschlagsmembran von der Innenseite des Plasmamantels trennen (WEIS 1926, p. 169). Zerplatzt ein Tonoplast, so bleiben von dessen Haut höchstens noch ganz wenige, außerordentlich kleine Tropfen zu erkennen (vgl. KLERCKER 1892, p. 473; CHAMBERS und HÖFLER 1931, p. 345).

Die Tonoplastenhaut in den ausgewachsenen Zellen von *Sphaeroplea annulina* ist auch unter natürlichen Verhältnissen zu beobachten (MOTHES 1934, p. 488).

Ihrer chemischen Natur nach besteht sie wahrscheinlich aus *Lipoiden* (vgl. CHAMBERS und HÖFLER 1931, p. 351), daneben dürfte auch eine *Eiweiß*komponente vorhanden sein (EICHBERGER 1934, p. 628; MOTHES 1934, p. 498f.; BOGEN 1938, p. 576; CHOLNOKY 1952c, p. 229).

Bei Aufquellung des Plasmas durch Vacuolenkontraktion oder Kappenplasmolyse (S. 88_8, 89_5) tritt die Tonoplastenhaut viel stärker hervor, rundet sich, und ihre *Dicke* nimmt zu (Lit.: CHOLNOKY und HÖFLER 1950, p. 177; S. 85_7, 91_6). Sie vermag ferner zu *erstarren* (vgl. DE VRIES 1885, p. 529/381f.; JANSE 1887b, p. 420; EICHBERGER 1934, p. 626; CHOLNOKY 1952b, p. 63) und *aufzuquellen* (MOTHES 1934, p. 506).

Bei Kappenplasmolyse werden die normalerweise sehr *hohe Wasserpermeabilität* (HUBER und HÖFLER 1930, p. 451) und die Durchlässigkeit für Salze herabgesetzt (HÖFLER 1939b, p. 552, 571; vgl. hingegen BOGEN 1951, p. 8; S. 85_2, 89_3), andererseits scheint diese bei Vacuolenkontraktion auch für Zellsaftbestandteile erhöht (vgl. HARTMAIR 1937, p. 586; HENNER 1934, p. 106; S. 91_2). Die *Dehnbarkeit, Festigkeit* und *Hitzeresistenz* der Tonoplastenhaut lassen sich ebenfalls experimentell beeinflussen (DÖRING 1933, p. 421f.).

C. Das Mesoplasma.

Als solches wird das von Plasmalemma und Tonoplastenhaut umgrenzte mehr oder weniger flüssige *Binnenplasma* bezeichnet (PLOWE 1931a, p. 202). Es besitzt eine gegenüber dem Plasmalemma niedrigere Elastizität (p. 208) und ist in besonders hohem Maße zur Wasseraufnahme und -abgabe fähig, was zu *Quellung* (S. 92) sowie *Entquellung* bzw. *Verfestigung* des Protoplasmas führen kann, wobei auch die Tonoplastenhaut Veränderungen zu zeigen vermag (S. 88_6).

Wegen der Umgrenzung durch die Membran ist eine Zunahme des Plasmavolumens stets an eine Verkleinerung der Vacuole gebunden, die entweder gleichzeitig erfolgt oder (z. B. als Reizplasmolyse, S. 83_1) vorausgehen kann. In ersterem Fall stammt das Quellungswasser wohl hauptsächlich aus der Vacuole (vgl. WEBER 1930e, p. 315; DE HAAN 1935, p. 395), in letzterem aus der Außenlösung, wobei sich noch Beziehungen zu anderen Kontraktionsvorgängen zeigen (S. 83_3), und auch nekrotische Bildungen entstehen können („*Doppelsaumplasmolyse*“, HÖFLER 1947, p. 598).

Eine solche Quellung vermag spontan oder nach bestimmten Einwirkungen, in Wasser oder hypotonischen, bzw. hypertonischen Außenmedien einzusetzen (*spontane oder aplasmolytische* bzw. *plasmolytische Vacuolenkontraktion*[1], WEBER 1932c, p. 518). In letzterem Fall kann die Quellung mitunter erst nach dem Erreichen der perfekten Plasmolyse beginnen, sich dann vorwiegend auf die freien Kuppen beschränken, was mit der Tendenz zur minimalen Vacuolenoberfläche

[1] Die Termini „*Tonoplastenplasmolyse*“ (vgl. STRUGGER 1931, p. 461f.) oder „*Vacuolenplasmolyse*“ [vgl. HÖFLER 1928, p. (79); FITTING 1919, p. 25f.] erscheinen ungünstig gewählt (vgl. KÜSTER 1951, p. 483). Gelegentlich wird auch eine als Zellsaftentmischung aufgefaßte Erscheinung als Vacuolenkontraktion betrachtet (S. 100).

zusammenhängen mag (vgl. KÜSTER 1938a, p. 153), und meist einen Vorraum bestehen lassen [*Kappenplasmolyse*, HÖFLER 1928, p. (74); vgl. BOGEN 1951, p. 2], doch sitzen bisweilen auch Bläschen und Schleifen gequollenen Plasmas der Vacuolenoberfläche auf (HÖFLER 1934b, p. 72).

a) Kappenplasmolyse.

Sie zeigt auch Beziehungen zu anderen Kontraktionserscheinungen (S. 83), gilt als echt vitale Veränderung des Plasmas und tritt an vielen Objekten in hypertonischen Lösungen von Alkalisalzen ein [1928, p. (78); 1939b, p. 548; s. Abb. 4; S. 93_1].

Die Salze gelangen dabei durch das Plasmalemma zwar leicht in das Mesoplasma (hohe *Intrabilität*), wo sie die Quellung bewirken, aber kaum in die Vacuole (sehr geringe *Permeabilität*, 1934b, p. 75; S. 85_5, 88_7).

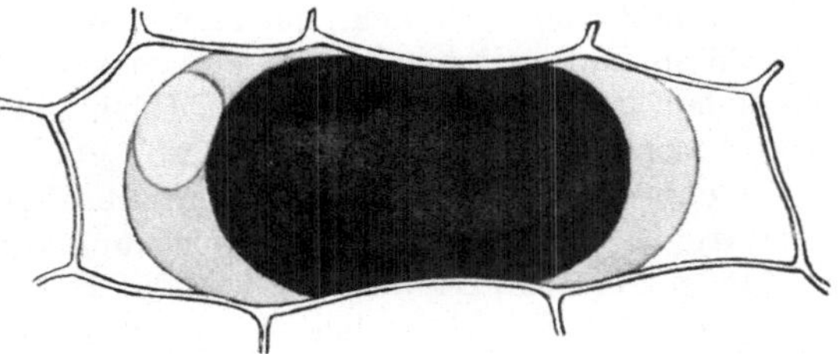

Abb. 4. Kappenplasmolyse einer Zelle der Außenepidermis der Zwiebelschuppe von *Allium Cepa* [nach HÖFLER 1928, p. (74)].

In bestimmten Fällen läßt sich Quellung hervorrufen auch durch Harnstoff, Sulfo- und Methylharnstoff (vgl. MODER 1932, p. 17; HÖFLER 1939a, p. 131f.; HOFMEISTER 1948a, p. 73; BIEBL 1948, p. 140), Zusatz von Neutralrot (BOGEN 1951, p. 20) oder $MgSO_4$ (SEEMANN 1952, p. 118) zum Plasmolytikum, Deplasmolyse in hypotonischem Milieu (vgl. DE HAAN 1935, p. 398; LANZ 1942, p. 398; KÜSTER 1942a, p. 140; S. 84), Wundreiz (STRUGGER 1931, p. 458), Einwirkung von α- oder UV-Strahlen (BIEBL 1941, p. 216f.; 1943, p. 6) oder Ca-Nitrat (LANZ 1942, p. 401).

Die Quellung, die durch Permeabilitätserhöhung des Plasmalemma (S. 86) beschleunigt wird und mit Veränderungen in der Tonoplastenhaut und im Plasma einhergeht (S. 88_6, 93_1), führt zu einer oft sehr erheblichen Vergrößerung des Plasmavolumens und stets zur Sistierung der Plasmaströmung (S. 93_4). Diese Volumzunahme ist während begrenzter Zeit durch Übertragen des Objektes in Ca- oder Sr-Salzlösungen oder Mischlösungen aus K- und Ca-Salzen rückgängig zu machen (HÖFLER 1940b, p. 299), wobei es mitunter zur Ausbildung von Teilvacuolen kommt (KAISERLEHNER 1939, p. 587; S. 100), und kann in solchen Lösungen auch überhaupt unterbleiben [HÖFLER 1928, p. (80)]. Bei Kappenplasmolyse zeigt sich in den Schnitten oft eine *Zonierung*, indem in der Schnittmitte normale, langsam zurückgehende Plasmolysen auftreten und die Kappenplasmolysen in der umgebenden Außenzone, die von einer Randzone mit durch den Schnittvorgang geschädigten Zellen begrenzt wird (1934b, p. 77f.; 1940b, p. 297f.; S. 104), zu finden sind.

Längeres Wässern erschwert die Kappenbildung (HOUSKA 1941, p. 15; S. 93_3), die bei Plasmasystrophe vorzugsweise in den Randzellen beobachtet wird (S. 99).

Quellung in LiCl ist reversibel und erreicht dort nach etwa 48 Std ihr Maximum (vgl. BOGEN 1951, p. 7f.). Bei Deplasmolyse (S. 85_2), die nach Kappenplasmolyse verlangsamt wird, kann es in den Plasmakuppen zu sekundären Zonungen kommen (KAISERLEHNER 1939, p. 594).

Das in K-Salz gequollene Kappenplasma ist gegenüber mechanischen Eingriffen, Wasser, hypotonischen und salzfreien isotonischen Lösungen weniger resistent als normales Plasma (HÖFLER 1940b, p. 295), wohingegen die Hitzekoagulation verzögert wird (DÖRING 1933, p. 425).

Mit der Ausbildung der Kuppen beginnen in diesen *Entmischungen* (S. 93_5), von welchen die frühen Stadien noch entquellbar sind (KAISERLEHNER 1939,

p. 593). Bei den anschließenden irreversiblen Entmischungen kommt es zur Trennung des Plasmas in einen flüssigen äußeren und einen als *Tonoplastenscheide* (S. 88) auftretenden koagulierten inneren Anteil. Bei Konzentrationswechsel treten in den Kappen deutliche Volumänderungen auf (HÖFLER 1940b, p. 295; BOGEN 1951, p. 10f.).

b) Spontane oder aplasmolytische Vacuolenkontraktion.

Sie führt entweder zu einer schwachen, oft nur an den Zellecken sichtbaren Quellung oder zu einer bisweilen sehr starken Volumzunahme des Plasmas („Allgemeinkontraktion" oder „Einzelkontraktion", HÖFLER 1947, p. 606; Lit.: SCHEIDL 1954, p. 645f.; S. 93_1), die in zylindrischen Zellen quantitativ bestimmbar ist (HENNER 1934, p. 87). Bei gewissen Objekten scheint sich aber die Tonoplastenhaut von dem nicht quellenden wandständig bleibenden übrigen Plasma abzulösen und zwischen beiden ein von farbloser Flüssigkeit erfüllter Raum zu entstehen (CHOLNOKY und HÖFLER 1950, p. 178). Die Vacuolenkontraktion vermag besonders bei hohem Kontraktionsgrad (KÜSTER 1938a, p. 151) *irreversibel* zu sein, und das Plasma kann von Zellsaftschläuchen durchsetzt werden oder erstarren (1926, p. 99), durch Entmischungsvorgänge Vacuolen ausbilden (WEBER 1930e, p. 314; vgl. HENNER 1934, p. 85; PARDATSCHER 1951b, p. 172f.; SCHEIDL 1954, p. 660; S. 93_5), wobei der Kern aufzuquellen vermag (p. 666; S. 97) oder grobschaumig werden (WEBER 1930d, p. 600), was gelegentlich die Vacuole deformiert (KÜSTER 1940, p. 415). Durch Plasmalamellen getrennte Teilvacuolen können weiter auseinanderrücken und sich dabei abrunden (KIERMAYER 1954, p. 218). Die Kontraktion ist bisweilen aber auch *reversibel*, und die Wiederausdehnung beginnt vielleicht unter Sprengung einer dünnen Oberflächenschicht (vgl. KÜSTER 1926, p. 100) spontan, in Wasser, in hypo- oder hypertonischen Lösungen oder solchen mit bestimmtem p_H (vgl. PARDATSCHER 1953, p. 27; 1951a, p. 105; b, p. 179; SCHEIDL 1954, p. 559f.; STRUGGER 1936, p. 58; HENNER 1934, p. 97f.; S. 84). Sie geht dann langsamer oder gleich schnell wie die Kontraktion vor sich (vgl. KEIL 1930, p. 570; HENNER 1934, p. 100) und ist manchmal von einer neuen Kontraktion mit Wiederausdehnung usw. gefolgt. Die Zelle nimmt dabei um so weniger Schaden, je langsamer die Kontraktion verläuft und je schwächer die Vacuole kontrahiert ist (SCHEIDL 1954, p. 659; S. 101). Bei starker Kontraktion und bei Wiederausdehnung platzt gelegentlich die Vacuole (vgl. p. 660; KEIL 1930, p. 578), die sich bisweilen auch entmischt (PARDATSCHER 1951a, p. 39, 60).

Meist besitzt die kontrahierte Vacuole eine konvexe Form (vgl. hingegen KÜSTER 1938b, p. 30; PARDATSCHER 1951a, p. 36), doch setzt die Quellung im Plasma mitunter nicht gleichmäßig ein und die Vacuole kann, besonders bei amphinekrotischen Zellen (KEIL 1930, p. 576; KÜSTER 1926, p. 100f.; S. 104_1), exzentrisch zu liegen kommen, in mehrere Teilvacuolen aufgelöst werden (p. 97; 1940, p. 414; CHOLNOKY und HÖFLER 1950, p. 148f.; SCHEIDL 1954, p. 657, 665; PARDATSCHER 1951a, p. 23; S. 100) sowie bei gefärbten Zellsäften Farbumschlag zeigen (Lit.: p. 37). Die Plasmaquellung vermag auch lokal auszusetzen, so daß krampfplasmolytische Formen der Vacuole entstehen (vgl. SCHEIDL 1955, p. 339; am Schnittrand liegende Zellen: KEIL 1930, p. 570; S. 104_3).

Bei Zusatz hypertonischer Lösungen erfolgt Plasmolyse und das gequollene Plasma hebt sich von der Wand ab, wobei die Vacuole unverändert bleibt, kleiner wird, sich deformiert, teilt (KÜSTER 1938b, p. 34), verlagert, vergrößert (siehe oben) oder unter Entleerung in das Plasma zerplatzt (PARDATSCHER 1953, p. 32).

Mitunter kommt es bei stark kontrahierten Zellen auch zur Ausdehnung der Vacuole (KÜSTER 1938a, p. 154; 1938b, p. 34). Diese Plasmolyse, welche dann sogleich als Kappenplasmolyse auftritt (WEBER 1930c, p. 113; HOFMEISTER 1948a, p. 60), und durch die sich eine Vacuolenkontraktion klar von einer Reizplasmolyse unterscheiden läßt, führt gelegentlich zu anfangs fädigen und reversiblen Entmischungen im Plasma (S. 93_5), wird durch Vorbehandlung mit Li-Carbonat unterbunden (WEBER 1930d, p. 603, 601) und ist in hypotonischen Lösungen oder Wasser meist reversibel (HENNER 1934, p. 95; HARTMAIR 1937, p. 589).

Vacuolenkontraktion, die offenbar mit Veränderungen der Permeabilität der Tonoplastenhaut einherzugehen (S. 88_7) und den osmotischen Wert der Zelle bis auf $^1/_{10}$ seiner ursprünglichen Größe zu erniedrigen vermag (HENNER 1934, p. 96; HARTMAIR 1937, p. 586; BANCHER 1938, p. 239; hingegen: HOFMEISTER 1948a, p. 73), kann auch in wasserfreiem Milieu einsetzen (HENNER 1934, p. 98; vgl. KEIL 1930, p. 582f.), tritt spontan bzw. nach Einwirkungen, denen wohl ein gewisser schädigender Charakter zuzusprechen ist, auf (*spontane bzw. aplasmolytische* Vacuolenkontraktion, KÜSTER 1938a, p. 150; S. 101) und scheint Anthocyan-führende Zellen zu bevorzugen (PARDATSCHER 1951a, p. 16f.).

In experimentell nicht alterierten Zellen kommt Vacuolenkontraktion als Alterserscheinung (WEBER 1930d, p. 599; 1930e, p. 313; vgl. BANCHER 1938, p. 239; PARDATSCHER 1951b, p. 182; S. 104), aber auch schon in früheren Entwicklungsstadien der Zelle vor (1951a, p. 34). In letzterem Falle können die Vacuolen unter gleichzeitiger Degeneration des Plasmas verfallen oder aufplatzen (1953, p. 26f.).

Andererseits bewirken bereits *mechanische Einwirkungen* (vgl. KEIL 1930, p. 596, 586; Bürstenbehandlung, KÜSTER 1929c, p. 168) sowie *UV-Bestrahlung* (BIEBL 1942, p. 500) oder *Kälte* (MEINDL 1934, p. 385) Vacuolenkontraktion.

Sie wird ferner durch bestimmte *Milieus* ausgelöst (vgl. ÅKERMAN 1917, p. 158f.). Als solche wirken destilliertes Wasser (HENNER 1934, p. 89; HOFMEISTER 1940, p. 84), Leitungswasser (HENNER 1934, p. 85f.; BANCHER 1938, p. 239; vgl. PARDATSCHER 1953, p. 26; WEBER 1930d, p. 600), hypotonische oder isotonische Plasmolyticis (FITTING 1919, p. 25, 158; HENNER 1934, p. 90; WEBER 1925e, p. 73), undissoziierte Basen (SCHEIDL 1954), stark verdünnte Lösungen von Neutralrot (KÜSTER 1926, p. 96f.; 1940, p. 413), wenn sich dieses zunächst diffus im Zellsaft löst (WEBER 1930c, p. 107), oder besonders stark bei Temperaturerhöhung (KÜSTER 1937a, p. 439), und andere Farbstoffe (1933b, p. 410; 1937a, p. 440f.; HÖFLER 1947, p. 609; HOFMEISTER 1948a, p. 60; STRUGGER 1949, p. 72; vgl. KIERMAYER 1954, p. 188f.; SCHEIDL 1954; 1955), wobei die Kontraktion mit der Stärke der Farbstoffspeicherung in der Vacuole (1954, p. 652f.), dem Farbton (vgl. DRAWERT 1938, p. 126) oder der Fluorescenzfarbe (HÖFLER 1947, p. 600f.) korreliert. Mitunter erfolgt dabei die Kontraktion sehr rasch (vgl. KÜSTER 1938b, p. 29) und es setzt zusätzlich Reizplasmolyse ein (S. 83).

Auch in verdünnten Säuren oder Basen (BRENNER 1918, p. 82f.), in hypotonischen Alkalisalzlösungen (außer KCNS, STRUGGER 1932, p. 30) und Pufferlösungen (PARDATSCHER 1951a, p. 83) vermag das Plasma aufzuquellen (vgl. HÖFLER 1940b, p. 294; BOGEN 1951, p. 18), wobei sich dieses und die Tonoplastenhaut bisweilen merklich verändern (S. 88_6, 93_1). Eisessigdämpfe bewirken Kontraktionen mit unregelmäßigen Vacuolenumrißformen (KEIL 1930, p. 595). In anderen Fällen ist gleichzeitiges oder aufeinanderfolgendes Einwirken von Neutralrot und Leitungswasser (WEBER 1930c, p. 116; KUNZE 1931, p. 162; KRESSIN 1935, p. 44; STRUGGER 1936, p. 58) bzw. ein bestimmter p_H-Wert der Neutralrotlösung (KUNZE 1931), *Wundreiz* (vgl. KÜSTER 1926, p. 101; 1929c, p. 168; 1937a, p. 439; 1938b, p. 28; 1939c, p. 45; KEIL 1930, p. 576, 591; STRUGGER 1949, p. 72; DRAWERT 1938, p. 128), Neutralrot und *Narkotica* (KÜSTER 1937a, p. 440), Mischlösungen zweier Vitalfarbstoffe und Wundreiz (1942b, p. 246) oder hypotonische Lösungen und *Erwärmung* (KEMMER 1928, p. 56) zur Auslösung der Vacuolenkontraktion nötig, wobei auffällt, daß die Kontraktion oft nur bei alkalischem Zellsaft auftritt (KÜSTER 1942b, p. 264).

c) *Plasmolytische Vacuolenkontraktion.*

Hierbei (vgl. FITTING 1919, p. 25, 158) bleibt der Protoplast entweder dauernd mit der Membran verbunden (STRUGGER 1931, p. 461; vgl. BANK 1935, p. 240), was mitunter mit seiner hohen Adhäsion zusammenhängt (vgl. TAKAMINE 1940b, p. 583; S. 73), oder eine Quellung beginnt während einer plasmolytischen

Kontraktion (vgl. CHOLNOKY 1950a, p. 381), oder es erfolgt zunächst Kappenplasmolyse (z.B. in KCNS, STRUGGER 1932, p. 24; KRESSIN 1935, p. 38f.). Auch hier scheint die Permeabilität des Plasmalemma durch die Versuchseinwirkungen gegenüber jener des Tonoplasten wesentlich erhöht (vgl. STRUGGER 1931, p. 463; S. 86), wobei die Zellkerne oft eine starke Quellung erfahren (S. 97_2).

d) *Plasmoschise.*

Sie stellt eine Spaltung bzw. Zerreißung des Plasmamantels dar, die oft so erfolgt, daß ein dicker Plasmabelag an der Membran zurückbleibt (ISRAEL 1897, p. 300; S. 72, 73), und kann bei Plasmolyse gleich am Beginn der Protoplastenkontraktion auftreten oder während der Kontraktion, wenn die äußere Plasmaschicht innehält und sich von den weiter kontrahierenden inneren Plasmateilen trennt (KÜSTER 1926, p. 88; 1929b, p. 29; vgl. STRUGGER 1931, p. 466, 473). Plasmoschise erfolgt auch bei Deplasmolyse, wenn sich die inneren Plasmapartien weniger stark ausdehnen (vgl. WEBER 1929b, p. 160; S. 85). An der Vacuole, die sich stark zu verformen vermag (S. 100), bleibt dann oft nur eine dünne Plasmaschicht, die durch Fäden mit dem Wandplasma verbunden sein kann (S. 94). Sie ist ferner eine oft durch Verfestigungen im Plasma ausgelöste Absterbeerscheinung, für die Tonoplastenbildung von Bedeutung (S. 96), zeigt zu anderen Kontraktionserscheinungen zahlreiche Übergänge (S. 83) und läßt sich nach geeigneter Vorbehandlung (vgl. GICKLHORN 1933, p. 574f.) und bei osmotischem Wasserentzug (vgl. WEBER 1933, p. 243; Plasmolyse geschädigter Zellen, vgl. KÜSTER 1926, p. 89f.; Zugabe von Netzmittel zum Plasmolytikum, HOFMEISTER 1954a, p. 306), aber auch durch Reizeinwirkungen (*Reizplasmoschise*, SCHÖNLEBER 1935) hervorrufen.

VIII. Die einzelnen Teile des Protoplasten bei Plasmolyse.

Je nach Art, Stärke und Dauer des plasmolytischen Eingriffes vermögen die Zellorgane ihr normales Verhalten und Aussehen in oft charakteristischer Weise abzuändern, es können ferner neue Differenzierungen entstehen, bereits vorhandene verschwinden oder modifiziert werden und Schädigungen auftreten (S. 101). Dies ist, soweit es allein eine Entquellung und Verformung betrifft, manchmal als unmittelbare Folge des osmotischen Wasserentzuges zu erkennen, meist aber steht die kausale Erklärung noch aus.

A. Das Cytoplasma.

a) *Allgemeines Verhalten bei Plasmolyse.*

Plasmolyse wirkt auf das Cytoplasma durch den Wasserentzug zunächst entquellend (vgl. KÜSTER 1951, p. 19; WALTER 1922, p. 164f.; RUGE 1940, p. 327; TAKAMINE 1940a, p. 310; S. 74). Dieser Effekt kann (z. B. bei Vacuolenkontraktion) durch zusätzliche Wirkungen überkompensiert, aber auch verstärkt werden, wobei es sich primär stets um Änderungen des *Hydratationsgrades* und damit auch der Viscosität handelt (vgl. BOGEN 1951, p. 30f.; S. 93).

Quellung. Eine leichte Quellung des Plasmas (vgl. Bd. I, Abschnitt II D; S. 88) ist oft bei Plasmolyse und besonders nach Alkalisalzeinwirkungen zu finden (vgl. HÖFLER 1939b, p. 553; S. 97_4, 102). Bei stärkerer Volumzunahme, die sich als eine Überschwemmung des Binnenplasmas mit Wasser bzw. Außenlösung auffassen läßt (HOFMEISTER 1948a, p. 74), sind wahrscheinlich zwei Arten solcher Mechanismen zu unterscheiden, von denen der eine unter intensiver Aufquellung bei starker Volumverkleinerung der Vacuole zu offenbar schwerwiegenden Veränderungen und baldigem Tod des Plasmas führt. Im anderen

Falle erfolgt eine begrenzte, manchmal reversible Volumzunahme („schwache Kontraktion“, HENNER 1934, p. 104f.; Kappenplasmolyse, S. 89_2; HÖFLER 1939b, p. 551; S. 90_2, 91_6). Das Plasma ist dann leichtflüssig (p. 548; KÜSTER 1938a, p. 153), hat geringe Viscosität, neigt mit der Zeit zu Entmischungen (l. c.; KAISERLEHNER 1939, p. 591f.; S. 89_5, 93_5) und zeigt mitunter Änderungen in der Plasmaströmung (S. 93_4).

Umgekehrt vermögen Ca-Salze zu einer *Verfestigung* und *Verhärtung* des Plasmas zu führen, bei langer Einwirkung selbst in hypotonischen Lösungen (vgl. SEEMANN 1951, p. 125). Ihre hydratationsherabsetzende Wirkung soll die gelartigen Außenschichten verfestigen (S. 86), während sie im solartigen Binnenplasma die Viscosität verringert (vgl. KESSLER und RUHLAND 1938, p. 194; S. 93_3). Auch Al-Salzplasmolyse wirkt verfestigend (vgl. KÜSTER 1929b, p. 116; MISSBACH 1928, p. 331f.; S. 73_4).

Viscosität. Die Viscosität (vgl. Bd. I, Abschn. II Dc), die mit dem Quellungszustand sowie den Verfestigungen eng zusammenhängt (S. 92) und sich z. B. auch in der Fähigkeit zum Fadenziehen manifestiert (vgl. WEBER 1921, p. 173f.; PFEIFFER 1935; S. 94_1), ist für die Plasmolyseform und Rundungszeit mitbestimmend (S. 75, 80_3, 81), spielt auch für die Protoplastenverschiebungen innerhalb des Zellumens eine Rolle (S. 82) und läßt sich aus ersteren sowie unter anderem durch Beobachtung der BROWNschen Molekularbewegung beurteilen (vgl. PEKAREK 1940; S. 80_1). Dabei entspricht oft einer höheren Viscosität auch eine hohe Adhäsion des Protoplasten, doch ist auch das umgekehrte Verhalten zu finden (HÖFLER 1932b, p. 211; S. 73_2). Die Viscosität kann mit steigendem Alter abnehmen (vgl. FISCHER 1948, p. 519f.) und scheint gelegentlich außer durch Plasmaquellung noch durch Temperaturerniedrigung, Narkotica (vgl. WEBER 1925b, p. 710; DERRY 1929, p. 15f.; SCHEITTERER 1930, p. 292), Zugabe von Netzmittel zum Plasmolytikum (besonders für die äußeren Plasmaschichten, HOFMEISTER 1954a, p. 304), bestimmte Plasmolytica (vgl. WEIS 1926, p. 156; S. 93_2), Vorbehandlung mit Alkohol, Änderung des p_H-Wertes der Außenlösung (STRUGGER 1934, p. 434), hypotonische Salzlösungen (WEBER 1924a, p. 263f.) und durch Vitalfärbung (vgl. HOFMEISTER 1948a, p. 60) variiert zu werden. Vorwässerung (S. 73_4, 85, 89_6, 98_8, 101_2, 101_4, 103) bewirkt Viscositätserniedrigung, doch scheint auch der gegensinnige Effekt möglich (S. 73_6).

Plasmaströmung, ein sicheres Indiz des Lebenszustandes (vgl. SCHNEIDER 1924, p. 32; S. 101_2), ist für die systrophischen Verlagerungen (S. 98_3) von Bedeutung und wird bisweilen durch den Plasmolysevorgang stimuliert (vgl. STRUGGER 1949, p. 97) oder behindert (vgl. ROMIJN 1931, p. 295; YOTSUYANAGI 1953, p. 150). Gequollenes Plasma vermag Glitschbewegungen zu zeigen (KÜSTER 1938a, p. 153), hingegen unterbleibt bei Kappenplasmolyse jede Strömung (vgl. HÖFLER 1940b, p. 294; S. 89_5), ohne daß die Vitalität der Zelle vermindert scheint. Änderungen der Strömungsintensität deuten auf Alterationen des Plasmas hin (PLOWE 1931a, p. 200; S. 93_1).

Doppelbrechung nach Plasmolyse wurde an *Allium*-Epidermen beobachtet (ULLRICH 1937, p. 316). *Entmischungen* im Plasma (in der Vacuole, S. 100) können bei Plasmolyse, nach Vacuolenkontraktion, Kappenplasmolyse und nach dem Auftreten besonderer Plasmabildungen in der Vacuole einsetzen (S. 89_9, 90_2, 91_1, 93_1, 94_7). Bisweilen entmischt sich auch der gesamte Zellinhalt, mitunter nach Entstehung von Plasmatentakeln in der Vacuole (GICKLHORN 1932b). Einen *netzartigen Zerfall* erleidet gelegentlich das Plasma von *Valonia*-Zellen in hypertonischen Salzlösungen (JOST 1929, p. 6f.). Eine *dichtwabige Plasmakonfiguration* entsteht nach Plasmolyse in jüngeren Epidermiszellen des Labellums von *Platanthera bifolia* (GERM 1947, p. 515).

b) Besondere plasmatische Elemente.

Die bei Plasmolyse auftretenden *Plasmafäden, -stränge, -lamellen* u. ä. gehen von der Plasmahaut *nach außen* zur Zellwand, *durch den Zellsaftraum hindurch*, verschiedene Stellen des Plasmamantels verbindend, oder *liegen* diesem unmittelbar *auf*. Sie können sich auch zwischen die durch Plasmoschise (S. 92) zerrissenen Plasmateile spannen, in stark gequollenem Plasma oder bei Vacuolenentmischung als unterscheidbare Elemente hervortreten (vgl. KÜSTER 1926, p. 96; WEBER 1930d, p. 601; S. 100_4), zum Tonoplasten führen (S. 96_3) und sind Ausdruck der Plasmaviscosität (S. 93_3).

Plasmafäden. Die vom Protoplasten nach außen abgehenden Plasmafäden (HECHT*sche Fäden*) sind zur Membran gerichtet, endigen an einer Regenerationsmembran oder einem Teilprotoplasten (vgl. CHODAT 1907, p. 19; S. 96_5). Sie sind oft sehr verschieden dick und verzweigt, wobei die stärkeren häufig die Fortsetzung von Plasmodesmen bilden (vgl. WEBER 1929d, p. 584; S. 73), bestehen aus Lipoiden wohl ohne Proteine (vgl. WEIS 1926, p. 178), sind mitunter außerordentlich zahlreich, was auch mit der Konzentration der Außenlösung zusammenhängen mag (vgl. WEBER 1921, p. 173) und werden manchmal nur in Elektrolytlösungen beobachtet (TAKAMINE 1940b, p. 585). *Plasmalamellen* zwischen Zellwand und Protoplast entstehen gelegentlich auch aus der Annäherung zweier konkaver Plasmaabhebungen und werden dann vom Protoplasten meist wieder eingezogen (vgl. KUCHAR 1950a, p. 58).

Die zunächst flüssigen Fäden (vgl. LEPESCHKIN 1925, p. 21) setzen an den Tüpfeln und dem wandständigen Plasma an und halten sich oft bis zu mehreren Stunden (HECHT 1912, p. 158f., 172, 180; S. 74). Sie können (besonders bei Deplasmolyse, BÖRGER 1926, p. 132) *reißen*, wobei der eine Rest meist vom Protoplasten aufgenommen wird, während der andere als Plasmatröpfchen an der Membran bleibt (HECHT 1912, p. 163; S. 95), oder es bilden sich später (bei Zugabe von Wasser sofort, p. 180) perlschnurartig aneinandergereihte Plasmatröpfchen, wahrscheinlich aus Mesoplasma mit einer Oberflächenschicht aus Plasmalemma bestehend (PLOWE 1931a, p. 203f., 216), die nur durch ganz dünne Fäden aus Plasmalemmasubstanz (p. 204) miteinander verbunden sind und schließlich *zerfallen* (HECHT 1912, p. 163f.; S. 86_2). Ferner vermögen die Fäden zu *erstarren* (vgl. LEPESCHKIN 1925, p. 25; KÜSTER 1951, p. 97) und manchmal zu *fusionieren*, was gelegentlich zu Vacuolen mit Außenlösung führt (DIANNELIDIS 1951, p. 37; S. 96_7).

Die *Fußpunkte* am Protoplasten bleiben fix, so daß ursprünglich senkrecht stehende Fäden bei seitlicher Verschiebung der Ansatzstelle ihre Richtung ändern (TIROLD 1932, p. 354; S. 85). Das wandständige Plasma, die Plasmaviscosität (vgl. WEBER 1921) und die Klebrigkeit des Plasmalemma scheinen für die Bildung der Fäden wichtig zu sein. Diese lassen sich nämlich nach beliebigem In-Berührung-Bringen von Membran oder mikrurgischer Nadel aus dem Protoplasten ausziehen (PLOWE 1931a, p. 210; S. 86_1), doch bleiben nach der einer Deplasmolyse folgenden zweiten Plasmolyse die dünnen Fäden aus (vgl. HECHT 1912, p. 181).

Plasmafäden, deren Spitze frei endigt, können bei Plasmorrhyse von Myxomyceten in Form feiner Nadeln oder Haare stehenbleiben (vgl. DE BARY 1864, p. 127, Fig. 11; BALBACH 1936, p. 198; S. 106) oder an nackten Protoplasten als feine cilienartige Bildungen auftreten (KLERCKER 1892, p. 470).

Plasmafäden und -lamellen längs des Plasmamantels entstehen als Verbindung lokaler Plasmaballungen, manchmal auch in stärkerer Ausbildung (KÜSTER 1910a, p. 279; GERM 1932, p. 610f.).

Plasmagebilde in der Vacuole ragen zuweilen als *Plasmazungen* (z. B. bei Systrophe, S. 98) in den Zellsaftraum hinein (KÜSTER 1926, p. 75f.; GICKELHORN 1932b), was mitunter zu Vacuolenteilung oder Entmischung führt (S. 93_5, 100_1).

Häufig treten auch Plasmafäden, die die Vacuole durchziehen (ÅKERMAN 1916, p. 60), als *Plasmanetzwerk* bei Systrophe auf (vgl. KLERCKER 1892, p. 471; GERM 1932, p. 573, 582). Nach langdauernder Plasmolyse entstehen aus den dickeren Plasmasträngen durch Querverbindungen *Plasmaraumgitter* (p. 582). *Plasmalamellen*, die den Zellsaftraum in voneinander getrennte Vacuolen zerteilen, können sich nach Systrophe, wahrscheinlich aus den Plasmafäden, bilden (vgl. HABERLANDT 1919a, p. 332; 1920, p. 327f.). Sie ziehen manchmal stellenweise die Plasmaoberfläche gegen das Zellinnere zu, so daß sich der Protoplast an ihrer Ansatzstelle einbiegt (*Protoplastenfurchung*, vgl. GERM 1932, p. 616; „*Kerbplasmolyse*", HÖFLER 1918a, p. 155; HABERLANDT 1919a, p. 337; S. 78, 95_3) und kommen auch in den kleinen Vacuolen plasmareicher Zellen vor, wo sie bei Plasmolyse gelegentlich reißen (vgl. KUCHAR 1950a, p. 59).

Plasmatröpfchen bilden sich beim Zerfall von Plasmafäden (S. 94) oder bisweilen als sphärische Ausgliederungen des Protoplasten nach länger dauernder Plasmolyse (YOTSUYANAGI 1953, p. 148). Sie können auch mit einer oder mehreren Vacuolen ausgestattet sein und so Übergänge zu den Teilprotoplasten darstellen. „*Plasmaamöben*" (KÜSTER 1910a, p. 279; vgl. GERM 1932, p. 614), d. h. kleine Plasmatropfen, die lebhafte Formänderungen zeigen, ohne sich von der Stelle zu bewegen, sind manchmal an der Oberfläche plasmolysierter Protoplasten zu beobachten.

c) Spezielle Neubildungen.

Teilprotoplasten. Die Entstehung von Teilprotoplasten in langgestreckten Zellen wird durch die Oberflächenspannung bedingt (BERTHOLD 1886, p. 86f.; vgl. PFEIFFER 1936, p. 530f.), hängt aber noch von der Viscosität besonders der äußeren Plasmaschichten (vgl. CHOLNOKY 1931a, p. 324; KÜSTER 1951, p. 24), der Adhäsion (S. 73) und gegebenenfalls von mechanischen Eigenschaften des Chromatophors ab (EIBL 1939a, p. 166). Der Protoplast wird zunächst je nach Zellänge ein- oder mehrmals sanduhrförmig oder asymmetrisch (vgl. KUCHAR 1950a, p. 56f.) eingeschnürt und schließlich vollends zerteilt, wobei die Teilprotoplasten zumindest anfangs meist mit Plasmafäden verbunden bleiben, die oft noch Plasmakügelchen tragen, Chromatophoren oder Teile von solchen beinhalten können (S. 97) und mitunter noch kleine runde Teilprotoplasten bilden (vgl. p. 57; s. Abb. 5b).

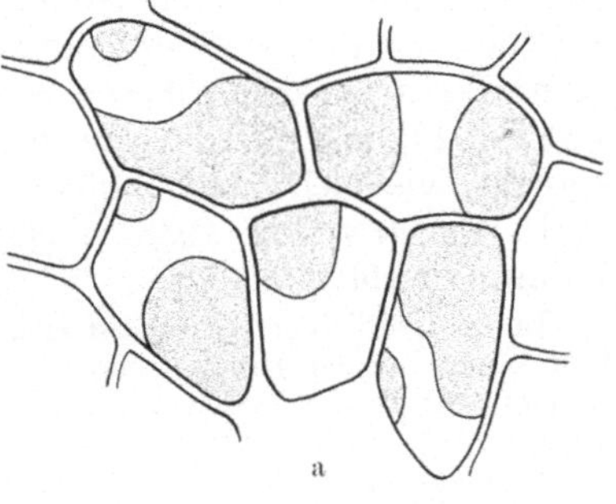

Abb. 5a u. b. Teilprotoplastenbildung. a Teilprotoplasten in isodiametrischen Zellen von *Cololejeunia calcarea* (grau: Protoplast mit Vacuole; nach WILL-RICHTER 1949, p. 517); b Einschnürung des Protoplasten in einer zylindrischen Zelle. Teil eines Wurzelhaares von *Trianea bogotensis* (nach BERTHOLD 1886, Tafel 1).

Manchmal wird auch durch die Faltung des Plasmaschlauches bei Kerbplasmolyse (S. 95_1) eine Protoplastenteilung ausgelöst (HABERLANDT 1919a, p. 337). Bei anderen Zellformen tritt eine solche gelegentlich nach mechanischen Einwirkungen (vgl. LOREY 1929, p. 173f.) oder durch sehr hohe Adhäsion eng begrenzter Membranstellen auf (vgl. KÜSTER-WINKELMANN 1938, p. 257; s. Abb. 5a). Teilprotoplasten werden schon in länglichen Zellen leicht gebildet (vgl. KÜSTER 1909, p. 590) und vermögen miteinander zu fusionieren oder nach längerer Zeit zu zerfallen (TIROLD 1932, p. 366; S. 87). Als Teilprotoplasten sind vielleicht auch die bei mechanischer Zerkleinerung von *Cladophora*-Zellen nach Vacuolenkontraktion übrigbleibenden, sehr widerstandsfähigen, isolierten Teilvacuolen anzusprechen (MOSER 1942, p. 152).

Tonoplasten, worunter ursprünglich die vermeintlichen Vacuolenbildner verstanden wurden (DE VRIES 1885, p. 469/325; S. 85_8), streben stets minimale Oberfläche an und können partiell durch *Fensterbildung* des Protoplasten freigelegt werden (HÖFLER 1932b, p. 207f.; TIROLD 1932, p. 364). Sie vermögen aber auch als Ganzes aus den Plasmakuppen auszuschlüpfen (p. 359f.; BECKEROWA 1935, p. 387), bei Deplasmolyse bruchsackartig auszutreten (EICHBERGER 1934, p. 609; S. 86) sowie zu mehreren in einer Zelle zu entstehen (DE VRIES 1885, p. 474/329). Sie bleiben allgemein nach mikrurgischer, plasmolytischer (WEBER 1934; LEDERER 1935, p. 215; BIEBL 1938a, p. 571) oder elektrischer *Zerstörung der anderen Plasmapartien* übrig (Lit.: BECKER und BECKEROWA 1934, p. 377; S. 101), wobei auch Plasmoschise eine Rolle zu spielen vermag (S. 92), werden aber bisweilen auch spontan gebildet (KÜSTER 1928, p. 225; MOTHES 1934, p. 494).

Die Tonoplasten sind ähnlich wie die Protoplasten zu Fusionen befähigt (JANSE 1887b, p. 419; MOTHES 1934, p. 496; S. 87_2), kontrahieren sich bei Konzentrationserhöhung(KEMMER 1928, p. 17), zeigen eine sehr geringe Oberflächenspannung (PIGOŃ 1952, p. 416), besitzen begrenzte Lebensdauer (vgl. EICHBERGER 1934) und können durch Fäden mit dem wandständigen Plasma verbunden sein (BIEBL 1938a, p. 572; S. 94_1). Sie platzen häufig bei Deplasmolyse (S. 84_5) und sehr leicht unter der Einwirkung des elektrischen Stromes (vgl. KEMMER 1928, p. 17; BECKER und BECKEROWA 1934, p. 377) sowie bei Verletzung durch mikrurgische Eingriffe (MOTHES 1934, p. 498), welche allerdings auch zur *Teilung der Tonoplasten* (vgl. KÜSTER 1928, p. 231; PLOWE 1931a, p. 212; CHAMBERS und HÖFLER 1931, p. 344) oder zu einem *Durchstich* (p. 343) führen können.

Regenerations- oder Vernarbungsmembranen (Lit.: WEISSENBÖCK 1938, p. 45f.; KÜSTER 1951, p. 767f.), die auch Übergänge zu den Verhärtungen der Plasmaoberfläche zeigen (S. 87_1), bilden sich an der Oberfläche von abgehobenen oder abgehoben gewesenen Stellen (vgl. MIEHE 1905, p. 259; HABERLANDT 1919a, p. 339; WEISSENBÖCK 1938, p. 85) oder am ganzen Umfang (GROHROCK 1935, p. 315; KANERT 1947, p. 10) des Protoplasten und selbst um dessen kernlose Anteile (vgl. PALLA 1890; CHOLNOKY 1930, p. 294) und treten bei seiner erneuten Kontraktion an der früheren Stelle der Plasmaoberfläche im Zellumen auf (vgl. LANZ 1942, p. 407). Sie werden bei Deplasmolyse (S. 85_1) nach Dehnung gesprengt, vielleicht auch aufgelöst (KAISERLEHNER 1939, p. 590), wachsen aus (vgl. WEISSENBÖCK 1938, p. 85) oder halten dem Turgordruck stand (vgl. KOBINGER 1953, p. 39).

In unschädlichen Plasmolyticis ist die junge Vernarbungsmembran zart und biegsam und bleibt bei der Kontraktion des Protoplasten mit diesem durch zahlreiche Fäden verbunden (vgl. KLEBS 1886/8, p. 527; HECHT 1912, p. 183f.; WEISSENBÖCK 1938, p. 61; KAMIYA 1939, p. 391; S. 94_2), die bei älteren verdickten und starren Membranen fehlen.

In chemischer Hinsicht bestehen die Regenerationsmembranen wahrscheinlich aus Cellulose (vgl. WEISSENBÖCK 1938, p. 59f.; KOBINGER 1953, p. 39). Sie kommen auch mehrschichtig vor (vgl. KLEMM 1894, p. 27; GROHROCK 1935, p. 321f.; KÜSTER 1939b, p. 64; KANERT 1947, p. 7; KOBINGER l. c.), fehlen bei *Desmidiaceen* (KLEBS 1886/8, p. 504; hingegen nicht bei *Diatomeen,* Lit.: KÜSTER 1951, p. 768), können gallertig degenerieren (KANERT 1947, p. 23f., 40), quellbar und färbbar sein und bei tiefen Temperaturen ausbleiben (WEISSENBÖCK 1938, p. 63f.). Bei Vorplasmolyse in K-Salzen und folgender stärkerer Plasmolyse in Ca-Salzen entsteht gelegentlich als besondere Membranbildung ein *Ca-Fällungshäutchen,* das aber eine Deplasmolyse nicht behindert (KAISERLEHNER 1939, p. 588f.). An der wachsenden Fadenspitze von *Vaucheria* bilden sich bisweilen bei Zusatz von Kongorot zum Plasmolytikum im Vorraum kristallähnliche rote Fäden (KANERT 1947, p. 41).

Vacuolen mit Außenlösung können durch partielle Verschmelzung von Plasmasträngen (S. 94_3) außerhalb oder, bei einer extrem ausgehöhlten Krampfplasmolyseform, innerhalb der Zentralvacuole (KUCHAR 1950a, p. 58) zu liegen kommen. Abnormes Wachstum (LOTHRING 1942, p. 618) oder *schlauchartige Auswüchse* plasmolysierter Protoplasten vermögen nach längerem Liegen im Plasmolytikum sowie nach langsamer Deplasmolyse aufzutreten (WEISSENBÖCK 1938, p. 84; TIROLD 1932, p. 367; S. 84_2, 102).

B. Der Zellkern.

Er kann bei Plasmolyse Veränderungen seiner Größe, Lage (vgl. HABERLANDT 1920, p. 327; S. 98_6) und Struktur (vgl. NEMEC 1899, p. 5f.) erleiden. Einer Volumverringerung, die wohl auf die osmotische Wirkung der Außenlösung zurückgeht (vgl. ABELE 1951; S. 115), steht eine Volumvergrößerung bei Vacuolenkontraktion (S. 90) und in Alkalisalzlösungen, welche auch einen Wechsel in der Gestalt zu bewirken vermögen (vgl. JANSE 1887b, p. 390), gegenüber (KÜSTER 1932, p. 357). Die derart gequollenen Kerne sind glasig, hyalin, prall gerundet und weisen eine deutlich erkennbare Kernmembran auf [HÖFLER 1928, p. (77)f.]. Aus dem Quellungszustand des Kernes wird auf die im Cytoplasma vorhandene Konzentration des betreffenden K-Salzes geschlossen (STRUGGER 1931, p. 463; 1932, p. 26f.). Die Kernform verändert sich manchmal auch durch den mechanischen Druck des Protoplasten (vgl. CHOLNOKY 1928, p. 456), und die Chromatinsubstanz zeigt bisweilen bei Plasmolyse eine gröbere Körnung (HABERLANDT 1919a, p. 339). Die Anzahl der *Kernteilungen* ist mitunter erhöht (PRÁT 1925, p. 7; S. 102) und ihr Ablauf wird modifiziert (S. 115).

In K-nitrat oder -bichromat vermag sich zwischen Cytoplasma und Kern eine semipermeable Niederschlagsmembran zu bilden, die in hypotonischen Lösungen platzt und den Kerninhalt explosiv ausschleudert (KÜSTER 1932, p. 356), was auch bei den in plasmolytischer Vacuolenkontraktion stark gequollenen Kernen gelegentlich vorkommt (STRUGGER 1931, p. 464; S. 92_1).

C. Die Plastiden.

Bei Plasmolyse erfahren die Plastiden durch die Kontraktion des Plasmamantels meist passiven Lagewechsel, gegebenenfalls auch Zerreißungen (vgl. CHOLNOKY 1928, p. 460), Formänderungen (zum Teil direkt auf der osmotischen Beeinflussung durch die Außenlösung beruhend, vgl. SENN 1908, p. 13; BAUER 1938, p. 288; KÜSTER 1937b, p. 113), Verklebungen (p. 77) und wohl auch andersartige Alterationen, die bei Deplasmolyse manchmal nur unvollständig zurückgehen (vgl. GICKLHORN 1933, p. 590; KÜSTER 1937b, p. 6f.; EIBL 1939b; S. 84). Doch sind auch Lageänderungen (z. B. bei Systrophe, S. 98_3) oder Gestaltwechsel, die nicht unmittelbar mit der Protoplastenkontraktion zusammenhängen, anzutreffen (vgl. PRÁT 1923b, p. 225f.; GICKLHORN 1932a; SEEMANN 1952, p. 118). Daneben sind Plastidenverlagerungen nach verschiedensten Reizen und schon in Wasser oder hypotonischen Konzentrationen (vgl. JANSE 1887b, p. 390; FRANK 1872, p. 221f.; KÜSTER 1905; SENN 1908; WALTER 1929, p. 130; PRÁT 1931, p. 428; GERM 1932, p. 591: Gruppenbildung; HELLWEGER 1935, p. 233; s. unten), aber auch bei Ausbleiben einer Plasmolyse in stark konzentrierten Lösungen, möglich (BRILLIANT 1927, p. 158f.). An Protoplastenteilstücken von *Vaucheria* kann ein rhythmischer Wechsel zwischen Ansammlung und Verteilung von Chloroplasten erfolgen (TIROLD 1932, p. 366). Zwischen Teilprotoplasten z. B. von *Spirogyra* vermag sich Plastidensubstanz als „*grüne Brücke*“ auszuziehen (EIBL 1939a, p. 165f.; S. 95).

Systrophe.

Als Systrophe (SCHIMPER 1885, p. 221) wird die lokale Anhäufung von Plasma und/oder Plastiden bezeichnet, wie sie vorzüglich um den Kern, aber auch an anderen Stellen an der Plasmaoberfläche und an kernlosen Teilprotoplasten auftritt (KÜSTER 1910a, p. 274; HÖFLER und WEBER 1926, p. 698; GERM 1932, p. 613) und wohl meist mit einer leichten Volumzunahme des Plasmas verbunden ist (HÖFLER 1940b, p. 293; S. 92_5).

Die Systrophe stellt eine *reversible Reizerscheinung* dar, die nach Erschütterung, mechanischem Druck (z. B. bei *Biddulphia* unter Verschwinden der Vacuole, BÜNNING 1934, p. 452), intensiver Belichtung (*Lichtsystrophe*, vgl. KÜSTER 1929b, p. 74f.; DIANNELIDIS 1951, p. 30), Verwelken (GERM 1933a, p. 521), Einwirkung von Tellursalzen (MODER 1932, p. 19f.), Vitalfärbung (BECKEROWA 1935, p. 388; KÜSTER 1942b, p. 252), Säurebehandlung (WEBER 1932e, p. 289), Ultraschalleinwirkung (KÜSTER 1952, p. 85), Verwundung von Nachbarzellen, Kälte, Verdunkelung, Plasmolyse und anderen Einwirkungen (vgl. 1929b, p. 72f.) oder spontan (S. 76) erfolgt und bisweilen auch Beziehungen zum Alter der Zelle zeigt (S. 104). Dabei handelt es sich primär meist um die Verlagerung des Protoplasmas (*Plasmasystrophe*, 1932, p. 569), welches dann die Inhaltskörper (Plastiden, Kristalle) mittransportiert (vgl. FRANK 1872, p. 228; KÜSTER 1910a, p. 282).

Auch im normalen Lebensablauf sind Plasmasystrophen (Lit.: KÜSTER 1951, p. 34f.) sowie auch die alleinige Ansammlung der Plastiden (*Plastidensystrophe*, vgl. p. 369f.) anzutreffen.

Plasmasystrophe, wie sie durch Plasmolyse ausgelöst wird, gilt als eine Lebensreaktion (S. 101), kann nur in völlig lebensfähigem Zustand auftreten (GERM 1933b, p. 261) und wird auch für nackte Protoplasten erwähnt (KLERCKER 1892, p. 469). Die Verlagerung des Plasmas und der Plastiden besorgt hier die Plasmaströmung (HÖFLER 1918a, p. 150; GERM 1932, p. 573; S. 93_4), die zunächst ruckweise die Plastiden erfaßt und in der Nähe des Kernes deponiert (HÖFLER und WEBER 1926, p. 698; S. 97_3). Bei Plastidensystrophe unterbleibt dabei eine Ansammlung des Plasmas selbst (GERM 1932, p. 594).

Sitzt alles Plasma an einer Stelle als sphärische Linse an der Innenseite eines kugeligen Protoplasten, so ist das Volumen des Plasmas berechenbar (KÜSTER 1935b). Die Plasmaballung zeigt aber auch Strömung und Gestaltsänderungen (1910a, p. 276, 278; BLAESS 1941), mitunter sogar amöboide Bewegungen mit Plasmazungen (KÜSTER 1935b, p. 431; S. 94). Manchmal liegen zahlreiche gestreckte Vacuolen in den dem Zellsaftraum aufsitzenden Plasmapartien, radial von der Hauptmasse der Plasmaansammlung ausgehend (*Rosettensystrophe*, HÖFLER zit.: SCHINDLER 1937, p. 197f.).

Als Reizanlaß für Systrophe fungiert offenbar nicht der osmotische Wasserentzug, sondern die Abhebung des Protoplasten von der Membran. Eine bestimmte Konzentration ist für die Ausbildung einer Systrophe optimal und in zu geringen oder zu hohen Konzentrationen werden nur Vorstufen erreicht (GERM 1932, p. 569; 1933a, p. 511f., 528f.).

In wenig hypertonischen Lösungen und bei viscosem Plasma wandert der Kern (S. 97_1) zunächst an Plasmafäden in die Vacuolenmitte (*Zentrierung*, 1932, p. 574) und dann gemeinsam mit Plastiden und Plasma, dessen Fäden hernach verschwinden, an die Peripherie, oder er verlagert sich direkt längs der Plasmaoberfläche und das Plasmanetzwerk wandert ihm durch den Zellsaftraum zu (vgl. Abb. 6). Die Zusammenballung ragt zuerst meist in die Vacuole hinein und wölbt sich später nach außen vor (KÜSTER 1929b, p. 74; *Aussackung*, GERM 1932, p. 576f.; *Prolaps*, LANZ 1942, p. 384, 389), wobei es im Endstadium mitunter zu Vacuolenteilungen kommt (S. 100).

Unterhalb einer bestimmten Konzentration wird die Systrophe nur selten perfekt, löst sich dann aber wieder auf, und der Kern gelangt an Plasmafäden erneut in die Zellmitte („*sekundäre Zentrierung*", GERM 1932, p. 581). In stärkeren Lösungen und stets bei wenig viscosem Plasma erfolgt die Zusammenballung direkt an der Oberfläche ohne Zentrierung und führt in langgestreckten Zellen zu einem Plasma- und Chloroplastengürtel (*Zonenbildung*, p. 586f.).

Die *Systrophezeit* (d. i. die Zeitspanne vom Übertragen des Schnittes in das Plasmolytikum bis zum Erreichen perfekter Systrophe, 1933a, p. 532) wird mit steigender Temperatur (p. 538) und bei Wässerung der Schnitte, offenbar wegen einer damit verbundenen Viscositätsherabsetzung, kürzer (vgl. p. 547; HÖFLER und WEBER 1926, p. 699; KAMIYA 1939, p. 383; S. 93_3).

Durch Deplasmolyse geht die Systrophe meist wieder zurück (KÜSTER 1906, p. 257; GERM 1932, p. 597; S. 84), doch vermag sie bei genügend langer Dauer des Plasmolysereizes, analog zu den Verhältnissen bei anderen Reizeinwirkungen, auch während der Deplasmolyse weiterzugehen und perfekt zu werden (1933a, p. 523; S. 84_3).

Bei der *Auflösung*, die zuweilen in Stadien stärkerer Zusammenballung zurückpendelt (1932, p. 607), zerteilt sich meist zuerst das Plasma an der Protoplastenoberfläche, während die Plastiden erst später nachfolgen.

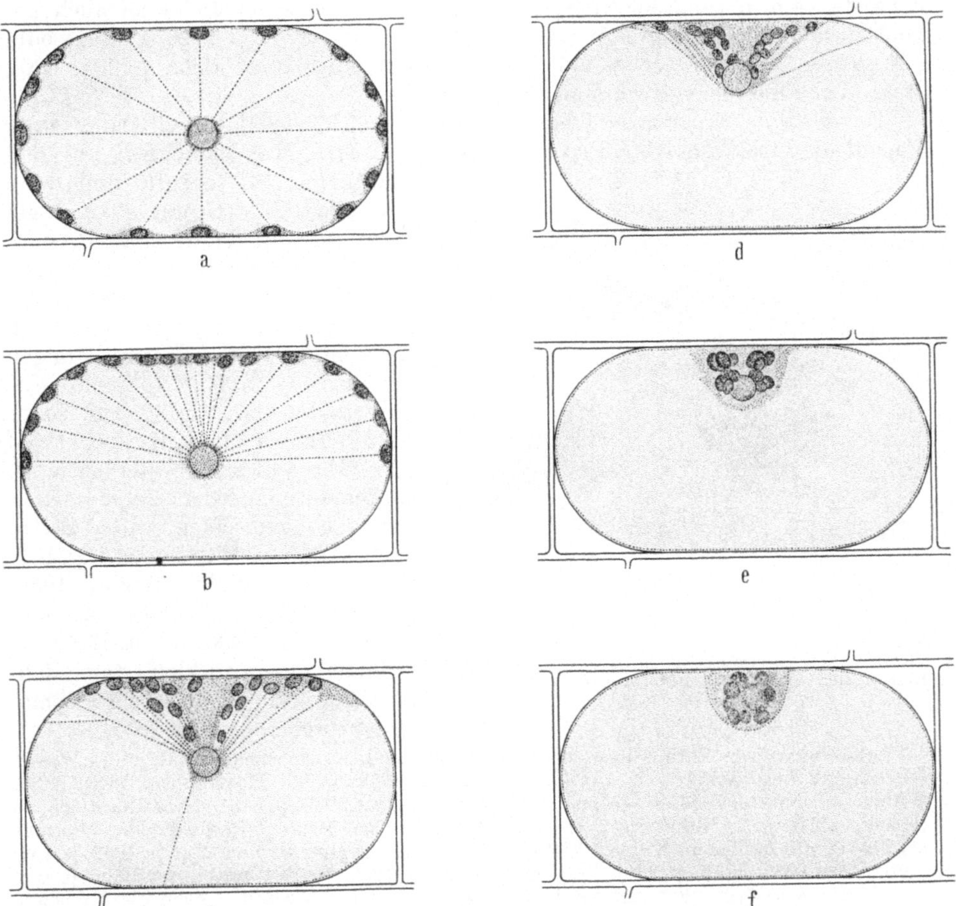

Abb. 6a—f. Schema des Verlaufes einer Plasmasystrophe: a „Zentrierung"; b „Flankenstellung"; c „Sammlung"; d „Häufung"; e „perfekte Systrophe"; f „perfekte Systrophe" (Endstadium). (Nach GERM 1932, p. 575.)

Die Systrophe wird durch Äthernarkose (HÖFLER und WEBER 1926, p. 699) und Kappenplasmolyse verhindert, wenn sich letztere rascher ausbildet, sonst quillt das um den Kern angesammelte Plasma, besonders in den Randzellen, auf (GERM 1932, p. 610; LANZ 1942, p. 400; S. 89, 104).

D. Die Vacuole.

Durch die Plasmolyse erleidet die Vacuole Veränderungen ihrer *Größe*, meist auch der *Gestalt* und manchmal selbst ihrer *Struktur*.

In plasmareichen Zellen können die dann nur kleinen Vacuolen bei fortschreitender Plasmolyse aufreißen sowie sich bis zur völligen Unsichtbarkeit reversibel kontrahieren (vgl. KUCHAR 1950a, p. 55f.; S. 84_3), oder große Vacuolen entstehen (NEMEC 1899, p. 14). Bei großer Zentralvacuole und leichtflüssigem

Zellsaft sucht die Vacuole bei ihrer Verkleinerung die Minimalform anzunehmen, doch treten bisweilen auch starke Verformungen auf (S. 92), und Plasmazungen oder Plasmalamellen führen oft zur *Zerklüftung* (vgl. KÜSTER 1926, p. 79; S. 94_7), *Furchung* oder *Teilung*, besonders nach Kappenplasmolyse, bei Systrophe oder Vacuolenkontraktion (vgl. 1918; 1929b, p. 30; S. 89, 98, 90). Dabei entstehen entweder am plasmareichen Teil der Zelle kleine gegeneinander ziemlich abgeplattete Vacuolen („grobschaumige" Plasmastruktur, vgl. 1918, p. 291) oder am ganzen Protoplasten ziemlich große Kammern (GERM 1932, p. 604, 620), stets aber ohne eigentliche Vacuolenneubildung.

Die *Zellsäfte* bestimmter Objekte vermögen sich nach längerer Dauer einer Vacuolenkontraktion oder Plasmolyse zu *verfestigen*, was manchmal nur den peripheren Teil betrifft und dann reversibel ist (HOFMEISTER 1940, p. 84) oder im Laufe einer Plasmolyse *Entmischungen* zu zeigen (vgl. PFEFFER 1886/8, p. 245; CHOLNOKY 1953, p. 523; *Myelinfiguren*, vgl. GICKLHORN 1932b; Entmischungen des Plasmas, S. 93). Sie sind mitunter fädig bzw. tropfig (vgl. BANK 1933; FLASCH und KINZEL 1954) oder führen in anderen Fällen (z. B. Blütenblattzellen der *Borraginoideen* unter Farbumschlag roter Zellen nach blau, vgl. WEBER 1936, p. 101; GICKLHORN und MÖSCHL 1930, p. 524; Involukralblätter von *Compositen*, SCHITTENGRUBER 1953) zu Gebilden von verkleinertem Zellumriß, wie dieser auch bei Plasmolyse bisweilen auftritt (S. 72, 81).

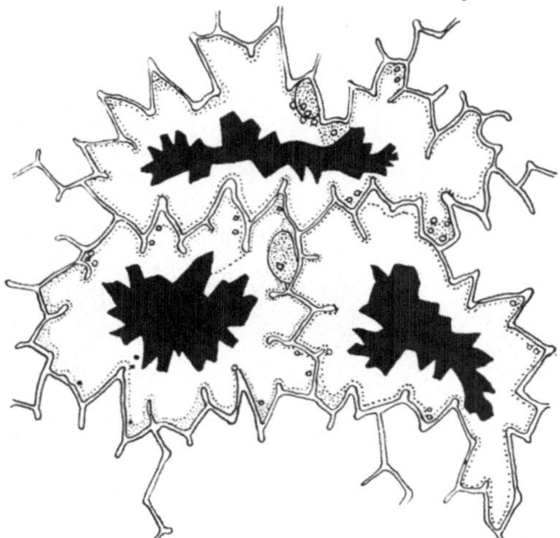

Abb. 7. Vacuolenentmischung in Mesophyllzellen der Korollblätter von *Anchusa off.* (Nach GICKLHORN und WEBER 1927 p. 429.)

Letztere Art von Entmischungen entsteht spontan bzw. schon nach Einlegen in Wasser (GICKLHORN und WEBER 1927, p. 428; vgl. KEIL 1930, p. 590; HOFMEISTER 1940, p. 80; WEBER und KENDA 1953, p. 315), nach Klopfreizen (KEIL 1930, p. 588f.), Neutralrotfärbung (WEBER 1935, p. 5; HOFMEISTER 1940, p. 74; vgl. MOSER 1942, p. 152; besondere Formen bei Zusatz von Al-Salzen, KÜSTER 1949), Säureeinwirkung (KENDA und WEBER 1952, p. 459; SCHITTENGRUBER 1953, p. 330), Verwundung (KEIL 1930, p. 589; STRUGGER 1935, p. 108) oder Plasmolyse (vgl. SCHARINGER 1936, p. 418), werden gelegentlich auch als Vacuolenkontraktionen bezeichnet und gedeutet (vgl. GICKLHORN und WEBER 1927; KÜSTER 1939c, p. 48; 1949, p. 21; S. 88), sind mitunter reversibel und wiederholbar (vgl. HENNER 1934, p. 102; HOFMEISTER 1940, p. 78; KENDA und WEBER 1952, p. 463) sowie auf Zellen bestimmten Alters beschränkt und zeigen Beziehungen zu anderen Kontraktionsvorgängen (S. 104, 83).

Es handelt sich dann wahrscheinlich um eine auf *Synaerese* (vgl. GICKLHORN und MÖSCHL 1930, p. 531; WEBER 1935) oder anderen Mechanismen beruhende Entquellung stark kolloider, vermutlich pectinreicher (vgl. SCHITTENGRUBER 1953, p. 332) Zellsäfte, die auch äußerst rasch erfolgen kann (KENDA und WEBER 1952, p. 463, 459), wobei ein zentral liegender, fest und elastisch werdender (HOFMEISTER 1940, p. 74) oder spröder (vgl. SCHITTENGRUBER 1953, p. 329) Gelteil sich von einem Solteil trennt, welcher den übrigen Vacuolenraum erfüllt und mitunter eine weitere Entmischung erfährt (WEBER 1935, p. 7; WEBER und KENDA 1953, p. 317; s. Abb. 7). Bisweilen zeigen sich Plasmafäden zwischen dem an der Wand verbliebenen Plasmamantel und dem Gelteil (GICKLHORN und WEBER 1927, p. 428; KÜSTER 1938a, p. 157; S. 94_1). In hypertonischen Lösungen erfolgt Plasmolyse, wobei die äußere, flüssige Komponente des Zellsaftes Wasser abgibt und sich der Plasmamantel schließlich dem festen Solteil anlegt (S. 77); bei darauffolgender Deplasmolyse setzen an ihr manchmal lokale Volumvergrößerungen ein (WEBER 1935, p. 12). Auf dem Vorhandensein der Zellsaftkolloide basiert auch die Annahme eines für osmotische Substanzen „nichtlösenden" Raumes (vgl. NETTER 1927; hingegen BOGEN und PRELL 1953, p. 464; S. 76).

IX. Der Einfluß von Plasmolyse und Deplasmolyse auf das Verhalten der Zelle.

A. Lebensdauer und Schädigung.

Bleibt bei Plasmolyse der Lebenszustand normalerweise wohl immer erhalten (vgl. URSPRUNG 1939, p. 1173), so zeigt sich doch oft ein schädigender Einfluß auf die Zelle als Ganzes sowie ihre Teile (S. 92), der mit Objekt und Bedingungen wechselt und an Veränderungen von Zelleigenschaften und -strukturen erkennbar ist (z. B. Permeabilität, S. 103; Tonoplastenbildung, S. 96; Lit.: KARZEL 1926, p. 551f.; KÜSTER 1929b, p. 8f.; DE HAAN 1933, p. 303f.). Die *Plasmolyseresistenz* (vgl. WEBER 1932a, p. 179) kann dann aus Lebensdauer bzw. *Lebensreaktionen*, wie Vitalfärbung (vgl. STRUGGER 1949, p. 168f.; SIMINOVITCH und BRIGGS 1953, p. 16), Deplasmolyse (S. 85), Plasmaströmung (S. 93_4) oder Systrophe (vgl. CHALAUPKA 1939, p. 18; S. 98) erschlossen werden, welch letztere zwei auch bei voller Vitalität mitunter ausbleiben. *Rotalgen* besitzen im Gegensatz zu *Vaucheria* meist nur sehr geringe Plasmolyseresistenz (HÖFLER 1930b, p. 573; TIROLD 1932, p. 387). Diese vermag ferner durch Kälte (WEBER 1932c, p. 518; vgl. MODER 1932, p. 33), Narkotica (p. 26) und Wässerung (vgl. KACZMAREK 1929, p. 288f.; ALBACH 1931, p. 262; S. 93_3) verändert zu werden und Relationen zum jahreszeitlichen Verlauf der Frostresistenz (vgl. SIMINOVITCH und BRIGGS 1953, p. 21) und zum osmotischen Grundwert der Zelle zu zeigen (ALBACH 1931, p. 263).

Schädigend wirken dabei, abgesehen von besonderen Effekten, wie z. B. bei einer Vacuolenkontraktion (S. 91), allgemein die Schnelligkeit des Konzentrationswechsels, die osmotische Wirksamkeit des Plasmolytikums, die Plasmolysedauer, bei Vacuolenkontraktion deren Schnelligkeit (S. 90) und gegebenenfalls die Qualitäten der betreffenden Außenlösung. Im übrigen können noch bei der Präparation durch Schneiden usw. hervorgerufene Verwundungen lokale Alterationen bedingen (S. 104).

Die schädigende Wirkung wird oft durch zusätzliche Maßnahmen (z. B. Narkose, WEBER 1932a, p. 181; Zugabe von Säuren, Laugen, Alkaloiden usw., vgl. BRENNER 1920, p. 278f.; HÖFLER 1938, p. 458f.; Wässern, ALBACH 1931, p. 263; S. 93_3) verändert und mitunter erst als *Nachwirkung* bei scheinbar intakt gebliebenen Objekten zu erkennen sein (KARZEL 1926, p. 560f.).

Der *Konzentrationswechsel* ist um so weniger alterierend, je langsamer das Objekt in die höhere Konzentration gelangt (vgl. DE VRIES 1885, p. 476/332; HÖFLER 1930b, p. 583; ALBACH 1931, p. 258; RAFFELT 1944, p. 21; SEEMANN 1952, p. 112); seine schädigende Wirkung erreicht beim *Gel-Auflageverfahren* (vgl. STRUGGER 1949, p. 94) ein Minimum und vermag eine p_H-Abhängigkeit zu zeigen (KACZMAREK 1928, p. 252). Allgemein gilt stets der WEBERsche Satz, daß die Schädigung um so geringer ist, je leichter sich der Protoplast von der Wand zu lösen vermag (WEBER 1932a, p. 185). Während so zu schneller Konzentrationswechsel meist stark schädigt, ertragen *Hefe*-Zellen selbst plötzliches Übertragen aus dem Kulturmedium in eine 4-molare Salzlösung (TAKADA 1953, p. 19, 23).

Eine Alteration durch die *osmotische Wirksamkeit* ist um so geringer, je niedriger die Plasmolytikumkonzentration liegt (ALBACH 1931, p. 259), kann durch Anpassung an den höheren osmotischen Wert überwunden werden (Osmoregulation, S. 84) und tritt vielfach überhaupt erst in hypertonischen Konzentrationen auf (vgl. HOFMEISTER 1937, p. 54).

Cyanophyceen vermögen hohe Zuckerkonzentrationen gut zu ertragen (PRÁT 1923a, p. 96); andererseits zeigt sich auch eine mit dem Grad der Hypotonie zunehmende Schädlichkeit des Milieus (SCHEITTERER und WEBER 1930; HÖFLER 1931b).

Bei zu langer *Plasmolysedauer* können selbst relativ unschädliche Außenlösungen den Protoplasten alterieren (vgl. DE VRIES 1885, p. 567/416f.).

Von den als Plasmolytica gebräuchlichen *Lösungen* schädigen Zucker (vgl. KEMMER 1928, p. 11) und äquilibrierte, d. h. balancierte Elektrolytlösungen (Seewasser, vgl. HÖFLER 1930b, p. 574; BRENNERsche Lösung, BRENNER 1920, p. 284; vgl. HOFMEISTER 1954a, p. 294, 296; KACZMAREK 1928, p. 278) oft nur wenig. Auch die übrigen Anelektrolyte bewirken vielfach nur geringe Alteration. Unter den Elektrolyt-Einzellösungen sind Ca-Salze am unschädlichsten (vgl. BRENNER 1920, p. 283; KEMMER 1928, p. 21), während Kationen, die eine Quellung fördern (S. 92), stärker einwirken (DE HAAN 1933, p. 306) und Anionen hinsichtlich der Stärke ihres Effektes eine Modifikation der lyotropen Reihe ergeben (vgl. KAHHO 1921, p. 132; PRÁT 1921); daneben ist auch der p_H-Wert der Lösung von Wichtigkeit (vgl. KACZMAREK 1928, p. 292).

Die auf die Wirkung der einzelnen Plasmolytica zurückgehenden Vitalitätsunterschiede, wie sie sich z. B. in der Lebensdauer zu erkennen geben, werden in den *Resistenzuntersuchungen* behandelt (Lit.: LINDNER 1950; STADELMANN 1952).

Die Schädigungen durch *Deplasmolyse* (vgl. DE VRIES 1885, p. 532/383; KAHHO 1921, p. 127; Lit.: KEMMER 1928, p. 17; URSPRUNG 1939, p. 1175; S. 84_2, 85), besonders in destilliertem Wasser (ALBACH 1931, p. 257; DE HAAN 1933, p. 307f.) sind meist stärker als jene durch Plasmolyse, können vornehmlich die älteren Zellen betreffen (ALBACH l. c.; S. 104) und sind um so geringer, je langsamer die Konzentrationsänderung erfolgt (vgl. REINHARDT 1899, p. 431).

B. Änderungen von Zellfunktionen durch Plasmolyse.

Lokomotion. *Bakterien* behalten ihre Bewegungsfähigkeit nach Plasmolyse oft noch lange bei (vgl. FISCHER 1895, p. 36f.; 1903, p. 25; RAICHEL 1928, p. 335; hingegen: GARBOWSKI 1907, p. 61); *Diatomeen* sistieren, mitunter reversibel durch Deplasmolyse (HÖFLER 1940a, p. 101, 111), je nach Spezies die Lokomotion schon in hypotonischen Lösungen oder erst bei stärkerer Plasmolyse (vgl. CHOLNOKY 1928, p. 481; BAUER 1938, p. 282f., 290; CHOLNOKY und HÖFLER 1943, p. 162f.), während *Chlamydomonas* ebenfalls in hypotonischen Lösungen die Bewegung einstellt (LOTHRING 1942, p. 623).

Die **Zellteilung** kann durch Plasmolyse je nach Außenbedingungen, Objekt und Jahreszeit zuweilen auch irreversibel beeinflußt werden (vgl. JANSE 1887a, p. 23; 1887b, p. 386f.; KLEBS 1886/8, p. 533, 536; LUNDEGÅRDH 1911, p. 50; PRÁT 1925, p. 7; S. 97).

Wachstum. Der plasmolysierte Protoplast bestimmter Objekte ist offenbar zum Wachstum befähigt (vgl. KLEBS 1886/8, p. 525f.) und vermag dann oder erst nach Deplasmolyse schlauchartige Auswüchse zu bilden oder abnormes Wachstum zu zeigen, was vielleicht auf eine desorganisierende Wirkung der Plasmolyse zurückgeht (vgl. MIEHE 1905, p. 259f.; ZEHETNER 1934, p. 535; REINHARDT 1899, p. 434f.; S. 84_2, 96). An Wurzeln scheint Nebenwurzelbildung bei Plasmolyse möglich (p. 431f.). Nach Deplasmolyse entstehen an *Farn*-Prothallien durch mehrfache Teilungen einzelner Zellen gelegentlich Adventivsprosse (ISABURO-NAGAI 1914, p. 305f.) und in *Elodea*-Blattzähnen kernlose Zellen (HABERLANDT 1919b, p. 724f.; Lit.: PFEIFFER 1930, p. 255f.). Im übrigen scheint das Wachstum durch Plasmolyse generell gehemmt (vgl. BÖRGER 1926,

p. 143f.; Lit.: KÜSTER 1929b, p. 9f.), was vielleicht auf dessen Sistierung bei Abhebung des Protoplasten von der Wand zurückführbar ist (REINHARDT 1899, p. 429; BURSTRÖM 1953, p. 275). Das Wachstum kann auch nach Deplasmolyse ganz unterbleiben (vgl. DETMER 1909, p. 201; S. 85_1).

Das Ergrünen etiolierter Pflanzenteile wird durch Plasmolyse verhindert (FÜRLINGER 1938).

Stoffwechsel. Die Atmung des Wurzelgewebes von *Beta* erfährt zunächst eine starke Intensivierung (etwa während der Plasmolysezeit; S. 79), die von einer durch Rückübertragen in Wasser reversiblen Depression gefolgt ist (BENNET-CLARK und BEXON 1943, p. 79, 86; vgl. ALBACH 1929, p. 414 für *Elodea*; PFEIFFER 1936, p. 536). Die CO_2-Assimilation scheint bei Plasmolyse eher nur indirekt durch den Wasserentzug gehemmt zu werden (vgl. WALTER 1929, p. 154; TREBOUX 1903, p. 53f.). Auch die Aufnahme von KCl ist verringert (SUTCLIFFE 1953), und die Tetrazoliumreaktion an der Zellwand bleibt bei Plasmolyse und Deplasmolyse aus (CURRIER und VAN DER ZWEEP 1955, p. 129).

Die **Resistenz** gegenüber Austrocknung wird durch Plasmolyse oft erhöht (vgl. ILJIN 1935, p. 748f.), diejenige gegen giftige Substanzen erniedrigt (vgl. HÖFLER 1934a, p. 218), wobei diese Einwirkungen wohl meist die „nicht umweltbezogene, konstitutionelle Resistenz“ betreffen (vgl. BIEBL 1947, p. 147).

C. Permeabilität.

Eine Durchlässigkeit des Protoplasmamantels für die osmotisch wirksamen Substanzen der Vacuole (*Exosmose zelleigener Stoffe*, S. 85_5) zeigt sich schon oft bei Wässerung (vgl. FITTING 1915, p. 11; KACZMAREK 1928, p. 288; ALBACH 1931, p. 261f.; JÄRVENKYLÄ 1937, p. 22; KAMIYA 1939, p. 379; S. 93) oder beim Einlegen der Objekte in hypotonische Lösungen (vgl. WÄCHTER 1905, p. 168f.; BROOKS 1916; ILJIN 1928, p. 562f.) und ist an einer langsamen Kontraktionszunahme in nicht-diosmierenden Außenlösungen zu erkennen (vgl. DE VRIES 1885, p. 567/416f.; S. 75, 83_3, 83_6). Eine Permeabilität für die gelösten Stoffe der Außenlösung manifestiert sich hingegen als Deplasmolyse (S. 84). Sie tritt entweder überhaupt erst als Folge einer Schädigung auf (S. 101), kann aber durch eine solche wohl ebenfalls erhöht (von der *Normalpermeabilität* zur *Plasmolysepermeabilität*, WEBER 1932a, p. 190, 181, 185; vgl. BÜNNING 1934, p. 446) oder erniedrigt werden (SCHMIDT 1936, p. 495), erfährt hingegen bei einer Reihe von Stoffen und Objekten durch Plasmolyse keine Änderung (vgl. HÖFLER 1934a, p. 218; MODER 1932, p. 17; Lit.: SCHMIDT 1936, p. 470). Die *Wasserpermeabilität* vermag sich mitunter bei Plasmolyse (MYERS 1951, p. 141) oder auch während der Deplasmolyse zu vergrößern (Lit.: URSPRUNG 1939, p. 1176).

X. Plasmolyseverhalten der Zelle in Relation zu ihrer Lage im Zellverband und zum Alter.

Die Lage im Zellverband.

Je nach ihrer Lokalisierung im Zellkomplex oder Präparat plasmolysieren mitunter an sich völlig gleichwertige Zellen verschieden (S. 81). Dabei vermögen für den Plasmolyseort *zelleigene Faktoren* eine Rolle zu spielen (vgl. SCHEITTERER 1930; CHOLNOKY 1931a, p. 327), wie sie sich gelegentlich bei den für Zellreihen charakteristischen Plasmolyseformen (S. 78) manifestieren oder bei Zellen, welche Wunden bzw. nekrotische Herde umgrenzen (*amphinekrotische Zellen*, MODER 1932, p. 42). Dort liegt der negative Plasmolyseort meist auf der dem nekrotischen Herd zugewandten Seite (WEBER 1929d, p. 588; MODER 1932, p. 44; S. 76_5; s. Abb. 8), oder es tritt bei Vacuolenkontraktion an diesen Orten eine stärkere Quellung auf (S. 90_3).

Bestimmen *Diffusionsverhältnisse* die Lage des positiven Plasmolyseortes (S. 72_9, 81_3), so liegt er stets in der Richtung des leichteren Plasmolytikumzutritts (vgl. BRILLIANT 1927, p. 156f.) und zeigt auch eine deutliche Orientierung um begrenzte Stellen erhöhter Membranpermeabilität (STRUGGER 1935, p. 28). Am *Schnittrand* findet sich oft ein gegenüber den zentralen Partien deutlich abweichendes Verhalten, das zum Teil wohl auch durch traumatische Einwirkungen verursacht ist (vgl. KÜSTER 1935b, p. 430; ALBACH 1931, p. 260; KAMIYA 1939, p. 375f.; S. 83, 90_3, 101). Die Plasmolysezeit kann dann auch durch den raschen Zutritt des Plasmolytikums verkürzt werden (S. 72_9).

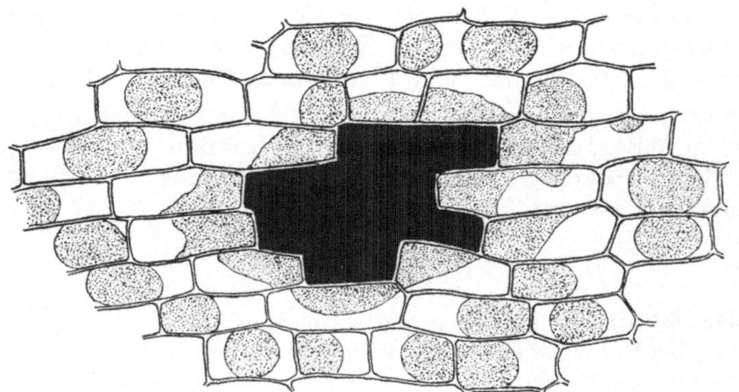

Abb. 8. Plasmolyseorte um einen nekrotischen Herd (tote Zellen schwarz). (Blattstück von *Elodea canadensis*.) (Nach WEBER 1929d, p. 588.)

Bei gleicher Umspülung vermögen noch Unterschiede in den Eigenschaften der Zellwand (Membranpermeabilität, S. 72_7), des Plasmamantels (Adhäsion, Viscosität, Permeabilität, S. 80) oder der Vacuole (z. B. Unterschiede im osmotischen Wert, MODER 1932, p. 13; S. 81_1) den Plasmolyseeintritt zu variieren, wie diese gelegentlich auch zwischen verschiedenen Zellsorten desselben Objektes bestehen (vgl. DIANNELIDIS 1951, p. 33). *Deplasmolyse* ist am Schnittrand mitunter verlangsamt (Lit.: URSPRUNG 1939, p. 1186; S. 85_2). Eine deutliche Zonierung zeigt meist die Kappenplasmolyse (S. 89), die nach Systrophe hauptsächlich in den Randzellen auftritt (S. 99).

Relationen zum Alter.

Solche können Plasmolyseort (S. 76_6), Plasmolyseform (FISCHER 1948; BANCHER 1949, p. 90; S. 77), Systrophe (vgl. 1941, p. 103; S. 98), Plasmolysierbarkeit (S. 82), Vacuolenkontraktion (S. 91), Resistenz gegen Deplasmolyseschädigung (S. 102), Deplasmolyseeinsatz (S. 85_1), sowie osmotischer Wert und Plasmolysehäufigkeit (Lit.: LOTHRING 1942, p. 600, vgl. p. 606; S. 76_2) zeigen, wobei die letztere auch mit den Milieubedingungen korreliert (vgl. p. 626). Entmischungen des Zellsaftes werden manchmal nur an alten Blüten beobachtet (KEIL 1930, p. 590; SCHARINGER 1936, p. 417; S. 100).

Gradienten für Plasmolysezeit und Plasmolyseform, bei ersterer wahrscheinlich auf Adhäsionsunterschiede zurückgehend (SCHAEFER 1955, p. 432), lassen sich zwischen Wachstums- und Dauerzonen von Organen feststellen (vgl. PIRSON und SEIDEL 1950, p. 437; S. 81_6).

XI. Plasmoptyse und Plasmorrhyse.

Plasmoptyse.

Als Plasmoptyse (FISCHER 1900, p. 5) wird der plötzliche Austritt des ganzen Zellinhaltes oder eines Teiles davon, der dann meist zugrunde geht, bezeichnet; die austretende Zellsubstanz wird dabei entweder zufolge starken Anwachsens des Innendruckes (*osmotische Plasmoptyse*, HOLDHEIDE 1930, p. 706; 1932, p. 244) oder nach Verringerung der mechanischen Membranfestigkeit (*Defektplasmoptyse*, p. 244) durch Membranporen *ausgepreßt* bzw. unter Sprengung der Zellwand, die stets an der schwächsten Stelle und oft in charakteristischer Weise erfolgt (vgl. p. 289f.; SCHULTE 1938, p. 93), *ausgeschleudert*, wobei die Membran geringe Schwellungen aufzufangen vermag (FÖRSTER 1933, p. 480), oder zufolge einer nach Innen gerichteten Quellung der Zellwand, die schließlich lokal aufreißt, *ausgequetscht* (*Quellungsplasmoptyse*, ÚLEHLA 1923, p. 248; 1926, p. 637). Bei *Diatomeen* erfolgt auch Auseinanderklappen der Theken (BAUER 1938, p. 278).

Plasmoptyse, die bei günstigen Objekten gut reproduzierbar ist, bei anderen aber ausbleibt (vgl. HOLDHEIDE 1930, p. 710), was mitunter von deren Zustand abhängt (ZEHETNER 1934, p. 528f.), tritt in bestimmten Zellen auch im normalen Lebensablauf ein und läßt sich gelegentlich bereits durch Erschütterungen herbeiführen (vgl. KÜSTER 1951, p. 129f.; PALLA 1890, p. 317). Die Menge und Art des ausgetretenen Plasmas ist oft recht verschieden (vgl. ÚLEHLA und MORÁVEK 1922, p. 12). Die Größe der zur Sprengung nötigen Kraft läßt sich dabei näherungsweise berechnen (FREY-WYSSLING 1952, p. 243; vgl. HOLDHEIDE 1932, p. 284).

Osmotische Plasmoptyse kann durch eine entsprechend erhöhte Saugkraft der Zellen (z. B. Übertragen in hypotonisches Milieu, vgl. NÄGELI 1855, p. 2; NOLL 1888, p. 522; SCHÜTT 1895, p. 108; KIENITZ-GERLOFF 1902, p. 112; GARBOWSKI 1907, p. 71; PRÁT 1925, p. 2; ÚLEHLA 1926, p. 635; RAICHEL 1928, p. 337f.; HÖFLER 1931b, p. 58; BAUER 1938, p. 273f.) oder eine andersartig verursachte Aufnahme von Wasser oder anderen Stoffen ausgelöst werden und ist durch eine Erhöhung des osmotischen Wertes der Außenlösung behindert (vgl. LOPRIORE 1895, p. 592f.; HOLDHEIDE 1930, p. 707; BONTE 1935, p. 232). Hierbei spielen für die Plasmoptysehäufigkeit (*Plasmoptysezahl*, HOLDHEIDE 1932, p. 255), Zellmaterial, Konzentration und Art der Außenlösung (vgl. p. 260; SCHULTE 1938, p. 66, 94), sowie für die Zeit vom Einlegen bis zum Platzen der ersten Zellen (*Plasmoptyseeintrittszeit*, HOLDHEIDE 1932, p. 254; ZEHETNER 1934, p. 508) mitunter auch noch Zellgröße, Temperatur und Vorbehandlung eine Rolle (vgl. HOLDHEIDE 1930, p. 709; 1932, p. 261f., 274f.).

Bei Einwirkung von Elektrolyten bzw. Säuren (vgl. KLEMM 1895, p. 660; LOPRIORE 1895, p. 591f.; GARBOWSKI 1907, p. 74f.; ÚLEHLA und MORÁVEK 1922; STRUGGER 1926, p. 460f.; BONTE 1935, p. 228f.; SCHULTE 1938, p. 66f.) handelt es sich offenbar um eine durch elektrische Kräfte bewirkte Wasseraufnahme (*Elektroosmose*, vgl. p. 61; HÖBER 1947, p. 668f.), die nur bei einer gegenüber dem Zellinhalt höheren Acidität der Außenlösung auftreten soll (SCHULTE 1938, p. 93). Die Plasmoptysezahl zeigt dann eine Abhängigkeit vom p_H-Wert (Lit.: KÜSTER 1929b, p. 88) in einer zweigipfeligen (STRUGGER 1926, p. 461; 1928, p. 158, 164) oder eingipfeligen Kurve (vgl. SCHULTE 1938, p. 67f., 91).

Bei organischen Substanzen (Alkohol, Aceton usw., vgl. HOLDHEIDE 1930, p. 706; 1932, p. 252; ZEHETNER 1934, p. 508f.) scheint ihre überaus rasche *Permeation* die Ursache der Plasmoptyse zu bilden.

Defektplasmoptyse läßt sich offenbar durch Störung der Membranbildung (HOLDHEIDE 1932, p. 245) oder Erniedrigung der Wandfestigkeit auslösen.

Bei nur kleinerer Wandöffnung kann durch deren zeitweiligen Verschluß der Austritt des Zellinhaltes periodisch in kleineren Portionen erfolgen (vgl. LINSBAUER 1929, p. 578) oder es tritt Wundverschluß ein. *Bakterien* behalten mitunter nach Plasmoptyse ihre Beweglichkeit noch bei (GARBOWSKI 1907, p. 68f.).

Plasmorrhyse.

Als Plasmorrhyse (Plasmorhyse) wird die Volumverringerung nackter Protoplasten durch Einwirkung hypertonischer Lösungen bezeichnet (BALBIANI 1898, p. 537), wobei das Protoplasma entquillt und sich allenfalls vorhandene Vacuolen verkleinern.

Es treten dann häufig Kräuselungen und Knitterungen der Oberfläche auf (vgl. PFEIFFER 1931, p. 28) und an den Plasmodien von *Myxomyceten*, die durch eine Plasmorrhyse geschädigt werden können (VOUK 1913, p. 685), nadelförmige Plasmafäden (S. 94) oder rhythmische Einschnürungen. An Zellen mit Zentralvacuole können anfangs Plasmaanhäufungen linsenförmig in diese hineinragen sowie Ballungen und Plasmastränge auftreten (PFEIFFER 1931, p. 30). Mit dem Grad der Plasmorrhyse ändert sich die Stärke der Adhäsion nackter Protoplasten an Oberflächen (1934a, p. 315). Durch Konzentrationsverringerung läßt sich Plasmorrhyse wieder rückgängig machen (1931, p. 31) und aus einer solchen *Deplasmorrhyse* eine *Deplasmorrhysegeschwindigkeit* bestimmen (1933, p. 61f.).

Literatur.

ABELE, K.: Über die Volumenabnahme des Zellkernes in der Plasmolyse und über das Zustandekommen der Kernplasmarelation. Protoplasma 40, 324—337 (1951). — ÅKERMAN, Å.: Studier över trådlika protoplasmabildningar i växtcellerna. (Studien über fadenförmige Protoplasmastrukturen in Pflanzenzellen.) Lunds Univ. Årsskr. N. F. 12, Nr 4, 1—64 (1916). — Untersuchungen über die Aggregation in den Tentakeln von *Drosera rotundifolia*. Bot. Not. (Lund) 1917, 145—192. — ALBACH, W.: Mikrorespirometrische Untersuchungen über den Einfluß der Vitalfärbung und der Plasmolyse auf die Atmung von Pflanzenzellen. Protoplasma 7, 395—422 (1929). — Über die schädigende Wirkung der Plasmolyse und der Deplasmolyse. Protoplasma 12, 255—267 (1931).

BALBACH, H.: Die Wirkung von Harnstoff auf Gymnoplasten. Protoplasma 26, 192—204 (1936). — BALBIANI, E. G.: Etudes sur l'action des sels sur les infusoires. Arch. Anat. microsc. (Paris) 2, 518—600 (1898). — BANCHER, E.: Zellphysiologische Untersuchungen über den Abblühvorgang bei *Iris* und *Gladiolus*. Österr. bot. Z. 87, 221—244 (1938). — Zellphysiologische Beobachtungen an *Iris Reichenbachii* während des Abblühens. Österr. bot. Z. 90, 97—106 (1941). — Zellphysiologische Untersuchung des Abblühens von *Delphinium orientale*. Vorläufige Mitteilung. Mitt. bot. Inst. techn. Hochsch. Wien 16, 85—91 (1949). — BANK, O.: Der Einfluß des Koffeins auf Plasmolyseform und -zeit bei *Allium Cepa*. Protoplasma 16, 452—453 (1932). — Die Entmischung des vitalgefärbten Zellsaftes der Zwiebelepidermis. Protoplasma 18, 620—627 (1933). — Zur Tonoplasten-Frage. Protoplasma 23, 239—249 (1935). — BARY, A. DE: Die *Mycetozoen* (Schleimpilze). Leipzig: Wilhelm Engelmann 1864, 132 S. — BAUER, L.: Über das Verhalten der *Diatomeen* in hyper- und hypotonischen Medien. Arch. Protistenkde 91, 267—291 (1938). — BECKER, W. A., u. Z. BECKEROWA: Über einige Desintegrationserscheinungen des Protoplasten. Acta Soc. Bot. Poloniae 11, 367—392 (1934). — BECKEROWA, Z.: Über Zellsaft und Tonoplasten von *Bryopsis*. Protoplasma 23, 384—392 (1935). — BEHRISCH, R.: Zur Kenntnis der Endodermiszelle. Ber. dtsch. bot. Ges. 44, 162—164 (1926). — BENECKE, W.: Über farblose *Diatomeen* in der Kieler Föhrde. Jb. wiss. Bot. 35, 535—572 (1900). — BENNET-CLARK, T. A., and D. BEXON: Water relations of plant cells. III. The respiration of plasmolysed tissues. New Phytologist 42, 65—92 (1943). — Water relations of plant cells. IV. Diffusion effects observed in plasmolysed tissues. New Phytologist 45, 5—17 (1946). — BERTHOLD, G.: Studien über die Protoplasmamechanik. Leipzig: A. Felix 1886, 332 S. — BIEBL, R.: Tonoplastenbildung bei *Heterosiphonia plumosa* Batt. Protoplasma 30, 570—576 (1938a). — Zur Frage der Salzpermeabilität bei *Braunalgen*. Protoplasma 31, 518—523 (1938b). — Protoplasmatische Ökologie der Meeresalgen. Ber. dtsch. bot. Ges. 57, (78)—(90) (1939). — Weitere Untersuchungen über die Wirkung der α-Strahlen auf die Pflanzenzelle. Protoplasma 35, 187 bis 236 (1941). — Wirkung der UV-Strahlung auf *Allium*-Zellen. Protoplasma 36, 491—513 (1942). — Wirkung der UV-Strahlen auf die Plasmapermeabilität. Protoplasma 37, 1—24 (1943). — Die Resistenz gegen Zink, Bor und Mangan als Mittel zur Kennzeichnung verschiedener pflanzlicher Plasmasorten. Sitzgsber. österr. Akad. Wiss., Math.-naturwiss. Kl., Abt. I 155, 145—157 (1947). — Permeabilitätsversuche an der *Kartoffel*pflanze. Österr. bot. Z. 95, 129—146 (1948). — BLAESS, H.: Bewegungen des systrophischen und intravacuolären Protoplasmas. Protoplasma 36, 177—180 (1941). — BOCHSLER, A.: Die Wasserpermeabilität des Protoplasmas auf Grund des FICKschen Diffusionsgesetzes. Ber. schweiz. bot. Ges. **58**, 73—122 (1948). — BÖRGER, H.: Über die Kultur von isolierten Zellen und Gewebsfragmenten. Arch. exper. Zellforsch. **2**, 123—190 (1926). — BOGEN, H. J.: Untersuchungen zu den „spezifischen Permeabilitätsreihen“ HÖFLERS. II. Harnstoff und Glycerin. Planta (Berl.) **28**, 535—581 (1938). — Permeabilitäts-Untersuchungen an polyploiden Zellen. Z. Abstammgslehre **83**, 93—105 (1949). — Über Kappenplasmolyse und Vakuolenkontraktion. I. Die Wirkung von LiCl und Neutralrot und ihre Abhängigkeit von der Konzentration und dem osmotischen Wert in der Außenlösung. Planta (Berl.) **39**, 1—35 (1951). — Beiträge zur Physiologie der nichtosmotischen Wasseraufnahme. Planta (Berl.) **42**, 140—155 (1953). — BOGEN, H. J., u. H. PRELL: Messung nichtosmotischer Wasseraufnahme an plasmolysierten Protoplasten. Planta (Berl.) **41**, 459—479 (1953). — BOKORNY, TH.: Die Vakuolenwand der Pflanzenzellen. Biol. Zbl. **13**, 271—275 (1893). — BONTE, H.: Vergleichende Permeabilitätsstudien an Pflanzenzellen. Protoplasma **22**, 209—242 (1953). — BORRISS, H.: Plasmolyseform und Streckungswachstum. Jb. wiss. Bot. **86**, 784—831 (1938). — BRAND, F.: Über das osmotische Verhalten der *Cyanophyceen*zelle. Ber. dtsch. bot. Ges. **21**, 302—309 (1903). — BRAUN, N.: Über die Richtungsverhältnisse der Saftströme in den Zellen der *Characeen.* Ber. Verh. kgl. preuß. Akad. Wiss. Berlin **1852**, 220—268. — BRENNER, W.: Studien über die Empfindlichkeit und Permeabilität pflanzlicher Protoplasten für Säuren und Basen. Öfvers. Finska Vet.-Soc. Förh. **60**, Afd. A, No. 4, 1—124 (1918). — Über die Wirkung von Neutralsalzen auf die Säureresistenz, Permeabilität und Lebensdauer der Protoplasten. Ber. dtsch. bot. Ges. **38**, 277—285 (1920). — BROOKS, S. C.: Studies on exosmosis. Amer. J. Bot. **3**, 483—492 (1916). — BRILLIANT, B.: Les formes de la plasmolyse produites par des solutions concentrées de sucres et de sels dans les cellules de *Mnium* et de *Catharinea.* C. R. Acad. Sci. URSS. **1927**, 155—160. — BRYANT, A. E.: A demonstration of the connection of the protoplasts of the endodermal cells with the Casparian strips in the roots of *barley.* New Phytologist **33**, 231 (1934). — BÜNNING, E.: Zellphysiologische Studien an Meeresalgen. Protoplasma **22**, 444—456 (1934). — BUGAEVSKY, M. F.: Dynamics of vegetable cell decay due to low temperature. C. R. (Doklady) Acad. Sci. URSS., N. S. **22**, 131—134. Ref.: Ber. wiss. Biol., Abt. A **51**, 397 (1939). — BUHMANN, A.: Kritische Untersuchungen über vergleichende plasmolytische und kryoskopische Bestimmungen des osmotischen Wertes bei Pflanzen. Protoplasma **23**, 579—612 (1935). — BURSTRÖM, H.: Studies on growth and metabolism of roots. IX. Cell elongation and water absorption. Physiol. Plantarum (Copenh.) **6**, 262—276 (1953).

CASARI, K.: Über den Plasmolytikumwechsel-Effekt. Protoplasma **42**, 427—447 (1953). — CHALAUPKA, I.: Permeabilitätsstudien an Meeresalgen vornehmlich an *Braunalgen.* Thalassia (Rovigno, Istrien) **3**, No 5, 1—36 (1939). — CHAMBERS, R., and H. P. HALE: The formation of ice in protoplasm. Proc. Roy. Soc. Lond., Ser. B **110**, 336—352 (1932). — CHAMBERS, R., u. K. HÖFLER: Micrurgical studies on the Tonoplast of *Allium Cepa.* Protoplasma **12**, 338—355 (1931). — CHATTAWAY, M.: Protoplasmic retractions in *Bryopsis plumosa.* New Phytologist **28**, 359—368 (1929). — CHAUDAT, R.: Principes de botanique. Genève: Georg et Cie. Libraires-éditeurs 1907, 744 S. — CHAYEN, J.: The structure of root meristem cells of *Vicia Faba.* Symposia Soc. Exper. Biol. **6**, 290—305 (1952). — CHOLNOKY, B. v.: Über die Wirkung von Hyper- und Hypotonischen Lösungen auf einige *Diatomeen.* Internat. Rev. Hydrobiol. **19**, 452—500 (1928). — Untersuchungen über den Plasmolyse-Ort der Algenzellen. I—II. Protoplasma **11**, 278—297 (1930). — Untersuchungen über den Plasmolyse-Ort der Algenzellen. III. Die Plasmolyse der ruhenden Zellen der fadenbildenden *Konjugaten.* Protoplasma **12**, 321—337 (1931a). — Untersuchungen über den Plasmolyse-Ort der Algenzellen. IV. Die Plasmolyse der Gattung *Oedogonium.* Protoplasma **12**, 510—523 (1931b). — Neue Beiträge zur Kenntnis der Plasmolyse bei den *Diatomeen.* Internat. Rev. Hydrobiol. **27**, 306—314 (1932). — Protoplasmatische Untersuchungen durch Lebendfärbung an den Epidermiszellen der *Senecio cruentus*-Blumenblätter. Österr. bot. Z. **97**, 380—390 (1950a). — Einfluß des Plasmolytikums auf den Verlauf der Plasmolyse bei den Epidermiszellen der Blumenblätter von *Senecio cruentus* („*Cineraria*“ hort.). Portugal. Acta Biol., Ser. A **3**, 69—89 (1950b). — Ein Beitrag zur Kenntnis des Plasmalemmas. Ber. dtsch. bot. Ges. **65**, 369—373 (1952a). — Beobachtungen über die Wirkung der Kalilauge auf das Protoplasma. Protoplasma **41**, 57—68 (1952b). — Ein Beitrag zur Kenntnis der Plasmolyse der *Oedogonium*-Zelle. Österr. bot. Z. **100**, 226—234 (1952c). — Beobachtungen über die Plasmolyse. II. Sitzgsber. österr. Akad. Wiss., Math-naturwiss. Kl., Abt. I **162**, 521—543 (1953). — CHOLNOKY, B. v., u. K. HÖFLER: Plasmolyse und Bewegungsvermögen der Diatomee *Amphiprora paludosa.* Protoplasma **38**, 155—164 (1943). — Vergleichende Vitalfärbungsversuche an Hochmooralgen. Sitzgsber. österr. Akad. Wiss., Math.-naturwiss. Kl., Abt. I **159**, 143 bis 182 (1950). — CHOLODNY, N.: Über Protoplasmaveränderungen bei Plasmolyse. Biochem. Z. **147**, 22—29 (1924). — CHOLODNY, N., u. E. SANKEWITSCH: Plasmolyseform und Ionenwirkung. Protoplasma **20**, 57—72 (1934). — COLLANDER, R.: Plasmolytische Beobachtungen an den Epidermiszellen von *Rhoeo discolor.* Protoplasma **21**, 226—234 (1934). — CRAFTS,

A. S.: The relation between structure and function of the phloem. Amer. J. Bot. **26**, 172—177 (1939). — CURRIER, H. B., and W. VAN DER ZWEEP: Plasmolysis and the tetrazolium reaction in *Anacharis canadensis*. Protoplasma **45**, 125—132 (1955). — CZAJA, A. TH.: Ein allseitig geschlossenes, selektiv-permeables System. Ber. dtsch. bot. Ges. **40**, 381—385 (1922).

DERRY, B. H. EL: Plasmolyseform- und Plasmolysezeit-Studien. Protoplasma **8**, 1—49 (1930). — DETMER, W.: Das kleine pflanzenphysiologische Praktikum, 3. Aufl. Jena: Gustav Fischer 1909. 319 S. — DIANNELIDIS, TH.: Zur protoplasmatischen Anatomie des Blattes von *Halophila stipulacea*. Phyton, Ann. rei botanicae **3**, 29—43 (1951). — DÖRING, H.: Beiträge zur Frage der Hitzeresistenz pflanzlicher Zellen. Planta (Berl.) **18**, 405—434 (1933). — DRAWERT, H.: Untersuchungen über den Erregungs- und den Erholungsvorgang in pflanzlichen Geweben nach elektrischer und mechanischer Reizung. Planta (Berl.) **26**, 391—419 (1937). — Beiträge zur Entstehung der Vakuolenkontraktion nach Vitalfärbung mit Neutralrot. Ber. dtsch. bot. Ges. **56**, 123—131 (1938).

EIBL, K.: Plasmolytische Untersuchungen an den Plastiden von *Spirogyra*. Protoplasma **33**, 161—199 (1939a). — Studien über das Plasmolyseverhalten der *Desmidiaceen*-Chromatophoren. Protoplasma **33**, 531—544 (1939b). — EICHBERGER, R.: Über die „Lebensdauer" isolierter Tonoplasten. Protoplasma **20**, 606—632 (1934). — ELO, J. E.: Vergleichende Permeabilitätsstudien, besonders an niederen Pflanzen. Ann. bot. Soc. zool.-bot. fenn. „Vanamo" **8**, Nr 6, 1—108 (1937). — ESCHRICH, W.: Beiträge zur Kenntnis der Wundsiebröhrenentwicklung bei *Impatiens Holsti*. Planta (Berl.) **43**, 37—74 (1953). — ETZ, K.: Über die Wirkung des Austrocknens auf den Inhalt lebender Pflanzenzellen. Protoplasma **33**, 481—511 (1939).

FISCHER, A.: Die Plasmolyse der *Bacterien*. Ber. kgl. sächs. Ges. Wiss., Math.-phys. Cl. **43**, 52—74 (1891). — Untersuchungen über *Bakterien*. Jb. wiss. Bot. **27**, 1—163 (1895). — Die Empfindlichkeit der *Bakterien*zelle und das baktericide Serum. Z. Hyg. **35**, 1—58 (1900). — Vorlesungen über *Bakterien*, 2. Aufl. Jena: Gustav Fischer 1903. 374 S. — FISCHER, H.: Plasmolyseform und Mineralsalzgehalt in alternden Blättern. I. Untersuchungen an *Helodea* und *Fontinalis*. Planta **35**, 513—527 (1948). — FITTING, H.: Untersuchungen über die Aufnahme von Salzen in die lebende Zelle. Jb. wiss. Bot., PFEFFER-Festschr. **56**, 1—64 (1915). — Untersuchungen über die Aufnahme und über anomale osmotische Koeffizienten von Glycerin und Harnstoff. Jb. wiss. Bot. **59**, 1—170 (1919). — FLASCH, A., u. H. KINZEL: Rasche Bildung von Entmischungskörpern in Zellsäften von *Buxus sempervirens*. Protoplasma **44**, 266—271 (1954). — FLURI, M.: Der Einfluß von Aluminiumsalzen auf das Protoplasma. Flora (Jena) **99**, 81—126 (1909). — FÖRSTER, K.: Quellung und Permeabilität der Zellwand von *Rhizoclonium*. Planta (Berl.) **20**, 476—505 (1933). — FRANK, B.: Über Veränderungen der Lage der Chlorophyllkörner und des Protoplasmas in der Zelle, und deren innere und äußere Ursachen. Jb. wiss. Bot. **8**, 216—303 (1872). — FREY-WYSSLING, A.: Deformation and flow in biological systems. Amsterdam: North-Holland Publ. Comp. 1952. 564 S. — Submicroscopic morphology of protoplasm, 2nd english edit. New York: Elsevier Publ. Comp. 1953. 411 p. — Die submikroskopische Struktur des Cytoplasmas. In Protoplasmatologia, Bd. II/A 2. Wien: Springer 1955. 244 S. — FRITSCH, F. E., and F. M. HAINES: The moisture-relations of terrestrial algae. II. The changes during exposure to drought and treatment with hypertonic solutions. Ann. of Bot. **37**, 683—728 (1923). — FÜRLINGER, H.: Plasmolyse verhindert Ergrünen. Protoplasma **30**, 328—333 (1938).

GAIDUKOV, N.: Ultramikroskopische Untersuchungen der Stärkekörner, Zellmembranen und Protoplasten. Ber. dtsch. bot. Ges. **24**, 581—590 (1906). — GARBOWSKI, L.: Gestaltsänderung und Plasmoptyse. Arch. Protistenkde **9**, 53—83 (1907). — GEITLER, L.: Zur Morphologie der Pollenkörner von *Clarkia elegans*. Planta (Berl.) **27**, 426—431 (1938). — GERM, H.: Untersuchungen über die systrophische Inhaltsverlagerung in Pflanzenzellen nach Plasmolyse. I. Protoplasma **14**, 566—621 (1932). — Untersuchungen über die systrophische Inhaltsverlagerung in Pflanzenzellen nach Plasmolyse. II. Protoplasma **17**, 509—547 (1933a). — Untersuchungen über die systrophische Inhaltsverlagerung in Pflanzenzellen nach Plasmolyse. III. Protoplasma **18**, 260—280 (1933b). — Plasmakonfigurationen und Inhaltskörper in Blütenzellen von *Platanthera bifolia* (L.) Rich. Sitzgsber. österr. Akad. Wiss., Math.-naturwiss. Kl., I. Abt. **156**, 509—520 (1947). — Ein Fall spontaner Plasmasystrophe. Protoplasma **39**, 483—492 (1950). — GICKLHORN, J.: Plasmolyse-Orte verschiedener Entwicklungsstadien einer Zelle. Protoplasma **12**, 79—85 (1931). — Vorübergehende Formänderungen von Plastiden während der Plasmolyse. Protoplasma **15**, 71—89 (1932a). — Intracelluläre Myelinfiguren und ähnliche Bildungen bei der reversiblen Entmischung des Protoplasmas. Protoplasma **15**, 90—109 (1932b). — Über aktive Chloroplasten-Kontraktion bei *Spirogyra* und dem Aggregatzustand der Spiralbänder. Protoplasma **17**, 571—592 (1933). — GICKLHORN, J., u. L. MÖSCHL: Vitalfärbung und Vakuolenkontraktion an Zellen mit stabilem Plasmaschaum. Protoplasma **9**, 521—535 (1930). — GICKLHORN, J., u. FR. WEBER: Über Vakuolenkontraktion und Plasmolyseform. Protoplasma **1**, 427—432 (1927). — GREELEY, A. W.: On the analogy between the effects of loss of water and lowering of temperature.

Amer. J. Physiol. **6**, 122—128 (1902). — GROHROCK, E.: Über die Umhäutung isolierter Protoplasmastücke. Untersuchungen an *Saprolegnia*. Planta (Berl.) **23**, 313—339 (1935).
HAAN, Iz. DE: Protoplasmaquellung und Wasserpermeabilität. Rec. Trav. bot. néerl. **30**, 234—335 (1933). — Kappenplasmolyse und Wasserpermeabilität. Protoplasma **22**, 395—404 (1935). — HABERLANDT, G.: Zur Physiologie der Zellteilung. III. Über Zellteilungen nach Plasmolyse. Sitzgsber. preuß. Akad. Wiss., Physik.-math. Kl. **1919**a, Nr 20, 322—348. — Zur Physiologie der Zellteilung. IV. Über Zellteilungen in *Elodea*-Blättern nach Plasmolyse. Sitzgsber. preuß. Akad. Wiss., Physik.-math. Kl. **1919**b, Nr 39, 721—733. — Zur Physiologie der Zellteilung. 5. Mitt. Über das Wesen des plasmolytischen Reizes bei Zellteilungen nach Plasmolyse. Sitzgsber. preuß. Akad. Wiss., Physik.-math. Kl. **1920**, Nr 11, 323—338. — HANSTEEN CRANNER, B.: Zur Biochemie und Physiologie der Grenzschichten lebender Pflanzenzellen. Meld. fra Norges Landbruksh. **2**, 1—160 (1922). — HANSTEIN, J.: Einige Züge aus der Biologie des Protoplasmas. Bot. Abh. Morph. u. Physiol. **4**, H. 2, 1—56 (1880). — HARTMAIR, V.: Über Vakuolenkontraktion in Pflanzenzellen. Protoplasma **28**, 582—592 (1937). — HECHT, K.: Studien über den Vorgang der Plasmolyse. Beitr. Biol. Pflanz. **11**, 137—192 (1912). — HEILBRUNN, L. V.: The colloid chemistry of protoplasm. V. A preliminary study of the surface precipitation reaction of living cells. Arch. exper. Zellforsch. **4**, 246—263 (1927). — HELLWEGER, H.: Über Plasmolyseorte, -form und -zeit im Zusammenhang mit der Chloroplastenstellung (Untersuchungen an *Farn*prothallien). Protoplasma **23**, 221—238 (1935). — HENNER, J.: Untersuchungen über Spontankontraktion der Vakuolen. Protoplasma **21**, 81—111 (1934). — HILMBAUER, K.: Zellphysiologische Studien an *Euglenaceen*, besonders an *Trachelomonas*. Protoplasma **43**, 192—227 (1954). — HÖBER, R.: Physikalische Chemie der Zellen und Gewebe. Bern: Stämpfli & Cie. 1947. 717 S. — HÖFLER, K.: Die plasmolytisch-volumetrische Methode und ihre Anwendbarkeit zur Messung des osmotischen Wertes lebender Pflanzenzellen. Ber. dtsch. bot. Ges. **35**, 706—726 (1917). — Eine plasmometrisch-volumetrische Methode zur Bestimmung des osmotischen Wertes von Pflanzenzellen. Denkschr. Akad. Wiss. Wien, Math.-naturwiss. Kl. **95**, 99—170 (1918a). — Permeabilitätsbestimmung nach der plasmometrischen Methode. Ber. dtsch. bot. Ges. **36**, 414—422 (1918b). — Über die Permeabilität der Stengelzellen von *Tradescantia elongata* für Kalisalpeter. Ber. dtsch. bot. Ges. **36**, 423—442 (1918c). — Über Kappenplasmolyse. Ber. dtsch. bot. Ges. **46**, (73)—(82) (1928). — Über Eintritts- und Rückgangsgeschwindigkeit der Plasmolyse und eine Methode zur Bestimmung der Wasserpermeabilität des Protoplasten. Jb. wiss. Bot. **73**, 300—350 (1930a). — Das Plasmolyse-Verhalten der *Rotalgen*. Z. Bot. **23**, 570—588 (1930b). — Das Permeabilitätsproblem und seine anatomischen Grundlagen. Ber. dtsch. bot. Ges. **49**, (79)—(95) (1931a). — Hypotonietod und osmotische Resistenz einiger *Rotalgen*. Österr. bot. Z. **80**, 51—71 (1931b). — Zur Tonoplastenfrage. Protoplasma **15**, 462—477 (1932a). — Plasmolyseformen bei *Chaetomorpha* und *Cladophora*. Protoplasma **16**, 189—214 (1932b). — Vergleichende Protoplasmatik. Ber. dtsch. bot. Ges. **50**, (53)—(67) (1932c). — Permeabilitätsstudien an Stengelzellen von *Majanthemum bifolium*. Zur Kenntnis spezifischer Permeabilitätsreihen. I. Sitzgsber. Akad. Wiss. Wien, Math.-naturwiss. Kl., I. Abt. **143**, 213—264 (1934a). — Kappenplasmolyse und Salzpermeabilität. Z. wiss. Mikrosk. (KÜSTER-Festschr.) **51**, 70—87 (1934b). — Nekroseformen pflanzlicher Zellen. Ber. dtsch. bot. Ges. **56**, 451—473 (1938). — Nekrose in Sulfoharnstoff. Flora (Jena) **133**, 131—142 (1939a). — Kappenplasmolyse und Ionenantagonismus. Protoplasma **33**, 545—578 (1939b). — Aus der Protoplasmatik der *Diatomeen*. Ber. dtsch. bot. Ges. **58**, 97—120 (1940a). — Salzquellung des Protoplasmas und Ionenantagonismus. Ber. dtsch. bot. Ges. **58**, 292—305 (1940b). — Einige Nekrosen bei Färbung mit Akridinorange. Sitzgsber. österr. Akad. Wiss., Math.-naturwiss. Kl., Abt. I **156**, 585 bis 644 (1947). — Plasmolyse mit Natriumkarbonat. Protoplasma **40**, 426—460 (1951). — Über die Farbionenpermeabilität der Tonoplastenmembran. Ber. dtsch. bot. Ges. **65**, 183 bis 187 (1952). — HÖFLER, K., u. FR. WEBER: Die Wirkung der Äthernarkose auf die Harnstoffpermeabilität von Pflanzenzellen. Jb. wiss. Bot. **65**, 643—737 (1926). — HOFMEISTER, L.: Die Wirkung von Äthylenglykol auf die Plastiden von *Spirogyra*. Protoplasma **28**, 48—65 (1937). — Mikrurgische Studien an *Borraginoideen*-Zellen. I. Mikrodissektion. Protoplasma **35**, 65—94 (1940). — Vitalfärbungsstudien mit Chrysoidin. Sitzgsber. österr. Akad. Wiss., Math.-naturwiss. Kl., Abt. I **157**, 55—82 (1948a). — Über die Permeabilitätsbestimmung nach der Deplasmolysezeit. Sitzgsber. österr. Akad. Wiss., Math.-naturwiss. Kl., Abt. I **157**, 83—95 (1948b). — Mikrurgische Untersuchung über die geringe Fusionsneigung plasmolysierter, nackter Pflanzenprotoplasten. Protoplasma **43**, 278—326 (1954a). — Persönliche Mitteilung 1954b. — HOFMEISTER, W.: Handbuch der physiologischen Botanik, Bd. 1, 1. Abt., Die Lehre von der Pflanzenzelle. Leipzig: Wilhelm Engelmann 1867. 404 S. — HOLDHEIDE, W.: Über einen eigenartigen Fall von Plasmoptyse (vorläufige Mitteilung). Biol. Zbl. **50**, 705—712 (1930). — Über Plasmoptyse bei *Hydrodictyon utriculatum*. Planta (Berl.) **15**, 244—298 (1932). — HOUSKA, H.: Beiträge zur Kenntnis der Kappenplasmolyse. Zur Ätiologie und protoplasmatischen Anatomie der Kappenplasmolyse bei *Allium Cepa*.

Protoplasma **36**, 11—51 (1941). — HUBER, B., u. K. HÖFLER: Die Wasserpermeabilität des Protoplasmas. Jb. wiss. Bot. **73**, 351—511 (1930).

ILJIN, W. S.: Die Durchlässigkeit des Protoplasmas, ihre quantitative Bestimmung und ihre Beeinflussung durch Salze und durch die Wasserstoffionenkonzentration. Protoplasma **3**, 558—602 (1928). — Über Absterben der Pflanzengewebe durch Austrocknung und über ihre Bewahrung vor dem Trockentode. Protoplasma **19**, 414—442 (1933). — Über den Kältetod der Pflanzen und seine Ursachen. Protoplasma **20**, 105—124 (1934). — Lebensfähigkeit der Pflanzenzellen in trockenem Zustand. Planta (Berl.) **24**, 742—754 (1935). — ISABURO-NAGAI: Physiologische Untersuchungen über *Farn*prothallien. Flora (Jena) **106**, 281—330 (1914). — ISRAEL, O.: Biologische Studien mit Rücksicht auf die Biologie. III. ISRAEL und KLINGMANN: Oligodynamische Erscheinungen (v. NÄGELI) an pflanzlichen und tierischen Zellen. Virchows Arch. **147**, 293—340 (1897).

JÄRVENKYLÄ, Y. T.: Über den Einfluß des Lichtes auf die Permeabilität pflanzlicher Protoplasten. Ann. bot. Soc. zool.-bot. fenn. „Vanamo“ **9**, Nr 3, 1—99 (1937). — JANSE, J. M.: Plasmolytische Versuche an Algen. Bot. Zbl. **32**, 21—26 (1887a). — Die Permeabilität des Protoplasma. Versl. Meded. Kon. Akad. Wet. Naturk. **4**, 332—436 (1887b). — JOST, L.: Einige physikalische Eigenschaften des Protoplasmas von *Valonia* und *Chara*. Protoplasma **7**, 1—22 (1929). — JUNGERS, V.: Die Verlagerungsfähigkeit des Zellinhaltes der Zwiebelschuppen von *Allium Cepa* durch Zentrifugierung. Protoplasma **21**, 351—361 (1934).

KACZMAREK, A.: Untersuchungen über Plasmolyse und Deplasmolyse in Abhängigkeit von der Wasserstoffionenkonzentration. Protoplasma **6**, 209—301 (1928). — KAHHO, H.: Zur Kenntnis der Neutralsalzwirkungen auf das Pflanzenplasma. Biochem. Z. **120**, 125 bis 142 (1921). — KAISERLEHNER, E.: Über Kappenplasmolyse und Entmischungsvorgänge im Kappenplasma. Zugleich ein Beitrag zur Kenntnis der Salznekrose des Cytoplasmas. Protoplasma **33**, 579—601 (1939). — KAMIYA, N.: Über Doppelschalen bei *Melosira*. Arch. Protistenkde **91**, 324—342 (1938). — Zytomorphologische Plasmolysestudien an *Allium*-Epidermen. Protoplasma **32**, 373—396 (1939). — KANERT, M.: Über Restitutionsmembranen an Algenzellen, insbesondere die der *Konjugaten*. Diss. Univ. Gießen **1947**. **45** S. — KARSTEN, G.: Die *Diatomeen* der Kieler Bucht. Wiss. Meeresunters., Abt. Kiel, N. F. **4**, 17—205 (1899). — KARZEL, R.: Über die Nachwirkungen der Plasmolyse. Jb. wiss. Bot. **65**, 551—591 (1926). — KEIL, R.: Über systolische und diastolische Veränderungen der Vakuole in den Zellen höherer Pflanzen. Protoplasma **10**, 568—597 (1930). — KEMMER, E.: Beobachtungen über die Lebensdauer isolierter Epidermen. Arch. exper. Zellforsch. **7**, 1—68 (1928). — KENDA, G., u. FR. WEBER: Rasche Vakuolen-Kontraktion in *Cerinthe*-Blütenzellen. Protoplasma **41**, 458—466 (1952). — KESSLER, W., u. W. RUHLAND: Weitere Untersuchungen über die inneren Ursachen der Kälteresistenz. Planta (Berl.) **28**, 159—204 (1938). — KIENITZ-GERLOFF, F.: Neue Studien über Plasmodesmen. Ber. dtsch. bot. Ges. **20**, 93—117 (1902). — KIERMAYER, O.: Die Vakuolen der *Desmidiaceen*, ihr Verhalten bei Vitalfärbe- und Zentrifugierungsversuchen. Sitzgsber. österr. Akad. Wiss., Math.-naturwiss. Kl., Abt. I **163**, 175—222 (1954). — KLEBS, G.: Beiträge zur Physiologie der Pflanzenzelle. Unters. bot. Inst. Tübingen **2**, 489—568 (1886/88). — Ueber die Bildung der Fortpflanzungszellen bei *Hydrodictyon utriculatum* ROTH (Fortsetzung). Bot. Ztg **49**, 805—817 (1891). — KLEMM, P.: Über Regenerationsvorgänge bei den *Siphonaceen*. Ein Beitrag zur Erkenntnis der Mechanik der Protoplasmabewegungen. Flora (Jena) **78**, 19—41 (1894). — Desorganisationserscheinungen der Zelle. Jb. wiss. Bot. **28**, 627—700 (1895). — KLERCKER, J. AF: Eine Methode zur Isolierung lebender Protoplasten. Pflanzenphysiologische Mitteilungen 3. Öfvers. Kongl. Vet. Akad. Förh. **49**, Nr 9, 463—474 (1892). — KOBINGER, I.: Die Vernarbungsmembran an plasmolysierten Protoplasten. Phyton, Ann. rei botanicae **5**, 38—40 (1953). — KOLKWITZ, R.: Untersuchungen über Plasmolyse, Elastizität, Dehnung und Wachstum an lebendem Markgewebe. Beitr. wiss. Bot., Abt. 2 **1**, 221—254 (1896). — KREBS, I.: Beiträge zur Kenntnis des *Desmidiaceen*-Protoplasten. I. Osmotische Werte. II. Plastidenkonsistenz. Sitzgsber. Österr. Akad. Wiss., Math.-naturwiss. Kl., Abt. I **160**, 579—613 (1951). — Beobachtungen über das Plasmolyseverhalten von *Spirotaenia condensata* BRÉB. Protoplasma **44**, 106—108 (1953). — KRESSIN, G.: Beiträge zur vergleichenden Protoplasmatik der *Moos*zelle. Diss. Univ. Greifswald 1935. 64 S. — KUCHAR, K.: Plasmolyseverhalten von *Oospora lactis*. Sydowia, Ann. Mycolog., Ser. II **4**, 53—65 (1950a). — Plasmolyseformverlauf und Trichomzerfall bei zwei *Oscillatorien*. Phyton, Ann. rei botanicae **2**, 213—222 (1950b). — KÜSTER, E.: Über den Einfluß von Lösungen verschiedener Konzentration auf die Orientierungsbewegungen der Chromatophoren. Ber. dtsch. bot. Ges. **23**, 254—256 (1905). — Über den Einfluß wasserentziehender Lösungen auf die Lage der Chromatophoren. Ber. dtsch. bot. Ges. **24**, 255—259 (1906). — Über die Verschmelzung nackter Protoplasten. Ber. dtsch. bot. Ges. **27**, 589—598 (1909). — Über Inhaltsverlagerungen in plasmolysierten Zellen. Flora (Jena) **100**, 267—287 (1910a). — Eine Methode zur Gewinnung abnorm großer Protoplasten. Arch. Entw.mechan. **30**, I, 351—355 (1910b). — Über Veränderungen der Plasmaoberfläche bei Plasmolyse. Z. Bot. **2**, 689—717 (1910c). — Über Vakuolenteilung und grobschaumige

Protoplasten. Ber. dtsch. bot. Ges. **36**, 283—292 (1918). — Experimentelle Physiologie der Pflanzenzelle. In ABDERHALDENS Handbuch der biologischen Arbeitsmethoden, Abt. 11, Teil 1, S. 961—1058. 1924. — Beiträge zur Kenntnis der Plasmolyse. Protoplasma **1**, 73 bis 104 (1926). — Über die Gewinnung nackter Protoplasten. Protoplasma **3**, 223—233 (1928). — Beiträge zur zellphysiologischen Methodik. I u. II. Protoplasma **5**, 191—200 (1929a). — Pathologie der Pflanzenzelle. I. In Protoplasma-Monographien, Bd. 3. Berlin: Gebrüder Borntraeger 1929(b). 200 S. — Beobachtungen an verwundeten Zellen. Protoplasma **7**, 150—170 (1929c). — Über die Bildung semipermeabler Kernmembranen. (Beitrag zur Pathologie des Zellkerns.) Ber. dtsch. bot. Ges. **50**, 350—358 (1932). — Dellen und Löcher im Protoplasma lebender Pflanzenzellen. Beiträge zur Pathologie des Protoplasmas. Protoplasma **19**, 443—451 (1933a). — Über Färbung lebenden Protoplasmas von Pflanzenzellen mit Prune pure. Beiträge zur zellenphysiologischen Methodik IV. Z. wiss. Mikrosk. **50**, 409 bis 418 (1933b). — Über die Bildung plasmatischer Scheinwände nach Plasmolyse. Ber. dtsch. bot. Ges. **53**, 823—833 (1935a). — Systrophe und Messung des Protoplasmas. Beiträge zur zellenphysiologischen Methodik. Z. wiss. Mikrosk. **52**, 427—432 (1935b). — Über Vakuolenkontraktion und Membranfärbung bei *Helodea* nach Behandlung mit Vitalfärbemitteln. Beiträge zur zellenphysiologischen Methodik. V. Z. wiss. Mikrosk. **54**, 433—444 (1937a). — Pathologie der Pflanzenzelle. II. Pathologie der Plastiden. In Protoplasma-Monographien, Bd. 13. Berlin: Gebrüder Borntraeger 1937(b). 152 S. — Über Vakuolenkontraktion. Ber. oberhess. Ges. Nat.- u. Heilk., Naturwiss. Abt. 18, 148—158 (1938a). — Über *Hyacinthus*, ein neues zur Untersuchung der Vakuolenkontraktion geeignetes Objekt. Beiträge zur zellphysiologischen Methodik VI. Z. wiss. Mikrosk. **55**, 26—40 (1938b). — Über Plasmapfropfungen. In Beiträge zur entwicklungsmechanischen Anatomie der Pflanzen, H. 2. Jena: Gustav Fischer 1939(a). 80 S. — Über Membranbildung an kontrahierten Protoplasten höherer Pflanzen. Z. wiss. Mikrosk. **56**, 63—65 (1939b). — Über Vakuolenkontraktion und Anthozyanophoren bei *Pulmonaria*. Cytologia **10**, 44—50 (1939c). — Neue Objekte für die Untersuchung der Vakuolenkontraktion. Ber. dtsch. bot. Ges. **58**, 413—416 (1940). — Über Plasmolyse- und Deplasmolyseformen pflanzlicher Protoplasten. Protoplasma **36**, 134—146 (1942a). — Vitalfärbung und Vakuolenkontraktion. Z. wiss. Mikrosk. **58**, 245—268 (1942b). — Über Vakuolenkontraktion in gegerbten Zellen. Protoplasma **39**, 14—22 (1949). — Die Pflanzenzelle, 2. Aufl. Jena: Gustav Fischer 1951. 866 S. — Beobachtungen über die Wirkung des Ultraschalls auf lebende Pflanzenzellen. Sitzgsber. österr. Akad. Wiss., Math.-naturwiss. Kl., Abt. I **161**, 79—92 (1952). — KÜSTER-WINKELMANN, G.: Über die Doppelschalen der *Diatomeen*. Arch. Protistenkde **91**, 237—266 (1938). — KUNZE, R.: Der Einfluß der Wasserstoffionen-Konzentration auf die Vakuolenkontraktion vital-gefärbter *Elodea*-Zellen. Protoplasma **12**, 161—166 (1931).

LAMBERTZ, P.: Untersuchungen über das Vorkommen von Plasmodesmen in den Epidermisaußenwänden. Planta (Berl.) **44**, 147—190 (1954). — LANZ, I.: Neue Beiträge zur Kenntnis der Systrophe des Protoplasmas. Protoplasma **36**, 381—409 (1942). — LEDERER, B.: Färbungs-, Fixierungs- und mikrochirurgische Studien an *Spirogyra*-Tonoplasten. Protoplasma **22**, 405—430 (1934). — Färbung, Fixierung und Mikrodissektion von Tonoplasten. Biol. generalis (Wien) **11**, 211—242 (1935). — LEPESCHKIN, W. W.: Über den Turgordruck der vakuolisierten Zellen. Ber. dtsch. bot. Ges. **26a**, 198—214 (1908). — Über den Aggregatzustand der protoplasmatischen Fäden und Stränge der Pflanzenzellen. Ber. dtsch. bot. Ges. **43**, 21—26 (1925). — LEVITT, J.: Further remarks on the thermodynamics of active (non-osmotic) water absorption. Physiol. Plantarum (Copenh.) **6**, 240—252 (1953). — LINDNER, E.: Zellphysiologische Resistenzuntersuchungen an Ruderal-, Wiesen- und Kulturpflanzen. Protoplasma **39**, 507—534 (1950). — LINSBAUER, K.: Untersuchungen über Plasma und Plasmaströmung an *Chara*-Zellen. I. Beobachtungen an mechanisch und operativ beeinflußten Zellen. Protoplasma **5**, 563—621 (1929). — LOPRIORE, G.: Über die Einwirkung der Kohlensäure auf das Protoplasma der lebenden Pflanzenzelle. Jb. wiss. Bot. **28**, 531—626 (1895). — LOREY, E.: Mikrochirurgische Untersuchungen über die Viskosität des Protoplasmas. Protoplasma **7**, 171—203 (1929). — LOTHRING, H.: Beiträge zur Biologie der Plasmolyse. Planta (Berl.) **32**, 600—629 (1942). — LUNDEGÅRDH, H.: Über die Permeabilität der Wurzelspitzen von *Vicia Faba* unter verschiedenen äußeren Bedingungen. Kgl. Sv. Vetenskapsakad. Hdl. **47** (3), 1—254 (1911).

MARKLUND, G.: Vergleichende Permeabilitätsstudien an pflanzlichen Protoplasten. Acta bot. fenn. **18**, 1—110 (1936). — MEINDL, T.: Weitere Beiträge zur protoplasmatischen Anatomie des *Helodea*-Blattes. Protoplasma **21**, 362—393 (1934). — MICHEL, W.: Über die experimentelle Fusion pflanzlicher Protoplasten. Arch. exper. Zellforsch. **20**, 230—252 (1937). — MIEHE, H.: Wachstum, Regeneration und Polarität isolierter Zellen. Ber. dtsch. bot. Ges. **23**, 257—264 (1905). — MISSBACH, G.: Versuche zur Prüfung der Plasmaviskosität. Protoplasma **3**, 327—344 (1928). — MODER, A.: Beiträge zur protoplasmatischen Anatomie des *Helodea*-Blattes. Protoplasma **16**, 1—55 (1932). — MOSER, L.: Zellphysiologische Untersuchungen an *Cladophora fracta*. Österr. bot. Z. **91**, 131—167 (1942). — MOTHES, K.: Der

Tonoplast von *Sphaeroplea*. Planta (Berl.) **21**, 486—510 (1934). — MYERS, G. M. P.: The water permeability of unplasmolysed tissues. J. of Exper. Bot. **2**, 129—144 (1951).

NÄGELI, C.: In C. NÄGELI u. C. CRAMER, Pflanzenphysiologische Untersuchungen. 1. H., Zürich: Fr. Schulthess 1855. 120 S. — NEMEC, B.: Über Ausgabe ungelöster Körper in hautumkleideten Zellen. Sitzgsber. böhm. Ges. Wiss., Math.-naturwiss. Kl. **1899**, Nr 42, 1—15. — NETTER, H.: Über den nichtlösenden Raum (sog. disperse Phase) und seine Bedeutung für zellphysiologische Probleme. Protoplasma **2**, 554—563 (1927). — NOLL, F.: Beitrag zur Kenntnis der physikalischen Vorgänge, welche den Reizkrümmungen zu Grunde liegen. Arb. bot. Inst. Würzburg **3**, 496—533 (1888).

OSTERHOUT, W. J. V.: On plasmolysis. Bot. Gaz. **46**, 53—55 (1908). — Protoplasmic contractions resembling plasmolysis which are caused by pure distilled water. Bot. Gaz. **55**, 446—451 (1913).

PALLA, E.: Beobachtungen über Zellhautbildung an des Zellkernes beraubten Protoplasten. Flora (Jena) **73**, 314—330 (1890). — PANTANELLI, E.: Zur Kenntnis der Turgorregulationen bei *Schimmelpilzen*. Jb. wiss. Bot. **40**, 303—367 (1904). — PARDATSCHER, G.: Beiträge zur Kenntnis der Vakuolenkontraktion. Diss. Univ. Wien 1951 (a). 121 S. — Protoplasmatische Studien an Blütenzellen von *Dahlia*. Portugal. Acta Biol., Sér. A **3**, 171—186 (1951 b). — Vakuolenkontraktion in *Iris*-Blüten. Phyton, Ann. rei botanicae **5**, 26—33 (1953). — PECKSIEDER, E.: Permeabilitätsstudien an *Lebermoosen*. Sitzgsber. österr. Akad. Wiss., Math.-naturwiss. Kl., Abt. I **156**, 521—584 (1947). — PEKAREK, J.: Absolute Viskositätsmessungen mit Hilfe der BROWNschen Molekularbewegung. IX. Die Viskosität des Protoplasmas bei Kappenplasmolyse. Protoplasma **34**, 177—187 (1940). — PFEFFER, W.: Osmotische Untersuchungen. Studien zur Zellmechanik. Leipzig: Wilhelm Engelmann 1877. 236 S. — Über Aufnahme von Anilinfarben in lebende Zellen. Unters. bot. Inst. Tübingen **2**, 179—332 (1886/88). — PFEIFFER, H.: Experimentelle und theoretische Untersuchungen über die Entdifferenzierung und Teilung pflanzlicher Dauerzellen. II. Protoplasma **10**, 253—288 (1930). — Über die Plasmorrhyse nackter Protoplasten in hypertonischen Medien. I. Cytologia **3**, 26—35 (1931). — Über die Plasmorrhyse nackter Protoplasten. II. Cytologia **4**, 52—67 (1933). — Über die Plasmorrhyse nackter Protoplasten. III. Cytologia **5**, 308—316 (1934 a). — Über die Plasmorrhyse nackter Protoplasten. IV. Cytologia **5**, 507—516 (1934 b). — Versuche über die Beeinflussung von Form und Adhäsion nackter Protoplasten. Arch. exper. Zellforsch. **15**, 203—212 (1934 c). — Versuche über das Fadenziehvermögen isolierter pflanzlicher Protoplasten. Kolloid-Z. **70**, 26—31 (1935). — Beiträge zur physikalischen Analyse der plasmolytischen Zerschnürung langgestreckter Protoplasten. Protoplasma **25**, 528—545 (1936). — PIGOŃ, A.: The tension at the surface of dissected vacuoles. Part II. Plant vacuoles. Bull. internat. Acad. pol. Sci. Lettres, Cl. math. et nat. Sér. B **1951**, 403—418. — PIRSON, A., u. E. GÖLLNER: Beobachtungen zur Entwicklungsphysiologie der *Lemna minor* L. Flora (Jena) **140**, 485—498 (1953). — PIRSON, A., u. F. SEIDEL: Zell- und stoffwechselphysiologische Untersuchungen an der Wurzel von *Lemna minor* L. unter besonderer Berücksichtigung von Kalium- und Calciummangel. Planta (Berl.) **38**, 431—473 (1950). — PLOWE, J. Q.: Membranes in the plant cell. I. Morphological membranes at protoplasmic surfaces. Protoplasma **12**, 196—220 (1931 a). — Membranes in the plant cell. II. Localization of differential permeability in the plant protoplast. Protoplasma **12**, 221—240 (1931 b). — PRÁT, S.: Plasmolyse und Permeabilität. Biochem. Z. **128**, 557—567 (1921). — Plasmolyse des *Cyanophycées*. Bull. internat. Classe Sci. Math. nat. méd., Acad. Tchèque Sci. **23**, 96—97 (1923 a). — Plasmolyse und Permeabilität. III. Ber. dtsch. bot. Ges. **41**, 225—227 (1923 b). — Weiteres über Plasmolyse und Permeabilität. Mém. Soc. Roy. Sci. Bohême, Cl. Sci. **1925**, Nr 1, 1—12. — Wasserstoffionenkonzentration und Plasmolyse. Kolloid-Z. **40**, 248—251 (1926). — Plasmolysis and swelling in plant cells. Fifth Int. Bot. Congr. Rep. Proc. Cambridge 1931, S. 428—430. 1931. — Stimulation plasmolysis on marine algae. Acta adriatica inst. oceanogr. Split. **1934**, Nr 4, 1—20. — PRELL, H.: Untersuchungen über die Aufnahme von Anelektrolyten in Zellen di- und polyploider Pflanzen. Planta (Berl.) **41**, 480—508 (1953). — PRINGSHEIM, N.: Untersuchungen über den Bau und die Bildung der Pflanzenzelle. 1. Abt. Grundlinien einer Theorie der Pflanzenzelle. Berlin: August Hirschwald 1854. 90 S. — PRUD'HOMME VAN REINE jr., W. J.: Versuche über die Konsistenz des Protoplasmas. Rec. Trav. bot. néerl. **32**, 467—515 (1935). Siehe auch: Über Plasmolyse und Deplasmolyse. Proc. Kon. Akad. Wetensch. **38**, 199—209 (1935).

RAFFELT, G.: Untersuchungen über den Einfluß des Milieuwechsels auf die *Bakterien*zelle. Diss. tierärztl. Hochsch. Hannover 1944. 34 S. — RAICHEL, B.: Über den Einfluß osmotisch wirksamer Mittel auf die *Bakterien*zelle. Arch. Protistenkde **63**, 333—361 (1928). — REINHARDT, M. O.: Plasmolytische Studien zur Kenntniss des Wachsthums der Zellmembran, S. 425—463. Botanische Untersuchungen S. SCHWENDENER zum 10. Februar 1899 dargebracht. SCHWENDENER-Festschr. Berlin: Gebrüder Borntraeger 1899. — REUTER, L.: Zur protoplasmatischen Anatomie des Keimblattes von *Soja hispida*. Ein Beitrag zur Protoplasmatik ernährungsphysiologisch differenter Zustände. Österr. bot. Z. **95**, 373—421 (1948). — ROMIJN, C.: Über den Einfluß der Temperatur auf die Protoplasmaströmung bei *Nitella flexilis*. Proc. Sect. Sci. Kon. Akad. Wetensch. **34**, 289—296 (1931). — RUGE, U.: Untersuchungen über den Einfluß des Hetero-Auxins auf das Streckungswachstum des Hypocotyls von *Helianthus annuus*. Z. Bot. **31**, 1—56 (1937). — Kritische zell- und entwicklungsphysiologische Untersuchungen an den Blattzähnen von *Helodea densa*. Flora (Jena) **134**, 311—376 (1940). — RUHLAND, W., u. C. HOFFMANN: Die Permeabilität von *Beggiatoa mirabilis*. Ein Beitrag zur Ultrafiltertheorie des Plasmas. Planta (Berl.) **1**, 1—83 (1926). — RUHLAND, W., H. ULLRICH u. S. ENDO: Untersuchungen zu den „spezifischen Permeabilitätsreihen" HÖFLERS. I. Zur Frage der Alkoholpermeabilität von Pflanzenzellen unter verschiedenen Versuchsbedingungen. Planta (Berl.) **27**, 650—668 (1938).

SCARTH, G. W.: Colloidal changes associated with protoplasmic contraction. Quart. J. Exper. Physiol. **14**, 99—113 (1924). — The influence of external osmotic pressure and of disturbance of the cell surface on the permeability of *Spirogyra* for acid dyes. Protoplasma **1**, 204—213 (1926). — SCHAEFER, G.: Ein Versuch zur quantitativen Auswertung der Plasmolyseform- und -zeitmethode. Protoplasma **44**, 422—436 (1955). — SCHARINGER, W.: Cytologische Beobachtungen an *Ranunculaceen*-Blüten. Protoplasma **25**, 404—426 (1936). — SCHEIDL, W.: Auslösung von Vakuolenkontraktion durch undissoziierte Basen. Sitzgsber. österr. Akad. Wiss., Math.-naturwiss. Kl., Abt. I **163**, 645—688 (1954). — Vakuolenkontraktion bei vollen Zellsäften an Zwiebelzellen von *Tulipa silvestris* und *Colchicum speciosum*. Protoplasma **44**, 336—341 (1955). — SCHEITTERER, H.: Plasmolyse-Ort der Blatt-Palisaden-Zellen. Protoplasma **10**, 289—293 (1930). — SCHEITTERER, H., u. FR. WEBER: Hypotonie-Tod von Pflanzenzellen. Protoplasma **10**, 474—477 (1930). — SCHIMPER, A. F. W.: Untersuchungen über die Chlorophyllkörper und die ihnen homologen Gebilde. Jb. wiss. Bot. **16**, 1—247 (1885). — SCHINDLER, H.: Tötungsart und Absterbebild. I. Der Alkalitod der Pflanzenzelle. Protoplasma **30**, 186—205 (1937). — SCHITTENGRUBER, B.: Kontraktion Anthoorphnin-haltiger Vakuolen. Protoplasma **42**, 328—333 (1953). — SCHMIDT, H.: Plasmolyse und Permeabilität. Jb. wiss. Bot. **83**, 470—512 (1936). — SCHMIDT, H., K. DIWALD u. O. STOCKER: Plasmatische Untersuchungen an dürreempfindlichen und dürreresistenten Sorten landwirtschaftlicher Kulturpflanzen. Planta (Berl.) **31**, 559—596 (1940). — SCHNEE, L.: Bandplasmolyse der Endodermiszellen von *Cobaea scandens*. Protoplasma **26**, 97—99 (1936). — SCHNEIDER, E.: Über die Plasmolyse als Kennzeichen lebender Zellen. Z. wiss. Mikrosk. **42**, 32—54 (1924). — SCHOCH-BODMER, H.: Zur Physiologie der Pollenkeimung bei *Corylus Avellana:* Pollen- und Narbensaugkräfte, Quellungserscheinungen der Kolloide des Pollens. Protoplasma **25**, 337—371 (1936). — SCHÖNLEBER, K.: Reizplasmoschise bei *Spirogyra*. Planta (Berl.) **24**, 387—401 (1935). — SCHÜTT, F.: Die *Peridineen* der Plankton-Expedition. I. Ergebnisse der Plankton-Expedition der HUMBOLDT-Stiftung **4**. M. 2. A. Kiel u. Leipzig: Lipsius & Tischer 1895. 170 S. — SCHULTE, E.: Untersuchungen über das Auftreten von anomaler (negativer) Osmose im Pflanzenkörper. Protoplasma **29**, 60—98 (1938). — SCHUMACHER, W.: Über die Plasmolysierbarkeit der Siebröhren. Jb. wiss. Bot. **88**, 545—553 (1939). — SEEMANN, F.: Der Einfluß von Neutralsalzen und Nichtleitern auf die Wasserpermeabilität des Protoplasmas. Protoplasma **42**, 109—132 (1952). — SENN, G.: Die Gestalts- und Lageveränderung der Pflanzen-Chromatophoren. Leipzig: Wilhelm Engelmann 1908. 397 S. — SIMINOVITCH, D., and D. R. BRIGGS: Studies on the chemistry of the living bark of the *black locust* in relation to its frost hardiness. III. The validity of plasmolysis and desiccation tests for determining the frost hardiness of bark tissue. Plant Physiol. **28**, 15—34 (1953). — STADELMANN, ED.: Zur Messung der Stoffpermeabilität pflanzlicher Protoplasten. II. Sitzgsber. österr. Akad. Wiss., Math.-naturwiss. Kl., Abt. I **161**, 373—408 (1952). — Sur la plasmolyse des plantes sahariennes par des electrolytes (note préliminaire). In: [M. MALABAN:] Desert Res. Jerusalem 1952: Research Council of Israel. 641 S.; p. 332—334. — STRASBURGER, E.: Über Plasmaverbindungen pflanzlicher Zellen. Jb. wiss. Bot. **36**, 493—610 (1901). — STRUGGER, S.: Untersuchungen über den Einfluß der Wasserstoffionen auf das Protoplasma der Wurzelhaare von *Hordeum vulgare* L. I. Sitzgsber. österr. Akad. Wiss., Math.-naturwiss. Kl., Abt. I **135**, 453—477 (1926). — Untersuchungen über den Einfluß der Wasserstoffionen auf das Protoplasma der Wurzelhaare von *Hordeum vulgare* L. II. Sitzgsber. österr. Akad. Wiss., Math.-naturwiss. Kl., Abt. I **137**, 143—169 (1928). — Zur Analyse der Vitalfärbung pflanzlicher Zellen mit Erythrosin. Ber. dtsch. bot. Ges. **49**, 453—476 (1931). — Über Plasmolyse mit Kaliumrhodanid. Ber. dtsch. bot. Ges. **50**, 24—31 (1932). — Beiträge zur Physiologie des Wachstums. I. Zur protoplasma-physiologischen Kausalanalyse des Streckungswachstums. Jb. wiss. Bot. **79**, 406—471 (1934). — Praktikum der Zell- und Gewebephysiologie der Pflanze. Berlin: Gebrüder Borntraeger 1935. 181 S. — Beiträge zur Analyse der Vitalfärbung pflanzlicher Zellen mit Neutralrot. Protoplasma **26**, 56—69 (1936). — Praktikum der Zell- und Gewebephysiologie der Pflanze, 2. Aufl. Berlin-Göttingen-Heidelberg: Springer 1949. 225 S. — SUTCLIFFE, J. F.: The absorption of potassium ions by plasmolysed cells. J. of Exper. Bot. **5**, 215—231 (1953). — SWELLENGREBEL, N. H.: Über

Plasmolyse und Turgorregulation der Preßhefe. Zbl. Bakter. II 14, 374—388, 481—492 (1905). — SZÜCS, J.: Studien über Protoplasmapermeabilität. Sitzgsber. ksl. Akad. Wiss., Math.-naturwiss. Kl., Abt. I 119, 737—773 (1910).

TAKADA, H.: Protoplasmological studies on the *Yeast* cell adapted to high osmotic environement. J. Inst. Polytechn. Univ. Osaka (Japan), Ser. D 4, 17—26 (1953). — TAKAMINE, N.: On the plasmolysis form in *Allium Cepa* with special reference to the influence of potassium ion upon it. Cytologia 10, 302—323 (1940a). — Studies in the influence of ultraviolet rays on the cytoplasm, tested by the plasmolysis-form method. Cytologia 10, 577—587 (1940b). — TAMMANN, G.: Über Osmose durch Niederschlagsmembranen. Ann. Physik u. Chemie, N. F. 34, 299—315 (1888). — TIROLD, M.: Untersuchungen über das Plasmolyseverhalten von *Vaucheria*. Protoplasma 18, 345—389 (1932). — TÖRNÄVÄ, S. R.: Expansion-capacity of naked plant protoplasts. Protoplasma 32, 329—341 (1939). — TREBOUX, O.: Einige stoffliche Einflüsse auf die Kohlensäureassimilation bei submersen Pflanzen. Flora 92, 49—76 (1903). — TRÖNDLE, A.: Die Aufnahme von Salzen in die Pflanzenzelle. Denkschr. schweiz. naturforsch. Ges., Abh. 1 58, 1—59 (1922).

ÚLEHLA, V.: Jak působí vodíkové ionty na některé nižší rostliny. Über den Einfluß der Wasserstoffionen auf einige niedere Pflanzen. Studia Mendeliana, Brunae 1923, 229 bis 253. — Die Quellungsgeschwindigkeit der Zellkolloide als gemeinschaftlicher Faktor in Plasmolyse, Plasmoptyse und ähnlichen Veränderungen des Zellvolumens. Planta (Berl.) 2, 618—639 (1926). — Über einen natürlich isolierten und überlebenden Ektoplasten in den „leeren“ Zellen von *Basidiobolus ranarum* und seine Semipermeabilität. Arch. exper. Zellforsch. 22, 501—514 (1939). — ÚLEHLA, V., u. V. MORÀVEK: Über die Wirkung von Säuren und Salzen auf *Basidiobolus ranarum* Eid. Ber. dtsch. bot. Ges. 40, 8—20 (1922). — ULLRICH, H.: Einige Beobachtungen über Doppelbrechung am lebenden Protoplasten, an verschiedenen Zellorganellen sowie der Zellwand. Planta (Berl.) 26, 311—318 (1937). — UMRATH, K.: Die Bildung von Plasmalemma (Plasmahaut) bei *Nitella mucronata*. Protoplasma 16, 173—188 (1932). — UNGER, F.: Anatomie und Physiologie der Pflanzen. Pest, Wien u. Leipzig: C. A. Hartleben 1855. 461 S. — URSPRUNG, A.: Über die gegenseitigen Beziehungen der osmotischen Zustandsgrößen. Planta (Berl.) 2, 640—660 (1926). — Zur Terminologie und Analyse der osmotischen Zustandsgrößen. Z. Bot. 23, 183—202 (1930). — Die Messung der osmotischen Zustandsgrößen pflanzlicher Zellen und Gewebe. Die Messung des Widerstandes, den das Substrat (Boden, Lösung, Luft) dem Wasserentzug durch die Pflanze entgegensetzt. In ABDERHALDENS Handbuch der biologischen Arbeitsmethoden, Abt. 11, Teil 4, 2. Hälfte, S. 1109—1572. 1939. — URSPRUNG, A., u. G. BLUM: Dürfen wir die Ausdrücke osmotischer Wert, osmotischer Druck, Turgordruck, Saugkraft synonym gebrauchen? Biol. Zbl. 40, 193—216 (1920). — Eine Methode zur Messung polarer Saugkraftdifferenzen. Jb. wiss. Bot. 65, 1—27 (1926).

VRIES, H. DE: Untersuchungen über die mechanischen Ursachen der Zellstreckung ausgehend von der Einwirkung von Salzlösungen auf den Turgor wachsender Pflanzenzellen. Leipzig: Wilhelm Engelmann 1877. 120 S. — Opera coll. 1, 357—486. — Eine Methode zur Analyse der Turgorkraft. Jb. wiss. Bot. 14, 427—601. — Opera coll. 2, 137—296 (1884). — Plasmolytische Studien über die Wand der Vacuolen. Jb. wiss. Bot. 16, 465—598. — Opera coll. 2, 321—446 (1885). — Isotonische Koeffizienten einiger Salze. Z. physikal. Chem. 3, 103—109. — Opera coll. 2, 536—542 (1889). — VOUK, V.: Untersuchungen über die Bewegung der Plasmodien. II. Teil. Studien über die Protoplasmaströmung. Denkschr. ksl. Akad. Wiss., Math.-naturwiss. Kl. 88, 653—692 (1913).

WÄCHTER, W.: Untersuchungen über den Austritt von Zucker aus den Zellen der Speicherorgane von *Allium Cepa* und *Beta vulgaris*. Jb. wiss. Bot. 41, 165—220 (1905). — WALTER, H.: Protoplasma- und Membranquellung bei Plasmolyse. Jb. wiss. Bot. 62, 145—213 (1923). — Plasmaquellung und Assimilation. Protoplasma 6, 113—156 (1929). — WEBER, FR.: Das Fadenziehen und die Viskosität des Protoplasmas. Österr. bot. Z. 70, 172—180 (1921). — Plasmolyseform und Protoplasmaviskosität. Österr. bot. Z. 73, 261—266 (1924a). — Krampf-Plasmolyse bei *Spirogyra*. Pflügers Arch. 206, 629—634 (1924b). — Über die Beurteilung der Plasmaviskosität nach der Plasmolyseform. Untersuchungen an *Spirogyra*. Z. wiss. Mikrosk. 42, 146—156 (1924c). — Schrauben-Plasmolyse bei *Spirogyra*. Ber. dtsch. bot. Ges. 43, 217—223 (1925a). — Plasmolyseform und Ätherwirkung. Pflügers Arch. 208, 705 bis 717 (1925b). — Physiologische Ungleichheit bei morphologischer Gleichheit. Österr. bot. Z. 74, 256—261 (1925c). — Plasmolyseform und Kernform funktionierender Schließzellen. Jb. wiss. Bot. 64, 687—701 (1925d). — Experimentelle Physiologie der Pflanzenzelle. Arch. exper. Zellforsch. 2, 67—92 (1925e). — Plasmolyse-Zeit-Methode. Protoplasma 5, 622—624 (1929a). — Fadenziehen des Endoplasmas bei *Spirogyra*. Protoplasma 6, 159—161 (1929b). — Plasmolysezeit und Lichtwirkung. Protoplasma 7, 256—258 (1929c). — Plasmolyse-Ort. Protoplasma 7, 583—601 (1929d). — Protoplasmatische Pflanzenanatomie. Protoplasma 8, 291—306 (1930a). — Plasmolyse in verdünntem Gewebesaft. Protoplasma 8, 437—439

(1930b). — Vakuolen-Kontraktion vitalgefärbter *Elodea*-Zellen. Protoplasma **9**, 106—119 (1930c). — Vakuolenkontraktion und Protoplasmaentmischung in Blütenblattzellen. Protoplasma **10**, 598—607 (1930d). — Vakuolenkontraktion und Vitalfärbung in Blütenzellen. Protoplasma **11**, 312—316 (1930e). — Harnstoff-Permeabilität ungleich alter *Spirogyra*-Zellen. Protoplasma **12**, 129—140 (1931). — Plasmolyse-Resistenz und -Permeabilität bei Narkose. Protoplasma **14**, 179—191 (1932a). — Plasmalemma oder Tonoplast? Protoplasma **15**, 453—461 (1932b). — Plasmolyse-Permeabilität bei Kälte. Protoplasma **15**, 517—521 (1932c). — Plasmolyse und „surface precipitation reaction". Protoplasma **15**, 522—531 (1932d). — Unterschiede in der Säureresistenz der *Helodea*-Blattzellen. Protoplasma **16**, 287—290 (1932e). — Plasmalemma-Zerstörung und Tonoplasten-Bildung. Protoplasma **21**, 424—426 (1934). — Vakuolen-Kontraktion der *Borraginaceen*-Blütenzellen als Synärese. Protoplasma **22**, 4—16 (1935). — Vakuolenkontraktion und Anthocyanophoren in *Pulmonaria*-Blütenzellen. Protoplasma **26**, 100—107 (1936). — WEBER, FR., u. G. KENDA: Zweimalige Vakuolenkontraktion in *Cerinthe*-Zellen. Phyton, Ann. rei botanicae **4**, 315—318 (1953). — WEBER, R.: Plasmolyse und Vakuolenkontraktion bei *Antithamnion plumula*. Protoplasma **19**, 242—248 (1933). — WEIS, A.: Beiträge zur Kenntnis der Plasmahaut. Planta (Berl.) **1**, 145—186 (1926). — WEISSENBÖCK, K.: Membranregeneration plasmolysierter *Vaucheria*-Protoplasten. Protoplasma **32**, 44—91 (1938). — WILL-RICHTER, G.: Der osmotische Wert der *Lebermoose*. Sitzgsber. österr. Akad. Wiss., Math.-naturwiss. Kl., Abt. I **158**, 431—542 (1949). — WULFF, H. D.: Lebendfärbungen mit Chrysoidin. Planta (Berl.) **22**, 70—79 (1934).

YAMAHA, G., u. R. UEDA: Über den Einfluß der Ultraschallwellen auf die Wurzelspitzenzellen von *Vicia Faba* L. Ein Orientierungsversuch. Cytologia **9**, 524—532 (1939). — YOTSUYANAGI, Y.: Recherches sur les phénomènes moteurs dans les fragments de protoplasme isolés. I. Mouvement rotatoire et le processus de son apparition. Cytologia **18**, 146 bis 156 (1953).

ZEHETNER, H.: Untersuchungen über die Alkoholpermeabilität des Protoplasmas. Jb. wiss. Bot. **80**, 505—566 (1934).

Nachtrag.

Die folgenden Arbeiten sind an der durch Seitenzahl und Absatz (s. Anmerkung 3 auf S. 71) gekennzeichneten Stelle, an welcher stets auf die Seitenzahl dieses Nachtrages rückverwiesen ist, zusätzlich zu berücksichtigen.

S. 76_1: PRELL, H.: Die Wasserbilanz plasmolysierter Protoplasten. Planta (Berl.) **46**, 272—285 (1955).

S. 82_5: CURRIER, H. B., K. ESAU and V. I. CHEADLE: Plasmolytic studies of phloem. Amer. J. Bot. **42**, 68—81 (1955).

S. 83_2: MODLIBOWSKA, I., and W. S. ROGERS: Freezing of plant tissues under the microscope. J. of Exper. Bot. **6**, 384—391 (1955).

S. 97_1: WADA, B.: Mikrurgische Untersuchungen lebender Zellen in der Teilung. III. Die Einwirkung der Plasmolyse auf die Mitose bei Staubfadenhaarzellen von *Tradescantia reflexa*. Cytologia **7**, 198—212 (1936).

S. 97_1: STROHMEYER, G.: Beiträge zur experimentellen Zytologie. Planta (Berl.) **24**, 470—509 (1935).

S. 97_1: BĚLAŘ, K.: Beiträge zur Kausalanalyse der Mitose. III. Untersuchungen an den Staubfadenhaar- und Blattmeristemzellen von *Tradescantia virginica*. Z. Zellforsch. **10**, 73—134 (1930).

Stoffaustausch.
Einführung, Begriffsbestimmung: Permeabilität, nichtdiosmotische Stoffaufnahme und -abgabe.

Von

Hans Joachim Bogen.

Der Stoffaustausch der Zelle, gemessen als Stoffaufnahme, als Stoffabgabe und als Bilanz der beiden gegenläufigen Prozesse, ist eines der ältesten Arbeitsgebiete der Zellphysiologie; er wurde schon von W. PFEFFER (1877) und H. DE VRIES (1888) theoretisch und experimentell behandelt. Das erklärt sich zunächst aus der Möglichkeit, mit verhältnismäßig einfachen Methoden verläßliche Meßdaten zu erhalten (gravimetrisch, mit speziellen chemischen Nachweisreaktionen, osmotisch), zum anderen aber aus der Einsicht, daß der Stoffaustausch in engster Beziehung zum Stoffwechsel steht, einem Grundphänomen der lebenden Zelle also: soweit man ihn nicht überhaupt als einen Teil des Stoffwechsels auffassen will, ist die Stoffaufnahme eine Voraussetzung für dessen Ablauf, die Stoffabgabe Folge bzw. Ausdruck des Stoffwechselgeschehens. Da aber der Stoffwechsel seinerseits Voraussetzung für alle anderen Lebensäußerungen der Zelle wie Wachstum, Reizbarkeit, Fortpflanzung, Vererbung, Adaptation usw. ist, erscheint es geboten, dem Stoffaustausch ein umfangreiches und zentral stehendes Kapitel des 2. Bandes zu widmen.

Die unübersehbare Fülle der Arbeiten, die in den letzten 75 Jahren experimentelle und theoretische Beiträge insbesondere zum Problem der Stoff*aufnahme* geliefert haben, ist der Herausbildung präziser Begriffsbestimmungen weniger förderlich gewesen, als man erwarten durfte; ein Blick auf die Publikationen lehrt, daß oftmals so verschiedenartige Begriffe wie Permeabilität, Aufnahme und Akkumulation bzw. Exosmose, Sekretion, loss und leakage synonym verwendet wurden. Diese Verwirrung kann aus der Geschichte der Zellphysiologie verstanden werden; wie die jüngsten Veröffentlichungen zeigen, ist sie aber inzwischen endgültig überwunden worden (vgl. ROBERTSON 1951).

1. Stoffaustauschwerte M.

Jeder Stoffaustausch zwischen der Zelle und ihrer Umgebung setzt voraus, daß die äußersten Schichten, mit denen der Protoplast an seine Umgebung grenzt, für die in Rede stehenden Stoffe durchlässig *(„permeabel“)* sind (von der Durchlässigkeit der Zellwand, die in der Regel nicht der begrenzende Faktor ist, wird hier abgesehen — vgl. hierzu III E). Die Durchlässigkeit („Permeabilität“) der Grenzschichten ist in dieser Betrachtungsweise lediglich eine *Eigenschaft*, die noch nichts aussagt über die Mechanismen, nach denen der Stoffaustausch erfolgt; sie *ermöglicht* nur den Durchtritt von Substanz in der einen oder anderen oder in beiden Richtungen. Die experimentelle Analyse mißt zunächst die Stoffmenge M, die unter definierten Bedingungen eine Translozierung erfährt. Durch die Indices i und a kann dabei die Richtung der Translozierung bezeichnet werden: M_i ist die nach innen, M_a die nach außen bewegte Stoffmenge. Die Gesamtverschiebung ist dann

$$M = M_i - M_a. \tag{1}$$

Um verschiedene M-Werte untereinander vergleichen zu können, bedarf es folgender Zusätze:

M ist die Substanzmenge, die in der Zeit t (weil die Stoffbewegung ein Vorgang, also zeitabhängig ist) durch die Oberfläche A bewegt wird (weil der Durchtritt durch die Zellgrenzfläche erfolgen *muß*, diese aber für verschiedene Zellen verschieden groß ist):

$$M = f(t, A). \tag{2}$$

Die so definierten M-Werte sind identisch mit den in der ausländischen Literatur oft verwendeten „flux"-Werten:

net flux = influx — outflux.

2. Osmotische Stoffaufnahme und -abgabe I: Der Konzentrationsgradient.

Die empirisch ermittelten M-Werte (über Meßmethoden vgl. IV C) sind hingegen *nicht identisch mit den Permeabilitätskonstanten P*. Bei der Berechnung von P wird die Stoffmenge M nämlich fernerhin in Beziehung gesetzt zum Konzentrationsgefälle C für den betreffenden Stoff:

$$P = f(t, A, C). \tag{3}$$

Meist wird folgende Formel angewendet:

$$dM = P \cdot dt \cdot A\,(C_a - C_i), \tag{4}$$

die einen Spezialfall der Fickschen Diffusionsgleichung darstellt

$$\frac{dM}{dt} = -D \cdot \frac{dc}{dx}. \tag{5}$$

Hierin ist

$\frac{dc}{dx}$ = Konzentrationsgradient und

D = Diffusionskonstante.

Mit der Einbeziehung des Konzentrationsgefälles als einem modifizierenden Faktor wird ein sehr weitgehendes Postulat erhoben, nämlich daß die gesamte Stoffbewegung eine *behinderte Diffusion* ist, die in Richtung und Ausmaß allein durch das Konzentrationsgefälle bestimmt wird, und die beendet ist, wenn das *Gleichgewicht* zwischen Innen- und Außenkonzentration erreicht ist. Das System Zelle/Umgebung folgt somit dem Gleichverteilungssatz; die treibende Kraft der translozierten Partikel ist deren thermische bzw. kinetische Energie und die Stoffverschiebung innerhalb des Systems eine Folge des Konzentrationsgefälles.

In den Ausgleich dieses Konzentrationsgradienten sind die (permeablen) Zellgrenzschichten als Diffusionshindernis eingeschaltet. (Wir dürfen annehmen, daß es sich nicht um ein *mechanisches* Hindernis handelt, sondern daß die zwischen Grenzschichtmolekülen und durchtretenden Molekülen wirksamen zwischenmolekularen Kräfte hemmend wirken — vgl. IV G.) Dadurch wird das Gleichgewicht nicht etwa geändert, sondern lediglich seine Einstellung zeitlich verschoben. Die Stoffbewegung ist damit als osmotisch = diosmotisch gekennzeichnet, wenn man unter Diosmose den („diffundorischen") Stoffdurchtritt durch eine begrenzt permeable Membran versteht. Sie erfordert keinerlei Leistung von seiten der Zelle — es sei denn die Aufrechterhaltung der Lebensfunktionen überhaupt — und erweist sich als „passiv".

Es ist charakteristisch, daß diese Auffassung von der diosmotischen Natur der Stoffaustauschvorgänge älter ist als die Anwendung der M-Werte, mit der man sich ja jeglicher vorwegnehmenden Hypothese über die Mechanik der Stoffbewegung entschlägt. Das hängt mit der frühzeitigen Entdeckung der osmotischen Gesetze durch W. PFEFFER zusammen. Diese zeigten nicht allein vollkommene Übereinstimmung mit der kinetischen Theorie der Gase und Lösungen, sondern ergänzten sie noch in ungeahnter Weise. Die von PFEFFER genial konstruierte und nach ihm benannte Zelle demonstrierte am unbelebten Modell die ausschließliche Gültigkeit der Formel 3 und der Theorie. So war die Übertragung dieser Vorstellung auf die lebende Zelle das Gebot der Stunde, und es gelang denn auch, mit Hilfe der Plasmolyseversuche hier die gleichen Gesetzmäßigkeiten in erster Annäherung wiederzufinden, zunächst für Wasser als Lösungsmittel, späterhin auch für gelöste Stoffe („Diosmotika"). Damit schien bewiesen, daß die Gesamtaufnahme der lebenden Zelle ebenso wie der PFEFFERschen Zelle als diosmotische Aufnahme = Permeation aufzufassen sei, daß die Zellgrenzschichten lediglich als Diffusionshindernis fungieren — wenigstens unter den hier verwendeten experimentellen Bedingungen, im plasmolytischen Versuch also bei verhältnismäßig hohen Konzentrationsdifferenzen usw. —, und daß deren Durchlässigkeit durch die *Permeabilitäts*konstanten definierbar war. Dem entsprach es, daß nunmehr die Permeabilitätsbestimmungen als Mittel zur Analyse des Grenzschichtfeinbaues eingesetzt wurden.

Die folgenden Jahrzehnte standen völlig unter dem Zeichen der Permeabilitätskonstanten und der diosmotischen Stoffaufnahme — nicht ganz zu Recht, wie wir heute sagen können. Wenn die Annahme richtig ist, daß die Gesamtaufnahme gleich der diosmotischen Aufnahme ist, so müssen die berechneten Permeabilitätskonstanten unabhängig vom Zeitpunkt der Messung, von der Zellgröße und von der Höhe des angelegten Konzentrationsgefälles für ein Objekt eine genaue *Konstanz der Werte* liefern. Es stellte sich indessen bald heraus, daß das keineswegs die Regel war. So unterschieden sich z. B. Permeabilitätskonstanten eines Objekts oft um mehr als eine Zehnerpotenz, je nachdem ob sie im plasmolytischen Versuch nach der Total- oder der Partialmethode ermittelt wurden (HOFMEISTER 1935). Auch die Konstanz über längere Versuchszeiten ließ sich (soweit man überhaupt geneigt war, mehr als 2 Messungen zu machen, vgl. HÖFLER 1934) nur selten verifizieren: entweder stiegen sie immer mehr an, oder sie zeigten einen oft außerordentlich starken Abfall (vgl. z. B. BOGEN und FOLLMANN 1955). Noch bedenklicher hätte es stimmen müssen, daß in vielen Fällen die Zellen imstande waren, Stoffe, vor allem Ionen, über den Ausgleich des Konzentrationsgefälles hinaus aufzunehmen und ein solches (oder entgegengesetzt gerichtetes) Ungleichgewicht zum Teil über Jahre hinaus aufrechtzuerhalten (z. B. STEWARD 1932, OSTERHOUT 1936).

Es würde den Gebräuchen wissenschaftlicher Erkenntnis entsprochen haben, nach dem Ausfall solcher Versuche die ursprüngliche Hypothese zu revidieren, d. h. entweder die Beziehung zwischen Stoffmenge M und Konzentrationspotential C (Formel 3) gänzlich zu lösen, nach einem anderweitigen Faktor zu suchen, und damit die Ausschließlichkeit der diosmotischen Stoffbewegung aufzugeben, oder aber wenigstens eine Überlagerung der diosmotischen Prozesse durch andersartige Vorgänge in Betracht zu ziehen. Unter der Suggestionskraft der Versuche an der PFEFFERschen Zelle war das jedoch für lange Zeit bzw. für eine große Zahl von Forschern nicht möglich. Vielmehr wurde an dem oben angegebenen Postulat festgehalten; man projizierte die gefundenen Abweichungen gewissermaßen in die Struktur der Grenzschichten hinein und konstruierte besondere und sehr komplizierte *Permeabilitäts*verhältnisse dieser

Grenzschichten. Ein erstes Mißverständnis führte zu dem bekannten Widerspruch, in den die Lipoidtheorie zur Ultrafiltertheorie geriet — ein Problem, das inzwischen als Scheinproblem erkannt worden ist (vgl. IV G). Weiterhin ergab sich die Notwendigkeit, besondere permeabilitätsregulierende Mechanismen einzuführen („physiologische Permeabilität", Höber 1926). In extremster Ausprägung wurden dann die „spezifischen" Permeabilitätsreihen und -typen aufgestellt, nach denen für jedes Objekt besondere Durchlässigkeitseigenschaften und (diosmotische!) Permeationsmechanismen anzunehmen waren (Höfler 1942), ja sogar mit einem spezifischen Wechsel der Permeabilitätsverhältnisse im Laufe eines Jahres gerechnet wurde (Hofmeister 1938). Das Ergebnis war eine solche Fülle von theoretischen Zusätzen bzw. Hilfsannahmen, Permeationsmechanismen, Permeabilitätstypen und Strukturdetails, daß von durchgängig gültigen Permeabilitätsregeln nicht mehr gesprochen werden konnte und die Mehrzahl der Forscher sich von diesem Gebiet abwandte. Es ist gewiß kein Zufall, daß diese Abkehr mit dem Höhepunkt der „vergleichenden Permeabilitäts-Forschung" zusammenfiel.

3. Osmotische Stoffaufnahme und -abgabe II: Der Gradient des elektrochemischen Potentials.

Schon frühzeitig hatte sich eine Gruppe von Forschern speziell der Aufnahme und Abgabe von *Ionen* gewidmet (z. B. Osterhout, vgl. hierzu Hoagland 1948). Ausgangspunkt war die unbestrittene, mit den einfachen Osmosegesetzen nicht zu vereinbarende Tatsache der Ionen-*Akkumulation*, d. h. der Anreicherung von Ionen im Zellinnern über den Ausgleich des Konzentrationsgefälles hinaus. Dieser Effekt wurde nunmehr zum *elektrochemischen Potential* in Beziehung gesetzt, welches der bloßen („anschaulichen") Konzentration vorzuziehen ist, denn in jedem Falle (auch bei den Anelektrolyten) ist allein schon aus thermodynamischen Gründen die Einsetzung des elektrochemischen Potentials korrekt (Broyer 1951).

Die Beobachtung, daß in der Zelle eine Anisotropie der Ladungen besteht (Membranpotentiale, Überschuß nichtdiffusibler Ionen des einen Ladungssinnes im Binnenplasma oder in der Vacuole, Donnan-Potentiale) ließ erwarten, 1. daß die Grenzschichten des Protoplasten der Ionendiffusion einen qualitativ anderen Widerstand entgegensetzen als der Anelektrolytdiffusion, und 2. daß im Zellinnern sich Donnan-Gleichgewichte einstellen, d. h. bei Gleichheit der Ionenprodukte Konzentrationsungleichgewichte auftreten. Die konsequente Ausarbeitung dieses Gedankens führte schließlich zu den Formeln (Teorell 1953)

$$M = - K \cdot u\,(c_2 \cdot \xi - c_1) \tag{6}$$

bzw.

$$M = - K \cdot u\,(a_2 \cdot r_2 \cdot \xi - a_1 \cdot r_1). \tag{7}$$

In ihnen ist

M = Stoff-(Ionen-)Menge; entspricht „flux" bzw. „net flux" aus Formel (3),
u = Ionenbeweglichkeit,
c_1, c_2 = Ionenkonzentration in der Membran,
a_1, a_2 = Ionenkonzentration in der äußeren Lösung,
r_1, r_2 = Donnan-Verteilung,
$\xi = \exp(F\varphi/RT)$, entspricht dem Diffusionspotential im Innern der Grenzschicht gemäß der Planck-Verteilung,
K = *Proportionalitäts*konstante.

Gl. (6) kann als eine verallgemeinerte Form der 2. FICKschen Gleichung für den molekularen Transport angesehen werden (TEORELL), wobei $\xi = 1$ ist. Damit wird auch die DONNAN-Verteilung = 1, und die Gleichung geht über in die FICKsche Gleichung

$$M = \frac{R \cdot T}{\delta} \cdot u \, (a_2 - a_1).$$

Die *Permeabilitäts*konstante wird darin durch $\frac{R \cdot T}{\delta} \cdot u$ wiedergegeben.

Wesentlich ist, daß M (und also auch K) proportional dem elektrochemischen Potentialgradienten sind. Die Stoffbewegungen sind somit gleichfalls diosmotischer Art, d.h. *Permeations*vorgänge; sie können wegen der beteiligten Potentiale zu einer Akkumulation führen, die nunmehr quantitativ berechnet werden kann. — Solche Stoffbewegung wird von einigen Autoren auch als *Elektroosmose* bezeichnet.

Zugleich wird eine weitere Beobachtung erklärt, die der *selektiven Permeabilität*, d. h. die bevorzugte (diosmotische) Aufnahme von Ionen des einen Ladungssinnes gegenüber den entgegengesetzt geladenen.

4. Nicht-diosmotische Stoffaufnahme und -abgabe.

Wie TEORELL mehrfach betont, reicht die Formel (6) nicht aus, um *alle* Akkumulationserscheinungen lediglich auf Permeabilitätseffekte zurückzuführen. Es kann danach kein Zweifel mehr bestehen, daß wenigstens bei den darüber hinausgehenden Akkumulationserscheinungen *nichtosmotische* (= nichtdiosmotische) Mechanismen beteiligt sind.

Wichtige Hinweise hierfür waren die Beobachtungen, daß das Ausmaß der Stoffaufnahme und -abgabe von der Intensität gewisser Stoffwechselprozesse, vor allem der aeroben *Atmung*, abhängig ist (zuerst STEWARD *et al.*) und demzufolge unter anaeroben Bedingungen oder bei Vergiftungen des Stoffwechsels (Kaliumcyanid, Natriumazid, Dinitrophenol, Natriumarsenit, Fluoressigsäure u. a.) reduziert oder gleich Null wird. Diese Proportionalität zwischen Stoffmenge M und Stoffwechsel wurde nicht allein bei Ionen, sondern auch bei Anelektrolyten wie Glucose (vgl. ROSENBERG und WILBRANDT 1952), Wasser (HACKETT und THIMANN 1952, BOGEN 1953, BONNER und Mitarbeiter 1953) und den in plasmolytischen Versuchen verwendeten „Diosmoticis" wie Harnstoff, Glycerin usw. (BOGEN und FOLLMANN 1955) beobachtet. Für sie gelten unter anderem folgende Kriterien (nach ROSENBERG und WILBRANDT 1952 und ROSENBERG 1954):

1. Keine Übereinstimmung mit den Gesetzen der Osmose, statt dessen häufig sogar Konstanz der M-Werte, unabhängig vom Konzentrationspotential.
2. Konkurrenz gleichzeitig transportierter Substanzen (bei der Permeation bewegt sich jeder Stoff unabhängig von jedem anderen, allein gemäß seinem eigenen Konzentrationsgefälle).
3. Hohe Strukturspezifität, häufig sogar Stereospezifität.
4. Beeinflussung der M-Werte durch Enzyminhibitoren und -aktivatoren (soweit sich deren Wirkung nicht lediglich auf Veränderungen der Konzentrationsdifferenz der transportierten Substanzen erstreckt).

Welche Mechanismen den nichtosmotischen Transportvorgängen zugrunde liegen, ist gegenwärtig noch kontrovers; es darf jedoch vermutet werden, daß es deren verschiedene gibt, so etwa die Bildung eines spezifischen Komplexes aus Co-Enzym und durchtretender Substanz unter Vermittlung von Enzympaaren in der äußeren und inneren Grenzschicht und Bewegung dieses Komplexes durch das Plasma (ROSENBERG und WILBRANDT 1952), oder die Faltung und Entfaltung von Trägerproteinen (GOLDACRE 1952), oder die Wanderung entlang der „Elektronenleiter" der Cytochrome, entgegen dem bei intakter Atmung fließenden Elektronenstrom (Theorie der Anionenatmung LUNDEGÅRDHs 1951); sie werden in den folgenden Abschnitten des Handbuches besprochen. Ihnen allen gemeinsam ist jedoch, daß die Partikel der zu transportierenden Substanz im Plasma oder auch nur in seinen Grenzschichten an bestimmte carrier-Aggregate gebunden und mit diesen zusammen transportiert werden, oftmals entlang einer vorgebildeten Struktur. Für Bindung und/oder Transport und wohl auch für die Aufrechterhaltung der Struktur ist ein *beständiger* Energieaufwand von seiten der Zelle erforderlich, ein Energieaufwand, der entweder direkt aus der Atmung oder mittelbar aus dem ATP-Vorrat der Zelle gespeist wird, und dessen Verhinderung die gesamten nichtosmotischen Translozierungsprozesse zum Erliegen bringt. Damit ist die Bezeichnung „aktive Aufnahme" gerechtfertigt, und es darf die nichtosmotische Aufnahme der aktiven Aufnahme gleichgesetzt werden.

LEVITT hat in mehreren theoretischen Publikationen betont, daß eine aktive Stoffaufnahme thermodynamisch bzw. aus Gründen des Energiehaushaltes der Zelle nicht möglich sei; eine Zelle, die sich energetisch so kostspieliger Aufnahmemechanismen bediente, würde sich in Kürze zu Tode geatmet haben. Dem widerspricht indessen alle experimentelle Erfahrung. Man muß wohl annehmen, daß für die nichtosmotische Aufnahme nicht ein zusätzlicher Energiebetrag bereitgestellt werden muß, sondern daß bereits ein Teil der „Normalatmung" für die nichtosmotische Aufnahme zur Verfügung steht. Speziell für den Fall der Faltung und Entfaltung von Trägerproteinen (GOLDACRE 1952) läuft dieser Umfaltungsmechanismus beständig ab, und es ist energetisch ziemlich bedeutungslos, ob er für den Transport benutzt wird oder nicht.

Wenn ein Fall gegeben ist, wo ein Stoff ausschließlich nichtosmotisch aufgenommen oder abgegeben wird, so sind zweifellos die Grenzschichten des Protoplasten für diesen Stoff durchlässig, d. h. *permeabel* im Wortsinn. Ebenso steht aber außer Frage, daß eine *Permeation*, also eine osmotische Aufnahme, *nicht* stattfindet, mithin keine Permeabilitätskonstanten berechnet werden dürfen und können. Wenn man daher den Ausdruck „permeabel" verwendet, muß man jeweils angeben, ob er im weiteren oder im engeren Sinne verstanden werden soll, also als „durchlässig schlechthin" oder als „durchlässig für *permeierende* Stoffe". Zugleich erweist sich damit, daß die Bezeichnung *Permeabilitäts*konstante weniger glücklich gewählt ist, als es früher unter dem absoluten Primat der osmotischen Gesetze erschien; diese Konstanten würden besser als diosmotische Konstanten bzw. Koeffizienten benannt.

Um den Begriff „*Intrabilität*" ist es nicht besser bestellt. Ethymologisch müßte er, wie „Permeabilität", eine Eigenschaft der Grenzschichten bezeichnen; in praxi versteht man darunter jedoch den Vorgang der Intrameation, d. h., daß ein Stoff wohl die äußeren, nicht aber die inneren Grenzschichten (Tonoplast) passieren kann und also nur bis ins Binnenplasma gelangt. Das Phänomen Intrabilität besagt in Wirklichkeit aber nur, daß der Tonoplast diesen Stoff nicht durchläßt. Da es bei pflanzlichen Zellen kaum möglich ist, die Stoffmenge M, die sich im Binnenplasma befindet, exakt zu bestimmen, kann auf das Bestehen einer „Intrabilität" nur aus qualitativen Befunden (Anfärbung, Ausfällung, „Aufquellung") geschlossen werden; die Erscheinung hat daher für die quantitative Analyse der Zellvorgänge nur geringe Bedeutung.

Die hier beschriebenen nichtosmotischen Effekte dürfen auch nicht unter dem Terminus „physiologische Permeabilität" (HÖBER 1926) diskutiert werden, denn es handelt sich eben nicht um Permeationsvorgänge. Dieser Ausdruck muß jenen physiologischen Mechanismen vorbehalten werden, mit denen die Zelle ausschließlich die Permeabilität reguliert (echte Erhöhung oder Herabsetzung des Diffusionswiderstandes, was bekanntlich nur die Geschwindigkeit der Aufnahme, nicht aber das Ausmaß verändert).

Hingegen ist die von Overton (1899) und Collander und Holmström (1937) beobachtete *adenoide Tätigkeit* (Einfluß der Atmung auf die Aufnahme von Sulfosäurefarbstoffen) eine „aktive" und damit echte nichtosmotische Aufnahme (vgl. dagegen Drawert 1940).

Ebensowenig angezeigt erscheint es, die nichtosmotischen Effekte auf „irreziproke Permeabilität" zurückzuführen. Die unterschiedliche Durchlässigkeit der Grenzschichten in verschiedenen Richtungen ist augenscheinlich die Folge eines nichtosmotischen Transportes mittels gerichtet arbeitender carrier-Aggregate bzw. entlang einer gerichteten (anisotropen) Struktur (s. S. 121).

Nach den osmotischen und nichtosmotischen Aufnahme- und Abgabemechanismen ist noch kurz die „metaosmotische" Stoffaufnahme anzuführen (Bogen und Prell 1953). Dem Begriff liegt folgende Überlegung zugrunde: es wird häufig beobachtet, daß ein Stoff, nachdem er auf osmotischem Wege in die Vacuole einer Pflanzenzelle gelangt ist, dort nachträglich, etwa durch Adsorption an gewisse Inhaltsstoffe wie Kolloide u. a. aus dem Konzentrationsgefälle entfernt wird. Das ist, solange die Menge der absorbierenden Stoffe nicht verändert wird, ein Fall echter osmotischer Stoffbewegung, nur ist nicht das Konzentrations-, sondern das elektrochemische Potential maßgebend. (Das gleiche gilt übrigens für die Funktion der Vacuole als „Ionenfalle", wie sie aus Vitalfärbeversuchen bekannt ist, vgl. Drawert, Höfler.) Sobald jedoch im Gefolge der Stoffwechseltätigkeit des Protoplasten in der Vacuole adsorbierende Stoffe neu entstehen bzw. dort zusätzlich deponiert werden, sind die M-Werte mehr oder weniger proportional der Stoffwechselaktivität und unabhängig vom Konzentrationsgefälle, d. h. *einige* der Kriterien der nichtosmotischen Stoffaufnahme (Rosenberg und Wilbrandt, vgl. S. 120) sind erfüllt. Gleichwohl würde es nicht zutreffen, wollte man deshalb von einer nichtosmotischen Aufnahme sprechen, denn der Aufnahmemechanismus ist nach wie vor der osmotische. Lediglich nach stattgehabter Endosmose, also „metaosmotisch", unterliegt der Stoff metabolisch bedingten Veränderungen.

Die Entdeckung der nichtosmotischen Aufnahme und ihrer weiten Verbreitung haben das Pendel der Zellforschung nach der anderen Seite ausschlagen lassen. Nunmehr werden diese Vorgänge besonders hoch bewertet, und es gibt nicht wenige Autoren, die ihnen die Alleinherrschaft zuerkennen wollen und jegliche osmotische Stoffbewegung leugnen (vgl. hierzu die Ausführungen von Steinbach 1951, ferner Overstreet und Jacobson 1952).

Sicherlich kommt ihnen allein schon deshalb große Bedeutung zu, weil sie mit physiologischen Mitteln reguliert werden können und die Zelle nicht zum bloßen Spielball des Konzentrationsgradienten werden lassen. Auch vermögen sie uns z. B. zu erklären, wieso die lebenswichtigen Zucker trotz der Geringfügigkeit der osmotischen Zuckeraufnahme von Zelle zu Zelle gelangen können. Und endlich sind sie das Privileg, das die lebende Zelle der Pfefferschen Zelle voraus hat. Vielleicht sollte auch nicht unerwähnt bleiben, daß die osmotischen Gesetze oftmals um so stärker in Erscheinung treten, je „unphysiologischer" die experimentellen Bedingungen (Konzentrationen) sind. Dennoch — wer einmal einen Plasmolyseversuch gemacht hat, und wer die Einzelheiten des Wasserhaushaltes der Zellen und Pflanzen kennt, wird ohne Zögern zugeben, daß es ebenso einseitig sein würde, die osmotischen Prinzipien völlig aufzugeben, wie die früheren Permeabilitätsvorstellungen einseitig waren. Ohne Zweifel wirken beide Formen der Stoffbewegung neben-, mit- und gegeneinander, und wir müssen die Gesamtaufnahme wie folgt aufgliedern:

Gesamtaufnahme = osmotische (einschließlich metaosmotische) Aufnahme
+ nichtosmotische Aufnahme.

Es wird Aufgabe der künftigen Forschung sein, in jedem Einzelfalle die beiden Komponenten voneinander abzutrennen und ihre Anteile an der Gesamtaufnahme zu bestimmen (hierzu: Bogen und Follmann 1955).

5. Der Stoffaustausch.

Dieser wird gemessen nach der auf S. 116 gegebenen Formel

$$M = M_i - M_a. \tag{1}$$

Aus den Meßwerten kann nicht ohne weiteres auf die beteiligten Mechanismen geschlossen werden, da sie von den letzteren in ganz unterschiedlicher Weise abhängen können. Ihre Interpretierung erfordert besondere Sorgfalt.

a) Am einfachsten liegen die Verhältnisse, wenn nur ein einziger Stoff betrachtet wird und dieser sich gemäß dem *Konzentrationsgradienten* bewegt. Dann sind die Wege in beiden Richtungen gleichmäßig gangbar. Wenn wir zwei solcher Stoffe haben, können wir zwar voraussetzen, daß jeder von ihnen, unabhängig vom anderen, nur nach seinem eigenen Konzentrationsgradienten transportiert wird. Sobald indessen die Stoffmengen M_i und M_a nicht direkt bestimmt werden können, sondern lediglich „osmotisch" (plasmolytisch), so interferiert die Wasserpermeation. Sie folgt ebenfalls nur dem Konzentrationsgradienten für Wasser, aber dieser wird sowohl durch die Konzentration von Stoff 1 als auch von Stoff 2 beeinflußt, d. h. die plasmolytischen M-Bestimmungen gestatten nicht ohne weiteres eine getrennte Behandlung der Transportraten beider Stoffe.

b) Prinzipiell gleich können die Verhältnisse bei der Austauschrate von Stoffen sein, die sich gemäß dem *elektrochemischen* Gradienten („elektroosmotisch") bewegen. Hier muß jedoch berücksichtigt werden, daß der erste Schritt der Stoffaufnahme oftmals eine sog. *Austauschadsorption* ist, d. h. daß die aufzunehmenden Ionen in der Oberfläche der Grenzschicht gegen ionische Zellbestandteile ausgetauscht werden, die dann in die Außenlösung übertreten. Die Menge M' dieser letzteren muß dann ebenfalls ermittelt und in die Berechnung einbezogen werden.

Dabei kann der Fall eintreten, daß der Vorrat an austauschbaren Zellbestandteilen bzw. deren Nachlieferung aus dem Stoffwechsel zum begrenzenden Faktor der Aufnahme wird und so eine Abhängigkeit des Stoffaustausches vom Stoffwechsel anzeigt.

Wird der gleichzeitige Austausch mehrerer Stoffe betrachtet, so ist ferner mit einer Konkurrenz dieser Stoffe um die austauschfähigen Zellbestandteile zu rechnen (Punkt 2 der Kriterien der nichtosmotischen Stoffaufnahme! S. 120).

Generell ist schließlich zu sagen, daß eine solche Austauschadsorption in der Regel *nicht* reversibel ist, allein schon weil die durch den Austausch in die Umgebung gelangenden Zellbestandteile sich sehr bald aus dem Wirkungsbereich der Grenzschicht entfernen.

Wegen der Einzelheiten und methodischen Erfassung dieser Vorgänge sei auf USSING (1952) verwiesen.

Besonderer Erwähnung bedarf ein Austauschphänomen, das mit der oben genannten Austauschadsorption nicht identisch ist. Es wird meist als „*exchangeability*" bezeichnet und besagt, daß in der Zelle ein beständiger Austausch von Ionen oder Molekülen *der gleichen Art* stattfindet, also z. B. K^+ gegen K^+, Wasser gegen Wasser. Bekanntlich unterscheidet man zwischen „austauschbarem" und „nicht austauschbarem" Kalium, oder zwischen „gebundenem" und „freiem" Wasser. Das austauschbare K^+ (oder Wasser) kann fortlaufend gegen andere, *also auch markierte* Kaliumionen (bzw. Wassermoleküle) ausgetauscht werden, *ohne daß die Gesamtmengen* M bzw. M_i and M_a *eine Änderung erfahren.* Diese exchangeability muß bei der Anwendung von tracern für die Ermittlung der M-Werte zusätzlich in Rechnung gestellt werden (Formeln bei USSING 1952).

c) Bei der *nichtosmotischen Stoffbewegung* ist die Reversibilität gleichfalls nicht gegeben: hier liegt ein *gerichteter* Transport (irrtümlich „irreziproke Permeabilität", vgl. S. 122) entlang einer vorgebildeten Struktur oder eines inhärenten Gradienten vor, außerdem substratspezifische carrier-Substanzen und kompetitive Beeinflussung zwischen mehreren translozierten Stoffen (Punkt 3 und 2 der auf Seite 120 angeführten Kriterien). M_a und M_i unterliegen also *verschiedenen* Abhängigkeitsbeziehungen.

Aus dem Gesagten ergibt sich, daß die M-Werte in jedem Falle der eingehenden experimentellen Analyse bedürfen, bevor sie als Argumente für oder gegen eine Hypothese verwendet werden (RUHLAND, ULLRICH und ENDO 1938).

Literatur.

BONNER, J., R. S. BANDURSKI u. A. MILLERD: Linkage of respiration to auxin-induced water uptake. Physiol. Plantarum (Copenh.) **6**, 511—522 (1953). — BOGEN, H. J.: Beiträge zur Physiologie der nichtosmotischen Wasseraufnahme. Planta (Berl.) **42**, 140—155 (1953). — BOGEN, H. J., u. G. FOLLMANN: Osmotische und nichtosmotische Stoffaufnahme bei Diatomeen (Untersuchungen über die „spezifische Permeabilität" der Diatomeen). Planta (Berl.) (im Druck). — BOGEN, H. J., u. H. PRELL: Messung nichtosmotischer Wasseraufnahme an plasmolysierten Protoplasten. Planta (Berl.) **41**, 459—479 (1953). — BROYER, T. C.: An outline of energetics in relation to the movement of materials through an two-phased solution system. Plant Physiol. **26**, 598—610 (1951).

COLLANDER, R., u. A. HOLMSTRÖM: Die Aufnahme von Sulfosäurefarbstoffen seitens pflanzlicher Zellen — ein Beispiel der adenoiden Tätigkeit der Protoplasten. Acta Soc. Fauna et Flora fenn. **60**, 129—135 (1937).

DRAWERT, H.: Zur Frage der Stoffaufnahme durch die lebende Pflanzenzelle. II. Die Aufnahme basischer Farbstoffe und das Permeabilitätsproblem. Flora (Jena) **34**, 159—214 (1940). — DRAWERT, H.: Zur Theorie der Aufnahme basischer Stoffe. Z. Naturforsch. **3**b, 111—120 (1948).

GOLDACRE, R. J.: The folding and unfolding of protein molecules as a basis of osmotic work. Internat. Rev. Cytol. **1**, 135—164 (1952).

HACKETT, D. P., and K. V. THIMANN: The nature of the auxin-induced water uptake by potato tissue. Amer. J. Bot. **39**, 553—560 (1952). — HOAGLAND, D. R.: Lectures on the inorganic nutrition of plants. Waltham, Mass. 1948. — HÖBER, R.: Physikalische Chemie der Zelle und der Gewebe, 6. Aufl. Leipzig 1926. — HÖFLER, K.: Permeabilitätsstudien an Stengelzellen von *Majanthemum bifolium*. (Zur Kenntnis spezifischer Permeabilitätsreihen, I.) Sitzgsber. Akad. Wiss. Wien. Math.-naturwiss. Kl., Abt. 1 **143**, 213—264 (1934). — Unsere derzeitige Kenntnis von den spezifischen Permeabilitätsreihen. Ber. dtsch. bot. Ges. **60**, 179—200 (1942). — HOFMEISTER, L.: Vergleichende Untersuchungen über spezifische Permeabilitätsreihen. Bibl. botanica (Stuttgart) **113** (1935). — Verschiedene Permeabilitätsreihen bei ein und derselben Zellsorte von *Ranunculus repens*. Jb. Bot. **86**, 401—419 (1938).

LUNDEGÅRDH, H.: The translocation of salts and water through wheat roots. Physiol. Plantarum (Copenh.) **3**, 103—159 (1950).

OSTERHOUT, W. J. V.: The absorption of electrolytes in large plant cells. Bot. Review **2**, 283—315 (1936). — OVERSTREET, R., and L. JACOBSON: Mechanism of ion absorption by roots. Annual Rev. Plant Physiol. **3**, 189—206 (1952).

PFEFFER, W.: Osmotische Untersuchungen. Leipzig 1877.

ROBERTSON, R. N.: Mechanism of absorption and transport of inorganic nutrients in plants. Annual Rev. Plant Physiol. **2**, 1—24 (1951). — ROSENBERG, TH.: The concept and definition of active transport. Symposia Soc. Exper. Biol. 8, 27—41 (1954). — ROSENBERG, TH., and W. WILBRANDT: Enzymatic processes in cell membrane penetration. Internat. Rev. Cytol. **1**, 65—92 (1952). — RUHLAND, W., H. ULLRICH u. S. ENDO: Untersuchungen zu den „spezifischen Permeabilitätsreihen" HÖFLERS. I. Zur Frage der Alkoholpermeabilität von Pflanzenzellen unter verschiedenen Versuchsbedingungen. Planta (Berl.) **27**, 650—668 (1938).

STEINBACH, H. B.: Permeability. Annual Rev. Plant Physiol. **2**, 323—342 (1951). — STEWARD, F. C.: The absorption and accumulation of solutes by living plant cells. I. Experimental conditions which determine salt absorption by storage tissues. Protoplasma (Berl.) **15**, 29—58 (1932).

TEORELL, T.: Transport processes and electrical phenomena in ionic membranes. Progr. Biophysics a. Biophysical Chem. **3**, 305—319 (1953).

USSING, H. H.: Some aspects of the application of tracers in permeability studies. Adv. Enzymol. **13**, 21—65 (1952).

VRIES, H. DE: Über die Permeabilität der Protoplasten für Harnstoff. Bot. Ztg 1888, Nr. 19, 1.

Energetics and mathematical treatment of diffusion.

By

D. C. Spanner.

With 5 figures.

The concept of diffusion.

If a lump of sugar is placed in a glass of water it proceeds to dissolve, and in a short time has formed a layer of concentrated solution at the bottom. If the glass is now set aside where it will be free from disturbance it is a matter of common observation that the sugar will move in such a way that it will eventually reach a uniform concentration everywhere. We call this phenomenon *diffusion*. It is a very important example of what are termed transport processes, and is to be distinguished more or less sharply from another type of movement spoken of as *mass transport*. The distinction can be made from two standpoints, the thermodynamic or macroscopic, and the molecular. From the first viewpoint we can say that diffusion is a manifestation of thermal energy, that is, energy possessed by virtue of temperature. It would therefore be non-existent at the absolute zero. Mass transport, on the other hand, always signifies the doing of work: somewhere in the system there is an unbalanced force producing observable movement. Temperature here has no direct relevance, and mass transport can still be conceived as occurring at absolute zero. The thermodynamic distinction is therefore parallel to that between heat and work.

From the molecular point of view it can be said that diffusion depends essentially on the *random* thermal motions of the molecules, whereas mass transport depends on a motion possessed by them in common *i.e.* on *ordered* movements. Thus any agency which can exert a force simultaneously on each of a large number of molecules will impart to them a common motion and cause mass transport; such agencies are illustrated by a pressure-difference, a gradient of electrical potential, or a centrifugal field. It is hardly necessary to emphasise that the ideas of diffusion and mass transport are both statistical and that therefore the distinction between them progressively ceases to have meaning as assemblages of fewer and fewer molecules are considered. Further, the idea of diffusion is only strictly applicable to systems fairly close to internal equilibrium, since only in such cases can molecular motions adequately be described as random.

Net and gross diffusion rates.

The molecular picture of diffusion leads us to suppose that across any surface within a system there is, in fact, a two-way diffusion going on. Molecules are passing across it backwards as well as forwards in conformity with their random motions. However, unless the system is in thermodynamic equilibrium these two processes will be going on at different rates, and it is the difference between them which is the only rate we can actually observe. We call this the net diffusion rate. What we may call the gross diffusion rates are often important in theoretical studies. Although they are in essence impossible of observation, in practice they can be satisfactorily measured. This possibility arises from the fact that means now exist for labelling molecules strongly while leaving their diffusion properties, it is believed, virtually unaffected. Such means are of course the incorporation of radioactive atoms or of such heavy isotopes as deuterium.

FICK's law.

It will be found convenient to deal first with the problem of diffusion from the macroscopic angle, the complementary approach from the molecular point of view being reserved till later. In the example of diffusion quoted above two substances are involved, which for convenience we distinguish as solute and solvent. Across any given horizontal plane through the solution there will be during the progress to equilibrium a two-fold movement taking place, of sugar upwards and of water downwards. However, in a pure diffusion system these two flows are not independent, but are related by the requirement that no unbalanced forces should develop, that is, that the pressure should remain constant everywhere. What this means is that the interdiffusion of two substances is not two processes but one, and can therefore be described by a single equation. Were mass transport involved the case would in general be otherwise, and two equations would be necessary.

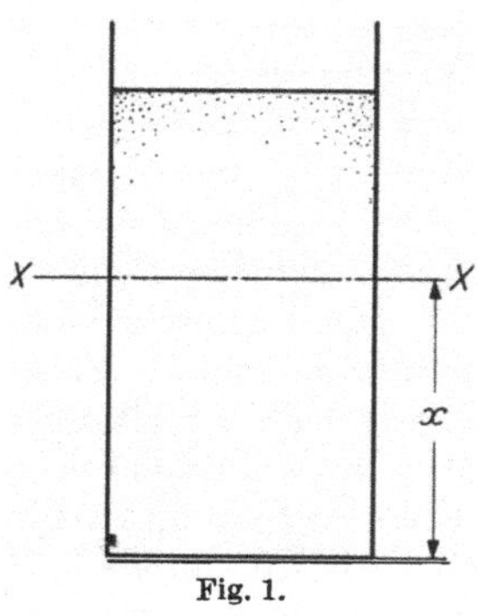

Fig. 1.

The mathematical formulation of simple diffusion was first given by FICK in 1855. He made the reasonable assumption that across any section normal to the concentration gradient there will be a net passage of solute proportional to that gradient, and of course depending directly on the area of the section.

Thus, referring to figure 1, let m be the mass of sugar in the system above the plane XX defined by its coordinate x measured in the direction of diffusion. Further, let c be the concentration per unit volume of sugar at the level XX and the time t. Then FICK's law states that the net rate of passage of sugar across the section XX can be written

$$\frac{\mathrm{d}m}{\mathrm{d}t} = -DA\frac{\mathrm{d}c}{\mathrm{d}x} \tag{1}$$

where A is the area of the section, and the negative sign is introduced because the flow is down the concentration gradient. Much of the value of this equation lies in the fact that the coefficient D proves to be approximately constant under a wide range of conditions, though experiment shows that it does in fact vary with both the concentration c and the gradient $\mathrm{d}c/\mathrm{d}x$. As is to be expected, it varies also with the temperature. It is called the coefficient of diffusion of sugar in water.

Comparison with mass flow.

It is instructive at this point to compare FICK's law with the law expressing the rate of mass flow of a fluid in response to a pressure gradient (POISEUILLE's law). For a fluid of viscosity η moving down a tube of radius r POISEUILLE's law can be expressed in the form

$$\frac{\mathrm{d}m}{\mathrm{d}t} = -\frac{\pi r^4 \varrho}{8\eta}\cdot\frac{\mathrm{d}p}{\mathrm{d}x}. \tag{2}$$

A comparison of this with (1) shows that while for a given gradient the diffusive flow depends simply on the area, and would therefore not be affected by merely subdividing the latter, for viscous flow it varies as the fourth power of the radius or as the square of the area. Subdividing a sectional area equally into n equal parts would therefore reduce it to $1/n$th. It is for these reasons that while a dilute gel such as 2% agar offers such a high resistance to the forcible passage of water, it provides nearly as free a path as plain water for the diffusion of solutes. The same consideration applies of course to the passage of materials across the cellulose wall of the cell.

Alternative form of FICK's law.

FICK's law can be put into a more convenient form by eliminating m in favour of c as follows.

Referring to figure 2, let XX have the same meaning as before, and let $X'X'$ be a second cross-section with the coordinate $(x + \delta x)$. For simplicity, let the gradient $\mathrm{d}c/\mathrm{d}x$ be denoted by g; then the gradient at $X'X'$ can be written $(g + \delta g)$.

Consider now the sugar content of the thin element of solution between the planes XX and $X'X'$. Sugar will be entering this element by diffusion across XX at the rate $(-DAg)$ as given by equation (1). Similarly it will be leaving the element across $X'X'$ at the rate $\{-DA(g+\delta g)\}$. The rate of accumulation will be the difference between these two, and clearly it may also be expressed in terms of the volume of the element and the rate of change of the concentration. Equating these expressions we find

$$-DAg + DA(g + \delta g) = A\,\delta x \cdot \frac{\mathrm{d}c}{\mathrm{d}t} \tag{3}$$

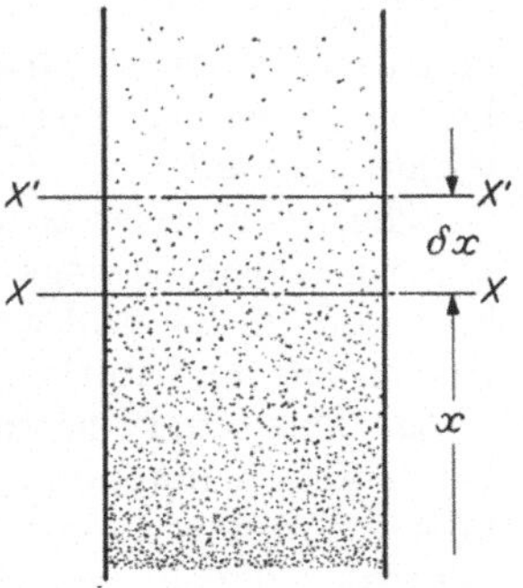

Fig. 2.

or dividing by $A\,\delta x$ and remembering that $g = \mathrm{d}c/\mathrm{d}x$, we have in the limit when δx is made infinitesimally small,

$$D\frac{\mathrm{d}g}{\mathrm{d}x} = D\frac{\mathrm{d}^2 c}{\mathrm{d}x^2} = \frac{\mathrm{d}c}{\mathrm{d}t}. \tag{4}$$

This equation is more unambiguously expressed in terms of partial derivatives, when we have as the general equation for one-dimensional flow

$$\frac{\partial c}{\partial t} = D\frac{\partial^2 c}{\partial x^2}. \tag{5}$$

Some general properties of the equation.

Some useful information follows at once from the form of FICK's equation just derived. In the first place, since the dimensions of the derivatives are respectively c/t and c/x^2, the dimensions of D will be simply x^2/t. Usually x is measured in centimetres and t in seconds, so that the units of D are centimetres^2 per second. If we then use the equation (1) to calculate actual diffusion rates we must be careful to express x in cm., A in cm.^{2}, c in mass units per cm.^{3} and t in secs. The mass units may be anything we choose. If we are dealing with a gaseous system a useful alternative exists, for c may be replaced with the partial pressure and then, without changing the units of D, the flow will be given directly in cm.^{3} of gas per sec. measured at unit pressure.

Scale effect.

Consider now the effect on diffusion phenomena of altering the linear dimensions of a system. Suppose the concentration at a point A be c_1, and at a point B, c_2. Without changing these concentrations or the physical properties of the system imagine the latter to be altered in scale by the factor λ. Then clearly the gradient $\partial c/\partial x$ will everywhere be multiplied by the factor $1/\lambda$, and the second derivative $\partial^2 c/\partial x^2$ by $1/\lambda^2$. Since D remains the same the rate of change $\partial c/\partial t$ will be generally reduced according to equation (5) in the ratio of the square of the scale factor. Even if the concentrations are not the same as originally, it still remains the case that their *relative* rates of change $\left(i.e.\ \frac{1}{\Delta c}\frac{\partial c}{\partial t}\right)$ will be modified by the factor $1/\lambda^2$. This has very important consequences for the biologist, for it means that however slowly solutes may diffuse *in vitro* where distances of the order of 1–10 cm. are involved, the evening out of concentration differences in the cell, where distances are rather 1–10 microns, is an exceedingly rapid process. With the dimensions quoted it is, in fact, just one hundred million times as fast. This emphasises, in particular, how immense is the contribution of cellular membranes to the maintenance of concentration differences within the cell.

The scale effect is also important in connection with external relations of the cells. This is very evidently so in the case of the stomata. When a leaf is placed in dry still air it begins to lose water by a process of diffusion, and if it is left undisturbed for long enough the diffusion pattern reaches a steady state. This steady pattern incidentally, is such that the rate of water loss from the leaf is a minimum; before the steady state supervenes it is higher. If now the leaf is subjected to variable winds the steady state pattern is never allowed to develop, and in accordance with the preceding remark the transpiration rate is higher. This is true both for the diffusion shells set up over the stomata and for those belonging to the leaf as a whole. However, the former are established very much more rapidly than the latter; in fact the 'square factor' again operates, and for a leaf 5 cm. diameter with stomata of 25 μ the stomatal shells arise $(5 \times 10^4/25)^2$ or four million times as fast as the leaf shells. They are much less affected by winds therefore, and it is obvious that the stomata will be the main factor controlling transpiration whenever air movement is sufficiently rapid.

One further interesting point emerges from considerations of scale. If the scale of the system is magnified λ times, the concentrations remaining unaffected, the gradients will be everywhere reduced by the factor $1/\lambda$. But equation (1) shows that the actual flow of material is also proportional to A, which has been increased λ^2 times. Hence the total flow rate across "corresponding areas" has increased by $\lambda^2 . 1/\lambda$ or λ times. This is the basis of the famous "diameter law" of STEFAN, which was used by BROWN and ESCOMBE to show why the effect of stomata in reducing the diffusive interchange of gases by leaves is so much less than the reduction of free area which they involve.

Integration of FICK's equation.

Integration of the diffusion equation requires rather more mathematics than most biologists have at their command, but methods for doing so can be found in textbooks dealing with the theory of heat conduction (see, for instance, INGERSOLL and ZOBELL, 1913). Two cases of interest to the physiologist will be briefly discussed here.

Case 1. Consider a long vertical column of liquid, at the base of which a given concentration c_0 of solute is maintained. the rest of the column being initially pure solvent. Such a system could be set up by having excess of solid solute at the foot of the column, when c_0 would correspond to saturation (fig. 3).

The solution of FICK's equation here has to comply with the conditions

$$c = 0 \text{ for all values of } x \text{ at } t = 0$$
$$c = c_0 \text{ for all values of } t \text{ at } x = 0.$$

The required solution is

$$c = c_0 \left[1 - \frac{2}{\sqrt{\pi}} \int_0^{x/2\sqrt{Dt}} e^{-\gamma^2} d\gamma\right] \tag{6}$$

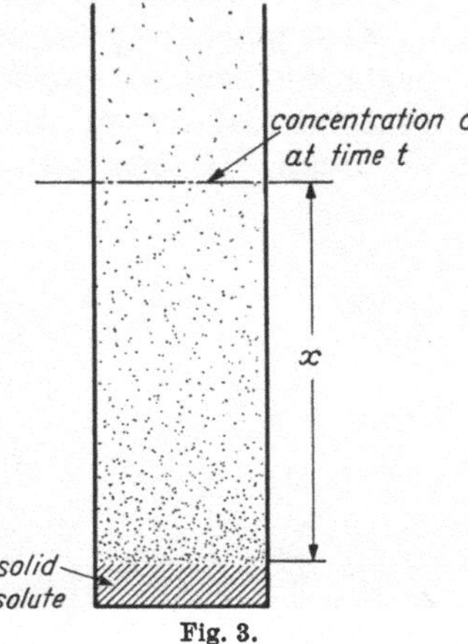

Fig. 3.

which clearly satisfies the above two boundary conditions, and can be shown by simple differentiation to satisfy equation (5). In this solution we have an illustration of the square law referred to earlier, for if two planes are distant x_1, x_2 from the base of the column the times t_1, t_2 for a given concentration to reach them are clearly proportional to x_1^2, x_2^2, since x and t occur together as the quotient $x/\sqrt{t}$. For any value of Dt the variation of c with x can be plotted quite easily from tables giving values of the "probability integral",

$$I = \frac{2}{\sqrt{\pi}} \int_0^{\gamma} e^{-\gamma^2} d\gamma. \tag{7}$$

By this means the mass of solute which has diffused beyond a specified plane can be calculated and compared with experiment. Such results enable the diffusion coefficient D to be found.

Case 2. Diffusion waves. A very interesting case occurs when the concentration c_0 at a given cross-section of a long column is made to vary harmonically. With a similar notation to the previous we may then write

$$c_0 = a + b \sin \frac{2\pi t}{\tau} \tag{8}$$

with the concentration varying between $(a - b)$ and $(a + b)$ with a period of τ. The solution of FICK's equation must therefore satisfy (8) when x is put equal to zero. With this condition the result turns out to be

$$c = a + b\, e^{-\sqrt{\frac{\pi}{D\tau}}\,x} \sin\left(\frac{2\pi t}{\tau} - \sqrt{\frac{\pi}{D\tau}}\,x\right) \tag{9}$$

for the case in which there is no steady overall gradient present.

Thus diffusion waves travel along the column (fig. 4) with an amplitude which decreases exponentially according to the factor $e^{-\sqrt{\frac{\pi}{D\tau}}\,x}$. The velocity of the waves is found by writing the phase-angle $\left(\frac{2\pi t}{\tau} - \sqrt{\frac{\pi}{D\tau}}\,x\right)$ constant and evaluating dx/dt: that is

$$v = \frac{dx}{dt} = 2\sqrt{\frac{\pi D}{\tau}}\,. \tag{10}$$

The wavelength is given by

$$\lambda = v\tau = 2\sqrt{\pi D \tau}. \tag{11}$$

Over a wavelength the exponential factor is $e^{-2\pi}$ or 1/540; thus successive crests fall off extremely rapidly in amplitude.

One potentially interesting application of this case lies in the question of sugar conduction from leaves. To a first approximation it might be imagined that the leaf concentration of sugar would vary harmonically with a period of 24 hours. If then translocation is governed by a process, such as diffusion,

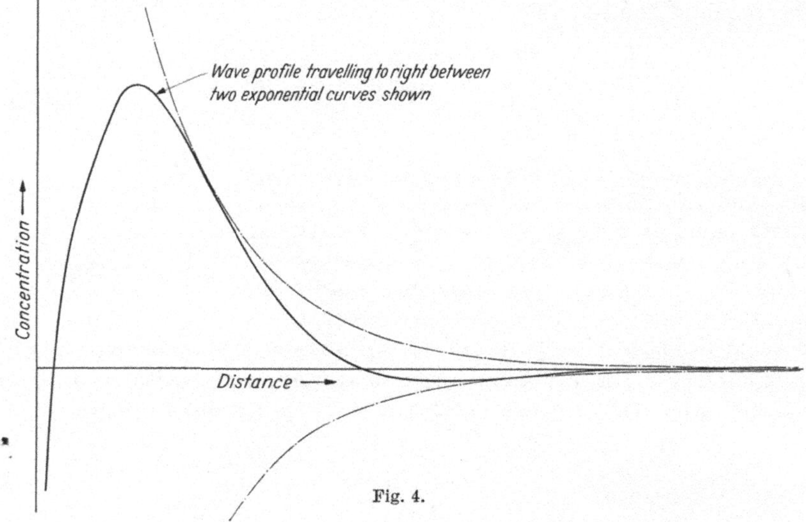

Fig. 4.

conforming to an equation of the type of (5) the above general features of the wave pattern ought to be manifest; for example, the rapid exponential drop in amplitude. On the other hand, if mass flow is the mechanism a different pattern should develop involving probably a much greater constancy of amplitude. In the case of straightforward diffusion of sugar it is easy to evaluate the velocity and wavelength of the movement. Taking D as 0.37 cm.² per day we have

$$v = 2\sqrt{\frac{\pi D}{\tau}} = 2\sqrt{\frac{\pi \times 0.37}{1}} = 2.2 \text{ cm. per day}$$

$$\lambda = 2\sqrt{\pi D \tau} = 2.2 \text{ cm.}$$

Beyond a centimetre or so from the start therefore the harmonic variation would be hardly apparent and the rate of passage of sugar would be constant day and night. There would, of course, probably be a steady overall gradient superimposed on the harmonic pattern, but this will not affect the above conclusions.

ONSAGER's generalisation of FICK's equation.

Suppose that in the diffusion system of sugar and water we place a third substance, say urea. Imagine further that initially the urea is everywhere at a uniform concentration, or rather, at a uniform activity. It is a very natural question to ask, is the distribution of the urea affected at all by the diffusion of the sugar? Instinctively we should imagine that it would be so, and experiment confirms that in general the movement of any one molecular species induces movements in every other species present, even if these latter already possess the distribution that will characterise them at final equilibrium. Into this picture the flow of heat enters in an analogous way.

ONSAGER in 1931 introduced the method of taking account of this general situation which can be illustrated as follows. For simplicity let us confine ourselves to an isothermal one-dimensional system containing three substances, two of which might be considered solutes and the third solvent, say water. Let these three be denoted by subscripts $_{1,2,3}$. Across any section such as XX (fig. 1) let the net rates of passage $\mathrm{d}m_{1,2,3}/\mathrm{d}t$ of the three substances be denoted by J_1, J_2 and J_3. It will simplify matters if the plane XX is supposed not to be fixed but to move with such a velocity that the resultant net flow of matter across it is zero; this will usually make a negligible difference in practice. Diffusion will occur in response to certain gradients, which in FICK's original equation were the concentration gradients; but ONSAGER specified them in a different way which introduced an important simplification. In the present case it amounts to taking them as the gradients of chemical potential, $\mathrm{d}\mu_{1,2,3}/\mathrm{d}x$. Using these we can then write down three generalised equations of diffusion thus:

$$\left.\begin{aligned} J_1 &= L_{11}\frac{\mathrm{d}\mu_1}{\mathrm{d}x} + L_{12}\frac{\mathrm{d}\mu_2}{\mathrm{d}x} + L_{13}\frac{\mathrm{d}\mu_3}{\mathrm{d}x} \\ J_2 &= L_{21}\frac{\mathrm{d}\mu_1}{\mathrm{d}x} + L_{22}\frac{\mathrm{d}\mu_2}{\mathrm{d}x} + L_{23}\frac{\mathrm{d}\mu_3}{\mathrm{d}x} \\ J_3 &= L_{31}\frac{\mathrm{d}\mu_1}{\mathrm{d}x} + L_{32}\frac{\mathrm{d}\mu_2}{\mathrm{d}x} + L_{33}\frac{\mathrm{d}\mu_3}{\mathrm{d}x}\,. \end{aligned}\right\} \tag{12}$$

These three equations contain nine "diffusion coefficients", but the latter are not all independent. As a consequence of the reference of the flow-rates to the moving plane it follows that

$$J_1 + J_2 + J_3 = 0 \tag{13}$$

identically. If equations (11) are inserted in this it can be seen at once that we must have

$$\left.\begin{aligned} L_{11} + L_{21} + L_{31} &= 0 \\ L_{12} + L_{22} + L_{32} &= 0 \\ L_{13} + L_{23} + L_{33} &= 0 \end{aligned}\right\} \tag{14}$$

since the result is an equation which is true for all values of the variables $(\mathrm{d}\mu_{1,2,3}/\mathrm{d}x)$. Further, it can be shown to be a consequence of the Second Law of Thermodynamics (that is, that all natural processes lead to an increase of entropy) that the L's in each horizontal row add up to zero. Thus

$$\left.\begin{aligned} L_{11} + L_{12} + L_{13} &= 0 \\ L_{21} + L_{22} + L_{23} &= 0 \\ L_{31} + L_{32} + L_{33} &= 0\,. \end{aligned}\right\} \tag{15}$$

To these relations there is a final set which ONSAGER showed was a consequence of the Principle of Microscopic Reversibility. It follows from this principle that provided the flows and gradients are specified in the way indicated, coefficients with the same numbers in their suffixes are equal. In symbols

$$L_{12} = L_{21}$$

or more generally,

$$L_{rs} = L_{sr}\,. \tag{16}$$

Broadly speaking therefore the "cross-coefficient" which describes how the gradient of solute 1 influences the movement of solute 2, describes also how the gradient of 2 influences the movement of 1.

The final result of the above considerations may be put thus; in an isothermal system of n different molecular species the description of transport[1] processes involves $\frac{1}{2}n(n-1)$ independent coefficients. Of these $(n-1)$ are "straight" transport coefficients, while the remaining $\frac{1}{2}(n-1)(n-2)$ are "cross-coefficients" describing how the gradient of one solute can influence the movement of another. The "straight" coefficients describe what is often spoken of as passive movement; the "cross-coefficients", active movement. The latter are usually small, but become of greater importance with electrolytes. Electro-osmosis is an important manifestation. For further information on this subject the reader is referred to DE GROOT, The Thermodynamics of Irreversible Processes.

Free energy of dilution.

Like all spontaneous processes occurring in systems whose temperature remains uniform and constant, diffusion involves a dissipation of free energy. It is often of interest to compute this, since the diffusive process may be linked energetically with other processes; that is, it may 'drive' or be 'driven by' them. A case in point is the electro-osmotic 'pumping' of water by the diffusion of electrolytes; the free energy of diffusion represents the maximum work that the mechanism can do.

It will be sufficient to state the relation in the simplest possible case. Imagine a large volume of solution at a concentration c_1, and a similar large volume at a lower concentration c_2. Both solutions are of course at the same temperature T, otherwise the concept of free energy is without relevance. Then if a relatively small quantity n moles of solute is transferred from the first solution to the second the loss in free energy of the system is given by

$$-\Delta G = nRT \ln \frac{c_1}{c_2}. \tag{17}$$

This assumes that the concentrations are low enough for the osmotic pressures to be calculable by the ideal gas laws; for stronger solutions concentrations must be replaced by activities.

The molecular picture of diffusion.

Hitherto we have followed the macroscopic approach to the problem of diffusion, confining our attention to characteristics of diffusion systems which can be actually observed and measured. Such an approach, while it is the obvious one for many purposes, gives little insight into the real nature of the processes involved. It leaves untouched, for example, the following questions: what makes one substance diffuse faster than another? has diffusion rate any connection with viscosity? what is the significance of the effect of temperature? These problems all call for an enquiry into the mechanism of diffusion, and for this purpose we have to turn to the molecular picture. In doing so, it will be convenient to start with gaseous systems, since in these many of the factors which complicate condensed systems are absent.

The law of equipartition of energy.

Let us fix our attention for a moment on a single molecule of gas moving freely in a containing vessel. Its translatory motion can be conveniently specified

[1] The case is rather more general than would be indicated if "diffusion" were put here.

by the three independent components of velocity v_x, v_y, v_z parallel to a suitable set of rectangular axes. If the mass of the molecule is m the total kinetic energy of translation can then be written $(\frac{1}{2} m v_x^2 + \frac{1}{2} m v_y^2 + \frac{1}{2} m v_z^2)$. But it possesses *thermal*[1] energy of kinds other than this; for instance, the whole molecule will be rotating, and also sustaining internal vibrations. If we desire a complete statement of its thermal energy we shall have to add other terms to take account of these forms; for example, rotational energy might be written as the sum of three independent terms $(\frac{1}{2} I_1 \omega_1^2 + \frac{1}{2} I_2 \omega_2^2 + \frac{1}{2} I_3 \omega_3^2)$, where the I's are moments of inertia and the ω's angular velocities about rectangular axes located in the molecule. The number of independent terms given to the vibratory energy will depend on the number of atoms in the molecule, and will be of two kinds: terms describing the kinetic energy of vibration (of the form of a mass times the square of a velocity) and terms describing the potential energy of elastic deformation (of the form of a "stiffness" times the square of a displacement). The final result is that the thermal energy of the molecule can be put into some such form as

$$E = \tfrac{1}{2} m v_x^2 + \tfrac{1}{2} m v_y^2 + \tfrac{1}{2} m v_z^2 + \tfrac{1}{2} I_1 \omega_1^2 + \tfrac{1}{2} I_2 \omega_2^2 + \tfrac{1}{2} I_3 \omega_3^2 + \Sigma \tfrac{1}{2} m' u^2 + \Sigma \tfrac{1}{2} \sigma a^2. \quad (18)$$

At this point a very important generalisation is applicable. Provided a gas is in internal equilibrium the energy represented by each one of the above independent "square terms", averaged over all the molecules in the gas, is the same. It is in fact, equal to $\frac{1}{2} kT$ per molecule where T is the absolute temperature and k is a constant called BOLTZMANN's constant. This generalisation, the Law of the Equipartition of Energy, is subject to certain limitations, especially at very low temperatures; but for present purposes it can be taken as exact for gases, and as applying rather more broadly to liquids and solids. Further it holds as between different sorts of molecules. To illustrate this latter point, consider two different gases having molecules of mass m_1 and m_2. If the two gases are at the same temperature T we can write

$$\tfrac{1}{2} m_1 \overline{v_1^2} = \tfrac{1}{2} m_2 \overline{v_2^2} = 3 \cdot \tfrac{1}{2} k T \quad (19)$$

where $\overline{v_1^2}$ and $\overline{v_2^2}$ are mean square molecular velocities. This equation leads at once to the relation[2]

$$\frac{\bar{v}_1}{\bar{v}_2} = \sqrt{\frac{m_2}{m_1}}\,. \quad (20)$$

It is reasonable to suppose that the rates of diffusion of gases through very fine pores are proportional to their molecular velocities; equation (20) can therefore be taken as an expression of GRAHAM's law, that the speed of diffusion is inversely proportional to the square root of the molecular weight.

The MAXWELLian distribution.

The law discussed above applies of course to averages; it goes without saying that at any given instant individual molecules may possess velocities differing widely from the average. Biologists are familiar with the state of affairs in which a population whose members differ among themselves has to be described by a frequency distribution. This sort of description was applied by CLERK MAXWELL in 1860 to the present problem. In fact, he showed that provided a gas was in internal equilibrium the distribution of molecular velocities among

[1] That is, loosely, energy associated with the idea of temperature.

[2] Strictly, this should be in terms of r.m.s. velocities, but the result holds also for mean velocities.

its molecules followed a quite definite pattern. Confining our attention to those components of velocity parallel to one given direction it proves to be the case that the fraction of molecules dn/N possessing velocities between v and $v+dv$ can be written

$$\frac{dn}{N} = \sqrt{\frac{m}{2\pi kT}}\, e^{-\frac{mv^2}{2kT}}\, dv \tag{21}$$

where N is the total number of molecules, m the mass of a molecule and k is BOLTZMANN's constant. In form this resembles the equation of the normal error curve so important in the theory of experiments. Its shape is shown in fig. 5. Since the gas is assumed at rest the mean velocity-component is zero, and the curve is in fact symmetrical about the vertical axis.

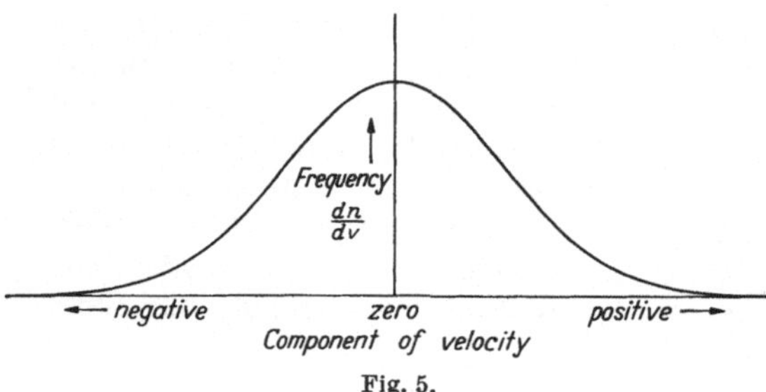

Fig. 5.

Rate of diffusion across a membrane.

The simple form of MAXWELL's equation just given enables us to calculate quite easily the rate at which molecules will diffuse across a plane membrane. Membranes as a rule offer a resistance to the penetration of molecules, and a very satisfactory way of taking account of this is to specify the minimum kinetic energy a molecule must possess in order to enable it to penetrate. The membrane is thus regarded as a "potential energy barrier", analogous to a ridge across a billiard table. Molecules possessing only a little kinetic energy penetrate only a short distance and are then thrown back.

Suppose that the value of the potential energy barrier is $\frac{1}{2} m v_0^2$; this will mean that only molecules with a velocity exceeding v_0 will penetrate. In order to calculate how many cross per second we make the simplifying assumption that we can neglect molecular collision, so that all molecules with a large enough component v perpendicular to the membrane and directed towards it will cross in the next second provided they are distant not more than v from it. If A is the area of the membrane and if V is the volume occupied by the N molecules on one side of the membrane the number $d\nu$ of molecules with components between v and $v+dv$ which cross in the next second will be

$$\left.\begin{aligned} d\nu &= \text{concentration of molecules of stated category times} \\ &\quad\ \text{volume of prism of base } A \text{ and height } v \\ &= \frac{dn}{V}\cdot vA \\ &= \frac{NA}{V}\sqrt{\frac{m}{2\pi kT}}\, v\, e^{-\frac{mv^2}{2kT}}\, dv \end{aligned}\right\} \tag{22}$$

introducing (21).

The total number crossing per second from one side is obtained by integrating this expression between the limits v_0 and ∞. Thus

Thus

$$\left.\begin{aligned} \nu &= \frac{NA}{V}\sqrt{\frac{m}{2\pi kT}}\int_{v_0}^{\infty} v\, e^{-\frac{mv^2}{2kT}}\, dv \\ &= \frac{NA}{V}\sqrt{\frac{kT}{2\pi m}}\, e^{-\frac{mv_0^2}{2kT}}. \end{aligned}\right\} \tag{23}$$

This equation is more convenient if it is put into molar units instead of molecular. If N is AVOGADRO's number, $\frac{v}{AN} = z$ is the rate of crossing in moles per second per square centimetre. Further, BOLTZMANN's constant k is equal to R/N, where R is the gas constant; and mN is the molecular weight, M. Finally, writing $N\frac{1}{2}mv_0^2 = E_0$, the energy barrier per mole, and $\frac{N}{NV} = c$ the molar concentration, we have

$$z = c\sqrt{\frac{RT}{2\pi M}}\,e^{-\frac{E_0}{RT}} \tag{24}$$

moles per unit area per second, or dividing by c,

$$\text{Permeability } \mu = \sqrt{\frac{RT}{2\pi M}}\,e^{-\frac{E_0}{RT}} \tag{25}$$

moles per unit area per second per unit concentration.

These equations express what was called above the *gross* rate of diffusion; to find the *net* rate the quantity c in (24) must be replaced by $(c_1 - c_2)$, the difference in concentration on the two sides of the membrane.

Effect of temperature and nature of gas.

The present interest of equation (24) lies in the fact that it gives us an insight into the effect of temperature and of the nature of the gas on the speed of diffusion across the membrane. As will appear later it applies *qualitatively* also to the ordinary process of diffusion in liquids, as well as to the penetration of membranes in liquid systems.

First, the nature of the diffusing molecule will obviously influence E_0 in more than one way. Increase of size will tend to raise E_0, as will "awkwardness" of shape. Chemical nature will be important too, since on this depend the forces of cohesion or adhesion which draw back the molecule into its own phase or pull it forward into the membrane. For these reasons the rate of diffusion, when any factor introducing an energy barrier is present, cannot be correlated simply with any one property of the molecular species as GRAHAM's law for gases is able to do on the basis of the molecular weight.

Secondly, consider how the rate of diffusion across the barrier varies with temperature. For two temperatures T_1 and T_2 it is easy to derive from (24) the relation

$$\ln\left(\frac{z_1}{z_2}\right) = \frac{1}{2}\ln\left(\frac{T_1}{T_2}\right) + \frac{E_0}{R}\left(\frac{1}{T_2} - \frac{1}{T_1}\right) \tag{26}$$

or if T_1 and T_2 differ by 10° C and E_0 is large compared to RT,

$$\ln Q_{10} \approx \frac{10\,E_0}{RT^2} \tag{27}$$

where Q_{10} is the ordinary temperature coefficient. This shows that the effect of temperature reflects principally the influence of E_0. If E_0 is large the rate of diffusion (from 24) tends to be small, owing to the influence of the exponential factor; but concurrently the effect of a temperature rise is great. The correlation of these two properties, low diffusion rate and high temperature sensitivity, has often been noted.

Diffusion in liquid systems.

When we turn from gases to liquids the molecular picture of diffusion is complicated by new factors. In liquids not only is the mean free path[1] very small, relative to molecular dimensions, but the molecules tend to "stick together", so that unlike the state of affairs in an ideal gas a molecule can only diffuse when its instantaneous share of thermal energy is sufficient to break the bonds attaching it to neighbouring solvent molecules and these to each other. It is, in fact, probably to be pictured as alternately vibrating erratically about a mean position and jumping to a new one. The energy for the jump may be considered as being contributed in varying degrees by the molecule itself and by the 'co-operation' of its neighbours, which may make the molecule's own requirement greater or less according as they crowd together in front of it or separate to form a hole. But in any case an "activation energy of diffusion" E_a will be attributable to the system, analogous to the energy of penetration previously discussed. This will be the case even when a long flexible molecule such as mannitol is involved, when of course a clean-cut jump in which the molecule moves as a whole becomes increasingly unlikely. It is the presence of the energy factor that makes the problem of diffusion in liquids rather like the one discussed above, where an ideal gas penetrates a membrane. In particular, the effect of temperature on the diffusion rate, and the magnitude of the diffusion coefficient itself, are very largely determined by the activation energy.

Origin of the activation energy.

The activation energy arises from the fact that molecules exert forces on one another, especially attractive ones. Sometimes these forces are merely the weak VAN DER WAALS' forces, as in the case of non-polar systems like bromoform diffusing in ether. Here we would expect E_a to be low and diffusion fairly rapid, and this is found broadly-speaking to be the case (Table 1).

Table 1. *Coefficients of diffusion.*

	Temp. °C	M. W. of solute	$D \times 10^5$ (cm.²/sec.)
Bromobenzene in ether	7.5	157	3.50
in benzene	7.5	157	1.45
Bromoform in ether	20	253	3.39
in benzene	20	253	1.69
in methanol	20	253	1.93
Urea in water	20	60	1.18
Glycerol in water	20	92	0.83
Glucose in water	18	180	0.57
NaCl in water	18		1.24

Where on the other hand there are stronger bonds present, particularly the hydrogen bond, the diffusion coefficient is smaller. This is of course the case with water, especially with *solutes* capable of forming hydrogen bonds (*e.g.* glucose). The rate of diffusion in such cases would be much smaller were it not for the operation of a second factor, the entropy of activation (see below). It is hardly necessary to add that the significance of the activation energy extends also to the question of viscosity. Thus there is a parallelism between the viscosity of a solvent and the rate at which solutes diffuse in it. Of course, the diffusion case must reflect also the properties of the solute.

The entropy of activation.

In deducing the rate of penetration of a membrane by molecules of a perfect gas it was assumed that no eligible molecules were impeded by finding others

[1] That is, the mean distance a molecule travels between collisions.

in their path. Actually of course molecular collisions do occur and complicate the actual course of events. In general, it does not follow that a molecule qualified so far as its energy requirements is concerned will therefore diffuse. One has first to enquire about the probability that the energy is suitably distributed between its different degrees of freedom, for instance. It may be that the requirements here are very stringent, in which case the diffusion rate will be greatly lowered; or they may be so easily met that the molecule is in this respect almost "encouraged" to move. Considerations such as these lead to the notion of an entropy of activation. In the case of solvents like water the breaking of hydrogen bonds, a necessary condition for diffusion, signifies an increase in entropy. As is the case in fixing the position of equilibria, the entropy factor operates in the reverse direction to the energy factor. Thus the high entropy of activation partially offsets the high energy of activation; and the diffusion of solutes in water is faster than one would expect if the latter alone were influential.

The nature of the diffusion coefficient.

The physiologist often needs to enquire into the nature of the factors which determine the rate of diffusion of a given solute, since often, as in the case of diffusion through the plasma membrane, such knowledge will increase his understanding of the molecular architecture of the cell. In the case of diffusing molecules, the most satisfactory approach to the calculation of the diffusion coefficient is that of the theory of absolute reaction rates (see GLASSTONE, LAIDLER and EYRING, 1941). This leads to the equation

$$D = \frac{\lambda^2}{v_f^{\frac{1}{3}}} \left(\frac{RT}{2\pi M}\right)^{\frac{1}{2}} e^{-\frac{L}{nRT}}. \tag{28}$$

Here λ is the average length of jump made by a diffusing molecule, M is the molecular weight, L is the latent heat of evaporation, n is a coefficient whose value is about $2\frac{1}{2}$, and v_f is the "free volume" per molecule; that is, the average volume of empty space which may be regarded as surrounding each molecule and belonging to it. The values of λ, M, L and v_f have to be chosen to take account of the properties of both solvent and solute, and this is a difficulty where these differ at all widely. But while the equation can hardly be used to give an accurate value of D, it does show the way in which different factors enter into its magnitude. For instance, no simple dependence of D on the inverse square root of the molecular weight is to be expected independently of other factors, as in the case of GRAHAM's law for gases. The chemical relationship between solute and solvent and the sizes and shapes of their molecules will certainly affect v_f, λ and L. Above all, the exponential factor will exert a pronounced effect both on the magnitude of D and on its temperature dependence.

These remarks apply qualitatively also to the permeability of the plasma membrane.

Diffusion of large molecules.

Where the solute particles are large (as for instance in the case of proteins) the assumptions on which the above theory is based no longer hold, and a different approach must be made. Using STOKE's law, which expresses the magnitude of the viscous drag on a spherical particle, EINSTEIN derived the equation

$$D = \frac{kT}{6\pi r \eta} \tag{29}$$

where k is BOLTZMANN's constant, r the radius of the diffusing particle, and η the viscosity of the solvent. With a suitable variation in the numerical factor this equation can be applied to such cases as the diffusion of rod-shaped or other particles. Assuming that particles of constant density and shape are compared, the equation leads to the expectation that D will vary inversely as the cube root of the molecular weight, and this is borne out by experiment. It also shows clearly the connection between diffusion rate and viscosity, a connection which will also hold broadly for the previously discussed case of the diffusion of small molecules.

Literature.

GLASSTONE, S., K. J. LAIDLER and H. EYRING: The theory of rate processes. New York: McGraw Hill Book Co. 1941. — GROOT, S. R. DE: Thermodynamics of irreversible processes. Amsterdam: North Holland Publishing Co. 1951.

INGESOLL, L. R., and O. J. ZOBEL: The mathematical theory of heat conduction. Boston 1913.

Mathematische Analyse experimenteller Ergebnisse: Gewinnung der Permeabilitätskonstanten, Stoffaufnahme- und -abgabewerte.

Von

Eduard Stadelmann.

Mit 3 Abbildungen.

I. Einleitung.

Jede Stoffwanderung in biologischen Objekten scheint allgemein an zwei Voraussetzungen gebunden: an den *Mechanismus*, welcher die Dislokation bewirkt, und an die *Wegsamkeit* der Strecke, längs welcher die Substanz transportiert werden soll (TEORELL 1953, p. 308). Die einzelnen Abschnitte eines solchen Transportweges, als welche bei den untersuchten Objekten besonders Membranen interessieren (z. B. die Protoplastenhaut bei Pflanzenzellen, bestimmte Zellschichten bei größeren Komplexen), setzen der Stoffwanderung meist einen sehr unterschiedlichen Widerstand entgegen.

Zur Charakterisierung solcher Transportvorgänge können speziell bei Membranen verschiedene Kriterien dienen, so z. B.

a) *Der Mechanismus, welcher die Substanzverlagerung bewerkstelligt.* Es handelt sich hier entweder um eine Stoffwanderung, die keine weitere Energiezufuhr benötigt *(passive*, auch als osmotisch bezeichnete *Stoffaufnahme und -abgabe*; *passive Permeabilität)*, oder um einen aktiven Vorgang (*nicht-diosmotische Stoffaufnahme- und -abgabe*; vgl. „*adenoide Tätigkeit der Zelle*", OVERTON 1899, p. 102), der in seiner Art noch wenig erkannt ist und offenbar eines dauernden Arbeitsaufwandes bedarf.

Im ersten Falle ist meist die *Diffusion* wirksam, welche eine statistische Gleichverteilung von Teilchen zur Folge hat und daher *ohne Energieumsetzungen* vor sich geht. Es können aber auch *physikalische Kräfte* (z. B. elektrostatische Anziehungskräfte) die Ursache der Substanzwanderung darstellen, wobei dann eine *beschränkte Energieumsetzung* erfolgt (z.B. die Umwandlung der elektrischen Feldenergie in die Bewegungsenergie der einzelnen Ionen). Diese beiden Phänomene können sich in ihrer Wirkung auch überlagern, und es läßt sich der resultierende passive Stofftransport dann aus ihrer jeweiligen Größe erklären.

Für Ionen ist die Entscheidung, ob es sich in einem bestimmten Fall um einen passiven oder aktiven Transportvorgang handelt, wohl dadurch möglich, daß außer den Werten für den Zufluß und Abfluß (S. 152) noch die elektrische Potentialdifferenz zwischen beiden Seiten der Membran gemessen wird. Für diese drei Größen muß bei rein passiver Stoffwanderung eine mathematische Beziehung erfüllt sein (USSING 1949, p. 44; 1951).

b) *Der Ausgangs- bzw. Endpunkt der Substanzverschiebung.* Betrachtet man die Verhältnisse, wie sie sich in der Zelle selbst ergeben, so kann bei Vorhandensein einer Vacuole der Zellsaftraum oder der Plasmamantel als Ausgangspunkt bzw. Ziel in Betracht kommen („*Permeabilität*" bzw. „*Intrabilität*", HÖFLER 1934a, p. 75). Aber der Stoffdurchtritt kann auch durch eine aus einer oder mehreren Zellschichten bestehende lebende Membran, die zwei Milieus voneinander trennt, welche ihrerseits dann Ausgangs- bzw. Endpunkt darstellen, untersucht werden.

Zu beachten ist hierbei noch, daß die Stoffwanderung die Resultierende zweier simultan, aber verschieden intensiv ablaufender gegensinniger Transportvorgänge darstellt.

c) *Die Art des Milieus.* Das Untersuchungsobjekt kann in seinem natürlichen bzw. einem diesem gleichwertigen Milieu beobachtet werden oder in einer anderen mehr oder weniger fremdartigen Umgebung, die sich von ersterer durch die Art ihrer stofflichen Zusammensetzung oder auch nur durch die Konzentration unterscheidet.

Die Verhältnisse des *passiven Stofftransportes* sind hauptsächlich in zweierlei Weise Gegenstand experimenteller und theoretischer Arbeiten:

Erstens wird versucht, aus der Intensität der Stoffwanderung durch die betreffende biologische Membran die Größe ihrer Wegsamkeit (als Permeabilität bzw. Durchlässigkeit bezeichnet) zu bestimmen *(Permeabilitätsmessung)*, wobei es zum Vergleich der Ergebnisse der verschiedenen Versuche erforderlich ist, geeignete Maße zu finden.

Zweitens trachtet man, diese Stoffwanderung aus theoretischen Annahmen und Modellvorstellungen über den Transportmechanismus, die Eigenschaften der zu prüfenden Substanz und die Membranstrukturen kausal zu erklären *(Permeabilitätstheorien*, vgl. Abschnitt IV G). Hierbei soll entweder die Größe der Permeabilität, wie sie aus den erstgenannten Untersuchungen folgt, quantitativ auf diese Faktoren zurückgeführt werden, oder es ist aus theoretischen Annahmen die Größe des im Experiment im Einzelfall zu erwartenden Stofftransportes zu berechnen und die Übereinstimmung mit dem experimentell gefundenen Wert zu prüfen (vgl. DAVSON und DANIELLI 1943, p. 341; Lit.: TEORELL 1953).

Bei der *aktiven Stoffaufnahme* interessiert, abgesehen von ihrem Mechanismus, hauptsächlich die Menge des transportierten Stoffes, das jeweilige Ausmaß der Speicherfähigkeit sowie ihre Korrelation mit Stoffwechselvorgängen und deren experimentelle Beeinflussung (Lit.: STEWARD 1937; LUNDEGÅRDH 1940; BURSTRÖM 1954; vgl. Abschnitt D d, H).

Für beide Mechanismen läßt sich die Größe des Stofftransportes für die transportierte Substanz meist auf einfache Weise ermitteln, da hierzu nähere Annahmen über dessen treibende Kraft nicht erforderlich sind.

Die folgenden Darlegungen beschränken sich allein auf die *Bestimmung der Durchlässigkeit biologischer Membranen beim passiven Stofftransport sowie die Ermittlung der Stoffaufnahme- und -abgabewerte*, wobei nur einige der zahlreichen Berechnungsverfahren als Beispiele angeführt werden können und von relativen, an spezielle Versuchsanordnungen und Methoden gebundenen Permeabilitätsmaßen abgesehen ist. Über die Theorie der Stoffwanderung und der Membrandurchlässigkeit ist in anderen Abschnitten berichtet (vgl. IV D—J).

Bevor nun einzelne dieser Berechnungsmethoden, mit deren Hilfe Zahlenwerte für die Transportgröße und die Permeabilität aus den verschiedenen experimentellen Daten abgeleitet werden, zu zeigen sind, sei kurz und soweit erforderlich auf die experimentelle Methodik, die im folgenden Abschnitt IV C noch ausführlich behandelt ist, verwiesen. Eingangs sind ferner jene Grundannahmen und Voraussetzungen, die bei solchen Ableitungen herangezogen werden, zusammengestellt. Welche dieser Postulate bei einem bestimmten Berechnungsverfahren erforderlich sind, läßt sich leicht aus der Versuchsanordnung sowie der Art der Ableitung ersehen und ist im betreffenden Text meist nicht speziell angeführt.

Da das Wasser in seiner Qualität als Lösungsmittel der anderen permeierenden Stoffe diesen gegenüber eine ausgezeichnete Stellung einnimmt, werden die den Durchtritt des Wassers betreffenden Betrachtungen von jenen der gelösten

Substanzen getrennt diskutiert. Schließlich ist noch die Möglichkeit eines Vergleiches der nach den verschiedenen Endformeln berechneten Permeabilitätsfaktoren und der Stofftransportwerte behandelt.

Während sich die Ableitung von Maßen für die Stoffaufnahme meist in einfacher Weise aus den experimentellen Befunden ergibt, beansprucht die Berechnung der Permeabilitätskonstanten, besonders aus Versuchen nach der plasmolytischen Methode, oft einen größeren mathematischen Aufwand. Es schien daher angebracht, spezielle Beispiele für solche Berechnungsgänge zu zeigen. Dabei wurden die Endgleichungen möglichst stets so weit ausgerechnet, daß die sich im Experiment ergebenden Meßgrößen (z. B. Konzentration, Zeit, Zellänge usw.) unmittelbar eingesetzt werden können und zur Kennzeichnung durch Fettdruck der Formelnummer hervorgehoben. Zum Vergleich mit der Originalformel ist diese meist in eckigen Klammern mit den dort benutzten Formelzeichen wiedergegeben. Die einzelnen Formelzeichen haben sonst überall dieselbe Bedeutung, die gelegentlich im zugehörigen Text, im übrigen aber vollständig und in tabellarischer Form im Anhang angeführt ist. Dort sind auch das ihnen in dem hier gewählten Dimensionssystem mit den Grunddimensionen Länge, Masse, Zeit, Temperatur (L, M, T, $\mathsf{\Theta}$) zugehörige Dimensionsprodukt (auch kurz als Dimension bezeichnet, vgl. STILLE 1955, p. 20), und die hier benutzten Maßeinheiten, die sich meist an das CGS-System anschließen, angegeben. Das Dimensionsprodukt der behandelten Permeabilitätsfaktoren und Stofftransportmaße wird unmittelbar bei der betreffenden Ausgangsgleichung erwähnt.

Zur Beurteilung der Vergleichbarkeit der Permeabilitätskonstanten und Stofftransportwerte kommt den Ausgangsgleichungen eine wichtige Rolle zu (vgl. S. 178f.). Diese Formeln werden daher stets angeführt und ihre Formelnummern unterstrichen. Für die Permeabilitätsfaktoren sind die Gleichungen auch textlich formuliert, wobei aber für denselben mathematischen Zusammenhang oft auch noch andere solcher Auslegungen möglich sind (vgl. BOCHSLER 1948, S. 81). In den hier gewählten Darstellungen wurde immer darnach getrachtet, möglichst elementare Größen miteinander zu verknüpfen (z. B. Volumen und Oberfläche statt Oberflächenentwicklung).

Die Beobachtung einer Durchlässigkeit des Plasmamantels pflanzlicher Zellen für gelöste Substanzen reicht weit zurück und erfolgte zuerst wohl mittels *Fällungs-* und *Farbreaktionen* (vgl. JANSE 1887, p. 345f.). Durch die Einführung *plasmolytischer Methoden* (vgl. DE VRIES 1889, p. 310/553) und besonders der *Plasmometrie* (HÖFLER 1918b, p. 441) erhielten die quantitativen Untersuchungen der Permeabilität einen starken Impuls und wurden in großer Zahl ausgeführt. Ein weiterer Aufschwung, besonders der Messung des Ionentransportes bei tierischen Objekten, ergab sich in letzter Zeit durch die Anwendung von *Isotopen* (vgl. KAMEN 1947, p. 118f.; BROOKS 1951; USSING 1953; HEVESY 1948, p. 191f.; SCHWIEGK 1953, p. 330f.).

II. Übersicht über die Grundannahmen und Voraussetzungen.

Um die Stoffwanderung durch biologische Membranen einer mathematischen Analyse zugänglich zu machen, ist es erforderlich, bestimmte vereinfachende Annahmen zu machen und deren Zulässigkeit a priori anzuerkennen. Ein solches Vorgehen ist bei der quantitativen Behandlung von andersartigen Phänomenen (z. B. im Gebiete der Chemie und Physik) durchaus gebräuchlich und führt wie dort zu Aussagen begrenzter Gültigkeit.

Diese Bedingungen lassen sich unterteilen in 1. Grundannahmen, welche den Mechanismus des Versuchsablaufes betreffen und somit meist den Ausgangspunkt der mathematischen Ableitungen bilden, 2. in Voraussetzungen über die Eigenschaften der Versuchsanordnung, 3. in solche, die gewisse Eigenschaften des Objekts betreffen, und 4. in einige zur Berechnung nötige Prämissen.

A. Allgemeine Grundannahmen für den Versuchsverlauf.

(*a*) *Die Wanderung eines gelösten Stoffes erfordert keine ständige Energiezufuhr* und beruht also auf der Wirkung der Diffusion und/oder elektrostatischer Kräfte. Das Ausmaß des Substanztransportes wird dann direkt proportional der Stärke dieser Effekte angenommen, die sich so aus einem aus der Diffusion stammenden und einem elektrischen Anteil zusammensetzen können. Dabei sind Konzentrationsdifferenz und/oder elektrisches Feld innerhalb der Membran zwischen deren Grenzflächen wirksam und vermögen durch letztere noch Modifikationen zu erfahren (vgl. TEORELL 1953, Fig. 2b). Es wird dann für den Transport einer bestimmten Stoffart im stationären Zustand die Gültigkeit einer auf NERNST (1888, 1889) und PLANCK (1890a, b) zurückgehenden Beziehung angenommen (vgl. TEORELL 1951, p. 461), die lautet:

$$M = u \cdot c_x^* \cdot \left(\frac{R \cdot T}{c_x^*} \cdot \frac{d c_x^*}{d x} \pm F \cdot \frac{d \varphi}{d x} \right)^{\dagger}. \qquad \text{(II, 1)}$$

Erfährt der transportierte Stoff auf jener Membranseite, zu welcher er hinwandert, keinerlei Änderungen (Fehlen von chemischen Reaktionen, Adsorption usw.), so reichert er sich dort an, was zu einer Verkleinerung des Gefälles führt, so daß sich ein *Endgleichgewicht* ausbilden kann. Besteht die betreffende Substanz aus ungeladenen Teilchen (Anelektrolyte), so fallen die elektrostatischen Kräfte weg, und als Endzustand stellt sich auf beiden Seiten der Membran die gleiche Konzentration ein.

Handelt es sich um Ionen, so können dann die beiderseitigen Konzentrationen verschieden sein, wobei die für dieses Konzentrationsungleichgewicht erforderliche osmotische Arbeit gleich sein muß der elektrischen Arbeit (TEORELL 1937, p. 941; 1949, p. 549):

$$\varepsilon \cdot F \cdot \pi = R \cdot T \cdot \ln \frac{c_i^*}{c_a^*}. \qquad \text{(II, 2)}$$

Konzentrationsungleichheit beiderseits der Membran weist in solchen Fällen daher noch nicht ohne weiteres auf das Vorhandensein eines aktiven Stofftransportes hin. Im Falle einer Strömung des Lösungsmittels (Wasser) oder der Mitwirkung noch anderer Phänomene (z. B. der Adsorption) kommen in (II, 2) weitere Glieder hinzu (vgl. 1953, p. 365[1]).

Treten während der betrachteten Versuchsdauer keinerlei Änderungen der Konzentrationen, der elektrischen Kräfte und der übrigen Faktoren auf (z. B. wenn die Menge des transportierten Stoffes an jener Membranseite, zu welcher er hinwandert, vernachlässigbar klein ist oder im Ausmaß des Zuflusses unwirksam gemacht wird), so bezeichnet man diesen Zustand als *stationäres Gleichgewicht (steady state)*.

Flußgleichgewichte (*flux equilibria*, 1953, p. 354) stellen sich bei Systemen mit mehreren Arten transportabler Partikel dann ein, wenn für einzelne Teilchenarten das Endgleichgewicht, für andere das stationäre Gleichgewicht vorliegt.

Verkleinert sich der Gesamtfluß mit fortschreitender Zeit und kehrt schließlich seine Richtung um, so entsteht im Umkehrpunkt ein *vorübergehendes Gleichgewicht* (*temporary distribution equilibrium*, 1953, p. 350f.).

Gleichung (II, 1) geht in Grenzfällen (vgl. 1953, p. 317) und für Anelektrolyte wegen des Verschwindens des elektrischen Potentials in die einfache Form

$$M = u \cdot c_x^* \cdot \frac{R \cdot T}{c_x^*} \cdot \frac{d c_x^*}{d x} \qquad \text{(II, 3)}$$

über. Führt man für $u \cdot R \cdot T$ die Diffusionskonstante D und für M nach (IV, 1) den Wert

$$M = \frac{d m}{A \cdot d t}$$

† Wobei vereinfachend angenommen wird, daß statt der Aktivitäten unmittelbar die Konzentrationen eingesetzt werden dürfen. Die hier als *Beweglichkeit* (Dim $u = \mathsf{M}^{-1}\mathsf{T}$) bezeichnete Größe ist nicht gleich jener der Elektrochemie, da oben die bei Vorliegen der Einheit der treibenden Kraft (Dim $= \mathsf{LMT}^{-2}$) auftretende *Geschwindigkeit* verstanden sein soll (TEORELL 1953, p. 308), während diese Größe dort auf die elektrische Feldstärke [Dim (im el.stat. LMT-System) $= \mathsf{L}^{-\frac{1}{2}} \mathsf{M}^{\frac{1}{2}} \mathsf{T}^{-1}$] und daher als Geschwindigkeit bei Einheit der Feldstärke aufzufassen ist. Die „elektrochemische" Beweglichkeit (Dim $= \mathsf{L}^{\frac{3}{2}} \mathsf{M}^{-\frac{1}{2}}$) ergibt sich durch Multiplikation von u mit der Ladung (Dim $= \mathsf{L}^{\frac{3}{2}} \mathsf{M}^{\frac{1}{2}} \mathsf{T}^{-1}$).

[1] Unterbleibt die Angabe von Name bzw. Jahreszahl, so ist stets der unmittelbar vorher angeführte Autor bzw. die zuletzt zitierte Publikation gemeint.

ein, wobei von den Differenzen zu den Differentialen übergegangen wird, so folgt unmittelbar das erste FICKsche Gesetz (FICK 1855, p. 66):

$$\frac{dm}{dt} = D \cdot A \cdot \frac{dc_x^*}{dx}\dagger. \qquad (II, 4)$$

Hierbei ist zu beachten, daß die stillschweigende Voraussetzung, die Diffusionskonstante D sei konzentrationsunabhängig, nur angenähert und innerhalb bestimmter Konzentrationsbereiche gilt [vgl. ZUBER 1932, p. 298 (Anelektrolyte), DEAN 1947; p. 505 (Elektrolyte)].

Aus Gleichung (II, 4), die hier die Transportvorgänge innerhalb der Membran beschreibt, wobei für deren Anwendung bei dünnen Membranen gewisse Bedenken bestehen (vgl. DAVSON und DANIELLI 1943, p. 342), läßt sich nun die vorzugsweise für die Permeabilitätsberechnungen benutzte Gleichung ableiten (vgl. BÄRLUND 1929, p. 65). Man führt zunächst unter den Voraussetzungen (u) und (v) ein:

$$\frac{dc_x^*}{dx} = \frac{C_a - C_i}{X}. \qquad (II, 5)$$

Mit der Annahme (t) läßt sich der Quotient $-\frac{D}{X}$ als *Permeabilitätskonstante* K einsetzen, und bei Gültigkeit der Postulate (i) und (ff) kann man für C_a und C_i die Konzentrationen der betreffenden Lösungen einführen, so daß (II, 4) schließlich

$$\frac{dm_s}{dt} = K_s \cdot A \cdot (C - k) \qquad (II, 6)$$

wird (vgl. COLLANDER und BÄRLUND 1933, p. 25).

Das Vorzeichen von K hängt von der relativen Größe von C und k sowie vom Standpunkt des Beobachters ab und soll hier stets positiv gewählt sein.

Auf ein Ausgleichsgesetz wies schon SEBOR (1904, p. 348) bei Betrachtungen über die Diffusionsgeschwindigkeit von Wasser durch semipermeable Membranen hin, und RUNNSTRÖM (1911, p. 14) schlug eine ähnliche Formel zur Permeabilitätsberechnung am Seeigelei vor.

Schon die Versuchsergebnisse mit *Spirogyra* und *Rhoeo discolor* legen die Gültigkeit dieses Gesetzes für die betrachteten Vorgänge nahe (SZÜCS 1910, p. 746; BÄRLUND 1929, p. 70), und diese wurde später mikrochemisch an den Riesenzellen von *Chara ceratophylla* bewiesen (COLLANDER und BÄRLUND 1933, p. 19, 25). Schließlich gelingt es für plasmometrische Versuche, rein formelmäßig über den Versuchsablauf gewisse Aussagen abzuleiten, die bei verschiedenen Objekten (z. B. *Taraxacum, Sonchus, Carduncellus*[1]) mit großer Genauigkeit erfüllt werden und damit indirekt die Richtigkeit der Annahmen bestätigen (STADELMANN 1952, p. 381, 386). Bei Beteiligung nicht-osmotischer Vorgänge (vgl. Abschnitt IV A) läßt sich natürlich keine solche Übereinstimmung erwarten.

Außer durch obige Ableitung folgt Gleichung (II, 6) noch aus der Überlegung, daß die eintretende Stoffmenge (Zufluß, influx) M_I proportional der Konzentration der betreffenden Substanz in der Außenlösung ist, während die austretende Stoffmenge (Abfluß, out-flux) Proportionalität zur Konzentration im Zellinnern zeigt:

$$M_I = \text{Const}_1 \cdot C \qquad M_A = \text{Const}_2 \cdot k.$$

Nimmt man bei diesen beiden einander überlagerten gegensinnigen Permeationsvorgängen an, daß die Membran in beiden Richtungen gleich gut durchlässig ist, so wird

$$\text{Const}_1 = \text{Const}_2 = \text{Const}$$

und der Gesamtfluß M, auf den sich die Messung einer Permeabilitätskonstanten ja meist bezieht:

$$M = M_I - M_A = \text{Const} \cdot (C - k).$$

† Das Vorzeichen von D, welches auch vom Standpunkt des Beobachters abhängt, wird hier stets als positiv angenommen.

[1] Bei letzterem in noch nicht veröffentlichten Versuchen.

Setzt man hierin Const $= K_s$ und nach Gleichung (IV, 1) für M ein, so folgt hieraus unmittelbar (II, 6):

$$M = \frac{dm}{A \cdot dt} = K_s \cdot (C - k).$$

Die Verhältnisse bei einer solchen durch Membranen behinderten Diffusion eines gelösten Stoffes sind von jenen für das Lösungsmittel selbst (*Wasserpermeation*) nicht grundsätzlich verschieden. Als Transportmechanismus kommen dabei neben der Diffusion bei gewissen Versuchsanordnungen noch osmotische Kräfte in Betracht. Es handelt sich aber in beiden Fällen immer um die quantitative Erfassung der Behinderung, die eine den Wassertransport hemmende Membran auf die Einstellung des Endzustandes ausübt.

(*b*) Analog dem DALTONschen Gesetz für Partialdrucke ist der osmotische Gesamtdruck einer Lösung stets gleich der Summe aller osmotischen Partialdrucke.

(*c*) Nach dem VAN'T HOFFschen Gesetz und dem BOYLE-MARIOTTEschen Gesetz bleibt bei konstanter Teilchenzahl des gelösten Stoffes sein osmotischer Druck in einer Lösung dem Lösungsvolumen umgekehrt proportional [vgl. Voraussetzung (*cc*)], wobei von einer VAN DER WAALSschen Korrektur abgesehen werden kann (vgl. RESÜHR 1935, p. 339; 1936, p. 437).

(*d*) Ein einmal erreichtes osmotisches Gleichgewicht zwischen Außenlösung und der Gesamtkonzentration im Zellsaftraum bleibt während des ganzen folgenden Versuchsverlaufes erhalten, sofern nicht elastische Kräfte der Wanddehnung in Erscheinung treten oder in geeigneter Weise Energie zugeführt wird (vgl. JACOBS und STEWART 1932, p. 72; STADELMANN 1951, p. 768).

B. Eigenschaften der Versuchsanordnung.

(*i*) Die Konzentration der Außenlösung ist während der Versuchsdauer in allen ihren Volumelementen dieselbe und bleibt konstant. Insbesondere soll dies für die dem Protoplasten unmittelbar anliegende Flüssigkeitsschicht in bezug auf den übrigen Raum der Außenlösung gelten (vgl. BÄRLUND 1929, p. 49; JACOBS und STEWART 1932, p. 73; RESÜHR 1936, p. 437).

(*j*) Auf einen allfälligen Einfluß von Licht und Temperatur ist stets Rücksicht zu nehmen; nötigenfalls sind die zu vergleichenden Versuche bei ähnlichen Temperatur- und Beleuchtungsverhältnissen auszuführen (Lit.: STADELMANN 1951, p. 768).

(*k*) Äquimolare Lösungen von Anelektrolyten gelten als einander genau isosmotisch (vgl. BÄRLUND 1929, p. 49; hingegen FITTING 1919).

(*l*) Ein höherer Gehalt an bestimmten Isotopen bewirkt keine Veränderungen der für den Substanztransport maßgebenden Verhältnisse und der diesbezüglichen Eigenschaften des betreffenden Stoffes (vgl. HEVESY und HOFER 1934, p. 879; für DHO: FLEXNER und Mitarbeiter 1942, p. 36; WARTIOVAARA 1944, p. 4; für D_2O: LUCKÉ und HARVEY 1935, p. 480).

C. Eigenschaften des Objekts.

(*p*) Das Objekt wird durch die benutzte Außenlösung nicht geschädigt und ändert seine Permeabilitätseigenschaften im Laufe des Versuches nicht (vgl. BÄRLUND 1929, p. 15).

(*q*) Die Durchlässigkeit des Protoplasten für Wasser ist um ein Mehrfaches höher als jene für die gelösten Substanzen (vgl. RESÜHR 1935, p. 440).

Diese Voraussetzung ist zur Bestimmung der Permeabilität für gelöste Stoffe nach plasmolytischen Methoden von Bedeutung, wenn der osmotische Gleichgewichtszustand erreicht ist und die Rückdehnungsphase (vgl. S. 148) zur Beobachtung ausgewertet wird. Eine zu geringe Wasserdurchlässigkeit des Protoplasten wirkte dann wegen Voraussetzung (*d*) bremsend auf den Eintritt permeierender Substanz in die Vacuole und würde dadurch für diese eine geringere Permeabilität des Protoplasmamantels vortäuschen (vgl. JACOBS und STEWART 1932, p. 72).

(*r*) Der Diffusionswiderstand der Zellwand fehlt oder ist vernachlässigbar klein gegenüber jenem des Plasmamantels (RESÜHR 1935, p. 339).

(*s*) Der Diffusionswiderstand für Wasser und gelöste Stoffe bleibt im wesentlichen in einer sehr dünnen Oberflächenschicht des Protoplasten lokalisiert (RESÜHR 1935, p. 339; 1936, p. 347).

(*t*) Die Dicke dieser sehr dünnen diffusionshemmenden Schicht erfährt während des Versuches keine Änderung (*ibid.;* JACOBS und STEWART 1932, p. 74).

(*u*) Für jede der zwei Grenzflächen der diffusionshemmenden Schicht soll gelten, daß die Konzentration des permeierenden Stoffes an der Innenseite gleich derjenigen an der Außenseite der Grenzschicht ist.

(*v*) Der Konzentrationsgradient des permeierenden Stoffes bleibt innerhalb der ganzen Dicke der diffusionshemmenden Schicht gleich groß $\left(\frac{d c^*}{dx} = \text{Const.};\right.$ vgl. JACOBS und STEWART 1932, p. 74$\Big)$.

Für Versuche nach osmotischen Methoden werden meist noch folgende Voraussetzungen erwähnt:

(*aa*) Die osmotische Wirkung der gelösten Substanz auf die Zelle hängt nur von der Konzentration ab und ist unabhängig von der Durchlässigkeit des Protoplasten für diesen Stoff (BÄRLUND 1929, p. 49).

(*bb*) Die Menge der in der Vacuole ursprünglich vorhandenen osmotisch wirksamen Substanz wird während des Versuches nicht verändert (keine Katatonose, Anatonose oder unbeabsichtigte Exosmose; vgl. *ibid.*). Es gilt daher für diese Stoffe die *Volumen-Konzentrationsbeziehung*, die besagt, daß die Konzentration c_1 (in mol/cm³), welche diese Substanzen beim Volumen V_1 (in cm³) des Zellsaftraumes besitzen, multipliziert mit diesem Volumen als Produkt stets die Menge m_u (in mol) dieser Stoffe angibt. Da diese nun konstant ist, so gilt allgemein:

$$m_u = V_1 \cdot c_1 = V_2 \cdot c_2 = V_n \cdot c_n .$$

Entsprechend ergibt sich mit Voraussetzung (*c*) für den osmotischen Druck:

$$\text{Const.} = V_1 \cdot p_1 = V_2 \cdot p_2 = V_n \cdot p_n ,$$

wobei nach dem VAN'T HOFFschen Gesetz [Voraussetzung (*c*)] gilt:

$$p = C \cdot R \cdot T . \qquad \text{(II, 7)}$$

Obige Beziehung besteht ebenso auch für die osmotischen Saugkräfte der Lösungen[1]:

$$\text{Const.} = V_1 \cdot S_1 = V_2 \cdot S_2 = V_n \cdot S_n .$$

Die Richtigkeit dieser Voraussetzung ist für die nach osmotischen (und innerhalb dieser speziell für die nach plasmolytischen) Methoden ausgeführten Versuche von besonderer Wichtigkeit und wurde erstmalig wohl von KLERCKER (1892, p. 472) durch Beobachtungen an kugeligen, isolierten Protoplasten von *Stratiotes aloides* bestätigt. Auch die Grundgewebszellen von *Tradescantia elongata* gehorchen sehr genau dieser Beziehung zwischen Protoplastenvolumen und Konzentration (HÖFLER 1918c, p. 123), und sie gilt auch für die Anthocyane der Epidermiszellen von *Rhoeo discolor* (ZEEUW und KUENEN 1935). In anderen Zellen

[1] Da als osmotischer Druck in der physikalischen Chemie der in der PFEFFERschen Zelle auftretende hydrostatische Maximaldruck definiert ist, so wird hier im Sinne von URSPRUNG (1930) besser von einer osmotischen Saugkraft (gemessen in Atmosphären) gesprochen, die bei Fehlen irgendwelcher elastischer Wanddrucke oder hydrostatischer Drucke auf das Lösungsvolumen stets dem osmotischen Druck, den diese Lösung in einem Osmometer entwickeln würde, numerisch gleich ist.

zeigen sich mitunter Abweichungen, die als durch das Plasmavolumen („Protoplasmakorrektur", HÖFLER 1918c, p. 113f.) oder einen „nicht-lösenden Raum" (vgl. RESÜHR 1935, p. 339; 1936, p. 437; Abschnitt III D, S. 101) verursacht erscheinen und sich dann rechnerisch berücksichtigen lassen, so daß die geforderte Gesetzmäßigkeit wieder beobachtet wird (vgl. 1935, *l. c.*). Andererseits treten aber auch Unterschiede auf, die vielleicht durch andere Mechanismen für den Stoff- oder Wassertransport hervorgerufen sind (BOGEN und PRELL 1953, p. 464f.).

(*cc*) Die osmotische Wirksamkeit der permeierten Substanz wird in der Vacuole nicht verändert (vgl. BÄRLUND 1929, p. 49).

(*dd*) Im Gleichgewichtszustand zwischen Außenkonzentration der diosmierenden Substanz und deren Konzentration im Zellsaftraum ist letztere gleich groß wie erstere, d. h. das *Verteilungsverhältnis* $\frac{C^*}{C}$ ist dann 1 (vgl. COLLANDER und BÄRLUND 1933, p. 19).

(*ee*) Der Diffusionswiderstand innerhalb der Vacuole ist vernachlässigbar klein und es bestehen dort in allen Volumelementen zu gleichen Zeiten dieselben Partialkonzentrationen der im Zellsaft gelösten osmotisch wirksamen Stoffe [vgl. p. 33f.; JACOBS und STEWART 1932, p. 74; die entgegengesetzte Annahme von BACHMANN (1939) gilt als wenig wahrscheinlich, vgl. HUBER 1943, p. 442f.; HÖFLER 1949, p. 117].

(*ff*) Bei Versuchen nach dem Partialverfahren (S. 147) besteht hinreichende Undurchlässigkeit des Plasmamantels für die als nichtpermeierende Komponente gewählten Stoffe (vgl. BÄRLUND 1929, p. 49).

Bei den plasmolytischen Methoden werden zur Auswertung meist noch die folgenden Voraussetzungen postuliert:

(*kk*) Der Protoplast ist derart leicht beweglich, daß er eine Volumänderung der Vacuole nicht verzögert oder erschwert (vgl. BÄRLUND 1929, p. 49).

(*ll*) Die Adhäsion des Protoplasten an der Membran bleibt so gering, daß bei der Ablösung hierdurch keine Störungen resultieren.

Schließlich wird bei Versuchen nach dem plasmometrischen Verfahren meist noch die folgende Annahme erforderlich:

(*mm*) Das Plasma zeigt keine wesentlichen Änderungen seines Quellungszustandes, und sein Volumen kann als Anteil am Gesamtvolumen des Protoplasten vernachlässigt werden (Wegfall einer Protoplasmakorrektur; vgl. HÖFLER 1918c, p. 113; 1920, p. 292; HÖFLER und WEBER 1926, p. 652).

D. Weitere Voraussetzungen zur Auswertung der Versuchsresultate.

(*uu*) Die Zellmembran übt keinen Wanddruck aus. Die Messungen erfolgen in plasmolysiertem Zustand, d.h. nur während der Kontraktions- und Rückdehnungsphase (vgl. RESÜHR 1935, p. 339).

(*vv*) Die Protoplastenoberfläche (bei nackten Zellen: Zelloberfläche) wird während des Versuches als konstant betrachtet (vgl. *ibid.*; 1936, p. 437; JACOBS und STEWART 1932, p. 78), ebenso auch das Protoplasten (bzw. Zell-)volumen (*ibid.*; BOCHSLER 1948, p. 80).

(*ww*) Die Partialkonzentration der diosmierenden Substanz in der Vacuole nimmt linear mit der Zeit zu (HÖFLER 1934b, p. 226).

(*xx*) Die aus dem Gestaltwechsel des sich rückdehnenden Protoplasten zylindrischer Zellen nach dem Berühren der Zellquerwand resultierenden Oberflächen- und Volumänderungen (Oberfläche bzw. Volumen einer Kugelschicht statt jener einer Halbkugel an den Menisken) sind vernachlässigbar.

III. Die experimentellen Methoden zur Feststellung von Stoffwanderung und Wasserverschiebung.

Die Versuchsverfahren sind für die Berechnung der Stofftransportwerte und Permeabilitätskonstanten von grundlegender Wichtigkeit, da die Meßresultate als Ausgangswerte der mathematischen Ableitungen dienen. Die Art der zu wählenden experimentellen Anordnung hängt von der Qualität der zu prüfenden Substanz (Kolloide, Anelektrolyte, Ionen, Farbstoffe), vom Versuchsobjekt und von der Problemstellung ab (ob z. B. die Stoffwanderung oder Wasserverschiebung bei Einzelzellen interessiert oder ob Durchschnittswerte von einem größeren Zellkomplex genügen, ob diese zwischen Objekt und Außenlösung oder durch als Membran dienende Zellschichten untersucht werden soll).

Je nach dem Nachweisprinzip können die Versuchsmethoden zunächst in physikalische und chemische eingeteilt werden (vgl. Wieringa 1930, p. 527). In der ersten Gruppe sind neben den physikalischen Eigenschaften der gelösten Stoffe auch jene Unterschiede von Bedeutung, die eine Lösung gegenüber dem reinen Lösungsmittel auszeichnen (z. B. Gefrierpunkterniedrigung, osmotische Erscheinungen). Um deutlich erkennbare Veränderungen hervorzurufen, muß allerdings eine genügend große Teilchenzahl vorhanden sein, was bei ionaler oder molekulardisperser Verteilung im Lösungsmittel (*Kristalloide*, vgl. Graham 1861, p. 2) bei den gebräuchlichen Konzentrationen meist erreicht wird.

Da Verfahren, welche *osmotische Effekte als Volumänderungen des Objektes direkt auswerten*, eine besondere Bedeutung erlangten, werden diese meist als *osmotische Methoden* zusammengefaßt und separat behandelt, so daß schließlich drei Gruppen experimenteller Nachweisverfahren resultieren: *1. Osmotische Methoden. 2. Übrige physikalische Methoden. 3. Chemische Methoden.*

Eine ausführliche Diskussion der einzelnen experimentellen Verfahren erfolgt an anderer Stelle (Abschn. IV C; vgl. Stadelmann 1956), und es wird daher hier nur insoweit auf die osmotischen Methoden eingegangen, als dies für einige der mathematischen Ableitungen erforderlich ist.

Bei den *osmotischen Methoden* wird eine *Wasser*verschiebung als Volumänderung am Objekt direkt festgestellt, während auf eine Endosmose oder Exosmose *gelöster Substanzen* nur indirekt mittels des damit verbundenen Wassertransportes geschlossen werden kann. Zur Untersuchung der Wasserpermeabilität können für die Objekte als *Außenmilieu* Lösungen nicht-endosmierender Substanzen geeigneter Konzentration und zur Bestimmung des Wassereintrittes in die Zellen auch reines Wasser dienen, während zur Erfassung der Permeabilität für gelöste Stoffe Milieus, welche diese allein enthalten *(Totalverfahren)* oder Mischlösungen mit einer nicht-endosmierenden Komponente (*Partialverfahren, Methode der Partialdrucke*, Overton 1895, p. 172; Hofmeister 1935, p. 6) gewählt werden und zur Exosmose von zuvor in die Vacuole permeierten Substanzen auch noch reines Wasser als Außenflüssigkeit geeignet ist.

Bei nackten Zellen betreffen die Volumänderungen stets die ganze Zelle, während die mit Membran und Zentralvacuole ausgestatteten Zellen ihr Volumen nur in hypotonischen Außenmedien variieren und in hypertonischen Konzentrationen Plasmolyse erfolgt, bei welcher dann nur noch der Protoplast Änderungen seiner Größe erfährt.

Natürlich durchläuft die Zelle auch in hypertonischen Lösungen Stadien, bei welchen der Protoplast der Membran noch anliegt und die Zelle als Ganzes ihr Volumen variiert. Eine genaue Analyse dieser Verhältnisse für den häufigen Fall, daß die Außenlösung allein die endosmierende Substanz enthält, läßt vier Phasen unterscheiden (Stadelmann 1951, p. 770):

a) Die *Anfangsphase*, welche vom Beginn der Einwirkung der hypertonischen Lösung bis zum Zustand der Grenzplasmolyse dauert.

b) Die *Kontraktionsphase*, welche sich vom Stadium der Grenzplasmolyse bis zum Zustand maximaler Kontraktion erstreckt, also der Plasmolyse entspricht.

c) Die *Rückdehnungsphase*, die mit der maximalen Kontraktion beginnt und mit dem Wiedererreichen der Grenzplasmolyse endet und demnach die Deplasmolyse erfaßt.

d) Die *Endphase*, welche von diesem Grenzplasmolysestadium bis zum Wiedererlangen der vollen Turgescenz reicht.

Andererseits wurde auch die Aufeinanderfolge verschiedenartiger Milieus, z. B. der alternierende Wechsel von solchen, welche kontrahierend wirken und anderen, die eine Wiederausdehnung des Protoplasten hervorrufen, für die Permeabilitätsbestimmung herangezogen (vgl. COLLANDER 1949, p. 306; STADELMANN 1952, p. 374).

Je nachdem, ob zur Erfassung der Volumänderungen Stadien, bei welchen der Protoplast der Wand anliegt oder von dieser abgehoben ist, herangezogen werden, unterscheidet man zwischen Turgescenzmethoden und plasmolytischen Methoden (vgl. ULLRICH 1948, p. 113).

Bei den *Turgescenzmethoden* handelt es sich um die unmittelbare Beobachtung der *Volumänderung* an einzelnen Zellen (vgl. RUHLAND und HOFFMANN 1926, p. 8) bzw. Zellkomplexen oder um die Auswertung von *Form-* und *Längenänderungen* (vgl. LUNDEGÅRDH 1911, p. 24f.), z. B. bei der Krümmung von Gewebestreifen (vgl. BROOKS 1916a, p. 563f.; b, p. 569f.; BOUILLENNE 1930, p. 1019f.), welche sich aus solchen Volumwechsel der einzelnen Zellen des Objektes summieren. Die experimentellen Resultate der Turgescenzmethoden sind nur in beschränktem Ausmaß zum Vergleich der Permeabilität verschiedener Objekte geeignet, worauf hier nicht näher eingegangen wird.

Die *plasmolytischen Methoden*, mittels welcher schon sehr bald die Endosmose gelöster Substanzen in die Vacuole festgestellt wurde (vgl. DE VRIES 1885, p. 581/429; 1888, p. 232/487; JANSE 1887; KLEBS 1887, p. 187), bieten besonders günstige Möglichkeiten zu einer genauen Auswertung von Volumänderungen des Protoplasten.

Unabhängig von der Form der Zellen des betreffenden Objekts läßt sich zunächst der Zeitpunkt feststellen, an welchem das Protoplastenvolumen das Zelllumen gerade erfüllt (Grenzplasmolyse). Dies kann zur Bestimmung der *Deplasmolysezeit*, als welche man die Zeitspanne von der maximalen Kontraktion bis zur vollständigen Deplasmolyse bezeichnet, dienen (vgl. Abschnitt III D, S. 84). Die Deplasmolysezeit wird mitunter auch schon vom Moment des Einlegens des Objekts in eine permeierende Lösung an gerechnet, wobei man die Ausfüllung letzter kleiner Volumanteile der Zelle oft nicht abwartet (vgl. OVERTON 1895, p. 182f.; HOFMEISTER 1948, p. 88, 86). Bei Versuchen nach der Partialmethode entspricht der Deplasmolysezeit sinngemäß die Zeitspanne bis zum Erreichen desjenigen Protoplastenvolumens, welches der Konzentration der nichtpermeierenden Komponente allein zugehört (vgl. p. 92).

Die Feststellung des Grenzplasmolysezustandes ist auch zur Beobachtung eines *Verzuges des Plasmolyseeintrittes* erforderlich, der in permeierenden Lösungen gegenüber solchen von nicht-permeierenden Stoffen zu beobachten ist. Er läßt sich insofern zur Bestimmung der Permeabilität auswerten, als die Konzentration bestimmt wird, die in beiden Milieus die Grenzplasmolyse hervorzurufen vermag (vgl. OVERTON 1895, p. 170).

In größeren Verbänden möglichst gleichartiger Zellen kann auch die zeitliche Änderung der *Plasmolysehäufigkeit* (prozentualer Anteil der plasmolysierten Zellen im Objekt) zur Permeabilitätsbestimmung dienen (*Grenzplasmolytische Methode*, FITTING 1915, 1919).

Bei weitergehender Abhebung des Protoplasten läßt sich die Stärke seiner Kontraktion *Plasmolysegrad*, vgl. Abschnitt III D, S. 75) entweder — bei unregelmäßig geformten Zellen — schätzen (vgl. DE VRIES 1885, p. 551/401) oder — bei geometrisch einfachen Zellformen (zylindrische oder zylindrisch-prismatische Gestalt) — aus dem Protoplastenvolumen genau berechnen (plasmolytisch-volumetrische = *plasmometrische Methode*, HÖFLER 1918b, p. 441). Der plasmolysierte Protoplast nimmt in letzterem Falle schließlich die Form eines Zylinders mit zwei ihm an den Stirnflächen aufsitzenden, mehr oder weniger halbkugeligen Menisken an (Form des *endplasmolysierten Protoplasten*), dessen Länge L gemessen werden kann.

Formeln für Protoplastenoberfläche und -volumen von zylindrischen und kugeligen Zellen.

a) Zylindrische Zellen. Bei halbkugeligen Menisken ergibt sich das *Volumen des endplasmolysierten Protoplasten* durch Addition der Rauminhalte des Zylinders und der beiden Halbkugeln (LEPESCHKIN 1908, p. 209; HÖFLER 1918c, p. 102; vgl. Abb. 1):

$$V_p = V_{\text{zyl}} + V_{\text{kugel}}$$

$$V_p = \left(\frac{b}{2}\right)^2 \cdot \pi \cdot (L - b) + \frac{4}{3} \cdot \left(\frac{b}{2}\right)^3 \cdot \pi$$

$$V_p = \frac{\pi b^2}{4} \cdot \left(L - \frac{b}{3}\right). \qquad \text{(III, 1)}$$

Bei kugelsegmentförmigen Menisken (vgl. Abb. 2) kann das Protoplastenvolumen nach der Formel

$$V_p = \frac{\pi b^2}{4} \cdot (L - 2 \cdot \lambda \cdot m) \qquad \text{(III, 2)}$$

bestimmt werden, wobei m die Höhe des Meniscus und λ den sog. Meniscusfaktor, welcher vom Verhältnis der Zellbreite b zur Meniscushöhe m abhängt und aus Tabelle 1 zu entnehmen ist, bedeutet (HÖFLER 1918c, p. 104f.).

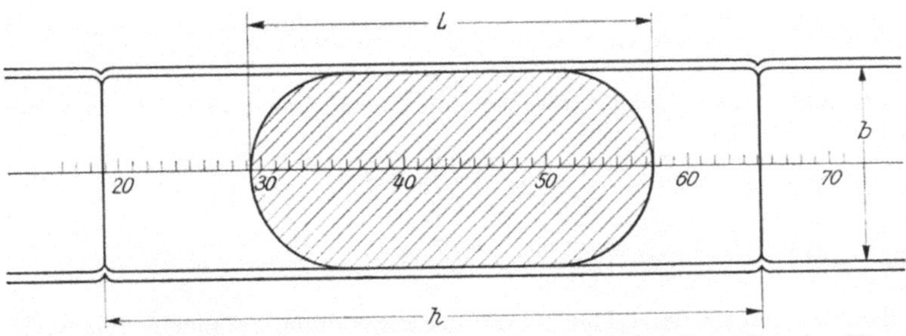

Abb. 1. Schema einer plasmolysierten zylindrischen Zelle mit halbkugeligen Protoplastenmenisken (Zustand der Endplasmolyse). L Länge des Protoplasten von Kuppe zu Kuppe; b innere Zellbreite; h innere Zellänge.

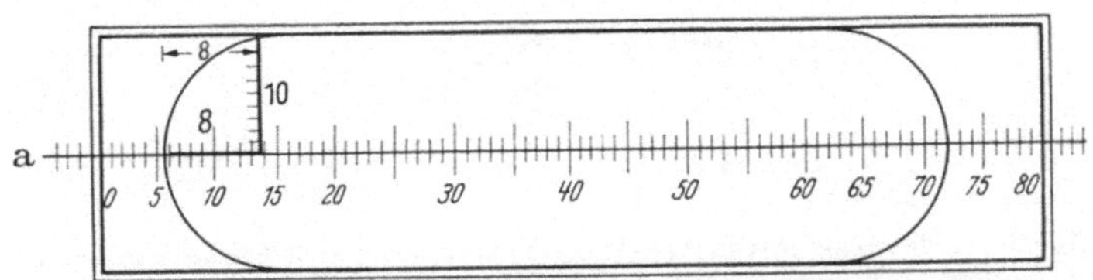

Abb. 2. Schema einer plasmolysierten zylindrischen Zelle mit kugelsegmentförmigen Protoplastenmenisken (Zustand der Endplasmolyse). m (Höhe des Meniscus) = 8′; b (innere Zellbreite) = 20′; L (Länge des Protoplasten von Kuppe zu Kuppe) = 72,0—5,5 = 66,5′; h = 80′. (Nach HÖFLER 1918c, p. 112.)

Tabelle 1. *Ermittlung des Meniscusfaktors λ aus dem Verhältnis von Meniscushöhe m zur inneren Zellbreite b.* (Nach HÖFLER 1918c, p. 105.)

$\frac{m}{b}$	λ	$\frac{m}{b}$	λ	$\frac{m}{b}$	λ	$\frac{m}{b}$	λ
0,00	0,5	0,15	0,485	0,30	0,44	0,45	0,365
0,05	0,4983	0,20	0,473	0,35	0,4183	0,50	0,3
0,10	0,493	0,25	0,4583	0,40	0,393		

Hebt sich der plasmolysierte Protoplast rotationssymmetrisch durch eine leichte Einschnürung in der Mitte etwas von der Zellängswand ab (vgl. Abb. 3), so läßt sich das fehlende

Volumen angenähert als Rotationsdreieck mit der Grundlinie $(L-b)$ und der Höhe s auffassen, welches sich durch Subtraktion des Volumens zweier Kegelstümpfe

$$2\,V = \frac{\pi}{3}\cdot(L-b)\cdot\left[\frac{b^2}{4}+\frac{b}{2}\cdot\left(\frac{b}{2}-s\right)+\left(\frac{b}{2}-s\right)^2\right]$$

vom Rauminhalt des Zylinders

$$V = \frac{\pi b^2}{4}\cdot(L-b)$$

zu

$$V_\nabla = \pi\cdot s\cdot(L-b)\left(\frac{b}{2}-\frac{s}{3}\right) \tag{III, 3}$$

ergibt, und um welchen Wert sich dann das Protoplastenvolumen (nach III, 1 oder III, 2 berechnet) verringert (BOCHSLER 1948, p. 95).

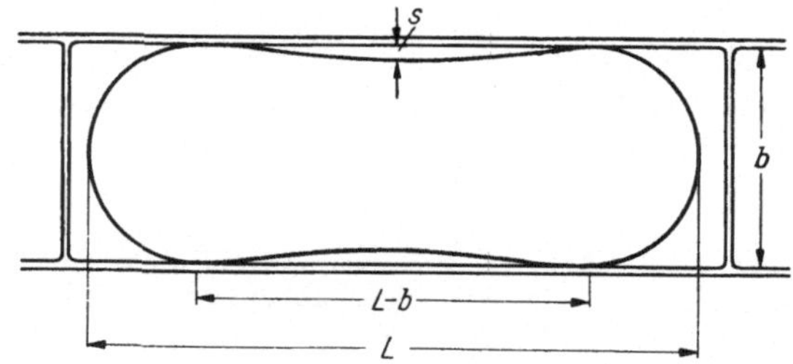

Abb. 3. Schema eines plasmolysierten Protoplasten mit leichter mittlerer Einschnürung. s Maximale Höhe der Abhebung. (Nach BOCHSLER 1948 p. 95.)

Für den unplasmolysierten Protoplasten einer zylindrischen Zelle ist das Volumen (vgl. HÖFLER 1918c, p. 102):

$$V_Z = \left(\frac{b}{2}\right)^2\cdot\pi\cdot h$$

$$V_Z = \frac{\pi b^2}{4}\cdot h\,. \tag{III, 4}$$

Soll das Volumen V_Z der zylindrischen Zelle nicht durch die innere Zellänge h, sondern durch eine fiktive Protoplastenlänge L^* für die Protoplastenform bei Endplasmolyse angegeben werden, so folgt für L^* nach (III, 4) und (III, 1):

$$V_Z = \frac{\pi b^2}{4}\cdot h = \frac{\pi b^2}{4}\cdot\left(L^*-\frac{b}{3}\right)$$

$$L^* = h+\frac{b}{3}\,. \tag{III, 5}$$

Die Oberfläche A_p des plasmolysierten Protoplasten setzt sich bei halbkugeligen Menisken aus jener der beiden Halbkugeln und des Zylindermantels zusammen (vgl. LEPESCHKIN 1908, p. 209; SCHMIDT 1936, p. 485):

$$A_p = A_{\text{Kugel}} + A_{\text{Zyl.mantel}}$$

$$A_p = 4\cdot\pi\left(\frac{b}{2}\right)^2 + 2\cdot\left(\frac{b}{2}\right)\cdot\pi\cdot(L-b)$$

$$A_p = \pi\cdot b\cdot L. \tag{III, 6}$$

Für den unplasmolysierten Protoplasten einer zylindrischen Zelle ergibt sich die Oberfläche zu (*ibid.*):

$$A_Z = 2\cdot\left(\frac{b}{2}\right)\cdot\pi\cdot h + 2\cdot\left(\frac{b}{2}\right)^2\cdot\pi$$

$$A_Z = \pi\cdot b\cdot\left(h+\frac{b}{2}\right). \tag{III, 7}$$

Als *Plasmolysegrad* G wird der Quotient $\frac{\text{Protoplastenvolumen}}{\text{Volumen der entspannten Zelle}}$ definiert (HÖFLER 1918c, p. 107). Er ist eine dimensionslose Zahl, die höchstens den Wert 1 erreichen kann und folgt aus (III, 1) und (III, 4) zu:

$$G = \frac{V_p}{V_Z} = \frac{L - \frac{b}{3}}{h}. \qquad \text{(III, 8)}$$

Die *Oberflächenentwicklung* Q $\left(\text{d. h. der Quotient } \frac{\text{Oberfläche}}{\text{Volumen}}\right)$ läßt sich aus (III, 1) und (III, 6) berechnen (vgl. BOCHSLER 1948, p. 80):

$$Q = \frac{A_p}{V_p} = \frac{4}{b} \cdot \frac{L}{L - \frac{b}{3}}. \qquad \text{(III, 9)}$$

b) Kugelige Zellen. Volumen V und Oberfläche A für kugelige Protoplasten oder nackte kugelige Zellen können leicht aus deren Durchmesser d ermittelt werden:

$$V = \frac{d^3}{6} \cdot \pi. \qquad \text{(III, 10)}$$

$$A = d^2 \cdot \pi. \qquad \text{(III, 11)}$$

IV. Maße des Transportes und der Permeation gelöster Stoffe.

Das einzelne Meßergebnis, welches aus einer der oben beschriebenen Methoden folgt, wird erst dadurch für die Ermittlung der Größe des Stofftransportes wertvoll, daß es die Grundlage zu einem Vergleich dieser Charakteristica für verschiedene Substanzen und Objekte liefert. Hierzu kann der direkt gemessene Zahlenwert selbst oder ein aus ihm erst abzuleitender Faktor als geeignet befunden werden. Solche Größen, die als zahlenmäßiger Ausdruck der Stoffaufnahme oder der Durchlässigkeitseigenschaften einer bestimmten Membran (z. B. Protoplasmamantel) dienen, sind hier als Maß für den *Stofftransport* bzw. für die *Permeabilität* (letztere meist auch *Permeabilitätskonstante* oder *Permeabilitätsfaktor* genannt) bezeichnet.

Eine weitere mathematische Behandlung der Meßresultate verfolgt das Ziel, einen Zahlenwert zu finden, der allein das Ausmaß des Stofftransportes oder der Permeabilität für die betreffende Membran repräsentiert und von den durch die Größe des Objekts und die Konzentrationsverhältnisse gegebenen experimentellen Bedingungen unabhängig ist. Hierbei sind oft noch andere Größen als die unmittelbar gemessenen Variablen einzuführen und über die herrschenden Verhältnisse bestimmte Annahmen und Voraussetzungen zu postulieren, wie sie im Kapitel III zusammengefaßt wurden.

A. Maße für den Stofftransport.

Da bei den Maßen für den Stofftransport bewußt auf eine Berücksichtigung der Konzentration der transportierten Substanz verzichtet wird, geben diese in jenen Fällen, wo der Einfluß der Konzentrationsverhältnisse bekannt ist, nur einen begrenzten Aufschluß über die Faktoren der betreffenden Stoffwanderung. Andererseits ermöglichen aber diese Maße in allen jenen Fällen, bei welchen der Mechanismus für die Substanzverschiebung noch nicht geklärt (z. B. aktive Stoffaufnahme) oder umstritten ist, die zahlenmäßige Angabe seiner Auswirkungen und damit die für solche Verhältnisse momentan einzig mögliche quantitative Aussage.

Als *Stoffaustauschwert* (*Stoffaufnahme- bzw. Stoffabgabewert, Stofftransportwert, Fluß, Zufluß, Abfluß, Gesamtfluß, influx, outflux, flux*; vgl. USSING 1951, p. 470) wird die auf Zeiteinheit und Oberflächeneinheit der Membran reduzierte durch diese durchtretende Substanzmenge bezeichnet:

$$M = \frac{m_2 - m_1}{A \cdot (t_2 - t_1)}\,. \qquad \text{(IV, 1)}$$

Schreibt man für

$$m_2 - m_1 = \Delta m, \qquad \text{(IV, 2)}$$

so wird (IV, 1):

$$M = \frac{\Delta m}{A \cdot (t_2 - t_1)}\,. \qquad \text{(IV, 3)}$$

Die Größe M kann je nach dem Standpunkt des Beobachters und der Wanderungsrichtung des Stoffes positiv oder negativ sein. Die Menge Δm wird dabei in Grammen, Molen, bei Ionen auch in Gramm-Äquivalenten oder durch deren Ladungsmenge in Cb gemessen.

Die Ermittlung der durchtretenden Menge kann je nach den Versuchsgegebenheiten z. B. durch Verfolgen der zeitlichen Konzentrationsänderungen auf einer oder beiden Seiten der Membran mit Hilfe chemischer Analysen geschehen. Hierdurch wird allerdings nur der resultierende Gesamtfluß (net-flux), der sich aus dem tatsächlichen Zufluß (influx, M_I) nach Abzug des Abflusses (out-flux, M_A) ergibt, erfaßt:

$$M = M_I - M_A\,. \qquad \text{(IV, 4)}$$

Um alle drei Größen zu ermitteln, kann man entweder zusätzlich zur chemischen Bestimmung des Gesamtflusses den Abfluß erfassen (z. B. wenn in der inneren Lösung ein radioaktives Isotop, dessen Auftreten im Außenmilieu zu messen ist, in der zu prüfenden Substanz eingebaut wurde; vgl. USSING 1951, p. 471), oder unter Verwendung zweier verschiedener Isotopen desselben Elementes Zufluß und Abfluß getrennt feststellen (*Doppel-tracer-Methode*, vgl. LEVI und USSING 1949, p. 929).

Ist der Transportmechanismus als Diffusion erkannt, so läßt sich für zylindrische Zellen im Zustand der Endplasmolyse der Stoffaustauschwert für eine gelöste Substanz mit Gleichung (III, 6) und (IV, 13) leicht berechnen, wenn die Messung während der Rückdehnungsphase (S. 148) und in der Außenkonzentration C erfolgt, für die Protoplastenoberfläche Mittelwerte und für die Menge Δm ihre Größe in Mol eingeführt werden:

$$\left.\begin{aligned} M_m = \frac{\Delta m_s}{A_m \cdot (t_2 - t_1)} = \frac{\frac{\pi b^2}{4} \cdot C \cdot (L_2 - L_1)}{\pi b \cdot \frac{L_1 + L_2}{2} (t_2 - t_1)} = \frac{b \cdot C}{2\,(L_1 + L_2)} \cdot \frac{L_2 - L_1}{t_2 - t_1} \\ [\mathrm{Dim}\, M_m = \mathsf{L}^{-2}\,\mathsf{T}^{-1}]. \end{aligned}\right\} \qquad \text{(IV, 5)}$$

Hierher ist auch ein Verfahren zu rechnen, mit dessen Hilfe die durch eine *Membran unbekannter Flächengröße* (Blutcapillarenwandung beim Meerschweinchen; MERELL und Mitarbeiter 1944) je Zeiteinheit durchtretende relative Substanzmenge bestimmt wird. Man benützt hierzu ein radioaktives Isotop (Na*), welches auf der einen Membranseite einzuführen ist (Injektion in den Blutkreislauf); Nach der sehr rasch erfolgenden Durchmischung im Blut wird unter Voraussetzung (l) angenommen, daß die zu beobachtende Abnahme des Na*-Gehaltes des Blutplasmas je Zeiteinheit $\left(\frac{dm}{dt}\right)$ die Differenz darstellt aus der tatsächlich je Zeiteinheit austretenden Na*-Menge und jener, welche von dem zuvor bereits in die Körperflüssigkeit ausgetretenen Na*-Anteil von dort wieder in das Blutplasma zurückwandert:

$$\frac{dm}{dt} = -\frac{dm_A}{dt} + \frac{dm_I}{dt}. \qquad \text{(IV, 6)}$$

Die Größe $\frac{dm_A}{dt}$ ermittelt sich als die je Zeiteinheit aus den Capillaren austretende Menge des Na (m_A') multipliziert mit dem Anteil des Na* an der gesamten Natriummenge des Blutplasmas $\left(\frac{m}{N_{pl}}\right)$. Analog folgt der Rückfluß $\frac{dm_I}{dt}$ von Na* als das Produkt aus der je Zeiteinheit aus der Körperflüssigkeit in das Blutplasma austretenden Na-Menge (welche ja, da im stationären Gleichgewicht untersucht wird, gleich der Größe m_A' ist) und dem Anteil von Na* in der Na-Menge der Körperflüssigkeit. Letzterer ergibt sich als der Quotient aus der in der Körperflüssigkeit vorhandenen Na*-Menge [das ist die Differenz aus der beim Versuchsbeginn in das Blutplasma injizierten Na*-Menge (m_B) und der dort im betreffenden Zeitpunkt befindlichen Quantität (m)] und ihrer Gesamt-Na-Menge. Damit wird die obige Gleichung:

$$\frac{dm}{dt} = -m_A' \cdot \frac{m}{N_{pl}} + m_A' \cdot \frac{m_B - m}{N_k} \qquad \text{(IV, 7)}$$

$$\left[(1)\ \frac{dN_p}{dt} = -r \cdot \frac{N_p}{Na_p} + r \cdot \frac{N_0 - N_p}{Na_E}\right].$$

Dividiert man beiderseits durch das Volumen des Plasmas V_{pl} und setzt $\frac{m}{V_{pl}} = c$ und $\frac{m_B}{V_{pl}} = c_B$ (Anfangskonzentration von Na* im Plasma), so folgt nach Integration der Gleichung, wobei zur Bestimmung der Integrationskonstanten für $t = 0$ $c = c_B$ eingeführt wurde:

$$\ln\left(c - c_B \cdot \frac{N_{pl}}{N_{pl} + N_k}\right) - \ln\left(c_B - c_B \cdot \frac{N_{pl}}{N_{pl} + N_k}\right) = -\frac{m_A'}{N_{pl}} \cdot \frac{N_{pl} + N_k}{N_k} \cdot t.$$

Das Produkt $c_B \cdot \frac{N_{pl}}{N_{pl} + N_k}$ stellt die Konzentration c_E von Na* im Endgleichgewichtszustand dar und kann als solche direkt gemessen werden. Aus ihr und c_B errechnet sich der Quotient $\frac{N_{pl}}{N_{pl} + N_k}$, mit welchem für

$$\frac{N_k}{N_{pl} + N_k} = 1 - \frac{N_{pl}}{N_{pl} + N_k}$$

schließlich

$$\frac{N_k}{N_{pl} + N_k} = 1 - \frac{c_E}{c_B} \qquad \text{(IV, 8)}$$

wird, womit sich der Quotient $\frac{m_A'}{N_{pl}}$, welcher den Bruchteil der Gesamt-Na-Menge im Blutplasma darstellt, der je Zeiteinheit in die Körperflüssigkeit übertritt, ergibt:

$$\frac{m_A'}{N_{pl}} = \frac{1 - \frac{c_E}{c_B}}{t} \cdot \ln \frac{c_B - c_E}{c - c_E} \qquad \left[\text{Dim}\left(\frac{m_A'}{N_{pl}}\right) = \mathrm{T}^{-1}\right]. \qquad \textbf{(IV, 9)}$$

$$[(2)\ q\,[\ln(c_p - c_{eq}) - \ln(c_0 - c_{eq})] = -R\,t].$$

B. Maße für die Permeabilität.

Die Vergleichsgrößen für die Durchlässigkeitseigenschaften der Protoplastenmembranen sind je nach den experimentellen Voraussetzungen und den theoretischen Annahmen oft sehr verschiedener Natur und vielfach nur durch komplizierte

Berechnungsverfahren zu ermitteln. Diese wurden bereits in großer Zahl ausgearbeitet, so daß es hier nur möglich ist, einige Beispiele anzuführen[1]. Die Maße lassen sich, je nachdem, ob zu ihrer Ableitung ein Ausgleichsgesetz angenommen wird oder nicht, in 2 Gruppen zusammenfassen. In ersterem Falle ist zur Berechnung meist ein hoher mathematischer Aufwand erforderlich.

a) Permeabilitätsmaße ohne Annahme eines Ausgleichsgesetzes.

Hierher gehören zunächst die ältesten zahlenmäßigen Angaben der Permeabilität, wobei, mangels geeigneter Formulierungen, die Meßresultate selbst oft unmittelbar zum Vergleich dienten. Hierher zählen vielfach auch jene Maße, die einer weiteren mathematischen Behandlung unzugänglich sind.

Als erste Meßwerte wurden bei den plasmolytischen Methoden die zeitliche *Änderung des Plasmolysegrades* beobachtet (DE VRIES 1885, p. 556/405f.), die *Deplasmolysezeit* festgestellt oder die Konzentrationen, bei welchen in den verschiedenen Lösungen permeierender Substanzen gerade noch Grenzplasmolyse erfolgt (*Plasmolyse-Eintritt*, OVERTON 1895, p. 182, 170), verglichen.

Die Methode der isotonischen Koeffizienten. Ohne den dynamischen Charakter der Stoffpermeation genügend zu berücksichtigen, versuchten LEPESCHKIN (1908, p. 207) und TRÖNDLE (1910, p. 181f.) die Permeabilität quantitativ festzustellen. Es wurde hierbei im Experiment diejenige Konzentration des Diosmoticums gesucht, die am Ende einer gewissen Versuchsdauer mit einer bestimmten Konzentration eines nichtpermeierenden Stoffes (Zucker) isotonisch ist. Anschließend wird die dieser Zuckerlösung isotonische KNO_3-Lösung aus physikalisch-chemischen Tabellen ermittelt und dort auch die dieser KNO_3-Konzentration isotonische Konzentration des Diosmoticums nachgeschlagen. Der Quotient der letzten beiden Größen gibt den theoretisch geforderten Wert des isotonischen Koeffizienten an, der mit seiner experimentell (aus der im Versuch isotonisch gefundenen Diosmoticumkonzentration und der obigen tabellarisch ermittelten KNO_3-Konzentration) sich ergebenden Größe verglichen wird. Der Unterschied dieser beiden isotonischen Koeffizienten führt schließlich zu einem Ausdruck, der als Maß für die Permeabilität dient.

Die grenzplasmolytische Methode (FITTING 1915, p. 19f.). Bei Verwendung von Teilen eines möglichst gleichartigen Materials, bei welchen der osmotische Grundwert der einzelnen Zellen stets um dasselbe Mittel in gleicher Weise statistisch streut, entsprechen den Unterschieden der *Plasmolysehäufigkeit* in Lösungsreihen fein abgestufter Konzentrationen nichtendosmierender Substanzen bestimmte Konzentrationsdifferenzen der Außenlösung. Eine solche Zuordnung ist auch noch kurz nach dem Einlegen in schwach endosmierende Lösungen möglich, in welchen dann die Plasmolyse durch die Permeation langsam zurückgeht. Die sich damit ändernden Plasmolysehäufigkeiten werden nun in bestimmten Zeitintervallen erfaßt, und mit Hilfe der anfangs gewonnenen Relation zwischen Unterschied der Plasmolysehäufigkeit und Konzentrationsdifferenz wird auf die durch die Permeation in der Vacuole hervorgerufene Konzentrationszunahme geschlossen, welche sich in $mol/cm^3 \cdot sec$ angeben läßt und als Maß für die Permeabilität dienen kann.

Mittels der Plasmolysehäufigkeit lassen sich ferner die *temporären plasmolytischen Koeffizienten* bestimmen, indem beobachtet wird, zu welchen Zeiten $t_1, t_2, t_3, \ldots$ in den verschiedenen Konzentrationen $C_1, C_2, C_3, \ldots$ der endosmierenden Substanz die Plasmolysehäufigkeit genau 50% beträgt. Als temporärer plasmolytischer Koeffizient π wird nun der Quotient aus dem ursprüng-

[1] Wegen weiterer Maße vgl. SZÜCS (1910), HEUSSER (1917), POIJÄRVI (1928), JACOBS (1933a, 1933b, 1934), HOLM-JENSEN und Mitarbeiter (1944), ÄYRÄPÄÄ (1950).

lichen osmotischen Wert O_g (meist bestimmt mit Hilfe von Konzentrationsreihen von Saccharose als jene Konzentration, in welcher die Plasmolysehäufigkeit 50% auftritt) und den Konzentrationen $C_1, C_2, C_3, \ldots$ bezeichnet $\left(\pi_{t_1} = \frac{O_g}{C_1}; \pi_{t_2} = \frac{O_g}{C_2}; \pi_{t_3} = \frac{O_g}{C_3}; \ldots\right)$. Er läßt sich in Abhängigkeit von der ihm zugehörigen Zeit t graphisch darstellen, und ein Vergleich solcher *Plasmolysekurven* für die verschiedenen Substanzen gibt Hinweise auf deren Permeiervermögen (BÄRLUND 1929).

Bei Versuchen nach der *plasmometrischen Methode* (S. 148) kann schon die *Plasmolysegradänderung, bezogen auf die Zeiteinheit* (ΔG), als Vergleichsmaß für die Permeabilität dienen (vgl. HÖFLER 1934b, p. 225; HOFMEISTER 1948, p. 87). Mit Gleichung (III, 7) ergibt sich schließlich für ΔG:

$$\Delta G = \frac{G_2 - G_1}{t_2 - t_1} = \frac{L_2 - L_1}{h\,(t_2 - t_1)}; \quad [\mathrm{Dim}\,\Delta G = \mathsf{T}^{-1}]. \tag{IV, 11}$$

Ferner wird noch das Produkt aus Außenkonzentration und Differenz des Plasmolysegrades (als Ausdruck für die aufgenommene Stoffmenge), reduziert auf die Zeiteinheit (auf die Stunde bezogen als „Stundenwert der Stoffaufnahme" oder die je Stunde aufgenommene Lösungsmenge bezeichnet), als Permeabilitätsmaß (M^*) benutzt (HÖFLER 1918a, p. 421; 1934b, p. 226):

$$\left.\begin{aligned} M^* = \Delta G \cdot C = \frac{(G_2 - G_1)\cdot C}{t_2 - t_1} = \frac{(V_2 - V_1)\cdot C}{V_Z \cdot (t_2 - t_1)} = \frac{(L_2 - L_1)\cdot C}{h \cdot (t_2 - t_1)}; \\ [\mathrm{Dim}\, M^* = \mathsf{L}^{-3}\mathsf{T}^{-1}]. \end{aligned}\right\} \tag{IV, 12}$$

Führt man in diese Gleichung noch die Beziehung ein, daß die Menge der permeierten Substanz im osmotischen Gleichgewicht der Rückdehnungsphase (S. 148) gleich dem Produkt aus der Differenz der Protoplastenvolumina $V_2 - V_1$ und der Außenkonzentration (vgl. STADELMANN 1951, p. 774), also

$$\Delta m_s = (V_2 - V_1) \cdot C \tag{IV, 13}$$

ist, so wird

$$M^* = \frac{\Delta m_s}{V_Z \cdot (t_2 - t_1)}, \tag{IV, 14}$$

womit sich M^* demnach auch als die auf die Einheit des Volumens der entspannten Zelle (1 ccm) und der Zeit (1 sec) reduzierte permeierende Substanzmenge (in mol) auffassen läßt.

b) Permeabilitätsmaße mit Annahme eines Ausgleichsgesetzes.

Die hierher gehörenden Maße gehen auf bestimmte Anfangsgleichungen zurück, welche die mathematische Formulierung des jeweils postulierten Ausgleichsgesetzes darstellen. Dieses wird aber von den Autoren nicht immer ausdrücklich angeführt, sondern ist bisweilen nur aus dem Berechnungsgang erkennbar. Die Permeabilitätsmaße sollen hier je nach der Anfangsgleichung, aus welcher sie sich ergeben, zusammengefaßt werden.

1. Permeabilitätsmaße, welche sich aus der Gleichung $\frac{dm_s}{dt} = K_s \cdot A(C - k)$ ableiten.

Mißt man in (II, 6) bei den Konzentrationsangaben C und k (als Menge je Volumeneinheit) die Mengen mit den gleichen Einheiten wie dm_s, so ist:

$$\frac{dm_s}{dt} = {}_1K_s \cdot A \cdot (C - k) \quad [\mathrm{Dim}\, {}_1K_s = \mathsf{L}\mathsf{T}^{-1}], \tag{IV, 21}$$

wobei als einfachere Gleichung für näherungsweise Berechnung die Differenzenform dienen kann, so daß die Integration umgangen wird:

$$\frac{m_{s\,2} - m_{s\,1}}{t_2 - t_1} = {}_1K_s \cdot A \cdot (C-k). \tag{IV, 22}$$

Da hieraus folgt:

$${}_1K_s = \frac{d\,m_s}{d\,t} \cdot \frac{1}{A \cdot (C-k)}$$

bzw. mit Gleichung (IV, 2):

$${}_1K_s = \frac{\Delta\,m_s}{t_2 - t_1} \cdot \frac{1}{A\,(C-k)}$$

so gibt, wählt man als Maß für die Menge das Mol, ${}_1K_s$ als Permeabilitätsmaß an, wie groß die Menge (in mol) der je Flächeneinheit (1 cm²) und Zeiteinheit (1 sec) durch die betreffende Membran diosmierenden Substanz ist, wenn das Konzentrationsgefälle zwischen Außenlösung und Vacuole für den permeierenden Stoff die Größe der Konzentrationseinheit (1 mol/ccm) besitzt.

Je nachdem, ob bei den einzelnen Objekten das Volumen der Vacuole (bzw. der Zelle) und die Fläche der permeationshemmenden Membran (bei ausgewachsenen Pflanzenzellen der Protoplasmamantel) während des Versuches seine Größe nicht ändert oder zumindest als konstant bleibend mit einem Mittelwert in die Formel eingeführt wird [Voraussetzung (*vv*)], ergeben sich verschiedene Wege zur Lösung der Ausgangsgleichung. Der Berechnungsgang hängt aber auch davon ab, welche Art von Meßresultaten zum Einsetzen in die Endformel vorliegen, was wieder durch die gewählte Versuchsmethode und das Objekt gegeben ist. Oft dienen hier osmotische Methoden (S. 147), bei welchen aus dem unmittelbar meßbaren Wassertransport auf die permeierte Menge der gelösten Substanz zu schließen ist, als experimentelle Grundlage, so daß die Beziehung zwischen diesen beiden Größen bekannt sein muß. Sieht man von Komplikationen durch eventuelle nicht-osmotische Vorgänge ab, so bleiben noch allfällig vorhandene andere osmotische Kräfte für die Wasserverschiebung zu berücksichtigen, sofern man nicht im Zustand des osmotischen Gleichgewichtes zwischen Außenlösung und Vacuole (Rückdehnungsphase, S. 148) mißt.

α) Berechnungen für konstant bleibende oder konstant angenommene Oberfläche A und/oder Volumen V des Protoplasten bzw. der Zelle.

Die ersten drei der hierher zu rechnenden Ableitungen gelten Versuchen nach der *chemischen Methode*. Die Änderungen von A und V bleiben dabei, weil mit hypotonischen Lösungen gearbeitet wird, vernachlässigbar.

a) Bei *Riesenzellen* von *Chara* u. ä. läßt sich das Volumen einfach bestimmen und auch die Konzentration einer endosmierenden Substanz im Zellsaft kann mikrochemisch erfaßt werden. COLLANDER und BÄRLUND (1933, p. 25f.) wenden darauf die analog Gleichung (IV, 21) lautende Formel an:

$$\frac{d\,m_s}{d\,t} = {}_{1a}K_s \cdot (C^* - k) \cdot A\,, \tag{IV, 23}$$

worin C^* die Konzentration der permeierenden Substanz im Zellsaft, welche im Gleichgewicht mit der Außenkonzentration C steht, bedeutet. Unter Voraussetzung (*dd*) geht (IV, 23) dann in Gleichung (IV, 21) über und führt man

$$m_s = k \cdot V \tag{IV, 24}$$

ein, so gilt bei konstantem Zellvolumen V:

$$d\,m_s = d\,k \cdot V, \tag{IV, 25}$$

womit folgt:

$$\frac{dk}{C-k} = {}_{1a}K_s \cdot \frac{A}{V} \cdot dt,$$

was für konstantes V und A zwischen den Grenzen 1 und 2 integriert werden kann:

$$ {}_{1a}K_s = \frac{V}{A} \cdot \frac{1}{t_2 - t_1} \cdot \ln \frac{C - k_1}{C - k_2} \qquad \textbf{(IV, 26)}$$

$$\left[k' = \frac{v}{q \cdot t} \cdot \ln \frac{C}{C - c} \quad (4)\right]$$

wobei für $t_1 = 0$ $k_1 = 0$ ist.

b) Für dasselbe Zellmaterial läßt sich auch die *Exosmosepermeabilität* auf einfache Weise messen, wenn die betreffende Substanz zuvor in genügendem Maße in die Vacuole endosmiert ist. Ohne dabei die Gleichgewichtskonzentration kennen zu müssen, kann man nach WARTIOVAARA (1942, p. 28) die Permeabilitätskonstante ${}_{1b}K_s$ errechnen. Man bestimmt zunächst die während zweier gleich langer und aufeinanderfolgender Zeitintervalle Δt_1 und Δt_2 exosmierende Substanzmenge $\Delta m_{s\,2-1}$ und $\Delta m_{s\,3-2}$, wobei als Außenmilieu reines Wasser gewählt wird. Da dann die Konzentrationsdifferenz, welche die Exosmose bewirkt, gleich der Partialkonzentration k ist, so gilt zunächst nach (IV, 22) mit (IV, 2):

$$\frac{\Delta m_{s\,2-1}}{t_2 - t_1} = {}_{1b}K_s \cdot A \cdot k_1 \qquad \frac{\Delta m_{s\,3-2}}{t_3 - t_2} = {}_{1b}K_s \cdot A \cdot k_2. \qquad \text{(IV, 27)}$$

Darin bedeuten k_1 und k_2 die Partialkonzentrationen der endosmierenden Substanz im Zellsaft zu den Zeiten der Intervallmitte von Δt_1 und Δt_2.

Da die Außenkonzentration Null ist, vereinfacht sich unter Nichtbeachtung des Vorzeichens die Ausgangsgleichung für diesen Fall zu:

$$\frac{dm_s}{dt} = {}_{1b}K_s \cdot k \cdot A. \qquad \text{(IV, 28)}$$

Durch Substitution von dm_s nach (IV, 25), Integration und Einführung der aus (IV, 27) resultierenden Werte für k_1 und k_2 folgt schließlich (WARTIOVAARA 1944, p. 7):

$$ {}_{1b}K_s = \frac{V}{A} \cdot \frac{1}{\Delta t} \cdot \ln \frac{m_{s\,3-2}}{m_{s\,2-1}}. \qquad \textbf{(IV, 29)}$$

$$\left[(2)\ P = 3{\cdot}45 \frac{d}{t} \log k\right].$$

c) Für *Valonia*-Zellen läßt sich bei Versuchen in isosmotischen Seewassergemischen ebenfalls annehmen, daß sie ihr Volumen nicht ändern, womit dann die Gleichung (IV, 21), wie für (IV, 23) gezeigt, integriert werden kann und die Endformel (IV, 26), welche hier zur Ermittlung des Anionendurchtritts dient, resultiert (ULLRICH 1935, p. 156f.).

Bei den folgenden Methoden sind Volumen und Oberfläche zwar als nicht konstantbleibend erkannt, aber deren Veränderungen werden nur durch Einführen eines Mittelwertes berücksichtigt. Bei den ersten drei Verfahren, die alle vom Differenzenansatz (IV, 22) ausgehen, bleibt auch die stetige Zunahme der Partialkonzentration der endosmierenden Substanz in der Vacuole nicht erfaßt, sondern es ist hierfür ebenfalls eine Konstante substituiert. Da beim osmotischen Gleichgewicht beobachtet wird, fallen die Komplikationen durch zusätzlichen Wassertransport weg (vgl. S. 156).

d) Für *zylindrische Zellfäden* läßt sich die Permeabilität aus Versuchen nach der *Turgescenzmethode* mit (IV, 22) berechnen, wobei folgendes Verfahren eingeschlagen wird (LEPESCHKIN 1909, p. 136): Zunächst kommt der Zellfaden aus Wasser in eine hypotonische Zuckerlösung C_{Z1}, worin man seine Länge H_{Z1} zur Zeit t_{Z1} im osmotischen Gleichgewicht bestimmt. Anschließend kommt er in eine andere, aber wiederum hypotonische Zuckerkonzentration C_{Z2} und man mißt dort zur Zeit t_{Z2} die entsprechende Länge H_{Z2}. Zu einer späteren Zeit t_{Z3} erfolgt zur Bestimmung des Wachstums eine neue Messung H_{Z3}. Hernach gelangt der Faden zur Zeit t_B in die mit der zweiten Zuckerlösung beinahe isotonische Konzentration C des Diosmoticums, worin die Fadenlänge H_1 und H_2 zu den Zeiten t_1 und t_2 gemessen wird. Anschließend überträgt man den Faden erneut in eine Zuckerkonzentration C_{Z4}, deren Höhe derart gewählt ist, daß der Faden beim Übertragen seine Länge möglichst nicht ändert. Nach einiger Zeit wird darin sein Längenwachstum durch Messung der Längen H_{Z4} und H_{Z5} zu den Zeiten t_{Z4} und t_{Z5} festgestellt.

Als mittlere Längenzunahme des Fadens ($\overline{\Delta H'_G}$) in der Zeiteinheit durch das Wachstum folgt dann der Wert:

$$\frac{\dfrac{H_{Z3}-H_{Z2}}{t_{Z3}-t_{Z2}}+\dfrac{H_{Z5}-H_{Z4}}{t_{Z5}-t_{Z4}}}{2}=\overline{\Delta H'_G}.$$

Diese Größe ist zu subtrahieren von der während der Endosmose eintretenden Verlängerung des Zellfadens je Zeiteinheit, damit die aus der Endosmose allein resultierende Größe der mittleren Längenzunahme je Zeiteinheit ($\overline{\Delta H'}$) berechnet wird:

$$\frac{H_2-H_1}{t_2-t_1}-\overline{\Delta H'_G}=\overline{\Delta H'}.$$

Um hieraus die Differenz der Partialkonzentrationen k_1 und k_2 der diosmierenden Substanz in den Zellen des Fadens zu ermitteln, bedient man sich der Längenbestimmung in den beiden Zuckerkonzentrationen C_{Z1} und C_{Z2}, unter der Annahme, daß das Verhältnis zwischen der Vergrößerung der Gesamtkonzentration Z in den Zellen und der Zunahme der Fadenlänge in beiden Fällen gleich sei, was dann ebenso auch für die Längenzunahme je Zeiteinheit gilt:

$$\frac{H_{Z2}-H_{Z1}}{C_{Z2}-C_{Z1}}=\frac{\overline{\Delta H'}}{\dfrac{Z_2-Z_1}{t_2-t_1}}.$$

Vernachlässigt man die durch die Vacuolenvergrößerung bei der Zelldehnung entstehende Veränderung der Partialkonzentration der ursprünglich vorhandenen Substanz, so wird

$$Z_1=c+k_1 \quad \text{und} \quad Z_2=c+k_2,$$

so daß folgt:

$$Z_2-Z_1=k_2-k_1.$$

Für die mittlere Partialkonzentrationszunahme der diosmierenden Substanz in der Vacuole je Zeiteinheit ($\overline{\Delta k'}$) ergibt sich hieraus:

$$\overline{\Delta k'}=\frac{k_2-k_1}{t_2-t_1}=\frac{C_{Z2}-C_{Z1}}{H_{Z2}-H_{Z1}}\cdot\overline{\Delta H'}.$$

Da der Faden zur Zeit t_B in das Diosmoticum gelangte, ist $k_1=\overline{\Delta k'}\cdot(t_1-t_B)$ und $k_2=\overline{\Delta k'}\cdot(t_2-t_B)$, womit $C-\frac{k_1+k_2}{2}$, die mittlere Konzentrationsdifferenz, wird:

$$C-\overline{\Delta k'}\cdot\frac{t_1+t_2-2t_B}{2}.$$

Mit diesen Größen sowie den Werten für das mittlere Volumen V_{F_m} und die mittlere Oberfläche A_{F_m} des Zellfadens $\left(V_{F_m} = \frac{\pi b^2}{4} \cdot \frac{H_1 + H_2}{2};\ A_{F_m} = \pi \cdot b \cdot \frac{H_1 + H_2}{2}\right)$ und

$$\frac{\Delta m_s}{t_2 - t_1} = \overline{\Delta k'} \cdot V_{F_m}$$

nach Gleichung (IV, 2) und (IV, 25) in (IV, 22) eingesetzt, folgt schließlich:

$${}_{1d}K_s = \frac{\frac{b}{4} \cdot \frac{C_{Z_2} - C_{Z_1}}{H_{Z_2} - H_{Z_1}} \cdot \left[\frac{H_2 - H_1}{t_2 - t_1} - \frac{\frac{H_{Z_3} - H_{Z_2}}{t_{Z_3} - t_{Z_2}} + \frac{H_{Z_5} - H_{Z_4}}{t_{Z_5} - t_{Z_4}}}{2}\right]}{C - \frac{C_{Z_2} - C_{Z_1}}{H_{Z_2} - H_{Z_1}} \cdot \left[\frac{H_2 - H_1}{t_2 - t_1} - \frac{\frac{H_{Z_3} - H_{Z_2}}{t_{Z_3} - t_{Z_2}} + \frac{H_{Z_5} - H_{Z_4}}{t_{Z_5} - t_{Z_4}}}{2}\right] \cdot \frac{t_1 + t_2 - 2\,t_B}{2}}. \quad \text{(IV, 30)}$$

e) Für *zylindrische Einzelzellen*, die in hypertonischen Lösungen plasmolysiert werden, kann man aus Versuchen nach der *plasmometrischen Methode* von Gleichung (IV, 22) ebenfalls eine Endformel zur Bestimmung der Permeabilität ableiten (LEPESCHKIN 1908, p. 209). Das Objekt wird zunächst annähernd isotonisch vorplasmolysiert und zur Zeit t_0 in das Diosmoticum, worin zur Zeit t_1 und t_2 die Messung der Protoplastenlängen L_1 und L_2 erfolgt, übertragen. Zur Berechnung von Δm_s dient wieder (IV, 13). Für die Partialkonzentration k ist der Mittelwert $\frac{k_1 + k_2}{2}$ eingeführt, für welchen sich k_1 und k_2 nach (IV, 24) ergeben:

$$k_1 = \frac{m_1}{V_1} = \frac{V_1 - V_0}{V_1} \cdot C; \qquad k_2 = \frac{m_2}{V_2} = \frac{V_2 - V_0}{V_2} \cdot C.$$

Die Ermittlung von V_0, dem Protoplastenvolumen zur Zeit t_0, erfolgt durch lineare Extrapolation aus der Rückdehnung:

$$\frac{V_2 - V_1}{t_2 - t_1} = \frac{V_1 - V_0}{t_1 - t_0} = \frac{V_2 - V_0}{t_2 - t_0},$$

womit

$$k_1 = \frac{C}{V_1} \cdot \frac{V_2 - V_1}{t_2 - t_1} \cdot (t_1 - t_0) \quad \text{und} \quad k_2 = \frac{C}{V_2} \cdot \frac{V_2 - V_1}{t_2 - t_1} \cdot (t_2 - t_0)$$

wird.

Für die Protoplastenoberfläche A ist der Mittelwert von A_1 und A_2 einzusetzen, so daß Gleichung (IV, 22) mit diesen Größen ergibt:

$${}_{1e}K_s = \frac{(V_2 - V_1) \cdot C}{\frac{A_1 + A_2}{2} \cdot (t_2 - t_1) \left[C - \frac{C}{2} \cdot \frac{V_2 - V_1}{t_2 - t_1} \cdot \left(\frac{t_1 - t_0}{V_1} + \frac{t_2 - t_0}{V_2}\right)\right]}.$$

Für A und V nach (III, 6) und (III, 1) substituiert, erhält man als Endformel:

$${}_{1e}K_s = \frac{b}{2} \cdot \frac{L_2 - L_1}{(L_1 + L_2)(t_2 - t_1) \cdot \left[1 - \frac{L_2 - L_1}{t_2 - t_1} \cdot \left(\frac{t_1 - t_0}{L_1 - \frac{b}{3}} + \frac{t_2 - t_0}{L_2 - \frac{b}{3}}\right)\right]} \quad \text{(IV, 31)}$$

$$\left[\beta = \frac{V_3 - V_2}{1000 \cdot P \left(1 - \frac{(V_3 - V_2)(V_3 + 4\,V_2)}{8\,V_2 \cdot V_3}\right)}\right]^{\dagger}.$$

† Bei der Berechnung der mittleren Partialkonzentration nach der Gleichung

$$k = \frac{1}{2} \cdot \frac{V_3 - V_2}{4} \cdot \frac{C_2}{V_2} + \frac{(V_3 - V_2) \cdot C_2}{V_3}$$

wurde von LEPESCHKIN beim zweiten Glied des Klammerausdruckes die Menge der zur Zeit t in der Vacuole vorhandenen permeierten Substanz $\frac{V_3 - V_2}{4} \cdot C_2$ vernachlässigt.

f) Ebenfalls für *zylindrische Zellen* und aus *plasmometrischen Messungen* in hypertonischen Lösungen des Diosmoticums allein *(Totalverfahren)* kann nach SCARTH (1939, p. 139) die Berechnung erfolgen, wenn man die Zelle in einer nicht-endosmierenden und gegenüber dem nachfolgenden Diosmoticum isotonischen Außenlösung vorplasmolysiert, wobei deren Konzentrationen so zu wählen sind, daß sie doppelt so hoch liegen als der osmotische Wert der Zelle. Aus der beobachteten *Deplasmolysezeit* T^* (Zeit vom Übertragen des Objektes in das Diosmoticum bis zum Ende der Rückdehnung), der Zellänge und Zellbreite wird dann die Permeabilitätskonstante berechnet.

Da während des osmotischen Gleichgewichtes der Rückdehnungsphase (S. 148) beobachtet wird, so gilt:

$$C = c + k, \tag{IV, 32}$$

woraus sich $C - k = c$, also gleich der Konzentration der ursprünglich vorhandenen osmotisch wirksamen Substanz der Vacuole, ergibt. Aus (IV, 24) folgt hier, weil sich das Vacuolenvolumen ändert und die Außenkonzentration konstant bleibt, die Gleichung (IV,13):

$$\Delta m_s = (V_2 - V_1) \cdot C.$$

Die Größen c und A in Formel (IV, 22) als Mittelwerte eingesetzt, führen zunächst zur Gleichung

$$\frac{(V_2 - V_1) \cdot C}{t_2 - t_1} = {}_{1f}K_s \cdot c_m \cdot A_m. \tag{IV, 33}$$

Für c_m und A_m werden die Werte beim mittleren Volumen $V_m = \frac{V_1 + V_2}{2}$ gewählt. Die mittlere Partialkonzentration c_m läßt sich aus der Volumen-Konzentrationsbeziehung [Voraussetzung (*bb*)] $c_m \cdot V_m = O_g \cdot V_Z$ ableiten:

$$c_m = \frac{V_Z}{V_m} \cdot O_g. \tag{IV, 34}$$

Wird nun V_1 gleich V_0 (Wert zu Beginn der Rückdehnung) und V_2 gleich V_Z (Protoplastenvolumen am Ende der Rückdehnungsphase) gesetzt und ist bei isotonischer Vorplasmolyse $C = 2 \cdot O_g$, so gilt wieder nach der Volumen-Konzentrationsbeziehung $C \cdot V_0 = O_g \cdot V_Z$. Aus diesen beiden Gleichungen ergibt sich $V_0 = \frac{V_Z}{2}$, und damit

$$V_m = \frac{3}{4} \cdot V_Z. \tag{IV, 35}$$

Für dieses Volumen wird nach (IV, 34):

$$c_m = \frac{4}{3} \cdot O_g. \tag{IV, 36}$$

Um A_m durch V_m auszudrücken, geht man von den Formeln (III, 1 und III, 6) aus, löst nach L auf und setzt sie gleich, wodurch sich ergibt:

$$A_m = \frac{V_m + \frac{\pi b^2}{4} \cdot \frac{b}{3}}{\frac{b}{4}}. \tag{IV, 37}$$

Mit (IV, 36), (IV, 37) und $(t_2 - t_1) = T^*$ folgt zunächst aus (IV, 33):

$$\frac{V_Z \cdot 2 \cdot O_g}{2 \cdot T^*} = {}_{1f}K_s \cdot \frac{4}{3} \cdot O_g \cdot \frac{\frac{3}{4} \cdot V_Z + \frac{\pi b^2}{4} \cdot \frac{b}{3}}{\frac{b}{4}}.$$

Diese Gleichung läßt sich mit Hilfe von Formel (III, 4) vereinfachen:

$$ {}_{1f}K_s = \frac{3}{4} \cdot \frac{h}{T^*} \cdot \frac{b}{3h + \frac{4}{3} \cdot b} \qquad \text{(IV, 38)} $$

$$ \left[P_s = \frac{3d\left(L - \frac{1}{3}d\right)}{4t\left(3L + \frac{1}{3}d\right)} \quad (19) \right]. $$

Die folgenden drei Ableitungen basieren auf der Differentialgleichung (IV, 21), wobei die stetigen Änderungen von Oberfläche und Volumen des Objektes nicht berücksichtigt werden, außer in der dritten Berechnungsmethode, bei welcher nur die Oberfläche mit einem konstanten Wert eingesetzt ist. Beziehen sich das zweite und dritte Verfahren auf Zellen, die im osmotischen Gleichgewicht mit der Außenlösung stehen, so betrifft die erste Ableitung jene Verhältnisse, bei welchen noch ein zusätzlicher von osmotischen Saugkräften verursachter Wassertransport durch die betrachtete Membran erfolgt.

g) Für zylindrische oder kugelige Zellen (z. B. Seeigeleier), die bei Versuchen nach der *Partialmethode* aus dem Kulturmedium mit der Konzentration E plötzlich in eine hypertonische Lösung, welche die permeierende Substanz mit der Partialkonzentration C enthält und die Gesamtkonzentration $C + E$ besitzt, übertragen werden, läßt sich (IV, 21) natürlich gleicherweise auf die Kontraktionsphase (S. 148), d. h. für den Zeitabschnitt bis zum Erreichen des osmotischen Gleichgewichtes zwischen Zelle und neuer Außenlösung anwenden (JACOBS und STEWART 1932, p. 78f.). Werden für Oberfläche und Volumen konstante Mittelwerte A_m und V_m eingeführt, und setzt man für $k = \frac{m_s}{V_m}$, so folgt zunächst:

$$ \frac{dm_s}{dt} = {}_{1g}K_s \cdot \frac{A_m}{V_m} \cdot (C \cdot V_m - m_s), $$

was integriert gibt:

$$ \ln \frac{C \cdot V_m - m_{s1}}{C \cdot V_m - m_{s2}} = {}_{1g}K_s \cdot \frac{A_m}{V_m} \cdot (t_2 - t_1). $$

Wählt man für $t_1 = t_B$, die Zeit des Einbringens der Zelle in die permeierende Lösung, so wird m_{s1} null und für t_2, den Endpunkt der Kontraktionsphase, der durch das Minimum des Zell- bzw. Protoplastenvolumens V_0 leicht erkannt wird und den Beginn des osmotischen Gleichgewichtes darstellt, so gilt bei und nach diesem Zeitpunkt:

$$ C + E = k + c. \qquad \text{(IV, 39)} $$

Aus dieser Gleichung kann man durch die Substitutionen

$$ k = \frac{m_{s0}}{V_0} \qquad c = \frac{m_u}{V_0} \qquad \text{[nach (IV, 24)]} $$

den Wert für m_{s0} errechnen:

$$ m_{s0} = (C + E) \cdot V_0 - m_u. $$

Die Menge der ursprünglich vorhandenen Substanz (m_u) läßt sich aus dem Volumen V_B der Zelle im isotonisch angenommenen Kulturmedium mit der Konzentration E nach der Gleichung $m_u = V_B \cdot E$ (JACOBS und STEWART 1932, p. 80) ermitteln.

Durch Einsetzen dieser Werte folgt schließlich:

$$_{1g}K_s = \frac{V_m}{A_m} \cdot \frac{1}{t_0 - t_B} \cdot \ln \frac{C \cdot V_m}{C \cdot V_m - [(C+E) \cdot V_0 - E \cdot V_B]} \qquad \textbf{(IV, 40)}$$

$$\left[K = \frac{V}{A \cdot t} \cdot \ln \frac{C \cdot V}{C \cdot V - S} \quad (7)\right].$$

Die Ausrechnung der integrierten Gleichung unter Verwendung von (IV, 39) scheint für die Verhältnisse der Kontraktionsphase allerdings nicht gegeben, da das mit (IV, 39) postulierte osmotische Gleichgewicht während der Dauer des betreffenden Meßintervalls ständig vorhanden sein müßte, um in der zwischen t und m_s für die Kontraktionsphase gefundenen Beziehung eingesetzt werden zu können. Dies trifft aber keinesfalls zu; denn es herrscht während der Kontraktion zweifellos kein osmotisches Gleichgewicht. Formel (IV, 40) kann aber ohne Bedenken für die Phase der Wiederausdehnung der Zelle verwendet werden, in welcher sich diese dauernd im osmotischen Gleichgewicht mit dem Außenmedium befindet.

h) Bei *zylindrischen Zellen*, die entsprechend der *Partialmethode* in eine Mischlösung einer endosmierenden und einer nicht-endosmierenden Komponente gelangen, läßt sich Formel (IV, 21) *nach Abwarten des osmotischen Gleichgewichtes* auf einfache Weise auswerten (FREY-WYSSLING 1945, p. 331; BOCHSLER 1948, p. 78). Man führt zunächst für m_s und A/V nach (IV, 24) und (III, 9) ein, wobei für die weitere Berechnung die Oberflächenentwicklung Q als konstant bleibend angesehen wird, und erhält zunächst:

$$\frac{dk}{dt} = {}_{1h}K_s \cdot Q \cdot (C - k). \qquad \text{(IV, 41)}$$

Durch Integration dieser Gleichung folgt:

$$\ln \frac{C - k_1}{C - k_2} = {}_{1h}K_s \cdot Q \cdot (t_2 - t_1). \qquad \text{(IV, 42)}$$

Da im osmotischen Gleichgewicht, also während der Rückdehnungsphase beobachtet wird, gelingt die weitere Umformung mittels der Gleichgewichtsbedingung (IV, 39).

Im Endgleichgewicht ist die Partialkonzentration k der permeierenden Komponente in der Vacuole, die dann das Volumen V_E besitzt, gleich groß wie in der Außenlösung ($k_E = C$), und jene der ursprünglich vorhandenen Substanz entspricht dann nach (IV, 39) der Partialkonzentration der nicht-permeierenden Komponente E in der Außenlösung ($c_E = E$). Für die ursprünglich vorhandene Substanz m_u gilt daher die Volumen-Konzentrationsbeziehung [Voraussetzung (bb)]:

$$V \cdot c = V_E \cdot c_E = V_E \cdot E.$$

Setzt man für c aus dieser Formel in (IV, 39) ein, so ergibt sich für die Konzentrationsdifferenz:

$$C - k = \frac{V_E}{V} \cdot E - E. \qquad \text{(IV, 43)}$$

Substituiert man hierin nach (III, 1) V durch die Protoplastenlänge L, führt diesen Ausdruck in (IV, 42) ein und ersetzt Q nach Formel (III,9), wobei der Mittelwert $L_m = \frac{L_1 + L_2}{2}$ eingeführt ist, so wird schließlich:

$$_{1h}K_s = \frac{1}{t_2 - t_1} \cdot \frac{b}{4} \cdot \frac{L_1 + L_2 - \frac{2b}{3}}{L_1 + L_2} \cdot \ln \frac{(L_E - L_1)\left(L_2 - \frac{b}{3}\right)}{(L_E - L_2)\left(L_1 - \frac{b}{3}\right)} \qquad \textbf{(IV, 44)}$$

$$\left[P_1 = \frac{1}{Q \cdot t_{2-1}} \cdot \ln \frac{\frac{1}{g_1} - 1}{\frac{1}{g_2} - 1} \qquad \text{p. 110} \quad (25)\right]$$

i) Für *Zellen mit meßbarem Volumen* (zylindrische oder kugelige Form) kann man bei Versuchen nach der *Partialmethode* unter Berücksichtigung eines nicht-lösenden Raumes x^* nach RESÜHR (1936) eine Endformel ableiten, in welcher zwar die stetigen Änderungen des Zell- bzw. Protoplastenvolumens berücksichtigt sind, aber die Oberfläche mit ihrem konstanten Anfangswert eingeführt wird. Man geht hierzu von Gleichung (II, 7) aus, welche auch für die osmotischen Saugkräfte gilt:

$$S_{sz} = k \cdot R \cdot T; \qquad S_{sl} = C \cdot R \cdot T. \tag{IV, 45}$$

Mit dieser Beziehung und (IV, 24) wird (IV, 21)

$$\frac{d\,(S_{sz} \cdot V)}{dt} = {}_{1i}K_s \cdot A \cdot (S_{sl} - S_{sz})\,. \tag{IV, 46}$$

Erfolgen die Beobachtungen im *Gleichgewichtszustand der osmotischen Gesamtsaugkräfte* innerhalb und außerhalb der Vacuole (d. h. für die Endosmose während der Rückdehnungsphase), so gilt mit (IV, 24) und (IV, 45) in (IV, 13) eingeführt:

$$d\,(S_{sz} \cdot V) = S_l \cdot dV,$$

wobei sich die Saugkraft S_l der Außenlösung aus den Partialsaugkräften der nicht-diosmierenden (Q^*) und der diosmierenden Komponente (S_{sl}) zusammensetzt. Berücksichtigt man ferner beim Volumen noch den nichtlösenden Raum x^*, so wird (IV, 46):

$$\frac{(Q^* + S_{sl}) \cdot d\,(V - x^*)}{dt} = {}_{1i}K_s \cdot A \cdot (S_{sl} - S_{sz})\,. \tag{IV, 47}$$

Ist V_E das Volumen der Zelle am Schluß der Endosmose, so folgt aus der Volumen-Saugkraftbeziehung [Voraussetzung (*bb*)] für die ursprünglich vorhandene Substanz m_u:

$$S_u \cdot (V - x^*) = S_{uE} \cdot (V_E - x^*)\,. \tag{IV, 48}$$

Es gilt nun analog (IV, 39) während des osmotischen Gleichgewichtes

$$S_{sl} + Q^* = S_u + S_{sz} \tag{IV, 49}$$

und am Versuchsende, wenn die Saugkraft der diosmierenden Substanz in der Vacuole gleich groß ist, wie jene im Außenmilieu ($S_{szE} = S_{sl}$):

$$Q^* = S_{uE}\,. \tag{IV, 50}$$

Durch Berechnung von $S_u = S_{sl} + Q^* - S_{sz}$ aus (IV, 49) und Substitution von S_{uE} nach (IV, 50) in Gleichung (IV, 48) folgt:

$$(S_{sl} + Q^* - S_{sz}) \cdot (V - x^*) = Q^* \cdot (V_E - x^*)$$

und hieraus:

$$S_{sl} - S_{sz} = Q^* \cdot \frac{(V_E - x^*) - (V - x^*)}{(V - x^*)}\,.$$

Die Einführung dieses Wertes in (IV, 47) und Trennung der Variablen, wobei für die Oberfläche A ein konstanter Anfangswert A_B eingesetzt wird, ergibt:

$$\frac{Q^* + S_{sl}}{Q^*} \cdot \frac{V - x^*}{(V_E - x^*) - (V - x^*)} \cdot d\,(V - x^*) = {}_{1i}K_s \cdot A_B \cdot dt\,.$$

Nach Integration zwischen den Grenzen 1 und 2, bei konstantem x^*, wird schließlich:

$$ {}_{1i}K_s = \frac{Q^* + S_{sl}}{Q^*} \cdot \frac{1}{A_B \cdot (t_2 - t_1)} \cdot \left[V_1 - V_2 + (V_E - x^*) \cdot \ln \frac{V_E - V_1}{V_E - V_2}\right] \tag{IV, 51}$$

$$\left[\begin{array}{ll} + P = \dfrac{A + B}{t \cdot q_a \cdot A} \cdot \left[(V_Z - x) \cdot \ln \dfrac{Z_g}{Z_g - Z_t} - Z_t\right] & (10) \\ - P = \dfrac{A + B}{t \cdot q_a \cdot A} \cdot \left[(V_a - x) \cdot \ln \dfrac{A_g}{A_g - A_t} + A_t\right] & (13) \end{array}\right]$$

Die Größe des nicht-lösenden Raumes x^* wird an ähnlichem Versuchsmaterial aus der Messung der Volumina V_1 und V_2 in den möglichst stark voneinander verschiedenen Konzentrationen C_1 und C_2 einer nicht-endosmierenden Lösung nach der Volumen-Konzentrationsbeziehung [Voraussetzung (*bb*)] aus der Gleichung

$$(V_1 - x^*) \cdot C_1 = (V_2 - x^*) \cdot C_2$$

bestimmt und in Prozenten des Zellvolumens V_Z ausgedrückt (RESÜHR 1936, p. 444):

$$x^{*\%} = \frac{V_1 \cdot C_1 - V_2 \cdot C_2}{C_1 - C_2} \cdot \frac{100}{V_Z}.$$

Mit dem Mittelwert $x_m^{*\%}$ ergibt sich dann der nicht-lösende Raum x^* einer Zelle mit dem Volumen V_Z:

$$x^* = \frac{x_m^{*\%} \cdot V_Z}{100}. \qquad \text{(IV, 52)}$$

Gleichung (IV, 51) ist, sofern man das Vorzeichen von K nicht beachtet und V_E das Volumen des Protoplasten am Ende des jeweiligen Versuches darstellt, sowohl für Endosmose- als auch für Exosmose-Versuche anwendbar.

β) Berechnungen, bei welchen die stetigen Änderungen von A und V berücksichtigt sind.

Nachdem zuerst ein *halb-graphisches Verfahren* zu zeigen ist, durch welches die Größenänderungen von Oberfläche und Volumen der Zelle bzw. des Protoplasten für die Berechnung erfaßbar sind, folgen zwei rein rechnerische Ableitungen, welche die Variabilität dieser Größen berücksichtigen.

j) Bei *nackten kugeligen Zellen* (z. B. Seeigeleiern) in *hypotonischen Lösungen* gelingt es nach JACOBS und STEWART (1932, p. 75f.) durch ein halbgraphisches Verfahren, die Permeabilität auch bei Vorliegen des aus der *Konzentrationsungleichheit* zwischen Zelle und Außenlösung resultierenden osmotisch verursachten zusätzlichen Wassertransportes zu erfassen.

Man geht dabei von einer Gleichung aus, welche analog (V, 17) den gesamten Wassertransport beschreibt:

$$\frac{dV}{dt} = K_w \cdot A \cdot (Z - U).$$

Besteht die Außenkonzentration U aus einer permeierenden Komponente C sowie einer nicht-permeierenden Komponente E, setzt man für Z die Partialkonzentrationen der ursprünglich vorhandenen (c) und der endosmierten Substanz (k) ein und drückt c und k durch die betreffenden Substanzmengen (m_u und m_s) sowie das Zellvolumen (V) aus, so geht obige Gleichung über in:

$$\frac{dV}{dt} = K_w \cdot A \cdot \left(\frac{m_u + m_s}{V} - C - E\right).$$

Hieraus folgt nun:

$$m_s = \frac{V}{K_w \cdot A} \cdot \frac{dV}{dt} + (C + E) \cdot V - m_u. \qquad \text{(IV, 53)}$$

Aus der Darstellung der Ausdehnung der Zelle im Zeit-Protoplastenlängen-Diagramm kann man graphisch für jede Zeit t die Steigung dV/dt bestimmen. Da als Versuchsobjekte kugelige Zellen gewählt wurden, so lassen sich aus dem Durchmesser leicht dieses Volumen sowie auch die Zelloberfläche berechnen. Die Menge der ursprünglich vorhandenen Substanz m_u vermag aus dem Volumen

V_B der Zelle in dem als isotonisch mit der Konzentration E gewählten Kulturmedium analog (IV, 24) ermittelt zu werden (JACOBS und STEWART 1932, p. 80):

$$m_u = V_B \cdot E.$$

Falls noch K_w aus Messungen an anderen Zellen des gleichen Versuchsmaterials bekannt ist, läßt sich mittels Gleichung (IV, 53) ein Zeit-m_s-Diagramm konstruieren.

Führt man in Formel (IV, 21) für die Partialkonzentration k der in die Zelle endosmierten Substanz aus Gleichung (IV, 24)

$$k = \frac{m_s}{V}$$

ein und löst die Gleichung ohne Integration nach ${}_{1j}K_s$ auf, so folgt:

$$ {}_{1j}K_s = \frac{d m_s}{d t} \cdot \frac{1}{A} \cdot \left(\frac{V}{C \cdot V - m_s} \right) \qquad \textbf{(IV, 54)}$$

$$\left[K_1 = \frac{1}{A} \cdot \frac{d S}{d t} \cdot \frac{V}{C_s \cdot V - S} \qquad (5) \right].$$

Aus dieser Formel kann nun mittels des Zeit-m_s-Diagrammes für einen Wert von V zur Zeit t leicht die Größe ${}_{1j}K_s$ berechnet werden: Durch das $m_s - t$-Diagramm läßt sich zum betreffenden Zeitpunkt die Menge der endosmierten Substanz m_s ermitteln und aus der Steigung der Tangente auch der Differentialquotient $d m_s/d t$ finden. Die Oberfläche A folgt schließlich aus dem zur Zeit t beobachteten Zelldurchmesser nach (III, 11). Die auf diese Weise gewonnene Zahlengröße von ${}_{1j}K_s$ besitzt wegen der zweimaligen Anwendung graphischer Methoden allerdings nur eine geringe Genauigkeit.

k) Die Integration von (IV, 21) ohne Einführung von Vereinfachungen gelang erstmalig SCARTH (1939) für *zylindrische Zellen* im *osmotischen Gleichgewicht* beim *Totalverfahren* und der *plasmolytischen Methode*, indem alle abhängigen Variablen durch das Volumen V des Protoplasten ausgedrückt werden.

Für die Rückdehnungsphase wird nämlich stets (IV, 13) gelten:

$$d m_s = d V \cdot C;$$

man berücksichtigt nun noch den als konstant bleibend angenommenen nichtlösenden Raum x^*, für welchen

$$V = x^* + y \quad \text{(und hieraus } d V = d y) \qquad \text{(IV, 55)}$$

gilt und drückt A nach (IV, 37) durch V aus. Ferner folgt für $C - k$ aus der Gleichgewichtsbedingung der Rückdehnungsphase nach (IV, 32):

$$C - k = c.$$

Ist der osmotische Grundwert O_g der entspannten Zelle bekannt, so gilt wegen der Volumen-Konzentrationsbeziehung [Voraussetzung (*bb*)] unter Berücksichtigung des nicht-lösenden Raumes x^* für die Menge m_u der ursprünglich vorhandenen Substanz:

$$(V_Z - x^*) \cdot O_g = (V - x^*) \cdot c,$$

woraus mit (IV, 55) folgt:

$$c = \frac{V_Z - x^*}{y} \cdot O_g.$$

Durch Einsetzen dieser Werte in Formel (IV, 21) folgt:

$$\frac{dV}{dt}\cdot C = {}_{1k}K_s \cdot \frac{x^* + y + \frac{\pi b^2}{4}\cdot\frac{b}{3}}{\frac{b}{4}} \cdot O_g \cdot \frac{V_Z - x^*}{y}. \qquad \text{(IV, 56)}$$

Substituiert man zur Übersichtlichkeit für die Integration die Konstanten:

$$\alpha = {}_{1k}K_s \cdot \frac{4}{b}\cdot\frac{O_g}{C}\cdot(V_Z - x^*), \qquad \beta = \alpha\cdot\left(x^* + \frac{\pi b^2}{4}\cdot\frac{b}{3}\right)$$

so geht die Differentialgleichung (IV, 56) über in

$$\frac{dy}{dt} = \alpha + \frac{\beta}{y}.$$

Man integriert zwischen den Grenzen 1 und 2:

$$\frac{y_2 - y_1}{\alpha} - \frac{\beta}{\alpha^2}\cdot\ln\frac{\beta + \alpha\cdot y_2}{\beta + \alpha\cdot y_1} = t_2 - t_1.$$

Nach Wiedereinführung der Werte von α und β sowie Umformung ergibt sich zunächst:

$${}_{1k}K_s = \frac{b}{4}\cdot\frac{C}{(V_Z - x^*)\cdot O_g\cdot(t_2 - t_1)}\cdot\left[y_2 - y_1 - \left(x^* + \frac{\pi b^2}{4}\cdot\frac{b}{3}\right)\cdot\ln\frac{x^* + \frac{\pi b^2}{4}\cdot\frac{b}{3} + y_2}{x^* + \frac{\pi b^2}{4}\cdot\frac{b}{3} + y_1}\right]$$

Vernachlässigt man den nicht-lösenden Raum und substituiert man nach Gleichung (IV, 55), (III, 1) und (III, 4), so folgt zunächst:

$${}_{1k}K_s = \frac{b}{4}\cdot\frac{C}{h\cdot O_g\cdot(t_2 - t_1)}\cdot\left(L_2 - L_1 - \frac{b}{3}\cdot\ln\frac{L_2}{L_1}\right)$$

und hieraus unter Vernachlässigung des Gliedes $\frac{b}{3}\cdot\ln\frac{L_2}{L_1}$

$${}_{1k}K_s = \frac{b}{4}\cdot\frac{C\,(L_2 - L_1)}{h\cdot O_g\cdot(t_2 - t_1)} \qquad \textbf{(IV, 57)}$$

$$\left[P_s = \frac{C\,(l_2 - l_1)\cdot d}{4\,t\,O\cdot\left(L - \frac{1}{3}\cdot d\right)} \qquad (20)\right]$$

Endformel (IV, 57) läßt sich noch für spezielle Versuchsverhältnisse weiter vereinfachen:

α) Wird die Konzentration C der Außenlösung doppelt so hoch als der osmotische Grundwert O_g der Zelle gewählt, so folgt aus (IV, 57):

$${}_{1l}K_s = \frac{b}{2}\cdot\frac{L_2 - L_1}{h\,(t_2 - t_1)} \qquad \textbf{(IV, 58)}$$

$$\left[P_s = \frac{d\,(l_2 - l_1)}{2\,t\left(L - \frac{1}{3}\,d\right)} \qquad (21)\right].$$

β) Plasmolysiert man zusätzlich mit einer Lösung der Konzentration $C_{\overline{p}} = 2\cdot O_g$ vor, überträgt die Zelle zur Zeit t_0 in die gleichstarke Konzentration des Diosmoticums ($C_{\overline{p}} = C$), besitzt der Protoplast bei der Vorplasmolyse die Länge L_0 und mißt man als Zeitdifferenz $t_2 - t_1$ die Deplasmolysezeit $T^* = t_E - t_0$, bei welcher dann der Protoplast wieder das Zellvolumen V_Z ausfüllt, so ist nach der Volumen-Konzentrationsbeziehung bei Substitution von V nach Gleichung (III, 1):

$$C \cdot \left(L_0 - \frac{b}{3}\right) = 2 \cdot O_g \cdot \left(L_0 - \frac{b}{3}\right) = O_g \cdot h.$$

Vernachlässigt man den Quotienten $\frac{b}{3}$, ergibt sich hieraus:

$$L_0 = \frac{h}{2}.$$

Dies, sowie die Beziehung $C = 2 \cdot O_g$ in (IV, 57) eingeführt, liefert schließlich die Gleichung

$$_{1m}K_s = \frac{b}{4} \cdot \frac{1}{T^*} \qquad \textbf{(IV, 59)}$$

$$\left[P_s = \frac{d}{4t} \quad (18)\right].$$

γ) Erfolgt keine Vorplasmolyse, sondern gelangt die Zelle gleich in das Diosmoticum, dann wird bei langsam permeierenden Substanzen das Volumen V_0 am Ende der Kontraktionsphase nur wenig von dem Volumen in einem isotonischen Plasmolyticum verschieden sein, so daß man mit guter Näherung setzen kann: $V_0 \cdot C = V_Z \cdot O_g$, woraus mit (III, 1) und (III, 4) folgt:

$$h = \frac{\left(L_0 - \frac{b}{3}\right) \cdot C}{O_g}. \qquad \text{(IV, 60)}$$

Dieser Wert in (IV, 57) eingesetzt führt zu folgender Schlußformel:

$$_{1n}K_s = \frac{b}{4} \cdot \frac{L_2 - L_1}{\left(L_0 - \frac{b}{3}\right)(t_2 - t_1)} \qquad \textbf{(IV, 61)}$$

$$\left[P_s = \frac{d\,(l_2 - l_1)}{4\,t\left(l_{\min} - \frac{1}{3}d\right)} \quad (22)\right].$$

l) Für *zylindrische Zellen*, bei welchen die Protoplastenlänge L_0 für isotonische Vorplasmolyse entweder durch eine solche bestimmt oder auf andere Weise erschlossen wird, gelingt es, für das *osmotische Gleichgewicht* mit der Außenlösung während der Rückdehnungsphase im *Totalverfahren* Gleichung (IV, 21) zu integrieren, wenn man (unter Vernachlässigung eines nicht-lösenden Raumes) die abhängigen Variablen durch die Protoplastenlänge L ersetzt (Stadelmann 1951). Man erhält dann mittels der Gleichungen (IV, 13), (III, 1) und (III, 6) sowie aus der Volumen-Konzentrationsbeziehung [Voraussetzung (*bb*)], wenn bei isotonischer Vorplasmolyse das Protoplastenvolumen V_0 ist und demnach $V_0 \cdot C = V \cdot c$ wird:

$$dm_s = dV \cdot C = dL \cdot \frac{\pi b^2}{4} \cdot C$$

$$C - k = c = C \cdot \frac{L_0 - \frac{b}{3}}{L - \frac{b}{3}}$$

$$A = \pi \cdot b \cdot L.$$

Durch Einsetzen dieser Werte in die Ausgangsgleichung (IV, 21) folgt:

$$\frac{dL}{dt} \cdot \frac{\pi b^2}{4} \cdot C = {}_{10}K_s \cdot \pi b L C \cdot \frac{L_0 - \frac{b}{3}}{L - \frac{b}{3}}$$

und nach Trennung der Variablen sowie Integration zwischen den Grenzen 1 und 2 ergibt sich schließlich:

$${}_{10}K_s = \frac{b}{4} \cdot \frac{L_2 - L_1 - \frac{b}{3} \cdot \ln \frac{L_2}{L_1}}{\left(L_0 - \frac{b}{3}\right)(t_2 - t_1)}. \qquad \textbf{(IV, 62)}$$

Diese Gleichung entspricht Formel (IV, 57), bei welcher allerdings das Glied $\frac{b}{3} \cdot \ln \frac{L_2}{L_1}$ vernachlässigt wurde, da nach (IV, 60)

$$\frac{C}{h \cdot O_g} = \frac{1}{L_0 - \frac{b}{3}}$$

substituiert werden kann. Eine nähere Prüfung von Formel (IV, 62) ergibt, daß deren Vereinfachung nicht angebracht ist (p. 782). Es gelingt ferner, die beiden aus der Gleichung theoretisch abgeleiteten Gesetzmäßigkeiten für den Verlauf der Rückdehnung (Zeitproportionalität und Konzentrationsabhängigkeit) experimentell zu bestätigen (1952, p. 381, 386).

2. Permeabilitätsmaße, welche auf andere Ausgangsgleichungen zurückgehen.

Neben Formel (IV, 21) fanden noch andere Gleichungen, von denen im folgenden einige erwähnt werden sollen, zur Ableitung von Permeabilitätsmaßen Anwendung. Die ersten beiden gehen dabei von einer Differenzenformel aus, während die dritte Gleichung als Differentialgleichung nach ihrer Integration ausgewertet wird.

a) Bei Versuchen nach der *plasmolytischen Methode* wird von HÖFLER (1934b, p. 226) als Permeabilitätsmaß vorgeschlagen:

$${}_2K_s = \frac{M^*}{C - k} \quad (\text{Dim } {}_2K_s = \mathsf{T}^{-1}) \qquad (\text{IV, 70})$$

$$\left[P' = \frac{M}{C - c}\right].$$

Dieser Formel entspricht als Ausgangsgleichung bei Einsetzen nach (IV, 12):

$$M^* = \Delta G \cdot C = \frac{(G_2 - G_1) \cdot C}{t_2 - t_1} = \frac{\Delta m_s}{V_Z \cdot (t_2 - t_1)} = {}_2K_s \cdot (C - k). \qquad \underline{(\text{IV, 71})}$$

Der Faktor $_2K_s$ gibt als Permeabilitätsmaß an, wie groß die Menge (in mol) der auf die Zeiteinheit (1 sec) und die Einheit des Volumens der entspannten Zelle (1 cm³) reduzierten, durch die Protoplastenhaut diosmierenden Substanz ist, wenn das Konzentrationsgefälle zwischen Außenlösung und Vacuole für den betreffen den Stoff die Größe der Konzentrationseinheit (1 mol/cm³) besitzt.

Die Ermittlung von M^* und $(C - k)$ kann entweder aus dem Plasmolysegrad oder aus der Deplasmolysezeit erfolgen.

α) Berechnung bei zylindrischen Zellen aus dem Plasmolysegrad. Im Totalverfahren läßt sich unter Voraussetzung (*ww*) die Partialkonzentration k für den Zeitpunkt t berechnen:

$$k = (t - t_B) \cdot M^*, \qquad (\text{IV, 72})$$

so daß mit Gleichung (IV, 12) aus (IV, 70) folgt:

$$ {}_{2a}K_s = \frac{\Delta G}{1 - \Delta G \cdot (t - t_B)} \qquad \text{(IV, 73)} $$

$$ \left[P' = \frac{\Delta G}{1 - \Delta G \cdot t} \right]. $$

Wählt man für t die Mitte des Intervalls $t_2 - t_1$, so wird schließlich

$$ {}_{2a}K_s = \frac{L_2 - L_1}{h\,(t_2 - t_1) - (L_2 - L_1) \cdot \left(\frac{t_1 + t_2}{2} - t_B \right)} \qquad \textbf{(IV, 74)} $$

Meist ist eine Messung einer größeren Zahl von Zellen möglich, wobei dann das Mittel der Plasmolysegrade in Gleichung (IV, 73) eingesetzt wird und für t_1 und t_2 die Mittelwerte des zur Protokollierung dieser Größen erforderlichen Zeitintervalls zu wählen sind.

Da sich k in Formel (IV, 72) auf den Innenraum der entspannten Zelle bezieht (ASHIDA; zit.: HOFMEISTER 1938, p. 404), so ist dieser Wert noch auf den mittleren Plasmolysegrad $\frac{G_1 + G_2}{2}$ des gemessenen Intervalls umzurechnen (ASHIDA-Korrektur):

$$ k \cdot V_Z = k_{\text{korr}} \cdot V_{pm}. $$

Daraus folgt mit (III, 8):

$$ k_{\text{korr}} = k \cdot \frac{V_Z}{V_{pm}} = k \cdot \frac{2}{G_1 + G_2} = \frac{2h \cdot k}{L_1 + L_2 - \frac{2b}{3}}. \qquad \text{(IV, 75)} $$

Somit nimmt Gleichung (IV, 73) die Form an:

$$ {}_{2b}K_s = \frac{\Delta G}{1 - \Delta G\,(t - t_B) \cdot \frac{2}{G_1 + G_2}} \qquad \text{(IV, 76)} $$

$$ \left[P' = \frac{M}{C - \frac{2c}{G_1 + G_2}} \right]. $$

Nach Substitution mit ΔG entsprechend (IV, 11) und Einsetzen für t wie oben ergibt sich schließlich:

$$ {}_{2b}K_s = \frac{L_2 - L_1}{h \cdot (t_2 - t_1) - (L_2 - L_1) \left(\frac{t_1 + t_2}{2} - t_B \right) \cdot \frac{2h}{L_1 + L_2 - \frac{2b}{3}}}. \qquad \textbf{(IV, 77)} $$

β) Zur Berechnung aus Versuchen nach dem *Partialverfahren* (HOFMEISTER 1935, p. 13f.) wird zunächst das Protoplastenvolumen V_Z bzw. die Protoplastenlänge L_Z beim osmotischen Gleichgewicht in der Lösung der nicht-diosmierenden Komponente (z. B. Rohrzucker) allein bestimmt. Hernach gelangt der Schnitt zur Zeit t_0 in die Mischlösung der Konzentration $C + E$, wo nach einer anfänglichen osmotisch bedingten Kontraktion die Wiederausdehnung, durch das Eindringen der diosmierenden Substanz verursacht, einsetzt. Zu Beginn dieser Volumvergrößerung werden zu den Zeiten t_1 und t_2 die Protoplastenlängen L_1 und L_2 gemessen.

Die je Zeiteinheit und Volumeinheit der entspannten Zelle eintretende Substanzmenge M^* wird wieder nach Gleichung (IV, 12) bestimmt. Zur Ermittlung von k dient die Proportion

$$k : C = \left(\frac{G_1 + G_2}{2} - G_0\right) : (G_Z - G_0),$$

wobei der Plasmolysegrad G_0 unter Voraussetzung (ww) durch lineare Extrapolation aus den Plasmolysegraden G_1 und G_2 zu den Zeiten t_1 und t_2 berechnet wird:

$$\frac{G_1 - G_0}{G_2 - G_1} = \frac{t_1 - t_0}{t_2 - t_1}; \quad G_0 = G_1 - \frac{t_1 - t_0}{t_2 - t_1} \cdot (G_2 - G_1).$$

Daraus ergibt sich für k:

$$k = \frac{\frac{G_1 + G_2}{2} - G_1 + \frac{t_1 - t_0}{t_2 - t_1} \cdot (G_2 - G_1)}{G_Z - G_1 + \frac{t_2 - t_0}{t_2 - t_1} \cdot (G_2 - G_1)} \cdot C. \tag{IV, 78}$$

Nach Substitution von G gemäß Formel (III, 8) und Einsetzen der Werte von k nach (IV, 78) und M^* nach (IV, 12), wobei als Konzentration die Gesamtkonzentration $(C + E)$ eingeführt wird, folgt aus Gleichung (IV, 70) zunächst:

$${}_{2c}K_s = \frac{L_2 - L_1}{h(t_2 - t_1)} \cdot \frac{1}{1 - \frac{\frac{L_1 + L_2 - \frac{2b}{3}}{2} - \left(L_1 - \frac{b}{3}\right) + \frac{t_1 - t_2}{t_2 - t_1} \cdot (L_2 - L_1)}{L_Z - L_1 + \frac{t_2 - t_0}{t_1 - t_2} \cdot (L_2 - L_1)}} \cdot \frac{C + E}{C}$$

womit schließlich wird:

$${}_{2c}K_s = \frac{L_2 - L_1}{h \cdot (t_2 - t_1)} \cdot \frac{L_Z - L_1 + \frac{t_2 - t_0}{t_2 - t_1} \cdot (L_2 - L_1)}{L_Z - \frac{L_1 + L_2}{2}} \cdot \frac{C + E}{C} \tag{IV, 79}$$

$$\left[P_p' = \frac{M}{D - c'} \quad (7)\right].$$

γ) *Bestimmung von M^* und $(C - k)$ aus der Deplasmolysezeit bei Zellen von beliebiger Form* (HOFMEISTER 1948). Da am Ende der Rückdehnungsphase (S. 148), d. h. also nach Deplasmolyse für das osmotische Gleichgewicht *beim Totalverfahren* gilt:

$$k_E + O_g = C, \tag{IV, 80}$$

so folgt unter Voraussetzung (ww) die während der Deplasmolysezeit T^* (gerechnet vom Versuchsbeginn an) je Zeiteinheit und Volumeinheit der entspannten Zelle durchschnittlich eintretende Substanzmenge M^*:

$$M^* = \frac{k_E}{T^*} = \frac{C - O_g}{T^*}.$$

Während dieser Zeit hat sich das Konzentrationsgefälle $C - k$ von C auf den Wert O_g verringert, so daß als mittlere Konzentrationsdifferenz während des Versuches $C - k_m = \frac{C + O_g}{2}$ als wirksam betrachtet werden kann. Damit berechnet sich nun nach (IV, 70):

$${}_{2d}K_s = \frac{2 \cdot (C - O_g)}{T^* \cdot (C + O_g)} \tag{IV, 81}$$

$$\left[P_d' = \frac{120\,(C - O)}{T\,(C + O)}\right].$$

δ) Wird beim *Partialverfahren* (S. 147) die Konzentration E der nicht-endosmierenden Komponente schwach hypotonisch gewählt, so daß Plasmolyse nur in der Konzentration $C+E$ eintritt, so wird am Ende der Deplasmolyse die Partialkonzentration k auf C angestiegen sein, so daß sich unter Voraussetzung (ww) ergibt: $M^* = \frac{C}{T^*}$. Ebenso ist die Konzentrationsdifferenz von C auf Null gesunken, womit sich als mittlerer Konzentrationsunterschied $C - k_m = \frac{C}{2}$ berechnet und schließlich zu folgender Gleichung führt:

$$_{2l}K_s = \frac{\frac{C}{T^*}}{\frac{C}{2}} = \frac{2}{T^*} \qquad \text{(IV, 82)}$$

$$\left[P'_{pd} = \frac{120}{T}\right].$$

b) Um für *zylindrische Zellen* und Versuche nach der *plasmolytischen Methode* beim Permeabilitätsfaktor $_2K_s$ die Oberfläche mit zu berücksichtigen, werden von HÖFLER (1934b, p. 234) diese $_2K_s$-Werte auf die Oberflächenentwicklung Q (III, 9) bezogen, so daß gilt:

$$_3K_s = \frac{M^*}{C-k} \cdot \frac{1}{Q} = \frac{M^* \cdot V_p}{(C-k) \cdot A_p} \quad (\text{Dim}\, _3K_s = \mathsf{L\,T^{-1}}), \qquad \text{(IV, 83)}$$

was der Ausgangsgleichung

$$M^* \cdot V_p = \frac{\Delta m_s}{t_2 - t_1} \cdot \frac{V_p}{V_Z} = \frac{\Delta m_s}{t_2 - t_1} \cdot \frac{1}{G} = {_3K_s} \cdot A_p \cdot (C-k) \qquad \underline{\text{(IV, 84)}}$$

entspricht.

Damit gibt der Faktor $_3K_s$ als Permeabilitätsmaß an, wie groß das Produkt aus der Menge der permeierenden Substanz (in mol) und dem Protoplastenvolumen (in cm^3), reduziert auf die Zeiteinheit (1 sec), die Oberflächeneinheit (1 cm^2) und die Einheit des Volumens der entspannten Zelle (1 cm^3) ist, wenn das Konzentrationsgefälle zwischen Außenlösung und Vacuole für den permeierenden Stoff die Größe der Konzentrationseinheit (1 mol/cm^3) besitzt.

Bei der Berechnung erfolgt die Ermittlung von M^* und $(C-k)$ in gleicher Weise wie bei $_2K_s$ gezeigt ist, und für Oberfläche und Volumen werden nach (III, 6) und (III, 1) Mittelwerte eingesetzt, was zur folgenden Formel führt:

$$\left.\begin{aligned} {_{3a-c}K_s} &= \frac{_{2a-c}K_s}{Q} = {_{2a-c}K_s} \cdot \frac{V_p}{A_p} = {_{2a-c}K_s} \cdot \frac{b}{4} \cdot \frac{L_m - \frac{b}{3}}{L_m} \\ &= \frac{b}{4} \cdot \frac{L_1 + L_2 - \frac{2b}{3}}{L_1 + L_2} \cdot {_{2a-c}K_s} \end{aligned}\right\} \qquad \text{(IV, 85)}$$

$$\left[P = \frac{\left(l - \frac{2}{3}r\right)r}{2l} \cdot P'\right].$$

c) Für Versuche nach der *plasmolytischen Methode* mit *Zellen beliebiger Form* wurde von BÄRLUND (1929, p. 65) folgende Gleichung eingeführt:

$$\frac{dk^1}{dt} = {_4K_s} \cdot (C-k). \qquad (\text{Dim}\, _4K_s = \mathsf{T^{-1}}). \qquad \underline{\text{(IV, 90)}}$$

Hierbei sind also Oberfläche und Volumen des Protoplasten nicht berücksichtigt.

[1] Ursprünglich als Menge bezeichnet, aber bei der Integration als Konzentration behandelt.

Der Faktor $_4K_s$ gibt als Permeabilitätsmaß an, wie groß die Konzentrationszunahme an permeierender Substanz (in mol/cm³) je Zeiteinheit (1 sec) im Zellsaftraum ist, wenn das Konzentrationsgefälle zwischen Außenlösung und Vacuole für den eindringenden Stoff die Größe der Konzentrationseinheit (1 mol/cm³) besitzt.

Formel (IV, 90) läßt sich ohne Umformung zwischen den Grenzen 1 und 2 integrieren:

$$_4K_s = \frac{1}{t_2 - t_1} \cdot \ln \frac{C - k_1}{C - k_2}. \qquad \textbf{(IV, 91)}$$

Wählt man für t_1 den Beginn der Endosmose, so wird $t_1 = 0$ und $k_1 = 0$, womit sich dann ergibt:

$$_4K_s = \frac{1}{t} \cdot \ln \frac{C}{C - k}, \qquad \text{(IV, 92)}$$

$$\left[P = \frac{1}{t} \cdot \ln \frac{C}{C - x} \qquad \text{p. 65}\right].$$

Zur Ermittlung von k aus dem Experiment geht man von gleichartigem Material aus, bestimmt bei einem Teil davon den osmotischen Grundwert O_g, während der Rest in die Konzentration C des Diosmoticums gelangt, worin nach der Zeit t durch die Deplasmolyse wieder der Grenzplasmolysezustand erreicht ist. Die Protoplasten haben dann während dieser Zeitspanne t die Partialkonzentration $k = C - O_g$ aufgenommen (p. 49).

V. Maße für Wassertransport und Wasserpermeabilität.

Die separate Behandlung des Wassertransportes ist durch die besondere Stellung des Wassers innerhalb der die Zelloberfläche passierenden Stoffe berechtigt. Sie liegt darin begründet, daß ihm als Solvens der häufig sehr verdünnten Lösungen von permeierenden Substanzen eine von diesen wohl verschiedene Rolle zukommt und es außerdem die Protoplastenmembran meist um ein Vielfaches leichter zu durchdringen vermag.

Die Berechnungsweise der Wassertransportwerte und -permeabilität lehnt sich gleichwohl an jene für die gelösten Stoffe an. Dabei ergeben sich in jenen Fällen, bei welchen die transportierte Wassermenge durch Volumenänderungen zu erkennen ist, meist einfache Formulierungen.

A. Maße für den Wassertransport.

Die Größe der Wasseraufnahme bzw. -abgabe folgt als Gesamtfluß aus Versuchen, bei welchen die verschobene Wassermenge meßbar ist. Durch Einsetzen ihres Werte z.B. als Volumen (es wäre aber auch eine Angabe der Wassermenge in Gramm oder Mol denkbar), des Zeitintervalls und der betreffenden Protoplasten- bzw. Zelloberfläche in die (IV, 1) analoge Gleichung

$$M_w = \frac{V_2 - V_1}{A \cdot (t_2 - t_1)} \qquad (\text{Dim } M_w = \mathsf{LT}^{-1}) \qquad \underline{\text{(V, 1)}}$$

berechnet sich dann die Wasseraufnahme bzw. -abgabe.

Ist die Fläche der betreffenden Membran unbekannt, z. B. im Falle der Blutcapillarenwandung, so läßt sich der je Zeiteinheit durch diese eindringende Bruchteil der Gesamtwassermenge als Quotient $\frac{V_w'}{V_{wb}}$ nach FLEXNER und Mitarbeiter (1942) unter Voraussetzung (l) ermitteln, wenn man eine bestimmte Menge m_{wB}

schweren Wassers ins Blut einführt. Man geht dann von der Formel (IV, 7) analogen Gleichung aus, wobei die gleichen Überlegungen gelten:

$$\frac{dm_w}{dt} = -V_w' \cdot \frac{m_w}{V_{wb}} + V_w' \cdot \frac{m_{wB} - m_w}{V_{wk}}. \qquad \text{(V, 2)}$$

Hieraus resultiert wieder:

$$\ln\left(c_w - c_{wB} \cdot \frac{V_{wb}}{V_{wk} + V_{wb}}\right) - \ln\left(c_{wB} - c_{wB} \cdot \frac{V_{wb}}{V_{wk} + V_{wb}}\right) = -\frac{V_w'}{V_{wB}} \cdot \frac{V_{wk} + V_{wb}}{V_{wk}} \cdot t.$$

Das Produkt $c_{wB} \cdot \frac{V_{wb}}{V_{wk} + V_{wb}}$ ist die Konzentration c_{wE} des schweren Wassers im Endgleichgewichtszustand im Blut und daher direkt meßbar. Aus ihr und c_{wB}, die ja ebenfalls direkt bestimmt werden kann, läßt sich der Quotient $\frac{V_{wb}}{V_{wk} + V_{wb}}$ errechnen, so daß schließlich

$$\frac{V_{wk}}{V_{wk} + V_{wb}} = 1 - \frac{V_{wb}}{V_{wk} + V_{wb}} = 1 - \frac{c_{wE}}{c_{wB}}$$

wird. Damit folgt nun für den Quotienten $\frac{V_w'}{V_{wb}}$:

$$\frac{V_w'}{V_{wb}} = \frac{1 - \frac{c_{wE}}{c_{wB}}}{t} \cdot \ln \frac{c_{wB} - c_{wE}}{c_w - c_{wE}} \qquad \left(\text{Dim}\ \frac{V_w'}{V_{wb}} = \mathsf{T}^{-1}\right) \qquad \text{(V, 3)}$$

$$\left[(3) \quad q\,[\ln(c_t - c_{eq}) - \ln(c_0 - c_{eq})] = -Rt\right].$$

B. Maße für die Wasserpermeabilität.

Als solche werden meist Größen vorgeschlagen, die sich aus Endformeln ergeben, welche ihrerseits auf bestimmte Ausgleichsgesetze als Ausgangsgleichungen zurückgehen, während unmittelbar im Versuch beobachtbare Veränderungen selbst oder Formulierungen, die nicht auf Ausgleichsgesetzen basieren, zur quantitativen Angabe der Wasserpermeabilität weniger Verwendung finden und hier nicht näher behandelt werden sollen.

Die entsprechenden Anfangsgleichungen sind zunächst grundsätzlich verschieden, je nachdem, ob bei der betreffenden Versuchsanordnung der Wassertransport durch Diffusion erfolgt, indem es sich allein um die statistische Gleichverteilung einer bestimmten Wassermenge handelt, oder ob besondere Kräfte (z.B. bei Vorliegen von Konzentrationsverschiedenheiten osmotische Saugkräfte) als wirksam postuliert werden. Analog Voraussetzung (*a*) und Gleichung (IV, 21) läßt sich Proportionalität zwischen der in der Zeiteinheit transportierten Wassermenge $\frac{dm_w}{dt}$ und der Größe der beteiligten Kraft sowie der durchdrungenen Fläche A (Oberfläche der Zelle bzw. des Protoplasten) annehmen, wobei dadurch, daß das transportierte Wasser einerseits als Masse bzw. Menge (in g oder mol), andererseits als Volumen (in cm³) in die Ausgangsgleichungen eingeführt werden kann, eine zusätzliche Variationsmöglichkeit entsteht.

Die aus der großen Zahl von Ableitungen für Maße der Wasserpermeabilität[1] hier zu zeigenden Beispiele sollen im folgenden nach dem als wirksam postulierten Mechanismus in Gruppen zusammengefaßt werden.

[1] Vgl. noch: Lillie (1916), de Haan (1930), Lucké und Mitarbeiter (1931), Manegold und Stüber (1933), McCutcheon und Lucké (1933), Jacobs (1933a, 1933b, 1934), Levitt und Mitarbeiter (1936), Bachmann (1939), Palva (1939), Lucké (1940), Frey-Wyssling (1945), Løvtrup und Pigoń (1951), Myers (1951), Kamiya und Tazawa (1956); Lit.: Jacobs (1933, p. 428).

1. Gleichungen für den Wassertransport durch Diffusion.

Ist allein die Diffusion wirksam und betrachtet man eine bestimmte Wassermenge m_w, die nur einen verhältnismäßig geringen Anteil der Gesamtwassermenge des Außenraumes darstellt, welche von dort bis zu ihrer Gleichverteilung in das Zellinnere diffundiert, so lassen sich die Verhältnisse völlig analog jenen, wie sie bei den gelösten Substanzen vorliegen, betrachten.

Die Möglichkeit der Auswertung solcher Wasserverschiebungen, die durch Diffusion hervorgerufen werden, ist grundsätzlich nur insoweit gegeben, als es tatsächlich gelingt, die Permeation einer bestimmten Wassermenge durch eine Membran messend zu verfolgen, wobei Volumänderungen, wie sie osmotische Kräfte meist verursachen, hier natürlich nicht vorkommen. Durch Verwendung von mit Isotopen gekennzeichnetem Wasser (besonders DHO) läßt sich unter Voraussetzung (*l*) die verschobene Wassermenge gut bestimmen, so daß die weitere Berechnung möglich wird.

In solchen Fällen können die für die Diffusion gelöster Substanzen aus (IV, 21) abgeleiteten Formeln ohne weiteres auch zur Bestimmung der Wasserpermeabilität herangezogen werden, außer jenen, die für Versuche nach osmotischen Methoden berechnet wurden, da bei diesen eine die Diffusion der gelösten Substanz begleitende Wasserverschiebung ausgewertet wird.

Um störende Einflüsse osmotisch verursachter Wasserverschiebungen auf die zu messende Diffusion des schweren Wassers zu vermeiden, darf diese nur bei Fehlen solcher Kräfte, also im osmotischen Gleichgewicht, erfolgen.

Es ergibt sich daher analog (II, 6) die allgemeine Formel:

$$\frac{d m_w}{d t} = K_w \cdot A \cdot (C_w - k_w). \qquad \text{(V, 11)}$$

Setzt man die Menge m_w in mol und die Konzentration dieser Menge im gesamten Wasservolumen wieder in mol/cm³ ein, so folgt die Ausgangsgleichung:

$$\frac{d m_w}{d t} = {}_1K_w \cdot A \cdot (C_w - k_w) \qquad (\mathrm{Dim}\,{}_1K_w = \mathsf{L}\,\mathsf{T}^{-1}) \qquad \underline{\text{(V, 12)}}$$

bzw. die Differenzenform:

$$\frac{m_{w2} - m_{w1}}{t_2 - t_1} = {}_1K_w \cdot A \cdot (C_w - k_w). \qquad \text{(V, 13)}$$

Es gibt dann der *Faktor* ${}_1K_w$ *als Maß für die Wasserpermeabilität an, wie groß die Menge (in mol) des je Zeiteinheit (1 sec) und Oberflächeneinheit (1 cm²) durch die betreffende Membran permeierenden Wassers ist, wenn für die betrachtete Wassermenge das Konzentrationsgefälle zwischen Außenraum und Vacuole die Größe der Konzentrationseinheit (1 mol/cm³) besitzt.*

α) Bei Zellen mit *gleichbleibender Oberfläche* und *konstantem Volumen*, welche in einem Vorversuch bereits schweres Wasser in die Vacuole aufnahmen, läßt sich nach WARTIOVAARA (1944, p. 7) der *Wasseraustritt* aus der (IV, 23) analogen Formel ermitteln:

$$\frac{d m_w}{d t} = {}_{1a}K_w \cdot (C_w^* - k_w) \cdot A. \qquad \text{(V, 14)}$$

Ihre Ausrechnung erfolgt in gleicher Weise, wie dort gezeigt (S. 157), und liefert bei Messung der ausgetretenen Menge schweren Wassers m_{w2-1} und m_{w3-2} für zwei aufeinanderfolgende gleich lange Zeitintervalle $\Delta t = t_2 - t_1 = t_3 - t_2$ unter Voraussetzung (*dd*) schließlich die (IV, 29) entsprechende Endgleichung:

$$ {}_{1a}K_w = \frac{V}{A} \cdot \frac{1}{\Delta t} \cdot \ln \frac{m_{w2-1}}{m_{w3-2}} \qquad \textbf{(V, 15)}$$

$$\left[(2) \quad P = 3{,}45 \cdot \frac{d}{t} \cdot \log k\right].$$

2. Gleichungen für osmotische Saugkräfte.

Werden allein die osmotischen Saugkräfte (vgl. S. 145), welche sich als Differenz ihrer innerhalb und außerhalb der Zelle herrschenden Größe ergeben, als für den Wassertransport maßgebend angenommen, so folgt die allgemeine Gleichung:

$$\frac{dm_w}{dt} = K_w \cdot A \cdot (S_l - Si). \qquad \text{(V, 16)}$$

Mißt man die Wassermenge dm_w als Volumen V, so ergibt sich schließlich (JACOBS 1935, p. 80; RESÜHR 1935, p. 340)

$$\frac{dV}{dt} = {}_2K_w \cdot A \cdot (S_l - Si) \qquad (\text{Dim}\ {}_2K_w = \mathsf{L^2 M T^{-1}}\dagger) \qquad \underline{\text{(V, 17)}}$$

bzw. deren Differenzenform

$$\frac{V_2 - V_1}{t_2 - t_1} = {}_2K_w \cdot A \cdot (S_l - Si). \qquad \text{(V, 18)}$$

Der Faktor ${}_2K_w$ *gibt hierin als Maß für die Wasserpermeabilität an, wie groß die Menge des durch den Protoplasten hindurchtretenden Wassers, gemessen als Volumen (in cm³) je Zeiteinheit (1 sec) und Oberflächeneinheit (1 cm²) ist, wenn der Unterschied der osmotischen Saugkräfte zwischen Vacuole und Außenlösung die Größe der Saugkrafteinheit (1 Atm) besitzt.*

Die Gleichungen (V, 17) und (V, 18) können nun für die verschiedenen Versuchsverhältnisse nach ${}_2K_w$ aufgelöst werden, wobei unter α) zunächst eine Ableitung folgt, welche ${}_2K_w$ bei Plasmolyse und auch bei Deplasmolyse zu bestimmen gestattet (Wasseraustritt bzw. -eintritt). Hierbei werden allerdings die Oberflächenänderungen vernachlässigt, während der Wechsel der Größe von Volumen und Konzentration voll beachtet ist.

Die daran anschließenden Ausrechnungen (β—δ) von ${}_2K_w$ berücksichtigen alle Variablen, gelten aber nur für die Deplasmolyse von zylindrischen bzw. kugeligen Zellen in Wasser oder teilweise auch in hypotonischen Milieus:

α) Bei Versuchen mit Zellen bzw. Protoplasten, deren Volumen meßbar ist *(kugelige oder zylindrische Form)*, läßt sich mittels der osmotischen bzw. *plasmolytischen Methode im Totalverfahren* nach RESÜHR (1935, p. 340) Gleichung (V, 17) unter *Berücksichtigung eines nicht-lösenden Raumes* x^* ausrechnen, wenn die *Oberfläche als konstant* betrachtet wird und mit ihrem Anfangswert A_B eingeführt wird.

Man geht dabei von der Volumen-Saugkraftbeziehung [Voraussetzung (*bb*)] aus, wobei V_E das Volumen des Protoplasten beim osmotischen Endgleichgewicht mit der Außenlösung bedeutet:

$$(V - x^*) \cdot Si = (V_E - x^*) \cdot S_l; \qquad Si = S_l \cdot \frac{V_E - x^*}{V - x^*}.$$

Setzt man dies in Formel (V, 17) ein:

$$\frac{d(V - x^*)}{dt} = {}_{2a}K_w \cdot A_B \cdot S_l \cdot \left(1 - \frac{V_E - x^*}{V - x^*}\right).$$

Nach Trennung der Variablen folgt zunächst:

$$dt = \frac{1}{{}_{2a}K_w \cdot A_B \cdot S_l} \cdot \left[1 + \frac{V_E - x^*}{(V - x^*) - (V_E - x^*)}\right] \cdot d(V - x^*).$$

Bleibt x^* konstant, so ergibt sich nach Integration zwischen den Grenzen 1 und 2:

$$_{2a}K_w = \frac{1}{A_B \cdot S_l (t_2 - t_1)} \cdot \left[V_2 - V_1 + (V_E - x^*) \cdot \ln \frac{V_2 - V_E}{V_1 - V_E}\right]. \qquad \textbf{(V, 19)}$$

† Diese folgt aus (V, 17), wenn man dort die Dimensionen einsetzt und dabei die Saugkraft als Druck (= Kraft je Fläche) angibt:

$$\frac{\mathsf{L^3}}{\mathsf{T}} = (\text{Dim}\ {}_2K_w) \cdot \mathsf{L^2} \cdot \frac{\mathsf{L M T^{-2}}}{\mathsf{L^2}}.$$

Für $t_1 = 0$ wird bei Versuchen nach der plasmolytischen Methode und Auswertung der Kontraktion (Plasmolyse) $V_1 = V_Z$, dem Volumen der entspannten Zelle.

$$\left[k = -\frac{1}{t \cdot q \cdot P_g} \cdot \left\{(V-x)_0 - (V-x)_g \cdot \ln\left[\frac{(V-x)_0 - (V-x)_g}{(V-x)_t - (V-x)_g}\right]\right\} \cdot \text{(p. 341)}\right].$$

Die osmotische Saugkraft der Außenlösung S_l, die ja numerisch gleich dem osmotischen Druck ist, läßt sich nach (II, 7) mittels der Gleichung

$$S_l = C \cdot R \cdot T$$

aus der Konzentration C des Außenmediums berechnen, und x^* bestimmt sich aus Versuchen an gleichartigem Material nach S. 164.

β) Überträgt man *vorplasmolysierte zylindrische Zellen in Wasser*, so lassen sich die Gleichungen (V, 17) und (V, 18) auf die Wiederausdehnung des Protoplasten in verschiedener Weise anwenden, wobei die Protoplastenoberfläche zum Teil als konstant bleibend mit einem Mittelwert eingeführt ist, zum Teil als Variable behandelt wird (Scarth 1939).

$\alpha\alpha$) Endformeln für *zwei beliebige Deplasmolysegrade* bei *Vernachlässigung eines nicht-lösenden Raumes.*

Formel (V, 17) vereinfacht sich, da für Wasser $S_l = 0$ ist, wenn man das Vorzeichen von ${}_{2b}K_w$ als positiv wählt, zu dem Ausdruck

$$\frac{dV}{dt} = {}_{2b}K_s \cdot A \cdot Si_{\overline{p}}. \tag{V, 20}$$

Man kann nun A nach (IV, 37) durch V ausdrücken und zunächst den nicht-lösenden Raum x^* nach (IV, 55) berücksichtigen, was zu der Gleichung

$$A = \frac{x^* + y + \frac{\pi b^2}{4} \cdot \frac{b}{3}}{\frac{b}{4}} \tag{V, 21}$$

führt. Nach der Volumen-Saugkraftbeziehung [Voraussetzung (bb)] gilt nun:

$$V_Z \cdot Si_g = Si_{\overline{p}} \cdot V,$$

bzw. unter Berücksichtigung des nicht-lösenden Raumes:

$$(V_Z - x^*) \cdot Si_g = (V - x^*) \cdot Si_{\overline{p}},$$

so daß wird:

$$Si_{\overline{p}} = \frac{V_Z - x^*}{y} \cdot Si_g.$$

Durch Einsetzen der Werte für A und $Si_{\overline{p}}$ in Gleichung (V, 20) ergibt sich:

$$\frac{dy}{dt} = {}_{2b}K_w \cdot \frac{x^* + y + \frac{\pi b^2}{4} \cdot \frac{b}{3}}{\frac{b}{4}} \cdot \frac{(V_Z - x^*)}{y} \cdot Si_g.$$

Diese Gleichung kann anlog (IV, 56), wie auf S. 166 angegeben, integriert werden, womit folgt:

$$\left.\begin{aligned} {}_{2b}K_w = \frac{b}{4} \cdot \frac{1}{(V_Z - x^*) \cdot Si_g (t_2 - t_1)} \times \\ \times \left[y_2 - y_1 - \left(x^* + \frac{\pi b^2}{4} \cdot \frac{b}{3}\right) \cdot \ln \frac{x^* + \frac{\pi b^2}{4} \cdot \frac{b}{3} + y_2}{x^* + \frac{\pi b^2}{4} \cdot \frac{b}{3} + y_1}\right] \end{aligned}\right\}. \tag{V, 22}$$

Substituiert man nach Formel (IV, 55), (III, 1) sowie (III, 4) und vernachlässigt man die Größe des nicht-lösenden Raumes, so errechnet sich schließlich:

$$ {}_{2b}K_w = \frac{b}{4} \cdot \frac{1}{h \cdot Si_g \cdot (t_2 - t_1)} \cdot \left[L_2 - L_1 - \frac{b}{3} \cdot \ln \frac{L_2}{L_1}\right] \qquad \text{(V, 23)} $$

$$ \left[P = \frac{d}{4tO} \cdot \left[\frac{l_2 - l_1}{L - \frac{1}{3} \cdot d} - \frac{d}{3\left(L - \frac{1}{3} d\right)} \cdot \log_e\left(\frac{l_2}{l_1}\right)\right] \quad \text{p. 135 (6)}\right]. $$

Die osmotische Saugkraft des Zellinhaltes bei Grenzplasmolyse Si_g kann aus Versuchen mit isotonischen Zuckerkonzentrationen leicht bestimmt werden (vgl. SCARTH 1939, p. 142).

Für Zellen, welche die Bedingung $\frac{L}{b} \geqq 4$ erfüllen, kann statt (V, 23) in erster Näherung die einfachere Beziehung

$$ {}_{2c}K_w = \frac{b}{4} \cdot \frac{L_2 - L_1}{h \cdot Si_g \cdot (t_2 - t_1)} \qquad \text{(V, 24)} $$

$$ \left[P = \frac{d\,(l_2 - l_1)}{4tOL} \quad (8)\right] $$

gesetzt werden (p. 136).

$\beta\beta$) Endformeln für Versuche, bei welchen die *Deplasmolysezeit* gemessen wird und ein *nicht-lösender Raum nicht berücksichtigt* ist.

Man geht hierbei von Gleichung (V, 23) aus, in welche man für $t_2 - t_1$ die Deplasmolysezeit $T^* = t_E - t_0$ einsetzt. Es wird dann $L_1 = L_0$, die Protoplastenlänge im Moment des Übertragens der Zelle aus dem Vorplasmolyticum in das Wasser, d. h. also am Ende der Vorplasmolyse. Für L_2 ist nach (III, 5) unter Voraussetzung (xx) zunächst die *fiktive Protoplastenlänge* L^* für die volle Rückdehnung einzuführen. Wählt man schließlich zur Vorplasmolyse eine Konzentration, die doppelt so hoch ist wie der osmotische Wert der betreffenden Zelle, so wird $S_l = 2 \cdot Si_g$, und es folgt aus der Volumen-Saugkraftbeziehung [Voraussetzung (bb)]:

$$ \left(L^* - \frac{b}{3}\right) \cdot Si_g = \left(L_0 - \frac{b}{3}\right) \cdot 2\,Si_g; \qquad L_0 - \frac{b}{3} = \frac{L^* - \frac{b}{3}}{2}. \qquad \text{(V, 25)} $$

Einsetzen dieser Werte in (V, 23) führt zu:

$$ {}_{2d}K_w = \frac{b}{4} \cdot \frac{1}{\left(L^* - \frac{b}{3}\right) \cdot Si_g \cdot T^*} \cdot \left[L^* - \left(\frac{L^* - \frac{b}{3}}{2} + \frac{b}{3}\right) - \frac{b}{3} \cdot \ln \frac{L^*}{\frac{L^* - \frac{b}{3}}{2} + \frac{b}{3}}\right]. $$

Schließlich folgt mit Formel (III, 5):

$$ {}_{2d}K_w = \frac{b}{8} \cdot \frac{1}{Si_g \cdot T^*} \cdot \left[1 - \frac{2}{3} \cdot \frac{b}{h} \cdot \ln \frac{2\left(h + \frac{b}{3}\right)}{h + \frac{2b}{3}}\right] \qquad \text{(V, 26)} $$

$$ \left[P = \frac{d}{8tO}\left(1 - \frac{2d}{3\left(L - \frac{1}{3} d\right)} \cdot \log_e \frac{2L}{L + \frac{1}{3} d}\right) \quad \text{p. 135 (5)}\right]. $$

Für Zellen, bei welchen die Bedingung $\frac{h}{b} \geq 4$ erfüllt ist, kann diese Formel in guter Näherung wieder vereinfacht werden:

$$_{2e}K_w = \frac{b}{8} \cdot \frac{1}{Si_g \cdot T^*} \qquad \textbf{(V, 27)}$$

$$\left[P = \frac{d}{8tO} \quad (7)\right].$$

Benützt man wegen der einfacheren Berechnungsweise den Differenzenansatz (V, 18), so vereinfacht er sich bei Deplasmolyse in Wasser ($S_l = 0$) unter Nichtbeachtung des Vorzeichens von $_2K_w$ zu der Formel:

$$\frac{V_2 - V_1}{t_2 - t_1} = {}_{2f}K_w \cdot Si \cdot A. \qquad (V, 28)$$

Man setzt nun für die Oberfläche A und die Saugkraft Si jene Mittelwerte A_m und Si_m, die sich beim mittleren Volumen $V_m = \frac{V_1 + V_2}{2}$ ergeben, ein, wählt als V_1 den Anfangswert V_0 zu Beginn der Deplasmolyse und als V_2 den Wert V_Z am Schluß der Deplasmolyse, entsprechend der Deplasmolysezeit $T^* = t_E - t_0$. Erfolgt die Vorplasmolyse wieder in einer doppelt so hohen Konzentration als der osmotische Grundwert der Zelle ($C = 2 \cdot O_g$), so wird das mittlere Volumen V_m nach (IV, 35):

$$V_m = \frac{3}{4} \cdot V_Z.$$

Die diesem Wert entsprechende mittlere Saugkraft Si_m ergibt sich damit aus der Volumen-Saugkraftbeziehung [Voraussetzung (bb)] $Si_g \cdot V_Z = Si_m \cdot V_m$ zu:

$$Si_m = \frac{4}{3} \cdot Si_g. \qquad (V, 29)$$

Für die mittlere Oberfläche A_m folgt nach (IV, 37) und (IV, 35):

$$A_m = \frac{\frac{3}{4} \cdot V_Z + \frac{\pi b^2}{4} \cdot \frac{b}{3}}{\frac{b}{4}}. \qquad (V, 30)$$

Einführen dieser Werte in Gleichung (V, 28) gibt zunächst

$$_{2f}K_w = \frac{V_Z - \frac{V_Z}{2}}{T^* \cdot \frac{4}{3} \cdot Si_g \cdot \frac{\frac{3}{4} \cdot V_Z + \frac{\pi b^2}{4} \cdot \frac{b}{3}}{\frac{b}{4}}}.$$

Daraus folgt schließlich mit Formel (III, 4) und unter Voraussetzung (xx):

$$_{2f}K_w = \frac{3}{8} \cdot \frac{b \cdot h}{T^* \cdot Si_g \cdot \left(3h + \frac{4}{3} \cdot b\right)} \qquad \textbf{(V, 31)}$$

$$\left[P = \frac{3d\left(L - \frac{1}{3}d\right)}{8tO\left(3L + \frac{1}{3}d\right)} \quad (10)\right].$$

$\gamma\gamma$) Endformel für *zwei beliebige Deplasmolysegrade* unter *Berücksichtigung eines nicht-lösenden Raumes.*

Wählt man zur Vorplasmolyse eine Konzentration, die doppelt so hoch ist wie der osmotischen Grundwert der betreffenden Zelle ($C = 2 \cdot O_g$), so wird $S_l = 2 \cdot Si_g$, und damit läßt sich aus der Volumen-Saugkraftbeziehung [Voraussetzung (bb)]

$$(V_Z - x^*) \cdot Si_g = (V_0 - x^*) \cdot 2\, Si_g$$

x^* ausrechnen:

$$x^* = 2 \cdot V_0 - V_Z.$$

Nach Einsetzen dieser Werte in Gleichung (V, 22) und Substitution von x^* nach Formel (IV, 55) gilt:

$${}_{2g}K_w = \frac{b}{4} \cdot \frac{1}{2 \cdot (V_Z - V_0) \cdot Si_g \cdot (t_2 - t_1)} \cdot \left[V_2 - V_1 - \left(2 V_0 - V_Z + \frac{\pi b^2}{4} \cdot \frac{b}{3}\right) \cdot \ln \frac{V_2 + \frac{\pi b^2}{4} \cdot \frac{b}{3}}{V_1 + \frac{\pi b^2}{4} \cdot \frac{b}{3}} \right],$$

woraus mit (III, 1) und (III, 4) wieder folgt:

$${}_{2g}K_w = \frac{b}{8} \cdot \frac{1}{\left(h + \frac{b}{3} - L_0\right) \cdot Si_g\,(t_2 - t_1)} \cdot \left[L_2 - L_1 - \left(2\, L_0 - h + \frac{b}{3}\right) \ln \frac{L_2}{L_1} \right] \quad \text{(V, 32)}$$

$$\left[P = \frac{d}{8\, t\, O} \cdot \left[\frac{l_2 - l_1}{L - l_{\min}} - \frac{2\, l_{\min} - L}{L - l_{\min}} \cdot \log_e \frac{l_2}{l_1} \right] \quad (4) \right].$$

$\delta\delta$) Endformel für Versuche, bei welchen die *Deplasmolysezeit* gemessen wird und ein *nicht-lösender Raum berücksichtigt* ist.

Erfolgt die Vorplasmolyse wieder in einer Außenlösung von der Konzentration $C = 2 \cdot O_g$, so kann in (V, 32) für $t_2 - t_1$ die Deplasmolysezeit T^* eingeführt werden, wenn $L_1 = L_0$ (die Länge des Protoplasten beim Deplasmolysebeginn) und $L_2 = L^*$ (die fiktive Protoplastenlänge; vgl. S. 177), wird. Es folgt dann aus Gleichung (V, 32) mit (III, 5):

$${}_{2h}K_w = \frac{b}{8} \cdot \frac{1}{Si_g \cdot T^*} \cdot \left[1 - \frac{2\, L_0 - \left(h + \frac{b}{3}\right)}{h + \frac{b}{3} - L_0} \cdot \ln \frac{h + \frac{b}{3}}{L_0} \right] \quad \text{(V, 33)}$$

$$\left[P = \frac{d}{8\, t\, O} \cdot \left[1 - \frac{2\, l_{\min} - L}{L - l_{\min}} \cdot \log_e \frac{L}{l_{\min}} \right] \quad (3) \right].$$

γ) Deplasmolyse *zylindrischer Zellen* in *hypotonischer Außenlösung.* Man geht vom Differenzenansatz (V, 18) aus und wählt die osmotische Saugkraft der hypotonischen Außenlösung halb so groß wie jene des Zellinhaltes bei Grenzplasmolyse $\left(S_l = \frac{Si_g}{2}\right)$, womit Formel (V, 18) in

$$\frac{V_2 - V_1}{t_2 - t_1} = {}_{2i}K_w \cdot \left(Si - \frac{Si_g}{2}\right) \cdot A \quad \text{(V, 34)}$$

übergeht.

Bestimmt man wieder die Deplasmolysezeit $T^* = t_E - t_0$, so werden die zugehörigen Vacuolenvolumina V_Z und V_0. Für die Saugkraft Si des plasmolysierten Protoplasten und dessen Oberfläche A werden die Werte beim mittleren Volumen $V_m = \frac{V_0 + V_Z}{2}$ eingesetzt. Hierzu wird noch festgelegt, daß die Vorplasmolyse in einer Konzentration $C = 2 \cdot O_g$ zu erfolgen habe, so daß V_m nach Formel (IV, 35) und dadurch die mittlere Saugkraft Si_m der Vacuole nach (V, 29) sowie die mittlere Oberfläche A_m nach (V, 30) bestimmt sind. Damit wird nun (V, 34):

$$\frac{V_Z - \frac{V_Z}{2}}{T^*} = {}_{2i}K_w \cdot Si_g \cdot \left(\frac{4}{3} - \frac{1}{2}\right) \cdot \frac{\frac{3}{4} \cdot V_Z + \frac{\pi b^2}{4} \cdot \frac{b}{3}}{\frac{b}{4}},$$

woraus sich mit Gleichung (III, 1) und (III, 4) errechnet:

$${}_{2i}K_w = \frac{3 \cdot b \cdot h}{5 \cdot T^* \cdot Si_g \cdot \left(3\,h - \frac{4b}{3}\right)} \qquad \text{(V, 35)}$$

$$\left[P = \frac{3\left(L - \frac{1}{3}d\right)d}{5\,t\,O\left(3\,L + \frac{1}{3}d\right)} \qquad \text{p. 137 approx (12)}\right].$$

δ) *Deplasmolyse kugeliger Protoplasten oder Zellen in Wasser bei Vernachlässigung eines nicht-lösenden Raumes.* Geht man von (V, 20) aus und ist für ein Volumen V_Z die Saugkraft Si_Z des Zellinhaltes bekannt, so kann mittels der Volumen-Saugkraftbeziehung für jedes beliebige Volumen die Saugkraft $S\,i_{\overline{p}}$ errechnet werden:

$$\frac{Si_Z \cdot V_Z}{V_1} = Si_1 .$$

Drückt man ferner die Kugeloberfläche nach Gleichung (III, 10) und (III, 11) durch ihr Volumen aus, nämlich

$$A = \sqrt[3]{36\,\pi\,V^2},$$

so wird Formel (V, 20):

$$\frac{dV}{dt} = {}_{2j}K_w \cdot \frac{S\,i_Z \cdot V_Z}{V} \cdot \sqrt[3]{36 \cdot \pi \cdot V^2}.$$

Durch Trennung der Variablen

$$dV \cdot V^{\frac{1}{3}} = {}_{2j}K_w \cdot S\,i_Z \cdot V_Z \cdot \sqrt[3]{36\,\pi} \cdot dt$$

und Integration zwischen den Grenzen 1 und 2 folgt:

$$\frac{3}{4} \cdot \left(V_2^{\frac{4}{3}} - V_1^{\frac{4}{3}}\right) = {}_{2j}K_w \cdot S\,i_Z \cdot V_Z \cdot \sqrt[3]{36\,\pi} \cdot (t_2 - t_1).$$

Einsetzen von V gemäß Gleichung (III, 10) gibt endlich:

$${}_{2j}K_w = \frac{d_2^4 - d_1^4}{8 \cdot Si_Z \cdot d_Z^3 \cdot (t_2 - t_1)} \qquad \text{(V, 36)}$$

$$\left[P = \frac{r^4 - r_0^4}{4\,t\,O \cdot R^3} \quad (15)\right].$$

Wird als d_2 der Wert d_E zur Berechnung benutzt, ist die osmotische Saugkraft der vorplasmolysierenden Lösung $S_l = 2 \cdot S\,i_E$ und wählt man für d_1 den Wert d_0, erfolgt also die Messung unmittelbar vor dem Beginn der Ausdehnung, so bestimmt sich nach der Volumen-Saugkraftbeziehung [Voraussetzung (bb)]

$$S_l \cdot V_0 = S\,i_E \cdot V_E$$

schließlich $2 \cdot V_0 = V_E$ und damit nach (III, 10) $2 \cdot d_0^3 = d_E^3$. Dies in (V, 36) eingesetzt, gibt:

$${}_{2k}K_w = \frac{\left[1 - \left(\frac{1}{2}\right)^{\frac{4}{3}}\right] \cdot d_E}{8 \cdot S\,i_E \cdot (t_E - t_0)},$$

woraus folgt:

$$_{2k}K_w = \frac{0{\cdot}075 \cdot d_E}{S i_E \cdot (t_E - t_0)} \qquad \text{(V, 37)}$$

$$\left[P = \frac{0{\cdot}15 \cdot R}{t \cdot O} \quad (16)\right].$$

3. Ausgangsgleichungen für Konzentrationsdifferenzen.

Als Konzentrationen wurden zunächst diejenigen der osmotisch wirksamen Substanzen, die ja die Saugkräfte unmittelbar hervorrufen und daher zweckmäßigerweise selbst in die Ausgangsgleichungen eingesetzt sind (vgl. FREY-WYSSLING und RECHENBERG-ERNST 1943, p. 213), vorgeschlagen. Andererseits lassen sich aber auch *Wasserkonzentrationen* ableiten und als treibendes Prinzip der Wasserverschiebung postulieren.

aa) Konzentrationen von osmotisch wirksamen Substanzen.

α) Unter *Vernachlässigung der Protoplastenoberfläche* kann nach HÖFLER (1930, p. 322f.) die *zeitliche Änderung des Plasmolysegrades* während Plasmolyse oder Deplasmolyse *zylindrischer Zellen* zur Ableitung eines Maßes für die Wasserpermeabilität herangezogen werden, indem man (ohne Berücksichtigung des jeweiligen Vorzeichens von $_3K_w$) als Ausgangsgleichung formuliert:

$$\frac{dG}{dt} = {}_3K_w \cdot (C - c) \quad (\text{Dim}\, {}_3K_w = \mathsf{L}^3\,\mathsf{T}^{-1}). \qquad \underline{\text{(V, 38)}}$$

Darin bedeutet C die Außenkonzentration des Plasmolyticums, in welches das Objekt zur Plasmolyse bzw. Deplasmolyse aus einem nichtisotonischen anderen Milieu plötzlich übertragen wird, und c stellt die momentane Gesamtkonzentration der in der Vacuole vorhandenen osmotisch wirksamen Substanz dar, deren Menge dort nach Voraussetzung (*bb*) konstant bleibt.

Der Faktor $_3K_w$ *gibt als Maß für die Wasserpermeabilität die Plasmolysegradänderung [d. h. die Größe der Volumänderung des Protoplasten (in cm³), als Maß für die transportierte Wassermenge, je Volumeinheit der entspannten Zelle (1 cm³)] als reine Verhältniszahl (dimensionslos), je Zeiteinheit (1 sec) an, wenn der Unterschied der Konzentrationen der osmotisch wirksamen Substanzen von Vacuole und Außenlösung die Größe der Konzentrationseinheit (1 mol/cm³) besitzt* (vgl. HUBER und HÖFLER 1930, p. 482).

Aus der Volumen-Konzentrationsbeziehung [Voraussetzung (*bb*)] für die in der Vacuole ursprünglich vorhandene osmotisch wirksame Substanz m_u, also

$$c \cdot V = O_g \cdot V_Z,$$

folgt zunächst mit Gleichung (III, 8):

$$c = O_g \cdot \frac{V_Z}{V} = \frac{O_g}{G}.$$

Damit wird (V, 38):

$$\frac{dG}{dt} = {}_3K_w \cdot \left(C - \frac{O_g}{G}\right),$$

was zur Integration umgeformt ergibt:

$$dG\left(1 + \frac{O_g}{C} \cdot \frac{1}{G - \frac{O_g}{C}}\right) = C \cdot {}_3K_w \cdot dt.$$

Man integriert zwischen den Grenzen 1 und 2 und führt dann den Endplasmolysegrad G^*, den der Protoplast im betreffenden Plasmolyticum schließlich erreicht, ein. Der Protoplast besitzt dann die Protoplastenlänge L_E, so daß aus der Volumen-Konzentrationsbeziehung und nach Formel (III, 8) gilt:

$$G^* = \frac{O_g}{C} = \frac{V_E}{V_Z} = \frac{L_E - \frac{b}{3}}{h}.$$

Damit wird schließlich:

$$_3K_w = \frac{1}{C \cdot h \cdot (t_2 - t_1)} \cdot \left[L_2 - L_1 + \left(L_E - \frac{b}{3}\right) \cdot \ln \frac{L_2 - L_E}{L_1 - L_E}\right] \qquad \text{(V, 39)}$$

$$\left[k = \frac{1}{C\,(t_2 - t_1)} \cdot \left\{y_1 - y_2 + G \cdot \ln \frac{y_1 - G}{y_2 - G}\right\} \quad \text{(6, II)}\right].$$

β) Bei *Berücksichtigung der Oberfläche* kommt man zu einer analogen Ausgangsgleichung:

$$\frac{dm_w}{dt} = K_w \cdot A \cdot (C - c). \qquad \text{(V, 40)}$$

Mißt man die Wassermenge als Volumen V, so folgt unmittelbar (vgl. JACOBS 1935, p. 80; RESÜHR 1935, p. 340):

$$\frac{dV}{dt} = {}_4K_w \cdot A \cdot (C - c) \quad (\text{Dim } {}_4K_w = \mathsf{L}^4\mathsf{T}^{-1}). \qquad \text{(V, 41)}$$

Darin sind wieder C und c die Konzentrationen der osmotisch wirksamen Substanzen außerhalb und innerhalb des Protoplasmamantels bzw. der Zelle. Der Faktor $_4K_w$ *gibt als Maß für die Wasserpermeabilität an, wie groß das Volumen (in cm³) des je Zeiteinheit (1 sec) und Oberflächeneinheit (1 cm²) durch die Zell- bzw. Protoplastenoberfläche permeierenden Wassers ist, wenn der Unterschied der Konzentration der osmotisch wirksamen Substanzen von Vacuole und Außenlösung die Größe der Konzentrationseinheit (1 mol/cm³) besitzt.*

αα) Nach JACOBS und STEWART (1932, p. 75) läßt sich für nackte *kugelige Zellen* aus Formel (V, 41) *auf halbgraphischem Wege* der Faktor $_4K_w$ berechnen. Ist das Volumen V_E im osmotischen Gleichgewichtszustand mit der Außenlösung C bekannt, so kann die Konzentration c nach Voraussetzung (*bb*) für jedes andere Volumen berechnet werden:

$$c = \frac{V_E \cdot C}{V}.$$

Man setzt diese Beziehung in Gleichung (V, 41) ein und formt um:

$$_{4a}K_w = \frac{dV}{dt} \cdot \frac{1}{A} \cdot C \cdot \left(1 - \frac{V_E}{V}\right). \qquad \text{(V, 42)}$$

Die Volumänderungen der Zelle während des Versuches werden zeitlich genau verfolgt und in ein Zeit-Zellvolumendiagramm eingetragen, mit dessen Hilfe sich für einen beliebigen Wert von V der Differentialquotient dV/dt aus der Tangentensteigung angeben läßt. Da sich aus V mittels des Zelldurchmessers leicht die Oberfläche A der kugeligen Zelle errechnet, so sind alle in (V, 42) vorkommenden Größen bekannt, und der Wert von $_{4a}K_w$ ist damit zahlenmäßig zu ermitteln.

ββ) Für *zylindrische Zellen* mit endplasmolysierten Protoplasten läßt sich bei als *konstant bleibend angenommener Protoplastenoberfläche* die Formel (V, 41) leicht integrieren (HUBER und HÖFLER 1930, p. 482). Man erweitert hierzu vorerst mit dem Volumen der Zelle V_Z:

$$\frac{d\left(\frac{V}{V_Z}\right)}{dt} = \frac{dG}{dt} = {}_{4b}K_w \cdot \frac{A}{V_Z} \cdot (C - c). \qquad \text{(V, 43)}$$

Da der Quotient A/V_Z eine Konstante darstellt, erfolgt die weitere Ausrechnung nach dem auf S. 181 angegebenen Wege, so daß zunächst gilt:

$$_{4b}K_w = {_3K_w} \cdot \frac{V_Z}{A}. \qquad \text{(V, 44)}$$

Wird als die für die Permeation maßgebende Protoplastenoberfläche nur die Oberfläche der beiden halbkugeligen Menisken betrachtet, so formt sich der Quotient $\frac{V_Z}{A}$ mit Formel (III, 4) und der Beziehung

$$A = b^2 \cdot \pi$$

schließlich um zu dem Ausdruck:

$$\frac{V_Z}{A} = \frac{\pi b^2 \cdot h}{4} \cdot \frac{1}{b^2 \cdot \pi} = \frac{h}{4}$$

$$\left[\frac{V_0}{O} = \frac{h}{4}\right],$$

womit endlich folgt:

$$_{4b}K_w = \frac{h}{4} \cdot {_3K_w} = \frac{1}{4\,C \cdot (t_2 - t_1)} \cdot \left[L_2 - L_1 + \left(L_E - \frac{b}{3}\right) \cdot \ln \frac{L_1 - L_E}{L_2 - L_E}\right]. \qquad \textbf{(V, 45)}$$

bb) Wasserkonzentrationen.

Ein Unterschied der Wasserkonzentrationen von Vacuole und Außenlösung wird von FREY-WYSSLING (1945, p. 333; 1946, p. 133; vgl. BOCHSLER 1948) als maßgebend für den Wassertransport angenommen. Unter Wasserkonzentration ist hierbei, besonders beim Auflösen flüssiger Substanzen (z. B. Glycerin) in Wasser, der für dieses resultierende Anteil im Lösungsvolumen zu verstehen. Diese Wasserkonzentration unterscheidet sich gegenüber jener, welche beim Diffusionspotential angenommen wurde, in zweierlei Hinsicht: Einmal handelt es sich dort nur um einen verhältnismäßig geringen Anteil der gesamten Wassermenge, während *hier der immer weit überwiegende Teil des Volumens vom Wasser erfüllt* ist, und zweitens *gehört hier das verbleibende Restvolumen einem gelösten Stoff*, während es dort ebenfalls von Wasser eingenommen wird.

Als Ausgangsgleichung kann die (IV, 21) bzw. (V, 11) analoge Formel postuliert werden, wobei wieder das Vorzeichen von K_w stets positiv gewählt wird:

$$\frac{dm_w}{dt} = K_w \cdot A \cdot (k_{\overline{w}} - C_{\overline{w}}). \qquad \text{(V, 46)}$$

Mißt man die transportierte Wassermenge in mol, so gilt speziell:

$$\frac{dm_w}{dt} = {_5K_w} \cdot A \cdot (k_{\overline{w}} - C_{\overline{w}}) \qquad (\text{Dim}\ {_5K_w} = \mathsf{L\,T^{-1}}). \qquad \underline{\text{(V, 47)}}$$

Der Faktor $_5K_w$ gibt als Maß für die Wasserpermeabilität an, wie groß die Menge (in mol) des je Zeiteinheit (1 sec) und Oberflächeneinheit (1 cm²) durch die Protoplastenhaut hindurchtretenden Wassers ist, wenn der Unterschied der Wasserkonzentration von Vacuole und Außenlösung die Größe der Konzentrationseinheit (1 mol/cm³) besitzt.

Diese Formel wird für *zylindrische Zellen* ausgewertet, indem man zunächst die *Oberflächenentwicklung* $Q = \frac{A}{V}$ *als Konstante* annimmt und damit zur Gleichung

$$\frac{dc_{\overline{w}}}{dt} = {_5K_w} \cdot Q \cdot (c_{\overline{w}} - C_{\overline{w}}) \qquad \text{(V, 48)}$$

$$[d\overline{x}_2 = P_2 \cdot Q \cdot (\overline{x}_2 - \overline{C}_2) \cdot dt \qquad \text{(5a)}]$$

gelangt. Die Wasserkonzentrationen $C_{\overline{w}}$ und $c_{\overline{w}}$ werden nun als relative Volumina V^{rel} der Außenlösung und v^{rel} der Vacuole (cm³ Wasser/cm³ Lösung) ausgedrückt, wobei zur Umrechnung gilt (Bochsler 1948, p. 79):

$$V^{\mathrm{rel}} = C \cdot \frac{M}{\varrho}. \qquad \text{(V, 49)}$$

Dies in (V, 48) eingesetzt, führt zur Formel:

$$\frac{d v_{\overline{w}}^{\mathrm{rel}}}{d t} = {}_5K_w \cdot Q \cdot (v_{\overline{w}}^{\mathrm{rel}} - V_{\overline{w}}^{\mathrm{rel}}). \qquad \text{(V, 50)}$$

Es läßt sich nun zeigen (p. 80f.), daß die Verschiedenheit der chemischen Zusammensetzung von Außenlösung und Zellsaft ohne Einfluß bleibt und im Gleichgewichtszustand am Ende der Wasserpermeation die Wasserkonzentration in der Vacuole ($v_{\overline{w}E}^{\mathrm{rel}}$) ebenso groß ist, wie diejenige der Außenlösung ($V_{\overline{w}}^{\mathrm{rel}}$). Damit wird Formel (V, 50) zu:

$$\frac{d v_{\overline{w}}^{\mathrm{rel}}}{d t} = {}_5K_w \cdot Q \cdot (v_{\overline{w}}^{\mathrm{rel}} - v_{\overline{w}E}^{\mathrm{rel}}),$$

woraus durch Integration zwischen den Grenzen 1 und 2 unmittelbar folgt:

$${}_5K_w = \frac{1}{Q\,(t_2 - t_1)} \cdot \ln \frac{v_{\overline{w}2}^{\mathrm{rel}} - v_{\overline{w}E}^{\mathrm{re}}}{v_{\overline{w}1}^{\mathrm{rel}} - v_{\overline{w}E}^{\mathrm{rel}}} \qquad \text{(V, 51)}$$

$$\left[P_2 = \frac{1}{Q\,t} \cdot \ln \frac{\overline{V}_2' - \overline{V}_2}{\overline{y}_2 - \overline{V}_2} \quad (6)\right]$$

Für $t_1 = 0$ wird $v_{\overline{w}1}^{\mathrm{rel}} = v_{\overline{w}B}^{\mathrm{rel}}$, gleich dem relativen Volumen des Wassers im Zellsaft bei Versuchsbeginn.

Da die Summe der relativen Volumina des Wassers und der gelösten Stoffe im Zellsaft stets Eins ist, so gilt:

$$v_{\overline{w}}^{\mathrm{rel}} = 1 - v_u^{\mathrm{rel}}.$$

Nach Einsetzen in (V, 51) ergibt sich damit

$${}_5K_w = \frac{1}{Q\,(t_2 - t_1)} \cdot \ln \frac{v_{uE}^{\mathrm{rel}} - v_{u2}}{v_{uE}^{\mathrm{rel}} - v_{u1}}.$$

Es soll nun auch für die im Zellsaft vorhandenen gelösten Substanzen die Beziehung (V, 49) anwendbar sein, womit wird:

$${}_5K_w = \frac{1}{Q\,(t_2 - t_1)} \cdot \ln \frac{c_E - c_2}{c_E - c_1}.$$

Unter der Annahme, daß das absolute Volumen der osmotisch wirksamen Substanz im Zellsaftraum konstant bleibt (p. 83), läßt sich die Volumen-Konzentrationsbeziehung [Voraussetzung (*bb*)] anwenden:

$$c_E \cdot V_E = c_1 \cdot V_1 = c_2 \cdot V_2; \qquad c_2 = \frac{c_E \cdot V_E}{V_2}; \qquad c_1 = \frac{c_E \cdot V_E}{V_1},$$

woraus sich ergibt:

$${}_5K_w = \frac{1}{Q\,(t_2 - t_1)} \cdot \ln \frac{1 - \frac{V_E}{V_2}}{1 - \frac{V_E}{V_1}}$$

$$\left[P_2 = \frac{1}{Q \cdot t_{2-1}} \cdot \ln \frac{1 - \frac{G}{g_1}}{1 - \frac{G}{g_2}} \quad (12)\right].$$

Setzt man für V und Q schließlich nach Formel (III, 1) und (III, 9) ein und wählt zur Berechnung der Oberflächenentwicklung den Mittelwert $\frac{L_1 + L_2}{2}$, so resultiert als Schlußgleichung:

$$_5K_w = \frac{1}{t_2 - t_1} \cdot \frac{b}{4} \cdot \frac{L_1 + L_2 - \frac{2b}{3}}{L_1 + L_2} \cdot \ln \frac{(L_2 - L_0)\left(L_1 - \frac{b}{3}\right)}{(L_1 - L_0)\left(L_2 - \frac{b}{3}\right)}. \quad \text{(V, 52)}$$

VI. Die gegenseitige Vergleichbarkeit der nach den verschiedenen Formeln berechneten Transportwerte und Permeabilitätskonstanten.

Die quantitative Angabe der Größe des Stofftransportes oder der Permeabilität hat zum Ziel, das Ergebnis eines bestimmten Experimentes mit Resultaten anderer Versuche vergleichen zu können, und es soll nun untersucht werden, unter welchen Bedingungen und innerhalb welcher Grenzen dies erfolgen kann. Unter „*Vergleichen*" sei dabei stets das Urteil „*Der Permeabilitätsfaktor K (oder der Stofftransportwert M), welcher im Experiment A gefunden wurde, ist um einen Faktor* α *größer (kleiner) als jener, der sich aus dem Versuch B ergab* ($K_{(A)} = \alpha \cdot K_{(B)}$)" verstanden. Die Aussage, daß ein Permeabilitätsfaktor (bzw. Stofftransportwert) um eine gewisse Anzahl (n) seiner Maßeinheiten (E) größer oder kleiner ist als ein anderer Permeabilitätsfaktor ($K_{(A)} = K_{(B)} \pm n \cdot E$), wird seltener anzutreffen sein und läßt sich stets leicht in die Form des erstgenannten Urteils bringen.

Die Prüfung der für einen solchen Vergleich maßgebenden Faktoren ist, wie bei rein physikalischen Problemen, auch hier durchaus angezeigt und um so wichtiger, als möglichst viele mit verschiedenen Substanzen, Objekten und Versuchsanordnungen gewonnene Ergebnisse zum Vergleich herangezogen werden sollen.

Aus der Gegenüberstellung der verschiedenen Permeabilitätskonstanten werden meist Schlüsse über Eigenschaften und Struktur der diffusionshemmenden Schicht gezogen, was zu einem tieferen Verständnis der kausalen Zusammenhänge zwischen Transportmechanismus und Membranqualitäten führen soll und die Wichtigkeit einer Untersuchung der Vergleichbarkeit unterstreicht.

A. Allgemeines.

Die aus den oben behandelten Formeln für Stofftransportwerte und Permeabilitätskonstanten errechneten *Größen* stellen stets das *Produkt aus Maßzahl und Einheit* dar. Das Gewicht, welches den betreffenden Werten bei der Beurteilung zugelegt werden kann, ist hierbei von der *Genauigkeit* der Messung und der Kenntnis der *Streuung* bei gleichartigem Versuchsmaterial abhängig.

Für die sich aus einer Endformel ergebende Größe ist die *Genauigkeit* stets durch diejenige der einzusetzenden Meßergebnisse bedingt und kann jene nicht übersteigen. Eine exakte Berücksichtigung des Einflusses des Fehlers der einzelnen Werte auf das Resultat ist mittels des GAUSSschen Fehlerfortpflanzungsgesetzes möglich, wobei man bei einmaliger Bestimmung der Meßgrößen (α, β, γ, ...) als absolute mittlere Fehler ε_α, ε_β, ε_γ, die aus der Ablesegenauigkeit geschätzten Fehler einsetzt (z. B. betrage der Fehler der Messung der Protoplastenlänge $L \pm 0{,}2$ Skalenteile, was aussagt, daß die Ablesung an der

Mikrometerskala nur auf 0,2 Teilstriche genau erfolgen konnte). Gilt für das Endresultat (z. B. die Permeabilitätskonstante K) die allgemeine Funktion

$$K = f(\alpha, \beta, \gamma, \ldots)$$

so wird der Fehler ε_K von K nach der GAUSSschen Formel

$$\varepsilon_K = \pm \sqrt{\left(\frac{\delta f}{\delta \alpha} \cdot \varepsilon_\alpha\right)^2 + \left(\frac{\delta f}{\delta \beta} \cdot \varepsilon_\beta\right)^2 + \left(\frac{\delta f}{\delta \gamma} \cdot \varepsilon_\gamma\right)^2 + \ldots}$$

berechnet (vgl. FREY-WYSSLING und v. RECHENBERG-ERNST 1943, p. 204)[1]. Die Angabe der Ungenauigkeit erfolgt dabei nach dem Schema $K \pm \varepsilon_K$ oder andernfalls auch durch besondere Kennzeichnung der letzten Stelle (vgl. STILLE 1955, p. 32f.).

Die so erzielbare exakte Angabe der Genauigkeit des Stofftransportwertes bzw. der Permeabilitätskonstanten wird allerdings in vielen Fällen wieder durch die relativ große Variationsbreite, wie sie bei selbst homogenem Material auftritt, kompensiert. Die meisten Autoren versuchen diese Streuungen durch Mittelwertbildung bestimmter Meßgrößen, Zwischen- oder Endresultate zu berücksichtigen [vgl. z. B. die Berechnungsweise wie sie für ${}_2K_s$ und ${}_3K_s$ üblich ist (HÖFLER 1934b; HOFMEISTER 1935 p. 15; u. a.)]. Eine genauere Erfassung der jeweils auftretenden Streuungen wurde bisher kaum angestrebt (vgl. BOGEN 1950, p. 66f.).

Die Richtigkeit des Endresultates kann auch durch die Art der mathematischen Ableitung der betreffenden Endformel beeinflußt werden (z. B. Berücksichtigung der stetigen Änderungen Variabler oder nur Einsetzen von Mittelwerten). Die auf diese Weise entstehenden Fehler lassen sich bisweilen aus einer

[1] Als Berechnungsbeispiel zur Ableitung der Formel für ε_K sei die verhältnismäßig einfache Gleichung (IV, 81) für ${}_{2d}K_s$ (HOFMEISTER 1948) gewählt:

$${}_{2d}K_s = \frac{2 \cdot (C - O_g)}{T^* (C + O_g)}.$$

Der Permeabilitätsfaktor ${}_{2d}K_s$ ist hierbei eine Funktion von C, O_g und T^*. Die partielle Differentiation obiger Gleichung nach diesen drei Größen ergibt:

$$\frac{\delta\, {}_{2d}K_s}{\delta C} = \frac{4 \cdot T^* \cdot O_g}{[T^* (C + O_g)]^2},$$

$$\frac{\delta\, {}_{2d}K_s}{\delta O_g} = \frac{-4\, T^* \cdot C}{[T^* (C + O_g)]^2},$$

$$\frac{\delta\, {}_{2d}K_s}{\delta T^*} = \frac{-2\,(C^2 + O_g^2)}{[T^* (C + O_g)]^2}.$$

Damit wird nun die Formel für den Fehler $\varepsilon_{{}_{2d}K_s}$:

$$\varepsilon_{{}_{2d}K_s} = \pm \frac{1}{[T^* (C + O_g)]^2} \cdot \sqrt{16\,(T^* \cdot O_g \cdot \varepsilon_C)^2 + 16\,(T^* \cdot C \cdot \varepsilon_{O_g})^2 + 4\,[(C^2 + O_g^2) \cdot \varepsilon_{T^*}]^2}$$

und hieraus folgt schließlich:

$$\varepsilon_{{}_{2d}K_s} = \pm \frac{2}{[T^* (C + O_g)]^2} \cdot \sqrt{4 \cdot T^{*2} \cdot (O_C^2 \cdot \varepsilon_c^2 + C^2 \cdot \varepsilon_{O_g}^2) + [(C^2 + O_g^2) \cdot \varepsilon_{T^*}]^2}.$$

Durch Einsetzen der für die betreffenden Versuche zu erwartenden Zahlenwerte von T^*, O_g, C, ε_{T^*}, ε_{O_g}, und ε_C läßt sich aus dieser Formel bereits erkennen, welcher der drei Meßfehler ε_{T^*}, ε_{O_g} und ε_C verhältnismäßig den größten Beitrag für $\varepsilon_{{}_{2d}K_s}$ liefert und daher möglichst klein zu halten ist.

genaueren Analyse der betreffenden Gleichungen erschließen (vgl. STADELMANN 1951, p. 782).

Die *Maßeinheiten* für die Stofftransportwerte und die Permeabilitätsfaktoren sind oft zusammengesetzter Natur und leiten sich meist von den Grundeinheiten für Länge, Masse und Zeit ab (cm, μ, Mikrometereinheiten; g; sec, min, Std). Die hier als dimensionslos betrachtete Menge wird vorwiegend in mol gemessen. Hinsichtlich der Konzentrationsangaben ist noch besonders zu beachten, daß diese besonders, wenn osmotische Phänomene wirksam werden, stets auf das Volumen und nicht auf das Gewicht der Lösung zu beziehen sind [volummolare Lösungen (d. h. n Grammoleküle Substanz in 1 Liter Lösung) statt gewichtsnormaler Lösungen].

Obwohl die Angabe der Maßeinheiten eine unerläßliche Voraussetzung für weitere Verwendbarkeit der zahlenmäßigen Resultate darstellt, werden sie in der Literatur häufig gar nicht, bisweilen auch falsch, angeführt. Oft sind die Grundeinheiten auch als Dimensionen bezeichnet, ein Vorgehen, welches zwar naheliegend, aber unrichtig ist, da als Dimensionen wohl Länge, Masse, Zeit, Temperatur usw. und deren Produkte (besser Dimensionsprodukte genannt), nicht aber die für sie eingeführten Maßeinheiten gelten[1]. Letztere entsprechen, soweit in obigen Kapiteln überhaupt angeführt, meist dem CGS-System, um hierdurch den Anschluß an die überwiegend auf diesem System basierenden anderen physikalischen Meßgrößen zu erleichtern.

Für den *unmittelbaren Vergleich zweier Permeabilitätsfaktoren oder Stofftransportwerte* sind stets zwei Bedingungen zu erfüllen: Die betreffenden Größen müssen sich, wie dies bekanntlich allgemein für physikalische Größen gilt, *von derselben Definitionsgleichung* (hier als Ausgangsgleichung bezeichnet) *ableiten*[2] und in den *gleichen Maßeinheiten gemessen* werden. Sind in der Definitionsgleichung bestimmte Parameter, die mit Objekt oder Versuch variieren können (z. B. die Zellgröße bzw. -oberfläche), nicht erfaßt, so ist ein Vergleich solcher Endresultate nur möglich, wenn diese Parameter dieselbe Größe besitzen (vgl. FREY-WYSSLING und RECHENBERG-ERNST 1943, p. 204f.), sofern man nicht bewußt auf deren Berücksichtigung verzichtet (Konzentration in der Definitionsgleichung für die Stofftransportwerte).

Die Übereinstimmung zweier K- bzw. M-Werte in ihrer Dimension ist zwar eine notwendige, aber nicht hinreichende Voraussetzung für ihre Vergleichbarkeit, wie auch aus der Physik Beispiele bekannt sind, daß zwei ihrer Natur nach völlig verschiedene Größen dieselbe Dimension besitzen können [z. B. Energie und Drehmoment, Länge und Kapazität (im elektrostatischen LMT-System)].

Da sich alle *Stofftransportwerte* [sowie auch die Wassertransportwerte mit Formel (V, 1)] nach derselben Definitionsgleichung (IV, 1) berechnen, so ist zur Ermöglichung ihrer Vergleichbarkeit allein zu beachten, daß die darin vorkommenden Größen mit denselben Maßeinheiten gemessen werden. Insbesondere ist dies für die Mengen m_1 und m_2 erforderlich, deren Angabe auf sehr verschiedene Weisen erfolgen kann (in mol, g, g-Äqu usw., bei Wasser auch als Volumen). Betrachtet man die Gesetzmäßigkeiten der Diffusion und Osmose während der Rückdehnungsphase als gültig, lassen sich auch die mittels Gleichung (IV, 5) ableitbaren Stofftransportwerte mit den aus (IV, 1) folgenden vergleichen.

Entstammen die beiden zu vergleichenden K-Werte verschiedenen Ausgangsgleichungen, so muß versucht werden, einen mittelbaren Vergleich zu ermöglichen.

---

[1] So ist z. B. das Dimensionsprodukt von ${}_1K_s$ der Quotient Länge/Zeit, als Maßeinheiten können aber verschiedene Größen gewählt werden, etwa cm/sec, cm/min, μ/sec, '/sec usw.

[2] Allen Formelzeichen für Permeabilitätskonstanten, welche von derselben Ausgangsgleichung abstammen, ist als Index die gleiche Ziffer vorangestellt. Die Ausrechnung dieser Gleichungen für die speziellen Versuchsverhältnisse führt dann zu verschiedenen Endformeln für K, die gegeneinander durch Kleinbuchstaben als zweiten vorangestellten Index unterschieden sind [z. B. bedeutet ${}_{1c}K_s$ die dritte (Buchstabe c) Permeabilitätskonstante für gelöste Substanzen (Index s), welche aus der Definitionsgleichung für ${}_1K_s$ abgeleitet wurde].

B. Der mittelbare Vergleich von Permeabilitätskonstanten verschiedener Ausgangsgleichungen.

Es erweist sich hierzu als zweckmäßig, die Ausgangsgleichungen nach Tabelle 2 in drei Kategorien einzuteilen. Die erste Gruppe enthält dabei jene Definitionsgleichungen, bei welchen die transportierte Stoffmenge bzw. die sie enthaltende abhängige Variable mit einem Gradienten ihrer eigenen Konzentration (bzw. ihres osmotischen Druckes usw.) in Beziehung gesetzt wird, wobei die Verhältnisse derart liegen sollen, daß die Gesetzmäßigkeiten, wie sie für verdünnte Lösungen gelten, zutreffen.

Die zweite Gruppe umfaßt Ausgangsgleichungen, bei welchen die transportierte Stoffmenge (es handelt sich hierbei stets um Wasser) mit einem Gradienten von Konzentration oder osmotischem Druck einer anderen Substanz korreliert,

Tabelle 2. *Gruppierung der Definitionsgleichungen der behandelten Permeabilitätsfaktoren (mit Angabe ihrer Dimension und derjenigen Autoren, deren hier angeführte Endformeln sich auf die betreffende Ausgangsgleichung zurückführen lassen).*

Gruppe 1:

$$ {}_1K_s = \frac{dm}{dt} \cdot \frac{1}{A \cdot (C - k)} \qquad (\mathrm{Dim}\ {}_1K_s = \mathsf{L\,T^{-1}}) \qquad \underline{(\mathrm{IV}, 21)} $$

(LEPESCHKIN 1908, 1909, JACOBS und STEWART 1932, COLLANDER und BÄRLUND 1933, ULLRICH 1935, RESÜHR 1936, SCARTH 1939, WARTIOVAARA 1942, BOCHSLER 1948, STADELMANN 1951).

$$ {}_2K_s = \frac{\Delta m_s}{V_Z \cdot (t_2 - t_1)} \cdot \frac{1}{C - k} \qquad (\mathrm{Dim}\ {}_2K_s = \mathsf{T^{-1}}) \qquad \underline{(\mathrm{IV}, 71)} $$

(HÖFLER 1934b, HOFMEISTER 1935, 1948).

$$ {}_3K_s = \frac{\Delta m_s}{t_2 - t_1} \cdot \frac{V_p}{V_Z} \cdot \frac{1}{A_p \cdot (C - k)} \qquad (\mathrm{Dim}\ {}_3K_s = \mathsf{L\,T^{-1}}) \qquad \underline{(\mathrm{IV}, 84)} $$

(HÖFLER 1934b).

$$ {}_4K_s = \frac{dk}{dt} \cdot \frac{1}{C - k} \qquad (\mathrm{Dim}\ {}_4K_s = \mathsf{T^{-1}}) \qquad \underline{(\mathrm{IV}, 90)} $$

(BÄRLUND 1929).

$$ {}_1K_w = \frac{dm_w}{dt} \cdot \frac{1}{A\,(C_w - k_w)} \qquad (\mathrm{Dim}\ {}_1K_w = \mathsf{L\,T^{-1}}) \qquad \underline{(\mathrm{V}, 12)} $$

(WARTIOVAARA 1944).

Gruppe 2:

$$ {}_2K_w = \frac{dV}{dt} \cdot \frac{1}{A\,(S_l - S_i)} \qquad (\mathrm{Dim}\ {}_2K_w = \mathsf{L^2\,M\,T^{-1}}) \qquad \underline{(\mathrm{V}, 17)} $$

(RESÜHR 1935, SCARTH 1939).

$$ {}_3K_w = \frac{dG}{dt} \cdot \frac{1}{C - c} \qquad (\mathrm{Dim}\ {}_3K_w = \mathsf{L^3\,T^{-1}}) \qquad \underline{(\mathrm{V}, 38)} $$

(HÖFLER 1930).

$$ {}_4K_w = \frac{dV}{dt} \cdot \frac{1}{A\,(C - c)} \qquad (\mathrm{Dim}\ {}_4K_w = \mathsf{L^4\,T^{-1}}) \qquad \underline{(\mathrm{V}, 41)} $$

(HUBER und HÖFLER 1930, JACOBS und STEWART 1932).

Gruppe 3:

$$ {}_5K_w = \frac{dm_w}{dt} \cdot \frac{1}{A\,(k_{\overline{w}} - C_{\overline{w}})} \qquad (\mathrm{Dim}\ {}_5K_w = \mathsf{L\,T^{-1}}) \qquad \underline{(\mathrm{V}, 47)} $$

(BOCHSLER 1948).

wobei ebenfalls die Gesetzmäßigkeiten verdünnter Lösungen angewandt werden können. Die dritte Kategorie schließlich enthält eine Definitionsgleichung, in welcher die transportierte Stoffmenge (Wasser) zu einem Gradienten ihrer eigenen Konzentration (Wasserkonzentration) in Beziehung gesetzt wird, wobei aber die Gesetzmäßigkeiten für verdünnte Lösungen nicht mehr gelten.

Es fällt auf, daß in allen hier gezeigten Definitionsgleichungen für Permeabilitätsfaktoren die Dicke der diffusionshemmenden Schicht nicht auftritt (vgl. hingegen MANEGOLD und STÜBER 1933, p. 316). Durch dieses Vorgehen, welches der Annahme konstanter Dicke der permeationshemmenden Schicht gleichkommt [Voraussetzung (*t*)] und durch die meist großen Schwierigkeiten der Bestimmung der Membrandicke bedingt ist, unterscheidet sich die Permeabilitätskonstante prinzipiell von der Diffusionskonstanten (vgl. FREY-WYSSLING und RECHENBERG-ERNST 1943, p. 206; S. 143).

Betrachtet man die Vergleichsmöglichkeiten der K-Werte innerhalb jeder der einzelnen Gruppen, so ist leicht zu erkennen, daß eine Umrechnung ohne Schwierigkeiten durchführbar ist. Hierzu sind lediglich die Faktoren, durch welche sich die betreffenden Ausgangsgleichungen voneinander unterscheiden, für die gewählten Maßeinheiten als zahlenmäßige und zum Teil dimensionsbehaftete Umrechnungsfaktoren anzugeben. Freilich können sich dabei insoweit Komplikationen ergeben, als die im Umrechnungsfaktor auftretenden Größen bei den betreffenden Experimenten vielleicht nicht ermittelt wurden, so daß durch deren nachträgliche Schätzung eine größere Ungenauigkeit entstehen kann. Diese Schwierigkeiten bleiben aber hier, wo es sich um die Diskussion der prinzipiellen Möglichkeit einer Umrechnung handelt, ohne Belang. Natürlich sind bei solchen Umrechnungen die oben erwähnten Bedingungen hinsichtlich Gleichheit der Maßeinheiten und Berücksichtigung eines variablen, nicht in der Ausgangsgleichung vorkommenden Parameters zu erfüllen.

Innerhalb Gruppe 1 ist der ${}_1K_w$-Wert überhaupt nur ein Sonderfall von ${}_1K_s$, bei welchem als geprüfte Substanz Wasser gewählt wird. Für ${}_4K_s$ läßt sich das Differential dk meist in einfacher Weise als $d(m/V)$ ausdrücken, womit sich bei bekanntem Volumen V sogleich die Beziehung zu ${}_2K_s$ herstellt und die Angabe des Umrechnungsfaktors für ${}_1K_s$ möglich ist (vgl. COLLANDER und BÄRLUND 1933, p. 54).

Auch innerhalb Gruppe 2 treten bei der Umrechnung keine Schwierigkeiten auf, da G nach Formel (III, 8) als V_p/V_Z und S nach Gleichung (II, 7) allgemein durch $C \cdot R \cdot T$ ersetzt werden kann.

Es verbleibt somit noch die Aufgabe, die Umrechnung von Permeabilitätsmaßen der ersten Gruppe in jene der zweiten Gruppe (und umgekehrt), durch welche sich auch die für Formel (V, 47) geeigneten Versuche beschreiben lassen, zu ermöglichen. Das Problem besteht dabei darin, eine Stoffbewegung, welche durch Diffusion verursacht ist, mit einer solchen, die aus der Wirkung von Kräften (osmotische Saugkraftdifferenzen), d. h. aus einer Filtration (vgl. FREY-WYSSLING 1946, p. 132f.) resultiert, zu vergleichen (vgl. WARTIOVAARA 1944, p. 1; LØVTRUP und PIGOŃ 1951, p. 3) und keineswegs nur in der rein formalen Bedingung, die beiden Permeabilitätskonstanten auf gleiche Dimension zu bringen, was ja meist durchführbar ist (vgl. DEAN 1947, p. 509). Angesichts der Wichtigkeit der Fragestellung, die zahlreichen Bestimmungen der Wasserpermeabilität aus den Beobachtungen einer osmotisch verursachten Wasserverschiebung mit den aus der Diffusion abgeleiteten Permeabilitätskonstanten für gelöste Substanzen in Beziehung zu setzen, wurden bereits mehrere Umrechnungsversuche unternommen, ohne daß diese offenbar eine endgültige Klärung des Problems zu bringen vermochten (vgl. RESÜHR 1936, p. 449f.; RASHEVSKY und LANDAHL 1940, p. 10f.; BOCHSLER 1948; LANDAHL 1948).

Bedeutung, Dimensionsprodukt und Maßeinheit der Formelzeichen.

a) Hauptsymbole.

	Dimensions-produkt[1]	Maßeinheit	
A	L^2	cm^2	Fläche der Membran
			Oberfläche
			Oberfläche des endplasmolysierten Protoplasten zylindrischer Zellen
b	L	cm	Innere Zellbreite
C	L^{-3}	mol/cm^3	Konzentration des permeierenden Stoffes
			Konzentration des permeierenden Stoffes in der Außenlösung
			In Formeln für die Wasserpermeabilität: Konzentration des (nicht permeierenden) Plasmolyticums
C^*	L^{-3}	mol/cm^3	Konzentration der permeierenden Substanz im Zellsaft, die im Gleichgewicht mit der Außenkonzentration C steht
c	L^{-3}	mol/cm^3	Konzentration der ursprünglich vorhandenen osmotisch wirksamen Substanz in der Vacuole
			Konzentration im Blutplasma
c^*	L^{-3}	mol/cm^3 $g\text{-Ion}/cm^3$	Konzentration des diffundierenden Stoffes innerhalb der Membran oder an deren Grenzflächen
	$\mathsf{L}^{-3}\,\mathsf{M}$	g/cm^3	
D	$\mathsf{L}^2\,\mathsf{T}^{-1}$	cm^2/sec	Diffusionskonstante
d	L	cm	Kugeldurchmesser
			Durchmesser einer kugeligen Zelle
E	L^{-3}	mol/cm^3	Konzentration eines nichtpermeierenden Stoffes in der Außenlösung
			Maßeinheit
F	$\mathsf{L}^{\frac{3}{2}}\,\mathsf{M}^{\frac{1}{2}}\,\mathsf{T}^{-1}$	Cb/g-Äqu Cb/g-Ion	FARADAYsche Konstante
G	0	—	Plasmolysegrad $= V_p/V_Z$
G^*	0	—	Endplasmolysegrad nach Einstellung des osmotischen Gleichgewichtes bei Plasmolyse oder Deplasmolyse in einer hypertonischen Außenlösung
H	L	cm	Länge eines Zellfadens
h	L	cm	Zellänge
			Innere Länge der entspannten Zelle
K	jeweils vermerkt		Permeabilitätskonstante (wegen der Indices s. S. 187, Anm. 2)
k	L^{-3}	mol/cm^3	Konzentration der permeierenden Substanz in der Vacuole
L	L	cm	Länge des plasmolysierten Protoplasten zylindrischer Zellen von Kuppe zu Kuppe
L^*	L	cm	Fiktive Protoplastenlänge für das Volumen der entspannten Zelle
L			Länge (als Dimension)
M	$\mathsf{L}^{-2}\,\mathsf{T}^{-1}$	$mol/cm^2 \cdot sec$ $g\text{-Ion}/cm^2 \cdot sec$ $g\text{-Äqu}/cm^2 \cdot sec$	Stoffaustauschwert (Flux)
	$\mathsf{L}^{-2}\,\mathsf{M}\,\mathsf{T}^{-1}$	$g/cm^2 \cdot sec$	
	$\mathsf{L}^{-\frac{1}{2}}\,\mathsf{M}^{\frac{1}{2}}\,\mathsf{T}^{-2}$	$Cb/cm^2 \cdot sec$	
	M	g/mol	Molekulargewicht
M^*	$\mathsf{L}^{-3}\,\mathsf{T}^{-1}$	$mol/cm^3 \cdot sec$	Die je Zeiteinheit und Einheit des Volumens der entspannten Zelle in die Vacuole eintretende Substanzmenge (Produkt aus Außenkonzentration und Differenz des Plasmolysegrades je Zeiteinheit. Auf die Stunde bezogen: „Stundenwert der Stoffaufnahme" oder „je Stunde aufgenommene Lösungsmenge")
M			Masse (als Dimension)

[1] Für das elektrostatische $\mathsf{LMT\Theta}$-(Länge-Masse-Zeit-Temperatur) System angegeben.

	Dimensionsprodukt[1]	Maßeinheit	
m	0	mol	Menge des zu transportierenden Stoffes auf der einen Membranseite
	$L^{3/2} M^{1/2} T^{-1}$	Cb	
	M	g	
	L	cm	Höhe des Meniscus des plasmolysierten Protoplasten zylindrischer Zellen
N	0	mol	Gesamtmenge
O	L^{-3}	mol/cm³	Osmotischer Wert
p	$L^{-1} M T^{-1}$	Atm	Osmotischer Druck des Zellsaftes
Q	L^{-1}	cm⁻¹	Oberflächenentwicklung
Q^*	$L^{-1} M T^{-1}$	Atm	Osmotische Saugkraft der nicht-permeierenden Komponente der Außenlösung
R	$L^2 M T^{-2} \Theta^{-1}$	erg/grad · mol erg/grad · g-Ion	Gaskonstante
S	$L^{-1} M T^{-1}$	Atm	Osmotische Saugkraft
Si	$L^{-1} M T^{-1}$	Atm	Osmotische Saugkraft des Zellinhaltes
s	L	cm	Maximale Höhe einer rotationssymmetrischen Einschnürung in der Mitte eines plasmolysierten Protoplasten zylindrischer Zellen
T	Θ	grad	Absolute Temperatur
T^*	T	sec	Deplasmolysezeit
T			Zeit (als Dimension)
t	T	sec	Zeit
U	L^{-3}	mol/cm³	Gesamtkonzentration in der Außenlösung
u	$M^{-1} T$	cm/Dyn · sec	Beweglichkeit
V	L^3	cm³	Volumen
			Volumen der Außenlösung
			Volumen der Zelle
			Volumen des Zellsaftraumes = (bei Vernachlässigung des Protoplasmavolumens) Protoplastenvolumen
v	L^3	cm³	Volumen des Zellsaftraumes
X	L	cm	Dicke der Membran
x	L	cm	Abstand einer zur Membrangrenzfläche parallelen Fläche, die innerhalb der Membran liegt, von dieser Grenzfläche
x^*	L^3	cm³	Volumen des nicht-lösenden Raumes der Vacuole
y	L^3	cm³	Als Lösungsraum verfügbares Volumen der Vacuole
Z	L^{-3}	mol/cm³	Gesamtkonzentration in der Vacuole
ΔG	T^{-1}	sec⁻¹	Plasmolysegradänderung bezogen auf die Zeiteinheit
Δ			Differenz von ...
ε	0		Stöchiometrische Wertigkeit eines Ions (unbenannte Zahl)
	—	—	Mittlerer Fehler
Θ			Temperatur (als Dimension)
λ	0		Meniscusfaktor (dimensionslos)
π	$L^{1/2} M^{1/2} T^{-1}$	Volt	Elektrische Potentialdifferenz an den beiden Membrangrenzflächen
	0		LUDOLPHsche Zahl (3,14...)
	0	—	Temporärer osmotischer Koeffizient
ϱ	$L^{-3} M$	g/cm³	Dichte
φ	$L^{1/2} M^{1/2} T^{-1}$	Volt	Elektrisches Potential innerhalb der Membran in der Entfernung x von der Grenzfläche (für $x = 0$ meist als Null angenommen)

b) Indices.

n_A n in der Richtung von innen nach außen
n_a n an der Außenseite der äußeren Membrangrenzfläche
n_B n zu Versuchsbeginn
n beim Einsetzen der Permeation

[1] Für das elektrostatische $L M T \Theta$-(Länge-Masse-Zeit-Temperatur) System angegeben.

n_b n im Blut
n_E n am Versuchsende
n im osmotischen Gleichgewicht zwischen Außenlösung und Vacuole
n_F n einen Zellfaden betreffend
n_G n das Wachstum betreffend
n_g n bei Grenzplasmolyse
n_I n in der Richtung von außen nach innen
n_i n an der Außenseite der inneren Membrangrenzfläche
n_k n in der Körperflüssigkeit
n_l n in der Außenlösung
n_m Mittelwert von n_1 und n_2
n_p n des Protoplasten
$n_{\bar{p}}$ n bei Plasmolyse
n_{pl} n im Blutplasma
n_s n der permeierenden gelösten Substanz
n_u n der ursprünglich in der Vacuole vorhandenen osmotisch wirksamen Substanz
n_w n des Wassers
$n_{\bar{w}}$ n des Wassers bei Vorhandensein einer darin verdünnt gelösten Substanz
n_x n im Abstand x von einer der Membrangrenzflächen innerhalb der Membran
n_Z n der (entspannten) Zelle $= n$ des unplasmolysierten Protoplasten
n bei Plasmolyse mit Zucker
n in einer bestimmten als Bezugslösung benutzten Außenkonzentration
n_z n in der Zelle
n_0 n zu Beginn des osmotischen Gleichgewichtszustandes zwischen Vacuole und Außenlösung
n bei isotonischer Vorplasmolyse
$n_1, n_2, n_3, \ldots$ n gemessen zur Zeit $t_1, t_2, t_3, \ldots$
$n_\triangledown$ n ein Rotationsdreieck betreffend
n' n je Zeiteinheit
n^{rel} n gemessen als relativer Anteil
$n^{\%}$ n gemessen als relativer Anteil in Prozenten
$\bar{n}$ Mittelwert von n
n^* Siehe unter dem betreffenden Buchstaben
$_1, _2, \ldots K$ Fortlaufende Numerierung der Permeabilitätskonstanten (vgl. S. 187, Anm. 2).

Literatur.

ÄYRÄPÄÄ, T.: On the base permeability of *yeast*. Physiol. Plantarum (Copenh.) **3**, 402 bis 429 (1950).

BACHMANN, FR.: Zur Analyse von Permeabilitätsmessungen. I. Wasserpermeabilität. Planta (Berl.) **30**, 224—251 (1939). — BÄRLUND, H.: Permeabilitätsstudien an Epidermiszellen von *Rhoeo discolor*. Acta bot. fenn. **5**, 1—117 (1929). — BOCHSLER, A.: Die Wasserpermeabilität des Protoplasmas auf Grund des FICKschen Diffusionsgesetzes. Ber. schweiz. bot. Ges. **58**, 73—122 (1948). — BOGEN, H. J.: Kritische Untersuchungen über Permeabilitätsreihen. Planta (Berl.) **38**, 65—90 (1950). — BOGEN, H. J., u. H. PRELL: Messung nichtosmotischer Wasseraufnahme an plasmolysierten Protoplasten. Planta (Berl.) **41**, 459—479 (1953). — BOUILLENNE, R.: Contribution à l'étude des phénomènes d'osmose dans les cellules végétales. Bull. Acad. roy. Belg., Cl. Sci., Sér. V **16**, 1017—1026 (1930). — BROOKS, S. C.: A study of permeability by the method of tissue tension. Amer. J. Bot. **3**, 562—570 (1916a). — New determinations of permeability. Proc. Nat. Acad. Sci. U.S.A. **2**, 569—574 (1916b). — Penetration of radioactive isotopes P^{32}, Na^{24} and K^{42} into *Nitella*. J. Cellul. a. Comp. Physiol. **38**, 83—93 (1951). — BURSTRÖM, H.: Mineralstoffwechsel. In Fortschritte der Botanik, Bd. 16, S. 269—291. 1954.

COLLANDER, R.: The permeability of plant protoplasts to small molecules. Physiol. Plantarum (Copenh.) **2**, 300—311 (1949). — COLLANDER, R., u. H. BÄRLUND: Permeabilitätsstudien an *Chara ceratophylla*. Acta bot. fenn. **11**, 1—114 (1933).

DAVSON, H., and J. F. DANIELLI: The permeability of natural membranes. Cambridge: University Press 1943. 361 S. — DEAN, R. B.: The effects produced by diffusion in aqueous systems containing membranes. Chem. Reviews **41**, 503—523 (1947).

FICK, A.: Über Diffusion. Poggendorffs Ann. **94**, 59—86 (1855). — FITTING, H.: Untersuchungen über die Aufnahme von Salzen in die lebende Zelle. Jb. wiss. Bot. **56**, 1—64 (1915). — Untersuchungen über die Aufnahme und über anomale osmotische Koeffizienten von Glycerin und Harnstoff. Jb. wiss. Bot. **59**, 1—170 (1919). — FLEXNER, L. B., A. GELLHORN and M. MERRELL: Studies on rates of exchange of substances between the blood and extravascular fluid. J. of Biol. Chem. **144**, 35—40 (1942). — FREY-WYSSLING, A.: Die Turgorschwankung bei Permeabilitätsversuchen. Verh. naturforsch. Ges. Basel **56**, II, 330—342 (1945). — Zur Wasserpermeabilität des Protoplasmas. Experientia (Basel) **2**, 132—137 (1946). — FREY-WYSSLING, A., u. V. v. RECHENBERG-ERNST: Über die Wasserpermeabilität der Epithemzellen von Hydathoden. Flora (Jena) **137**, 193—215 (1943).

GRAHAM, TH.: Anwendung der Diffusion der Flüssigkeiten zur Analyse. Ann. Chem. Pharm. **121** (n. R. **45**), 1—77 (1861).

HAAN, IZ DE: Protoplasmaquellung und Wasserpermeabilität. Rec. Trav. bot. néerl. **30**, 234—335 (1930). — HEUSSER, K.: Neue vergleichende Permeabilitätsmessungen zur Kenntnis der osmotischen Verhältnisse der Pflanzenzelle im kranken Zustande. Vjschr. naturforsch. Ges. Zürich **26**, 565—589 (1917). — HEVESY, G.: Radioactive indicators. Their application in biochemistry, animal physiology and pathology. New York: Interscience Publ. Inc. 1948. 556 S. — HEVESY, G., and E. HOFER: Elimination of water from the human body. Nature (Lond.) **134**, 879 (1934). — HÖFLER, K.: Permeabilitätsbestimmung nach der plasmometrischen Methode. Ber. dtsch. bot. Ges. **36**, 414—422 (1918a). — Über die Permeabilität der Stengelzellen von *Tradescantia elongata* für Kalisalpeter. Ber. dtsch. bot. Ges. **36**, 423—442 (1918b). — Eine plasmometrisch-volumetrische Methode zur Bestimmung des osmotischen Wertes von Pflanzenzellen. Denkschr. Akad. Wiss. Wien, Math.-naturwiss. Kl. **95**, 99—170 (1918c). — Über Eintritts- und Rückgangsgeschwindigkeit der Plasmolyse und eine Methode zur Bestimmung der Wasserpermeabilität des Protoplasten. Jb. wiss. Bot. **73**, 300—350 (1930). — Kappenplasmolyse und Salzpermeabilität. Z. wiss. Mikrosk. **51** (Küster-Festschr.), 70—87 (1934a). — Permeabilitätsstudien an Stengelzellen von *Majanthemum bifolium*. (Zur Kenntnis spezifischer Permeabilitätsreihen, I.) Sitzgsber. Akad. Wiss. Wien, Math.-naturwiss. Kl., Abt. I **143**, 213—264 (1934b). — Über Wasser- und Harnstoffpermeabilität des Protoplasmas. Phyton, Ann. rei botanicae **1**, 105—121 (1949). — HÖFLER, K., u. FR. WEBER: Die Wirkung der Äthernarkose auf die Harnstoffpermeabilität von Pflanzenzellen. Jb. wiss. Bot. **65**, 643—737 (1926). — HOFMEISTER, L.: Vergleichende Untersuchungen über spezifische Permeabilitätsreihen. Bibl. botanica **113**, 1—83 (1935). — Verschiedene Permeabilitätsreihen bei einer und derselben Zellsorte von *Ranunculus repens*. Jb. wiss. Bot. **86**, 401—419 (1938). — Über die Permeabilitätsbestimmung nach der Deplasmolysezeit. Sitzgsber. österr. Akad. Wiss., Math.-naturwiss. Kl., Abt. I **157**, 83—95 (1948). — HOLM-JENSEN, I., A. KROGH and V. WARTIOVAARA: Some experiments on the exchange of potassium and sodium between single cells of *Characeae* and the bathing fluid. Acta bot. fenn. **36**, 1—22 (1944). — HUBER, B.: Zur Theorie der spezifischen Permeabilitätsreihen. Protoplasma **37**, 439—444 (1943). — HUBER, B., u. K. HÖFLER: Die Wasserpermeabilität des Protoplasmas. Jb. wiss. Bot. **73**, 351—511 (1930).

JACOBS, M. H.: The simultaneous measurement of cell permeability to water and to dissolved substances. J. Cellul. a. Comp. Physiol. **2**, 427—444 (1933a). — Volume changes of cells in solutions containing both penetrating and non-penetrating solutes, and their relation to the "Permeability ratio". J. Cellul. a. Comp. Physiol. **3**, 121—129 (1933b). — The quantitative measurement of the permeability of the erythrocyte to water and to solutes by the hemolysis method. J. Cellul. a. Comp. Physiol. **4**, 161—183 (1934). — Diffusion processes. Erg. Biol. **12**, 1—160 (1935). — JACOBS, M. H., and D. R. STEWART: A simple method for the quantitative measurement of cell permeability. J. Cellul. a. Comp. Physiol. **1**, 71—82 (1932). — JANSE, J. M.: Die Permeabilität des Protoplasma. Versl. Meded. Kon. Akad. Wetensch. Natuurk. **4**, 332—436 (1887).

KAMEN, M. D.: Radioactive tracers in Biology. New York: Academic Press Inc. Publ. 1947. 281 S. — KAMIYA, N., u. M. TAZAWA: Studies on water permeability of a single plant cell by means of transcellular osmosis. Protoplasma **46** (WEBER-Festband), 394—422 (1956). — KLEBS, G.: Beiträge zur Physiologie der Pflanzenzelle. Ber. dtsch. bot. Ges. **5**, 181—188 (1887). — KLERCKER, J. AF: Eine Methode zur Isolierung lebender Protoplasten. (Pflanzenphysiologische Mitteilungen 3.) Oefvers. Vet. Akad. Förh. **49**, 463—474 (1892).

LANDAHL, H. D.: A note on the use of membrane permeability to water. Bull. Math. Biophysics **10**, 187—190 (1948). — LEPESCHKIN, W. W.: Über den Turgordruck der vakuolisierten Zellen. Ber. dtsch. bot. Ges. **26a**, 198—214 (1908). — Über die Permeabilitätsbestimmung der Plasmamembran für gelöste Stoffe. Ber. dtsch. bot. Ges. **27**, 129—142 (1909). — LEVI, H., and H. H. USSING: Resting potential and ion movements in the frog skin. Nature (Lond.) **164**, 928—929 (1949). — LEVITT, J., G. W. SCARTH and R. R. GIBBS: Water permeability of isolated protoplasts in relation to volume change. Protoplasma **26**, 237—248 (1936). — LILLIE, R. S.: Increase of permeability to water following normal and artificial activation in *Sea-Urchin*-eggs. Amer. J. Physiol. **40**, 249—266 (1916). — LØVTRUP, S., u. A. PIGOŃ: Diffusion and active transport of water in the amoeba *Chaos chaos*. C. r. Labor. Carlsberg, Sér. Chim. **28**, 1—36 (1951). — LUCKÉ, B.: The living cell as an osmotic system and its permeability to water. Cold Spring Harbor Symp. Quant. Biol.

8, 123—131 (1940). — LUCKÉ, B., H. K. HARTLINE and M. McCUTCHEON: Further studies on the kinetics of osmosis in living cells. J. Gen. Physiol. **14**, 405—419 (1930). — LUCKÉ, B., and E. N. HARVEY: The permeability of living cells to heavy water (Deuterium oxide). J. Cellul. a. Comp. Physiol. **5**, 473—482 (1935). — LUNDEGÅRDH, H.: Über die Permeabilität der Wurzelspitzen von *Vicia Faba* unter verschiedenen äußeren Bedingungen. Kgl. Sv. Vetenskapsakad. Hdl. **47** (3), 1—254 (1911). — Investigations as to the absorption and accumulation of inorganic ions. Lantbruks-Högskolans Ann. **8**, 233—404 (1940).

MANEGOLD, E., u. C. STÜBER: Über Kapillar-Systeme XIV (2). Zur Dynamik der Plasmolyse. Erster Teil. Die mathematische Behandlung semipermeabler Protoplasten. Kolloid-Z. **63**, 316—323 (1933). — McCUTCHEON, M., and B. LUCKÉ: The effect of temperature on permeability to water of resting and of activated cells (unfertilized and fertilized eggs of *Arbacia punctulata*). J. Cellul. a. Comp. Physiol. **2**, 11—26 (1933). — MERRELL, M., A. GELLHORN and L. B. FLEXNER: Studies on rates of exchange of substances between the blood and extravascular fluid. J. of Biol. Chem. **153**, 83—89 (1944). — MYERS, G. M. P.: The water permeability of unplasmolysed tissues. J. of Exper. Bot. **2**, 129—144 (1951).

NERNST, W.: Zur Kinetik der in Lösung befindlichen Körper. Z. physik. Chem. **2**, 613 bis 637 (1888). — Die elektromotorische Wirksamkeit der Ionen. Z. physik. Chem. **4**, 129—181 (1889).

OVERTON, E.: Über die osmotischen Eigenschaften der lebenden Pflanzen- und Tierzelle. Vjschr. naturforsch. Ges. Zürich **40**, 159—201 (1895). — Über die allgemeinen osmotischen Eigenschaften der Zelle, ihre vermutlichen Ursachen und ihre Bedeutung für die Physiologie. Vjschr. naturforsch. Ges. Zürich **44**, 88—135 (1899).

PALVA, P.: Die Wasserpermeabilität der Zellen von *Tolypellopsis stelligera*. Protoplasma **32**, 265—271 (1939). — PLANCK, M.: Über die Erregung von Electricität und Wärme in Electrolyten. Ann. Physik u. Chemie, N. F. **39**, 161—186 (1890a). — Über die Potentialdifferenz zwischen zwei verdünnten Lösungen binärer Electrolyte. Ann. Physik u. Chemie, N. F. **40**, 561—576 (1890b). — POIJÄRVI, L. A. P.: Über die Basenpermeabilität pflanzlicher Zellen. Acta bot. fenn. **4**, 1—102 (1928).

RASHEVSKY, N., and H. D. LANDAHL: Permeability of cells, its nature and measurement from the point of view of mathematical biophysics. Cold Spring Harbor Symp. Quant. Biol. **8**, 9—16 (1940). — RESÜHR, B.: Zur Theorie der Plasmolyse- und Deplasmolysedauer (Entquellungs- und Quellungsdauer) lebender Protoplasten. Protoplasma **23**, 337—348 (1935). — Zur mathematischen Behandlung der Stoffaufnahme lebender Protoplaste. Protoplasma **25**, 435—460 (1936). — RUHLAND, W., u. C. HOFFMANN: Die Permeabilität von *Beggiatoa mirabilis*. Ein Beitrag zur Ultrafiltertheorie des Plasmas. Planta (Berl.) **1**, 1—83 (1926). — RUNNSTRÖM, J.: Untersuchungen über die Permeabilität des Seeigeleies für Farbstoffe. Ark. Zool. (Stockh.) **7**, Nr 13, 1—17 (1911).

SCARTH, G. W.: Estimation of protoplasmic permeability from plasmolytic tests. Plant Physiol. **14**, 129—143 (1939). — SCHMIDT, H.: Plasmolyse und Permeabilität. Jb. wiss. Bot. **83**, 470—512 (1936). — SCHWIEGK, H.: Künstliche radioaktive Isotope in Physiologie, Diagnostik und Therapie. Berlin-Göttingen-Heidelberg: Springer 1953. 842 S. — ŠEBOR, J.: Über die Diffusionsgeschwindigkeit von Wasser durch eine halbdurchlässige Membran. Z. Elektrochem. **10**, 347—353 (1904). — STADELMANN, ED.: Zur Messung der Stoffpermeabilität pflanzlicher Protoplasten. I. Die mathematische Ableitung eines Permeabilitätsmaßes für Anelektrolyte. Sitzgsber. österr. Akad. Wiss., Math.-naturwiss. Kl., Abt. I **160**, 761—787 (1951). — Zur Messung der Stoffpermeabilität pflanzlicher Protoplasten. II. Sitzgsber. österr. Akad. Wiss., Math.-naturwiss. Kl., Abt. I **161**, 373—408 (1952). — Zu Versuchsmethodik und Berechnung der Permeabilität pflanzlicher Protoplasten. Protoplasma **46** (WEBER-Festband), 692—710 (1956). — STEWARD, F. C.: Salt accumulation by plants — the rôle of growth and metabolism. Trans. Faraday Soc. **33**, 1006—1016 (1937). — STILLE, U.: Messen und Rechnen in der Physik. Grundlagen der Größeneinführung und Einheitenfestlegung. Braunschweig: Vieweg & Sohn 1955. 416 S. — SZÜCS, J.: Studien über die Protoplasmapermeabilität. Über die Aufnahme der Anilinfarben durch die lebende Zelle und ihre Hemmung durch Elektrolyte. Sitzgsber. ksl. Akad. Wiss., Math.-naturwiss. Kl., Abt. I **119**, 737—773 (1910).

TEORELL, T.: In A. KEYS, The apparent permeability of the capillary membrane in man. Trans. Faraday Soc. **33**, 930—943 (1937). — Permeability. Annual Rev. Physiol. **11**, 545—564 (1949). — Zur quantitativen Behandlung der Membranpermeabilität. (Eine erweiterte Theorie.) Z. Elektrochem. **55**, 460—469 (1951). — Transport processes and electrical phenomena in ionic membranes. Progr. Biophysics a. Biophysical Chem. **3**, 305—369 (1953). — TRÖNDLE, A.: Der Einfluß des Lichtes auf die Permeabilität der Plasmahaut. Jb. wiss. Bot. **48**, 171—282 (1910).

ULLRICH, H.: Über den Anionendurchtritt bei *Valonia* sowie dessen Beziehungen zum Zellbau. Planta (Berl.) **23**, 146—176 (1935). — Die Protoplasmapermeabilität. Naturwiss.

35, 111—118 (1948). — URSPRUNG, A.: Zur Terminologie und Analyse der osmotischen Zustandsgrößen. Z. Bot. **23**, 183—202 (1930). — USSING, H. H.: The distinction by means of tracers between active transport and diffusion. The transfer of iodide across the isolated frog skin. Acta physiol. scand. (Stockh.) **19**, 43—56 (1949). — Unterscheidung zwischen aktivem Transport und Diffusion mit Hilfe von radioaktiven Isotopen. Z. Elektrochem. **55**, 470—475 (1951). — Transport through biological membranes. Annual Rev. Physiol. **15**, 1—20 (1953).

VRIES, H. DE: Plasmolytische Studien über die Wand der Vacuolen. Jb. wiss. Bot. **16**, 465—598. — Opera coll. **2**, 321—446 (1885). — Über den isotonischen Coefficient des Glycerins. Bot. Ztg **1888**, 229—253. — Opera coll. **2**, 484—497 (1888). — Über die Permeabilität der Protoplaste für Harnstoff. Bot. Ztg **1889**, 309—333. — Opera coll. **2**, 553—566 (1889).

WARTIOVAARA, V.: Über die Temperaturabhängigkeit der Protoplastenpermeabilität. Ann. bot. Soc. zool.-bot. fenn. „Vanamo" **16** (1), 1—111 (1942). — The permeability of *Tolypellopsis* cells for heavy water and methyl alcohol. Acta bot. fenn. **34**, 1—21 (1944). — WIERINGA, K. T.: Quantitative Permeabilitätsbestimmungen. (Mit einem kritischen Überblick über die Theorien der Zellpermeabilität und die bisherigen Untersuchungsmethoden.) Protoplasma **8**, 522—584 (1930).

ZEEUW, J. DE, u. D. J. KUENEN: Note on a micro-colorimeter and its possible applications. Protoplasma **23**, 626—629 (1935). — ZUBER, R.: Untersuchungen über Diffusion in Flüssigkeiten. II. Über einen Mikrodiffusionsapparat für ungefärbte Flüssigkeiten. III. Diffusionsmessungen an elektrisch neutralen Flüssigkeitsgemischen und Lösungen. Z. Physik **79**, 280—305 (1932).

Methoden zur Messung des Stoffaustausches.

Von

Runar Collander.

Einleitung.

Zwei Größen sind es hauptsächlich, deren Messung im folgenden behandelt werden soll.

1. Die Geschwindigkeit des Stoffaustausches zwischen lebenden Zellen und dem umgebenden Medium, d. h. die Größe der Stoffmenge M, die in der Zeiteinheit von den Zellen aufgenommen oder abgegeben wird. Die Messung dieser Größe, für die neuerdings die Benennung *Flux* vorgeschlagen worden ist (TEORELL 1949, USSING 1952), ist an sich unabhängig von allen Theorien über die Art des Stoffaustausches und über die Kräfte, die dabei wirksam sind.

2. Die Größe der Zellpermeabilität (im beschränkten Sinne dieses Wortes). Als deren Maß gilt die *Permeabilitäts-* oder *Permeationskonstante*[1] P, welche angibt, wieviel Substanz in der Zeiteinheit durch die Einheit der Zelloberfläche permeiert, wenn zwischen den beiden Seiten der Zellgrenzschicht (Protoplasmaschlauch, Plasmahaut) die Einheit der Konzentrations- oder richtiger Aktivitätsdifferenz besteht. (Über die Berechnung der Konstante P s. Bd. 2, IV B.) Es ist zu beachten, daß im Falle nichtosmotischer (aktiver, metabolischer, adenoider) Stofftransporte die Berechnung einer Permeabilitätskonstanten nicht in Frage kommen kann. Vielmehr sind nur rein osmotische sowie metaosmotische Stoffaustauschprozesse durch eine Permeabilitätskonstante zu charakterisieren (vgl. Bd. 2, IV A).

Zur Messung der Geschwindigkeit des Stoffaustausches und der Größe der Zellpermeabilität ist eine sehr große Zahl der verschiedenartigsten Methoden benutzt worden. Die meisten von ihnen sind aber mit Fehlerquellen behaftet, die es bedingen, daß die mit ihrer Hilfe erzielten Resultate in quantitativer Hinsicht oft nicht sehr zuverlässig sind. Dies gilt besonders für die Bestimmung der Permeabilitätskonstante. Ehe aber auf die einzelnen Meßmethoden und die ihnen anhaftenden Fehlerquellen eingegangen wird, soll zunächst auf zwei Fehlermöglichkeiten allgemeinerer Art hingewiesen werden.

Die eine liegt darin, daß die zu prüfenden Zellen bei der Ausführung der Messungen beschädigt werden können, falls sie allzu abnormen Verhältnissen ausgesetzt sind. Es müssen z. B. Gewebe zerschnitten werden, um die in Frage kommenden Zellen freizulegen, es werden Zellen von Landpflanzen in wäßrigen Lösungen eingetaucht gehalten, die Zellen werden der Einwirkung giftig wirkender Stoffe ausgesetzt usw. Es liegt auf der Hand, daß durch derartige Maßnahmen der physiologische Zustand der Zellen und damit unter anderem auch ihre Permeabilität und ihre Befähigung zu aktiven Transportleistungen unabsichtlich beeinflußt werden kann.

Um die hieraus entspringenden Fehler zu vermeiden, sollte man bereits bei der Planung derartiger Messungen immer im Auge behalten, daß den Zellen

[1] Wir sprechen von der Permeabilität (oder Durchlässigkeit) der Zelle, aber von dem Permeationsvermögen (oder Durchdringungsvermögen) der verschiedenen Stoffe. Daraus ergeben sich die zwei verschiedenen Benennungen für die Konstante P.

nicht mehr als eben nötig zugemutet wird. Wichtig ist es aber auch, durch besondere Versuche sich davon zu vergewissern, ob die Zellen während der Ausführung der Messungen noch „normal“ sind. Die Exosmose von Anthocyanfarbstoffen, das Eindringen von sonst nicht permeierenden, leichtdiffusiblen Sulfosäurefarbstoffen (z. B. Orange G), eine Herabsetzung der Turgescenz sowie die Aufhebung der Plasmolysierbarkeit oder der Deplasmolysierbarkeit sind im allgemeinen Zeichen einer tiefgreifenden Alterierung der normalen Permeabilität. Doch muß man auch mit dem Vorkommen kleinerer Permeabilitätsänderungen rechnen, die nicht so leicht festzustellen sind. Noch empfindlicher gegenüber störenden Außenfaktoren ist der aktive Stofftransport.

Eine zweite Fehlerquelle allgemeiner Art bei der Bestimmung der Protoplasmapermeabilität liegt darin, daß die Geschwindigkeit des Stoffaustausches oft nicht allein von der Durchlässigkeit der Protoplasten, sondern auch von derjenigen der Zellwand abhängt. (Über die Permeabilität der Zellwand vgl. dieses Handbuch Bd. 2, IV E.) Auch die Geschwindigkeit (oder vielmehr Langsamkeit), mit der die zu prüfende Substanz an die Zellen etwa im Innern eines Gewebeschnittes herandiffundiert, kann einen begrenzenden Faktor darstellen. Manche sonst mit großer Sorgfalt ausgeführten Messungen sind an der Nichtbeachtung dieses trivialen Umstandes gescheitert.

Die im folgenden zu besprechenden Meßmethoden können in 4 Gruppen eingeteilt werden: a) Optische Methoden, d. h. Verfahren, die auf sichtbaren Veränderungen (außer Plasmolyse und Deplasmolyse) innerhalb der Zellen basieren. b) Chemisch-analytische Methoden (im weitesten Sinne des Wortes). c) Osmotische Methoden. d) Andere Methoden.

Zu einer eingehenden Beschreibung der verschiedenen Meßmethoden reicht der zur Verfügung stehende Raum nicht aus. In dieser Hinsicht muß vielmehr auf die zitierte Literatur hingewiesen werden. Dabei werden in erster Linie Veröffentlichungen neueren Datums genannt, in denen auch die ältere Literatur erwähnt ist.

1. Optische Methoden.

a) Vitalfärbung.

Unter den optischen Methoden der Bestimmung des Stoffaustausches ist an erster Stelle die Vitalfärbung zu nennen, d. h. die Beobachtung des Eindringens von Farbstoffen in lebende Zellen. Wegen ihrer Anschaulichkeit und scheinbaren Einfachheit ist diese Methode seit den grundlegenden Untersuchungen Pfeffers (1886) in weitem Ausmaß zum Studium der Protoplasmapermeabilität herangezogen worden. Dabei sind jedoch die vielen gefährlichen Fehlerquellen dieser Methode bei weitem nicht immer gebührend berücksichtigt worden, was zu zahlreichen, höchst verwirrenden Widersprüchen in den erzielten Ergebnissen geführt hat (vgl. Bd. 2, IV D c).

Die zu Vitalfärbungsversuchen benutzten Farbstoffe sind großmolekulare Verbindungen komplizierter Struktur. In wäßriger Lösung sind sie oft kolloidal oder halbkolloidal (Robinson 1935, Lison und Fautrez 1939). Polymerisierung und Tautomerisierung kommt bei ihnen häufig vor. Außerdem bestehen die käuflichen Farbstoffpräparate oft aus Gemengen mehrerer Farbstoffe, die sich in ihrem physiologischen Verhalten deutlich voneinander unterscheiden können. Ein eklatantes Beispiel hierfür bietet der seit altersher viel benutzte Vitalfarbstoff Methylenblau. Nach Irwin (1928, 1930; vgl. auch Brooks und Brooks 1941, Drawert 1949) soll nämlich das eigentliche Methylenblau (= Tetramethylthionin) etwa in *Valonia*-Zellen fast gar nicht eindringen. Ihr Vitalfärbungsvermögen sollen die üblichen Methylenblaupräparate vielmehr dem Trimethylthionin verdanken, das in allen käuflichen Methylenblaupräparaten enthalten ist und außerdem leicht aus Tetramethylthionin entsteht. Auch enthalten die Farbstoffpräparate des Handels häufig bis über 50% an farblosen Beimischungen (Dextrin, Glaubersalz usw.). Man wird also gut tun, möglichst zuverlässige Farbstoffpräparate zu benutzen, wie etwa die von der Biological Stain Commission in USA geprüften (vgl. CONN 1953). Angaben über die wechselnden Handelsnamen der Farbstoffe findet man in SCHULTZ' Farbstofftabellen.

Mißlich ist es, daß die Aufnahme eines Farbstoffes bisweilen infolge intracellularer Reduktion desselben unsichtbar bleiben kann (M. M. BROOKS 1926, BETZ 1953).

Eine weitere Komplikation liegt in der wechselnden Ionisation der Farbstoffe. Die sog. basischen Farbstoffe sind größtenteils Salze schwacher, farbiger Basen mit starken, farblosen Säuren. Wenigstens in vielen Fällen permeiert dabei die durch Hydrolyse freigemachte Base (bzw. Pseudobase) unvergleichlich viel schneller als das Salz bzw. das Farbkation (HÖFLER und SCHINDLER 1951). Die sog. sauren Farbstoffe wiederum sind Salze teils starker, teils schwacher Farbsäuren (Sulfonsäuren, Carbonsäuren, Phenole usw.) mit farblosen starken Basen. Auch hier gilt der Satz, daß die Ionen höchstens nur sehr langsam aufgenommen werden. Es ist somit bei allen Vitalfärbungsversuchen wichtig, die Ionisationskonstante der Farbbase bzw. der Farbsäure sowie den p_H-Wert der Außenlösung und des Zellsaftes zu kennen, da diese Faktoren häufig für die Farbstoffaufnahme ausschlaggebend sind.

Eine lästige Komplikation ist auch die besonders innerhalb bestimmter p_H-Grenzen eintretende, oft sehr kräftige Adsorption des Farbstoffes in den Zellwänden (DRAWERT 1949, KINZEL 1953). Für hochkolloidale Farbstoffe sind die Zellwände undurchlässig.

Viele Farbstoffe, vor allem alle basischen, sind mehr oder weniger giftig. (Über verschiedene Grade toxischer Beeinflussung lebender Zellen bei der Vitalfärbung vgl. STRUGGER 1936.) Um schädliche Wirkungen auszuschließen, muß die Konzentration der Farblösung daher häufig so niedrig gewählt werden, daß die Lösung in dünner Schicht nicht mehr farbig erscheint und demgemäß ein Eindringen in die Zelle gar nicht zu erkennen ist, außer wenn eine intracellulare „Speicherung" des Farbstoffes etwa infolge seiner Überführung in nichtpermeierende Formen stattfindet. Eine derartige Speicherung — wobei keine „vitalen" Kräfte mitzuwirken brauchen — kommt, wie bereits PFEFFER feststellte, in sehr vielen pflanzlichen Zellen vor, nicht aber in allen. Im Falle der sog. „leeren" Zellsäfte läßt sich der theoretisch mögliche maximale Speicherungsgrad verhältnismäßig leicht berechnen (JACOBS 1940, KINZEL 1954). Bei den „vollen" ist dies dagegen einstweilen noch nicht möglich (HÖFLER 1954). Für die Größe der Farbstoffaufnahme kann das Speicherungsvermögen der Zellen geradezu ausschlaggebend sein, aber selbstverständlich nur bei vorhandener genügend großer Permeabilität für die betreffenden Farbstoffe. Das wechselnde Ausmaß der Vitalfärbung nur auf Differenzen des Speicherungsvermögens zurückführen zu wollen wäre also mindestens ebenso einseitig wie seine Erklärung unter alleiniger Berücksichtigung der Permeabilität.

Viele fluorescierende Farbstoffe haben den Vorteil, bereits in äußerst niedrigen Konzentrationen sichtbar zu sein. In den letzten Jahren hat daher die „Fluorochromierung" der lebenden Zellen eine immer wichtigere Rolle neben der gewöhnlichen Vitalfärbung errungen (HÖFLER 1954).

Infolge der langsamen Diffusion und oft sehr großen Permeationsfähigkeit der Farbstoffe ist bei Versuchen mit verdünnten Farbstofflösungen unbedingt für stetige Durchmischung der Lösung zu sorgen, sonst wird die Geschwindigkeit der Farbstoffaufnahme nicht von der Permeabilität der Zellen, sondern von dem Diffusionswiderstand des Außenmediums geregelt.

Die Bestimmung der aufgenommenen Farbstoffmengen bzw. der erreichten intracellulären Farbstoffkonzentration ist bei Verwendung der Riesenzellen der Valoniaceen oder Characeen als Versuchsobjekte leicht: man braucht nur den gefärbten Zellsaft abzuzapfen und ihn mit Standardlösungen bekannter Farbstoffkonzentration zu vergleichen (Irwin 1925, M. M. Brooks 1926, Collander und Mitarbeiter 1943). Anders im Falle mikroskopisch kleiner Zellen, sofern man sich nicht etwa mit ganz unbestimmten Ausdrücken (Färbung „nicht merkbar", „schwach", „mäßig" usw.) begnügt. Um die in ihnen erreichten Farbstoffkonzentrationen wenigstens annäherungsweise zu schätzen, kann man zwar nach dem Vorschlag von Overton (1907, S. 814; vgl. auch Collander und Mitarbeiter 1943) die Farbintensität der Zellen mit derjenigen von Glascapillaren vergleichen, die mit Farbstofflösungen bekannter Konzentration gefüllt sind. Oder aber man bringt — besonders im Falle langsam permeierender Farbstoffe — die gefärbten Zellen auf dem Objektglas in Farbstofflösungen abgestufter Konzentration und vergleicht so die Färbung des Zellsaftes mit demjenigen der die Zellen umspülenden Lösung (Collander 1921). Eine große Genauigkeit darf man sich aber von solchen Bestimmungen nicht erwarten. Weitaus leistungsfähiger ist dagegen, wie Bartels (1954) gezeigt hat, eine mikroabsorptionsspektroskopische Meßanordnung in Verbindung mit einem Elektronenvervielfacher.

Anstatt die von den Zellen aufgenommenen Farbstoffmengen zu bestimmen, kann man natürlich auch die Abnahme der Farbkonzentration der Außenlösung messen (Buchy 1941, Kölbel 1948). Dabei bleibt allerdings zunächst unentschieden, inwieweit der aus der Außenlösung verschwundene Farbstoff tatsächlich in die Zellen eingedrungen und inwieweit er nur von den Zellwänden gespeichert worden ist.

Hinsichtlich der technischen Ausführung von Vitalfärbungsversuchen vgl. Strugger (1949).

b) Die Indicatormethode.

Der Eintritt von Säuren und Basen in lebende Zellen kann leicht beobachtet werden, falls der Zellsaft von vornherein einen geeigneten Indicatorfarbstoff (z. B. Anthocyan) enthält oder aber falls er mit einem künstlichen Indicatorfarbstoff (etwa Neutralrot) vital gefärbt worden ist. Weitaus schwieriger ist es, auch in diesem Falle zu quantitativen Ergebnissen zu gelangen.

Erstens ist zu beachten, daß die Anthocyane recht wenig empfindliche Indicatoren sind (vgl. Smith 1933, Drawert 1954). Anthocyanhaltige Zellen eignen sich daher schon aus diesem Grunde hauptsächlich nur für qualitative oder halbquantitative Permeationsbestimmungen (Brenner 1918, Jacobs 1920, 1922, Molisch 1921, Poijärvi 1928). Dagegen geben mit Neutralrot gefärbte Zellen einen scharfen Farbumschlag beim Eindringen von Basen (Harvey 1911, Poijärvi 1928, Äyräpää 1950). Der Farbumschlag ist aber von einer Entfärbung der Zellen begleitet, denn die von der eingedrungenen Base freigemachten, nichtionisierten Moleküle der Neutralrotbase exosmieren schnell.

Die Säuren- oder Basenmenge, die nötig ist, um das Erreichen einer bestimmten Farbnuance des Indicators zu bewirken, hängt selbstverständlich davon ab, wie kräftig gepuffert der Zellsaft ist. Daneben spielt aber auch die Stärke der eindringenden Säure oder Base unter Umständen eine wichtige Rolle. So z. B. zeigt die folgende Zusammenstellung, wie große relative Mengen verschieden starker Basen nötig sind, um den p_H-Wert des Zellsaftes auf 10 zu erhöhen.

Ionisationskonstante der Base:	∞	10^{-2}	10^{-3}	10^{-4}	10^{-5}	10^{-6}
Erforderliche Basenmenge:	1,0	1,0	1,1	2,0	11	101

Soll die Durchlässigkeit der Protoplasten für Säuren oder Basen quantitativ bestimmt werden, muß auch die Konzentrationsdifferenz zwischen Außenlösung und Zellsaft in bezug auf die Säuren- bzw. Basenmoleküle bestimmt werden. Im allgemeinen sind nämlich die nichtionisierten Säuren- und Basenmoleküle allein permeationsfähig. Aus den Arbeiten von POIJÄRVI (1928) und ÄYRÄPÄÄ (1950) ist ersichtlich, wie die Permeationskonstante auf Grund derartiger Versuche berechnet werden kann.

c) Die Niederschlagsmethode.

OVERTON (1896, 1907) hat darauf hingewiesen, daß gerbstofführende Zellen, in verdünnte Lösungen von Ammoniak oder organischen Basen (einschließlich der Alkaloide) gebracht, in ihrem Zellsaft mikroskopisch leicht wahrnehmbare Niederschläge ergeben, die als Zeichen der stattgefundenen Permeation dienen können. Die Reaktion ist außerordentlich empfindlich. So z. B. erzeugt Strychnin noch in einer Verdünnung von 1:10000000 einen deutlichen Niederschlag in *Spirogyra*-Zellen. Die Grenzkonzentration, welche zur Entstehung des Niederschlags nötig ist, ist aber für jedes Alkaloid eine andere. Denn damit die Ausfällung eintrete, muß erst das Löslichkeitsprodukt aus der Alkaloid- und der Gerbsäurekonzentration überschritten werden, und dieses Produkt ist selbstverständlich bei den verschiedenen Alkaloiden verschieden groß.

Von OVERTON ist diese Methode in halbquantitativer Weise benutzt worden; sie dürfte aber weiter ausbaufähig sein (vgl. auch TRÖNDLE 1920 und DEYSSON 1952).

d) Weitere optische Methoden.

Seit langem ist es bekannt, daß in stärkefreigemachten Algen oder Blättern höherer Pflanzen, die in Lösungen geeigneter Zuckerarten oder Zuckeralkohole gelegt werden, allmählich — auch im Dunklen — eine durch die Jodprobe leicht nachzuweisende Stärkebildung in den Zellen eintritt. Offenbar ist also Zucker bzw. Zuckeralkohol von den Zellen aufgenommen worden, was von einem gewissen Interesse ist, da die Zellen sonst oft in hohem Grade undurchlässig für derartige Verbindungen zu sein scheinen (vgl. WEEVERS 1931). Bis jetzt dürfte diese Methode nicht für quantitative Bestimmungen benutzt worden sein.

Die Blattzellen von *Fontinalis* und vielen anderen Moosen enthalten in ihrem Saftraum eigenartige Fadenknäuel, die aus fettähnlichen Substanzen bestehen. Wie BORESCH (1919) fand, wirken viele Alkohole, Phenole und Amine auf die Knäuel emulgierend, so daß sie sich in feinste Tröpfchen auflösen. Da der Vorgang intravital abläuft, wurde er von BORESCH als Kriterium des Permeierens der betreffenden Substanzen benutzt (vgl. auch COLLANDER und SOMER 1931).

Auch Wasserstoffperoxyd (PFEFFER 1889), Kalium- und Natriumhydroxyd (CHOLNOKY 1953) sowie manche andere Substanzen bewirken in lebenden oder absterbenden Zellen verschiedene mikroskopisch sichtbare Veränderungen, die als Kriterien der Permeation dieser Substanzen benutzt worden sind. Quantitative Ergebnisse scheinen aber in dieser Weise noch nicht erzielt worden zu sein.

2. Chemisch-analytische Methoden.

Bei den unter dieser Rubrik vereinigten Methoden handelt es sich darum, daß entweder die von den Zellen aufgenommenen oder die von ihnen abgegebenen Stoffmengen quantitativ bestimmt werden, wobei außer eigentlichen chemisch-analytischen auch verschiedene physikalische Verfahren wie Colorimetrie, Kryoskopie, Messung der Leitfähigkeit, des p_H-Wertes oder der Radioaktivität in Frage kommen. Wie aus folgendem hervorgeht, hängt die Zuverlässigkeit der in dieser Weise durchgeführten Permeationsbestimmungen in entscheidendem Maße von der Beschaffenheit der benutzten Objekte ab.

a) Makroskopische Einzelzellen als Objekt.

Am einfachsten und zugleich am zuverlässigsten gestalten sich die Versuche mit Zellen, die so groß sind, daß sie einzeln erfaßt und behandelt werden können.

In bezug auf Größe stehen die Zellen der in warmen Meeren vorkommenden Valoniaceen an der Spitze. Eine einzige solche Zelle kann bis etwa 5 cm^3 Zellsaft enthalten. Der Zellsaft ist nur von einem ganz dünnen Plasmaschlauch und einer ebenfalls dünnen Zellwand umgeben und läßt sich somit leicht in unvermischtem Zustand abzapfen und analysieren. Er besteht aus einer wäßrigen Lösung von hauptsächlich nur anorganischen Salzen. Unter Umständen ist es sogar möglich, den Zellsaft einer solchen Zelle mittels zweier Glascapillaren durch eine andere Flüssigkeit zu ersetzen, ohne daß die Zelle dabei abstirbt (BLINKS 1935). Die Zellen der Valoniaceen scheinen somit geradezu ideale Objekte für Permeationsstudien zu sein. Seit dem Anfang der zwanziger Jahre sind sie denn auch ausgiebig für derartige Untersuchungen benutzt worden, besonders von den Amerikanern.

Seit etwa derselben Zeit hat man aber auch die wesentlich kleineren Zellen von *Nitella* und anderen Characeen zu ähnlichen Versuchen herangezogen. Allmählich hat es sich dann herausgestellt, daß die Characeen nicht nur als eine Art Ersatz für die schwerer erhältlichen Valoniaceen anzusehen sind, sondern daß sie ihnen gegenüber auch wenigstens einen klaren Vorzug aufzuweisen haben: bei den so sehr voluminösen Valoniaceenzellen kann nämlich die Diffusion durch den Zellsaftraum leicht die Rolle eines begrenzenden Faktors annehmen, während diese Gefahr bei Versuchen mit den so viel dünneren Characeenzellen bei weitem nicht so groß ist. Übrigens hat sich ja die Mikrochemie in den letzten Jahrzehnten so sehr entwickelt, daß die relative Kleinheit der Characeenzellen keinen so großen Übelstand mehr bedeutet.

Permeationsversuche mit Valoniaceen- und Characeenzellen sind meistens recht einfach. Man beläßt z. B. die Zellen eine bestimmte Zeit in einer Lösung der zu prüfenden Substanz und ermittelt dann in einer aus den Zellen isolierten Zellsaftprobe den Gehalt an dieser Substanz. Methodische Einzelheiten solcher Versuche ergeben sich z. B. aus den Veröffentlichungen von ALBAUM und Mitarbeitern (1937), S. C. BROOKS (1935), COLLANDER und BÄRLUND (1933), HOAGLAND, HIBBARD und DAVIS (1926), IRWIN (1925), JACQUES (1936), OSTERHOUT (1931, 1933, 1936) sowie STEWARD und MARTIN (1937). Eine andere Vorgangsweise ist die, daß eine Zelle zuerst mit dem zu prüfenden Stoff „gesättigt" wird, worauf man die Exosmose desselben verfolgt, indem man die Zelle in eine kleine Wassermenge bringt, die in passenden Zeitabschnitten erneuert und analysiert wird. Versuche dieser Art sind von WARTIOVAARA (1942, 1944, 1949) und COLLANDER (1954) ausgeführt worden. Infolge der relativen Eindeutigkeit ihrer Ergebnisse können beide Methoden fast als ideal bezeichnet werden. Nur schade, daß die Zahl der in Frage kommenden Objekte so beschränkt ist und daß man nicht im voraus wissen kann, inwieweit die an diesen Objekten gewonnenen Ergebnisse zu verallgemeinern sind.

Flächenhaft ausgebreitete Algen sowie Fadenalgen sind bisher nur wenig für Studien dieser Art benutzt worden, dürften aber manchmal fast ebenso günstige Versuchsobjekte darstellen wie die makroskopischen Einzelzellen. Vor allem geht aus den Studien von SCOTT und HAYWARD (1954) hervor, daß sich der aus nur zwei Zellschichten bestehende Thallus von *Ulva lactuca* zu Studien über die Kationenaufnahme ausgezeichnet eignet: jede Zelle steht ja in direktem Kontakt mit dem Außenmedium, und dabei ist das Verhältnis Zelloberfläche/Zellvolumen weit größer als bei den Valoniaceen und Characeen, wodurch die Geschwindigkeit des Stoffaustausches vergrößert wird. Auch Fadenalgen, z. B. Zygnemalen, scheinen nach orientierenden Versuchen im hiesigen Laboratorium recht geeignet für Studien über die Stoffaufnahme und -abgabe mittels analytischer Methoden.

b) Zellsuspensionen als Versuchsobjekte.

Versuche dieser Art sind besonders mit Hefezellen ausgeführt worden, die in der Tat ein in vieler Hinsicht ausgezeichnetes Versuchsmaterial darstellen. Eine Schwierigkeit liegt allerdings darin, daß in einer Hefesuspension das Gesamtvolumen der suspendierten Zellen meistens wesentlich kleiner ist als das Volumen des Suspensionsmittels. Bei Substanzen, die nicht in den Zellen gespeichert oder im Stoffwechsel verbraucht werden, erfolgt daher oft nur eine sehr geringe und somit schwer meßbare Abnahme der Außenkonzentration, auch wenn die betreffenden Substanzen in kurzer Zeit bis zum Konzentrationsausgleich in die Zellen eindringen.

Hinsichtlich der älteren Literatur über den Stoffaustausch der Hefezellen vergleiche man die Übersichten von MALM (1947) sowie von ROSENBERG und WILBRANDT (1952). Von neueren Untersuchungen seien genannt diejenige von CONWAY und Mitarbeitern (1954) über den Ionenaustausch sowie die von RIEDER (1951) und ROTHSTEIN (1954) über die Zuckeraufnahme der Hefe. Von Bedeutung ist auch die Beobachtung von CONWAY und DOWNEY (1950) sowie von ÄYRÄPÄÄ (1950, S. 407), daß sich die Aufnahme sowohl von Säuren wie Basen recht genau verfolgen läßt, wenn man sie zu einer Hefesuspension gibt und die hierdurch veranlaßten Änderungen des p_H-Wertes in der Suspensionsflüssigkeit potentiometrisch verfolgt.

Auch mit Bakterien (COWIE und Mitarbeiter 1949, EDDY und HINSHELWOOD 1951), Chlorellen (ÖSTERLIND 1952, KNAUSS und PORTER 1954) sowie anderen einzelligen Algen sind ähnliche Versuche wie mit Hefen ausgeführt worden.

c) Gewebescheiben als Versuchsobjekte.

Hier handelt es sich um Versuche, in denen Scheiben aus irgendwelchen fleischigen Speichergeweben (Kartoffeln, Rüben, Karotten usw.) in Wasser oder wäßriger Lösung gehalten und die dabei stattfindende Stoffabgabe oder Stoffaufnahme messend verfolgt wurde. Untersuchungen dieser Art gibt es in sehr großer Zahl. Daß diese Versuchsanordnung so sehr beliebt war, hängt wohl damit zusammen, daß sie gleichzeitig bequem und exakt anmutet: die Auswahl an Speichergeweben verschiedener Kulturpflanzen, die sich in gleichmäßige Scheiben zerlegen lassen, ist groß, und bei der Wahl geeigneter Analysenmethoden lassen sich die Bestimmungen schnell und mit großer Genauigkeit ausführen. Was dabei aber oft übersehen wurde, ist, daß noch so genaue Analysenresultate leider nicht dafür bürgen, daß die Ergebnisse auch in physiologischer Hinsicht eindeutig sind.

Unter Verzicht auf die ältere Literatur, betreffs deren auf die Zusammenstellung von STILES (1924) hingewiesen werde, mögen hier nur einige bemerkenswertere Untersuchungen dieser Art etwa aus den letzten 30 Jahren genannt werden. Sie können in 3 Gruppen eingeteilt werden.

Bei den Untersuchungen der 1. Gruppe wurde die *Exosmose* von normalen Zellbestandteilen wie Zuckern, Anthocyanen und Salzen studiert (DE VISSER SMITS 1926, STILES 1927, KRASSINSKY 1930, INGOLD 1931, BRAUNER und BRAUNER 1936, KAHO 1937, IRMAK 1938). Bei der Beurteilung der in dieser Weise gewonnenen Resultate darf vor allem nicht vergessen werden, daß die Durchlässigkeit der unversehrten Zellen für derartige Stoffe im allgemeinen nicht etwa hundert- oder tausendmal, sondern wohl eher hunderttausendmal kleiner ist als diejenige toter Zellen. Hierin liegt eine gefährliche Fehlerquelle verborgen. Denn bei Experimenten dieser Art ist es schwer mit voller Sicherheit zu vermeiden, daß einzelne Zellen aus irgendeinem Grunde während der Versuchsdauer absterben. Nur allzu leicht geschieht es dann, daß die festgestellte Exosmose hauptsächlich eben von diesen abgestorbenen Zellen herrührt und nicht oder nur zum geringsten Teil von den gesunden Zellen, deren Permeabilität der Experimentator zu messen wünscht und wohl auch zu messen glaubt.

Bei den Untersuchungen der 2. Gruppe handelt es sich dagegen um die *Aufnahme* von gelösten Stoffen, in den meisten Fällen von Salzen bzw. Ionen, seitens der Gewebescheiben. Die Arbeiten von HOMÈS (1933), ASPREY (1937), STEWARD und Mitarbeiter (1943, dort auch zahlreiche ältere Arbeiten derselben Verfasser), STILES und SKELDING (1944), STILES und DENT (1946), ROBERTSON und Mitarbeiter (1947), BRIGGS und ROBERTSON (1948), REES (1949), SUTCLIFFE (1952, 1954), STEWARD und MILLAR (1954), HANSON und BONNER (1954) können hier als Beispiele genannt werden. Auch die Resultate derartiger Versuche sind nicht immer so leicht zu beurteilen, wie es zunächst den Anschein hat. Besonders STEWARD hat mit großem Nachdruck darauf hingewiesen, daß die in den Gewebescheiben oberflächlich gelegenen Zellen wesentlich anderen Bedingungen ausgesetzt sind und deshalb auch wesentlich anders reagieren als die im Inneren befindlichen. Während diese in einem Zustand relativer Ruhe verharren, zeigen die oberflächlich gelegenen Zellen oft Zeichen einer auffallenden Verjüngung: ihre Atmungsintensität wird um ein Vielfaches gesteigert, es werden Zellteilungen eingeleitet, eine rege Proteinsynthese setzt ein, und im Zusammenhang hiermit findet eine effektive Ionenaufnahme ausgesprochen „aktiver" Art statt. Von einem homogenen, ruhenden Speichergewebe, mit dem man ursprünglich zu arbeiten glaubte, kann somit in derartigen Fällen keine Rede sein. Nach SUTCLIFFE (1952) stellen jedoch Schnitte der roten Rübe ein besonders günstiges Versuchsmaterial dar: obwohl die Zellen wochenlang am Leben bleiben und zu einer beträchtlichen Ionenaufnahme befähigt sind, werden keine Zellteilungen eingeleitet.

Eine 3. Gruppe bilden diejenigen Untersuchungen, bei denen der Ionen*austausch* zwischen den Gewebescheiben und der umgebenden Lösung studiert wird. Eine Arbeit von SUTCLIFFE (1954) kann als vorbildlich für derartige Untersuchungen bezeichnet werden.

Über Bestimmung der Wasseraufnahme von Gewebescheiben vgl. HACKETT und THIMANN (1952) sowie BURSTRÖM (1953) und die dort zitierte Literatur.

d) Wurzeln und andere Pflanzenorgane als Versuchsobjekte.

HOAGLAND und Mitarbeiter haben in einer Reihe von Untersuchungen (Zusammenfassungen: HOAGLAND und BROYER 1936, BROYER und HOAGLAND 1940; vgl. auch JACOBSON und Mitarbeiter 1950) gezeigt, daß die Wurzeln von in

Wasserkultur gezogenen höheren Pflanzen ein sehr geeignetes Objekt zur Untersuchung der Ionenaufnahme abgeben. Die Vorgeschichte des benutzten Pflanzenmaterials zeigt sich dabei als ausschlaggebend für den Erfolg. Pflanzen, die bei reichlicher Beleuchtung in salzarmer Nährlösung aufgewachsen sind, liefern nämlich Wurzeln, die zu einer sehr effektiven aktiven Ionenaufnahme befähigt sind, während Wurzeln, die aus Kulturen in schwachem Licht bei reichlicher Salzzufuhr stammen, hauptsächlich nur auf dem Wege des Ionenaustausches neue Ionen aufzunehmen vermögen. Auf eine Untersuchung von Butler (1953) über den „scheinbaren freien Raum" im Wurzelgewebe sei in diesem Zusammenhang hingewiesen. Auch die Zuckeraufnahme der Wurzeln ist studiert worden (Burström 1941).

Einigermaßen ähnlich verhalten sich, wie Arisz (1948) zeigte, Blattstücke von *Vallisneria*. Durch chemische Analyse ließ sich sowohl eine einfache Permeation bestimmter Stoffe (Coffein) wie eine aktive Aufnahme anderer (Asparagin, anorganische Ionen) nachweisen. Auch mit *Elodea*sprossen lassen sich ganz entsprechende Versuche durchführen (Warner 1932, Rosenfels 1935).

Durch Bestimmung der Zunahme des Trockengewichts von Blattstücken, die auf Zuckerlösung schwammen, konnte Weatherley (1954) deren Zuckeraufnahme verfolgen (vgl. Porter und May 1955).

Die Aufnahme von Heteroauxin seitens Erbsenepikotylen wurde unter anderen von Reinhold (1954) chemisch-analytisch studiert.

In diesem Zusammenhang sei auch die Versuchstechnik von Wiersum (1947) erwähnt. Er saugte Wasser oder wäßrige Lösungen bekannter Zusammensetzung durch die Gefäße junger Wurzeln und verfolgte analytisch den Stoffaustausch in radialer Richtung durch den Zentralzylinder, die Endodermis und die Wurzelrinde.

Die Methodik der Untersuchung der Stoffaufnahme und -abgabe intakter, bewurzelter Pflanzen gehört nicht zu unserem Thema.

e) Die Isotopenmethode.

Obwohl logisch zu den „chemisch-analytischen" Methoden (in der hier benutzten sehr weiten Fassung des Wortes) gehörig, bedarf die Isotopenmethode wegen ihrer besonderen Vorzüge und Fehlerquellen einer besonderen Erörterung (vgl. auch die Zusammenfassungen von Hevesy 1948, Martin und Scott Russel 1950, Kamen 1951, Ussing 1952 sowie die *Reports from the Second Isotope Conference*).

Die Bedeutung der Isotopenmethode, also der Verwendung von „markierten" Molekülen oder Ionen, liegt zunächst darin, daß sich besonders die radioaktiven Isotopen bereits in äußerst geringen Mengen relativ bequem und sehr exakt bestimmen lassen. Darüber hinaus hat die Isotopenmethode es ermöglicht, gewisse Probleme des Stoffaustausches anzugreifen und zu lösen, die auf anderen Wegen völlig unnahbar gewesen wären. Dies gilt vor allem für das Problem der Ionenpermeabilität. Infolge der ihm anhaftenden elektrischen Ladung kann ja im allgemeinen ein bestimmtes Ion nur entweder zusammen mit einem Ion entgegengesetzter Ladung oder aber unter Austausch gegen ein Ion gleicher Ladung durch die Zellgrenzschicht passieren. Es konnte daher bis vor kurzem nur der *gemeinsame* Permeationswiderstand für je 2 Ionenarten experimentell bestimmt werden, nicht aber der Widerstand für irgendeine Ionensorte für sich. Denn ein Austausch von, sagen wir, Kaliumionen gegen Kaliumionen oder von Chlorionen gegen Chlorionen war ja vor der Einführung der Isotopenmethode in keiner Weise nachzuweisen.

Man kann somit sagen, daß mit der Isotopenmethode eine neue Epoche in der Erforschung der Ionenumsetzung und der Ionenpermeabilität begonnen hat. Trotzdem hinterläßt die Mehrzahl der bisherigen Untersuchungen auf diesem Gebiet keinen ganz befriedigenden Eindruck. Oft ist nur irgendeine unbestimmte Aufnahme- oder Austauschgeschwindigkeit gemessen worden, von der man nicht recht weiß, was sie eigentlich physiologisch zu bedeuten hat (vgl. USSING 1952, S. 28). Inwieweit es sich im konkreten Fall um einen aktiven, gerichteten Ionentransport und inwieweit um einen einfachen Austausch radioaktiver Ionen gegen nichtradioaktive Ionen derselben oder einer anderen Ionenart handelt, derartige Fragen sollten nicht vergessen werden vor der Freude, die radioaktiven Impulse mittels des GEIGER-MÜLLER-Zählrohres exakt registrieren zu können. In vielen Fällen empfiehlt es sich, die radioaktiven Messungen mit gewöhnlichen chemischen Analysen zu kombinieren, um in dieser Weise sowohl den Netto- wie den Bruttoaustausch zu erfassen. Eine Arbeit von SUTCLIFFE (1954) ist ein Beispiel dafür, wie aufschlußreich eine solche Vorgangsweise sein kann.

Auch für die Bestimmung der Wasserpermeabilität hat die Isotopenmethode neue Wege geöffnet. Es gab schon vor der Einführung dieser Methode zahlreiche Untersuchungen über die *Filtrations*permeabilität der Protoplasten für Wasser, aber ein eindeutiger quantitativer Vergleich der so ermittelten Werte mit der *Diffusions*permeabilität für gelöste Stoffe war jedenfalls nicht ganz einfach (vgl. BOCHSLER 1948, USSING 1952). Diese Schwierigkeit kann jedoch leicht umgangen werden, indem man die Diffusionspermeabilität für schweres Wasser (eigentlich DHO), in gewöhnlichem Wasser gelöst, bestimmt (WARTIOVAARA 1944, COLLANDER 1954). Dabei ist allerdings zu berücksichtigen, daß DHO ein wenig langsamer als H_2O durch das Protoplasma permeiert.

3. Osmotische Methoden.

Die bisher besprochenen Meßmethoden können als direkte Methoden bezeichnet werden insofern, als bei ihnen das Eindringen oder Austreten der zu prüfenden Verbindung mehr oder weniger direkt zur Beobachtung gelangt. Den jetzt zu besprechenden osmotischen Methoden ist es dagegen gemeinsam, daß man auf indirektem Wege, und zwar auf Grund der Wasseraufnahme oder -abgabe der Zellen, Rückschlüsse hinsichtlich einer stattgefundenen Aufnahme oder Abgabe gelöster Stoffe zieht.

Der Gedanke, der allen diesen Methoden gemeinsam ist, ist etwa folgender: Die zu untersuchenden Zellen werden in eine ziemlich konzentrierte wäßrige Lösung des zu prüfenden Stoffes gebracht. Ist nun das Protoplasma (das als leicht durchlässig für Wasser vorausgesetzt wird) durchlässig für den betreffenden gelösten Stoff, so wirkt die Lösung schwächer wasserentziehend auf die Zelle, als wenn keine merkbare Permeabilität für ihn besteht. Und zwar ist die wasserentziehende Kraft der Außenlösung ceteris paribus um so geringer, je schneller der gelöste Stoff von den Protoplasten aufgenommen wird.

Die osmotischen Methoden zur Messung des Stoffaustausches sind meistens recht bequem anzuwenden. Sie haben ferner den großen Vorzug, daß mit ihrer Hilfe die Aufnahme oder Abgabe der verschiedenartigsten Stoffe untersucht werden kann, da ja das osmotische Wasseranziehungsvermögen eine gemeinsame Eigenschaft aller gelösten Stoffe ist. Es ist daher nicht verwunderlich, daß diese Methoden eine hervorragende Rolle in der Permeabilitätsforschung gespielt haben und immer noch spielen.

Andererseits aber darf nicht übersehen werden, daß die osmotischen Methoden zur Messung des Stoffaustausches auf gewissen Voraussetzungen basieren, deren tatsächliches Erfülltsein nicht ohne weiteres klar steht. Die wichtigsten dieser Voraussetzungen sind: 1. Der Wasseraustausch zwischen den Protoplasten und dem umgebenden Medium wird praktisch allein durch osmotische Kräfte geregelt, nicht dagegen in nennenswertem Grade durch Quellungs- oder Entquellungserscheinungen im Protoplasma und auch nicht durch Elektroosmose oder durch einen sonstigen nichtosmotischen Wassertransport. 2. Der in die Protoplasten eingedrungene Stoff behält dort seinen osmotischen Effekt unverändert bei. Er wird also z. B. weder ausgefällt, noch adsorbiert oder zersetzt. 3. Eine Exosmose normaler Inhaltsstoffe der Zellen geschieht während der Messung nicht. 4. Regulatorische Änderungen des osmotischen Wertes des Zellsaftes (Anatonose, Katatonose) kommen nicht vor.

Inwieweit diese Bedingungen tatsächlich erfüllt sind, ist eine Frage, die heute noch nicht definitiv beantwortet werden kann. Während z. B. BÄRLUND (1929) durch sorgfältige Versuche an Epidermiszellen von *Rhoeo discolor* zu dem Ergebnis gelangte, daß die erwähnten Voraussetzungen im Falle seines Objektes sehr weitgehend erfüllt sind, wird die Zuverlässigkeit der osmotischen Methoden der Permeationsmessung von mehreren späteren Forschern stark bezweifelt. So z. B. glauben unter anderen BENNET-CLARK und BEXON (1946) eine elektroosmotische Komponente der Wasseraufnahme der Protoplasten nachgewiesen zu haben, während BOGEN und PRELL (1953) sowie BOGEN und FOLLMANN (1955) das Eingreifen metabolischer Transportkräfte sowohl bei der Wasseraufnahme wie bei der Aufnahme gelöster Stoffe stark hervorheben. In der Tat zeigen ja Phänomene wie die „Reizplasmolyse" und die „Spontankontraktion" der Vacuolen, daß sich die Wasserbilanz der Zelle nicht immer auf ein einfaches osmotisches Schema zurückführen läßt. Permeationsmessungen mittels osmotischer Methoden dürfen daher nicht kritiklos ausgeführt werden. Bei geeigneten Objekten und mit der nötigen Umsicht verwendet dürften sie trotzdem von großem Wert sein.

Die osmotischen Methoden können in 2 Gruppen eingeteilt werden — je nachdem, ob sie eine Plasmolyse involvieren oder nicht. Wir beginnen mit den plasmolytischen Methoden.

Die Methode der isotonischen Koeffizienten oder der „Permeabilitätskoeffizienten" ist vornehmlich von LEPESCHKIN (1909) und TRÖNDLE (1910) ausgearbeitet. Sie gründet sich auf die Tatsache, daß von einem permeierenden Stoff eine höhere Konzentration (C') zur Erzielung der Plasmolyse erforderlich ist als von einem nichtpermeierenden (C). Der Ausdruck $\frac{C' - C}{C'}$ wird daher als ein Maß der Permeabilität benutzt. Von FITTING (1917, 1919) und STILES (1924) ist diese Methode scharf kritisiert worden (vgl. auch LEPESCHKIN 1923 und ZYCHA 1928). Der schwerwiegendste von diesen Einwänden ist wohl der, daß bei der Benutzung der in Rede stehenden Methode der Zeitfaktor fast ganz vernachlässigt wird. Wird er in Rechnung gestellt, dann gelangt man zu einer Methode, die eigentlich „die Methode der Grenzkonzentrationsverschiebung" heißen müßte. Der Kürze halber soll sie im folgenden einfach

Die grenzplasmolytische Methode genannt werden. Es ist dies die Methode, deren OVERTON sich in seinen grundlegenden Permeabilitätsstudien vornehmlich bedient hat. Später ist sie besonders von FITTING (1919) und BÄRLUND (1929) kritisch geprüft worden.

Bei der Ausführung von Messungen mittels dieser Methode werden möglichst gleichartige Zellen in Lösungen abgestufter Konzentration der zu prüfenden

Substanz eingetragen und hier von Zeit zu Zeit auf Plasmolyse hin untersucht. Falls die Konzentrationen richtig gewählt waren und der Stoff nicht gar zu schnell permeiert, so findet man nach einer gewissen Zeit in den konzentrierten Lösungen alle Zellen plasmolysiert, während in den verdünntesten Lösungen keine Plasmolyse eingetreten ist. Irgendwo dazwischen liegt die plasmolytische Grenzkonzentration, bei der die halbe Anzahl der Zellen eben merkliche Plasmolyse aufweist. (Einzelheiten betreffs der genauen Bestimmung der plasmolytischen Grenzkonzentration findet man bei URSPRUNG 1938.) Wenn nun der plasmolysierende Stoff nicht in die Protoplasten eindringt, bleibt der Zustand der Zellen unverändert, sobald der Zellsaft durch Wasserabgabe den gleichen osmotischen Wert wie die Außenlösung erreicht hat. Eine nachträgliche Verschiebung der plasmolytischen Grenzkonzentration findet in diesem Falle also nicht statt. Dringt der Stoff dagegen in die Protoplasten ein, so tritt, nachdem die Plasmolyse ihren Höhepunkt erreicht hat, ein allmählicher Rückgang der Plasmolyse ein. Hierdurch steigt die plasmolytische Grenzkonzentration immer mehr. Als Maß der Aufnahmegeschwindigkeit dient nun eben die in der Zeiteinheit stattfindende Erhöhung der Grenzkonzentration.

Wenn es sich um die Untersuchung schwerlöslicher, in größeren Konzentrationen giftig wirkender oder aber sehr schnell permeierender Verbindungen handelt, bedient man sich der von OVERTON (1896) angegebenen „Partialdruckmethode“ (auch „Partialmethode“ genannt): man versetzt die an sich nicht oder kaum plasmolysierende Lösung des zu prüfenden Stoffes mit einer bestimmten Menge einer nichtpermeierenden Verbindung, so daß beide Stoffe zusammen, gemäß der Summe ihrer osmotischen Partialdrucke, eine deutliche Plasmolyse verursachen, und verfolgt dann in gewöhnlicher Weise die zeitliche Verschiebung der plasmolytischen Grenkonzentration. Als nichtpermeierende Hilfssubstanz („osmotisches Widerlager“) benutzt man in solchen Versuchen entweder Zuckerarten (Saccharose, Glucose), vielwertige Alkohole (Mannit) oder ein physiologisch balanciertes Salzgemisch (z. B. Natriumchlorid + Calciumchlorid).

Um die Aufnahme extrem schnell permeierender schwacher Säuren und Basen auf plasmolytischem Wege zu verfolgen, kann man sich eines von JACOBS (1927, 1940) ersonnenen Kunstgriffes bedienen (vgl. auch BOUILLENNE 1940). Nehmen wir an, es handle sich um die Permeation von Essigsäure und Ammoniak. Dann plasmolysiert man die Zellen mit einer Lösung von Ammoniumacetat. Das Salz als solches (d. h. die Ammonium- und Acetationen) permeiert nur sehr langsam. Trotzdem deplasmolysieren sich die Zellen einigermaßen schnell, was zum größten Teil davon herrührt, daß die hydrolytisch abgespaltenen Essigsäure- und Ammoniakmoleküle äußerst schnell permeieren und im Zellsaft wieder zu Ammoniumacetat zusammentreten. Unter Umständen kann man so aus der Geschwindigkeit des Plasmolyserückganges die Permeabilität der Zellen für sowohl Essigsäure wie Ammoniak berechnen (COLLANDER und Mitarbeiter 1931).

Die Deplasmolysezeitmethode. Dies ist gewissermaßen eine vereinfachte und abgekürzte Modifikation der vorangehenden (oder der nachfolgenden) Methode. Sie besteht darin, daß man den osmotischen Wert der zu untersuchenden Zellen in Zuckerlösungen abgestufter Konzentration bestimmt und an streng vergleichbarem Material die Deplasmolysezeit in einer hypertonischen Lösung des zu prüfenden Stoffes feststellt (BÄRLUND 1929, S. 84, HOFMEISTER 1948). Die Methode ist einfach und bequem und daher in vielen Fällen empfehlenswert. Sie hat jedoch den Nachteil, daß eine eventuelle Änderung der Permeabilität während des Versuches unbeachtet bleiben kann.

Die plasmometrische Methode ist von HÖFLER (1918, 1934) ausgearbeitet worden und hat sich in den Untersuchungen zahlreicher Forscher gut bewährt. Sie unterscheidet sich von der grenzplasmolytischen Methode darin, daß man hier den zeitlichen Verlauf des Plasmolyserückganges an einzelnen, verhältnismäßig stark plasmolysierten Zellen verfolgt. Dazu muß man den Grad der Plasmolyse, d. h. die Maßzahl für das Volumenverhältnis zwischen dem plasmolysierten Protoplasten und dem Innenraum der turgorlosen Zelle genau bestimmen können, was durch direkte mikrometrische Messung geschieht. Aus der Größe des Plasmolysegrades G und der bekannten Konzentration C der plasmolysierenden Lösung wird dann der osmotische Wert O des Zellsaftes bei Grenzplasmolyse berechnet. Es gilt nämlich die Beziehung $O = C \cdot G$. Besteht nun Permeabilität für den plasmolysierenden Stoff, so bleibt die anfänglich beobachtete maximale Plasmolyse nicht bestehen, sondern der Protoplast dehnt sich aus, d. h. der Plasmolysegrad (G) nimmt dauernd ab. Der Ausdruck $O_2 - O_1 = C(G_2 - G_1)$, worin G_1 und G_2 die zu zwei verschiedenen Zeitpunkten gemessenen Plasmolysegrade und O_1 und O_2 die entsprechenden osmotischen Werte bei Grenzplasmolyse bezeichnen, gibt somit die im betrachteten Zeitabschnitt eingedrungene Substanzmenge an (vgl. auch STADELMANN 1952).

Zur Bestimmung der Wasserpermeabilität wurde die plasmometrische Methode unter anderen von HUBER und HÖFLER (1930), BOCHSLER (1948) und SEEMANN (1953) benutzt. Bei solchen Bestimmungen ist es vor allem wichtig, daß das Plasmolyticum möglichst freien Zutritt zu den zu untersuchenden Zellen hat. Sonst bekommt man viel zu kleine Werte der Wasserpermeabilität.

Die grenzplasmolytische und die plasmometrische Methode haben je ihre eigenen Vor- und Nachteile. Jene setzt (in ihrer hier geschilderten Form) voraus, daß man über eine größere Zahl von Zellen möglichst gleichmäßigen osmotischen Wertes verfügt, während bei der plasmometrischen Methode das Verhalten einzelner, individueller Protoplasten verfolgt wird und daher im Prinzip eine einzige Zelle für eine Bestimmung genügt. Eine Fehlerquelle, die bei der grenzplasmolytischen, nicht aber bei der plasmometrischen Methode zu berücksichtigen ist, ist die Adhäsion der Protoplasten an der Zellwand (BONTE 1934). Auf der anderen Seite müssen die plasmometrisch zu untersuchenden Zellen möglichst regelmäßiger Form sein, am besten zylindrisch, und ihre Protoplasten müssen bei der Plasmolyse oder wenigstens bei beginnender Deplasmolyse regelmäßige, genau meßbare Formen annehmen. (Eine Auswahl für plasmometrische Zwecke verwendbarer Zellsorten findet man z. B. in den Arbeiten von HUBER und HÖFLER 1930 und HOFMEISTER 1935 zusammengestellt.)

Das Partialdruckverfahren ist bei der plasmometrischen Methode ebenso wie bei der grenzplasmolytischen verwendbar (HOFMEISTER 1935). Auch die gleichzeitige Permeation zweier Verbindungen in entgegengesetzter Richtung läßt sich sowohl grenzplasmolytisch (BÄRLUND 1929, S. 77) wie plasmometrisch (COLLANDER 1949, URL 1952, vgl. auch CASARI 1953) verfolgen. Die Bedeutung dieser „Simultanmethode" liegt vor allem darin, daß sie es erlaubt, die Permeation zweier Stoffe unter absolut gleichen Bedingungen zu beobachten.

Ein Nachteil sowohl der plasmometrischen wie der grenzplasmolytischen Methode (in der Form, wie sie gewöhnlich verwendet wird) liegt darin, daß die zu untersuchenden Protoplasten ziemlich hohen Stoffkonzentrationen ausgesetzt werden, wozu noch kommt, daß die äußere Plasmahaut beim Losreißen von der Zellwand mehr oder weniger stark verletzt wird. A priori ist es daher durchaus nicht unwahrscheinlich, daß die Durchlässigkeit der plasmolysierten Protoplasten von der normalen Permeabilität stark abweichen würde. Die experimentelle Nachprüfung dieser naheliegenden Vermutung unter anderen durch SCHMIDT (1936; vgl. auch SUTCLIFFE 1954) zeigte jedoch, daß sich die „Plasmolysepermeabilität" in vielen Fällen nur sehr wenig von der Normalpermeabilität unterscheidet. Bei hochpermeablen Objekten wurde allerdings eine bedeutende Permeabilitätserniedrigung durch Zuckerplasmolyse festgestellt. Soll der eventuelle Einfluß der Plasmolyse auf die Permeabilität ganz vermieden werden, kann man so vorgehen, daß man die Zellen zuerst in einer hypotonischen Lösung der zu prüfenden Substanz liegen läßt und die aus *dieser* Lösung stattgefundene Stoffaufnahme hinterher plasmolytisch ermittelt.

Wir kommen jetzt zu den nichtplasmolytischen osmotischen Meßverfahren.

Die Knickungsmethode.. Bei *Beggiatoa* und *Oscillatoria* ist der Protoplast so innig mit der Zellwand verwachsen, daß es in hypertonischen Lösungen meistens zu keiner Plasmolyse kommt (vgl. DRAWERT 1949, KUCHAR 1950). Statt dessen findet ein Einknicken der Zellwand statt. Bei eintretender Isotonie gleichen sich die Knickungen wieder aus. Die Knickung und „Entknickung" der Fäden kann daher in ganz gleicher Weise wie die Plasmolyse und Deplasmolyse zur Messung der Stoffpermeation benutzt werden (RUHLAND und HOFFMANN 1925, ELO 1937, RUHLAND und HEILMANN 1951). Bei *Beggiatoa* sind die zur Erzeugung von Knickungen benötigten Stoffkonzentrationen außerordentlich gering, was einen großen Vorteil bedeutet, da hierdurch Experimente mit schwerlöslichen oder in höherer Konzentration giftig wirkenden Substanzen ermöglicht werden.

Die Plasmoptysemethode. Unter den Tierphysiologen gelten die roten Blutkörperchen als ein besonders günstiges Objekt für Permeabilitätsuntersuchungen. Die Methode, deren man sich dabei in erster Linie bedient, ist die Beobachtung der Hämolyse, d. h. einer osmotisch bedingten Plasmoptyse, welche eintritt, wenn die Zellen entweder mit hypotonischen Lösungen oder mit Lösungen einigermaßen schnell permeierender Substanzen behandelt werden. Auf botanischem Gebiet scheinen entsprechende Untersuchungen bis jetzt nur mit Leuchtbakterien ausgeführt worden zu sein. Dieselben sind größtenteils an hohe Salzkonzentrationen adaptiert. Ihre Zellen werden daher bei Verdünnung des normalen Suspensionsmediums, entweder mit Wasser oder mit der Lösung eines permeierenden Stoffes, osmotisch gesprengt, was makroskopisch an dem Erlöschen der Chemoluminescenz zu erkennen ist (HILL 1929, 1932 sowie unveröffentlichte Versuche im Helsingforser botanischen Institut).

Auch wenn lebende Zellen in konzentrierte wäßrige Lösungen von z. B. Methanol oder Aceton versetzt werden, kann Plasmoptyse eintreten. Dies zeigt, daß die betreffenden Substanzen unter Umständen schneller in die Zellen eindringen, als Wasser aus ihnen heraustritt (vgl. ZEHETNER 1933, RUHLAND und Mitarbeiter 1938).

Beobachtungen von Größenänderungen an Einzelzellen oder an vielzelligen Gewebestreifen. An Eizellen von Seeigeln und anderen Meerestieren ist die Wasserpermeabilität sehr eingehend und genau in der Weise untersucht worden, daß man die Eier in hypo- oder hypertonische Lösungen gebracht hat und die hierdurch veranlaßten Änderungen des Zellvolumens messend verfolgt hat (Zusammenfassungen bei LUCKÉ 1940 sowie DAVSON und DANIELLI 1943). Auch das Eindringen gelöster Stoffe in die Eizellen läßt sich in derselben Weise bequem messen. Bisher scheinen nur wenige pflanzliche Objekte in dieser Weise studiert worden zu sein, und zwar unter anderen die Eizellen von *Fucus* (RESÜHR 1935) sowie die aus dem plasmolysierten Zwiebelparenchym von *Allium* freipräparierten kugeligen Protoplasten (LEVITT und Mitarbeiter 1936). Ein beachtenswerter Vorzug dieser Methode liegt darin, daß sie die diffusionshemmende Wirkung der Zellwand ganz ausschaltet.

Auch mit behäuteten Einzelzellen können ähnliche Versuche ausgeführt werden. So hat LAIBACH (1932) eine interferometrische Methode zur Messung sehr geringer Größenänderungen an *Valonia*zellen ausgearbeitet, während v. HOFE (1933) und COLLANDER (1950) die osmotisch bedingten Längenänderungen von Pilzhyphen bzw. *Nitella*zellen mikrometrisch verfolgten. Im letzteren Fall konnte der osmotische Effekt von Substanzen, die nur wenig langsamer als das Wasser permeieren, immer noch festgestellt werden.

Schon lange vor diesen Untersuchungen hat man es versucht, aus den Längenänderungen vielzelliger Gewebestreifen auf osmotisch bedingte Wasseraufnahme oder -abgabe zu schließen. Versuche dieser Art wurden unter anderen von DE VRIES (1884), LUNDEGÅRDH (1911), DELF (1916), S. C. BROOKS (1916), BOUILLENNE (1930) und ELO (1937, S. 81) ausgeführt. Bei der Verwendung von Gewebestreifen anstatt einzelner Zellen kompliziert sich die Situation insofern, als nun nicht alle Zellen gleichzeitig mit dem Osmoticum in Berührung kommen. Die Zellen der verschiedenen Gewebeschichten verkürzen und verlängern sich daher auch nicht gleichzeitig, sondern sukzessiv. Dies erschwert die quantitative Auswertung der erhaltenen Versuchsdaten so sehr, daß diese Methode der Permeationsbestimmung wohl nur als ein Notbehelf für spezielle Fälle gelten kann. Am ehesten wird man auf diesem Wege zu anwendbaren Ergebnissen gelangen können, wenn man mit möglichst dünnen Objekten arbeitet.

Die Wasseraufnahme und -abgabe von Gewebestücken läßt sich nicht nur auf Grund ihrer Dimensionsänderungen, sondern auch gravimetrisch feststellen (MYERS 1951, DRAWERT 1952, HACKETT und THIMANN 1952). Die Fehlerquellen bleiben dabei wohl ungefähr dieselben.

Lebende Zellen als Osmometer. KORNMANN (1934) hat Steigrohre auf lebende *Valonia*zellen aufgesetzt und konnte so die in die Zelle eingedrungenen oder von der Zelle abgegebenen Flüssigkeitsmengen direkt bestimmen. Durch Kontrollversuche mit abgetöteten Zellen ließ sich der Einfluß der Zellwand ermitteln. Es fragt sich jedoch, ob nicht die normale Permeabilität der Protoplasten durch den in Rede stehenden ziemlich groben Eingriff alteriert wird.

OSTERHOUT (1950) wiederum befestigte eine lebende *Nitella*zelle so, daß ihre entgegengesetzten Enden mit verschiedenen Lösungen in Berührung standen, und maß dann direkt auf Grund der Bewegungen des Flüssigkeitsmeniskus in einer Meßcapillare die durch die Zelle hindurchgesogenen Flüssigkeitsmengen.

4. Andere Methoden.

a) Elektrische Messungen.

In dem elektrischen Widerstand der Protoplasten haben wir ein Maß für den Widerstand, den sie dem Durchtritt von Ionen entgegensetzen, und somit ein reziprokes Maß der Ionenpermeabilität. Die Zahl der pflanzlichen Organismen, an denen eine exakte Bestimmung des elektrischen Widerstandes der Protoplasten möglich ist, ist indessen nicht groß. In vielzelligen Geweben werden im allgemeinen die Zellwände den Strom wesentlich besser leiten als die Protoplasten. Aus Messungen des Gesamtwiderstandes von Gewebescheiben (OSTERHOUT 1922) ist es daher nicht leicht, den Widerstand der Protoplasten zu berechnen. Bei weitem aufschlußreicher gestalten sich dagegen Widerstandsmessungen an einzelnen makroskopischen Zellen, vor allem an solchen der Valoniaceen und der Characeen, an denen auch Potential- und Kapazitätsmessungen verhältnismäßig leicht auszuführen sind. Die hier in Frage kommende Methodik ist besonders von BLINKS (1937, 1940) und COLE (1940) ausgearbeitet worden. Die genaue Interpretierung der gewonnenen Meßdaten scheint aber noch immer nicht ganz leicht.

b) Bestimmung des spezifischen Gewichts.

Nach COOPER und Mitarbeitern (1929) kann man durch Zufügen von Caesiumchlorid zum Seewasser sein spezifisches Gewicht so erhöhen, daß *Valonia*zellen, die in normalem Seewasser sofort untersinken, mehr als ein Jahr an der Oberfläche schwimmen bleiben. Dies deutet darauf hin, daß die Caesiumionen nur sehr langsam von den Zellen aufgenommen werden.

Die Wasserpermeabilität der Zellen von *Nitellopsis (Tolypellopsis)* hat PALVA (1939) in der Weise bestimmt, daß er die Zunahme des spezifischen Gewichts der Zelle, wenn ihr Wasser auf osmotischem Wege entzogen wird, messend verfolgte und zwar so, daß die Zelle in einem Gefäß schwebend gehalten wurde, das mit Lösungen abgestuften spezifischen Gewichts gefüllt war. Analog hiermit haben PIGÓN und ZEUTHEN (1951) die Wasseraufnahme und -abgabe einer Amöbe mittels des cartesischen Tauchers bestimmt.

BUFFEL (1952) hat in sorgfältig durchgeführten Versuchen diejenigen Änderungen des spezifischen Gewichts von *Avena*-Koleoptilen bestimmt, die beim gegenseitigen Austausch von H_2O und D_2O zustandekommen. Sein Ergebnis, wonach die Wasserpermeabilität dieser Zellen sehr viel kleiner wäre als bei anderen Pflanzenzellen, deutet jedoch darauf hin, daß in seiner Methodik irgendein Fehler steckt. In der Tat muß es eine ungemein schwierige Aufgabe sein, die Wasserpermeabilität der Protoplasten auf Grund von Versuchen mit einem so kompliziert gebauten Organ, wie es die Koleoptilen sind, zu bestimmen.

c) Diffusionsversuche mit membranartigen Geweben.

In der Tierphysiologie spielen Untersuchungen über die Durchlässigkeit solcher membranartiger Gewebe wie etwa der Froschhaut, der Darmwand oder der Kiemenepithelien eine wichtige Rolle. Die Frage, inwieweit die Permeation dabei durch die Protoplasten selbst geschieht, und inwieweit es sich um eine Diffusion zwischen den Protoplasten handelt, ist wohl nicht immer leicht zu entscheiden. Noch viel schlimmer steht es in dieser Hinsicht mit den entsprechenden pflanzlichen Objekten. Das Vorkommen von verhältnismäßig leicht durchlässigen Zellwänden zwischen den Protoplasten läßt es von vornherein aussichtslos erscheinen, aus Diffusionsversuchen mit derartigen Gebilden quantitative Aufschlüsse hinsichtlich der Durchlässigkeitseigenschaften der Protoplasten zu gewinnen (vgl. STEWARD 1930). STEEMANN NIELSEN (1947) vertrat allerdings in dieser Hinsicht eine mehr optimistische Auffassung.

d) Physiologische und toxikologische Effekte als Kriterien der Permeation.

Wenn die Substanz A einen gewissen physiologischen Effekt hat, den die analoge Verbindung B vermissen läßt, dann kann dies davon herrühren, daß wohl A, nicht aber B zum Zellinnern Zutritt hat. Wenn wiederum eine Substanz unter gewissen Bedingungen toxisch wirkt, unter anderen aber nicht, liegt es nahe, diesen Befund so zu erklären, daß die Substanz unter den erstgenannten Bedingungen in die Zellen eindringt, unter den anderen dagegen nicht.

In der Literatur findet man eine Unzahl derartiger „Erklärungen". Zum großen Teil handelt es sich jedoch dabei um unbewiesene Vermutungen, die einer Bestätigung auf anderem Wege bedürfen.

In einigen Fällen kann man immerhin auf Grund von Erfahrungen der angedeuteten Art zu recht sicher begründeten Vorstellungen betreffs des Permeationsvermögens gewisser Substanzen gelangen. So z. B. ist es bekannt, daß manche Schimmelpilze einerseits in schwefelsäurehaltigen Lösungen, deren p_H-Wert etwa 2 beträgt, gedeihen, andererseits aber auch in alkalischen Lösungen mit einem p_H-Wert von etwa 10. Dies ist ein überzeugender Beweis für die fast vollständige Undurchlässigkeit des Plasmalemmas dieser Pilze für diejenigen Säuren und Basen, die die p_H-Werte der Außenlösung bedingen. Wenn nun andererseits festgestellt wird, daß etwa Fettsäuren und Amine bereits bei viel weniger extremen p_H-Werten giftig wirken, so deutet dies darauf hin, daß diese Säuren und Basen viel leichter als die erstgenannten permeieren.

Ein zweites Beispiel dafür, wie man auf Grund der Beeinflussung gewisser physiologischer Prozesse der Zellen Schlüsse hinsichtlich ihrer Stoffaufnahme ziehen kann, bietet eine Arbeit von Krech (1954). Nach ihr ruft Infiltration von Blattstücken mit Saccharose- oder Glucose-1-phosphat-Lösung eine deutliche Atmungssteigerung hervor, woraus zu schließen ist, daß jene Substanzen in die Blattzellen eindringen. Citronensäure verändert außerdem den respiratorischen Quotienten der Blattstücke. Dadurch ist die Aufnahme dieser Substanz noch eindeutiger erwiesen.

Zu quantitativen Bestimmungen der Permeabilität oder der Aufnahmegeschwindigkeit wird man durch derartige Versuche allerdings nicht leicht gelangen.

5. Wahl der geeignetsten Methode.

Bereits ein flüchtiger Blick auf die im Vorangehenden erwähnten Methoden zur Messung des Stoffaustausches und der Permeabilität lebender Zellen läßt erkennen, daß sie nicht nur sehr verschiedener Art, sondern auch von sehr verschiedenem Wert sind. Zunächst in bezug auf die Amplitude ihrer Anwendbarkeit: es gibt Methoden, die nur für ganz bestimmte Stoffe oder bei ganz bestimmten Zellen in Frage kommen, während andere sehr vielseitig verwendbar sind. Eine wirkliche Universalmethode, die in allen Fällen anwendbar wäre, gibt es jedoch nicht.

Die einzelnen Methoden unterscheiden sich ferner hinsichtlich ihrer Zuverlässigkeit und Genauigkeit. Trotzdem kann man sie nicht einfach in genaue und ungenaue Methoden einteilen. Vielmehr verhält es sich so, daß eine Methode, die in bestimmten Fällen Ausgezeichnetes leistet, in anderen Fällen ganz versagen kann. Es muß somit in jedem konkreten Einzelfall erwogen werden, welche Methode eben hier die besten Resultate ergeben wird. Wenn es sich z. B. darum handelt, das Permeationsvermögen sehr schnell permeierender Stoffe zu messen, ist es vor allem wichtig, die Versuchsbedingungen so zu gestalten, daß der zu prüfende Stoff möglichst freien Zutritt zu den zu untersuchenden Zellen hat. Sonst kann es einem leicht passieren, daß das Meßergebnis eher ein Ausdruck ist für die Geschwindigkeit, mit der der Stoff an die Zellen herantritt, als für sein Permeationsvermögen. Viele Bestimmungen etwa der Wasserpermeabilität auf plasmolytischem Wege sind gerade wegen der Nichtbeachtung dieser Fehlerquelle ziemlich wertlos. Im Falle sehr langsam permeierender Stoffe wird man wiederum in erster Linie danach streben, die Versuche so zu gestalten, daß keine Zellen dabei beschädigt werden, da in diesem Falle bereits eine geringe Anzahl pathologisch veränderter Zellen das Gesamtergebnis stark beeinflussen kann. Handelt es sich schließlich um die Verfolgung eines aktiven Stofftransportes der Protoplasten, so wird man vor allem bemüht sein, diejenigen Außenfaktoren (Sauerstoffversorgung u. dgl.), welche auf den Prozeß einen entscheidenden Einfluß haben, möglichst günstig zu gestalten.

Trotz aller Sorgfalt können Messungen der Stoffaustauschgeschwindigkeit der Zellen und besonders auch solche der Zellpermeabilität selten oder nie mit demselben Grad der Exaktheit ausgeführt werden wie etwa entsprechende Messungen an leblosen Systemen. Oft muß man sich bewußt mit Annäherungswerten begnügen. Dabei sollte man aber wenigstens nicht versäumen, die Fehlergrenzen der benutzten Bestimmungsmethode so realistisch wie nur irgendmöglich abzuschätzen. In dieser Hinsicht ist viel gesündigt worden, indem man Fehlerquellen außer acht gelassen hat, die die erzielten Resultate in ganz entscheidender Weise beeinflußt haben.

Literatur.

ÄYRÄPÄÄ, T.: On the base permeability of yeast. Physiol. Plantar. **3**, 402—429 (1950). — ALBAUM, H. G., S. KAISER and H. A. NESTLER: The relation of hydrogen-ion concentration to the penetration of 3-indole acetic acid into Nitella cells. Amer. J. Bot. **24**, 513—518 (1937). — ARISZ, W. H.: Uptake and transport of chlorine by parenchymatic tissue of leaves of Vallisneria spiralis. Proc. Kon. Ned. Akad. v. Wetensch. **51**, 25—36 (1948). — ASPREY, G. F.: Some observations on the absorption and exosmosis of electrolytes by storage organs. Protoplasma (Berl.) **27**, 153—168 (1937).

BÄRLUND, H.: Permeabilitätsstudien an Epidermiszellen von Rhoeo discolor. Acta bot. fenn. **5**, 1—117 (1929). — BARTELS, P.: Quantitative mikrospektroskopische Untersuchung der Speicherungs- und Permeabilitätsverhältnisse akridinorangegefärbter Zellen. Planta (Berl.) **44**, 341—369 (1954). — BENNET-CLARK, T. A., and D. BEXON: Water relations of plant cells. New Phytologist **45**, 5—17 (1946). — BETZ, A.: Untersuchungen über Verhalten und Wirkung der Vitalfarbstoffe Prune pure und Akridinorange sowie Beobachtungen über das Reduktions-Oxydationspotential in Zellen höherer Pflanzen. Planta (Berl.) **41**, 323—357 (1953). — BLINKS, L. R.: Protoplasmic potentials in Halicystis. J. Gen. Physiol. **18**, 409—420 (1935). — Electrical evidence on the nature and alterations of membranes in large plant cells. Trans. Faraday Soc. **33**, 991—997 (1937). — The relations of bioelectric phenomena to ionic permeability and to metabolism in large plant cells. Cold Spring Harbor Symp. Quant. Biol. 8, 204—215 (1940). — BOCHSLER, A.: Die Wasserpermeabilität des Protoplasmas auf Grund des FICKschen Diffusionsgesetzes. Ber. schweiz. bot. Ges. **58**, 73—122 (1948). — BOGEN, H. J., u. G. FOLLMANN: Osmotische und nichtosmotische Stoffaufnahme bei Diatomeen. Planta (Berl.) **45**, 125—146 (1955). — BOGEN, H. J., u. H. PRELL: Messung nichtosmotischer Wasseraufnahme an plasmolysierten Protoplasten. Planta (Berl.) **41**, 459—479 (1953). — BONTE, H.: Vergleichende Permeabilitätsstudien an Pflanzenzellen. Protoplasma (Berl.) **22**, 209—242 (1934). — BORESCH, K.: Über den Eintritt und die emulgierende Wirkung verschiedener Stoffe in Blattzellen von Fontinalis antipyretica. Biochem. Z. **101**, 110—158 (1919). — BOUILLENNE, R.: Un appareil nouveau pour la mesure des vitesses de pénétration des solutions salines dans le protoplasme végétal. Bull. Acad. roy. Belg., Cl. d. sc. V. s. **16**, 1017—1026 (1930). — Pénétration dans des cellules d'Allium cepa des sels ammoniques d'acids gras à differente p_H. Bull. Soc. roy. sci. Liège **1940**, 11—26. — BRAUNER, L., u. M. BRAUNER: Untersuchungen über den Einfluß des Lichtes auf die Zuckerpermeabilität lebenden Pflanzengewebes. Rev. Fac. Sci. Univ. Istanbul, N. S. **1**, 58—73 (1936). — BRENNER, W.: Studien über die Empfindlichkeit und Permeabilität pflanzlicher Protoplasten für Säuren und Basen. Öfvers. Finska Vetensk.-Soc. Förh. A **60** Nr. 4 (1918). — BRIGGS, G. E., and R. N. ROBERTSON: Diffusion and absorption in disks of plant tissue. New Phytologist **47**, 265—283 (1948). — BROOKS, M. M.: The penetration of certain oxidation-reduction indicators as influenced by p_H. Amer. J. Physiol. **76**, 360—379 (1925). — The influence of experimental conditions upon the penetration of methylene blue and trimethyl thionine. Protoplasma (Berl.) **7**, 46—61 (1929). — BROOKS, S. C.: A study of permeability by the method of tissue tension. Amer. J. Bot. **3**, 562—570 (1916). — The accumulation of ions: relations between protoplasm and sap in Valonia. J. Cellul. a. Comp. Physiol. **6**, 169—180 (1935). — BROOKS, S. C., u. M. M. BROOKS: The permeability of living cells. Berlin: Gebr. Borntraeger 1941. — BROYER, T. C., and D. R. HOAGLAND: Methods of sap expression from plant tissues with special reference to studies on salt accumulation by excised barley roots. Amer. J. Bot. **27**, 501—511 (1940). — BUCHY, A.: Recherches quantitatives sur la coloration vitale des cellules de levure. Rev. cytol. et cytophysiol. végét. **5**, 265—299 (1941). — BUFFEL, K.: New techniques for comparative permeability studies on the oat coleoptile. Meded. Kon. Vlaamse Akad., Kl. Wetensch. **14**, Nr 7 (1952). — BURSTRÖM, H.: Studies on the carbohydrate nutrition of roots. Ann. Agric. Coll. Sweden **9**, 264—284 (1941). — Growth and water absorption of Helianthus tuber tissue. Physiol. Plantar. **6**, 685—691 (1953). — BUTLER, G. W.: The "apparent free space" of wheat roots. Physiol. Plantar. **6**, 617—635 (1953).

CASARI, K.: Über den Plasmolytikumwechsel-Effekt. Protoplasma (Berl.) **42**, 427—447 (1953). — CHOLNOKY, B. J.: Ein Beitrag zur Kenntnis des Plasmalemmas. Ber. dtsch. bot. Ges. **65**, 369—373 (1953). — COLE, K. S.: Permeability and impermeability of cell membranes for ions. Cold Spring Harbor Symp. Quant. Biol. **8**, 110—121 (1940). — COLLANDER, R.: Über die Permeabilität pflanzlicher Protoplasten für Sulfosäurefarbstoffe. Jb. wiss. Bot. **60**, 354—410 (1921). — The permeability of plant protoplasts to small molecules. Physiol. Plantar. **2**, 300—311 (1949). — The permeability of Nitella cells to rapidly penetrating non-electrolytes. Physiol. Plantar. **3**, 45—57 (1950). — The permeability of Nitella cells to non-electrolytes. Physiol. Plantar. **7**, 420—445 (1954). — COLLANDER, R., u. H. BÄRLUND: Permeabilitätsstudien an Chara ceratophylla. II. Acta bot. fenn. **11**, 1—114 (1933). — COLLANDER, R., H. LÖNEGREN u. E. ARHIMO: Das Permeationsvermögen eines basischen Farbstoffes mit demjenigen einiger Anelektrolyte verglichen. Protoplasma (Berl.) **37**, 527—537 (1943). — COLLANDER, R., u. K. SOMER: Über die angebliche Permeabilität der Fontinalis-Zellen für Alkaloidkationen. Protoplasma (Berl.) **14**, 1—10 (1931). — COLLANDER, R., O. TURPEINEN u. E. FABRITIUS: Die Permeabilität der Rhoeo-Zellen für Ammoniak und Essigsäure. Protoplasma (Berl.) **13**, 348—362 (1931). — CONN, H. J.: Biological stains, 6. Aufl. Geneva, N.Y. 1953. — CONWAY, E. J., and M. DOWNEY: An outer metabolic region of the yeast cell. Biochemic. J. **47**, 347—355 (1950). — CONWAY, E. J., H. RYAN and E. CARTON: Active transport of sodium ions from the yeast cell. Biochemic. J. **58**, 158—167 (1954). — COOPER, W. C., M. J. DORCAS and W. J. V. OSTERHOUT: The penetration of strong electrolytes. J. Gen. Physiol. **12**, 427—433 (1929). — CORSON, S. A.: An optical torsion microlever for measuring cell permeability. Proc. Oklahoma Acad. Sci. **23**, 32—33 (1943). — COWIE, D. B., R. B. ROBERTS and I. Z. ROBERTS: Potassium metabolism in Escherichia coli. J. Cellul. a. Comp. Physiol. **34**, 243—258 (1949).

DAVSON, H., and J. F. DANIELLI: The permeability of natural membranes, 2. Aufl. Cambridge: University Press 1952. — DELF, E. M.: Studies of protoplasmic permeability by measurement of rate of shrinkage of turgid tissues. Ann. of Bot. **30**, 283—310 (1916). — DEYSSON, G.: Recherches sur la perméabilité des cellules végétales. Rev. cytol. et biol. végét. **13**, 153—315 (1952). — DRAWERT, H.: Zur Frage der Methylenblauspeicherung in Pflanzenzellen. Z. Naturforsch. 4b, 35—37 (1949). — Zellmorphologische und zellphysiologische Studien an Cyanophycen. Planta (Berl.) **37**, 161—209 (1949). — Kritische Untersuchungen zur gravimetrischen Bestimmung der Wasserpermeabilität. Planta (Berl.) **41**, 65—82 (1952). — Über die Eignung zelleigener Anthozyane zur p_H-Wert-Bestimmung des Zellsaftes. Protoplasma (Berl.) **44**, 370—373 (1954).

EDDY, A. A., and C. HINSHELWOOD: Alkali-metal ions in the metabolism of Bact. lactis aerogenes. Proc. Roy. Soc. Lond., Ser. B **138**, 237—240 (1951). — ELO, J. E.: Vergleichende Permeabilitätsstudien, besonders an niederen Pflanzen. Ann. bot. Soc. zool.-bot. fenn. „Vanamo" 8, Nr 6 (1937).

FITTING, H.: Untersuchungen über isotonische Koeffizienten und ihren Nutzen für Permeabilitätsbestimmungen. Jb. wiss. Bot. **57**, 553—612 (1917). — Untersuchungen über die Aufnahme und über anomale osmotische Koeffizienten von Glycerin und Harnstoff. Jb. wiss. Bot. **59**, 1—170 (1919).

HACKETT, D. P., and K. V. THIMANN: The nature of the auxin-induced water uptake by potato tissue. Amer. J. Bot. **39**, 553—560 (1952). — HANSON, J. B., and J. BONNER: The relationship between salt and water uptake in Jerusalem artichoke tuber tissue. Amer. J. Bot. **41**, 702—710 (1954). — HARVEY, E. N.: Studies on the permeability of cells. J. of Exper. Zool. **10**, 507—556 (1911). — HEVESY, G.: Radioactive indicators. New York: Interscience 1948. — HILL, S. E.: The penetration of luminous bacteria by the ammonium salts of the lower fatty acids. J. Gen. Physiol. **12**, 863—872 (1929). — The effects of ammonia, of the fatty acids, and of their salts, on the luminescence of Bacillus Fisheri. J. Cellul. a. Comp. Physiol. **1**, 145—159 (1932). — HOAGLAND, D. R., and T. C. BROYER: General nature of the process of salt accumulation by roots with description of experimental methods. Plant Physiol. **11**, 471—507 (1936). — HOAGLAND, D. R., P. L. HIBBARD and A. R. DAVIS: The influence of light, temperature, and other conditions on the ability of Nitella cells to concentrate halogens in the sap. J. Gen. Physiol. **10**, 121—146 (1926). — HÖFLER, K.: Eine plasmolytisch-volumetrische Methode zur Bestimmung des osmotischen Wertes von Pflanzenzellen. Denkschr. ksl. Akad. Wiss. Wien, Math.-naturwiss. Kl., Abt. I **95**, 99—168 (1918). — Permeabilitätsbestimmungen nach der plasmometrischen Methode. Ber. dtsch. bot. Ges. **36**, 414—422 (1918). — Permeabilitätsstudien an Stengelzellen von Majanthemum bifolium. Sitzgsber. Akad. Wiss. Wien, Math.-naturwiss. Kl., Abt. I **143**, 213—264 (1934). — Zur Vital- und Fluoreszenzfärbung. Ber. bot. Ges. **66**, 453—467 (1954). — HÖFLER, K., u. H. SCHINDLER: Vitalfärbung von Algenzellen mit Toluidinblaulösungen gestufter Wasserstoffionenkonzentration. Protoplasma (Berl.) **40**, 137—151 (1951). — HOFE, F. v.: Permeabilitätsuntersuchungen an Psalliota campestris. Planta (Berl.) **20**, 354—390 (1933). — HOF-

MEISTER, L.: Vergleichende Untersuchungen über spezifische Permeabilitätsreihen. Bibl. Bot. 113 (1935). — Über die Permeabilitätsbestimmung nach der Deplasmolysezeit. Sitzgsber. österr. Akad. Wiss., Math.-naturwiss. Kl., Abt. I 157, 83—95 (1948). — HOLDHEIDE, W.: Über Plasmoptyse bei Hydrodictyon utriculatum. Planta (Berl.) 15, 244—298 (1931). — HOMÈS, M. V.: La pénétration des chlorures dans les cellules de la racine charnue du navet. Protoplasma (Berl.) 18, 161—193 (1933). — HUBER, B., u. K. HÖFLER: Die Wasserpermeabilität des Protoplasmas. Jb. wiss. Bot. 73, 351—511 (1930).

INGOLD, C. T.: On the effect of previous treatment with salt solutions on the subsequent outward diffusion of electrolytes from plant tissue. Ann. of Bot. 45, 709—716 (1931). — IRMAK, L. R.: The lyotropic effect of ions on the sugar permeability of living plant cells. Rev. Fac. Sci. Univ. Istanbul, N.S. 3, 1—39 (1938). — IRWIN, M.: On the accumulation of dye in Nitella. J. Gen. Physiol. 8, 147—182 (1925). — Penetration of trimethylthionine into Nitella and Valonia from methylene blue. J. Gen. Physiol. 12, 147—165 (1928). — Why does azure B penetrate more readily than methylene blue or crystal violet? J. Gen. Physiol. 14, 19—29 (1930).

JACOBS, M. H.: The production of intracellular acidity by neutral and alkaline solutions containing carbon dioxide. Amer. J. Physiol. 53, 457—463 (1920). — The influence of ammonium salts on cell reaction. J. Gen. Physiol. 5, 181—188 (1922). — The exchange of material between the erythrocyte and its surroundings. Harvey Lectures 22, 146—164 (1927). — Some aspects of cell permeability to weak electrolytes. Cold Spring Harbor Symp. Quant. Biol. 8, 30—39 (1940). — JACOBSON, L., R. OVERSTREET, H. M. KING and R. HANDLEY: A study of potassium absorption by barley roots. Plant Physiol. 25, 639—647 (1950). — JACQUES, A. G.: The kinetics of penetration. XII. Hydrogen sulfide. J. Gen. Physiol. 19, 397—418 (1936).

KAHO, H.: Über den Einfluß der Kohlensäure auf die Exosmose von Elektrolyten aus Stengelzellen. Protoplasma (Berl.) 27, 502—522 (1937). — KAMEN, M. D.: Radioactive tracers in biology, 2. Aufl. New York: Academic Press Inc. 1951. — KINZEL, H.: Die Bedeutung der Pektin- und Zellulosekomponente für die Lage des Entladungspunktes pflanzlicher Zellwände. Protoplasma (Berl.) 42, 209—226 (1953). — Theoretische Betrachtungen zur Ionenspeicherung basischer Vitalfarbstoffe in leeren Zellsäften. Protoplasma (Berl.) 44, 52—72 (1954). — KNAUSS, H. J., and J. W. PORTER: The absorption of inorganic ions by Chlorella pyrenoidosa. Plant Physiol. 29, 229—234 (1954). — KÖLBEL, H.: Quantitative Untersuchungen über den Speicherungsmechanismus von Rhodamin B, Eosin und Neutralrot in Hefezellen. Z. Naturforsch. 3b, 442—453 (1948). — KORNMANN, P.: Osmometer aus lebenden Valonia-Zellen und ihre Verwendbarkeit zu Permeabilitätsbestimmungen. Protoplasma (Berl.) 21, 340—350 (1934). — KRASSINSKY, N.: Über jahreszeitliche Änderungen der Permeabilität des Protoplasmas. Protoplasma (Berl.) 9, 622—631 (1930). — KRECH, E.: Über die Phosphorylase in höheren Pflanzen. Beitr. Biol. Pflanz. 30, 379—405 (1954). — KUCHAR, K.: Plasmolyseformverlauf und Trichomzerfall bei zwei Oscillatorien. Phyton 2, 213—222 (1950).

LAIBACH, F.: Interferometrische Untersuchungen an Pflanzen. Jb. wiss. Bot. 76, 218—282 (1932). — LEPESCHKIN, W. W.: Über die Permeabilitätsbestimmung der Plasmamembran für gelöste Stoffe. Ber. dtsch. bot. Ges. 27, 129—142 (1909). — Permeabilitätsänderungen des Protoplasmas nach der Methode der isotonischen Koeffizienten. Biochem. Z. 142, 291—307 (1923). — LEVITT, J., G. W. SCARTH u. R. D. GIBBS: Water permeability of isolated protoplasts in relation to volume change. Protoplasma (Berl.) 26, 237—248 (1936). — LISON, L., u. J. FAUTREZ: L'étude physicochimique des colorants dans ses applications biologiques. Protoplasma (Berl.) 33, 116—151 (1939). — LUCKÉ, B.: The living cell as an osmotic system and its permeability to water. Cold Spring Harbor Symp. Quant. Biol. 8, 123—132 (1940). — LUNDEGÅRDH, H.: Über die Permeabilität der Wurzelspitzen von Vicia faba unter verschiedenen äußeren Bedingungen. Kgl. Sv. Vetenskapsakad. Hdl. 47, Nr 3 (1911).

MALM, M.: Über die Permeabilität der Hefezellen und die von den permeierenden Stoffen, insbesondere Fluorwasserstoff, bedingten Plasmaveränderungen. Ark. Kemi, Mineral., Geol. (Stockh.) A 25, Nr 1 (1947). — MARTIN, R. P., and R. SCOTT RUSSEL: Studies with radioactive tracers in plant nutrition. J. of Exper. Bot. 1, 141—158 (1950). — MEDWEDEW, G.: Kinetische Theorie der alkoholischen Hefegärung. Protoplasma (Berl.) 27, 242—263 (1937). — MOLISCH, H.: Über eine auffallende Farbenänderung einer Blüte durch Wassertröpfchen und Kohlensäure. Ber. dtsch. bot. Ges. 39, 57—62 (1921). — MYERS, G. M. P.: The water permeability of unplasmolysed tissues. J. of Exper. Bot. 2, 129—144 (1951).

ÖSTERLIND, S.: Inhibition of respiration and nitrate absorption in green algae by enzyme poisons. Physiol. Plantar. 5, 292—297 (1952). — OSTERHOUT, W. J. V.: Injury, recovery, and death, in relation to conductivity and permeability. Philadelphia: J. B. Lippincott Company 1922. — Physiological studies of single plant cells. Biol. Rev. 6, 369—411 (1931). — Permeability in large plant cells and in models. Erg. Physiol. 35, 967—1021 (1933). — The absorption of electrolytes in large plant cells. Bot. Rev. 2, 283—315 (1936). — Higher permeability for water than for ethyl alcohol in Nitella. J. Gen. Physiol. 33, 275—284 (1950). — OVERTON, E.: Über die osmotischen Eigenschaften der Zelle in ihrer Bedeutung für die Toxikologie und Pharmakologie. Vjschr. naturforsch. Ges. Zürich 41, 383—406 (1896). — Über den Mechanismus der Resorption und der Sekretion. In Handbuch der Physiologie des Menschen (herausgeg. von W. NAGEL), Bd. 2. Braunschweig: Vieweg & Sohn 1907.

PALVA, P.: Die Wasserpermeabilität der Zellen von Tolypellopsis stelligera. Protoplasma (Berl.) 32, 265—271 (1939). — PFEFFER, W.: Über Aufnahme von Anilinfarben in lebende Zellen. Unters. bot. Inst. Tübingen 2, 179—332 (1886). — Beiträge zur Kenntnis der Oxydationsvorgänge in lebenden Zellen. Abh. kgl.-sächs. Ges. Wiss., Math.-phys. Cl. 15, Nr 5 (1889). — PIGÓN, A., u. E. ZEUTHEN: Cartesian diver balance in permeability studies. Experientia (Basel) 7, 455—456 (1951). — POIJÄRVI, L. A. P.: Über die Basenpermeabilität pflanzlicher Zellen. Acta bot. fenn. 4, 1—102 (1928). — PORTER, H. K., and L. H. MAY: Metabolism of radioactive sugars by tobacco leaf tissue. J. of Exper. Bot. 6, 43—63 (1955).

REES, W. J.: Some observations on the effect of the preparation of storage tissue on its subsequent absorption of manganese chloride. Ann. of Bot., N. S. 13, 29—51 (1949). — REINHOLD, L.: The uptake of indole-3-acetic acid by pea epicotyle segments and carrot discs. New Phytologist 53, 217—239 (1954). — *Reports from the Second Isotope Conference Oxford.* London: Butterworth's Scient. Publ. 1954. — RESÜHR, B.: Hydratations- und Permeabilitätsstudien an unbefruchteten Fucus-Eiern. Protoplasma (Berl.) 24, 531—586 (1935). — RIEDER, H. P.: Über die Zuckeraufnahme von Hefezellen. Ber. schweiz. bot. Ges. 61, 539—621 (1951). — ROBERTSON, R. N., J. S. TURNER and M. J. WILKINS: Salt respiration and accumulation in red beet tissue. Austral. J. of Exper. Biol. a. Med. 25, 1—8 (1947). — ROBINSON, C.: The nature of the aqueous solutions of dyes. Trans. Faraday Soc. 31, 245—261 (1935). — ROSENBERG, TH., and W. WILBRANDT: Enzymatic processes in cell membrane penetration. Internat. Rev. Cytology 1, 65—92 (1952). — ROSENFELS, R. S.: The absorption and accumulation of potassium bromide by Elodea as related to transpiration. Protoplasma (Berl.) 23, 503—519 (1935). — ROTHSTEIN, A.: Enzyme systems of the cell surface involved in the uptake of sugars by yeast. Symposia Soc. Exper. Biol. 8, 165—201 (1954). — RUHLAND, W., u. U. HEILMANN: Über die Permeabilität von Beggiatoa mirabilis für Anelektrolyte bei Narkose mit den homologen Alkoholen C_1—C_9. Planta (Berl.) 39, 91—120 (1950). — RUHLAND, W., u. C. HOFFMANN: Die Permeabilität von Beggiatoa mirabilis. Planta (Berl.) 1, 1—83 (1925). — RUHLAND, W., H. ULLRICH u. S. ENDO: Zur Frage der Alkoholpermeabilität von Pflanzenzellen unter verschiedenen Versuchsbedingungen. Planta (Berl.) 27, 650—668 (1938).

SCARTH, G. W.: Estimation of protoplasmic permeability from plasmolytic tests. Plant Physiol. 14, 129—143 (1939). — SCHMIDT, H.: Plasmolyse und Permeabilität. Jb. wiss. Bot. 83, 470—512 (1936). — SCHULTZ, G.: Farbstofftabellen, 7. Aufl., Bd. I—II. Leipzig: Akademische Verlagsgesellschaft 1931/32 sowie Ergänzungsbände. — SCOTT, G. T., and H. R. HAYWARD: Evidence for the presence of separate mechanisms regulating potassium and sodium distribution in Ulva lactuca. J. Gen. Physiol. 37, 601—620 (1954). — SEEMANN, F.: Der Einfluß von Neutralsalzen und Nichtleitern auf die Wasserpermeabilität des Protoplasmas. Protoplasma (Berl.) 42, 109—132 (1953). — SMITH, E. P.: The calibration of flower colour indicators. Protoplasma (Berl.) 18, 112—125 (1933). — STADELMANN, E.: Zur Messung der Stoffpermeabilität pflanzlicher Protoplasten. Sitzgsber. österr. Akad. Wiss., Math.-naturwiss. Kl., Abt. I 161, 373—408 (1952). — STEEMANN NIELSEN, E.: Diffusion of dissolved substances through thalli and leaves of aquatic plants. Nature (Lond.) 160, 376 (1947). — STEWARD, F. C.: Diffusion of certain solutes through membranes of living plant cells. Protoplasma (Berl.) 11, 521—557 (1930). — STEWARD, F. C., W. E. BERRY, C. PRESTON and T. K. RAMAMURTI: The absorption and accumulation of solutes by living plant cells. Ann. Bot., N. S. 7, 221—260 (1943). — STEWARD, F. C., and J. C. MARTIN: The distribution and physiology of Valonia at the Dry Tortugas, with special reference to the problem of salt accumulation in plants. Carnegie Inst. Wash. Publ. 475, 87—150 (1937). — STEWARD, F. C., and F. K. MILLAR: Salt accumulation in plants. Symposia Soc. Exper. Biol. 8, 367—406 (1954). — STILES, W.: Permeability. New Phytologist Reprint Nr 13. London 1924. — The exosmosis of dissolved substance from storage tissue into water. Protoplasma (Berl.) 2, 577—601 (1927). — STILES, W., and K. W. DENT: Further observations on the absorption of manganese chlorid by storage tissue. Ann. of Bot., N. S. 10, 203—222 (1946). — STILES, W., and A. D. SKELDING: The scope, technique, and interpretation of the results of experiments on absorption of salts by storage tissue. Ann. of Bot., N. S. 8, 149—156 (1944). — STRUGGER, S.: Die Vitalfärbung der Chloroplasten von Helodea mit Rhodaminen. Flora, N. F. 31, 113—128 (1936). — Praktikum der Zell- und Gewebephysiologie der Pflanze, 2. Aufl. Berlin-Göttingen-Heidelberg: Springer 1949. — SUTCLIFFE, J. F.: The influence of internal ion concentration on potassium absorption and salt respiration of red beetroot tissue. J. of Exper. Bot. 3, 59—76 (1952). — The absorption of potassium ions by plasmolyzed cells. J. of Exper. Bot. **5**, 215—231 (1954). — The exchangeability of potassium and bromide ions in cells of red beetroot tissue. J. of Exper. Bot. **5**, 313—326 (1954).

TEORELL, T.: Permeability. Ann. Rev. of Physiol. **11**, 545—564 (1949). — TRÖNDLE, A.: Der Einfluß des Lichtes auf die Permeabilität der Pflanzenzelle. Jb. wiss. Bot. **48**, 171—282 (1910). — Der Einfluß des Lichtes auf die Permeabilität der Plasmahaut und die Methode der Permeabilitätskoeffizienten. Vjschr. naturforsch. Ges. Zürich **63**, 187—213 (1918). — Neue Untersuchungen über die Aufnahme von Stoffen in die Zelle. Biochem. Z. **112**, 259—285 (1920).

URL, W.: Permeabilitätsstudien mit Fettsäureamiden. Protoplasma (Berl.) **41**, 287—301 (1952). — URSPRUNG, A.: Die Messung der osmotischen Zustandsgrößen pflanzlicher Zellen und Gewebe. In ABDERHALDENS Handbuch der biologischen Arbeitsmethoden, Abt. XI, Teil 4, H. 7, Liefg 469. 1938. — USSING, H. H.: Some aspects of the applications of tracers in permeability studies. Adv. Enzymol. **13**, 21—65 (1952).

VISSER SMITS, D. DE: Einfluß der Temperatur auf die Permeabilität des Protoplasmas bei Beta vulgaris L. Rec. Trav. bot. néerl. **23**, 104—199 (1926). — VRIES, H. DE: Eine Methode zur Analyse der Turgorkraft. Jb. wiss. Bot. **14**, 427—601 (1884). — Über die Permeabilität der Protoplaste für Harnstoff. Bot. Ztg **47**, 309—315, 325—333 (1889).

WARNER, T.: Zur Aufnahme von Zucker aus hypotonischen Lösungen durch Helodea. Planta (Berl.) **15**, 739—751 (1932). — WARTIOVAARA, V.: Über die Temperaturabhängigkeit der Protoplasmapermeabilität. Ann. bot. Soc. zool.-bot. fenn. „Vanamo“ **16**, Nr 1 (1942). — The permeability of Tolypellopsis cells for heavy water and methyl alcohol. Acta bot. fenn. **34**, 1—22 (1944). — The permeability of the plasma membranes of Nitella to normal primary alcohols at low and intermediate temperatures. Physiol. Plantar. **2**, 184—196 (1949). — WEATHERLEY, P. E.: Preliminary investigations into the uptake of sugars by floating leaf discs. New Phytologist **53**, 204—216 (1954). — WEEVERS, T.: Aufnahme, Verarbeitung und Transport der Zucker im Blattgewebe. Rec. Trav. bot. néerl. **28**, 400—420 (1931). — WEICHHERZ, J., u. F. F. NORD: Kinetische und moleküldynamische Betrachtungen zum Gärungsproblem. Protoplasma (Berl.) **10**, 41—52 (1930). — WIERSUM, L. K.: Transfer of solutes across the young root. Rec. Trav. bot. néerl. **41**, 1—79 (1947).

ZEHETNER, H.: Untersuchungen über die Alkoholpermeabilität des Protoplasmas. Jb. wiss. Bot. **80**, 505—566 (1934). — ZYCHA, H.: Über den Einfluß des Lichtes auf die Permeabilität von Blattzellen für Salze. Jb. wiss. Bot. **68**, 499—548 (1928).

Der Ort des Penetrationswiderstandes.

Von

Runar Collander.

Einleitung.

Wie bereits im Abschnitt IV A ausgeführt, besteht der Stoffaustausch zwischen den lebenden Protoplasten untereinander sowie der zwischen den Protoplasten und dem umgebenden Medium aus wenigstens zwei prinzipiell verschiedenartigen Vorgängen.

In dem einen Falle handelt es sich um eine Arbeitsleistung seitens des Protoplasten auf Kosten etwa der bei der Atmung freigemachten Energie. Man spricht in solchen Fällen von einer aktiven, metabolischen oder nichtosmotischen Stoffbeförderung oder von einer adenoiden (d. h. drüsenartigen) Tätigkeit des Protoplasten.

In anderen Fällen handelt es sich dagegen um verhältnismäßig einfache Diffusions- (oder Filtrations-)prozesse, wobei der Protoplast nur die Rolle eines Penetrationswiderstandes spielt. Den reziproken Wert dieses Widerstandes nennt man die Durchlässigkeit oder die (passive) Permeabilität des Protoplasten, ebenso wie in der Elektrizitätslehre Widerstand und Leitfähigkeit reziproke Größen darstellen.

Daß diese beiden Arten des Stoffaustausches von grundlegender Bedeutung für das Leben der Organismen sind, wird wohl heute von niemandem bestritten. Die nachfolgenden Erörterungen beziehen sich jedoch allein auf den einfachen Diffusions- bzw. Filtrationswiderstand, denn die Widerstände, die bei dem aktiven Stofftransport zu überwinden sind, lassen sich vorläufig nicht angeben.

Und zwar befassen wir uns hier nur mit der Frage, in welchen Teilen des Protoplasten dieser oft sehr große Widerstand lokalisiert ist. (Über den in der Zellwand gelegenen Penetrationswiderstand vergleiche man Abschnitt IV E.)

Geschichtliches.

Die Frage nach dem Sitz des Penetrationswiderstandes der Protoplasten wurde bereits von Pfeffer in seinen „Osmotischen Untersuchungen" aus dem Jahre 1877 aufgeworfen und vorläufig beantwortet. In seiner Studie „Zur Kenntnis der Plasmahaut und der Vakuolen" (1890) behandelt er diese Frage eingehender. Pfeffer nahm an, daß sich an der Oberfläche der Protoplasten bei der Berührung derselben mit Wasser eine Art Niederschlagsmembran bilde, welche ähnlich den künstlichen Niederschlagsmembranen für Wasser permeabel, für viele gelöste Stoffe dagegen mehr oder weniger impermeabel sei. Eine ähnliche Membran bildet sich nach Pfeffer auch an der Grenze zwischen dem Cytoplasma und der wäßrigen Lösung, welche die Vacuole erfüllt. Das wasserreiche Binnenplasma, das sich zwischen diesen beiden „Plasmahäuten" befindet, soll dagegen ähnlich wie andere „gelatinöse Körper" für gelöste Stoffe leicht durchlässig sein.

Als Stütze seiner Auffassung führt Pfeffer in erster Linie Beobachtungen an vorsichtig getöteten Protoplasten an. So z. B. fand er, daß plasmolysierte Protoplasten, die mit einer säurehaltigen Zuckerlösung behandelt worden waren, immer noch undurchlässig für gewisse von außen gebotene Farbstoffe waren. Erst wenn in der coagulierten Plasmamembran ein Riß entstanden war, breitete sich der Farbstoff von dieser Rißstelle aus in dem zwischen der äußeren und inneren Plasmahaut eingeschlossenen toten Protoplasma aus. Die plasmatischen „Grenzhäute" hatten also trotz der Coagulation ihre Semipermeabilität im wesentlichen beibehalten, bis sie irgendwie mechanisch verletzt wurden.

Wertvoller, weil an ganz intakten Protoplasten durchgeführt, sind wohl die folgenden Beobachtungen (Pfeffer 1890): — Myxomyceten-Plasmodien, die sich in einer gesättigten Asparaginlösung befinden, werden zur Aufnahme von Asparaginkriställchen veranlaßt. Wird hiernach die umgebende Lösung verdünnt, so hat dies eine Auflösung der im Plasma liegenden Kriställchen zur Folge. Dabei entsteht in der Regel um jedes Kriställchen eine Vacuole, die auf osmotischem Wege sich vergrößert. Dies deutet darauf hin, daß die Vacuolenwandung für Asparagin sehr viel schwieriger durchlässig ist als für Wasser. Unter Umständen aber können sich die Kriställchen auflösen, ohne daß es zur Bildung einer Vacuole kommt. In den zuletzt genannten Fällen muß sich das gelöste Asparagin recht schnell im Cytoplasma verbreiten, woraus nach Pfeffer zu schließen ist, daß das Cytoplasma ganz im Gegensatz zur Vacuolenwandung für Asparagin leicht durchlässig ist.

Nach Pfeffer ist die Plasmahaut im allgemeinen nicht mit der mikroskopisch sichtbaren Hyaloplasmaschicht identisch, sondern wesentlich dünner. Er erwägt sogar die Möglichkeit, die Plasmahaut könne „auf eine Molekularschicht" reduziert sein (1890, S. 235). Als „wesentliches Baumaterial" der Plasmahäute nimmt er Proteine an.

Im Jahre 1885 zeigte de Vries, daß es gelingt, den größten Teil des Protoplasten in solcher Weise abzutöten, daß dabei die innere Plasmahaut, d. h. die Vacuolenwandung oder der Tonoplast zunächst noch, wenigstens in großen Zügen, dieselben Durchlässigkeitseigenschaften wie der ursprüngliche Protoplast zeigt: sie ist für Anthozyan undurchlässig und kann sich verkleinern oder vergrößern, je nachdem, ob der osmotische Wert der umgebenden Lösung steigt oder fällt.

Overton (1899) übernahm die Pfeffersche Lehre von den Plasmahäuten und bildete sie weiter aus, indem er auf Grund des von ihm aufgedeckten Parallelismus zwischen Permeationsvermögen und Lipoidlöslichkeit annahm, daß die Plasmahäute „mit Lipoiden imprägniert" seien.

Lange Zeit hindurch behielt die Vorstellung, daß der Penetrationswiderstand der Protoplasten fast allein in den Plasmahäuten lokalisiert sei, ihre Stellung als ein gewissermaßen „offiziell" sanktioniertes Dogma, das in allen Lehrbüchern unter Hinweis auf Pfeffers klassische Untersuchungen mit aller Reverenz, aber ohne eingehendere Auseinandersetzungen vorgeführt wurde.

In den zwanziger Jahren ließen sich jedoch einige oppositionelle Stimmen vernehmen. So z. B. behauptete Lepeschkin in seiner „Kolloidchemie des Protoplasmas" (1924), daß die Beobachtungen Pfeffers, weil an abgetöteten Protoplasten gemacht, nicht das Vorkommen einer osmotisch maßgebenden Plasmahaut bei den lebenden Protoplasten beweisen. Lepeschkin ging so weit, die Existenz der Plasmahäute zu bestreiten, indem er annahm, daß das Dispersionsmittel des Protoplasmas nicht Wasser sei, wie sonst allgemein angenommen worden ist, sondern „eine Lösung von Wasser in einer organischen Flüssigkeit" (l. c. S. 143), bzw. eine innige Mischung oder eine labile chemische Verbindung von Wasser, Lipoiden und Eiweißkörpern (S. 151, 153).

Eine neue Epoche in der Erforschung der plasmatischen Membranen eröffneten die Methoden der Mikrurgie einschließlich der Mikroinjektion. Die ersten bedeutsamen Erfolge dieser Arbeitsrichtung wurden an tierischen Zellen erreicht

(Chambers 1922). Bei pflanzlichen Zellen aber bildeten die starren Zellwände ein zunächst nicht ganz leicht zu überwindendes Hindernis für derartige Operationen. Doch im Jahre 1931 erschienen drei ebenso geschickt ausgeführte wie erfolgreiche Untersuchungen über plasmatische Membranen bei Pflanzenzellen.

Zwei dieser Untersuchungen stammten von Janet Plowe, einer Schülerin von Seifriz. Ihre erste Arbeit (Plowe 1931a) ist dem anatomischen Nachweis differenzierter Plasmamembranen gewidmet. Zieht man mit der Mikronadel aus dem Plasma eines plasmolysierten Protoplasten einen Strang aus, so sondert sich seine Substanz alsbald in abgerundete, trübe Tröpfchen, eingebettet in einen feinen, aus einer optisch homogenen Substanz bestehenden Faden. Hieraus ist zu schließen, daß die äußere Plasmahaut, für die Plowe die Benennung Plasmalemma einführt, mit dem Binnenplasma oder Mesoplasma nicht identisch ist, denn eine homogene, viscöse Flüssigkeit (etwa Teer oder Leimlösung) könnte unter ähnlichen Bedingungen zwar zu einem langen Faden ausgezogen werden, zur Bildung abgerundeter Tröpfchen in dem Faden würde es aber nicht kommen. In ähnlicher Weise konnte Plowe auch einen stofflich differenzierten Tonoplasten nachweisen. Derselbe wurde in mehrfach verschiedener Weise „überlebend" isoliert. Die Verfasserin schließt: "Certainly a layer which appears in both plasmolyzed and unplasmolyzed cells, whether the death occurs slowly or suddenly, whether death is caused by mechanical injury or by toxic products ..., which is invariably distinct from the remainder of the protoplasm and which always possesses the same properties, must represent a layer distinct from the remainder of the protoplast in the living condition."

Diese Beobachtungen wurden, soweit sie sich auf den Tonoplasten beziehen, in schöner Weise durch die etwa gleichzeitig erschienene Arbeit von Chambers und Höfler (1931) ergänzt. Es gelang diesen Forschern, die überlebenden Tonoplasten aus getötetem Protoplasma freizupräparieren und von den letzten Spuren von Plasmacoagulum zu säubern. Wird ein solcher Tonoplast mit einer Nadel angestochen, dellt sich seine Oberfläche zunächst nach einwärts, dann dringt die Nadel ein, ohne daß der Tonoplast dabei zerreißt. Wird die Nadel wieder herausgezogen, so bleibt kein Loch, der Tonoplast schließt sich glatt und sieht unversehrt aus wie zuvor. Der Tonoplast erweist sich somit als ein Häutchen aus einer mit Wasser nicht mischbaren Flüssigkeit, die starke Oberflächenspannung gegen wäßrige Lösungen besitzt. Er ist in frisch isoliertem Zustand semipermeabel und nach dem physikalischen Verhalten (Benetzbarkeit mit Paraffinöl und Olivenöl, Löslichkeit in Chloroform) höchstwahrscheinlich lipoider Natur (vgl. Weber 1932).

In der zweiten Mitteilung von Plowe (1931b) wurde die Durchlässigkeit des Plasmalemmas, des Mesoplasmas und des Tonoplasten gegenüber Farbstoffen getrennt studiert. Als Versuchsobjekte dienten Wurzelhaare von *Trianea*, Epidermiszellen von *Allium* sowie Zellen der Rotalge *Griffithsia*. Die verwendeten Farbstoffe (Anilinblau, Säurefuchsin, Bromkresolpurpur und Phenolrot) waren solche, die, von außen dargeboten, nicht in die Protoplasten einzudringen vermögen. Es zeigte sich, daß sie, in die zentrale Vacuole injiziert, sich in dieser ausbreiteten, aber nicht in das Mesoplasma übertraten. Wurde die Farbstofflösung dagegen in das Mesoplasma injiziert, breitete sich der Farbstoff schnell in diesem aus, drang aber nicht in die Vacuole und exosmierte auch nicht aus dem Protoplasten heraus. Diese Versuche scheinen einwandfrei zu beweisen, daß beide Plasmahäute wesentlich schwerer durchlässig als das Mesoplasma für die geprüften Farbstoffe sind. (Einige Ergebnisse von Chambers und Kerr 1932 an Wurzelhaaren von *Limnobium* stimmen allerdings nicht ganz mit denen von Plowe überein.)

Um diese Zeit, d. h. Anfang der dreißiger Jahre, hat HÖFLER (1931, 1932) durch eine Reihe von Vorträgen und Veröffentlichungen wirksam dazu beigetragen, die Aufmerksamkeit der Zellphysiologen auf das Hautschichtproblem zu lenken. Unter anderem hat er wohl als erster das uns hier interessierende Problem ganz scharf und zugleich eindrucksvoll formuliert, indem er hervorhob, daß der Penetrationswiderstand des ganzen Protoplasten (R) sich additiv aus 3 Komponenten zusammensetzt: dem Widerstand des Plasmalemmas (R_L), demjenigen des Mesoplasmas (R_M) und demjenigen des Tonoplasten (R_T), also

$$R = R_L + R_M + R_T,$$

und daß es eine wichtige Aufgabe der Permeabilitätsforschung sein muß, wenigstens die relative Größe der 3 Teilwiderstände festzustellen.

Wir reißen hier den Faden unserer geschichtlichen Darstellung ab, um uns der Frage zuzuwenden, was man heutzutage von diesen 3 Penetrationswiderständen weiß. Die in der Literatur repräsentierten diesbezüglichen Ansichten können, etwas schematisiert, in der folgenden Weise gruppiert werden:

I. R_M ist so groß, daß er gegenüber $R_L + R_T$ nicht vernachlässigt werden darf.

II. R_M ist so viel kleiner als $R_L + R_T$, daß er gegenüber diesen kaum ins Gewicht fällt.

A. R_L ist viel kleiner als R_T.

B. R_T ist bedeutend kleiner als R_L.

C. R_L und R_T sind etwa von derselben Größenordnung.

Wir wollen nun diese Alternativen der Reihe nach einer Prüfung unterziehen.

Der Penetrationswiderstand des Mesoplasmas verglichen mit demjenigen der Plasmagrenzschichten.

Wie bereits erwähnt, hat LEPESCHKIN (1924) die Existenz der Plasmahäute in Zweifel gezogen und somit den ganzen Penetrationswiderstand des Protoplasten hypothetisch in das Mesoplasma verlegt. In einer so extremen Fassung ist seine These kaum länger aufrechtzuerhalten. Doch haben auch in neuerer Zeit namhafte Forscher die Meinung vertreten, daß der Penetrationswiderstand des Mesoplasmas gegenüber demjenigen der Plasmagrenzschichten durchaus nicht zu vernachlässigen sei.

So nimmt HÖFLER (1949, 1951, 1953) an, daß das Plasma „durch und durch aus einer zusammenhängenden Phase mit lipoiden Lösungsmitteleigenschaften" bestehe, „die von der Außenfläche bis an die Innenoberfläche des Plasmaschlauches reicht" (1953, S. 393). In erster Linie stützt sich diese Stellungnahme auf die von STRUGGER (1938, 1949) gemachten fluorescenzmikroskopischen Beobachtungen an mit den lipoidlöslichen Farbstoffen Rhodamin und Neutralrot vitalgefärbten Protoplasten. Es zeigte sich nämlich hierbei das ganze Binnenplasma gefärbt, nicht nur die Hautschichten. Bei der Verwendung von Neutralrot fiel es außerdem auf, daß der Farbstoff im Binnenplasma als fluorescierende, undissoziierte Neutralrotmoleküle vorhanden war, also in einer Form, die hauptsächlich in organischen, nichtwäßrigen Lösungsmitteln vorkommt. Unserer Meinung nach besagt aber dies alles nur, daß das Binnenplasma *auch* Lipoide enthält. Ob aber die lipoide oder die wäßrige Phase dominiert, ist damit nicht entschieden. Und noch weniger läßt sich aus diesen Beobachtungen schließen, daß die Lipoide eine zusammenhängende Phase bilden, die von der Außenfläche bis an die Innengrenzfläche des Plasmaschlauches reicht. Übrigens gibt auch HÖFLER (1953, S. 393) zu, daß neben der lipoiden auch eine zusammenhängende wäßrige Phase im Binnenplasma vorhanden sein kann.

Eine weitere Stütze für seine Auffassung, daß der Penetrationswiderstand des Mesoplasmas nicht zu vernachlässigen sei, sieht HÖFLER (1949) in dem Umstand, daß Harnstoff- und Wasserpermeabilität verschiedener Protoplasten durchaus nicht parallel variieren, obwohl ja sowohl Harnstoff wie Wasser kleinmolekulare, sehr wenig lipoidlösliche Substanzen sind, die daher nach HÖFLER beide auf dem „Porenweg" permeieren müßten. Die einfachste Erklärung hierfür sieht er in der Annahme, daß der maßgebende Permeationswiderstand für den Wasserdurchtritt nicht in den Hautschichten, sondern im Binnenplasma seinen Sitz hat, während die Permeation des Harnstoffs ihrem größten Widerstand in den Plasmagrenzschichten begegnet. Gegen diese Argumentierung kann eingewendet werden, daß unsere Kenntnisse von dem Permeationsmechanismus verschiedener Stoffe noch allzu lückenhaft sind, um die Möglichkeit eindeutig auszuschließen, daß es Plasmahäute geben kann, bei denen die Durchlässigkeit für Wasser einerseits und für Harnstoff andererseits weitgehend unabhängig voneinander variieren könnten. Trotzdem ist der Gedanke, daß die Wasserdurchlässigkeit der Plasmagrenzschichten ebenso groß oder noch größer als diejenige des Mesoplasmas sein könnte, durchaus plausibel. In diesem Sinne sprechen nämlich, wie HÖFLER betont, die Beobachtungen von HUBER und HÖFLER (1930), daß frisch isolierte Tonoplasten für Wasser viel durchlässiger sind als ganze Protoplasten[1].

Übrigens ergibt schon eine theoretische Überlegung, daß die Permeation des Wassers und anderer etwa ebenso schnell permeierender Stoffe in *relativ* bedeutendem Maße vom Mesoplasma gehemmt sein dürfte. Wir brauchen uns nur zu vergegenwärtigen, daß das Mesoplasma meistens wohl rund 1000mal dicker als die Plasmahäute ist. Auch wenn der Penetrationswiderstand des Mesoplasmas, je Dickeneinheit berechnet, 1000mal kleiner als derjenige der Plasmahäute wäre, kann somit der Gesamtwiderstand des Mesoplasmas demjenigen der Plasmahäute gleichkommen und vielleicht sogar ein wenig überschreiten.

Eine solche Überlegung gilt selbstverständlich nur für Stoffe, die sehr schnell durch die Plasmahäute dringen. Dagegen glaubt CHOLNOKY (1952, 1953) nachgewiesen zu haben, daß auch langsam permeierende Stoffe wie KOH und NaOH sehr stark vom Mesoplasma in ihrer Diffusion gehemmt werden, nachdem das Plasmalemma von ihnen zerstört worden ist. Die Befunde von CHOLNOKY sind in der Tat teilweise recht beachtlich. Es fragt sich aber, ob sie nicht so gedeutet werden können, daß, nachdem das ursprüngliche Plasmalemma zerstört worden ist, eine neue Grenzschicht mit ähnlichen Durchlässigkeitseigenschaften entsteht.

Jedenfalls scheint es, daß wirklich überzeugende Gründe noch nicht beigebracht worden sind dafür, daß die von PFEFFER begründete und später durch mikrurgische Versuche bestätigte Auffassung von der Lokalisation des *hauptsächlichen* Penetrationswiderstandes in den Plasmagrenzschichten aufgegeben werden müßte.

Ist der Penetrationswiderstand des Plasmalemmas viel kleiner als derjenige des Tonoplasten?

Viele mikrurgische Beobachtungen sprechen dafür, daß zwischen Plasmalemma und Tonoplast deutliche stoffliche und/oder strukturelle Unterschiede bestehen (z. B. CHAMBERS und HÖFLER 1931, HÖFLER 1931, 1932, 1953, LEDERER

[1] Auch für Farbstoffkationen ist der Tonoplast nach HÖFLER (1952) viel durchlässiger als der Gesamtprotoplast. Eine systematische Prüfung der Permeabilität isolierter Tonoplasten für verschiedenartige Stoffe steht noch aus, wäre aber gewiß von großem Interesse.

1935, PLOWE 1931a; vgl. dagegen auch EICHBERGER 1934.) An sich wäre es daher durchaus nicht unerwartet, wenn es sich zeigte, daß sich die beiden Plasmahäute auch hinsichtlich ihrer Durchlässigkeit bedeutend unterscheiden würden. In der Tat scheint es eine recht verbreitete Ansicht zu sein, daß die in der Überschrift enthaltene Frage zu bejahen sei.

Ein terminologischer Vorschlag HÖFLERS (1932) hängt mit solchen Vorstellungen zusammen. Er schlug nämlich vor, daß die Benennung *Permeabilität* in erster Linie „auf die Eignung des Protoplasmaschlauches der vacuolisierten Zelle, den Durchtritt von Stoffen von außen bis in den Zellsaft oder umgekehrt vom Zellsaft nach außen" einzuschränken sei, wogegen die Fähigkeit, „den Eintritt von Stoffen durch die äußeren Plasmagrenzschichten bis in das Cytoplasma zuzulassen" als *Intrabilität* zu bezeichnen sei. Das Wort Intrabilität hat in der deutschsprachigen pflanzenphysiologischen Literatur eine weite Verbreitung gefunden. Ganz glücklich gewählt ist diese Benennung aber kaum, denn nur zu leicht verleitet sie zu der Annahme, daß der Penetrationswiderstand des Plasmalemmas in der Richtung von außen nach innen kleiner als in der entgegengesetzten Richtung sei — obwohl HÖFLER selbst dies nie behauptet hat.

Welche Beweise gibt es nun dafür, daß das Plasmalemma wesentlich durchlässiger als der Tonoplast sei?

a) Bereits DE VRIES hat 1885 gezeigt, daß bei rascher Plasmolyse mit Alkalisalzen der Großteil des Cytoplasmas manchmal getötet werden kann, während nur die Vacuolenwandung „am Leben" bleibt, d. h. ihre Semipermeabilität bewahrt. Analoge Erfahrungen sind später von zahlreichen anderen Forschern (z. B. CHAMBERS und HÖFLER 1931, CHOLNOKY 1952a, b, c, 1953, HÖFLER 1931, 1951, 1952, 1953, PLOWE 1931a) gemacht worden unter anderem bei der Behandlung der Protoplasten mit Alkalien oder Säuren. Es liegt nahe, aus derartigen Beobachtungen den Schluß zu ziehen, daß das Plasmalemma leichter durchlässig und daher weniger resistent gegenüber den benutzten Giften sei. Tatsächlich sind diese Erfahrungen wohl kaum ganz eindeutig. Erstens ist nämlich zu beachten, daß ein von außen einwirkendes Agens — wie etwa eine nicht oder nur langsam permeierende Base oder Säure — ja zunächst allein mit dem Plasmalemma in Berührung kommt und vielleicht erst viel später den Tonoplasten erreicht. Wenn also nach einer bestimmten Zeit festgestellt wird, daß das Plasmalemma und das Mesoplasma jetzt getötet, der Tonoplast dagegen noch am Leben ist, so besagt eine solche Feststellung an sich noch gar nichts in bezug auf die relative Permeabilität oder Resistenz der beiden Plasmagrenzschichten. Bei Plasmolyseversuchen ist außerdem zu beachten, daß das Plasmalemma ja häufig mit der Zellwand innig verklebt ist und somit bei der Loslösung des Protoplasten von der Zellwand mechanisch unvergleichlich schlimmer als der Tonoplast insultiert wird.

b) Bekanntlich zeigen sich die lebenden Protoplasten, z. B. plasmolytisch untersucht, in hohem Grade undurchlässig für Zuckerarten, Aminosäuren und die meisten anorganischen Salze, d. h. also für zahlreiche Stoffe, die für die Ernährung der Zellen unentbehrlich sind. Um dieses zunächst geradezu paradox anmutende Ergebnis zu erklären, hat man sich vielfach gezwungen gesehen, anzunehmen, daß, obwohl der Tonoplast für diese Stoffe undurchlässig ist, sie trotzdem mehr oder weniger leicht durch das Plasmalemma diffundieren sollen und so also doch ohne Schwierigkeit von den Protoplasten ausgenützt werden können. Eine solche Annahme ist jedoch keineswegs zwingend, denn die Aufnahme der betreffenden Stoffe kann sehr wohl das Ergebnis einer besonderen aktiven Transportleistung der Protoplasten sein — und ist es tatsächlich nachgewiesenerweise in vielen Fällen (vgl. z. B. ARISZ 1948).

c) WEEVERS wies darauf hin, daß Zuckerarten dauernde Plasmolyse bewirken, während in ihren Lösungen doch Stärkebildung beobachtet wird, was darauf hindeute, daß wohl das Plasmalemma, nicht aber der Tonoplast für Zucker durchlässig sei. Beweisend für eine bevorzugte Permeabilität des Plasmalemmas ist dieser Befund aber nicht, denn erstens ist es denkbar, daß die Zucker eben deshalb nicht (oder kaum) ihren Weg in den Zellsaft finden, weil sie unterwegs von den Plastiden aufgefangen und in Stärke umgewandelt werden, und zweitens ist doch ein gewisser, allerdings sehr langsamer Durchtritt des Zuckers bis in den Zellsaft plasmolytisch nachgewiesen (z. B. HÖFLER 1926, BÄRLUND 1929). Außerdem besteht ja die Möglichkeit, daß die Zuckeraufnahme ein nichtosmotischer Vorgang ist (RIEDER 1951, ROTHSTEIN 1954, WEATHERLEY 1954).

d) Das als „Vacuolenplasmolyse" (FITTING 1920) bzw. „Kappenplasmolyse" (HÖFLER 1928, 1939) bezeichnete Phänomen wird häufig als ein Beweis dafür angesehen, daß das Plasmalemma wesentlich durchlässiger als der Tonoplast für bestimmte Stoffe (Glycerin, KCl u. a.) sei (Literatur bei HOUSKA 1941). Eine solche Deutung hat in der Tat vieles für sich. Es sind jedoch auch ganz andersartige Vorstellungen von dem Mechanismus dieses Vorganges entwickelt worden (BOGEN 1951). Unabhängig hiervon aber läßt sich die Kappenplasmolyse nicht als Beweis dafür verwenden, daß das Plasmalemma *im allgemeinen* bedeutend durchlässiger ist als der Tonoplast. Eher könnte man wohl umgekehrt aus dem Umstande, daß die Kappenplasmolyse nur unter ganz speziellen Bedingungen zustandekommt, den Schluß ziehen, daß das Plasmalemma meistens nicht den Tonoplasten an Durchlässigkeit stark übertreffen würde, denn in diesem Falle müßte wohl die Kappenplasmolyse eine noch viel häufigere Erscheinung sein.

e) S. C. BROOKS (1940, 1953) gibt an, daß radioaktive Natrium- und Kaliumionen äußerst schnell in das Protoplasma von *Nitella*- und *Valonia*-Zellen aufgenommen werden, daß sie aber dann sehr viel langsamer aus dem Protoplasma in den Zellsaft hinübertreten. Es scheint jedoch, daß diese Befunde kaum noch endgültig feststehen (vgl. BROOKS 1951).

f) SUTCLIFFE (1954) stellte in offenbar sehr umsichtig ausgeführten Versuchen fest, daß nur ein recht geringer Teil der in Scheiben der roten Rübe enthaltenen Kalium- und Bromidionen während 6 Std gegen die entsprechenden radioaktiven Ionen austauschbar sind, während der Großteil dieser Ionen auch nicht in 24 Std in merkbarem Grade ausgetauscht werden. Er vermutet hypothetisch, daß die leicht austauschbaren Ionen in den Intercellularen, den Zellwänden und "in parts of the protoplasm" enthalten sind, während die nicht leicht austauschbaren Ionen "may be situated in the cell vacuoles, or else strongly associated with protoplasmic constituents" (vgl. HOPE 1953 und BUTLER 1953). Er meint ferner, daß "the results suggest that a considerable barrier to the free diffusion and exchange of ions is located in the region of the tonoplast". Einen entscheidenden Beweis dafür, daß das Plasmalemma für Ionen leicht durchlässig sei, wird man aber in diesen Befunden kaum sehen.

Zusammenfassend müssen wir somit feststellen: Es gibt zwar Beobachtungen, die vielleicht darauf hindeuten, daß das Plasmalemma nicht ganz so schwer durchlässig sei wie der Tonoplast. Die vielfach vertretene Annahme aber, daß das Plasmalemma ganz wesentlich durchlässiger für gelöste Stoffe sei als der Tonoplast, findet in den bisherigen Erfahrungen keine sichere Stütze.

Dagegen gibt es, wie im folgenden gezeigt werden soll, Umstände, die entschieden dafür sprechen, daß auch das Plasmalemma für viele Stoffe in hohem Grade schwer- oder sogar undurchlässig ist.

Erstens haben wir da die Gesamtheit der an tierischen Zellen gesammelten Erfahrungen, wonach auch diese Zellen eine ausgesprochen selektive Permeabilität besitzen — im großen ganzen recht ähnlich derjenigen der Pflanzenzellen —, obwohl ihnen ja im allgemeinen eine große, zentrale Vacuole fehlt und der Penetrationswiderstand eines Tonoplasten also meist nicht in Frage kommt. Auf mikrurgischem Wege ist zudem in durchaus überzeugender Weise dargetan, daß der große Penetrationswiderstand z. B. bei den Seeigeleiern seinen Sitz eben in der Außengrenzfläche des Protoplasten hat (Zusammenfassung bei CHAMBERS 1949). Liegt es da nicht überaus nahe anzunehmen, daß ähnliches auch für pflanzliche Zellen gilt?

In der Tat ist es ja auch bekannt, daß z. B. viele Schimmelpilze, wie etwa *Penicillium* oder *Aspergillus*, sowohl in stark sauren wie in schwach alkalischen Lösungen, d.h. in einem p_H-Bereich, das sich etwa von 2 bis 10 erstreckt, wachsen können. Wenn wir uns nun andererseits vergegenwärtigen, innerhalb wie enger p_H-Grenzen die einzelnen Enzyme wirksam sind, so bleibt wohl nur die Annahme übrig, daß die erstaunliche Unabhängigkeit dieser Organismen von der Wasserstoffionenkonzentration des Milieus davon herrühren muß, daß das Binnenplasma so effektiv gegen die Umwelt abgeschlossen ist, daß es von den p_H-Schwankungen des Mediums praktisch nicht beeinflußt wird. Ähnliches gilt übrigens auch für die meisten anderen Pflanzenzellen: ihre Widerstandsfähigkeit gegenüber zahlreichen (nicht allen) Säuren und Basen ist so groß, daß man annehmen muß, daß ihr Plasmalemma fast undurchlässig für sie ist (vgl. z. B. BRENNER 1918, HÖFLER 1951, STEEMANN NIELSEN 1955).

Auch in konzentrierten Lösungen von gewissen Schwermetallsalzen wie Kupfersulfat können einige Schimmelpilze wachsen (PFEFFER 1904, S. 334, STARKEY und WAKSMAN 1943). Da diese Salze bekanntlich bereits in sehr niedrigen Konzentrationen auf Proteine denaturierend wirken, muß man wohl schließen, daß das Plasmalemma der betreffenden Pilze in höchstem Grade undurchlässig für diese Salze sein muß. PRINGSHEIM (1924) sowie BIEBL und ROSSI-PILLHOFER (1954) rechnen allerdings auch mit der Möglichkeit, daß die Schwermetalle erst bei der Berührung mit der Plasmaoberfläche dort eine irreversibel koagulierte Oberflächenschicht erzeugen, die das Eindringen des Salzes verhindert und so den Großteil des Plasmas vor Abtötung schützt.

Auf eine ausgesprochen selektive Durchlässigkeit des Plasmalemmas deutet auch die längst bekannte Tatsache, daß sich zahlreiche physiologisch aktive schwache Säuren und Basen (Inhibitoren, Wuchsstoffe usw.) weitaus wirksamer zeigen in Form undissoziierter Moleküle als in Ionenform. Es liegt nämlich jedenfalls sehr nahe anzunehmen, dies rühre davon her, daß das Plasmalemma die Ionen viel stärker zurückhält als die entsprechenden undissoziierten Moleküle.

Natürlich gestatten die hier genannten Beobachtungen keinen quantitativen Vergleich zwischen dem Penetrationswiderstand des Plasmalemmas und dem des Tonoplasten, wohl aber zeigen sie, daß bereits das Plasmalemma für viele Substanzen äußerst schwerdurchlässig ist.

Ist der Penetrationswiderstand des Plasmalemmas wesentlich größer als derjenige des Tonoplasten?

Diese Möglichkeit scheint nur in einem Falle ernstlich ins Auge gefaßt zu sein. HOLM-JENSEN und Mitarbeiter (1944) studierten die Aufnahme von radioaktivem Na und K in Zellen von *Nitellopsis (Tolypellopsis)* und *Nitella*, wobei Radioaktivität des Plasmas und des Zellsaftes getrennt gemessen wurde. Die Berechnungen ergaben, daß der Penetrationswiderstand des Plasmalemmas erheblich größer als derjenige des Tonoplasten sein müsse. In Anbetracht der vielen Fehlerquellen, sowohl bei der Ausführung der Versuche wie bei deren Deutung, bedarf dieses Ergebnis noch der Bestätigung.

Zusammenfassendes über den Penetrationswiderstand des Plasmalemmas, des Mesoplasmas und des Tonoplasten.

Das Hauptergebnis der obigen Ausführungen kann in 4 Sätzen zusammengefaßt werden:

1. Das Binnenplasma scheint ziemlich frei permeabel für beliebige Stoffe zu sein.
2. Der selektive Penetrationswiderstand der Protoplasten scheint somit seinen Sitz ganz überwiegend in den Plasmagrenzschichten zu haben.
3. Eine auch nur halbwegs quantitative Schätzung der relativen Größen der 3 Addenden R_L, R_M und R_T, die gemeinsam den Gesamtwiderstand des Protoplasmaschlauches ausmachen, ist noch nicht möglich.
4. Ein wesentlicher Unterschied zwischen den Penetrationswiderständen in der äußeren und inneren Plasmagrenzschicht ist zwar oft angenommen, aber bisher nicht einwandfrei nachgewiesen worden.

Der Penetrationswiderstand in der Grenzschicht Cytoplasma/Zellkern.

Über die Struktur der Kernumhüllung liegen bereits mehrere Untersuchungen vor (z. B. WILKINS 1951, CALLAN 1952, BAIRATI und LEHMANN 1952, WATSON 1954). Nach ihnen zu schließen besteht die Kernumhüllung aus 2 Schichten: einer inneren, die allein aus Proteinen zusammengesetzt ist, und einer äußeren, die aus Proteinen und Lipoiden aufgebaut ist. Bezüglich der Durchlässigkeit der Kernumhüllung gibt es dagegen zwei stark divergierende Ansichten.

Auf der einen Seite deuten zahlreiche Beobachtungen darauf hin, daß die Kernmembran ziemlich frei durchlässig ist auch für äußerst wenig lipoidlösliche Stoffe wie für Zucker und anorganische Salze (CALLAN und TOMLIN 1950, GOLDSTEIN und HARDING 1950, ABELE 1951, HOGEBOOM und SCHNEIDER 1953). Auch solche Farbstoffe, für die die Plasmahäute undurchlässig sind, dringen in den Zellkern hinein, wenn sie in das Cytoplasma, in die Nähe des Kerns, injiziert werden (CHAMBERS und POLLACK 1927, MONNÉ 1935). Nach diesen Autoren ist die Kernmembran allein für hochkolloidale Stoffe wie Albumin, Glykogen und Acacia-Gummi undurchlässig.

Ganz andersartige Vorstellungen von der Durchlässigkeit der Grenzschicht Cytoplasma/Kern hat dagegen KIHLMAN (1951, 1952a, b) entwickelt. Er studierte Chromosomenveränderungen, hervorgerufen durch Behandlung der Zellen (besonders *Allium*) mit verschiedenen Purinderivaten. Dabei fand er, daß bestimmte Purinverbindungen nur dann die Zellkerne beeinflussen, wenn diese sich teilen, während andere Purine auch während der Interphase wirksam sind. Aus diesen Befunden schloß er, daß, obwohl alle diese Substanzen durch die äußere Plasmahaut hindurch in das Cytoplasma eindringen, nur diejenigen der zweiten Gruppe imstande sind durch die Grenzschicht Cytoplasma/Kern zu permeieren. Da es sich weiter zeigte, daß die Purine dieser zweiten Gruppe im allgemeinen eine größere relative Ätherlöslichkeit haben als die der ersten Gruppe, nahm er dementsprechend an, daß die permeationshemmende Schicht lipoider Natur sei. Die perinucleare Lipoidschicht soll sogar eine noch viel effektivere Abdichtung gegenüber lipoidunlöslichen Stoffen zustandebringen als etwa Plasmalemma und

Tonoplast zusammen. Die Angaben früherer Autoren von der freien Durchlässigkeit der Kernumhüllung erklärt KIHLMAN damit, daß die innere Schicht der Kernumhüllung tatsächlich beliebige nichtkolloidale Stoffe durchlasse, wogegen die äußere, in intaktem Zustande selektiv durchlässige Schicht in jenen Versuchen alteriert worden sei.

Es leuchtet ein, daß die Befunde, auf die KIHLMAN seine Hypothese aufbaut, recht vieldeutig sind. Einstweilen muß wohl somit die Frage von dem Penetrationswiderstand in der Grenzschicht Cytoplasma/Kern als unentschieden bezeichnet werden.

Die Penetrationswiderstände in den Grenzschichten der Plastiden und der Mitochondrien.

McCLENDON (1954) hat neuerdings beobachtet, daß die isolierten „Rhodoplasten" mariner Rotalgen sich schnell deformieren, nicht nur in Wasser, sondern auch in Lösungen von beliebigen niedermolekularen Stoffen wie Kaliumchlorid oder Saccharose. Die Plastiden verquellen dabei, oft unter deutlicher Vacuolisierung, und geben ihren roten Farbstoff an das Medium ab. Anders dagegen in genügend konzentrierten Lösungen von Dextrin, Protein oder solchen Polyäthylenglykolen („Carbowax"), deren Molekulargewicht wenigstens etwa 2400 beträgt. In solchen Lösungen bleiben die Rhodophyceen-Plastiden stunden- und sogar tagelang anscheinend intakt. In Lösungen von niedermolekularen Polyäthylenglykolen (Molekulargewicht 1500) behalten sie dagegen ihren natürlichen Zustand nur kurzfristig bei. Es scheint somit, daß die betreffenden Plastiden für beliebige niedrigmolekulare Stoffe permeabel, für Stoffe mit einem Molekulargewicht über etwa 2400 dagegen impermeabel sind. Auch Chloroplasten von *Nitella* und *Elodea* blieben in den Versuchen von McCLENDON länger intakt in Lösungen hochmolekularer Polyäthylenglykole als etwa in Saccharoselösungen.

Isolierte Mitochondrien sind sehr empfindlich gegen Hypotonie des Aufbewahrungsmediums. In Saccharoselösungen geeigneter Konzentration und ebenso in entsprechenden Salzlösungen bleiben sie dagegen verhältnismäßig lange intakt. Die Annahme erscheint daher naheliegend, daß sie von einer selektiv permeablen Membran umgeben sind, die unter anderem für Saccharose mehr oder weniger undurchlässig ist (DUVE und Mitarbeiter 1951, CLELAND 1952, MILLERD und BONNER 1953, DIANZANI 1953, LATIES 1953, 1954).

Literatur.

ABELE, K.: Über die Volumenabnahme des Zellkerns in der Plasmolyse und über das Zustandekommen der Kernplasmarelation. Protoplasma (Berl.) **40**, 324—337 (1951). — ARISZ, W. H.: Active uptake, vacuole-secretion and plasmatic transport of chlorine-ions in leaves of Vallisneria spiralis. Acta bot. néerl. **1**, 506—515 (1953).

BAIRATI, A., u. F. E. LEHMANN: Über die submikroskopische Struktur der Kernmembran bei Amoeba proteus. Experientia (Basel) **8**, 60—61 (1952). — BÄRLUND, H.: Permeabilitätsstudien an Epidermiszellen von Rhoeo discolor. Acta bot. fenn. **5**, 1—117 (1929). — BIEBL, R., u. W. ROSSI-PILLHOFER: Die Änderung der chemischen Resistenz pflanzlicher Plasmen mit dem Entwicklungszustand. Protoplasma (Berl.) **44**, 113—135 (1954). — BOGEN, H. J.: Über Kappenplasmolyse und Vakuolenkontraktion. I. Planta (Berl.) **39**, 1—35 (1951). — BRENNER, W.: Studien über die Empfindlichkeit und Permeabilität pflanzlicher Protoplasten für Säuren und Basen. Öfvers. Finska Vet. Soc. Förh. A **60**, Nr 4 (1918). — BROOKS, S. C.: The intake of radioactive isotopes by living cells. Cold Spring Harbor Symp. Quant. Biol. **8**, 171—180 (1940). — Penetration of radioactive isotopes P^{32}, Na^{24} and K^{42} into Nitella. J. Cellul. a. Comp. Physiol. **38**, 83—94 (1951). — The penetration of radioactive sodium into Valonia and Halicystis. Protoplasma (Berl.) **42**, 63—68 (1953). — BUTLER, G. W.: The "apparent free space" of wheat roots. Physiol. Plantarum (Copenh.) **6**, 617—635 (1953).

CALLAN, H. G.: A general account of experimental work on amphibian oocyte nuclei. Symposia Soc. f. Exper. Biol. 6, 243—255 (1952). — CALLAN, H. G., and S. G. TOMLIN: Experimental studies on amphibian oocyte nuclei. I. Proc. Roy. Soc. Lond., Ser. B 137, 367—378 (1950). — CHAMBERS, R.: A micro injection study on the permeability of the starfish egg. J. Gen. Physiol. 5, 189—193 (1922). — Micrurgical studies on protoplasm. Biol. Rev. 24, 246—265 (1949). — CHAMBERS, R., u. K. HÖFLER: Micrurgical studies on the tonoplast of Allium cepa. Protoplasma (Berl.) 12, 338—355 (1931). — CHAMBERS, R., and T. KERR: Intracellular hydrion concentration studies. J. Cellul. a. Comp. Physiol. 2, 105—119 (1932). — CHAMBERS, R., and H. POLLACK: Colorimetric determination of the nuclear and cytoplasmic p_H in the starfish egg. J. Gen. Physiol. 10, 739—755 (1927). — CHOLNOKY, B. J.: Beobachtungen über die Wirkung der Kalilauge auf das Protoplasma. Protoplasma (Berl.) 41, 57—68 (1952). — Beobachtungen über die Plasmolyse. I. Sitzgsber. österr. Akad. Wiss., Math.-naturwiss. Kl., Abt. I 161, 539—557 (1952). — Ein Beitrag zur Kenntnis des Plasmalemmas. Ber. dtsch. bot. Ges. 65, 369—373 (1952). — Ein Beitrag zur Kenntnis der Plasmolyse der Oedogonium-Zellen. Österr. bot. Z. 100, 226—234 (1953). — CLELAND, K. W.: Permeability of isolated rat heart sarcosomes. Nature (Lond.) 170, 497—499 (1952).

DIANZANI, M. U.: On osmotic behaviour of mitochondria. Biochim. et Biophysica Acta 11, 353—367 (1953). — DUVE, C. DE, J. BERTHET, L. BERTHET and F. APPELMANS: Permeability of mitochondria. Nature (Lond.) 167, 389—390 (1951).

EICHBERGER, R.: Über die „Lebensdauer" isolierter Tonoplasten. Protoplasma (Berl.) 20, 606—632 (1934).

FITTING, H.: Untersuchungen über die Aufnahme und über normale osmotische Koeffizienten von Glyzerin und Harnstoff. Jb. wiss. Bot. 59, 1—170 (1920).

GOLDSTEIN, L., and C. V. HARDING: Osmotic behavior of isolated nuclei. Federat. Proc. 9, 48—49 (1950).

HÖFLER, K.: Über die Zuckerpermeabilität plasmolysierter Protoplaste. Planta (Berl.) 2, 454—475 (1926). — Über Kappenplasmolyse. Ber. dtsch. bot. Ges. 46, (73)—(81) (1928). — Das Permeabilitätsproblem und seine anatomischen Grundlagen. Ber. dtsch. bot. Ges. 49, (79)—(95) (1931). — Zur Tonoplastenfrage. Protoplasma (Berl.) 15, 462—477 (1932). — Kappenplasmolyse und Ionenantagonismus. Protoplasma (Berl.) 33, 545—578 (1939). — Über Wasser- und Harnstoffpermeabilität des Protoplasmas. Phyton 1, 105—121 (1949). — Plasmolyse mit Natriumkarbonat. Protoplasma (Berl.) 60, 426—460 (1951). — Über die Farbionenpermeabilität der Tonoplastenmembran. Ber. dtsch. bot. Ges. 65, 183—187 (1952). — Zur Kenntnis der Plasmahautschichten. Ber. dtsch. bot. Ges. 65, 391—399 (1953).— HOGEBOOM, G. H., and W. C. SCHNEIDER: On the nuclear envelope. Science (Lancaster, Pa.) 118, 419 (1953). — HOLM-JENSEN, I., A. KROGH u. V. WARTIOVAARA: Some experiments on the exchange of potassium and sodium between single cells of Characeae and the bathing fluid. Acta bot. fenn. 36, 1—22 (1944). — HOPE, A. B.: Salt uptake by root tissue cytoplasm. Austral. J. Biol. Sci. 6, 396—409 (1953). — HOUSKA, H.: Beiträge zur Kenntnis der Kappenplasmolyse. Protoplasma (Berl.) 36, 11—51 (1941). — HUBER, B., u. K. HÖFLER: Die Wasserpermeabilität des Protoplasmas. Jb. wiss. Bot. 73, 351—511 (1930).

KIHLMAN, B.: The permeability of the nuclear envelope and the mode of action of purine derivatives on chromosomes. Symb. bot. Upsal. 11, H. 2 (1951). — Induction of chromosome changes with purine derivatives. Symb. bot. Upsal. 11, H. 4 (1952). — A survey of purine derivatives as inducers of chromosome changes. Hereditas (Lund) 38, 115—127 (1952).

LATIES, G. G.: The physical environment and oxidative and phosphorylative capacities of higher plant mitochondria. Plant Physiol. 28, 557—575 (1953). — The osmotic inactivation *in situ* of plant mitochondrial enzymes. J. of Exper. Bot. 5, 49—70 (1954). — LEDERER, B.: Färbungs-, Fixierungs- und mikrochirurgische Studien an Spirogyra-Protoplasten. Protoplasma (Berl.) 22, 405—430 (1934). — LEPESCHKIN, W.: Kolloidchemie des Protoplasmas. Berlin: Springer 1924.

MCCLENDON, J. H.: The physical environment of chloroplasts as related to their morphology and activity in vitro. Plant Physiol. 29, 448—458 (1954). — MILLERD, A., and J. BONNER: The biology of plant mitochondria. J. Histochem. a. Cytochem. 1, 254—264 (1953). — MONNÉ, L.: Permeability of the nuclear membrane to vital stains. Proc. Soc. Exper. Biol. a. Med. 32, 1197—1199 (1935).

OVERTON, E.: Über die allgemeinen osmotischen Eigenschaften der Zelle, ihre vermutlichen Ursachen und ihre Bedeutung für die Physiologie. Vjschr. naturforsch. Ges. Zürich 44, 88—135 (1899).

PFEFFER, W.: Osmotische Untersuchungen. Leipzig: Wilhelm Engelmann 1877. — Zur Kenntnis der Plasmahaut und der Vakuolen. Abh. kgl.-sächs. Ges. Wiss. 28, 187—344 (1890). — Pflanzenphysiologie, 2. Aufl., Teil 2. Leipzig: Wilhelm Engelmann 1904. — PLOWE, J. Q.: Membranes in the plant cell. I. Protoplasma (Berl.) 12, 196—220 (1931). —

Membranes in the plant cell. II. Protoplasma (Berl.) **12**, 221—240 (1931). — PRINGSHEIM, E. G.: Über Plasmolyse durch Schwermetallsalze. Beih. bot. Zbl., Abt. I **41**, 1—14 (1924).

RIEDER, H. P.: Über die Zuckeraufnahme von Hefezellen. Ber. schweiz. bot. Ges. **61**, 539—621 (1951). — ROTHSTEIN, A.: Enzyme systems of the cell surface involved in the uptake of sugars by yeast. Symposia Soc. Exper. Biol. **8**, 165—201 (1954).

STARKEY, R. L., u. S. A. WAKSMAN: Fungi tolerant to extreme acidity and high concentrations of copper sulfate. J. Bacter. **45**, 196 (1943). — STEEMANN NIELSEN, E.: Influence of p_H on the respiration in Chlorella pyrenoidosa. Physiol. Plantarum (Copenh.) **8**, 106—114 (1955). — STRUGGER, S.: Die Vitalfärbung des Protoplasmas mit Rhodamin B und 6 G. Protoplasma (Berl.) **30**, 85—100 (1938). — Praktikum der Zell- und Gewebephysiologie der Pflanze, 2. Aufl. Berlin-Göttingen-Heidelberg: Springer 1949. — SUTCLIFFE, J. F.: The exchangeability of potassium and bromide ions in cells of red beetroot tissue. J. of Exper. Bot. **5**, 313—326 (1954).

VRIES, H. DE: Plasmolytische Studien über die Wand der Vakuolen. Jb. wiss. Bot. **16**, 465—598 (1885).

WATSON, M. L.: Pores in the mammalian nuclear membrane. Biochim. et Biophysica Acta **15**, 475—479 (1954). — WEATHERLEY, P. E.: Preliminary investigations into the uptake of sugars by floating leaf discs. New Phytologist **53**, 204—216 (1954). — WEBER, F.: Plasmalemma oder Tonoplast? Protoplasma (Berl.) **15**, 453—461 (1932). — WEEVERS, T.: Aufnahme, Verarbeitung und Transport der Zucker im Blattgewebe. Rec. Trav. bot. néerl. **28**, 400—420 (1931). — WILKINS, M. H. F.: Electron microscopy of nuclear membranes. Pubbl. Staz. zool. Napoli **23**, Suppl. 104—114 (1951).

Die Aufnahme der Anelektrolyte.

Von

H. J. Bogen.

Mit 5 Abbildungen.

Unter Anelektrolyten sollen solche Stoffe verstanden werden, deren Dissoziation sich im physiologischen p_H-Bereich in der Größenordnung der Dissoziation des Wassers oder darunter bewegt. Sie sind also praktisch nicht ionisiert und werden auch in dieser Form aufgenommen.

Im Sinne der klassischen kinetischen Theorie der Gase gelten sie als indifferente, „inerte" Stoffe, die keine Hauptvalenzkräfte besitzen und daher nicht befähigt sind, größere Wechselwirkungen zu entfalten, sei es untereinander, sei es gegenüber anderen Substanzen der Umgebung. Dieser Auffassung ist es zu danken, daß sich schon die ersten Untersuchungen über die Aufnahme von Stoffen in die lebende Zelle auf solche Anelektrolyte erstreckten:

W. Pfeffer: Osmotische Untersuchungen, 1877,

H. de Vries: Über die Permeabilität der Protoplasten für Harnstoff, 1888,

E. Overton: Über die osmotischen Eigenschaften der lebenden Pflanzen- und Tierzelle, 1895.

Es fällt auf, daß hier einhellig die Gesamtaufnahme mit der diosmotischen Aufnahme gleichgesetzt wird. Diese Gleichsetzung lag nahe, eben weil die Anelektrolyte als indifferente Stoffe galten, für die das elektrochemische Potential mit dem (leichter zu handhabenden) Konzentrationspotential zusammenfällt und Störungen des Ausgleichs dieses Konzentrationspotentials nicht befürchtet wurden. (Es verdient indessen festgehalten zu werden, daß bereits Pfeffer die Beteiligung von Molekularkräften auch beim Durchtreten von Anelektrolyten in Rechnung stellte).

Die Anelektrolyte erscheinen somit für die Analyse der osmotischen Stoffaufnahme besonders gut geeignet.

Alle diosmotischen Untersuchungen verfolgen den Zweck, zunächst an einem Objekt die Aufnahmeraten möglichst vieler Stoffe zu bestimmen und sie zu physikalischen Konstanten dieser Stoffe (z. B. Molekularvolumen, Löslichkeit, Grenzflächenaktivität) in Beziehung zu setzen. Praktisch geht man dabei so vor, daß man die Stoffe nach ihrer Aufnahmerate ordnet („Permeabilitätsreihe") und prüft, ob eine Anordnung der gleichen Stoffe nach einer der genannten Eigenschaften eine ähnliche Reihe ergibt. Werden Parallelen gefunden, so gestattet das in zweierlei Richtung Schlüsse zu ziehen: 1. auf die Aufnahmemechanismen („Permeationsprinzipien"), 2. auf die Art des Widerstandes, den das Plasma bzw. seine Grenzschichten dem Stoffdurchtritt entgegensetzen, und wie die Grenzschichten gebaut sein müssen, damit sie derartige Widerstände entfalten können („Strukturprinzipien"). Der Vergleich verschiedener Objekte untereinander wird dann ergeben, ob die Permeationsprinzipien und/oder die Strukturprinzipien generell gültig oder aber objektgebunden sind.

Bekanntlich gibt es eine ganze Anzahl von Objekten, an denen ausgedehnte, sorgfältige Messungen der Aufnahmeraten von Anelektrolyten ausgeführt und Permeabilitätsreihen (PR) aufgestellt wurden: *Beggiatoa*, *Chara*, *Rhoeo* und zahlreiche weitere Objekte des Pflanzen- und Tierreiches. Des weiteren besteht eine umfangreiche Literatur, die sich nicht allein mit den dort jeweils ermittelten Gesetzmäßigkeiten befaßt, sondern die vor allem die verschiedenen Objekte im Hinblick auf ihre Permeabilität miteinander vergleicht.

Wir wissen heute indessen, daß die Gleichsetzung von Gesamtaufnahme und diosmotischer Aufnahme nicht ohne weiteres statthaft ist; die Gesamtaufnahme setzt sich meist zusammen aus osmotischer (einschließlich metaosmotischer) und nichtosmotischer Aufnahme (vgl. III A). Die Berechnung von Permeationskonstanten (PK) aus der *Gesamt*aufnahme kann somit zu erheblichen Fehlschlüssen führen, *auch bei der Anelektrolytaufnahme* (s. unten). Die Fehler werden jedoch je nach der Methode, die zur Bestimmung der PK angewendet wurde, verschieden groß ausfallen. Es wird somit Aufgabe des Abschnittes sein, bereits bei jedem Objekt die Stichhaltigkeit der Daten und Argumente zu prüfen. Daraus wird sich ergeben, inwieweit es überhaupt noch ratsam ist, die aus dem Vergleich *verschiedener* Objekte auf deren Permeabilität gezogenen Folgerungen („vergleichende Permeabilitätsforschung") zu berücksichtigen.

A. Osmotische Aufnahme von Anelektrolyten.

1. Die Untersuchungen Overtons; die Lipoidtheorie.

Die ersten Untersuchungen auf breiter Basis, d. h. mit zahlreichen Stoffen und an verschiedenen Objekten, verdanken wir Overton (1897). Leider hat — was schon seit je beklagt wurde — Overton keinerlei Meßzahlen veröffentlicht, sondern sich darauf beschränkt, relative Angaben über die „Permeationsgeschwindigkeit" zu machen: „recht schnell", „mäßig", „langsam" u. ä. Auch über die untersuchten Objekte und die angewendeten Methoden finden sich nur gelegentliche Aussagen, und die Frage nach der Schädlichkeit der geprüften Diosmotica bleibt völlig ungeklärt.

Bekanntlich gelangte Overton zu zwei grundsätzlichen Einsichten: 1. Die eindringenden Stoffe (und zwar nicht nur die Anelektrolyte unter ihnen) würden um so schneller aufgenommen, je stärker sie sich in „organischen Lösungsmitteln" wie Äther, Chloroform, Öl usw., d. h. in „Lipoiden", lösen; bei der Permeation spiele sich mithin ein Lösen in lipoiden Medien ab, und die Grenzschicht müsse als lipoidhaltig aufgefaßt werden: Die Lipoidtheorie der Permeabilität war begründet. 2. Alle lebenden Objekte verhielten sich in den wichtigen Grundzügen gleichartig; Overtons PR gelte in gleicher Weise für jede lebende Zelle.

Das Fehlen quantitativer experimenteller Angaben (vgl. hierzu Ruhland und Hoffmann 1925) verhindert eine kritische Stellungnahme zu diesen Folgerungen. Da insbesondere niemals der Nachweis geführt wurde, daß die (mit unterschiedlicher Methode gemessene) Stoffaufnahme ausschließlich osmotischer Art ist, kann nicht ausgeschlossen werden, daß nichtosmotische Mechanismen beteiligt waren und die Aufnahmeraten höher ausfallen ließen, als es den osmotischen Gesetzen entsprach. Es dürfte daher gerechtfertigt sein, den summarischen Angaben Overtons und seiner Lipoidlöslichkeitshypothese mit großer Zurückhaltung zu begegnen, um so mehr, als die zweite Folgerung Overtons, nämlich die Gleichartigkeit im diosmotischen Verhalten der verschiedenen Zellen, in der Folgezeit *von allen Seiten* gründlich widerlegt wurde.

2. Beggiatoa mirabilis; das Ultrafilterprinzip; die Turgescenzmethode.

(Ruhland und Hoffmann 1925, Schönfelder 1930.)

An diesem Objekt wurde das Eindringungsvermögen von 68 verschiedenen Anelektrolyten quantitativ bestimmt und damit die erste zuverlässige PR aufgestellt.

Beggiatoa mirabilis, früher zu den Fadenbakterien gestellt, ist eine farblose Cyanophycee, die in Form vielzelliger Fäden auftritt. Die einzelnen Zellen besitzen eine große Zentralvacuole, die von einer dünnen Plasmaschicht und einer ebenfalls sehr dünnen, gleichwohl zweischichtigen Membran umgeben ist. Die Eigenart der Membran und die feste Verwachsung mit dem Plasma verhindert eine Plasmolyse; in hypertonischen Lösungen findet daher nur ein Einknicken der Fäden statt, das nach einiger Zeit wieder zurückgeht. Diese Zeit wird als Maß für das Eindringen der gelösten Moleküle verwendet, und zwar, indem diejenigen geringen osmotischen „Überkonzentrationen" ermittelt werden, die entweder eine gerade 5 min anhaltende Einknickung bewirken („5 min-Versuche"), oder aber nur eine „momentane" Knickung mit sofort anschließender Streckung hervorrufen („Momentanversuche"). Die „plasmolytischen" Grenzkonzentrationen sind mit ungewöhnlicher Genauigkeit erfaßbar; es handelt sich indessen immer um *Relativwerte*, und es verdient hervorgehoben zu werden, daß weder RUHLAND und HOFFMANN noch SCHÖNFELDER daraus absolute PK berechnet haben.

Gleichwohl ist diese Reihe der Aufnahmeraten in der Folgezeit stets als Reihe der diosmotischen Anelektrolytaufnahme, d. h. als *Permeabilitäts*reihe s. str. aufgefaßt worden; es bedarf der Erörterung, in welchem Maße diese Gleichsetzung statthaft ist.

Aus der Bestimmung der Ausgleichszeiten in Lösungen verschiedener Konzentrationen (Salze, aber auch Anelektrolyte wie Harnstoff, Arabinose, Rhamnose) ergibt sich, daß *mit steigendem Konzentrationsgefälle die in der Zeiteinheit aufgenommene Stoffmenge abnimmt.* Bereits RUHLAND und HOFFMANN betonen nachdrücklich, daß in solchen Fällen „das Permeieren nicht in einer einfachen Diosmose bestehen" kann (S. 32). Diese Abweichung vom FICKschen Diffusionsgesetz kann als Permeabilitätsherabsetzung unter dem Einfluß hochkonzentrierter Lösungen gedeutet werden, sie kann aber auch als Anhaltspunkt für das Interferieren nichtosmotischer Prozesse gelten, die dann in Abzug gebracht werden müßten, wenn man reine diosmotische Koeffizienten bzw. Konstanten erhalten und auswerten will.

Eine Entscheidung konnte aus methodischen Gründen nicht getroffen werden; es scheint indessen, als ob, wiederum aus methodischen Gründen, derartige Abweichungen als unerheblich betrachtet werden dürfen: 1. Der außerordentlich geringe „Überwert" der Zellen gegenüber dem umgebenden Solwasser (0,00015 M Raffinose = 0,00036 atm !) gestattet, mit besonders niedrigen Versuchskonzentrationen der Diosmotica auszukommen, in der Regel 0,1 mol. 2. Die Versuche, insbesondere die Momentanversuche, dauern nur wenige Sekunden oder Minuten, daher können nichtosmotische Prozesse sich nur beschränkt auswirken. Da die Turgescenz ein osmotisches Phänomen ist, kann angenommen werden, daß die Methode im wesentlichen osmotisch wirksame Substanzen erfaßt (vgl. hierzu S. 38ff.).

Das Ergebnis der Untersuchungen ist allgemein bekannt und kann daher in gedrängter Form wiedergegeben werden: Die nach steigender Aufnahmegeschwindigkeit geordneten Stoffe ergeben eine PR, die eine bemerkenswerte Übereinstimmung mit derjenigen Reihe zeigt, die aus der Anordnung der gleichen Stoffe nach fallendem Molekulargewicht M bzw. fallendem Molekularvolumen MV (berechnet nach der KOPPschen Formel) folgt. Von einzelnen Ausnahmen abgesehen, folgt daraus, daß die Permeabilität umgekehrt proportional ist der Raumerfüllung der permeierenden Moleküle

$$P = \frac{1}{f(M)} \tag{1}$$

bzw.

$$P \cdot f(M) = \text{const.}$$

Ähnliches ist von der freien Diffusion bekannt, von der die Beziehung gilt:

$$P \cdot M^{\frac{1}{2}} = \text{const.} \tag{2}$$

Mithin besteht eine weitgehende Übereinstimmung zwischen freier Diffusion („Hydrodiffusion") und Permeation. Da indessen, wie gleichfalls experimentell nachweisbar ist, die Permeabilitätskoeffizienten mit steigender Molekülgröße

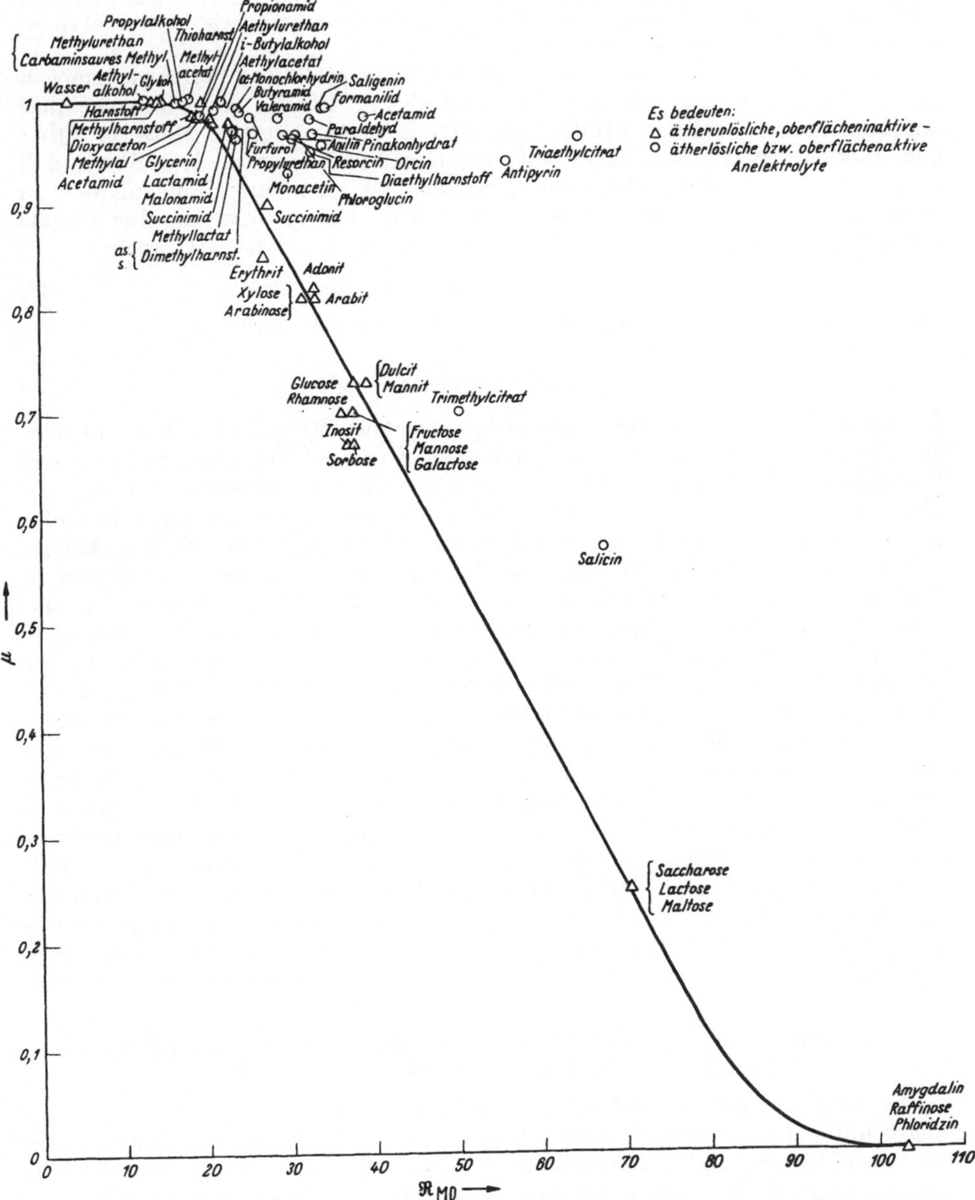

Abb. 1. *Beggiatoa mirabilis*, „μ-Werte" für verschiedene Anelektrolyte, aufgetragen gegen die Molekularrefraktion R_{MD} (Abszisse). Während die Werte der „lipoidunlöslichen Stoffe (△) auf einer S-förmigen Kurve liegen, sind die Werte der „lipoidlöslichen" Stoffe (○) zu hoch. (Nach SCHÖNFELDER 1930.)

viel schneller abnehmen als die Diffusionskoeffizienten, ist zu folgern, daß „beim Permeieren keine freie Diffusion ... stattfindet, sondern daß ... in erster Linie die *behindernde* Wirkung der Micellarinterstitien zum Ausdruck gelangt ...“. Die Permeation ist also eine *Ultrafiltration*, und der Plasmamembran kommt Struktur und Funktion eines Ultrafilters zu.

Die bis in die neueste Zeit geübte synonyme Anwendung des Ausdruckes „Molekülsieb“ die dann meistens eine Ablehnung des Ultrafilterprinzips rechtfertigen sollte (vgl. DAVSON und DANIELLI 1943), wird bereits von RUHLAND und HOFFMANN verworfen mit dem Hinweis, daß ja ein Sieb lediglich über Passieren oder Zurückhalten entscheidet. Eine „kritische Teilchengröße“ besteht zumindest bei *Beggiatoa* offenbar nicht (möglicherweise aber bei Erythrocyten, vgl. MOND und HOFFMANN 1928).

Die hier erkennbare Beziehung zwischen Teilchengröße und Permeiervermögen gilt streng genommen nur für „lipoidunlösliche“ Stoffe; sie wird scheinbar durchbrochen bei solchen Stoffen, die sich in nennenswertem Maße in Äther und anderen organischen Lösungsmitteln lösen, und zwar in dem Sinne, daß solche Stoffe schneller aufgenommen werden, als nach ihrer Molekülgröße zu erwarten ist (Abb. 1, nach SCHÖNFELDER 1930). Nimmt man als Maß für die Lipoidlöslichkeit den sog. Verteilungskoeffizienten K, so ist also auch

$$P = f(K). \tag{3}$$

Die Formel 1 muß also dahingehend erweitert werden, daß

$$P = \frac{f'(K)}{f''(M)} \tag{4}$$

ist. Mit anderen Worten: von zwei gleichgroßen Anelektrolytmolekülen wird stets das „lipoidlöslichere“ schneller aufgenommen, von zwei Molekülen mit gleichem Verteilungskoeffizienten das kleinere. Hierin erblicken zahlreiche Autoren eine Bestätigung der alten OVERTONschen Löslichkeitshypothese; SCHÖNFELDER hingegen, die sich besonders der lipoidlöslichen Stoffe angenommen hat, konnte zeigen, daß die Parallelität zwischen P und dem Verteilungskoeffizienten in Wirklichkeit eine Parallelität zwischen P und der *Adsorbierbarkeit* der eindringenden Moleküle im Sinne der Adsorptionstheorie WARBURGs bedeutet, insofern als der HENRYsche Verteilungssatz nur ein Spezialfall der Adsorptionsgesetze ist (ABDERHALDEN und FODOR 1919, 1920). Somit stellt sich der Permeationsvorgang nach RUHLAND und HOFFMANN sowie SCHÖNFELDER dar als behinderte Diffusion „durch die zwischen den einzelnen Kolloidpartikeln der Membran befindlichen (wassererfüllten) Interstitien“. Wie freilich diese „molekulare Osmose“ in den Einzelheiten ablaufe, mußte so lange offen bleiben, als die zwischen den Filterporen und den permeierenden Molekülen wirksamen Wechselbeziehungen noch unbekannt waren und auch die molekularphysikalischen Grundlagen der Diffusion noch der Klärung bedurften. Der Gewinn der Ultrafiltertheorie war, daß sie einen für alle Stoffe gültigen Permeationsmechanismus angeben konnte, einen Mechanismus, der den PFEFFERschen Gesetzen entsprach, und dessen Funktionieren in Modellversuchen bewiesen wurde (COLLANDER 1926).

3. Characeen; Lipoidfilterhypothese; chemische Zellsaftanalyse.

(COLLANDER und BÄRLUND 1933, COLLANDER 1954.)

Es könnte zunächst anfechtbar erscheinen, daß die Untersuchungen an *Chara* unter den *osmotischen* Untersuchungen aufgeführt werden, denn mit den Methoden der chemischen Zellsaftanalyse wird fraglos die *Gesamt*menge des aufgenommenen Stoffes (also osmotische Aufnahme + nichtosmotische Aufnahme) bestimmt. Nach COLLANDER und BÄRLUND geht jedoch aus den Analysendaten hervor, daß die geprüften Stoffe ausschließlich diosmotisch aufgenommen werden;

die Verfasser interpretieren ihre Befunde daher unter dem Gesichtspunkt des *Permeabilitäts*problems. Wegen dieser Schlußfolgerung erscheint es angezeigt, die Ergebnisse bereits hier zu behandeln.

Die untersuchten Characeen *Chara*, *Nitella* und *Nitellopsis* haben mit *Beggiatoa* gemein, daß sie wasserbewohnende Algen sind. Daneben freilich bestehen erhebliche Unterschiede nicht allein gegenüber *Beggiatoa*, sondern auch gegenüber den Zellen höherer Pflanzen; vor allem sind die untersuchten Internodialzellen vielkernige Riesenzellen mit ungewöhnlich großer Zentralvacuole. Dadurch wird es möglich, aus verhältnismäßig wenig Zellen soviel Vacuolenflüssigkeit zu gewinnen, daß eine quantitative chemische Analyse durchgeführt werden kann.

Solche Internodialzellen werden eine gewisse Zeit lang in einer wäßrigen Lösung des Diosmoticums belassen, dann entfernt und angestochen. Der Zellsaft wird in eine Capillare eingesogen, und aus 6—12 Zellen können auf diese Weise 0,05 bis 0,1 ml Zellsaft gesammelt werden. Die Menge der in den Zellsaft gelangten („endosmierten") Substanz wird bei N-haltigen Verbindungen mittels Mikro-Kjeldahl, bei den übrigen an Hand des Reduktionsvermögens (Chromsäuremethode nach Bang 1927) bestimmt. In beiden Fällen muß von den Versuchswerten der Kjeldahl-N bzw. das Reduktionsvermögen der Kontroll-(„Normal"-) Zellen abgezogen werden.

Auf diese Weise gelingt es, die Konzentration der Stoffe im Zellsaft, die bei anderen Methoden der Permeabilitätsmessung immer nur indirekt ermittelt werden kann, *direkt* zu bestimmen. Collander und Bärlund betrachten sie als „osmotische" bzw. „osmotisch wirksame" Binnenkonzentration c und gelangen so über die Formel

$$\frac{dm}{dt} = k \cdot q \cdot (C - c)$$

(vgl. S. 117) zu Permeationskonstanten.

Die aus diesen PK aufgestellte PR zeigt nun eine bemerkenswerte Beziehung zum Verteilungskoeffizienten K der Diosmotica: je höher der $K_{\text{Äther/Wasser}}$ bzw. $K_{\text{Olivenöl/Wasser}}$ ist, um so höher ist auch die PK. In der graphischen Darstellung (Abb. 2), in der die PK gegen K aufgetragen sind, läßt sich durch zwei Diagonalen ein Bereich abgrenzen, in dem die Mehrzahl der PK zu finden ist. Eine Ausnahme machen vor allem kleinmolekulare Stoffe, die *über* diesem Bereich liegen, d. h. höhere PK besitzen, als nach ihrer Lipoidlöslichkeit zu erwarten wäre. Für die 3 Characeen ergeben sich nach Collander (1954) folgende (formale) Abhängigkeitsbeziehungen:

$$\textit{Chara:}\ P = \frac{K_{\text{Öl}}}{M^{1,5}}, \quad \textit{Nitellopsis:}\ P = \frac{(K_{\text{Öl}})^{1,15}}{M^{1,5}}, \quad \textit{Nitella:}\ P = \frac{(K_{\text{Öl}})^{1,3}}{M^{1,5}}.$$

Es ist ohne weiteres ersichtlich, daß diese Formeln Spezialfälle der allgemeinen Gleichung

$$P = \frac{f'(K)}{f''(M)}$$

von S. 234 sind (vgl. Bogen 1950).

Die Folgerungen, die Collander und Bärlund aus den so formulierten Abhängigkeitsbeziehungen gezogen haben, sind bekanntlich folgende: Der Permeationsvorgang ist als Lösen bzw. „Hindurchlösen" (Diasolyse, vgl. Brintzinger und Beier 1937) durch Grenzschichtlipoide der Zelle aufzufassen, wobei deren Lösungsvermögen durch die Lösungseigenschaften des Olivenöls besser als durch die des Äthers ($K_{\text{äther}}$) charakterisiert wird. Daneben hat aber auch die Molekülgröße des Diosmoticums einen gewissen Einfluß auf dessen Permeiervermögen. Damit die Schwierigkeit umgangen werden kann, zwei verschiedene

Permeationsmechanismen nebeneinander annehmen zu müssen (Diasolyse und Ultrafiltration), vermutet COLLANDER (1954), daß das Molekularvolumen bzw. Molekulargewicht nicht eigentlich als *permeabilitäts*bestimmender Faktor einwirke; vielmehr manifestiere sich sein Einfluß über einen „Orientierungseffekt" (WARTIOVAARA 1949, 1950). Insbesondere dürfe aus der Tatsache, daß P umgekehrt proportional der 1,5. Potenz von M, aber proportional der 1,3. Potenz

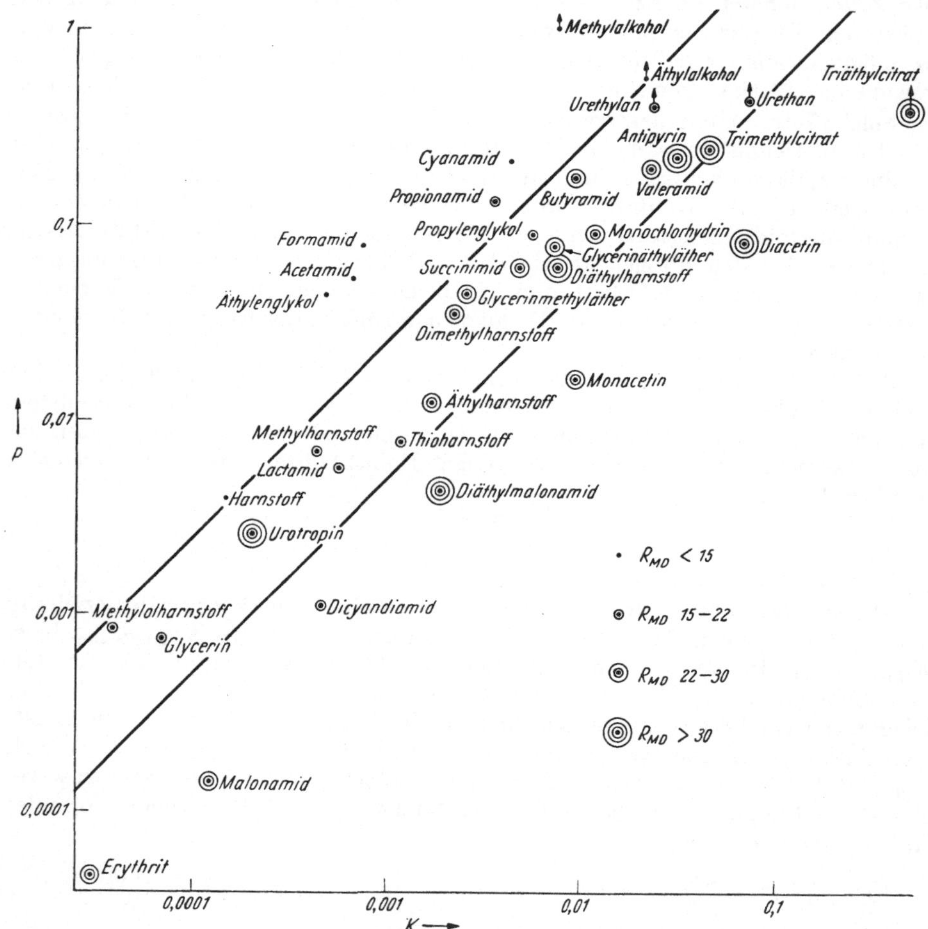

Abb. 2. *Chara ceratophylla*, Permeationskonstanten P (Ordinate), aufgetragen gegen den Verteilungskoeffizienten $K_{\text{Olivenöl/Wasser}}$. Die meisten P-Werte liegen zwischen zwei etwa diagonalen Geraden; die Werte der kleinmolekularen Stoffe jedoch sind im Verhältnis zu hoch. (Nach COLLANDER und BÄRLUND 1933.)

von K *(Nitella)* oder gar nur K selbst *(Chara)* sei, nicht geschlossen werden, daß die Molekülgröße wichtiger sei als die Lipoidlöslichkeit. Dieser Annahme widerspräche es, daß K in erheblich weiteren Grenzen variiere als M.

Ohne auf die Einzelheiten in der Theorie der Anelektrolytpermeation hier schon näher eingehen zu wollen, sei verzeichnet, daß die Schlußfolgerungen COLLANDERs (1954) und RUHLANDs, SCHÖNFELDERs und auch BOGENs (1950) formal völlig gleich sind. Diskrepanzen bestehen zunächst nur in der *Bewertung* der Anteile von M bzw. K: COLLANDER sieht die Molekülgröße als zweitrangig an, da sie den Permeationsvorgang nur mittelbar beeinflußt, und da die VK stärker variieren als die Molekulargewichte. Umgekehrt ist nach RUHLAND (1925) und

BOGEN (1950) das Molekulargewicht M bzw. die Molekularrefraktion R_{MD} der entscheidende Faktor, wohingegen der Verteilungskoeffizient lediglich modifizierend eingreift; insbesondere beweist der Einfluß von K nicht die Mitwirkung eines *Lösungs*vorganges, und K ist nur ein stellvertretendes Maß für die zwischenmolekularen Kräfte (vgl. Abschnitt V G).

Schließlich wäre noch folgendes anzumerken: Wenn K in erheblich weiteren Grenzen variiert als M, so kann man daraus auch das Gegenteil der Ansicht COLLANDERs ableiten: Es genügen oftmals geringe Unterschiede von M, um eine Änderung der PK herbeizuführen, während andererseits VK sehr stark wechseln kann, ohne daß die PK nennenswert betroffen werden. Damit müßte man wohl der Molekülgröße den größeren Einfluß zuschreiben (wie das ja auch aus der Tatsache hervorgeht, daß P umgekehrt proportional $M^{1,5}$ ist).

Die Diskussion hierüber ist über lange Zeit hinweg fruchtlos geblieben; sie konnte auch durch Herbeitragen weiteren experimentellen Materials nicht zu einer Lösung geführt werden, die alle Teilnehmer überzeugt hätte. Einer der wichtigsten Gründe hierfür liegt offenbar tiefer, denn er betrifft nicht so sehr die Bewertung der einzelnen PK, sondern überhaupt die Berechtigung, die Meßdaten COLLANDERs als *Permeations*konstanten, d. i. als *diosmotische* Koeffizienten oder Konstanten aufzufassen.

Ganz augenscheinlich wird, wie bereits erwähnt, mit der chemischen Zellsaftanalyse die *gesamte* eingedrungene Stoffmenge erfaßt. Diese mit der osmotisch aufgenommenen gleichzusetzen, ist nur erlaubt, wenn nachgewiesen wird, daß keinerlei nichtosmotische Stoffaufnahme stattfindet — der exakte Beweis steht noch aus —, oder wenn die Kriterien der osmotischen Stoffaufnahme eindeutig erfüllt sind. Als solche haben (unter anderen) zu gelten:

1. der Nachweis der „Gleichverteilung", d. h. daß das Diosmoticum genau bis zum Konzentrationsausgleich eintritt, der „Verteilungskoeffizient" Medium/Zellsaft also = 1 ist, und
2. daß die „Zeitproportionalität" dem Diffusionsgesetz entspricht und wohl die aufgenommene Menge, nicht aber die PK vom Konzentrationspotential abhängig ist. Dazu kommt
3. daß die aufnehmende Zelle nicht zwischen verschiedenen Diosmoticis unterscheidet, soweit diese gleiche Molekülgröße und gleiche Lipoidlöslichkeit aufweisen, d. h. keine Spezifität gegenüber dem aufzunehmenden Substrat besteht.

COLLANDER und BÄRLUND bemühen sich nachzuweisen, daß die Bedingungen 1 und 2 erfüllt sind. Es finden sich indessen mehrere bemerkenswerte Abweichungen:

Zu 1. Grundsätzlich ist der Konzentrationsausgleich erst nach unendlich langer Zeit erreicht; alle in endlicher Zeit abgebrochenen Versuche müssen zu „Verteilungskoeffizienten" < 1 bzw. zu „Gleichgewichtskonzentrationen" < 100 (%) führen. Um so auffälliger ist, daß die Hälfte aller Verbindungen zu Mittelwerten von 98—102 führen, für Glycerin z. B. 85—119. Besonders kraß ist der Fall des Methylharnstoffes: nach 24 Std beträgt die Gleichgewichtskonzentration in 0,1 mol und 0,02 mol Lösung ziemlich genau 100, in 0,002 mol hingegen 281 und 300. Hier scheint eine erhebliche Speicherung stattgefunden zu haben, doch wird diese Deutung von COLLANDER und BÄRLUND aus methodischen Gründen abgelehnt: „Angenommen, daß bei der Titration ... etwa 0,06 cm³ mehr n/100 NaOH-Lösung verbraucht worden wäre, würde man das Verteilungsverhältnis 1 bekommen haben". Mit anderen Worten, die Abweichung vom „theoretischen" Wert liegt im Bereich der methodischen Fehler. Somit ist gewiß der Schluß berechtigt, daß innerhalb der Versuchsfehler „eine Änderung des Verteilungsverhältnisses zwischen Zellsaft und Außenlösung ... in keinem Falle nachzuweisen" war, es liegt aber ebenso auf der Hand, daß die Genauigkeit der Methode nicht ausreicht, um bei geringen Außenkonzentrationen eine Speicherung endosmierter Substanz *selbst auf das 3fache der Außenkonzentration* mit

Sicherheit nachzuweisen. Es muß zwar zugegeben werden, daß bei höheren Versuchskonzentrationen die Fehler geringer werden, aber selbst eine Speicherung von 10 oder 20% würde, wenn man eine Methode besäße, sie sicher nachzuweisen, genügen, um eine Überlagerung der osmotischen Stoffaufnahme sicherzustellen.

Andere Autoren haben, zum Teil mit genaueren Methoden, am selben Objekt wie COLLANDER in der Tat Speicherung von Anelektrolyten nachweisen können. So fanden HOAGLAND und DAVIS (1929), daß *Nitella* Harnstoff speichert; BAZIN (1947) konnte bei *Chara* mit spezifischen Fällungsreaktionen (Kieselwolframsäure) zeigen, daß Aminoalkohole zum Teil bis zum 20fachen der Außenkonzentration (Cocain) akkumuliert werden. Somit muß in der Tat mit erheblichen Speicherungsvorgängen gerechnet werden. Diese täuschen dann eine viel zu hohe PK vor, weil auch die gespeicherte Substanz bei der Zellsaftanalyse miterfaßt wird und in die Formel für die Berechnung der PK eingeht.

Zu 2. Um die Gültigkeit des Diffusionsgesetzes nachzuweisen, wurden in zahlreichen Versuchen die „Sättigungsgrade" zu verschiedenen Zeiten analytisch bestimmt und verglichen mit den Werten, die nach der Theorie zu berechnen waren. Die Abweichungen sind, numerisch betrachtet, nicht sehr erheblich (meist zwischen +0,05 und —0,07). Prozentual jedoch macht das bis zu 30% aus (z. B. Urethylan). Dabei sind die Abweichungen zu Versuchsbeginn meist stärker als zu Versuchsende, und häufig läßt sich überdies ein „Gang" erkennen insofern, als die Differenzen (gefunden/berechnet) zuerst positiv, zuletzt negativ sind, d. h. zu Versuchsbeginn wird mehr, gegen Versuchsende weniger aufgenommen, als dem Konzentrationsgefälle entspricht. Die von COLLANDER und BÄRLUND geübte Methode der unspezifischen Zellsaftanalyse erlaubt also auch nicht, Abweichungen vom Diffusionsgesetz sicher nachzuweisen.

Man darf danach annehmen, daß bei den Characeen neben der osmotischen Stoffaufnahme auch meta- und nichtosmotische Aufnahmevorgänge ablaufen. Die Gleichsetzung von (analytisch erfaßbarer) Gesamtaufnahme und osmotischer Aufnahme ist augenscheinlich nicht immer statthaft. Setzt man trotzdem die Analysendaten in die Permeationsgleichung ein, so müssen die daraus errechneten PK zu hoch ausfallen. Sehr wahrscheinlich ist diese Überhöhung nicht für alle Stoffe gleichgroß (vgl. die Versuche BAZINs); es müssen daher unterschiedliche Korrekturen angebracht werden, und die „wirkliche" PR für *Chara* wird vermutlich etwas anders aussehen. Damit verlieren aber viele der Argumente, die auf Grund der *Chara*-Ergebnisse gegen Permeabilitätstheorien ins Feld geführt worden sind, merklich an Gewicht, und wir müssen zugeben, daß die unkorrigierten *Chara*-Werte nur sehr begrenzt Aufschluß geben können über die *osmotische* Anelektrolytaufnahme.

4. Rhoeo discolor; die grenzplasmolytische Methode.

(COLLANDER und BÄRLUND 1927, BÄRLUND 1929.)

Die gleichen Autoren, denen wir die quantitative Bestimmung der (Gesamt-) Anelektrolytaufnahme der Characeen verdanken, haben sich auch der Untersuchung von Zellen höherer Pflanzen gewidmet. Das Objekt — Epidermiszellen der Blattunterseite von *Rhoeo discolor* — gestattet keine chemische Analyse des Zellsaftes; daher griffen COLLANDER und BÄRLUND zu einer der plasmolytischen Methoden, nämlich der von FITTING ausgearbeiteten *grenzplasmolytischen Methode*. Ihr Prinzip ist, verschiedene Schnitte desselben Blattes in passend abgestufte hypertonische Lösungen des auf seine Permeationsgeschwindigkeit zu untersuchenden Stoffes einzulegen und in bestimmten Zeitabständen diejenige Konzentration zu ermitteln, bei der 50% der Zellen Grenzplasmolyse zeigen und 50% bereits deplasmolysiert sind. (Es wird also nicht mit Einzelzellen gearbeitet, sondern der Durchschnittswert eines Kollektivs von 100—200 Zellen bestimmt.) Diese „temporären osmotischen Werte" können als Maß für das Anwachsen der Konzentration osmotischer wirksamer Substanz im Zellsaft gelten.

COLLANDER und BÄRLUND haben in ihrer vorläufigen Mitteilung keine PK berechnet, sondern lediglich die temporären osmotischen *Koeffizienten* (vgl. STADELMANN):

$$\omega = \frac{C}{C_1},$$

worin C die plasmolytische Grenzkonzentration für die nichtpermeierende Saccharose, C_1 die jeweilige plasmolytische Grenzkonzentration des permeierenden Stoffes ist.

Der temporäre osmotische Koeffizient wird von COLLANDER und BÄRLUND (1927), MARKLUND (1936) und ELO (1937) als ω-Wert, von BÄRLUND (1929) als π-Wert bezeichnet. COLLANDER und BÄRLUND errechnen die Zunahme des osmotischen Wertes aus der Abnahme der ω-Werte: „Wenn z. B. der osmotische Koeffizient des Methylharnstoffes 1 Std nach Versuchsbeginn 0,80 beträgt, so ist die einfachste Deutung hierfür die, daß zu jener Zeit Methylharnstoff in einer Menge, die 20% der Außenkonzentration entspricht, in die Zelle eingedrungen, wodurch der ω-Wert des Methylharnstoffs vom theoretischen Maximalwert 1,00 um 20%, d. h. auf 0,80 herabgesetzt worden ist." Die Autoren wollen also keine quantitativen Angaben machen, sondern, ähnlich wie RUHLAND und HOFFMANN an Beggiatoa, nur Relativwerte ermitteln, die die *Reihenfolge* der Permeierfähigkeit der verschiedenen Diosmotica angeben. MARKLUND, ELO und andere Schüler COLLANDERs haben dann später, obgleich COLLANDER die Bedenken FITTINGs gegen die quantitative Auswertung der Grenzplasmolysekurven teilte, die ω-Werte zur Berechnung der Binnenkonzentration in der Vacuole verwendet:

$$p' = \frac{1}{t} \cdot \ln \frac{C}{(C - x)},$$

$$= \frac{1}{t} \cdot \ln \frac{1}{\omega},$$

$$= \frac{1}{t} \cdot \ln \frac{C_1}{C},$$

und so PK errechnet.

COLLANDER und BÄRLUND haben die gleichen Stoffe verwendet, wie später für die *Chara*-Versuche; sie stellen einen Teil der Stoffe dar, die auch RUHLAND und HOFFMANN sowie SCHÖNFELDER bei *Beggiatoa* benutzten. Die „qualitative" PR von *Rhoeo discolor* zeigt nun eine überraschende Ähnlichkeit mit der „quantitativen" Charareihe: auch hier sind die „lipoidlöslichen" Diosmotica stark gefördert, desgleichen die kleinmolekularen, d. h. die PK ist proportional der Lipoidlöslichkeit (repräsentiert durch den Verteilungskoeffizienten $K_{\text{Öl}}$) und umgekehrt proportional der Molekülgröße (Molekularrefraktion, Molvolumen). Unterschiede finden sich einmal in der „absoluten", d. h. auf die Flächeneinheit berechneten Permeabilität, die bei *Chara* etwa 40—50mal größer ist als bei *Rhoeo*, zum anderen bei den Amiden (und dem noch stärker basischen Urotropin): Diese Stoffe dringen bei *Chara* im Verhältnis zu den nichtbasischen Verbindungen schneller ein als bei *Rhoeo*. Vor allem die letztgenannte Differenz wird auf eine andersartige Zusammensetzung der Grenzschichtlipoide zurückgeführt; sie sollen bei *Chara* saurer sein als bei *Rhoeo*.

(Dagegen wäre einzuwenden, daß das Fehlen einer Übereinstimmung zwischen Hypothese — Lipoidlöslichkeit als permeabilitätsbestimmender Faktor — und Befund — PK bei *Rhoeo* bzw. *Chara* — ebensogut zur Einführung einer neuen Hilfshypothese — stärker saure Lipoide — wie zur Ablehnung der ursprünglichen Hypothese Veranlassung geben kann.)

Hier wird zum ersten Male bewußt jener Weg eingeschlagen, der später von seiten der Schule Höflers energisch beschritten wird, der Weg des Vergleichs von PR, der dann zur Aufstellung von „Permeabilitätstypen“ führen sollte (s. unten). Die Resultate dieser Forschungsrichtung werden unter III F h — Objekttypen der Permeabilität — behandelt. Bereits hier sei aber darauf eingegangen, daß gegen diesen Vergleich wiederum ernste Bedenken anzumelden sind:

Die grenzplasmolytische Untersuchungsweise ist eine *osmotische* Methode. Es scheint zunächst so, als ob sie gerade deshalb besonders gut geeignet sei, die osmotischen Gesetzmäßigkeiten, d. h. die Permeation als osmotische Stoffaufnahme, rein zu erfassen, unbeschadet aller etwa ablaufenden meta- und nichtosmotischen Prozesse. Das ist indessen nicht der Fall. Hierzu sei auf folgende Tatsachen verwiesen:

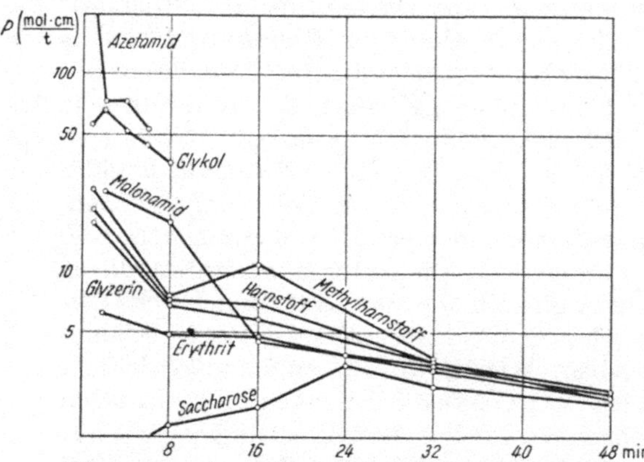

Abb. 3. *Licmophora oedipus*, Permeationskonstanten *P*, berechnet nach den Angaben Elos (1937) von Bogen und Follmann (1955). Die Konstanten fallen mit fortschreitender Versuchsdauer erheblich ab (Ausnahme: Saccharose).

Aus der finnischen Schule hat Elo (1937) mit der grenzplasmolytischen Methode die Anelektrolytaufnahme der Diatomee *Licmophora oedipus* untersucht. Berechnet man aus seinen Versuchsdaten die absoluten PK, so ergibt sich für die geprüften Stoffe ein eigenartiges Bild (Abb. 3): Die PK sind keineswegs konstant, sondern fallen *zumeist* im Verlaufe des Versuches sehr stark ab, zum Teil auf $^1/_{10}$ des anfänglichen Wertes. Vereinzelt ist aber auch ein Ansteigen zu verzeichnen (z. B. für Saccharose). Wäre die grenzplasmolytisch erfaßte Anelektrolytaufnahme ausschließlich osmotischer Art, so würden die PK zu allen Meßzeiten konstant sein müssen. Das gegenteilige Verhalten zwingt zu der Annahme, daß entweder die Permeabilität während des Versuches verändert wird, oder aber, daß nichtosmotische Prozesse mitwirken.

Die erste Annahme erscheint plausibel, wenn man berücksichtigt, daß starke Plasmolyse permeabilitätsherabsetzend wirken soll (Schmidt 1936), und gerade für die später bestimmten PK müssen ja die Schnitte in stark hypertonische Lösungen eingelegt werden. Indessen versagt sie im Falle der Saccharose. Andererseits haben Bogen (1953), Bogen und Prell (1953), Bogen und Follmann (1955) zeigen können, daß in der Tat nichtosmotische Aufnahmevorgänge so ablaufen können, daß sie auch mit plasmolytischen Methoden faßbar sind: Bei *Rhoeo* und anderen höheren Pflanzen findet im plasmolytischen Versuch eine nichtosmotische Wasseraufnahme statt, die durch Stoffwechselgifte verhindert, durch Stoffwechselanreger beschleunigt werden kann. Sie führt zu einer Protoplastenausdehnung, ohne daß osmotisch wirksame Substanz aus der plasmolysierenden Außenlösung in die Vacuole tritt (Bogen 1953).

Bei den Diatomeen *Melosira arenaria* und *varians* läßt sich das gleiche Verhalten grenzplasmolytisch nachweisen; hinzu tritt eine nichtosmotische Aufnahme gelöster Anelektrolyte (Einzelheiten s. unten).

Damit ist sichergestellt, daß selbst die grenzplasmolytische Methode keine unverfälschten PK liefert, sondern zu hohe Werte, wobei die Überhöhung bei verschiedenen Diosmoticis, je nach deren osmotischem und nichtosmotischem Eindringungsvermögen, verschieden stark ist. Auch grenzplasmolytisch ermittelte PR haben nur begrenzten Anspruch auf Gültigkeit, und der Vergleich solcher Reihen untereinander wie auch der Vergleich mit PR, die an Hand anderer Methoden *(Chara, Beggiatoa)* erhalten wurden, kann zu verhängnisvollen Mißdeutungen etwa aufgefundener Unterschiede führen.

5. Majanthemum bifolium; plasmometrische Methode.

(HÖFLER 1934ff.)

Gegenüber der grenzplasmolytischen Methode werden hier Einzelzellen untersucht: diese werden stark plasmolysiert und in verschiedenen Zeitabständen auf den Grad der Rückdehnung vermessen. Der Grad der Protoplastenkontraktion („Plasmolysegrad“, vgl. diesen Band S. 207f.) wird berechnet aus dem Verhältnis Volumen der plasmolysierten Zelle zu Volumen der entspannten (deplasmolysierten) Zelle. Das erfordert einfache geometrische Formen der Zellen und schränkt so den Anwendungsbereich der Methode ein.

Aus dem Plasmolysegrad kann der „osmotische Wert bei Grenzplasmolyse“ (O_g-Wert) berechnet werden; sein Anstieg im Laufe des Permeationsversuches dient dann als Maß für die eingedrungene Menge osmotisch wirksamer Substanz („Binnenkonzentration“), was wiederum eine Berechnung von PK gestattet.

HÖFLER hat auf diese Weise die PK für 12 verschiedene Anelektrolyte bestimmt und aus ihnen die „*Majanthemum*-Permeabilitätsreihe“ zusammengestellt. Die Ähnlichkeit mit der *Chara*- und *Rhoeo*-Reihe ist überraschend groß; HÖFLER erblickt hierin mit COLLANDER „einen kräftigen Wahrscheinlichkeitsbeweis für die Anwendbarkeit und Zuverlässigkeit der von manchen Seiten stark angezweifelten plasmolytischen Methode der Permeabilitätsmessung“; des weiteren betrachtet er den „*Majanthemum*-Typus“ als den „Normaltypus“ einer PR.

Kleinere Abweichungen, die gegenüber *Chara* nur quantitativer, gegenüber *Rhoeo* in einem Falle auch qualitativer Art sind (bei *Majanthemum* ist $PK_{\text{Harnstoff}} > PK_{\text{Glycerin}}$, bei *Rhoeo* umgekehrt) fallen demgegenüber wenig ins Gewicht.

Das Ergebnis wird, wie bei *Chara* und *Rhoeo*, im Sinne der Lipoidfilterhypothese ausgewertet; neu kommt hinzu, daß die Besonderheiten verschiedener PR als *spezifisch* bezeichnet und als solche ausdrücklich auf das „anatomische Substrat des Permeationsvorganges“ bezogen werden.

Mit analoger Methode wurden dann in schneller Folge zahlreiche weitere Objekte aus allen Teilen des Pflanzenreiches untersucht, und heute liegen dank den Bemühungen der Schulen HÖFLERs und COLLANDERs etwa 30 verschiedene spezifische PR vor. Sie sind vielfältig voneinander unterschieden, wenigstens hinsichtlich einzelner Glieder der Reihe (hauptsächlich Harnstoff und Glycerin), und eine umfängliche Literatur befaßt sich mit dem Vergleich der Reihen und der Deutung der Differenzen („vergleichende Permeabilitätsforschung“). Über die „Objekttypen der Permeabilität“ wird gesondert berichtet (III F h). Hier ist lediglich die Frage zu stellen, inwieweit es berechtigt ist, die plasmometrisch bestimmten PK bzw. PR als *quantitativen* Ausdruck der osmotischen Anelektrolytaufnahme zu werten.

Die plasmometrische Methode arbeitet ebenso wie die grenzplasmolytische an *plasmolysierten* Zellen und wird somit von den gleichen Einwänden betroffen, die bereits im vorigen Abschnitt gegen die grenzplasmolytische Arbeitsweise geltend gemacht wurden. Sie fallen um so mehr ins Gewicht, als hier nicht mehr lediglich die Rückdehnungsgeschwindigkeiten *verglichen*, also *Relativ*maße erhalten werden, sondern der *absolute* osmotische Wert (bei Grenzplasmolyse) in die Formel zur Berechnung von PK eingesetzt wird. Dazu muß aber der Plasmolysegrad *ausschließlich* eine Funktion der Konzentration an osmotisch wirksamer Substanz in der Vacuole sein. Wie in den letzten Jahren eingehend nachgewiesen werden konnte, ist diese Voraussetzung nicht gegeben.

1. Zellen, die in hypertonischen Lösungen von Saccharose (die während kurzfristiger Versuche nicht nennenswert aufgenommen wird) plasmolysiert werden, verändern ihr Protoplastenvolumen nach Übertragung in schwächer oder auch stärker hypertonische Lösungen *nicht proportional* der Veränderung des Plasmolyticums („osmotisch nicht veränderlicher Volumanteil" in der Vacuole; Bogen und Prell 1953 an *Oenothera*, Neuber unveröffentlicht an *Taraxacum*).

2. In Saccharoselösung plasmolysierte Zellen dehnen sich aus, obwohl keine Saccharose endosmiert. Die Protoplastenausdehnung, die plasmometrisch als Erhöhung des O_g-Wertes in Erscheinung tritt, beruht vielmehr auf einer nichtosmotischen (stoffwechselabhängigen) *Wasser*aufnahme (Bogen 1953 *plasmometrisch* an *Rhoeo* und *Taraxacum*, Bogen und Follmann 1955 *grenzplasmolytisch* und mittels Mikroosmometer bestimmt an *Melosira*).

3. Bereits die unplasmolysierte Zelle enthält einen gewissen, ansehnlichen Betrag an Wasser, das mit nichtosmotischen Kräften in der Vacuole festgehalten wird (Bogen 1953 an *Rhoeo*, Bogen und Follmann 1955 an *Melosira*, Neuber unveröffentlicht an *Taraxacum*). In Extremfällen werden plasmometrisch 0,37 mol, im Osmometer nur 0,21 mol gemessen.

Aus diesen experimentellen Fakten geht klar hervor, daß der grenzplasmolytisch oder plasmometrisch ermittelte osmotische Wert durchaus kein verläßliches Maß für die im Zellsaft befindliche Menge osmotisch wirksamer Substanz ist, und daß die Ausdehnung plasmolysierter Protoplasten in der Regel ganz erheblich stärker ist als der Endosmose entspricht (vgl. hierzu die „extrem hohe Zuckerpermeabilität" der Diatomeen, die zu $^9/_{10}$ durch nichtosmotische Wasseraufnahme vorgetäuscht wird, Bogen und Follmann 1955).

Demgegenüber hat Höfler schon frühzeitig Wert auf den Nachweis gelegt, daß die gemessene Erhöhung des O_g-Wertes dem Diffusionsgesetz folgt („Zeitproportionalität"). Dieser Nachweis erscheint dann als geführt, wenn die zu verschiedenen Zeiten an der gleichen Zelle gemessenen PK von gleicher Größe sind, und Höfler hat demgemäß nur solche Versuche ausgewertet, in denen diese Forderung realisiert war. Umfangreiche statistische Untersuchungen (die in den Einzelheiten noch nicht publiziert sind, vgl. Prell 1953) haben indessen ergeben, daß in wenigstens 50% aller Fälle die zweite gemessene PK signifikant verschieden ist von der zuerst gemessenen. Wenn daher die 1. und 2. PK manchmal übereinstimmen, so schließt das ein Interferieren nichtosmotischer Prozesse und damit eine Überhöhung der PK nicht aus.

Daß diese Veränderung der PK nur in den seltensten Fällen auf echter *Permeabilitäts*erhöhung beruht, geht bereits aus Höflers Untersuchungen hervor; besonders deutlich wird es in den „Partialversuchen" Hofmeisters (1935): die so erhaltenen PK sind zum Teil (aber keineswegs regelmäßig) um eine Zehnerpotenz verschieden von denen der Totalversuche.

Damit wird aber zumindest die quantitative Seite der PK und PR ernstlich in Frage gestellt, und alle Folgerungen, die aus quantitativen Differenzen hinsichtlich der Permeationsmechanismen wie des anatomischen Substrates des Permeationsvorganges gezogen worden sind, können nur mit Vorbehalten zur Kenntnis genommen werden.

Zugleich wird ersichtlich, daß wir, entgegen den oft als selbstverständlich geäußerten Ansichten zahlreicher Autoren, bisher noch *keine Methode besitzen, die es uns erlaubt, die osmotische Aufnahme der Anelektrolyte allein und exakt zu messen.* Am ehesten wird noch die an *Beggiatoa* geübte Turgescenzmethode strengen Anforderungen gerecht, obgleich auch hier schon einige jener Bedenken ins Feld geführt werden können, die dann bei *Chara*, *Rhoeo* und *Majanthemum* so schwerwiegend werden. In vielen Fällen messen wir mit den genannten Methoden die *Summe* aus osmotischer und nichtosmotischer Aufnahme bzw. einen wechselnden Bruchteil dieser Summe. Die — für die hier diskutierten Fragen unerwünschte — nichtosmotische Aufnahme kann aber nicht *generell* abgezogen werden; vielmehr bedarf es in jedem Einzelfalle experimenteller Prüfung, ob und welche nichtosmotischen Vorgänge mitwirken, und in welchem Maße sie an der Gesamtaufnahme beteiligt sind.

6. Die Mechanismen der osmotischen Aufnahme von Anelektrolyten.

Alle Versuche, auf Grund experimenteller Daten Aussagen über die Mechanismen der osmotischen Anelektrolytaufnahme und die beteiligten Grenzschichtstrukturen zu machen, müssen von der Tatsache ausgehen, daß allenfalls *Näherungswerte* bekannt sind, die in vielen Fällen und in durchaus wechselndem Ausmaß zu hoch sind. Wenn daher von seiten der Schule HÖFLERs die Meßwerte als echte, unverfälschte PK betrachtet werden, kann es nicht wundernehmen, daß zahlreiche Differenzen in den Einzelheiten die gemeinsamen Prinzipien, die der osmotischen Aufnahme und dem Grenzschichtbau fraglos zugrunde liegen, überdeckt werden. Wenn gar noch nach Verschiedenheiten „planmäßig Ausschau gehalten" wird (HÖFLER und STIEGLER 1921, 1930) und der Vergleich der verschiedenen PR auf die PK gerade jener Diosmotica gegründet wird, die am stärksten variieren, so ist vorauszusehen, daß schließlich nur noch „Permeabilitätstypen" gegeneinandergestellt und Differenzen diskutiert werden, ein Unterfangen, das sich der kausalanalytischen Lösung des Problems als nicht sehr förderlich erwiesen hat.

Demgegenüber hat BOGEN (1950) versucht, in Anrechnung der (prinzipiellen) Ungenauigkeit der PK nicht die quantitativen Differenzen der einzelnen PR in den Vordergrund zu stellen, sondern die Gemeinsamkeiten aufzuzeigen. Als Grundlage dienten hierbei die gleichen experimentellen Daten, mit denen die vergleichende Permeabilitätsforschung ihre Diskussionen führte, d. h. die plasmolytisch ermittelten PR (die *Beggiatoa*- und die *Chara*-Reihe wurden wegen der andersartigen Methode nicht mit einbezogen).

Diese PR enthalten freilich nur 7 Glieder, nämlich die PK von Erythrit, Malonamid, Glycerin, Harnstoff, Methylharnstoff, Äthylenglykol und Acetamid. 31 solcher PR wurden geordnet, hier aber nicht nach steigenden Glycerinwerten (wie z. B. ELO 1937), sondern nach den Malonamidwerten, in der Annahme, daß beim Malonamid weniger mit „Sekundäreffekten" aller Art zu rechnen sei als beim Glycerin (eine Annahme, die freilich strenger Prüfung auch nicht standhalten wird).

Während so die „Bezugsgrößen" $PK_{Malonamid}$ im logarithmischen Netz auf der Diagonalen angeordnet werden, zeigt sich, daß auch die PK der übrigen Stoffe auf ähnliche Geraden zu liegen kommen; diese laufen indessen nicht parallel, sondern konvergieren nach den Bereichen höherer Permeabilität (hierzu Abb. 4 und 5; bezüglich der Einzelheiten wird auf III F h verwiesen). Naturgemäß fallen eine Anzahl der PK nicht genau auf „ihre" Gerade; dabei ist aber zu berücksichtigen, daß ja die PK in Wirklichkeit durch nichtosmotische und andere Vorgänge (wie z. B. Permeabilitätsveränderung unter dem Einfluß der Diosmotica,

Bogen 1938, 1940, 1941) erheblich verfälscht, und zwar unterschiedlich verfälscht, sein dürften. Die gezogenen Geraden müssen also ebenso wie die plasmolytisch bestimmten PK als Darstellungen des durchschnittlichen Verhaltens genommen werden.

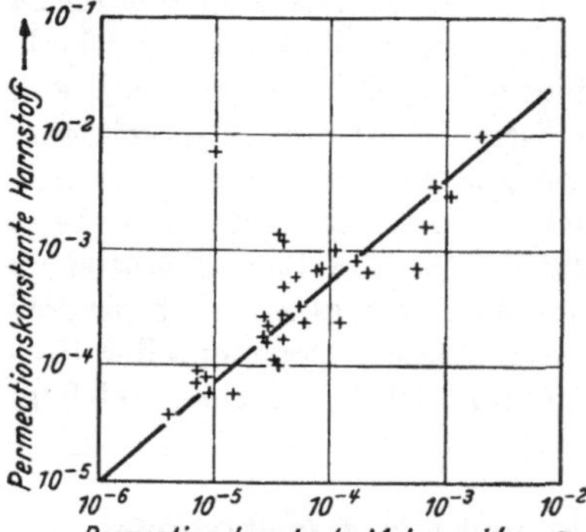

Abb. 4. Permeationskonstanten des Harnstoffs von 32 Objekten (Ordinate), dargestellt in Abhängigkeit von den entsprechenden Permeationskonstanten des Malonamids (Abszisse). Logarithmischer Maßstab. (Aus Bogen 1950.)

Wie unter III F h genauer ausgeführt ist, spiegeln diese Geraden 2 Abhängigkeitsbeziehungen wieder:

1. Das Konvergieren der Geraden gegen die Bereiche höherer Permeabilität repräsentiert das Ultrafilterprinzip, wie theoretisch zu fordern und experimentell (Collander 1926 an Kollodiummembranen) erhärtet worden ist.

2. Die Reihenfolge der Geraden hingegen paßt nicht zum Ultrafilterprinzip: Sie lautet Erythrit $<$ Malonamid $<$ Glycerin $<$ Harnstoff $<$ Methylharnstoff $<$ Glykol Acetamid (I); nach der Molekülgröße bzw. R_{MD} der Diosmotica geordnet müßte sie lauten Erythrit $<$ Malonamid $<$ Glycerin $<$ Methylharnstoff $<$ Acetamid $<$ Glykol $<$ Harnstoff (II). Die Abweichung zeigt, daß die sog. lipoidlöslichen Stoffe in der Aufnahme gefördert sind; die Reihenfolge II drückt mithin den Einfluß des Verteilungskoeffizienten $K_{\text{Öl}}$ aus.

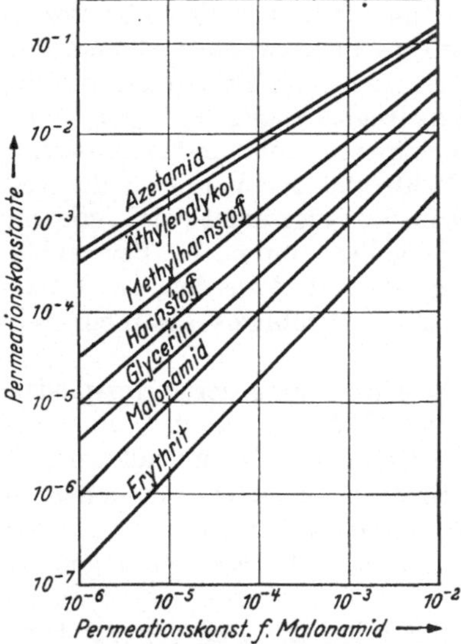

Abb. 5. Permeationsgeraden für die Stoffreihe Acetamid bis Erythrit, bezogen auf die Permeationsgerade für Malonamid. Die Geraden wurden aus den einzelnen PK der betreffenden Stoffe von 31 Objekten erhalten (vgl. Abb. 4). Abszisse: $PK_{\text{Malonamid}}$. Ordinate: PK der übrigen Stoffe. Logarithmischer Maßstab. (Aus Bogen 1950.)

Formal lassen sich beide Beziehungen durch die Formel

$$P = \frac{f'(K_{\text{Öl}})}{f''(R_{MD})}$$

erfassen, d. h. die gleiche Formel, die aus den Untersuchungen von Ruhland und Hoffmann sowie Schönfelder und den Daten von Collander und Bärlund 1929 und Collander 1954 abgeleitet wurde (S. 234).

Höfler schließt aus diesen beiden Abhängigkeitsbeziehungen, daß zwei verschiedene Permeabilitätsprinzipien nebeneinander bestehen, nämlich „Lipoidpermeabilität“ und „Porenpermeabilität“. Collander erkennt als direkt wirksam allein das Löslichkeitsprinzip an und sieht im Ultrafiltereffekt lediglich die Auswirkung eines Orientierungsfaktors. Ruhland und Hoffmann hingegen, desgleichen Schönfelder, stellen das Ultrafilterprinzip in den Vordergrund und sehen in der Förderung der lipoidlöslichen Stoffe nur einen Spezialfall der Ultrafiltration, hervorgerufen durch Adsorptionsphänomene. Bogen geht noch einen Schritt weiter, indem er sowohl den Verteilungskoeffizienten $K_{\text{Öl}}$ als auch die Adsorptionsphänomene als Auswirkungen der *zwischenmolekularen Kräfte* betrachtet.

Seit dem Erscheinen der ersten Arbeiten über das Permeabilitätsproblem konnten vor allem von seiten der Physikochemiker die Vorstellungen über die

Diffusion im molekularen Bereich sehr viel weiter entwickelt und präzisiert werden. Das neue Bild haben DAVSON und DANIELLI (1944) sowie SPANNER in diesem Band (S. 125ff.) beschrieben. Danach ist in Lösung die (thermische) Beweglichkeit der Moleküle, die ja die Grundlage der Diffusion ist, gegenüber der Diffusion in Gasen wesentlich beeinträchtigt durch die „short-range"-Kräfte, die als „zwischenmolekulare Kräfte", „VAN DER WAALSsche Kräfte", „Kohäsionskräfte", Wasserstoffbrücken usw. bekannt sind. Die „Hydrodiffusion" solcher Moleküle ist besonders behindert, wenn die zwischenmolekularen Kräfte (ZMK) groß sind, weniger bei geringen ZMK. Nun sind aber die ZMK nicht nur um so stärker, je größer das Molekül ist, sondern auch, je *mehr hydrophile Gruppen* es enthält, während der Energiegehalt hydrophober Gruppen geringer ist. Daraus erklärt sich die stärkere Hemmung der hydrophilen Stoffe bei der Aufnahme gegenüber den hydrophoben bzw. die *relative* Förderung hydrophober („lipoidlöslicher") Stoffe. Des weiteren wirken sich die geringeren ZMK lipoidlöslicher Stoffe in höheren Verteilungskoeffizienten und gesteigerter Adsorbierbarkeit aus. Die scheinbare Abhängigkeit der PK vom Verteilungskoeffizienten K spiegelt also nur die Abhängigkeit vom Energiegehalt, von den ZMK wieder; der experimentell bestimmte Verteilungskoeffizient darf nur *stellvertretend* für eine theoretische Größe verwendet werden (Ausmaß und Anordnung der ZMK) und beweist durchaus nicht die Mitwirkung von Lösungsvorgängen bei der Permeation.

Diese Vorstellungen erlauben es, den Permeationsvorgang einheitlich auf das *Prinzip der räumlichen Durchdringung* zurückzuführen, das nunmehr auch die schnellere Permeation der lipoidlöslichen Stoffe einschließt. Zugleich damit entfällt eine Schwierigkeit, deren die Lipoid- wie die Lipoidfilterhypothese bislang nicht gut Herr werden konnte, nämlich die Existenz von Lipoiden *als Lösungsmittel* nachzuweisen. Wohl hatte COLLANDER (1937) versucht, die Grenzschicht als lipoidhaltigen Coacervatfilm vorzustellen, wobei die Penetration durch die Poren zwischen den Filmpalisaden erfolgen sollte und zwischen echter Lösung und Adsorption nicht mehr streng unterschieden werden könne. Indessen würden solche „kondensierten" Filme Grenzflächenspannungen von 35—55 dyn/cm zeigen, während alle verläßlichen Messungen der Grenzflächenspannung lebender Zellen zu Werten unter 1 dyn/cm geführt haben (HARVEY 1954). Die Grenzflächen können mithin allenfalls aus „gasförmigen" Filmen bestehen, und bei solchen kann von echter Lösung durchaus nicht mehr gesprochen werden. Wahrscheinlich sind die Grenzschichten aus Filmen von Proteinen mit hydrophilen und hydrophoben Gruppen aufgebaut, doch über Einzelheiten vermögen PK keine Hinweise mehr zu geben.

Die Gesetzmäßigkeiten, die sich im molekularen Bereich bei der Permeation abspielen, hatte bereits PFEFFER 1877 erkannt; ihre Neuformulierung bzw. Präzisierung war freilich nur möglich auf Grund der neuen Vorstellungen, die die Physikochemie uns in die Hand geben konnte, und weiterhin nur, nachdem klar geworden war, daß die sog. PK lediglich Näherungswerte darstellen und somit von vornherein nur als korrekturbedürftig in Rechnung gestellt werden können. Sie erklären zweifellos nicht alle Details der Differenzen in den PR; wir sind aber heute berechtigt, sie als Ausfluß des Interferierens nichtosmotischer und anderer Prozesse anzusehen, nicht jedoch als Absonderlichkeiten der *Permeationsmechanismen*.

B. Nichtosmotische Aufnahme von Anelektrolyten.

Die auf S. 237 genannten Kriterien der osmotischen Aufnahme seien — in veränderter Reihenfolge — nochmals aufgeführt:

1. Die aufgenommene Stoffmenge ist proportional dem Konzentrationsgefälle.

2. Die Diosmotica dringen bis zum Konzentrationsausgleich ein, so daß „Gleichverteilung“ zwischen Außenmedium und Zellsaft hergestellt wird.

3. Die Aufnahme ist „unspezifisch“, d. h. die Zelle kann nicht zwischen verschiedenen Diosmotici unterscheiden, soweit diese gleiche Molekülgröße und gleiche Lipoidlöslichkeit aufweisen.

Demgegenüber ist die nichtosmotische Aufnahme durch das Fehlen dieser 3 Kriterien gekennzeichnet. Gleichwohl sind damit nicht lediglich negative Merkmale gegeben, denn man kann sie auch wie folgt formulieren:

1. Die Aufnahmerate ist unabhängig vom Konzentrationsgefälle und unter konstanten Bedingungen konstant.

2. Die Diosmotica können — wenn die nichtosmotischen Prozesse genügend lange ablaufen — über den Ausgleich des Konzentrationsgefälles hinaus aufgenommen werden. Das führt zu einer *Akkumulation* und Aufrechterhaltung eines Ungleichgewichtes entgegen dem Konzentrationsgradienten. Dies wiederum erfordert von seiten der Zelle einen Aufwand von Energie, die aus dem Stoffwechsel bezogen werden muß. Die nichtosmotische Stoffaufnahme ist also stoffwechselabhängig und wird durch stoffwechselbeeinflussende Agentien verändert.

3. Die Aufnahmemechanismen sind spezifisch auf das aufzunehmende Substrat eingestellt, oftmals wird Stereospezifität beobachtet.

Diesem dritten Kriterium kann nach anderweitigen Erfahrungen noch ein viertes angefügt werden: Im Gegensatz zur osmotischen Aufnahme besteht hier oftmals eine ausgeprägte Konkurrenz gleichzeitig eindringender Stoffe.

Die 4 Punkte sind identisch mit den 4 Kriterien, die ROSENBERG und WILBRANDT (1952) für die nichtosmotische Aufnahme aufgestellt haben[1]. Die Autoren haben lediglich Punkt 2 allgemeiner und zugleich schärfer gefaßt:

2. Beeinflussung der Aufnahmerate durch Enzymeffektoren — soweit deren Wirkung nicht lediglich in einer Veränderung der Konzentrationsdifferenz für die eindringenden Stoffe besteht. (Diese Einschränkung ist erforderlich z. B. für den Fall, daß aufgenommener Zucker unter Mitwirkung hemmbarer Enzyme verarbeitet und so aus dem Konzentrationsgefälle wieder entfernt wird, wodurch neuer Nachschub *osmotisch* erfolgen kann.)

Damit eröffnen sich mehrere Möglichkeiten, die Auswirkung nichtosmotischer Aufnahmemechanismen experimentell zu erfassen, von denen der Nachweis der Stoffakkumulation am einfachsten zu führen ist (a). Wenn in der Versuchszeit eine solche nicht eintritt, ist die Situation schwieriger, weil sich nunmehr das Nebeneinander von osmotischer und nichtosmotischer Aufnahme störend bemerkbar macht. Überdies hat sich herausgestellt, daß die letztere wesentlich empfindlicher gegen Eingriffe aller Art ist; die Zelle nimmt noch auf osmotischem Wege Stoffe auf, wenn ihre nichtosmotische Aufnahme schon längst zum Erliegen gekommen ist. Daher muß man entweder die Aufnahme solcher Stoffe untersuchen, die auf osmotischem Wege nicht oder nur sehr wenig aufgenommen werden, insbesondere Zucker (b), oder man vernichtet die nichtosmotische Aufnahme durch Blockierung des Stoffwechsels mit den modernen, oft recht spezifisch wirkenden Stoffwechsel- bzw. Koppelungsgiften; man erhält dann, falls die osmotische Aufnahme unbeeinflußt bleibt, deren Anteil an der Gesamtaufnahme gesondert und kann die nichtosmotische Aufnahmerate daraus berechnen (c). Die in der Literatur vorliegenden Befunde sind im folgenden nach diesen Gesichtspunkten geordnet.

[1] Nach ROSENBERG (1954) kommen noch folgende hinzu: 5. Der Temperaturkoeffizient ist ebenso hoch wie bei enzymatischen Reaktionen; 6. Sättigungsphänomene.

1. Experimentelle Befunde.

a) Akkumulation.

Die ersten Hinweise ergaben sich schon verhältnismäßig frühzeitig aus vereinzelten, manchmal mehr beiläufigen Beobachtungen. So fanden z. B. STEWARD und DAVIS (1929), daß *Chara*-Zellen Harnstoff zu speichern vermögen; analoges wies BAZIN (1949) für Aminoalkohole nach, gleichfalls an *Chara*.

b) Zuckeraufnahme.

Diese hat schon immer größeres Interesse gefunden, denn man versprach sich von der Analyse eine Auflösung jenes Paradoxons, auf das schon viele Permeabilitätsforscher immer wieder hingewiesen haben, daß nämlich die für den Stoffwechsel so überaus wichtigen Zucker osmotisch kaum aufgenommen werden. PHILLIS und MASON (1937) konnten an *Gossypium*-Blattscheibchen zeigen, daß die Glucoseaufnahme von der Atmung bzw. der O_2-Gegenwart abhängt; ähnlich verhalten sich *Atropa belladonna* (WEATHERLEY 1954), *Scenedesmus* (TAYLOR 1950) u. a.

Genauer hat RIEDER (1951) die Glucoseaufnahme durch Hefezellen untersucht; er wies nach, daß 75—80% der gesamten Zuckeraufnahme nichtosmotisch erfolgten, „aktiv" unter Aufwand von Energie, die aus der Atmung bzw. Gärung stammt. Demgemäß hat eine Hemmung des Energiestoffwechsels die Verhinderung der nichtosmotischen Glucoseaufnahme zur Folge.

Diese Nachweise sind freilich mehr allgemeiner Art, indem die oben angeführten Forderungen 1—4 nicht sämtlich streng erfüllt sind. Insbesondere ist bei den C-autotrophen Zellen mit dem Interferieren eines lichtabhängigen, über Phosphorylierungen laufenden Glucoseeinbaues zu rechnen (vgl. KANDLER 1954). Aber in den letzten Jahren hat sich ROTHSTEIN mit seinen Mitarbeitern in systematischen Untersuchungen dem Problem der Glucoseaufnahme in Hefezellen gewidmet und zeigen können, daß in der Tat alle 4 Kriterien gegeben sind.

Im Anschluß an ROTHSTEINs zusammenfassende Darstellungen (1954a, b) ergibt sich folgendes:

Hefezellen nehmen Hexosen in sehr unterschiedlichem Maße auf: Glucose, Mannose und Fructose außerordentlich rasch, Galaktose, Sorbose und Arabinose hingegen, trotz weitgehender Übereinstimmung in Molekulargewicht und Verteilungskoeffizient, nur zu $^1/_{1000}$ der Glucoserate. Diese „Substratspezifität" stimmt überraschend genau überein mit der Substratspezifität der Hexokinase; mehr noch: Hefe kann an Galaktose adaptiert werden (SPIEGELMANN, REINER und COHNBERG 1947), wobei sie das Enzym Galaktokinase ausbildet — zugleich wird die vordem galaktose-impermeable Zelloberfläche für Galaktose durchlässig. Durch Uranylionen wird die Zuckeraufnahme und die Zuckerverarbeitung blockiert bzw. die Überführung der Glucose in den Tricarbonsäurecyclus verhindert, ohne daß jedoch der TSC selbst beeinträchtigt wird. Als uranylionenbindende Komplexe haben Polyphosphate (Nucleinsäuren, ATP) der Zelloberfläche zu gelten (bei der Gärung; bei der Atmung wirken ferner Carboxylgruppen von Oberflächenproteinen mit). Der erste Schritt der Glucoseaufnahme besteht also offensichtlich in einer enzymatischen, ATP-bedürftigen Oberflächenreaktion, an der die Hexokinase maßgeblich beteiligt ist. Auf weitere Einzelheiten soll an dieser Stelle nicht eingegangen werden, zumal die Untersuchungen noch in vollem Fluß sind.

c) Anwendung von Stoffwechselgiften.

Hatte bereits ROTHSTEIN bei seinen biochemischen Untersuchungen Enzyminhibitoren verwendet, so versuchten BOGEN und FOLLMANN (1955) das Nebeneinander von osmotischer und nichtosmotischer Anelektrolytaufnahme unter dem

Einfluß anderer Gifte mit *osmotischen* Methoden zu analysieren. Ausgangspunkt waren die Beobachtungen über die nichtosmotische Wasseraufnahme bei *Rhoeo* und *Taraxacum* (Bogen 1953), aber auch bei Diatomeen. Unter dem Einfluß von Enzymeffektoren (Atmungs- bzw. Koppelungsgiften) wie Natriumazid, Dinitrophenol (DNP) u. a. wird sie unterbunden, ohne daß indessen die osmotische Wasseraufnahme beeinträchtigt wird. Bogen und Follmann konnten bei analoger Versuchsanstellung zeigen, daß das gleiche auch für solche Diosmotica wie Acetamid, Harnstoff, Glycerin u. a. m. gilt: Bereits 10^{-3} mol Azid bzw. 10^{-4} mol DNP setzen die — grenzplasmolytisch gemessene — Gesamtaufnahme der genannten Stoffe erheblich herab. Durch diese Dosen wird indessen nur die gesamte nichtosmotische Aufnahme verhindert (sowie der größte Teil der nichtosmotischen Wasseraufnahme). Eine zehnmal höhere Dosis ändert daran nichts mehr (lediglich der Rest der nichtosmotischen Wasseraufnahme — der in allen Versuchen gleich groß ist — wird noch vernichtet). Es bleibt aber noch eine ansehnliche osmotische Anelektrolytaufnahme bestehen, die selbst durch noch höhere Dosen der Stoffwechselgifte nicht beeinflußt wird. Daraus kann geschlossen werden 1. daß die Permeabilität durch die Enzymeffektoren nicht herabgesetzt wird, 2. daß die osmotische Aufnahme von einer nichtosmotischen, stoffwechselabhängigen, konzentrationsunabhängigen Aufnahme solcher Anelektrolyte wie Harnstoff und Glycerin u. a. überlagert wird. Berechnungen ergaben, daß diese 30—60% der Gesamtaufnahme ausmacht, womit sie nahe an das Ausmaß der nichtosmotischen Glucoseaufnahme der Hefe heranreicht (Rieder 1951). Diese Zahlen machen es besonders deutlich, wie wenig berechtigt es ist, die plasmolytisch (oder anderweitig) bestimmte Gesamtaufnahme mit der osmotischen gleichzusetzen, und wie hinfällig eine ganze Reihe von Folgerungen sind, die aus scheinbaren Permeabilitätsdifferenzen gezogen worden sind.

2. Mechanismen der nichtosmotischen Anelektrolytaufnahme.

Das bisher vorliegende Versuchsmaterial beweist ausreichend, daß auch Anelektrolyte der verschiedensten Art unter Energieaufwand nichtosmotisch in das Plasma und die Vacuole der Pflanzenzelle aufgenommen werden können. Es genügt indessen bei weitem nicht, um begründete Vorstellungen über die Mechanismen zu entwickeln, die zu dieser Art der Aufnahme führen. Hierzu sind wir auf die Untersuchungen über Ionen- und Farbstoffaufnahme und vor allem auf Untersuchungen an tierischen Objekten angewiesen.

Da die Anelektrolyte nicht ionisiert sind, können alle jene Reaktionen außer Betracht bleiben, die unter Mitwirkung von Hauptvalenzkräften ablaufen. Die daneben verbleibenden Möglichkeiten zu Wechselwirkungen zwischen den aufzunehmenden Molekülen und den Molekülen der Grenzschicht erstrecken sich daher in der Hauptsache auf Enzymwirkungen und Adsorptionsvorgänge. Den Vorstellungen, die in letzter Zeit hierüber entwickelt worden sind, ist gemeinsam, daß die Moleküle des eindringenden Substrates an zelleigene Substanzen gebunden werden, die dann als „Träger“ (carrier) eine Translocierung erfahren; sie weichen jedoch voneinander ab hinsichtlich der Bindungsart und der Form der Translocierung.

Beim „enzymatisch kontrollierten“ Transport (Rosenberg 1948, Rosenberg und Wilbrandt 1952) wirken zwei Enzyme („ein Enzympaar“) mit, die durch die Zellgrenzschicht getrennt sind. Das äußere Enzym bindet das Substrat an den Träger (vermutlich ein Coenzym). Dadurch wird es in eine „membranlösliche Transportform“ überführt. Der Substrat-Carrier-Komplex bewegt sich dann im Inneren der Grenzschicht (oder auch im Binnenplasma) gemäß seinen *Konzentrationsgradienten*. An der Innenseite wird dann der Komplex durch das zweite Enzym wieder aufgespalten, und der Carrier diffundiert zurück. Damit dieser allgemein formulierte Carriermechanismus zu einem „thermodynamisch aktiven Transport" führt, müssen die Diffusionsrichtungen für Substrat (allein) und Carrier (allein) entgegengesetzt sein; nur dann wird das Substrat aus einem Bereich geringen in einen Bereich höheren chemischen bzw. elektrochemischen Potentials überführt. — Hier ist es also die *Bindung* des Substrates, die Energie erfordert, während der Transport des Komplexes eine gewöhnliche Diffusion darstellt.

Gerade umgekehrt verhält es sich nach der Theorie, die GOLDACRE über die Akkumulierung gebildet hat. Nach dieser fungieren als Trägersubstanzen Proteine, die in verschiedenen Faltungszuständen vorliegen: In den äußeren Grenzschichten sind sie gestreckt und haben, da alle Seitengruppen freiliegen, die Möglichkeit, Substrat absorptiv an sich zu binden. So beladen werden sie mit der Plasmaströmung ins Zellinnere geführt und dort — unter Energieaufwand — gefaltet, wobei das Substrat desorbiert werden muß. Die entladenen Trägerproteine (nach DANIELLI 1952 Phosphatasen) gelangen dann wieder in die Zelloberfläche, wo sie unter dem Einfluß der Grenzflächenkräfte erneut gestreckt werden und neues Substrat adsorbieren können. Das Prinzip ist also ein Faltungs-/Entfaltungscyclus der Trägerproteine; die Orte der Auffaltung und der Entfaltung sind getrennt; zwischen ihnen vermittelt die Plasmaströmung. Das heißt, nach GOLDACRE ist hier der *Transport* des Substrat-Carrierkomplexes (sowie die Auffaltung der Trägerproteine) der energiebedürftige Teil der nichtosmotischen Aufnahme.

Die Wirksamkeit dieses Mechanismus konnte GOLDACRE nicht allein an tierischen Zellen, sondern auch an *Neurospora*-Hyphen und an Wurzelhaaren demonstrieren, hier allerdings für Farbstoffe wie Neutralrot; es besteht indessen keine prinzipielle Schwierigkeit, sich auf diese Weise auch den Transport von Anelektrolyten vorzustellen.

Eine Aufnahme von Anelektrolyten nach dem Enzym- und nach dem Faltungsmechanismus erfüllt die 4 auf S. 246 genannten Kriterien und muß daher als nichtosmotisch bezeichnet werden.

Es erscheint ohne Schwierigkeiten möglich, die Anschauungen ROSENBERGS und GOLDACRES zu vereinigen zu einem Schema, in dem die Bindung an das Carriersystem enzymatisch oder adsorptiv erfolgt, der Komplex selbst aber contractil ist oder diffundieren kann. Die Selektivität solcher Systeme und ihr Arbeiten in verschiedenen Richtungen (Aufnahme, Abgabe und Translocierung innerhalb der Zelle, etwa zwischen Oberfläche und Mitochondrien oder Zellkern) ist ohne weiteres verständlich; die Koppelung mit dem Stoffwechsel würde freilich an verschiedenen Stellen stattfinden.

Nach BOGEN (1956) und NEUBER (1956, im Druck) ist damit zu rechnen, daß Anelektrolyte auch metaosmotisch aufgenommen werden: in der Vacuole befinden sich labile Kolloide, die ansehnliche Mengen Wasser und gelöster Anelektrolyte zu binden vermögen. Diese Vacuolenkolloide — vermutlich Glykoproteide — zerfallen, wenn die Atmung unterbunden wird (sowie bei der Preßsaftherstellung). Dabei werden die vordem gebundenen Stoffe frei und verlassen die Vacuole auf osmotischem Wege. Bei Wiedereinsetzen der Atmung werden sie erneut aufgebaut; sie binden einen Teil des Vacuolenwassers bzw. der darin gelösten Stoffe, und neues Wasser usw. strömt osmotisch nach. Danach wäre der Aufnahmevorgang selbst osmotisch; lediglich nach stattgehabter Osmose, *meta*osmotisch, wird in der Vacuole osmotisch wirksame Substanz gebunden und aus den Konzentrationsgradienten entfernt. Energiebedürftig wären somit die Vacuolenkolloide bzw. deren Zustand, der über ihr Bindungsvermögen entscheidet.

Wahrscheinlich existieren auch noch anderweitige, vorerst unbekannte Formen der meta- und nichtosmotischen Aufnahme — so finden sich Hinweise dafür, daß ganze Mitochondrien als Transportaggregate fungieren (vgl. ROBERTSON, dieser Band, S. 449) —, doch ist es schon heute klar, daß die nichtosmotische Anelektrolytaufnahme wesentlich vielgestaltiger ist als die osmotische; es wird in jedem einzelnen Falle sorgfältiger Analyse bedürfen, um ihren Anteil und die beteiligten Vorgänge aufzuklären.

Die Beeinflussung der Anelektrolytaufnahme durch Ionen ist vor etwa 15 Jahren Gegenstand lebhafter Diskussionen gewesen. Ihnen lagen experimentelle Daten über die Veränderung der Aufnahmeraten („Permeationskonstanten") in Gegenwart verschiedener Ionen zugrunde. Da sie zumeist plasmometrisch bzw. grenzplasmolytisch erhalten wurden (z. B. DE HAAN 1935, BOGEN 1940, 1941, KREUZ 1941), wurden die Befunde als *Permeabilitäts*änderungen diskutiert. Es fanden sich in der Tat sehr erhebliche Veränderungen, meist Herabsetzung der Konstanten. In einigen Fällen ließen sich dabei Ionenreihen (lyotrope Reihen, Adsorptionsreihen usw.) erkennen (DE HAAN, BOGEN), doch waren diese nicht nur von Objekt zu Objekt, sondern auch bei ein und demselben Objekt für die Diosmotica verschieden. Daher ließen sich einheitliche Gesichtspunkte nur selten gewinnen. HÖFLER wertete sie aus vom Standpunkt der „spezifischen Permeabilität". BOGEN hingegen schloß auf eine Beeinflussung des Hydratationszustandes und damit der Wegsamkeit der Plasmahäute durch die Ionen, aber auch, und mit diesen konkurrierend, der permeierenden Anelektrolyte.

Berücksichtigt man, daß die Deplasmolysegeschwindigkeit der Protoplasten nicht nur durch osmotische, sondern auch durch nichtosmotische Stoffaufnahme zustande kommt, so ergibt sich heute ein anderes Bild. Es ist anzunehmen, daß neben der Permeabilität auch der Stoffwechsel und, von ihm abhängig, die nichtosmotische Anelektrolytaufnahme betroffen wird. In der damaligen Diskussion konnte dieser Gesichtspunkt nicht berücksichtigt werden; ihre Wiederaufnahme erscheint erst dann erfolgversprechend, wenn es gelingt, die Ioneneffekte auf die osmotische und nichtosmotische Aufnahme zu trennen. Von einer Aufführung der Befunde kann daher an dieser Stelle abgesehen werden.

Literatur.

ABDERHALDEN, E., u. H. FODOR: Forschungen über Fermentwirkung. II. Studien über die Adsorption von Aminosäuren und Polypeptiden durch Tierkohle. Fermentforsch. **2**, 74 (1919). — Forschungen über Fermentwirkung. III. Weitere Studien über die Adsorption von Aminosäuren und Polypeptiden durch Tierkohle. Fermentforsch. **2**, 151 (1919). — Studien über die Adsorption von Aminosäuren, Polypeptiden und Eiweißkörpern durch Tierkohle. Beziehungen zwischen Adsorbierbarkeit und gelöstem Zustand. I. Kolloid-Z. **27**, 49 (1920).

BÄRLUND, H.: Permeabilitätsstudien an Epidermiszellen von *Rhoeo discolor*. Acta bot. fenn. **5**, 1 (1929). — BANG, L.: Mikromethoden zur Blutuntersuchung, 6. Aufl. München 1927. — BAZIN, S.: Pénétration de quelques esters d'amino-alcohols dans le suc vacuolaire de *Chara*. C. r. Soc. Biol. Paris **141**, 224—226 (1947). — BOGEN, H. J.: Untersuchungen zu den „spezifischen Permeabilitätsreihen" HÖFLERS. II. Harnstoff und Glycerin. Planta (Berl.) **28**, 535—581 (1938). — Untersuchungen über den Quellungseffekt permeierender Anelektrolyte. I. Ionenwirkung auf die Permeabilität von *Rhoeo discolor*. Z. Bot. **36**, 65 bis 106 (1940). — Untersuchungen über den Quellungseffekt permeierender Anelektrolyte. II. Ionenwirkung auf die Permeabilität von *Gentiana cruciata*. Planta (Berl.) **32**, 150—175 (1941). — Kritische Untersuchungen über Permeabilitätsreihen. Planta (Berl.) **38**, 65—90 (1950). — Beiträge zur Physiologie der nichtosmotischen Wasseraufnahme. Planta (Berl.) **42**, 140—155 (1953). — Nichtosmotische Aufnahme von Wasser und gelösten Anelektrolyten. Ber. dtsch. bot. Ges. **20** (1956). — BOGEN, H. J., u. G. FOLLMANN: Osmotische und nichtosmotische Stoffaufnahme bei Diatomeen. Planta (Berl.) **45**, 125—146 (1955). — BOGEN, H. J., u. H. PRELL: Messung nichtosmotischer Wasseraufnahme an plasmolysierten Protoplasten. Planta (Berl.) **41**, 459—479 (1953). — BRINTZINGER, H., u. H. BEIER: Die Diasolyse. Kolloid-Z. **79**, 324—331 (1937).

COLLANDER, R.: Über die Permeabilität von Kollodiummembranen. Soc. Sci. Fenn. Comment. Biol. II **6**, 1—48 (1926). — Einige neuere Ergebnisse und Probleme der botanischen Permeabilitätsforschung. Phys.-ökon. Ges. Königsberg (Pr.) **69**, 251—266 (1937). — The permeability of *Nitella* cells to non-electrolytes. Physiol. Plantarum (Copenh.) **7**, 420—445 (1954). — COLLANDER, R., u. H. BÄRLUND: Über die Protoplasmapermeabilität von *Rhoeo discolor*. Soc. Sci. Fenn. Comment. Biol. II **9**, 1—13 (1926). — Permeabilitätsstudien an *Chara ceratophylla*. II. Acta bot. fenn. **11**, 1—114 (1933).

DANIELLI, J. F.: Structural factors in cell permeability and in secretion. Symposia Soc. Exper. Biol. **6**, 1—15 (1952). — DAVSON, H., and J. F. DANIELLI: The permeability of natural membranes. Cambridge 1943.

ELO, J. E.: Vergleichende Permeabilitätsstudien. Ann. bot. Soc. zool.-bot fenn. „Vanamo" **8**, Nr 6 (1937).

FITTING, H.: Untersuchungen über die Aufnahme von Salzen in die lebende Zelle. Jb. wiss. Bot. **56**, 1—64 (1915). — Untersuchungen über isotonische Koeffizienten und ihren Nutzen für Permeabilitätsbestimmungen. Jb. wiss. Bot. **57**, 553—612 (1917). — Untersuchungen über die Aufnahme und anomale isotonische Koeffizienten von Glycerin und Harnstoff. Jb. wiss. Bot. **59**, 1—170 (1919).

GOLDACRE, R. J.: The folding and unfolding of protein molecules as a basis of osmotic work. Internat. Rev. Cytology **1**, 135—164 (1952).

HAAN, J. DE: Ionenwirkung und Wasserpermeabilität. Ein Beitrag zur Koazervattheorie der Plasmagrenzschichten. Protoplasma (Berl.) **24**, 188—197 (1935). — HARVEY, E. N.: Tension at the cell surface. Protoplasmatologia (Wien) II E **5**, 1—30 (1954). — HOAGLAND, D. R., u. A. R. DAVIS: The intake and accumulation of electrolytes by plant cells. Protoplasma **6**, 610—626 (1929). — HÖFLER, K.: Permeabilitätsstudien an Stengelzellen von *Majanthemum bifolium*. Sitzgsber. Akad. Wiss. Wien, Math.-naturwiss. Kl., Abt. I **143**, 213—264 (1934). — Unsere derzeitige Kenntnis von den spezifischen Permeabilitätsreihen. Ber. dtsch. bot. Ges. **60**, 179—200 (1942). — HÖFLER, K., u. A. STIEGLER: Ein auffälliger Permeabilitätsversuch in Harnstofflösung. Ber. dtsch. bot. Ges. **39**, 157—164 (1921). — Permeabilitätsverteilung in verschiedenen Geweben der Pflanze. Protoplasma **9**, 469—512 (1930). — HOFMEISTER, L.: Vergleichende Untersuchungen über spezifische Permeabilitätsreihen. Bibl. Bot. **113** (1935).

KANDLER, O.: Über die Beziehungen zwischen Phosphathaushalt und Photosynthese. I. Gesteigerter Glukoseeinbau im Licht als Indikator einer lichtabhängigen Phosphorylierung. Z. Naturforsch. **9**b, 625—644 (1954). — KREUZ, J.: Der Einfluß von Calcium- und Kalium-Salzen auf die Permeabilität des Protoplasmas für Harnstoff und Glycerin. Österr. bot. Z. **90**, 1—30 (1941).

MARKLUND, G.: Vergleichende Permeabilitätsstudien an pflanzlichen Protoplasten. Acta bot. fenn. **18**, 1—110 (1936). — MOND, H., u. F. HOFFMANN: Weitere Untersuchungen über die Membranstruktur der roten Blutkörperchen: Beziehungen zwischen Durchlässigkeit und Molekularvolumen. Pflügers Arch. **219**, 467 (1928).

PHILLIS, E., and T. G. MASON: The effect of light and of oxygen on the uptake of sugar by foliage leaf. Ann. of Bot., N. S. **1**, 231—237 (1937). — PRELL, H.: Untersuchungen über die Aufnahme von Anelektrolyten in Zellen di- und polyploider Pflanzen. Planta (Berl.) **41**, 480—508 (1953).

RIEDER, P.: Über die Zuckeraufnahme von Hefezellen. Ber. schweiz. bot. Ges. **61**, 539 bis 621 (1951). — ROSENBERG, T.: On accumulation and active transport in biological systems. I. Thermodynamic considerations. Acta chem. scand. (Copenh.) **2**, 14—33 (1948). — The concept and definition of active transport. Symposia Soc. Exper. Biol. 8, 27—41 (1954). — ROSENBERG, T., u. W. WILBRANDT: Enzymatic processes in cell membrane penetration. Internat. Rev. Cytology **1**, 65—92 (1952). — ROTHSTEIN, A.: The enzymology of the cell surface. Protoplasmatologia (Wien) II E **4**, 1—86 (1954). — Enzyme systems of the cell surface involved in the uptake of sugars bei yeast. Symposia Soc. Exper. Biol. 8 (1954). — RUHLAND, W., u. C. HOFFMANN: Die Permeabilität von *Beggiatoa mirabilis*. Ein Beitrag zur Ultrafiltertheorie des Plasmas. Planta (Berl.) **1**, 1—83 (1925).

SCHMIDT, H.: Plasmolyse und Permeabilität. Jb. wiss. Bot. **83**, 470—512 (1936). — SCHÖNFELDER, S.: Weitere Untersuchungen über die Permeabilität von *Beggiatoa mirabilis*, nebst kritischen Ausführungen zum Gesamtproblem der Permeabilität. Planta (Berl.) **12**, 414—504 (1931). — SPIEGELMANN, S., J. M. REINER and R. COHNBERG: The relation of enzymatic adaptation to the metabolism of endogeneous and exogeneous substrates. J. Gen. Physiol. **31**, 27—49 (1947). — STEWARD, F. C.: The absorption and accumulation of solutes by living plant cells. I. Protoplasma (Berl.) **15**, 29—58 (1932).

TAYLOR, F. J.: Effect of aeration on the absorption of glucose by a green alga. Nature (Lond.) **166**, 519—520 (1950).

WARTIOVAARA, V.: The permeability of the plasma membranes of *Nitella* to normal primary alcohols. Physiol. Plantarum (Copenh.) **2**, 184—196 (1949). — Zur Erklärung der Ultrafilterwirkung der Plasmahaut. Physiol. Plantarum (Copenh). **3**, 462 (1950). — WEATHERLEY, P. E.: Preliminary investigations into the uptake of sugars by floating leaf discs. New Phytologist **53**, 204—216 (1954).

Die Aufnahme der Farbstoffe. Vitalfärbung.

Von

H. Drawert.

Mit 10 Abbildungen.

I. Einleitung.

Für das Studium der Stoffaufnahme durch die ganze Pflanze oder durch die einzelne Zelle besitzen die Farbstoffe den großen Vorteil, daß man ihren Verbleib in der Pflanze bzw. in der Zelle visuell verfolgen kann. Es ist daher verständlich, daß die Farbstoffe bei den Untersuchungen über die Stoffaufnahme eine besondere Rolle gespielt haben. Man muß allerdings auch einige Nachteile in Kauf nehmen. Es handelt sich bei den meisten zur Vitalfärbung benutzten Farbstoffen um Substanzen, die im natürlichen Stoffaustausch der Pflanze nicht vorkommen. Ferner sind viele Farbstoffe so giftig, daß sie nur in sehr geringen Konzentrationen angewandt werden können, so daß sie in der Zelle nur zu sehen sind, wenn sie von einem Bestandteil der Zelle gespeichert werden, falls es sich nicht um fluorescierende Substanzen handelt, die man unter Umständen noch in einer Verdünnung von $1:10^{11}$ (Uranin) fluorescenzoptisch nachweisen kann. Aber auch bei der Anwendung sehr geringer Konzentrationen können Farbstoffe noch sehr giftig sein, wenn die Speicherung in lebenswichtigen Bestandteilen der Zelle erfolgt. Ein nicht zu vernachlässigender Faktor ist der geringe Reinheitsgrad der handelsüblichen Farbstoffpräparate (vgl. S. 258). Trotz dieser Nachteile werden die Farbstoffe auch heute noch wegen der Sichtbarkeit ihrer Lokalisationsorte in der Zelle oder bei Fluorochromen auch wegen der Sichtbarkeit ihrer Wanderwege in der Pflanze in der Pflanzenphysiologie viel benutzt.

Zusammenfassende Darstellungen über die Vitalfärbung finden sich bei Höber (1926), Gellhorn (1929), Gellhorn und Régnier (1936), Becker (1936), Zeiger (1938), Kiyono, Sugiyama und Amano (1938), Küster (1939), Guilliermond und Gautheret (1940), Brooks und Moldenhauer-Brooks (1941), Guilliermond (1941), Strugger (1949a, b), Milovidov (1949, 1954).

Die bereits erwähnte Giftigkeit der Farbstoffe macht es zur Pflicht, den Vitalitätsgrad der Zelle während des Versuchs zu prüfen. Der Grad der Giftigkeit einer Substanz kann von Objekt zu Objekt und auch beim selben Objekt, je nach dem physiologischen Zustand sehr verschieden sein. Bei einer „Vitalfärbung" müssen wir verschiedene Stufen der Verträglichkeit unterscheiden. Strugger (1936b) schlägt folgende Nomenklatur vor: Eine Färbung ist inturbant, wenn weder unmittelbar nach der Farbstoffbehandlung noch bei Weiterkultur der Pflanze eine sichtbare Schädigung der gefärbten Zellen eintritt. Zeigen sich bei der Weiterkultur der Pflanze allmählich Schädigungen, so wird die Färbung als turbant bezeichnet. Bei einer perturbanten Färbung sind bereits während der Färbung und unmittelbar danach pathologische Veränderungen der Zellen zu beobachten. Ist die Giftwirkung so stark, daß die Zellen bereits während des Färbeprozesses absterben, so spricht Strugger von einer disturbanten Färbung.

Die erste eingehende Beschreibung eines Vitalfärbungsversuches finden wir bei Unger (1848). Mit der Transpirationsmethode läßt Unger den angesäuerten Saft aus den roten Früchten von *Phytolacca decandra* in weißblütigen Hyacinthen aufsteigen und erhält so eine rote Vacuolenfärbung in den Parenchymzellen längs der Schraubengefäße. Der älteste Anilinfarbstoff, das Mauvein, wurde 1854 entdeckt, aber erst 1886 benutzt PFEFFER Anilinfarbstoffe zum Studium der Stoffaufnahme und der Stoffspeicherung. Vor allem bei den Untersuchungen über die Permeabilitätseigenschaften der Zelle haben die Anilinfarben in der Folgezeit eine große Rolle gespielt. Es sei nur an die Arbeiten von OVERTON, RUHLAND, HÖBER, BETHE, NIRENSTEIN und COLLANDER erinnert.

Zum Verständnis der Vitalfärbungen müssen wir uns zunächst kurz mit den physikalisch-chemischen Eigenschaften der Anilinfarben befassen.

II. Die physikalisch-chemischen Eigenschaften der Farbstoffe.

A. Einteilung der Farbstoffe.

Nach ihrer chemischen Zusammensetzung kann man die Anilinfarbstoffe in Azo-, Triphenylmethan-, Azin-, Oxazin-, Thiazinfarbstoffe u. a. einteilen. Für die Physiologie ist aber die Einteilung nach physikalisch-chemischen Gesichtspunkten von größerer Bedeutung. Je nachdem, ob der die chromophore Gruppe führende Bestandteil eine Säure oder eine Base ist, unterscheiden wir saure oder anionische und basische oder kationische Farbstoffe und eine dritte Gruppe wird von den amphoteren Farbstoffen gebildet. Als Beispiele seien der Triphenylmethanfarbstoff Säurefuchsin (I), der Azinfarbstoff Neutralrot (II) und der Oxazinfarbstoff Prune pure (III) gebracht.

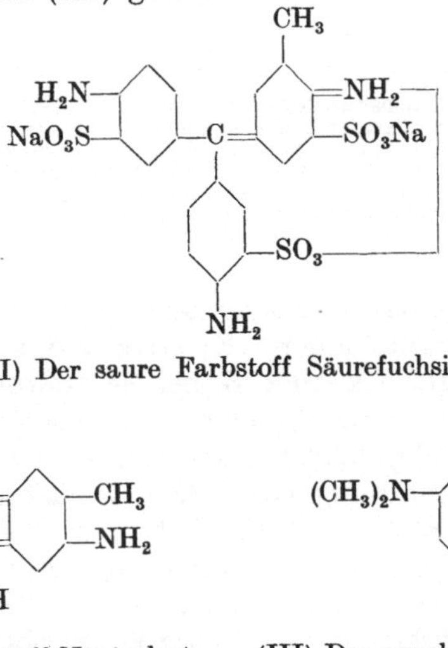

(I) Der saure Farbstoff Säurefuchsin.

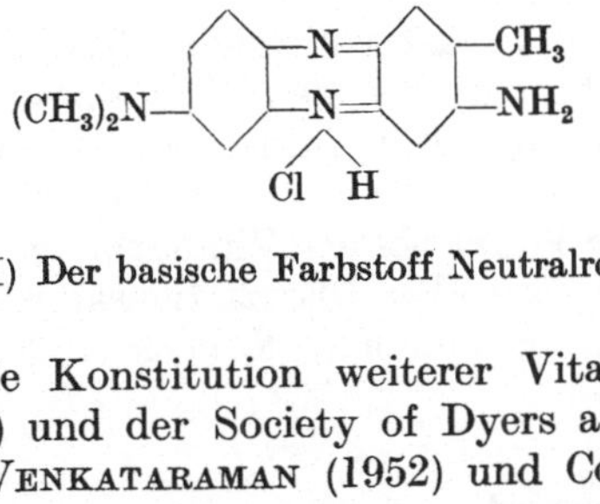

(II) Der basische Farbstoff Neutralrot.

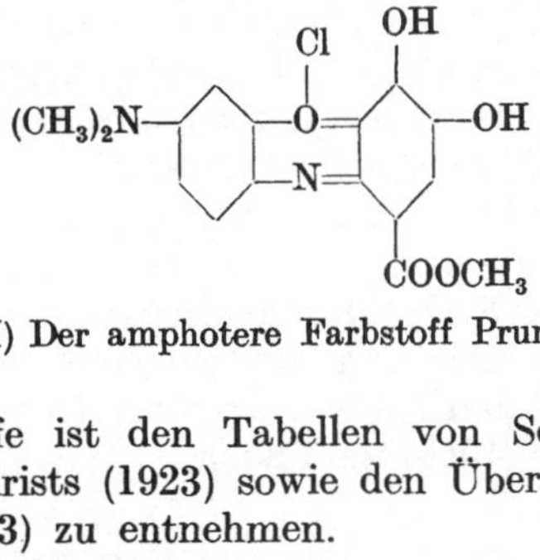

(III) Der amphotere Farbstoff Prune pure.

Die Konstitution weiterer Vitalfarbstoffe ist den Tabellen von SCHULTZ (1931) und der Society of Dyers and Colourists (1923) sowie den Übersichten von VENKATARAMAN (1952) und CONN (1953) zu entnehmen.

Bei Farbstoffen unbekannter Konstitution genügt in vielen Fällen bereits eine Capillaranalyse zur Bestimmung des elektrischen Ladungssinns. Läßt man nach RUHLAND (1912) einen Tropfen einer wäßrigen Farbstofflösung auf einem Stück Filtrierpapier sich ausbreiten, dann wandert ein negativ geladener saurer Farbstoff gleichzeitig mit dem Lösungsmittel, bei einem positiv geladenen basischen Farbstoff wandert das Lösungsmittel dagegen voraus. Genauer sind Bestimmungen der Wanderungsrichtung mit Hilfe der Elektrophorese mit 220 V Gleichstrom. Hierbei benutzt man entweder Filtrierpapierstreifen als Elektroden (FÜRTH 1929, DRAWERT 1938b) und vergleicht die Steighöhen in den beiden Papierstreifen, oder man bestimmt die Wanderungsrichtung und die Geschwindigkeit im Lösungsmittel mit einer Apparatur, wie sie BETHE (1920) beschreibt.

B. Dissoziationsverhältnisse.

Der Dissoziationsgrad eines Farbstoffes kann nicht nur für seine Löslichkeit in hydrophilen und hydrophoben Medien, sondern auch für die Aufnahme und Speicherung durch die lebende Zelle von ausschlaggebender Bedeutung sein. Die basischen Farbstoffe setzen sich im allgemeinen aus einer schwachen Farbbase und einer starken Säure — meist Salzsäure — zusammen, so daß wir in den Lösungen, neben der elektrolytischen Dissoziation im sauren Bereich, im alkalischen Gebiet eine starke hydrolytische Spaltung antreffen (IRWIN 1926, CZAJA 1934, DRAWERT 1938 u. f.). In Abb. 1 sind die wahrscheinlichen Dissoziationsverhältnisse für den häufig benutzten Vitalfarbstoff Neutralrot in Anlehnung an ein Schema von DRAWERT (1940 und 1948) und Hinweise von KÖLBEL (1948) schematisch dargestellt. Bei vielen Farbstoffen mit Indicatoreigenschaften kann

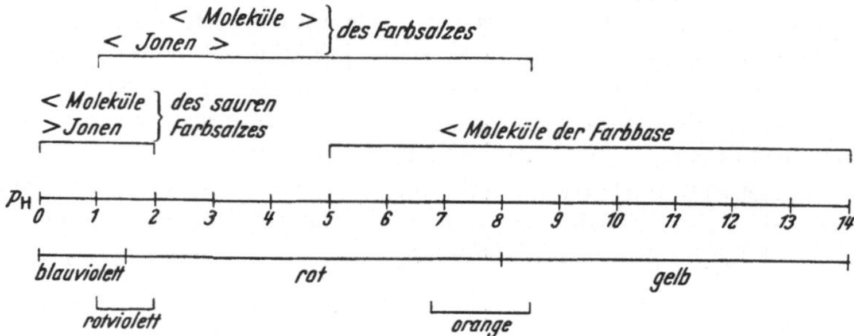

Abb. 1. Die wahrscheinlichen Dissoziationsverhältnisse des Neutralrots. Unter p_H 1 dürften nur Ionen des sauren Farbsalzes und über p_H 8,5 nur undissoziierte Moleküle der Farbbase vorliegen. Unter p_H 1 ist der Farbton blau bis blauviolett. Über p_H 8,5 sind die Lösungen gelb und unbeständig. In dem zwischen den beiden Umschlagspunkten liegenden p_H-Bereich sind neben Farbkationen und Molekülen der beiden Farbsalze auch Moleküle der Farbbase vorhanden, deren Anteil mit steigendem p_H-Wert zunimmt. Der Farbton ändert sich zwischen p_H 1,0 und 8,5 je nach dem Mischungsverhältnis der einzelnen Anteile von rotviolett (sauer) über rot nach orange (alkalisch).

die in Abhängigkeit vom p_H-Wert auftretende Farbtonänderung damit zusammenhängen, daß sich Farbkation (-anion), Farbsalz und Farbbase (-säure) in ihrer Färbung unterscheiden. Dazu kann z. B. bei Neutralrot noch die Bildung eines abweichend gefärbten sauren Salzes kommen. Die sauren Farbstoffe, besonders die sulfosauren, sind im allgemeinen stärker dissoziiert als die basischen. Sie dürften in einer wäßrigen Lösung vorwiegend in fast vollständig dissoziiertem Zustand vorliegen (COLLANDER 1942). Faktoren, die den Dissoziationsgrad ändern, beeinflussen auch häufig die Vitalfärbung. Eine Änderung der cH oder der Zusatz von Elektrolyten werden sich in vielen Fällen auf den Dissoziationsgrad des Farbstoffes auswirken, so daß dieser Punkt bei der Vitalfärbung besonders berücksichtigt werden muß. Einige Angaben über Dissoziationskonstanten von Vitalfarbstoffen sind bei IRWIN (1926a), BRUCH und NETTER (1930) und KINZEL (1954) zu finden.

C. Dispersitätsgrad.

Änderung der cH, Zusatz von Elektrolyten können sich aber auch auf den Dispersitätsgrad als weiteren Faktor, der eine Vitalfärbung beeinflussen kann, auswirken. Die Teilchengröße hängt außerdem häufig von der Farbstoffkonzentration, dem Alter der Lösung und der Temperatur ab. Mit dem Dispersitätsgrad ändert sich nicht selten der Farbton, die Lage der Absorptionsbanden verschiebt sich (SCHEIBE 1938). Für den wichtigen Vitalfarbstoff Acridinorange hat BARTELS (1954) die mit steigender Farbstoffkonzentration auftretende Bandenverschiebung untersucht. Auch in diesem Fall beruht wohl die Bandenverschiebung auf einer Assoziation der Farbstoffkationen.

Auf den Dispersitätsgrad eines Farbstoffes hat man vorwiegend aus Diffusionsversuchen geschlossen. In erster Linie wurden dafür 5—20%ige Gelatinesäulen benutzt (SCHULEMANN 1917, COLLANDER 1921, TRAUBE und SHIKATA 1923, GELLHORN 1928, DRAWERT 1937a, 1941, 1951, STRUGGER 1939a, RUGE 1940) oder man bestimmte die freie Diffusion in Wasser (FÜRTH 1927, NISTLER 1929, 1930, 1931, AXMACHER 1933, HANUT und FAUTREZ 1935). Ferner wurde mit Hilfe der Dialyse (RUHLAND 1908, 1912, HÖBER 1909, v. MOELLENDORFF 1916, BENOIST und Mitarbeiter 1929) und der Ultrafiltration (FISCHER 1929, CZAJA 1930, 1934) versucht, Auskunft über die Teilchengröße zu erhalten. Alle diese Methoden sind aber mit vielen Fehlerquellen behaftet, so daß sie uns bestenfalls nur relative Aussagen in vergleichenden Untersuchungen liefern können. So zeigen die für eine Anzahl von Farbstoffen in Gelatinegel und in reinem Wasser bestimmten Diffusionskoeffizienten große Abweichungen. Das Verhältnis der beiden Koeffizienten schwankt zwischen 2 und 10 (HERZOG und POLOTZKY 1914). Eine eingehende Kritik der aufgeführten Methoden findet sich bei FAUTREZ und LISON (1937) und LISON und FAUTREZ (1939). Vergleichende Diffusionskurven für die wichtigsten Vitalfarbstoffe in einem 10%igen Gelatinegel sind bei DRAWERT (1941, 1951) zu entnehmen.

D. Oberflächenaktivität.

Da eine Änderung der Oberflächenspannung sich auch auf die Vitalfärbung auswirken kann, ist des öfteren die Oberflächenaktivität der Vitalfarbstoffe bestimmt worden. Ein zunächst vermuteter Zusammenhang zwischen Oberflächenaktivität und Giftwirkung hat sich nicht bestätigt (TSCHERNORUTZKY 1912, HÖBER 1914). Die meisten Farbstoffe zeigen nur eine geringe Oberflächenaktivität; bei den basischen ist sie meist stärker ausgeprägt als bei den sauren (TSCHERNORUTZKY 1912, TRAUBE 1912, AXMACHER 1933). Neuere Untersuchungen zu dieser Frage liegen von GOLDACRE (1952) vor.

E. Löslichkeitsverhältnisse.

Von großer Wichtigkeit für die Vitalfärbung sind die Löslichkeitsverhältnisse der Farbstoffe in hydrophilen und hydrophoben Lösungsmitteln. Im allgemeinen sind bisher nur solche Farbstoffe für Vitalfärbungen benutzt worden, die bis zu einem gewissen Grade in Wasser löslich sind.

Wenn die Zelle für einen Stoff ausgesprochene Speicherfähigkeit besitzt, genügen schon Spuren davon im umgebenden Medium, um eine sichtbare Vitalfärbung zu erzielen, wie DRAWERT (1955a) für ein aliphatisch N-substituiertes Aminopyren zeigen konnte, dessen Wasserlöslichkeit so gering ist, daß die Lösung im normalen Licht völlig farblos bleibt und im UV nur eine äußerst schwache bläulichweiße Fluorescenz gerade noch zu erkennen ist.

Nach GRAFFI (1940) kann man zur Fluorochromierung geeignete Kohlenwasserstoffe durch Lösung in Glycerin und Zugabe von Serum in einer wäßrigen Lösung erhalten und so zur vitalen Fluorochromierung benutzen. Auch bei Gegenwart geringer Konzentrationen von Seifen oder kolloidaler Elektrolyte gehen normalerweise wasserunlösliche Substanzen in Lösung (ROSS 1951).

Von den basischen Farbstoffen sind im allgemeinen die Farbsalze gut wasserlöslich, die Farbbasen dagegen nur schlecht oder gar nicht, so daß der cH des wäßrigen Lösungsmittels und damit dem elektrolytischen und noch mehr dem hydrolytischen Dissoziationsgrad des Farbstoffes für die Löslichkeit eine Bedeutung zukommt. Bei der geringeren hydrolytischen Spaltbarkeit der meist stärker elektrolytisch dissoziierten sauren Farbstoffe spielt die cH — von wenigen Ausnahmen wie Eosin und Erythrosin abgesehen — keine entsprechende Rolle (FAURÉ-FREMIET 1912, DRAWERT 1940, 1941, SPEK 1944). Umgekehrt verhält sich dementsprechend die Löslichkeit in hydrophoben Medien, worauf bereits OVERTON (1899) hingewiesen hat.

Mit Farbstofflösungen in hydrophoben Medien sind bisher kaum Vitalfärbungen durchgeführt worden. Nur YAMAHA (1937) bringt Angaben über

Paraffinöl als Lösungsmittel. Danach dürfte durchaus die Möglichkeit bestehen, auch mit Farbstoffen, die nicht in Wasser löslich sind, bei Benutzung eines unschädlichen hydrophoben Lösungsmittels Vitalfärbungen zu erzielen.

Verteilungskoeffizient.

Neben den Untersuchungen über den Dispersitätsgrad eines Farbstoffes in wäßriger Lösung spielte in der Diskussion über die Permeabilitätstheorien die „Lipoidlöslichkeit" eines Farbstoffes eine große Rolle, so daß auch auf diesem Gebiet eine umfangreiche Literatur vorliegt. Besonders häufig ist der Verteilungskoeffizient eines Farbstoffes zwischen einer hydrophoben und einer hydrophilen Phase und seine Beeinflussung durch andere Faktoren in Ausschüttelversuchen bestimmt worden.

v. MOELLENDORFF (1918) weist darauf hin, daß die Löslichkeit eines basischen Farbstoffes in Lecithinxylol beträchtlich größer ist, wenn der Farbstoff nicht in Substanz, sondern aus einer wäßrigen Lösung geboten wird. Diese Erscheinung ist ohne Zweifel auf die hydrolytische Spaltung zurückzuführen. Die dadurch freiwerdende Farbbase ist bedeutend lipophiler als das Farbsalz (SPEK 1944, 1951). Daraus geht weiter hervor, daß die cH der wäßrigen Phase einen großen Einfluß auf den Verteilungskoeffizienten haben muß. Viele in der älteren Literatur anzutreffenden Widersprüche über die „Lipoidlöslichkeit" eines Farbstoffes sind auf die Nichtbeachtung dieser Tatsache zurückzuführen. Die lipophilen Eigenschaften eines basischen Farbstoffes nehmen mit steigender cH ab, die eines sauren Farbstoffes zu. Dementsprechend wird auch der Verteilungskoeffizient bei basischen Farbstoffen sich mit fallender cH zugunsten der hydrophoben Phase verschieben und sich bei den sauren Farbstoffen in Abhängigkeit von der cH entgegengesetzt verhalten (ROBERTSON 1908, REINDERS 1913, HERZFELD 1917, McCUTCHEON und LUCKÉ 1924, IRWIN 1925/28, 1927, BAILEY und ZIRKLE 1931, DRAWERT 1938 u. f., STRUGGER 1939 u. f.).

Die amphoteren Farbstoffe wie das Prune pure weisen im mittleren p_H-Bereich eine stärkere Lipophilie auf, die sowohl nach der sauren wie nach der basischen Seite hin abnimmt (DRAWERT 1938b, BETHE 1950).

Wie die cH wirken sich auch andere Faktoren, die die Dissoziation beeinflussen, auf den Verteilungskoeffizienten aus. So kann ein Salzzusatz den Verteilungskoeffizienten zugunsten der hydrophilen Phase (IRWIN 1926c, 1930) oder auch der hydrophoben Phase (BUNGENBERG DE JONG und BANK 1940) verschieben bzw. auch ohne Einfluß sein (CZAJA 1936). Wie das Salz wirkt, hängt von der Dissoziationskonstante und dem Charakter — ob sauer oder basisch — des benutzten Farbstoffes sowie von der Wertigkeit besonders des Kations und der Art des Anions des geprüften Salzes ab (DRAWERT 1938 u. f., BUNGENBERG DE JONG und BANK 1940).

Für die Deutung von Vitalfärbungsversuchen ist es auch noch von Interesse, daß ein Zusatz von Proteinen (ROBERTSON 1908) oder von Alkaloiden (RÜTER und BORNSTEIN 1925, WASSILJEWA 1938) den Verteilungskoeffizienten zugunsten der hydrophilen Phase verschieben kann.

Der Verteilungskoeffizient hängt aber auch von der Art der hydrophoben Phase ab. Das Lösungsvermögen völlig hydrophober Medien wie Benzol, Xylol mit oder ohne Cholesterin wird durch Zusatz von lipophilen Substanzen, die aber einige hydrophile Gruppen enthalten, wie Lecithin, Ölsäure, Diamylamin, unter Umständen ganz beträchtlich erhöht (NIRENSTEIN 1920). Dies tritt besonders deutlich hervor, wenn man den Verteilungskoeffizienten in Abhängigkeit von der cH der wäßrigen Phase untersucht (DRAWERT 1938 u. f.). Bei Ölsäure und

Diamylamin kann natürlich eine Salzbildung eine höhere „Löslichkeit" vortäuschen (BEUTNER und Mitarbeiter 1930); aber überblickt man die vorliegende Literatur, so kommt man doch zu dem Schluß, daß eine gewisse Hydrophilie der organischen Phase sich auf alle Fälle auf den Verteilungskoeffizienten so auswirkt, daß er zugunsten der natürlich vorwiegend noch hydrophoben Phase verschoben wird. So nimmt das ganz schwach hydrophile Chloroform im Ausschüttelversuch mehr Farbstoff aus der wäßrigen Phase auf als die völlig hydrophoben Flüssigkeiten Xylol oder Toluol. Nach BUNGENBERG DE JONG und BANK (1940) lassen sich die Farbstoffe mit Äthylacetat leichter ausschütteln als mit Benzol oder Toluol. Ein Zusatz von Substanzen mit polaren Gruppen macht auch das Benzol geeigneter zum Ausschütteln. Nach ROSS (1951) kommt auch der Oberflächenaktivität der zugesetzten Stoffe eine gewisse Bedeutung für den Verteilungskoeffizienten zu.

Vor allem bei den hydrophoben Medien mit schwach hydrophilen Zusätzen scheint die Verteilung des Farbstoffes zwischen den beiden Phasen nicht auf reinen Lösungseffekten zu beruhen, da es sich immer wieder zeigt, daß in diesem Fall die Verteilung nicht dem HENRYschen Satz folgt. Eher treffen die Gesetzmäßigkeiten der Adsorptionsvorgänge zu (LOEWE 1912, REINDERS 1913, v. MOELLENDORFF 1918, NIRENSTEIN 1920, ZIPF 1927, HÖBER und PUPILLI 1931).

F. Adsorbierbarkeit.

Ein Vorgang, der besonders für die technischen Färbeprozesse eine Bedeutung hat, ist die Adsorption von Farbstoffen an andere Körper. Aber auch bei der Vitalfärbung spielt die Adsorbierbarkeit nicht nur für die Färbung der Zellwände eine Rolle, sondern ist ohne Zweifel auch an Speicherungsvorgängen in der Zelle beteiligt, nur daß heute die in der Zelle ablaufenden Prozesse noch nicht so klar zu übersehen sind wie etwa die Färbung der Zellwand. Entsprechend der technischen Bedeutung der Adsorbierbarkeit liegt auch auf diesem Gebiet eine kaum noch zu übersehende Literatur vor. Im Rahmen dieses Referates können aber nur einige Grundphänomene zur Sprache kommen.

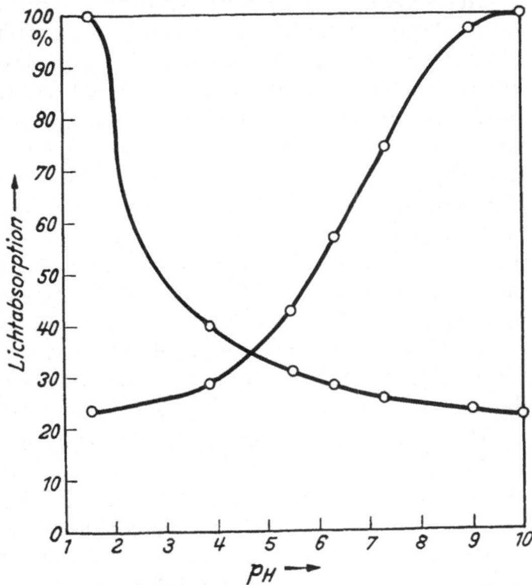

Abb. 2. Farbstoffspeicherung von Gelatine in Abhängigkeit vom p_H-Wert der Farblösungen. Abszisse: p_H-Werte der Farbstofflösungen. Ordinate: Lichtabsorption als Maß der Speicherungsintensität. Die oben links beginnende Kurve bezieht sich auf Säurefuchsin (saurer Farbstoff) und die unten links beginnende Kurve auf Toluidinblau (basischer Farbstoff). Der Schnittpunkt beider Kurven entspricht annähernd dem IEP der Gelatine. (Nach DRAWERT 1937.)

Für die Adsorptionsvorgänge sind folgende Faktoren von grundlegender Bedeutung: elektrische Ladung und Dichte des Adsorbens sowie elektrische Ladung, Dissoziationsgrad und Dispersitätsgrad des Farbstoffes. Negativ geladene Körper wie Kieselsäure (BERL und PFANNMÜLLER 1924, SEKI 1932), Kollodiumfolie (SEKI 1933) färben sich mit basischen Farbstoffen, positiv geladene dagegen mit sauren (RHEINBOLDT und WEDEKIND 1923, hier weitere Literatur, DOLADILHE 1932). Nach SPEK (1939) soll allerdings Kieselsäuregel außer basischen auch saure Farbstoffe gut adsorbieren. Amphotere Körper wie Proteine adsorbieren je nach der cH des umgebenden Mediums sowohl basische wie saure Farbstoffe. Diese Fähigkeit benutzt man zur Bestimmung des isoelektrischen Punktes amphoterer Substanzen (Abb. 2, vgl. auch MESZ 1956).

Für die Klärung mancher Vitalfärbungseffekte kann es von Bedeutung sein, daß das Farbbindungsvermögen der Proteine von ihrer Zustandsform abhängt. Durch Erwärmen oder Vergiften geschädigtes Eiweiß (Hühnereiweiß, Pferdeserum) nimmt durch einen dünnwandigen Kollodiumschlauch viel mehr Methylenblau auf als ungeschädigtes (ALEXANDROW und NASSONOV 1939). Durch Hitze (90° C) denaturiertes Tabakmosaikvirus-Eiweiß vergrößert seine Farbstoffaufnahmefähigkeit wahrscheinlich durch das Freiwerden von Nucleinsäuren um das 300fache (OSTER und GRIMSSON 1949). Eialbumin adsorbiert in monomolekularer Schicht 8mal soviel Farbstoff als in globulärer Form, da ungefaltete Proteinmoleküle eine viel größere Oberfläche zur Adsorption anderer Moleküle besitzen als gefaltete (GOLDACRE 1952).

Wie die Löslichkeit der Farbstoffe in Lösungsmitteln, so wird auch die Adsorbierbarkeit nicht nur durch die cH, sondern auch durch Salze und andere Stoffe stark beeinflußt. Hierbei handelt es sich wohl allgemein um eine Adsorptionsverdrängung. Doch kann hier nicht näher darauf eingegangen werden. Es sei auf folgende Literatur verwiesen: ZEIGER (1938), BERSIN (1946), SINGER (1952).

G. Farbton und Fluorescenzerscheinungen.

In den Abschnitten über Dissoziationsverhältnisse und Dispersitätsgrad wurde schon darauf hingewiesen, daß die elektrolytische und hydrolytische Spaltung sowie die Teilchengröße den Farbton einer Lösung bestimmen können. Auch viele Farbtonänderungen, die durch Wechsel des Lösungsmittels, Änderung der Farbstoffkonzentration, Adsorptionsvorgänge hervorgerufen werden, lassen sich wenigstens zum Teil auf die oben erwähnten Faktoren zurückführen. Für die Vitalfärbung wichtig ist z. B. der Konzentrationseffekt des Acridinorange, dessen Fluorescenz in der wäßrigen Lösung mit zunehmender Konzentration von Grün nach Kupferrot umschlägt (STRUGGER 1940b). Diesen Konzentrationseffekt zeigt das Acridinorange aber nur im dissoziierten Zustand. Die nicht dissoziierte Farbbase weist ihn nicht auf (STRUGGER 1947). Andere Farbstoffe wie Berberinsulfat und Auramin fluorescieren überhaupt erst im adsorbierten Zustand stärker.

Der Farbton kann auch durch Reduktions- und Oxydationsvorgänge geändert werden (Redoxindikatoren). Eins der bekanntesten Beispiele ist das Janusgrün B, das bei Reduktion von Blau über Rot nach farblos umschlägt. Die farblose Stufe weist aber eine grünliche Fluorescenz auf (DRAWERT 1953).

Eine viel diskutierte Erscheinung ist die sog. „Metachromasie". Darunter verstehen wir seit EHRLICH (1877) die Fähigkeit eines Farbstoffes, verschiedene Zell- und Gewebselemente in verschiedenen Farbtönen anzufärben. Toluidinblau ist der in dieser Richtung bisher am eingehendsten untersuchte Farbstoff. Es kann hier aber nur auf die Literatur verwiesen werden: LISON und FAUTREZ (1939), BANK und BUNGENBERG DE JONG (1939), SPEK (1940), LISON und MATSAARS (1950), GILLISSEN (1953).

H. Reinheitsgrad der handelsüblichen Farbstoffe.

Für die Reproduzierbarkeit von Vitalfärbungsversuchen ist es unbedingt notwendig, auf den Reinheitsgrad der benutzten Präparate zu achten. Derselbe Farbstoff kann sich je nach Herkunft auf Grund der unterschiedlichen Beimengungen sehr verschieden verhalten. So war nach IRWIN (1925/28) die Aufnahmegeschwindigkeit von Brillantkresylblau durch *Nitella* für ein Präparat von Grübler größer als für ein Präparat der National Aniline Chemical Company. Nach DRAWERT und METZNER (1955) enthalten mehrere geprüfte Brillantkresylblauproben Nilblau als Verunreinigung, und zwar in wechselnder Menge, so daß man mit ihnen unterschiedliche Ergebnisse erhält. Ebenso verhalten sich Toluidinblaupräparate, besonders in bezug auf ihre metachromatischen Eigenschaften sehr unterschiedlich (BALL und JACKSON 1953). Ein handelsübliches Janusgrün B-Präparat hatte nach BRENNER (1953) einen Reinheitsgrad von 72%. Diese wenigen Beispiele mögen genügen, um darzulegen, wie wichtig eine chromatographische Analyse der benutzten Farbstoffpräparate ist (vgl. Abb. 3). Häufig trifft man bei den handelsüblichen Präparaten auch unter demselben Namen chemisch ganz verschiedene Stoffe an (DRAWERT und GUTZ 1953, DRAWERT 1955b), oder das im Handelsnamen angegebene Anion ist gar nicht darin enthalten. So konnten nach GUTZ (1955) in keinem der geprüften „Nilblausulfat"-Präparate Sulfationen in einer entsprechenden Menge nachgewiesen werden. In Abb. 3 ist ein Papierchromatogramm von 5 Pyroninpräparaten verschiedener Herkunft und Bezeichnung, sowie von zwei mit Pyronin chemisch verwandten Farbstoffen (Rhodamin S und Acridinrot) wiedergegeben. Die Aufnahme ist mit UV gemacht. Aus dem Chromatogramm geht deutlich hervor, wie wichtig eine genaue Bezeichnung des für Vitalfärbungen benutzten Farbstoffes, besonders auch die Angabe der Herkunft ist.

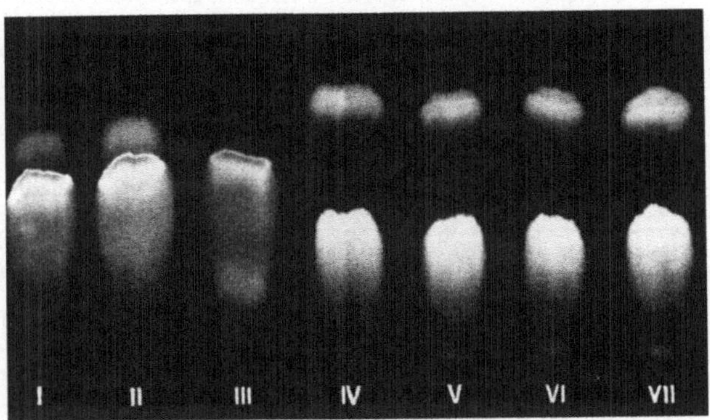

Abb. 3. Ein Papierchromatogramm von 5 verschiedenen Pyroninpräparaten, einem Rhodamin S- und einem Acridinrotpräparat. Aufnahme im UV. I Pyronin (Grübler); II Pyronin extra (Grübler); III Pyronin B echt stand. (Bayer); IV Pyronin stand. (Bayer); V Pyronin (Merck); VI Rhodamin S (Grübler); VII Acridinrot (Grübler).

III. Methoden der Vitalfärbung.

In dem Kapitel „Der p_H-Wert des Zellsaftes" im 1. Band sind bereits zwei Methoden der Vitalfärbung erwähnt worden, so daß wir uns hier kurz fassen können. Im einzelnen muß auf das Praktikum von STRUGGER (1949b) verwiesen werden. Im wesentlichen handelt es sich um 3 Methoden: Das Immersionsverfahren, das Injektionsverfahren und die Transpirationsmethode.

A. Immersionsverfahren.

Seit PFEFFER (1886) kommen die Farbstoffe in 0,001 oder 0,0001 %igen wäßrigen Lösungen zur Anwendung. Als Lösungsmittel wird destilliertes, Regen- oder Leitungswasser angegeben, die sich in ihren p_H-Werten beträchtlich unterscheiden können. Viele der sich widersprechenden Angaben in der älteren Literatur sind wohl auf die Nichtbeachtung der cH der Farbstofflösungen zurückzuführen. Für die Reproduzierbarkeit von Vitalfärbungsversuchen ist für viele Farbstoffe die Angabe des p_H-Wertes der benutzten Lösungen unerläßlich. Zur genauen Einstellung des p_H-Wertes werden häufig verdünnte Phosphatpuffer benutzt, deren Zusammensetzung bei STRUGGER (1949b, S. 134) zu entnehmen ist. Bei längerer Versuchsdauer darf aber nicht außer acht gelassen werden, daß die Phosphate — besonders das tertiäre — allmählich in die Zelle eindringen und den p_H-Wert des Zellinneren verschieben.

Die Objekte werden für eine bestimmte Zeit in oder auf der Farblösung schwimmen gelassen und dann in dem farblosen Lösungsmittel mit entsprechendem p_H-Wert beobachtet. Der p_H-Wert ist hierbei wieder besonders zu beachten, da eine Änderung der Außen-cH, auch die eines farblosen Mediums, die Farbstoffverteilung innerhalb der vorher vitalgefärbten Zelle in kürzester Zeit verändern kann (STRUGGER 1935, 1936; DRAWERT 1940). Selbst der Abschluß mit einem Deckglas kann die Farbstoffverteilung wahrscheinlich über eine durch den Stoffwechsel bedingte cH-Verschiebung oder auch Reduktion des Farbstoffes beeinflussen (STRUGGER 1936, DRAWERT 1953, 1954).

Bei ganzen Gewebestücken empfiehlt es sich, die Farblösung in die Objekte mit der Zentrifugen- (WEBER 1927) oder der Vakuummethode (KELLER und GICKLHORN 1928) zu infiltrieren.

Die Beurteilung der Färbung erfolgt bei Diachromen im Mikroskop bei Tageslicht, größter Blendenöffnung und gehobenem Kondensor. Nur bei „Überstrahlung" des Objektes kann man einwandfrei entscheiden, was in der Zelle gefärbt ist. Bei Benutzung von Lampenlicht empfiehlt es sich, ein Tageslichtfilter zwischenzuschalten, um den Farbton richtig und vergleichbar beurteilen zu können. Bei Fluorochromen benutzt man als Lichtquelle eine Kohlenbogenlampe oder eine Quecksilberhöchstdrucklampe unter Zwischenschaltung von UV- oder Blaufiltern und Anwendung entsprechender Ocularsperrfilter (Näheres s. STRUGGER 1949a). Bei Beschreibung des Farbtons des Fluorescenzlichtes muß das benutzte Ocularsperrfilter mit angegeben werden.

Tabelle 1. *Pinnularia viridis mit Toluidinblau 1:10000 in Phosphatpuffern vitalgefärbt. Prozentsatz der Zellen mit Inhaltsfärbung bei verschiedenen p_H-Werten nach verschiedenen Färbezeiten.* (Nach HIRN 1953.)

Färbedauer	pH 6,0	pH 7,1	pH 8,0	pH 9,1	pH 11
2 min . .	0	0	6	50	100
15 min . .	0	8	45	—	—
45 min . .	50	—	—	—	—

Ein nicht zu vernachlässigender Faktor ist die Färbungszeit. Bei den mehr oder weniger giftigen basischen und lipoidlöslichen sauren Farbstoffen werden zwar im allgemeinen nur Kurzfärbungen bis zu 15 min durchgeführt. Die nichtlipoidlöslichen Sulfosäurefarbstoffe benötigen aber mehrere Stunden, bis es zu einer sichtbaren Färbung der Zellen kommt. Besonders bei den basischen Farbstoffen kann es je nach der Färbezeit zu ganz verschiedenen Färbungsbildern kommen. Bei Diatomeen tritt hier schon innerhalb kürzester Zeit unter Umständen eine Änderung des Färbungsbildes auf, wie aus Tabelle 1 hervorgeht.

Bei Untersuchungen mit Mikroorganismen kann der Farbstoff auch den festen Nährböden zugesetzt werden (DIETRICH 1929).

B. Injektionsverfahren.

Farbstoffe, die im Immersionsverfahren von der lebenden Zelle nicht bis zur Sichtbarkeit aufgenommen werden, können mit Hilfe des Mikromanipulators injiziert werden. Bei der Pflanzenzelle erweist sich allerdings die starre Zellwand als großes Hindernis für Injektionsversuche. An Pflanzenzellen sind solche Versuche vor allem von PLOWE (1931) und HOFMEISTER (1940a—c) durchgeführt worden.

C. Transpirationsmethode.

Farbstoffe, die mit dem Immersionsverfahren sich nicht in der Vacuole anreichern lassen, können aber unter Umständen mit der Transpirationsmethode in den Zellen zur Speicherung gebracht werden. Bereits UNGER (1848) hat mit dieser Methode gearbeitet, später ist sie von KÜSTER (1911) wieder in die Vitalfärbungstechnik eingeführt worden. Abgeschnittene Organe werden mit der Schnittfläche in Farbstofflösungen gestellt, so daß der Farbstoff mit dem Transpirationsstrom in das Pflanzeninnere eindringen kann. Nach 12—24 Std werden dann die Organe zerschnitten und die Färbung der Gewebe und Zellen mikroskopisch geprüft. Diese Methode ist von den älteren Autoren besonders zur Erforschung der Wanderwege in der Pflanze viel benutzt worden (vgl. WIELER 1888), aber mit Ausnahme von UNGER hat man vor KÜSTER (1911) nur die Färbung der Membranen und nicht die des Zellinhaltes beachtet.

IV. Verteilung der Farbstoffe in der Zelle.

Wie aus dem Abschnitt über die Methoden hervorgeht, werden die Farbstoffe im allgemeinen in einer solchen Verdünnung angewandt, daß sie in der Schichtdicke eines mikroskopischen Präparates — zwischen Objektträger und Deckglas — häufig unter der Sichtbarkeitsgrenze liegen. Eine Ausnahme davon machen viele Fluorochrome. Wenn der Farbstoff in der Zelle also sichtbar werden soll, muß er gespeichert werden, sich demnach in der Zelle in höherer Konzentration befinden als im umgebenden Medium. Je nachdem, welcher Teil der Zelle den Farbstoff akkumuliert, können wir eine Färbung der Membran, der Vacuole, des Kernes, des Cytoplasmas, der Plastiden, der Mitochondrien oder der Sphärosomen (= Mikrosomen) erhalten. Die Farbstoffe werden zum Teil ganz spezifisch von den einzelnen Zellbestandteilen gespeichert, so daß man elektive Vitalfärbungen erhalten kann. Allerdings haben auch noch eine Reihe anderer Faktoren auf die Farbstoffverteilung in der Zelle einen Einfluß, so daß ein und derselbe Farbstoff unter verschiedenen Bedingungen verschiedene Zellbestandteile mehr oder weniger elektiv färbt. Wenn in den folgenden Abschnitten eine Übersicht über die Vitalfärbung der einzelnen Zellbestandteile mit den gebräuchlichsten Vitalfarbstoffen gebracht wird, so beziehen sich die Angaben — wenn nicht anders vermerkt — auf Kurzfärbungen mit Farbstofflösungen in Aqua dest. oder Leitungswasser und geben das unter diesen Bedingungen am häufigsten anzutreffende Färbungsbild an. Es ist damit also nicht gesagt, daß man bei jedem Objekt mit einer wäßrigen Farbstofflösung die hier angeführte Färbung erhalten wird. Wenn z. B. Neutralrot unter den Farbstoffen, die vorwiegend die Vacuole färben, angegeben wird, so besteht durchaus die Möglichkeit, daß unter gleichen Bedingungen bei bestimmten Objekten — etwa bei der Oberepidermis der Schuppenblätter von *Allium cepa* — unter Umständen eine reine Membranfärbung anzutreffen ist. In den anschließenden Abschnitten werden dann die Einflüsse von inneren und äußeren Faktoren auf das „normale" Färbungsbild und seine Veränderung besprochen werden.

A. Membranfärbung.

Zu den guten Membranfärbern können wir folgende basische Farbstoffe zählen: Azur I, Kresylechtviolett, Pyronin, Methylenblau, Safranin, Thionin, Toluidinblau. Diese Farbstoffe zeichnen sich durch einen recht hohen Dissoziationsgrad (Dissoziationskonstante der Farbbase um 10^{-3}) und damit zusammenhängend durch eine schlechte Lipoidlöslichkeit aus. Saure Farbstoffe färben nur in Ausnahmefällen die Zellwände.

B. Vacuolenfärbung.

Unter den basischen Farbstoffen färben Neutralrot, Nilblau, Brillantkresylblau und Acridinorange vorwiegend die Vacuole. Diese Farbstoffe sind schwächer dissoziiert als die unter A. genannten und besitzen auch eine bessere Lipoidlöslichkeit (Dissoziationskonstante der Farbbase um 10^{-6}). Es soll aber hier schon darauf hingewiesen werden, daß auch die unter A. sowie C. und D. aufgeführten basischen Farbstoffe die Vacuole färben können, sobald der Zellsaft besondere Speicherstoffe, z. B. Phenolabkömmlinge oder Kolloide, besitzt (Abb. 4—6). Die Färbung des Zellsaftes kann diffus sein (Abb. 7b), oder es entstehen tröpfchen- bzw. körnchenartige Entmischungen, die zu großen Tropfen zusammenfließen können (Abb. 4) oder auch größere Aggregate bilden (Abb. 5). Auch kristalline Farbstoffausfällungen kommen vor (Abb. 6). Nicht selten zeigen die Zellen eine sog. Vacuolenkontraktion, d. h. eine Ansammlung und wohl auch Quellung des Plasmas in den Zellzwickeln (Abb. 7b). Zu einer Vacuolenkontraktion scheinen vor allem Zellen zu neigen, deren Zellsaft sich mit Neutralrot erdbeerfarbig oder zinnoberrot anfärbt, also wahrscheinlich arm an Speicherstoffen ist (DRAWERT 1938a).

Zu einer diffusen Vacuolenfärbung — allerdings nur in bestimmten Zelltypen — sind aber auch stark dissoziierte sulfosaure Farbstoffe befähigt. Dazu gehören Cyanol, Orange G, Prontosil, Säurefuchsin und viele andere (vgl. KÜSTER 1911, 1921; RUHLAND 1912, COLLANDER 1921, DRAWERT 1941, 1950). Zu den ausgezeichneten Zellen gehören nach COLLANDER (1921): 1. Blumenblattzellen, 2. jugendliche, noch nicht ganz ausgewachsene Zellen und 3. das Leitbündelparenchym; aber auch bei diesen Zelltypen kann eine Vacuolenfärbung mit den sulfosauren Farbstoffen erst nach stundenlanger Farbstoffeinwirkung erzielt werden. Es muß also ein ganz anderer Speicherungsmechanismus als bei der Färbung mit basischen und lipoidlöslichen sauren Farbstoffen vorliegen. Auffällig ist auch, daß fast nie Entmischungserscheinungen oder Niederschläge in den mit sulfosauren Farbstoffen gefärbten Vacuolen beobachtet worden sind.

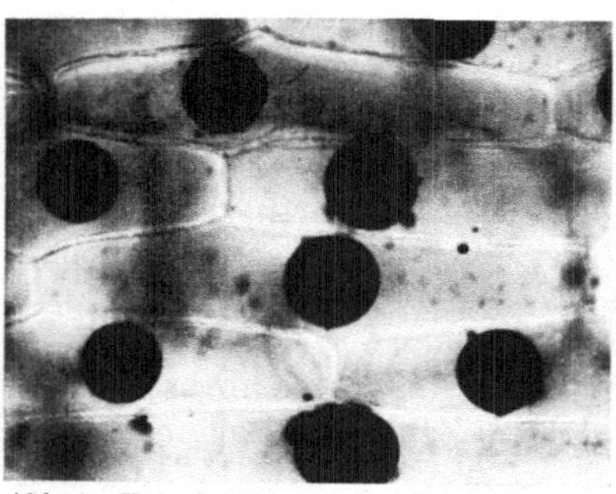

Abb. 4. Vacuolenfärbung der Unterepidermiszellen der Schuppenblätter von *Allium cepa* mit Rhodamin B. Bildung großer Entmischungstropfen. (Nach DRAWERT 1939.)

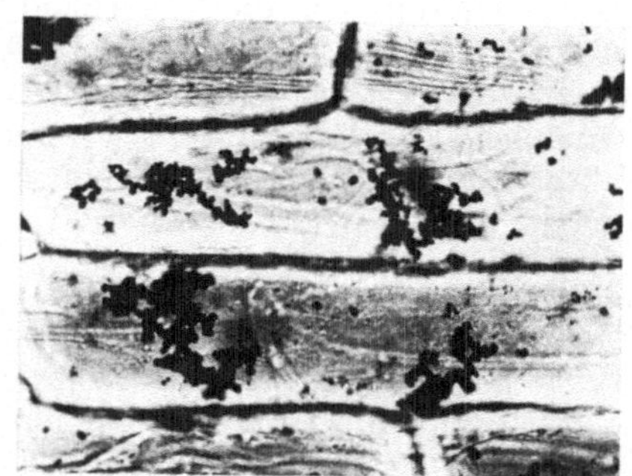

Abb. 5. Vacuolenfärbung der Oberepidermiszellen der Schuppenblätter angetriebener Zwiebeln von *Allium cepa* mit Rhodamin B. Bildung von Entmischungsaggregaten. (Nach DRAWERT 1939.)

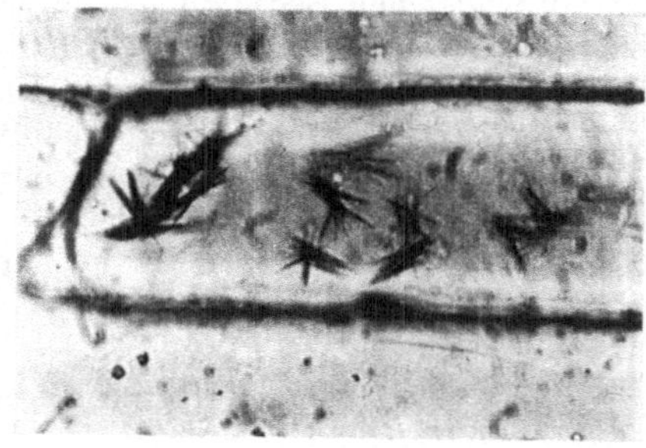

Abb. 6. Vacuolenfärbung der Oberepidermiszellen der Schuppenblätter angetriebener Zwiebeln von *Allium cepa* mit Methylenblau. Auftreten von kristallinen Farbstoffausfällungen. (Nach DRAWERT und STRUGGER 1938.)

C. Kern- und Cytoplasmafärbung.

Farbstoffe, die vom Kern vital gespeichert werden, färben im allgemeinen auch das Cytoplasma mehr oder weniger gut und umgekehrt. Eine wirklich elektive Kern- oder Cytoplasmafärbung tritt nur selten auf. In den meisten Fällen — von einigen Fluorochromen abgesehen — führen außerdem Kern- und Plasmafärbungen zu einer baldigen Schädigung der Zelle. Es handelt sich also in der Terminologie von STRUGGER (1936b) vorwiegend um perturbante Färbungen. Zur Kern- und Plasmafärbung eignen sich unter anderen folgende basische Farbstoffe: Anilingrün, Berberinsulfat, Bismarckbraun (= Vesuvin), Chrysoidin, Dahlia (= meist Methylviolett), Gentianaviolett (Methylviolett + Kristallviolett), Kristallviolett, Methylviolett (enthält wohl meist Kristallviolett), Prune pure, Rhodamin B. Bei den Farbstoffen Dahlia, Gentianaviolett, Kristallviolett und Methylviolett handelt es sich um Mischungen verschiedener violetter Rosaniline, so daß sehr wahrscheinlich die vitalfärbende Eigenschaft in allen 4 Farbstoffen auf dieselbe gemeinsame Komponente zurückzuführen ist.

Zu den Kern- und Plasmafärbern gehören aber auch einige saure Farbstoffe, die sich durch eine stärkere Lipophilie auszeichnen. Es sind dies vor allem

Eosin, Erythrosin, K-Fluorescein und Uranin (= Na-Fluorescein), also alles Oxyfluoronfarbstoffe.

Verglichen mit den Membran- und Vacuolenfärbern, handelt es sich bei den Kern- und Plasmafärbern vorwiegend um relativ schwach dissoziierte und stark lipoidlösliche Farbstoffe (Dissoziationskonstante der Farbbase um 10^{-9}).

D. Plastidenfärbung.

Lassen sich Membran, Vacuole und Cytoplasma bei Verwendung der entsprechenden Farbstoffe in der Regel bei allen Objekten vital färben, so ist dies bei den Einschlüssen des Plasmas nicht immer so ohne weiteres möglich. Diese Färbungen sind in viel höherem Grade von äußeren und inneren Faktoren abhängig. Hierbei spielt der „physiologische Zustand" der Zelle eine große Rolle, so daß in der Literatur recht widersprechende Angaben zu finden sind. Es soll im folgenden vorwiegend nur über positive Ergebnisse berichtet werden.

Über eine Vitalfärbung der Chloroplasten mit Neutralrot berichten WEBER (1930) und DIANNELIDIS (1951). Aus den Angaben der Autoren zu schließen, scheint es sich aber ähnlich der Plastidenfärbung mit Rhodamin 6 G um perturbante Färbungen zu handeln, während Plastidenfärbungen mit Rhodamin B als inturbant zu bezeichnen sind (STRUGGER 1936b, WEBER 1937, LÄRZ 1942, HÖFLER 1951, SCHMIDT 1951). Ferner sind noch Vitalfärbungen der Chloroplasten mit Prune pure (DRAWERT 1938b, LÄRZ 1942), Janusgrün, Mauvein, Vesuvin (LÄRZ 1942), Fluorescein (FREUDENBERGER 1941) und Cölestinblau (DRAWERT 1954) erzielt worden. In den Chloroplasten werden die Farbstoffe vorwiegend in den Grana gespeichert, das Stroma bleibt meist farblos. Vitalfärbungen von Leukoplasten geben STRUGGER (1937) und DRAWERT (1955b) für Rhodamin B, GUILLIERMOND (1937) und GUILLIERMOND und GAUTHERET (1940) für Janusgrün B und DRAWERT (1954) für Coelestinblau an. Eine Vitalfärbung der Chromoplasten scheint bisher noch nicht durchgeführt worden zu sein.

E. Mitochondrienfärbung.

Wie bei der tierischen Zelle hat sich auch bei der Pflanzenzelle für eine elektive Färbung der Mitochondrien bisher das Janusgrün B am besten bewährt (GUILLIERMOND 1923, GUILLIERMOND und GAUTHERET 1940, TARWIDOWA 1938, SOROKIN 1938, 1941, BHARGAVA 1951, DRAWERT 1953, LAZAROW und COOPERSTEIN 1953, COOPERSTEIN und Mitarbeiter 1953). Die Färbung gelingt aber nur bei sehr guter Sauerstoffversorgung der Objekte. Ebenso ist eine vitale Fluorochromierung der Mitochondrien mit Berberinsulfat nur bei Sauerstoffgegenwart möglich (DRAWERT 1953). Ferner wurden noch Mitochondrienfärbungen mit Methylviolettpräparaten (Dahlia) (GUILLIERMOND 1923, 1937b, BHARGAVA 1951), Phenosafranin (BECKER 1933, SIWIKKA-TARWIDOWA 1934) und Rhodamin B (JOHANNES 1941, AGTHE 1951, PERNER und PFEFFERKORN 1953) erzielt.

F. Sphärosomen-(Mikrosomen-)färbung.

Unter dem von HANSTEIN (1880) eingeführten Begriff „Mikrosomen" verstand man in der Botanik bisher die mikroskopisch sichtbaren granulaartigen Einschlüsse des Cytoplasmas, außer den Mitochondrien und den Plastiden in rein morphologischem Sinn, ohne damit über die physiologische Bedeutung dieser Granula etwas auszusagen. Leider sind dann von CLAUDE (1943) submikroskopische Teilchen des Plasmas als „Mikrosomen" bezeichnet worden, und in dieser Definition hat sich der Begriff in den letzten Jahren so eingebürgert, daß man den alten Begriff „Mikrosomen" wohl fallen lassen muß. Im Anschluß an PERNER (1952, 1953) soll dafür der von DANGEARD (1919) geprägte Begriff „Sphärosomen" benutzt werden, allerdings zum Unterschied von PERNER in einem rein morphologischen Sinn (DRAWERT 1955a, DRAWERT und METZNER 1955, vgl. auch SOROKIN 1955).

Eine einwandfreie vitale Färbung der Sphärosomen ist bis jetzt nur fluorescenzoptisch gelungen. Nach Strugger (1939) und Perner (1952) eignet sich Berberinsulfat dazu. Drawert (1953) konnte zeigen, daß besonders bei Sauerstoffmangel die Sphärosomen sich mit Berberinsulfat fluorochromieren lassen. Dasselbe ist der Fall bei Janusgrün B und reinem Nilblau. Diese Farbstoffe müssen erst bei Sauerstoffmangel durch die lebende Zelle reduziert werden, ehe sie von den Sphärosomen mit intensiver Fluorescenz gespeichert werden (Drawert 1953, Gutz 1955). Bei Zutritt von Sauerstoff wird diese Sphärosomenfluorescenz sofort gelöscht. Mit älteren Nilblaulösungen kann man aber auch eine von Sauerstoff unbeeinflußte Sphärosomenfluorescenz erhalten (Drawert 1952). Nach Gutz (1955) ist diese auf Nilrot zurückzuführen, das in den Nilblaupräparaten bereits als Verunreinigung enthalten ist bzw. in der wäßrigen Lösung durch Oxydation entsteht.

Es muß noch kurz auf das Verhalten der Granula in Pilzzellen hingewiesen werden, die wenigstens zum Teil in ihrem färberischen Verhalten eine Mittelstellung zwischen Mitochondrien und Sphärosomen einzunehmen scheinen. Einige der im Abschnitt über Mitochondrienfärbung zitierten Arbeiten sind mit Pilzen, besonders Saprolegniaceen, durchgeführt worden. Die sich mit Janusgrün B färbenden Plasmaeinschlüsse der Pilzzellen wurden den Mitochondrien der höheren Pflanzen gleichgesetzt. Neuere Ergebnisse lassen es aber doch nachprüfenswert erscheinen, ob es sich hier wirklich immer um identische Gebilde handelt (Graffi 1941, Ritchie und Hazeltine 1953, Bautz 1954, Steiner und Heinemann 1954, Gutz 1955).

V. Faktoren, die die Aufnahme der Farbstoffe und ihre Verteilung in der Zelle beeinflussen.

Die im Abschnitt IV geschilderten Färbungsbilder können durch verschiedene Innen- und Außenfaktoren verändert werden. Einer der wirksamsten Faktoren ist die cH der Farbstofflösung.

A. cH des Außenmediums.

Bereits aus der Angabe von Pfeffer (1886), daß Zellen, die mit Methylenblau vital gefärbt waren, in Citronensäure den Farbstoff wieder abgaben, war zu vermuten, daß die cH des Außenmediums die Aufnahme und Abgabe von Farbstoffen beeinflussen muß. In vielen folgenden Arbeiten wurde dann bestätigt, daß die Aufnahme basischer Farbstoffe mit fallender cH gefördert wird und daß auf die Abgabe von bereits aufgenommenem Farbstoff die cH eines farblosen Außenmediums gerade umgekehrt wirkt (Endler 1912b, Collander 1921, McCutcheon und Lucké 1924, Albach 1928). Für die Aufnahmegeschwindigkeit z. B. von Brillantkresylblau (basischer Farbstoff) durch *Nitella* in Abhängigkeit von der Außen-cH führt Irwin (1923a) folgende Vergleichszahlen an: p_H 7,38 = 5; p_H 9 = 350 und p_H 10 = 910. Es gibt allerdings auch Farbstoffe, deren Aufnehmbarkeit durch die lebende Zelle im physiologischen p_H-Bereich kaum von der cH beeinflußt wird; dazu gehört z. B. der amphotere Farbstoff Rhodamin B (Bailey und Zirkle 1931, Drawert 1937c, 1939, Strugger 1938a, Yamaha 1938, Johannes 1941, Kölbel 1948). Auch auf die Aufnahme der im allgemeinen stark dissoziierten sulfosauren Farbstoffe mit Hilfe der Transpirationsmethode hat die cH der Farblösung kaum einen Einfluß (Albach 1928, Drawert 1941). Dies steht im Einklang mit den Angaben von Lundegårdh

(1940), daß ganz allgemein die Anionenaufnahme zum Unterschied von der Kationenaufnahme weitgehend unabhängig von der Außen-cH ist (vgl. aber COLLANDER 1921 und PERNER 1950a).

Die cH der Farbstofflösung bestimmt aber nicht nur die Geschwindigkeit der Farbstoffaufnahme, sondern auch die Verteilung des Farbstoffes in der Zelle, bei den basischen Farbstoffen vor allem die Verteilung zwischen Zellwand und Vacuole. Im sauren Bereich färben die meisten basischen Vitalfarbstoffe die Zellwand, im neutralen oder alkalischen Bereich aber die Vacuole (Abb. 7a und b) (RUHLAND 1908, 1912, IRWIN 1923b, GUILLIERMOND, DUFRÉNOY, LABROUSSE 1930, PRÀT 1931, DÖRING 1935). RUHLAND (1908, 1912) beobachtete mit Neutralrot, daß die Verteilung des Farbstoffes zwischen Membran und Vacuole auch geändert werden kann, wenn die vitalgefärbte Zelle in farblose Medien mit entsprechender cH übertragen wird. Diese Erscheinung wurde dann von STRUGGER

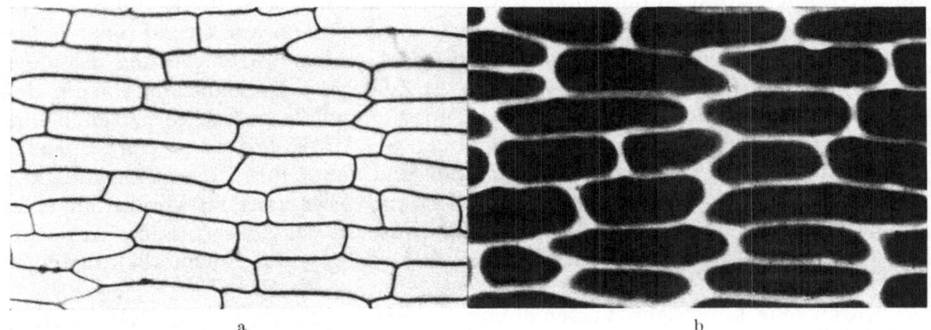

a b

Abb. 7a u. b. a Membranfärbung der Oberepidermiszellen der Schuppenblätter von *Allium cepa* mit Neutralrot 1:10000 in dest. Wasser gelöst (p_H 4,7). b Vacuolenfärbung der Oberepidermiszellen der Schuppenblätter von *Allium cepa* mit Neutralrot 1:10000 in Leitungswasser gelöst (p_H 7,2). Diffuse Zellsaftfärbung, Zellen mit Vacuolenkontraktion.

(1935, 1936) eingehend analysiert; und durch die Verwendung von abgestuften Pufferreihen konnte STRUGGER den p_H-Wert bestimmen, bei dem ein Umschlag von einer Membran- zu einer Vacuolenfärbung auftritt. Bei den Wurzelhaarzellen von *Trianea bogotensis* — dem auch von RUHLAND (1908) benutzten Objekt — lag er für Neutralrot um p_H 6,4 und für die Oberepidermiszellen der Schuppenblätter von *Allium cepa* bei $p_H \sim 7{,}1$. Danach ist die Lage des Umschlagpunktes Membran-/Vacuolenfärbung einerseits vom Objekt abhängig, andererseits wird er aber auch vom Farbstoff mitbestimmt; denn für Methylenblau liegt der Umschlagspunkt bei den *Allium*-Oberepidermen zwischen $p_H \sim 11$ und 11,5 (DRAWERT und STRUGGER 1938). Für Neutralrot schwankt der Umschlagspunkt je nach Objekt und auch nach dem physiologischen Zustand desselben Objektes zwischen p_H 3 und 7,6 (DRAWERT 1937b, STRUGGER 1937b). Angaben über die p_H-Werte für die Umschlagsbereiche weiterer basischer Farbstoffe finden sich in folgenden Arbeiten: DRAWERT (1940, 1951), STRUGGER (1940, 1941), HÖFLER (1947a, b), STIEGLER (1950), WIESNER (1951).

Der Einfluß der Außen-cH scheint sich nach SCARTH (1926) und BROOKS (1927, 1929, 1931) nur auf die Aufnahmegeschwindigkeit auszuwirken, aber nicht auf die Endkonzentration des Farbstoffes im Zellsaft im Gleichgewichtszustand. Nach IRWIN (1923b, 1926d) soll dagegen auch die Endkonzentration von der cH der Farblösung abhängen, dasselbe erwähnt auch BROOKS (1933) für ein Indophenol. Allem Anschein nach kommt sowohl das eine wie das andere vor, dies wird wohl von der Art des jeweiligen Speichermechanismus in der Vacuole abhängen.

B. cH des Zellsaftes und Speicherstoffe.

Der Umschlagspunkt von einer Membran- zu einer Vacuolenfärbung kann für denselben Farbstoff, je nach dem benutzten Objekt bei einem unterschiedlichen pH-Wert der Außenlösung liegen. Außer den physikalisch-chemischen Eigenschaften des Farbstoffes muß demnach auch dem physiologischen Zustand des untersuchten Objektes eine Bedeutung für die Aufnahme und die Verteilung eines basischen Farbstoffes zukommen. In erster Linie werden hierbei die Speicherfähigkeiten des Zellsaftes beteiligt sein.

Die Speicherfähigkeit des Zellsaftes und auch der anderen Zellbestandteile kann auf 4 Faktoren zurückgeführt werden: a) Zelleigene Stoffe reagieren chemisch mit dem eindringenden Farbstoff und überführen ihn in eine unlösliche Verbindung, so daß das Diffusionsgefälle aufrechterhalten bleibt. Für die basischen Farbstoffe kommt den häufig im Zellsaft anzutreffenden Gerbstoffen und anderen Phenolabkömmlingen in dieser Hinsicht eine besondere Rolle zu, wie bereits PFEFFER (1886) erkannt hat. b) Lipoide können auf Grund einer hohen Lipoidlöslichkeit des eindringenden Farbstoffes zu einer Anreicherung desselben in bestimmten Teilen der Zelle führen. c) Zelleigene Kolloide adsorbieren den Farbstoff, so daß auf diese Weise das Diffusionsgefälle aufrechterhalten bleibt. d) Die cH des Zellsaftes kann in Verbindung mit den Kolloiden zu einer Speicherung beitragen, da die Ladung der Kolloide und damit ihre Adsorptionsfähigkeit von der cH des umgebenden Mediums abhängt. Aber auch im Verein mit einer Impermeabilität des Plasmas für Ionen kann die cH des Zellsaftes eine Anreicherung des Farbstoffes in der Vacuole nach dem Prinzip der „Ionenfalle" bedingen (vgl. S. 278).

Durch eine künstliche Verschiebung der cH des Zellsaftes mit NH_3, H_2CO_3, CH_3COOH ist öfter eine Änderung der Speicherfähigkeit erzielt worden (McCUTCHEON und LUCKÉ 1924, IRWIN 1925, 1926, SCARTH 1926, HÖFLER 1947b). Im allgemeinen fördert eine Ansäuerung des Zellsaftes die Aufnahmegeschwindigkeit und die Speicherung der basischen Farbstoffe.

Häufig ist aus dem Farbton gespeicherter p_H-Indicatoren auf die cH des Zellsaftes und deren Einfluß auf die Vitalfärbung geschlossen worden (BAILEY und ZIRKLE 1931, CHADEFAUD 1933, DRAWERT 1937b u. f., PRESCOTT 1953). Es haben sich aber immer mehr Bedenken ergeben, ob man aus einer Vitalfärbung mit Indicatoren wirklich auf die cH der wäßrigen Phase schließen kann (vgl. DRAWERT 1955c). Selbst wenn wir von diesen strittigen Fällen absehen, hat es sich aber doch immer wieder bestätigt, daß ein basischer Farbstoff mittleren oder stärkeren Dissoziationsgrades in der Zelle sich in einem Bereich höherer Acidität anreichert (DRAWERT 1948, BARTELS 1954). Die gegenteilige Schlußfolgerung von ROHDE (1917) hat sich nicht aufrechthalten lassen (vgl. DRAWERT 1941).

Wie sich die cH des Zellsaftes auf den Beginn der Vacuolenfärbung in Abhängigkeit von der Außen-cH auswirken kann, geht aus den in Tabelle 2 wiedergegebenen Versuchen mit *Bryophyllum* hervor. Die Bryophyllen zeigen wie alle

Tabelle 2. *Beginn der Vacuolenfärbung mit Neutralrot im Mesophyll von Bryophyllum-Blättern in Abhängigkeit von der* cH *der Farbstofflösung und der* cH *des Preßsaftes.* (Nach OVERBECK, unveröffentlicht.)

	B. tubiflorum		*B. daigremontianum*	
	Tageszeit		Tageszeit	
	morgens	abends	morgens	abends
p_H-Wert des Preßsaftes	4,44	5,82	4,49	5,74
Beginn der Vacuolenfärbung bei p_H	4,36 ± 0,04	6,20 ± 0,05	4,53 ± 0,1	5,76 ± 0,06

Succulenten in Abhängigkeit von der Beleuchtung einen täglichen Rhythmus in der Acidität des Zellsaftes (vgl. DRAWERT 1955c, S. 642). Morgens besitzt der Zellsaft eine höhere Acidität als abends, dementsprechend beginnt auch die Vacuolenfärbung mit Neutralrot morgens weiter im sauren Bereich als abends. Hier dürfte wohl kein Zweifel vorliegen, daß es sich um eine Aciditätsänderung der wäßrigen Phase des Zellsaftes handelt.

C. Salze.

Wie die Wasserstoffionen wirken sich auch die anderen Ionen auf die Farbstoffaufnahme aus. So können Kationen die Färbung der Zellwände mit basischen Farbstoffen blockieren oder bereits gefärbte Zellwände entfärben (PFEFFER 1886, SCARTH 1926, GENEVOIS 1930, BRAUNER 1933, CZAJA 1936, BORRISS 1937a, b, DRAWERT 1937b, KERSTING 1937, PEKAREK 1938, HÖFLER und STIEGLER 1947). Der Wirkungsgrad der Salze nimmt mit der Wertigkeit der Kationen zu. Nach KERSTING (1937) erhält man für die Entfärbungsgeschwindigkeit in 0,1 n Salzlösungen folgende Reihe: $AlCl_3 > CaCl_2 > MgCl_2 > KCl > NaCl$, und bei p_H 7,4 liegt die Konzentration, die gerade eine Membranfärbung mit 0,02% Methylenblau verhindert, für $K^{\cdot}$ und $Na^{\cdot}$ bei 0,2 Mol und für $Ca^{\cdot\cdot}$ bei 0,05 Mol (BORRISS 1937a).

Für den Einfluß der Salze auf die Plasma-, Kern- und Vacuolenfärbung ist das aus der Literatur zu entnehmende Bild nicht so einheitlich wie in bezug auf die Zellwandfärbung. Aber für die Beurteilung z. B. der cH-Wirkung wäre es wichtig, den Einfluß der anderen Ionen der benutzten Puffer genau zu kennen. So hängt die im Gleichgewichtszustand erreichte Endkonzentration von Indophenolen im Zellsaft von *Valonia*-Arten von der benutzten Pufferlösung ab. Bei der Verwendung von Phosphatpuffern beträgt sie bei einem bestimmten p_H-Wert $^1/_4$ der Außenkonzentration, bei Boratpuffern dagegen bei demselben p_H-Wert nur $^1/_8$ (BROOKS 1932). Bei konstanter cH (p_H 6,8) wird die Aufnahme von Dahlia durch *Nitella* von Chloriden in folgender Reihe gehemmt: Na < K < Ca < Mg. Ca und Na wirken dabei antagonistisch und der stärkste antagonistische Effekt wird bei einem Verhältnis von $NaCl:CaCl_2 = 98:2$ erzielt (BROOKS 1927b). Nach IRWIN (1926c) soll eine Vorbehandlung von *Nitella*-Zellen mit NaCl die Aufnahme von Brillantkresylblau stark hemmen, bei gleichzeitiger Gabe von Salz + Farbstoff dagegen fördern. Ca-, Mg- und La-Salze sollen immer die Aufnahmegeschwindigkeit für den Farbstoff etwas erhöhen. Ebenso beobachten BAILEY und ZIRKLE (1931) eine Beschleunigung der Aufnahme basischer Farbstoffe durch die Cambiumzellen von *Pinus* durch Ca-Salze. HOMÈS (1930) stellt dagegen eine Herabsetzung der Aufnahmegeschwindigkeit von Methylenblau durch die Zellen von *Helodea canadensis* nach Zusatz von NaCl, KCl oder $CaCl_2$ fest. Ein unterschiedliches Verhalten einzelner Farbstoffe gibt BORRISS (1937b) an. Während die Aufnahme von Neutralrot aus Acetatpufferlösungen durch $CaCl_2$ stark gefördert wird, hemmt dasselbe Salz die Aufnahme von Methylenblau. In ähnlicher Richtung liegen auch die Beobachtungen von HÖFLER und SCHINDLER (1953). Diese Autoren finden einen Unterschied im Verhalten von basischen Farbstoffen, je nachdem, ob sie molekular gelöst und lipoidlöslich sind, oder ob sie im ionisierten Zustand vorliegen. Auf die Aufnahme der ersten Gruppe haben Neutralsalze selbst in hohen Konzentrationen keinen Einfluß, die Aufnahme der zweiten Gruppe wird durch Salzzusatz stark gehemmt. Nach den älteren Untersuchungen von ENDLER (1912a, b) kann ein Salz auch je nach der benutzten Konzentration auf die Farbstoffaufnahme hemmend oder fördernd wirken.

Nach den geschilderten Befunden scheint die Wirkung der Salze auf die Zellsaftfärbung mit basischen Farbstoffen recht komplexer Natur zu sein. Etwas einheitlicher sind die Ergebnisse über die Salzwirkung auf die Plasmafärbungen mit Prune pure und besonders mit den sauren Farbstoffen Eosin und Erythrosin. Hier überwiegen die Angaben über eine Förderung durch Salzzusatz (ALBACH 1928, SCHÖNLEBER 1936, DRAWERT 1938, YAMAHA und NOMURA 1939). Selbst Aluminiumsalze begünstigen die Plasmafärbung mit den sauren Farbstoffen, während bei Prune pure dagegen nach DRAWERT (1938) eine Hemmung zu beobachten ist.

D. Anelektrolyte.

Harnstoff, Saccharose, Glucose, Fructose, Lactose scheinen die Aufnahme basischer Farbstoffe zu hemmen (ENDLER 1912b, BAILEY und ZIRKLE 1931). Auf eine Vitalfärbung mit dem sauren Farbstoff Eosin soll dagegen Rohrzucker in hypertonischer Konzentration nach SCARTH (1926b) fördernd wirken, während GELLHORN (1927) an tierischen Eiern auch für den sauren Farbstoff Erythrosin wie für die basischen Farbstoffe eine Hemmung durch Zucker beobachtet hat. Die nach einer Vitalfärbung mit Rhodamin B in den vollen Zellsäften (s. S. 277) von Blumenblattepidermen auftretenden Entmischungskugeln lösen sich bei einer Plasmolyse in stark hypertonischen Harnstofflösungen auf (DRAWERT 1955b). Noch rascher lassen sich diese Entmischungstropfen mit 1% Coffein zur Lösung bringen (FLASCH 1955). Das Coffein bedingt auch eine sehr rasche Entfärbung der mit Rhodamin B diffus gefärbten Zellsäfte und führt bei Vacuolen, die nach einer Acridinorange-Färbung rot fluorescieren, zu einem Farbumschlag nach Grün (FLASCH). An tierischen Objekten und zum Teil auch an Blumenblättern, die mit Neutralrot oder Nilblau vital gefärbt waren, beobachteten bereits BORNSTEIN und RÜTER (1925) sowie LAUER (1930) und LEWIN (1933) eine Entfärbung in Lösungen von Coffein und anderen Alkaloiden.

Sehr widersprechend sind die Angaben über die Wirkung von Alkohol, Äther und Chloroform. Eine Förderung wird besonders für Kernfärbungen mit Erythrosin, Prune pure und mit basischen Farbstoffen beschrieben (PALTAUF 1928, SCHÖNLEBER 1936, BANK 1938). Nach ALBACH (1928) hemmt eine Vorbehandlung mit Ätherwasser die Vitalfärbung mit Eosin und nach HOMÈS (1933) haben Alkohol und Äther keinen Einfluß auf die Methylenblauaufnahme, während sie nach LEPESCHKIN (1911a, b) hemmend wirken. So manche Diskrepanz wird sich vielleicht aus der Beobachtung von LEPESCHKIN (1913, 1932) erklären lassen, daß die Aufnahme der äther- und chloroformlöslichen Farbstoffe durch diese Narkotica gefördert, die der vorwiegend wasserlöslichen aber gehemmt wird.

E. Stoffwechselvorgänge.

Der Einfluß von Narkotica auf die Aufnahme verschiedener Farbstoffe weist schon darauf hin, daß auch dem Stoffwechsel der Zelle eine gewisse Bedeutung für die Vitalfärbung zukommt, sei es in Form eines aktiven Aufnahmevorganges, sei es durch die Bildung von Stoffwechselprodukten, die für die Farbstoffspeicherung eine Rolle spielen; aber auch der aufgenommene Farbstoff selber kann in den Stoffwechsel einbezogen und chemisch so verändert werden, daß es zu einer Verschiebung des Verteilungsbildes kommt.

Stoffwechselveränderungen finden wir im normalen Entwicklungscyclus der Pflanze, und dementsprechend können sich die Zellen in ihren verschiedenen Entwicklungszuständen ganz unterschiedlich gegenüber einem Farbstoff verhalten. Die Aktivität einer Zelle kann von ausschlaggebender Bedeutung für die Aufnahme und Speicherung sein. So speichern nach BAILEY (1930) z. B. meristematische Zellen nur im Winter Neutralrot, aber nicht in der Hauptwachstumsperiode. Oder nach GUILLIERMOND und GAUTHERET (1937a, b) ist die Neutralrotfärbung bei Keimwurzeln und Pilzmycelien abhängig von der Wachstumsgeschwindigkeit. Die einzelnen Objekte verhalten sich aber verschieden. Während Getreidekeimlinge und Saprolegniaceen Neutralrot auch in wachsenden Zellen speichern, nehmen andere Pilze wie *Saccharomyces*-Arten und *Oidium lactis* den Farbstoff nur im Ruhezustand aus der Kulturflüssigkeit auf und entfärben sich mit der Wiederaufnahme des Wachstums. Acridin färbt in den Oberepidermiszellen der Schuppenblätter von *Allium cepa* bei Sommerzwiebeln den Zellsaft, während bei ruhenden Winterzwiebeln statt einer Zellsaftfärbung manchmal eine Fluorochromierung des Cytoplasmas zu beobachten ist. Auch durch Ultraviolettbestrahlung kann eine Vacuolenfluorescenz hier in eine Plasmafluorescenz übergeführt werden (STRUGGER 1941).

Die Erscheinung, daß Erdalgen und Moosprotonemen Eosin und Erythrosin nach einer Austrocknungsperiode schneller aufnehmen (FRITSCH und HAINES 1923) als vorher, ist wohl ebenfalls auf eine Änderung des Stoffwechsels und damit der Speicherfähigkeit der Zellen zurückzuführen.

Nach PFEFFER (1886) hat ein Sauerstoffentzug keinen Einfluß auf die Speicherung von Methylenblau. Bei einer Vacuolenfärbung mit Neutralrot kann sich aber ein Deckglasabschluß und damit ein Sauerstoffmangel auf die Verteilung des Farbstoffes auswirken. Der Farbstoff verläßt die Vacuole und färbt die Membran (STRUGGER 1936a). Die mit Brillantkresylblau gefärbten Vacuolen der Raphidenzellen von *Haemaria discolor* entfärben sich ebenfalls bei längerem Deckglasabschluß (DISKUS und KIERMAYER 1954). Sehr wahrscheinlich beruhen diese Effekte auf einer Ansäuerung des umgebenden Mediums durch CO_2-Ausscheidung. Auch die Aufnahme von Acridinorange durch Hefezellen ist nach BOGEN (1953) von der Durchlüftung abhängig.

Allem Anschein nach wird aber die Aufnahme und Speicherung der meisten basischen und der lipoidlöslichen sauren Farbstoffe nicht direkt von der Sauerstoffgegenwart beeinflußt. Anders liegt es aber mit der Mitochondrienfärbung durch Janusgrün B. Diese Färbung verläuft nur positiv bei ausreichender Sauerstoffversorgung (SOROKIN 1938, 1941, 1955, DRAWERT 1953, LAZAROW und COOPERSTEIN 1953). Bei Sauerstoffmangel wird der Farbstoff durch Reduktion in die im weißen Licht farblose, aber im Blaulicht oder UV. grünlich fluorescierende Leukoform umgewandelt, die jetzt auf Grund ihrer höheren Lipoidlöslichkeit von den Sphärosomen gespeichert wird (DRAWERT 1953). Ähnlich liegen die Verhältnisse beim Berberinsulfat.

Auch noch andere basische Farbstoffe werden bei Sauerstoffmangel von der lebenden Zelle reduziert und ändern dabei ihre physikalisch-chemischen Eigenschaften und damit ihre Verteilung in der Zelle. So wandert das im oxydierten Zustand aufgenommene Methylenblau nach der Reduktion durch die lebende Zelle aus der Vacuole in das Plasma (FRITZ 1951, DRAWERT 1953), das reine Nilblau aus der Vacuole in die Sphärosomen (DRAWERT und GUTZ 1953, GUTZ 1955), das Coelestinblau aus den Leukoplasten oder auch aus Plasma, Kern und Mitochondrien in die Vacuole, und das Prune pure bleibt im Plasma, oder falls es sich in der oxydierten Stufe im Zellsaft befindet, geht es in das Plasma (FRITZ 1951, BETZ 1953, DRAWERT 1953). Nach einer Reoxydation in der Zelle reichern sich die meisten Farbstoffe in der Vacuole an.

Aus diesen Befunden geht hervor, daß es nicht möglich ist, aus den Speicherungsorten der reduzierten oder der oxydierten Stufe eines Redoxindicators etwas Näheres über Reduktions- oder Oxydationsorte in der Zelle auszusagen. Dies trifft auch für Formazan und Indophenol zu.

Für die Aufnahme und Speicherung der sulfosauren Farbstoffe scheint dem Sauerstoff eine viel größere Bedeutung zuzukommen als für die entsprechenden

Vorgänge bei den basischen Farbstoffen (Collander 1921, Collander und Holmström 1937, Drawert 1941, Drawert und Endlich 1956).

Ein bisher sehr wenig beachteter Faktor, der sich vor allem bei längerer Farbstoffeinwirkung bemerkbar machen muß, ist die Änderung des Stoffwechsels durch den aufgenommenen Farbstoff. Dadurch wird der physiologische Zustand der Zelle geändert, und damit kann sich auch sekundär das Färbungsbild ändern, wie nach einem Deckglasabschluß oder nach einer Reduktion des Farbstoffes durch die Zelle. Im einzelnen kann hier nicht darauf eingegangen werden. Es soll aber darauf hingewiesen werden, daß nach der überwiegenden Zahl der vorliegenden Angaben basische Farbstoffe, die wie Methylenblau in der Vacuole gespeichert werden, die O_2-Aufnahme und die CO_2-Abgabe fördern. Die Förderung soll um so stärker sein, je höher der rH-Wert des betreffenden Farbstoffes ist (Genevois 1928). Allerdings fehlen auch nicht Angaben, nach denen selbst Methylenblau die Atmung hemmt (Massart und Mitarbeiter 1947). Basische und saure Farbstoffe, die das Plasma und seine Bestandteile färben, setzen dagegen ganz allgemein die Atmungsintensität herab. Für Berberinsulfat liegen aus neuerer Zeit darüber Angaben von Perner (1952) vor und für Acridinorange von Bogen und Elste (1955). Bei photodynamisch wirksamen Farbstoffen kann dabei allerdings auch noch das Licht von ausschlaggebender Bedeutung sein, so hemmt Eosin die Atmung im Dunkeln und fördert sie im Licht (Ziegler 1950). Auf die zahlreichen Untersuchungen über die Änderung der Wachstumsgeschwindigkeit und der geo- und phototropischen Empfindlichkeit durch fluorescierende Farbstoffe wie Eosin kann hier auch nur hingewiesen werden (Schweighart 1935). Mit dem Einfluß von Acridinorange auf Wachstum, Proteinumsatz und Gaswechsel befassen sich neuere Arbeiten von Bogen und Keser (1954) sowie Bogen und Elste (1955).

VI. Theorien der Vitalfärbung.

„Es gibt sicher nicht viele technische Disziplinen, über deren theoretische Grundlagen derartig viel und von verschiedenster Seite gearbeitet worden ist, wie über die Vorgänge bei der Färberei. Kaum ein anderes Gebiet hat auch derartig verschiedene, zum Teil schroff einander entgegengesetzte Meinungen hervorgebracht wie das vorliegende" (Auerbach 1923). Diese für den technischen Färbeprozeß geprägten Sätze gelten heute noch in vollem Umfang für die Vitalfärbung. Bei den folgenden Betrachtungen über die Vitalfärbungstheorien steht das Problem der Farbstoffspeicherung im Vordergrund. Die Intrabilitäts- und Permeabilitätsfragen werden an anderen Stellen für die Stoffaufnahme im allgemeinen besprochen. Die dort behandelten Punkte gelten natürlich für die Farbstoffaufnahme genau so, wie für die Aufnahme der anderen Stoffe. Die Fragen der Speicherung treten dagegen bei der Vitalfärbung am augenfälligsten in Erscheinung. Intrabilität und Permeabilität sollen deshalb nur dort gestreift werden, wo sie mit der Speicherung in direktem Zusammenhang stehen.

A. Theorien der Membranfärbung.

Am klarsten liegen wohl noch die Vorgänge bei der Zellwandfärbung, bei der wir ihrer Mechanik nach drei Färbungstypen unterscheiden können (Höfler 1948).

Auf Grund ihres hohen Kohlenhydratgehaltes ist die Zellwand im physiologischen p_H-Bereich negativ geladen. Sie kann also in diesem Bereich Farbkationen adsorptiv binden (Austauschadsorption). Für diesen ersten Färbungstyp sprechen folgende Tatsachen: 1. Eine Membranfärbung kann — wenn wir von den Fällen des dritten Färbungstyps absehen — nur mit basischen Farbstoffen erzielt werden. 2. Die Färbung kann nur erfolgen, wenn der Farbstoff in dissoziiertem Zustand vorliegt; schwach dissoziierte basische Farbstoffe wie Methylviolett oder Rhodamin B färben die Zellwand nicht im physiologischen p_H-Bereich (vgl. aber den zweiten Färbungstyp). 3. Die Membranen besitzen einen isoelektrischen Punkt (Drawert 1937c) oder Entladungspunkt (Höfler 1946). Im sauren Bereich unterhalb des IEP findet auch mit den stark dissoziierten basischen Farbstoffen keine Zellwandfärbung mehr statt. Je nach der Zusammensetzung der Zellwände liegt ihr IEP zwischen p_H 2 und 3,5, bei den Chitinwänden der Pilze nach Höfler und Pecksieder (1947) sogar um p_H 4. Liegt der IEP sehr weit im sauren Bereich, so ist es möglich, auch mit den basischen Farbstoffen, die erst im stark sauren Bereich dissoziieren, wie Methylviolett, bei entsprechender cH der Farbstofflösung eine Membranfärbung zu erzielen (Drawert 1937c). Für die negative Ladung der Zellwände sind wohl in erster Linie Pektine verantwortlich zu machen, wie aus Untersuchungen von Kinzel (1953) hervorgeht.

Neben der rein adsorptiven Zellwandfärbung finden wir als zweiten Färbungstyp eine chemische Bindung basischer Farbstoffe an primär vorhandene Membranstoffe oder auch sekundär in die Membran eingewanderte Substanzen. Diese Art der Färbung wurde von Drawert (1937c) in Anlehnung an Ausführungen von Brauner (1933) als „Niederschlagsfärbung“ bezeichnet. Es entsteht ein „Farblack“. Gerbstoffführende Zellwände, wie die Membranen der Blättchen von *Funaria hygrometrica*, tingieren sich mit basischen Farbstoffen auf Grund einer Niederschlagsfärbung. Sobald die in der Membran vorhandenen Gerbstoffe abgesättigt sind, wird, falls der Farbstoff in dissoziiertem Zustand vorliegt, nach der alkalischen Seite vom IEP die Niederschlagsfärbung von einer Adsorptionsfärbung nach dem ersten Färbungstyp überdeckt (Drawert 1937c). Dies tritt besonders augenfällig in Erscheinung bei metachromatischen Farbstoffen. Beim Toluidinblau z. B. ist die auf Grund einer Farblackbildung gefärbte Zellwand blau, und sie ist violett, wenn eine Adsorptionsfärbung vorliegt. Die Niederschlagsfärbung ist nicht wie die Adsorptionsfärbung vom IEP der Zellwand und vom Dissoziationszustand des Farbstoffes und damit vom p_H-Wert des umgebenden Mediums abhängig. Nach diesem zweiten Färbungstyp können auch schwach oder nicht dissoziierte basische oder amphotere Farbstoffe wie Rhodamin B entsprechende Zellwände färben. Die auf S. 267 beschriebene Blockierung der Membranfärbung oder auch Entfärbung bereits gefärbter Membranen durch Salze trifft nur für den ersten Färbungstyp (Austauschadsorption) zu. Eine Niederschlagsfärbung ist salzfest. So kann man durch Zugabe von $CaCl_2$-Lösungen entscheiden, ob eine Adsorptions- oder eine Niederschlagsfärbung vorliegt (Höfler und Stiegler 1947, Höfler 1948, 1951, Stiegler 1950).

Bei postmortalen Membranfärbungen handelt es sich meist um Niederschlagsfärbungen. Aus dem absterbenden Protoplasten treten Stoffe aus, die die Zellwand durchtränken und in der Zellwand mit basischen Farbstoffen reagieren und so Inkrusten bilden (Pfeffer 1886, Brauner 1933). Brauner bezeichnet diese Erscheinung als „sekundäre Induktion“. Wie die primäre Niederschlagsfärbung, ist auch die sekundäre salzfest.

Ein dritter Färbungstyp beruht auf einer Einlagerung saurer Farbstoffe mit hoher Teilchengröße in die submikroskopischen Capillaren oder auch Intermicellarräume der Zellwände. Die bekannte Membranfärbung mit Kongorot dürfte wohl unter diesen Färbungstyp einzuordnen sein. Dafür sprechen die Befunde von Klebs (1919) und Dorner (1922). Die „lockerer“ gebauten Zellwände der Rhizoiden von Farnprothallien färben sich mit Kongorot, die „dichteren" Wände der Prothalliumzellen selber aber nicht. Erst nach der Behandlung mit fettlösenden Substanzen färben sich auch die Prothalliumzellwände, da jetzt die verstopfenden Inkrusten herausgelöst worden sind oder vielleicht auch eine Cuticula aufgelöst worden ist (BRAUNER 1933). Zum Unterschied von der bei basischen Farbstoffen auftretenden „Sekundärinduktion" bezeichnet BRAUNER diese Erscheinung als „Parallelinduktion". Mit polydispersen kolloidalen sauren Farbstoffen, deren Farbtöne von der Teilchengröße abhängen, kann man mit verschieden dichten Zellwänden metachromatische Färbungen erhalten, wie sie von SCHWARZ (1924) und CZAJA (1930) beschrieben werden. In der Terminologie der technischen Färberei könnte man diesen dritten Färbungstyp als eine „Echtfärbung" mit substantiven Farbstoffen bezeichnen. Diese Färbungen sind unabhängig vom IEP der Wand und von der cH der Lösungen (falls nicht in Grenzfällen durch Quellungserscheinungen doch eine Farbstoffeinlagerung nur bei bestimmten p_H-Werten ermöglicht werden sollte) und absolut salzfest. Für eine gerichtete Einlagerung oder auch Anlagerung spricht das Auftreten eines positiven Dichroismus, z. B. bei der Färbung von Pflanzenfasern (ZIEGENSPECK 1941).

Zusammenfassend können wir über die Theorien der Membranfärbung sagen, daß der häufigste Typ die Adsorptionsfärbung ist. Deshalb sind auch die im physiologischen p_H-Bereich stärker dissoziierten basischen Farbstoffe vorwiegend Membranfärber. Diese Farbstoffe können aber auch beim Vorliegen der dafür notwendigen Voraussetzungen auf dem Wege einer Niederschlagsfärbung die Membranen tingieren und bei entsprechender Teilchengröße und -form auch eine Echtfärbung der Zellwände bewirken. Das letzte scheint bis zu einem gewissen Grade, bei der Membranfärbung mit Coriphosphin HK vorzuliegen (HÖFLER und MÜLLER-HAITINGER 1949). Die im physiologischen p_H-Bereich schwach oder gar nicht dissoziierten basischen oder amphoteren Farbstoffe können die Zellwände nur mit Hilfe der Niederschlagsfärbung tingieren. Das klassische Beispiel für diesen Typ ist das Rhodamin B. Eine Adsorptionsfärbung ist bei einem Teil dieser Farbstoffgruppe nur im stark sauren — also im unphysiologischen p_H-Bereich — möglich, wenn der IEP der Zellwand weit genug im sauren Gebiet liegt, wie bei *Helodea canadensis* und Methylviolett. Theoretisch wäre auch eine Echtfärbung möglich, falls die Farbstoffe eine entsprechende Teilchengröße besitzen. Coriphosphin liegt z. B. auf der Grenze zwischen den stärker und den schwächer dissoziierten Farbstoffen. Die sauren Farbstoffe können dagegen allem Anschein nach nur auf dem Wege einer Echtfärbung die Zellwände tingieren. Dazu sind aber eine bestimmte Größe und wahrscheinlich auch eine bestimmte Gestalt der Teilchen Voraussetzung, so daß im allgemeinen nur kolloidal gelöste Farbstoffe dazu befähigt sind.

B. Theorien der Cytoplasmafärbung.

Voraussetzung für das Zustandekommen einer Plasmafärbung ist das Intrameieren des Farbstoffes. Aus dem Ausbleiben einer Plasmafärbung dürfen wir aber nicht schließen, daß der untersuchte Farbstoff nicht intrameieren kann, denn zur Färbung ist außer dem Intrameieren eine Speicherung im Plasma notwendig. Dafür spricht die Tatsache, daß bei kräftiger Färbung des Zellsaftes — was ja ein Permeieren und damit auch Intrameieren des Farbstoffes zur Voraussetzung hat — das Plasma völlig farblos erscheinen kann, auch während des Färbevorganges, da die Konzentration des Farbstoffes im Cytoplasma unter der Sichtbarkeitsgrenze bleibt. Die Theorien der Speicherung sollen deshalb auch bei der Besprechung der Plasmafärbung im Vordergrund stehen.

Theoretisch sind folgende Möglichkeiten einer Farbstoffspeicherung im Cytoplasma und seinen Einschlüssen denkbar: 1. Die Farbionen werden von der hydrophilen Eiweißkomponente des Plasmas adsorbiert. 2. Die Farbbasen- oder Farbsäuremoleküle werden auf Grund ihrer Lipophilie in den lipoiden Komponenten — etwa Phosphatiden — gespeichert. 3. Die Farbstoffe reagieren chemisch mit bestimmten Plasmabestandteilen; die basischen Farbstoffe z. B. mit Nucleinsäuren.

Alle drei Möglichkeiten werden auch in der Literatur diskutiert. Die völlige Ablehnung einer vitalen Farbstoffspeicherung im Plasma, wie sie neuerdings wieder JOHANNES (1954) ausspricht, wird heute kaum noch geteilt, wenn auch zugegeben werden muß, daß die Zelle gegen Giftwirkungen im Plasma naturgemäß bedeutend empfindlicher ist als bei einer Lokalisierung des Giftes im Zellsaft.

Eine Farbstoffspeicherung auf Grund einer Adsorption der Farbkationen bzw. -anionen an die Eiweißkomponente wird von STRUGGER (1947) und besonders von seinem Schüler KÖLBEL (1947, 1948, 1949) vertreten. Außerdem wird zur Voraussetzung gemacht, daß das Farbstoffmolekül in seiner Gestalt den submikroskopischen Plasmastrukturen sterisch adäquat sein muß. Früher nahm STRUGGER wohl für K-Fluorescein (1938b) und Acridinorange (1941) eine Anionen- bzw. Kationenadsorption an, neigte dann aber dazu, die Cytoplasmafluorochromierung mit Neutralrot (1940a), Benzoflavin (1941) und Auramin (1943) auf eine Lösung in den Plasmalipoiden zurückzuführen. Die Fluorochromierung mit Pyronin soll nach beiden Mechanismen erfolgen (STRUGGER 1941). KÖLBEL (1948) betont dann aber, daß auch die Fluorochromierung mit Neutralrot auf einer Kationenadsorption beruhen soll, und er erklärt die Fluorescenzerscheinung mit einem Konzentrationseffekt. Die Farbkationen zeigen, an Gelatine adsorbiert, nur in geringer Konzentration eine gelbe Fluorescenz. Nach BETZ (1953) soll auch die Plasmafärbung mit Prune pure auf einer Ionenadsorption beruhen. Nach seiner Auffassung ist eine Dissoziation des Farbstoffes im Plasma eine Voraussetzung für seine Speicherung. Auch die Auffassung von GOLDACRE (1952) über die Farbstoffaufnahme und den Farbstofftransport bei Amöben könnte man hier anführen. Danach sind die Proteinmoleküle im entfalteten Zustand fähig, Farbstoffe zu adsorbieren, im gefalteten Zustand dagegen nicht. Doch ist diese Färbung des Amöbenplasmas nicht ohne weiteres mit der Färbung des Plasmas einer normalen Pflanzenzelle zu vergleichen. Das Amöbenplasma verhält sich wie anderes tierisches Plasma, bei dem man eine Diffus- und eine Granulafärbung unterscheiden muß (v. MOELLENDORFF 1918). Nur die Diffusfärbung entspricht in ihrer Mechanik der Cytoplasmafärbung einer Pflanzenzelle, die Granulafärbung weist dagegen vielmehr Parallelen zu der Vacuolenfärbung auf (DRAWERT 1948, vgl. auch LEPESCHKIN 1936).

Die vorliegenden, experimentell gewonnenen Ergebnisse sprechen aber vielmehr für die Auffassung, daß die Plasmafärbung der Pflanzenzelle ebenso wie die diffuse Färbung der tierischen Zelle auf einer Farbstoffspeicherung in den lipoiden Phasen des Plasmas beruht. Diese Auffassung wurde bereits von LEPESCHKIN (1913, vgl. auch 1950) und von v. MOELLENDORFF (1918, 1920) vertreten. Neuerdings setzen sich dafür vor allem DRAWERT (1938b, 1951), HÖFLER (1948, 1949, 1951) und HOFMEISTER (1948) ein, für die tierische Zelle vergleiche man MONNÉ (1942). Gegen die Vorstellung von STRUGGER (1947) und KÖLBEL (1947), daß das Plasma eine Intrabilität für Farbionen besitzt und die Plasmafärbung durch eine Ionenadsorption zustande kommt, wendet sich besonders HÖFLER.

Nach DRAWERT (1948 u. f.) sprechen folgende Punkte nicht nur für eine Speicherung der Farbstoffe in der lipoiden Phase des Plasmas, sondern auch für eine Abschirmung der meisten freien Carboxyl- und Aminogruppen der Plasmaproteine und -proteide, wahrscheinlich durch Lipoide:

1. Im allgemeinen ist eine Färbung nur mit schwach dissoziierten Farbstoffen, unabhängig von ihrer Ladung, nur auf Grund ihrer Lipoidlöslichkeit möglich.

2. Der Farbton der vom Cytoplasma gespeicherten basischen Farbstoffe entspricht immer dem Farbton der in neutralen Lipoiden gelösten Farbbase.

Aus diesem Farbton wurde bisher fälschlicherweise auf eine alkalische Reaktion des Plasmas geschlossen. — Die Tatsache, daß sich das Plasma mit dem sauren Indicatorfarbstoff Methylrot gelb ($= p_H \sim 7$) anfärbt (SCHAEDE 1924, DRAWERT 1941), ist kein stichhaltiges Gegenargument, da der aus einer sauren, roten, wäßrigen Lösung mit hydrophoben Medien ausgeschüttelte Farbstoff der lipophilen Phase merkwürdigerweise einen gelben bis orangefarbigen Farbton verleiht. Man sollte eher den roten Farbton der Farbsäure vermuten.

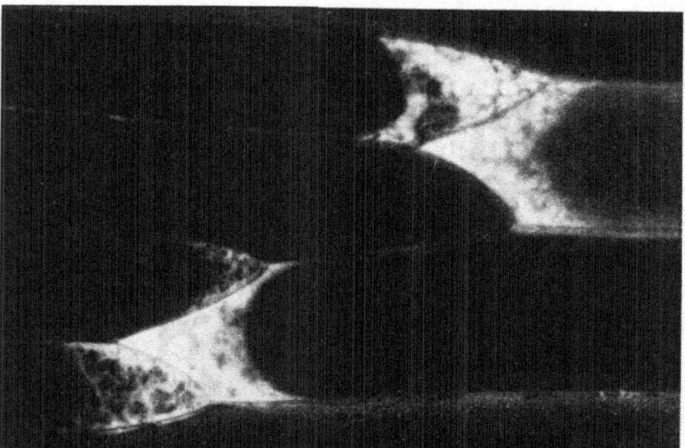

Abb. 8. Oberepidermiszellen eines Schuppenblattes von *Allium cepa* nach 24stündigem Aufenthalt in einer Neutralrotlösung 1:10000 in Leitungswasser. Die Vacuolen sind sehr intensiv gefärbt, ebenso zeigen die Membranen eine Färbung, das in den Zellecken systrophisch zusammengezogene Cytoplasma (Vacuolenkontraktion) beginnt, als erstes Anzeichen des Absterbevorganges, sich zu färben. (Nach DRAWERT 1948.)

3. Besonders instruktiv sind fluorescenzoptische Versuchsergebnisse von STRUGGER (1940a) und TOTH (1952) mit Neutralrot. Hier zeigt das Cytoplasma die Fluorescenzfarbe der in neutralen hydrophoben Medien gelösten Farbbase, nicht die der wäßrigen Farbbasenlösung. Die Möglichkeit einer Reduktion des Neutralrots im Plasma, was ebenfalls zu einer Fluorescenz führen müßte, kann ausgeschlossen werden, da entsprechend dem Redoxpotential des Neutralrots die lebende Zelle nach den bisherigen Erfahrungen zu einer Reduktion des Neutralrots, selbst bei Sauerstoffmangel, nicht befähigt ist. Die auf S. 273 geschilderte Annahme von KÖLBEL (1948), daß auch bei der Neutralrotfluorochromierung eine Kationenadsorption vorliegt, läßt sich nicht aufrechthalten, da nach unveröffentlichten eigenen Untersuchungen eine künstliche Alkalinisierung des Plasmas keine Fluorescenzlöschung, sondern im Gegenteil eine Verstärkung bedingt.

4. Mit den stärker dissoziierten Farbstoffen konnte bisher im Hellfeld nie eine vitale Cytoplasmafärbung beobachtet werden, die aber bei dem Vorhandensein freier kationischer Gruppen unbedingt bei einer dafür günstigen p_H-Konstellation Außenmedium/Zellsaft zu fordern wäre. Das lebende Cytoplasma bleibt im Gegensatz zum reaktionsfähigen toten immer farblos bzw. zeigt nie den Farbton des Farbsalzes oder den des Farbkations, was besonders gut bei der Erscheinung der Vacuolenkontraktion zu beobachten ist.

5. Die unter Punkt 4 geforderte Anfärbung des Cytoplasmas tritt aber beim Absterbevorgang als postvitale Erscheinung auf, was man wieder gut an Zellen mit Vacuolenkontraktion erkennen kann, wie Abb. 8 belegt. Erst beim Absterbevorgang werden die Verbindungen Eiweiß-Lipoid gelöst, das Cytoplasma wird labil und auch nach außen reaktionsfähig, es färbt sich im Tone des Farbsalzes. Die stärkere Färbbarkeit des absterbenden Plasmas kann auch mit einer Entfaltung der vorher gefalteten oder aufgerollten Proteinmoleküle zusammen-

hängen. Dies unterschiedliche Verhalten von lebendem und absterbendem bzw. totem Plasma gegenüber den stärker dissoziierten Farbstoffen ist eine bekannte Erscheinung (DEVAUX 1930). Für die hier vorgetragene Auffassung sprechen auch die Befunde an tierischen Zellen. Nach SPEK (1951) zeigt das mit Safraninen vital gefärbte tierische Plasma ein Spektrum, das mit dem einer Safraninlösung in Phosphatiden übereinstimmt. Erst bei einer Zellschädigung treten auch die Banden einer wäßrigen Safraninlösung auf als Anzeichen dafür, daß sich jetzt auch hydro phile Substanzen (wohl der Eiweißkörper) färben (vgl. auch LEPESCHKIN 1913)

Damit soll nicht die Möglichkeit einer Adsorptionsfärbung des lebenden Cytoplasmas grundsätzlich geleugnet werden, doch scheint diesem Färbungsmechanismus nur eine sehr untergeordnete Bedeutung zuzukommen.

Eine dritte Möglichkeit besteht in der chemischen Bindung der Farbstoffe an Plasmakomponenten. Dies ist vor allem für basische Farbstoffe postuliert worden und trifft auch wohl zu, wenn das Plasma z. B. kleine Gerbstoffbläschen führt. Doch handelt es sich hierbei nicht um die Vitalfärbung des eigentlichen Cytoplasmas. Neuerdings wird auch mit Bindungen an Nucleinsäuren gerechnet. Nach ZEIGER, HARDERS und MÜLLER (1951) fallen im Cytoplasma der Nervenzellen von Frosch und Regenwurm die Areale stärkster Acridinorangespeicherung mit den Orten höchster Ribonucleinsäurekonzentration zusammen, und STICH (1951) schließt aus Wachstumshemmungen bei *Acetabularia*-Fragmenten auf eine Affinität des Trypaflavins zu Nucleinsäuren auch im Cytoplasma. Wieweit den Nucleinsäuren wirklich eine Bedeutung für die Cytoplasmafärbung zukommt, müssen erst zukünftige Untersuchungen klären. Nach dem oben über die „Lösungsfärbung" Gesagten, scheint aber auch der chemischen Bindung — jedenfalls nach unseren bisherigen Kenntnissen — höchstens eine untergeordnete Bedeutung zuzukommen.

C. Theorien der Kernfärbung.

Nach BEUTNER, CAYWOOD, LOZNER und DOUTHITT (1930) sollen nur die in Olivenöl + Ölsäure löslichen basischen Farbstoffe den Kern färben und die in Olivenöl + Amin löslichen sauren Farbstoffe das Cytoplasma. GICKLHORN (1930) und KELLER (1931) halten dagegen nur mit sauren Farbstoffen eine vitale Kernfärbung für möglich, STRUGGER (1932) erhielt sowohl mit sauren wie mit basischen Farbstoffen Kernfärbungen und zwar färbte sich mit den sauren die Karyolymphe und mit den basischen das Karyotin. Allerdings dürfte es sich hierbei nicht mehr um wirkliche Vitalfärbungen gehandelt haben, der Autor spricht auch zum Teil von „überlebenden" Kernen. In der Folgezeit hat es sich dann aber bestätigt, daß sowohl mit sauren wie mit basischen Farbstoffen Vitalfärbungen des Kernes erhalten werden können. Nach STRUGGER (1939, 1943) soll es sich bei den Kernfärbungen um eine Adsorption von Anionen bzw. Kationen handeln. Diese Vorstellung wird von HÖFLER, TOTH und LUHAN (1949) abgelehnt. Aus Fluorochromierungsversuchen mit Acridinorange auch an Kernen, die auf verschiedene Weise abgetötet waren, schließen die Autoren, daß es sich bei den basischen Farbstoffen um eine chemische Bindung des Farbstoffs sehr wahrscheinlich mit Thymonucleinsäuren handelt. Auf eine mögliche Bindung mit Nucleinsäuren hat auch schon KIESEL (1930) hingewiesen.

Was für ein Mechanismus der vitalen Kernfärbung zugrunde liegt, läßt sich heute noch nicht entscheiden. Man sollte bei den theoretischen Betrachtungen aber nicht außer acht lassen, daß die meisten Plasmafärber auch gute Kernfärber sind und umgekehrt und daß ferner bei einer gleichzeitigen Kern- und Plasmafärbung im allgemeinen keine größeren Unterschiede im Farbton auftreten. Vielleicht liegt beiden Färbungen ein gleicher Mechanismus zugrunde.

Es soll auch noch auf die bis zu einem gewissen Grade reversiblen Kernentmischungen hingewiesen werden, die man bei längerer Einwirkung des amphoteren Farbstoffes Cölestinblau erhält (DRAWERT 1954). Diese Entmischungserscheinungen sprechen wohl eher für eine Adsorptions- oder auch chemische Bindung des Farbstoffes an kerneigenen Substanzen, andererseits kann aber auch eine Entmischung der lipoiden Bestandteile des Kernes vorliegen, so daß sich aus dieser Erscheinung auch noch keine Schlußfolgerungen über die Mechanik der Kernfärbung ableiten lassen. Es kann sich hier aber auch um eine spezifische Reaktion des Cölestinblaus handeln, die sich nicht verallgemeinern läßt, zumal mit anderen Farbstoffen entsprechende Entmischungserscheinungen bisher nicht beobachtet worden sind.

D. Theorien der Plastidenfärbung.

Die wenigen vorhandenen Angaben über eine Vitalfärbung der Plastiden sprechen alle dafür, daß eine Lösung des Farbstoffes in Lipoiden vorliegt. Besonders Rhodamin B läßt sich leicht wieder auswaschen, dies ist selbst mit einer verdünnten Rhodamin B-Lösung möglich (STRUGGER 1937b). Der Farbstoff scheint sich also nur auf Grund seiner Lipophilie entsprechend dem HENRY-NERNSTschen Verteilungssatz in den Plastiden anzureichern, eine festere Bindung kann nicht vorliegen. Dasselbe trifft sehr wahrscheinlich auch bei der Färbung von *Allium*-Leukoplasten mit Cölestinblau zu. Sobald der Farbstoff in der Zelle durch Reduktion seine Lipophilie verliert, verläßt er die Plastiden und reichert sich in der Vacuole an (DRAWERT 1954). MENKE (1938) schließt aus dem Dichroismus der mit Rhodamin B vital gefärbten Chloroplasten auf die Art der Anordnung der Lipoidmoleküle.

E. Theorien der Mitochondrienfärbung.

Nach MONNÉ (1942) können sich die Mitochondrien der tierischen Zelle entweder mit stärker lipoidlöslichen Farbstoffen zusammen mit dem Grundplasma oder elektiv mit nicht lipoidlöslichen Farbstoffen färben. Demnach beruht die Färbung in dem einen Fall auf einer Lösung der Farbbasenmoleküle in der lipoiden Phase der Mitochondrien und im anderen Fall auf einer Adsorption an die Eiweißphase. In der Pflanzenzelle soll die Vitalfärbung der Mitochondrien mit Rhodamin B auch auf einer Lösung des Farbstoffes in Lipoiden zurückzuführen sein (STRUGGER 1938a, JOHANNES 1941). Die für die Unterscheidung der Plasmaeinschlüsse so wichtige elektive Färbung mit Janusgrün B ist dagegen noch recht ungeklärt. Nach MARSTON (1923) wird der Farbstoff durch den Stickstoff des Diazinringes an die Mitochondrien gebunden, und nach BRENNER (1953) ist der Basizitätsgrad eines Farbstoffes für sein Vitalfärbungsvermögen gegenüber Mitochondrien verantwortlich. Je basischer der Farbstoff, desto geringer ist die zur Vitalfärbung notwendige Konzentration. Eine Verbindung des Farbstoffes mit Cholinesterase soll nach ZACKS und WELSCH (1951) eine Voraussetzung für die Vitalfärbung sein. LAZAROW und COOPERSTEIN (1953a) sind der Meinung, daß es nicht notwendig ist, eine stärkere Bindung des Farbstoffs an das Mitochondrienprotein als an die übrigen Proteine der Zelle anzunehmen. Wahrscheinlich sind alle Teile der lebenden Zelle befähigt, das Janusgrün B zu binden und zu reduzieren. Die Mitochondrien sind aber der Sitz des Cytochrom-Oxydasesystems und aus diesem Grunde bleibt der Farbstoff in den Mitochondrien in der oxydierten Form erhalten und bedingt so, gekoppelt mit der Bildung einer unlöslichen Verbindung, ihre elektive Färbung. Diese Auffassung von LAZAROW und COOPERSTEIN kann aber nicht zutreffen. Dagegen spricht das Auftreten einer Rotfärbung von Kern und Plasma nach einem Deckglasabschluß. Das blaue Janusgrün B wird bei Sauerstoffmangel zunächst zur ersten roten Reduktionsstufe reduziert, die sich sowohl im Kern wie im Plasma, aber vorwiegend in der

Vacuole anreichert. Dann geht die Reduktion unter starker Zunahme der lipophilen Eigenschaften weiter bis zur im normalen Licht farblosen Stufe, die nach DRAWERT (1953) grünlich fluoresciert und elektiv die Sphärosomen fluorochromiert. Würde die Vorstellung von LAZAROW und COOPERSTEIN richtig sein, dürften sich Kern und Plasma bei Sauerstoffmangel nicht rot färben, und außerdem müßten alle Zellbestandteile außer den Mitochondrien von vornherein eine grünliche Fluorescenz zeigen, das ist aber nicht der Fall. Eine Lösung von Janusgrün B in der lipoiden Phase, entsprechend dem Rhodamin B, liegt aber auch wohl kaum vor, da mit Zunahme der Lipophilie der Farbstoff sich nicht in den Mitochondrien, sondern in den Sphärosomen anreichert. Ähnlich verhält sich auch das Berberinsulfat (DRAWERT 1953).

F. Theorien der Sphärosomenfärbung.

Zum Unterschied von der Mitochondrienfärbung scheint die elektive Sphärosomenfärbung nach den bisher vorliegenden Untersuchungen eindeutig auf einer Lösung in der lipoiden Phase dieser Plasmaeinschlüsse zu beruhen. Das nimmt STRUGGER (1936a) für die Fluorochromierung mit Berberinsulfat an, und dafür sprechen vor allem die Versuche von DRAWERT (1952, 1953, 1955) mit Nilblau, Berberinsulfat, Janusgrün B und Aminopyren. Allerdings dürfte man auch die Möglichkeit einer Oberflächenadsorption nicht außer acht lassen.

G. Theorien der Vacuolenfärbung.

Die weitaus größte Zahl der Arbeiten befaßt sich mit der Theorie der Zellsaftfärbung, zumal in den älteren Arbeiten eine Klärung der Permeabilitätsverhältnisse der Zelle von den Untersuchungen der vitalen Vacuolenfärbung erhofft wurde. Man hat dann aber erkannt, daß auch die Zellsaftfärbung mehr ein Problem der Speicherung als der Permeabilität darstellt, wenn natürlich auch die Permeabilitätsverhältnisse für die Speicherung von Bedeutung sein können.

Für die Mechanik der Farbstoffspeicherung in der Vacuole lassen sich folgende sechs Möglichkeiten postulieren:

1. Der eindringende Farbstoff wird von Zellsaftsubstanzen chemisch gebunden und fällt aus; oder die Verbindung bleibt in Lösung, ist aber so groß molekular oder kolloidal, daß sie nicht exosmieren kann. In allen Fällen wird das Konzentrationsgefälle solange aufrechterhalten, bis alle zelleigenen Substanzen, die mit dem Farbstoff reagieren, abgesättigt sind.

Eine Anreicherung auf Grund chemischer Bindung trifft wohl besonders für die basischen Farbstoffe beim Vorliegen von Phenolabkömmlingen im Zellsaft zu. Bereits PFEFFER (1886) nahm eine chemische Bindung an Gerbstoffe oder andere zelleigene Stoffe an. LEPESCHKIN (1911a) und RUHLAND (1912) denken an salzartige Bindungen der Farbbase an hochmolekulare Säuren und MCCUTCHEON und LUCKÉ (1924) sprechen von einer Verbindung mit sauren Zellsaftsubstanzen. Eine chemische Bindung an großmolekulare Inhaltsstoffe liegt auch der Vorstellung von DEVAUX (1930) zugrunde. HÖFLER (1947a) bezeichnet die Zellsäfte mit chemischen Affinitäten zu den Farbstoffen als „volle Zellsäfte", sie zeigen mit Acridinorange eine grüne Fluorescenz und färben sich mit Neutralrot violett.

2. Entsprechend dem ersten Mechanismus muß es auch zu einer Speicherung kommen, wenn der Farbstoff in der Vacuole an zelleigene Kolloide adsorbiert wird. Mit dieser Möglichkeit rechnen ebenfalls viele Autoren, so SCARTH (1926a), GUILLIERMOND und OBATON (1934), GUILLIERMOND (1937a), BÜNNING (1936),

DRAWERT (1948), BETHE (1950), BARTELS (1954). PERNER (1950a) nimmt auch für sulfosaure Farbstoffe eine Bindung an positiv geladene Vacuolenkolloide an.

3. Der Farbstoff kann sich aber auch bei ausreichender Lipophilie in lipoiden Phasen des Zellsaftes anreichern. Doch scheint diese Möglichkeit im Gegensatz zur Cytoplasma-, Plastiden- und Sphärosomenfärbung für die Vacuolenfärbung nur von untergeordneter Bedeutung zu sein; jedenfalls soweit es sich um neutrale Lipoide handelt. Saure Lipoide können für die Speicherung basischer Farbstoffe eine größere Rolle spielen, aber dann erfolgt die Speicherung wohl eher nach den unter 1. und 2. aufgeführten Mechanismen.

4. Setzen wir eine Permeation der Farbstoffe als Ionen voraus, so kann es zu einer Anreicherung in der Vacuole auf Grund eines DONNAN-Gleichgewichtes kommen (AXMACHER und NARATH 1934 und vor allem WILBRANDT 1938). Die p_H-Abhängigkeit der Speicherung von stärker dissoziierten Farbstoffen würde zugunsten eines DONNAN-Gleichgewichtes sprechen; denn danach könnte bei gleicher Innen- und Außen-cH und erst recht bei höherer Alkalinität des Zellsaftes keine Speicherung basischer Farbstoffe mehr stattfinden.

5. Die Abhängigkeit der Speicherung vom p_H-Wert der Außenlösung und des Zellsaftes läßt sich aber auch sehr gut mit der Annahme erklären, daß das Plasma zwar für Moleküle, aber nicht für Ionen permeabel ist. In diesem Fall muß es zu einer Speicherung kommen, wenn die cH-Differenz Außen/Innen so liegt, daß der Farbstoff im Zellsaft stärker dissoziiert ist als im Außenmedium. Bei umgekehrtem cH-Verhältnis muß eine Färbung der Vacuole ausbleiben. Für die basischen Farbstoffe liegen mehrere Angaben dafür vor, daß eine Speicherung im Zellsaft nur stattfindet, wenn die Innen-cH größer ist als die Außen-cH (IRWIN 1923a, MCCUTCHEON und LUCKÉ 1924, GUTSTEIN 1932, OSTERHOUT 1933, DRAWERT 1937b, 1948, COLLANDER, LÖNEGREN und ARHIMO 1943, HÖFLER 1947a, BARTELS 1954, vgl. auch diese Arbeit S. 266, Tabelle 2). Die gegenteiligen Angaben von ROHDE (1917) und BETHE (1922) haben sich nicht bestätigen lassen. HÖFLER bezeichnet das auf der Annahme einer Impermeabilität des Plasmas für Ionen beruhende Speicherungsprinzip als „Ionenfalle", und die Vacuolen, die sich nach diesem Mechanismus mit basischen Farbstoffen färben, sollen „leere Zellsäfte" besitzen. Dieser Speicherungsmechanismus wird eingehend von IRWIN diskutiert.

6. Bei den bisher betrachteten Möglichkeiten der Farbstoffspeicherung handelt es sich um passive Vorgänge, die nicht direkt mit der Lebenstätigkeit der Zelle zusammenhängen, sondern höchstens indirekt, sei es in Aufrechterhaltung der Semipermeabilität des Plasmas oder der cH des Zellsaftes u. a. Besonders für die Aufnahme und Speicherung der sulfosauren Farbstoffe wird aber eine aktive, energieverbrauchende Tätigkeit der Zelle angenommen, da hier diese Vorgänge von der Atmung der Zelle abhängen (COLLANDER und HOLMSTRÖM 1937, DRAWERT und ENDLICH 1955). Es fragt sich nur, ob die energetischen Vorgänge direkt mit der Aufnahme zusammenhängen (COLLANDER), oder ob bei diesen Prozessen Stoffwechselprodukte entstehen, die für die Speicherung der sulfosauren Farbstoffe und damit für ihr Sichtbarwerden in der Zelle wichtig sind (DRAWERT). Dafür, daß auch in diesem Fall Fragen der Speicherung und weniger der Permeabilität im Vordergrund stehen, spricht die antagonistische Färbung der Wundränder von Blumenblattstücken mit sauren und basischen Farbstoffen, wie sie in Abb. 9a und b wiedergegeben ist (DRAWERT 1941). Eine Parallele dazu bieten die jungen Blättchen von *Helodea densa*. Mit dem basischen Fluorochrom Acridinorange färben sich nur die ausgewachsenen Zellen, die embryonalen Zonen bleiben farblos. Mit dem sulfosauren Fluorochrom Brillantsulfoflavin FF erhält man das

umgekehrte Bild. Die mit Acridinorange und die mit Brillantsulfoflavin FF gefärbten Zonen schließen sich gesetzmäßig aus (PERNER 1950b).

Alle sechs der hier betrachteten Möglichkeiten können der vitalen Zellsaftfärbung zugrunde liegen. Je nach den Eigenschaften des Farbstoffes und dem physiologischen Zustand der Zelle wird die Färbung nach dem einen oder dem anderen Mechanismus erfolgen. Vor allem müssen wir uns vor Verallgemeinerungen hüten. Früher wurde z. B. der Einfluß der cH oder der Salze auf die Vitalfärbung einseitig von den Permeabilitätseigenschaften des Plasmas her betrachtet. Heute wissen wir, daß die sich hier abspielenden Vorgänge viel komplizierterer Natur sind. cH und Salze wirken nicht nur auf die Permeabilität der Plasmagrenzflächen, sondern ändern auch den Dissoziationsgrad des benutzten

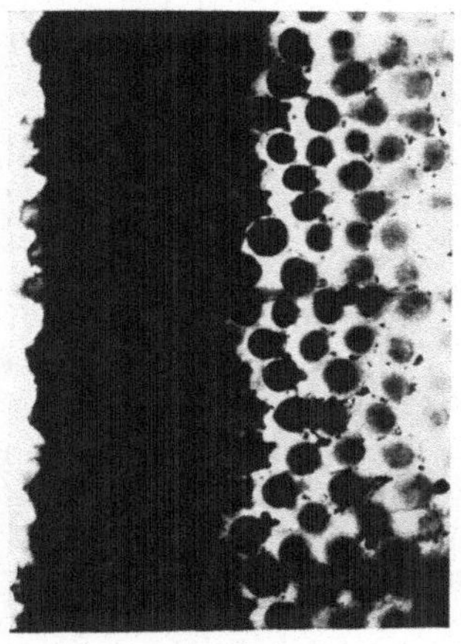

a

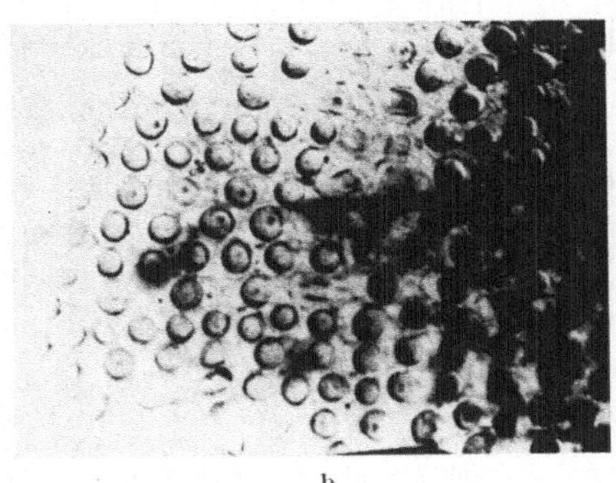

b

Abb. 9a u. b. Schnittrand (links im Bild) eines Blumenblattes von *Chrysanthemum leucanthemum* a mit Neutralrot gefärbt in 1 Mol Traubenzucker plasmolysiert. In den toten Schnittrandzellen sind alle Zellbestandteile, auch die Membranen, so stark gefärbt, daß der Schnittrand im Bild nur als undifferenzierte dunkle Zone erscheint. Die dann folgenden lebenden Zellen haben das Neutralrot sehr intensiv in den Vacuolen gespeichert. Die Vacuolenfärbung nimmt aber mit der Entfernung der Zellen vom Schnittrand an Intensität ab. b Mit dem sauren Farbstoff Ponceau gefärbt und plasmolysiert. In den toten Schnittrandzellen sind alle Zellbestandteile, ebenso wie die dann zunächst folgenden lebenden Zellen, farblos. — Die in dem plasmolysierten Protoplasten erkennbaren dunklen Körper (meist nur einer) sind gelbliche Entmischungstropfen des Zellsaftes, wie sie in den *Chrysanthemum*-Epidermen nach einer Plasmolyse manchmal erscheinen. — Je weiter die Zellen vom Schnittrand entfernt liegen, desto stärker speichern sie den Farbstoff diffus in der Vacuole. (Nach DRAWERT 1941.)

Farbstoffes, die Ladung der Zellwände und unter Umständen den Stoffwechsel der Zelle, und alle diese Faktoren wirken sich auf die Verteilung eines Farbstoffes zwischen Außenmedium und den einzelnen Bestandteilen der Zelle aus.

Wenn wir von den basischen Farbstoffen aus den 3 Gruppen der Membran-, Plasma- und Vacuolenfärber je einen Vertreter in bezug auf das Verhalten bei der Vitalfärbung „voller“ und „leerer“ Zellsäfte, auf die Verteilung zwischen hydrophober und hydrophiler Phase im Modellversuch und auf die Wanderung in einem elektrischen Feld in Abhängigkeit von der cH der Farblösung betrachten, so erhalten wir das in Tabelle 3 wiedergegebene Bild.

In Abb. 10 ist die Farbstoffverteilung an einer Pflanzenzelle in Abhängigkeit von der cH der Farbstofflösung und des Zellsaftes für Zellen mit „leeren“ Zellsäften nach einer Kurzfärbung schematisch dargestellt. Aus diesen Übersichten geht hervor, wie verschieden das Verteilungsbild ein und desselben Farbstoffes in Abhängigkeit allein von der cH aussehen kann. So erklären sich wohl viele der in der Vitalfärbungsliteratur anzutreffenden Widersprüche.

Folgende Punkte können wir aus Tabelle 3 und Abb. 10 als Faustregeln für die Verteilung eines basischen Farbstoffes nach einer Kurzfärbung auf die

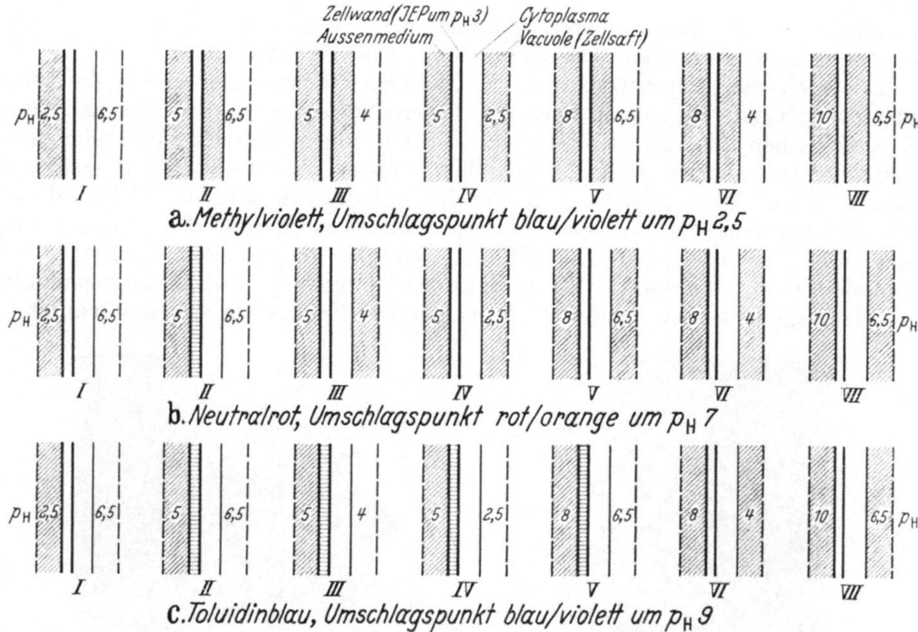

Abb. 10a—c. Schematische Darstellung der Farbstoffverteilung in einer Zelle, die arm an Speicherstoffen ist (Vacuole mit „leerem" Zellsaft), in Abhängigkeit von den p_H-Werten des Außenmediums und des Zellsaftes für die basischen Farbstoffe Methylviolett (a), Neutralrot (b) und Toluidinblau (c). (Nach DRAWERT 1948.)

Tabelle 3. *Die Verteilung von Farbstoffen in der lebenden Zelle (I—III),*
Es bedeuten: O.E. und U.E. = Ober- bzw. Unterepidermiszellen der Schuppenblätter der oberen. Die Anzahl der Kreuze gibt die Färbungsintensität an,

Farbstoff	p_H	I Membran		II Vacuole		III Cytoplasma
		O.E.	U.E.	O.E.	U.E.	O.E.
Methylviolett	2,0	—	—	—	—	—
	3,0	+	—	—	+++	++++
	5,0	—	—	—	++++	++++
	7,0	—	—	—	++++	++++
	8,5	—	—	—	++++	++++
	10,0	—	—	—	++++	++++
	11,5	—	—	—	++++	++++
Neutralrot	2,0	—	—	—	—	—
	3,0	++++	++++	—	—	—
	5,0	++	—	++	++++	—
	7,0	—	—	++++	++++	—(+)
	8,5	—	—	++++	++++	—(+)
	10,0	—	—	++++	++++	—(+)
	11,5	—	—	++++	++++	—(+)
Toluidinblau	2,0	—	—	—	—	—
	3,0	++++	++++	—	—	—
	5,0	++++	++++	—	—	—
	7,0	++++	++++	—	—	—
	8,5	++++	++	—	++	—
	10,0	—	—	++++	++++	—
	11,5	—	—	++++	++++	—

einzelnen Zellbestandteile ableiten: 1. Liegt der Farbstoff oberhalb des IEP (nach der alkalischen Seite) der Zellwand in einer nichtdissoziierten lipophilen Form vor, so wird er das Plasma färben, vorausgesetzt, daß der Zellsaft nicht so sauer ist, daß der Farbstoff in der Vacuole dissoziieren kann und daß der Zellsaft arm an Speicherstoffen ist. Trifft einer der beiden letzten Punkte nicht zu, dann wird der Farbstoff statt vom Plasma von der Vacuole gespeichert. 2. Liegt der Farbstoff oberhalb des IEP der Zellwand in dissoziiertem Zustand vor, dann färbt er die Zellwand. Mit abnehmender Außen-cH wird aber die Dissoziation des Farbstoffes zurückgedrängt, und es tritt eine hydrolytische Spaltung auf. Die Farbbasenmoleküle dringen in die Vacuole ein und werden dort chemisch gebunden oder bei entsprechender Aciditätslage des Zellsaftes dissoziiert und von Vacuolenkolloiden adsorbiert bzw. nach dem Prinzip der „Ionenfalle" gespeichert. Dadurch tritt die Vacuole als Konkurrent zur Zellwand auf, und wir erhalten in Abhängigkeit von der Außen-cH bei einem bestimmten p_H-Wert einen Umschlag von einer Membran- zu einer Vacuolenfärbung. Der p_H-Wert, bei dem dieser Färbungsumschlag auftritt, wird um so kleiner sein, je weiter im sauren Bereich eine hydrolytische Spaltung des Farbstoffes einsetzt und je saurer der Zellsaft bzw. je reicher die Vacuole an farbstoffbindenden Substanzen chemischer und adsorptiver Art ist. Die von BARTELS (1954) geäußerte Ansicht, daß der p_H-Wert der Außenlösung, bei dem der Umschlag von der Membran- zur Vacuolenfärbung erfolgt, mit der aktuellen cH der Vacuole identisch sein soll, kann zufällig für einen Farbstoff mit bestimmter Dissoziationskonstante zutreffen (vgl. z. B. S. 266, Tabelle 2), darf aber nicht verallgemeinert werden. Je nach der Dissoziationskonstante des benutzten Farbstoffes liegt dieser Umschlagspunkt

ihre Verteilung im Zweiphasenmodell (IV) und ihr Verhalten im elektrischen Feld (V).
ruhender Zwiebeln von *Allium Cepa*. Der Zellsaft der unteren Epidermis ist saurer als der
— = ungefärbt, ? = Farbstoff entfärbt sich. (Nach DRAWERT 1948.)

IV Verteilung zwischen						V Wanderung	
a		b		c		Kathode	Anode
Wasser	Toluol	Wasser	Chloroform	Wasser	Toluol + Ölsäure		
++++	—	+	+++	+	+++	++++	—
++++	—	—	++++	—	++++	+++	+
++++	—	—	++++	—	++++	+++	+
++++	—	—	++++	—	++++	+++	+
++	+	—	++++	—	++++	++	++
?	?	—	++++	—	++++	++	++
?	?	—	++++	—	++++	++	++
++++	—	++++	—	++++	—(+)	++++	—
++++	—	+++	+	++	++	++++	—
+	+++	++	++	+	+++	+++	+
—	++++	+	+++	+	+++	++(+)	++
—	++++	—	++++	—	++++	++	++
—	++++	—	++++	—	++++	++	++
—	++++	—	++++	—	++++	++	++
++++	—	++++	—	++++	—	++++	—
++++	—	++++	—	++++	—(+)	++++	—
++++	—	++++	—	+++	+	++++	—(+)
++++	—	++++	—(+)	++	++	+++	+
++++	—(+)	++	++	+	+++	+++	+
+	+++	—	++++	—	++++	++	++
—(+)	++++	—	++++	—	++++	++	++

für ein und dieselbe Zelle bei einem anderen p_H-Wert. Außerdem verhalten sich „leere" und „volle" Zellsäfte ganz verschieden. Beginnt die hydrolytische Spaltung eines Farbstoffes bereits unterhalb des Neutralpunktes im sauren Bereich, dann kann es außer zu einer Membran- oder Vacuolenfärbung bei Zellen mit „leeren" und nicht zu sauren Zellsäften zu einer Plasmafärbung kommen.

Eine Änderung des Stoffwechsels wird sich auf die cH des Zellsaftes, den Gehalt der Vacuole an Speicherstoffen und auf noch andere Faktoren der Zelle auswirken; ferner kann auch der Farbstoff durch Reduktion oder Oxydation in seinen physikalisch-chemischen Eigenschaften beeinflußt werden. Es ist daher aus dem oben Gesagten zu schließen, daß Stoffwechseländerungen auch eine Änderung der Verteilung des Farbstoffes in der Zelle bewirken. In der Tat kann ja bereits ein Deckglasabschluß und der damit verbundene allmählich auftretende Sauerstoffmangel das Verteilungsbild völlig ändern. Wir dürfen dabei auch nicht vergessen, daß der aufgenommene Farbstoff als Fremdkörper in der Zelle den Stoffwechsel beeinflussen wird.

In der Behandlung der Vitalfärbung stand das Problem der Speicherung im Vordergrund. Einer der Hauptfaktoren, das cH-Gefälle Außen/Innen, wird sich aber nicht nur auf die Speicherung auswirken, sondern auch für die Aufnahme von Bedeutung sein, besonders wenn es sich um eine Ionenaufnahme handelt. Es sei nur an den Austausch etwa von $Na^{\cdot}$ gegen $H^{\cdot}$ durch Kationenaustauscherharze erinnert und die Übertragung dieses Prinzips auf die Stoffaufnahme der Pflanze, wie es z. B. von Reichenberg und Sutcliffe (1954) durchgeführt wird, oder an die Trägertheorie, die wohl weniger für die Aufnahme der lipoidlöslichen basischen Farbstoffe, aber für die der sulfosauren Farbstoffe von Bedeutung sein kann. Doch werden diese Fragen an anderer Stelle eingehender behandelt. Aus dem Dargestellten geht aber wohl deutlich hervor, daß es sich bei der Aufnahme und der Speicherung von Farbstoffen durch die Zelle um sehr komplexe Vorgänge handelt, die sich nicht mit einer allgemein gültigen Theorie erklären lassen.

Literatur.

Agthe, C.: Über die physiologische Herkunft des Pflanzennektars. Ber. schweiz. bot. Ges. **61**, 240—273 (1951). — Albach, W.: Zellphysiologische Untersuchungen über vitale Protoplasmafärbung. Protoplasma **5**, 412—442 (1928). — Alexandrow, W. J., u. D. N. Nassonov: On the cause of colloidal changes in the protoplasm and the augmentation of its affinity to dye substances, as produced by injurious influences. Arch. f. Anat. **22**, 11—43 (1939). — Auerbach, R.: Zur Kolloidchemie der Färbevorgänge. Kolloid-Z. **33**, 262—264 (1923). — Axmacher, Fr.: Die Beeinflussung von Zell- bzw. Organfunktionen durch organische Farbstoffe. IV. Zur Frage der Farbstoffaufnahme durch die lebende Zelle (Hefe). Arch. exper. Path. u. Pharmakol. **171**, 289—310 (1933). — Axmacher, Fr., u. H. Narath: Die Beeinflussung von Zell- bzw. Organfunktionen durch organische Farbstoffe. VI. Weitere Beiträge zum Mechanismus der Farbstoffaufnahme durch Hefezellen. Arch. exper. Path. u. Pharmakol. **175**, 293—306 (1934).

Bailey, J.: The cambium and its derivative tissues. V. A reconnaissance of the vacuome in living cells. Z. Zellforsch. **10**, 651—682 (1930). — Bailey, J., and C. Zirkle: The cambium and its derivative tissues. VI. The effects of hydrogen ion concentration in vital staining. J. Gen. Physiol. **14**, 363—383 (1931). — Ball, J., and D. S. Jackson: Histological, chromatographic and spectrophotometric studies of toluidine blue. Stain Technol. **28**, 33—40 (1953). — Bank, O.: Die Vitalfärbung des Zellkernes mit basischen Farbstoffen. Protoplasma **29**, 587—594 (1938). — Bank, O., u. H. G. Bungenberg de Jong: Untersuchungen über Metachromasie. Protoplasma (Berl.) **32**, 489—516 (1939). — Bartels, P.: Quantitative mikrospektroskopische Untersuchung der Speicherungs- und Permeabilitätsverhältnisse akridinorange gefärbter Zellen. Planta (Berl.) **44**, 341—369 (1954). — Bautz, E.: Untersuchungen über die Mitochondrien von Hefen. Ber. dtsch. bot. Ges. **67**, 281—288 (1954). — Becker, W. A.: Application de la coloration vitale à l'étude de la cytodiérèse. C. r. Acad. Sci. Paris **196**, 2022—2023 (1933). — Vitale Cytoplasma- und Kernfärbungen. (Sammelreferat.) Protoplasma **26**, 439—487 (1936). — Benoist, H., V. Golblinu. W. Kopaczewski:

Études sur les phénomènes electrocapillaires. VIII. Coloration vitale. Protoplasma 5, 481—510 (1929). — BERL, E., u. W. PFANNMÜLLER: Über das Verhalten von organischen Farbstoffen zu Kieselsäure. Kolloid-Z. 35, 166—169 (1924). — BERSIN, TH.: Die Bedeutung der Austauschadsorption für das biochemische Geschehen. Naturwiss. 33, 108—111 (1946). — BETHE, A.: Ladung und Umladung organischer Farbstoffe. Kolloid-Z. 27, 11—17 (1920). — Der Stoffaustausch zwischen Zellen und Umgebung vom Standpunkt der Ladungshypothese und der Austauschadsorption. Naturwiss. 37, 177—182 (1950). — BETZ, A.: Untersuchungen über Verhalten und Wirkung der Vitalfarbstoffe Prune pure und Akridinorange sowie Beobachtungen über das Reduktions-Oxydationspotential in Zellen höherer Pflanzen. Planta (Berl.) 41, 323—357 (1953). — BEUTNER, R., B. E. CAYWOOD, J. LOZNER u. H. M. DOUTHITT: The relation of life to electricity. I. Stainability and electromotive forces of artificial systems which reproduce conditions in living tissues in accordance with the experimental work of G. W. CRILE. Protoplasma 10, 1—23 (1930). — BHARGAVA, K. S.: Cytological studies of some members of the family Saprolegniaceae. II. The chondriome. Cytologia 16, 84—94 (1951). — BOGEN, H. J.: Anfärbung, Schädigung und Abtötung von Hefezellen durch Acridinorange. Arch. Mikrobiol. 18, 170—197 (1953). — BOGEN, H. J., u. U. ELSTE: Zur Physiologie der Acridinorange-Wirkung. Untersuchungen an Hefezellen. Planta (Berl.) 45, 325—375 (1955). — BOGEN, H. J., u. M. KESER: Eiweiß-Abbau durch Acridinorange bei Hefezellen. Physiol. Plantarum (Copenh.) 7, 446—462 (1954). — BORNSTEIN, A., u. E. RÜTER: Über den Einfluß von Alkaloiden und Salzen auf die Vitalfärbung. I. Versuche an lebendem Gewebe. Pflügers Arch. 207, 596—613 (1925). — BORRISS, H.: Beiträge zur Kenntnis der Wirkung von Elektrolyten auf die Färbung pflanzlicher Zellmembranen mit Thioninfarbstoffen. Protoplasma 28, 23—47 (1937a). — Die Abhängigkeit der Aufnahme und Speicherung basischer Farbstoffe durch Pflanzenzellen von inneren und äußeren Faktoren. Ber. dtsch. bot. Ges. 55, 584—597 (1937b). — BRAUNER, L.: Zur Frage der postmortalen Farbstoffaufnahme von Pflanzenzellwänden. Flora (Jena) 127, 190—214 (1933). — BRENNER, S.: Supravital staining of mitochondria with phenosafranin dyes. Biochem. et Biophysica Acta 11, 480—486 (1953). — BROOKS, M. M.: Does methylene blue itself penetrate? Univ. California Publ. Zool. 31, 79—92 (1927a). — Studies on the permeability of living cells. VIII. The effect of chlorides upon the penetration of Dahlia into Nitella. Protoplasma 2, 420—427 (1927b). — Studies on the permeability of living cells. X. The influence of experimental conditions upon the penetration of methylene blue and trimethyl thionine. Protoplasma 7, 46—61 (1929). — The penetration of 1-naphthol-2-sulphonate indophenol and o-cresol indophenol in Valonia. Proc. Nat. Acad. Sci. U.S.A. 17, 1—3 (1931). — Studies on the permeability of living cells. XIV. The penetration of certain oxidation-reduction indicators with different species of Valonia. Protoplasma 17, 89—96 (1932). — Studies on the permeability of living cells. XV. A review of the penetration into Valonia ventricosa of oxidation-reduction indicators including m-bromo phenol indophenol and guaiacol indophenol. J. Cellul. a. Comp. Physiol. 3, 61—73 (1933). — BROOKS, S. C., u. M. MOLDENHAUER-BROOKS: The permeability of living cells. Berlin-Zehlendorf: Gebrüder Bornträger 1941. — BRUCH, H., u. H. NETTER: Über die Gesetze der Farbstoffverteilung an einfachen biologischen Systemen. Pflügers Arch. 225, 403—415 (1930). — BÜNNING, E.: Über die Farbstoff- und Nitrataufnahme bei Aspergillus niger. Flora (Jena) 131, 86—112 (1936). — BUNGENBERG DE JONG, H. G., u. O. BANK: Mechanismen der Farbstoffaufnahme. II. Farbstoffspeicherung, elektrische Bindung, Ausschüttelung. Protoplasma 34, 1—21 (1940).

CHADEFAUD, M.: Les colorations vitales chez les algues. C. r. Acad. Sci. Paris 197, 90—92 (1933). — CLAUDE, A.: The constitution of protoplasm. Science (Lancaster, Pa.) 97, 451—456 (1943). — COLLANDER, R.: Über die Permeabilität pflanzlicher Protoplasten für Sulfosäurefarbstoffe. Jb. wiss. Bot. 60, 354—410 (1921). — Diffusion und adenoide Tätigkeit bei der Ionenaufnahme pflanzlicher Zellen. Naturwiss. 30, 484—489 (1942). — COLLANDER, R., u. A. HOLMSTRÖM: Die Aufnahme von Sulfosäurefarbstoffen seitens pflanzlicher Zellen — ein Beispiel der adenoiden Tätigkeit der Protoplasten. Acta Soc. Fauna et Flora Fenn. 60, 129—135 (1937). — COLLANDER, R., H. LÖNEGREN u. E. ARHIMO: Das Permeationsvermögen eines basischen Farbstoffes mit demjenigen einiger Anelektrolyte verglichen. Protoplasma 37, 527—537 (1943). — CONN, H. J.: Biological stains, 6. Aufl. Geneva, N. Y. 1953. — COOPERSTEIN, S. J., A. LAZAROW and J. W. PATTERSON: Studies on the mechanism of Janusgreen B staining of mitochondria. II. Reactions and properties of Janusgreen B and its derivatives. Exper. Cell Res. 5, 69—82 (1953). — CZAJA, A. TH.: Über das Verhalten der Membranen von Pflanzenhaaren zu organischen Farbstoffen. Planta (Berl.) 10, 424—455 (1930a). — Untersuchungen über metachromatische Färbungen von Pflanzengeweben. I. Substantive Farbstoffe. Planta (Berl.) 11, 582—626 (1930b). — Untersuchungen über metachromatische Färbungen von Pflanzengeweben. II. Basische Farbstoffe. Planta (Berl.) 21, 531—601 (1934). — Untersuchungen über den Membraneffekt des Absorptionsgewebes und über die Farbstoffaufnahme in die lebende Zelle. Planta (Berl.) 26, 90—119 (1936)

DANGEARD, P. A.: Sur la distinction du chondriome des auteurs en vacuome, plastidome et sphérome. C. r. Acad. Sci. **169**, 1005—1010 (1919). — DEVAUX, A.: Les affinités cellulaires. Bull. Soc. bot. France **77**, 144—159 (1930). — DIANNELIDIS, TH.: Zur protoplasmatischen Anatomie des Blattes von Halophila stipulacea. Phyton **3**, 29—43 (1951). — DIETRICH, F.: Beobachtungen über Stoffwanderung in lebendigen Zellen. Protoplasma 8, 161—198 (1929). — DISKUS, A., u. O. KIERMAYER: Die Raphidenzellen von Haemaria discolor bei Vitalfärbung. Protoplasma (Wien) **43**, 450—454 (1954). — DÖRING, H.: Versuche über die Aufnahme fluoreszierender Stoffe in lebenden Pflanzenzellen. Ber. dtsch. bot. Ges. **53**, 415—437 (1935). — DORNER, A.: Über das Verhalten der Zellwand zu Kongorot, insbesondere bei Farnprothallien. Zbl. Bakter. II **56**, 14—27 (1922). — DOLADILHE, M.: Influence exercée par un électrolyte sur la fixation des matières colorantes colloidales par les granules d'un hydrosol. C. r. Acad. Sci. Paris **194**, 1934—1936 (1932). — DRAWERT, H.: Untersuchungen über die p_H-Abhängigkeit der Plastidenfärbung mit Säurefuchsin und Toluidinblau in fixierten pflanzlichen Zellen. Flora (Jena) **131**, 341—354 (1937a). — Der Einfluß anorganischer Salze auf die Aufnahme und Abgabe von Farbstoffen durch die pflanzliche Zelle. Ber. dtsch. bot. Ges. **55**, 380—390 (1937b). — Das Verhalten der einzelnen Zellbestandteile fixierter pflanzlicher Gewebe gegen saure und basische Farbstoffe bei verschiedener Wasserstoffionenkonzentration. Flora (Jena) **132**, 91—124 (1937c). — Beiträge zur Entstehung der Vakuolenkontraktion nach Vitalfärbung mit Neutralrot. Ber. dtsch. Bot. Ges. **56**, 123 bis 131 (1938a). — Über die Aufnahme und Speicherung von Prune pure durch die pflanzliche Zelle. Planta (Berl.) **29**, 179—214 (1938b). — Zur Frage der Stoffaufnahme durch die lebende pflanzliche Zelle. I. Versuche mit Rhodaminen. Planta (Berl.) **29**, 376—391 (1939). — Zur Frage der Stoffaufnahme durch die lebende pflanzliche Zelle. II. Die Aufnahme basischer Farbstoffe und das Permeabilitätsproblem. Flora (Jena) **134**, 159—214 (1940). — Zur Frage der Stoffaufnahme durch die lebende pflanzliche Zelle. III. Die Aufnahme saurer Farbstoffe und das Permeabilitätsproblem. Flora (Jena) **135**, 21—64 (1941). — Zur Frage der Stoffaufnahme durch die lebende pflanzliche Zelle. V. Zur Theorie der Aufnahme basischer Stoffe. Z. Naturforsch. **3**b, 111—120 (1948). — Das Sulfonamid Prontosil solubile als Vitalfarbstoff. Protoplasma **39**, 686—692 (1950). — Beiträge zur Vitalfärbung pflanzlicher Zellen. Protoplasma **40**, 85—106 (1951). — Vitale Fluorochromierung der Mikrosomen mit Nilblausulfat. Ber. dtsch. bot. Ges. **65**, 263—271 (1952). — Vitale Fluorochromierung der Mikrosomen mit Janusgrün, Nilblausulfat und Berberinsulfat. Ber. dtsch. bot. Ges. **66**, 135—151 (1953). — Vitalfärbung der Plastiden von Allium cepa mit Coelestinblau. Ber. dtsch. bot. Ges. **67**, 33—42 (1954). — Die vitale Fluorochromierung der Sphärosomen (Mikrosomen) mit einem aliphatisch im N substituierten Aminopyren. Naturwiss. **42**, 419 (1955a). — Vitalfluorochromierungen mit angeblichem „Magdalarot". Flora (Jena) **142**, 479—488 (1955b). — Der p_H-Wert des Zellsaftes. In Handbuch der Pflanzenphysiologie, Bd. 1, S. 627—648, 1955c. — DRAWERT, H., u. B. ENDLICH: Der Einfluß des Sauerstoffs auf die Aufnahme und Speicherung saurer und basischer Farbstoffe durch die lebende Pflanzenzelle. Protoplasma **46**, 170—183 (1956). — DRAWERT, H., u. H. GUTZ: Zur vitalen Fluorochromierung der Mikrosomen mit Nilblau. Naturwiss. **40**, 512 (1953). — DRAWERT, H., u. I. METZNER: Vitalfluorochromierungen mit Brillantkresylblaupräparaten verschiedener Herkunft. Ber. dtsch. bot. Ges. **68** 385—394 (1955). — DRAWERT, H., u. S. STRUGGER: Zur Frage der Methylenblauspeicherung in Pflanzenzellen. I. Ber. dtsch. bot. Ges. **56**, 43—54 (1938).

EHRLICH, P.: Beiträge zur Kenntnis der Anilinfärbungen und ihrer Verwendung in der mikroskopischen Technik. Arch. mikrosk. Anat. **13**, 263 (1877). — ENDLER, J.: Über den Durchtritt von Salzen durch das Protoplasma. Über die Beeinflussung der Farbstoffaufnahme in die lebende Zelle durch Salze. Biochem. Z. **42**, 440—469 (1912a). — Über den Durchtritt von Salzen durch das Protoplasma. Über eine Methode zur Bestimmung des isoelektrischen Punktes des Protoplasmas auf Grund der Beeinflussung des Durchtrittes von Farbstoffen durch OH- und H-Ionen. Biochem. Z. **45**, 359—411 (1912b).

FAURÉ-FREMIET, E.: Sur la valeur des indications microchimiques fournies par quelques colorants vitaux. Anat. Anz. **40**, 378—380 (1912). — FAUTREZ, J., et L. LISON: Etudes sur la diffusibilité des colorants. Protoplasma **27**, 169—189 (1937). — FISCHER, E. P.: Über die Permeabilität der Hornhaut und über Vitalfärbungen des vorderen Bulbusabschnittes mit Bemerkungen über die Vitalfärbung des Plexus chorioideus. Arch. Augenheilk. **101**, 480—555 (1929). — FLASCH, A.: Die Wirkung von Coffein auf vitalgefärbte Pflanzenzellen. Protoplasma **44**, 412—421 (1955). — FREUDENBERGER, H.: Die Reaktion der Schließzellen auf Kohlensäure und Sauerstoffentzug. Protoplasma **35**, 15—54 (1941). — FRITSCH, F. E., and F. M. HAINES: The moisture-relations of terrestrial Algae. II. The changes during exposure to drought and treatment with hypertonic solutions. Ann. of Bot. **37**, 683—728 (1923). — FRITZ, A.: Veränderungen von Plasmaeigenschaften durch Vitalfarbstoffe. Nach Beobachtungen im Hellfeld und im Fluorescenzlicht. Sitzgsber. österr. Akad. Wiss., Math.-naturwiss. Kl., Abt. I **160**, 789—828 (1951). — FÜRTH, R.: Zur physikalischen Chemie der Farbstoffe. IV. Eine neue Methode zur exakten Bestimmung des Dispersitionsgrades der

Farbstofflösungen. Kolloid-Z. **41**, 300—304 (1927). — Die elektrische Charakteristik der Lösungen, Farbstoffe und Biokolloide. Kolloidchem. Beih. **28**, 285—292 (1929).

GELLHORN, E.: Studien zur vergleichenden Physiologie der Permeabilität. II. Vitalfärbung und Permeabilität, nach Versuchen an den Eiern von Meerestieren. Pflügers Arch. **216**, 234—248 (1927). — Über die Permeabilität tierischer Membranen für Farbstoffe. Pflügers Arch. **221**, 230—246 (1928). — Das Permeabilitätsproblem. Berlin: Springer 1929. — GELLHORN, E., et J. RÉGNIER: La perméabilité en physiologie et en pathologie générale. Paris: Masson & Cie. 1936. — GENEVOIS, L.: Coloration vitale et respiration. Protoplasma **4**, 67—87 (1928). — Les échanges d'ions dans les tissus végétaux. Protoplasma **10**, 478—502 (1930). — GICKLHORN, J.: Zur Frage der Lebensbeobachtung und Vitalfärbung von Chromosomen pflanzlicher Zellen. Protoplasma **10**, 345—355 (1930). — GILLISSEN, G.: Die Färbungsmetachromasie. (Sammelreferat.) Protoplasma **42**, 448—457 (1953). — GOLDACRE, R. J.: The folding and unfolding of protein molecules as a basis of osmotic work. Internat. Rev. Cytology **1**, 135—164 (1952). — GRAFFI, A.: Zelluläre Speicherung cancerogener Kohlenwasserstoffe. Z. Krebsforsch. **49**, 477—495 (1940). — Benzpyrenspeicherungsversuche an Hefezellen. Z. Krebsforsch. **52**, 234—239 (1941). — GUILLIERMOND, A.: Nouvelles observations cytologiques sur les Saprolégniacées. Cellule **32**, 431—454 (1923). — Sur la présence de lipides (phospholipides et stéroides) dans les vacuoles de certaines cellules végétales. Protoplasma **27**, 290—307 (1937a). — Rémarques sur la coloration vitale des cellules épidermiques des écailles bulbaires d'Allium cepa par le vert janus et par violets de dahlia et de methyle. Hommage Prof. Teodoresco Bucuresti 1937b. — The cytoplasma of the plant cell. Waltham, Mass. U.S.A. 1941. — GUILLIERMOND, A., J. DUFRÉNOY et F. LABROUSSE: Germination des graines de tabac en milieux additionés de rouge neutre et coloration vitale de leur vacuome pendant la croissance. C. r. Acad. Sci. Paris **190**, 1439—1441 (1930). — GUILLIERMOND, A., et R. GAUTHERET: Sur les conditions dans lesquelles se produit la coloration vitale des vacuoles par le rouge neutre. C. r. Acad. Sci. Paris **204**, 1377—1381 (1937a). — Sur la propriété des cellules végétales d'excréter le rouge neutre après l'avoir accumulé dans leur vacuoles. C. r. Acad. Sci. Paris **204**, 1520—1523 (1937b). — Recherches sur la coloration vitale des cellules végétales. Rev. gén. Bot. **52**, 5 (1940). — GUILLIERMOND, A., et F. OBATON: Sur l'action de p_H du milieu dans la coloration vitale des cellules végétales. C. r. Soc. Biol. Paris **116**, 984—988 (1934). — GUTSTEIN, M.: Farbstoffspeicherung durch lebende Bakterien. Zbl. Bakter. I Orig. **105**, 478—480 (1932). — GUTZ, H.: Zur Analyse der Granula-Fluorochromierung mit Nilblau in den Hyphen von Mucor racemosus Fres. Planta (Berl.) **46**, 481—511 (1956).

HANSTEIN, J. v.: Einige Züge aus der Biologie des Protoplasmas. Bot. Abh. **4**, H. 2 (1880). — HANUT, CH., u. J. FAUTREZ: Etudes sur les variations de dispeısion des colorants. Protoplasma **23**, 93—108 (1935). — HERZFELD, E.: Über die Natur der am lebenden Tier erhaltenen granulären Färbungen bei Verwendung basischer und saurer Farbstoffe. Anat. H., I. Abt. **54**, 450—523 (1917). — HERZOG, R. O., u. A. POLOTZKY: Die Diffusion einiger Farbstoffe. Z. physik. Chem. **87**, 449—489 (1914). — HIRN, I.: Vitalfärbung von Diatomeen mit basischen Farbstoffen. Sitzgsber. österr. Akad. Wiss., Math.-naturwiss. Kl., Abt. I **162**, 571—595 (1953). — HÖBER, R.: Die Durchlässigkeit der Zellen für Farbstoffe. Biochem. Z. **20**, 56—99 (1909). — Beiträge zur physikalischen Chemie der Vitalfärbung. Biochem. Z. **67**, 420—430 (1914). — Physikalische Chemie der Zelle und der Gewebe. 6. Aufl. Leipzig: Wilhelm Engelmann 1926. — HÖBER, R., u. G. PUPILLI: Neue Versuche über die Aufnahme von Farbstoffen durch die roten Blutkörperchen. Pflügers Arch. **226**, 585—599 (1931). — HÖFLER, K.: Über den isoelektrischen Punkt natürlicher Zellulosemembranen und deren Färbbarkeit mit Fluorochromen. Anz. Akad. Wiss. Wien, Math.-naturwiss. Kl. **1946**, Nr 7, 41—48. — Was lehrt die Fluorescenzmikroskopie von der Plasmapermeabilität und Stoffspeicherung? Mikroskopie (Wien) **2**, 13—29 (1947a). — Einige Nekrosen bei Färbung mit Akridinorange. Sitzgsber. österr. Akad. Wiss., Math.-naturwiss. Kl., Abt. I **156**, 585—644 (1947b). — Neue Ergebnisse der Vital- und Fluorescenzfärbung. Wiss. Mitt. Pharmaz. Forsch.-Inst. Österr. Apothekerver. **1**, 23—27 (1948). — Fluoreszenzmikroskopie und Zellphysiologie. Biol. generalis (Wien) **19**, 90—113 (1949). — Fluorochromfärbung am lebenden Protoplasten. Mikrochem. **36/37**, 1146—1157 (1951). — HÖFLER, K., u. TH. MÜLLNER-HAITINGER: Zur Wirkung des Coriphosphins auf die Pflanzenzelle. Mikroskopie (Wien) 1. Sonderbd. „Beiträge zur Fluoreszenzmikroskopie" 119—125 (1949). — HÖFLER, K., u. E. PECKSIEDER: Fluoreszenzmikroskopische Beobachtungen an höheren Pilzen. Österr. bot. Z. **94**, 99—127 (1947). — HÖFLER, K., u. H. SCHINDLER: Vitalfärbbarkeit verschiedener Closterien. Protoplasma **42**, 296—311 (1953). — HÖFLER, K., u. A. STIEGLER: Cresylechtviolett als Vitalfarbstoff. Mikroskopie (Wien) **2**, 250—258 (1947). — HÖFLER, K., A. TOTH u. M. LUHAN: Beruht die Fluorochromfärbung von Zellkernen auf Elektroadsorption an der Eiweißphase? Protoplasma **39**, 62—78 (1949). — HOFMEISTER, L.: Mikrurgische Studien an Borraginoiden-Zellen. I. Mikrodissektion. Protoplasma **35**, 65—94 (1940a). — Mikrurgische Studien an Borraginoiden-Zellen. II. Mikroinjektion und mikrochemische Untersuchungen. Protoplasma **35**, 161—186 (1940b). — Studien über Mikroinjektion in Pflanzenzellen. Z. wiss. Mikrosk. 57, 274—290 (1940c). — Vitalfärbungsstudien mit Chrysoidin. Sitzgsber. österr. Akad. Wiss., Math.-naturwiss. Kl., Abt. I 157, 55—82 (1948). — Homès, M.: La pénétration du bleu de méthylène dans les cellules d'Elodea canadensis, Rich. C. r. Congrès National de Sciences, Bruxelles 1930. — Recherches sur la perméabilité cellulaire des algues marines. Archives de Zool. 75, 75—101 (1933).

Irwin, M.: The permeability of living cells to dyes as affected by hydrogen ion concentration. J. Gen. Physiol. 5, 223—224 (1923a). — The penetration of dyes as influenced by hydrogen ion concentration. J. Gen. Physiol. 5, 727—740 (1923b). — On the accumulation of dye in Nitella. J. Gen. Physiol. 8, 147—182 (1925/28). — Mechanism of the accumulation of dye in Nitella on the basis of the entrance of the dye as undissociated molecules. J. Gen. Physiol. 9, 561—573 (1926a). — Exit of dye from living cells of Nitella at different p_H values. J. Gen. Physiol. 10, 75—102 (1926b). — Certain effects of salts on the penetration of brilliant cresyl blue into Nitella. J. Gen. Physiol. 10, 425—436 (1926c). — Accumulation of dye in Nitella as related to dissociation. Proc. Soc. Exper. Biol. Med. 23, 251—253 (1926d). — On the nature of the dye penetrating the vacuole of Valonia from solutions of methylene blue. J. Gen. Physiol. 10, 927—947 (1927). — Studies on penetration of dyes with glass electrode. IV. Penetration of brilliant cresyl blue into Nitella flexilis. J. Gen. Physiol. 14, 1—17 (1930).

Johannes, H.: Beiträge zur Vitalfärbung von Pilzmyzelien. II. Die Inturbanz der Färbungen mit Rhodaminen. Protoplasma 36, 181—194 (1941). — Beiträge zur Vitalfärbung von Pilzmycelien. IV. Die Vitalfärbung der Achlya racemosa (Hildebrand) Pringsheim mit den basischen Farbstoffen Neutralrot und Akridinorange. Protoplasma 44, 165—191 (1954).

Keller, R.: Lebende Zellen als Säure-Basenkette? Protoplasma 13, 402—404 (1931). — Keller, R., u. J. Gicklhorn: Methoden der Bioelektrostatik. In Abderhaldens Handbuch der biologischen Arbeitsmethoden, Abt. V, Teil 2, S. 1189—1280. 1928. — Kersting, F.: Über Adsorption von Farbstoffen an Zellwänden und ihre Verdrängung durch anorganische Salze. Ber. dtsch. bot. Ges. 55, 329—337 (1937). — Kiesel, A.: Chemie des Protoplasmas. Berlin: Gebrüder Bornträger 1930. — Kinzel, H.: Die Bedeutung der Pektin- und Zellulosekomponente für die Lage des Entladungspunktes pflanzlicher Zellwände. Protoplasma 42, 208—226 (1953). — Theoretische Betrachtungen zur Ionenspeicherung basischer Vitalfarbstoffe in leeren Zellsäften. Protoplasma 44, 52—72 (1954). — Kiyono, K., S. Sugiyama u. S. Amano: Die Lehre von der Vitalfärbung. Kyoto 1938. — Klebs, G.: Über das Verhalten der Farnprothallien gegenüber Anilinfarben. Sitzgsber. Heidelberg. Akad. Wiss., Math.-naturwiss. Kl., Abt. B, Abh. 18 (1919). — Kölbel, H.: Quantitative Untersuchungen über die Farbstoffspeicherung von Akridinorange in lebenden und toten Hefezellen und ihre Beziehung zu den elektrischen Verhältnissen der Zelle. Z. Naturforsch. 2b, 382—392 (1947). — Quantitative Untersuchungen über den Speicherungsmechanismus von Rhodamin B, Eosin und Neutralrot in Hefezellen. Z. Naturforsch. 3b, 442—453 (1948). — Zur Kenntnis und kataphoretischen Bestimmung des isoelektrischen Punktes einiger Bakterien. Z. Naturforsch. 4b, 145—150 (1949). — Küster, E.: Über die Aufnahme von Anilinfarben in lebende Pflanzenzellen. Jb. wiss. Bot. 50, 261—288 (1911). — Über die Vitalfärbung von Pflanzenzellen. II—IV. Z. wiss. Mikrosk. 38, 280—292 (1921). — Vital-staining of plant cells. Bot. Review 5, 351—370 (1939).

Lärz, H.: Beiträge zur Pathologie der Chloroplasten. Flora (Jena) 135, 319—355 (1942). — Lauer, A.: Über den Einfluß der Alkaloide auf die vitale Färbung mit basischen Farbstoffen. Pflügers Arch. 224, 462—470 (1930). — Lazarow, A., and S. J. Cooperstein: Studies on the mechanism of Janus Green B staining of mitochondria. I. Review of the literature. Exper. Cell. Res. 5, 56—69 (1953). — Lepeschkin, W. W.: Zur Kenntnis der chemischen Zusammensetzung der Plasmamembran. Ber. dtsch. bot. Ges. 29, 247—261 (1911a). — Über die Einwirkung anästhesierender Stoffe auf die osmotischen Eigenschaften der Plasmamembran. Ber. dtsch. bot. Ges. 29, 349—355 (1911b). — Über die kolloidchemische Beschaffenheit der lebenden Substanz und über einige Kolloidzustände, die für dieselbe eigentümlich sind. Kolloid-Z. 13, 181—192 (1913). — The influence of narcotica, mechanical agents and light on the permeability of protoplasm. Amer. J. Bot. 19, 568—580 (1932). — Fortschritte der Kolloidchemie des Protoplasmas in den letzten zehn Jahren. Protoplasma 25, 124—149 (1936). — Über die Struktur und den molekularen Bau der lebenden Materie. Protoplasma 39, 222—243 (1950). — Levin, B. S.: L'action de quelques alcaloides sur les tissus colorés in vivo. C. r. Soc. Biol. Paris 113, 1042—1044 (1933). — Lison, L., et J. Fautrez: L'étude physicochimique des colorants dans ses applications biologiques. — Etude critique. Protoplasma 33, 116—151 (1939). — Lison, L., et W. Matsaars: Metachromasy of nucleic acids. Quart. J. Microsc. Sci. 91, 309—313 (1950). — Loewe, S.: Zur physikalischen Chemie der Lipoide. I. Beziehungen der Lipoide zu den Farbstoffen. Biochem. Z. 42, 150—189 (1912). — Lundegårdh, H.: Investigations as to the absorption and accumulation of inorganic ions. Annals agricult. College Sweden 8, 233—404 (1940).

MARSTON, H. R.: Azine n-Azonium compounds of the proteolyte enzyme. Biochemic. J. 17, 851—859 (1923). — MASSART, L., G. PEETERS, J. DE LEY u. R. VERCAUTEREN: The influence of dyes on the respiration of bakersyeast. Experientia (Basel) 3, 119 (1947). — McCUTCHEON, M., and B. LUCKÉ: The mechanism of vital staining with basic dyes. J. Gen. Physiol. 6, 501—507 (1924). — MENKE, W.: Über den Feinbau der Chloroplasten. Kolloid-Z. 85, 256—259 (1938). — MESZ, E.: Kritische Untersuchungen zur kolorimetrischen Bestimmung der isoelektrischen Punkte einzelner Zellbestandteile fixierter Gewebe. Protoplasma (Wien) 1956. — MILOVIDOV, P. F.: Physik und Chemie des Zellkernes. Teil I und II. Berlin: Gebrüder Bornträger 1949 u. 1954. — MOELLENDORFF, W. v.: Die Speicherung saurer Farben im Tierkörper, ein physikalischer Vorgang. Kolloid-Z. 18, 81—90 (1916). — Die Bedeutung von sauren Kolloiden und Lipoiden für die vitale Farbstoffbindung in den Zellen. Arch. mikrosk. Anat. 90, 503—542 (1918). — Vitale Färbungen an tierischen Zellen. Erg. Physiol. 18, 141—306 (1920). — MONNÉ, L.: Über die elektive Vitalfärbung des Vakuoms und der Mitochondrien sowie über die diffuse Vitalfärbung des gesamten Cytoplasmas. Arch. exper. Zellforsch. 24, 373—393 (1942).

NIRENSTEIN, E.: Über das Wesen der Vitalfärbung. Pflügers Arch. 179, 233—337 (1920). — NISTLER, A.: Dispersoidanalyse mittels eines neuen Diffusionsapparates. Kolloidchem. Beih. 28, 296—313 (1929). — Dispersitätsuntersuchungen an Farbstoffen. Kolloidchem. Beih. 31, 1—58 (1930). — Über die Bedeutung der Dispersoidanalyse und eine Neukonstruktion des Diffusions-Mikroskopes. Protoplasma 13, 517—545 (1931).

OSTER, G., and H. GRIMSSON: The interaction of tobacco mosaic virus and of its degradation products with dyes. Arch. of Biochem. 24, 119—128 (1949). — OSTERHOUT, W. J. V Permeability in large plant cells and in models. Erg. Physiol. 35, 967—1021 (1933). — OVERTON, E.: Über die allgemeinen osmotischen Eigenschaften der Zelle, ihre vermutlichen Ursachen und ihre Bedeutung für die Physiologie. Vjschr. naturforsch. Ges. Zürich 44, 88—135 (1899).

PALTAUF, A.: Die Lebendfärbung von Zellkernen. Sitzgsber. Akad. Wiss. Wien, Math.-naturwiss. Kl., Abt. I, 137, 691—716 (1928). — PEKAREK, J.: Über die Wirkung von Nitraten auf die Färbung pflanzlicher Zellmembranen und Zellsäfte mit Azur I. Protoplasma 30, 161—185 (1938). — PERNER, E. S.: Die intravitale Aufnahme und Speicherung sulfosaurer Fluorochrome durch die obere Zwiebelschuppenepidermis von Allium cepa. Protoplasma 39, 180—205 (1950a). — Die intravitale Fluorochromierung junger Blätter von Helodea densa. Protoplasma 39, 400—422 (1950b). — Die Vitalfärbung mit Berberinsulfat und ihre physiologische Wirkung auf Zellen höherer Pflanzen. Ber. dtsch. bot. Ges. 65, 52—59 (1952). — Die Sphärosomen (Mikrosomen) pflanzlicher Zellen. (Sammelreferat.) Protoplasma 42, 457—481 (1953). — PERNER, E. S., u. G. PFEFFERKORN: Pflanzliche Chondriosomen im Licht- und Elektronenmikroskop unter Berücksichtigung ihrer morphologischen Veränderungen bei der Isolierung, Flora (Jena) 140, 98—129 (1953). — PFEFFER, W.: Über Aufnahme von Anilinfarben in lebende Zellen. Unters. bot. Inst. Tübingen 2, 179 (1886). — PLOWE, J. A.: Membranes in the plant cell. II. Localization of differential permeability in the plant protoplast. Protoplasma 12, 221—240 (1931). — PRESCOTT, D. M.: Relation between dye uptake and cytoplasmic streaming in Amoeba proteus. Nature (Lond.) 172, 593 (1953).

REICHENBERG, D., and J. F. SUTCLIFFE: Absorption of ions by plants. Nature (Lond.) 174, 1047—1048 (1954). — REINDERS, W.: Zur Theorie der Färbung. Die Verteilung von Farbstoffen zwischen zwei Lösungsmitteln. Kolloid-Z. 13, 96—105 (1913). — RHEINBOLDT, H., u. E. WEDEKIND: Über die Bindung organischer Farbstoffe durch anorganische Substrate. Kolloidchem. Beih. 17, 115—188 (1923). — RITCHIE, D., and P. HAZELTINE: Mitochondria in Allomyces under experimental conditions. Exper. Cell. Res. 5, 261—274 (1953). — ROBERTSON, T. B.: On the nature of the superficial layer in cells and its relation to their permeability and to the staining of tissues by dyes. J. of Biol. Chem. 4, 1—34 (1908). — ROHDE, K.: Untersuchungen über den Einfluß der freien H-Ionen im Innern lebender Zellen auf den Vorgang der vitalen Färbung. Pflügers Arch. 168, 411—433 (1917). — ROSS, S.: Solubilization of dyes in mineral oil and its application to a model biological cell membrane. J. Colloid Sci. 6, 497—507 (1951). — RÜTER, E., u. A. BORNSTEIN: Über den Einfluß von Alkaloiden und Salzen auf die Vitalfärbung. Pflügers Arch. 207, 614—623 (1925). — RUGE, U.: Kritische zell- und entwicklungsphysiologische Untersuchungen an den Blattzähnen von Helodea densa. Flora (Jena) 134, 311—376 (1940). — RUHLAND, W.: Die Bedeutung der Kolloidnatur wässeriger Farbstofflösungen für ihr Eindringen in lebende Zellen. Ber. dtsch. bot. Ges. 26a, 772—782 (1908). — Studien über die Aufnahme von Kolloiden durch die pflanzliche Plasmahaut. Jb. wiss. Bot. 51, 376—431 (1912).

SCARTH, G. W.: The mechanism of accumulation of dyes by living cells. Plant Physiol. 1, 215—229 (1926a). — The influence of external osmotic pressure and of disturbance of the cell surface on the permeability of Spirogyra for acid dyes. Protoplasma 1, 204—213 (1926b). — SCHAEDE, R.: Über die Reaktion des lebenden Plasmas. Ber. dtsch. bot. Ges.

42, 219—224 (1924). — SCHEIBE, G.: Reversible Polymerisation als Ursache neuartiger Absorptionsbanden von Farbstoffen. Kolloid-Z. 82, 1—14 (1938). — SCHMIDT, H.H.: Studien zur experimentellen Pathologie der Chloroplasten. II. Untersuchungen über die Streifen- und Näpfchenbildung der Chloroplasten. Protoplasma 40, 506—525 (1951). — SCHÖNLEBER, K.: Über Prune pure und seine Verwendung als Protoplasma-Vitalfärbemittel. Z. wiss. Mikrosk. 53, 303—321 (1936). — SCHULEMANN, W.: Die vitale Färbung mit saueren Farbstoffen in ihrer Bedeutung für Anatomie, Physiologie und Pharmakologie. Biochem. Z. 80, 1—142 (1917). — SCHULTZ, G.: Farbstofftabellen, 7. Aufl., Bd. I. Leipzig: Akademische Verlagsgesellschaft 1931. — SCHWARZ, FR.: Metachromatische Färbungen pflanzlicher Zellwände durch substantive Farbstoffe. I. u. II. Ber. dtsch. bot. Ges. 42, 21—38 (1924). — SCHWEIGHART, O.: Eosin und Keimpflanzen. Beih. bot. Zbl., Abt. A 53, 217—292 (1935). — SEKI, M.: Zur physikalischen Chemie der histologischen Färbung. I. Elektrische Ladung von Farbstoffen und ihre wichtige Rolle bei der Färbung des Siliciumdioxyds (SiO_2). Fol. anat. jap. 10, 621—634 (1932). — Zur physikalischen Chemie der histologischen Färbung. VIII. Z. Zellforsch. u. mikr. Anat. 18, 1—55 (1933). — SINGER, M.: Factors which control the staining of tissue sections with acid and basic dyes. Internat. Rev. Cytology 1, 211 bis 255 (1952). — SIWICKA-TARWIDOWA, H.: Sur l'évolution du chondriome pendant le développement du sac embryonnaire de l'Orchis latifolius. Acta Soc. Bot. Poloniae 11, 511—539 (1934). — *Society of Dyers and Colourists:* Colour Index. Bradford: Yorkshire 1923. — SOROKIN, H.: Mitochondria and plastids in living cells of Allium cepa. Amer. J. Bot. 25, 28—33 (1938). — The distinction between mitochondria and plastids in living epidermal cells. Amer. J. Bot. 28, 476—485 (1941). — Mitochondria and spherosomes in the living epidermal cell. Amer. J. Bot. 42, 225—231 (1955). — SPEK, J.: Ein Schlußwort zu dem vorstehenden Artikel von E. RIES über die Differenzierung der Eizelle. (Mit besonderer Berücksichtigung des Metachromasie-Problems.) Protoplasma 33, 608—621 (1939). — Metachromasie und Vitalfärbung mit p_H-Indikatoren. Protoplasma 34, 533—584 (1940). — Optische Analysen von Vitalfärbungen. Jena. Z. Naturwiss. 77, 48—67 (1944). — Über das optische Verhalten von Safraninen und Aposafraninen in den verschiedenen Komponenten des Protoplasmas. Protoplasma 40, 239—255 (1951). — STEINER, M., u. H. HEINEMANN: Über die Beziehungen zwischen den fettbildenden Grana und typischen Mitochondrien in den Zellen von Oospora lactis. Naturwiss. 41, 90 (1954). — STICH, H.: Trypaflavin und Ribonucleinsäure. Untersucht an Mäusegeweben, Condylostoma spec. und Acetabularia mediterranea. Naturwiss. 38, 435—436 (1951). — STIEGLER, A.: Vitalfärbungen an Pflanzenzellen mit Cresylechtviolett. Protoplasma 39, 493—506 (1950). — STRUGGER, S.: Über das Verhalten des pflanzlichen Zellkernes gegenüber Anilinfarbstoffen. Planta 18, 561—570 (1932). — Beiträge zur Gewebephysiologie der Wurzel. Zur Analyse und Methodik der Vitalfärbung pflanzlicher Zellen mit Neutralrot. Protoplasma 24, 108—127 (1935). — Beiträge zur Analyse der Vitalfärbung pflanzlicher Zellen mit Neutralrot. Protoplasma 26, 56—69 (1936a). — Die Vitalfärbung der Chloroplasten von Helodea mit Rhodaminen. Flora (Jena) 131, 113—128 (1936b). — Die Vitalfärbung als gewebsanalytische Untersuchungsmethode. Arch. exper. Zellforsch. 19, 199—208 (1937a). — Weitere Untersuchungen über die Vitalfärbung der Plastiden mit Rhodaminen. Flora (Jena) 131, 324—340 (1937b). — Die Vitalfärbung des Protoplasmas mit Rhodamin B und 6 G. Protoplasma 30, 85—100 (1938a). — Fluoreszenzmikroskopische Untersuchungen über die Speicherung und Wanderung des Fluoreszein-Kaliums in pflanzlichen Geweben. Flora (Jena) 132, 253—304 (1938b). — Die lumineszenzmikroskopische Analyse des Transpirationsstromes in Parenchymen. II. Die Eigenschaften des Berberinsulfats und seine Speicherung durch lebende Zellen. Biol. Zbl. 59, 274—288 (1939a). — Die lumineszenzmikroskopische Analyse des Transpirationsstromes an Helxine Soleirolii Reg. Biol. Zbl. 59, 409—442 (1939b). — Neues über die Vitalfärbung pflanzlicher Zellen mit Neutralrot. Protoplasma 34, 601—608 (1940a). — Fluoreszenzmikroskopische Untersuchungen über die Aufnahme und Speicherung des Akridinorange durch lebende und tote Pflanzenzellen. Jena. Z. Naturwiss. 73, 97—136 (1940b). — Zellphysiologische Studien mit Fluoreszenzindikatoren. I. Basische, zweifarbige Indikatoren. Flora (Jena) 135, 101—134 (1941). — Aufnahme und Speicherung des Auramins durch lebende Pflanzenzellen. Protoplasma 37, 429—438 (1943). — Die Vitalfluorochromierung des Protoplasmas. Naturwiss. 34, 267—273 (1947). — Fluoreszenzmikroskopie und Mikrobiologie. Hannover: M. u. H. Schaper 1949a. — Praktikum der Zell- und Gewebephysiologie der Pflanze, 2. Aufl. Berlin-Göttingen-Heidelberg: Springer 1949b.

TARWIDOWA, H.: Über die Entstehung der Lipoidtröpfchen bei Basidiobolus ranarum. Cellule 47, 203—216 (1938). — TOTH, A.: Neutralrotfärbung im Fluoreszenzlicht. Protoplasma 41, 103—110 (1952). — TRAUBE, J.: Über Oberflächenspannung und Flockung kolloidaler Systeme. Beitrag zur Theorie der Gifte, Arzneimittel und Farbstoffe. Kolloidchem. Beih. 3, 237—336 (1912). — TRAUBE, J., u. M. SHIKATA: Diffusion von Farbstoffen in Gele. Kolloid. Z. 32, 313—316 (1923). — TSCHERNORUTZKY, H.: Über die Wirkung von Natriumcarbonat auf einige Alkaloidsalze und Farbstoffe. Biochem. Z. 46, 112—120 (1912).

UNGER, F.: Über Aufnahme von Farbstoffen bei Pflanzen. Sitzgsber. ksl. Akad. Wiss. Wien, Math.-naturwiss. Kl. 1848.

VENKATARAMAN, K.: The chemistry of synthetic dyes, Bd. I u. II. New York: Academic Press 1952.

WASSILJEWA, N. E.: Wirkung verschiedener Substanzen auf die Verteilung von Vitalfarbstoffen zwischen wässeriger und lipoider Phase. Biol. Z. 7, 131—142 (1938).—WEBER, F.: Vitale Blattinfiltration (eine zellphysiologische Hilfsmethode). Protoplasma 1, 581—588 (1927). — Vakuolen-Kontraktion vital gefärbter Elodeazellen. Protoplasma 9, 106—119 (1930). — Assimilationsfähigkeit und Doppelbrechung der Chloroplasten. Protoplasma 27, 460—461 (1937). — WIELER, A.: Über den Anteil des sekundären Holzes der dicotyledonen Gewächse an der Saftleitung und über die Bedeutung der Anastomosen für die Wasserversorgung der transpirierenden Flächen. Jb. wiss. Bot. 19, 82—137 (1888). — WIESNER, G.: Untersuchungen über Vitalfärbung von Alliumzellen mit basischen Hellfeldfarbstoffen. Protoplasma 40, 405—425 (1951). — WILBRANDT, W.: Die Permeabilität der Zelle. Erg. Physiol. 40, 204—291 (1938).

YAMAHA, G.: Zur Methodik und Theorie der Vitalfärbung pflanzlicher Protoplasten. Botanic. Mag. 51, 533—538 (1937). — Vitalfärbung an den Pollenmutterzellen von Lilium speciosum. Cytologia 9, 193—202 (1938). — YAMAHA, G., u. K. NOMURA: Über den Einfluß der Wasserstoffionenkonzentration und der Neutralsalze auf die Vitalfärbung pflanzlicher Protoplasten mit sauren Farbstoffen. Sci. Rep. Tokyo Bunrika Daigaku, Sect. B 4, 27—42 (1939).

ZACKS, S. I., and J. H. WELSH: Cholinesterases in rat liver mitochondria. Amer. J. Physiol. 165, 620—623 (1951). — ZEIGER, K.: Physikochemische Grundlagen der histologischen Methodik. Dresden u. Leipzig: Theodor Steinkopff 1938. — ZEIGER, K., H. HARDERS, u. W. MÜLLER: Der STRUGGER-Effekt an der Nervenzelle. Protoplasma 40, 76—84 (1951). — ZIEGENSPECK, H.: Dichroskopie und Metachroskopie. Der Dichroismus und Metachroismus besonders substantiv gefärbter natürlicher Pflanzenmembranen als ein Hilfsmittel zum Erforschen der physikalisch-chemischen Beschaffenheit derselben. Protoplasma 35, 235—264 (1941). — ZIEGLER, H.: Die Beeinflussung der Atmungsintensität pflanzlicher Gewebe durch eine Belichtung in fluorescierenden Farbstofflösungen. Z. Naturforsch. 5b, 345—350 (1950). — ZIPF, K.: Die Austauschbedingungen als Grundlage der Aufnahme basischer und saurer Fremdsubstanzen in die Zelle. I. u. II. Arch. exper. Path. u. Pharmakol. 124, 259—325 (1927).

The uptake of salts by plant cells.

By

Paul J. Kramer.

With 10 figures.

Introduction.

The uptake of materials from their environment is one of the most characteristic and important processes of living organisms. The plant as a whole absorbs water, gases and solutes from the soil, water and air around it. The organs of the plant absorb from each other and the cells within each organ are continually involved in the exchange of a variety of materials. Some of this movement is by mass flow, some by diffusion, and some involves active transport and accumulation against concentration gradients. Most of the uptake of solutes seems to occur by the activities of individual cells and the basic processes can therefore best be discussed in terms of cells and tissues rather than of organs or entire plants. Relatively little attention will be given to absorption by intact, growing plants because that topic will be discussed in detail in other chapters of this series.

It is possible to mention in this chapter only a few of the numerous publications dealing with accumulation. These of course contain numerous references to other work dealing with almost every phase of this field. Readers also are referred to the Annual Review of Plant Physiology and the Fortschritte der Botanik for reviews of current literature. The monograph entitled Active Transport and Secretion, which was published in 1954 as Symposium VIII by the Society for Experimental Biology, contains a particularly valuable group of papers dealing with accumulation.

Most of our information concerning the kinds and amounts of salts absorbed and the nature of the absorption process has been obtained by studies of salt uptake by slices of storage tissue such as beet, carrot, and potato, and excised roots, especially those of barley. Some studies were made on *Chara*, *Nitella* and *Tollypellopsis* because their large cells yield sufficiently large volumes of sap for chemical analysis (COLLANDER 1936, HOAGLAND and DAVIS 1929, OSTERHOUT 1922). Among those who have worked extensively with slices of storage tissue are STEWARD, STILES and ROBERTSON. HOAGLAND and his colleagues at the University of California used barley roots, and LUNDEGÅRDH used wheat roots for the most part. The availability of radioactive tracers has increased the ease with which exchange and accumulation phenomena can be studied and work on a wide variety of plant materials is now under way in various laboratories.

The elements absorbed by plants. Examination of the list of elements which have been found in plants suggests that they are able to absorb almost any element occurring in their environment. Over 40 different elements have been detected in plants, although some of these have been found in only a few samples, presumably because they seldom occur in the plant environment. JACOBSON and OVERSTREET (1948) found that plants even could absorb plutonium and other products of nuclear fission which are produced only in the laboratory and were never previously encountered by plants.

The kinds and amounts of mineral elements absorbed are affected by various climatic and soil factors, including the kind and concentration of elements in the soil or root environment, and also vary widely among species. The extensive data on mineral uptake by crop plants were summarized by MILLER (1938). An extreme example of species differences in ability to accumulate silicon is shown in table 1. Another excellent example of species differences in ability to accumulate sodium is found in the paper by COLLANDER (1941). The effect of differences in concentration of elements in the environment on uptake is shown for *Chara* growing in brackish and salt water in table 2, taken from data of COLLANDER (1936).

Table 1. *Differences in the mineral composition of four species of plants growing in the same environment.* (From RICHARDSON 1920.)

Constituent	Percentage of soil	Percentage of ash			
		Prunus pumila	*Artemisia caudata*	*Andropogon*	*Equisetum hiemale*
Silicon	92.0	1.5	12.1	65.4	49.4
Fe_2O_3	1.0	0.7	1.7	2.5	1.0
Al_2O_3	3.5	0.02	0.4	2.6	2.3
CaO	1.6	44.1	35.5	10.2	13.4
K_2O	1.0	10.9	11.6	6.7	6.0
MgO	0.7	4.2	6.4	3.2	3.7

Table 2. *Effects of differences in concentration of the surrounding solution on composition of vacuolar sap of Chara.* (From COLLANDER 1936.)

	Cl	Na	K	Ca	Mg
Chara ceratophylla in brackish water	232	148	69	11	26
Composition of water	80	68	1.4	3.8	14
Concentration factor	2.9	2.2	49	2.9	1.9
Chara ceratophylla in fresh water	194	99	85	21	32
Composition of water	0.78	0.81	0.13	2.2	0.6
Concentration factor	294	122	650	10	53

The numerous data of this type which are available show that plants exercise considerable control over the amounts of various elements which they absorb. Not only are some species able to accumulate certain elements until the concentration inside is much higher than that in the environment, but they are able to exclude other elements and maintain an internal concentration which is lower than that in their root environment. The mechanism by which such effects are produced constitutes the principal theme of this chapter.

Terminology. Before discussing processes it seems desirable to define some of the commonly used terms because they are not always used in the same way by different writers. For the most part the usage of ROBERTSON (1951) will be followed.

Absorption and uptake are used as general terms for the entrance of a substance into a cell, tissue, organ, or organism by any process or combination of processes. Accumulation refers to entrance of a substance into a cell or tissue against a concentration or activity gradient or gradient of chemical potential. Generally, it requires the expenditure of metabolic energy by the cells or tissues in which it occurs but nonmetabolic accumulation by adsorption also occurs. The term accumulation is most often used in connection with the uptake of ions against a concentration gradient, but it can be applied to the uptake of dyes and other organic compounds if they occur in higher concentration within the cells than outside. If nonosmotic uptake of water occurs, as claimed by some writers, it would constitute an example of accumulation against an activity gradient (see chapter on water uptake in this volume). The terms nonmetabolic and metabolic accumulation are used by some writers (OVERSTREET and JACOBSON 1946). Nonmetabolic absorption refers to the first step in ion uptake which is an adsorption exchange process occurring independently of metabolism, while metabolic accumulation refers to the active transport of the ions into the vacuoles, which is closely related to metabolic processes.

The term secretion is sometimes used by physiologists, especially animal physiologists, with the same meaning as accumulation. For example BARTLEY and DAVIES (1952) wrote of the secretory activity of mitochondria when they were really discussing the accumulatory activity of mitochondria, and SUTCLIFFE's (1953) excellent review of ion accumulation is entitled, "Ion Secretion in Plants". Plant physiologists also sometimes speak of secretion of substances into the vacuole, but in most instances accumulation would be a better term. The term secretion is better reserved for use in connection with the emission of substances from cells, especially glandular cells or structures such as the nectaries. Secretion in this sense is treated in another chapter of this volume.

Movement of substances along concentration gradients caused by the kinetic energy of the moving molecules or ions is termed diffusion. If the movement is caused by hydrostatic pressure it is mass flow. If movement occurs against a concentration or activity gradient by the expenditure of respiratory energy it is often termed active transport. Active transport is responsible for accumulation and presumably also for secretion. ROSENBERG (1954) has an interesting discussion of theories of active transport.

A membrane, according to USSING (1954), is a boundary that presents a higher resistance (is less permeable) to the movement of a substance than the phases separated by it. Membranes can vary from monomolecular layers at interfaces to multicellular structures such as the epidermis, endodermis, and the coats of various seeds. The important cell membranes will be discussed in the next section. The extent to which substances can pass through a membrane is termed its permeability. Various aspects of permeability have been discussed in other chapters of this volume, but it is evident that the subjects of permeability and uptake of materials are so interrelated that one cannot be considered without the other. Membranes which exhibit differences in permeability to different substances are often termed semipermeable or selectively permeable, but the writer prefers to term them differentially permeable because it is more descriptive of their actual behavior.

It seems that a distinction must be made between permeability in the classical sense, that is permeability to substances diffusing along a concentration or free energy gradient and substances moved by active transport against a concentration gradient. Cells obviously must be relatively impermeable to substances which accumulate in them. If they were not relatively impermeable accumulation would be impossible because the substances would leak out by diffusion as rapidly as they were transported in. Accumulation in the vacuoles is the result of active transport across a relatively impermeable membrane. The location of this impermeable membrane will be discussed in the next section.

Cell membranes in relation to uptake of salt.

The principal membranes of a plant cell are the wall, the plasmalemma at the outer surface of the protoplast, and the tonoplast at the inner surface between the cytoplasm and the vacuole. Primary cell walls which have not become suberized, cutinized, or lignified contain large amounts of water in the spaces between the cellulose microfibrils and are relatively permeable to water and solutes. Generally the walls are so much more permeable than the cytoplasmic membranes that they are disregarded in discussions of differential permeability. CASARI (1953) claims, however, that cell walls are measurably less permeable to large molecules than to small ones. The decrease in permeability of the surface layers of maturing roots caused by suberization of the surface cells greatly decreases uptake of both water and salts. For the present discussion it will be assumed that the walls of the cells under discussion are relatively so permeable that they are not an important hindrance to the uptake of salts.

The differential permeability which is such an essential characteristic of living cells obviously is located in the layer of cytoplasm lying between the wall and the vacuole. It is not unanimously agreed, however, whether it is at the outer surface of this layer, the plasmalemma, in the body of the cytoplasm, or in the tonoplast which forms the boundary next to the vacuole. HÖFLER (1950) has summarized considerable evidence indicating that the entire layer of cytoplasm is involved in permeability to water and SEEMAN (1953) holds the same view.

Most people probably have assumed that the plasmalemma or outer surface of the protoplast is the layer impermeable to salts, but there is reason to doubt if this is true. BROOKS (1940) for example, found that a high concentration of radioactive sodium and potassium occurred in the cytoplasm of *Nitella* cells considerably before it occurred in the vacuolar sap. He concluded that the cytoplasm is relatively permeable to potassium and that the impermeable layer must be the tonoplast rather than the plasmalemma. HOAGLAND and BROYER (1942) found the same situation, but HOLM-JENSEN, KROGH and WARTIOVAARA (1944) found radioactive sodium and potassium evenly distributed between cytoplasm and vacuole in *Nitella* and *Tollypellopsis*. Perhaps these discrepancies are related to the length of the experiment and variations in metabolic activity as HOAGLAND and BROYER found that the amount of Rb in the vacuole increased with time. BROOKS (1953) later reported that the tonoplast of *Valonia* is relatively impermeable to radioactive sodium, but not that of *Halicystis*, where the plasmalemma seemed to be the impermeable layer. ARISZ (1945, 1948) also concluded that the plasmalemma must be permeable to ions and the tonoplast impermeable.

HOPE and STEVEN (1952) claim that ions move in and out of the cytoplasm freely by diffusion and that the differentially permeable membrane involved in accumulation must be located at the inner surface. They term the total cell space which can be occupied reversibly by diffusing ions the "apparent free space", often abbreviated as A. F. S. BUTLER (1953a) measured the apparent free space of wheat roots by several methods and concluded that it included the cytoplasm and cell walls plus whatever intercellular space is filled with water. HOPE and STEVEN (1952) estimated the A. F. S. of bean root tips to be about 13% by volume, BUTLER estimated it to amount to 24 to 33% in wheat roots, but HYLMÖ (1953) found only 8 to 9% in pea roots. HOPE (1953) found that the A. F. S. increased with increase in concentration of the external solution and suggested the existence of a DONNAN equilibrium between the solution and root cell cytoplasm. SUTCLIFFE (1954b, 1954c) concluded that measurable amounts of easily exchangeable ions occur in intercellular spaces, cell walls, and the cytoplasm of beet tissue. He suggested that the larger part of the ions not easily exchangeable occur in the vacuole and a smaller part associated or bound strongly in the cytoplasm. It seems reasonable to suppose that part of the ions which diffuse into the cytoplasm would be adsorbed so firmly that they would not easily exchange, just as part of the water molecules are so firmly bound that they do not move easily. Perhaps the concept of apparent free space must be considered in terms of the amount of material free to diffuse in and out, rather than in terms of volume occupied.

The concept of diffusion into apparent free space has been criticized by RUSSELL and AYLAND (1955) because they found the absorption of Rb and Cs to be reduced by the presence of other monovalent ions, even during the first five minutes of an experiment. They reason that diffusion of one ion into the cytoplasm ought not be hindered by the presence of another ion at very low concentrations, but that adsorption at the cell surface would be if there were only a limited number of sites on which adsorption could occur. On the other hand SUTCLIFFE (1954c) thinks it is unlikely that the surface of the protoplast is the sole region of adsorption and he, like several other writers, appears to favor the tonoplast as the chief barrier to inward movement of ions.

HOPE and ROBERTSON (1953) summarized considerable experimental data, beginning with the microinjection experiments of PLOWE (1931), indicating that the tonoplast is relatively less permeable to solutes than the plasmalemma. They attribute the resistance to entrance of ions partly to the DONNAN effect of non-diffusible anions in the body of the cytoplasm and partly to the impermeability of the tonoplast, the latter being much the most important factor. In their opinion the tonoplast is the chief barrier to be penetrated by the mechanism of active transport which causes accumulation of ions in the vacuole.

Before leaving this topic it should be mentioned that there is some evidence that the impermeability to sugars is localized at the surface of the protoplast, in the plasmalemma. STREET and LOWE (1950) proposed that sucrose absorption depends on phosphorylation occurring at the external surface of the protoplasts. BROWN (1952) also suggested that the experimental evidence on fructose uptake by root segments could best be explained by assuming that it is controlled by enzyme reactions occurring at the surface of the protoplasts. Absorption of sugar by growing root cells was not increased by increase in surface area of the cells or decrease in thickness of the cytoplasm caused by cell growth. This suggests that absorption probably occurs at specific sites on the surface which do not increase in number during cell enlargement, and these points are either on the inner or the outer surface of the cytoplasm. BROWN thinks they are on the outer surface because respiration does not increase immediately when sugar is added to the external solution, but only after the sugar concentration inside the cell has increased. Perhaps the location of the differentially permeable membrane in the cytoplasm is different for different kinds of solutes. The reader is referred to a paper by ROTHSTEIN (1954) for further discussion of the role of enzymes at the cell surface in the uptake of solutes.

According to BROWN and CARTWRIGHT (1953) the absorption of potassium by enlarging root cells decreases per unit of surface area as the roots enlarge, probably because the protein content decreases per unit of cell surface during enlargement. They regarded salt absorption as closely related to protein content and further concluded that during cell growth the cytoplasm differentiates in such a way that absorption of ions per unit of protein increases. SUTCLIFFE (1954c) also found that the rate of accumulation was not increased by a large increase in root surface. Further careful study of ion accumulation will be necessary before it is known with certainty whether the principal barrier to diffusion of ions occurs at the outer or inner surface of the cytoplasm of vacuolated cells.

In this discussion it has been assumed that the principal site of ion accumulation in plant cells is the central vacuole. It should be mentioned, however, that LUNDEGÅRDH (1954, pp. 266–267) regards the vacuole as a transitory storage place which is ordinarily in equilibrium with the cytoplasm, and thinks that ione move from cytoplasm to vacuole by diffusion. The preponderance of evidence certainly supports the view that the tonoplast is most likely the differentially permeable membrane which controls accumulation in the vacuole.

Forces concerned in salt uptake.

The nature of the forces bringing about the uptake of ions by plant cells has long been a topic of discussion. Final conclusions concerning the details of the process are not yet possible and some revision of what is presented in this section is likely to be necessary in the future as our information grows, but the main outline seems reasonably well established.

It was naturally supposed by the earlier physiologists that various solutes found in plant cells entered by diffusion, but this belief had to be abandoned when it was found that some substances occur in cells in higher concentration and others in lower concentration than they occur in the environment. The importance of diffusion as the initial step in salt uptake has been partly reestablished by observations such as those of HOPE and STEVENS (1952), HYLMÖ (1953), BUTLER (1953a), and SUTCLIFFE (1954b). Their results suggest that considerable amounts of ions diffuse into the roots along cell walls and even in the cytoplasm. This is a revival of the suggestion made by SCOTT and PRIESTLEY (1928) that the soil solution diffuses into the roots and ions are accumulated in the vacuoles from this solution. BROYER (1950) also supports this view, which is discussed in more detail in the last section of this chapter.

According to ROBERTSON (1951) diffusion of ions through the cytoplasm probably is slowed down by the presence of fixed ions. Experiments of HOPE (1953) indicate the presence of immobile ions (probably anions) in bean and maize roots equivalent in concentration to 0.01 M per liter. Mature cells seem to have a higher concentration than immature cells. It can be seen that electrical charges of membranes and moving particles and the presence of electrical potentials affect movement into cells, but such phenomena receive less attention than formerly because they seem so much less important than the process of active transport and accumulation.

Accumulation processes.

Any diffusion which occurs through the walls and into the cytoplasm can be regarded as an incidental or preliminary process, rather than directly related to accumulation. Accumulation is concerned with the transport of various substances, especially ions, into cells against a concentration or activity gradient until the concentration in the cells (usually in the vacuoles) becomes higher than the concentration in the environment.

Many and perhaps most investigators regard accumulation of ions as occurring in two stages (BROYER and OVERSTREET 1940, EPSTEIN and LEGGETT 1954) and this seems to be a convenient way to divide the discussion. The first step is a physical or nonmetabolic process involving exchange adsorption at the cell surface, the second step is a metabolic process in which active transport of ions occurs across the differentially permeable membrane of the cell.

Exchange adsorption. The first step was termed nonmetabolic absorption by OVERSTREET and JACOBSON (1946) because it occurs at 0° C. and is independent of metabolic activity. It also is an accumulation process in the sense that the concentration of ions at the adsorbing surface is much higher than in the surrounding solution. Unlike the later stage, however, it usually is easily reversible because the adsorbed ions can be removed by exchange for other ions. It also attains equilibrium very rapidly, often within less than an hour, as shown in Fig. 1.

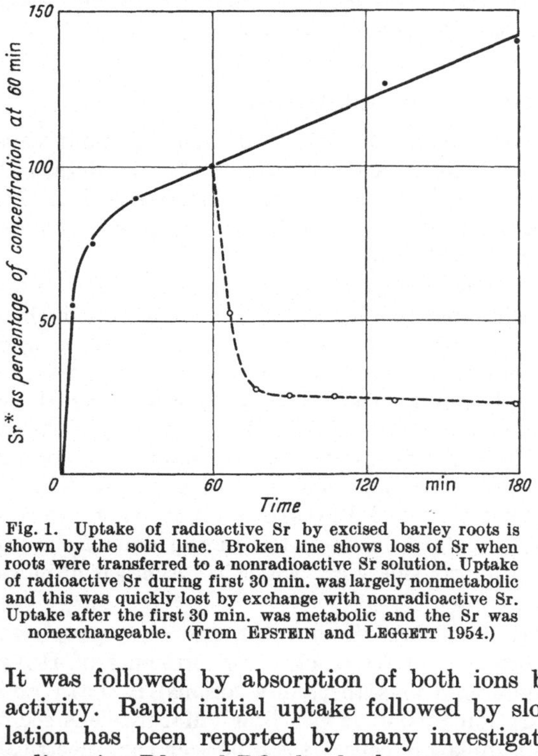

Fig. 1. Uptake of radioactive Sr by excised barley roots is shown by the solid line. Broken line shows loss of Sr when roots were transferred to a nonradioactive Sr solution. Uptake of radioactive Sr during first 30 min. was largely nonmetabolic and this was quickly lost by exchange with nonradioactive Sr. Uptake after the first 30 min. was metabolic and the Sr was nonexchangeable. (From EPSTEIN and LEGGETT 1954.)

According to ROBERTSON (1951) ion exchange is affected by the concentration of the adsorbed ion, the presence of competing ions of similar charge, and the relative affinities of adsorbing and competing ions. Numerous papers have been written on exchange phenomena between roots and soil and the reader is referred to a review by STOUT and OVERSTREET (1950) and an article by WILLIAMS and COLEMAN (1950) on roots as adsorbing surfaces.

STEWARD and HARRISON (1939) observed a rapid initial uptake of Rb by potato discs from RbBr solutions which was not related to metabolic activity. It was followed by absorption of both ions by a process related to metabolic activity. Rapid initial uptake followed by slower, but long continued accumulation has been reported by many investigators. Nonmetabolic absorption of radioactive Rb and PO_4 by barley roots was observed to occur even at 0° C by OVERSTREET and JACOBSON (1946). Uptake was largely restricted to the apical few millimeters of the roots. The Rb was easily removed by exchange for nonradioactive isotope, but the PO_4 was difficult to remove. JACOBSON and OVERSTREET (1947) found that maximum nonmetabolic uptake of radioactive Sr and I occurred in the root tip and observed a similar pattern of accumulation in dead roots, although less was absorbed than in living roots.

EPSTEIN and LEGGETT (1954) have also distinguished exchange adsorption from active uptake. Results of an experiment with radioactive Sr are shown in Fig. 1. Very rapid uptake occurred during the first few minutes, followed by a long period of slower uptake at a constant rate. Uptake during the first 30 min. was largely by exchange adsorption and much of the radioactive Sr was removed when the roots were removed to a nonradioactive solution of Sr. Uptake after the first 30 min. was largely by active transport and the Sr could not be removed by exchange. About 25% of the Sr accumulated during the first 60 min. could not be removed by exchange and was assumed to have been accumulated in the cells. The fraction not readily exchanged increased slowly with time. Ca and Mg were exchanged as rapidly with radioactive Sr as was nonradioactive Sr. A much larger fraction of Sr than of Rb was found to be readily exchangeable. In earlier experiments (EPSTEIN and HAGEN 1952) a very small fraction of the total Rb was found to be easily exchanged and exchange adsorption appeared to be a minor factor in Rb accumulation. The reason for this is unknown, but perhaps there are fewer sites capable of firmly binding Sr than Rb.

RUSSELL and AYLAND (1955) claim that the initial step in ion uptake is not passive or nonmetabolic, as claimed by OVERSTREET and JACOBSON (1946), because they were able to inhibit it with dinitrophenol and azide. They consider the process of ion accumulation to be more complex than was supposed by most writers. It is difficult to explain these discrepant results with respect to the role of metabolic processes in the initial stage or stages of ion accumulatiom. Possibly some workers have not always distinguished clearly between the initial exchange adsorption at the surface and active transport into the cell, and perhaps the two processes overlap more in some tissues than in others.

Active transport. The really critical problem in respect to the uptake of ions is the means by which they are moved across an impermeable membrane located somewhere in the cytoplasm and caused to accumulate in the vacuole against a concentration gradient. The theoretical aspects of this problem will be discussed in more detail by ROBERTSON in another chapter, so relatively brief mention of them will be made here. FRANCK and MAYER (1947), ROSENBERG (1948), and CONWAY (1953) have all contributed theoretical explanations of how metabolic energy may be used to bring about active transport of substances into cells. DAVIES and OGSTON (1950) dealt with the situation in multicellular animal membranes and OVERSTREET and JACOBSON (1952) summarized much of the theoretical and experimental work on accumulation in plant cells. SPIEGELMAN and REINER (1942) analyzed the problem and decided that differences in ionic mobility, DONNAN equilibria and other physical and electrical properties of ions and membranes could not explain ion accumulation, but that chemical forces probably are involved.

The LUNDEGÅRDH theory. One of the best known theories of salt accumulation is that first proposed by LUNDEGÅRDH and BURSTRÖM (1933) and developed by LUNDEGÅRDH (1945, 1946, 1947, 1950, 1954). LUNDEGÅRDH has modified some of the details in successive papers and probably the best exposition of his views are those of BURSTRÖM (1951) and ROBERTSON and WILKINS (1948). LUNDEGÅRDH assumes that the absorption of anions occurs independently of the absorption of cations and by a different mechanism.

LUNDEGÅRDH developed his theory to explain the transport of ions across roots, but it is also applied to uptake by cells and tissues. A basic assumption of this theory is the existence of a gradient in oxygen concentration from outside to inside, such that oxidation is favored at the outer surface and reduction at the inner end of the path of transport. Anions are supposed to move inward on an electron carrier, probably cytochrome, while electrons are moving outward. When reduced iron of the cytochrome is oxidized at the outer surface it loses an electron and can pick up an anion which moves along a cytochrome bridge across the membrane to the inner surface. Here the iron is reduced, gaining an electron and losing the anion which it carried. Thus streams of electrons and anions are continually moving in opposite directions along the cytochrome bridge. The accumulation of anions produces a potential gradient along which cations move passively by exchange for H ions (LUNDEGÅRDH 1946).

Salt respiration. Another basic assumption of the LUNDEGÅRDH theory is that it is possible to distinguish between ground respiration which is not directly related to salt uptake and "anion respiration". The rate of anion respiration is supposed to be related to the rate of anion absorption and to supply energy for it. This assumption is based on observations that respiration increases when plant tissue is transferred from water to dilute salt solutions. This increased respiration is more sensitive to cyanide, carbon monoxide, and azide than ground respiration, indicating that it involves a cytochrome-cytochrome oxidase system. WANNER (1948) also reported that the Q_{10} for the uptake of anions is considerably higher than for cations, suggesting that the former may be more closely related to respiration than the latter.

ROBERTSON (1951) seemed to accept the major premises of the LUNDEGÅRDH theory. The only important change was to suggest that the ions are carried across the membrane by cytochrome molecules instead of jumping from one to another along a chain of cytochromes. He also prefers to call the increased respiration sometimes associated with salt uptake salt respiration rather than anion respiration.

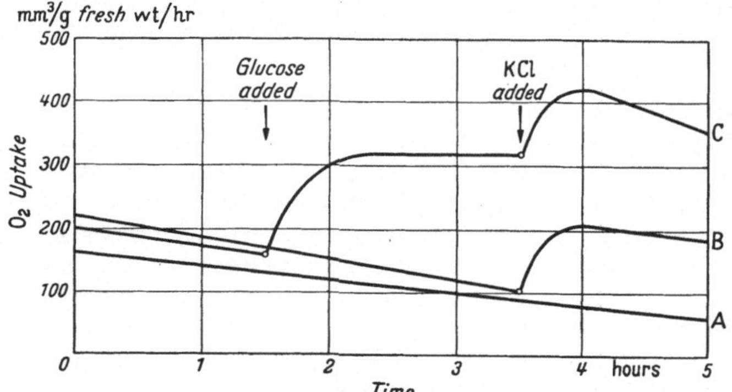

Fig. 2. Salt respiration in barley roots. Curve A for roots in water, curve B for roots in water to which 0.01 M KCl was added. Curve C is for roots in water to which 2% of glucose was first added, and later KCl. Addition of KCN reduced respiration of all roots below the control rate. (From MILTHORPE and ROBERTSON 1948.)

The relation of so-called salt respiration to uptake of salt has been debated vigorously. LUNDEGÅRDH regards anion uptake as directly dependent on salt respiration, but this is questioned by some writers (HUMPHRIES 1952, SUTCLIFFE 1952, EPSTEIN 1953, 1954). There is no doubt that when well washed plant tissue which has stood for some time (many hours to several days) in distilled water is transferred to a dilute salt solution respiration often increases immediately and salt uptake also occurs (see Figs. 2 and 6). Furthermore, inhibitors such as azide, cyanide and carbon monoxide which affect the cytochrome system reduce respiration and salt uptake (LUNDEGÅRDH 1945, MACHLIS 1944a, MILTHORPE and ROBERTSON 1948, WEEKS and ROBERTSON 1950). On the other hand ROBERTSON, WILKINS and WEEKS (1951) found that 2,4-dinitrophenol increases respiration, but decreases uptake of ions by wheat roots. They state, as is obviously true, that this requires some modification of the LUNDEGÅRDH theory to allow the participation of phosphorylation in ion accumulation. HUMPHRIES (1952) was unable to find any evidence of salt respiration in pea roots, just as he had been unable to find it earlier in wheat roots. SUTCLIFFE (1954a) found that although glucose and sucrose increased both respiration and uptake of K in beet roots, Ca and Mg increased K uptake without any increase in respiration. Thus there is abundant evidence that there is no close proportionality between respiration and salt accumulation. EPSTEIN (1954) recently reported that an increase in respiration of barley roots can also be produced by cations which suggests that salt respiration represents a general effect of salt on respiration. ROBERTSON (1941) calculated that less than one percent of the energy released by respiration was needed for salt uptake in one of his experiments. The direct need for energy seems so low that it is difficult to measure it accurately and most of the energy must be expended indirectly in some essential subsidiary process such as the synthesis of binding or carrier compounds. It seems more likely that salt uptake causes an increase in respiration than that increase in respiration is necessary for salt uptake.

A puzzling fact is that maximum salt uptake and respiration are often obtained only after the tissue has been soaked in aerated distilled water for some time. SUTCLIFFE (1954b) found that soaking beet tissue in aerated water up to a week causes increased salt uptake (see Fig. 6) and suggested that perhaps new absorption centers are produced during this time. Perhaps the increase in respiration obtained after this preconditioning is connected with effects of ions on enzymes. LATIES (1954) states that exposure of cauliflower tissue to hypotonic solutions of sucrose reduces its enzyme activity and presents some evidence suggesting that soaking effects are exerted through changes in respiratory enzyme activity. ORDIN and JACOBSON (1955) suggested that absorption of ions could stimulate respiration because the ions might act as cofactors in oxidative processes.

One of the most serious objections to the LUNDEGÅRDH theory is the observation of EPSTEIN (1953) that the Br anion appears to be absorbed in the same manner as cations. His results lead him to conclude that both cations and anions are accumulated in the same manner. Another serious objection is that it provides no means of explaining differential uptake of various ions. This is better explained by the theory to be discussed next.

Carrier or ion binding systems. At the present time the most popular theory of active transport of ions is some form of the carrier theory. The usefulness of this theory was well summarized by OSTERHOUT (1952) who was one of the first workers to propose it. The carrier theory has been presented in different forms by various investigators. JACOBSON et al. (1950) applied the carrier theory to ion absorption by roots as follows. They assume that the first step consists of exchange reactions as follows:

$$Z^+ + HR \rightarrow ZR + H^+$$

$$A^- + ROH \rightarrow RA + OH^-$$

In this scheme Z^+ represents a cation and A^- an anion. It is assumed that the carrier substances HR and ROH are produced on the outer side of the membrane and that the membrane is permeable to the complexes ZR and RA, but not to Z^+, A^-, R^- or R^+. The complexes ZR and RA are carried across the membrane and break down on the inner side, releasing the ions Z^+ and A^-. EPSTEIN and HAGEN (1952) expanded this theory somewhat. They regard the process as similar to the reaction between enzyme and substrate to form a temporary complex in enzyme catalyzed reactions. They suggest that the carriers possess several reactive centers having different affinities for different ions or groups of ions. This provides a chemical basis for differences in uptake of various ions. They concluded that Ca, K, and Rb are bound on the same sites because they compete with each other, but Na is bound on a different site. EPSTEIN and LEGGETT (1954) found that Ba, Ca, and Sr compete for identical binding sites, but Mg does not compete with them for this site.

Development of the carrier theory does not explain how the temporary complexes between ions and carrier molecules are transported across cell membranes. GOLDACRE (1952) suggested that they are moved by cytoplasmic streaming. Their nature and origin also is uncertain. BROOKS (1937) suggested that amino acids might serve as carriers, and STEWARD and STREET (1947) suggested that high energy phosphorylated nitrogen compounds might be involved. HUMPHRIES (1952) suggested that the binding compounds are formed from sugars because he found a relationship between sugar content and ion uptake, but this may be because sugar is needed as a substrate for respiration. The close relationship which exists between respiration, protein metabolism, and salt uptake may be related in part to the synthesis of carrier compounds and in part to the energy required to transport them across membranes and split off the attached ions. According to ROSENBERG and WILBRANDT (1952) an essential feature of this theory of active transport is the maintenance of a gradient of carrier molecules across the membrane. This, they suggest, probably involves formation of carrier on the outer side and destruction on the inner side by enzymatic action. Under these conditions the carrier itself might move passively along a gradient of decreasing concentration, and at the same time bring about active transport of ions or other solutes against a concentration gradient (ROSENBERG 1954). WILBRANDT (1954) has a discussion of carrier systems, principally in animal tissues.

It should be remembered that the carrier theory of transport is at present only a theory. It is attractive because it explains a variety of observations and because it has a close analogy in enzyme action, also because its general action is easy to show. A variation of this theory is the suggestion of GOLDACRE and LORCH (1950) and GOLDACRE (1952) that expanded protein molecules adsorb materials at the outer surface of the cytoplasm and then are carried across it by cytoplasmic streaming and undergo desorption at the inner surface by contraction. Some years ago ARISZ (1942) proposed that organic substances such as asparagine are translocated by binding to protoplasmic particles. He later extended this to include ions (ARISZ 1945) and suggested that movement might occur in either of two ways. The ions might be carried on particles transported by protoplasmic streaming, or they might move from particle to particle across the protoplasm. This could occur by exchange which might be facilitated by rapid synthesis and breakdown of the binding compounds. It also has been proposed that the mitochondria themselves are the accumulatory structures (BARTLEY and DAVIES 1952). In view of the manner in which they are moved about in the cytoplasm they might serve as carriers, also. More research is needed before the details of active transport can be explained satisfactorily.

CONWAY (1953) has proposed that a redox system acts both as a carrier and an energy donator for active transport. He proposed what he termed a redox pump in which energy released by electrons jumping from one redox system to another at a lower level provide energy for active transport of ions against concentration gradients. In this system, as in that of LUNDEGÅRDH, ion accumulation is directly linked to respiration.

Respiration and salt accumulation. As pointed out by RUSSELL (1954) the two important questions in connection with this problem are, (a) is the cytochrome system the only respiratory enzyme system which makes energy available for salt accumulation, and (b) is salt accumulation linked directly to electron transfer in respiration or is it related indirectly through products of metabolism which act as binding sites or carrier molecules?

LUNDEGÅRDH (1954) continues to contend vigorously that the accumulation of anions is directly linked to respiration through the cytochrome system. It is by no means certain, however, that cytochrome oxidase is the only important terminal oxidase in plant tissue. JAMES and BOULTER (1955) reported that although the cytochrome system accounts for over 80 per cent of electron transfer in barley embryos it is replaced gradually by ascorbic acid oxidase during the first week of growth of the roots. STEWARD and MILLAR (1954) found that cyanide inhibited uptake of Cs^{137} by nondividing carrot tissue, but did not seriously inhibit the growth or Cs uptake of dividing carrot cells. This suggests that cytochrome is not an essential part of the respiratory mechanism involved in salt uptake by growing carrot tissue.

LUNDEGÅRDH and CONWAY assume the existence of a direct linkage between electron transfer and salt absorption. It seems unlikely that such a direct linkage exists because dinitrophenol, which stimulates electron transfer but hinders phosphorylation, greatly decreases salt accumulation. This situation suggests that some product of respiration associated with phosphorylation, perhaps a high energy phosphate compound, is concerned in salt accumulation. In general the evidence available at this time suggests that salt absorption is not linked directly to respiration, but is indirectly related through the energy required for synthesis of binding sites or carrier molecules in the cytoplasm. This type of relationship has been discussed by several writers, including RUSSELL (1954), STEWARD and MILLAR (1954), and SUTCLIFFE 1954). The existence of such a relationship is supported by the observation that apparently roots (RUSSELL 1954) and beet tissue (SUTCLIFFE 1954) can continue to absorb ions for a short time after respiration is inhibited. ORDIN and JACOBSON (1955) concluded from inhibitor studies that both the KREBS cycle and phosphorylation are involved in ion uptake, the latter supplying energy in the form of high energy phosphate and the former intermediates to function as carriers.

Factors affecting salt accumulation.

Some of the internal and environmental factors affecting salt accumulation will be discussed separately, although most of them at least have been mentioned in connection with theories of transport. Attention will be centered on factors affecting salt uptake by cells and tissues because factors affecting salt uptake by entire plants will be discussed in detail in volume 4 of the Handbook. Readers also are referred to BROYER's (1951) well organized discussion of salt absorption by plants.

Metabolic activity. The earlier studies by STEWARD and others brought out the fact that much more accumulation occurs in growing tissue than in mature tissue. This is particularly noticeable in roots where the meristematic region and adjacent cells often accumulate more salt than the older tissues (PREVOT and STEWARD 1936, STEWARD, PREVOT and HARRISON 1942, OVERSTREET and JACOBSON 1946, BROWN and CARTWRIGHT 1953). Recent studies with longer root segments and with attached roots indicate, however, that considerable accumulation of ions occurs in the root hair zone, several centimeters behind the root apex (KRAMER and WIEBE 1952, WIEBE and KRAMER 1954).

The high accumulation of ions near the root tips is assumed to be related to the higher metabolic activity of that region, but the relationship is complex and not fully explained. GREGORY and WOODFORD (1939), MACHLIS (1944b), and BERRY and BROCK (1946) observed a decreasing gradient of respiration from apex toward the base in roots. The gradients in respiration and P^{32} accumulation in barley root tips are shown in Fig. 3. Synthetic processes obviously are occurring more rapidly near the tip than in the mature region (BROWN and BROADBENT 1950, BROWN and CARTWRIGHT 1953, ROBINSON and BROWN 1952). The production of carrier or binding compounds would naturally be intimately related to cell metabolism. Organic acid metabolism may also be important in ion accumulation by providing organic anions to balance the excess of inorganic cations which sometimes accumulate in plant cells (ULRICH 1941, 1942). ULRICH found that when roots absorbed an excess of cations organic acids were formed, and when they absorbed an excess of anions the organic acid anions tended to disappear. More recent work by JACOBSON and ORDIN (1954) indicates that ionic balance during ion absorption is maintained primarily by metabolic gain or loss of malic acid anion in young barley roots, and in older roots by exchange of previously adsorbed ions.

The nature of the relationship between ion accumulation and respiration has been discussed in an earlier section of this chapter. The direct energy requirements for ion uptake are probably relatively small, but it has been shown many times that treatments which lower respiration and general metabolic activity, such

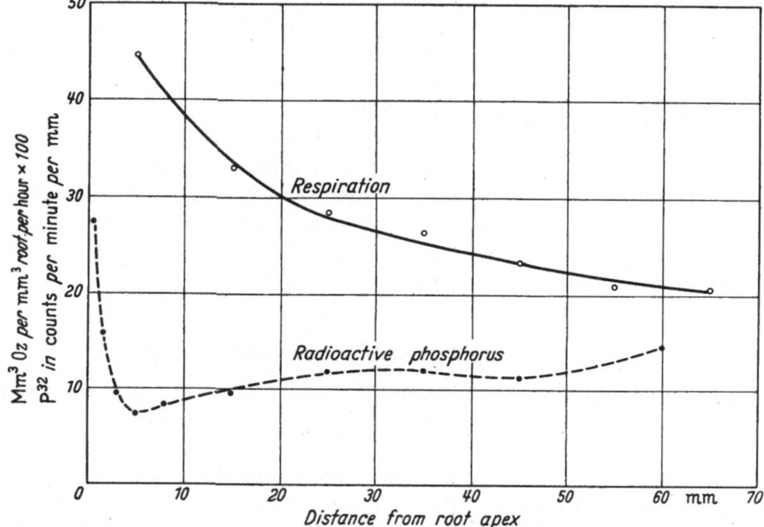

Fig. 3. Respiration and accumulation of P^{32} at various distances behind the apex of barley roots. Data for respiration are from MACHLIS (1944b). Data for accumulation of P^{32} are for 6-day old barley roots grown in dilute nutrient solution containing 20 μc per liter of P^{32}. (From unpublished data of WIEBE.)

as inhibitors, oxygen deficiency, sugar deficiency, and low temperature also reduce or prevent accumulation of ions. It appears that the indirect energy requirements for synthesis of binding sites or carrier molecules may be rather high. The effect of a respiration inhibitor, KCN, on salt uptake is shown in Fig. 4.

For over 20 years STEWARD and his colleagues have emphasized the relationship between aerobic respiration, growth, protein synthesis, and salt accumulation. On the basis of these observations it was concluded that meristematic tissue ought to be able to accumulate more salt than nonmeristematic tissue, but as mentioned earlier in this paper this does not seem to be true. Recent studies with radioactive tracers indicate that surprisingly great accumulation can occur in the root hair zone of roots, where cell division is no longer occurring. Even more surprising is the discovery that nongrowing carrot tissue absorbs Cs^{137} more rapidly than growing dividing cells of carrot tissue treated with coconut milk (STEWARD, CAPLIN and MILLAR 1952, STEWARD and MILLAR 1954). Addition of cyanide did not seriously inhibit growth, respiration, or Cs uptake of dividing cells, but it almost completely inhibited Cs uptake and greatly reduced respiration of nondividing cells. The same situation was found in artichoke tissue.

These observations led STEWARD and MILLAR (1954) to suggest that in growing plant tissue salt absorption occurs by two different mechanisms at different stages in growth. In actively dividing cells ion uptake appears to involve binding in certain specific sites produced continously during cell growth. This uptake mechanism is closely related to protein synthesis and phosphorylation. Apparently formation of binding sites or carrier molecules can also occur in nongrowing cells such as the beet tissue studied by SUTCLIFFE (1954b). In cells which are no longer dividing, but enlarging with consequent increase in size of the vacuole most of the accumulation occurs in the vacuoles. STEWARD and MILLAR suggest that development of the tonoplast around the central vacuole is chiefly responsible for the increasing importance of the second type of accumulation. This chiefly involves removal of ions from their binding sites and secretion of them into the vacuole. This type of accumulation is regarded as chiefly responsible for accumulation from dilute solutions, and it must be responsible for most of the salt absorption by roots because of the relatively large proportion of vacuolated cells in growing roots. Uptake of ions is not restricted to cells which are enlarging as shown by the accumulation which occurs in the root hair zone of roots and in mature beet tissue.

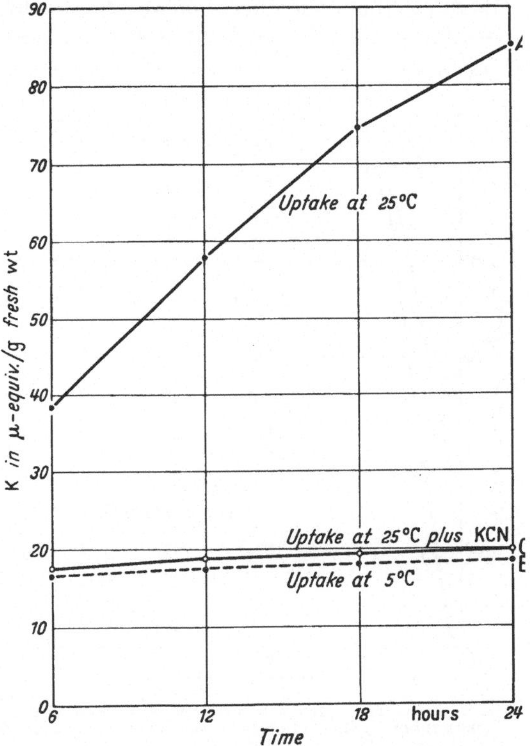

Fig. 4. Effects of low temperature and a respiration inhibitor on uptake of K from KBr by slices of washed beet root. Curve A shows uptake of K at 25° C., curve B uptake at 5° C., and curve C uptake at 25° C. in the presence of 0.001 M KCN. Uptake at 5° C. and in the presence of KCN may be regarded as nonmetabolic. (From SUTCLIFFE 1954b.)

STEWARD believes that salt accumulation is related to respiration principally through its dependence on protein synthesis. According to his view the kind of mechanism which is responsible for salt uptake depends on whether the cells are dividing or not.

Temperature. Interdependence between salt accumulation and metabolism naturally makes the process react strongly to environmental factors which reduce metabolism, such as low temperature. A typical example of the relation between ion uptake and temperature is shown in Fig. 4. It should be recalled that the initial adsorption exchange stage of uptake is little affected by temperature, but the second step which consists of transport into the vacuole is greatly affected.

Oxygen. Many investigators have found that reduction in oxygen supply results in reduced uptake of salt (HOAGLAND and BROYER 1936, STEWARD, BERRY and BROYER 1936, VLAMIS and DAVIS 1944). In general the effects do not become serious until the oxygen concentration has been reduced far below normal. An example of the reduction in ion uptake and respiration with decreasing oxygen supply is shown in Fig. 5.

Preconditioning. An interesting and puzzling feature of salt uptake is the fact that some kinds of tissue show a marked increase in salt uptake and respiration only if they are first washed in distilled water for a time. SUTCLIFFE (1954a) found it necessary to wash slices of beet tissue in aerated distilled water for 7 days in order to get maximum salt intake and increase in respiration when transferred to dilute salt solution. HUMPHRIES (1952) suggested that the cytochrome or some other component of the accumulation system must be regenerated. SUTCLIFFE (1954b) suggested that during the washing process carrier or binding molecules or centers are developed in the cytoplasm, while SKELDING and REES (1952) attribute the effects of washing to removal of an inhibitor which interferes with ion uptake. The effects of washing beet tissue for various periods of time on K content, K uptake, and amount of exchangeable potassium are shown in Fig. 6.

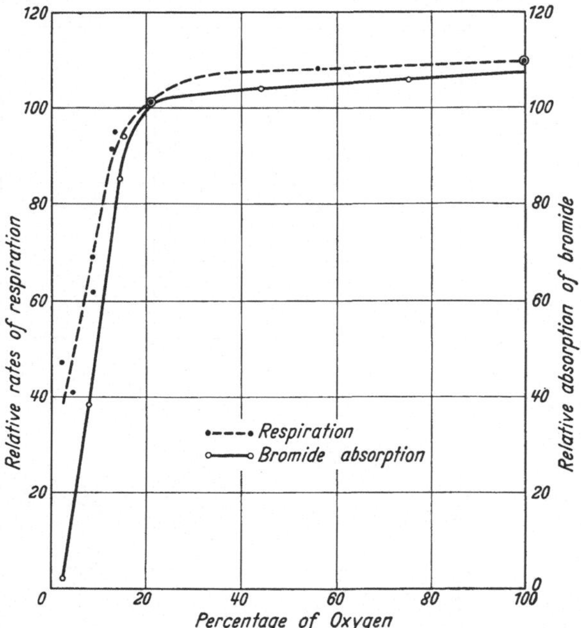

Fig. 5. Effect of oxygen concentration on respiration and bromide absorption by slices of potato tissue. (From STEWARD 1933.)

Kind and concentration of ions. It is well known that the uptake of a particular ion often is affected by the presence of other ions, but the manner of this interaction is complex and difficult to summarize. VIETS (1944) reported that the rate of absorption of KBr by excised barley roots is affected by the concentration of Ca, Mg, and other polyvalent ions in the external solution. To be effective these ions must be present in the external solution, as previous accumulation in the cells does not affect subsequent absorption of KBr from a solution in which they are absent. This suggests that these ions have important effects on the permeability of protoplasmic membranes, but as they only affect uptake under aerobic conditions the effect must be linked in some way with respiration. OVERSTREET, JACOBSON, and HANDLEY (1952) found that a given concentration of Ca both depresses and stimulates K uptake. The depression is attributed to competition for the same binding compound, the stimulation to increased utilization of the potassium complex formed during absorption. The effects of various concentrations of Ca on uptake of Ca, K and Br are shown in Fig. 7.

EPSTEIN and HAGEN (1952) explain interaction of K, Rb, Cs, Na, and Li in terms of competition for binding sites. K and Cs interfere with the absorption of Rb because they compete for the same sites, but Li and Na do not interfere competitively because they are not bound on the same sites. At high concentrations they might spill over on these sites and interfere in a noncompetitive manner. EPSTEIN and LEGGETT (1954) later reported that Ba, Ca, and Sr compete for the same sites, but Mg does not compete for this site. RUSSELL (1954) claims that no reliable conclusions can be drawn from observing the effect of varying the concentration of one ion on the absorption of another ion, but this seems to be an unduly pessimistic view.

If cells contain a high concentration of salt this reduces their capacity to absorb more, as shown by the tendency for absorption curves to flatten out when plotted over time (see Fig. 7, for example). This has been noted repeatedly by HOAGLAND, STEWARD, and other workers who made a distinction between the behavior of low-salt and high-salt tissues and plants. This is being reexamined by various investigators in an attempt to explain it in terms of current theories of accumulation. SUTCLIFFE (1954a) found that the uptake of K by beet root tissue stops at an internal concentration of 0.15 to 0.20 molar K. HUMPHRIES (1952) found that the uptake of N, P, and K by excised roots depended primarily on the concentration of the ion already present in the roots. There was little evidence that one ion affected the uptake of another ion under the conditions of his experiments.

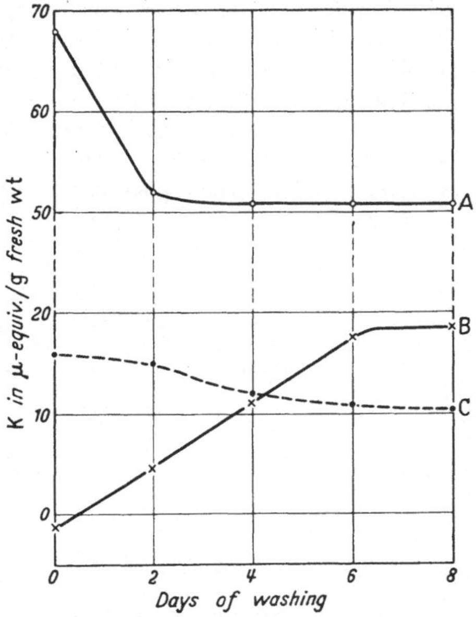

Fig. 6. Effect of length of preliminary period of washing of beet root tissue on, (A) the amount of K initially present in the tissue, (B) the amount absorbed from 0.02 M KCl in 4 hrs. at 5° C., and (C) the amount of K subsequently exchanged. (From data of SUTCLIFFE 1954.)

It has been suggested that salt accumulation represents a balance between accumulation and outward diffusion, or leakage, and that at high internal concentrations an increase in loss by leakage causes decreased net absorption even though gross absorption continues at a high level (KROGH 1946, LUNDEGÅRDH 1946, BROWN and CARTWRIGHT 1954). SUTCLIFFE (1954b) found that K and Br did not diffuse out of salt-saturated tissue, indicating that decreased absorption at high concentrations is not the result of increased leakage. He suggests that at high internal concentrations the free energy barrier to accumulation becomes so high that the carrier complex cannot break down and release ions, hence the supply of free carrier becomes depleted and uptake ceases. In connection with leakage it appears that there is more loss of salts by leakage or outward diffusion from excised roots than from storage tissue such as beet or carrot. This apparent leakage may really represent salt secreted into the xylem which escapes through the open ends of excised root segments.

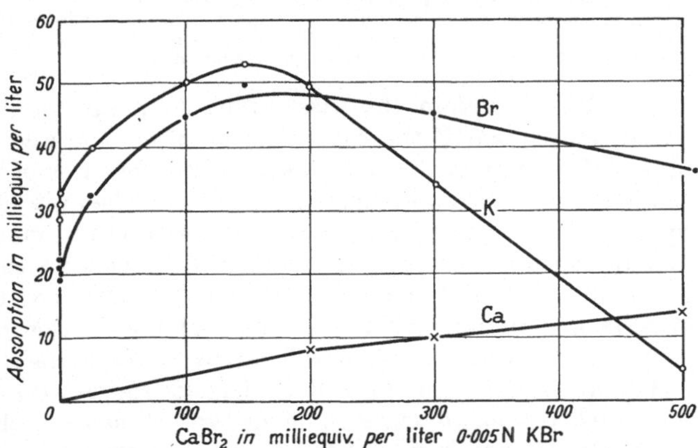

Fig. 7. The effect of various concentrations of Ca on uptake of Ca, K and Br by excised barley roots. (From VIETS 1944.)

JACQUES (1938) observed that if *Valonia* cells were impaled on a capillary tube which prevented increase in turgor pressure by allowing escape of surplus sap, ion intake occurred for at least two days at a much higher rate than in turgid cells. He suggested that in intact cells entrance of solutes increases the diffusion pressure deficit of the cell sap and causes dehydration of the protoplasm, reducing its permeability to ions. According to MYERS (1951) and BROUWER (1954) decrease in turgidity of plant tissue short of plasmolysis may increase its permeability to water and solutes, and perhaps the lesser turgidity of the impaled cells may be the significant factor. It seems that the method of JACQUES might be useful for other studies of the accumulation process in large cells. SUTCLIFFE (1954a) reported that plasmolysis decreases uptake of K by beet tissue by decreasing gross uptake and increasing outward leakage. This might be interpreted as resulting from increased permeability to K accompanying a major decrease in active transport to the vacuole. He states that among the factors involved in the effects of plasmolysis on ion uptake are change in surface area, thickness or density of the cytoplasm, concentration of ions in the vacuole, and respiration.

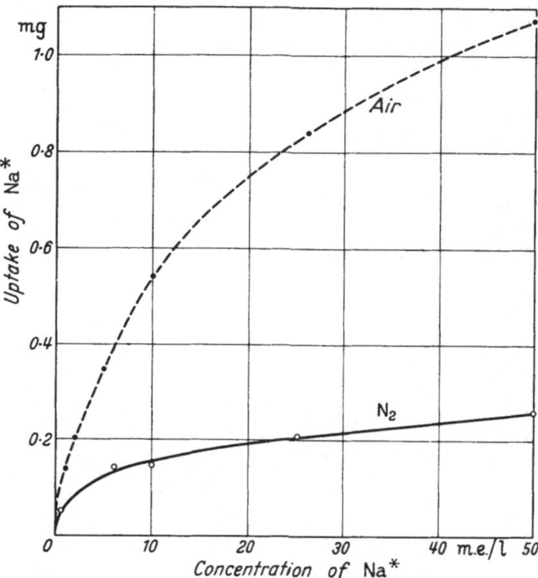

Fig. 8. Effects of concentration and oxygen deficiency on uptake of radioactive Na by excised barley roots. Uptake in nitrogen can be regarded as nonmetabolic, that in air as largely metabolic accumulation. (From EPSTEIN and HAGEN 1952.)

Views concerning the effect of changes in concentration of the external solution on salt uptake seem to vary. OLSEN (1950) claimed that the rate of absorption of a given ion from a solution is independent of the concentration of that ion over a wide range, provided that the relative concentrations of the various ions in the solution are kept constant. This is seldom the case, either in nature or in experimental solutions. OLSEN found that if the concentration of one ion was increased relative to the others, then more of that ion was absorbed, but less of the others so the sum of the ions remained approximately constant. The quite different results obtained by other investigators were attributed by OLSEN to faulty techniques, especially lack of constant stirring. HYLMÖ (1953) found that the external concentration materially affected the amount of Ca and Cl absorbed by pea plants. RUSSELL, MARTIN and BISHOP (1954) reported that the relationship between external concentration and uptake of phosphate by excised pea roots over the range from 0.0003 to 30.0 p.p.m. formed a sigmoid curve with the steepest slope from 0.1 to 0.3 p.p.m. A more common type of curve is shown in Fig. 8.

Auxin. It has been claimed that auxin stimulates the uptake of both salts and water, but the evidence is somewhat contradictory. The writer has indicated in the chapter on uptake of water that socalled auxin-induced uptake of water appears to result from effects of auxin on cell wall growth rather than from non-osmotic uptake of water. It seems possible that some of the data indicating that auxin causes increased uptake of minerals can be explained through the effects of auxin on metabolic processes. HIGINBOTHAM, LATIMER, and EPPLEY (1953) found an increase in uptake of Rb by pea epicotyl and slices of rutabaga when indoleacetic acid was added, which seemed to be relatively independent of water uptake. LUNDEGÅRDH (1954) found that low concentration of indoleacetic acid did not affect uptake of chloride ions by wheat roots over a period of one hour, but a concentration of 10^{-5} M increased uptake slightly and 10^{-3} M prevented it. HANSON and BONNER (1954) decided that 2,4-D has no direct effect on salt uptake by slices of Jerusalem artichoke tuber *(Helianthus tuberosus)*, but at least three indirect effects. One effect is increased cation exchange capacity resulting from increased growth, another is competition between water and salt uptake for a common energy suply, and a third is increased hydration. The first and third effects cause increased salt uptake, the second decreases it. In contrast HIGINBOTHAM *et al.* did not think there was competition between water and salt uptake for a common source of energy. Obviously no such competition occurs unless active uptake of water actually occurs. Evidently this problem deserves further study.

Accumulation in roots in relation to absorption through roots.

The behavior of the shoot has important effects on the amount of salt accumulation in the roots. This presumably will be discussed in detail in Volume 4 which deals with mineral nutrition, but it seems desirable to discuss the relation between accumulation of salts in the cells of roots and absorption of salts through the roots by growing shoots. Many valuable studies of salt accumulation have been made on excised roots, and these have contributed materially to our knowledge of salt uptake by intact plants. It seems possible, however, that concentration of attention on the behavior of storage tissue and excised roots has resulted in a rather inadequate picture of salt uptake through the roots of growing plants.

For example, studies with excised roots indicate that the greatest accumulation of ions occurs usually near the root tip. It has therefore been assumed that this also is the principal region of mineral intake by the plant as a whole. Studies with potometers on roots of intact plants indicate that the greatest absorption of water occurs somewhat behind the tip, where the xylem has matured, but suberization has not progressed far enough to reduce permeability (HAYWARD and SPURR 1943, ROSENE 1937). These observations led to the conclusion that water and salts are not absorbed through the same regions of roots. This conclusion has re-enforced the view that absorption of salts by plants is quite independent of the absorption of water. It seems to the writer that both of these assumptions need re-examination in view of more recent work.

In the first place it seems theoretically unlikely that a tissue capable of accumulating large amounts of ions would release these ions to the shoot. It is true that STEWARD, PREVOT, and HARRISON (1942) reported that when barley roots were placed in water, ions were first removed from the tips where they had been accumulated first. This almost certainly was abnormal behavior, however, which may have resulted from injury to root tips by prolonged immersion in water or some other cause. WIEBE (1953) found that barley root tips always maintained a high concentration of radioactive phosphorus and sulphur when the plants were transferred from a radioactive solution to nonradioactive, dilute HOAGLAND solution and kept there for up to 60 hrs.

In the second place actual measurements of translocation of several radioactive isotopes out of the root tips indicate that very little translocation occurs from the terminal 3 or 4 mm. where accumulation is so great. The results of a series of such measurements from a study by WIEBE and KRAMER (1954) are summarized in table 3. These data show that although considerable translocation of isotopes occurred when they were supplied to regions 10 to 60 mm. behind the tips, practically no ions were translocated out of the meristematic region. Fig. 9 shows accumulation in and translocation from various distances behind the apex of barley roots.

In view of these data it seems probable that the region of a root through which most of the salts are absorbed also is the region through which most of

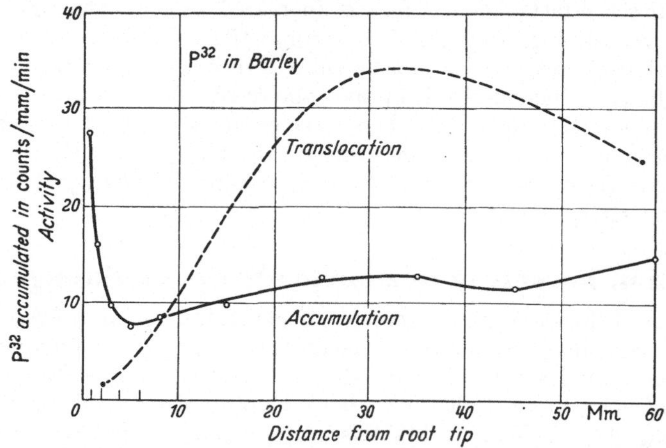

Fig. 9. Accumulation of P^{32} in roots of barley seedlings grown for 6 days in dilute nutrient solution containing 20 μc per liter of P^{32}. (From unpublished data of WIEBE.) The curve for translocation shows the percentage of P^{32} translocated more than 2 cm. from the point at which it was supplied to the root. See also table 3. (From WIEBE and KRAMER 1954.)

the water is absorbed. It further seems probable that if both water and salt are absorbed through the same regions of the roots that salt absorption by transpiring plants cannot be entirely independent of water absorption. HOAGLAND (1944) and his colleagues minimized the relation between water and salt absorption and emphasized the relationship of salt absorption by the intact plant to accumulation in the roots and the general metabolic condition of the roots and the plant. There is no doubt about the importance of accumulation in the roots or the effects of the general level of metabolism of the plant on absorption of salts (ALBERDA 1948, HELDER 1952, VAN ANDEL, ARIZS and HELDER 1950). It seems probable, however, that because of excessive preoccupation with accumulation certain aspects of absorption by entire plants have been overlooked or underemphasized.

Table 3. *Data on translocation of radioactive isotopes from various regions of barley roots. Each value represents the amount translocated as a percentage of the total amount absorbed and is the average of two to four replications.* (From WIEBE and KRAMER 1954.)

Approximate distance from root tip to point at which isotope was supplied	Percentage of isotope translocated away from region at which it was supplied			
	P^{32}	Rb^{86}	I^{131}	S^{35}
0–4 mm.	1.3	4.2	1.0	1.7
7–10 mm.	8.5	14.3	28.3	5.2
27–30 mm.	34.4	14.7	28.9	11.8
57–60 mm.	24.9	9.4	22.7	9.2

The early studies on the relation between water and salt absorption were subject to criticism (KRAMER 1949), but FREELAND (1937) and WRIGHT (1939) avoided most of the defects in technique. Both investigators found that the plants which absorbed the most water also absorbed the most salt, although the increase in salt absorption was not proportional to the increase in water absorption of the more rapidly transpiring plants. HYLMÖ (1952) reported recently that the rate of uptake of calcium and chloride ion by young pea plants was related to the rate of water uptake when water uptake was varied by four different methods. The ion uptake by excised roots was the same as the calculated ion uptake for plants with zero transpiration. HYLMÖ suggested that there might be three phases in ion uptake by and through roots. The first step is diffusion of ions into the roots along the walls and probably in the cytoplasm of the cells. This is accompanied by exchange adsorption on walls and in cytoplasm. The second phase is accumulation of ions in the vacuoles. In transpiring plants there is a third phase, the movement of ions into the roots carried by the entering water. Some of these ions also are adsorbed and accumulated until a steady state has been attained, after which they are carried along toward the xylem by the water moving in the cell walls, cytoplasm, and possibly the intercellular spaces (the apparent free space mentioned earlier). BUTLER (1953b) found that more K was retained in the roots of slowly transpiring plants than in roots of rapidly transpiring plants (see Fig. 10). On the other hand BROUWER (1954) found no direct effect of transpiration on salt absorption, but he did find evidence that a high rate of transpiration increases the permeability of the roots to water and salts. HANSON and BIDDULPH (1953) found diurnal variations in movement of Rb and P from roots to shoots, the maximum translocation occurring about midday and the minimum about midnight. They attributed it to some obscure property of the root cytoplasm, but it may well have been caused by differences in rate of water movement through the roots. The difficulties in obtaining definite results from experiments on this problem have been well presented by VAN DEN HONERT, HOOYMANS and VOLKERS (1955).

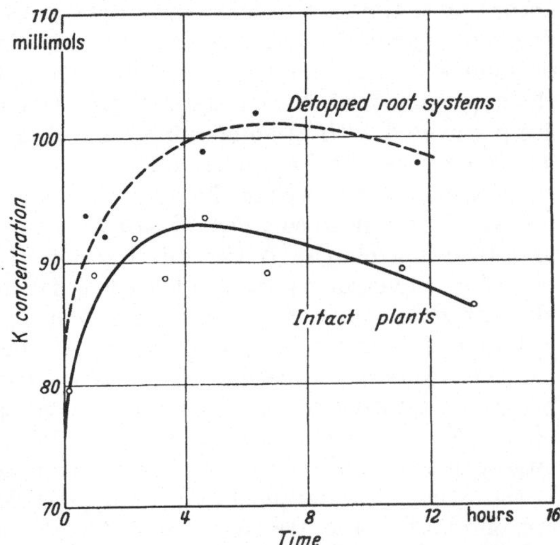

Fig. 10. Difference in K content of roots of intact, transpiring plants and detopped root systems. The ower K content in the roots of transpiring plants is attributed to greater translocation to the shoot in the transpiration stream. (From BUTLER 1953b.)

In a plant having roots low in salt, accumulation by the root cells will remove more of the entering ions than if the roots are already high in salt. RUSSELL and his colleagues have considerable data on the effect of conditions in the roots on salt transfer to the shoot. If the roots are pretreated with 10 p.p.m. of P more P will be translocated to the shoot from a very dilute solution of P than if they are not pretreated (RUSSELL, MARTIN, and BISHOP 1953). According to ALBERDA (1948), in high-salt plants some of the ions carried to the shoot in the transpiration stream return to the roots. Most workers agree that the metabolism of the shoot has important effects on salt uptake by the roots. The relation between water and salt uptake obviously is therefore a complex one, and it is not surprising that various investigators have come to somewhat different conclusions concerning the nature of the relationship.

There is increasing evidence that the transport of ions across a root need not occur through the vacuoles of the cortical cells and therefore can occur somewhat independently of accumulation, although the vacuoles of course accumulate ions from the surrounding solution (BROYER 1950, HYLMÖ 1953).

This is essentially the situation proposed by SCOTT and PRIESTLEY (1928). The possibility of movement across the root through the plasmodesmata or strands of cytoplasm which connect the protoplasts of adjoining cells has been emphasized by ARISZ (1945, 1947, 1953, 1954). ARISZ terms the interconnected protoplasm the symplast and has shown that chloride is transported through it in leaves of *Vallisneria*. If there are considerable numbers of mobile ions in the roots they are likely to be carried inward in the water which moves inward passively toward the xylem. This passive transport might be stopped at the endodermis according to the view of SCOTT and PRIESTLEY (1928) and ARNOLD (1952). HYLMÖ does not regard the endodermis as an important barrier to the entrance of water or solutes and it certainly did not prevent the entrance of radioactive ions in the experiments of WIEBE and KRAMER (1954). In at least some species its walls are permeated with plasmodesmata and it is often pierced by branch root initials which decrease its impermeability.

It seems to the writer that although accumulation of ions in roots is an important factor in absorption of salt by plants it does not completely control it in transpiring plants. A variable fraction of the total salt which enters a plant probably moves in passively at least as far as the endodermis, and probably, in at least some roots, as far as the xylem. Thus under some circumstances the uptake of water can considerably modify the uptake of salts through the roots.

A characteristic feature of detopped root systems of many plants is the presence of sufficient solutes in the xylem sap to produce an osmotic pressure of 1 or 2 atm. (KRAMER 1949, pp. 176–179). In herbaceous plants this consists largely of salts, indicating that ions are transported across the cortex and released in the xylem. In detopped plants the possible effects of the transpiration stream are eliminated and transport across the root seems to resemble accumulation in cells in several respects, including its sensitivity to oxygen deficiency and respiration inhibitors. The theories of active transport proposed for cells have therefore quite naturally been applied to explain transport of ions into roots. As mentioned previously, some investigators regard the endodermis as the differentially permeable membrane controlling movement of ions into the roots, but the writer is inclined to think that the importance of the endodermis has been overemphasized.

ARISZ and his coworkers regard the protoplasts of the root cells as forming a unit, the symplast, through which ions move across the root in the same manner as they move across the cytoplasm in an individual cell. LUNDEGÅRDH applies his anion respiration theory to transport across the root, and like CRAFTS and BROYER (1938), WIERSUM (1947) and others, emphasizes the effects of a gradient of decreasing oxygen concentration from epidermis to stele. According to BROWN (1947) and WOODFORD and GREGORY (1948) it is doubtful if an important oxygen deficit exists in the stele. If a deficit does not exist considerable revision of certain theories of salt transport into the xylem will be necessary.

There seems to be no really satisfactory explanation for the accumulation of salts in the xylem sap. HYLMÖ (1953) proposed a novel mechanism which eliminates the necessity of assuming that ions are "secreted" into the xylem. He suggested that the protoplasts of xylem initials continue to accumulate ions even after their upper transverse walls are ruptured. The ions accumulated during the formation of xylem vessels are released into the sap as the protoplasts disintegrate, providing a continuous supply of solutes so long as differentiation is occurring. This hypothesis is consistent with what is known about the factors affecting root pressure and with the general failure of most gymnosperms to develop root pressure, but there are no experimental data as yet to support it.

Discussion.

It must not be supposed that accumulation is restricted to cells of roots and slices of beet, carrot, or potato tissue. Although most experiments deal with accumulation in these tissues, the process occurs in the cells of all of the growing regions of plants. Expanding leaves absorb minerals from older leaves and from the roots, and growing buds, cambial cells, developing ovules, and other metabolically active regions are fully as capable of accumulating ions as are root tips. Some data on relative amounts of accumulation by various parts of the plant are presented by STEWARD and MILLAR (1954). It should be mentioned also that accumulation is not restricted to ions. Organic compounds, particularly sugars, are transported to and accumulated in fruits, seeds, tubers, and other storage organs. Often such movement is along concentration gradients because the soluble compounds are converted into insoluble storage forms, as sugar into starch, but in some instances active transport of organic compounds against concentration gradients occurs.

It is somewhat unfortunate that the chapter on salt accumulation had to be written at this time because we are uncertain about so many details of the process. Although numerous contributions have been made to our knowledge of the process in recent years, many of them are so contradictory that they often add to our perplexity rather than increase our understanding. As a result it has not been possible to give complete or final explanations for the phenomena discussed and it probably will be necessary to modify some of our conclusions in the light of information yet to be obtained.

Summary.

One of the outstanding characteristics of living plant cells is their ability to accumulate certain ions until the concentration in the vacuole is much higher than that in the environment. This is brought about by the active transport of ions across a differentially permeable membrane located in the layer of cytoplasm lying between the cell wall and the vacuole. In many instances this membrane appears to be the tonoplast, lying next to the vacuole, but it is possible that the location of the impermeable layer varies in different kinds of cells and for different solutes.

Uptake of salts appears to occur in two steps. The first step is nonmetabolic uptake by exchange adsorption. This process attains equilibrium rapidly, is reversible, and largely independent of temperature, oxygen and metabolic activity. The second step, sometimes termed metabolic uptake, is active transport across the impermeable layer into the vacuole. This process is slow, long continued, and affected by temperature, oxygen, and metabolic activity. One of the best known theories of active transport is that of LUNDEGÅRDH who proposed a close relationship between a special fraction of the respiration, called anion respiration, and anion absorption. Anion respiration involves the cytochrome-cytochrome oxidase system and LUNDEGÅRDH proposed that anions are carried across the cell membrane along a chain of cytochrome molecules. Cations then move in along a potential gradient caused by the accumulation of anions. This theory requires some modification because ion uptake is not directly proportional to respiration and because it does not explain differential uptake of various ions. Furthermore it now appears that cations are accumulated in the same manner as anions.

Some of these difficulties are answered by the increasingly popular carrier theory of active transport. According to this theory ions form loose combinations with specific binding compounds or carrier molecules, probably similar to those formed between an enzyme and its substrate. The complex then moves across the membrane and breaks up on the inner side, setting the ion free. The exact nature of the binding compounds are unknown although proteins, amino acids, high energy phosphorylated compounds and mitochondria have been suggested as possibilities. The close relationship which exists between respiration, protein metabolism, and salt uptake may be related in part to the synthesis of carrier or binding compounds and in part to their transport across the membranes. The method by which such complexes are moved across a membrane is unknown, although protoplasmic streaming may be involved.

Accumulation of ions is not always highest in meristematic vegions, and otten is guite high in en larging cells. A general correlation between salt uptake, respiration, protein synthesis and other metabolic processes seems to exist. The correlation between salt uptake and respiration probably is not as close or as direct as it once was supposed to be, however. Low oxygen and low temperature greatly decrease accumulation and a low sugar supply in the tissue also results in less accumulation. Within a wide range the process is little affected by p_H. The effects of auxin on ion uptake are complex, but it is said to increase accumulation in some tissues.

The rate and amount of accumulation are affected by the concentration of ions in the cells and by the kinds and concentration of ions in the environment. At least part of the effect of one ion on the absorption of another ion is supposed to be caused by competition for binding sites on carrier molecules, but effects on permeability and metabolic processes probably also are involved.

Uptake of salt by entire plants probably depends primarily on accumulation in the root cells and active transport to the xylem, but it seems likely that considerable salt moves to the xylem without entering the vacuoles of the cortical cells. Furthermore, much of the salt absorbed by plants probably enters through the root hair zone rather than the meristematic region where accumulation is heaviest. In view of these facts it seems probable that salt intake by intact plants is measurably affected by the rate of water intake.

Literature.

ALBERDA, TH.: The influence of some external factors on growth and phosphate uptake of maize plants of different salt conditions. Rec. Trav. bot. néerl. **41**, 541–602 (1948). — ANDEL, O. M. VAN, W. H. ARISZ and R. J. HELDER: Influence of light and sugar on growth and salt intake by maize roots. Proc. Kon. Ned. Akad. Wetensch. **53**, 159–171 (1950). — ARISZ, W. H.: Absorption and transport by the tentacles of *Drosera capensis*. I., II. Proc. Kon. Ned. Akad. Wetensch. **45**, 2–8, 794–801 (1942). — Contribution to a theory on the absorption of salts by the plant and their transport in parenchymatous tissue. Proc. Kon. Ned. Akad. Wetensch. **48**, 420–446 (1945). — Uptake and transport of chlorine by parenchymatic tissue of leaves of *Vallisneria spiralis*. III. Proc. Kon. Ned. Akad. Wetensch. **51**, 28–36 (1948). — Active uptake, vacuole-secretion and plasmatic transport of chloride-ions in leaves of *Vallisneria spiralis*. Acta bot. néerl. **1**, 506–515 (1953). — Transport of chloride in the "symplasm" of *Vallisneria* leaves. Nature (Lond.) **174**, 223–224 (1954). — ARNOLD, A.: Über den Funktionsmechanismus der Endodermiszellen der Wurzeln. Protoplasma **41**, 189–211 (1952).

BARTLEY, W., and R. E. DAVIES: Secretory activity of mitochondria. Biochemic. J. **52**, XX (1952). — BERRY, L. J., and M. J. BROCK: Polar distribution of respiratory rate in the onion root tip. Plant Physiol. **21**, 542–549 (1946). — BROOKS, S. C.: Selective accumulation with reference to ion exchange by the protoplasm. Trans. Faraday Soc. **33**, 1002–1006 (1937). — The intake of radioactive isotopes by living cells. Cold Spring Harbor Symp. Quant. Biol. 8, 171–177 (1940). — The penetration of radioactive sodium into *Valonia* and *Halicystis*. Protoplasma (Berl.) **42**, 63–68 (1953). — BROUWER, R.: The regulating influence of transpiration and suction tension on the water and salt uptake by the roots of intact *Vicia faba* plants. Acta bot. néerl. **3**, 264–312 (1954). — BROWN, R.: The gaseous exchange between the root and the shoot of the seedling of Cucurbita pepo. Ann. of Bot. **11**, 417–437 (1947). — Protoplast surface enzymes and absorption of sugar. Internat. Rev. Cytology **1**, 107–118 (1952). — BROWN, R., and D. BROADBENT: The development of cells in the growing zones of the roots. J. of Exper. Bot. **1**, 249–263 (1950). — BROWN, R., and P. M. CARTWRIGHT: The absorption of potassium by cells in the apex of the root. J. of Exper. Bot. **4**, 197–221 (1953). — BROYER, T. C.: Further observations on the absorption and translocation of inorganic solutes using radioactive isotopes with plants. Plant Physiol. **25**, 367–377 (1950). — The nature of the process of inorganic solute accumulation in roots. In Mineral Nutrition of Plants, p. 187–249. Ed. by E. TRUOG. Univ. Wisconsin Press 1951. — BROYER, T. C., and R. OVERSTREET: Cation exchange in plant roots in relation to metabolic factors. Amer. J. Bot. **27**, 425–430 (1940). — BURSTRÖM, H.: The mechanism of ion absorption. In Mineral Nutrition of Plants, p. 251–260. Ed. by E. TRUOG. Univ. Wisconsin Press 1951. — BUTLER, G. W.: Ion uptake by young wheat plants. II. The "apparent free space" of wheat roots. Physiol. Plantarum (Copenh.) **6**, 617–635 (1953a). — Ion uptake by young wheat plants. I. Time course of the absorption of potassium and chloride. Physiol. Plantarum (Copenh.) **6**, 594–616 (1953b).

CASARI, K.: Über den Plasmolytikum-Wechsel-Effekt. Protoplasma **42**, 427–447 (1953). — COLLANDER, R.: Der Zellsaft der Characeen. Protoplasma **25**, 201–210 (1936). — Selective absorption of cations by higher plants. Plant Physiol. **16**, 691–720 (1941). — CONWAY, E. J.: A redox pump for the biological performance of osmotic work and its relation to the kinetics of free ion diffusion across membranes. Internat. Rev. Cytology **2**, 419–445 (1953).—CRAFTS, A. S., and T. C. BROYER: Migration of salts and water into xylem of the roots of higher plants. Amer. J. Bot. **25**, 529–535 (1938).

DAVIES, R. E., and A. G. OGSTON: On the mechanism of secretion of ions by gastric mucosa and by other tissues. Biochemic. J. **46**, 324–333 (1950).

EPSTEIN, E.: Mechanism of ion absorption by roots. Nature (Lond.) **171**, 83–84 (1953). — Cation-induced respiration in barley roots. Science (Lancaster, Pa.) **120**, 987–988 (1954). — EPSTEIN, E., and C. E. HAGEN: A kinetic study of the absorption of alkali cations by barley roots. Plant Physiol. **27**, 457–474 (1952). — EPSTEIN, E., and J. L. LEGGETT: The absorption of alkaline earth cations by barley roots: kinetics and mechanism. Amer. J. Bot. **41**, 785–792 (1954).

FRANCK, J., and J. E. MAYER: An osmotic diffusion pump. Arch. of Biochem. **14**, 297–313 (1947). — FREELAND, R. O.: Effect of transpiration upon the absorption of mineral salts. Amer. J. Bot. **24**, 373–374 (1937).

GOLDACRE, R. J.: The folding and unfolding of protein molecules as a basis of osmotic work. Internat. Rev. Cytology **1**, 135–164 (1952). — GOLDACRE, R. J., and I. J. LORCH: Folding and unfolding of protein molecules in relation to cytoplasmic streaming, amoeboid movement and osmotic work. Nature (Lond.) **166**, 497–500 (1950). — GREGORY, F. G., and H. K. WOODFORD: An apparatus for the study of the oxygen, salt, and water uptake of various zones of the root, with some preliminary results with *Vicia faba*. Ann. of Bot., N. S. **3**, 147–154 (1939).

HANSON, J. B., and O. BIDDULPH: The diurnal variation in the translocation of minerals across bean roots. Plant Physiol. **28**, 356–370 (1953). — HANSON, J. B., and J. BONNER: The relationship between salt and water uptake in Jerusalem artichoke tuber tissue. Amer. J. Bot. **41**, 702–710 (1954). — HAYWARD, H. E., and W. B. SPURR: Effects of osmotic concentration of substrate on the entry of water into corn roots. Bot. Gaz. **105**, 152–164 (1943). — HIGINBOTHAM, N., H. LATIMER and R. EPPLEY: Stimulation of rubidium absorption by auxins. Science (Lancaster, Pa.) **118**, 243–245 (1954). — HELDER, R. J.: Analysis of the process of anion uptake of intact maize plants. Acta bot. néerl. **1**, 361–434 (1952). — HOAGLAND, D. R.: The inorganic nutrition of plants. Waltham, Mass.: Chronica Botanica Co. 1944. — HOAGLAND, D. R., and T. C. BROYER: General nature of the process of salt accumulation by roots with description of experimental methods. Plant Physiol. **11**, 471–507 (1936).— Accumulation of salt and permeability in plant cells. J. Gen. Physiol. **25**, 865–880 (1942). — HOAGLAND, D. R., and A. R. DAVIS: The intake and accumulation of electrolytes by plant cells. Protoplasma (Berl.) **6**, 610–626 (1929). — HÖFLER, K.: New facts on water permeability. Protoplasma (Wien) **39**, 677–683 (1950). — HOLM-JENSEN, I., A. KROGH and V. WARTIOVAARA: V. Some experiments on the exchange of potassium and sodium between single cells of *Characeae* and the bathing fluid. Acta bot. fenn. **36**, 1 (1944). — HONERT, T. H. VAN DEN, J. J. HOOYMANS and W. S. VOLKERS: Experiments on the relation between water absorption and mineral uptake by plant roots. Acta bot. néerl. **4**, 139–155 (1955). — HOPE, A. B.: Salt uptake by root tissue cytoplasm: the relation between uptake and external concentration. Austral. J. Biol. Sci. **6**, 396–409 (1953). — HOPE, A. B., and R. N. ROBERTSON: Bioelectric experiments and the properties of plant protoplasm. Austral. J. Sci. **15**, 197–203 (1953). — HOPE, A. B., and P. G. STEVENS: Electrical potential differences in bean roots and their relation to salt uptake. Austral. J. Sci. Res. B **5**, 335–343 (1952). —

HUMPHRIES, E. C.: III. Observations on roots of pea plants grown in solutions deficient in phosphorus, nitrogen or potassium. J. of Exper. Bot. 3, 291–309 (1952). — HYLMÖ, B.: Transpiration and ion absorption. Physiol. Plantarum (Copenh.) 6, 333–405 (1953).

JACOBSON, L., and L. ORDIN: Organic acid metabolism and ion absorption in roots. Plant Physiol. 29, 70–75 (1954). — JACOBSON, L., and R. OVERSTREET: A study of the mechanism of ion absorption by plant roots using radioactive elements. Amer. J. Bot. 34, 415–420 (1947). — The uptake by plants of plutonium and some products of nuclear fission adsorbed on soil colloids. Soil Sci. 65, 129–134 (1948). — JACOBSON, L., R. OVERSTREET, H. M. KING and R. HANDLEY: A study of potassium absorption by barley roots. Plant Physiol. 25, 639–647 (1950). — JACQUES, A. G.: Kinetics of penetration XV. The restriction of the cellulose wall. J. Gen. Physiol. 22, 147–163 (1938). — JAMES, W. O., and D. BOULTER: Further studies of the terminal oxidases in the embryos and young roots of barley. New Phytologist 54, 1–12 (1955).

KRAMER, P. J.: Plant and soil water relationships. New York: McGraw-Hill Book Co. 1949. — KRAMER, P. J., and H. H. WIEBE: Longitudinal gradients of P^{32} absorption in roots. Plant Physiol. 27, 661–674 (1952). — KROGH, A.: The active and passive exchanges of inorganic ions through the surfaces of living cells and through living membranes generally. Proc. Roy. Soc. Lond., Ser. B 133, 140–200 (1946).

LATIES, G. G.: The osmotic inactivation in situ of plant mitochondrial enzymes. J. of Exper. Bot. 5, 49–70 (1954). — LUNDEGÅRDH, H.: Absorption, transport and exudation of inorganic ions by the roots. Ark. Bot. (Stockh.) A 32 (12), 1–139 (1945). — Transport of water and salts through plant tissues. Nature (Lond.) 157, 575–577 (1946). — Mineral nutrition of plants. Annual Rev. Biochem. 16, 503–528 (1947). — Translocation of salt and water through wheat roots. Physiol. Plantarum (Copenh.) 3, 103–151 (1950). — Anion respiration: the experimental basis of a theory of absorption, transport and exudation of electrolytes by living cells and tissues. Soc. Exper. Biol. Symp. 8, 262–296 (1954). — LUNDEGÅRDH, H., and H. BURSTRÖM: Atmung und Ionenaufnahme. Planta (Berl.) 18, 683–699 (1933).

MACHLIS, L.: The influence of some respiratory inhibitors and intermediates on respiration and salt accumulation of excised barley roots. Amer. J. Bot. 31, 183–192 (1944a). — The respiratory gradient in barley roots. Amer. J. Bot. 31, 281–282 (1944b). — MILLER, E. C.: Plant physiology. Second edition. New York: McGraw-Hill Book Co. 1938. — MILTHORPE, J., and R. N. ROBERTSON: 6. Salt respiration and accumulation in barley roots. Austral. J. Exper. Biol. a. Med. 26, 191–197 (1948). — MYERS, G. M. P.: The water permeability of unplasmolyzed tissues. J. of Exper. Bot. 2, 129–144 (1951).

OLSEN, C.: The significance of concentration for the rate of ion absorption by higher plants in water culture. Physiol. Plantarum (Copenh.) 3, 152–164 (1950). — ORDIN, L., and L. JACOBSON: Inhibition of ion absorption and respiration in barley roots. Plant Physiol. 30, 21–27 (1955). — OSTERHOUT, W. J. V.: Some aspects of selective absorption. J. Gen. Physiol. 5, 225–231 (1922). — The mechanism of accumulation in living cells. J. Gen. Physiol. 35, 519–594 (1952). — OVERSTREET, R., and L. JACOBSON: The absorption by roots of rubidium and phosphate ions at extremely small concentrations as revealed by experiments with Rb^{86} and P^{32} prepared without inert carrier. Amer. J. Bot. 33, 107–112 (1946). — Mechanisms of ion absorption by roots. Annual Rev. Plant Physiol. 3, 189–206 (1952). — OVERSTREET, R., L. JACOBSON and R. HANDLEY: The effect of calcium on the absorption of potassium by barley roots. Plant Physiol. 27, 583–590 (1952).

PLOWE, J.: Membranes in the plant cells. Protoplasma (Berl.) 12, 196–240 (1931). — PREVOT, P., and F. C. STEWARD: Salient features of the root system relative to the problem of salt absorption. Plant Physiol. 11, 509–534 (1936).

ROTHSTEIN, A.: Enzyme systems of the cell surface involved in the uptake of sugars by yeast. Soc. Exper. Biol. Symp. 8, 165—201 (1954). — RUSSELL, R. S.: The relationship between metabolism and the accumulation of ions by plants. Soc. Exper. Biol. Symp. 8, 343–366 (1954). — RUSSELL, R. S., and M. J. AYLAND: Exchange reactions in the entry of cations into plant tissues. Nature (Lond.) 175, 204–205 (1955). — RUSSELL, R. S., R. P. MARTIN and O. N. BISHOP: II. The effect of phosphate status and root metabolism on the distribution of absorbed phosphate between roots and shoots. J. of Exper. Bot. 4, 136–156 (1953). — III. The relationship between the external concentration and the absorption of phosphate. J. of Exper. Bot. 5, 327–342 (1954).

SCOTT, L. I., and J. H. PRIESTLEY: A reconsideration of the entry of water and salts in the absorbing region. New Phytologist 27, 125–140 (1928). — SEEMANN, F.: Der Einfluß von Neutralsalzen und Nichtleitern auf die Wasserpermeabilität des Protoplasmas. Protoplasma (Wien) 42 (3), 109–132 (1953). — SKELDING, A. D., and W. J. REES: An inhibitor of salt absorption in the root tissue of red beet. Ann. of Bot. 16, 513–529 (1952). — SPIEGELMAN, S., and J. M. REINER: A kinetic analysis of potassium accumulation and sodium exclusion. Growth 6, 367–389 (1942). — STEWARD, F. C.: V. Observations upon the effects of time oxygen and salt concentration upon absorption and respiration by storage tissue. Protoplasma

(Berl.) **18**, 208–242 (1933). — STEWARD, F. C., W. E. BERRY and T. C. BROYER: VIII. The effect of oxygen upon respiration and salt accumulation. Ann. of Bot. **50**, 345–366 (1936). — STEWARD, F. C., S. M. CAPLIN and F. K. MILLAR: New techniques for the investigation of metabolism, nutrition and growth in undifferentiated cells. Ann. of Bot. **16**, 57–77 (1952). — STEWARD, F. C., and J. A. HARRISON: IX. The absorption of rubidium bromide by potato discs. Ann. of Bot., N. S. **3**, 427–453 (1939). — STEWARD, F. C., and F. K. MILLAR: Salt accumulation in plants: a reconsideration of the role of growth and metabolism. Soc. Exper. Biol. Symp. **8**, 367–406 (1954). — STEWARD, F. C., P. PREVOT and J. A. HARRISON: Absorption and accumulation of rubidium bromide by barley plants. Localization in the root of cation accumulation and of transfer to the shoot. Plant Physiol. **17**, 411–421 (1942). — STEWARD, F. C., and H. E. STREET: The nitrogenous constituents of plants. Annual Rev. Biochem. **16**, 471–502 (1947). — STOUT, P. R., and R. OVERSTREET: Soil chemistry in relation to inorganic nutrition of plants. Annual Rev. Plant Physiol. **1**, 305–342 (1950). — STREET, H. E., and J. S. LOWE: The carbohydrate nutrition of tomato roots. II. The mechanism of sucrose absorption by excised roots. Ann. of Bot. **14**, 307–329 (1950). — SUTCLIFFE, J. F.: The influence of internal ion concentration on potassium accumulation and salt respiration of red beet root tissue. J. of Exper. Bot. **3**, 59–76 (1952). — Ion secretion in plants. Internat. Rev. Cytology **2**, 179–200 (1953). — The absorption of potassium ions by plasmolyzed cells. J. of Exper. Bot. **5**, 215–231 (1954a). — The exchangeability of potassium and bromide ions in cells of red beet root tissue. J. of Exper. Bot. **5**, 313–326 (1954b). — Cation absorption by non-growing plant cells. Soc. Exper. Biol. Symp. **8**, 325–341 (1954c).

ULRICH, A.: Metabolism of non-volatile organic acids in excised barley roots as related to cation-anion balance during salt absorption. Amer. J. Bot. **28**, 526–537 (1941). — Metabolism of organic acids in excised barley roots as influenced by temperature, oxygen tension and salt concentration. Amer. J. Bot. **29**, 220–227 (1942). — USSING, H. H.: Ion transport across biological membranes. In Ion Transport across Membranes. Ed. by H. T. CLARK. New York: Academic Press 1954.

VIETS, F. G.: Calcium and other polyvalent cations as accelerators of ion accumulation by excised barley roots. Plant Physiol. **19**, 466–480 (1949). — VLAMIS, J., and A. R. DAVIS: Effects of oxygen tension on certain physiological responses of rice, barley and tomato. Plant Physiol. **19**, 33–51 (1944).

WANNER, H.: Die relative Größe der Temperaturkoeffizienten von Kationen- und Anionenaufnahme. Ber. schweiz. bot. Ges. **58**, 123–130 (1948). — WEEKS, D. C., and R. N. ROBERTSON: VIII. Dependence of salt accumulation and salt respiration upon the cytochrome system. Austral. J. Sci. Res. B **3**, 487–500 (1950). — WIEBE, H. H.: A study of absorption and translocation of radioactive isotopes in various regions of barley roots. Ph. D. Diss. Duke University 1953. — WIEBE, H. H., and P. J. KRAMER: Translocation of radioactive isotopes from various regions of roots of barley seedlings. Plant Physiol. **29**, 342–348 (1954). — WIERSUM, L. K.: Transfer of solutes across young roots. Rev. Trav. bot. néerl. **41**, 1–79 (1947). — WILBRANDT, W.: Secretion and transport of nonelectrolytes. Soc. Exper. Biol. Symp. **8**, 136–162 (1954). — WILLIAMS, D. E., and N. T. COLEMAN: Cation exchange properties of plant root surfaces. Plant a. Soil **2**, 243–256 (1950). — WOODFORD, E. K., and F. G. GREGORY: Preliminary results obtained with an apparatus for the study of salt uptake and root respiration of whole plants. Ann. of Bot., N. S. **12**, 335–370 (1948). — WRIGHT K. E.: Transpiration and the absorption of mineral salts. Plant Physiol. **14**, 171–174 (1939).

The uptake of water by plant cells.

By

Paul J. Kramer.

Introduction.

The importance of water in the physiology of plants makes it desirable to have an understanding of the nature and origin of the forces responsible for the movement of water into and out of plant cells and tissues. Only incidental attention will be given to the absorption and translocation of water in the plant as a whole in this chapter because these topics are discussed in detail in Volume 3 of this series. It may seem that the uptake of water by cells must be fully understood by this time, but those most familiar with the literature of this subject know that this is not true. There are differences in opinion concerning the relative importance of osmotic and nonosmotic forces in water uptake, concerning the relation of water uptake to growth and cell enlargement, and even concerning the best terminology to use in discussing the problems.

It is impossible to mention all of the important literature in the space available and the reader is referred to the book by CRAFTS, CURRIER, and STOCKING (1949), the review by KRAMER (1955), and the reviews by BOGEN in recent volumes of the Fortschritte der Botanik for additional literature and other viewpoints. Many of the topics which are touched on in this chapter are treated in more detail in other chapters of this volume, but some repetition seems unavoidable.

The cells of growing plants undergo almost continual changes in water content. Some of these changes result from the competition for water which occurs among the various organs and tissues, other changes occur because of changes in amount and kind of cell constituents, and still others result from growth or responses to various stimuli. The water in a plant can be regarded as forming essentially a continuous system, the hydrostatic system of the plant, in which the cells of the various tissues are connected through their water saturated walls and the organs are connected to each other through the xylem. As a result of the continuity of the hydrostatic system any significant change in water content in one part of the plant is likely to produce changes in water content of other parts of the plant, and the hydrostatic system of a plant is therefore in a constant state of flux. The hydrostatic system is discussed in detail by MEYER in Volume 3 of this series.

Changes in rates of transpiration and absorption produce diurnal variations in water content of leaves (KRAMER 1937, ACKLEY 1954) and if rapid transpiration results in development of a large water deficit there may be movement of water into the leaves from other organs such as from mature fruits (BARTHOLOMEW 1926), and probably from stems (WILSON, BOGGESS, and KRAMER 1952). On the other hand young growing cotton bolls (ANDERSON and KERR 1943) and growing stem tips (STOCKING 1945, WILSON 1948) seem to be able to compete successfully for water with wilting leaves, at least during short periods of severe moisture stress. Absorption of sufficient water to maintain a more or less turgid condition is essential for growth and for normal occurrence of various processes such as photosynthesis and respiration. Certain plant processes such as stomatal movements, the opening and closing of some flowers, the rolling and unrolling of leaves, pulvinal movements, and other socalled turgor movements, are caused by changes in water content of cells. For a more detailed discussion of changes in water content of plant tissues the reader is referred to other volumes of this series, for example the chapter on Water Content and Water Turnover in Cells in Volume 1, and to the book by CRAFTS, CURRIER, and STOCKING (1949).

When DUTROCHET (1837) proposed the osmotic theory he provided for the first time a workable explanation of water uptake based on physical principles to replace the vague, vitalistic explanations previously proposed. The classical osmotic theory of water movement and water uptake has been accepted generally by botanists for over a century with only minor modifications. According to this theory the cell may be regarded as an osmometer in which the cytoplasm functions as a differentially permeable membrane, separating the solution in the vacuole from the external solution. For many decades water movement in and out of cells was assumed to occur along osmotic gradients from regions of lower to regions of higher osmotic pressure. The only important modification of this simple concept of water movement was the realization during the second and third decades of this century that water movement from cell to cell and in and out of cells occurs along diffusion pressure gradients (see next section for definition) instead of along gradients of osmotic pressure. During recent years questions have begun to arise concerning the adequacy of osmotic movement to explain the uptake of water by cells and it has been suggested that perhaps at least part of the uptake of water by plant cells is brought about by nonosmotic forces which are dependent on energy released by respiration.

Terminology.

The problem of terminology has been treated in detail in the introductory chapter of this volume by BOGEN and is touched on by other contributors. It seems necessary, however, to define some of the terms and concepts used in this chapter. The basic fact in cell water relations is that water tends to move along gradients of decreasing free energy from regions of higher to regions of lower free energy, activity, or diffusion pressure. The free energy or diffusion pressure of water can be reduced by the addition of solutes, the action of surface forces, decrease in temperature, decrease in pressure, or the application of tension, while its free energy or diffusion pressure is increased by increase in temperature or the application of pressure. Although it is difficult to measure the diffusion pressure of water or of a solution it is readily possible to measure the difference in diffusion pressure between pure water and a solution in an osmotic system because the addition of solutes lowers the diffusion pressure by an amount equal to the osmotic pressure of the solution. The difference between the diffusion pressure of pure water and of water in a solution at the same temperature is termed the diffusion pressure deficit (DPD) by MEYER (1945) and MEYER and ANDERSON (1952), and the suction force, suction tension, net osmotic pressure, and other names by other writers. Regardless of the name applied it is a measure of the pressure with which water tends to diffuse into a cell along a free energy gradient.

MEYER (1945) attempted to clarify the terminology relating to osmotic movement of water and his terminology will be used in this discussion, with the minor exception that WP (wall pressure) will be used instead of TP (turgor pressure) in the basic equation.

It seems more logical to use WP than TP in the equation for water uptake because the DPD and the turgor pressure of a cell are so much affected by the properties of the cell wall. For example, in growing tissue water uptake occurs over a long period of time because continued deposition of new wall material permits extension of the walls and prevents development of sufficient wall pressure to eliminate the DPD of the cells. In mature tissue the amount of cell wall extension is quite limited and resistance to expansion soon results in development of wall pressure, with consequent cessation of water intake. The increased uptake of water in the presence of added auxin also can be attributed largely, or perhaps even entirely, to increased extensibility of the cell walls. The wide differences in volume change of different kinds of mature cells over the range from incipient plasmolysis to full turgor also are caused by differences in extensibility of the cell walls. In view of these relationships it seems best to relate the changes in DPD to changes in wall pressure rather than to changes in turgor pressure, even though the two pressures are numerically equal. According to this terminology the relationships among the various osmotic quantities can be represented as follows:

$$\mathrm{DPD} = \mathrm{OP} - \mathrm{WP}\ (\mathrm{WP} = \mathrm{TP})$$

The diffusion pressure deficit or DPD represents the force with which water tends to diffuse into a cell and it varies from a value equal to the osmotic pressure of the cell sap at incipient plasmolysis to zero in a fully turgid cell where the wall pressure has become equal to the osmotic pressure. The wall pressure (WP) results from the expansion of the protoplast against the wall, caused by the entrance of water, and is equal numerically but exerted in the opposite direction to the turgor pressure which is the pressure developed in the protoplast and against the wall during the entrance of water. In an open beaker of solution DPD = OP because no wall pressure exists, but in cells the DPD decreases as the WP increases because application of pressure increases the diffusion pressure of water in the vacuole.

This terminology has been critized by various writers. SPANNER (1952) claims that if active forces are involved in water uptake then the total "water absorbing effort" of the cell cannot be adequately described in terms of pressure. He would restrict such terms as "diffusion pressure deficit" or "suction potential" to the passive osmotic component of water uptake and treat the active or nonosmotic component separately. As will be seen later the existence of a nonosmotic component is still uncertain. LEVITT (1951) and WALTER (1952) object to the use of 2 or 3 symbols to represent a single physical quantity and BROYER (1951a) prefers to use "specific free energy" rather than diffusion pressure deficit. These suggestions have their merits, but the terminology of MEYER will be used in this article because of its simplicity and because it is familiar to many readers.

Forces concerned in water uptake.

There have been a number of attempts in recent years to distinguish between osmotic and nonosmotic forces in connection with water uptake by cells. BOGEN and PRELL (1953) proposed that water uptake might occur by one or more of the following ways, 1. osmotic uptake of water, caused by diffusion and depending on the osmotic pressure of the cell sap, 2. metaosmotic uptake, also by diffusion, but dependent on the binding of water by adsorptive forces in the cells, and 3. nonosmotic uptake where water movement is caused by energy released by respiration. BOGEN (1953, 1954) suggests that several mechanisms may be involved in nonosmotic water uptake. Probably no distinction need be made between osmotic and metaosmotic water uptake because in both instances water movement occurs along a diffusion pressure gradient which is produced by purely physical forces. BROYER (1951a, 1951b) has interesting discussions of the complex of factors involved in the water relations of plant cells, also BOGEN (1954).

It would be highly desirable to distinguish between water uptake caused by the expenditure of metabolic energy and water uptake caused by diffusion along a DPD gradient produced by physical forces, but as SPANNER (1952) states it is sometimes difficult to distinguish between them. Such terms as absorption, intake, and uptake, are general and do not indicate the nature of the forces involved. Osmotic and imbibitional forces are obviously of a physical nature, but electro-osmotic forces are more difficult to classify because electrical potentials originate as the result of both physical and metabolic processes. Even the participation of enzymes in transport across membranes would not prove the occurrence of active uptake according to ROSENBERG and WILBRANDT (1952). Nonosmotic movement or nonosmotic uptake has been used to designate water uptake against a diffusion pressure gradient, caused by the expenditure of metabolically released energy. Nonosmotic uptake of water would be analogous in most respects to the accumulation of solutes by cells.

The osmotic uptake of water.

In mature tissue it seems probable that osmotic forces are responsible for most or perhaps all of the uptake of water. Imbibitional forces will also be treated in this section because they usually tend to come into equilibrium with osmotic forces. Osmotic movement of water occurs because the presence of solutes in water reduces the concentration of water molecules, decreasing the activity or free energy per unit of volume. The attractive forces of solute molecules and ions for water molecules also decreases the free energy of water. This causes a decrease in specific free energy or diffusion pressure and increase in the DPD of the water in a solution as compared with pure water. Water movement along osmotic gradients causes the development of hydrostatic pressure (called turgor pressure in cells) if a solution is separated from pure water or from a solution of different concentration by a membrane permeable to water, but not to the solute. In plant cells the cytoplasm acts as the differentially permeable membrane and the solution is the cell sap in the vacuoles.

Imbibitional forces are responsible for a limited amount of water uptake, especially in germinating seeds and in growing regions where the vacuoles are small and constitute a small percentage of the total cell volume. Imbibitional uptake of water also occurs because of movement along a DPD gradient produced by the adsorption of water molecules on surfaces and in microcapillaries where they are held so firmly that their free energy is greatly reduced. Much water is held in cell walls by imbibitional forces, mostly in the tiny spaces between the micelles and fibrils. Most of the water in the larger spaces between the cellulose fibrils is held rather loosely and can move freely, hence cellulose walls are very permeable to water. BUTLER (1953) estimated that 70 to 75% by weight of the water in young wheat roots occurs in the vacuoles, 5% in the cytoplasm, and 20 to 25% in the walls.

The imbibitional forces of the walls and protoplasm and the osmotic forces of the vacuoles tend to come into equilibrium with each other, and collectively determine the ability of a cell to absorb water. In most mature cells the volume of vacuolar sap so greatly exceeds the volume of imbibed water that the osmotic forces predominate over the imbibitional forces and the DPD of a cell is controlled largely by the osmotic pressure of the vacuolar sap. Some exceptions to this occur, however. Kerr and Anderson (1944) found in growing cotton seeds *(Gossypium hirsutum)* from 24 to 36 days old that the DPD exceeded the osmotic pressure of the expressed cell sap, although in younger seeds the reverse had occurred. Their explanation was that as the seeds mature there is a large increase in amount of meristematic tissue in the embryos consisting of small cells with small vacuoles, and an accompanying decrease in endosperm which consists of large cells with very large vacuoles. Water absorption in the older seeds is therefore largely controlled by imbibitional forces and the osmotic pressure of the expressed sap does not indicate the potential DPD of the mature seeds. In these experiments the vacuolar sap probably was diluted by imbibed water released from the hydrophilic colloids when the tissue was frozen prior to expressing the sap, further reducing its apparent osmotic pressure. Stocking (1945) observed a similar situation in the tip leaves of squash *(Cucurbita pepo)* which remained turgid and enlarged after the older, lower leaves had wilted, although they possessed no higher osmotic pressure than the older leaves. Probably the high imbibitional forces of the young cells of these leaves enabled them to develop higher DPDs than the older leaves where the DPD was controlled by the osmotic pressure of the vacuolar sap.

In plant cells the diffusion pressure deficit of the cell sap is greatly modified by the wall pressure, and to a lesser extent by the pressure of surrounding cells and the existence of tension in the hydrostatic system. The relationships among the various osmotic quantities of a cell in the range from incipient plasmolysis to full turgidity can be shown by the following equation, the symbols of which were defined earlier.

$$\text{Cell DPD} = \text{OP} - \text{WP} \quad (\text{WP} = \text{TP})$$

The surrounding tissues often exert pressure on cells in addition to their own wall pressure, causing a corresponding increase in the diffusion pressure of the water in the cell. This is usually included with the wall pressure, but can be shown separately if desired, as follows:

$$\text{Cell DPD} = \text{OP} - \text{WP} - \text{Pressure of surrounding cells}$$

In rapidly transpiring plants the pressure on the water in the hydrostatic system of the plant often falls below zero and becomes a negative pressure or tension. If the tension becomes greater than the osmotic pressure of the cells they too are subjected to tension. Just as imposition of a positive pressure decreases the DPD of a solution so imposition of a negative pressure increases the DPD. If water diffuses out of a cell until the protoplast and wall are pulled inward the resistance of the wall causes the water to be subjected to tension and the turgor pressure and wall pressure become negative values. This can be shown as follows.

$$\text{Cell DPD} = \text{OP} - (-\text{WP})$$

The essential fact in this modernized form of the osmotic theory is that water movement occurs along gradients of diffusion pressure deficit (suction force, suction tension of some authors), rather than along gradients of osmotic pressure as was once supposed. This has made it possible to understand certain processes previously difficult to understand. For example, it formerly was difficult to understand how water moves from the very dilute soil solution across the root parenchyma with an osmotic pressure of several atmospheres into the xylem with an osmotic pressure of less than 2 atmospheres. Following the work of URSPRUNG and BLUM (1916a), THODAY (1918), HÖFLER (1920), and others it became clear that water might move from cells or regions of high osmotic pressure to cells of low osmotic pressure if the cell or region of lower osmotic pressure had the higher diffusion pressure deficit (DPD).

Perhaps the application of the diffusion pressure theory to uptake of water can be made clearer by an illustration. Suppose that a cell at incipient plasmolysis is placed in pure water. In such a cell the DPD is equal to the OP and the WP is zero, but as water diffuses in and the volume of the vacuole increases hydrostatic pressure or turgor pressure develops because the wall resists expansion. The development of turgor pressure in the protoplast is accompanied by the development of an equal, but inwardly directed wall pressure. Imposition of this wall pressure on the protoplast tends to increase the diffusion pressure and decrease the DPD of the water in the vacuole until finally no more water diffuses into the cell because there is no longer a diffusion pressure gradient from the outside to the inside of the cell. If any appreciable increase in volume of the cell occurs during this process there will also be a small decrease in OP of the cell contents because of dilution. The changes in the various osmotic quantities can be shown as follows.

	Cell at incipient plasmolysis atmos.	Cell partly turgid atmos.	Cell fully turgid atmos.
OP	10	9.5	9.0
WP	0	5.0	9.0
DPD	10	4.5	0.0

In general it may be concluded that any cell or tissue in which a DPD exists can absorb water from any other cell or tissue possessing a lower DPD regardless of their relative osmotic pressures. As an illustration suppose that cell A with an OP of 10 atmos. is placed in a solution with an OP of 2 atmos. At equilibrium with this solution it will have a DPD of 2 atmos. because the DPD of a cell tends to come into equilibrium with the DPD of its environment. If cell B with an OP of only 5 atmos. is placed in a solution with an OP of 4 atmos. it will possess a DPD of 4 atmos. at equilibrium. If the two cells are then brought into contact with each other water will move from A to B because cell B has the higher DPD although it has the lower OP. This explains how a transpiring leaf in which a considerable DPD has developed can absorb water from a turgid fruit of low DPD, but of higher OP. If transpiration produces a DPD of only 1 or 2 atmos. in the xylem of a turgid root, water will move from soil to xylem in spite of the fact that the OP of the intervening cortical cells may be 5 or 6 atmos.

It should now be clear that although the OP of a cell or tissue varies somewhat with changes in water content and amount and kind of solutes, the DPD which is the real measure of its ability to absorb water, varies over a much wider range, from zero in a fully turgid cell to a value equal to the osmotic pressure of the cell sap at incipient plasmolysis. The DPD may even exceed the osmotic pressure of the cell sap if the water in the cell is subjected to tension. As mentioned earlier, in certain instances where the quantity of imbibed water is very large in comparison to the amount of vacuolar sap the DPD may exceed the OP because imbibitional forces exceed osmotic forces (KERR and ANDERSON 1944, STOCKING 1954).

Numerous studies of the osmotic pressure and some studies of the diffusion pressure deficit of various plant tissues have been made. The difficulties inherent in measuring the DPD of plant tissue have limited the number of measurements and the accuracy of some of these is questionable. Limitations of space permit presentation of only enough data to illustrate the range of variation in these values which is ordinarily encountered. Readers are referred to CRAFTS, CURRIER, and STOCKING (1949) for a review of the extensive literature dealing with osmotic pressures, diffusion pressure deficits, and methods of measuring them. Information on these topics also will be found in other chapters of this volume. Some data also are presented in Volume 1 of this Handbook and the reader is referred to extensive collections of data on osmotic pressures by HARRIS (1934) and WALTER (1951).

Among the studies which have been made of the effect of environmental factors on the DPD of plant tissue are those of URSPRUNG and BLUM (1916b), MOLZ (1926), LI (1929), HERRICK (1933), CHU (1936), and STOCKING (1945). LI (1929) attempted to analyze the factors responsible for diurnal variation in the DPD of leaves of *Syringa oblata*. He found the lowest values early in the morning and the highest values at 12 to 2 p.m., the range being from 10.8 to 16.3 atmos., when the rate of transpiration was moderate and 11.0 to 19.2 atmos. when evaporation was high. There was a close correlation between evaporation and DPD, good correlation with relative humidity and light, but a low correlation with temperature. HERRICK (1933) measured the OP of expressed leaf sap cryoscopically and the DPD of strips of leaves of *Ambrosia trifida* subjected to severe moisture stress. He found the OP of the expressed sap to vary from 10.1 to 17.2 atmos., while the DPD varied from 6.5 to 17.5 atmos. STOCKING (1945) found the DPD of petioles of squash leaves *(Cucurbita pepo)* to vary from 0.8 to 11.0 atmos., depending principally on the rate of transpiration and water supply. The highest DPDs were found in wilted leaves and the lowest ones at night.

Development of high DPDs is accompanied by development of tensions in the hydrostatic system of the plant. STOCKING calculated that the xylem sap of the squash plants which he studied was subjected to tensions of up to 9.2 atmos. Much higher tensions have been reported to occur in plants. STOCKER (1928) reported osmotic pressures and DPDs of 35 to 48 atmos. in the roots of plants growing in the Egyptian desert and during droughts he found DPDs in excess of 40 atmospheres in plants of the Hungarian steppes (STOCKER 1930). ARCICHOVSKIJ and OSSIPOV (1931) measured a DPD of 142.9 atmos. in a desert shrub. GREENIDGE (1954) doubts if tensions in transpiring plants often exceed 30 atmos. because the average tensile strength of distilled water under field conditions probably is only about 30 atmos. Whenever a high DPD is developed in one part of a plant all of the cells of the plant tend to be subjected to the same high DPD because it is transmitted through the water which forms a continuous system throughout the entire plant. When the tension on the water in the xylem and the hydrostatic system in general exceeds the OP of the cell contents the cells are also subjected to tension. CHU (1936) concluded that during times of high transpiration the tension in the leaf parenchyma greatly exceeded the OP in the trees studied by him and therefore was the dominant factor in determining the magnitude of their DPD.

Factors affecting osmotic uptake of water.

BROYER (1951) made a detailed analysis of the forces involved in the movement of water and solutes into and out of cells. He concluded that in living systems the important factors affecting water movement are solute concentration and pressure. Temperature and surface forces do not exert much influence on water movement, and the importance of electrical forces is doubtful. Experience and theory agree in indicating that water movement in and out of cells is controlled primarily by the osmotic pressures on the two sides of the membranes and the pressure exerted on the contents of the cells. The combined effects of these two forces are exerted through their effects on the specific free energy of water, which we have chosen to express as its diffusion pressure deficit or DPD. BOGEN in an earlier chapter of this volume has discussed this concept in more general terms with respect to the exchange of materials.

According to this view osmotic uptake of water should depend almost entirely on the steepness of the DPD gradient from cells to external solution. It has been demonstrated many times that even small increases in the OP of the external solution reduce or prevent the absorption of water. ROSENE (1941) found that water uptake by excised roots of *Allium cepa* was stopped by immersion in a solution having an OP of 1.8 to 3.3 atmos. and HAYWARD and SPURR (1943) reported that water uptake by detached roots of *Zea mays* was greatly reduced in a solution having an OP of 4.8 atmos. The report of HUBER and MERKENSCHLAGER (1951) that seed germination is slowed by a water tension or DPD as low as 0.00525 atm. seems improbable because KAUSCH (1952) found that the differences in germination in the apparatus used by HUBER and MERKENSCHLAGER apparently were caused by differences in water supply rather than by differences in DPD. Furthermore OWEN (1952) found that some wheat seeds germinated at a DPD of over 30 atmos.

Although the absorption of water is affected by the DPD gradient it is not always proportional to it. Change in weight or size of plant tissue plotted over the OP of the solution in which it is immersed is not always a straight line (THIMANN, SLATER and CHRISTIANSEN 1950). Although osmotic effects predominate, ionic effects may be of importance because of modifications in hydration of cell contents, in permeability, and in extensibility of the walls. EATON (1941) found that when root systems of corn and tomato were divided between solutions having OPs of 0.3 and 1.8 atmos. that 1.8 times as much water was absorbed from the more dilute as from the concentrated solution. Growth of root systems left in the more concentrated solution was greatly reduced. This probably was not because of a decreased DPD gradient from solution to plant because when the OP of the environment is increased there usually also is an increase in the OP of the plant sap.

The OP of the sap of plants growing in saline habitats usually is abnormally high and even a small increase in the OP of the substrate usually is accompanied by increase in the OP of the plant sap, especially the roots (see McCOOL and MILLAR 1917, for example). ROBERTS (1916) found that as she increased the concentration of the sucrose solution in which radish roots *(Raphanus sativus)* were immersed the concentration of the cell sap of the root hairs increased so that there was always a difference of 4 to 6 atmos. over a range of external concentrations from 0.02 M to 0.65 M. EATON (1942) found that the average difference between the OP of the expressed sap and that of the external solution in which the plants were grown averaged 11 atmos. for 5 species of plants growing in solutions with OPs of 0.72 and 6.0 atmos. Although both groups of plants maintained the same DPD gradient from nutrient solution to plant the plants in the more concentrated solution made less growth, apparently because they were suffering from a water deficit.

This is perhaps a good place to emphasize that although plants and plant cells are governed by physical laws they do not behave so simply as physical systems. In terms of a physical system water uptake ought to occur just as readily from a solution with an OP of 4 atmos. to a cell with an OP of 14 atmos. as from a solution with an OP of 1 atm. to a cell with an OP of 11 atmos. The effect of the higher external OP on the growth of cells is quite marked, however. In general plant growth is considerably reduced by an OP in excess of 2 atmos. in the soil or nutrient solution (MAGISTAD and REITEMEIER 1943), even though the DPD gradient from soil to plant may be as great in the more concentrated solution as in the dilute solution.

The reduction in growth occurs because even a small external DPD reduces cell enlargement in most tissues. BURSTRÖM (1953a) observed that elongation of wheat roots was decreased by immersion in 0.03 M mannitol which has an OP of slightly over 1 atm. Growth was further decreased in more concentrated solutions and stopped at incipient plasmolysis of the growing tissue. The growth of oat coleoptiles (THIMANN and SCHNEIDER 1938) and of segments of pea stem (THIMANN, SLATER and CHRISTIANSEN 1950) is reduced by low concentrations of mannitol and salts. FREY-WYSSLING (1952) has recently stated that turgor pressure plays no important role in growth. It appears, however, that although turgor is very low in growing cells, plasmolyzed cells do not grow. Although the absorption of water is not the cause of growth, water absorption appears essential to continued growth. Probably this is because as new wall material is deposited the cell wall tends to become separated from the cytoplasm unless there is enough turgor to keep the protoplasts pressed against the actively extending walls.

Active or nonosmotic uptake of water.

By nonosmotic uptake of water is meant the movement of water into cells and tissues against a diffusion pressure or DPD gradient. This is sometimes termed "active" uptake because movement of water against a DPD gradient could only be brought about by the expenditure of metabolic energy, whereas osmotic uptake is a passive physical process requiring no metabolic energy. The term "secretion" is occasionally applied to water movement supposed to occur against a diffusion pressure gradient and it has also been termed "anomalous osmosis" (BRAUNER 1945, for example). Neither of these terms specifies any particular mechanism for the movement of water. Active or nonosmotic uptake of water presumably is analogous to accumulation of salts by cells, although its occurrence is not as generally accepted as is salt accumulation.

It seems that it should be easy to distinguish between osmotic and nonosmotic uptake of water, but this is not the case. Claims for the occurrence of nonosmotic or active uptake of water are based principally on discrepancies between the OP of cell sap and the DPD of the cells and on the effects of auxin and respiration inhibitors on water uptake. There is a very extensive literature on this subject, some of which is discussed in Volume I of the Handbook, in a review by KRAMER and CURRIER (1950), in the book by CRAFTS, CURRIER, and STOCKING (1949) and in reviews by BOGEN (1951, 1953, 1954). Although nonosmotic water uptake probably could at most account for only a very small percentage of the total water uptake of plant tissue, it receives a disproportionately large amount of space in discussions of water relations such as this because of the controversial nature of the evidence for its existence and the varied mechanisms proposed to explain its occurrence. The literature will be discussed under several headings for convenience and clarity.

Discrepancies between cryoscopic and plasmolytic measurements.

If the expression DPD = OP—WP is strictly true then the osmotic pressure of a solution which prevents water uptake at incipient plasmolysis (a measure of DPD) ought to be equal to the OP of the expressed sap. Although this is often found to be approximately true some notable exceptions have been observed. BENNET-CLARK, GREENWOOD, and BARKER (1936) found the cryoscopically determined OP of sap expressed from roots of *Beta vulgaris* and *Brassica napabrassica*, and from petioles of *Begonia rex* was several atmospheres lower than the DPD as determined plasmolytically. This difference was not found in *Rheum* or *Caladium*. They proposed that the conventional formula be rewritten to read DPD = OP—WP + X, X standing for the unknown force responsible for the nonosmotic fraction of water uptake. Similar discrepancies have been reported by other investigators, including BUHMANN (1935), MASON and PHILLIS (1939), ROBERTS and STYLES (1939), BENNET-CLARK and BEXON (1940), LYON (1942), and CURRIER (1944).

These observations have been variously interpreted by the workers who observed them. OPPENHEIMER (1932) attributed many discrepancies to difficulties inherent in the methods used, such as adhesion of protoplasts to walls, inelastic walls, a large volume of protoplasm in the cells, and failure to allow for volume changes. BUHMANN (1935) attributed most of her discrepancies to adhesion of the protoplasm to the wall, causing an overestimate of the OP required to produce incipient plasmolysis, but CURRIER (1944) could detect no adhesion in *Beta vulgaris*. He attributed the discrepancies observed in some samples in his work to contamination of the vacuolar sap by liquid expressed from the cytoplasm.

As mentioned earlier, KERR and ANDERSON (1944) attributed the large excess of DPD over OP in developing cotton seeds to dilution of the vacuolar sap by imbibitionally held water released during expression of the sap from frozen tissue. EATON (1943) attributed many of the discrepancies to penetration of the plasmolyzing solute into the vacuole. It is agreed by most workers that such penetration is common, but it is usually small in amount and could scarcely explain some of the differences observed. LEVITT (1947) believed that most of the discrepancies arise from errors in technique or faulty interpretation of the data. There is no doubt that considerable errors are likely to occur in both of these methods. Sap expressed from the vacuole is contaminated to a varying extent with sap from the cytoplasm and walls, and plasmolytic determinations are often unreliable because of the occurrence of anomalous plasmolysis, vacuolar contraction, and possible variations in permeability of the cell membranes during changes in turgidity and hydration. GASSER (1947) has discussed some of these difficulties.

Because of the difficulties in making osmotic measurements it seems doubtful if the existence of discrepancies of the sort described constitute adequate evidence for the occurrence of nonosmotic water uptake. CURRIER (1944) for example, was unable to relate the discrepancies to age or condition of the tissue, although they would be expected to be greatest in growing tissue.

Plasmometric measurements.

BOGEN (1953) and BOGEN and PRELL (1953) have made the most recent plasmolytic studies. BOGEN and PRELL measured the uptake of water by epidermal cells of *Oenothera franciscana* by plasmometric methods. Their work was based on the assumption that it should be possible, by applying the principles governing osmotic phenomena, to calculate the volume change resulting from a given change in concentration of the plasmolyzing solution. They found that the cells consistently were plasmolyzed to a smaller extent than expected on a theoretical basis. This difference between the expected and the actual change in volume of the cells could be corrected by assuming the existence of "an osmotically inactive volume" in the cells. This might be colloidal material which contains nonosmotically held water, thereby increasing the concentration of the solutes in the vacuole to a value higher than expected. BOGEN (1953) studied the behavior of epidermal cells of *Taraxacum* and *Rhoeo* which were subjected to long continued plasmolysis. In *Rhoeo* the protoplasts continued to expand, and they also expanded in *Taraxacum* if the tissue was kept above 25° C. The enlargement was stopped by NaN_3, and DNP, but not by KCN or indoleacetic acid, and it was increased by glucose-1-phosphate. BOGEN regarded the continued increase in volume observed in his experiments as caused by nonosmotic uptake of water different from that previously reported because it was not affected by indoleacetic acid or cyanide. He thought that it could not be directly related to respiration because it was not immediately reduced by KCN, but it must be related to general cell metabolism because it was reduced by DNP.

Respiration and water uptake.

If the absorption of water occurs against a free energy or DPD gradient, energy to move the water must be supplied by respiration. It has been shown by several investigators that treatments which affect respiration usually also affect water uptake. An adequate supply of oxygen was found essential for maximum water uptake by potato tissue (REINDERS 1938, 1942, BRAUNER, BRAUNER, and HASMAN 1940), oat coleoptiles (KELLY 1947), and root hairs (ROSENE 1950, ROSENE and BARTLETT 1950). HACKETT and THIMANN (1950) found much greater water uptake by discs of potato tissue arranged so they broke the surface and were well aerated than by discs allowed to become completely submerged, and they later reported (HACKETT and THIMANN 1952) that water uptake was inhibited about 50% in 7% oxygen.

Respiration inhibitors have also been observed to reduce or prevent the absorption of water by various kinds of tissue, including tomato roots (VAN OVERBEEK 1942), onion roots (ROSENE 1944), oat coleoptiles (KELLY 1947), and potato tuber (HACKETT and THIMANN 1952). As mentioned previously, BOGEN (1953) observed a type of water uptake in epidermal cells of *Rhoeo* and *Taraxacum* which was inhibited by azide and arsenite, but not by KCN, and which was stimulated by glucose-1-phosphate. LEVITT (1948) reported that KCN did not inhibit auxin-induced water uptake by potato tissue.

There seems to be little doubt that there is some kind of relationship between water uptake and respiration, but the nature of the relationship is more debatable. The general relationship was recognized some years ago when STEWARD, STOUT and PRESTON (1940) wrote that results of studies of salt and water uptake by discs of potato tissue "suggest that aerobic respiration, protein synthesis, water absorption, and salt accumulation are all mutually dependent processes which occur in all cells which are not subject to equilibrium conditions". More recently STEWARD, CAPLIN and MILLER (1952) have again stressed the interrelationship of these processes. To this list of related processes might be added growth in the sense of permanent cell enlargement (THIMANN 1954).

BONNER, BANDURSKI, and MILLERD (1953) proposed that respiration and water intake are linked directly by transfer of energy through adenosinetriphosphate, possibly in a system in which the auxin molecule serves as a carrier. On the other hand HACKETT and THIMANN (1953) concluded that although water uptake is related to the general level of metabolism it cannot be directly geared to respiration because when both processes are inhibited, water uptake is inhibited more than respiration, when both processes are stimulated by naphthalenacetic acid water uptake is stimulated more than respiration, and the two processes can be caused to vary oppositely by DNP and arsenite which increase respiration, but decrease water uptake in the absence of auxin. Later studies indicate that the relation is probably through effects of metabolism on cell wall formation (THIMANN 1954) rather than directly on water uptake.

The fact that respiration and water intake are affected in the same way by certain treatments does not prove that water intake is directly dependent on respiration. Respiration might affect water uptake indirectly by affecting permeability, the synthesis of substances which modify the degree of hydration of the tissues or the extension of the cell wall, or directly by supplying energy for nonosmotic uptake of water. BRAUNER, BRAUNER, and HASMAN ((1940) observed that permeability of potato tissue was decreased under anaerobic conditions to 63% of permeability of tissue under aerobic conditions. BRAUNER and BRAUNER ((1943) also observed that the extensibility of the cell walls of potato tissue is considerably greater in aerobic than in anaerobic conditions. They attributed at least part of the increased water uptake of well aerated tissue to the decreased wall pressure resulting from increased extensibility of the walls. The role of cell wall extensibility will be discussed again in connection with the effects of auxin on water uptake.

The structure of protoplasmic membranes is complex and probably can be maintained only by a constant expenditure of energy. Thus any serious changes in rate of respiration might be expected to cause modifications in permeability of cell membranes. It should be remembered, however, as BOGEN (1951) pointed out, that change in permeability to water does not change the total amount of water absorbed at equilibrium, but merely changes the time required for cells to come to equilibrium. BOGEN (1953, 1954) seems to think that water uptake is related directly to respiration through the synthesis of carrier molecules and streaming of the cytoplasm. He favors the theory of GOLDACRE (1952) who proposed that protein carrier molecules unfold at the outer surface of the cell membranes, exposing side chains on which are adsorbed water molecules and ions. These carrier molecules are then transported across the membrane, perhaps aided by protoplasmic streaming, and contract, shedding whatever materials they are carrying on the other side of the membrane. BOGEN believed that this theory is supported by the observation of ALLEN and PRICE (1950) that KCN inhibited respiration, but not protoplasmic streaming of a slime mold, while DNP inhibited streaming, but increased respiration. BOGEN found that KCN did not inhibit water uptake, but DNP did under the conditions of his experiments.

LUNDEGARDH (1950) related water uptake by roots to respiration through salt uptake. According to his theory, salt uptake is linked to respiration and the accumulation of ions is followed by the osmotic uptake of water. While intake of solutes certainly will be accompanied by intake of water, in some instances uptake of water has been demonstrated where no uptake of salt could have occurred. Metabolic activity might also cause loss of water if the concentration of solutes is decreased as when nitrate is converted into protein or sugar is converted to starch or used in respiration.

LEVITT (1947) calculated the energy required to maintain water uptake against a diffusion gradient and concluded that maintenance of a difference of more than one or two atmos. would require an impossibly large expenditure of energy by the cells. LEVITT's calculations were criticized by BENNET-CLARK (1948) and MYERS (1951) because he used a permeability value which they regarded as too high. SPANNER (1952) claimed that LEVITT used too large a cell area, that his permeability factor was too high, and that he calculated the energy requirement improperly. LEVITT (1953) seems to have answered these objections satisfactorily. SPANNER (1954) has recently presented new calculations supporting the possibility of nonosmotic gradients of some magnitude, but LEVITT (1954) claims that SPANNER used an equation which is not applicable to analyses of the "steady state" which would exist in a cell absorbing water nonosmotically.

LOVTRUP and PIGON (1952) estimated that the ameba studied by them absorbed actively a volume of water equal to 2 to 4% of its total volume per hour. The calculated energy requirement for this active water movement was regarded as negligible compared to the total metabolism of the organism. BOGEN (1954) suggested that if water uptake occurred by means of carrier molecules very little additional energy would be required if these carrier molecules are normal products of cell metabolism.

These contradictory conclusions indicate the unsatisfactory results obtained by applying a rigorous mathematical analysis to a biological process which is not completely understood. Until the process is well enough understood to indicate the proper type of equation to use and the proper values to insert in the equation, nothing but confusion will result from such analyses. Obviously more laboratory research is needed to provide reliable data before mathematical analyses will yield useful explanations.

Auxin and water uptake.

It has been known for several years that an external supply of auxin increases water uptake by certain kinds of plant tissue. Among those who have reported this are REINDERS (1938, 1942), COMMONER and MAZIA (1942), COMMONER, FOGEL, and MULLER (1943), VAN OVERBEEK (1944), LEVITT (1947), HACKETT (1952), HACKETT and THIMANN (1952), and BRAUNER and HASMAN (1952). HACKETT observed that naphthaleneacetic acid causes much greater water uptake than indoleacetic acid in potato tissue, but BONNER, BANDURSKI, and MILLERD (1953) found that with good aeration indoleacetic acid caused a large increase in water uptake by slices of Jerusalem artichoke *(Helianthus tuberosus)*. On the other hand BOGEN (1953) observed what he regarded as nonosmotic water uptake by epidermal cells of *Taraxacum* and *Rhoeo* which was not affected by indoleacetic acid.

TS'O and STEINBAUER (1953) found that indoleacetic acid greatly increases water uptake of pea segments, but maleic hydrazide reduces the auxin effect about 50%. Addition of fumaric acid greatly reduces maleic hydrazide inhibition, perhaps because it bypasses the inhibiting action of maleic hydrazide on respiration. HASMAN (1953) found that di-n-amylacetic acid increases aerobic water uptake by potato tissue, possibly by activating auxin, but 6-aminoundecane inhibits it, apparently by affecting permeability of cell membranes.

Although there is general agreement that auxin causes an increase in the uptake of water there is no agreement on the means by which the increase is produced. Among the reasons which have been proposed are increased uptake of solutes by the cells, increased permeability of cell membranes, increased extensibility of cell walls, and direct stimulation of active uptake of water. Any of these causes might be related to the increase in respiration which usually occurs when auxin is supplied. Each of these possibilities will be discussed separately.

Increased uptake of solutes. REINDERS (1942) thought the increased uptake of water occurred because starch was hydrolyzed to sugar, increasing the concentration of the cell sap. COMMONER and MAZIA (1942) and COMMONER, FOGEL, and MULLER (1943) thought that added auxin caused increased uptake of salt which was accompanied by increased uptake of water. This explanation seems to be incorrect because VAN OVERBEEK (1944), LEVITT (1948), HACKETT (1952), and BRAUNER and HASMAN (1952) all found that there is a decrease in the OP of the cell sap at the same time that there is an increase in water content of the tissue. HANSON and BONNER (1954) concluded that auxin (2,4-D) has no direct effects on uptake of salts. They state that salt uptake is increased by increased hydration and increased growth, but decreased by competition for energy with other processes associated with growth.

Permeability changes. Auxin seems to increase the permeability of cells to water. BRAUNER and HASMAN (1949) found that indoleacetic acid increased the permeability of potato tissue to water within 6 hours, and VON GUTTENBERG and his colleagues (1951, 1952) found that permeability was almost doubled in several kinds of tissue by 10^{-7} g. per liter of indoleacetic acid. POHL (1953) reported that auxin increases the permeability and the DPD of oat coleoptiles. During the first 4 to 7 hrs. elongation of oat coleoptiles is proportional to the DPD gradient from external solution to coleoptiles because the only effect of auxin is on permeability, but after that time the proportionality disappears because auxin causes an increase in their DPD. Later POHL (1954) reported that the effect of auxin on permeability decreases from tip to base of oat coleoptiles, but there is no definite longitudinal gradient with respect to the effect of auxin on DPD. POHL suggested that the WENT Avena test measures effects of auxin on permeability and the cylinder and pea tests the effects of auxin on DPD. POHL apparently ascribes the increased DPD in the presence of auxin to increased nonosmotic uptake of water, but it might be caused by increased extensibility of the cell wall.

Extensibility of cell walls. If the cell wall were to be so modified that it would stretch more easily, increase in volume caused by intake of water might occur without a proportional increase in wall pressure. So long as the wall continues to stretch plastically water intake can continue because the DPD of the cell is not reduced to zero. LEVITT (1948) claimed that cells of potato tissue stretch well beyond their elastic limits in water, and attributed increased water uptake in the presence of auxin to increased plasticity of the cell walls, a view maintained in his later papers (LEVITT 1953). BRAUNER and BRAUNER (1943) found that indoleacetic acid increased plasticity of cell walls and BRAUNER and HASMAN (1949) concluded that increased water uptake in the presence of auxin was caused during the first day by increased permeability, but after 24 to 48 hours it resulted chiefly from increased extensibility of the cell walls. In a later paper BRAUNER and HASMAN (1952) decided that decreased wall pressure, resulting from increased extensibility of the walls, is the principal cause of auxin-induced water uptake in potato tissue. They also observed increased permeability and a considerable increase in DPD with 10 mg. per liter of indoleacetic acid. These effects were not obtained under anaerobic conditions. The increase in DPD was attributed to increased extensibility of the cell walls because there was no increase in OP and no evidence of nonosmotic uptake of water. THIMANN (1954) has now decided that auxin-induced uptake of water probably occurs because of increased extension of cell walls.

According to current views of cell wall structure it seems doubtful if auxin would have much effect on extensibility of mature cell walls because they are composed of firmly interwoven cellulose fibrils (FREY-WYSSLING 1952), but it might have considerable effect on the extensibility of walls of cells which are still growing or which are capable of resuming growth, as is the case with potato tuber. It should be noted that BRAUNER and HASMAN ((1949) thought auxin did not greatly affect wall extensibility until after 24 hours, and HACKETT and THIMANN (1952, 1953) did not observe much effect of auxin on water uptake by potato tissue during the first 48 hours. The potato tissue was not growing at the beginning of the experiments, but it certainly began to grow during the experiments. It is not certain whether growth occurred in the artichoke tissue used by BONNER, BANDURSKI, and MILLERD ((1953), but it may have done so. Oat coleoptiles show immediate reaction to auxin because they are actively growing.

Stimulation of nonosmotic water uptake. The evidence for direct "secretion" of water into the vacuole is exceedingly difficult to evaluate. THIMANN (1951) proposed that auxin causes "the direct metabolic pumping of water" into the vacuoles of cells, but HACKETT and THIMANN (1953) later concluded that although there is some relation between respiration and water uptake the two processes are not directly linked. THIMANN (1954) has now abandoned the idea that auxin directly causes nonosmotic uptake of water and suggests that the increased uptake of water in the presence of auxin is caused by increased extensibility of cell walls resulting from increased synthesis and deposition of cell wall material.

In contrast, BONNER, BANDURSKI, and MILLERD (1953) proposed a direct linkage between respiration and water uptake in which auxin might even function as a carrier molecule. They concluded that auxin increased respiration of their artichoke tissue only if water was being absorbed, suggesting the existence of a "water respiration" analogous to the "salt respiration" mentioned in connection with salt accumulation. BURSTRÖM (1953) repeated some of their experiments and concluded that they have not demonstrated the operation of a nonosmotic force in water uptake. He attributed part of the supposed nonosmotic uptake of water to unavoidable errors in measuring the osmotic pressure of the artichoke tissue at incipient plasmolysis and part to the absorption of mannitol. BURSTRÖM claimed that all of the water uptake by artichoke tissue, both with and without auxin, can be accounted for by osmosis and a nonosmotic force need not be postulated to explain the behavior of the tissue. BONNER and his colleagues also have decided recently that auxin-induced water uptake is related to increased extensibility of the cell walls and is not a nonosmotic process.

Conclusions. Although a relationship between auxin, respiration, and water uptake has been demonstrated, the occurrence of auxin-induced nonosmotic uptake of water has not been demonstrated. The increased uptake of water observed in the presence of auxin is more probably related to the increased extension of cell walls which occurs in the presence of added auxin. Apparently auxin stimulates metabolic processes which result in deposition of new cell wall material and during this process wall pressure is greatly reduced, permitting the continued absorption of water by osmosis. Thus in a certain sense absorption of water is the result of growth, but it is also essential to continued growth because unless the protoplasts are kept expanded against the walls, cell wall growth will cease (BURSTRÖM 1953, THIMANN 1954). According to this scheme water uptake occurs along a DPD gradient which is maintained as a result of growth, but the writer would not regard it as nonosmotic water uptake because it does not involve movement against a free energy gradient.

It seems appropriate to mention the comment of AUDUS (1949) that auxin probably acts through some master system rather than individually on all of the various processes in which it appears to be involved. In accord with this view, it seems more probable that auxin acts directly on a single key metabolic process rather than individually on permeability, wall extensibility, respiration, and other specific processes.

Electroosmosis.

It has been proposed by various investigators that intake of water may occur because of the difference in electrical potential which often occurs across cell membranes. It is well known that water will move across membranes from the positive to the negative pole. The mechanism has been discussed in detail by CRAFTS, CURRIER, and STOCKING (1949) and by SPANNER (1952). It is uncertain whether electroosmotic movement of water should be classified as active or passive. Even if the movement of ions responsible for producing a difference in potential is dependent on metabolic energy the movement of water which results might be regarded as a physical process.

Among the investigators who have proposed electroosmotic theories of water uptake are BENNET-CLARK and BEXON (1946), BRAUNER and HASMAN (1946, 1947), STUDENER (1948), ARENS (1949), and BAUER (1952). Several of these investigators proposed that there is an electroosmotic component in water uptake because they observed that a higher osmotic pressure is required with sugar than with salt solution to prevent uptake of water by plant tissue. BRAUNER and HASMAN (1947) for example, found the apparent DPD of beet tissue was about 1.15 atmos. higher in sucrose than in $CaCl_2$ solution. This difference was ascribed to elimination of the electroosmotic component of water uptake by addition of an electrolyte, but it seems to the writer that it might more reasonably be ascribed to the effects of ions on hydration and permeability of the protoplasm. Although in one paper BRAUNER and HASMAN (1946) attributed about 10% of the total water uptake by potato tissue to electroosmosis they have disregarded it in their most recent papers.

LUNDEGARDH (1944) observed a potential difference of about 100 millivolts across the cortex of wheat roots, but he did not believe that it has much effect on water movement. BLINKS (1940) measured potential differences of 70 to 80 millivolts between the exterior and the interior of cells of *Halicystis*, the outer surface being positive to the interior. The potential was evidently related to metabolic activity as it was modified by oxygen supply, temperature, respiration inhibitors, and salt concentration. BLINKS and AIRTH (1951) found that applied potential differences of up to 1500 millivolts produced no appreciable water movement into *Nitella* cells.

In conclusion, it appears that although a limited amount of electroosmotic movement of water may occur it is a negligible factor in the uptake of water by plant cells and tissues.

Discussion.

There naturally is a tendency to attribute any discrepancies in cell water relations which cannot be explained readily on a physical basis to nonosmotic uptake of water. The idea of active or nonosmotic uptake of water is a very attractive one. It might almost be described as "a la mode", because the active accumulation of solutes has been demonstrated and it seems reasonable to suppose that active accumulation of water also occurs. The writer does not deny the possible occurrence of nonosmotic uptake of water. It seems quite possible that some active or nonosmotic uptake of water does occur, but it is doubtful if its occurrence has yet been demonstrated.

It is exceedingly difficult to prove the occurrence of nonosmotic water uptake because in mature tissue it would constitute a very small percentage of the total volume, and in growing tissue water uptake is so intimately associated with growth that it is affected by factors which modify cell metabolism and growth. For example it has been repeatedly claimed that auxin-induced water uptake is an example of nonosmotic water uptake, but it now appears probable that auxin affects cell wall growth and this in turn results in increased water uptake. The reduction in water intake caused by oxygen deficiency and the action of various respiration inhibitors may also be the indirect result of their effects on metabolic processes rather than any direct effect on active water uptake. The difficulties inherent in plasmolytic or plasmometric measurements may also result in considerable discrepancies quite unrelated to the occurrence of nonosmotic water uptake.

It is evident that large deficiencies exist in our knowledge of cell water relations. For example we have no satisfactory mechanism to explain nonosmotic uptake of water. The idea of "carrier molecules" is popular at present (GOLDACRE 1952, ROSENBERG and WILBRANDT 1952), but their role in water transport is still hypothetical. It would be expected that some useful information on mechanisms for active water transport might be obtained from the animal physiologists, but unfortunately they seem to have no better theories than the plant physiologists. According to recent reviews by KEITH (1953), ROBINSON and MCCANCE (1952) and ROBINSON (1953) there is increasing doubt that water movement in animals can be explained by osmosis. ROBINSON (1953) suggested that kidney cells might secrete water on the basal margin because of higher metabolic activity in that region, but he did not attempt to explain how higher metabolic activity caused water movement. OSTERHOUT (1947) suggested that if a plant cell maintains a higher concentration of solutes at one end than at the other by expenditure of metabolic energy, it might absorb water at that end and expel it at the other end. USSING (1953) seems to think that some of the supposed instances of nonosmotic water movement in animals result from failure to distinguish between filtration or osmotic permeability and diffusion permeability. He cites data indicating that permeability of membranes to movement along an osmotic gradient is much greater than permeability as measured by diffusion of heavy water (D_2O) through the membrane. Failure to realize this, he thinks, lead to assumptions that water movement was occurring by active transport when it actually may have been moving by osmosis.

Among the factors given little attention up to the present time is the effect of plasmodesmata in the cell walls. If they are as abundant as claimed by some investigators (MEEUSE 1941) they form numerous connections between cells and require modification of the classical picture of a cell as an osmotic unit. LAMBERTZ (1954) recently reported that plasmodesmata even occur in the outer walls of epidermal cells and suggested that their presence might affect the intake and loss of materials.

It is obvious that much more research is needed and that discovery of new facts will require modification of some of our present views on cell water relations. It is necessary, therefore, that plant physiologists maintain an open mind which is receptive to new ideas while at the same time they maintain a reasonable degree of scepticism to prevent too ready acceptance of attractive but unproven ideas.

Summary.

The uptake of water by plant cells occurs mostly by osmosis along diffusion pressure gradients, but it is claimed that nonosmotic water uptake also occurs against diffusion pressure gradients. Such active uptake of water would be analogous to accumulation of salts by plant cells. The discussion may be simplified by differentiating between water uptake by mature nongrowing tissue and tissue which is either growing such as oat coleoptiles and stem tips or like potato is capable of resuming growth, especially when supplied with auxin.

In mature, nongrowing tissue rapid, reversible changes in water content of considerable magnitude occur. Changes of 10% or more may occur daily in leaves and other parenchymatous tissue under field conditions or within less than an hour in the laboratory. These changes are caused largely or entirely by osmotic movement of water along diffusion pressure gradients and the amount and direction of movement are determined largely by the osmotic pressure of the cell sap and the amount of pressure exerted on it by the wall. The water relations of such cells can be expressed by the equation DPD = OP—WP. In this equation DPD (an abbreviation for diffusion pressure deficit) is a measure of the force with which the cell can absorb water expressed as the difference between the diffusion pressure of pure water and that of the water in the vacuole. The DPD of the sap in the vacuole is equal to the OP (osmotic pressure) of the cell sap at incipent plasmolysis because the wall pressure (WP) equals zero. As water enters WP increases and DPD decreases until when WP = OP, DPD falls to zero and water uptake ceases. In some instances the DPD of cells seems to exceed their OP and this difference is claimed by some investigators to be caused by a nonosmotic component in water uptake. It seems probable that many of the supposed instances of nonosmotic uptake of water by plant cells result from unavoidable errors in measurement of osmotic quantities of the tissue studied.

In growing tissue the uptake of water dealt with experimentally usually is slow, long continued, and largely irreversible. In such tissue there is a close relationship between respiration, water uptake, and cell enlargement, and all of these processes are stimulated by added auxin. It has been claimed, therefore, that addition of auxin causes nonosmotic uptake of water to occur in such tissue, but it now appears more probable that added auxin increases metabolic processes concerned with cell wall growth. Increased cell wall extension reduces wall pressure, causing the entrance of more water along a DPD gradient and this permits continued cell enlargement or growth to occur.

Although it is possible that limited nonosmotic uptake of water occurs, it certainly is not a quantitatively important factor in cell water relations. It is not even certain whether or not its occurrence has been demonstrated. The mechanisms proposed to explain nonosmotic uptake of water are not very satisfactory.

It is obvious that more research is needed on cell water relations to evaluate the importance of non-osmotic uptake of water and to explore further the nature of the mechanism or mechanisms which might bring it about. It is possible that the results of such research might require considerable revision of the views set forth in this paper.

Literature.

ACKLEY, W. B.: Seasonal and diurnal changes in the water contents and water deficits of BARTLETT pear leaves. Plant Physiol. 29, 445–448 (1954). — ALLEN, P. J., and W. H. PRICE: The relation between respiration and protoplasmic flow in the slime mold, Physarum polycephalum. Amer. J. Bot. 37, 393–402 (1950). — ANDERSON, D. B., and T. KERR: A note on the growth behavior of cotton bolls. Plant Physiol. 18, 261–269 (1943). — ARCICHOVSKIJ, V., and A. OSSIPOV: Die Saugkraft der baumartigen Pflanzen der zentralasiatischen Wüsten nebst Transpirationsmessungen am Saxaul (Arthrophytum haloxylon LITW.). Planta (Berl.) 14, 552–565 (1931).

BARTHOLOMEW, E. T.: Internal decline of lemons. III. Water deficit in lemon fruits caused by excessive leaf evaporation. Amer. J. Bot. 13, 102–117 (1926). — BAUER, L.: Über den Wasserhaushalt der Submersen. I. Zur Frage der Saugkräfte der Submersen. Protoplasma 41, 178–188 (1952). — BENNET-CLARK, T. A.: Non-osmotic water movement in plant cells. Discussion of the Faraday Soc. 3, 134–139 (1948). — BENNET-CLARK, T. A., and D. BEXON: Water relations of plant cells. III. The respiration of plasmolyzed tissues. New Phytologist 42, 65–92 (1943). — BENNET-CLARK, T. A., A. D. GREENWOOD and J. W. BARKER: Water relations and osmotic pressures of plant cells. New Phytologist 35, 277–291 (1936). — BLINKS, L. R.: The relations of bioelectric phenomena to ionic permeability and to metabolism in large plant cells. Cold Spring Harbor Symp. Quant. Biol. 8, 204–215 (1940). — BLINKS, L. R., and R. L. AIRTH: The rôle of electroosmosis in living cells. Science (Lancaster, Pa.) 113, 474–475 (1951). — BOGEN, H. J.: Zellphysiologie und Protoplasmatik. Fortschr. Bot. 13, 192–226 (1951). — Zellphysiologie und Protoplasmatik. Fortschr. Bot. 14, 256–288 (1953). — Beiträge zur Physiologie der nichtosmotischen Wasseraufnahme. Planta (Berl.) 42, 140–155 (1953). — Zellphysiologie und Protoplasmatik. Fortschr. Bot. 15, 212–258 (1954). — BOGEN, H. J., and H. PRELL: Messung nichtosmotischer Wasseraufnahme an plasmolysierten Protoplasten. Planta (Berl.) 41, 459–479 (1953). — BONNER, J., R. S. BANDURSKI and A. MILLERD: Linkage of respiration to auxin-induced water uptake. Physiol. Plantarum (Copenh.) 6, 511–522 (1953). — BRAUNER, L.: Experiments on anomalous osmosis. Rev. Fac. Sci. Univ. Istanbul, Sér. B 10, 1–59 (1945). — BRAUNER, L., and M. BRAUNER: The relations between water-intake and oxybiosis in living plant-tissues. II. The tensility of the cell wall. Rev. Fac. Sci. Univ. Istanbul, Sér. B 8, 30–75 (1943). — BRAUNER, L., M. BRAUNER and M. HASMAN: The relation between water-intake and oxybiosis in living plant-tissues. Rev. Fac. Sci. Univ. Istanbul 5, 266–309 (1940). — BRAUNER, L., and M. HASMAN: Über den Mechanismus der Heteroauxinwirkung auf die Wasseraufnahme von pflanzlichem Speichergewebe. Bull. Fac. Méd. Istanbul 12, 57–71 (1949). — Weitere Untersuchungen über den Wirkungsmechanismus des Heteroauxins bei der Wasseraufnahme von Pflanzenparenchymen. Protoplasma 41, 302–326 (1952). — BROYER, T. C.: Further theoretical considerations of modes of expression and factors possibly concerned in the movement of materials through a two-phased system. Plant Physiol. 26, 655–676 (1951a). — Experiments on imbibition and other factors concerned in the water relations of plant tissues. Amer. J. Bot. 38, 485–495 (1951b). — BUHMANN, A.: Kritische Untersuchungen über vergleichende plasmolytische und kryoskopische Bestimmung des osmotischen Wertes bei Pflanzen. Protoplasma 23, 579–612 (1935). — BURSTRÖM, H.: Studies on growth and metabolism of roots. IX. Cell elongation and water absorption. Physiol. Plantarum (Copenh.) 6, 262–276 (1953a). — Growth and water absorption of Helianthus tuber tissue. Physiol. Plantarum (Copenh.) 6, 685–691 (1953b). — BUTLER, G. W.: Ion uptake by young wheat plants. I. Time course of the absorption of potassium and chloride ions. Physiol. Plantarum (Copenh.) 6, 594–616 (1953).

CHU, C. R.: Der Einfluß des Wassergehaltes der Blätter der Waldbäume auf ihre Lebensfähigkeit, ihre Saugkräfte und ihren Turgor. Flora (Jena) 130, 384–437 (1936). — COMMONER, B., and D. MAZIA: The mechanism of auxin action. Plant Physiol. 17, 682–685 (1942). — COMMONER, B., S. FOGEL and W. H. MULLER: The mechanism of auxin action. The effect of auxin on water absorption by potato tissue. Amer. J. Bot. 30, 23–28 (1943). — CRAFTS, A. S., H. B. CURRIER and C. R. STOCKING: Water in the physiology of plants. Waltham, Mass.: Chronica Botanica 1949. — CURRIER, H. B.: Water relations of root cells of Beta vulgaris. Amer. J. Bot. 31, 378–387 (1944).

DUTROCHET, H. J.: Memoires pour servir a l'histoire anatomique et physiologique des vegetaux et des animaux. Paris: J. B. Baillière 1837.

EATON, F. M.: Water uptake and root growth as influenced by inequalities in the concentration of the substrate. Plant Physiol. 16, 545–564 (1941). — Toxicity and accumulation of chloride and sulfate salts in plants. J. Agricult. Res. 64, 357–399 (1942). — The osmotic and vitalistic interpretations of exudation. Amer. J. Bot. 30, 663–674 (1943).

FREY-WYSSLING, A.: Growth of plant cell walls. Symposium Soc. Exp. Biol. 6, 320–328 (1952).

GOLDACRE, R. J.: The folding and unfolding of protein molecules as a basis of osmotic work. Internat. Rev. Cytology **1**, 135–164 (1952). — GREENIDGE, K. N. H.: Studies in the physiology of forest trees. I. Physical factors affecting the movement of water. Amer. J. Bot. **41**, 807–811 (1954). — GUTTENBERG, H. v., and A. BEYTHIEN: Über den Einfluß von Wuchsstoffen auf die Wasserpermeabilität des Protoplasmas. Planta (Berl.) **40**, 36–69 (1951). — GUTTENBERG, H. v., and G. MEINL: Über den Einfluß von Wirkstoffen auf die Wasserpermeabilität des Protoplasmas. II. Über den Einfluß des p_H-Wertes und der Temperatur auf die durch Heteroauxin bedingten Veränderungen der Wasserpermeabilität. Planta (Berl.) **40**, 431–442 (1952).

HACKETT, D. P.: The osmotic change during auxin-induced water uptake by potato tissue. Plant Physiol. **27**, 279–284 (1952). — HACKETT, D. P., and K. V. THIMANN: The action of inhibitors on water uptake by potato tissue. Plant Physiol. **25**, 648–652 (1950). — The nature of the auxin-induced water uptake by potato tissue. Amer. J. Bot. **39**, 553–560 (1952). — The nature of the auxin-induced water uptake by potato tissue. II. The relation between respiration and water absorption. Amer. J. Bot. **40**, 183–188 (1953). — HANSON, J. B., and J. BONNER: The relationship between salt and water uptake in Jerusalem artichoke tuber tissue. Amer. J. Bot. **41**, 702–710 (1954). — HARRIS, J. A.: The physico-chemical properties of plant saps in relation to phytogeography. Minneapolis: Univ. Minnesota Press 1934. — HASMAN, M.: Investigations of the water exchange of potato tissue under the effect of 6-aminoundecane and di-n-amylacetic acid. Physiol. Plantarum (Copenh.) **6**, 187–198 (1953). — HAYWARD, H. E., and W. B. SPURR: Effects of osmotic concentration of substrate on the entry of water into corn roots. Bot. Gaz. **105**, 152–164 (1943). — HERRICK, E. H.: Seasonal and diurnal variations in the osmotic values, and suction tension values in the aerial portions of Ambrosia trifida. Amer. J. Bot. **20**, 18–34 (1933). — HÖFLER, K.: Ein Schema für die osmotische Leistung der Pflanzenzelle. Ber. dtsch. bot. Ges. **38**, 288–298 (1920). — HUBER, B., and G. MERKENSCHLAGER: Über die Wirkung kleinster Saugkräfte auf die Samenkeimung. Planta (Berl.) **40**, 112–120 (1951).

KAUSCH, W.: Physiologische Wirkung kleinster Saugkräfte. Planta (Berl.) **41**, 59–63 (1952). — KEITH, N. M.: Water metabolism. Ann. Rev. Physiol. **15**, 63–84 (1953). — KELLY, S.: The relationship between respiration and water uptake in the oat coleoptile. Amer. J. Bot. **34**, 521–526 (1947). — KERR, T., and D. B. ANDERSON: Osmotic quantities in growing cotton bolls. Plant Physiol. **19**, 338–349 (1944). — KRAMER, P. J.: The relation between rate of transpiration and rate of absorption of water in plants. Amer. J. Bot. **24**, 10–15 (1937). — Water relations of plant cells and tissues. Ann. Rev. Plant Physiol. **5**, 253–272 (1955). — KRAMER, P. J., and H. B. CURRIER: Water relations of plant cells and tissues. Ann. Rev. Plant Physiol. **1**, 265–284 (1950).

LAMBERTZ, P.: Untersuchungen über das Vorkommen von Plasmodesmen in den Epidermisaußenwänden. Planta (Berl.) **44**, 147–190 (1954). — LEVITT, J.: The thermodynamics of active (non-osmotic) water absorption. Plant Physiol. **22**, 514–525 (1947). — The role of active water absorption in auxin-induced water uptake by aerated potato discs. Plant Physiol. **23**, 505–515 (1948). — Toward a clearer concept of osmotic quantities in plant cells. Science (Lancaster, Pa.) **113**, 228–231 (1951). — Further remarks on the thermodynamics of active (non-osmotic) water absorption. Physiol. Plantarum (Copenh.) **6**, 240–252 (1953). — Steady state versus equilibrium thermodynamics in the concept of "active" water absorption. Physiol. Plantarum (Copenh.) **7**, 592–594 (1954). — LI, T. T.: Effect of climatic factors on suction force. Quart. Rev. Biol. **4**, 401–414 (1929). — LOVTRUP, S., and A. PIGON: Diffusion and active transport of water in the ameba Chaos chaos. C. r. Trav. Labor. Carlsberg, Ser. Chim. **28**, 1–36 (1951). — LUNDEGARDH, H.: Bleeding and sap movement. Ark. Bot., Ser. A **31** (2), 1–56 (1944). — The translocation of salts and water through wheat roots. Physiol. Plantarum (Copenh.) **3**, 103–151 (1950). — LYON, C. J.: A non-osmotic force in the water relations of potato tubers during storage. Plant Physiol. **17**, 250–266 (1942).

MAGISTAD, O. C., and R. F. REITEMEIER: Soil solution concentrations at the wilting point and their correlation with plant growth. Soil Sci. **55**, 351–360 (1943). — MASON, T. G., and E. PHILLIS: Experiments on the extraction of sap from the vacuole of the cotton plant and their bearing on the osmotic theory of water absorption by the cell. Ann. of Bot. **3**, 531–544 (1939). — MCCOOL, M. M., and C. E. MILLAR: The water content of the soil and the composition and concentration of the soil solution as indicated by the freezing point lowerings of the roots and tops of plants. Soil Sci. **3**, 113–138 (1917). — MEEUSE, A. D. J.: Plasmodesmata. Bot. Review **7**, 249–262 (1941). — MEYER, B. S.: A critical evaluation of the terminology of diffusion phenomena. Plant Physiol. **20**, 142–164 (1945). — MEYER, B. S., and D. B. ANDERSON: Plant physiology. New York: D. van Nostrand 1952. — MOLZ, F. J.: A study of suction force by the simplified method. II. Periodic variations and the influence of habitat. Amer. J. Bot. **13**, 465–501 (1926). — MYERS, G. M. P.: The water permeability of unplasmolyzed tissues. J. of Exper. Bot. **2**, 129–144 (1951).

OPPENHEIMER, H. R.: Über Zuverlässigkeit und Anwendungsgrenzen der üblichsten Methoden zur Bestimmung der osmotischen Konzentration pflanzlicher Zellsäfte. Planta (Berl.) **16**, 467–517 (1932). — OSTERHOUT, W. J. V.: Some aspects of secretion. I. Secretion of water. J. Gen. Physiol. **30**, 439–447 (1947). — OVERBEEK, J. VAN: Water uptake by excised root systems of the tomato due to non-osmotic forces. Amer. J. Bot. **29**, 677–683 (1942). — Auxin, water uptake and osmotic pressure in potato tissue. Amer. J. Bot. **31**, 265–269 (1944). — OWEN, P. C.: The relation of germination of wheat to water potential. J. of Exper. Bot. **3**, 188–203 (1952).

POHL, R.: Zur Reaktionsweise des Wuchsstoffes bei der Zellstreckung. Z. Bot. **41**, 343–372 (1953). — Die Ursache der Aktivitätsunterschiede von Wuchsstoffen im Test und im Zylindertest. Planta (Berl.) **44**, 191–202 (1954).

REINDERS, D. E.: The process of water-intake by discs of potato tuber tissue. Proc. Roy. Acad. Sci. Amsterdam **41**, 820–831 (1938). — Intake of water by parenchymatic tissue. Rec. Trav. bot. néerl. **39**, 1–140 (1942). — ROBERTS, E. A.: The epidermal cells of roots. Bot. Gaz. **62**, 488–506 (1916). — ROBERTS, O., and S. A. STYLES: An apparent connection between the presence of colloids and the osmotic pressure of conifer leaves. Sci. Proc. Roy. Dublin Soc. **22**, 119–125 (1939). — ROBINSON, J. R.: The active transport of water in living systems. Biol. Rev. Cambridge Philos. Soc. **28**, 158–192 (1953). — ROBINSON, J. R., and R. A. MCCANCE: Water metabolism. Ann. Rev. Physiol. **14**, 115–142 (1952). — ROSENBERG, T., and W. WILBRANDT: Enzymatic processes in cell membrane penetration. Internat. Rev. Cytology **1**, 65–92 (1952). — ROSENE, H. F.: Control of water transport in local root regions of attached and isolated roots by means of the osmotic pressure of the external solution. Amer. J. Bot. **28**, 402–410 (1941). — Effect of cyanide on rate of exudation in excised onion roots. Amer. J. Bot. **31**, 172–174 (1944). — The effect of anoxia on water exchange and oxygen consumption of onion root tissues. J. Cellul. a. Comp. Physiol. **35**, 179–193 (1950). — ROSENE, H. F., and L. E. BARTLETT: Effect of anoxia on water influx of individual radish root hair cells. J. Cellul. a. Comp. Physiol. **36**, 83–96 (1950).

SPANNER, D. C.: The suction potential of cells and some related topics. Ann. of Bot. **16**, 379–407 (1952). — The thermodynamics of actively-maintained turgor pressure, with a note on the idea of permeability. Physiol. Plantarum (Copenh.) **7**, 278–282 (1954). — STEWARD, F. C., S. M. CAPLIN and F. K. MILLAR: I. New techniques for the investigation of metabolism, nutrition and growth in undifferentiated cells. Ann. of Bot. **16**, 57–77 (1952). — STEWARD, F. C., P. R. STOUT and C. PRESTON: The balance sheet of metabolites for potato discs showing the effect of salts and dissolved oxygen on metabolism at 23° C. Plant Physiol. **15**, 409–447 (1940). — STOCKER, O.: Über die Messung von Bodensaugkräften und ihrem Verhältnis zu den Wurzelsaugkräften. Z. Bot. **23**, 27–56 (1930). — STOCKING, C. R.: The calculation of tensions in Cucurbita pepo. Amer. J. Bot. **32**, 126–134 (1945).

THIMANN, K. V.: Studies on the physiology of cell enlargement. Growth Symposium **10**, 5–22 (1951). — The physiology of growth in plant tissues. Amer. Scientist **42**, 589–606 (1954). — THIMANN, K. V., and C. L. SCHNEIDER: The interdependence of auxin and sugar for growth. Amer. J. Bot. **25**, 270–280 (1938). — THIMANN, K. V., R. R. SLATER and G. S. CHRISTIANSEN: The metabolism of stem tissue during growth and its inhibition. IV. Growth inhibition without enzyme poisoning. Arch. of Biochem. **28**, 130–137 (1950). — THODAY, D.: On turgescence and the absorption of water by the cells of plants. New Phytologist **17**, 108–113 (1918). — TS'O, P., and G. P. STEINBAUER: Effect of maleic hydrazide on auxin-induced water uptake by pea stem segments. Science (Lancaster, Pa.) **118**, 193–194 (1953).

URSPRUNG, A., and G. BLUM: Zur Methode der Saugkraftmessung. Ber. dtsch. bot. Ges. **34**, 525–539 (1916a). — Über den Einfluß der Außenbedingungen auf den osmotischen Wert. Ber. dtsch. bot. Ges. **34**, 123–142 (1916b). — USSING, H. H.: Transport through biological membranes. Ann. Rev. Physiol. **15**, 1–20 (1953).

WALTER, H.: Die Grundlagen der Pflanzenverbreitung. Stuttgart 1949. — Kritisches zur Darstellung der osmotischen Zustandsgrößen in den verschiedenen Lehrbüchern der Botanik. Planta (Berl.) **40**, 550–554 (1952). — WILSON, C. C.: Diurnal fluctuations in growth and length of tomato stems. Plant Physiol. **23**, 156–157 (1948). — WILSON, C. C., W. R. BOGGESS and P. J. KRAMER: Diurnal fluctuations in the moisture content of some herbaceous plants. Amer. J. Bot. **40**, 97–100 (1953).

Die Permeabilität der Zellwand.

Von

L. Brauner.

Mit 11 Abbildungen.

I. Die Struktur der Zellwand.

A. Mikroskopischer Aufbau.

Junge, noch wachsende Pflanzenzellen sind nur von der plastisch dehnbaren primären Zellwand umgeben. Im Gewebeverband werden die einzelnen Zellen durch die amorphe intercellulare Kittsubstanz der Mittellamelle untereinander verbunden. Nach der Erreichung der Endgröße wird auf die Primärwand in der Regel eine Sekundärwand aufgelagert. Diese Membran kann bei starker Verdickung mehrschichtig werden. In vielen Fällen erscheint sie dreischichtig und besteht dann aus je einer dünnen Innen- und Außenlamelle, die eine viel dickere Mittelschicht einschließen. Diese Mittelschicht kann ihrerseits lamelliert sein.

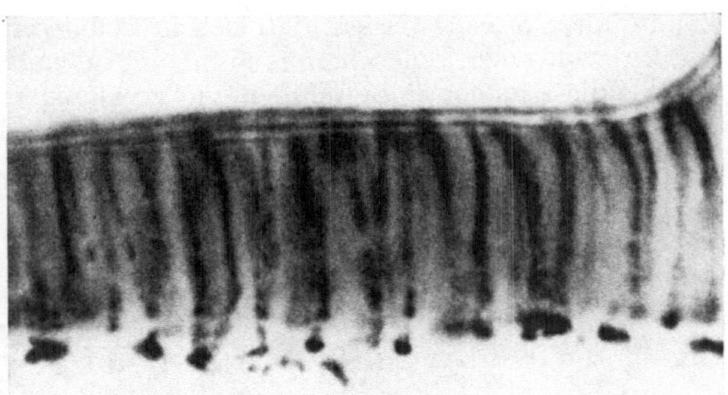

Abb. 1. *Primula acaulis.* Epidermisaußenwand mit zahlreichen Plasmodesmen. Vergr. 1800×. (Nach LAMBERTZ.)

Zunehmende Verdickung der sekundären Membranen muß den Stoffaustausch zwischen Nachbarzellen erschweren. Zur Erniedrigung der Diffusionswiderstände bleiben jedoch einzelne Partien der trennenden Wände als Tüpfel beiderseits unverdickt. Ihre Schließhaut besteht demnach aus der Mittellamelle und den beiderseits aufgelagerten, hier in der Regel besonders dünnen Primärwänden. Die Wegsamkeit dieser zarten Felder wird noch weiterhin dadurch erhöht, daß sie von zahlreichen, meist submikroskopisch feinen Poren durchsetzt sind (vgl. Abb. 5).

Zur Erleichterung des Stoffaustausches durch die Zellwand dienen neben den Tüpfeln auch die Plasmodesmen, die als zarte, plasmatische Verbindungsfäden die Membranstruktur senkrecht durchqueren. Sie finden sich besonders gehäuft in den Schließhäuten der Tüpfel, wo sie ihren Weg durch die erwähnten Poren nehmen, doch lassen sie sich meistens auch in den verdickten Partien der Zellwand nachweisen. Neuerdings konnten SCHUMACHER (1942) und LAMBERTZ (1954) zeigen, daß solche Strukturen selbst in den Epidermisaußenwänden vieler höherer Pflanzen vorkommen (Abb. 1).

B. Feinstruktur.

Schon die indirekten polarisationsoptischen und röntgenologischen Untersuchungsmethoden haben erkennen lassen, daß die Feinstruktur der Zellwand aus parallelgelagerten Strängen von kristallinischen Fadenmolekülen aufgebaut ist. Diese Micellarstränge, die sich in Cellulosemembranen aus etwa 100 Cellulosemolekülketten zusammensetzen, besitzen einen Durchmesser von 50—60 Å. 25—30 dieser Elemente treten ihrerseits zu größeren Bündeln, den *Mikrofibrillen* zusammen, wobei zwischen ihnen enge Spalten von etwa 10 Å Durchmesser bestehen bleiben. Diese *Intermicellarräume* spielen bei der Quellung der Zellwand eine entscheidende Rolle, da sie Wassermolekülen und gewissen anderen Quellungsmitteln den Eintritt gestatten. Dabei werden die Cellulosemicellen unter Vergrößerung des Gesamtvolumens der Mikrofibrillen auseinandergedrängt. Nun konnte aber gezeigt werden, daß sich in Cellulosefasern kolloidale Metalle (Au, Ag, Cu) von einer Teilchengröße bis zu 100 Å einlagern lassen. Diese Tatsache beweist die Existenz eines weiteren, viel geräumigeren Capillarsystems im Gefüge der Zellwand. Es wurde daher angenommen, daß die Mikrofibrillen sich so aneinander lagern, daß zwischen ihnen submikroskopische Interstitien von etwa 100 Å Durchmesser freibleiben (Abb. 2). Diese weiteren *Interfibrillarräume*, die mit dem engen Intermicellarsystem in Verbindung stehen, sind in der natürlichen Membran häufig der Ort der Einlagerung von Imprägnierungsstoffen wie Lignin, Cutin u. a.

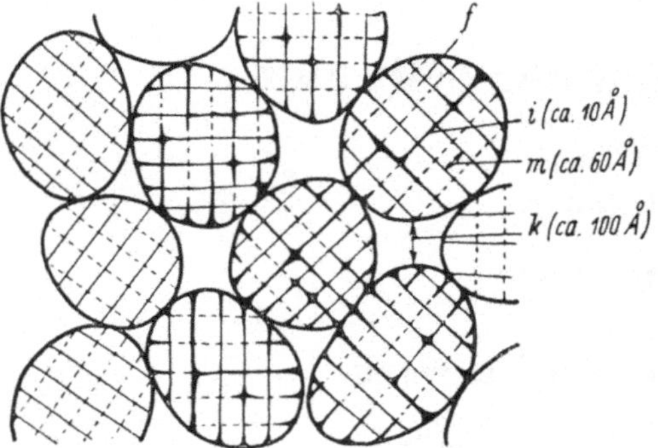

Abb. 2. Schematischer Querschnitt durch das Fibrillensystem der sekundären Zellwand einer Bastfaser. *f* Mikrofibrille; *m* Micellarstrang; *i* intermicellare Spalte; *k* Interfibrillarraum. (Nach FREY-WYSSLING.)

Abb. 3. Elektronenbild der Zellwand von *Valonia ventricosa*. Fibrillen in Paralleltextur. (Nach PRESTON.)

Der direkte Nachweis der Realität dieser Vorstellung gelang FREY-WYSSLING, MÜHLETHALER und WYCKOFF (1948) mit Hilfe des Elektronenmikroskops. Die gewonnenen Aufnahmen lassen aufs deutlichste eine Textur von feinen Fasern erkennen, die bei allen untersuchten Objekten bemerkenswerterweise annähernd den gleichen Durchmesser, 200—300 Å besitzen. Die abbildbaren Strukturen müssen also Aggregaten von je 25—30 Micellarsträngen entsprechen und stellen demnach die postulierten Mikrofibrillen dar. Diese Fasern finden sich in den einzelnen Schichten der Zellwand in typisch verschiedener Anordnung: In der sekundären Membran verlaufen sie als dichte, annähernd parallele Strähnen *(Paralleltextur)*, deren Richtung sich in den aufeinanderfolgenden Lamellen jedoch oft überkreuzt (Abb. 3, PRESTON 1952). In der Primärwand dagegen bildet das Cellulosegerüst eine viel lockerere und ungerichtetere Textur *(Streuungstextur)*, wobei die einzelnen Fibrillen wie in einem Textilgewebe miteinander verwoben sind (Abb. 4, FREY-WYSSLING 1949). — Ähnliche Texturunterschiede wurden auch in der Primär- und Sekundärmembran von Chitinzellwänden gefunden (FREY-WYSSLING und MÜHLETHALER 1950).

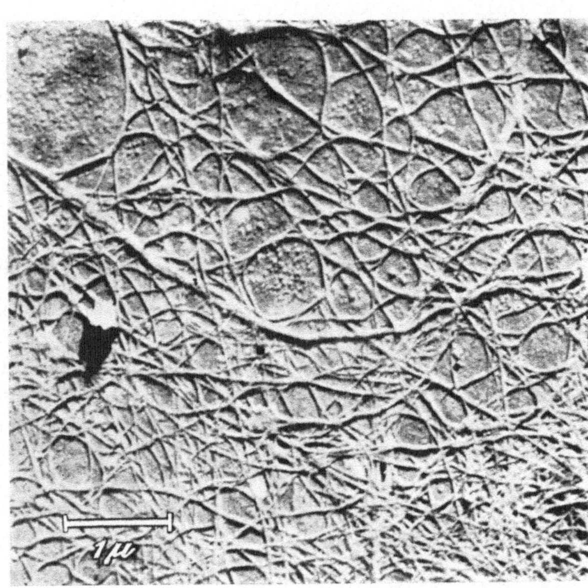

Abb. 4. Elektronenbild der wachsenden Primärwand in einer Maiscoleoptile. Fibrillen in Streuungstextur. (Nach FREY-WYSSLING.)

In noch wachsenden Primärwänden ist der Volumanteil der Cellulosefibrillen sehr gering, nur etwa 2,5%. Es bleiben daher weite Maschen frei (vgl. Abb. 4), die reichlich Raum für die Einlagerung von lebendem Plasma, Pektinen und Hemicellulosen lassen. Im weiteren Verlauf werden diese Maschen dann durch Einbau neuer Mikrofibrillen erheblich verdichtet.

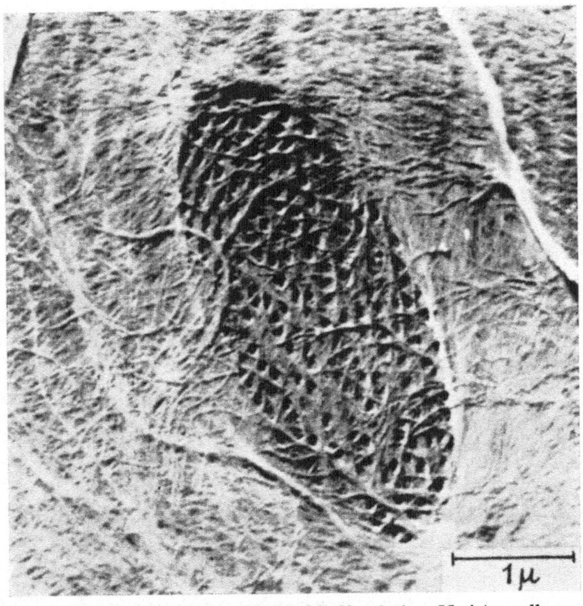

Abb. 5. Elektronenbild der Tüpfelschließhaut einer Meristemzelle aus der Keimwurzel von *Zea mays*. Man erkennt die Durchtrittsstellen der Plasmodesmen. (Nach FREY-WYSSLING.)

Die Elektronenmikroskopie gibt auch Auskunft über den Feinbau der Tüpfelschließhaut, die im Lichtmikroskop meist homogen erscheint. Es konnte gezeigt werden, daß sie tatsächlich von zahlreichen submikroskopischen Löchern durchbrochen ist, um die herum sich die Fibrillen der Membran tangential lagern (Abb. 5). Nach einer besonders klaren Aufnahme FREY-WYSSLINGS (1949) scheint der Durchmesser dieser Poren 80—90 mμ zu betragen. HUBER und KOLBE (1948) geben für die Plasmodesmentüpfel in Siebröhren Werte von 100 mμ an. — Nach neueren Beobachtungen soll im Koniferenholz der Margo der Hoftüpfel-Schließhaut von sehr viel größeren, lichtmikroskopisch sichtbaren Poren durchbrochen sein (LIESE und HARTMANN-FAHNENBROCK 1953, EICKE 1954). Diesen Angaben stehen allerdings die letzten Befunde von FREY-WYSSLING und BOSSHARD (1953) entgegen, die auch in den Hoftüpfel-Schließhäuten von *Abies* und *Picea* nur submikroskopische Poren entdecken konnten.

C. Chemischer Aufbau.

Das Grundgerüst der Pflanzenzellwand besteht aus Kohlenhydraten, vornehmlich Cellulose und Hemicellulosen; bei vielen Pilzen findet sich an deren Stelle das N-haltige Chitin, ein Polymerisationsprodukt des Acetylglucosamins. Diese Skeletsubstanzen bauen das fibrilläre Rahmenwerk auf, in dessen Hohlräume verschiedene Füllsubstanzen eingelagert sein können, vor allem Pektine, Lignin, Pflanzenwachse, Suberin, Cutin und, in lebenden Zellen, stets erhebliche Mengen von Wasser. In älteren Membranen finden sich gelegentlich Gerbstoffderivate, häufig auch Mineralstoffe, vor allem SiO_2, $CaCO_3$ und Ca-Oxalat als Inkrustierungen. Es wird sich zeigen, daß die Permeabilitätseigenschaften der Zellwände weitgehend von der Natur und Menge dieser interfibrillären Füllsubstanzen bestimmt werden.

Für die Verteilung der wesentlichen Bauelemente in den einzelnen Schichten der Zellwand gilt grundsätzlich folgendes:

a) Die Mittellamelle besteht zur Hauptsache aus amorphen Pektinverbindungen (Mg- und Ca-Pektat); in verholzten Geweben enthält sie ferner auch noch Lignin.

b) Die Primärwand besitzt bereits ein Cellulosegerüst, in dessen Interstitien erhebliche Mengen von Protopektinen und Hemicellulosen eingelagert sind. In älteren Entwicklungsstadien kann sie verholzen. Wichtig ist ferner der Einschluß von hydrophoben, wachsartigen Substanzen und Cutin in Außenwänden, wodurch eine drastische Herabsetzung der Wasserpermeabilität verursacht wird. An der Oberfläche von Zellwänden, die an die freie Atmosphäre oder an innere Lufträume im Gewebe grenzen (ARZT 1933, HÄUSERMANN 1944), bildet Cutin als Cuticula meist einen homogenen Überzug. Bei der Entwicklung der Samenschalen aus den Integumenten kommt es meistens zur Ausbildung von inneren Cuticulen, die die Permeabilitätseigenschaften der reifen Testa maßgebend bestimmen.

c) Die Sekundärwand enthält den Hauptanteil der Cellulose, die hier ein wesentlich dichteres Gerüst bildet als in der Primärmembran. Als Einlagerungen finden sich wieder Protopektine, vor allem aber Lignin, das bei Holzfasern bis zu 30% der Gesamtstruktur ausmachen kann. In verkorkten Geweben enthält die Sekundärwand als Füllsubstanz Suberin und in Borken, älterem Holz und in manchen Samenschalen Gerbstoffderivate (Phlobaphene).

D. Quellungseigenschaften der Zellwandbausteine.

Für die Beurteilung der Permeabilitätseigenschaften von Zellwänden ist ein Vergleich der relativen Hydrophilie ihrer Bauelemente erforderlich.

Reine Cellulose ist in Wasser begrenzt quellbar, das Maximum der Gewichtszunahme beträgt etwa 25%. Verantwortlich für die Wasserbindung sind die zahlreichen hydrophilen OH-Gruppen des Kohlenhydratgerüstes, deren hohes Dipolmoment der Oberfläche der Micelle eine ansehnliche negative Ladung verleiht. Dieser Effekt kann noch durch gelegentlich vorkommende freie COOH-Gruppen verstärkt werden. — Die Hemicellulosen, die in ihrem Molekulargerüst neben der Zuckerkomponente (Arabinose, Xylose) bereits regelmäßig einen bestimmten Anteil von Säureresten der Urongruppe enthalten, übertreffen die Cellulose dementsprechend in ihrem Wasserbindungsvermögen. Noch erheblich stärker ist die Quellbarkeit der Pektine, in deren Makromolekülen der Anteil an Polygalakturonsäureresten weit überwiegt. Die auffällige Hydrophilie dieser Körper wird wieder durch die Häufung von OH- und COOH-Gruppen bedingt, welch letztere hier häufig mit Methanol verestert sind. In dieser Bindung gewinnen die sonst lipophilen CH_3-Gruppen bemerkenswerterweise hydrophilen Charakter (FREY-WYSSLING 1953, S. 60). — Lignin ist ein komplexes Kondensationsprodukt aromatischer Kerne von noch nicht völlig aufgeklärter Struktur (vgl. Band 1, S. 708ff.). In verholzten Wänden erfüllt es die submikroskopischen Räume zwischen den Kohlenhydratfibrillen in regellosen Partikeln unter erheblicher Ausweitung der Struktur („Ligninquellung", FREY-WYSSLING 1928). Diese Imprägnation verringert das Quellvermögen verholzter Membranen in Wasser durch die teilweise Besetzung des sonst verfügbaren Interfibrillarraums mit einer wenig hydrophilen Füllmasse. Merkwürdigerweise wird dadurch aber die Permeabilität der Struktur für Wasser und kleinmolekulare wasserlösliche Verbindungen nur wenig beeinträchtigt. Offenbar bleibt also *zwischen* den Ligninpartikeln noch genügend freier Diffusionsweg offen (FREY-WYSSLING 1935).

Pflanzenwachse, Cutin und Suberin sind hydrophobe Einlagerungssubstanzen fettartiger Konstitution, deren Anwesenheit im Zellwandgerüst die Permeabilität der Membranen für Wasser, wasserlösliche Moleküle und Ionen drastisch herabsetzt. Die Wachse stellen im freien Zustand Ester zwischen Fettsäuren und langkettigen aliphatischen Alkoholen dar. Ihre Anlagerung an die hydrophilen Cellulosefibrillen soll nach FREY-WYSSLING (1953, S. 293) so zustande kommen, daß innerhalb der Wandstruktur die Säure- und Alkoholkomponenten nicht direkt verestert vorliegen, sondern mittels ihrer hydrophilen COOH- bzw. OH-Gruppen die hydrophilen Gruppen des Cellulosegerüstes besetzen.

Cutin und Suberin sind hochpolymere Ester von Fettsäuren mit zwei oder mehr reaktiven Gruppen, z. B. Suberinsäure $COOH—(CH_2)_6—COOH$, Phellonsäure $C_{21}H_{42}(OH) \cdot COOH$. Während die Häufung von $—CH_2—$-Gruppen die Hydrophobie des Systems bedingt, vermitteln die hydrophilen OH- und COOH-Gruppen einerseits die Bildung komplexer Ester- bzw. Ätherbindungen zwischen den Säuremolekülen, andererseits aber auch die Anheftung der Komplexe an die hydrophilen Gruppen der Cellulosemicelle. Da jedoch ein Teil der Carboxylgruppen offenbar frei dissoziiert bleibt, gewinnt die cutinisierte Membran eine weitere für ihr Permeabilitätsverhalten entscheidende Eigenschaft: die gegenüber wäßrigen Medien nur schwach negative Eigenladung reiner Cellulose wird durch die Imprägnierung mit dem stark acidoiden Cutin so sehr erhöht, daß der Komplex nunmehr ausgesprochen ionenselektiven Charakter gewinnt (relativ stärkere Herabsetzung der Anionenbeweglichkeit, vgl. BRAUNER 1930b).

II. Permeabilitätstypen der Zellwände.

A. Durchlässige hydrophile Membranen.

Unter dieser Kategorie sollen in erster Linie Zellmembranen verstanden werden, die in ihren Interfibrillarräumen vorwiegend Pektine, Hemicellulosen und Plasmareste enthalten. Solche „unveränderte" Cellulosewände wurden lange

Zeit für unbeschränkt permeabel gehalten, vor allem wohl auf Grund der Tatsache, daß sie in lebenden Zellen den Ablauf der Plasmolyse selbst bei Einwirkung großmolekularer Osmotica (Raffinose!) nicht verhindern. Die fortschreitende Kontraktion des Protoplasten innerhalb des Zellgehäuses beweist ja den Durchtritt der Lösung durch die Zellwand und ihre Ansammlung im entstehenden „Vorraum".

Nun ist es aber schon älteren Beobachtern aufgefallen, daß die *Geschwindigkeit* des Plasmolyseeintritts in vielen Fällen vom Teilchenvolumen des verwendeten Osmotikums abhängt. FITTING (1917) fand bei *Rhoeo discolor*-Epidermen Erreichung des Plasmolysegleichgewichts in KNO_3 bereits nach 15—30 min, in isosmotischer Rohrzuckerlösung dagegen erst nach $1^1/_2$—2 Std. HÖFLER (1930) und HUBER und HÖFLER (1930) studierten diese Erscheinung eingehender und kamen dabei zum bemerkenswerten Ergebnis, daß der Permeationswiderstand einer Zellwand um so auffälliger in Erscheinung tritt, je größer die *Wasser*permeabilität des von ihr umschlossenen Protoplasten ist, je weniger also die Geschwindigkeit des Wasseraustausches von den Plasmagrenzschichten selbst limitiert wird. Es konnte gezeigt werden, daß bei Objekten, in denen Plasmolyse an sich langsam verläuft (*Maianthemum, Salvinia, Spirogyra*), kleinmolekulare Osmotica wie KCl oder Malonamid nicht merklich schneller wirken als der größervolumige Rohrzucker. Dagegen ergaben sich bei Geweben mit besonders kurzer Plasmolysezeit (*Vallisneria, Zygnema*) sehr deutliche Unterschiede zugunsten der kleinmolekularen Plasmolytica, die das Verhältnis ihrer Diffusionskonstanten im freien Lösungsmittel weit übersteigen. So verhielten sich etwa bei *Vallisneria* die Plasmolysegeschwindigkeiten in Rohrzucker, Malonamid und KNO_3 wie 1,0:2,9:4,5.

Auch bei der viel untersuchten Zwiebelepidermis hatte sich gezeigt, daß die Zellwand die Permeation von Rohrzucker merklich verlangsamt. Als Maß diente die Geschwindigkeit des Ablaufs der Deplasmolyse unter folgenden Bedingungen: Das vorplasmolysierte Gewebe wurde nach dem KÜSTERschen Verfahren quer durchschnitten und sodann in eine verdünntere Stufe des verwendeten Osmotikums übertragen. In diesen Präparaten wurde nun die Expansionsgeschwindigkeit unversehrt gebliebener Protoplasten geöffneter Zellen des Schnittrandes mit dem Deplasmolyseablauf in umwandeten Zellen verglichen. Nach Rohrzuckerplasmolyse reagierten geöffnete Zellen merklich schneller als geschlossene; bei Verwendung diffusibler Salzlösungen dagegen waren die Deplasmolysegeschwindigkeiten in beiden Fällen praktisch gleich. Da der Ausgleich zwischen der höheren Vorraumkonzentration und dem verdünnteren Außenmedium in der geöffneten Zelle direkt, in den intakten jedoch nur durch die Zellwand möglich ist, beweist dieses Ergebnis deren beschränkte Permeabilität für den großmolekularen Zucker.

DE HAAN (1933) prüfte die Rohrzuckerpermeabilität der Zellwände von Zwiebelepidermen nach einer anderen, bereits von HÖFLER (1930) angegebenen Methode: das vorplasmolysierte Gewebe wird verschieden lang in Wasser gespült und sodann in Paraffinöl übertragen. Vergleicht man nun den Fortgang der Deplasmolyse in diesem isolierenden Medium, so läßt sich die Zeit ermitteln, die bei der vorherigen Wässerung zur völligen Beseitigung des Osmotikums aus den „Vorräumen" erforderlich war. Es zeigte sich, daß bereits 30 sec langes Spülen genügte, um die Deplasmolyse im Paraffinöl ebenso schnell ablaufen zu lassen wie bei fortgesetztem Aufenthalt in Wasser. Dieser Befund läßt also auf eine erstaunlich hohe Zuckerpermeabilität der Zwiebelzellwand schließen. Der Gegensatz zu den vorerwähnten Beobachtungen erklärt sich möglicherweise aus der Verschiedenheit des Objekts: DE HAAN untersuchte die Innenepidermis, HUBER und HÖFLER die Außenepidermis.

Neuerdings hat CASARI (1953) interessante Beobachtungen über das Verhalten plasmolysierter Protoplasten beim Wechsel des Plasmolytikums mitgeteilt, die wichtige Rückschlüsse auf den Permeationswiderstand der Zellwand gestatten. Werden Zwiebelepidermiszellen mit einem kleinmolekularen Osmotikum (z. B. Glycerin) vorplasmolysiert und nach Eintritt des Gleichgewichts in eine isotonische Lösung eines größermolekularen Osmotikums (z. B. Saccharose) übertragen, so dehnen sich die Protoplasten vorübergehend wieder aus. Bei Darbietung der beiden Plasmolytica in umgekehrter Reihenfolge ruft die Übertragung in die kleinermolekulare Lösung vorübergehende Verstärkung der Plasmolyse hervor. Das gleiche Verhalten ist auch beim Wechsel zwischen einem leicht diffusiblen Elektrolyten (z. B. KCl) und einem schwer diffusiblen Anelektrolyten (z. B. Zucker) zu beobachten (Abb. 6). Da dieser Effekt nur bei umwandeten Zellen, nicht aber bei isolierten Protoplasten auftritt, glaubt CASARI eine von

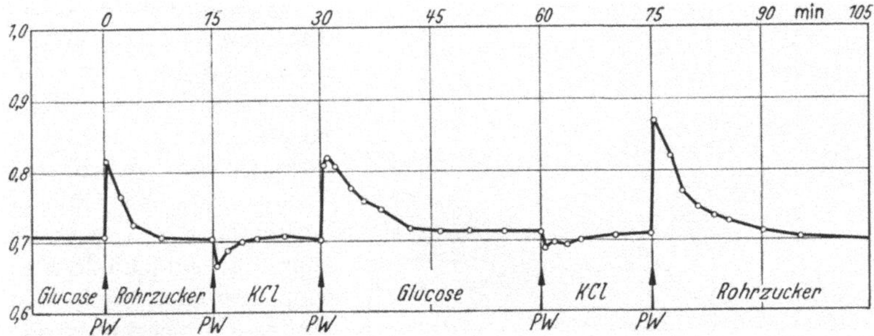

Abb. 6. Plasmolytikumwechseleffekt bei der Innenepidermis von *Allium cepa*. Ordinate: Plasmolysegrad, Abszisse: Zeit. *PW* Wechsel des Plasmolytikums. (Nach CASARI.)

anderen Autoren (BENNET-CLARK und BEXON 1946) vermutete Mitwirkung elektroosmotischer Faktoren ausschließen zu können. Er sieht die Erklärung des „Plasmolytikumwechseleffekts" ganz allgemein in einer vorübergehenden Entstehung von Konzentrationsdifferenzen im Vorraum, bedingt durch die beschränkte Permeabilität der umgebenden Zellwand: Enthält der Vorraum ursprünglich das größermolekulare Osmotikum, so wird beim Wechsel des Plasmolytikums der Einstrom der kleineren Außenmoleküle schneller erfolgen als der gleichzeitige Austritt der voluminöseren Innenmoleküle. Der osmotische Wert der Vorraumlösung steigt infolgedessen vorübergehend an. Erfolgt der Wechsel im umgekehrten Sinn, so muß die Konzentration im Vorraum dementsprechend sinken. — Ähnliche Effekte wurden auch an Epidermis- und Mesophyllzellen von *Vallisneria spiralis* beobachtet.

Der Diffusionswiderstand der Zellwand tritt auch bei den üblichen Bestimmungsmethoden der Gewebesaugkraft auffällig in Erscheinung. E. G. PRINGSHEIM (1931) hat wohl als erster darauf hingewiesen, daß bei derartigen Messungen an Parenchymgeweben das Volumen der Proben mit steigender Außenkonzentration bis zu einer bestimmten Stufe annähernd linear abnimmt, bei weiterer Konzentrationssteigerung sich jedoch nicht mehr verändert. Diese Erscheinung wurde später von HASMAN (1943) und von SKENE (1943) näher analysiert. Gravimetrische Untersuchungen an verschiedenen Speichergeweben (Kartoffel, Rübe, Karotte) ergaben, daß bei kürzeren Expositionszeiten (1—2 Std) nur leicht diffusible Osmotica (KNO_3) den erwähnten abrupten Übergang des Kurvenabfalls in die horizontale Endstrecke hervorrufen, nicht aber Rohrzuckerlösungen, in denen die Gewichtsabnahme sich bis in die höchsten Konzentrationsstufen fortsetzte (Abb. 7). Dieses Verhalten erklärt sich daraus, daß der Knickpunkt der Kurve im KNO_3-Versuch bei der Konzentration der Grenzplasmolyse liegt, die sich in diesem Osmotikum unbehindert entwickeln kann. Ein weiteres Fortschreiten der Protoplastenkontraktion führt dann zu keinem neuen Gewichtsverlust des Gewebes mehr, da der Wasserverlust der Vacuolen durch eine Auffüllung der „Vorräume“ mit der Außenlösung ausgeglichen wird. Anders liegen die Verhältnisse im Rohrzuckerversuch, indem der Durchtritt des größermolekularen Osmotikums durch die Zellwand bereits merklich erschwert ist. Dadurch verzögert sich der Eintritt der Plasmolyse, das Gewebe beginnt unter weiterem Gewichtsverlust zu schrumpfen. Läßt man die Zuckerlösung jedoch länger einwirken, so entwickelt sich nach etwa einem Tag auch hier bei der Konzentration der Grenzplasmolyse der charakteristische Knick in der Kurve, da die Zuckermoleküle unterdessen die Zellwände durchqueren konnten (vgl. Abb. 7).

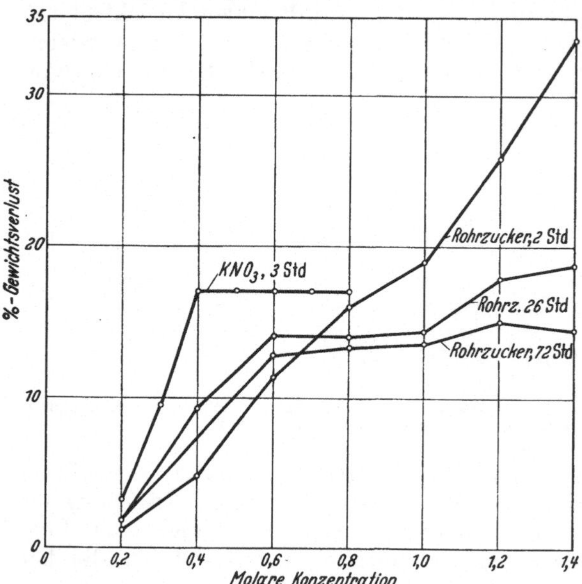

Abb. 7. Gewichtsverlust von Rübengewebe in steigenden Konzentrationen von KNO_3 und Rohrzucker. (Nach SKENE.)

SKENE führte ferner an durch Chloroform abgetötetem Rübengewebe auch quantitative Messungen der Zellwandpermeabilität aus. Durch gleichzeitige Bestimmung der Volum- und der Gewichtsänderungen der Gewebescheibchen in verschiedenen Versuchslösungen ließ sich die Menge der permeierten Substanz berechnen. Bei Verwendung 1-molarer Konzentrationen wurde der Diffusionswiderstand in der Zellwand für Rohrzucker etwa 80mal, für Glucose 60mal, für Glycerin 50mal und für KNO_3 40mal so hoch gefunden wie im freien Lösungsmittel. Diese Reihe bringt die Ultrafiltereigenschaft der untersuchten Membran deutlich zum Ausdruck. Verglichen mit dem enormen Permeationswiderstand des lebenden Protoplasten fällt aber dennoch der Widerstand der Wand für den Stoffaustausch kaum ins Gewicht. Denn trotz ihrer beschränkten Durchlässigkeit läßt die Wand Saccharose immer noch 80000mal, Glucose 50000mal und KNO_3 14000mal leichter passieren als die Plasmahaut!

B. Beschränkt durchlässige hydrophile Membranen.

Daß der Diffusionswiderstand der Zellwand in Plasmolyseversuchen zu einer Schrumpfung der ganzen Zelle führen kann, ist schon bei Parenchymgeweben zu beobachten gewesen. Noch viel auffälliger tritt dieser Effekt bei vielen Fadenalgen in Erscheinung, wo großmolekulare Osmotica, je nach dem Dichtheitsgrad der Wände, vorübergehenden oder dauernden Kollaps der Zellen (Cytorrhyse, Schrumpfelung) hervorrufen können. FÖRSTER (1933) untersuchte die Situation eingehend bei der Cladophoracee *Rhizoclonium*. Hier rufen leicht hypertonische Lösungen noch normale Plasmolyse hervor; in hohen Konzentrationsstufen dagegen treten mehr oder weniger starke Eindellungen der ganzen Zellen auf, die sich im weiteren Verlauf wieder ausgleichen, wobei der Protoplast kontrahiert bleibt. Daß es sich beim anfänglichen Kollaps nicht etwa um eine passive Deformation der Wand durch die Kontraktion des anhaftenden Plasmaschlauchs handelt, geht schon daraus hervor, daß der gleiche Effekt auch an vorplasmolysierten Fäden hervorzurufen ist. Aus der Intensität und Dauer der Cytorrhyse lassen sich demnach Rückschlüsse auf die Durchlässigkeit der Zellwand für das einwirkende Osmotikum ziehen. Als entscheidend erwies sich wieder dessen Teilchengröße, ferner aber auch sein Einfluß auf den Quellungszustand der Membran. Begünstigt wird das Zustandekommen der Eindellung durch eine Verdichtung der Wand im Osmotikum, während andererseits selbst großmolekulare Verbindungen keine Cytorrhyse hervorrufen, wenn sie den Quellungsgrad der Membran erhöhen (Buttersäure!). Diese Wechselwirkung beeinträchtigt gelegentlich die Gültigkeit des Ultrafilterprinzips, da die Porenweite der Membran in den einzelnen Medien erheblich zu variieren scheint. Immerhin läßt sich für indifferente Nichtelektrolyte die Permeationsreihe:

$$\text{Glycerin, Rohrzucker} < \text{Äthylalkohol} < \text{Methylalkohol}$$

aufstellen, während das Durchtrittsvermögen der Ionen durch ihre lyotropen Eigenschaften bestimmt wird:

$$Mg < Ca < K,\ NH_4 \text{ und } PO_4 < CO_3 < SO_4 < Cl,\ NO_3.$$

Nicht ganz klar ist die Rolle der zarten Cuticula, die die Längswände der Zellen umgibt. FÖRSTER hält ihre Filterwirkung für unbedeutend, doch bedürfen seine Argumente wohl noch der Nachprüfung.

An einer anderen Cladophoracee, der marinen *Chaetomorpha aerea* studierte HOFFMANN (1936) die Zellwandpermeabilität abgetöteter Fäden. Die offenbar höhere Selektivität dieses Objekts gestattete die Anwendung sehr viel niedrigerer Konzentrationen, die den Quellungszustand der Membran noch nicht entscheidend beeinflussen. So bewirkte hier Raffinose schon in 0,04 m-Lösung eben erkennbare Cytorrhyse. Bei kleinermolekularen Osmoticis wurde die erforderliche Grenzkonzentration (G.K.) entsprechend dem zunehmenden Permeiervermögen größer gefunden. Die von HOFFMANN mit einer großen Anzahl von Testsubstanzen ermittelte Wirkungsreihe spricht eindeutig für die Gültigkeit des Ultrafilterprinzips. Nur sind die *Chaetomorpha*-Zellwände immerhin noch so permeabel, daß die Deformation der Fäden in allen geprüften Lösungen nach einer bestimmten Frist wieder vollständig zurückgeht. In diesen, für jedes Osmotikum charakteristischen Ausgleichszeiten (t) wird die Beziehung zwischen Permeiervermögen und Molekularvolumen (M.V.) besonders eindrucksvoll erkennbar.

Tabelle 1. *Chaetomorpha. Permeabilität der Zellwand für Anelektrolyte von verschiedenem Molekularvolumen.*

Stoff	M.V.	G.K.	t in 0,5 m-Lösungen sec
Saccharose	345,8	8×10^{-2} m	270
Antipyrin	213,2	15×10^{-2} m	105
Glucose	183,2	16 10^{-2} m	75
Dimethylharnstoff	147,2	21×10^{-2} m	60
Erythrit	130,2	27×10^{-2} m	45
Glycerin	87,8	41×10^{-2} m	20

Sehr interessant ist auch die Beobachtung, daß in toten *Chaetomorpha*-Zellen selbst stark lipophile Stoffe (Antipyrin, Dimethylharnstoff) entsprechend ihrem Molekularvolumen permeieren, während sie von den lebenden Fäden bevorzugt schnell aufgenommen werden. Ähnliche Erfahrungen machte KORNMANN (1934) auch an *Valonia*-Zellen, die er als natürliche Osmometer untersuchte. Die lebende Alge erwies sich dabei als auffällig permeabel für Alkohole. Nach mechanischer Entfernung des Plasmaschlauchs wird die isolierte Zellwand für die gleichen Verbindungen jedoch sehr viel weniger durchlässig. Während beim lebenden System das Permeiervermögen aliphatischer Alkohole mit steigendem Molekulargewicht zunimmt, gilt für die tote Wand dabei die umgekehrte Reihenfolge:

—Permeationsgeschwindigkeit→

Lebende Zelle: Methylalkohol<Äthylalkohol<Propylalkohol

Tote Zellwand: Propylalkohol<Äthylalkohol<Methylalkohol

Der Komplex Zellwand/Plasmabelag wirkt also als Lipoidfilter, die Zellwand allein als Ultrafilter..

Auch bei der von RUHLAND und seiner Schule eingehend studierten farblosen Cyanophycee *Beggiatoa mirabilis* ist die Zellwand selektiv permeabel. Die Membran dieser Alge besteht im wesentlichen aus Pektinstoffen; Cellulose scheint in ihrem Gerüst zu fehlen. An abgetöteten Fäden lassen sich hier durch noch viel schwächere Lösungen als bei den vorerwähnten Grünalgen typische Cytorrhysen (Knickung der Wände) erzielen (RUHLAND und HOFFMANN 1925). Die hierzu erforderlichen Minimalkonzentrationen stehen wieder im umgekehrten Verhältnis zur Teilchengröße der einwirkenden Osmotica, die Permeationsraten folgen also auch hier weitgehend dem Ultrafilterprinzip. Dabei zeigt es sich abermals, daß diese Regel für tote Membranen viel strenger gilt als für die lebende Zelle.

In den bisher besprochenen Fällen war die beobachtete Permeabilitätserniedrigung der Zellwände wohl in erster Linie ihrem Gehalt an Pektinstoffen zuzuschreiben. In solchen Membranen scheint die Erfüllung der interfibrillaren Capillarräume mit einem extrem hydrophilen Körper die Beweglichkeit wasserlöslicher Stoffe durch die Bindung von Wassermolekülen an die heteropolaren Gruppen der Füllsubstanz zu behindern. Die dadurch bedingte Verengerung des Diffusionsweges führt aber selbst bei den voluminösesten der geprüften Osmotica nur zu einer Einschränkung, nie jedoch zu einer völligen Sistierung der Permeation.

C. Hydrophobe Zellwände.

Sehr viel wirkungsvoller ist die Abdichtung von Zellwänden mit hydrophoben Füllstoffen, Cutinen und Wachsen, die in manchen Fällen zu vollkommener Semipermeabilität und selbst zu einer drastischen Reduktion der Wasserdurchlässigkeit führen kann. Solche Wände kommen demnach in ihren diosmotischen Eigenschaften schon den Plasmagrenzschichten nahe. Eindrucksvolle Beispiele für diesen Typus finden wir unter den Trichomen, den Annuluszellen der Farnsporangien, den Blättchen vieler Laubmoose und den Testen mancher Samen.

a) Trichome, Annuluszellen, Moosblättchen, Epidermen.

Das Verhalten von Trichomen und Annuluszellen wurde von FRENZEL (1929) näher studiert. Intakte Haare von *Fittonia verschaffeltii* zeigen in 1 Mol Lösungen von Rohrzucker und Glucose bald ausgeprägte Cytorrhyse und bleiben dann, im Gegensatz zu den früher besprochenen Fadenalgen, dauernd deformiert (Abb. 8). Eine isosmotische Glycerinlösung dagegen ruft schon nach wenigen Minuten normale Plasmolyse hervor. — In Annuluszellen von *Blechnum brasiliense* bewirken Glycerin und Weinsäure nach etwa 30 min Plasmolyse. Die Ablösung des Plasmaschlauchs beginnt auffälligerweise stets von der Innenwand her, die trotz ihrer erheblichen Dicke durchlässiger ist als die dünne Außenmembran (Abb. 9). Für Glucose, Mannit und Rohrzucker sind dagegen sämtliche Wände des Annulus praktisch impermeabel: in diesen Osmoticis entwickelt sich Cytorrhyse (Eindellung der Außenwand), die im Falle von Rohrzucker selbst nach Tagen nicht zurückgeht (RENNER 1915). Auf Grund des Molekularvolumens der Testsubstanzen wurde von RENNER (1925) und FRENZEL geschlossen, daß in diesen Membranen die für hydrophile Körper verfügbare Porenweite unter 10 Å liegen

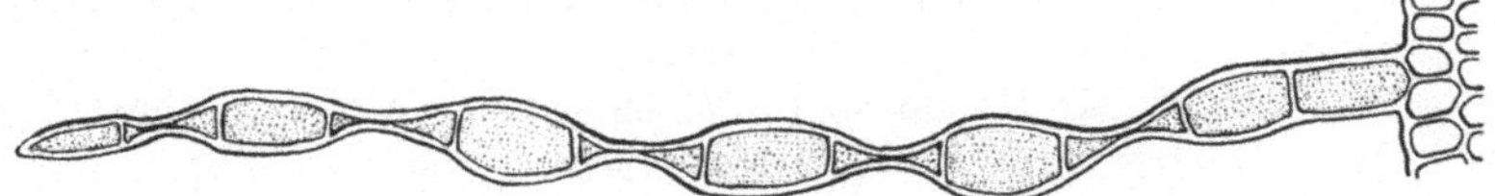

Abb. 8. Haar von *Fittonia verschaffeltii*. Cytorrhyse nach Behandlung mit Rohrzuckerlösung. (Nach FRENZEL.)

muß. Da andererseits lipophile basische Farbstoffe von weit größerem Molekularvolumen (Methylgrün, Jodgrün) in rohrzuckerimpermeable Haare einzudringen vermögen, ist zu vermuten, daß in den Interfibrillarräumen der Membran Cutin einen erheblich größeren Anteil des Volumens erfüllt als das Quellungswasser.

Daß die Blättchen vieler Laubmoose sich wegen der Dichte ihrer Außenwände mit Rohrzucker nur schwer plasmolysieren lassen, ist schon lange bekannt (BRILLIANT 1927, RENNER 1932, KRESSIN 1935). Neuerdings hat BIEBL (1954) diese älteren Befunde nachgeprüft und erweitert. Das Grundphänomen besteht hier wieder darin, daß intakte Blättchen, die zusammen mit ihrem ebenfalls unverletzten Tragsproß in 1—1,5 Mol Rohrzuckerlösung eingelegt werden, selbst nach langer Expositionszeit keinerlei Plasmolyse erkennen lassen. Statt dessen zeigt sich schon innerhalb der ersten Stunde starke Cytorrhyse, die oft mehrere Tage erhalten bleibt. In Osmoticis von kleinem Teilchenvolumen (NaCl, KNO_3) entwickelt sich dagegen normale Plasmolyse. Wird der Versuch mit abgeschnittenen Blättchen angestellt, so ruft auch Rohrzucker Plasmolyse hervor, die sich jetzt von der Wunde her, den Zellen der Mittelrippe folgend, über die Blattfläche ausbreitet. Dieses Verhalten läßt deutlich erkennen, daß die Seitenwände der Blattzellen und vor allem die

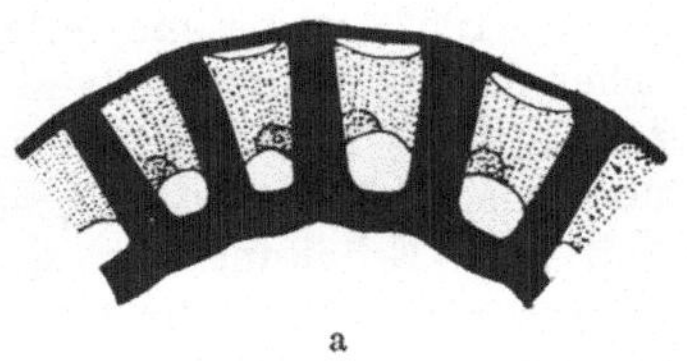

a

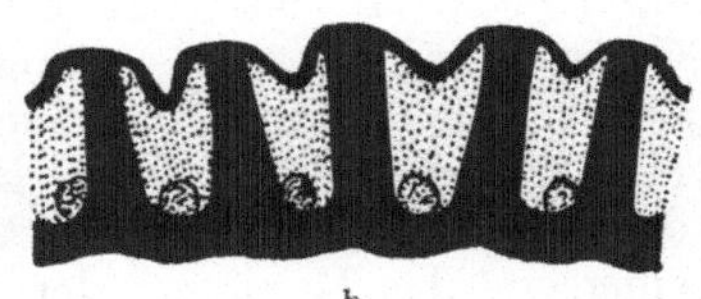

b

Abb. 9a u. b. Selektive Permeabilität der Zellwände des Farnannulus. a Plasmolyse in Glycerin, von der dickeren Innenwand her beginnend; b Cytorrhyse in Rohrzuckerlösung. (Nach FRENZEL.)

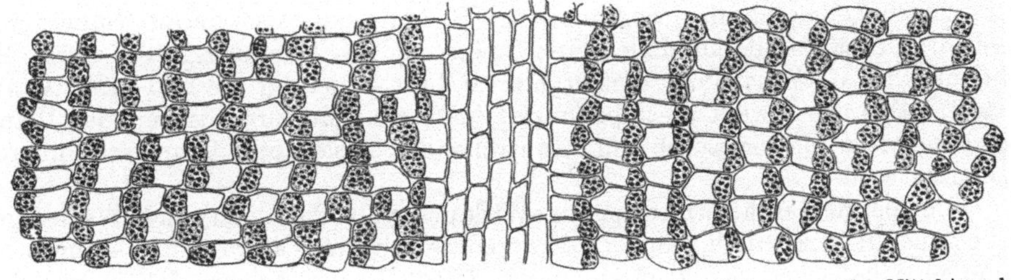

Abb. 10. Einseitige Plasmolyse in einem Blättchen von *Catharinea* in 1 m Maltose, von der Mittelrippe her beginnend. (Nach BRILLIANT.)

der Mittelrippe wesentlich durchlässiger für das Osmotikum sind als die Außenwände (Abb. 10). Durch die hohe Permeabilität ihrer Quer- und Seitenwände stellt die Mittelrippe zweifellos die leitende Verbindung zwischen Sproß und Blattfläche her. BIEBL konnte zeigen, daß bei Präparaten, die aus einem gestutzten Sproßstück mit einem ansitzenden, intakten Blatt bestehen, Rohrzucker nach längerer Zeit wieder von der Mittelrippe her in den Blattzellen Plasmolyse hervorruft. Der Zucker ist also offenbar durch die Schnittfläche vom Sproß über die „Leitzellen" der Rippe von innen her in die Blattfläche gelangt.

Der Permeabilitätsgrad der Blattaußenwände variiert innerhalb der einzelnen Laubmoosgattungen erheblich, und selbst zwischen den Arten der gleichen Gattung bestehen gelegentlich die erstaunlichsten Unterschiede. So ist *Mnium rostratum* selbst noch für Raffinose (M.V. 498,8) durchlässig, während *Mnium serratum* schon Harnstoff (M.V. 59,2) kaum mehr passieren läßt.

Über die Natur der Zellwandabdichtung in Moosblättchen besteht noch keine volle Klarheit. Die Existenz einer Cuticula scheint fraglich zu sein (KRESSIN 1935), doch ist, wenigstens bei den weniger durchlässigen Typen, Cutinisierung der Außenwände wahrscheinlich.

Schließlich sei hier auch noch auf die Bedingungen der Wasserpermeabilität bei Epidermen und Peridermen der höheren Pflanzen eingegangen, deren Kenntnis für das Verständnis der cuticularen Transpiration wesentlich ist. O. HÄRTEL (1947—1952) und seine Mitarbeiter untersuchten die Wechselbeziehungen zwischen dem Quellungszustand der Grenzschichten und ihrem Diffusionswiderstand für flüssiges und dampfförmiges Wasser. Wenn auch die bisherigen Befunde noch nicht völlig eindeutig sind, so scheint sich doch die Grundregel zu ergeben, daß jede Erhöhung der Quellungskapazität in cutinisierten Membranen die Wasserpermeabilität herabsetzt. Dies ließ sich in erster Linie aus dem Verhalten der Gewebe nach Behandlung mit abgestuften Pufferlösungen entnehmen. Dabei ergab sich zunächst, daß Cutin, vermutlich wegen des Vorhandenseins freier Carboxylgruppen und organischer Basenreste, ampholytische Eigenschaften besitzt, ein Befund, der auch durch das färberische Verhalten cutinisierter Membranen bestätigt wird (HÄRTEL 1952). In der isoelektrischen Zone, die in der Regel nahe am Neutralitätspunkt zu liegen scheint (p_H 6—7,2), wurde nun einerseits das Maximum der Wasserdurchlässigkeit, andererseits das Minimum der Quellung beobachtet (Befunde an Laubblättern, Coniferennadeln, Peridermen verholzter Triebe). Eine ähnliche Korrelation zeigte sich unter bestimmten Versuchsbedingungen auch bei der Einwirkung lyotroper Ionenreihen.

HÄRTEL interpretiert diese Beziehung durch die Annahme, daß im verhältnismäßig starren Gefüge der Membran jede Zunahme der Grenzflächenladung zu einer steigenden Immobilisierung des Füllwassers in den Poren führen muß. Der gleiche Effekt ist auch bei adsorptiver Einlagerung stark hydratisierbarer Ionen (Lithium!) zu erwarten. — Einer ähnlichen Situation sind wir bereits bei sehr pektinreichen Zellmembranen begegnet (vgl. S. 346), und es sei daran erinnert, daß selbst in reinen Cellophanfolien eine Steigerung des Quellungsgrades durch Erhöhung des p_H-Wertes die Wasserpermeabilität erniedrigt (L. und M. BRAUNER 1943). HÄRTEL erwägt die Möglichkeit, daß diese Beeinflußbarkeit der Wandeigenschaften physiologisch zur Kontrolle der cuticularen Transpiration ausgenutzt wird.

Die permeabilitätserniedrigende Wirkung von Cuticulen und Wachsüberzügen hängt vermutlich weniger von deren Schichtdicke als von ihrer chemischen und physikalischen Struktur ab. KAMP (1930) konnte zeigen, daß in manchen Fällen die cuticulare Transpiration bei jungen Blättern mit sehr zarter Cuticula stärker eingeschränkt ist als bei alten Blättern der gleichen Species mit viel derberer Cuticula. Offenbar führen Alterungserscheinungen zu Strukturänderungen im Cutinüberzug, die seine Wasserdurchlässigkeit merklich erhöhen. Daß gutentwickelte Cuticulen selbst für größere Moleküle erstaunlich permeabel sein können, geht aus den Angaben von ARENS (1929) hervor, der bei Blättern von *Vitis, Syringa* u. a. cuticulare Exosmose von Calciumsalzen und selbst von Phosphatiden chemisch nachweisen konnte. Besonders wegsam scheinen die

Bezirke über den pektinreichen Antiklinalen der Epidermis zu sein, ein Umstand, der vermutlich eine wichtige Rolle bei der Leitung des „extrafascicularen" Transpirationsstroms spielt (STRUGGER 1938, 1949).

In anderen Fällen dagegen können Cuticula- und Wachsüberzüge die Epidermisaußenwand bis zu hohen Graden von Ionenselektivität verdichten. Ein bekanntes Beispiel hierfür ist die Schale des Apfels, deren Wasserundurchlässigkeit schon von HABERLANDT (1918) studiert worden ist. In einem Vergleichsversuch mit intakten und geschälten Äpfeln verloren 100 cm^2 der epidermisfreien Oberfläche in 3 Std 385 mg, eine gleichgroße epidermisbedeckte Fläche nur 15 mg. Später konnte BEUTNER (1920) zeigen, daß Apfelschalen als Membran in Konzentrationsketten von Alkalisalzen das entstehende elektrische Diffusionspotential so sehr erhöhen, daß daraus auf eine praktisch vollkommene Immobilisierung des Anions im Gewebe geschlossen werden muß. So entstand etwa in einer $^1/_{250}$ m/$^1/_{1250}$ m-NaCl-Kette eine Potentialdifferenz von 38—41 mV mit dem positiven Pol auf der Seite der verdünnteren Stufe. Der theoretische Maximalwert dieser Kette bei der Anionenbeweglichkeit Null beträgt 40 mV (20° C).

Für diese extreme Anionenundurchlässigkeit ist neben der geringen Porenweite der Cuticularstruktur zweifellos auch ihre hohe negative Eigenladung verantwortlich (vgl. S. 341).

b) Samenschalen.

1. Permeabilität für Wasser und hydrophile Stoffe.

Die Membranverhältnisse bei dichten Fruchtschalen leiten über zu denen bei vielen Samenhüllen. Auch hier begegnen wir Permeabilitätstypen, die in den extremsten Fällen bereits manche Eigenschaften der Plasmagrenzschichten aufweisen. Da sich an dichten Testen alle wesentlichen Durchlässigkeitswerte nach exakten physikalischen Methoden meist zuverlässiger ermitteln lassen als an lebenden Protoplasten, stellen solche natürliche Membranen wertvolle Modelle für das Studium von Permeabilitätsproblemen dar. Zur Messung der Durchlässigkeit für Wasser und hydrophile Nichtelektrolyte kann bei mechanisch resistenten Samenschalen Druckfiltration unter definierten Bedingungen dienen. Die Elektrolytpermeabilität läßt sich durch Bestimmung der Leitfähigkeit der Lösungen in den Membranen ermitteln und die relative Selektivität für Anionen und Kationen durch die schon erwähnte Messung der Beeinflussung von Diffusionspotentialen (BRAUNER 1928, 1930a, b). In günstigen Fällen lassen sich Samenschalen auch als Membranen in Mikroosmometern untersuchen (DENNY 1917a, b, v. DELLINGSHAUSEN 1933, BRAUNER 1935).

Tabelle 2. *H_2O- und KCl-Permeabilität einiger Samenschalen.*

Objekt	Dicke (μ)	Pw	λ	Pw/λ
Pisum sativum	115	11,0	2,60	4,23
Phaseolus multiflorus .	250	3,1	0,75	4,13
Vicia faba	315	3,2	0,43	7,44
Aesculus hippocastanum	550	0,86	0,15	5,73
Xanthium macrocarpum	84	1,02	0,015	68,0
Triticum vulgare . . .	92	0,51	0,0017	300,0

Um zunächst eine Vorstellung davon zu geben, in wie weiten Grenzen der Dichtheitsgrad der Samenschalen bei verschiedenen Pflanzen variieren kann, sind in der nebenstehenden Tabelle (BRAUNER 1928) für einige charakteristische Typen Vergleichszahlen der Wasserpermeabilität (Pw) und der spezifischen Leitfähigkeit (λ) für 1 Mol KCl zusammengestellt. Die Pw-Werte geben die Strömungsgeschwindigkeit (μ/min) von dest. H_2O in der Membran bei 1 atm. Druckgefälle und 22° C an, die λ-Werte den reziproken Widerstand Ω^{-1} einer Leitfläche von 1 cm^2.

In Tabelle 2 sind die geprüften Objekte nach fallender KCl-Leitfähigkeit angeordnet. Dabei zeigt sich nun, daß die entsprechenden Werte der Wasserdurchlässigkeit keineswegs immer im gleichen Verhältnis absinken. Verglichen mit dem weitaus permeabelsten Typ der Reihe, *Pisum*, erscheint zwar bei *Phaseolus* und bei *Aesculus* Pw und λ annähernd proportional reduziert. Dagegen ist schon bei *Vicia* und noch viel auffälliger bei *Xanthium* und *Triticum* die Permeabilität für Ionen sehr viel stärker beeinträchtigt als die Wasserdurchlässigkeit. Während also bei *Phaseolus* und *Aesculus* nur eine allgemeine *Verdichtung* in Erscheinung tritt, erweisen sich *Xanthium* und vor allem *Triticum* bereits als ausgesprochen *semipermeabel*. In der letzten Spalte der Tabelle sind als Maß der Selektivität die Quotienten Pw/λ eingetragen. Diese Zahlen zeigen deutlich, daß der Semipermeabilitätsgrad einer Membran keineswegs in direkter Korrelation zu ihrer Wasserdichtheit steht: die Samenschale von *Aesculus*, die Wasser 16% schlechter leitet als die *Xanthium*-Testa, ist dieser in ihrer KCl-Permeabilität 10mal überlegen. — Bei weitem am vollkommensten ist die Elektrolyt-Semipermeabilität jedoch bei der Hülle des Weizenkorns ausgebildet, bei der allerdings auch die Wasserdurchlässigkeit den niedersten Absolutwert besitzt. Dieser erreicht hier bemerkenswerterweise schon die Größenordnung der Filtrationswiderstände lebender Protoplasten. HUBER und HÖFLER (1930) fanden beim Plasmahäutchen von *Salvinia* *Pw*-Werte von 0,55 μ/atm./min.

Für den Dichtheitsgrad von Samenschalen ist in erster Linie wohl die Gegenwart von cutinisierten Membranen und von Cuticularüberzügen verantwortlich. Dies läßt sich eindrucksvoll an der *Aesculus*-Testa zeigen, bei der nur die äußersten 3—4 Zellagen cutinisiert sind. Diese periphere Schicht macht etwa $^1/_{10}$ der Gesamtdicke aus. Wird sie abpräpariert, so steigt die Wasserpermeabilität der Schale um 280%! — Neben der Dicke der cutinisierten Schichten spielt aber offenbar auch die chemische Natur und die Dispersionsform der imprägnierenden Cutinart eine entscheidende Rolle. Diesen letzteren Faktoren ist es wohl zuzuschreiben, daß die zarte Hülle des Weizenkorns mit nur einer einzigen 4—5 μ dicken cutinisierten Doppelmembran (das innere Integument) einen fast 90mal höheren Elektrolytwiderstand besitzt als die mächtige *Aesculus*-Testa mit ihrer derbwandigen cutinisierten Außenschicht.

Andererseits läßt sich feststellen, daß Cutinisierung die Elektrolytpermeabilität in der Regel stärker beeinträchtigt als die Wasserdurchlässigkeit. Dies ist vermutlich der schon mehrfach erwähnten Erhöhung der negativen Eigenladung der Membran durch die Cutineinlagerung zuzuschreiben, deren Ventilwirkung sich Ionen gegenüber stärker auswirkt als gegenüber den Dipolen des Wassers.

Von anderen permeabilitätserniedrigenden Imprägnationsstoffen finden sich in Samenschalen häufig Gerbstoffe, Lipoide und Pektine. Ihre Wirkung läßt sich dadurch nachweisen, daß man die Membranen vor und nach Extraktion mit spezifischen Lösungsmitteln auf ihre Durchlässigkeit prüft. So fand DENNY (1917b) in Osmometerversuchen mit Samenschalen von *Arachis hypogaea*, daß Extraktion der Testalipoide mit kaltem Aceton die Geschwindigkeit des Wasserstroms um mehrere 100% erhöht. Dabei ist anscheinend die Semipermeabilität der Membran gegenüber dem Osmoticum (0,5 m NaCl) unbeeinträchtigt geblieben.

Bei manchen Samenschalen (z. B. *Amygdalus communis*) wird die Wasserpermeabilität schon durch Kochen in Wasser enorm erhöht (DENNY 1917b). Hier ist der Effekt offenbar auf die Entfernung von in heißem Wasser löslichen Pektinkörpern zurückzuführen. — In gerbstoffreichen Testen (*Aesculus!*) führt Extraktion mit Tanninlösungsmitteln (40% Aceton, KOH) zu einer ansehnlichen Steigerung der Elektrolytpermeabilität (BRAUNER 1930b). Eine Untersuchung der Wirkung dieser Behandlung auf die Wasserdurchlässigkeit steht noch aus.

In Ausnahmefällen scheinen selbst reine Cellulosemembranen bei besonders dichter Packung der Mikrofibrillen sehr undurchlässig zu werden. CAVAZZA (1950) hat einen solchen Fall in der Testa von „harten“ *Gleditschia*-Samen beobachtet, deren „Lichtlinie“ ihre geringe Permeabilität einer weitgehenden Verengerung der Interfibrillarräume verdankt.

Wie bei allen hydrophilen Membranen hängt der Permeabilitätsgrad der Samenschalen weitgehend von ihrem Hydratationszustand ab. Steht nun eine solche beschränkt durchlässige Membran in direktem Kontakt mit einer Lösung, so wird ihr Quellungsgrad herabgesetzt, und zwar um so mehr, je höher die Saugkraft und je geringer das Permeiervermögen des einwirkenden Osmotikums ist. Diese Entquellung äußert sich im Osmometerversuch darin, daß die Einströmungsgeschwindigkeit „E“ des Wassers nicht proportional mit dem osmotischen Wert der Füllung ansteigt. Dieses Phänomen hat schon DENNY (1917a) an Samenschalen von *Arachis* beobachtet und zwar sowohl bei Einschaltung der Membran zwischen Wasser und Rohrzuckerlösungen verschiedener Saugkraft, wie auch bei Anwendung eines konstanten Saugkraftgefälles, hergestellt durch verschiedene Konzentrationspaare. So erzeugt etwa der Gradient 10,3 atm. zwischen Zuckerlösung und destilliertem H_2O einen Wasserstrom von 20,5 E; dagegen wurde beim Gradienten 10,95 atm. zwischen 21,25 atm. und 10,30 atm. Zuckerlösung nur eine Geschwindigkeit von 10,8 E erreicht. — BRAUNER (1935) ermittelte in Osmometerversuchen mit *Aesculus*-Samenschalen den Einfluß der Zuckerkonzentration auf den Quellungsgrad und die Wasserpermeabilität der Membran (Tab. 3).

Tabelle 3. *Aesculus. Einfluß der Konzentration des Osmotikums auf Quellung und Pw.*

Saccharose-konzentration g-mol	Saugkraftgradient atm.	Quellung in % des Wasserwertes	Strömungsgeschwindigkeit μ/h	Pw μ/h/atm.
0,57	14,6/0,0	98,3	114	7,8
1,1	29,2/0,0	97,0	162	5,6
2,0	58,4/0,0	94,7	218	3,7

Der hohe Semipermeabilitätsgrad der Gramineen-Fruchthülle ist schon seit den Untersuchungen von A. J. BROWN (1907, 1909) bekannt. Hier sei nur an ein eindrucksvolles Experiment dieses Autors (1907) mit Körnern einer Gerstensorte, *Hordeum vulgare*, var. *coerulescens* erinnert, die in ihrer Aleuronschicht einen blauen Anthocyanfarbstoff enthalten. Werden solche Körner in verdünnte Schwefelsäure eingebracht, so nehmen sie daraus ansehnliche Mengen Wasser auf. Dabei behält jedoch der natürliche Indicator seine blaue Farbe bei und beweist damit die Impermeabilität der Hüllschicht für die Säure (in Schnitten färbt sich die anthocyanhaltige Zone bei Säurezusatz sofort rot!).

Durch Anwendung der empfindlicheren Leitfähigkeitsmethode konnte später GUREWITSCH (1929) Vergleichswerte für die Beweglichkeit verschiedener Ionen in der Hülle von *Triticum vulgare* gewinnen. Der Eigenwiderstand der abpräparierten Schale war in allen Fällen sehr hoch, bei einer Leitfläche von 1,9 mm^2 und einem spezifischen Widerstand der Versuchslösungen von 10 Ω betrug er stets einige zehntausend Ohm. Dabei zeigten sich jedoch zwischen verschiedenen Elektrolyten charakteristische Unterschiede, die Beziehungen zu den lyotropen Eigenschaften und vor allem der Wertigkeit ihrer Ionen erkennen ließen. In einer Anionenreihe von Ammoniumsalzen stieg der Membranwiderstand in folgendem Sinn an:

	$J' < NO_3' < Cl' < SO_4''$
$10^3\,\Omega$:	21 23 31 51

(Spezifischer Widerstand aller Lösungen: 10 Ω).

In einer entsprechenden Kationenreihe mit Cl′ als Anion zeigte sich dagegen nur zwischen den 1- und 2-wertigen Metallen ein signifikanter Unterschied:

	$Na^{\cdot}, K^{\cdot}, NH_4^{\cdot}$	$<$	$Mg^{\cdot\cdot}, Ca^{\cdot\cdot}$
$10^3\,\Omega$:	30—31		38—39

Eine weitere, für die Theorie der Ionenpermeation wichtige Beobachtung betrifft die Beziehung zwischen Membranleitfähigkeit und Konzentration der einwirkenden Lösungen. GUREWITSCH konnte zeigen, daß der Widerstand in der Membran bei steigender Leitfähigkeit der „freien" Lösung nach einem anfänglich annähernd proportionalen Abfall schließlich asymptotisch einem konstanten Minimalwert zustrebt. Dieses dem FICKschen Diffusionsgesetz widersprechende Verhalten wird durch die Annahme erklärt, daß durch den beschränkten Querschnitt der engen Membranporen nur eine limitierte Menge von Ionen zu wandern vermag. Ist diese Kapazität erreicht, so kann eine weitere Steigerung der Außenkonzentration die Stärke des Diffusionsstroms nicht mehr erhöhen. Ein zusätzlicher Hemmfaktor ist die schon früher diskutierte Verdichtung der Membran durch den osmotischen Entzug von Quellungswasser.

Die weitgehende Annäherung an den Zustand idealer Halbdurchlässigkeit begrenzt beim Weizenkorn die Möglichkeit einer differenzierteren Analyse der Ionenpermeation. Viel günstiger liegen die Verhältnisse bei der wesentlich leitfähigeren *Aesculus*-Testa. Hier zeigt sich bei einer Ausmessung von Konzentrationsreihen, daß die spezifische Leitfähigkeit der Membran $\Lambda = \Omega^{-1}/\text{cm}^2$ (Membran): Ω^{-1} (freie Lösung) in den verdünntesten Stufen am höchsten ist und dann steil bis in den Bereich mittlerer Konzentrationen ($\sim$ 0,25 n) abfällt. Im weiteren Verlauf bleibt Λ dann innerhalb eines weiten Konzentrationsgebietes fast unverändert, d. h. der Membranwiderstand nimmt hier proportional mit der Leitfähigkeit der Lösung ab. In diesem Bereich gilt also das FICKsche Gesetz. Erst in den höchsten Stufen treten schließlich neue Abweichungen auf. Diese Situation läßt sich folgendermaßen interpretieren: Kommt die ursprünglich elektrolytfreie Membran in Kontakt mit einer verdünnten Elektrolytlösung, so entsteht an den Porenwänden eine Adsorptionsschicht von Ionen, deren Ausbildung bei Steigerung der Konzentration bis zur Absättigung der verfügbaren Adsorptionsflächen fortschreitet. Während dieser Phase verengt sich der freie Diffusionsweg bis auf einen bestimmten Endwert, der dann in noch höheren Konzentrationen annähernd unverändert bleibt. Die dabei in Erscheinung tretende Verringerung der Ionenbeweglichkeit ist vermutlich der elektrostatischen Bindung eines Teils des Füllwassers durch die adsorbierte Ionenschicht zuzuschreiben. Der wesentliche Unterschied im Verhalten der *Triticum*- und der *Aesculus*-Testa würde demnach darin bestehen, daß in der dichteren Membran praktisch der ganze Poreninhalt den Feldkräften der Wandung unterliegt. Für die Richtigkeit dieser Vorstellung spricht, daß in der weiterporigen *Aesculus*-Testa das Permeiervermögen der Kationen gleicher Wertigkeit viel mehr von ihren lyotropen Eigenschaften bestimmt wird als bei *Triticum*. Wir finden hier im ganzen Konzentrationsbereich Permeationsreihen, die umgekehrt zum Wasserbindungsvermögen der Ionen ansteigen (BRAUNER 1930b):

—Permeiervermögen→

	$Li^{\cdot}$	$Na^{\cdot}$	$K^{\cdot}$	$NH_4^{\cdot}$
$\Lambda_{0,25n}$:	1,14	1,43	1,66	1,89

	$Mg^{\cdot\cdot}$	$Ca^{\cdot\cdot}$
$\Lambda_{0,25n}$:	0,77	0,94

←Hydratationskraft—

Im Gegensatz dazu scheinen in den engen Poren der Weizenhülle bereits die schwächst hydratisierbaren Kationen den Diffusionsweg maximal zu blockieren. — Da beide Membrantypen negative Eigenladung besitzen, äußert sich in ihnen die Wirkung der nicht direkt adsorbierten Anionen in prinzipiell anderer Weise.

Sofern diese nicht mit den Strukturelementen des Membrangefüges chemisch reagieren und auf diese Weise verdichtend oder peptisierend wirken. beeinflussen sie den Permeabilitätsgrad nur sekundär, als Konkurrenten um das Hydratationswasser. Da das verfügbare Wasser mit fallendem Porenareal in steigendem Maß zum limitierenden Faktor der Durchlässigkeit wird, tritt der Anioneneffekt bei *Triticum* sehr auffällig in Erscheinung. Bei der beträchtlich permeableren *Aesculus*-Testa dagegen wird das Durchtrittsvermögen eines Elektrolyten weit mehr von seinem Kation als von seinem Anion bestimmt (BRAUNER 1930b):

	K_2SO_4	KCl	KNO_3	KJ
$\Delta_{0,25n}$:	1,52	1,66	1,73	1,52

2. Permeabilität für lipophile Stoffe.

Schon A. J. BROWN (1907) war es aufgefallen, daß die für Ionen fast undurchlässige Hülle des Getreidekorns in LUGOLscher Lösung gebotenes Jod leicht passieren läßt. Unerwartet gut permeieren noch zwei weitere anorganische Stoffe: Sublimat ($HgCl_2$) und Osmiumsäure (OsO_4) (SCHROEDER 1911 und GUREWITSCH 1929). In freier Lösung diffundieren alle diese Substanzen beträchtlich langsamer als Alkalisalze, für die die Gramineen-Testa, wie erwähnt, praktisch impermeabel ist. Es lag daher nahe, die Sonderstellung der permeierenden Stoffe mit ihrer beträchtlichen Lipoidlöslichkeit in Beziehung zu bringen. In der kontrollierenden Schicht der Kornhülle, die ihre selektiven Eigenschaften der Imprägnierung mit lipoidem Cutin verdankt, steht lipophilen Substanzen als zusätzlicher Permeationsweg die Cutinfüllung der Interfibrillarräume zur Verfügung.

Sehr interessant ist in diesem Zusammenhang das Verhalten einiger organischer Farbstoffe (GUREWITSCH 1929): Aus wäßrigen Lösungen vermögen nur besonders kleinmolekulare basische Verbindungen zu permeieren, die sich durch große Lipoidlöslichkeit auszeichnen (Chrysoidin, Anilingelb, o-Nitranilin). Aber bereits das etwas umfangreichere Methylenblaumolekül kann trotz seiner ebenfalls guten Lipoidlöslichkeit die Weizentesta nicht mehr passieren. Werden die Farbstoffe dagegen in 50% Alkohol gelöst geboten, so permeiert nicht nur Methylenblau, sondern selbst das erheblich voluminösere Anilinblau. Das gleiche Verhalten hat SCHROEDER (1911, 1922) auch bei zwei anorganischen Salzen beobachtet, $AgNO_3$ und NaCl, die in wäßriger Lösung die Membran kaum zu durchdringen vermögen. Dieses Phänomen läßt sich vermutlich dadurch erklären, daß der Alkohol im Cutin löslich ist und durch dessen Verquellung die Interfibrillarräume vergrößert. Auf diese Weise wird für alkohol- und lipoidlösliche Stoffe nunmehr ein größerer Diffusionsquerschnitt verfügbar.

3. Polare Permeabilität.

Daß tote Samenschalen nach Richtungen verschieden durchlässig sein können, ist zuerst von DENNY (1917a) beobachtet worden: Untersucht man das Verhalten von *Arachis*-Samenschalen im Osmometer, so zeigt sich, daß die Geschwindigkeit des Wassereinstroms weitgehend von der Orientierung der Membran abhängt. Im gleichen Saugkraftgefälle bewegt sich das Wasser stets erheblich schneller von der Außen- nach der Innenseite der Testa als umgekehrt. Die Differenz beträgt im Mittel $+52\%$. DENNY führt diesen Effekt auf eine asymmetrische Schichtung der Samenschale zurück, die er als eine „Doppelmembran" im Sinne HAMBURGERS (1908) auffaßt.

Ein analoges Phänomen konnte BRAUNER (1930a, b) an der Samenschale von *Aesculus* beobachten. Läßt man Wasser unter hydrostatischem Druck durch die Membran filtrieren, so ist die Strömungsgeschwindigkeit in der „normalen"

Richtung, von außen nach innen, um 52—66% größer als im inversen Sinn. Eine Analyse dieser auffälligen Erscheinung führte zum Ergebnis, daß die Asymmetrie des Filtrationswiderstands durch elektroosmotische Vorgänge verursacht wird. Aus potentiometrischen Messungen ging nämlich hervor, daß bei symmetrischer Ableitung mit Leitungswasser die Außenwand der Testa mehrere Millivolt positiv gegenüber der Innenwand geladen ist. Da nun das Füllwasser in den Poren der acidoiden Membran sich elektropositiv verhalten muß, wird eine Verschiebung der Wasserfäden nach dem negativen Innenpol erleichtert und nach dem positiven Außenpol erschwert sein. Die Potentialdifferenz zwischen den beiden Flanken der Testa kommt dadurch zustande, daß ihr Gewebe natürlicherweise einen Vorrat an diffusiblen Elektrolyten enthält, und daß ihre Außenhälfte wesentlich dichter ist als die Innenhälfte. Nun beeinträchtigt in acidoiden Membranen zunehmende Verdichtung die Beweglichkeit der Anionen weit mehr als die der Kationen. Befindet sich die Samenschale in Wasser, so werden sich demnach nach beiden Seiten hin Diffusionspotentiale verschiedener Größe entwickeln.

Für die Richtigkeit dieser Vorstellung spricht die folgende Erfahrung: Durch schrittweise Erhöhung der Ionenkonzentration im Medium läßt sich die Potentialbildung an den Membranflächen reduzieren und schließlich aufheben. Führt man nun Filtrationsversuche mit Lösungen eines geeigneten Elektrolyten, z. B. K_2SO_4 aus, so verschwindet bei Steigerung der Konzentration tatsächlich auch der polare Permeabilitätsunterschied. Die Strömungsgeschwindigkeit in „inverser“ Richtung wird zunächst erheblich über den Wasserwert erhöht, die Bewegung in „normaler“ Richtung dagegen etwas verlangsamt. Bei $^1/_8$ n treffen sich die beiden Kurven. Wird die Konzentration schließlich noch weiter gesteigert, so nimmt die Strömungsgeschwindigkeit infolge des wachsenden osmotischen Widerstandes in beiden Richtungen gleichmäßig ab (BRAUNER 1930b).

Tabelle 4. *Aesculus-Testa. Filtrationsgeschwindigkeit „v“ bei 70 cm Hg Überdruck in μ/10′. Temperatur 25° C.*

K_2SO_4-konz.	dest. H_2O	n/100	n/50	n/16	n/8	n/4
„v“ normal	6,30	—	—	6,20	5,80	4,70
„v“ invers	4,15	5,26	5,70	6,05	5,80	4,80

Die Erscheinung der polaren Wasserpermeabilität ist in diesem Fall also die Folge einer gerichteten Triebkraft, deren Arbeitsfähigkeit von der Aufrechterhaltung des notwendigen Ionenkonzentrationsgefälles abhängt.

Nun gibt es aber auch Bedingungen, unter denen an asymmetrischen Membranen eine rein passive Ventilwirkung in Erscheinung tritt. Schaltet man eine *Aesculus*-Testa als Diaphragma in einen elektrolytischen Stromkreis ein, so hängt der Gleichstromwiderstand der Membran von der Richtung des Stromes ab. Die Samenschale erscheint in der gleichen Lösung etwa doppelt so leitfähig, wenn ihre Außenseite der positiven Elektrode zugekehrt ist wie bei umgekehrter Polung (BRAUNER 1930b). Dieser Effekt läßt sich folgendermaßen verstehen: Infolge der beschränkten Anionenpermeabilität der Membran ruft der Stromdurchgang an deren Anodenseite Alkalisierung hervor und säuert die Kathodenseite an (BETHE und TOROPOFF 1914, 1915). Die Schaffung einer solchen Säure/Alkalikette produziert ein Gegenpotential, das den Gleichstromwiderstand des Systems erhöht. Da die Außenseite der Testa weit selektiver ist als die Innenseite, muß hier jede Art von Neutralitätsstörung ein höheres Maß erreichen. Nun ist unter den gegebenen Verhältnissen die Wirkung des Gegenpotentials bei Ansäuerung größer als bei Alkalisierung. Daher wird die Widerstandserhöhung bei derjenigen Orientierung der Membran größer sein, bei der ihre Außenflanke der Kathode zugekehrt ist und damit die Säurebildung begünstigt.

Diese Leitfähigkeitsasymmetrie geschichteter Membranen bedingt in Wechselstromkreisen einen merklichen Gleichrichtereffekt, wie METZNER (1930) an verschiedenen Samen- und Fruchtschalen zeigen konnte.

D. Verholzte Membranen.

Einlagerung von Lignin in die Interfibrillarräume beeinträchtigt die Permeabilitätseigenschaften von Zellwänden in ganz anderer Weise als Cutinisierung. Wie schon in einem früheren Abschnitt (S. 341) ausgeführt worden ist, setzt die teilweise Erfüllung des submikroskopischen Capillarsystems mit der wenig hydrophilen Ligninsubstanz die Quellungskapazität der Wand in Wasser merklich herab. Die Permeabilität wird dadurch jedoch unerwartet wenig beeinträchtigt. JACCARD und FREY (1928) prüften gelegentlich einer Studie der physiologischen Eigenschaften von Zug- und Druckholz auch die Rohrzuckerpermeabilität 100 μ dicker Tangentialschnitte von *Populus*- und *Pseudotsuga*-Zweigen. Obwohl solche Holzscheibchen als Membranen in Diffusionshülsen einen starken osmotischen Wasserstrom verursachen, lassen sie doch erhebliche Saccharosemengen passieren. Dabei erweist sich bei *Populus* das Druckholz noch permeabler als das Zugholz.

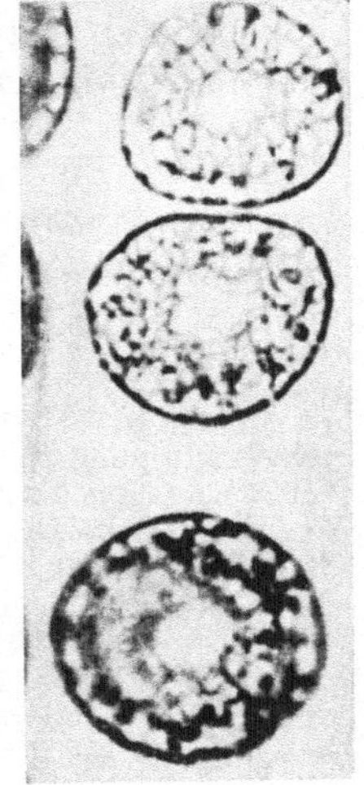

Abb. 11. Hoftüpfel von *Pinus* nach Durchsaugung einer kolloidalen Goldlösung von 1440 Å Teilchengröße. Die Goldpartikeln sind durch die Poren des Margo durchgetreten und haben sich auf der Innenseite der Tüpfelwand festgesetzt. Vergr. 1500 ×. (Nach FRENZEL.)

FRENZEL (1929) untersuchte die Durchlässigkeit von *Ginkgo*- und *Pinus*-Holz für verschiedene Farbstoffe und Kolloide. Läßt man die Testlösungen in beblätterten Sprossen mit dem Transpirationsstrom oder in Zweigstücken durch Pumpensaugung hochsteigen, so ergibt sich, daß saure Farbstoffe selbst von erheblichem Molekularvolumen (Eosin!) viele Tracheidengrenzen durchqueren können (basische Farbstoffe lassen sich nach dieser Methode schlecht prüfen, weil sie schon an den Kontaktflächen adsorbiert werden). Sogar verdünnte Tusche dringt über 10 Tracheidenlängen vor. Dabei werden die Schließhäute der Hoftüpfel als Diffusionsweg deutlich erkennbar. Genauere Maße ließen sich durch Filtration von kolloidalem Gold bekannter Partikelgröße gewinnen. Es zeigte sich, daß tüpfelfreie Membranen Teilchen bis zu einem Durchmesser von etwa 90 Å eindringen lassen, was ungefähr den Dimensionen der Interfibrillarräume entspricht. Dagegen können durch die Hoftüpfel-Schließhäute selbst noch 1440 Å große Goldteilchen passieren (Abb. 11). Dieser Wert steht in befriedigender Übereinstimmung mit neueren elektronenmikroskopischen Messungen der Porendimensionen bei ähnlichen Objekten.

ULLRICH (1953) untersuchte die Frage, ob bei verholzten Membranen das Ultrafilterprinzip auch für die Permeation lipophiler Stoffe gilt. Als Versuchsobjekte dienten tangential geschnittene dünne Furniere von *Fagus silvatica* und *Abies alba*, als Testsubstanzen hauptsächlich eine Reihe einwertiger aliphatischer Alkohole von zunehmender Kettenlänge (C_1—C_5). Das Eindringungsvermögen in die *Mikrofibrillen* wurde aus dem longimetrisch verfolgten Verlauf der Quellung erschlossen. Am Splintholz von *Fagus silvatica* fand ULLRICH die folgenden Endwerte für die tangentiale Dilatation:

Tabelle 5. *Quellung von Buchenholz in Alkoholen verschiedener Kettenlänge (C_1—C_5).*

Quellmittel	H_2O	C_1	C_2	C_3	C_4	C_5
Maximale Verlängerung in %	10,5	10,0	8,1	7,1	4,8	0,7

Die Maximalwerte der Volumzunahme (und auch deren Geschwindigkeit!) fallen also bei steigender Molekülgröße des Quellungsmittels stetig ab. Noch höhere Homologe verursachen schließlich überhaupt keine Quellung mehr. Interessant ist ferner, daß bei den noch passierenden Alkoholen die sperrigere Isoform schwerer eindringt als die Normalform. Diese Befunde sind von erheblicher Tragweite. Sie zeigen, daß in einer verhältnismäßig starren hydrophilen Membran das Permeiervermögen ausgesprochen lipophiler Moleküle ebenso von deren Dimensionen bestimmt wird wie bei hydrophilen Körpern. Die scheinbaren Abweichungen vom Ultrafilterprinzip, die unpolare Moleküle in den engeren, aber dilatationsfähigeren Poren lipoider Membranen zeigen, verschwinden hier vollkommen. Es ist schließlich noch bemerkenswert, daß bei der Holzmembran die Grenze des Eindringungsvermögens in die Mikrofibrillen beim Hexanol liegt, dessen Kettenlänge, nach RUHLAND und HEILMANN (1951) ~7,7 Å, bereits dem Durchmesser der Intermicellarräume nahekommt.

Literatur.

ARENS, K.: Physiologische Untersuchungen an Plasmopara viticola, unter besonderer Berücksichtigung der Infektionsbedingungen. Jb. wiss. Bot. **70**, 93—157 (1929). — ARZT, TH.: Untersuchungen über das Vorkommen einer Kutikula in den Blättern dikotyler Pflanzen. Ber. dtsch. bot. Ges. **51**, 470—500 (1933).

BENNET-CLARK, T. A., and D. BEXON: Water relations of plant cells. IV. Diffusion effects observed in plasmolysed tissues. New Phytologist **45**, 5—17 (1946). — BETHE, A., u. TH. TOROPOFF: Über elektrische Vorgänge an Diaphragmen. I. Die Neutralitätsstörung. Z. physik. Chem. **88**, 686 (1914). — Über elektrische Vorgänge an Diaphragmen. II. Die Abhängigkeit der Größe und Richtung der Konzentrationsveränderungen und der Wasserbewegung von der H-Ionenkonzentration. Z. physik. Chem. **89**, 597 (1915). — BEUTNER, R.: Die Entstehung elektrischer Ströme in lebenden Geweben usw. Stuttgart: Ferdinand Enke 1920. — BIEBL, R.: Zellwandpermeabilität einiger Moose. Protoplasma (Wien) **44**, 73—88 (1954). — BRAUNER, L.: Untersuchungen über das geoelektrische Phänomen. II. Membranstruktur und geoelektrischer Effekt. Jb. wiss. Bot. **68**, 711—770 (1928). — Über polare Permeabilität. Ber. dtsch. bot. Ges. **48**, 109—118 (1930a). — Untersuchungen über die Elektrolyt-Permeabilität und Quellung einer leblosen natürlichen Membran. Jb. wiss. Bot. **73**, 513—632 (1930b). — Über den Einfluß der Saugspannung auf die Wasserpermeabilität toter und lebender Gewebe. Protoplasma **22**, 539—552 (1935). — BRAUNER, L., and M. BRAUNER: Studies in the relations between water permeability and electric charge in membrane models and in living plant cells. Rev. Fac. Sci. Univ. Istanbul, Ser. B **8**, 264—310 (1943). — BRILLIANT, W.: Les formes de la plasmolyse produit par des solutions concentrées de sucres et de sels dans les cellules de Mnium et de Catharinea. C. r. Acad. Sci. USSR. **1927**, 155—160. — BROWN, A. J.: On the existence of a semi-permeable membrane enclosing the seeds of some of the Gramineae. Ann. of Bot. **21**, 79—87 (1907). — The selective permeability of the coverings of the seeds of Hordeum vulgare. Proc. Roy. Soc. Lond., Ser. B **81**, 82—93 (1909).

CASARI, K.: Über den Plasmolytikumwechsel-Effekt. Protoplasma (Wien) **42**, 427—447 (1953). — CAVAZZA, L.: Recherches sur l'imperméabilité des graines dûres chez les légumineuses. Bull. Soc. Bot. Suisse **60**, 596—610 (1950).

DELLINGSHAUSEN, M. v.: Untersuchungen über die Wechselbeziehungen zwischen Quellwirkung und Permeiervermögen der Elektrolyte. Planta (Berl.) **21**, 51—97 (1933). — DENNY, F. E.: Permeability of certain plant membranes to water. Bot. Gaz. **63**, 373—397 (1917a). — Permeability of membranes as related to their composition. Bot. Gaz. **63**, 468—485 (1917b).

EICKE, R.: Beitrag zur Frage des Hoftüpfelbaues der Koniferen. Ber. dtsch. bot. Ges. **67**, 213—217 (1954).

FITTING, H.: Untersuchungen über isotonische Koeffizienten etc. Jb. wiss. Bot. **57**, 553—612 (1917). — FÖRSTER, K.: Quellung und Permeabilität der Zellwand von Rhizoclonium. Planta (Berl.) **20**, 476—505 (1933). — FRENZEL, P.: Über die Porengrößen einiger pflanzlicher Zellmembranen. Planta (Berl.) **8**, 642—665 (1929). — FREY, A.: Über die Intermicellar-Räume der Zellmembranen. Ber. dtsch. bot. Ges. **46**, 444—456 (1928). — FREY-

WYSSLING, A.: Die Stoffausscheidung der höheren Pflanzen. Berlin: Springer 1935. — Formgestaltung im sublichtmikroskopischen Gebiet. Ber. schweiz. bot. Ges. **59**, 5—22 (1949). — Submicroscopic morphology of protoplasm, 2. Aufl. New York u. Amsterdam: Elsevier Publ. Co. 1953. — FREY-WYSSLING, A., u. H. H. BOSSHARD: Über den Feinbau der Schließhäute in Hoftüpfeln. Holz **11**, 417—420 (1953). — FREY-WYSSLING, A., u. K. MÜHLETHALER: Der submikroskopische Feinbau von Chitinzellwänden. Vjschr. naturforsch. Ges. Zürich **95**, 45—52 (1950). — FREY-WYSSLING, A., K. MÜHLETHALER u. R. W. G. WYCKOFF: Mikrofibrillenbau der pflanzlichen Zellwände. Experientia (Basel) **4**, 475—476 (1948).

GUREWITSCH, A.: Untersuchungen über die Permeabilität der Hülle des Weizenkorns. Jb. wiss. Bot. **70**, 657—706 (1929).

HAAN, Iz. DE: Protoplasmaquellung und Wasserpermeabilität. Rec. Trav. bot. néerl. **30**, 234—335 (1933). — HABERLANDT, G.: Physiologische Pflanzenanatomie, 5. Aufl., S. 101. Leipzig: Wilhelm Engelmann 1918. — HÄRTEL, O.: Über pflanzliche Kutikulartranspiration und ihre Beziehungen zur Membranquellbarkeit. Sitzgsber. Akad. Wiss. (Wien), Math.-naturwiss. Kl. I **156**, 57—86 (1947). — Wirkungen von Ionen auf die Wasserdurchlässigkeit des primären und sekundären Hautgewebes pflanzlicher Organe. Protoplasma (Wien) **39**, 364—385 (1950). — Ionenwirkung auf die Kutikulartranspiration von Blättern. Protoplasma (Wien) **40**, 107—136 (1951). — Färbungsstudien an der pflanzlichen Kutikula. Protoplasma (Wien) **41**, 1—14 (1952). — HÄUSERMANN, E.: Über die Benetzungsgröße der Mesophyllinterzellularen. Ber. schweiz. bot. Ges. **54**, 541—578 (1944). — HAMBURGER, H. J.: Permeabilität von Membranen in zwei entgegengesetzten Richtungen. Biochem. Z. **11**, 443—480 (1908). — HASMAN, M.: A study of the shape of the determinant curve in measurements of the suction potential in plant tissues. Rev. Fac. Sci. Univ. Istanbul, Sér. B **8**, 167—200 (1943). — HÖFLER, K.: Über Eintritts- und Rückgangsgeschwindigkeit der Plasmolyse etc. Jb. wiss. Bot. **73**, 300—350 (1930). — HOFFMANN, C.: Beiträge zur Physiologie der Meeresalgen. I. Permeabilitätsuntersuchungen an der Grünalge *Chaetomorpha aerea*. Kiel. Meeresforsch. **1**, 125—166 (1936). — HUBER, B., u. K. HÖFLER: Die Wasserpermeabilität des Protoplasmas. Jb. wiss. Bot. **73**, 351—511 (1930). — HUBER, B., u. R. W. KOLBE: Elektronenmikroskopische Untersuchungen an Siebröhren. Sv. bot. Tidskr. **42**, 364—371 (1948).

JACCARD, P., u. A. FREY: Quellung, Permeabilität und Filtrationswiderstand des Zug- und Druckholzes von Laub- und Nadelbäumen. Jb. wiss. Bot. **69**, 549—571 (1928).

KAMP, H.: Untersuchungen über Kutikularbau und kutikuläre Transpiration von Blättern. Jb. wiss. Bot. **72**, 403—465 (1930). — KORNMANN, P.: Osmometer aus lebenden *Valonia*-zellen und ihre Verwendung zu Permeabilitätsbestimmungen. Protoplasma (Berl.) **21**, 340—350 (1934). — KRESSIN, G.: Beiträge zur vergleichenden Protoplasmatik der Mooszelle. Diss. Greifswald 1935, 64 S.

LAMBERTZ, P.: Untersuchungen über das Vorkommen von Plasmodesmen in den Epidermisaußenwänden. Planta (Berl.) **44**, 147—190 (1954). — LIESE, W., and M. HARTMANN-FAHNENBROCK: Elektronenmikroskopische Untersuchungen über die Hoftüpfel der Nadelhölzer. Biochim. et Biophysica Acta **11**, 190 (1953).

METZNER, P.: Über polare Leitfähigkeit lebender und toter Membranen. Ber. dtsch. bot. Ges. **48**, 207—211 (1930).

PRESTON, R. D.: Biological units of cellulose structure. Symposia Soc. Exper. Biol. **6**, 348—357 (1952). — PRINGSHEIM, E. G.: Untersuchungen über Turgordehnung und Membranbeschaffenheit. Jb. wiss. Bot. **74**, 749—796 (1931).

RENNER, O.: Theoretisches und Experimentelles zur Kohäsionstheorie der Wasserbewegung. Jb. wiss. Bot. **56**, 617—667 (1915). — Die Porenweite der Zellhäute und ihre Beziehung zum Saftsteigen. Ber. dtsch. bot Ges. **43**, 207—211 (1925). — Zur Kenntnis des Wasserhaushalts javanischer Kleinepiphyten. Planta (Berl.) **18**, 215—287 (1932). — RUHLAND, W., u. U. HEILMANN: Über die Permeabilität von Beggiatoa mirabilis für Anelektrolyte bei Narkose mit den homologen Alkoholen C_1—C_9, als Beitrag zur Ultrafiltertheorie. Planta (Berl.) **39**, 91—120 (1951). — RUHLAND, W., u. C. HOFFMANN: Die Permeabilität von Beggiatoa mirabilis. Planta (Berl.) **1**, 1—83 (1925).

SCHROEDER, H.: Über die selektiv permeable Hülle des Weizenkorns. Flora (Jena) N. F. **2**, 186—208 (1911). — Über die Semipermeabilität von Zellwänden. Biol. Zbl. **42**, 172—188 (1922). — SCHUMACHER, W.: Über plasmodesmenartige Strukturen in den Epidermisaußenwänden. Jb. wiss. Bot. **90**, 530—545 (1942). — SKENE, M.: The permeability of the cellulose cell wall. Ann. of Bot., N. S. **7**, 261—273 (1943). — STRUGGER, S.: Die lumineszenzmikroskopische Analyse des Transpirationsstromes in Parenchymen. I. Flora (Jena) N. F. **33**, 56—68 (1938). — Praktikum der Zell- und Gewebephysiologie der Pflanze. Berlin-Göttingen-Heidelberg: Springer 1949.

ULLRICH, H.: Über das Quellungsverhalten von Holz als Substanz mit Ultrafilterstruktur in Abhängigkeit von Größe und Bau der Quellungsmittelmoleküle. Planta (Berl.) **42**, 129 bis 139 (1953).

Permeability in relation to respiration.

By

Paul J. Kramer.

Introduction.

In recent years it has become increasingly difficult to define or describe permeability with any degree of precision. In general terms permeability is a property of membranes and refers to the extent to which various substances can pass through them. In order to make comparisons of permeability it is necessary also to describe the forces causing movement across the membrane. BROOKS and BROOKS (1941) defined permeability as the rate of movement of a substance through a permeable layer under a given driving force. This, as they state, involves two concepts which should be distinguished. One is permeability in its narrowest sense, a property of a membrane, and the other is the driving force which causes movement across the membrane. In the simpler instances of movement by diffusion along activity gradients the driving force may be quite distinct from any property of the membrane, but where "active transport" or accumulation is involved it may be related to certain properties of the membrane itself. TEORELL (1949, 1953) also distinguishes between permeability in its narrow sense as a property of the membrane, and the forces causing movement across the membrane. He reminds his readers that failure of a substance to enter a cell can result either from inability to pass through the membrane (impermeability) or from lack of a driving force to cause movement across the membrane.

There has been a marked change in viewpoint recently concerning the nature and role of cell membranes. In the early discussions of permeability cell membranes often were regarded as static structures which could be treated in the same manner as nonliving membranes of porcelain or collodion. Thus permeability was treated as a physical phenomenon which could be described by relatively simple equations such as FICK's law.

Although this view explains many permeability phenomena fairly adequately it does not explain some of the most important examples of the movement of materials across cell membranes. In the first place it has become increasingly clear that protoplasmic membranes are not static structures, but often vary considerably over a period of time in permeability to a particular substance. Permeability is affected by many environmental factors such as gases, light, and various chemicals, and treatments which affect respiration are likely also to affect permeability.

There also has been a change in view concerning the nature of the forces causing movement across membranes, which further involves respiration in any consideration of permeability. Research during the past three decades has shown that a considerable amount of movement occurs against concentration or activity gradients and cannot be explained in terms of diffusion along gradients of decreasing free energy. As a result, interest has been shifting from study of the nature and extent of permeability of cell membranes to study of the nature of the forces causing movement across membranes, especially movement against concentration gradients. Thus much present day research deals with active transport and accumulation of substances by cells rather than with cell permeability in the classical sense. This shift in viewpoint has been discussed by STEINBACH (1951) who also points out the increasing difficulty in defining permeability caused by this new concept of the exchange of materials between cells and their environment. It often is difficult to know whether we are dealing with permeability, active transport and accumulation, or a combination of the two processes. Permeability in the narrow sense sometimes is termed "passive permeability" while if both the properties of the membrane and active transport are involved it is termed "physiological permeability".

It is because of the increased complexity of present day theories that a chapter dealing with respiration and permeability is desirable. According to STERN *et al.* (1949) "the dominating factor in cell permeability is not any inherent and constant physical property of the cell wall, like porosity or a lipoid-sieve structure, but a mechanism dependent on the supply of energy". In this chapter we will deal with both the relation between respiration and the structure and consequent permeability of cell membranes, and with the relation of respiration to the driving forces causing movement across the membranes. The relation of respiration to the accumulation of salts and the uptake of water is discussed in other chapters, hence this chapter will deal with permeability and respiration in a general manner, but some repetition seems to be unavoidable.

Cell membranes in relation to permeability.

Modern discussions of permeability assume that protoplasmic membranes are complex systems of oriented molecules maintained in a rather definite structural arrangement. Maintenance of such a complex structure requires the constant expenditure of energy supplied by respiration. When protoplasm is killed it immediately loses its differential permeability because the complex structure collapses. Even treatments which merely modify respiration such as the addition of respiration inhibitors often produce marked changes in permeability of plant cells to water and to solutes. The two principle theories of permeability, the sieve or pore theory and the lipoid-solubility theory are discussed in detail elsewhere in this volume. Another interesting concept of permeability is that of USSING (1953) who attempts to distinguish between diffusion permeability and filtration permeability.

USSING (1953) and his colleagues have attempted to study the nature of permeability in frog skin and other animal membranes by comparing their permeability to heavy water (D_2O) as a solute in ordinary water with permeability to water as a solvent in osmotic systems. They assume that if water forms a continuous phase through the pores of a membrane water movement under pressure or along an osmotic gradient occurs by flow through the larger pores, but as pore size is reduced toward molecular dimensions resistance to flow increases and more and more water moves by diffusion as individual molecules through the smaller pores. When heavy water is supplied to one side of a membrane it must move by diffusion as individual molecules. USSING therefore distinguishes between 1. diffusion permeability as measured by diffusion of heavy water along a concentration gradient and 2. filtration permeability as measured by the rate of movement of water along hydrostatic or osmotic gradients. Movement by diffusion should be controlled by the total area of pores, but is little affected by their size or shape so long as they are large enough to permit water molecules to pass through. Filtration or laminar flow will decrease with decrease in size of pores even though increased number of pores results in maintenance of the same total pore surface. If filtration and diffusion permeability are equal then water movement across a membrane is occurring by diffusion through very small pores, but if filtration permeability is much larger than diffusion permeability then mass flow is occurring through large pores. According to this idea comparison of diffusion and filtration permeability ought to indicate something about the pore size in a membrane.

Several investigators have found the filtration permeability of frog skin greater than the diffusion permeability. KOFOED-JOHNSON and USSING (1953) observed that neurohypophyseal hormone caused a large increase in filtration permeability of toad skin, but no change in diffusion permeability. They suggest that the hormone causes an increase in size of individual pores without any change in total pore area. PRESCOTT and ZEUTHEN (1953) found that the filtration permeability was higher than the diffusion permeability in the eggs of several fish and amphibians studied by them and that the difference was greater in ovarian eggs than in fertilized eggs. They therefore concluded that ovarian eggs have fewer but larger pores than fertilized eggs. GARBY and LINDERHOLM (1953) decided that permeability to water and to Cl ions are independent of each other in frog skin because the diuretic, amonophylline, has no effect on permeability to heavy water, but increases permeability to Cl ions fourfold. They suggest that this increase occurs because increase in pore size opens up additional paths to the hydrated Cl ions which were previously too small for them. It was assumed that Cl ions move across frog skin by diffusion, as claimed by USSING (1953). JACOBS (1952) calculated that the filtration permeability of *Tollypellopsis* is about 8 times the diffusion permeability.

These observations support the concept of pore permeability. They also indicate that even the pore size of a sieve-type membrane is not fixed, but can be varied by addition of drugs and as a result of fertilization. No doubt many factors which affect cell metabolism result in both structural and chemical changes in cell membranes. If the proportion of lipid at the cell surface were changed, for example, this would almost certainly affect permeability, especially to nonpolar compounds.

Although the lipoid-sieve theory of permeability is the most commonly accepted theory among physiologists at this time (USSING 1954) it must be regarded as a workable theory rather than a proven fact. DAVSON and DANIELLI (1952) showed that theoretically the permeability data of COLLANDER could be explained as well by means of a homogenous, nonporous membrane in which both ions and nonpolar molecules dissolve as by a membrane in which ions and polar molecules pass through pores and nonpolar molecules dissolve. DANIELLI (1954) regards the plasma membranes of most cells as basically lipoid layers about 50 Ångström units in thickness, with protein layers on both sides of them. He suggests that the pores of COLLANDER and USSING are channels formed by polypeptide chains of the protein lattice extending through the lipoid layers. These chains would constitute areas of high permeability to polar compounds as compared with the remainder of the membrane. DANIELLI also suggests that the protein lattice of the cell membrane might contract and expand, thereby varying its permeability.

COLLANDER (1949) and WARTIOVAARA (1949, 1950) also warn against accepting a sieve theory of permeability too literally. COLLANDER states that because the plasma membrane of a cell exerts a sieve effect it should not be assumed that it must contain water-filled pores. WARTIOVAARA suggests that the apparent sieve action of the plasma membrane of *Nitella* really depends on orientation of the molecules passing through it rather than on the existence of actual pores. On the other hand USSING seems to accept the pore theory of permeability rather literally and to regard certain multicellular animal membranes as containing water-filled pores of varying size. OSTERHOUT (1952) thinks that the protoplasmic membranes cannot be regarded as a system of electrically charged pores because there is no consistent difference in behavior of anions and cations. In *Valonia*, for example, K ions have a higher and Na ions a lower mobility than Cl ions, and in *Nitella* K has a higher mobility than OH ions which in turn have a higher mobility than Na ions. OSTERHOUT thinks that if electrical charge were an important factor then all cations should behave differently from all anions.

In contrast to the movement of materials by diffusion across relatively permeable membranes which has been discussed, there is a large amount of "active transport" of physiologically important substances across relatively impermeable membranes against concentration or activity gradients. This active transport is dependent largely on metabolic energy released by respiration. The primary problem in active transport is not that of membrane permeability, but that of a mechanism by which substances are transported across membranes through which they cannot pass, or pass very slowly, by diffusion. Active transport must be fully as dependent on certain properties of protoplasmic membranes as diffusion through them but it is even more dependent on respiration because of its energy requirement and the probability that transport is related to the synthesis of carrier compounds. It will be discussed in the section dealing with the nature of the forces causing movement across membranes. It also is discussed in the chapter on uptake of salts.

Forces causing movement across membranes.

Attempts to classify the driving forces responsible for movement of substances across membranes have proven difficult because too little is known about the various mechanisms to place some of them in their proper categories. The writer would prefer to separate the various kinds of forces into those responsible for passive movement, controlled by the principles governing diffusion in physical systems, and those responsible for active transport against activity gradients, which require the expenditure of energy released by respiration. Unfortunately, although certain examples of these two types of movement are easily distinguished, it is often difficult to decide whether certain phenomena should be classified as passive movement or active transport. This has been recognized by several writers, including DANIELLI (1954), ROSENBERG (1954), and STEINBACH (1951). The following classification of transport mechanisms attempts to place them in three categories.

1. Passive movement, mostly by diffusion through pores or in solution in membranes. Movement is from regions of higher to regions of lower concentration, activity, or chemical potential of the diffusing substance, and it is controlled by the laws which govern diffusion in physical systems. Most or perhaps all water movement occurs in this way, also a limited amount of movement of solutes.

2. Active transport by systems not requiring the direct expenditure of metabolic energy. For example the formation of a complex between the substance being transported and a carrier molecule might be a physical process. If the complex is formed at the outer and destroyed at the inner surface of the membrane, then

the carrier probably moves by diffusion along a concentration gradient although the substance carried is moved against a concentration gradient. Another example difficult to classify is "facilitated diffusion" which refers to diffusion through restricted areas of unusually high permeability in a cell membrane (DANIELLI 1954). The actual movement may be by diffusion, although metabolic energy may be expended in maintaining the membrane structure favorable to high permeability. Most theories of transport by carrier mechanisms fall into this category. It is debatable whether movement of substances across membranes by some of these mechanisms should be regarded as active or passive movement, but the writer prefers to classify any movement against an activity gradient as active transport, whether metabolic energy is expended directly or indirectly in causing the movement.

3. Active transport requiring the direct expenditure of energy released in respiration. Not only is energy required for the formation of the carrier, if such exists, but it is also required for the formation of and/or the breakdown of the complex between the carrier being transported, as well as for the movement of the complex. LUNDEGÅRDH's anion respiration theory and the redox pump mechanism proposed by CONWAY (1953) seem to fit into this category.

Respiration, permeability and active transport.

Discussion of passive movement supplies a good opportunity for further consideration of the relation of respiration to permeability, because the driving force is the kinetic energy of the molecules which are moving and is therefore entirely independent of respiration, but the rate of transfer depends on the permeability of the membrane. Changes in respiration doubtless produce changes in the structure and possibly in the composition of the cytoplasmic membranes which affect their permeability. According to DANIELLI (1954) expansion and contraction of the lattice structure of the membrane might produce important changes in permeability, and such changes in structure certainly require metabolic energy. DANIELLI's reasoning concerning the nature of the forces involved in binding ions and molecules to cytoplasmic constituents suggests that differential permeability depends more on the orientation of the components of a membrane than on its chemical composition. This seems to be the view of WARTIOVAARA (1949, 1950), also. Their orientation certainly is controlled by, among other things, the supply of metabolic energy. The importance of metabolic processes in the maintenance of differential permeability also is demonstrated by the loss of differential permeability and outward diffusion of contents from cells after they are killed or their metabolism is severely disturbed. Respiration might also indirèctly affect passive movement across the cytoplasm through its effects on streaming, adding the effects of mass flow to this diffusion.

In active transport there is not only the effect of respiration on permeability of the membrane itself, but also the relationship between respiration and the forces causing movement of substances across the membrane against a concentration or activity gradient. In such movement respiration must supply either directly or indirectly the energy to move substances against the activity gradient. Indirectly, metabolic energy is necessary for the synthesis of the hypothetical carrier molecules which are supposed to form complexes with the solutes being transported. Even if these carrier complexes diffuse across the membrane their movement will be speeded up by cytoplasmic streaming (GOLDACRE 1952). Possibly energy is required for formation of the complex between the substance transported and the carrier molecule, and it almost certainly is required to break the complex and release the substance carried into a region of higher concentration than that on the other side of the membrane. Even if the carrier is conceived to be a rotating molecule or part of a molecule, energy is required to make it operate, and the folding and unfolding protein molecule proposed by GOLDACRE (1952) requires energy, possibly supplied by adenosinetriphosphate, to make it operate. The LUNDEGÅRDH theory of anion transport along cytochrome bridges from a region of high to a region of low oxidative activity also requires a direct supply of energy from respiration, as does the CONWAY redox pump.

Some of these mechanisms of movement across membranes will be discussed in more detail in other chapters, but it is clear from what has been said that permeability, even in the narrowest sense, is definitely affected by the respiratory activity of the cell.

Active transport. In contrast to the passive movement of materials by diffusion along activity or concentration gradients is the "active transport" of materials against concentration or activity gradients. This results in accumulation of substances in cells and tissues in higher concentrations than exists outside of them. This occurs with respect to various ions and certain organic compounds. Such an accumulation against a free energy gradient can take place only by expenditure of energy released in respiration. Theories dealing with active transport therefore mostly concentrate on explaining how respiratory energy is used to move materials and largely disregard the "permeability" of the cell membranes as measured by diffusion. It may be said, however, that in general cytoplasmic membranes are relatively impermeable to the substances accumulated in them by active transport. If roots or pieces of plant tissue are allowed to accumulate ions and then are placed in water surprisingly little of the accumulated ion leaks out, although it may escape by exchange with similar ions outside, as when tissue containing a radioactive isotope is placed in a solution containing a nonradioactive isotope. Thus we must explain transport across relatively impermeable membranes and furthermore we must explain why this is selective so that some substances are accumulated and others are not. For example certain marine algae accumulate K, but exclude Na so the concentration of K in the vacuolar sap is much higher and the concentration of Na is lower than in the sea water in which the algae grow.

After the demonstration of ion accumulation in plant cells in the 1920's numerous theories of active transport and accumulation were proposed, but only a few have merited much attention. The reader is referred to a review by OVERSTREET and JACOBSON (1952) for a good discussion of various theories of ion accumulation by roots and it also is discussed in some detail in other chapters of this volume. There also is a recent monograph entitled Active Transport and Secretion, published as Symposium VIII by the Society for Experimental Biology which contains a series of excellent articles on this somewhat controversial problem.

In considering active transport across cell membranes there seem to be two principal problems. These are, (a) how to explain differential permeability to various substances, and (b) how the energy released in respiration is used to move ions and molecules against a concentration gradient. At present most investigators believe ion accumulation consists of two stages. In the first stage ions are adsorbed by exchange on the surface of cells. This process usually is regarded as nonselective, reversible, and probably not dependent on metabolic energy, although RUSSELL and AYLAND (1955) claim that it is related to metabolism. The second stage is transport across the cytoplasm into the vacuole.

This is highly selective with respect to various ions, relatively nonreversible, and dependent on energy released by respiration (EPSTEIN and LEGGETT 1954) and it is this active transport with which we now are concerned.

Although there are considerable differences in details, all the important transport theories in vogue at the present time are "carrier" theories in the sense that some temporary combination is assumed to occur between the substance being transported and an organic compound produced in the cytoplasm. The principal difference is that in certain theories it is assumed that the energy released in respiration is used directly to bring about active transport, while other theories assume that it is used indirectly. For example, the redox pump theory of CONWAY (1953) assumes that a single substance is used as an energy donor and carrier and the energy required is released when electrons jump from one redox system to another. LUNDEGÅRDH'S anion respiration theory assumes a direct connection between respiration and transport of anions, but only an indirect relation to cation transport. Anions are supposed to move along a bridge of cytochrome molecules from a region of high to a region of lower redox potential, as from the surface to the interior of the root. Accumulation of anions produces an electrical potential along which cations move in exchange for hydrogen ions (LUNDEGÅRDH 1946, 1954). ROBERTSON and his colleagues propose the same theory with the minor change that instead of proposing exchange of anions for electrons along a cytochrome bridge they suggest that as cytochrome molecules are oxidized at the outer surface they pick up anions and the entire complex is carried across the membrane (ROBERTSON and WILKINS 1948). This mechanism of transport is so closely linked to respiration that a special type of respiration, "anion respiration" or "salt respiration" has been proposed as the source of energy.

Various difficulties have arisen in connection with the attempts to link active transport directly to respiration. A serious one is the observation that DNP (2,4-dinitrophenol) inhibits salt accumulation at the same time that it increases respiration as measured by oxygen consumption (ROBERTSON, WILKINS and WEEKS 1951). As the increased respiration is sensitive to cyanide and carbon monoxide it seems that DNP inhibits salt absorption at the same time that it increases the activity of the cytochrome-cytochrome oxidase system. DNP prevents phosphorylation, hence it appears that respiration is related to active transport by a mechanism involving phosphorylation. ORDIN and JACOBSON (1955) concluded from studies with respiration inhibitors that both the Krebs cycle and phosphorylation are involved in ion uptake by barley roots. They also concluded that separate, but similar mechanisms are involved in the uptake of K and Br ions. EPSTEIN (1953, 1954) also has evidence that cations and anions are accumulated by the same type of mechanism. It may even be questioned whether the cytochrome system is always the principal terminal oxidase in plants. JAMES and BOULTER (1955) reported that in barley roots a week old most of the cytochrome system has been replaced by ascorbic acid oxidase as the terminal oxidase in respiration. RUSSELL (1954) and STEWARD and MILLAR (1954) offer other evidence against a direct relationship between respiration and active transport. Nevertheless LUNDEGÅRDH (1954) believes that most of these difficulties can be explained without abandoning the essential features of his theory.

Because of the objections to a theory involving direct use of respiratory energy for active transport there has been a general trend toward theories which assume a less direct linkage between respiration and active transport. Many people have contributed to the development of the present carrier theory, and some of these workers will be mentioned later, but special mention should be made of OSTERHOUT whose work on permeability evolved into the development of one of the earliest concepts of a carrier system. Most of his important work is cited and summarized in a recent paper (OSTERHOUT 1952).

In brief, current theories propose that ions or molecules moved by active transport combine with organic molecules known as "binding compounds", or "carriers" at the outer surface of a membrane impermeable to the ions, but permeable to the complex of carrier plus ion. The complex is broken up at the inner surface of the membrane, releasing the ions or molecules on the inner side of the membrane. The substances released cannot escape because the membrane is relatively impermeable to them. EPSTEIN (1953) and EPSTEIN and HAGEN (1952) regard the kinetics of this process as analogous to that of enzyme action. Differences in permeability depend principally on the extent to which various substances form complexes with the carrier molecules. The carrier molecules possess reactive centers specific for certain ions and both competitive and noncompetitive interference with binding is shown by various ions. For example Ca, Ba, and Sr ions compete for the same site on the carrier, but Mg does not compete for this site (EPSTEIN and LEGGETT 1954). Also Rb, K and Cs compete for identical sites, but Na and Li do not compete for the sites used by the other three ions (EPSTEIN and HAGEN 1952).

The role of respiration is fully as important in this type of theory as in the anion respiration theory, but it is more indirect. Energy is necessary for the synthesis of the carrier and possibly also for the splitting of the complex, but not necessarily for the movement of the carrier complex across a membrane. As stated earlier, if the carrier complex is formed on the outer side and destroyed on the inner side, then it probably moves across the gradient by diffusion and little or no energy is required for actual transport. It may be that as much or more energy is required to separate the substance transported from the carrier as to form the carrier, but this is uncertain. Various investigators have proposed various substances as carriers, but none have been specifically identified. They probably are proteins, possibly phosphorylated, high energy proteins (STEWARD and STREET 1947). ORDIN and JACOBSON (1955) found evidence that both the Krebs cycle and phosphorylation are involved in ion absorption. Possibly intermediates produced in the Krebs cycle serve as carriers, but perhaps a high energy compound such as adenosinetriphosphate controls their functioning in some manner. GOLDACRE (1952) thought that ATP might control the contraction and expansion of protein molecules, exposing binding sites when unfolded and rendering them unavailable when contracted. In any carrier mechanism enzymes probably also play an important role in the synthesis of the carrier, and possibly in the splitting of the carrier complex. According to this view active transport and accumulation are related to the general metabolism of the cell rather than to a specific fraction of total respiration (STEWARD and MILLAR 1954).

Factors affecting permeability.

Ordinarily a paper on permeability would include a discussion of various environmental factors which affect the permeability of cells, but this is unnecessary because most of these factors are discussed in later chapters of this volume. It seems desirable, however, to consider briefly how any environmental factor might affect permeability. The simplest way would be by producing a direct change in the properties of the membrane, for example, by changing its hydration or the electrical charge on it. Lowering the temperature increases the viscosity and decreases protoplasmic streaming, thereby decreasing the rate of movement across protoplasmic membranes. Changes in environmental factors may affect permeability indirectly by affecting respiration and other metabolic processes which in turn modify the structure and the permeability of cell membranes. They may also affect the active transport mechanism, either by limiting the synthesis of carrier molecules or binding sites, the formation of complexes with binding sites or carrier molecules, the transport of these complexes, or the supply of energy required to split off the substance carried from the carrier molecule.

These various possibilities illustrate the complex relationship between cell metabolism in general and permeability and active transport. It is not surprising, therefore, that, some differences of opinion exist concerning the relation between respiration and permeability and related processes. It seems likely that much more research will be necessary before these relationships are fully understood. In conclusion it may be said with respect to permeability that there is a large body of data, a number of interesting theories but only a small number of generally accepted generalizations.

Summary.

Respiration is related to permeability in two general ways, first because the maintenance of the structure of protoplasmic membranes is dependent on energy supplied by respiration, and second because the energy required for active transport across membranes is supplied either directly or indirectly by respiration. Permeability has been defined as the rate of movement of a substance through a membrane under a given during force. Permeability in its classical sense is a property of membranes, but it cannot be discussed without consideration of the driving force which causes movement across the membrane, because failure of a substance to pass through a membrane can result either from impermeability or lack of a driving force to cause movement across the membrane. In recent years interest in the accumulation of ions and other substances by cells has shifted the emphasis from permeability of cell membranes to the nature of the driving force.

Various theories have been developed to explain the differences in permeability of cells to various substances. Although both the sieve and solubility theories explain some instances of differential permeability, the lipoid-sieve theory seems to have more general applicability. Perhaps the permeability of a membrane depends largely on the orientation of its components. Maintenance of such an oriented structure requires a continuous supply of respiratory energy, and death or serious interference with respiration results in loss of differential permeability.

The forces causing movement across a membrane can be described as passive and active. Passive movement is mostly by diffusion and occurs along gradients of decreasing concentration, activity, or chemical potential of the diffusing substance and the driving force is not related to respiration. Most water movement occurs in this way, also some movement of solutes. Active transport involves movement of a substance against a concentration or activity gradient by the expenditure of energy released in respiration. According to the theory of LUNDEGÅRDH respiratory energy is used directly to move anions while cations move in passively along an electrical gradient produced by the entrance of anions. There is increasing reason to believe, however, that respiratory energy is used indirectly to bring about active transport and that anions and cations are transported into cells by the same or similar mechanisms.

Various carrier theories of active transport are popular at present. These assume that the substance being moved forms a complex with an organic carrier molecule, perhaps a high energy phosphorylated compound, at the outer surfaces of the membrane. The cell membranes are impermeable to the carrier molecule and to ions but permeable to the complex which moves across the membrane and the substance carried is then split off on the inner side where it is accumulated. Metabolic energy is used in the synthesis of the carrier and possibly in splitting the complex, but not necessarily in moving it across the membrane.

It is clear that permeability and active transport are related to respiration and general metabolic activity in several ways. Changes in metabolic activity can affect the composition and structure of cell membranes, thereby affecting their permeability. They also affect the synthesis of carrier molecules, and the supply of energy required for active transport.

Literature.

Brooks, S. C., u. M. M. Brooks: The permeability of living cells. Protoplasma-Monogr. **19** (1941).

Collander, R.: The permeability of plant protoplasts to small molecules. Physiol. Plantarum (Copenh.) **2**, 300–311 (1949). — Conway, E. J.: A redox pump for the biological performance of osmotic work and its relation to the kinetics of free ion diffusion across membranes. Internat. Rev. Cytology **2**, 419–445 (1953).

Danielli, J. F.: Morphological and molecular aspects of active transport. Symposia Soc. Exper. Biol. 8, 502–516 (1954). — Davson, H., and J. F. Danielli: The permeability of natural membranes, 2d ed. Cambridge: Cambridge University Press 1952.

Epstein, E.: Mechanism of ion absorption by roots. Nature (Lond.) **171**, 83–84 (1953). — Cation-induced respiration in barley roots. Science (Lancaster, Pa.) **120**, 987–988 (1954). — Epstein, E., and C. E. Hagen: A kinetic study of the absorption of alkali cations by barley roots. Plant Physiol. **27**, 457–474 (1952). — Epstein, E., and J. L. Leggett: The absorption of alkaline earth cations by barley roots: Kinetics and mechanism. Amer. J. Bot. **41** 785–792 (1954).

Garby, L., u. H. Linderholm: The permeability of frog skin to heavy water and to ions, with special reference to the effect of some diuretics. Acta physiol. scand. (Stockh.) **28**, 336–346 (1953). — Goldacre, R. J.: The folding and unfolding of protein molecules as a basis of osmotic work. Internat. Rev. Cytology **1**, 135–164 (1952).

Jacobs, M. H.: The measurement of cell permeability with particular reference to the erythrocyte. In Trends in Physiology and Biochemistry, pp. 149–171. Ed. E. S. G. Barron, New York: Academic Press 1952. — James, W. O., and D. Boulter: Further studies of the terminal oxidases in the embryos and young roots of barley. New Phytologist **45**, 1–12 (1955).

Kofoed-Johnson, V., u. H. H. Ussing: The contributions of diffusion and flow to the passage of D_2O through living membranes. Acta physiol. scand .(Stockh.) **28**, 60–76 (1953).

Lundegårdh, H.: Transport of water and salts through plant tissues. Nature (Lond.) **157**, 575–577 (1946). — Anion respiration: the experimental basis of a theory of absorption, transport and exudation of electrolytes by living cells and tissues. Symposia Soc. Exper. Biol. 8, 262–296 (1954).

Ordin, L., and L. Jacobson: Inhibition of ion absorption and respiration in barley roots. Plant Physiol. **30**, 21–27 (1955). — Osterhout, W. J. V.: The mechanism of accumulation in living cells. J. Gen. Physiol. **35**, 579–594 (1952). — Overstreet, R., and L. Jacobson: Mechanisms of ion absorption by roots. Annual Rev. Plant Physiol. **3**, 189–206 (1952).

Prescott, D. M., u. E. Zeuthen: Comparison of water diffusion and water filtration across cell surfaces. Acta physiol. scand. (Stockh.) **28**, 77–94 (1953).

Robertson, R. N., and M. J. Wilkins: VII. The quantitative relation between salt accumulation and salt respiration. Austral. J. Sci. Res. B **1**, 17–37 (1948). — Robertson, R. N., M. J. Wilkins and D. C. Weeks: IX. The effects of 2,4-dinitrophenol on salt accumulation and salt respiration. Austral. J. Sci. Res. B **4**, 248–264 (1951). — Rosenberg, T.: The concept and definition of active transport. Symposia Soc. Exper. Biol. 8, 27–41 (1954). — Russell, R. S.: The relationship between metabolism and the accumualtion of ions by plants. Symposia Soc. Exper. Biol. 8, 343–366 (1954). — Russell, R. S., and M. J. Ayland: Exchange reactions in the entry of cations into plant tissues. Nature (Lond.) **175**, 204–205 (1955).

STEINBACH, H. B.: Permeability. Annual Rev. Plant Physiol. 2, 323–342 (1951). — STERN, J. R., L. B. EGGLESTON, R. HEMS and H. A. KREBS: Accumulation of glutamic acid in isolated brain tissue. Biochemic. J. 44, 410–418 (1949). — STEWARD, F. C., and F. K. MILLAR: Salt accumulation in plants: a reconsideration of the role of growth and metabolism. Symposia Soc. Exper. Biol. 8, 367–406 (1954). — STEWARD, F. C., and H. E. STREET: The nitrogenous constituents of plants. Annual Rev. Biochem. 16, 471–502 (1947).

TEORELL, T.: Permeability. Annual Rev. Physiol. 11, 545–564 (1949). — Transport processes and electrical phenomena in ionic membranes. Progr. in Biophysics a. Biophysical Chem. 3, 305–319 (1953).

USSING, H. H.: Transport of ions across cellular membranes. Physiologic. Rev. 29, 127–155 (1949). — Transport through biological membranes. Annual Rev. Physiol. 15, 1–20 (1953). — Ion transport across biological membranes. In Ion Transport across Membranes, pp. 3–22. Ed. by H. T. CLARK, New York: Academic Press 1954.

WARTIOVAARA, V.: The permeability of the plasma membranes of Nitella to normal primary alcohols at low and intermediate temperatures. Physiol. Plantarum (Copenh.) 2, 184–196 (1949). — Zur Erklärung der Ultrafilterwirkung der Plasmahaut. Physiol. Plantarum (Copenh.) 3, 462–478 (1950).

Die Abhängigkeit des Stoffaustausches von der Temperatur.

Von

Veijo Wartiovaara.

Mit 2 Abbildungen.

Physikalische Grundlagen.

Die Stoffaufnahme setzt eine Eigenbewegung der aufzunehmenden Moleküle oder Ionen voraus, denn von elektrischen Ladungen abgesehen verfügt die Zelle über keine nach einwärts gerichtete Fernkräfte, die die Aufnahme erzwingen könnten. Falls die Aufnahme mit chemischen Reaktionen verknüpft ist, erfolgen auch diese nur bei einem genügend kräftigen Zusammenstoß der Reaktionspartner. Wie die Stoffaufnahme auch geschieht, setzt sie somit eine thermische Molekularbewegung voraus. Die Intensität derselben nimmt, der absoluten Temperatur (K^0) proportional, mit steigender Temperatur zu, was die positive Temperaturabhängigkeit der Stoffaufnahme verständlich macht.

Es sei der Einfachheit halber angenommen, daß wir nur je einen Teilprozeß, den entschieden langsamsten und daher maßgebenden zu berücksichtigen brauchen, was auch in der Praxis oft zutrifft. Eine nähere Definition des begrenzenden Teilprozesses ist in diesem Zusammenhang nicht nötig, da in bezug auf die Temperatureinwirkung zwischen physikalischen und chemischen Vorgängen kein grundsätzlicher Unterschied besteht. Wird nun die Temperatur erhöht, so zeigt sich, daß die Geschwindigkeit der Stoffaufnahme nicht linear, proportional der absoluten Temperatur, sondern Grad für Grad immer steiler zunimmt. Die Beschleunigung erfolgt exponential und kann durch den Temperaturkoeffizient Q_{10} zum Ausdruck gebracht werden.

Es besteht somit ein scheinbarer Widerspruch zwischen dem linearen Anstieg der Wärmeenergie und der exponentiellen Temperaturabhängigkeit der Stoffaufnahmegeschwindigkeit. Dieser Widerspruch wird jedoch bei näherer Betrachtung aufgehoben. Die Temperatur ist nämlich ein Maß der *mittleren* Bewegungsenergie der Moleküle. Im biologischen Temperaturbereich würde diese aber zur Aufrechterhaltung der Lebensvorgänge gar nicht ausreichen. Zum Vergleich sei erwähnt, daß der Anteil der translatorischen Bewegung — in einer wäßrigen Lösung handelt es sich vornehmlich um ein Hin- und Herpendeln des Schwerpunktes in dem knapp bemessenen Raum zwischen den Wassermolekülen — bei Zimmertemperatur rund 900 cal Mol^{-1} ausmacht, während die Festigkeit der Wasserstoffbrücken zwischen Wassermolekülen 4500 und die einer Einfachbindung zwischen Kohlenstoffatomen 58600 cal Mol^{-1} entspricht. Die Wärmeenergie ist aber weder zeitlich noch räumlich gleichmäßig verteilt. Es entstehen vorübergehende Energieanhäufungen, welche dem Platzwechsel der Moleküle bei der Diffusion und der Umgruppierung von Atomen bei chemischen Reaktionen den Anstoß geben und so ihre exponentielle Temperaturabhängigkeit bedingen. Die statistische Wahrscheinlichkeit der verschieden hohen Gipfelwerte nimmt nämlich mit steigender Temperatur exponentiell zu.

Die durchschnittliche Wartezeit, bis ein gegebenes Molekül zufälligerweise das für den betreffenden Prozeß charakteristische Mindestmaß an Energie empfängt, wird durch einen Temperaturanstieg verhältnismäßig um so mehr verkürzt, je höher der zu überschreitende Schwellenwert liegt. Gerade hierauf beruht der geläufige Unterschied zwischen den Temperaturkoeffizienten von Diffusion und chemischen Reaktionen. (Es sei auf die relative Festigkeit der eben angeführten Bindungsarten hingewiesen.) Man darf aber nicht vergessen, daß die bei der Diffusion gegebenenfalls zu überwindende Gesamtwirkung zahlreicher, einzeln noch so schwacher Anheftungsstellen zwischen großen Molekülen (Lipoide!) einer starken Kovalenzbindung gleichkommen kann, und daß andererseits manche chemische Reaktionen, zumal katalysierte (Enzyme!), schrittweise über mehrere aufeinanderfolgende niedrige Energieschwellen verlaufen. Dies kann sogar zu einer Umkehrung der gewöhnten Reihenfolge der Temperaturkoeffizienten führen.

Die Stärke des erforderlichen Anstoßes, die Aktivierungsenergie des Prozesses, läßt sich nach der bekannten Formel von ARRHENIUS aus der Temperaturabhängigkeit berechnen. In cal Mol^{-1} ausgedrückt und mit dem Symbol μ bezeichnet, wird sie oft, besonders bei mehr physikalisch betonten Untersuchungen dem Q_{10} vorgezogen. Beide Ausdrücke haben ihre eigenen Vor- und Nachteile (BĚLEHRÁDEK 1935, BULL 1951, NETTER 1951). Während μ einen mehr unmittelbaren Einblick in die Energiebedingungen gestattet und im Gegensatz zu Q_{10} bis zu den höchsten hier in Frage kommenden Temperaturen praktisch konstant bleibt, ist Q_{10} dafür als eine rein empirische Größe allgemeiner verwendbar — sogar in den Fällen, wo thermische Reize die Stoffaufnahme regulatorisch beeinflussen, was die Berechnung eines μ-Wertes nicht nur aussichtslos, sondern schwer irreführend machen würde. Im folgenden soll unter dem Temperaturkoeffizient das Q_{10} verstanden werden.

Die Temperatur wirkt auf alle Teilprozesse der Aufnahme, von dem Herankommen an die Zelle bis zur Ausbreitung in der Vacuole. Die Vorgänge an oder in den Grenzschichten des Protoplasmas, weil der Regel nach die weitaus langsamsten, pflegen jedoch für die Höhe des Temperaturkoeffizienten ausschlaggebend zu sein, so daß wir die anderen Schritte oft vernachlässigen können. Leider ist die relative Bedeutung des Plasmalemmas und des Tonoplasten mangelhaft bekannt. Im folgenden müssen sie daher oft einfach als „die Grenzschicht" zusammengefaßt werden. Und wenn von der „Aufnahme" die Rede ist, kann möglicherweise bisweilen die aktive Absonderung in die Vacuole die entscheidende sein.

Im einfachsten Fall einer unkomplizierten osmotischen Aufnahme, wobei die Permeabilität der Grenzschicht die Geschwindigkeit allein bestimmt, wirkt die Temperatur unmittelbar auf die Beweglichkeit der eindringenden Moleküle. In einem solchen Fall sind, ähnlich wie bei anderen Diffusionserscheinungen, ziemlich einfache Gesetzmäßigkeiten hinsichtlich des Temperatureffektes zu erwarten, die zu einer rechnerischen Analyse der physikalischen Eigenschaften der Grenzschicht verwertet werden können.

Weniger durchsichtig ist es, wenn die aufgenommenen Moleküle mit irgendwelchen Bestandteilen des Zellsaftes reagieren. Eine im Verhältnis zur Permeation schnell und quantitativ zu Ende verlaufende Reaktion macht zwar keinen so großen Unterschied: sie erhöht nur die Aufnahmekapazität. Sonst ist mit der Beteiligung zweier zusätzlicher, von der Temperatur abhängiger Variablen, der Geschwindigkeit und dem Gleichgewicht der Reaktion, zu rechnen.

Unter den auf den chemischen Reaktionen (im weitesten Sinne) beruhenden Aufnahmemechanismen gibt es eine unübersehbare Fülle der verschiedenartigsten Übergänge zwischen einer geringfügigen, der Einverleibung durch Diffusion vorangehenden reversiblen Umwandlung und den komplizierten Fällen, wo die Reaktionsfolge an einer spezifischen Empfangsstelle beginnt und in der Folge von energieliefernden Lebensvorgängen direkt abhängig ist. Die Wirkungsweise der Temperatur gestaltet sich entsprechend mannigfaltig.

Es bietet schon ein scheinbar so einfacher Fall wie eine hypothetische Komplexbildung mit einem zelleigenen Träger mehrere Angriffspunkte für die Einwirkung der Temperatur. Die Geschwindigkeit und das Gleichgewicht des Primärprozesses, die Durchwanderung des Komplexes und die Freisetzung bzw. Nachlieferung des Trägers können jede für sich ausschlaggebend sein. Genügend große Unterschiede in den Temperaturkoeffizienten vorausgesetzt, mögen bei niedriger und hoher Temperatur verschiedene Teilprozesse limitierend wirken, da ein Prozeß mit höherem Q_{10}-Wert den mit niedrigerem beim Ansteigen der Temperatur schließlich überholen muß. Ist die Nachlieferung des Trägers mit aktiver Lebenstätigkeit verbunden, so kann ein Temperaturoptimum auftreten und der Prozeß leitet über zu den noch mehr verwickelten, vom Gesamtstoffwechsel intim abhängigen Aufnahmeprozessen, deren Geschwindigkeit der Atmungsintensität mehr oder weniger genau parallel variiert, was auf eine vorwiegend indirekte Temperaturabhängigkeit hindeutet.

Bisher wurde stillschweigend vorausgesetzt, daß die Eigenschaften der Grenzschicht beim Temperaturwechsel unverändert bleiben. Streng genommen verhält es sich wohl nie so. Da aber alle realen Systeme von der Temperatur beeinflußt werden — kaltes und heißes Wasser sind in mancher Hinsicht verschiedenartige Stoffe —, können die unmittelbar temperaturbedingten Unterschiede in der Festigkeit des Feinbaues („Viscosität") u. dgl. zu der schon besprochenen Primärwirkung gerechnet werden. Unter den mehr speziellen Veränderungen verdienen die des Wassergehaltes (Quellungsgrad) und mögliche Umlagerungen der Bausteine eine Erwähnung. In welchem Ausmaß sie etwa zu der Temperaturabhängigkeit der osmotischen Aufnahme beitragen können, ist nicht näher bekannt. Es ließe sich jedenfalls die in einigen Fällen beobachtete, theoretisch etwas unerwartete Erhöhung des Q_{10}-Wertes mit steigender Temperatur durch solche qualitative Veränderungen erklären, obwohl auch andere Möglichkeiten gut denkbar sind. Diese unmittelbaren Veränderungen sind von einer allmählichen, mehr aktiven Einstellung zu unterscheiden, die später besprochen werden soll.

Die nichtosmotische Aufnahme braucht nicht unbedingt von den mit der Temperatur sich passiv verändernden Eigenschaften der Grenzschicht direkt abhängig zu sein. Es läßt sich nämlich vorstellen, daß die vor allem als eine Isoliereinrichtung anzusehende Grenzflächenstruktur unter Umständen nur reversible, aus physikalischen Gründen unvermeidliche Veränderungen erleidet, während die Anzahl, die Bereitschaft und die Energieversorgung der spezifischen Aufnahmestellen physiologisch reguliert werden.

So gut wie alle nicht überaus großmolekularen Stoffe dürften wenigstens einigermaßen zum osmotischen Durchdringen befähigt sein, weshalb bei einer nichtosmotischen Aufnahme die passive Durchlässigkeit immer mehr oder minder stark mitwirkt. Diese kann je nach den Umständen den Temperaturkoeffizient der Gesamtaufnahme in verschiedener Weise beeinflussen. Eine eingehendere mathematische Analyse unter Berücksichtigung der Gleichgewichte, Energiequellen und anderer mit der Temperatur sich verändernden Faktoren würde zu weit führen und wäre auch wegen der nur spärlich aufgeklärten Variabilität dieser Faktoren ziemlich aussichtslos. Dafür soll ein theoretisch konstruiertes Beispiel von dem Zusammenspiel zweier Austauschvorgänge betrachtet werden. Es sei angenommen, daß der relative Anteil der osmotischen Komponente bei 20^0 ein Zehntel von der nichtosmotischen ist und daß im Falle I der Temperaturkoeffizient des aktiven Stofftransportes 2, derjenige der passiven Permeabilität dagegen 5 beträgt, während im Falle II das umgekehrte zutrifft. Die Temperaturkoeffizienten der berechneten Gesamtleistung sind in Tabelle 1 zusammengestellt.

Tabelle 1. *Theoretisch konstruiertes Beispiel von dem Zusammenspiel zweier Austauschvorgänge.* A = aktive Stoffaufnahme, O = osmotischer Austausch.

Temperatur	Fall I				Fall II			
	A	O	A+O	A—O	A	O	A+O	A—O
0—10	5	2	3,9	10,0	2	5	2,0	1,9
10—20	5	2	4,3	6,0	2	5	2,1	1,8
20—30	5	2	4,7	5,3	2	5	2,3	1,7

Die Zunahme von Q_{10} mit steigender Temperatur, wenn die beiden Aufnahmemechanismen gleichsinnig wirken, verdient eine besondere Aufmerksamkeit. Sie ist ein häufig auftretendes Merkmal uneinheitlicher Stoffaufnahmeprozesse (vgl. S. 376). Bemerkenswert ist auch, daß eine bei Zimmertemperatur geringfügige osmotische Abgabe eine beträchtliche Variationsbreite der auf den Gesamtprozeß sich beziehenden Q_{10}-Werte verursachen kann.

Der Zeitfaktor.

Die Geschwindigkeit der gewöhnlichen Diffusion folgt momentan und exakt allen Temperaturschwankungen, wie dies in Anbetracht des auf der Wärmebewegung unmittelbar beruhenden Mechanismus der Diffusion ja selbstverständlich ist. Dasselbe sollte für die als Diffusion in lebendem Substrat aufzufassende osmotische Stoffaufnahme gelten. In der Tat hat sich die Einwirkung der Temperatur auf die passive Permeabilität bei kurzfristigen Versuchen als momentan und reversibel erwiesen (WARTIOVAARA 1942, SEEMANN 1950).

Wird eine Nachwirkung des Temperaturwechsels bei deutlich osmotischen Stoffaustauschvorgängen gefunden, so ist dies zunächst auf 3 Ursachen zurückzuführen:

1. Einstellung eines neuen Konzentrationsgefälles in verschiedenen Schichten, deren Durchlässigkeit in ungleichem Maße von der Temperatur beeinflußt wird. Liegt ein Objekt mit einem erheblichen extravacuolaren Raum, etwa ein dicker, stark plasmolysierter Gewebeschnitt vor, so mag die Einstellung eine beträchtliche Zeit beanspruchen und eine allmählich abklingende physiologische Nachwirkung vortäuschen.

2. Träge verlaufende passive Zustandsänderungen des Protoplasmas. Beweise für solche fehlen jedoch einstweilen.

3. Umgestaltung des Baues und Veränderung der Zusammensetzung der Grenzschichten durch aktive Tätigkeit des Protoplasmas. Die Temperatur wirkt dabei als ein auslösender Faktor, d. h. als Reiz, und die Veränderung der passiven Permeabilität dürfte eine sekundäre Begleiterscheinung einer allgemeinen physiologischen Umstimmung darstellen. Wird z. B. nach einem thermischen Reiz erhöhte Permeabilität und angeregter Stoffwechsel gefunden, so beweist dies in erster Linie nur, daß eine aktive Zelle auf den möglichst vollkommenen Abschluß gegen die Umwelt verzichten kann und daß eine Erhöhung der Zellaktivität leicht eine gewisse Auflockerung des Gefüges der Plasmagrenzschicht nach sich zieht. Dagegen erscheint die häufig angenommene „Nützlichkeit" der Permeabilitätserhöhung recht zweifelhaft. Denn die Permeabilität z. B. für Wasser und Sauerstoff ist wohl immer hinreichend groß, während wiederum andere lebenswichtige Stoffe wie etwa Zucker und Mineralsalze so langsam permeieren, daß die Aufnahme derselben durch aktive Transportmechanismen geschehen muß. Die Erhöhung der passiven Permeabilität bei der Aktivierung des Stoffwechsels ist also eher eine Folge als etwa eine Voraussetzung der Beschleunigung des Stoffwechsels.

Die unmittelbare Wirkung der Temperatur auf die nichtosmotische Aufnahme ist ebenfalls momentan und reversibel, da die chemischen Teilprozesse in ähnlicher Weise wie die Diffusion von der Molekularbewegung abhängig sind. Auch die Ursachen einer Nachwirkung sind grundsätzlich dieselben wie oben, mit dem Unterschied, daß bei der Einstellung des Gradienten (gemäß Punkt 1) chemische Gleichgewichte neben den osmotischen und Akkumulation in das Protoplasma als zusätzliche verzögernde Faktoren zu berücksichtigen sind. Die auf aktiven Umänderungen beruhenden Nachwirkungen (Punkt 3) dürften, wenigstens unter einigermaßen normalen Lebensbedingungen, im Gegensatz zu den Verhältnissen bei dem osmotischen Stoffaustausch meistens in mehr oder minder innigem Zusammenhang mit den Änderungen des Stoffwechsels stehen.

Die während bzw. nach einer Kälteperiode häufig eintretenden Erscheinungen wie Erhöhung der Frosthärte, Aufhebung eines Ruhezustandes oder Übergang in eine fertile Phase lassen kaum die aktive oder die osmotische Aufnahme unberührt (vgl. Scarth und Levitt 1937) und dasselbe gilt wohl für weniger auffällige Folgen thermischer Reize, so daß immer mit einer möglichen Nachwirkung zu rechnen ist. Damit ist natürlich nicht gesagt, daß z. B. die täglichen Temperaturschwankungen unter stabilen klimatischen Verhältnissen tiefgreifendere Änderungen hervorrufen müssen. Es ist eher anzunehmen, daß jeweils eine bestimmte Schwelle in bezug auf die Temperatur und ihre Wirkungszeit überschritten werden muß.

Temperaturabhängigkeit der Anelektrolytaufnahme.

Das Eindringen der Anelektrolyte ist meistens eine direkte Folge der passiven Permeabilität. Speicherung durch chemische Bindung oder Ausfällung kommt zwar vor, der Regel nach stellt aber die Permeation einen unkomplizierten Diffusionsprozeß dar, der zu einem Ausgleich der Konzentrationsunterschiede führt. Diesem rein physikalischen Charakter entsprechend sind keine die eigentlichen Lebensvorgänge kennzeichnende Kardinalpunkte der Temperatur zu erwarten. Solche sind auch nicht gefunden worden. Es läßt sich die Aufnahme durch Abkühlung mehr oder weniger stark verlangsamen, aber nie ganz unterdrücken. Bei Temperaturerhöhung wiederum nimmt die Permeationsgeschwindigkeit kontinuierlich zu, bis irreversible Veränderungen und Wärmetod eintreten. Es besteht also in bezug auf die Temperaturabhängigkeit der Permeation eine qualitative Übereinstimmung mit anderen Diffusionserscheinungen, indem die Geschwindigkeit sich stetig und gleichsinnig mit der Temperatur verändert.

In quantitativer Hinsicht aber pflegt sich die Permeation von den geläufigen Erscheinungen der Hydrodiffusion durch ihren erheblich höheren Temperaturkoeffizienten zu unterscheiden. Soweit bekannt, liegen die Q_{10}-Werte ganz allgemein etwa zwischen 2 und 4, wenn der Diffusionswiderstand des Protoplasmas (d. h. der Plasmamembranen) entscheidend ist (Tabelle 2 und 3). Niedrigere Werte kommen leicht dadurch zustande, daß der weniger temperaturabhängige Widerstand der Zellwand, sich dem des Protoplasmas überlagernd, das Q_{10} des ganzen Vorganges herabsetzt. Alle Angaben über niedrige Temperaturkoeffizienten der Anelektrolytpermeabilität dürfen jedoch nicht ohne weiteres diesem Umstand zugeschrieben werden. Wenn aber das Objekt aus mehreren Zellschichten besteht und die Permeabilität des Protoplasmas groß ist, so liegt die Gefahr nahe, daß zumal bei höheren Temperaturen die Diffusionswiderstände nichtplasmatischer Teile als begrenzende Faktoren wirken können.

Die hohen Temperaturkoeffizienten der Anelektrolytpermeabilität haben viel Kopfzerbrechen und Diskussion verursacht. Die noch heutzutage ab und zu

Tabelle 2. *Temperaturkoeffizienten der mikrochemisch bestimmten Anelektrolytenpermeabilität bei Characeen.*

<table>
<tr><th></th><th>0—10</th><th>10—20</th><th>20—30°</th><th></th><th></th></tr>
<tr><td>Harnstoff . . . . .</td><td>2,5</td><td>2,6</td><td>2,8</td><td>Chara</td><td>WARTIOVAARA (1942)</td></tr>
<tr><td>Methanol . . . . .</td><td colspan="2">2,6</td><td>—</td><td>Nitella</td><td>WARTIOVAARA (1949)</td></tr>
<tr><td>Äthanol . . . . . .</td><td colspan="2">2,7</td><td>—</td><td>Nitella</td><td>WARTIOVAARA (1949)</td></tr>
<tr><td>Propanol . . . . .</td><td colspan="2">2,7</td><td>—</td><td>Nitella</td><td>WARTIOVAARA (1949)</td></tr>
<tr><td>Äthylenglykol . . .</td><td>2,3</td><td>2,9</td><td>3,5</td><td>Nitellopsis</td><td>WARTIOVAARA (1942)</td></tr>
<tr><td>Urethylan . . . . .</td><td>3,2</td><td>2,9</td><td>—</td><td>Nitellopsis</td><td>WARTIOVAARA (1942)</td></tr>
<tr><td>Glycerin. . . . . .</td><td>—</td><td>3,5</td><td>8</td><td>Nitellopsis</td><td>WARTIOVAARA (1942)</td></tr>
<tr><td>Tetraäthylenglykol .</td><td>—</td><td>4,9</td><td>—</td><td>Nitellopsis</td><td>WARTIOVAARA (1942)</td></tr>
<tr><td>Trimethylcitrat. . .</td><td>—</td><td>5,5</td><td>—</td><td>Nitellopsis</td><td>WARTIOVAARA (1942)</td></tr>
<tr><td>2,3-Butylenglykol. .</td><td>7,4</td><td>5,9</td><td>4,8</td><td>Nitellopsis</td><td>WARTIOVAARA (1942)</td></tr>
</table>

Tabelle 3. *Temperaturkoeffizienten verschiedener Objekte. Bestimmungen auf osmotischem Wege.*

Harnstoff	1,6	*Elodea* 10—20°	WARTIOVAARA (1942)
Harnstoff	1,8	*Tradescantia* 7,5—30°	HOFMEISTER (1935)
Harnstoff	2,6	*Rhoeo* 10—20°	WARTIOVAARA (1942)
Glycerin	1,8	*Spirogyra* 10—22°	HOFFMANN (1927)
Glycerin	2,3	*Tradescantia* 18—30°	HOFMEISTER (1935)
Glycerin	2,5	*Rhoeo* 10—20°	WARTIOVAARA (1942)
Malonamid . . .	2,5	*Tradescantia* 18—30°	HOFMEISTER (1935)

auftauchende Meinung, daß die Größe dieser Koeffizienten für die „chemische" Natur der Permeationsvorgänge spräche, ist schon längst als unstatthaft verworfen worden. Zwar sind die Temperaturkoeffizienten der Diffusion in Wasser und der Permeabilität wasserimbibierter Membranen der Regel nach niedriger — das Q_{10} pflegt unter 2 zu bleiben —, dafür sind aber Diffusionsprozesse in hochviscösen Flüssigkeiten und halbfesten Medien durch bedeutend höhere Temperaturkoeffizienten gekennzeichnet. Die Aktivierungsenergie der Diffusion, die hauptsächlich zu der notwendigen Lochbildung verbraucht werden soll (JOST 1952), ist nämlich in solchen Medien wesentlich größer als etwa im leichtbeweglichen Wasser, woraus folgt, daß das Q_{10} entsprechend größer sein muß (vgl. S. 370). Dies paßt mit der Annahme einer dichteren Grenzschicht der Protoplasten gut zusammen und würde die Höhe der Temperaturkoeffizienten der Protoplasmapermeabilität ohne weiteres erklären.

Bemerkenswert ist immerhin, daß die Koeffizienten verschiedener, auch recht ähnlicher Substanzen so große und anscheinend regellose Unterschiede aufzeigen (vgl. Tabelle 2).

DANIELLI (DAVSON und DANIELLI 1943) nimmt an, daß die hohen Energiewerte zum Durchbrechen der Phasengrenze zwischen Wasser und Plasmahaut vonnöten sind. Lipoidlösliche Stoffe, deren Moleküle den Sprung unter minderem Energieaufwand durchführen können, sollen folglich niedrigere Temperaturkoeffizienten aufweisen als weniger lipoidlösliche, für welche die Energieschwelle an der Grenze höher ist. Gemäß der Berechnungen von ZWOLINSKI, EYRING und REESE (1949) ist eine solche Abhängigkeit der Temperaturkoeffizienten von der Lipoidlöslichkeit nur in gewissen Fällen angedeutet, weshalb die genannten Verfasser die hohe Aktivierungsenergie der Diffusion einer aus dichtgepackten Lipoidmolekülen bestehenden Plasmahaut zuschreiben. Erklärungsversuche auf Grund der eigentlichen Ultrafiltertheorie scheinen nicht vorzuliegen. Welche die Zusammensetzung der Plasmahaut auch sein mag, sie muß ein sehr dichtes Gefüge besitzen. Sie wird nur durch die äußerst seltenen starken Energieanhäufungen hier und da vorübergehend soviel aufgelockert, daß ein fremdes

Molekül sich einfügen bzw. einen Schritt weiter bewegen kann. Da die kleinsten Moleküle dabei, ihrem geringeren Raumbedarf entsprechend, mit den häufiger auftretenden mäßigen Energiegipfeln auskommen dürften, möchte man wegen der Beziehung zwischen der Aktivierungsenergie und der Temperaturabhängigkeit bei ihnen verhältnismäßig niedrige Temperaturkoeffizienten erwarten. So verhält es sich auch wenigstens insofern, als solche kleinmolekulare und schnell permeierende Stoffe wie Methanol, Harnstoff und Äthylenglykol zu denen mit den niedrigsten Q_{10}-Werten gehören.

Bei näherer Betrachtung erweist sich aber das Molekularvolumen an sich ebensowenig entscheidend wie die Lipoidlöslichkeit. Die Temperaturkoeffizienten der höheren Homologen des Methanols sind nicht nachweisbar größer als der des Methanols (WARTIOVAARA 1949); entsprechendes gilt für Glykole mit endständigen Hydroxylgruppen, verglichen mit Äthylenglykol (unveröffentlicht). Der größermolekulare Thioharnstoff hat etwa denselben Temperaturkoeffizienten wie Harnstoff. Dagegen fallen die Alkylharnstoffe durch ihre im Verhältnis zum Harnstoff höheren Temperaturkoeffizienten auf (Tabelle 4), und 2 Kohlenstoffatome, nicht zwischen, sondern neben den hydroxyltragenden des Äthylenglykols eingeführt (2,3-Butylenglykol), üben eine außergewöhnlich starke Wirkung aus. Man bekommt somit den Eindruck, als ob eine gewisse Sperrigkeit oder „Dicke" der Moleküle ihre Diffusion in der Zellgrenzschicht besonders erschwere, so daß die Aktivierungsenergie und Q_{10} in solchen Fällen besonders hoch ausfallen. Diese Wirkung des sperrigen Molekülbaus kann sich übrigens bereits bei gewöhnlicher Zimmertemperatur durch ein relativ schwaches Permationsvermögen äußern: Methylharnstoff permeiert bisweilen, trotz größerer Lipoidlöslichkeit, langsamer als Harnstoff (COLLANDER und WIKSTRÖM 1949); ebenso scheint Pentaerythrit langsamer als Erythrit zu permeieren (COLLANDER, private Mitteilung).

Zugunsten der Auffassung, daß es sich in derartigen Fällen um eine unmittelbare Wirkung der Moleküldimensionen handelt, sprechen Beobachtungen über die Diffusion von Paraffinkohlenwasserstoffen in Polyisobutylen (PRAGER, BAGLEY und LONG 1953). Die Länge einer unverzweigten Kohlenstoffkette bedingt hier einen wesentlich kleineren Unterschied als das Vorhandensein seitlicher Methylgruppen, welche die Diffusiongeschwindigkeit stark erniedrigen. Das Medium in diesen Versuchen bestand, in Übereinstimmung mit dem wahrscheinlichen Bau der Plasmahaut, aus dicht gepackten, beweglichen, aber nicht wanderungsfähigen großen Molekülen, weshalb die ähnliche Wirkung der Sperrigkeit kein Zufall sein dürfte.

Immerhin muß wohl die Sperrigkeit nur als ein Faktor unter anderen, vorläufig noch unbekannten betrachtet werden. Ob z. B. Versteifungen der Kohlenstoffkette durch Doppelbindungen oder Ringschlüsse den Temperaturkoeffizient merklich beeinflussen und in welcher Richtung, bleibt noch zu untersuchen. Ferner wissen wir nicht, in welchem Ausmaß das Hydratationswasser polarer Atomgruppen jeweils in die Plasmahaut mitgeschleppt wird. Da hohe Temperatur die Desolvatation begünstigt, so besteht die Möglichkeit, daß die durchgelassenen Teilchen in der Kälte Hydratkomplexe, bei höherer Temperatur dagegen vorwiegend Einzelmoleküle sind, was den Temperaturkoeffizienten in unvoraussehbarer Weise beeinflussen könnte. Der Temperaturkoeffizient der Permeation ist eben eine von dem Zusammenspiel mehrerer schwer zu trennender Faktoren abhängige Größe.

Aus thermodynamischen Gründen wäre eine geringe, aber immerhin deutliche Abnahme der Q_{10}-Werte mit steigender Temperatur zu erwarten. Soweit die spärlichen bisher vorliegenden Versuchsreihen (Tabelle 2), die mehrere Temperaturintervalle umfassen, beurteilen lassen, trifft aber eine solche Erwartung nicht immer zu. Die Ursache hierfür ist unbekannt. Man möchte darin etwa die Wirkung der eben besprochenen, mit der Temperatur sich verändernden Beständigkeit der Hydrate vermuten oder es werden vielleicht bei höherer Temperatur neue Diffusionswege geöffnet (vgl. auch S. 372).

Über artspezifische Unterschiede hinsichtlich der Temperaturabhängigkeit weiß man nicht viel. Der Temperaturkoeffizient eines gegebenen Stoffes ist natürlich von dem Objekt abhängig; die bisher untersuchten Pflanzenzellen scheinen aber sämtlich Werte etwa von derselben Größe ergeben zu haben, so daß man zur Zeit nicht sagen kann, ob möglicherweise zwischen den verschiedenen Permeabilitätstypen und der Einwirkung der Temperatur irgendwelche regelmäßige Beziehungen bestehen (vgl. Tabelle 4). Es verdient immerhin

Tabelle 4. *Temperaturkoeffizienten des Harnstoffs und einiger Harnstoffderivate bei hoch- und niedrigpermeablen Protoplasten. Die Permeabilität P ausgedrückt in cm/sec · 10^7.* (Nach WARTIOVAARA und TIKKANEN 1951.)

	Harnstoff		Thioharnstoff		Methylthioharnstoff		Dimethylharnstoff	
	P	Q_{10}	P	Q_{10}	P	Q_{10}	P	Q_{10}
Rhoeo	0,25	2,3	0,51	2,3	0,73	3,4	3,5	3,6
Spirogyra . . .	0,64	2,3	1,3	2,4	1,4	3,6	5,4	3,8
Elodea	1,9	1,9	3,8	1,9	1,7	3,4	9,5	3,6
Gentiana	1,3	2,2	4,2	2,2	4,0	3,3	23	3,6
Lamium	2,8	2,3	12	2,4	13	3,2	54	3,4

Aufmerksamkeit, daß die bekanntlich hohe Harnstoffpermeabilität von *Elodea* ungewöhnlich wenig temperaturabhängig ist (WARTIOVAARA und TIKKANEN 1951).

Zuckerarten und Aminosäuren. Das oben Gesagte bezieht sich in erster Linie auf wenigstens einigermaßen leicht permeierende Anelektrolyte. Als stark hydrophile, nur in beschränktem Maße osmotisch aufnehmbare, dabei aber physiologisch wichtige Stoffe verdienen die Zucker und Aminosäuren eine besondere Behandlung.

Eine ausgiebigere Aufnahme der Zucker setzt wohl im allgemeinen die Mitbeteiligung nichtosmotischer Prozesse voraus. Da alle Kombinationen von osmotischen und nichtosmotischen Aufnahmemechanismen nicht nur denkbar, sondern vermutlich in der Natur realisiert sind, dürfte sich die Einwirkung der Temperatur hier entsprechend mannigfaltig gestalten. Daß die Aufnahme von organischer Nahrung bei den Heterotrophen innerhalb weiter Temperaturgrenzen mit den dissimilatorischen Prozessen und dem Wachstum gleichen Schritt hält, ist seit langem bekannt. Es handelt sich aber dabei wohl meist um eine sozusagen gezwungene, mit derjenigen der anderen Lebensvorgänge koordinierte Temperaturabhängigkeit.

Nach RIEDER (1951) ist die bereits bei 0^0 recht beträchtliche Glucoseaufnahme der Hefe auf die „passive" Permeabilität zurückzuführen, während bei 25^0 der Anteil „aktiver" Vorgänge etwa 75—80% ausmachen soll. Wenn dies zuträfe, würde sich also beim Einsetzen eines regeren Stoffwechsels zu einem ziemlich effektiven Aufnahmemechanismus ein zweiter, noch effektiverer gesellen. Unter solchen Umständen wäre wohl eine besonders große Temperaturabhängigkeit zu erwarten. In Wirklichkeit bleibt aber nach RIEDER der Temperaturkoeffizient auffallend niedrig (< 2). Es liegt daher nahe anzunehmen, daß beide Aufnahmemechanismen von einem dritten Vorgang (oder von einem den beiden Aufnahmemechanismen gemeinsamen Teilvorgang) mit niedrigen Temperaturkoeffizienten abhängig sind.

Auch bei der Zuckeraufnahme von Blattgewebestücken wurden sehr niedrige Q_{10}-Werte gefunden (WEATHERLEY 1954). Es mag sein, daß Diffusionsvorgänge außerhalb der Protoplasten oder sonstige, bei derartig kompliziert gebauten Objekten schwer auszuschaltende Faktoren das Ergebnis wesentlich beeinflußt haben.

Bei der Aufnahme von Glucose seitens roter Blutkörperchen haben sich dagegen große Q_{10}-Werte ergeben (MASING 1914).

Bei der Bestimmung der Aufnahme von Lysin seitens *Streptococcus faecalis* fand GALE (1947) einen Temperaturkoeffizient von 1,4.

Wasser. Auch die Temperaturabhängigkeit der Wasseraufnahme und -abgabe verdient eine besondere Besprechung. In Anbetracht der geringen Größe des Wassermoleküls ist zu erwarten, daß die Temperaturabhängigkeit der Wasserpermeabilität verhältnismäßig gering sei. Die zur Verfügung stehenden experimentellen Daten (Tabelle 5) bestätigen dies im großen ganzen, wobei allerdings

Tabelle 5. *Temperaturkoeffizienten der Wasseraufnahme und -abgabe.*

	10—20°	20—30 °C	
Allium, Blätter	1,5	2,6	DELF (1916)
Taraxacum, Blütenstengel	2,3	3,8	DELF (1916)
Solanum, Knollenparenchym	3,0	2,7	STILES und JØRGENSEN (1917)
Daucus, Wurzelparenchym	1,3	1,6	STILES und JØRGENSEN (1917)
Daucus, Wurzelparenchym	1,1	1,2	BRAUNER und BRAUNER (1940)
Allium, Epidermis	2,7	—	DE HAAN (1933)

zu beachten ist, daß die empirisch beobachtete Geschwindigkeit der osmotischen Wasseraufnahme und -abgabe meistens eine sehr komplexe Größe darstellt, die auch von allerhand nichtphysiologischen Faktoren beeinflußt sein kann.

Dem Einfluß niedriger Temperaturen auf die Wasseraufnahme von Wurzeln ist viel Aufmerksamkeit geschenkt worden, unter anderem wohl wegen der großen ökologischen und praktischen Bedeutung der Frage (vgl. DÖRING 1935, ROUSCHAL 1935, KÖHNLEIN 1930). Dabei hat sich herausgestellt, daß einige, zumal tropische Pflanzen bei Abkühlung der Wurzeln sofort welken, während andere eine solche Behandlung ohne weiteres vertragen. Die Ursache des unterschiedlichen Verhaltens ist ohne Zweifel in dem lebenden Gewebe der Wurzel (und nicht etwa in der Leistungsfähigkeit der Gefäße) zu suchen, die physiologische Grundlage derselben ist aber noch nicht endgültig aufgeklärt. Vermutlich handelt es sich um Besonderheiten im Protoplasma selbst. Der Ausdruck „erhöhte Wasserpermeabilität" besagt in diesem Zusammenhang nicht viel, solange nicht bekannt ist, wo und wie die größere Wegsamkeit in dem sowohl anatomisch wie physiologisch komplizierten System zustandekommt. Nach ROUSCHAL (1935) beansprucht die Einstellung auf eine veränderte Temperatur bis 2 Std, was auf vitale Vorgänge bei der Umstellung hinweist.

Die Temperaturabhängigkeit der Ionenaufnahme und -abgabe.

Bei der Ionenaufnahme pflegt die einfache Diffusion eine ganz untergeordnete Rolle zu spielen. Es überwiegen vielmehr Austausch und aktive Speicherung (vgl. in diesem Band, S. 230). Das weitere Schicksal der aufgenommenen Ionen ist verschieden. Alkalikationen sowie Halogene scheinen unverändert bis in die Vacuole zu gelangen, während z. B. Nitrate bei verschiedenen Synthesen verbraucht werden. Wieder andere, z. B. Ca und Mg, werden teils in freier und teils in gebundener Form gespeichert. Dies alles macht die Einwirkung der Temperatur auf die Ionenaufnahme kompliziert. Es mag die Temperatur an einer oder mehreren Stellen der Reaktionsfolge und ihrer Verzweigungen der maßgebende Faktor werden und zwar an verschiedenen, je nach der Höhe der Temperatur, dem physiologischen Zustand der Zellen sowie der Konzentration und Zusammensetzung der Außenlösung (SUTCLIFFE 1954).

Der Austausch gegen in der Zelle vorhandene freie Ionen dürfte eine verhältnismäßig einfache Temperaturabhängigkeit zeigen: eine mäßige Geschwindigkeit in der Kälte und eine mehr oder weniger regelmäßige Beschleunigung beim Temperaturanstieg. Klare experimentelle Beweise für die Richtigkeit dieser theoretischen Vermutung scheinen aber auffallenderweise nicht vorzuliegen.

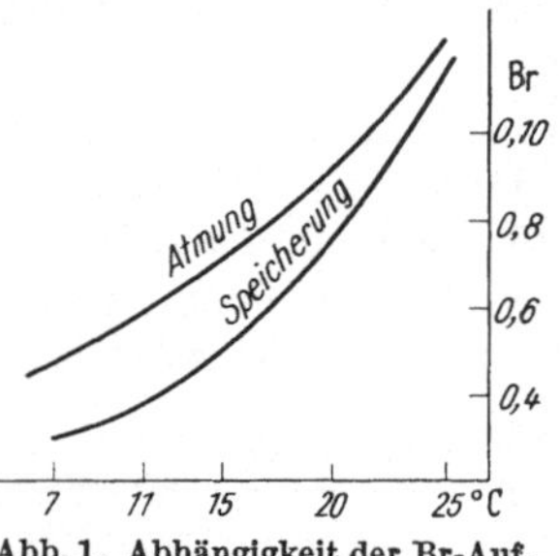

Abb. 1. Abhängigkeit der Br-Aufnahme von der Atmung bei *Elodea*. (Nach ROSENFELS 1935.)

Die Ionenspeicherung ist außer von der Atmung (Abb. 1) auch von solchen Lebensäußerungen wie Wachstum und Kohlensäureassimilation abhängig, deren temperaturbedingte Intensitätsveränderungen sich in der Ionenaufnahme widerspiegeln. Ungleiche Temperaturkoeffizienten dieser Vorgänge und spezifische Ansprüche (NO_3 für die Eiweißsynthese, K für die Vacuolenausdehnung) mögen dabei beträchtliche qualitative und quantitative Unterschiede zwischen verschiedenen Objekten verursachen. Näheres ist hierüber aber nicht bekannt, unter anderem weil die allermeisten Forscher mit recht ähnlichen Objekten, nämlich entweder mit Wurzeln (WANNER 1948, LUNDEGÅRDH 1950, KRAMER und WIEBE 1952, OVERSTREET und JACOBSON 1946) oder Schnitten aus parenchymatischen Pflanzenteilen gearbeitet haben. Außerdem ist die Bewertung der Versuchsergebnisse an solchen Objekten schwierig. Manche wichtige, aber schwer kontrollierbare Faktoren wie die Sauerstoffversorgung der Zellen, die Diffusion außerhalb der Protoplasten, Wundreize u. dgl. erschweren den Vergleich zwischen den Angaben verschiedener Autoren. Als ein Beispiel von dem Effekt solcher Einflüsse sei auf die auffallende Einwirkung der Belichtung auf den Temperaturkoeffizient des Kaliumaustausches bei *Ulva* hingewiesen (SCOTT und HAYWARD 1953). Der Q_{10}-Wert ist im Lichte etwa doppelt so groß wie im Dunkeln.

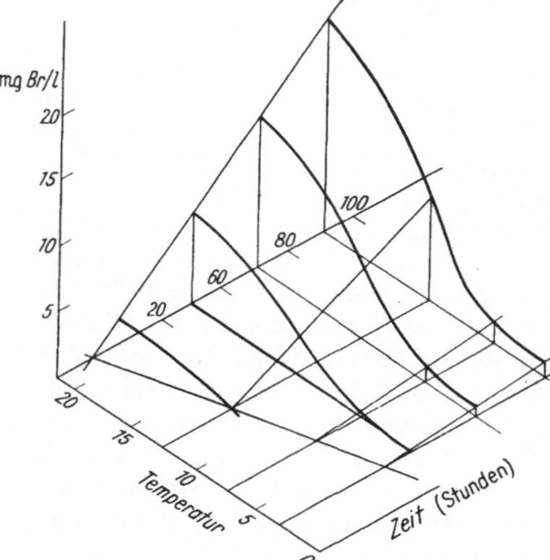

Abb. 2. Verlauf der Br-Aufnahme seitens Kartoffelscheiben bei verschiedenen Temperaturen. (Nach STEWARD, BERRY und RAMAMURTI 1943.)

Im Bereich der normalen Lebenstemperatur der betreffenden Objekte scheint der Temperaturkoeffizient der Ionenaufnahme im allgemeinen zwischen etwa 2 und 3 zu liegen, kann aber unter Umständen beträchtlich höher sein (STEWARD und Mitarbeiter 1943). Gegen höhere und niedrigere Temperatur pflegt der Q_{10}-Wert abzunehmen, so daß die Temperaturabhängigkeit graphisch dargestellt die für manche Lebensvorgänge charakteristische S-förmige Kurve ergibt (Abb. 2). Es kann also für die Ionenaufnahme kein über dem ganzen verträglichen Temperaturbereich gültiger ungefährer Temperaturkoeffizient angegeben werden (STEWARD 1932). Der Kurvenverlauf zeigt, daß mehrere bei mittlerer Temperatur gegenseitig ausbalancierte Teilvorgänge die Geschwindigkeit der Aufnahme und ihren Temperaturkoeffizient bestimmen. Nach BUTLER (1953) soll in der Kälte Adsorption und Austausch vorherrschen, während die aktive Speicherung erst bei höheren Temperaturen ausschlaggebend wird. Welche Faktoren beim weiteren Temperaturanstieg beschränkend werden, ist noch eine ungelöste Frage. Es mögen unter anderem Beschleunigung der Gegenreaktionen, Exosmose oder Sättigung der Transportmechanismen sein.

Über die Phosphoraufnahme der Hefe und der Bakterien ist viel gearbeitet worden (KAMEN und SPIEGELMAN 1948). Der Einfluß der Temperatur auf diesen Vorgang ist aber noch recht mangelhaft aufgeklärt.

Farbstoffe. Gegenüber den meisten anderen bei Permeationsversuchen benutzten Substanzen zeichnen sich die Farbstoffe unter anderem durch bedeutende Dimensionen aus sowie durch gedrungenen Bau ihrer Moleküle, die mehrere verschiedenartige funktionale Gruppen enthalten. Es ist vorauszusehen, daß dies eine besondere Wirkung auf die Temperaturabhängigkeit der Farbstoffaufnahme haben muß. Diese Behauptung kann leider nicht durch Literaturhinweise belegt werden, denn die Anzahl der Arbeiten, welche die Einwirkung der Temperatur auf die Farbstoffaufnahme berühren, ist im Verhältnis zu der großen Zahl der Vitalfärbungsstudien auffallend gering.

Die als undissoziierte Moleküle leicht eindringenden Farbbasen (Neutralrot, Methylenblau) werden von den Zellen stark gespeichert und kommen üblicherweise als äußerst verdünnte Lösungen zur Verwendung. Unter solchen Umständen können die Nachlieferung aus der Umgebung und die Diffusion in der Zellwand leicht beschränkend werden, was natürlich niedrige Temperaturkoeffizienten ergibt. Dies war offenbar der Fall in den Versuchen von PFEFFER (1886) und ENDLER (1912). Die anscheinend einzige eingehendere Arbeit über dieses Thema ist die von IRWIN (1928) über die Akkumulation von Brillant-Kresylblau seitens Nitella. Der Temperaturkoeffizient hatte den Wert von 4,9 bei 20—25°. Die Höhe dieses Wertes wurde von der Verfasserin auf die Beteiligung chemischer Reaktionen an dem Aufnahmevorgang zurückgeführt. Im Lichte späterer Erfahrungen an Anelektrolyten erscheint ihr Befund schon in Anbetracht der Größe bzw. Sperrigkeit des Farbstoffmoleküls einleuchtend.

Wie die Speicherung von Farbanionen von der Temperatur beeinflußt wird, ist nicht näher bekannt. Nach den Angaben von COLLANDER (1921) zu urteilen, muß der Temperaturkoeffizient jedenfalls sehr groß sein.

Literatur.

BĚLEHRÁDEK, J.: Temperature and living matter. Berlin: Gebrüder Bornträger 1935. — BRAUNER, L., and M. BRAUNER: Further studies of the influence of light upon the water intake and output of living plant cells. New Phytologist **39**, 104—128 (1940). — BULL, H. B.: Physical biochemistry. New York: John Wiley & Sons, London: Chapman & Hall 1951. — BUTLER, G. W.: Ion uptake by young wheat plants. III. Phosphate absorption by excised roots. Physiol. Plantarum (Copenh.) **6**, 637—661 (1953).

COLLANDER, R.: Über die Permeabilität pflanzlicher Protoplasten für Sulfosäurefarbstoffe. Jb. wiss. Bot. **60**, 354—410 (1921). — COLLANDER, R., u. B. WIKSTRÖM: Die Permeabilität pflanzlicher Protoplasten für Harnstoff und Alkylharnstoffe. Physiol. Plantarum (Copenh.) **2**, 235—246 (1949).

DAVSON, H., and J. F. DANIELLI: The permeability of natural membranes. Cambridge 1943. — DELF, E. M.: Studies of protoplasmic permeability by measurement of rate of shrinkage of turgid tissues. I. The influence of temperature on the permeability of protoplasm to water. Ann. of Bot. **30**, 283—310 (1916). — DÖRING, B.: Die Temperaturabhängigkeit der Wasseraufnahme und ihre ökologische Bedeutung. Z. Bot. **28**, 305—383 (1935).

HAAN, I. DE: Protoplasmaquellung und Wasserpermeabilität. Rec. Trav. bot. néerl. **30**, 234—235 (1933). — HARLEY, J. L., and J. K. BRIERLEY: The uptake of phosphate by excised mycorrhizal roots of the beech. VI. Active transport of phosphorous from the fungal sheat into the host tissue. New Phytologist 53, 240—252 (1954). — HOAGLAND, D. R., and T. C. BROYER: General nature of the process of salt accumulation by roots with description of experimental methods. Plant Physiol. 11, 475—507 (1936). — HOFFMANN, C.: Über die Durchlässigkeit kernloser Zellen. Planta (Berl.) 4, 584—605 (1927).

IRWIN, M.: On the accumulation of dye in Nitella. J. Gen. Physiol. 8, 147—182 (1928).

JOST, W.: Diffusion in solids, liquids, gases. New York 1952.

KAMEN, M. D., and S. SPIEGELMAN: Studies on the phosphate metabolism of some unicellular organisms. Cold Spring Harbor Symp. Quant. Biol. 8, 151—163 (1948). — KÖHNLEIN, E.: Untersuchungen über die Höhe des Wurzelwiderstandes und die Bedeutung aktiver Wurzeltätigkeit für die Wasserversorgung der Pflanzen. Planta (Berl.) 10, 381—423 (1930). — KRAMER, P. J., and H. H. WIEBE: Longitudinal gradients of P^{32} absorption in roots. Plant Physiol. 27, 661—674 (1952).

LUNDEGÅRDH, H.: The translocation of salts and water through wheat roots. Physiol. Plantarum (Copenh.) 3, 103—151 (1950).

MASING, E.: Über die Verteilung von Traubenzucker in Menschenblut und ihre Abhängigkeit von der Temperatur. Pflügers Arch. 156, 401—425 (1914).

NETTER, H.: Biologische Physikochemie. Potsdam: Akademische Verlagsgesellschaft Athenaion 1951.

OVERSTREET, R., and L. JACOBSON: The absorption of rubidium and phosphate ions at extremely small concentrations as revealed by experiments with Rb^{86} and P^{32} prepared without inert carrier. Amer. J. Bot. 33, 107—112 (1946).

PFEFFER, W.: Über Aufnahme von Anilinfarben in lebende Zellen. Unters. bot. Inst. Tübingen 2, 179—332 (1886). — PRAGER, S., E. BAGLEY and F. A. LONG: Diffusion of hydrocarbon vapors into polyisobutylene. II. J. Amer. Chem. Soc. 75, 1255—1256 (1953).

RIEDER, H. P.: Über die Zuckeraufnahme von Hefezellen. Ber. schweiz. bot. Ges. 61, 539—622 (1951). — ROBERTSON, R. N.: Studies in the metabolism of plant cells. II. Effects of temperature on accumulation of potassium chloride and on respiration. Austral. J. Exper. Biol. a. Med. Sci. 22, 237—245 (1944). — ROSENFELS, R. S.: The absorption and accumulation of potassium bromide by Elodea as related to respiration. Protoplasma 23, 503—519 (1935). — ROUSCHAL, E.: Untersuchungen über die Temperaturabhängigkeit der Wasseraufnahme ganzer Pflanzen. Sitzgsber. Akad. Wiss. Wien, Math.-naturwiss. Kl., Abt. I 144, 313—348 (1935).

SCARTH, G. W., and J. LEVITT: The frost-hardening mechanism of plant cells. Plant Physiol. 12, 51—78 (1937). — SCOTT, G. T., and H. R. HAYWARD: The influence of temperature and illumination on the exchange of potassium ion in Ulva lactuca. Biochim. et Biophysica Acta 12, 401—404 (1953). — SEEMANN, F.: Der Einfluß der Wärme und UV-Bestrahlung auf die Wasserpermeabilität des Protoplasmas. Protoplasma 39, 535—566 (1950). — STEWARD, F. C.: The absorption and accumulation of solutes by living plant cells. I. Experimental conditions which determine salt absorption by storage tissue. Protoplasma 15, 29—58 (1932). — STEWARD, F. C., W. E. BERRY and T. K. RAMAMURTI: The absorption and accumulation of solutes by living plant cells. X. Time and temperature effects on salt uptake by potato discs and the influence of the storage conditions of the tuber on metabolism and other properties. Ann. of Bot., N. S. 7, 221—260 (1943). — STILES, W., and I. JØRGENSEN: Studies in permeability. V. The swelling of plant tissue in water and its relation to temperature and various dissolved substances. Ann. of Bot. 31, 417—434 (1917). — SUTCLIFFE, J. F.: The exchangeability of potassium and bromide ions in cells of red beetroot tissue. J. of Exper. Bot. 5, 313—326 (1954).

WANNER, H.: Untersuchungen über die Temperaturabhängigkeit der Salzaufnahme durch Pflanzenwurzeln. I. Die relative Größe der Temperaturkoeffizienten (Q_{10}) von Kationen- und Anionenaufnahme. Ber. schweiz. bot. Ges. 58, 123—130 (1948). — WARTIOVAARA, V.: Über die Temperaturabhängigkeit der Protoplasmapermeabilität. Ann. bot. Soc. zool.-bot. fenn. „Vanamo" 16, 1—111 (1942). — The permeability of the plasma membranes of Nitella to normal primary alcohols at low and intermediate temperatures. Physiol. Plantarum 2, 184—196 (1949). — WARTIOVAARA, V., u. R. TIKKANEN: Zur Permeation des Harnstoffs in Pflanzenzellen. Arch. Soc. zool.-bot. fenn. „Vanamo" 6, 19—24 (1951). — WEATHERLEY, P. E.: Preliminary investigations into the uptake of sugars by floating leaf disks. New Phytologist 53, 204—216 (1954).

ZWOLINSKI, B. J., H. EYRING and C. E. REESE: Diffusion and membrane permeability. I. J. Physic. a. Colloid Chem. 53, 1426—1453 (1949).

Die Beeinflussung des Stoffaustausches durch das Licht.

Von

Leo Brauner.

Mit 4 Abbildungen.

I. Das Problem.

Bei einer Untersuchung der Wirkung von Außenfaktoren auf den Stoffaustausch lebender Zellen ist stets zu bedenken, daß das Auftreten von Veränderungen der Aufnahme- oder Abgabegeschwindigkeit prinzipiell von zwei Seiten her verursacht sein kann: durch eine Beeinflussung der Diffusionswiderstände im System oder durch eine Veränderung der treibenden Kräfte. Bei den in diesem Kapitel zu behandelnden Wirkungen des Lichts auf den Stoffaustausch ist die Entscheidung, welcher dieser beiden Faktoren in einem gegebenen Fall den Ausschlag gibt, bei grünen Geweben noch besonders erschwert. Denn die gleichzeitige Induktion von Photosynthese muß hier zu Umsetzungen führen, die den Mechanismus der Stoffbewegung sekundär beeinflussen können. So wird ein Überblick über die Fülle des bereits vorliegenden Beobachtungsmaterials zeigen, daß die Erforschung des Problems bisher noch kaum über ihre deskriptive Phase hinausgekommen ist. Nur in manchen weniger komplexen Reaktionstypen deuten sich bereits einige Grundeffekte an, die in einer allgemeinen Theorie der zusammenwirkenden Faktoren zu berücksichtigen sein werden.

Um unter den zahlreichen einander scheinbar widersprechenden Reaktionstypen gemeinsame Gesetzmäßigkeiten erkennen zu können, sollen in den folgenden Abschnitten die bisher bekanntgewordenen Lichteffekte nach den Hauptgruppen der ausgetauschten Stoffe zusammengestellt werden: Elektrolyte, Wasser und Nichtelektrolyte. Ferner scheint es mit Hinblick auf die polare Natur der zu untersuchenden Systeme zweckmäßig, Stoffaufnahme und -abgabe getrennt zu behandeln.

II. Austausch von Elektrolyten.

Die ersten ernsthaften Untersuchungen von „Lichtpermeabilitätsreaktionen" haben annähernd gleichzeitig Lepeschkin (1908, 1909) und Tröndle (1909, 1910) ausgeführt. Lepeschkin wurde durch seine Studien der Lichtturgorreaktionen von Blattgelenken dazu veranlaßt, den Ursachen der Druckänderungen bei Beleuchtungswechsel nachzugehen. Zur Analyse des Effekts ermittelte er unter anderem den isotonischen Koeffizienten von KNO_3 im Hellen und im Dunkeln. Das Ergebnis war, daß belichtete Zellen je Zeiteinheit erheblich mehr Salz eindringen lassen als verdunkelte. Ganz ähnlich verhielten sich in Lepeschkins Versuchen auch *Spirogyra*-Zellen und die Epidermis von *Rhoeo discolor*.

Tröndle prüfte diese Befunde an den Mesophyllzellen von *Tilia*- und *Buxus*-Blättern nach. Ein Vergleich der plasmolytischen Grenzkonzentrationen von NaCl und Saccharose führte auch ihn zum Schluß, daß in chlorophyllhaltigen Zellen „die Permeabilität für Kochsalz und vermutlich auch für andere Elektrolyte mit steigender Beleuchtung zu-, mit fallender Beleuchtung abnimmt".

Gegen das von diesen beiden Autoren angewandte Prinzip der Bestimmung der Stoffaufnahme sind in der Folgezeit ernsthafte methodische Bedenken vorgebracht worden, vor allem von FITTING (1917) und ZYCHA (1928); doch sind diese wohl nur soweit berechtigt, als sie sich auf die quantitative Auswertung der Resultate beziehen. Der hier wesentlichere qualitative Befund einer Begünstigung der Salzaufnahme durch das Licht bleibt jedoch von dieser Kritik unberührt. Es sei nur abermals betont, daß die aufgefundenen Effekte vor einer Analyse ihrer Faktoren noch nicht ohne weiteres als „Permeabilitäts"reaktionen interpretiert werden dürfen. — Wichtig und durch spätere Untersuchungen bestätigt ist ferner TRÖNDLEs Beobachtung (1910), daß die Erhöhung der Salzaufnahme im Tageslicht periodischen Schwankungen unterworfen ist.

In diesem Zusammenhang sei eine kleine Studie von BRAUNER (1922) erwähnt, der auf der Suche nach primären Wirkungen des Lichts bei den Photowachstums- und -krümmungsreaktionen der Haferkoleoptile auch die Beeinflußbarkeit der Elektrolyt„permeabilität" des Gewebes prüfte. Durch Messung der Zunahme der grenzplasmolytischen Konzentration von $NaNO_3$ während eines 10 min langen Aufenthalts im Osmotikum konnte die jeweils aufgenommene Salzmenge ermittelt werden. Vorbelichtung der Keimlinge mit 50000 MKS Glühlicht erhöht zunächst die Absorption gegenüber den Dunkelkontrollen. Im weiteren Verlauf dagegen kehrt sich dieses Verhältnis wieder um, wie dies bei einem periodischen Verlauf der Lichtreaktion zu erwarten war.

Die bisher besprochenen plasmolytischen Messungen machen jedoch nur die Tatsache wahrscheinlich, daß Licht in lebenden Pflanzenzellen die Salz*aufnahme* fördert, ohne Aufschluß über den Mechanismus der Reaktion zu geben. In einer späteren Arbeit wurde daher von BRAUNER (1924) der Wechselstromwiderstand des verdunkelten und belichteten Koleoptilgewebes verglichen. Bestrahlung mit Glühlampenlicht rief stets eine ansehnliche Steigerung der Leitfähigkeit hervor, die unmittelbar mit der Belichtung einsetzte und im Dauerlicht bis zum Versuchsende erhalten blieb. Diese Widerstandsänderungen lassen also bereits erkennen, daß an der Photoreaktion tatsächlich eine Erhöhung der Permeabilität für die zelleigenen Ionen beteiligt ist. Bemerkenswerterweise läßt der Ablauf dieses Effekts im Gegensatz zur Salzaufnahme keinerlei Periodik erkennen.

Eine weitere sehr wesentliche Wirkung des Lichts ist die Absorption von Ionen durch das bestrahlte Gewebe *gegen* den Konzentrationsgradienten. Dieses Phänomen wurde zuerst von HOAGLAND und DAVIS (1924) beobachtet, als sie die Chloridaufnahme von *Nitella clavata* im Tageslicht und im Dunkeln durch Messung des Konzentrationsverlustes im Außenmedium verfolgten. In einem Versuch mit einem $KCl + KH_2PO_4$-Gemisch absorbierte die Alge das als 0,001 Mol-Lösung gebotene KCl nach 4 Tagen im Licht vollständig, während im Dunkeln noch über 87% der Ausgangskonzentration im Medium verblieben war. — In einer späteren Arbeit untersuchten HOAGLAND, HIBBARD und DAVIS (1927) noch eingehender die Absorptionsbedingungen beim Bromid. Es konnte nunmehr eindeutig nachgewiesen werden, daß Belichtung der Kultur zu einer Speicherung von KBr in den Algenzellen auf über das Doppelte der Außenkonzentration führte, wenn diese ursprünglich 0,005 Mol betrug. Im Dunkeln dagegen blieb die Ionenkonzentration stets unter diesem Wert.

Tabelle 1. *Nitella.*

	Licht	Dunkel
Außenmedium 5×10^{-3} Mol KBr . . .		
Innenkonzentration nach 50 Std . . .	12,3	$3,3 \times 10^{-3}$ Mol KBr

Zur Erklärung dieses Lichteffekts ist die Annahme einer einfachen Photopermeabilitätsreaktion nicht ausreichend, da die Absorption über das Konzentrationsgleichgewicht hinaus ja einen zusätzlichen Energieaufwand erfordert.

Wahrscheinlicher dürfte die Auslösung einer aktiven Ionenaufnahme sein, bei der das Licht als Energiequelle wirkt. Über die Natur dieses Mechanismus konnten die Autoren jedoch keine konkreten Vorstellungen entwickeln.

Dies versuchen zuerst JACQUES und OSTERHOUT (1934), die bei einer anderen grünen Alge, *Valonia macrophysa*, eine Speicherung von Kalium im Licht feststellen konnten. Sie gehen von der Annahme aus, daß Kalium vorwiegend in Form von undissoziiertem KOH aus der Außenlösung in die Vacuole gelangt. Demnach müßte der Vorgang durch eine Steigerung der OH-Konzentration außerhalb der Zelle gefördert werden. Diese Bedingung soll nach der Vorstellung der beiden Forscher das Licht auf dem Wege über die Photosynthese schaffen, die durch Verbrauch von Kohlensäure den p_H-Wert im Medium erhöht.

INGOLD (1936) untersuchte an *Helodea*-Sprößchen durch Titration der Außenlösung den Einfluß von Glühlampenlicht auf die Aufnahme von KCl und KH_2PO_4. In beiden Fällen ließ sich eine merkliche Förderung der Absorption im Licht feststellen. Dabei kam es aber, entgegen den Erfahrungen von JACQUES und OSTERHOUT, zu einer leichten Ansäuerung des Mediums der Lichtprobe, die das oben angedeutete Erklärungsprinzip der K-Aufnahme hier unanwendbar macht. Für die Interpretation des Effekts ist nun eine weitere Beobachtung INGOLDS wichtig: werden *Helodea*-Sprosse in destilliertes H_2O eingebracht, so läßt sich durch Bestimmung der elektrischen Leitfähigkeit des Außenmediums die Exosmose von Elektrolyten aus dem Gewebe ermitteln. Im Gegensatz zur Förderung der Salz*aufnahme* wird nun die Ionen*abgabe* durch das Licht erheblich *gehemmt*. Dieses Ergebnis, das einen ähnlichen Befund DILLEWIJNS (1927) an *Helianthus*-Hypokotylen bestätigt, schließt wiederum das Vorliegen einer einfachen Permeabilitätsreaktion aus. INGOLD hält daher, wie schon die vorerwähnten Autoren, eine Verursachung aktiver Stoffaufnahme durch das Licht für wahrscheinlich, wobei der Photosynthese eine entscheidende Rolle zufallen soll. Die Hemmung der Exosmose wäre dann möglicherweise die Folge einer Bindung der Zellelektrolyte an bestimmte Produkte der Photosynthese. — Es muß allerdings betont werden, daß die Frage des Lichteffektes auf die Ionenabgabe nach den bisher vorliegenden Angaben noch einigermaßen kontrovers ist. So haben BLACKMAN und PAINE (1918) bei Belichtung von *Mimosa*-Blattgelenken in Wasser gerade die entgegengesetzte Reaktion beobachtet, nämlich eine *Förderung* der Ionen*abgabe*. Die Autoren schließen aus diesem Befund, daß am Mechanismus der normalen Lichtturgorreaktion des Gelenks eine Erhöhung der Inhaltspermeabilität in der bestrahlten Flanke beteiligt sei. — Später hat LEPESCHKIN (1940) an einem chlorophyllfreien Objekt, dem Speichergewebe der Kartoffelknolle, je nach Art der Belichtung den einen oder den anderen Effekt beobachten können: direktes Sonnenlicht und UV-Strahlung bestimmter Wellenlänge verstärken die Ionenabgabe, während andere Bereiche des UV-Spektrums Salzspeicherung begünstigen. Offenbar interferieren also in allen diesen Versuchen die Wirkungen einer Permeabilitätserhöhung mit einer lichtinduzierten aktiven Stoffaufnahme.

KAHO (1937) fand schließlich bei etiolierten Lupinenhypokotylen überhaupt keinen Einfluß von Glühlampenlicht (300 bzw. 835 Lux) auf die Ionenexosmose. Der Autor erwägt als Ursachen für das Ausbleiben jeden Lichteffekts einerseits das Fehlen von Chlorophyll im Gewebe und andererseits die Armut des Glühlampenlichts an wirksamen kurzwelligen Strahlen.

In der Folgezeit mehrten sich andererseits zuverlässige Beobachtungen, daß bei höheren Pflanzen Belichtung des beblätterten Sprosses zu einer Steigerung der Ionenabsorption durch das Wurzelsystem führt. O. SCHMIDT (1936) untersuchte die Salzaufnahme von *Sanchezia nobilis* in Wasserkulturen, wobei der Transpirationsstrom durch automatische Regulierung der Luftfeuchtigkeit im

Licht und im Dunkeln konstant gehalten wurde. Bei starker Belichtung des Sproßteils absorbierte die verdunkelte Wurzel mehr K, als dem aufgenommenen Lösungsvolumen entsprach. Dies spricht also wieder für aktive Ionenaufnahme. In anderen Fällen ergaben sich allerdings erhebliche Komplikationen. So zeigte die Wirkungskurve der Lichtintensität bei der Absorption von NO_3, PO_4 und Ca in ihrem mittleren Bereich ein deutliches Minimum, und bei bestimmten Beleuchtungsstärken wurden Ca und PO_4 sogar wieder an die Außenlösung abgegeben. Eine Interferenz von aktiver Stoffaufnahme mit einer Permeabilitätsreaktion ist diesmal schwer vorstellbar, da das Wurzelsystem ja dauernd verdunkelt geblieben war.

Einheitlichere Ergebnisse erzielten LUTTKUS und BÖTTICHER (1939) bei ihren Experimenten mit Mais-Wasserkulturen. Das Wurzelsystem nahm hier im Tageslicht aus der Nährlösung Kalium auf und schied es bei Verdunklung wieder aus. Die beiden Autoren nehmen an, daß Kalium unter dem Einfluß der Belichtung mit irgendwelchen Plasmaelementen in den Zellen eine labile Bindung eingeht. Die Förderung der Kaliumaufnahme sei demnach durch die Festlegung des permeierten Anteils, also durch Aufrechterhaltung des Konzentrationsgefälles verursacht. Bei Verdunklung würde diese Bindung wieder gelöst und die Ionen könnten im jetzt umgekehrten Konzentrationsgefälle zurückdiffundieren. Da das ausgeschiedene Kalium nachweislich aus dem Sproßteil der Pflanze stammt, ist anzunehmen, daß sich die entscheidende Lichtreaktion im assimilierenden Gewebe abspielt.

Mit dem gleichen Problem haben sich seit mehreren Jahren ARISZ und seine Mitarbeiter beschäftigt. Im Mittelpunkt dieser Studien steht die Frage nach der Beteiligung der Photosynthese an der aktiven Aufnahme von Ionen. ARISZ (1947/1948) leitet die Untersuchung mit der Analyse eines verhältnismäßig übersichtlichen Vorgangs ein, der Absorption von Cl-Ionen durch definierte Blattstreifen von *Vallisneria spiralis*. Starkes Glühlampenlicht (12900 Lux) steigert die Chloridaufnahme aus 10^{-3} Mol KCl-Lösung auf das 3—5fache des Dunkelwertes. Dieser Effekt entwickelt sich bemerkenswerterweise auch bei Ausschluß von CO_2. ARISZ schließt daher, daß das Licht in diesem System *nicht* über den Mechanismus der Photosynthese, sondern durch eine direkte Beeinflussung der aufnehmenden Protoplasten wirkt. In zwei späteren Arbeiten entwickelt ARISZ (1948, 1954) diese Vorstellung weiter, vorwiegend auf Grund seiner Erfahrungen über die Weiterleitung des Cl-Ions im Blatt. Wird ein Blattstreifen mit seinem Ende in verdünnte KCl-Lösung getaucht, so begünstigt Belichtung des eingetauchten Teils zunächst die Salzaufnahme in diese „Kontaktzone". Wird der Streifen sodann im Dunkeln in reines Wasser übertragen, so verursacht lokale Belichtung der angrenzenden Blattpartien Überwandern von Cl' aus der ursprünglichen Kontaktzone („re-distribution"). Auf Grund dieser Beobachtung stellte ARISZ folgende Hypothese auf: das von außen in das Cytoplasma der Kontaktzone diffundierte Cl' wird bei Belichtung aktiv in die Vacuole sezerniert. Bei nachfolgender Verdunklung kehrt das Chlorid wieder in das Cytoplasma zurück und wird über Plasmaverbindungen in die jetzt belichteten Nachbarzellen geleitet, die es ihrerseits aktiv in ihren Vacuolen speichern. Bemerkenswerterweise wird einmal in den Zellverband aufgenommenes Chlorid nicht wieder nach außen verloren.

Die übrigen Arbeiten des ARISZschen Laboratoriums beschäftigen sich mit den viel komplexeren Verhältnissen bei der Salzaufnahme ganzer Maispflanzen. ALBERDA (1948) stellte fest, daß Phosphate von der Wurzel nur bei Belichtung des Sproßteils absorbiert werden können, während im Dunkeln sogar ein schwacher PO_4-Verlust zu beobachten war. Hier tritt wieder der Gedanke auf, daß das Licht die Ionenaufnahme sekundär über die Photosynthese fördert, vermutlich durch Bildung gewisser Zwischenprodukte, die als Substrat der Ionenbildung dienen können.

Es folgen zwei wichtige Mitteilungen von VAN ANDEL, ARISZ und HELDER (1950) und von HELDER (1952), in denen zunächst nachgewiesen wird, daß die Phosphataufnahme im Licht wie im Dunkeln bei Zufuhr von Glucose ansteigt. Daraus läßt sich schließen, daß die Förderwirkung des Lichts in der photosynthetischen Bereitstellung von Glucose besteht, die das permeierte Phosphat vermutlich in organischer Bindung (Phosphorylierung!) fixiert. Die Mitwirkung der Photosynthese ließ sich hier auch direkt nachweisen: im Gegensatz zu den Bedingungen der Cl-Absorption bei *Vallisneria* verursacht beim Mais der Entzug von CO_2 eine erhebliche Hemmung der PO_4-Aufnahme im Licht. — Schließlich konnte VAN DER BURG (1952) zeigen, daß auch die Förderung der Kaliumabsorption im Licht ganz entsprechend durch CO_2-Ausschluß reduziert wird. Erneute Zufuhr von CO_2 setzt die Salzaufnahme unmittelbar wieder in Gang.

Es ist schon mehrfach betont worden, daß bei Untersuchungen der Lichtwirkung auf den Ionenaustausch in lebenden Geweben einer Isolierung des Permeabilitätsfaktors erhebliche Schwierigkeiten im Wege stehen. L. und M. BRAUNER (1937, 1938) versuchten nun, von einer anderen Seite her Einblick in das Permeabilitätsverhalten belichteter Membranen zu gewinnen. Den Ausgangspunkt bildete ein Studium der photoelektrischen Reaktionen grüner Blätter. Da sich bald Anzeichen dafür ergaben, daß die bei Belichtung auftretenden Potentialänderungen auf einer Beeinflussung präexistierender Membranpotentiale beruhen, wurde versucht, entsprechende Effekte an übersichtlicheren, leblosen Membranmodellen zu rekonstruieren. Als geeignet erwies sich gewöhnliches Pergamentpapier nach Behandlung mit bestimmten Sensibilisatoren: Chlorophyll, Carotinoide, Methylenblau und vor allem Graphit. Als Diaphragma in eine Konzentrationskette eines Elektrolyten eingeschaltet, produzieren solche Membranen infolge ihrer Ionenselektivität ein Membran-Diffusionspotential, dessen positiver Pol in der verdünnten Lösung liegt. Im Dunkeln fällt dieses Potential mit dem allmählichen Ausgleich des Konzentrationssprunges ganz langsam ab. Wird die Membran jedoch belichtet, so kommt es augenblicklich zu einer drastischen Veränderung des Dunkelpotentials, und zwar in den meisten Fällen zu einem Abfall. Diese Reaktion erweist sich bei nachfolgender Wiederverdunkelung als streng reversibel (Abb. 1, a). Da ihr Wirkungsspektrum nach dem kurzwelligen Gebiet steil ansteigt, handelt es sich offenbar um einen echten Photoeffekt. Zur Analyse des Mechanismus der Reaktion wurde deren Abhängigkeit von den Eigenschaften der potentialbildenden Ionen untersucht. Dabei ergab sich die allgemeine Regel, daß die Spannung im Licht um so stärker abfällt, je unbeweglicher die beteiligten Kationen und je beweglicher die Anionen sind:

Photoreaktion (P. D.-Abfall) →

Kationenreihe: $KCl < NaCl < LiCl$ bzw. $BaCl_2 < CaCl_2 < MgCl_2$

Anionenreihe: $KJO_3 < K_2SO_4 < KCl < KJ < KBr$

Das Licht löst also in der empfindlichen Membran offenbar einen Doppeleffekt aus: eine Verringerung der Kationenbeweglichkeit und eine gleichzeitige Erhöhung der Anionenbeweglichkeit. Diesen Veränderungen scheint das gleiche Grundphänomen zugrunde zu liegen: Belichtung verringert wie beim HALLWACHS-Effekt an Metalloberflächen die negative Eigenladung der Membran durch Abtrennung von Photoelektronen. Dies führt zunächst zu einer Verringerung der elektrischen Ventilwirkung gegenüber den gleichsinnig geladenen

Anionen (primärer Photoeffekt). Gleichzeitig wird aber der Ladungsverlust der Membranmicelle deren Quellungskapazität herabsetzen und dadurch eine mechanische Verdichtung des Systems hervorrufen (sekundärer Photoeffekt). Durch die so verursachte Verengung des Porenquerschnittes muß die Beweglichkeit der Kationen ihrem Volumen entsprechend reduziert werden. Beide Effekte bewirken gemeinsam eine Verringerung des relativen Abstandes zwischen Kation und Anion im Diffusionsstrom und damit eine Reduktion des Diffusionspotentials.

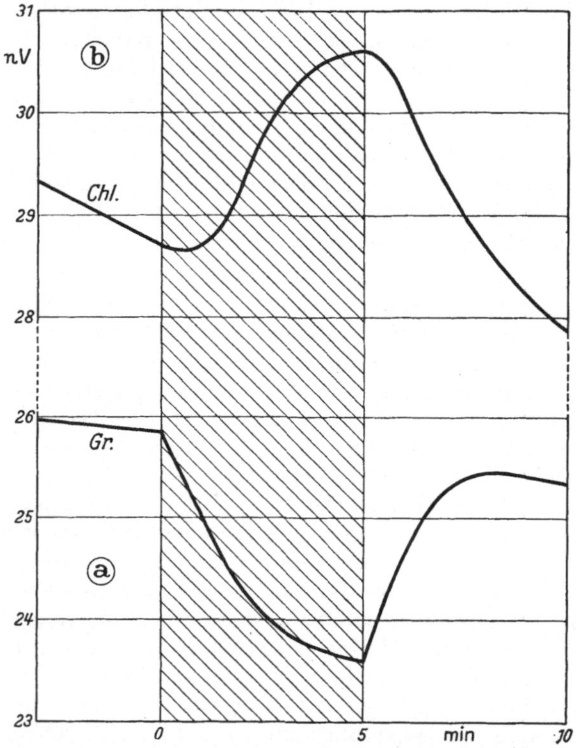

Abb. 1. Photoelektrischer Effekt in Pergamentpapiermembranen. Sensibilisator in „a“ Graphit in „b“ Chlorophyll. Konzentrationskette: $10^{-2}/10^{-3}$ Mol KCl. Bestrahlung mit 20×10^{-3} cal./cm²/sec wärmefreiem Weißlicht. Schraffiertes Feld: Lichtperiode. (Nach L. und M. Brauner 1938.)

Gelegentlich kommt es allerdings bei besonders hochsensibilisierten Membranen und in Gegenwart von sehr voluminösen Anionen (PO_4!) zu einer Umkehr der Reaktion in eine Potentialsteigerung (Abb. 1, b). In diesem Fall überwiegt der sekundäre Effekt der Porenverengerung so sehr, daß nunmehr die Verringerung der elektrostatischen Ventilwirkung in den Hintergrund tritt und die Beweglichkeit der Anionen in der Membran stärker beeinträchtigt wird als die kleinerer Kationen.

Aus dieser elektrometrischen Analyse ergibt sich also, daß Belichtung die Durchlässigkeit der geprüften Membranen für Kationen stets verringert. Bei den Anionen dagegen kann es je nach dem Überwiegen des primären oder des sekundären Photoeffektes zu einer Erhöhung oder Erniedrigung ihrer Beweglichkeit in der Membran kommen. Es hat sich nun gezeigt, daß auch die in lebenden Blättern (*Helodea*) zu beobachtenden photoelektrischen Effekte auf dem gleichen Reaktionsmechanismus beruhen. Hier kommt allerdings als komplizierender Faktor Produktion und Verbrauch von Ionen durch die gleichzeitig ausgelöste Photosynthese hinzu. Doch läßt sich deren Wirkung meist leicht übersehen und berücksichtigen.

Für das Problem der Lichtpermeabilitätsreaktionen folgt aus diesen Erfahrungen, daß der Durchtritt eines dissoziierten Elektrolyten von der Resultante des Verhaltens seiner Ionenpartner bestimmt wird. Besitzen beide Ionenarten kleines Volumen, so überwiegt in der Regel die permeabilitätserhöhende Wirkung des primären Photoeffekts. Ist andererseits das Anion voluminös und die Membran engporig, dann kann der sekundäre Photoeffekt eine Herabsetzung der Permeationsgeschwindigkeit bewirken.

III. Wasseraustausch.

Die Wirkung des Lichts auf den Wasseraustausch von Pflanzengeweben ist am eingehendsten an den Blattgelenken der Leguminosen untersucht worden. Der Pulvinus steht normalerweise in dynamischem Gleichgewicht mit der umgebenden Atmosphäre. Er transpiriert und deckt den entstehenden Wasserverlust laufend aus dem Sproß. Wird das Organ nun in Luft einseitig mit weißem Licht bestrahlt, so fällt der Turgor auf der Lichtflanke und es entsteht eine positive Krümmung. Als Ursache einer solchen Reaktion waren von vornherein a) eine Erniedrigung des osmotischen Wertes, b) eine Beeinträchtigung des Semipermeabilitätsgrades, oder aber c) eine Erhöhung der Wasserpermeabilität in Betracht zu ziehen. a) und b) würden die Saugkraft des belichteten Gewebes statisch erniedrigen, c) dagegen die Wasserbilanz durch eine Verminderung der Abgabewiderstände negativ werden lassen. Eine Entscheidung der Frage war durch einseitige Beleuchtung des Gelenks in Wasser oder Paraffinöl möglich (M. BRAUNER 1932, L. und M. BRAUNER 1947, L. BRAUNER 1948, Versuche an *Phaseolus* und *Robinia*). In beiden Medien rufen nun mittlere Lichtstärken, die in Luft kräftigen positiven Phototropismus erzeugen, negative Krümmungen hervor (Abb. 2). Da die Umkehrung der Reaktionsrichtung durch eine Sistierung der Transpiration bzw. durch Wasserzufuhr von außen bewirkt worden ist, scheidet eine photische Saugkrafterniedrigung als Ursache aus. Denn eine solche Veränderung müßte unter allen Umständen zu einer positiven Krümmung führen. Dagegen läßt sich das Negativwerden der Reaktion befriedigend durch die Annahme einer Erhöhung der Wasserpermeabilität in der Lichtflanke erklären. Die Verringerung des Diffusionswiderstandes, die in Luft eine Steigerung des Transpirations- und Turgorverlustes der Lichtflanke verursacht, muß im submersen Gelenk zu deren schnellerer Absättigung führen. Die resultierende Krümmung stellt jedoch im Gegensatz zum positiven Tropismus in Luft nur eine Übergangsreaktion dar, die schließlich zu einem statischen Gleichgewicht führt, sobald die Schattenflanke trotz ihrer geringeren Permeabilität ihr Turgordefizit aufgeholt hat.

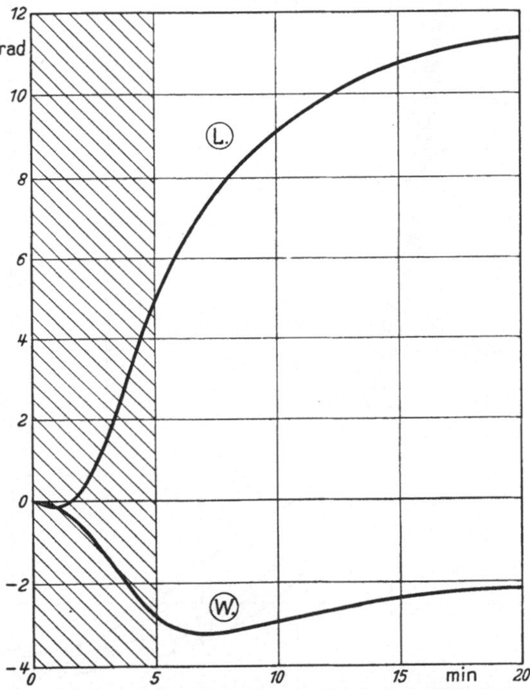

Abb. 2. Phototropische Reaktionen des *Robinia*-Fiederblattgelenks in Luft (L) und in Wasser (W). Bestrahlung von oben mit $1{,}44 \times 10^{-3}$ cal./cm²/sec wärmefreiem Weißlicht. Schraffiertes Feld: Lichtperiode. (Nach L. und M. BRAUNER 1947.)

Die Richtigkeit dieser Interpretation wird dadurch bestätigt, daß durch entsprechende Veränderungen der Belichtungsweise auch die anderen denkmöglichen Reaktionsmechanismen induziert werden können, die sich dann in der zu erwartenden Weise manifestieren. So ruft Weißlicht bei hinreichender Steigerung seiner Intensität auch an submersen Gelenken schließlich positive Krümmungen hervor (vgl. S. 394, Abb. 4). Diese Reaktionsform läßt erkennen, daß die Permeabilitätssteigerung jetzt tatsächlich bis zu einer Beeinträchtigung der Selektivität fortgeschritten ist. — Andererseits verursacht schwaches Rotlicht nicht nur in Wasser, sondern auch in Luft negative Krümmungen. Ein solcher Effekt kann nur die Folge einer Erhöhung der Saugkraft in der bestrahlten Gelenkhälfte sein. In diesem Wellenbereich kommen bei nicht zu hohen Intensitäten die Folgen der Photosynthese offenbar stärker zur Auswirkung als die Permeabilitätsreaktion. Bei einer Steigerung der Lichtstärke erscheint jedoch auch im Rotlicht schließlich wieder der Reaktionstypus des Weißlichts: positive Krümmungen in Luft, negative in Wasser. Der Anatonoseeffekt wird jetzt wieder von einer Erhöhung der Wasserpermeabilität verdeckt, die, wie dieser Versuch zeigt, also auch vom langwelligen Bereich des Spektrums hervorgerufen werden kann.

Eine Erklärung des Mechanismus der Wasserpermeabilitätserhöhung wurde auf Grund der gleichen Voraussetzungen versucht, die eine befriedigende Deutung der Wirkung des Lichts auf die Ionenpermeation ermöglicht hatten. Wenn eine Erniedrigung der negativen Eigenladung der Membranmicelle den Diffusionswiderstand für viele Anionen herabsetzen konnte, so ist eine analoge Verringerung der elektrostatischen Bremswirkung auch gegenüber den viel kleineren Dipolen des Wassers sehr wahrscheinlich.

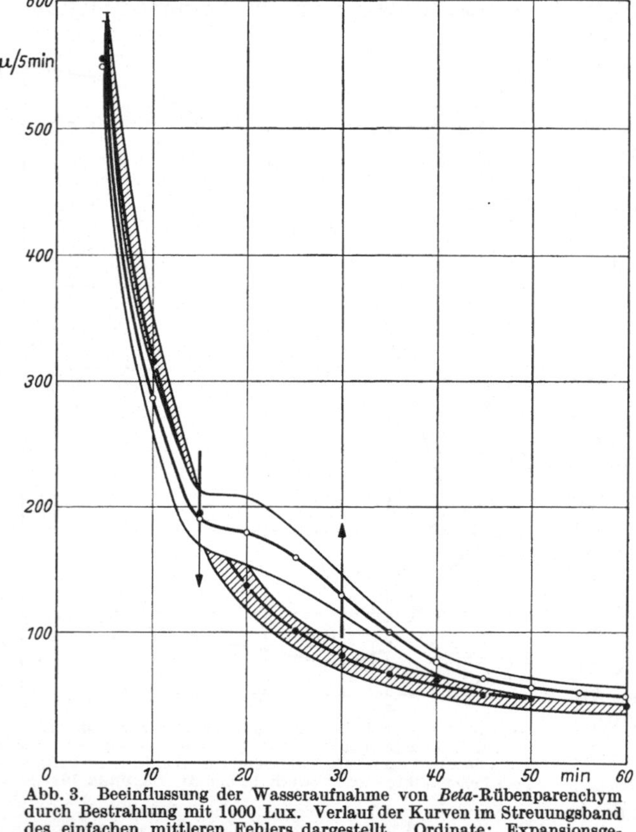

Abb. 3. Beeinflussung der Wasseraufnahme von *Beta*-Rübenparenchym durch Bestrahlung mit 1000 Lux. Verlauf der Kurven im Streuungsband des einfachen mittleren Fehlers dargestellt. Ordinate: Expansionsgeschwindigkeit; Abszisse: Zeit in Minuten. ↓ Beginn, ↑ Ende der Belichtung. ● Dunkelkontrolle. (Nach L. BRAUNER 1935.)

Während die Wirkung des Lichts in Blattgelenken schon deshalb als echter Permeabilitätseffekt gewertet werden kann, weil sie den Wasseraustausch *in beiden Richtungen* erleichtert, ist bei der Beurteilung des Verhaltens wasserabsorbierender Speichergewebe einige Vorsicht geboten. L. BRAUNER (1935) verglich den Verlauf der Absättigung angewelkter Gewebestreifen von Runkelrüben *(Beta vulgaris*, var. *rapacea)* im Dunkeln und im Licht durch Messung der Expansionsgeschwindigkeit mittels eines empfindlichen Spiegelauxanometers. Im Dunkelversuch verlief die Zeitkurve der Verlängerungsgeschwindigkeiten wie eine regelmäßig absinkende Hyperbel. Wurden die Proben nach einer Anlaufszeit im Dunkeln nunmehr mit wärmefreiem Glühlampenlicht (1000 Lux) symmetrisch bestrahlt, so ergab sich fast augenblicklich eine drastische Verzögerung dieses Abfalls (Abb. 3). Das Gewebe nahm also während der Lichtperiode erheblich mehr Wasser auf als die Dunkelkontrollen. Nach dem Löschen der Lampen begannen die Differenzen sich unmittelbar wieder auszugleichen.

In anderen Versuchen wurde die Wasseraufnahme von Karottenparenchym im Licht und im Dunkeln gravimetrisch verfolgt (L. und M. BRAUNER 1936, 1940). Auch hier war eine signifikante Förderung der Wasseraufnahme im Licht (maximal $8{,}3 \pm 1{,}06\%$) zu beobachten. Die weitere Analyse des Effekts zeigte nun, daß Belichtung, im Gegensatz zu seiner symmetrischen Wirkung im Pulvinus, beim *Daucus*-Gewebe nur die Aufnahme, nicht aber die Abgabe von Wasser beeinflußt. Wurden die Proben in hypertonische Glucoselösung übertragen, so war der induzierte osmotische Wasserverlust im Licht und Dunkeln praktisch gleich groß. Möglicherweise hat hier die einsetzende Plasmolyse die Lichtempfindlichkeit des Systems zerstört, während das *Phaseolus*-Gelenk diesen Eingriff zu ertragen scheint (M. BRAUNER 1932). Andererseits ließ sich in diesen Versuchen ein Einblick in den Mechanismus der Lichtreaktion durch einen Vergleich der Wasseraufnahme in hypotonischen Elektrolytlösungen gewinnen. Ausgehend von den im vorigen Abschnitt behandelten photoelektrischen Effekten in lichtempfindlichen Membranen lassen sich die folgenden Bedingungen erwarten: Befindet sich das Gewebe in destilliertem Wasser, so entwickelt sich zwangsläufig ein Membrandiffusionspotential, bei dem die vorauseilenden Gewebekationen das Medium positiv aufladen. Damit sind die Voraussetzungen für die Entstehung einer elektroosmotischen Triebkraft gegeben, die ihrer Richtung nach die normale osmotische Wasseraufnahme unterstützen muß. Belichtung erhöht in diesem System das Diffusionspotential, da die voluminösen Anionen des Gewebes in den engen Membranporen bereits dem sekundären Photoeffekt unterliegen. Dadurch verstärkt sich die elektroosmotische Komponente der Saugkraft über ihren Dunkelwert. Dieser Potentialgewinn muß sich zu der Wirkung der gleichzeitigen Herabsetzung des elektrostatischen Reibungswiderstandes addieren und als Steigerung der Wasseraufnahme zum Ausdruck kommen.

Ist diese Interpretation richtig, so sollte sich die Lichtwirkung auf den Wasseraustausch durch eine Änderung des äußeren Ionenmilieus beeinflussen lassen. Tatsächlich konnten mit zwei membranelektrisch verschieden wirkenden Elektrolyten, KCl und einem Na-Phosphatpuffergemisch vom p_H 6,8 die zu erwartenden Effekte induziert werden. Beide Salze erzeugen in der gewählten Konzentration (0,1 n) Membranpotentiale, die, ihrer Richtung entsprechend, der osmotischen Wasseraufnahme entgegenwirken müssen. Im Falle von KCl wird nun dieses Potential durch den Photoeffekt reduziert, beim Phosphat dagegen gesteigert (L. und M. BRAUNER 1938). Dementsprechend erhöht Belichtung die Wasseraufnahme in der KCl-Lösung und hemmt sie im Phosphatgemisch.

Tabelle 2. *Daucus carota*. (36×10^{-3} cal./cm²/sec; Weißlicht; 30 min-Werte.)

Außenmedium	Destilliertes H_2O	n/10 KCl	n/10 Phosphat
%-Veränderung der Wasseraufnahme im Licht .	$+ 8,3 \pm 1,06$	$+ 7,1 \pm 1,13$	$- 6,8 \pm 0,96$

Aus diesen Erfahrungen folgt, daß bei den bisher studierten Photoreaktionen der Wasserbewegung beide Bestimmungsfaktoren der Strömungsgeschwindigkeit, Widerstand und treibende Kraft lichtempfindlich sind.

Dafür, daß die modifizierbare Komponente des Saugpotentials, wie hier angenommen, elektroosmotischer Natur ist, spricht auch das fast latenzzeitlose Einsetzen der Reaktion, eine Situation, die bei osmotischer Anatonose kaum verständlich wäre. — Der relative Anteil des Permeabilitäts- und des Triebkraftfaktors am gesamten Lichteffekt scheint von Fall zu Fall beträchtlich zu variieren. Während bei den Speichergeweben eine entscheidende Mitwirkung einer gerichteten elektroosmotischen Komponente kaum zu bezweifeln ist, steht bei den Blattgelenken die apolare Permeabilitätsreaktion im Vordergrund; es sei in diesem Zusammenhang daran erinnert, daß Gelenke zu den wenigen Organen gehören, bei denen auch die Abgabe von Zellelektrolyten durch Belichtung gefördert wird (vgl. S. 383).

Nun finden sich in der neueren Literatur auch einige Angaben über das plasmolytische Verhalten belichteter und verdunkelter Gewebe. Dabei wurde stets die Eintrittszeit der Plasmolyse, also die Geschwindigkeit des Wasser*austritts* aus dem Protoplasten in einem definierten osmotischen Konzentrationsgefälle ermittelt. WAHRY (1936) untersuchte nach dieser Methode mit Glucose als Plasmolyticum Luft- und Wasserblätter von *Hippuris vulgaris*, die 24 Std vor der Messung entweder unter natürlichen Tageslichtbedingungen oder im Dunkeln gehalten worden waren. Bei Luftblättern wurde die Austrittsgeschwindigkeit des Wassers im Licht etwa 13% höher gefunden als im Dunkeln, bei den Wasserblättern dagegen war der Lichtwert um 29% niedriger als der Dunkelwert.

Über ähnlich gegensätzliche Effekte, diesmal im gleichen Organ, berichten BIEBL (1942) und seine Mitarbeiterin TOTH (1949). Kurzwelliges UV-Licht (>360 mμ) ruft in der Innenepidermis von *Allium cepa*-Zwiebelschuppen eine deutliche Beschleunigung des Plasmolyseeintritts hervor; die Außenepidermis dagegen reagiert gerade umgekehrt mit einer Verzögerung der Plasmolysegeschwindigkeit.

Einheitlich fördernde Wirkungen des UV-Lichts beobachtete andererseits SEEMANN (1950), der die Untersuchung auf eine Reihe für plasmometrische Messungen besonders geeignete Objekte ausdehnte: Stengelparenchymzellen von *Majanthemum bifolium* und *Convallaria majalis*, Wasserblätter von *Salvinia natans*. In allen diesen Geweben verursachte 2—4 min lange Bestrahlung mit kurzwelligem UV-Licht eine deutliche Erhöhung der Plasmolysegeschwindigkeit. Dabei ist es interessant, daß in vielen Fällen die cuticularisierte Epidermisaußenwand als UV-Schutzfilter für die darunterliegenden Protoplasten wirkt: Bestrahlung von außen her löst meistens viel schwächere Reaktionen aus als gleichstarke Lichtgaben von der Schnittfläche her.

Als Ursache der Förderwirkung nimmt BIEBL eine Porenerweiterung in der bestrahlten Plasmamembran an; der Mechanismus der Hemmwirkung, deren Erscheinungsformen vor allem von TOTH eingehend studiert worden sind, bleibt noch undiskutiert.

Eine befriedigende Interpretation der beiden Reaktionsformen wird wieder von der Entscheidung der Frage abhängen, ob das Licht neben der Wasserpermeabilität auch eine Komponente der Saugkraft beeinflußt. Wenn eine Erhöhung der Durchlässigkeit von der Verstärkung einer nach innen gerichteten elektroosmotischen Triebkraft begleitet ist, so würde dies dem osmotischen Wasserentzug entgegenwirken und wie eine Verminderung der Austrittspermeabilität in Erscheinung treten. Es müßte dann angenommen werden, daß in solchen Geweben, in denen Bestrahlung die Plasmolysegeschwindigkeit erniedrigt, die hemmende Triebkraftkomponente den Ausschlag gibt, in den umgekehrten Fällen dagegen die Permeabilitätserhöhung.

IV. Austausch von Anelektrolyten.

Innerhalb der heterogenen Gruppe der Nichtelektrolyte wird der Vorgang des Stoffaustausches durch das Licht in so komplexer Weise beeinflußt, daß es nicht immer leicht fällt, die wirksamen Mechanismen von einander zu trennen. Diese verwickelten Verhältnisse werden vermutlich dadurch verursacht, daß sich hier rein osmotische und nichtosmotische Faktoren noch mehr überlagern als bei den bisher behandelten Körpern. In der folgenden Darstellung wird versucht werden, durch Zusammenfassung der Stoffe nach bestimmten Eigenschaftskategorien einige allgemeine Gesetzmäßigkeiten abzuleiten.

1. Zucker.

WARNER (1932) kultivierte *Helodea*-Sprosse in hypotonischen Glucose- bzw. Saccharoselösungen bei künstlicher Beleuchtung und im Dunkeln und bestimmte nach 2 Tagen den Zuckergehalt des Gewebes. Die gewonnenen Werte machten es sehr wahrscheinlich, daß Beleuchtung die Aufnahme beider Zuckerarten, vor allem aber die von Glucose, erheblich steigert (+321%!). Leider ist jedoch aus den angegebenen Werten nicht eindeutig zu ersehen, wie groß der Anteil der Photosynthese an der gemessenen Konzentrationsdifferenz war.

Phillis und Mason (1937) studierten das Verhalten von *Gossypium*-Blattscheibchen in Rohrzuckerlösungen. Es ergab sich zunächst, daß vorher entstärktes Gewebe nur im Licht Stärke bildet, und zwar auch bei Ausschluß der Photosynthese in CO_2-freier Kultur. Da die Stärkesynthese dabei vom Schnittrand nach der Mitte der Scheibchen fortschritt, war schon zu vermuten, daß der Vorgang mit der Aufnahme von Zucker aus der Außenlösung zusammenhängt. In weiteren Versuchen wurde die Aufnahme von Zucker aus 0,1% Lösung in Licht- und Dunkelserien direkt bestimmt. Auch hier wurde durch Beseitigung der Kohlensäure für eine Inhibierung der Photosynthese gesorgt. Belichtung verursachte wieder eine erhebliche Anreicherung von Zucker im Gewebe.

Tabelle 3. *Gossypium*. Medium 0,1 % Saccharose.

	Saccharose	Reduzierender Zucker
Ausgangsgehalt (mg/100 Scheibchen)	0,18	3,75
Gehalt nach 12 Std Dunkelheit	0,69	3,11
Gehalt nach 12 Std Tageslicht	3,36	4,91

Weiter zeigte sich, daß dieser Effekt bis zu einem gewissen Grade von der Gegenwart von Sauerstoff abhängig ist. Dies legt den Gedanken an die Mitwirkung eines aktiven Aufnahmemechanismus nahe. Doch konnte die wichtige Frage, ob die Absorption nur mit oder auch gegen den Konzentrationsgradienten verläuft, nicht eindeutig entschieden werden. Phillis und Mason vermuten, daß das Licht die Zuckerpermeabilität im Gewebe erhöht und daß die Wirkung des Sauerstoffs in einer Steigerung der Lösungskapazität („solvent capacity") des Cytoplasmas bestehe.

Ganz kürzlich berichtete Weatherley (1954) über ähnliche Untersuchungen an *Atropa belladonna*. Läßt man Blattscheibchen auf Saccharoselösung schwimmen, so nehmen sie, offenbar aktiv, Zucker auf. Auch in diesen Versuchen förderte Licht die Absorption sehr deutlich (l. c., Tabelle 6), und zwar auch in Abwesenheit von CO_2. Dagegen scheint Dampfsättigung der umgebenden Atmosphäre den Effekt stark zu reduzieren. Weatherley vermutet daher, daß das Licht die Zuckeraufnahme nur indirekt, durch eine Veränderung der Dampfspannung und des Zustands der Stomata beeinflußt. Gegen diese Vorstellung spricht allerdings, daß die oben mitgeteilten positiven Ergebnisse von Phillis und Mason an *submersem* Material gewonnen worden sind.

Biebl (1942) fand in seiner schon erwähnten Untersuchung der Wirkungen des UV-Lichts, daß Rohrzucker in das Plasma der bestrahlten Innenepidermis von Zwiebelschuppen eindringt und es zum Quellen bringt (Kappenplasmolyse!).

In einem soeben erschienenen Bericht teilt Kandler (1954) bemerkenswerte Beobachtungen über den „Glucoseeinbau" in *Chlorella*-Zellen mit. Es zeigte sich, daß vor allem Hungerkulturen bei Belichtung bis zu 25% mehr Zucker aus dem Außenmedium aufnehmen als die Dunkelkontrollen. Der Autor vermutet, daß die eigentliche Lichtreaktion in der Bereitstellung von verfügbarem $\sim$ph besteht, das dann die eingedrungene Glucose phosphoryliert und so ihre Festlegung in der Zelle ermöglicht.

In scheinbarem Gegensatz zu allen diesen Ergebnissen stehen die Befunde von L. und M. Brauner (1936) am Wurzelparenchym von *Daucus carota*. Hier wurde nicht die Zuckeraufnahme aus dem Medium, sondern die Abgabe von zelleigenen Zuckern an Leitungswasser im Dunkeln und bei verschieden starker Glühlampenbeleuchtung (156—5000 Lux) verglichen. Dabei ergab sich nun stets eine signifikante Verringerung der Exosmose durch das Licht.

Diese Reaktion wurde seinerzeit dem sekundären Photoeffekt (vgl. S. 386) zugeschrieben, also einer Porenverengung in der bestrahlten Membran. Unterdessen ist jedoch eine Reihe von Tatsachen bekanntgeworden, die sich mit dieser Interpretation nur schwer vereinbaren lassen: der Nachweis der aktiven, nicht osmotischen Akkumulierbarkeit der Zucker (vgl. BROWN 1952), die eben besprochene Steigerung der Zucker*aufnahme* im Licht und das analoge Verhalten anderer großmolekularer Nichtelektrolyte (vgl. JÄRVENKYLÄ 1937). Nun hatte sich im Falle des Salzaustausches gezeigt, daß K- und PO_4-Ionen, die von Wurzeln im Licht aktiv gespeichert werden, im Dunkeln wieder freigegeben werden und exosmieren (vgl. S. 384). Es liegt daher nahe, einen ähnlichen Mechanismus auch für den Zuckeraustausch anzunehmen. Demnach würde im Licht die Exosmose der Gewebezucker durch eine nach innen gerichtete Triebkraft eingeschränkt, die osmotischer oder nichtosmotischer Natur sein kann; im Dunkeln dagegen könnte infolge der Abschwächung dieser Triebkraft ein entsprechend größerer Teil des Zuckers mit dem Konzentrationsgefälle nach außen diffundieren. Nach dieser Auffassung wäre also die beobachtete Abgabeeinschränkung im Licht nicht die Folge einer Permeabilitätserniedrigung, sondern der Ausdruck einer die Exosmose verhindernden Gegenkraft.

Tabelle 4. *Daucus carota. Zuckerabgabe je Gramm Frischgewicht während 20 Std im Dunkeln (D) und bei 1250 Lux Dauerlicht (L).*

	Monosaccharide	Disaccharide
D	122	121
L	95	93

2. Zellfremde Testsubstanzen.

JÄRVENKYLÄ (1937) prüfte nach plasmolytischen Methoden und durch direkte Analyse des Zellsafts den Einfluß von elektrischer Beleuchtung und von Sonnenlicht auf das Eindringen einer Reihe von Testsubstanzen in Blattzellen von *Helodea* und *Rhoeo* und in den Thallus von *Chara*. Es zeigte sich zunächst wieder, daß Bestrahlung mit starkem elektrischem Licht die Geschwindigkeit der Stoffaufnahme bei *Chara* und *Helodea* grundsätzlich erhöht. Die *Rhoeo*-Blattepidermis erwies sich als wesentlich unempfindlicher und reagierte erst im direkten Sonnenlicht. In allen Fällen war bei gleichem Energiegehalt der kurzwellige Bereich des Spektrums (400—600 mμ) wirksamer als der langwellige (600—1100 mμ). Das erste prinzipiell wichtige Ergebnis dieser Versuche war, daß der fördernde Einfluß der Belichtung um so *weniger* in Erscheinung tritt, je *hydrophober* die

Tabelle 5. *Chara*.

	Trimethylcitrat	Äthylenglykol	Harnstoff	Glycerin	Hexamethylentetramin
Mol.-Gewicht	234,2	62	60,1	92,1	140
Verteilungskoeffizient Äthyläther/H_2O = k_m .	0,43	0,0068	0,00047	0,00066	0,00026
Aufnahmeförderung im Licht	12%	32%	61%	65%	83%

Tabelle 6. *Helodea*.

	Harnstoff	Glycerin	Malonamid
k_m	0,00047	0,00066	0,00030
Mol.-Vol.	59,2	87,8	104,4
Permeationsförderung im Licht (300 W/41 cm)	97%	45%	34%

aufzunehmende Substanz ist. So ergaben Zellsaftanalysen in einer Versuchsreihe mit 5 Verbindungen von abgestuft verschiedener Lipophilie bei *Chara* die folgenden prozentualen Erhöhungen der Stoffaufnahme durch Tageslicht.

Andererseits besteht aber auch eine Beziehung zwischen Lichtwirkung und Molekülgröße. Dies ging aus Versuchen mit *Helodea*-Blättern hervor, bei denen die Aufnahme von drei annähernd gleich lipoidlöslichen Verbindungen verschiedenen Molekularvolumens plasmolytisch gemessen wurde.

Die Förderung der Absorption durch das Licht nimmt also mit fallender Teilchengröße erheblich zu.

Vor einer Diskussion der Deutungsmöglichkeiten dieser wichtigen Befunde sei noch über die Erfahrungen anderer Autoren mit den gleichen Testsubstanzen berichtet. Eine Steigerung der Glycerinaufnahme im Licht konnte schon HOFFMANN (1927) an *Spirogyra*-Zellen beobachten. Bei Bestrahlung mit 50 HK/50 cm Glühlampenlicht wurde in 3stündigen Versuchen plasmometrisch eine Absorptionszunahme von 90% gegenüber den Dunkelkontrollen gefunden. — MEINDL (1934) stellte an *Helodea*-Blättern plasmolytisch eine erhebliche Förderung der Harnstoffpermeation durch UV-Bestrahlung und durch direktes Sonnenlicht fest. Interessanterweise konnte dieser Effekt durch Zusatz von $CaCl_2$ weitgehend aufgehoben werden. — WAHRY (1936) fand bei Messungen an Luftblättern von *Hippuris* im Tageslicht und im Dunkeln folgende Verhältnisse der „Permeationskonstanten" P.

Tabelle 7. *Hippuris*.

	Harnstoff	Glycerin	Malonamid
P_{hell}/P_{dunkel}	6,03	2,12	0,25

Die Stufung der Effekte entspricht also den Befunden JÄRVENKYLÄs, nur zeigt sie sich hier so übersteigert, daß beim Malonamid an Stelle einer bloßen Abschwächung der Lichtförderung bereits eine Lichthemmung der Absorption in Erscheinung tritt.

ELO (1939) ist offenbar der einzige Beobachter, der überhaupt keinen Einfluß des Tageslichts auf die Aufnahme von Harnstoff und Glycerin auffinden konnte. Die Ursachen für das unterschiedliche Verhalten des gleichen Versuchsobjekts, *Hippuris*, in ELOs und in WAHRYs Messungen lassen sich aus der Darstellung der Methoden nicht ersehen.

Schließlich liegen noch Angaben von TOTH (1949) über die Wirkung kurzfristiger UV-Bestrahlung auf die Aufnahme von Harnstoff, Glycerin, Äthylenglykol, Sulfo- und Methylharnstoff vor. Das Versuchsobjekt war, wie schon in den früher erwähnten Messungen der Autorin, die rote Außenepidermis von *Allium cepa*-Zwiebeln. Kurz nach der Bestrahlung erschien der Eintritt von Glykol, Harnstoff, Methylharnstoff und Sulfoharnstoff gefördert, während Glycerin langsamer aufgenommen wurde. Es fällt nicht ganz leicht, in diesem Ergebnis eine gesetzmäßige Beziehung zu erkennen.

Die Verschiedenheiten der respektiven Lipoidlöslichkeiten geben hier keinen Hinweis auf die Natur des Reaktionsprinzips. Andererseits besitzt Glycerin unter den geprüften Verbindungen allerdings das größte Molekularvolumen, was daran denken ließe, daß hier der sekundäre Photoeffekt die Reaktion entscheidet. Dem widerspricht jedoch, daß UV-Bestrahlung im gleichen Gewebe auch den osmotischen Austritt des kleinmolekularen Wassers verlangsamt (vgl. S. 390).

Wenn wir von diesen noch ungeklärten Beobachtungen absehen dürfen, scheinen die Befunde JÄRVENKYLÄs schon soweit gesichert, daß sie als Basis für eine allgemeine Interpretation des Mechanismus der Lichtwirkung bei der Permeation der Nichtelektrolyte dienen können. Da die Förderung der Aufnahme mit der Hydrophilie der permeierenden Substanz zunimmt, liegt es nahe, den lichtempfindlichen Faktor in der elektrostatischen Beziehung zwischen den heteropolaren Gruppen des Membrangerüsts und der diffundierenden Moleküle zu suchen. Wenn die Bestrahlung die negative Eigenladung der Membran herabsetzt, so muß sich die resultierende Verringerung der elektrostatischen Reibung wiederum als Erhöhung der Permeabilität auswirken (vgl. S. 386, 388). Da aber auf den primären der sekundäre Photoeffekt folgt, ist auch bei hydrophilen Nichtelektrolyten zu erwarten, daß die Förderung der Diffusion um so mehr durch die Verdichtung der Membran kompensiert wird, je größer die permeierenden Teilchen sind. Bei einem bestimmten Verhältnis zwischen Molekülgröße und ursprünglicher Porenweite muß der Effekt schließlich in eine Diffusions*hemmung* übergehen (vgl. das Verhalten des Malonamids in WAHRYs Versuchen!).

Mit zunehmender Lipophilie der diffundierenden Substanzen verliert der primäre Photoeffekt selbst kleinmolekularen Körpern gegenüber graduell seine permeationsfördernde Wirkung, da bei abnehmender Wasseraffinität die photoelektrisch beeinflußbaren heteropolaren Kohäsionsbindungen immer mehr in den Hintergrund treten.

V. Beeinträchtigung der Semipermeabilität.

In einer bereits besprochenen Untersuchung hatten L. und M. BRAUNER (1936) beobachtet, daß Glühlampenlicht die Exosmose von zelleigenen Zuckern aus Karottengewebe herabsetzt (vgl. S. 391). Nun gibt es aber auch mehrere Hinweise dafür, daß starke Bestrahlung in manchen Geweben zu einer mehr oder weniger weitgehenden Beeinträchtigung der Semipermeabilität führen kann. LEPESCHKIN wies schon 1908 durch gravimetrische Bestimmungen nach, daß *Phaseolus*-Gelenke im Licht etwa $1^1/_2$mal mehr gelöste Zellinhaltsstoffe an Wasser abgeben als im Dunkeln. Für den Sonderfall der Elektrolytexosmose konnte ein solcher Effekt des Lichts, wie schon früher erwähnt, von BLACKMAN und PAINE (1918) an *Mimosa*-Gelenken, und in einer späteren Arbeit von LEPESCHKIN (1940) selbst an Kartoffelgewebe bestätigt werden. In diesem chlorophyllfreien Parenchym war eine Steigerung der Elektrolytabgabe allerdings erst durch direktes Sonnenlicht oder UV-Bestrahlung zu erreichen. BIEBL (1942) stellte in der Zwiebelinnenepidermis nach kurzer Einwirkung von UV-Licht plasmometrisch eine signifikante Erniedrigung des osmotischen Wertes fest. Eine 30 sec lange Bestrahlung hatte zur Folge, daß die Zellsaftkonzentration nach 6 Std gegenüber den Kontrollen um 0,08 Mol gefallen war. Dieser Effekt wird von BIEBL auf eine Abgabe von Osmoticis zurückgeführt.

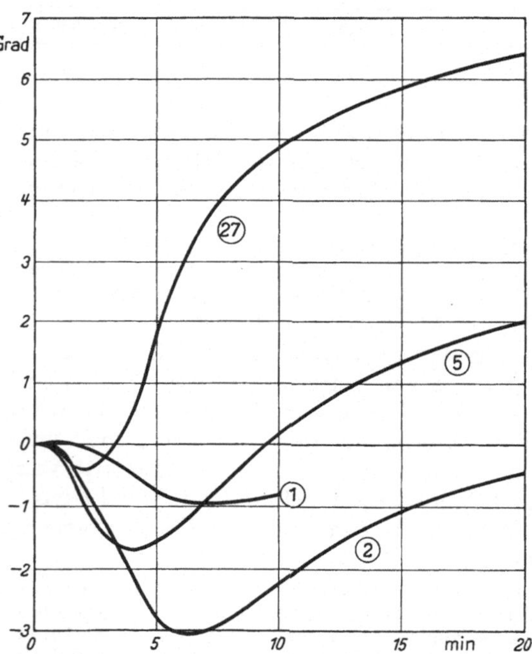

Abb. 4. Phototropische Reaktionen submerser *Robinia*-Fiederblattgelenke bei Bestrahlung von oben mit den relativen Intensitäten 1, 2, 5 und 27. ($1 = 0{,}29 \times 10^{-3}$ cal./cm²/sec.) — Belichtungszeit stets 5 min. (Nach L. und M. BRAUNER 1947.)

Über die Natur dieser extremen Formen der Strahlenwirkung auf den Stoffaustausch läßt sich einiges aus dem phototropischen Verhalten der Leguminosen-pulvini entnehmen. Wie erinnerlich, verursachen schwache und mittlere Lichtstärken in den Gelenken von *Robinia* und *Phaseolus* nur eine Erhöhung der Wasserpermeabilität. Dies wurde aus der Richtungsumkehr der phototropischen Reaktion bei Belichtung des Organs unter Wasser erschlossen (vgl. S. 387). Bei weiterer Steigerung der Lichtintensität wird nun die anfänglich immer noch negative Bewegung von einer positiven Krümmung abgelöst, und bei den höchsten Reizstärken dominiert diese positive Reaktionsphase schließlich vollkommen (L. und M. BRAUNER 1947, Versuche an *Robinia*, Abb. 4). Daraus ist zu schließen, daß die für die Bewegung verantwortliche Photopermeabilitätsreaktion je nach der Reizintensität verschiedene Stufen erreichen kann. Die Entstehung einer positiven Krümmung bringt unter den gegebenen Bedingungen ein Absinken des Turgordrucks in der Lichtflanke zum Ausdruck. In submersen Organen ist nun eine derartige Veränderung in Anbetracht ihrer kurzen Induktionszeit nur durch eine Beeinträchtigung der Semipermeabilität in den bestrahlten Zellen zu erklären. Vermutlich kommt ein solches „Undichtwerden" des Systems bei sehr starker Bestrahlung dadurch zustande, daß der sekundäre Photoeffekt in den äußerst labilen Plasmagrenzschichten zu einer weitgehenden Kontraktion der Micellarstruktur führt. Infolge der resultierenden Spannungen dürften an Stellen geringeren mechanischen Widerstandes „Risse" entstehen, die auch größeren Molekülen bereits den Durchtritt gestatten.

Es ist sehr wahrscheinlich, daß entsprechende Photopermeabilitätseffekte auch an der Entstehung der Lichtturgorreaktionen in Spaltöffnungen mitbeteiligt sind. Auch hier wird man entsprechend der einwirkenden Lichtintensität mit einer bloßen Erhöhung der Wasserpermeabilität oder mit einem Verlust der Semipermeabilität zu rechnen haben (vgl. BÜNNING 1953, S. 429).

VI. Die Strahlenaufnahme bei den Permeabilitätsreaktionen.

Über die sehr wesentliche Frage, was für Acceptoren für die Absorption der für die Auslösung von Permeabilitätsreaktionen nötigen Lichtenergie in der Zelle verfügbar sind, liegen bisher noch keine direkten Untersuchungen vor. Doch lassen die aufgefundenen Wirkungsspektren immerhin schon einige Rückschlüsse zu. Die sehr starken Effekte im UV-Licht sind offenbar nicht an die Gegenwart eines besonderen Photosensibilisators gebunden, da das Absorptionsvermögen des Plasmas selbst in diesem Wellenbereich hinreichend groß ist. Auch in den Fällen, in denen Lampen- oder diffuses Tageslicht noch wirkungslos ist, direktes Sonnenlicht dagegen bereits Reaktionen hervorbringt, dürfte der Effekt auf einer Absorption des mit eingestrahlten langwelligen Ultravioletts durch die Plasmakolloide selbst beruhen.

Für den sichtbaren Wellenbereich dagegen ist das farblose Protoplasma bereits so durchlässig, daß das Zustandekommen von Lichtreaktionen nur bei Gegenwart sensibilisierender Pigmente möglich wird. Aus den meisten der bisher vorliegenden Untersuchungen hat sich nun ergeben, daß der kurzwellige Teil des sichtbaren Spektrums weit wirksamer ist als der langwellige (JÄRVENKYLÄ 1937, L. und M. BRAUNER 1938, 1947, L. BRAUNER 1948); demnach kommen als Sensibilisatoren der Photopermeabilitätsreaktion wohl die gleichen gelben Pigmente in Betracht, die auch für die photochemischen Umsetzungen der Wuchsstoffe verantwortlich sind, also vor allem Riboflavin, möglicherweise aber auch Carotinoide (vgl. L. und M. BRAUNER 1938).

Nun werden aber in grünen Organen (Pulvini!) auch durch rote Strahlung genügender Intensität eindeutige Permeabilitätserhöhungen hervorgerufen. Da in diesem Bereich nur der Chlorophyllkomplex wirksame Absorptionsbanden aufweist, muß hier also das grüne Pigment als Sensibilisator fungieren. Ob das

Chlorophyll dabei direkt wirkt, indem es einen Teil der aufgenommenen Energie für nichtphotosynthetische Umsetzungen verwendet, muß noch dahingestellt bleiben. Eine alternative Möglichkeit wäre die Ausnutzung von Sekundäreffekten der primär induzierten Photosynthese für die Permeabilitätsreaktion, etwa die Entstehung von p_H-Änderungen.

Literatur.

ALBERDA, TH.: The influence of some external factors on growth and phosphate uptake of maize plants of different salt conditions. Rec. Trav. bot. néerl. **41**, 541—601 (1948). — ANDEL, O. M. VAN, W. H. ARISZ and R. J. HELDER: Influence of light and sugar on growth and salt intake by maize. Proc. Kon. Nederl. Akad. v. Wetensch. **53**, 159—171 (1950). — ARISZ, W. H.: Uptake and transport of chlorine by parenchymatic tissue of leaves of *Vallisneria spiralis*. Proc. Kon. Nederl. Akad. v. Wetensch. **50**, 1019—1032, 1235—1245 (1947); **51**, 25—36 (1948). — Transport of chloride in the „symplasm" of *Vallisneria* leaves. Nature (Lond.) **174**, 223—224 (1954).

BIEBL, R.: Wirkung der UV-Strahlen auf die Plasmapermeabilität. Protoplasma **37**, 1—24 (1942). — BLACKMAN, V. H., and S. G. PAINE: Studies in the permeability of the pulvinus of *Mimosa pudica*. Ann. of Bot. **32**, 69—85 (1918). — BRAUNER, L.: Lichtkrümmung und Lichtwachstumsreaktion. Z. Bot. **14**, 497—547 (1922). — Permeabilität und Phototropismus. Z. Bot. **16**, 113—132 (1924). — Über den Einfluß des Lichtes auf die Wasserpermeabilität lebender Pflanzenzellen. Rev. Fac. Sci. Univ. Istanbul **1**, 50—55 (1935). — Untersuchungen über die phototropischen Reaktionen des Primärblattgelenks von *Phaseolus multiflorus* in weißem und in farbigem Licht. Rev. Fac. Sci. Univ. Istanbul, Ser. B **13**, 211—267 (1948). — BRAUNER, L. u. M.: Untersuchungen über den Einfluß des Lichtes auf die Zuckerpermeabilität lebenden Pflanzengewebes. Rev. Fac. Sci. Univ. Istanbul **1**, 58 bis 73 (1936). — Untersuchungen über den photoelektrischen Effekt in Membranen. I. Weitere Beiträge zum Problem der Lichtpermeabilitätsreaktionen. Protoplasma **28**, 230—261 (1937). — Untersuchungen über den photoelektrischen Effekt in Membranen. II. Rev. Fac. Sci. Univ. Istanbul **3**, 1—66 (1938). — Further studies of the influence of light upon the water intake and output of living plant cells. New Phytologist **39**, 104—128 (1940). — Untersuchungen über den Mechanismus der phototropischen Reaktion der Blattfiedern von *Robinia pseudacacia*. Rev. Fac. Sci. Univ. Instanbul, Ser. B **12**, 35—79 (1947). — BRAUNER, M.: Untersuchungen über die Lichtturgorreaktionen des Primärblattgelenkes von *Phaseolus multiflorus*. Planta (Berl.) **18**, 288—337 (1932). — BROWN, R.: Protoplast surface enzymes and absorption of sugar. Intern. Rev. Cytology **1**, 107—118 (1952). — BÜNNING, E.: Entwicklungs- und Bewegungsphysiologie der Pflanzen, 3. Aufl. Berlin-Göttingen-Heidelberg 1953. — BURG, A. H. VAN DER: Influence of light on the absorption of potassium by maize plants in carbon dioxide free air. Proc. Kon. Nederl. Akad. v. Wetensch. C **55**, 279—281 (1952).

DILLEWIJN, C. VAN: Die Lichtwachstumsreaktionen von *Avena*. Rec. Trav. bot. néerl. **24**, 307—581 (1927).

ELO, J. E.: Zur Kenntnis der Permeabilitätseigenschaften von *Hippuris vulgaris* L. Protoplasma **32**, 423—442 (1939).

FITTING, H.: Untersuchungen über isotonische Koeffizienten und ihren Nutzen für Permeabilitätsbestimmungen. Jb. wiss. Bot. **57**, 553—612 (1917).

HELDER, R. J.: Analysis of the process of anion uptake of intact maize plants. Acta bot. néerl. **1**, 361—434 (1952). — HOAGLAND, D. R., and A. R. DAVIS: Further experiments on the absorption of ions by plants, including observations on the effect of light. J. Gen. Physiol. **6**, 47—62 (1924). — HOAGLAND, D. R., P. L. HIBBARD and A. R. DAVIS: The influence of light, temperature, and other conditions on the ability of *Nitella* cells to concentrate halogens in the cell sap. J. Gen. Physiol. **10**, 121—146 (1927). — HOFFMANN, C.: Über die Durchlässigkeit kernloser Zellen. Planta (Berl.) **4**, 584—605 (1927).

INGOLD, C. T.: The effect of light on the absorption of salts by *Elodea canadensis*. New Phytologist **35**, 132—141 (1936).

JACQUES, A. G., and W. J. V. OSTERHOUT: The accumulation of electrolytes. VI. The effect of external p_H. J. Gen. Physiol. **17**, 727—750 (1934). — JÄRVENKYLÄ, Y. T.: Über den Einfluß des Lichtes auf die Permeabilität pflanzlicher Protoplasten. Ann. bot. Soc. zool.-bot. Fenn. „Vanamo" **9**, 1—99 (1937).

KAHO, H.: Über den Einfluß künstlicher Belichtung auf die Exosmose von Elektrolyten aus Stengelzellen. Protoplasma **27**, 453—455 (1937). — KANDLER, O.: Über die Beziehungen zwischen Phosphathaushalt und Photosynthese. II. Gesteigerter Glukoseeinbau im Licht als Indikator einer lichtabhängigen Phosphorylierung. Z. Naturforsch. **9** b, 625—644 (1954).

LEPESCHKIN, W. W.: Zur Kenntnis des Mechanismus der Variationsbewegungen. Ber. dtsch. bot. Ges. **26** a, 724—735 (1908). — Zur Kenntnis des Mechanismus der photonastischen Variationsbewegungen und der Einwirkung des Beleuchtungswechsels auf die Plasmamembran. Beih. bot. Cbl., I. Abt. **24**, 308—356 (1909). — Über den Einfluß des Lichtes auf Exosmose und Speicherung von Salzen im Kartoffelknollengewebe. Protoplasma **34**, 55—69 (1940). — LUTTKUS, K., u. B. BÖTTICHER: Über die Ausscheidung von Aschenstoffen durch die Wurzeln. I. Planta (Berl.) **29**, 325—340 (1939).

MEINDL, T.: Weitere Beiträge zur protoplasmatischen Anatomie des *Helodea*-Blattes. Protoplasma **21**, 362—393 (1934).

PHILLIS, E., and T. G. MASON: The effect of light and of oxygen on the uptake of sugar by the foliage leaf. Ann. of Bot., N. S. **1**, 231—237 (1937).

SCHMIDT, O.: Die Mineralstoffaufnahme der höheren Pflanze als Funktion einer Wechselbeziehung zwischen inneren und äußeren Faktoren. Z. Bot. **30**, 289—334 (1936). — SEEMANN, F.: Der Einfluß der Wärme und UV-Bestrahlung auf die Wasserpermeabilität des Protoplasmas. Protoplasma **39**, 535—566 (1950).

TOTH, A.: Quantitative Untersuchungen über die Wirkungen der UV-Bestrahlung auf die Plasmapermeabilität. Österr. bot. Z. **96**, 161—195 (1949). — TRÖNDLE, A.: Permeabilitätsveränderung und osmotischer Druck in den assimilierenden Zellen des Laubblattes. Ber. dtsch. bot. Ges. **27**, 71—78 (1909). — Der Einfluß des Lichtes auf die Permeabilität der Plasmahaut. Jb. wiss. Bot. **48**, 171—282 (1910).

WAHRY, E.: Permeabilitätsstudien an *Hippuris*. Jb. wiss. Bot. **83**, 657—705 (1936). — WARNER, TH.: Zur Aufnahme von Zucker aus hypotonischen Lösungen durch *Helodea*. Planta (Berl.) **15**, 739—751 (1932). — WEATHERLEY, P. E.: Preliminary investigations into the uptake of sugars by floating leaf discs. New Phytologist **53**, 204—216 (1954).

ZYCHA, H.: Über den Einfluß des Lichtes auf die Permeabilität von Blattzellen für Salze. Jb. wiss. Bot. **68**, 499—548 (1928).

Uptake and ionic environment (including external p_H).

By

Emanuel Epstein.

The preceding sections (IV A–E) have demonstrated that plant cells take up a variety of substances, and that there is a number of mechanisms whereby solutes may enter cells and tissues. A given type of solute may be capable of entering by more than one of the possible mechanisms. For example, inorganic ions may reach a certain fraction of the total volume of the cells or tissue by simple diffusion. Simultaneously, cation exchange may take place along electrochemical potential gradients because of the presence of immobile anions acting as cation exchange surfaces, resulting in the establishment of DONNAN equilibria. And concomitantly, ions may be actively (non-osmotically) transported and concentrated within the cells. A given factor in the ionic environment will not be expected to affect uptake by these diverse mechanisms in parallel ways. In some instances, a certain factor in the ionic environment may increase uptake of some ion by one of these mechanisms, and greatly reduce the uptake of the same ion by another mechanism. In studies on the effects of the ionic environment on absorption it is therefore mandatory to discriminate between the various mechanisms of uptake, and to identify the particular mechanism through which a given ionic feature of the medium affects the absorption of the substance under investigation.

I. Osmotic uptake.

A. Diffusion of nonelectrolytes.

Diffusion of nonelectrolytes into cells and tissues occurs according to the prevailing concentration gradients, and involves their passage through diffusion barriers offering various degrees of resistance to penetration. The nature of the barriers and the uptake of nonelectrolytes have been discussed (IV D a and b), and in particular, the effect of inorganic ions on the uptake of nonelectrolytes has been examined elsewhere (see IV D b). In addition to osmotic uptake by diffusion, nonelectrolytes may be absorbed non-osmotically, by the mechanism of active transport. Ionic effects on both types of uptake are in general non-specific, although ionic series have been recognized (see IV D b).

A special case is represented by the effect of external p_H upon the uptake of weak organic acids and bases. These substances enter more readily in the undissociated form than in the form of ions. As a result, weak acids enter readily at low p_H values, and to a much lesser extent at high p_H values, where they are largely ionic (BEEVERS, GOLDSCHMIDT, and KOFFLER 1952). The reverse holds for weak bases. The effect of p_H on the dissociation of cytoplasmic anions may also be a factor (BRIAN and RIDEAL 1952, REINHOLD 1953).

B. Diffusion of electrolytes.

A fraction of the total volume of a plant tissue is accessible to electrolytes by diffusion (see III H I 2 a). In plant tissue, estimates of this space are of the order of 13–25 per cent of the total tissue volume (HOPE and STEVENS 1952, BUTLER 1953). In microorganisms, on the other hand, the "water space" may be far greater: a value of approximately 72 per cent was reported for *Escherichia coli* (ROBERTS, ABELSON, COWIE, BOLTON, and BRITTEN 1955).

In BUTLER's experiments with wheat roots, the magnitude of the measured space was independent of the concentration of the substances used in measuring it. HOPE (1953), working with broad bean roots, found an increase in the space with increasing KCl concentration, a finding consistent with the view that a DONNAN equilibrium was established between the ambient solution and immobile anions within the tissue. EPSTEIN (1955), by way of contrast, found for barley roots a decrease in the apparent space with increasing ambient concentration of sulfate. This suggests a labile binding of sulfate by the cytoplasm, resulting in high values for the measured "space" at low sulfate concentrations. At the higher sulfate concentrations, this factor becomes negligible, and the space remains constant at approximately 23 per cent of the tissue volume. The space, measured by means of a given salt, is not appreciably affected by the presence of a second salt in the solution (EPSTEIN 1955).

C. Ion exchange.

Most plant tissues, at physiological p_H values, contain immobile anions which give rise to DONNAN equilibria upon immersion of the tissue in salt solutions (see III A 3 II and III H I 2b). Such systems act as cation exchangers, much as do clays and other cation exchange materials. The exchanges are rapid, reversible, and stoichiometric. In experiments with barley roots, EPSTEIN and LEGGETT (1954) found that equilibrium was approached in approximately 30 minutes. Competition between cations for the available exchange spots suggested the lyotropic series commonly encountered in exchange chemistry. Competition between cations for the exchange spots bore no consistent relation to mutual interference between the same ions in the process of active ion transport. VERVELDE (1953) has discussed the DONNAN principle in the ionic relations of plant roots with respect to salt concentration and p_H.

MASSART and VAN DER STOCK (1950) have studied the uptake of, and inhibition of respiration by, trypaflavine and other organic cations. These processes, in yeast and *E. coli*, can be reversed by different mono-, di-, and trivalent cations (K^+, Na^+, Mg^{++}, Ca^{++}, La^{+++}), and by non-inhibiting organic cations such as spermine. They conclude that the reversing ions compete with trypaflavine for negatively charged groups on the cell surface, adsorption at these groups being the condition of their entrance into the cell. Quarternary ammonium salts also antagonized the uptake and respiratory effect of trypaflavine. These findings indicate a complete absence of specificity in the surface binding reaction, which is, in fact, characterized as an adsorption due to COULOMB forces. Their results stand in interesting contrast to those of EPSTEIN and LEGGETT (1954), who in their studies on cation uptake by barley roots concluded that reversible, non-selective adsorptive binding due to COULOMB charges on exchange surfaces was not a rate-limiting step in the process of active absorption.

REINHOLD (1954) studied the uptake of indole-3-acetic acid by pea epicotyls and carrot discs. As in the cases discussed above, two processes were distinguished, one physical (adsorptive), the other metabolic. Uptake by the former process was indifferent to metabolic poisons but decreased with increasing p_H of the medium; the latter was sensitive to poisons but indifferent to p_H.

II. Non-osmotic uptake.

A. Nonelectrolytes.

Osmotic uptake by diffusion predominates in the case of most nonelectrolytes, but in the case of sucrose and some other nonelectrolytes active transport is implicated also (see IV D b). Information on ionic effects is meager. STREET and LOWE (1950) concluded that the absorption of sucrose by excised tomato roots proceeds by a mechanism involving phosphorolysis at the surface of the absorbing cells. When phosphorus-deficient roots were supplied with phosphate, their previously impaired ability to absorb sucrose was rapidly restored.

GAUCH and DUGGER (1953) have demonstrated a striking dependence of the rate of absorption and translocation of sucrose in tomato and bean plants upon the boron concentration. When sucrose was applied to the leaves in the presence of 10 p.p.m. boron, sucrose absorption and translocation was increased several hundred per cent over that occurring in the absence of boron. The authors postulate that boron may enter into the reactions whereby sucrose is moved across a diffusion barrier by a transport mechanism such as that proposed by BROWN (1952). The effect seems to be specific to boron: a number of other elements tested failed to produce the response (DUGGER, GAUCH, and MITCHELL, unpublished data).

B. Electrolytes.

Non-osmotic uptake of electrolytes has been studied far more intensely than osmotic uptake of these substances. The process is metabolic and it is recognized that it involves chemical combination of the substances transported with carrier molecules (IV H). If a given ion species combines with a carrier, other ion species in the medium may or may not compete with the former for identical binding sites or reactive centers. Competition is the most clear-cut mechanism of mutual interference. Relative entrance of two competing ion species depends on the ratio of their concentrations in the medium and their relative association values for the common binding sites. Non-competitive and other types of interference have been recognized. In some instances, active transport of an ion is entirely unaffected by the presence of certain other ions. And finally, absorption of a given ion species may be promoted by the presence of other ion species in the medium ("synergism").

It is pertinent briefly to examine the effect of experimental conditions on the results obtained. During prolonged periods of exposure to inorganic ions, plant roots exhibit growth responses which are a function of the specific ions in question and their concentrations. HASSAN and OVERSTREET (1952) observed pronounced effects of single salts on the elongation of radish radicles during 24-hour exposures. Similar observations were made by LIBBERT (1953) who studied the effect of various salts on the elongation of the root of *Lepidium sativum* during 27-hour periods. HASSAN and OVERSTREET (1952) measured the absorption of the individual ions as well as their effect on elongation and found that when comparing different ions, there was no parallelism between the extent of their absorption and their simultaneous effect on the rate of elongation. The depressing effect of the salts was not due primarily to increased osmotic pressure of the medium, but rather to inhibitory effects of the several ions on metabolic processes. Exposure to solutes evidently may have profound metabolic consequences which in turn will affect the state and functioning of the absorbing membranes themselves. Hence, prolonged exposure of cells and tissues to substances designed to measure penetration of these substances may itself alter the very processes investigated. For this reason it is desirable to restrict absorption experiments to short periods in order to minimize such secondary metabolic feedback effects.

a) The effect of concentration.

Other conditions being equal, the rate of absorption of an ion is a function of its concentration. This applies to storage tissue like carrot and potato (STEWARD 1932), to the roots of cereals such as barley (OVERSTREET, BROYER, ISAACS, and DELWICHE 1942, EPSTEIN and HAGEN 1952), and other higher plants (VAN DEN HONERT 1937, ALBERDA 1949, HELDER 1952), unicellular autotrophs (HOAGLAND, DAVIS, and HIBBARD 1928, GOLDBERG, WALKER, and WHISENAND 1951, and KNAUSS and PORTER 1954) and heterotrophs (ROBERTS, ROBERTS, and COWIE 1949).

Of great theoretical interest is the quantitative relation between the concentration of an ion and its rate of penetration. When these two quantities are plotted, the concentration as abscissa and rate of uptake as ordinate, the resulting curve is often linear or nearly so at low concentrations, whereas at the higher concentrations the curve flattens out, the rate approaching a maximal value. HOAGLAND, DAVIS, and HIBBARD (1928) showed that bromide absorption by *Nitella* approached a limiting value at bromide concentrations of 20 m.e./l. VAN DEN HONERT (1937) and HELDER (1952) made similar observations in regard to phosphate uptake by sugarcane and corn, respectively. Using excised barley roots, EPSTEIN and HAGEN (1952) found the same relationship between concentration and penetration for rubidium and potassium, EPSTEIN (1953) for bromide, and EPSTEIN and LEGGETT (1954) for strontium. This typical connection between concentration and penetration is not limited to inorganic ions. JACQUES (1935) found precisely the same relation between the concentration of the strong base, guanidine $\left(\begin{matrix} NH_2 \\ NH_2 \end{matrix}\!\!>\!C=NH\right)$, and its rate of entrance into *Valonia macrophysa*.

The response of the rate of penetration to the concentration of the substance permits conclusions as to the mechanism involved. The curve in question should be essentially linear if the mechanism of penetration were simple diffusion of the substances through some limiting membrane. EPSTEIN and co-workers (1952, 1953, and 1954) plotted the reciprocal of the absorption rate of an ion $1/v$, against the reciprocal of its external concentration $(1/S)$, and obtained linear plots. Their treatment is based on the assumption that concentration and rate of penetration are related according to the well-known equation relating the rate of an enzymatic reaction to the concentration of substrate (MICHAELIS and MENTEN 1913, LINEWEAVER and BURK 1934):

$$v = \frac{V(S)}{Ks + (S)}, \tag{1}$$

where

v = the rate of reaction (penetration),

V = the maximal rate theoretically attainable at infinite substrate concentration,

(S) = the concentration of substrate (penetrating substance),

Ks = the "MICHAELIS constant", the concentration at which half the maximal velocity is attained.

The implication is that ion transport, like enzymatic catalysis, involves the formation of intermediate, labile complexes between the ions and protoplasmic constituents ("carriers"). The concentration of the carriers being finite, and small in relation to the concentration of the substances transported, a maximal rate of transport is approached at concentrations resulting in complete saturation of the carriers. So far as the overall transport is concerned, this does not represent an equilibrium condition. The carriers are held to operate in a turn-over manner, like enzymes, so that relatively few carrier molecules can transport large numbers of substrate ions. The carriers operate within, or across, the membranes which constitute the barriers to free diffusion. Following the transfer, the transported substances appear, apparently unaltered, within the cells or tissues.

b) The effect of counter-ions.

It is impossible to expose cells or tissues to given cations or anions without at the same time exposing them to counter-ions of opposite sign. The rate of penetration of the counter-ions may vary from nil to values far in excess of the rate for the given ion. In cases of unequal rates for the cation and anion of a salt, the general experience is that the rate of absorption of an ion is positively correlated with that of its counter-ion. Thus more potassium was accumulated by barley roots from the chloride than from the sulfate, and halide absorption was more rapid from the potassium salt than from one of calcium or magnesium, as though the rate of absorption of the anion were accelerated by a rapidly moving cation (HOAGLAND 1940). STEWARD and PRESTON (1941) found that a hundredfold increase in the external concentration produced approximately a tenfold increase in bromide absorption by potato discs, when the potassium salt was used. A hundredfold increase in the concentration of calcium bromide, on the other hand, only increased bromide uptake by 3.5 times. Similar observations were made by ULRICH (1942) in regard to bromide uptake by barley roots.

Under conditions of unequal penetration of the cations and anions of a salt, electrical neutrality is maintained within the tissue by the appearance or disappearance of organic acids (ULRICH 1941, 1942, JACOBSON and ORDIN 1954). When cations were absorbed in excess of anions, organic acids were formed, with a resulting decrease in the respiratory quotient. Organic acids diminished, and the respiratory quotient increased, when anions were absorbed in excess of cations, in ULRICH's experiments.

c) The effect of other ions of the same sign.

Effects of cations upon the rate of penetration of other cations, and of anions on the rate of transfer of other anions, are frequently discussed under the heading of "antagonism" (STILES 1950). The term implies that the presence of a given ion interferes with the absorption of another ion of the same sign, or decreases the "permeability" of the cells to that ion. However, frequently an ion is without effect on the rate of penetration of another ion of the same sign, and often it has the result of increasing rather than diminishing the rate of absorption of the other ion ("synergism"). A few characteristic instances of each type of behavior —retardation, lack of effect, and acceleration—will be given.

HOAGLAND, DAVIS and HIBBARD (1928) studied the penetration of bromide into *Nitella*. They found that chloride markedly decreased bromide penetration; nitrate and sulfate, on the other hand, had no such effect. Similar observations were made by EPSTEIN (1953) in regard to bromide uptake by excised barley roots. ALBERDA (1949) found that phosphate absorption by corn was quite independent of the concentration of nitrate, sulfate, and chloride. COLLANDER (1941) showed that rubidium and potassium absorption by *Avena*, *Helianthus*, and *Pisum* was essentially unaffected by variations of the lithium, sodium, calcium, and strontium concentrations. For other ion pairs, however, COLLANDER

(1941) discovered highly specific mutual relations between the members of each pair in regard to their absorption by numerous species of higher plants. There was a marked parallelism between potassium and rubidium, and between calcium and strontium, in that both members of each pair were absorbed to practically the same extent, if present in equivalent concentration in the solution. Moreover, each ion interfered with the absorption of the other member of its pair as though the plants were unable to distinguish between calcium and strontium, and between potassium and rubidium. COLLANDER's conclusions are borne out by the findings of EPSTEIN and HAGEN (1952) and EPSTEIN and LEGGETT (1954). In short-term experiments with barley roots, potassium and cesium competitively interfered with the absorption of rubidium, in the former paper, and in the latter it was found that calcium and barium competed with strontium. Sodium and lithium, on the other hand, did not effectively compete with rubidium, nor magnesium with strontium. The interpretation of these findings was that potassium, rubidium, and cesium are bound by identical reactive centers or "sites" on the carriers effecting the ion transport, whereas sodium and lithium are bound by other sites. Similarly, calcium, strontium, and barium were held to compete for identical sites for which magnesium has but little affinity. Interference by sodium with rubidium uptake, and by magnesium with strontium uptake, was quantitatively slight, and qualitatively not in the nature of a competition. These findings stress the impropriety of applying the term "competition" to *any* interference of a substance with the uptake of another. The term should only be used if the *mode* of interference has actually been established.

Instances of such specific mutual interference and non-interference are not restricted to inorganic ions. ARISZ (1952), in studying the transport of substances through the tentacles of leaves of *Drosera capensis* L. found that phloridzin inhibits the transport of phosphate but not that of asparagine, whereas penicillin inhibits the transport of asparagine but not that of phosphate. The mutual effects of L-methionine and seleno-methionine on the growth of *Chlorella* observed by SHRIFT (1954) also seem to reflect specific competition in uptake between the amino acid and its analog.

Some instances will now be given of "synergistic" effects in which an ion exhibits an accelerating influence upon the rate of penetration of another ion of like sign. ASPREY (1933) found that the uptake of ammonium ions by potato tuber tissue was greatly increased by the presence of lithium ions, and EPSTEIN and HAGEN (1952) found that absorption of rubidium by excised barley roots was increased by lithium. The effect was maximal at a lithium concentration of 10 m.e./l.; at higher concentrations, lithium depressed rubidium uptake. The accelerating effect of lithium was more pronounced the lower the rubidium concentration.

VIETS (1944) and OVERSTREET, JACOBSON and HANDLEY (1952) studied the effect of calcium on the uptake of potassium by excised barley roots. With increasing calcium concentration, up to 0.1 N, potassium absorption was greatly accelerated over that from solutions lacking calcium. The effect is only apparent under conditions of aerobic metabolism (VIETS 1944), and exchange of absorbed potassium with ambient potassium is less in the presence of calcium than in its absence (OVERSTREET, JACOBSON, and HANDLEY 1952). Magnesium and other di- and polyvalent cations produce similar effects, and the absorption of anions as well as of cations is accelerated by these ions (VIETS 1944). This is an instance in which the effect of an ion on the uptake of another by one mechanism is different from its effect on the uptake by another mechanism. In cation exchange, calcium effectively competes with other cations, especially monovalent ones like potassium (EPSTEIN and LEGGETT 1954). Active transport of potassium, on the other hand, is promoted by calcium.

TANADA (1955) studied the effects of calcium and magnesium on the uptake of rubidium and phosphate by excised mung bean roots during short periods. Calcium produced increases in the rate of absorption of both ions which were considerably larger than those observed for barley roots in the experiments of VIETS (1944) and OVERSTREET, JACOBSON, and HANDLEY (1952). There was no demonstrable time lag in the manifestation of the accelerating effect of calcium on phosphate uptake; the effect was apparent during the first minute of absorption. Magnesium brought about essentially similar responses. The responses of rubidium and phosphate absorption to calcium were affected by previous ultraviolet irradiation of the tissue. Irradiation increased rubidium uptake in the absence of calcium, but in the presence of calcium rubidium absorption by irradiated tissue was severely reduced. Phosphate absorption, on the other hand, was reduced by irradiation, both in the absence, and somewhat more, in the presence of calcium. These results were interpreted as indicating the involvement of surface-localized ribonucleo-proteins in the first step of salt absorption (cf. LANSING and ROSENTHAL 1952).

STOUT, MEAGHER, PEARSON, and JOHNSON (1951) found that absorption and translocation of molybdate by tomato plants during short periods was enhanced in the presence of phosphate and depressed by sulfate. The increase due to phosphate was particularly impressive in view of the extremely wide ratios of phosphate to molybdate which they employed. Phosphate increased molybdate uptake when the molar phosphate/molybdate ratio was over 100,000/1; the effect was less when the phosphate concentration (and hence the ratio) was lower.

d) The effect of other ions of opposite sign.

The most obvious manifestation of the effect of ions on the rate of uptake of other ions of opposite sign is revealed in experiments with single salts. This case has already been discussed (p. 402) under the heading "The effect of counter-ions". As a rule, the rate of absorption of a given ion is positively correlated with the rate of absorption of its counter-ion. When more than one salt is present, large effects of cations or anions on the rate of penetration of ions of the opposite sign have frequently been observed. It is essential in such cases to take into account also the influence of the accompanying ion of like sign (p. 402). In the experiments with *Nitella* discussed earlier (p. 402), HOAGLAND, DAVIS, and HIBBARD (1928) found that while chloride interfered with the absorption of bromide, sulfate and nitrate (as the potassium salts) increased bromide uptake. They attributed the result to the increased potassium concentration. Very pronounced positive effects of potassium on phosphate uptake by *Escherichia coli* were observed by ROBERTS and ROBERTS (1950). At high potassium concentrations, a saturation effect was evident in that further increases in the potassium concentration produced progressively smaller responses in terms of increased phosphate uptake. Phosphorus taken up in the trichloroacetic acid-insoluble fraction was proportional to the content of bound potassium in the cell. Sulfur uptake (from sulfate) was also increased by potassium. EDDY, CARROLL, DANBY, and HINSHELWOOD (1951) found the movement of potassium or rubidium into *Bact. lactis aerogenes* to be parallel to that of phosphate, and concluded that in the early stages of metabolism the functions of these ions are closely linked.

Mention has already been made (p. 403) of VIETS' observation that calcium and other polyvalent cations increased potassium absorption by barley roots. The uptake of the accompanying anion, bromide, was likewise increased by these ions. A feature of particular interest in VIETS' experiments was the observation that the calcium effect bore no direct relation to the calcium status of the tissue. Roots varying in initial calcium content from 2.6 m.e./l. expressed sap to 13.5 m.e./l. responded in quantitatively identical manner to calcium present during accumulation of potassium bromide. In other words, calcium, to exert this synergistic effect, must be present simultaneously with the salt being absorbed; pretreatment of the tissue with calcium is without effect. At very high calcium concentrations (above 0.2 N), absorption of potassium bromide declined again, but this was attributed to non-physiological osmotic pressures. BURSTRÖM (1954) observed that wheat seedlings absorbed practically no nitrate from solutions lacking calcium. As the calcium concentration was increased from 10^{-6} M to 10^{-3} M, nitrate uptake increased by a factor of 3.5. The effect could not be accounted for by the extension in root length due to the added calcium because this increased by a factor of only 1.5. The experiments of TANADA (1955) on the effect of calcium on phosphate (and rubidium) uptake by mung bean roots have been discussed above (p. 404).

e) The effect of external p_H.

General ecological observation shows that plants grow in, and hence, absorb ions from soils ranging in reaction from quite acid (p_H 4 and below) to alkaline (p_H 8 and above). Although many species are restricted to fairly narrow regions of the p_H spectrum, others are broadly tolerant of hydrogen ion concentrations varying by factors of a thousand or more (SMALL 1954). These observations suggest that for many species, the p_H of the medium, within rather wide limits, is unlikely to be a paramount factor in the absorption of ions. This conclusion is borne out by experimental evidence.

HOAGLAND and BROYER (1940) studied the absorption by barley roots of potassium, halide, and nitrate as a function of the p_H of the external solution in short-term experiments. In some of their experiments, the p_H of the solutions was adjusted by means of phosphate buffer, ranging from below 4 to approximately 8. The accumulation of potassium, bromide, and halide was not profoundly affected by the p_H of the culture solution, although there was a rather slight, general trend toward higher uptake at higher p_H values. The same conclusion was reached for potassium, when absorption from very dilute (10^{-4} M) unbuffered solutions of potassium bromide was determined. Bromide accumulation was unaffected by the p_H in these experiments. When the CO_3^{--}—HCO_3^-—CO_2 system was used to maintain the desired p_H values, essentially similar results were obtained.

ARNON, FRATZKE, and JOHNSON (1942) studied the uptake of potassium, calcium, magnesium, phosphate, and nitrogen by tomato, lettuce, and Bermuda grass as a function of the p_H of the nutrient solution in experiments lasting 96 hours. Within a p_H range of about 4 to 8, there was no profound effect of p_H on the absorption of these ions, a finding the authors considered in essential harmony with the conclusions of HOAGLAND and BROYER (1940). KYLIN (1953) found that varying the p_H of the solution from 5.0 to 7.2 had little effect on sulfate uptake by wheat seedlings in experiments lasting one hour.

The absorption by higher plants of ions normally present, and taken up, in relatively small quantities ("microelements") is in general more severely affected by variations in p_H than is the uptake of macroelements. PEARSON (1951) found zinc uptake by barley roots to increase markedly with increasing

p_H of the solution, between the values 4 and 8, approximately, while STOUT, MEAGHER, PEARSON and JOHNSON (1951) found the opposite trend with regard to molybdate uptake by tomato plants from solutions containing phosphate. It was doubtful, however, whether the increased uptake at low p_H values was due to the p_H, per se, or was the result of the fact that at those p_H values, phosphate is predominantly present as $H_2PO_4^-$. This may have affected the results since the authors showed that phosphate specifically increased molybdate uptake (see p. 404), an effect possibly due predominantly to the monobasic ion.

Phosphate is in a class by itself. It might be grouped with the micronutrients in view of the low concentrations at which it is normally present in the substrate, and in view, also, of the low concentrations at which maximal rates of uptake are attained (VAN DEN HONERT 1937). However, the amounts of phosphate taken up over extended periods are such as to place phosphate in the traditional category of macronutrients. In regard to the p_H dependence of ion absorption, phosphate is also unique among the macronutrients in that over the physiological range of p_H values, the ionic form shifts from predominantly $H_2PO_4^-$ (p_H 4–5) to predominantly $HPO_4^=$ (p_H 8), a factor which must be considered in studies of the effect of p_H on phosphate uptake (VAN DEN HONERT 1937, HAGEN and HOPKINS 1955). VAN DEN HONERT (1937) inferred that the response of the rate of phosphate uptake to changes in the p_H could be accounted for by the shift in ionic species, on the assumption that only the monobasic ion is taken up. However, HAGEN and HOPKINS (1955), on the basis of a kinetic analysis, came to the conclusion that both ionic species are absorbed, and that the uptake of each is interfered with by hydroxyl ions.

Ion absorption by the roots of higher plants may be considerably more sensitive to low p_H values of the ambient solution when the latter contains only a single salt at low concentration, than in cases in which salt, and especially calcium, is present in substantial concentrations. This observation parallels another, viz., that the injurious effect of low p_H values upon *growth* is likewise mitigated by the presence of moderate concentrations of salt, and especially of calcium (ÅSLANDER 1931, ARNON and JOHNSON 1942). In regard to absorption, FAWZY, OVERSTREET, and JACOBSON (1954) found a marked p_H dependence of potassium uptake by barley roots, from 10^{-3} M KCl. Uptake was maximal, or nearly so, at p_H 6 and dropped to nil at about p_H 4. Below p_H 4, potassium was lost from the roots. The presence of di- and trivalent cations, at 10^{-3} N, largely reversed this p_H effect. In the presence of these ions, some uptake took place at p_H 3, and loss occurred only at p_H values below 2.5, approximately. These results indicate that conflicting conclusions concerning the p_H sensitivity of salt uptake may have their basis in different experimental conditions employed, very dilute, unbuffered solutions giving results quite different from those obtained by the use of solutions containing more salt, and particularly calcium ions. However, HOAGLAND and BROYER (1940) found no marked p_H effect even in the case of dilute, unbuffered solutions.

For algae, as for the roots of higher plants, instances have been reported in which absorption of inorganic solutes was independent of external p_H over rather wide p_H ranges. JACQUES and OSTERHOUT (1935) found that the rate of penetration of potassium into *Nitella* was independent of p_H between p_H 6 and 8, and COLLANDER (1939) found no very marked response of the rate of lithium absorption by *Chara ceratophylla* to changes in the p_H between 5 and 8.4. In the experiments of STEWARD and MARTIN (1937), the absorption of potassium chloride by *Valonia ventricosa* was not significantly affected by varying the external p_H between 6.2 and 8.7. However, HOAGLAND and DAVIS (1923) found that penetration of nitrate into the sap of *Nitella* decreased with increasing p_H, over the range of 5.0 to 8.5. The hydrogen ion concentration within healthy cells was found to be approximately constant, at p_H 5.2, even when the external solution varied in p_H from 5.0 to 9.0.

Literature.

ALBERDA, T.: The influence of some external factors on growth and phosphate uptake of maize plants of different salt conditions. Rec. Trav. bot. néerl. **41**, 541–601 (1949). — ARISZ, W. H.: Transport of substances through the tentacles of leaves of *Drosera capensis* L. Nature (Lond.) **170**, 932–933 (1952). — ARNON, D. I., W. E. FRATZKE and C. M. JOHNSON: Hydrogen ion concentration in relation to absorption of inorganic nutrients by higher plants. Plant Physiol. **17**, 515–524 (1942). — ARNON, D. I., and C. M. JOHNSON: Influence of hydrogen ion concentration on the growth of higher plants under controlled conditions. Plant Physiol. **17**, 525–539 (1942). — ÅSLANDER, A.: Studies on antagonism in acid nutrient solutions. Sv. bot. Tidskr. **25**, 77–107 (1931). — ASPREY, G. F.: Studies on antagonism. I. The effect of the presence of salts of monovalent, divalent, and trivalent cations on the intake of calcium and ammonium ions by potato tuber tissue. Proc. Roy. Soc. Lond., Ser. B **112**, 451–472 (1933).

BEEVERS, H. E., P. GOLDSCHMIDT and H. KOFFLER: The use of esters of biologically active weak acids in overcoming permeability difficulties. Arch. of Biochem. a. Biophysics **39**, 236–238 (1952). — BRIAN, R. C., and E. K. RIDEAL: On the action of plant growth regulators. Biochem. et Biophysica Acta **9**, 1–18 (1952). — BROWN, R.: Protoplast surface enzymes and absorption of sugar. Internat. Rev. Cytology **1**, 107–118 (1952). — BURSTRÖM, H.: Studies on growth and metabolism of roots. X. Investigations of the calcium effect. Physiol. Plantarum (Copenh.) **7**, 332–342 (1954). — BUTLER, G. W.: Ion uptake by young wheat plants. II. The "apparent free space" of wheat roots. Physiol. Plantarum (Copenh.) **6**, 617–635 (1953).

COLLANDER, R.: Permeabilitätsstudien an Characeen. III. Die Aufnahme und Abgabe von Kationen. Protoplasma **33**, 215–257 (1939). — Selective absorption of cations by higher plants. Plant Physiol. **16**, 691–720 (1941).

EDDY, A. A., T. C. N. CARROLL, C. J. DANBY and C. HINSHELWOOD: Alkali metal ions in the metabolism of *Bact. lactis aerogenes*. I. Experiments on the uptake of radioactive potassium, rubidium and phosphorus. Proc. Roy. Soc. Lond., Ser. B **138**, 219–228 (1951). — EPSTEIN, E.: Mechanism of ion absorption by roots. Nature (Lond.) **171**, 83–84 (1953). — Passive permeation and active transport of ions in plant roots. Plant Physiol. **30**, 529 bis 535 (1955). — EPSTEIN, E., and C. E. HAGEN: A kinetic study of the absorption of alkali cations by barley roots. Plant Physiol. **27**, 457–474 (1952). — EPSTEIN, E., and J. E. LEGGETT: The absorption of alkaline earth cations by barley roots: kinetics and mechanism. Amer. J. Bot. **41**, 785–791 (1954).

FAWZY, H., R. OVERSTREET and L. JACOBSON: The influence of hydrogen ion concentration on cation absorption by barley roots. Plant Physiol. **29**, 234–237 (1954).

GAUCH, H. G., and W. M. DUGGER jr.: The role of boron in the translocation of sucrose. Plant Physiol. **28**, 457–466 (1953). — GOLDBERG, E. D., T. J. WALKER and A. WHISENAND: Phosphate utilization by diatoms. Biol. Bull. **101**, 274–284 (1951).

HAGEN, C. E., and H. T. HOPKINS: Ionic species in orthophosphate absorption by barley roots. Plant Physiol. **30**, 193–199 (1955). — HASSAN, M. N., and R. OVERSTREET: Elongation of seedlings as a biological test of alkali soils. I. Effects of ions on elongation. Soil Sci. **73**, 315–326 (1952). — HELDER, R. J.: Analysis of the process of anion uptake of intact maize plants. Acta bot. néerl. **1**, 361–434 (1952). — HOAGLAND, D. R.: Salt accumulation by plant cells, with special reference to metabolism and experiments on barley roots. Cold Spring Harbor Symp. Quant. Biol. **8**, 181–194 (1940). — HOAGLAND, D. R., and T. C. BROYER: Hydrogen-ion effects and the accumulation of salt by barley roots as influenced by metabolism. Amer. J. Bot. **27**, 173–185 (1940). — HOAGLAND, D. R., and A. R. DAVIS: Composition of cell-sap of the plant in relation to absorption of ions. J. Gen. Physiol. **5**, 629–646 (1923). — HOAGLAND, D. R., A. R. DAVIS and P. L. HIBBARD: The influence of one ion on the accumulation of another by plant cells with special reference to experiments with *Nitella*. Plant Physiol. **3**, 473–486 (1928). — HONERT, T. H. VAN DEN: Over eigenschappen van plantenwortels, welke een rol spelen bij de opname van voedingszouten. Natuurk. Tijdschr. v. Nederl.-Ind. **97**, 150–162 (1937). — HOPE, A. B.: Salt uptake by root tissue cytoplasm: The relation between uptake and external concentration. Austral. J. Biol. Sci. **6**, 396–409 (1953). — HOPE, A. B., and P. G. STEVENS: Electric potential differences in bean roots and their relation to salt uptake. Austral. J. Sci. Res. **5**, 335–343 (1952).

JACOBSON, L., and L. ORDIN: Organic acid metabolism and ion absorption in roots. Plant Physiol. **29**, 70–75 (1954). — JACQUES, A. G.: The kinetics of penetration. X. Guanidine. Proc. Nat. Acad. Sci. U.S. **21**, 488–492 (1935). — JACQUES, A. G., and W. J. V. OSTERHOUT: The kinetics of penetration. XI. Entrance of potassium into Nitella. J. Gen. Physiol. **18**, 967–985 (1935).

KNAUSS, H. J., and J. W. PORTER: The absorption of inorganic ions by *Chlorella pyrenoidosa*. Plant Physiol. **29**, 229–234 (1954). — KYLIN, A.: The uptake and metabolism of sulphate by deseeded wheat plants. Physiol. Plantarum (Copenh.) **6**, 775–795 (1953).

LANSING, A. I., and T. B. ROSENTHAL: The relation between ribonucleic acid and ionic transport across the cell surface. J. Cellul. a. Comp. Physiol. **40**, 337–345 (1952). — LIBBERT, E. Die Wirkung der Alkali- und Erdalkaliionen auf das Wurzelwachstum unter besonderer Berücksichtigung des Ionenantagonismus und seiner Abhängigkeit von Milieufaktoren. Planta (Berl.) **41**, 396–435 (1953). — LINEWEAVER, H., and D. BURK: The determination of enzyme dissociation constants. J. Amer. Chem. Soc. **56**, 658–666 (1934).

MASSART, L., and J. VAN DER STOCK: Antagonism between trypaflavine and cations. Nature (Lond.) **165**, 852–853 (1950). — MICHAELIS, L., and M. L. MENTEN: Die Kinetik der Invertinwirkung. Biochem. Z. **49**, 333–369 (1913).

OVERSTREET, R., T. C. BROYER, T. L. ISAACS and C. C. DELWICHE: Additional studies regarding the cation absorption mechanism of plants in soil. Amer. J. Bot. **29**, 227–231 (1942). — OVERSTREET, R., L. JACOBSON and R. HANDLEY: The effect of calcium on the absorption of potassium by barley roots. Plant Physiol. **27**, 583–590 (1952).

PEARSON, G. A.: Some factors influencing absorption of zinc by roots from single salt systems. Thesis (Ph. D.), University of California, Berkeley 1951.

REINHOLD, L.: The uptake of indole-3-acetic acid by pea epicotyl segments and carrot disks. New Phytologist **53**, 217–239 (1954). — ROBERTS, R. B., P. H. ABELSON, D. B. COWIE, E. T. BOLTON and R. J. BRITTEN: Studies of biosynthesis in *Escherichia coli*. Carnegie Instn. Publ. No 607, **1955**. — ROBERTS, R. B., and I. Z. ROBERTS: Potassium metabolism in *Escherichia coli*. III. Interrelationship of potassium and phosphorus metabolism. J. Cellul. a. Comp. Physiol. **36**, 15–39 (1950). — ROBERTS, R. B., I. Z. ROBERTS and D. B. COWIE: Potassium metabolism in *Escherichia coli*. II. Metabolism in the presence of carbohydrates and their metabolic derivatives. J. Cellul. a. Comp. Physiol. **34**, 259–291 (1949).

SHRIFT, A.: Sulfur-selenium antagonism. II. Antimetabolite action of seleno-methionine on the growth of *Chlorella vulgaris*. Amer. J. Bot. **41**, 345–352 (1954). — SMALL, J.: Modern aspects of p_H, with special reference to plants and soils. New York: Van Nostrand Co. 1954. — STEWARD, F. C.: The absorption and accumulation of solutes by living plant cells. I. Experimental conditions which determine salt absorption by storage tissue. Protoplasma **15**, 29–58 (1932). — STEWARD, F. C., and J. C. MARTIN: The distribution and physiology of *Valonia* at the Dry Tortugas, with special reference to the problem of salt accumulation in plants. Carnegie Instn. Washington, Papers from the Tortugas Laboratory **31**, 87–170 (1937). — STEWARD, F. C., and C. PRESTON: The effect of salt concentration upon the metabolism of potato discs and the contrasted effect of potassium and calcium salts which have a common ion. Plant Physiol. **16**, 85–116 (1941). — STILES, W.: An introduction to the principles of plant physiology, 2nd ed. London: Methuen 1950. — STOUT, P. R., W. R. MEAGHER, G. A. PEARSON and C. M. JOHNSON: Molybdenum nutrition of crop plants. I. The influence of phosphate and sulfate on the absorption of molybdenum from soils and solution cultures. Plant a. Soil **3**, 51–87 (1951). — STREET, H. E., and J. S. LOWE: The carbohydrate nutrition of tomato roots. II. The mechanism of sucrose absorption by excised roots. Ann. of Bot., N. S. **14**, 307–329 (1950).

TANADA, T.: Effect of ultraviolet radiation and calcium and their interaction on salt absorption by excised mung bean roots. Plant Physiol. **30**, 221–225 (1955).

ULRICH, A.: Metabolism of non-volatile organic acids in excised barley roots as related to cation-anion balance during salt accumulation. Amer. J. Bot. **28**, 526–537 (1941). — Metabolism of organic acids in excised barley roots as influenced by temperature, oxygen tension and salt concentration. Amer. J. Bot. **29**, 220–227 (1942).

VERVELDE, G. J.: The Donnan-principle in the ionic relations of plant roots. Plant a. Soil **4**, 309–322 (1953). — VIETS, F. G.: Calcium and other polyvalent cations as accelerators of ion accumulation by excised barley roots. Plant Physiol. **19**, 466–480 (1944).

Stoffaufnahme und innerer p_H-Wert.

Von

H. Drawert.

Mit 3 Abbildungen.

A. Einleitung.

Die p_H-Werte des Zellinnern werden ebenso wie die des Außenmediums sich auf die Stoffaufnahme durch die Zelle oder die ganze Pflanze auswirken. Dabei können der p_H-Wert der Plasmagrenzflächen, der des Binnenplasmas und der des Zellsaftes von Bedeutung sein. Zur Klärung der jeweils vorliegenden Verhältnisse ist es aber notwendig, daß wir den p_H-Wert dieser Zellorte kennen. Eine Bestimmung des realen p_H-Wertes des Zellsaftes einer lebenden Zelle stößt bereits auf große technische Schwierigkeiten (Drawert 1955); noch viel schwieriger, ja in den meisten Fällen noch unmöglich, ist es, etwas über den wirklichen p_H-Wert des Binnenplasmas oder gar der für die Stoffaufnahme besonders wichtigen Plasmagrenzflächen auszusagen. Daraus folgt, daß alle Ausführungen über die Rolle des intracellularen p_H-Wertes für die Stoffaufnahme notgedrungen vorwiegend hypothetischer Natur sein müssen. Experimentell gefundene p_H-Werte können uns bestenfalls Näherungswerte über den Aciditätsgrad des Zellsaftes geben.

Die Innen-cH kann auf verschiedene Weise in den Prozeß der Stoffaufnahme regulierend eingreifen. Sie kann direkt — etwa durch eine Änderung des Quellungsgrades — die Permeabilitätseigenschaften des Plasmas beeinflussen oder für die Aufnahme der Kationen Austauschionen liefern. Die cH-Lage bestimmt ferner Richtung und Ausmaß der Enzymwirkungen, so daß durch Auf- oder Abbau osmotisch wirksamer Substanzen die Aufnahme des Wassers beeinflußt werden wird. In diesem Zusammenhang sei z. B. an die Theorien der Spaltöffnungsbewegungen erinnert, die sich auf eine p_H-abhängige Kohlehydratumwandlung begründen (vgl. Small 1954, S. 165f.). Daneben treten natürlich auch kolloidale Änderungen in den Schließzellen auf, die sich auf den Turgor auswirken. Die Innen-cH wird aber auch die elektrische Ladung der Plasmagrenzflächen mitbestimmen und so eine der Ursachen für eine elektroosmotische Wasseraufnahme sein. Vor allem wird aber die cH des Zellsaftes auf die Speicherung eines Stoffes in der Vacuole einen Einfluß ausüben, sei es über die Ladung der adsorbierenden Zellsaftkolloide oder sei es über die Dissoziation der aufgenommenen Stoffe. Die verschiedenen Möglichkeiten der Wirkungsmechanismen der Innen-cH sind zur Erklärung von Phänomenen der Stoffaufnahme herangezogen worden und einige davon sollen hier näher betrachtet werden.

Es wurde bereits darauf hingewiesen, daß die Unkenntnis des natürlichen p_H-Wertes in den einzelnen Teilen der Zelle der schwächste Punkt aller mit der Innen-cH rechnenden Hypothesen ist. Man hat deshalb häufig versucht, die Reaktion des Zellinnern durch CO_2- oder NH_3-Behandlung künstlich nach der sauren oder alkalischen Seite hin zu verschieben. Man darf bei diesen Versuchen aber nicht übersehen, daß — abgesehen von den damit verbundenen mehr oder weniger pathologischen Veränderungen der Zelle — CO_2 und NH_3 die Zelle ebensoschnell wieder verlassen wie sie eingedrungen sind. Alle Stoffaufnahmeversuche müssen sich über eine gewisse Zeit erstrecken, ehe man zu faßbaren Ergebnissen kommt; Zellen, deren Innen-cH man vorher durch CO_2 bzw. NH_3 verändert hat, werden während dieser Zeit aber durch Abgabe der vorher aufgenommenen Gase nicht nur ihre Innen-cH wieder verändern, sondern auch den p_H-Wert des Außenmediums verschieben, so daß es in diesem Fall schwer zu entscheiden sein wird, ob ein beobachteter Effekt von der Innen- oder von der Außen-cH verursacht worden ist.

B. Permeabilitätseigenschaften des Plasmas.

Auf eine Änderung der Permeabilitätseigenschaften des Plasmas führt BOGEN (1938) die Abhängigkeit der Aufnahme von Harnstoff und Glycerin von der Zellsaft-cH zurück. Die cH der verschiedenen Gewebearten wurde kolorimetrisch im Abdruckverfahren (s. DRAWERT 1955, S. 635, Tab. 4) ermittelt. Die erhaltenen Zahlen geben also bestenfalls Näherungswerte für die natürliche Zellsaft-cH der lebenden Zellen. Für Harnstoff und Glycerin wurden die Permeationskonstanten P' plasmometrisch oder auch mit Hilfe der Deplasmolysemethode bestimmt und der Quotient $Q \frac{P'\text{-Harnstoff}}{P'\text{-Glycerin}}$ ($= Q_{H/G}$) berechnet. Für 19 verschiedene Gewebearten ergab sich in Abhängigkeit von der Preßsaft-cH für den Quotienten $Q_H/_G$ die in Abb. 1 wiedergegebene mehrgipfelige Kurve. Künstliche (CO_2- und NH_3-Gaben) und natürliche (Altern, Frosthärtung) cH-Verschiebungen in den betreffenden Geweben bewirkten eine Änderung des Quotienten streng im Sinne der Kurve, so daß ein kausaler Zusammenhang zwischen Quotient und cH vermutet wird. Diese komplizierte cH-Abhängigkeit der Permeabilität führt BOGEN auf Quellungserscheinungen zurück. In Modellversuchen mit Gelatine konnte nachgewiesen werden, daß Harnstoff eine zusätzliche Quellung, Glycerin dagegen eine Entquellung bewirkt und daß das Ausmaß dieser Quellungserscheinungen vom p_H-Wert der Lösungen abhängt. In gewissen p_H-Bereichen kann allerdings auch der Harnstoff zu einer Entquellung führen.

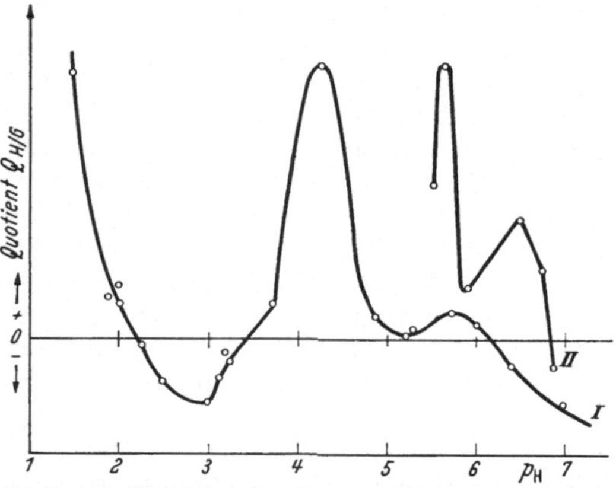

Abb. 1. Die Verteilung der Quotienten $Q_{H/G}$ verschiedener Gewebearten in Beziehung zum p_H-Wert des Zellsaftes. Abszisse: p_H-Wert; Ordinate: Quotient $Q_{H/G}$. Kurve I nach BOGEN (1938), Kurve II nach ROTTENBURG (1943).

HÖFLER (1942) lehnt diese Deutung der Ergebnisse durch BOGEN ganz entschieden ab, zumal ROTTENBURG (1943) an anderen Objekten in dem von ihr untersuchten p_H-Bereich 5,5—6,9 für den Quotienten $Q_H/_G$ ganz andere Werte erhalten hat, wie aus Kurve II in Abb. 1 hervorgeht. HÖFLER wendet sich vor allem auch gegen die Übertragung der an Gelatine gefundenen Verhältnisse auf das Plasma und den Zellsaft. Nach HÖFLER und ROTTENBURG soll zwischen dem Quotienten $Q_H/_G$ und dem p_H-Wert des Zellsaftes keine gesetzmäßige Beziehung bestehen. Andererseits geht aber aus Beobachtungen von DRAWERT (1948) und ALCER (1956) hervor, daß ohne Zweifel die Deplasmolyse-Geschwindigkeit in hypertonischen Harnstoff- und Glycerinlösungen von der cH des Zellsaftes abhängt.

Nach DRAWERT wird bei gleicher Außen-cH Harnstoff von einer Zelle um so schneller aufgenommen, je saurer der Zellsaft ist. Dasselbe kann ALCER für Harnstoff bestätigen und auch für Glycerin nachweisen. Abb. 2 zeigt die in Abhängigkeit von der Außen-cH erhaltenen Befunde an *Bryophyllum daigremontianum*, das als Succulente einen natürlichen täglichen Wechsel in der Zellsaft-Acidität aufweist. Über die Möglichkeit einer nicht osmotischen Wasseraufnahme in diesen Versuchen vgl. man S. 415, 416.

Es soll auch an die älteren Versuche von M. M. BROOKS (1924, 1925) mit *Valonia* über die Abhängigkeit der Aufnahmegeschwindigkeit von As_2O_3 und As_2O_5 von der Reaktion des Zellsaftes erinnert werden. Mit steigendem CO_2-Gehalt wird das fünfwertige As schneller aufgenommen. Gegen die künstliche p_H-Wertänderung müssen aber hier wie auch in den später noch zu schildernden Versuchen anderer Autoren die auf S. 410 erwähnten Bedenken erhoben werden.

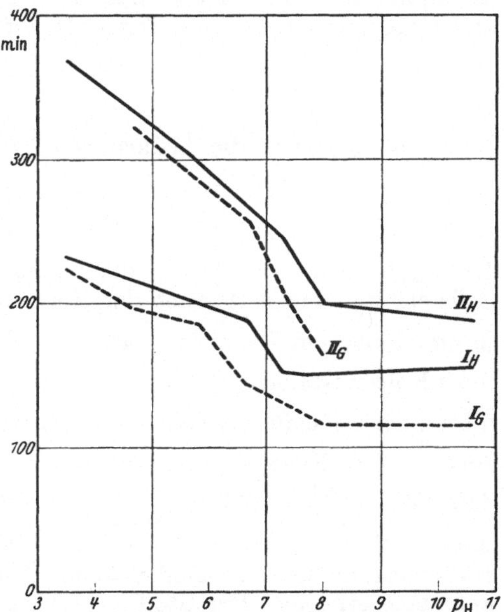

Abb. 2. Deplasmolysezeiten für die Oberepidermis von *Bryophyllum daigremontianum* in hypertonischen Harnstoff- (———) und Glycerinlösungen (- - - - -). Ordinate: Deplasmolysezeit in Minuten; Abszisse: p_H-Werte der Außenlösungen. Kurven I_G und I_H für Blätter mit Preßsaft-p_H 4 und Kurven II_G und II_H für Blätter mit Preßsaft-p_H 6. (Nach ALCER 1956.)

Für die Frage der Wasserpermeabilität ist es vielleicht noch von Interesse, daß nach gravimetrischen Bestimmungen von DRAWERT (1952) in einigen Fällen die Wasserabgabe von Gewebestücken in Abhängigkeit von der Außen-cH ein Minimum aufweist, wenn der p_H-Wert des Außenmediums übereinstimmt mit dem p_H-Wert des Preßsaftes. Für die Kartoffelknolle lag dieser Wert um p_H 6,3 und für den Blattstiel vom Rhabarber um p_H 3,2. Wieweit es sich hierbei um eine Änderung der Permeabilitätseigenschaft des Plasmas, um ein Minimum elektroosmotischer Wasserwanderung oder um andere Faktoren handelt, läßt sich noch nicht absehen. DRAWERT (1954) nimmt als Arbeitshypothese an, daß bei einer Verschiebung der Außen-cH in Abhängigkeit von der Innen-cH eine Änderung der elektrischen Verhältnisse in dem als Membran zwischen Zellsaft und Außenmedium liegenden Plasmaschlauch auftreten wird. Je nach der Richtung der sich einstellenden Potentialdifferenz wird diese die osmotische Wasseraufnahme und -abgabe hemmen oder fördern.

C. Ionenaufnahme durch Austauschadsorption.

An dieser Stelle soll nicht auf das Problem der Ionenaufnahme im allgemeinen eingegangen werden, sondern nur kurz auf die Möglichkeiten einer Beteiligung der Innen-cH, die vor allem für die Kationenaufnahme in Frage kommt, hingewiesen werden.

Die einseitige Wanderung einer Ionenart durch das Plasma ist wegen der dabei auftretenden hohen elektrostatischen Ladungen unmöglich. Mit LUNDEGÅRDH

(1945) kann man sich deshalb als ersten Schritt einer Ionenaufnahme nur einen Ionenaustausch in der Plasmaoberfläche vorstellen. Kationen werden dabei in erster Linie gegen H^+-Ionen ausgetauscht werden. Von den Vorgängen an Kationenaustauscherharzen ausgehend, entwickeln REICHENBERG und SUTCLIFFE folgende Vorstellung: Die Aktivität von Na^+ und H^+ wäre in der Außenlösung $[a_{Na}]_A$ sowie $[a_H]_A$ und in der Zelle $[a_{Na}]_I$ sowie $[a_H]_I$. Natrium wird durch die trennende Membran von außen nach innen diffundieren, vorausgesetzt, daß die Na-Konzentration in der Membran an der Seite der Außenlösung größer ist. Da die Aufnahme von Na^+ durch einen Kationenaustauscher von dem Verhältnis a_{Na}/a_H abhängt, wird eine Wanderung von Na^+ aus der Außenlösung in die Zelle stattfinden, wenn das Verhältnis (1) vorliegt.

$$\frac{[a_{Na}]_A}{[a_H]_A} > \frac{[a_{Na}]_I}{[a_H]_I}. \qquad (1)$$

Das trifft zu, wenn die Bedingung (2) erfüllt ist.

$$\frac{[a_H]_I}{[a_H]_A} > \frac{[a_{Na}]_I}{[a_{Na}]_A}. \qquad (2)$$

Ist $[a_{Na}]_A$ kleiner als $[a_{Na}]_I$, dann muß bei einer Natriumaufnahme durch die Zelle ein Na-Transport gegen das Konzentrationsgefälle stattfinden. Dazu wäre je gm. Ionen ein Energieaufwand von $RT \ln \frac{[a_{Na}]_I}{[a_{Na}]_A}$ erforderlich. Auf Grund ihres Konzentrationsgefälles werden die H^+-Ionen in entgegengesetzter Richtung diffundieren, das führt zu einer Freisetzung einer Energie von $RT \ln \frac{[a_H]_I}{[a_H]_A}$ je gm. Ionen. Diese Energie würde ausreichen für die Na^+-Wanderung, vorausgesetzt, daß $RT \ln \frac{[a_H]_I}{[a_H]_A} - RT \ln \frac{[a_{Na}]_I}{[a_{Na}]_A} > 0$. Die Zelle muß demnach durch den Stoffwechsel die Innen-cH so steigern, daß die Bedingung (2) erfüllt ist, damit eine Kationenaufnahme stattfinden kann. Entsprechend kann auch eine Anionenaufnahme erfolgen. Der Stoffwechsel wird nach dieser Vorstellung nur für die Aufrechterhaltung eines Gradienten austauschbarer Ionen beansprucht, bei der Kationenaufnahme also für die Aufrechterhaltung des cH-Gefälles. Es erhebt sich nur die Frage, wo die Austauschermembran in der vacuolisierten Zelle lokalisiert ist. Da die Stoffwechselprozesse sich außerhalb dieser Membran im Plasma abspielen, kann nach Meinung der Autoren der für einen Ionenaustausch notwendige H^+- und Anionengradient wohl zwischen Außenmedium und Plasma bestehen, aber nicht zwischen Plasma und Vacuole. Für die Aufnahme in die Vacuole müßte ein anderer Mechanismus als ein Ionenaustausch angenommen werden. Dieser Einschränkung ist aber entgegenzuhalten, daß, aus den bisherigen Ergebnissen der Vitalfärbung zu schließen, im allgemeinen die Vacuole der Ort höchster cH in der Zelle ist, so daß jedenfalls für die Kationenaufnahme die vorgetragene Vorstellung auch für die Vacuole zu diskutieren wäre. Dafür spricht ferner die Erfahrung, daß sich basische Farbstoffe und auch gerade die stärker dissoziierten immer an den Orten höchster Acidität in der Zelle anreichern, und das ist im allgemeinen die Vacuole (DRAWERT 1948, BARTELS 1954).

Die Speicherung eines Stoffes in der Vacuole kann auf der Bindung der Ionen an die Zellsaftkolloide beruhen. Für diesen Vorgang kommt dem p_H-Wert des Zellsaftes in zweierlei Hinsicht eine Bedeutung zu; denn sowohl die Ladung der Kolloide wie der Dissoziationsgrad des zu bindenden Stoffes sind von der cH abhängig. Auf dieser Tatsache beruht die vor etwa 40 Jahren von BETHE aufgestellte „Reaktions- oder Ladungshypothese" der Stoffaufnahme (vgl. BETHE

1950, 1952). Liegt die cH der wäßrigen Phase so, daß die amphoteren Zellsaftkolloide positiv geladen sind, müßten die Anionen gespeichert werden. Das wird um so eher der Fall sein, je saurer der Zellsaft ist, da der isoelektrische Punkt (IEP) der meisten zelleigenen Eiweißkörper im sauren Bereich liegt. Kationen werden dagegen vorwiegend gespeichert werden, wenn die Zellsaftkolloide negativ geladen sind, der Zellsaft also neutral oder alkalisch reagiert. Diese Voraussetzungen stehen aber im Widerspruch zu den Erfahrungen aus der Vitalfärbung, daß allgemein basische Farbstoffe von der Vacuole um so besser gespeichert werden, je höher die cH des Zellsaftes ist (DRAWERT 1940, 1948). Diese Tatsache läßt sich auch schlecht mit der Vorstellung von BARTELS (1954) vereinen, jedenfalls soweit sie der Auffassung von BETHE sehr nahe kommt. Andererseits weist sie aber auch Ähnlichkeiten mit den Ausführungen von REICHENBERG und SUTCLIFFE (1954) auf, wonach eine Kationenhäufung an den Orten höchster Acidität zu erwarten ist.

D. Das Prinzip der „Ionenfalle".

Auf die Diskrepanz zwischen der Forderung der Hypothese von BETHE in Bezug auf die Innen-cH für die Kationenspeicherung und den mit basischen Farbstoffen empirisch gefundenen Tatsachen weisen besonders McCUTCHEON und LUCKÉ (1924) hin. Werden z. B. *Nitella*-Zellen mit Brillantkresylblau bei gleichem p_H-Wert der Farbstofflösung vital gefärbt, die cH jedoch einmal mit NaOH und ein andermal mit NH_4OH eingestellt, so ist die vom Zellsaft aufgenommene Farbstoffmenge in der NH_4OH-haltigen Lösung geringer als in der NaOH-haltigen (Tabelle 1).

Tabelle 1. *Zwei Versuche mit Nitella, die den Einfluß von NaOH und NH_4OH bei gleicher Außen-cH auf den p_H-Wert des Zellsaftes und die Speicherung von Brillantkresylblau wiedergeben. Behandlungszeit 1 Std. Der p_H-Wert des Zellsaftes bezieht sich auf ungefärbte Zellen.* (Nach McCUTCHEON und LUCKÉ 1924.)

Farbstoff in der Außenlösung in %	Farbstoff im Zellsaft in %		p_H-Wert des Zellsaftes	
	NaOH	NH_4OH	NaOH	NH_4OH
0,0006	0,003	<0,001	5,2	6,6
0,0006	0,020	0,006	5,8	6,3

Werden die gefärbten Zellen nachträglich von der NaOH-haltigen Farblösung in die NH_4OH-haltige übertragen, so geben die Zellen Farbstoff ab und nach einer Übertragung von NH_4OH in NaOH vertieft sich die Färbung. Bei konstanter Außen-cH wird die Speicherung basischer Farbstoffe durch die Erhöhung des Zellsaft-p_H-Wertes gehemmt; denn NH_4OH dringt im Gegensatz zu NaOH in kurzer Zeit in die Zelle ein und verschiebt den p_H-Wert des Zellsaftes nach der alkalischen Seite (s. Tabelle 1). Durch entsprechende Versuche mit HCl und CO_2 konnte bestätigt werden, daß basische Farbstoffe um so stärker gespeichert werden, je saurer der Zellsaft ist. Die Autoren ziehen den Schluß, daß diese Tatsache gegen eine Adsorption der Farbstoffe, etwa an die Proteine des Zellsaftes, spricht. Es liegt vielmehr eine Salzbildung mit Säuren des Zellsaftes vor. Die Farbstoffe permeieren als Moleküle der Farbbase in die Vacuole, verbinden sich dort mit Säuren zu einem Farbsalz, das dissoziiert und die Farbkationen können nicht mehr extrameieren, da der Protoplast für Ionen impermeabel ist. Dadurch wird das Konzentrationsgefälle für die Farbbasenmoleküle aufrechterhalten und es kommt zu einer Speicherung in der Vacuole. Die Diffusion wird solange weitergehen, bis die permeierfähigen Farbbasenmoleküle im Zellsaft dieselbe Konzentration erreicht haben, wie in der Außenlösung. Die Diffusion ist bis zu einem gewissen Grade unabhängig von dem Verhältnis der Gesamtkonzentration des Farbstoffes im Zellsaft zu derjenigen in der Außenlösung. Da die Dissoziation im Zellsaft um so stärker sein muß je saurer dieser ist, folgt, daß dementsprechend auch die Gesamtfarbstoffmenge in der Zelle mit steigender Acidität zunehmen muß. Für die basischen Farbstoffe wird — bei einer entsprechenden Dissoziationskonstanten — die Gesamtkonzentration im Zellsaft immer größer sein als im umgebenden Medium, wenn die Innen-cH größer ist als die Außen-cH (vgl. KINZEL 1954).

Nachdem IRWIN (1923) in ihren ersten Arbeiten noch eine Speicherung durch Adsorption an Zellsaftkolloide annahm, schloß sie sich später (1926) mehr der Anschauung von MCCUTCHEON und LUCKÉ an (vgl. auch OSTERHOUT 1933). In neuerer Zeit wird dieses Speicherungsprinzip vor allem von HÖFLER (1947) für die Aufnahme der basischen Farbstoffe vertreten (vgl. auch COLLANDER, LÖNEGREN und ARHIMO 1943). In der Literatur sind dafür die anschaulichen Begriffe „trap mechanism" (BROOKS 1941) oder „Ionenfalle" (HÖFLER 1947) geprägt worden.

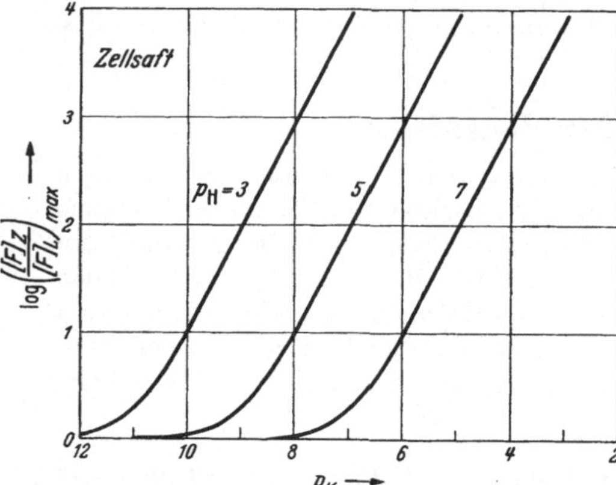

Abb. 3. Die Speichermaxima verschiedener basischer Farbstoffe in ideal gepufferten, kolloidfreien Zellsäften mit den p_H-Werten 3, 5 und 7. Abszisse: pK = der negative Logarithmus der Dissoziationskonstanten des Farbstoffes. Ordinate: Logarithmus des Speichermaximums. (Nach KINZEL 1954.)

Mit den theoretischen Grundlagen der Ionenspeicherung nach dem Prinzip der „Ionenfalle" setzt sich KINZEL (1954) auseinander. Danach ist für den Idealfall, daß der Farbstoff nicht an Zellsaftkolloide adsorbiert wird und der Zellsaft so gut gepuffert ist, daß der aufgenommene Farbstoff den p_H-Wert nicht verändert, der Speicherungsgrad lediglich eine Funktion der Dissoziationskonstanten der Farbbase und der Innen- und der Außen-cH. Für den Fall, daß der p_H-Wert der Außenlösung eine Größenordnung hat, bei der die Farbstoffe in völlig undissoziierter Form vorliegen, können wir die Außen-cH vernachlässigen und erhalten dann in Abhängigkeit von der Dissoziationskonstanten des Farbstoffes sowie dem p_H-Wert des Zellsaftes die in Abb. 3 für 3 Zellsäfte unterschiedlicher Acidität graphisch dargestellten Speichermaxima. Aus der Abbildung geht hervor, wie stark der Speicherungsgrad von der Dissoziationskonstanten K_B eines Farbstoffes und der cH des Zellsaftes abhängt. So ist ein hier angenommener idealer Zellsaft (s. oben) mit p_H 5 nicht in der Lage, einen basischen Farbstoff zu speichern, dessen K_B kleiner als 10^{-8} ist. Für einen Zellsaft mit p_H 7 liegt die Grenze für $K_B = 10^{-6}$.

E. Schlußbetrachtung.

Die meisten Untersuchungen über einen Einfluß der Innen-cH auf die Stoffaufnahme stammen aus dem Bereich der Vitalfärbung. Wenn wir von BETHE und seiner Schule absehen, können wir die sowohl aus der botanischen wie aus der zoologischen Literatur vorliegenden Ergebnisse auf folgenden Nenner bringen: Steigende Acidität des Zellinnern begünstigt die Aufnahme der basischen Farbstoffe und hemmt die der sauren. Diesen Schluß dürfen wir aber nicht auf die Aufnahme aller Stoffe verallgemeinern; denn erstens besteht die Stoffaufnahme aus mehreren Schritten und zweitens ist es nicht anzunehmen, daß für die Aufnahme von starken und schwachen Elektrolyten sowie von Anelektrolyten dieselben Gesetzmäßigkeiten gelten.

Über die Aufnahme eines Stoffes entscheiden — wenn wir die Möglichkeit einer sog. „aktiven Aufnahme" außer Betracht lassen — zwei Faktoren und zwar die Permeabilitätsverhältnisse des Protoplasten und die Speicherfähigkeiten des Zellinnern (DRAWERT 1940). Die Innen-cH kann sich auf beide Faktoren auswirken und beide Möglichkeiten sind auch in der Literatur diskutiert worden. Bei der Vitalfärbung steht aber das Problem der Speicherung im Vordergrund, so daß dementsprechend die hier erhaltenen Befunde nicht ohne weiteres auf farblose Stoffe übertragen werden können, deren Aufnahme mit plasmolytischen Methoden untersucht worden ist, bei denen naturgemäß den Permeabilitätseigenschaften die größere Bedeutung zukommt. Ferner sind die meisten Vitalfarbstoffe schwache Elektrolyte. Daß die Verhältnisse bei der Aufnahme der starken Elektrolyte und Anelektrolyte anders liegen dürften, geht schon daraus hervor, daß allem Anschein nach die stärker dissoziierten Sulfosäurefarbstoffe einerseits und einige nur sehr schwach dissoziierte basische oder amphotere Farbstoffe andererseits weder eine ausgesprochene Abhängigkeit von der Außen-cH noch von der Innen-cH aufweisen.

Bei den starken Elektrolyten wird die Innen-cH für die Austausch-Adsorption von Bedeutung sein, etwa im Sinne der von REICHENBERG und SUTCLIFFE (1954) entwickelten Vorstellung. Auch der Einfluß der cH auf DONNAN-Gleichgewichte darf nicht übersehen werden. Für die Speicherung der starken Elektrolyte etwa im Zellsaft wird auch die Adsorption an Zellsaftkolloide von Bedeutung sein, deren Ladung und damit Adsorptionsfähigkeit unter anderem vom p_H-Wert des Zellsaftes abhängen. Das Prinzip der Ionenfalle kommt für die Aufnahme starker Elektrolyte dagegen nicht in Frage.

Für die schwachen Elektrolyte und damit für die meisten Vitalfarbstoffe kann bei entsprechender Lage der Außen-cH die Innen-cH wie bei den starken Elektrolyten für eine Austauschadsorption von Bedeutung sein. Hier werden aber in erster Linie Moleküle permeieren, die im Zellinnern dissoziieren und so wird — eine Impermeabilität des Protoplasten für Ionen vorausgesetzt — die Innen-cH von ausschlaggebender Bedeutung für eine Speicherung nach dem Prinzip der Ionenfalle sein. Wie bei den starken Elektrolyten wird sich der p_H-Wert des Zellsaftes auch auf eine Adsorption an Zellsaftkolloide auswirken. Daneben kann aber auch eine Permeabilitätsänderung der Plasmagrenzflächen durch Quellungseffekte im Sinne von BOGEN die Aufnahmegeschwindigkeit für die Moleküle beeinflussen.

Am wenigsten klar liegen die Verhältnisse bei den Anelektrolyten, wo auf den ersten Blick eigentlich nur eine Änderung der Permeabilitätsverhältnisse im klassischen Sinne durch die Innen-cH in Betracht käme (BOGEN). Ein Ionenaustausch fällt fort. Den nicht wegzuleugnenden Einfluß der Innen-cH auf die Deplasmolysegeschwindigkeit in hypertonischer Harnstofflösung hat DRAWERT (1948) versucht, in Parallele zu der Aufnahme der basischen Farbstoffe zu setzen. Mit Recht hat BOGEN (1949) auf die sehr niedrige Dissoziationskonstante des Harnstoffes hingewiesen. Wir wissen aber nicht, was mit dem Harnstoff in der Zelle geschieht. Eine Salzbildung zwischen Harnstoff und Säuren in der Zelle muß immer mit in Betracht gezogen werden; mit einer bloßen Beeinflussung der Permeabilitätseigenschaften des Plasmas durch die Innen-cH kommen wir jedenfalls beim Harnstoff nicht aus. Da in den meisten Fällen die Aufnahme der Anelektrolyte mit plasmolytischen Methoden verfolgt worden ist, müssen wir mit BOGEN und PRELL (1953) aber auch eine nichtosmotische Wasseraufnahme in

Betracht ziehen. Es wäre also durchaus denkbar, daß die höhere Deplasmolysegeschwindigkeit der Zellen mit stärker saurem Zellsaft in hypertonischen Harnstofflösungen gar nicht auf eine schnellere Harnstoffaufnahme zurückzuführen ist, sondern diese nur durch eine von der Innen-cH abhängige nichtosmotische Wasseraufnahme vorgetäuscht wird. In diesem Zusammenhang sei auch nochmals auf die Befunde von DRAWERT (1952) hingewiesen, daß bei einigen Geweben die Wasserabgabe ein Minimum aufwies, wenn der p_H-Wert der Außenlösung mit dem p_H-Wert des Preßsaftes übereinstimmte. Die plasmolytischen Methoden zur Untersuchung der Stoffaufnahme bedürfen einer eingehenden Nachprüfung.

Auf alle Fälle wird aus den angeführten Beispielen hervorgehen, daß der Innen-cH der Zelle bei den Fragen der Stoffaufnahme eine größere Beachtung geschenkt werden muß, als es bisher geschehen ist. Die Wirkung der Innen-cH auf den Prozeß der Stoffaufnahme läßt sich nicht auf einen allgemein gültigen Nenner bringen, sondern von Fall zu Fall werden die verschiedensten Vorgänge beeinflußt und Überschneidungen treten auf. So wird bei der Aufnahme der schwachen Elektrolyte wohl kaum einmal das reine Ionenfallenprinzip vorliegen, sondern Austauschadsorption oder auch chemische Bindung werden bei den Vorgängen mehr oder weniger stark ausgeprägt beteiligt sein.

Die Innen-cH hängt weitgehend vom Stoffwechsel der Zelle ab, und so kann der Stoffwechsel über den Aciditätsgrad etwa des Zellsaftes auch die passive Stoffaufnahme beeinflussen und auf diesem Wege unter Umständen eine direkte aktive Beteiligung vortäuschen.

Literatur.

ALCER, G.: Die Aufnahme von Harnstoff und Glycerin in Abhängigkeit von der Acidität der Außenlösung und des Zellsaftes. Protoplasma **1956**.

BARTELS, P.: Quantitative mikrospektroskopische Untersuchung der Speicherungs- und Permeabilitätsverhältnisse akridinorangegefärbter Zellen. Planta (Berl.) **44**, 341—369 (1954). — BETHE, A.: Der Stoffaustausch zwischen Zelle und Umgebung vom Standpunkt der Ladungshypothese und der Austauschadsorption. Naturwiss. **37**, 177—182 (1950). — Allgemeine Physiologie. Berlin-Göttingen-Heidelberg: Springer 1952. — BOGEN, H. J.: Untersuchungen zu den „spezifischen Permeabilitätsreihen" HÖFLERS. II. Harnstoff und Glycerin. Planta (Berl.) **28**, 535—581 (1938). — Besprechung der Arbeit von DRAWERT (1948). Ber. wiss. Biol. **60**, 200—201 (1949). — Beiträge zur Physiologie der nichtosmotischen Wasseraufnahme. Planta (Berl.) **42**, 140—155 (1953). — BOGEN, H. J., u. H. PRELL: Messung nichtosmotischer Wasseraufnahme an plasmolysierten Protoplasten. Planta (Berl.) **41**, 459—479 (1953). — BROOKS, M. M.: The effects of varying the internal and external pH of *Valonia* upon penetration of arsenic. Proc. Soc. Exper. Biol. a. Med. **22**, 148—150 (1924). — Studies on the permeability of living and dead cells. V. The effect of $NaHCO_3$ and NH_4Cl upon the penetration into *Valonia* of trivalent and pentavalent arsenic at various H˙ ion concentrations. Publ. Health Rep. **40**, 139—161 (1925). — BROOKS, S. C., and M. M. BROOKS: The permeability of living cells. Protoplasma-Monogr. **19** (1941).

COLLANDER, R., H. LÖNEGREN u. E. ARHIMO: Das Permeationsvermögen eines basischen Farbstoffes mit demjenigen einiger Anelektrolyte verglichen. Protoplasma (Berl.) **37**, 527 bis 537 (1943).

DRAWERT, H.: Zur Frage der Stoffaufnahme durch die lebende pflanzliche Zelle. II. Die Aufnahme basischer Farbstoffe und das Permeabilitätsproblem. Flora (Jena) **134**, 159—214 (1940). — Zur Frage der Stoffaufnahme durch die lebende pflanzliche Zelle. IV. Der Einfluß der Wasserstoffionenkonzentration des Zellsaftes und des Außenmediums auf die Harnstoffaufnahme. Planta (Berl.) **35**, 579—600 (1948). — Zur Frage der Stoffaufnahme durch die lebende pflanzliche Zelle. V. Zur Theorie der Aufnahme basischer Stoffe. Z. Naturforsch. **3**b, 111—120 (1948). — Kritische Untersuchungen zur gravimetrischen Bestimmung der Wasserpermeabilität. Ein Beitrag zum Einfluß der cH auf die Wasseraufnahme und -abgabe. Planta (Berl.) **41**, 65—82 (1952). — Untersuchungen zur plasmolytischen Bestimmug der Wasserpermeabilität. Ein Beitrag zum Einfluß der cH auf die Wasseraufnahme. Planta (Berl.) **44**, 1—8 (1954). — Der p_H-Wert des Zellsaftes. In W. RUHLAND, Handbuch der Pflanzenphysiologie, Bd. I. 1955.

HÖFLER, H.: Unsere derzeitige Kenntnis von den spezifischen Permeabilitätsreihen. Ber. dtsch. bot. Ges. **60**, 179—200 (1942). — Was lehrt die Fluoreszenzmikroskopie von der Plasmapermeabilität und Stoffspeicherung? Mikroskopie (Wien) **2**, 13—29 (1947).

IRWIN, M.: The penetration of dyes as influenced by hydrogen ion concentration. J. Gen. Physiol. **5**, 727—740 (1923). — Accumulation of dye in *Nitella* as related to dissociation. Proc. Soc. Exper. Biol. a. Med. **23**, 251—253 (1926).

KINZEL, H.: Theoretische Betrachtungen zur Ionenspeicherung basischer Vitalfarbstoffe in leeren Zellsäften. Protoplasma **44**, 52—72 (1954).

LUNDEGÅRDH, H.: Die Blattanalyse. Jena: Gustav Fischer 1945.

MCCUTCHEON, M., and B. LUCKÉ: The mechanism of vital staining with basic dyes. J. Gen. Physiol. **6**, 501—507 (1924).

OSTERHOUT, W. J. V.: Permeability in large plant cells and in models. Erg. Physiol. **35**, 967—1021 (1933).

REICHENBERG, D., and J. F. SUTCLIFFE: Absorption of ions by plants. Nature (Lond.) **174**, 1047—1048 (1954). — ROTTENBURG, W.: Die Plasmapermeabilität für Harnstoff und Glycerin in ihrer Abhängigkeit von der Wasserstoffionenkonzentration. Flora (Jena) **137**, 230—264 (1943).

SMALL, J.: Modern aspects of p_H with special reference to plants and soil. London: Baillière, Tindall a. Cox 1954.

Die Beeinflussung der Stoffaufnahme durch den physiologischen Zustand.

Von

E. Bünning.

Mit 5 Abbildungen.

A. Einleitung.

Es gibt zahlreiche in der Literatur verstreute Angaben über die Beziehung der Permeabilität pflanzlicher Protoplasten zum physiologischen Zustand, also etwa zum Alterszustand, zu Ruheperioden usw. Systematische Untersuchungen über diesen Gegenstand aber sind selten. Oft haben sich, namentlich in der Frühzeit der Permeabilitätsforschung, die Autoren von der irrigen Vorstellung leiten lassen, am wahrscheinlichsten sei eine Parallelität in der Höhe der physiologischen Aktivität und der der Permeabilität. Später aber hat sich gezeigt, daß hoher Aktivität keineswegs auch hohe Permeabilität zu entsprechen braucht. Ein anderer Mangel vieler der veröffentlichten Hinweise besteht darin, daß die angewandten Methoden oft auch andere Deutungen zulassen.

Im übrigen soll hier nur ein kurzer Überblick gegeben werden, da in vielen Teilen dieses Handbuches auf Beziehungen bestimmter physiologischer Zustände zur Permeabilität ausführlicher eingegangen werden muß.

B. Beziehungen zum Entwicklungszustand.

Permeabilitätsverschiedenheiten unterschiedlich alter Zellen sind oft gefunden worden. WEBER (1931) stellte bei jungen *Spirogyra*-Zellen Impermeabilität für Harnstoff fest (Abb. 1). Auch bei Schließzellen von Spaltöffnungen konnte er eine Zunahme der Permeabilität mit dem Älterwerden der Zellen ermitteln. Bei

Abb. 1. Nach der Übertragung von Spirogyrafäden in hypertonische Harnstofflösung plasmolysieren die jungen, eben durch Teilung entstandenen Zellen, die älteren jedoch infolge ihrer höheren Permeabilität nicht. (Nach WEBER.)

Spaltöffnungen hat REUTER (1942) diese Beziehungen näher untersucht. Die Permeabilität der Schließzellen von *Polypodium vulgare* nimmt mit fortschreitender Ausdifferenzierung der Spaltöffnungen zu. Dabei zeichnen sich die Schließzellen auch noch gegenüber den gewöhnlichen Epidermiszellen durch eine besonders hohe Permeabilität aus. Ermittelt wurde das durch Messung der Deplasmolysezeiten bei Verwendung verschiedener Stoffe (Acetamid, Äthylenglykol, Glycerin, Harnstoff, Malonamid, KNO_3 und Erythrit).

Nach den Untersuchungen von MARKLUND an *Helodea* und *Taraxacum* ist die Permeabilität bei jungen Blättern höher als bei alten. In diesen Arbeiten zeigte sich zugleich, daß sich die Permeabilität für die einzelnen Substanzen nicht in gleicher Weise ändert; die „Permeabilitätsreihen" werden also modifiziert. Namentlich zeigte sich, daß die Harnstoffpermeabilität stärker vermindert wird als die Permeabilität für Glycerin. Bei *Helodea* unterschied MARKLUND 3 Zonen. In den jüngsten untersuchten Blättern erwies sich die Permeabilität als niedrig (während zugleich der Protoplast fest an der Wand haftete); Harnstoff permeierte langsamer als Methylharnstoff. In den älteren Blättern bzw. Blatteilen dringt Harnstoff schneller in die Zellen ein als Methylharnstoff; die Permeabilität ist im ganzen höher als in den jüngsten Zellen. In einem noch älteren Stadium ist die Permeabilität wieder niedriger, nämlich intermediär zwischen der der jüngsten und der der etwas älteren Zellen. Die Permeabilität für Harnstoff ist wieder geringer als die für Methylharnstoff (Tabelle 1).

Die Blätter von *Helodea canadensis* hatte mit etwas anderen Methoden und etwas anderer Fragestellung 1934 schon GAHLEN untersucht. Sie fand, daß die Wasserpermeabilität der ganz jungen Blätter erheblich größer ist als die der älteren. Bei jungen Zellen war im Plasmolyseversuch mit Rohrzucker die Hälfte der erreichbaren Volumenkontraktion schon in 15 min erreicht, in älteren Zellen aber erst nach 55 min. Bei Verwendung von NaCl-Lösungen waren die Unterschiede noch größer. Hierbei wird aber nicht einfach die normale Permeabilität gemessen, sondern es kommen in den Werten auch Schädigungen durch die verwendeten Lösungen zum Ausdruck.

Tabelle 1. *Deplasmolysezeiten in Minuten für die Epidermiszellen der Mittelrippe von Taraxacum-Blättern verschiedenen Alters.* (Nach MARKLUND.)

Plasmolyticum	Junges Blatt	Altes Blatt
Glycerin	$43^1/_3$	$17^1/_3$
Harnstoff. . . .	4	$37^2/_3$
Methylharnstoff .	$1^1/_2$	7
Malonamid . . .	145	370

Bei Epidermiszellen von *Allium cepa* und *Tradescantia virginica* hat BOUILLENNE gefunden, daß der Plasmolyserückgang in jungen Blättern schneller erfolgt als in älteren. ALBACH machte an *Rhoeo*-Schnitten Beobachtungen, die für ein schnelleres Exosmieren aus jungen Zellen sprechen (vgl. hierzu weiterhin auch die Zusammenfassung von FISCHER).

Für Moosblätter liegen Untersuchungen von KRESSIN vor. Bei der Verwendung von KNO_3 als Plasmolyticum zeigte sich an mehreren Moosen für junge Zellen eine langsamere Plasmolyse, also wohl geringere Wasserpermeabilität als für alte Zellen. Jedoch besteht diese allmähliche Permeabilitätszunahme mit dem Alter offenbar nur für einige Moosarten, während die Verhältnisse bei anderen umgekehrt liegen. OCHI fand, daß die Protoplasten bei alten Moospflanzen permeabler sind als bei jungen (vgl. auch MENDER).

Auch die Angaben von BENDER lassen das unterschiedliche Verhalten einzelner Moosarten erkennen. Aus dem Plasmolyseverhalten in KNO_3-Lösungen ergab sich, daß zumeist ältere Zellen eine höhere Permeabilität haben.

Bei Farnprothallien fand REUTER (1953), daß die Plasmolysezeit mit zunehmendem Alter abnimmt. In teilungsfähigen Zellen ist diese Zeit groß, während der Differenzierung mittelmäßig und in ausdifferenzierten Zellen niedrig. So wie sich die Wasserpermeabilität ändert, ändert sich bei diesem Objekt auch die Permeabilität für Harnstoff und Glycerin sowie das Verhältnis der Permeabilitäten für diese Substanzen. In teilungsfähigen Zellen ist die Permeabilität für Glycerin größer als für Harnstoff, während der Differenzierung sind die Permeabilitätswerte für beide Substanzen ungefähr gleich, in ausdifferenzierten Zellen dagegen ist die Permeabilität für Harnstoff größer als für Glycerin.

Recht gute Einblicke ermöglichen neuere Untersuchungen von PIRSON und GÖLLNER an Wurzeln von *Lemna minor*. Es ergab sich ein ausgeprägter Gradient der Plasmolysezeiten. Die längsten Plasmolysezeiten wurden bei eben ausgewachsenen Zellen gemessen, so daß das Maximum der Plasmolysezeit 5—8 mm von der Spitze entfernt lag (Abb. 2). Die Kurve der Plasmolysezeiten kann als

Ausdruck dafür gewertet werden, daß die Wasserpermeabilität in 5—8 mm Spitzenentfernung am geringsten ist. Zugleich wurde auch, und zwar durch Bestimmung der Deplasmolysezeiten, die Harnstoffpermeabilität gemessen. Auch das Minimum der Harnstoffpermeabilität liegt genau am Ende der Streckungszone (Abb. 3). Zur Basis hin kann die Harnstoffpermeabilität wieder abnehmen.

ROSENE gibt für Wurzelhaarzellen auf Grund mikrophotometrischer Messungen an, daß die Wasserpermeabilität mit zunehmendem Alter der Zellen sinkt.

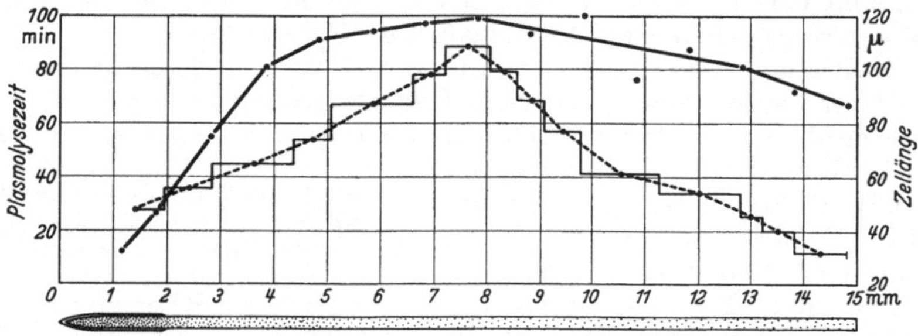

Abb. 2. Plasmolysezeitgradient in der Wurzel von Lemna minor. Die diskontinuierliche Darstellung (Treppenkurve) erläutert das Meßverfahren, bei dem die Wurzel laufend von der Spitze zur Basis verschoben werden mußte, um die beiderseits erfolgende Abrundung zu erfassen. Diese war jeweils um die durch die Stufenbreite gekennzeichnete Wurzelstrecke vorgerückt. Wurzel an der Basis fixiert, Verschiebung durch Kreuztisch. Objektiv: Wasserimmersion Leitz 1/7 W. Die Meßpunkte der Zellängen geben die Mittelwerte von 15 Zellen der vermessenen Wurzelzone wieder. (Nach PIRSON und SEIDEL.)

In diesem Zusammenhang sei darauf hingewiesen, daß aus wachstumsphysiologischen Untersuchungen oft die Folgerung abgeleitet worden ist, daß in der Streckungszone ein Maximum der Wasserpermeabilität besteht. Für diese Annahme spricht, daß diese Zone eben die lebhafteste Wasseraufnahme zeigt, obwohl sie durchweg keine höheren Saugkräfte besitzt als die anderen Zonen,

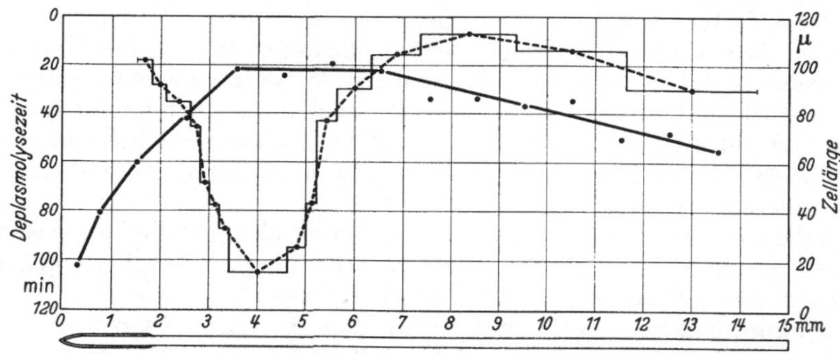

Abb. 3. Gradient der Harnstoffpermeabilität (Deplasmolysezeit in 0,4 Mo Harnstoff) in der Lemna-Wurzel. (Erläuterung der Treppenkurve vgl. Abb. 1.) (Nach PIRSON und SEIDEL.)

in der Regel in ihr offenbar sogar ein Saugkraftminimum besteht. Auch kann darauf hingewiesen werden, daß die Wirkungsweise der Wuchsstoffe gelegentlich durch die permeabilitätserhöhende Wirkung von Auxin gedeutet wird.

Nach diesen verschiedenen Angaben läßt sich feststellen, daß sehr wohl klare Beziehungen zwischen der Höhe der Permeabilität und dem physiologischen Zustand der Zellen mit der Teilungs- oder Wachstumsbereitschaft bestehen. Diese Beziehungen sind aber, soweit die bisher vorliegenden Untersuchungen ein Urteil zulassen, an den verschiedenen Objekten nicht übereinstimmend. So muß wohl der Schluß gezogen werden, daß kein unmittelbarer Kausalzusammenhang im Sinne älterer Vorstellungen besteht; es ist nicht etwa erhöhte

Permeabilität für erhöhte Aktivität charakteristisch. Vielmehr dürften die Permeabilitätsschwankungen in erster Linie als Ausdruck der protoplasmatischen Zustandsänderung aufzufassen sein, die für die wechselnden physiologischen Leistungen verantwortlich sind, uns aber noch im einzelnen weitgehend unbekannt bleiben. Vor allem ist bei der Beurteilung der Ergebnisse auch immer zu beachten, daß die Plasmolysepermeabilität nicht notwendig der Normalpermeabilität zu entsprechen braucht. In unterschiedlicher Plasmolysepermeabilität kommt also unter Umständen nur eine unterschiedlich leichte Beeinflussung der normalen Permeabilität durch das Plasmolyticum zum Ausdruck. Wie wichtig die Berücksichtigung dieses Umstandes ist, ergibt sich aus den Untersuchungen von LENK (1953), die an *Spirogyra* mit der plasmometrischen Methode Permeabilitätsmessungen durchführte, aber zur Schonung der Zellen nur schwach hypertonische Lösungen benutzte. Diese Autorin fand dabei, daß die Permeabilität der *Spirogyra*-Zellen für Anelektrolyte während der Teilungs- und Streckungsphase gleich hoch ist. Nur selten waren die jüngeren Zellen etwas permeabler. Danach kommt LENK zu dem Ergebnis, daß bei den erwähnten Untersuchungen WEBERS (1930 sowie 1933) nur eine ungleiche Intrabilität bzw. ungleiche Neigung, unter dem Einfluß der gebotenen Lösungen die ursprüngliche Intrabilität pathologisch zu erhöhen, gemessen wurde. Auch Ergebnisse HÖFLERS (1952) an *Spirogyra* wären so zu deuten.

Auch eine neue Untersuchung ZICKELS an Stengelepidermiszellen von *Bryophyllum daigremontanum* ergab mit dem Altern verbundene Permeabilitätsänderungen. Die nur kurz mitgeteilten Resultate schließen mit der Feststellung: „Vorläufig ist es nicht möglich, diese Differenzen auf bestimmte Struktur- bzw. Stoffwechseleigentümlichkeiten zurückzuführen, einmal, weil die verschiedenen Reaktionen in vorerst unübersehbarer Weise miteinander verknüpft sind, zum anderen, weil die plasmometrisch ermittelten Permeationskonstanten nicht ausschließlich auf der Basis der Permeabilität interpretiert werden dürfen: sie sind das Resultat mehrerer Prozesse, unter denen die osmotische Stoffaufnahme (= Permeation s. str.) wahrscheinlich nicht immer die wichtigste Rolle spielt, sondern durch nichtosmotische, stoffwechselabhängige Aufnahmemechanismen in erheblichem Ausmaß überdeckt werden kann."

Schließlich sei noch erwähnt, daß während der Annäherung an den Alterstod erhebliche Permeabilitätsänderungen auftreten. REUTER (1937) stellte fest, daß die Permeabilität für Harnstoff, Glycerin, KNO_3 und CaCl bei Protoplasten vergilbter Blätter geringer ist als bei Protoplasten grüner Blätter. Bei alternden Blüten fand BANCHER starke Erhöhungen der Permeabilität.

C. Beziehungen zum jahresperiodischen Aktivitätswechsel.

Über die Beziehungen der Permeabilität zum Aktivitätszustand der Zellen können wir einen Einblick am besten durch das vergleichende Studium der Permeabilitätsverhältnisse während der sommerlichen Aktivität und der winterlichen Ruheperiode gewinnen. Auch hierüber liegen etliche Angaben vor.

MARKLUND fand, daß bei *Taraxacum* im Hochsommer Glycerin schneller eindringt als Harnstoff, während die Verhältnisse im Frühjahr umgekehrt sind. Also auch hierbei zeigte sich wieder, daß sich offenbar nicht nur die Permeabilität ändert, sondern auch das Verhältnis des Permeiervermögens verschiedener Substanzen.

Änderungen der Permeabilitätsreihen fand auch HOFMEISTER (1938) an *Ranunculus repens*. Die von ihm festgestellten Verschiedenheiten entsprechen den von LENK an *Spirogyra* ermittelten. „Im Winter vertritt die *Spirogyra condensata* einen Glycerintyp, im März gehört sie einem schwachen Harnstofftypus an; nachdem sie sich im April durch Unplasmolysierbarkeit auszeichnete, zeigte sie im Juni wieder einen starken Harnstofftyp" (vgl. Abb. 4). Auch RUGE (1943) hatte, und zwar an Zellen der *Rhoeo*-Blätter, Änderungen der spezifischen Permeabilitätsreihen während der Entwicklung festgestellt.

Jahreszeitliche Schwankungen der Permeabilität haben z. B. TRÖNDLE (1910), FITTING (1915) und KRASSINSKI (1930) gefunden; die Permeabilität soll nach diesen älteren Angaben im Winter niedriger sein, oft soll dann sogar eine völlige Undurchlässigkeit bestehen. Diesen Angaben liegen ganz unterschiedliche Methoden zugrunde. KRASSINSKI z. B. prüfte die Exosmose von Zuckern. Im Winter erniedrigte Permeabilität fanden auch FITTING (1920) und FÜRLINGER (1938). Für Moose berichtet OCHI über jahreszeitliche Schwankungen.

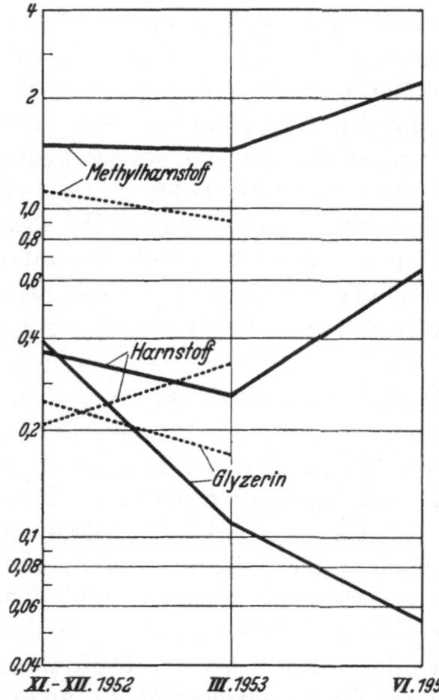

Abb. 4. Die Permeabilität von Spirogyra condensata zu verschiedenen Jahreszeiten. Ausgezogene Linien geteilte Zellen, punktierte Linien = gestreckte Zellen. Die Ordinatenwerte sind Mittelwerte von Permeabilitätskonstanten. (Nach LENK.)

Ausführlicher sind die Ausführungen von GAHLEN über das Verhalten der Blattzellen von *Helodea*. Diese Autorin bestimmte die Wasserpermeabilität aus der Geschwindigkeit der Erreichung des Plasmolyseendgrades. Es ergaben sich folgende Zahlen (Tabelle 2).

Tabelle 2. *Mittelwert der Zeit, in der der Plasmolyseendgrad erreicht wird.*

Monat	Rohrzuckerlösung min	NaCl-Lösung min
Mai	47	—
Juni	145	30
Juli	—	20
September	100	48
Oktober	292	60

Diese Zellen lassen etwas kompliziertere Verhältnisse erkennen als die älteren Angaben. Erwähnenswert ist dabei noch, daß nach den Ausführungen GAHLENs eine Beziehung zwischen Permeabilität und osmotischem Wert besteht. Sie meint sogar, für die Änderung der osmotischen Werte sei in erster Linie die veränderte Permeabilität verantwortlich zu machen (verringerte Permeabilität soll Assimilatstauungen, im Winter Ansammlung der durch Enzyme gelösten Kohlenhydrate bedingen). Besonders wertvoll sind die Mitteilungen von PIRSON und GÖLLNER über das Verhalten von *Lemna minor* (Abb. 5). Nach der Bestimmung der Plasmolysezeit zu urteilen, ist die Harnstoffpermeabilität im Sommer niedrig, im Winter hoch. Das harmoniert mit den Angaben von SCARTH (1936), nach denen ein Minimum der Permeabilität für Harnstoff, Wasser und KNO_3 im Sommer, ein Maximum im Winter besteht.

Auf jahreszeitliche Schwankungen in der Wasserpermeabilität der Epidermiszellen von *Allium cepa* deuten Versuche von LOEVEN. Der Einfluß von Salzen auf die Wasserpermeabilität nimmt vom Herbst zum Winter ab, vom Winter zum Frühjahr wieder zu. Bei anderen Geweben des gleichen Objekts wurden entsprechende Schwankungen gefunden (LOEVEN und BEETS). Folgende Zahlen können die Verhältnisse für die Innenepidermis veranschaulichen (Plasmolyse in 0,6 Mol Saccharose, Deplasmolyse in 0,2 Mol Saccharose).

Monat	Deplasmolyezeit (sec)	Monat	Deplasmolyezeit (sec)
Oktober	372 ± 21	März	274 ± 32
November	329 ± 51	Juli	224 ± 7
Dezember	330 ± 19		

Endlich kann hier noch auf die von HENCKEL (1953) mitgeteilten Untersuchungen russischer Autoren hingewiesen werden, nach denen sich der Protoplast während der Ruheperiode mit einer sichtbaren Lipoidhülle umgibt, die für die Herabsetzung der Permeabilität für Wasser und Sauerstoff verantwortlich

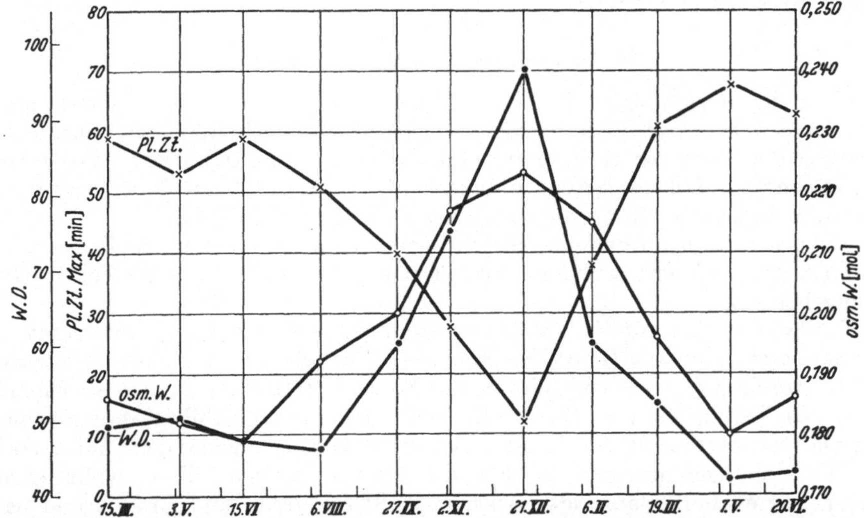

Abb. 5. Plasmolysezeit (in 0,4 mol Glucose), osmotischer Wert (mol) und Wachstum (Zeitdauer WD für die Verlängerung von 1 auf 20 mm) in der Wurzel von Lemna minor zu verschiedenen Jahreszeiten (Jahresperiodik). Die Plasmolysezeiten in Minuten beziehen sich auf die jeweiligen Maxima (Ende der Streckungszone). Die osmotischen Werte geben den Durchschnitt der jeweils ausgewachsenen Zellen (Dauerzone) wieder. Ordinaten für WD und osmotischen Wert gekürzt. (Nach PIRSON und GÖLLNER.)

gemacht wird. Auch hinsichtlich der jahresperiodischen Permeabilitätsschwankungen muß auf die Warnungen hingewiesen werden, die wir bei der Diskussion der mit fortschreitender Entwicklung auftretenden Permeabilitätsänderungen ausgesprochen haben. Es ist nicht sicher, ob die im Plasmolyseversuch ermittelten Schwankungen der Plasmolysepermeabilität immer mit Schwankungen der Normalpermeabilität parallel gehen oder nicht teilweise einfach dadurch bedingt sind, daß das Protoplasma zu den einzelnen Zeiten unter dem Einfluß des Plasmolyticums seine Permeabilität unterschiedlich stark verändert. Es kann dazu noch einmal auf die Arbeit HOFMEISTERs verwiesen werden, in der sich noch zeigte, daß Permeabilitätsänderungen der Art, wie sie beim Übergang von einer Jahreszeit zur anderen auftreten, auch schon durch geringfügige experimentelle Einwirkungen, etwa durch Wässerung der Schnitte, erzielt werden können. Unter solchen Einflüssen wird die Harnstoffpermeation stärker gefördert als die des Glycerins. Es kann sich also schon durch geringfügige Faktoren ein „Glycerintyp“ in einen „Harnstofftyp“ umwandeln.

D. Beziehungen zu anderen Typen des Aktivitätswechsels.

Nur ganz kurz sei hier darauf hingewiesen, daß auch die Ruheperioden bei Samen und Sporen nicht selten mit dem Permeabilitätsverhalten in Zusammenhang gebracht worden sind. Für mehrere Objekte liegen Angaben vor, nach denen die allmähliche Aufhebung der Ruheperiode mit einer allmählichen Erhöhung der Permeabilität zusammenhängt (vgl. etwa SEILER 1936, WEMMER 1954). Dabei handelt es sich aber wohl in der Regel dann nicht um Änderungen in der Permeabilität der Protoplasten, sondern solche in der Durchlässigkeit der Sporenwandung oder Samenschale.

Bei Kartoffelknollen fanden v. GUTTENBERG und MEINL vom Herbst zum Frühjahr hin eine ansteigende Wasserpermeabilität. Diese Autoren sehen in der Permeabilitätserhöhung, zumal sie sich durch das entwicklungshemmende Cumarin unterdrücken ließ, einen wesentlichen Faktor für die allmähliche Beendigung der Ruheperiode.

E. Beziehungen zur Resistenz.

Über die Beziehungen der Permeabilität zur Resistenz ist insofern schon indirekt etwas gesagt worden, als ja bekanntlich regelmäßig eine Parallelität zwischen der Tiefe der Ruhe und der Höhe der Resistenz besteht. Wir dürfen also annehmen, daß wenigstens in gewissen Grenzen die Permeabilitätseigentümlichkeiten resistenter Zellen denen ruhender Zellen entsprechen.

Außerdem wird an anderen Stellen dieses Handbuches auf die Beziehungen zwischen der Resistenz und den Protoplasmaeigentümlichkeiten der Zellen näher einzugehen sein.

Nur kurz sei hier auf die Untersuchungen von SCHMIDT (1939) sowie SCHMIDT, DIWALD und STOCKER (1940) hingewiesen. Danach zeichnen sich resistentere Sorten durch höhere Permeabilität für Wasser, Harnstoff und Glycerin aus. Auch das Verhältnis von Harnstoff- und Glycerinpermeabilität unterscheidet sich bei resistenteren und weniger resistenten Sorten voneinander. Wird Hafer bei Wassermangel gezogen, so fällt die Harnstoffpermeabilität, während die Glycerinpermeabilität ansteigt. Auch nach SIMINOVITCH und BRIGGS und nach LEVITT zeichnen sich resistente Zellen durch eine erhöhte Permeabilität aus.

Für Moose liegen einige Angaben von OCHI vor.

Literatur.

ALBACH, W.: Über die schädigende Wirkung der Plasmolyse und Deplasmolyse. Protoplasma **12**, 255 (1931).

BANCHER, E.: Zellphysiologische Untersuchungen über den Abblühvorgang bei Iris und Gladiolus. Österr. bot. Z. **87**, 221—244 (1938). — BENDER, F.: Der osmotische Druck in den Zellen der Moose. Diss. Münster 1916. — BOUILLENNE, R.: Études sur la perméabilité des cellules de Tradescantia virginica et d'Allium cepa. Bull. Acad. roy. Sci. Belg., V. s. **16**, 102 (1930).

FISCHER, H.: Über protoplasmatische Veränderungen beim Altern von Pflanzenzellen. Protoplasma **39**, 661—676 (1950). — FITTING, H.: Untersuchungen über die Aufnahme von Salzen in die lebende Zelle. Jb. wiss. Bot. **56**, 1—64 (1915). — Untersuchung über die Aufnahme und über anormale osmotische Koeffizienten von Glyzerin und Harnstoff. Jb. wiss. Bot. **59**, 1—170 (1920). — FÜRLINGER, H.: Harnstoffpermeabilität der Epidermis etiolierter und ergrünender Blätter. Protoplasma **31**, 277—285 (1938).

GAHLEN, K.: Beiträge zur Physiologie der Blattzellen von Helodea canadensis. Protoplasma **22**, 337—361 (1934). — GUTTENBERG, H. v., u. G. MEINL: Über die Veränderung der Wasserpermeabilität von Kartoffelknollen während der Lagerzeit und durch Cumarin. Planta (Berl.) **43**, 571—575 (1954).

HENCKEL, P. A.: The adaptative significance of dormancy in plants. (Presented at 7th Intern. Botan. Congr. Stockholm, July 1950.) Proc. Soc. Internat. Botan. Congr. Almqvist a. Wiksells Uppsala 1953. — HÖFLER, K.: Zur Kenntnis der Plasmahautschichten. Ber. dtsch. bot. Ges. 65, 391—399 (1952). — HOFMEISTER, L.: Verschiedene Permeabilitätsreihen bei einer und derselben Zellsorte von Ranunculus repens. Jb. wiss. Bot. 86, 401—419. (1938).

KRASSINSKY, N.: Über jahreszeitliche Änderungen der Permeabilität des Protoplasmas. Protoplasma 9, 622—631 (1930). — KRESSIN, G.: Beiträge zur vergleichenden Protoplasmatik der Mooszelle. Diss. Greifswald 1935.

LENK, I.: Über die Plasmapermeabilität einer Spirogyra in verschiedenen Entwicklungsstadien und zu verschiedener Jahreszeit. Sitzgsber. österr. Akad. Wiss., Math.-naturwiss. Kl., Abt. I 162, 235—271 (1953). — LEVITT, J.: Frost killing and hardiness of plants. Minneapolis: Burgess Publishing Co. 1941. — LOEVEN, W. A.: Seasonal variations in the water permeability of Allium cells. Proc. Kon. Ned. Akad. Wetensch., Ser. C 51, 411—420 (1951). — LOEVEN, W. A., u. H. P. BEETS: Seasonal variations in the water permeability of Allium cells. Proc. Kon. Ned. Akad. Wetensch., Ser. C 53, 453—461 (1955).

MARKLUND, G.: Vergleichende Permeabilitätsstudien an pflanzlichen Protoplasten. Acta bot. fenn. 18, 1—110 (1936). — MENDER, G.: Protoplasmatische Anatomie des Laubmooses Bryum capillare. I. Protoplasma 30, 373—400 (1938).

OCHI, H.: The preliminary report on the osmotic value, permeability, drought and cold resistance of mosses. Bot. Mag. (Tokyo) 65, 10—12 (1952).

PIRSON, A., u. E. GÖLLNER: Beobachtungen zur Entwicklungsphysiologie der Lemna minor L. Flora (Jena) 140, 485—498 (1953). — PIRSON, A., u. F. SEIDEL: Zell- und stoffwechselphysiologische Untersuchungen an der Wurzel von Lemna minor L. unter besonderer Berücksichtigung von Kalium- und Kalziummangel. Planta (Berl.) 38, 431—473 (1950).

REUTER, L.: Protoplasmatik vergilbender Blätter. Protoplasma 27, 270—279 (1937). — Beobachtungen an den Spaltöffnungen von Polypodium vulgare in verschiedenen Entwicklungsstadien. Ein Beitrag zur protoplasmatischen Anatomie. Protoplasma 36, 321—344 (1942). — A contribution to the cell-physiologic analysis of growth and morphogenesis in Fern Prothallia. Protoplasma 42, 1—29 (1953). — ROSENE, H. F.: Ageing and the influx of water into radish roothair cells. J. Gen. Physiol. 34, 65—73 (1951). — RUGE, U.: Permeabilitätsstudien an jungen und ausdifferenzierten Zellen des Rhoeoblattes. Planta (Berl.) 33, 589—599 (1943).

SCARTH, G. W.: The yearly cycle in the physiology of trees. Trans. Roy. Soc. Canada, Biol. Sci. 30 (1936). — SCHMIDT, H.: Plasmolysezustand und Wasserhaushalt bei Laminum maculatum. Protoplasma 33, 25—43 (1939). — SCHMIDT, H., K. DIWALD u. O. STOCKER: Plasmatische Untersuchungen an dürreempfindlichen und dürreresistenten Sorten landwirtschaftlicher Kulturpflanzen. Planta (Berl.) 31, 559—596 (1940). — SEILER, F.: Untersuchungen über Zygotenbildung und Zygotenkeimung bei Sporodinia grandis. Beih. bot. Zbl. Abt. A 54, 235—291 (1936). — SIMINOVITCH, D., and D. R. BRIGGS: The chemistry of the living bark of the black locust tree in relation to frost hardiness. I. Seasonal variations in protein content. Arch. of Biochem. 23, 8—17 (1949).

TRÖNDLE, A.: Einfluß des Lichtes auf die Permeabilität der Plasmahaut. Jb. wiss. Bot. 48, 171—282 (1910).

WEBER, F.: Harnstoff-Permeabilität ungleich alter Spirogyra-Zellen. Protoplasma 12, 129—140 (1931a). — Harnstoff-Permeabilität ungleich alter Spirogyra-Zellen. Protoplasma 14, 75—82 (1931b). — Alkohol-Resistenz ungleich alter Spirogyra-Zellen. Protoplasma 20, 15—19 (1933). — Kurzzellen-Schließzellen von Iris japonica. Protoplasma 35, 140—142 (1940). — WEMMER, L.: Über die Ruheperiode der Zygosporen von Saprolegnia. Arch. Mikrobiol. 21, 217—229 (1954).

ZICKEL, I.: Untersuchungen über zellphysiologische Reaktionen verschieden alter Zellen. Z. Bot. 43, 73—78 (1955).

Objekttypen der Permeabilität.

Von

Hans Joachim Bogen.

Mit 4 Abbildungen.

I. Einleitung.

Die ersten Untersuchungen über die Stoffaufnahme, die sich auf mehrere Objekte erstreckten, stammen von OVERTON (1897); sie führten zu der Auffassung, daß sich alle lebenden Objekte hinsichtlich der Stoffaufnahme in den wichtigen Grundzügen *gleichartig* verhielten. OVERTON bezog seine Aussage auf die Permeabilität (s. str., vgl. S. 117) bzw. die osmotische Stoffaufnahme, nicht aber auf die „adenoide Tätigkeit" (nichtosmotische Aufnahme); daher fand die Gleichheit im Verhalten ihren Ausdruck in der Gleichheit der „Permeabilitätsreihen", d. h. der nach ihrer Durchtrittsgeschwindigkeit geordneten Reihe der Diosmotika. Die Permeabilitätsreihen zeigen überdies recht gute Übereinstimmung mit derjenigen Reihe, die sich aus der Anordnung der gleichen Diosmotika nach ihrer Lipoidlöslichkeit ergab (soweit das aus den nicht eben quantitativen Angaben abzulesen war). Somit konnte gefolgert werden, daß sowohl die Mechanismen der osmotischen Aufnahme (Löslichkeitshypothese) als auch die permeabilitätsregelnden Substrate (Grenzschichtlipoide) und deren Durchlässigkeit bei allen Objekten gleich seien.

Die fortschreitende Verbesserung der Untersuchungsmethoden gestattete alsbald quantitative Messungen. Deren Ergebnisse mußten anders gedeutet werden; sie wiesen, generell betrachtet, auf so erhebliche Unterschiede im Aufnahmevermögen verschiedener Objekte hin, daß eine Revision der Auffassung OVERTONs unumgänglich war. In den Jahren zwischen 1925 und 1940 (letzte Zusammenfassung HÖFLER 1942) ist, vor allem dank der Breitenarbeit der Schulen COLLANDERs und HÖFLERs, eine Vielzahl von Permeabilitätsreihen ermittelt worden, die eigentlich niemals volle Kongruenz erkennen lassen und die Widerlegung der Gleichartigkeit nachdrücklich rechtfertigten.

Andererseits sind die Differenzen doch keineswegs so radikaler Art, daß jedem Objekt seine eigene, einmalige und unverwechselbare Permeabilitätsreihe hätte zugeschrieben werden können. Ähnlichkeiten und Verwandtschaften deuteten sich an und ermöglichten eine Zusammenfassung zu „Permeabilitätstypen" („die Typenbildung bedeutet ein Stadium, das überall dort durchlaufen zu werden pflegt, wo sich die Wissenschaft beschreibend einer neuen Mannigfaltigkeit von Erscheinungen bemächtigt", HÖFLER 1942). So etablierte sich, von HÖFLER (1930) inauguriert, eine eigene Disziplin unter dem Namen „Vergleichende Permeabilitätsforschung"; sie war bestrebt, Typen aufzustellen und aus ihrem Vergleich weitere Aufschlüsse zu gewinnen. In diesem Handbuch wird es die Aufgabe des vorliegenden Abschnittes sein, über diese Permeabilitätstypen und ihre Bedeutung für die kausalanalytische Betrachtungsweise zu berichten.

II. Permeabilitätstypen.

a) Allgemeines.

In den sog. historischen Disziplinen der Biologie, z. B. der Phylogenetik, ist der Vergleich des Tatsachenmaterials oftmals das einzige methodische Mittel, um den Ablauf unwiederholbar abgelaufener Vorgänge zu rekonstruieren; in den Experimentaldisziplinen hingegen tritt das Vergleichen zurück hinter dem Experiment, d. h. der Kausalanalyse wiederholbarer Vorgänge, im vorliegenden Falle der Stoffaufnahme. Es würde verfehlt sein, innerhalb der Pflanzenphysiologie den Vergleichen jede Berechtigung abzusprechen — um so mehr, weil ja am Prozeß der Stoffaufnahme auch Strukturen beteiligt sind, die im Laufe der Stammesgeschichte „phylogenetisch" erworben worden sind; es bedarf indessen sorgfältiger Prüfung, inwieweit die vergleichenden Fakten auch wirklich verglichen werden dürfen, und inwieweit die Fakten bzw. etwaige Differenzen *signifikant* sind für die (Details der) involvierten Mechanismen und Strukturen.

Die Frage nach der Vergleichbarkeit ist immer dann mit besonderem Nachdruck zu stellen, wenn die *Gesamt*aufnahme zweier Objekte in Rede steht. Die Zellsäfte von *Nitella*- und *Valonia*-Zellen unterscheiden sich in ihrer Zusammensetzung erheblich voneinander (Abb. 1). Der Unterschied fällt noch mehr ins Gewicht, wenn man die Zusammensetzung des umgebenden Mediums (Süß- bzw. Seewasser) berücksichtigt: dann ergibt sich, daß *Nitella* nicht nur Kalium-, sondern auch Natrium-Ionen akkumuliert hat, *Valonia* hingegen Natriumionen weitgehend von der Aufnahme ausschließen konnte. Unter dem Aspekt der Permeabilität s. str. (osmotische Stoffaufnahme) würde man der *Nitella*-Zelle eine Permeabilität für Natriumionen zuschreiben, der *Valonia*-Zelle aber nicht. Andererseits läßt die Aufrechterhaltung des Natriumungleichgewichtes auch auf Impermeabilität der *Nitella*-Zelle für Natriumionen schließen. Endlich muß wegen der Akkumulation von Natriumionen bei *Nitella* eine nichtosmotische Natriumaufnahme angenommen werden. Allein aus dem Vergleich, ohne weitere Kausalanalyse, ist es nicht möglich, zu entscheiden, ob die beiden Algen sich unterscheiden

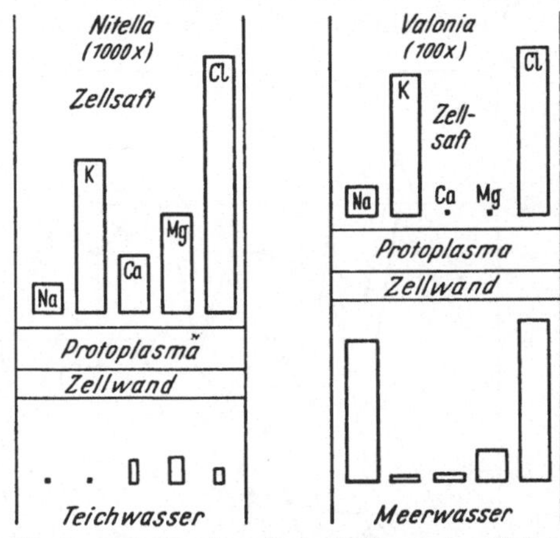

Abb. 1. Relative Ionenkonzentrationen im Kulturmedium und in den Zellsäften von *Nitella*- und *Valonia*-Zellen in Blockdarstellung. (Nach HOAGLAND 1948.)

a) in der osmotischen Stoffaufnahme (bzw. der Permeabilität),
b) der nichtosmotischen Aufnahme,
c) dem Verhältnis beider Mechanismen zueinander,
d) in der Reaktion beider Mechanismen auf die jeweilige Umgebung usw.

Da die Ungleichgewichte nicht nur für Natrium-, sondern auch für Kalium-, Calcium-, Magnesium- und Chlor-Ionen beständig aufrechterhalten werden, darf man wohl vermuten, daß die entscheidenden Vorgänge nichtosmotischer Art sind, denen gegenüber die osmotischen Vorgänge an Bedeutung stark zurücktreten. Man wird also hier nicht von einem *Nitella*- und einem *Valonia*-Typ *der Permeabilität* sprechen wollen, sondern die Differenzen den nichtosmotischen Prozessen zuschreiben bzw. sie auf Stoffwechseleigentümlichkeiten zurückführen.

Wenn *Permeabilitäts*typen diskutiert werden sollen, ist es mithin unerläßlich, nur solche Objekte miteinander zu vergleichen, bei denen möglichst ausschließlich die rein osmotische Aufnahme gemessen wird oder doch die nichtosmotische Aufnahme erheblich weniger ins Gewicht fällt. Damit wird der Umfang des Vergleichsmaterials eingeengt a) auf (meist) plasmolytische Untersuchungen und b) auf Anelektrolyte — wobei zu berücksichtigen ist, daß selbst dann nichtosmotische Prozesse interferieren und zu Verfälschungen führen können (vgl. diesen Band, S. 245ff.).

b) Experimentelle Ergebnisse.

Das gesamte bisher vorliegende, vergleichbare Versuchsmaterial sei in einem Diagramm zusammengestellt, das einer der letzten Arbeiten aus dem Gebiete der Vergleichenden Permeabilitätsforschung entnommen ist (Abb. 2, nach ELO

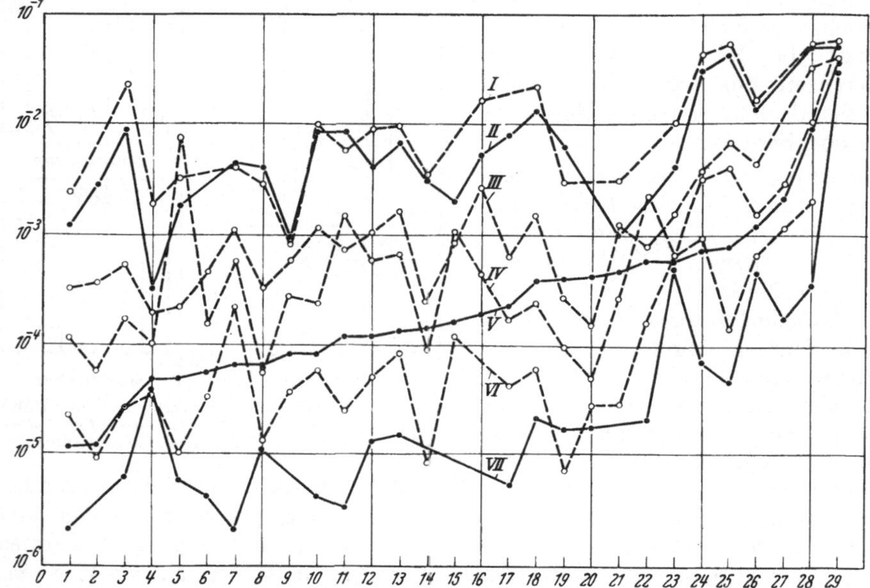

Abb. 2. Vergleichende Darstellung der Permeabilitätsreihen für verschiedene pflanzliche Objekte, geordnet nach steigenden Glycerinwerten. — 1. *Plagiothecium*, 2. *Anemone*, 3. *Zygnema*, 4. *Hippuris* (Wasserblatt), 5. *Hippuris* (Luftblatt), 6. *Maianthemum*, 7. *Oedogonium*, 8. *Curcuma*, 9. *Potamogeton*, 10. *Pylaiella*, 11. *Lemna*, 12. *Spirogyra*, 13. *Elodea*, 14. *Caltha*, 15. *Taraxacum*, 16. *Fucus*, 17. *Zygnema*, 18. *Taraxacum*, 19. *Rhoeo*, 20. *Gentiana* (Corollzellen), 21. *Muscari*, 22. *Gentiana* (Stengelepidermiszellen), 23. *Licmophora*, 24. *Ceramium*, 25. *Chara*, 26. *Melosira*, 27. *Bacterium*, 28. *Oscillatoria*, 29. *Beggiatoa*. (Nach ELO 1937.)

1937). Es handelt sich um nicht weniger als 29 Permeabilitätsreihen. Jede der Reihen ist als eine Vertikale wiedergegeben, auf der die einzelnen (absoluten) Permeabilitätskonstanten in logarithmischer Maßeinteilung abgetragen sind. Es sind hier jeweils „verkürzte" Reihen, die die Konstanten von nur 7 Stoffen enthalten: Der Alkohole Äthylenglykol, Glycerin und Erythrit, der Amide Acetamid und Malonamid sowie des Harnstoffs und des Methylharnstoffs. In den Originalarbeiten besitzen die Permeabilitätsreihen in der Regel sehr viel mehr Glieder; die genannte Auswahl wurde getroffen einmal aus Gründen der Übersichtlichkeit in der Darstellung, zum anderen, weil nicht alle Diosmotika an allen Objekten geprüft wurden. — Die 29 verschiedenen Vertikalen („Permeabilitätsreihen") sind in gleichen Abständen nebeneinandergestellt worden und zwar nach steigenden Glycerinwerten geordnet. Ferner wurden die Permeabilitätskonstanten der 7 Stoffe jeweils von Objekt zu Objekt durch Linien verbunden. Dem gewählten Bezugssystem entsprechend zeigt die „Glycerin-Linie“ einen annähernd stetigen, nach rechts ansteigenden Verlauf.

Zunächst fallen erhebliche quantitative Differenzen auf: *Plagiothecium* ist das Objekt mit den niedrigsten (zwischen 10^{-6} und 10^{-3}), *Beggiatoa* das mit den höchsten absoluten Konstanten (zwischen 10^{-2} und 10^{-1}). Auch die Ausdehnung der einzelnen Reihen ist verschieden groß: Bei *Zygnema* liegen die Konstanten der 7 Stoffe weit auseinander (von etwa $5 \cdot 10^{-6}$ bis $5 \cdot 10^{-2}$), bei *Melosira* zwischen $5 \cdot 10^{-4}$ und $5 \cdot 10^{-2}$, und bei *Beggiatoa* sind sie eng zusammengedrängt in einen Bereich von $5 \cdot 10^{-2}$ bis $8 \cdot 10^{-2}$.

Genauere Betrachtung lehrt ferner, daß auch *qualitative* Unterschiede vorliegen: die Verbindungslinien der Diosmotika (außer Glycerin) sind vielfältig gebrochen, mit steilen Anstiegen und Abfällen. Da diese nicht für alle Diosmotika gleichsinnig auftreten, kommt es häufig zu Überschneidungen der einzelnen Linien, d. h. nicht nur die absolute Größe der Permeabilitätskonstanten, sondern auch die Reihenfolge der Konstanten innerhalb der Permeabilitätsreihen wechselt mit dem Objekt.

Demgegenüber sind auch durchgängig gültige Tatsachen zu verzeichnen: bei allen Objekten ist die Reihenfolge der Alkohole: Glykol $>$ Glycerin $>$ Erythrit; stets liegen Acetamid und Glykol nahe beieinander; immer dringt der Methylharnstoff langsamer ein als Acetamid und Glykol, aber schneller als Malonamid und Glycerin (bei letzterem 2 Ausnahmen). Aber diese Beobachtungen reichen doch nicht aus, um allen Objekten im Sinne OVERTONs gleiche Permeabilitätseigenschaften zuzusprechen.

c) Objekttypen.

Aus dieser zunächst verwirrenden Fülle lassen sich auch einzelne Permeabilitätsreihen herausfinden, die einander mehr oder weniger ähnlich sind. Es war vor allem HÖFLERs Bemühen, solche Ähnlichkeiten zusammenzufassen zu „Typen“, die sich gegen stärker differierende Reihen abgrenzen lassen. Er gelangte bisher zu 5 solcher Typen, die von ihm, aber auch von seiten der finnischen Schule anerkannt und in einer Vielzahl von Publikationen diskutiert wurden:

1. *Chara-Maianthemum*-Typ,
2. *Gentiana sturmiana*-Typ,
3. *Rhoeo*-Typ,
4. *Diatomeen*-Typ,
5. *Beggiatoa*-Typ.

Der *Chara-Maianthemum*-Typ (1, COLLANDER und BÄRLUND 1933, HÖFLER 1934, MARKLUND 1936, WAHRY 1936, ELO 1937) ist recht häufig zu finden und wird als Haupttyp oder „Normaltyp“ bezeichnet; sein Auftreten reicht von *Bacterium coli* über *Chara ceratophylla* und *Plagiothecium denticulatum* bis zu Mono- und Dikotylen.

Der *Gentiana sturmiana*-Typ (2, HÖFLER und STIEGLER 1921, 1930, HÖFLER 1934, 1939), benannt nach dem Verhalten der anthocyanhaltigen Stengelepidermiszellen, ist demgegenüber nur vereinzelt gefunden worden; er läßt sich beschreiben durch besonders hohe Permeabilitätskonstanten für Harnstoff („rapider Harnstofftyp“ nach HOFMEISTER 1935). Die Zucker hingegen permeieren sehr langsam, und da die übrigen Verbindungen etwa in der Mitte liegen, erscheint hier die Permeabilitätsreihe besonders weit auseinandergezogen.

Beim *Rhoeo*-Typ (3, COLLANDER und BÄRLUND 1927) ist die Permeabilitätskonstante des Glycerins größer als die des Harnstoffes, eine Eigenart, die bei anderen Objekten kaum zu bemerken ist. Da auch andere Amide verhältnismäßig langsam eindringen, wird der Typ als „extrem amidophob“ bezeichnet.

Der *Diatomeen*-Typ (4, MARKLUND 1936, ELO 1937, HÖFLER und LEGLER 1940, HÖFLER 1940) fällt auf durch die hohen Permeabilitätskonstanten für Malonamid, Erythrit und insbesondere Zucker, d. h. für die großmolekularen Verbindungen. Hier rückt folglich die ganze Permeabilitätsreihe zusammen. Der *Diatomeen*-Typ wird auch als „Zuckertyp" bezeichnet.

Der *Beggiatoa*-Typ (5), bisher nur an *Beggiatoa mirabilis* gefunden (RUHLAND und HOFFMANN 1926, SCHÖNFELDER 1930, RUHLAND, ULLRICH und YAMAHA 1932), weist extrem hohe Permeabilitätskonstanten für alle untersuchten Verbindungen auf; hier wird eine strenge Parallelität zum Molekulargewicht bzw. zur Molekularrefraktion offenbar.

d) Permeabilitätstypen.

Diese — oder eine andere — Zusammenfassung von Typen ist sehr zu begrüßen, sofern sie ausschließlich als formales Prinzip zur übersichtlichen Ordnung eines großen Tatsachenmaterials dient; in diesem Sinne hat sie sich überdies als heuristisches Prinzip mehrfach bewährt. Sie läßt indessen das Streben nach Einsicht in die *Ursachen* der Tatsachen unbefriedigt. So hat denn HÖFLER versucht, diese Typen als *Permeabilitäts*typen zu interpretieren und angenommen, daß sich in ihnen besondere Eigenschaften widerspiegeln, die sich sowohl auf den Vorgang der Permeation als auch auf den Grenzschichtbau erstrecken sollen:

Beim *Chara-Maianthemum*-Typ folgt die Permeabilität vorwiegend dem Prinzip der Lipoidlöslichkeit; daneben ist auch mit einer gewissen Ultrafilterwirkung zu rechnen, vor allem gegenüber kleinmolekularen Substanzen, da diese schneller permeieren als ihrer Lipoidlöslichkeit entspricht (nach COLLANDER 1954 erstreckt sich jedoch die Ultrafilterwirkung auf alle Stoffe, auch die großmolekularen). Demgemäß sei bei *Chara* und *Maianthemum* mit einer „Porenphase" und einer „lösenden Phase" zu rechnen; es sollen zwei grundverschiedene Permeationsmechanismen existieren, und die Grenzschicht müsse nicht nur Porenstruktur aufweisen, sondern auch Lipoide mit bestimmten Lösungseigenschaften besitzen. Das Übergewicht habe die lösende Phase und die Lipoidlöslichkeit; bei *Maianthemum* gilt die „Porenphase" als noch schwächer entwickelt als bei *Chara*.

Die *Gentiana sturmiana*-Reihe wird so verstanden, daß das Plasma (der Grenzschichten) „stark amidophil im Sinne der Löslichkeit, ... außerdem hoch permeabel im Sinne einer Filterwirkung" ist. „Amidophilie und Ausbildung der Porenphase scheinen ... unabhängig voneinander in den Reihen variieren zu können" (Beispiel: *Plagiothecium*).

Beim *Rhoeo*-Typ (= Glycerintyp) ist die Permeationsgeschwindigkeit aller Amide herabgedrückt, die der Alkohole erhöht. Die Porenpermeation tritt zurück, und das Plasma wird als amidophob bezeichnet (nach dem Prinzip der gruppengebundenen Löslichkeit — HÖBER — ist das Lösungsvermögen „basischer" Lipoide für Amide in Modellversuchen geringer als das „saurer" Lipoide).

Der *Diatomeen*-Typ wird ebenfalls als „Löslichkeitstyp" aufgefaßt, weil groß- und kleinmolekulare Verbindungen wie Saccharose und Harnstoff nur geringe Unterschiede im Permeationsvermögen haben; ELO nimmt an, daß hier besonders hydrophile Grenzschichten mit hohem Lösungsvermögen für Zucker usw. vorliegen

Der *Beggiatoa-Oscillatoria*-Typ schließlich ist nach HÖFLER ein „reiner Ultrafiltertyp".

e) Beziehungen zwischen den Permeabilitätstypen.

Die genannten 5 wichtigsten Permeabilitätstypen stehen nach HÖFLER nicht völlig beziehungslos nebeneinander, vielmehr kann der *(Chara-)Maianthemum*-Typ nicht allein wegen der Häufigkeit, sondern auch in seiner Eigenschaft als „amidophiler Lösungstyp" in den Mittelpunkt gestellt werden, während die

4 folgenden Typen als Abänderungen dieses Grundtypus nach verschiedenen Richtungen aufzufassen sind. Diese Ableitungen sind in Abb. 3 (Permeabilitätstypen) und 4 (Objekttypen) in weitgehend kongruenten Diagrammen wiedergegeben. Die 4 Abzweigungen sind wie folgt zu charakterisieren: Der amidophobe Lösungstyp (unten) entsteht aus dem amidophilen Lösungstyp, wenn die Grenzschichtlipoide basischer werden; als Permeationsprinzip bleibt der Lösungsvorgang erhalten. Anders beim amidophilen Porentyp (oben), wo neben dem Löslichkeitsprinzip das Ultrafilterprinzip stärker zur Geltung kommt. *Chara* kann in dieser Betrachtung als vermittelndes Glied zwischen dem Haupttypus (*Maianthemum*) und dem extremen *Gentiana eturmiana*-Typ gelten. Der Zuckertyp (links) entsteht, wenn die Grenzschichten besonders hydrophil sind, gleichwohl aber als Lösungsmittel fungieren. Der Ultrafiltertyp (rechts) wird von HÖFLER abseits gestellt, da bei ihm das Löslichkeitsprinzip unwirksam geworden ist. Indessen kann *Oscillatoria* als Zwischenglied gelten.

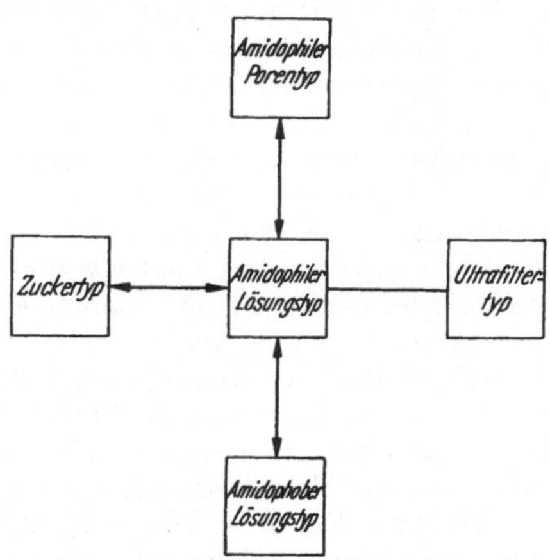

Abb. 3. Permeabilitätstypen. (Nach HÖFLER 1950).

Die Pfeile, mit denen HÖFLER drei der Strahlen seines Diagrammes (Abb. 3) ausgestattet hat, sollen zunächst andeuten, daß es Permeabilitätsreihen gibt, die zwischen dem Zentraltypus und den extremen „Abweichern" vermitteln. Nachdem dann die *Veränderlichkeit* von Permeabilitätsreihen „unter natürlichen Bedingungen" (z. B. Abhängigkeit von der Jahreszeit, HOFMEISTER 1938) oder durch experimentelle Eingriffe (SCHMIDT 1936, BOGEN 1938, 1940, 1941, KREUZ 1941) beobachtet war, fanden sich ferner Hinweise dafür, daß auch diese Variabilität sich den durch die Pfeile bezeichneten Richtungen einfügt und zwar offenbar im Sinne einer Reversibilität.

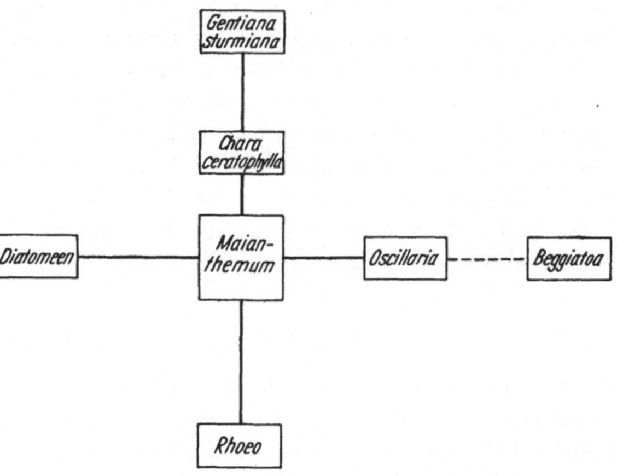

Abb. 4. Objekttypen. (Nach HÖFLER 1950.)

Für die experimentelle Veränderung von Permeabilitätsreihen seien einige Beispiele angeführt. *Rhoeo* ist als „amidophober" Typ unter anderem dadurch ausgezeichnet, daß das Glycerin erheblich schneller eindringt als der Harnstoff; der Quotient der betreffenden Permeabilitätskonstanten $Q_{Gl/Ha}$ liegt zwischen 4 und 6. BOGEN konnte durch Zugabe von $MgSO_4$ und $Al(NO_3)_3$ zur Lösung der Diosmotika diesen Quotienten auf 2,9 bzw. 1,1 erniedrigen und damit dem „amidophilen" Typus annähern. KREUZ erzielte am entgegengesetzten Typ (hochpermeable Harnstofftypen, bei denen der Harnstoff ganz erheblich schneller eindringt als das Glycerin) mit Ca-Salzen entsprechende Effekte, d. h. ebenfalls Annäherung an den „Zentraltyp".

Über *Ranunculus repens* (subepidermale Blattscheidenzellen) berichtete HOFMEISTER (1938), daß hier in der ersten Hälfte der Vegetationsperiode der Harnstofftyp, im Herbst der Glycerintyp in Erscheinung tritt.

Demgegenüber muß freilich betont werden, daß bisher keine Meßdaten vorliegen, die die Annahme rechtfertigen, jede der zahlreichen Permeabilitätsreihen könne experimentell in jeden beliebigen anderen Typ überführt werden.

Nach dem Gesagten ergibt sich, daß HÖFLER bei der Interpretierung der Permeabilitätstypen mehrere permeationsbestimmende Prinzipien in Betracht zieht: Ultrafiltration und Lösung in Lipoiden, entweder jedes für sich oder aber beide wechselnd kombiniert, daneben hydrophile und hydrophobe Grenzschichten sowie saure, neutrale und basische Grenzschichtlipoide mit vielfach wechselnden Anteilen. Die hieraus sich ergebenden Kombinationsmöglichkeiten sind zahlreich genug, um alle beobachteten Permeabilitätsunterschiede deuten und die Vielfalt der Erscheinungen in einem Gedankengebäude zusammenfassen zu können, dessen tragende Pfeiler die 5 genannten Permeabilitätstypen sind. Die Argumentationen konnten hier nur andeutungsweise wiedergegeben werden; aus den zahlreichen Diskussionen, die vor allem HÖFLER, aber auch MARKLUND, HOFMEISTER, ELO u. a. mit großem Scharfsinn geführt haben, geht hervor, daß das Gedankengebäude in sich weitgehend widerspruchsfrei erscheint.

Es soll freilich nicht verschwiegen werden, daß auch gegenteilige Ansichten laut geworden sind. So fand WAHRY (1936) bei ihren Untersuchungen an Luft- und Wasserblättern von *Hippuris* grundverschiedene Permeabilitätsreihen; durch Beleuchtung werden wiederum neue Reihen erhalten, wobei die einzelnen Permeabilitätskonstanten weder gleichsinnig noch im gleichen Maße verschoben werden. Wenn die Veränderung in den Zellen des Wasserblattes noch auf eine Erhöhung des Anteils saurer Lipoide in der Plasmahaut zurückgeführt werden kann, so führt eine ähnliche Diskussion der Permeabilitätsänderungen in den Zellen des Luftblattes zu keinem brauchbaren Ergebnis.

Solche und andere Widersprüche, die sich aus physiologischen Bedingtheiten ergeben, haben offenbar tiefere Gründe: Es muß mit allem Nachdruck betont werden, daß die oben angeführte Widerspruchsfreiheit nur besteht, sofern man die Meßdaten als wirkliche unverfälschte Permeabilitätskonstanten auffassen darf, und sofern man zugesteht, daß ein Stoff auf wenigstens zwei gänzlich verschiedenen Wegen *endosmieren* kann.

f) Kritik der Permeabilitätstypen.

In den letzten Jahren haben sich die Beweise gehäuft, daß beide Voraussetzungen *nicht* zutreffen. In einem vorangestellten Abschnitte des vorliegenden Bandes (Die Aufnahme der Anelektrolyte, S. 230) sind die Argumente ausführlich besprochen, die es verbieten, experimentell gefundene Meßwerte unbesehen in die Formeln zur Berechnung von Permeabilitätskonstanten einzusetzen, so daß es hier genügt, die wesentlichen Folgerungen herauszustellen: Mit plasmolytischen Methoden und mit Hilfe der chemischen Zellsaftanalyse wird stets ein mehr oder weniger großer Teil der gesamten Stoffaufnahme erfaßt, die sich aus osmotischer und nichtosmotischer Aufnahme zusammensetzt. Da für die Beurteilung der Permeabilitätsverhältnisse ausschließlich die osmotische Aufnahme herangezogen werden darf, sind die berechneten „Konstanten" in der Regel zu hoch. Einen großen Anteil an der Verfälschung hat die nicht- bzw. meta-osmotische *Wasser*aufnahme, die mit plasmolytischen Methoden gleichfalls erfaßt wird, doch ist auch mit nichtosmotischer Aufnahme gelöster Diosmotika zu rechnen. BOGEN und FOLLMANN (1955) konnten die Bedeutung dieser Effekte insbesondere für den sog. „Zuckertyp der Permeabilität" (*Diatomeen*-Typ, Nr. 4 in der Aufstellung HÖFLERS) aufzeigen. Die schnelle Deplasmolyse, die bei *Melosira*-Arten in Saccharoselösungen zu beobachten ist, und die bisher stets als (extrem schnelle) Saccharose-*Permeation* gedeutet worden ist, beruht in Wirklichkeit auf einer ansehnlichen nichtosmotischen Wasseraufnahme. Diese kann durch Stoffwechselgifte (Natriumazid, Dinitrophenol) völlig vernichtet werden, ohne daß die Permeabilität nachweisbar betroffen würde. Die in solchen Versuchen verbleibende Deplasmolysegeschwindigkeit, die nunmehr wirklich auf Saccharose-Endosmose beruhen dürfte, beträgt nur etwa $^1/_{10}$ bis $^1/_5$ des Kontrollwertes; die daraus berechneten Permeabilitätskonstanten liegen durchaus in der Größenordnung der Zuckerkonstanten anderer Objekte, z. B. des „amidophilen Lösungstyps". Bei weniger langsam eindringenden Stoffen, wie etwa Erythrit, fällt der Anteil der nichtosmotischen Wasseraufnahme schon geringer aus, und bei den am schnellsten aufgenommenen Stoffen (Acetamid, Glykol) spielt sie kaum noch eine Rolle. Allein damit ist der „Spezialtyp" der Diatomeen-Permeabilität zurückgeführt auf das Normalverhalten: Hinsichtlich ihrer Permeabilität weichen die Diatomeen von anderen Objekten nicht ab; was sie von diesen unterscheidet, sind gewisse Stoffwechseleigentümlichkeiten, die sich vor allem in einer erheblichen metaosmotischen Wasseraufnahme (und daneben auch in einer nichtosmotischen Aufnahme gelöster Anelektrolyte) äußern.

Was hier an einem Einzelbeispiel besonders deutlich gemacht werden konnte, gilt sinngemäß auch für andere Objekte (*Rhoeo*, *Taraxacum*, *Oenothera*, vgl. BOGEN 1953, BOGEN und PRELL 1953 u. a.) und spricht eindeutig gegen die bisher vertretene Anschauung, jeder Unterschied zwischen Permeabilitätsreihen sei als reeller Unterschied der *Permeabilität* anzusehen. Hierfür lassen sich auch noch andersartige Gründe aufführen. So wies z. B. BOGEN (1938, 1940, 1941) nach, daß gewisse Anelektrolyte wie Glycerin und Harnstoff die Permeabilität der Grenzschichten für sich selbst (und andere, gleichzeitig gebotene Stoffe) verändern können. Der Stoffdurchtritt durch eine solchermaßen veränderte Grenzschicht folgt nun ebenfalls dem FICKschen Diffusionsgesetz (und äußert sich daher auch nicht in einem Abfallen oder Ansteigen der Konstanten während des Versuchs), aber mit veränderter Geschwindigkeit; die Permeabilitätskonstanten für Harnstoff und Glycerin sind also an einem System gewonnen, das sich in einem anderen Zustand befand als bei Diosmoseversuchen etwa mit Glykol oder Malonamid und sind daher mit den Konstanten der letztgenannten Stoffe nicht vergleichbar. Diesem für zwei Diosmotika nachgewiesenen „Quellungseffekt" kommt für die gesamte Permeabilitätsreihe des betreffenden Objektes größere Bedeutung zu als es zunächst den Anschein hat, denn ein großer Teil der Diskussion über Permeabilitätsunterschiede basiert auf der Anordnung der Permeabilitätsreihe gerade nach den Konstanten dieser beiden, nachgewiesenermaßen *nicht inerten* Stoffe. Dadurch werden oftmals geringfügige Differenzen verzerrt und vergröbert (vgl. Abb. 1).

Die auf S. 427 erhobene Forderung, daß ein Vergleich von Permeabilitätsreihen nur dann statthaft sei, wenn die Vergleichbarkeit auch wirklich gewährleistet ist, erweist sich somit als unerfüllbar, soweit sie den größten Teil des bisher vorliegenden Versuchsmaterials betrifft, und soweit Differenzen als *Permeabilitäts*differenzen interpretiert werden sollen: die zu vergleichenden Objekte befinden sich physikalisch und physiologisch gesehen in ganz verschiedenen Zuständen, die mit der eigentlichen Permeabilität gar nichts zu tun haben, sondern sich auf die nichtosmotische Stoffaufnahme u. a. erstrecken. Damit sind aber dem HÖFLERschen Schema der 5 Permeabilitätstypen wesentliche Stützen entzogen.

Eine andere Art der Typenbildung sei noch kurz erwähnt, die ausschließlich auf Grund der Alkoholaufnahme vorgenommen wurde: ZEHETNER, ein Schüler HÖFLERs, berichtete 1934, daß einige Objekte (z. B. *Vallisneria*) nach Vorplasmolyse in Saccharoselösung bei anschließender Alkoholzufuhr Protoplastenkontraktion zeigen („Zellen leichter für Wasser durchlässig"), andere (z. B. *Cladophora*) hingegen Ausdehnung („Zellen leichter für Alkohol durchlässig"). Nach dem Ausfall dieser Versuche wurde ein Kontraktions- und ein Ausdehnungstyp aufgestellt und beide als Permeabilitätstypen im Sinne der Vergleichenden Permeabilitätsforschung diskutiert. RUHLAND, ULLRICH und ENDO (1937) konnten alsbald zeigen, daß sich die Reaktionsweisen jeweils umkehren lassen, je nach den experimentellen Bedingungen (insbesondere der Versuchstemperatur). Da es im wesentlichen physiologische Faktoren sind, die über den Ausfall der Reaktion im einen oder anderen Sinne entscheiden, lehnen die Autoren die von ZEHETNER gewählte Typenbildung mit Recht ab.

g) Über die Einheitlichkeit der Permeabilitätsreihen.

Es bleibt noch zu prüfen, ob die zweite der auf S. 432 genannten Voraussetzungen erfüllt ist, unter denen die HÖFLERsche Theorie der Permeabilitätstypen in sich widerspruchsfrei ist, d. h. also, ob ein Stoff auf wenigstens zwei gänzlich verschiedenen Wegen *endosmieren* kann (daß er osmotisch *und* nichtosmotisch aufgenommen werden kann, wurde bereits erwähnt). Hierzu ist ein Versuch BOGENs (1950) zu nennen, der im Gegensatz zu HÖFLER die Herausarbeitung der allen Permeabilitätsreihen *gemeinsamen* Eigenschaften zum Ziele hat. Auch hierüber ist bereits im Abschnitt über Anelektrolytaufnahme berichtet worden (S. 230). Als Basis dienten folgende Tatsachen und Überlegungen:

1. Es sollten die gleichen Permeabilitätsreihen verwendet werden, die der Vergleichenden Permeabilitätsforschung als Diskussionsgrundlage dienten (mit Ausnahme von *Beggiatoa* und *Chara*, da diese nicht mit plasmolytischen Methoden untersucht worden sind). Dabei ist die gleiche Verfälschung und Verzerrung in Kauf zu nehmen, die bereits oben herausgestellt wurde.

2. Die Anordnung der (insgesamt 31 bzw. 32) Permeabilitätsreihen sollte nicht an Hand steigender Glycerinwerte (wie bei ELO, vgl. Abb. 1) erfolgen, weil dieser Stoff Sekundäreffekte zeigt und damit doppelt verfälschte Konstanten liefert (BOGEN 1940). Stattdessen wurde die Konstante des Malonamids als Bezugsgröße gewählt, das zwar auch nicht als völlig inert gelten dürfte (BOGEN und FOLLMANN 1955), aber doch erheblich weniger kräftige Sekundäreffekte hervorruft.

3. Die experimentellen Daten (Konstanten) sind allein aus methodischen Gründen mit großen Fehlern behaftet; mittlere Fehler von $\pm$ 50% sind durchaus keine Seltenheit (vgl. z. B. die Meßwerte HOFMEISTERs 1935). Wenn daher bei graphischen Darstellungen die an verschiedenen Objekten ermittelten Permeabilitätskonstanten ein und desselben Stoffes verbunden werden sollen, so brauchen die Verbindungslinien nicht genau durch die Punkte zu laufen, die die Lage der Konstanten angeben.

Trägt man so die Permeabilitätskonstanten der 31 Reihen im doppelt logarithmischen Netz ein, so läßt sich durch die Punktdarstellung der Konstanten eines Stoffes eine Gerade legen, von der die einzelnen Punkte nur geringfügige Abstände haben, und es ergeben sich in grober Annäherung für die 7 Diosmotika 7 Gerade, die nach Bereichen (Objekten) höherer Permeabilität konvergieren.

Wie auf S. 243 abgeleitet und im nachfolgenden Abschnitt (Theorie der Permeabilität) näher begründet, drücken die „Permeabilitätsgeraden" folgende Abhängigkeitsbeziehung aus:

$$P = \frac{f'(K)}{f''(R_{MD})};$$

die Permeabilitätskonstante ist proportional dem Verteilungskoeffizienten K und umgekehrt proportional der Molekularrefraktion R_{MD}. Diese Abhängigkeitsbeziehung ist so zu interpretieren, daß der Permeationsvorgang für alle Stoffe und Objekte einheitlich eine Ultrafiltration ist bzw. eine durch eine „Durchdringungsstruktur" gehemmte Hydrodiffusion, wobei als hemmend die „zwischenmolekularen Kräfte" wirken. Die Proportionalität zum Verteilungskoeffizienten zeigt nicht etwa die Mitwirkung eines Lösungsvorganges an, sondern drückt lediglich aus, daß die sog. lipoidlöslichen Stoffe geringere zwischenmolekulare Kräfte entfalten und deshalb gegenüber den lipoidunlöslichen Verbindungen in der Permeation weniger gehemmt bzw. relativ gefördert sind.

Wenn man nur die von den endosmierenden Molekülen ausgehenden zwischenmolekularen Kräfte in Rechnung stellt, bleibt nach der graphischen Darstellung kein Raum für eigentliche Objekttypen der Permeabilität, denn die einzelnen Objekte sind nur quantitativ unterschieden, wobei zwischen extrem durchlässigen Objekten (z. B. *Oscillatoria princeps*) und solchen mit besonders geringer Permeabilität (z. B. *Rhoeo discolor* oder *Hippuris*, Wasserblatt bei Dunkelheit) alle Zwischenstufen gegeben sind.

Es ist indessen auch damit zu rechnen, daß die Grenzschichten der verschiedenen Objekte z. B. in ihrem Lipoidgehalt unterschieden sein können, oder daß die Lipoide bzw. die hydrophoben Gruppen des Proteins selbst differieren. Somit sind auch Unterschiede im Ausmaß jener zwischenmolekularen Kräfte zu erwarten, die von der Grenzschicht ausgehen und mit den endosmierenden Molekülen in Wechselwirkung treten. Sie können bei entsprechendem Ausmaß die Permeabilitätskonstanten in quantitativer Hinsicht modifizieren, doch bleibt der Permeationsvorgang immer der gleiche, nämlich eine Ultrafiltration.

Beim gegenwärtigen Stand der Meßmethoden erscheint es wenig aussichtsreich, solche Differenzen exakt zu erfassen, bzw. aus den Permeabilitätsreihen erschließen zu wollen, weil die Auswirkungen der gleichzeitig ablaufenden nichtosmotischen Prozesse erheblich stärker sind und daher feinere Unterschiede völlig zu überdecken vermögen.

Wie Bogen (1950) bereits betonte, darf die Formulierung von einheitlichen Permeabilitätseigenschaften aller Objekte nicht überschätzt werden. Ohne Zweifel gibt es einige Permeabilitätsreihen, die nicht in allen Einzelheiten mit der theoretisch vorausberechneten übereinstimmen. Solche Differenzen (so z. B. daß bei *Rhoeo* das Glycerin langsamer eindringt als der Harnstoff) können nicht auf Meßfehler zurückgeführt werden. Aber hier dürften entweder sog. Sekundäreffekte (z. B. Bogen 1940 an *Rhoeo*) mitwirken, oder aber es interferieren nichtosmotische Aufnahmevorgänge, wie das für den Fall der „extremen Zuckerpermeabilität" der Diatomeen nachgewiesen werden konnte. Das aber bestätigt den Schluß, daß es nicht gerechtfertigt ist, von *Permeabilitäts*typen zu sprechen, wo lediglich Stoffwechseleigentümlichkeiten in Rede stehen.

III. Die spezifischen Permeabilitätsreihen.

Höfler hat die Reihen seiner Permeabilitätskonstanten nicht allein Permeabilitätsreihen genannt, sondern darüber hinaus noch als „spezifisch" bezeichnet, um ihnen eine weitere Bedeutung beizulegen (Höfler und Stiegler

1930). Über diesen Begriff „spezifisch" hat lange Zeit Unklarheit geherrscht. Er wurde zunächst von einigen Autoren (z. B. WAHRY 1936) im taxonomischen bzw. genetischen Sinne aufgefaßt. Auch bei HÖFLER finden sich Formulierungen, die in diesem Sinne zu verstehen sind, so z. B. wenn er sagt: „verschiedene Zellsorten ... haben eine gewisse Plasmapermeabilität und diese ist in gewissen Grenzen autonom und erblich festgelegt, so gut wie die morphologische Gestalt und der spezifische Chemismus". Andererseits spricht HÖFLER, besonders nach seinen Erfahrungen an *Gentiana sturmiana*, von „objektspezifisch", indem er diesen Ausdruck nicht auf die Species, sondern auf „Zellsorten" bezieht. Objektspezifisch heißt hier aber „determiniert und dem Objekt eigentümlich". Diese vorsichtige Formulierung besagt mit Recht nichts anderes, als daß die (plasmometrisch bestimmten) Permeationskonstanten Besonderheiten des untersuchten Objekts widerspiegeln, und daß Differenzen in den Reihen der Ausdruck von irgendwelchen, zunächst nicht näher definierten Unterschieden dieser Objekte sind. Durch die Bezeichnung „*Permeabilitäts*"reihen bzw. -typen wird freilich bereits ausgesprochen, daß diese Differenzen als Permeabilitätsdifferenzen zu gelten haben — ein Postulat, das nach den oben angeführten Erörterungen nicht mehr aufrechterhalten werden kann.

Darüber hinaus aber sollen die spezifischen Permeabilitätsreihen ein *Charakteristikum* sein, d. h. sie sollen das Plasma der betreffenden Zellsorte *kennzeichnen*. Wenn man indessen unter „kennzeichnen" mehr verstehen will als das bloße Konstatieren von Details unbekannter Signifikanz, nämlich das Erfassen der *wesentlichen, unverwechselbaren und entscheidenden Merkmale des Protoplasmas*, so ist leicht einzusehen, daß die Reihenaufstellung dieser Aufgabe nicht gerecht werden kann.

1. Eine Überraschung war der Befund HOFMEISTERs (1938), daß die spezifische Permeabilitätsreihe der Blattscheidensubepidermis von *Ranunculus repens* nicht konstant ist, sondern, offenbar periodisch, einen Wechsel vom „Harnstofftyp" zum „Glycerintyp" und wieder zurück durchmacht. Um den Spezifitätsbegriff in der hier gebrauchten Fassung aufrechterhalten zu können, erweitert ihn HOFMEISTER (mit Zustimmung HÖFLERS) wie folgt: „Auch die Eigenheit der einzelnen Reihen, bestimmte Veränderlichkeit zu zeigen, ist spezifisch und als vollwertiges protoplasmatisches Merkmal den übrigen an die Seite zu stellen." Dem ist durchaus beizupflichten, wenn man den Wechsel der Reihen als Eigentümlichkeit der Zelle bezeichnen will, hingegen dürfte es kaum Zustimmung finden, wollte man ihn als „charakteristischen Wesenszug des Protoplasmas" der Blattscheidensubepidermiszellen deklarieren. Es handelt sich bestenfalls um eines von vielen Merkmalen, nicht jedoch um das oder auch nur ein Charakteristikum.

2. Überblickt man das Auftreten der verschiedenen spezifischen Reihen im Pflanzenreich, so bemerkt man den „amidophilen Lösungstyp" (*Chara-Maianthemum*-Typ) unter anderen bei *Bacterium coli* (ELO 1937), *Chara ceratophylla* (COLLANDER und BÄRLUND 1933), dem Laubmoos *Plagiothecium denticulatum* (MARKLUND 1936) und der Liliacee *Maianthemum bifolium* (HÖFLER 1934). Da sich „die Plasmen" der vier Pflanzen wegen ihrer extrem entfernten Positionen im Stammbaum sowohl phylogenetisch als auch hinsichtlich ihres Cytoplasma-Idiotypus und -Phänotypus nachdrücklich unterscheiden dürften, ergibt sich gerade aus der nahezu vollständigen Übereinstimmung ihrer Permeabilitätsreihen, daß diese nicht geeignet sind, die charakteristischen Merkmale bzw. Differenzen dieser Plasmen auch nur anzudeuten.

Ein Bild des anderen Extrems bietet *Gentiana sturmiana* (HÖFLER und STIEGLER 1921, 1930, HÖFLER 1936). Hier gehören „die Plasmen der Stengelhautzellen dem amidophilen Typ und zwar einem extremen Harnstofftyp an", die der Corollzellen der gleichen Pflanze hingegen dem Glycerintyp. Hier sind, am selben Pflanzenexemplar, die beiden Zellsorten eindeutig der gleichen idiotypischen Konstitution und lediglich im Verlaufe der Ontogenese verschieden determiniert worden — trotzdem sind ihre spezifischen Permeabilitätsreihen verschieden und charakterisieren somit zwei verschiedene Plasmen. Das würde zu der merkwürdigen Folgerung führen, daß die Plasmen von *Bacterium coli* und der Stengelepidermis von *Gentiana sturmiana* auf der einen Seite und die von *Rhoeo discolor* und den Corollzellen von *Gentiana sturmiana* andererseits einander plasmatisch viel mehr gleichen als Stengelepidermis und Corollzellen von *Gentiana sturmiana* untereinander.

Konstruktionen dieser Art erscheinen abwegig, und so muß die Koppelung zwischen den Begriffen „spezifisch" und „Permeabilitätsreihen" (gegen den letzteren wurden bereits die oben angeführten Bedenken geltend gemacht) aufgegeben werden, sobald ihr der Sinn unterlegt wird, man könne mit Hilfe der spezifischen Reihen verschiedene Plasmen *charakterisieren*. Sie kann hingegen bestehenbleiben, wenn mit ihr lediglich gewisse Eigentümlichkeiten der untersuchten Objekte zahlenmäßig erfaßt werden sollen. Eine Aufklärung über die Ursachen dieser Eigentümlichkeiten freilich vermag uns nur die eingehende experimentelle Analyse zu liefern. Sie kann allein durch den Vergleich oder gar durch die bloße Benennung „spezifisch" in keinem Falle ersetzt werden. Andererseits sind die Reihen ohne Zweifel ein wertvolles Mittel, um Differenzen physiologischer, konstitutiver und/oder anderer Art aufzuzeigen, gerade dort, wo sie mit anderen Mitteln nicht entdeckt werden können.

Literatur.

Bogen, H. J.: Untersuchungen zu den „spezifischen Permeabilitätsreihen" Höflers. II. Harnstoff und Glycerin. Planta (Berl.) **28**, 535—581 (1938). — Bogen, H. J., u. G. Follmann: Osmotische und nichtosmotische Stoffaufnahme bei Diatomeen. Planta (Berl.) **45**, 125—146 (1955). — Bogen, H. J., u. H. Prell: Messung nichtosmotischer Wasseraufnahme an plasmolysierten Protoplasten. Planta (Berl.) **41**, 459—479 (1953).

Collander, R.: The permeability of Nitella cells to non-electrolytes. Physiol. Plantarum (Copenh.) **7**, 420—445 (1954). — Collander, R., u. H. Bärlund: Über die Protoplasmapermeabilität von Rhoeo discolor. Soc. Sci. fenn., Comment. Biol. II **6**, 1—13 (1927). — Permeabilitätsstudien an Chara ceratophylla. II. Acta bot. fenn. **11**, 1—114 (1933).

Elo, J. E.: Vergleichende Permeabilitätsstudien. Ann. bot. Soc. zool.-bot. fenn. „Vanamo" **8**, Nr 6 (1937).

Hoagland, D. R.: Lectures on the inorganic nutrition of plants, 2. Aufl. Waltham, Mass. 1948. — Höber, R., u. H. Ulrich: Über das Eindringen lipoidunlöslicher Nichtelektrolyte mit relativ großem Molekularvolumen in Säugetiererythrocyten. Klin. Wschr. **1934**, 63. — Höfler, K.: Permeabilitätsstudien an Stengelzellen von Majanthemum bifolium. Sitzgsber. Akad. Wiss. Wien, Math.-naturwiss. Kl., Abt. I **143**, 213—264 (1934). — Nekrose in Sulfoharnstoff. Flora (Jena), N. F. **33**, 131—142 (1939). — Unsere derzeitige Kenntnis von den spezifischen Permeabilitätsreihen. Ber. dtsch. bot. Ges. **60**, 179—200 (1942). — New facts on water permeability. Protoplasma **39**, 677—683 (1950). — Höfler, K., u. F. Legler: Über die Salzresistenz einiger Diatomeen in dem Franzensbader Mineralmoore. Beih. bot. Zbl., Abt. A **60**, 327—342 (1940). — Höfler, K., u. A. Stiegler: Ein auffälliger Permeabilitätsversuch in Harnstofflösung. Ber. dtsch. bot. Ges. **39**, 157—164 (1921). — Permeabilitätsverteilung in verschiedenen Geweben einer Pflanze. Protoplasma **9**, 469—512 (1930). — Hofmeister, L.: Vergleichende Untersuchungen über spezifische Permeabilitätsreihen. Bibl. Bot. **1935**, H. 113. — Verschiedene Permeabilitätsreihen bei einer und derselben Zellsorte von Ranunculus repens. Jb. wiss. Bot. **1938**, 401—419.

Kreuz, J.: Der Einfluß von Kalzium- und Kaliumsalzen auf die Permeabilität des Protoplasmas für Harnstoff und Glycerin. Österr. bot. Z. **90**, 1—30 (1941).

Marklund, G.: Vergleichende Permeabilitätsstudien an pflanzlichen Protoplasten. Acta bot. fenn. **18**, 1—110 (1936).

OVERTON, E.: Über die allgemeinen osmotischen Eigenschaften der Zelle, ihre vermutlichen Ursachen und ihre Bedeutung für die Physiologie. Vjschr. naturforsch. Ges. Zürich **44**, 88—135 (1899).

RUHLAND, W., u. C. HOFFMANN: Die Permeabilität von Beggiatoa mirabilis. Ein Beitrag zur Ultrafiltertheorie des Plasmas. Planta (Berl.) **1**, 1—83 (1925). — RUHLAND, W., H. ULLRICH u. G. YAMAHA: Über den Durchtritt von Elektrolyten mit organischem Anion und einwertigem Kation in die Zellen von Beggiatoa mirabilis. Planta (Berl.) **18**, 338—382 (1932). — RUHLAND, W., H. ULLRICH u. S. ENDO: Untersuchungen zu den „spezifischen Permeabilitätsreihen" HÖFLERS. I. Zur Frage der Alkoholpermeabilität von Pflanzenzellen unter verschiedenen Versuchsbedingungen. Planta (Berl.) **27**, 650—668 (1938).

SCHMIDT, H.: Plasmolyse und Permeabilität. Jb. wiss. Bot. **86**, 470 (1936). — SCHÖNFELDER, S.: Weitere Beobachtungen über die Permeabilität von Beggiatoa mirabilis, nebst kritischen Ausführungen zum Gesamtproblem der Permeabilität. Planta (Berl.) **12**, 414—504 (1931).

WAHRY, E.: Permeabilitätsstudien an Hippuris. Jb. wiss. Bot. **83**, 657—705 (1936).

ZEHETNER, H.: Untersuchungen über die Alkoholpermeabilität des Protoplasmas. Jb. wiss. Bot. **80**, 505—566 (1934).

Die Theorie der Permeabilität.

Von

H. J. Bogen.

In diesem Abschnitt soll, in Übereinstimmung mit den auf S. 116ff. gegebenen Begriffsbestimmungen, unter „Permeabilität" jene Eigenschaft der Zelle bzw. ihrer Grenzschichten verstanden werden, die den osmotischen Durchtritt von gelösten Substanzen (und Lösungsmitteln) zur Folge hat. Sie findet ihren Ausdruck in der *Permeation* dieser Stoffe.

In den einschlägigen Publikationen ist immer wieder betont worden, daß sich die Permeabilitätsforschung zwei Aufgaben gestellt hat: 1. die Aufklärung der Permeationsmechanismen, 2. die Analyse der Grenzschichtstrukturen, die die Permeation in der jeweils beobachteten Weise ablaufen lassen. Eine Darstellung der Theorie der Permeabilität hat demzufolge die gleiche doppelte Zielsetzung. Wenn sie indessen unter Theorie nicht die Vielzahl der Hypothesen und Vermutungen begreifen will, wie sie im Überfluß geäußert worden sind, sondern — im Sinne der theoretischen Physik — die theoretischen physikalischen Grundlagen aufzudecken versucht, so wird sie sich vorwiegend mit dem ersten Punkt, den Permeationsmechanismen, zu befassen haben, weil hier empirische Versuchsdaten mit theoretisch wohlfundierten Vorstellungen verglichen werden können, während die Analyse des Grenzschichtbaues, von wenigen Ausnahmen abgesehen, noch im Stadium der morphologischen und physiologischen Beobachtungen steht und von einer eigentlichen Theorie der Grenzschichtstruktur einstweilen nicht gesprochen werden kann.

I. Permeationsmechanismen.

Permeation ist die Stoffbewegung entlang eines Gradienten der Konzentration oder des elektrochemischen Potentials; in diesen Gradienten eingeschaltet bzw. ein Teil von ihm ist die Zellgrenzschicht („Membran"), durch die hindurch die Permeation stattfindet. Sie ist beendet, wenn zu beiden Seiten der Membran Gleichheit der Konzentration („Gleichverteilung") bzw. des elektrochemischen Potentials erreicht wird. Das gleiche gilt, von geringfügigen Ausnahmen abgesehen, die überdies keine grundsätzliche Bedeutung haben, von der Diffusion, und es ist schon früh erkannt worden, daß die Permeation im Grunde nur ein Spezialfall der Diffusion ist — sowohl thermodynamisch als auch kinetisch betrachtet: thermodynamisch insofern, als in beiden Fällen die Endzustände übereinstimmen (Gleichheit der freien bzw. spezifischen Energie); kinetisch, indem die mathematische Behandlung der Permeation auf dem Fickschen Diffusionsgesetz basiert.

Der Spezialfall Permeation wird in der Regel als *behinderte Diffusion* aufgefaßt, und demgemäß hat jede theoretische Betrachtung von der Diffusion auszugehen. Dazu gibt es zwei Wege: 1: die statistische = formale Behandlung, die sich der Diffusionskonstanten bedient. Nach Graham und später Maxwell sind die Diffusionskonstanten D verschiedener Stoffe umgekehrt proportional der Quadratwurzel aus ihren Molekulargewichten M:

$$D = \text{const} \cdot M^{-\frac{1}{2}}; \tag{1}$$

dann ist zu überprüfen, inwieweit auch für die Permeationskonstanten P die Gleichung

$$P = \text{const} \cdot M^{-\frac{1}{2}} \tag{2}$$

gilt, bzw. jede Abweichung von dieser Gleichung zu interpretieren; 2. die *molekularkinetische*, die die von den *einzelnen* Molekülen des Systems ausgehenden Kräfte untersucht und aus diesen Daten das Verhalten der diffundierenden bzw. permeierenden Moleküle ableitet bzw. erklärt.

A. Die formale Behandlung der Diffusion und Permeation.

Ihre Grundlage ist die Diffusionsgleichung von FICK (1855):

$$\frac{ds}{di} = D \cdot \frac{dc}{dx} \tag{3}$$

worin ds diejenige Stoffmenge ist, die durch 1 cm² Grenz- bzw. Oberfläche in der Zeit dt bei einem Konzentrationsgradienten dc/dx bewegt wird. D ist die Diffusionskonstante und gibt die Stoffmenge an, die je Flächen- und Zeiteinheit bei der Einheit des Konzentrationsgefälles ($dc/dx = 1$) diffundiert. Sie ist ein Maß für den *Diffusionswiderstand* (Näheres s. SPANNER, diesen Band, S. 125).

Der Vergleich der Diffusionskonstanten *verschiedener* Stoffe vermag nähere Aufschlüsse zu geben über die diffusionsbestimmenden Faktoren. Schon GRAHAM konnte feststellen, daß die Diffusionskonstante umgekehrt proportional der Quadratwurzel aus den Molekulargewichten ist (Gleichung 1). Die theoretischen Grundlagen für diese Gleichung lieferte später MAXWELL, und THOVERT (1910) konnte zeigen, daß die MAXWELLsche Gleichung ziemlich genau gilt für die Diffusion in Flüssigkeiten wie Wasser und Alkohol und zwar vor allem für Moleküle bis zur Größe der Hexosen. Von besonderer Bedeutung ist es nun, daß es wenigstens einen Fall gibt, bei dem eine analoge Beziehung für die Permeationskonstanten bzw. Permeabilitätskoeffizienten P gilt:

$$P = \text{const} \cdot M^{-\frac{1}{2}}.$$

Er ist gegeben bei den künstlichen Kollodiummembranen, *Poren*membranen also, die COLLANDER (1926) auf ihre Permeabilität untersucht hat. Bekanntlich ist es möglich, solche Membranen mit verschiedenen Porenweiten herzustellen, und es zeigte sich, daß bei Membranen mit der größten Porenweite die obige Beziehung realisiert ist. Die Permeation durch solche Membranen darf also als *ungehinderte Diffusion* betrachtet werden. Bei „dichteren“ Membranen (mit engeren Poren) ergeben sich bemerkenswerte Abweichungen: die Produkte $P \cdot M^{\frac{1}{2}}$ sind nicht allein niedriger, sondern sie fallen auch mit steigendem M, anders ausgedrückt: die P-Werte fallen mit wachsendem Molekulargewicht schneller als die Diffusionskoeffizienten (RUHLAND und HOFFMANN 1926).

Der erste Befund, daß nämlich das Produkt $P \cdot M^{\frac{1}{2}}$ niedriger ist als bei freier Diffusion, kann mit der Verringerung des Diffusionsquerschnittes befriedigend erklärt werden: als Diffusionswege stehen nurmehr die Membranporen zur Verfügung. Wenn so die Permeation als *Poren*permeation in Erscheinung tritt, so muß doch berücksichtigt werden, daß, wie der zweite Befund zeigt, verschiedenartige Moleküle von der Diffusionshemmung nicht gleichmäßig betroffen werden. Offenbar sind die Membranporen unterhalb eines gewissen

Durchmessers nicht frei wegsam (daher ist bei dichten Kollodiummembranen z. B. $P_{\text{Glucose}} = 0$); dies wirkt sich bei den künstlichen Membranen jedoch nur in *quantitativer* Hinsicht aus, und eine (modifizierte) Beziehung zwischen P und M bleibt noch klar erkennbar.

Anders ist es in der lebenden Zelle. Hier gibt es, der Vielzahl der untersuchten Objekte gemäß, eine große Mannigfaltigkeit der Größenordnung von P und ihren Relationen zu M. Selbst bei den höchstpermeablen Objekten (z. B. *Beggiatoa*, vgl. S. 231 ff.) findet keine freie Diffusion statt, sondern die Permeationskonstanten fallen, wie bei den dichteren Kollodiummembranen, mit steigendem Molekulargewicht schneller als die entsprechenden Diffusionskoeffizienten; die Gleichung 2 ist nicht erfüllt. Geht man zu niedriger permeablen Objekten über, so wird diese Diskrepanz immer stärker. Daneben liegen zahlreiche Angaben vor, nach denen auch die gleichwohl noch erkennbare Beziehung

$$P \cdot f(M) = \text{const} \tag{4}$$

unterbrochen wird. Man ist daher davon abgegangen, das Molekulargewicht M zu den Permeationskonstanten in Beziehung zu setzen und hat statt dessen das Molekularvolumen bzw. die Molekularrefraktion R_{MD} (die ja gleichfalls die Dimension eines Volumens hat) gewählt:

$$P \cdot f(R_{MD}) = \text{const.} \tag{5}$$

Indessen ist auch diese Gleichung nur in erster Annäherung gültig, und es finden sich beinahe zu allen Objekten eine oder mehrere Abweichungen. Von diesen kann, wie auf S. 243 ausgeführt, ein großer Teil aus methodischen Gründen unberücksichtigt bleiben, vor allem, weil die experimentellen P-Werte in der Regel gar keine echten Permeationskonstanten sind, sondern durch das Interferieren nichtosmotischer Aufnahmeprozesse verfälscht werden und somit erst nach Korrektur in die oben angeführte Gleichung eingesetzt werden dürfen.

Aber auch bei Berücksichtigung dieser Einwände bleiben zahlreiche Fälle übrig, in denen die P-Werte bestimmter Stoffe eindeutig höher liegen, als nach Gleichung (5) zu erwarten ist. Bemerkenswerterweise sind es die sog. *lipoidlöslichen* Stoffe, die sich so verhalten. Das führte zur Aufdeckung folgender Abhängigkeitsbeziehungen:

$$P = \frac{f'(K)}{f''(R_{MD})}, \quad \text{(Bogen 1950)} \tag{6}$$

worin K der Verteilungskoeffizient ist, der angibt, in welchem Verhältnis sich der betreffende Stoff in den beiden Phasen des Systems Wasser/Äther bzw. Wasser/Öl verteilt (vgl. diesen Band S. 231).

Für die Characeen hat Collander kürzlich diese Beziehung auch zahlenmäßig erfassen können; es gilt z. B. für *Nitella*

$$P = b\,\frac{K^{1,32}}{M^{1,5}} \tag{7}$$

bzw. $\log(P \cdot M^{1,5}) = 1{,}32 \cdot \log K_{\text{oel}} + \log 6{,}5$ (Collander 1954).

Die Beziehung zwischen P und R_{MD} wird also überlagert durch eine zweite Beziehung zwischen P und K, und zwar in dem Sinne, daß von 2 Stoffen mit gleicher R_{MD} der „lipoidlösliche" die höhere Permeationskonstante hat, von 2 Stoffen mit gleichem K aber derjenige mit dem kleineren Molekularvolumen.

Wie bereits auf S. 231 erwähnt, kannte Overton (1899) nur die Beziehung zwischen P und K; er interpretierte sie so, daß die Permeation keine Ultrafiltration sei, sondern ein Lösungsvorgang, der sich in den Grenzschichtlipoiden abspiele, daß also der Permeationsvorgang grundsätzlich anderer Art sei als der Vorgang bei der freien Diffusion.

Die doppelte Abhängigkeitsbeziehung (zu K und R_{MD}) hingegen wird im Sinne der Lipoidfilterhypothese COLLANDERs so erklärt, daß *zwei* Permeationswege bestehen: der Porenweg, den hauptsächlich die lipoidunlöslichen Diosmotica beschreiten, und die Grenzschichtlipoide, in denen sich die lipoidlöslichen Diosmotica *lösen*.

Bei solchen Überlegungen ist indessen oftmals übersehen worden, daß, wie bereits NERNST (1926) betont, das FICKsche Diffusionsgesetz lediglich *formaler Natur* ist und nichts über die „treibenden Kräfte" aussagt. Es erscheint daher zweckmäßig, folgende Formulierungen zu bevorzugen:

Der Permeationsvorgang läßt sich, *jeweils in erster Annäherung*, sowohl als Lösungsvorgang wie als Ultrafiltration *umschreiben*, oder:

Der Permeationsvorgang kommt einem Lösungsvorgang und einer Ultrafiltration gleich.

Bindende Aussagen hingegen, ob er auch wirklich eine Lösungsvorgang bzw. eine Ultrafiltration *sei*, bzw. welcher der beiden *einander mehr oder weniger ausschließenden* Mechanismen der entscheidende sei, können auf Grund der hier aufgeführten formalen Abhängigkeitsbeziehungen *nicht* gemacht werden. Es darf lediglich gefolgert werden, daß durch die Einschaltung einer lebenden Grenzschicht in den Konzentrationsgradienten die Verhältnisse bei der Diffusion nicht nur quantitativ (wie bei den dichten Kollodiummembranen), sondern auch *qualitativ* abgeändert werden und daß möglicherweise „Lipoide" in irgendeiner Form beteiligt sind — wobei aus der Einbeziehung von Lipoiden keineswegs auf die Mitwirkung echter Lösungsvorgänge geschlossen werden *muß*.

Der Dualismus Lipoidtheorie/Ultrafiltertheorie bleibt solange bestehen, als nicht die „treibenden (bzw. hemmenden) Kräfte" bei der Diffusion berücksichtigt werden. Über diese vermag aber die „formale" Analyse keine Aufschlüsse zu geben; dazu bedarf es der molekularkinetischen Betrachtungsweise, die die zwischen den diffundierenden (permeierenden) Molekülen, den Lösungsmittelmolekülen und den Grenzschichtmolekülen wirksamen Kräfte berücksichtigt.

B. Die molekularkinetische Behandlung der Diffusion und Permeation.

Während die Grundlagen der Diffusion bereits mathematisch und quantitativ behandelt werden können (vgl. SPANNER, diesen Band, S. 125), sind für ein analoges Vorgehen bei den Permeationsvorgängen wichtige Parameter noch nicht ausreichend bekannt; es erscheint daher geboten, sich vorläufig auf qualitative Betrachtungen zu beschränken.

Diffusion ist der ungerichtete Transport einzelner Moleküle bzw. Ionen und beruht auf der kinetischen Energie dieser Moleküle. Grenzen zwei Volumina („Phasen") mit verschiedener Teilchenkonzentration aneinander, so ist die Zahl der Teilchen, die trotz „Rückdiffusion" aus dem Volumen mit höherer Konzentration in das von geringerer Konzentration übertreten, größer als in umgekehrter Richtung; *statistisch* folgt daraus eine in das Volumen von geringerer Konzentration gerichtete Stoffbewegung. Gegenüber der statistischen Betrachtungsweise des voraufgehenden Abschnittes muß nunmehr jedoch *das einzelne Molekül* betrachtet werden.

In *Gasen* besitzen die diffundierenden Teilchen eine große freie Weglänge; gleichwohl sind sie nicht völlig frei, sondern kollidieren gelegentlich mit benachbarten Teilchen. Eine solche Kollision bedingt nicht allein einen Austausch von kinetischer Energie, sondern es machen sich auch im Wirkungsbereich des benachbarten Teilchens die „short range" VAN DER WAALSschen Kräfte bemerbkar, die überwunden werden müssen, wenn das Teilchen dem Wirkungsbereich eines anderen Teilchens entfliehen soll. Entsprechend der Größe der freien Weglänge befinden sich die Teilchen jedoch zumeist außerhalb des Wirkungsbereiches der „short range"-Kräfte, so daß diese nur eine sehr untergeordnete Rolle spielen können.

Anders ist es bei den *Flüssigkeiten*. Hier ist die freie Weglänge der Teilchen erheblich geringer; sie befinden sich zumeist oder beständig innerhalb des Wirkungsbereiches der „short range"-Kräfte, wodurch die kinetische Energie jedes Teilchens stark herabgesetzt wird. Das findet einen anschaulichen Ausdruck im sog. „Binnendruck" der Flüssigkeiten, der mehrere Tausend Atmosphären betragen kann und als „Kohäsion" („Kohäsionsdruck") in Erscheinung tritt. Das bedeutet, daß jeweils nur ein Teil aller Moleküle genügend kinetische Energie besitzt, um die zwischenmolekularen Kräfte (ZMK) zu überwinden; die übrigen werden durch die VAN DER WAALSschen Kräfte an der Translokation verhindert, entweder weil sie von einem benachbarten Teilchen festgehalten werden, oder weil die umgebenden Teilchen untereinander in einem mehr oder minder festen Verband stehen, der dem in Rede stehenden Teilchen keine Bewegung erlaubt. Nach der anschaulichen Darstellung von DANIELLI (DAVSON und DANIELLI 1943) gibt es in nichtpolaren Flüssigkeiten folgende Möglichkeiten der Bewegung:

1. a) Wenn ein Teilchen genügend kinetische Energie besitzt, um die zwischenmolekularen Kräfte (ZMK) zu überwinden, die es an seine Nachbarn heften,

b) wenn es weiterhin genügend kinetische Energie besitzt, um die benachbarten Moleküle beiseitezustoßen,

2. wenn die umgebenden Moleküle sich so gegeneinander bewegen, daß ein „Loch" entsteht, in das das Teilchen hineingestoßen wird, obgleich es selbst nicht genügend kinetische Energie besitzt.

Modus 1 betrifft hauptsächlich kleine Moleküle, während Modus 2 überwiegt, wenn größere Moleküle in einem Lösungsmittel von kleinen Molekülen gelöst sind.

Daraus folgt, daß die ZMK die freie Diffusion behindern (gegenüber der Diffusion in Gasen). Es ist vorauszusagen, daß solche Teilchen, die nur geringe ZMK entfalten, nur wenig behindert werden, während große Moleküle, weil ja die kinetische Energie gleich $\frac{m}{2} v^2$ ist, sehr stark behindert werden — wie das Experiment bestätigt.

Ein weiterer diffusionsbehindernder Faktor wird wirksam bei der „*Hydrodiffusion*", der Stoffbewegung in *wäßrigen Lösungen*, wie sie ja in der lebenden Zelle vorliegen. Die Wassermoleküle sind polar gebaut (vgl. den Aufsatz von ERICHSEN, dieses Handbuch Band 1, S. 168—193) und vermögen *Wasserstoffbrücken* zu bilden. Wasserstoffbrücken (-bindungen) bestehen zwischen hauptvalenzmäßig abgesättigten Molekülen (BRIEGLEB 1949) und können daher ebenfalls als ZMK bezeichnet werden (gelegentlich werden sie als „chemische Bindung" angesprochen: DANIELLI 1943). Ihre Bindungsenergie ist wesentlich höher als die der VAN DER WAALSschen Kräfte (etwa 4 kcal je Mol.).

Die Wasserstoffbrücken schränken die freie Beweglichkeit der Wassermoleküle somit stärker ein als die VAN DER WAALSschen Kräfte; Wassermoleküle bedürfen einer höheren „Aktivierungsenergie" (Aktivierungsenergie E_a der Diffusion; DANIELLI 1943, SPANNER, dieses Handbuch, S. 136), um aus dem Wirkungsbereich der umgebenden Moleküle zu entfliehen. Das drückt sich nicht allein in einer geringeren Diffusionskonstanten aus, sondern auch in dem höheren

Siedepunkt und einem geringeren Dampfdruck, wodurch das Wasser von ähnlich gebauten Stoffen wie z. B. H_2S, die keine Wasserstoffbrücken zu bilden vermögen, unterschieden ist.

Im gleichen Maße benötigen auch in Wasser gelöste, stark polare Moleküle, wie z. B. Glycerin, eine höhere Aktivierungsenergie, da sie gleichfalls Wasserstoffbrücken bilden, nämlich sowohl untereinander als auch mit den umgebenden Wassermolekülen (Modus 1 der Diffusionsmöglichkeiten, S. 443). Weniger stark polare und unpolare („hydrophobe") Stoffe hingegen werden, wenn sie wasserlöslich sind, weniger stark von der Diffusionshemmung betroffen, denn sie neigen nicht zur Ausbildung von Wasserstoffbrücken. Soweit sie sich indessen vorwiegend nach dem Modus 2 bewegen (also z. B. größere Moleküle), wird ihre Translokation ebenfalls eingeschränkt, denn die umgebenden Wassermoleküle werden untereinander durch Wasserstoffbrücken zusammengehalten, und es bedarf längerer Zeit bzw. höherer Aktivierungsenergie, um ein „Loch" entstehen zu lassen, in das die unpolaren Moleküle hineingestoßen werden.

Der bei polaren und unpolaren Molekülen unterschiedliche Energiegehalt der ZMK ist die Ursache eines weiteren Phänomens, das in der Diskussion der Permeabilität eine hervorragende Rolle spielt: der sog. Lipoidlöslichkeit. Grenzt eine wäßrige (hydrophile) Phase an eine hydrophobe („organisches Lösungsmittel" z. B. Äther, Benzol, Öl, „Lipoide"), so verteilen sich Stoffe, die in der wäßrigen Phase gelöst waren, ganz unterschiedlich auf die beiden Phasen: Stoffe, die selbst hydrophil sind, also polare Moleküle, die Wasserstoffbrücken bilden, werden mit hohen ZMK in der wäßrigen Phase festgehalten und nur selten in die hydrophobe Phase übertreten. Sie benötigen mithin zum Übertritt eine hohe Aktivierungsenergie, um die Energie der Wasserstoffbrücken zu überwinden, denn von seiten der hydrophoben Phase werden ja keine Wasserstoffbrücken entwickelt, die sie anziehen könnten. Und wenn sie gleichwohl in geringem Ausmaße in die hydrophobe Phase gelangen, so werden sie, sobald sie wieder in den Wirkungsbereich der ZMK der wäßrigen Phase geraten, wieder in diese hineingezogen. Hydrophobe Stoffe dagegen bilden in der wäßrigen Phase keine Wasserstoffbrücken und brauchen daher nur eine geringe Aktivierungsenergie E_a, um in die hydrophobe Phase überzutreten, und sie verbleiben auch in dieser, weil sie nicht „zurückgezogen" werden.

Unter diesem Gesichtspunkt ist der sog. Verteilungskoeffizient K ($K_{\text{Äther}}$, $K_{\text{Öl}}$ usw.) nicht so sehr ein Maß für die Lipoidlöslichkeit, als vielmehr ein Maß für die Kräfte, mit denen die Moleküle in der hydrophilen Phase festgehalten werden, also ein (empirisches) *Maß für die ZMK*, insbesondere für die Fähigkeit der betreffenden Stoffe, Wasserstoffbrücken auszubilden.

Für die *Permeation* ergibt sich daraus folgendes: Bei der lebenden Zelle ist eine Membran in ein Konzentrationsgefälle eingeschaltet, das zwischen der hydrophilen Umgebung der Zelle und dem ähnlich hydrophilen Zellinneren (Binnenplasma, Vacuole) besteht. Durch diese Zwischenschaltung werden die bei der Hydrodiffusion gefundenen Verhältnisse sowohl quantitativ *als auch qualitativ* geändert:

a) quantitativ, indem die Permeationskonstanten bzw. Permeabilitätskoeffizienten allgemein niedriger sind als die Diffusionskonstanten, und indem sie bei steigender Molekularrefraktion schneller fallen als diese. Dieser Befund ließe sich noch interpretieren unter der Annahme, daß die Grenzschichten aus rein hydrophilen Molekülen aufgebaut sind, die demgemäß mit hohem ZMK, vor allem durch zahlreiche Wasserstoffbrücken, untereinander festgehalten werden, die aber auch hohe ZMK gegenüber allen permeierenden Molekülen entfalten. Zur Permeation ist daher eine höhere Aktivierungsenergie E_a vonnöten als für die Hydrodiffusion. Außerdem entfällt der Diffusionsmodus 2 („Lochbildung") fast vollständig, denn die Beweglichkeit der Grenzschichtmoleküle ist nur gering (größere Moleküle permeieren daher überhaupt nicht).

b) Der qualitative Unterschied zwischen Permeations- und Diffusionskonstanten besteht darin, daß bei den sog. lipoidlöslichen Substanzen die Translokationshemmung *geringer* ist als bei den lipoidunlöslichen, d. h. daß von zwei Stoffen mit gleicher Molekularrefraktion der lipoidunlösliche, von zwei Stoffen mit gleichem Verteilungskoeffizienten derjenige mit den größten Molekülen langsamer permeiert.

α) Rein hydrophile Membranen.

Die daraus abzuleitende „*Diskontinuität der Diffusionswege*" spricht eindeutig gegen die Auffassung, daß die Grenzschichten aus rein hydrophilen Molekülen aufgebaut seien, und daß ihre Wegsamkeit sich nur quantitativ von der Wegsamkeit der wäßrigen Lösung unterschiede.

β) Rein hydrophobe Membranen.

Es ist indessen leicht einzusehen, daß die entgegengesetzte (und meist stillschweigend als bewiesen angesehene) Auffassung, wonach die Grenzschichten ausschließlich auf Lipoiden aufgebaut seien, ebensowenig stichhaltig ist:

Beim Durchtritt durch eine solche Membran, die sich zwischen dem hydrophilen Außenmedium und dem hydrophilen Zellinneren befindet, sind folgende Vorgänge zu unterscheiden: a) der Eintritt des permeierenden Moleküls in die Grenzschicht, b) die Diffusion innerhalb der Grenzschicht, c) der Austritt aus der Membran.

Davon können a) und c) gemeinsam behandelt werden:

a) und c) Eintritt und Austritt. Die Phänomene sind nahezu identisch mit denen der „Verteilung". Hydrophile Stoffe, die Wasserstoffbrücken bilden, benötigen eine hohe Aktivierungsenergie E_{a1}, um in die Grenzschicht überzutreten, weil sie im hydrophilen Außenmedium mit Wasserstoffbrücken festgehalten werden, während sie in der hydrophoben Grenzschicht keine nennenswerten Wechselwirkungen mit deren Molekülen entfalten. Demgegenüber ist die Aktivierungsenergie E_{a2} für den Austritt aus der lipoiden Membran gering, denn die permeierenden hydrophilen Moleküle werden förmlich in das hydrophile Medium des Zellinneren „hineingesaugt": $E_{a1} > E_{a2}$. Stoffe, die wenige oder keine Wasserstoffbrücken bilden und auch sonst nur geringe ZMK besitzen (im folgenden der Einfachheit halber, wenn auch nicht streng korrekt, als „hydrophobe" Moleküle bezeichnet), gelangen sehr leicht in die Grenzschicht, benötigen aber eine hohe Aktivierungsenergie, um jenseits der Grenzschicht wieder ins hydrophile Zellinnere zu kommen, d. h. $E_{a1} < E_{a2}$. Die Summe aus beiden dürfte für hydrophile und hydrophobe Diosmotica nahezu die gleiche sein. Eintritt und Austritt vermögen also keine qualitative Veränderung der Permeationskonstanten gegenüber dem Diffusionskonstanten herbeizuführen (wohl aber quantitative, da ja infolge des schnelleren Eintretens der hydrophoben Moleküle diese in der Membran in höherer Konzentration vorhanden sind).

b) Die Diffusion im Inneren der hydrophoben Membranen. Hydrophobe Moleküle bewegen sich hier zumindest ebenso schnell wie im wäßrigen Medium, weil sie in beiden keine Wasserstoffbrücken entfalten. Wahrscheinlich erfolgt die Diffusion sogar etwas schneller, denn der Diffusionsmodus 2 („Lochbildung") kann stärker in Erscheinung treten, weil die hydrophoben Grenzschichtmoleküle wegen des Fehlens von Wasserstoffbrücken untereinander leichter beweglich sind als die Wassermoleküle, die durch Wasserstoffbrücken in ihrer Beweglichkeit gehindert sind.

Hydrophile Moleküle dagegen wandern in der rein hydrophoben Membran wesentlich *schneller als in Wasser*, weil sie nur untereinander, nicht aber gegen die Grenzschichtmoleküle diffusionshemmende Wasserstoffbrücken ausbilden, während im wäßrigen Medium Wasserstoffbrücken sowohl untereinander als auch zum Lösungsmittel bestehen.

In einer rein hydrophoben Grenzschicht müßten also die hydrophilen Stoffe in der Permeation gegenüber den hydrophoben Diosmotica (gleicher Molekularrefraktion) gefördert sein — im Gegensatz zum empirischen Befund.

Das gilt allerdings zunächst nur für das *einzelne* hydrophile Molekül. Bei der (statistischen) Behandlung der transportierten Stoff*menge* müssen noch die Konzentrationsgradienten in Rechnung gestellt werden. Diese sind bei hydrophilen Stoffen kleiner als bei hydrophoben. Dadurch wird der „Beschleunigungseffekt", der beim einzelnen Molekül sichtbar wird, teilweise kompensiert. Ob er freilich völlig aufgehoben wird, läßt sich gegenwärtig nicht angeben.

γ) Membranen mit hydrophoben und hydrophilen Anteilen.

Hingegen läßt eine Membran, die aus hydrophoben und hydrophilen Teilen besteht, sowohl quantitative als auch qualitative Differenzen solcher Art erwarten, wie sie im Experiment beobachtet werden:

Hydrophobe Moleküle werden wegen der hydrophoben Membrananteile schneller aufgenommen und befinden sich in der Membran in höherer Konzentration als hydrophile Moleküle. Ihre Diffusion im Membraninneren wird durch den hydrophoben Membrananteil nicht, durch den hydrophilen Membrananteil wenig gehemmt (Diffusionsmodus 2).

Hydrophile Moleküle hingegen benötigen eine hohe E_{a1}, um in die Membran überzutreten; im Membraninneren werden sie in der Diffusion durch den hydrophoben Membrananteil nicht gehemmt (eher etwas gefördert), durch die hydrophile Komponente jedoch, die Wasserstoffbrücken bildet, stark gehemmt. Die Summe aus Hemmung und Förderung ist für hydrophile und hydrophobe Diosmotica *nicht* gleich, und so erscheinen die hydrophoben Stoffe in der Permeation gefördert gegenüber den hydrophilen.

Die hier anschaulich und daher nicht völlig zutreffend geschilderten Verhältnisse bedürfen zur theoretischen Fundierung der quantitativen Behandlung. Eine solche ist aber vorerst noch nicht möglich. Wohl kann der empirische Verteilungskoeffizient in erster Annäherung als Maßstab für das Ausmaß der ZMK verwendet werden, aber es müßten, weil ja die Permeation einzelner Moleküle betrachtet werden muß, die je Einheit der Moleküloberfläche wirkenden ZMK in die Formeln eingesetzt werden. Eine solche Rechnung kann bisher nicht durchgeführt werden, weil *bereits bei Wassermolekülen* eine Anisotropie hinsichtlich der Bindungsenergien festzustellen ist. Wichtiger aber noch wird es sein, die von den Grenzschichtmolekülen ausgehenden ZMK quantitativ zu erfassen, die gleichfalls maßgebend für die Diffusionshemmung sind. Dazu sind jedoch unsere Kenntnisse von der chemischen Zusammensetzung der Grenzschichten noch zu ungenau. Wir müssen uns also vorläufig mit der Formulierung begnügen, daß die Permeation eine durch zwischenmolekulare Wechselwirkung behinderte Diffusion ist (Ultrafiltertheorie, RUHLAND und HOFFMANN 1926); die Diffusionshemmung ist um so größer, je größer die ZMK sind. Dieses aber ist der Fall sowohl bei höherem Molekulargewicht bzw. bei höherer Molekularrefraktion als auch bei geringerer „Lipoidlöslichkeit". Die Diffusionshemmung geht aus von den ZMK der Grenzschichten, die aus hydrophoben und hydrophilen Anteilen zusammengesetzt sein muß, wobei die hydrophoben als (relative) „Diffusionsbeschleuniger" zu gelten haben.

II. Grenzschichtfeinbau.

Die hier vertretene Ansicht, daß die Grenzschichten sowohl aus hydrophilen als auch aus hydrophoben Anteilen bestehen, entspricht der Auffassung der Mehrzahl aller Autoren, die sich mit den Strukturen der Zelle beschäftigt haben: chemische und physikalische Untersuchungen sprechen regelmäßig dafür, daß zumindest in der Grenzschicht Lipoide als hydrophobe Bestandteile in größeren Mengen auftreten, daß aber daneben hydrophile bzw. partiell hydrophobe Proteine vorliegen (neuerdings wird sogar mit dem Vorhandensein von Nucleinsäuren in der Grenzschicht gerechnet: LANSING).

In welcher Form sie allerdings vorliegen, und wie sie miteinander vergesellschaftet sind, bedarf noch durchaus der Klärung. Ein „Mosaik" aus nebeneinanderliegenden verschiedenen Arealen (NATHANSOHN) dürfte unwahrscheinlich sein; mehr Zustimmung hat die Vorstellung von einem *Schichten*aufbau gefunden, mit einem tieferliegenden oligomolekularen Lipoidfilm und einem aufgelagerten Proteinfilm. Elektronenmikroskopische Untersuchungen haben wenigstens bei dem Zellkern, bei den Mitochondrien und bei den Chloroplasten Mehrfachschichten nachweisen können, mit einer zentralen Lipoidschicht und beiderseits aufgelagerten Proteinschichten (letztere etwa 55 Å dick, d. h. gerade so viel, wie der Durchmesser globulärer Enzyme beträgt). Doch fehlt dieser Nachweis noch für die äußeren Zellhäute, und es ist nicht unwahrscheinlich, daß sich in diesen die Protein- und Lipoidkomponenten durchdringen. Das Vorhandensein kompakter („condensed") Lipoidfilme jedenfalls ist zweifelhaft, nachdem alle neueren Untersuchungen über die Grenzflächenspannung lebender Zellen zu Werten < 1 dyn/cm geführt haben (vgl. HARVEY 1954), kondensierte Lipoidfilme aber durch eine Grenzflächenspannung von etwa 35—55 dyn/cm charakterisiert sind. Selbst die Werte expandierter Lipoidfilme liegen noch um ein Vielfaches zu hoch.

Wenn hier noch zahlreiche Einzelheiten der Aufklärung harren, so ist doch soviel gewiß, daß die Lipoide der Grenzschicht *nicht als Lösungsmittel* im makroskopischen Sinne fungieren können; die oben angeführte Parallelität zwischen Permeabilität und Lipoidlöslichkeit ist kein Beweis für die Mitwirkung eines echten Lösungsvorganges, sondern demonstriert lediglich die überragende Rolle der ZMK, die ja ihrerseits auch die Ursache der Lipoidlöslichkeit sind. Der Verteilungskoeffizient darf also nur als stellvertretend für das Ausmaß der ZMK verwendet werden, nicht als Beweismittel für die „Löslichkeitshypothese". Trotzdem läßt sich natürlich der Permeationsvorgang in erster Annäherung auch als ein Lösungsvorgang *beschreiben.*

Inwieweit es ähnlich um den Vorgang der Ultrafiltration bestellt ist, ist noch nicht zu entscheiden. Zweifellos können wir nicht mit distinkten Poren rechnen, etwa in Form der Poren zwischen den Palisadenmolekülen eines kondensierten Filmes, die die Grenzschicht oder gar das gesamte Binnenplasma (DRAWERT 1948) durchziehen. Wohl aber dürfte ein „Maschenwerk" aus Makromolekülen (fibrillär oder globulär) vorliegen, das anastomosierende intramolekulare Räume freiläßt, die nur an ihren Grenzen von den „short range"-ZMK beherrscht werden. So läßt sich der Vorgang der Permeation mindestens ebenso richtig als Ultrafiltration beschreiben. Erst die Anerkennung der überragenden Rolle der ZMK und die Untersuchung der Permeationsvorgänge im molekularen

Bereich ermöglichen es, den anfänglich unüberbrückbar erscheinenden Widerspruch zwischen Löslichkeitshypothese und Ultrafiltertheorie aufzuhellen und das ,,Prinzip der räumlichen Durchdringung" als übergeordnet zu erkennen.

Literatur.

BOGEN, H. J.: Kritische Untersuchungen über Permeabilitätsreihen. Planta (Berl.) **38**, 65—90 (1950). — BRIEGLEB, G.: In Zwischenmolekulare Kräfte, herausgeg. von H. FRIEDRICH-FREKSA, B. RAJEWSKI u. M. SCHÖN. Karlsruhe: G. Braun 1948.

COLLANDER, R.: Über die Permeabilität der Kollodiummembranen. Soc. Sci. fenn., Comment. Biol. II **6**, 1—48 (1926). — The permeability of Nitella cells to non-electrolytes. Physiol. Plantarum (Copenh.) **7**, 420—445 (1954).

DAVSON, H., and J. F. DANIELLI: The permeability of natural membranes. Cambridge 1943. — DRAWERT, H.: Zur Frage der Stoffaufnahme durch die lebende pflanzliche Zelle. V. Zur Theorie der Aufnahme basischer Stoffe. Z. Naturforsch. **3** b, 111—120 (1948).

HARVEY, E. N.: Tension at the cell surface. Protoplasmatologia (Wien) II E **5**, 1—30 (1954).

LANSING, A. I.: The relation between RNS and ionic transport across the cell surface. J. Cellul. a. Comp. Physiol. **40**, 337—345 (1952).

NATHANSOHN. A,: Stoffwechselphysiologie der Pflanzen. Leipzig 1910. — NERNST, W.: Theoretische Chemie. Stuttgart 1926.

OVERTON, E.: Über die allgemeinen osmotischen Eigenschaften der Zelle, ihre vermutlichen Ursachen und ihre Bedeutung für die Physiologie. Vjschr. naturforsch. Ges. Zrich **44**, 88—135 (1899).

RUHLAND, W., u. C. HOFFMANN: Die Permeabilität von Beggiatoa mirabilis. Ein Beitrag zur Ultrafiltertheorie des Plasmas. Planta (Berl.) **1**, 1—83 (1925).

THOVERT, J.: C. r. Acad. Sci. Paris **150**, 270 (1910). Zit. nach DAVSON u. DANIELLI 1943.

The mechanism of absorption.

By

R. N. Robertson.

With 1 figure.

As has been seen in the preceding sections (IV A–G), living cells and tissues show a great capacity for absorbing a variety of substances from the surrounding medium. Large molecules, small molecules, electrolytes and nonelectrolytes can all be absorbed. Sometimes substances are absorbed by diffusion as long as there is a concentration difference between outside and inside; sometimes, as with ions, because an electrochemical potential difference has been established between inside and outside, ions can enter in exchange for other ions; sometimes a non-osmotic process of active transport, working against both concentration and electric gradients, results in internal concentrations greater than external. No single mechanism can account for the absorption processes which differ in different cells and for the different types of substances absorbed. Underlying these mechanisms is the highly complex and little understood structure of the cytoplasm. While we expect to find that the cytoplasm obeys physico-chemical laws, the history of discussions on the mechanism of absorption shows that over-simplification of the physical system involved has led to erroneous hypotheses. The mechanism of absorption is one aspect of cell physiology and our knowledge advances hand in hand with our general knowledge of cell physiology. For example, though a few years ago the functions of mitochondria were almost unknown, we now know that no hypothesis of absorption mechanism can neglect these particles and their important physiological role.

I. Osmotic uptake.

1. The concentration gradient.

Since all substances in solution tend to distribute themselves until their concentrations (or activities) in all parts of the system are equal, cells may acquire solutes by simple diffusion. The energy comes from the kinetic energy of the dissolved substances and absorption continues until the concentrations in cell and medium are equal. In this way, by simple diffusion, many substances enter cells and, as STEINBACH (1951) has pointed out, it is necessary to show that entry is not by this mechanism before invoking more complex mechanisms. Diffusion will also be responsible for movement of substances up to the surface of the cytoplasm through wet cell walls and water-filled intercellular spaces. The rate of entry of a substance being absorbed depends upon the diffusion constant or permeability constant of the substance in the cell or tissue and the two basic equations have been given (see IV A and B).

$$\frac{dM}{dt} = PA(C_a - C_i) \tag{1}$$

and

$$\frac{dM}{dt} = -D\frac{dc}{dx} \tag{2}$$

where M is the amount of substance entering, C_a and C_i are the external and internal concentration, A is the area of the cell or tissue, P is the permeability constant, defined as the amount of substance diffusing across unit area with unit concentration difference in unit time, and D is the diffusion constant defined as the amount of substance diffusing across unit area with unit concentration gradient in unit time. Thus, the permeability or diffusion constant summarises the properties of the cell surface or tissue through which the substance must diffuse and these properties govern the rate; for example, the diffusion constant of KCl in water is 1.67×10^{-5} cm²/sec at 20° C while in the plant mitochondrial membrane it is only 1×10^{-13} cm²/sec.

Any factor within the cell or tissue which alters either the permeability constant or the gradient will alter the rate of absorption. The entry of dissolved substances by simple diffusion is facilitated by the combination or change of the substance within the cell in such a way as to decrease its activity, thus increasing the concentration difference. Thus the utilization of a metabolite supplied from outside a cell results in continued diffusion or the combination of an ion (such as Fe^{++} or Mg^{++}), with a complexing or chelating compound, increases the gradient for the free ion.

Diffusion of nonelectrolytes.

Many investigations carried out with nonelectrolytes (see IV D b) have shown that the penetration depends on diffusion of the substances in solution, tending towards equality of concentration within and without but that different molecules differ greatly in their ability to penetrate the cells. These differences depend upon the properties of the cytoplasm and are thought to be due to specific membranes in the cell (see IV G). Understanding the mechanism of absorption of nonelectrolytes, moving according to their kinetic energy, involves study of the penetration of barriers to diffusion in the cytoplasm; as this has been discussed elsewhere it will not be discussed further here. Much work remains to be carried out on the precise structure and location of these barriers. Their existence depends on the continued activity of the cell and they cannot be regarded as static structures. Further, we need to know more about the influence of the complexity of the cytoplasm, with its different phases of plastids, mitochondria and microsomes and its continual movement in streaming which may affect the penetration of the nonelectrolyte; it cannot necessarily be assumed that the penetration can be reduced to that of a single membrane, a few molecules thick. Though most nonelectrolytes are absorbed under the influence of the concentration gradient, there may be times when some are moved by an active transport mechanism (see p. 456).

2. The electrochemical potential gradient.

The movement of electrolytes through intercellular spaces and wet cell walls and up to cytoplasmic surfaces takes place by simple diffusion. Once within the cell wall, the process of diffusion is not simple because the cells consist of immobile or indiffusible ions which influence the movement of the mobile ions. Even the cell wall contains ionizable substances of the pectate type and some interference with simple diffusion can be expected. Electric potential differences are known to exist between the medium and cytoplasmic surface (Lundegårdh 1940, Tendeloo, Vervelde and Zwart Voorspuy 1944, and Hope 1951) and probably exist across the tonoplast. They are also known to exist between different places on tissue surfaces (Lund *et al.* 1947).

a) Diffusion of electrolytes.

Not very much work on diffusion of electrolytes into plant cells or through plant tissue has been carried out systematically and few values are available. The general evidence supports the view that movement of ions through plant tissue is restricted to the wet cell walls, injected intercellular spaces and to some parts of the cytoplasm, *i.e.* that ions do not move readily into and out of the vacuoles. This has been demonstrated by various workers, *e.g.* by ARISZ (1947, 1948) for *Vallisneria* leaves, by BRIGGS and ROBERTSON (1948) for carrot tissue, by HOPE and STEVENS (1952) and BUTLER (1953) for roots. When plant material is placed in a solution, there is a rapid diffusion of the solute into those regions readily accessible. This entry of ions by diffusion and exchange is referred to conveniently as the *initial uptake* because the adjustment is complete in a short time. In carrot discs (1 mm thick) the initial uptake is 95 per cent complete in 25 minutes and in bean roots it is 95 per cent complete in 20 minutes. The initial uptake is reversible; on transferring the tissue to water, the ions taken up during initial uptake diffuse out again in about the same time. During this period, too, the root acts as a cation exchanger (EPSTEIN and LEGGETT 1954). The volume of tissue into which the electrolyte apparently moves by free diffusion, has been termed the *apparent free space* (a term due to BRIGGS, adopted by HOPE and STEVENS 1952), which may be as much as 25 per cent of the tissue volume.

BRIGGS and ROBERTSON (1948) and HOPE (1953) have shown that this diffusion through tissue follows FICK's law. Table 1 shows the apparent diffusion constant obtained for KCl in two different tissues.

Table 1. *Apparent diffusion constant.*

Carrot discs	BRIGGS and ROBERTSON 1948	0.03×10^{-5} cm²/sec.
Bean roots	HOPE 1953	0.07×10^{-5} cm²/sec.

These apparent diffusion constants are very much less than that for KCl in water (at 20° C, 1.67×10^{-5} cm²/sec) due to the restricted diffusion path within the tissue. There is much accumulated evidence that the cytoplasm contains predominantly indiffusible anions in quantity but the actual effective concentration is difficult to determine. BRIGGS and ROBERTSON (1948) have discussed the influence of indiffusible anions in a phase (or tissue) on the apparent diffusion constant of a salt diffusing through the phase and have shown that the constant of diffusion in such a phase is

$$u\,R\,T\,\frac{2v\,p - v\,a}{(u+v)\,p - v\,a}$$

as compared with $uRT\frac{2v}{u+v}$ for the constant of diffusion in water, where u is the mobility of the cation within the phase, v is the mobility of the anion, p is the activity of the cation in the phase, and a is the activity of the immobile anions, R is the gas constant and T the absolute temperature. Thus the diffusion constant in the phase may be different from that in water and depends upon the concentration of the salt, the concentration of immobile anions and the mobility of the anions and cations. The difficulty of estimating the concentration of a has been discussed by BRIGGS and ROBERTSON. Not only is it unlikely to be uniform throughout the tissue but it is also likely to change with time and conditions, especially as it may be largely due to weak acids. BRIGGS and ROBERTSON conclude that the concentration of a in carrot discs is not less than 0.1 M.

VERVELDE (1948) with barley roots concluded that a is about 0.02 M and with wheat 0.04 M, while HOPE (1953) with bean and maize roots concluded it is of the order of 0.01 M. We cannot assess a precisely but an estimate of its effects is important. The apparent diffusion constant falls with decrease in the ratio of concentration of mobile anions to immobile anions. The difficulties of applying the theoretical considerations to experimental material have been discussed by BRIGGS and ROBERTSON (1948) and the fundamental equations have been developed by TEORELL (1953).

Further difficulties would be introduced if, as USSING (1954) suggests, ions cannot always be considered as moving independently of each other. HODGKIN and KEYNES (quoted by USSING) propose that in the cephalopod nerve, the only paths where K^+ ions can pass the membrane have 4 sites each of which has to be occupied by a K^+ ion so no K^+ can leave at one boundary unless another one is taken in at the other. This means that all four of them will have to move as a unit if they are to move at all.

b) The DONNAN equilibrium.

When there are two aqueous phases in contact, one of which contains immobile ions, at equilibrium there is inequality of concentration of the mobile ions—the DONNAN *equilibrium*. In most biological systems, the DONNAN equilibrium results from a predominance of immobile anions in the cell and the distribution of cations (*e.g.* K^+) and anions (*e.g.* Cl^-) at equilibrium is given by the following:

$$[Cl^-]_i = \frac{\sqrt{a^2 + 4C_a^2} - a}{2} \qquad (3)$$

$$[K^+]_i = \frac{a + \sqrt{a^2 + 4C_a^2}}{2} \qquad (4)$$

where C_a is the external concentration of KCl (assumed constant and a is the concentration of immobile anions).

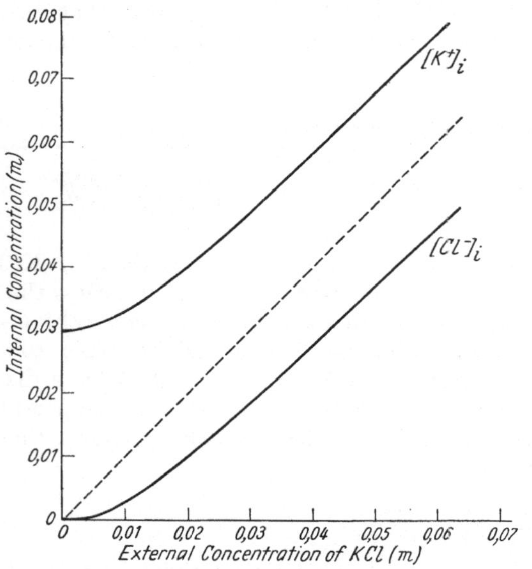

Fig. 1. Equilibrium concentration of K^+ and Cl^- in a phase containing immobile anions ($a = 0.03$ M) plotted against concentration in the external solution.

α) The immobile anions.

In such a simple system, taking a value for a such as 0.03 M, the expected concentration of mobile cations and anions in the phase containing immobile anions would be as illustrated by Fig. 1. Of course, the system of ions in the cytoplasm is not as simple as this but this figure serves to illustrate two important points: (1) that the concentration of immobile anions results in a higher concentration of mobile cations inside than in the external solution, and a lower concentration of mobile anions and (2) that the concentration ratio (inside to outside) differs at different concentrations. Such a system can therefore account for a high concentration of mobile cations in the cytoplasm but not for the accumulation of both mobile cations and mobile anions. An apparent accumulation of one of the ions from the external solution is not evidence for an active transport system. Such a movement does not require energy expenditure on the part of the cell while the ions are moving; energy is required only to establish the original electrochemical gradient.

The immobile anion concentration will always be difficult to assess (BRIGGS and ROBERTSON 1948). The value of the anion concentration in relation to the potential difference between cell surface and the external solution has been discussed by VERVELDE (1948) and HOPE (1953). The approximate concentration of immobile anions represents only an average concentration in a mixture containing a large number of different anions. Some of these will be the proteins of the cytoplasmic framework, some the proteins of the mitochondria and some the ionic intermediates of metabolism such as the organic acids. Further, the cytoplasm in which the anions occur is a polyphase system in which some of these anions are separated from other parts of the cell by membranes such as those of the mitochondria and that surrounding the vacuole. These complications affect the interpretation of the results which have been used to assess the average concentration of the anions. Two methods have been used: the first is the variation in the interfacial potential difference (used by VERVELDE 1948), and the second is the change in internal concentration of the mobile ions with different concentrations of external solution (HOPE 1953).

β) The effect of p_H.

The concentration of immobile anions will not be expected to remain constant under different conditions. There is now additional evidence from ion exchange studies that cations can penetrate the cytoplasm very readily (EPSTEIN and LEGGETT 1954). This means that hydrogen ions from the external solution enter readily. Since many of the anions in the cytoplasm are weak electrolytes, increase in hydrogen ion concentration results in suppression of the ionization of these weak anions and decrease in a. This results in an increase in apparent free space because the concentration of mobile anions in the cytoplasm is no longer depressed to the same extent (see Fig. 1) and more total salt can enter. This has been demonstrated by ROBERTSON and WILKINS (unpublished data) for the entry of KCl into discs of carrot tissue.

The effect of p_H on the ionization of the cytoplasm was largely neglected in earlier work on the absorption of weak electrolytes by the plant cell. For some time physiologists have realised that lowering the p_H in the solution external to a cell and increasing the concentration of undissociated molecules in a weak electrolyte would promote entry. Since some biological membranes were known to be more readily permeable to molecules than to ions, it was assumed that decrease in external p_H operated by increasing the number of undissociated molecules in the external solution. Such an approach neglected the possibility that the p_H was also affecting the ionization of the cytoplasm. The implications of this have been discussed by SIMON and BEEVERS (1952), and BRIGGS (1954).

γ) Metabolic production of ions.

Not only will the immobile anions of the cytoplasm affect the Donnan equilibrium but adjustment in the cell as a whole will be affected by the total organic acid content, some of which may be in mitochondria in the cytoplasm but much of which is probably in the vacuole. This content of organic acid for a number of plants is summarised in Table 2.

The principal acids occurring in quantity are citric, malic and oxalic but the percentages in which they occur and the amounts of other acids accompanying them is very variable. Alterations in the ionic environment of the plant affect the metabolism and hence slowly influence the content of organic anions. This

Table 2. *Some organic acids in plants.*

Plant	Organ	Investigator	Total acids mg. equiv./gm sap.
Barley	Root	ULRICH (1942)	0.038
Wheat	Root	BURSTRÖM (1945)	0.062
Wheat	Leaf	BURSTRÖM (1945)	0.108
Kleinia	Leaf	THODAY and JONES (1939)	0.045–0.086
Kleinia	Stem pith	THODAY and JONES (1939)	0.124–0.130
Apple.	Fruit	J. F. TURNER (1949)	0.22
Orange	Fruit	SINCLAIR, BARTHOLOMEW and RAMSEY (1945)	0.129–0.398

is strikingly illustrated by ULRICH (1942) who has shown that when roots of barley are placed in a solution of potassium bicarbonate, the considerable amount of K^+ entering the cell in exchange for hydrogen ions does not increase the internal p_H. The excess potassium ions are balanced by an increase in organic acid anions which results in the p_H remaining approximately constant. Again when the roots are immersed in divalent salts where the uptake of anions exceeds uptake of cations, organic acid anions in the tissue decrease and the p_H remains about the same. Further, ULRICH (1942) has shown that the respiratory quotient changes in such a way as to be consistent with the formation or utilization of organic acids. This has been confirmed by BURSTRÖM (1945) using wheat roots where the principal organic acid is malic. BURSTRÖM has shown that malic acid is formed or destroyed so that its content always balances an excess or deficiency of cations in the cell whether due to unequal absorption or to the formation or destruction of cations internally (*e.g.*, in reduction of nitrate to ammonium). This adjustment of the ionic balance by alteration in the metabolic balance will be slower and continue for longer periods than the relatively rapid Donnan adjustment in the apparent free space including the cytoplasm. Slow metabolic adjustments are probably associated with the continued difference between cation and anion uptake over several days as reported by STILES and SKELDING (1940).

Recent work has suggested that the ions available for exchange in the cytoplasm are also likely to be affected by the metabolism. The isolation of mitochondria and study of their ionic properties has shown that they adjust themselves to changes in the concentration of salt in the external solution to maintain both mobile anions and mobile cations in higher concentration than the supernatant. The work on animal mitochondria indicates that the concentrations of these ions in the particles are dependent upon the metabolic activity of the mitochondria. ROBERTSON, WILKINS, HOPE and NESTEL (1955) have shown that plant mitochondria contain concentrations of sodium and potassium in excess of those in the external solution and suggest that this effect can be explained by a Donnan equilibrium between immobile anions of the particles and mobile cations but that some additional mechanism must be postulated for the high internal concentration of anions. This will be referred to again in discussion of the accumulation mechanism. These observations however, indicate the complexities in attempting explanation of the Donnan adjustment when there is involved separate adjustment in cytoplasm, in mitochondria and in vacuole.

c) Combination and complex formation.

The entry of ions by diffusion is facilitated if the ions are combined within the cell so that the activity of free ions is reduced and the concentration gradient from outside to inside is maintained. Metallic ions may enter into the formation of complexes such as the iron porphyrins and chlorophylls. Other compounds such as nitrate, entering the cell as ions, may be removed from solution in the formation of organic molecules. Phosphate, which occupies a particularly important position in metabolism, is continually taken up in inorganic form and removed from solution as the free ion by the process of respiration, to form the variety of phosphorylated compounds associated with the synthetic reactions of the cell.

We require more detailed experimental evidence of the types of compound into which nutrient constituents such as iron, manganese, boron, aluminium, copper, zinc and cobalt enter. The chemistry of metal organic complexes requires further study with more detailed knowledge of the stability constants of compounds which complex with metals; further, little is known about the specific compounds into which these ions enter.

Special discussion is required of the possible combination of sodium and potassium with cell constituents. The marked difference in sodium and potassium concentrations in biological systems has been difficult to explain, since these two ions are thought to have rather similar chemical properties. Perhaps too much emphasis has been placed on the similarity between the ions rather than on the differences which may be quite sufficient to account for the different quantities combined with certain types of molecules. The work of ROBERTS, ROBERTS and COWIE (1949) on *E. coli* showed a marked difference in the combining power with potassium as compared with sodium which they associated with phosphorylated compounds in the cell. The importance of a difference in combination with phosphorylated compounds has been developed by LING (1952) who has pointed out that if there were a series of fixed negative charges in the cytoplasm, *i.e.* negative charges such as would be due to a phosphate group on a fixed structure, then it might be expected that the potassium ion with its much smaller hydration shell might approach adequately the fixed negative charge to be held by the Coulomb forces. On the other hand, the heavily hydrated sodium would be less capable of approaching closely the fixed negative charge and would be only weakly held. The net result would be that with sodium and potassium ions in competition a greater absorption of potassium could be expected. So far there is no experimental evidence on plant material for LING's hypothesis for the difference between sodium and potassium but clearly investigations are required. There may be some plants in which the compounds which can preferentially hold potassium predominate over those responsible for holding sodium.

As will be seen in the section dealing with active transport and accumulation some type of combination is accepted as being part of the mechanism. Carriers which combine with the ions have not yet been identified.

d) Exchange ratios.

Whether we are dealing with the rapid Donnan adjustment of cytoplasm or with the slower Donnan adjustments resulting from metabolic changes or complex formation, the ultimate composition of the cell and external solution approaching equilibrium will be complicated by the number of ion species in the two systems. In a mixture of ions, not only are the relative concentrations important, but also the competing power of the different ions for the anions or adsorption sites in the cytoplasm. In these complicated adjustments lie the explanations of the exchange ratios which have been observed in a variety of plants.

Valuable experimental evidence to sort out the relative importance of these adjustments has been obtained by EPSTEIN and LEGGETT (1954). EPSTEIN and LEGGETT, distinguishing clearly, as other workers have done, between the ion exchange and the active transport, have shown that ions supplied externally compete with each other differently in ion exchange from the way in which they compete in salt accumulation. As might be expected from the variety of ions which give the DONNAN adjustment, the competition between cations is much less selective in DONNAN adjustment than in active transport. The ions entering by exchange are rapidly exchangeable and the equilibrium is approached in about 30 minutes (corresponding to initial uptake). In such a system the competing power of polyvalent cations will be great compared with that of monovalent cations and will result in relatively high amounts of exchange for lower concentrations in the external solution.

II. Non-osmotic uptake.

In non-osmotic movement of dissolved substances, the cell or tissue must do work in moving the substances—the process referred to as *active transport*. The substances may be moved against a concentration gradient *(accumulation)* or may be moved more rapidly than they could move by diffusion. The mechanisms involved are still obscure. Where accumulation against a concentration gradient occurs it is generally agreed that (1) the substance must be moved across a region or membrane of resistance to its normal diffusion and (2) some chemical combination with a carrier is involved.

1. Nonelectrolytes.

Most nonelectrolytes which have been investigated move by diffusion to equality of concentration but not all compounds which the cell contains move along the concentration gradient. The cell frequently, for instance, contains a higher concentration of different sugars than that of the external solution and sugar is increased in the cell against a concentration gradient (GAWADI 1935, quoted by SAID 1941). It has been known for many years that leaf discs in a solution of sugar increase their total sugar concentration. Similar results have been obtained in cut tissue from carrot roots and in wheat roots. There seems to be some doubt as to whether sucrose can enter cells as an unaltered molecule or whether the sucrose is transformed by an invertase activity which has been shown to be present at the surfaces of a number of cells (SAID 1941, BURSTRÖM 1941). The ability of cells to concentrate sugar raises the question of what sort of mechanism might operate and though there have been speculations as to the part that phosphorylation plays in some such accumulatory mechanism (similar to the phosphorylation which is thought to be responsible for the intestinal absorption of glucose), no experimental evidence as to the mechanism has yet been produced. Suggestions about the mechanism require revision in the light of recent work on sucrose syntheses in higher plants (LELOIR and CARDINI 1953, TURNER 1954). Sugar absorption is therefore an example of accumulation which is probably dependent on an enzymatic transformation; how many other substances are accumulated in similar fashion is at present unknown. Enzymatic processes in cell membrane penetration, particularly in animal cells, were reviewed by ROSENBERG and WILLBRANDT (1952).

2. Large molecules.

Very little work has been done on the penetration of larger molecules, many of which are electrolytes but have special properties by virtue of their size. The cell is certainly not impermeable to molecules as large as proteins as shown by the penetration of virus and by the effect of molecules such as cytochrome C on the respiration. How such molecules affect the surface of the cell and enter subsequently will not be clear until we have a much better understanding of cytoplasmic organization. It is known that surface-active substances bring about rapid disorganization of the cells and result in a loss of compounds, such as the coloured sap of beet, to the external solution. What membranes are affected by such molecules is unknown. These observations, however, suggest that molecules of large size such as proteins, in approaching the living surface of the cytoplasm in the cell, may have profound local effects.

3. Electrolytes.

In considering non-osmotic uptake or accumulation of electrolytes, it is essential to distinguish this process from the electrochemical effect. If one mobile ion of a salt is in the cell or tissue in higher concentration than in the external solution, it may be due to ion exchange along the electrochemical potential gradient; but if both mobile ions of the salt enter in higher concentration than in the external solution, accumulation resulting from active transport has taken place.

Any hypothesis for the mechanism of salt accumulation must explain the following characteristics:

(a) Both cations and anions accumulate in the cells and no corresponding ions appear in the external solution.

(b) Efficiency, judged by concentration ratios is very high. For example, aerated carrot tissue in a 0.005 M solution of KCl will attain an internal concentration of 0.085 M while reducing the external concentration to 0.0001 M (ROBERTSON 1941). Such tissue will maintain a very low external concentration of electrolytes, shown by the observations that the electrical conductivity of the external solution will remain at 2×10^{-6} mho/cm.—the value of good distilled water.

(c) The rate of accumulation (of the order of 4×10^{-6} gm mol/gm fresh weight/hr) is limited by low external concentrations and approaches an asymptote as the concentration increases. Some tissues show a depressed rate at very high concentrations (greater than 0.1 M) (ROBERTSON and WILKINS 1948).

(d) The rate of accumulation changes slowly with time apparently in relation to the increasing internal concentration (ROBERTSON 1941).

(e) Ions from different salts are accumulated at different rates by any given tissue: K^+, Na^+, NH_4^+, Rb^+ and Cs^+ are accumulated rapidly, Li^+ less rapidly and Ca^{++}, Mg^{++}, Ba^{++} only slowly; NO_3^-, Br^- and Cl^- are accumulated rapidly while the divalent SO_4^{--} is accumulated very slowly.

(f) Different species accumulate different ions selectively. This is well illustrated by comparison of the internal concentrations of sodium and potassium in *Valonia* and *Halicystis* when both are grown in sea water.

Table 3. *Chemical analysis (mol.) of the sap and the milieu of cells.* (From HÖBER 1946.)

	Sap of *Valonia macrophysa* mol.	Sap of *Halicystis* mol.	Sea water mol.
Cl	0.597	0.603	0.580
Na	0.09	0.557	0.498
K	0.5	0.0064	0.012
Ca	0.0017	0.008	0.012
Mg	trace?	0.0167	0.057
SO_4	trace?	trace	0.036

(g) Ions which have been accumulated are free in solution in the vacuoles of *Valonia* and *Nitella* and presumably also in higher plant tissues.

(h) Different ions show some degree of competition during the process of accumulation (EPSTEIN and HAGEN 1952, EPSTEIN 1953).

Accumulation is related to the respiration. Sometimes the accumulation rate is clearly related to a special component of the respiration, the *salt respiration* (also known as the anion respiration). The term "anion respiration" was introduced by LUNDEGÅRDH to describe this component of respiration because it was found to be proportional to anion uptake rather than to uptake of cations which are taken up by ion exchange as well as by active transport. The term anion respiration should not be taken as meaning that this component of the respiration is caused by the anions alone (ROBERTSON 1941, EPSTEIN 1955). With some tissues (*e.g.* carrot discs), the salt respiration can be recognised as an increase in the rate of oxygen uptake and carbon dioxide output when the tissue is transferred from water or very dilute solution to salt solution. In other tissues the salt respiration can be recognised only with the use of inhibitors, cyanide or carbon monoxide, to which salt respiration is sensitive. The basal respiration is apparently not sensitive to these inhibitors. The fact that both salt respiration and accumulation are sensitive to carbon monoxide in the dark but not in the light suggests that the two processes are dependent on the cytochrome oxidase system (WEEKS and ROBERTSON 1950, SUTTER 1950). Whether other oxidases may also be involved is not certain; a recent report indicates that inhibition of ascorbic oxidase results in inhibition of accumulation but further information is required (JAMES and RUSSELL, unpublished, quoted by JAMES 1954).

Theories of mechanism.

Most plant physiologists have now come to the general hypothesis that the accumulation process requires the combination of the ion being accumulated, with a suitable carrier substance in the cells, a suggestion made originally by PFEFFER (1890). Models which would be suitable for accumulation have been suggested by various workers and discussed in detail by some (see the discussion of the permeability of the cell of *Beggiatoa mirabilis* by RUHLAND, ULLRICH and YAMAHA 1932). Particularly useful theoretical discussions have been given by FRANCK and MAYER (1947), ROSENBERG (1948) and OVERSTREET and JACOBSON (1952). Physical models have been constructed, particularly by OSTERHOUT (1940), but these are of limited value because of the probable complexity of the biological system, and examination of the biological active transport is of greater value. While the evidence is not yet sufficient to describe the mechanism of accumulation in detail, all valid hypotheses have certain features in common: (1) Combination of the ions to be accumulated with a suitable carrier, (2) the continuous liberation of the ions in the appropriate accumulation region and regeneration of the carrier, (3) a region of the cell in which the carrier-ion complex can move but which is largely impermeable to the ions, so back diffusion or leakage of the accumulated ions is limited.

While it is necessary to place constraint on only one ion of a salt to bring about the accumulation of both, the constraint must be of such a kind that it temporarily removes the ion from solution as an ion. This presents special difficulties in postulating the nature of the combination of monovalent ions with the carrier molecule. All preceding hypotheses have tended to simplify the cell to a relatively simple physical system. Recent work has shown that, whatever the ultimate fate of ions being accumulated, the cell has organization in the mitochondria which enables these bodies to hold ions in very different concentrations from the surrounding medium.

α) Hypotheses on the carrier.

The early experiments led to a clear recognition that the carrier was regenerated by some process associated with the respiration, but various indirect relationships have been suggested. Thus BROOKS (1937) suggested that the amino-acids might serve as the carriers, exchanging a H^+ ion from the carboxyl group for a cation and a OH^- from the hydrated amino group for an anion. Two such ionisable carriers would not result in accumulation (BRIGGS 1930). STEWARD and STREET (1947) suggested that energy-rich phosphorylated nitrogen compounds might function as the carriers since protein synthesis frequently accompanies ion accumulation, but no evidence of suitable compounds has been obtained. OVERSTREET and JACOBSON (1952) agree to the suggestion of carrier systems for cations and anions but attribute the absence of positive evidence in their favour to their great lability on the destruction of the cells and probably too, to their low concentration. An ingenious and plausible hypothesis has been advanced by GOLDACRE (1952) based on observations on the accumulation of dyes in root hairs and in *Amoeba*. GOLDACRE suggests that the cyclical folding and unfolding of proteins creates a succession of positive and negative charges which cause local concentrations of anions and cations respectively. When the protein is in the extended state, ions can be adsorbed, anions on the positive groups and cations on the negative groups. When the protein passes to the contracted state, the positive and negative groups compensate each other intra-molecularly and the mobile ions are desorbed. No direct evidence for this hypothesis has been obtained.

The most specific suggestion was made by LUNDEGÅRDH (1945), based upon the evidence that the accumulation mechanism in his experiments was closely dependent on a cytochrome respiration, and further that the anions which entered the cell mainly by accumulation were proportional to the respiration, whereas the cations entering by exchange as well as accompanying the anions in accumulation, bore no such clear relation to the oxygen uptake. LUNDEGÅRDH therefore suggested that the cytochrome which acts as an electron carrier in one direction might act as an anion carrier in the other direction. Simultaneously the H^+ ions liberated by the dehydrogenases were thought to exchange with the cations. In recent years, LUNDEGÅRDH (1951, 1952, 1953) has elaborated this hypothesis and produced spectroscopic evidence for the participation of the cytochrome system in the salt respiration of wheat and maize roots. By direct observation of the absorption bands of cytochrome *a*, *b* and *c* in intact roots, LUNDEGÅRDH concludes that the presence of neutral salt activates the oxidation, and that electron transference takes place through the cytochromes. LUNDEGÅRDH regards the cytochromes as an "electron ladder" so that as an electron is passed from the dehydrogenase to cytochrome *b*, to cytochrome *c*, to cytochrome *a* and to cytochrome oxidase, an anion can be passed back. LUNDEGÅRDH's scheme is summarised in the following diagram:

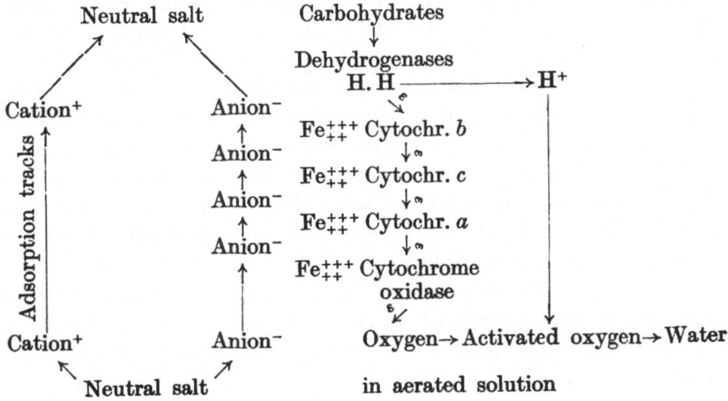

ROBERTSON and WILKINS (1948) using carrot tissue, which has the advantage (unlike roots) of accumulating ions in the cells without secreting them again, examined the possibility that the amount accumulated in unit time did not exceed the rate at which the electron-carrying cytochromes could act as anion carriers. If an anion were accumulated as each electron traversed the system, the maximum rate of accumulation would be four molecules of salt for each molecule of oxygen absorbed in the salt respiration, since each molecule of oxygen absorbed in respiration requires four electrons and four hydrogen ions for the formation of water. The experimental results were approximately equal to the hypothetical ratio of four. Thus LUNDEGÅRDH's hypothesis is possible but not proved by these data.

It is important to note that recent work on the secretion of HCl by the gastric mucosa suggests a system in which the hydrogen ion is liberated from the respiration and an equivalent number of hydroxyl ions pass to the blood. Most experiments suggest that the ratio of this hydrogen and hydroxyl ion formation to the oxygen uptake is that to be expected if the hydrogen ions and electrons of respiration were being separated. As with the plant accumulatory cells, inhibitor evidence suggests that cytochrome is involved (DAVIES 1951, CONWAY 1953, ROBERTSON 1955).

While this suggests the possible fundamental process responsible for secretion and accumulation of ions in living cells, many gaps in our knowledge must be filled before the hypothesis could be accepted. The separation of a hydrogen ion and an electron could be the first step in any type of ion secretion. This is equivalent to putting constraint on one ion, but the movement may not be of an ion but of an electron through a suitable resonance system which may be located in a region which is not permeable to cations.

Thus the separation of positive and negative ions, hydrogen and hydroxyl respectively, can be achieved without the movement of an ion. If now the anions (*e.g.* chloride) can enter from the external solution, electrostatic neutrality would be achieved by the formation of HCl on the inside and the increase of hydroxyl ions to accompany the cations on the outside. In such a system it is the primary separation of H^+ and e^- which leads to the secondary secretion of the anion to equality of concentration with the H^+ ions. In such a mechanism, the cytochrome system would not act as the anion carrier but the number of anions entering would be stoichiometrically related to the cytochrome electron transport. If now there were a change in permeability, so this part of the cell ceased to be permeable to anions and became permeable to cations, the H^+ ions could exchange for the cations in the external solution. Alternatively if there were two carriers, one for anions and the other for cations, accumulation could take place.

Specificity of carriers is suggested by the work of EPSTEIN and HAGEN (1952) and EPSTEIN and LEGGETT (1954) for cations, and of EPSTEIN (1953) for anions, absorbed by barley roots. Similar ions seemed to compete for the combining centres. K^+, Rb^+ and Cs^+ seemed to compete for identical binding sites whereas Na^+ and Li^+ did not compete for the binding sites common to the other three cations. Ca^{++}, Sr^{++} and Ba^{++} compete for sites which have little affinity for Mg^{++}. Among the anions, halides appeared to compete for the same sites but NO_3^- did not. The work of EPSTEIN and LEGGETT is particularly valuable as it suggests that the ion-combining centres concerned with active transport, *i.e.* the carriers, are more selective than those concerned with the Donnan absorption. Since the carriers of active transport are likely to be present only in small quantity in the cells, it is not surprising that nothing is known about their nature. They are not necessarily related in any way to compounds holding ions in the Donnan effect. The accumulation of some ions, *e.g.* phosphate, may be dependent on special mechanisms associated with the metabolism which may not reflect the ion accumulation mechanism of the other inorganic ions (SCOTT RUSSELL, MARTIN and BISHOP 1953).

Little is known about the accumulation of nitrogenous compounds. ARISZ and VAN DIJK (1939) showed that in *Vallisneria* leaves asparagine is accumulated by a process resulting in a high osmotic value in the cell sap and that passive diffusion of asparagine cannot be demonstrated. The accumulated asparagine is not lost under anaerobic conditions though further uptake is inhibited. These results suggest that the cell is not fully permeable to this small molecule and that the accumulation mechanism is dependent on active transport.

A considerable amount of work on *Drosera capensis*, carried out by ARISZ and his collaborators, was summarised recently by ARISZ (1953). Substances transported from tentacles to leaves give concentrations greater than those of the substances in the medium. The cells do not distinguish between the different amino acids, nor between the amino acids and the amides, asparagine and glutamine, which suggests that the mechanism may be similar to that for inorganic electrolytes. The active transport of these substances is different from that of phosphate however, emphasising that different compounds may depend on different carriers.

Whatever the nature of the carriers, they probably operate across a region of high resistance and are unlikely to be free in the cytoplasm. In this connection the behaviour of the mitochondria to different ions (in extension of the work on sodium, potassium and chloride, which will be discussed later), requires further investigation. Specific substances generated by the mitochondrial metabolism may be responsible for the carrier function. LING (1952) has suggested how a phosphorylated organic compound may be able to distinguish between sodium and potassium because of their difference in ionic radii but more work will be necessary on the accumulation carriers of plant cells.

β) Hypotheses on the diffusion barrier.

Since the ions are free in solution, the accumulation must take place across an area of low permeability to the two ions. This is shown by the following calculations: In a typical experiment on carrot tissue, the difference between internal and external concentrations $(c_i - c_a)$ was shown to be 0.075 M = 7.5×10^{-5} gm mol/g., the mean radius of each cell was 3.4×10^{-3} cm and the net leakage rate was apparently zero. If it is assumed that the actual leakage was equal to the maximum rate of accumulation, calculated from the salt respiration, which returned exactly the amount of salt leaking to the surface, actual leakage was 5×10^{-6} gm mol/g./hr.

From the equation

$$\frac{dc}{dt} = \frac{3D}{r^2}(c_i - c_a) \tag{5}$$

which represents diffusion from a sphere of uniform diffusion constant, D can be calculated as being 7×10^{-11} cm²/sec. This is much less than the diffusion constant of KCl in water and suggests that considerable resistance to diffusion must be encountered somewhere in the cell. The diffusion constant of salt in a high resistance membrane surrounding the cell could be calculated only if a thickness of membrane were assumed, but not only are we ignorant of the thickness of the membrane which has this resistance, we also have inadequate information on its location in the cell. A variety of experimental evidence, *e.g.*

the rapid exchange of H^+ ions and cations, and initial uptake, suggests that there is not much resistance at the outside of the cytoplasm but there might be resistance at the tonoplast. If the resistance were in a single membrane (*e.g.* the tonoplast) which had the thickness of many other biological membranes, 10^{-6} cm could be assumed. This hypothetical thickness aids calculation but it is not essential to development of the constants since "permeability" (which neglects thickness and number of membranes) could be used.

The method of determining the diffusion constant for the salt in the membranes rests on two observations: (1) the leakage of salt already accumulated is quite low when the salt respiration is inhibited by cyanide and (2) when salt is removed from the external solution, the salt respiration decreases gradually to a low value but no salt is lost to the external solution (ROBERTSON and THORN 1945). If it is assumed that this low salt respiration of 2.85×10^{-7} gm mol O_2/gm/hr is equal (in hydrogen ions and electrons) to the actual leakage rate (in cations and anions), the leakage rate would be 11.4×10^{-7} gm mol salt/gm/hr. If each cell is treated as a sphere surrounded by a membrane of high resistance, the equation:

$$\frac{dc}{dt} = \frac{3D}{rl}(c_i - c_a) \tag{6}$$

can be used to calculate D assuming that l, the thickness of the membrane is 10^{-6} cm. From experiments of ROBERTSON and THORN (1945) the mean diffusion constant of five observations in a membrane of this hypothetical thickness is 5.4×10^{-15} cm²/sec which is of the order of magnitude for some membranes of this thickness, composed of lipo-protein aggregations.

Two consequences are apparent in calculating the resistance to diffusion this way: (1) using the equation for half-time of adjustment, $\frac{1}{2} = e^{-\frac{3Dt}{lr}}$, it can be shown that when the tissue is transferred from one salt concentration to another, adjustment of concentration throughout by diffusion would be very slow (half time about 35 hours) and (2) the final steady state concentration within the tissue for a given salt respiration can be calculated. If the maximum salt respiration were maintained (11×10^{-7} gm mol O_2/gm/hr) throughout the accumulation period, the internal concentration could go as high as 0.26 M. Accumulation certainly continues until considerable internal concentrations have been established but not as high as this—perhaps due to the fall in respiration. More work is necessary to explain fully the course of the rate of salt accumulation in time with increasing internal concentration (ROBERTSON 1941).

While insufficient information on the nature of the resistance to diffusion in the cytoplasm is available, some resistance must be postulated. This resistance could be in the surface of the cytoplasm, through the bulk of the cytoplasm or at the tonoplast. For reasons already discussed it appears unlikely that the resistance lies at the surface of the cytoplasm. The view that the resistance is at the surface has been held by LUNDEGÅRDH for some time, but would seem improbable in the light of recent work on the apparent free space and the rapidity of exchange. LUNDEGÅRDH's theory (1942) was based largely upon the rapidity of adjustment of the potential differences at the surface of roots when the concentrations of the external media were changed. These data, however, were open to the interpretation that some change in the cytoplasm to some depth in the cell was taking place. Further, these views preceded the recent work which has demonstrated so clearly that the cytochrome system is associated with the mitochondria in plant cells (STAFFORD 1951, MILLERD 1951, and MILLERD, BONNER, AXELROD and BANDURSKI 1951). Recent work has shown not only that the mitochondria of storage tissue of carrot and beet (ROBERTSON, WILKINS, HOPE and NESTEL 1955) carry the cytochrome but also the mitochondria of young (2 day) and old (10 day) barley roots (HONDA 1955).

γ) The role of mitochondria.

The strong evidence for the association between the cytochrome respiration and salt accumulation led to the suggestion (ROBERTSON 1951) that the mitochondria may be responsible for the salt accumulation. Some evidence for this has been available for some time in the work of MULLINS (1940) who showed that radioactive potassium and phosphate supplied to cells of *Nitella*, became concentrated in granules. Accumulation of cations in animal mitochondria has recently been demonstrated by BARTLEY and DAVIES (1952, 1954), MACFARLANE and SPENCER (1953) and STANBURY and MUDGE (1953); BARTLEY and DAVIES also demonstrated that phosphate and organic acid anions were concentrated in mitochondria.

The ionic relations of mitochondria extracted from beet and carrot have been studied by ROBERTSON, WILKINS, HOPE and NESTEL (1955). The concentrations of sodium, potassium and chloride in mitochondrial pellets were analysed with different concentrations in the supernatants. The mitochondria hold both cations and anions in concentrations greater than those in the supernatant. The results are summarised in Table 4.

Table 4. *Concentrations of sodium, potassium and chloride in beet mitochondrial pellets in mM/l; means with standard errors of the number of observations given in brackets.*

Na^+-concentration		K^+-concentration		Cl^--concentration	
External	Internal	External	Internal	External	Internal
0– 6	22.8 ± 2.6 (12)	0– 6	11.0 ± 0.8 (17)	0– 2	4.9 ± 0.6 (11)
7–13	39.6 ± 3.9 (16)	7–13	20.6 ± 1.9 (5)	5–12	11.7 ± 0.7 (11)
		14–20	21.3 ± 1.6 (5)	13–20	16.9 ± 0.6 (18)
79–85	88.1 ± 2.6 (4)			45–47	38.7

Experiments on the time of adjustment to a new concentration of chloride were used to calculate the diffusion constant of salt in the mitochondria. On the assumption that most of the resistance to diffusion lay in the surface membrane of thickness about 200 Å (FARRANT, ROBERTSON and WILKINS 1953), the apparent diffusion constant of chloride in the membrane was shown to be of the order of 10^{-13} cm²/sec. This agrees with that found for heart muscle sarcosomes prepared under similar conditions. The concentrations of mobile cations (Na^+ and K^+) in the mitochondria, which are considerably greater than those in the supernatant, can be explained as being due to a Donnan equilibrium based on the immobile anions (proteins, phosphates, organic acid anions) of the particle. No simple Donnan equilibrium will account for the simultaneous concentrating of both mobile cations and mobile anions. Three explanations are possible: (i) the mobile ion of one charge is bound and removed from the aqueous phase, while the mobile ion of the other charge is held by immobile ions of opposite charge; (ii) there are two phases in the particle, one containing immobile anions and binding mobile cations and the other containing immobile cations and binding mobile anions, but the two phases must be quite distinct; (iii) the cations could be held by a simple Donnan equilibrium while the anions are maintained by some accumulating mechanism against the concentration gradient. ROBERTSON, WILKINS, HOPE and NESTEL (1955) examined the relations of concentration of chloride in the particles to the oxygen uptake and found that the internal concentration was highly correlated with oxygen uptake when the particles were using endogenous substrate. Further, the rate of oxygen uptake is of the right order of magnitude to maintain the concentration differences observed, assuming that there is leakage through a surface membrane with a diffusion constant of 10^{-13} cm²/sec and that the electron carrier of the cytochrome system acts as an anion carrier to accumulate the chloride in the particle.

This work does not prove the LUNDEGÅRDH hypothesis because the quantitative relations might be a coincidence and the real mechanism connecting oxygen uptake to the accumulation of chloride in the particles, might be indirect. The new facts of observation, however, are (1) that the cytochrome system of accumulating plant cells is definitely associated with the mitochondria and (2) that the mitochondria are capable of maintaining ions at greater concentration than the supernatants.

These observations can be related to other work. Thus ROBERTSON, WILKINS and WEEKS (1951) showed that 2,4-dinitrophenol which increased the oxygen uptake via the cytochrome system in cut carrot tissue, prevented the accumulation of salts. Simultaneously, electrolytes including phosphate began to leak from the tissue. The uncoupling nitrophenols are now known (SLATER 1953) to act by uncoupling the oxidative phosphorylations, *i.e.* they allow the formation of water by uptake of oxygen without the simultaneous phosphorylations accompanying the hydrogen and electron transfers. It has been suggested that the lack of transfer of phosphate would be sufficient to make the mitochondria leak and prevent them from being ion accumulators—the phosphorylations having an indirect effect on the mitochondrial organization. However, we do not know enough of the way in which the phosphorylation is related to the hydrogen and electron transfer. It is significant that 2,4-dinitrophenol and other uncoupling agents prevent the secretion of H^+ by the gastric mucosa. The possibility is open that the prevention of simultaneous phosphorylation results in the transfer of hydrogen and electrons to the oxygen of respiration without the separation of H^+ and e^- which is essential to accumulation and secretion.

If the mitochondria act as the temporary accumulators of electrolytes their behaviour in relation to the ultimate accumulation of ions in the vacuole must be examined. Mitochondria, consisting of a high content of lipoid and protein, would be suitable to transport ions through regions of the cell which are predominantly composed of lipoids, *e.g.* through the tonoplast. Mitochondria are moved in the protoplasmic streaming and have been frequently observed to come into contact with the tonoplast boundary in such cells as *Nitella* (MERCER, HODGE, HOPE and McLEAN 1955). From the point of view of comparative physiology it is interesting that USSING (1954) quotes WIGGLESWORTH, mentioning that, in the salt-reabsorbing part of the Malpighian tubule in insects, the giant mitochondria actually pierce the luminal cell boundary, waving with one end in the pre-urine, while having the base well within the cell body.

The accumulation mechanism suggested for plant cells implies that the mitochondria, which are presumably changing continuously during their metabolism in different parts of the cell, would pick up ions in one part and liberate them in another. Thus a mitochondrion actively metabolizing in one part of the cell may increase its content of mobile ions and then, in cytoplasmic streaming, be moved to another part of the cell, where if its oxidative activity decreases, the ion concentration in the particle will also decrease. If the movement of the particles has been through an area of the cell which has a low permeability to the free ions, the liberated ions will not diffuse back easily.

III. Comparative physiology.

The problem of absorption by plant cells is not merely a plant physiological problem but is one aspect of the general biological problem of how cells in all living systems maintain their contents and hold these contents against a concentration gradient. It seems probable that the plant cell is, in this respect as in many other physiological properties, rather less specialised than the highly developed tissues of the specific organs in higher animals. Thus for instance, the oxyntic cells of the gastric mucosa are specialised for the production of hydrochloric acid though the mechanism may be very similar to that which is responsible for accumulation of salts in plant tissues. Again, the red blood cell represents a specialised cell with a particular type of Donnan equilibrium which accounts for the maintenance of certain of its concentrations in relation to the blood plasma. The cells of the intestinal wall are specialised to function as sugar absorbing tissues. The plant cell, on the other hand, shows all these properties and, while differences between plants can be marked, the specialisation of function is much less than in animal tissues. Further work will show how far the same basic mechanisms underly the processes which differ merely in degree of specialisation from species to species and from tissue to tissue.

Literature.

ARISZ, W. H.: Uptake and transport of chlorine by parenchymatic tissue of leaves of *Vallisneria spiralis*. I. The active uptake of chlorine. Proc. Kon. Ned. Akad. v. Wetensch. **50**, 1019–1032 (1947). — II. Analysis of the transport of chlorine. Proc. Kon. Nederl. Akad. v. Wetensch. **50**, 1235–1245 (1947). — III. Discussion of the transport and uptake. Vacuole secretion theory. Proc. Kon. Nederl. Akad. v. Wetensch. **51**, 25–35 (1948). — Absorption and transport by the tentacles of *Drosera capensis*. V. Influence on the transport of substances inhibiting enzymatic processes. Acta bot. Nederl. **2**, 74–106 (1953). — ARISZ, W. H., u. P. J. S. VAN DIJK: Value of plasmolytic method for the demonstration of the active asparagine intake by *Vallisneria* leaves. Proc. Kon. Nederl. Akad. v. Wetensch. **42**, 820–831 (1939).

BARTLEY, W., and R. E. DAVIES: Secretory activity of mitochondria. Biochemic. J. **52**, XX (1952). — Active transport of ions by sub-cellular particles. Biochemic. J. **57**, 37–49 (1954). — BRIGGS, G. E.: The accumulation of electrolytes in plant cells. A suggested mechanism. Proc. Roy. Soc. Lond., Ser. B **107**, 248–269 (1930). — The relation between concentration of hydrogen ions and the effect of weak acids or bases on metabolic activity. J. of Exper. Bot. **5**, 263–268 (1954). — BRIGGS, G. E., and R. N. ROBERTSON: Diffusion and absorption in disks of plant tissue. New Phytologist **47**, 265–283 (1948). — BROOKS, S. C.: Selective accumulation with reference to the ion exchange by the protoplasm. Trans. Faraday Soc. **33**, 1002–1006 (1937). — BURSTRÖM, H.: Studies on the carbohydrate nutrition of roots. Ann. Agric. Coll. Sweden **9**, 264–284 (1941). — Studies on the buffer systems of cells. Ark. Bot. (Stockh.) A **32**, 1–18 (1945). — BUTLER, G. W.: Ion uptake by young wheat plants. 11. The "Apparent Free Space" of wheat roots. Physiol. Plantarum (Copenh.) **6**, 617–635 (1953).

CONWAY, E. J.: The biochemistry of gastric acid secretion. Springfield: Ch. C. Thomas 1953.

DAVIES, R. E.: The mechanism of hydrochloric acid production by the stomach. Biol. Rev. **26**, 87–120 (1951).

EPSTEIN, E.: Mechanism of ion absorption by roots. Nature (Lond.) **171**, 83–84 (1953). — EPSTEIN, E., and C. E. HAGEN: A kinetic study of the absorption of alkali cations by barley roots. Plant Physiol. **27**, 457–474 (1952). — EPSTEIN, E., and J. E. LEGGETT: The absorption of alkaline earth cations by barley roots. Kinetics and mechanism. Amer. J. Bot. **41**, 785–791 (1954).

FARRANT, J. L., R. N. ROBERTSON, and M. J. WILKINS: The mitochondrial membrane. Nature (Lond.) **171**, 401 (1953). — FRANCK, J., and J. E. MAYER: An osmotic diffusion pump. Arch. of Biochem. **14**, 297–313 (1947).

GOLDACRE, R. J.: The folding and unfolding of protein molecules as a basis of osmotic work. Internat. Rev. Cytol. **1**, 135–164 (1952).

HÖBER, R.: Physical chemistry of cells and tissues. London: J. & A. Churchill 1946. — HONDA, S. I.: Succinoxidase and cytochrome oxidase in barley roots. Plant Physiol. **1955**. —

HOPE, A. B.: Membrane potential differences in bean roots. Austral. J. Sci. Res. B 4, 265–274 (1951). — Salt uptake by root tissue cytoplasm; the relation between uptake and external concentration. Austral. J. Biol. Sci. 6, 396–409 (1953). — HOPE, A. B., and P. G. STEVENS: Electric potential differences in bean roots and their relation to salt uptake. Austral. J. Sci. Res. B 5, 335–343 (1952).

JAMES, W. O.: Plant respiration. London: Oxford Univ. Press 1954.

LELOIR, L. F., and C. E. CARDINI: The biosynthesis of sucrose. J. Amer. Chem. Soc. 75, 6084 (1953). — LING, G. N.: The role of phosphate in the maintenance of the resting potential and selective ionic accumulation in frog muscle cells. Phosphorus Metabolism. Vol. 2 (edit. W. D. McElroy a. B. Glass). Baltimore: The Johns Hopkins Press 1952. — LUND, E. J., and Collaborators: Bioelectric fields and growth. Austin: Univ. Texas Press 1947. — LUNDEGÅRDH, H.: Investigations as to the absorption and accumulation of inorganic ions. Ann. Agric. Coll. Sweden 8, 234–404 (1940). — Absorption, transport and exudation of inorganic ions by the roots. Ark. Bot. (Stockh.) A 32, 1–139 (1945). — Spectroscopic evidence of the participation of the cytochrome-cytochrome oxidase system in the active transport of salts. Ark. Kemi. (Stockh.) 3, 69–79 (1951). — The cytochrome-cytochrome oxidase system of living roots of wheat and corn. Ark. Kemi. (Stockh.) 5, 97–146 (1953).

MACFARLANE, M. G., and A. G. SPENCER: Changes in the water, sodium and potassium content of rat-liver mitochondria during metabolism. Biochemic. J. 54, 569–575 (1953). — MERCER, F. V., A. J. HODGE, A. B. HOPE, and J. D. MCLEAN: The structure and swelling properties of *Nitella* chloroplasts. Austral. J. Biol. Sci. 1955. — MILLERD, A.: Succinoxidase of potato tuber. Proc. Linnean. Soc. N.S.Wales 76, 123–132 (1951). — MILLERD, A., J. BONNER, B. AXELROD and R. BANDURSKI: Oxidative and phosphorylative activity of plant mitochondria. Proc. Nat. Acad. Sci. U.S.A. 37, 855–862 (1951). — MULLINS, L. J.: Radioactive ion distribution in protoplasmic granules. Proc. Soc. Exper. Biol. a. Med. 45, 856–858 (1940).

OVERSTREET, R., and L. JACOBSON: Mechanisms of ion absorption by roots. Annual. Rev. Plant Physiol. 3, 189–206 (1952). — OSTERHOUT, W. J. V.: Some models of protoplasmic surfaces. Cold Spring Harbor Symp. Quant. Biol. 8, 51–62 (1940).

PFEFFER, W.: Zur Kenntnis der Plasmahaut und der Vacuolen. Abh. kgl. sächs. Akad. Wiss., Math.-physik. Kl. 16, 185–344 (1890).

ROBERTS, R. B., I. Z. ROBERTS and D. B. COWIE: Potassium metabolism in *Escherichia coli*. II. Metabolism in the presence of carbohydrates and their metabolic derivatives. J. Cellul. a. Comp. Physiol. 34, 259–292 (1949). — ROBERTSON, R. N.: Studies in the metabolism of plant cells. I. Accumulation of chlorides by plant cells and its relation to respiration. Austral. J. Exper. Biol. a. Med. Sci. 19, 265–278 (1941). — Mechanism of absorption and transport of inorganic nutrients in plants. Annual. Rev. Plant Physiol. 2, 1–24 (1951). — The struggle against equilibrium—a physico-chemical problem in biology. Proc. Roy. Soc. New Zealand 1955. — ROBERTSON, R. N., and M. THORN: Studies in the metabolism of plant cells. IV. The reversibility of the salt respiration. Austral. J. Exper. Biol. a. Med. Sci. 23, 305–309 (1945). — ROBERTSON, R. N., and M. J. WILKINS: Studies in the metabolism of plant cells. VII. The quantitative relation between the salt accumulation and salt respiration. Austral. J. Sci. Res. B 1, 17–37 (1948). — ROBERTSON, R. N., M. J. WILKINS and D. C. WEEKS: Studies in the metabolism of plant cells. IX. The effects of 2,4-dinitrophenol on salt accumulation and salt respiration. Austral. J. Sci. Res. B 4, 248–264 (1951). — ROBERTSON, R. N., M. J. WILKINS, A. B. HOPE and L. NESTEL: Plant mitochondria and salt accumulation. Nature (Lond.) 1955. — Studies in the metabolism of plant cells. X. Respiratory activity and ionic relations of plant mitochondria. Austral. J. Biol. Sci. 1955. — ROSENBERG, TH.: On accumulation and active transport in biological systems. I. Thermodynamic considerations. Acta chem. scand. (Copenh.) 2, 14–33 (1948). — ROSENBERG, TH., and W. WILLBRANDT: Enzymatic processes in cell membrane penetration. Internat. Rev. Cytol. 1, 65–89 (1952). — RUHLAND, W., H. ULLRICH u. G. YAMAHA: Über den Durchtritt von Elektrolyten mit organischem Anion und einwertigem Kation in die Zellen von *Beggiatoa mirabilis* nebst allgemeinen Bemerkungen zum Problem der Salzpermeabilität. Planta (Berl.) 18, 338–382 (1932). — RUSSELL, R. SCOTT, R. P. MARTIN and O. N. BISHOP: A study of the absorption and utilisation of phosphate by young barley plants. II. The effect of phosphate status and root metabolism on the distribution of absorbed phosphate between roots and shoots. J. of Exper. Bot. 4, 136–156 (1953).

SAID, H.: Researches on plant metabolism. I. Respiration and sugar absorption by storage organs in relation to time and thickness of tissue slices. Bull. Fac. Sci. Univ. Cairo 24, 31–60 (1941). — SIMON, E. W., and H. BEEVERS: The effect of p_H on the biological activities of weak acids and bases. I. The most usual relationship between p_H and activity. New Phytologist 51, 163–190 (1952). — SINCLAIR, W. B., E. T. BARTHOLOMEW and R. C. RAMSEY: Analysis of the organic acids of orange juice. Plant Physiol. 20, 3–18 (1945). —

SLATER, E. C.: Mechanism of phosphorylation in the respiratory chain. Nature (Lond.) **172**, 975–978 (1953). — STAFFORD, H. A.: Intracellular localization of enzymes in pea seedlings. Physiol. Plantarum (Copenh.) **4**, 696–741 (1951). — STANBURY, S. W., and G. H. MUDGE: Potassium metabolism of liver mitochondria. Proc. Soc. Exper. Biol. a. Med. **82**, 675–681 (1953). — STEINBACH, H. B.: Permeability. Annual. Rev. Plant Physiol. **2**, 323–342 (1951). — STEWARD, F. C., and H. E. STREET: The nitrogenous constituents of plants. Annual. Rev. Biochem. **16**, 471–502 (1947). — STILES, W., and A. D. SKELDING: The salt relations of plant tissues. I. The absorption of potassium salts by storage tissue. Ann. of Bot., N.S. **4**, 329–364 (1940). — SUTTER, E.: Über Wirkung des Kohlenoxyds auf Atmung und Ionenaufnahme von Weizenwurzeln. Experientia (Basel) **6**, 264–265 (1950).

TENDELOO, H. J. C., G. J. VERVELDE and A. J. ZWART VOORSPUY: Electrochemical behaviour of ion-exchanging substances. I. Potential measurements on plant roots. Rec. Trav. Chim. **63**, 97–104 (1944). — TEORELL, T.: Transport processes and electrical phenomena in ionic membranes. Progr. Biophysics a. Biophysical Chem. **3**, 305–319 (1953). — THODAY, D., and K. M. JONES: Acid metabolism and respiration in succulent compositae. I. Malic acid and respiration during starvation in *Kleinia articulata*. Ann. of Bot., N.S. **3**, 677–698 (1939). — TURNER, J. F.: The metabolism of the apple during storage. Austral. J. Sci. Res. B **2**, 138–153 (1949). — Synthesis of sucrose by partially purified plant tissue extract. Nature (Lond.) **174**, 692–693 (1954).

ULRICH, A.: Metabolism of organic acids in excised barley roots as influenced by temperature, oxygen tension, and salt concentration. Amer. J. Bot. **29**, 220–227 (1942). — USSING, H. H.: Ion transport across biological membranes. Ion transport across membranes (edit. H. T. Clarke). New York: Academic Press 1954.

VERVELDE, G. J.: Electrochemical behaviour of ion-exchanging substances. Potential measurements on plant roots. Proc. Kon. Nederl. Acad. v. Wetensch. **51**, 308–313 (1948).

WEEKS, D. C., and R. N. ROBERTSON: Studies in the metabolism of plant cells. VIII. Dependence of salt accumulation and salt respiration upon the cytochrome system. Austral. J. Sci. Res. B **3**, 487–500 (1950).

The loss of substances by cells and tissues (salt glands).

By

R. J. Helder.

With 6 figures.

General introduction.

Undoubtedly the loss of substances by individual cells or tissues is an important aspect of the physiology of the plant. For instance, if we study the uptake of salts by intact plants and their subsequent distribution within the plant, we meet a good many processes which involve either a passive leakage or an active secretion. Exchange, *i.e.* a coupled binding and release of ions, is often the very first step in ion absorption. Accumulation of the ions in the vacuoles requires an active secretion mechanism which brings ions from the cytoplasm into the vacuoles (Arisz 1948). The transfer of ions across the root tissue, from the medium to the xylem vessels also involves either an alternating absorption and release by the individual cells or, if we accept the symplasm theory, at least one absorption process and another process by which ions are released from the stele tissue to the xylem vessels. Moreover, we know that, once the salts have become distributed among the different parts of the plant, redistribution may take place, which again involves a loss of ions by the different plant parts and which may lead to a loss from the roots back to the medium.

However, this process of loss of substances may be discussed more appropriately under headings like exchange, accumulation, transport, distribution and redistribution etc. This in fact, is done in this Encyclopedia, not only for salts but also for organic substances[1]. And what has been said with regard to the processes mentioned also applies to other types of processes like respiration and photosynthesis. This being so, interesting data may well be scattered over a great variety of physiological papers. In the present review, therefore, discussion has to be restricted in a somewhat artificial and arbitrary way, first of all by selecting those papers dealing exclusively or almost exclusively with loss of substances by individual cells or isolated tissues. These will be discussed very briefly in relation to two fundamental aspects, *viz.* the structural features and the metabolism of the cells involved. Secondly there are many interesting facts concerning the loss of substances by intact plants to their surroundings. Substances taken up by the plant during the earlier stages of its development may eventually be given of by the roots back to the medium. Evidence for this can be found in agricultural literature. It has always been a matter of doubt whether the losses found are the result of a normal physiological activity of the roots or whether they are the result of a dying off of the root systems. Achromeiko (1936) first demonstrated that normal healthy plants can excrete large quantities of mineral salts which can be taken up by plants growing in a salt deficient medium. He also reviewed much of the older literature concerning this problem. Very impressive amounts of potassium were

[1] See this volume p. 116: Der Stoffaustausch der Zelle (und der Gewebe). Vol. 4 (IV) The absorption of minerals, (V) The translocation and distribution of minerals in the plant, (VI) The physiological role of the minerals. Vol. 6 (V) The accumulation of carbohydrates, (VII) The secretion of carbohydrates.

found by LUTTKUS and BÖTTICHER (1939) to be given off by the roots of intact maize plants if the shoots were kept in the dark. The greater part of the potassium found in the culture solution was derived from the shoots. VAN ANDEL, ARISZ and HELDER (1950) were able to corroborate this, although the amounts of potassium given off in their experiments appeared to be much smaller than in those of LUTTKUS and BÖTTICHER. A loss of phosphorus and nitrogen can also be demonstrated provided the plants are kept in the dark for some days (HELDER 1952).

Mineral salts are not only lost by the roots but also by the leaves. This also applies to organic substances synthesized within the plants, in particular, sugars. It is a well-known fact that appreciable quantities of these substances can be leached out of the leaves by rain (MES 1954). Special precautions are therefore taken in agricultural research to prevent rain from doing so.

Most workers in this field agree that apart from the substances mentioned more complicated and specific compounds must be assumed to be released by the plant roots. These compounds affect the bacterial flora of the soil and are responsible for the fact that the presence of one species may inhibit or stimulate the growth of another species in its vicinity. Very little is known, however, about the chemical nature of these interesting substances.

The most impressive way in which substances are released from the plant to medium is by means of specialized tissues and organs. The more general aspects of glandular activity are discussed by STOCKING, whilst the sugar secretion of nectaries is discussed by the present author[1]. Here special emphasis will be laid on the structure and functioning of salt glands.

Loss of substances by individual cells and isolated tissues.

Exosmosis and permeability.

From the earliest investigations in which the penetration of substances into living cells was traced with the aid of plasmolytic methods, the reverse *i.e.* a loss of substances, often called exosmosis, has been observed. Thus, much incidental data can be found among the results of classic investigations into the problem of the permeability of the protoplasm. However, sometimes attention was particularly paid to this problem of exosmosis. And, whereas the results concerning permeability gave an impetus to speculations on the structure of cytoplasm and cytoplasmatic membranes, the data on exosmosis have especially been used to demonstrate the condition of the cytoplasm in a more general way. As a rule, healthy cells and tissues show little or no exosmosis, whilst all factors which decrease the vitality of the protoplasm, tend to destroy its semipermeability. A familiar object for demonstrating this are discs of red beet root tissue, in which loss of semipermeability becomes apparent at once, by the release of the red anthocyan in the vacuoles. Temperatures higher than 40° C are injurious, the injury increasing with time of exposure. The same applies to low temperature, in particular to temperatures below zero. Light causes the permeability of the cytoplasm to increase and also stimulates exosmosis (ILJIN 1928).

The influence of toxic substances has also been studied: Ethyl alcohol, octyl alcohol, mercury cyanide, acetaldehyde, strong alkali or acid solutions etc., all change the coagulation condition of the cytoplasm irreversibly, so that its impermeability is lost (STILES and JØRGENSEN 1917, THODAY 1918). From this it appears that exosmosis is a reliable test of the health of the protoplasm. It can be used to show the harmfulness of certain factors often present in physiological experiments, *e.g.* the use of distilled water or unbalanced solutions (BROOKS 1916). In the latter cases small amounts of calcium ions appeared to have beneficial effect.

[1] See Vol. 6 (VII) The secretion of carbohydrates.

It should be realized though, that it is not always easy to distinguish between harmful and harmless factors. Thus, MAXIMOW (1951) indicated that withdrawal of water from leaf tissue, even to a very slight extent, may have a profound influence on the permeability of the tissue. Wilted leaves lose considerable amounts of organic substances and electrolytes, even when the wilting is completely reversible.

With regard to the theory of permeability CZAPEK's ideas, which were based on the influence of surface active substances on exosmosis, are of some importance. He studied the exosmosis of tannins from the leaves of *Echeveria*, and claimed that substances will cause an exosmosis if they reduce the surface tension of the interface of solution and air to 68% of the value for water-air. Starting from this he arrived at the conclusion that the protoplasmic membrane consists mainly of neutral fats and lipoids. Similar ideas were put forward by KISCH (1912), who observed that a decrease of 50% is necessary to cause an exosmosis of invertase from yeast cells.

These ideas have been severely critisized (KOLTZOFF 1912, VERNON 1913, STILES and JØRGENSEN 1917). The principal objection was a theoretical one. The surface tension reducing action of a substance not only depends on its own nature but also on the nature of the interface involved. That is why the water-air interface ist not always a good measure of the solution-protoplasm interface. Of course, this criticism, which is mainly concerned with the theoretical conclusions, does not imply that the surface tension is of no importance to the permeability of the protoplasm.

The data so far discussed are related to the system medium-cytoplasm and cytoplasm-vacuole. Now whilst it is obvious that the structure of the tonoplast plays an important rôle, substances other than those resident in the vacuoles may also be given off. ROBERTSON (Vol. II) has discussed the evidence that there may be a free diffusion of electrolytes through the wet cell walls and the injected intercellular spaces into some parts of the cytoplasm. As a matter of fact, one way of measuring the volume of this "apparent free space" has been by determining the back leakage of ions out of the tissue after it has been allowed to absorb the ions for some time, and then transfered to distilled water or a more dilute solution (HOPE and STEVENS 1952, BUTLER 1953). It should be realized, however, that the "apparent free space" is a formal concept. If adsorption of the ions plays a dominant part, as is no doubt the case with phosphorus, the "true free space" will be much lower than the "apparent free space", which can be as high as 30% of the volume of the tissue.

Most tissues, when isolated from the intact plant and suspended in distilled water or dilute salt solutions, tend to lose some of their substances. The loss may be considerable and may involve both inorganic and organic substances. Thus ILJIN (1928) found with onion tissue a loss of sugar was as high as 1–2% of the initial fresh weight.

The release of electrolytes can easily be traced by measuring the electric conductivity of the bathing medium. This is often done with tissues well-known from the studies on active salt absorption *i.e.* the discs of storage tissue of red beet root, carrot root, potato and artichoke tissue etc. If these discs are immersed in distilled water, the conductivity will increase during the first few days. Then this increase will gradually turn into a decrease because of reabsorption of the ions, previously given off, provided the solution is shaken in such a way that a satisfactory respiration is made possible. It was found (STILES 1927, ASPREY 1937) that, if the discs were kept immersed under water without continuous movement of the medium, they lost their vitality and exosmosis continued until after 2–3 days the cells died.

This exosmosis by freshly cut tissue discs is mostly thought to be due to a release from the cut and other superficial cells, while the active absorption is assumed to be brought about by the inner living cells. On the other hand it appeared that exosmosis can be increased by immersing the discs in monosalt solutions (INGOLD 1931, ASPREY 1935). Pretreatment with sodium, potassium and lithium chlorides increased the subsequent exosmosis. As the capacity for absorbing ammonium ions rose simultaneously ASPREY assumed that physiological and biochemical changes can explain the increases of both the subsequent absorption and the exosmosis of electrolytes into distilled water.

Of particular interest are the substances given off by plant roots. As early as 1919 HANSTEEN-CRANNER stated that cells contain large amounts of phosphatides, which may be given off to the medium. BROWN *and collaborators* (1949) studied the stimulating effect of root exudates on the germination of seeds and on the extension growth of pea and maize root sections. Seeds of parasites showed a ready germination only when they had been exposed to substances released by the roots of their hosts. Thus, seeds of *Striga* plants would only germinate if the water which was applied to them, had been in contact with roots of *Sorghum vulgare* seedlings for a few days. A concentrate could be made from this water by shaking it with charcoal and subsequent elution of the charcoal with 70% acetone. This concentrate was found to contain sugars and in particular pentoses. By testing a large variety of sugars they were able to show that a powerful effect similar to that of the concentrate was given by D-xyloketose. Moreover, both the concentrate and the D-xyloketose stimulated the extension growth of root cells even at the concentration characteristic of hormone activity. Heating the solutions destroyed their activity, whilst simple storage also resulted in a decrease of their activity.

A thorough analysis of the nature of the substances given off by excised roots of young pea and wheat seedlings was made by LUNDEGÅRDH and STENLID (1944). They found that 200–300 young actively growing roots exuded about 100 mg substance per liter solution per 24 hours. Presence of potassium salts favoured this exudation, whereas calcium salt reduced it. Reducing sugars were always present, as were the ions potassium, calcium, phosphate and nitrate. However, apart from these common substances highly specific organic substances were exuded as well. Spectrographical analysis revealed that nucleotides were present, most likely adenosine monophosphoric acid. In the case of wheat roots still other compounds were detected. For example the absorption spectrum of the medium surrounding wheat roots appeared to be very similar to that obtained with a solution of flavone. These results were discussed in relation to the structure of the protoplasmic membrane and the conclusion arrived at by these authors was that nucleotides must be present in the membrane which are responsible for both its ability to form cellulose walls and to absorb ions actively from the medium. Also the significance of the results as regards the relationship of the root system with its environment was stressed. Clearly more data of this kind are highly desirable in order to get a better insight into the many ecological factors which make various plant species grow together or avoid each other.

Excretion and metabolism.

The data on the excretion of substances by young roots as discussed in the previous section, are also interesting in that they provide data which may give a clue to the metabolic processes going on simultaneously with the process of excretion. Thus, the appearance of flavone in the medium of young roots is a fact of importance when considering the respiratory mechanism of wheat roots.

On the other hand, the nucleotides, which were also found in the medium, were formed at an early stage of the development of the cells. In fact, loss of substances seems to be particularly associated with the younger, actively growing plant parts, so that one might ask whether there exists an even more intimate connection between metabolism and excretion. LUNDEGÅRDH and STENLID considered the excretion they found to be the result of a simple leakage of substances from the cells. But generally speaking, it is often difficult to say to what extent an excretion is connected with metabolism or not. This applies especially to the results obtained in experiments where the effects of inhibitors are studied. These substances may be said to be toxic and for that reason able to destroy the fragil structure of the protoplasmic membranes, as a result of which semipermeability is lost so that an exosmosis of cell constituents can occur. But we have learned from modern biochemical research that "toxic" may mean the inhibition of a definite metabolic step, which, of course, will eventually lead to irreversible injurious effects. In practice, however, it is not always possible to discern whether the exosmosis found is due to such injurious effects or not.

STENLID (1948) studied the influence of cyanide and azide on the excretion by young excised roots of wheat, barley, rye, oats and pea plants. Both inhibitors increased the exosmosis, the latter especially at p_H 4.5. The effect of iodoacetate was studied by CHRISTIANSEN (1950) in an investigation, which was a part of an extensive research into the metabolism of rapidly growing stem tissue (THIMANN and CHRISTIANSEN 1950). The growth of isolated sections of etiolated pea stems was stimulated by indoleacetic acid at a concentration of 1 mg/L. This stimulated growth could be blocked by adding either iodoacetate, arsenite or fluoride to auxin solution on which the stem sections were allowed to float. But only under the condition of a simultaneous stimulation by auxin and inhibition by iodoacetate a marked leakage of cell contents, up to 14% of the initial dry weight of the sections, became apparent.

An analysis of the exudate revealed that it contained 32% sugars (mainly glucose), 8% lipides, 8% organic acids, 15% asparagine, 26% amino acids (*i.a.* aspartic acid, α-alanine, histidine, tryptophane, valine, leucine) and 1% phosphate. Other inorganic salts may have been present in the remaining 9%, which was not accounted for.

As this composition corresponded roughly to the easily diffusible part of the cell constituents, iodoacetate was thought to increase the permeability of the cell membrane, as a result of which cell components simply leak out of the cells. This explanation was supported by the fact that in the case of the auxin-iodoacetate treatment the initial increase in fresh-weight changed into a rapid fall of the fresh-weight at the end of the first 12-hours period. The dry-weight continued to fall at a constant rate throughout the experimental period, whereas with the other inhibitors this fall slowed down regularly. Thus, there was a loss of water and dry matter during the second half of the 24-hours experiment. On the other hand it is a remarkable fact that this leakage only occurred if iodoacetate was present with the auxin. Moreover, this leakage could easily be stopped simply by transferring the sections to solutions lacking iodoacetate.

A loss of asparagine from leaf strips of *Vallisneria* was found by ARISZ (1943). These strips accumulate asparagine actively. However, if the strips were placed in a fresh solution at the end of a 24-hours absorption period an exosmosis occurred for a few hours, which then changed into a reabsorption of the asparagine given off. The reason for this tempory effect of changing the solution on the behaviour of the protoplast towards asparagine is not clear.

The loss of amino acids as well as the uptake of these substances by micro-organisms has been extensively studied by GALE and his colleagues (DAVIES, FOLKES, GALE and BIGGER 1953, GALE 1947, 1948, NAJJAR and GALE 1950). Gram-positive organisms, yeasts as well as bacteria, appeared to contain large amounts of free amino acids within their cells. With bacteria *(Streptococcus faecalis, Staphylococcus aureus)* it is particularly lysine and glutamic acid which are derived from the medium and concentrated in the cells. There appeared to exist a marked contrast between these two amino acids in the ways they are taken up and given off by yeasts *(Saccharomyces cerevisiae)* and the bacteria just mentioned.

With *Strept. faecalis* and *Strept. aureus* the uptake of lysine was proportional to the external concentration, it was only slightly dependent on temperature (Q_{10} is about 1.4), and it was uneffected by inhibitors. The reverse applied to glutamic acid. It needed the presence of glucose or casein hydrolysate in the medium for its being absorbed. The relationship between the rate of uptake and the external concentration could be represented by the well-known "adsorption curve" and the Q_{10} was about 1,9. Inhibitors like cyanide and iodo-acetate, which stop fermentation also stopped the uptake of glutamic acid. But even oxin, which does not inhibit fermentation, reduced the rate of uptake, indicating that metallic ions are important for the absorption mechanism.

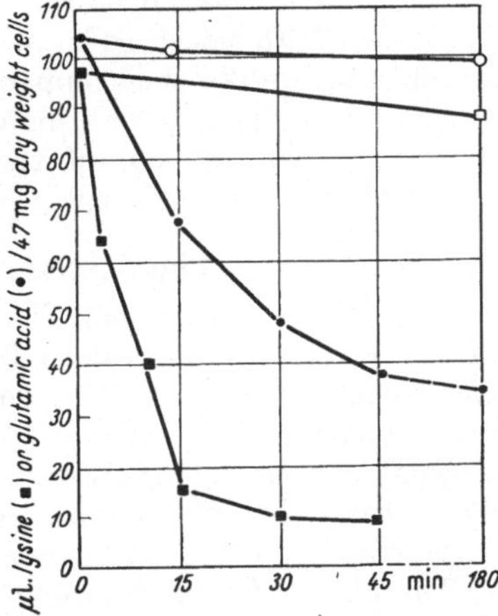

Fig. 1. The influence of 1% glucose on the loss of lysine (■——■) and glutamic acid (●——●) from cells of Streptococcus faecalis incubated at 37° C in salt solutions. In absence of glucose (○——○ and □——□), no loss occurs. (After GALE 1947.)

These data suggested that, whereas uptake of glutamic acid is an active process, lysine is taken up by simple diffusion, leading to an equilibrium gouverned by electrochemical gradients between medium and cell interior. This means that lysine, being an organic cation, tends to be held within the cells at a concentration higher than that of the medium by virtue of a DONNAN-equilibrium, which is reached after about 15 minutes.

If, on the other hand, bacterial cells rich in amino acids were suspended in an amino acid-free salt medium no loss, or only a very slight one, of either lysine or glutamic acid could be observed. But as soon as some glucose was added to *Strept. faecalis* suspensions a ready loss of both lysine and glutamic acid occurred. This suggested that not only the outward migration of glutamic acid but also that of lysine depends on metabolism and active processes (Fig. 1).

Other results, however, made it very difficult to interpret the exact relationship between this excretion process and metabolism. The presence of other amino acids in the cells appeared to have an influence; the retention ability for lysine for instance was enhanced. If no other amino acids were present, so that the cells contain only lysine in appreciable amounts, lysine was readily given off also in the absence of glucose in the medium. Thus, presence of glucose in the medium and absence of other amino acids in the cells had the same effect.

Similar results were obtained with *Strept. aureus* suspensions, except for the fact that addition of glucose completely suppressed the slight loss of glutamic acid from amino acid rich cells.

It will be clear that it is hard to decide whether there exists a direct relationship between metabolism and the transport mechanisms by which amino acids are moved either into or out of the cell, or a more indirect relationship, uptake and loss being passive processes governed by the metabolic state within the cells. For more details concerning the effect of orthophosphate, metallic ions, various amino acid mixtures the reader must refer to the original papers. The results are complex and an explanation is far from simple at this stage.

Of some interest are the changes in distribution of potassium and sodium, going along with the uptake of amino acids. The active uptake of glutamic acid was always accompanied by a simultaneous gain in cellular potassium. This also occurred with the passive diffusion of lysine into cells of *Strept. faecalis*. With *Strept. aureus* no effect on either potassium or sodium content could be observed. With yeast cells, however, which showed an active uptake of lysine as well, the uptake is accompanied by a loss of sodium. The uptake of glutamic acid was accompanied by a slight uptake of sodium. These results are reminiscent of the general problem of the potassium-sodium distribution in living cells. This problem deserves a special discussion, which will be given in the next section.

The excretion of sodium (sodium-pump).

It is a well-known fact that living cells, plant cells as well as animal cells display a characteristic cation distribution with respect to their milieu. Potassium concentration in the cell is always much higher than the external concentration, whereas the reverse applies to the sodium distribution. Hence, even at high external sodium concentration the internal potassium concentration is often higher than that of sodium. As a rule the sum of the sodium and potassium concentrations tends to be constant for any particular type of cells.

The ability for maintaining this characteristic ionic distribution will be lost under the influence of a great variety of factors, the common effect of which is a tendency to lower the metabolic activity of the cells and tissues involved. At first sight it may seem that these effects might well be explained in terms of selective permeability of cytoplasmic membranes, specific binding capacity for potassium of some cell constituents etc. and the loss of these features under the influence of unfavourable conditions. In fact, for many years physiologists have been discussing this problem along these lines. However, during the past fifteen years careful quantitative analysis of the movements of the monovalent cations has affected a profound change in attitude with respect to the interpretation of these ionic distributions. It is now realized that, although alterations in selective permeability may have a bearing on physiological processes of limited duration, the ultimate ionic distribution representing steady state conditions can never be explained by them, but should be considered as being the result of active processes by which ions are moved either inwards or outwards. Moreover, as with most cells and tissues the distribution of potassium seems to be in accordance with DONNAN-equilibrium, attention has moved from the "active" potassium uptake to the extrusion of sodium. DEAN (1941) was first in mentioning the possibility of a so-called sodium-pump being present in muscle cells as a result of which a low internal sodium level, and consequently a high potassium level is affected. The significance of this modern trend in physiology is perhaps best realized by considering that such an active excretion mechanism is also assumed to be present in such a senescent structure as the human erythrocyte has always been regarded to be.

At least in the opinion of the present author the significance of this recent development is such that it seems justified to discuss briefly some of the more important results from the appropriate literature, although up to now most work has been done in the field of animal physiology (erythrocytes, MAIZELS

1954, HARRIS and MAIZELS 1952, HARRIS 1954; muscle, STEINBACH 1954; nerve tissue, HODGKIN and KEYNES 1954; frog skin, USSING 1954). It is obvious that for the study of the excretion of sodium, cells and tissues enriched in sodium in one of the ways indicated above will be used. Such objects will excrete sodium when brought under the appropriate conditions. This process occurs even against an electrochemical gradient. It is almost independent of the external sodium concentration but dependent on the internal one, at least if the latter is below a definite value. At low internal sodium concentrations the rate of excretion increases with increasing concentration. At higher internal concentrations the rate becomes independent of it. Finally, the process is highly temperature sensitive, most metabolic inhibitors reducing the rate of excretion.

All these features are reminiscent of the factors and conditions involved in active uptake processes. As can be deduced from the differential effects of various inhibitors, the required energy may be derived from either glycolysis (human erythrocytes) or respiration (most other types of objects). Hence, a good supply of substrate (sugars, pyruvate etc.) is often an essential prerequisit for obtaining optimal results.

In addition it should be mentioned that there is also a passive movement of sodium ions as can easily be demonstrated by using tracers. As far as this passive movement is concerned, sodium appears to be replaceable by choline. Consequently, if with sodium excreting cells suspended in a sodium containing medium, choline diffuses into the cells in stead of sodium ions, the net loss of sodium by the cells is increased.

As a rule potassium is distributed according to the electrochemical gradients, *i.e.* the ratio of the internal and external concentration is often in accordance with a DONNAN-equilibrium. In these cases no active absorption mechanism needs to be postulated, but, of course, it does by no means exclude the existence of such a mechanism. The uptake of potassium in these cases can be regarded as being a direct consequence of a potential difference set up by the active extrusion of sodium.

With the human erythrocytes, however, the distribution ratio is stated to be much higher than can be explained by a DONNAN-equilibrium (30 in stead of 1.4). Thus, in this case potassium is also actively transported. But even is these cases both processes must be assumed to be intimately linked. This assumption is supported by numerous experimental data. If the potassium concentration of the external solution is reduced below a critical level, the rate of entrance is reduced but so also is the simultaneous excretion of sodium. This reduction is by no means due to the slight degree of the electrochemical gradient against which sodium has to be expelled. On the other hand, with low sodium cells, the sodium excretion of which is limited by their low internal concentration, the active potassium absorption is reduced. The same applies to cells in which much of the sodium has been substituted by choline.

Little is known about the transport mechanism involved. The idea most commonly accepted is that of a carrier system, which functions in a way similar to that in all other types of active transport mechanisms. The linkage between the movements of sodium and potassium is supposed to be due to the fact that the carrier which transports sodium from the cell interior to the medium is, after releasing the sodium ion, converted into a carrier which brings a potassium ion from the medium into the cell. This potassium carrier is subsequently converted into a sodium carrier.

This attractive hypothesis may explain most of the results so far discussed, but it raises a number of difficulties as well, in particular in connection with the quantitative data of erythrocyte studies concerning the interaction of the potassium influx and the sodium efflux (MAIZELS 1954).

In the field of plant physiology far less studies on this interesting topic have been made. With yeast cells it has been tackled chiefly by CONWAY *and his colleagues* (1954). It was found that yeast cells took up potassium actively both under conditions of rapid fermentation and of respiration. Under conditions allowing fermentation H ions are given off in quantities equal to the amounts of potassium taken up. The p_H of the medium tended to run very low, especially when the ratio of cells to suspending medium was low. The H ions were accompanied by organic anions, chiefly succinic acid. The active potassium uptake could be inhibited by azide but not by cyanide. But cyanide did inhibit the potassium uptake which occurred in aerated solutions.

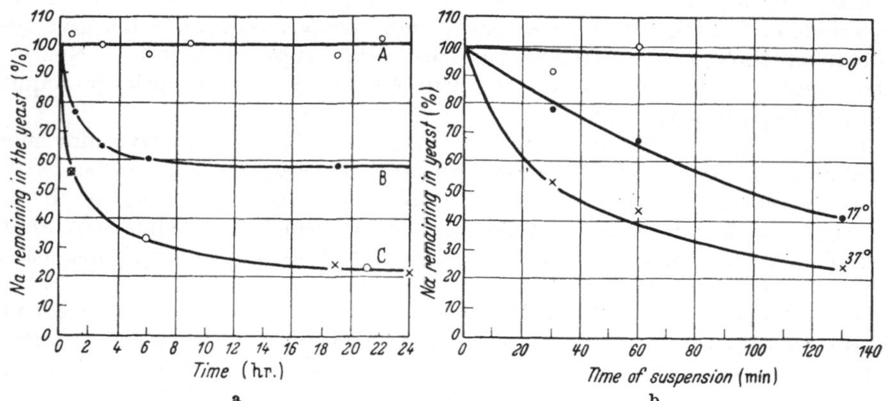

Fig. 2a and b. a The course of the excretion of sodium by Na-yeast. A: 0.1 M NaCl, B: tapwater, C: 0.1 M KCl (circles) and 0.1 M KCl plus 0.1 M NaCl (crosses). b The influence of temperature on the excretion of sodium by Na-yeast. (After CONWAY, RYAN and CARTON 1954.)

The excretion of sodium was also an active process. Na-yeast *i.e.* yeast the sodium concentration of which was increased, was prepared by suspending yeast cells in twenty times their volume of a solution containing 0.1 mol sodium citrate and 5% glucose. By repeating this procedure practically the entire amount of potassium present in the cells could be replaced by sodium. However, very low amounts of potassium in the solution sufficed to prevent sodium from being taken up by the cells, indicating that the sodium was absorbed by the potassium absorption mechanism.

Cells, forced to increase their sodium content in this way, would excrete sodium actively if brought under the appropriate conditions. They did so even in aerated tap water. But addition of potassium chloride to the medium enhanced the rate of excretion (Fig. 2a). But if azide was added simultaneously, the rate was reduced to its original level in tap water.

If on the other hand, Na-yeast was suspended in a 0.1 M sodium chloride solution no net loss of sodium was observed. Apparently there was under this condition a simultaneous transport into the cells. This could be proved by experiments in which either the internal or the external sodium was labelled. Consequently if the inward movement of sodium was inhibited by the presence of azide, a marked net excretion of sodium was obtained, the rate of which was equal to that in tap water.

Fig. 2b shows that the excretion was highly dependent on temperature. Obviously there were two active transport mechanisms: a mechanism that conveyed potassium into, and sodium out of, the cell. Both mechanisms were thought to be highly specific, although the potassium mechanism was able to transport sodium into the cell, provided the external concentration was high. If the sodium concentration was twenty times the potassium concentration both types of ions were absorbed at about the same rate. On the other hand, although in Na-yeast the internal concentrations of potassium and sodium are about the same (50–100 mmol/kg yeast) hardly any potassium is excreted under conditions favourable to the sodium excretion. Moreover, whereas the active potassium uptake is sensitive to azide, 2,4 dinitrophenol, cyanide and anoxia, the sodium excretion is only sensitive to cyanide and anoxia.

The results were considered to support CONWAY'S redox pump theory on active transport, in which a respiratory enzyme in its reduced state acts as a carrier for the sodium ions. By its oxidation, *i.e.* by transfering its electron to another enzyme system of higher redox potential the sodium ion is excreted. The enzym system that is now in its oxidized state is able to bind a potassium ion and to release it subsequently on its being reduced. In this way also the stimulating effect of the presence of potassium in the medium on the sodium excretion could be made understood. If no potassium is present in the medium, the sodium excreted has to be accompanied by anions such as bicarbonate and succinate, the excretion of which, being difficult, limits the excretion of the sodium.

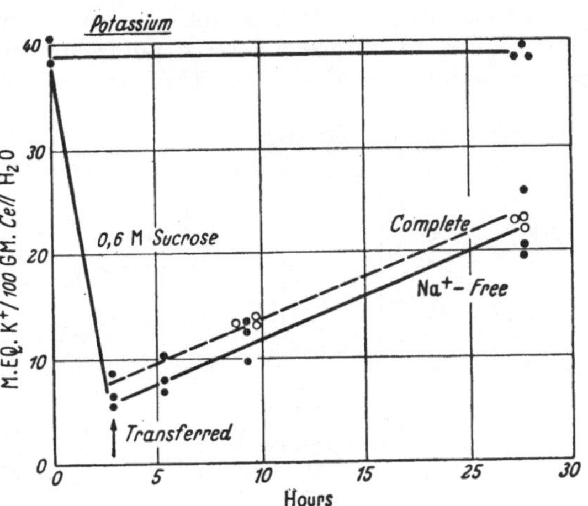

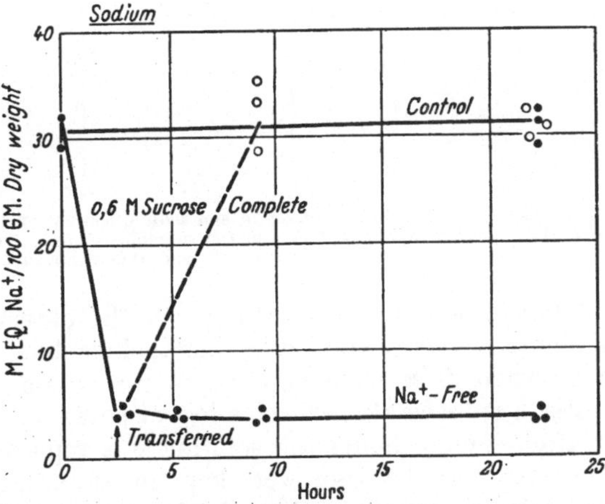

Fig. 3. The accumulation of sodium and potassium by tissue discs of Ulva lactuca in complete natural sea water and sodium free ALLEN'S artificial sea water. (After SCOTT and HAYWARD 1954.)

It cannot be denied that the potassium and chloride concentrations of the media used in these experiments were fairly high if compared with concentrations one comes across normally in transport studies. Moreover, it may be doubted whether it is permissible to generalize from the results obtained with such highly specialized cells as those of yeast. It is therefore important that SCOTT and HAYWARD (1954) were able to show that similar results may be obtained with tissue discs punched from the common sea lettuce *(Ulva lactuca)*. By washing these discs in isotonic sucrose solutions both the cellular potassium and sodium could be leached from the cells. If these potassium and sodium deficient tissue discs were incubated in natural sea water they showed a rapid uptake of potassium and an even faster uptake of sodium. If sodium free artificial sea water was used in stead, no uptake of sodium was possible of course. But in spite of that the reaccumulation of potassium proceeded at a rate equal to that in natural sea water (Fig. 3). This proved that the uptake of potassium was independent of the presence and simultaneous uptake of sodium.

Various substances which inhibit metabolism, photosynthesis as well as respiration, tended to induce a loss of potassium from the cells and a simultaneous gain in sodium by the cells. This applied to phenylurethane, iodoacetate and

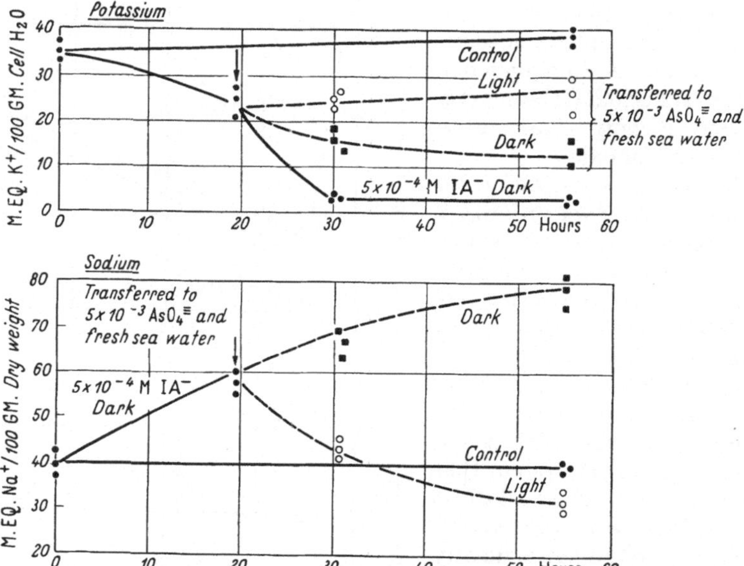

Fig. 4. The shifts in cellular potassium and sodium content of discs of Ulva lactuca as a result of a 20 hours iodoacetate treatment in the dark and the subsequent protective effects of arsenate and light. (After SCOTT and HAYWARD 1954.)

4,6-dinitro-o-cresol. Interesting are the observations on the protective effects of light, arsenate and some metabolites like pyruvate, phosphoglycerate and ATP when discs were exposed to the action of iodoacetate. As may be seen in Fig. 4 iodoacetate induced a loss of potassium and a gain in sodium in the dark. If after twenty hours the solution was replaced by fresh sea water containing arsenate, and the discs were kept in the dark, both the loss of potassium and the gain of sodium continued. But the rate of the potassium loss was much reduced whereas the rate of the gain in sodium was hardly affected by the absence of iodoacetate and the presence of arsenate. On the other hand, if the discs were illuminated while in this fresh solution, the loss of potassium was completely stopped, but all the sodium which was previously taken up under the influence of the iodoacetate was rapidly excreted. This iodoacetate induced increase in cellular sodium could be prevented by adding pyruvate.

For further details the reader should refer to the original paper. The authors arrived at the conclusion that there are two mechanisms regulating the normal cation distribution. There is a mechanism for moving potassium inwards, energetically coupled to the break-down of the sugars, and a mechanism for the active transport of sodium outwards, this mechanism being more directly coupled to photosynthetic processes.

For the moment it seems unwise to enter too much into the details of the speculations in the mechanisms involved. It seems safe to assume that considerable attention will be paid to this interesting problem of the cation distribution in the future, and that important contributions can be expected to be made in this field of permeability and metabolism.

Active loss of salts by salt glands.

Introduction.

Although salt glands were discovered and described as early as the middle of the nineteenth century, it was not until after 1910 that their real physiologic function and biologic meaning was fully appreciated by SCHTSCHERBACK. In addition to the preliminary communication by this author RUHLAND (1915) made a detailed examination of the structure and function and biologic significance of these glands, particularly in connection with the halophyte-problem. He remarks on the striking fact that plant physiologists have paid little attention to this unique process of salt secretion.

The same is true today, for, apart from a single paper on the anatomical features of the glands of *Spartina Townsendii* (SKELDING and WINTERBOTHAM 1939) no other investigations into this problem seem to have been made. This is the more striking as so much research has been done on the processes of active uptake and transport of substances since 1920. It may well be asked why this aspect of salt transport has been so completely neglected by plant physiologists.

One reason may be the fact that these glands occur only within a limited number of plant species, the distribution of which is mainly restricted to the saline regions of the earth. As a result of this, most species belong to the halophytes and this stimulates ecological research rather than causal physiological analysis. Moreover, as will be shown presently, such an analysis is very difficult because of the enormous variability of the experimental results.

In the author's opinion, the main reason for this negligence is, that it has never been sufficiently appreciated by plant physiologists how intimately this problem of glandular activity is connected with the general problems of active transport and, in particular, how much our insight into the process of transfer of substances by plant roots may be benefited by results obtained in a study of these salt glands.

During the last few years some research on these glands has, however, been done in the Botanical Laboratory in Groningen, the results of which have not yet been published. I am therefore grateful to Prof. Dr. W. H. ARISZ for allowing me to make use of some of these results and concepts based upon them in the following brief description of the structure and function of salt glands.

Distribution, morphologic and anatomic features of the glands.

Salt glands are common only within two families: the *Plumbaginaceae* and the *Frankeniaceae*. Outside these families they are found only in isolated plant species *e.g.*: *Acanthus*, *Avicennia*, *Tamarix*, *Glaux*, *Aegiceras*, *Spartina*, *Aeluropus*. Most species are able to tolerate high salinities. As a rule their distribution is limited to the saline margins of the oceans and the salt-lakes.

As to their structural features there exists a large similarity among the glands of the various plant species. Therefore, a more detailed description may be restricted to two representative species: *Statice* and *Spartina*.

Glands are found on almost every aerial part of the plant, but most abundantly on the leaves. According to RUHLAND there are 722 glands per cm^2 on the upperside and 644 glands per cm² on the underside of the leaves of *Statice Gmelini*. He also gives the numbers of stomata, which amount to 8890 and 5413 per cm² respectively. So the numbers on the upperside surpass the numbers on the underside to a slight extent. With *Spartina* there are 42 on the adaxial surface to 27 on the abaxial surface of the leaf.

The glands are initiated at a very early stage in the development of the leaf. And their differentiation is completed much more rapidly than the differentiation of the other leaf tissues. This stresses the significance of their special function. This first impression is supported by a closer examination of the anatomy of the gland.

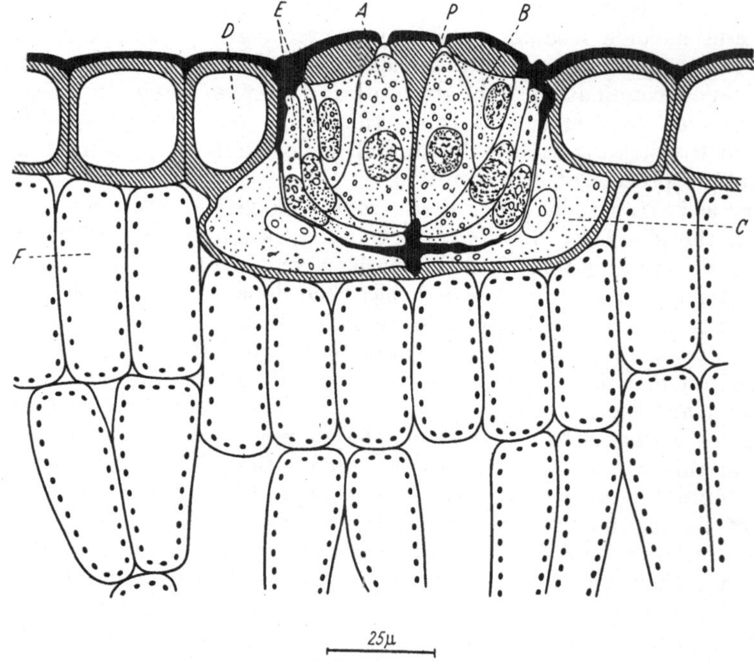

Fig. 5. Cross section of the gland of Statice Gmelini; A: secretion cells, B: small adjacent cells, C: collecting cells outside the gland proper, D: epidermal cells, E: outer cup-shaped layers of glandular cells, F: mesophyll, P: the narrow pores, a little less than 1 micron in diameter, through which the whole amount of the saline fluid has to pass (After RUHLAND 1915.)

With *Statice* the gland consists of a complex of 16 cells. There are 4 secretion-cells proper, arranged in a cross. Each of these cells is accompanied at the outer side by a much smaller cell, which does not reach the bottom of the gland. Both the secretion-cells and these adjacent cells are surrounded by two layers, each consisting of four flat shaped cells, the arrangement of which corresponds with the arrangement of the secretion-cells. These two layers are more or less cup-shaped (Fig. 5).

The top of the gland as well as the epidermis is covered by a thick cuticula. But the other outer walls of the gland, *i.e.* the outer walls of the outermost cup-shaped layer of cells are also strongly cutinized, as a result of which the gland is surrounded by a short of armouring. This process of cutinization is not exclusively restricted to these membranes, but extends, although to a slight degree, also to the adjacent membranes which touch them. In this way a very rigid structure is formed, by which the gland is firmly inserted in the ground tissue of the leaf and protected against mechanical injury (RUHLAND 1915).

These walls are pitted at a few places. In the first place there is only one very small pore, a little less than one micron in diameter, in the cuticula covering the top of each secretion-cell. Thus the whole amount of secretion produced by a gland has to pass through four small pores. The presence of these pores does not mean that the cytoplasm of the secretion cells is in direct contact with the atmosphere. The cell wall is still present, the chemical composition of the cellulose near the pores is however different from that of the bulk of the cellulose.

A large pit is found in the walls of the cells adjacent to the assimilating tissue of the leaves. In this way contact is provided with four big cells: the collecting cells. These cells are situated outside the gland, but differ from the bulk of the mesophyll cells by their extension. Each collecting cell is in contact with

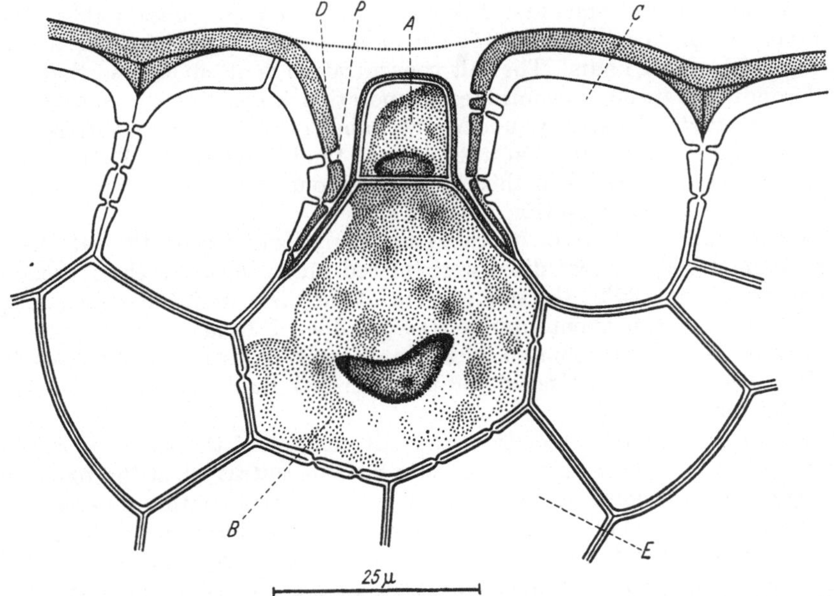

Fig. 6. Cross section of the salt gland of Spartina Townsendii. A: cap cell, B: basal cell, C: the pitted epidermal cell, D: the central well, E: mesophyll. (After SKELDING and WINTERBOTHAM 1949.)

a large number of ordinary mesophyll cells. Presumably their main task is to transfer substances from the assimilating tissue to the gland tissue.

Every cell of the gland proper possesses characteristic contents, consisting of dark granular cytoplasm with a large nucleus, and very thin membranes. There is no large vacuole. Instead a great number of very small vacuoles may be present. This applies particularly to the four secretion-cells. The nuclei are often found near the pits, *i.e.* near the places, where a rapid transfer of substances can be expected to occur. Chloroplasts are lacking both in the glandular cells and the collecting cells outside the gland. In this respect they are similar to the epidermal cells. As a result of this the whole complex of cells involved in the secretion process can easily be distinguished from the ground tissue.

It is tempting to speculate about the secretion mechanism on the basis of the anatomical data so far obtained. But we prefer to postpone doing so, until more has been said about the structure of the salt glands of *Spartina* and some more physiological data on the secretion activity have been given.

The gland of *Spartina Townsendii* is much simpler in structure than the gland of *Statice*. Instead of a 16-cell complex there are only two typical secretion-cells; the basal cell and the cap cell (Fig. 6). The wall of the cap cell is not pitted.

The basal cell is much larger and is connected by pits on one side with the cells of the assimilating tissue, on the other side with the four epidermal cells by which the well-like opening of the gland is bounded. These epidermal cells are perforated by numerous pits. So there is an important difference between the two types of the salt glands. Here the salt solution is forced to pass from the basal through the adjacent epidermal cells to the wall. Reduced turgor of the leaf will cause the well to close by bringing the walls of these cells into contact with the cap cell.

The influence of internal and external factors on the secretion process.

Even simple observation reveals that the secretion process must be a very important one, for the amounts of secreted fluid may be considerable. Under appropriate conditions leaf discs 150 mg in fresh weight may secrete as much as 70 mg fluid per 24 hours. The salt concentration is often high so that under normal conditions owing to continued evaporation leaves soon become covered with a salt crust. Especially under dry conditions this encrustation occurs to a considerable extent so that the original fluid is masked. With *Spartina* scaly structures are formed in this way. Thus one is forced to agree that the secretion is an efficient desalting process.

This first impression is corroborated by analytical data on the salt content of the leaf material. If leaf discs were placed on distilled water or weak salt solutions *e.g.* up to 1,85%, the salt content of the discs would decrease, in spite of the fact that a concurrent uptake was possible. During a 3-days period the content could be lowered from an initial value of 2% on fresh weight basis down to 0,5%. The salt concentration of the excreted fluid decreased also under these conditions.

Owing to this vigorous secretion, the salt content of the leaves is as a rule equal to, or only slightly higher than, the salt concentration in the root tissue. This might be considered to be the biological meaning of this process, but it should be borne in mind that these halophytes can tolerate very high salt concentrations. By gradual increasing the salt level of the nutrient solutions, RUHLAND found salt concentrations in the leaf tissue as high as 10–12%. The osmotic value, as measured by the plasmolytic method, amounted to 165 Atm. However, these teleological aspects of this problem will not be entered into. It will be clear that a physiological study may contribute important information on the time-honoured halophyte-problem.

As a rule NaCl is the most important salt present in the fluid, but apart from this salt all other kinds of inorganic substances, in particular ions like K, Mg, Ca, nitrate and sulphate may be present. Also phosphates are mentioned to occur in the fluid. We have not been able, however, to detect any phosphate. So, further research on the composition seems desirable.

In order to get an idea about the quantities of substances other than NaCl, present in the fluid, it is advisable to compare the total osmotic concentration with the total chloride content. According to RUHLAND's data there is a discrepancy of about 10%; the chloride content can only account for about 90% of the total osmotic value. However, if leaf discs were placed on a 2.5% sugar solution, a fluid was secreted which was isotonic to the sugar solution. But in this case the chloride could only account for about 50% of the total osmotic value. So, a fairly high concentration of other inorganic and perhaps organic substances must have been present.

In experiments performed in our laboratory a significant difference between total osmotic concentration and chloride content has never been found. It should be noticed however that material used in these experiments was, as a result of a NaCl pretreatment, in a high NaCl status. Moreover, during the experiments the leaf discs were also placed on a NaCl solution. And it is obvious that the salt status and the composition of the medium will have a direct influence on the composition of the secretion.

No doubt, there is always some calcium bicarbonate present in the fluid. This is responsible for the chalk incrustations which made some older authors regard the glands to be chalk glands (VOLKENS 1884). It is possible that this concept is the right one as far as the glands of *Armeria vulgaris* are concerned. This species is very sensitive to saline soils. RUHLAND claimes to have found that although oxalic acid is present in the tissue, no calcium oxalate cristals can be observed.

Of primary importance is the fact that secretion occurs not only with intact plants but also with detached leaves and even with almost isolated glands. From an experimental point of view this offers the possibility to use large numbers of randomised leaf discs, so that the enormous variability can be reduced to some extent. Otherwise this secretion process appears to depend on such a variety of internal yet unknown factors, that even after the process of randomisation there is often a considerable variability as to the experimental results of various sets of leaf discs. The reason why the secretory activity falls down rather rapidly during the winter months is still an unsolved problem.

A second difficulty often met with, is the small quantities of secretion available for analytical purposes. In this respect the thermo-electric osmometer according to BALDES and JONES has proved to be of invaluable importance (VAN ANDEL 1953). It needs only a single droplet for a fairly accurate determination of the osmotic value. Chloride determinations are made most satisfactorily by an electrometric method.

Finally, there is always the danger of secondary changes in the composition and osmotic value of the fluid, once it has been secreted. Owing to vapour pressure differences there will always be a tendency to a levelling in osmotic pressure between secretion and the medium on which the leaf discs are placed. This makes it necessary to mount small chambers on the leaf discs. Then placing the leaf discs on filter paper soaked by a solution of known composition, an experimental arrangement is obtained which works fairly well and which permits the study of the influence of various external and internal factors on the process of secretion.

It is almost self-evident that a good salt condition of the leaf tissue will promote secretion. It is wise, therefore, to give the leaves a 1—2 days pretreatment by allowing the detached leaves to absorb salts from a salt solution while kept under continuous light. If the leaf discs are then placed on a 2% NaCl solution, a constant rate of secretion occurs for at least 1 or 2 days. On the third day there is as a rule a slight slowing down of the rate although the concentration of the fluid remains at the initial level.

With plant material treated in this way the influence of temperature on the secretion process could clearly be demonstrated. By lowering the temperature from 20° C down to 5° C, the rate of secretion was decreased by about 60%. The amount of Cl given off was decreased to the same extent. So, the concentration being the same, only the total amount of the fluid was reduced by the temperature. Similar results could be obtained if the results of a dark-experiment were compared with the results of a light-experiment.

If the secretion process is an active one, the oxygen supply can be expected to have a direct influence on this process. In fact, anaerobic conditions were found to have a strong influence; the secretion was stopped completely.

RUHLAND found that toxic and narcotic substances either stimulated or inhibited the process. In order to see whether these processes are similar to the active processes of salt uptake and transport experiments were performed to trace the influence of inhibitors like potassium cyanide, iodoacetate, 2,4-dinitrophenol. The draw-back of this kind of experiments was that the results tended to be so variable that a very large number of experiments had to be done in order to obtain exact and reliable quantitative data. As a rule there was a slight stimulating effect at the lower concentrations of the inhibitors, whereas a 50% inhibition occurred at a concentration of about 10^{-3} molar. Also in these cases the composition of the secreted fluid, its osmotic value and Cl concentration, remained about constant. It was only the amount of the fluid, which was either increased or decreased.

Changes in the composition were mainly the result of factors, which interfered with osmotic relations and salt concentrations of the leaf tissue. Thus the loss was very much reduced either by extremely dry conditions or by placing intact plants on a concentrated salt solution. At the same time the concentration of the secretion was increased.

More detailed information about the influence of osmotic factors could be obtained by putting leaf discs on solutions with a constant Cl concentration but varying osmotic concentrations owing to the presence of an osmotic active substance. It was found that small increases in osmotic value of the medium reduced the amount of fluid secreted immediately. The amount of secreted Cl being almost the same, however, a more concentrated fluid was produced as a result of this. At much higher osmotic concentrations (10–20 Atm) the Cl secretion was also reduced considerably, sometimes to the same extent as the fluid was.

The influence of the salt status of the leaf tissue appeared from the experiments, mentioned above, in which leaf discs were placed on distilled water. It was found that, owing to the continuous secretion of a salt fluid, the Cl content of the leaf tissue decreased considerably. In accordance with this result the Cl secretion was also found to decrease more and more, so that at the end of a three days period a final concentration could be reached which is only one third of the initial one. In contrast with this the total amount of secretion tended to remain at about the same level. So the concentration showed a gradual decrease.

It might be asked in view of these results what the relationship is between the salt concentration of the leaf tissue and the concentration of the secretion. RUHLAND stated that there is hardly any difference between the concentrations mentioned. In our experiments however, we were able to show that the concentration of the secretion is always significantly higher than the average concentration of the leaf tissue. This difference may amount to 0.5% NaCl.

More complicated results can be expected to occur in those cases in which leaf discs were placed on solutions of various NaCl concentrations. At the lower concentrations increasing the NaCl concentration of the medium promoted both the rate of the secretion and the concentration of the secreted fluid. Further increase of the outer concentration reduced the amount of secretion for osmotic reasons. Finally, also the amount of chloride was reduced. But in spite of that the secretion process continued and a very highly concentrated fluid could be produced, even at very high medium concentrations. RUHLAND found that secretion was possible on a 16–18% sucrose solution.

Although it has been possible to trace the influence of a few factors quantitatively in this way, it should be stressed once more, that the results of this type of experiments are as a rule very variable. This must be due for the greater part to unknown internal factors involved in the process of secretion.

The secretion mechanism.

It follows from the observed effects of oxygen withdrawal and inhibitors that the secretion process as a whole is an active process, depending on metabolism. Moreover, it must be situated within the glandular tissue itself, for even more or less isolated glands are able to go on secreting a salt fluid. Of course, this does not exclude participation by the other leaf tissues under normal conditions. All substances, which can easily penetrate into the cytoplasm *e.g.* the ordinary nutrients and organic substances like mannitol, asparagine, urea, may be found in the secreted fluid (SCHTSCHERBACK 1910). A substance like ferro-ferricyanide, however, which is transported only in the cell walls, does not penetrate into the glandular tissue. To what extent active parenchymatic transport, that has been proved to be very important in modern investigations (ARISZ 1955), has any bearing on the total desalting process remains to be studied. The importance of the salt status of the leaf is not only a priori understandable but could experimentally be demonstrated.

For further insight it now seems appropriate to remember the striking fact that the concentration of the secretion tended to remain fairly constant even under conditions which strongly changed the total amount of secretion. Only factors, influencing the osmotic relations of the leaf tissue appeared to change also the concentration of the fluid. This gives the impression that the secretion process consists in a loss of a salt solution. This might involve an active secretion of ions out of the glandular cells into the outer medium, coupled with an osmotically controlled water loss. However, the anatomical data of *Spartina* are in discordance with this view. We found that with this species, the fluid has to pass through the adjacent epidermal cells, which do not show a glandular structure. Thus we are forced to conclude that the withdrawal of ions and the accumulation of these ions within the glandular cells constitute the active part of the secretion process. This also explains the similarity in behaviour between this process and the active uptake processes, so thoroughly studied with other tissues and organs. This accumulation of ions by the cytoplasm-rich glandular cells will lead to an osmotic withdrawal of water from the leaf tissue. The total result will be an unilateral transport of a salt solution out off the tissue into the salt gland. This involves a tendency for the glandular cells to increase in size, but this is prevented by the presence of the rigid wall structure. An enormous pressure will be created instead and the salt solution of high concentration will leak away passively through the small pores at the top of the secretion cells.

Unfortunately, determination of the osmotic pressure with the aid of the plasmolytic method has proved to be impossible. But that the fluid is given off under a high pressure could be demonstrated by RUHLAND by applying a mechanical counter-pressure direct to the glands: the secretion process continued even against a pressure of 22 cm mercury.

For osmotic reasons a large concentration difference between the secretion given off and the content of the glandular cells can hardly be expected. It was found that the concentration of the secretion tended to be higher than the average concentration in the tissue.

It should be stressed, however, that, though it is possible to distinguish between salt and water transport in the way indicated, this does not mean that both types of transport are completely independent of each other. The fact that factors, hampering the water transport, also inhibited the salt transport, points to such a connection. As in the case of the bleeding phenomenon with excised root systems, there may be an intimate mutual relation between ion and water transport, although ultimately the amounts of secretion are governed by osmotic rules (ARISZ, HELDER and VAN NIE 1951).

At any rate, there is undoubtedly a great similarity between the active processes which play a role in these salt glands and the transport processes in other tissues and organs. In particular, it is tempting to compare the process of secretion into the vessels of the root and the secretion processes of the salt glands, especially in view of the similarities in anatomical features. For instance, the endodermal cells and their Casparian strips might be compared with the basal cell in *Spartina*. Their function would be an active accumulation of ions, derived from the cortical cells. As a result of the high pressure created in this way within the endodermal cells, a salt solution is forced to pass through the cells of the stele and given off to the xylem vessels. The well-known impermeable walls of the endodermal cells would serve in this case as a barrier to a leakage of the secreted salt solution to the medium through the cell walls of the cortex.

How far this analogy is justified, can only be found out by further investigations into the secretion process of both the root and the salt gland. For the moment it may serve to illustrate the importance of this salt secretion process not only for the salt relations of plants but also for our insight into the general active transport processes of ions.

Literature.

ACHROMEIKO, A. J.: Über die Ausscheidung mineralischer Stoffe durch Pflanzenwurzeln. Z. Pflanzenernähr., Düngung u. Bodenkde **42**, 156–186 (1936). — ANDEL, O. M. VAN: Determinations of the osmotic value of exudation sap by means of the thermo-electric method of BALDES and JOHNSON. Proc. Kon. Ned. Akad. Wetensch., Ser. C **55**, 40–48 (1952). — ANDEL, O. M. VAN, W. H. ARISZ and R. J. HELDER: Influence of light and sugar on growth and salt uptake by maize. Proc. Kon. Ned. Akad. Wetensch. **53**, 159–171 (1950). — ARENS, K.: Die kutikuläre Exkretion des Laubblattes. Jb. wiss. Bot. **80**, 248–300 (1934). — ARISZ, W. H.: Het actief en passief opnemen van stoffen door Vallisneria. Versl. Ned. Akad. Wetensch. **52**, 639–645 (1943). — Uptake and transport of chlorine by parenchymatic tissue of leaves of Vallisneria spiralis. III. Discussion of the transport and the uptake. Vacuole secretion theory. Proc. Kon. Ned. Akad. Wetensch. **51**, 25–33 (1948). — Significance of the symplasm theory for the uptake of substances into the cells. In the press 1955. — ARISZ, W. H., I. J. CAMPHUIS, H. HEIKENS and A. J. VAN TOOREN: The secretion of the salt glands of Limonium latifolium. In the press 1955. — ASPREY, G. F.: On the relationship between exosmosis and salt absorption by potato tuber tissue previously treated with various salt solutions. Protoplasma **24**, 497–504 (1935).

BROOKS, S. C.: Studies on exosmosis. Amer. J. Bot. **3**, 483–492 (1916). — BROWN, R., A. W. JOHNSON, E. ROBINSON and A. R. TODD: The stimulant involved in the germination of Striga hermonthica. Proc. Roy. Soc. Lond., Ser. B **136**, 1–12 (1949). — BROWN, R., and E. ROBINSON: The effect of D-xyloketose and certain root exudates in extension growth. Proc. Roy. Soc. Lond., Ser. B **136**, 577–591 (1949). — BUTLER, G. W.: Ion uptake by young wheat plants. II. The "apparent free space" of wheat roots. Physiol. Plantarum (Copenh.) **6**, 617–635 (1953).

CHRISTIANSEN, G. S.: The metabolism of stem tissue during growth and its inhibition. V. Nature and significance of the exudate. Arch. of Biochem. **29**, 357–368 (1950). — CONWAY, E. J.: Some aspects of ion transport through membranes. Symposia Soc. Exper. Biol. **8**, 297–324 (1954). — CONWAY, E. J., H. RYAN and E. CARTON: Active transport of sodium ions from yeast cell. Biochemic. J. **58**, 158–167 (1954). — CZAPEK, F.: Versuche über Exosmose aus Pflanzenzellen. Ber. dtsch. bot. Ges. **28**, 159–169 (1910). — Über die Oberflächenspannung der Plasmahaut von Pflanzenzellen. Ber. dtsch. bot. Ges. **28**, 480–487 (1910). — Über eine Methode zur direkten Bestimmung der Oberflächenspannung der Plasmahaut von Pflanzenzellen. Jena 1910.

DAVIS, R., J. P. FOLKES, E. F. GALE and L. C. BIGGER: The assimilation of amino-acids by micro-organisms. 16. Changes in sodium and potassium accompanying the accumulation of glutamic acid or lysine by bacteria and yeast. Biochemic. J. **54**, 430–437 (1953). — DEAN, R. B.: Biol. Symp. **3**, 331 (1941).

ENGEL, H.: Das Verhalten der Blätter bei Benetzung mit Wasser. Jb. wiss. Bot. **88**, 816–861 (1939).

FITTING, H.: Die Wasserversorgung und die osmotischen Druckverhältnisse der Wüstenpflanzen. Z. Bot. **3**, 209–275 (1911).

GALE, E. F.: The assimilation of amino-acids by bacteria. 1. The passage of certain amino-acids across the cell wall and their concentration in the internal environment of Streptococcus faecalis. J. Gen. Microbiol. **1**, 53–76 (1947).

HANSTEEN-CRANNER, B.: Beiträge zur Biochemie und Physiologie der Zellwand und der plasmatischen Grenzschichten. Ber. dtsch. bot. Ges. **37**, 380–391 (1919). — HARRIS, E. J.: Linkage of sodium- and potassium-active transport in human erythrocytes. Symposia Soc. Exper. Biol. **8**, 228–241 (1954). — HELDER, R. J.: Analysis of the process of anion uptake of intact maize plants. Acta bot. néerl. **1**, 361–434 (1952). — HODGKIN, A. L., and R. D. KEYNES: Movements of cations during recovery in nerve. Symposia Soc. Exper. Biol. **8**, 423–437 (1954). — HOPE, A. B., and P. G. STEVENS: Electrical potential differences in bean roots and their relation to salt uptake. Austral. J. Sci. Res. B **5**, 335–343 (1952).

ILJIN, W. S.: Die Durchlässigkeit des Protoplasmas, ihre quantitative Bestimmung und ihre Beeinflussung durch Salze und durch die Wasserstoffionenkonzentration. Protoplasma **3**, 558–602 (1928). — INGOLD, C. T.: On the effect of previous treatment with salt solutions on the subsequent outward diffusion of electrolytes from plant tissue. Ann. of Bot. **45**, 709–716 (1931).

KISCH, B.: Über die Oberflächenspannung der lebenden Plasmahaut bei Hefe und Schimmelpilzen. Biochem. Z. **40**, 152–188 (1912). — KOLTZOFF, N. K.: Zur Frage der Zellgestalt. Anat. Anz. **41**, 183–207 (1912).

LAUSBERG, T.: Quantitative Untersuchungen über die kutikuläre Exkretion des Laubblattes. Jb. wiss. Bot. **81**, 769–806 (1935). — LECLERC, J. A., and J. F. BREAZEALE: Plant food removed from growing plants by rain or dew. U.S. Agric. Dept. Yearbook **1908**, 389. — LEPESCHKIN, W. W.: Untersuchungen über die Ausscheidungen wässeriger Lösungen durch die Pflanzen. Mem. Acad. St. Petersburg **1904**. — LUNDEGÅRDH, H., u. G. STENLID: On the exudation of nucleotides and flavanone from living roots. Ark. Bot. (Stockh.) Ser. A **31**, 1–27 (1944). — LUTTKUS, K., u. R. BÖTTICHER: Über die Ausscheidung von Aschenstoffen durch die Wurzeln. I. Planta (Berl.) **29**, 325–340 (1939).

MAIZELS, M.: Active cation transport in erythrocytes. Symposia Soc. Exper. Biol. **8**, 202–227 (1954). — MAURY, M. P.: Étude sur l'organisation et la distribution géographique des Plombaginacëes. Ann. des Sci. natur., Sér. VII **4**, 1–134 (1886). — MAXIMOW, N. A.: Kurzes Lehrbuch der Pflanzenphysiologie. Berlin: 1951 Kultur u. Fortschritt. — MES, M. G.: Excretion (recretion) of phosphorus and other mineral elements under the influence of rain. S. Afric. J. Sci. **50**, 167–172 (1954). — METTENIUS, G.: Filices horti botanici Lipsiensis, 10. 1856.

NAJJAR, V. A., and E. F. GALE: The assimilation of amino-acids by bacteria. IX. The passage of lysine across the cell wall of Streptococcus faecalis. Biochemic. J. **46**, 91–95 (1950).

RUHLAND, W.: Untersuchungen über die Hautdrüsen der Plumbaginaceen. Ein Beitrag zur Biologie der Halophyten. Jb. wiss. Bot. **55**, 409–498 (1915).

SCHTSCHERBACK, J.: Über die Salzausscheidung durch die Blätter von Statice Gmelini. (Vorläufige Mitteilung.) Ber. dtsch. bot. Ges. **28**, 30–34 (1910). — SCOTT, G., and H. R. HAYWARD: The influence of iodoacetate on the sodium and potassium content of Ulva lactuca and the prevention of its influence by light. Science (Lancaster, Pa.) **117**, 719 (1953). — Evidence for the presence of separate mechanisms regulating potassium and sodium distribution in Ulva lactuca. J. Gen. Physiol. **37**, 601–620 (1954). — SKELDING, A. D., and J. WINTERBOTHAM: The structure and development of the hydathodes of Spartina Townsendii Groves. New Phytologist **38**, 69–79 (1939). — STEINBACH, H. B.: The regulation of sodium and potassium in muscle fibres. Symposia Soc. Exper. Biol. **8**, 438–452 (1954). — STENLID, G.: The effect of 2,4-dinitrophenol upon oxygen consumption and glucose uptake in young wheat roots. Physiol. Plantarum (Copenh.) **2**, 350–355 (1949). — STILES, W.: The exosmosis of dissolved substance from storage tissue into water. Protoplasma **2**, 577–601 (1927). — STILES, W., and I. JØRGENSEN: Studies in permeability. I. The exosmosis of electrolytes as a criterion of antagonistic ion-action. Ann. of Bot. **29**, 349–367 (1915). — Studies in permeability. IV. The action of various organic substances on the permeability of the plant cell, and its bearing on CZAPEK's theory of the plasma membrane. Ann. of Bot. **31**, 47–76 (1917). — SUTHERLAND, G. H., and A. EASTWOOD: The physiological anatomy of Spartina Townsendii. Ann. of Bot. **30**, 333–351 (1916).

TAMM, C. O.: Removal of plant nutrients from tree crowns by rain. Physiol. Plantarum (Copenh.) **4**, 184–188 (1951). — THODAY, D.: Some observations on the behaviour of turgescent tissue in solutions of cane sugar and of certain toxic substances. New Phytologist **17**, 57–68 (1918). — TRUE, R. H., and H. H. BARTLETT: Absorption and excretion of salts in roots as influenced by concentration and composition of culture solutions. U.S. Dept. Agric. Bur. Plant Ind., Bull. 231, **1912**.

USSING, H.: Active transport of inorganic ions. Symposia Soc. Exper. Biol. **8**, 407–422 (1954).

VERNON, H. M.: Die Rolle der Oberflächenspannung und der Lipoide für die lebenden Zellen. Biochem. Z. **51**, 1–25 (1913). — VOLKENS, G.: Die Kalkdrüsen der Plumbagineen. Ber. dtsch. bot. Ges. **2**, 334–342 (1884). — VUILLEMIN, M. P.: Recherches sur quelques glandes épidermiques. Ann. des Sci. natur., Sér. VII, **5**, 152–177 (1877).

WÄCHTER, W.: Untersuchungen über den Austritt von Zucker aus den Zellen der Speicherorgane von Allium cepa und Beta vulgaris. Pringsh. Jb. wiss. Bot. **41**, 165 (1905). — WILSON, J.: The mucilage- and other glands of the Plumbagineae. Ann. of Bot. **4**, 231–258 (1889–1891).

Die Mitwirkung von Enzymen im Zellstoffwechsel. Konstitution, Kinetik und cytochemische Grundlagen.

Von

Franz Duspiva.

Mit 6 Abbildungen.

I. Einleitung.

Im Protoplasma laufen chemische Umsetzungen unaufhörlich nebeneinander ab und bilden untereinander ein äußerst komplexes Netzwerk. *In diesem ständigen Stoffumsatz,* der durch Reaktionen unterhalten wird, deren Aufeinanderfolge einer gewissen Ordnung unterliegt und deren Umfang gegenseitig fein abgestimmt ist, *offenbart sich das Leben.* Das *Material,* aus dem die Zell- und Organstrukturen bestehen, ist das Produkt dieser chemischen Umsetzungen; die *Energie,* die zum Aufbau der Strukturen benötigt und den errichteten Strukturen zur Ausübung ihrer Funktionen zugeführt werden muß, wird durch einen Oxydations-Reduktionsmechanismus bereitgestellt. Alle diese Prozesse gehen direkt oder indirekt auf die *Wirkung von Enzymen* zurück. Es ist die Leistung der Enzyme, den chemischen Umsatz mit ausreichender Geschwindigkeit zu erhalten, ohne daß Milieufaktoren, wie Temperatur, Wasserstoffionenkonzentration und Druck, wesentlich von der an der Erdoberfläche natürlicherweise vorkommenden Größenordnung abzuweichen brauchen. Ein Charakteristikum der biologischen Katalyse ist ihre *kontrollierte Wirkung.* Die Aktivität der Enzyme in einem gegebenen Reaktionssystem hängt nicht nur von physiko-chemischen Faktoren, wie Temperatur, Säuregrad und Druck, sondern vor allem auch von der Substratkonzentration, der Konzentration der Reaktionsprodukte und verschiedenartiger Begleitstoffe ab, die eine hemmende oder aktivierende Wirkung auf die katalytische Reaktion ausüben.

Man pflegt die an dem chemischen Umsatz in der Zelle beteiligten Stoffe in die Gruppen der *Bau-* und *Betriebsstoffe* sowie der *Wirkstoffe* aufzuteilen und zählt zu letzterer alle Stoffe, die in der Zelle nur in relativ geringer Konzentration anzutreffen sind, aber die wichtige Funktion erfüllen, die chemischen Umwandlungen der Baustoffe zu ordnen, zu steuern oder überhaupt erst zu ermöglichen. Außer den Enzymen wirken in diesem Sinne Gene, Hormone und Vitamine sowie gewisse zellfremde Stoffe, wie Viren, Pharmaka, Gifte und Antibiotica. Nach dem heutigen Stand unserer Einsicht läßt sich diese Aufteilung nicht mehr in vollem Umfang aufrechterhalten, da die *biologischen Wirkstoffe,* ebenso wie auch die Bau- und Betriebsstoffe Produkte des Stoffumsatzes sind und allesamt dem in der lebenden Zelle immerwährenden Werden und Vergehen unterliegen. Vielleicht bilden die Gene in dieser Hinsicht eine gewisse Ausnahme.

Daß in der belebten Natur weitaus die meisten Umsetzungen durch eine *katalytische Kraft* der lebenden Substanz hervorgebracht werden, hat als erster Berzelius (1837) erkannt. Seine Auffassung von der Bedeutung der Enzyme wurde seither in vollem Umfang bestätigt. Wir wissen, daß die Mehrzahl aller chemischen Umsetzungen in der Zelle nur unter der Mitwirkung von Katalysatoren mit meßbarer Geschwindigkeit abläuft. Infolge der Verquickung aller Reaktionen in der Zelle werden auch die zahlenmäßig in den Hintergrund tretenden nichtkatalytischen Reaktionen sekundär von katalytisch gesteuerten beeinflußt. Nach W. OSTWALD bewirkt ein Katalysator eine Änderung des zeitlichen Verlaufes einer chemischen Reaktion, wobei der Katalysator am Ende dieser Reaktion im selben Zustand vorliegt wie am Anfang. Aber praktisch gesehen wird eine chemische Reaktion durch den Katalysator nicht nur beschleunigt, sondern überhaupt erst ermöglicht, da der spontane Umsatz zumeist nur mit unmeßbar geringer Geschwindigkeit abläuft. Auch dirigiert der Katalysator die Reaktion sehr häufig in eine bestimmte Richtung. Glucose wird bekanntlich durch verschiedene Arten von Mikroorganismen in sehr unterschiedliche Gärungsprodukte verwandelt. Man versteht daher heute nach MITTASCH unter einem Katalysator einen Stoff, der eine Reaktion hervorrufen, beschleunigen und in bestimmte Bahnen lenken kann, obwohl er selbst an dieser nicht unmittelbar beteiligt ist. LIEBIG und WÖHLER vertraten mit aller Entschiedenheit den Standpunkt, daß die aus lebenden Zellen extrahierbaren Enzyme unter den Katalysatorbegriff fallen. Als aber SCHWANN, KÜTZING und CAYNARD-LATOUR erkannten, daß Gärungsprozesse an die Anwesenheit lebender Zellen gebunden sind, eine Erkenntnis, die PASTEUR eindeutig beweisen konnte, glaubte man zwischen den in zellfreier Form gewinnbaren *Enzymen* und den an den lebenden Organismus gebundenen *Fermenten* unterscheiden zu müssen. Mit der Gewinnung zellfreier Hefepräparate durch BUCHNER, die zur alkoholischen Gärung befähigt waren, ist diese Scheidung gegenstandslos geworden. Heute werden die Bezeichnungen „Enzyme“ und „Fermente“ gleichwertig gebraucht; man versteht unter diesen Begriffen *organische Katalysatoren*, die in lebenden Zellen gebildet werden und ihre katalytischen Eigenschaften auch in dem Zustand nach der Isolierung aus den Zellstrukturen beibehalten. Es bleibt jedoch die Frage offen, ob alle nach den heute geübten Methoden isolierten und rein dargestellten Enzyme tatsächlich auch ihrem in der Zelle vorliegenden, nativen Zustand entsprechen oder Bruchstücke komplexer Proteinstrukturen darstellen, die das durch sie in der lebenden Zelle katalysierte chemische Geschehen nach ihrer Isolierung im Probeglas nicht mehr adäquat, sondern nur noch modellartig vorführen können.

Die zentrale Bedeutung der Enzyme im Stoffwechsel der Zellen macht verständlich, daß man Enzyme in allen pflanzlichen und tierischen Zellen sowie in Mikroorganismen findet. Enzyme kommen im Cytoplasma und Zellkern, im Zellsaft, im Blutplasma und anderen Körperflüssigkeiten, in Sekreten und Exkreten vor. Enzyme sind an allen wichtigen Lebensvorgängen, wie Wachstum, Vermehrung, Differenzierung, an der Resorption und Exkretion, aber auch an der Erregungsleitung beteiligt. Die Wirkungsweise von Hormonen, Vitaminen, aber auch von zahlreichen Giften und therapeutisch wirkenden Stoffen kann man in vielen Fällen als aufgeklärt betrachten, wenn man den Einfluß dieser Stoffe auf die Aktivität von Enzymen erkannt hat. Das Leben beruht in letzter Linie auf einem geordneten Zusammenspiel der Enzyme; jedwede Störung des normalen Verhältnisses der enzymatischen Aktivitäten zueinander — Hemmung oder Überfunktion einzelner Enzyme — führt zu pathologischen Prozessen und in weiterer Folge zum Tode. Ein Satz von Enzymen wird bereits der befruchteten Eizelle mitgegeben. Mit der morphologischen Differenzierung der Keimareale im Verlaufe des Entwicklungsprozesses geht ein zunehmend kompliziertes Enzymmuster Hand in Hand. Der Genbestand des Zellkernes übt hierbei entscheidende Funktionen aus. Aufschlußreich für ein Verständnis der Enzymneubildung in der Zelle sind die Untersuchungen über die adaptive Enzymbildung bei Mikroorganismen (MONOD und COHN) sowie die Ergebnisse der biochemischen Genetik (BUTENANDT, HALDANE).

II. Die Konstitution der Fermente.

A. Allgemeines.

Schon lange bevor man reine Enzympräparate besaß, schloß man aus dem Verhalten der Enzyme gegen Lösungsmittel, gegen thermische Einwirkung und auf Grund der kolloiden Eigenschaften auf ihre Eiweißnatur.

E. FISCHER hat die Anschauung vertreten, daß die Proteine nicht nur einen ganz erheblichen Teil des Protoplasmas bilden, sondern auch das Material zur Herstellung von Enzymen liefern. Die hochgereinigten Enzympräparate R. WILLSTÄTTERS aber haben zur Stärkung der Proteintheorie nicht beigetragen. Als dann später eine Auftrennung von Enzymen (Holoferment) in einen niedermolekularen Bestandteil (prosthetische Gruppe, Coferment) und ein Trägerprotein (Apoferment) gelang, die sich getrennt als inaktiv erwiesen, nach ihrer Wiedervereinigung aber die enzymatische Wirksamkeit zurückerlangen, schien es sogar sehr unwahrscheinlich, daß die Proteintheorie geeignet sei, ein Licht auf den Mechanismus der enzymatischen Katalyse zu werfen. So blieb die Frage nach der chemischen Natur der Fermente lange ungeklärt. Einen Markstein auf dem Wege der Erforschung der stofflichen Natur der Enzyme bildete die Darstellung eines kristallisierten globulinartigen Proteins aus Jack-beans (Schwertbohnen) durch J. SUMNER (1926, 1932), das Harnstoff mit einer enormen Aktivität spaltet. Nach einer gründlichen Untersuchung dieses Stoffes war SUMNER von der Identität dieses Proteins mit dem Enzym Urease überzeugt. Nachdem in der Folgezeit die Kristallisation zahlreicher weiterer Enzyme, wie z. B. Pepsin[1], Trypsin[2], Chymotrypsin, Katalase, Peroxydase u. a. glückte[3] und alle diese rein dargestellten Enzyme (heute schon mehr als 50) die chemische Natur von Proteinen haben, setzt sich die Anschauung langsam durch, daß die Enzyme Eiweißkörper sind. Die Diskussion über diese Frage ist aber auch heute noch nicht vollständig abgeschlossen. Eine wichtige Frage ist, ob die Kristallisation tatsächlich die sichere Gewähr für die Reinheit eines Proteins gibt. In Anbetracht der Größe des Molekulargewichtes isolierter Enzyme und der Vielzahl ihrer Bausteine liegen Zweifel auf der Hand, ob die Elementarteilchen eines solchen Proteinkristalles untereinander völlig identisch sein müssen, ob nicht Mischkristalle vorliegen können, so daß allein die Tatsache der Bildung von Kristallen noch kein ausreichender Beweis für die Einheitlichkeit des Präparates zu sein braucht[4]. Die bisherigen Erfahrungen haben aber gezeigt, wie BÜCHER[4] kürzlich feststellte, daß die Kristallisation immer noch das beste Verfahren darstellt, einzelne Enzymarten aus einem Fermentgemisch zu isolieren. Proteinfremde Begleitstoffe lassen sich bei der Kristallisation abtrennen, wenn sie nicht zu fest am Enzym haften; im anderen Fall verhindern sie meistens die Kristallisation. Durch planmäßige Waschungen können fremde Enzymspuren weitgehend entfernt werden. Für die Brauchbarkeit der Methode spricht, daß bis heute noch kein Enzym in Kristallform abgeschieden wurde, das mit Sicherheit mehr als *eine* spezifische Reaktion katalysiert[4].

---

[1] NORTHROP (1929/30a, b).

[2] NORTHROP und KUNITZ (1931).

[3] Tabellarische Zusammenstellung kristallischer Enzyme in B. FLASCHENTRÄGER und E. LEHNARTZ, „Physiologische Chemie“, Bd. 1, S. 1034. Berlin-Göttingen-Heidelberg 1951.

[4] BÜCHER (1953a).

B. Enzyme als Proteine.

a) Löslichkeit.

Die Enzyme verhalten sich im isolierten Zustand wie lyophile Kolloide. Sie sind in Wasser löslich, aber in typischen organischen Lösungsmitteln unlöslich. Enzyme von Globulinnatur fallen bei der Dialyse aus und sind nur in Gegenwart einer minimalen Elektrolytkonzentration in Lösung zu halten. Aber größere Mengen an Ammonium-, Natrium- oder Magnesiumsulfat fällen Enzyme aus ihrer Lösung aus. Die Bedingungen zur Fällung (Salzkonzentration, Acidität) sind für jedes Ferment verschieden. Multivalente Ionen wie z. B. Hg oder Pb sowie Phosphorwolframsäure oder Tannin wirken ebenfalls fällend. Gereinigte Enzyme besitzen bei einer bestimmten H^+-Konzentration (isoelektrischer Punkt) ein Löslichkeitsminimum; sie lassen sich bei diesem Säuregrad in Gegenwart einer passend gewählten Salzkonzentration besonders leicht fällen.

Organische Lösungsmittel wie Aceton, Alkohol, Dioxan scheiden, in ausreichender Menge zugesetzt, Enzyme aus ihren Lösungen ab. Stark polare Lösungsmittel wie Glycerin oder Glykol wirken dagegen lösend. Das Lösungsverhalten der Enzyme deutet auf ihre Proteinnatur. Die Löslichkeit eines Fermentproteins unter wohldefinierten Bedingungen bezüglich Salzgehalt, p_H, Temperatur u. a. ist eine charakteristische, konstante Eigenschaft. Setzt man steigende Mengen eines einheitlichen Proteins einem definierten Medium zu, so ist die in Lösung gehende Proteinmenge bis zum Sättigungspunkt eine Funktion der eingetragenen Proteinmenge und nimmt bei weiterem Proteinzusatz nicht mehr zu. Trägt man den löslichen Protein-N gegen den gesamten suspendierten Protein-N in einem Koordinatensystem auf, so zeigt die resultierende Kurve im Falle eines reinen Proteins einen scharfen Knick im Sättigungspunkt und läuft dann parallel zur Abszisse weiter.

b) Stabilität.

Enzyme in Lösung werden bei höheren Temperaturen meist rasch inaktiviert. In vielen Fällen genügt bereits eine Erwärmung auf 60°. Trockene Enzympräparate sind resistenter. Die Hitzeinaktivierung der Enzyme hat mit der Hitzedenaturierung der Proteine viel Ähnlichkeit; Aktivitätsverlust und Denaturierung gehen streng parallel. In beiden Fällen wird ein außerordentlich großer Temperaturkoeffizient beobachtet. Dadurch wird der Eindruck erweckt, als ob ein ganz bestimmter *Inaktivierungspunkt* existiere. Mit „kritischer Inaktivierungstemperatur" eines Enzymes bezeichnet man die Temperatur, bei der das Enzym innerhalb 1 Std die Hälfte seiner Aktivität verliert.

So wie es kochbeständige Proteine gibt — wie beispielsweise die Proteine der Milch —, so widersteht auch eine Anzahl von Enzymen dem Erhitzen. Der beliebte Kochtest zur Unterscheidung von enzymatischen und nicht-enzymatischen Reaktionen ist daher *nicht* allgemein gültig. Durch eine hohe Temperaturbeständigkeit zeichnen sich z. B. Ribonuclease, kristallisiertes Trypsin bei p_H 1 und Lysozym sowie hochgereinigte Meerrettichperoxydase aus. Die Hitzedenaturierung der Proteine verläuft nicht unbedingt irreversibel. Nach ANSON und MIRSKY besteht bei Trypsin ein temperaturabhängiges Gleichgewicht zwischen einer nativen (= aktiven) und denaturierten (= inaktiven) Form des Moleküls. Das Verschwinden und Wiederauftreten der enzymatischen Aktivität erweist sich hier als ein außerordentlich empfindlicher Test für die Intaktheit eines Proteinmoleküls.

Die *Aktivierungswärme* der Hitzeinaktivierung (aus dem Temperaturkoeffizienten berechnet) ist bei Proteinen und in analoger Weise auch bei Enzymen sehr hoch. Mit der Hitzeinaktivierung ist daher eine sehr große Zunahme der Entropie verbunden, wohl im Zusammenhang mit dem Lösen einer großen Zahl von Bindungen im Proteinmolekül (STEARN 1949).

c) Verhalten im elektrischen Feld.

Enzyme verhalten sich im elektrischen Feld wie amphotere Kolloide, die je nach Ladung anodisch oder kathodisch wandern. Die Richtung ihrer Wanderung und die Wanderungsgeschwindigkeit sind vom p_H-Wert der Enzymlösung abhängig. Im *isoelektrischen Punkt* geht der Ladungsüberschuß auf Null zurück; in diesem Zustand sind daher Enzympartikel im elektrischen Feld bewegungslos. Da die Proteine sich durch die Art und Zahl der Ladungen je Molekül sowie durch die Lage des isoelektrischen Punktes stark unterscheiden, ist die *Kataphorese* eine beliebte und bewährte Methode zur Auftrennung von Fermentgemischen und Prüfung von Enzympräparaten auf Einheitlichkeit. Bei unreinen Präparaten kann der Ladungssinn durch anhaftende Begleitkolloide oder Ionen mitunter stark verändert werden; es treten daher im Verlauf der Reinigung Verschiebungen im kataphoretischen Verhalten auf.

d) Das Molekulargewicht.

Zuverlässige Bestimmungen des Molekulargewichtes konnten erst an hochgereinigten, kristallisierten Enzymen durchgeführt werden. Neben *osmotischen Methoden* und *Diffusionsmessungen* werden mit besonderem Erfolg Bestimmungen der *Sedimentationsgeschwindigkeit* und des *Sedimentationsgleichgewichtes* in der Ultrazentrifuge zur Ermittlung der Teilchengrößen herangezogen. Am Beispiel des Pepsins kann gezeigt werden, daß die nach verschiedenen Methoden erhaltenen Werte des Molekulargewichtes untereinander übereinstimmen und auch mit chemischen Daten in gutem Einklang stehen (MOELWYN-HUGHES).

Molekulargewicht des Pepsins.

Verteilung im Zentrifugalfeld	39200	
Aus osmotischen Daten	35000	
Aus dem Diffusionskoeffizienten	36000	
Aus dem P-Gehalt	39800	(1 Atom P/Mol)
Aus dem Cl-Gehalt	35000	(2 Atome Cl/Mol)
Aus dem S-Gehalt	36000	(10 Atome S/Mol)
Aus der Sedimentationsgeschwindigkeit im Zentrifugalfeld	35500	
Mittel	36800 ± 3000	

Aus allen bisherigen Untersuchungen ergibt sich, daß der Teilchengröße ein auch chemisch wohldefiniertes Molekulargewicht entspricht. Die gewonnenen Werte umspannen den weiten Bereich zwischen 14000 (Ribonucleodepolymerase) und 483000 (Urease); sie stimmen in der Größenordnung mit den Molekulargewichten von Eiweißkörpern überein.

Infolge der erheblichen Teilchengrößen ist es möglich, sich der *Dialyse* oder *Ultrafiltration* zur Abtrennung von niedermolekularen Stoffen aus Enzympräparaten zu bedienen. In besonderen Fällen gelingt auf diese Weise auch die Abtrennung der Cofermente aus dem Enzymkomplex.

e) Artspezifität.

Ein bis in Einzelheiten gehender Vergleich zwischen gleichnamigen Enzymen, die mit der gleichen reaktiven Gruppe des Substrates reagieren, aber verschiedenen Arten von Organismen entstammen, bringt stets deutliche Unterschiede zutage. Solche Unterschiede zeigt z. B. ein Vergleich der *Pepsine* verschiedener Wirbeltierarten.

Sehr deutliche artspezifische Unterschiede zeigen sich ferner bei Hemmversuchen. Diäthyl-p-nitrophenylphosphat ist ein sehr starker Inhibitor der Acetylcholinesterase. Dieser organische Phosphorsäureester hemmt die Acetylcholinesterase aus dem Gehirn von Bienen und Stubenfliegen ungefähr gleich stark. Dasselbe gilt für Diäthyl-p-nitrophenylthiophosphat. Prüft man aber Di-isopropyl-p-nitrophenylthiophosphat, so wird die aus Fliegengehirn stammende Acetylcholinesterase weit wirksamer gehemmt als die Bienencholinesterase[1]. BEARD zeigte in einfachsten Versuchsanordnungen, wie ausgeprägt und verbreitet artspezifische Unterschiede gegenüber Enzyminhibitoren nicht nur bei der Cholinesterase, sondern auch bei anderen Fermenten sind. Unterschiede zwischen artlich verschiedenen Fermenten von gleicher Grundspezifität äußern sich auch bei der Bestimmung des Verhältnisses der Spaltungsgeschwindigkeit mehrerer Substrate des gleichnamigen Enzyms. ZELLER (1955) schlägt vor, die einander entsprechenden Enzyme aus verschiedenen Arten von Organismen unter Anlehnung an den in der vergleichenden Anatomie gebräuchlichen Begriff als „*homologe Enzyme*" zu bezeichnen. Eine Erklärung für das unterschiedliche Verhalten homologer Enzyme ist in ihrer Proteinnatur zu suchen. Es ist bekannt, daß auch in der Zusammensetzung nicht-enzymatischer Proteine deutlich artliche Unterschiede bestehen. So unterscheiden sich die Hämoglobine verschiedener Tierarten charakteristisch in Kristallform und Löslichkeit, im Molekulargewicht, elektro-phoretischem Verhalten, immunologischem Verhalten, Absorptionsspektrum und in ihrer Aminosäurezusammensetzung (Tabelle 1). Die Unterschiede betreffen die S-haltigen Aminosäuren, aber auch den Gehalt an Leucin, Isoleucin, Valin und Alanin. Da die Hämoglobine immer dasselbe „Häm" enthalten, gehen die Unterschiede ausschließlich auf einen etwas verschiedenen Aufbau des Globinanteils zurück[2].

Tabelle 1. *Artliche Unterschiede in der Aminosäurezusammensetzung der Globine.*

	Gehalt in Prozent	
	Cystin	Methionin
Menschen	1,2	1,45 (1,32)
Affen	1,2	1,45
Rind	0,55	1,8
Pferd	0,8	1,0
Caniden (Hund, Schakal, Fuchs)	1,6	0,6

Bei Pferd und Esel werden die endständigen Aminogruppen in den Hämoglobinen ausschließlich von Valinresten, bei Kuh, Schaf und Ziege zur Hälfte von Valin und Methionin gebildet. Die Analysenergebnisse deuten auf eine allgemeingültige Beziehung zwischen dem Aufbau von Proteinen und der systematischen Stellung der Organismen hin. Artliche Unterschiede sind auch für das *Metakentrin* (Luteinisierungshormon des Hypophysenvorderlappens) bekanntgeworden.

Ein Vergleich der α-Amylasen zwischen Organismen sehr verschiedener systematischer Zugehörigkeit zeigt bedeutende Unterschiede in zahlreichen Eigenschaften[3]. Fraglich bleibt allerdings, ob diese als α-Amylase funktionierenden Proteine in Anbetracht so großer Verschiedenheit im Bau und Verhalten gegen Aktivatoren als homolog gelten können. Enzyme, die bei gleicher Wirkung einen zweifellos verschiedenen Mechanismus haben, sind die *Aldolase* aus Hefe,

[1] METCALF und MARCH (1949).

[2] Eine eingehende Besprechung der Artspezifität des Globulinanteiles im Hämoglobin verschiedener Tiere findet sich in B. FLASCHENTRÄGER und E. LEHNARTZ, „Physiologische Chemie", Bd. 1, S. 708ff. und S. 853ff.; hier auch weitere Literaturangaben.

[3] Vgl. Tabelle 3 in J. B. SUMNER und K. MYRBÄCK, „The Enzymes", Bd. 1, Teil 1, S. 685, New York 1950.

Tabelle 2. *Aminosäuregehalt einiger kristallisierter Enzyme.* (Aus DESNUELLE.)

Lfd. Nr.	Aminosäuren	Chymotrypsinogen (Rinderpankreas) Mg 25000			Ribonuclease (Rinderpankreas) Mg 15000			Aldolase (Skeletmuskulatur vom Kaninchen) Mg 140000			Triosephosphat-dehydrogenase (Skeletmuskulatur vom Kaninchen) Mg 140000			Phosphorylase (Skeletmuskulatur vom Kaninchen)	
		I	II	III	I	II	III	I	II	III	I	II	III	I	II
1	Alanin	7,6	7,4	21	—	—	—	8,6	8,0	135	6,7	6,4	75	4,8	45
2	Arginin	2,8	5,6	4	5,2	10,2	5	6,3	12,1	51	5,2	10,2	30	11,6	22,6
3	Asparaginsäure . .	11,3	7,3	21	14,2	9,1	16	9,7	6,1	102	12,4	8,0	95	9,3	5,9
4	Glutaminsäure . .	9,0	5,3	15	13,0	7,5	13	11,4	6,5	109	6,8	3,9	46	13,4	7,8
5	Cystein	1,3	0,9	3	0,6	0,4	0,7								
6	Cystin/2	3,3	2,4	7	6,5	4,6	8	1,1	0,8	13	1,1	0,8	9	0,4	0,4
7	Glykokoll	5,3	6,1	18	1,3	1,5	3	5,6	6,2	105	6,0	6,9	80	3,8	4,3
8	Histidin	1,2	2,0	2	4,2	6,9	4	4,2	6,8	38	5,0	8,3	32	3,3	5,4
9	Isoleucin	5,7	3,8	11	3,1	2,0	4	7,9	5,0	84	9,1	5,9	69	6,5	4,2
10	Leucin	10,4	6,9	20	0	0	0	11,5	7,3	123	6,8	4,4	51	10,5	6,8
11	Lysin	8,0	9,5	14	10,4	12,1	11	9,5	10,9	91	9,4	11,0	64	7,2	8,4
12	Methionin	1,2	0,7	2	4,4	2,5	5	1,2	0,6	11	2,7	1,7	18	2,7	1,5
13	Phenylalanin . .	3,6	1,9	5	3,6	1,8	3	3,1	1,5	26	5,5	2,9	33	6,2	3,2
14	Prolin	5,9	4,4	13	3,6	2,7	5	5,7	4,1	69	3,7	2,7	32	4,7	3,5
15	Serin	11,4	9,4	27	12,0	9,7	17	7,3	5,8	97	8,5	6,9	81	3,1	2,4
16	Threonin	11,4	8,3	24	9,0	6,4	11	7,4	4,8	87	7,6	5,5	63	4,2	3,0
17	Tryptophan . . .	5,6	4,7	7	0	0	0	2,3	1,9	16	2,0	1,7	10	2,0	1,6
18	Tyrosin	3,0	1,4	4	7,9	3,7	7	5,3	2,4	41	4,6	2,2	25	5,9	2,8
19	Valin	10,1	7,4	22	7,3	5,3	9	7,4	5,3	88	12,0	9,0	105	7,3	5,3
20	Amid (NH_3) . . .	1,86	9,5	(27)	2,5	12,5	(22)	1,1	5,4	(91)	1,2	5,8	(67)	1,5	7,3
	Summe:	119,96	104,9	240	108,8	98,9	121	116,6	101,5	1286	116,3	104,2	916	108,4	100,9

I g Aminosäuren je 100 g Proteine; II g Stickstoff je 100 g Gesamtstickstoff; III Anzahl des Restes je Mol.

die der Mithilfe zweiwertiger Metallionen bedarf, und die Aldolase aus Muskeln, die ohne solche auskommt[1]. Im Gegensatz zu Muskelaldolase wird die Hefealdolase durch Metallkomplexbildner, wie Phosphat, Cystein und Glutathion inaktiviert, durch Fe^{++}, Cu^{++} und Zn^{++} reaktiviert. Bücher[1], der zu diesen Fragen eingehend Stellung nimmt, meint daß das „*Gebiet der vergleichenden Molekular-Morphologie der Fermente* noch wenig bearbeitet ist".

f) Bausteine der Enzyme.

Nach den heute vorliegenden Unterlagen unterscheiden sich die Enzyme untereinander in ihrer *Aminosäurezusammensetzung* ganz bedeutend, auch dann, wenn diese Enzyme in der gleichen Zelle entstehen wie z. B. Aldolase, Triosephosphatdehydrogenase und Phosphorylase aus dem Skeletmuskel des Kaninchens oder Chymotrypsinogen und Ribunoclease aus Rinderpankreas (Tabelle 2). Es

Tabelle 2. (Fortsetzung).

Seitenketten	Chymotrypsinogen	Ribonuclease	Aldolase	Triosephosphatdehydrogenase	Phosphorylase
Polare	50,4	69,5	49,0	48,4	50,6
a) ionisierte	13,7	28,0	26,6	24,4	30,3
kationische	8,3	16,5	14,0	13,8	17,0
anionische	5,4	11,5	12,6	10,6	13,3
b) nicht-ionisierte	36,7	41,5	22,5	24,0	20,3
Apolare	49,6	30,5	51,0	51,6	49,4
Aliphatische	86,6	84,3	85,4	85,6	82,3
Aromatische	6,7	8,3	6,5	7,4	10,0
Heterocyclische	6,7	7,4	8,1	7,0	7,7
Zahl der Aminogruppen je Mol					
total	17	14	—	—	—
α-Amino (Total-Lysin)	3	3	—	—	—

wäre aber interessant zu wissen, ob und inwieweit Enzyme mit verwandter Wirkung im Aufbau ihres Moleküls übereinstimmen. Wir wissen, wie Bücher[1] schreibt, heute noch nicht, wie exakt die Natur beim Bau der Fermentmoleküle vorgeht, und ob „bestimmten Fermenten eine bestimmte bis auf die letzte Bindung und Aminosäure festgelegte Konstruktion" zukommt. Martin und Porter konnten auf chromatographischem Wege Ribonuclease in zwei enzymatisch aktive Fraktionen aufteilen, was von Hirs, Moore und Stein bestätigt werden konnte. Neilands gelang es, die Milchsäuredehydrogenase elektrophoretisch in zwei Komponenten aufzuteilen, die sich auch bei der Aussalzung sowie bei der spektralen und kinetischen Analyse unterschieden. Man kann aber heute noch nicht mit Sicherheit sagen, ob es sich bei den Begleitproteinen um native Proteine handelt, die schon in der lebenden Zelle enthalten sind, oder um Derivate der Hauptkomponente, die bei der Isolation entstanden sind. Die beobachteten Veränderungen können auch auf ganz geringfügigen Unterschieden in den Oberflächengruppierungen der Enzymproteine beruhen. Im anderen Fall müßte man annehmen, daß in der Zelle aktive Proteine in mehreren nahe verwandten Typen produziert werden können. Daß in *heterozygoten Zellen* spezifische Proteine in *zwei Typen* nebeneinander vorkommen, ist in einigen Fällen erwiesen[2]. Faßt man alle bisher gesammelten Tatsachen zusammen, so ergibt sich, daß ein *spezifisches, natives Protein* der Träger der enzymatischen Wirkung ist.

[1] Bücher (1953a).
[2] Pauling, Itano, Singer und Wells.

Die eingehende Untersuchung der Enzyme zeigt, daß nicht alle Enzyme ausschließlich Aminosäuren als Bausteine enthalten, wie es z. B. für Chymotrypsin, Pepsin, Ribonuclease zutrifft; manche Enzyme enthalten außerdem noch andere Gruppen. Ein großes Tatsachenmaterial spricht heute dafür, daß solche Gruppen Träger von Zwischenreaktionen im Verlaufe der katalytischen Reaktion sind. Die Grundlagen zu dieser Kenntnis hat O. WARBURG (1938, 1953) in seinen berühmten Arbeiten über die Fermente der Zellatmung gelegt. Die Beobachtung, daß Kohlenoxyd im Dunkeln die Zellatmung hemmt, und daß Licht diese Hemmung wieder aufhebt, führte zur Entdeckung des *sauerstoffübertragenden Eisens*. In der lebenden Zelle reagiert der Sauerstoff mit komplexem Ferroeisen, das zu Ferrieisen oxydiert wird. Durch reduzierende Systeme der Zelle zu Ferroeisen rückverwandelt, kann es durch neuen Sauerstoff oxydiert werden. Später gelang die Isolierung des sauerstoffübertragenden Eisens; es erwies sich als eine *Häminverbindung*. Im „sauerstoffübertragenden Ferment der Zellatmung" ist das Hämin als Wirkungsgruppe an ein spezifisches Protein gebunden (WARBURG und NEGELEIN). An spezifische Proteine ist das Hämin ferner in der Katalase (ZEILE und HELLSTRÖM, STERN), der Peroxydase (R. KUHN und Mitarbeiter, KEILIN und Mitarbeiter 1937, 1951, ELLIOT und KEILIN) und den Cytochromen (THEORELL 1935, 1936) verkettet. Eine andere Wirkungsgruppe ist das *Nicotinsäureamid*. Diese Pyridinverbindung kann bei der Atmung durch partielle Hydrierung Wasserstoff übertragen. Bei der Aufnahme von Wasserstoff bildet sich im langwelligen UV-Licht eine Bande aus und es zeigt sich zugleich eine bläulich-weiße Fluorescenz. Bei den Tieren spielt das Nicotinsäureamid als Vitamin eine Rolle. In der Zelle kommt es als Komponente zweiter Nucleotide vor, die als Diphosphopyridinnucleotid (DPN, Co I, Cozymase) (WARBURG und CHRISTIAN 1936a, b, c; MYRBÄCK 1933; EULER und Mitarbeiter 1935a, b, 1936) und Triphosphopyridinnucleotid (TPN, Co II) (WARBURG und CHRISTIAN 1934, 1935a, b, c) bekannt sind und in den Geweben von Pflanzen, Tieren und Mikroorganismen gefunden werden.

Die reversible Reduktion spielt sich am C-Atom 4 des Pyridinringes ab (PULLMAN, GOTTSCHALK). In der Zelle pendelt das Coenzym I zwischen mehreren Apoenzymen und vollführt dabei jedesmal einen bestimmten Oxydationsreduktionsschritt (WARBURG und CHRISTIAN 1939; WARBURG, CHRISTIAN und GRIESE 1935). Es ist beispielsweise an folgender enzymatischer Reaktion beteiligt:

$$\text{Acetaldehyd} + \text{Co I} - H_2 \rightleftharpoons \text{Äthanol} + \text{Co I}.$$

Bei dieser Reaktion tritt intermediär ein ternärer Komplex zwischen Apoenzym, Coenzym und Substrat auf, wobei Substrat und Coenzym an der Proteinoberfläche in engem Kontakt stehen. Für die Existenz eines solchen Verbandes spricht, daß der Komplex zwischen Apoenzym und Co I—H_2 ein Absorptionsspektrum besitzt, das etwas von dem Absorptionsspektrum des freien Co I—H_2 abweicht (THEORELL und CHANCE). Die direkte Übertragung von Wasserstoff vom Äthanol auf Co I konnte mittels 1,1-Dideuteroäthanol erwiesen werden (FISHER, CONN, VENNESLAND und WESTHEIMER). Es resultiert hierbei ein Co I—DH mit einem neuen Asymmetriezentrum. Wenn Co I auf chemischem Wege in Gegenwart

$$CH_3CD_2OH + \text{[Pyridinring, } CONH_2\text{, } N^+\text{–R]} \xrightleftharpoons[\text{der Alkohol-dehydrogenase}]{\text{Apoenzym}} \text{[Dihydropyridinring, C-4: D, H; } CONH_2\text{; N–R]} + CH_3CDO + H^+$$

von D_2O reduziert wird (z. B. durch Na-hydrosulfit), werden beide Lagemöglichkeiten von H und D am Kohlenstoffatom 4 realisiert, aber die Alkoholdehydrogenase reagiert nur mit einer bestimmten sterischen Konfiguration. Co I verbindet sich als dissoziables Coenzym auch mit verschiedenen anderen Apoenzymen in reversibler Weise.

Wenn chemisch reduziertes Monodeutero-Co I (Mischung beider stereoisomerer Konfigurationen bezüglich C-4) in Gegenwart von Milchsäuredehydrogenase mit Pyruvat reagiert, so bildet sich Lactat mit etwa 0,58 Atomen Deuterium je Molekül Lactat. Diese Beobachtungen sprechen für eine strenge dreidimensionale Orientierung des Coenzyms auf der Oberfläche des Apoenzyms, die zur Folge hat, daß nur auf *einer* Seite des Pyridinringes die Wasserstoffübertragung erfolgt. In diesem Beispiel wird Wasserstoff auf das Carbonyl-C der Brenztraubensäure übertragen, wobei es zur Bildung der l-Form von Milchsäure kommt (LOEWUS, OFNER, FISHER, WESTHEIMER und VENNESLAND).

Eine weitere Wirkungsgruppe ist das *Alloxazin*, welches ebenfalls in reversibler Weise Wasserstoff aufnehmen und abgeben kann.

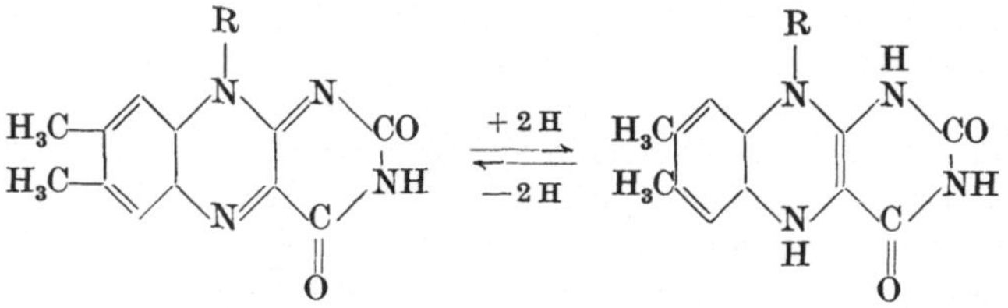

$$\text{Alloxazin (R; } H_3C, H_3C; \text{N, N, CO, NH, C=O)} \underset{-2\,H}{\overset{+2\,H}{\rightleftharpoons}} \text{Dihydroalloxazin (R; } H_3C, H_3C; \text{N, N-H, CO, NH, N-H, C=O)}$$

Diese Wirkungsgruppe hat eine gelbe Farbe und fluoresciert in blauem Licht gelbgrün; die reduzierte Wirkungsgruppe ist farblos und fluoresciert nicht. In Verbindung mit Ribose (Lactoflavin, Riboflavin) ist es als Wachstumsfaktor (Vitamin B_2) bekannt (KUHN, RUDY und WAGNER-JAUREGG). In den Zellen erscheint Alloxazin als Riboflavin-phosphorsäure und als Alloxazin-Adenin-Dinucleotid und gibt in Verbindung mit spezifischen Proteinen mehrere Fermente, die bei der Zellatmung den durch die hydrierten Pyridine transportierten Wasserstoff (Elektronen) an das katalytisch wirksame Eisen weitergeben (Flavinenzyme) (WARBURG 1953). Weitere mit Fermentproteinen vergesellschaftete Cofaktoren sind Aneurinpyrophosphorsäure (AUHAGEN, LOHMANN und SCHUSTER) als Bestandteil der Carboxylase und Transketolase, Pyridoxalpyrophosphorsäure (GUNSALUS und BELLAMY) als Cofaktor einer Anzahl in Bakterien gefundener Decarboxylasen und Transaminasen.

Die „*Transketolase*“ enthält Aneurinpyrophosphat als prothetische Gruppe (RACKER, DE LA HABA und LEDER; HORECKER, SMYRNIOTIS und KLENOW). Sie spaltet Ribulosephosphat in Glycerinaldehyd-3-phosphat und „aktiven Glykolaldehyd“. Dieser kann dann mit Ribose-5-phosphat zu Sedoheptulose-7-

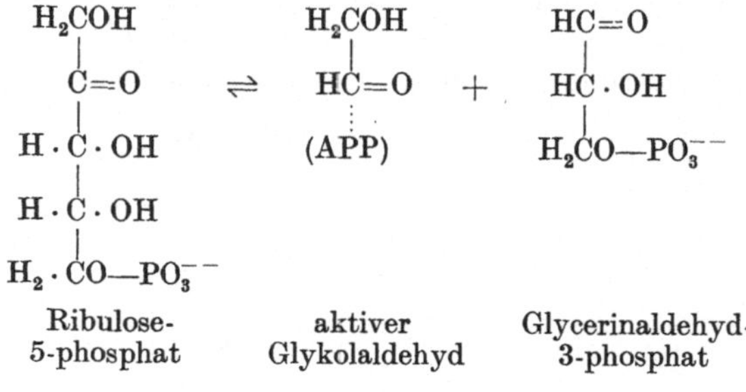

$$\begin{array}{ccccc}
H_2COH & & H_2COH & & HC{=}O \\
| & & | & & | \\
C{=}O & \rightleftharpoons & HC{=}O & + & HC\cdot OH \\
| & & \vdots & & | \\
H\cdot C\cdot OH & & (APP) & & H_2CO{-}PO_3^{--} \\
| & & & & \\
H\cdot C\cdot OH & & & & \\
| & & & & \\
H_2\cdot CO{-}PO_3^{--} & & & & \\
\text{Ribulose-} & & \text{aktiver} & & \text{Glycerinaldehyd-} \\
\text{5-phosphat} & & \text{Glykolaldehyd} & & \text{3-phosphat}
\end{array}$$

phosphat kondensieren (HORECKER, GIBBS, KLENOW und SMYRNIOTIS; BERNSTEIN, LENTZ, MALM, SCHAMBYE und WOOD).

$$\begin{array}{ccccc} H\cdot C{=}O & & H_2\cdot C\cdot OH & & H_2COH \\ | & & | & & | \\ H\cdot C\cdot OH & + & H\cdot C{=}O & \rightleftharpoons & C{=}O \\ | & & \vdots & & | \\ H\cdot C\cdot OH & & (APP) & & HO\cdot C\cdot H \\ | & & & & | \\ H\cdot C\cdot OH & & & & H\cdot C\cdot OH \\ | & & & & | \\ H_2CO{-}PO_3^{--} & & & & H\cdot C\cdot OH \\ & & & & | \\ & & & & H\cdot C\cdot OH \\ & & & & | \\ & & & & H_2CO{-}PO_3^{--} \end{array}$$

Ribose-5-phosphat — aktiver Glykolaldehyd — Sedoheptulose-7-phosphat

Pyridoxal und Pyridoxamin lassen sich ineinander überführen:

$$\text{Pyridoxal (4-CHO, 3-HO, 5-}CH_2OH\text{, 2-}H_3C\text{-Pyridin)} \underset{H_2O,\ O}{\overset{NH_3,\ H_2}{\rightleftharpoons}} \text{Pyridoxamin (4-}CH_2NH_2\text{, 3-HO, 5-}CH_2OH\text{, 2-}H_3C\text{-Pyridin)}$$

Pyridoxal — Pyridoxamin

Auf diesem Übergang beruht wahrscheinlich der Wirkungsmechanismus der Transaminasen.

$$\begin{array}{l} COOH \\ | \\ C{=}O \\ | \\ R \end{array} + H_2N\cdot CH_2\cdot X \rightleftharpoons \begin{array}{l} COOH \\ | \\ C{=}N\cdot CH_2X \\ | \\ R \end{array} \rightleftharpoons \begin{array}{l} COOH \\ | \\ CHN{=}CHX \\ | \\ R \end{array} \rightleftharpoons \begin{array}{l} COOH \\ | \\ CH\cdot NH_2 \\ | \\ R \end{array} + OHC\cdot X$$

α-Keto-glutarat — Pyridoxamin-pyrophosphat — Glutamat — Pyridoxal-pyrophosphat

Aneurin spielt im Tierreich und bei Pilzen als Vitamin B_1 bzw. Wuchsstoff, *Pyridoxal* als Derivat von 3-Oxy-4,5-di[oxymethyl]-2-methylpyridin eine Rolle, welches als *Adermin* ein Schutzstoff der Rattenpelagra ist. Zur Bildung von Citronensäure ist Oxalessigsäure und eine aktivierte Form der Essigsäure nötig. Eine solche Verbindung kann aus Zuckern über die Brenztraubensäure, aus der Oxydation von Fettsäuren, aus β-Ketosäuren wie Acetessigsäure, sowie aus Essigsäure entstehen; letztere kann durch eine Reaktion mit *Adenosintriphosphat* aktiviert werden. Das an diesen für den Zellstoffwechsel fundamentalen Umsetzungen beteiligte Coenzym wurde durch die Arbeiten von LIPMANN und Mitarbeitern, sowie SNELL und Mitarbeitern bekannt und führt den Namen *Coenzym A*. Es ist aufgebaut wie auch andere Cofermente vom Adenindinucleotidtypus, enthält aber die als Wuchsstoff bekannte Pantothensäure als Baustein, an der Thioäthanolamin, eine sulfhydrylhaltige Gruppe, gebunden ist. Co A wird leicht zu einem entsprechenden Disulfid oxydiert, oder es bildet mit anderen Thiolen Disulfide. In der oxydierten Form ist Co A biologisch unwirksam. Die „aktivierte Essigsäure" erwies sich nach LYNEN, REICHERT und RUEFF als ein energiereicher Thioester, das Acetylderivat von Co A, in dem der Acetylrest an die SH-Gruppe von Thioäthanolamin gebunden ist: $R{-}NH{-}CH_2\cdot CH_2{-}S{-}CO\cdot CH_3$. Co A ist bei Mikroorganismen und tierischen Organen weit verbreitet und ist diejenige Form, in der im Körper die Hauptmenge des Wuchsstoffes Pantothensäure untergebracht ist.

Am oxydativen Pyruvatabbau sind außer Co A noch drei weitere Cofaktoren beteiligt: Diphosphopyridinnucleotid (DPN), Aneurinpyrophosphat und ein weiterer S-haltiger Faktor, genannt „*α-lipoic-acid*" (Thioctic acid, Protogen A, Liponsäure). Dieser als Coenzym der Dehydrierungsreaktion funktionierende Faktor wurde als Wachstumsfaktor von Milchsäurebakterien entdeckt, ist aber in einer Vielzahl von tierischen und pflanzlichen Geweben, in Hefe- und Bakterienzellen gefunden worden, also weit verbreitet. Er ist kristallisierbar und stellt ein cyclisches Disulfid von 6,8-Di-mercapto-octansäure dar. Auch das Reduktionsprodukt 6,8-Di-mercaptooctansäure ist biologisch aktiv. Beobachtungen von REED und DE BUSK sprechen dafür, daß Liponsäure nicht in freier Form, sondern als Konjugat mit Aneurinpyrophosphat wirksam ist (Lipothiamid), wobei jene durch eine Säureamidbindung mit der NH_3-Gruppe des Aneurin verkettet ist. Die oxydative Decarboxylierung von Brenztraubensäure erfolgt unter Anteilnahme spezifischer Fermentproteine, wobei an den prosthetischen Gruppen folgender Umsatz abläuft[1]:

$$\underbrace{CH_3{-}CO{-}COO^-}_{\text{Brenztraubensäure}} + \underbrace{H_2C{-}(CH_2)_2{-}CH{-}(CH_2)_3{-}COO^-}_{\text{(S—S) α-Liponsäure — Aneurinpyrophosphat}}$$

$$\downarrow$$

$$H_2C{-}(CH_2)_2{-}CH{-}(CH_2)_3{-}COO^- \quad (\text{S—C(=O)—}CH_3;\ \text{S—}COO^-)$$

$$\xleftarrow{+\ \text{Co A—SH}}$$

$$H_2C{-}(CH_2)_2{-}CH{-}(CH_2)_3{-}COO^- \quad (\text{SH},\ \text{SH}) + \underbrace{H_3C{-}CO{-}S{-}CoA}_{\text{Acetyl-Coenzym A}} + CO_2$$

$$\text{DPN}^+ \longrightarrow \text{DPN}\cdot\text{H}$$

Der enzymatische Austausch zwischen dem Formylrest und C-Atom 2 der Inosinsäure wird durch den „*Citrovorum-factor*" (= Formyl-5,6,7,8-tetrahydrofolinsäure) gefördert. Dieser Faktor wirkt als Coenzym bei der Transformylierungsreaktion (GREENBERG).

Daß Fermente zu ihrer Wirkung Cofaktoren benötigen, ist schon lange bekannt. Die Bezeichnung Coferment stammt von BERTRAND; er fand, daß zur Aktivierung von Laccase Mn-Salze notwendig sind. Die klassischen Untersuchungen von HARDEN und YOUNG haben gezeigt, daß sich das von BUCHNER aus Hefezellen isolierte, katalytisch wirksame Prinzip der alkoholischen Gärung in eine niedermolekulare, als „Co-Zymase" bezeichnete, und eine hochmolekulare Komponente zerlegen läßt, die beide für sich allein unwirksam, vereint aber wieder wirksam werden. In der Zwischenzeit wurde aufgeklärt, daß das komplette System der Gärung einer Hefezelle aus 13 Fermentproteinen und 7 verschiedenartigen Cofaktoren besteht.

Wie es in der Wissenschaft bisweilen vorkommt, wurden diese Befunde zu frühzeitig verallgemeinert und die Fermente als „Symplexe" angesprochen, die aus einem hochmolekularem Träger und einer niedermolekularen prosthetischen Gruppe bestehen *(dualistische Fermenttheorie)*. Man ist sogar so weit gegangen, daß man die prosthetische Gruppe als das eigentliche fermentative Prinzip (Agon) und das hochmolekulare Protein als Träger (Pheron), also als etwas verhältnismäßig Nebensächliches betrachtet. Sobald aber eingehende Informationen vorlagen, die für die Proteinnatur der Fermente sprachen, billigte man allmählich auch dem Fermentprotein eine wichtige Rolle zu, hielt aber an dem dualistischen Prinzip fest. Aber heute stößt eine Verallgemeinerung des dualisti-

[1] Aus FLASCHENTRÄGER-LEHNARTZ, Bd. II/1b, S. 1039; dort weitere Literatur.

schen Prinzips bereits auf Schwierigkeiten. Bei einer Anzahl sehr gründlich untersuchter Fermente hat man keine besonderen prosthetischen Gruppen finden können. In Anbetracht der Vielfalt der Bausteine in einem Proteinmolekül bietet zur Untersuchung dieser Frage die Analyse von Fermentproteinen mit niederem Molekulargewicht, wie z. B. der Ribonuclease von KUNITZ, besondere Vorteile. Der Aufbau dieses Moleküls ist heute bereits so gut bekannt, daß das generelle Vorkommen von prosthetischen Gruppen sehr unwahrscheinlich geworden ist. Man kann also schon mit gutem Grund behaupten, daß es neben Fermenten mit prosthetischen Gruppen auch solche ohne diese gibt. Wenn auch den Cofermenten bei der Übertragung von Elektronen, Wasserstoffatomen und Acetyl- oder Phosphatgruppen zweifellos eine wichtige Rolle zukommt, so ist doch ihre spezifische Reaktionsfähigkeit nur durch ihre Bindung an ein natives und spezifisches Zellprotein ermöglicht.

III. Kinetik der Fermentreaktionen.

Kinetischen Studien an biologischen Objekten steht in den weitaus meisten Fällen die hochgradige Komplikation dieses Materials im Wege. Das hauptsächliche Problem liegt dann in der Frage, wie weit man den vorliegenden Fall vereinfachen darf, ohne den engeren Zusammenhang mit dem natürlichen, nicht vereinfachten Prozeß zu verlieren, wie STEARN (1949) in einem Referat über Kinetik äußerte. Heute wird die Kinetik auch auf dem Gebiete der Enzymologie hauptsächlich an zwei verschiedenen Typen von biologischem Material untersucht: 1. an *isolierten und weitgehend gereinigten Enzymen*; dabei werden Fragen wie z. B. die katalytische Reaktion der Enzyme, die Inaktivierung und Denaturierung von Enzymproteinen u. dgl. bearbeitet. 2. am *lebenden Organismus* selbst; Beispiele hierfür sind Untersuchungen über die stationäre Konzentration von Metaboliten und zahlreiche Arbeiten über die Prozesse der Atmung und der Photosynthese, der Morphogenese und Regeneration. Man hat mitunter versucht, solche komplexe Vorgänge wie einfache chemische Reaktionen zu behandeln; offenbar sind derartige Studien durchaus geeignet, konkretere Vorstellungen über den Ablauf solcher Prozesse zu bahnen.

A. Thermodynamische Voraussetzungen.

Nach dem *Satz von der Erhaltung der Energie* (I. R. MAYER 1842) bleibt die Gesamtenergie in einem geschlossenen System bei allen Energieumwandlungen konstant. Der Satz gilt auch für den Ablauf chemischer Reaktionen sowie für alle Stoffumsetzungen im lebenden Körper.

Man kann schreiben:

$$\underset{\substack{\text{Zufuhr an}\\ \text{gesamter Energie}}}{\Delta U} = \underset{\substack{\text{Zufuhr an}\\ \text{freier Energie}}}{\Delta F} + \underset{\substack{\text{Zufuhr an}\\ \text{thermischer Energie}}}{\Delta Q} \tag{1}$$

ΔF und ΔQ werden nach Übereinkommen positiv gezählt, wenn die Energiebeträge dem System zugeführt werden, und negativ, wenn das System mechanische oder thermische Energie nach außen abgibt. Auch bei ΔU wählt man ein positives Vorzeichen, wenn der Energiebetrag dem System zugeführt wird, und ein negatives, wenn er entnommen werden kann.

Nach dem 2. Hauptsatz der Thermodynamik kann zwar Arbeit stets in Wärme verwandelt werden, aber von der Wärme nur ein Teil in Arbeit überführt werden. Es gilt

$$dF = \Delta Q \frac{dT}{T}. \tag{2}$$

Aus (1) und (2) ergibt sich die thermodynamische Grundgleichung

$$\Delta F = \Delta U - T\frac{dF}{dT} = \Delta U - T \cdot \Delta S, \tag{3}$$

wobei $\frac{dF}{dT}$ den Temperaturkoeffizienten der maximalen Arbeit (= Entropie, S) bedeutet.

Damit aus einem Vorgang die *maximale Arbeit* gewonnen werden kann, sind 2 Bedingungen erforderlich: Der Vorgang muß *isotherm* verlaufen und muß *reversibel* geleitet werden, damit jeder Teilschritt der Reaktion rückgängig und der ganze Prozeß als Kreisprozeß verlaufen kann. Die zunächst gewonnene maximale Arbeit ist es, die den Prozeß auf späterer Phase wieder in den Ausgangszustand zurückbringen kann. Wenn Energie aus dem System nach außerhalb abfließt, so verläuft der Prozeß irreversibel und bedarf zu seiner Rückführung zusätzlicher Arbeit.

B. Reversibilität chemischer Reaktionen, Gleichgewicht.

Reversible Reaktionen können in der Richtung sowohl von rechts nach links, als auch von links nach rechts verlaufen.

$$A + B \underset{k_2}{\overset{k_1}{\rightleftharpoons}} C + D. \tag{4}$$

Bei konstanter Temperatur ist die Geschwindigkeit einer chemischen Reaktion durch die gegebene Konzentration der Partner festgelegt. Auch die Richtung der Reaktion hängt von der Konzentration der reagierenden Komponenten ab (Massenwirkungsgesetz). Wenn n_1 Moleküle des Stoffes A mit n_2 Molekülen des Stoffes B unter Bildung von n_1' Molekülen des Stoffes C und n_2' Molekülen des Stoffes D reagieren und die entsprechenden Konzentrationen C_1, C_2 bzw. C_1', C_2' sind, so herrscht Gleichgewicht, sofern die Geschwindigkeiten der beiden, in entgegengesetzte Richtung enlaufenden Reaktionen gleich geworden sind. Es gilt dann:

$$\frac{C_1'^{n_1'} \cdot C_2'^{n_2'}}{C_1^{n_1} \cdot C_2^{n_2}} = \frac{k_1}{k_2} = K, \tag{5}$$

wobei K die Gleichgewichtskonstante bei gegebener Temperatur darstellt. Da diese Konstante stets einen bestimmten endlichen Wert hat, läuft in der Natur keine Reaktion vollständig nach einer bestimmten Richtung ab. Je nach Wahl der Konzentration der Teilglieder läuft jede reversible Reaktion nach Gleichung (4) entweder von rechts nach links oder umgekehrt ab, bis die jeweilige Gleichgewichtslage eingestellt ist.

C. Maximale Arbeit chemischer Reaktionen.

Eine Reaktion läuft freiwillig nur in Richtung nach dem thermodynamischen Gleichgewicht ab, wobei freie Energie verfügbar wird.

Bringt man bei isothermer und reversibler Leitung des Vorganges 1 Mol einer Substanz A von der Konzentration C_1 auf die Konzentration C_2, so ist der maximale Arbeitsgewinn $\Delta F = -RT \ln \frac{C_1}{C_2}$. Der reversible Umsatz an freier Energie in einem System nach Gleichung (4) ist von der Konzentration der Reaktionsteilnehmer abhängig und errechnet sich nach VAN'T HOFF zu

$$\Delta F = -RT\left[\ln \frac{C_1'^{n_1'} \cdot C_2'^{n_2'}}{C_1^{n_1} \cdot C_2^{n_2}} - \ln K\right]. \tag{6}$$

K bedeutet hierbei die Gleichgewichtskonstante, R die Gaskonstante und T die absolute Temperatur. Mit $C_1'^{n_1}$, $C_2'^{n_2}$ usw. ist die Konzentration der Reaktionsteilnehmer zu Beginn der Umsetzung gemeint.

Wir betrachten die Umwandlung eines Mols des Stoffes A in der Konzentration C_1 in 1 Mol eines Stoffes B in der Konzentration C_2. Wenn das Reaktionssystem in den Gleichgewichtszustand kommt, sollen die Konzentrationen C_1' und C_2' betragen. Setzt man das Verhältnis der Anfangskonzentration $\frac{C_1}{C_2} = x$, das Verhältnis der Endkonzentration $\frac{C_1'}{C_2'} = K$, so gilt

$$\Delta F = -RT(\ln K - \ln x) = -RT \ln \frac{K}{x}. \tag{7}$$

Aus (7) ersieht man, daß die maximale Arbeit, die bei einer freiwillig ablaufenden Reaktion anfällt, um so größer sein wird, je weiter das System vom Gleichgewichtszustand entfernt wird. Im Gleichgewicht selbst ist die Arbeitsleistung gleich Null. L. v. BERTALANFFY hat darauf hingewiesen, daß die chemischen Systeme im Organismus, der ständig arbeitsfähig sein muß, nicht im Gleichgewichtszustand verharren können. ,,Das scheinbare Gleichgewicht, das wir im Organismus vorfinden, ist daher nicht ein echtes und daher arbeitsunfähiges, sondern ein *Fließgleichgewicht*, das in einem gewissen Abstand vom wahren konstant erhalten wird, daher arbeitsfähig ist, andererseits aber zur Aufrechterhaltung der Distanz vom wahren Gleichgewicht ständig neu zugeführter Energien bedarf" (p. 28). Führt man die Umsetzung mit je 1 Mol der beteiligten Reaktionskomponenten durch, so geht Gleichung (6) in folgende sehr einfache Form über:

$$\Delta F = -RT \ln K = -4{,}58 \cdot T \cdot \log K. \tag{8}$$

Ist zu Beginn der Umsetzung die Konzentration der Reaktionsteilnehmer nur wenig von der Konzentration im Gleichgewicht verschieden, so ist die freie Energieänderung nach (6) klein; eine solche Reaktion ist durch Konzentrationsveränderungen der Reaktionsteilnehmer leicht umkehrbar zu gestalten. Die Umkehrung wird aber um so schwieriger, je größer die freie Energie des Umsatzes ist, bzw. je stärker die Konzentration der Teilnehmer zu Reaktionsbeginn von den Gleichgewichtskonzentrationen verschieden ist. Das folgende Beispiel (aus FLASCHENTRÄGER-LEHNARTZ, Bd. I, S. 993) erläutert eindrucksvoll, welchen Einfluß die energetischen Verhältnisse auf die Lage eines chemischen Gleichgewichtes nehmen.

Wir betrachten die hydrolytische Reaktion:

$$A + H_2O \rightleftharpoons C + D; \qquad \frac{(C)(D)}{(A)(H_2O)} = K. \tag{9}$$

Wählt man für A einen energiearmen Phosphorsäureester, wie z. B. Hexosephosphat oder Glycerinphosphat, so wird bei der Hydrolyse eine Energie ($-\Delta F$) von etwa 3000 cal/Mol frei. Aus Gleichung (8) erhält man für 25° C: $-3000 = -4{,}58 \times 298 \times \log K$. Es folgt $K = 158$. Da ein Liter Wasser etwa 56 Mole H_2O enthält und $(C) = (D)$ ist, ergibt sich aus Gleichung (9)

$$K = \frac{(C)^2}{56\,(A)},$$

oder

$$(C) = \sqrt{56 \cdot K \cdot (A)} = 7{,}5 \cdot 12{,}6 \cdot \sqrt{(A)}.$$

Setzt man für $(A) = 0{,}001$ Mol/Liter, so erhält man für $(C) = 3{,}45$ Mole/Liter. Das bedeutet aber, daß im Gleichgewicht die Konzentration der Spaltstücke etwa 3000mal höher ist als die von (A). Man müßte schon sehr hohe Konzentrationen von C und D anwenden, wenn man auf enzymatischem Wege eine praktisch ins Gewicht fallende Synthese von (A) erzielen wollte.

Handelt es sich bei A jedoch um einen energiereichen Phosphorsäureester, so beträgt die bei der Hydrolyse freiwerdende Energie 10000—12000 cal/Mol. Setzt man diesen hohen Betrag in Gleichung (8) ein, so kann man sich überzeugen, daß das Gleichgewicht noch sehr viel stärker nach rechts verschoben wird. Reaktionen dieser Art sind daher unter biologischen Bedingungen praktisch nicht mehr umkehrbar. Der Organismus verwendet zur Synthese von Estern im allgemeinen nicht die einfache Umkehrung der Hydrolyse, sondern einen grundsätzlich anderen Weg.

Eine umkehrbare Reaktion läßt sich auch in die thermodynamisch nicht begünstigte Richtung lenken, wenn man das entstehende Produkt laufend aus dem System entfernt. Man macht von dieser Möglichkeit z. B. bei der Bestimmung von Metaboliten mittels des „optischen Testes" Gebrauch.

Die Alkoholdehydrogenase katalysiert die Reaktion:

$$C_2H_5OH + DPN^+ \xrightleftharpoons{\text{Alkoholdehydrogenase}} CH_3CHO + DPN \cdot H + H^+.$$

Obwohl das Gleichgewicht der Reaktion natürlicherweise sehr stark zugunsten des Äthanols verschoben ist, kann man dennoch diese Umsetzung im Laboratorium z. B. zur Blutalkoholbestimmung gebrauchen. Durch Abfangen des entstehenden Acetaldehyds mit Semicarbazid verlagert sich das Gleichgewicht praktisch vollständig nach rechts, so daß in Gegenwart eines Enzymüberschusses auch sehr kleine Mengen an Äthanol quantitativ zu Acetaldehyd dehydriert werden. Das im Verlaufe der Reaktion stöchiometrisch gebildete DPN · H kann infolge seiner selektiven Absorption bei 340 mμ photometrisch bestimmt werden.

Auch im Organismus laufen Umsetzungen in der Gegenrichtung ab, wenn die Reaktionsprodukte schon im Moment ihres Entstehens weiter reagieren. Wie schon bemerkt, werden sich im allgemeinen in der lebenden Zelle echte Gleichgewichte wohl nur sehr selten vorfinden (s. S. 503).

D. Die Reaktionsgeschwindigkeit.

Ein reagierendes System von Substanzen muß auf seinem Wege vom Beginn der Reaktion bis zum Auftreten der Reaktionsprodukte mindestens einen gewissen Energiebetrag E aufnehmen, um den Energieberg zu überwinden, der der Reaktion entgegensteht. Diese freie Energie der Aktivierung ist ein entscheidender Faktor der Reaktionsgeschwindigkeit. Nach L. MICHAELIS (1949) entspricht die Aktivierungsenergie bei den meisten chemischen Reaktionen der zur Ausbildung radikalartiger Zwischenstadien nötigen Energie (Resonanzenergie). Oxydoreduktionen verlaufen z. B. stets unter Aufnahme bzw. Abgabe von zwei Elektronen, die nur nacheinander, also in zwei Phasen, ausgetauscht werden können. Der *aktivierte Zustand* entspricht der radikalartigen energiereichen Zwischenver-

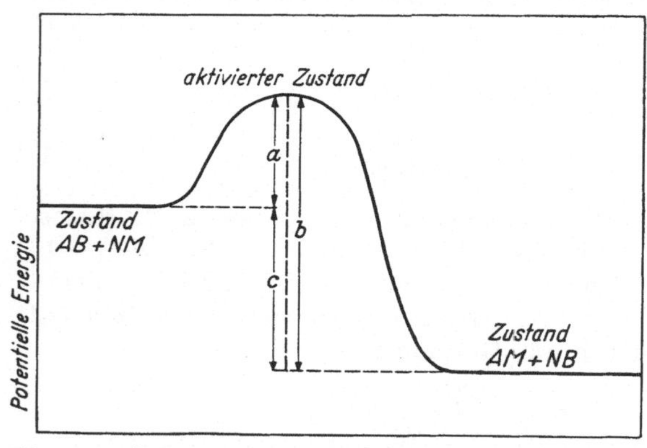

Abb. 1. Die Änderung der potentiellen Energie im Verlaufe einer chemischen Reaktion. (Nach BERSIN 1938.) a Aktivierungsenergie für die exotherme Reaktion $AB + NM = AM + NB$. b Reaktionswärme.

bindung nach Austausch *eines* Elektrons. Die Abb. 1 zeigt die freien Energien einer Reihe von Konfigurationsstufen des reagierenden Systems zwischen dem Anfangs- und Endstadium. Die Ordinate des Diagramms bedeutet den Betrag der Aktivierungsenergie. Das System in seinem Zustand am Gipfel des Energieberges repräsentiert den „*aktivierten Komplex*". Ungeachtet der Art der zugrunde liegenden Reaktion hängt die spezifische Abbaugeschwindigkeit des aktivierten Komplexes in die Endprodukte der Reaktion nur von der Temperatur ab.

a) Einfluß der Temperatur.

Die Beziehungen zwischen der Reaktionsgeschwindigkeit, Aktivierungsenergie und Temperatur beschreibt die Gleichung von ARRHENIUS:

$$\frac{d \ln k}{d T} = \frac{E}{R T^2}, \qquad (10)$$

k Konstante der Reaktionsgeschwindigkeit; T absolute Temperatur; R Gaskonstante; E Aktivierungsenergie der Moleküle, d. h. das Minimum an kinetischer Energie, die ein Molekül haben muß, um den Energieberg zu überwinden und die Reaktion zu ermöglichen. Dieser Wert wird häufig auch mit „μ" bezeichnet. Die Gleichung besagt, daß die Veränderung des natürlichen Logarithmus der Reaktionskonstanten dem Quadrat der absoluten Temperatur umgekehrt proportional ist. Bezüglich einer Ableitung der Gleichung aus der statistischen Mechanik sei auf STEARN (1938, 1949) verwiesen; daselbst findet sich auch die ältere Literatur.

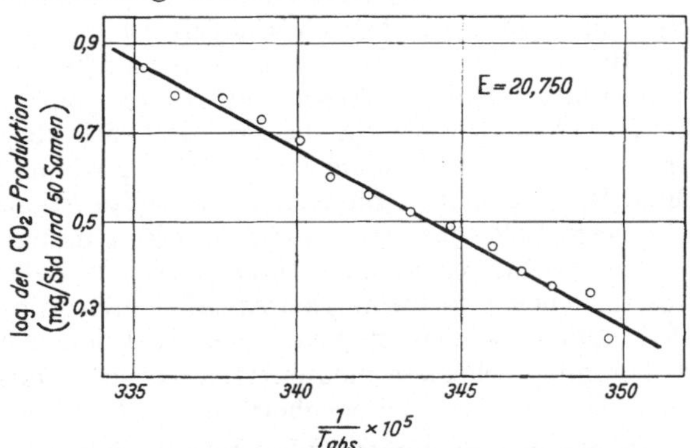

Abb. 2. Anwendung der ARRHENIUSschen Formel auf die CO_2-Produktion durch keimende Maissamen. (Nach TANG aus v. BERTALANFFY 1942.)

Die Integration ergibt:

$$k = C\, e^{-E/RT}, \qquad (11)$$

wobei C ein Maß der Wahrscheinlichkeit ist, daß aktive Moleküle Endprodukte der Reaktion ergeben und $e^{-E/RT}$ den Anteil der Moleküle angibt, die einen Energieüberschuß von E oder mehr besitzen. Man kann die Gleichung auch schreiben:

$$\log k = \frac{-E}{2{,}303\, R} \frac{1}{T} + \log C. \qquad (12)$$

Zur Ermittlung der Größe der Aktivierungsenergie (E) trägt man die Werte von $\log k$ gegen $1/T$ auf; die Neigung (tg α) der statistisch ermittelten „besten" Geraden durch die Versuchspunkte gleicht dem Wert von $\frac{-E}{2{,}303\, R}$; daraus läßt sich E berechnen (Abb. 2). Die Konstante E hat einen spezifischen Charakter, der sich darin äußert, daß sich die bei sehr verschiedenartigen Reaktionsprozessen ermittelten Werte von E um ganz bestimmte Zahlenwerte gruppieren. So ist z. B. für die Oxydation von Fe^{++} zu Fe^{+++} ein Wert von 16000—16500 charakteristisch. Biologische Vorgänge sind stets komplizierte Kettenreaktionen, an denen eine sehr große Zahl gleichzeitig ablaufender Reaktionen beteiligt ist. Wenn sich daher bei verschiedenen Vorgängen am häufigsten Werte wie 8000, 11000 und 16000 feststellen lassen (CROZIER; HOAGLAND; HADIDIAN und HOAGLAND), so bedeutet das, daß diese Zahlen auf die jeweils langsamste Reaktion zu beziehen sind, die sich in der betreffenden Kettenreaktion als „Schrittmacher“ auswirkt.

Aus Gleichung (12) folgt:

$$\log \frac{k_{2(T_2)}}{k_{1(T_1)}} = \frac{E}{2{,}303\,R} \cdot \frac{T_2 - T_1}{T_1\,T_2}. \qquad (13)$$

Setzt man E gleich 12000 cal — einem sehr häufig vorkommenden Wert — und T_1, T_2 gleich 295° bzw. 305° absolut, dann ist

$$\log \frac{k_2}{k_1} = \frac{12000}{4{,}6} \cdot \frac{10}{295 \cdot 305} = 0{,}29.$$

Wie man sieht, kommt das Verhältnis der beiden für den Temperaturunterschied von 10° zutreffenden Reaktionsgeschwindigkeiten (k_2/k_1) dem Wert 2 sehr nahe. Man erhält auf diesem Wege eine Bestätigung der bekannten VAN'T HOFF*schen Regel* (R.G.T.-Regel), die besagt, daß sich die Reaktionsgeschwindigkeit bei einer Temperaturerhöhung von 10° für gewöhnlich verdoppelt bis verdreifacht. Diese Regel gilt auch für enzymatische Reaktionen und zahlreiche enzymatisch gesteuerte Lebensprozesse, wie z. B. die Atmung oder das Wachstum. Dagegen besitzen Diffusionsvorgänge, die ebenfalls im Zellstoffwechsel eine wichtige Rolle spielen, einen Geschwindigkeitskoeffizienten von nur 1 bis höchstens 1,5. Die Aktivierungswärmen enzymatischer Reaktionen liegen im allgemeinen niedriger als bei entsprechenden nichtenzymatischen Reaktionen, wie Tabelle 3 zeigt, was ein Licht auf die Überlegenheit der biologischen Katalyse wirft.

In manchen Systemen ändert sich E mit der Temperatur. So hat E im System Urease-Harnstoff in einem schwach oxydierenden Milieu über den ganzen

Tabelle 3. *Aktivierungsenergien einiger enzymatischer und nicht enzymatisch katalysierter Reaktionen.* (Aus FLASCHENTRÄGER-LEHNARTZ, Bd. 1, S. 998.)

Reaktion	Katalysator	E cal/Mol
H_2O_2-Zersetzung	ohne	18000
	kolloidales Platin	11000
	Leberkatalase	5000
Rohrzuckerinversion. . .	Wasserstoffionen	26000
	Hefeinvertase	11500
	Malzinvertase	13000
Caseinhydrolyse.	HCl	20600
	kristallisiertes Trypsin	12000
	kristallisiertes Chymotrypsin	12000
Äthylbutyrat-Hydrolyse .	Wasserstoffionen	13200
	Pankreaslipase	4200

Temperaturbereich den Wert 11700, in einem schwach reduzierenden Milieu hingegen den Wert von 8700. Bei mittlerem Oxydations-Reduktionspotential hat E bei tiefen Temperaturen den Wert 11700, aber bei 22° zeigt die Kurve einen Knick, um im höheren Temperaturbereich den Wert 8700 zu zeigen. Die Ursache hierfür dürften Modifikationen des Enzymmoleküls sein (SIZER; STEARN 1949). Auch bei anderen enzymatischen Umsetzungen beobachtete man bei der graphischen Ermittlung von E einen scharfen Knick der Linie bei einer ganz bestimmten Temperatur, der 2 Gebiete trennt, die bedeutend verschiedenen E-Werten entsprechen. STEARN (1949) vermutet, daß bei der Übergangstemperatur ein Wechsel in der Schrittmacherfunktion zwischen der Geschwindigkeitskonstante der Bildungsreaktion des Enzym-Substrat-Komplexes und der Geschwindigkeitskonstante des Zerfalles desselben in die Produkte eintritt.

Außer Temperatur und Redoxpotential können in bestimmten Fällen auch spezifische Reagentien einen Einfluß auf den Wert von E nehmen, ohne die Größenordnung der Reaktionsgeschwindigkeit zu verändern. STEARN (1949) bespricht z. B. die Beobachtung von HADIDIAN und HOAGLAND, daß die Zugabe einer kleinen Menge von Cyanid den E-Wert bezüglich des O_2-Verbrauches in einen Succinat-Succinodehydrogenase-Cytochrom-Cytochromoxydase-System des Rinderherzens nicht verändert. Aber ein etwas größeres Quantum an Cyanid ändert den E-Wert von 11200 in 16000. Cyanid hemmt die Wirkung der Cytochromoxydase.

$$\text{Bernsteinsäure} \xrightarrow[\text{Succinodehydrogenase}]{\overset{\text{A}}{E\,=\,11200}} \text{Fumarsäure} + 2\,\text{H} + \text{O} \xrightarrow[\text{Cytochrom-Cytochromoxydase}]{\overset{\text{B}}{E\,=\,16000}} H_2O\,.$$

Im normalen System ist die Reaktion A langsamer; sie gibt den Schrittmacher, und ihr E gilt für den ganzen Ablauf. Wenn aber ausreichend Cyanid gegeben wird, um die Reaktion B zum Schrittmacher zu verlangsamen, verändert sich das E des ganzen Prozesses auf 16000. Eine nachträgliche Gabe von Selenit, welches die Dehydrogenase selektiv hemmt, bringt wieder den E-Wert von 11200 zutage.

In der Regel ist aber der Wert von E für Enzym-Substrat-Systeme von vielen Umweltfaktoren wie p_H, Ionenstärke usw. unabhängig, welche die Reaktionsgeschwindigkeit deutlich beeinflussen. Die Wirkung dieser Faktoren erstreckt sich hauptsächlich auf die Aktivierungsentropie (s. STEARN 1949). So dürfte auch eine Verunreinigung der Enzympräparate den E-Wert nicht beeinflussen, es sei denn, daß es sich um spezifische Gifte handelt.

CROZIER dachte erstmalig daran, daß ein bestimmter E-Wert einem bestimmten Enzymsystem zugeordnet sein könnte, und erkannte die sich hieraus ergebende Möglichkeit, in einem physiologischen Prozeß dasjenige Enzymsystem zu erkennen, das die Gesamtgeschwindigkeit dieses Vorgangs bestimmt. Er hat zahlreiche Prozesse daraufhin studiert und zumeist E-Werte gefunden, die den bei oxydativen Reaktionen gefundenen zahlenmäßig gleichen. Fraglich bleibt hierbei allerdings, ob der E-Wert eines komplexen physiologischen Prozesses mit dem einer einfachen katalytischen Reaktion verglichen werden darf (STEARN 1949; BOOIJ und WOLVERKAMP; BURTON).

b) Einfluß der Substratkonzentration.

Variiert man bei gegebener Enzymmenge die Substratkonzentration unter gleichbleibenden sonstigen Reaktionsbedingungen, so findet man, daß die Initialgeschwindigkeit der enzymatischen Umsetzung in der Regel mit steigender Substratkonzentration (S) ansteigt bis zu einem Grenzwert, dessen Größe von der gewählten Enzymkonzentration abhängt und dieser proportional ist (Abb. 3). Trägt man die Initialgeschwindigkeit gegen log (S) auf, so resultiert eine S-förmige Kurve, die eine gewisse Ähnlichkeit mit der Dissoziationskurve eines organischen Elektrolyten hat (Abb. 4).

Diese Beobachtungen sind die Grundlage der Hypothese von MICHAELIS und MENTEN, die wie folgt lautet: Jeder enzymatische Prozeß durchläuft 3 Phasen. In der 1. Phase bildet das Enzymmolekül mit dem Substratmolekül durch direkten Kontakt einen dissoziablen, instabilen *Enzym-Substrat-Komplex*. In der 2. Phase kommt es zu einer Veränderung der Bindungsverhältnisse im Substratmolekül, wobei der spezifischen Natur des Enzyms und dem speziellen Mechanismus der Anheftung des Substratmoleküls an das Enzym bei der Bildung des Enzym-

Substrat-Komplexes eine entscheidende Rolle zukommt[1]. Infolge dieser Bindungslockerung im Substratmolekül läuft die Umsetzung ab. In der 3. Phase erfolgt die Abwanderung der Spaltprodukte. Man kann den Prozeß wie folgt skizzieren:

$$\underset{(E)-(ES)}{\text{Freies Enzym}} + \underset{(S)}{\text{Substrat}} \underset{k_2}{\overset{A \atop k_1}{\rightleftharpoons}} \underset{(ES)}{\text{Enzym-Substrat-Komplex}} \overset{B \atop k_3}{\longrightarrow} \text{freies Enzym} + \text{Produkte.}$$

Bezeichnet man mit (E), (S), $(E\,S)$ die Konzentration (in Mole je Liter) an *totalem Enzym*, Substrat bzw. Enzym-Substrat-Komplex, dann ist die Konzentration an freiem Enzym $[(E)-(E\,S)]$. Für die Bildungsgeschwindigkeit des Enzym-Substrat-Komplexes gilt:

$$\frac{d\,(E\,S)}{d\,t} = k_1\,(S)\,[(E)-(E\,S)] - (k_2+k_3)\,(E\,S). \tag{14}$$

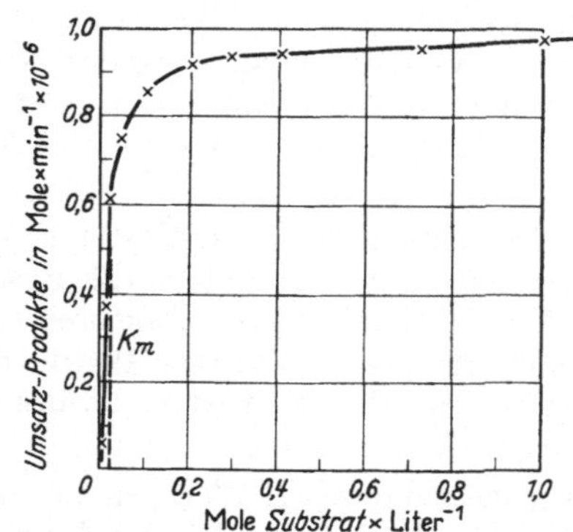

Abb. 3. Die Abhängigkeit der Anfangsgeschwindigkeit einer enzymatischen Reaktion von der Substratkonzentration. Die Substratkonzentration wird linear aufgetragen. (Nach Daten von MICHAELIS und MENTEN.)

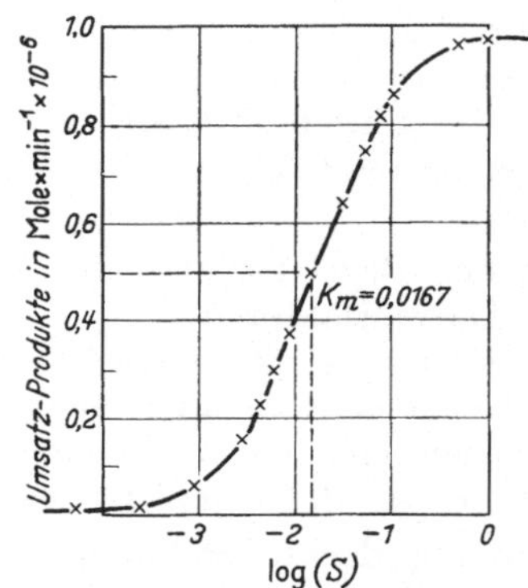

Abb. 4. Die Abhängigkeit der Anfangsgeschwindigkeit einer enzymatischen Reaktion von der Substratkonzentration. Die Substratkonzentration wird logarithmisch aufgetragen.

Für das Verschwinden von Substrat gilt:

$$\frac{d\,(S)}{d\,t} = -\,k_1\,(S)\,[(E)-(E\,S)] + k_2\,(E\,S). \tag{15}$$

Wie die Skizze des Vorganges zeigt, bildet sich bei der enzymatischen Aufspaltung des Substrates ein *Fließgleichgewicht* aus, das wie folgt charakterisiert werden kann:

$$(E\,S) = (E\,S)_{\max}; \quad \frac{d\,(ES)}{dt} = 0.$$

So ergibt sich aus (14), daß $(E\,S)_{\max} = \frac{(E)\,(S)}{K_m + (S)}$, wobei

$$K_m = \frac{(k_2+k_3)}{k_1} = \frac{(S)\,[(E)-(E\,S)_{\max}]}{(E\,S)_{\max}}. \tag{16}$$

Durch Addition der Gleichungen (14) und (15) erhält man für das Fließgleichgewicht:

$$\frac{d\,(E)}{dt} = -\,k_3\,(E\,S)_{\max}. \tag{17}$$

Aus (17) kann man k_3 erhalten.

[1] Es handelt sich dabei um eine echte chemische Wechselwirkung zwischen funktionellen Gruppen des Substrates und der prosthetischen Gruppe des Enzyms (Näheres bei GOTTSCHALK).

K_m wurde ursprünglich von MICHAELIS und MENTEN als Dissoziationskonstante des Enzym-Substrat-Komplexes betrachtet; sie wird MICHAELIS-MENTEN-*Konstante* genannt. Sie drückt aber nur formal ein Gleichgewicht aus, in Wirklichkeit ist sie dem Fließgleichgewicht der beiden aufeinanderfolgenden Reaktionen A und B zugeordnet, von denen die erste reversibel ist (s. Schema). Nur wenn $k_3 \gg k_2$ ist, gleicht K_m dem Quotienten k_2/k_1, in den übrigen Fällen entspricht K_m der komplizierteren Funktion (16). So erklärt sich auch der Befund, daß K_m mit der Temperatur ansteigen und mit dem p_H-Wert variieren kann, mithin Eigenschaften des ganzen Systems trägt.

Da nur in verhältnismäßig seltenen Fällen die molaren Konzentrationen des im Ansatz enthaltenen Enzyms und des Enzym-Substrat-Komplexes bekannt sein dürften oder ermittelt werden können, müssen diese Glieder zur Berechnung von K_m durch bekannte Größen ersetzt werden.

Bezeichnet man mit (v) die Reaktionsgeschwindigkeit (Anfangsgeschwindigkeit), mit welcher der Enzym-Substrat-Komplex aufgespalten wird, so gilt

$$v = k_3\,(E\,S)_{\max}. \qquad (18)$$

Bei großem Substratüberschuß (Sättigung der Enzymoberfläche mit Substrat) werden praktisch alle Enzymmoleküle im Enzym-Substrat-Komplex eingeschlossen sein; (ES) erreicht daher unter diesen Bedingungen den Wert von (E) und die Reaktionsgeschwindigkeit (v) einen Grenzwert (V). Bei großem Substratüberschuß gilt also:

$$V = k_3\,(E) \quad \text{und} \quad \frac{v}{V} = \frac{(E\,S)}{(E)}. \qquad (19)$$

Die Gleichung (16) läßt sich wie folgt umformen:

$$(S)\,[(E) - (E\,S)] = (E\,S)\cdot K_m,$$

$$(S)\,(E) = (S)\,(E\,S) + (E\,S)\,K_m = (E\,S)\,[(S) + K_m],$$

$$(S) = \frac{(E\,S)}{(E)}\,[(S) + K_m].$$

Setzt man für $(ES)/(E)$ den Wert v/V aus Gleichung (19) ein, so kommt man zur MICHAELIS-MENTEN-*Gleichung:*

$$(S) = \frac{v}{V}\,[(S) + K_m] \qquad (20a)$$

oder

$$v = \frac{V\,(S)}{(S) + K_m} \quad (\text{Mole} \times \text{Liter}^{-1} \times \text{min}^{-1}). \qquad (20b)$$

Die MICHAELIS-MENTEN-Gleichung gibt die Abhängigkeit der Reaktionsgeschwindigkeit von der Substratkonzentration zahlenmäßig wieder; sie entspricht formal der Gleichung einer rechtwinkeligen Hyperbel mit folgenden Eigenschaften:

a) Die Grenzgeschwindigkeit (V) der enzymatischen Reaktion entspricht der Asymptote, nach der die Reaktionsgeschwindigkeit (v) strebt, wenn die Substratkonzentration wächst.

b) Die MICHAELIS-Konstante (K_m) gleicht zahlenmäßig der Substratkonzentration (Mole × Liter^{-1}), bei welcher die halbe Grenzgeschwindigkeit erreicht wird.

Die Voraussetzungen für die Gültigkeit der Gleichung von MICHAELIS und MENTEN sind die Bildung einer dem Massenwirkungsgesetz gehorchenden Enzym-Substrat-Verbindung, sowie einer Weiterreaktion des Enzym-Substrat-Komplexes zu den Endprodukten, die für den Gesamtverlauf der Reaktion geschwindigkeitsbestimmend ist. Soviel heute bekannt, dürften diese Voraussetzungen für die meisten Enzymreaktionen gelten. Wenn Abweichungen auftreten, so sind sie durch das Eingreifen zusätzlicher Faktoren, wie z. B. einer Hemmung der katalytischen Reaktion durch Spaltprodukte oder durch den Überschuß von Substrat bedingt (Abb. 5).

Zur Anwendung der MICHAELIS-MENTEN-Gleichung wählt man am besten experimentelle Bedingungen, bei denen der Quotient $(E)/K_m$ klein gehalten wird. In diesem Falle werden praktisch alle Enzymmoleküle im Substrat-Enzym-Komplex aufgenommen sein. Bei niederer Substratkonzentration und verhältnismäßig hohem K_m nähert sich die Kinetik einer Reaktion 1. Ordnung.

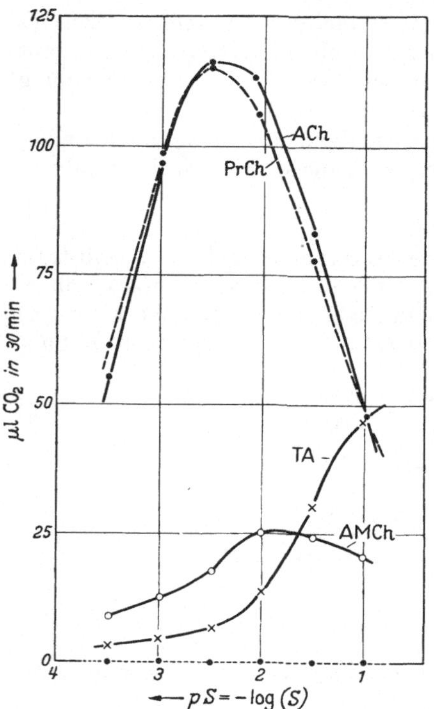

Abb. 5. Aktivitäts-pS-Kurven der enzymatischen Spaltung verschiedener Ester durch die Acetylcholinesterase des elektrischen Organs von *Electrophorus*. *ACh* Acetylcholin; *PrCh* Propionylcholin; *BCh* Butyrylcholin; *AMCh* Acetyl-β-methylcholin; *TA* Triacetin. (Nach AUGUSTINSSON.)

Enzyme als Proteine besitzen im allgemeinen ein Molekulargewicht zwischen 10000 bis einigen 100000; das Molekulargewicht der meisten Substrate beträgt nur etwa den 1000. Teil. Vereint man in einem Spaltungsansatz gleiche Gewichtsteile von Enzym und Substrat, so kommen etwa 1000 Substratmoleküle auf 1 Enzymmolekül. Man kann annehmen, daß in einem solchen Ansatz für gewöhnlich schon das Gros der Enzymmoleküle mit einem Substratmolekül vereint sein wird (BALDWIN 1949).

Bei definiertem Ionenmilieu und Temperatur, sowie geeignet gewählten Reaktionsbedingungen, die eine Rückreaktion möglichst ausschließen, gilt unter der Voraussetzung einer Substratsättigung des Fermentes:

$$v = W \cdot (E) \quad (\text{Mole} \times \text{Liter}^{-1} \times \text{min}^{-1}). \tag{21}$$

Das Produkt ist ein Ausdruck für die *Wirkungsgröße eines Enzyms* unter den gewählten Bedingungen; v ist die Reaktionsgeschwindigkeit; W bedeutet hier die bei Substratsättigung je Mol Ferment und je Minute zum Umsatz gelangenden Mole Substrat und wird nach O. WARBURG *Wechselzahl* (turnover number) eines Ferments genannt. Soll die Gleichung auch für den Bereich niederer Substratkonzentrationen Geltung haben, erweitert man sie durch Multiplikation mit dem Wert für v/V aus der MICHAELIS-MENTEN-Gleichung:

$$v = W \cdot (E) \frac{(S)}{(S) + K_m} \quad (\text{Mole} \times \text{Liter}^{-1} \times \text{min}^{-1}). \tag{22}$$

W und K_m sind wichtige Konstanten zur Charakterisierung von Fermenten; sie können auch zur Bestimmung der relativen Affinität eines Fermentes (oder anderen aktiven Proteins) zu zwei verschiedenen Substraten herangezogen werden. Je größer W und je kleiner K_m, um so größer ist die Affinität des Enzyms zu dem betreffenden Substrat (HOLZER).

c) Ermittlung der Reaktionskonstanten.

Die Bestimmung von K_m nach Abb. 3 oder 4 ist ungenau. LINEWEAVER und BURK haben vorgeschlagen, die MICHAELIS-MENTEN-Gleichung in ihrer Form (20b) reziprok zu nehmen; es gilt:

$$\frac{1}{v} = \frac{(S) + K_m}{V(S)} = \frac{K_m}{V}\left[\frac{1}{(S)}\right] + \frac{1}{V}. \tag{23}$$

Wenn die experimentell gewonnenen Daten der MICHAELIS-MENTEN-Gleichung gehorchen, so resultiert eine gerade Linie, wenn man $1/v$ gegen $1/(S)$ aufträgt. Die Neigung der Geraden entspricht K_m/V, der Ordinatenabschnitt $1/V$. Beide Größen lassen sich auf diese Weise bestimmen. Eine noch genauere Ermittlung des Wertes der Konstanten wird erreicht, wenn man alle Glieder der Gleichung (23) mit (S) multipliziert.

$$\frac{(S)}{v} = \frac{(S)}{V} + \frac{K_m}{V}. \tag{24}$$

Trägt man $(S)/v$ gegen (S) auf, so ist die Neigung der resultierenden Geraden $1/V$ und der Ordinatenabschnitt K_m/V. Aus V kann k_3 errechnet werden nach $V = k_3(E)$ (vgl. P. W. WILSON).

In modifizierter Form erlaubt die Formulierung von LINEWEAVER und BURK eine Bestimmung der Zahl der aktiven Arealen der Enzymoberfläche, die zur Substratbindung befähigt sind (LOEWUS und BRIGGS).

Die Bestimmung der Geschwindigkeitskonstanten k_1, k_2 und k_3 auf Grund von kinetischen Messungen war bis vor kurzem noch eim ungelöstes Problem, da die MICHEALIS-MENTEN-Gleichung diese Größen nicht direkt zu bestimmen gestattet. Eine Möglichkeit dazu gaben die spektrophotometrischen Arbeiten von CHANCE. In neuerer Zeit haben SLATER und BONNER eine Methode zur Bestimmung dieser Konstanten entwickelt, die auf kinetischen Messungen in Gegenwart eines kompetitiven Inhibitors beruht, der schnell und reversibel mit dem Enzym reagiert. SLATER (1953) gibt noch weitere Möglichkeiten zur Berechnung dieser Konstanten an. Man kann unter anderem von der MICHAELIS-Konstanten (K_m) ausgehen. Wenn V die Reaktionsgeschwindigkeit bei Substratsättigung und (E) die totale Enzymkonzentration ist, so gilt:

$$K_m = \frac{k_2(E) + k_3(E)}{k_1} = \frac{k_2(E) + V}{k_1(E)}. \tag{25}$$

Tabelle 4. *Geschwindigkeitskonstanten von Enzymen der Atmungskette.* (Nach SLATER 1953.)

Reaktion	Temperatur °C	k_1 (Mole$^{-1}\cdot$sec^{-1})	k_2 (sec^{-1})	k_3 (sec^{-1})
DPN · H + Fl $\xrightarrow{k_1}$ DPN + FlH_2	20	2,2 × 10^5	—	—
Succinat + E $\underset{k_2}{\overset{k_1}{\rightleftharpoons}}$ Succinat · E	38	0,7 × 10^5	2	—
Succinat · E $\xrightarrow{k_3}$ Fumarat + E (aerobe Oxydation) .	38	—	—	32
C^{++} + Cyt. Ox.$^{+++}$ $\xrightarrow{k_1}$ C^{+++} + Cyt. Ox.$^{++}$ (lösl.) .	38	40 × 10^5	—	—
Cyt. Ox.$^{++}$ $\xrightarrow{k_3}$ Cyt. Ox.$^{+++}$ (aerobe Oxydation) .	38	—	—	216

Fl = Flavoprotein im Herzmuskel.
C^{++}, C^{+++} = Cytochrom reduziert bzw. oxydiert.
Cyt.Ox.$^{++}$, Cyt.Ox.$^{+++}$. = Cytochromoxydase reduziert bzw. oxydiert.
E = Bernsteinsäure-Dehydrogenase.

K_m ist proportional zu V, die Reaktion kann aber bei Dehydrogenasen durch die Natur und Konzentration von H-Acceptoren verändert werden. Trägt man auf diese Art gewonnene K_m-Werte gegen V auf, so erhält man eine Gerade, aus der sich die Konstanten ermitteln lassen. Im Schnittpunkt der Ordinate $V = 0$ ist $K_m = k_2(E)/k_1(E) = k_2/k_1$. Im Schnittpunkt mit der Abszisse ist $K_m = 0$ und $V = -k_2(E)$. Wenn (E) bekannt ist, kann k_2 berechnet und aus dem Wert k_2/k_1 kann k_1 erhalten werden (Tabelle 4).

d) Der Enzym-Substrat-Komplex.

Die Hypothesen von MICHAELIS und MENTEN beruhen ursprünglich auf der Beobachtung, daß die Geschwindigkeit einer enzymatischen Reaktion sehr häufig linear zur Enzymkonzentration und hyperbolisch zur Substratkonzentration verläuft. Obwohl die fundamentale Basis der Hypothese, die Bildung eines Enzym-Substrat-Komplexes zunächst rein hypothetisch war, hat sich diese Annahme in der Folgezeit doch als sehr fruchtbar erwiesen. Die photoelektrische Spektrophotometrie ermöglichte es aber, intermediäre Enzym-Substratverbindungen bei solchen Enzymen zu entdecken, die ihrer Natur nach Hämoproteine sind, wie z. B. die Peroxydasen und die Katalase. KEILIN und MANN fanden, daß sich die charakteristischen Absorptionsbanden der Enzyme ändern, wenn der Eisen-Porphyrinkern dieser Enzyme Wasserstoffperoxyd bindet. Dieser Komplex verschwindet in Gegenwart eines Wasserstoff-(oder Elektronen-)donators (Pyrogallol), wobei letzterer oxydiert wird. Es sind bisher 3 Peroxydase-(bzw. Katalase-)peroxyde bekannt, ein grüner (Komplex I) und 2 rote Komplexe (Komplex II und III) (KEILIN und HARTREE 1951). *Katalase* leitet die Spaltung von H_2O_2 in Wasser und Sauerstoff durch die Bildung der grünen Zwischenverbindung (Komplex I) ein. Dieser Komplex reagiert mit einem anderen H_2O_2-Molekül unter Bildung der Spaltprodukte und Regeneration des Enzyms (CHANCE 1951, CHANCE, GREENSTEIN und ROUGHTON).

$$\text{Katalase} + H_2O_2 \underset{k_2}{\overset{k_1}{\rightleftharpoons}} \text{Komplex I}$$

$$\text{Komplex I} + H_2O_2 \xrightarrow{k_3} \text{Katalase} + O_2 + 2\,H_2O.$$

In Gegenwart von geeigneten Wasserstoffdonatoren [wie niedere primäre Alkohole (KEILIN und HARTREE 1936), Formaldehyd, Ameisensäure (CHANCE 1947, 1948, 1949) und Nitrit (HEPPEL und PORTERFIELD), auch H_2O_2 (CHANCE 1947, 1948, 1949)] bewirkt Katalase eine Oxydoreduktion nach Art einer Peroxydase, eine Reaktion, die bei den in der lebenden Zelle vorliegenden sehr niederen Konzentration an H_2O_2 eine wichtigere Rolle spielt, als die Zersetzung von H_2O_2 (KEILIN und HARTREE 1936).

$$AH_2 + H_2O_2 \rightarrow A + 2\,H_2O.$$

Diese Reaktion geht auch bei niederer Konzentration von H_2O_2 vor sich, bei der fast keine Aufspaltung von H_2O_2 in O_2 und Wasser stattfindet (GOTTSCHALK).

Gegen die Notwendigkeit, die Bildung eines Enzym-Substrat-Komplexes anzunehmen, sprechen Versuche von ROTHÉN. Die Beweiskraft dieser Versuche ist jedoch sehr umstritten.

e) Die Hemmstoffwirkung.

Um zu einem grundsätzlichen Verständnis für das Zustandekommen des biologischen Wirkstoff-Hemmstoff-Antagonismus zu gelangen, muß man von der Vorstellung eines räumlichen Kontaktes zwischen einem aktiven Protein und einem Metaboliten ausgehen, der durch einen Hemmstoff verdrängt werden kann. Bei Zugrundelegung dieser Vorstellung wird es verständlich, daß eine Aktivitätshemmung von Enzymen stets dann eintreten wird, wenn irgendwelche Substanzen (Antimetaboliten) mit dem Substrat (Metaboliten) in einen Wettbewerb um den Platz am reagierenden System treten. Dieser Wettbewerb kann auf drei verschiedenen Wegen erfolgen: Bei der *kompetitiven Hemmung* konkurriert der Hemmstoff mit dem Substrat ausschließlich um spezifische Gruppen der Enzymoberfläche; die meßbare Abnahme der Enzymaktivität hängt dabei sowohl von der Substrat- als auch der Inhibitorkonzentration ab. Man hat deshalb die Möglichkeit, die Hemmungsintensität kompetitiv wirkender Antimetabolite durch den „Hemmungsindex" zu präzisieren. Dieser Index zeigt an, bei welchem molaren Mengenverhältnis sich die Wirkungen von Inhibitor und Metabolit genau kompensieren. Bei der *nicht kompetitiven Hemmung* wird der Antimetabolit an solche Gruppen der Enzymoberfläche gebunden, die an der Bindung des Substratmoleküls nicht beteiligt sind. Daher gibt es hier keine Verdrängung oder Konkurrenz mit dem Substrat um den Platz an der Fermentoberfläche. Das Ausmaß der Inaktivierung hängt deshalb in diesem Fall allein von der Konzentration des Inhibitors ab; ein Hemmungsindex ist nicht meßbar. Man spricht schließlich von einer *unkompetitiven Hemmung*, wenn der Antimetabolit nur mit dem Enzym-Substrat-Komplex, nicht aber mit dem Enzymmolekül selbst, zusammentritt (KÜHNAU).

Fußend auf den klassischen Anschauungen von MICHAELIS und MENTEN von der Bildung dissoziabler Enzym-Substrat-Komplexe bei jeglicher Art von Enzymwirkung kann man den *kompetitiven Wirkstoff-Hemmstoff-Antagonismus* wie folgt formulieren:

$$E + S \underset{k_2}{\overset{k_1}{\rightleftharpoons}} ES \xrightarrow{k_3} E + P, \tag{26}$$

$$E + I \underset{k_2'}{\overset{k_1'}{\rightleftharpoons}} EI \quad \text{(inaktiv mit Dissoziationskonstante } K_i\text{)}, \tag{27}$$

wobei I die Konzentration des Inhibitors bedeutet und die übrigen Symbole in dem bisher gebrauchten Sinn verwendet werden. Ebenso wie k_3 und K_m zur Ermittlung der relativen Affinität eines Enzyms gegenüber zwei verschiedenen Substraten herangezogen wurden, können analoge Konstanten (K_i, k_3') zur Ermittlung der relativen Affinität des Enzyms gegenüber Inhibitoren verwendet werden.

Nach den mathematischen Formulierungen von LINEWEAVER und BURK ist die Konzentration an freiem Enzym:

$$(E) - (E\,S) - (E\,I)$$

$$K_m = \frac{[(E) - (E\,S) - (E\,I)]\,(S)}{(E\,S)}$$

und

$$K_i = \frac{[(E) - (E\,S) - (E\,I)]\,(I)}{(E\,I)}.$$

Nach einer einfachen mathematischen Ableitung (siehe P. W. WILSON) erhält man

$$(E\,S) = \frac{(E)\,(S)\,K_i}{K_m K_i + K_m (I) + K_i (S)}.$$

Setzt man wie auf S. 509 $v = k_3\,(E\,S)$, wobei k_3 die Geschwindigkeitskonstante der Weiterreaktion nach Gleichung (26) ist, so erhält man:

$$v = \frac{k_3\,(E)\,(S)\,K_i}{K_m K_i + K_m (I) + K_i (S)}.$$

Setzt man für $k_3(E) = V$, wobei V die maximale Reaktionsgeschwindigkeit bei Sättigung des Enzyms mit Substrat ist, so ergibt sich:

$$v = \frac{V(S)K_i}{K_m K_i + K_m(I) + K_i(S)}.$$

Zur graphischen Darstellung empfiehlt es sich, nach dem Verfahren von LINEWEAVER und BURK den reziproken Wert der Gleichung zu nehmen

$$\frac{1}{v} = \frac{K_m K_i + K_m(I) + K_i(S)}{V(S)K_i} \tag{28}$$

$$= \frac{1}{V(S)}\left[K_m + \frac{K_m(I)}{K_i}\right] + \frac{1}{V} \tag{29}$$

oder

$$\frac{1}{v_i} = \frac{1}{V^*}\left[K_m + \frac{K_m(I)}{K_i}\right]\frac{1}{(S)} + \frac{1}{V^*},$$

wobei v_i die meßbaren Geschwindigkeiten und V^* die maximale Reaktionsgeschwindigkeit eines Inhibitors bedeuten.

Man erhält eine gerade Linie, wenn man die experimentell gewonnenen Werte für $1/v_i$ gegen zugehörende Werte von $1/(S)$ aufträgt. Ein Vergleich von Gleichung (29) mit Gleichung (23) auf S. 511 zeigt, daß sich die Zugabe eines kompetitiven Inhibitors in einer bestimmten Konzentration zu den Reaktionsansätzen in einer Vergrößerung der Neigung einer derartig ermittelten Geraden um den Betrag von $\frac{K_m(I)}{K_i}$ ändert. Prüft man nach dieser graphischen Methode den Einfluß eines kompetitiven Inhibitors in mehreren Konzentrationsstufen, so erhält man eine ebenso große Anzahl von Geraden, die sich untereinander durch ihre Neigung unterscheiden, aber die Ordinate fast an dem gleichen Punkt durchstoßen. Der steilere Anstieg solcher aus den Versuchspunkten statistisch ermittelten „besten Geraden" ohne signifikante Änderung des Ordinatenabschnittes ist ein Charakteristikum des kompetitiven Hemmungstyps. Der Zahlenwert der Dissoziationskonstante (K_i) der Enzym-Hemmstoffverbindung läßt sich aus der Gleichung

$$K_m + \frac{K_m(I)}{K_i} = V^* \times (\text{Neigung der Geraden})$$

berechnen.

Aus Gleichung (23) und (29) erhält man:

$$\frac{v}{v_i} = 1 + \frac{K_m}{K_i}\left[\frac{(I)}{K_m + (S)}\right].$$

Trägt man v/v_i gegen verschiedene Werte der Inhibitorkonzentration (I) auf, so bekommt man ebenfalls gerade Linien mit einem gemeinsamen Ordinatenabschnitt. Die Neigung dieser Linien ist hier mehr abhängig von der den Bestimmungen von v bzw. v_i zugrunde liegenden Substratkonzentration (S).

Bei der kompetitiven Hemmung konkurriert der Antagonist mit dem Substrat um den katalytisch aktiven Enzymbereich; das kann auf verschiedene Art geschehen.

1. Der Antagonist ist dem Metaboliten (Substrat) in seiner Struktur zwar so ähnlich, daß er sich mit den aktiven Gruppen verbindet, der gebildete Enzym-Inhibitor-Komplex reagiert aber äußerst langsam, d. h. praktisch überhaupt nicht weiter. Auf diese Weise bringt der Inhibitor die Reaktion durch Verdrängung des Substrates zum Stehen.

$$E + I \underset{k_2}{\overset{k'}{\rightleftharpoons}} EI \qquad K_i = \frac{k_2'}{k_1'}.$$

Nach diesem Hemmtyp wirken zahlreiche Strukturanaloge der Substrate und Cofermente. So wird z. B. die Succinodehydrogenase durch Malonat gehemmt. Eserin geht mit Cholinesterase einen Enzym-Substrat-Komplex ein, der äußerst langsam weiterreagiert und dadurch das Ferment für Acetylcholin blockiert (I. B. WILSON 1951 c).

2. Es kann aber auch eine Hemmung dadurch hervorgebracht werden, daß der Antimetabolit von dem Enzym schneller umgesetzt wird als das Substrat; dann wird das Enzym vom Substrat auf den Antagonisten abgelenkt. Es sind ferner Fälle bekannt, wo das Reaktionsprodukt (P') einer enzymatischen Reaktion

$$E + I \underset{k_2'}{\overset{k_1'}{\rightleftharpoons}} E\,I \xrightarrow{k_3'} E + P' \qquad K_i = \frac{k_2' + k_3'}{k_1'}$$

hemmend in eine weitere Reaktion eingreift; ein derartiges Beispiel ist die an sich unschädliche *Fluoressigsäure*, die im Stoffwechsel offenbar in Fluorcitronensäure umgewandelt wird und durch Hemmung der Aconitase toxisch wirkt (LOTSPEICH, PETERS und WILSON). Ähnlich verhält sich Schradan (Octamethyltetramidopyrophosphat), welches Cholinesterase in vitro nicht hemmt, im Gewebe aber enzymatisch in eine sehr aktive Anticholinesterase verwandelt wird (CASIDA und STAHMANN).

Eine kompetitive Hemmung wird nicht nur durch einen Wettbewerb mit Substraten, sondern auch durch Antimetabolite hervorgerufen, die gewisse Analogien zu *Cofermenten* aufweisen. So kennt man die Ausschaltung von Flavin-Adenin-Dinucleotid durch Adenin (WILLIAMS) und von Pyridinnucleotiden durch Nicotinsäureamid (ALIVASATOS und DENSTEDT). Man kennt auch Beispiele einer Hemmung, die durch Antagonisten einer Coenzymvorstufe, eines freien Vitamins, hervorgerufen werden. Aminopterin hemmt beispielsweise die Umwandlung der ähnlich gebauten Folsäure in den als Coenzym wirkenden Citrovorumfaktor (Folinsäure) (NICHOL und WELCH).

Bei der *nicht kompetitiven Hemmung* wird der Inhibitor an katalytisch unwirksame Teile der Oberfläche des Fermentproteins gebunden. Es gilt daher

$$E + S \underset{k_2}{\overset{k_1}{\rightleftharpoons}} E\,S \xrightarrow{k_3} E + P,$$

$$E + I \underset{k_2'}{\overset{k'}{\rightleftharpoons}} E\,I \quad \text{(inaktiv mit Dissoziationskonstante } K_i)$$

$$E\,S + I \rightarrow E\,S\,I \quad \text{(inaktiv mit Dissoziationskonstante } K_{esi} = K_i).$$

Nach einer ähnlichen Ableitung wie auf S. 513 erhält man (siehe P. W. WILSON):

$$\frac{1}{v_i} = \left[1 + \frac{(I)}{K_i}\right]\left[\frac{1}{V^*} + \left(\frac{K_m}{V^*}\right)\frac{1}{(S)}\right] \tag{30}$$

und

$$\frac{v}{v_i} = 1 + \frac{(I)}{K_i}. \tag{31}$$

Der Vergleich von Gleichung (30) mit Gleichung (23) auf S. 511 zeigt, daß ein nicht kompetitiver Inhibitor sowohl die Neigung als auch den Ordinatenabschnitt einer Geraden um den Faktor $\left[1 + \frac{(I)}{K_i}\right]$ vergrößert. Man erhält diese Gerade, wenn man die ermittelten Werte für $1/v_i$ gegen $1/(S)$ in ein Koordinatensystem einträgt.

Unter diesen Typ der Hemmwirkung fallen die Wirkungen mancher anorganischer Ionen, von Reaktionsprodukten und hohen Substratkonzentrationen. Ein Beispiel hierfür ist die von AUGUSTINSSON (1949) untersuchte Hemmung der

Acetylcholinesterase durch Acetylcholin (Abb. 5). Ein ähnlicher Fall ist auch die Hemmung der Urease durch Harnstoff. Im aktiven Areal der Cholinesterase befinden sich zwei für die katalytische Wirkung wesentliche Strukturen, genannt „anionic site" und „esteratic site". Das Substrat Acetylcholin muß erst mit dem Cholin-N am „anionic site", mit der Esterverbindung am „esteratic site" verankert werden, wenn die Bedingungen zur Spaltung erfüllt sein sollen. Bei hoher Substratkonzentration ist die Möglichkeit gegeben, daß die beiden aktiven Stellen im gleichen aktiven Areal eines Fermentmoleküls von zwei verschiedenen Acetylcholinmolekülen besetzt werden. Unter diesen Umständen kann keine Spaltung eintreten; die Stellen bleiben daher blockiert (ZELLER und BISEGGER; WILSON 1951a, b; AUGUSTINSSON und NACHMANSOHN; WILSON und BERGMANN). Bei hohen Harnstoffkonzentrationen wird der Harnstoff auch an Oberflächenarealen gebunden, die normalerweise mit Wasser beladen werden. Dadurch wird die Spaltungsreaktion, zu der Wasser benötigt wird, verhindert. Die Katalasehemmung durch HCN ist nicht kompetitiv, dagegen ist die Peroxydasehemmung kompetitiv. Die Erklärung hierfür liegt in dem Besitz von 4 Fe-Atomen der Katalase, von denen nur eines vom Substrat beansprucht wird, während Peroxydase nur ein Eisen-Atom enthält (LAIDLER und HOARE).

Bei der *unkompetitiven Hemmung* kombiniert sich der Hemmstoff nur mit der Enzym-Substrat-Verbindung, nicht mit dem Enzym. Ein Beispiel hierfür ist die Azidhemmung der oxydierten Form des Atmungsfermentes (WINZLER). Es wurden Übergänge zwischen den einzelnen Hemmungstypen beobachtet, besonders im Zusammenhang mit Änderungen der Inhibitorkonzentration.

f) Die Umsatzgeschwindigkeit im Verlauf der Spaltung.

Im Verlaufe der enzymatischen Reaktion ist meist eine Abnahme der Umsatzgeschwindigkeit zu merken, die auf die allmählich abnehmende Substratkonzentration und den hemmenden Einfluß der gebildeten Produkte der Umsetzung beruht. Bei der verschiedenen Affinität der einzelnen Enzyme zu ihren Substraten und den Spaltprodukten ist es zu verstehen, daß alle Übergänge von einem streng linearen Reaktionsverlauf bis zu einer starken Abnahme der Reaktionsgeschwindigkeit im Verlaufe der Umsetzung vorkommen werden.

Bei *hoher Affinität des Enzyms zum Substrat und geringer Affinität zu den Reaktionsprodukten* ist ein Grenzfall gegeben, der durch $(S) \gg K_m$ charakterisiert werden kann. Da K_m neben (S) vernachlässigt werden kann, nähert sich das Glied $\frac{(S)}{(S)+K_m}$ aus Gleichung (15) auf S. 508 dem Wert 1. Die gesamte, im Ansatz enthaltene Enzymmenge ist an das Substrat gebunden $(ES) = (E)$ und die Reaktionsgeschwindigkeit (v) gleicht der Maximalgeschwindigkeit (V). Das Substrat wird mit konstanter Geschwindigkeit abgebaut, solange die Bedingung $(S) \gg K_m$ noch Geltung hat. Man spricht von linearem Reaktionsverlauf oder einer Reaktion 0. Ordnung.

Bei *geringer Affinität des Enzyms zum Substrat und den Spaltprodukten* oder *geringer Substratkonzentration* oder *sehr schnellem Zerfall des Enzym-Substrat-Komplexes* gilt $K_m \gg (S)$. Die Reaktionsgeschwindigkeit ist unter diesen Umständen $(S)/K_m$ proportional und damit der Substratkonzentration. Das Ergebnis ist ein monomolekularer Reaktionsverlauf (Reaktion 1. Ordnung),

$$-\frac{d(S)}{dt} = k(S)(E) = k(a-x)(E),$$

da (E) im Verlauf der Reaktion annähernd konstant bleibt, gilt nach Integration:

$$k(E) = K' = \frac{1}{t}\ln\frac{a}{a-x} = \frac{2{,}303}{t}\log\frac{a}{a-x}$$

a Substratkonzentration zu Reaktionsbeginn; x umgesetzte Substratmenge; K' Konstante der monomolekularen Reaktion.

Wenn die Affinität zu den Spaltprodukten groß ist bei geringer Affinität zum Substrat, so entwickelt sich im Verlauf der Spaltung eine zunehmende Hemmung der Reaktionsgeschwindigkeit, was man an dem Fallen der berechneten Konstanten der monomolekularen Reaktion erkennen kann. Der Wert $\log \frac{a}{a-x}$ gibt eine Maßzahl für die Initialgeschwindigkeit (v) der enzymatischen Reaktion. Nach Zusammenziehung der Konstanten wird die Gleichung gewöhnlich $K = \frac{1}{t} \log \frac{a}{a-x}$ geschrieben; es ist aber nicht exakt, dieses K als monomolekulare Reaktionskonstante anzusprechen.

g) Der Einfluß der H+-Konzentration.

Die von SØRENSEN entdeckte p_H-Abhängigkeit der enzymatischen Aktivität ist ein sehr charakteristisches Merkmal enzymatischer Reaktionen; durch ein p_H-Optimum unterscheiden sie sich von andersartigen Katalysen in wäßrigen Medien.

Enzyme werden durch p_H in zweifacher Hinsicht betroffen, in ihrer *Stabilität* sowie in ihrer *Aktivität*. Mit einigen Ausnahmen werden die Enzyme von stärker sauren wie auch stärker basischen Lösungen irreversibel denaturiert. Die p_H-Werte, bei denen dieser Prozeß einsetzt, sind von Enzym zu Enzym verschieden. In den meisten Fällen ist das Enzym im neutralen Teil des p_H-Gebietes stabil, und die p_H-Kurve der Stabilität hat eine typische Glockenform. Es gibt aber auch Fälle eines sehr komplizierten Verhaltens. Ein Beispiel hierfür ist die p_H-Stabilität des kristallisierten Trypsins. Unter p_H 2 erfolgt eine irreversible Destruktion. Bei höheren p_H-Werten steht Trypsin im Gleichgewicht mit einer inaktiven Form, die durch die aktive hydrolysiert wird. Zwischen p_H 12 und 13 wird der aktive Anteil sehr klein, daher fällt die Inaktivierungsgeschwindigkeit stark ab. Oberhalb p_H 13 setzt eine irreversible Inaktivierung ein (KUNITZ und NORTHROP 1934).

Der p_H-Einfluß auf die enzymatische Aktivität kann auf einer Wirkung von H^+-Ionen auf das Enzymprotein selbst, auf das Substrat, auf Coenzyme und auch auf Aktivatoren beruhen. Da die Enzymproteine Ampholyte sind, und da unter allen möglichen Ionenformen offensichtlich nur *eine* existiert, die katalytisch wirksam ist und im Optimum überwiegen wird, hat man die beiden Hälften einer p_H-Aktivitätskurve als Dissoziationskurven einer sauren und basischen Gruppe zu deuten versucht. Man hat auch darauf hingewiesen, daß zwischen dem IEP des Enzymproteins und dem p_H der optimalen katalytischen Aktivität auffallende Parallelen bestehen. MICHAELIS und DAVIDSOHN führten die p_H-Kurve der Saccharase auf die amphotere Natur des Enzyms zurück und betrachten die elektrisch neutralen Enzymmoleküle zur Verbindung mit der Saccharase für geeignet, deren positive und negative Ladungen sich die Waage halten. MYRBÄCK (1947) bestimmte die Dissoziationskonstanten der amphoteren Saccharasemoleküle zu $K_a \sim 10^{-6,6}$ und $K_b \sim 10^{-11}$. In der Annahme, daß nur die undissoziierte Enzym-Substrat-Verbindung an der Spaltung beteiligt ist, stellte er eine Formel auf

$$[(ES)_{\text{undiss.}}]/(E) = 1/C_1 + K_a/[H^+]\{1 + (K_m/(S))\,(1 + K_b\,[OH])\},$$

die in guter Übereinstimmung mit den experimentellen Befunden über den Einfluß von p_H und (S) auf die Aktivität der Saccharase steht. Obwohl man sich vorstellen kann, daß in der Nähe des IEP die Anlagerung polarer Wassermoleküle am geringsten ist und dieser Zustand zur Bindung von Substratmolekülen besonders geeignet ist, ist es doch ungewiß, ob solche Vorstellungen eine allgemeine Gültigkeit besitzen. Es gibt Beispiele für die alleinige Wirksamkeit der sauren Form eines Enzyms. Das p_H-Optimum stimmt auch nicht immer mit dem p_H-Wert überein, den das Enzym in seinem biologischen Milieu vorfindet. Ist das Substrat selbst auch ionisiert, wie z. B. Polypeptide, so wirkt das Enzym gewöhnlich nur auf eine ganz bestimmte Form des Substrates ein, z. B. das Zwitterion. Daher ändert sich bei den Proteinasen nach einem Wechsel des Substrates auch die Form und Lage der p_H-Kurve.

Die p_H-Aktivitätskurve ist daher in den meisten Fällen eine Resultierende zahlreicher, zum Teil wenig oder unbekannter Faktoren, da oft sogar die Löslichkeit von Aktivatoren und die Stabilität von Cofaktoren mit eingeschlossen sind. Wenn die p_H-Effekte dagegen nur durch die Dissoziationsverhältnisse von Substrat und Enzym hervorgerufen werden, die bekannt sind oder ermittelt werden können, kann der Versuch einer rechnerischen Behandlung der p_H-Kurve von Erfolg begleitet sein. So gelang es JOHNSON, JOHNSON und PETERSON, die p_H-Abhängigkeit der Triglycinspaltung durch die Darm-Aminopolypeptidase mit einer theoretisch errechneten Kurve für die Reaktionsgeschwindigkeit in gute Übereinstimmung zu bringen.

h) Das Zusammenwirken von Enzymen
("Multi-Enzyme-Systems").

Am Stoffwechsel der Zelle ist eine große Zahl von Fermenten beteiligt. Sie unterhalten den Abbau, Aufbau und Umbau von Metaboliten und Zwischenstoffen. Seit Jahrzehnten ist man bestrebt, einzelne Enzyme aus Extrakten und Homogenisaten von Zellen anzureichern, zu isolieren und rein darzustellen. Manche der am Stoffwechsel beteiligten Fermente sind daher bereits auch in ihrem Wirkungsmechanismus aufgeklärt worden. Im Zellstoffwechsel gibt es aber keine isolierten enzymatischen Umsetzungen; die einzelnen Fermentreaktionen sind Glieder eines reich verzweigten Reaktionsnetzes, in dem sich alle Komponenten gegenseitig beeinflussen, wenn ein stationärer Stoffdurchsatz läuft. Aber eine noch so genaue Kenntnis der Spezifität und Kinetik von Einzelfermenten gibt noch keine Aufklärung darüber, welche Ursachen der Verteilung von Metaboliten auf verschiedene Reaktionswege zugrunde liegen und in welchem Mengenverhältnis diese Aufteilung stattfindet. Die Organismen besitzen besondere Regulationsmechanismen, die dafür sorgen, daß in einem Fall Zwischenstoffe des Kohlenhydratstoffwechsels in Fett, in einem anderen Fall in Eiweißstoffe oder Milchsäure verwandelt oder aber zu H_2O und CO_2 verbrannt werden. Eine ausgezeichnete Diskussion dieser Frage verdanken wir KREBS. Es ist bekannt, daß Hormone den Stoffwechsel höherer Organismen steuern. Aber der Stoffwechsel einzelliger, hormonloser Organismen (Bakterien, Hefen) unterliegt ebenfalls einer Regulation. Es existieren Mechanismen, die den Zellstoffwechsel entsprechend den Milieufaktoren abwandeln. Bei ausreichender Sauerstoffzufuhr wird Energie von der Atmung, unter anaeroben Bedingungen von der Gärung geliefert. Bei ausreichender Stickstoff- und Sauerstoffzufuhr gehen Hefezellen aus der Ruhe zu Wachstum über. Wir haben es hierbei nach KREBS mit elementaren Mechanismen der Zelle zu tun, auf denen die Wirkung der Hormone aufbaut. Welche Struktur haben diese cellulären Steuerungsmechanismen?

Die Basis der Stoffwechselreaktionen ist das Ferment-Substrat-System. Bisher wurden die Reaktionen isolierter Enzyme behandelt, nun sollen die Verhältnisse betrachtet werden, die durch das Zusammenspiel mehrerer nebeneinander ablaufender enzymatischer Prozesse gegeben sind.

Es hat als erster DIXON verschiedene Verknüpfungstypen unterschieden, nach denen Fermente zu einer Wirkungskette zusammengeschlossen werden können.

1. Nach dem ersten Typ können beliebig lange Fermentketten dadurch entstehen, daß das Reaktionsprodukt des einen Gliedes das Substrat des folgenden Gliedes bildet.

$$\mathrm{A} \xrightarrow[E_1]{} \mathrm{B} \xrightarrow[E_2]{} \mathrm{C} \xrightarrow[E_3]{} .$$

Enzymketten dieser Art sind im Organismus sehr häufig. Der Abbau eines Proteins durch ein Gemisch aus Proteinasen und Peptidasen ist ein Beispiel dafür. Man benützt diesen Reaktionstyp gegenwärtig oft zur Bestimmung von enzymatischen Aktivitäten und Metaboliten in der Form des „zusammengesetzten optischen Testes" (WARBURG und CHRISTIAN 1936b).

2. Nach dem zweiten Typ von Fermentketten wird ein Reaktionsprodukt von E_1 in der folgenden Reaktion durch E_2 in den Ausgangsstoff rückverwandelt und verhält sich dadurch in der Kette wie ein Katalysator.

$$AH + X \xrightarrow[E_1]{} XH + A$$

$$XH + B \xrightarrow[E_2]{} X + BH.$$

Diese Funktion kommt noch klarer bei folgender Schreibweise zum Ausdruck. Dadurch, daß die Substanz (X) bei einem stationären Durchsatz von H ständig regeneriert wird, erfüllt sie die Rolle eines Katalysators. Es genügt bereits eine geringe Menge von (X) im Zellplasma, um große Mengen von (H) durch die Fermentkette von $E_1\,E_2$ hindurchzuschleusen.

$$\begin{array}{ccccc} AH & \rangle\langle & X & \rangle\langle & BH \\ A & & XH & & B \\ & \uparrow & & \uparrow & \\ & E_1 & & E_2 & \end{array}$$

Man kennt eine Reihe von Substanzen, die in der Zelle eine ähnliche Rolle spielen wie (X). Hierher zählt z. B. die Glutaminsäure, die in einem System aus Transaminase und Glutaminsäuredehydrase, sowie DPN · H und NH_3, Oxalessigsäure zu Asparaginsäure aminiert, selbst aber immer wieder regeneriert wird. Auch das hierbei entstandene DPN^+ kann wieder reduziert werden, wenn ein passendes Enzym und ein H-Donator dem System zusätzlich beigefügt werden. Glutaminsäure und Cozymase I und II haben eine ausgesprochene Doppelfunktion. Vom Enzym aus betrachtet spielen sie die Rolle von Substraten. Das geht aus folgender Überlegung hervor: das „oxydierende Gärungsferment" stellt, wie bekannt, ein Gleichgewicht zwischen folgenden Reaktionskomponenten her:

$$\begin{array}{l} DPN^+ + \text{Phosphoglycerinaldehyd} + \text{Phosphat} \\ \quad \rightleftarrows \ \textit{spezifisches Ferment} \\ DPN \cdot H + \text{Diphosphoglycerinsäure} + H^+. \end{array}$$

Da nach W. OSTWALDS Definition ein katalytischer Vorgang daran zu erkennen ist, daß „eine für sich in einer bestimmten Zeit verlaufende chemische Reaktion durch die Gegenwart eines fremden Stoffes, der am Ende der Reaktion im selben Zustand ist, wie am Anfang, eine Änderung ihres zeitlichen Verlaufes erfährt", so kann in diesem System nur das Fermentprotein den Katalysator darstellen (BÜCHER 1953a, b). Eine Prüfung des Reaktionsgleichgewichtes hat ergeben, daß die beiden Formen der Cozymase, vom Massenwirkungsgesetz aus betrachtet, allen anderen Reaktionspartnern, auch den H-Ionen, gleichwertig sind.

Auch bei einer noch eingehenderen Betrachtung der Wirkungsweise des isolierten „oxydierenden Gärungsfermentes“, wie sie Bücher (1947, 1953a) anstellt, ergeben sich keine Anhaltspunkte, die Cozymase für etwas anderes als ein Substrat anzusprechen, denn sie erscheint in der Bilanz der katalytischen Reaktion.

Als Bestandteil einer Fermentkette und im Rahmen eines umfassenderen Stoffwechselprozesses besitzt das Pyridinnucleotidsystem ($DPN^+ \leftrightarrow DPN \cdot H$) jedoch die Funktion, Wasserstoff von einem wasserstoffübertragenden Enzymsystem auf ein anderes zu transportieren. Seine Aufgabe in der lebenden Zelle ist daher, zahlreiche Fermentreaktionen, auch wenn sie ganz verschiedenen Fermentketten angehören, miteinander zu verbinden; das Pyridinnucleotidsystem hat daher eine Schlüsselfunktion im Stoffwechsel. In der Bilanz eines komplexen Stoffwechselprozesses erscheint das DPN-System nicht mehr. Da es immer wieder regeneriert wird, kann es in kleinen Mengen große Umsätze erzielen, erhält aber erst durch das Zusammenwirken von mindestens zwei Fermentreaktionen katalytische Eigenschaften. Eine solche Transportfunktion besitzen auch andere Nucleotide. An der Übertragung der energiereichen Phosphatbindung ist das Adenosinsystem $ADP \leftrightarrow ATP$ maßgeblich beteiligt. Den Transport von Acetylgruppen bewirkt das pantothensäurehaltige Nucleotid Coenzym A. Der katalytischen Funktion wegen wurden die genannten Systeme „cyclische Substrate“ genannt; Bücher (1953a) schlägt die Bezeichnung „Transportmetabolite“ vor.

In den Zellen sind Fermentketten nach Typ (1) aufgebaut, die durch Wasserstoff- oder Phosphatübertragungen nach Typ (2) Nebenschlüsse zeigen. Fermentketten können auch einen Ringschluß besitzen. Der Citronensäurecyclus und der Harnstoffcyclus spielen vielfach im Zellstoffwechsel eine besondere Rolle.

Ein heute schon weitgehend aufgeklärter Reaktionsweg im Kohlenhydratstoffwechsel der Zellen ist die Gärung bzw. Glykolyse. Die an diesem Vorgang beteiligte Fermentkette entspricht dem Typus (1), eine Verbindung zwischen einzelnen Fermentproteinen nach Typus (2) wird von DPN^+—$DPN \cdot H$, sowie ADP—ATP-System bewirkt. An dem Abtransport der gewonnenen Energie an andere Prozesse des Zellgeschehens ist das ATP—ADP-System beteiligt.

Die *quantitative Behandlung eines Multienzymsystems* ist ein sehr schwieriges Problem. Unter günstigen Bedingungen herrscht in einem solchen System ein lebhafter Stoffumsatz, es bilden sich aber stationäre Konzentrationen von Zwischenprodukten aus, die einer Analyse zugänglich sind. Man kann diesen Sachverhalt besonders klar durch Versuche „in vitro“ mit reinen Fermenten und Substraten demonstrieren (Warburg und Christian 1939). Holzer hat folgendes Beispiel beschrieben:

$$1.\ \text{Triosephosphat} + \text{DPN} \xrightarrow[\text{Arsenat}]{\text{Triosephosphatdehydrase}} \text{Phosphoglycerinsäure} + \text{DPN} \cdot \text{H}_2$$

$$2.\ \text{Acetaldehyd} + \text{DPN} \cdot \text{H}_2 \xrightarrow{\text{Alkoholdehydrase}} \text{Alkohol} + \text{DPN}.$$

Das Zusammenwirken dieser beiden wasserstoffübertragenden Fermente der Gärung läßt sich spektrophotometrisch verfolgen, da $DPN \cdot H$ im Gegensatz zu DPN^+ das Licht der Wellenlänge im Bereich um 340 $m\mu$ stark absorbiert (Warburg und Christian 1936b). Bei dieser Reaktion handelt es sich um eine Übertragung von Wasserstoff durch das DPN-System von einem Enzymsystem auf das andere nach Fermentkettentyp (2) von Dixon. In der lebenden Zelle verbindet außerdem eine Kette nach Typ (1) beide Fermentreaktionen, im Holzerschen Modell hingegen wurde bereits zu Versuchsbeginn Acetaldehyd im Überschuß zugegeben. Arsenat wurde zugefügt, um die Acceptorfrage auszuschalten; es gibt mit Triosephosphat eine der 1,3-Diphosphoglycerinsäure entsprechende Intermediärverbindung, die aber bereits spontan zerfällt. Wird Glycerinaldehyd mit DPN, Arsenat und Acetaldehyd vereint und dem Ansatz Triosephosphatdehydrase zugegeben, so erfolgt nach Abb. 6 eine rasche Reduktion von DPN. Fügt man ein wenig Alkoholdehydrase zu, so startet zusätzlich Reaktion (2), und es stellt sich ein stationäres DPN · H/DPN$^+$-Gleichgewicht ein, „das dem dynamischen Gleichgewicht der DPN-Hydrierung an der Triosephosphatdehydrase und der DPN—H-Dehydrierung an der Alkoholdehydrase entspricht" (HOLZER).

Ein wichtiger Faktor, welcher die Lage des Gleichgewichtes bedingt, ist die *Wirkungsgröße* (Produkt aus Konzentration und Wechselzahl) der beteiligten Fermente. Erhöht man z. B. im Modell nach Abb. 6 die Konzentration der Alkoholdehydrase, so verschiebt sich das Gleichgewicht in Richtung zu DPN. Herrscht in einem Multifermentsystem für alle Komponenten der Zustand der Substratsättigung, so begrenzt das Enzym mit der geringsten Wirkungsgröße die Reaktionsgeschwindigkeit der gesamten Enzymkette. Aber auch der Substratmangel an einer Stelle des Reaktionssystems kann geschwindigkeitsbestimmend wirken. Starke Veränderungen der Substratkonzentration können sogar den Richtungssinn enzymatisch katalysierter Gleichgewichte umkehren (s. S. 502). Physiologisch gesehen kommt diesem Faktor die höhere Bedeutung zu. In tierischen Organen ist die Aktivität der meisten Enzyme im allgemeinen um eine Zehnerpotenz höher, als der tatsächlich stattfindende Umsatz erwarten läßt. Dieser Überschuß an katalytischer Potenz ermöglicht der Zelle die Bewältigung fallweise auftretender Spitzenkonzentrationen an Substrat. Für gewöhnlich aber begrenzt die unter der Sättigung liegende Substratkonzentration die Geschwindigkeit des betreffenden Stoffwechselprozesses (LANG). Von nicht zu unterschätzender Bedeutung ist ferner die Konzentration an Cofermenten, aktivierenden (Me^{++}) und hemmenden Ionen (Schwermetalle), SH-Verbindungen und anderen Effektoren der Enzymaktivität. Untersucht man den Stoffumsatz *lebender Zellen*, so tritt die Permeabilität der Zellmembran und Plasmapartikel als sehr wichtiger Faktor in Erscheinung. Es ist durch CONWAY und DOWNEY bekannt, daß Glucose leicht in Hefezellen eindringt; das gilt aber nicht für Pyruvat und noch weniger für Fructosediphosphat. Für die Geschwindigkeit und die stationären Verhältnisse der alkoholischen Gärung kann ein Abdiffundieren von ATP, ADP und Phosphat in Cytoplasmapartikel regulative Bedeutung haben. Aus Zellhomogenisaten isolierte Mitochondrien sind weder für DPN und TPN, noch für ATP und ADP unbeschränkt durchlässig.

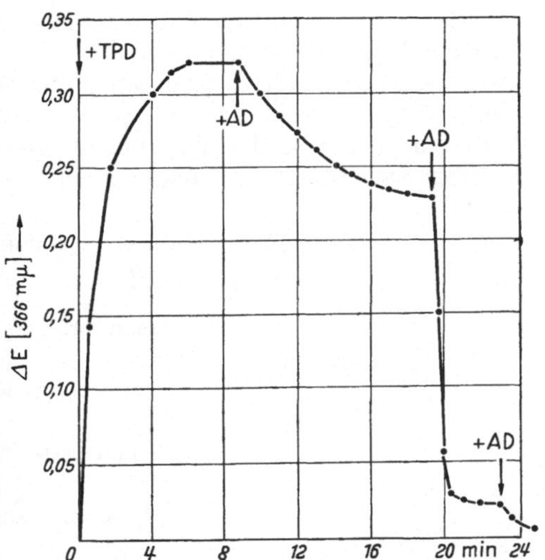

Abb. 6. Stationäre Konzentration an DPN—H im System Triosephosphatdehydrase-Alkoholdehydrase. *TPD* Triosephosphatdehydrase; *AD* Alkoholdehydrase. (Nach HOLZER 1953.)

Im allgemeinen wird — wie bereits betont — die Geschwindigkeit von Zwischenreaktionen des Stoffwechsels von der Substratmenge gesteuert. KREBS macht aber darauf aufmerksam, daß es Reaktionen gibt — verhältnismäßig wenige an Zahl —, deren Geschwindigkeit von anderen Faktoren abhängt. Er schreibt diesen Reaktionen, die er „Schrittmacher" des Stoffwechsels nennt, eine ganz besondere regulatorische Bedeutung zu. Eine solche Rolle spielt z. B.

die Hexokinasereaktion; sie bestimmt den Glucoseverbrauch der Zelle. Ein Schrittmacher im Gärungsablauf ist ferner die komplexe Dehydrierung von Triosephosphat, welche DPN, ADP und anorganisches Phosphat als Cofaktoren benötigt. Auch bei der Zellatmung kann nur eine kleine Anzahl der zahlreichen Reaktionsfolgen zu den Schrittmachern zählen. Die Zwischenstufen des Citronensäurecyclus, des Fettsäureabbaues und des Aminosäurestoffwechsels werden im allgemeinen so schnell durchlaufen, daß es zu keiner Anhäufung von Metaboliten kommt. KREBS folgert daraus, daß es die *einleitenden Stufen* der Substratoxydation sein müssen, die als Schrittmacher der Atmung funktionieren. Sie bestimmen darüber, ob Zucker, Fett- oder Aminosäuren in einer gegebenen Situation vorrangsweise als Energiequelle dienen.

Alternative Reaktionen dieser Art sind jedoch nicht voneinander unabhängig. In einer längeren Reaktionsfolge ist oft nur die 1. Stufe substratabhängig (substratspezifische Dehydrogenasen); die nachfolgenden Zwischenstufen sind auf einer langen Strecke für eine größere Anzahl von Substraten gemeinsam. So läuft beispielsweise der Weg des Wasserstoffes von den Substraten zum molekularen Sauerstoff auf einer gemeinsamen Endstrecke (nach KREBS):

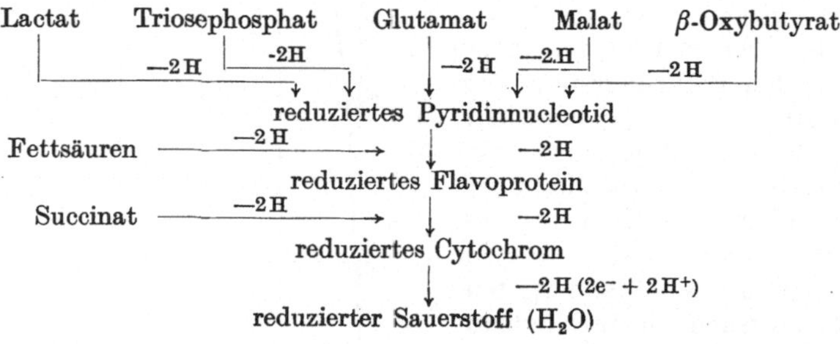

Das Problem der Wahl unter mehreren Substraten bedeutet nichts anderes als eine Konkurrenz verschiedener Stoffe um den gemeinsamen Katalysator. Welches Substrat zum Zuge kommt, ist nach KREBS allein eine Frage der physikochemischen Bedingungen und der Eigenschaften der Partner, die sich bestimmen lassen.

Wenn die Alternative zwei gegenläufige Richtungen derselben Reaktion betrifft, wie z. B. Synthese oder Abbau, so ist die Rückreaktion, falls alle Teilschritte reversibel verlaufen, die Folge einer Veränderung der Gleichgewichtslage. Zu den über die Richtung entscheidenden Faktoren gehört der Oxydoreduktionszustand der beteiligten Coenzyme. Der Fettsäureabbau wird durch die Verfügbarkeit von Coenzym A und den oxydierten Formen von Flavoprotein und DPN, die Synthese hingegen durch die reduzierte Form der Cofaktoren, ferner durch Acetyl-Coenzym A und Abwesenheit von freiem Coenzym A gefördert. Da Zwischenstufen des Kohlenhydratstoffwechsels DPN und Flavoprotein schneller reduzieren als Fettsäuren, und Brenztraubensäure auch mit Coenzym A schneller als Fettsäure reagiert, führt reichlich verfügbares Kohlenhydrat zu einem Anstieg der Quotienten:

$$\frac{\text{Acetyl-CoA}}{\text{freies CoA}}, \quad \frac{\text{red. Flavoprotein}}{\text{oxyd. Flavoprotein}}, \quad \frac{\text{DPN}\cdot\text{H}}{\text{DPN}^+}.$$

So kann man verstehen, daß Anwesenheit von Zucker die Synthese, Abwesenheit den Abbau von Fettsäuren begünstigt. Von entscheidender Bedeutung sind nach KREBS die drei obengenannten Quotienten.

IV. Cytochemische Grundlagen der Enzymwirkung.

A. Allgemeines.

Unklare Vorstellungen über eine „chemische Organisation der Zelle" bestehen schon lange. Anfangs dachte man sich Substrate und Katalysatoren in der Zelle räumlich getrennt gehalten und nach Maßgabe einer „regulierenden Wirkung" des Protoplasmas zur Reaktion gebracht (HOFMEISTER). Es bestanden aber damals noch keine experimentellen Möglichkeiten, diese Frage zu bearbeiten. Als man sich bemühte, Fermente aus den Zellen zu isolieren und von Begleitstoffen zu reinigen, entwickelte man Verfahren zum Aufschluß von Zellen und Ablösung von Enzymen aus der organisierten Struktur. Man fand hierbei, daß manche Enzyme verhältnismäßig leicht aus aufgebrochenen Zellen in Lösung zu bringen (Lyoenzyme), andere nur sehr schwer oder gar nicht vom Zellgerüst abzutrennen sind (Desmoenzyme) (BAMANN und MUKHERJEE; WILLSTÄTTER und ROHDEWALD). Es wurde aber damals noch kein Versuch gemacht, diese Beobachtungen mit cytologischen Gegebenheiten zu verbinden. O. WARBURG gebührt das Verdienst, bereits 1913 auf Grund experimenteller Befunde die Folgerung gezogen zu haben, daß der Stoffwechsel mit strukturellen Einrichtungen der Zellkomponenten innig verknüpft sein muß. Bezüglich der Natur dieser „strukturellen Anordnungen" bestanden damals allerdings noch keine konkreteren Vorstellungen. Seit ungefähr 15 Jahren wird die Frage nach der Lokalisation der Enzyme in der Zelle auf mehreren Wegen experimentell bearbeitet. Die Möglichkeit hierzu bot die Entwicklung neuer Methoden. Eine ausgezeichnete Diskussion der hierbei aufgetretenen Probleme hat kürzlich HOLTER (1952) gebracht.

Der eine Weg führte zur Ausarbeitung der *Homogenisattechnik* (HOGEBOOM 1951). Man versucht die Zellen mechanisch aufzubrechen unter Schonung der Zellbestandteile. Das Prinzip dieser Methode ist schon lange bekannt. WARBURG hat die Technik bereits vor 1913 angewandt. Die Verbesserung beruht auf der *Entwicklung geeigneterer Homogenisatoren*, unter denen der von POTTER-ELVEHJEM sich trotz mancher Unzulänglichkeiten besonders eingeführt hat, und der *Erprobung von Medien*, die sich zur Suspension der Zelltrümmer und Zellbestandteile eignen. Da Wasser nicht befriedigte und Salzlösungen meist zu einer Verklumpung granulärer Zellbestantdeile führen, verwendet man heute elektrolytfreie Medien. Am besten geeignet scheinen Rohrzuckerlösungen zu sein, von denen eine isotonische (0,25 molare) (SCHNEIDER 1948) und eine hypertonische (0,88 molare) (HOGEBOOM, SCHNEIDER und PALADE 1947, 1948) gebraucht werden; sie dürften auch zur Stabilisierung der suspendierten enzymtragenden Partikel beitragen. Die Suspension wird dann einer *fraktionierten Zentrifugierung* (CLAUDE 1940—1946, GLICK 1949, DUVE und BERTHET 1954) zugeführt. Die gewonnenen Sedimente werden für gewöhnlich in einem geeigneten Medium resuspendiert und abermals einer Zentrifugierung unterworfen mit der Absicht, durch wiederholte Waschungen zu möglichst homogenen Suspensionen bestimmter Zellbestandteile, wie Zellkernen, Mitochondrien, Mikrosomen zu gelangen, die von anhaftendem Grundplasma und Trümmern fremder Zellkomponenten weitgehend befreit werden sollten. Die auch bei hoher Zentrifugalbeschleunigung nicht sedimentierbare Fraktion wird als Suspension des Grundcytoplasmas betrachtet. Diese Methodik wurde neben einigen Versuchen auf botanischem Gebiet hauptsächlich zur Untersuchung parenchymatöser Organe von Säugetieren verwendet, da nur sie eine zur Anwendung dieser Technik geeignete Beschaffenheit besitzen. Die Homogenisattechnik ist in den letzten Jahren von sehr vielen

Autoren in den verschiedensten Laboratorien benutzt worden. Sie erfreut sich einer so großen Beliebtheit, weil sie auf verhältnismäßig bequeme Weise genügende Mengen an Material liefert, so daß sich die Anwendung besonderer Mikromethoden zur Untersuchung der enzymatischen Aktivität der Fraktionen erübrigt. Die Methodik schließt aber eine Reihe von Gefahren (BRADFIELD) ein, darunter das Auftreten von *Autolyse* nach mechanischer Zerstörung der gröberen Zellstruktur und die Möglichkeit einer *Elution* von Enzymen, Coenzymen, Aktivatoren u. a. aus dem Sediment im Verlauf der nötigen Waschungen und *einer Readsorption* von Enzymen und anderen Bestandteilen des Grundcytoplasmas, z. B. an nucleinsäurehaltige Areale der isolierten Zellpartikel. Die fraktionierte Zentrifugierung ist ferner wenig geeignet, eine *scharfe* Trennung verschiedenartiger Zellpartikel herbeizuführen. Eine teilweise Überlappung der einzelnen Fraktionen ist nur sehr schwierig zu vermeiden. Eingehende mikroskopische Kontrollen auf Einheitlichkeit des Sedimentes unter Benützung geeigneter färberischer Verfahren sowie des Phasenkontrastmikroskopes sind dringend erforderlich, genügen aber in den meisten Fällen nicht, da die morphologischen Kriterien zur Charakterisierung isolierter Zellbestandteile im allgemeinen unzureichend sind. Das ist in noch gesteigertem Maße der Fall, wenn nicht Leberzellen von Säugetieren, sondern Zellen niederer Tiere (Protisten) oder von Pflanzen untersucht werden. Die Abgrenzung der einzelnen Fraktionen unterliegt daher aus allen diesen Gründen einer gewissen Willkür.

Die andere heute übliche Methode geht von der lebenden Zelle aus. HARVEY hat gezeigt, daß man durch Zentrifugieren intakter Eizellen des Seeigels eine Schichtung des Zellinhaltes erreicht und in besonders günstigen Fällen sogar eine spontane Abtrennung geschichteter Teilstücke erzielt, wenn man die Zellen vor dem Zentrifugieren in ein Medium einbettet, welches einen Dichtegradienten besitzt. In weniger günstigen Fällen muß die Aufteilung mit Hilfe des Mikromanipulators durchgeführt werden. Da man als Medien Kulturflüssigkeit mit indifferenten Zusätzen verwendet, bleiben die Zellen während der Stratifizierung in der Zentrifuge am Leben. Auch die Stratifizierung selbst ist meist kein schwerer Eingriff; nach Aufhören der Zentrifugalkraft setzt schon in kurzer Zeit spontan eine Durchmischung der separierten Zellkomponenten ein und die Zellen überleben den Eingriff. Da diese Methodik im allgemeinen nur wenig Material liefert, ist man auf Mikromethoden zur Untersuchung der enzymatischen Aktivität angewiesen. Hier haben sich die im Carlsberg-Laboratorium von K. LINDERSTRØM-LANG und H. HOLTER entwickelten maßanalytischen, dilatometrischen und gasometrischen (Cartesianische Taucher) Analysenverfahren sehr bewährt. Zur Untersuchung gelangten bisher mit dieser Methode fast ausschließlich Gewebe und größere Zellen, wie z. B. Protisten, Oocyten und Eizellen wirbelloser Tiere, sowie niederer Wirbeltiere, nicht dagegen einzelne Gewebezellen aus den Organen von höheren Wirbeltieren und des Menschen, sowie von Pflanzenzellen. Die Empfindlichkeit der meisten heute vorliegenden analytischen Verfahren ist der Größenordnung dieser Objekte noch nicht adäquat. Auch stößt ihre Isolierung auf sehr große methodische Schwierigkeiten. Da auch hier zur Stratifizierung der Zellen Zentrifugalkräfte gebraucht werden, ist die komplette Isolierung eines bestimmten Zellbestandteiles nicht erreichbar, weil zumindest das Grundcytoplasma sich zwischen die separierten Teilchen gewissermaßen als Verunreinigung einschiebt. Aber der große Vorteil dieses Verfahrens ist die Anreicherung von Plasmapartikeln in ihrem natürlichen Medium, also unter physiologischen Bedingungen. Da die Objekte überleben, ist nicht anzunehmen, daß schwerwiegende Verlagerungen von Enzymen während der Stratifizierung zwischen Partikel und Grundplasma stattfinden. Das äußert sich z. B. darin, daß die Summe der enzymatischen Aktivitäten der Teilfraktionen mit der Gesamtaktivität der intakten Zellen im allgemeinen sehr gut ist (Andresen, Engel und Holter, Holter 1936, Holter und Kopac, Holter und Løvtrup).

B. Verteilung der Enzyme in der Zelle.

a) Grundcytoplasma und Plasmapartikel.

Es ist verständlich, daß zur Zeit das Interesse der Cytochemiker hauptsächlich auf die partikulären Zellkomponenten gerichtet ist. Die Homogenisattechnik ermöglicht die Bereitstellung ausreichender Mengen dieses Materials. Es wurde bald erkannt, daß insbesondere die Mitochondrien als Träger wichtiger Enzyme eine entscheidende Rolle im Stoffwechsel der Zelle spielen dürften. Über das Grundcytoplasma ist hingegen sehr viel weniger gearbeitet worden. Der Überstand nach der Aktivierung aller in der Zentrifuge sedimentierbarer Teilchen enthält im allgemeinen eine Anzahl von Enzymen, die dem Grundcytoplasma der lebenden Zelle zugeschrieben werden. Hierzu zählen zahlreiche Fermente der Glykolyse, darunter Aldolase, Phosphoglucomutase und das oxydierende Gärungsferment, dann Isocitronensäuredehydrogenase, Hexosediphosphatase, Katalase, Peptidasen, saure und alkalische Phosphatase, sowie Diaminoxydase. Wie aber Holter (1952) in einem sehr kritischen Referat betont, ist heute noch kaum bekannt, wieviel von den in den Zellpartikeln lokalisierten Enzymen sich im Verlaufe der Homogenisierung aus diesen herauslöst oder durch Autolyse freigesetzt wird. Solche Enzyme werden, falls sie ihre Aktivität nicht einbüßen, sich ebenfalls im Überstand einfinden. Man ist im allgemeinen der Meinung, daß die Enzyme im Grundcytoplasma nicht strukturgebunden sind. *Das Cytoplasma könnte demnach als ein ungeordnetes Multienzymsystem betrachtet werden.* Alle abdissoziierbaren Cofermente könnten das Enzymprotein verlassen und zusammen mit Substratmolekülen mehr oder weniger ungehindert weite Wegstrecken in der Zelle zurücklegen, um an einer ganz anderen Stelle zur Reaktion zu kommen. Wenn diese Vorstellung richtig ist, müßte sich der glykolytische Abbau der Hexosen nach diesem Mechanismus vollziehen und die entstandene Brenztraubensäure schließlich in den Mitochondrien verbrannt werden.

In letzter Zeit ist es aber fraglich geworden, ob das Cytoplasma mit Recht als ein ungeordnetes Multienzymsystem aufgefaßt werden darf. Die schönen elektronenmikroskopischen Aufnahmen ultradünner Schnitte durch das Mäusepankreas von Sjöstrand und Hanzon zeigen eine reiche Strukturierung des Cytoplasmas, wobei dicht gepackte Systeme konzentrisch angeordneter, etwa 40 Å dicker Membranen auffallen, denen einseitig winzige Partikelchen angeheftet sind. Palade (1955) bestätigt mit ausgezeichneten Aufnahmen die lamellare Struktur des Grundcytoplasmas und bringt eine genaue Beschreibung dieser winzigen partikulären Komponente. Die meisten der granulären Gebilde haben einen Durchmesser von 100—150 Å und fallen daher in die Dimensionen der Makromoleküle. Diese Granula kommen am reichlichsten im Cytoplasma embryonaler Zellen, in rasch wachsenden Zellen adulter Gewebe (Matrix der Epidermis, Epithel der Darmkrypten) und vor allem in den Drüsenzellen des Pankreas und der Speicheldrüse vor, alles Zellen, die durch einen hohen Gehalt an Ribonucleinsäure ausgezeichnet sind. Am wenigsten reichlich fand Palade diese Körnchen im Cytoplasma von Granulocyten, sie fehlten in adulten Erythrocyten. Es ist zunächst nicht möglich, zu entscheiden, ob die winzigen Teilchen schon von vornherein oder erst durch den Prozeß der Fixierung mit Osmiumsäure an die Membranstrukturen des Cytoplasmas verankert wurden. Aber Palade macht darauf aufmerksam, daß die Assoziation der kleinen Körnchen an die Membranen des endoplasmatischen Reticulums hauptsächlich in verschiedenen Zelltypen beobachtet werden kann, die durch einen hohen Differenzierungsgrad charakterisiert sind, während die Körnchen bei anderen Zelltypen, die sich durch eine rasche Proliferation auszeichnen, mehr oder weniger frei in der Grundsubstanz des Cytoplasmas vorkommen.

Es ist wenig darüber bekannt, ob und inwieweit synthetische Prozesse im Grundcytoplasma ablaufen. Nach MÜLLER und LEUTHARDT kommen die Enzyme zur Synthese von Glutamin und Arginin im Grundcytoplasma vor, benötigen aber die Gegenwart von Mitochondrien; ähnliches gilt für das Transaminasesystem (HIRD und ROWSELL). Die Kohlensäureanhydrase der höheren Pflanzen wurde ebenfalls im Grundplasma gefunden (WAYGOOD und CLENDENNING). Die Lokalisation der Dipeptidase und Katalase im Grundplasma konnte mittels der Methode der Zentrifugierung von lebenden Zellen (vgl. S. 524) sichergestellt werden (HOLTER 1937, 1939, 1952 und 1954; HOLTER und DOYLE; HOLTER und KOPAC; HOLTER und LØVTRUP).

Eine ganze Reihe von Enzymen kommt sicherlich sowohl im Grundplasma als auch in geformten Zellbestandteilen vor.

Man darf nicht außer acht lassen, daß Partikelchen mit einem Durchmesser von weniger als 50 mμ von den üblicherweise verwendeten Zentrifugen nicht abgeschleudert werden. Nach den von FREY-WYSSLING (1938, 1947) entwickelten, neuerdings revidierten Vorstellungen (FREY-WYSSLING 1955), sowie den Arbeiten von MONNÉ und LEHMANN ist das Grundplasma von einem Fasergerüst durchzogen, in welches subvisibles, partikuläres Material eingeschlossen ist. Es ist heute noch unbekannt, inwieweit die zur Zeit im Grundplasma nachgewiesenen Enzyme an solche geformte Strukturen verankert sind.

Bald nachdem eine systematische Bearbeitung der Plasmapartikel in Angriff genommen wurde, erkannte man, daß es sich hierbei um eine Vielfalt von Strukturen handelt. CLAUDE (1944) unterschied 2 Kategorien von Teilchen:

a) *Große Teilchen* mit einem Durchmesser von etwa 0,5 μ und einer erheblich schwankenden Länge. Zu dieser Gruppe zählen kugelige oder kurze Teilchen von 0,5—1,5 μ Länge oder fadenförmige, bis zu 4 μ lange Gebilde.

b) *Kleine, submikroskopische Teilchen* von 50—200 mμ Durchmesser (CLAUDE 1940). Beide Gruppen von Teilchen unterscheiden sich in zahlreichen Merkmalen, auch in chemischer Hinsicht. Die kleinen Teilchen sind durch einen hohen Gehalt an Lipoiden und Ribonucleotiden ausgezeichnet. Da die großen Teilchen gegen Änderung der osmotischen Verhältnisse sehr empfindlich sind, aber in Rohrzuckerlösungen vitalen Mitochondrien weitgehend gleichen, beispielsweise supravital eine Speicherung von Janusgrün zeigen, wurden sie mit diesen identifiziert. Als aber CLAUDE (1943) die submikroskopischen Teilchen als „Mikrosomen“ ansprach, entstand eine gewisse Verwirrung, zumal diese Teilchen anfangs auch von Mitochondrien nicht eindeutig unterschieden wurden. Unter Mikrosomen verstand man bisher, vor allem in der Botanik, mikroskopisch sichtbare körnchenartige Plasmaeinschlüsse ohne genauere Festlegung auf eine bestimmte chemische Zusammensetzung. Die Existenz sehr kleiner Teilchen, die sich mit Kernfarbstoffen leicht anfärben lassen, war schon lange vor CLAUDE bekannt. Auf Grund ihrer Färbbarkeit nahm man ihre Herkunft aus dem Zellkern an und nannte sie in Anlehnung an das Chromatin des Zellkernes „Chromidien“. Diese Ableitung wurde nicht allgemein anerkannt und geriet bei den Cytologen später in Vergessenheit. Da man aber heute weiß, daß diese winzigen Teilchen reich an RNS sind, ist zumindest ihre gute Färbbarkeit mit Kernfarbstoffen durchaus erklärlich. MONNÉ verwendete diesen längst eingeführten Terminus zur Charakterisierung der submikroskopischen Partikelchen, die sich nach Behandlung von

Seeigeleiern mit Natriumazid zu lichtmikroskopisch sichtbaren, stark färbbaren, granulären Gebilden aggregieren, die wie die Perlen einer Halskette an nicht färbbaren Fäden aufgereiht erscheinen. Diese grobe Struktur ist zwar ein Kunstprodukt, geht aber wahrscheinlich auf ein submikroskopisches, plasmatisches Reticulum zurück, dessen Interstitium von einer proteinhaltigen Flüssigkeit, dem Enchylema ausgefüllt wird und in dessen Maschenwerk „Chromidien" ein- und angelagert sind (MONNÉ). Elektronenmikroskopische Aufnahmen von LEHMANN konnten eine solche Struktur im Plasma der Amöben nachweisen. In der biochemischen Literatur versteht man heute unter Mikrosomen subvisible, reichlich RNS-haltige Plasmateilchen, die sich aus Homogenisaten nur bei Anwendung sehr hoher Zentrifugalkräfte abschleudern lassen. Nach HOLTER, OTTESEN und WEBER (1953) sind sie die dichtesten Gebilde im Cytoplasma, ihre Dichte kann bis zu 1,29 betragen.

Man kennt heute noch nicht viele Enzyme, von denen man behaupten könnte, daß sie sich größtenteils in Mikrosomen vorfinden. Erschwerend wirkt der Umstand, daß man nicht scharf definieren kann, was man unter Mikrosomen versteht. Es ist heute sogar sehr unwahrscheinlich geworden, daß es sich um native Teilchen des Cytoplasmas handelt.

Wahrscheinlich sind die Mikrosomen keine einheitliche Population von Plasmateilchen. Sie unterscheiden sich in ihrer Größe untereinander ganz bedeutend. Aus Homogenisaten der Mäuseleber lassen sich Mikrosomen isolieren, die 3 Gruppen angehören: große (125—130 mμ), mittlere (65—70 mμ) und kleine (23—25 mμ). Es ist aber nicht bekannt, ob diese Verteilung nativ ist oder ein Kunstprodukt darstellt (SLAUTTERBACK). Nach heutiger Ansicht sind folgende Enzyme zum überwiegenden Teil in Mikrosomen lokalisiert: DPN-Cytochrom-Reductase, Glucose-6-phosphatase der Leber, Niere und Gehirn, alkalische Phosphatase der Darmwand, das Enzym zur Umwandlung von Cortison in 17-Oxycorticosteron in der Leber, eine Esterase der Leber, die Vitamin A-Esterase und Gewebsthrombokinase (LANG und SIEBERT). Daß der RNS bei der Bindung von Enzymen an diese Teilchen eine Rolle zukommt, geht aus Beobachtungen von LUNDBLAD und HULTIN am Seeigelei hervor. In unbefruchteten Eiern haben bei neutraler Reaktion optimal spaltende proteolytische Enzyme eine nur geringe und schwankende Aktivität, aber in befruchteten Eiern wird diese bedeutend und konstant. Setzt man aber Extrakten aus unbefruchteten Eiern Ribonuclease zu, so steigt die proteolytische Aktivität um fast eine Zehnerpotenz an. Offenbar inaktiviert die Ribonucleinsäure in unbefruchteten Eiern diese Enzyme, verliert aber ihre Wirkung nach der Befruchtung infolge eines enzymatischen Abbaues zu nicht mehr hemmenden Spaltstücken. Es ist heute noch nicht entschieden, ob die Mikrosomen eine Partikelgruppe sui generis sind oder in Beziehung zu Mitochondrien stehen. Für eine Sonderstellung sprechen die charakteristische chemische Zusammensetzung, der Enzymgehalt und die erheblichen synthetischen Leistungen. Wie Versuche mit radioaktiven Aminosäuren gezeigt haben, sind die Mikrosomen allen anderen Zellelementen bezüglich der Einbaurate von Aminosäuren überlegen (Literatur bei LANG und SIEBERT). Für einen Zusammenhang zwischen Mikrosomen und Mitochondrien sprechen hingegen Beobachtungen von CHANTRENNE, daß die Granula aus dem Homogenisat der Mäuseleber eine kontinuierliche Reihe von Übergangsstufen bilden, nicht nur hinsichtlich der Größe, sondern auch der chemischen Zusammensetzung und des Enzymgehaltes. Sie lassen sich nicht einfach in 2 Klassen aufteilen, die großen und kleinen Granula, wie CLAUDE (1943) es vorschlug. Mit steigender Größe fällt der RNS-Gehalt, die Aktivität an alkalischer Phosphatase und ATP-ase nimmt hingegen zu. JEENER behandelte große Granula mit starken Salzlösungen und fand, daß sie hierbei in kleinere Elemente zerfallen von höherem RNS-Gehalt, aber niedrigerem Gehalt an Cytochromoxydase. Jeener knüpfte an diese Beobachtungen die Hypothese, daß die kleinen RNS-reichen Partikel aggregieren und daß das Innere dieser Teilchen der Ort der Enzymsynthese sei. Die Synthese geht unter Verlust von RNS vonstatten, wobei laufend neue RNS-reiche Teilchen an der Peripherie hinzukommen. Das Endstadium dieser Entwicklung sollen Mitochondrien sein. Weitere Beispiele, die für einen Übergang von Mikrosomen von einer Größenklasse in eine andere sprachen, sind die Beobachtungen von Lehmann am Ei von *Tubifex*, sowie die älteren Versuche von E. B. Harvey am Seeigelei. Die durch Zentrifugalkraft abgetrennten hellen Fragmente des Eies von *Arbacia* sind frei von groben granulären Einschlüssen wie Mitochondrien, Dotter und Pigment. Nach der Befruchtung entwickeln sie sich aber trotzdem zu fast normalen, wenn auch kleineren Plutei weiter, die mit Mitochondrien ausgestattet sind. Hutchens, Kopac und Krahl finden die Cytochromoxydase bei *Arbacia*-Eiern mehr an kleinere Granula gebunden, als an solche, die ihrer Größe und Färbbarkeit nach beurteilt als Mitochondrien angesprochen werden müßten. Dieser Befund zeigt besonders eindrucksvoll, wie wenig abgeklärt unsere gegenwärtigen Vorstellungen von der Natur der Mikrosomen sind. Eine rein morphologische Charakterisierung der Zellpartikel bringt von vornherein die Gefahr, ein starres, statisches Moment in die Betrachtungen über die Natur dieser Teilchen einzuführen, die dem dynamischen Verhalten dieser Zellelemente in der lebenden Zelle nicht gerecht wird. Man sieht an diesem Beispiel besonders klar, welche Bedeutung der Forschung über die Lokalisation der Enzyme in den Zellkomponenten zukommt, da die berechtigte Hoffnung besteht, aus dem enzymatischen Verhalten dieser Teilchen einen Rückschluß auf ihre Funktion ziehen zu können. Man muß Holter (1952) recht geben, wenn er fragt: "But if is so (gemeint ist, die Eigenschaften der Partikel auf Grund ihres enzymatischen Verhaltens zu erhellen), has not the time come to rely less upon morphological terms, to assume that the existence of a great variety of more or less closely related granular structures is highly probable, and to try to characterize and designate the single members of the family by their chemistry and function?" Allerdings darf sich die Forschung nicht auf wenige, aus methodischen Gründen besonders dazu geeignete Gewebe von Säugetieren beschränken, wie das heute im großen und ganzen der Fall ist, sondern danach trachten, die Zellen in ihrer ganzen Fülle von Differenzierungshöhen und Anpassungsformen an spezielle physiologische Leistungen mit in die Betrachtung einzuschließen.

Über *Mitochondrien* ist in der letzten Zeit viel geschrieben worden. Das Elektronenmikroskop zeigt eine charakteristische Feinstruktur. Die Teilchen besitzen eine 7—8 mμ dicke Membran, von der aus parallele Leisten (cristae mitochondriales) nach innen ausgehen. Zwischen der Anzahl dieser Leisten und der Atmungsintensität scheint eine engere Beziehung zu herrschen (Palade 1952, 1953, 1955). Die Struktur der Mitochondrien dürfte im ganzen Organismenreich einheitlich sein, da beispielsweise die Mitochondrien des Flagellaten *Euglena* denen aus der Säugerleber im Bau durchaus gleichen (Wolken und Palade). Auf Zeitrafferaufnahmen zeigen Mitochondrien lebhafte Bewegungen, die den Eindruck einer Eigenbeweglichkeit machen (Gey, Sapranaukas und Bang). Man weiß auch, daß Mitochondrien im allgemeinen aus etwa 60% Protein und etwa 30% Lipiden bestehen, wozu noch kleinere Mengen an RNS kommen (Lang und Siebert); in diesen Partikeln wurde überdies eine große Anzahl biochemisch bedeutsamer Verbindungen, wie Cytochrom, DPN, TPN, Coenzym A, Flavin-Adenin-Dinucleotid, Diphosphothiamin, α-Liponsäure, Pyridoxalphosphat,

Folinsäure, Biotin und B_{12} gefunden, was allein schon für ihre Bedeutung im Stoffwechsel der Zellen spricht. Nach dem Aufbrechen der Mitochondrienmembran durch Ultraschall oder mechanische Kräfte gehen etwa 60% ihres Proteingehaltes in Lösung und mit diesen Proteinen auch manche Enzyme, wie z. B. die saure Phosphatase (BERTHET und DE DUVE), während andere Enzyme, zu denen die Succinodehydrogenase und die Cytochromoxydase zählt (HOGEBOOM 1951), fest an dem Gerüst dieser Partikel verankert bleiben. In der Mitochondrienmembran liegen Proteine. Nach der Behandlung von Mitochondrien aus der Rattenleber mit gereinigtem Pepsin oder Trypsin, frei von Nucleasen, Lecithinasen und Lipase, wurde eine Aktivitätsverminderung der Cytochromoxydase, d-Aminosäureoxydase und Milchsäureoxydase beobachtet (DIANZANI 1953b). Man hat eine Reihe von Enzymen und Enzymsystemen hauptsächlich in Mitochondrien lokalisiert gefunden, so z. B. Cytochromoxydase, Succinoxydase, TPN-Cytochrom-Reductase, Octansäureoxydase und ATP-ase (vgl. Tabellen von LANG und SIEBERT). Um den Anteil, den ein Zellbestandteil am Stoffwechsel der ganzen Zelle nimmt, in adäquater Weise zu erfassen, wäre zu fordern, daß nach erfolgter Homogenisation und Auftrennung der Fraktionen eine Bilanz der Aktivitäten aufgestellt wird, und zwar sollte als Kriterium die Aktivität der Zellen und nicht die des Homogenisates gewählt werden, um den Einfluß der Homogenisation in halbphysiologischen Medien auf die Aktivität der Enzyme mit zu erfassen und nicht nur den der späteren Schritte bei der Aufarbeitung, wie BRADFIELD und HOLTER forderten. Das ist jedoch bisher kaum in ausreichender Weise geschehen; so fehlen heute noch genügend gesicherte Zahlen über die Verteilung von Enzymen auf Mitochondrien und die übrigen Zellkomponenten. Aber es ist wohl sehr wahrscheinlich, daß die Systeme des Citronensäurecyclus (mit Ausnahme der Isocitronensäuredehydrogenase), der Fettsäureoxydation, des Harnstoffcyclus und der Synthese von Hippursäure und Glutamin hauptsächlich in den Mitochondrien lokalisiert sind, obwohl in vielen Fällen die volle Aktivität erst nach der Vereinigung mehrerer Fraktionen des Homogenisates erreicht wird (KENNEDY und LEHNINGER; POTTER, RECKNAGEL und HURLBERT; DE DUVE und BERTHET).

Alle diese Befunde sind an parenchymatösen Organen von Säugetieren erhoben worden, aber viele Beobachtungen sprechen dafür, daß sie allgemeinere Gültigkeit haben. So fanden HOLTER und DOYLE an der Amöbe, daß die Amylase an mitochondrienartige Teilchen gebunden ist. HOLTER und LØVTRUP konnten den gleichen Befund für die Proteinase und ANDRESEN, ENGEL und HOLTER für die Bernsteinsäuredehydrogenase der Amöbe erhalten. Auch die Mitochondrien der Pflanzen scheinen sich ähnlich zu verhalten. Eine Mitochondrienfraktion aus etiolierten Keimlingen der Lupine verarbeitete ein Gemisch aus Brenztraubensäure (2-C^{14}) und Apfelsäure in Gegenwart anderer Säuren des Citronensäurecyclus vollständig zu $C^{14}O_2$ und Wasser (BRUMMOND und BURRIS). Eine Zusammenfassung der botanischen Befunde gab NEWCOMER. Protisten, wie *Chaos chaos* und *Tetrahymena geleii*, besitzen jedoch an Stelle des Cytochrom c-Systems der Zellen höherer Organismen Cytochrom e und eine terminale Cytochrom e-Oxydase (MØLLER und PRESCOTT).

Eine für den Zellstoffwechsel sehr wichtige Frage ist, ob die Mitochondrien alle von einheitlicher chemischer Zusammensetzung sind, oder ob sie eine Differenzierung in Abhängigkeit von der Zellfunktion erkennen lassen (PAIGEN; KUFF und SCHNEIDER). Untersuchungen von DE DUVE, GIANETTO, APPELMANS und WATTIAUX sprechen dafür, daß in den Zellen der Rattenleber mindestens 2 Klassen von Mitochondrien vorkommen, die eine untereinander verschiedene, aber innerhalb einer Klasse einheitliche Enzymausrüstung haben. Die eine Klasse enthält neben Cytochromoxydase auch Rhodanese, die andere neben saurer Phosphatase auch α-Glucuronidase und Kathepsin. Untersuchungen von FELIX, DECKER und ROKA haben ergeben, daß die Mitochondrien aus den homologen Organen verschiedener Individuen derselben Tierart die gleiche Aktivitätsverteilung der Enzyme zeigen, daß aber die Mitochondrien der einzelnen Organe dasselbe Substrat verschieden schnell abbauen. So zeigen z. B. die Mitochondrien der Nebenniere des Rindes eine höhere Aktivität an Apfelsäuredehydrogenase und Fumarase als die der Niere. In den Mitochondrien der Leber fehlte die Fumarase. Die Mitochondrien der Milz oxydieren d,l-Alanin nicht, haben aber die höchste Aktivität an Diaphorase.

Zusammenfassend kann man heute wohl behaupten, daß den Mitochondrien eine sehr wichtige Rolle im Stoffwechsel der Zellen zukommt. Man darf aber dabei nicht außer acht lassen, daß es viele Enzyme gibt, die weitverbreitet sind, in keiner wachsenden oder irgendwie am Stoffwechsel aktiv beteiligten Zelle fehlen, wie z. B. die Peptidasen, ja sogar in ihrer Aktivität im allgemeinen dem Gehalt der Zellen an RNS parallel gehen, der geradezu als Charakteristikum einer wachsenden Zelle gilt (BOTTELIER, HOLTER und LINDERSTRØM-LANG; DUSPIVA 1942; HOLTER 1952; PATTERSON, DACKERMAN und SCHULTZ; URBANI). Diese Enzyme sind nicht vorzugsweise an Mitochondrien gebunden. Ihre allgemeine Verbreitung spricht aber für eine wichtige Funktion im Zellstoffwechsel, obwohl man heute noch nichts Konkretes darüber aussagen kann. Andererseits zeigen die Befunde von KRAHL (1950), daß in den Frühstadien der Embryonalentwicklung manche Enzyme im Grundcytoplasma lokalisiert sind, die man auf späteren Stadien, wenn die Differenzierung der Zellen eingesetzt hat, an Granula gebunden findet. In neuester Zeit haben GUSTAFSON und LENIQUE an verschiedenen Stadien des Seeigelkeimes durch Zählung zu zeigen versucht, daß Partikelpopulationen eine Verteilung aufweisen, die dem Gefälle der morphogenetischen Leistung entspricht. All das deutet darauf hin, worin wir uns HOLTER (1952) anschließen möchten, daß die Mitochondrien spezialisierte Strukturen sind, welche für gewisse Funktionen der differenzierten Zelle notwendig sind, daß aber die elementaren Funktionen des Lebens, Wachstums und der Zellteilung nicht von ihnen getragen werden.

b) Zellkern.

Nach einer älteren, auf J. LOEB und UNNA zurückgehenden Anschauung sollte dem *Zellkern* eine führende Rolle bei den Oxydationsvorgängen der Zelle zukommen. Am isolierten Keimbläschen des Amphibieneies ließ sich aber weder eine Peroxydase noch eine Indophenoloxydase nachweisen. Isolierte Keimbläschen zeigten einen sehr geringen Sauerstoffverbrauch (BRACHET). Es konnte also biochemisch keine Stütze für obengenannte Hypothese gewonnen werden. Mehr Interesse wurde der Frage nach dem Enzymgehalt des Zellkernes und seiner Rolle im Stoffwechsel der Zelle entgegengebracht, als die moderne Homogenisattechnik die Isolierung von Kernen aus den Organen von Säugetieren in ausreichender Menge ermöglichte. Die Ergebnisse dieser Forschungsepoche sind erst kürzlich von LANG und SIEBERT tabellarisch zusammengefaßt worden, welche auch selbst zahlreiche Beiträge zur Frage nach dem Enzymgehalt des Zellkernes lieferten. Es hat sich gezeigt, daß der Enzymgehalt der isolierten Kerne im allgemeinen nicht höher ist als im Cytoplasma, meist sogar niedriger. Es konnte aber das Vorkommen zahlreicher Enzyme im Kern nachgewiesen werden. LANG und SIEBERT kommen zu dem Schluß, daß Zellkerne zwar reichlich mit Fermenten versehen sind, die zur Aufspaltung von Eiweiß, Nucleotiden und Lipoiden befähigt sind, also aller Stoffklassen, aus denen sich der Kern selbst zusammensetzt, daß der Kern aber aller Enzyme der biologischen Oxydation entbehrt. In den isolierten Zellkernen konnte weder das Vorkommen des WARBURG-KEILINschen Systems, noch von Flavinproteinen, Bernsteinsäureoxydase oder anderer Oxydasen gesichert werden. LANG und SIEBERT machten ferner die überraschende Beobachtung, daß der isolierte Kern mancher Zellen gewisse Enzyme in hoher Konzentration enthalten kann, für die man im Kern keine Funktion angeben könnte und die hier nur in einer mehr oder weniger inaktiven Form enthalten sein dürften. So wurde z. B. in Leberzellkernen reichlich Arginase angetroffen und in isolierten Pankreaszellkernen die mehr als 5fache Konzentration an Trypsin als im Cytoplasma der betreffenden Zellen. Nach Ansicht der Autoren deutet die hohe Konzentration der betreffenden Enzyme im Kern ohne eine ersichtliche funktionelle Leistung auf eine synthetische Neubildung dieser Fermente im Zellkern hin. Bei der Frage, wieweit die Ergebnisse der Enzymanalyse an isolierten Kernen heute bereits zu einem zuverlässigen Bild über die Stoffwechselleistung des Zellkernes beitragen können, muß man die Schwierigkeiten kritisch in Erwägung ziehen, die bei der Isolierung von Kernen — ebenso wie auch bei anderen Zellbestandteilen — aus Zellhomogenisaten in semiphysiologischen Medien zweifellos bestehen. DOUNCE, der selbst viel zu unserer Kenntnis vom Enzymgehalt des Zellkernes beigetragen hat, nahm auch kürzlich zu dieser Frage Stellung und meint, daß es zur Zeit wichtiger sei, *einen* positiven Enzymbefund im Zellkern nach allen Seiten zu sichern, anstatt einen nach einer bestimmten Methode isolierten Kerntyp auf eine große Zahl von Enzymen hin zu untersuchen. Man kann heute noch nicht absehen, ob und in welchem Maße die unverletzte Kernmembran im Suspensionsmedium während der Manipulationen bei der Isolierung für Proteine und damit auch für Enzyme durchlässig ist. Isolierte Kerne sind ferner dazu befähigt, fremde Proteine aus dem Suspensionsmedium aufzunehmen, was man z. B. mit Hämoglobin demonstrieren kann; sie geben diese bei gründlichem Waschen wieder ab. Viele Enzyme haben ein kleineres Molekulargewicht als Hämoglobin, und nichts spricht dagegen, daß sie eventuell noch leichter eindringen könnten. Zur Zeit unterscheiden sich die mit verschiedenartiger Methodik gewonnenen Resultate über den Enzymgehalt des Zellkernes noch so sehr, daß man nicht imstande ist, etwas Sicheres über den Enzymgehalt dieses Zellbestandteiles auszusagen. Die Analysenergebnisse bezüglich alkalischer Phosphatase, ATP-ase und Milchsäuredehydrogenase weichen ganz bedeutend voneinander ab, je nachdem, ob die Zellkerne in wäßrigen Medien oder in nichtwäßrigen Medien nach der Methode von BEHRENS (1939) oder einer Modifikation derselben gewonnen wurden. Es ist doch sehr verdächtig, daß der isolierte Kern ein Enzym besonders reichlich dann enthält, wenn es auch im Cytoplasma der zugehörigen Zelle entsprechend stark vertreten ist. MARSHALL behandelte Gefrierschnitte durch Pankreasgewebe mit Antikörpern gegen Chymotrypsinogen und Procarboxypeptidase, die mit Fluoresceinisocyanat fluorescierend gemacht wurden. Nach gründlichem Auswaschen der Schnitte fluorescierten wohl die Zymogengranula im Apex der Zellen, ein Zeichen, daß dort die Proenzyme gespeichert liegen, wie schon seit langem vermutet wurde, aber die Zellkerne, wie auch die Mitochondrien und das Grundplasma blieben dunkel. Markierte Antisera für Ribonuclease und Desoxyribonuclease färbten Granulen im Cytoplasma, nicht aber den Zellkern und Mitochondrien. Das Resultat paßt nicht zu den Befunden der Homogenisattechnik, da SCHNEIDER und HOGEBOOM Ribo- und Desoxyribonuclease in Mitochondrien aus Leberhomogenisaten, LANG, SIEBERT, BALDUS und CORBET hauptsächlich nur in Zellkernen gefunden haben. Es ist zwar nicht daran zu zweifeln, daß Zellkerne im allgemeinen keine führende Rolle bei den Oxydationsprozessen der

Zelle spielen, aber es ist doch fraglich, ob sie gänzlich frei von Oxydationsfermenten sind. Kerne, die aus Zellen gewonnen werden, die reich an Mitochondrien sind, lassen stets eine gewisse Aktivität an Oxydationsfermenten erkennen; sie wird gewöhnlich auf eine nur schwer vermeidbare Verunreinigung der Kernfraktion durch Mitochondrien zurückgeführt. Warburg beobachtete 1911, daß bei Vogelerythrocyten, die keine oder nur wenige Mitochondrien enthalten, fast das gesamte Atmungsvermögen mit den Zellkernen abzentrifugiert werden kann. Bei diesem Objekt ist nach Rubinstein und Denstedt (1953, 1954) die gesamte Cytochromoxydase- und Succinoxydaseaktivität der Zelle im Kern lokalisiert. In ihrer neuesten Arbeit bringen diese Autoren Indizien, die dafür sprechen, daß auch die Leberzellkerne über Oxydationsfermente verfügen, wenn auch der Gehalt an Kerncytochromoxydase an der gesamten Atmungskapazität dieser Zelle einen relativ geringen Anteil hat.

Untersuchungen über die Einbaurate markierter Substanzen sprechen für bedeutende synthetische Fähigkeiten des Zellkernes. P^{32} wird sogar schneller in die RNS der Rattenleberzellkerne eingebaut als in die des Cytoplasmas (Jeener und Szafarz; Barnum und Huseby; McIndoe und Davidson; Smellie, McIndoe, Logan, Davidson und Dawson; Davidson, McIndoe und Smellie; Anderson und Åqvist). Als weitere wichtige synthetische Leistung wird dem Kern die Bildung von DPN aus Nicotinsäureamidmononucleotid und ATP zugeschrieben. In den Leberzellen soll diese Synthese ausschließlich im Zellkern stattfinden (Hogeboom und Schneider 1952). Eine Bestätigung dieser Befunde durch Untersuchungen, die sich einer anderen Methodik bedienen, wäre sehr erwünscht.

Untersuchungen zur Frage, welche Rolle der Kern bei der Atmung der Zelle spielt, haben Brachet zusammen mit Chantrenne und Chantrenne-van Halteren an Amöben und *Acetabularia* durchgeführt. In entkernten Teilstücken von Amöben bleibt die Atmungsgröße über geraume Zeit konstant. Der O_2-Verbrauch kernloser *Acetabularia*-Zellen selber bleibt sogar über 3 Monate lang praktisch gleich hoch und erweist sich dennoch von der Gegenwart eines Kernes weitgehend unabhängig. Der Einbau von C^{14} in die Carboxylgruppen der Aminosäuren bleibt in kernlosen Teilstücken von *Acetabularia* erst 14 Tage nach der Entkernung den kernhaltigen Teilen gegenüber etwas zurück, ist aber selbst nach 5 Wochen noch nicht gänzlich unbedeutend geworden. Demnach dürfte auch die Eiweißsynthese ohne Zellkern eine beträchtliche Zeit fortgesetzt werden können. Vorläufig noch unerklärbar ist der Befund von Mazia und Hirshfield, daß die innerhalb von 24 Std in Amöben aufgenommene Menge an P^{32} bei den kernfreien Hälften um $^1/_3$ tiefer liegt als bei den kernhaltigen.

c) Golgi-Körper und Plasmamembran.

Über den Enzymgehalt der übrigen Zellkomponenten ist sehr wenig bekannt. Hier müßte eine Besprechung der Golgi-Körper, der pflanzlichen Plastiden und der Plasmamembran anschließen. Da die Golgi-Körper sich infolge ihrer außerordentlichen Empfindlichkeit zur Isolierung aus Homogenisaten nicht eignen, ist über ihren Anteil am Enzymgehalt der Zelle nichts bekannt. Trümmer dieser Zellorganelle finden sich im Überstand nach erfolgter Abscheidung der Mikrosomen. Typische pflanzliche Zellen mit reifen und funktionellen Chloroplasten sind nach der Homogenisierungstechnik äußerst schwer zu bearbeiten. Da hier nur ein allgemeines Bild von dem gegenwärtigen Stand unserer Kenntnisse über die Enzymverteilung in der Zelle entworfen werden soll, wird bezüglich des Enzymgehaltes der Plastiden auf die speziellen Kapitel des Handbuches verwiesen.

Eine besondere Bedeutung im Zellstoffwechsel kommt der Plasmamembran zu. Umfassendere Untersuchungen über deren Enzymologie liegen zur Zeit noch nicht vor. Bei Hefezellen konnten mehrere Phosphatasen in der Zellhaut lokalisiert werden, die phosphorylierte Zucker und andere Phosphatester spalten. Es ist sehr wahrscheinlich, daß diesen Enzymen eine Bedeutung im Zuckerhaushalt bzw. bei der Aufnahme von Zuckern zukommt (ROTHSTEIN, MEIER und HURWITZ). Für den Sitz der Saccharase an der Außenseite der Zellmembran bei der Bäckerhefe spricht die starke Abhängigkeit der Aktivität des Enzyms vom p_H des Mediums, sowie die Reaktivierung nach Uranylhemmung durch niedere Phosphatkonzentrationen. Der Befund, daß die Aktivität der Saccharase nach Behandlung der Hefe mit verdünnter Salzsäure sehr viel rascher zerstört wird als die Gärung, läßt sich in gleicher Weise deuten. Das native Enzym ist ein unlösliches Protein, das sehr fest an die Zellmembran gebunden ist. Durch Einwirkung von aktiviertem Papain bildet sich ein wasserlösliches Teilstück des nativen Saccharasemoleküls, das noch die volle enzymatische Aktivität besitzt (MYRBÄCK und WILLSTAEDT).

Es gibt eine Reihe von Untersuchungen, die eine Lokalisation der Acetylcholinesterase in der Plasmahaut der Neuronen und in der Membran der Erythrocyten wahrscheinlich macht (NACHMANSOHN und WILSON). Auch dieses Enzym steht mit Stoffverschiebungen in der Zellmembran und indirekt mit der Erregungsleitung in Zusammenhang. Das gilt auch für niedere Organismen, wie z. B. Protisten. Bei *Tetrahymena gelei* findet sich die gesamte Acetylcholinesteraseaktivität in der Membran (SEAMAN).

V. Die Organisation des enzymatischen Reaktionsapparates der Zelle.

Während in einem homogenen Reaktionsansatz der durch ein Enzym bewirkte Umsatz außer durch Milieufaktoren, wie Temperatur, p_H, r_H, Ionendichte und Aktivatoren, hauptsächlich durch die Enzym- und Substratkonzentration bestimmt wird, haben in der lebenden Zelle noch weitere Faktoren einen entscheidenden Einfluß auf den enzymatischen Umsatz; sie beruhen auf der ungleichmäßigen Verteilung von Enzym und Substrat im Protoplasma, die auf eine submikroskopische strukturelle Organisation der Zelle zurückgeht.

Man kann heute bereits eine Reihe von Beobachtungen heranziehen, die für einen Einfluß der Zellstruktur auf die Aktivität der Enzyme sprechen. So ist die ATP-ase-Aktivität frisch isolierter Mitochondrien sehr niedrig; mit dem Alter des Präparates steigt die Aktivität (HOGEBOOM und SCHNEIDER 1950). Aber parallel zu diesem Aktivitätsanstieg verliert das Präparat seine Fähigkeit, Phosphorylierungen auszuführen. Bei schonender Aufarbeitung von Rattenleberhomogenisaten liegt auch die saure Phosphatase weitgehend inaktiv vor; auch die Ribonuclease ist in intakten Teilchen unwirksam. Alterung des Präparates und viele andere Maßnahmen aktivieren solche Präparate. Die Anwesenheit von Rohrzucker und Glycerophosphat (Substrat) schützt die Phosphatase vor Aktivierung (BERTHET, BERTHET, APPELMANS und DE DUVE; BERTHET und DE DUVE; DE DUVE, BERTHET, BERTHET und APPELMANS).

Die stäbchenförmigen Mitochondrien der Leber-, Nieren- und Herzzellen behalten ihre Gestalt, wenn man sie in einer 17—30%igen Rohrzuckerlösung homogenisiert, nehmen aber Kugelform an, wenn das Medium nur 8,5% Zucker oder 0,9% KCl enthält (HOGEBOOM, SCHNEIDER und PALADE 1948). Die Veränderungen in der äußeren Form gehen mit inneren Strukturveränderungen einher: Struktur und Funktion stehen in enger Beziehung. „In vitro“ haben die kugelförmigen Mitochondrien einen höheren O_2-Verbrauch als die stäbchenförmigen (Harman 1950a, b).

In hypotonischen Zuckerlösungen oder in verdünnten Salzlösungen schwellen Mitochondrien; das gleiche tritt ein, wenn diese Plasmapartikel in isotonischem Medium nur ein paar Minuten bei Zimmertemperatur gehalten werden, es sei denn, daß ein Ca^{++}-bindendes Reagens, wie z. B. Äthylendiamintetraessigsäure zugegeben wird. Ein Zusatz von Ca^{++} beschleunigt die Schwellung. ATP oder Substrate der Enzyme des Citronensäurecyclus hemmen dagegen diese Quellung (Slater und Cleland).

Bei der Schwellung von Mitochondrien wird zuerst der Zentralkörper undeutlich, dann bleibt nur noch eine von der Membran des ursprünglichen Mitochondriums umgebene Blase übrig, die schließlich platzt, besonders bei mechanischer Belastung. Als sichtbare Bestandteile bleiben nur noch Fetzen der Mitochondrienmembran übrig. Mit zunehmender Quellung nehmen die Stoffwechselleistungen der Mitochondrien im allgemeinen ab, so z. B. die Fähigkeit, Glieder des Citronensäurecyclus zu oxydieren (Raaflaub 1952, 1953a, b). Die Mitochondrien geben hierbei Nucleotide ab (Dianzani 1953a). Es ist heute noch nicht genauer bekannt, welches die Ursachen des funktionellen Zusammenbruches sind. Denkbar wäre ein Verlust an Coenzymen und anderen Hilfsfaktoren enzymatischer Aktivität, die aus den Mitochondrien herausdiffundieren oder durch Autolyse zerstört werden.

Die einzelnen Enzyme der Mitochondrien verhalten sich bei dem osmotischen Abbau der Mitochondrienstruktur verschieden. Eine Gruppe von Enzymen, die dem Citronensäurecyclus angehört, zu der α-Ketoglutarsäure-Dehydrogenase, Fumarase und das „kondensierende Enzym“ Ochoas gehören, geht in Gegenwart von Salzen in Lösung (Slater). Eine andere Gruppe von Enzymen, zu der Milchsäuredehydrogenase und Aminosäureoxydase zählt, verliert in destilliertem Wasser an Aktivität. Die Enzymsysteme der Succinoxydase und α-Ketoglutarsäureoxydase der Sarkosomen des Herzmuskels gewinnen dabei an Aktivität (Slater; Slater und Cleland). An den Membranen der Sarkosomen bleiben Flavoprotein, Succinodehydrogenase und das ganze Cytochromsystem verankert, mithin wahrscheinlich die ganze Kette der Atmungsenzyme, die am Elektronen- oder H-Transport vom Succinat oder reduziertem DPN zu molekularem Sauerstoff beteiligt ist.

Die Sarkosomen, die Mitochondrien des Herzmuskels, reagieren leicht mit reduziertem oder oxydiertem Cytochrom c, das im Medium gelöst ist. Dieser Befund ist ungewöhnlich, da man weiß, daß Cytochrom c ein Protein vom Molekulargewicht 12500 ist, von dem man nicht gut erwarten kann, daß es durch die Membran der Sarkosomen hindurchtritt. Wenn aber die Enzymkette der Atmung in der Membran der Mitochondrien lokalisiert ist, wird dieser Befund verständlich. Im Zellkern liegen die Verhältnisse anders. Die Kerne der Vogelerythrocyten enthalten eine schwache Cytochromoxydase. Dem Suspensionsmedium zugeführtes Cytochrom c reagiert aber nicht mit den Enzymen der kerneigenen Atmung. Das kerneigene Cytochrom c läßt sich auch durch Ascorbinsäure nicht reduzieren, sondern allein durch p-Phenylendiamin. Die Kernmembran ist für Cytochrom c des Mediums undurchlässig und die Enzymkette der Atmung ist nicht in der Membran, sondern im Inneren des Kernes lokalisiert (Rubinstein und Denstedt).

Nach dem heute vorliegenden Beobachtungsmaterial wird man nicht daran zweifeln können, daß die Mitochondrien ein geordnetes Multienzymsystem enthalten. Das Cytophorasesystem von Green ist eine gelartige Fraktion, die hauptsächlich aus Mitochondriensubstanz gebildet wird. Dieses System verhält sich nicht wie ein einzelnes Ferment, sondern wie eine Hierarchie von Enzymen und die chemische Organisation, durch welche die einzelnen Enzymindividuen zu einem System integriert werden, bedingt bei den einzelnen Enzymen Eigenschaften, die ihnen fehlen, wenn sie aus dem Komplex herausgesprengt werden und als einzelne Enzymmoleküle vorliegen (Tabelle 5). Beispiele hierfür sind das Verhalten der Apfelsäuredehydrogenase und Aconitase.

In den Mitochondrien sind nicht nur Enzyme, sondern auch Coenzyme fest gebunden. Der Umsatz von Metaboliten geht so rasch vor sich, daß sich keine größeren Mengen davon anhäufen.

Ebenso wahrscheinlich ist heute, daß das ordnende Prinzip, welches die Einzelenzyme in ein System zusammenfügt, die Feinstruktur der Cytoplasmapartikel ist.

Tabelle 5. *Eigenschaften der Apfelsäuredehydrogenase im Verband und nach Isolierung.* (Nach HUENNEKENS.)

	Im Verband	Frei
Bedarf an Co-Zymase	—	+
Hemmung durch Dinitrophenol	+	—
Hemmung durch Oxalessigsäure	—	+
p_H-Optimum	7—8	9,5

Komplizierte Leistungen, wie z. B. die Konzentration und Anreicherung von anorganischen Ionen und organischen Stoffen im Inneren des Mitochondriums (BARTLEY und DAVIES) oder die oxydative Phosphorylierung, wenn diese mit Substrat versorgt werden und lebhaft oxydieren, kommen bei den geringsten Schädigungen zum Erliegen, wie z. B. durch Frost (PORTER, DEMING, WRIGHT und SCOTT) oder Fehlen des Adenosinphosphorsäuresystems (McFARLANE und SPENCER). Es ist weiterhin wahrscheinlich, daß RNS und Lipoide am Aufbau der für die planmäßige Anordnung der Enzyme entscheidenden submikroskopischen Strukturen einen wesentlichen Anteil haben.

Durch elektrische Reizung läßt sich die O_2-Aufnahme der Mitochondrien auf das Doppelte und mehr steigern. Bei Gehirnmitochondrien erfolgt dabei eine Entkoppelung zwischen Atmung und Phosphorylierung. Bei Lebermitochondrien ist das aber nicht der Fall. Lösliche Enzyme, wie z. B. die Enzyme der Glykolyse, werden durch einen solchen Eingriff nicht beeinflußt (ABOOD, GERARD und OCHS). Die Angriffspunkte der elektrischen Reizung sind Grenzflächen an den Strukturen der Mitochondrien. Wahrscheinlich spielen hier Phosphatide eine Rolle. Es wurde von NYGAARD und SUMNER (1953) gefunden, daß die Einwirkung von kristallisierter Schlangengiftlecithinase auf Lebermitochondrien zu einer perfekten Hemmung der Oxydation von Bernsteinsäure führt, ohne daß irgendein Teilenzym der Mitochondrien (Bernsteinsäure-Dehydrogenase oder DPN-Cytochrom c-Reductase) an Aktivität einbüßt. Der gegen Lecithinase empfindliche Faktor liegt vermutlich zwischen Cytochrom b und Cytochrom c und ist möglicherweise ein Phosphatid (McFARLANE).

Die Ursache für den optimalen Gesamteffekt ist wahrscheinlich darin zu suchen, daß die Fermente des Citronensäurecyclus und der Atmungskette in einem räumlich engen und passenden Abstand stehen. Hierfür sind 2 Möglichkeiten gegeben:

1. Die Wirkgruppen der Enzyme stehen in direktem Kontakt,
2. die Verbindung der Wirkgruppen wird durch Systeme der Energieübertragung besorgt.

Es läßt sich heute noch nicht entscheiden, wie eine solche Enzymkette im Mitochondrium funktioniert, aber offensichtlich ist im Mitochondrium eine ganz andere Situation gegeben als in einem unstrukturiertem System, wo Metaboliten und Coenzyme von den Apofermenten abdissoziieren und an anderer Stelle weiterreagieren können. In einem organisierten System ist das Zufallsmoment des Zusammentreffens passender Komponenten und die Länge der Wegstrecken für den Transport der beweglichen Reaktionspartner auf ein Minimum reduziert.

Während die Umsatzgeschwindigkeit der RNS im Mitochondrium, verglichen mit anderen Zellkomponenten, sehr gering ist, ist die Umsatzgeschwindigkeit der Phosphatide im Mitochondrium der Leberzelle am höchsten. In fallender Intensität folgen Mikrosomen, Cytoplasma und Zellkerne (ADA). Jedes dieser Strukturelemente baut sein eigenes Phosphatid auf und übernimmt nicht fertige Phosphatide aus anderen Zellelementen (vgl. LANG und SIEBERT). Der Mangel an essentiellen Fettsäuren setzt den Gehalt der Lebermitochondrien an aktiver Bernsteinsäureoxydase und Glutaminsäuredehydrogenase deutlich herab. Auch steht die Aktivität mancher Leberenzyme mit der Jodzahl der Leberlipoide in einer deutlichen Beziehung (TULPULE und PATWARDAHN). Dies wie auch der zerstörende Einfluß von α-Toxin (Lecithinase aus *Cl. welchii*) auf die Succinoxydaseaktivität deuten auf einen engeren Zusammenhang zwischen Funktion und Feinbau der Strukturelemente in der Zelle hin (MCFARLANE).

Literatur.

Zusammenfassende Darstellungen und Bücher.

AMMON, R., u. W. DIRSCHERL: Fermente, Hormone, Vitamine und die Beziehungen dieser Wirkstoffe zueinander, 2. Aufl. Leipzig 1948.

BALDWIN, E.: Dynamic aspects of biochemistry. Cambridge 1949. — BERTALANFFY, L. v.: Theoretische Biologie. Bd. 2 Stoffwechsel, Wachstum. Berlin 1942. — BERSIN, TH.: Kurzes Lehrbuch der Enzymologie, 2. Aufl. Leipzig 1939. — BERZELIUS, J.: Lehrbuch der Chemie, 3. Aufl. Übersetzt von L. WÖHLER. Dresden u. Leipzig 1837.

DIXON, M.: Multi enzyme systems. Cambridge: Univ. Press 1951.

EDSALL, J. T.: Enzymes and enzyme systems. Cambridge, Mass.: Harvard Univ. Press 1951. — EULER, H. v.: Chemie der Enzyme. München 1925—1934.

FLASCHENTRÄGER, B., u. E. LEHNARTZ: Physiologische Chemie. Berlin-Göttingen-Heidelberg: Springer. Bd. 1. 1951, Bd. 2. 1954.

GLICK, D.: Techniques of histo- and cytochemistry. New York: Interscience 1949.

HALDANE, J. B. S.: Enzymes. London 1930. — HALDANE, J. B. S., u. K. G. STERN: Allgemeine Chemie der Enzyme. Dresden u. Leipzig 1932. — HOFFMANN-OSTENHOF, O.: Enzymologie. Wien: Springer 1954.

LANGENBECK, W.: Die organischen Katalysatoren, 2. Aufl. Berlin-Göttingen-Heidelberg: Springer 1949.

MITTASCH, A.: Über Katalyse und Katalysatoren in Chemie und Biologie. Berlin 1936.

NORD, F., u. R. WEIDENHAGEN: Ergebnisse der Enzymforschung, Bd. 1ff. Leipzig 1932. — NORD, F. F., u. C. H. WERKMAN: Advances in enzymology, Bd. 1ff. New York 1941. — NORTHROP, J. H.: Crystalline enzymes; the chemistry of pepsin, trypsin and bacteriophage. New York 1939.

OPPENHEIMER, C.: Die Fermente und ihre Wirkungen, 5. Aufl. Leipzig: Bd. 1—4, 1924—1929; Suppl. 2 Bde. 1936—1938.

SUMNER, J. B., u. K. MYRBÄCK: The enzymes. 2 Bde. in 4 Teilen. New York 1951/52. — SUMNER, J. B., u. G. F. SOMERS: Chemistry and methods of enzymes. 2. Aufl. New York 1947.

WAKSMAN, S. A., u. W. C. DAVISON: Enzymes, properties distribution, methods and applications. Baltimore 1926. — WALDSCHMIDT-LEITZ, E.: Die Enzyme, Wirkungen und Eigenschaften. Braunschweig 1926. — WARBURG, O.: Schwermetalle als Wirkungsgruppen von Fermenten. Berlin 1946. — WILLSTÄTTER, R.: Untersuchungen über Enzyme, 2 Bde. Berlin 1928.

Originalaufsätze.

ABOOD, L. G., R. W. GERARD and S. OCHS: Electrical stimulation of metabolism of homogenates and particulates. Amer. J. Physiol. **171**, 134—139 (1952). — ADA, G. L.: Phospholipid metabolism in rabbit-liver cytoplasm. Biochemic. J. **45**, 422—428 (1949). — ALIVISATOS, S. G., and O. F. DENSTEDT: Nicotinamide inhibition of tri- and diphosphopyridine nucleotide-linked dehydrogenases. J. of Biol. Chem. **199**, 493—504 (1952). — ANDERSON, E. P., and S. E. G. ÅQVIST: A double precursor study of nucleic acid turnover in normal and regenerating liver. J. of Biol. Chem. **202**, 513—520 (1953). — ANDRESEN, N., F. ENGEL u. H. HOLTER: Succinic dehydrogenase and cytochrome oxidase in Chaos chaos. C. r. Trav. Labor. Carlsberg **27**, 408—420 (1951). — ANSON, M. L., and A. E. MIRSKY: The equilibrium between active native trypsin and inactive denatured trypsin. J. Gen. Physiol. **17**, 393—398 (1934). — AUGUSTINSSON, K.-B.: Substrate concentration and specificity of choline ester-splitting enzymes. Arch. of Biochem. **23/24**, 111—126 (1949). — AUGUSTINSSON, K.-B., and D. NACHMANSOHN: Studies on cholinesterase. VI. Kinetics of the inhibition of acetylcholine esterase. J. of Biol. Chem. **179**, 543—559 (1949). — AUHAGEN, E.: Co-Carboxylase, ein neues Co-Enzym der alkoholischen Gärung. Z. physiol. Chem. **204**, 149—167 (1932).

BAMANN, E., u. J. N. MUKHERJEE: Über die protoplasmatische Verankerung der Leberesterase. (Zur Kenntnis der Zellverankerung der Enzyme.) Z. physiol. Chem. **229**, 1—14 (1934). — BARNUM, C. P., and R. A. HUSEBY: The intracellular heterogeneity of pentose nucleic acid as evidenced by the incorporation of radiophosphorus. Arch. of Biochem. **29**, 7—26 (1950). — BARTLEY, W., and R. E. DAVIES: Secretory activity of mitochondria. Biochemic. J. **52**, 20 (1952). — BEARD, R. L.: Chemical activity ratios in relation to species-specifity. J. Econ. Entomol. **44**, 469—471 (1951). — BEHRENS, M.: Über die Verteilung der Lipase und Arginase zwischen Zellkern und Protoplasma der Leber. Z. physiol. Chem. **258**, 27 (1939). — BERNSTEIN, I. A., K. LENTZ, M. MALM, P. SCHAMBYE and H. G. WOOD: Degradation of glucose-C^{14} with Leuconostoc mesenteroides; alternate pathways and tracer patterns. J. of Biol. Chem. **215**, 137—152 (1955). — BERTHET, J., L. BERTHET, F. APPELMANS and C. DE DUVE: Tissue fractionation studies. II. The nature of the linkage between acid phosphatase and mitochondria in rat-liver. Biochemic. J. **50**, 182—189 (1951). — BERTHET, J., and C. DE DUVE: Tissue fractionation studies. I. The existence of a mitochondria-linked enzymically inactive form of acid phosphatase in rat-liver tissue. Biochemic. J. **50**, 174—181 (1951). — BERTRAND, G.: Chimie organique. — Sur l'intervention du manganèse dans les oxydations provoquées par la laccase. C. r. Acad. Sci. Paris **124**, 1032—1035 (1897). — BOOIJ, H. L., u. H. P. WOLVERKAMP: On the concepts „limiting factor", "master reaction" and "temperature analysis". Rec. Trav. chim. Pays-Bas (Amsterd.) **64**, 316—317 (1945). — BOTTELIER, H. P., H. HOLTER u. K. LINDERSTRØM-LANG: Studies on enzymatic histochemistry. XXXVI. Determination of peptidase activity, nitrogen content and reduced weight in roots of the barley, hordeum vulgare. C. r. Trav. Labor. Carlsberg, Sér. Chim. **24**, 289—313 (1943). — BRACHET, J.: Oxygen uptake of nucleated and non-nucleated halves of Amoeba proteus. Nature (Lond.) **168**, 205 (1951). — Chemical Embryology. New York: Interscience 1950. — BRACHET, J., and H. CHANTRENNE: Protein synthesis in nucleated and non-nucleated halves of Acetabularia mediterranea studied with carbon-14 dioxide. Nature (Lond.) **168**, 950 (1951). — BRADFIELD, J. R. G.: The localization of enzymes in cells. Biol. Rev. **25**, 113—157 (1950). — BRUMMOND, D. O., u. R. H. BURRIS: Transfer of C^{14} by lupine mitochondria through reactions of the tricarboxylic acid cycle. Proc. Nat. Acad. Sci. USA. **39**, 754—759 (1953). — BUCHNER, E.: Alkoholische Gärung ohne Hefezellen. Ber. dtsch. chem. Ges. **30**, 117—124, 1110—1113 (1897). — BÜCHER, TH.: Über ein phosphatübertragendes Gärungsferment. Biochim. et Biophysica Acta **1**, 292—314 (1947). — Proteine als Träger der Fermentwirkung. In Biologie und Wirkung der Fermente. Berlin-Göttingen-Heidelberg 1953a. — Probleme des Energietransports innerhalb lebender Zellen. Adv. Enzymol. **14**, 1—42 (1953b). — BURTON, A. C.: The properties of the steady state compared to those of equilibrium as shown in characteristic biological behavior. J. Cellul. a. Comp. Physiol. **14**, 327—349 (1939). — BUTENANDT, A.: Biochemie der Gene und Genwirkungen. Verh. Ges. dtsch. Naturforsch. (97. Verslg) **1952**.

CAGNIARD-LATOUR, M.: Mémoire sur la fermentation vineuse. Ann. Chim. Physique **68**, 206—222 (1838). — CASIDA, J. E., and M. A. STAHMANN: Metabolism and mode of action of schradan. J. Agricult. a. Food. Chemistry **1**, 883—888 (1953). — CHANCE, B.: The kinetics of the enzyme-substrate compound of peroxydase. J. of Biol. Chem. **151**, 553—577 (1943). — An intermediate compound in the catalase-hydrogen peroxide reaction. Acta chem. scand. (Copenh.) **1**, 236—267 (1947). — The enzyme-substrate compounds of catalase and peroxides. Nature (Lond.) **161**, 914—917 (1948). — The primary and secondary compounds of catalase and methyl or ethyl hydrogen peroxide. J. of Biol. Chem. **180**, 947—959 (1949). — Enzyme-substrate compounds. Adv. Enzymol. **12**, 153—190 (1951). — The kinetics of the complexes of peroxidase formed in the presence of chlorite or hypochlorite. Arch. of Biochem. a. Biophysics **41**, 425—431 (1952). — CHANCE, B., D. S. GREENSTEIN and F. J. W. ROUGHTON: The mechanism of catalase action. I. Steady-state analysis. Arch. of Biochem. a. Biophysics **37**, 301—321 (1952). — CHANCE, B., and J. HIGGINS: Peroxidase kinetics in coupled oxidation; an experimental and theoretical study. Arch. of Biochem. a. Biophysics **41**, 432—441 (1952). — CHANTRENNE, H.: Hétérogénéité des granules cytoplasmiques du fois de souris. Biochim. et Biophysica Acta **1**, 437 (1947). — CHANTRENNE-VAN HALTEREN, M. B., et J. BRACHET: La respiration de fragments nucléés et énucléés d' "Acetabularia mediterranea". Arch. internat. Physiol. **60**, 187—188 (1952). — CLAUDE, A.: Particulate compounds of normal and tumor cells. Science (Lancaster, Pa.) **91**, 77—78 (1940). — The constitution of protoplasm. Science (Lancaster, Pa.) **97**, 451—456 (1943). — The constitution of mitochondria and microsomes, and the distribution of nucleic acid in the cytoplasm of a leukemic cell. J. of Exper. Med. **80**, 19—29 (1944). — Fractionation of mammalian liver cells by differential centrifugation. J. of Exper. Med. **84**, 51—59 (1946). — CONWAY, E. J., and M. DOWNEY: An outer metabolic region of the yeast cell. Biochemic. J. **47**, 347—355 (1950). — CROZIER, W. J.: Zahlreiche Arbeiten in J. Gen. Physiol. **7**, **9**, **10**, **13**.

DAVIDSON, J. N., W. M. MCINDOE and R. M. S. SMELLIE: The uptake of ^{32}P by ribonucleotides in liver-cell fractions. Biochemic. J. **49**, 36 (1951). — DESNUELLE, P.: Quelques techniques nouvelles pour l'étude de la structure des proteines. Adv. Enzymol. **14**, 261—310 (1953). — DIANZANI, M. U.: On the osmotic behaviour of mitochondria. Biochim. et biophysica Acta **11**, 353—367 (1953a). — Action of papain and of trypsin on the morphology and some enzymatic activities of isolated mitochondria. Experientia (Basel) **9**, 343—345 (1953b). — DOUNCE, A. L.: The significance of enzyme studies on isolated nuclei. Internat. Rev. Cytology **3**, 199—223 (1954). — DUSPIVA, F.: Die Verteilung der Peptidase auf Kern und Plasma bei Froschoocyten im Verlauf der zweiten Wachstumsperiode. Biol. Zbl. **62**, 403—431 (1942). — Enzymatische Histo- und Cytochemie. In HOPPE-SEYLER-THIERFELDER's Handbuch der physiol. und pathol.-chem. Analyse, 10. Aufl., Bd. 2. Berlin-Göttingen-Heidelberg 1955. — DUVE, CHR. DE, and J. BERTHET: The use of differential centrifugation in the study of tissue enzymes. Internat. Rev. Cytology **3**, 225—275 (1954). — DUVE, CHR. DE, J. BERTHET, L. BERTHET and F. APPELMANS: Permeability of mitochondria. Nature (Lond.) **167**, 389—390 (1951). — DUVE, CHR. DE, R. GIANETTO, F. APPELMANS and R. WATTIAUX: Enzymic content of the mitochondria fraction. Nature (Lond.) **172**, 1143—1144 (1953).

ELLIOT, K. A. C., and D. KEILIN: The haematin content of horseradish peroxidase. Proc. Roy. Soc. Lond., (Ser. B) **114**, 210—222 (1934). — EULER, H. v., H. ALBERS u. F. SCHLENK: Hochgereinigte Co-Zymase. Z. physiol. Chem. **234**, 1 (1935a). — Chemische Untersuchungen an hochgereinigter Co-Zymase. Z. physiol. Chem. **240**, 113—126 (1936). — EULER, H. v., u. R. VESTIN: Zur Kenntnis der Wirkungen der Co-Zymase. Z. physiol. Chem. **237**, 1—7 (1935b).

FELIX, K., M. DECKER u. L. ROKA: Dehydrogenase-Aktivität der Mitochondrien verschiedener Organe. Hoppe-Seylers Z. **294**, 79—85 (1953). — FISHER, H. F., E. E. CONN, B. VENNESLAND and F. H. WESTHEIMER: The enzymatic transfer of hydrogen. I. The reaction catalyzed by alcohol dehydrogenase. J. of Biol. Chem. **202**, 687—697 (1953). — FREY-WYSSLING, A.: Submicroscopic morphology of protoplasm and its derivatives, 2. Aufl. New York u. Amsterdam 1947. 1. Aufl. i. dtsch. Berlin 1938. — Die submikroskopische Struktur des Cytoplasmas. In L. V. HEILBRUNN u. F. WEBER, Protoplasmatologia, Bd. 2, A 2. Wien 1955.

GEY, G., P. SAPRANAUKAS and F. B. BANG: Cine-phase microscope studies of the behavior of Mitochondria in living cells and of actors affecting their morphology. 4. Annual meeting of the histochemical society, Chicago 1953. — GOTTSCHALK, A.: On the mechanism of enzyme action. Rev. Pure a. Appl. Chem. **3**, 179—206 (1953). — GRASSMANN, W.: Neue Methoden und Ergebnisse der Enzymforschung. Erg. Physiol. **27**, 407—551 (1928). — GREEN, D. E.: The cyclophorase system. In J. T. EDSALL, Enzymes and enzyme systems. Cambridge, Mass. 1951. — GREENBERG, G. R.: De novo synthesis of hypoxanthine via inosine-5-phosphate and inosine. J. of Biol. Chem. **190**, 611—631 (1951). — GUNSALUS, I. C., and W. D. BELLAMY: A function of Pyridoxal. J. of Biol. Chem. **155**, 357 (1944). — GUSTAFSON, T., and P. LENICQUE: Studies on mitochondria in the developing sea urchin egg. Exper. Cell Res. **3**, 251—274 (1952).

HADIDIAN, Z., and H. HOAGLAND: Chemical pacemakers. I. Catalytic brain iron. II. Activation energies of chemical pacemakers. J. Gen. Physiol. **23**, 81—99 (1939). — HALDANE, J. B. S.: The biochemistry of genetics. London 1954. — HARMAN, J. W.: Studies on mitochondria: I. The association of cyclophorase with mitochondria. Exper. Cell Res. **1**, 382—393 (1950a). — Studies on mitochondria: II. The structure of mitochondria in relation to enzymatic activity. Exper. Cell Res. **1**, 394—402 (1950b). — HARVEY, E. B.: The development of half and quarter eggs of arbacia punctulata and of strongly centrifuged whole eggs. Biol. Bull. **62**, 155—167 (1932). — Development of the parts of sea urchin eggs separated by centrifugal force. Biol. Bull. **64**, 125 (1933). — HARVEY, E. N.: The tension at the surface of marine eggs, especially those of the sea urchin, arbacia. Biol. Bull. **61**, 273—279 (1931). — HEPPEL, L. A., and V. T. PORTERFIELD: Metabolism of inorganic nitrite and nitrate esters. I. The coupled oxidation of nitrite by peroxideforming systems and catalase. J. of Biol. Chem. **178**, 549—556 (1949). — HIRD, F. J. R., and E. V. ROWSELL: Additional transaminations by insoluble particle preparations of rat liver. Nature (Lond.) **166**, 517—518 (1950). — HIRS, C. H. W., ST. MOORE and W. H. STEIN: A chromatographic investigation of pancreatic ribonuclease. J. of Biol. Chem. **200**, 493—506 (1953). — HOAGLAND, H.: Pacemaker aspects of nervous activity. Cold Spring Harbor Symp. Quant. Biol. **4** (1936). —

HOGEBOOM, G. H.: Separation and properties of cell components. Federat. Proc. **10**, 640—645 (1951). — HOGEBOOM, G. H., and W. C. SCHNEIDER: Cytochemical studies of mammalian Tissues. III. Isocitric dehydrogenase and triphosphopyridine nucleotide cytochrome c reductase of mouse liver. J. of Biol. Chem. **186**, 417—427 (1950). — Cytochemical studies. VI. The synthesis of diphosphopyridinnucleotide by liver cell nuclei. J. of Biol. Chem. **197**, 611—620 (1952). — HOGEBOOM, G. H., W. C. SCHNEIDER and G. E. PALADE: The isolation of morphologically intact mitochondria from rat liver. Proc. Soc. Exper. Biol. a. Med. **65**, 320—321 (1947). — Cytochemical studies of mammalian tissues. I. Isolation of intact mitochondria from rat liver; some biochemical properties of mitochondria and submicroscopic particulate material. J. of Biol. Chem. **172**, 619—635 (1948). — HOLTER, H.: Studies on enzymatic histochemistry. XVIII. Localization of peptidase in marine ova. J. Cellul. a. Comp. Physiol. **8**, 179—200 (1936). — Enzymverteilung im Protoplasma. Arch. exper. Zellforsch. **19**, 232—237 (1937). — Zur Chemie einiger Zellstrukturen. Arch. exper. Zellforsch. **22**, 534—540 (1939). — Localization of enzymes in cytoplasm. Adv. Enzymol. **13**, 1—20 (1952). — Distribution of some enzymes in the cytoplasma of amoebae. Proc. Roy. Soc. Lond., Ser. B **142**, 140—146 (1954). — HOLTER, H., and W. L. DOYLE: Studies on enzymatic histochemistry. XXVIII. Enzymatic studies on protozoa. J. Cellul. a. Comp. Physiol. **12**, 295—308 (1938). — HOLTER, H., and M. J. KOPAC: Studies on enzymatic histochemistry. XXIV. Localization of peptidase in thc ameba. J. Cellul. a. Comp. Physiol. **10**, 423—437 (1937). — HOLTER, H., and S. LØVTRUP: Proteolytic enzymes in Chaos chaos. C. r. Trav. Labor. Carlsberg, Sér. Chim. **27**, 27—62 (1949). — HOLTER, H., M. OTTESEN and R. WEBER: Separation of cytoplasmic particles by centrifugation in a density-gradient. Experentia (Basel) **9**, 346—348 (1953). — HOLZER, H.: Über Fermentketten und ihre Bedeutung für die Regulation des Kohlenhydratstoffwechsels in lebenden Zellen. In Biologie und Wirkung der Fermente. Berlin-Göttingen-Heidelberg 1953. — HORECKER, B. L., M. GIBBS, H. KLENOW and P. Z. SMYRNIOTIS: The mechanism of pentose phosphate conversion to hexose monophosphate. I. With a liver enzyme preparation. J. of Biol. Chem. **207**, 393—403 (1954). — HORECKER, B. L., P. Z. SMYRNIOTIS and H. KLENOW: The formation of sedoheptulose phosphate from pentose phosphate. J. of biol. Chem. **205**, 661—682 (1953). — HUENNEKENS, F. M.: Studies on the cyclophorase system. 15. The malic oxidase. Exper. Cell Res. **2**, 115—125 (1951). — HUTCHENS, J. O., M. J. KOPAC and M. E. KRAHL: The cytochrome oxidase content of centrifugally separated fractions of unfertilized arbacia eggs. J. Cellul. a. Comp. Physiol. **20**, 113—116 (1942).

JEENER, R.: L'Héterogénéité des granules cytoplasmiques: Données complémentaires fournies par leur fractionnement en solution saline concentrée. Biochim. et biophysica Acta **2**, 633—641 (1948). — JEENER, R., and D. SZAFARZ: Relations between the rate of renewal and the intracellular localization of ribonucleic acid. Arch. of Biochem. **26**, 54—67 (1950). — JOHNSON, M. J., G. H. JOHNSON and W. H. PETERSON: The magnesium-activated leucyl peptidase of animal erepsin. J. of Biol. Chem. **116**, 515—526 (1936).

KEILIN, D., and E. F. HARTREE: Coupled oxidation of alcohol. Proc. Roy. Soc. Lond., Ser. B **119**, 141—159 (1936). — Purification of horse-radish peroxidase and comparison of its properties with those of catalase and methaemoglobin. Biochemic. J. **49**, 88—104 (1951). — KEILIN, D., and T. MANN: On the haematin compound of peroxidase. Proc. Roy. Soc. Lond., Ser. B **122**, 119—133 (1937). — KENNEDY, E. P., and A. L. LEHNINGER: Oxidation of fatty acids and tricarboxylic acid cycle intermediates by isolated rat liver mitochondria. J. of Biol. Chem. **179**, 957—972 (1949). — KRAHL, M. E.: Metabolic activities and cleavage of eggs of the sea urchin, Arbacia punctulata. Biol. Bull. **98**, 175—217 (1950). — KREBS, H. A.: Die Steuerung der Stoffwechselvorgänge. Dtsch. med. Wschr. **1956**, Nr 1, 4—8. — KÜHNAU, J.: Antagonismen und Konkurrenzen um den Platz am Ferment. In Biologie und Wirkung der Fermente. Berlin-Göttingen-Heidelberg 1953. — KÜTZING, F.: Mikroskopische Untersuchungen über Hefe und Essigmutter, nebst mehreren anderen dazu gehörigen vegetabilischen Gebilden. J. prakt. Chem. **11**, 385—409 (1837). — KUFF, E. L., and W. C. SCHNEIDER: Intracellular distribution of enzymes. XII. Biochemical heterogeneity of mitochondria. J. of Biol. Chem. **206**, 677—685 (1954). — KUHN, R., D. B. HAND u. M. FLORKIN: Über die Natur der Peroxydase. Z. physiol. Chem. **201**, 255—266 (1931). — KUHN, R., H. RUDY u. TH. WAGNER-JAUREGG: Über Lactoflavin (Vitamin B_2). Ber. dtsch. Chem. Ges. **66**, 1950 bis 1956 (1933). — KUNITZ, M., and J. H. NORTHROP: Isolation of a crystalline protein from pancreas and its conversion into a new crystalline proteolytic enzyme by trypsin. Science (Lancaster, Pa.) **78**, 558—559 (1933). — Inactivation of crystalline trypsin. J. Gen. Physiol. **17**, 591—615 (1934).

LAIDLER, K. J., and J. P. HOARE: The molecular kinetics of the urea-urease system. I. The kinetic laws. J. Amer. Chem. Soc. **71**, 2699—2702 (1949). — LANG, K.: Die Biologie der Enzyme. In Biologie und Wirkung der Enzyme. 4. Kolloquium der Ges. für physiol. Chem., Mosbach. Berlin-Göttingen-Heidelberg 1953. — LANG, K., u. G. SIEBERT: Die chemischen Leistungen der morphologischen Zellelemente. In B. FLASCHENTRÄGER u.

E. LEHNARTZ, Physiologische Chemie, Bd. 2, Teil B. Berlin-Göttingen-Heidelberg 1954. — LANG, K., G. SIEBERT, I. BALDUS u. A. CORBET: Über das Vorkommen von Desoxyribonuclease und Kathepsin in Zellkernen aus Nieren. Experientia (Basel) **6**, 59—60 (1950). — LARDY, H. A.: Respiratory enzymes. Minneapolis: Burgess Publ. Comp. 1950. — LEHMANN, F. E.: Die submikroskopische Organisation der Zelle. Verh. Ges. dtsch. Naturforsch. **1955**. — LEHMANN, F. E., u. R. BISS: Elektronenoptische Untersuchungen an Plasmastrukturen des Tubifex-Eies. Rev. suisse Zool. **56**, 264 (1949). — LEHNINGER, A. L.: The organised respiratory activity of isolated rat-liver mitochondria. In J. T. EDSALL, Enzymes and enzyme systems. Cambridge, Mass. 1951. — LIEBIG, J. v.: Über die Erscheinungen der Gährung, Fäulnisz und Verwesung und ihre Ursachen. Ann. Pharmazie **30**, 250—287 (1839).— LINDERSTRØM-LANG, K., u. H. HOLTER: Die enzymatische Histochemie. In Methoden der Fermentforschung von BAMANN u. MYRBAECK, S. 1132—1162. 1941. — LINEWEAVER, H., and D. BURK: The determination of enzyme dissociation constants. J. Amer. Chem. Soc. **56**, 658—666 (1934). — LIPMANN, F.: Acetylation of sulfanilamide by liver homogenates and extracts. J. of Biol. Chem. **160**, 173—190 (1945). — On chemistry and function of coenzyme A. Bacter. Rev. **17**, 1—16 (1953). — LIPMANN, F., N. O. KAPLAN, G. D. NOVELLI, L. C. TUTTLE and B. M. GUIRARD: Isolation of Coenzyme A. J. of Biol. Chem. **186**, 235—243 (1950). — LOEWUS, F. A., P. OFNER, H. F. FISHER, F. H. WESTHEIMER and B. VENNESLAND: The enzymatic transfer of hydrogen. II. The reaction catalysed by lactic dehydrogenase. J. of Biol. Chem. **202**, 699—704 (1953). — LOEWUS, M. W., and D. R. BRIGGS: The number of catalytically active sites present on the chymotrypsin molecule. J. of Biol. Chem. **199**, 857—864 (1952). — LOHMANN, K., u. PH. SCHUSTER: Über die Co-Carboxylase. Naturwiss. **25**, 26 (1937). — Untersuchungen über die Co-Carboxylase. Biochem. Z. **294**, 188—214 (1937). — LOTSPEICH, W. D., R. A. PETERS u. T. H. WILSON: The inhibition of aconitase by "inhibitor fractions" isolated from tissues poisoned with fluoroacetate. Biochemic. J. **51**, 20—25 (1952). — LUNDBLAD, G., and E. HULTIN: Liberation of proteolytic enzymes of the sea urchin egg by ribonuclease. Exper. Cell Res. **6**, 249—250 (1954). — LYNEN, F., u. E. REICHERT: Zur chemischen Struktur der „aktivierten Essigsäure". Angew. Chem. **63**, 47—48 (1951). — LYNEN, F., E. REICHERT u. L. RUEFF: Zum biologischen Abbau der Essigsäure. VI. „Aktivierte Essigsäure", ihre Isolierung aus Hefe und ihre chemische Natur. Liebigs Ann. **574**, 1—32 (1951).

MACFARLANE, M. G.: Inhibition of succinoxidase activity of mitochondria by Clostridium welchii toxin. Biochemic. J. **47**, 29—30 (1950). — MACFARLANE, M. G., and A. G. SPENCER: Changes in the water, sodium and potassium content of rat-liver mitochondria during metabolism. Biochemic. J. **54**, 569—575 (1953). — MARSHALL jr., J. M.: Distributions of chymotrypsinogen, procarboxypeptidase, desoxyribonuclease and ribonuclease in bovine pankreas. Exper. Cell. Res. **6**, 210—212 (1954). — MARTIN, A. J. P., and R. R. PORTER: The chromatographic fractionation of ribonuclease. Biochemic. J. **49**, 215—218 (1951). — MAZIA, D., and H. I. HIRSHFIELD: The nucleus-dependence of P^{32} uptake by the cell. Science (Lancaster, Pa.) **112**, 297—299 (1950). — MCINDOE, W. M., and J. N. DAVIDSON: The phosphorus compounds of the cell nucleus. Brit. J. Cancer **6**, 200—214 (1952). — METCALF, R. L., and R. B. MARCH: Studies of the mode of action of parathion and its derivatives and their toxicity to insects. J. Econ. Entomol. **42**, 721—728 (1949). — MICHAELIS, L.: Fundamental principles in oxidation-reduction. Biol. Bull. **96**, 293—295 (1949). — MICHAELIS, L., u. H. DAVIDSOHN: Die Wirkung der Wasserstoffionen auf das Invertin. Biochem. Z. **35**, 386—412 (1911). — MICHAELIS, L., u. M. L. MENTEN: Die Kinetik der Invertinwirkung. Biochem. Z. **49**, 333—369 (1913). — MITTASCH, A.: Katalyse und Determinismus. Erg. Enzymforsch. **7**, 377—417 (1938). — MØLLER, K. M., and D. M. PRESCOTT: Observations on the cytochromes of Amoeba proteus, Chaos chaos and Tetrahymena geleii. Exper. Cell Res. **9**, 375—377 (1955). — MOELWYN-HUGHES, E. A.: Physical chemistry and chemical kinetics of enzymes. In J. B. SUMNER u. K. MYRBÄCK, The enzymes, Bd. 1, Teil 1, S. 28. 1950. — MONNÉ, L.: Functioning of the cytoplasm. Adv. Enzymol. 8, 1—69 (1948). — On the induced formation of chromosomelike struktures within the cytoplasm of mature sea urchin eggs. Ark. Zool. (Stockh.) **42**a, 1—11 (1949). — MONOD, J., and M. COHN: La biosynthèse induite des enzymes. Adv. Enzymol. **13**, 67—116 (1952). — MÜLLER, A. F., u. F. LEUTHARDT: Oxydative Phosphorylierung und Citrullinsynthese in den Lebermitochondrien. Helvet. chim. Acta **32**, 2349—2356 (1949). — MYRBÄCK, K.: Co-Zymase. Erg. Enzymforsch. **2**, 139—168 (1933). — Zur Wirkungsweise der Hydrolasen. Acta chem. scand. (Copenh.) **1**, 142—148 (1947). — MYRBÄCK, K., and E. WILLSTAEDT: Studies on yeast invertase (saccharase). Localization of the enzyme in the cell and its liberation. Ark. Kemi (Stockh.) **8**, 367—374 (1955).

NACHMANSOHN, D., and I. B. WILSON: The enzymic hydrolysis and synthesis of acetylcholine. Adv. Enzymol. **12**, 259—339 (1951). — NEILANDS, J. B.: Studies on lactic dehydrogenase of heart. I. Purity, kinetics and equilibria. J. of Biol. Chem. **199**, 373—381 (1952). — NEWCOMER, E. H.: Mitochondria in Plants. Bot. Rev. **17**, 53—89 (1951). — NICHOL, C. A., and A. D. WELCH: Synthesis of citrovorum factor from folic acid by liver slices; augmentation by ascorbic acid. Proc. Soc. Exper. Biol. a. Med. **74**, 52—55 (1950). — NORTHROP, J. H.: Crystalline pepsin. I. Isolation and tests of purity. J. Gen. Physiol. **13**, 739—766 (1929 bis 1930a). — Crystalline pepsin. II. General properties and experimental methods. J. Gen. Physiol. **13**, 767—791 (1929/30b). — NORTHROP, J. H., and M. KUNITZ: Isolation of protein crystals possessing tryptic activity. Science (Lancaster, Pa.) **73**, 262—263 (1931). — NYGAARD, A. P.: Factors involved in the enzymatic reduction of cytochrome c. J. of Biol. Chem. **204**, 655—663 (1953). — NYGAARD, A. P., and J. B. SUMNER: The effect of lecithinase A on the succinoxydase system. J. of Biol. Chem. **200**, 723—729 (1953).

PAIGEN, K.: The occurence of several biochemically distinct types of mitochondria in rat liver. J. of Biol. Chem. **206**, 945—957 (1954). — PALADE, G. E.: The fine structure of mitochondria. Anat. rec. **114**, 427—452 (1952). — An electron microscope study of the mitochondrial structure. J. Histochem. a. Cytochem. **1**, 188—211 (1953). — A small particulate component of the cytoplasm. J. Biophys. a. Biochem. Cytology **1**, 59—68 (1955). — PASTEUR, L.: Mémoire sur la fermentation alcoolique. Ann. Chim. Physique (3) **58**, 323—426 (1860). — PATTERSON, E. K., M. E. DACKERMAN and J. SCHULTZ: Peptidase increase accompanying growth of the larval salivary gland of Drosophila melanogaster. J. Gen. Physiol. **32**, 623—645 (1949). — PAULING, L., H. A. Itano, S. J. SINGER and I. C. WELLS: Sickle cell anemia, a molecular disease. Science (Lancaster, Pa.) **110**, 543—548 (1949). — PORTER, V. S., N. P. DEMING, R. C. WRIGHT and E. M. SCOTT: Effects of freezing on particulate enzymes of rat liver. J. of Biol. Chem. **205**, 883—891 (1953). — POTTER, V. R., and C. A. ELVEHJEM: A modified method for the study of tissue oxidations. J. of Biol. Chem. **114**, 495—504 (1936). — POTTER, V. R., R. O. RECKNAGEL and R. B. HURLBERT: Intracellular enzyme distribution; interpretations and significance. Federat. Proc. **10**, 646—653 (1951). — PULLMAN, M. E.: Structure of reduced pyridine nucleotides. Federat. Proc. **12**, 255 (1955).

RAAFLAUB, J.: Die Korrelation zwischen Struktur und Aktivität von isolierten Leberzellmitochondrien. Helvet. physiol. Acta **10**, C22—C24 (1952). — Die Schwellung isolierter Leberzellmitochondrien und ihre physikalisch-chemische Beeinflußbarkeit. Helvet. physiol. Acta **11**, 142—156 (1953a). — Über den Wirkungsmechanismus von Adenosintriphosphat (ATP) als Cofaktor isolierter Mitochondrien. Helvet. physiol. Acta **11**, 157—165 (1953b). — RACKER, E., G. DE LA HABA and I. G. LEDER: Thiamine pyrophosphate, a coenzyme of transketolase. J. Amer. Chem. Soc. **75**, 1010—1011 (1953). — REED, L. J., and B. G. DE BUSK: Lipothiamide and its relation to a thiamin coenzyme required for oxidative decarboxylation of α-keto acids. J. Amer. Chem. Soc. **74**, 3457 (1952). — Lipothiamide pyrophosphate: Coenzyme for oxidative decarboxylation of α-keto acids. J. Amer. Chem. Soc. **74**, 3964—3965 (1952). — Lipoic acid conjugase. J. Amer. Chem. Soc. **74**, 4727—4728 (1952). — Mechanism of enzymatic oxidative decarboxylation of pyruvate. J. Amer. Chem. Soc. **75**, 1261—1262 (1953). — REED, L. J., B. G. DE BUSK, I. C. GUNSALUS and C. S. HORNBERGER jr.: Crystalline α-lipoic acid: A catalytic agent associated with pyruvate dehydrogenase. Science (Lancaster, Pa.) **114**, 93—94 (1951). — REED, L. J., B. G. DE BUSK, I. C. GUNSALUS and G. H. F. SCHNAKENBERG: Chemical nature of α-lipoic acid. J. Amer. Chem. Soc. **73**, 5920 (1951). — ROTHEN, A.: On the mechanism of enzymatic activity. J. of Biol. Chem. **163**, 345—346 (1946). — Immunological reactions between films of antigen and antibody molecules. J. of Biol. Chem. **168**, 75—97 (1947). — ROTHSTEIN, A., R. MEIER and L. HURWITZ: The relationship of the cell surface to metabolism. V. The role of uranium-complexing loci of yeast in metabolism. J. Cellul. a. Comp. Physiol. **37**, 57—81 (1951). — RUBINSTEIN, D., and O. F. DENSTEDT: The metabolism of the erythrocyte. III. The tricarboxylic acid cycle in the avian erythrocyte. J. of Biol. Chem. **204**, 623—637 (1953). — Cytochrome oxydase activity of cell nuclei. Canad. J. Biochem. a. Physiol. **32**, 548—552 (1954).

SEAMAN, G. R.: Localization of acetylcholinesterase activity in the Protozoan, Tetrahymena geleii S. Proc. Soc. Exper. Biol. a. Med. **76**, 169—170 (1951). — SIZER, I. W.: Temperature activation of the urease-urea system using crude and crystalline urease. J. Gen. Physiol. **22**, 719—741 (1939). — SJÖSTRAND, F. S., und V. HANZON: Membrane structures of cytoplasm and mitochondria in exocrine cells of mouse pancreas as revealed by high resolution electron microscopy. Exper. Cell Res. **7**, 393—414 (1954). — SLATER, E. C.: Structurally-bound enzymes. In Biologie und Wirkung der Enzyme. 4. Kolloquium der Ges. für physiol. Chem. Mosbach. Berlin-Göttingen-Heidelberg 1953. — SLATER, E. C., and W. D. BONNER jr.: Effect of fluoride on the succinic oxidase system. Biochemic. J. **52**, 185—196 (1952). — SLATER, E. C., and K. W. CLELAND: The effect of tonicity of the medium on the respiratory and phosphorylative activity of heart muscle sarcosomes. Biochemic. J. **53**, 557—567 (1953). — SLAUTTERBACK, D. B.: Elektron microscopic studies of small cytoplasmic particles (microsomes). Exper. Cell Res. **5**, 173—186 (1953). — SMELLIE, R. M. S., W. M. MCINDOE, R. LOGAN, J. N. DAVIDSON and I. M. DAWSON: Phosphorus compounds in the cell. 4. The incorporation of radioactive phosphorus into liver cell fractions. Biochemic. J. **54**, 280—286 (1953). — SNELL, E. E., G. M. BROWN, V. J. PETERS, J. A. CRAIG, E. L. WITTLE, J. A. MOORE, V. M. MCGLOHON and O. D. BIRD: Chemical nature and synthesis of the Lactobacillus bulgaricus factor. J. Amer. Chem. Soc. **72**, 5349—5350 (1950). — SØRENSEN, S. P. L.: Enzymstudien. II. Mitt. Über die Messung und die Bedeutung der Wasserstoffionenkonzentration bei enzymatischen Prozessen. Biochem. Z. **21**, 131—200, 201—304 (1909). — SUMNER, J. B.: The isolation and crystallisation of the enzyme urease. J. of Biol. Chem. **69**, 435—441 (1926). — Crystalline urease. Erg. Enzymforsch. **1**, 295—301 (1932). — SUMNER, J. B., and A. L. DOUNCE: Crystalline catalase. J. of Biol. Chem. **121**, 417—424 (1937). — SCHNEIDER, W. C.: Intracellular distribution of enzymes. III. The oxidation of octanoic acid by rat liver fractions. J. of Biol. Chem. **176**, 259—266 (1948). — SCHNEIDER, W. C. and G. H. Hogeboom: Intracellular distribution of enzymes. X. Desoxyribonuclease and ribonuclease. J. of Biol. Chem. **198**, 155—163 (1952). — SCHWANN, TH.: Vorläufige Mittheilung betreffend Versuche über die Weingährung und Fäulnisz. Ann. Physik **41**, 184—193 (1837). — STEARN, A. E.: The theory of absolute reaction rates applied to enzyme catalysis. Erg. Enzymforsch. **7**, 1—27 (1938). — Kinetics of biological reactions with special reference to enzymic processes. Adv. Enzymol. **9**, 25—74 (1949). — STERN, K. G.: The constitution of the prosthetic group of catalase. J. of Biol. Chem. **112**, 661—669 (1935 bis 1936).

THEORELL, H.: Reines Cytochrom c. Biochem. Z. **279**, 463—464 (1935). — Reines Cytochrom c. II. Mitt. Darstellung, Eigenschaften, Ionenbeweglichkeit, Diffusion und Absorptionsspektrum des Cytochrom c. Biochem. Z. **285**, 207—218 (1936). — Reversible Spaltung einer Peroxydase. Ark. Kem., Mineral. Geol., Ser. B **14**, Nr 20, 1—3 (1940). — THEORELL, H., u. B. CHANCE: Studies on liver alcohol dehydrogenase. II. The Kinetics of the compound of horse liver alcohol dehydrogenase and reduced diphosphopyridine nucleotide. Acta chem. scand (Copenh.) **5**, 1127—1144 (1951). — TULPULE, P. G., and V. N. PATWARDHAN: The effect of fat and pyridoxine deficiencies on rat liver dehydrogenases. Arch. of Biochem. **39**, 450—456 (1952).

URBANI, E.: Gli enzimi proteolitici nella cellula e nell'embrione. Experientia (Basel) **9**, 209—218 (1955).

WARBURG, O.: Chemische Konstitution von Fermenten. Erg. Enzymforsch. **7**, 210—245 (1938). — Über die Wirkungsgruppen der oxydierenden und reduzierenden Fermente. Naturwiss. **40**, 493—496 (1953). — WARBURG, O., u. W. CHRISTIAN: Co-Fermentprobleme. Biochem. Z. **274**, 112—116 (1934). — Co-Fermentprobleme. Biochem. Z. **275**, 112—113 (1935a).— Co-Fermentprobleme. Biochem. Z. **275**, 464 (1935b). — Zerstörung des wasserstoffübertragenden Co-Ferments durch ultraviolettes Licht. Biochem. Z. **282**, 221—223 (1935). — Gärungs-Co-Ferment. Biochem. Z. **285**, 156—158 (1936a). — Optischer Nachweis der Hydrierung und Dehydrierung des Pyridins im Gärungs-Co-Ferment. Biochem. Z. **286**, 81—82 (1936b). — Pyridin, der wasserstoffübertragende Bestandteil von Gärungsfermenten. (Pyridin-Nucleotide.) Biochem. Z. **287**, 291—328 (1936c). — Isolierung und Kristallisation des Proteins des oxydierenden Gärungsferments. Biochem. Z. **303**, 40—68 (1939). — WARBURG, O., W. CHRISTIAN u. A. Griese: Wasserstoffübertragendes Co-Ferment, seine Zusammensetzung und Wirkungsweise. Biochem. Z. **282**, 157—205 (1935). — WARBURG, O., u. E. NEGELEIN: Über die photochemische Dissoziation bei intermittierender Belichtung und das absolute Absorptionsspektrum des Atmungsferments. Biochem. Z. **202**, 202—228 (1928). — WAYGOOD, E. R., and K. A. CLENDENNING: Intracellular localization and distribution of carbonic anhydrase in plants. Science (Lancaster, Pa.) **113**, 177—179 (1951). — WILLIAMS jr., J. N.: Inhibition of coenzyme I-requiring enzyme by adenine and adenyl metabolities in vitro. J. of Biol. Chem. **195**, 629—635 (1952). — WILLSTÄTTER, R., u. M. ROHDEWALD: Zur enzymatischen Methodik. Z. physiol. Chem. **229**, 241—268 (1934). — WILSON, I. B.: Mechanism of enzymic hydrolysis. I. Role of the acidic group in the esteratic site of acetylcholinesterase. Biochim. et Biophysica Acta **7**, 466—470 (1951a). — Mechanism of hydrolysis. II. New evidence for an acylated enzyme as intermediate. Biochim. et Biophysica Acta **7**, 520—525 (1951b). — Acetylcholinesterase. XI. Reversibility of tetraethyl pyrophosphate inhibition. J. of Biol. Chem. **190**, 111—117 (1951b). — WILSON, I. B., and F. BERGMANN: Studies on cholinesterase. VII. The active surface of acetylcholine esterase derived from effects of p_H on inhibitors. J. of Biol. Chem. **185**, 479—489 (1950). — WILSON, P. W.: Kinetics and mechanisms of enzyme reactions. In LARDY, Respiratory enzymes. Minneapolis 1950. — WINZLER, R. J.: A comperative study of the effects of cyanide, acide and carbon monoxide on the respiration of bakers yeast. J. Cellul. a. Comp. Physiol. **21**, 229—252 (1943). — WOLKEN, J. J., and G. E. PALADE: An electron microscope study of two flagellates. Chloroplast structure and variation. Ann. New York Acad. Sci. **56**, 873—889 (1953).

ZEILE, K., u. H. HELLSTRÖM: Über die aktive Gruppe der Leberkatalase. Hoppe-Seylers Z. **192**, 171—192 (1930). — ZELLER, E. A.: Allgemeine Physiologie und Pathologie der Enzyme. Handbuch der allgemeinen Pathologie, Bd. 2. Heidelberg: Springer 1955. — ZELLER, E. A., u. A. BISSEGGER: Über die Cholinesterase des Gehirns und der Erythrocyten. Helvet. chim. Acta **26**, 1619—1630 (1943).

Der Zellkern.

Von

G. F. Bahr.

Unter Mitwirkung von

T. Caspersson und G. Klein.

Unsere Kenntnis der Funktion der einzelnen Zellbestandteile ist im Augenblick in rascher Entwicklung begriffen. Das wird hauptsächlich durch die Ausarbeitung neuer Methoden zur Analyse der chemischen Zusammensetzung der einzelnen Zellabschnitte bedingt. Zwei Gruppen solcher Methoden zum Studium von Zellumsatzfragen müssen dabei unterschieden werden. Die erste hat die Analyse des Zellkernes als Ganzem zum Ziel, die andere die Analyse einzelner Organellen im einzelnen Zellkern. Die Größenordnungen, die die beiden Methodengruppen bearbeiten, liegen im ersten Falle bei 500—1000 μ^3 und im zweiten Falle bei 0,5—1 μ^3, d. h. das „Auflösungsvermögen" ist ungefähr tausendmal größer. Dieser Größenordnungssprung trennt etwa auch die konventionellen Makromethoden der Chemie von denen der Mikromethoden. Die Arbeiten der erstgenannten Art, welche also die eigenen Strukturen des Zellkernes außer acht lassen, werden in hohem Maße durch die Möglichkeit erleichtert, in großem Umfange Zellkerne vom Cytoplasma zu trennen und so ein einheitliches Material zu gewinnen, an welchem übliche mikrochemische Methoden zur Anwendung kommen können. Die zweite Gruppe der Methoden, die mit größerer Auflösung als 1 μ arbeitenden Verfahren, fordert ein recht kompliziertes Instrumentarium. Die Entwicklung auf beiden Gebieten ist in den letzten Jahren rasch vorangeschritten. Ein wesentlicher Teil unserer heutigen Kenntnisse um die Mitwirkung des Zellkernes im Zellstoffwechsel stammt von der erstgenannten Arbeitsform, die vor allem von dem Interesse an der speziell leicht zugänglichen Substanzklasse der Enzyme dominiert wird, aber auch durch rein chemische Analysen der Kernsubstanzen und Untersuchungen mit radioaktiven Isotopen gefördert wurde.

Der folgende Abschnitt behandelt diese Gebiete, in denen das Hauptgewicht der Betrachtungen auf die Arbeiten der letzten 5 Jahre gelegt wurde, und weist gleichzeitig auf umfassende Darstellungen hin, die auch früheren und über den Rahmen des behandelten Stoffes hinausgehenden Arbeiten Rechnung tragen. Allgemeinere Fragen betreffend, wird auf den ersten Teil des Handbuches hingewiesen, der unter anderem die Chemie des Zellkernes behandelt, wohin auch solche Fragen, wie die primäre Synthese der Kernbestandteile und besonders der gentragenden Desoxyribonucleoproteine gehören.

Die Hochauflösungsmethoden, die in starker Entwicklung begriffen sind, sind bisher hauptsächlich verwandt worden, um den „Eiweißwandel in der Zelle" zu studieren und werden unter diesem Abschnitt in Band VIII behandelt.

Die Diskussion aller mit Zell*teilungs*prozessen verbundener Fragen liegt außerhalb des Rahmens der vorliegenden Darstellung (dieselben werden ausführlich an anderen Stellen dieses Handbuches abgehandelt). Folglich wird im wesentlichen der Interphasenkern Gegenstand unserer Betrachtungen sein.

Der Interphasenkern erscheint morphologisch als ein abgeschlossenes System im Systemkomplex der cytoplasmatischen Organisation mit ihren hochorganisierten Bestandteilen wie Mitochondrien usw. Er ist durch eine wohldefinierte *Kernmembran* vom Cytoplasma getrennt, deren Existenz für das physiologische Verhalten des Kernes von größter Bedeutung sein muß. Kernmembranen konnten in allen bisher elektronenoptisch untersuchten Objekten demonstriert werden, nachdem schon seit einigen Dezennien lichtmikroskopische und mikromanipulatorische Studien eine solche Struktur hatten zeigen können. Siehe dazu vor allem die polarisationsoptischen Untersuchungen SCHMIDTS (1939). Bei einer Reihe von Objekten konnten elektronenoptisch Poren in der Membran nachgewiesen werden, deren Größe zwischen 200 und 3000 Å für die jeweilige Species variieren. (CALLAN und TOMLIN 1950, *Triturus*, *Xenopus*. HARRIS und JAMES 1954, BAIRATI und LEHMANN 1952, *Amoeba proteus*. BAHR und BEERMANN 1954, *Chironomus*. RHODIN (1954), Niere der Maus. WATSON 1954, Pankreas der Maus. GALL 1954, *Triturus*. AFZELIUS 1954, an 4 Arten schwedischer Seeigel.) Die Untersuchungen von GALL und AFZELIUS sind von besonderem Interesse, da diese Autoren in ihren Objekten nachweisen konnten, daß jede Pore der Kernmembran aus einer Pyramide globulärer Körperchen (8—10) mit etwa 400 Å Durchmesser aufgebaut ist.

Durch die Auffindung von Poren in der Kernmembran und deren eigenartigem Aufbau werden auch die Permeabilitätsangaben von ANDERSON (1952, 1953), WADDINGTON (1952) und von HOLTFRETER (1954) für die Kernmembran erklärlicher. Nach diesen Autoren ist die Kernmembran leicht für Saccharose, Dextrin, Gummi arabicum, depolymerisierte Nucleinsäuren und Rinderhämoglobin passierbar, während die Durchlässigkeit für Plasmaalbumin und Glykogen umstritten ist. Auch anorganische Ionen und Wasser können ungehindert durch die Kernmembran diffundieren. Die Kernmembran ist also nach Obengesagtem nicht semipermeabel, während allerdings CHURNEY (1942), HOGEBOOM und SCHNEIDER (1952) sie als semipermeabel betrachten.

Die Permeabilität der Kernmembran ist besonders dann interessant, wenn die Frage nach der Molekülgröße physiologisch auf das Cytoplasma wirkender Substanzen aus dem Zellkern gestellt wird. Es ist jedenfalls unwahrscheinlich, daß ein nucleo-cytoplasmatischer Transport großer Moleküle nur während der Mitose, d. h. dem Verschwinden der Kernmembran möglich ist. Besonders lebhaft wird jedoch die Frage von allen Laboratorien diskutiert, die mit isolierten Kernen arbeiten. Dabei wird oft — und vor allem trifft das für Enzyme zu — das Zellinnere mit Reaktionssystemen vom Lösungstyp gleichgesetzt, die man auch in vitro zu verwenden gewohnt ist. Wahrscheinlich ist aber ein großer Teil komplizierter Prozesse strukturgebunden; d. h. die Frage nach dem Wanderungs- und Diffusionsvermögen einzelner Substanzen in der Zelle, besonders aber der durch entsprechende Präparationsmethoden aufgebrochenen Zellen, ist von ihrer wechselnden Bindung an bestimmte funktionelle Struktureinheiten abhängig.

Über morphologische Organisationen des Kerninnern existieren bisher nur Hypothesen, von denen allerdings die Chromosomen der Schleifenkerne und die Nucleolen (s. die Übersicht STOCKINGERS 1953) ausgenommen sind. Auch die sich rasch entwickelnde deskriptive Elektronenmikroskopie hat bisher nichts zu dieser Frage beitragen können.

Doch liegen von seiten der jüngeren morphologischen Literatur der letzten Jahre eine Reihe von Betrachtungen vor, die in dem hier behandelten Zusammenhang Beachtung fordern. Zellkerne verschiedener Gewebe zeigten unter normalen und pathologischen Bedingungen Einschlußbildungen, die in Form von Cysten oder Kristallen als Ganzes in das Cytoplasma ausgestoßen werden und dort zur Auflösung kommen (SCHARRER 1934, APITZ 1937, RAVEN 1946, ORTMANN 1948). Solche Ausstoßungsprozesse geschehen mit einer Art Schleusenmechanismus (SCHILLER 1949), ohne die Kernmembran zu unterbrechen. Übertritte ganzer Nucleoli in das Cytoplasma mit darauffolgender Entleerung desselben wurden ebenfalls beschrieben (HETT 1927, GRAUPNER und FISCHER 1934, ALTMANN 1952, HERTL 1955). Vgl. besonders auch die umfangreiche Darstellung von DITTUS (1940), der gleich ALTMANN (1952) eine Extrusions- und eine Restitutionsphase aus seinen Beobachtungen herleitet. Auch werden rotierende Bewegungen des Zellkernes gegenüber dem Cytoplasma beschrieben, die in bestimmten Zeiten die Richtung ändern (POMERAT 1953).

Die genannten morphologischen Beobachtungen lassen jedoch nur bedingt Schlüsse auf normale Aktivitäten des Zellkernes während der Ruhepause zu.

Unsere allgemeinere Auffassung der physiologischen Funktion des Zellkernes umfaßt:

1. Die Rolle der in ihm enthaltenen Chromosomen bei genetischen Prozessen. Das heißt der Zellkern bestimmt die Arteigenschaften der Zelle.

2. Den Ausgangspunkt für eine andere Seite der funktionellen Beurteilung des Kernes, nämlich die Kenntnis seiner chemischen Zusammensetzung. Darauf aufbauend kann der Umsatz sowie der Stoffwechsel einzelner Substanzen im Kern weitere Aufschlüsse geben, wie Cytoplasma oder auch der Trägerorganismus mit dem einzelnen Kern in Wechselbeziehung stehen.

3. Schließlich können die Relationen des Zellkernes zu seiner Umgebung dadurch verdeutlicht werden, daß man ihn aus der Zelle entfernt und das Verhalten des enucleierten Zellrestes verfolgt.

Durch mannigfache Untersuchungen ist die Tatsache gesichert, daß der Kern die Hauptmenge der Erbfaktoren enthält, und daß diese Faktoren kontinuierlich an den physiologischen und morphogenetischen Prozessen der Zelle teilnehmen. Wenn auch im Rahmen dieser Darstellung nicht auf Einzelheiten eingegangen werden kann, sollen doch einige der bestbearbeiteten Beispiele genannt werden. So die klassischen Untersuchungen von KÜHN, BECKER und BUTENANDT an der Mehlmotte *Ephestia kühniella*, BEADLE, TATUM und EPHRUSSI an dem Pilz *Neurospora crassa*. Aber auch an zahlreichen anderen Objekten, einschließlich des Menschen, konnte die Genabhängigkeit einzelner physiologisch-chemischer Prozesse nachgewiesen werden (s. hierzu die Übersichten von HOROWITZ 1950, PONTECORVO 1952 und BARATT und Mitarbeitern 1954).

Durch cytochemische und biochemische Untersuchungen konnte ein grobes *Bild der ungefähren Zusammensetzung des Zellkernes* gewonnen werden, das hier besonders in seinen neueren Aspekten umrissen werden soll. Danach sind die Hauptbestandteile des Zellkernes: Desoxyribonucleinsäure (DNS), Ribonucleinsäure (RNS), ein basisches Protein (Histon oder Protamin), andere Proteine, die nicht dem Histontyp angehören, auch manchmal als Restprotein bezeichnet, und eine Reihe weiterer Substanzen, unter denen besonders die in letzter Zeit intensiv bearbeiteten Enzyme zu nennen sind.

Wenn auch bisher sicherlich nur ein Teil der betroffenen Nucleinsäuren untersucht wurde, so wächst doch die Anzahl der Angaben stetig, daß der Zucker in den verschiedenen Pentosenucleinsäuren (PNS) Ribose ist. Wir werden darum hier die weniger konservative Bezeichnung DNS und RNS verwenden, für die auch die Gründe der sprachlichen Unterscheidung sprechen. ALBAUM und OGUR (1947) isolierten allerdings ein Adenin-Pentose-Triphosphat, das wahrscheinlich

Arabinose als Zuckerkomponente enthält und in Pflanzen häufig vorkommen soll. Es liegen aber unseres Wissens noch keine Befunde über Nucleinsäuren mit dieser Komponente vor.

Als Trägersubstanz genetischer Faktoren wird die DNS angesehen, die, von einigen noch weiter unten zu diskutierenden Ausnahmen abgesehen, nicht außerhalb des Kernes angetroffen worden ist, also quasi ein Kriterium des Kernes darstellt. An Froscheiern konnten HOFF-JØRGENSEN und ZEUTHEN (1952) mit Hilfe einer bakteriellen Bestimmungsmethode Desoxyriboside nachweisen und schlossen aus den Daten ihrer Untersuchung auf eine Verteilung von DNS in Kern und Cytoplasma von 1:5000. Früher hatten schon FRAENKEL-CONRAT und DUCAY (1951) Avidin aus dem Weißen des Eies isoliert und es als DNS bezeichnet. Diese Befunde dürfen nicht erstaunen, da ja der ganze Organismus, der aus der betreffenden Eizelle entstehen soll, aus einer Vielzahl von Zellen zusammengesetzt sein wird und man annehmen darf, daß zumindest ein Teil der benötigten Bausteine in der Eizelle gespeichert ist. Die Bestimmungsmethode nach HOFF-JØRGENSEN (1952) schließt Nucleoside, Nucleotide und Nucleinsäuren ein. Das heißt, es ist nicht eindeutig möglich, die gefundenen Werte auf vorhandene, wohl aber auf synthetisierbare DNS zu beziehen. Im Falle des Vorkommens von DNS außerhalb des Kernes ist es wichtig, deren Polymerisationsgrad zu beachten, besonders auch, wenn deren genetische Funktion betrachtet wird, womit die Diskussion an der unentschiedenen Frage der Definition der DNS stehenbleibt. FRAENKEL-CONRAT und DUCAYS Befunde lassen das in Eiern vorkommende Avidin als DNS-Protein erscheinen. Es ist als wasserlösliches Nucleoprotein vorhanden und hat spezifisch Biotin-bindende Eigenschaften, wird von DNase abgebaut und von RNase nicht angegriffen. Der Gehalt ist für verschiedene Eiweißpräparationen abweichend. Auch in diesem Falle dürfte die Definition entscheiden, die einerseits der DNS des Kernes mit ihrer genetischen Funktion zukommt, auf der anderen Seite die große Zahl möglicher Vorläufer und Stoffwechselbruchstücke der Nucleinsäuren, auch im Cytoplasma, in einen sinnvollen Zusammenhang bringt. MARSHAK und MARSHAK (1953) untersuchten *Arbacia*eier und -spermien und errechneten den DNS-Gehalt aus Thymin-Isotopverdünnungsversuchen. Der Kern des reifen Eies zeigte eine negative FEULGEN-Reaktion, woraus geschlossen wurde, daß die DNS ein sekundäres Kernprodukt sei und nicht die eigentliche gentragende Substanz. MARRIAN (1954) konnte durch Versuche an regenerierender Rattenleber mit markiertem Adenin wahrscheinlich machen, daß zwar die DNS des Interphasenkernes kaum RNS und lösliche Nucleotide aufnimmt. Stimuliert man aber zu vermehrter Mitosefrequenz, so nimmt die sich neubildende DNS aus dem vorhandenen Vorrat an Nucleotiden auf, wobei dann unter anderem auch eine Konvertierbarkeit RNS-DNS wahrscheinlich sein soll. Solche cytoplasmatischen Nucleotide dürfen, wie weiter unten noch besprochen wird, auch nicht als einfache Speichersubstanzen für eventuellen DNS-Bedarf betrachtet werden, sondern erfüllen gewisse Aufgaben in anderen chemischen Prozessen der Zelle. Daran sind auch Desoxyriboside beteiligt, wie es für das Adenindesoxyribosid-triphosphat, dem Analogon des ATP nachgewiesen werden konnte. Durch die Tatsache, daß hochpolymerisierte DNS als transformierendes Prinzip auf Bakterienkulturen wirken kann (AVERY, MACLEOD und MACCARTY (1944), werden die Zusammenhänge allerdings sehr kompliziert.

Die DNS ist die konstanteste bekannte Größe in der Zelle. Alle anderen Größen wie Trockengewicht (Masse), Stickstoffgehalt, Größe usw sind erheblichen physiologischen Schwankungen unterworfen. Das wurde unter anderem durch die umfangreichen Untersuchungen von THOMSON und Mitarbeitern (1953) festgestellt, die DNS per Kern unter allen untersuchten Bedingungen konstant fanden, während RNS, Lipoide, Stickstoff und Phosphor stark schwankten. Die Behandlung der DNS-Konstanz in der Zelle und damit zusammenhängender Fragen sowie die umfangreiche Literatur dazu wird an anderer Stelle dieses Handbuches vorgenommen werden. Die Festlegung des DNS-Gehaltes für die betreffende Zelle als Bezugssystem cytochemischer und cytophysiologischer Studien ist von größter Bedeutung. Versuche, DNS als Referenzsystem zu benutzen, waren darum naheliegend (s. z. B. die Arbeiten von PATTERSON und DACKERMANN 1952, DAVIDSON und LESLIE 1951, LOWE und SALMON 1951). In Tabelle 1 finden sich einige Angaben über den Gehalt an DNS in Zellkernen verschiedener Herkunft.

Große prinzipielle Bedeutung muß dem Umstand zugemessen werden, daß DNS mit verschiedenen Zusammensetzungen und physikalisch-chemischen Eigenschaften aus gleichem Ausgangsmaterial isoliert werden konnten (REDDI 1954). BENDICH, RUSSEL und BROWN (1953) fanden in Lebern erwachsener Ratten nur eine DNS-Fraktion, während regenerierende Leber zwei Fraktionen enthielt, die auch durch unterschiedlichen Umsatz von C^{14}-Formiat gekennzeichnet waren (s. auch FURST und BROWN 1951). Diese zwei Fraktionen variieren von Organ zu Organ in der Ratte. Drei Fraktionen isolierten CRAMPTON, LIPSCHÜTZ und CHARGAFF (1954) aus Kalbsthymus durch fraktionierte Fällung mit basischen Eiweißen, betonten aber, daß eine solche Aufteilung nicht in unverändertem „intakten" Nucleohiston vorliegt. Bemerkenswerterweise waren in allen Fraktionen aller Ausgangsmaterialien die Verhältnisse von Adenin zu Thymin, Guanin zu Cytosin und Purinen zu Pyrimidinen konstant. In Bakterien fanden CHARGAFF und Mitarbeiter (1950) ungewöhnliche Proportionen der Nucleotide zueinander. PRICE und Mitarbeiter (1949) konnten aus Kalbsthymus allerdings immer nur eine Fraktion DNS durch ein basisches Polypeptid fällen. Die DNS verschiedener Ausgangsmaterialien unterschieden sich jedoch deutlich durch charakteristische Variationen in der Basenzusammensetzung (CHARGAFF und Mitarbeiter 1949, 1950, 1951, 1952 u. a.), ohne daß die genannten Proportionen verändert wurden (VISCHER, ZAMENHOF und CHARGAFF 1949, GANDELMAN, ZAMENHOF, CHARGAFF 1952, LALAND, OVEREND und WEBB 1952). ZAMENHOF, BRAWERMAN und CHARGAFF (1952), HURST, MARKO und BUTLER (1953), WYATT (1950, 1952), KHOUVINE und GREGOIRE (1953) konnten die DNS verschiedener Herkunft durch ihre Basenzusammensetzung charakterisieren. Nicht nur Proportionsunterschiede in der Zusammensetzung und unterschiedliche enzymatische Suszeptibilität, sondern auch direkte Differenzen der chemischen Komponenten wurden gefunden. So konnten durch Untersuchungen an Säugern und Pflanzen, an Hühnereiern und Phagen artspezifische kleine Mengen von 5-Methylcytosin als Komponente der DNS nachgewiesen werden, während die DNS von Protozoen, Rikettsien, Bakterien und Viren diese Komponente nicht enthielt (ELMES, SMITH und WHITE 1952, WYATT 1951, FRAENKEL-CONRAT, SNELL und DUCAY 1952, WYATT und COHEN 1952). Außerordentlich wenig kann jedoch bis jetzt über den Wirkungsmechanismus der DNS gesagt werden, obwohl über die prinzipielle Rolle dieser im Kern lokalisierten Substanz kaum ein Zweifel besteht. Auf direkte Aktionen im Kernbereich selber weisen einige Arbeiten hin, andere lassen auf größere Zusammenhänge mit dem ganzen Organismus schließen. ALLFREY (1954) fand, daß der Einbau von C^{14}-Alanin in nucleäres Protein direkt von der „Funktionstüchtigkeit" der DNS abhängt. Mit DNase konnte er die Synthese unterbinden, nicht dagegen mit RNase. Auf Zugabe von DNS zu Systemen, in denen DNS enzymatisch zerstört war, konnte die Synthese wieder merkbar gehoben werden.

In diesem Zusammenhang sind auch die Versuche von LU und WINNICK (1954) bedeutungsvoll, die eine Aufnahme von Bausteinen markierter DNS in sowohl DNS als auch RNS der Chorioallantoismembran von bebrüteten Hühnereiern feststellten. Ebenso verhielt sich markierte RNS. Daraus schließend wird angenommen, daß in der Zelle vollständige Enzymsysteme zur Überführung des Desoxyribosezuckers in Ribosezucker vorhanden seien. Da bisher keine direkt transformierenden Systeme gefunden werden konnten, wäre diese Umformung nur über den Weg eines vollständigen Abbaues einschließlich des Pentosezuckers zu Di- und Triosen zu denken. Damit scheint aber auch den beiden wichtigen Bestandteilen des Kernes die Möglichkeit einer gewissen, bedeutungsvollen funktionellen Plastizität gegeben zu sein.

Es war zu vermuten, daß zu den aus anderen Gründen schon geforderten verschiedenen DNS entsprechende Desoxyribonucleasen (DNasen) existieren würden. Solche konnten dann auch nachgewiesen werden (MAVER und GRECO 1949, SIEBERT, LANG und CORBERT 1950, LASKOWSKY 1951, ALLFREY und MIRSKY 1952), ALLFREY, STERN und MIRSKY 1952, WEBB 1953, BROWN, JACOBS und LASKOWSKY 1952).

Nach WEBB kommen DNasen jedoch nicht im Kern vor; die gefundenen Enzymaktivitäten für diesen Zellbestandteil sollen Artefakte sein. BROWN und Mitarbeiter (1952) fanden DNase außerordentlich fest an den Kern gebunden. Siehe auch die Tabelle 2. Nach ALLFREY und MIRSKY variiert der Enzymgehalt der Kerne von Gewebe zu Gewebe. Ihr mengenmäßiges Vorkommen im Gewebe steht in direktem Zusammenhang mit der Aufnahmegeschwindigkeit der DNS für ^{15}N-Glykokoll. Die Gewebe-DNasen sind vor allem durch ihre p_H-abhängigen Wirkungsoptima unterschieden (SIEBERT, LANG und CORBERT 1950) und liegen in aktiver Form vor. Die gewöhnlich verwandte DNase aus Pankreas ist ein aktivierbares Verdauungsferment (ALLFREY und MIRSKY 1952).

Aus Untersuchungen über den Effekt dieser pankreatischen DNase auf DNS konnten ZAMENHOF und CHARGAFF (1949, 1950) schließen, daß die DNS heterogen abgebaut werden, d. h. einen nur langsam angreifbaren „Kern" besitzen, der mit fortschreitendem Abbau Adenin- und Thymin-reicher wird.

RNS wurde sowohl im Cytoplasma als auch im Kern angetroffen (CASPERSSON 1936, 1950) und wurde dort besonders im Nucleolus lokalisiert. Dieser Zellbaustein, ist verglichen mit der DNS, viel stärkeren Veränderungen des physiologischen und pathologischen Geschehens in der Zelle unterworfen. Lokale Konzentrationen und chemische Zusammensetzungen schwanken, während allerdings die molaren Proportionen der Bausteine zueinander, ähnlich der DNS, von physiologischen Änderungen nicht beeinflußt werden sollen (CROSBIE, SMELLIE und DAVIDSON 1953). Die Zusammensetzung ändert sich dagegen von Kerntyp zu Kerntyp verschiedener Gewebe desselben Organismus (McINDOE und DAVIDSON 1952). Solche Unterschiede sind vor allem auch für die zwei prinzipiellen RNS, die nucleäre und die cytoplasmatische RNS, gefunden worden. Nucleäre RNS (nucRNS) enthält im allgemeinen sehr viel weniger Guanin als die RNS des Cytoplasmas (cytRNS) (CHARGAFF 1950, ELSON und CHARGAFF 1952), ja dieser Baustein fehlt bei gleichzeitig hohen Werten im Cytoplasma ganz in der RNS der Kerne von gewissen Tumoren und Kükenlebern (BEALE, HARRIS und ROE 1950).

Eine Reihe von Arbeiten beschreiben die nucRNS mit erheblich höherem Umsatz radioaktiver Isotope als die cytRNS. Diese vermehrten Umsatzwerte nucRNS gelten für alle vier Nucleotide, die praktisch gleichen Umsatz zeigen (McINDOE und DAVIDSON 1952, MARSHAK 1951, SMITH und STOKER 1951, SMITH und WYATT 1951, WHITFELD 1952). Seitdem es möglich wurde, die Bausteine und denkbaren Vorläufer der Nucleinsäuren radioaktiv zu markieren, widmeten sich eine Reihe von Untersuchern Stoffwechselproblemen dieser Kernsubstanzen. Es stellte sich bald heraus, daß die einzelnen Bausteine charakteristische Aufnahmequoten zeigen, sowohl ihrem Verbleib in den verschiedenen Typen von Nucleinsäuren nach, als auch differenziert für die einzelnen Purin- und Pyrimidinbasen der Nucleinsäuren und in Abhängigkeit des angebotenen Vorläufers. Dazu zeigten sich Unterschiede für die Nucleinsäuren verschiedener Herkunft in allen genannten Punkten. Gleichzeitig konnte durch chemische Basenanalysen wie durch Untersuchungen mit markierten Substanzen gezeigt werden, daß der aus ultraviolett-mikrospektrographischen Untersuchungen her bekannten Rolle der nucRNS (CASPERSSON) eine chemische und stoffwechselmäßige Differenzierung zugrunde liegt. Die nucRNS hat eine andere Zusammensetzung als die cytoplasmatische und vor allem bedeutend höhere Umsatzquoten. In diesem Zusammenhang zeigten HURLBERT und POTTER (1952) mit radioaktiven Isotopen an Rattenlebern, daß das Umsatzmaximum sehr schnell die lösliche Phase des Cytoplasmas passiert, um dann im Kern nach etwa 4—6 Std aufzutreten. NucRNS baut z. B. c^{14}-Orotsäure mit einem Maximum bei 4—8 Std ein, während das erste cytoplasmatische Maximum bei 2 Std liegt. Nach etwa 16—20 Std hat sich das Maximum wieder in das Cytoplasma, überwiegend in die kleingranulären Fraktionen, verschoben, wo es mit dem Maximum der Eiweißsynthese zusammenfällt.

Als ein bedeutungsvolles Ergebnis solcher Untersuchungen darf die relative biologische Stabilität der DNS angesehen werden. Von allen dargebotenen Vorläufern wird in der Interphase nur sehr wenig aufgenommen und lange retiniert, und da diese Untersuchungen an großen Organen vorgenommen wurden, muß damit gerechnet werden, daß ein Teil der Aufnahme durch vereinzelte Mitosen in dem Untersuchungsobjekt bedingt ist, z. B. Endothelzellen usw.

Dazu sollte beachtet werden, daß solche Tracerstudien oft unter den geringen Aktivitäten leiden, mit denen teilweise gearbeitet werden muß, und daß auch unterschiedliche Isolierungsverfahren der untersuchten Substanzen gewisse Fehlerquellen darstellen. Auch haben solche Untersuchungen nur relativ wenig Rücksicht auf histologische Gesichtspunkte genommen und auf Grund der verwandten Methoden nicht nehmen können. Bei der Isolation verschiedener Fraktionen DNS und RNS mit verschiedenen metabolischen Charakteristiken von Gewebe zu Gewebe ist es sehr wohl möglich, daß gewebliche Heterogenitäten einen wesentlichen Einfluß auf die Ergebnisse ausüben. Eine Arbeit der Autoren FICQU und ERRERA (1955), die die intracelluläre Verteilung radioaktiver Vorläufer von Nucleinsäuren und Proteinen in Autoradiographien zum Gegenstand hatte, zeigt unter anderem, daß das seiner relativen Einheitlichkeit wegen häufig für biochemische Studien verwandte Gewebe der Leber kräftige Aufnahme der Substanzen in die Kerne aufweist (4mal mehr als das Cytoplasma), daß aber die KUPFFERschen Sternzellen, die dem sog. reticuloendothelialen System des Körpers angehören, noch wesentlich mehr aufnehmen.

In Tabelle 1 sind einige Angaben über den Gehalt verschiedener Kerne an **Proteinen** eingetragen. Kernproteine sind, von der Unterscheidung in Histone, Protamine und Restproteine abgesehen, bis jetzt noch recht wenig charakterisiert. Aus verschiedenen Geweben isolierten STEDMAN und STEDMAN (1951) Zellkerne, deren basische Proteine sie untersuchten. Es konnten zwei Histone und zwei Protaminfraktionen isoliert werden, von denen drei sich als zellspezifisch erwiesen. Auch HAMER (1953) und KHOUVINE, sowie GREGOIRE und ZALTA (1953) konnten zwei Proteinfraktionen aus verschiedenen Materialien isolieren. LAMIRANDE, ALLARD und CANTERO (1953) isolierten 5 Proteinfraktionen aus Zellkernen. Ein

Tabelle 1.

Substanz	Menge	Bezugssystem	Gewebe
N	13,8	% des Trockengewichts	Lunge
N	17	% des Trockengewichts	Thymus
N	19,8	% des Trockengewichts	Sperma
P	3,1	% des Trockengewichts	Thymus
P	1,5	% des Trockengewichts	Lunge
Lipoidphosphor . . .	30	% des Homogenats	Gehirn
Lipoidphosphor . . .	0,2	% des Trockengewichts	Leukocyten
Lipoidphosphor . . .	0,309	% des Trockengewichts	Leukocyten
DNS	30—33	% des Trockengewichts	Thymus
DNS	26	% des Trockengewichts	Thymus
DNS	15	% des Trockengewichts	Leber
DNS	17	% des Trockengewichts	Niere
DNS	19	% des Trockengewichts	Herz
DNS	23,5	% des Trockengewichts	Knochenmark
DNS	16,5	% des Trockengewichts	Herz
DNS	17	% des Trockengewichts	Pankreas
DNS	22	% des Trockengewichts	Pankreas
DNS	12	% des Trockengewichts	Leber
DNS	18	% des Trockengewichts	Leber
DNS	27	% des Trockengewichts	Leber
DNS	25	% des Trockengewichts	Erythrocyten
DNS	50,5	% des Kerns	Thyreoidea
DNS	2,52	mg/g Frischgewicht	Leber
DNS	3,26	mg/g Frischgewicht	Leber
DNS	2,19—2,70	mg/g Frischgewicht	Leber
DNS	1,75—1,93	mg/g Frischgewicht	Leber
DNS	0,9—1,9	% der Nucleinsäure	Leber
RNS	0,6—1,3	% der Nucleinsäure	Thymus
RNS	0,7—1,1	% der Nucleinsäure	Milz
RNS	2,0	% der Nucleinsäure	Leber
RNS	1,2	% der Nucleinsäure	Niere
RNS	2,9—4,5	% der Nucleinsäure	Leber
RNS	4,0—5,6	% der Nucleinsäure	Leber
RNS	2,0—2,2	% der Nucleinsäure	Leber
RNS	1,3—1,4	% der Nucleinsäure	Thymus
RNS	0,7—2,5	% der Nucleinsäure	Erythrocyten
RNS	0,3	% der Nucleinsäure	Sperma
RNS	0,1	% der Nucleinsäure	Sperma
RNS	0,0—0,2	% der Nucleinsäure	Sperma
RNS	3,4	% des Kerns	Leber
RNS	7,5	% des Kerns	Thyreoidea
RNS	1,13	mg/g Frischgewicht	Leber
RNS	0,96	mg/g Frischgewicht	Leber
RNS	0,43—1,75	mg/g Frischgewicht	Leber
RNS	1,19	mg/g Frischgewicht	Leber
Protein	88	% des Trockengewichts	Leber
Protein	76	% des Trockengewichts	Pankreas
Protein	81,5	% des Trockengewichts	Pankreas
Protein	83,5	% des Trockengewichts	Herz
Protein	80	% des Trockengewichts	Herz
Protein	85	% des Trockengewichts	Leber
Protein	74	% des Trockengewichts	Thymus
Protein	79	% des Trockengewichts	Niere
Protein	76,5	% des Trockengewichts	Knochenmark
Protein	74,5	% des Trockengewichts	Erythrocyten
Protein	66	% des Kerns	Leber
Protein	31,34	mg/g Frischgewicht	Leber
Protein	18,5	% des Homogenats	Leber
Gesamtfettsäuren . .	5,7	% des Trockengewichts	Leukocyten

Substanzen im Zellkern.

Tier	Bemerkungen	Autor	Jahr
Rind		Chargaff	1945
Kalb		v. Euler und Mitarbeiter	1945
Fisch		Felix	1952
Kalb		v. Euler und Mitarbeiter	1945
Rind		Chargaff	1945
Mensch	Cortex	Tyrell, Richter	1951
Pferd		Pollie, Ratti	1953
Mensch		Pollie, Ratti	1953
Kalb		v. Euler und Mitarbeiter	1945
Kalb	Prot./DNS = 2,85	Allfrey und Mitarbeiter	1951
Kalb	Prot./DNS = 5,7	Allfrey und Mitarbeiter	1951
Kalb	Prot./DNS = 4,6	Allfrey und Mitarbeiter	1951
Kalb	Prot./DNS = 4,2	Allfrey und Mitarbeiter	1951
Kalb	Prot./DNS = 3,3	Allfrey und Mitarbeiter	1951
Rind	Prot./DNS = 5,1	Allfrey und Mitarbeiter	1951
Rind	Prot./DNS = 4,8	Allfrey und Mitarbeiter	1951
Pferd	Prot./DNS = 3,45	Allfrey und Mitarbeiter	1951
Pferd	Prot./DNS = 7,3	Allfrey und Mitarbeiter	1951
Pferd	fastend, Prot./DNS = 4,55	Allfrey und Mitarbeiter	1951
Maus	ZBC-Linie	Barnum und Mitarbeiter	1950
Hühnchen	Prot./DNS = 2,9	Allfrey und Mitarbeiter	1951
		Rerabeck	1952
Ratte		Muntwyler, Seifter, Harkness	1950
Ratte	proteinarme Nahrung	Muntwyler, Seifter, Harkness	1950
Maus	C^3H	Price, Miller, Miller	1951
Hamster		Price, Miller, Miller	1951
Rind		Mauritzen und Mitarbeiter	1952
Rind		Mauritzen und Mitarbeiter	1952
Rind		Mauritzen und Mitarbeiter	1952
Kaninchen		Mauritzen und Mitarbeiter	1952
Kaninchen		Mauritzen und Mitarbeiter	1952
Ratte		Mauritzen und Mitarbeiter	1952
Ratte	regenerierend	Mauritzen und Mitarbeiter	1952
Huhn		Mauritzen und Mitarbeiter	1952
Huhn		Mauritzen und Mitarbeiter	1952
Huhn		Mauritzen und Mitarbeiter	1952
Dorsch		Mauritzen und Mitarbeiter	1952
Lachs		Mauritzen und Mitarbeiter	1952
Hering		Mauritzen und Mitarbeiter	1952
Maus	ZBC-Linie	Barnum und Mitarbeiter	1950
		Rerabeck	1952
Ratte		Muntwyler, Seifter, Harkness	1950
Ratte	proteinarme Nahrung	Muntwyler, Seifter, Harkness	1950
Maus		Price, Miller, Miller	1951
Hamster		Price, Miller, Miller	1951
Pferd	lipoidfrei	Allfrey und Mitarbeiter	1952
Pferd	lipoidfrei	Allfrey und Mitarbeiter	1952
Rind	lipoidfrei	Allfrey und Mitarbeiter	1952
Rind	lipoidfrei	Allfrey und Mitarbeiter	1952
Kalb	lipoidfrei	Allfrey und Mitarbeiter	1952
Kalb	lipoidfrei	Allfrey und Mitarbeiter	1952
Kalb	lipoidfrei	Allfrey und Mitarbeiter	1952
Kalb	lipoidfrei	Allfrey und Mitarbeiter	1952
Kalb	lipoidfrei	Allfrey und Mitarbeiter	1952
Huhn	lipoidfrei	Allfrey und Mitarbeiter	1952
Maus	ZBC-Linie	Barnum und Mitarbeiter	1950
Maus		Price, Miller, Miller	1951
Ratte		Lee, Williams	1953
Mensch		Pollie, Ratti	1953

Tabelle 1.

Substanz	Menge	Bezugssystem	Gewebe
Gesamtfettsäuren . .	6,3	% des Trockengewichts	Leukocyten
Gesamtcholesterin . .	~40	% des Homogenats	Gehirn
Gesamtcholesterin . .	1,39	% des Trockengewichts	Leukocyten
Gesamtcholesterin . .	1,47	% des Trockengewichts	Leukocyten
Cholesterinester . . .	0,32	% des Trockengewichts	Leukocyten
Cholesterinester . . .	0,285	% des Trockengewichts	Leukocyten
Phosphatide	10,9	% des Trockengewichts	Leber
Phosphorlipoide . . .	3,4	% des Kerns	Leber
Monoaminophosphor-lipoide	~35	% des Homogenats	Gehirn
Lecithine	~45	% des Homogenats	Gehirn
Cerebroside	~60	% des Homogenats	Gehirn
Cephaline	~30	% des Homogenats	Gehirn
Sphingomyeline . . .	~40	% des Homogenats	Gehirn
Riboflavin	4fache Konzentration des gesamten Gewebes		Herz
Riboflavin	2,1—3,8	mg/g Frischgewicht	Leber
Riboflavin	4,2—6,8	% des Homogenats	Leber
Riboflavin	4,5—6,3	% des Homogenats	Leber
Riboflavin	11—13	% des Homogenats	Leber
Riboflavin	4,0	mg/g Frischgewicht	Leber
Vitamin A	23,6	% des Homogenats	Leber
Vitamin A	16	% des Homogenats	Leber
Vitamin A-Ester . .	0,5—6,9	% des Homogenats	Leber
Vitamin A-Alkohol .	0—8,8	% des Homogenats	Leber
Pyridoxin	gleiche Konzentration wie gesamtes Gewebe		Herz
Nicotinsäure	2,7fache Konzentration wie gesamtes Gewebe		Herz
Pantothensäure . . .	3,5fache Konzentration wie gesamtes Gewebe		Herz
Inositol	4,8fache Konzentration wie gesamtes Gewebe		Herz
Biotin	0,5fache Konzentration wie gesamtes Gewebe		Herz
Folsäure	3fache Konzentration wie gesamtes Gewebe		Herz
Folinsäure	7,3—8,2	% des Homogenats	Leber
Thiamin	3fache Konzentration des gesamten Gewebes		Herz
Thiaminpyrophosphat	7	% des Homogenats	Leber
Vitamin B 12	8,6	% des Homogenats	Leber
Pantothensäure . . .	43,5	% des Homogenats	Leber
Co-Enzym A	23,5	% des Homogenats	Leber
Na	0,044	mg/g Frischgewicht	Leber
K	0,21	mg/g Frischgewicht	Leber
J	0,0		Thyreoidea
J	0,0		Thyreoidea
Dijodtyrosin	4,7	% des Homogenats	Thyreoidea
Thyroxin	2,7	% des Homogenats	Thyreoidea
Acetat-Ersatz-Faktor	17,3	% des Homogenats	Leber

Lipoprotein, das leicht alkalisch löslich ist und 50% des Trockengewichtes des Kernes ausmacht, ist von WANG, MAYER und THOMAS (1953) beschrieben worden. Es ist unklar, ob die ebenfalls gefundenen Lipoide einer bestimmten Struktur angehören, z. B. der Kernmembran, oder ob das von WANG und Mitarbeitern gefundene Lipoprotein diffus im Interphasenkern verteilt ist, wie es die Osmiumaufnahme solcher Kerne auf elektronenoptischen Bildern vermuten läßt. Nach VINCENT (1952) besteht der Nucleolus überwiegend aus einem Phosphorprotein, das mit einer signifikanten Menge RNS verknüpft ist. Außerdem ließ sich nur saure Phosphatase als Enzym dieser Kernstruktur entdecken. Globuline wurden von KIRKHAM und THOMAS (1953) aus Zellkernen des Kalbsthymus und der Kalbsleber isoliert. Zusammen mit DNS als Desoxynucleoproteine konnten drei elektrophoretisch einheitliche Fraktionen isoliert werden (ROTHERHAM, IRVIN und IRVIN 1954). Solche Kernproteine weisen einen regen Stoffaustausch auf,

(Fortsetzung.)

Tier	Bemerkungen	Autor	Jahr
Pferd		Pollie, Ratti	1953
Mensch	Cortex	Tyrell, Richter	1951
Mensch		Pollie, Ratti	1953
Pferd		Pollie, Ratti	1953
Mensch		Pollie, Ratti	1953
Pferd		Pollie, Ratti	1953
Ratte		Levine, Chargaff	1952
Maus	ZBC-Linie	Barnum und Mitarbeiter	1950
Mensch	Cortex	Tyrell, Richter	1951
Mensch	Cortex	Tyrell, Richter	1951
Mensch	Cortex	Tyrell, Richter	1951
Mensch	Cortex	Tyrell, Richter	1951
Mensch	Cortex	Tyrell, Richter	1951
Rind		Isbell und Mitarbeiter	1942
Hamster		Price, Miller, Miller	1951
Hamster		Price, Miller, Miller	1951
Maus		Price, Miller, Miller	1951
Ratte		Muntwyler, Seifters, Harkness	1951
Maus		Price, Miller, Miller	1951
Ratte	mit A gefüttert	Powell, Krause	1953
Ratte		Powell, Krause	1953
Ratte		Krinsky, Ganguly	1953
Ratte		Krinsky, Ganguly	1953
Rind		Isbell und Mitarbeiter	1942
Rind		Isbell und Mitarbeiter	1942
Rind		Isbell und Mitarbeiter	1942
Rind		Isbell und Mitarbeiter	1942
Rind		Isbell und Mitarbeiter	1942
Rind		Isbell und Mitarbeiter	1942
Maus		Swendseid, Bethele, Ackermann	1951
Rind		Isbell und Mitarbeiter	1942
Ratte	Co-Enzym	Goethart	1952
Maus		Swendseid, Bethele, Ackermann	1951
Ratte		Higgins, Miller, Price, Strong	1950
Ratte		Higgins, Miller, Price, Strong	1950
Ratte		McFarlane, Spencer	1953
Ratte		McFarlane, Spencer	1953
Rind		Graff, Hebekerl	1952
Schwein		Graff, Hebekerl	1952
		Rerabeck	1952
		Rerabeck	1952
Ratte		Reed, Cornier	1951

wie Smellie (1951) mit ^{32}P und Daly, Allfrey und Mirsky (1952) mit ^{15}N-Glykokoll zeigen konnten. Histon und noch mehr das Restprotein waren beteiligt. Der Einbau von ^{14}C-Glykokoll kann auch in vitro von isolierten Kernen vollzogen werden (Lang und Mitarbeiter 1953), gleichzeitig mit der Aufnahme in Lipoide und Nucleotide. Die aus verschiedenen Ausgangsmaterialien gewonnenen Histone zeichnen sich in der Regel durch einen hohen Argingehalt aus (10—30%), Tyrosingehalt wird zwischen 1,5 und 5,5% angegeben. Tryptophan konnte nicht oder nur in Spuren gefunden werden, Phenylalanin und Prolin sind zu 1,5—2,5% enthalten, siehe z. B. Fischer und Kreuzer (1953). Der schon genannten unterschiedlichen Präparationsverfahren wegen ist es im Augenblick noch wenig sinnvoll, weiter auf Einzelwerte der Proteinanalysen einzugehen, da diese auf die hier betrachtete Rolle des Zellkerns keinen unmittelbaren Bezug haben. Solche Analysen sind jedoch bei Davidson und Lawrie (1948),

BRUNISH, FAIRLEY und LUCK (1951), KHOUVINE und BARON (1951), STEDMAN und STEDMAN (1951), HARPER und MORRIS (1953), HAMER (1953), KHOUVINE und GREGOIRE (1953) u. a. zu finden.

Aus den schon genannten Studien mit markierten Vorläufern geht hervor, daß die größte Aktivität bei den Kernproteinen zu finden ist, wenn man sie mit den Nucleinsäuren vergleicht. Die Kernproteine verschiedener Herkunft waren Gegenstand der Untersuchungen von PRICE, MILLER und MILLER (1951), POLLISTER (1952) und ENGBRING und LASKOWSKY (1953). Die in dieser und der voraufgenannten Literatur veröffentlichten Aminosäureanalysen isolierter Kernproteine zeigen durch ihr ziffernmäßiges Abweichen voneinander doch sehr deutlich den Umfang und die Komplexität des Problemkreises nucleärer Proteine.

In Tabelle 1 sind auch eine Anzahl von Substanzen aufgeführt, deren allgemeine, physiologisch celluläre oder humorale Wirkungen lange bekannt sind. Sie sind in ihrem Vorkommen entweder ganz auf den Kern beschränkt oder haben andere Zellstrukturen zur Unterlage. In den meisten der genannten Fälle darf wohl angenommen werden, daß der Ort hoher Konzentration entweder der Syntheseort ist oder die Aktion der betreffenden Substanz sich an diesen Plätzen vollzieht. Wenn allerdings, wie im Falle von Tyrosin und Dijodtyrosin, die celluläre Konzentration fast ganz auf den Nucleus beschränkt ist, während die Hauptmenge der nicht strukturgebundenen Wirkstoffe im Kolloid gefunden wird (RERABECK 1952), so ist wohl kaum an einen in den Kern der Thyreoideazellen lokalisierten Wirkungsort dieser beiden Substanzen zu denken, sondern vielmehr an Synthese. Nach spektrographischen und histochemischen Untersuchungen von CHAYEN (1953) sollen im Zellkern größere Mengen Vitamin C angehäuft sein. Radioaktiv markiertes Sulfat (^{35}S) wird vom Kern umgesetzt (ODEBLAD und BOSTRÖM 1953), was auf einen besonderen Schwefelstoffwechsel in den Kernen verschiedener Gewebe der Maus hinweist. Von vielen anderen in isolierten Kernen gefundenen Substanzen, z. B. freien Aminosäuren (DOUNCE und Mitarbeiter 1950, 1954), ist der Motor vieler physiologischer Bewegungserscheinungen, das ATP, besonders interessant. Sein Einfluß auf intranucleäre Tonuszustände, Chromosomenbewegungen und Spindelfaserkontraktionen wird von DANIELLI und CATCHSIDE (1945), LETTRÉ (1952), HOFFMANN-BERLING (1954), LETTRÉ (1954) diskutiert. Nach MAZIA (1952) soll dieses ATP aus dem Cytoplasma an den Kern abgegeben werden, da der ATP-Gehalt isolierter Kerne rasch fällt und gleichzeitig der normale Stoffwechsel aufhört. Wie weiter unten noch deutlich gemacht werden wird, scheinen aber Nucleotide aus dem Kern in das Cytoplasma zu wandern, um dort teilweise aus anderen energieliefernden Prozessen Phosphor — sowohl speichernd, als auch zur Weitergabe an den Kern — aufzunehmen. Doch ist auch der Kern selbst zu Phosphorylierungen fähig, wie deutlich aus seiner Enzymausrüstung hervorgeht.

Die Phosphatide und anderen Lipoide, die aus Zellkernen isoliert wurden, sind in Tabelle 1 zusammengestellt.

Auf die raschen Fortschritte der Chemie enzymatisch auf- und umgebauter Kohlenhydrate, die als Vorläufer der Nucleinsäure-Zuckerkomponente denkbar sind, kann hier nicht eingegangen werden. Auch können aus den bisherigen Ergebnissen noch keine Schlüsse auf die Wege des Einbaues gezogen werden. Das gleiche gilt für den Problemkreis der intracellulären Synthese der Purine und Pyrimidine bzw. deren Konvertibilität, soweit nicht unmittelbare Vorstellungen mit der Tätigkeit des Kernes damit verbunden sind.

Freie Nucleotide.

Biochemische Extraktionsstudien und eine größere Zahl Untersuchungen mit radioaktiven Substanzen haben die Bedeutung der *freien Nucleotide* und *Nucleoside* in der Zelle und teilweise auch ihren Zusammenhang mit den Nucleinsäuren zu erläutern versucht. Diesen Verbindungen wurde aus drei Gründen besonderes Interesse gewidmet. Sie sind, wie für einzelne Fälle bekannt ist, *wichtige Co-Enzyme*, sie sind *Speicher und Überträger von Energie des Phosphorstoffwechsels*, und sie dürfen als die wesentlichen *Vorläufer oder Vorstufen der Nucleinsäuren* betrachtet werden. Ein enger Zusammenhang mit der Tätigkeit des Kernes ist in der folgenden Darstellung entweder nachgewiesen oder kann begründet angenommen werden.

Freie säurelösliche Mono-, Di- und Triphosphate von Adenosin, Guanosin, Cytidin und Uridin, Thymin, sowie verschiedene Uridindiphosphorsäurederivate konnten aus Geweben isoliert werden (Bergquist und Deutsch 1953, Schmitz 1954, Smith und Mills 1954, Singer und Kearney 1954, Berg und Joklik 1954, Hurlbert, Schmitz, Barum und Potter 1954, Schmitz, Hurlbert und Potter 1954, Marshak und Marshak 1954).

Die Nucleotide können ineinander umgewandelt werden, z. B. ehe sie zum Einbau in Nucleinsäuren aufgenommen werden (Hammarsten, Reichard und Saluste 1950, Reichard und Estborn 1951, Brown, Roll und Weinfeld 1952, Lu und Winnik 1954). Auch scheinen die Synthesewege nicht an bestimmte „Kanäle“ gebunden zu sein, sondern haben eine ganze Reihe von Kombinationen zur Verfügung. Diese Wege können in gewissen Abschnitten artspezifisch sein, wie die Arbeit von Rabinowitz und Mitarbeitern (1954) wahrscheinlich macht. Der aufbauende Zellorganismus kann dabei auch wählerisch gegenüber den angebotenen Bausteinen sein (Rabinowitz 1954). Teilweise werden die aufgenommenen markierten Substanzen zunächst vermehrt im Plasma aufgenommen, um dann eventuell an den Kern abgegeben zu werden (Edmonds und le Page 1953). Steinert (1951, 1952) konnte sogar direkt zeigen, daß die Mengen Purine, die während der Gastrulation verschwinden, sofort in den Purinen der Nucleinsäuren wiedergefunden werden können. Hypoxanthin- und Guaninmengen nehmen zum gleichen Zeitpunkt ab, an dem die Nucleinsäuresynthese beginnt.

Von den erwähnten Umbauprozessen der Nucleotide und vor allem auch den Phosphorylierungsprozessen des Energiehaushaltes mit Hilfe von Enzymen konnten einige erfaßt werden. Da manche solcher Vorgänge nur in Gegenwart isolierter Kerne ablaufen, ist anzunehmen, daß entsprechende Enzyme im Kern lokalisiert und demnach auch dort wirksam sind (Leloir 1953, Smith und Mills 1954, Berg und Joklik 1954, Hecht, Potter und Herbert 1954, Schmitz und Mitarbeiter 1954). Brachet (1951) hatte schon vermutet, daß die Co-Enzymsynthese kernkontrolliert sei, und Hogeboom und Schneider fanden in mehreren Untersuchungen das Enzym, das an der Reaktion:

$$\text{Nicotinamidnucleotid} + \text{ATP} = \text{DPN} + \text{Pyrophosphat}$$

beteiligt ist, zu 85% im Kern akkumuliert (s. hierzu auch die Tabelle 2). Nucleotide und Basen sind aber nicht nur Substrate für Phosphatasen und Phosphorylasen im Kern (Kalckar 1947, Friedkin und Kalckar 1950, Wajzer und Baron 1949), sie können, wie Lang und Mitarbeiter es (1953) begründet annehmen, auch über die Totalsynthese des Purinringes im Kern selbst gewonnen werden. Ein weiterer Hinweis auf die enge Verknüpfung der freien Nucleotide mit dem Kern ist durch die Isolierung von Desoxyadenosindi- und -triphosphat als einer der Energieüberträger in der Zelle gegeben (Sable und Mitarbeiter 1954). Die

Cohydrogenasen (WARBURG und CHRISTIAN 1931, SCHLENK und v. EULER 1938) enthalten Adenylsäuren und Nicotinsäureamid als Dinucleotide. Auch das Coenzym A enthält Adeninnucleotid (LIPMAN und Mitarbeiter 1950). Ein Alloxazin-adenin-dinucleotid ist als wirksame Gruppe in Glucoseoxydase (KEILIN und HARTREE 1946, SMITH und MILLS 1954), in Glykokolloxydase (RATNER, NOCITO und GREEN 1944), in Luciferase (JOHNSON und EYRING 1944) enthalten. In den Diaphorasen (v. EULER und HASSE 1938, v. EULER und GÜNTHER 1938, MAHLER und Mitarbeiter 1953, GREEN und Mitarbeiter 1953) kommt dieses Dinucleotid ebenfalls vor. Auch Fumarathydrase (FISCHER und Mitarbeiter 1939), Xanthinoxydase (BALL 1939), l- und d-aminooxydase (STRAUB und Mitarbeiter 1939), sowie Cytochrom-c-reductase (HAAS und Mitarbeiter 1942) bedürfen dieses Co-Enzyms. Ein Diphosphopyridin ist für die Aktivität der Cholindehydrogenase erforderlich (STRENGTH und Mitarbeiter 1953). Zu den Nucleotiden siehe auch SINGER und KEARNAY (1954) sowie KORNBERG (1951), die die Rolle der Nucleotide im Phosphorstoffwechsel der Nucleotide selbst behandeln. Die vorstehende, nicht einmal erschöpfende Aufreihung von ganz allgemein nucleotidabhängigen Enzymaktionen läßt gut verstehen, daß Mangel sowohl als auch Anhäufung gewisser Nucleotide die cellulären Stoffwechselprozesse empfindlich beeinflussen kann. An Mitochondrien konnten MCFARLANE und SPENCER (1953) zeigen, daß eine normale Tätigkeit dieser Zellorganellen nicht ohne Zufuhr von Adenosin-mono-5-Phosphat oder anderen Nucleotiden möglich ist. Vielleicht noch klarer geht die Bedeutung der Nucleotide, und in diesem Zusammenhang auch sehr deutlich die des Kernes, aus Untersuchungen hervor, die das Verhalten entkernter Zellen zum Gegenstand hatten, wie es weiter unten noch ausführlicher beschrieben wird.

Enzymatische Aktivitäten des Zellkernes.

Verbesserte und leichter zugängliche Meßmethoden enzymatischer Aktivität sowie die Ausarbeitung geeigneter Verfahren zur Isolierung der corpusculären Zellelemente gaben die Basis zu einem intensiven Studium *enzymatischer Aktivitäten im Zellkern* und den anderen Zellfraktionen. Wenn auch immer wieder aus den Untersuchungen klar wurde, daß jede der verwandten Isolierungsmethoden nachteilige Einflüsse auf irgendeinen Teil der untersuchten Prozesse ausübte, wenn auch Diffusions- und Oxydoreduktionsprozesse während der Zerstörung der Zellintegrität auftraten, so zeichnen sich doch ganz allgemein die Grundzüge der enzymatischen Aktionen in der Zelle ab. Soviel darf heute jedenfalls angenommen werden, daß die oxydativen Prozesse sich ausschließlich im Cytoplasma (Mitochondrien) abspielen, daß die glykolytischen Vorgänge ebenfalls im wesentlichen in das Cytoplasma lokalisiert werden dürfen, und daß damit für den Zellkern die wesentlichen Energiequellen außerhalb seiner Membran liegen. Ihm wird die relativ geringe, benötigte Energie vom Cytoplasma aus in Form energiereicher Phosphate zugeführt. Doch ist der Zellkern selbst, wenn auch nur in begrenztem Umfange, in der Lage, Energie aus glykolytischen Prozessen zu gewinnen.

Aus der Tabelle 2 geht hervor, daß einige Phosphatasen und Phosphorylasen im Zellkern auftreten, in manchen Fällen sogar in beträchtlichen Mengen. Das Vorkommen von Phosphatasen im Zellkern war von histochemischer Seite lange diskutiert worden, besonders im Hinblick auf Adsorptionsphänomene (DANIELLI 1945, CHÉVREMONT und FIRQUET 1953). STEDMAN und STEDMAN (1951) zeigten unter anderem, daß saure Phosphatase aus Prostata reversibel und inaktivierend an isolierte Zellkerne gebunden werden kann. Die Bindung kann durch Proteine wieder gelöst werden. METZGER-FREED (1953) fand, daß die Phosphoproteinphosphatase in der Entwicklung von *Rana pipiens* unabhängig von der normalen oder haploiden Chromosomenzahl der Zellen sei; diese Angabe soll auf eine Unabhängigkeit der Menge des genannten Enzymes von der genetisch wirksamen DNS-Menge hinweisen. WACHSTEIN und MEISEL bekräftigen (1954) das Vorkommen von 5-Nucleotidase auf histochemischem Wege. Adenosintriphosphatase hatten LANG und SIEBERT und deren Mitarbeiter (s. die Publikationen unter diesen Namen) in hoher Konzentration im Zellkern gefunden. HOGEBOOM und SCHNEIDER (1952) hatten ein Enzym, das Diphosphonucleotid (DPN) aus ATP und Nicotinamidnucleotid synthetisiert, fast ausschließlich in den Kern lokalisieren können. DPN wird dann wahrscheinlich an das Cytoplasma abgegeben, um als Co-Enzym an glykolytischen Prozessen teilzunehmen. Wird diese Nucleotidase gehemmt, kommt es zu einer Akkumulation von ATP im Cytoplasma. Anaerobe Bedingungen setzen umgekehrt die Gesamt-ATP-Produktion erheblich herab (BRACHET 1954). ROWEN und KORNBERG (1951) interpretieren ihre Befunde über die phosphorylasegesteuerte Bildung von Nicotinamid-ribosid im Sinne der Bildung von Co-Enzymen für cytoplasmatische Prozesse durch den Kern, während ALLFREY, STERN und MIRSKY (1952) nur geringe Mengen Nucleotidphosphatasen im Zellkern fanden. Aber hohe Werte für Adenosindesaminase, Nucleosidphosphorylase und Guanase (4mal der Gewebewert) wurden gefunden. Die Konzentrationen variieren von Gewebe zu Gewebe und mit dem Alter des Objektes. SCHNEIDER (1946) hatte die größte Menge ATPase in den Mitochondrien gefunden. Die genannten Enzyme dürften auch an Transphosphorylierungen beteiligt sein.

Wie schon erwähnt, scheinen keine oxydativen Prozesse im Kern stattzufinden, der KREBS-Cyclus fehlt, also keine katabolischen Prozesse und keine Atmungsvorgänge konnten gezeigt werden. Die Arbeitsgruppe SCHNEIDER und HOGEBOOM und auch andere betrachten heute die früher gefundenen Aktivitäten entsprechender Enzyme als unsignifikant oder als Artefakt (DOUNCE 1952, 1954). Angaben über diese Enzyme finden sich in Tabelle 2. LANG, SIEBERT und Mitarbeiter konnten auch gewisse Aminosäureoxydasen im isolierten Kern nicht entdecken.

Je nach seiner geweblichen Herkunft ist der Zellkern zu geringeren oder größeren glykolytischen Leistungen befähigt. DOUNCE (1954) meint allerdings, daß das glykolytische System unvollständig sei, da bisher noch keine Hexokinase entdeckt werden konnte. Die Glykolyse ginge nur bis zum Lactat, das in das Cytoplasma diffundiere. STERN und MIRSKY (1952) fanden DPN vermehrt in Kernen von Weizenkeimen, ebenso eine Reihe glykolytischer Enzyme und glauben, daß der Kern eine außerordentlich lebhafte glykolytische Aktivität besitzt.

Amylase findet man nicht im Kern (STERN und MIRSKY 1952); sie ist nicht an granuliertes Material gebunden, sondern diffus im Plasma verteilt (HOLTER und LØVTRUP 1949); auch Protease und Dipeptidase fehlen (HOLTER und POLLOCK 1952), ebenfalls wird die Lipase vermißt (ALLFREY und Mitarbeiter 1952). Gewisse Enzyme werden in hoher Konzentration im Zellkern gefunden, so z. B. Arginase zu 80% (ALLFREY und Mitarbeiter 1952), ohne daß eine andere Vorstellung als die der Synthese dieses Enzymes damit verknüpft werden kann (LANG, SIEBERT und Mitarbeiter 1951). Ähnliches gilt für das durch Enterokinase aktivierbare reichliche Trypsin wie auch das Katepsin, das LANG, SIEBERT und Mitarbeiter (1953) im Zellkern fanden. Das beobachtete Vorkommen von DNasen im Kern (unter anderen LANG, SIEBERT und Mitarbeiter) wird von WEBB (1953) als ein Artefakt bezeichnet. Ebenso soll die Sulfatase im Kern (ROY 1954) nach DODGSON und Mitarbeitern (1954) ein Artefakt sein, d. h. auf geringe Verunreinigungen der Kernfraktion mit Resten anderer Fraktionen zurückzuführen sein.

Tabelle 2[1].

Enzym	Menge	Bezugsgröße	Gewebe
Esterase	14—19	% des Homogenats	Leber
Esterase	6,5	% des Homogenats	Leber
Esterase	50	% des Homogenats	Leber
Monobutyrase (Monoesterase)	4,5	% des Homogenats	Leber
Alkali-Phosphatase	193	% des Homogenats	Leber
Alkali-Phosphatase	40,1	% des Homogenats	Leber
Alkali-Phosphatase	30	% des Homogenats	Nerv, peripher
Phosphomonoesterase	10,4	% des Homogenats	Leber
Phosphomonoesterase	20,7	% des Homogenats	Leber
Saure Phosphatase	4,8	% des Homogenats	Leber
Saure Phosphatase	25—30	% des Homogenats	Leber
Glucose-6-phosphatphosphatase	5—25	% des Homogenats	Leber
Hexosediphosphatase	8	% des Homogenats	Leber
RNase	49	% des Homogenats	Thymus
RNase	10,2	% des Homogenats	Leber
RNase	11—28	% des Homogekats	Leber
RNase	44	% des Homogenats	Herz
DNase	10,3	% des Homogenats	Leber
DNase	37	% des Homogenats	Thymus
DNase	0		Thymus
Aryl-Sulfatase	15 (9—20)	% des Homogenats	Thymus
Aryl-Sulfatase	0		
Aryl-Sulfatase	(9,1—12,5)	% des Homogenats	Leber
β-Amylase	17	% des Homogenats	Kerne von
Guanase	0		Nucleolus
Adenosindesaminase	0		Nucleolus
Glutaminase I	10,3—14,5	% des Homogenats	Leber
Arginase	113	% des Homogenats	Leber
Arginase	33,6	% des Homogenats	Leber
Arginase	8	% des Homogenats	Leber
Arginase	36	% des Homogenats	Leber
Arginase	80	% des Homogenats	Leber
DPN-Nucleosidase	37—38	% des Homogenats	Leber
Adenosintriphosphatase	31,3	% des Homogenats	Leber
Adenosintriphosphatase	37,8	% des Homogenats	Hepatom
Adenosintriphosphatase	18,7	% des Homogenats	Leber
Adenylpyrophosphatase	20; 8; 10—16	% des Homogenats	Nerv, peripher
DPN-Synthese	61—101	% des Homogenats	Leber
Nucleosidphosphorylase	30,6	Aktivität des Nucleolus : Aktivität	
Phosphorylase	66	% des Homogenats	Leber
Pyruvat-kinase	62	% des Homogenats	Kerne von
Rhodanese	15,4	% des Homogenats	Leber
Rhodanese	44	% des Homogenats	Leber
Alkoholdehydrogenase	19	% des Homogenats	Leber
Alkoholdehydrogenase	6,2	% des Homogenats	Leber
Alkoholdehydrogenase	14,7	% des Homogenats	Leber
α-Glycerophosphatdehydrogenase	16,7	% des Homogenats	Leber
Phosphoglyceraldehyddehydrase	48	% des Homogenats	Kerne von
Glutaminsäuredehydrogenase	0	% des Homogenats	Leber
Betain-aldehydoxydase	17; 21	% des Homogenats	Leber
Milchsäuredehydrogenase	40	% des Homogenats	Leber
Milchsäuredehydrogenase	28,7	% des Homogenats	Leber
Milchsäuredehydrogenase	27,4	% des Homogenats	Niere
Succinatdehydrogenase	0	% des Homogenats	Leber
Succinatdehydrogenase	10	% des Homogenats	Leber
Glucosedehydrogenase	0	% des Homogenats	Leber
d-Aminosäureoxydase	100	% des Homogenats	Leber
l-Aminosäureoxydase	0		Leber

[1] Zusammengestellt nach den Richtlinien von O. HOFFMANN-OSTENHOF (1954).

Enzyme.

Organismus	Bemerkung	Autoren	Jahr
Maus		Omachi, Barnum, Glick	1948
Ratte		Ludewig, Chauntin	1950
Ratte		Dounce	1950
Ratte		Heller, Bargoni	1949
Ratte		Dounce	1950
Ratte		Ludewig, Chanutin	1950
Ratte		Abood, Gerard	1954
Maus		Tsuboi	1952
Maus	regeneriert	Tsuboi	1952
Ratte		Palade	1951
Ratte		Dounce	1950
Ratte		Hers, Berthet, Berthet, de Duve	1951
Ratte		Hers, Berthet, Berthet, de Duve	1951
Kalb		Brown, Jacobs, Laskowski	1952
Maus	C_3H	Schneider, Hogeboom	1952
verschieden		Pirotte, Desreux	1952
Kalb		Moyle	1953
Maus	C_3H	Schneider, Hogeboom	1952
Kalb		Brown, Jacobs, Laskowski	1952
Kalb	DNase im Kernartefakt	Webb	1953
Ratte		Roy	1954
Ratte	Vorkommen = Artefakt	Dodgson, Spencer, Thomas	1954
Ratte	nicht signifikant	Dodgson, Spencer, Thomas	1955
Weizensamen		Stern, Mirsky	1952
Seeigelei		Baltus	1954
Seeigelei		Baltus	1954
Ratte		Shepherd, Kalnitzky	1951
Ratte		Dounce	1950
Ratte		Ludewig, Chauntin	1950
Ratte		Lang, Siebert, Lucius, Lang	1951
Ratte		Schein, Young	1952
Ratte		Allfrey, Stern, Mirsky	1952
Ratte		Sung, Williams	1952
Maus		Schneider, Hogeboom	1950
Maus	98/15	Schneider, Hogeboom	1950
Ratte		Allard, Cantero	1952
Ratte	AS → ADP → ATP	Abood, Gerard	1954
Maus	C_3H	Hogeboom, Schneider	1952
des übrigen Seeigeleies		Baltus	1954
Ratte, nach Zerkleinern		Dounce	1950
Weizensamen		Stern, Mirsky	1952
Ratte		Ludewig, Chanutin	1950
Ratte		Moyle	1953
Ratte		Nyberg, Schuberth, Änggård	1953
Pferd		Nyberg, Schuberth, Änggård	1953
Ratte		Dianzani	1951
Ratte		Dianzani	1951
Weizensamen		Stern, Mirsky	1952
Ratte	Vorkommen im Kernartefakt	Hogeboom, Schneider	1953
Ratte		Williams	1952
Ratte		Dounce	1950
Ratte		Dianzani	1951
Ratte		Dianzani	1951
Ratte		Dounce	1950
Ratte		Strittmaker, Ball	1954
Ratte	Dia	Dianzani	1951
Ratte		Dounce	1950
Ratte	nicht FAD aktievierbar	Lang, Siebert	1950

Tabelle 2.

Enzym	Menge	Bezugsgröße	Gewebe
l-Aminosäureoxydase	0		Niere
Oxalessigsäureoxydase	10,5	% des Homogenats	Niere
Oxalessigsäureoxydase	9	% des Homogenats	Leber
Xanthinoxydase	0		Leber
Xanthinoxydase	0		Niere
l-Prolinoxydase	0		Niere
l-Prolinoxydase	0		Leber
Uricase	1	% des Homogenats	Leber
Cytochrom C	10; 11	% des Homogenats	Leber
Cytochrom C	24,9—26,3	% des Homogenats	Leber
Succinoxydase	10	% des Homogenats	Leber
Succinoxydase	11,1	% des Homogenats	Leber
Succinoxydase	10	% des Homogenats	Leber
Cholinoxydase	0	% des Homogenats	Leber
Cholinoxydase	9	% des Homogenats	Leber
Cholinoxydase	24	% des Homogenats	Leber
Cytochromoxydase	10	% des Homogenats	Leber
Cytochromoxydase	0,3—1,5	% des Homogenats	Leber
DPN-Cytochromoxydase	9,1	% des Homogenats	Leber
DPN-Cytochromoxydase	14,8	% des Homogenats	Hepatom
DPN-Cytochromreduktase . . .	9,1	% des Homogenats	Leber
DPN-Cytochromreduktase . . .	14,8	% des Homogenats	Hepatom
DPN-Cytochromreduktase . . .	12	% des Homogenats	Leber
Katalase	0	% des Homogenats	Leber
Katalase	4,5	% des Homogenats	Leber
Katalase	0,05—0,1	% des Homogenats	Leber
Katalase	18,6	% des Homogenats	Leber
Katalase	1,2	% des Homogenats	Leber
Katalase	6,9	% des Homogenats	Leber
Aldolase	40	% des Homogenats	Leber
Aldolase	67	% des Homogenats	Kerne von
Enolase	64	% des Homogenats	Kerne von
Enolase	50	% des Homogenats	Leber
Cozymase-Synthese-Enzym . .	48,1	Aktivität des Nucleolus : Aktivität	

LANG und Mitarbeiter haben in einer Reihe von Untersuchungen proteolytische Enzyme und Enzyme des Aminosäurestoffwechsels in verschiedenen Zellkernen finden können, sowie eine Reihe von hydrolytischen Enzymen.

Weiter oben wurde näher auf Untersuchungen mit radioaktiven Substanzen eingegangen. Die Ergebnisse dieser Studien ließen schon sehr stark vermuten, daß eine Reihe von Enzymen innerhalb des Kernes die schließliche Synthese hochmolekularer Substanzen vollziehen würde. Als Energielieferant dieser komplizierten Prozesse für Proteine, Nucleinsäuren, Lipoide usw. müssen hauptsächlich energiereiche Phosphate in Frage kommen, da die Energiequellen des Kernes selbst begrenzt erscheinen.

Die in diesem Artikel wiedergegebenen Tabellen über die chemischen Bestandteile des Kernes sind zum Teil in den Werten umgerechnet, um einen Vergleich der untersuchten Materialien untereinander zuzulassen, wie auch der Werte verschiedener Autoren an gleichen Objekten. So wurde unter anderem immer der totale Gehalt eines Gewebes mit dem Gehalt des entsprechenden Homogenats gleichgesetzt. Alle Substanzen, deren Vorkommen im Zellkern etwa 8—12% des gesamten Vorkommens ausmachen, müssen als insignifikant angesehen werden, insofern damit eine besondere Konzentration innerhalb des Zellkernes gemeint ist. Das Volumen des Zellkernes ist etwa 8—12% der Durchschnittszelle,

(Fortsetzung.)

Organismus	Bemerkung	Autoren	Jahr
Schwein	nicht FAD aktivierbar	Lang, Siebert	1950
Ratte		Schneider, Potter	1949
Ratte		Schneider, Potter	1949
Ratte	nicht FAD aktivierbar	Lang, Siebert	1950
Schwein	nicht FAD aktivierbar	Lang, Siebert	1950
Schwein	nicht FAD aktivierbar	Lang, Siebert	1950
Ratte	nicht FAD aktivierbar	Lang, Siebert	1950
Ratte		Hogeboom, Schneider, Striebich	1952
Ratte		Schneider, Hogeboom	1950
Ratte	Artefakt?	Dianzani, Viti	1955
Ratte		Schneider, Hogeboom	1950
Ratte		Shepherd, Kalnitzky	1951
Ratte		Strittmaker, Ball	1954
Ratte		Dounce	1950
Ratte		Kensler, Langemann	1951
Ratte		Williams	1952
Ratte		Strittmaker, Ball	1954
Ratte		Hogeboom, Schneider, Striebich	1952
Maus		Hogeboom, Schneider	1950
Maus		Hogeboom, Schneider	1950
Maus	C^3H	Schneider, Hogeboom	1950
Maus		Schneider, Hogeboom	1950
Ratte		Strittmaker, Ball	1954
Ratte	ebenso für Sarkomfragmente	v. Euler, Heller	1949
Ratte		Ludewig, Chanutin	1950
Ratte		Dounce	1950
Ratte		Nyberg, Schuberth, Änggård	1953
Pferd		Nyberg, Schuberth, Änggård	1953
Meerschweinchen		Nyberg, Schuberth, Änggård	1953
			1953
Ratte		Dounce	1950
Weizensamen		Stern, Mirsky	1952
Weizensamen		Stern, Mirsky	1952
Ratte		Dounce	1950
des übrigen Seeigeleies		Baltus	1954

d. h. bei unspezifischer Verteilung der betreffenden Substanz fallen 8—12% auf den Zellkern (Marshak 1941, Harrison 1951). Aber wie auch für die nucleäre und cytoplasmatische RNS klar wurde, ist die gefundene Konzentration einer Substanz in einem Zellpartikel niemals Ausdruck für ihre biologische Aktivität. Dazu kommt, daß für die Enzyme Phänomene der Hemmung und Aktivierung eine entscheidende Rolle spielen.

Enucleationsstudien.

Ein besonders wertvoller Beitrag zum Verständnis der Rolle des Zellkernes in der Zelle konnte durch Versuche mit enucleierten Zellen gewonnen werden. Solche Versuche sind allerdings auf große Einzeller beschränkt, womit sich die Frage erhebt, ob die zellphysiologischen Prozesse dieser oft autotrophen, selbständigen Organismen denen von Gewebezellen gleichgesetzt werden dürfen, wie umgekehrt natürlich eine Allgemeingültigkeit von Untersuchungen an Gewebezellen diskutabel ist. Ohne Zweifel sind jedoch eine Reihe von Erkenntnissen gewonnen worden, die auch über die Grenzen des jeweils untersuchten Objektes hinaus Gültigkeit haben dürften.

Entkernte Zellen können noch erhebliche Zeiten „leben", also auch Stoffwechselerscheinungen zeigen. Sie können Baumaterialien synthetisieren und damit eine begonnene Entwicklung ihres Zelleibes fortsetzen. Fragmente lassen ihre Cilien noch eine Weile nach der Abtrennung schlagen. Amöben nehmen noch Nahrung auf, ohne daß es jedoch zur Verdauung kommt (HOLTER 1952), andere Fragmente können noch assimilieren und Stärke aufbauen bzw. eine Cellulosemembran bilden, wie es ältere Untersuchungen von KLEBS (1896) zeigen. Die Permeabilität enucleierter *Arbacia*-Eizellen gegenüber Wasser (LUCKÉ 1932) oder von *Spirogyra*-Zellen gegenüber Glycerin (HOFFMANN 1927) ändert sich nicht. Ebenfalls wird die Permeabilität von *Amoeba proteus* gegenüber D_2O während 38 Std nach der Enucleation nicht beeinflußt (PRESCOTT und MAZIA 1954). BRACHET und SZAFARZ (1953) konnten an kernhaltigen und kernlosen Zellen von *Acetabularia mediterranea* mit radioaktiver Orotsäure zeigen, daß die RNS auch nach 2 Monaten nach der Enucleation ihren *Umsatz* nicht veränderte. Die RNS und die Proteinfraktionen nehmen nach MALKIN (1954) noch Glycin-C^{14} auf, aber es findet in den untersuchten Seeigeleiern, *Strongylocentrotus purpuratus*, keine Mehrproduktion von RNS oder Proteinen mehr statt.

LINET und BRACHET (1951) beobachteten jedoch, daß entkernte Amöben rasch ihren Gehalt an RNS verlieren, während die Glykogenmenge konstant bleibt, woraus die Kontrolle des Kernes über den RNS-Gehalt der Zelle hergeleitet wurde. Diese Versuche wurden von JAMES (1954) bestätigt. Er konnte durch Vergleich der Ergebnisse der Literatur mit eigenen zeigen, daß die Widersprüche zwischen RNS-Gehalt der kernhaltigen Fragmente, der entkernten und der ganzen Zellen eine einfache Erklärung zu finden scheinen. Danach ist der RNS-Verbrauch in hungernden Amöben eine direkte Funktion des Zellvolumens (s. auch ANDRESEN und HOLTER 1945), d. h. eine cytoplasmatische Eigenschaft. Der Abfall des RNS-Gehaltes in entkernten, also durch verhinderte Nahrungsaufnahme hungernden Fragmenten ist ähnlich, aber etwas steiler. Dieser Mehrabfall der RNS-Verbrauchskurve der entkernten Hälfte entspricht genau der Mehrproduktion an RNS, den kernhaltige Fragmente gegenüber ganzen Zellen aufweisen. Kern und Cytoplasma sind in bezug auf RNS also Antagonisten. Wie in voraufgehenden Abschnitten schon beschrieben worden ist, zeigt die geringe Menge nucRNS dementsprechend sehr kräftige Umsätze oder Aufnahme radioaktiver Substanzen, die dann später im Cytoplasma wiedergefunden werden.

Die Sauerstoffaufnahme in enucleierten Amöbenhälften änderte sich in den auf den Eingriff folgenden 9 Tagen nicht (BRACHET 1951), womit der Hinweis gegeben ist, daß die respiratorischen Enzyme der Zelle (überwiegend in den Mitochondrien) in ihrer Tätigkeit also weitgehend vom Kern unabhängig sind. Ja, für *Acetabularia*-Algen wurden sogar unveränderte O_2-Aufnahmewerte über 3 Monate gefunden (BRACHET 1952). Dagegen fallen die Werte für saure Phosphatase und die Phosphoraufnahme (P^{32}) in entkernten Acetabularien (BRACHET 1952) und für die P^{32}-Aufnahme in Amöben (MAZIA und HIRSCHFIELD 1950) rasch. Sie fallen aber auch während der Mitose (MAZIA und PRESCOTT 1954). Auch in *Amoeba proteus* fallen Esterase und saure Phosphatase rapid, die Dipeptidaseaktivität fällt auf ein Drittel, um sich dann konstant zu halten (BRACHET 1954). Der Kern übt also hauptsächlich seinen Einfluß auf die Phosphorylierungsprozesse aus, während oxydative Vorgänge nicht davon berührt werden. Energie*speicher*, wie Adenosin-Triphosphorsäure und das entsprechende Enzym ATPase halten sich 12 Tage konstant. HÄMMERLING und STICH (1954) fanden allerdings die P^{32}-Aufnahme in entkernter *Acetabularia* unverändert, und CLARK (1942) beschrieb Beobachtungen an Amöben, nach denen der Sauerstoffverbrauch in kernlosen Zellen um 70% fiel. Die restlichen 30% erwiesen sich als Kaliumcyanid-resistent.

Die $C^{14}O_2$-Aufnahme fällt bei *Acetabularia* erst nach 2 Wochen, d. h. solange geht die Proteinsynthese ungehindert weiter. Danach tritt langsam Abfall der Aufnahmewerte ein, doch ist auch nach 5 Wochen die Aufnahme noch deutlich (BRACHET und CHANTRENNE 1951). Diese Daten stimmen recht gut mit den Werten für die Enzyme überein, deren Aktivität unverändert erhalten bleibt: Amylase, Protease, Enolase und die schon genannte Adenosintriphosphatase (BRACHET 1952). URBANI (1952) fand die Amylaseaktivität nach der Enucleation sogar enorm erhöht. Die Werte für Proteinstickstoffentwicklung und Gewicht an enucleierten Acetabularien gehen mit den genannten Werten etwa parallel. Bis zum 15. Tage keine Differenzen zwischen kernhaltigen und kernlosen, dann rascher Abfall im kernlosen Teil, zunächst in den Chloroplasten, während die Mikrosomen unberührt bleiben. Es wird daher vermutet, daß schließlicher Mangel an Nucleotid-Coenzymen, die vom Kern geliefert werden, die Verwendung der Stärkereserven verhindert (VANDERHAEGHE 1954).

Die kernunabhängige Proteinsynthese der ersten Tage nach der Enucleation kann bei *Acetabularia* mit Trypaflavin gestört werden. Dieser reversible Störungseffekt ist unabhängig von der Gegenwart des Kernes (STICH 1951).

Der Kern übt also, wie aus den genannten Stoffwechsel- und Enzymstudien geschlossen werden kann, seine Wirkung hauptsächlich auf die kleinen Granula und das Hyaloplasma aus, während die Mitochondrien relativ unabhängig erscheinen; ihr Bestand an oxydativen Enzymen wird nicht verändert. Mit anderen Worten, der Kern hat primär Einfluß auf anabolische Prozesse, während die Mitochondrien primär katabolische Funktionen haben, die sie, zunächst kernunabhängig, noch eine Weile nach der Kernabtrennung fortsetzen können. Dann aber scheinen Stoffe, unter anderen auch gewisse Nucleotide, aus dem Kern zu fehlen, während es gleichzeitig im Cytoplasma zur Anhäufung von Stoffwechselprodukten kommt, die nicht weiterverarbeitet werden können. Eine gewisse Parallelität zu dem genannten Verhalten entkernter Einzeller ist in den Beobachtungen von LONDON, SHEMIN und RITTENBERG (1950) gegeben, die Syntheseprozesse des Blutfarbstoffes an unreifen, kernlosen Erythrocyten *in vitro* beschrieben. Im Prinzip ähnliche Problemstellungen sind an strahlengeschädigten Bakterien studiert worden, können aber an dieser Stelle nicht eingehender behandelt werden.

Solche Überlegungen bringen die Gefahr mit sich, die morphologischen Zellbestandteile getrennt in ihren Funktionen beurteilen zu wollen. Die Integrität der Zelle ist aber eine *conditio sine qua non* für jeden ihrer Konstituenten.

So ist natürlich auch der Zellkern nicht autonom. Das Plasma übt seinerseits vielmehr bestimmte Wirkungen auf den Kern aus. Zum Beispiel kann die Teilung des Zellkernes, die bei *Acetabularia* erst in älteren Stadien des Algenwachstums auftritt, durch Pfropfung eines reifen, kernlosen Teiles der Alge auf einen kernhaltigen, jungen Teil induziert werden (HÄMMERLING). Die Induktionen solcher Teilungen sind nicht artspezifisch, da sie auch von Teilen anderer Algenarten veranlaßt werden können (HÄMMERLING 1953). Nach HÄMMERLING wird der Kern, der während der Entwicklung der Zelle die bestimmte Rolle für die Differenzierung der Zelle spielte, also seinerseits vom differenzierten Cytoplasma zur Teilung angeregt. Die Wirkung des Cytoplasmas auf den Kern wird auch durch folgende Versuche STICHs (1951) demonstriert: Wird die Pflanze ins Dunkle gebracht, verliert der Kern an Größe, und die zahlreichen Nucleoli verschwinden. Im Licht aber gewinnt der Kern sein altes Aussehen zurück. Dieser reversible Prozeß kann beliebig wiederholt werden. Der Gedanke an einen Substanzgradienten zwischen Kern und Cytoplasma, dessen chemische Gleichgewichtslage lichtabhängig ist, liegt daher nahe.

HÄMMERLING (1953) konnte in einer Reihe von Untersuchungen zeigen, daß der vom Kern abgetrennte Teil des riesigen Zelleibes der Alge *Acetabularia* noch erhebliches Differenzierungs- und Wachstumsvermögen besitzt, das noch über Monate wirken kann. Regelmäßig tritt aber früher oder später Absterben des Objektes ein. Die Differenzierungspotenz ist auch vom Alter der Alge abhängig, von der der Kern abgetrennt wurde; HÄMMERLING nimmt daher auf Grund seiner Versuche kernkontrollierte „morphogenetische Substanzen" an, deren Menge und Wirkung von der Algenspecies und dem jeweiligen Differenzierungsstadium der Zelle abhängen.

Der Teilungsmechanismus des Zellkörpers ist bei *Arbacia*-Eiern kernunabhängig (HARVEY 1936). Kernlose Eier können durch parthenogenetische Agentien zur Teilung angeregt werden. Dabei entstehen bis zu 500 Zellen mit einem Monat Lebensdauer.

Daß die heutige Technik der Zellentkernung und Transplantation sehr weit entwickelt ist, geht aus den Untersuchungen von BRIGGS und KING (1952) hervor, die Blastulakerne in enucleierte Eier von Fröschen übertrugen und die Entwicklung von Embryonen erzielten. Hybridtransplantate blieben erfolglos. Der Kern ist also artspezifisch für die Entwicklung und Differenzierung der Zelle notwendig, was auch die folgenden Autoren klar zeigen konnten. LORCH, DANIELLI und HÖRSTADIUS (1953) enucleierten aus sich entwickelnden Seeigellarven im 2., 4. und 8. Zellenstadium eine Zelle. Es traten außer Störungen des räumlichen Wachstumsmusters durch das Ausfallen einer Zelle keine anderen Veränderungen auf. Die enucleierte Zelle kann sich in viele Zellen verschiedenster Größe teilen und gleicht den übrigen Zellen optisch vollkommen. Sie nimmt aber nie an der Differenzierung von Geweben und Organen teil. Auch übt offenbar die Nachbarschaft sich differenzierender Zellen keinen Einfluß aus. Im Larvenstadium kann jedoch der animale Teil der Larve auf den im 8. Zellenstadium enucleierten vegetativen Teil differenzierend einwirken. Das ist umgekehrt nicht der Fall, wenn der animale Teil enucleiert worden war.

Literatur.

ABOOD, L. G., and R. W. GERARD: Enzyme distribution in isolated particulates of rat peripherical nerve. J. Cellul. a. Comp. Physiol. **43**, 379 (1954). — ABOOD, L. G., R. W. GERARD, J. BANKS and R. D. TSCHIRGI: Substrate and enzyme distribution in cells. Amer. J. Physiol. **168**, 728 (1952). — ABRAMS, R.: Some factors influencing nucleic acid purine renewal in the rat. Arch. of Biochem. **33**, 436 (1951). — AFZELIUS, B. A.: The ultrastructure of the nuclear membrane of the sea urchin oocyte as studied with the electron microscope. Exper. Cell Res. 8, 147 (1955). — ALBAUM, H. G., and M. OGUR: An adenine-pentose-pyrophosphate from plant tissue. Arch. of Biochem. **15**, 158 (1947). — ALLARD, C., and A. CANTERO: Adenosine triphosphatase study during rat liver damage. I. Adenosine-trophosphatase activity of rat liver during regeneration after partial hepatectomy. Canad. J. Med. Sci. **30**, 295 (1952). — ALLFREY, A. H., H. STERN, A. E. MIRSKY and H. SAETREN: The isolation of cell nuclei in non-aqueous media. J. Gen. Physiol. **35**, 529 (1951). — ALLFREY, V.: Amino acid incorporation by isolated thymus nuclei. I. The role of Desoxyribonucleic acid in protein synthesis. Proc. Nat. Acad. Sci. **40**, 881 (1954). — ALLFREY, V., and A. E. MIRSKY: Some aspects of the Desoxyribonuclease activities of animal tissues. J. Gen. Physiol. **36**, 227 (1952). — ALLFREY, V., H. STERN and A. E. MIRSKY: Some enzymes of isolated cell nuclei. Nature (Lond.) **169**, 128 (1952). — ALTMANN, H. W.: Über die Abgabe von Kernstoffen in das Protoplasma der menschlichen Leberzelle. Z. Naturforsch. **4b**, 138 (1949). — Morphologische Bemerkungen zur Funktion des Ganglienzellkernes. Naturwiss. **39**, 348 (1952). — ALTMANN, H. W., u. R. MENY: Der Funktionswechsel des Zellkernes im exokrinen Pankreasgewebe. Naturwiss. **39**, 138 (1952). — ANDERSON, N. G.: On the nuclear envelope. Science (Lancaster, Pa.) **117**, 53 (1953). — ANDERSON, N. G., and K. M. WILBUR: Studies on isolated cell components. IV. The effect of various solutions on the isolated rat liver nucleus. J. Gen. Physiol. **35**, 781 (1952). — ANDRESEN, N., u. H. HOLTER: Cytoplasmic changes during starvation of the amoeba chaos chaos L. C. r. Trav. Labor. Carlsberg, Ser. chim. **25**, 107

(1944/47). — APITZ, K.: Über Pigmentbildungen in den Zellkernen melanotischer Geschwülste. Virchows Arch. 300, 89 (1937). — Über die Bildung RUSSELscher Körperchen in den Plasmazellen multipler Myelome. Virchows Arch. 300, 113 (1937). — AVERY, O. T., C. M. MACLEOD and M. MACCARTY: Studies on the chemical nature of the pneumococcal types. Induction of transformation by a desoxyribonucleic acid fraction isolated from pneumococcus type III. J. of Exper. Med. 79, 137 (1944).

BAHR, G. F., and W. BEERMANN: The fine structure of the nuclear membrane in the larval salivary gland and midgut of chironomus. Exper. Cell Res. 6, 519 (1954). — BAIRATI, A., u. F. LEHMANN: Über die submikroskopische Struktur der Kernmembran bei Amoeba proteus. Experientia (Basel) 8, 60 (1952). — BALL, E. G.: Xanthine oxydase: purification and properties. J. of Biol. Chem. 128, 51 (1939). — BALTUS, E.: Observations sur le rôle biochimique du nucléole. Biochim. et Biophysica Acta 15, 263 (1954). — BARNUM, C. P., C. W. NASH, E. JENNINGS, O. NYGAARD and H. VERMUND: The separation of pentose and desoxypentose nucleic acid from isolated mouse liver cell nuclei. Arch. of Biochem. 25, 376 (1950). — BARRATT, R. W., D. NEWMEYER, D. D. PERKINS and L. GARNJOBST: Map construction in neurospora crassa. Adv. Genet. 6, 1 (1954). — BEALE, R. N., R. J. C. HARRIS and E. M. F. ROE: The nucleic acid of normal and tumour tissues. The preparation and composition of a pentose nucleic acid from the fowl sarcoma G. R. C. H. 15. J. Chem. Soc. 1950, 1397. — BENDICH, A., P. J. RUSSEL jr. and G. B. BROWN: On the heterogeneity of the desoxyribonucleic acids. J. of Biol. Chem. 203, 305 (1953). — BERG, P., and W. K. JOKLIK: Enzymatic phosphorylation of nucleoside diphosphates. J. of Biol. Chem. 210, 657 (1954). — BERGQUIST, R., u. A. DEUTSCH: Guanosine triphosphate and uridine triphosphate from muscle. Acta chem. scand. (Copenh.) 7, 1307 (1953). — BRACHET, J.: Oxygen uptake of nucleated and non-nucleated halves of Amoeba proteus. Nature (Lond.) 168, 205 (1951). — Le rôle noyau cellulaire dans les oxydations et les phosphorylations. Biochim. et Biophysica Acta 9, 221 (1952). — La composition enzymatique de fragments nucléés et énucléés d'amœbes. Biochim. et Biophysica Acta 14, 449 (1954). — Constitution anormale du noyau et teneur en acide adenosine-triphosphorique de la cellule. Experientia (Basel) 10, 492 (1954). — BRACHET, J., and H. CHANTRENNE: Protein synthesis in nucleated and non-nucleated halves of Acetabularia mediterranea studied with carbon-14 dioxide. Nature (Lond.) 168, 950 (1951). — BRACHET, J., and D. SZAFARZ: L'incorporation d'acide orotique radioactif dans des fragments nucléés et Anucléés d'écetabularia mediterranea. Biochim. et Biophysica Acta 12, 589 (1953). — BRIGGS, R., and T. J. KING: Transplantation of living nuclei from blastula cells into enucleated frog's eggs. Proc. Nat. Acad. Sci. 38, 455 (1952). — BROWN, G. B., P. M. ROLL and H. WEINFELD: Biosynthesis of nucleic acids. Phosphorus Metabolism 2, 385 (1952). — BROWN, K. D., G. JACOBS and M. LASKOWSKI: The distribution of nucleodepolymerases in calf thymus fractions. J. of Biol. Chem. 194, 445 (1952). — BRUES, A. M., M. M. TRACY and W. E. COHN: Nucleic acids of rat liver and hepatoma: their metabolic turnover in relation to growth. J. of Biol. Chem. 155, 619 (1944). — BRUNISH, R., D. L. FAIRLEY and J. M. LUCK: Composition of histone prepared from rat liver desoxypentose nucleoprotein. Nature (Lond.) 168, 82 (1951). — BULLOUGH, W. S.: Zusammenhang zwischen Mitose und energieliefernden Reaktionen. Biol. Rev. 27, 133 (1952).

CALLAN, H. G., and S. G. TOMLIN: Experimental studies on amphibian oocyte nuclei. I. Investigation of the structure of the nuclear membrane by means of the electron microscope. Proc. Roy. Soc. Lond., Ser. B 137, 367 (1950). — CARVER, M. J., and L. E. THOMAS: Lipoproteins from chicken erythrocytes with isoelectric points of 4,65—4,85. Arch. of Biochem. 40, 342 (1952). — CASPERSSON, T.: Über den chemischen Aufbau der Strukturen des Zellkernes. Skand. Arch. Physiol. (Berl. u. Lpz.) 73, Suppl. 8 (1936). — Cell growth and cell function. New York: Norton & Co. 1950. — CHARGAFF, E.: Cell structure and the problem of blood coagulation. J. of Biol. Chem. 160, 351 (1945). — Chemical specificity of nucleic acids and mechanism of their enzymatic degradation. Experientia (Basel) 6, 201 (1950). — CHARGAFF, E., R. LIPSHITZ and CH. GREEN: Composition of the desoxypentose nucleic acids of four genera of sea-urchin. J. of Biol. Chem. 195, 155 (1952). — CHARGAFF, E., R. LIPSHITZ, CH. GREEN and M. E. HODES: The composition of the desoxyribonucleic acid of salmon sperm. J. of Biol. Chem. 192, 223 (1951). — CHARGAFF, E., E. VISCHER, R. DONIGER, CH. GREEN and F. MISANI: The composition of the desoxypentose nucleic acids of thymus and spleen. J. of Biol. Chem. 177, 405 (1949). — CHARGAFF, E., S. ZAMENHOF, G. BRAWERMAN and L. KERIN: Bacterial desoxypentose nucleic acids of unusual composition. J. Amer. Chem. Soc. 72, 3825 (1950). — CHARGAFF, E., S. ZAMENHOF and CH. GREEN: Composition of human desoxypentose nucleic acid. Nature (Lond.) 165, 756 (1950). — CHAYEN, J.: Ascorbic acid and its intracellular localisation, with special reference to plants. Internat. Rev. Cytology 2, 78 (1953). — CHAYEN, J., and K. P. NORRIS: Cytoplasmatic localization of nucleic acids in plant cells. Nature (Lond.) 171, 472 (1953). — CHÈVREMONT, M., and H. FIRKET: Alkaline phosphatase of the nucleus. Internat. Rev. Cytology 2, 261 (1953). — CHURNEY, L.: The osmotic properties of the nucleus. Biol. Bull. 82, 52 (1942). — CLARK,

A. M.: Some effects of removing the nucleus from amoeba. Austral. J. Exper. Biol. a. Med. Sci. **20**, 241 (1942). — CRAMPTON, C. F., R. LIPSHITZ and E. CHARGAFF: Studies on nucleoproteins. II. Fractionation of desoxyribonucleic acid through fractional dissication of their complexes with basic proteins. J. of Biol. Chem. **211**, 125 (1954). — CROSBIE, G. W., R. M. S. SMELLIE and J. N. DAVIDSON: V. The composition of the cytoplasmatic and nuclear ribonucleic acid of the liver cell. Biochemic. J. **54**, 287 (1953).

DALY, M. M., V. G. ALLFREY and A. E. MIRSKY: Purine and pyrimidine contents of some desoxypentose nucleic acids. J. Gen. Physiol. **33**, 497 (1949/50). — Uptake of glycine-N^{15} by components of cell nuclei. J. Gen. Physiol. **36**, 173 (1952). — DALY, M. M., and A. E. MIRSKY: Histones with high lysine content. J. Gen. Physiol. **38**, 405 (1955). — DANIELLI, J. F.: A critical study of techniques for determining the cytological position of alkaline phosphatase. J. of Exper. Biol. **22**, 110 (1945). — DANIELLI, J. F., and D. G. CATCHSIDE: Phosphatase on chromosomes. Nature (Lond.) **156**, 294 (1945). — DAVIDSON, J. N., and R. A. LAWRIE: Amino acids in nuclear proteins. Biochemic. J. **43**, proc. XXIX (1948). — DAVIDSON, J. N., and I. LESLIE: The changing cell number and composition of chick heart explants growing in vitro. Exper. Cell Res. **2**, 366 (1951). — DIANZANI, M. U.: La ripartizione della alcooldeidrogenasi e della glucosideidrogenasi nelle cellule epatiche di ratti normale e di ratti con degenerazione grassa del fegato. Arch. di Fisiol. **49/50**, 175 (1949/51). — La ripartizione del sistema ossidante acido lattico nelle cellule del fegato e del vene di ratti normali e di ratti con degenerazione grassa del fegato. Arch. di Fisiol. **49/50**, 181 (1949/51). — La ripartizione della alfa-glicerofosfatodeidrogenasi nelle cellule epatiche di ratti normali e di ratti con degenerazione grassa, del fegato. Arch. di Fisiol. **49/50**, 187 (1949/51). — DIANZANI, M. U., and I. VITI: The content and distribution of cytochrome c in the fatty liver of rats. Biochemic. J. **59**, 141 (1955). — DITTUS, P.: Histologie und Cytologie des Interrenalorganes der Selachier unter normalen und experimentellen Bedingungen. Z. wiss. Zool. **154**, 40 (1940). — DODGSON, K. S., B. SPENCER and J. THOMAS: Studies on sulphatases. 6. The localization of arylsulphatase in the rat liver cell. Biochemic. J. **56**, 177 (1954). — Studies on sulphatases. 9. The arylsulphatases of mammalian liver. Biochemic. J. **59**, 29 (1955). — DOUNCE, A. L.: Cytochemical foundations of enzyme chemistry. In SUMMER and MYRBÄCK: The Enzymes. New York: Acad. Press 1950. — Enzyme systems of isolated cell nuclei. Ann. New York Acad. Sci. **50**, 982 (1950). — The significance of enzyme studies on isolated cell nuclei. Internat. Rev. Cytology **3**, 199 (1954). — DOUNCE, A. L., G. H. TISHKOFF, S. R. BARNETT and R. M. FREER: Free amino acids and nucleic content of cell nuclei isolated by a modification of BEHRENS' technique. J. Gen. Physiol. **33**, 629 (1950).

EDMONDS, M., and G. A. LEPAGE: In vivo studies on acid-soluble precursors of nucleic acid purines. Federat. Proc. **12**, 199 (1953). — ELMES, P. C., J. D. SMITH et J. C. WHITE: 2nd Internat. Congr. Biochem. Abstr. of Communications, Paris, France. S. 7. 1952. — ELSON, D., and E. CHARGAFF: Observations on pentose nucleic acid composition in sea urchin embryos and in mammalian cell fractions. Phosphorus Metabolism **2**, 329 (1952). — ENGBRING, U. K., and M. LASKOWSKI: Protein components of chicken erythrocyte nuclei. Biochim. et Biophysica Acta **11**, 244 (1953). — EULER, H. v., I. FISCHER, H. HASSELQUIST u. M. JAARMA: Stability of isolated cell nuclei in different media. Ark. Kemi, Mineral. Geol., Ser. A **21**, Nr 12 (1945). — EULER, H. v., u. L. HELLER: Katalaseaktivität in Leberfraktionen normaler und sarkomtragender Ratten. Z. Krebsforsch. **56**, 393 (1949).

FELIX, K.: Zur Chemie des Zellkernes. Experientia (Basel) **8**, 312 (1952). — FIGQU, A., and M. ERRERA: Étude autoradiographique de l'incorporation dans le foie de souris de précurseurs des acides nucléiques et des protéines. Biochim. et Biophysica Acta **16**, 45 (1955). — FISCHER, H., u. L. KREUZER: Über Gallin. Hoppe-Seylers Z. **293**, 176 (1953). — FRAENKEL-CONRAT, H., and E. D. DUCAY: The nucleic acid of egg white. Biochemic. J. **49**, proc. XXIX (1951). — FRAENKEL-CONRAT, H., N. S. SNELL and E. D. DUCAY: Avidin. Isolation and characterization of the protein and nucleic acid. Arch. of Biochem. a. Biophysics **39**, 80 (1952). — FRIEDKIN, M., and H. M. KALCKAR: Desoxyribose-l-phosphate. I. Phosphorolysis and resynthesis of purine desoxyribose nucleoside. J. of Biol. Chem. **184**, 437 (1950). — FURST, S. S., and G. B. BROWN: On the role of glycine and adenine as precursors of nucleic acid purines. J. of Biol. Chem. **191**, 239 (1951).

GALL, J. G.: Observations on the nuclear membrane with the electron microscipe. Exper. Cell Res. **7**, 197 (1954). — GANDELMAN, B., S. ZAMENHOF and E. CHARGAFF: The desoxypentose nucleic acid of three strains of escheria coli. Biochim. et Biophysica Acta **9**, 399 (1952). — GOETHART, G.: The thiamine pyrophosphate content of centrifugally prepared fractions of rat liver homogenate. Biochim. et Biophysica Acta **8**, 479 (1952). — GRAFF, A., u. W. HEBEKERL: Untersuchungen über die chemische Zusammensetzung und das biochemische Verhalten der präformierten Zellbestandteile der Schilddrüse. Arch. Geschwulstforsch. **4**, 349 (1952). — GRAUPNER, H., u. I. FISCHER: Das Tintendrüsenepithel von Sepia vor, während und nach der Pigmentbildung. Z. Zellforsch. **21**, 329 (1934).

HAAS, E., C. J. HARRER and T. R. HOGNESS: Cytochrome reductase. II. Improved method of isolation; inhibition and inactivation; reaction with oxygen. J. of Biol. Chem. 143, 341 (1942). — HÄMMERLING, J.: Nucleo-cytoplasmic relationships in the development of Acetabularia. Internat. Rev. Cytology 2, 475 (1953). — HÄMMERLING, J., u. H. STICH: Über die Aufnahme von ^{32}P in kernhaltige und kernlose Acetabularien. I. Z. Naturforsch. 9b, 149 (1954). — HAMER, D.: A comparison of different protein fractions obtained from thymus nuclei isolated in an aqueous medium. Brit. J. Canc. 7, 151 (1953). — HAMMARSTEN, E., P. REICHARD and E. SALUSTE: Pyrimidine nucleosides as precursors of pyrimidines and polynucleotides. J. of Biol. Chem. 183, 105 (1950). — HARPER, H. H., and M. D. MORRIS: The amino acid composition of chick erythrocyte nucleohiston. Arch. of Biochem. 42, 61 (1953). — HARRIS, P., u. T. JAMES: Electron microscope study of the nuclear membrane of Amoeba proteus in thin section. Experientia (Basel) 8, 384 (1954). — HARRISON, M. F.: Composition of the liver cell. Proc. Roy. Soc. Lond., Ser. B 141, 203 (1951). — HARVEY, E. B.: Parthenogenetic merogony or cleavage without nuclei in Arbacia punctata. Biol. Bull. 71, 101 (1936). — HECHT, L. I., R. VAN POTTER and E. HERBERT: In vitro phosphorylation of pyrimidine desoxyribosemononucleotides. Biochim. et Biophysica Acta 15, 135 (1954). — HELLER, L., u. N. BARGONI: Studien über die intrazelluläre Verteilung der Enzyme. III. Die intrazelluläre Verteilung der Monobutyrase in der Leber normaler und sarkomatöser Ratten. Ark. Kemi (Stockh.) 1, 447 (1949). — HERTL, M.: Über den Nucleolarapparat der Nervenzellen im Hypothalamus der weißen Maus. Z. Zellforsch. 41, 207 (1955). — HERS, H. G., J. BERTHET, L. BERTHET et CH. DE DUVE: Le système hexose phosphatasique. III. Localisation intracellulaire des ferments par centrifugation fractionnée. Bull. Soc. Chim. biol. Paris 33, 21 (1951). — HETT, J.: Weitere Befunde über den Austritt von Kernsubstanzen in das Protoplasma. Z. Zellforsch. 26, 473 (1927). — HIGGINS, H., J. A. MILLER, J. M. PRICE and F. M. STRONG: Levels and intracellular distribution of coenzyme A and pantothenic acid in rat liver and tumors. Proc. Soc. Exper. Biol. a. Med. 75, 462 (1950). — HOFF-JØRGENSEN, E.: A microbiological assay of desoxyribonucleosides and desoxyribonucleic acid. Biochemic. J. 50, 400 (1952). — HOFF-JØRGENSEN, E., u. E. ZEUTHEN: Evidence of cytoplasmic desoxyribosides in the frog's egg. Nature (Lond.) 169, 245 (1952). — HOFFMANN, C.: Über die Durchlässigkeit kernloser Zellen. Planta (Berl.) 4, 584 (1927). — HOFFMANN-BERLING, H.: Adenosintriphosphat als Betriebsstoff von Zellbewegungen. Biochim. et Biophysica Acta 14, 182 (1954). — Die Bedeutung des Adenosintriphosphat für die Zell- und Kernteilungsbewegungen in der Anaphase. Biochim. et Biophysica Acta 15, 332 (1954). — HOFFMANN-OSTENHOF, O.: Enzymologie. Wien: Springer 1954. — HOGEBOOM, G. H., and W. C. SCHNEIDER: Intracellular distribution of enzymes. VIII. The distribution of diphosphopyridine nucleotide-cytochrome C reductase in normal mouse liver and mouse hepatoma. J. Nat. Canc. Inst. 10, 983 (1950). — HOGEBOOM, G. H., and W. C. SCHNEIDER: Cytochemical studies. IV. The synthesis of diphosphopyridin nucleotide by liver cell nuclei. J. of Biol. Chem. 197, 611 (1952). — Intracellular distribution of enzymes. XI. Glutamic dehydrogenase. J. of Biol. Chem. 204, 233 (1953). — HOGEBOOM, G. H., W. C. SCHNEIDER and M. J. STRIEBICH: Cytochemical studies. V. On the isolation and biochemical properties of liver cell nuclei. J. of Biol. Chem. 196, 111 (1952). — HOLTER, H.: Localization of enzymes in cytoplasm. Adv. Enzymol. 13, 1 (1952). — HOLTER, H., u. S. LØVTRUP: C. r. Trav. Labor. Carlsberg 27, 27 (1949). — HOLTER, H., u. B. POLLOCK: C. r. Trav. Labor. Carlsberg 28, 221 (1952). — HOLTFRETER, J.: Observations on the physico-chemical properties of isolated nuclei. Exper. Cell Res. 7, 95 (1954). — HOROWITZ, N. H.: Biochemical genetics of Neurospora. Adv. Genet. 3 (1950). — HURLBERT, R. B., and V. R. POTTER: A survey of the metabolism of orotic acid in the rat. J. of Biol. Chem. 195, 257 (1952). — Nucleotid metabolism. I. The conversion of orotic acid-6-C^{14} to uridine nucleotides. J. of Biol. Chem. 209, 1 (1954). — HURLBERT, R. B., H. SCHMITZ, A. F. BARNUM and V. R. POTTER: Nucleotid metabolism. II. Chromatographic separation of acid soluble nucleotides. J. of Biol. Chem. 209, 23 (1954). — HURST, R. O., A. M. MARKO and G. C. BUTLER: The mononucleotide content of some desoxyribonucleic acids. J. of Biol. Chem. 204, 847 (1953).

ISBELL, E. R., H. K. MITCHELL, A. TAYLOR and R. J. WILLIAMS: A preliminary study of B vitamins in cell nuclei. Univ. Texas Publ. 1942, No 4237, 81.

JAMES, T. W.: The role of the nucleus in the maintenance of ribonucleic acid in Amoeba proteus. Biochim. et Biophysica Acta 15, 367 (1954). — JOHNSON, F. H., and H. EYRING: The nature of the luciferin-luciferase system. J. Amer. Chem. Soc. 66, 848 (1944).

KALCKAR, H. M.: The enzymatic synthesis of purine ribosides. J. of Biol. Chem. 167, 377 (1947). — Differential spectrophotometry of purine compounds by means of specific enzymes. I. Determination of hydroxypurine compounds. J. of Biol. Chem. 167, 429 (1947). — KEILIN, D., and E. F. HARTREE: Prosthetic group of glucose oxidase (notatin). Nature (Lond.) 157, 801 (1946). — KENSLER, G. J., and H. LANGEMANN: The distribution of choline oxidase activity in rat liver. Nature (Lond.) 192, 551 (1951). — KHOUVINE, Y., et E. BARON: Influence du calcium sur la composition des histones de l'épithélioma atypique du rat. Bull. Soc. Chim. biol. Paris 33, 229 (1951). — KHOUVINE, Y., et J. GREGOIRE: Désoxyribonucléoproteides de l'épithelioma atypique du rat. II. Acides désoxyribonucléiques. Bull. Soc. Chim. biol. Paris 35, 603 (1953). — KHOUVINE, Y., J. GREGOIRE et J. P. ZALTA: Désoxyribonucléoprotéides de l'épithélioma atypiques du rat. I. Histones et protéines non basiques. Bull. Soc. Chim. biol. Paris 35, 244 (1953). — KIRKHAM, W. R., and L. E. THOMAS: The isolation of globulins from cellular nuclei. J. of Biol. Chem. 200, 53 (1953). — KLEBS, G.: Die Bedingungen der Fortpflanzung bei einigen Algen und Pilzen. Jena: Gustav Fischer 1896. — KORNBERG, A.: The metabolism of phosphorus-containing coenzymes. Phosphorus Metabolism 392 (1951). — KRINSKY, N. I., and J. GANGULY: Intracellular distribution of vitamin A ester and vitamin A alcohol in rat liver. J. of Biol. Chem. 202, 227 (1953).

LALAND, S. G., W. G. OVEREND and M. WEBB: Desoxypentose nucleic acids. IV. The properties and composition of the desoxypentose nucleic acids from certain animal, plant and bacterial sources. J. Chem. Soc. Lond. 1952, 3324. — LAMIRANDE, G. DE, C. ALLARD and A. CANTERO: Cancer (N. Y.) 6, 179 (1953). — LANG, K., H. LANG, G. SIEBERT u. S. LUCIUS: Über den Einbau von C^{14}-Glycokoll durch Zellkerne in vitro. Biochem. Z. 324, 217 (1953). — LANG, K., u. G. SIEBERT: Untersuchungen über Stoffwechselvorgänge in Zellkernen. 2. Das Fehlen von Oxydationsfermenten in Zellkernen aus Rattenleber und Schweineniere. Biochem. Z. 320, 402 (1950). — Untersuchungen über Stoffwechselvorgänge in isolierten Zellkernen. 5. Über Energielieferung durch Glycolyse und Adenosin-triphosphat-Spaltung sowie über das Vorkommen von Dehydrasen in isolierten Zellkernen. Biochem. Z. 322, 196 (1951). — Untersuchungen über Stoffwechselprozesse in isolierten Zellkernen. 6. Methodik der Gewinnung reiner intakter Zellkerne in beliebigem Maßstab. Biochem. Z. 322, 360 (1952). — LANG, K., G. SIEBERT, I. BALDUS u. A. CORBERT: Über das Vorkommen von Desoxyribonuclease und Kathepsin in Zellkernen aus Nieren. Experientia (Basel) 6, 59 (1950). — LANG, K., G. SIEBERT u. F. FISCHER: Untersuchungen über Stoffwechselprozesse in isolierten Zellkernen. Biochem. Z. 324, 1 (1953). — LANG, K., G. SIEBERT u. S. LUCIUS: Über die Hemmung der Arginase durch Stickstofflost. Experientia (Basel) 8, 228 (1952). — LANG, K., G. SIEBERT, S. LUCIUS u. H. LANG: Untersuchungen über Stoffwechselvorgänge in Zellkernen. 3. Über Arginase und Mangan in isolierten Zellkernen der Leber. Biochem. Z. 321, 538 (1951). — LASKOWSKI, M.: Nucleolytic enzymes. In SUMNER and MYRBÄCK: The Enzymes, Vol. I/2, p. 956. New York: Acad. Press 1951. — LEE, N. D., and R. H. WILLIAMS: Protein turnover and enzymatic adaptation. J. of Biol. Chem. 204, 477 (1953). — LELOIR, L. F.: Enzymic isomerization and related processes. Adv. Enzymol. 14, 193 (1953). — LETTRÉ, H.: Some investigations on cell behaviour under various conditions: a review. Cancer Res. 12, 847 (1952). — Zur Anwendbarkeit des Zellmodells nach HOFFMANN-BERLING für die Analyse der Zellbewegungen. Biochim. et Biophysica Acta 14, 584 (1954). — LEVINE, C., and E. CHARGAFF: Phosphatide composition in different liver cell fractions. Exper. Cell Res. 3, 154 (1952). — LINET, N., and J. BRACHET: L'évolution de l'acide ribonucléique et du glycogène dans des fragments nucléés et énucléés d'amœbes. Biochim. et Biophysica Acta 7, 607 (1951). — LIPMANN, F., N. O. KAPLAN, G. D. NOVELLI, L. C. TUTTLE and B. M. GUIRARD: Isolation of Co-Enzyme A. J. of Biol. Chem. 186, 235 (1950). — LONDON, I. M., D. SHEMIN and D. RITTENBERG: Synthesis of heme in vitro by the immature non nucleated mammalian erythrocyte. J. of Biol. Chem. 183, 749 (1950). — LORCH, I. J., J. F. DANIELLI and S. HÖRSTADIUS: The effect of enucleation on the development of sea urchin eggs. Exper. Cell Res. 4, 253 (1953). — LOWE, C. U., and R. J. SALMON: Sex variation in alkaline phosphatase activity and desoxyribonucleic acid content of rat liver. Arch. of Biochem. a. Biophysics 34, 481 (1951). — LU, K. H., and T. WINNICK: Studies of nucleic acid metabolism in embryonic tissue culture with the aid of C^{14}-labeled purines. Exper. Cell Res. 6, 345 (1954). — Biosynthesis of nucleic acids in embryonic chick heart tissue cultures. Federat. Proc. 13, 255 (1954). — LUCKÉ, B.: On osmotic behaviour of living cell fragments. J. Cellul. a. Comp. Physiol. 2, 193 (1932/33). — LUDEWIG, S., and A. CHANUTIN: Distribution of enzymes in the livers of control and X irradiated rats. Arch. of Biochem. 29, 8 (1950).

MALKIN, H. M.: Synthesis of ribonucleic acid purines and protein in enucleated and nucleated sea urchin eggs. J. Cellul. a. Comp. Physiol. 44, 105 (1954). — MARRIAN, D. H.: Concerning an aspect of desoxyribonucleic acid synthesis in regenerating rat liver. Biochim. et Biophysica Acta 14, 502 (1954). — MARSHAK, A.: Uptake of radioactive phosphorus by nuclei of liver and tumors. Science (Lancaster, Pa.) 92, 460 (1940). — Purine and pyrimidine content of the nucleic acids of nuclei and cytoplasma. J. of Biol. Chem. 189, 607 (1951). — MARSHAK, A., and C. MARSHAK: Desoxyribonucleic acid in Arbacia eggs. Exper. Cell Res. 5, 288 (1953). —Thymine in the acid-soluble fraction of Arbacia eggs. Biochim. of Biophysica Acta 15, 584 (1954). — MAURITZEN, C. M., A. B. ROY and E. STEDMAN: The ribonucleic acid content of isolated cell nuclei. Proc. Roy. Soc. Lond., Ser. B 140, 18 (1952). — MAVER, M. E., and A. E. GRECO: The hydrolysis of nucleoproteins by cathepsins from calf thymus. J. of Biol. Chem. 181, 853 (1949). — The nuclease activities of cathepsin preparations from calf spleen and thymus. J. of Biol. Chem. 181, 861 (1949). — MAVER, M. E., A. E. GRECO,

E. LøVTRUP and A. I. DALTON: Catheptic activities of the nuclei of normal, regenerating and neoplastic tissues of the rat. J. Nat. Canc. Inst. **13**, 687 (1952/53). — MAZIA, D.: Modern trends in physiology and biochemistry, p. 77. New York: Acad. Press. 1952. — MAZIA, D., and H. I. HIRSHFIELD: The nucleus-dependence of P^{32} uptake by the cell. Science (Lancaster, Pa.) **112**, 297 (1950). — MAZIA, D., and D. M. PRESCOTT: Nuclear function and mitosis. Science (Lancaster, Pa.) **120**, 120 (1954). — MCFARLANE, M. G., and A. G. SPENCER: Changes in the water, sodium and potassium content of rat liver mitochondria during metabolism. Biochemic. J. **54**, 569 (1953). — MCINDOE, W. M., and J. N. DAVIDSON: The phosphorus compounds of the cell nucleus. Brit. J. Canc. **6**, 200 (1952). — METZGER-FREED, L.: Phosphoprotein phosphatase activity in normal haploid and hybrid amphibian development. J. Cellul. a. Comp. Physiol. **41**, 493 (1953). — MICHAELIS, P.: Cytoplasmic inheritance in Epilobium and its theoretical significance. Adv. Genet. **5**, 288 (1954). — MOYLE, J. F.: Intracellular distribution of rhodanese in cardiac muscle. Nature (Lond.) **172**, 508 (1953). — MUNTWYLER, E., S. SEIFTER and D. M. HARKNESS: Some effects of restriction of dietary protein on the intracellular components of liver. J. of Biol. Chem. **184**, 181 (1950).

NYBERG, A., J. SCHUBERTH u. L. ÄNGGÅRD: On the intracellular distribution of catalase and alcohol dehydrogenase in horse, guinea pig and rat liver tissues. Acta chem. scand. (Copenh.) **7**, 1170 (1953).

ODEBLAD, E., and H. BOSTRÖM: Observations supporting the presence of a nuclear uptake of S^{35}-labelled sulfate in the mouse. Exper. Cell Res. **4**, 482 (1953). — OMACHI, A., C. P. BARNUM and D. GLICK: Ouantitative distribution of an esterase among cytoplasmic components of mouse liver cells. Proc. Soc. Exper. Biol. Med. **67**, 133 (1948). — ORTMANN, R.: Über Kernsekretion, Kolloid und Vakuolenbildung in Beziehung zum Nucleinsäuregehalt in Trophoblast-Riesenzellen der menschlichen Placenta. Z. Zellforsch. **34**, 562 (1948).

PALADE, G. E.: Intracellular distribution of acid phosphatase in rat liver cells. Arch. of Biochem. **30**, 144 (1951). — PATTERSON, E. K., and M. E. DACKERMAN: Nucleic acid content in relation to cell size in the mature larval salivary gland of Drosophila melanogaster. Arch. of Biochem. **36**, 97 (1952). — PIROTTE, M., et V. DESREUX: Distribution de la ribonucléase dans les extraits de granules cellulaires du foie. Bull. Soc. chim. Belg. **61**, 167 (1952). — POLLI, E., u. G. RATTI: Studien über Leucozyten. I. Die Fettfraktionen in intakten Zellen und in isolierten Zellkernen bei normalen und pathologischen Zuständen. Biochem. Z. **323**, 546 (1953). — POLLISTER, A. W.: Nucleoproteins of the nucleus. Exper. Cell Res. Suppl. **2**, 59, 70 (1952). — POMERAT, C. M.: Rotating nuclei in tissue cultures of adult human nasal mucosa. Exper. Cell Res. **5**, 191 (1953). — PONTECORVO, G.: Genetic formulation of gene structure and gene action. Adv. Enzymol. **13**, 121 (1952). — POWELL, L. T., and R. F. KRAUSE: Vitamin A distribution in the rat liver cell. Arch. of Biochem. a. Biophysics **44**, 102 (1953). — PRESCOTT, D. M., and D. MAZIA: The permeability of nucleated and enucleated fragments of Amoeba proteus to D_2O. Exper. Cell Res. **6**, 117 (1954). — PRICE, J. M., E. C. MILLER and J. A. MILLER: Intracellular distribution of vitamin B_6 in rat and mouse livers and induced rat liver tumors. Proc. Soc. Exper. a Med. **71**, 575 (1949). — PRICE, J. M., J. A. MILLER and E. C. MILLER: The intracellular distribution of protein, nucleic acids and riboflavin in the livers of mice and hamsters fed 4-dimethyl-aminoazobenzene. Cancer Res. **11**, 523 (1951). — PRICE, J. M., J. A. MILLER, E. C. MILLER and G. M. WEBER: Studies on the intracellular composition of liver and liver tumor from rats fed with 4-dimethyl-aminobenzene. Cancer Res. **9**, 96 (1949).

RABINOWITZ, J. C., G. JACOBS, L. J. TEPLY, V. H. CHELDELIN, E. DIAMANT, H. R. MAHLER and F. M. HUENNEKENS: The incorporation of ^{32}P-labelled orthophosphate into nucleotides. Biochim. et Biophysica Acta **13**, 413 (1954). — RATNER, S.: Conversion of d-glutamic acid to pyrrolidonecarboxylic acid by the rat. J. of Biol. Chem. **152**, 559 (1944). — RATNER, S., V. NOCITO and D. E. GREEN: Glycine oxidase. J. of Biol. Chem. **152**, 119 (1944). — RAVEN, CH. P.: In M. W. WOERDEMANN u. CH. P. RAVEN, Monographs on the progress of research in Holland during the war. New York u. Amsterdam 1946. — REDDI, K. K.: Desoxyribonucleic acid from pseudomonas hydrophila. Biochim. et Biophysica Acta **15**, 585 (1954). — REED, L. J., and M. J. CORMIER: Acetate-replacing factors for lactic acid bacteria: Intracellular distribution in liver. Proc. Soc. Exper. Biol. a. Med. **77**, 724 (1951). — REICHARD, P., and B. ESTBORN: Utilization of desoxyribosides in the synthesis of polynucleotides. J. of Biol. Chem. 188, 839 (1951). — RERABEK, J.: Fractionation of thyroid cells. Quantitative distribution of thyroxine, diiodtyrosine and nucleic acids in isolated cell fractions. Biochim. et Biophysica Acta 8, 389 (1952). — RHODIN, J.: Correlation of ultrastructural organisation and function in normal and experimentally changed proximal convoluted tubule cells of the mouse kidney. Diss. Karol. Inst. Stockholm 1954. — ROTHERHAM, J., J. L. IRVIN and E. M. IRVIN: Desoxyribonucleoproteins of the nuclei of rat liver cells. Federat. Proc. **13**, 285 (1954). — ROWEN, J. W., and A. KORNBERG: The phosphorolysis of nicotinamide riboside. J. of Biol. Chem. **193**, 497 (1951). — ROY, A. B.: The intracellular distribution of sulphatase. Biochim. et Biophysica Acta **14**, 149 (1954).

Sable, H. Z., P. B. Wilber, A. E. Cohen and M. R. Kane: Desoxyadenylic acid as acceptor for high energy phosphate. Biochim. et Biophysica Acta **13**, 156 (1954). — Scharrer, E.: Über die Beteiligung des Zellkernes an sekretorischen Vorgängen in Nervenzellen. Frankf. Z. Path. **47**, 143 (1934). — Schein, A. H., and E. Young: Intracellular localization of arginase in homogenates of rat liver suspended in distilled water. Exper. Cell Res. **3**, 383 (1952). — Schiller, E.: Variationsstatistische Untersuchungen über Kerneinschlüsse und Kristalle in der menschlichen Leber. Z. Zellforsch. **34**, 335 (1949). — Schlenk, F.. u. H. v. Euler: Desaminocozymase. Ark. Kemi, Mineral. Geol., Ser. B **12**, 8 (1938) cf. C. A. 6277, 8 (1938). — Schmidt, W. J.: Über Doppelbrechung und Feinbau der Kernmembran. Protoplasma **32**, 193 (1939). — Schmitz, H.: Isolierung von freien Nucleotiden aus verschiedenen Geweben. Biochem. Z. **325**, 555 (1954). — Schmitz, H., R. B. Hurlbert and V. R. Potter: Nucleotid metabolism. III. Mono, di- and triphosphates of cytidine, guanosine and uridine. J. of Biol. Chem. **209**, 41 (1954). — Schneider, W. C.: II. The distribution of succinic dehydrogenase, cytochrome oxidase, adenosintriphosphatase and phosphorus compounds in normal rat liver and in rat hepatomas. Cancer Res. **6**, 685 (1946). — Intracellular distribution of enzymes. III. The oxydation of octanonic acid by rat liver fractions. J. of Biol. Chem. **176**, 259 (1948). — Schneider, W. C., and G. H. Hogeboom: Intracellular distribution of enzymes. V. Further studies on the distribution of cytochrome C in rat liver homogenates. J. of Biol. Chem. **183**, 123 (1950). — Intracellular distribution of enzymes. VI. The distribution of succinoxidase and cytochrome oxidase activities in normal mouse liver and mouse hepatoma. — Intracellular distribution of enzymes. VII. The distribution of nucleic acids and adenosintriphosphatase in normal mouse liver and mouse hepatoma. J. Nat. Canc. Inst. **10**, 977 (1950). — Cytochemical studies of mammalian tissues: the isolation of cell components by differential centrifugation: A review. Cancer Res. **11**, 1 (1951). — Intracellular distribution of enzymes. IX. Certain purine-metabolizing enzymes. J. of Biol. Chem. **195**, 161 (1952). — Intracellular distribution of enzymes. X. Desoxyribonuclease and ribonuclease. J. of Biol. Chem. **198**, 155 (1952). — Schneider, W. C., and V. R. Potter: Intracellular distribution of enzymes. IV. Distribution of oxalacetic oxidase activity in rat liver and rat kidney fractions. J. of Biol. Chem. **177**, 893 (1949). — Shepherd, J. A., and G. Kalnitsky: Intracellular distribution of the phosphateactivated glutaminase of rat liver. J. of Biol. Chem. **192**, 1 (1951). — Siebert, G., K. Lang u. A. Corbert: Über das p_H-Optimum der Desoxyribonuclease verschiedener Tierarten. Biochem. Z. **320**, 418 (1950). — Siebert, G., K. Lang u. H. Lang: Über Stoffwechselvorgänge in Zellkernen. 4. Über Vitamin B_{12} und Kobalt in Zellkernen. Biochem. Z. **321**, 543 (1951). — Siebert, G., K. Lang, S. Lucius u. G. Rossmüller: Untersuchungen über Stoffwechselprozesse an isolierten Zellkernen: Über den Einbau von anorganischem Phosphat (P^{32}) in isolierte Zellkerne in vitro. Biochem. Z. **324**, 311 (1953). — Siebert, G., K. Lang, L. Müller, S. Lucius, E. Müller u. E. Kühle: Untersuchungen über Stoffwechselprozesse in isolierten Zellkernen. 7. Fermente des Aminosäure- und Peptidumsatzes in Zellkernen. Biochem. Z. **323**, 532 (1953). — Singer, T. P., and E. B. Kearney: Chemistry, metabolism and scope of action of the pyridine nucleotide. Adv. Enzymol. **15**, 79 (1954). — Smellie, R. M. S.: The uptake of ^{32}P by ribonucleotides in liver cell fractions. Biochemic. J. **49**, proc. XXXIV (1951). — Smellie, R. M. S., and W. M. McIndoe: The incorporation of ^{32}P and 14C into the nucleic acids of liver cell fractions. Biochemic. J. **52**, proc. XXII (1952). — Smith, E. E. B., and G. T. Mills: Uridine nucleotide compounds of liver. Biochim. et Biophysica Acta **13**, 386 (1954). — Smith, J. D., and M. G. P. Stoker: The nucleic acids of rickettsia burneti. Brit. J. Exper. Path. **32**, 433 (1951). — Smith, J. D., and G. R. Wyatt: The composition of some microbial desoxypentose nucleic acids. Biochemic. J. **49**, 144 (1951). — Stedman, E., and E. Stedman: The basic proteins of cell nuclei. Trans. Roy. Soc. Lond., Ser. B **235**, 565 (1951). — The chemical nature and functions of the components of cell nuclei. Cold Spring Harbor Symp. Quant. Biol. **12**, 224 (1951). — Steinert, M.: La synthèse de l'acide ribonucléique au cours du développement embryonnaire des batraciens. Bull. Soc. Chim. biol. Paris **33**, 549 (1951). — Sur la présence de bases puriques libres dans les œfs de batraciens. Arch. internat. Physiol. **60**, 192 (1952). — Stern, H., V. Allfrey, A. E. Mirsky and H. Saetren: Some enzymes of isolated nuclei. J. Gen. Physiol. **35**, 559 (1951). — Stern, H., and A. E. Mirsky: The isolation of wheat germ nuclei and some aspects of their glycolytic mechanism. J. Gen. Physiol. **36**, 181 (1952). — Soluble enzymes of nuclei isolated in sucrose and non-aqueous media. J. Gen. Physiol. **37**, 177 (1953). — Stich, H.: Trypaflavin und Ribonucleinsäure. Untersucht an Mäusegeweben, Condylostoma spec. und Acetabularia mediterranea. Naturwiss. **38**, 435 (1951). — Experimentelle karyologische und cytochemische Untersuchungen an Acetabularia mediterranea. Z. Naturforsch. **6**b, 319 (1951). — Stokkinger, L.: Das Kernkörperchen. Protoplasma **42**, 365 (1953). — Straub, F. B., H. S. Corran and D. E. Green: Mechanism of the oxidation of reduced coenzyme. I. Nature (Lond.) **143**, 119 (1939). — Strength, D. R., J. R. Christensen and L. J. Daniel: The requirement of diphosphopyridine nucleotide for choline dehydrogenase activity. J. of

Biol. Chem. **203**, 63 (1953). — STRITTMATTER, C. F., and E. G. BALL: The intracellular distribution of cytochrome compounds and of oxidative enzyme activity in rat liver. J. Cellul. a. Comp. Physiol. **43**, 57 (1954). — SUNG, S. CH., and J. N. WILLIAMS: The intracellular distribution of diphosphopyridine nucleotide nucleosidase in rat liver. J. of Biol. Chem. **197**, 175 (1952). — SWENDSEID, M. E., F. H. BETHELL and W. W. ACKERMANN: The intracellular distribution of vitamin B_{12} and folinic acid in mouse liver. J. of Biol. Chem. **190**, 791 (1951).

THOMSON, R. Y., F. C. HEAGY, W. C. HUTCHISON and J. N. DAVIDSON: The desoxyribonucleic acid content of the rat cell nucleus and its use in expressing the results of tissue analysis, with particular reference to the composition of liver tissue. Biochemic. J. **53**, 460 (1953). — TSUBOI, K. K.: Phosphomonoesterase activity in hepatic tissues of the mouse. Biochim. et Biophysica Acta 8, 173 (1952). — TYRELL, L. W., and D. RICHTER: The lipids of cell nuclei isolated from human brain cortex. Biochemic. J. **49**, proc. LI (1951).

URBANI, E.: Sur la teneur en amylase de fragments nucléés et énucléés d'Amoeba proteus. Biochim. et Biophysica Acta **9**, 108 (1952).

VANDERHAEGHE, F.: Les effects de l'énucléation sur la synthese des protéines chez Acetabularia mediterranea. Biochim. et Biophysica Acta **15**, 281 (1954). — VINCENT, W. S.: The isolation and chemical properties of the nucleoli of starfish oocytes. Proc. Nat. Acad. Sci. **38**, 139 (1952). — VISCHER, E., S. ZAMENHOF and E. CHARGAFF: Microbial nucleic acids: The desoxypentose nucleic acids of avian tubercle bacilli and yeast. J. of Biol. Chem. **177**, 429 (1949).

WACHSTEIN, M., and E. MEISEL: The histochemical distribution of 5-nucleotidase and unspecific alkaline phosphatase in the testicle of various species and in two human seminomas. J. Histochem. a. Cytochem. **2**, 137 (1954). — WADDINGTON, C. H.: Quinquennial report 1947—1951 of the institute of Animal Genetics, Univ. of Edinburgh 1952. — WAJZER, J., and F. BARON: Enzyme-promoted reactions between ribose phosphoric esters and purine bases. Bull. Soc. Chim. Biol. **31**, 750 (1949). — WANG, T. Y., D. T. MAYER and L. E. THOMAS: A lipoprotein of rat liver nuclei. Exper. Cell Res. **4**, 102 (1953). — WARBURG, O., u. W. CHRISTIAN: Über Aktivierung der ROBINSONschen Hexose-mono-Phosphorsäure in roten Blutzellen und die Gewinnung aktivierender Fermentlösungen. Biochem. Z. **242**, 206 (1931). — WATSON, M. L.: Pores in the mammalian nuclear membrane. Biochim. et Biophysica Acta **15**, 475 (1954). — WEBB, M.: The intracellular distribution of the desoxyribonucleases of mammalian tissues. Exper. Cell Res. **5**, 16 (1953). — WHITFELD, P. R.: Nucleic acids in erythrocytic stages of a malaria parasite. Nature (Lond.) **169**, 751 (1952). — WILLIAMS, J. N.: Intracellular distribution of choline oxidase. J. of Biol. Chem. **194**, 139 (1952). — Intracellular distribution of betaine aldehyde oxidase. J. of Biol. Chem. **195**, 37 (1952). — WYATT, G. R.: The quantitative analysis of desoxypentose nucleic acids. Biochemic. J. **47**, proc. VII (1950). — Recognition and estimation of 5-methylcytosine in nucleic acids. Biochemic. J. **48**, 581 (1951). — The purine and pyrimidine composition of desoxyribonucleic acid. Biochemic. J. **48**, 584 (1951). — The nucleic acids of some insect viruses. J. Gen. Physiol. **36**, 201 (1952). — WYATT, G. R., and S. S. COHEN: Nucleic acids of rickettsiae. Nature (Lond.) **170**, 846 (1952).

ZAMENHOF, S., G. BRAWERMAN and E. CHARGAFF: On the desoxypentose nucleic acid from several microorganisms. Biochim. et Biophysica Acta **9**, 402 (1952). — ZAMENHOF, S., and E. CHARGAFF: Evidence of the existence of a core in desoxyribonucleic acids. J. of Biol. Chem. **178**, 531 (1948). — Dissymmetry in nucleotide sequence of desoxypentose nucleic acids. J. of Biol. Chem. **187**, 1 (1950).

Zusammenfassende Darstellungen,

die ältere Literatur berücksichtigen und über den Rahmen des vorliegenden Abschnittes hinaus Probleme des Zellkernes behandeln.

Bücher:

SHARP, L.: Fundamentals of Cytology. New York: McGraw-Hill Book Co. 1943.

GRESSON, R.: Essentials of general cytology (mit vier Kapiteln Pflanzencytologie von H. HESLOP CLARK). Edinburgh: Oliver & Boyd 1948.

ROBERTIS, E., W. NOWINSKI and F. SAEZ: General cytology. Philadelphia: W. B. Saunders & Company 1948.

BRACHET, J.: Chemical embryology. New York: Interscience Publ. 1950.

CASPERSSON, T.: Cell growth and cell function. New York: Norton & Co. 1950.

BADDILLY, J.: Nucleic acids, purines and pyrimidines. Annual Rev. Biochem. **20**, 149 bis 178 (1951).

DAVIDSON, J.: The biochemistry of the nucleic acids. New York: J. Willey & Sons 1952.

SCHRADER, F.: Mitosis, the movements of chromosomes in cell division. New York: Columbia Univ. Press 1953.

TISCHLER, G.: Allgemeine Pflanzenkaryologie. In LINSBAUERS Handbuch der Pflanzenanatomie, S. 385—1040. 1921—1953.

Einzelpublikationen:

GREENBERG, D. M.: Chemical pathways of metabolism, Bd. 1. New York: Academic Press 1954.

MILOVIDOV, P.: Physik und Chemie des Zellkernes. Berlin: Gebrüder Bornträger, Teil I 1949; Teil II 1954.

JORDAN, D. O.: Physicochemical properties of the nucleic acids. Progr. in Biophysics a. Biophysical Chem. **2**, 51—89 (1951).

JORDAN, D. O.: Nucleic acids, purines and pyrimidines. Annual Rev. Biochem. **21**, 209—244 (1952).

BROWN, G. B.: Nucleic acids, purines and pyrimidines. Annual Rev. Biochem. **22**, 141 bis 178 (1953).

BÜCHER, T.: Probleme des Energietransportes innerhalb lebender Zellen. Adv. Enzymol. **14**, 1—48 (1953).

ALFERT, M.: Composition and structure of giant chromosomes. Internat. Rev. Cytology **3**, 131—176 (1954).

ALLEN, F. W.: Nucleic acids. Annual Rev. Biochem. **23**, 99—124 (1954).

DAVISON, P. F., B. E. CONWAY and I. A. V. BUTLER: The nucleoprotein complex of the cell nucleus and its reactions. Progr. Biophysics a. Biophysical Chem. **4**, 148—194 (1954).

DOUNCE, A.: The significance of enzyme studies on isolated cell nuclei. Internat. Rev. Cytology **3**, 199—224 (1954).

SINGER, T. P., and E. B. KEARNEY: Chemistry, metabolism and scope of action of the pyridine nucleotide coenzymes. Adv. Enzymol. **15**, 79—140 (1954).

Symposia:

Genes and chromosomes, structure and organization. Cold Spring Harbor, Symp. Ouant. Biol. **1941**.

Nucleic Acid. Symposia Soc. Exper. Biol. **1** (1947).

The Chemistry and physiology of the nucleus. Exper. Cell Res. Suppl. **2** (1952).

Structural aspects of cell physiology. Symposia Soc. Exper. Biol. **6** (1952).

Während der Drucklegung dieses Bandes erschien das Handbuch der Physiologischen Chemie, FLASCHENTRÄGER und LEHNARTZ, in dem K. LANG und G. SIEBERT das Kapitel „Chemische Leistungen der morphologischen Zellelemente“ schrieben, auf das besonders hingewiesen wird. Bd. 2, S. 1064—1156. Berlin: Springer 1954.

Mitochondria and microsomes.

By

Adèle Millerd.

A. Mitochondria.

a) Introduction.

The spatial relationships of the myriad enzymes concerned in cell metabolism have assumed great significance in our efforts to understand the intergrated economy of the plant cell. During investigations into the intracellular distribution of enzymes, the biochemical activities of the mitochondria have been widely studied.

In this present discussion the term mitochondrion is used in accordance with the combined biochemical and cytological definition of these bodies. Such a definition is proffered by GODDARD and STAFFORD (1954). "... the most appropriate definition applicable to both plant and animal tissue would be to define mitochondria as cellular particles associated with enzymes of the cytochrome system, the KREBS cycle, fatty acid oxidation, and with oxidative phosphorylation. These lipo-protein and pentose nucleic acid-containing particles may vary in size from the limit of the light microscope, about 0.1 μ to 6.0 μ in diameter, and range in shape from spheres to long rods." Preparations for enzymatic studies are obtained by differential centrifugation and hence would be contaminated by any other cytoplasmic particles of mitochondrial size.

The metabolic significance of plant mitochondria has been the subject of several reviews (MILLERD and BONNER 1953, GODDARD and STAFFORD 1954).

b) Mitochondria and organised respiratory activity.

The first awareness of the enzymatic importance of the insoluble components of plant cytoplasm appeared in the study by HILL and BHAGVAT (1939) on the isolation of particulate succinoxidase from germinating seeds and other plant tissues. OKUNUKI in the same year (1939) separated and examined an insoluble succinoxidase preparation from the pollen of *Lilium auratum*. A similar preparation from *Phaseolus mungo* was examined in detail by DAMODARAN and VENKATESEN (1941). The significance of the insoluble (particulate) nature of this enzyme was not stressed by these authors nor was the mitochondrion recognised as the bearer of the activity.

Though the important nature of the enzymes present in plant cell cytoplasm in insoluble form was thus clearly established, the association of such insoluble enzymes with specific cytoplasmic bodies (mitochondria) was made only after the identification of mitochondria of animal cells as the locus of the enzymes responsible for the respiratory synthesis of adenosine triphosphate (ATP), recently reviewed by DOUNCE (1950), GREEN (1951), SCHNEIDER and HOGEBOOM (1951), HOGEBOOM *et al.* (1953).

DU BUY *et al.* (1951) in a study of cytochrome oxidase of leaves, suggested that the mitochondria might be the bearers of this enzyme system. A similar suggestion was made by STAFFORD (1951) in her work on succinoxidase activity of pea seedlings, and by MILLERD (1951) on the basis of studies on the succinoxidase of potato tuber. In none of these cases was the identity of the particles as mitochondria established; the particles corresponded in size with mitochondria but did not stain vitally with Janus Green B, a property considered characteristic of mitochondria. This lack of staining was presumably a result of damage to the particles during isolation.

The early work on plant mitochondria was all concerned with single enzymes or simple enzyme systems, although the particulate preparations from potato (MILLERD 1951) were found to possess fumarase and malic dehydrogenase in addition to succinoxidase activity.

The first demonstration of organised respiratory activity by particles isolated from a higher plant was that of MILLERD *et al.* (1951). Cytoplasmic particles were isolated by differential centrifugation from seedlings of the mung bean, *Phaseolus aureus*. These particles, in the presence of phosphate and a catalytic amount of one of the acids of the KREBS cycle were found to be capable of oxidising pyruvate to carbon dioxide and water, and in so doing, to conserve the energy liberated in the form of ATP. These oxidations were apparently mediated by the cytochrome system. The active components of the preparation were identified as mitochondria on the basis of their size and their ability to stain vitally with Janus Green B. These observations showed then that the role of mitochondria in higher plants is basically similar to that which they play in animal tissues.

The work of MILLERD *et al.* indicated that oxidation of pyruvate takes place by way of the KREBS cycle. Thus pyruvate is oxidised only with the concomitant oxidation of a small amount of any one of the KREBS cycle acids. The subsequent work of BRUMMOND and BURRIS (1953) using mitochondria isolated from lupine cotyledons provided confirmation of the operation of a KREBS cycle in plants. Mitochondria were incubated with C^{14} carbonyl-labelled pyruvate and a small amount of malic acid. In separate experiments each of the KREBS cycle acids was added as trapping acid. After a suitable period of incubation the trapping acids were isolated. In each instance the C^{14} initially given as pyruvate had been incorporated. These facts together with the demonstration by LATIES (1953a) of the conversion of the varied plant acids to succinate by cauliflower mitochondria provide rigorous and complete evidence for the operation of the KREBS cycle in plant metabolism.

Of major significance in the operation of the KREBS cycle in animal tissues, are the phosphorylations which accompany each oxidative step of the cycle. Prior to the work just described there was good but indirect evidence that a tricarboxylic acid cycle, similar to that first postulated for animal tissues by KREBS and JOHNSON (1937) functioned in the respiration of higher plants. It had been suggested (KREBS 1950) that the cycle did operate in plant tissues but with a purpose different from that in animal tissues and that perhaps its primary role was to supply carbon skeletons rather than to be intimately concerned in energy conservation during the aerobic phases of carbohydrate breakdown. With the demonstration that phosphorylations do accompany KREBS cycle oxidations, it is apparent that the cycle has the same metabolic significance in both tissue types.

It has proved possible to isolate successfully from a variety of plant tissues mitochondria capable of oxidising pyruvic acid by way of the KREBS cycle. Metabolically active mitochondria have been isolated from pea seedlings (DAVIES 1953, PRICE and THIMANN 1954, SMILLIE 1955). Similar cytoplasmic particles have been isolated from fruits (from avocado, MILLERD *et al.* 1953, from apple, PEARSON and ROBERTSON 1954), storage tissue (Jerusalem artichoke, BONNER *et al.* 1953, potato, SHARPENSTEEN and CONN 1954, carrot and beetroot, ROBERTSON *et al.* 1954, chicory root, LATIES 1954). In addition they have been isolated from young roots (SMILLIE 1955).

In the leaf the task of demonstrating the biochemical significance of the mitochondria is made more difficult by the presence of large numbers by chloroplasts, but it has nonetheless been successfully accomplished (McCLENDON 1952, 1953; WEIER 1952 and JAGENDORF and WILDMAN 1954).

BRUMMOND and BURRIS (1954) have suggested that the enzymes of the tricarboxylic acid cycle do not exist as an integrated particulate system in the leaves of young lupine plants. The enzymes responsible for the reactions of the cycle were found to be present, but they appeared to have an intracellular localisation different from those of the mitochondrial complex of etiolated plants. Fumarase, aconitase, and α-ketoglutaric dehydrogenase were associated exclusively with the particulate fraction from green leaves, whereas the isocitric, succinic, and malic dehydrogenases were in the supernatant fraction. The supernatant fraction would contain the enzymes of the cytoplasm.

SMILLIE (1955) using pea seedlings and somewhat different isolation procedures obtained active mitochondria from all major organs of both green and etiolated pea plants. Particulate preparations from green leaves and green stems showed high activity and pyruvic, isocitric, succinic, malic and glutamic acids were rapidly oxidised. The oxidation in all cases was linear for at least one hour. The most pure mitochondrial preparations did contain small amounts of chlorophyll but the enzymatic activity was not correlated with chlorophyll content. Thus in the green leaves of the pea plant the KREBS cycle enzymes apparently exist in a particulate integrated complex as found in other plant tissues.

c) The relationship of mitochondria with terminal oxidases.

It has been possible, then, to isolate active mitochondria from numerous plant tissues. The fundamental biochemical role of mitochrondria in plant metabolism is indicated by the demonstration (BHAGVAT and HILL 1951, DUCET and ROSENBERG 1951, WEBSTER 1952) that cytochrome oxidase may be isolated on particles from a wide variety of leaves and other organs. Cytochrome oxidase is the terminal oxidase of the mitochondrial oxidations so far investigated.

The isolation (HACKETT and SIMON 1954) from the spadix of *Arum maculatum* of particulate preparations of a very high degree of activity in the reactions of the KREBS cycle is particularly interesting. The respiration of either slices or homogenates of this aroid tissue appears to be cyanide resistant and evidence has been presented (JAMES and BEEVERS 1950) that the terminal oxidase is a flavoprotein rather than a metallo-enzyme. The particulate preparations of HACKETT and SIMON respire at a rate sufficient to account for at least a major part of the oxygen consumption of the homogenate from which they were prepared. It will be of great interest therefore to determine whether the terminal oxidase operating in these particles is cytochrome oxidase. In the avocado fruit a similar problem arises. The respiration of tissue slices is cyanide resistant but that of the isolated cytoplasmic particles is not always so (BIALE 1953). In the avocado the problem is complicated by the high fat content of the tissue and the consequent penetration problems, but apparently the respiration of the isolated particles, under correctly chosen conditions, is substantially cyanide and azide sensitive (HATCH and MILLERD 1954).

Ascorbic acid oxidase has been postulated as the terminal oxidase functioning in the respiration of barley roots (JAMES 1953), However, as is shown by the recent work of HONDA (1955), cytochrome oxidase is also present in this tissue and may be isolated in particulate form in amounts sufficient to account for all of the respiration of the tissue. It is perhaps of interest, then, that ascorbic acid oxidase has been obtained attached to mitochondria isolated from avocado (CONN and YOUNG 1954). Other investigations indicate, however, that the localisation of ascorbic acid oxidase in the cell is not exclusively mitochondrial (WAYGOOD 1950, STAFFORD 1951, NEWCOMB 1951).

A possible link between ascorbic acid oxidase and mitochondrial oxidations is suggested by the finding that glutathione reductase is also attached to isolated avocado mitochondria (CONN and YOUNG 1954). A system which would oxidise reduced triphosphopyridine nucleotide (TPNH) and involving these two enzymes and dehydroascorbic acid reductase is illustrated below. (GSH and GSSG signify respectively reduced and oxidised glutathione, AA and DHA ascorbic and dehydroascorbic acids.)

$$\mathrm{TPNH} + \mathrm{H}^+ + \mathrm{GSSG} \underset{\text{reductase}}{\overset{\text{glutathione}}{\rightleftharpoons}} 2\,\mathrm{GSH} + \mathrm{TPN}$$

$$2\,\mathrm{GSH} + \mathrm{DHA} \underset{\text{reductase}}{\overset{\text{dehydroascorbic acid}}{\rightleftharpoons}} \mathrm{GSSG} + \mathrm{AA}$$

$$\mathrm{AA} + \mathrm{O}_2 \underset{\text{oxidase}}{\overset{\text{ascorbic acid}}{\rightleftharpoons}} \mathrm{DHA} + \mathrm{H}_2\mathrm{O}_2.$$

Tyrosinase (polyphenoloxidase) activity was found by McCLENDON (1953) associated with all cellular fractions obtained from tobacco leaf homogenates. The intracellular distribution of tyrosinase is reviewed by WEIER and STOCKING (1952).

d) Energy conservation during mitochondrial oxidations.

That plant mitochondria utilise the energy liberated in substrate oxidation for the formation of energy-rich phosphate bonds of ATP was first demonstrated with preparations from the mung bean (MILLERD *et al.* 1951). In this work, the mitochondria were allowed to oxidise substrate in the presence of adenylic acid and isotopically labelled inorganic phosphate, ATP containing the labelled phosphate was isolated from the reaction mixture after a suitable incubation period.

Broadly speaking, the principles of oxidative phosphorylation and energy transfer in the higher plant appear to resemble the system present in animal tissues.

The efficiency of conversion of oxidative energy to phosphate bond energy is conventionally expressed in the P:O ratio — the ratio of atoms of phosphorus esterified to atoms of oxygen consumed. The mung bean system first studied (BONNER and MILLERD 1953) had one apparent peculiarity, it was significantly less efficient in oxidative phosphorylation than are mammalian mitochondria. Each step in the oxidation of pyruvate to carbon dioxide and water by the mung bean system involved the esterification and transfer of a single molecule of phosphate per atom of oxygen consumed rather than 2 to 4 as in corresponding mammalian preparations. Whether this is a real property of the mung bean or whether the phosphorylative capacity of mung bean particles is adversely affected by some aspect of the preparative procedure is uncertain. This low efficiency is not however a general feature of isolated plant mitochondria. The oxidation of α-ketoglutarate by mitochondria from cauliflower (LATIES 1953b) and from avocado (BIALE 1953) is accompanied by P:O ratios approaching 3, and the oxidation of this acid by pea mitochondria may show a P:O ratio in excess of 3 (SMILLIE 1955). These values approach the maximum suggested for the oxidation of α-ketoglutarate in animal tissues (HUNTER 1951).

The locus of phosphate esterifications occurring during the transfer of electrons from substrate to oxygen is an important problem. SMILLIE (1955) has shown, with pea mitochondria, that some phosphate esterification does take place during the reoxidation of reduced diphosphopyridine and triphosphopyridine nucleotides.

e) Metabolic spectrum of plant mitochondria.

Carbohydrate metabolism.

The role of mitochondria in the aerobic phases of carbohydrate metabolism has already been discussed. A link between the glycolytic phase, catalysed by enzymes present in the cytoplasm, and the aerobic phase, catalysed by mitochondrial enzymes, is provided by the finding (SALTMAN 1953) that hexokinase is attached to the mitochrondrion. Hexokinase is found exclusively in the mitochrondria when the enzyme is carefully and suitably prepared. The linkage of this enzyme to the mitochondrion is however a labile one and any damage to the mitochondria results in liberation of soluble hexokinase and the ratio of particle-bound to soluble hexokinase is thus dependent on the methods of preparation used.

Fat metabolism.

Contrary to the situation with animal tissues (reviewed by GREEN 1954), the relationship of the mitochondria with fatty acid metabolism is not yet clear. It has been suggested that a particulate fraction other than the mitochondria is concerned (see Section B of this article). However mitochondria have been shown to oxidise butyric acid (SMILLIE 1955) and the acetate activating enzyme is mitochondrial (MILLERD and BONNER 1954).

It has been shown by appropriate experiments in which isotopically labelled acetate is supplied to intact plant tissues that several plant materials become labeled and with a specific activity approximating that of the acetate given. Thus it has been shown (BONNER 1949a, ARREGUIN *et al.* 1951) that the carbon atoms which go to make up the isoprenoids, certain amino acids, waxes, and fatty acids (NEWCOMB and STUMPF 1953) are derived principally or entirely from acetate. Acetyl coenzyme A has been shown to be the form in which acetate is incorporated into the isoprenoid rubber (BONNER *et al.* 1954) and is believed to be the form in which acetate is involved in other synthetic reactions. It is worthy of note then that the synthesis of acetyl coenzyme A is associated with the mitochondria.

Nitrogen metabolism.

1. Amino acid oxidation. l-Glutamic acid is rapidly metabolised by pea mitochondria (SMILLIE 1955), this would indicate the mitochondrion as the locus of glutamic dehydrogenase.

2. Amide bond synthesis. The enzyme system responsible for synthesis of glutamine has been studied in a wide variety of tissues by WEBSTER (1953a) who has shown that the system is not only widely distributed but also that it is largely (and probably entirely) mitochondrial. As with hexokinase, the enzyme is bound in the mitochondrial unit by a somewhat labile linkage and is readily liberated as a soluble enzyme.

3. Peptide bond synthesis. The first direct demonstration of peptide bond synthesis by plant preparations has been achieved by WEBSTER (1953b) who has shown that homogenates of bean seedlings synthesise glutathione from glutamate, cysteine and glycine using ATP as an energy source. This reaction involves, as in animal systems, the formation of glutamyl-cysteine from glutamate and cysteine (WEBSTER 1953c, WEBSTER and VARNER 1954) and the subsequent addition of glycine to the glutamyl-cysteine. The entire system is associated with the mitochondria.

4. Protein metabolism. The metabolically active nature of the mitochondrial proteins is shown by the *in vitro* experiments of WEBSTER (1954a). Mitochondria isolated from bean seedling homogenates showed an active incorporation of C^{14} labeled glycine, glutamic and aspartic acids into the mitochondrial protein. This incorporation process resembled that of peptide bond synthesis in that it was energy dependent. The system utilised either the energy of added ATP, or energy produced during respiration.

f) Problems of inactivation associated with isolation of plant mitochondria.

The preparational procedures necessary for isolation of active intact mitochondria have been investigated with various tissues (for example see LATIES 1953b, MILLERD 1953). In general it may be said that with the present established procedures, it is possible to isolate active respiratory particles from tissues of a wide variety of higher plants.

Plant mitochondria have perhaps not been more studied because of their lability and the extreme care which must be exercised if they are to be isolated with their vital properties unimpaired. There are several ways in which the integrity of plant mitochondria may be damaged. LATIES (1953b) has made a detailed study of the effects of physical environment on the oxidative and phosphorylative capacities of mitochondria from immature inflorescences of *Brassica oleracea.* These experiments show that mitochondria are severely damaged by swelling in hypotonic solution if this occurs in the presence of certain components of the protoplast. Such damage, which is rapid and irreversible, is apparently enzymatic. The malic, α-ketoglutaric and succinoxidase systems are all affected by this type of damage, which may be avoided by the preparation of mitochondria in cold hypertonic solution.

When mitochondria are suspended in distilled water, in the absence of other cytoplasmic components, the succinoxidase system remains intact, whereas the malic and α-ketoglutaric oxidase systems suffer considerable injury. This type of injury is minimised or precluded by the presence of dilute solutions of neutral salt. It is avoided also by the use of slightly hypertonic solutions (sucrose or mannitol).

The phosphorylative system which is associated with the succinoxidase complex of plant mitochondria is completely destroyed by exposure to a hypotonic environment even though the capacity to oxidise succinate is retained. On the other hand, the phosphorylative machinery which is associated with malate or with α-ketoglutarate oxidation is lost by exposure of the mitochondria to hypotonic solution and the capacity to oxidise these substrates is lost to approximately the same extent. Both the oxidative and phosphorylative systems are protected by neutral salt in these instances.

LATIES (1953c) has also investigated the dual nature of the requirement for adenylate by these mitochondria. Thus the presence of adenylate in the reaction mixture is essential to maintenance of the integrity of the phosphorylative system of cauliflower mitochondria at the substrate level. In addition adenylate is required as an acceptor for energy rich phosphate formed by oxidative phosphorylations. Indeed it is well established that the rate of mitochondrial oxydations is limited by the amount of acceptor available.

In addition LATIES (1954b) has indicated that the integrity of mitochondria can be damaged *in situ*. Mitochondria were prepared from tissue previously exposed to an environment of low osmotic pressure. The reduced enzymatic activity associated with such preparations suggested that the mitochondria, while *in situ* suffered fundamental transformation.

Another intriguing hazard in the successful isolation of plant mitochondria is indicated by GOODWIN and WAYGOOD (1954). These workers were studying the enzymatic activity of mitochondria isolated from barley seedlings. They observed that fractions from 4–5 day old seedlings apparently lost the ability to oxidise succinic acid, whereas mitochondria from embryonic tissue actively oxidised this substrate. The falling off in extractable succinoxidase activity was correlated with increase in lecithinase C activity. Lecithinase C hydrolyses phospholipids at the phosphate-glycerol linkage. Phospholipids are known to occur in high concentration in plant mitochondria (LEVITT 1954) and are known to play an important structural part in particulate enzyme systems as is shown for heart succinoxidase by MORTON (1950). The authors make the reasonable suggestion that exposure of the mitochondria to lecithinase during isolation could well bring about a breakdown in the functional integrity of succinoxidase.

g) Mitochondrial regulation of physiological activity.

Our growing knowledge of the functions of plant mitochondria has served to give us a deeper insight into the behaviour of the plant as a whole. This has been particularly the case in relation to those physiological processes which are dependent on or intimately associated with plant respiration. It has long been known that plant tissues or organs which are doing biological work respire more rapidly than do similar tissues or organs at rest. Thus increased rates of respiration attend the performance of osmotic work such as the accumulation of water or ions. Such increases in rate of respiration now appear to be the result of and to be brought forth by the work which the tissue is called upon to do. In addition, it is now possible to outline in some detail the mechanism by which respiratory rate is geared to and regulated by the energy requirement of the tissue.

As has been discussed the respiration of mitochondria is obligatorily coupled to the concomitant phosphorylations and their respiratory rate is limited by the supply of phosphate acceptor. Mitochondrial oxidations may be emancipated from these coupled phosphorylations by the use of uncoupling agents such as dinitrophenol (LATIES 1953c).

The rate of respiration of resting intact plant tissue is also increased by dinitrophenol, and here too the effect is mediated through an uncoupling of the phosphorylative system (BONNER 1949b). Dinitrophenol may then be used as a reagent for ascertaining the extent to which the respiratory rate is limited *in vivo* by the capacity or rate of turnover of the phosphate accepting system. In the cases which have thus far been studied (Table 1) it is clear that even though dinitrophenol causes very considerable increases in respiratory rate of resting plant tissues, it elicits little or no increase in rate in similar tissues which are carrying on active growth or osmotic work. In all of the instances summarised in

Table 1. *Rates of respiration of active and resting tissue as affected by 2,4-dinitrophenol (DNP)*[1].

Species	Organ or tissue	Rate of oxygen uptake mm³/gm/hr				
		resting tissue		active tissue		tissue activated by
		control	DNP	control	DNP	
Jerusalem artichoke	tubers	39	77	77	84	Auxin-induced water uptake
Oat	coleoptile	38	62	50	56	Auxin induced growth
Avocado	fruit	50	100	100	100	Normal fruit ripening
Carrot	root	40	135	75	140	Salt accumulation

[1] Taken from MILLERD and BONNER (1953).

Table 1, respiratory rate is increased by the act of doing biological work and this rate is not further increased by dinitrophenol. It may be concluded therefore that the rate of respiration of plant tissues is frequently regulated by the capacity of the mitochondrial phosphorylative system, and that this is in turn regulated by the rate at which the phosphate acceptor is renewed through the utilisation of ATP. It may be further concluded that for these several types of biological work energy is supplied often in the form of ATP. Rate at which active ATP-using work is carried on, is then, in the last analysis, a factor primarily responsible for regulation of plant respiration rate. However, as shown by HONDA (1955b), this is not the complete explanation for respiratory control.

Also among the physiological processes of the plant which are controlled by the coupled phosphorylation of the mitochondria is the process of fruit ripening as studied in the avocado (MILLERD *et al.* (1953).

The probable direct role of mitochondria in salt accumulation has been suggested (ROBERTSON 1951). ROBERTSON pointed out that if the mitochondrion were capable of accumulating ions, it would be a suitable body to transfer ions across the cytoplasm and into lipoid regions because not only do mitochondria have a lipo-protein structure but they also move effectively in the protoplasmic streaming. Recent work (ROBERTSON *et al.* 1955) has shown that mitochondria can indeed hold cations and anions in concentrations greater than those of the surrounding medium. This work emphasises the possibility that a definite ion transport mechanism exists in the mitochondria. This mechanism is related to the respiratory activity of the mitochondria and is made possible by the membrane which apparently surrounds most mitochondria. This role of the mitochondria in the physiological activity of the cell is discussed in detail by ROBERTSON in Section 3 of this volume.

h) Summary.

It is clear then that the mitochondria of the plant cell have associated with them the enzymatic machinery for the complete oxidation of pyruvate to carbon dioxide and water via the Krebs cycle. In addition they effect the conservation of the energy liberated in oxidation in a generally useable form, that of ATP. Although the ATP thus formed is used by the plant cell for the conduct of energy consuming processes at sites remote from the mitochondrion, still the latter can utilise the energy of ATP for the conduct of important synthetic reactions including the synthesis of peptide bonds, hexose phosphate and acetyl coenzyme A. In addition to the fundamental importance of these cellular entities in respiratory and synthetic mechanisms, the mitochondria are also concerned in other physiological activities such as control of respiration rate, fruit ripening and salt accumulation.

B. Microsomes.

a) Introduction.

In contrast with work on mitochondria, the biochemical significance of the microsomes of the plant cell is not well documented. Microsomes may be defined as cytoplasmic particles of submicroscopic size (less than 0.1 mμ) and possessing high nucleic acid content (SCHNEIDER and HOGEBOOM 1951).

b) Fatty acid metabolism.

NEWCOMB and STUMPF (1953) and HUMPHREYS *et al.* (1954) have reported a fatty acid oxidation system associated with a microsomal fraction isolated from germinating peanut cotyledons. This oxidation system is quite different from that of animal tissues (GREEN 1954) and is active towards the long chain fatty acids, stearic, palmitic and myristic. It is inactive towards short chain fatty acids. What relationship this microsomal system may have with the mitochondrial fatty acid oxidase system is not yet apparent.

c) Protein metabolism.

The relationship of microsomes with protein metabolism is shown by WEBSTER (1954b) in a study of the incorporation of C^{14} labelled glutamate into the proteins of etiolated bean hypocotyls. After a period of incubation, the distribution of the incorporated glutamate among the cellular fractions was determined. The cellular proteins examined possessed varied capacities for incorporation, thus giving some indication of the relative activities of protein turnover in different parts of the cell. The microsomes exhibited marked incorporation, suggestive of the possible significance of this fraction in the protein metabolism of the cell.

The significance of microsomes in protein metabolism has been further stressed by the finding of an apparent *in vitro* synthesis of protein by an isolated microsomal fraction of pea seedlings (WEBSTER 1955a, b). For synthesis to proceed, the system requires the microsomal preparation, ATP, an ATP regenerating system, ribonucleic acid, and a mixture of the 17 amino acids normally found in protein. If the particles are deprived of their ribonucleic acid by treatment with crystalline ribonuclease, the activity is greatly inhibited, but can be almost completely restored by addition of a ribonucleic acid preparation. This evidence strongly suggests that the ribonucleic acid generally found in microsomal preparations has a function in protein formation.

Literature.

ARREGUIN, B., J. BONNER and B. J. WOOD: Studies on the mechanism of rubber formation in the guayule. III. Experiments with isotopic carbon. Arch. of Biochem. a. Biophysics **31**, 234–247 (1951).

BHAGVAT, K., and R. HILL: Cytochrome oxidase in higher plants. New Phytologist **50**, 112–120 (1951). — BIALE, J. B.: Metabolism in the avocado fruit at different levels of organisation. Abstr. Amer. Soc. Plant Physiol., p. 21, 1953. — BONNER, J.: Synthesis of isoprenoid compounds in plants. J. Chem. Educat. **26**, 628–631 (1949a). — Relations of respiration and growth in the Avena coleoptile. Amer. J. Bot. **36**, 429–436 (1949b). — BONNER, J., R. S. BANDURSKI and A. MILLERD: Linkage of respiration to auxin-induced water uptake. Physiol. Plantarum (Copenh.) **6**, 511–522 (1953). — BONNER, J., and A. MILLERD: Oxidative phosphorylation by plant mitochondria. Arch. of Biochem. a. Biophysics **42**, 135–148 (1953).— BONNER, J., M. W. PARKER and J. C. MONTERMOSO: Biosynthesis of rubber. Science (Lancaster, Pa.) **120**, 549–551 (1954). — BRUMMOND, D. O., and R. H. BURRIS: Transfer of C^{14} by lupine mitochondria through the reactions of the tricarboxylic acid cycle. Proc. Nat. Acad. Sci. U.S.A. **39**, 754–759 (1953). — Reactions of the tricarboxylic acid cycle in green leaves. J. of Biol. Chem. **209**, 755–765 (1954).

CONN, E. E., and L. C. T. YOUNG: Reduction of oxidised glutathione by plant particles. Federat. Proc. **13**, 194 (1954).

DAMODARAN, M., and T. R. VENKATESEN: Amide synthesis in plants. I. The succinoxidase system in plants. Proc. Indian Acad. Sci. Sect. B **13**, 345–359 (1941). — DAVIES, D. D.: The Krebs cycle enzyme system of pea seedlings. J. of Exper. Bot. **4**, 173–183 (1953). — DOUNCE, A. L.: Cytochemical foundations of enzyme chemistry. The Enzymes. Edit. J. B. SUMNER and K. MYRBÄCK, Vol. 1, Part 1, p. 187–266. New York: Academic Press 1950. — DU BUY, H. G., M. W. WOODS and M. D. LACKEY: Enzymatic activities of isolated normal and mutant mitochondria and plastids of higher plants. Science (Lancaster, Pa.) **111**, 572–574 (1950). — DUCET, G., and A. ROSENBERG: Activité respiratoire chez les végétaux supérieurs. II. Activités cytochrome oxydasique et polyphénoloxydasique chez quelques végétaux supérieurs. Bull. Soc. Chim. Biol. Paris **33**, 321–336 (1951).

GODDARD, D. R., and H. A. STAFFORD: Localization of enzymes in the cells of higher plants. Annual Rev. Plant Physiol. **5**, 115–132 (1954). — GOODWIN, B. C., and E. R. WAYGOOD: Succinoxidase inactivation by a lecithinase of barley seedlings. Nature (Lond.) **174**, 517–518 (1954). — GREEN, D. E.: The cyclophorase complex of enzymes. Biol. Rev. **26**, 410–455 (1951). — Fatty acid oxidation in soluble systems of animal tissue. Biol. Rev. **29**, 330–366 (1954).

HACKETT, D. P., and E. W. SIMON: Oxidative activity of particles prepared from the spadix of *Arum maculatum*. Nature (Lond.) **173**, 162–163 (1954). — HATCH, M., and A. MILLERD: Unpublished data. 1954. — HILL, R., and K. BHAGVAT: Cytochrome oxidase in flowering plants. Nature (Lond.) **143**, 726 (1939). — HOGEBOOM, G. H., W. C. SCHNEIDER and M. J. STRIEBICH: Localization and integration of cellular function. Cancer Res. **13**, 617–632 (1953). — HONDA, S. I.: Succinoxidase and cytochrome oxidase in barley roots. Plant Physiol. **1955**a. — The salt respiration of barley roots. Plant Physiol. **1955**b. — HUMPHREYS, T. E., E. H. NEWCOMB, A. H. BOKMAN and P. K. STUMPF: Fat metabolism in higher plants. II. Oxidation of palmitate by a peanut particulate system. J. of Biol. Chem. **210**, 941–948 (1954). — HUNTER jr., F. E.: Oxidative phosphorylation during electron transport. Phosphorus Metabolism. Edit. W. D. MCELROY and B. GLASS, Vol. 1, p. 297–330. Baltimore: John Hopkins Press 1951.

JAGENDORF, A. T., and S. G. WILDMAN: The proteins of green leaves. VI. Centrifugal fractionation of tobacco leaf homogenates and some properties of isolated chloroplasts. Plant Physiol. **29**, 270–279 (1954). — JAMES, W. O.: The terminal oxidases in the respiration of the embryos and young roots of barley. Proc. Roy. Soc. Lond., Ser. B **141**, 289—299 (1953). — JAMES, W. O., and H. BEEVERS: The respiration of Arum spadix. A rapid respiration, resistant to cyanide. New Phytologist **49**, 353—374 (1950).

KREBS, H. A.: The tricarboxylic acid cycle. Harvey Lect. **44**, 165–199 (1950). — KREBS, H. A., and W. A. JOHNSON: The role of citric acid in intermediate metabolism in animal tissues. Enzymologia **4**, 148–156 (1937).

LATIES, G. G.: Transphosphorylating systems as a controlling factor in mitochondrial respiration. Physiol. Plantarum (Copenh.) **6**, 215–225 (1953a). — The physical environment and oxidative and phosphorylative capacities of higher plant mitochondria. Plant Physiol. **28**, 557–575 (1953b). — The dual role of adenylate in the mitochondrial oxidations of a higher plant. Physiol. Plantarum (Copenh.) **6**, 199–214 (1953c). — The nature of the respiratory rise in sliced tuberous root tissue. Abstr. Amer. Soc. Plant Physiol., p. 36–37, 1954a. — The osmotic inactivation *in situ* of plant mitochondrial enzymes. J. of Exper. Bot. **5**, 49–70 (1954b). — LEVITT, J.: Investigations of the cytoplasmic particles and proteins of potato tubers. I. Bound water and lipid contents. Physiol. Plantarum (Copenh.) **7**, 109–116 (1954).

MCCLENDON, J. H.: The intracellular localization of enzymes in tobacco leaves. I. Identification of components of the homogenates. Amer. J. Bot. **39**, 275–282 (1952). — The intracellular localization of enzymes in tobacco leaves. II. Cytochrome oxidase, catalase, and polyphenol oxidase. Amer. J. Bot. **40**, 260–266 (1953). — MILLERD, A.: Succinoxidase of potato tuber. Proc. Linnean Soc. N. S. Wales **76**, 123–132 (1951). — Respiratory oxidation of pyruvate by plant mitochondria. Arch. of Biochem. a. Biophysics **42**, 149–163 (1953). — MILLERD, A., and J. BONNER: The biology of plant mitochondria. J. Histochem. a. Cytochem. **1**, 254–264 (1953). — Acetate activation and acetoacetate formation in plant systems. Arch. of Biochem. a. Biophysics **49**, 343–355 (1954). — MILLERD, A., J. BONNER, B. AXELROD and R. S. BANDURSKI: Oxidative and phosphorylative activity of plant mitochondria. Proc. Nat. Acad. Sci. U.S.A. **37**, 855–862 (1951). — MILLERD, A., J. BONNER and J. B BIALE: The climacteric rise in fruit respiration as controlled by phosphorylative coupling. Plant Physiol. **28**, 521–531 (1953). — MORTON, R. K.: Separation and purification of enzymes associated with insoluble particles. Nature (Lond.) **166**, 1092–1095 (1950).

NEWCOMB, E. H.: Effect of auxin on ascorbic oxidase activity in tobacco pith cells. Proc. Soc. Exper. Biol. a. Med. **76**, 504–509 (1951). — NEWCOMB, E. H., and P. K. STUMPF:

Fat metabolism in higher plants. I. Biogenisis of higher fatty acids by slices of peanut cotyledons in vitro. J. of Biol. Chem. **200**, 233–239 (1953).

OKUNUKI, K.: Über den Gaswechsel der Pollen. 3. Weitere Untersuchungen über die Dehydrasen aus den Pollenkörnern. Acta phytochim. (Tokyo) **11**, 65–80 (1939).

PEARSON, J. A., and R. N. ROBERTSON: The physiology of growth in apple fruits. VI. The control of respiration rate and synthesis. Austral. J. Biol. Sci. **7**, 1–17 (1954). — PRICE, C. A., and K. V. THIMANN: The estimation of dehydrogenases in plant tissue. Plant Physiol. **29**, 113–124 (1954).

ROBERTSON, R. N.: Mechanism of absorption and transport of inorganic nutrients in plants. Annual. Rev. Plant Physiol. **2**, 1–24 (1951). — ROBERTSON, R. N., M. J. WILKINS, A. B. HOPE and L. NESTEL: Studies in the metabolism of plant cells. X. Respiratory activity and ionic relations of plant mitochondria. Austral. J. Biol. Sci. **8**, 164–185 (1955).

SALTMAN, P.: Hexokinase in higher plants. J. of Biol. Chem. **200**, 145–154 (1953). — SCHNEIDER, W. C., and G. H. HOGEBOOM: Cytochemical studies of mammalian tissues. The isolation of cell components by differential centrifugation. A review. Cancer Res. **11**, 1–22 (1951). — SHARPENSTEEN, H., and E. E. CONN: Preparation and properties of potato mitochondria. Abstr. Amer. Soc. Plant Physiol., p. 37, 1954. — SMILLIE, R. M.: Enzymic activity of particles isolated from various tissues of the pea plant. Austral. J. Biol. Sci. **8**, 186–195 (1955). — STAFFORD, H. A.: Intracellular localization of enzymes in pea seedlings. Physiol. Plantarum (Copenh.) **4**, 696–741 (1951).

WAYGOOD, E. R.: Physiological and biochemical studies in plant metabolism. II. Respiratory enzymes in wheat. Canad. J. Res., Sect. C **28**, 7–62 (1950). — WEBSTER, G. C.: The occurrence of a cytochrome oxidase in the tissues of higher plants. Amer. J. Bot. **39**, 739–745 (1952). — Enzymatic synthesis of glutamine in higher plants. Plant Physiol. **28**, 724–727 (1953a). — Peptide bond synthesis in higher plants. I. The synthesis of glutathione. Arch. of Biochem. a. Biophysics **47**, 241–250 (1953b). — Enzymatic synthesis of gamma-glutamylcysteine in higher plants. Plant Physiol. **28**, 728–730 (1953c). — An energy dependent incorporation of amino acids into the protein of mitochondria. Plant Physiol. **29**, 202–203 (1954a). — Incorporation of radioactive glutamic acid into the proteins of higher plants. Plant Physiol. **29**, 382–385 (1954b). — Incorporation of radioactive amino acids into the protein of plant tissue homogenates. Plant Physiol. **1955**a. — Unpublished experiments. **1955**b. — WEBSTER, G. C., and J. E. VARNER: Peptide bond synthesis in higher plants. II. Studies on the mechanism of synthesis of γ-glutamylcysteine. Arch. of Biochem. a. Biophysics **52**, 22–32 (1954). — WEIER, T. E.: The cytology of leaf homogenates. Abstr. Amer. Soc. Plant Physiol., Western Sect., p. 11, 1952. — WEIER, T. E., and C. R. STOCKING: The chloroplast: structure, inheritance and enzymology. Bot. Review **18**, 14–75 (1952).

Hydration.

By

J. Levitt.

A. Meaning.

The significance of the term hydration depends on whether the quantity factor (water content), or the intensity factor (water activity) is involved. The former is usually expressed as a percent of the fresh weight, or sometimes of the dry weight. Another measure of the quantity factor is saturation deficit (LEVITT 1951). This gives a value for the water content relative to that in the saturated state, which is taken as 100. Water activity, on the other hand, is usually obtained from one of the colligative properties of the cell solution—*i.e.* its vapor pressure, osmotic pressure, or freezing point lowering. The first of these is measured indirectly, by determining the relative humidity in equilibrium with the cell. By analogy with temperature, WALTER (1931) has coined the term "hydrature" for the intensity factor. He expresses it quantitatively in atmospheres of osmotic pressure or in percent of relative humidity. It is essential to realize, however, that water activity is inversely related to osmotic pressure, directly related to relative humidity.

Of the two quantities, water activity is usually far more useful, because it can be easily measured osmotically even in a single cell, and because it reveals whether water will move into or out of the cell when the water activity of its environment is known. Water content, however, is the value that must be known if cell volume changes due to hydration or dehydration are to be estimated, for instance when dehydration injury is investigated.

B. Hydration of cell components.

The hydration of the cell, of course, depends on that of its environment, and may therefore be governed by many factors external to and beyond the control of the cell. As might be expected, cell hydration in most cases decreases with environmental hydration. In succulents, however, the reverse occurs—under dry conditions water content actually increases due to an increased succulence (WERK 1954). Within the cell itself, forces exist that markedly affect both its water content and water activity. The intracellular forces controlling hydration of the wall and protoplasm are mainly imbibitional; those controlling the hydration of the vacuole are mainly osmotic. Changes in hydration of each component frequently occur. In the wall, this may be due to the deposition of new particles having hydrophilic properties different from those of the older ones—*e.g.* by the addition of hemicelluloses to a cellulose wall (CLEMENTS 1937). In the vacuole, it may be due to 1. a change in salt content due to uptake or loss, 2. the accumulation or utilization of organic acids, or 3. a change in sugar content. Sugars may accumulate either due to photosynthesis or to starch hydrolysis. They may decrease due to respiration or starch synthesis. The most rapid and common changes in vacuole hydration result from these starch ⇌ sugar changes. Significant changes may occur within 2 hours in response to wilting (AHRENS 1924) or within minutes in guard cells in response to light (SCARTH 1932). In some exceptional cases, the hydration of the vacuole may be largely controlled by imbibitional forces (*e.g.* ILJIN 1931).

The importance of these forces becomes clear when the cell is subjected to dehydration. Due to the large size of the vacuole in most mature cells of higher plants, the largest part of the cell's water resides there. Dehydration is, therefore, most obvious in the reduced size of the vacuole. The quantity of dehydration — *i.e.* the amount of water removed by a specific dehydrating force—depends on the intracellular hydrating force. Frost dehydration at any one freezing temperature will therefore remove much more water from cells with low original sap concentration than from these with high sap concentration (table 1). At the same time as the vacuole is dehydrated, protoplasm also loses water, and sometimes even to the same degree (LEVITT and SCARTH 1936).

Table 1. *Relation between amount of ice formed at any one freezing temperature and original concentration of solution.* (From ÅKERMAN 1927.)

Temperature (°C)	Concentration of solution in equilibrium with ice (M)	Percent of water frozen in solution with the original concentration of:			
		0.25	0.5	1.0	2.0
		molar			
—0.46	0.25	trace			
—0.93	0.5	50	trace		
—1.86	1.0	75	50	trace	
—3.72	2.0	87.5	75	50	trace
—7.44	4.0	93.8	87.5	75	50
—14.88	8.0	96.9	93.8	87.5	75

Though internally produced changes in the hydration of protoplasm are well known (LEVITT 1941), the reasons for them are not so well understood. It is usually assumed that they are mainly due to changes in the number of hydrophilic polar groups available on the protoplasmic proteins (*e.g.* CHRISTOPHERSEN and PRECHT 1953), due to the breaking or formation of junctures between them. This may be caused by hydrolysis or condensation, or by changes in the ionic content of the protoplasm (SCHMIDT *et al.* 1940, BOGEN 1948), or perhaps even by sulfhydryl substances (EWART *et al.* 1953).

Due to differences in the nature of the hydrophilic forces, the water contents of the three cell components are not likely to be the same. As a rule, it is highest in the vacuole, lowest in the wall, though exceptions may occur (see above). But due to the very large specific surface of each component, and the high permeability of the cell to water, the water activity at any one time must be essentially identical in all three (LEVITT 1947, 1953). Nevertheless, it is not correct to assume that the water activity of the cell as a whole is identical with that of each component, if the component is removed and its hydration then measured (*e.g.* by freezing point determinations on the vacuole sap). This is due to the fact that if the living cell is turgid, the positive wall pressure raises the water activity, if it is flaccid, the negative wall pressure (or tension) lowers it. For this reason, in order to measure the activity of the cell's water, it is necessary to find a solution with the same water activity—*i.e.* one that the cell may be placed in without causing any water movement into or out of it.

Due to the equilibrium between the water activities of the cell components, any change in one results in a change in the others. Thus, an accumulation of sugars in the cell sap lowers the activity of its water. As a result, water quickly moves from the protoplasm and wall into the vacuole and equilibrium between the three is almost instantaneously achieved. When changes in hydration occur slowly, they may be more complex. LE CHATELIER's principle appears to apply. Any factor that tends to cause a change in the water content of the cell is opposed by a change in the activity of this water. Thus, cells of fresh water plants are exposed to conditions inducing maximum water content. But they have the lowest cell sap concentration and therefore absorb less water than would terrestrial plants under the same conditions. On the other hand, cells exposed to drought and frost are subjected to a water loss. Due to an increase in cell sap concentration, this water loss is not so great. Similarly, the very increase in cell sap concentration dehydrates the protoplasm. Perhaps as a result of this, the imbibitional forces in the protoplasm increase and less water is lost from it (LEVITT 1941). The best example of the action of this principle is the increase in "bound water" of *Aspergillus niger* with the concentration of the nutrient medium in which it is grown (table 2).

Apparent exceptions occur in the case of succulents. But the cells of these plants are not really exposed to drought. On the contrary, their cells are so well protected against water loss that excess water accumulates. It must be admitted, however, that the mechanisms by means of which these changes are brought about—*i.e.* the starch → sugar conversion, and the change in protoplasmic imbibition—are not understood.

Table 2. *Bound water in the mycelium of Aspergillus niger when grown in media of different concentrations.* (From TODD and LEVITT 1951.)

Osmotic potential of medium (atms.)	Bound water in mycelium (%)
7.7	7.8
15.5	6.8
30.6	7.7
52.7	14.1
83.7	26.5
132.2	28.5

It has long been known that ions affect the hydration of protoplasm. The best example is "vacuole contraction" or "cap plasmolysis" (STRUGGER 1949), both of which involve an increased protoplasmic hydration at the expense of the vacuole. Monovalent cations such as K^+ and Na^+ are particularly effective, as might be expected from their high hydration and relatively ready penetration into the protoplasm. According to BOGEN (1948), it is not simply the hydration of the protoplasm as a whole that is affected. Instead, new dipole moments are induced under the influence of the ion, and new hydration centers are created. In this way, the arrangement of the hydration centers in the protoplasm may be altered. In spite of this creation of new hydration centers, BOGEN believes that the total water content of the protoplasm may actually increase, decrease, or remain unchanged, depending on the conditions.

C. Extremes of hydration.

The degree of dehydration tolerated by plants varies markedly with the species, the plant part, and the state of the cells. Some plants or plant parts may survive drying over concentrated H_2SO_4 for weeks or even months. This is true, not only in the case of resting plant parts such as spores and seeds, but also in the case of seedlings, sporelings, mosses, and even some mature higher plants (RABE 1905, MAXIMOV 1929). No evidence, however, has been produced that any cell can survive complete dehydration, for about 5% water always remains (SCHRÖDER 1886). Only by raising its temperature above the boiling point can the last traces of water be removed (TODD and LEVITT 1951).

Most growing plants cannot survive such extreme dehydration as mentioned above. They are usually killed by a loss of 40–90% of their normal water content

(SCHRÖDER 1886), or when they come to equilibrium with relative humidities of 92–97% (ILJIN 1931). Survival of relative humidities as low as 88% is not uncommon; but only exceptional plants survive 85%. Different methods of dehydration all appear to be injurious to about the same degree (SCARTH 1941). The three main methods are by drought, frost, and osmotic water removal.

The ability of a plant to survive dehydration by any one of these methods is called its hardiness. The term resistance is sometimes also used in this sense, but it is ambiguous, since an ability to prevent rather than survive dehydration is implied. The hardiness of a plant can be increased in several ways. Repeated wilting has been reported to produce a marked increase (TUMANOV 1927). Exposure to low temperature has the same effect (LEVITT 1941). Even simply starving the plant for nutrients such as nitrogen may be effective (LEVITT 1951). Hardiness to all forms of dehydration increases simultaneously (SIMINOVITCH and BRIGGS 1953). Even hardiness to non-dehydrating injurious agents may be induced in the same way—*e.g.* hardiness to heat (SAPPER 1935).

Table 3. *Degree of hydration survived by different plants.* (From ILJIN 1934.)

Species	Habitat	Osmotic value (moles sucrose)	Conc. difference required to kill cells (moles)
Sempervivum soboliferum	rock	0.22	0.35–0.40
Sedum purpureum	rock	0.20	0.35–0.40
Bidens tripartitis	swamp	0.28	0.6–0.7
Diplotaxis tenuifolia	field	0.44	0.9–1.0
Cirsium canum	meadow	0.36–0.43	0.9–1.1
Ranunculus repens	meadow	0.45	1.8–2.0
Anthericum ramosum	steppe	0.5–0.6	1.5–2.3
Centaurea scabiosa	steppe	0.55	1.5–2.1
Lycium barbatum	steppe	—	2.1–2.2
Poa badensis	steppe	0.49–0.62	2.7–3.3
Festuca glauca	rock	0.6	3.0–3.2
Bupleurum falcatum	rock	0.85	3.2
Buxus sempervirens	garden	0.8	4.5–5.5

The hydration of the cell changes markedly during this hardening period. A decrease in total water content occurs (LEVITT 1941), as well as in activity of the remaining water. The latter may be produced by an accumulation of carbohydrates in winter annuals due to an excess of photosynthesis over respiration (ANDERSSON 1944). In perennials, however, this excess photosynthate usually accumulates during summer in the form of starch, which is converted to sugars during the hardening period (LEVITT 1941). The increased cell sap concentration dehydrates the protoplasm, leading to an increased protoplasmic viscosity (LEVITT 1951). But the most important changes occur in the protoplasm itself. Its increased hydrophily on hardening can be shown in many ways—by the lower refractive index of the cytoplasmic strands, making them more difficult to see, by the inability of osmotic dehydration to make them stiff enough to break, etc.

Though the above changes have been most intensively investigated in relation to frost hardening, those that have been observed in relation to other methods of hardening have shown similar responses (LEVITT 1951). LUNDEGÅRDH (1914) showed that even osmotic dehydration produces the same hydrolysis of starch to sugar as does wilting. Osmotic rehydration led to the reformation of starch. ILJIN (1930) found that the degree of osmotic dehydration needed to produce the starch → sugar conversion in leaves floated on solutions differs with the plant. Aquatic plants show it at the lowest concentrations, rock plants require the highest.

Extremes of hydration have not received as much attention as extremes of dehydration. ILJIN (1934) increased the cell sap concentration of different plants by allowing them to take up large quantities of glycerine. He then exposed them to different degrees of hydration by transferring them to hypotonic solutions or to water. The degree of hydration survived, varied markedly with the plant (table 3).

D. Hydration and metabolism.

Besides its effect on hardiness, hydration may also control the metabolic rate of the plant. As a general (though not inviolable) principle, high cell hydration is associated with high metabolic rate and low hardiness; low cell hydration with low metabolic rate and high hardiness. In applying this principle, it is necessary to distinguish between rapidly produced and slowly produced changes in hydration. Only a slow dehydration is, in general, accompanied by increased hardiness. Conversely, a rapid hydration may not, at first, be accompanied by an increased metabolic rate.

This concept has been frequently reiterated by various investigators, in one form or another, usually on the basis of the forces with which the plant's water is held. Most recently, CHRISTOPHERSEN and PRECHT (1953) have reemphasized the importance of free and bound water, the latter being due to the so-called hydration coats of the proteins. The boundary between the two is, of course, changeable according to the dehydrating force the cell is subjected to. The free water is important in metabolism because of its ability to act as a solvent. The bound water is important in hardiness because of its ability to prevent denaturation of the proteins (see GORTNER 1938). But CHRISTOPHERSEN and PRECHT point out that free water and solvent space are not strictly equal. The solvent space is somewhat larger, since some soluble substances penetrate the hydration coat, the degree of penetration depending on the molecular size. The relation of solvent to non-solvent space is decisive for the metabolic rate, since chemical reactions depend on the transport of substances in the water. An increase in solvent space or a reduction in non-solvent space favor metabolism and vice versa. Since solvent space and free water, though not identical, are both altered in the same way by the same factors, either quantity may be related to metabolic rate.

From this simplified concept, one might expect all metabolic processes to be affected by dehydration to the same extent. But complications occur. According to ILJIN (1923), the first effect of water reduction in the leaves is stomatal closure. This markedly slows up the movement of carbon dioxide into assimilating leaves, reducing the photosynthetic rate 2–10×, according to the amount of water removal and the sensitivity of the plant. Others, however, have shown a marked reduction in assimilation rate even in the absence of stomatal closure (WALTER 1929). In very sensitive plants, such as *Helodea canadensis*, the rate is markedly reduced at relatively slight degrees of dehydration. In sucrose solutions above 0.5 M, CO_2 is actually evolved in full sunlight (WALTER 1929). This is obviously not due to lack of free water in the plant.

Respiration is also affected. According to SMITH (1915), respiration increases gradually, up to a 30% loss of water. This maximum rate is maintained up to a loss of 50–60% of the plant's water. Any further loss is accompanied by a progressive decrease in respiration rate. Others have obtained similar results (*e.g.* DOMIEN 1949). Obviously, dehydration markedly affects respiration long before solvent space is an important factor, and in the opposite direction to what would be expected. It is apparently only when 50–60% of the plant's water is removed that solvent space becomes a factor. The lack of a uniform effect of dehydration on all phases of metabolism is shown most clearly in the case of *Helodea canadensis* (WALTER 1929). No change in respiration rate occurs at a degree of dehydration that is sufficient to stop photosynthesis completely.

This discrepancy is not surprising, since photosynthesis takes place in the chloroplasts, respiration in the mitochondria. The responses of these two processes to dehydration must, therefore, depend not solely on the "free solvent space" of the protoplasm, but also on the sensitivity of these two protoplasmic structures to dehydration injury. Chloroplasts are the first structures to show heat injury (DÖRING 1933), and may therefore perhaps be more sensitive to dehydration, too. This would explain WALTER's results.

These direct effects of water loss on assimilation and respiration rates may not be the same as the indirect effects of long-continued growth under conditions of partial dehydration. SIMONIS (1952) showed that this results in a marked increase in rate of photosynthesis, though respiration rate sometimes increased and sometimes decreased.

Perhaps the most striking case of a dehydration induced metabolic change that appears to occur in the wrong direction, is the starch → sugar hydrolysis. From the law of mass action, increased hydration would be expected to displace the equilibrium in the direction of hydrolysis. In actual fact, the reverse occurs. The metabolic response to dehydration varies with the plant part. MOTHES (1928) showed that asparagine or glutamine moved from the lower leaves of slightly wilted plants to the upper ones, resulting in protein deficiency and ultimate death of the lower leaves. As in the cases of the starch → sugar change, the hydrolysis of the proteins is in the wrong direction from the point of view of mass action. MOTHES suggests that the increased concentration of the enzymes in the dehydrated cells may be responsible for the hydrolysis. In favor of this explanation, osmotic dehydration of detached leaves produced the same result. According to him, wilting causes proteolysis, the amino acids activate diastase, and this produces an increase in soluble carbohydrates leading to increased respiration. But this does not agree with his statement that starch hydrolysis occurs before proteolysis. All these metabolic responses to hydration changes indicate that any concept of the cell as a simple solution in which chemical reactions occur is a gross oversimplification. There are, undoubtedly, many other effects on protoplasmic structure as well as on activity of enzymes that must play a significant role. CHRISTOPHERSEN and PRECHT (1952), themselves give evidence of this.

Literature.

AHRENS, W.: Weitere Untersuchungen über die Abhängigkeit des gegenseitigen Mengenverhältnisses der Kohlenhydrate im Laubblatt vom Wassergehalt. Bot. Archiv **5**, 234–259 (1924). — ÅKERMAN, Å.: Studien über den Kältetod und die Kälteresistenz der Pflanzen, S. 1–232. Lund 1927. — ANDERSSON, GÖSTA: Gas change and frost hardening studies in winter cereals, S. 5–163. Lund 1944.

BOGEN, HANS J.: Untersuchungen über Hitzetod und Hitzeresistenz pflanzlicher Protoplasten. Planta (Berl.) **36**, 298–340 (1948).

CHRISTOPHERSEN, J., u. H. PRECHT: Untersuchungen zum Problem der Hitzeresistenz. II. Untersuchungen an Hefezellen. Biol. Zbl. **71**, 585–601 (1952). — Die Bedeutung des Wassergehaltes der Zelle für Temperaturanpassungen. Biol. Zbl. **72**, 104–119 (1953). — CLEMENTS, H. F.: Studies in drought resistance of the soy bean. Res. Stud. State Coll. Wash. **5**, 1–16 (1937).

DÖRING, H.: Beiträge zur Frage der Hitzeresistenz pflanzlicher Zellen. Planta (Berl.) **18**, 405–434 (1933). — DOMIEN, F.: Influence de la déshydratation sur la respiration des feuilles de végétaux aériens. Rev. Bot. **56**, 285–317 (1949).

EWART, M. H., D. SIMINOVITCH and D. R. BRIGGS: Studies on the chemistry of the living bark of the black locust tree in relation to frost hardiness. VI. Amylase and phosphorylase systems of the bark tissues. Plant Physiol. **28**, 629–644 (1953).

GORTNER, R. A.: Outlines of biochemistry. New York: John Wiley & Sons 1938.

ILJIN, W. S.: Der Einfluß des Wassermangels auf die Kohlenstoffassimilation durch die Pflanzen. Flora (Jena) **116**, 360–378 (1923). — Der Einfluß des Welkens auf den Ab- und Aufbau der Stärke in der Pflanze. Planta (Berl.) **10**, 170–184 (1930). — Austrocknungsresistenz des Farnes *Notochlaena Marantae* R. Br. Protoplasma (Berl.) **13**, 322–330 (1931). — Über Absterben der Pflanzengewebe durch Austrocknung und über ihre Bewahrung vor dem Trockentode. Protoplasma (Berl.) **19**, 414–442 (1933). — Kann das Protoplasma durch den osmotischen Druck des Zellsaftes zerdrückt werden? Protoplasma (Berl.) **20**, 570–585 (1934).

LEVITT, J.: Frost killing and hardiness of plants. Minneapolis: Burgess Publ. Co. 1941. — The thermodynamics of active (non-osmotic) water absorption. Plant Physiol. **22**, 514–525 (1947). — Frost, drought, and heat resistance. Ann. Rev. Plant Physiol. **2**, 245–268 (1951). — Further remarks on the thermodynamics of active (non-osmotic) water absorption. Physiol. Plantarum (Copenh.) **6**, 240–252 (1953). — LEVITT, J., and G. W. SCARTH: Frost-hardening studies with living cells. Canad. J. Res., Sect. C **14**, 267–305 (1936). — LUNDEGÅRDH, H.: Einige Bedingungen der Bildung und Auflösung der Stärke. Pringsh. Jb. **53**, 421–463 (1914).

MAXIMOV, N. A.: The plant in relation to water. London: Allen & Unwin 1929. — MOTHES, K.: Die Wirkung des Wassermangels auf den Eiweißumsatz in höheren Pflanzen. Ber. dtsch. bot. Ges. **46**, (59)–(67) (1928).

RABE, FRANZ: Über die Austrocknungsfähigkeit gekeimter Samen und Sporen. Flora (Jena) **95**, 253–324 (1905).

SAPPER, I.: Versuche zur Hitzeresistenz der Pflanzen. Planta (Berl.) **23**, 518–556 (1935). — SCARTH, G. W.: Mechanism of the action of light and other factors on stomatal movement. Plant Physiol. **7**, 481–504 (1932). — Dehydration injury and resistance. Plant Physiol. **16**, 171–179 (1941). — SCHMIDT, H., K. DEWALD u. O. STOCKER: Plasmatische Untersuchungen. an dürreempfindlichen und dürreresistenten Sorten landwirtschaftlicher Kulturpflanzen Planta (Berl.) **31**, 559–596 (1940). — SCHRÖDER, G.: Über die Austrocknungsfähigkeit der Pflanzen. Inaug.-Diss. Tübingen 1886. S. 1–51. — SIMINOVITCH, D., and D. R. BRIGGS: Studies on the chemistry of the living bark of the black locust in relation to its frost hardiness. III. The validity of plasmolysis and desiccation tests for determining the frost hardiness of bark tissue. Plant Physiol. **28**, 15–34 (1953). — SIMONIS, W.: Untersuchungen zum Dürreeffekt. I. Morphologische Struktur. Wasserhaushalt, Atmung und Photosynthese feucht und trocken gezogener Pflanzen. Planta (Berl.) **40**, 313–332 (1952). — SMITH, A., and J. MALINS: The respiration of partly dried plant organs. Rep. Brit. Assoc. Adv. Sci. **1915**, 725. — STRUGGER, S.: Praktikum der Zell- und Gewebephysiologie der Pflanze. Berlin 1949.

TODD, GLENN W., and J. LEVITT: Bound water in *Aspergillus niger*. Plant Physiol. **26**, 331–336 (1951). — TUMANOV, I. I.: Ungenügende Wasserversorgung und das Welken der Pflanzen als Mittel zur Erhöhung ihrer Dürreresistenz. Planta (Berl.) **3**, 391–480 (1927).

WALTER, H.: Plasmaquellung und Assimilation. Protoplasma (Berl.) **6**, 113–156 (1929). — Hydratur der Pflanze und ihre physiologisch-ökologische Bedeutung. Jena 1931. — WERK, O.: Untersuchungen zum Dürreeffekt. 2. Über den Kalium- und Calciumgehalt feucht und trocken gezogener Pflanzen. Flora (Jena) **141**, 312–355 (1954).

Viscosität.

Von

M. G. Stålfelt.

Unter Viscosität versteht man die innere Reibung eines fließenden Körpers. Die Reibung setzt sich zusammen aus der eigenen Reibung des Körpers sowie der Reibung von dispergierten Partikeln, gelösten wie suspendierten.

Bei gewöhnlichen molekulardispersen Lösungen und reinen Flüssigkeiten steigt die Viscosität mit zunehmendem Molekulargewicht (über Ausnahmen von der Regel s. z. B. PHILIPOFF 1942, S. 337). Suspendierte Kolloidpartikel üben prinzipiell die gleiche Wirkung aus: jedes Teilchen bietet einen gewissen Widerstand, stört also die Strömungsbewegung. Der Widerstand steigt proportional der Partikelanzahl, solange die Partikel nicht aufeinander einwirken. Aber oberhalb einer bestimmten Grenzkonzentration stoßen die Partikel zusammen und stören sich gegenseitig in ihren Bewegungen, was eine Vergrößerung des Strömungswiderstandes verursacht, so daß dieser jetzt schneller zunimmt als die Konzentration. Die Wirkung bleibt dieselbe, wenn die Moleküle in einen gewissen Kontakt miteinander treten, z. B. indem die Kettenmoleküle sich ineinander verhaken (PHILIPOFF 1942, S. 160).

Solange die suspendierten oder gelösten Kolloidpartikel sich gegenseitig nicht stören, ist die Viscosität der Lösung oder der Suspension konstant. Diese unterliegt dann dem HAGEN-POISEUILLEschen Gesetz für eine durch eine Capillare fließende Flüssigkeit, d. h. die Viscosität ist unabhängig von dem Druck, welcher die Flüssigkeit vorwärts treibt. Treten die Partikel jedoch miteinander in Verbindung und werden sie dadurch zu einer bestimmten Orientierung gezwungen, so wird die Viscosität vom Druck abhängig — diese wird geringer, wenn der Druck zunimmt —, und die Flüssigkeit folgt nicht mehr dem vorgenannten Gesetz. Da die Abweichung als Folge einer mehr oder weniger festen Orientierung der Partikel im Raum entstanden ist, d. h. auf Grund des Vorhandenseins einer gewissen Struktur, wurde diese Form von Viscosität von OSTWALD (1925) als Strukturviscosität bezeichnet (s. FREY-WYSSLING 1953, S. 64 und SEIFRIZ 1952, S. 28). SEIFRIZ (1952, S. 9) betont, daß Viscosität nicht mit anderen nahestehenden physikalischen Eigenschaften verwechselt werden darf, welche auch für das Plasma charakteristisch sind, wie Kleistrigkeit (glutinosity), Klebrigkeit (takkiness), Elastizität oder Zerreißfestigkeit (tensile strength).

Daß das Protoplasma Strukturviscosität besitzt, ist unter anderen von PFEIFFER gezeigt worden (1930). Er bestimmte die Viscosität eines Zellplasmas, welches durch ein Rohr gesaugt wurde, und stellte fest, daß die Viscosität abnahm mit Zunahme des Druckes.

Viele Faktoren tragen zur Entstehung der Strukturviscosität bei. SCHMIDT, DIWALD und STOCKER (1940) haben über bisher Bekanntes eine Übersicht gegeben, wobei sie von der Annahme ausgingen, daß das Plasma aus zwei Phasen aufgebaut ist: einem festen, aus Polypeptidketten der Eiweißmoleküle zusammengesetzten Gerüst und einer die Maschenräume ausfüllenden Interstitialflüssigkeit, die zum großen Teil aus Wasser mit gelösten Stoffen, zum kleineren aus Lipoiden besteht. SEIFRIZ (1952, S. 110) charakterisiert die Interstitialflüssigkeit als eine aus globulären Nahrungsproteinen zusammengesetzte Lösung.

Die folgende Übersicht wird hauptsächlich nach SCHMIDT, DIWALD und STOCKER wiedergegeben.

Die Viscosität des Cytoplasmas dürfte von 2 Hauptkomponenten abhängig sein, und zwar der inneren Reibung der Zwischenflüssigkeit und dem Reibungswiderstand des Gerüstes.

1. Für die innere Reibung der Zwischenflüssigkeit sind bestimmend:

a) Die Konzentration der Lösung, die ihrerseits nicht nur vom Verhältnis Wasser/gelöste Substanz, sondern auch von der Hydratur des Plasmas und der Hydratation seiner Bestandteile abhängt; die Hydratur ändert sich mit Trockenheit, Bewässerung und durch Änderung der Hydratation des Gerüstes und der Kolloide.

b) Die Zusammensetzung der dispergierten Teile, die sowohl durch Zusammenfügung als Aufspaltung der Moleküle verändert werden kann. Eine Änderung wird besonders wirksam beim Auftreten großer und stark hydratisierter Moleküle z. B. von Zucker und von aus dem Gerüst abgesprengten Teilen.

2. Der Reibungswiderstand des Gerüstes ist vor allem von der Anordnung der Kettenmoleküle und der Zahl und Festigkeit der Bindungen abhängig. Darauf wirkt außerdem sowohl Hydratation der Fäden (weil durch die Wasserhülle der freie Maschenraum verengt wird) als auch Ausstreckung bzw. Zusammenrollung der Kolloidmoleküle, wodurch auch ihre wasserbindende Oberfläche und ihr Volumen erheblich verändert wird (BOGEN 1948, S. 318, 1950).

Meßmethoden.

Für eine physikalisch definierbare Flüssigkeit kann — mit Hilfe der STOKESschen Formel — die Viscosität gemessen und berechnet werden (s. z. B. STARLING-WOODALL 1952). Laut dieser ist der Widerstand (F) gegen einen sphärischen Körper, welcher sich langsam in einer Flüssigkeit bewegt (deren äußere Grenze auf den Körper nicht einwirkt), gleich dem Produkt $6\pi\eta \cdot av$, wobei a = den Radius der Kugel, v = ihre Geschwindigkeit, η = die Viscosität der Flüssigkeit bedeuten.

Für einen fallenden Körper heißt die Formel:

$$\eta = \frac{2a^2(\delta - p)}{9v} g$$

(δ = Dichte des Körpers; p = Dichte der Flüssigkeit; g = Beschleunigung auf Grund der Gravitation).

Der Gebrauch dieser Methode (*Fallmethode*) für die Messung der Viscosität des Zellplasmas wird dadurch erschwert, daß freibewegliche sphärische Körper den Zellen fehlen. In der Wurzelhaube und in der Stärkescheide der Stengel gewisser Pflanzen gibt es aber bewegliche Stärkekörner (Statolithenstärke), die durch die Zellen fallen können, wenn die Lage des Objekts verändert wird. Durch Messungen solcher Bewegungen hat HEILBRONN (1914) den relativen Viscositätswert berechnet, der wertvolle Aufschlüsse über den Zustand des Plasmas ergab. HEILBRONN (1923) benutzte für denselben Zweck auch Eisenfeilspäne, die er in das Plasma von Myxomyceten einführte. Ein solches Verfahren ist aber nur möglich, wenn die Zellen keine feste Membran haben (übrige Literatur s. SEIFRIZ 1952).

Eine Modifikation der Fallmethode ist die *Zentrifugierungsmethode*. Wenn man Zellen einer Zentrifugierung aussetzt, werden gewisse feste, halbflüssige und flüssige Körper, z. B. Chloroplasten, in Bewegung gesetzt. Indem man deren Bewegungen mißt und diese mit der angewandten Zentrifugalkraft in Beziehung setzt oder die Formel von STOKES anwendet, kann man den relativen Wert der Plasmaviscosität bestimmen. Die Anwendbarkeit dieser Methode ist jedoch durch die nachstehenden Mängel begrenzt:

Das spezifische Gewicht des Plasmas und der zentrifugierten Körper konnte nur in Ausnahmefällen gemessen werden (HEILBRONN 1952, S. 81). Es kann auch nicht als konstant betrachtet werden, z. B. bei Chloroplasten infolge des Vorkommens von Stärke. Wird das Objekt eine Zeitlang vor dem Versuch im Dunkeln gehalten, so daß die Chloroplastenstärke verschwindet, so kann dieser Fehler allerdings ausgeschaltet werden (MODER 1932, KESSLER 1935).

Die Teilchen besitzen keine Kugelform, und ihr Volumen kann nicht exakt bestimmt werden. Jedoch variiert das Volumen z. B. von Chloroplasten bei den einzelnen Arten nur in geringen Grenzen, wie dies Messungen der Durchmesser von Chloroplasten gezeigt haben (SCHRATZ 1927, SCHMIDT 1939, STÅLFELT 1954).

Eine wiederholte und starke Zentrifugierung und vor allem eine wiederholte Umkehr der Schleuderrichtung (LANZ 1939, KÜSTER 1939) ändert die Plasmaviscosität — das Plasma ist thixotropisch —, jedoch nicht eine schwache und kurzdauernde Zentrifugierung (zur Literatur s. RUGE 1940, S. 342, BRECKHEIMER-BEYRICH 1949).

Die Plasmolysemethode.

WEBER (1924, 1925) deutete den Wechsel des Plasmolysebildes als Ausdruck der verschiedenen Viscosität des Plasmas: konvexe Konturen als Zeichen einer geringen, konkave als Zeichen einer hohen Plasmaviscosität. Neben dieser „Plasmolyseformmethode“ entwickelte er später seine sog. „Plasmolysezeitmethode“, die auf den Unterschieden der Plasmolysezeit beruht: eine langsame Plasmolyse wird als Zeichen einer hohen Viscosität, Adhäsion und Klebrigkeit betrachtet. WEBER (1929a, b) verstand dabei als Plasmolysezeit diejenige Zeitdauer, „die verstreicht vom Moment des Einlegens der Zelle in das Plasmolyticum bis zur Erreichung der konvexen Plasmolyseform“ (1929, S. 623).

Wie WEBER (1925) betont, wird das Plasmolysebild und die Plasmolysezeit nicht nur durch die Viscosität, sondern auch durch Adhäsion und Klebrigkeit bedingt. In der Folge haben einige Forscher (Zusammenstellung bei RUGE 1940, S. 327, FISCHER 1948) auf die Kohäsion, Oberflächenspannung, Intrabilität und Permeabilität des Protoplasmas hingewiesen und hervorgehoben, daß der Plasmolyseverlauf durch diese Faktoren beeinflußt wird, und daß die Plasmolysezeit auch vom Diffusionswiderstand im Gewebe, dem osmotischen Wert des Zellsaftes, der Konzentration des Plasmolyticums und der elastischen Dehnung der Membranen abhängig ist.

Die Anwendbarkeit der Methode beruht mithin auf der relativen Bedeutung der Viscosität für die Entstehung des Plasmolysebildes. Die Anwendbarkeit wird ferner durch die Schäden beschränkt, die die Plasmolyse hervorruft, wie dies bereits WEBER (1932) hervorhob. Je schwerer ein Protoplast sich plasmolytisch von der Zellwand löst, desto stärker schädigt die Plasmolyse. HÖFLER (1936, S. 384) bestätigt dies und bemerkt, daß die Protoplasten manchmal so fest an den Wänden haften, daß da und dort eher Rißwunden eintreten, als daß die Loslösung gelingt. Ferner wird die Hydratur durch die Plasmolyse geändert. Bei der Dehydrierung wird das Ektoplasma verdichtet und gelartig (SCARTH 1941, SIMINOVITCH 1941, JUNGERS 1934). Die Wirkung hängt aber auch von der Natur des Plasmolyticums ab (TIMMEL 1927).

PEKAREK (1931) verwandte die BROWN*sche Molekularbewegung* zur Messung der Viscosität des Zellplasmas (betr. Methodik und Berechnung s. STRUGGER

1949). Bei dieser Bewegung ist die Geschwindigkeit bedingt durch die Teilchengröße, die Viscosität der Flüssigkeit und die Temperatur. Die Methode hat, besonders bei tierischen Objekten, wertvolle Resultate ergeben (s. z. B. HEILBRUNN 1952, S. 84). In der Pflanzenphysiologie ist sie jedoch in der von PEKAREK angegebenen Form nur selten zur Anwendung gekommen. Die Erklärung ist wohl darin zu suchen, daß die pflanzlichen Objekte nur selten die Bedingungen erfüllen, die die Methode erfordert. SCARTH und LEVITT (1932, S. 68) fanden z. B., daß die Bewegungsamplitude bei bestimmten Objekten zu gering war, um gemessen werden zu können. Bei den Untersuchungen von KESSLER (1935) sowie KESSLER und RUHLAND (1938) wurde die Erfahrung gemacht, daß der Plasmabelag der Zellen des verwendeten Pflanzenmaterials zu dünn war, um eine freie BROWNsche Bewegung der noch sichtbaren Microsomen zu gestatten. Die gleiche Erfahrung machte HIRAOKA (1951, S. 172). KATO (1933) maß dagegen die Amplitude der mittelgroßen Microsomen (standard granules) und wandte den ermittelten Wert als Ausdruck der Viscosität an. Ältere Untersuchungen nach dieser Methode werden von KATO (S. 201) angeführt.

Vergleich der Methoden.

Von den genannten Methoden scheint diejenige von PEKAREK die exakteste zu sein, obgleich sie zum Teil auf Schätzungen beruht, insbesondere bei der Messung von Teilchengrößen (PEKAREK 1931, S. 647). Andererseits ist ihre Anwendbarkeit begrenzt, sie erfaßt nur kleine, lokale Bezirke des Zellplasmas und scheint nur in Ausnahmefällen für pflanzliche Objekte geeignet zu sein. Die beiden anderen Methoden haben größere Anwendungsmöglichkeit, sind jedoch mit einigen Unsicherheitsmomenten behaftet. Bei den Plasmolysenmethoden kommen hauptsächlich die Eigenschaften des Ektoplasmas zum Ausdruck (WEBER 1924, 1925, KESSLER 1935). Im Gegensatz hierzu dürfte bei den durch die Zentrifugierungsmethode bedingten Bewegungen das gesamte Plasma oder dessen größter Teil herangezogen werden. Diese Methode ist auch empfindlicher als die Plasmolysemethoden (RUGE 1940, STÅLFELT 1954), läßt sich aber nur bei solchen Objekten anwenden, die zentrifugierbare Teilchen in den Zellen enthalten. Die Plasmolysemethoden können bei den meisten pflanzlichen Objekten verwendet werden.

Trotz der vielen theoretisch denkbaren Unsicherheitsquellen der Plasmolyse- und Zentrifugierungsmethoden haben sie — wie viele Untersuchungen zeigten — im großen und ganzen übereinstimmende Resultate ergeben (WEBER 1924, SCHMIDT 1939, SCHMIDT, DIWALD und STOCKER 1940, S. 572; Literatur s. auch RUGE 1940, S. 342). Mangelnde Übereinstimmung fand TIMMEL (1927) für eine seiner Versuchspflanzen *(Caltha)*, es scheint aber, daß er ein ungeeignetes Plasmolyticum (KNO_3) benutzt hat. Im übrigen deuten die gleichlautenden Resultate darauf, daß unter allen besprochenen Faktoren, die das Meßergebnis beeinflussen, der Viscosität eine dominierende Rolle zukommt.

Die Plasmaviscosität kann innerhalb weiter Grenzen schwanken. Gewöhnlich etwa 20mal so hoch wie die des Wassers, nur selten niedriger als der obige Wert, kann sie in bestimmten Fällen bedeutend höhere Werte annehmen, bis zum 800fachen Wert des Wassers (Literatur s. SEIFRIZ 1952, S. 9, FREY-WYSSLING 1953, S. 169).

Die Werte der Plasmaviscosität sind verschieden, sowohl innerhalb der einzelnen Zelle als auch in den Zellen eines oder verschiedener Organe. Laut SCARTH (1941) und SIMINOVITCH (1941) kommen solche Unterschiede bei der durch die Plasmolyse bewirkten Dehydratisierung des Plasmas zum Ausdruck. Hierbei erleidet das Ektoplasma eine größere Veränderung als das Endoplasma: das erstere wird viscoser und schließlich gelartig. Es ist möglich, daß derartige Unterschiede bereits bei turgescenten Zellen vorhanden sind. Auch im Zellinneren ist die Viscosität ungleich. KATO (1933) maß die höchsten Werte in der Nähe des Zellkerns und fand eine Abnahme gegen die Peripherie hin. Im Zellsaft war die Viscosität niedriger als im Plasma. Verschiedene Werte in verschiedenen Teilen der Zelle wurden auch von SCARTH (1942) und HIRAOKA (1951) nachgewiesen. Auf Grund derartiger lokaler Differenzen können Messungen mittels der BROWNschen Bewegungsmethode andere Werte ergeben als beispielsweise solche, die nach der Plasmolyseformmethode ermittelt wurden. SCARTH und LEVITT (1927), die beide Methoden anwandten, konnten zuweilen Viscositätsunterschiede bei Objekten nachweisen, wenn sie die Plasmolysemethode anwandten, nicht aber, wenn sie die BMB-Methode benutzten. Von der Hypothese von den 2 Phasen des Plasmas betrachtet, deutet ein solches Resultat auf Viscositätsunterschiede in dem festeren Gerüstteil, aber nicht in der Interstitialsubstanz.

In einigen Fällen wurden Vergleiche zwischen den Zellen eines Organs angestellt. An Algen mit flächigem Thallus fand BIEBL (1932), daß die Viscosität der Randzellen höher war als die der zentralen Zone. Blätter zeigen auch bestimmte Unterschiede. Bei den höheren Pflanzen ist die Viscosität gewöhnlich an der Blattbasis am niedrigsten, sie steigt nach der Spitze zu und steigt ferner von der Mittelrippe nach den Blattkanten hin (REUTER 1946, STÅLFELT 1954). Bei *Bryum capillare* fand MENDER (1938) einen Gradienten in umgekehrter Richtung, indem die Viscosität bei den basalen Zellen am höchsten war.

Entwicklung und Insertion finden ihren Ausdruck in Viscositätsunterschieden. Die Viscosität ist am niedrigsten bei dem vollentwickelten und noch nicht beschädigten oder zerstörten Blatt und steigt sowohl in der Richtung nach den jungen, noch unentwickelten Blättern an der Sproßspitze, als nach solchen Basalblättern, die durch Insekten oder Pilze beschädigt wurden oder andere Zeichen einer Zerstörung aufwiesen (STÅLFELT 1954). Mit dem Entwicklungszustand ändert sich auch die chemische Resistenz der pflanzlichen Plasmen (BIEBL und ROSSI-PILLHOFER (1954).

Über die Größenordnung der Viscosität bei den verschiedenen Arten gibt es nur wenige Angaben. CHOLODNY (1924) gibt an, daß die Landpflanzen eine höhere Viscosität besitzen als Wasserpflanzen. Laut HENKEL (1949) haben Succulenten eine höhere Viscosität als andere Xerophyten.

Der Zusammenhang zwischen Blattinsertion und Viscosität ist zum Teil von der gleichen Art wie derjenige zwischen Plasmaviscosität und dem Alter der Zellen bzw. der Organe. FISCHER (1950) faßte die Resultate der zahlreichen Untersuchungen, die über diesen Gegenstand vorliegen, zusammen und zeigte, daß die Abhängigkeit der Viscosität mit dem Alter des Organismus variiert. Vergleiche zwischen Viscosität und dem K/Ca-Quotienten, der mit dem Alter des Blattes kleiner wird, haben ebenfalls kein eindeutiges Resultat ergeben (FISCHER 1949).

Einige Forscher stellten die Resistenz der Pflanzen gegen Kälte und Trockenheit sowie die Veränderungen, die dabei hervorgerufen werden, in Zusammenhang mit Änderungen der Plasmaviscosität sowie der Faktoren, von denen die Viscosität abhängt.

KESSLER (1935) und KESSLER und RUHLAND (1938) verglichen den Plasmazustand von frostbeständigen und frostempfindlichen Pflanzen (*Sempervivum*, *Saxifraga* und andere Arten) und fanden eine Korrelation zwischen Kälteresistenz und Plasmaviscosität. Im ruhenden Zustand sind die Pflanzen mehr kälteresistent, ihr Plasma ist dann auch stärker viscos, als im treibenden Zustand. Mit beginnender Entwicklung sinkt die Resistenz sofort, und die Viscosität wird vermindert. Bei den Messungen wurden sowohl die Plasmolysemethode als auch die Zentrifugierungsmethode verwandt; die Resultate waren übereinstimmend. Ungefähr gleichzeitig wurden ähnliche Versuche von SCARTH und LEVITT (1937) und SIMINOVITCH und LEVITT (1941) ausgeführt. Deren Resultate waren jedoch entgegengesetzter Art als bei den erstgenannten Forschern. Bei der Plasmolyse nahm das Plasma bei den resistenten *(Catalpa)* eine konvexe, bei den empfindlichen dagegen eine konkave Form an. KESSLER und RUHLAND (1938) nehmen an, daß der Widerspruch im Meßverfahren liegt, indem die Unterschiede zwischen den Zellen und den hypertonischen Lösungen in den beiden Fällen nicht die gleichen waren. Jedoch ist es denkbar, daß die Ursache in der komplizierten Art des Phänomens selbst liegt. Die Kälteresistenz ist nämlich mit verschiedenen Plasmaeigenschaften gekoppelt — SCARTH und LEVITT (1932) betonen besonders den osmotischen Wert, die Permeabilität und die Menge der Kolloide —, welche auf die Viscosität auf verschiedene Art einwirken. Einen direkten Zusammenhang zwischen Resistenz und Viscosität sehen jedoch weder die genannten deutschen, noch die kanadischen Forscher.

Bei *Avena*-Coleoptilen, die einige Tage lang der Trockenheit ausgesetzt wurden, ist die Beweglichkeit der Chloroplasten bei Zentrifugierung geringer als bei gewässerten Pflanzen, wie dies die Untersuchungen von NORTHEN (1938) zeigten. SCHMIDT (1939) studierte den Zusammenhang zwischen Plasmaviscosität und Trockenresistenz und fand, daß Pflanzen *(Lamium maculatum)* in Trockenkultur ein viscoseres Plasma hatten, als Pflanzen in Feuchtkultur. Später haben SCHMIDT, DIWALD und STOCKER (1940) diese Untersuchungen auf mehrere Ackergewächse erweitert und denselben Schluß ziehen können. Sie finden es daher wahrscheinlich, daß höhere Viscosität ein spezifisches Kennzeichen höherer Dürreresistenz ist. Von diesen Ergebnissen ausgehend und gestützt auf die thixotropische Wirkung, die ein langes und starkes Schütteln auf das Plasma ausübt, hat dann STOCKER (1946) eine Theorie der Dürreresistenz aufgestellt. LEVITT (1951, S. 266) kritisiert aber diese Theorie und verteidigt die Meinung, daß Dürreresistenz nicht mit einer Verfestigung, sondern mit einer Auflockerung des Plasmas verbunden ist[1].

Unter den Plasmakomponenten, welche zusammen die Plasmaviscosität bestimmen — einige derselben waren oben angeführt —, befinden sich auch solche, die auf dem osmotischen Wert basieren. Eine Änderung der osmotisch wirksamen Faktoren (Natur der gelösten Stoffe, Konzentration, Hydratisierung) kann daher auch die Viscosität ändern. Da die Viscosität aber außerdem von anderen Bedingungen abhängt, wie die obige Zusammenstellung zeigt, kann ein deutlicher Zusammenhang zwischen dem osmotischen Wert und der Plasmaviscosität kaum erwartet werden, wie dies auch aus den Versuchen nicht hervorgeht (SCHMIDT, DIWALD und STOCKER 1940; übrige Literatur s. STÅLFELT 1954).

Mit dem Alter der Organe und der Zellen werden manche Eigenschaften und Prozesse verändert — Zusammensetzung, Funktion, Differenzierung, Saugkraft, Wasserdefizit, Zerstörungsprozesse usw. — die, jedes für sich, auf die Viscosität einwirken und sie entweder erhöhen oder mindern. Je nach dem Verhältnis zwischen der Stärke und der Bedeutung der einzelnen Faktoren wird die Gesamtwirkung auf die Plasmaviscosität verschieden sein. Die Beziehungen zwischen Viscosität und Alter sind daher vielgestaltig (FISCHER 1950). Ein regelmäßiger Zusammenhang kann in einzelnen Fällen gewiß konstatiert werden, doch scheint er bei verschiedenen Objekten von verschiedener Art zu sein.

Die Plasmaviscosität wird von verschiedenartigen äußeren Faktoren beeinflußt.

[1] Vgl. Band III: STOCKER, O., Die Dürreresistenz, S. 696ff.

Wasserbilanz.

Unter der Einwirkung eines langdauernden Wassermangels treten Verschiebungen im chemischen Gleichgewicht der Zellen ein, indem Konzentrationsänderungen von Stoffen entstehen. Bereits nach 36 Std kann z. B. der Zuckergehalt um 49% steigen bei einem Wasserdefizit von 21% (ILJIN 1929). Viscositätsänderungen, die mit solchen Folgen eines wirklichen Wassermangels (Dürre, „Trockenkultur") zusammenhängen, sind bereits behandelt worden. Jedoch kann man Änderungen der Plasmaviscosität bereits dann erwarten, wenn die Turgescenz der Zellen sinkt, da der Wasserverlust eine Konzentration des Zellinhalts mit sich bringt. Solche Änderungen sind auch nachgewiesen worden. FRITSCH und HAINES (1923) fanden, daß die Plasmaviscosität bei Erdalgen und Moosprotonemen höher war, wenn sie Trockenheit oder der Behandlung mit einer hypertonischen Lösung ausgesetzt wurden: ihre Neigung zur Plasmolyse wurde nämlich geringer, und die Beweglichkeit der Teilchen bei der Zentrifugierung wurde verringert. Viscositätsänderungen der gleichen Art wurden auch für höhere Pflanzen nachgewiesen. Bei *Vicia sepium* und *Achillea millefolium* nimmt die Viscosität sofort ab, wenn ein Wassermangel eintritt. Ein Defizit von 5% verursacht bereits eine Abnahme, die eine deutliche Verschiebung des Zentrifugierungswerts der Chloroplasten mit sich führt (STÅLFELT 1954). Bei *Mnium*- und *Bryum*-Arten fand NORTHEN (1943), daß die Beweglichkeit der Chloroplasten bei Zentrifugierung erst zunahm, wenn das Plasma einer 10—50 min langen Austrocknung ausgesetzt worden war, danach aber abnahm.

Die durch eine Dehydrierung hervorgerufenen Veränderungen, die anfangs reversibel sind, werden irreversibel, wenn der Wasserverlust eine bestimmte Grenze überschreitet. Das Plasma wird dann gelatinös und kann nach SCARTH (1941) bei Wasseraufnahme zerrissen werden.

Temperatur.

WEBER (1921) maß die Fallgeschwindigkeit von Calciumoxalatkristallen in den Zellen von *Callisia repens* und fand, daß die Geschwindigkeit mit der Temperatur zunahm. Nach den Untersuchungen, die HEILBRONN (1924) mit seiner Magnetmethode (Eisenfeilspäne) an Plasmodien von verschiedenen Pilzen vornahm, dürfte der Zusammenhang jedoch recht kompliziert sein. Die Bewegungsgeschwindigkeit der Teilchen nahm sowohl bei einer Temperaturerhöhung, wie bei einer Temperaturabnahme zu, auf die Änderung folgte jedoch eine Gegenreaktion, indem die Beweglichkeit wieder geringer wurde. In den Versuchen, die NORTHEN (1940) mit *Spirogyra* anstellte, nahm die Viscosität bei Temperaturerhöhung ab, wenn die Zellen vorher 10—20 min auf —3° abgekühlt worden waren. Ähnliche Resultate ergaben andere Versuche an *Spirogyra* (BAAS-BECKING und Mitarb. 1928) sowie Versuche mit Eiern von *Cumingia* (HEILBRONN 1924) und *Amoeba* (THORNTON 1935, übrige Literatur s. SEIFRIZ 1952). Es hat sich im allgemeinen gezeigt, daß die Viscosität bei Temperaturerhöhung sowohl zu- als abnehmen kann; es scheint mithin, daß Temperaturänderungen eine Serie von Viscositätsänderungen hervorrufen. Durch Änderungen dieser Art (Abhängigkeit von der Temperatur) unterscheidet sich das Protoplasma von leblosen Proteinsystemen. SEIFRIZ (1952, S. 23) deutet die Änderungen als Ausdruck eines viscositätskontrollierenden Mechanismus. Die gleiche Auffassung wurde früher von HEILBRONN (1922) zum Ausdruck gebracht.

Licht.

HEILBRONN (1922) zeigte, daß die Plasmodien von *Reticularia* lichtempfindlich sind. Direktes Sonnenlicht von 1 und $1^1/_2$ min Dauer ruft zuerst Steigerung, dann

Herabsetzung der Viscosität hervor. Nach einigen Stunden klingt die Zustandsänderung wieder ab, und zwar um so später, je intensiver der Reiz war. HEILBRONN vermutet, daß es sich um eine Wirkung der ultravioletten Strahlen handelt.

Bei plasmolytischen Untersuchungen fand WEBER (1929) konkave Plasmolyseform bei *Ranunculus*-Blättern, die vorher 1—2 Std dem Sonnenlicht ausgesetzt waren, aber konvexe Plasmolyseform nach derselben Zeit im Dunkeln. Nach MACKE (1939) besteht eine gewisse Beziehung zwischen Borwirkung und Lichtwirkung. Nach einer Vorbehandlung von *Helodea*-Pflanzen mit Borax und Borsäure zeigten die Blattzellen verminderte Plasmaviscosität (Plasmolysemethode), aber nur im Licht. Im Dunkeln konnte sogar eine leichte Erhöhung eintreten.

In den erwähnten Fällen handelte es sich um Wirkungen des direkten Sonnenlichts, d.h. um Prozesse, bei denen auch ultraviolettes Licht beteiligt war. Daß aber nicht nur ultraviolettes Licht, sondern auch langwelligere Strahlen wirksam sind, geht daraus hervor, daß auch Lampenlicht Viscositätsänderungen auslöst. Bei Blättern von *Helodea densa*, die 2—3 Tage im Dunkeln standen, zeigt die Plasmaviscosität eine große Lichtempfindlichkeit (STÅLFELT 1945, 1946). Eine Belichtung löst unmittelbar Viscositätsschwankungen aus, die solange anhalten, wie die Belichtung andauert; sie wirken einige Stunden nach. Deren Amplitude nimmt im Anfang zu, geht dann zurück, um später wieder anzusteigen. Die Dauer der Nachwirkung bringt es mit sich, daß die Schwankungen auch nachts nicht aufhören, sondern ständig fortdauern. Ein Objekt, das direkt von seinem Standort kommt, zeigt daher wechselnde Viscositätsänderungen des Zellplasmas. Nur bei starkem Licht (32000—48000 Lux) hören die Schwankungen auf, beginnen jedoch wieder, sobald die Lichtstärke abnimmt. Die Ursache ist nicht in der Photosynthese zu suchen, denn bereits eine 15 sec dauernde Belichtung mit 100 Lux reicht aus, um die Änderungen auszulösen. VIRGIN (1947, 1950) fand noch niedrigere Schwellenwerte, nämlich 1 Lux und im Herbst 0,001 Lux. Er wies ferner nach, daß *Spirogyra* auf die gleiche Weise reagiert, daß das wirksame Licht innerhalb der Wellenlängen von 390—500 mμ liegt, und daß der Lichtreiz lokal wirkt. Schließlich wurde diese Erscheinung auch bei Landpflanzen festgestellt (STÅLFELT, im Druck). DUGGAR hat in seiner Arbeit Biological Effects of Radiation (1954) eine Zusammenstellung der Lichtwirkungen auf die Plasmaviscosität gegeben.

Die Abhängigkeit der Viscosität vom Licht erinnert in manchen Beziehungen an den von FITTING (1925) festgestellten Zusammenhang zwischen Plasmarotation und Licht. Nach seinen Untersuchungen wird die Plasmarotation durch den Übergang von schwachem Licht zum stärkeren ausgelöst. Der auslösende Faktor war nicht das Licht als solches, sondern der Wechsel zum stärkeren Licht. Bei einem an Dunkelheit adaptierten Material rief bereits eine wenige Minuten dauernde Belichtung eine Rotation hervor. Je stärker das Licht, desto stärker war dessen Nachwirkung. Im Sommer hört infolge der Stärke der Lichteinwirkung die Plasmarotation auch nachts nicht auf. In welchem Maße die Plasmaströmung und deren Änderungen von der Plasmaviscosität oder von der treibenden Kraft abhängen, ist allerdings noch nicht bekannt. Trotzdem kann mit LINSBAUER (1928, S. 565) und RUGE (1940, S. 344) ausgesprochen werden, daß „jeder Faktor, der das Ausmaß der Viscosität verändert, ... auch auf die Plasmaströmung von Einfluß ist".

Ultraviolette, Röntgen- und Radiumstrahlen, die im allgemeinen tiefergehende Wirkungen auf das Plasma ausüben, ändern auch dessen Viscosität. Auch in diesen Fällen können Viscositätsänderungen in der Form von Schwankungen auftreten (NORTHEN und VICAR 1940). Eine Übersicht der Arbeiten, die dieses Thema behandeln, ist von HEILBRUNN (1952, S. 96) gegeben worden[1].

[1] Vgl. auch in diesem Band: SIMONIS, W., Die Wirkung von Licht und Strahlung auf die Zelle, S. 690ff.

Strahlen anderer Art. Eine vorübergehende Verminderung der Plasmaviscosität *(Helodea)* durch die Einwirkung von α- und β-Strahlen radioaktiver Isotope haben VIRGIN und EHRENBERG (1953) festgestellt.

Elektrizität.

Durch den elektrischen Strom wird die Plasmaviscosität geändert. Ein Strom von 2 mA während 2 min ruft z. B. laut NORTHEN (1940a) bei *Spirogyra* erst eine Minderung und danach eine Steigerung der Plasmaviscosität hervor. Bei *Amoeba* löst der Strom eine reversible Abnahme der Viscosität aus (ANGERER und WILBUR 1943), ebenso bei Plasmodien der Myxomyceten (angelegte Spannung 64 V, SEIFRIZ 1950). Höhere Spannung (16—100 V) bedingt in Blattzellen von *Helodea* anodische Verlagerung der Chloroplasten und des Cytoplasmas und gleichzeitig Viscositätserhöhung (TOBIAS und SALOMON 1951).

Mechanische Einwirkung.

Stöße und Druck setzen bei *Spirogyra* die Plasmaviscosität herab (NORTHEN 1939, 1940a), ebenso langdauerndes und starkes Schütteln (KAHL 1951). Dagegen steigt die Viscosität bei dem Plasma von Myxomyceten *(Physarum polycephalum)* bei mechanischer Einwirkung (SEIFRIZ 1950). Bei Berührung von *Amoeba*-Zellen mit einem Mikromanipulator tritt je nach Reaktionslage und Berührungsstelle eine Verflüssigung oder Gelbildung auf (GOLDACRE 1952). Die Wirkung des Zentrifugierens ist bereits im vorhergehenden behandelt worden.

Elektrolyte.

Gemäß dem entworfenen Bild von der Zusammensetzung des Plasmas besteht dessen Interstitialphase unter anderem aus Proteinen und Lipoiden. Zusammen mit dem Plasmaskelet bieten diese Bestandteile den eindringenden Elektrolyten viele Möglichkeiten, in den Plasmazustand einzugreifen. Aus den zahlreichen Untersuchungen über die Wirkung von Elektrolyten auf das Plasma geht hervor, daß das Plasma in vielen Fällen auf die gleiche Weise reagiert, wie Kolloide in vitro. Eine solche Übereinstimmung ist ja auch zu erwarten, solange die Elektrolyte nur die Komponenten der Interstitialphase angreifen. Oft jedoch treten Abweichungen von diesem Bilde auf. Man kann in solchen Fällen vermuten, daß das Plasmaskelet in Mitleidenschaft gezogen wurde.

Die Wirkung der Elektrolyte auf die Plasmaviscosität wird durch deren Konzentration und Art bestimmt, insbesondere durch die Kationen und deren Valenz, des weiteren durch den p_H-Wert des Systems sowie durch die Hydrophilie oder Hydrophobie der Kolloide (s. z. B. PHILIPPOFF 1942, S. 293), möglicherweise auch durch antagonistische Wirkungen der einzelnen Stoffe.

In einer Untersuchung über die Wirkung verschiedener Salze auf die Hitzeresistenz des Protoplasmas der Pflanzen *(Rhoeo, Gentiana)* hat BOGEN (1948) gezeigt, daß die Wirkungsreihe der Ionen mit der HOFMEISTERschen lyotropen Reihe übereinstimmt. Ergebnisse derselben Art hat SEIFRIZ (1950) mitgeteilt. Bei einer Untersuchung über den toxischen Effekt verschiedener Salze auf das Plasma von Myxomyceten fand er dieselbe Übereinstimmung. Als Folge der Giftwirkung wurde das Plasma verdichtet und in Gel-Zustand übergeführt.

Das Interesse war vorzugsweise auf die beiden gewöhnlichsten Kationen in den Pflanzenzellen, Kalium und Calcium gerichtet, weil diese bei vielen physiologischen Prozessen antagonistische Wirkungen zeigen. In manchen Versuchen sowohl mit der Plasmolyse- als auch der Zentrifugierungsmethode ist eine Viscositätszunahme des Plasmas durch Ca- und eine Viscositätsabnahme durch K-Behandlung gezeigt worden (Literatur s. FISCHER 1948, PIRSON und SEIDEL 1950). Jedoch liegen auch andere Angaben vor. HEILBRUNN (1923, 1952, S. 97, hier auch Literaturangaben) fand, daß Na und K die Viscosität bei Protozoen erhöhen, während Mg und Ca diese herabsetzen und weiter, daß dies nur das Innenplasma betrifft und nur für solche Versuche gilt, bei denen die Ionen langsam in die Zellen eindringen. Bei *Amoeba* ruft Ca eine Versteifung (stiffening) des Außenplasmas hervor, wobei K, Na und Mg antagonistisch wirken (vgl. auch FISCHER 1949, S. 285). Ein Mangel an den genannten Ionen scheint ebenfalls verschiedenartige Wirkungen hervorzurufen. KALCHHOFER (1936) hatte bei *Helodea, Tradescantia* und *Vicia faba* gefunden, daß Kalimangel die Plasmaviscosität erhöht, Calciummangel dagegen herabsetzt. PIRSON und SEIDEL (1950) zeigten aber, daß Wurzelzellen von *Lemna minor* gleichmäßig reagieren, K- und Ca-Mangel führen beide zu einer Verkürzung der Plasmolysezeit (Glucose) und Deplasmolysezeit (Harnstoff). Eine generelle Bedeutung der Quotienten K/Ca für die Plasmaviscosität dürfte daher kaum vorliegen.

Die osmotische Wirkung der Elektrolyte äußert sich unter anderem in einer Entquellung des Plasmas, aber nicht immer. Versuche, die HÖFLER (1939) mit *Allium*-Zellen anstellte, haben gezeigt, daß das Plasma entgegen dem ansteigenden osmotischen Wert des Zellsaftes aufquellen kann. Lösungen von K, Na- und Li-Salzen bewirkten eine mächtige Aufquellung des Cytoplasmas und Kappenplasmolyse. Dabei wirkte Ca antagonistisch und ein geringer Ca-Zusatz verhinderte die Aufquellung. Mit der Aufquellung ist eine Herabsetzung der Plasmaviscosität verbunden, wie PEKAREK (1940) nachgewiesen hat.

Wie WALTER (1924) klarlegte, wirken die Quellungsänderungen auf die Plasmaviscosität verschieden ein, je nachdem, ob das System Wasser aufnehmen kann oder nicht. Wird ein quellungsfähiger Stoff hinzugefügt, so werden die einzelnen Teilchen stärker quellen, mehr Wasser aufnehmen und ihr Volumen vergrößern. Infolgedessen wird der Zwischenraum zwischen den Teilchen kleiner, die Viscosität muß also steigen. Wenn aber das System Wasser durch eine semipermeable Membran aufnehmen kann, so wird neues Wasser nachgesaugt — die Saugkraft ist durch die Quellung gesteigert worden —, bis wiederum die frühere Saugkraft erreicht worden ist. Die Lösung wird dadurch mehr verdünnt und die Viscosität geringer.

Infolge der verschiedenen Wirkungsweise der Ionen und der wechselnden Umstände, unter denen ihre Wirkungen im Plasma zustande kommen, dürfte eine einfache Beziehung zwischen Elektrolyten und Plasmaviscosität kaum vorliegen. Die zahlreichen Angaben über die Ionenwirkung auf die Plasmaviscosität unterscheiden sich auch stark in ihren Ergebnissen, wie FISCHER (1948) bemerkt. Nach ihm kann die Ursache darin liegen, daß die Ionen im normalen Stoffwechsel vielfach anders wirken, als wenn sie von außen geboten werden. Es ist wohl auch möglich, daß die Hydratation auf die verschiedenen Plasmakomponenten verschieden einwirkt, wie BOGEN (1948) meint.

In Anbetracht der verschiedenartigen Wirkungen, welche K und Ca ausüben, scheint es schwer, an der oft vorgebrachten Hypothese festzuhalten, daß nämlich diese beiden Kationen und deren Gleichgewicht einen Regulator für die Plasmakolloide bilden, welche durch ihn in einem für die Zellfunktionen optimalen Zustand erhalten werden sollen. Diese Hypothese stößt auch auf Schwierigkeiten angesichts der Resultate der Untersuchungen von PIRSON und SEIDEL (1950) über den Zusammenhang zwischen gewissen zell- und stoffwechselphysiologischen Faktoren — Plasmolysezeit, Deplasmolysezeit, osmotischer Wert und respiratorischer O_2-Wert — und dem K- und Ca-Mangel.

p_H-Wert. Obgleich die Wasserstoffionen von ausschlaggebender Bedeutung für die Sol-Gel-Bilanz der Kolloide sind, sind unsere Kenntnisse von dem p_H-Wert des Plasmas und den Schwankungen dieses Wertes gering; dies ist bedingt durch die Schwierigkeit, Messungen im Zellinneren vorzunehmen. Daß die Viscosität (Plasmolysezeit) des Cytoplasmas durch den p_H-Wert des Plasmolyticums beeinflußbar ist, hat STRUGGER (1934) gezeigt. Die Acidität des Außenmediums wirkt auch unter normaleren Umständen. MINSHALL und SCARTH (1952) zeigten, daß die Viscosität des Cytoplasmas (Strömungsgeschwindigkeit der proplasmatischen Teilchen) bei *Hydrocharis* in stark saurer ($p_H = 3{,}5$) Kultur etwa doppelt so hoch war wie bei $p_H = 5$.

Fettlösende Stoffe.

Fettlösende Stoffe wirken auf die Viscosität des Zellplasmas ein, was wohl auf die Reaktionen der lipoidartigen Bestandteile des Plasmas zurückzuführen ist. KESSLER (1935) wies eine Minderung der Plasmaviscosität bei *Sempervivum*, das mit Chloroform behandelt wurde, nach. NORTHEN (1939 und 1946) prüfte ähnliche Substanzen (2% Äthylenchlorhydrin, 4% Äther, 6% Äthylalkohol, 3,5% Propylalkohol) bei Versuchen mit *Helodea*. Nach einer Behandlungszeit von 2 Std verursachten alle die genannten Stoffe eine verminderte Viscosität. Auch bei einer Herabsetzung der genannten Konzentrationen um die Hälfte war eine verminderte Viscosität zu konstatieren. Bei einer Erhöhung der Konzentrationen trat dagegen — nach einer vorangehenden Herabsetzung — eine Steigerung der Viscosität ein (übrige Literatur s. NORTHEN 1946). Umfassende Untersuchungen über diese Fragen wurden von SEIFRIZ (1951, 1952, hier auch Literatur) ausgeführt, der die Einwirkung einiger organischer Substanzen, darunter fettlösender Stoffe, auf die Plasmaviscosität und andere plasmatische Zustände und Prozesse nachwies. Die Resultate dieser Arbeiten deuten auf einen allgemeinen Zusammenhang zwischen Viscositätsänderungen und dem physiologischen Zustand einer Zelle. Stoffe mit stimulierender Wirkung setzen die Plasmaviscosität herab, narkotische Substanzen erhöhen sie. SEIFRIZ (1951) deutet diese Erscheinung auf die Weise, daß der narkotische Effekt durch eine Verdichtung des Plasmas bestimmt ist, der stimulierende dagegen durch eine Auflockerung. Der Mechanismus dieser Prozesse besteht mithin in einer Änderung des Zusammenhangs zwischen den Molekülen (molecular cohesion). Die Untersuchungen über den Zusammenhang zwischen Plasmaviscosität und Anaestheticis sind vorwiegend an tierischen Objekten durchgeführt worden (HEILBRUNN 1952).

Zelleigene Substanzen.

Die Wirkung einiger zelleigener Substanzen und ihnen nahestehender Stoffe zeigte sich unter anderem in der Veränderung der Viscosität. Die chemische Konstitution einiger dieser Stoffe ist noch unbekannt.

a) Zelleigene Stoffe bekannter Konstitution.

Adenosintriphosphat (ATP). KRISZAT (1950) hat die Wirkung von ATP auf Amoeben untersucht und gefunden, daß Konzentrationen von 0,002—0,006 Mol nach 10—15 min Wirkungsdauer die Plasmaviscosität zuerst vorübergehend erniedrigen, dann aber irreversibel erhöhen. Mit steigender Konzentration und verlängerter Einwirkungsdauer nimmt diese Veränderung zu. Noch wirkungsvoller war ATP in Ca-freiem Kulturmedium.

Heteroauxin (*β-Indolylessigsäure*) und *Colchicin*, die beide mitotisch aktiv sind — Heteroauxin außerdem aktiv als Wuchsstoff —, verursachen eine Änderung der Plasmaviscosität (STÅLFELT 1947). Noch bei einer Verdünnung von

0,0000001 mg/l verursacht Heteroauxin eine Viscositätssteigerung, die 60 Std und mehr fortdauern kann. Bei höheren Konzentrationen erfolgt die Steigerung rascher, wird aber nach ungefähr 30 Std durch eine Viscositätsminderung abgelöst. Es handelt sich also in Wirklichkeit um einen Wechsel der Viscosität. Wahrscheinlich gilt dies auch für die niederen Konzentrationen, was jedoch noch nicht gezeigt werden konnte, da die Objekte die lange Versuchsdauer nicht ertragen, die für die Entstehung der Gegenreaktion notwendig ist. NORTHEN (1942), der Parenchymzellen von Blättern und Blattstielen untersuchte, bemerkte eine Herabsetzung der Plasmaviscosität nach Zusatz von Heteroauxin in Lanolinpasta. Eine Steigerung der Viscosität (Blatt von *Rhoeo*) als Folge eines Zusatzes von Heteroauxin fanden dagegen GUTTENBERG und BEYTHIEN (1952). Dieser Widerspruch kann auf dem langsam vor sich gehenden Wechsel in der Viscosität beruhen.

Colchicin wirkt auf die gleiche Weise, jedoch erst bei höheren Konzentrationen — genauer gesagt bei 0,001 mg/l und darüber *(Helodea)* —, ebenso Indol-β-Propionsäure (geringste wirksame Konzentration 0,01 mg/l) und das laut TJIO und LEVAN (1950) mitotisch wirksame Oxychinolin (geringste wirksame Konzentration 0,001—0,002 mol/l; STÅLFELT 1947, 1950).

Die beiden viscositätsaktiven Stoffe, Colchicin und Oxychinolin, sind, wie gesagt, mitotisch aktiv, und die Zellteilungsprozesse, in welche sie eingreifen, sind unter normalen Bedingungen mit vorübergehenden Änderungen der Plasmaviscosität gekoppelt. Die Änderungen sind verschieden bei den morphologisch unterscheidbaren, bei der Mitose beteiligten Plasmakomponenten (HIRAOKA 1951, Literatur bei SEIFRIZ 1952, S. 19). Wahrscheinlich üben Colchicin und Oxychinolin gewisse mitosestörende Wirkungen aus, indem sie Viscositätsänderungen bei den an der Mitose beteiligten Plasmabestandteilen hervorrufen (TJIO und LEVAN 1950). Zur Zellteilung gehören nämlich auch gewisse Bewegungserscheinungen, bei welchen Plasmakomponenten verlagert werden. Hierbei ergeben sich infolge des Fließwiderstandes, d. h. der Viscosität, Folgeerscheinungen verschiedener Art.

b) Zelleigene Stoffe unbekannter Art.

BÜNNING (1926) beobachtete in einigen Fällen, daß Verwundungen eine Steigerung der Plasmaviscosität bei den benachbarten Zellen hervorriefen. JUNGERS (1934) fand, daß Zellen (Zwiebelschuppen von *Allium cepa*), welche unter dem Einfluß traumatischer Reize gestanden hatten, z. B. Zellen in der Nähe von verwundeten Zellen, durch Zentrifugenbehandlung stark beeinflußt wurden; ihr Inhalt erwies sich als leicht verlagerungsfähig. Besonders groß war diese Veränderung bei Schnittpräparaten von Epidermen, während Zellen von Gewebeprismen erst nach Stunden reagierten. LANZ (1939) konnte bei *Cladophora* keinen solchen „Wundreiz" feststellen, aber UMRATH (1942) hat die Erscheinung eingehend an *Spirogyra* studiert. In solchen *Spirogyra*-Zellen, die einer durchschnittenen Zelle im Faden benachbart waren, waren die Chromatophoren bei Zentrifugierung leichter verlagerbar. Die Beobachtungen sprechen dafür, daß es sich nicht um eine Erregungsleitung handelt, sondern um eine aus der durchschnittenen Zelle freiwerdende Substanz, die sich ausbreitet und die intakten Zellen beeinflußt. Dies wird auch dadurch gestützt, daß ein Kochextrakt aus fast reiner *Spirogyra*, der in geringer Menge dem Wasser zugesetzt worden war, ebenfalls die Verlagerbarkeit der Chromatophoren erhöhte. Ein Kochextrakt aus anderen Algen zeigte nicht diese, sondern eher die entgegengesetzte Wirkung.

STÅLFELT (1947, 1948, 1949) fand, daß die Viscosität (Zentrifugierung) der Blattzellen von *Helodea densa* sich ändert, wenn die Blätter aus dem Standorts-

oder Leitungswasser in destilliertes Wasser übergeführt werden und daß dieses Verhalten durch Substanzen hervorgerufen wird, die im Leitungswasser vorhanden sind, im destillierten oder Regenwasser dagegen fehlen. Die Stoffe kommen in der Erde und wahrscheinlich in lebendem Pflanzenmaterial und frischer Streu (schwedisch: förna) vor. Ein Extrakt, sogar ein Kochextrakt aus frischer Streu liefert Lösungen von hoher Aktivität. Selbst eine Verdünnung von $1:10^9$ (das Gewicht der extrahierten Streu gleich 1 gesetzt) ruft eine Änderung der Viscosität hervor. Wahrscheinlich handelt es sich um 2 Stoffe, von denen der eine Steigerung und der andere eine Minderung der Plasmaviscosität verursacht. Es sind mithin Substanzen, die von den Pflanzen selbst erzeugt, bei der Humifizierung frei werden und infolge ihrer relativ großen Beständigkeit in Erde und Wasser vorkommen.

In diesem Zusammenhang kann noch erwähnt werden, daß HARDING (1951) ein ähnliches Verhalten bei tierischem Material nachwies; er beobachtete, daß ein Extrakt von Froschmuskeln, die durch Hitze abgetötet wurden, die Viscosität von Seeigeleiern erhöht und eine Teilung und Pathogenese hervorruft, und daß die Viscositätsänderung der Zellteilung vorausgeht.

Ein bestimmter spezieller Viscositätswert ist eine der charakteristischen Eigenschaften, die den Unterschied zwischen lebendem und totem Protoplasma ausmachen. Der Viscositätswert des lebenden Plasmas ist niedrig und labil, im Gegensatz zu demjenigen der toten Zelle. Durch ihre Labilität ist die Viscosität, wie bereits erwähnt, empfindlich gegenüber einer Reihe chemischer, physikalischer und biotischer Faktoren. Bezeichnend für die Änderungen der Plasmaviscosität (solange die physiologischen Prozesse noch nicht die Grenzen zum Destruktiven überschritten haben) ist deren Reversibilität, die es mit sich bringt, daß die Änderungen in der Form von Schwankungen um einen für eine Art oder ein Organ charakteristischen Wert auftreten. Die Viscosität scheint mithin durch eine Art physiologischen Regulators gesteuert zu werden. Durch diesen Regulator ist das Plasma vor den Folgen unberechenbarer Veränderungen gesichert, die ständig durch allerlei äußere Faktoren ausgelöst werden. Da die oben besprochenen viscositätsaktiven Stoffe ganz allgemein bei den Pflanzen vorkommen, ist es durchaus denkbar, daß sie auch zu den Bestandteilen eines solchen Regulators gehören (STÅLFELT 1949). Die ältere Hypothese von der Bedeutung des K/Ca-Quotienten für den Zustand der Plasmakolloide dürfte daher nicht mehr haltbar sein, wie dies bereits bemerkt wurde (s. S. 600).

Die Annahme viscositätsregulierender Substanzen erfährt eine Stütze durch die Tatsache, daß ein bestimmter Zusammenhang zwischen der Viscosität und der Aktivität des Plasmas herrscht (SEIFRIZ 1952, S. 18, dort auch Literatur). Die Viscosität ist relativ niedrig, wenn die Aktivität — die physiologische wie auch die physikalische — lebhaft ist, steigt aber, wenn die Aktivität abnimmt. Die Annahme wird weiter dadurch gestützt, daß das Plasma bei einigen Prozessen mit viscositätsaktiven Stoffen arbeitet (Wundstoffe, ATP, Heteroauxin) und daß ein Zusatz von viscositätsaktiven Substanzen (Colchicin, Oxychinolin) den normalen Verlauf dieser Prozesse stört (Literatur bei HIRAOKA 1951, S. 175, SEIFRIZ 1952, S. 18).

Literatur.

ANGERER, C. A., and K. M. WILBUR: On the action of various types of electric fields on the relative viscosity of the plasmagel of *Amoeba proteus*. Physiologic. Zool. **16**, 84 (1943).

BAAS-BECKING, L. G. M. u. Mitarb.: The physical state of protoplasm. Verh. Kon. Ned. Akad. v. Wetensch., Sect. B **25**, 1 (1929). — BAJER, A.: Absolute viscosity and living mitotic spindle structure. Acta Soc. Bot. Poloniae **22**, 331 (1953). — BIEBL, R.: Zur protoplasmatischen Anatomie der Rotalgen. Protoplasma **28**, 582 (1937). — BIEBL, R., u. W. ROSSI-

PILLHOFER: Die Änderung der chemischen Resistenz pflanzlicher Plasmen mit dem Entwicklungszustand. Protoplasma (Berl.) **44**, 9 (1954). — BOGEN, H. J.: Vergleichende Untersuchungen über Ionenreihen an pflanzlichen Protoplasten. Biol. Zbl. **67**, 490—503 (1948). — Permeabilitätsuntersuchungen an polyploiden Zellen. Z. Abstammgslehre **83**, 93—105 (1949).— BORRISS, H.: Plasmolyseform und Streckungswachstum. Jb. wiss. Bot. **86**, 784 (1938). — BRECKHEIMER-BEYRICH, HELGA: Über die Wirkung zentrifugaler Kräfte auf das Protoplasma von *Nitella flexilis*. Ber. dtsch. bot. Ges. **62**, 55 (1949). — BÜNNING, E.: Untersuchungen über Reizleitung und Reizreaktion bei traumatischer Reizung von Pflanzen. Bot. Archiv **14**, 138; **15**, 4 (1926).

DELLINGSHAUSEN, MARGARETE v.: Entwicklungsgeschichtlich-genetische Untersuchungen an *Epilobium*. VIII. Planta (Berl.) **25**, 282 (1936). — DIANNELIDIS, T.: Plastiden-Rückverlagerung nach Zentrifugierung und Narkose. Protoplasma **39**, 244 (1950).

FISCHER, HERMANN: Plasmolyseform und Mineralsalzgehalt in alternden Blättern. Planta (Berl.) **35**, 513—527 (1948). — Plasmolyseform und Mineralsalzgehalt in alternden Blättern. II. Planta (Berl.) **37**, 244—292 (1949). — Über protoplasmatische Veränderungen beim Altern von Pflanzenzellen. Protoplasma **39**, 661 (1950). — FREY-WYSSLING, A.: Submicroscopic. morphology of protoplasm. Amsterdam-Houston-London-New York 1953. — FRITSCH, F. E., and F. M. HAINES: The moisture relations of terrestrial algae. II. The changes during exposure to drought and treatment with hypertonic solution. Ann. Bot. **37**, 683—728 (1923).

GOLDACRE, R. J.: The action of general anaestetics on Amoebae and the mechanism of the response to touch. Symposia Soc. Exper. Biol. **6**, 128—144 (1952). — GUTTENBERG, H. v., and A. BEYTHIEN: Über den Einfluß von Wirkstoffen auf die Wasserpermeabilität des Protoplasmas. Planta (Berl.) **40**, 36 (1952).

HEILBRONN, A.: Eine neue Methode zur Bestimmung der Viskosität lebender Protoplasten. Jb. wiss. Bot. **68**, 284 (1922). — HEILBRUNN, L. V.: The viscosity of protoplasm at various temperatures. Amer. J. Physiol. **68**, 645 (1924).—Colloid Chemistry of Protoplasm. Berlin 1928. — An Outline of General Physiology. Philadelphia 1952. — HENKEL, P. A.: Über Ursachen der Dürrebeständigkeit einiger Xerophyten und Halophyten. Bot. Ž. **34**, 461—473 (1949) [Russisch]. Ref. Ber. wiss. Biol. **69** (1950). — HIRAOKA, TOSISUKE: Viscosity change of cytoplasm in spore mother cells undergoing meiosis. Mem. Coll. Sci. Univ. Kyoto, Ser. B **20**, 171 (1953). — HÖFLER, KARL: Vertragen Rotalgen das Zentrifugieren? Protoplasma **26**, 377 (1936). — Kappenplasmolyse und Ionenantagonismus. Protoplasma **33**, 545 (1939).

JUNGERS, W.: Die Verlagerungsfähigkeit des Zellinhaltes der Zwiebelschuppen von *Allium cepa* durch Zentrifugierung. Protoplasma **21**, 351 (1934).

KAHL, H.: Über den Einfluß von Schüttelbewegungen auf Struktur und Funktion des pflanzlichen Plasmas. Diss. Darmstadt 1951. — KALCHHOFER, Z.: Protoplasmazustand nährsalzmangelkranker Pflanzen. Protoplasma **26**, 249 (1936). — KATO, KAZUO: Viscosity changes in the cytoplasm during mitosis as indicated by Brownian movement. Mem. Coll. Sci. Univ. Kyoto, Ser. B **8**, 201 (1933). — KESSLER, W.: Über die inneren Ursachen der Kälteresistenz der Pflanzen. Planta (Berl.) **24**, 312 (1935). — KESSLER, W., u. W. RUHLAND: Weitere Untersuchungen über die inneren Ursachen der Kälteresistenz. Planta (Berl.) **28**, 159 (1938). — KRISZAT, GEORG: Die Wirkung von Adenosintriphosphat und Calcium auf Amöben. *(Chaos Chaos.)* Ark. Zool. (Stockh.) Ser. II **1**, 477—485 (1950). — KÜSTER, E.: Über die Wirkung des Zentrifugierens auf die Viskosität des lebenden Protoplasmas. Kolloid-Z. **89**, 237 (1939).

LANZ, I.: Über die Wirkung der Zentrifugenbehandlung auf den lebendigen Zellinhalt. Untersuchungen an *Cladophora*. Arch. exper. Zellforsch. **23**, 220 (1939). — LEVITT, J.: Frost, drought and heat resistance. Annual Rev. Plant Physiol. **2**, 245 (1951). — LINSBAUER, K.: Untersuchungen über Plasma und Plasmaströmung an Chara-Zellen. Protoplasma **5**, 563 (1928).

MACKE, W.: Untersuchungen über die Wirkung des Bors auf *Helodea canadensis*. Z. Bot. **34**, 241, 267 (1939). — MENDER, GRETE: Protoplasmatische Anatomie des Laubmooses *Bryum capillare*. Protoplasma **30**, 373 (1938). — MINSHALL, WM. HAROLD, and G. W. SCARTH: Effect of growth in acid media on the morphology, hydrogen-ion concentration, viscosity and permeability of water hyacinth and frogbit rootcells. Canad. J. Bot. **30**, 188—208 (1952). — MOLÈ-BAJER, J.: Influence of hydration and dehydration on mitosia. II. Acta Soc. Bot. Poloniae **22**, 33 (1953).

NORTHEN, H. T.: Effect of drought on protoplasmic elasticity. Plant physiol. **13**, 658 bis 660 (1938). — Studies on the protoplasmic nature of stimulation and anesthesia. II.

Cytologia **10**, 105 (1939). — Studies on the protoplasmic nature of stimulation and anaestetica. II. Plant. Physiol. **15**, 645—659 (1940). — Relationship of dissociation of cellular proteins by auxins to growth. Bot. Gaz. **103**, 668—683 (1942). — Relationship of dissociation of cellular proteins by incipient drought to physiological process. Bot. Gaz. **104**, 480—486 (1943). — Effects of various agents on the structural viscosity of *Elodea* protoplasm. Plant physiology **21**, 148 (1946). — NORTHEN, H. T., and R. MACVICAR: Effects of x-rays on the structural viscosity of protoplasm. Biodynamica (Normandy) **3**, 28 (1940).

OSTWALD, W.: Über die Geschwindigkeitsfunktion der Viskosität disperser Systeme. I. Kolloid-Z. **36**, 99 (1925).

PAECH, K.: Veränderungen des Plasmas während des Alterns pflanzlicher Zellen. Planta (Berl.) **31**, 295 (1940). — PEKAREK, J.: Absolute Viskositätsmessungen mit Hilfe der BROWNschen Molekularbewegung. Protoplasma **13**, 632—665 (1931). — Absolute Viskositätsmessungen mit Hilfe BROWNscher Molekularbewegung. IX. Mitt. Die Viskosität des Plasmas nach Kappenplasmolyse. Protoplasma **34** (1940). — PFEIFFER, HANS: Versuche über das Fadenziehvermögen isolierter pflanzlicher Protoplasten. Kolloid-Z. **70**, 26 (1935). — PHILIPPOFF, W.: Viskosität der Kolloide. Dresden u. Leipzig 1942. — PIRSON, A., u. F. SEIDEL: Zell- und Stoffwechselphysiologische Untersuchungen an der Wurzel von *Lemna minor* L. unter besonderer Berücksichtigung von K- und Ca-Mangel. Planta (Berl.) **38**, 431 (1950).

REUTER, LOTTE: Über die Salzresistenz der Epidermiszellen des Blattes von *Pisum sativum*. Protoplasma **35**, 329 (1941). — Zur protoplasmatischen Anatomie des Keimblattes von *Soja hispida*. Österr. bot. Z. **95**, 374 (1948). — RUGE, ULRICH: Kritische Zell- und entwicklungsphysiologische Untersuchungen an den Blattzähnen von *Helodea densa*. Flora (Jena) **34**, 311 (1940). — RUHLAND, W.: Zur Kälteresistenz der Pflanzen. Ber. sächs. Akad. Wiss., Math.-physik. Kl. **87**, 37 (1935).

SCARTH, G. W.: Dehydration injury and resistance. Plant Physiol. **16**, 171 (1941). — Structural differentiation of cytoplasm, pp. 109—126. In: The Structure of Protoplasm edit. by W. SEIFRIZ, Iowa State College Press 1942. — SCARTH, G. W., and J. LEVITT: The frost-hardening mechanism of plant cells. Plant Physiol. **12**, 51—78 (1937). — SCHMIDT, HELMUT: Plasmazustand und Wasserhaushalt bei *Lamium maculatum*. Protoplasma **33**, 25 (1939). — SCHMIDT, H., K. DIWALD u. O. STOCKER: Plasmatische Untersuchungen an dürreempfindlichen und dürreresistenten Sorten landwirtschaftlicher Kulturpflanzen. Planta (Berl.) **31**, 559 (1940). — SEIFRIZ, W.: Les transformations sol-gel du protoplasme. Rev. d'Hématol. **5**, 191—602 (1950). — A molecular interpretation of toxicity. Protoplasma **60**, 313 (1951). — The rheological properties of protoplasm. In FREY-WYSSLING, Deformation and flow in biological systems. Amsterdam 1952. — SEIFRIZ, W., and H. L. POLLACK: A colloidal interpretation of biological stimulation and depression. J. Colloid. Sci. **4**, 19 (1949). — SIMINOVITCH, D., and J. LEVITT: The relation between frost resistance and the physical state of protoplasm. Canad. J. Res. Sect. C **19**, 9—20 (1941). — STÅLFELT, M. G.: Über die lichtbedingten Hemmungsvorgänge in der Kohlensäureassimilation. Sv. bot. Tidskr. **39**, 365 (1945). — The influence of light upon the viscosity of protoplasm. Ark. Bot. (Stockh.) **33**, 17 (1946). — Influence of litter extract on the viscosity of protoplasm. Sv. bot. Tidskr. **41** (1946). — Effect of heteroauxin and colchicine on protoplasmic viscosity. Proc. of the Sixth Internat. Congr. of Exp. Cytology, Stockholm. Exper. Cell Res. Suppl. **1** 63—78. — Soil substances affecting the viscosity of the protoplasm. Sv. bot. Tidskr. **42**, 17—33 (1948). — The lability of the protoplasmic viscosity. Physiol. Plantarum (Copenh.) **2**, 341—349 (1949). — The effect of oxyquinoline on protoplasmic viscosity. Ann. Estacion exper. Aula Dei **2** (1950). — STOCKER, O.: Beiträge zu einer Theorie der Dürreresistenz. Planta (Berl.) **35**, 445—466 (1948). — STARLING, S. G., and A. J. WOODALL: Physics. London 1952. — STRUGGER, S.: Zur protoplasmatischen Kausalanalyse des Streckungswachstums. Jb. wiss. Bot. **79**, 406 (1934). — Praktikum der Zell- und Gewebephysiologie der Pflanze. Berlin 1949.

THORNTON, F. E.: The action of radium, potassium, calcium, and magnesium ions on the plasmagel. Physiologic. Zool. 8, 246 (1935). — TIMMEL, H.: Zentrifugierversuche über die Wirkung chemischer Agentien insbesondere des Kalciums auf die Viskosität der Protoplasten. Protoplasma **3**, 127 (1922/28). — TJIO, J. H., and A. LEVAN: The use of oxyquinoline in chromosome analysis. An. Estacion exper. Aula Dei **2**, 21 (1950). — TOBIAS, JULAIN M., and S. SALOMON: Electrically induced polar changes in viscosity in the hyaline protoplasm of *Elodea* with observations on streaming and plastid charge. J. Cellul. a. Comp. Physiol. **35**, 1—9 (1950).

UMRATH, K.: Über die Ausbreitung der durch Verwandlung bedingten Viskositätsveränderung bei *Spirogyra*. Protoplasma **36**, 410 (1942).

Virgin, H. I.: Changes in the viscosity of *Spirogyra* cytoplasm under the influence of light. Proc. 6th. Internat. Congr. of Exper. Cytol., Stockholm 1947. — A localized effect of light on the protoplasmic viscosity of plant cells. Nature (Lond.) **166**, 485 (1950). — The effect of light on the protoplasmic viscosity. Physiol. Plantarum (Copenh.) **4**, 255 (1951). — An action spectrum for the light induced changes in the viscosity of plant protoplasm. Physiol. Plantarum (Copenh.) **5**, 575—582 (1952). — Physical properties of protoplasm. Annual Rev. Plant Physiol. **4**, 363 (1953). — Virgin, H., and L. Ehrenberg: Effects of β- and γ-rays on the protoplasmatic viscosity of *Helodea* cells. Physiol. Plantarum (Copenh.) **6**, 159—165 (1953).

Walter, H.: Plasmaquellung und Wachstum. Z. Bot. **16**, 353 (1924). — Weber, Friedl: Die Zellsaftviskosität lebender Pflanzenzellen. Ber. dtsch. bot. Ges. **39**, 188 (1921). — Plasmolyseform und Protoplasmaviskosität. Österr. Bot. Z. **73**, 261 (1924). — Die Beurteilung der Plasmaviskosität nach der Plasmolyseform. Z. Mikrosk. **42**, 146 (1925). — Plasmolyse-Zeit-Methode. Protoplasma **5**, 622 (1929a). — Plasmolysezeit und Lichtwirkung. Protoplasma **7**, 256 (1929b).

Die Koordination der Reaktionssysteme.

Von

H. J. Bogen.

Mit 2 Abbildungen.

Einleitung.

Mit vollem Recht bezeichnet EPHRUSSI (1953) die Formulierung der Zelltheorie durch VIRCHOW (1858) als den entscheidenden Schritt, durch den die Biologie zu einer wissenschaftlichen Disziplin wurde. VIRCHOWs Aussage, daß die Zelle die letzte lebende Einheit sei, die mit sämtlichen Charakteristicis des Lebens ausgestattet ist, daß jede Zelle „durch legitime Succession auf eine Ursprungszelle zurück"gehe, daß jedes Leben Zellenleben und außerhalb der Zelle kein Leben sei — diese Aussagen sind in der Tat die fundamentalen Sätze nicht nur der Zellphysiologie, sondern der Biologie überhaupt. Nach ihnen stellt sich die Zelle gewissermaßen als ein Elementarorganismus dar: wohl ist sie, wie das Element, zerlegbar und damit der Analyse zugänglich, aber durch die Zerlegung verliert sie zugleich das Leben, d. h. die Fähigkeit, sämtliche Charakteristika des Lebens zu zeigen — was nicht ausschließt, daß einzelne der Zellbruchstücke das eine oder andere der Charakteristika noch lange Zeit beibehalten können.

In der Folgezeit tauchten bei manchen Forschern Zweifel an der Gültigkeit der Zelltheorie auf. Die Fülle der licht- und elektronenmikroskopisch erkennbaren Feinstrukturen in der Zelle schien die Möglichkeit offen zu lassen, daß es allerletzte Einheiten „unterhalb" der Zelle gäbe, die das eigentlich Lebendige seien und die die Zelle nicht allein aufbauten, sondern auch beherrschten. Biosomen, Bioplasten, Chondriosomen u. ä. wurden diese Steuerungszentren benannt. Auch die Viren erschienen eine Zeitlang als letzte Einheiten, die dem untergeordneten Substrat „Zelle" ihren Stempel aufdrückten. In gewisser Hinsicht war sogar die Genetik an dieser Abwendung von der Zelltheorie beteiligt: Die Erforschung der karyotischen Vererbung machte die überragende Rolle des Zellkerns im Zellgeschehen deutlich.

Seit einiger Zeit mehren sich jedoch die Stimmen, die denen beipflichten, die allen Gegenargumenten zum Trotz an der Zelltheorie festgehalten haben. Die Entwicklung der makromolekularen Chemie, die Bereicherung unserer Kenntnisse über die außerkaryotische Vererbung, die reichen Ergebnisse aus der Biochemie und Physiologie der „partikulären Zellfraktionen" — um einige der wichtigsten Disziplinen zu nennen — haben entscheidend dazu beigetragen, daß wir heute *beweisen* können, was VIRCHOW formulierte: die Zellkonstituenten, mögen sie auch diese oder jene Lebensäußerung zeigen, sind unselbständige Mitspieler in der Zelle; erst ihr hochorganisiertes Zusammenwirken macht die Zelle aus, die selbständig ist und *alle* Charakteristika des Lebens aufweist.

Der Ausdruck „Organisation der Zelle" besagt hier nicht etwa das unterschiedslose Erfassen und Organisieren „großer Massen", sondern im Gegenteil das im Endeffekt *harmonische* Zusammenspiel — darin liegt seine Rechtfertigung. Im übrigen freilich bemäntelt er nur unser Unvermögen, die ungeheuere Vielfalt der Einzelbeziehungen zu konkretisieren. Es kann nicht Aufgabe des vorliegenden Artikels sein, aus den spärlichen Daten ein Bild vom Leben der Zelle zu entwerfen, denn für den Fall, daß wir alle Details kennten, würde das einer „allgemeinen Biologie" gleichkommen und unübersehbares Ausmaß annehmen müssen. Ebensowenig soll hier ein bloßer Katalog derjenigen Einzelreaktionen gegeben werden, die hier und da aufgefunden wurden. Es kann allenfalls versucht werden, an Hand gewisser Beispiele zu zeigen, mit welchen *Möglichkeiten* der Wechselbeziehung überhaupt zu rechnen ist; ob, wann und wo sie im Einzelfall realisiert sind, muß vielfach noch offen bleiben.

Bei der Behandlung des Stoffes werden von zahlreichen anderen Möglichkeiten 2 Gesichtspunkte zugrunde gelegt:

1. Das Zusammenwirken wird diskutiert unter dem Gesichtspunkt der Zusammenarbeit und der gegenseitigen Konkurrenz der Reaktionsteilnehmer und zwar als Konkurrenz um Substrate im weitesten Sinne und als Konkurrenz um Positionen (Zellorte). Das setzt, in Übereinstimmung mit anderer Erfahrung, zunächst voraus, daß in der Zelle ein Gleichgewicht im thermodynamischen Sinne nicht existiert (v. BERTALANFFY 1953); die Mehrzahl der Reaktionsteilnehmer ist nicht etwa abgesättigt, sondern befindet sich in einem „Sättigungsdefizit". Nur so kann ein Fließgleichgewicht entstehen (und sich automatisch einregulieren). Es setzt aber, im Hinblick auf die Harmonie des Zusammenspiels, weiter voraus, daß die Sättigungsdefizite der Reaktionspartner sich innerhalb gewisser Grenzen halten und aufeinander abgestimmt sind. Gesellt sich dem Kreis der Partner ein neuer hinzu, der die Grenzen nicht einhält, so können irreversible Störungen auftreten. In diesem Sinne kann formal, d. h. ohne Kenntnis bzw. Berücksichtigung der Detailvorgänge, z. B. das Ausbleiben der Ergrünung von Chloroplasten in einem nicht passenden Plasmon verstanden werden, in dem die neu hinzugekommenen (Pro-) Plastiden entweder ein so geringes Potential haben, daß es zu genügender Substratbeschaffung nicht ausreicht, oder in dem das Defizit so groß ist, daß andere Reaktionspartner beeinträchtigt werden. Möglicherweise liegen bei einer Virusinfektion mit „Umschaltung" der Proteinproduktion ähnliche Verhältnisse vor.

2. Als Prinzip der Stoffgliederung bietet sich zunächst die Einteilung nach den Lebenserscheinungen an, also Stoffwechsel, Fortpflanzung, Vererbung, Differenzierung, Entwicklung, Reizbarkeit usw. Da wir indessen über fast alle diese Phänomene noch völlig im Ungewissen sind und allenfalls beim Stoffwechsel und der Differenzierung erste Ansatzpunkte haben, soll statt dessen nach den Reaktionspartnern der Zelle unterteilt werden:

1. Nieder- und makromolekularer Bereich,
2. Enzyme und Enzymsysteme (Multienzyme),
3. partikuläre Gebilde (Kern, Plastiden, Chondriosomen, Mikrosomen sowie Cytoplasma).

Darauf folgen Abschnitte über das Schicksal dieser Partner als Objekt des Stoffwechsels (4) und bei der Entstehung (5). Einige Anmerkungen über Fragen der Differenzierung (6) bilden den Abschluß.

1. Molekularer Bereich.

a) Wasserbindung.

Bereits am Beispiel des Wassers lassen sich einige Möglichkeiten der Wechselwirkung demonstrieren. Gemeinhin wird den verschiedenen Zellbestandteilen der gleiche Dampfdruck zugeschrieben. Das trifft zu, insofern das Wasser „als Masse" auftritt, z. B. in Vacuolen, als Wasser in mikrocapillaren Räumen usw., und daher der statistischen Erfassung zugänglich ist. Treten jedoch die Wassermoleküle mit Zellbestandteilen in (chemische oder physikalische) Wechselwirkung, so müssen sie als diskrete Partikel betrachtet werden. Als solche besitzen sie eine Anisotropie der Ladung, die als Dipolmoment gemessen wird. Eine analoge Anisotropie der Ladung besteht aber auch bei anderen Partnern, sei es im niedermolekularen Stoffwechsel, bei Makromolekülen oder bei Enzymen. Hier gibt es apolare Atomgruppen bzw. Molekülbereiche, wie z. B. Kohlenwasserstoffketten, die überhaupt kein Wasser binden, ferner Gruppen mit ansehnlichen Dipolmomenten (OH-Gruppen, undissoziierte COOH- oder NH_2-Gruppen) und ionisierte Gruppen. Alle diese Gruppen eines einzigen Makromoleküls werden sich in ihrer Wasserbindung zunächst untereinander und ferner insgesamt mit dem Wassergehalt bzw. der Wasserkonzentration des umgebenden Mediums abgleichen. Der resultierende Wassergehalt des Makromoleküls — der keineswegs maximal zu sein braucht, da ja das umgebende Medium in der Regel „Saugkräfte" entwickelt — vermag dann natürlich nur etwas über die Verteilung des Wassers zwischen Makromolekül und Medium auszusagen, nicht aber über die Verteilung innerhalb des Makromoleküls, und erst recht nicht über die Absättigungsdefizite der verschiedenen wasserbindenden Zentren. Gerade die Kenntnis der letzteren wäre aber vonnöten, denn eine Verlagerung von Wassermolekülen zieht ja nicht nur quantitative Veränderungen in Makromolekülen nach sich, sondern auch qualitative; es sei nur an die Beeinflussung der Molekülgestalt durch unterschiedliche Hydratation gedacht, die z. B. bei der Aktivität von Enzymen, Antigenen usw. eine erhebliche Rolle spielt.

Mit solchen Veränderungen ist in der Pflanzenzelle stets zu rechnen, da sich der Wassergehalt der Zelle aus Wasseraufnahme und Wasserabgabe (an Nachbarzellen oder an intercelluläre Lufträume) ergibt.

Aber auch anderweitig kommt es zu Änderungen des Wassergehaltes: Bei der Synthese von Makromolekülen wird Wasser gebildet, bei der Hydrolyse verschwindet Wasser ebenso wie bei der Photolyse. Wenn auch diese Veränderungen angesichts der gesamten Wassermenge geringfügig erscheinen mögen, am Reaktionsort selbst können sie erhebliche Ausmaße annehmen, und dank der gegenüber der Umsatzrate mancher Enzyme geringen Diffusionsgeschwindigkeit braucht sich ein solcher stoffwechselbedingter Überschuß oder Mangel nicht „schlagartig" auszubreiten, sondern wird langsam ausgeglichen und die den Reaktionen benachbarten Stellen früher betreffen als entferntere (Diffusionsfaktor, vgl. S. 624).

Die Vernichtung des gebildeten Potentials durch Diffusion kann aber auch verhindert oder wenigstens weit hinausgeschoben werden, wenn der Reaktionsort innerhalb eines höher organisierten Systems liegt, das durch Membranen von der Umgebung abgegrenzt ist, wie z. B. im Kern, in den Chondriosomen usw.

In anderen Fällen ist das Wasserbindungsvermögen stärker als die „osmotische Barriere". Die Verarmung von Chondriosomen an ATP (und/oder anderen Cofaktoren) führt zu einem Anschwellen, das im wesentlichen die Folge einer Wasseraufnahme ist (RAAFLAUB 1953). Um eine osmotische Aufnahme scheint es sich dabei nicht zu handeln (HARMAN 1952); es ist freilich noch unbekannt, welche Chondriosomenbestandteile ihr Wasserbindungsvermögen steigern, und welches die Ursachen sind. Noch beträchtlicher ist die Wasserverschiebung, die bei der sog. Vacuolenkontraktion von der Vacuole in „das" Cytoplasma gerichtet ist. Dabei ändert sich das Volumen des Protoplasten insgesamt nicht (gleiches ereignet sich bei plasmolysierten Zellen: Kappenplasmolyse). Hier dürften umfangreiche, durch unspezifische Reize ausgelöste Form- und Faltungsänderungen von Plasma- und Vacuolenkolloiden beteiligt sein (BOGEN 1951).

Diese Beispiele zeigen, daß selbst das Wasser, ein Stoff also, der in größten Mengen in der Zelle vorhanden ist und biochemisch nur eine bescheidene Rolle spielt, innerhalb der Zelle höchst unterschiedlich verteilt ist und zwar auch in Bereichen so verschiedener Organisationshöhe wie Makromolekül, Enzym, Chondriosom und Cytoplasma. Diese Verteilung resultiert aus dem Zusammenspiel aller Reaktionspartner und ist keineswegs konstant. Bereits kleine Veränderungen im Ladungsmuster eines Makromoleküls, im Faltungszustand eines Enzymproteins, im Stoffwechsel eines Chondriosoms usw., wie sie sich aus geringfügigen Stoffumsätzen oder bei der Reizaufnahme ergeben können, vermögen die Wasserverteilung ganz grundlegend umzugestalten und damit zugleich den Anstoß zu einem veränderten Zusammenspiel zu geben. Das kann in extremen Fällen zum Zelltod führen, ebensogut aber, auf unbegreiflichen Wegen, auf eine Wiederherstellung des ursprünglichen Zustandes gerichtet sein. In keinem Fall aber ist die Wasserverteilung mit den Begriffen Gleichgewicht der Dampfdrucke, Isotonie u. dgl. ausreichend zu beschreiben.

Tabelle 1. *Höchstzahlen der ionenbindenden Loci bei Serumalbumin.* (Nach KLOTZ 1949.)

Ion	*n*
Chlorid	10
Dodecylsulfat. . .	14
Phenylbutyrat . .	24
o-Nitrophenolat. .	6
m-Nitrophenolat .	24
p-Nitrophenolat. .	25
Methylorange. . .	22
Cu^{++}	16

b) Ionenbindung.

Gegenüber der Wasserbindung erfolgt die Bindung von Ionen durch Makromoleküle im stöchiometrischen Verhältnis (nur Detergents können anscheinend als Micellen gebunden werden — HAUROWITZ 1950); sie ist eine Hauptvalenzbindung zwischen kationischen Gruppen und Anionen bzw. umgekehrt, während VAN DER WAALSsche Kräfte nur eine untergeordnete Rolle spielen (KLOTZ 1949). Dabei ist die Zahl der ionenbindenden „Ladungszentren" ziemlich gering; sie liegt, wie Tabelle 1 für Serumalbumin ausweist, zwischen 6 und 25, wobei so verschiedene Ionen wie die anorganischen Ionen Kupfer und Chlor, Detergents, Nitrophenol und Farbstoffe berücksichtigt sind.

Da die Ionenbindung von der Ionen*konzentration* abhängig ist, liegen die Werte in der Regel niedriger, in 10^{-2} mol Cu^{++} z. B. bei 13. Diese Zahl hat nur statistischen Aussagewert und bedeutet, daß neben 13 Cu-haltigen Molekülen auch solche mit $\gtrless$ 13 Cu-Ionen vorliegen. Bereits hierin äußert sich eine Konkurrenz gleicher Makromoleküle auch um die Ionen der Umgebung. Die Konkurrenzfähigkeit der Makromoleküle („Ionenaffinität") läßt sich nach KLOTZ (1949) durch den sog. Bindungs*index* ausdrücken:

$$\sum(\equiv NH^+)/|\sum(-OH) - \sum(-COO^-)|.$$

Die Menge gebundener Ionen ergibt sich aus dem Zusammenspiel zwischen Bindungs*index* und Bindungs*energie*. Die Bindungsenergie ihrerseits wächst, wenigstens innerhalb homologer Reihen, mit dem Molekulargewicht; einige für Serumalbumin geltende Beispiele sind in Tabelle 2 aufgeführt.

Daraus ergibt sich, daß Ionen mit geringer Bindungsenergie (z. B. Chlorid) auch nur in geringem *Ausmaße* gebunden werden; Ionen mit energiereicher Bindung hingegen (z. B. Dodecylsulfat) bilden *zahlreiche* Ion-Protein-Komplexe, selbst mit Proteinen geringer Bindungsindices. In den letzteren Fällen können sogar ursprünglich vorhandene Ion-Protein-Komplexe gelöst werden; das hat weitreichende Folgen, wenn sog. funktionelle Gruppen betroffen werden.

Ferner nimmt die Bindungsenergie immer mehr ab, wenn vom Protein mehr Ionen gebunden werden. Hierfür ist zunächst ein statistischer Faktor verantwortlich zu machen· die Wahrscheinlichkeit für die Heranführung eines Ions an eine noch freie Gruppe im Makromolekül sinkt mit fortschreitender Besetzung dieser Stellen. Dazu kommt die „elektrostatische Abstoßung“, die von bereits gebundenen Ionen auf die übrigen Ionen ausgeübt wird. Gleich*artige* Ionenbindungen dürfen somit hinsichtlich der Bindungs*energie* nicht gleich bewertet werden.

Haben wir an Stelle eines einheitlichen makromolekularen Proteins ein Gemisch aus mehreren Proteinen oder gar aus Proteinen, Nucleinsäuren, Lipoiden und höheren Komplexverbindungen, so wird sofort die Unmöglichkeit offenbar, die ungeheuere Vielfalt der Konkurrenzeffekte mit lediglich statistischen Angaben zu erfassen. Das Bild wird noch verwickelter, weil die Effekte solcher Komplexe nicht auf den Bindungsort beschränkt zu sein brauchen. Eine Ionenbindung (unter „Entladung“) verändert das Ladungsmuster, das Wasserbindungsvermögen und die Gestalt des *gesamten* Makromoleküls; ähnliche Folgen mögen die sog. Chelatbildungen haben; Schwermetalle können Proteine ausfällen; andere Metallproteide (Cu, Fe) haben katalytische Wirkungen usw. Alle diese Faktoren fallen besonders dann ins Gewicht, wenn Makromoleküle als Bestandteile von Enzymen auftreten; in solchen Fällen sind nachhaltige Änderungen der enzymatischen Aktivität zu erwarten.

Tabelle 2. *Bindungsenergie einiger Serumalbuminkomplexe.* (Nach KLOTZ 1949.)

Ion	ΔF_1^0
Chlorid	— 3,700
p-Nitrophenolat. .	— 4,865
Methylorange. . .	— 6,050
Dodecylsulfat. . .	—11,000

Unter den Ionen spielt das bisher nicht ausdrücklich genannte Wasserstoffion wie auch das OH-Ion eine besondere Rolle, denn vor allem Proteine haben ein ansehnliches Protonenbindungsvermögen. Dessen besondere Bedeutung liegt darin, daß die Wasserstoffionenkonzentration der Umgebung im Zusammenwirken mit dem inhärenten Faktor „IEP“ den Ladungssinn und den Überschuß an positiver bzw. negativer Ladung bestimmt. Gestalt, Stabilität, Wasser- und Ionenbindungsvermögen usw. werden auf diese Weise weit nachdrücklicher beeinflußt als durch Wasser allein oder durch anorganische Ionen in geringen Konzentrationen.

Das unterschiedliche Protonenbindungsvermögen führt zu neuen Konkurrenzbeziehungen. Als Beispiel sind in Tabelle 3 die Prozentsätze derjenigen Hämoglobinmoleküle aufgeführt, die im IEP des Hämoglobins (p_H 6,9) wirklich isoelektrisch sind bzw. mit einer bis mehreren positiven bzw. negativen Überschußladungen versehen sind; die einzelnen gleichartigen Makromoleküle binden die Wasserstoffionen in recht verschiedenem Ausmaß.

Tabelle 3. *Isoelektrisches Hämoglobin bei p_H 6,9; Anteil der wirklich isoelektrischen Moleküle (freie Ladung 0) und der Moleküle mit 1—3 und mehr positiven bzw. negativen Ladungen.* (Nach HAUROWITZ 1950.)

Freie Ladung	> —3	—3	—2	—1	0	+1	+2	+3	> +3
% der Moleküle . . .	2,5	3,0	14,2	21,2	22,4	17	9,4	3,9	1,5

Sind neben den Wasserstoffionen noch andere anorganische Kationen vorhanden, so treten diese gleichfalls als Konkurrenten auf, um so mehr, als der Energiegehalt der Metallionenbindung größer ist als der der Protonenbindung. Diese Verhältnisse sind bei der Auswertung von Versuchen mit Pufferlösungen zu berücksichtigen.

In solchen Versuchen werden indessen nicht allein die Wasserstoffionen mit den vorhandenen Kationen, sondern auch die Anionen des Puffers mit den vorgegebenen Anionen in Konkurrenz treten. Abb. 1 zeigt das Ausmaß der

Bindung von Methylorange durch Serumalbumin in Abhängigkeit von der Pufferlösung; je länger der Balken, desto mehr Methylorange wird gebunden, d. h. desto größer ist seine Überlegenheit über das konkurrierende Anion des Puffers.

Somit ist die Voraussage über den Effekt von Ionen, die einem Gemisch von Makromolekülen von außen zugeführt werden, vorerst nur dann möglich, wenn sowohl die einzelnen Bindungsenergien der verschiedenen Ionenkomplexe als auch die Bindungsindices der Makromoleküle bekannt sind.

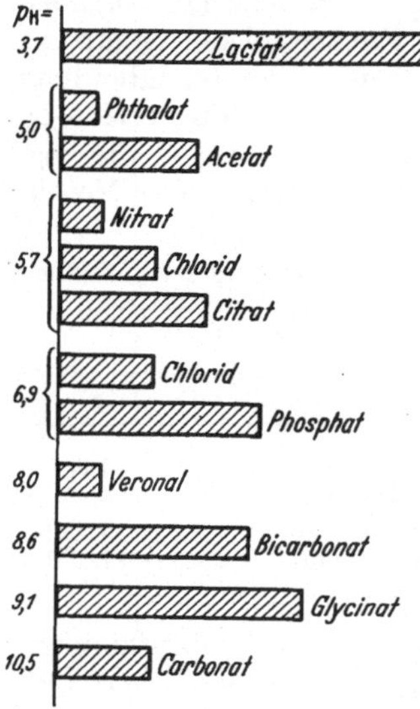

Abb. 1. Einfluß verschiedener Pufferlösungen auf die Bindung von Methylorange an Serumalbumin. Die Länge des Balkens gibt die relative Menge an gebundenem Farbstoff an. (Nach Klotz 1949.)

In der lebenden Zelle haben wir ferner damit zu rechnen, daß im Stoffwechsel selbst Ionen erzeugt werden — hauptsächlich Wasserstoffionen und organische Anionen. Diese sind gleicherweise an der Konkurrenz beteiligt, nur sind hier Produktion oder Verschwinden von Ionen an die Reaktionsorte gebunden; die Konzentrationsunterschiede gegenüber anderen Zellorten können, wie beim Wasser, längere Zeit bestehenbleiben und werden in besonderen Fällen womöglich überhaupt nicht ausgeglichen. Statistische Zahlen wie „der" p_H-Wert des Cytoplasmas oder „die" Calciumkonzentration in der Zelle sind also angesichts der Aufteilung der Zelle in Mikroreaktionsräume mit besonderen Konzentrations- und Konkurrenzverhältnissen wenig ergiebig.

Für Lipoide können insofern noch besondere Bedingungen gelten, als sie, oder wenigstens die Phosphatide unter ihnen, mono- und bimolekulare Filme zu bilden vermögen. Bei monomolekularen Filmen ist die Dichte und die Durchlässigkeit von der Dissoziation der Phosphorsäure abhängig; bimolekulare Filme benötigen zur Aufrechterhaltung ihrer Stabilität ansehnliche Mengen von Kationen, die sie in sich einschließen (Davson und Danielli 1943). Daher treten auch solche Filme, die überdies für die Abgrenzung von Reaktionsräumen gegen die Umgebung von Bedeutung sein mögen, als Konkurrenten um die Ionen auf, während zugleich ihre Stabilität von Ionenart und -menge abhängt.

2. Enzyme und Multienzyme.

Enzyme können als makromolekulare Proteine bzw. Komplexe (Symplexe) aus makromolekularen Proteinen und Nichtproteinanteilen aufgefaßt werden. Von denjenigen Proteinen, die lediglich oder vorwiegend als („inerte") Bausteine dienen, sind sie durch ihre sehr aktive Rolle als Katalysatoren des Stoffwechsels unterschieden. Für sie gelten zunächst die gleichen Abhängigkeitsbeziehungen, die im vorangehenden Abschnitt diskutiert wurden; die enzymatische Aktivität wird daher maßgeblich von der Wasserbindung, der Ionenbindung usw. beeinflußt. Darüber hinaus muß jedoch mit einer neuen Art von Relationen gerechnet werden, die im übrigen gleichfalls unter dem Gesichtspunkt Kooperation/Konkurrenz betrachtet werden darf. Sie ergibt sich letztlich aus der Tatsache, daß die meisten Enzyme aus einem Proteinanteil und einer prosthetischen Gruppe bestehen, die im einfachsten Falle ein metallisches Ion, in anderen Fällen eine niedermolekulare organische Verbindung ist („Vitamin").

Es besteht weitgehende Übereinstimmung in der Auffassung, daß der Proteinanteil eines Enzyms, das Apoenzym, Träger der Substratspezifität ist und wahrscheinlich eine Enzym-Substrat-Verbindung eingeht[1]; das Coenzym hingegen ist maßgebend für die Wirkungsspezifität, d. h. für die Art der enzymatischen Veränderung, der das Substrat unterliegt. In der Mehrzahl der Fälle arbeiten zahlreiche Enzyme in der Weise zusammen, daß das erste das Ausgangssubstrat in bestimmter Weise verändert und es damit zum Substrat für das nachfolgende Enzym aufbereitet usw. So können ganze Enzymketten nach Art eines Fließbandes Hand in Hand arbeiten und das Prinzip der Kooperation nahezu vollkommen verwirklichen (z. B. im Tricarbonsäurecyclus). Solche Kettenprozesse und Cyclen bestehen aber natürlich nicht unabhängig nebeneinander, sondern berühren bzw. überschneiden sich an bestimmten Stellen. Derartige Stellen können z. B. dort gegeben sein, wo ein und dasselbe Substrat in verschiedener Weise verändert wird, wie etwa die α-Ketoglutarsäure, die entweder zu Bernsteinsäure decarboxyliert und dehydriert wird oder aber unter Hydrierung und Aminierung zur Glutaminsäure wird. Hier konkurrieren zwei Enzyme um das gleiche Substrat, wobei die Verteilung des Substrates nicht allein durch die „Affinitäten" der beiden Enzyme geregelt wird, sondern auch nach Maßgabe der Konzentration an $DPNH_2$, NH_3 usw. (vgl. hierzu S. 616).

Das Entgegengesetzte tritt ein, wenn verschiedene Stoffe der gleichen Verarbeitung (z. B. der Dehydrierung) unterworfen werden. In einem solchen Falle ist dasselbe Coenzym wirksam (DPN bzw. $DPNH_2$) und es ist bekannt, daß es zwischen verschiedenen Apoenzymen hin- und herpendeln kann, und daß verschiedene Apoenzyme um das Coenzym konkurrieren. Der Übertragung von Wasserstoff („Transhydrierung") entsprechen die Vorgänge der Transaminierung, Transmethylierung, Transglykosidierung, Transphosphorylierung usw. Auch hier wird der erreichte Effekt sowohl von der Konkurrenzfähigkeit („Affinität") der Apoenzyme als auch von der Konzentration der zu aminierenden usw. Substrate abhängen.

Noch andere Phänomene sind zu beobachten, wenn dem normalerweise zu verarbeitenden Substrat bei seiner Bindung an das Enzym ein Konkurrent entgegentritt, der die Bindungsstellen blockieren kann und vielleicht sogar das Substrat aus seiner ursprünglichen Bindung verdrängt (z. B. die Hemmung der Bernsteinsäuredehydrierung durch Malonsäure). Solche kompetitiven Hemmungen können an verschiedenen Stellen des Stoffwechselgeschehens entweder durch eigene Stoffwechselprodukte oder aber auch durch künstlich zugeführte Substanzen (Inhibitoren wie Natriumazid u. ä.) bewirkt werden; sie spielen bei der Analyse von Stoffwechselketten eine hervorragende Rolle.

Zu der Vielfalt der Faktoren, die den metabolischen Effekt eines Enzyms bestimmen, kommen noch jene, die mit dem Zellort des Enzyms zusammenhängen. Es ist schon längere Zeit bekannt, daß lösliche Enzyme bzw. Enzyme *in vitro* sich anders verhalten können als gebundene Enzyme *(in situ)*. Die Unterschiede beziehen sich unter anderem auf das p_H-Optimum, die Hemmbarkeit durch Stoffwechselprodukte und zellfremde Stoffwechselgifte und den Bedarf an Cofaktoren (HUENNEKENS 1951), aber auch auf die Spezifität (d- bzw. l-Milchsäure; MAHLER und HUENNEKENS 1952).

In letzter Zeit sind schließlich sog. Multienzyme genauer untersucht worden (Zusammenstellung bei MAHLER 1953); es handelt sich dabei zumeist um partikuläre Gebilde, die bei der Zerschlagung von Chondriosomen erhalten wurden. Bei ihnen ist vor allem bemerkenswert, daß sie einen besonders niedrigen Bedarf an Cofaktoren haben; das wird anscheinend ermöglicht durch die räumliche Nähe der verschiedenen hintereinandergeschalteten Enzyme.

[1] Vgl. diesen Band: DUSPIVA, F., Die Mitwirkung von Enzymen im Zellstoffwechsel, S. 489ff.

3. Die mikroskopisch sichtbaren Komponenten des Protoplasten.

a) Allgemeines.

Unter diese sind zunächst die sog. partikulären Gebilde zu rechnen: der Zellkern, die Plastiden, die Chondriosomen und die Mikrosomen (obgleich letztere oft nur submikroskopische Ausmaße erreichen). Dazu gesellen sich das Cytoplasma i. e. S. sowie vielleicht die Zellgrenzschicht und die Vacuole. Im vorliegenden Abschnitt soll lediglich ihre Anwesenheit konstatiert und ihre Funktionen und Strukturen analysiert werden, ohne Rücksicht auf die Genese, d. h. ob sie Gebilde *sui generis* sind oder nicht, ob sie autoreproduktive Einheiten darstellen oder nicht; den letzteren Fragen ist ein eigener Abschnitt gewidmet.

Sicherlich handelt es sich in jedem Fall um subcelluläre (Mikro-) Reaktionsräume, in denen verschiedene Prozesse ablaufen. In erster Annäherung könnte man differenzieren zwischen stoffwechselphysiologischen Aufgaben (Chloroplasten: Photosynthese; Chondriosomen: Kohlenhydratabbau, ATP-Gewinnung, Lipoidsynthese) und entwicklungsphysiologischen Funktionen (Zellkern: „Genstoffe"). Dann bleibt freilich die Rolle des Cytoplasmas, in das alle diese Partikel eingebettet sind, unklar, ebenso die Rolle der Vacuole, die als Abfallstätte zweifellos nicht ausreichend beschrieben ist. Und die Mikrosomen scheinen auf beiden Gebieten tätig zu sein: als Orte der Proteinsynthese und — wenigstens zum Teil — als Vorstufen für Chondriosomen. Aber es gibt stichhaltigere Gründe für die Ansicht, daß die oben getroffene Unterscheidung nicht streng durchführbar ist; betrachten wir die Enzymaktivität als Indicator für die metabolische Aktivität, so sind alle Komponenten, wenn auch unterschiedlich, am Stoffwechsel beteiligt — auch im Zellkern und in der Vacuole konnten Enzymaktivitäten gemessen werden —, andererseits ist die Rolle der Plastiden, der Chondriosomen (EPHRUSSI 1953), gewisser Mikrosomen (SONNEBORN 1943, 1946) und des Cytoplasmas (MICHAELIS 1947, 1951, OEHLKERS 1952, WARIS 1950) bei den Vererbungs- und Entwicklungsvorgängen, ihrer Mutabilität usw. bekannt. Es muß daher vorausgesetzt werden, daß *alle* Komponenten sowohl im stoffwechselphysiologischen Geschehen der ausdifferenzierten Zelle als auch entwicklungsphysiologisch zusammenwirken.

Da alle Komponenten aus makromolekularen Bausteinen bestehen, gelten für sie die gleichen, im Abschnitt 1 behandelten Gesetzmäßigkeiten: Konkurrenz um Wasser, Ionen, zu verarbeitende Substrate usw., wobei die in der Mehrzahl vorhandenen Partikel wie Plastiden, Chondriosomen und Mikrosomen untereinander nicht *direkt* konkurrieren können, sondern nur auf einem Umweg über das Cytoplasma. Gleiches gilt für die Kerne polyenergider Zellen, für Heterokaryen usw.

Da die stoffwechselphysiologische Aktivität im wesentlichen eine *enzymatische* Aktivität ist, gelten ferner die im 2. Abschnitt diskutierten Gesichtspunkte. Wegen der Spezialisierung der Enzyme auf ganz bestimmte Substrate kann man in vielen Fällen erwarten, daß außer einer Konkurrenz auch eine Kooperation sichtbar wird. Die Chloroplasten benötigen Licht und CO_2 und produzieren O_2 und Kohlenhydrate, die Chondriosomen verarbeiten Kohlenhydrate, benötigen O_2 und liefern ATP, Aminosäuren, Phosphatide und wahrscheinlich Purin- und Pyrimidinbasen bzw. deren Vorstufen, die Mikrosomen verwenden Aminosäuren und liefern Proteine, der Kern bedarf der Nucleotide und Proteine usw.

Neu hinzu treten Faktoren, die mit der gegenüber den einfachen Enzymen höheren Organisation der partikulären Gebilde zusammenhängen. Die Tatsache, daß sie von Membranen umgeben sind (vielleicht mit Ausnahme der Mikrosomen), bringt es mit sich, daß die Konzentrationsverhältnisse im Inneren anders und für den Stoffwechsel durchaus nicht immer günstiger sind als im umgebenden Cytoplasma. So ist nach HARMAN (1952) die Kaliumkonzentration in den Mitochondrien normalerweise infraoptimal, kann aber unter bestimmten Bedingungen optimal werden. Die Abgeschlossenheit der Reaktionsräume verschärft aber auch die Konkurrenz zwischen den Makromolekülen bzw. Enzymen ein und derselben Partikel.

Diese hier skizzierten Relationen sind lediglich allgemeiner Art; es wird im einzelnen zu prüfen sein, inwieweit die Möglichkeiten realisiert sind und welches Ausmaß und welche Bedeutung sie annehmen können.

b) Die Verteilung der Enzyme in der Zelle.

Zur Lokalisierung der Enzyme bzw. Enzymaktivitäten innerhalb der Zelle werden vorwiegend 2 Methoden benutzt: mikrochemische Fällungsreaktionen und fraktionierte Zentrifugierung eines Zellhomogenats mit anschließender Bestimmung der Enzymaktivität in den einzelnen Fraktionen. Beide Methoden sind mit erheblichen Fehlern behaftet: bei 1 ist es nicht sicher, ob das sichtbare Fällungsprodukt auch wirklich am Ort seiner Entstehung verbleibt, bei 2 ist damit zu rechnen, daß bereits während der Homogenisierung erhebliche Schädigungen der partikulären Gebilde eintreten, die zu einer (unterschiedlichen!) Abgabe von Enzymen in die Umgebung führen (vgl. NOSSAL an Hefe). Aus diesen Gründen sind alle Lokalisationsangaben, insbesondere solche quantitativer Art, mit großer Vorsicht zu bewerten.

Es ist nicht überraschend, daß diese Angaben einander oft widersprechen. Sehr selten ist darin Übereinstimmung zu finden, daß bestimmte Enzyme *ausschließlich* in einer bestimmten Zellkomponente auftreten bzw. niemals auftreten; meist berichten die Autoren nur, daß die *Hauptmenge* eines Enzyms hier oder dort zu finden ist, z. B. in den Chondriosomen, während in den Mikrosomen, dem Plasma, dem Zellkern usw. nur wesentlich geringere Mengen vorkommen. Man wird sich daher darauf beschränken müssen, nur drastische Aktivitätsunterschiede als real anzuerkennen, vor allem dann, wenn sie übereinstimmend nach Anwendung beider Methoden erhalten wurden. Überblickt man die Fülle der Angaben (WILDMAN und COHEN, SERRA, GRANICK, STEFFEN, s. Band 1 dieses Handbuches) unter diesem Vorbehalt, so ergibt sich etwa folgendes: die Chondriosomen sind besonders reich mit Enzymen ausgestattet. Wir finden hier sämtliche glykolytischen Enzyme, die des Tricarbonsäurecyclus, der Endatmung bis zur Cytochromoxydase, ferner Transaminasen, Fettsäureoxydasen und ATPase. Bei den Mikrosomen scheinen die Esterasen zu überwiegen; das Cytoplasma ist reich an Katalasen, einigen Dehydrasen und Tyrosinasen, während die Angaben über die Chloroplasten noch spärlich und widerspruchsvoll sind. Sicher scheint hier nur das Auftreten von Phosphorylasen und Polyphenoloxydasen zu sein. Der Zellkern schließlich[1] besitzt eine Reihe glykolytischer Enzyme in geringer Menge, ferner die Hauptmasse der Phosphatasen und (in tierischen Zellen) Enzyme des Eiweiß- und Aminosäurestoffwechsels.

Danach bestätigt sich in großen Zügen das bekannte Bild, nach dem die Chloroplasten mit Hilfe der Lichtenergie Kohlenhydrate aufbauen, die Chondriosomen hingegen durch Glucoseabbau (im Unterschied zu den tierischen Zellen setzt die Verarbeitung nicht erst bei der Brenztraubensäure ein) energiereiche Phosphate herstellen, die für anderweitige Synthesen benötigt werden. Einige von ihnen laufen in den Chondriosomen selbst ab (z. B. Aminierungen und Transaminierungen, vielleicht auch der Fettstoffwechsel; vgl. STEINER 1954), andere in den Mikrosomen (Proteinsynthese) und im Kern (Proteinsynthese?,

[1] Vgl. diesen Band: BAHR, G. F., Der Zellkern, S. 543 ff.

Nucleinsäuresynthese); letzterer selbst ist nur in begrenztem Maße fähig, durch Glykolyse energiereiche Verbindungen zu gewinnen.

Diese Übersicht läßt erkennen, daß eine nahezu reibungslose „Arbeitsteilung" erzielt ist; ähnlich wie in einem zusammengeschalteten Enzymsystem produziert hier die jeweilige partikuläre Fraktion „im Überschuß" gerade diejenigen Substrate bzw. Verbindungen, die die anderen benötigen. Ebenso deutlich ist aber, daß diese Koordination nur recht vergröberte Beziehungen zwischen einigen großen Bereichen des Stoffwechselgebietes erfaßt. Weite Areale bleiben noch völlig rätselhaft, so z. B. das Cytoplasma, dessen Tätigkeit sich mit der Rolle eines bloßen Vermittlers zwischen den partikulären Komponenten keineswegs erschöpft. Vor allem aber läßt diese Übersicht die Vielzahl der Einzelbeziehungen außer acht, die erst die „Organisation der Zelle" ausmachen. Über diese ist erst recht wenig bekannt; augenblicklich beginnen wir erst bei den Chondriosomen, die ja in den letzten 5 Jahren zum umfangreichsten Feld der Experimentalforschung geworden sind, derartige Zusammenhänge zu ahnen.

c) Chondriosomen.

Die Prüfung der enzymatischen Ausstattung von Chondriosomen richtete sich zunächst auf die Glieder des Citronensäurecyclus (z. B. SCHNEIDER und HOGEBOOM 1951), nachdem schon aus GREENS (1947) Untersuchungen das sog. Cyclophorasesystem bekannt geworden war. Die Chondriosomen besitzen die gesamte Enzymgarnitur des Citronensäurecyclus, und sie vermögen diesen Cyclus auch *in vitro* durchzuführen. Selbst ein Gemisch aus den entsprechenden löslichen Enzymen kann (Glucose bzw.) Brenztraubensäure zu CO_2 und H_2O abbauen. Gibt man hierzu einige Cofaktoren, wie vor allem Co-A, so führt das Gemisch auch die oxydative Phosphorylierung durch. Der Unterschied gegenüber den intakten Chondriosomen äußert sich zunächst lediglich darin, daß der Bedarf an Cofaktoren erheblich größer ist. Die submikroskopische Struktur der Chondriosomen, nämlich die Ausbildung der Cristae und vor allem einer Membran (vgl. STEFFEN, Band 1 dieses Handbuches), scheint also geeignet zu sein, die oxydativen Prozesse unter geringstem Aufwand mit maximaler Ausbeute ablaufen zu lassen.

Indessen kann sich die Bedeutung der Struktur darin nicht erschöpfen. Das wurde offenbar, als die entscheidende Entdeckung gemacht wurde, daß in den Chondriosomen auch energieverbrauchende Syntheseprozesse ablaufen, vor allem die Bildung von Aminosäuren (Transaminierung) und vielleicht von Fettsäuren („Transacylierungen"). Das bedeutet, daß das Substrat nur zum Teil abgebaut wird, zum anderen Teil aber für Syntheseprozesse bereitgehalten werden muß, die oxydative Kapazität also gar nicht voll ausgenutzt werden darf. LINDBERG und ERNSTER (1954) entwarfen hierfür ein sehr instruktives Bild, wobei sie von folgenden zwei Gleichungen ausgingen:

$$\text{(Abbau)} \quad \alpha\text{-keto-Glutarsäure} + \text{DPN} \rightarrow \text{Bernsteinsäure} + \text{DPNH}_2 + \text{CO}_2 \quad (1)$$

$$\text{(Synthese)} \quad \alpha\text{-keto-Glutarsäure} + \text{DPNH}_2 + \text{NH}_3 \rightleftharpoons \text{Glutaminsäure} + \text{DPN} + \text{H}_2\text{O} \quad (2)$$

$$2\,\alpha\text{-keto-Glutarsäure} + \text{NH}_3 \rightleftharpoons \text{Bernsteinsäure} + \text{Glutaminsäure} + \text{CO}_2 + \text{H}_2\text{O}$$

Unter anaeroben Bedingungen wird das Gleichgewicht (2) bestimmt durch die Konzentration von NH_3; bei O_2-Zufuhr jedoch bewerben sich um den Wasserstoff des $DPNH_2$ *zwei* Konkurrenten: α-Ketoglutarsäure + NH_3 (→ Glutaminsäure) und (letztlich) O_2 (→ H_2O).

Die Chondriosomen *in situ* sind imstande, das Gleichgewicht zwischen Verbrennung und Synthese herzustellen bzw. einzuregulieren, während *in vitro* meist ausschließlich Verbrennung stattfindet. Diese Regulation spielt sich offenbar unter der Mitwirkung zahlreicher Faktoren ab: zunächst dem in den Chondriosomen gebildeten $DPNH_2$, ferner dem ebendort entstandenen Substrat α-Ketoglutarsäure, weiterhin den Enzymen der Atmungskette, deren Anspruch wiederum durch die O_2-Zufuhr geregelt wird, und schließlich dem (wahrscheinlich von außen zugeführten) NH_3; d. h. die Menge der Reaktionspartner am Reaktionsort und die Sättigungsdefizite der Enzyme sind mitbestimmend.

Das errichtete Gleichgewicht ist seinerseits bestimmender Faktor für eine Reihe weiterer Reaktionen: die Synthese von Proteinen aus Aminosäuren, die Gewinnung energiereicher Phosphate aus der Atmungskettenphosphorylierung usw. Diese Zusammenhänge können heute allenfalls qualitativ betrachtet werden, da die Bedingungen für die vielfältig verflochtenen, meist außerordentlich schnell ablaufenden Reaktionen noch nicht meßbar sind. Wir können nur hoffen, daß es eines Tages gelingen wird, die Reaktionskonstanten exakt zu erfassen.

Ebensowenig wissen wir über die spezielle Bedeutung der Chondriosomenfeinstruktur. Können wir die Funktion der Membran noch darin sehen, daß Reaktionsprodukte, Cofaktoren u. ä. im Reaktionsraum gehalten und am Abdiffundieren in das umgebende Cytoplasma gehindert werden, so bleibt der Bau der Cristae noch rätselhaft. Wohl dürfen wir vermuten (vgl. die Multienzyme), daß auf ihnen die kooperierenden Enzyme nebeneinander gelagert sind, und daß die Transportwege der Substrate möglichst klein gehalten werden — experimentelle Beweise dafür sind kaum vorhanden. Hingegen wissen wir, *daß* die Struktur erforderlich ist. Beeinträchtigen wir sie durch mechanische oder osmotische Kräfte (HARMAN 1952, LATIES 1954), so sinkt die Leistung der Chondriosomen allmählich ab. Aber auch das Umgekehrte ist der Fall: blockieren wir den Stoffwechsel etwa durch DNP oder Natriumazid, so verschwindet die Struktur, und es konnte bereits nachgewiesen werden, daß ein Teil der in den Chondriosomen produzierten energiereichen Phosphate, insbesondere ATP, in den Chondriosomen selbst wieder verbraucht wird *zur Aufrechterhaltung der Struktur* (RAAFLAUB 1953).

Hier ist der Zusammenhang — fast möchte man sagen die Identität — von Stoffwechsel und Struktur besonders klar erkennbar.

Weitere Beziehungen zwischen der Chondriosomenaktivität und der Tätigkeit der gesamte Zelle scheinen bei der Stoffaufnahme zu bestehen. Nach LUNDEGÅRDHs (1950) Theorie der Anionenatmung erfolgt die Aufnahme der Anionen über die „Elektronenleiter" der Cytochromkette; der Anionenstrom ist um so stärker, je lebhafter der Elektronentransport erfolgt, ja die Anionen sollen sogar einen Teil dieser Atmung hervorrufen (ROBERTSON 1955). Allerdings sind auch gegenteilige Stimmen laut geworden, und es ist nicht sicher, ob diese Beziehungen zwischen Stoffwechsel (= Chondriosomenaktivität) und Stoffaufnahme wirklich so unmittelbar sind. Immerhin kann an einer wenigstens *mittelbaren* Beteiligung der Chondriosomen an der nichtosmotischen Stoffaufnahme wohl nicht mehr gezweifelt werden, selbst wenn sie sich nur insofern äußert, als energiereiche Verbindungen für den Transport oder die Aufrechterhaltung labiler, wasserbindender Vacuolenkolloide bereitgestellt werden. Das betrifft auch die nichtosmotische Wasseraufnahme (BOGEN 1953, BOGEN und FOLLMANN 1955, NEUBER 1956, im Druck).

Die Vorstellung über den Anteil der Chondriosomen am Leben der Zelle bedarf noch einer wichtigen Ergänzung, deren Bedeutung freilich noch nicht exakt faßbar ist. Bisher wurde so argumentiert, als seien die Chondriosomen einer Zelle untereinander gleich. Das scheint nach mehreren neuen, voneinander unabhängig erhobenen Befunden nicht der Fall zu sein. Die Chondriosomen-fraktion läßt sich ihrerseits noch unterteilen in Subfraktionen mit recht unterschiedlichem Enzymgehalt und daher wohl unterschiedlicher Aktivität (z. B. KUFF und SCHNEIDER 1954, PAIGEN 1954). Selbst wenn ein Teil dieser Differenzen auf unterschiedliches Herauslösen chondriosomeneigener Enzyme infolge von Präparationsschäden zurückgeführt werden kann, so bleiben doch genügend Hinweise, die die Auffassung rechtfertigen, die Chondriosomengarnitur einer Zelle müsse als *Population* aufgefaßt werden (cytologische Befunde stützen diese Ansicht, vgl. die verschiedenen nadipositiven Körper der Hefezelle nach BAUTZ 1956 u. a. m.). Damit wird es aber nötig, weitere Konkurrenzerscheinungen in Rechnung zu stellen, nämlich sowohl zwischen den einzelnen Chondriosomen als auch zwischen verschiedenen Chondriosomen und denjenigen Zellkomponenten, die, wie das Cytoplasma, die Mikrosomen und der Zellkern, Chondriosomenprodukte benötigen bzw. weiterverarbeiten.

d) Mikrosomen.

Unter Mikrosomen sollen hier im Sinne CLAUDEs (1946) "small granules" verstanden werden, Partikel, deren Größe submikroskopisch oder an der Grenze der mikroskopischen Sichtbarkeit ist, und die im Elektronenmikroskop keine Strukturen erkennen lassen. Cytochemisch können sie nur selten nachgewiesen, wohl aber aus Zellhomogenaten als Reinfraktionen gewonnen werden. Sie sind durch ihren hohen Gehalt an Ribonucleinsäure (etwa 50% der Gesamt-RNS der Zelle) und Lipoiden (30%) ausgezeichnet. Daneben treten Proteine auf, vor allem die Enzyme Glucose-6-Phosphatase (bis zu 85%), Cytochromreduktase und Fettsäureoxydase, während andere Enzyme nur in so geringen Mengen gefunden werden, daß sie als Verunreinigungen infolge Adsorption oder ähnlichem angesehen werden können. Ihre physiologische Funktion scheint in der Proteinsynthese zu bestehen; dafür spricht zunächst die ausnahmslos aufgefundene Proportionalität zwischen RNS-Gehalt und Ausmaß der Proteinsynthese, ferner die Tatsache, daß der Einbau markierter Aminosäuren fast ausschließlich in den Mikrosomen stattfindet. Die gebildeten Proteine verbleiben dabei in den Mikrosomen. Die Versuche gelingen auch *in vitro* (z. B. mit Erbsenkeimlingen, WEBSTER 1954), wenn ATP, ein ATP regenerierendes System und ein Gemisch aus 17 verschiedenen Aminosäuren zugegeben wird. Umgekehrt unterbleibt die Proteinsynthese, wenn die RNS durch Ribonuclease abgebaut wird.

In situ scheint freilich die Zufuhr freier Aminosäuren (aus den Chondriosomen) nicht immer zu genügen. Schon 1952 konnte SIEKEWITZ zeigen, daß der Einbau markierten Glykokolls in die Mikrosomenproteine der isolierten Fraktion schneller erfolgt, wenn gleichzeitig Chondriosomen vorhanden sind. Die Chondriosomen können durch einen makromolekularen Faktor ersetzt werden, der wahrscheinlich ein γ-Glutamylpeptid ist. Dieses wird in den Chondriosomen synthetisiert. Danach bestände die Rolle der Mikrosomen bei der Proteinsynthese in der Transpeptidierung.

Besteht so ein Zusammenwirken von Chondriosomen und Mikrosomen (unter Vermittlung des Cytoplasmas, in das beide Partikelgruppen eingebettet sind), so scheint die Proteinsynthese von der Anwesenheit des Kerns weitgehend unabhängig zu sein (bei Amöben und *Acetabularia:* BRACHET 1952, HÄMMERLING 1954). Das Verbleiben der neu synthetisierten Proteine in den Mikrosomen, zum Teil als spezifische Enzymproteine, hat zu der Ansicht geführt, die Mikrosomen seien Vorstufen von Chondriosomen. Dafür sprechen erste elektronenmikroskopische Aufnahmen von EICHENBERGER (1953), aber auch die Entwicklung normaler, chondriosomenhaltiger Zellen aus Zellen, die durch

Zentrifugierung chondriosomenfrei gemacht wurden (HARVEY 1946 an Seeigeleiern). Diese Frage gehört jedoch nicht mehr in den biochemischen Bereich und soll deshalb im entwicklungsphysiologischen Teil abgehandelt werden.

e) Zellkern.

Gegenüber der großen Bedeutung, die der Kern in der Vererbung und in der Entwicklung spielt, ist seine Rolle im Stoffwechsel der ausdifferenzierten Zelle, wie wir sie bisher betrachtet haben, verhältnismäßig geringfügig. Der Kern hat keine oxydative Aktivität, er vermag den Citronensäurecyclus nicht durchzuführen, der Proteinanteil (Histon und „Nichthiston") besitzt keine nennenswerte enzymatische Aktivität, der DNS-Gehalt ist bemerkenswert konstant und stabil usw. Er ist lediglich in begrenztem Umfang imstande, aus glykolytischen Prozessen Energie zu gewinnen. Damit vermag er vielleicht den Energiebedarf des Interphasenkerns zu decken — für die wichtigen Aufgaben bei der Zellteilung und -differenzierung ist er auf die Anlieferung energiereicher Verbindungen aus Cytoplasma und Chondriosomen angewiesen.

In der Tat zeigt sich, daß bei der Entkernung der normale Stoffwechsel der Zelle noch eine lange Zeit unverändert weiterlaufen kann: die Sauerstoffaufnahme bei Amöben 9 Tage, bei *Acetabularia* sogar über 3 Monate (BRACHET 1952, 1954); die Proteinsynthese 14 Tage, und der ATP-Gehalt bleibt 12 Tage konstant. Eine Ausnahme machen die Phosphataufnahme und die Aktivität der sauren Phosphatase, die durch Entkernung rasch absinken (MAZIA und HIRSHFIELD 1950, BRACHET 1952 — vgl. aber STICH und HÄMMERLING 1953), ein Zeichen, daß der Kern in die Phosphorylierungsprozesse eingreift, wahrscheinlich durch Produktion von Co-Faktoren. Auch der Einbau von C^{14}-Orotsäure in die RNS des *Acetabularia*-Plasmas ist unabhängig von der Anwesenheit des Zellkerns (SZAFARZ und BRACHET 1954).

Anders ist es mit dem Kohlenhydratstoffwechsel: der ATP-Gehalt kernloser und kernhaltiger Zellhälften von *Amöba proteus* bleibt unter aeroben Bedingungen unverändert, während in der Anaerobiose die kernlosen Fragmente einen großen Teil ihres ATP-Gehaltes einbüßen. Die Aufrechterhaltung des hohen ATP-Gehaltes kernhaltiger Hälften dürfte mit der Glykogenolyse zusammenhängen. In der Tat verbrauchen die kernhaltigen Hälften ihren Glykogenbestand sehr schnell, während er in den kernlosen unverändert bleibt. Danach dürfte der Zellkern auch den anaeroben KH-Stoffwechsel kontrollieren, vermutlich gleichfalls über sog. Cofaktoren (BRACHET 1954).

Besondere Verhältnisse können bei mehrkernigen Zellen vorliegen, vor allem, wenn die Kerne verschiedenwertig sind. Zeigt schon die Kerndifferenzierung bei der Pollenschlauchmitose, daß es sich dabei keineswegs um seltene Ausnahmen handelt, so verdient sie besonderes Interesse, wenn es gelingt, verschiedene, wohldefinierte Kerne in ein gemeinsames Plasma zu bringen, so bei den heterokaryotischen Pilzen. Die Kombination genetischer und physiologischer Methoden hat hier wertvolle Aufschlüsse geliefert. Sie betreffen vor allem das Zusammenwirken bestimmter Gene, die zum Teil in demselben Chromosom, zum Teil nur im selben Kern, in anderen Fällen aber auch in verschiedenen Kernen liegen können bzw. müssen. Im letzteren Fall müssen die „Genprodukte" ihren Weg ins Plasma bzw. durchs Plasma in den anderen Kern finden.

Eine sehr interessante Form der Wechselwirkung beschreibt PONTECORVO (1952):

Unter den Mutanten von *Aspergillus nidulans* fand PONTECORVO einen argininbedürftigen Stamm, der durch Lysin kompetitiv gehemmt wird, sowie einen weiteren Stamm, bei dem die Verhältnisse umgekehrt lagen (lysinbedürftig, durch Arginin gehemmt); keine der beiden Substanzen hemmt den nichtbedürftigen Stamm. Durch Kreuzung wurde ein Stamm hergestellt, der beide Stoffe benötigt.

Heterozygote Diploide dieses Stammes, die die beiden Allele („Arginin" und „Lysin") in Cis- und Transkombination auf homologen Chromosomen *innerhalb eines Kernes* enthalten, sind in beiden Typen nicht bedürftig und werden auch nicht gehemmt. Anders bei Heterokaryen, die die beiden Allele in cis- und trans-Kombination *in verschiedenen Kernen* (im gemeinsamen Plasma) tragen. Diese sind sehr empfindlich gegenüber den Hemmstoffen; vor allem aber zeigen sie einen rhythmischen Wechsel zwischen gutem Wachstum und beeinträchtigtem Wachstum (erkennbar an der Kolonieform). Daraus folgt zunächst, daß die kompetitive Hemmung zwischen Arginin und Lysin innerhalb der Zelle, nicht aber innerhalb des Kernes stattfindet, ferner, daß sich eine Art Schaukelgleichgewicht (see-saw-equilibrium) einstellt, in dem ein Kern die Synthese eines Stoffwechselproduktes anregt, das die Aktivität eben dieses Kernes hemmt, das aber von dem anderen Kern benötigt wird, und umgekehrt (Abb. 2).

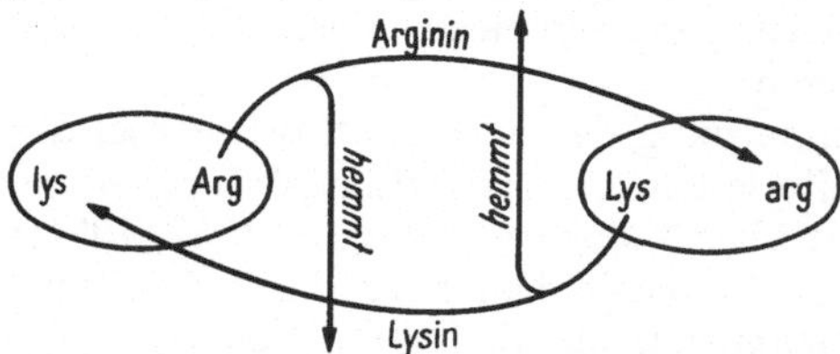

Abb. 2. Wechselseitige Hemmung zweier Kerne in einem Heterokaryon von *Aspergillus nidulans*. (Nach PONTECORVO 1952.)

Ähnlich wie bei den Chondriosomen- und Mikrosomenpopulationen ist hier mit einer Vervielfachung der Konkurrenzbeziehungen zu rechnen. Es bewerben sich nicht allein verschiedene plasmatische Fraktionen bzw. Partikel um die Kernstoffe, sondern umgekehrt konkurrieren auch die verschiedenen Kerne um die vom Plasma zu beziehenden Substrate. Es können experimentelle Bedingungen gefunden werden, unter denen die Konkurrenzfähigkeit der einen Kernsorte zu stark absinkt und das Heterokaryon zusammenbricht, ohne daß freilich vorerst zu entscheiden wäre, welcher biochemische Faktor letztlich den Ausschlag gibt.

f) Cytoplasma.

Das Bild, das wir uns vom Aufbau und der Funktion der partikulären Zellkomponenten machen können, ist reich an Einzelheiten, wenn wir es mit dem des Cytoplasmas vergleichen. Die ältesten Cytoplasmatheorien stammen aus der Zeit, in der die Zelltheorie begründet wurde; seither ist ihre Zahl ins Ungemessene gestiegen, aber nicht eine hat den fortschreitenden Erkenntnissen über die Biochemie und Physiologie der Zelle standhalten können. Das Cytoplasma ist noch immer das große Rätsel in der Biologie. Die älteren Vorstellungen über die submikroskopische Struktur des Cytoplasmas müssen sämtlich abgelehnt werden; es ist wohl kein Zufall, daß FREY-WYSSLING (1955) in seiner neuesten Darstellung dem Cytoplasma je nach Zustand und Funktion verschiedene „Aspekte" zuschreibt: den reticularen, granularen, fibrillaren und lamellaren Aspekt. Elektronenmikroskopische Belege lassen sich für jeden Spezialfall erbringen. Berücksichtigt man indessen, daß reticulare, granulare und fibrillare Strukturen auch in unbelebten Modellsubstanzen, z. B. Natriumoleat (HAUSER und LE BEAU 1950) gefunden werden und daher weder als typisch noch als „nativ" zu gelten brauchen, so bleibt für vorerst allein überzeugend der lamellare Aspekt. Die ganz vorzüglichen Aufnahmen von SJÖSTRAND und HANZON (1954), PALADE (1955), WATSON (1955) u. a. schließen jeden Verdacht auf Fixierungsartefakte aus. Sie zeigen im Cytoplasma verschiedenster Zellen einen lamellaren Aufbau:

ziemlich streng parallel verlaufende Membranen umschließen „leere" Hohlräume (cavities) mit der cytoplasmatischen Matrix; die zwischen den membranbegrenzten Hohlräumen liegenden Partien hingegen enthalten zahlreiche partikuläre Komponenten von etwa 100—150 Å Durchmesser (es gibt Hinweise für die Annahme, daß die — perforierte — Kernmembran nichts anderes ist als eine solche flachgedrückte Kaverne — perinuclear cisterna, WATSON 1955 — und somit cytoplasmatischer Herkunft). Neuerdings sind solche lamellaren Plasmabezirke auch im Pflanzenreich aufgefunden worden, nämlich in *Nitella*-Zellen (MERCER und Mitarbeiter 1955).

Muß die Deutung dieses Strukturfaktors an Hand elektronenmikroskopischer Aufnahmen noch offenbleiben, so ergibt sich aus Homogenisierungsversuchen ein wichtiger Anhaltspunkt. WILDMAN und COHEN (vgl. Band 1 dieses Handbuches) gelang es, aus Tabakblättern eine Fraktion abzutrennen (fraction I), die chemisch und elektrophoretisch weitgehend einheitlich ist. Es handelt sich dabei um ein Nucleoprotein, dessen Phosphorgehalt in Abhängigkeit vom physiologischen Alter und von der stoffwechselphysiologischen Aktivität zwischen 0,15 und 1,5% wechselt. Das Elektronenmikroskop weist sie aus als globuläre Partikel von der Größe 100—200 Å — in bester Übereinstimmung mit den Befunden *in situ*. Ultrazentrifugenversuche lassen auf die gleiche Größe schließen (EGGMAN 1953). Diese fraction I scheint nichtenzymatischen Charakter zu haben, zugleich aber ein Ort der Proteinsynthese im Cytoplasma zu sein. Zu analogen Schlüssen kam PALADE (1955).

Fraction I macht indessen nur 35% der gesamten cytoplasmatischen Proteinmenge aus; in dem Restprotein scheinen Enzyme verschiedenster Art vorzuliegen: Cytochrom c, Katalase, Dehydrasen und Aldolase (bei den tierischen Zellen), ferner saure und alkalische Phosphatasen. Das Bild, das man sich hieraus von der physiologischen Aktivität des Cytoplasmas (auch „Grundplasma" genannt) machen kann, bleibt noch außerordentlich lückenhaft. Es ist schwer vorzustellen, daß sich die Rolle des Cytoplasmas mit der Proteinsynthese und einigen Teilschritten der Oxydation erschöpfen sollte, zumal die Anwesenheit von Phosphatasen auf einen starken Verbrauch energiereicher Phosphate schließen läßt. Es muß indessen berücksichtigt werden, daß ja dem Cytoplasma auch die Rolle eines stoffaustauschvermittelnden Agens zukommt — Plasmaströmung, aktive Stoffaufnahme und die Zu- und Abfuhr von Reaktionsprodukten von und zu den Reaktionsorten müssen erhebliche Energiemengen erfordern. Auf eine sich hieraus ergebende Konsequenz für die Konkurrenz der Transportaggregate (Carrier-Systeme, GOLDACRE 1952) um die Substrate und der Abnehmer um die Transportsysteme sei hier nur hingewiesen.

g) Zelloberfläche.

Der Vollständigkeit halber sollen die Untersuchungen ROTHSTEINS über die Enzymologie der Zelloberfläche, insbesondere der Hefezellen, angeführt werden. ROTHSTEIN gelangt zu der Ansicht, daß die Zelloberfläche nicht eine bloße Ansammlung von Enzymen darstellt, die an der Stoffaufnahme, dem Stofftransport und der Stoffverarbeitung beteiligt sind, sondern vielmehr ein organisiertes Gebilde, das mit der Fähigkeit zur Selbstreproduktion verbunden sei, ein Zellorganell also, das den Chondriosomen und Mikrosomen gleichzustellen wäre. In der Tat lassen sich mancherlei Hinweise für die Berechtigung dieser Auffassung finden, so z. B., daß insbesondere bei Adaptationserscheinungen die Substrate, für die die Zelle adaptiv Enzyme bildet, gar nicht in die Zelle eindringen, sondern an der Oberfläche ihre Wirkung entfalten bzw. dort verarbeitet werden.

Ganz zweifellos liegen in der Grenzfläche besondere Strukturen und Enzymgarnituren vor, die von denen des Binnenplasmas sehr verschieden sind (vgl. z. B. HILLIER und HOFFMANN 1953 an Erythrocyten, HOUWINK 1955 an Spirillen). Gleichwohl sind es chemisch die gleichen Konstituenten, die wir im Plasma vorfinden: Proteine, Lipoide, Nucleinsäuren (?) und Polysaccharide, so daß die Möglichkeit nicht ausgeschlossen werden kann, es handle sich bei den Oberflächenbausteinen eben doch um Elemente binnenplasmatischer Herkunft, die *nicht* in der Grenzschicht selbst hergestellt („reproduziert") werden. Solange unwiderlegliche Beweise ausstehen, muß die oben angeführte Vorstellung als Hypothese gewertet werden, die freilich der Diskussion wert ist.

4. Makromolekularer Stoffwechsel.

In den vorangegangenen Abschnitten wurden einige Möglichkeiten erörtert, die für das Zusammenwirken wichtiger Zellkomponenten, nämlich von makromolekularen Stoffen, von Enzymen und von partikulären Gebilden, in der fertig ausgebildeten Zelle bestehen. Sie betrafen in der Hauptsache den *niedermolekularen* Stoffwechsel und zeigten, wie dieser von der Existenz bestimmter Strukturen und deren Ordnung bzw. Organisation abhängig ist. Nun ist Ordnung zunächst nichts anderes als die Bezeichnung für eine Abweichung von der statistisch zu erwartenden Verteilung. Das könnte den Anschein erwecken, als seien die Strukturen im weitesten Sinne, die „vorgegebenen Anisotropien", das entscheidende Prinzip, das dem niedermolekularen Stoffwechsel „übergeordnet" (FREY-WYSSLING 1955) ist und dessen Ablauf „steuert". Das ist nur bedingt richtig. In der fertig ausgebildeten Zelle sind zumindest die makromolekularen und enzymatischen Komponenten nicht statisch aufzufassen, nicht dem Stoffwechsel *vorgesetzt*, sondern zugleich auch *Objekt des Stoffwechsels*.

Hierüber können wenigstens einige grundsätzliche Anmerkungen gemacht werden. Die von SCHÖNHEIMER bereits 1941 getroffene Feststellung, daß makromolekulare Stoffe, z. B. Proteine, einem beständigen Abbau und Wiederaufbau unterworfen sind, ist seither, vor allem durch die Anwendung der Isotopentechnik, in überwältigendem Maße bestätigt worden. Für die noch immer in Rede stehende fertig ausgebildete Zelle besagt das, daß bei Proteinen, Nucleinsäuren, Lipoiden und zum Teil wohl auch Polysacchariden beständig Abbau- und Syntheseprozesse ablaufen müssen, Zerlegung und Neubildung von Makromolekülen aus niedermolekularen Bausteinen. Dabei stehen z. B. die Proteine im Gleichgewicht mit dem „Aminosäurepool". Es kann zunächst gleichgültig bleiben, ob jeweils nur einzelne Aminosäuren bzw. größere Bruchstücke der Polypeptidkette ausgewechselt werden, oder ob Makromoleküle *in toto* neu hergestellt werden. Trotz dieses beständigen Wechsels bleibt der Bestand an Makromolekülen im großen und ganzen erhalten (vgl. aber die Adaptationserscheinungen, SŁONIMSKI 1954) und damit auch die Ordnung der makromolekularen Zellkomponenten in einem Fließgleichgewicht besonderer Art. Es steht zu erwarten, daß auch die verschiedenen Proteine und Enzyme um die auszutauschenden Aminosäuren des Pools konkurrieren.

Der *Abbau* makromolekularer Stoffe erscheint zunächst verständlich, wenn man sich vergegenwärtigt, daß diese ja thermodynamisch instabil sind und nur aus energiereichen Verbindungen entstehen können, die dem niedermolekularen Stoffwechsel entstammen (Hexosephosphate, Acylphosphate, Aminosäurephosphate usw.); indessen können auch chemische Systeme, die von einem Gleichgewicht weit entfernt sind, ihren Zustand sehr lange aufrechterhalten. Mit Recht führt G. V. SCHULTZ (1950) das Beispiel des Serumalbumins an, das, in reinster Form in Wasser gelöst, jahrelang ohne merkliche Veränderung haltbar ist — das "turnover" hingegen geht sehr viel schneller vonstatten, wobei wahrscheinlich Enzyme mitwirken. Der beständige Umbau scheint vielmehr erforderlich zu sein, um die Makromoleküle untereinander als auch mit den Bausteinen und energieliefernden niedermolekularen Stoffwechselcyclen im Fließgleichgewicht zu halten.

Noch ein weiterer Gesichtspunkt verdient in diesem Zusammenhang erwähnt zu werden. Bei enzymatischen Umsetzungen können die Strukturänderungen in der prosthetischen Gruppe Rückwirkungen auf das Enzymprotein haben. Diese wirken sich, da es ja thermodynamisch instabil ist, in einer partiellen Denaturierung aus (Jung 1954). In der Tat konnte Jung zeigen, daß bei erzwungenem schnellen Wechsel des Hämoglobins zwischen oxydierter und reduzierter Stufe diesem Wechsel parallel denaturiertes Protein entsteht. Damit kann die Lebensdauer eines Enzyms und also der Zeitpunkt, zu dem seine Erneuerung notwendig wird, von seiner Aktivität bzw. Beanspruchung abhängen. Eine derart begründete Notwendigkeit des turnover mag ein weiterer Schlüssel für das Verständnis der Tatsache sein, daß der Organismus nicht die Tendenz hat, seine makromolekularen Stoffe zu stabilisieren (G. V. Schultz 1950). Auf die Beziehung des turnover zum Problem des Alterns soll hier nicht eingegangen werden.

5. Die Entstehung der Reaktionssysteme.

Neben dem bisher behandelten Zusammenwirken der *fertig ausgebildeten* Zellkomponenten kann deren Entstehung im Laufe der ontogenetischen Entwicklung mindestens gleich großes Interesse des Zellphysiologen beanspruchen. Freilich sind hier die verfügbaren physiologischen Daten gegenüber den cytomorphologischen Befunden in der Minderzahl. Daraus ergibt sich die Unmöglichkeit, die verstreuten Fakten zu einem geschlossenen Bild zusammenzufügen. Ein solches soll hier allerdings auch nicht angestrebt werden, sondern den Bänden 14—19 dieses Handbuches vorbehalten bleiben. Es kann lediglich versucht werden, auch für das Gebiet der Zellentwicklung einige Koordinationsprinzipien herauszustellen, die zellphysiologisch von Bedeutung sind. Es ist damit zu rechnen, daß diese etwas andersartig sind als im bisher bekannten Bereich des Stoffwechsels. Sie werden aber gerade deshalb geeignet sein, neues Licht auch auf stoffwechselphysiologische Fragen zu werfen.

Das Problem der Entwicklung von Zellkomponenten ist zuerst bei den partikulären Fraktionen behandelt worden, nämlich dem Zellkern, den Plastiden und den Chondriosomen. Da bei ihnen niemals Neuentstehung beobachtet worden ist, gelten sie als Gebilde *sui generis* oder, wie man heute vielfach sagt, als „autoreproduktive Gebilde“. Der modernere Ausdruck bedarf eines entsprechenden Zusatzes: diese Gebilde sind nicht imstande, vollständig sich selbst zu reproduzieren; es besteht nicht "self-sufficiency", sondern nur "self-dependence" (Lederberg 1952); sie bedürfen der Mitwirkung anderer Reaktionssysteme, um sich vollständig auszubilden.

a) Der Zellkern.

Vom Zellkern ist eine Entstehung aus Vorstufen nicht bekannt; ein neuer Kern entsteht stets durch exakte Teilung eines bereits vorhandenen Kernes. Die Einzelheiten der Kernteilung werden in Band 14 behandelt. Hier sollen nur folgende Punkte betrachtet werden: Im Zellkern befinden sich die „Anlagen“ (Gene, Allele) nicht frei an beliebigen Orten, sondern sie sind streng (linear) geordnet auf Gebilden höherer Organisation, den Chromosomen. Bei der der Teilung voraufgehenden Verdoppelung der Substanz werden nicht allein die Gene verdoppelt („identisch reproduziert"), sondern auch die Chromosomen. Offenbar muß nicht allein das Genmaterial, sondern auch seine Anordnung auf die Tochterkerne weitergegeben werden (vgl. den position-effect).

Die Verdoppelung der Gene ist heute leicht vorzustellen, wenn wir den neuen Ansichten von WATSON und CRICK (1953), DELBRÜCK (1954) u. a. folgen. Danach wird das Gen-„Material" durch DNS-Makromoleküle dargestellt, die in zwei komplementären Schrauben umeinandergewunden sind. Sie tragen in Gestalt wechselnder Kombinationen von Purin- bzw. Pyrimidinbasen die gesamte „Information", die weitergegeben werden soll. Sie vermögen sich zu „entwinden" und zu trennen. Jede Hälfte erzeugt durch Anlagerung das Komplement, und es sind dann 2 Doppelschrauben entstanden.

Die DNS-Doppelschrauben fungieren wahrscheinlich als „Matrizen" für Proteinkörper, deren Spezifität sich aus der durch die Information vorgeschriebenen Anordnung der Aminosäuren (und damit der Auffaltungsmöglichkeiten der Polypeptidkette) ergibt. Diese spezifischen Proteine sind freilich keineswegs die spezifischen Enzyme selbst, die zur Ausbildung des betreffenden „Merkmals" führen bzw. benötigt werden; soweit wirklich die Beziehung 1 Gen = 1 Enzym bestehen sollte, ist sie wesentlich komplizierter. Wahrscheinlich sind noch RNS-Moleküle (des Nucleolus?) dazwischengeschaltet, die entweder selbst, als Cofaktoren, ins Plasma übertreten, oder aber im Plasma die Ausbildung äquivalenter Cofaktoren veranlassen. Erst die „genkontrollierten" Faktoren verbinden sich mit einem "inducer" für die Ausbildung eines Enzyms A zu einem Komplex, der als Co-Ferment eines Enzyms wirkt, das die Synthese des Enzyms A katalysiert (POLLOCK 1952, PORTER, HOLMES und CROCKER 1953). In anderen Fällen können die vom Kern ausgehenden „Befehlsübermittler" auch andere Wirkungen haben, so z. B. bestimmte Veränderungen in der Membran auslösen und dergleichen mehr (s. S. 625).

Mechanismen dieser Art schließen zunächst eine Reihe von bereits besprochenen Konkurrenzphänomenen ein: da der Kern nur eine geringe Stoffwechselaktivität aufweist, ist er für die Reproduktionsvorgänge auf die Anlieferung von Purin- und Pyrimidinbasen, Pentosen und Phosphaten sowie energiereichen Verbindungen aus dem niedermolekularen Stoffwechsel angewiesen, für die Ausbildung spezifischer Proteine bzw. Oligopeptide auf die Zufuhr von Aminosäuren. Nur bei ausreichender Anlieferung, und nur, wenn die einzelnen DNS-Schrauben in ihren Ansprüchen aufeinander abgestimmt sind, wird gewährleistet, daß alle Gene reproduziert werden, bzw. erst wenn das letztere der Fall ist, kann die Verteilung erfolgreich vonstatten gehen. Dabei muß sich der Kern als Ganzes z. B. mit dem Aminosäurebedürfnis der übrigen Zellkomponenten auseinandersetzen. Hier ist folglich eine quantitative Grundlage der sog. Kernplasmarelation bzw. der Kern-Plasma-Chondriosomen-Relation (LETTRÉ und LETTRÉ 1953) gegeben.

Andersartige Möglichkeiten des Zusammenspiels bestehen für die vom Kern ausgehenden Substanzen. In vielen Fällen ist dieser Transport eine *Diffusion*. Da diese verhältnismäßig langsam erfolgt, kommt hier ein Zeitfaktor ins Spiel, vor allem dann, wenn am Reaktionsort erst eine gewisse Konzentration des genkontrollierten Stoffes erreicht werden muß, ehe die Reaktion in Gang kommt. Vielleicht hat hier die zeitliche Aufeinanderfolge gewisser Genauswirkungen eine ihrer Ursachen. Aber auch räumliche Faktoren spielen eine Rolle, denn bei größeren Zellvolumen wird eine Schwellenkonzentration später erreicht als in einer kleinen Zelle. Sogar die Selbstregulierung des Wachstums kann auf diese

Weise „quantitativ" gedeutet werden (vgl. WENT, Band 1 dieses Handbuches). In einem anderen Falle konnte MITCHISON (1952) das Phänomen der ersten Furchung und Teilung des Seeigeleies auf einen Diffusionsgradienten zurückführen: von den beiden Astern diffundieren gewisse Kernsubstanzen (Nucleotide ?) in alle Richtungen; wenn sie auf die charakteristisch strukturierten Proteinunterlagen der Zellmembran auftreffen, bewirken sie deren Ausdehnung. Da sie zuerst an der polaren Zone der Membran ankommen, findet hier auch die erste Ausdehnung statt. Das führt, da das Volumen des Eis etwas abnimmt, zu einer Streckung des Eies in polarer Richtung. Später werden auch polfernere Membranareale von der diffundierenden Substanz erreicht und strecken sich. So kommt es zu einer Einfaltung in der Äquatorialzone (Furchung). Zuletzt strecken sich die Membranproteine dieser Furche und vollenden so die Durchschnürung. Nach BRØNDSTEDT (1954) ist auch die harmonische Regeneration bei Planarien und sogar die bilaterale Symmetrie die Folge eines Zeitfaktors (vgl. hierzu auch die diffusion-reaction theory der Differenzierung nach TURING 1952). Hierbei erscheint besonders bemerkenswert, daß quantitative Faktoren an der Ausbildung von „Qualitäten" wie Furchung und Teilung, Bilateralsymmetrie usw. maßgeblich beteiligt sein können. Gleichzeitig muß aber berücksichtigt werden, daß solche Diffusions- und Zeitfaktoren nunmehr auch für den niedermolekularen und makromolekularen Stoffwechsel in Betracht gezogen werden müssen.

b) Chondriosomen.

Bei den Chondriosomen liegen die Verhältnisse etwas anders als im Kern Unzweifelhaft sind echte Chondriosomenteilungen (nicht nur Aufteilung von Chondriosomenaggregaten) beobachtet worden (FRÉDÉRIC und CHÈVREMONT, vgl. STEFFEN, Band 1 dieses Handbuches). In den meisten Fällen jedoch wird die mikroskopisch sichtbare Struktur (Cristae, vielleicht auch die Membran) *nicht* „identisch reproduziert". Für die Chondriosomen genügt es offenbar, wenn *Vorstufen* identisch reproduziert werden, „Prochondriosomen", vielleicht Mikrosomen (HARVEY 1946, HÖRSTADIUS 1953). Mikrosomen aber sind extranucleäre Gebilde, die die Aufgabe der Proteinsynthese ohne direkte Beteiligung des Kerns durchführen können (wobei freilich noch offenbleiben muß, ob ihre RNS-Matrize nicht ein Kernprodukt ist, vgl. LINDBERG und ERNSTER 1954). Die Mikrosomen nehmen Lipoide und Aminosäuren aus den Chondriosomen auf, bauen daraus Makromoleküle bzw. Enzyme und wachsen selbst zu Chondriosomen heran.

Wir wissen nicht, wie sich die Makromoleküle bzw. Enzyme dann zu der charakteristischen Struktur zusammenordnen; vielleicht genügen Ladungsanisotropie und/oder eine Polarität der einzelnen Makromoleküle hinsichtlich der Hydrophilie und Hydrophobie schon, um eine solche Ordnung „automatisch" herbeizuführen, so wie auch aus der Polarität der einzelnen Komponenten des Seeigeleies die Anisotropie des gesamten Eies resultieren kann (HARRISON 1945): "The anisotropy of the egg, sometimes revealed by the distribution of cytoplasmic materials, is therefore only a reflection of another more fundamental and deeper asymmetry, presumably seated on the molecular level" (EPHRUSSI 1953). Gewisse räumliche Anordnungsweisen könnten also der Ausdruck der makromolekularen Konstitution und einer daraus resultierenden *Konkurrenz um Zellorte* sein.

c) Chloroplasten.

Die Chloroplasten nehmen eine Zwischenstellung ein: in den meisten Algen werden die fertigen Chloroplasten ebenso wie der Zellkern geteilt und also die ausgebildeten mikroskpoisch sichtbaren Strukturen identisch reproduziert und auf die Tochterzelle übertragen. Bei den höheren Pflanzen jedoch sind Eizelle und meristematische Zelle plastidenfrei; hier werden, wie vielfach bei den Chondriosomen, *Vorstufen* (Proplastiden) übertragen und zu Chloroplasten weiterentwickelt. Die Proplastiden besitzen ein Primärgranum (STRUGGER 1950), das im Elektronenmikroskop eine kristallgitterähnliche Ordnung von Makromolekülen zeigt. Es liefert Produkte, die sich außerhalb des Granums sofort zur Lamellenstruktur ordnen (HEITZ 1954, LEYON 1954). Auch hier kann also die mikroskopisch sichtbare Struktur nachgebildet werden, ohne daß sie an Hand eines entsprechenden Musters von mikroskopischer Sichtbarkeit geformt würde.

Im übrigen gilt für den Stoffwechsel bzw. den Bedarf an niedermolekularen Substanzen das gleiche wie für die Chondriosomen.

Besonders aufschlußreich sind in diesem Zusammenhang die Untersuchungen STEFFENs über die Umwandlung von Chloroplasten in Chromoplasten (*Solanum capsicastrum*). Sie ist, wie die elektronenmikroskopischen Aufnahmen ausweisen, mit einer durchgreifenden Veränderung der Struktur verbunden: die Lamellen werden eingeschmolzen und bilden vorübergehend kugelige Tropfen; diese werden dann zu langen, mehr oder weniger parallel liegenden *Fibrillen* ausgezogen. Hier kann der gleiche Ausgangskörper, das Proplastid, zwei grundsätzlich verschiedene Strukturprinzipien verwirklichen und sogar ineinander überführen, wenigstens in *einer* Richtung. Das zeigt, daß auch so charakteristische Strukturen wie die Chloroplastenlamellen sich ohne unmittelbares Muster bilden und ihre Elemente nach Einschmelzung anders anordnen können, wiederum ohne unmittelbares Muster. Zugleich wird offenbar, daß diese Strukturausbildung nur innerhalb gewisser Grenzen autonom ist, denn sie hängt augenscheinlich auch von Faktoren ab, die außerhalb des Chloroplasten liegen, im Cytoplasma, und die, wenigstens zum Teil, vom Kern ausgegangen sein können.

d) Das Cytoplasma.

Über die Neubildung des Cytoplasmas sind wir fast völlig auf Vermutungen angewiesen. Als sicher kann gelten, daß der Kern- und Zellteilung eine Vermehrung der Plasmamasse vorausgeht. Es ist möglich, aber nicht sehr wahrscheinlich, daß es sich dabei lediglich um einen „unspezifischen" Zuwachs an Masse handelt. Wir müssen jedoch berücksichtigen, daß das Cytoplasma seine Komponenten, wenigstens vorübergehend, in Form des endoplasmatischen Reticulums anordnen kann (Lamellierung, Kavernenbildung), und daß es außer der Matrix offenbar eine Population von Nucleoproteinpartikeln enthält (fraction I, WILDMAN und COHEN). Diese Partikel besitzen aller Wahrscheinlichkeit nach eine mehr oder weniger ausgeprägte Spezifität, und es kann nicht von vornherein ausgeschlossen werden, daß sie, wie die Plastiden, Chondriosomen und der Zellkern, „autoreproduktive" Gebilde sind, zwar nicht unabhängig von den übrigen Zellkomponenten, aber in der Vermehrung "self-dependent". Bei der Zellteilung erfolgt dann die Verteilung der Partikel wohl ebenso zufällig wie die der Mikrosomen und Chondriosomen (vgl. EPHRUSSI) und Plastiden; wir dürfen annehmen, daß das endoplasmatische Reticulum vorher eingeschmolzen wird, um die Verteilung zu erleichtern — wie ja überhaupt zu erwarten ist, daß die Anordnung im endoplasmatischen Reticulum nur eine von mehreren Möglichkeiten darstellt und nur unter bestimmten physiologischen Bedingungen verwirklicht wird. Analoges ist beim Übergang der Chloroplasten in Chromoplasten beobachtet worden.

Bei der Befruchtung einer Eizelle liegen die Verhältnisse insofern anders, als aus dem Pollenschlauch der männliche Kern, Proplastiden und (Pro-)Chondriosomen übertreten, dagegen so gut wie kein Cytoplasma. Eine Entmischung cytoplasmatischer Komponenten analog der Entmischung der veränderten Plastiden und Chondriosomenpopulation ist daher wenig wahrscheinlich. Hingegen wissen wir, daß zur Ausprägung eines Merkmals nicht nur die Kerngene, sondern auch „plasmatische" Strukturen erforderlich sind (Kappa bei *Paramaecium*, SONNEBORN; Chondriosomen bei Hefe, EPHRUSSI). Es ist bisher nicht erwiesen, daß alle Plasmagene Partikel von der Größenordnung bzw. Organisationsstufe der Mikrosomen, Chondriosomen und Plastiden sind; es könnten auch Partikel der Nucleoproteinfraktion I sein. Dann aber können nicht nur den Genen des männlichen Kerns die cytoplasmatischen Gegenspieler fehlen, sondern umgekehrt auch den cytoplasmatischen Faktoren gewisse „Genstoffe", so daß erstere degenerieren und die Konkurrenzbedingungen der Population verändert werden.

Das hier auftretende Problem der „Unverträglichkeit" soll indessen an dieser Stelle nicht erörtert werden, doch sei in diesem Zusammenhang auf die von *Acetabularia*-Kernen ausgehenden Stoffe hingewiesen, die teils speciesspezifisch, teils speciesunspezifisch die Morphogenese nicht nur auslösen, sondern sogar in ihrer Richtung bestimmen (HÄMMERLING 1953).

e) Vacuolen.

In Eizellen und meristematischen Zellen findet man in der Regel keinerlei Vacuolen; daher herrscht die Ansicht vor, Vacuolen entstünden jeweils *de novo*. Das braucht indessen nicht auszuschließen, daß sie gleichwohl Gebilde *sui generis* sind, wenn sie nämlich ihren Ausgang nehmen von submikroskopischen Vorstufen, die partikulärer Natur sind und spezifische Proteine, Nucleoproteine u. dgl. enthalten. Zwar liegt insbesondere bei wäßrigen Vacuolen die Vermutung nahe, daß sie infolge „unspezifischer" *Entmischung* (Koazervation) entstehen (vgl. GUILLERMOND, Zusammenfassung DRAWERT, Band I dieses Handbuches). Zudem konnte PFEFFER (1890) durch Einbringen von Asparaginkriställchen in Plasmodien künstliche Vacuolen erzeugen. Danach wäre die Vacuolenbildung im wesentlichen ein Phänomen, das aus der wechselnden Konkurrenzfähigkeit gewisser Makromoleküle um Hydratationswasser resultierte.

Andererseits wissen wir, daß in einer Zelle mehrere Vacuolen verschiedener Art mit unterschiedlichen Inhaltsstoffen sich ausbilden können. Ferner ist die Grenze zwischen Eiweißvacuolen und sog. Proteinoplasten nicht scharf zu ziehen; auch die Existenz von Tanninoplasten wird diskutiert. Diese Benennungen deuten an, daß wenigstens gewissen Vacuolen Plastennatur zukommt, und daß sie als Zellorganelle *sui generis* zu gelten hätten. Das entspricht der Auffassung DANGEARDs, der die Ausgangssubstanz der Vacuolen in Metachromatinkörnchen (Metaphosphate ?) sieht, die ihrerseits Reduplikanten seien. Falls diese Ansicht zutreffen sollte, wäre die Vacuole auch in physiologischer und genetischer Hinsicht aktiver Teilnehmer am Zusammenwirken innerhalb der Zelle.

6. Differenzierung.

Das Problem der Differenzierung, d. h. daß Abkömmlinge einer einzigen Zelle verschieden sein können in der Entwicklungsrichtung, der Stoffwechselleistung und der Gestalt, bedarf noch sehr der experimentellen Bearbeitung. Immerhin gestatten es die bisherigen Ergebnisse sowie einige der hier dargelegten Erkenntnisse, uns wenigstens einen Teil der dabei beteiligten Mechanismen vorzustellen. Wir wissen heute, daß die einzelnen Zellkomponenten nicht nur um die Substrate, sondern auch um die Zellorte konkurrieren, und daß sich bereits in der befruchteten Eizelle eine räumliche Anordnung einstellt, die keineswegs zufällig ist, sondern ihre Ursache in der Anisotropie (Asymmetrie) der Makromoleküle, Enzyme und „organisierten" Partikel hat; Polarität, Symmetrieverhältnisse u. a. m. sind ihrerseits Folgen dieser Anordnung. Jede Teilung dieser Zelle, die nicht die Verdopplung einer jeden Komponente und die genaue Verteilung der Paarlinge auf die beiden Tochterzellen erzwingt, so wie das beim Zellkern der Fall ist, muß zu Ungleichheit der Tochterzellen führen. Erfahrungsgemäß werden aber in der Tat die Proplastiden wie die Prochondriosomen, Mikrosomen und das Cytoplasma zufallsgemäß auf die Tochterzellen verteilt.

Wenn alle Proplastiden, alle Prochondriosomen, alle Mikrosomen und alle cytoplasmatischen Partikel jeweils untereinander gleich wären, würde das wenig Bedeutung haben, denn alsbald müßten sich die ursprünglichen Verhältnisse wieder einstellen. Aber die genannten Fraktionen, vielleicht mit Ausnahme der Plastiden, stellen *Populationen* dar; bei einer Aufteilung werden die beiden Teilpopulationen auch qualitativ verschieden sein. Damit wird aber das bisherige Gefüge der Korrelation gestört, die Konkurrenz verändert und ein neues Gleichgewicht eingestellt: beide Tochterzellen haben nunmehr verschiedene Eigenschaften. Dieses allgemeine Schema ließe sich um zahlreiche Einzelzüge bereichern, doch soll hier davon abgesehen werden, da diese Prinzipien bereits in den vorhergehenden Abschnitten dargelegt worden sind. Lediglich auf die Beziehung des Volumenfaktors soll hingewiesen werden. Wenn durch innere oder äußere Faktoren (z. B. Lage der Zellen im Meristem) eine sog. inäquale Teilung erzwungen wird, müssen sich beispielsweise die vom Kern ausgehenden Substanzen auf verschieden große Plasmavolumina verteilen, so daß die kleinere Zelle zunächst im Vorteil sein kann. Umgekehrt aber kann die kleinere Zelle ihren Kern nicht so reichlich mit Substraten versorgen wie die größere — es ist nicht generell vorauszusehen, ob sich die Wirkungen immer aufheben oder in welcher Richtung sie das Zusammenwirken innerhalb der Zelle verändern (WENT, Band 1 dieses Handbuches). Völlig unbekannt bleibt schließlich, wieso eine ausdifferenzierte Zelle oder ein solches Organ bei der Regeneration sich wieder „ent"differenzieren und eine neue vollständige Pflanze bilden kann.

Abschließende Bemerkungen.

Bereits die wenigen hier aufgeführten Fakten beweisen, daß es innerhalb der Zelle keine autonomen Untereinheiten gibt, keine „lebendigen Moleküle" und keine „Lebenseinheiten kleiner als die Zelle". Das Leben resultiert aus dem Zusammenwirken einer Vielzahl von Komponenten, die sich auf verschiedener Organisationshöhe befinden: der Region des niedermolekularen Stoffwechsels, der Makromoleküle, der Enzyme und der organisierten partikulären Gebilde. Auf jeder dieser Organisationsstufen ist das Zusammenwirken zunächst als gegenseitige Konkurrenz verwirklicht, Konkurrenz um Substrate, Positionen usw. Gleichwohl ist es keine hemmungslose Konkurrenz, sondern die Partner sind in ihren Ansprüchen aufeinander abgestimmt — zweifellos eine Erwerbung, die im Laufe der Phylogenese gemacht wurde. Aber auch die verschiedenen Organisationsstufen, die ihrerseits geordnet zu sein scheinen, konkurrieren miteinander und wirken zusammen, und die niederste Ebene ist von der höchsten ebenso abhängig wie umgekehrt. Es existiert also kein Steuerungszentrum in der Zelle, das die einzige Befehlsstelle für eine Vielzahl ausführender Organe ist, sondern jedes Gebilde steuert und wird zugleich gesteuert.

Viele dieser Wechselbeziehungen sind offenbar rein quantitativer Art, aber manche Beispiele zeigen, daß quantitative Differenzen zu Qualitätsunterschieden führen. Es könnte daher verlockend erscheinen, das Problem nicht nur der Koordinierung, sondern das des Lebens überhaupt lediglich an Hand der Quantifizierung von Einzelbeziehungen zu interpretieren. Dem ist jedoch entgegenzuhalten, daß gerade die genetische Kontinuität auf der Weitergabe *qualitativer* Faktoren beruht. So tragen die DNS-Doppelschrauben die (verschlüsselte) Information für die Sequenz der Aminosäuren in der Polypeptidkette — ein qualitativer Faktor, der die Faltung, Spezifität und Aktivität der Eiweißkörper usw. kontrolliert.

Somit erweisen sich die Lebenserscheinungen der Zelle als ein Netz aus einer wahrhaft ungeheuerlichen Vielfalt von Einzelrelationen, dessen Ausmaß wir gerade erst zu ahnen beginnen. Es erscheint recht zweifelhaft, ob es jemals möglich sein wird, mehr als eine verhältnismäßig kleine Zahl solcher Einzelrelationen zu analysieren, ohne die übrigen zu beeinträchtigen oder auch nur vorübergehend außer acht zu lassen. Vielleicht sind hier dem Physiologen Grenzen gesetzt, die denen des Physikers entsprechen. Andererseits gibt es für ihn keinen anderen Weg als den, die Einzelreaktionen konsequent zu verfolgen und zu versuchen, sie untereinander in Zusammenhang zu bringen. Das bisher Erreichte dürfte ermutigend genug sein, diesen Weg weiter zu verfolgen.

Literatur.

Bautz, E.: Die Mitochondrien und Sphärosomen der Pflanzenzelle. Z. Bot. **44**, 109—136 (1956). — Bertalanffy, L. v.: Biophysik des Fließgleichgewichtes. Braunschweig 1953. — Bogen, H. J.: Über Kappenplasmolyse und Vakuolenkontraktion. Planta (Berl.) **39**, 1—35 (1951). — Beiträge zur Physiologie der nichtosmotischen Wasseraufnahme. Planta (Berl.) **42**, 140—155 (1953). — Bogen, H. J., u. G. Follmann: Osmotische und nichtosmotische Stoffaufnahme bei Diatomeen. Planta (Berl.) **45**, 125—146 (1955). — Brachet, J.: The role of the nucleus and the cytoplasm in synthesis and morphogenesis. Symposia Soc. Exper. Biol. **6**, 172—200 (1952). — Influence of the nucleus on anaerobic breakdown of adenosine triphosphate. Nature (Lond.) **173**, 725 (1954). — Brøndstedt, H. V.: The time-graded regeneration field in planarians and some of its cytophysiological implications. Recent developments in cell physiology, S. 121—138. London 1954.

Claude, A.: The constitution of mitochondria and microsomes and the distribution of nucleic acid in the cytoplasm of a leukemic cell. J. of Exper. Med. **80**, 19—29 (1944).

Dangeard, P.: Recherche sur l'appareil vacuolaire dans les végétaux. Botaniste **15**, 1—267 (1923). — Davson, H., and J. D. Danielli: The permeability of natural membranes. Cambridge 1943. — Delbrück, M.: Wie vermehrt sich ein Bakteriophage? Behringwerk Mitt. **29**, 113—125 (1954).

Eggmann, L., S. J. Singer and S. G. Wildman: The proteins of green leaves. V. A cytoplasmic nucleoprotein from spinach and tobacco leaves. J. of Biol. Chem. **205**, 969—983 (1953). — Eichenberger, M.: Elektronenmikroskopische Beobachtungen über die Entstehung der Mitochondrien aus Mikrosomen. Exper. Cell Res. **4**, 275—282 (1953). — Ephrussi, B.: Nucleo-cytoplasmic relations in micro-organisms. Oxford 1953.

Frédéric, J., et M. Chèvremont: Recherches sur les chondriosomes de cellules vivantes par la microscopie et la microcinématographie en contraste de phase (1. partie). Archives de Biol. **63**, 109—131 (1952). — Frey-Wyssling, A.: Die submikroskopische Struktur des Cytoplasmas. Protoplasmatologia II, A 2. Wien 1955.

Goldacre, R. J.: The folding and unfolding of protein molecules as a basis of osmotic work. Internat. Rev. Cytology **1**, 135—164 (1952). — Green, D. E., W. F. Loomis, S. R. Dickman, V. H. Auerbach and B. Noyce: Cyclophorase system of kidney for complete oxidation of pyruvic acid. Abstracts of papers, 111th Meeting Amer. Chem. Soc. Atlantic City, 26 B, 1947. — Guilliermond, A.: The cytoplasm of the plant cell. Waltham, Mass. 1941.

Hämmerling, J.: Nucleo-cytoplasmic relationships in the development of *Acetabularia*. Internat. Rev. Cytology **2**, 475—498 (1953). — Hämmerling, J., u. H. Stich: Über die Aufnahme von P^{32} in kernhaltige und kernlose Acetabularien. I. Z. Naturforsch. **9b**, 149—155 (1954). — Harman, J. W., and M. Feigelson: Studies on mitochondria. V. The relationship of structure and oxidative phosphorylation on mitochondria of heart muscle. Exper. Cell Res. **3**, 509—525 (1952). — Harrison, R. G.: Relations of symmetry in the developing embryo. Trans. Connecticut Acad. Arts a. Sci. **36**, 277—330 (1945). — Harvey, E. B.:

Structure and development of the clear quarter of the *Arbacia punctulata* egg. J. of Exper. Zool. **102**, 253—271 (1946). — HAUROWITZ, F.: Chemistry and biology of proteins. New York 1950. — HAUSER, E. A., and D. S. LE BEAU: Lyogels. Adv. Colloid. Sci. **3**, 137—160 (1950). — HEITZ, E.: Kristallgitterstruktur des Granum junger Chloroplasten von *Chlorophytum*. Exper. Cell Res. **7**, 606—608 (1954). — HILLIER, J., and J. F. HOFFMANN: On the ultrastructure of the plasma membrane as determined by the electron microscope. J. Cellul. a. Comp. Physiol. **42**, 203—247 (1953). — HÖRSTADIUS, S.: Influence of implanted micromeres on reduction gradients and mitochondrial distribution in developing see urchin eggs. J. Embryol. a. Exper. Morph. **1**, 257—259 (1953). — HOUWINK, A. L.: A macromolecular mono-layer in the cell wall of *Spirillum* spec. Biochim. et Biophysica Acta **10**, 360—367 (1953). — HUENNEKENS jr., F. N.: Studies on the cyclophorase system. XV. The malic oxidase. Exper. Cell Res. **2**, 115—125 (1951).

JUNG, F.: Zur Thermodynamik biologischer Systeme. Naturwiss. **41**, 120—121 (1954).

KLOTZ, J. M.: The nature of some ion-protein complexes. Cold Spring Harbor Symp. Quant. Biol. **14**, 97—112 (1949). — KUFF, E. L., and W. C. SCHNEIDER: Intracellular distribution of enzymes. XII. Biochemical heterogeneity of mitochondria. J. of Biol. Chem. **206**, 677—685 (1954).

LATIES, D. G.: The osmotic inactivation in situ of plant mitochondrial enzymes. J. of Exper. Bot. **5**, 49—70 (1954). — LEDERBERG, J.: Cell genetics and hereditary symbiosis. Physiologic. Rev. **32**, 403—430 (1952). — LETTRÉ, H., u. R. LETTRÉ: Kern-Plasma-Mitochondrien-Relation als Zellcharakteristikum. Naturwiss. **40**, 203 (1953). — LEYON, H.: The structure of the chloroplasts. V. The origin of the chloroplast laminae. Exper. Cell Res. **7**, 609—611 (1954). — LINDBERG, O., and L. ERNSTER: Chemistry and physiology of mitochondria and microsomes. Protoplasmatologia III, A 4. Wien 1954. — LUNDEGÅRDH, H.: The translocation of salts and water through wheat roots. Physiol. Plantarum (Copenh.) **3**, 103—151 (1950).

MAHLER, H. R., u. F. M. HUENNEKENS jr.: Zit. nach MAHLER 1953. — Multienzyme sequences in soluble extracts. Internat. Rev. Cytology **2**, 201—230 (1953). — MAZIA, D., and H. HIRSHFIELD: The nucleus dependence of P^{32} uptake by the cell. Science (Lancaster, Pa.) **112**, 297—299 (1950). — MERCER, F. V., A. J. HODGE, A. B. HOPE and J. D. MCLEAN: The structure and swelling properties of *Nitella* chloroplasts. Austral. J. Biol. Sci. 8, 1—18 (1955). — MICHAELIS, P.: Über das genetische System der Zelle. Naturwiss. **34**, 18—22 (1947). — Interactions between genes and cytoplasm in *Epilobium*. Cold Spring Harbor Symp. Quant. Biol. **16**, 121—129 (1951). — MITCHISON, J. M.: Cell membranes and cell division. Symposia Soc. Exper. Biol. **6**, 105—127 (1952).

NOSSAL, P. M.: A mechanical cell desintegrator. Austral. J. Exper. Biol. a. Med. Sci. **31**, 583—590 (1953).

OEHLKERS, F.: Neue Überlegungen zum Problem der außerkaryotischen Vererbung. Z. Abstammungslehre **84**, 213—250 (1952).

PAIGEN, K.: The occurrence of several biochemically distinct types of mitochondria in rat liver. J. of Biol. Chem. **206**, 945—957 (1954). — PALADE, G. E.: A small particulate component of the cytoplasm. J. Biophys. a. Biochem. Cytology **1**, 59—68 (1955). — PFEFFER, W.: Pflanzenphysiologie, 2. Aufl. Leipzig 1897. — POLLOCK, M. R.: Stages in enzyme formation. 3. Symp. Soc. Gen. Microbiol., S. 150—177, 1953. — PONTECORVO, G.: Genetical analysis of cell organization. Symposia Soc. Exper. Biol. **6**, 218—229 (1952). — PORTER, C. J., R. HOLMES and B. F. CROCKER: The mechanism of the synthesis of enzymes. II. Further observations with particular reference to the linear nature of the course of enzyme formation. J. Gen. Physiol. **37**, 271—289 (1953).

RAAFLAUB, J.: Die Schwellung isolierter Leberzellenmitochondrien und ihre physikalisch-chemische Beeinflußbarkeit. Helvet. physiol. Acta **11**, 142—156 (1953). — ROBERTSON, R. N., M. J. WILKINS and A. B. HOKE: Plant mitochondria and salt accumulation. Nature (Lond.) **175**, 640—641 (1955). — ROTHSTEIN, A.: The enzymology of the cell surface. Protoplasmatologia II, E 4. Wien 1954.

SCHNEIDER, W. C., and G. H. HOGEBOOM: Cytochemical studies of mammalian tissues: The isolation of cell components by differential centrifugation: a review. Cancer Res. **11**, 1—22 (1951). — SCHÖNHEIMER, R.: The dynamic state of body constituents. Cambridge (Mass.) 1941. — SCHULTZ, G. V.: Über den makromolekularen Stoffwechsel der Organismen. Naturwiss. **37**, 196—200 u. 223—229 (1950). — SIEKEWITZ, P.: Uptake of radioactive alanine in vitro into the protein of rat liver fraction. J. of Biol. Chem. **195**, 549 (1952). — SJÖSTRAND, F. S., and V. HANZON: Membrane structures of cytoplasm and mitochondria in exocrine cells of mouse pancreas as revealed by high resolution electron microscopy. Exper. Cell Res. **7**, 393—414 (1954). — SŁONIMSKI, P.: La formation des enzymes respiratoires chez la levure. Paris 1955. — SONNEBORN, T. M.: Gene and cytoplasm. Proc. Nat. Acad. Sci. U.S.A. **29**, 329—343 (1943). — Experimental control of the concentration of cytoplasmic genetic factors in *Paramecium*. Cold Spring Harbor Symp. Quant. Biol. **11**, 236—248 (1946).—

STEFFEN, K.: Vortrag auf der Tagg. der Dtsch. Bot. Ges. 1955. — STEINER, M., u. H. HEINEMANN: Über die Beziehungen zwischen fettbildenden Grana und typischen Mitochondrien in den Zellen von *Oospora lactis*. Naturwiss. **41**, 90 (1954). — STICH, H., u. J. HÄMMERLING: Der Einbau von P^{32} in die Nucleolarsubstanz des Zellkernes von *Acetabularia mediterranea*. Z. Naturforsch. **8b**, 329—333 (1953). — STRUGGER, S.: Über den Bau der Proplastiden und Chloroplasten. Naturwiss. **37**, 166—167 (1950). — SZAFARZ, D., et J. BRACHET: Le rôle du noyeau dans le métabolisme de l'acide ribonucléique chez „*Acetabularia mediterranea*". Arch. internat. Physiol. **62**, 154—155 (1954).

TURING, A. M.: The chemical basis of morphogenesis. Philosophic. Trans. Roy. Soc. Lond., Ser. B **237**, 37—47 (1952).

VIRCHOW, R.: Die Cellularpathologie in ihrer Begründung auf physiologische und pathologische Gewebelehre. Berlin 1858.

WARIS, H.: Cytophysiological studies on *Micrasterias*. II. The cytoplasmic framework and its mutation. Physiol. Plantarum (Copenh.) **3**, 236—246 (1950). — WATSON, J. D., and F. H. C. CRICK: A structure for desoxyribose nucleic acid. Nature (Lond.) **171**, 737 (1953). — Genetical implications of the structure of desoxyribose-nucleic acid. Nature (Lond.) **171**, 964 (1953). — WATSON, M. L.: The nuclear envelope. Its structure and relation to cytoplasmic membranes. J. Biophys. a. Biochem. Cytology **1**, 257—270 (1955). — WEBSTER, G. C.: Incorporation of radioactive glutamic acid into the proteins of higher plants. Plant Physiol. **29**, 382—385 (1954).

Temperature (heat and cold resistance, frost hardening).

By

J. Levitt.

The literature on these subjects is so extensive that it cannot begin to be covered in the following brief account. For further details see LEVITT (1956).

Cell activity is markedly affected by temperature in many ways. In general, as is commonly true of all physical and chemical processes, cell activity increases with rise in temperature. But there is a definite limit to this relation, and the actual effect varies with the plant, the temperature range, and the environmental conditions (BELEHRADEK 1935). There are also more or less definite upper and lower temperature limits to cell activity—*i.e.* temperatures at which all activity stops and death sets in. But this varies markedly with the plant. The lower limit may be anywhere from about $+5^0$ C to the lowest temperatures obtainable. The upper limit may be from as low as 20^0 C to as high as 120^0 C though no one exact temperature can be given, since it varies so much with exposure time (table 1).

Table 1. *Heating time and killing temperature for Tradescantia discolor.* (From LEPESCHKIN 1910.)

Heating time (mins.)	Killing temperature (°C)
4	72.1
10	69.6
25	63.2
60	57.0
80	55.7
100	54.1
150	52.0

These upper and lower extremes, however, are true of resting organs in the dehydrated state—*e.g.* seeds and spores. Merely allowing these to take up small quantities of water markedly lowers the upper limit and even more markedly raises the lower limit (ROBBINS and PETSCH 1932). In the case of normally hydrated tissues, the upper limit is 85^0 C, or according to one unconfirmed observation 93^0 C (VOUK 1923), the lower limit under natural conditions is about -60^0 C (SCHOLANDER *et al.* 1953). The importance of hydration as distinct from the resting stage can be shown in another way. If the cells are dehydrated osmotically, by means of noninjurious solutions such as sucrose, they can survive heating to higher temperatures and freezing at lower temperatures than when in the normally hydrated state. There are exceptions that cannot be protected in this way, due to the fact that the plasmolysis itself produces injury (SCHEIBMAIR 1938). It should be pointed out that the upper limits of life for hydrated tissues cannot be raised appreciably above the limits mentioned. The lower limit of life for hydrated tissues can, however, be dropped probably to the lowest temperature obtainable, if a sufficiently rapid temperature drop and rise is used to prevent the formation of ice crystals of microscopically visible size (LUYET 1954).

In the case of any one plant, the lower or the higher the cooling or the heating limits are, the greater its cold resistance or heat resistance respectively. The term resistance, however, is an ambiguous one. It may imply the ability of the plant to resist a change in temperature. And, indeed, mechanisms are available for preventing the plant's temperature from reaching that of its environment—*e.g.* respiration may raise the plant's temperature, transpiration may lower it. But these are not very effective methods of preventing either cold or heat injury, for at the low temperatures when the heat of respiration is needed, the process is too slow, and at the high temperatures when cooling by transpiration is needed, insolation raises the plant's temperature above atmospheric. In most cases, the "resistance" of the plant refers to its ability to survive high or low temperatures in its tissues. The term "hardiness", used commonly by practical men, describes this state somewhat better and will be used here. In the same way, a plant is said to undergo hardening, when its ability to survive temperature extremes increases.

Heat hardiness.

(BELEHRADEK 1935, LEPESCHKIN 1937, LEVITT 1951.)

The upper temperature limit varies somewhat with the habitat of the species (SAPPER 1935), from 38–40° C for aquatic plants, to 50° C or higher for plants of dry and hot sunny regions. It also may vary with the conditions the individual plant is exposed to. Cultures of microorganisms may have their heat hardiness increased simply by growing them at moderately high temperatures. But this involves the formation of new individuals at the higher temperatures. Individual higher plants do not show this adaptability. On the contrary, their heat hardiness may actually decrease if they are grown at high temperatures. Yet, if exposed to drought or low temperatures or even to deficiencies of mineral elements (*e.g.* Na, K) their heat hardiness does increase (table 2) though the effect is far less pronounced than in the case of frost hardiness. Younger tissues may be more hardy than older ones. Some evidence indicates a relation between hardiness and cell sap concentration (table 2), though this does not always hold.

Table 2. *Effect of dry cultivation on heat hardiness.* (From SAPPER 1935.)

Species	Heat killing temperature (°C)		Osmotic value at incip. plas. (M. sucrose)	
	sparingly watered	kept saturated	sparingly watered	kept saturated
Melilotus officinalis	46.5	45	0.55–0.60	0.45–0.50
Taraxacum officinale	46	44.5	0.65	0.45
Avena sativa	44.5	< 43.5	0.65	0.40–0.45
Hordeum distichum	46–46.5	< 44	0.80	0.45–0.50
Hieracium pilosella	> 50.5	48.5	0.65–0.75	0.35–0.55
Ceterach officinarum	100–120	47		

The nature of heat hardiness is not too well understood. Heat injury is probably due to denaturation of the protoplasmic proteins. Evidence of this is the similarity between the temperatures for the two, the high Q_{10}s for both, and the similar effects of acids, alkalis, narcotics and salts on both processes. Nearly all the theories proposed to explain heat hardiness, in fact, assume that the injury is due to protein denaturation. Hardiness may therefore be due either to 1. the higher heat denaturation temperatures of the proteins, or 2. the ability to resynthesize the proteins more rapidly than they are denatured (VAN HALTEREN 1950). Though the latter theory has been proposed primarily for microorganisms, there is no reason why it should not also apply to higher plants. However, insufficient evidence is as yet available for adopting it as a general rule.

Other factors besides heat denaturation of proteins may also play a part in heat hardiness. HEILBRUNN suggested that the liquefaction of the protoplasmic lipids may be involved, though LEPESCHKIN concludes that this does not agree with the high Q_{10} of the process. When heat killing is slow, the balance between photosynthesis and respiration may be important, since the maximum temperature for the former (sometimes 40° C) is much lower than that for respiration in some plants. As a result, the plant may starve to death at a temperature not quite high enough to cause protein denaturation. HARDER *et. al.* (1932) showed that plants adapted to high temperatures have compensation points much higher than the above, and therefore continue to assimilate at temperatures as high as 45–53° C. These plants are never in danger of starvation due to high temperature, since their tissues never go above 45–53° C for long enough periods of time to suffer an appreciable loss of reserves.

Many investigators have shown that the heat resistance of sections of tissues is markedly affected by the nature of the ions in the solution in which they are heated. Since the ions may be arranged in lyotropic series according to their effectiveness, BOGEN (1948) concluded that heat resistance is not simply due to dehydration but to the order of charges and the hydration centers on individual molecules. Any disturbance of this order may lead to death, since the molecule is thereby deformed and the system of molecular bonds exposed to tensions.

Cold resistance.

(LEVITT 1941, 1951.)

In accord with what was pointed out above, cold resistance is a general term that includes many things. Plants have little ability to maintain their temperatures above that of the air when near the freezing point. Consequently, those that survive exposure to low temperatures must be able to tolerate cooling of their tissues. Some tropical plants are very sensitive and may be killed if kept at temperatures of 0–5° C for days or even hours (table 3). Many fruit show this same sensitivity (PENTZER and HEINZE 1954). The nature of the injury is not well understood. Unlike high temperatures, low temperatures cannot lead to danger of starvation, since the respiration process is markedly slowed up. Death by starvation would therefore require much longer than a few days. It might, however, involve a disturbance of the metabolism, leading to the formation of toxic substances (SMITH 1954), though even this would probably require long periods of time. Suggestions that N-metabolism may produce the injury have not been upheld (SEIBLE 1939).

Table 3. *Chilling injury in plants from warm climates. Exposed continuously to 1.4 to 3.7° C in diffuse light and covered to prevent transpiration. From 28 species listed by* MOLISCH (1897).

Species	Time for first injury to appear	Time for complete killing (days)
Episcia bicolor Hook.	18 hrs.	5
Sciadocalyx Warsewitzii Regel	24 hrs.	5
Eranthemum tricolor Nichols	48 hrs.	4–5
Eranthemum Couperi Hook.	3–5 days	10
Boehmeria argentea Linden	8 days	20
Iresine acuminata	11 days	19
Uhdea bipinnatifida Kunth.	15 days	16
Eranthemum nervosum R. Br.	20 days	30–35

The most difficult cases to explain are those that occur as a result of as little as a few minutes' exposure to low temperature. The only available hints at an explanation are to be found in some old observations of pseudoplasmolysis in cells suddenly cooled to temperatures of 0–5° C. This has been explained by a sudden, injurious increase in permeability; though why such a change should occur, no one has tried to explain. It follows from all these uncertainties that no reason can be given for the hardiness of most plants to this kind of cold injury.

The above type of cold injury that occurs at supra-freezing temperatures has been called "catching cold" (Erkältung—MOLISCH 1897). A far more prevalent kind of cold injury occurs at freezing temperatures and is known as frost injury (Erfrieren). That injury is due to an actual freezing of the plant is easily shown; for if it is undercooled, it can survive temperatures far below its frost killing point (*i.e.* the temperature at which the frozen plant dies). The ice does not have to form inside the cell to cause injury. In fact, under natural conditions it always or nearly always occurs in the intercellular spaces.

All gradations occur between plants that are killed by the first touch of frost and those that survive freezing solid at the lowest temperatures. A single plant can show these same gradations from completely tender (killed by —5° C) to completely hardy in mid-winter (not killed by —50° C). Hardening is therefore continuous and quantitative, rather than a discontinuous, step-by-step process. It may be produced in several ways. The maximum hardiness results from exposure to low temperatures (*e.g.* 0–5° C) for some days or weeks. It has even been stated that full hardening requires a second exposure to subfreezing temperatures. Winter annuals harden under these conditions only if light is adequate. Simply withholding water for some time, or starving the plant of nitrogen, may induce a significant hardening, though not as much as the low temperature treatment. A combination of treatments may be necessary to obtain maximum hardiness.

Frost hardiness also varies with the method of freezing and thawing. If the rate of freezing is too rapid, not as low a temperature can be survived. Rate of thawing may also have this effect. Even the treatment after thawing may affect survival, and the length of time frozen, as well as the number of times the plant is frozen and thawed. In order to obtain duplicable results, a standardized method is usually adopted, involving a single slow freezing and thawing, followed by recovery for a day at low (suprafreezing) temperatures.

Many changes occur in the plant during the hardening period. In general, the methods of hardening all reduce growth and therefore lead to a small cell size. But even the cells that were already mature before the hardening treatment show increased hardening without any morphological change. Consequently, physiological factors must form the basis of frost hardening. Many such factors have, indeed, been shown to change during the hardening period. As a rule, the moisture content of the plant decreases and its cell sap concentration increases. This is primarily due to the accumulation of sugars. In the case of winter annuals that harden at low temperatures in the light, the accumulation is directly due to an excess of photosynthesis over respiration at these low temperatures, and the absence of sufficient growth to use up these reserves. In the case of perennials, the carbohydrates accumulate in the form of starch during the summer, when growth has stopped but photosynthesis continues at a high rate. When these plants are exposed to the hardening low temperatures of autumn, the starch is hydrolyzed to sugar.

The importance of sugars in frost hardiness is so pronounced, that if varieties of a species, such as wheat, are arranged according to their sugar contents, they are found to be also in the order of their hardiness (table 4). But this parallelism holds true only if the plants have been hardened. Even then, in many cases the relation does not hold. Many of the hardiest evergreens, especially among the conifers, show little parallelism between frost hardiness and sugar content.

The reason for these exceptions is the existence of protoplasmic factors that are essential to frost hardiness. Plants without these protoplasmic properties may have the highest sugar contents without being able to develop any frost hardiness (*e.g.* sugar cane). On the other hand, some plants that have these protoplasmic properties may be very hardy, even though their sugar content is low (*e.g.* some succulents). The nature of these protoplasmic properties is not fully understood, though their existence can be readily demonstrated. During hardening, there is a marked increase in protoplasmic permeability to polar substances, and a marked change in other physical properties of protoplasm. Some controversy at first existed as to just what these changes were. According to Kessler and Ruhland (1938) protoplasmic viscosity increased on hardening. According to Scarth and Levitt (1937) it decreased on hardening. Later work (Siminovitch and Levitt 1941, Levitt and Siminovitch 1940), however, established the fact that the change is not readily detected in fully hydrated cells but becomes easily recognized when the protoplasm is partially dehydrated. In this state, frost hardy protoplasm shows an ability to tolerate all kinds of rough handling that frost tender protoplasm subjected to the same dehydrating forces quickly succumbs to. Pressure, stretching of the cytoplasmic strands, pushing the protoplasm through pits in the wall, plasmolysis and deplasmolysis, all fail to injure frost hardy protoplasm though readily killing the frost tender. This ability of frost hardy protoplasm to survive rough handling is associated with its ability to remain ductile even when subjected to powerful dehydrating forces. Judging from this fact, from its higher water permeability, and from its lower refractive index, the frost hardy protoplasm is apparently able to hold on to more water when subjected to dehydration than can frost tender protoplasm.

Table 4. *Relation between sugar content and frost hardiness in wheat.* (From Åkerman 1927.)

Variety	Relative sugar content	Relative frost hardiness (I. = highest)
Sammet	100	I
Svea II	87	II
Thule II	67	IV
Standard	66	IV
Sol II	65	IV
Pansar II	48	V
Extra Squarehead II .	44	VI
Danish small wheat .	41	VII
Wilhelmina	39	VIII
Perl summer-wheat .	29	IX
Halland summer-wheat	22	X

The basis for these protoplasmic properties must be sought in changes that occur in the protoplasmic proteins during the hardening period. A series of investigations by Siminovitch and his coworkers has shown that one soluble protein component undergoes a marked seasonal increase prior to the seasonal rise in frost hardiness. Its quantity, in fact, always paralleled frost hardiness very closely, even under conditions that prevented a similar parallel between sugars and frost hardiness. The significance of this protein in frost hardiness cannot, as yet be explained.

It is a striking fact that frost hardening increases the plant's ability to tolerate dehydration whether it is due to extracellular ice formation in the frozen plant, to drought, or to osmotic water removal. Thus, the seasonal changes in frost hardiness of evergreens, rising to a maximum in mid-winter, are paralleled by similar seasonal changes in drought hardiness (Pisek and Larcher 1954). These plants are actually more drought hardy during mid-winter than in mid-summer. The same holds true for osmotic dehydration. It is, in fact, possible to dispense with freezing tests for determining frost hardiness, and to replace them by osmotic dehydration tests (SIMINOVITCH and BRIGGS 1953).

The roles of all these hardiness changes can be explained only if the mechanism of frost injury is understood. All the evidence indicates that the injury is due to dehydration, though many other explanations have been proposed from time to time. This is in accord with the following facts: 1. Frost hardiness is developed toward extracellular freezing but not toward intracellular freezing; the latter cannot be tolerated but can be avoided by hardy plants. 2. Undercooling below the frost killing point fails to injure. 3. Frost hardiness parallels hardiness to other forms of dehydration. 4. Hardy protoplasm appears able to hold more water against dehydrating forces than can frost tender protoplasm. All the known facts about frost hardiness can, in fact, be fit into this concept. But this only pushes the question back a step. If frost injures due to dehydration, why does dehydration produce injury?

To date, the only accepted explanation of dehydration injury is still the mechanical theory proposed by ILJIN, or modifications of it. When ice forms outside the cell, the water removal is accompanied by cell collapse. The greater the amount of extracellular ice, the greater this cell collapse. The cell wall is pulled in with the collapsing cell, and exerts a spring-like tension on the protoplasm it adheres to. At some point, this tension mechanically injures the protoplasm, perhaps by inducing a tear in the collapsed cell; or by leading to a rupture during thawing, when the cell wall quickly snaps back into place while the less water-permeable (and now somewhat stiffened) protoplasm "deplasmolyzes" more slowly.

On this basis, the above frost hardiness factors can be easily explained. Small cell size would reduce the danger of injury because of the greater specific surface and more nearly spherical shape. Reduced water content and increased sugar content would both insure a less severe cell collapse, the former because less water is available to be removed, the latter because of osmotic forces opposing water removal. A cell with a molar sugar solution in its vacuole would lose only three quarters of its water at —7.5° C, a cell with only a half molar sugar solution would lose seven eighths. But if the protoplasm is easily dehydrated, such high sugar concentrations would fail to confer any hardiness on the cell, because the protoplasm would be injured by far less than a cell collapse to nearly one quarter its normal size. On the other hand, protoplasm with a marked ability to hold water against dehydrating forces would not be injured by severe collapse, and would be hardy even in the absence of high sugar concentrations. Maximum hardiness, however, would require the development of both factors.

Much remains to be explained, but the above theory of frost hardiness agrees with the known facts and therefore serves as a satisfactory working hypothesis. Indirectly, it also leads to an explanation of heat hardiness, since the two kinds of hardiness have been found to parallel each other in so many cases. As mentioned, both are inversely related to the hydration of the cell and those environmental conditions that increase frost hardiness also result in increased heat hardiness. Even when different cells of a tissue are compared, the order of hardiness is the same for both frost and heat. This parallelism is far more surprising than that between frost and drought hardiness, since both frost and drought dehydrate the cell. True heat injury, on the other hand, does not involve dehydration. The only logical explanation of the correlation seems to be on the basis of the first theory of heat hardiness given above—*i.e.* that it is due to a greater ability of the protein to survive high temperatures without denaturation. The water-binding properties of the protoplasmic proteins that enable protoplasm to resist injury due to dehydration also, presumably, help the proteins to tolerate heat without being denatured. In agreement with this explanation, and with the relation of heat hardiness to cell hydration, heat denaturation of proteins is known to occur more readily in dilute than in concentrated solutions. From the limited evidence available, however, it seems probable that the correlation between heat and frost hardiness does not always hold. If so, this might be due to heat hardiness of the second kind—the ability to resynthesize proteins more rapidly than heat tender plants at high temperatures.

Literature.

ÅKERMAN, Å.: Studien über den Kältetod und die Kälteresistenz der Pflanzen, S. 1–232. Lund 1927.

BELEHRADEK, J.: Temperature and living matter. Protoplasma monograph 8. Berlin: Gebrüder Bornträger 1935. — BOGEN, H. J.: Untersuchungen über Hitzetod und Hitzeresistenz pflanzlicher Protoplaste. Planta (Berl.) **36**, 298–340 (1948).

HALTEREN, P. VAN: Effets d'un choc thermique sur le métabolisme des levures. Bull. Soc. Chim. Biol. **32**, 458–463 (1950). — HARDER, RICHARD, PAUL FILZER u. ALFRED LORENZ: Über Versuche zur Bestimmung der Kohlensäureassimilation immergrüner Wüstenpflanzen während der Trockenzeit in Beni Unif (algerische Sahara). Jb. wiss. Bot. **75**, 45–194 (1932). — HUBER, B.: Der Wärmehaushalt der Pflanzen. Naturwiss. u. Landwirtsch. **17**, 148 (1935).

KESSLER, W., u. W. RUHLAND: Weitere Untersuchungen über die inneren Ursachen der Kälteresistenz. Planta (Berl.) **28**, 159–204 (1938).

LEPESCHKIN, W. W.: Zell-Nekrobiose und Protoplasma-Tod. Protoplasma (Berl.) **12** (1937). — LEVITT, J.: Frost killing and hardiness of plants. Minneapolis: Burgess Publ. Co. 1941. — Frost, drought, and heat resistance. Annual Rev. Plant Physiol. **2**, 245–268 (1951). — The hardiness of plants. New York: Academic Press 1956. — LEVITT, J., and D. SIMINOVITCH: The relation between frost resistance and the physical state of protoplasm. I. Canad. J. Res. C **18**, 550–561 (1940). — LUYET, B. J.: Le mécanisme du gel et la résistance au froid. VIII. Congr. Internat. de Botanique, 1954, p. 259–267.

MOLISCH, HANS: Untersuchungen über das Erfrieren der Pflanzen. Jena 1897. S. 1–73.

PENTZER, W. T., and P. H. HEINZE: Post harvest physiology of fruits and vegetables. Annual Rev. Plant Physiol. **5**, 205–224 (1954). — PISEK, A., u. W. LARCHER: Zusammenhang zwischen Austrocknungsresistenz und Frosthärte bei Immergrünen. Protoplasma (Wien) **44**, 30–46 (1954).

ROBBINS, W. J., and K. F. PETSCH: Moisture content and high temperature in relation to the germination of corn and wheat grains. Bot. Gaz. **93**, 85–92 (1932).

SAPPER, I.: Versuche zur Hitzeresistenz der Pflanzen. Planta (Berl.) **23**, 518–556 (1935). — SCARTH, G. W., and J. LEVITT: The frost-hardening mechanism of plant cells. Plant Physiol. **12**, 51–78 (1937). — SCHEIBMAIR, G.: Hitzeresistenz-Studien an Moos-Zellen. Protoplasma (Berl.) **29**, 394–424 (1937). — SCHOLANDER, P. F., WALTER FLAGG, R. J. HOCK and LAURENCE IRVING: Studies on the physiology of frozen plants and animals in the Arctic. J. Cellul. a. Comp. Physiol. **42**, 1–56 (1953). — SEIBLE, DORA: Ein Beitrag zur Frage der Kälteschäden an Pflanzen bei Temperaturen über dem Gefrierpunkt. Beitr. Biol. Pflanz. **26**, 289–330 (1939). — SIMINOVITCH, D., and D. R. BRIGGS: Studies on the chemistry of the living bark of the black locust in relation to its frost hardiness. III. The validity of plasmolysis and desiccation tests for determining the frost hardiness of bark tissue. Plant Physiol. **28**, 15–34 (1953). — SIMINOVITCH, D., and J. LEVITT: The relation between frost resistance and the physical state of protoplasm. II. Canad. J. Res. C **19**, 9–20 (1941). — SMITH, W. HUGH: Non-freezing injury in plant tissues with particular reference to the detached fruit. VIII. Congr. Internat. de Botanique, 1954, p. 280–285.

VOUK, V.: Die Probleme der Biologie der Thermen. Internat. Rev. d. Hydrobiol. **11**, 89–99 (1923).

Wassermangel und Zellaktivität.

Von

O. Stocker.

Mit 12 Abbildungen.

Es ist nicht die Absicht, an dieser Stelle eine vollständige und eingehende Darstellung des Themas zu geben. Das wird in Band 3, Abschnitt VI, „Wasserzustand und Wasserbilanz" geschehen, namentlich in dem Artikel „Die Dürreresistenz". Über die Beeinflussung einzelner Aktivitäten wird man nähere Angaben in den sie behandelnden Bänden finden.

I. Einführung.

Wassermangel beeinflußt die Aktivität der Zelle *direkt* über die Verschlechterung des *chemischen Potentials* des Wassers und *indirekt* über die Deformation und Destruktion der submikroskopischen *Struktur des Plasmas.*

Das *chemische Potential* bestimmt den *thermodynamischen Zustand* des Wassers. Es bedeutet ein *Arbeitsvermögen*, und dieses erfährt eine Minderung, wenn die Moleküle des Wassers in Lösungen und Quellkörpern durch Ionen, Moleküle und kolloidale Teilchen auseinandergedrängt oder mittels elektrischer und adsorptiver bzw. VAN DER WAALSscher Kräfte gebunden werden.

Die submikroskopische *Struktur des Plasmas*[1] wird über Quellungs- und Hydratationsänderungen der Eiweißmicellen durch Wasserentzug beeinflußt. Mit der Deformation der micellaren Struktur ändern sich nicht nur die physikalisch-physiologischen *Zustandsgrößen* wie Viscosität, Permeabilität usw., sondern auch die chemischen *Reaktionsbedingungen* vieler Prozesse, vor allem fermentativer, weil die Aktivität vieler Fermente von dem micellaren Zustand ihrer Trägergruppen und ihrer Bindung an Plasmastrukturen abhängig ist.

II. Die Beeinflussung des chemischen Potentials des Wassers durch Wassermangel.

A. Physikalisch-chemische Grundlagen.

Das *chemische Potential* μ eines durch die absolute Temperatur T und den (Sättigungs-) Dampfdruck p charakterisierten Wasserzustandes[2] ist durch die Gleichung definiert:

$$\mu = \mu_0 + R \cdot T \cdot \ln \frac{p}{p_0}. \qquad (1)$$

In ihr bedeutet μ_0 das als absolute Größe unbekannte chemische Potential eines Standardzustandes des Wassers bei T Grad absolut und 1 at mit dem (Sättigungs-) Dampfdruck p_0 (STERN, S. 256, dort auch die Gleichung bei verschiedener Temperatur der beiden Zustände).

[1] Vgl. Band I, S. 301ff.: W. SEIFRIZ, Microscopic and submicroscopic structure of cytoplasm.

[2] Vgl. dazu Band I, S. 194ff.: P. J. KRAMER: Water content and water turnover in plant cells.

Für den *thermodynamischen Zustand* des Wassers in der *Zelle* ist maßgebend die isotherme Nutzarbeit $\mu_0 - \mu$, die beim Überführen reinen Wassers in den Zustand des Zellwassers maximal gewonnen werden kann. Sie bestimmt sich nach obiger Gleichung zu

$$\mu_0 - \mu = -R \cdot T \cdot \ln \frac{p}{p_0} \tag{1a}$$

Durch Entwicklung von $\ln \frac{p}{p_0}$ in logarithmischer Reihe und Vernachlässigung der höheren Glieder derselben ergibt sich:

$$\mu_0 - \mu = -R \cdot T \cdot \ln\left(1 + \frac{p - p_0}{p_0}\right) = -R \cdot T \cdot \frac{p - p_0}{p_0}$$

$$\mu_0 - \mu = R \cdot T \cdot \frac{p_0 - p}{p_0} \quad \text{(STERN, S. 263),} \tag{2}$$

d. h. *die thermodynamisch formulierte Aktivität des Wassers fällt bei gegebener Temperatur annähernd proportional der relativen Dampfdruckerniedrigung.*

Das so entwickelte chemische Potential des Wassers entspricht der von WALTER (1931) als *Hydratur* bezeichneten Größe in ihrer strengen Bedeutung (vgl. STERN, S. 273). Wir werden diese Bezeichnung im folgenden aber deshalb vermeiden, weil sie oft in einer lässigen, nicht mehr auf die korrekte Definition beschränkten Weise gebraucht wird. Dagegen werden wir statt der etwas umständlichen Bezeichnung „chemisches Potential des Wassers" im folgenden auch den abgekürzten Ausdruck „Wasserpotential" verwenden.

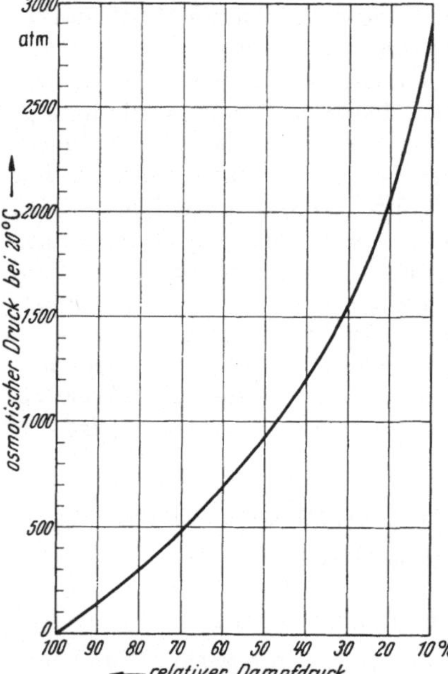

Abb. 1. Zusammenhang zwischen osmotischem Druck (20° C) und relativem Dampfdruck einer Lösung. (Nach Angaben von WALTER 1931.)

Als relatives Maß des chemischen Potentials des Wassers in einer Lösung kann man statt ihrer relativen Dampfdruckerniedrigung auch den *osmotischen Druck O* benutzen. Er berechnet sich unter vereinfachenden Annahmen ebenfalls aus den Dampfdrucken der beiden Zustände nach der Formel

$$O = \frac{R \cdot T}{V_0} \cdot \ln \frac{p_0}{p} \tag{3}$$

worin V_0 das Molvolumen des Wassers ist[1].

Der Zusammenhang der Formel (3) mit der Formel (1a) ergibt sich, wenn man durch Herüberbringen von V_0 auf die linke Seite mit $O \cdot V_0$ zur Dimension der Energie übergeht.

Die zahlenmäßige Beziehung des *osmotischen Druckes* von Lösungen zu ihrem *relativen Dampfdruck* ergibt sich aus der Abb. 1. Bei Quellkörpern tritt an Stelle des osmotischen Druckes der *Quellungsdruck.*

Im osmotisch-semipermeablen System der lebenden Zelle wird der *osmotische* bzw. der mit ihm stets im Gleichgewicht stehende *Quellungsdruck* (WALTER 1923) nur insoweit *real*, als er nicht durch den *Wanddruck* der elastisch gespannten Zellmembran aufgehoben wird. Die durch Verdünnung und Adsorption verursachte thermodynamische Zustandsänderung und biologische „Entwertung"

[1] Die Formel ist nur bei Dampfdruckerniedrigungen bis etwa 80% relativer Feuchtigkeit anwendbar (WALTER 1931).

der Wassermoleküle im Plasma und Zellsaft wird dabei durch Kompression des Zellinhaltes ausgeglichen; es ist derselbe Vorgang, der im Osmometer durch den hydrostatischen Druck der Wassersäule bewirkt wird. Allgemein gilt für das *tatsächliche* Saugpotential der Zelle, das man als *Saugkraft* zu bezeichnen pflegt, die bekannte osmotische Zustandsgleichung

$$S = O - W$$

[S = Saugkraft der Zelle, O = osmotischer Druck (osmotischer Wert), bzw. dessen Differenz zum osmotischen Druck der Außenlösung, W = Wanddruck.]

Für den *thermodynamischen Zustand des Wassers* ist also nicht der kryoskopisch oder grenzplasmolytisch bestimmte Wert des osmotischen Druckes (O), sondern allein die Größe der *Saugkraft* (S) charakteristisch. O kann nur dann als Annäherung angesehen werden, wenn sich W der Null nähert, d. h. die Pflanze beim Welken ihren Turgor zu verlieren beginnt. Es wird dann zum oberen Grenzwert von S, das im allgemeinen kleiner als O ist. Jedoch kann auch der Fall vorkommen, daß S größer als O wird, nämlich dann, wenn die Wand einer Zelle im Gewebeverband durch Zug der Nachbarzellen eine Dehnung von außen her erfährt, W also negativ wird. Es können so sehr erhebliche Saugkräfte zustande kommen.

B. Die zellphysiologischen Verhältnisse.

a) Die Regulation der Saugkraft.

In der Zelle ist das chemische Potential des Wassers durch den Gehalt an *gelösten Kristalloiden* und *solvatisierten Kolloiden* bestimmt. Die ersteren bestimmen den *osmotischen Druck*, die letzteren den *Quellungsdruck*. Im Zustand des thermodynamischen Gleichgewichtes sind diese beiden Drucke gleich. Das bedeutet, daß jeder Wasserentzug im Zellsaft eine entsprechende Entquellung des Plasmas zur Folge hat.

Über den Anstieg des *osmotischen Druckes* bei zunehmender *Konzentration von Lösungen* orientiert die Abb. 2, die sich auf Saccharose und Kochsalz bezieht. Man ersieht aus ihr, daß der osmotische Druck bei niederen Konzentrationen etwa proportional der Konzentration linear ansteigt, um dann bei höheren schneller zu wachsen. Die Beziehung zwischen *Quellungsdruck* und *Volumen von Quellkörpern* gibt die Abb. 3, in die neben Gelatine, Casein und Stärke auch das Plasma von *Lemanea*-Karposporen aufgenommen ist (Walter 1923, 1925). Auch hier wächst bei niederen Werten der Quellungsdruck proportional dem Wasserentzug, um dann mit einem Knick bei etwa 95% relativem Dampfdruck im höheren Bereich außerordentlich steil anzusteigen.

Wenn eine Zelle aus dem Zustand der *Wassersättigung* in den eines *Wassermangels (Wasserdefizit)* übergeht, so erhöht sich durch den Entzug von Wasser der *osmotische* und der *Quellungsdruck*, und der *Wanddruck* der Membran vermindert sich durch die Schrumpfung des Volumens. Da die Entspannung größer ist als die Drucksteigerung, tritt gemäß der Gleichung $S = O - W$ eine *Saugkraft* auf, die mit völliger Entspannung der Zellwand im Stadium der Grenzplasmolyse gleich dem osmotischen und Quellungsdruck wird.

Die Tabelle 1 erläutert diesen Vorgang zahlenmäßig am Beispeil der Markzellen von *Impatiens noli tangere* (Ursprung 1932). Diese Zellen haben einen nur dünnen Plasmabelag, so daß der Quellung des Plasmas keine Bedeutung zukommt und die Vorgänge überwiegend durch das *osmotische Verhalten des Zellsaftes* bestimmt sind. Das *Quellungsverhalten des Plasmas* wird in vacuolenlosen Zellen entscheidend, wie etwa den Karposporen von *Lemanea* in Abb. 4; auch hier erfolgt im Bereich geringer Drucke der Anstieg etwa umgekehrt proportional dem Volumen.

Wir untersuchen jetzt den Einfluß der Dehnbarkeit, d. h. des *Elastizitätskoeffizienten* der Zellwand (WALTER 1924).

In Abb. 5 sind schematisch 3 Fälle A, B und C der Volumenverkleinerung einer wassergesättigten Zelle gezeichnet, deren osmotischer Druck O zu 14 at angenommen wird. Jedesmal soll eine Überführung der Zelle in ein Medium mit der Saugkraft 16 at stattfinden, wobei aber in A, B und C verschiedene Elastizitätskoeffizienten der Wand und verschiedene Anteile von Zellsaft und Plasma vorausgesetzt werden.

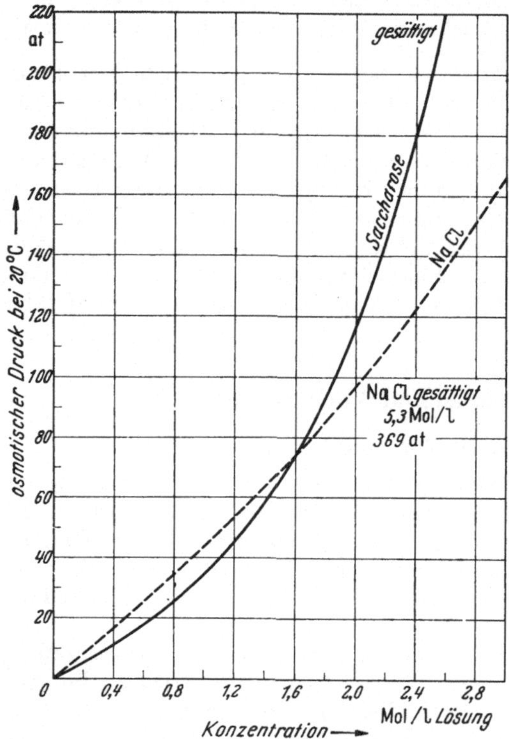

Abb. 2. Osmotischer Druck von Saccharose und NaCl-Lösungen bei 20° C. (Nach Angaben von WALTER 1931.)

Im Falle A soll in einer *vacuolisierten Zelle* der Elastizitätskoeffizient der Membran so gering sein, daß schon nach einer *Volumschrumpfung um etwa* $^1/_8$ die Wand völlig entspannt ist. Da sich bei dieser Schrumpfung der osmotische Druck um etwa $^1/_8$ auf 16 at erhöht, ist nach ihrem Vollzug die Saugkraft der Zelle gleich der des Mediums, das Gleichgewicht also erreicht.

Bei der ebenfalls vacuolisierten Zelle B dagegen soll erst nach einer *Schrumpfung auf die Hälfte* die Wand entspannt sein. Infolgedessen ist nach einer $^1/_8$-Schrumpfung zwar wie bei A $O = 16$, W aber nur auf $^3/_4$ seines Wertes von 14 at, also auf 10,5 at reduziert, so daß noch eine Saugkraft von 5,5 at besteht. Erst eine Volumverkleinerung auf etwa $^2/_3$ erreicht das Gleichgewicht mit der Außenlösung von 16 at, indem dann $O = 21$ und $W = 4{,}7$ at ist.

Tabelle 1. *Osmotische Zustandsgrößen einer Markzelle von Impatiens noli tangere.* (Nach URSPRUNG 1932.)

Zustand der Zelle	Relatives Zellvolumen	Osmotischer Druck at	Wanddruck at	Saugkraft at
Wassersättigung	148	9,3	9,3	0,0
Normaler Zustand . . .	141	9,7	5,4	4,3
Grenzplasmolyse	132	10,5	0,0	10,5

Hinsichtlich des *chemischen Potentials* besteht kein Unterschied zwischen dem Fall A und B — es ist beide Male auf 16 at verschlechtert — aber die verschobene Wassermenge ist bei B fast 3mal so groß als bei A; solche stark dehnbaren Zellen sind denn auch charakteristisch für Wassergewebe (RENNER 1915, WALTER 1924).

Nehmen wir nun als dritten Fall *C* eine *vacuolenfreie Zelle*, deren Inhalt nur aus *Plasma* besteht. Einem Anstieg des Quellungsdruckes um 1 at ent-

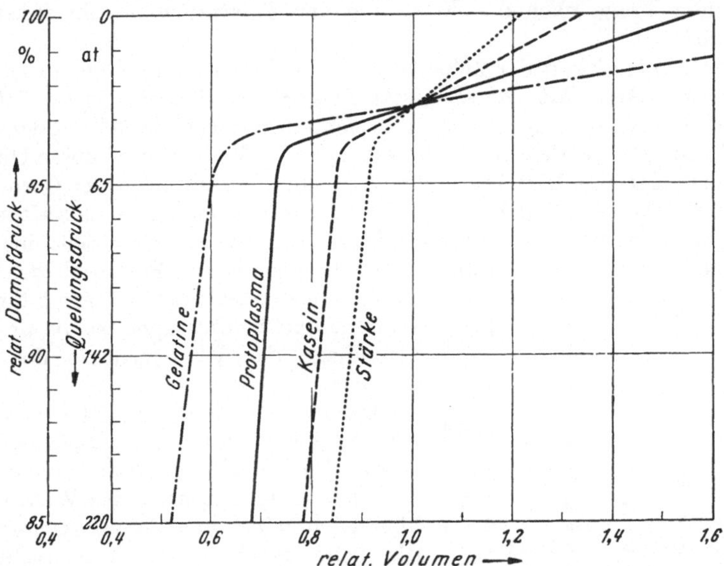

Abb. 3. Quellungsdruck von Quellkörpern. Die Protoplasmakurve bezieht sich auf *Lemanea*-Karposporen (Nach WALTER 1923 und 1925.)

spricht eine Volumabnahme um etwa 1% (Abb. 4, vgl. WALTER 1924, S. 377). Bei einer *Wanddehnbarkeit* wie im *Fall A* wird das Gleichgewicht mit 16 at bei einer Schrumpfung um etwa 7,5% erreicht. Dann ist nämlich der Quellungsdruck von 14 at um 7,5 at auf 21,5 at gestiegen und der Wanddruck, der gemäß unserer Annahme bei einer Schrumpfung um $^1/_8$ um 14 at abnehmen würde, um $\frac{14 \cdot 8 \cdot 7{,}5}{100} = 8{,}4$ at vermindert, so daß er noch 5,6 at beträgt. Als Saugkraft der Zelle bleiben also 15,9 at.

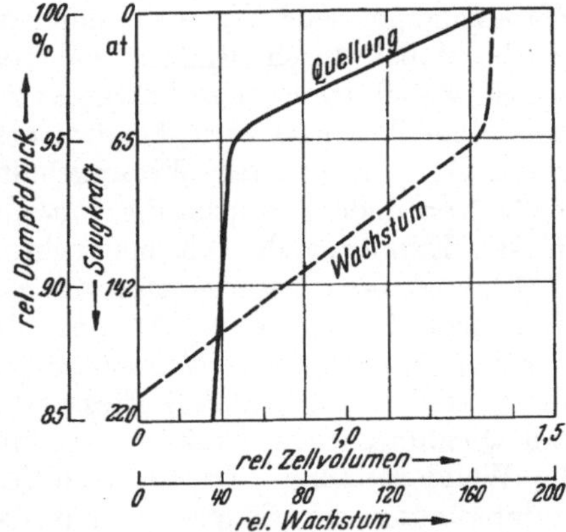

Abb. 4. Protoplasmaquellung (*Lemanea*-Karposporen) und Wachstum (*Aspergillus glaucus*) in Abhängigkeit vom Wasserpotential des Mediums. Die punktierte Kurve bezieht sich auf entspannte Zellwände. (Nach Angaben von WALTER 1923 und 1924.)

So verschieden die 3 Fälle hinsichtlich der Zellschrumpfung und des osmotischen bzw. Quellungsdruckes sind, so ist der Effekt hinsichtlich der *Zellaktivität* derselbe: das durch die Saugkraft gemessene chemische Potential des Wassers hat sich jedesmal von 0 auf 16 at verschlechtert.

Nach Eintritt der *Plasmolyse* ist bei weiterem Wasserverlust die Saugkraft gleich dem osmotischen Wert des Zellsaftes und dem Quellungsdruck des Plasmas. Dabei hat die Zelle bis zum Erreichen des Knickpunktes der Quellungskurve, der im allgemeinen bei etwa 65 at oder 95% relativem Dampfdruck liegt, den größten Teil ihres Wassers (Abb. 4) verloren. Der Rest des Wassers wird in dem nun beginnenden Gebiet der Lufttrockenheit sehr zäh festgehalten und nur gegenüber stark ansteigenden Saugkräften des Mediums abgegeben (Abb. 4).

In dieser Phase besteht nur noch eine sehr begrenzte *aktive* Zellvitalität; sie interessiert uns deshalb hier wenig, wird aber wichtig sein für die später zu diskutierende Frage nach der Erhaltung der *Plasmastruktur* im inaktivierten Lebensbezirk.

Für die Abhängigkeit der Zellaktivität vom Wassermangel gilt in jedem Fall eines *Gleichgewichtes, daß das chemische Potential des Wassers in der Zelle gleich dem des Mediums sein muß.* Handelt es sich um freiliegende Zellen oder Gewebe, wie im allgemeinen bei den Thallophyten, so ist als Medium die äußere Umgebung des Wassers oder der Luft bestimmend. In der Organisation der Cormophyten wird es der Pflanze möglich, in gewissem Umfang eigene Saugkraftpotentiale der Gewebe aufrechtzuerhalten, mit denen sich die einzelne Zelle ins Gleichgewicht setzen muß. Diese Stabilisierung des chemischen Potentials des Wassers und damit der Zellaktivität gegen den Wechsel der Außenbedingungen ist die wesentliche Grundlage der Überlegenheit des Cormophyten[1].

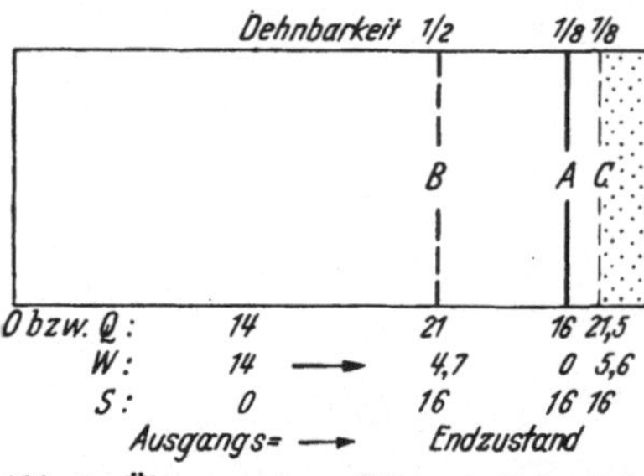

Abb. 5. Übergang einer Zelle vom Potential 0 at auf das von 16 at bei verschiedener Dehnbarkeit der Zellwand und verschiedenem Zellsaft-Plasmaverhältnis. Nähere Erklärung s. Text. (Nach Angaben von WALTER 1923).

Das in der pflanzlichen Zelle verwirklichte System der *Gegenwirkung zwischen osmotischem Druck und Wanddruck* verbessert also das chemische Potential des Wassers nicht, aber es mindert die *Volumenschwankungen*, die beim Wechsel der Außenbedingungen auftreten, und das ist, wie wir in der Folge sehen werden, wahrscheinlich wichtig für die Schonung der submikroskopischen *Feinstruktur des Plasmas.*

Neben diesem elastischen Gegendrucksystem geht die pflanzliche Zelle bei Wassermangel sehr allgemein noch den anderen Weg der *Osmoregulation.* Sie erhöht dabei durch Bildung osmotisch wirksamer Stoffe ihren osmotischen Druck, wobei das System Stärke $\rightleftharpoons$ Zucker eine besondere Rolle spielt. Für das chemische Potential wird dabei nichts gewonnen, aber es wird der Eintritt der Plasmolyse auf niederere Wassergehalte hinausgeschoben und damit die Gefahrenquelle vermindert, welche die *Plasmolyse* mit ihren Spannungen und Abrissen für die Erhaltung der Plasmastruktur bildet.

Eine *Verbesserung des chemischen Potentials des Wassers* in der *Zelle* gegenüber dem der *Umgebung* ist nur unter dauerndem Energieaufwand möglich. Der Weg dabei ist die *nichtosmotische (adenoide) Wasseraufnahme.* Sie setzt einmal durch Verdünnung des Zellsaftes und Aufquellung des Plasmas den osmotischen und Quellungsdruck herab und erhöht zweitens durch die Volumenvergrößerung die Wandspannung. Daraus resultiert gemäß der Gleichung $S = O - W$ eine Herabsetzung der Saugkraft und damit eine Verbesserung der Zellaktivität gegenüber dem rein osmotischen Saugkraftgleichgewicht mit dem Medium. *Adenoide Salzausscheidung* wird diesen Effekt erhöhen. Wir sind noch nicht näher orientiert, inwieweit diese Möglichkeit in der Pflanzenwelt ausgenützt wird; wahrscheinlich wird man unter anderem auch die contractilen Vacuolen unter diesem Gesichtspunkt betrachten müssen.

b) Die Zellaktivität.

Die *Wachstums- bzw. Zellteilungsgeschwindigkeit* ist wohl der beste Ausdruck für die in der „Zellaktivität" zusammengefaßten Einzelfaktoren. Ihre Abhängig-

[1] Vgl. dazu in Band III, S. 1ff.: O. STOCKER, Einführung und Übersicht, und S. 511ff.: B. HUBER, Allgemeine Grundlagen der Wasserleitung.

keit vom Wasserpotential des Mediums kann am leichtesten bei Thallophyten untersucht werden, wenn man sie in Luft von bekanntem relativen Dampfdruck bringt. Man benützt dazu Lösungen bestimmter Konzentration von Zucker, H_2SO_4, NaCl usw., deren Dampfdruck man kennt (Abb. 2). Mit ihnen werden kleine Schälchen gefüllt, die über der Flüssigkeit einen nur kleinen, rasch ins Dampfdruckgleichgewicht kommenden Luftraum lassen; in diesem können auf der Unterseite des Deckels auf Gelatine Pilzmycele, Bakterien usw. kultiviert werden (WALTER 1924). Man wird in der Beurteilung der so gewonnenen *absoluten* Werte freilich vorsichtig sein müssen, weil es einerseits sehr schwierig ist, in dem Luftraum konstante Feuchtigkeitsbedingungen zu halten, und weil andererseits bisher nicht die Möglichkeit untersucht worden ist, daß die Zellen durch aktive adenoide Aufnahme von Wasserdampf bzw. Wasser ihr chemisches Potential des Wassers über das der Umgebung verbessern könnten.

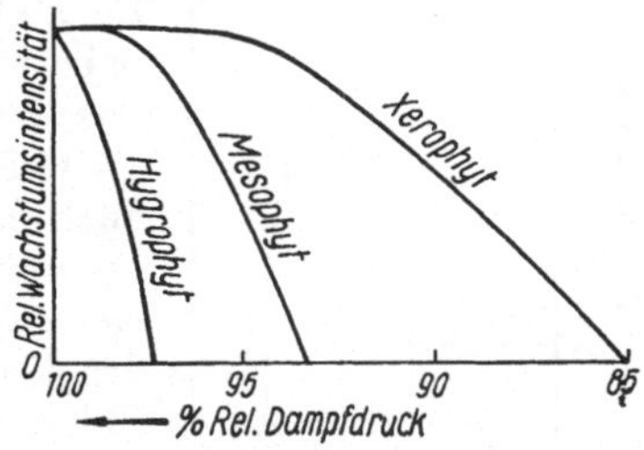

Abb. 6. Aktiver Lebensbereich von Schimmelpilzen. (Nach WALTER 1931 verändert.)

Um einen ersten Einblick in die Abhängigkeit der Zellaktivität von dem chemischen Potential des Zellwassers zu vermitteln, ist in Abb. 4 der *Quellungskurve* der *Lemanea*-Karpospore die *Wachstumskurve* eines Schimmelpilzes beigezeichnet worden. Diese Darstellung hat natürlich nur den Wert eines Schemas, weil es sich um zwei ganz verschiedene, nicht ohne weiteres vergleichbare Objekte handelt, aber sie zeigt doch das Grundsätzliche: Die Wachstumsaktivität bleibt durch Saugkraftschwankungen fast unbeeinflußt vom Sättigungspunkt bis zu etwa 95% relativem Dampfdruck bzw. 65 at Saugkraft, also in dem Bereich, in welchem große Wassermengen bei geringen Potentialdifferenzen verschoben werden können. Bei weiterer Verschlechterung des Potentials aber fällt dann die Wachstumsgeschwindigkeit schnell ab bis zum Nullpunkt.

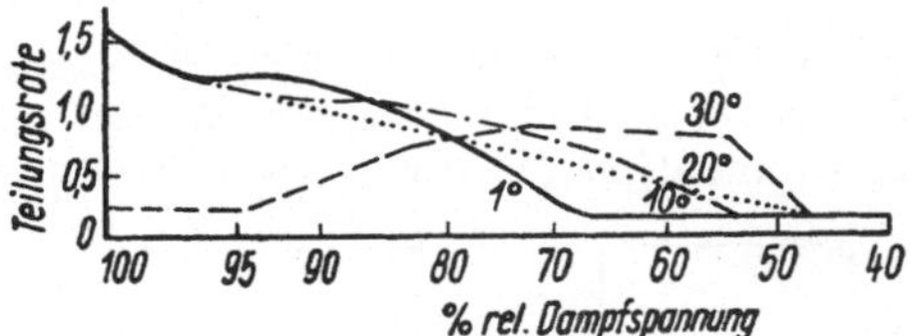

Abb. 7. Aktiver Lebensbereich von *Pleurococcus vulgaris* bei verschiedener Temperatur.

$$\text{Teilungsrate} = \frac{\text{Anzahl Zellteilungen}}{\text{Anfängliche Anzahl der Zellen.}}$$

Die Teilungsrate 0,16 ist als Nullinie zu betrachten (Teilungen vor Einstellung des Potentialgleichgewichtes). (Nach ZEUCH.)

Sowohl die Ausdehnung des *Optimumbereiches* wie die Lage des *Minimums* sind *artspezifisch* sehr verschieden. Die Abb. 6 veranschaulicht das für Pilze nach WALTER (1924, 1926, 1931), der zahlreiche Arten untersucht hat. Bei 100% relativem Dampfdruck beobachtete WALTER ein Optimum des Wachstums. Nach BAVENDAMM und REICHELT ist das aber zumindest nicht immer zutreffend; bei holzzersetzenden Pilzen und auch bei *Penicillium* kann das Optimum tiefer, etwa erst bei 95% liegen und nach dem Sättigungspunkt ein wenn auch nur schwacher Abfall des Wachstums und der holzzersetzenden Wirkung vorliegen. Dies ist von allgemeinem Interesse insofern, als hier für die biologische Wirkung des chemischen Potentials des Wassers auch eine obere Begrenzung angedeutet ist, wie sie für andere lebenswichtige Stoffe die Regel bildet. Nach unten hin kann sich das Optimum über einen mehr oder weniger großen Bereich erhalten, bevor der Rückgang der Wachstumsgeschwindigkeit bis zum Erliegen einsetzt. WALTER (1931) unterscheidet dabei 3 Verhaltensgruppen (Abb. 6). Bei den *Hygrophyten* führt eine Verschlechterung des chemischen Potentials des Wassers sofort nach Unterschreitung von 100% relativem Dampfdruck zur Verminderung des Wachstums, das schon bei 95% oder darüber völlig aufhört; so verhalten sich einige Pilze und die meisten Bakterien. Die *Xerophyten* dagegen werden bis etwa 95% nicht merklich gehemmt und haben ihr Maximum erst bei 90—85%; *Penicillium*- und *Aspergillus*-Arten gehören zu diesen Trockenresistenten. Zwischen beiden Gruppen vermitteln die *Mesophyten*, zu denen beispielsweise *Phycomyces* und die Hefen zählen. Die bisher beobachteten absoluten Minima für dauernde Entwicklung liegen für Schimmelpilze bei etwa 73% relativem Dampfdruck (entsprechend 419 at) (HEINTZELER), für Hefen bei etwa 88% (171 at) (BURCIK), für Bakterien bei etwa 90% (140 at) (BURCIK), für die Luftalge *Prasiola crispa* bei etwa 88% (171 at) (ITZEROTT).

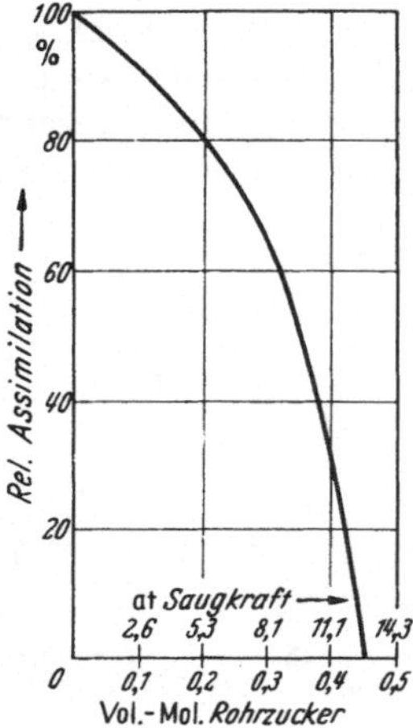

Abb. 8. Assimilation von *Helodea* in Abhängigkeit von Konzentration und Saugkraft von Rohrzuckerlösungen. (Nach WALTER 1929.)

Bei manchen Objekten ist aktives Wachstum aber auch noch bei viel *tieferen chemischen Potentialen des Wassers* beobachtet worden. Die Luftalge *Pleurococcus vulgaris* z. B. teilt sich noch bei 47% relativem Dampfdruck (ZEUCH). Die Vermutung liegt nahe, daß in solchen Fällen durch adenoide Wasseraufnahme das Wasserpotential der Zelle höher gelegt wird; darauf deutet auch die Temperaturabhängigkeit des Minimums hin, das sich von 68% relativem Dampfdruck bei 1° auf 47% bei 20—30° erniedrigt (Abb. 7); eine solche Temperaturbeziehung ist auch bei *Aspergillus Sydowi* gefunden worden (TAMMES).

Für die Potentialabhängigkeit eines physiologischen *Einzelprozesses* gibt Abb. 8 einen Beleg. Er bezieht sich auf die *Assimilation* von *Helodea*-Sprossen in reinem Wasser und in Rohrzuckerlösungen verschiedener Konzentration (WALTER 1928, 1929). Wir sehen einen sehr potentialempfindlichen Prozeß, der schon in 0,45 molarer Rohrzuckerlösung, also bei nur 13 at Saugkraft zum Stillstand kommt. Ihm gegenüber ist die *Atmung* allgemein viel weniger wasserpotentialempfindlich gefunden worden. Nach den Untersuchungen verschiedener Autoren (z.B. KOLKWITZ 1901, QVAM 1906, JAUERKA 1912, BAILEY und GURJAR 1918, LÖFFLER 1953) bleibt sie in Samen bis in unmittelbare Nähe der Lufttrockenheit bestehen (Abb. 9). Die Atmungskurve der Abb. 9 zeigt außerdem, daß das Optimum schon unterhalb der Wassersättigung erreicht wird, und daß nach dieser hin die Atmung wieder abfällt. Diese Erscheinung ist mehrfach gefunden worden, ohne daß bisher klargestellt ist, ob es sich um eine Abhängigkeit vom chemischen Potential oder nur um physikalische Diffusionshemmungen handelt.

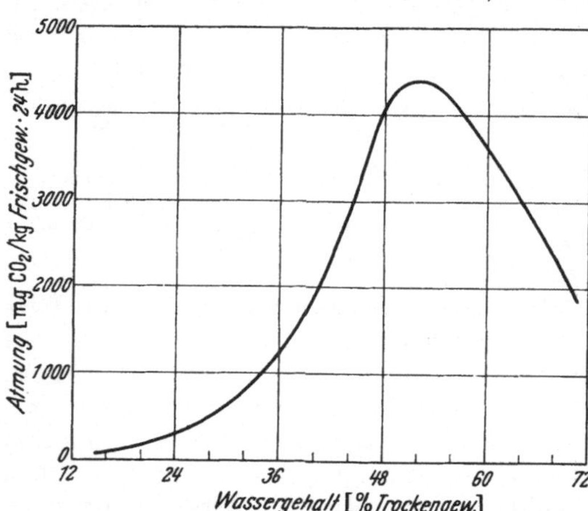

Abb. 9. Atmung von Weizensamen (Lohmanns Weender-Winterweizen II) in Abhängigkeit vom Wassergehalt bei 25°. Lufttrockene Körner haben 10% Wassergehalt. (Nach LÖFFLER.)

Die Beispiele der Assimilation und Atmung zeigen, daß *verschiedene Funktionen* vom Wassermangel in *verschiedener Weise* betroffen werden, so daß die Gesamtaktivität der Zelle bei einem bestimmten Wasserpotential aus der relativen Wirkung der verschieden stark und verschieden gerichtet beeinflußten Komponenten hervorgeht. Weiterhin hat der Fortschritt der Untersuchungen zu der wichtigen Erkenntnis geführt, daß das physiologische Geschehen in jedem Potentialpunkt nicht nur von der Größe des Potentials, sondern auch von dem *Vorleben der Zelle* abhängt. Die Reaktion kann verschieden sein, je nach der Zugehörigkeit des untersuchten Wasserzustandes zu einem Austrocknungs- oder Wiederbefeuchtungsvorgang, je nach der Schnelligkeit des Ablaufs der Vorgänge und je nach der Dauer der Zustände. Das weist darauf hin, daß neben der unmittelbaren Wirkung des chemischen Potentials noch ein mittelbar nachwirkender Einfluß des Wassermangels vorhanden ist, den wir in einem Angriff auf die Feinstruktur des Plasmas zu suchen haben.

III. Der Angriff des Wassermangels auf die Plasmastruktur.

A. Die letale Zerstörung der Plasmastruktur.

Viele Pflanzen ertragen ohne sichtbaren Schaden völliges *Austrocknen* für längere Zeit. Das gilt vor allem sehr allgemein für reproduktive Dauerzustände wie Samen und Sporen. Aber auch in der vegetativen Phase sind viele Bakterien, Pilze, Luftalgen, Flechten und Moose austrocknungsresistent. Auch bei Farnen kommt diese Fähigkeit noch öfters vor, nur noch selten aber bei Samenpflanzen[1].

Die Austrocknungsresistenz geht vielfach bis zum Ertragen von Exsiccatortrockenheit. So erträgt in der von HÖFLER (1942—1954) eingehend untersuchten Gruppe der *Lebermoose* die zarte kleine *Cololejeunea calcarea*, die an sonnigen Kalkfelsen zu Hause ist, mehr als einen Monat lang den Aufenthalt über konzentrierter Schwefelsäure. Solchen *hochresistenten* Formen stehen außerordentlich *empfindliche* gegenüber, wie die im Gischt von Wasserfällen lebende *Haplozia sphaerocarpa*, die noch nicht einmal einen 24stündigen Aufenthalt in Luft von 95% relativem Dampfdruck lebend zu überstehen vermag. Diese Letalpunkte sind aber keine Konstanten, sondern hängen vom *Vorleben*, von *Abhärtung* oder *Verweichlichung* ab. So fand HÖFLER bei dem Lebermoos *Metzgeria conjugata* in einem feuchten Monat die Absterbegrenze bei 60%, in einer sehr trockenen Föhnperiode aber erst bei 6% relativer Feuchtigkeit, und bei ein und demselben Individuum von *Lophozia porphyroleuca* konnte er durch langsames Überführen in immer trockenere Luft den Letalpunkt von 78% auf 37% relativer Feuchtigkeit verschieben.

Alle diese Beobachtungen lassen sich nur durch die Annahme erklären, daß der Trockentod nicht durch eine einfache Austrocknung des Plasmas zustande kommt, sondern auf eine irreversible Zerstörung der submikroskopischen *Plasmastruktur* zurückgeht, und daß die Widerstandsfähigkeit dieser Struktur variabel ist, je nachdem durch die vorausgegangenen Umweltbedingungen das Plasma abgehärtet oder verweichlicht ist.

Den klarsten Beweis für die entscheidende Rolle der *Plasmastruktur* beim Austrocknungstod hat uns die Experimentierkunst ILJINS (1927, 1930, 1935) gegeben. Er arbeitete mit *Zellen von Samenpflanzen*, die er in Gewebeschnitten im üblichen Kurzversuch gegen Antrocknung äußerst *empfindlich* fand; so starben die großen Zellen von Mark- und Wassergeweben schon in Luft von 99% relativer Feuchtigkeit, Mesophyll- und Epidermiszellen von Hygrophyten bei 97—96%, solche von Xerophyten teilweise erst bei 85%. Bei der weiteren Vertiefung in das Problem aber konnte er zeigen, daß man alle untersuchten

[1] Nähere Einzelheiten findet man in Band III in den Artikeln: F. GESSNER, Die Wasseraufnahme durch Blätter und Samen, S. 234ff., und O. STOCKER, Die Dürreresistenz, S. 696.

Zellen mindestens 2—3 Tage lang in *vollständig angetrocknetem Zustand* in Zimmerluft lebend erhalten kann und daß einzelne Zellarten, wie z. B. die Epidermiszellen von Rotkohlblättern, sogar einen wochenlangen Aufenthalt im Exsiccator ertragen, wenn man nur ihre Plasmastruktur sehr schonend behandelt. Dazu ist notwendig, die Austrocknung langsam über vielfach abgestufte Lösungen von H_2SO_4 oder $CaCl_2$ oder unter stufenweiser Plasmolysierung mit Rohrzucker vorzunehmen und nicht weniger vorsichtig die Wiederaufsättigung durchzuführen. Unter diesen Bedingungen besteht die Aussicht, irreversible Zerreißungen des Plasmas zu vermeiden. Der trockene Zustand ist aber labil, da Spannungen der Plasmastruktur vorhanden sind, die mit zunehmender Dauer der Entquellung oder bei wiederholter Austrocknung und Einquellung schließlich doch zu letalen Zerreißungen führen.

Neben der artspezifischen Resistenz der Plasmastruktur spielen nach ILJIN auch Form- und Baueigentümlichkeiten der Zelle eine Rolle bei der Austrocknungsresistenz. Günstig ist alles, was die *Schrumpfung und Deformation des Plasmas* beim Austrocknen einschränkt, also Kleinheit der Zellen und Kleinheit oder Fehlen der Vacuolen. Das ist besonders in meristematischen Zellen verwirklicht, die sich allgemein durch große Resistenz auszeichnen. Es gibt auch Fälle, in denen gelartiges Erstarren des Zellsaftes mit hoher Austrocknungsresistenz zusammenfällt (ILJIN 1931, vgl. auch LEPESCHKIN).

B. Vitale Änderungen der Plasmastruktur.

a) Plasmatische Zustandsgrößen.

Der letale Zusammenbruch der Plasmastruktur ist nur die letzte Phase von *reversiblen* und *irreversiblen Veränderungen des plasmatischen Strukturgefüges*, die mit dem Entzug von Wasser *von Anfang an* verbunden sind. Diese Veränderungen haben den Charakter von Reizerscheinungen: Eine durch Wasserentzug hervorgerufene destruktive *Reaktionsphase* ist gefolgt von einer konstruktiven *Restitutionsphase*, die bei Fortdauer des Wassermangels über den Ausgangszustand hinaus zu einem *Abhärtungszustand* führt (STOCKER 1947, 1948, 1953, 1954).

Am besten sind diese Vorgänge bei der *Viscosität* des Plasmas bekannt (STOCKER und ROSS, vgl. auch STÅLFELT 1954 und die dort besprochene spärliche frühere Literatur). Wenn man Pflanzen von *Lamium maculatum*, die bisher bei maximaler Feuchtigkeit der Topferde kultiviert wurden, dadurch einem Trockenreiz aussetzt, daß man die Erde innerhalb von 26 Std auf 22% der Wasserkapazität austrocknen läßt und sie dann auf diesem Wassergehalt konstant hält, so beobachtet man die in Abb. 10 dargestellten Änderungen der Viscosität und des osmotischen Druckes; die erstere wurde in Halbwerten der Plasmolysezeitmethode (SCHMIDT, DIWALD und STOCKER), der letztere plasmometrisch in Rohrzucker gemessen.

In der *Reaktionsphase* erfolgt eine Herabsetzung der *Viscosität* auf die Hälfte ihres ursprünglichen Wertes; es muß also eine Auflockerung der Struktur des Plasmas stattgefunden haben. Dann aber tritt, nachdem die Austrocknung des Bodens auf 22% der Wasserkapazität stabilisiert ist, die *Restitutionsphase* in Erscheinung. Die Viscosität steigt wieder an, d. h. das Plasma strebt in eine festere Bindung zurück. Diese Tendenz überpendelt am 3. Tag den Ausgangswert und erreicht am 9. Tag mit dem $3^1/_2$fachen desselben einen Gleichgewichtszustand, den wir als *Abhärtungszustand* bezeichnen können. Wird der Boden optimal durchfeuchtet, so geht die Abhärtung innerhalb 48 Std wieder verloren.

Reaktion und Restitution sind in Wirklichkeit keine zeitlich aufeinanderfolgenden Vorgänge, sondern wirken als *Gegenspieler* während der ganzen Dauer des Dürrereizes, jedoch mit wechselnden Intensitäten: Zunächst setzt die Reaktion voll ein, die Restitution aber wird erst allmählich von der Reaktion ausgelöst. Das wird deutlich, wenn man *langsam*, über 6 und mehr Tage hinweg, austrocknet. Dann hat die Restitution Zeit, von Anfang an der Reaktion entgegenzuwirken und der Abfall der Viscosität in der Reaktionsphase tritt nicht in Erscheinung; die Abhärtung dagegen wird auch hier sichtbar.

Aus der Abb. 10 geht hervor, daß der *osmotische Druck* bei den Viscositätsänderungen keine unmittelbare Rolle spielt. Sein Anstieg in der Reaktionsphase ist durch Entwässerung des Plasmas und Osmoregulation bedingt; während der Restitutions- und Abhärtungsphase bleibt er konstant.

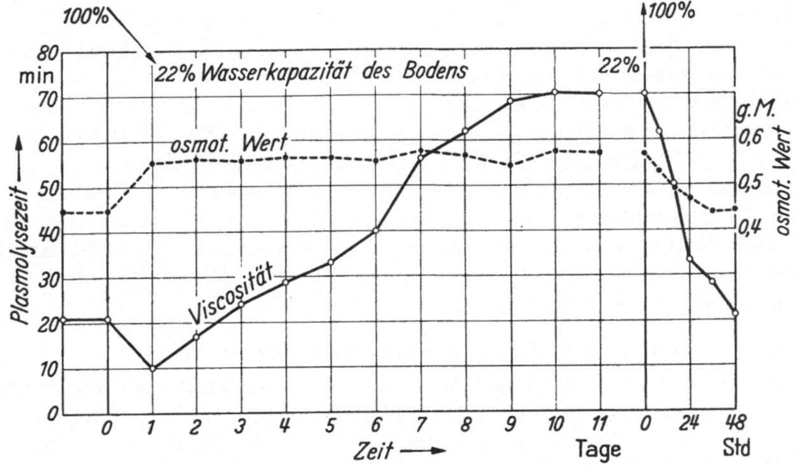

Abb. 10. Reaktions- und Restitutionsphase der Plasmaviscosität von *Lamium maculatum* bei Dürreeinwirkung. Am Tag 0 wird der Boden innerhalb 26 Std von 100% der Wasserkapazität auf 22% ausgetrocknet und auf diesem Stand bis zum 11. Tag gehalten. Im Punkt 0 h erfolgt Wiederbewässerung auf 100%, 36°, 40—45% relativer Luftfeuchtigkeit. (Nach STOCKER und ROSS.)

Die *plasmatischen Strukturänderungen* kann man am sinnfälligsten nach den Vorstellungen von FREY-WYSSLING als Lösung von Haftpunkten und Änderung der Molekülfaltung auffassen. Sie könnten aber auch auf der Basis anderer kolloidchemischer Auffassungen erklärt werden, z. B. der Vitaide nach LEPESCHKIN. Jedenfalls hat der Verlauf der Dürreeffekte eine bemerkenswerte Parallele in den Vorgängen, die bei Pflanzen durch *Schüttelreize* ausgelöst werden (STOCKER 1948, KAHL 1951, vgl. auch BÜNNING 1953, S. 353). Im nichtlebenden Bereich treten Viscositätsänderungen beim Schütteln gewisser Kolloide als *thixotroper Effekt* auf, wobei Vernetzungen von Molekülen gelockert und dann in der Ruhe wiederhergestellt werden.

Lockerungen und Festigungen der Plasmastruktur hängen sicher mit *Entquellungs-* und *Quellungsvorgängen* an hydrophilen Gruppen des Eiweißgerüstes zusammen. Nach den von BOGEN entwickelten Anschauungen muß man dabei an Angriffe auf spezifische Stellen des „*Hydratationsmusters*" des Plasmanetzes denken. Solche partiellen Entquellungen führen dann zu inneren mechanischen Spannungen, die mit der Zeit oder bei Verschärfung des Wassermangels zu Lockerungen und schließlich Brüchen führen. Umgekehrt wird die Quellung bestimmter Stellen bei der Restitution die Stabilität der Struktur fördern. Die zuerst von RUHLAND für die Kälteresistenz betonte Wichtigkeit einer verstärkten *Hydratation* gliedert sich damit in die Strukturtheorie ein, und die mehrfachen Beobachtungen einer Vermehrung des „*gebundenen Wassers*" bei

Wassermangel (vgl. MIGAHID 1938, LEVITT 1951, HENCKEL 1954) werden verständlich. Da auch bei anderen Plasmaschädigungen, wie vor allem Hitze und Kälte, Plasmastruktur und Hydratation von entscheidender Bedeutung sind, können gewisse Parallelen in den Resistenzerscheinungen nicht überraschen. Eine gegenüber allen Angriffen gleiche *allgemeine* „*physiologische Resistenz*" der Zelle ist aber bei dem spezifischen Eingriff der einzelnen Schädigungen an differenten Plasmaorten entgegen der Ansicht von LEVITT ebenso unwahrscheinlich wie unbewiesen.

Mit der Viscosität ist natürlich nur eine Teilerscheinung des Komplexes herausgegriffen, der den *plasmatischen Zustand* ausmacht. Unsere Kenntnisse über weitere Teileigenschaften desselben sind noch gering, und die Beurteilung der ermittelten Tatsachen wird dadurch erschwert, daß man bisher nicht zwischen Reaktions- und Restitutionsphase unterschieden hat und sich die Ergebnisse deshalb oft scheinbar widersprechen.

Ziemlich gesichert sind die Änderungen der *Porenpermeabilität* auf Grund von Untersuchungen verschiedener Autoren (z. B. SCHMIDT 1939, SCHMIDT, DIWALD und STOCKER 1940, LEVITT 1951). Für *Harnstoff* ist sie in der Reaktionsphase erhöht, in der Restitutionsphase erniedrigt, was unseren Vorstellungen der Plasmastruktur entspricht. Die *Wasserpermeabilität* verhält sich wahrscheinlich ebenso. Bei Stoffen vom „*Glycerintypus*" (Äthylenglykol, Glycerin, Erythrit) liegen die Verhältnisse anders, wobei vielleicht kolloidchemische Änderungen der Plasmalipoide eine Rolle spielen (SCHMIDT 1939).

b) Die Beeinflussung physiologischer Prozesse.

Änderungen der Plasmastruktur bedeuten nicht nur andere Zustandsgrößen, sondern beeinflussen auch den *Ablauf physiologischer Vorgänge* in der Zelle. Das kann als *unmittelbare Folge* des plasmatischen Zustandes geschehen, z. B. bei der *Transpiration*, insoweit sie von der Wasserpermeabilität abhängig ist; experimentell Gesichertes können wir darüber freilich noch nicht sagen.

Von viel weitreichenderer Bedeutung ist der *indirekte* Eingriff in das Zellgeschehen über die *Aktivität der Fermente*, die oft von ihrem Bindungszustand abhängt (KURSANOW 1940, 1946, SISSAKIAN 1940, 1954, vgl. auch BÜNNING 1953, S. 47)[1]. Die Fermente können nämlich frei oder an den Lipoideiweißkomplex mehr oder weniger stabil oder instabil gebunden vorkommen. Dabei bedeutet *Freisetzung* der Fermente eine Steigerung ihrer *hydrolytischen Aktivität*, während ihre *Bindung* das Reaktionsgleichgewicht nach der Seite der *Synthese* verschiebt. Mit der Lockerung der Plasmastruktur wird also in der *Reaktionsphase* eine Tendenz des Stoffwechsels nach *hydrolytisch-oxydativen Abbauprozessen*, in der *Restitutionsphase* aber umgekehrt in *synthetisch-reduktiver Aufbaurichtung* zu erwarten sein. Das hat sich denn auch in den bisherigen Erfahrungen überall bestätigt.

Als Beispiel wollen wir auf die Beeinflussung von *Atmung* und *Assimilation* durch Wassermangel kurz eingehen, wobei hinsichtlich weiterer Einzelheiten und Literaturangaben wie immer auf meinen Artikel „Dürreresistenz" in Band III dieses Handbuches verwiesen werden muß. Die *Atmung* zeigt bei fallendem Wassergehalt des Bodens, also in der Reaktionsphase, eine starke Steigerung, in der Restitutionsphase einen Wiederabfall, der im Abhärtungszustand, also bei Kultur in dauernd trockenem Boden, zu Werten unterhalb denen der Feuchtpflanzen führt (KRAFT 1944, vgl. STOCKER 1948). Umgekehrt erfährt die *Assimilation* bei Wassermangel einen sehr starken und schnellen Abfall bis zum völligen

[1] Vgl. weiter in Band III: K. MOTHES, Der Einfluß des Wasserzustandes auf Fermentprozesse und Stoffumsatz, S. 661.

Zusammenbruch (vgl. dazu Abb. 8), während sie, wie namentlich SIMONIS (1947, 1952) gezeigt hat, in der Abhärtung bei Trockenkultur potentiell, d. h. bei Darbietung eines optimalen Wasserpotentials während des Versuches, über die Fähigkeit von Feuchtpflanzen gesteigert ist.

Auch für das physiologische Verhalten spielt das *Vorleben* eine wesentliche Rolle. Der bei einem bestimmten Wasserdefizit vorhandene *Strukturzustand* des Plasmas ist, wie schon betont, *instabil*, mit noch unausgelösten inneren Spannungen. Je *schneller* die *Austrocknung* vor sich geht, um so stärker wirken die zerreißenden Kräfte, und die tiefer greifende Störung der Struktur wird in der weitergehenden Beeinflussung der physiologischen Reaktionen, Assimilation und Atmung bemerkbar (ENSGRABER 1954). Aber auch das *anhaltende* Defizit wirkt

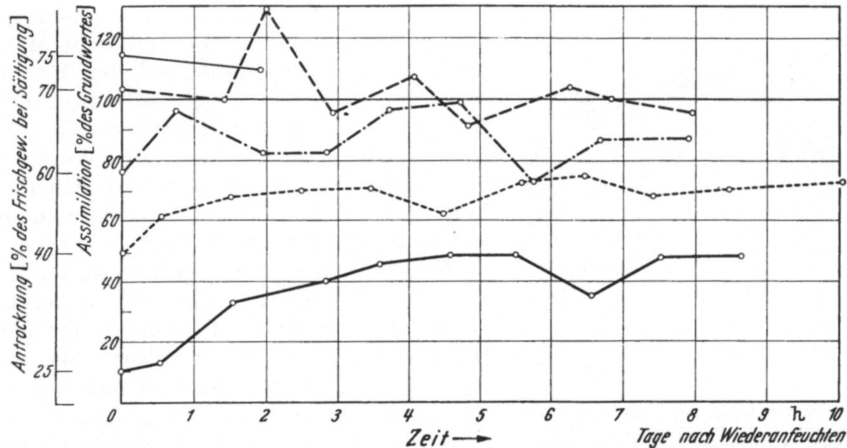

Abb. 11. Reaktivierung der Assimilation bei dem Lebermoos *Conocephalum conicum* nach eintägiger verschieden starker Antrocknung. Assimilationswerte der wieder wassergesättigten Thalli in Prozenten der Versuchswerte vor der Antrocknung. Der Grad der Antrocknung der einzelnen Versuche ist auf der Achse links der Figur verzeichnet. (Nach ENSGRABER.)

sich störend aus, weil mit ihm die Wahrscheinlichkeit von Brüchen an Spannungsstellen steigt. Ein weiterer schädlicher Faktor ist eine vorangegangene *Verweichlichung* durch dauernden optimalen Wassergehalt, während *Abhärtung* durch periodisches Welken eine besonders stabile Struktur erzeugt (TUMANOW 1927, 1930). Eine solche Abhärtung ist in Trockengebieten auch von praktischer Wichtigkeit. Nach HENCKEL kann man sie schon im *Samen* induzieren, wenn man durch vorübergehende Anfeuchtung die ersten Keimungsstadien anregt, durch Austrocknen aber sogleich wieder unterbricht.

Wenn die Austrocknung einen gewissen Grad erreicht, erfolgt eine *Inaktivierung* der Lebensvorgänge, besonders früh für die Assimilation. Die mögliche Spanne eines so latenten Lebens ist, wie schon gesagt, artspezifisch sehr verschieden. Sie kann bis zur völligen Austrocknungsfähigkeit über längere Zeit gehen. Der inaktivierte Zustand bedeutet aber immer eine *Schädigung der Plasmastruktur*, und es wird bei der *Reaktivierung*, d. h. der Wiederherstellung günstiger Wasserpotentiale, eine mehr oder weniger lange Zeit benötigt, um die normale Struktur wiederherzustellen. Dieser Vorgang erfordert Energieaufwand, und man beobachtet eine Steigerung der Atmung. Ist der Reaktionsschaden *reversibel*, so erreicht die physiologische Funktion wieder ihre frühere Leistung, um so schneller, je geringer der Schaden war. Ging dieser tiefer bis zu teilweiser oder ganzer *Irreversibilität*, so bleibt eine dauernde Schwächung, oder die Zelle geht über kurz oder lang zugrunde (Abb. 11). Dabei kann die irreversible Beschädigung nur einzelne Funktionen betreffen; besonders empfindlich erweist sich dabei wieder die Assimilation, was auf eine besonders empfindliche Struktur der Chloroplasten hinweist (ENSGRABER). Oft gehen in solchen Fällen zwar die ausgebildeten vegetativen Organe zugrunde, aber es bleiben dürreresistentere *meristematische Gewebe* erhalten und setzen dann durch Austrieb der Knospen das Leben des Individuums fort. Auch in *austrocknungsresistenten Zellen* befindet sich das Plasma in einem instabilen Spannungszustand, welcher mit fortschreitender Zeit immer mehr Brüche erfolgen läßt; die Reaktivierungszeiten werden damit länger, und die Irreversibilität der Schäden nimmt zu (Abb. 12).

Von weiteren fermentativen Prozessen sind die *Kohlenhydratumsetzungen* am eingehendsten untersucht (vgl. vor allem ILJIN 1929, 1930, VASSILJEW 1931, 1936, SISSAKIAN 1940, HENRICI 1946). In der Reaktionsphase verschiebt sich z. B. bei der Luzerne (HENRICI 1946) zunächst kurze Zeit das Gleichgewicht Monosaccharide ⇌ Disaccharide nach der Seite der letzteren. Dann aber erfolgt der hydrolytische Abbau von Stärke und Rohrzucker zu reduzierenden Monosacchariden. Beim Weizen ist der Verlauf ähnlich (VASSILJEW). Über die Vorgänge in der Restitutionsphase haben wir noch kein gesichertes Bild.

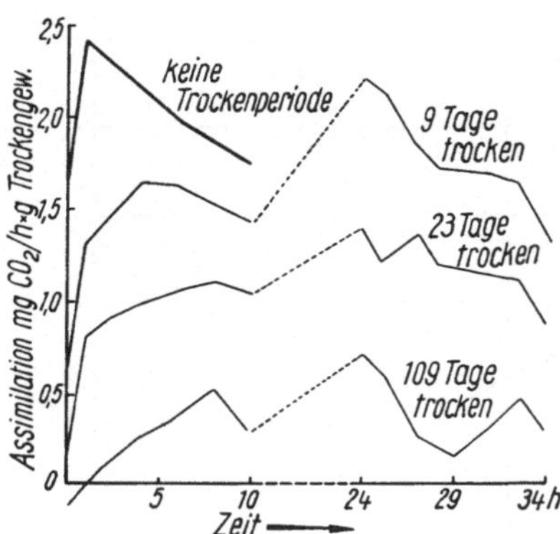

Abb. 12. Reaktivierung der Assimilation einer Flechte (*Ramalina farinacea*) nach vorangegangenen Trockenperioden verschiedener Dauer. (Nach STÅLFELT 1939.)

Der Einfluß des Wassermangels auf den *Eiweißgehalt* (MOTHES 1928, 1931, SISSAKIAN 1940) folgt derselben allgemeinen Tendenz: In der Reaktionsphase resultiert aus dem geschwächten synthetischen Vermögen der Proteasen ein Überwiegen des proteolytischen Abbaues, während für die Restitutionsphase Angaben über eine Eiweißerhöhung vorliegen (SCHARRER, DUISBERG).

Wir haben hier das Problem der Wirkung von Wassermangel auf die Zellaktivität nur insoweit angeschnitten, als es sich auf das aktuelle physiologische Zellgeschehen bezieht. Das ist eine hier notwendige Beschränkung, die aber nicht darüber täuschen darf, daß das Problem noch weitere wesentliche Seiten hat, so die verschiedene Dürreresistenz der einzelnen *Entwicklungsstadien* (SKAZKINE), die *genotypische Resistenzveranlagung*, die *formative Wirkung* von Wassermangel auf morphologische und anatomische Strukturen und das Zusammenwirken der physiologischen Einzelprozesse in einer dürreresistenten „*Konstitution*". Darüber vergleiche man in Band III den Artikel Dürreresistenz.

Literatur.

BAILEY, C. H., and A. M. GURJAR: Respiration of stored wheat. J. Agricult. Res. **12**, 685—713 (1918). — BAVENDAMM, W., u. H. REICHELT: Die Abhängigkeit des Wachstums holzzersetzender Pilze vom Wassergehalt des Nährsubstrates. Arch. Mikrobiol. **9**, 486—544 (1938). — BOGEN, H. J.: Untersuchungen über Hitzetod und Hitzeresistenz pflanzlicher Protoplaste. Planta (Berl.) **36**, 298—340 (1948). — BÜNNING, E.: Entwicklungs- und Bewegungsphysiologie der Pflanze, 3. Aufl. Berlin-Göttingen-Heidelberg: Springer 1953. — BURCIK, E.: Über die Beziehungen zwischen Hydratur und Wachstum bei Bakterien und Hefen. Arch. Mikrobiol. **15**, 203—235 (1950).

DUISBERG, P. C.: Some relationships between xerophytism and the content of resin, nordihydroguaiaretic acid and protein of Larrea divaricata Cav. Plant Physiol. **27**, 769—777 (1953).

ENSGRABER, A.: Über den Einfluß der Antrocknung auf die Assimilation und Atmung von Moosen und Flechten. Flora (Jena) **141**, 432—475 (1954).

FREY-WYSSLING, A.: Submicroscopic morphology of protoplasm. Sec. ed. Amsterdam-Houston-London-New York, N. Y.: Elsevier Publishing Co. 1953.

HEINTZELER, J.: Das Wachstum der Schimmelpilze in Abhängigkeit von den Hydraturverhältnissen unter verschiedenen Außenbedingungen. Arch. Mikrobiol. **10**, 92—132 (1939). — HENCKEL, P. A.: Sur la résistance des plantes à la sécheresse et les moyens de la diagnostiquer et de l'augmenter. Acad. Sci. 'URSS, Essais Bot. **2**, 436—456, Moscou 1954. — HENRICI, M.: Effect of excessive water loss and wilting on the life of plants. Union South Africa, Dep. Agric. Forestry, Bull. **1946**, 256. — Comparative study of the content of starch and sugars of Tribulus terrestris, Lucerne, some Gramineae and Pentzia incana under different meteorological conditions. Paper 2. Carbohydrate nutrition. Onderstepoort J. Veterin. Res. **25**, 45—92 (1952). — HÖFLER, K.: Über die Austrocknungsfähigkeit des Protoplasmas. Ber. dtsch. bot. Ges. **60** (94)—(107) (1942). — Über Trockenhärtung des Protoplasmas. Ber. dtsch. bot. Ges. **63**, 3—10 (1950). — Durch plasmatische Trockengrenzen bedingte Lebermoosvereine. Proceedings VII. Internat. Bot. Congr. Stockholm 1950, S. 621—623. 1953. — La résistance du protoplasme à la sécheresse. VIII. Congr. Internat. Bot. Paris, Rapp. et Communic. Sect. 11 et 12, S. 239—241. Paris 1954.

ILJIN, W. S.: Über die Austrocknungsfähigkeit des lebenden Protoplasmas der vegetativen Pflanzenzellen. Jb. wiss. Bot. **66**, 947—964 (1927). — Standortsfeuchtigkeit und der Zuckergehalt in den Pflanzen. Planta (Berl.) **7**, 59—71 (1929). — Die Ursache der Resistenz von Pflanzenzellen gegen Austrocknen. Protoplasma **10**, 379—414 (1930). — Der Einfluß des Welkens auf den Ab- und Aufbau der Stärke in der Pflanze. Planta (Berl.) **10**, 170—184 (1930). — Austrocknungsresistenz des Farnes Notochlaena Marantae R. B. Protoplasma **13**, 322—330 (1931). — Die Veränderung des Turgors der Pflanzenzellen als Ursache ihres Todes. Protoplasma **22**, 299—311 (1935). — Die Lebensfähigkeit der Pflanzenzellen in trockenem Zustand. Planta (Berl.) **24**, 742—754 (1935). — ITZEROTT, H.: Untersuchungen zum Wasserhaushalt von Prasiola crispa. Jb. wiss. Bot. **84**, 254—275 (1936).

JAUERKA, O.: Die ersten Stadien der Kohlensäureausscheidung bei quellenden Samen. Beitr. Biol. Pflanz. **11**, 193—248 (1912).

KAHL, H.: Über den Einfluß von Schüttelbewegungen auf Struktur und Funktion des pflanzlichen Plasmas. Planta (Berl.) **39**, 346—376 (1951). — KOLKWITZ, R.: Über die Atmung ruhender Samen. Ber. dtsch. bot. Ges. **19**, 285—287 (1901). — KRAFT, M.: Die Atmung dürreempfindlicher und dürreresistenter Getreidesorten bei verschiedener Wasserversorgung. Diss. Darmstadt 1944 (auszugsweise in STOCKER 1948). — KURSANOW, A. L.: Die reversible Fermentwirkung in der lebenden Pflanzenzelle. Moskau u. Leningrad 1940. — Die Fermentadsorption durch die Gewebe höherer Pflanzen. Biochimija **11**, 333 (1946).

LEPESCHKIN, W. W.: Zur Kenntnis des Absterbens der Hefe beim Austrocknen. Protoplasma **37**, 404—414 (1942). — LEVITT, J.: Frost, drought and heat resistance. Annual Rev. Plant Phys. **2**, 245—268 (1951). — LÖFFLER, M.: Die Atmung von lagernden Weizenkörnern in Abhängigkeit von ihrem Feuchtigkeitsgrad. Diplomarbeit Darmstadt 1953.

MIGAHID, A. M.: Binding of water in relation to drought resistance. Bull. Fac. Sci. Fouad I Univ. Cairo **1938**, Nr 18. — MOTHES, K.: Die Wirkung des Wassermangels auf den Eiweißumsatz in höheren Pflanzen. Ber. dtsch. bot. Ges. **46**, (59)—(67) (1928). — Zur Kenntnis des N-Stoffwechsels höherer Pflanzen. 3. Beitrag unter besonderer Berücksichtigung des Blattalters und des Wasserhaushaltes. Planta (Berl.) **12**, 686—731 (1931).

QVAM, O.: Zur Atmung des Getreides. Jb. Ver. angew. Bot. **4**, 70 (1906).

RENNER, O.: Theoretisches und Experimentelles zur Kohäsionstheorie der Wasserbewegung. Jb. wiss. Bot. **56**, 617—667 (1915). — RUHLAND, W.: Zur Kälteresistenz der Pflanzen. Ber. sächs. Akad. Wiss. 87, 37—40 (1935).

SCHARRER, K.: Der Einfluß verschiedener Wasser- und Stickstoffversorgung auf den Eiweißgehalt der Gerste. Forschungsdienst **7**, 127—140 (1939). — SCHMIDT, H.: Plasmazustand und Wasserhaushalt bei Lamium maculatum. Protoplasma **33**, 25—43 (1939). — SCHMIDT, H., K. DIWALD u. O. STOCKER: Plasmatische Untersuchungen an dürreempfindlichen und dürreresistenten Sorten landwirtschaftlicher Kulturpflanzen. Planta (Berl.) **31**, 559—596 (1940). — SIMONIS, W.: CO_2-Assimilation und Stoffproduktion trocken gezogener Pflanzen. Planta (Berl.) **35**, 188—224 (1947). — Untersuchungen zum Dürreeffekt. I. Morphologische Struktur, Wasserhaushalt, Atmung und Photosynthese feucht und trocken gezogener Pflanzen. Planta (Berl.) **40**, 313—332 (1952). — SISSAKIAN, N. M.: The biochimical character of drought-resistant plants. Moskau u. Leningrad 1940. — Die fermentative Aktivität der protoplasmatischen Struktur. Berlin: Akademie-Verlag 1954a. — Biochimie des plastides. Acad. Sci. URSS. Essais Bot. **1**, 224—258 (1954b). — SISSAKIAN, N. M., i A. M. KOBIAKOWA: Die Tendenz der fermentativen Wirkung als Kennzeichen der Dürrefestigkeit von Kulturpflanzen. VI. Mitt. Die Adsorption der Invertase durch pflanzliche Gewebe beim Verwelken. Biochimija **12**, 377 (1947). — SKAZKINE, F. D.: La résistance des céréales au manque d'eau dans le sol durant différentes périodes de leur développement. Acad. Sci. URSS., Essai Bot. **2**, 751—762 (1954). — STÅLFELT, M. G.: Vom System der

Wasserversorgung abhängige Stoffwechselcharaktere. Bot. Not. (Lund) **1939**, 176—192. — The relation between the fluidity of the protoplasm and the insertion and function of the leaves. Physiol. Plantarum (Copenh.) **7**, 354—374 (1954). — Stern, K.: Pflanzen-Thermodynamik. Monographien Physiol. 30. Berlin: Springer 1933. — Stocker, O.: Probleme der pflanzlichen Dürreresistenz. Naturwiss. **34**, 362—371 (1947). — Beiträge zu einer Theorie der Dürreresistenz. Planta (Berl.) **35**, 445—466 (1948). — Begriff und Wesen der Dürreresistenz. Proc. VII. Internat. Bot. Congr. 1950, S. 232—233, Stockholm 1953. — Die Trockenresistenz der Pflanzen. VIII. Congr. Internat. de Botanique Paris 1954. Rapports et Communic. Sect. 11/12, 223 bis 232. Paris 1954. — Stocker, O., u. H. Ross: Reaktions- und Restitutionsphase der Viskosität bei Dürrereizen. Naturwiss. **43**, 283 (1956).

Tammes, P. M. L.: On the influence of temperature upon the growth of Aspergillus Sydowi Sartory in concentrated brines Rec. Trav. bot. néerl. **34**, 610—614 (1937). — Tumanow, J. J.: Ungenügende Wasserversorgung und das Welken der Pflanzen als Mittel zur Erhöhung ihrer Dürreresistenz. Planta (Berl.) **3**, 391—480 (1927). — Welken und Dürreresistenz. Wiss. Arch. Landwirtsch., Abt. A **3**, 389—419 (1930).

Ursprung, A.: Osmotische Zustandsgrößen. In Handwörterbuch der Naturwissenschaften 2. Aufl., Bd. 7, S. 493—510. Jena: Gustav Fischer 1932.

Vassiljew, J. M.: Untersuchungen über die Dynamik der Kohlehydrate bei den Weizen. Über den Einfluß der Wasserbewegung auf die Umwandlung der Kohlehydrate. Arch. Pflanzenbau **7**, 126—146 (1931). — Vassiljew, J. M., and M. G. Vassiljew: Changes in carbohydrate content in wheat plants during the process of hardening for drought resistance. Plant Physiol. **2**, 115—125 (1936).

Walter, H.: Protoplasma- und Membranquellung bei Plasmolyse. Jb. wiss. Bot. **62**, 145—213 (1923). — Plasmaquellung und Wachstum. Z. Bot. **16**, 353—417 (1924). — Der Wasserhaushalt der Pflanze in quantitativer Betrachtung. Naturwissenschaft und Landwirtschaft, Heft 6. Freising u. München 1925. — Die Wassersättigung der Pflanze und ihre Bedeutung für das Pflanzenwachstum. Z. Pflanzenernährg. A **6**, 65—88 (1926). — Die Bedeutung des Wassersättigungszustandes für die CO_2-Assimilation der Pflanzen. Ber. dtsch. bot. Ges. **46**, 530—539 (1928). — Plasmaquellung und Assimilation. Protoplasma **6**, 113—156 (1929). — Die Hydratur der Pflanze. Jena: Gustav Fischer 1931.

Zeuch, L.: Untersuchungen zum Wasserhaushalt von Pleurococcus vulgaris. Planta (Berl.) **22**, 614—643 (1934).

Die Wirkung von Licht und Strahlung auf die Zelle.

Von

W. Simonis.

Mit 27 Abbildungen.

I. Allgemeine Hinweise.

In dem vorliegenden Abschnitt sollen die Grundlagen der Wirkungen von Licht und Strahlung auf biologische Objekte, besonders auf die einzelne Zelle, dargestellt werden. Hierüber gibt es eine höchst ausgedehnte Literatur, die insgesamt zu behandeln völlig unmöglich ist. Abgesehen von den später erwähnten Originalarbeiten und Zusammenfassungen einzelner Gebiete sei auf die folgenden, grundlegenden Werke hingewiesen: DESSAUER [2], HOLLAENDER [2, 3], LEA, NICKSON, RABINOWITCH [1, 2], RAJEWSKY, TIMOFÉEFF-RESSOVSKY und ZIMMER, SOMMERMEYER. Allgemeine Zusammenfassungen finden sich besonders in BUTLER und RANDALL (Progress in Biophysics and Biophysical Chemistry), sowie in den Fortschritten der Botanik (SIMONIS).

a) Die verschiedenen Strahlensorten.

Für eine Wechselwirkung mit biologischen Objekten kommt sowohl die *elektromagnetische* als auch die *Corpuscularstrahlung* in Frage. Zur erstgenannten gehören die zur Nachrichtenübertragung dienenden Strahlen (Kilometer-, Kurz-, Ultrakurz-, Dezimeterwellen), ferner die Wärmestrahlen, das ultrarote, das sichtbare und das ultraviolette Licht, im folgenden kurz UV genannt, die Röntgen- und die γ-Strahlen. Zur Corpuscularstrahlung rechnet man die aus + und — geladenen Partikeln gebildeten Elektronenstrahlen (β-Strahlen), ferner die α-Strahlen (positiv geladene He-Kerne), die Protonen, Deuteronen, sowie die ungeladenen Neutronen. Die Energie dieser Strahlung ist äußerst verschieden. Unter dem Gesichtspunkt, daß die Energie (E) der elektromagnetischen Strahlung in Form einzelner Energiequanten Q ($E = h\nu$) und die der Corpuscularstrahlung auf einzelne bewegte Korpuskel lokalisiert weitergegeben wird, kann der Energiegehalt der Quanten bzw. der Korpuskel in Elektronenvolt (eV) gemessen werden. 1 eV ist dabei der kinetischen Energie gleichzusetzen, die eine Elementarladung bei dem Durchlaufen einer Potentialdifferenz von 1 V erhält. Die Tabellen 1—3 geben die notwendigsten Daten an.

Infolge der unterschiedlichen Energie der genannten Strahlenarten ist ihre Wirkung auf Atome, mit denen sie in Wechselwirkung treten, verschieden. Bei geringster Energie ist, außer besonderen, auf S. 695 besprochenen Wirkungen, mit einer *Wärmebewegung* zu rechnen. Ist die Energie höher (0,5—6 eV), so entstehen vorwiegend *Anregungen* von Elektronen in der an der Oberfläche der Atome befindlichen Elektronenhülle. Ab 6 eV kommt es zusätzlich und in steigendem Maße zur Abtrennung von Elektronen vom Atom *(Ionisation)*. Bei sehr hoher Energie treten weitere, hier nicht näher erörterte Prozesse auf. Obwohl eine prinzipielle Trennung zwischen den Strahlen, die eine Ionisation hervorrufen und jenen, deren Wirkung lediglich auf einer Anregung beruht, schon allein deswegen nicht möglich ist, weil ionisierende Strahlen auch Anregungen verursachen, sollen im folgenden die vorwiegend *ionisierend* wirkenden Strahlen im zweiten Abschnitt, die durch *Anregung* von Elektronen wirkenden Strahlen (UV und sichtbares Licht) dagegen im dritten Abschnitt behandelt werden. Im Abschnitt IV schließlich wird auf die Bedeutung sehr *langwelliger* elektromagnetischer Strahlung eingegangen.

b) Übersicht über die verschiedenen Strahlenwirkungen.

Manche Strahlensorten verhalten sich in der Zelle weitgehend *indifferent*, andere sind für den normalen Ablauf der Lebensvorgänge völlig *unentbehrlich*, manche *fördern* bestimmte Zellvorgänge oder wirken *hemmend*. Bei elektromagnetischer Strahlung, deren Quantenenergie größer als etwa 6 eV ist, also bei Wellenlängen unter 200 mμ (Tabelle 1), treten Ionisationen auf, die *Schädigungen* der Zelle hervorrufen; bereits 1 Energiequant ist hierzu unter Umständen ausreichend. Corpuscularstrahlen wirken in gleicher Weise. Auch bei elektromagnetischer Strahlung kleinerer Quantenenergie (bis zu etwa 4 eV; 300 mμ) überwiegen noch die Schädigungen, obwohl in diesem UV-Bereich fast nur noch eine elektronische Anregung vorhanden ist. Zu den „Schädigungen" sind hier

Tabelle 1. *Übersicht über die elektromagnetische Strahlung.*

Strahlenart	Strahlenquelle	Wellenlänge		Energie der Quanten (Elektronen-Volt) [1]	Wirkungsweise
Radiowellen	Rundfunksender	10^6 cm	10 km		vgl. S. 694.
		10^5 cm			
Kurzwellen		10^4 cm	100 m		
UK-Wellen	Dipolsender	10^3 cm	10 m		
		10^2 cm	1 m		
Dezimeterwellen		10^1 cm	10 cm		
Zentimeterwellen		10^0 cm	1 cm		
Ultrarot	Ultrarotstrahler	10^{-1} cm	1 mm		Wärmewirkungen
		10^{-2} cm	0,1 mm		
	Sonnenlicht	10^{-3} cm	10 μ		
		10^{-4} cm	1 μ		
Sichtbares Licht	Glühlampen Sonnenlicht		700 mμ	1,8 eV	elektronische Anregung
			400 mμ	3,1 eV	
Ultraviolett (UV)	Sonnenlicht Quecksilberlampen		300 mμ	4 eV	Anregung
			2000 Å	6 eV	Anregung und (Ionisation)
		10^{-5} cm	1000 Å	13 eV	
		10^{-6} cm			
Röntgenstrahlen	Röntgenröhren		20 Å	0,6 keV	Ionisation und (Anregung)
		10^{-7} cm	10 Å	1,3 keV	
			8,3 Å	1,5 keV	
		10^{-8} cm	1 Å	12,5 keV	
		10^{-9} cm	0,1 Å	125 keV	
γ-Strahlen			0,04 Å	300 keV	
	^{131}J			0,6 MeV	
	^{60}Co			1,1 MeV	
				1,3 MeV	
	^{226}Ra			2 MeV	
	Elektronenschleuder	10^{-14} cm		>100 MeV	

[1] 1 keV = 10^3 eV; 1 MeV = 10^6 eV.

auch diejenigen *bleibenden* Veränderungen, vor allem im Zellkern, zu rechnen, die sich als *Mutationen* manifestieren. Bei noch niedrigerer Anregungsenergie (im langwelligen UV und im sichtbaren Spektralbereich) kommt es zu *photochemischen Umsetzungen* mit den mannigfaltigsten Folgeerscheinungen. Es treten, wie später im einzelnen ausgeführt wird, sowohl Änderungen von *Plasmaeigenschaften* und *stoffwechselphysiologische* Vorgänge als auch *Wachstumsprozesse* aller Art auf. Das Licht wirkt dabei entweder in der Weise, daß die auf die Zelle auftreffenden Lichtquanten mit Hilfe ihrer eingestrahlten Energie die genannten Vorgänge *vollständig* ermöglichen, wie dies etwa bei der Photosynthese der Fall ist, oder aber sie wirken als *Auslöser*. Hier wird dann, wie bei den durch Licht hervorgerufenen *reizphysiologischen* Vorgängen eine Reaktion durch oft nur relativ wenige Energiequanten in Gang gebracht. Die für die Reaktion benötigte Energie selbst wird jedoch von der Zelle bzw. dem Organismus zur Verfügung gestellt. Von manchen der genannten Wirkungen erfolgt eine ausführliche Darstellung in anderen Abschnitten des Handbuches. Im folgenden behandle ich lediglich die *Grundlagen* der Strahlenwirkungen und einige *Wirkungen auf die Zelle*, einschließlich der durch Strahlung hervorgerufenen *Zellschädigungen*.

Tabelle 2. *Übersicht über einige Corpuscularstrahlen*[1].

Strahlensorte	Strahlenquelle	Korpuskelenergie
Elektronen	Elektronenschleuder künstliche radioaktive β-Strahler z. B. ^{32}P	1—100 MeV 1,69 MeV (max)
α-Strahlen ($^4_2$He-Kerne)	Teilchen-beschleuniger (z. B. Synchro-cyclotron) — natürliche radioaktive Elemente ^{210}Po ^{214}Po — künstliche radioaktive α-Strahler	1—400 MeV (>1000 MeV) 5,3 MeV 7,7 MeV
Neutronen	Kernreaktor — Deuteronenbeschuß α-Beschuß ^{2}D(d, n)^{3}He ^{9}Be(α, n)^{12}C	 2,5 MeV 13,7 MeV
Deuteronen (schwere Wasserstoffkerne, ^{2}H)	Teilchen-beschleuniger (z. B. Synchro-cyclotron)	1—1000 MeV
Protonen (Wasserstoffkerne, ^{1}H)	Neutronenbeschuß Deuteronenbeschuß α-Strahlenbeschuß (vgl. Erzeugung dieser)	0,1—1000 MeV

[1] Vgl. z. B. W. RIEZLER: Einführung in die Kernphysik, 5. Aufl. Berlin 1954. — H. SCHWIEGK: Künstliche radioaktive Isotope in Physiologie, Diagnostik und Therapie. Berlin: Springer 1953.

II. Die Wirkungen ionisierender Strahlen.

a) Grundlagen.

1. Die Energieübertragung durch Ionisation und Anregung.

Die physikalischen Voraussetzungen zum Verständnis der Primärwirkungen ionisierender Strahlen sind unter anderen ausführlich von LEA und von SOMMERMEYER, ferner in den Zusammenfassungen von FANO, FRANCK und PLATZMANN, LIVINGSTON [1] und von PLATZMANN behandelt worden. Durch die *elektro-*

magnetische Strahlung (Röntgen- und γ-Strahlung) treten in den Atomen des durchstrahlten Mediums Ionisationen und Anregungen der Elektronen in den äußeren Elektronenschalen auf. Beträgt die Quantenenergie der Photonen ($E = h\nu_1$) weniger als 30 keV ($\lambda > 0{,}4$ Å), so entstehen bei der Ionisation mehr oder weniger schnelle Elektronen (Photoelektronen), die ihrerseits Sekundärionisationen bzw. Anregungen hervorrufen können. Liegt die Quantenenergie ($h\nu_2$) der eingestrahlten Photonen aber höher ($E > 30$ keV), so wird in steigendem Maß nur ein Teil der eingestrahlten Energie auf Elektronen (COMPTON-Elektronen) übertragen; der andere Teil der Energie wird als Quant verminderter Energie ($h\nu'$), also als elektromagnetische Strahlung von längerer Wellenlänge als der eingestrahlten, weitergegeben (Abb. 1). Da die erzeugten Elektronen weitere Sekundärionisationen hervorzurufen vermögen, wirkt die energiereiche elektromagnetische Strahlung also vorwiegend als Elektronenstrahlung.

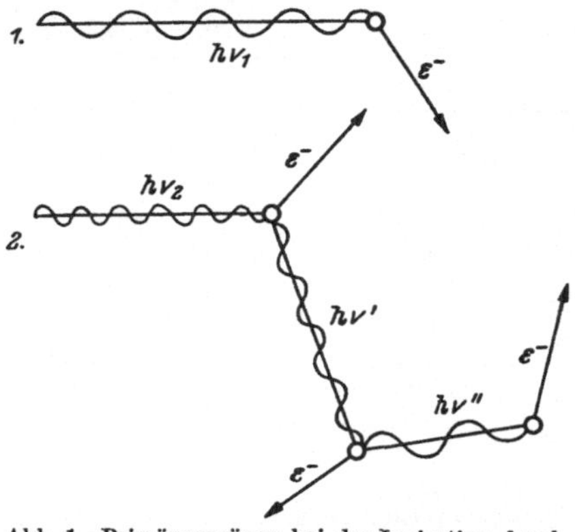

Abb. 1. Primärvorgänge bei der Ionisation durch Röntgen- bzw. γ-Strahlen. *1.* Die Quantenenergie der Photonen ist kleiner als etwa 30 keV. Bei der Einstrahlung werden Photoelektronen freigesetzt (Photoionisation). *2.* Die Quantenenergie der Photonen ist größer als 30 keV. Nur ein Teil der einfallenden Strahlung dient zur Freisetzung eines Elektrons (Comptonelektron, Comptonionisation), der Rest wird als Photon verminderter Energie abgestreut.

Die Größe der von der Strahlung abgegebenen Energie wird durch den sog. *linearen Energieübertragungsfaktor* (LET, Linear Energy Transfer), den je Weglänge aufgetretenen Energieverlust (keV/μ), dividiert durch die Dichte des durchstrahlten Mediums, gemessen. Die in der Raumeinheit absorbierte Strahlenenergie bezeichnet man als Strahlendosis. Als Einheit der Dosis wurde zunächst für Röntgen- und γ-Strahlen mittlerer Energie das „Röntgen" ($1\,\text{r} = 1{,}65 \times 10^{12}$ Ionen/cm^3) gewählt; um aber eine von der Strahlenart unabhängige Dosiseinheit zur Verfügung zu haben, wird in steigendem Ausmaß 1 rep bzw. 1 rad als Einheit für eine absorbierte Dosis von 83 bzw. 100 erg/g in Luft benutzt. 1 r entspricht 83,8 erg/g in Luft.

Tabelle 3. *Mittlere Reichweite und mittlere spezifische Ionisation verschiedener Corpuscularstrahlen*[1].

Strahlensorte	Energie	Reichweite	Mittlere spezifische Ionisation je μ
Elektronen	1 keV	0,053 μ	190
Elektronen	100 keV	141 μ	7,2
Elektronen	5 MeV	2,5 cm	2
Elektronen	100 MeV	30,4 cm	3,3
Protonen	2 MeV	73 μ	260
Protonen	10 MeV	1211 μ	78,5
Deuteronen	3,7 MeV		250
Deuteronen	190 MeV	14 cm	15
α-Strahlen	2 MeV	10 μ	$1{,}9 \times 10^3$
α-Strahlen	10 MeV	108 μ	$0{,}8 \times 10^3$

[1] Nach LEA, TOBIAS, SETLOW und DOYLE, SOMMERMEYER.

Bei der *Corpuscularstrahlung* ist die Größe der Wechselwirkung, also Zahl und Ausmaß der Anregungen und Ionisationen, durch den *Abstand* der fliegenden Korpuskel von den Atomen des Mediums sowie durch die *Geschwindigkeit* der an den Atomen vorbeifliegenden Partikel gegeben (vgl. POLLARD). Kernzusammenstöße und Übertragungen der Anregungsenergie auf innere Elektronen sind dabei relativ selten zu erwarten.

Mit *wachsender* Geschwindigkeit der Teilchen *sinkt* die Zahl der Anregungen bzw. der Ionisationen; entsprechend sinkt der lineare Energieübertragungsfaktor.

Je größer die *Masse* des Korpuskels, desto kleiner ist bei gleicher kinetischer Energie die Geschwindigkeit, und desto größer dann die Ionisation bzw. LET (vgl. Tabelle 3). Am Anfang der Bahn, bei hoher Partikelgeschwindigkeit, ereignen sich demnach viel weniger Anregungen und Ionisationen als am Ende der Bahn.

Die von den Primärionisationen herrührenden Elektronen rufen *Sekundärionisationen* hervor, die aber infolge der geringen Energie dieser Sekundärelektronen nur in der Nachbarschaft der Primärionisationen wirksam werden können. Bei *niedrigem* LET entstehen so Ionisationshaufen, bei *hohem* LET, z. B. bei α-Strahlen, dagegen Ionisationssäulen.

2. Die Treffertheorie der Strahlenwirkung.

Zur Erklärung der Strahlenwirkung in biologischen Objekten ist die *Treffertheorie* der Strahlenwirkung (zuerst BLAU und ALTENBURGER, DESSAUER [1], später CROWTHER u. a.) entwickelt worden. Sie gründet sich auf die Beobachtung, daß auch bei der Verwendung höherer Strahlungsintensitäten (Dosis) nur ein bestimmter Bruchteil der vorhandenen Zellen verändert wird, im Gegensatz zu der bei der Einwirkung von Giften vielfach vorhandenen Wirkungsweise, bei der bei schwachen Dosen noch kein Effekt auftritt, mit steigender Konzentration aber sehr bald alle Zellen vergiftet werden. Es wurde deshalb die wesentliche Annahme gemacht, daß der z. B. bei Bestrahlung von Einzellern gefundene Verlauf der Wirkung auf Grund einer *nicht kontinuierlichen*, sondern *statistischen* Absorption der Strahlungsenergie erklärt werden könne (vgl. z. B. GLOCKER, LEA, TIMOFÉEFF-RESSOWSKY und ZIMMER, WYCKOFF sowie DESSAUER [2] und SOMMERMEYER). Im einzelnen wird ein strahlenbiologisch wirksamer Elementarakt als *Treffer* bezeichnet, der z. B. im UV und sichtbaren Licht in der Absorption eines Lichtquants, bei γ- und Röntgenstrahlen in der Auslösung eines Elektrons im biologischen Objekt und bei Corpuscularstrahlen in dem wirksamen Durchgang eines Korpuskels durch das biologische Material besteht. Ferner wird die Zahl der zur Wirkung nötigen Treffer als *Trefferzahl* bezeichnet. Die Orte der Wirksamkeit der Treffer nennt man *Treffbereiche* bzw. Treffervolumen. Aus Wahrscheinlichkeitsbetrachtungen läßt sich dann die relative Anzahl der biologischen Objekte (N/N_0), die mindestens n Treffer erhalten haben (N_0 = Gesamtzahl der Objekte; N = Anzahl der getroffenen), berechnen: Wird *ein* empfindlicher Treffbereich angenommen, ergibt sich $N/N_0 = 1 - e^{-VD} \cdot \sum_{K=0}^{n-1} \frac{(VD)^K}{K!} = 1 - B$.

V = Volumen/cm³ (Treffbereich); D = Strahlendosis; K = eine Zahl, die entsprechend der Anzahl der Treffer von 0 bis $n-1$ läuft. B bedeutet den zweiten Ausdruck in der angegebenen Gleichung. Für $n = 1$ Treffer vereinfacht sich diese Gleichung zu $N/N_0 = 1 - e^{-VD}$. Stellt man diese Gleichung als Kurve für N/N_0 in Abhängigkeit von D dar *(Dosiseffektkurve)*, erhält man eine Kurve mit exponentiellem Verlauf; in halblogarithmischer Darstellung demnach eine Gerade. Werden weiterhin m *Treffbereiche* angenommen, wobei jeder von ihnen n Treffer erhalten muß, so ergibt sich nach GLOCKER $N/N_0 = (1 - B)^m$. Genügt es auch hier zur Auslösung der Wirkung, daß jeder der m Bereiche nur *einen Treffer* erhält, so vereinfacht sich die Gleichung zu $N/N_0 = (1 - e^{-VD})^m$ (Multitarget theory). Eine andere Form der Gleichung ergibt sich schließlich noch dann, wenn m empfindliche Bereiche angenommen werden, von denen *einer* n Treffer erhalten muß: $N/N_0 = 1 - B^m$; ist $n = 1$, vereinfacht sich hier die Gleichung zu $N/N_0 = 1 - e^{-VDm}$. Für weitere Einzelheiten muß auf die genannte Literatur verwiesen werden.

An Hand vieler Beispiele, wie etwa der Inaktivierung von *Escherichia coli* durch Elektronen, Röntgenstrahlen verschiedener Wellenlänge und auch durch

UV (WYCKOFF, Abb. 2), ließ sich nachweisen, daß durch Bestrahlung bedingte Vorgänge in der Tat vielfach durch Dosiseffektkurven vom 1-Treffertyp (also solche mit exponentiellem Verlauf) beschrieben werden können. Durch weitere Untersuchungen zeigte sich dann allerdings, daß das Medium und die unmittelbare *Umgebung* des bestrahlten biologischen Objekts eine sehr wesentliche Bedeutung für die Strahlenwirkung haben. Zu den in der ursprünglichen Theorie vorwiegend behandelten *direkten* Treffern kommen damit die *indirekten Strahlenwirkungen*, weil durch die Bestrahlung auch in der Nachbarschaft der biologischen Objekte Treffer erzielt werden, deren Folgen erst die biologischen Objekte selbst erfassen. So können etwa durch Bestrahlung im umgebenden Wasser Radikale gebildet werden, die zur Bildung von schädigenden, diffundierenden Verbindungen führen. Auch für diese indirekte Wirkung der Bestrahlung ist inzwischen eine theoretische Formulierung auf statistischer Grundlage gefunden worden, die im einfachsten Fall, wenn nämlich *eine* wesentliche Stelle eines wichtigen Makromoleküls durch ein Molekül eines wirksamen Agens getroffen wird, zu einer 1-Trefferkurve führt. Dieses *Diffusionsmodell* der Strahlenwirkung schließt die direkten Treffer mit ein und ermöglicht die zur Zeit beste Deutung der Strahlenwirkung (SIX, TOBIAS, ZIRKLE [1]). Im folgenden werden durch direkte Treffer hervorgerufene Strahlenwirkungen und durch indirekte Effekte verursachte, soweit es geht, getrennt voneinander behandelt, obwohl eine solche Trennung in der Praxis nur in seltenen Fällen exakt möglich ist.

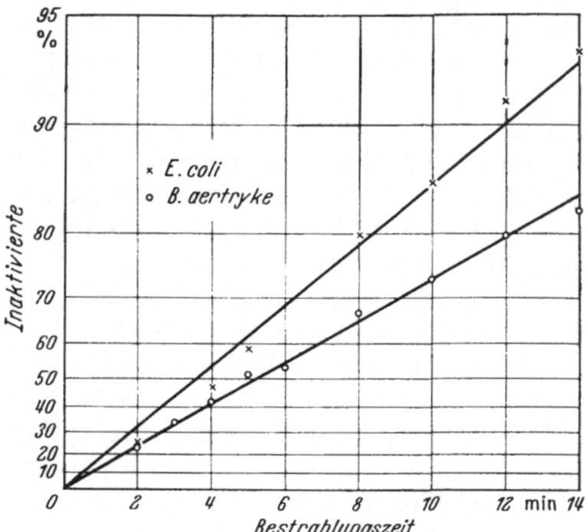

Abb. 2. Inaktivierung von *Escherichia coli* und *Bacterium aertryke* durch 12 kV-Röntgenstrahlen. (Nach WYCKOFF, aus TIMOFÉEFF-RESSOVSKY u. ZIMMER, Leipzig 1947.)

Gelegentlich wird die Meinung vertreten, daß der Verlauf der Strahlenwirkung, etwa bei Mikroorganismen, *nicht* durch die *statistische* Verteilung der auftreffenden Strahlungsenergie zustande kommt, sondern, ähnlich der oben genannten Giftwirkung, vor allem durch die *Variabilität der Empfindlichkeit* der Organismen (Mikroorganismen, Zellen, Zellbestandteile) hervorgerufen wird. Diese Ansicht scheint sich in Einzelfällen neuerdings durch die genaueste Untersuchung des Verlaufs der Schädigungskurve bei jungen Bakterienkulturen stützen zu lassen (HOUTERMANS [3]). Die Bedeutung dieser Befunde für die gesamte Theorie der Strahlenwirkung bleibt jedoch abzuwarten.

3. Die radiochemischen Vorgänge bei der ionisierenden Strahlung und ihre Bedeutung für die Strahlenwirkungen.

Die im biologisch wichtigen Bereich erfolgten „direkten Treffer“ können mancherlei Veränderungen des betroffenen Moleküls zur Folge haben: die durch Ionisation gebildeten Radikale werden *oxydiert* oder *reduziert* und weiter umgebaut; durch Stabilisierung des Radikals kommt es zur *Lösung schwächerer Bindungen* im Molekül (z. B. Wasserstoffbrücken im Eiweiß) oder auch zum *Bruch stärkerer* Bindungen (z. B. Peptidbindungen); es können *Depolymerisierungen*, aber ebenso auch *Polymerisierungen* auftreten.

Zu dem Verständnis der „*indirekten Strahlenwirkungen*“ hat die in den letzten Jahren intensiv betriebene Untersuchung der *Radiochemie des Wassers*, besonders die Feststellung der unterschiedlichen Wirkung verschiedener Strahlensorten im

Zusammenhang mit dem unterschiedlichen Einfluß des Sauerstoffs, wesentlich beigetragen.

Auf Grund der Untersuchungen von WEISS und der vor allem von LEA entwickelten Vorstellungen kommt es bei der Bestrahlung des Wassers zur Ionisation und unter Dissoziation zur Bildung freier *Radikale:*

$$H_2O^+ \longrightarrow OH + H^+ \qquad R_{(1)} \tag{1}$$

$$\varepsilon^- + H_2O \longrightarrow OH^- + H \qquad R_{(2)} \tag{2}$$

Neuerdings stellten die Untersuchungen von ALLEN außerdem fest, daß bei der Wasserzersetzung auch die *Molekularprodukte* H_2 und H_2O_2 auftreten, deren Ausbeute mit steigender Ionisationsdichte der Strahlung ansteigt. Demnach läßt sich die radiochemische Wirkung von Strahlen auf Wasser durch die folgenden beiden Gleichungen darstellen:

$$H_2O \longrightarrow H + OH \qquad (R) \tag{3}$$

$$2\,H_2O \longrightarrow H_2 + H_2O_2 \qquad (M) \tag{4}$$

Die Molekularprodukte (M) werden durch die H-Atome und OH-Radikale aus (R) in einer Rückreaktion zum Teil in Wasser zurückverwandelt (O. A. ALLEN und Mitarbeiter).

Die vorliegenden Ergebnisse der Radiochemie des Wassers haben SAMUEL und MAGEE sowie DEWHURST, SAMUEL und MAGEE in einer plausiblen *Modellvorstellung* vereinigt, wodurch die unterschiedlichen Reaktionen für verschiedene Strahlensorten bereits recht gut verständlich werden. Hiernach kollidiert unter anderem ein Teil der gebildeten Radikale; es kommt zu *Rekombinationen,* und nur der Rest erscheint in Gleichung (3). Wegen der viel größeren Ionisationsdichte bei α-Strahlen sind hier im Gegensatz zu Röntgen-, γ- und β-Strahlen die Rekombinationen viel häufiger; dadurch verliert hier Gleichung (3) gegenüber Gleichung (4) an Bedeutung. HART konnte die genannte Modellvorstellung in seinen Versuchen im wesentlichen bestätigen (vgl. Tabelle 4). Das Modell erklärt aber nicht nur die wechselnden Anteile der Reaktion (R) und (M) für die unterschiedlichen Strahlensorten, sondern auch den verschiedenartigen Einfluß von *Sauerstoff* auf die Wirkungen der genannten Strahlen. Sauerstoff verringert danach vor allem die Aktivität des H-Atoms, greift also in die Radikalreaktion (R) ein:

Tabelle 4. *Die Ausbeute an Radikalpaaren* (R) *und Molekularprodukten* (M) *in 0,8 n* H_2SO_4. (*Nach* HART.)

Strahlensorten		Reaktion (R) %	Reaktion (M) %
UV < 2000 Å	(UV)	100	0
^{60}Co γ	(γ)	74,4	22,6
^{3}H β	(β)	70	30
^{210}Po α	(α)	12	88
^{10}B(n, α)^{7}Li	(α)	8	92

$$2\,H + 2\,O_2 \longrightarrow 2\,HO_2 \tag{5}$$

$$HO_2 + HO_2 \longrightarrow H_2O_2 + O_2 \tag{6}$$

Dabei wird das reaktive HO_2-Radikal gebildet, und es entsteht im Gefolge der Reaktion (R) Peroxyd. Demnach ist also vor allem bei den Strahlenarten ein O_2-Effekt zu erwarten, bei denen die Radikalreaktion (R) eine Rolle spielt. Bei Abwesenheit von Sauerstoff zeigt sich, entsprechend dem Anteil der Reaktionen (R) und (M), eine Abhängigkeit des gebildeten Peroxyds von der Strahlensorte sowie von der Masse und Energie der ionisierenden Corpuscularstrahlung. Das Modell deutet also sowohl den fehlenden O_2-Einfluß bei α-Strahlen aus dem geringen Anteil der Radikalreaktion (*R*), die verschiedene Beteiligung von Peroxyd bei unterschiedlichen O_2-Gaben, als auch die unterschiedliche Peroxydempfindlichkeit nach Bestrahlung mit verschiedenen Strahlensorten bei O_2-Ausschluß. Im einzelnen besteht allerdings noch eine Reihe von Diskrepanzen (ALLEN, GHORMLEY und HOCHANADEL).

b) Die Strahlenwirkungen.

1. Die Wirkung der direkten Treffer.

Zur Untersuchung der direkten Wirkungen ionisierender Strahlen müssen alle sekundären Einflüsse ausgeschaltet werden, die durch Strahlenwirkungen im umgebenden Medium erfolgen können. Die Objekte müssen deshalb getrocknet oder im gefrorenen Zustand bzw. im Vakuum bestrahlt werden.

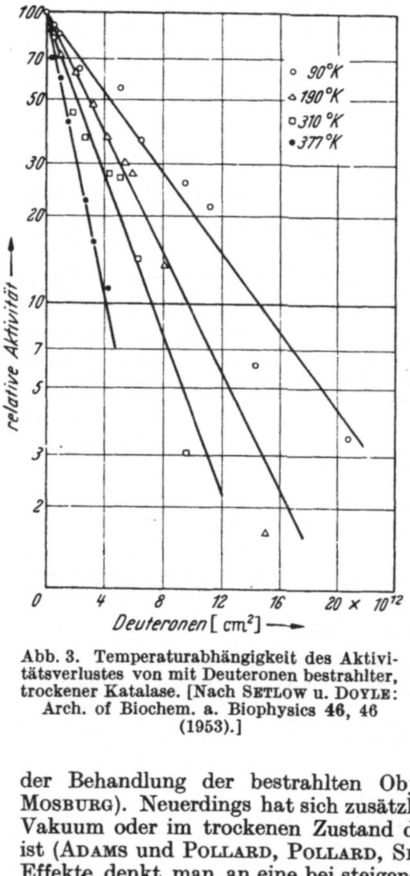

Abb. 3. Temperaturabhängigkeit des Aktivitätsverlustes von mit Deuteronen bestrahlter, trockener Katalase. [Nach SETLOW u. DOYLE: Arch. of Biochem. a. Biophysics 46, 46 (1953).]

Modellsubstanzen.

Als Modellsubstanzen wurden Proteine und Enzyme benutzt, die nach trockener Bestrahlung erneut gelöst und dann auf den durch die Bestrahlung hervorgerufenen Verlust ihrer serologischen Eigenschaften oder ihrer Enzymaktivität hin geprüft werden (POLLARD). Die *Empfindlichkeit* solcher *trocken* bestrahlten Eiweiße und Enzyme ist *wesentlich geringer* als bei Bestrahlung im gelösten Zustand (APPLEYARD [1, 2], BARRON und FINKELSTEIN, BIER und NORD u. v. a.). Bei steigender Strahlendosis ergibt sich vielfach ein exponentieller Verlauf der Inaktivierung. Die Schädigungen können also unter dem Gesichtspunkt einer Eintrefferwirkung behandelt werden und es läßt sich der Treffbereich bzw. das Treffvolumen berechnen (vgl. oben, S. 659). Beim gleichen Objekt ist aber die Größe der Schädigung verschiedener Eigenschaften (z. B. der Enzymaktivität und der serologischen Aktivität) durchaus ungleich. Nach POLLARD kann man unter Berücksichtigung der Wirkung verschiedener Strahlenarten (z. B. Deuteronen und Elektronen) und unter Beachtung des Schädigungsverlaufs neben der Bestimmung des Treffbereichs auch Rückschlüsse auf den Molekülaufbau und das Molekulargewicht der bestrahlten Substanzen vornehmen. Die Größe des berechneten Treffbereichs hängt dabei aber vielfach von der Art der Behandlung der bestrahlten Objekte nach der Bestrahlung ab (HUTCHINSON und MOSBURG). Neuerdings hat sich zusätzlich auch noch gezeigt, daß selbst bei Bestrahlung im Vakuum oder im trockenen Zustand der Bestrahlungseffekt von der *Temperatur* abhängig ist (ADAMS und POLLARD, POLLARD, SETLOW und DOYLE [1]) (Abb. 3). Zur Erklärung dieser Effekte denkt man an eine bei steigenden Temperaturen (0—200° K) langsam abnehmende Molekülstabilität, bei noch höherer Temperatur an Anregungen und nicht nur an Ionisationen im getroffenen Molekül, die die Schädigungen hervorrufen.

Biologische Objekte.

Manche biologische Objekte (Viren, Bakterien, Samen) lassen sich ohne Verlust ihrer Aktivität weitgehend austrocknen, so daß im *trockenen* Zustand hier die *direkte* Strahlenwirkung untersucht werden kann. Es zeigen sich die gleichen Effekte *(Reduktion* der Strahlenwirkung und *Temperaturabhängigkeit)* wie bei den Modellsubstanzen (ADAMS und POLLARD, NYBOOM und Mitarbeiter, RAJEWSKY [2], REINHOLZ u. a.). Bei solchen Objekten ist das Treffvolumen und vielleicht auch die Molekularorganisation der getroffenen Bereiche ebenso wie bei den Modellsubstanzen einer genaueren Untersuchung zugänglich. Der Wirkungsquerschnitt der Inaktivierung von Phagen T_1 erwies sich z. B. auf $^1/_{10}$ des Phagenvolumens beschränkt. Der Wirkungsbereich für die serologische Aktivität der gleichen Phagen war noch um ein Vielfaches kleiner, so daß die Annahme gerechtfertigt ist, daß die serologisch wichtigen Einheiten in vielen kleinen Bezirken an der Oberfläche des Virus liegen (POLLARD). In allen anderen Fällen, in denen eine Austrocknung nicht ohne Schädigung vertragen wird, lassen sich direkte Treffer aber nicht für sich allein untersuchen. Meistens sind dann indirekte Strahlungseffekte an der Gesamtwirkung beteiligt.

2. Indirekte Strahlenwirkungen.

Abbau von Modellsubstanzen in Lösungen.

Indirekte Strahlenwirkungen auf biologische Objekte lassen sich mit Hilfe von gelösten *Modellsubstanzen* (Proteine, Enzyme, Nucleinsäuren (NS), einfache organische Verbindungen)

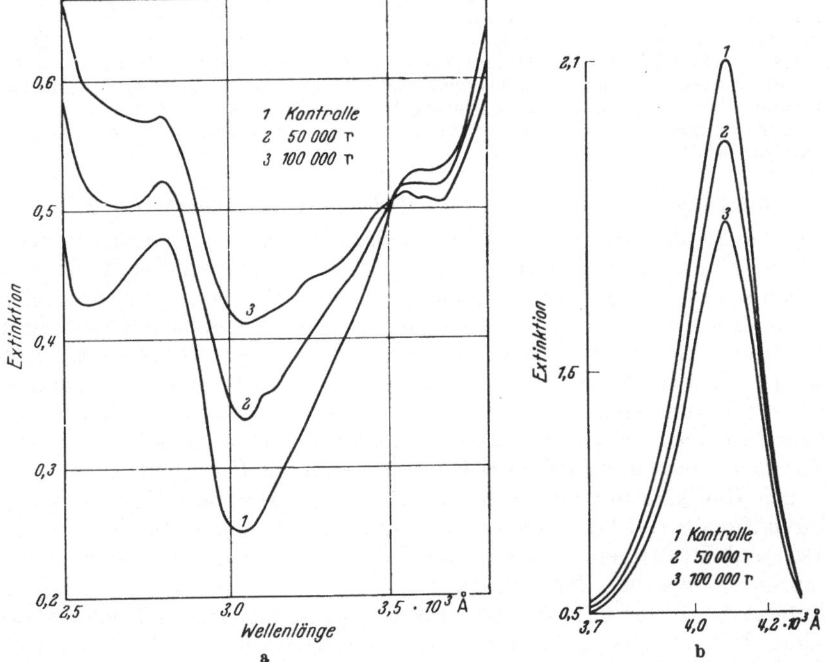

Abb. 4a u. b. Der Einfluß von Röntgenbestrahlung auf das UV-Absorptionsspektrum von Ferricytochrom c. p_H 7,7. a Die Veränderung des Proteinanteils (Zunahme der Extinktion zwischen 250 und 350 mμ). b Die Veränderung des Coenzyms (Abnahme der Extinktion bei 410 mμ). (Nach E. S. G. BARRON: NICKSON, Symposium, S. 216. 1952.)

untersuchen, die in mannigfaltigster Weise durch ionisierende Strahlen abgebaut und verändert werden können (BARRON [1—3], DALE [1—3], GRAY [1, 2], HEVESY [1], SPARROW und RUBIN).

Es kommt zu einem *Abbau* mit vielfältigen Spaltprodukten: Ameisensäureabbau, HART, Essigsäureabbau, GARRISON und Mitarbeiter, Abbau von Aminosäuren, DALE [2], und von Nucleinsäuren, SCHOLES und WEISS [1, 2]. Beim aeroben Abbau von Glycerinlösungen mit Röntgenstrahlen wurden z. B. 9 verschiedene Produkte gefunden (MAXWELL und Mitarbeiter). Es treten *Oxydationen* auf, gelegentlich auch Reduktionen (BARRON und Mitarbeiter, SWALLOW). Bei Proteinen und Enzymen ist vor allem an die Oxydation von SH-Gruppen, von freien Aminogruppen, von OH-Gruppen und an die Auftrennung von Doppelbindungen zu denken; durch die Zerstörung von H-Brücken kommt es zur Entfaltung von Peptidketten. Damit werden die *physikalisch-chemischen* Eigenschaften der bestrahlten Substanzen verändert. Es wurde sowohl ein Viscositätsabfall (Nucleinsäuren, BERNSTEIN, DANIELS und

Mitarbeiter) als auch ein Anstieg der Viscosität z. B. bei Eiweißen gefunden (ALEXANDER, BARRON und FINKELSTEIN, PATRICK und BURTON, SCHERAGA und NIMS u. a.). Die *UV-Absorption* von bestrahlten Proteinen, Enzymen und NS wird verändert. Bei steigender Dosis findet sich neben gelegentlich beobachtetem Anstieg (z. B. BARRON [1]) vor allem ein Abfall der Absorption, also ein Zeichen für die Zerstörung der lichtabsorbierenden (chromophoren) Gruppen (NS, Purine, Pyrimidine, DPN, z. B. BARRON und Mitarbeiter) (Abb. 4). Die verschiedenen Modellsubstanzen zeigen gegenüber der Bestrahlung zum Teil eine recht erhebliche, wenn auch unterschiedliche Resistenz. Proteine sind vielfach strahlenresistenter als manche Enzyme, die gelegentlich aber auch eine große *Resistenz* aufweisen (BARRON und JOHNSON, SMITH u. v. a.). NS werden proportional zur eingestrahlten relativ hohen Dosis zu vielfältigen Abbauprodukten verändert (z. B. SCHOLES und WEISS [1, 2]).

Die Strahlenwirkung kann in verschiedener Weise beeinflußt werden. Besonders wirksam erwies sich der Einfluß von *Sauerstoff*, der sich bei α-Strahlen am wenigsten, stärker bei Neutronen und am stärksten bei Röntgen- und γ-Strahlen sowie bei Elektronen bemerkbar macht (DALE [1, 3], GRAY [2, 4], HART, PATT). Das Vorhandensein von Sauerstoff führt zur Bildung sekundärer Radikale, besonders von HO_2, und damit zur Entstehung verschiedenster Folgereaktionen, wie sich aus dem S. 661 Gesagten ergibt.

Auch *Chemikalien* können die Strahlenwirkung erheblich verändern. Viele Substanzen, die den bestrahlten Lösungen zugefügt werden, setzen die Strahlenwirkung herab. Sie wirken als *Schutzstoffe*. Die verschiedensten Modellsubstanzen (Proteine, Enzyme, Nucleoproteide, Farbstoffe) können durch solche Schutzstoffe vor der Strahlenwirkung geschützt werden (BACQ, DALE [1—3], HERVE und Mitarbeiter, STAPLETON und Mitarbeiter [1]). Als Schutzstoffe wirken z. B. neben Allylthioharnstoff, Thioharnstoff, Gluthathion und Cystein auch Äthanol, Glycin, Glycol usw. (ALEXANDER, BARRON und JOHNSON, MINDER und SCHOEN, PATT). Zur Erklärung dieser Schutzwirkung wurden verschiedene Theorien aufgestellt; wir kommen auf S. 666 darauf zurück.

Wirkungen von vorbestrahltem Medium auf biologische Objekte.

Daß die Strahlenwirkung bei biologischen Objekten zu einem großen Teil sekundärer Natur ist und durch Veränderungen der bestrahlten Flüssigkeit in den Zellen selbst oder in ihrer Umgebung hervorgerufen ist, läßt sich durch *Vorbestrahlung* des Kulturmediums nachweisen, denn nach dessen Bestrahlung überimpfte Viren, Zellen und Gewebe werden geschädigt (GRAY [2, 4], LURIA). Gab man z. B. nicht bestrahlte Hefezellen zu einer vorbestrahlten Pufferlösung, so trat eine Hemmung der Gärung auf (SHERMAN und CHASE). Entsprechende Ergebnisse zeigten sich bei der Inaktivierung von Viren (ALPER [3, 4], WATSON). Offensichtlich treten dabei neben H_2O_2 auch organische Peroxyde bzw. möglicherweise auch Radikale mit langer Lebensdauer auf. Der Bestrahlungseffekt läßt sich durch Zugabe von *Schutzstoffen* oder durch Nährbouillon zum Teil eliminieren (DANIEL und PARK); er ist *temperaturabhängig*, und die gebildeten instabilen Substanzen *nehmen* mit der *Zeit ab*.

Ähnliche Wirkungen lassen sich erzielen, wenn nur ein *Teil* eines Gewebes oder einer Zellkultur bestrahlt wird und dabei auch die Strahlenwirkungen auf die *nicht* bestrahlte Kulturhälfte beobachtet werden. So wurde die *nicht* bestrahlte Hälfte einer Hühnerherzfibroblastenkultur erheblich *geschädigt* (Untersuchung von Teilungsstadien). Die nicht bestrahlte Hälfte setzte jedoch gleichzeitig die Wirkung der Strahlung im bestrahlten Abschnitt herab (PETERS [1, 2]). Auch die Partialbestrahlung von Zellen gehört hierher. So können z. B. durch alleinige Bestrahlung des Cytoplasmas genetische Effekte ausgelöst werden (GRAY [4], NAKAO).

3. Wirkungen ionisierender Strahlung auf Zellen und Gewebe.

Allgemeines.

Da die Wirkungen ionisierender Strahlung im lebenden Organismus außerordentlich zahlreich sind (Zusammenfassungen: CARLSON, DALE [1, 3], DE PLAEN, ERRERA [1], HEVESY [1, 2], ORD und STOCKEN, PATT und BRUES, sowie die eingangs genannten Bücher), entsteht die Frage, welche dieser Wirkungen unmittelbar durch die Bestrahlung selbst erfolgen, und welche Wirkungen nur Folgeprozesse darstellen. Man hat deshalb schon seit längerem nach Vorgängen gesucht, die gegen eine Bestrahlung besonders empfindlich sind. Wie schon oben betont, sind die besonders wichtigen Zellbestandteile, nämlich die Proteine, Enzyme und Nucleinsäuren, relativ strahlenresistent. Die lebende Zelle ist aber bereits gegen *geringe* Strahlendosen (50—100 r) so *empfindlich*, daß Letalfolgen auftreten. Es müssen also in der Zelle hoch empfindliche Vorgänge inaktiviert oder Zellbestandteile durch die Bestrahlung zerstört werden. Man hat in diesem Zusammenhang daran gedacht, daß etwa die für Wachstum und Teilung wichtigen Desoxyribosenucleinsäuren (DNS) zerstört würden. Es stellte sich aber heraus, daß die DNS bzw. die Ribosenucleinsäuren (RNS) durch die (geringen) Bestrahlungsdosen nicht unmittelbar abgebaut werden (GROSS und MANDEL, HARRINGTON und KOZA, v. SALLMANN u. a.). Dagegen konnte vielfach festgestellt werden, daß die *Synthese* der NS bzw. der *Einbau* markierter Substanzen in die NS nach Bestrahlung sehr rasch intensiv *abnimmt* (GROSS und MANDEL, HEVESY [1, 2], HOWARD und PELC, KELLY und JONES, KLEIN und FORSSBERG, LAVIK und BUCKALOW, WEYMUTH und KAPLAN). Es ist aber durchaus noch fraglich, ob die hier herausgegriffene Hemmung der NS-Systeme in besonders engem Zusammenhang mit den primären zellphysiologischen Strahlenwirkungen steht, da ein völlig paralleler Zusammenhang zwischen NS-Synthese und dem besonders strahlenempfindlichen Stadium der Zelle bei der Zellteilung bisher *nicht* nachgewiesen werden konnte (DONIACH und Mitarbeiter, HOWARD und PELC, SPARROW und Mitarbeiter). Es kann sogar nach Bestrahlung zu einer erhöhten Einlagerung z. B. von ^{14}C in NS kommen, wie Untersuchungen an *Chlorella* von PORTER und KNAUSS zeigen. Auf der anderen Seite mehren sich die Arbeiten, die eine *allgemeine Hemmung der Synthesefähigkeit* der Zelle nach Bestrahlung feststellen und die Aktivitätshemmung der Enzyme als den Haupteffekt der Bestrahlung ansehen. Besonders Thiolgruppen enthaltende Enzyme werden für strahlenempfindlich gehalten (BARRON [1], BARRON und SEKI, HEVESY, POWELL und POLLARD). Da zudem die normalen Enzymwirkungen vielfach als Radikalreaktionen angesehen werden können (MICHAELIS, WATERS), wird die genannte Theorie besonders wahrscheinlich, weil die vor allem durch die indirekten Strahlenwirkungen gebildeten Radikale mit dem normalen Enzymprozeß konkurrieren können und ihn inaktivieren (z. B. BARRON und LEVINE, SCHOLES und WEISS [1]).

Wahrscheinlich sind aber sogar *verschiedene* Primärwirkungen der Bestrahlung an der Schädigung beteiligt. Dazu dürften gehören: Direkte Treffer in empfindlichen Enzymsystemen, durch Strahlung entstandene Gifte, Synthesen hemmende Radikale und die von diesen Schäden ausgehenden physiologischen Folgeprozesse (DALE [3], HEVESY [1, 2]).

Die Beeinflussung der Strahlenwirkung durch die Umwelt.

Wäßriges Medium in der Umgebung der Zellen und die Zellflüssigkeit selbst lassen die indirekten Strahlenwirkungen ansteigen. Sowohl die Höhe des *Wassergehalts* der Umgebung als auch seine Erhöhung in den zu bestrahlenden biologischen Objekten selbst ist auf die Strahleneinwirkung von größtem Einfluß, wie Untersuchungen von Pilzsporen oder von Samenkörnern zeigen (z. B. CALDECOTT [1, 2], ENGEL, KAPLAN [4, 5], KONZAK, STAPLETON und HOLLAENDER) (Abb. 5).

Ferner ist der die Diffusionsvorgänge ändernde *Temperatureinfluß* auf die Bestrahlungsvorgänge und ihre Folgen zu erwähnen (GILES, HOUTERMANS [2], RAJEWSKY [2], STAPLETON und EDINGTON, WOOD) (Abb. 6). Viele Temperatureffekte sind aber nicht so einfach zu deuten, da bereits trockenbestrahlte Objekte eine Temperaturabhängigkeit der hier im wesentlichen direkten Strahlenwirkung zeigen (vgl. S. 662). Auch ist die Temperaturwirkung vom O_2-Gehalt während der Bestrahlung (z. B. Chromatidbrüche bei *Tradescantia*; GILES) abhängig. Und schließlich verlaufen im Gefolge der Bestrahlung auftretende *Erholungsvorgänge* je nach der Temperatur verschieden schnell (HOLLAENDER und STAPLETON, LANGENDORFF und SOMMERMEYER, STAPLETON und Mitarbeiter [2]).

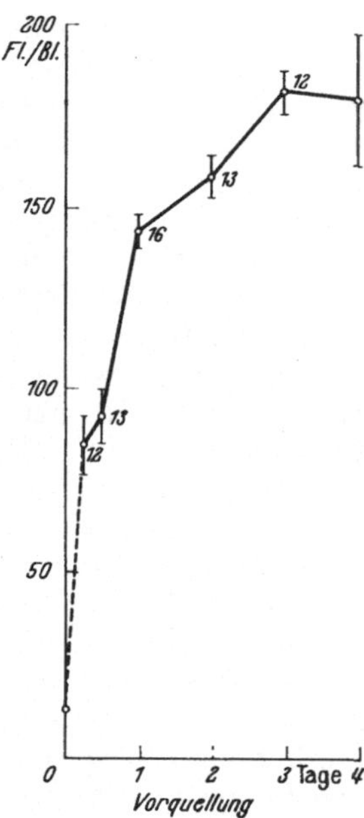

Abb. 5. Der Einfluß des Wassergehaltes auf die Wirkung von Röntgenstrahlen. Durch die Bestrahlung von Samen mit 2000 r induzierte und durch die Vorquellzeit beeinflußte Fleckenmutationen von *Soja hispida*. [Nach R. W. KAPLAN: Strahlenther. **94**, 106 (1954).]

Weiter macht sich die bereits oben, S. 661 erörterte Abhängigkeit der Bestrahlungsprodukte des Wassers vom *Sauerstoffgehalt*, entsprechend der Theorie vor allem bei Röntgenstrahlen, auch bei den Strahlenwirkungen auf die Zelle selbst bemerkbar, wie sich aus Untersuchungen über die Strahlenresistenz von Bakterien oder von Sporen sowie aus der Beeinflussung von Samen und Keimlingen und an Hand von genetischen Veränderungen durch Bestrahlung ergibt (EHRENBERG, GUSTAVSON und Mitarbeiter, GILES und RILAY, HOLLAENDER [1], NYBOM und Mitarbeiter, READ [1, 2], STAPLETON und HOLLAENDER, THODAY und READ). Auf die zahlreichen Untersuchungen über den vielfach verstärkend wirkenden O_2-Einfluß bei durch Bestrahlung hervorgerufenen Chromosomenaberrationen kann hier nur hingewiesen werden (unter anderen GILES, GRAY, LÜNING, MULLER) (Abb. 7). Es ist bei diesen Untersuchungen aber nicht immer leicht zu entscheiden, ob die Beeinflußbarkeit der Röntgenbestrahlung durch Sauerstoff wirklich durch erhöhte Radikalbildung im wäßrigen Medium zustande kommt, ob sie auf Erholungsvorgänge zurückzuführen ist oder aber von der Veränderung von Stoffwechselprozessen herrührt, die durch die Sauerstoffzufuhr in andere Bahnen geleitet werden (z. B. LASER).

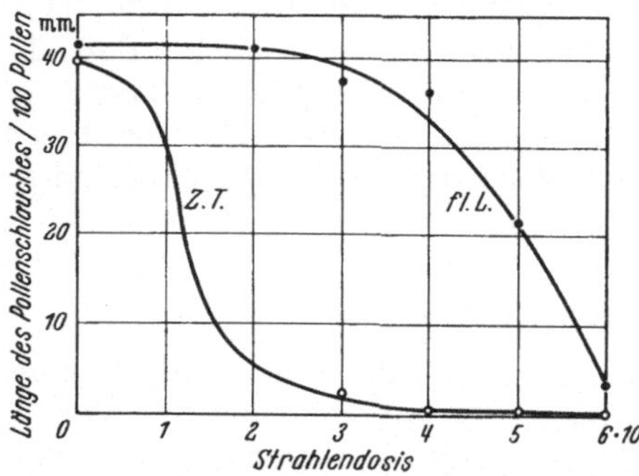

Abb. 6. Die Abhängigkeit der Röntgenstrahlenwirkung von der Temperatur. Bestrahlung von Pollenkörnern von *Digitalis* bei Zimmertemperatur und bei der Temperatur der flüssigen Luft. Zur Messung der Strahlenempfindlichkeit diente die Länge des Pollenschlauchwachstums. [Nach RAJEWSKY: Brit. J. Radiol. **25**, 350 (1952).]

Von weiteren die Strahlenwirkung beeinflussenden Faktoren sei die recht verschieden beurteilte *H_2O_2-Wirkung* wenigstens erwähnt. Zugabe von H_2O_2 abbauender Katalase vermag gelegentlich die Strahlenwirkung herabzusetzen (ALPER [1], HOWARD, KIMBALL und GAITHER [2]). Auch konnte H_2O_2 verschiedentlich nachgewiesen und eine Nachwirkung der Bestrahlung auf einen H_2O_2-Effekt zurückgeführt werden (ALPER [2—4]). In anderen Fällen konnte aber durch H_2O_2-Zugabe keine Steigerung der Strahlenschäden beobachtet werden (KIMBALL, BARRON [3]).

Vor allem sind es nun aber eine Reihe von Chemikalien, sog. *Schutzstoffe*, die, in kleiner Menge *vor* der Bestrahlung geboten, die Strahlungsfolgen mindern.

Die Literatur über derartige Stoffe ist bereits außerordentlich angewachsen (Zusammenfassungen: BACQ, BACQ und HERVE, DALE, FORSSBERG, HOLLAENDER und STAPLETON, LANGENDORFF und Mitarbeiter [1, 2]). Wirksam sind besonders Verbindungen mit SH-Gruppen wie Cystein, Glutathion, Cysteamin, ferner auch Thioharnstoff und Dithiophosphonat. SH-Gruppen blockierende Substanzen wie Monojodacetat erhöhten die Strahlenschäden. Weiterhin sind aber auch Stoffwechselprodukte wichtig, wie Brenztraubensäure oder Milchsäure (HOLLAENDER und STAPLETON, STAPLETON und Mitarbeiter [2], TROWELL).

Über die Ursachen dieser Schutzwirkung liegen zwei Theorien vor: bei der *radiochemischen* Theorie der Schutzwirkung wird angenommen, daß die Schutzstoffe, insbesondere die Sulfhydrylverbindungen, durch ihre Konkurrenz mit den

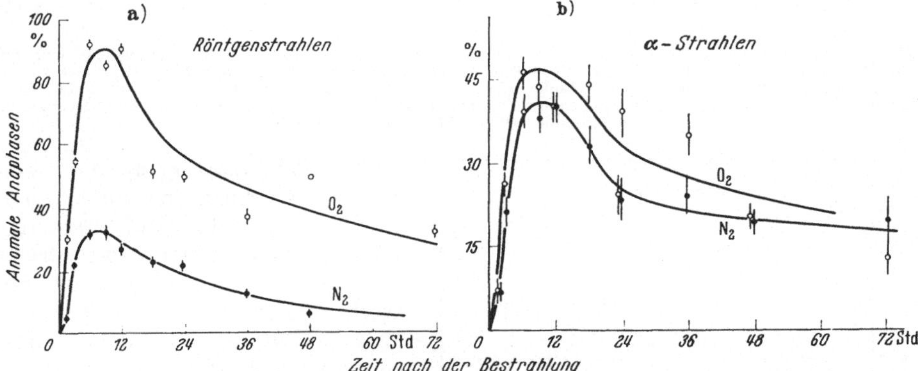

Abb. 7a u. b. Der Einfluß von Sauerstoff auf die Strahlenwirkung bei Röntgen- und α-Strahlen, gemessen an Chromosomenaberrationen in Wurzelspitzen von *Vicia faba*. a Erheblicher O_2-Einfluß bei Röntgenstrahlen. b Geringer O_2-Einfluß bei α-Strahlen. [Nach READ u. THODAY, aus GRAY: Progr. Biophysics a. Biophysical Chem. 2, 240 (1951).]

in der Lösung gebildeten Radikalen bzw. dem in der Lösung vorhandenen, die Strahlenschäden erhöhenden Sauerstoff wirksam sind. Bei der *biologischen Theorie* der Schutzwirkung ist man dagegen der Auffassung, daß durch die Schutzstoffe Oxydationen an den biologisch wichtigen Molekülen, besonders an den Enzymen mit SH-Gruppen, verhindert werden, daß die Schutzstoffe mit geschädigten Teilen von Enzymen austauschen können, und daß manche Schutzstoffe auf dem Weg über Stoffwechselabläufe wirksam sind.

Für beide Theorien lassen sich, wie sich auch schon aus den obigen Hinweisen ergibt, experimentelle Stützen anführen. Besonders sei darauf hingewiesen, daß bereits durch Zugabe der genannten Schutzstoffe zu Modellsubstanzen in vitro (vgl. S. 664) Schutzwirkungen zu erzielen sind. Bei Anaerobiose ist die Schutzwirkung oft geringer, auch ist die Schutzwirkung bei α-Strahlen und bei Neutronen *geringer* als bei Röntgen- und γ-Strahlen (FORSSBERG und NYBOM, PATT und Mitarbeiter). Wirkungen von vorbestrahltem Medium können durch Glutathiongaben ausgeschaltet werden (DANIEL und PARK). Dies alles spricht für die radiochemische Theorie. Andere Ergebnisse, wie der Schutz durch Stoffwechselprodukte, die Hemmung der Schutzwirkung durch bestimmte Stoffwechselhemmstoffe, sowie der Befund, daß nur in Enzyme einbaufähige SH-Körper einen Schutz auszuüben vermögen, sprechen für die biologische Theorie (vgl. HOLLAENDER und STAPLETON, LANGENDORFF und Mitarbeiter [1,2]).

Von Umwelteinflüssen auf die Wirkung ionisierender Strahlen sei schließlich noch die *Zeit* genannt. Wird eine bestimmte Strahlendosis nicht unmittelbar hintereinander, sondern in mehreren Abschnitten geboten (protrahierte Gabe), so treten besonders bei Röntgen- und γ-Strahlung, weniger offenbar bei Neutronen,

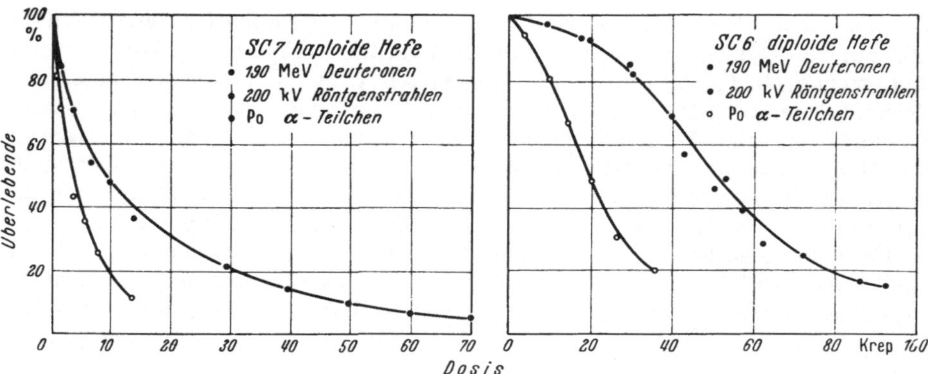

Abb. 8. Die unterschiedliche Wirksamkeit verschiedener Strahlenarten bei der Inaktivierung haploider und diploider Hefezellen. Die Bestrahlung mit α-Teilchen vergleichbarer Dosis ist wesentlich wirksamer als die mit Deuteronen und Röntgenstrahlen. Links: Haploide Hefe; die Inaktivierung folgt einer Eintrefferkurve. Rechts: Diploide Hefe; die Inaktivierung entspricht einer Mehrtrefferkurve. (Nach TOBIAS: NICKSON, Symposium on Radiobiology, S. 357. New York 1952.)

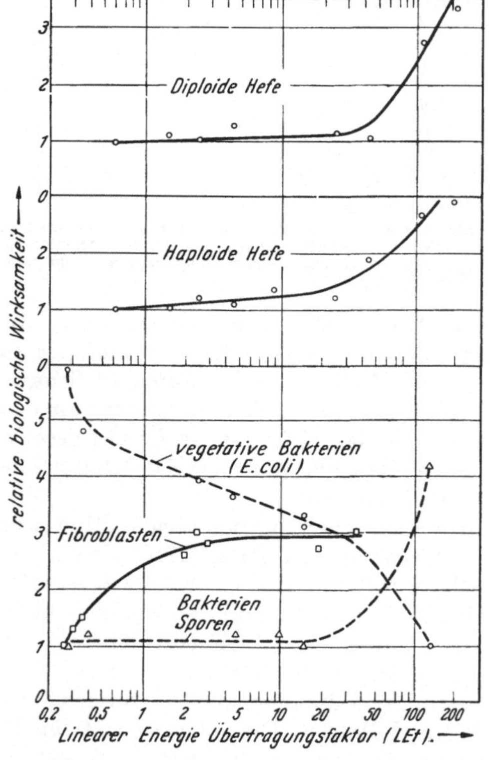

Abb. 9. Relative biologische Wirksamkeit (*RBW*) in Beziehung zum linearen Energieübertragungsfaktor (*LET*; kV/μ) bei verschiedenen Objekten. Nach E. R. ZIRKLE, in HOLLAENDER, Radiation biology, Bd. I, Teil 1, S. 315. New York 1954.)

allerlei *Erholungsvorgänge* und sonstige Unterschiede gegenüber einer einzigen Bestrahlung auf. Derartige Effekte sind auch dann zu beobachten, wenn schnelle Elektronen in rasch hintereinander erfolgenden, aber getrennten Fraktionen abgegeben werden. Besonders bei cytologischen Vorgängen sind Einflüsse einer derartigen *Fraktionierung* der Strahlendosis festgestellt worden (BORA, GÄRTNER und PETERS, GILES, SERRES und BEATTY, HAAS und Mitarbeiter [2, 3], SERRES und GILES, WOLFF und ATWOOD).

Die unterschiedliche Wirkung verschiedener Strahlensorten.

Bei der Zusammenstellung der Wirkung verschiedener Strahlensorten auf das gleiche Material ist trotz vergleichbarer Strahlendosen vielfach eine *unterschiedliche* Schädigung festgestellt worden. So erwiesen sich α-Partikel gegenüber Röntgenstrahlen und Deuteronen bei der Bestrahlung von Hefe, abgesehen von dem unterschiedlichen Schädigungsverlauf bei haploiden und diploiden Zellen, als wirksamer (Abb. 8).

Zum Vergleich der Wirkung verschiedener Strahlensorten untereinander wird neuerdings vorwiegend das Verhältnis der Wirkung der untersuchten Strahlensorte zu der der Röntgenstrahlen benutzt (RAJEWSKY [3]). Dieses Zahlenverhältnis heißt „*relative biologische Wirksamkeit*“ (RBW), entsprechend in englisch geschriebenen Arbeiten RBE (Relative Biological Effectiveness). Die RBW läßt sich zu der Ionisationsdichte, besser zu dem linearen Energieübertragungsfaktor (LET), in Beziehung setzen (vgl. S. 658). Für verschiedene Röntgen- und γ-Strahlen liegt die RBW, auch bei der Untersuchung pflanzlichen Materials, im Bereich von 0,5—1 (Dittrich und Schubert, Fritz-Niggli, Gärtner und Peters, Kirby-Smith und Daniels, Lindemann, Heidenthal und Mitarbeiter, Moos u. a.). Bei Neutronen ist dagegen die RBW wesentlich höher, sie liegt vielfach zwischen 7—13 (Baker und v. Halle, Kirby-Smith und Swanson, Rajewsky [3], Yost und Mitarbeiter).

Entsprechend zeigt sich bei einem Vergleich der RBW mit dem LET, vielfach bei steigendem LET, also steigender Ionisationsdichte, ein Anstieg der RBW. Es gibt aber, wie die Abb. 9 mit von Zirkle [2] zusammengestellten vergleichbaren Daten zeigt, auch andersartige Ergebnisse. In der Abb. 9 befinden sich links am Anfang sehr schwach ionisierende Elektronenstrahlen von mehr als 1 MeV Energie; rechts am Ende sind die besonders stark ionisierenden α-Strahlen aufgetragen; die schnellen Neutronen liegen bei 30—60 kV/μ, 200 kV-Röntgenstrahlen bei 1,5 kV/μ.

III. Strahlungswirkungen von UV und sichtbarem Licht.

A. Allgemeine Grundlagen.

a) Chromophore Gruppen und Strahlung absorbierende Substanzen.

Elektromagnetische Strahlung, deren Quanten nicht mehr zur Ionisation genügen, sondern nur noch zur elektronischen Anregung von Molekülen ($\lambda > 190$ mμ) ausreichen, kann in der Zelle nur dann wirksam werden, wenn die fragliche Strahlung in der Zelle von einem Molekül absorbiert wird (Grothus-Drapersches Gesetz). Für die Absorption sind bestimmte „chromophore Gruppen" in den Molekülen von Bedeutung, die durch Strahlen leicht anregbare Elektronen besitzen (Bowen [1], Dannenberg, Förster [1, 2], Mohler, Sinsheimer [2]). Hierfür kommen einmal die sog. *Chromophore* 1. *Ordnung* in Frage; dazu rechnen Atomgruppen mit Doppelbindungen, die π-Elektronen besitzen. Solche Gruppen sind z. B.: $>C=C<$, $>C=N-$, $>C=O$, $-N{\genfrac{}{}{0pt}{}{\nearrow O}{\searrow O}}$. Eine wesentliche Erhöhung der Absorption, verbunden mit einer Verschiebung zum Rot hin, tritt ein, wenn mehrere solcher Gruppen benachbart liegen und Wechselwirkungen zwischen ihnen möglich sind (Ringsysteme; Systeme konjugierter Doppelbindungen). *Chromophore* 2. *Ordnung*, auch auxochrome Gruppen genannt, zum anderen sind durch einsame Elektronenpaare in der äußeren Elektronenschale ausgezeichnet. Hierzu rechnen unter anderem die OH-Gruppe, in stärkerem Maße die NH_2-Gruppe, besonders $N(CH_3)_2$, ferner die SH-Gruppe, auch Halogene in organischer Bindung.

Durch Substitution geeigneter Ausgangsstoffe mit mehreren solcher „auxochromer" Gruppen ergeben sich auch die zur Photosensibilisierung (vgl. S. 687 f.) benutzten Farbstoffe. Diese besitzen besonders dann starke Absorptionsmöglichkeiten, wenn es sich um Ionen mit „Ladungsresonanz" handelt, deren Ladung also nicht an einer Stelle des Moleküls festgelegt ist (z. B. Acridinorange, Eosin oder Methylenblau). Sind nämlich durchlaufend konjugierte Doppelbindungen vorhanden, so kommt es im Anregungszustand zusätzlich zu einer Bewegung von Ladungen, und es entsteht ein Ladungsresonanzspektrum.

Aus dem Gesagten geht hervor, daß die Zelle zahlreiche in den verschiedensten Spektralbereichen absorbierende Stoffe mit chromophoren Gruppen enthält. Einzelheiten vgl. man unten, S. 672 und 683.

b) Absorptionsspektrum, Wirkungsspektrum, Quantenausbeute.

Aus der Gesamtheit der absorbierenden Substanzen der Zellen ergibt sich bei Bestrahlung ihr *Absorptionsspektrum* (z. B. für UV Abb. 16, S. 674). Werden nun die Folgen einer Bestrahlung in Abhängigkeit von der Wellenlänge untersucht, so ist die Wirkung je nach der Wellenlänge sehr verschieden. Dieses „*Wirkungsspektrum*" entspricht vielfach weitgehend dem Absorptionsspektrum *eines* Stoffes, der dann für die Absorption der in diesem Fall wirksamen Strahlen verantwortlich zu machen ist. Die Übereinstimmung zwischen Wirkungsspektrum und Absorptionsspektrum ist allerdings meistens nicht vollständig, weil für einen exakten Vergleich im untersuchten Bereich eine konstante Quantenausbeute erforderlich ist, eine nur selten zu verwirklichende Bedingung (ERRERA [3], MCLAREN, PUTNAM, UBER, ZELLE und HOLLAENDER).

Die *Quantenausbeute* (Φ) einer Reaktion bedeutet das Verhältnis der in einer Reaktion verschwindenden bzw. entstehenden Molekeln zur Anzahl der absorbierten Quanten. Der Quantenbedarf ist der reziproke Wert dieser Größe. Der Zahlenwert von Φ variiert bei den verschiedenen Reaktionen sehr erheblich und ist bei manchen Vorgängen sehr niedrig.

c) Das Schicksal der Anregungsenergie.

Durch die Einwirkung der Lichtenergie und ihrer Absorption in den chromophoren Gruppen der Moleküle (M_0) wird die Energie des bestrahlten Moleküls vom stabilen Grundzustand auf ein höheres Energieniveau gebracht (M_1^*, Gl. 1 sowie Abb. 10). Dabei kommt es vorwiegend zur Entstehung eines elektronischen, aber nur sehr kurze Zeit währenden (gegen 10^{-9} bis 10^{-7} sec) Anregungszustandes, der sofort in verschiedenster Weise auch im biologischen Material umgewandelt wird (BOWEN [2], FÖRSTER [2], PRINGSHEIM, RABINOWITCH [1, 2], ferner die Zusammenfassungen von FRANCK und PLATZMANN, LIVINGSTON [2], SCHENCK [1, 2], WEISS [3]).

1. $M_0 \xrightarrow{h\nu} M_1^*$ Absorption
2. $M_1^* \longrightarrow M_0$ intramolekulare Wärmebewegung
3. $M_1^* \longrightarrow M_2^*$; $M_2^* \longrightarrow M_3^{**}$; $M_1^* \longrightarrow M_3^{**}$ strahlungslose Übergänge
4. $M_2^* \xrightarrow{h\nu'} M_0$ Fluorescenz 5. $M_3^{**} \xrightarrow{h\nu''} M_0$ Phosphorescenz
6. $M_1^* + L \longrightarrow M_0 + L^*$ Übertragung der Anregungsenergie z. B. durch Resonanz
7. $M_0 \xrightarrow{h\nu} M_1^* \longrightarrow M_3^{**}$ Entstehung des metastabilen Anregungszustandes

7a. $M_3^{**} + O_2 \longrightarrow MO_2$ Photooxydation

7b. $M_3^{**} + L \longrightarrow M + L_1$ Photosensibilisierende Wirkung von M auf L

Häufig kommt es zu einer *inneren Umwandlung* der Anregungsenergie in Oscillationsenergie der aufbauenden Atome des Systems *(intramolekulare Wärmebewegung)* (Gl. 2, Abb. 10, 2), oder es treten *strahlungslose Übergänge* zum nächst niederen, stabileren Anregungszustand auf (Gl. 3). Derartige Zustände heißen *metastabil* (M^{**}). Ihre Lebensdauer beträgt 10^{-3} sec oder mehr. Der strahlungslose Übergang erfolgt dabei vielfach so rasch, daß eine Wiederabgabe der eingestrahlten Energie durch Strahlung hier *nicht* möglich ist. Bei bestimmter Molekülkonfiguration vermag aber das angeregte Molekül einen Teil der aufgenommenen Energie durch Strahlung unmittelbar wieder abzugeben. Diese *Fluorescenzstrahlung* (z. B. bei Riboflavin oder bei Chlorophyll) ist höchstens energiegleich, meistens aber energieärmer als die eingestrahlte Energie (STOKESsche Regel) und besitzt eine größere Wellenlänge als das eingestrahlte Licht (Gl. 4, Abb. 10,

4 $h\nu'$). Aus dem fluorescenzfähigen Zustand kann es noch zu einem weiteren Umbau in den metastabilen Zustand kommen, der damit in Konkurrenz um die Fluorescenzausbeute tritt; die Fluorescenz wird gelöscht. Die Anregungsenergie des metastabilen, sekundären Anregungszustandes kann nun wiederum durch innere Umwandlung in *Wärme* überführt werden und damit verlorengehen, oder es treten verschiedene weitere Folgeprozesse auf. So kann die Energie durch Strahlung wieder abgegeben werden; es kommt zum langsamen Energieverlust durch Phosphorescenz (Gl. 5, Abb. 10, 5 $h\nu''$) (vgl. z. B. das Auftreten eines derartigen phosphorescenzfähigen Zustandes beim Chlorophyll!).

Unter besonderen Umständen kann es zu einer *Energiewanderung*, z. B. durch Resonanzanregung anderer strahlungsfähiger Moleküle oder auch durch elektronische Weiterleitung der aufgenommenen Energie kommen; im biologischen Material z. B. zu einem elektronischen Transport auf dem Weg über Polypeptidketten (z. B. BOWEN [2], BÜCHER). In diesen Fällen dient das absorbierende Molekül also lediglich als *Absorptionsort* der Strahlung. Das „Sensibilisator"-Molekül erfährt hier *keine* (vorübergehende) chemische Veränderung (Gl. 6).

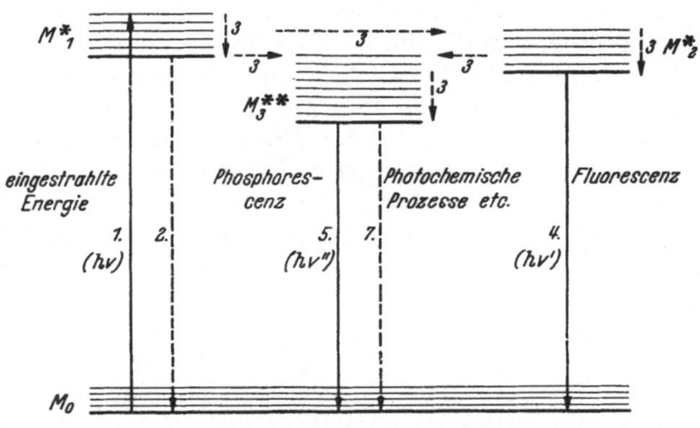

Abb. 10. Schema der Energieübertragung (vgl. Text sowie Gleichung 1—7).

Geht die Anregungsenergie nicht auf einem der genannten Wege verloren, so kann sie für photochemische Reaktionen verwendet werden; wegen seiner Langlebigkeit, etwa das 10^6—10^8fache der Anregungszeit, ist der metastabile Zustand (M^{**}) besonders für derartige chemische Reaktionen geeignet. Atomphysikalisch handelt es sich dabei wahrscheinlich um einen sog. Triplettzustand (FÖRSTER [2], FRANCK, LEWIS und KASHA, LIVINGSTON [2]).

Chemisch handelt es sich vermutlich um den durch Umlagerung von Elektronen hervorgerufenen Übergang des angeregten Moleküls in ein Radikal, bzw. nach G. O. SCHENCK [1, 2] in ein *Diradikal*. Als Beispiel sei der wahrscheinliche Übergang des vielfach bei photobiologischen Versuchen benutzten Eosin in ein derart phototrop entstandenes isomeres Diradikal dargestellt:

HO, Br, O, Br, O; Br, Br; C; COONa $\xrightarrow{h\nu}$ HO, Br, O, Br, $\overline{\underline{O}}$•; Br; •C; COONa

Die Punkte (•) bedeuten die Radikalstellen. Die Radikale sind nun zu weiteren Folgereaktionen bereit, die zu photochemischen Umwandlungen führen. Sie können im Verlauf der Reaktion entweder selbst *irreversibel* verändert werden (z. B. durch Autoxydation) oder aber durch Wechselwirkung mit anderen Molekülen in einem Kreisprozeß, und damit *reversibel*, wieder in den nichtangeregten

Grundzustand übergehen. Ein zu solcher reversiblen Radikalbildung befähigter Stoff heißt ebenfalls (vgl. oben) *Photosensibilisator*, weil durch seine Gegenwart andere, sonst gegen Belichtung inaktive Substanzen zu chemischen Reaktionen veranlaßt werden können. Hierzu gehören z. B. viele, vorwiegend fluorescierende Farbstoffe wie das genannte Eosin. Beide Vorgänge sind für die Lebensvorgänge von grundsätzlicher Bedeutung. Der *irreversible* Verlauf spielt z. B. bei Schädigungen durch UV-Bestrahlung eine wichtige Rolle; der *reversible* Verlauf, also die Beteiligung von Photosensibilisierungsvorgängen, ist für die in vivo normalerweise durch Licht in Gang gesetzten Prozesse wichtig (z. B. die Photosynthese).

B. Bestrahlung mit kurzwelligem ultraviolettem Licht (UV < 310 mμ).

a) Allgemeines.

In dem hier zu besprechenden Wellenlängenbereich ist die Energie der Quanten noch so groß, daß die die Lebensvorgänge zerstörende Wirkung der Strahlen bei weitem überwiegt. Da der Anteil dieser Strahlen im Sonnenlicht in der untersten Schicht der Atmosphäre aber nur noch sehr gering ist oder überhaupt fehlt, ist in der *natürlichen Umwelt* der Pflanzen auf der Erdoberfläche, abgesehen vom höheren Gebirge, durch diesen Strahlenbereich nur eine relativ geringe Wirkung zu erwarten (PIRSCHLE, WERFFT). Hinzu kommt, daß besonders die höheren Pflanzen durch ihre Epidermis, durch kutinisierte Wände und durch Wachsschichten vor diesem Strahlenbereich geschützt sind (WUHRMANN-MEYER). Bereits eine verdickte Zellmembran hält einen wesentlichen Anteil der UV-Strahlung zurück (BIEBL [1]). Auch Gerbstoffe, Anthozyane, Flavone und Flavonole, die in diesem Spektralbereich stark (Abb. 14, S. 674) absorbieren, sind oft im Zellsaft der Epidermiszellen zu finden und erhöhen so den Strahlenschutz (vgl. BÜNNING [2], S. 386).

Nichtsdestoweniger erfordert dieser Spektralbereich über die natürlich vorhandene UV-Strahlenwirkung hinaus eine genaue Beachtung, weil durch *künstliche UV-Bestrahlung* die verschiedensten Wirkungen auf Zellen und Gewebe erzielt worden sind.

b) Absorptions- und Wirkungsspektrum.

1. Absorption im UV unter 300 mμ.

Zahlreiche Substanzen in der Zelle besitzen in diesem Spektralbereich Absorptionsbanden. Hierzu gehören vor allem die Aminosäuren und Proteine, die Nucleinsäuren und Nucleotide, ferner Sterine, Flavine, verschiedene Coenzyme und außerdem zahlreiche einfacher gebaute Substanzen (vgl. BEAVEN und HOLIDAY, CASPERSSON, DANNENBERG, DAVIES und WALKER, DOTY und GEIDUSCHEK, MEYER-SEITZ, SINSHEIMER [2]).

Das UV-Absorptionsspektrum der Proteine und Aminosäuren.

Für die Absorption des ultravioletten Lichtes sind in den Proteinen vorwiegend die in der Polypeptidkette auftretenden, vor allem unter 220 mμ absorbierenden —Co—NH-Gruppen und ferner die aromatischen *Aminosäuren* in den Seitengruppen der Polypeptidkette der Eiweiße verantwortlich zu machen. Zu den letztgenannten gehören Tyrosin, Tryptophan, Phenylalanin und Histidin. Von diesen Aminosäuren absorbieren Histidin, außerdem noch Cystin, im Bereich von 220—250 mμ und überlappen sich daher teilweise mit der Absorption der CONH-Gruppen. Tryptophan, Tyrosin und Phenylalanin haben dagegen ihr Absorptionsmaximum im Bereich von 280, 275 und 258 mμ. Aus den Verhältnissen der Absorption dieser drei Aminosäuren 27:7:1 ergibt sich die Bedeutung des Tryptophan für die Proteinabsorption (Abb. 11a).

Aus der Summierung der Absorption der genannten Systeme entsteht wegen ihrer additiven Wirkung das Absorptionsspektrum des *Eiweiß*. Es beginnt etwa bei 310 mμ, hat ein erstes Maximum bei etwa 280 mμ (Tryptophan), um nach einem Minimum (etwa 240 mμ) erneut zum kurzen Wellenbereich hin anzusteigen (Globuline und Albumine, Abb. 11b). Bei globulärem Eiweiß vom Histontyp liegt das Maximum bereits bei 290 mμ, eine durch die Gruppierung von Diaminosäuren und Tyrosin erklärbare Verschiebung der Absorption.

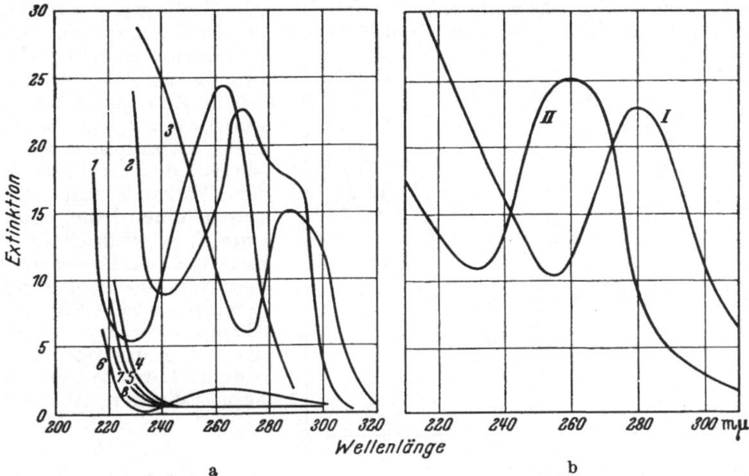

Abb. 11a u. b. Absorptionsspektrum von Aminosäuren, Proteinen und Nucleotiden. a Die Absorptionsspektra von *1* Adenin; *2* Tryptophan; *3* Tyrosin; *4* Histidin; *5* Arginin; *6* Phenylalanin; *7* Leucylglycin; *8* Prolin. b *I* Serumglobulin; *II* Polynucleotid. (Nach CASPERSSON: Cell growth and cell function. New York 1950.)

Die UV-Absorption der Nucleinsäuren, Purin- und Pyrimidinverbindungen.

Von besonderer Bedeutung für die UV-Absorption in der Zelle erweisen sich die Purin- und Pyrimidinringsysteme, deren Absorptionsmaximum zwischen 250 und 280 mμ, vor allem aber bei 260 mμ liegt. Die einzelnen Systeme unterscheiden sich durch wenig auseinanderliegende, vom p_H abhängige Absorptionsmaxima (HOTCHKISS, SINSHEIMER). In Abb. 11a ist der Absorptionsverlauf für Adenin und in Abb. 11b für ein Polynucleotid dargestellt. Das Absorptionsspektrum von Desoxy-

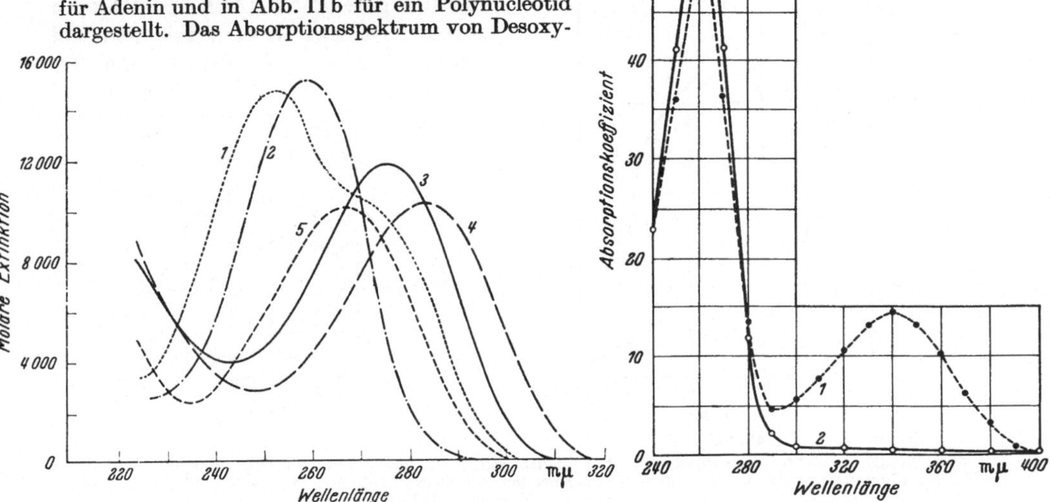

Abb. 12. Absorptionsspektren von Desoxyribonucleotiden bei p_H 4,3. *1* Desoxyguanylsäure; *2* Desoxyadenylsäure; *3* Desoxycytidylsäure; *4* Desoxy-5-methylcytidylsäure; *5* Thymidylsäure. (Nach SINSHEIMER, aus HOLLAENDER, Radiation biology, Bd. II, S. 165. 1955.)

Abb. 13. Absorption der Codehydrase (Diphosphopyridinnucleotid). p_H 7,4. *1* hydriert; *2* dehydriert. (Nach WARBURG: Wasserstoffübertragende Fermente. Freiburg 1949.)

ribonucleotiden ist aus Abb. 12 zu ersehen; wegen der Teilnahme von Purin- und Pyrimidinsystemen an der prosthetischen Gruppe von Fermenten ist für derartige Coenzyme im

UV-Bereich ebenfalls eine typische Absorption zu erwarten. Als Beispiel sei auf das Absorptionsspektrum von Diphosphopyridinnucleotid (Abb. 13) verwiesen.

Sonstige Substanzen.

Schließlich muß eine Reihe weiterer im UV absorbierender Stoffe genannt werden (BRODE, SINSHEIMER [2]), wie die *Sterine* (z. B. Ergosterin) und die *Flavine* (darunter vor allem das Riboflavin (Abb. 21, S. 684). Von Interesse sind auch noch bei pflanzlichem Material vorkommende Vitamine, wie die *Ascorbinsäure* mit einem Absorptionsmaximum bei 265 mμ, das zur Verwechslung mit dem Vorhandensein von Nucleinsäuren führen kann, sowie das *Vitamin K*, ein vielleicht in Zusammenhang mit der Photosynthese wichtiges Naphthochinonderivat, das im Bereich von 240—270 mμ sowie bei 320 mμ absorbiert. Schließlich besitzen auch die *Anthocyane, Flavone* und ihre Derivate deutliche VU-Absorptionsbanden (Abb. 14).

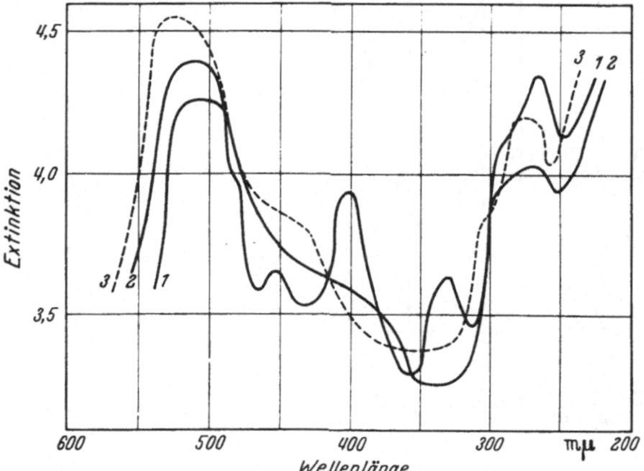

Abb. 14. Absorptionsspektren von Anthocyanen in saurer Lösung. *1* Pellargonidin; *2* Cyanidin; *3* Delphinidin. (Nach SCHOU, aus SINSHEIMER, in HOLLAENDER, Radiation Biology, Bd. II, S. 165, 1955.)

UV-Absorption in der lebenden Zelle.

In der lebenden Zelle überlagern sich diese verschiedenen Absorptionsspektren. Da aber die Mengen der Proteine und der Purin- bzw. Pyrimidinsysteme (Nucleinsäuren) andere absorbierende Systeme in der Zelle weit übertreffen, bestimmen deren Spektren den Absorptionsverlauf von Zellen im UV, wie CASPERSSON an fixiertem und lebendem Material ausführlich gezeigt hat (Abb. 15). Der Absorptionsverlauf in verschiedenen Teilen der Zelle ist recht verschieden. Die

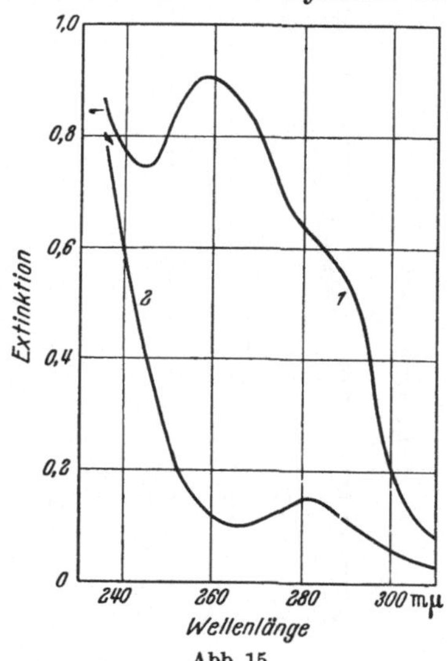

Abb. 15.

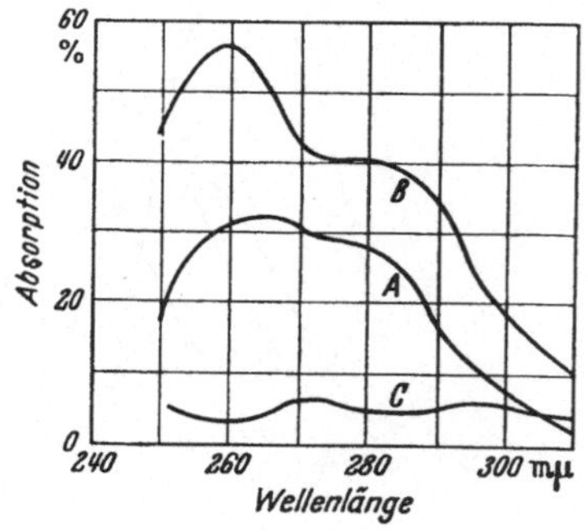

Abb. 16.

Abb. 15. Die UV-Absorption in lebenden Pollenmutterzellen von *Tradescantia*. *1* Die Absorption von Stellen in den Chromosomen; *2* die Absorption im umgebenden Cytoplasma. (Nach CASPERSSON: Cell growth and cell function. New York 1950.)

Abb. 16. Die UV-Absorption in lebenden Epidermiszellen von *Allium cepa*. *A* Die Absorption des Cytoplasmas; *B* die Absorption des Zellkerns (einschließlich der umhüllenden Cytoplasmaschicht); *C* die Absorption der Zellmembran allein in der Umgebung der für die Messung von *A* und *B* benutzten Zellen. (Nach von GLUBRECHT freundlicherweise zur Verfügung gestellten, noch unveröffentlichten Messungen.)

Untersuchung wurde durch die Entwicklung der Ultraviolettmikroskopie, besonders aber durch das Spiegelmikroskop am lebenden Objekt vereinfacht (BLOUT,

MELLORS und Mitarbeiter). An pflanzlichen Objekten hat sich kürzlich GLUBRECHT (sowie persönliche Mitteilung) mit der UV-Absorption von Zellteilen befaßt. Abb. 16 zeigt derartige Messungen an lebenden Epidermiszellen der Zwiebelschuppen von *Allium cepa*. Die Kombination der Eiweiß- (Aminosäuren) und Nucleinsäuren-Absorption ist auch hier deutlich.

2. Wirkungsspektrum im UV-Bereich unter 300 mμ.

Im Bereich des UV gibt es mehrere Typen von Wirkungsspektren (vgl. S. 670) für verschiedene untersuchte Strahlenfolgen. Der erste Typ zeigt ein der *Protein*-absorption entsprechendes Wirkungsspektrum, so z. B. bei der Beeinflussung der Geißelbeweglichkeit (GIESE, SCHREIBER); auch bestimmte Mutationen folgen nach R. W. KAPLAN [3] diesem Typ.

Bei dem zweiten Typ gleicht das Wirkungsspektrum dem Absorptionsspektrum der *Nucleinsäuren*. Dies gilt z. B. für die Bacteriophagen-Inaktivierung (ZELLE und HOLLAENDER, Abb. 17), die Inaktivierung von Viren, mancher Bakterien und Pilze (vgl. GIESE [1]) und ebenfalls für das Auftreten mancher Mutationen (vgl. ERRERA [3]).

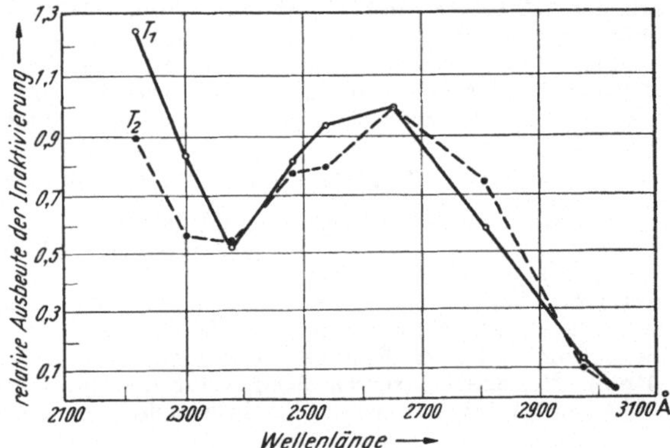

Abb. 17. UV-Wirkungsspektrum für die Inaktivierung von T_1 und T_2 Bakteriophagen nach UV-Bestrahlung. [Nach ZELLE u. HOLLAENDER: J. Bacter. 68, 210 (1954).]

Gelegentlich kommen Wirkungsspektren vor, deren Verlauf auf eine *kombinierte NS*- und *Protein*-Absorption deutet: Das Wirkungsspektrum von bestimmten Chromosomenbrüchen stimmt z. B. mit einem Nucleoproteinspektrum überein (GRAY [3]).

Besonders zu betonen ist, daß verschiedene durch UV beeinflußbare Zelleigenschaften bei dem *gleichen* Organismus *verschiedene* Wirkungsspektren besitzen (z. B. GLUBRECHT, JAKOB).

Hieraus geht hervor, daß für die UV-Wirkung in der lebenden Zelle *verschiedene* Orte der UV-Absorption wichtig sind. Ob dieser primäre Absorptionsort nun aber mit dem späteren Angriffspunkt der absorbierten Energie übereinstimmt, ist durchaus nicht gesagt (vgl. die folgenden Abschnitte).

c) Die Wirkungen der UV-Bestrahlung.

1. Abbau von Modellsubstanzen. Photochemische Primärwirkungen.

Zur Feststellung der nach UV-Absorption in den betreffenden Molekülen ablaufenden Reaktionen sind ebenso wie bei der ionisierenden Strahlung verschiedene Modellsubstanzen geprüft worden. Besonders zahlreich wurden Aminosäuren, Peptide, Eiweiße und Enzyme, aber ebenso auch Nucleinsäuren, Purine und Pyrimidine untersucht.

Aminosäuren, Eiweiße und Enzyme.

Auf Grund vieler Versuche über die nach UV-Bestrahlung ablaufenden Reaktionen ergab sich, daß verschiedene Veränderungen am Proteinmolekül eintreten (vgl. z. B. MCLAREN, sowie DOTY und GEIDUSCHEK). Vielfach scheinen *Oxydationsprozesse* abzulaufen. Hierfür spricht die O_2-Abhängigkeit der Abbauvorgänge und die beobachtete Erhöhung der UV-Absorption nach Bestrahlung (GIESE [1], MCLEAN und GIESE, SHUGAR [3]). Oft findet eine Oxydation der aromatischen Ringe der beteiligten Aminosäuren statt, oder es kommt zu einer

Oxydation von SH-Gruppen; auch eine Auftrennung von Wasserstoffbrücken wurde beobachtet. Vor allem aber tritt eine *Zerstörung von Peptidbindungen* auf, wie sich aus dem Anwachsen freier Aminogruppen und dem Auftreten von abgespaltenen Aminosäuren folgern läßt.

Nicht völlig geklärt ist die Frage, in welcher Weise die Spaltung der Peptidbindung vor sich geht. Entweder wird die Peptidbindung durch Absorption an dieser Stelle *selbst* zerstört, oder aber es muß wegen der vorwiegenden UV-Absorption in einem chromophoren Ring der aromatischen Aminosäuren eine *Energieübertragung* zum Ort der Zerstörung der Peptidbindung vorhanden sein. Nun *wächst* aber nach Untersuchungen von MANDL und Mitarbeitern sowie McLAREN und Mitarbeitern die Quantenausbeute mit der Zahl der C-Atome zwischen der zerstörten Peptidbindung und dem aromatischen Ring. Ein Ergebnis, das nicht ganz leicht mit der Theorie der Energieübertragung in Übereinstimmung zu bringen ist. Unter Umständen ist an eine Faltung der Peptidkette gegen den Benzolring hin und damit an einen Kontakt mit einer Peptidbindung zu denken.

Neuere Beobachtungen wie die, daß bei dem Abbau von Enzymen (s. MANDL und Mitarbeiter, McLAREN und Mitarbeiter, SHUGAR [3]) die Quantenausbeute mit steigendem Molekulargewicht, also mit steigender Größe fällt, sowie weitere Untersuchungen über den Protein- und Enzymabbau durch Untersuchung der Spaltprodukte mit Hilfe der Papierchromatographie und schließlich eine Berücksichtigung der auch hier möglichen indirekten Strahlenwirkungen dürften in naher Zukunft eine weitere Aufklärung der noch offenen Fragen bringen.

Nucleinsäuren. Purin- und Pyrimidinabbau.

Bei der Untersuchung des durch UV bedingten Abbaus von Nucleinsäuren ergab sich, daß die Purin- und Pyrimidinbasen gegenüber der UV-Bestrahlung am empfindlichsten sind (CANZANELLI und Mitarbeiter, ERRERA [2]). Hierfür spricht auch die Abnahme der UV-Absorption bei 260 mμ nach Bestrahlung, die auf eine Ringöffnung deutet (CONRAD, RAPPORT und CANZANELLI, SINSHEIMER [1]). Eine Depolymerisierung der NS scheint nur eine geringe Bedeutung zu haben (ERRERA). Gelegentlich wird aber auch der Pentoseanteil angegriffen (RICE). Bei einer stärkeren Bestrahlung der Purin- und Pyrimidinbasen treten zahlreiche Abbausubstanzen auf, wie z. B. Untersuchungen am Uracil gezeigt haben (CONRAD).

2. Direkte UV-Strahlenwirkungen.

Zur Untersuchung der direkten Strahlenwirkungen müssen auch im UV die Möglichkeiten eines Energietransportes verringert und vor allem die Diffusionsvorgänge ausgeschaltet werden. Die Bestrahlung ist also entsprechend den vergleichbaren Versuchen mit ionisierender Strahlung (S. 662) in trockenem oder gefrorenem Zustand vorzunehmen. Zur Untersuchung werden wiederum Modellsubstanzen, z. B. getrocknete Enzyme wie Katalase oder Trypsin, herangezogen, deren Inaktivierung nach der UV-Bestrahlung gemessen wird. Hierbei ergeben sich häufig exponentiell verlaufende Dosiseffektkurven. Die Steilheit dieser Kurven erweist sich auch bei Trockenbestrahlung als temperaturabhängig (SETLOW und DOYLE [2—4]). Die Inaktivierungsrate steigt im allgemeinen bei gleicher Dosis mit steigender Temperatur. Im Bereich der Zimmertemperatur ist die Temperaturabhängigkeit allerdings gering und entging so früher der Beobachtung. Zur Erklärung wird angenommen, daß bei der niedrigen Temperatur von 290° K ein großer Teil der in den Enzymen absorbierten Energie in metastabilen Zuständen der Proteinmoleküle festgelegt ist, die als langlebige Phosphorescenzstrahlung abgegeben wird (DEBYE und EDWARDS). Sie kann daher zur Inaktivierung nicht benutzt werden. Bei niedrigen Temperaturen ist außerdem möglicherweise die Rekombination zerstörter Bindungen erhöht, auch könnte der Absorptionskoeffizient der UV-Strahlung temperaturabhängig sein.

Neue Untersuchungen von HILL und ROSSI [2, 3] über die Abhängigkeit der Inaktivierung von Phagen bei abnehmendem Wassergehalt sowie in trockenem oder gefrorenem Zustand zeigen überdies, daß die Verhältnisse selbst der direkten Strahlenwirkungen noch komplexer liegen als bisher vermutet wurde: Das umgebende Medium ändert nämlich durch seine unterschiedliche Hydratation die Strahlenempfindlichkeit der Phagen, obgleich hier direkte Treffer vorzuliegen scheinen.

3. Indirekte UV-Strahlenwirkungen.

Indirekte UV-Wirkungen werden sich besonders dann nachweisen lassen, wenn das Medium, in das die zu untersuchenden Objekte übertragen werden, vor der Übertragung bestrahlt wird.

HAAS und Mitarbeiter [1], STONE und Mitarbeiter und WYSS und Mitarbeiter konnten bei Streptokokken auf diese Weise eine deutliche, spontane Erhöhung der Mutationsrate feststellen, die durch Katalasegaben eliminiert werden konnte. Es wurde deshalb angenommen, daß durch in der Nährlösung vorhandene Proteine oder Nucleoproteide Reaktionen sensibilisiert werden, bei denen Peroxyde entstehen. WAGNER und Mitarbeiter zeigten dann allerdings, daß das Wirkungsspektrum für diese indirekten Wirkungen bei Wellenlängen < 200 mμ wesentlich ansteigt, also im Wasser gebildete Radikale beteiligt sein können. HARM und STEIN [1, 2] und STEIN und HARM [1—3] konnten aber solche indirekten UV-Wirkungen bei der Inaktivierung von *E. coli* durch vorbestrahlte Bouillonnährböden bestätigen, wobei die Art des Nährbodens, aber auch das Milieu von erheblichem Einfluß auf die Strahlenwirkung war. LATARJET und MILÉTIC fanden, daß selbst noch längere Wellen (> 340 mμ) in Bouillon durch Katalase hemmbare Substanzen erzeugen, die auf T_2-Bacteriophagen toxisch wirken. In reinem Wasser ist bei der Benutzung von UV-Wellenlängen über 200 mμ bei T_1 Phagen aber nicht mit indirekten Effekten zu rechnen (HILL und ROSSI [2, 3]).

Aus den genannten Arbeiten geht also hervor, daß bei UV-Bestrahlung mit Wellenlängen *unter* 200 mμ auch im *rein* wäßrigen Medium infolge der hier zunehmenden Radikalbildung indirekte Strahlenwirkungen auftreten können. Bei *längeren* Wellenlängen sind dagegen indirekte Wirkungen nur dann möglich, wenn im bestrahlten Medium in dem fraglichen Wellenbereich *absorbierende* Substanzen (Aminosäuren, Proteine, NS usw.) vorhanden sind, die durch UV verändert werden (Bildung von Radikalen oder von organischen Peroxyden), und die dann auf andere Substanzen bzw. auf die biologischen Objekte selbst einzuwirken vermögen.

Wie groß der Anteil der hier besprochenen indirekten UV-Strahlenwirkung an der Gesamtwirkung des UV in den Zellen selbst ist, läßt sich auf Grund der bisherigen Untersuchungen noch nicht sicher abschätzen (GIESE [1]). Ihr Anteil wird teilweise als groß (STEIN und HARM [3]), teilweise aber auch noch als relativ gering erachtet (R. W. KAPLAN [3]).

4. Beeinflussung der UV-Strahlenwirkung.

Die Entfernung von *Sauerstoff* während der UV-Bestrahlung verringert die Wirkung auf die Zelle wegen der Verminderung der Oxydationsprozesse (MEFFERD und CAMPBELL, MEFFERD und MATNEY). Auch bestimmte, unmittelbar vor der Bestrahlung gebotene Chemikalien sind in der Lage, die UV-Wirkung abzuschwächen; sie wirken als *Schutzstoffe* gegenüber UV-Strahlenschäden (vgl. S. 666f.). An erster Stelle stehen SH-Gruppen tragende Verbindungen, wie Cystein und Glutathion (CALCUTT, GIESE und Mitarbeiter, SHUGAR [1], WELS, WHITEHEAD). Durch Zugabe von Stoffwechselzwischenprodukten, wie z. B. Brenztraubensäure, kann ebenfalls eine erhebliche Verringerung der Strahlenwirkung eintreten (HEINMETS, HEINMETS und Mitarbeiter [1, 2]). Auch *Enzymhemmstoffe* wie Natriumazid verringern die Strahlenwirkung (BERGER und Mitarbeiter). Endlich waren *Katalasegaben* in der Lage, eine wesentliche Verringerung der durch UV hervorgerufenen Einflüsse auf die Zelle herbeizuführen (LATARJET und MILÉTIC, MILÉTIC und MORENNE, PRATT und Mitarbeiter, WYSS und Mitarbeiter).

Die Beeinflussung der UV-Wirkung durch die genannten Substanzen und Enzyme kommt sehr wahrscheinlich auf recht verschiedene Weise zustande.

Im einfachsten Fall kann es sich um eine Verringerung der Wirkung durch *konkurrierende Absorption* handeln: Zugegebene Substanzen absorbieren einen Teil der sonst auf die zelleigenen Stoffe wirkenden UV-Strahlen. Die genannten Effekte sind aber nicht allein auf diese Weise zu erklären (vgl. GIESE und Mitarbeiter [3]). Es wird vielmehr angenommen, daß durch die obengenannten Sulfhydrylgruppen enthaltenden Substanzen die oxydierten SH-Enzyme oder andere Proteine mit SH-Gruppen wieder reduziert werden. Es handelt sich dann also um eine *Umkehr* von *Oxydationsprozessen* (CALCUTT, WELS). Die Wirkung der Katalase könnte darin bestehen (BACQ, PRATT und Mitarbeiter, sowie die oben angegebenen Arbeiten), daß sie die durch indirekte UV-Strahlenwirkung in der Nährlösung oder in der Zelle gebildeten *organischen Peroxyde* wieder abbaut. Und schließlich könnten durch Zugabe bestimmter Substanzen *Stoffwechselvorgänge* — vgl. die Verringerung der Strahlenwirkung durch Stoffwechselzwischenprodukte und Enzymhemmstoffe — so geändert werden, daß sie eine Verringerung der Strahlenwirkung zur Folge haben.

5. Cytologische und stoffwechselphysiologische Wirkungen durch UV.

Cytologische Veränderungen.

Durch UV-Bestrahlung treten vielfältige cytologische Veränderungen auf (ERRERA [3]). An erster Stelle betreffen diese den *Zellkern* und auch den Nucleolus, was schon deswegen verständlich ist, weil das Wirkungsspektrum der UV-Bestrahlung vor allem der NS- und Nucleoproteinabsorption folgt (vgl. oben S. 675). Eine bestimmte UV-Dosis zeigt z. B. bei kernlosen Seeigeleiern noch keine Beeinflussung der untersuchten Teilungsprozesse, während kernhaltige durch die gleiche Dosis schon erheblich abzuändern sind (BLUM und Mitarbeiter [3]). Die Cytoplasmaströmung von *Allium*-Epidermiszellen wird bei Bestrahlung des Kerns (und des umgebenden Cytoplasmaanteils) viel stärker retardiert als bei Cytoplasmabestrahlung allein (GLUBRECHT). Die Empfindlichkeit hängt dabei von der *Polyploidie* des Kerns ab. Zellen mit haploidem Kern sind empfindlicher als die von Diploiden oder gar von Polyploiden (CALDAS und CONSTANTIN, SARACHEK und LUCKE).

Die *Teilungsphase*, in der sich die Zellen befinden, ist für die UV-Wirkung ebenfalls von Bedeutung (BLUM und Mitarbeiter [2]), Bakterien scheinen kurz nach der Teilung gegen UV besonders empfindlich zu sein (HOUTERMANS [2]). Die an *Chromosomen* eintretenden Veränderungen sind besonders viel untersucht worden. Beobachtet wurden vorwiegend Deletionen am Ende der Chromosomen (Chromatid-Deletionen), sehr viel weniger oder gar keine Translokationen (*Tradescantia*-Pollen, Tomate, *Anthirrhinum*). Auch die Frage des Auftretens von Chromosomenbrüchen durch UV und ihre Heilung ist Gegenstand zahlreicher Arbeiten. Hier muß auf die Literatur verwiesen werden (Zusammenfassung: SWANSON und STADLER). Im *Cytoplasma* zeigt sich bei Bestrahlung mit geeigneten UV-Wellenlängen u. a. ein Aufhören der sehr empfindlich reagierenden Protoplasmaströmung, eine erhöhte Anfärbbarkeit bestrahlter Zellen, eine Erhöhung der Permeabilität, ein Aufhören der Plasolysierbarkeit und der BROWNschen Molekularbewegung (GLUBRECHT). In den zuletzt genannten Fällen tritt die Wirkung meistens erst nach Verlauf einer *Latenzperiode* sichtbar in Erscheinung. Die oben genannte Beeinflussung der Plasmaströmung und auch der Teilungsprozesse durch das Cytoplasma allein zeigt, daß die Strahlenwirkung bestimmt nicht nur durch eine Vermittlung des Kerns erzielt wird, sondern daß an der Primärwirkung auch das Cytoplasma beteiligt ist.

Zellphysiologische Veränderungen.

Durch bestimmte UV-Dosen werden ganz bestimmte, zum Teil sehr spezifische Vorgänge beeinflußt; es kommt keineswegs zu einer allgemeinen, destruktiven Wirkung auf die zellphysiologischen Eigenschaften der Zelle. Die *Atmung* wird, zunächst jedenfalls, bei *E. coli nicht* gehemmt (KELNER [4]). Auch bei Rikettsien ging die Atmung nur wenig zurück, während gleichzeitig die Infektivität gegenüber Hühnerembryonen bereits auf 0,001 % herabgesetzt war (E. G. ALLEN und Mitarbeiter). Bei *Chlorella* wurde bei einer bestimmten UV-Dosis die Atmung noch nicht geschädigt, obwohl die Photosynthese bereits abgenommen hatte (HOLT und Mitarbeiter, REDFORD und MEYERS). Auch das *Zellwachstum* wird nach KELNER zunächst *nicht* gehemmt. Ebenso bleibt die Aktivität einer Reihe von Enzymsystemen auch nach stärkerer Bestrahlung fast völlig erhalten. So wurden bei *Chlorella* Polyphenoloxydasen, Cytochromoxydase, Katalase (HOLT und Mitarbeiter), aber auch Alkoholdehydrase und Milchsäuredehydrase, ebenso wie Katalase bei der Hefe nicht geschädigt (ALDOUS und STEWART).

Dagegen werden andere Enzyme, wahrscheinlich eher aber die durch sie hervorgerufenen *Synthesen*, stark und auch sehr bald nach der Bestrahlung angegriffen: Eine Galaktosidasehemmung *(E. coli)* trat ein (JAKOB und Mitarbeiter). Die *Adaptation* an Galaktose wurde nach S. KAPLAN und ebenso nach NORMAN [2] bei der Hefe sehr stark gehemmt. Bei der Hefe wurde weiter die Myokinase, Carboxylase und Zymase inaktiviert (ALDOUS und STEWART). Der Glucoseabbau nahm entsprechend ab (REDFORD und MYERS). Auch die *Photosynthese* wurde bei *Scenedesmus* und *Chlorella* stark inaktiviert (HOLT und Mitarbeiter, REDFORD und MYERS).

Demnach werden Atmung und Atmungsenzyme wenig betroffen, der anfängliche Glucoseabbau, die Photosynthese und die Adaptation an bestimmte Substrate dagegen sehr stark gehemmt.

Besonders hat sich gezeigt, daß die *Synthese* von *Desoxyribosenucleinsäure* nach UV-Bestrahlung sehr rasch gehemmt wird, während die Ribose-Nucleinsäure-Synthese kaum beeinflußt wird (COURCY und Mitarbeiter, KANAZIR und ERRERA, KELNER [4], SIMINOWITCH und RAPTINE). Da ATP nicht angegriffen wird, auch Purin- und Pyrimidinbasen nach Bestrahlung angehäuft werden, ist mit KANAZIER und ERRERA zu vermuten, daß die UV-Strahlung erst auf späte Abschnitte der DNS-Synthese einwirkt. KELNER [4] nimmt an, daß die schnelle und intensive Beeinflussung der DNS-Synthese ein sehr wichtiger Ansatzpunkt für die bereits durch niedrige UV-Dosen zu erzielende Inaktivierung („Abtötung“) von Zellen darstellt (vgl. auch Photoreaktivierung S. 680).

d) Erholungsvorgänge nach UV-Bestrahlung.

1. Erholungsvorgänge durch Chemikalien und Temperatur.

Chemikalien.

Hier ist besonders auf die erholende Wirkung durch während und nach der Bestrahlung gegebene Stoffwechselprodukte hinzuweisen. HEINMETS nimmt an, daß durch bestimmte Stoffwechselzwischenprodukte, z. B. Brenztraubensäure bei *E. coli*, eine Erholung von photochemisch leicht angreifbaren Zwischengliedern in der komplexen Kette von Folgereaktionen nach UV-Bestrahlung erfolgen kann. In ähnlicher Weise sind Oxalessigsäure und andere Glieder des Citronensäurecyclus wirksam (HEINMETS und Mitarbeiter [1, 2]). Auch bei NaF-Gaben, durch die Wachstum und Vermehrung von *E. coli* sistiert wurden, konnte nach Zugabe neuer Kulturlösung eine Erholung beobachtet werden. Da hier die zugesetzten Substanzen im Gegensatz zu den oben S. 677 besprochenen „Schutzstoffen" auch nach Bestrahlung wirksam sind, können diese Erholungsvorgänge nicht mit jenen unmittelbar während der Bestrahlung ablaufenden Prozessen verglichen werden.

Temperatur.

Durch die Erhöhung der Temperatur nach UV-Bestrahlung können ebenfalls Erholungsprozesse in Gang gesetzt werden. Bereits bestrahlte Enzyme weisen eine solche „Erholung" auf (RAPKINE, SHUGAR [1], SIMINOWITCH). Auch Bakteriophagen (BRESCH) und Bakterien sind durch Temperaturerhöhung nach sonst inaktivierender UV-Bestrahlung reaktivierbar (ANDERSON, HARM und STEIN [1], HOLLAENDER [1], STEIN und MEUTZNER). Die Erholungsfähigkeit ist von den benutzten Bakterienstämmen und vom gegebenen Nährmedium abhängig. Schließlich zeigte sich, daß auch bei vor der Beimpfung bestrahlten Kulturen, bei denen also indirekte Strahlenwirkungen vorliegen, eine Erholung möglich ist (STEIN und HARM [2]). Welche besonderen Vorgänge durch diese Temperatureinflüsse verändert werden, ist noch nicht bekannt.

2. Erholungsvorgänge nach UV-Bestrahlung durch langwelliges UV und Licht.

Allgemeine Beobachtungen.

DULBECCO und KELNER stellten gleichzeitig erneut fest (zuerst GIESE 1938), daß eine „*Belichtung*" mit langwelligem UV und auch mit blauem Licht Zellen und Organismen nach tödlich wirkender UV-„*Bestrahlung*" teilweise zu reaktivieren vermag (Zusammenfassung DULBECCO [4]). Derartige Erholungsvorgänge wurden seither bei vielen biologischen Objekten festgestellt. Genannt seien Phagen (DULBECCO [1—3]), Bakterien (KELNER [2—4], (Abb. 18), NOVIK und SZILARD), Protozoen (GIESE und Mitarbeiter [1—3], KIMBALL und GAITHER [1], Pilze (KELNER [1—3], NORMAN [1], SWENSON und GIESE) und Blätter höherer Pflanzen (BAWDEN und KLECZKOWSKI, TANADA und HENDRICKS). Selbst durch UV induzierte erhöhte Mutationsraten können durch Behandlung mit sichtbarem Licht verringert werden (ALTENBURG und ALTENBURG, AUERBACH). Ebenso sind durch UV inaktivierte Enzymsysteme durch Licht zu reaktivieren (SHUGAR [1, 2]). Als erholungsfähig erwiesen sich im übrigen z. B. die Vermehrung von Bakterien, die Überlebensrate, die Teilungsfähigkeit, die Enzymaktivität von Enzymsystemen, die Adaptation an verschiedene Substrate (Glucosevergärung bei Hefen, SWENSON und GIESE) und die Synthese bestimmter Zellbestandteile, vor allem die DNS-Synthese (KELNER [4]).

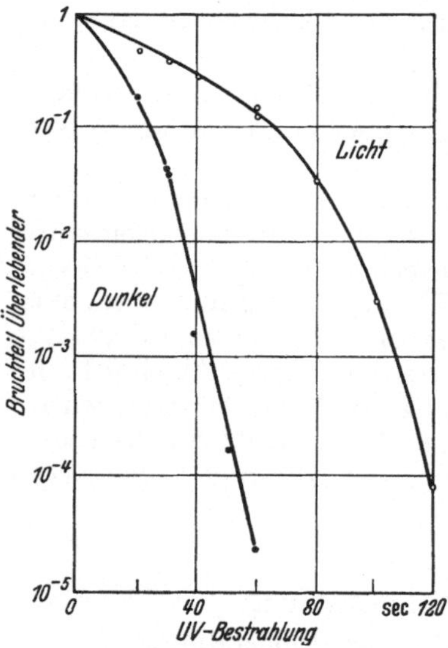

Abb. 18. Reaktivierung von *Escherichia coli* nach verschieden langer UV-Bestrahlung durch Belichtung. [Nach KELNER: Proc. Nat. Acad. Sci. U.S.A. **35**, 73 (1949).]

Das *Ausmaß der Erholung* ist bei verschiedenen Objekten und sogar bei verschiedenen Stämmen des gleichen Objekts unter Umständen verschieden groß (Phagen: DULBECCO [2, 3], HILL und ROSSI [2, 3]; Bakterien: LATARJET und CALDAS, LATARJET und MILÉTIC, HARM und STEIN [2]). Auf alle Fälle wird nur ein bestimmter, jedoch von verschiedenen Bedingungen abhängiger und oft wechselnder Prozentsatz bestrahlter Zellen durch Belichtung in den ursprünglichen Zustand zurückversetzt (unter anderen HILL und ROSSI [2, 3], NOVIK und SCILLARD). Auch unterliegen die durch UV-Bestrahlung hervorgerufenen Veränderungen durchaus nicht alle in gleichem Ausmaß der Erholung. Überhaupt sind nicht alle durch UV-Strahlung hervorgerufenen Vorgänge durch nachfolgende Belichtung reaktivierbar (BLUM und Mitarbeiter [1], HIRSHFIELD und GIESE). Ob die für die Bestrahlung mit UV benutzten Wellenlängen für das Ausmaß der späteren Erholung von maßgeblicher Bedeutung sind, steht noch nicht sicher fest (DULBECCO [2, 3], KIMBALL und GAITHER [1]).

Das Wirkungsspektrum der Belichtung. Quantenbedarf.

DULBECCO [2] stellte das Wirkungsspektrum der Reaktivierung von Bacteriophagen fest. Hiernach trat eine Reaktivierung durch Licht mit Wellenlängen von 315 mμ bis zu 480 mμ auf; das Maximum lag bei 365 mμ (Abb. 19). KELNER [2] fand ein Maximum der Erholung bestrahlter *E. coli* bei 375 mμ. Als Erholungsbereich bei bestrahltem Sperma von Seeigeln und Seeigeleiern waren Wellenlängen von 300—500 mμ (BLUM und Mitarbeiter [3]) wirksam. Aus den Untersuchungen geht einstweilen nicht mit Sicherheit hervor, ob die erholende Wirkung überall durch das gleiche lichtabsorbierende System hervorgerufen wird. KELNER [2] diskutiert z. B. die Möglichkeit eines Porphyrinkörpers als Absorbens. Da bei der Reaktivierung von Triosephosphatdehydrase eine Erholung durch Belichtung mit Wellenlängen von 310—380 mμ nur bei Gegenwart von reduzierter Co-Dehydrase möglich war, die bei 340 mμ ihr Absorptionsmaximum besitzt (SHUGAR [2]) (vgl. Abb. 13, S. 673), können ähnliche Wirkungen in lebenden Zellen erwartet werden (Weiteres vgl. S. 683).

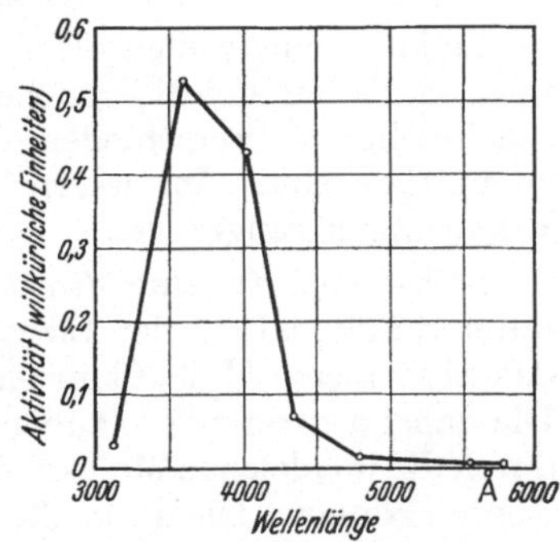

Abb. 19. Das Wirkungsspektrum der Photoreaktivierung von Bakteriophagen (T_2). Nach 20 sec UV-Bestrahlung wurden die Phagen an Bakterien adsorbiert und in Lösung bei 37° C belichtet. [Nach DULBECCO: J. Bacter. 59, 329 (1950).]

Der Quantenbedarf für die Erholungsvorgänge ist erheblich. GIESE und Mitarbeiter [2, 3] stellten fest, daß etwa 100 Quanten Blaulicht (435 mμ) zur Erholung von einem durch ein Quant UV (2654 Å) hervorgerufenen Effekt bei *Colpidium* nötig waren. Außerdem hängt der Quantenbedarf aber auch noch vom Ernährungszustand ab: Bei hungernden Colpidien war er wesentlich größer (800 Quanten). Verschiedene Wellenlängen reaktivierenden Lichtes (366 bis 546 mμ) erwiesen sich im übrigen als gleich wirksam.

Abhängigkeit der Erholung von äußeren und inneren Faktoren.

Die Höhe der Erholung durch Nachbelichtung ist von verschiedenen äußeren und inneren Faktoren abhängig. Soeben wurde bereits erwähnt, daß der *Ernährungszustand* die lichtbedingte Erholung nach Bestrahlung wesentlich beeinflußt (GIESE und Mitarbeiter). Besonders scheint aber das Vorhandensein von *Cytoplasma* eine notwendige Bedingung für die Möglichkeit von Reaktivierungsvorgängen durch Belichtung zu sein. So machten BLUM und Mitarbeiter [3] sowie GIESE [1] darauf aufmerksam, daß UV-bestrahltes und vor der Befruchtung belichtetes Sperma später zu keiner Furchung des Seeigeleies führt, während UV-bestrahlte Eier oder Eier, die mit UV-bestrahltem Sperma befruchtet wurden, durch Belichtung sehr wohl reaktivierbar waren. In ähnlicher Weise erwiesen sich

Bakteriophagen nach UV-Bestrahlung nur dann als reaktivierbar, wenn die adsorbierten Phagen zusammen mit den Bakterien belichtet wurden (DULBECCO [2, 3], KLECZKOWSKI und KLEZKOWSKI, LATARJET und CALDAS).

Von äußeren Faktoren erwiesen sich sowohl die *Lichtintensität* (DULBECCO [2]) als auch die *Belichtungszeit* nach der Bestrahlung (KELNER [3]) als wichtig. Auch die *Temperatur* während der Belichtung ist für die lichtabhängige Erholung von Bedeutung (DULBECCO [3]), wobei die Temperatureffekte noch durch eine von der Temperatur selbst beeinflußte Erholung nach UV überlagert werden. Der *Sauerstoffgehalt* während der Erholung scheint keine Bedeutung zu haben (JOHNSON und Mitarbeiter). Jedoch spielt der *Wassergehalt* der bestrahlten Objekte für die Erholung offenbar eine erhebliche Rolle. Trocken bestrahlte Phagen sind nicht reaktivierbar (HILL und ROSSI [1]). Verringerung hohen Wassergehalts kann aber zunächst die Reaktivierung steigern (HILL und ROSSI [2, 3]).

Ursachen der lichtbedingten Erholung.

Es kann einstweilen nicht entschieden werden, ob die lichtbedingte Erholung in allen Fällen auf die gleichen Ursachen zurückzuführen ist, oder ob — wahrscheinlicher — verschiedene Effekte für ihr Zustandekommen verantwortlich zu machen sind. Im wesentlichen werden *drei* verschiedene *Möglichkeiten* zur Erklärung herangezogen.

1. Es wird an eine *direkte Reparatur* von entstandenen Strahlungsschäden unter Zuhilfenahme der Lichtenergie gedacht: Durch die UV-Bestrahlung wird das absorbierte Molekül in einen angeregten, höheren Energiezustand versetzt: Die dabei auftretenden, zunächst labilen Veränderungen des Moleküls sollen nun durch die im gleichen Molekül absorbierte Lichtenergie wieder rückgängig gemacht werden können. Die durch die UV-Bestrahlung erfolgte Umwandlung von Pyridin in NH_4-Glutaronaldehydenolat und die Rückverwandlung in Pyridin durch sichtbares Licht kann als Modellbeispiel für diese Lichtwirkung angeführt werden (FREYTAG). Ein ähnliches, auf den Erfahrungen der Färbung und Entfärbung von Alkalihalogeniden aufgebautes, theoretisch recht plausibles elektronisches Modell legte kürzlich ROTTGARDT vor. Auch HILL und ROSSI [3] vermuten, daß die vom Wassergehalt abhängige Reaktivierung von Phagen auf eine reversible Änderung der durch direkte UV-Treffer veränderten Molekülkonfiguration zurückzuführen ist. Gegen die allgemeine Anwendbarkeit dieser in speziellen Fällen recht plausiblen Erklärung sprechen mehrere Ergebnisse, von denen die Notwendigkeit des Cytoplasmas für manche Reaktivierungsvorgänge sowie die Erfahrungen der lichtbedingten Erholung von Strahlenschäden nach Alleinbestrahlung des Kulturmediums durch UV genannt seien.

2. Die weiteste Anerkennung hat die Annahme eines durch Belichtung hervorgerufenen *Abbaus* von durch UV entstandenen *toxischen Substanzen* gefunden, wie sie zuerst von NOVIK und SCILLARD vorgeschlagen wurde. Ein Teil der durch UV gebildeten giftigen Verbindungen soll danach lichtempfindlich, ein anderer lichtunempfindlich sein. Sie basiert auf der in der Folgezeit aber durchaus nicht immer verwirklicht gefundenen Erfahrung, daß jeweils ein bestimmter Prozentsatz der Schädigungen durch Licht rückgängig gemacht werden kann: Das Licht setzt nämlich die Wirkung einer bestimmten UV-Dosis vielfach ohne Änderung der Form der Dosiseffektkurve herab (vgl. Abb. 18, S. 680). Eine Weiterentwicklung dieser Theorie unter Verknüpfung mit den Erfahrungen der indirekten UV-Strahlenwirkung und den Ergebnissen der oben besprochenen Erholung durch Wärme wird von STEIN und HARM [3] erörtert. Hiernach soll die durch Belichtung bedingte Erholung erst an den Sekundärprozessen der durch indirekte Strahlen-

wirkung (toxische Substanzen) zustande gekommenen UV-Strahlenfolgen angreifen. Besonders wird hier gegenüber der unter 1. genannten Vorstellung darauf hingewiesen, daß der Absorptionsort der UV-Strahlung durchaus nicht der Wirkungsort der UV-Strahlung zu sein braucht. Über den eigentlichen Mechanismus der Lichtwirkung wird hier aber nichts ausgesagt.

3. An dritter Stelle sei die notwendige *Mitwirkung* von *Stoffwechselvorgängen* an den lichtbedingten Erholungsprozessen diskutiert. Hierfür spricht die Notwendigkeit von Cytoplasma für die Erholung (Blum und Mitarbeiter [3], Giese [1], der unterschiedliche Quantenbedarf bei hungernden und gut ernährten Zellen (Giese und Mitarbeiter [3]), sowie der Hinweis auf Erholungsvorgänge nach UV durch Zugabe von Stoffwechselzwischenprodukten (Heinmets und Mitarbeiter [1, 2]). Auch hier konnten aber über den wirkenden Mechanismus der Erholung noch keine Angaben gemacht werden. Es kann etwa an Wechselwirkungen von lichtempfindlichen, im Stoffwechsel gebildeten Substanzen mit der durch UV geschädigten Reaktionskette gedacht werden, wofür als Modell die Reaktivierung von Triosephosphatdehydrase durch Licht bei Gegenwart von reduzierter, im fraglichen Wellenbereich absorbierender Co-Dehydrase herangezogen werden mag (Shugar [1, 2]). Ob die von Kelner [4] beobachtete schnelle Erholung der durch UV sehr rasch intensiv gehemmten DNS-Synthese bei Belichtung eine frühe Zwischenstufe oder aber bereits eine Spätfolge solcher Prozesse darstellt, müssen weitere Versuche klären.

C. Die Wirkungen von langwelligem UV und sichtbarem Licht.

a) Primärvorgänge der Lichtwirkung.

1. Allgemeines.

Die Wirkungen des sichtbaren Lichtes auf die Zelle kommen einmal dadurch zustande, daß verschiedene als *Photoreceptoren* (Sensibilisatoren) dienende lichtabsorbierende Substanzen *reversibel* verändert werden. Sie übertragen die absorbierte Lichtenergie auf andere Stoffe, die dann weiter reagieren. In gleicher Weise wirken bestimmte, in die Zellen gebrachte *Sensibilisatorfarbstoffe*. Zum andern können manche Substanzen durch das von ihnen absorbierte Licht *direkt* und dann meistens *irreversibel* verändert werden; z. B. werden sie im Gefolge der Strahleneinwirkung oxydiert. Sind diese Stoffe lebenswichtig, treten Schädigungen ein.

Die Lichtwirkung ist bei diesen Vorgängen entweder um so größer, je mehr Energie, also eine je größere Lichtmenge auf die Photoreceptoren trifft. Eine solche Wirkung findet z. B. bei der Photosynthese statt; auch Photooxydationen kommen so zustande (vgl. S. 693). Oder aber die schließlich hervorgerufene Lichtwirkung steht in keiner engen Beziehung zur auftreffenden Energiemenge. Das einfallende Licht hat dann nur eine auslösende Funktion, wirkt also als Reiz. Beispiele finden sich sehr reichlich; die meisten lichtabhängigen Reaktionen der Zelle und auch die formativen Lichtwirkungen gehören hierher.

Für das Studium von durch Licht induzierten Lebensvorgängen sei auf die Bücher von Bünning [1], Duggar und Rabinowitch [1, 2] verwiesen. Viele Hinweise auf die ältere Literatur finden sich z. B. bei du Buy und Nuernbergk.

2. Die lichtabsorbierenden Stoffe in der Zelle.

Das auf die Pflanzenzelle auftretende langwellige ultraviolette und sichtbare Licht kann in zahlreichen Substanzen absorbiert werden. Auf verschiedene dieser Stoffe wird an anderer Stelle des Handbuches ausführlich eingegangen, so daß sie hier nur erwähnt zu werden brauchen. An erster Stelle sollen die in den Chromatophoren lokalisierten Chlorophylle, Carotinoide und Phycobiline genannt werden, die, abgesehen von ihrem Zusammenhang mit der Photosynthese, die Acceptoren für bestimmte strahlungsabhängige Vorgänge in den Zellen darstellen dürften. An zweiter Stelle sind die im Zellsaft gelösten Stoffe wie Anthocyane, Flavonole und Flavone zu erwähnen (Abb. 14, S. 674). Weiter müssen die im Cytoplasma oder in besonderen Zellorganellen lokalisierten absorbierenden Substanzen genannt werden, zu denen

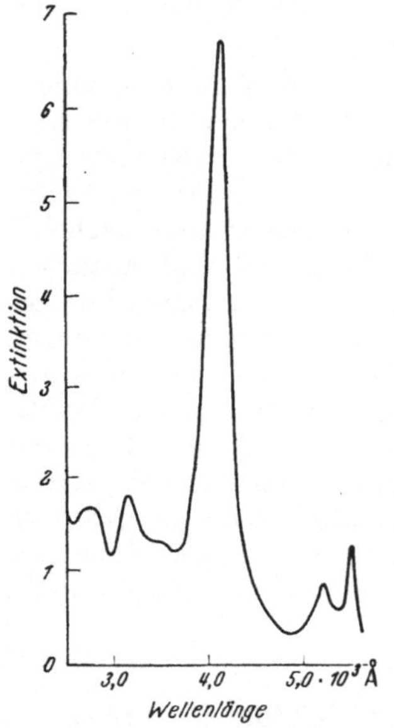

Abb. 20. Das Absorptionsspektrum von Ferrocytochrom c. (Nach BARRON, in NICKSON, Symposium, S. 216, 1952)

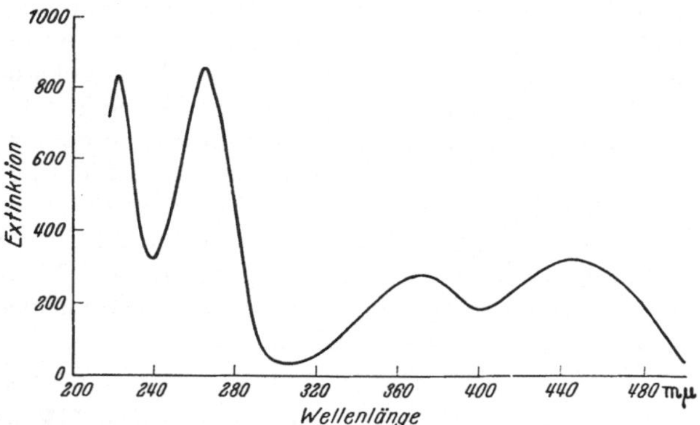

Abb. 21. Das Absorptionsspektrum von Riboflavin bei p_H 2,5. [Nach DAGLISH u. Mitarb., aus SINSHEIMER, in HOLLAENDER, Radiation biology, Bd. II, S. 165, 1955.)

vor allem Co-Enzyme gehören; z. B. die Co-Dehydrase (Abb. 13, S. 673), das Cytochromsystem (Abb. 20) und das gelbe Atmungsferment. Ein wesentlicher chromophorer Bestandteil des zuletzt genannten Co-Enzyms ist das auch sonst in den Zellen vorkommende Vitamin B_2 (Lactoflavin; Riboflavin), das im langwelligen UV und im blauen Licht eine deutliche Absorption aufweist (Abb. 21).

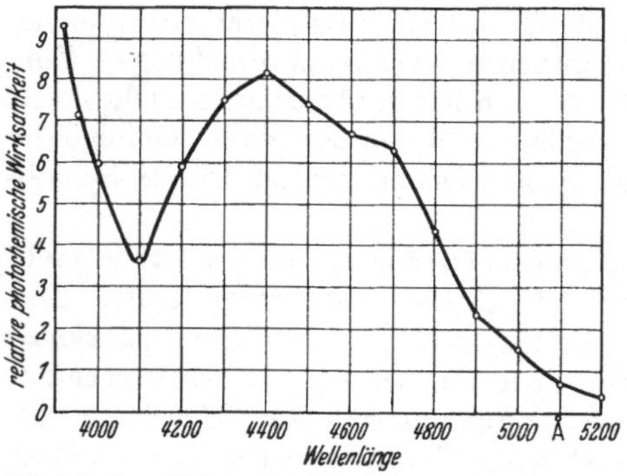

Abb. 22. Das Wirkungsspektrum für die Photoinaktivierung von β-Indolylessigsäure durch einen Brei von etioliertem Erbsenepikotyl. Das Spektrum gleicht einem Flavinabsorptionsspektrum. [Nach GALSTON: Science (Lancaster, Pa.) 111, 619 (1950.)]

Aus der Untersuchung der verschiedensten photosensibilisierten Prozesse (s. unten) geht hervor, daß, neben dem immer wieder als wichtig angesehenen *Chlorophyll* in den letzten Jahren dem *Riboflavin* für allein im kurzwelligen, sichtbaren Bereich photosensibilisierte Vorgänge als Lichtacceptor eine zur Zeit noch wachsende Bedeutung beigelegt wird [vgl. z. B. das Wirkungsspektrum der Indolessigsäure-(IES-)Inaktivierung durch Pflanzenextrakte im Vergleich zum Riboflavinspektrum (Abb. 22)], während bis vor kurzem nur dem *Carotin* diese Aufgabe zugeschrieben wurde (unter anderen BRAUNER, BRAUNER und BRAUNER [2], GALSTON [1, 2], GALSTON und Mitarbeiter [2, 3]).

Eine besondere und anscheinend zentrale Bedeutung scheint neuerdings schließlich ein lichtabsorbierendes System zu gewinnen, das sowohl für das Auftreten der in verschiedenen Spektralbereichen unterschiedlichen Samenkeimung (BORTHWICK und Mitarbeiter [1, 2]) wie auch für das Internodialwachstum, für die Förderung des Wachstums von *Avena*-Koleoptilen (LIVERMANN und BONNER), für die unterschiedliche Nebenwurzelbildung durch verschiedenfarbiges Licht (TORREY) und vor allem für die photoperiodische Reaktion von Kurztagspflanzen und eines bestimmten Typs von Langtagspflanzen verantwortlich gemacht wird (BORTHWICK und Mitarbeiter (1, 2), LANG [1, 2], WASSINK und STOLWIGK). Hiernach handelt es sich vielleicht um ein *Porphyrinpigment* oder eine ganze Gruppe derartiger Pigmente, die in zwei ineinander überführbaren Formen mit verschiedenem Absorptionsspektrum vorliegen sollen. Das Maximum der Absorption des einen Pigmentes liegt im Rot, ein schwaches 2. Maximum findet sich im Blau. Die Wirkungen des roten Lichtes können durch Infrarotabsorption des zweiten Pigments reversibel aufgehoben werden. Ob das kürzlich von TODD und GALSTON in nicht chlorophyllhaltigen Pflanzenzellen festgestellte Methylpyrophaeophorbid a (ein metallfreies Porphyrin) an diesem die formativen Wachstumsvorgänge beeinflussenden Pigmentsystem beteiligt ist, steht nicht fest.

3. Die photochemischen Wirkungen des Lichtes.

Über die photochemische Umsetzung der absorbierten Anregungsenergie (vgl. S. 670) sind die verschiedensten Hypothesen entwickelt worden (E. J. BOWEN [1], FÖRSTER [2], FRANCK, FRANCK und PLATZMANN, PRINGSHEIM, RABINOWITCH [1, 2], G. O. SCHENCK [1, 2], J. WEISS [1, 3]). Ich beschränke mich hier auf die Darstellung einiger neuerdings entwickelter Vorstellungen.

Übertragung der Anregungsenergie durch photosensibilisierende Substanzen.

1. Vielfach ist angenommen worden, daß der Photosensibilisator (S) die Anregungsenergie *ohne* chemische Reaktion auf einen Acceptor (A) überträgt, wie es bei sensibilisierter Fluorescenz beobachtet werden kann:

$$S \xrightarrow{h\nu} S^* \qquad 1a$$

$$S^* + A \longrightarrow S + A^* \qquad 1b$$

$$A^* + O_2 \longrightarrow AO_2 \qquad 1c$$

Der angeregte Acceptor kann dann weiter reagieren, etwa mit O_2 (vgl. GAFFRON, TAPPEINER, WARBURG und SCHOCKEN). Eine solche Übertragung der Anregungsenergie ist aber nur dann wahrscheinlich, wenn Sensibilisator und Acceptor sich überlappende Absorptionsspektren besitzen (vgl. z. B. die von FRENCH und YOUNG untersuchte Übertragung der Anregungsenergie von Phycocyan auf Chlorophyll bei Rotalgen; s. auch BANNISTER).

2. Wie schon oben S. 669 diskutiert, ist aus energetischen Gründen eine primäre Dissoziation der als Sensibilisatoren dienenden Stoffe (Farbstoffe) im langwelligen UV und sichtbaren Licht unwahrscheinlich. Dagegen ist es denkbar, daß ein elektronisch angeregtes Molekül oder ein Molekülkomplex erhöhte Neigung zeigt, ein einzelnes *Elektron* und im Gefolge davon ein Proton abzugeben oder aufzunehmen, also univalente Oxydationen bzw. Reduktionen durchzuführen (EVANS und URI, LEWIS und LIPKIN, WEISS [2, 3], WEISS und FISCHGOLD, vgl. RABINOWITCH).

$$S \xrightarrow{h\nu} S^* \qquad 2a$$

$$S^* + A \longrightarrow S^- + A^+ \qquad 2b$$

oder

$$S^* + A \longrightarrow S^+ + A^- \qquad 2c$$

Als Elektronen- bzw. Wasserstoffdonator wird bei diesen Reaktionen verschiedentlich sogar das Wasser in Erwägung gezogen. Als Beispiel einer solchen Vorstellung sei die durch Farbstoffe (Methylenblau, Thionin) photosensibilisierte Reaktion einer $Fe^{2+} \rightarrow Fe^{3+}$-Oxydation bei Wellenlängen von 650 mμ angeführt, die unter Beteiligung von H_2O vor sich gehen soll. Die Reaktion läuft nach WEISS in folgender Weise ab:

$$S \xrightarrow{h\nu} S^* \qquad \text{(Lichtabsorption)} \qquad 3a$$

$$S^* + H_2O\,Fe^{2+} \longrightarrow S^* \cdot H_2O \cdot Fe^{2+} \qquad \text{(Komplexbildung)} \qquad 3b$$

$$S^* \cdot H_2O \cdot Fe^{2+} \longrightarrow SH + OH^- + Fe^{3+} \qquad \text{(Elektronenübertragung, Hydrierung)} \qquad 3c$$

$$SH + O_2 \longrightarrow SH^+ + O_2^- \qquad \text{(Reoxydation des Sensibilisators)} \qquad 3d$$

$$\downarrow$$

$$S + H^+$$

$$2SH \longrightarrow SH_2 + S \qquad \text{(Semichinongleichgewicht)} \qquad 3e$$

Der Farbstoff S geht dabei also unter Reduktion durch den Wasserstoff des Wassers in den Leukofarbstoff über (vorübergehendes Ausbleichen) und wird vom Luftsauerstoff wieder oxydiert. Das Endergebnis ist dann die Oxydation des Eisens. Der Farbstoff selbst dient nur als Photokatalysator und bleibt bei O_2-Gegenwart unverändert.

Die biologische Bedeutung solcher Vorstellungen wurde neuerdings bei Untersuchungen von BRAUNER, BRAUNER und BRAUNER [2] (vgl. LIVERMANN und BONNER, GALSTON [2], GALSTON und Mitarbeiter [3]) über den photosensibilisierten Abbau der β-Indolylessigsäure durch verschiedene Farbstoffe, besonders durch Riboflavin (vgl. Abb. 23, S. 688), diskutiert. Als H-Donatoren sollen nach BRAUNER auch hierbei das Wasser bzw. im Gefolge der vermuteten, in einfachen Systemen energetisch allerdings unwahrscheinlichen Photolyse des Wassers gebildete hydrierende Radikale dienen.

3. Durch ein großes Beobachtungsmaterial gestützt, hat in den letzten Jahren G. O. SCHENCK ein Schema entworfen, nach dem photosensibilisierende Farbstoffe wie Eosin, Methylenblau und auch Chlorophyll nur bei O_2-Abwesenheit dehydrierend wirken. Bei Sauerstoffgegenwart wird dagegen zwischen O_2 und dem Sensibilisator eine lockere, transitorische chemische Bindung hergestellt (Zwischenreaktionskatalyse) und der Sauerstoff (O_2) auf den Acceptor übertragen, der so oxydiert wird. Ausgehend von der oben S. 670 besprochenen Erkenntnis, daß der durch Bestrahlung hervorgerufene elektronische Anregungszustand wahrscheinlich sehr rasch zu einem metastabilen Molekülzustand führt, der chemisch einem Diradikal entspricht (vgl. S. 671), stellt SCHENCK folgendes allgemeine Schema einer photosensibilisierten Reaktion auf:

$$+O_2\colon \quad S \xrightarrow{h\nu} S^* \qquad 4a$$

$$S^* + O_2 \longrightarrow S^* \ldots O_2 \qquad 4b$$

$$S^* \ldots O_2 + A \longrightarrow S + AO_2 \qquad 4c$$

$$S^* + B \longrightarrow S + B + \text{Wärme} \qquad 4a'$$

$$-O_2\colon \quad S^* + RH \longrightarrow SH + R \qquad 4b'$$

Nach der Überführung von S in das Diradikal S* (4a) wird bei O_2-Gegenwart an S* O_2 angelagert (4b) — Bildung eines instabilen Peroxyds — und auf einen Acceptor (A) übertragen (4c). Sind Löschsubstanzen oder Inhibitoren (B) vorhanden, kann der angeregte Zustand in den unangeregten zurückgeführt werden

(4a′). Nur unter O_2-Ausschluß kommt es zur Hydrierung von S durch einen H-Donator RH (4b′). Dieser Vorgang läuft wegen der großen Affinität von S* zu O_2 bei Sauerstoffgegenwart nur in ganz geringem Umfang ab. Ist keine O_2-Affinität vorhanden (z. B. ortho- und para-Chinone), erfolgt die Sensibilisierung allein nach dem RH-Schema (4b′).

Es sei noch erwähnt, daß Beziehungen der photosensibilisierenden Wirkung von Farbstoffen zur Höhe ihrer *Fluorescenz* verschiedentlich erörtert wurden. Die Fluorescenzintensität läuft jedoch keineswegs mit der sensibilisierenden Wirkung parallel (BRAUNER, DREBINGER [1, 2], RABINOWITCH [1]). Eher ist die sensibilisierende Wirkung dann besonders groß, wenn der direkt zum metastabilen Zustand führende Anteil der Anregungsenergie relativ hoch und damit die Fluorescenzfähigkeit niedrig ist (z. B. DREBINGER [2]). Auch die Lebensdauer des metastabilen Zustandes ist wichtig (RABINOWITCH). Der zur Fluorescenz führende Anteil der Anregungsenergie kann durch sekundäre Übergänge in den metastabilen Zustand weiter geschwächt (gelöscht) werden, und erst diese sekundären Übergänge sind es vor allem, die mit der Fluorescenzintensität in Konkurrenz treten (vgl. das Schema S. 671 sowie FÖRSTER [2], FRANCK, RABINOWITCH [1, 2]).

Inhibitoren der Strahlenwirkung.

Ebenso wie im UV gibt es im sichtbaren Licht Stoffe, die imstande sind, durch photosensibilisierende Substanzen hervorgerufene Strahlenwirkungen zu unterdrücken. Hierbei kann es sich um Stoffe handeln, durch die die Anregungsenergie des Photosensibilisators in *Wärme* überführt wird. Ferner kann der die Anregungsenergie löschende Stoff zum Photosensibilisator eine größere Affinität haben als der Acceptor; es handelt sich dann um eine *konkurrierende Hemmung*. Schließlich ist es möglich, daß durch den zugefügten Inhibitor eine eingeleitete *Folgereaktion*, z. B. eine Oxydation durch chemische Veränderung des Hemmstoffes rückgängig gemacht wird.

Als Beispiele von in vitro-Versuchen seien genannt: Zugabe von Cyclooctatetraen (C_8H_8) zu einer durch Methylenblau photosensibilisierten Autoxydation hemmt die Reaktion. Das Cyclooctatetraen bleibt dabei chemisch unverändert (G. O. SCHENCK [1]). Eine Konkurrenzhemmung scheint bei der Schutzwirkung des Crocins auf die Photoinaktivierung der IES bei Riboflavingegenwart vorzuliegen; das Crocin bleicht dabei aus (BRAUNER), Guajacol oder Ascorbinsäure können den Effekt verhindern. Auch reduzierte Ascorbinsäure selbst verringert den durch Riboflavin photosensibilisierten Abbau. Die Reaktion läßt sich ebenfalls als Konkurrenzhemmung auffassen. Auch Polyphenole (o- und p-Phenole), Flavone und Flavonole können den photosensibilisierten Abbau von IES zum Teil hemmen (BRAUNER und BRAUNER [2]). In der Zelle sollen derartige Reaktionen z. B. bei dem lichtbedingten Abbau von Indolylessigsäure, der durch Carotinoide gehemmt zu werden scheint, eine Rolle spielen (BRAUNER, BÜNNING [2], REINERT). Auch die soeben erörterten Versuche in vitro sprechen hierfür. In der Zelle kommt aber noch hinzu, daß Enzymsysteme solche Vorgänge in mannigfaltiger Weise abändern dürften. Hinweise dafür bieten etwa die Untersuchungen des IES-Abbaus in enzymhaltigen Pflanzenextrakten durch Licht (vgl. GALSTON und BAKER [3], GALSTON und Mitarbeiter).

4. Photosensibilisierter Abbau von biologischen Modellsubstanzen.

Bereits an biologischen Modellsubstanzen konnten in vitro vielfach photochemische Wirkungen durch sichtbares Licht nach Zugabe von photosensibilisierenden Stoffen festgestellt werden. (Ältere Arbeiten sind der Zusammenfassung von MCLAREN zu entnehmen.)

So tritt durch Riboflavin ein Abbau von Urease, Tyrosinase, Amylase und Invertase auf (GALSTON und BAKER [1, 2] bzw. MANDELS). Die Inaktivierung von Galaktozymase durch Rose bengale beobachteten BRANDT und Mitarbeiter, die von Aminosäuren durch Methylenblau und Riboflavin WEIL und Mitarbeiter sowie GIRI und Mitarbeiter. Es entstehen mannigfaltige Abbauprodukte, die erst in den wenigsten Fällen genauer analysiert worden sind, z. B. wird Tryptophan unter O_2-Aufnahme und Abgabe von CO_2 abgebaut; das sich ändernde

UV-Absorptionsspektrum legt außerdem die Aufspaltung des Indolringes in den Benzol- und Pyrrolanteil nahe (WEIL und Mitarbeiter). Es können aber auch Polymerisationen auftreten (EVANS und URI).

Der Abbau von IES wurde wegen ihrer Bedeutung für die Lichtwachstumsreaktionen im Modellversuch mit vielen verschiedenen Farbstoffen (Riboflavin, Methylenblau, Phthaleinfarbstoffe usw. (Abb. 23) gründlicher untersucht (BRAUNER, FERRI, GALSTON); auf das von BRAUNER letzthin vorgeschlagene Schema des Primärabbaus wurde bereits oben hingewiesen (S. 686). Die Spaltung der IES verläuft vermutlich unter Sauerstoffaufnahme über β-Indolylglykolsäure zur β-Indolylglyoxylsäure. Wahrscheinlich ist an der weiteren Photolyse eine oxydative Carboxylierung beteiligt. Zur Frage der Photolyse der IES in der Zelle selbst durch ein Flavoprotein bzw. durch Riboflavin vgl. S. 687.

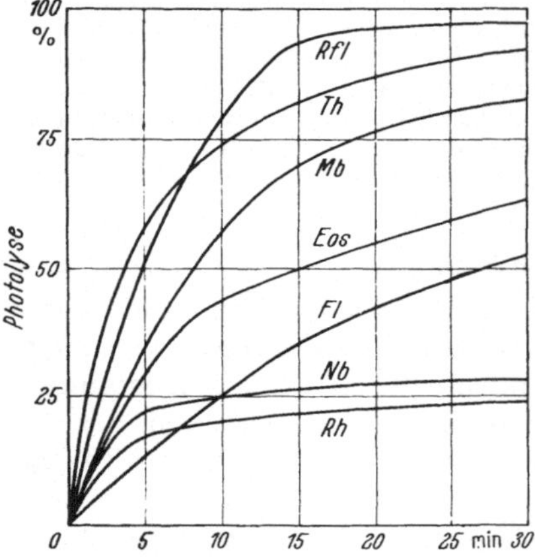

Abb. 23. Photosensibilisierter Abbau von β-Indolessigsäure bei Gegenwart verschiedener Farbstoffe. *Rfl* Riboflavin; *Th* Thiamin; *Mb* Methylenblau; *Eos* Eosin; *Fl* Fluorescein; *Nb* Nilblau; *Rh* Rhodamin B. Beleuchtungsstärke 2000 Lux. [Nach L. BRAUNER: Z. Bot. 41, 291 (1953).]

b) Lichtwirkungen auf die Zelle.

1. Photodynamische Wirkungen.

Durch die Einlagerung von photosensibilisierenden Substanzen, also teils im langwelligen UV, vor allem aber im sichtbaren Bereich absorbierenden Farbstoffen, in lebende Zellen lassen sich die verschiedensten Wirkungen erzielen. Sie gestatten Hinweise auf den Ablauf derjenigen Vorgänge, die in der Zelle unter dem Einfluß von Belichtung unter normalen Bedingungen ablaufen. TAPPEINER und JODLBAUER gebrauchten für derartige Effekte bereits den Ausdruck „Photodynamische Wirkungen", der sich in der Folgezeit eingebürgert hat und hier dementsprechend beibehalten wird. Über die Vorstellungen von ihrer Wirkungsweise vergleiche das soeben (S. 685) Erörterte. Allgemeine Hinweise auf photodynamische Wirkungen finden sich bei BOWEN [1], METZNER, RABINOWITCH [1]. Die erzielten Wirkungen seien an Hand einiger neuerer Untersuchungen besprochen. Vielfach wurde eine *Inaktivierung* bzw. eine Abtötung oder „Schädigung" festgestellt: so von *Viren* durch Acriflavin (OSTER und MCLAREN), von *Phagen* durch Methylenblau (WELSH und ADAMS) und Riboflavin (GALSTON [1]), von *Bacterien* und *Pilzen* einschließlich Hefen durch Erythrosin (R. W. KAPLAN [1,2] (Abb. 24) und durch Rose bengale (FREEMAN und GIESE), von Wurzeln und Blättern *höherer Pflanzen* durch Phtaleinfarbstoffe (ZIEGLER) und Rhodamin (GESSNER). Ferner wird mehrfach über eine Änderung der *Atmungsintensität*, also der O_2-Aufnahme, berichtet. Während die exogene Atmung der Hefe erniedrigt wird, zeigt die endogene O_2-Aufnahme bei Rose bengale-Behandlung eine deutliche Erhöhung (FREEMAN und GIESE). Bei Wurzeln (Phtaleinfarbstoffe) wurde die Atmung ebenfalls erhöht (ZIEGLER). Die Photosynthese dagegen (z. B. bei *Helodea* mit Rhodamin B) wird erniedrigt (GESSNER, PIRSON und ALBERTS).

Die Abbauvorgänge sind, wie nicht anders zu erwarten, vielfach vom *Sauerstoffgehalt* der Umgebung abhängig. Genannt seien die durch Methylenblau induzierte O_2-abhängige Inaktivierung von Phagen (WELSH und ADAMS); die O_2-abhängige Abtötung von *Blepharisma undulans* im Starklicht nach Bildung eines zelleigenen Pigments im Dunkeln (GIESE [2]). Das gleiche gilt für die im langwelligen UV-Licht (300—400 mμ) in Gegenwart von Benzpyren entstehende Inaktivierung von Hefe (WINDISCH und Mitarbeiter, GRAFFI und Mitarbeiter). Im Stickstoffstrom waren nach 32 min UV noch 82% überlebende Hefezellen vorhanden, bei O_2-Gegenwart nach der gleichen Zeit aber nur noch 0,3%.

Für die genannten photodynamischen Wirkungen lassen sich *Dosiseffektkurven* aufstellen. Der Kurvenverlauf hat teilweise eine exponentielle Form, teilweise erhält man Kurven, die den untersuchten Vorgang als „Mehrtreffervorgang" zu beschreiben gestatten (z. B. R. W. KAPLAN [1—3]) (Abb. 24). Eine genauere Analyse der photodynamischen Effekte steckt erst in den Anfängen.

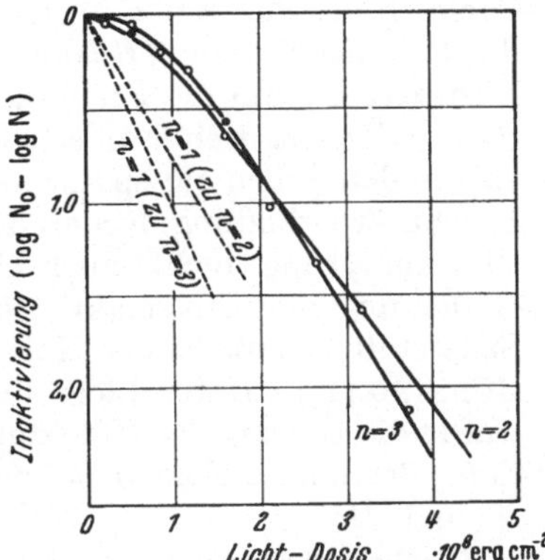

Abb. 24. Photodynamische Wirkung von Erythrosin. Die Inaktivierung von Sporen von *Penicillium notatum* nach Anfärbung mit Erythrosin durch Belichtung. Eine 2- oder 3-Trefferkurve ($n = 2$ bzw. $n = 3$) bei verschiedenem Verlauf der zugehörigen 1-Trefferkurven entspricht dem experimentell gefundenem Inaktivierungsverlauf. [Nach R. W. KAPLAN: Planta (Berl.) **38**, 1 (1950).]

2. Beeinflussung zellphysiologischer Vorgänge durch Belichtung.

Ob es sich bei den durch Belichtung hervorgerufenen zellphysiologischen Vorgängen um primär mit der Lichtwirkung in Zusammenhang stehende Prozesse handelt, oder ob es erst infolge sekundärer Abläufe, etwa infolge des Photosynthesebeginns oder der lichtbedingten Umstimmung des Stoffwechsels aus dem Ruhezustand in Aktivität usw. zu Veränderungen des Stoffwechsels kommt, bedarf im einzelnen genauer Untersuchung und ist vielfach erst wenig bekannt.

Welche Teile der Zelle für die in Frage kommenden lichtempfindlichen Prozesse verantwortlich zu machen sind, hängt ganz davon ab, an welcher Stelle die die Strahlung absorbierenden Substanzen lokalisiert sind, welche mit den Strahlungsfolgen in Korrelation stehenden Prozesse betroffen sind, und ob Energietransportmechanismen eine Rolle spielen. So können kernhaltige und kernfreie Teile von *Acetabularia*-Zellen, obwohl die absorbierenden Substanzen offenbar *nicht* im Zellkern liegen, doch recht verschieden auf gleiche Lichteinwirkung reagieren (BETH).

Stoffwechselaktivität, Enzymaktivität, Wuchsstoffe.

Durch Belichtung wird die *Stoffwechselaktivität* chlorophyllhaltiger Zellen infolge der einsetzenden Photosynthese sehr wesentlich verändert (vgl. Handbuch Bd. 5). Aber auch chlorophyllfreie Zellen und Gewebe können stark betroffen werden, wie die unterschiedliche Abhängigkeit der Keimung vom Licht bei verschiedenen Samen zeigt. Auf die diesbezüglichen Arbeiten von BORTHWICK und Mitarbeiter [1,2] sei wenigstens hingewiesen. Das Wirkungsspektrum folgt dem oben S. 685 angegebenen Modus. Weiter sei die Aktivierung der *Chlorophyllbildung* aus seinen Vorstufen durch Belichtung erwähnt (GRANICK, RABINOWITCH, SEYBOLD). Das Wirkungsspektrum folgt, wie KOSKI, FRENCH und SMITH feststellten, dem Absorptionsspektrum des Protochlorophylls; dabei wird Protochlorophyll zu Chlorophyll a hydriert.

Auch sonst kann die *Pigmentbildung* der Zellen durch Licht angeregt werden; so etwa die Bildung von Anthocyan in Blüten (HARDER, SCHRÖDER). Jüngst untersuchten PIRINGER und HEINZE die bei geringer Lichtintensität vor sich gehende Farbstoffausbildung (wahrscheinlich ein Flavonoid) in den Epidermiszellen der Tomatenfruchtschale. Hier wurde die Pigmentbildung durch gelbes Licht (580 mμ) gefördert, durch rotes Licht (von 695 mμ) aber rückgängig gemacht. Damit ähnelt das Wirkungsspektrum dem der lichtabhängigen Keimung von *Lactuca*-Samen (vgl. S. 685).

Die Beeinflussung der *Enzymaktivität* zeigten in einem besonderen Fall TOLBERT und BURRIS. Durch Licht geringer Intensität wurde bei etiolierten Weizenblättern unabhängig von der Chlorophyllbildung und ohne Vorhandensein von Chlorophyll ein Enzym aktiviert, das Glykolsäure zu Glyoxylsäure oxydierte. Daß durch Licht die Bildung besonderer Stoffe in Gang gebracht wird, ergeben auch die Untersuchungen der lichtabhängigen Bildung von Fruchtkörpern und Conidien bei Pilzen und Myxomyceten. Teils fördert nach SAGROMSKY das Licht die Conidienbildung *(Sclerotinia)*, teils wird sie gehemmt *(Penicillium)*. Die Lichtwirkung auf die Fruchtkörperbildung von *Didymium* läßt sich durch abgetötetes Plasma weiter übertragen. Durch Belichtung der Plasmodien entsteht also in den Zellen ein später nach Abtötung und Plasmadurchmischung noch die Fruchtkörperbildung fördernder Stoff (STRAUB). Rotes und blaues Licht ist wirksam, grünes nicht; auch UV von 350—390 mμ wirkt induzierend. Eine Beeinflussung der Enzymaktivität durch Licht geht auch aus der Änderung des Sauerstoffverbrauchs der Hefe bei der *Atmung* im Licht hervor (GUERRINI, vgl. BÜNNING [1]). Rotes Licht fördert, blaues Licht hemmt die Atmung. Ebenso wurde die Gärung der Hefe durch Blaulicht gehemmt, rotes Licht war ohne Einfluß. Bestimmte Enzyme könnten als Photoreceptoren dienen (vgl. z. B. Abb. 13, 20 und 21).

Die Beeinflussung der *Wuchsstoffaktivität* durch Licht wird an anderer Stelle des Handbuchs besprochen (vgl. Bd. 16). Als Photoreceptoren kommen für den durch Auxin beeinflußten Phototropismus das Carotin und möglicherweise Riboflavin, für die Lichtwachstumsreaktion das Riboflavin bzw. für die durch Wuchsstoffbeteiligung gekennzeichneten formativen Lichtwirkungen das noch nicht genauer bekannte, aber wohl sehr verbreitete Porphyrinsystem (vgl. S. 685) in Frage.

Viscosität.

Viscositätsänderungen bei Belichtung von mehr oder weniger lange vorher verdunkelten Zellen sind verschiedentlich beobachtet worden. WEBER fand bei belichteten Blättern von *Ranunculus ficaria* eine längere Plasmolysezeit als bei Dunkelkontrollen. STÅLFELT stellte rasche, zum Teil fluktuierende Viscositätsschwankungen in den Zellen der Blätter von *Helodea densa* fest, die je nach der Lichtintensität zu einer vorübergehenden Erhöhung oder Senkung der Viscosität führten (Abb. 25). VIRGIN [1, 2] bestätigte diese Beobachtungen. Danach wird bei sehr hoher und bei niedriger Beleuchtungsstärke, auch nach kurzfristigem Einwirken die Viscosität herabgesetzt, bei mittleren Beleuchtungsstärken dagegen erhöht. Nur direkt bestrahlte Zellen werden betroffen. Außerdem ist die Vorbehandlung der Pflanzen wichtig. Es zeigt sich, daß die Viscositätsänderung besonders intensiv im Bereich von 400—490 mμ, mit einem schnellen Abfall zum roten Bereich auftritt (Abb. 26). Interessant ist, daß durch Bestrahlung mit Infrarot die Verlagerungsfähigkeit der Chloroplasten erniedrigt werden kann. Dies läßt auf eine Erhöhung der Plasmaviscosität schließen (VOERKEL). Es liegt nahe, hier an eine Strahlenabsorption zu denken, die ebenso erfolgt wie die bei Beeinflussung formativer Prozesse (vgl. S. 685). Andererseits wird auf Grund des Wirkungsspektrums eine Absorption im Carotin bzw. im Riboflavin für wahrscheinlich gehalten (VIRGIN [1, 2]). Da die Viscositätsänderung eine Abhängig-

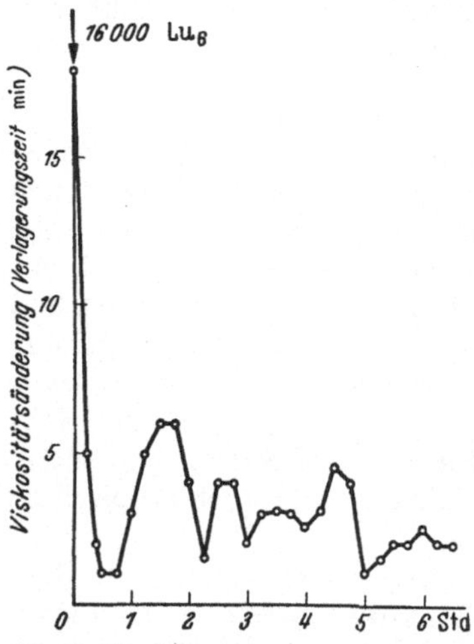

Abb. 25. Viscositätsänderung in den Blattzellen von *Helodea densa* nach Belichtung. Die Viscosität wurde durch die Verlagerungsgeschwindigkeit der Chloroplasten beim Zentrifugieren festgestellt. [Nach STÅLFELT: Ark. Bot. A 33 (1946).]

keit von der Beleuchtungsstärke und nicht eine solche von der Lichtmenge zeigt, wird ein Reizvorgang vermutet (vgl. S. 683). Die weitere Übertragung der Lichtwirkung soll mit Hilfe des Auxins über Änderungen der Plasmastruktur vor sich gehen, wobei aber zu berücksichtigen ist, daß die Plasmaviscosität durch die verschiedensten Stoffe beeinflußt werden kann.

Es muß nun auch noch daran erinnert werden, daß eine Steigerung der Viscosität ebenfalls durch Belichtung von *Modellsubstanzen* (z. B. β-Lactoglobulin) in Gegenwart von sensibilisierenden Stoffen (Methylenblau) möglich ist (WEIL und BUCHERT), auch sind gelegentlich photosensibilisierte Polymerisationen in vitro beobachtet worden, z. B. von EVANS und URI (vgl. S. 688), so daß Viscositätsänderungen nicht nur durch Reizvorgänge in Gang gesetzt zu werden brauchen.

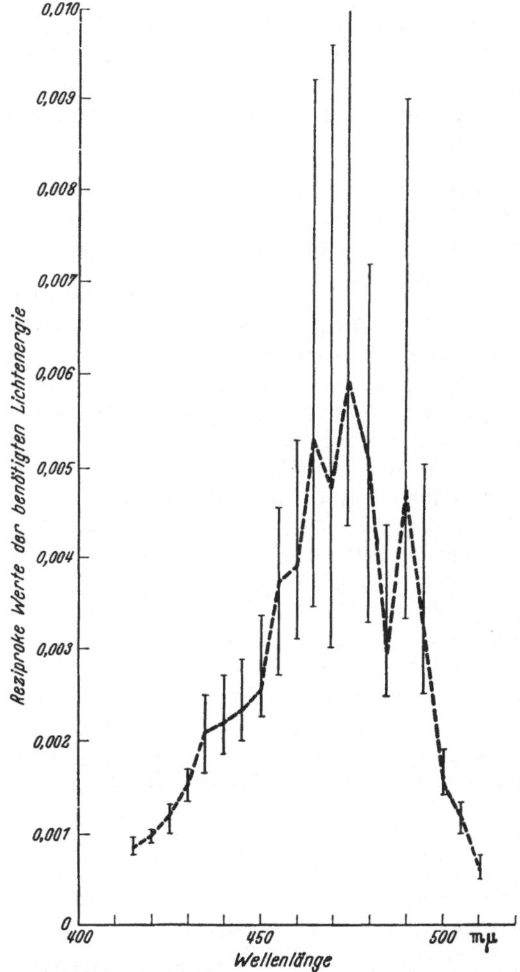

Abb. 26. Das Wirkungsspektrum der Viscositätsänderung von *Helodea densa*-Zellen nach Belichtung. Aufgetragen sind die reziproken Werte der für die Testreaktion erforderlichen Lichtenergie. [Nach VIRGIN: Physiol. Plantarum (Copenh.) 7, 343 (1954).]

Plasmaströmung.

Da durch Belichtung im Plasma Viscositätsänderungen hervorgerufen werden, ist es verständlich, wenn auch die Plasmaströmung durch Licht beeinflußt wird. Nach BOTELLIER wird durch intensive Beleuchtung mit Blaulicht und langwelligem UV, besonders durch die Wellenlänge von 436 mμ eine *Hemmung* der Plasmaströmung verursacht (*Avena*-Koleoptilzellen, Abb. 27). Durch andere Wellenlängen, besonders durch Rotlicht, weniger durch blaue und nicht durch grüne Strahlen, kann es aber auch zu einer *Auslösung* der Plasmaströmung kommen (vgl. BÜNNING [2]). Die angegebenen Wellenlängen lassen die Vermutung zu, daß diese Auslösungsreaktion durch Vermittlung des als Sensibilisator wirksamen Chlorophylls zustande kommen könnte. Da aber die Plasmaströmung

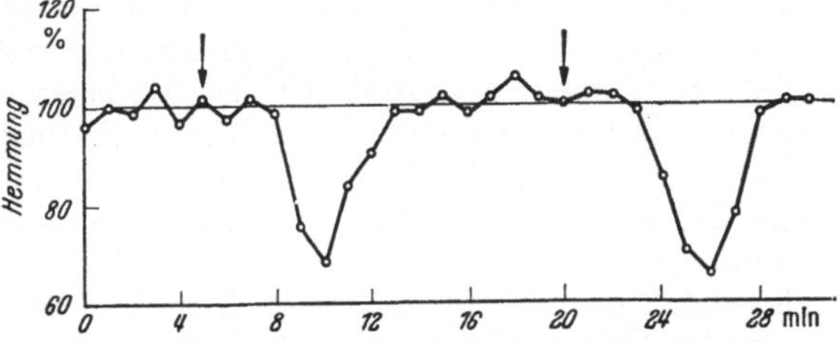

Abb. 27. Hemmung der Protoplasmaströmung in den Zellen der *Avena*-Koleoptile durch zweimalige (Pfeile) Gaben von Licht (436 mμ) für je 8 sec (190 Erg/cm²). [Nach BOTELLIER, aus DU BUY u. NUERNBERGK, Erg. Biol. 12, 325 (1935).]

möglicherweise auch durch Infrarot beeinflußt wird, ist es nicht ausgeschlossen, daß auch hier das gleiche Receptorsystem wie bei Lichtwachstumsreaktionen eine Rolle spielt. Eine kausale Deutung der Lichtwirkung und ein Anschluß an die verschiedenen Theorien der Plasmaströmung ist bisher nicht möglich.

Permeabilität.

Der Einfluß des Lichtes auf die Permeabilität wird im Zusammenhang mit dem Lichteinfluß auf die Stoffaufnahme an anderer Stelle dieses Buches[1] besprochen, auf den deshalb hier verwiesen werden kann.

3. Polarität und Licht.

Bei der Induktion der Polarität handelt es sich offenbar um die Herstellung einer protoplasmatischen Asymmetrie, die ihren Sitz in Oberflächenschichten des Protoplasmas hat (BÜNNING und VON WETTSTEIN). Diese Polarität kann bei jungen Zellen durch Licht und UV induziert werden (KNAPP, MOSEBACH [2], dort weitere Literatur). So ließ sich bei befruchteten Eiern von *Cystosira* durch Licht, besonders der Wellenlängen unter 490—520 mμ, eine Polarität der Eier dieser Braunalgen hervorrufen (MOSEBACH [1]). Bei *Equisetum*-Sporen konnte die Polarität sogar durch eine partielle Bestrahlung des oberflächlich gelegenen Cytoplasmas ohne Bestrahlung des Kerns oder der Plastiden induziert werden (MOSEBACH [2]). Infrarotes bis grünes Licht ist wirkungslos, erst durch Blaulicht und UV wird die Polarisierung ermöglicht; besonders stark wirken Wellenlängen von 235—280 mμ, weniger jedoch solche von 313—366 mμ (MOSEBACH [1], WHITAKER).

Aus diesen Beobachtungen ergibt sich, daß, wie zunächst angenommen, Carotinoide wohl kaum als Receptoren der Strahlenwirkung in Frage kommen. Wegen der Absorption allein im Cytoplasma, der Wirkung im kurzwelligen UV und im blauen Licht könnte an Riboflavin als Lichtacceptor gedacht werden (BÜNNING [2]). Auch hier ist es noch unbekannt, wie durch das Licht die Strukturänderung des Plasmas schließlich erreicht wird. WHITAKER denkt an eine die Wirkung vermittelnde Auxininaktivierung. In ähnlicher Weise wie die Polarität kann auch die Dorsiventralität bei Brutkörpern von *Marchantia* durch Licht induziert werden (FITTING).

4. Photoelektrische Effekte.

Bei Belichtung lassen sich an den belichteten Pflanzenteilen und Zellen Potentialänderungen nachweisen; so beobachteten L. und M. BRAUNER [1], daß bei Belichtung der Blättchen von *Helodea densa* eine vorübergehende Negativität auftritt, die später von einer starken Positivität abgelöst wird. Die Potentialänderungen könnten nach der Meinung der Autoren vielleicht darauf zurückgeführt werden, daß zunächst die Kationenpermeabilität, später die Anionenpermeabilität verändert wird. Dieses „Diffusionspotential" ließ sich besonders durch kürzere Wellenlängen beeinflussen, so daß an Lichtwirkungen gedacht wurde, die durch von den Plasmakolloiden selbst absorbierte Strahlen zustande kommen. BROWN beobachtete ebenso Potentialänderungen bei *Chara* und fand hier, daß durch rotes und auch blaues Licht, nicht jedoch durch grünes Licht, eine Wirkung erzielt wurde; als Photosensibilisator wurde daher das Chlorophyll in Erwägung gezogen. Auch BACKUS und SCHRANK beobachteten unterschiedliche Potentiale bei Avenakoleoptilen. Wurde mit niedriger, zu positiver Krümmung führender Lichtintensität bestrahlt, erhielten sie auf der Lichtseite ein negatives Potential, bei hoher Lichtintensität (die zu negativer Krümmung

[1] Vgl. diesen Band S. 381: BRAUNER, L., Die Beeinflussung des Stoffaustausches durch das Licht.

führt) dagegen ein positives Potential. Wichtig ist hier, daß dieser transversale elektrische Effekt sich bereits vor der Krümmung der Koleoptile zeigt. Er muß also durch Vorgänge verursacht sein, die im Gefolge der primären Lichtwirkung auftreten. Ob die verschiedenen genannten Beobachtungen von Potentialänderungen auf die gleiche Ursache zurückzuführen sind, ist unwahrscheinlich und nicht bekannt.

5. Irreversible Veränderungen durch sichtbares Licht.

Letalwirkungen.

Durch sichtbares Licht können unter besonderen Umständen Schädigungen und Inaktivierungen hervorgerufen werden. So beobachteten WAHL und LATARJET unter Verwendung hoher Lichtintensitäten Inaktivierungen bei *Bactericphagen.* Auch langwelliges UV und Blaulicht geeigneter Intensität führte bei Phagen zu Inaktivierungen (LATARJET und MILÉTIC). Bei verschiedenen tierischen *Viren* konnten SKINNER und BRADISH kürzlich die erhebliche inaktivierende Wirkung von Tageslicht (4 Std), aber auch die von künstlichem Licht feststellen. In bestimmten Fällen vermochte eine Zugabe von 10% Kaninchenserum zum Suspensionsmittel die Inaktivierung zu reduzieren. JONES und HOLLAENDER fanden in solchen Versuchen eine Abtötung von *Bakterien (Escherichia coli).* Auf frühere entsprechende Untersuchungen von NOETHLING und STUBBE, sowie LIECHTI und FEISTMANN über Inaktivierungen durch Starklicht sei verwiesen. SWART-FÜCHTBAUER und RIPPEL-BALDES untersuchten die bactericide Wirkung des Sonnenlichtes bei *Bacterium prodigiosum* genauer. Es ließ sich auch hier feststellen, daß Licht der Wellenlängen >313 mμ ohne Verwendung von Sensibilisatoren die Bakterien schädigt. Ein Gipfel der Schädigung scheint zwischen 366—405 mμ zu liegen. Die Autoren vermuten, daß als wirksame Photoreceptoren hier gewisse Enzyme eine Rolle spielen, wie etwa das Cytochromsystem (vgl. Abb. 20, S. 684). Die oben S. 690 erwähnte Hemmung der Hefeatmung durch Blaulicht würde damit übereinstimmen. Auch hier hängt übrigens die Größe der Wirkung von der Art des verwendeten Agarnährbodens ab (vgl. Photoreaktivierung, S. 681 u. 677).

Bei niederen Organismen können Schädigungen schließlich auch dadurch entstehen, daß durch Änderung der Ernährungsbedingungen oder durch Mutation in den Zellen photosensibilisierende Substanzen gebildet werden, die Schädigungen der Organismen herbeiführen und diese bei Belichtung zum Absterben bringen (CLAES, GIESE [2]).

Photooxydation.

Besonders müssen auch noch jene Photooxydationen erörtert werden, die bei chlorophyllhaltigen Pflanzen in Gegenwart von sehr hoher Lichtintensität, bei Entzug des CO_2 während der Photosynthese, bei Erhöhung des O_2 oder bei Hemmung der Photosynthese durch Enzyminhibitoren auftreten. Diese in engem Zusammenhang mit der Photosynthese stehenden Reaktionen sind viel untersucht worden (NOACK, WARBURG, vgl. auch RABINOWITCH [1]); sie werden in Bd. 5 des Handbuches ausführlich besprochen. Die Effekte beruhen darauf, daß die im Chlorophyll absorbierte Lichtenergie nicht allein den Umsetzungen im Gang der Photosynthese dient, sondern auch zur Oxydation von zur Verfügung stehendem Substrat verwendet werden kann und schließlich zum Abbau des Chlorophylls selbst (Ausbleichen) und damit zu irreversiblen Schädigungen führt. RABINOWITCH nimmt im Zusammenhang mit Messungen von GAFFRON, FRANCK und FRENCH, MYERS und BURR an, daß der primäre photochemische Prozeß der Photooxydation in vivo bei chlorophyllhaltigen Pflanzen identisch ist mit dem primären photochemischen Prozeß der Photosynthese. Irreversibles Ausbleichen des Chlorophylls und vorhergehende Photooxydation wurde in gleicher Weise unter anderen bei *Algen* (BIEBL [2], MONTFORT, MYERS und BURR), bei *Farnen* (FÖCKLER) und in Blättern *höherer Pflanzen* (FRANCK und FRENCH, MONTFORT und Mitarbeiter u. a.) beobachtet.

IV. Die Wirkungen von Hochfrequenzfeldern.

(Zentimeter-, Ultrakurz- und Langwellen.)

Die bisherigen Erfahrungen über die Wirkung von verschiedenen Hochfrequenzfeldern auf Zellen und Gewebe sind äußerst widerspruchsvoll, obwohl eine Klärung dieses Problems, abgesehen von der Aufhellung des Fragenkreises selbst, für die Ursachen der Korrelation zwischen Sonnenfleckenperiodizität und biologischer Rhythmik und für eine eventuelle Beziehung biologischer Vorgänge zu Wetterfronten und anderen meteorologischen Vorgängen von erheblicher Bedeutung sein könnte. Die unterschiedlichen Ergebnisse liegen möglicherweise zum Teil an der komplexen Materie selbst und sind sicher zum Teil durch sehr unterschiedliche Versuchsanordnungen bedingt. Außerdem ist der für die Untersuchung in Frage kommende Wellenlängenbereich sehr ausgedehnt, so daß, falls eine Strahlenwirkung in einem Bereich bei bestimmter Feldstärke vorliegt, eine solche in einem anderen Wellenbereich völlig fehlen kann. Die Versuchsergebnisse sind aber selbst bei vergleichbaren Fragestellungen durchaus unterschiedlich. Hinweise auf einen Teil der vorliegenden älteren Literatur sind in der Arbeit von RAMSHORN [2] zu finden.

Neue Untersuchungen liegen zunächst bei relativ *niedrigen* Frequenzen bis zu 20 kHz (also Wellenlängen von 15 km) vor: REITER fand bei 7 kHz im Wechselfeld bei tierischen Zellen p_H-Wert Änderungen um 0,02—0,1 Einheiten vom Ruhewert. Im Gegensatz dazu stehen die Versuche von KAUDEWITZ, der die Generationsdauer von Amöben und Protozoen bei 10 und 15 kHz und etwa 1,2 V/cm Feldstärke im Kondensatorfeld untersuchte und *keinen* Einfluß dieser Wellenlängen feststellen konnte, obwohl die Genauigkeit der Methode gestattete, Änderungen der Generationsdauer bei Temperaturunterschieden von 0,3° C eindeutig zu erkennen.

Bei *hohen* Frequenzen über 30 MHz (also Wellenlängen unter 10 m) liegen relativ viele, aber auch widerspruchsvolle Ergebnisse vor. JONAS fand bei ölhaltigen Samen eine Erhöhung der Keimungsrate; ebenso KÖHLER bei Samen von *Lepidium, Triticum, Helianthus* und *Solanum lycopersicum*. RAMSHORN [1,2] bestätigte diese Befunde an 24 Std vorgequollenen Samen von *Solanum lycopersicum* und konnte darüber hinaus bei verschieden langer Bestrahlung (von 5 sec bis zu 960 sec) bei 75 MHz 2 Maxima der Förderung, dagegen bei 150 MHz nur 1 Maximum der Förderung und dazu bei längerer Bestrahlung eine statistisch gesicherte Keimungshemmung feststellen. Bei 200 MHz (1,5 m) fanden KIEPENHEUER, BRAUER und HARTE sowie BRAUER eine Erhöhung der Mitosen in Wurzelspitzen von *Vicia faba*; höhere Feldstärken ergaben Wachstumshemmungen. Nach HARTE läßt sich auch die Meiose von *Oenothera* durch relativ niedrige Strahlungsintensitäten beeinflussen. RAMSHORN fand außerdem eine Atmungssteigerung bei Weizenkörnern und vermehrte Zellteilung bei der Hefe (37,5 und 75 MHz).

Demgegenüber konnten ELWERT, REINERT und MITLACHER weder eine Erhöhung der Keimungsgeschwindigkeit bei Samen *(Digitalis)* noch eine Wachstumsbeschleunigung bei Hefe nachweisen (200 MHz). Auch KAUDEWITZ vermochte mit 166 MHz mit der oben genannten Technik und bei gleichen Objekten keine Änderung der Generationsdauer zu erzielen.

Im Bereich von Zentimeterwellen haben BOYLE und Mitarbeiter am Hautbindegewebe von Ratten deutliche Veränderungen feststellen können.

Durch die genannten Strahlungen können grundsätzlich *verschiedene* physikalische *Primärwirkungen* hervorgerufen werden. Zunächst ist an eine Erhöhung der *Temperatur* durch die Wechselfelder zu denken, die in verschiedenen Untersuchungen auch festgestellt wurde (vgl. JONAS, RAMSHORN [2]) und die in der Medizin für die wesentliche Wirkung der Ultrakurzwellenbehandlung gehalten wird (RAJEWSKY [1]). Ungenügende Beachtung der Temperaturwirkungen könnte somit allein schon manche Widersprüche in den Befunden älterer Arbeiten erklären. RAMSHORN und auch JONAS halten aber in ihren Untersuchungen eine solche allgemeine und dazu relativ geringe Temperaturerhöhung als Erklärung ihrer Befunde für unzureichend. Für wichtiger wird in diesen Arbeiten eine mögliche, stärkere Erhöhung der *Punktwärme* in bestimmten Zellen bzw. Zellteilen gehalten (vgl. auch SAUTER und SCHWARTZ). Von RAMSHORN [2] wird schließlich noch darauf aufmerksam gemacht, daß durch die Wechselfelder bei einem elektrisch neutralen Molekül eine *elektrische Polarisation* hervorgerufen werden kann, und daß bei Molekülen mit Dipolcharakter *Molekülschwingungen* auftreten können, die möglicherweise zu Änderungen des Zustandes der Zellen führen.

Die Übersicht ergibt also, daß die bisherigen Erfahrungen über die Wirkung von Hochfrequenzfeldern nicht eindeutig sind, und daß die durch die Wechselfelder hervorgerufenen Wirkungen im einzelnen im biologischen Material noch gar nicht genauer untersucht wurden. Das Problem erfordert noch gründliche Prüfung; selbst die positiven Befunde bedürfen der weiteren Bestätigung.

Literatur.

ADAMS, W. R., and E. POLLARD: Combined thermal and primary ionization effects on a bacterial virus. Arch. of Biochem. a. Biophysics **36**, 311—322 (1952). — ALDOUS, J. G., and D. K. R. STEWART: The effect of ultraviolet radiation upon enzymatic activity and viability of the yeast cell. Canad. J. Med. Sci. **30**, 561—570 (1952). — ALEXANDER, P.: A physicochemical method of testing the protective action of chemical compounds against the lethal effects of ionizing radiations. Brit. J. Radiol. **26**, 413—416 (1953). — ALLEN, A. O.: The yields of free H and OH in the irradiation of water. Radiation Res. **1**, 85—96 (1954). — ALLEN, A. O., C. J. HOCHANADEL, I. A. GHORMLEY and T. W. DAVIS: Decomposition of water and aqueous solutions under mixed fast neutron and gamma radiation. J. Physic. Chem. **56**, 575—586 (1952). — ALLEN, E. G., M. R. BOVARNICK and J. C. SNYDER: The effect of irradiation with ultraviolet light on various properties of typhus rickettsiae. J. Bacter. **67**, 718—723 (1954). — ALPER, T.: [1] Hydrogen peroxide and the indirect effect of ionizing radiations. Nature (Lond.) **162**, 615—616 (1948). — [2] A new after-effect of X-rays on dilute aqueous suspensions of bacteriophage. Nature (Lond.) **169**, 964—965 (1952). — [3] The indirect inactivation of bacteriophage S 13 during and after exposure to ionizing radiations. Nature (Lond.) **169**, 183—184 (1952. — [4] The inactivation of free bacteriophage by irradiation and by chemical agents. J. Gen. Microbiol. **11**, 313—324 (1954). — ALTENBURG, L. S., and E. ALTENBURG: The lowering of the mutagenic effectiveness of ultraviolet by photoreactivating light in Drosophila. Genetics **37**, 545—553 (1952). — ANDERSON, E. H.: Heat reactivation of ultraviolet inactivated bacteria. J. Bacter. **61**, 389—394 (1951). — APPLEYARD, R. K.: [1] The inactivation of dried hemoglobins by fast charged particles. Arch. of Biochem. a. Biophysics **35**, 121—131 (1952). — [2] The irradiation of dried hemoglobins by fast charged particles. II. Arch. of Biochem. a. Biophysics **40**, 111—126 (1952). — AUERBACH, C.: The chemistry of biological after-effects of ultraviolet and ionizing radiations. Brit. J. Radiol. **26**, 212—213 (1953).

BACHOFER, C. S., C. F. EHRET, S. MAYER and E. L. POWERS: The influence of temperature upon the inactivation of a bacterial virus by X-rays. Proc. Nat. Acad. Sci. U.S.A. **39**, 744—750 (1953). — BACKUS, G. E., and A. R. SCHRANK: Electrical and curvature responses of the Avena coleoptile to unilateral illumination. Plant Physiol. **27**, 251—262 (1952). —

Bacq, Z. M.: Die indirekte Wirkung von Röntgen- und ultravioletten Strahlen. Experientia (Basel) **7**, 11—19 (1951). — Bacq, Z. M., u. A. Herve: Ein chemischer Schutz gegen Röntgenstrahlen. Strahlenther. **95**, 215—237 (1954). — Baker, W. K., and E. v. Halle: The production of dominant lethals in Drosophila by fast neutrons from cyclotron irradiation and nuclear detonations. Science (Lancaster, Pa.) **119**, 46—49 (1954). — Bannister, T. T.: Energy transfer between chromophore and protein in phycocyanin. Arch. of Biochem. a. Biophysics **49**, 222—233 (1954). — Barron, E. S. G.: [1] The effect of ionizing radiations on some systems of biological importance. Symposium on Radiobiology, ed. by J. J. Nickson. New York: John Wiley & Sons Inc. 1952. — [2] The effect of X-rays on systems of biological importance. In A. Hollaender, Radiation biology, Bd. I, Teil 1, S. 283—313. New York u. London 1954. — [3] The role of free radicals and oxygen in reactions produced by ionizing radiations. Radiation Res. **1**, 109—124 (1954). — Barron, E. S. G., and P. Finkelstein: Studies on the mechanism of action of ionizing radiations. X. Effect of X-rays on some physicochemical properties of proteins. Arch. of Biochem. a. Biophysics **41**, 212—232 (1952). — Barron, E. S. G., and Ph. Johnson: Studies on the mechanism of action of ionizing radiations. XI. Inactivation of yeast alcohol dehydrogenase by X-irradation. Arch. of Biochem. a. Biophysics **48**, 149—153 (1954). — Barron, E. S. G., and S. Levine: Oxydation of alcohols by yeast alcohol dehydrogenase and by the living cell. The thiol groups of the enzyme. Arch. of Biochem. a. Biophysics **41**, 175—187 (1952). — Barron, E. S. G., and S. L. Seki: Studies on the mechanism of action of ionizing radiations. VII. Cellular respiration, cell division and ionizing radiations. J. Gen. Physiol. **35**, 865—872 (1952). — Barron, E. S. G., Ph. Johnson and A. Cobure: Effect of X-irradiation on the absorption of purines and pyrimidines. Radiation Res. **1**, 410—425 (1954). — Bawden, F. C., and A. Kleczkowski: Ultraviolet injury to higher plants counteracted by visible light. Nature (Lond.) **169**, 90—91 (1952). — Beaven, G. H., and E. R. Holiday: Ultraviolet absorption spectra of proteins and amino acids. Adv. Protein Chem. **7**, 319—386 (1952). — Berger, H., F. L. Haas, O. Wyss and W. S. Stone: Effect of sodium azide on radiation damage and photoreactivation. J. Bacter. **65**, 538—543 (1953). — Bernstein, M. H.: Deoxyribonucleoproteins of cell nuclei: Sensitivity to X-rays. Nature (Lond.) **174**, 463 (1954). — Beth, K.: Experimentelle Untersuchungen über die Wirkung des Lichtes auf die Formbildung von kernhaltigen und kernlosen Acetabularia-Zellen. Z. Naturforsch. 8b, 334—342 (1953). — Biebl, R.: [1] Wirkung der UV-Strahlung auf Allium-Zellen. Protoplasma **36**, 491—513 (1942). — [2] Resistenz der Meeresalgen gegen sichtbares Licht und gegen kurzwellige UV-Strahlen. Protoplasma **41**, 353—377 (1952). — Bier, M., and F. F. Nord: On the mechanism of enzyme action XLVII. Effects of high intensity electron bombardment on crystalline trypsin. Arch. of Biochem. a. Biophysics **35**, 204—215 (1952). — Bora, K. C.: Delayed effects in chromosome breakage by X-rays in Tradescantia bracteata. J. Genet. **52**, 140—151 (1954). — Blau, M., u. K. Altenburger: Über einige Wirkungen von Strahlen. II. Z. Physik **12**, 315—329 (1922). — Blout, E. R.: Ultraviolet microscope and ultraviolet microspectroscopy. Adv. Biol. a. Med. Physics **3**, 285—336 (1953). — Blum, H. F., J. S. Cook and G. M. Loos: [1] A comparison of five effects of ultraviolet light on the Arbacia egg. J. Gen. Physiol. **37**, 313—324 (1954). — Blum, H. F., E. F. Kauzmann and G. B. Chapman: [2] Ultraviolet light and the mitotic cycle in the Sea Urchins egg. J. Gen. Physiol. **37**, 325—333 (1954). — Blum, H. F., J. C. Robinson and G. M. Loos: [3] The loci of action of ultraviolet and X-radiation and of photorecovery in the egg and sperm of the Sea Urchin Arbacia punctulata. J. Gen. Physiol. **35**, 323—342 (1951). — Borthwick, H. A., M. W. Parker and S. B. Hendricks: [1] Recent developments in the control of flowering by photoperiod. Amer. Naturalist **84**, 117—134 (1950). — Borthwick, H. A., S. B. Hendricks and M. W. Parker: [2] Action spectrum for inhibition of stem growth in dark grown seedlings of albino and nonalbino barley. Bot. Gaz. **113**, 95—105 (1952). — Bowen, E. J.: [1] The chemical aspects of light, 3. Aufl. Oxford: Clarendon Press 1949. — [2] Resonance transfer of energy between molecules. Symposia Soc. f. Exper. Biol. **5**, 152—159 (1951). — Bottelier, H. B.: Über den Einfluß äußerer Faktoren auf die Protoplasmaströmung in der Avena-Koleoptile. Rec. Trav. Bot. néerl. **31**, 474—582 (1934). — Boyle, A. C., H. F. Cook and T. J. Buchanan: The effects of microwaves. A preliminary investigation. Brit. J. Physic. Med., N. S. **13**, 2—9 (1950). — Brandt, C. L., P. J. Freemann and P. A. Swenson: The effect of radiations on galactocymase formation in yeast. Science (Lancaster, Pa.) (N. Y.) **113**, 383—384 (1950). — Brauer, I.: Experimentelle Untersuchungen über die Wirkung von Meterwellen verschiedener Feldstärke auf das Teilungswachstum der Pflanzen. Chromosoma (Wien) **3**, 483—509 (1950). — Brauner, L.: Untersuchungen über die Photolyse des Heteroauxins I. Z. Bot. **41**, 291—342 (1953). — Brauner, L., u. M. Brauner: [1] Untersuchungen über den photoelektrischen Effekt in Membranen I. Protoplasma **28**, 230—261 (1937). — [2] Untersuchungen über die Photolyse des Heteroauxins II. Z. Bot. **42**, 83—124 (1954). — Bresch, C.: Zur Reaktivierung von Bacteriophagen. 1. Mitteilung. Z. Naturforsch. 5b, 420—422 (1950). — Brode, W. R.: The absorption spectra of vitamins, hormones and enzymes. Adv. Enzymol. **4**, 269—311

(1946). — BROWN, S. O.: Relation between light and the electric polarity of Chara. Plant. Physiol. **13**, 713—736 (1938). — BÜCHER, TH.: Probleme des Energietransportes innerhalb lebender Zellen. Adv. Enzymol. **14**, 1—48 (1953). — BÜNNING, E.: [1] Physikalisch-chemische Grundlagen der biologischen Vorgänge. Fortschr. Bot. **11**, 128—145 (1944). — [2] Entwicklungs- und Bewegungsphysiologie der Pflanze, 3. Aufl. Berlin: Springer 1953. — BÜNNING, E., u. D. v. WETTSTEIN: Polarität und Differenzierung an Mooskeimen. Naturwiss. **40**, 147—148 (1953). — BUTLER, J. A. V., and J. T. RANDALL: Progress in biophysics and biophysical chemistry, Bd. I und folgende. 1949—1955. — BUY, H. G. DU, and E. L. NUERNBERGK: Phototropismus und Wachstum der Pflanzen III. Erg. Biol. **12**, 323—543 (1935).

CALCUTT, G.: A factor influencing the intracellular exposure of sulphhydryl groups. Nature (Lond.) **166**, 443—444 (1950). — CALDAS, L. R., et TH. CONSTANTIN: Courbes de survie de levures haploides et de diploides soumises aux rayons ultraviolets. C. r. Acad. Sci. (Paris) **232**, 2356—2358 (1951). — CALDECOTT, R. S.: [1] Inverse relationship between the water content of seeds and their sensitivity of X-rays. Science (Lancaster, Pa.) **120**, 809—810 (1954). — [2] Effect of ionizing radiation on seeds of barley. Radiation Res. **1**, 490 (1954). — CANZANELLI, A., R. GUILD and D. RAPPORT: Ammonia and urea production and changes in absorption spectra of nucleic acid derivatives following ultraviolet irradiation. Amer. J. Physiol. **167**, 364—374 (1951). — CARLSON, J. G.: Immediate effects on divisions, morphology and viability of the cell. In HOLLAENDER, Radiation Biology, Bd. I, Teil 2, S. 763—824. 1954. — CASPERSSON, T. O.: Cell growth and cell function. New York: W. W. Norton & Comp. Inc. 1950. — CLAES, H.: Analyse der biochemischen Synthesekette für Carotinoide mit Hilfe von Chlorella-Mutanten. Z. Naturforsch. **9**b, 461—469 (1954). — CONRAD, W. E.: The isolation of oxamide and parabanic acid from the products of ultraviolet irradiated uracil. Radiation Res. **1**, 523—529 (1954). — COURCY jr., S. J. DE, J. O. ELY and M. H. ROSS: Effect of ultraviolet light on deoxyribonucleic acid in rat thymocytes. Nature (Lond.) **172**, 119 (1953). — CROWTHER, J. A.: Some considerations relative to the action of X-rays on tissue cells. Proc. Roy. Soc. Lond. **96**, 207 (1924).

DALE, W. M.: [1] Some aspects of the biochemical effects of ionizing radiations. Symposium on Radiobiology, ed. by J. J. NICKSON, S. 177—188. New York: John Wiley & Sons Inc. 1952. — [2] The indirect action of ionizing radiations on aqueous solutions and its dependance on the chemical structure of the substrate. J. Cellul. a. Comp. Physiol. **39**, Suppl. 1, 39—56 (1952). — [3] Basic radiation biochemistry. In A. HOLLAENDER, Radiation biology, Bd. I, Teil 1, S. 255—281. New York u. London: McGraw Hill Book Comp. Inc. 1954. — DANIEL, G., and H. PARK: Glutathion and X-ray injury in hydra and paramaecium. J. Cellul. a. Comp. Physiol. **42**, 359—367 (1953). — DANIELS, M., G. SCHOLES and J. WEISS: After-effect in aqueous solutions of deoxyribonucleic acid irradiated with X-rays. Nature (Lond.) **171**, 1153 (1953). — DANNENBERG, M.: Zur Systematik der Ultraviolettabsorption I. Über substituierte Acetophenone, Benzoesäuren und Zimtsäuren. Z. Naturforsch. **4**b, 327—344 (1949). — DAVIES, H. G., and P. M. B. WALKER: Microspectrometry of living and fixed cells. Progr. Biophysics a. Biophysical Chem. **3**, 195—236 (1953). — DEBYE, P., and J. V. EDWARDS: A note on the phosphorescence of proteins. Science (Lancaster, Pa.) **116**, 143—144 (1952). — DENFFER, D. v., u. A. FISCHER: Papierchromatographischer Nachweis des β-Indolaldehyds in photolytisch zersetzter IES-Lösung. Naturwiss. **39**, 549—550 (1952). — DESSAUER, F.: [1] Über einige Wirkungen von Strahlen I. Z. Physik **12**, 38—47 (1922). — [2] Quantenbiologie. Berlin: Springer 1954. — DEWHURST, H. A., A. H. SAMUEL and J. L. MAGEE: A theoretical survey of the radiation chemistry of water and aqueous solution. Radiation Res. **1**, 62—84 (1954). — DITTRICH, W., u. G. SCHUBERT: Der Wirkungsmechanismus schneller Elektronen in biologischer Materie. Strahlenther. **92**, 532—554 (1953). — DONIACH, J., A. HOWARD and S. R. PELC: Autoradiography. Progr. Biophysics a. Biophysical Chem. **3**, 1—26 (1953). — DOTY, B., and E. P. GEIDUSCHEK: Optical properties of proteins. In H. NEURATH and K. BAILEY, The proteins, S. 393—460. New York 1953. — DREBINGER, K.: [1] Kerngifte und Lichtstrahlung. Eine Studie an Froschspermien zur Wirkungsanalyse der Kerngifte. Roux' Arch. **145**, 174—204 (1951). — [2] Die an das Licht gebundene Kernschädigung der Kerngifte des Trypoflavintyps. Roux' Arch. **147**, 128—130 (1954). — DUGGAR, B. M.: Biological effects of radiation. New York u. London 1936. — DULBECCO, R.: [1] Reactivation of ultraviolet inactivated bacteriophages by visible light. Nature (Lond.) **163**, 949—950 (1949). — [2] Experiments on photoreactivation of bacteriophages inactivated with ultraviolet radiation. J. Bacter. **59**, 329—347 (1950). — [3] Experiments on photoreactivation of inactive bacteriophages. J. Cellul. a. Comp. Physiol. **39**, Suppl. 1, 125—128 (1952). — [4] Photoreactivation. In A. HOLLAENDER, Radiation biology, Bd. II, S. 455—486. New York: McGraw Hill Book Comp. Inc. 1955.

ELWERT, G., J. REINERT u. W. MITLACHER: Zur Frage des Einflusses von Meterwellen auf Keimungs- und Wachstumsvorgänge bei Pflanzen. Z. Naturforsch. **7**b, 109—112 (1952). — ENGEL, O. S.: Veränderungen der Strahlenempfindlichkeit der Weizensamen. Dokl. Akad. Nauk SSSR., N.S. **85**, 229—231 (1952). — EHREBERG, L., Å. GUSTAFSSON, U. LUNDQUIST and N. NYBOM: Irradiation effects, seed soaking and oxygen pressure in barley. Hereditas (Lund) **39**, 493—504 (1953). — ERRERA, M.: [1] Mécanismes de l'action des radiations sur le noyau cellulaire. Ann. Soc. roy. Sci. méd. et natur. Brux. **5**, 65—176 (1952). — [2] Etude photochimique de l'acide désoxyribonucleique. I. Measures énergetiques. II. Etude des produites de la photolyse. Biochim. et Biophysica Acta **8**, 30—37, 115—124 (1952). — [3] Mechanism of biological action of ultraviolet and visible radiations. Progr. Biophysics a. Biophysical Chem. **3**, 88—130 (1953). — ERRERA, M., et A. HERVE: Méchanismes de l'action biologique des radiations. Liège: Desoer; Paris: Masson & Cie. 1951. — EVANS, M. G., and N. URI: The photochemical formation and reactions of atoms and radicals in aqueous systems. Symposia Soc. f. Exper. Biol. **5**, 130—140 (1952).

FANO, U.: Principles of radiological physics. In A. HOLLAENDER, Radiation biology, Bd I, Teil 1, S. 1—144. New York u. London: McGraw Hill Book Comp. Inc. 1954. — FERRI, M. G.: Fluorescence and photoinactivation of indolacetic acid. Arch. Biochim. **31**, 127—131 (1951). — FITTING, H.: Untersuchungen über die Induktion der Dorsiventralität bei den Brutkörperkeimlingen der Marchantieen. III. Jb. wiss. Bot. **85**, 169—242 (1937). — FÖCKLER, H.: Über den Einfluß des Lichtes auf die Atmung farbloser und assimilierender Gewebe und seine Rolle beim „funktionellen Sonnenstich". Jb. wiss. Bot. **87**, 45—92 (1938). — FÖRSTER, TH.: [1] Farbe und Konstitution organischer Verbindungen vom Standpunkt der modernen physikalischen Theorie. Z. Elektrochem. **45**, 548—573 (1939). — [2] Fluoreszenz organischer Verbindungen. Göttingen: Vandenhoeck & Ruprecht 1951. — FORSSBERG, A.: On the possibility of protecting the living organism against roentgen rays by chemical means. Acta radiol. (Stockh.) **33**, 296—304 (1950).— FORSSBERG, A., and N. NYBOM: Combined effects of cystein and irradiation on growth and cytology of Allium cepa roots. Physiol. Plantarum (Copenh.) **6**, 78—95 (1953). — FRANCK, J.: A critical survey of the physical background of photosynthesis. Annual Rev. Plant Physiol. **2**, 53—86 (1951). — FRANCK, J., and C. S. FRENCH: Photooxydation processes in plants. J. Gen. Physiol. **25**, 309—324 (1941). — FRANCK, J., and R. PLATZMAN: Physical principles underlying photochemical radiation. Chemical and radiobiological reactions. In A. HOLLAENDER, Radiation biology, Bd. I, Teil 1, S. 191—253. New York u. London: McGraw Hill Book Comp. Inc. 1954. — FREEMANN, P. J., and A. C. GIESE: Photodynamic effects on metabolism and reproduction in yeast. J. Cellul. a. Comp. Physiol. **39**, 301—322 (1952). — FRENCH, C. S., and V. K. YOUNG: The fluorescence spectra of red algae and the transfer of energy from phycoerythrin to phycocyanin and chlorophyll. J. Gen. Physiol. **35**, 873—890 (1952). — FREYTAG, H.: Möglichkeit einer Photosynthese von Pyridin-Derivaten. Z. Naturforsch. **3**b, 465—466 (1948). — FRITZ NIGGLI, H.: Unterschiedliche Wirksamkeit der 180 keV- und 31 MeV-Strahlen auf die Letalität von C 57-Mäusen. Experientia (Basel) **10**, 209—210 (1954).

GÄRTNER, H., u. K. PETERS: Einfluß der Dosis und Dosisleistung von Röntgenstrahlen und schnellen Elektronen auf die Mitosehäufigkeit in Gewebekulturen. Strahlenther. **92**, 555—562 (1953). — GAFFRON, H.: Über den Mechanismus der Sauerstoffaktivierung durch belichtete Farbstoffe. Biochem. Z. **264**, 251—271 (1933). — GALSTON, A. W.: [1] Riboflavin-sensitiv photooxydation of indol-acetic acid and related compounds. Proc. Nat. Acad. Sci. U.S.A. **35**, 10—17 (1949). — [2] Riboflavin, light and the growth of plants. Science (Lancaster, Pa.) **111**, 619—624 (1950). — GALSTON, A. W., and R. S. BAKER: [1] Inactivation of enzymes by visible light in the presence of riboflavin. Science (Lancaster, Pa.) **109**, 485—486 (1949). — [2] Studies on the physiology of light action II. The photodynamic action of riboflavin. Amer. J. Bot. **36**, 773—780 (1949). — [3] Studies on the physiology of light action. V. Photoinductive alteration of auxin metabolism in etiolated peas. Amer. J. Bot. **40**, 512—516 (1953). — GALSTON, A .W., J. BONNER and R. S. BAKER: Flavoprotein and peroxidase as components of the indol-acetic acid oxydase system of peas. Arch. of Biochem. a. Biophysics **42**, 456—469 (1953). — GARRISON, V. M., H. R. HAYMOND and B. M. WEEKS: Some affects of heavy-particle irradiation of aqueous acetic acid. Radiation Res. **1**, 97—108 (1954). — GESSNER, F.: Die Assimilation vitalgefärbter Chloroplasten. Planta (Berl.) **32**, 1—5 (1941). — GHORMLEY, J. A., and C. J. HOCHANADEL: The yields of hydrogen and hydrogen peroxide in the irradiation of oxygen saturated water with cobalt gamma-rays. J. Amer. Chem. Soc. **76**, 3351—3352 (1954). — GILES, N. H.: Radiation induced chromosome aberrations in Tradescantia. In A. HOLLAENDER, Radiation biology, Bd. I, Teil 2, S. 713—762. New York u. London: McGraw Hill Book Comp. Inc. 1954. — GILES jr., N. H., and H. PH. RILEY: Studies on the mechanism of the oxygen effect on the radiosensitivity of Tradescantia chromosomes. Proc. Nat. Acad. Sci. U.S.A. **36**, 337—344 (1950). — GILES, N. H., F. J. DE SERRES and A. V. BEATTY: The effect of radiation dose fractionation on chromosome aberration, frequencies in Tradescantia microspores. II. Studies with fast neutrons. Genetics **38**, 416—420 (1953). — GIESE, A. C.: [1] Action of ultraviolet radiation on protoplasm. Physiologic. Rev. **30**, 431—458 (1950). — [2] Some properties of a photodynamic pigment from Blepharisma. J. Gen. Physiol. **37**, 259—269 (1954). — GIESE, A. C., C. S. BRANDT, R. IVERSON and P. H. WELLS: [1] Photoreactivation in Colpidium colpoda. Biol. Bull. **103**, 336—344 (1952). —

GIESE, A. C., R. M. IVERSON, D. C. SHEPARD, C. JACOBSON and C. L. BRANDT: [2] Quantum relations in photoreactivation of Colpidium. J. Gen. Physiol. **37**, 249—258 (1953). — GIESE, A. C., C. L. BRANDT, C. JACOBSON, D. C. SHEPARD and R. T. SANDERS: [3] The effect of starvation on photoreactivation in Colpidium colpoda. Physiologic. Zool. **27**, 71—78 (1954). — GIRI, K. V., G. D. KALYANKAR and C. S. VAIDYANATHAN: Photolysis of metionine in presence of photocatalysis. Naturwiss. **41**, 88 (1954). — GLOCKER, R.: Quantenphysik der biologischen Röntgenstrahlenwirkung. Z. Physik **77**, 653—675 (1932). — GLUBRECHT, H.: Über die Wirkung von UV-Strahlen in somatischen Zellen. Z. Naturforsch. 8b, 17—27 (1953). — GRAFFI, A., H. KRIEGEL, H. SCHREIBER u. F. WINDISCH: Die photosensibilisierende Wirkung verschiedener cancerogener und nicht cancerogener Kohlenwasserstoffe auf die Hefezellen. Z. Naturforsch. 8b, 142—145 (1953). — GRANICK, S.: Biosynthesis of chlorophyll and related pigments. Ann. Rev. Plant Physiol. **2**, 115—144 (1951). — GRAY, L. H.: [1] Biological actions of ionizing radiations. Progr. in Biophysics **2**, 240—305 (1951). — [2] The initiation and development of cellular damage by ionizing radiations. Brit. J. Radiol. **26**, 609—618 (1953). — [3] Characteristics of chromosome breakage by different agents. Heredity (Lond.) **6** Suppl. 311—315 (1953). — [4] Some characteristics of biological damage induced by ionizing radiations. Radiation Res. **1**, 189—213 (1954). — GROS, CH. M., et P. MANDEL: Action d'irradiation totale du rat sur les acides nucleiques de la rate. J. belge Radiol. **35**, 357—359 (1952). — GUERRINI, G.: Vgl. E. BÜNNING [1].

HAAS, F., J. B. CLARK, O. WYSS and W. S. STONE: [1] Mutations and mutagenic agents in bacteria. Amer. Naturalist **84**, 261—275 (1950). — HAAS, F. L., E. DUDGEON, F. E. CLAYTON and W. S. STONE: [2] Frequency of chromosomal aberrations as related to rate of irradiation, temperature and gases. Genetics **37**, 589—590 (1952). — [3] Measurement and control of some direct and indirect effects of X-radiation. Genetics **39**, 453—471 (1954). — HARDER, R.: Über Farb- und Musteränderungen bei Blüten. Naturwiss. **26**, 713—722 (1938). — HARM, W., u. W. STEIN: [1] Beeinflussung der UV-Inaktivierung von Coli-Bakterien durch Bebrütungstemperatur und Nährboden. Z. Naturforsch. 8b, 123—133 (1953). — [2] Weitere Experimente zur Reaktivierung von ultraviolett- und peroxyd-inaktivierten Coli-Stämmen. Z. Naturforsch. 8b, 729—741 (1953). — HARRINGTON, N. J., and R. W. KOZA: Effect of X-radiation on the desoxyribonucleic acid and on the size of grasshopper embryonic nuclei. Biol. Bull. **101**, 138—150 (1951). — HART, E. J.: Molecular product and free radical yields of ionizing radiations in aqueous solutions. Radiation Res. **1**, 53—61 (1954). — HARTE, C.: Mutationsauslösung durch Ultrakurzwellen. Chromosoma **3**, 440—447 (1949). — HEIDENTHAL, G., L. B. CLARK and J. W. GOWEN: Comparative effectiveness of X-rays of 124 KV and 50 MeV on Habrobracon eggs. Radiation Res. **1**, 499 (1954). — HEINMETS, F.: Reactivation of ultraviolet inactivated Escherichia coli by pyruvate. J. Bacter. **66**, 455—457 (1953). — HEINMETS, F., W. W. TAYLOR and I. I. LEHMAN: [1] The use of metabolites in the restoration of the vitability of heat and chemically inactivated Escherichia coli. J. Bacter. **67**, 5—12 (1954). — HEINMETS, F., J. J. LEHMANN, W. W. TAYLOR and R. H. KATHAN: [2] The study of factors with influence metabolic reactivation of the ultraviolet inactivated Escherichia coli. J. Bacter. **67**, 511—522 (1954). — HERVE, A., Z. M. BACQ und H. BETZ: Schutzwirkung von Natriumcyanid und -azid gegen letale Röntgenbestrahlung. J. chim. physique etc. **48**, 256—257 (1951). — HEVESY, G. v.: [1] Ionizing radiation and cellular metabolism. Symposium on Radiobiology, ed. by J. J. NICKSON, S. 189—213. New York: John Wiley & Sons Inc. 1952. — [2] Die Anwendung der radioaktiven Indikatoren in der Radiobiologie. Strahlenther. **93**, 325—348 (1954). — HILL, R. F., and H. H. ROSSI: [1] Abscence of photoreactivation in T_1 bacteriophage irradiated with ultraviolet in the dry state. Science (Lancaster, Pa.) **116**, 424—425 (1952). — [2] The ultraviolet sensitivity and photoreactivability of T_1 bacteriophage. I. Effect of irradiation conditions upon survival curves. Radiation Res. **1**, 282—293 (1954). — [3] The ultraviolet sensitivity and photoreactivability of T_1 bacteriophage. II. Interpretation of the survival curves. Radiation Res. **1**, 358—368 (1954). — HIRSHFIELD, H., and A. C. GIESE: Ultraviolet radiation effects on growth processes of Blepharisma undulans. Exper. Cell Res. **4**, 283—294 (1953). — HOLLAENDER, A.: [1] Physical and chemical factors modifying the sensitivity of cells to high energy and ultraviolet radiation. U.S.A. E. C. Document No ORNL 844, 1950. — [2] Radiation Biology, vol. I, High energy radiation. Part 1 u. part 2. New York: McGraw Hill Book Comp. 1954. — [3] Radiation Biology Vol. II. Ultraviolet and related radiations. New York: McGraw Hill Book Comp. 1955. — [4] Effect of long ultraviolet and short visible radiation (3500—4900 Å) on Escherichia coli. J. Bacter. **46**, 531—541 (1943). — HOLLAENDER, A., and G. E. STAPLETON: Fundamental aspects of radiation protection from a microbiological point of view. Physiologic. Rev. **33**, 77—84 (1953). — HOLT, A. S., I. A. BROOKS and W. A. ARNOLD: Some effects of 2537 Å on green algae and chloroplast preparations. J. Gen. Physiol. **34**, 627—645 (1951). — HOTCHKISS, R. D.: The quantitative separation of purines, pyrimidines and nucleosides by paper chromatography. J. of Biol. Chem. **175**, 315—332 (1948). — HOUTERMANS, TH.: [1] Über den Einfluß des Wachstumszustandes eines Bacteriums auf seine Strahlenempfindlichkeit. Z. Naturforsch. 8b, 767—771 (1953). — [2] Über den Einfluß der Temperatur auf biologische Strahlenwirkungen. Z. Naturforsch. 9b, 600—602 (1954). — [3] Über den Verlauf der Inaktivierungskurven für E. coli bei sehr kleinen Bestrahlungsdosen. Strahlenther. 93, 130—137 (1954). — HOWARD, A.: Report of "Symposium on mode of action of ionizing radiations". Nucleonics 7, 26—30 (1950). — HOWARD, A., and S. R. PELC: Synthesis of desoxyribonucleic acid in normal and irradiated cells and its relation to chromosome breakage. Heredity (Lond.) 6 Suppl. 261 (1953). — HUTCHINSON, F., and E. R. MOSBURG jr.: Deuteron inactivation of adsorbed monolayers of bovine serum albumin. Arch. of Biochem. a. Biophysics 51, 436 (1954).

JACOB, F.: Effets de la carence glucidique sur l'induction d'un Pseudomonas pyocyanea lysogène. Ann. Inst. Pasteur 82, 433—456 (1952). — JACOB, F., A. M. TORRIANI and J. MONOD: L'effet du rayonnement ultraviolet sur la biosynthèse de la β-galactosidase et sur la multiplication du bactériophage T_2 chez Escherichia coli. C. r. Acad. Sci. (Paris) 233, 1230—1232 (1951). — JOHNSON, F. H., E. A. FLAGLER and H. F. BAUM: Relation of oxygen to photoreactivation of bacteria after ultraviolet radiation. Proc. Soc. Exper. Biol. a. Med. 74, 32—35 (1950). — JONAS, H.: Some effects of radio frequency irradiations on small oil-bearing seeds. Physiol. Plantarum (Copenh.) 5, 41—51 (1952).

KANAZIR, D., et M. ERRERA: Métabolism des acides nucleiques chez Escherichia coli B après irradiation ultraviolette. Biochim. et biophysica Acta (Amsterd.) 14, 62 (1954). — KAPLAN, R.: [1] Photodynamische Auslösung von Mutationen in den Sporen von Penicillium notatum. Planta (Berl.) 38, 1—11 (1950). — KAPLAN, R. W.: [2] Auslösung von Phagenresistenzmutationen bei Bacterium coli durch Erythrosin mit und ohne Belichtung. Naturwiss. 37, 308 (1950). — [3] Genetische Wirkungen durch UV und Licht. Strahlenther. 86, 157—163 (1952). — [4] Über Möglichkeiten der Mutationsauslösung in der Pflanzenzüchtung. Z. Pflanzenzücht. 32, 121—131 (1953). — [5] Beeinflussung des durch Röntgenstrahlen induzierten mutativen Fleckenmosaiks. Strahlenther. 94, 106—118 (1954). — KAPLAN, S., E. DE ROSENBLUM and V. BRYSON: Adoptive enzyme formation in radiation sensitive and radiation resistant Escherichia coli following exposure to ultraviolet. J. Cellul. a. Comp. Physiol. 41, 153—162 (1953). — KAUDEWITZ, F.: Untersuchung des Einflusses von Meter- und Kilometerwellen auf die Generationsdauer einiger Protozoen. Z. Naturforsch. 9b, 145—148 (1954). — KELLY, S. L., and H. B JONES: Effects of irradiation on nucleic acid formation. Proc. Soc. Exper. Biol. a. Med. 74, 493—497 (1950). — KELNER, A.: [1] Effect of visible light on the recovery of Streptomyces griseus conidia from ultraviolet irradiation injury. Proc. Nat. Acad. Sci. U.S.A. 35, 73—79 (1949). — [2] Action spectra for photoreactivation of ultraviolet-irradiated Escherichia coli and Streptomyces griseus. J. Gen. Physiol. 34, 835—852 (1951). — [3] Experiments on photoreactivations with bacteria and other microorganisms. J. Cellul. a. Comp. Physiol. 39, Suppl. 1, 115—118 (1952). — [4] Growth, respiration and nucleic acid synthesis in ultraviolet irradiated and in photoreactivated Escherichia coli. J. Bacter. 65, 252—262 (1953). — KIEPENHEUER, K. O., J. BRAUER u. C. HARTE: Über die Wirkung von Meterwellen auf das Teilungswachstum der Pflanzen. Naturwiss. 36, 27 (1949). — KIMBALL, R. F.: The influence of H_2O_2 on mutation production by X-rays in Paramaecium aurelia. Radiation Res. 1, 501 (1954). — KIMBALL, R. F., and N. GAITHER: [1] The influence of light upon the action of ultraviolet on Paramaecium aurelia. J. Cellul. a. Comp. Physiol. 37, 211—233 (1951). — [2] Role of externally produced hydrogen peroxide in damage to Paramaecium aurelia by X-rays. Proc. Soc. Exper. Biol. a. Med. 80, 525—529 (1952). — KIRBY-SMITH, J. S., and D. S. DANIELS: The relative effects of X-rays, gamma rays and beta rays on chromosomal breakage in Tradescantia. Genetics 38, 375—388 (1953). — KIRBY-SMITH J. S., and C. P. SWANSON: The effects of fast neutrons from a nuclear detonation on chromosome breakage in Tradescantia. Science (Lancaster, Pa.) 119, 42—45 (1954). — KLECZKOWSKI, J., and A. KLECZKOWSKI: The behaviour of Rhizobium bacteriophages during and after exposure to ultraviolet radiation. J. Gen. Microbiol. 8, 135—144 (1953). — KLEIN, G., and A. FORSSBERG: Studies in the effect of X-rays on the biochemistry and cellular composition of ascites tumors. I. Effect on growth, cell volumes, nucleic acid and nitrogen synthesis in the Ehrlich ascites tumor. Exper. Cell Res. 6, 211—220 (1954). — KNAPP, E.: Entwicklungsphysiologische Untersuchungen an Fucaceen-Eiern. I. Planta (Berl.) 14, 731—751 (1931). — KÖHLER, H.: Untersuchungen über den Einfluß von Kurzwellen auf Keimfähigkeit und Wachstum von Pflanzen. Diss. Greifswald 1944. — KONZAK, C. F.: Differential sensitivity of soaked barley seeds to X-rays and thermal neutrons. Radiation Res. 1, 220 (1954). — KOSKI, V. M., C. S. FRENCH and I. H. C. SMITH: The action spectrum for the transformation of protochlorophyll to chlorophyll a in normal and albino corn seedlings. Arch. of Biochem. a. Biophysics 31, 1—17 (1951).

LANG, A.: [1] Entwicklungsphysiologie. Fortschr. Bot. 15, 400—475 (1952). — [2] Entwicklungsphysiologie. Fortschr. Bot. 16, 342—376 (1954). — LANGENDORFF, H. u. M., u. K. SOMMERMEYER: Die Deutung des LD 50-Anstieges mit steigender spezifischer Ionisation und die Reaktivierung sowie Sensibilisierung durch Wärme bei E. choli. Naturwiss. 41,

189—190 (1954). — LANGENDORFF, H., R. KOCH u. H. SAUER: [1] Untersuchungen über einen biologischen Strahlenschutz. IV. Die Bedeutung Sulfhydrylgruppen tragender Verbindungen für den biologischen Strahlenschutz. Strahlenther. **93**, 281—288 (1954). — LANGENDORFF, H., R. KOCH u. U. HAGEN: [2] Untersuchungen über einen biologischen Strahlenschutz. VIII. Mitt. Zur Spezifität des Zystein und verwandter Sulfhydrylkörper beim Strahlenschutz. Strahlenther. **95**, 238—250 (1954). — LASER, H.: The oxygen-effect in ionizing irradiation. Nature (Lond.) **174**, 753 (1954). — LATARJET, R., and L. R. CALDAS: Restoration induced by catalase in irradiated microorganisms. J. Gen. Physiol. **35**, 455—470 (1952). — LATARJET, R., et B. MILÉTIC: Actions des ultra-violets longs et des visibles courts (3400—5500 Å) sur les complexes bacterie bacteriophage. Ann. Inst. Pasteur **84**, 205—217 (1953). — LAVIK, P. S., and G. W. BUCKALOO: Nucleic acid synthesis in X-irradiated chick embryos. Radiation Res. **1**, 221 (1954). — LEA, D. E.: Action of radiation of living cells. Cambridge 1947. — LEWIS, G. N., and M. CALVIN: Paramagnetism of the phosphorescent state. J. Amer. Chem. Soc. **67**, 1232—1233 (1945). — LEWIS, G. N., and M. KASHA: Phosphorescence and the triplet state. J. Amer. Chem. Soc. **66**, 2100—2116 (1944). — LEWIS, G. N., and D. L. LIPKIN: Reversible photochemical processes in rigid media: The dissociation of organic molecules into radicals and ions. J. Amer. Chem. Soc. **64**, 2801—2808 (1942). — LIECHTI, A., u. E. FEISTMANN: Über die Empfindlickkeit von Einzellern auf ultraviolettes ind sichtbares Licht. Strahlenther. **62**, 393—405 (1938). — LINDEMANN, J.: Die Röntgenschädigung von Escherichia coli bei 180 keV und 31 MeV. Experientia (Basel) **9**, 22—23 (1953). — LIVERMAN, I. L., and J. BONNER: Biochemistry of the photoperiodic response. The high intensity light reaction. Bot. Gaz. **115**, 121—128 (1954). — LIVINGSTON, R.: [1] General statements about chemical reactions induced by ionizing radiation. Symposium on Radiobiology, ed. by J. J. NICKSON, S. 56—69. New York 1952. — [2] Photochemistry. In A. HOLLAENDER, Radiation Biology, Bd. II, S. 1—40. New York: McGraw Hill Book Comp. 1955. — LÜNING, K. G.: Effect of oxygen on irradiated males and females of Drosophila. Hereditas (Lund) **40**, 295—312 (1954). — LURIA, S. E., and F. M. EXNER: The inactivation of bacteriophages by X-rays influence of the medium. Proc. Nat. Acad. Sci. U.S.A. **27**, 370—375 (1941).

MANDELS, G. R.: The photoinactivation of enzymes by riboflavin. Plant Physiol. **25**, 763—766 (1950). — MANDL, I., B. LEVY and A. D. MCLAREN: The photochemistry of proteins. IX. J. Amer. Chem. Soc. **72**, 1790—1792 (1950). — MAXWELL, C. R., D. C. PETERSON and N. E. SHARPLESS: The effect of ionizing radiation on amino acids. I. The effect of X-rays on aqueous solutions of glycine. Radiation Res. **1**, 530—545 (1954). — MCLAREN, A. D.: Photochemistry of enzymes, proteins and viruses. Adv. Enzymol. **9**, 75—170 (1949). — MCLAREN, A. D., P. Genitle, D. C. KIRK jr. and N. A. Levin: Photochemistry of proteins. XVII. Inactivation of enzymes with UV light and photolysis of the peptide bond. J. Polymer. Sci. **10**, 333—344 (1953). — MCLEAN, A. D., and A. C. GIESE: Absorption spectra of proteins and aminoacids after ultraviolet irradiation. J. of Biol. Chem. **187**, 537—542 (1950). — MEFFERD jr., R. B., and L. L. COMPBELL jr.: Influence of temperature upon radiation sensitivity of thermophilic and mesophilic bacteria. Proc. Soc. Exper. Biol. a. Med. **79**, 12—16 (1952). — MEFFERD jr., R. B., and T. S. MATNEY: Protection of Escherichia coli against ultraviolet radiation by pretreatment with carbon monoxide. Science (Lancaster, Pa.) **115**, 116—117 (1952). — MELLORS, R. C., R. E. BERGER and H. G. STREIM: Ultraviolet microscopy and microspectroscopy of resting and dividing cells: studies with a reflecting microscope. Science (Lancaster, Pa.) **111**, 627—632 (1950). — METZNER, P.: Zur Kenntnis der photodynamischen Erscheinung. III. Mitteilung: Über die Bindung der wirksamen Farbstoffe in der Zelle. Biochem. Z. **148**, 498—523 (1924). — MEYER, A. E., u. E. O. SEITZ: Ultraviolette Strahlen. Berlin 1942. — MICHAELIS, L.: Fundamentals of oxidation and respiration. Amer. Scientist **34**, 573—596 (1946). — MILÉTIC, B., et P. MORENNE: Nouvelles recherches sur la restauration induite par la catalase chez des Bact. irradiées. Ann. Inst. Pasteur **83**, 515—527 (1952). — MINDER, W., u. D. SCHÖN: Vergleichende Untersuchungen über den Schutzeffekt bei Bestrahlung definierter Systeme. Strahlenther. **91**, 126—134 (1953). — MOHLER, H.: Das Absorptionsspektrum der chemischen Bindung. Jena 1943. — MONTFORT, C.: Lichtlähmung und Lichtbleichung bei Wasserpflanzen. Planta (Berl.) **32**, 121—149 (1941). — MONTFORT, C., I. FELGNER u. L. MÜLLER: Zeitphasen im Jahreslauf des lichtökologischen Chlorophyllspiegels beim photostabilen Laubblatt. Beitr. Biol. Pflanz. **29**, 106—128 (1952). — MOOS, W. S.: How biological effectiveness varies with X-ray energy. Nucleonics **12**, 46—49 (1954). — MOROWITZ, H. J.: The action of ultraviolet light and ionizing radiation on spores of Bacillus subtilis. I. The ultraviolet lethal action, mutation action and absorption spectra. Arch. of Biochem. a. Biophysics **47**, 325—337 (1953). — MOSEBACH, G.: [1] Über den Einfluß des Lichtes auf die Polarisierung des befruchteten Eies von Cystosira barbata Ag. Ber. dtsch. bot. Ges. **56**, 210—225 (1938). — [2] Über die Polarisierung der Equisetum-Spore durch das Licht. Planta (Berl.) **33**, 340—387 (1943). — MULLER, H. J.: The nature of the genetic effects produced by radiation. The manner of production of mutations by radiation. In A. HOLLAENDER, Radiation biology

Bd. I, Teil 1, S. 351—473 u. 475—626. New York u. London 1954. — MYERS, J., and G. O. BURR: Studies on photosynthesis. Some effects of light of high intensity in Chlorella. J. Gen. Physiol. **24**, 45—67 (1940).

NAKAO, Y.: Action of irradiated cytoplasm on untreated chromosomes of the silkworm. Nature (Lond.) **172**, 625—626 (1953). — NICKSON, J. L.: Symposium on radiobiology. The basic aspect of radiation effects on living systems. New York: John Wiley & Sons, Inc. 1952. — NOACK, K.: Photochemische Wirkungen des Chlorophylls und ihre Bedeutung für die Kohlensäureassimilation. Z. Bot. **17**, 481—548 (1925). — NOETHLING, W., u. H. STUBBE: Neuere botanische Untersuchungen über die Beziehung von Genmutabilität zur Quantität und Qualität kurzwelliger Strahlung. Strahlenther. **61**, 622—630 (1938). — NORMAN, A.: [1] Inactivation of Neurospora conidia by ultraviolet radiation. Exper. Cell. Res. **2**, 454—473 (1951). — [2] Production of phenocopies in aerobacter aerogenes by ultraviolet radiation. J. Bacter. **65**, 151—156 (1953). — NOVICK, A., and L. SZILARD: Experiments on light-reaction of ultraviolet inactivated bacteria. Proc. Nat. Acad. Sci. U.S.A. **35**, 591—600 (1949). — NYBOM, N., A. GUSTAFSSON and L. EHRENBERG: On the injurious action of ionizing radiations in plants. Bot. Not. (Lund) **1952**, 343—365.

ORD, M. G., and L. A. STOCKEN: Biochemical aspects of the radiation syndrome. Physiologic. Rev. **33**, 356—386 (1953). — OSTER, G., and A. D. MCLAREN: The ultraviolet light and photosensitived inactivation of tobacco mozaic virus. J. Gen. Physiol. **33**, 215—228 (1949/50).

PATRICK, W. N., and M. BURTON: Polymer production in radiolysis of benzene. J. Amer. Chem. Soc. **76**, 2626—2629 (1954). — PATT, H. M.: Protective mechanism in ionizing radiation injury. Physiologic. Rev. **33**, 35—76 (1953). — PATT, H. M., and A. M. BRUES: The pathological physiology of radiation injury in the mammal. I. Physical and biological factors in radiation action. In A. HOLLAENDER, Radiation biology, Bd. I, Teil 2, S. 919—958. New York u. London: McGraw Hill Book Comp. Inc. 1954. — PATT, H. M., J. W. CLARK and H. H. VOGEL jr.: Comparative protective effect of cysteine against fast neutron and γ-irradiation in mice. Proc. Soc. Exper. Biol. a. Med. **84**, 189—193 (1953). — PETERS, K.: [1] Stoffwechselbeziehungen zwischen bestrahltem und unbestrahltem Gewebe in ihrem Einfluß auf die Mitosenhäufigkeit in vitro. Z. Zellforsch. **39**, 203—211 (1953). — [2] Über die Bedeutung des Mediums für die Wirkung sekundärer Stoffwechselprodukte auf die Mitosehäufigkeit in halbbestrahlten Gewebekulturen. Z. Zellforsch. **40**, 510—518 (1954). — PIRINGER, A. A. and P. H. HEINZE: Effect of light on the formation of a pigment in the tomato fruit cuticle. Plant Physiol. **29**, 467—472 (1954). — PIRSCHLE, K.: Weitere Beobachtungen über den Einfluß von langwelliger und mittelwelliger UV-Strahlung auf höhere Pflanzen. Biol. Zbl. **61**, 452—473 (1941). — PIRSON, A., u. F. ALBERTS: Über die Assimilation von Helodeablättchen nach Vitalfärbung mit Vitamin B. Protoplasma **35**, 131—136 (1940). — PLATZMANN, R. L.: On the primary processes in radiation chemistry and biology. Symp. on radiobiol., edited by J. J. NICKSON, S. 97—115. New York: John Wiley & Sons, Inc. 1952. — PLAEN, P. DE: Tendances actuelles dans l'interprétation de la radiolésion cellulaire. J. belge Radiol. **35**, 113—129 (1953). — POLLARD, E.: Primary ionisation as a test of molecular organisation. Adv. Biol. a. Med. Physics **3**, 153—190 (1953). — POLLARD, E., and J. SETLOW: Effect of ionizing radiation on the serological affinity of T_1-bacteriophage. Arch. of Biochem. a. Biophysics **50**, 376—382 (1954). — PORTER, J. W., and H. J. KNAUSS: Inhibition of growth of chlorella pyrenoidosa by β-rays emitting radioisotopes. Plant Physiol. **29**, 60—63 (1954). — POWELL, W. F., and E. POLLARD: Radiation sensitivity of enzymes in intact cells. Bull. Amer. Phys. Soc. **28**, 70 (1953). — PRATT, R., J. DUFRENOY and G. GARDNER: Effect of ultraviolet irradiation on Escherichia coli and its reversal. Pharmaceut. Assoc. **39**, 496—500 (1950). — PRINGSHEIM, P.: Fluorescence and phosphorescence. New York: Intersci. Publ. 1949. — PUTNAM, F. W.: Bacteriophages: Nature and reproduction. Adv. Protein Chem. **8**, 175—284 (1953).

RABINOWITCH, E.: [1] Photosynthesis and related Processes. I. New York 1945. — [2] Photosynthesis and related processes. II. Teil 1. New York 1951. — RAJEWSKY, B.: [1] Biophysikalische Grundlagen der Ultrakurzwellenbehandlung im lebenden Gewebe. In Ergebnisse der biophysikalischen Forschung. Leipzig: Georg Thieme 1938. — [2] The limits of the target theory of the biological action of radiation. Brit. J. Radiol. **25**, 550—552 (1952). — [3] Strahlendosis und Strahlenwirkung. Stuttgart: Georg Thieme 1954. — RAMSHORN, K.: [1] Einige Beobachtungen über den Einfluß von Hochfrequenzfeldern auf die pflanzliche Entwicklung. Ber. dtsch. bot. Ges. **64**, 24—25 (1952). — [2] Über den Einfluß von Hochfrequenzfeldern auf den pflanzlichen Organismus. I. Die Wirkung auf die Entwicklung von Tomatenpflanzen bei Frequenzen von 75 und 150 Mhz. Die Kulturpflanze **1**, 79—110 (1953). — RAPKINE, S., D. SHUGAR and L. SIMINOWITCH: The activation by heat of triose phosphate dehydrogenase. Arch. of Biochem. **26**, 33—49 (1950). — RAPPORT, D., and A. CANZANELLI: The photochemical action of ultraviolet light on the absorption spectra of nucleic acid and related substances. Science (N. Y.) **112**, 469—471 (1950). — READ, J.:

[1] The effect of ionizing radiations on the broad bean root. Part X. The dependence of the X-ray sensitivity on dissolved oxygen. Brit. J. Radiol. **25**, 89—99, 154—160 (1952). — [2] Mode of addition of X-ray doses given with different oxygen concentrations. Brit. J. Radiol. **25**, 336—338 (1952). — REDFORD, E. L., and J. MYERS: Some effects of ultraviolet radiations on the metabolism of Chlorella. J. Cellul. a. Comp. Physiol. **38**, 217—243 (1951). — REINERT, J.: Über die Wirkung von Riboflavin und Carotin beim Phototropismus von Avena-Koleoptilen und bei anderen pflanzlichen Lichtreizreaktionen. Z. Bot. **41**, 103—121 (1953). — REINHOLZ, E.: Beiträge zur Kenntnis der indirekten Strahlenwirkung. I. Röntgenbestrahlung biologischer Objekte in fester Phase. Strahlenther. **95**, 131—147 (1954). — REITER, R.: Die Bedeutung extrem langwelliger elektromagnetischer Strahlungen in der Bioklimatologie. Strahlenther. **89**, 628—633 (1953). — RICE, E. W.: The action of ultraviolet light on the pentose moiety of nucleic acids and related compounds. Science (Lancaster, Pa.) **115**, 92—93 (1952). — ROTTGARDT, K. H. J.: Versuch zu einer Erklärung der Photoreaktivierung von Bacteriophagen. Naturwiss. **40**, 169 (1953).

SAGROMSKY, H.: Lichtinduzierte Ringbildung bei Pilzen. II. Flora (Jena) **139**, 560—564 (1952). — SALLMANN, L. v.: The effects of radiation on the cytology of the eye. J. Cellul. a. Comp. Physiol. **39**, Suppl. 2, 217—233 (1952). — SAMUEL, A. H., and J. L. MAGEE: Theory of radiation chemistry. II. Track effects in radiolysis of water. J. Chem. Phys. **21**, 1080—1087 (1953). — SARACHEK, A.: Ultraviolett inactivation of Saccharomyces during the budding cycle. Exper. Cell Res. **6**, 45—55 (1954). — SARACHEK, A., and W. H. LUCKE: Ultraviolet inactivation of polyploid Saccharomyces. Arch. of Biochem. a. Biophysics **44**, 271—279 (1953). — SAUTER, E., u. W. SCHWARTZ: Untersuchungen über die Wirkung ultrakurzer Wellen auf die lebende Bakterienzelle. Arch. Mikrobiol. **10**, 189—225 (1939). — SCHENCK, G. O.: [1] Über den photochemischen Primärakt und die anschließende erste Dunkelreaktion der Photosynthese. Naturwiss. **40**, 205—212, 229—238 (1953). — [2] Reaction phototropisomerer Radikale in Natur und Technik. Z. Elektrochem. **56**, 855—868 (1952). — SCHERAGA, H. A., and L. F. NIMS: The action of X-rays on fibrinogen solutions. Arch. of Biochem. a. Biophysics **36**, 336—344 (1952). — SCHOLES, G., and J. WEISS: [1] Chemical action of X-rays on nucleic acids and related substances in aqueous systems. 1. Degradation of nucleic acids and nucleotides by X-rays and by free radicals produced chemically. Biochemic. J. **53**, 567—578 (1953). — [2] Chemical action of X-rays on nucleic acids and related substances in aqueous systems. 2. The mechanism of the action of X-rays on nucleic acids in aqueous systems. Biochemic. J. **56**, 65—72 (1954). — SCHREIBER, H.: Die Wellenlängenabhängigkeit des lichtbiologischen Effektes. Strahlenther. **77**, 243—258 (1948). — SCHRÖDER, H.: Untersuchungen über die Beeinflussung des Blütenfarbmusters von Petunia hybrida grandiflora hort. Jb. Bot. **79**, 714—752 (1934). — SERRES, F. J. DE, and N. H. GILES: The effect of radiation dose fractionation on chromosome aberration frequencies in Tradescantia microspores. I. Studies with X-rays. Genetics **38**, 407—415 (1953). — SETLOW, R., and B. DOYLE: [1] The effect of temperature on the direct action of ionizing radiation on catalase. Arch. of Biochem. a. Biophysics **46**, 46—52 (1953). — [2] The effect of temperature and ultraviolet radiation on dry catalase. Bull. Amer. Phys. Soc. **28**, 70 (1953). — [3] The combined effect of temperature and ultraviolet radiation on dry catalase. Arch. of Biochem. a. Biophysics **46**, 31—38 (1953). — [4] The effect of temperature on the ultraviolet light inactivation of trypsin. Arch. of Biochem. a Biophysics **48**, 441—447 (1954). — SHERMAN, F. G., and H. B. CHASE: Effects of ionizing radiations on enzyme activities of yeast cells. II. Influence of dilution on X-ray induced inhibition of anaerobic CO_2 production and colony formation. J. Cellul. a. Comp. Physiol. **34**, 207—219 (1949). — SHUGAR, D.: [1] Ultra-violet irradiation of triosephosphate dehydrogenase. Biochim. et Biophysica Acta **6**, 548—561 (1951). — [2] Photoreactivation in the near ultraviolett of D-Glyceraldehyde-3-phosphatedehydrogenase. Experientia (Basel) **7**, 26—28 (1951). — [3] The measurement of lysozyme activity and the ultraviolet inactivation of lysozyme. Biochim. et Biophysica Acta 8, 302—309 (1952). — SIMINOVITCH, L. et S. RAPKINE: Modifications biochimiques au cours de développement des bactériophages chez une bactérie lysogène. C. r. Acad Sci (Paris) **232**, 1603—1605 (1951). — SIMONIS, W.: Physikalisch-chemische Grundlagen der Lebensprozesse (Strahlenbiologie). Fortschr. Bot. **14**, (1953 ff.). — SINSHEIMER, R. L.: [1] The photochemistry of uridylic acid. Radiation Res. **1**, 505—513 (1954). — [2] Ultraviolet absorption spectra. In A. HOLLAENDER, Radiation biology, Bd. II, S. 165—201. New York: McGraw Hill Book Comp. 1955. — SIX, E.: Zur Treffertheorie der indirekten Strahlenwirkung. Z. Naturforsch. **9b**, 265—273 (1954). — SKINNER, H. H., and C. J. BRADISH: Exposure to light as a source of error in the estimation of the infectivity of virus suspensions. J. Gen. Microbiol. **10**, 377—397 (1954). — SMITH, C. L.: The inactivation of monomolekular films of protein and its relation to the lifetime of active radicals formed in water by X-radiations. Arch. of Biochem. a. Biophysics **50**, 322—336 (1954). — SOMMERMEYER, K.: Quantenphysik der Strahlenwirkung in Biologie und Medizin. Leipzig: Grest & Portig K.-G. 1952. — SPARROW, A. H., and B. A. RUBIN: Effects of radiations on biological systems. In: Survey of Biological

Progress, Bd. II, S. 1—52. New York: Acad. Press. Inc. Publ. 1952. — SPARROW, A. H., M. J. Moses and R. STEELE: A cytological and cytochemical approach to an understanding of radiation damage in dividing cells. Brit. J. Radiol. **25**, 182—189 (1952). — STÅLFELT, M. L.: The influence of light upon the viscosity of protoplasm. Ark. Bot. **33** A, 1—17 (1947). — STAPLETON, G. E. and C. W. EDINGTON: Temperature dependence of bacterial inactivation by X-rays. Radiation Res. **1**, 229 (1954). — STAPLETON, G. E., and A. HOLLAENDER: Mechanism of lethal and mutagenic action of ionizing radiations on Aspergillus terreus. II. Use of modifying agents and conditions. J. Cellul. a. Comp. Physiol. **39**, 101—114 (1952). — STAPLETON, G. E., D. BILLEN and A. HOLLAENDER: [1] The role of enzymatic oxygen removal in chemical protection against X-ray inactivation of bacteria. J. Bacter. **63**, 805—812 (1952). — [2] Recovery of X-irradiated bacteria at suboptimal incubation temperatures. J. Cellul. a. Comp. Physiol. **41**, 345—357 (1953). — STEIN, W., u. W. HARM: [1] Wärmeaktivierung, „spontan"-inaktiver und UV-inaktivierter Colibakterien. Naturwiss. **39**, 113 (1952). — [2] Wärmeaktivierung von Coli-Bakterien nach Inaktivierung durch UV-bestrahlten Bouillon-Agar. Naturwiss. **39**, 383 (1952). — [3] Modellvorstellungen zur UV-Inaktivierung von Colibakterien im Licht von Reaktivierungseffekten. Z. Naturforsch. 8b, 742—754 (1953). — STEIN, W., u. I. MEUTZNER: Reaktivierung von UV-inaktiviertem Bacterium coli durch Wärme. Naturwiss. **37**, 167—168 (1950). — STONE, W. S., F. Haas, J. B. CLARK and O. WYSS: The role of mutation and of selection in the frequency of mutants among microorganisms grown on irradiated substrate. Proc. nat. Acad. Sci. U.S.A. **34**, 142—149 (1948). — STRAUB J.: Das Licht bei der Auslösung der Fruchtkörperbildung von Didymium eunigripes und die Übertragung der Lichtwirkung durch das tote Plasma. Naturwiss. **41**, 219—220 (1954). — SWALLOW, A. J.: The radiation chemistry of ethanol and diphosphopyridinnucleotide and its bearing in dehydrogenase action. Biochemic. J. **54**, 253—257 (1953). — SWANSON, C. P., and L. J. STADLER: The effect of ultraviolet radiation on the genes and chromosomes of higher organisms. In: A. HOLLAENDER, Radiation biology, Bd. II, S. 249—284. New York: McGraw Hill Book Comp. 1955. — SWART-FÜCHTBAUER, H., u. A. RIPPEL-BALDES: Die bactericide Wirkung des Sonnenlichtes. Arch. Mikrobiol. **16**, 358—362 (1951). — SWENSON, P. A.: The action spectrum of the inhibition of galactozymase production by ultraviolet light. Proc. Nat. Acad. Sci. U.S.A. **36**, 699—703 (1950). — SWENSON, P. A., and A. C. GIESE: Photoreactivation of galactocymase formation in yeast. J. Cellul. a. Comp. Physiol. **36**, 369—380 (1950).

TANADA, T., and S. B. HENDRICKS: Photoreversal of ultraviolet effects in soybean leaves. Amer. J. Bot. **40**, 634—637 (1953). — TAPPEINER, H. v.: Die photodynamische Erscheinung (Sensibilisierung durch fluoreszierende Stoffe). Erg. Physiol. **8**, 698—741 (1909). — TAPPEINER, H. v., u. A. JODLBAUER: Die sensibilisierende Wirkung fluoreszierender Substanzen. Leipzig: F. C. W. Vogel 1907. — THODAY, J. M., and J. READ: Effect of oxygen on the frequency of chromosome aberrations produced by α-rays. Nature (Lond.) **163**, 133 (1949). — TIMOFÉEFF-RESSOVSKY, N. W., u. K. G. ZIMMER: Das Trefferprinzip in der Biologie. Leipzig: S. HIRZEL 1947. — TOBIAS, C. A.: The dependence of some biological effects of radiation on the rate of energy loss. Symposium on Radiobiology, ed. by NICKSON, S. 357—384. New York: John Wiley & Sons, Inc. 1952. — TODD, G. W., and A. W. GALSTON: A porphyrin pigment from photosensitive non chlorophyllous plant tissues. Plant Physiol. **29**, 311—318 (1954). — TOLBERT, N. E., and R. H. BURRIS: Light motivation of the plant enzyme which oxidises glycolic acid. J. of Biol. Chem. **186**, 791—804 (1950). — TORREY, J. G.: Effects of light on elongation and branching in pea roots. Plant Physiol. **27**, 591—602 (1952). — TROWELL, O. A.: The effect of environmental factors on the radiosensitivity of lymph nodes cultured in vitro. Brit. J. Radiol. **26**, 302—309 (1953).

UBER F. M.: Biophysical Research methods. New York: Intersci. Publ. 1950.

VIRGIN, H. J.: [1] The effect of light on the protoplasmatic viscosity. Physiol. Plantarum (Copenh.) **4**, 255—357 (1951). — [2] Further studies of the action spectrum for light induced changes in the protoplasmic viscosity of Helodea densa. Physiol. Plantarum (Copenh.) **7**, 343—353 (1954). — VOERKEL, H.: Untersuchungen über die Phototaxis der Chloroplasten. Planta (Berl.) **21**, 156—205 (1934).

WAGNER, R., C. H. HADDOX, R. FÜRST and W. S. STONE: The effect of irradiated medium, cyanide and peroxide on the mutation rate in Neurospora. Genetics **35**, 237—248 (1950). — WAHL, R. und R. LATARJET: Inactivation de bactériophages par les radiations de grande longueurs d'onde (3,600—6,000 Å) Ann. Inst. Pasteur **73**, 957—971 (1947). — WARBURG, O.: Über die Geschwindigkeit der photochemischen Kohlensäurezersetzung in lebenden Zellen. I und II. Biochem. Z. **100**, 230—270 (1919); **103**, 188—217 (1920). — WARBURG, O., and V. SCHOKken: A manometric actinometer for the visible spectrum. Arch. of Biochem. a. Biophysics **21**, 363—369 (1949). — WASSINK, E. C., and J. A. STOLWIJK: Effects of light narrow spectral regions on growth and development of plants. I und II. Proc. Kon. Ned. Akad. v. Wetensch., Sect. Sci (C) **55**, 471—480, 481—488 (1952). — WATERS, W. A.: Chemistry of free radicals. Oxford 1948. — WATSON, J. D.: The properties of X-ray inactivated bacteriophage. II.

Inactivation by indirect effects. J Bacter. **63**, 473—485 (1952). — WEBER, F.: Plasmolysezeit und Lichtwirkung. Protoplasma **7**, 256—258 (1929). — WEIL, L., u. A. R. BUCHERT: Photooxydation of crystalline β-lactoglobuline in the presence of methylene blue. Arch. of Biochem. a. Biophysics **34**, 1—15 (1951). — WEIL, L., W. G. GORDON and A. R. BUCHERT: Photooxidation of amino acids in the presence of methylene blue. Arch. of Biochem. a. Biophysics **33**, 90—109 (1951). — WEISS, J.: [1] Über das Auftreten eines metastabilen, aktiven Sauerstoffmoleküls bei sensibilisierten Photooxydationen. Naturwiss. **23**, 610 (1935). — [2] Photochemical reactions of SH-Compounds in solutions. Nature (Lond.) **137**, 71—72 (1936). — [3] Photochemical oxydation-reduction processes in aqueous systems. Symposia Soc. Exper. Biol. **5**, 141—151 (1951). — [4] Possible biological significance of the action of ionizing radiations on nucleic acids. Nature (Lond.) **169**, 460—461 (1952). — WELS, P.: Grundlagen der biologischen Strahlenwirkung. Arch. exper. Path. u. Pharmakol. **208**, 116 — 133 (1949). — WELSH, J. N. and M. H. ADAMS: Photodynamic inactivation of bacteriophage. J. Bacter. **68**, 122—127 (1954). — WERFFT, R.: Über die Lebensdauer der Pollenkörner in der freien Atmosphäre. Biol. Zbl. **70**, 354—367 (1951). — WEYMOUTH, P. P. and H. S. KAPLAN: Effect of X-radiation on lymphoid tissue nucleic acids in C57 black mice. Cancer Res. **12**, 307 (1952). — WHITAKER, D. M.: Physical factors of growth. Growth, Suppl. **1940**, 75—90. — WHITEHEAD, H. A.: The protection of bacteria against radiation effects. Science (Lancaster, Pa.) **116**, 459—460 (1952). — WINDISCH, F., W. HEUMANN, H. KRIEGEL u. A. GRAFFI: Untersuchungen an Hefezellen über die Abhängigkeit des photodynamischen Effektes vom molekularen Sauerstoff. Z. Naturforsch. 8b, 673—675 (1953). — WOLFF, S., and K. C. ATWOOD: Independent X-ray effects on chromosome breakage and reunion. Proc. Nat. Acad. Sci. U.S.A. **40**, 187—192 (1954). — WOOD, TH. H.: Influence of low temperature and phase states on X-ray sensitivity of yeast. Radiation Res. **1**, 234 (1954). — WUHRMANN-MEYER, K., u. W.: Untersuchungen über die Absorption ultravioletter Strahlen durch Cuticular- und Wachsschichten von Blättern. Planta (Berl.) **32**, 43—50 (1941). — WYCKOFF, R. M. G.: The killing of certein bacteria by X-rays. J. of Exper. Med. **52**, 435 (1930). — WYSS, O., F. HAAS, J. B. CLARK and W. S. STONE: Some effects of ultraviolet irradiation on microorganisms. J. Cellul. Physiol. **35** Suppl. 1, 133—140 (1950). —

YOST jr., H. T., J. CUMMINGS and A. F. BLAKESLEE: The effect of fast neutron radiation from a nuclear detonation on chromosome aberration in Datura. Proc. Nat. Acad. Sci. U.S.A. **40**, 447—451 (1954).

ZELLE, M. R., and A. HOLLAENDER: Monochromatic ultraviolett action spectra and quantum yields for inactivation of T_1 and T_2 E. coli bacteriophages. J. Bacter. **68**, 210—215 (1954). — ZIEGLER, H.: Beeinflussung der Atmungsintensität pflanzlicher Gewebe durch eine Belichtung in fluoreszierenden Farbstofflösungen. Z. Naturforsch. **5**b, 345—350 (1950). — ZIRKLE, R. E.: [1] Speculations on cellular actions of radiations. Symposium on Radiobiology ed. by NICKSON, S. 333—356. New York 1952. — [2] The radiological importance of linear energy transfer. In A. HOLLAENDER, Radiation biology, Bd. I, Teil 1, S. 315—350. New York u. London: McGraw Hill Book Comp. Inc. 1954. — ZIRKLE, R. E., and W. BLOOM: Irradiation of parts of individual cells, Science (Lancaster, Pa.) **117**, 487—493 (1953). — ZIRKLE, R. E., and C. A. TOBIAS: Effects of ploidy and linear energy transfer on radiobiological survival curves. Arch. of Biochem. a. Biophysics **47**, 282—306 (1953).

Ionenwirkungen.

Von

Hermann Fischer.

Mit 8 Abbildungen.

I. Einleitung.

Der Verknüpfung des Lebens mit dem Wasser als dem „Dispersionsmittel" der „lebenden Substanz" und als dem notwendigen Milieu für die chemischen Reaktionen, die sich an ihr abspielen, entspricht eine ebenso enge Beziehung zu Elektrolyten, speziell den Salzen, die in Gewässern und im Wasser der Böden gelöst sind. Ebenso wie Leben ohne ein Minimum an Wasser nicht bestehen kann, ist es auch in reinem Wasser höchstens zeitweilig möglich; mit dem Wasser, das jede lebende Zelle in allen ihren Bestandteilen durchtränkt, durchsetzen auch die wasserlöslichen Salze die Zelle; allerdings ist ihre Verteilung innerhalb der Zelle im einzelnen bisher experimentell kaum erfaßt worden. Dagegen mußte die Beeinflussung der Lebenserscheinungen durch Elektrolyte aus dem Medium, das die Zellen umgibt, immer wieder registriert werden, seitdem eine physiologische Forschung besteht; jede Verwendung einer Lösung zur Kultur von Organismen oder zur Erhaltung des Lebens eines Organs oder Gewebes führt zur Beobachtung von Ionenwirkungen auf alle Arten von Lebensäußerungen. Dabei stellte sich sehr bald die Schädlichkeit von Lösungen nur *eines* Salzes und die Notwendigkeit von Salzgemischen für physiologische Bedingungen heraus. Auf Grund solcher Erfahrungen wurden einerseits „physiologische" Lösungen, wie die RINGERsche, andererseits die große Zahl von Nährmedien für niedere Organismen und höhere Pflanzen entwickelt.

Daß ein Salz*gemisch* ein optimales Medium darstellt, gleichgültig ob freilebende Organismen darin suspendiert sind, oder ob Gewebe mit ihm in Gestalt einer Körperflüssigkeit in Berührung kommen, ist ökologisch und phylogenetisch verständlich. In der Frage nach der Wirkungsweise der Elektrolyte auf die Zellen und ihrer Bedeutung in ihnen spiegelt sich die nach Bau und Wirkungsweise des Protoplasmas, und es sind gerade Ionen als wirksame und leicht zu variierende experimentelle Faktoren immer wieder in allen Zweigen der Physiologie verwendet worden. Dabei wurden Gesetzmäßigkeiten gefunden, die in den letzten 50 Jahren bei der Entwicklung unserer Vorstellung von der Protoplasmastruktur entscheidend mitgewirkt haben. Unter den fast unübersehbaren einschlägigen Arbeiten sind besonders die aus den Schulen von JACQUES LOEB und RUDOLF HÖBER neben wertvollen pflanzenphysiologischen Arbeiten der grundsätzlichen Analyse der Ionenwirkungen gewidmet (Zusammenfassung bei HÖBER 1945, 1947; ausführlicher bei HÖBER 1926). Sie fallen zeitlich mit der raschen Entwicklung der Kolloidchemie, speziell mit der zunehmenden Kenntnis von den hydrophilen Kolloiden, zusammen; es gelang, Salzwirkungen auf Eiweiß, Lecithin und andere Kolloide mit solchen auf Lebensvorgänge zu parallelisieren, woraus umgekehrt Schlüsse auf Zellstrukturen gezogen wurden.

In dem hier zu gebenden Überblick werden zunächst unter diesem Gesichtspunkt die Ionenwirkungen auf die Zelle im allgemeinen erörtert, danach wird versucht, an einer Reihe ausgewählter Phänomene herauszuarbeiten, wieweit die zu ihrer Untersuchung verwendeten

Methoden grundsätzliche Folgerungen über die Ionenwirkungen im Sinne der Aufgabenstellung dieses Bandes zulassen. Hierdurch wird auch die Auswahl bestimmt, die sich bewußt auf möglichst einfache Erscheinungen beschränkt und keine Vollständigkeit anstrebt. Sie berücksichtigt hauptsächlich pflanzenphysiologische Arbeiten; wegen der umfangreichen human- und tierphysiologischen Literatur muß auf HÖBER (1926, 1945, 1947) verwiesen werden. Schließlich werden auch Wirkungen des Ionenmangels erörtert, soweit sie von allgemeiner Bedeutung sind, dagegen wird auf die Darstellung der Salzwirkungen auf Stoffproduktion und morphologische Ausbildung und auf die der Rolle einzelner Elemente im Stoffwechsel verzichtet, soweit sich nicht Hinweise auf die strukturellen Grundlagen des Zellstoffwechsels dabei ergeben. Auch wegen der Ionenaufnahme und der Beeinflussung der Permeabilität für Elektrolyte und Anelektrolyte durch Ionen wird auf die entsprechenden Abschnitte dieses Handbuches verwiesen. Für $K^{\cdot}$ und Cl' liegen monographische Bearbeitungen von SCHMALFUSS (1936) und ARNOLD (1955) vor, welche neben grundsätzlichen Fragen des Ionenhaushaltes besonders solche der landwirtschaftlichen Praxis und der Ökologie berücksichtigen.

II. Allgemeine Gesetzmäßigkeiten.

A. Wirkungen von Einzelsalzen.

Schon die älteren Arbeiten ergaben bei Verwendung von Lösungen nur *eines* Salzes, daß ganz verschiedene Vorgänge wie Hämolyse, Phagocytose, Erregbarkeit von Muskeln, Bewegungen von Flimmerepithelien und Spermatozoen, Wachstum von Pflanzen, Entwicklung von marinen Tieren, Permeabilitätserscheinungen u. a. durch Einzelsalze in grundsätzlich gleicher Art beeinflußt werden. Dabei stört eine Einzelsalzlösung in der Regel die normalen physiologischen Abläufe. Bei systematischer Untersuchung einer größeren Anzahl von Salzen mit gleichen Kationen bzw. Anionen lassen sich quantitative Unterschiede in deren Wirksamkeit herausstellen, die auch bei sehr heterogenen Lebensvorgängen immer wiederkehren. Ordnet man die Ionen nach ihrer Wirksamkeit, so erhält man Anordnungen, die den aus der Kolloidchemie bekannten lyotropen Reihen über die Wirkung der Neutralsalze auf die Löslichkeit von Eiweißkörpern mehr oder weniger vollständig entsprechen (FREUNDLICH 1930, PAULI und VALKÓ 1933):

$$SO_4'',\ Cl',\ Br',\ NO_3',\ J',\ SCN'$$

$$Li^{\cdot},\ Na^{\cdot},\ K^{\cdot},\ Rb^{\cdot},\ Cs^{\cdot}$$

$$Mg^{\cdot\cdot},\ Ca^{\cdot\cdot},\ Ba^{\cdot\cdot},\ Sr^{\cdot\cdot}.$$

Diese Reihen können beim gleichen physiologischen Vorgang im einzelnen verschieden oder sogar umgekehrt sein; dies zeigen sehr schön die Versuche von HÖBER und NAST (1914, vgl. HÖBER 1926, S. 623) über die Wirkung der Alkalisalze auf die Saponinhämolyse (Tabelle 1). Entsprechende Reihen und Gegenreihen, zwischen denen Übergangsreihen stehen, lassen sich bei der Ionenwirkung auf die Löslichkeitseigenschaften von Eiweiß wiederfinden, wenn die Acidität

Tabelle 1. *Wirkung der Alkalikationen auf die Hämolyse mit Saponin bei verschiedenen Tieren.* (Nach HÖBER und NAST 1914.)

Tier	Reihe
Pferd	$Li^{\cdot} < Na^{\cdot} < Rb^{\cdot} < K^{\cdot}$
Schwein	$Na^{\cdot} < Li^{\cdot} < Rb^{\cdot} < K^{\cdot}$
Kaninchen . . .	$Na^{\cdot} < Li^{\cdot} < Rb^{\cdot} < K^{\cdot}$
Meerschweinchen	$Li^{\cdot} < Na^{\cdot} < Rb^{\cdot} < K^{\cdot}$
Hund	$Li^{\cdot} < Na^{\cdot} < K^{\cdot} < Rb^{\cdot}$ und $Li^{\cdot} < K^{\cdot} < Na^{\cdot} < Rb^{\cdot}$
Katze	$Li^{\cdot} = Na^{\cdot} = Rb^{\cdot} = K^{\cdot}$
Ziege.	$Li^{\cdot} > Na^{\cdot} > Rb^{\cdot} > K^{\cdot}$
Mensch.	$Li^{\cdot} > Na^{\cdot} > Rb^{\cdot} > K^{\cdot}$
Rind.	$Li^{\cdot} > Na^{\cdot} > Rb^{\cdot} > K^{\cdot}$
Hammel	$Li^{\cdot} > Na^{\cdot} > Rb^{\cdot} > K^{\cdot}$

variiert wird (Beispiele bei Höber 1926, S. 233). In den Reihen und Gegenreihen drücken sich physikalische Eigenschaften der Ionen aus. So ist die Hydratationsreihe der Alkalikationen

$$Cs^{\cdot} < Rb^{\cdot} < K^{\cdot} < Na^{\cdot} < Li^{\cdot}$$

nach fallender Größe der im Kristallgitter gemessenen Ionenradien geordnet und damit nach steigender Feldstärke, die mit $1/r^2$ dem Ionenradius umgekehrt proportional ist. Die höhere Feldstärke bedingt eine stärkere Orientierung von Wasserdipolen in der Umgebung der Ionen und damit eine stärkere Wasserbindung; die angegebene Reihe stellt also die von $Cs^{\cdot}$ zu $Li^{\cdot}$ wachsende Ionenhydratation dar.

Für die Kolloidionen bedeutet nach der Hydratationstheorie (Pauli und Valkó 1933) hohe Ionisation auch starke Hydratation; daher wird z. B. die Quellung eines Ampholyten wie Eiweiß durch eine Entladung herabgesetzt, wie sie beim Übergang in den isoelektrischen Zustand oder durch Ionen, die schwächer dissoziierende Salze mit dem Kolloid bilden, eintritt. Die Folge der Entladung ist die Schwächung zweier Faktoren, welche die kolloidale Verteilung stabilisieren, der elektrostatischen Abstoßung und der Hydrathülle, welche die Teilchen mechanisch trennt, so daß im Extremfall Flockung eintritt. Bei hydrophoben Systemen beherrscht die Ladung, bei hydrophilen die Hydratation die Vorgänge. Die Ladung, speziell die der hydrophilen Kolloide, wird nicht nur durch ihre Ionisation, sondern auch durch polare Gruppen bestimmt, die auf Grund ihrer Dipolkräfte Wassermoleküle in ihrer Umgebung orientieren, also Hydrathüllen ausbilden. Solche Gruppen adsorbieren Salze, wobei deren eines Ion auf Grund seines Ladungssinnes bevorzugt festgehalten, aber auch das Gegenion des Salzes an die adsorbierende Gruppe herangezogen wird. Für die Ionenwirkung auf Eiweißkörper sind beide geschilderten Formen der Ladung bei der Wechselwirkung mit Ionen zu berücksichtigen: Die Ionisation, die zu Salzbildung und Ionenaustausch auf Grund von Hauptvalenzbindungen führt und die speziell bei Ampholyten p_H-abhängigen Quellungseffekte zeigt, auf der anderen Seite die Affinität polarer Gruppen zu Ionen durch Nebenvalenzen, welche zu einer Adsorption von Ionenpaaren führt. Dabei resultiert aus der Annäherung der Salzionen an das Kolloid eine gegenseitige Störung der Felder und damit der Orientierung der Wassermoleküle, die um so größer sein wird, je näher die Ionen an die polarisierten Gruppen herantreten können. Daher bestimmt neben der Konzentration der Partner deren spezifische Hydratation (Feldstärke) die auftretenden Effekte.

Wird z. B. das Kation eines Neutralsalzes durch ein negativ geladenes Eiweiß adsorbiert, so wird die Orientierung der Wassermoleküle um die polare Gruppe um so weniger gestört werden, je größer die Wasserhülle des Kations ist, da diese der Annäherung des Kations an das Kolloid entgegenwirkt. Die Adsorbierbarkeiten der Kationen verhalten sich also umgekehrt wie die Größen ihrer Hydrathüllen; infolgedessen wirkt bei den Alkalikationen das kleinste Ion ($Li^{\cdot}$) auf Grund seiner stärksten Hydratation am schwächsten dehydratisierend auf das Kolloid. Die Adsorptionsreihe

$$Li^{\cdot} < Na^{\cdot} < K^{\cdot} < Rb^{\cdot} < Cs^{\cdot}$$

drückt die Entladung (Dehydratation), die Hydratationsreihe

$$Cs^{\cdot} < Rb^{\cdot} < K^{\cdot} < Na^{\cdot} < Li^{\cdot}$$

die relative Quellung des Kolloids aus. An dessen Gesamthydratation ist aber auch das Anion des Salzes beteiligt; es wirkt im angenommenen Falle der Entladung des negativen Kolloids durch das Kation in dem Maße entgegen, als es selbst das Kation elektrostatisch bindet, was um so mehr der Fall ist, je kleiner die Wasserhülle des Anions ist, d. h. entgegengesetzt der Hydratationsreihe der Anionen

$$SCN' < J' < Br' < Cl' < SO_4''.$$

Die Anionen fördern also die Hydratation gemäß ihrer Adsorptionsreihe

$$SO_4'' < Cl' < Br' < J' < SCN'.$$

Die beschriebenen Wechselwirkungen zwischen den elektrischen Feldern sind für niedrigere Ionenkonzentrationen charakteristisch (*direkter* Ioneneffekt nach Michaelis 1925). Bei höheren Konzentrationen führt die Affinität der Ionen zum Lösungsmittel zu einer Konkurrenz um das Wasser zwischen dem Kolloidsystem und den nicht adsorbierten Ionen (*indirekter* Ioneneffekt, lyotroper Effekt), d. h. zu einem Wasserentzug vom Kolloid durch die Salzionen, der bei Kationen *und* Anionen entsprechend der Hydratationsreihe steigt:

$$Cs^{\cdot} < Rb^{\cdot} < K^{\cdot} < Na^{\cdot} < Li^{\cdot}$$

$$SCN' < J' < Br' < Cl' < SO_4''.$$

Auch hier wirken beide Ionen des Salzes additiv. Im Extremfalle kann das Kolloid einerseits in einer Salzlösung über den Wert im reinen Wasser hinaus quellen (z. B. in KCNS, einem Salz mit zwei schwach hydratisierten Ionen), andererseits ausgesalzen werden.

Für ein positiv geladenes Kolloid kehren sich alle Reihen sinngemäß um; auch die Umkehr von Ionenreihen bei p_H-Änderungen erklärt sich aus den Beziehungen zwischen Ladung und Hydratation. Nach Durchschreiten des isoelektrischen Zustandes verläuft mit der Umladung eines Ampholyten zu einem Kolloidkation im mehr sauren Bereich die Hydratationsförderung durch Anionen nach der Hydratationsreihe

$$SCN' < J' < Br < Cl' < SO_4'',$$

durch Kationen nach der Adsorptionsreihe

$$Li^{\cdot} < Na^{\cdot} < K^{\cdot} < Rb^{\cdot} < Cs^{\cdot}.$$

Brauner (1930) hat an der Quellung der *Aesculus*-Testa den direkten und indirekten Ioneneffekt klar herausgestellt. Nach dem Gesagten werden auch die physiologischen Ionenwirkungen, soweit sie auf Quellungsbeeinflussung beruhen, je nach den verwendeten Konzentrationen nicht die gleichen sein. Als weiterer konzentrationsabhängiger Effekt kann auf eine anfängliche Entladung eine Umladung mit erneuter Quellungsförderung bei höherer Konzentration folgen; dabei werden die Verhältnisse im einzelnen von der Feldstärke der Salzionen, aber auch von ihrem Polarisierungsvermögen und der Polarisierbarkeit der mit ihnen reagierenden Gruppen des Kolloidions abhängen (Teunissen 1938). Mehrwertige Ionen wirken einerseits stärker entladend, andererseits auch stärker quellungsfördernd, wenn sie selbst stark hydratisiert sind. Schließlich ist bei niedrigen Konzentrationen eine Ionenadsorption auch ohne Entladung des Kolloids (auf Grund der Induktion von Dipolmomenten im Kolloid durch adsorbierte Ionen?) mit dem Ergebnis einer Quellungsförderung zu berücksichtigen (Bogen 1948a). Aus dem Gesagten geht hervor, daß schon auf Grund des einen Wirkungsmechanismus (Hydratation) eine Vielzahl von Effekten zu erwarten ist, deren Analyse an biologischem Material noch schwieriger sein wird als an Kolloidmodellen und daß sich im Experiment die Bearbeitung eines größeren Konzentrationsbereiches empfiehlt.

Die Bedeutung atomarer Größen für die Ionenreihen diskutiert SEIFRIZ (1949) auf Grund der Giftwirkung auf Myxomyceten-Plasmodien (Tötungszeit, Sistierung der Plasmaströmung u. a.), für welche er die folgenden Reihen fand, wobei die Staffelung der drei Gruppen nach der Giftigkeit erfolgt[1]

$$Li^{\cdot} < Na^{\cdot} < K^{\cdot} < Rb^{\cdot} = Cs^{\cdot}$$
$$Ca^{\cdot\cdot} < Mg^{\cdot\cdot} < Sr^{\cdot\cdot} < Ba^{\cdot\cdot}$$
$$La^{\cdot\cdot\cdot} < Pb^{\cdot\cdot} < Au^{\cdot\cdot\cdot} < Ag^{\cdot} < Th^{\cdot\cdot\cdot\cdot}.$$

In den beiden ersten Reihen wächst die Giftwirkung mit dem Ionenradius und der Ionenbeweglichkeit, fällt aber mit der Feldstärke und der Hydratation. Der Einfluß der Valenz wird aus dem Vergleich der drei Reihen klar, die Verschiebung gegeneinander (nach der z. B. die Giftigkeit für $Ca^{\cdot\cdot} < Cs^{\cdot}$ und für $La^{\cdot\cdot\cdot} < Ba^{\cdot\cdot}$ ist) zeigt ihr Zusammenwirken mit anderen Faktoren. SEIFRIZ (1949) betont die stärkere Hydratation des dreiwertigen $La^{\cdot\cdot\cdot}$ im Vergleich zum zweiwertigen $Ba^{\cdot\cdot}$, die an eine Schutzwirkung der Wasserhülle denken läßt; danach muß unter den dreiwertigen Ionen das schwächer hydratisierte $Au^{\cdot\cdot\cdot}$ giftiger sein als $La^{\cdot\cdot\cdot}$, was auch der Fall ist. Da die Reihe der Ionenbeweglichkeit in umgekehrter Richtung verläuft wie die der Hydratation, denkt SEIFRIZ (1949) auch an einen leichteren Zutritt der beweglicheren Ionen zu gewissen Plasmastrukturen, wodurch deren Potential leichter erniedrigt werden könnte. Eine Beziehung zur Oberflächenenergie besteht insofern, als die Giftwirkung weitgehend der Adsorbierbarkeit der Ionen entspricht. TEUNISSEN (1938) betont das im Vergleich zu denen der Hauptgruppen stärkere Polarisierungsvermögen der Ionen aus den Nebengruppen des periodischen Systems ($Zn^{\cdot\cdot}$, $Cd^{\cdot\cdot}$, $Cu^{\cdot\cdot}$, $Pb^{\cdot\cdot}$, $Mn^{\cdot\cdot}$, $Ni^{\cdot\cdot}$) und bringt ihre kräftige biologische Wirkung mit einer stärkeren Umladung ionogener Gruppen in den Protoplasmakolloiden in Zusammenhang. Es gelingt demnach nicht ohne weiteres, die physiologischen Wirkungen auf *eine* physikalische Größe zurückzuführen; in den Ionenreihen drücken sich verschiedene Größen aus, die voneinander mehr oder weniger abhängig sind. In der dritten der angegebenen Reihen reichen die bisher diskutierten Prinzipien vor allem zur Erklärung der $Ag^{\cdot}$-Wirkung nicht aus; bei den Schwermetallen scheinen Hydratation und Beweglichkeit hinter anderen Eigenschaften zurückzutreten, während sich die lyotrope *Anionen*reihe auch bei der Giftwirkung von Schwermetallsalzen finden läßt (KAHO 1933). Die starke physiologische Wirkung der Edelmetalle kommt nach SCARTH (1924a, b) auch in der zur Kontraktion der Plastiden von *Spirogyra* notwendigen Minimalkonzentration neben der Valenz zum Ausdruck (vgl. WEBER 1924a); sie läßt sich mit ihrer Stellung in der Spannungsreihe (s. S. 711) in Verbindung bringen, aber auch mit einer Fähigkeit zur chemischen Reaktion mit dem Protoplasma bzw. bestimmten Gruppen darin, wobei in erster Linie an die Bildung von Eiweiß-Schwermetallkomplexen zu denken ist (PAULI und VALKÓ 1933, S. 100). Über die Beeinflussung von Enzymen s. S. 739. Bei der starken Giftigkeit mancher Schwermetalle („oligodynamische Wirkung“) ist ihre Anreicherung durch die Zelle zu berücksichtigen (Literatur bei CLARK 1937, S. 44ff.).

B. Wirkungen von Salzgemischen. Ionenantagonismus.

Zwischen den beiden Extremen (Einzelsalzlösung und balancierte Salzlösung) liegen Salzkombinationen, bei denen sich die schädlichen Wirkungen der Einzelsalze mehr oder weniger weitgehend aufheben. Dabei kann schon die Kombination

[1] Vgl. dieses Handbuch, Band I, S. 395ff.

zweier Salze optimal wirken, meist sind aber Gemische mehrerer Salze überlegen, vor allen Dingen solche, die in ihrer Zusammensetzung dem Seewasser ähnlich sind. Besonders sind solche antagonistischen Wirkungen von zwei- und dreiwertigen gegenüber einwertigen Kationen studiert worden (Höber 1926, S. 666ff.; 1947, S. 321 und 335). Aus den Untersuchungen Loebs (1902 u. a.), die als klassisch auf diesem Gebiet gelten können, sind die an befruchteten *Fundulus*-Eiern hervorzuheben, bei denen die durch NaCl oder ein anderes Alkalisalz verursachte Verhinderung der Entwicklung durch zahlreiche zweiwertige Kationen kompensiert werden kann, wobei ein Konzentrationsoptimum auftritt. Seit Benecke (1907) und Osterhout (1906, 1907) die Notwendigkeit von Salzgemischen für das Gedeihen von Algen, *Lunularia*-Brutknospen, Wurzeln und für die Keimung der Sporen von *Vaucheria* und *Equisetum* bewiesen, ist bis in die neueste Zeit (z. B. v. Eiselsberg 1937, 1938, MacLeod und Snell 1950, Libbert 1953) die Gültigkeit dieses Prinzips in der ganzen belebten Welt immer wieder nachgewiesen worden. In der Regel ruft ein Ion die gleiche antagonistische Wirkung in einer um so geringeren Konzentration hervor, je höher seine Wertigkeit ist. Allerdings sind nicht alle mehrwertigen Ionen als Antagonisten brauchbar, wobei Unterschiede je nach dem Vorgang, der untersucht wird, auftreten.

In den Versuchen von Loeb (1902) und in denen von Lillie (1904) an *Arenicola*-Larven (deren Cilien werden in reinen NaCl-Lösungen verflüssigt, mehrwertige Ionen wirken antagonistisch) zeigen sich weitere Analogien zwischen Ionenwirkungen, denen mikroskopisch sichtbare Strukturänderungen parallel laufen, und kolloidchemischen Gesetzmäßigkeiten. Die Entgiftung kann nicht darauf beruhen, daß das „Gift" durch ein „Gegengift" gebunden wird, da es sich bei beiden um Ionen gleichen Ladungssinnes handelt (Loeb 1902); dagegen findet der starke Anstieg der antagonistischen Kationenwirkung mit der Wertigkeit ein Modell in der Fällung von Kolloiden durch entgegengesetzt geladene Ionen, deren Fällungskraft mit steigender Wertigkeit annähernd in geometrischer Progression zunimmt (Schulze-Hardysche Regel). Der Einfluß der elektrischen Ladung (Wertigkeit) pflegt bei mehr hydrophoben Kolloiden gegenüber der lyotropen Wirkung der Ionen hervorzutreten, ohne daß diese ganz zu verschwinden braucht. In den Versuchen von Lillie (1904) zeigt sich neben dem Valenzeffekt eine deutliche Abstufung der Wirksamkeit innerhalb der zweiwertigen Ionen: Der Verflüssigung von *Arenicola*-Cilien durch m/2 NaCl wirken sie z. B. in m/200-Lösungen in der folgenden Reihe von links nach rechts in abnehmendem Maße entgegen.

$$Mg^{\cdot\cdot} > Ba^{\cdot\cdot} > Ca^{\cdot\cdot} > Sr^{\cdot\cdot} > Mn^{\cdot\cdot} > Fe^{\cdot\cdot} > Co^{\cdot\cdot} > Ni^{\cdot\cdot} > Cd^{\cdot\cdot} > Pb^{\cdot\cdot} > Zn^{\cdot\cdot} > Cu^{\cdot\cdot} > UO_2^{\cdot\cdot}.$$

Diese Reihe entspricht angenähert der des elektrolytischen Lösungsdruckes (Standard-Elektroden-Potential), welcher wiederum ein Maß für die Lösungsmittelaffinität der Metalle darstellt. Danach finden sich bei den antagonistischen ebenso wie bei den Einzelsalzwirkungen Beziehungen zu bestimmten physiologischen Größen und zur Beeinflussung von Kolloiden *in vitro*. Im einzelnen wird der Vergleich erschwert, weil die Eignung der zwei- und mehrwertigen Ionen als Antagonisten sehr wechselnd ist; auch können Folgewirkungen auftreten, wie z. B. in der genannten Reihe von Lillie (1904), bei deren letzten Gliedern nach einiger Zeit im Protoplasma der Cilien Flockungserscheinungen sichtbar werden, die zum Tode führen. Solche Wirkungen können offenbar je nach dem Vorgang und der zugrunde liegenden Struktur sehr verschieden sein, wie der Vergleich einer Anzahl biologischer Phänomene zeigt (s. Höber 1926, S. 671ff.). Bei den Schwermetallen wird man wieder mit ihren besonderen chemischen Eigenschaften rechnen müssen, durch welche es zu Reaktionen mit dem Substrat kommt, die für die Organismen und ihre Teile im einzelnen spezifisch sind.

Neben antagonistischen Erscheinungen zwischen verschiedenwertigen Ionen treten auch solche zwischen gleichwertigen auf; so ist ein ausgesprochener Antagonismus zwischen Mg˙˙ und den übrigen Erdalkalimetallen vielfach beobachtet worden (Literatur bei WIECHMANN 1920, HÖBER 1926, S. 692, LIBBERT 1953) aber auch zwischen Alkaliionen (HÖBER 1926, S. 683). Nach OSTERHOUT (1909) erreichen Weizenwurzeln in Gemischen von 0,1 m KCl und NaCl etwa die dreifache Länge wie in reinen Lösungen der beiden Salze (weitere Beispiele bei HÖBER 1926, S. 681ff.). Zur Erklärung könnte hier die Stellung innerhalb der lyotropen Reihe in Betracht kommen. Nach RUBINSTEIN (1928), der sich auch um die begriffliche Klärung der antagonistischen Ionenwirkungen bemüht hat, bestehen aber weder mit der Hydratation noch mit dem elektrolytischen Lösungsdruck durchgehende Zusammenhänge, die in allen Fällen von Ionenantagonismus gelten. Eine allgemeine Theorie ist schon deshalb nicht zu erwarten, weil bisher in der Physiologie eine Anzahl heterogener Erscheinungen als antagonistische bezeichnet worden sind, darunter auch solche, bei denen die „Antagonisten" sicherlich nicht die gleichen Angriffspunkte in der Zelle haben.

Es ist hier unter Hinweis auf die entsprechenden Abschnitte in Band 4, 5 und 12 dieses Handbuchs nur kurz zu erwähnen, daß sich auch bei Stoffwechselvorgängen die besprochenen Einzelsalzwirkungen und antagonistischen Erscheinungen wiederfinden. Als Beispiele seien Salzwirkungen auf die Atmung von *Helodea* genannt (LYON 1921); recht klare Verhältnisse ergeben sich bei Mikroorganismen: So steigt nach BOAS (1921) die Hefegärung nach der Adsorptionsreihe der Kationen und nach der Hydratationsreihe der Anionen (vgl. auch PIRSCHLE 1930). Vergleichende Untersuchungen über hemmende und fördernde Wirkung von Salzen verschiedener Konzentration und ihrer Kombination auf die Atmung finden sich in mehreren Arbeiten von BROOKS (1919, 1920, 1921) für *Bacillus subtilis* und von GUSTAFSON (1919) für *Aspergillus niger*, umfangreiches Material über Wachstum und Alkoholoxydation von *Bacterium acetigenoideum* bringt KREHAN (1932). Für die Atmung von *Azotobacter chroococcum* besteht nach ENDRES (1934) ein deutlicher Antagonismus von Ca˙˙ und Mg˙˙ gegenüber Alkaliionen. Die Rolle der Ionen im Stoffwechsel wird im Zusammenhang mit Mangelerscheinungen auf S. 735ff. noch zu diskutieren sein.

C. Die Kolloidtheorie der Ionenwirkungen.

Das hier gezeichnete Bild der physiologischen Ionenwirkungen als Beeinflussung von Protoplasmakolloiden war mit dem ersten Viertel unseres Jahrhunderts im wesentlichen abgeschlossen. Es ist dadurch gekennzeichnet, daß man nicht so sehr bestimmte Phänomene im einzelnen zu erklären suchte, sondern sich bemühte, „bei geeigneter Wahl der Bedingungen für den physiologischen Versuch die einfachen Gesetze der Kolloidchemie im Organischen wiederzufinden" (HÖBER 1926, S. 699), wobei man sich durchaus bewußt blieb, daß im Organismus „ein Gemisch vieler Kolloide" vorhanden sei, „die mit fluktuierenden Grenzen teils dem Bereich der Suspension, teils dem der hydrophilen Kolloide angehören, und deren Dispersions-, Lösungs- und Ladungszustand von zahlreichen Variablen abhängig ist; die Kolloidchemie gibt uns aber zum Vergleich bisher fast nur ganz einfache Systeme" (HÖBER l. c.). Dieser Vorbehalt ist in der folgenden Zeit oft allzu sehr zurückgetreten, obwohl er mit der fortschreitenden Kenntnis der Makromoleküle und des submikroskopischen Baues des Substrates für die biologischen Ionenwirkungen stärker zu betonen gewesen wäre. Ferner sind in der biologischen Literatur die Voraussetzungen für das Zustandekommen der „Quellung" oft unzureichend berücksichtigt worden, wenn man diesen Komplex zur Deutung von Beobachtungen heranzog, und es muß demgegenüber betont werden, daß — abgesehen von dem Fehlen einer quantitativen Theorie — Ioneneffekte in vitro schon an einfachen Kolloiden wie Gelatine recht kompliziert sein können (vgl. PAULI und VALKÓ 1933); selbst bei Annahme einer weitgehenden Vergleichbarkeit des Protoplasmas mit einem Gemisch hydrophiler Kolloide müßten die Verhältnisse sehr unübersichtlich sein.

Eine Möglichkeit zu feinerer Differenzierung von Hydratationsphänomenen an Systemen in der lebenden Zelle bietet der Vergleich von Ionenreihen für verschiedene Vorgänge am gleichen Material, wie ihn BOGEN (1948b) durchführt (Tabelle 2), wobei er Ergebnisse mehrerer Arbeiten verwendet (BOGEN 1940, 1941, 1948a), in denen die Zusammenhänge zwischen Hydratation und Permeabilität bzw. Hitzeresistenz und Kernvolumen im einzelnen diskutiert werden. Je nach dem Test, der verwendet wird, treten also unterschiedliche Reihen auf, während der gleiche Test bei verschiedenen Objekten übereinstimmende Reihen ergibt (Tabelle 3). Danach lassen sich die Einzeleffekte als Beeinflussung von *Komponenten* der Gesamthydratation deuten, die für die getestete Erscheinung auch bei verschiedenen Objekten übereinstimmen. Eine zunächst ganz allgemeine Vorstellung struktureller Art vermitteln hierzu die neueren Anschauungen über den Aufbau der Proteine, speziell derjenigen des Protoplasmas (HÖBER 1945, 1947; FREY-WYSSLING 1953; vgl. dieses Handbuch Band I, S. 371ff.), nach denen auch die Form der Einzelmoleküle und ihre Verknüpfung die physiologischen Erscheinungen bestimmt. Eine festgestellte Beeinflussung der Gesamthydratation, etwa durch Ionen, darf als die Summe der nicht

Tabelle 2. *Ionenreihen für die Beeinflussung verschiedener Erscheinungen an den Epidermiszellen von Rhoeo discolor.* (Nach BOGEN 1948b.)

Effekt	Ionenreihe	
Hitzeresistenz . . .	Kontr. $< Mg^{\cdot\cdot} < Cs^{\cdot} < K^{\cdot} < Na^{\cdot} < Li^{\cdot} < Ca^{\cdot\cdot} < Al^{\cdot\cdot\cdot}$ Resistenzabnahme, Hydratationsförderung →	(Chloride)
	Kontr. $< SO_4'' < C_2O_4'' < Cl' < Br' < NO_3' < SCN'$ Resistenzabnahme, Hydratationsförderung →	(K-Salze)
Harnstoffpermeation	Kontr., $Na^{\cdot} < K^{\cdot} < Mg^{\cdot\cdot} < Ca^{\cdot\cdot} < Al^{\cdot\cdot\cdot}$ Permeabilitätsverringerung, Hydratationsherabsetzung →	(Nitrate)
	Kontr. $< SO_4'' < Cl' < NO_3'$ Permeabilitätssteigerung, Hydratationsförderung →	(Na-Salze)
Glycerinpermeation .	Kontr. $< K^{\cdot} < Na^{\cdot} < Ca^{\cdot\cdot} < Mg^{\cdot\cdot} < Al^{\cdot\cdot\cdot}$ Permeabilitätsverringerung, Hydratationsherabsetzung →	(Nitrate)
	Kontr., $SO_4'' < Cl' < NO_3'$ Permeabilitätsverringerung, Hydratationsherabsetzung →	(Na-Salze)
Kernvolumen . . .	Kontr., $Al^{\cdot\cdot\cdot} < Ca^{\cdot\cdot} < Mg^{\cdot\cdot} < Li^{\cdot} < Na^{\cdot} < K^{\cdot} < Rb^{\cdot} < Cs^{\cdot}$ Volumenzunahme, Hydratationsförderung →	(Chloride)
	Kontr. $< SCN' < Br' < Cl' < SO_4'' < NO_3'$ Volumenzunahme, Hydratationsförderung →	(K-Salze)

Tabelle 3. *Ionenreihen für Hitzeresistenz und Harnstoffpermeabilität bei Rhoeo discolor (Blattepidermis) und Gentiana cruciata (Stengelepidermis).* (Nach BOGEN 1948b.)

	Hitzeresistenz	
Rhoeo	Kontr. $< Cs^{\cdot} < K^{\cdot} < Na^{\cdot} < Li^{\cdot}$	Kontr. $< SO_4'' < C_2O_4'' < Cl' < Br' < NO_3' < SCN'$
Gentiana	Kontr. $< Rb^{\cdot} < K^{\cdot} < Na^{\cdot} < Li^{\cdot}$	Kontr. $< SO_4'' < Cl' < Br' < NO_3' < SCN'$
		Hydratationsförderung →
	Harnstoffpermeation	
Rhoeo	Kontr., $Na^{\cdot} < K^{\cdot} < Mg^{\cdot\cdot} < Ca^{\cdot\cdot} < Al^{\cdot\cdot\cdot}$	Kontr. $< SO_4'' < Cl' < NO_3'$
Gentiana	Kontr., $Na^{\cdot}$, $K^{\cdot} < Mg^{\cdot\cdot} < Ca^{\cdot\cdot}$	Kontr. $< SO_4'' < Cl' < NO_3'$
	Hydratationsherabsetzung →	Hydratationsförderung →

notwendig gleichsinnigen Wirkungen auf verschiedenartige hydrophile Gruppen angesehen werden. Die mehr morphologische Betrachtung der Molekülstruktur hebt aber auch die Formveränderung des Einzelmoleküls hervor; Ladungsänderungen werden Modifikationen des „Ladungsmusters" und damit der Hydratationszentren hervorrufen (vgl. auch PAULI und VALKÓ 1933, S. 262ff.), die, wenn sie in stärkerem Maße auftreten, zu Spannungen im Molekülnetz und damit zur Störung seiner Struktur führen, wie BOGEN (1948a) am Beispiel der Hitzeresistenz diskutiert. Der im Zusammenhang mit der Quellung und Entquellung des Protoplasmas durch Ionen oft betonte „optimale Quellungszustand" erscheint damit nicht als durch die Gesamtheit des von den Kolloiden gebundenen Wassers bestimmt, sondern als differenzierte Ordnung von Ladungszentren, über die im einzelnen noch nichts ausgesagt werden kann, die aber wenigstens grundsätzlich sehr empfindliche Mechanismen für die physiologische Regulation ermöglicht; es bleibt abzuwarten, welche Rolle außer Polypeptidketten anderen gesetzmäßig im Protoplasma angeordneten Bauelementen wie Lipoiden oder labilen Lipoproteinen zukommt. Die Ergebnisse kritischer Prüfung der Ionenwirkung auf die lebende Zelle befindet sich somit in Übereinstimmung mit der neueren Entwicklung der Morphologie der Protoplasmakolloide, so wie sie in der „klassischen Periode" der Theorie von den hydrophilen Kolloiden als dem Substrat der Lebensvorgänge entscheidende Impulse gegeben hat.

Die neuere Entwicklung zeigt schon der Vergleich zwischen den beiden Darstellungen HÖBERs (1926, 1945). Es muß fast wundernehmen, wenn sich angesichts eines so komplexen Substrats bei den Lebenserscheinungen einfache Beziehungen, wie die lyotropen Reihen, wiederfinden; gerade daraus erwächst aber die Berechtigung, Ionenwirkungen auf einfache physikalisch-chemische Größen zurückzuführen. Für Abweichungen im einzelnen wird die Mannigfaltigkeit des Substrats ebenso verantwortlich sein können wie die differenzierten Reaktionsmöglichkeiten der Ionen in ihm (z. B. Quellungsbeeinflussung, Metall-Proteinverbindungen); darüber hinaus wird aber eine primäre Ionenwirkung durch von ihr selbst ausgelöste Reaktionsketten verschleiert sein können. Durch experimentelle Auflösung der Vorgänge und ihrer Verknüpfung wird im Einzelfall festgestellt werden müssen, ob die besprochenen Gesetzmäßigkeiten zur Erklärung der Ionenwirkung ausreichen, oder ob noch andere Prinzipien beteiligt sind.

III. Spezielle Ionenwirkungen.

A. Höhere Pflanzen.

Unter den Organen der Landpflanzen scheint die Wurzel am besten zur Feststellung von Salzwirkungen geeignet zu sein, weil sie unmittelbar mit Lösungen in Kontakt gebracht werden kann. Schon aus älteren ernährungsphysiologischen

Arbeiten geht hervor, daß Lösungen einzelner Salze, auch in schwachen Konzentrationen, das Wurzelwachstum verhindern, die Wurzeln schädigen und zum Absterben bringen (Literatur bei HANSTEEN-CRANNER 1922, KAHO 1926a, MEVIUS 1927, KERSTING 1939). Eine Ausnahme machen Ca-Salze, die wachstumsfördernd wirken und hervorragend geeignet sind, die nachteilige Wirkung anderer Salze aufzuheben. KERSTING (1939) bestätigt HANSTEEN-CRANNERs (1922) Auffassung, daß die $Ca^{\cdot\cdot}$-Wirkung auf das Wurzelwachstum exogen und vom $Ca^{\cdot\cdot}$-Gehalt der Wurzel unabhängig sei; er führt weiter die Befunde von SOROKIN und SOMMER (1929), nach welchen in Einzelsalzlösung als Folge von Ca-Mangel Mitoseanomalien eintreten, auf Fixierungsartefakte zurück. Nach OSTERHOUT (1907, 1909) gelten bei Wurzeln die gleichen antagonistischen Erscheinungen wie auch sonst im Tier- und Pflanzenreich (vgl. S. 711), eine „physiologisch balancierte" Salzlösung wirkt optimal; am empfindlichsten gegen Einzelsalze ist nach HANSTEEN (1910, dort auch quantitative Daten über den Antagonismus von $K^{\cdot}$, $Na^{\cdot}$, $Mg^{\cdot\cdot}$, $Ca^{\cdot\cdot}$ bei Wurzeln) die Streckungszone (s. S. 719). Wurzelhaarbildung unterbleibt nach KERR (1933) bei *Limnobium Spongia* in Lösung von NaCl, KCl und $MgCl_2$, die bereits vorhandenen Haare sterben ab, während in reiner $CaCl_2$-Lösung die Bildung neuer Wurzelhaare vorübergehend möglich ist; gegenüber dem hemmenden Einfluß von $K^{\cdot}$ und $Mg^{\cdot\cdot}$ sind $Ca^{\cdot\cdot}$, $Sr^{\cdot\cdot}$, $Ba^{\cdot\cdot}$, $Co^{\cdot\cdot}$, $Mn^{\cdot\cdot}$ und $Ni^{\cdot\cdot}$ bei Wurzelhaaren Antagonisten (WIECHMANN 1920). Die Unentbehrlichkeit von $Ca^{\cdot\cdot}$ für die Wurzelhaarbildung geht aus einer großen Anzahl von Arbeiten hervor, die CORMACK (1949) zusammengefaßt hat; die Acidität des Milieus wirkt sich nach diesem Autor dahingehend aus, daß höhere Wasserstoffionenkonzentration durch Verhinderung der Calciumpectatbildung in den Membranen die Wurzelhaare pathologisch verändert, im extremen Falle unterbleibt die Haarentwicklung ganz (vgl. auch S. 730).

Die schädliche Wirkung von Einzelsalzen auf Wurzel- und Sproßwachstum ist unter den Alkali-, Erdalkali- und Halogenionen am geringsten bei $K^{\cdot}$, $Ca^{\cdot\cdot}$ und Cl' (PIRSCHLE 1930, 1932). Eine Beziehung zur Anordnung im periodischen System ergibt sich dahin, daß nicht Atomgewicht oder Ordnungszahl unmittelbar mit der Giftigkeit in Verbindung zu bringen sind, sondern der Atombau insgesamt: Die am wenigsten giftigen Ionen gehören dem Argon-Typ an, und die Giftigkeit nimmt beiderseits zum Helium- und Xenon-Typ zu:

	F^{9}	Cl^{17}	Br^{35}	J^{59}	VII
Li^{3}	Na^{11}	K^{19}	Rb^{37}	Cs^{55}	I
Be^{4}	Mg^{12}	Ca^{20}	Sr^{38}	Ba^{56}	II
He^{2}	Ne^{10}	Ar^{18}	Kr^{36}	X^{54}	0

Mit der geochemischen Verteilung besteht insofern ein Zusammenhang, als die häufigsten Ionen am wenigsten giftig wirken. Für $Cr^{\cdot\cdot\cdot}$, $Mn^{\cdot\cdot}$, $Fe^{\cdot\cdot\cdot}$, $Co^{\cdot\cdot}$, $Ni^{\cdot\cdot}$, $Cu^{\cdot\cdot}$, $Zn^{\cdot\cdot}$ ließen sich derart einfache Beziehungen zwischen Atombau und physiologischer Wirkung nicht aufzeigen. PIRSCHLE (1932) betont, daß seine Darstellung andere, ebenfalls aus dem Atombau ableitbare Schlußfolgerungen über die physiologische Rolle der Ionen ergänze. Die Giftwirkung der Halogenide

$$J' > Br' > Cl'$$

entspricht ihrer Adsorption (s. S. 709), ebenso die der Erdalkaliionen

$$Be^{\cdot\cdot} > Ba^{\cdot\cdot} > Mg^{\cdot\cdot} > Sr^{\cdot\cdot} > Ca^{\cdot\cdot}$$

bis auf die Stellung von $Mg^{\cdot\cdot}$. Für die Reihenfolge der Alkaliionen

$$Li^{\cdot} > Cs^{\cdot} > Na^{\cdot} > Rb^{\cdot} > K^{\cdot}$$

ist die Übereinstimmung mit der Übergangsreihe (s. S. 707) der Eiweißfällung und der Hämolyse bemerkenswert.

Während die bisher genannten Versuche sich meist über Wochen ausdehnten, untersuchte Libbert (1953) den Einfluß von Salzen sämtlicher Alkali- und Erdalkalimetalle und ihrer Kombinationen auf den Zuwachs von *Lepidium*- und *Agrostemma*-Wurzeln in kurzdauernden Messungen. Abb. 1 und 2 zeigen die Bedeutung der Konzentration für die Stellung der einzelnen Ionen in den Reihen,

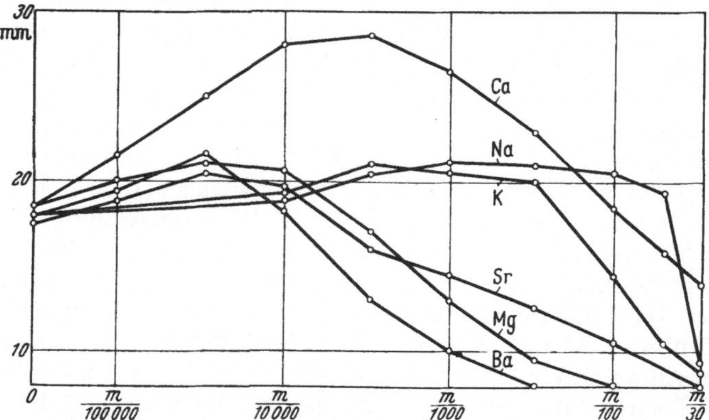

Abb. 1. Wachstum von *Lepidium*-Wurzeln in Einsalzlösungen (Chloride, Versuchsdauer 27 Std). (Nach Libbert 1953.)

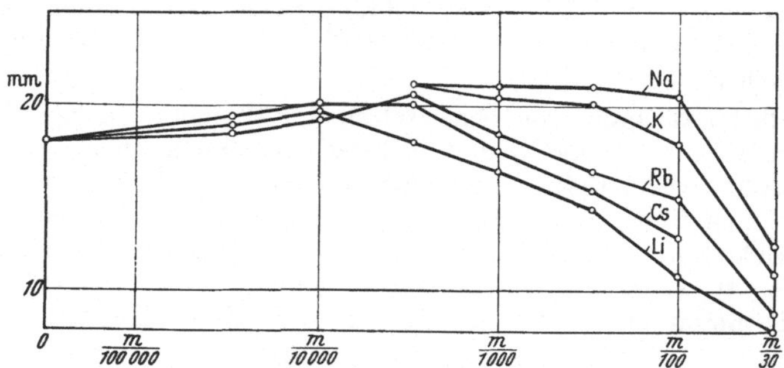

Abb. 2. Wachstum von *Lepidium*-Wurzeln in Einsalzlösungen (Chloride, Versuchsdauer 27 Std). (Nach Libbert 1953.)

wie sie auch aus ähnlichen Versuchen von Eisenmenger (1928) mit Mischungen von KH_2PO_4, $Ca(NO_3)_2$ und $MgSO_4$ bei verschiedenen Gesamtkonzentrationen hervorgeht. Entsprechend der Valenzregel tritt die Schädigung bei den Erdalkaliionen schon bei geringeren Konzentrationen auf als bei den Alkaliionen; dies gilt sogar für $Ca^{··}$ neben dessen spezifischer fördernder Wirkung auf das Wurzelwachstum. Eine besondere Bedeutung dieses Ions drückt sich auch in der Reihe der Giftigkeit für die Wurzeln aus:

$$Ba^{··} > Sr^{··} > Mg^{··} > Li^{·} > Cs^{·} > Rb^{·} > K^{·}, Na^{·} > Ca^{··};$$

in der fast gleichen Reihe Borowikows (1913, 1916) an ganzen *Helianthus*-Keimlingen steht aber $Ca^{··}$ bei den Erdalkaliionen. Die Reihen können nicht nur bei Wurzeln und Sproß, sondern auch bei verschiedenen Objekten ungleich sein (Marcozzi 1937, Borriss 1937, 1939); aus der Untersuchung von Kombinationen der sämtlichen Ionen in Zweisalzgemischen (Libbert 1953) ergibt sich wiederum die Bedeutung der Konzentration beider Partner. Die breit angelegten Versuche LIBBERTS (1953) an vergleichbarem Material gestatten einen Überblick über mögliche Effekte von „Entgiftung" und „Antagonismus", aber keine grundsätzliche Einsicht in das Antagonismusproblem, das als „außerordentlich verwickelt" bezeichnet wird; sie erreichen wohl die Grenze der Leistungsfähigkeit der Methode, die mit ganzen Organen arbeitet, in denen selbst innerhalb relativ kurzer Versuchszeiten „außerordentlich verwickelte" Vorgänge ablaufen, wobei die „Nebenwirkungen" (LIBBERT 1953, S. 433) dominieren können.

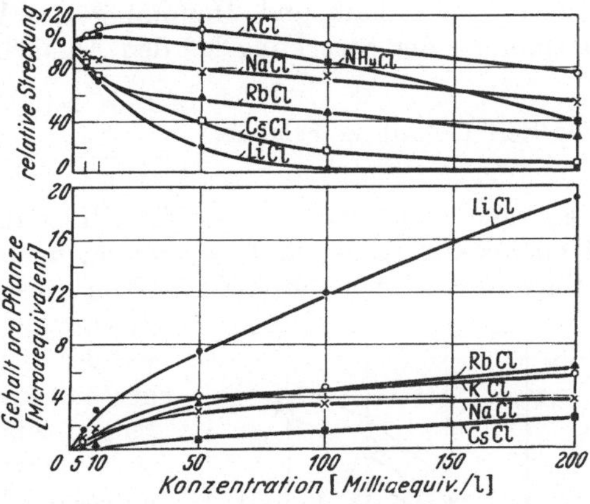

Abb. 3. Oben: Prozentuale Streckung (aq. dest = 100) von *Raphanus*-Keimlingen in Alkalichloridlösungen verschiedener Konzentration (Versuchsdauer 24 Std). Unten: Gehalt der Asche an dem geprüften Kation am Ende des Versuches. (Nach HASSAN und OVERSTREET 1952.)

Die Schwierigkeit, daß Versuchsanordnungen, die durch Messung der Länge ganzer Organe die Wirkung eines Faktors zu erfassen suchen, im allgemeinen keine Primärwirkungen ermitteln, sucht BORRISS (1937, 1939) durch Versuchsobjekte zu umgehen, deren Zuwachs in hohem Maße durch *einen* Faktor, nämlich das Streckungswachstum, beherrscht wird (als Test dient die Zunahme der Hypokotyllänge etiolierter Keimlinge von *Nicotiana rustica* und verschiedenen Caryophyllaceen, die auf Filterpapier wachsen, das mit Salzlösungen getränkt wird; die Länge wird innerhalb maximal 160 Std mehrmals gemessen). Allgemein fördert hier $K^{\cdot}$ und zum Teil $Na^{\cdot}$, im Gegensatz zum Wurzelwachstum hemmt $Ca^{\cdot\cdot}$ deutlich (vgl. BOROWIKOW 1913, 1916). Für die methodische Beurteilung ist wichtig, daß die gegenüber Aqua bidestillata festgestellten fördernden und hemmenden Wirkungen je nach Species und Änderungen in der Versuchsanordnung verschoben sein können. Lediglich die $K^{\cdot}$-Förderung wird stets deutlich, der eine analytisch festgestellte Erhöhung des $K^{\cdot}$-Gehalts der Hypokotyle entspricht (BORRISS 1939, vgl. aber HASSAN und OVERSTREET 1952). Übereinstimmend damit hatten schon ARNAUDOW und POPOW (1936) gegenüber salzfrei angezogenen Tabakkeimlingen eine Förderung durch KCl und eine Hemmung durch $CaCl_2$ festgestellt (Gesamtlänge von Wurzel + Hypokotyl 10 Tage nach dem Auslegen auf Filterpapier). Auch HASSAN und OVERSTREET (1952) finden in 24 Std dauernden Versuchen an Keimlingen von *Raphanus sativus* (Abb. 3 und 4) auf feuchtem Sand mit Einzelsalzlösungen bei geringen Konzentrationen eine $K^{\cdot}$-Förderung des

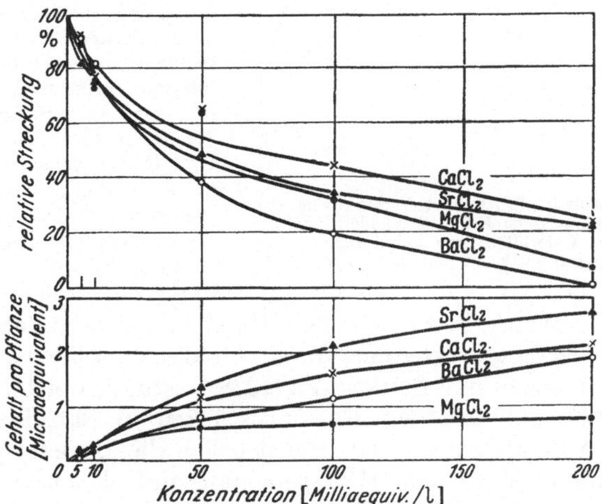

Abb. 4. Oben: Prozentuale Streckung (aq. dest = 100) von *Raphanus*-Keimlingen in Erdalkalichloridlösungen verschiedener Konzentration (Versuchsdauer 24 Std.) Unten: Gehalt der Asche an dem geprüften Kation am Ende des Versuches. (Nach HASSAN und OVERSTREET 1952.)

Wachstums (Längenzunahme von der Wurzelspitze bis zur Krümmungsstelle am Hypokotyl). Die quantitative Bestimmung des spezifischen Kations in den Keimlingen nach den Versuchen ergibt hier aber keine einfache Beziehung zwischen Ionengehalt und Hemmung des Wachstums; so ist die Reihe für die Hemmung der Streckung bei den Alkalichloriden

$$Li^{\cdot} > Cs^{\cdot} > Rb^{\cdot} > Na^{\cdot} > K^{\cdot}$$

für den Gehalt je Pflanze

$$Li^{\cdot} > Rb^{\cdot}, K^{\cdot} > Na^{\cdot} > Cs^{\cdot}.$$

Die bevorzugte Stellung von $K^{\cdot}$ (PIRSCHLE 1930) tritt auch nach HASSAN und OVERSTREET (1952) in Erscheinung, nicht aber eine spezifische Förderung des Wurzelwachstums durch $Ca^{\cdot\cdot}$. Von einem Test, der die Summe der Längenzunahmen von Wurzeln und Sproßteilen verwendet, wie es in den zuletzt genannten Arbeiten geschah, die von praktischen Fragestellungen ausgingen, sind kaum physiologisch klare Ergebnisse zu erwarten; immerhin zeigt sich die relativ günstige Wirkung von $K^{\cdot}$ und zum Teil von $Na^{\cdot}$, die bei geeigneter Versuchsanordnung und Konzentration zu einer Förderung der Längenzunahme im Vergleich zur Wasserkontrolle führen kann.

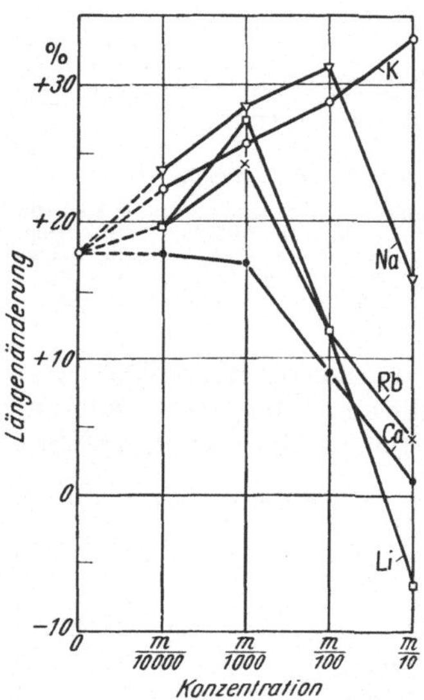

Abb. 5. Prozentuale Verlängerung von 3 mm langen Koleoptilzylindern (*Avena*) in Chloridlösungen mit 1 p.p.m. IES. (Nach THIMANN und SCHNEIDER 1938.)

Chemische Analysen des Pflanzenmaterials vermögen zwar Anhaltspunkte über die aufgenommene Salzmenge, über Mangelerscheinungen usw. zu geben, sind aber für die zellphysiologische Ausdeutung nur von beschränktem Wert, weil sie nichts über die Konzentration der Ionen an deren Reaktionsort aussagen; es muß mit einer ungleichen Ionenverteilung innerhalb der Zelle und mit lokalen Differenzen innerhalb des Protoplasmas gerechnet werden. Speziell bei Pflanzenzellen mit großen Vacuolen wird das Analysenergebnis in erster Linie den Gehalt des Zellsaftes repräsentieren, so daß nur unter der Annahme unbekannter Gleichgewichte auf das Protoplasma geschlossen werden kann; auch die Bindung von Aschenbestandteilen in den Zellwänden und die Entfernung aus dem Stoffwechsel — z. B. als Calciumoxalat — ist zu berücksichtigen.

Wenn BORRISS (1937, 1939) aus seinen Versuchen auf eine Förderung des *Streckungs*wachstums durch Alkalisalze schließt, so wird diese Auffassung durch THIMANN und SCHNEIDER (1938) gestützt, welche die Beeinflussung der Streckung von zylindrischen Koleoptilstücken in Salzlösung untersuchen. Ihre Kurven (Abb. 5) zeigen wiederum deutlich die Bedeutung der Konzentration. Dagegen findet WUHRMANN (1938) ebenfalls an Koleoptilzylindern durchweg eine Hemmung entsprechend der Reihe

$$Ba^{\cdot\cdot} > Sr^{\cdot\cdot} > Ca^{\cdot\cdot} > Mg^{\cdot\cdot} > Li^{\cdot} > Cs^{\cdot} > Na^{\cdot} > Rb^{\cdot} > K^{\cdot} > \text{Kontrolle}.$$

Die Unterschiede in den Ergebnissen mögen in solchen des Versuchsmaterials begründet sein; die unphysiologischen Bedingungen schließen auch bei den Koleoptilzylindern neben Schädigungen innerhalb der kurzen Versuchszeit von 9—24 Std (WUHRMANN 1938) Regulationen nicht aus. Wegen solcher Regulationen wird aber gerade der intakte, aktiv wachsende Keimling als Testobjekt für *primäre* Ionenwirkung nicht unbedingt geeignet sein. Überdies wirkt auf die Zellen intakter Hypokotyle und Wurzeln keine definierte Salzlösung ein, sondern ein Gemisch aus den von außen zugeführten Salzen und denen des Samens; die beobachteten Wirkungen sind mithin als Folgen einer Bilanzverschiebung zu werten. Für Hypokotyle gilt dieser Gesichtspunkt sicher noch mehr als für Wurzeln im Kontakt mit der Versuchslösung, bei welchen er aber auch nicht übersehen werden darf (Kersting 1939).

Die Analyse der $Ca^{\cdot\cdot}$-Förderung bei Wurzeln wurde neuerdings weiter vorangetrieben; dabei ließen sich Beziehungen zu Wuchs- und Hemmstoffen aufzeigen. Burström (1952) stellt durch Trennung der Beeinflussung von Zuwachs, Zellgröße und -länge bei verschiedenen p_H-Werten an Weizen- und Gerstenwurzeln drei $Ca^{\cdot\cdot}$-Effekte heraus: Für die Zellteilung (bei p_H 5—6) genügen schon 10^{-6} m $Ca^{\cdot\cdot}$ im Außenmedium, 10^{-5} m $Ca^{\cdot\cdot}$ fördern die Zellstreckung. Diese spezifische Förderung der Zellstreckung wird durch die Annahme verständlich, daß $Ca^{\cdot\cdot}$ bei den Wurzeln als Wuchsstoffantagonist wirkt: Die Förderung läßt sich durch 1-Naphthylessigsäure aufheben, andererseits durch den Hemmstoff α-p-Chlorphenoxyisobuttersäure ersetzen. Eine dritte Funktion kommt $Ca^{\cdot\cdot}$ in höheren Konzentrationen als $H^{\cdot}$-Entgifter zu. Die Verhältnisse werden durch eine $Ca^{\cdot\cdot}$-Förderung der Nitrataufnahme und die im einzelnen undurchsichtigen Beziehungen zwischen $Ca^{\cdot\cdot}$, N und Wuchsstoff kompliziert (Burström 1954). Für die Wurzelhaarbildung scheint (jedenfalls quantitativ) eine andere $Ca^{\cdot\cdot}$-Abhängigkeit zu bestehen (Burström 1952). Auf diese Frage wird bei der Besprechung der Ca-Rolle für die Zellwandbildung noch einmal einzugehen sein (s. S. 730). Nach Brink (1924) wird das Wachstum von Pollenschläuchen durch $Ca^{\cdot\cdot}$ gefördert; in Salzgemischen kompensiert $Ca^{\cdot\cdot}$ die Giftwirkung von $Na^{\cdot}$, $K^{\cdot}$, $Li^{\cdot}$, $Mg^{\cdot\cdot}$, $Sr^{\cdot\cdot}$ und $Ba^{\cdot\cdot}$ auf Pollenkeimung und Pollenschlauchwachstum. Entsprechende Ergebnisse teilen Bungenberg de Jong und Hennemann (1934) mit, ebenso Booij (1940), der auch untersucht, wieweit Salze mit ein- und mehrwertigem Kation das Platzen von *Lathyrus*-Pollen in verdünnten Zuckerlösungen verhindern. Die für diesen Schutzeffekt geltenden Ionenreihen sind andere als die für die Keimungshemmung am gleichen Pollenmaterial.

B. Mikroorganismen.

Einfacher scheinen die Verhältnisse bei Mikroorganismen zu liegen, die zum Wachstum ebenfalls ein ausgeglichenes Ionenmilieu verlangen. Ihr Verhalten gegenüber Einzelsalzen und Salzgemischen entspricht den geschilderten Gesetzmäßigkeiten (Literatur bei Pirschle 1935a, b). Phänomene, die in der Ausbildung der Zellen und Kulturen auftreten, lassen sich in Ionenreihen ordnen, bleiben aber ebenfalls der kausalen Erklärung vorerst unzugänglich, wie es einige Beispiele aus der neueren Literatur zeigen.

Nach Voss (1952) gilt für die Grenzkonzentrationen, die zur Sistierung des Wachstums von Mikroorganismen erforderlich sind, bei verschiedenen Stämmen (Kokken, *Mycobacterium*, *Rhodotorula*) grundsätzlich die gleiche Reihe der Toxicität, lediglich $Li^{\cdot}$ und $Cs^{\cdot}$ stehen bei den verschiedenen Organismen an verschiedener Stelle.

$$K^{\cdot} < Na^{\cdot} < Rb^{\cdot} < Mg^{\cdot\cdot} < Ca^{\cdot\cdot} < Sr^{\cdot\cdot} < Ba^{\cdot\cdot}$$

$\leftarrow NH_4^{\cdot} \rightarrow$

$\longleftarrow$ $Li^{\cdot}$ $\longrightarrow$

$\leftarrow Cs^{\cdot\cdot} \rightarrow$

Ordnet man nach dem periodischen System, so zeigt sich wieder die bevorzugte Stellung der Ionen vom mittleren Radius (Abb. 6); diese sind auch bei den von Voss (1952) eingehend untersuchten Veränderungen der Zellgröße und -form am wenigsten wirksam. Die Anionenwirkung entspricht der lyotropen Reihe.

Nach HEINZEL (1950) gelten bei *Bacterium prodigiosum* auch für den Übergang von der beweglichen Einzelzelle zu Ketten und Fäden, wie sie bei Kultur mit Salzlösung auftreten, Reihen, nach denen die schwach hydratisierten Ionen die Kettenbildung fördern.

$$Cs^{\cdot} < Rb^{\cdot} < NH_4^{\cdot} < Na^{\cdot} < K^{\cdot} < Li^{\cdot} < (Mg^{\cdot\cdot}, Sr^{\cdot\cdot}, Ca^{\cdot\cdot}, Ba^{\cdot\cdot})$$

Fäden, Ketten ⟵⟶ Kokken

$$NO_3' < J', Br' < Cl' < \text{Benzoat} < \text{Formiat} < \text{Tartrat} < SO_4'' < \text{Citrat}$$

Dagegen folgt die Aufhebung der Beweglichkeit der Bakterien in den Salzkulturen, die auf eine Schädigung des Geißelapparats zurückgeführt wird, einer Übergangsreihe mit $K^{\cdot}$ als dem harmlosesten Ion. Die Salze üben ihre chemomorphotische Wirkung nur in der logarithmischen Wachstumsphase aus.

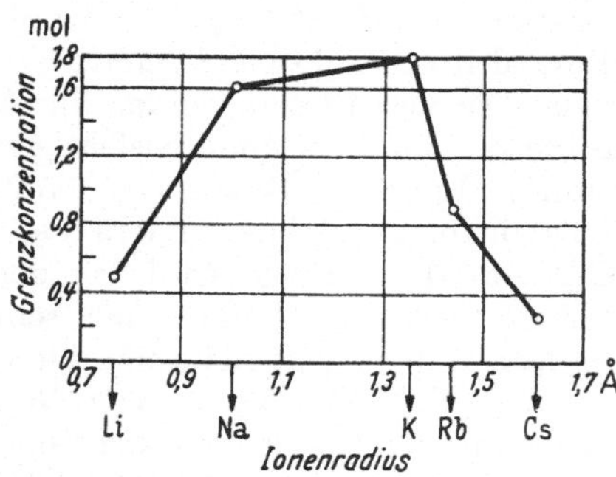

Abb. 6. Grenzkonzentrationen (Chloride) für das Wachstum von *Micrococcus flavus*. (Nach VOSS 1952.)

Bei den blasenförmigen Anschwellungen, wie sie durch Salze, besonders $MgSO_4$, an Bakterien hervorgerufen werden können (BOAS 1928, KATTERMANN 1930, STAPP und ZYCHA 1931) handelt es sich um pathologische Bildungen, die vielleicht sogar postmortal entstehen (VOSS 1952).

Die Salzempfindlichkeit von Pilzen und Bakterien steigt nach KATTERMANN (1930) entsprechend der Adsorptionsreihe der Anionen

$$SO_4'' < Cl' < Br' < NO_3' < SCN'.$$

Gegen SCN′ sind Bakterien empfindlicher als Pilze (BOAS 1927, 1928), unter den von GISTL (1932, 1933) untersuchten Erdalgen vor allem die Cyanophyceen. BOAS (1927) sieht hierin einen Ausdruck phylogenetischer Verschiedenheit, „Rhodan ist ein protoplasmatisches, ein stammesgeschichtliches (phyletisches) Reagens ersten Ranges; ein Reagens für Stammestrennung, für Großprotoplasmatik ...“ (BOAS 1942, S. 162). Die hohe SCN′-Empfindlichkeit der Cyanophyceen würde ihre Verwandtschaft zu den Bakterien nahelegen (GISTL 1933); die relativ rhodanfesten Purpurbakterien müßten eine Sonderstellung einnehmen (BOAS 1934). Der Physiologe wird sich vorerst mit der Vorstellung einer spezifischen SCN′-Wirkung, etwa im Stoffwechsel, begnügen; BOAS (1934) selbst nimmt auf Grund einiger Versuche eine Katalasehemmung durch SCN′ an, betont aber andererseits die Wirkung auf den „Kolloidzustand der Zelle“, wobei das schnell eindringende Rhodanidion diesen zuerst stören, das Sulfation ihn optimal halten soll. Umfangreiche Versuche über stimulierende und toxische Wirkungen einer großen Anzahl von Ionen auf das Wachstum von *Aspergillus niger* bringt PIRSCHLE (1935a, b).

C. Zellphysiologische Erscheinungen.

a) Plasmolyseverhalten.

Plasmolyse mit Salzlösungen. Der Verlauf der Plasmolyse wird unter anderem durch die Art der plasmolysierenden Lösung beeinflußt; bei Verwendung von Neutralsalzen (speziell von $K^{\cdot}$ und $Ca^{\cdot\cdot}$) mit einwertigen Kationen findet sich vielfach eine mehr konvexe Plasmolyseform bzw. kürzere Plasmolysezeit im Sinne WEBERs (1924a, 1928) im Vergleich mit zweiwertigen, bei denen eine Abrundung des abgelösten Protoplasten langsamer oder gar nicht eintritt. Diese Regel gilt aber keineswegs immer, wie sich bei vergleichenden Plasmolysestudien und bei Durchsicht der Literatur feststellen läßt.

In sehr vielen Fällen sind Plasmolyseform und Plasmolysezeit bei $K^{\cdot}$- und $Ca^{\cdot\cdot}$-Salzen gleich (z. B. PRÁT 1922a, CHOLODNY 1924, BORRISS 1938). Die gleichen Arbeiten geben zum Teil eine glatte Abhebung bei $K^{\cdot}$- und $Ca^{\cdot\cdot}$-Plasmolyse (PRÁT 1922a), zum Teil bei Salzen eine gegenüber Zuckern und anderen Nichtelektrolyten mehr konkave Plasmolyseform an (CHOLODNY 1924). Überwiegend wird aber eine „verflüssigende" Wirkung (Viscositätsherabsetzung) durch einwertige, eine „verfestigende" (Viscositätserhöhung) durch zweiwertige Kationen berichtet: HANSTEEN-CRANNERs (1919, 1922) Dunkelfeldaufnahmen zeigen bei Plasmolyse mit KNO_3 abgerundete, durch zahlreiche feine Fäden mit der Wand verbundene Protoplasten, mit $CaCl_2$ Krampfplasmolyse, bei welcher breite Partien des Cytoplasmas mit der Wand verbunden bleiben. Seine Versuche wurden von WEIS (1926) erweitert, der eine zunehmende Neigung zur Krampfplasmolyse in reinen Chloridlösungen angibt, entsprechend der Reihe

$$NH_4^{\cdot} < K^{\cdot} < Mg^{\cdot\cdot} < Ca^{\cdot\cdot}.$$

Auch die Stärke der Verbindungsfäden zwischen der Zellwand und dem von ihr getrennten Protoplasten wächst in der Reihe von links nach rechts. Die BROWNsche Molekularbewegung wird dagegen von $NH_4^{\cdot}$ zu $Mg^{\cdot\cdot}$ schwächer, ebenso ein im Dunkelfeld zu beobachtendes Zittern der Plasmafäden; bei $CaCl_2$ fehlen solche Bewegungen ganz. Veränderung des Anions ergibt die Reihe

$$NO_3' < Cl' < SO_4''$$

für zunehmende „Verfestigung". Übereinstimmend damit gelten nach RUGE (1940) an *Helodea*-Blattzähnen bei Plasmolyse mit 0,25 m Chloriden bzw. K-Salzen für die Plasmolysezeit die Reihe

$$Li^{\cdot} < NH_4^{\cdot} < Na^{\cdot} < K^{\cdot} < Mg^{\cdot\cdot} < Ca^{\cdot\cdot}$$

$$SCN' < J' < NO_3' < Cl' < SO_4'' < HPO_4'' < \text{Citrat}.$$

Die genannten Autoren verwendeten zum Teil stark hypertonische Lösungen gleicher Molarität bzw. Normalität, also nicht gleicher osmotischer Wirksamkeit. Durch Unterschiede im Grade der Hypertonie kann aber der Plasmolyseverlauf in dem Sinne verändert werden, daß bei starker Hypertonie die Plasmolysezeit verlängert und die Plasmolyseform mehr konkav wird (DERRY 1929, PRUD'HOMME VAN REINE 1935). Bei starker Hypertonie können Plasmolyseformunterschiede auftreten, wenn sie bei Außenkonzentrationen, die sich nicht stark vom osmotischen Wert der Zelle unterscheiden, fehlen; BORRISS (1938) betont daher die Forderung zum Einhalten vergleichbarer Konzentrationen bei vergleichenden Plasmolyseuntersuchungen besonders. Aus seinen Versuchen geht weiter hervor, daß auch durch beginnende Deplasmolyse auf Grund einer Permeabilität für K-Salze bei längerer Beobachtungsdauer eine Abrundung der Protoplastenoberfläche einsetzen kann, die bei den nicht permeierenden Ca-Salzen ausbleibt. Bei Plasmolyse mit höheren Salzkonzentrationen zeigen sich zum Teil pathologische Effekte, die durch die üblichen Bezeichnungen der Plasmolyseformen nicht ausreichend gekennzeichnet werden (SAKAMURA 1933, TAKAMINE 1940).

Kritische Versuche von TAKAMINE (1940) bestätigen im ganzen die verflüssigende Wirkung von $K^{\cdot}$ im Vergleich zur Plasmolyse mit Glucose; $Ca^{\cdot\cdot}$ und $Al^{\cdot\cdot\cdot}$ wirken antagonistisch. Die Ionenwirkungen werden aber oft durch den Zustand oder die Vorbehandlung des Versuchsmaterials verdeckt und zum Teil sogar umgekehrt. Das Wässern von Schnitten, wie es vielfach als methodischer Kunstgriff zum Erzielen von Konvexplasmolyse bei Permeabilitätsuntersuchungen verwendet wird, ist ein weiterer wichtiger Faktor für das Plasmolysebild, der mit dem Reinheitsgrad des verwendeten Wassers wechselt, nach TAKAMINE (1941) aber bei verschiedenen Objekten nicht zum gleichen Effekt zu führen braucht. Wässern mit Aqua dest. fördert bei *Spirogyra* Krampfplasmolyse, bei Zwiebelschuppen und *Rhoeo* Konvexplasmolyse (CHOLODNY und SANKEWITSCH 1933), nach TAKAMINE (1941) bei Zwiebelschuppen Konvexplasmolyse, bei *Rhoeo* Konkavplasmolyse; WEBER (1924a) gibt für *Spirogyra* nach Wässern in Leitungswasser „eckige" Plasmolyse an (vgl. auch KAMIYA 1939). Die Wirkung von Aqua dest. kann auf H˙ (s. S. 734) oder auf dem Auswaschen von Ionen und damit auf einer Verschiebung der Ionenbilanz beruhen; die von Leitungswasser wird auch durch die Befunde STÅLFELTs (1948) verständlich, daß Bodenextrakte viscositätsbeeinflussende Substanzen aus zersetzter Pflanzensubstanz enthalten, es kommen viscositätssteigernde und viscositätsherabsetzende Substanzen vor, so daß verschiedene Effekte zu erwarten sind.

Zusammenfassend ergibt sich für den Plasmolyseverlauf die Bedingtheit durch zahlreiche Faktoren, die selten ausreichend beachtet worden sind, so daß die Wirkung eines Faktors, wie des Ionenmilieus, nicht klar hervortreten konnte. Ein großer Teil der Widersprüche zwischen den Beobachtungen dürfte auf Unterschieden in Versuchsmaterial und -durchführung beruhen, auch die Versuchsdauer ist meist nicht vergleichbar. Der Zeitfaktor ist um so wichtiger, weil durch Lichtwirkung über längere Zeit fortdauernde Viscositätsschwankungen induziert werden (STÅLFELT 1946, VIRGIN 1951); auch durch mehrfache Beobachtung mit starker Mikroskopbeleuchtung können die Ergebnisse verändert werden (TAKAMINE 1940). Im ganzen überwiegen aber die Beobachtungen über eine in Alkalisalzen raschere Abrundung des Protoplasten als in Erdalkalisalzen.

Wegen der Kritik der Plasmolyseform- und Plasmolysezeitmethode der Viscositätsbestimmung wird auf den Abschnitt „Viscosität" in diesem Bande verwiesen[1]. Zu den ionenabhängigen Faktoren, die den Plasmolyseverlauf bestimmen gehört neben der Konsistenz des Cytoplasmas (Viscosität, Adhäsion usw.) auch seine Wasserpermeabilität; es ist vielfach beobachtet worden, daß die Plasmolyse-Eintrittsgeschwindigkeit, die nicht mit der Plasmolysezeit im Sinne WEBERs (1928) zu verwechseln ist, durch Alkalisalze gegenüber Erdalkalisalzen gefördert wird. Die Messung der Wasserpermeabilität durch die Deplasmolysegeschwindigkeit unter dem Einfluß verschiedener Salze ergibt Ionenreihen, die auch hier vom Gegenion (KAHO 1937) und von der Salzkonzentration (DE HAAN 1935) abhängen. Nach DE HAAN (1935) erhöhen einwertige Kationen stets die Wasserpermeabilität, bei mehrwertigen ergibt sich ein Minimum in mittleren Konzentrationen, das unter dem Wert der Vergleichslösung (Saccharose) liegt. Die Widersprüche am gleichen Objekt (Zwiebelschuppen) zwischen den Kurven DE HAANs (1935) und KLOMPs (unveröffentlicht), welcher auch bei mehrwertigen Kationen stets eine Erhöhung der Wasserpermeabilität feststellt, klärt LOEVEN (1951) als jahreszeitlich bedingt und bestätigt damit, daß ein Vergleich der Ionenwirkungen unter verschiedenen Bedingungen nur im beschränkten Maße zulässig ist und daß die Einzelbeobachtungen in ihrer Bedeutung für die primären Reaktionen im Protoplasma nicht überbewertet werden dürfen. "The factor so often neglected in such discussions is the protoplasm itself. Protoplasm is in a constant state of change; no two masses of it are ever identical, nor is one mass quite the same during any two periods of time." Diese Bemerkung von SEIFRIZ (1936, S. 445) drückt keine Resignation gegenüber der Mannigfaltigkeit der Phänomene aus, sondern kennzeichnet „das" Protoplasma als bestimmt durch den ständigen Wechsel seiner Komponenten, der besonders dann auftreten wird, wenn in seine Struktur von außen eingegriffen wird.

[1] Vgl. diesen Band, S. 591, STÅLFELT, M. G.: Viscosität.

Plasmolyse nach Vorbehandlung mit Salzen. Die Wirkung einer Vorbehandlung mit Salzen auf eine nachfolgende Plasmolyse in Lösungen von Nichtelektrolyten entspricht weitgehend der Plasmolyse mit den Salzen selbst. Während WEBER (1924a) bei derartigen Versuchen mit $Ca^{\cdot\cdot}$ Konvex-, mit $Na^{\cdot}$ Krampfplasmolyse erhält, geben CHOLODNY und SANKEWITSCH (1933) Beispiele für eine Förderung der Konvexplasmolyse durch $NH_4^{\cdot}$, $Na^{\cdot}$ und $K^{\cdot}$, der Konkavplasmolyse durch $Ca^{\cdot\cdot}$, $Ba^{\cdot\cdot}$ und $Mg^{\cdot\cdot}$. Die Befunde dieser Autoren sind keineswegs so eindeutig, wie man nach ihrer Wiedergabe in mehreren Lehrbüchern und anderen zusammenfassenden Darstellungen annehmen könnte: So konnten sie nach längerer Vorbehandlung (72 statt 24 und 48 Std) mit allen Salzen nur Konvexplasmolyse erzielen. Die von ihnen gefundene Reihe, die mit der von WEIS (1926, s. S. 721) übereinstimmt,

$$NH_4^{\cdot} < Na^{\cdot} < K^{\cdot} < Mg^{\cdot\cdot} < Ca^{\cdot\cdot}$$

gilt nur bei Vorbehandlung mit hypotonischen Lösungen; nach Einlegen der Schnitte in isotonische Lösungen bewirken die einwertigen Kationen konkave, die zweiwertigen konvexe Plasmolyse. Diesen Umkehreffekt, den TAKAMINE (1940) im Gegensatz zu der Wirkung der hypotonischen Lösungen nicht bestätigt, erzielen CHOLODNY und SANKEWITSCH (1933) nur bei Zwiebelschuppen, nicht bei *Rhoeo* und *Spirogyra*.

SCARTH (1923) ermittelt an *Spirogyra*-Material, das sich in m KNO_3 (bei Zusatz von etwas $CaCl_2$) konvex plasmolysieren läßt, für eine Reihe von Salzen diejenige Konzentration, bei der ein Haften des Protoplasten an der Zellwand eintritt, wenn die Algen vor der Plasmolyse eine Stunde lang mit Lösungen dieser Salze behandelt worden waren. Entsprechend S. 710 sind die Edelmetalle ($Hg^{\cdot\cdot}$, $Cu^{\cdot\cdot}$, $Ru^{\cdot\cdot}$) am wirksamsten, ihnen folgen eine Reihe vier- und dreiwertiger Ionen [$Fe^{\cdot\cdot\cdot}$, $Al^{\cdot\cdot\cdot}$, $Th^{\cdot\cdot\cdot\cdot}$, $Zr^{\cdot\cdot\cdot}$, $Sn^{\cdot\cdot\cdot}$, $La^{\cdot\cdot\cdot}$, $Ce^{\cdot\cdot\cdot}$, $Pr^{\cdot\cdot\cdot}$, $Sa^{\cdot\cdot\cdot}$, $Y^{\cdot\cdot\cdot}$, $Co(NH_3)_6^{\cdot\cdot\cdot}$], ferner $UO_2^{\cdot\cdot}$, $CrCl^{\cdot\cdot}$ und $Pb^{\cdot\cdot}$ Noch höhere Konzentrationen erfordern die übrigen zweiwertigen Ionen, bei Alkaliionen tritt glatte Ablösung ein, unter den Anionen wirken nur Chromat und Dichromat. Weitere Konzentrationssteigerung führt zur Verfestigung und weiter zum Tode des Plasmas. Die ähnlichen Feststellungen von FLURI (1909) und SZÜCS (1913) über eine Erstarrung des Cytoplasmas und Aufhebung der Plasmolysierbarkeit nach Behandlung mit $Al^{\cdot\cdot\cdot}$- (auch $Y^{\cdot\cdot\cdot}$- und $La^{\cdot\cdot\cdot}$-)Salzen sind vor allen Dingen von BRAMBRING (1930) als letale Erscheinungen kritisiert worden, nach WEBER (1933) liegt ihnen eine erhöhte Plasmolysepermeabilität zugrunde, die zur Ausbildung von Tonoplastenplasmolysen und zum Absterben des Protoplasten führt. Nach kurzer Einwirkung schwacher $Al^{\cdot\cdot\cdot}$-Konzentration tritt an Stelle dieser stark pathologischen Erscheinungen stärkeres Haften des Protoplasten an der Wand und Krampfplasmolyse (WEBER 1924b, übereinstimmend mit SCARTH 1923; vgl. auch PRINGSHEIM 1925). Die Reversibilität der Erscheinungen (SZÜCS 1913) wird im Einzelfalle zu untersuchen sein, die Aufhebung der $Al^{\cdot\cdot\cdot}$-Schädigung durch Zucker, Isodulcit, Glycerin und Harnstoff bleibt ebenso aufzuklären wie das Ausbleiben der $Al^{\cdot\cdot\cdot}$-Wirkung in anthocyanführenden Zellen (SZÜCS 1913).

Bei der Auswertung von Ioneneffekten muß die Dissoziation der Salze berücksichtigt werden, die auch bei gleicher molarer Salzkonzentration zu verschiedener *Ionen*konzentration führt, ebenso die Hydrolyse vieler Salze. Da Aluminiumsalzlösungen stark sauer reagieren, ist ihre Wirkung oft auf das H-Ion zurückgeführt worden (LEPESCHKIN 1927, ALBACH 1929, Literatur bei LEPESCHKIN 1938 und SCHARRER 1955). Es besteht aber eine spezifische $Al^{\cdot\cdot\cdot}$-Wirkung, die sich besonders nach TRÉNEL und ALTEN (1934) und EISENMENGER (1936) an Wurzeln und Keimlingen von der Säurewirkung trennen läßt.

b) Verhalten beim Zentrifugieren.

Die in den vorhergehenden Abschnitten zusammengefaßten Daten sprechen für eine Ionenwirkung auf die Viscosität des Cytoplasmas oder wenigstens seiner Oberfläche. WEBER (1924a) und WEIS (1928) stützen das Vorliegen einer Beeinflussung des gesamten Protoplasten durch Zentrifugen- neben den Plasmolyseversuchen, ebenso SZÜCS (1913) für die $Al^{\cdot\cdot\cdot}$-Wirkung. Auch TIMMEL (1928) gibt an, daß nach Behandlung mit hypo-und hypertonischen Lösungen von K- (nicht von Na- und Li-) Salzen der Zellinhalt sich leichter verlagern läßt; die $K^{\cdot}$-Wirkung kann durch $Ca^{\cdot\cdot}$-Zusatz aufgehoben werden (zum Teil auch durch $Mg^{\cdot\cdot}$, aber nicht durch $Fe^{\cdot\cdot\cdot}$ und $Mn^{\cdot\cdot}$), ebenso durch Wässern der Schnitte. TIMMELs (1928) Ergebnisse sind durch jahreszeitliche Veränderungen des Materials im einzelnen undurchsichtig. Nach JUNGERS (1934) ist der Protoplast in K-Salzen leichter verschiebbar als in Na- und Ca-Salzen. Die von NORTHEN und NORTHEN (1939) aufgestellten Ionenreihen sind wenig überzeugend: Es wurde die Prozentzahl der *Spirogyra*-Fäden festgestellt, in denen sich nach Zentrifugieren von nur 30 sec die Chloroplasten verlagert hatten.

Die Versuche von HEILBRUNN (1928, S. 146ff.), nach denen bei Seeigeleiern *(Arbacia)* und Protozoen *(Stentor)* eine Viscositätserhöhung durch $K^{\cdot}$, $Na^{\cdot}$ und $NH_4^{\cdot}$, eine Viscositätsherabsetzung durch $Ca^{\cdot\cdot}$ und $Mg^{\cdot\cdot}$ eintritt, werden auf S. 725 diskutiert. Nach HEILBRUNN und DAUGHERTY (1931, 1932) sind die aus Zentrifugenversuchen erschlossenen Viscositätsänderungen durch $K^{\cdot}$ und $Ca^{\cdot\cdot}$ im gelartigen Rindenplasma von Amöben denen im Binnenplasma entgegengesetzt. Die Angaben BRINLEYs (1928), nach welchen durch Einlegen von Amöben in NaCl und KCl die Plasmaviscosität (aus der BROWNschen Molekularbewegung erschlossen) gesenkt, durch $MgCl_2$ und $CaCl_2$ gesteigert wird, werden dadurch beeinträchtigt, daß ungleiche, zum Teil hypertonische und letale Konzentrationen verwendet wurden.

c) Injektion von Salzlösungen.

Um die Einflüsse auf die äußeren Plasmaschichten von denen auf das Binnenplasma zu trennen, wurden vielfach Salzlösungen nicht nur von außen an die Zelle herangebracht, sondern auch mit Mikropipetten in die Zellen injiziert. Die in Wurzelhaaren von *Limnobium Spongia* auf diese Weise mit NaCl und KCl hervorgerufene Verflüssigung des Cytoplasmas beruht nach KERR (1933) auf osmotischer Wasseraufnahme, da sie auch bei Injektionen von Zuckerlösungen auftritt; Voraussetzung ist die Injektion einer genügenden Menge osmotisch wirksamer Substanz. Die durch Salze mit zweiwertigen Kationen hervorgerufenen Wirkungen bestehen umgekehrt in einer „Verfestigung" des Cytoplasmas, welche bei $MgCl_2$ sich in der Zelle ausbreitet und reversibel ist, während bei $CaCl_2$ ein um die Einstichstelle lokal koagulierter Plasmaanteil auch nach Wiederherstellung des normalen Zellzustandes (Plasmaströmung) erhalten bleibt. Vergleichbar sind die Versuche der Schule von CHAMBERS an Amöben (CHAMBERS und REZNIKOFF 1926, REZNIKOFF und CHAMBERS 1927): KCl und NaCl verflüssigen injiziert das Cytoplasma (noch mehr LiCl nach REZNIKOFF 1928), daneben wird Zusammenklumpen der Granula beobachtet, das HEILBRUNN (1928, S. 152) als Koagulation deutet. Die verfestigende Wirkung ist ebenfalls vorhanden, bei $CaCl_2$ mehr lokalisiert, bei $MgCl_2$ über die ganze Zelle fortschreitend. Neben interessanten Einzelbeobachtungen über Lebensdauer der Amöben in Salzlösungen, Abschnürung von Zellteilen, die injiziert wurden, und antagonistischen Erscheinungen, die REZNIKOFF (1928) eingehender studiert hat, ist wesentlich, daß Alkalisalze bei der Injektion in die Zelle grundsätzlich ebenso wie von außen wirken, während $MgCl_2$ und $CaCl_2$ von außen keine Veränderungen im Zellinnern hervorrufen. Ähnlich verhalten sich Seeigeleier, in die Salze injiziert werden (CHAMBERS 1925), bei *Actinosphaerium* lassen sich Abweichungen im einzelnen nach CHAMBERS und HOWLAND (1930) auf die Baueigentümlichkeiten (stark vacuolisiertes Plasma) dieses Tieres zurückführen. Nach SPEK (1921) entspricht die Fällung der Plasmakolloide von *Actinosphaerium* durch Salze in der Kulturflüssigkeit ihrer Permeabilität ($K^{\cdot} > Na^{\cdot} > Li^{\cdot} > Ca^{\cdot\cdot}$); die Wirkung steigt proportional der Konzentration, außer bei dem nicht permeierenden $Ca^{\cdot\cdot}$, das erst bei einer kritischen Konzentration durch Risse eindringt und dann stark koagulierend wirkt. Neuerdings berichten KASSEL und KOPAC (1953) über Injektionsversuche mit balancierten Salzlösungen an Amöben, welche die von CHAMBERS und REZNIKOFF (1926) ergänzen. Neben $K^{\cdot}$ und $Na^{\cdot}$ wirkt auch $Mn^{\cdot\cdot}$ als $Ca^{\cdot\cdot}$-Antagonist, $Rb^{\cdot}$ und $Sr^{\cdot\cdot}$ verhalten sich ähnlich wie $K^{\cdot}$ bzw. $Ca^{\cdot\cdot}$.

Nach SPEK (1928) entsprechen die an *Opalina ranarum* auftretenden Erscheinungen, wenn die Tiere in Salzlösungen eingelegt werden, grundsätzlich der Auffassung, daß die leicht eindringenden Salze eine Wirkung im Zellinnern entfalten, wie sie sich z. B. an einer Trübung im Plasma erkennen läßt und in ähnlicher Weise auch an zerschnittenen Zellen auftritt. Überraschend ist aber die Wirkung von $CaCl_2$: Während dieses Salz wie bei CHAMBERS und REZNIKOFF (1926) gewöhnlich nur auf die Außenschicht zu wirken scheint (erst höhere Konzentrationen führen zum Tode), werden angeschnittene Zellen sofort desorganisiert, wobei auch die Pellicula vom Schnittrand her aufgelöst wird, so daß von ihr nur die Cilien übrigbleiben. $Ca^{\cdot\cdot}$ wirkt also auf die Pellicula von innen („verflüssigend") entgegengesetzt wie von außen („verfestigend"), ferner auf das Innenplasma bei *Opalina* überhaupt entgegengesetzt wie bei anderen Zellen. Über ein vergleichbares Verhalten von *Spirostomum* gegenüber Zuführung von außen und Injektion von $Ca^{\cdot\cdot}$ berichten EPHRUSSI und RAPKINE (1928).

Aus Befunden wie diesen ergeben sich Hinweise, wie sich die scheinbar extremen Widersprüche in den optisch feststellbaren Reaktionen von Zellbestandteilen auf Salze auflösen lassen: Einmal können sie in den Versuchsbedingungen und im Objekt liegen, wie besonders am Beispiel der Plasmolyseform gezeigt wurde und, wie ILJIN (1935a, b) betont, auch in Unterschieden der Auffassung und Terminologie. Eindringen von KCl und NaCl kann einerseits zur „Verflüssigung" des Cytoplasmas führen, andererseits aber im gleichen Versuch zur Bildung von Plasmagerinnseln, wie KERR (1933) in Übereinstimmung mit CHOLODNY (1923) an Wurzelhaaren zeigt: An der Spitze findet eine lokale Plasmafällung statt, von der Basis zuströmendes Plasma reichert sich dort an. Für das Erscheinungsbild ist bestimmend, ob eine sichtbare Plasmafällung ausgelöst werden kann. Auch die scheinbar HEILBRUNN und seiner Schule widersprechenden Versuche von CHAMBERS u. a. (1926—1928) bewirken eine Koagulation des Amöbenprotoplasmas, bevor die Tiere in Alkalichloridlösung absterben. Grundsätzlich sind bei höheren Salzkonzentrationen am Reaktionsort andere Phänomene zu erwarten; die Gerinnung, die den Tod kennzeichnet, mag wiederum anderer Art sein. Die unterschiedliche Reaktion des Außen- und Innenplasmas von Amöben (HEILBRUNN und DAUGHERTY 1931, 1932) beweist, daß es mehr als nur eine Wirkung des gleichen Salzes auf „das Protoplasma" gibt, wie auch LINSBAUER (1933) und SPEK (1928) auf Grund seiner *Opalina*-Versuche betont. Bei einer Reihe aufeinanderfolgender Zustände wird man verschiedene Erscheinungen beobachten: So geht der Koagulation des Kernsaftes von Amphibienoocyten durch $Mg^{\cdot\cdot}$ und $Ba^{\cdot\cdot}$ eine „Verflüssigung" voraus (CALLAN 1952). Schließlich lassen sich aus dem mikroskopischen Bild Änderungen des Kolloidzustandes, wie sie durch Bezeichnungen wie „Verflüssigung", „Koagulation", „Quellung" usw. charakterisiert werden sollen, nur unzureichend erschließen; selbst wenn das Protoplasma mit einem einfachen Kolloidsystem verglichen wird, wie dies vielfach geschehen ist, sind die sichtbaren Veränderungen nicht ohne weiteres bestimmten physikalisch-chemischen Wirkungen gleichzusetzen. Ihre Beobachtung ist die Grundlage, auf der eine Kausalanalyse erst aufzubauen hat.

d) Zellkern.

Für den Zellkern werden die Vorbehalte bei der Deutung optischer Effekte eher in noch höherem Maße gelten als für das Cytoplasma. Über die Veränderung der Kernstruktur bei Behandlung mit Salzlösungen berichten STRUGGER (1929, 1930) und BANK (1939). Mit KNO_3 lassen sich, besonders im Dunkelfeld, konzentrationsabhängige Strukturveränderungen des Karyotins zeigen, die reversibel sind; die Konzentrationsabhängigkeit gilt für Kerne in den Zellen verschiedener Pflanzen ebenso wie für isolierte Kerne aus *Chara*-Internodien. Bei Plasmolyse mit „quellenden" Salzen, besonders wenn diese zur Kappenplasmolyse führt, kann der Kern zu einer optisch homogenen „Solblase" aufquellen; STRUGGER (1931) schließt aus den sichtbaren Änderungen der Kernstruktur auf die in die Protoplasten eingedrungene Salzmenge. CHAMBERS und BLACK (1941) beobachten Veränderungen des optischen Erscheinungsbildes der Kerne von *Allium* und *Tradescantia* nur dann, wenn die geprüften Ionen als Bestandteil hypertonischer Lösungen geboten werden und schließen daher ebenfalls, daß die Plasmolyse erst das Eindringen der Ionen im erforderlichen Maße ermöglicht. Ionenreihen für die Volumenänderungen von Zellkernen bringen BOGEN (1948b, s. S. 713) für *Rhoeo* und CALLAN (1952, vgl. dieses Handbuch Bd. I, S. 474) für Amphibienoocyten; nach diesen Arbeiten treten am Kern grundsätzlich die gleichen Phänomene auf wie am Cytoplasma, auch hinsichtlich der Konzentrationsabhängigkeit der Volumenänderungen (CALLAN 1952). Nach YAMAHA und ISHII (1932) ist es außer vom p_H-Wert auch von der Art und der Konzentration der in der Beobachtungsflüssigkeit vorhandenen Salzionen abhängig, ob die Chromosomen in den Pollenmutterzellen von *Tradescantia* in vivo sichtbar sind.

Über den Ionengehalt von Zellkernen, über die Wirkung von Salzen bei der Mitose und auf die Chromosomen vgl. MILOVIDOV (1949—1954, I S. 356—368, II S. 157—158) und SERRA (dieses Handbuch Bd. I, S. 419 und 486), auch SOROKIN und SOMMER (1929, 1940) und KERSTING (1939; s. S. 715). Angaben über die Beeinflussung der Zellteilung, besonders bei Mangel an bestimmten Ionen, finden sich vielfach und zerstreut in der Literatur, ohne daß spezifische Effekte erfaßt zu sein scheinen; allein eine mögliche Viscositätsbeeinflussung durch Ionen bei der Mitose dürfte sehr verwickelte Folgen haben, da ohnehin während der Mitose regelmäßige Viscositätsveränderungen stattfinden (FRY und PARKS 1934). Nach CHÈVREMONT und FIRKET (1952) hemmt $Be^{\cdot\cdot}$ spezifisch die Mitose in Gewebekulturen durch Hemmung der alkalischen Phosphatase im Zellkern; $Mg^{\cdot\cdot}$ wirkt antagonistisch zu $Be^{\cdot\cdot}$; dagegen scheint bei der Hemmung von Wachstum und Mitosen in Gewebekulturen durch $Li^{\cdot}$ eine unspezifische Wirkung auf Stoffwechsel und Viscosität vorzuliegen (CHÈVREMONT-COMHAIRE 1953).

D. Ionenwirkung und Ionenaufnahme.

Eine direkte Wirkung wird ein Ion nur an solchen Zellstrukturen entfalten können, zu denen es Zutritt hat. Vergleicht man die Veränderungen, wie sie beim Einlegen der Zelle in eine Salzlösung auftreten, mit den durch Injektion des gleichen Salzes hervorgerufenen (s. S. 724), so scheinen die Erdalkalisalze im Gegensatz zu den Alkalisalzen bei Zufuhr von außen nur auf die Außenschicht einzuwirken, weil sie nicht tiefer eindringen können (vgl. auch WEBER 1933 über den Zusammenhang zwischen $Al^{\cdot\cdot\cdot}$-Wirkung und Plasmolysepermeabilität, s. S. 723). Die Bedeutung der Permeabilität für die Salzwirkung, speziell für die Giftigkeit von Einsalzlösungen und für antagonistische Erscheinungen ist besonders von KAHO (1926a, dort Literaturzusammenfassung) herausgestellt worden. Mit Salzen plasmolysierte Zellen zeigen durch allmähliche Deplasmolyse eine Permeabilität für Salze an, die stets bei einwertigen Kationen größer ist als bei zweiwertigen, wie PANTANELLI (1904) bei *Aspergillus*, RUHLAND und HOFFMANN (1926) bei *Beggiatoa*, PRÁT (1922a) bei *Spirogyra*, PRÁT (1923) bei Cyanophyceen, FITTING (1915) und TRÖNDLE (1918, 1922) an Geweben höherer

Pflanzen grundsätzlich übereinstimmend angeben; auch das Anion ist von Bedeutung. Entsprechende Feststellungen macht LUNDEGÅRDH (1911) an Wurzeln mit seiner Gewebespannungsmethode (Messung der Wiederausdehnung eines plasmolysierten Gewebes), die auch KAHO (1926a) benutzt. Die mit den verschiedenen Methoden gefundenen Reihen für die Permeabilität stimmen (bei Abweichung im einzelnen) im großen überein:

$$K^{\cdot} > Na^{\cdot} > Li^{\cdot} > Mg^{\cdot\cdot} > Ba^{\cdot\cdot} > Ca^{\cdot\cdot}$$

$$J' > Br' > NO_3' > Cl' > \text{Tartrat} > SO_4'' > \text{Citrat.}$$

Die Reihen entsprechen weitgehend denen von KAHO (1921a, 1921b, 1926a) für die Giftwirkung auf Blattepidermiszellen, z. B.

$$CNS' > J' > Br' > NO_3' > CH_3COO' > Cl' > \text{Tartrat} > \text{Citrat} > SO_4''$$

und von LEPESCHKIN (1927) auf *Spirogyra*. Die Alkalikationen sind giftiger als die Erdalkalikationen, $Mg^{\cdot\cdot}$ kann eine Mittelstellung einnehmen oder in seiner Wirkung den Alkaliionen gleichen, wie z. B. nach v. EISELSBERG (1937) beim Absterben von *Spirogyra* in hypertonischen Lösungen; aus dieser Arbeit, die artspezifische Eigenheiten betont, wäre eine von der Regel abweichende hohe Mortalität der Zellen in $SrCl_2$ hervorzuheben. Überwiegend findet sich aber die Reihe

$$K^{\cdot} > Na^{\cdot} > Mg^{\cdot\cdot} > Ba^{\cdot\cdot} > Ca^{\cdot\cdot}$$

für die Giftwirkung, wie nach KAHO, FITTING und TRÖNDLE für die Permeabilität.

KAHO (1926a) sieht die Giftigkeit als Folge einer durch das Salz selbst abnorm gesteigerten Permeabilität an, wobei die Ionen das Salzes additiv wirken, die zum Tode führende Störung (Koagulation) im Plasmainneren als eine Folge eines durch einseitige Salzanreicherung gestörten Gleichgewichts. Die Hemmung der Aufnahme von Alkalisalzen durch andere Salze, besonders durch solche mit mehrwertigen Kationen (NETTER 1923, KAHO 1921a, b, 1926a, b), legt es nahe, deren antagonistische Wirkung ebenfalls mit Permeabilitätsphänomenen in Verbindung zu bringen; nach KAHO (1926a) hemmt jedes Kation der Reihe

$$K^{\cdot} > Na^{\cdot} > Li^{\cdot} > Mg^{\cdot\cdot} > Ba^{\cdot\cdot} > Ca^{\cdot\cdot}$$

die Aufnahme eines links von ihm stehenden und zwar um so mehr, je größer der Abstand zweier Ionen in dieser Reihe ist. Dabei wird die Wirkung beider Ionen eines Salzes kombiniert, so daß durch Mischung sowohl eine Steigerung wie eine Herabsetzung von Permeabilität und Salzwirkung resultieren kann (vgl. auch SPEK 1928).

Für die Annahme, daß die beobachtete Salzaufnahme durch das unphysiologische Milieu hervorgerufen werden kann, sprechen verschiedene Arbeiten (z. B. OSTERHOUT 1922, KACZMAREK 1929; vgl. aber WEIXL-HOFFMANN 1930); besonders HÖBER (1947, Kapitel 11 und 17, dort auch entsprechende Literatur über Tierzellen) betont die Aufhebung der selektiven Ionenpermeabilität als Folge einer Veränderung der Plasmaoberfläche (Quellungsförderung, Auflockerung) durch Kationen und Anionen der Alkalisalze. Die lyotropen Reihen und die antagonistischen Effekte bei Giftwirkung und Permeabilität fügen sich in den Rahmen der Kolloidtheorie ein; allerdings bedeutet die Parallelität nicht unbedingt eine feste kausale Bindung von Stoffaufnahme und Ionenwirkung. KAHO (1926a) selbst vermutet auf Grund der Versuche von BOROWIKOW (1913), nach denen bei der Hemmung des Keimlingswachstums die Ionenreihen umgekehrt sind, daß außer dem Eindringen der Salze auch eine Quellungsbeeinflussung der jungen Zellwände in der Wachstumszone das Bild der Ionenwirkung bestimme. Sicherlich ist das Eindringen von Salzen in die Zelle eine Voraussetzung für ihre Wirkung im Zellinneren; es wird aber mehr als früher Gewicht auf eine mikroskopische oder submikroskopische Heterogenität der Ionenverteilung zu legen sein, um solche Wirkungen zu kennzeichnen. Mindestens bei Permeabilitätsversuchen mit osmotischen Methoden ist auch eine durch diese nicht erfaßbare Intrabilität im Auge zu behalten.

Zum Beweis einer Salzintrabilität hat man die als Kappenplasmolyse bezeichnete Aufquellung des Cytoplasmas herangezogen, wie sie bei Plasmolyse, vor allem mit Alkalisalzen, aber auch mit anderen Lösungen, auftritt (vgl. dieses Handbuch, Band I, S. 385—386). Nach STRUGGER (1932) und HÖFLER (1934) geht das Aufquellen des Cytoplasmas auf die Anreicherung von Salzen zurück, die infolge einer bei der Plasmolyse erhöhten Intrabilität (bei verringerter Permeabilität) aufgenommen werden, wofür auch die Reversibilität der Erscheinung durch $CaCl_2$ und $SrCl_2$ zu sprechen scheint (HÖFLER 1939; weitere Literatur bei HOUSKA 1942 und BOGEN 1951). Schwierigkeiten entstanden dieser Deutung schon dadurch, daß Kappenplasmolyse mit sehr verschiedenen Stoffen, darunter auch Farbstoffen und Nichtelektrolyten, und sogar mit $Ca(NO_3)_2$-Lösung (LANZ 1942) ausgelöst werden, andererseits auch in LiCl ohne Mediumwechsel rückläufig sein kann (BOGEN 1951). Dieses Beispiel zeigt zugleich, daß manche „Ioneneffekte" ebenso durch andere Einwirkungen hervorgerufen werden können. In einer kritischen Untersuchung über die Ätiologie der Kappenplasmolyse und der mit ihr verwandten Vacuolenkontraktion zeigt neuerdings BOGEN (1951), daß beide Phänomene nicht im genannten Sinn zu deuten sind, sondern eher als eine Wechselwirkung zwischen quellbaren Kolloiden im Cytoplasma und in der Vacuole, wobei möglicherweise der Vacuole eine größere Aktivität zukommt, als man bisher annahm. Danach lösen die wirksamen Stoffe (auch in kleinsten Mengen) die Vorgänge nur aus. Als maßgebend werden zunächst ladungsbedingte Hydratationsänderungen angenommen, ohne daß über die Rolle von Ionen in diesem Zusammenhang einzelnes ausgesagt werden kann.

E. $Ca^{\cdot\cdot}$ als Beispiel spezifischer Ionenwirkungen.

An die Beschränkung der Wirkung zweiwertiger Ionen auf die äußeren Plasmaschichten lassen sich Beobachtungen anschließen, die über die bei der Permeabilitätsregulierung angenommene „Abdichtung" des Plasmolemmas hinaus eine Abdichtung verletzter Stellen des Protoplasten anzeigen. Hierher gehört die Bildung von Plasmapfropfen beim Einstich von Mikroelektroden (UMRATH 1932) und Mikropipetten (CHAMBERS und REZNIKOFF 1926, KERR 1933) um die verletzte Stelle, wenn Ca zugeführt wird, das eine lokale Koagulation im Plasma hervorruft, und zahlreiche Fälle von Membranbildung an verletzten und plasmolysierten Protoplasten. HEILBRUNN (1928, S. 215ff.) bezeichnet die Ausbildung eines Membranfilmes, wie er beim Ausfließen des Plasmas aus zerquetschten Seeigeleiern an dessen Oberfläche gebildet wird, als *surface precipitation reaction*; sie findet nur statt, wenn Ca in der umgebenden Flüssigkeit vorhanden ist. Entsprechende Bildungen finden sich beim Ausfließen des Plasmas aus vielen verletzten Tier- und Pflanzenzellen (Literatur bei HEILBRUNN 1928, COSTELLO 1933, TERRY 1950). Aus zerschnittenen *Chara*-Internodialzellen austretende Plasmakugeln umgeben sich mit einem Film, der erhärtet und brüchig wird, wenn Ca vorhanden ist; fließt der Zellinhalt in eine Kaliumsalzlösung aus, so zerfließt das Protoplasma ohne eine solche Hautbildung. LINSBAUER (1933), der dieses „Grenzhäutchen" eingehend studiert hat, sieht es als ausgesprochenen Grenzflächenfilm (Haptogenmembran) mit semipermeablen Eigenschaften, nicht wie HEILBRUNN als Niederschlagsmembran an; er betont, daß die Ähnlichkeit mit der normalen Plasmahaut nicht die Identität des Aufbaues beweise, wie überhaupt eine Reihe ähnlicher Bildungen verschiedene Eigenschaften haben dürfte, vielleicht erklären sich hieraus die Widersprüche in der Literatur. Die Untersuchungen des ausfließenden *Chara*-Protoplasmas erweitert DIETZEL (1953)[1] mit K-, Na-, Mg-, Ca-, Sr-, Ba-, Zn- und Al-Salzen: $SrCl_2$ und

---

[1] Herrn Professor Dr. ULLRICH danke ich dafür, daß er mir das Manuskript dieser Arbeit zur Verfügung stellte.

$BaCl_2$ sind im Gegensatz zu $MgCl_2$ ebenfalls zur Bildung des Grenzhäutchens geeignet, dabei aber weniger wirksam: die minimale Konzentration, die erforderlich ist, liegt bei $CaCl_2$ niedriger (n/5000, dagegen n/100 $SrCl_2$, n/10 $BaCl_2$); im Antagonismus gegenüber $K^{\cdot}$ hat $Ca^{\cdot\cdot}$ ebenfalls die vielfach gefundene bevorzugte Stellung.

Nach WEBER (1932, 1934) wird von Pflanzenzellen Plasmolyse mit Harnstoff nach Vorbehandlung mit Kaliumoxalat nicht vertragen, weil die Entfernung von $Ca^{\cdot\cdot}$ aus den äußeren Plasmaschichten die Bildung einer neuen Plasmahaut verhindert und damit die Voraussetzung für eine abnorm hohe Plasmolysepermeabilität schafft, die zu einer Überschwemmung des Protoplasten mit Harnstoff führt (Quellungsdegeneration); bei Zwischenbehandlung mit $Ca^{\cdot\cdot}$ verläuft die Plasmolyse normal. Desmidiaceen vertragen nach HÖFLER (1951) auch Plasmolyse mit Kaliumoxalat (Anpassung an das $Ca^{\cdot\cdot}$-arme Hochmoorwasser?). EICHBERGER (1934, 1935) bestreitet die spezifische $Ca^{\cdot\cdot}$-Wirkung bei der Hautbildung, da sich der Oxalateffekt auch mit anderen Salzen aufheben läßt und führt den heilenden Einfluß auf das Fällungsvermögen der Salze für Eiweiß und Lipoide zurück. Die Behäutung von plasmolysierten Protoplasten verläuft bei $Ca^{\cdot\cdot}$ schneller als bei $K^{\cdot}$ und Na im Plasmolyticum, wie MISSBACH (1928) und LOREY (1929) daraus schließen, daß die Fähigkeit zur Fusion von Teilprotoplasten, die nicht durch Fäden verbunden sind, unter dem Einfluß von $Ca^{\cdot\cdot}$ schneller verlorengeht. JOSTS (1929) Beobachtungen über Wundheilung und Plasmolyseschäden bei *Valonia* und *Chara* unterstreichen die Bedeutung von $Ca^{\cdot\cdot}$ für Wundverschluß und Zellwandregeneration und zugleich die Unterschiede im Verlauf der Ionenwirkung auf zwei Protoplasten ungleicher Struktur. Nach KOBINGER (1953) tritt die Bildung einer Vernarbungsmembran (Cellulose) an plasmolysierten *Helodea*-Blattzellen nur in Zuckerlösung ein, wird aber durch $Ca^{\cdot\cdot}$ gefördert.

Die zuletzt geschilderten Vorgänge fügen sich in den Rahmen der Kolloidtheorie der Ionenwirkung ein, und sie unterstreichen mit der Veränderung der äußeren Plasmaschichten durch mehrwertige Ionen besonders die Beziehung zwischen Permeabilität und Salzwirkung. Es wird aber hier auch die Frage gestellt, ob die Kolloidaktivität der Ionen zur Erklärung ausreicht oder nicht vielmehr spezifische Effekte hinzukommen, wie sie speziell für $Ca^{\cdot\cdot}$ durch seine Stellung in vielen physiologischen Ionenreihen und durch die besondere Förderung des Wurzelwachstums nahegelegt werden. $Ca^{\cdot\cdot}$ läßt sich je nach dem untersuchten System oder Vorgang mehr oder weniger durch andere mehrwertige Ionen ersetzen, wofür außer den genannten Beispielen bei HÖBER (1926, S. 688ff.; 1947, S. 335ff.) auch eine größere Anzahl von Beispielen aus dem Tierreich zusammengefaßt wird. Danach kann $Ca^{\cdot\cdot}$ bei den Zell- und Organfunktionen in manchen Fällen gar nicht, in anderen nur durch $Sr^{\cdot\cdot}$ oder auch durch $Ba^{\cdot\cdot}$, aber meist nicht durch $Mg^{\cdot\cdot}$, in wieder anderen durch andere mehrwertige Kationen vertreten werden. Nach LOEB (vgl. HÖBER 1947, S. 338) wird die schädigende Wirkung einer reinen NaCl-Lösung auf *Fundulus*-Embryonen, die in einer Auflockerung der Eihülle besteht, durch zweiwertige Kationen ganz allgemein behoben, dagegen können die Embryonen nach Entfernung der Eihülle außer durch $Ca^{\cdot\cdot}$ nur durch $Sr^{\cdot\cdot}$ und $Mg^{\cdot\cdot}$ vor der $Na^{\cdot}$-Schädigung geschützt werden. Besonders eindeutig tritt eine spezifische $Ca^{\cdot\cdot}$-Förderung beim Wurzelwachstum hervor (s. S. 715). Die Vertretbarkeit durch $Sr^{\cdot\cdot}$, die gelegentlich behauptet wurde (MEVIUS 1926) ist mindestens nicht in allen Fällen gesichert (MEVIUS 1928, LIBBERT 1953). Man hat neben der kolloidchemischen Wirkung auch eine chemische Reaktion zur Erklärung der verfestigenden Wirkung von Ca angenommen; Grundlage könnte die Bildung von schwerlöslichen Ca-Salzen einer schwachen Säure sein (Proteinate,

Phosphatidsalze, Seifen usw.), welche Plasmakomponenten auf Grund der Zweiwertigkeit von Ca aneinander binden; beim Fehlen von $Ca^{\cdot\cdot}$ würde durch Bildung von Alkalisalzen diese Bindung aufgehoben, wie folgendes Schema nach HÖBER (1947, S. 337) zeigt:

$$R{-}C({=}O){-}O{-}Ca{-}O{-}C({=}O){-}R \rightleftharpoons R{-}C({=}O){-}O{-}Na + Na{-}O{-}C({=}O){-}R$$

Für eine solche Kittsubstanz spricht z. B., daß nach HERBST (1900) Blastomeren von Seeigeleiern in Ca-freiem Meerwasser reversibel getrennt werden. OVERTON (1905), der auch bei Epithelien eine Verkittung durch ein wenig gequollenes Ca-Salz einer schwachen Säure annimmt, weist besonders auf die entsprechende Rolle des Ca-Pectats in den Mittellamellen der Pflanzenzellen hin; nach BENECKE (1898) können die Zellen von *Spirogyra*-Fäden bei ungünstigen Bedingungen, unter anderem Ca-Mangel, leichter getrennt werden. Neuerdings gibt LUNDEGÅRDH (1946) für die Zellwand, speziell von Wurzelhaaren, ein Schema der Verbindung anionischer Makromoleküle durch $Ca^{\cdot\cdot}$ zur Erklärung der Verfestigung durch $Ca^{\cdot\cdot}$ und Verflüssigung durch $K^{\cdot}$ (Abb. 7). Danach ist bei $Ca^{\cdot\cdot}$ der erreichte Effekt der günstigste, $Mg^{\cdot\cdot}$ wird zu schwach gebunden, die Schwermetallionen dagegen zu stark, so daß sie dehydratisierend wirken und die Moleküle der Membran in einem physiologisch nicht mehr günstigem Maße einander nähern. CORMACK (1949) sieht vor allem auf Grund seiner Versuche bei verschiedener Acidität in einer Verfestigung der Zellwand durch Calciumpectat in den Wurzelhaarinitialen sogar den Grund für die Bildung der Wurzelhaare selbst: Die Zellen bleiben infolge bevorzugter Verfestigung kurz, und wachsender Turgordruck führt zur Ausstülpung der Zellwand nach außen am Orte des geringsten Widerstandes. Daß diese Vorstellung zu einfach ist, geht aus Befunden von BURSTRÖM (1952) hervor, durch welche drei Ca-Effekte auf die Wurzeln getrennt werden und eine davon unabhängige Förderung der Wurzelhaarbildung wahrscheinlich gemacht wird (s. S. 719). Nach BURSTRÖMs (1952, 1954) Hypothese ist $Ca^{\cdot\cdot}$ für die Einlagerung von elastischem Material in die schon gestreckte Wand unentbehrlich, während die zu $Ca^{\cdot\cdot}$ antagonistische Wuchsstoffwirkung, die in einer Lockerung der Wand besteht, sich in der ersten Phase der Zellstreckung auswirkt, in der unabhängig von $Ca^{\cdot\cdot}$ die Dehnbarkeit der Zellwand ohne Bildung von neuem Wandmaterial erhöht wird. BURSTRÖM (1952) diskutiert in diesem Sinne die Auffassung von CORMACK (1949); die Analyse der Faktoren (Wuchsstoff, $Ca^{\cdot\cdot}$ und $H^{\cdot}$) legt schon im gegenwärtigen Stadium der Kenntnisse, welches die Art der $Ca^{\cdot\cdot}$-Wirkung auf die Zellwand im einzelnen noch offen läßt, nahe, zur Erklärung der Beobachtungen nicht nur chemische Reaktionen und kolloidale Zustandsänderungen, deren Beteiligung nicht bezweifelt werden soll, heranzuziehen, sondern den Ablauf von Reaktions*ketten* und damit die Verbindung zum Stoffwechsel zu betonen. BURSTRÖM (1954) schneidet die Frage des Zusammenhangs von Streckungswachstum, Ca- und Eiweißstoffwechsel an.

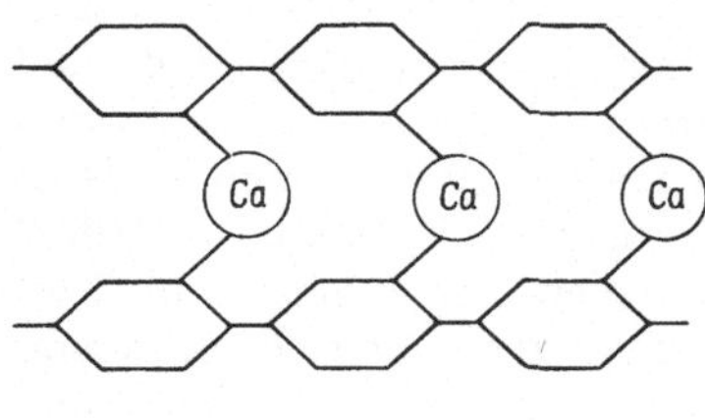

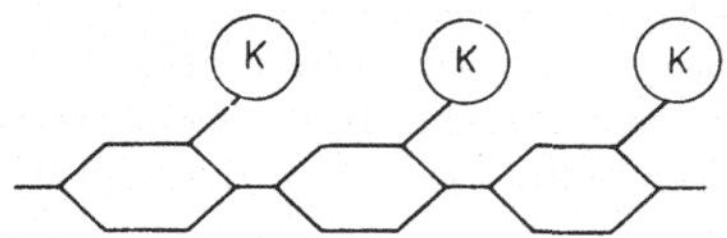

Abb. 7. Hypothetisches Schema der Bindung von anionischen Makromolekülen durch Ca. (Nach LUNDEGÅRDH 1946.)

Speziell für $Ca^{\cdot\cdot}$ ergeben sich Anhaltspunkte für eine Verknüpfung mit dem Stoffwechsel auch aus der Entwicklungsphysiologie der Tiere, die hier nur kurz gestreift werden sollen. Die von HEILBRUNN (1928) gezogene Parallele zwischen surface precipitation reaction, Blutgerinnung und den Vorgängen bei der Weiterentwicklung des befruchteten oder künstlich aktivierten Seeigeleies bringt $Ca^{\cdot\cdot}$ mit enzymatischen Prozessen in Verbindung; auch für die Struktur der Rindenschicht ist $Ca^{\cdot\cdot}$ von Bedeutung (Zusammenfassung bei RUNNSTRÖM 1949), ATP, ATPase und $Ca^{\cdot\cdot}$ als deren Aktivator scheinen zusammen zu wirken. Über den Einfluß von ATP und $Ca^{\cdot\cdot}$ auf Seeigeleier berichteten RUNNSTRÖM und KRISZAT (1950), auf Amöben KRISZAT (1950). Möglicherweise bestehen Zusammenhänge zwischen der $Ca^{\cdot\cdot}$-Wirkung auf die ATPase und der raschen Steigerung der O_2-Aufnahme, die nach HULTIN (1949, 1950a, b) unmittelbar auftritt, wenn $Ca^{\cdot\cdot}$ zum Homogenisat von Seeigeleiern aus $Ca^{\cdot\cdot}$-freiem Milieu gegeben wird.

F. Die Ionen des Wassers.

Verglichen mit den Salzionen nehmen die $H^{\cdot}$- ($H_3O^{\cdot}$-) und OH'-Ionen schon dadurch eine Sonderstellung ein, daß sie in wäßrigen Lösungen stets und in ganz bestimmten durch die Dissoziation des Wassers gegebenen Verhältnissen vorhanden sind (vgl. dieses Handbuch, Band I, S. 184ff.). Unter ihren physikalischen Eigenschaften ist ihre Beweglichkeit bemerkenswert, welche die der anderen Kationen und Anionen um ein Vielfaches übertrifft; es liegt daher nahe, mit dieser Eigenschaft die starke biologische Wirkung der Ionen des Wassers in Verbindung zu bringen, welche so auffallend ist, daß die Wasserstoffionenkonzentration, obwohl sie in Zellen und Körperflüssigkeiten im allgemeinen gering ist, zu den am meisten beachteten physiologischen und ökologischen Faktoren gehört. Die Erscheinungen werden im einzelnen bei verschiedenen Abschnitten dieses Handbuchs behandelt; an dieser Stelle sollen nur einige grundsätzliche Fragen gestreift werden.

Die Wasserstoffionenkonzentration wird einerseits durch puffernde Substanzen beeinflußt, andererseits ist die Dissoziation anderer Elektrolyte von ihr abhängig; die $H^{\cdot}$ und OH' werden in der Zelle und auf die Zelle einmal direkt wirken können, andererseits auch indirekt durch Beeinflussung des gesamten Ionenmilieus. So beeinflussen sie auch Dissoziation und Ladungszustand der Makromoleküle; auf diesem Wege modifizieren sie unter anderem die Effekte der Mineralsalzionen (s. S. 709). Ein so auffallendes Merkmal der Lebensprozesse wie die p_H-Abhängigkeit der Fermenttätigkeit wird zu einem wesentlichen Teil auf einer derartigen Strukturbeeinflussung von Ferment und Substrat neben einem Zusammenspiel mit Cofermenten und Aktivatoren beruhen[1].

Man könnte erwarten, daß Aciditätsverschiebungen im Außenmedium einen tiefgreifenden Einfluß auf die Plasmastruktur haben, so wie sie an Bodenkolloiden beim Austausch von metallischen Kationen gegen $H^{\cdot}$ auftreten; vielfache Beobachtungen haben aber gezeigt, daß die in der Zelle gemessenen Werte von den p_H-Veränderungen in der Umgebung weitgehend unabhängig sind, soweit sich diese im Rahmen physiologischer Bedingungen halten. Dies ist häufig an Zellsäften festgestellt worden (Literatur bei DRAWERT, dieses Handbuch, Band I, S. 642). RIETSEMA (1949) berichtet, daß er bei Mikroinjektion von Indicatoren in Zellen von *Avena*-Koleoptilen in manchen Fällen neben der Vacuolenfärbung auch eine Färbung im Cytoplasma beobachten konnte; der p_H-Wert blieb nach 24stündigem Aufenthalt der Koleoptilenstücke in Pufferlösungen von p_H 4,2, 5,8 und 8,3 unverändert (5,6—5,9 in der Vacuole wie im Cytoplasma); erst

[1] Vgl. diesen Band, S. 489, DUSPIVA, F.: Die Mitwirkung von Enzymen im Zellstoffwechsel.

durch die letal wirkende Lösung von p_H 3,4 sank auch der p_H-Wert in der Zelle auf 4,4. Auch Amöben sind, solange sie leben, von der Reaktion des Außenmediums unabhängig, wie SPEK und CHAMBERS (1934) durch Injektion von Indicatoren zeigen, durch welches allerdings nur einzelne Plasmakomponenten, die zum Teil wohl auch Vacuolen sind, nicht aber die „Grundmasse“ des Cytoplasmas gefärbt werden. Bei Injektion von Säuren (HCl, Buttersäure, Milchsäure) oder NaOH mit dem Indicator verändert sich der um 7,3 liegende p_H-Wert des Amöbenplasmas nicht, dagegen lassen sich im hängenden Tropfen mit CO_2 und NH_3 reversible Umfärbungen in der ganzen Zelle erzielen.

Gelegentlich können auch plasmaeigene Indicatoren zur intracellulären p_H-Messung dienen; der Myxomycet *Physarum polycephalum* enthält ein gelbes Pigment, das bei alkalischer Reaktion nach Grün, bei saurer nach Rot umschlägt, wodurch sich das Eindringen von NaOH, HN_4OH und HCl (nicht von Phosphorsäure und Essigsäure) zeigen läßt (SEIFRIZ 1935). Durchleiten eines Gleichstroms führt an der Kathode zu alkalischer Reaktion und faseriger elastischer Plasmastruktur, an der Anode zur Ansäuerung und butterartig-plastischer Konsistenz.

Bei der Ausdeutung solcher Befunde muß das Fehlen einer zuverlässigen Bestimmungsmethode für intracelluläre p_H-Werte berücksichtigt werden; die damit zusammenhängenden methodischen Fragen werden von DRAWERT in diesem Handbuch (Band I, S. 626ff.) ausführlich dargestellt. Die Bedenken gegenüber p_H-Messungen im Protoplasma sind noch schwerwiegender als gegenüber solchen in der relativ homogenen Vacuolenflüssigkeit. Bei der für Bestimmungen im Protoplasma allein in Betracht kommenden colorimetrischen Methode färben sich im allgemeinen nur gewisse plasmatische Komponenten und diese durchaus nicht übereinstimmend; ferner können die Färbungen in verschiedenen Zellbezirken differieren. Eine Störung der Plasmastruktur durch die Indicatoren selbst muß berücksichtigt werden; bei Injektionsversuchen an *Asterias*-Eiern findet nach CHAMBERS und POLLACK (1927) um die Einstichstelle herum eine Ansäuerung (von p_H 6,7 auf 5,6 bis 5,4) statt, die sich mehr oder weniger weit im Cytoplasma ausbreiten kann. Auf Grund einer größeren Anzahl von Untersuchungen an Eiern mariner Tiere und an Protozoen hat besonders SPEK (1938, 1937) die Heterogeneität des Plasmas hinsichtlich der p_H-Verteilung betont, die es nicht zuläßt, von einem p_H-Wert des Protoplasmas oder gar der Zelle zu sprechen, die aber andererseits eine feinere Regulation, etwa der Enzymtätigkeit, durch lokale p_H-Differenzen ermöglicht; er betont besonders die räumliche Trennung der gegenüber Indicatoren verschieden reagierenden Strukturen. Physiologisch wäre eine solche Vorstellung naheliegend, auch wenn die (in ihrem Werte ja zweifelhaften) p_H-Messungen an plasmatischen Strukturen nicht vorlägen; auf Grund unserer heutigen morphologischen Vorstellungen über den Plasmafeinbau ist zu erwägen, wie weit die Vorstellung eines p_H-Wertes bestimmter „disperser Phasen“ des Protoplasmas überhaupt sinnvoll ist.

Zur Erklärung der weitgehenden Unabhängigkeit der Reaktion in der intakten Zelle von der Umwelt könnte die Pufferung des Protoplasmas und seine mangelnde Permeabilität für $H^{\cdot}$ und $OH^{\cdot}$ dienen; für die zweite Deutung spricht, daß eine Reaktionsveränderung durch Kohlensäure und Ammoniak, die wegen ihrer relativ schwachen Dissoziation in wäßriger Lösung leicht eindringen können, leicht zu erreichen ist. Darüber hinaus bleiben aber gerade die exogenen p_H-Einflüsse im physiologischen Bereich zu erklären, die entweder im Binnenplasma in einer Weise wirken, die nicht mit p_H-Indicatoren erkennbar wird, oder die von den äußeren Plasmagrenzschichten aus die Tätigkeit der ganzen Zelle beeinflussen müßten. Man wird solche Überlegungen auch für andere Ionen anstellen dürfen, die ebenfalls nicht unmittelbar im Plasma nachweisbar sind. Für die Pflanzenzelle ergeben sich unter diesem Gesichtspunkt noch besondere Fragen hinsichtlich der Grenzschicht zwischen dem meist sauren Zellsaft und dem Protoplasma mit Reaktionsverhältnissen um den Neutralpunkt, die vielleicht auch im Cytoplasma der Pflanzen vielfach gegeben sind, so wie sie nach SPEK (1937) im ausfließenden Plasma aus Tierzellen für das „Dispersionsmittel der Plasmakolloide" charakteristisch sind.

Zur Beurteilung der $H^{\cdot}$ -und $OH^{\cdot}$-Wirkung ist wesentlich, daß der Physiologe sie im allgemeinen unter Verwendung von Pufferlösungen zu erfassen sucht und damit andere, zum Teil sehr wirksame Ionen zuführt. Bei Verwendung von Puffern, in denen die Konzentration solcher Ionen möglichst konstant ist, läßt sich die Rolle von $H^{\cdot}$ und $OH^{\cdot}$ herausschälen, ebenso wenn die Ergebnisse mit verschiedenen Puffern und möglichst auch mit reinen Säuren und Basen übereinstimmen. Von den einzelnen Pufferbestandteilen sind spezifische physiologische Wirkungen (auch Zellschädigungen) zu erwarten, viele von ihnen, wie Acetate, Citrate, Carbonate und besonders Phosphate, werden leicht in den Stoffwechsel einbezogen.

Es geht aus einer großen Anzahl von Arbeiten hervor, daß Art und Konzentration des Puffers bei zellphysiologischen Erscheinungen von Bedeutung sind, wofür nur einige Beispiele genannt seien. YAMAHA und SUITA (1940) verglichen vier verschiedene Pufferlösungen hinsichtlich ihrer Fähigkeit, Strukturenanomalien an Chloroplasten von *Spirogyra* hervorzurufen, und fanden, daß bei gleichem p_H-Wert die auftretenden Erscheinungen je nach Zusammensetzung und Konzentration des verwendeten Puffers verschieden ausfallen; entsprechende Feststellungen machte STRUGGER (1928) an Wurzelhaaren. SAKAMURA betont besonders die Wirkung des Acetations für die in saurer Lösung eintretenden Schädigungen an *Spirogyra*-Zellen (SAKAMURA 1934) und Wurzelhaaren (SAKAMURA und KANAMORI 1945). Wenn auch derartige Beobachtungen starken subjektiven Fehlern unterliegen, so zeigen sie doch die Bedeutung insbesondere der Anionen der Puffergemische. Auch SCHULTE (1938), der bei seinen Untersuchungen über die Säureplasmoptyse die geplatzten Wurzelhaare aus*zählte*, fand ein verschiedenes p_H-Optimum für diese Erscheinung, je nachdem, ob er Acetat-, Phosphatpuffer oder reine Säuren verwendete. ILJIN (1935a), der die Bedingungen prüfte, unter denen plasmolysierte Protoplasten längere Zeit am Leben bleiben, hebt hervor, daß die optimalen p_H-Werte artspezifisch sind und daß Phosphatpuffer relativ schädlich wirken. Über das Verhalten von Amöben gegenüber verschiedenen injizierten oder von außen wirkenden Säuren vgl. die schon genannten Arbeiten von REZNIKOFF und CHAMBERS (1927) sowie SPEK und CHAMBERS (1934); die von BRINLEY (1938) am Amöbenplasma erzielten Effekte („Verfestigung" mit HCl, „Verflüssigung" mit NaOH) scheinen stark pathologisch zu sein. Nach BARTH (1929) koagulieren im Seewasser gelöste organische Säuren das Cytoplasma von *Arbacia*-Eiern schon bei höherem p_H-Wert als anorganische; in isosmotischer NaCl-Lösung wirken Säuren und Laugen stärker (Permeabilitätserhöhung durch die nicht balancierte Salzlösung?).

Aus den genannten methodischen und grundsätzlichen Erwägungen ist es verständlich, wenn die Einzelbeobachtungen über p_H-abhängige zellmorphologische und -physiologische Erscheinungen voneinander abweichen; daß sie nur beschränkt für bestimmte physikalisch-chemische Zustände auszuwerten sind, wurde schon betont (S. 725—726). Ein großer Teil der älteren Literatur ist bei STRUGGER (1934) und SCHULTE (1938) zusammengefaßt; es finden sich vielfach komplizierte Abhängigkeiten, wie z. B. mehrgipfelige Kurven für „Verflüssigung" und „Verfestigung" des *Spirogyra*-Protoplasten in den Zentrifugenversuchen von SAKAMURA und LOO (1925). Relativ einheitlich sind die wenigen Daten über die Plasmolysezeit, für die mehrfach angegeben wird, daß sie bei alkalischer und neutraler Reaktion meßbar lang sei gegenüber sofortiger Abrundung des plasmolysierten Protoplasten im sauren Bereich (PRÁT 1926, DERRY 1929); die umgekehrten Verhältnisse fand TAKAMINE (1940) bei Plasmolyse mit KCl anstatt mit Zucker. Ihre Feststellung, daß die Plasmolysezeit im stärker saurem Gebiet wieder zunehme, werten PFEIFFER (1932a) und STRUGGER (1934) im folgenden Sinne aus: Im schwachsauren Gebiet zeigt das Plasmolysezeitminimum ein Viscositätsminimum an, wie es im isoelektrischen Punkt eines Eiweißsols auftritt („elektroviscoser Effekt"); umgekehrt ließe sich aus der Plasmolysezeit auf den isoelektrischen Punkt schließen, wofür nach PFEIFFER (1932b) auch sprechen würde, daß bei *Nitella* nach Vorbehandlung mit Pufferlösungen höchste Geschwindigkeit der Plasmaströmung und geringster Widerstand des Zellinhaltes gegen Verlagerung beim Zentrifugieren im gleichen p_H-Bereich (um 5) liegen wie die kürzeste Plasmolysezeit bei Epidermisschnitten von Blütenpflanzen. Für das Gedeihen der meisten Pflanzen liegt die optimale Reaktion des Substrats im schwachsauren Bereich; ob dabei eine Kausalverbindung zum isoelektrischen Punkt einer Plasmakomponente besteht, wäre noch zu beweisen. Gegen die Annahme eines bestimmten isoelektrischen Punktes eines Protoplasten gelten ähnliche Einwände wie gegen die eines bestimmten p_H-Wertes; schon SAKAMURA und LOO (1925), STRUGGER (1928) und PFEIFFER (1932b) kommen mit der Vorstellung *eines* isoelektrischen Punktes nicht aus, sondern deuten die erwähnten mehrgipfligen Kurven mit dem Vorhandensein mehrerer Kolloide von unterschiedlichen elektrischen Eigenschaften.

Die Folgerungen, die aus den Beobachtungen über die p_H-Abhängigkeit der Plasmolysezeit hinsichtlich der Mechanik des Streckungswachstums gezogen wurden (STRUGGER 1934), sind später aufgegeben worden (RUGE 1940); wegen neuerer Auffassungen über die Rolle der Wasserstoffionen bei der Zellstreckung vgl. SCHULTE (1938), POHL (1948), RIETSEMA (1949) u. a.

Vielfach ist der Antagonismus anderer Ionen gegenüber $H^{\cdot}$ festgestellt worden (BRENNER 1920; weitere Literatur bei MEVIUS 1927 und KACZMAREK 1929). Besonders wirksam ist $Ca^{\cdot\cdot}$ bei der Entgiftung schädlicher $H^{\cdot}$-Konzentrationen (auch ohne Veränderung des p_H-Wertes wie bei Aufhebung von Säureschäden durch $CaCO_3$); wie schon erwähnt wurde (s. S. 719), besteht auch ein Teil der $Ca^{\cdot\cdot}$-Wirkung auf das Wurzelwachstum in einer $H^{\cdot}$-Entgiftung (BURSTRÖM 1952). Andere Ionen können schon in sehr niedrigen Konzentrationen mit der $H^{\cdot}$- und OH'-Wirkung interferieren, so daß der Ionengehalt des Wassers einer Versuchslösung von Bedeutung sein kann.

Die toxischen Wirkungen, die schwermetallfreies destilliertes Wasser ausübt, beruhen mindestens in vielen Fällen nicht auf Verunreinigungen, wie schon daraus hervorgeht, daß wiederholte Destillation die Giftigkeit steigert (MEVIUS 1928, EHRING 1934, KERSTING 1939, LIBBERT 1953; in diesen Arbeiten auch weitere Literatur zu dieser Frage), obwohl mangelnde Reinheit des Wassers der Grund für manche Widersprüche in den Befunden sein dürfte (FITTING 1928). Als wirksamer Bestandteil kommen im reinen Wasser vor allem die $H^{\cdot}$-Ionen in Betracht (SCARTH 1924c, MEVIUS 1928), deren Giftwirkung infolge Erhöhung ihrer Konzentration durch ausgeschiedene Atmungskohlensäure noch gesteigert werden kann (SAKAMURA 1922, BODE 1926). Nach SCARTH (1924c) kann die Giftwirkung auf *Spirogyra* außer durch Neutralisieren des Wassers auch durch eine große Zahl anderer Ionen kompensiert werden; im einzelnen abweichende Ergebnisse hatten bei anderen Objekten MEVIUS (1928) und EHRING (1934). FOX (1933) vermutet eine Schädigung durch CO_2, nicht durch $H^{\cdot}$, da er eine solche mit HCl und Essigsäure bei p_H 4,0 bei Nitella nicht beobachtet, wohl aber mit CO_2-haltigem Wasser von p_H 4,5. Unabhängig von solchen Giftwirkungen ist im Auge zu behalten, daß reines Wasser als „unbalanciertes" Milieu mindestens auf die äußeren Plasmagrenzschichten wirken und von dort aus — z. B. durch Störung der normalen Permeabilitätsverhältnisse — auch das Zellinnere beeinflussen kann.

Zusammengefaßt ergibt sich eine große Variabilität der Lebenserscheinungen bei verschiedenen p_H-Werten, aber auch mehr noch als bei den Salzionen das

Problem, die Beobachtungen als Wirkungen der Ionen des Wassers zu erkennen und deren Weg und Angriffspunkt in der Zelle festzulegen. Ihre Eigenart besteht darin, daß sie ständig im Stoffwechsel gebildet und verbraucht werden, der umgekehrt selbst von ihnen abhängig ist und zwar in einer für die einzelnen Fermente spezifischen Weise. Daraus ergibt sich die Vorstellung eines Wechselspieles zwischen Stoffwechselleistungen, innerem und äußerem Ionenmilieu mit der Möglichkeit zu differenzierten Steuerungen der Lebensvorgänge.

IV. Ionen im Stoffwechsel.

A. Quellungsbeeinflussung als Grundlage von Stoffwechselregulationen.

In den vergangenen Jahrzehnten ist die Kolloidaktivität von Ionen auch zur Erklärung der Erscheinungen herangezogen worden, die bei Mangel oder Überschuß der für die Ernährung der Pflanzen notwendigen Elemente auftreten. Besonders den Stoffen, deren Beteiligung am Aufbau organischer Verbindungen in der Zelle nicht nachzuweisen war, sollte eine solche „überwiegende Ionenwirkung“ zukommen, die vor allem in einer Regulierung der Hydratation im Sinne eines für den Ablauf der physiologischen Vorgänge optimalen Quellungszustandes gesehen wurde, wobei eine Veränderung des Ionengleichgewichtes auch eine Lenkung der Lebensvorgänge zuließe. Es wurde vor allem das Antagonistenpaar $K^{\cdot}$ und $Ca^{\cdot\cdot}$ berücksichtigt, daneben auch $Mg^{\cdot\cdot}$ und (vor allem bei Tieren) $Na^{\cdot}$. Die Vorstellung wurde durch die Erfolge der Kolloidtheorie nahegelegt, und es besteht kein Grund, eine solche Rolle neben eventuell vorhandenen spezifischen Wirkungen gerade für die in größerer Menge in der Zelle vorhandenen Ionen zu bezweifeln.

Schmalfuss (1936) hat die zell- und ernährungsphysiologische Literatur zur Frage der Ionenwirkung monographisch zusammengefaßt und in diesem Sinne ausgedeutet. „Grundlage für die Lebens- und Stoffwechselvorgänge der Pflanze ist ihr Wasserhaushalt und damit im Zusammenhang ihr Quellungszustand“ (Schmalfuss 1936, S. 43). Besonders die vielfach festgestellte Erhöhung des Wassergehaltes von Blättern bei guter K-Versorgung bzw. der höhere Gehalt an Trockensubstanz bei K-Mangel (relativer Ca-Überschuß) wird oft zur Bestätigung einer Quellungsförderung durch $K^{\cdot}$ und einer Quellungsherabsetzung durch $Ca^{\cdot\cdot}$ herangezogen. Abgesehen von der Problematik, die ein Schluß vom Wassergehalt eines Pflanzengewebes auf den Quellungszustand des Protoplasmas mit sich bringt, gibt es von dieser Regel manche Ausnahmen: Für Gerste geben Gregory und Richards (1929) und Richards (1932) höhere Succulenz bei K-Mangel an; nach Nightingale, Schermerhorn und Robbins (1930) ist bei K-Mangel an Tomaten der Wassergehalt zunächst geringer, in älteren Entwicklungsstadien aber höher als bei den Kontrollpflanzen. Gregory (1937) klärt die Widersprüche auf Grund ausgedehnter Versuche seines Schülers Shih: Für das Verhältnis von Wasser und Trockensubstanz ist das Verhältnis der Nährstoffe zueinander von Bedeutung; je nach der Gabe von Ca, Na und P kann der Wassergehalt bei K-Mangel gesteigert oder herabgesetzt sein (Richards und Shih 1940). Nach Böning und Böning-Seubert (1940) ergibt die getrennte Bestimmung des Wassergehaltes im Preßsaft und -rückstand Verschiebungen im Verhältnis der beiden Wasserfraktionen in Abhängigkeit von allen Ernährungsfaktoren (auch von P und besonders von N); die Autoren betonen selbst die Argumente (abgetötetes Gewebe, hoher Anteil der Wandsubstanz), die dem Schluß vom Wassergehalt des Preßrückstandes auf die Quellfähigkeit „der

Kolloide“ entgegenstehen. Nach ARNOLD (1955) bestehen Beziehungen zwischen dem Wassergehalt und dem Verhältnis von adsorbierten und freien Ionen. Entsprechend uneinheitlich sind auch die Angaben über die K- und Ca-Wirkung auf die Wasseraufnahme und -abgabe höherer Pflanzen (Zusammenfassung bei WÖSTMANN 1942, BAUMEISTER 1954). Es entsteht hier aus den unterschiedlichen Befunden unter verschiedenen Versuchsbedingungen nicht notwendig ein Widerspruch zu einer grundlegenden Ionenwirkung, wie sie oben skizziert wurde; vielmehr drückt sich darin das Prinzip der „Harmonie der Nährstoffe“ aus, indem deren Kombination über die Größe z. B. der Transpiration entscheidet (ALTEN, GOEZE und FISCHER 1937). Gerade am Beispiel der Transpiration braucht nicht weiter ausgeführt zu werden, wie sehr die abgegebene Wassermenge von einer großen Anzahl von Faktoren morphologischer und funktioneller Art abhängt; es ist charakteristisch, daß für einen der transpirationsbestimmenden Faktoren, die Stomataweite, nach DESAI (1937) nicht bestimmte Wirkungen einzelner Nährstoffe auftreten, sondern daß die Spalten der vollernährten Pflanzen am stärksten auf die Außenbedingungen reagieren und ihre Apertur ungefähr der durch den Ernährungszustand bedingten Größe der Pflanzen entspricht. Der komplexe Charakter der Stomatareaktion macht es zweifelhaft, daß aus ihr oder aus der Transpirationsgröße entscheidende Argumente für oder gegen eine bestimmte Art der Ionenwirkung abgeleitet werden können.

Entsprechend ist bei anderen Vorgängen wie Atmung, Photosynthese oder Aktivität bestimmter Fermente die Kausalkette, die von der (angenommenen) Quellungsbeeinflussung bis zu den beobachteten Mangel- und Überschußwirkungen führt, kaum zu erfassen. Auch SCHMALFUSS (1936, z. B. S. 52) hat diese Schwierigkeit hervorgehoben. Morphologische und biochemische Symptome, wie Dimensionsquotienten, Chlorose, Kohlenhydrat- und N-Fraktionen, sind als Teste für *Ionen*wirkungen nicht unmittelbar geeignet.

B. Beziehungen zwischen protoplasmatischen und Stoffwechselerscheinungen.

Die Entwicklung der protoplasmatischen Methoden veranlaßte Versuche, die unter Mangelbedingungen auftretenden Plasmaveränderungen unmittelbar zu beobachten. Der ersten Arbeit in dieser Richtung (KALCHHOFER 1936), nach der Konkavplasmolyse überwiegt, wenn Pflanzen K-frei, Konvexplasmolyse, wenn sie Ca-frei angezogen wurden, stehen die Angaben von SCHMIDT, DIWALD und STOCKER (1940) gegenüber, nach denen bei Getreidepflanzen Faktoren, die zu einer Erhöhung des K-Gehalts im Vergleich zum Ca-Gehalt im Gewebe führen (z. B. N-Mangel im Material dieser Autoren), die Viscosität steigern, während solche, die den Quotienten K:Ca erniedrigen (K-Mangel) auch die Viscosität herabsetzen. FISCHERs (1948, 1949) Untersuchungen gingen von der Frage aus, ob die im Laufe der Entwicklung eines Blattes auftretende Veränderung im K-Gehalt verglichen mit dem Ca-Gehalt das Altern der Blätter als Ionenwirkung charakterisiere; der Vergleich des Plasmolyseverlaufs in verschiedenen Blattentwicklungsstadien bis zum Vergilben und die Bestimmung von K und Ca im gleichen Material ergibt folgendes Bild: Wo aus der Plasmolyseform zu erschließende Viscositätsunterschiede bemerkbar werden, haben die jüngeren Blätter bei höherem Quotienten K:Ca das viscösere Cytoplasma. Die Literatur über die Viscositätsveränderungen mit dem Alter ist nicht ohne Widersprüche (vgl. FISCHER 1950 und STÅLFELT 1954, nach dessen Zentrifugenversuchen unter anderem die Wasserbilanz von Bedeutung ist). FISCHERs (1948, 1949) Ergebnisse stimmen hinsichtlich der gleichsinnigen Abnahme des Quotienten K:Ca und der Viscosität mit denen von SCHMIDT, DIWALD und STOCKER (1940) grundsätzlich überein. Der Schluß, daß K˙ auf das Protoplasma hydratationsfördernd und viscositätserhöhend, Ca˙˙ umgekehrt wirkt, entspricht den Vorstellungen, die außer SCHMIDT, DIWALD und STOCKER (1940) auch KESSLER und RUHLAND (1938) über Zusammenhänge zwischen Plasmastruktur und Resistenz gegen Dürre bzw. Kälte entwickeln. Hierzu scheint außer den schon genannten Befunden von KALCHHOFER (1936) im Widerspruch zu stehen, daß im Plasmolyse- und Injektionsversuch im allgemeinen gerade eine Viscositätserhöhung durch Ca˙˙ eine Viscositätsherabsetzung durch K˙ angezeigt wurde, so auch nach PIRSON und SEIDEL (1950) an der *Lemna*-Wurzel. Die Arbeit dieser Autoren zeigt eindeutig, daß die auftretenden Effekte am gleichen Objekt beim Entzug eines Ions im Mangelversuch die entgegengesetzten sein können wie bei Zufuhr des gleichen Ions von außen, besonders bei Ionenüberschuß; sie erlaubt ferner, an einem einheitlichen Versuchsmaterial protoplasmatische und Stoffwechselgrößen zu vergleichen und ihre Veränderungen als Folge von K- und Ca-Mangel zeitlich und zum Teil kausal zu trennen. Dabei tragen die Beeinflussungen der Gradienten der Einzelgrößen von der Spitze zur Basis der Wurzeln zur Analyse der Erscheinungen bei.

Tabelle 4. *Vergleich der Mangeleffekte auf die Wurzelzellen von Lemna minor.* (Nach PIRSON und SEIDEL 1950.)

	K-Mangel	Ca-Mangel	Alternde Zellen
Plasmolysezeit in Glucose .	verkürzt	verkürzt	verkürzt
Deplasmolysezeit in Harnstoff	verkürzt	verkürzt	verkürzt
Osmotischer Wert	erniedrigt	erhöht	erhöht
Respiratorischer O_2-Verbrauch.	erhöht	erhöht	erniedrigt

Die Ergebnisse, die hier in ihrer grundsätzlichen Bedeutung etwas eingehender besprochen werden sollen, lassen sich vereinfacht folgendermaßen darstellen (Tabelle 4).

Für die Auswertung ist die zeitliche Folge der Einzelwirkungen wichtig: Zuerst tritt von der Dauerzone zur Wurzelspitze fortschreitend die Veränderung des osmotischen Wertes auf, die für K und Ca antagonistisch ist; die im Gegensatz zu ihr bei verschiedenem Ionenmilieu gleichsinnigen Veränderungen von Plasmolysezeit und Harnstoffpermeabilität machen sich erst später bemerkbar, wenn das Wachstum der Mangelwurzeln gegenüber den vollernährten gehemmt ist. Der in der Wurzel von der Spitze zur Basis normalerweise vorhandene Gradient von Viscosität und Harnstoffpermeabilität, der in der Tabelle (ohne Berücksichtigung von Einzelheiten) durch die Spalte „alternde Zellen" gekennzeichnet ist, wird unter K- wie unter Ca-Mangel flacher, das Maximum der Viscosität und das Minimum der Permeabilität wird zur Wurzelspitze hin verschoben. Die Beeinflussung dieser beiden Größen ist somit keine spezifische Wirkung des K- oder Ca-Mangels, sondern drückt eine unspezifische Wachstumshemmung aus, die eine vorzeitige Veränderung der Zellen im gleichen Sinne bewirkt, wie sie beim normalen Altern der Zellen erst später eintritt.

Ein vorzeitiges Nachlassen der Funktionstätigkeit gehört zu den (im einzelnen mannigfaltigen) Symptomen des Nährstoffmangels bei Blättern; auch bei der *Lemna*-Wurzel beschleunigen andere Wachstumsbedingungen das „Altern". Wenn KALCHHOFER (1936) die Plasmolysezeit bei K-Mangel gegenüber Ca-Mangel (nicht gegenüber vollständiger Nährlösung) als erhöht angibt, so läßt sich dieser scheinbare Widerspruch zu den Ergebnissen von PIRSON und SEIDEL (1950) auflösen: Bei der *Lemna*-Wurzel tritt die Wachstumshemmung bei Ca-Mangel schon vor der K-Hemmung auf (Abb. 8); die kürzere Plasmolysezeit in C gegenüber B entspricht den Ergebnissen KALCHHOFERS (1936), bedeutet aber keine spezifische Ca-Wirkung, sondern nur eine weiter fortgeschrittene plasmatische Veränderung, wie sie bei K- und Ca-Mangel in gleicher Weise, nur zeitlich verschoben, als „Altern“ auftritt.

Die spezifische Herabsetzung der Zellsaftkonzentration bei K-Mangel deckt sich mit entsprechenden Angaben von SCHMIDT, DIWALD und STOCKER (1940). Sind schon diese Tatsachen überraschend, so gibt der respiratorische O_2-Verbrauch grundsätzlich wichtige Hinweise für die kausale Verknüpfung der Mangelerscheinungen: Bei den älteren Teilen normaler Wurzeln ist die Atmung ebenso wie die Plasmolysezeit herabgesetzt, dagegen läuft bei K- wie bei Ca-Mangel der Plasmolysezeitherabsetzung ein erhöhter O_2-Verbrauch parallel; die Atmung zeigt also nicht wie Viscosität und Permeabilität ein Altern der Mangelzellen an, sondern scheinbar das Gegenteil. Eine Unabhängigkeit der Atmung — mindestens unter den geprüften Mangelbedingungen — von den gemessenen protoplasmatischen Größen geht auch daraus hervor, daß bei K-Mangel die Atmungssteigerung *vor* der Plasmolysezeitverkürzung (Wachstumshemmung) auftritt.

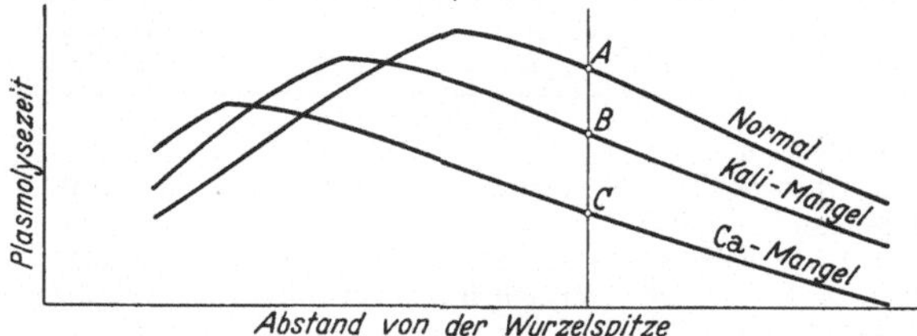

Abb. 8. Plasmolysezeit-Gradient bei normal ernährten und Mangelwurzeln von *Lemna minor* (schematisch). (Nach PIRSON und SEIDEL 1950.)

Bei Mangel von N und P, die im herkömmlichen Sinne als Plasmabausteine im Gegensatz zu den in erster Linie kolloidaktiven Ionen betrachtet werden, treten vergleichbare Effekte an der *Lemna*-Wurzel auf (PIRSON und GÖLLNER 1953). Dabei reagiert ebenfalls die Atmungsgröße empfindlicher als die Plasmolyseform, und zwar im Einsetzen der Mangelwirkung ebenso wie in deren Reversibilität bei Zugabe von N bzw. P zu den Mangelwurzeln. Die bei Mangel gehemmte Atmung wird innerhalb weniger Stunden weitgehend normalisiert, ohne daß die mangelbedingten Plasmolysezeitveränderungen, die außer der wachstumsbedingten bei N und P auch eine mangelspezifische Komponente zu enthalten scheinen, ebenfalls aufgehoben würden. Entsprechendes gilt für die Reversibilität bei K-Mangel. Der Ausgleich der Mangeleffekte im Stoffwechsel kann demnach ebensowenig wie diese selbst in einer Hydratationsbeeinflussung des Gesamtcytoplasmas bestehen, jedenfalls nicht in einer solchen, wie sie in anderen Fällen (entwicklungsbedingt oder durch exogene Faktoren wie Veränderung des Ionenmilieus) an der Plasmolyseform abzulesen sein mag. Falls im Mangelversuch bei der *Lemna*-Wurzel ein Kausalzusammenhang zu diesen Größen überhaupt besteht, so können die mit der Plasmolysezeit und der Deplasmolyse im Harnstoff gemessenen Veränderungen nicht die Ursachen der Atmungsbeeinflussung ausdrücken, sondern es kann höchstens der veränderte Stoffwechsel zu einer mit den protoplasmatischen Methoden erfaßbaren Strukturverschiebung führen. Auch CHÈVREMONT-COMHAIRE (1953) denkt bei der auf S. 726 geschilderten $Li^{\cdot}$-Wirkung an sekundäre Viscositätsänderungen. „Damit soll experimentell begründeten Angaben nicht widersprochen werden, wonach außerhalb des Bereiches physiologischer Ionenwirkung liegende Veränderungen im kolloidchemischen Allgemeinverhalten des Cytoplasmas vielfach mit dem Abweichen der Stoffwechselleistung von der Norm zusammengehen. Auch in solchen Fällen erübrigt sich allerdings die kritische Prüfung der Frage nicht, ob unmittelbare Kausalbeziehungen oder parallele Symptome vorliegen“ (PIRSON und GÖLLNER 1953).

C. Mittelbare und unmittelbare Ionenwirkungen im Stoffwechsel.

Weitere Ergebnisse über die Wirkung von Mineralstoffen auf den Stoffwechsel sind durch PIRSON und seine Schüler an Grünalgen erarbeitet worden (PIRSON 1937, 1939, PIRSON und WILHELMI 1950, PIRSON, TICHY und WILHELMI 1952, NEEB 1953, PIRSON und BERGMANN 1955, BERGMANN 1955, zusammenfassende Diskussion im Hinblick auf allgemeine Fragen des Mineralstoffwechsels grüner Pflanzen bei PIRSON 1955). Diese Arbeiten sollen hier unter Hinweis auf andere Bände dieses Handbuches, in denen die Rolle der Mineralstoffe im Stoffwechsel behandelt wird, nur kurz erwähnt werden. Die Versuchstechnik, bei der Stoffwechselmessungen an Mangelkulturen unter definierten Bedingungen vor und nach der Zugabe des in der Nährlösung fehlenden Stoffes vorgenommen werden, ergibt z. B. für K-Mangel eine kurz befristete Reversibilität der Mangelwirkung (Erholung der herabgesetzten Photosynthese, Senkung der gesteigerten Atmung) nach K-Zusatz. Da der Photosyntheseanstieg ohne Chlorophyllzunahme in den chlorotischen Zellen erfolgt, liegt eine direkte Beeinflussung der photosynthetischen Strukturen vor. Die Analyse von Photosynthese, Dunkelatmung, oxydativer Glucoseassimilation und Nitratreduktion liefert eine Fülle von Effekten, die je nach dem fehlenden Element und dem Grad des Mangelzustandes variieren; sie zeigen zum Teil spezifische Wirkungen auf einzelne Vorgänge, zum Teil Stoffwechselbeeinflussungen auf allgemeinerer struktureller Grundlage an. Im ersten Fall wird vor allem an eine spezifische Beeinflussung von Enzymen zu denken sein, im zweiten an eine Wirkung auf andere Reaktionsorte im Plasma, die eine differenzierte Steuerung zuläßt; eine mehr einheitliche Veränderung der „Plasmakolloide" ist bei der großen Anzahl offenbar selbständiger Effekte hier ebensowenig anzunehmen wie bei der *Lemna*-Wurzel. Wie am Beispiel des photosynthetischen Apparats dargelegt wird (PIRSON, TICHY und WILHELMI 1952), ist neben einer *unmittelbaren* Ionenwirkung, die sich als kurzfristige Veränderung der Leistung äußert, eine *mittelbare* beim Aufbau (Eiweißsynthese, Chlorophyllbildung usw.) möglich, die bei Mangel wie bei der Erholung langsamer in Erscheinung tritt; schließlich kann die durch Ionenmangel herabgesetzte Photosynthese *sekundär* durch Materialmangel den Chloroplastenaufbau und damit die Photosynthese beeinträchtigen. Während die Algenversuche die Erfassung unmittelbarer Ioneneffekte gestatten und die der mittelbaren erleichtern, wird man bei höheren Pflanzen vielfach höchstens die sekundären Wirkungen feststellen können, wenn man nicht einfache Systeme wie die *Lemna*-Wurzel unter entsprechenden Versuchsbedingungen verwendet.

Einen Angriffspunkt bei Ionenwirkungen auf Stoffwechselerscheinungen wird man in der Beeinflussung von Enzymen zu suchen haben. Für eine große Anzahl von Ionen ist die Beteiligung oder Notwendigkeit bei Enzymreaktionen nachgewiesen worden und unsere Kenntnisse hierüber wachsen ständig. Vor allen Dingen gilt dies für Schwermetalle, aber auch für die Hauptnährstoffe ($K^{\cdot}$, $Ca^{\cdot\cdot}$ und besonders $Mg^{\cdot\cdot}$); wegen der Einzelheiten muß auf die Lehr- und Handbücher der Enzymologie und auf einige neuere Zusammenfassungen (FOSTER 1949, MCELROY und NASON 1954, SCHARRER 1955, PIRSON 1955) verwiesen werden. Grundsätzlich ist eine doppelte Wirkung der Ionen anzunehmen: Über die Hydratation des Enzymeiweißes oder als Bestandteil aktiver Gruppen als Kofermente und Aktivatoren, wobei sich vor allem strukturell ähnliche Ionen gegenseitig verdrängen können. Unabhängig davon, wie im Einzelfall beobachtete Erscheinungen zu erklären sein werden, ist das Bild einer Beeinflussung spezieller Reaktionszentren durch Ionen und andere Einwirkungen physiologisch befriedigender als die Annahme einer allgemeinen Strukturveränderung des

Plasmas. Damit führt die allgemeine enzymologische Betrachtung zu den gleichen Schlußfolgerungen wie die kritische Auswertung von Ionenreihen und der Vergleich von Stoffwechsel und Protoplasmatik bei Mangelzuständen.

Nach MAC LEOD und SNELL (1950) lassen sich viele Fälle von Ionenantagonismus darauf zurückführen, daß ein für die Aktivität eines Enzyms wichtiges Ion (wie $K^{\cdot}$) durch ein anderes (z. B. $Na^{\cdot}$) ersetzt wird; die Autoren betonen dabei besonders die Bildung von enzymatisch unwirksamen Metallproteinen durch reversible oder irreversible Verdrängung des enzymaktivierenden Ions. Sie diskutieren am Beispiel des Wachstums von Milchsäurebakterien die Hemmwirkung anderer Ionen gegenüber den unentbehrlichen $K^{\cdot}$ und $Mn^{\cdot\cdot}$, die Ersetzbarkeit von $K^{\cdot}$ durch $Rb^{\cdot}$ und die gegenseitige Vertretbarkeit von Ionen als Antagonisten. Die Grenze zwischen Nährstoffwirkung und Kolloidwirkung wird undeutlich, wenn das gleiche Ion spezifisch als Enzymaktivator oder -bestandteil und allgemein kolloidaktiv wirken kann und wenn die Tätigkeit des gleichen Enzyms spezifisch (etwa über besondere metallische Gruppen) oder unspezifisch über seinen Hydratations- und Ladungszustand beeinflußt wird. Im einzelnen hierüber detaillierte Aussagen zu machen, ist noch nicht möglich. Die Frage mündet in die allgemeinere nach den Beziehungen von Struktur und physiologischer Funktion ein, die heute mehr und mehr als Wechselwirkungen beider Größen erkannt werden. Auch das Wesen der Ionenwirkungen wächst damit über eine kolloidale Strukturbeeinflussung als Voraussetzung von Stoffwechselregulationen hinaus, wie sie allgemein der Ausgangspunkt der bisherigen Arbeiten gewesen ist, deren Ergebnisse den einen Aspekt des Problems heute recht gut charakterisieren und auf den anderen hingeführt haben.

Literatur.

ALBACH, W.: Zellphysiologische Untersuchungen über vitale Protoplasmafärbung. Protoplasma **5**, 412—443 (1929). — ALTEN, F., G. GOEZE u. H. FISCHER: Kohlensäureassimilation und Stickstoffhaushalt bei gestaffelter Kaligabe. Bodenkde u. Pflanzenernähr. **5**, 259—289 (1937). — ARNAUDOW, N., u. K. J. POPOW: Über das Verhalten von Tabakkeimlingen zu Salzen. Jb. Univ. Sofia, Physiko-math. Fak. **3**, 1—50 (1936). Ref. Bot. Zbl. **30**, 222 (1938). — ARNOLD, A.: Die Bedeutung der Chlorionen für die Pflanze, insbesondere deren physiologische Wirksamkeit. Botanische Studien, H. 2. Jena: Gustav Fischer 1955.

BANK, O.: Abhängigkeit der Kernstruktur von der Ionenkonzentration. Protoplasma **32**, 20—30 (1939). — BARTH, L. G.: The effects of acids and alcalies on the viscosity of protoplasm. Protoplasma **7**, 505—533 (1929). — BAUMEISTER, W.: Mineralstoffe und Pflanzenwachstum, 2. Aufl. Stuttgart: Gustav Fischer 1954. — BENECKE, W.: Mechanismus und Biologie des Zerfalls der Conjugatenfäden in einzelne Zellen. Jb. wiss. Bot. **32**, 453—476 (1898). — Über die Giftwirkung verschiedener Salze auf *Spirogyra* und ihre Entgiftung durch Calciumsalze. Ber. dtsch. bot. Ges. **25**, 322—337 (1907). — BERGMANN, L.: Stoffwechsel und Mineralsalzernährung einzelliger Grünalgen. II. Vergleichende Untersuchungen über den Einfluß mineralischer Faktoren bei heterotropher und mixotropher Ernährung. Flora (Jena) **142**, 493—539 (1955). — BOAS, F.: Untersuchungen über die Mitwirkung der Lipoide beim Stoffaustausch der pflanzlichen Zelle. Biochem. Z. **117**, 166—214 (1921). — Das phyletische Anionenphänomen. Ein Beitrag zur Hylergographie. Jena: Gustav Fischer 1927. — Die Pflanze als kolloides System. In Naturwissenschaft und Landwirtschaft, H. 14. Freising u. München: F. P. Datterer & Cie. 1928. — Über Ionenwirkungen, besonders über das Anionenphänomen. Planta (Berl.) **22**, 445—461 (1934). — Dynamische Botanik, 2. Aufl. München: J. F. Lehmann 1942. — BODE, H. R.: Untersuchungen über die Abhängigkeit der Atmungsgröße von der $H^{\cdot}$-Ionenkonzentration bei einigen *Spirogyra*-Arten. Jb. wiss. Bot. **65**, 352—387 (1926). — BÖNING, K., u. E. BÖNING-SEUBERT: Der Einfluß der Mineralsalzernährung der Pflanze auf den Gehalt an Saft, Kolloiden und Wasser sowie auf die Wasserverteilung im Blattgewebe. Bodenkde u. Pflanzenernähr. **16**, 260—326 (1940). — BOGEN, H. J.: Untersuchungen über den Quellungseffekt permeierender Anelektrolyte. I. Ionenwirkung auf die Permeabilität von *Rhoeo discolor*. Z. Bot. **36**, 65—106 (1940). — Untersuchungen über den Quellungseffekt permeierender Anelektrolyte. II. Ionenwirkung auf die Permeabilität von *Gentiana cruciata*. Planta (Berl.) **32**, 150—175 (1941). — Untersuchungen über den Hitzetod und die Hitzeresistenz pflanzlicher Protoplaste. Planta (Berl.) **36**, 298—340 (1948a). — Vergleichende Untersuchungen über Ionenreihen an pflanzlichen Protoplasten. Biol. Zbl. **67**, 490—503 (1948b). — Über Kappenplasmolyse und Vacuolenkontraktion. I. Mitt. Die Wirkung vom LiCl und Neutralrot und ihre Abhängigkeit von der Konzentration und dem osmotischen Wert in der Außenlösung. Planta (Berl.) **39**, 1—35 (1951). — Booij, H. L.: The protoplasmic membrane regarded as a complex system. Rec. Trav. bot. néerl. **37**, 1—77 (1940). — Borowikow, G. A.: Über die Ursachen des Wachstums der Pflanzen. I. Mitt. Biochem. Z. **48**, 230—246 (1913). — Über die Einwirkung von salzartigen Stoffen auf die Wachstumsgeschwindigkeit der Pflanzen. Mém. Soc. Natural. Nouv. Russie Odessa **41**, 1 (1916). Zit. nach Kaho 1926a. — Borriss, H.: Die Beeinflussung des Streckungswachstums durch Salze. I. Jb. wiss. Bot. **85**, 732—769 (1937). — Plasmolyseform und Streckungswachstum. Jb. wiss. Bot. **86**, 784—831 (1938). — Die Beeinflussung des pflanzlichen Wachstums durch die Neutralsalze der Alkali- und Erdalkalimetalle. Ernähr. d. Pflanze **35**, 289—297 (1939). — Brambring, F.: Untersuchungen über die Wirkung des Aluminiums auf Wasserpflanzen. Jb. wiss. Bot. **73**, 241—299 (1930). — Brauner, L.: Untersuchungen über die Elektrolyt-Permeabilität einer leblosen natürlichen Membran. Jb. wiss. Bot. **73**, 513—632 (1930). — Brenner, W.: Über die Wirkung von Neutralsalzen auf die Säureresistenz, Permeabilität und Lebensdauer der Protoplasten. Ber. dtsch. bot. Ges. **38**, 277—285 (1920). — Brink, R. A.: Preliminary report of the role of salts in pollen tube growth. Bot. Gaz. **78**, 361—377 (1924). — Brinley, F. J.: The effect of chemicals on viscosity of protoplasm of *Amoeba* as indicated by Brownian movement. Protoplasma **4**, 177—182 (1928). — Brooks, M. M.: Comparative studies on respiration. VIII. The respiration of *Bacillus subtilis* in relation to antagonism. J. Gen. Physiol. **2**, 5—15 (1919). — Comparative studies on respiration. X. Toxic and antagonistic effects of magnesium in relation to the respiration of *Bacillus subtilis*. J. Gen. Physiol. **2**, 331—336 (1920). — Comparative studies on respiration. XIV. Antagonistic effect of lanthanum as related to respiration. J. Gen. Physiol. **3**, 337—342 (1921). — Bungenberg de Jong, H. G., and J. P. Hennemann: Preliminary experiments on the influence of neutral salts on the germination of *Lathyrus*-pollen. Rec. Trav. bot. néerl. **31**, 743—750 (1934). — Burström, H.: Studies on growth and metabolism of roots. VIII. Calcium as a growth factor. Physiol. Plantarum (Copenh.) **5**, 391—402 (1952). — Studies on growth and metabolism of roots. X. Investigations of the calcium effects. Physiol. Plantarum (Copenh.) **7**, 332—342 (1954).

Callan, H. G.: A general account of experimental work on amphibian oocyte nuclei. Symposia Soc. Exper. Biol. **6**, 243—255 (1952). — Chambers, R.: Microdissection and injection studies on the antagonistic action of salts upon protoplasm. Amer. J. Physiol. **72**, 210 (1925). — Chambers, R., and M. Black: Electrolytes and nuclear structure of the cells of the onion bulb epidermis. Amer. J. Bot. **28**, 364—371 (1941). — Chambers, R., and R. B. Howland: Micrurgical studies in cell physiology. VII. The action of the chlorides of Na, K, Ca and Mg on vacuolated protoplasm. Protoplasma **11**, 1—18 (1930). — Chambers, R., and H. Pollack: Micurgical studies in cell physiology. IV. Colorimetric determination of the nuclear and cytoplasmic p_H in the starfish egg. J. Gen. Physiol. **10**, 739—755 (1927). — Chambers, R., and P. Reznikoff: Micrurgical studies in cell physiology. I. The action of the chlorides of Na, K, Ca and Mg on the protoplasm of *Amoeba proteus*. J. Gen. Physiol. **8**, 369—401 (1926). — Chèvremont, M., et H. Firket: Action du béryllium en culture de tissus. II. Étude histochimique. Mécanisme d'action du Be et rôle de la phosphatase alcaline nucléaire dans la mitose. Archives de Biol. **63**, 515—535 (1952). — Chèvremont-Comhaire, S.: Action du lithium sur la croissance et la mitose dans les cultures des fibroblastes et myoblastes. Archives de Biol. **64**, 295—310 (1952). — Cholodny, N.: Zur Frage der Beeinflussung des Protoplasmas durch mono- und bivalente Metallionen. Beih. bot. Zbl. **39** (I), 231—238 (1923). — Über Protoplasmaveränderungen bei Plasmolyse. Biochem. Z. **147**, 22—29 (1924). — Cholodny, N., u. E. Sankewitsch: Plasmolyseform und Ionenwirkung. Protoplasma **20**, 57—72 (1934). — Clark, A. J.: General Pharmacology. In Handbuch der experimentellen Pharmakologie, Erg.-Werk, Bd. 4. Berlin: Springer 1937. — Cormack, R. G. H.: The development of root hairs in Angiosperms. Bot. Review **15**, 583—612 (1949). — Costello, D. P.: The surface precipitation reaction in marine eggs. Protoplasma **17**, 239—257 (1933).

Desai, M. C.: Effect of certain nutrient deficiencies on stomatal behavior. Plant Physiol. **12**, 253—283 (1937). — Derry, B. H.: Plasmolyseform- und Plasmolysezeitstudien. Protoplasma **8**, 1—49 (1929). — Dietzel, W.: Untersuchungen an aus *Chara*-Zellen austretendem Plasma unter besonderer Berücksichtigung vom Ionen-Milieu, insonderheit der Calcium-Ionen. Diss. Stuttgart 1953. (Unveröffentlicht.)

Ehring, L.: Untersuchungen über die Wirkung des destillierten Wassers und verschiedener Salze auf das Wachstum von Lebermoosen. Diss. Münster 1934. — Eichberger, R.: Über die Lebensdauer „isolierter" Tonoplasten. Protoplasma **20**, 606—632 (1934). —

Über die Wirkung von Kaliumoxalat auf den lebenden Protoplasten von *Allium cepa*. Planta (Berl.) **23**, 479—485 (1935). — EISELSBERG, C. v.: Ionenantagonismus und Giftwirkungen an *Spirogyra*. I. Über Ionenantagonismus ein- und zweiwertiger Kationen und deren Giftwirkung auf Spirogyrazellen. Biol. generalis (Wien) **13**, 529—560 (1937). — Ionenantagonismus und Giftwirkungen an *Spirogyra*. II. Über die Wirkung von Kaliumarsenit, Akonitin und Wasserstoffperoxyd. Biol. generalis (Wien) **14**, 21—46 (1938). — EISENMENGER, W. S.: Toxicity, additive effects, and antagonism of salt solutions as indicated by growth of wheat roots. Bull. Torrey Bot. Club **55**, 261—304 (1928). — Toxicity of aluminium on seedlings and action of certain ions in the elimination of the toxic effects. Plant Physiol. **10**, 1—25 (1936). — ENDRES, G.: Zur Kenntnis der stickstoffassimilierenden Bakterien. I. Liebigs Ann. **512**, 54—80 (1934). — EPHRUSSI, B., et L. RAPKINE: Action des differents sels sur le *Spirostomum*. Protoplasma **5**, 35—40 (1928).

FISCHER, H.: Plasmolyseform und Mineralsalzgehalt in alternden Blättern. I. Untersuchungen an *Helodea* und *Fontinalis*. Planta (Berl.) **35**, 513—527 (1948). — Plasmolyseform und Mineralsalzgehalt in alternden Blättern. II. Untersuchungen an Land- und Schwimmpflanzen. Planta (Berl.) **37**, 244—292 (1949). — Über protoplasmatische Veränderungen beim Altern von Pflanzenzellen. Protoplasma **39**, 661—676 (1950). — FITTING, H.: Untersuchungen über die Aufnahme von Salzen in die lebende Zelle. Jb. wiss. Bot. **56**, 1—64 (1915). — Untersuchungen über die Chemodinese bei *Vallisneria*. Jb. wiss. Bot. **67**, 427—596 (1928). — FLURI, M.: Der Einfluß von Aluminiumsalzen auf das Protoplasma. Flora (Jena) **99**, 81—126 (1909). — FOSTER, J. W.: Chemical activities of fungi. New York: Academic Press 1949. — FOX, D. L.: Carbon dioxide narcosis. J. Cellul. a. Comp. Physiol. **3**, 75—100 (1933). — FREUNDLICH, H.: Kapillarchemie, Bd. 1. Leipzig: Akademische Verlagsgesellschaft 1930. — FREY-WYSSLING, A.: Submicroscopic morphology of protoplasm, 2. Aufl. Amsterdam: Elsevier Publ. Comp. 1953. — FRY, H. J., and M. E. PARKS: Mitotic changes and viscosity changes in the eggs of *Arbacia*. *Cumingia*, and *Nereis*. Protoplasma **21**, 479—499 (1934).

GISTL, R.: Zur Kenntnis der Erdalgen. Arch. Mikrobiol. **3**, 634—649 (1932). — Erdalgen und Düngung. Erdalgen und Anionen. Arch. Mikrobiol. **4**, 348—378 (1933). — GREGORY, F. G.: Mineral nutrition of plants. Annual Rev. Biochem. **6**, 557—578 (1937). — GREGORY, F. G., and F. J. RICHARDS: Physiological studies in plant nutrition. I. The effect of manurial deficiency on the respiration and assimilation in barley. Ann. of Bot. **43**, 119—161 (1929). — GUSTAFSON, F. G.: Comparative studies on respiration. IX. The effects of antagonistic salts on the respiration of *Aspergillus niger*. J. Gen. Physiol. **2**, 17—24 (1919).

HAAN, I. DE: Ionenwirkung und Wasserpermeabilität. Ein Beitrag zur Koazervattheorie der Grenzschichten. Protoplasma **24**, 186—197 (1935). — HANSTEEN, B.: Über das Verhalten der Kulturpflanzen zu den Bodensalzen. I. u. II. Jb. wiss. Bot. **47**, 289—376 (1910). — HANSTEEN-CRANNER, B.: Beiträge zur Biochemie und Physiologie der Zellwand und der plasmatischen Grenzschichten. Ber. dtsch. bot. Ges. **37**, 380—391 (1919). — Zur Biochemie und Physiologie der Grenzschichten lebender Pflanzenzellen. Meld. Norges Landbrukshøiskole **1922**, 1—60. — HASSAN, M. N., and R. OVERSTREET: Elongation of seedlings as a biological test of alcali soils. I. Effects of ions on elongation. Soil Sci. **73**, 315—326 (1952). — HEILBRUNN, L. V.: The colloid chemistry of protoplasm. Protoplasma-Monogr. **1** (1928). — HEILBRUNN, L. V., and K. DAUGHERTY: The action of chlorides of sodium, potassium, calcium, and magnesium on the protoplasm of *Amoeba dubia*. Physiologic. Zool. **4**, 635—651 (1931). — The action of sodium, potassium, calcium, and magnesium on the plasmagel of *Amoeba proteus*. Physiologic. Zool. **5**, 254—274 (1932). — HEINZEL, E.: Experimentelle Untersuchungen zur Ausbildung von Faden- und Kettenformen bei *Bacterium prodigiosum*. Arch. Mikrobiol. **15**, 119—136 (1950). — HERBST, C.: Über das Auseinandergehen von Furchungs- und Gewebezellen in kalkfreiem Medium. Arch. Entw.mechan. **9**, 424—463 (1900). — HÖBER, R.: Physikalische Chemie der Zelle und der Gewebe, 6. Aufl. Leipzig: Wilhelm Engelmann 1926. — Physical chemistry of cells and tissues. Philadelphia u. Toronto: Blakiston Comp. 1945. Deutsche Übersetzung: Physikalische Chemie der Zellen und Gewebe. Bern: Stämpfli & Cie. 1947. — HÖBER, R., u. O. NAST: Beiträge zum arteigenen Verhalten der roten Blutkörperchen. I. Hämolyse bei gleichzeitiger Einwirkung von Neutralsalzen und anderen cytolysierenden Stoffen. Biochem. Z. **60**, 131—145 (1914). — HÖFLER, K.: Kappenplasmolyse und Salzpermeabilität. Z. wiss. Mikrosk. **51**, 70—87 (1934). — Kappenplasmolyse und Ionenantagonismus. Protoplasma **33**, 544—579 (1939). — Plasmolyse mit Natriumkarbonat. Protoplasma **40**, 426—460 (1951). — HOUSKA, H.: Beiträge zur Kenntnis der Kappenplasmolyse. Protoplasma **36**, 11—51 (1942). — HULTIN, T.: The effect of calcium on respiration and acid formation in homogenates of sea-urchin eggs. Ark. Kemi., Mineral. Geol., Ser. A **26**, Nr 27, 1—10 (1949). — On the oxygen uptake of *Paracentrotus lividus* egg homogenates after the addition of calcium. Exper. Cell Res. **1**, 159—168 (1950). — On the acid formation, breakdown of cytoplasmic inclusions, and increased viscosity in

Paracentrotus egg homogenates after the addition af calcium. Exper. Cell Res. **1**, 273—283 (1950).

ILJIN, W. S.: Plasmolyse und Deplasmolyse und ihre Beeinflussung durch Salze und durch die Wasserstoffionenkonzentration. Protoplasma **24**, 296—318 (1935a). — Das Absterben der Pflanzenzellen in reinen und balancierten Salzlösungen. Protoplasma **24**, 409—430 (1935b).

JOST, L.: Einige physikalische Eigenschaften des Protoplasmas von *Valonia* und *Chara*. Protoplasma **7**, 1—22 (1929). — JUNGERS, V.: Die Verlagerungsfähigkeit des Zellinhaltes der Zwiebelschuppen von *Allium cepa* durch Zentrifugieren. Protoplasma **21**, 351—361 (1934).

KACZMAREK, A.: Untersuchungen über Plasmolyse und Deplasmolyse in Abhängigkeit von der Wasserstoffionenkonzentration. Protoplasma **6**, 209—301 (1929). — KAHO, H.: Zur Kenntnis der Neutralsalzwirkungen auf das Pflanzenplasma. II. Mitt. Biochem. Z. **120**, 125—142 (1921a). — Ein Beitrag zur Permeabilität des Pflanzenplasmas für Neutralsalze. IV. Mitt. Biochem. Z. **123**, 284—303 (1921b). — Das Verhalten der Pflanzenzelle gegen Salze. Erg. Biol. **1**, 380—406 (1926a). — Ein Beitrag zur Theorie der antagonistischen Ionenwirkungen auf das Pflanzenplasma. VII. Biochem. Z. **167**, 25—37 (1926b). — Das Verhalten der Pflanzenzelle gegen Schwermetallsalze. Planta (Berl.) **18**, 664—682 (1933). — Ein Beitrag zur Kenntnis der Wasserpermeabilität des Protoplasmas. Cytologia (Tokyo), FUJII-Jub.-Bd. **1937**, 129—148. — KALCHHOFER, Z.: Protoplasmazustand nährsalzmangelkranker Pflanzen. Protoplasma **26**, 249—281 (1936). — KAMIYA, N.: Zytomorphologische Plasmolysestudien an *Allium*-Epidermen. Protoplasma **32**, 373—396 (1939). — KASSEL, R., and M. J. KOPAC: Experimental approaches to the evaluation of fractionation media. I. The action of ions on the protoplasm of *Amoeba proteus* and *Pelomyxa carolinensis*. J. of Exper. Zool. **124**, 279—301 (1953). — KATTERMANN, G.: Über Neutralsalzwirkungen bei Pilzen und Bakterien. Ein Beitrag zum phyletischen Anionenphänomen. Bot. Archiv **28**, 73—176 (1930). — KERR, T. A.: The injection of salts into the protoplasm and vacuoles of the root hairs of *Limnobium Spongia*. Protoplasma **18**, 420—440 (1933). — KERSTING, F.: Zur Frage der Kationenwirkung auf die lebende Pflanzenzelle. Jb. wiss. Bot. **87**, 706—749 (1939). — KESSLER, W., u. W. RUHLAND: Weitere Untersuchungen über die inneren Ursachen der Kälteresistenz. Planta (Berl.) **28**, 157—204 (1938). — KOBINGER, I.: Die Vernarbungsmembran an plasmolysierten Protoplasten. Phyton (Horn, N.-Oe.) **5**, 38—40 (1953). — KREHAN, M.: Beiträge zur Physiologie und Systematik der Essigbakterien. II. Teil. Der Einfluß von Salzen auf die Gärtätigkeit (Alkoholoxydation) und das Wachstum von *Bacterium acetigenoideum* in Nährlösungen ohne Zuckerzusatz. Arch. Mikrobiol. **3**, 277—321 (1932). — KRISZAT, G.: Die Wirkung von Adenosintriphosphat und Calcium auf Amöben (*Chaos chaos*). Ark. Zool. (Stockh.) **1**, 477—485 (1950).

LANZ, I.: Neue Beiträge zur Kenntnis der Systrophe des Protoplasmas. Protoplasma **36**, 380—409 (1942). — LEPESCHKIN, W. W.: Über den Zusammenhang zwischen mechanischen und chemischen Schädigungen des Protoplasmas und der Wirkung einiger Schutzstoffe. Protoplasma **2**, 239—269 (1927). — Kolloidchemie des Protoplasmas, 2. Aufl. Dresden u. Leipzig: Theodor Steinkopff 1938. — LIBBERT, E.: Die Wirkung der Alkali- und Erdalkaliionen auf das Wurzelwachstum unter besonderer Berücksichtigung des Ionenantagonismus und seiner Abhängigkeit von Milieufaktoren. Planta (Berl.) **41**, 396—435 (1953). — LILLIE, R. S.: The relation of ions to ciliary movement. Amer. J. Physiol. **10**, 419—443 (1904). — LINSBAUER, K.: Untersuchungen über Plasma und Plasmaströmung an *Chara*-Zellen. Protoplasma **18**, 554—593 (1933). — LOEB, J.: Über den Einfluß der Werthigkeit und möglicherweise der elektrischen Ladung von Ionen auf die antitoxische Wirkung. Arch. ges. Physiol. **88**, 68—78 (1902). — LOEVEN, W. A.: Seasonal variations in the water permeability of *Allium* epidermal cells. Proc., Kon. Ned. Akad. v. Wetensch. C **54**, 411—420 (1951). — LOREY, E.: Mikrochirurgische Untersuchungen über die Viskosität des Protoplasmas. Protoplasma **7**, 171—203 (1929). — LUNDEGÅRDH, H.: Über die Permeabilität der Wurzelspitzen von *Vicia Faba* unter verschiedenen äußeren Bedingungen. Kgl. Sv. Vetenskapsakad. Hdl. **47**, Nr 3, 1—254 (1911). — The growth of root hairs. Ark. Bot. (Stockh.) A **33**, Nr 5, 1—19 (1946). — LYON, C. J.: Comparative studies on respiration. XVIII. Respiration and antagonism in *Elodea*. Amer. J. Bot. **8**, 458—463 (1921).

MACLEOD, R. A., and E. E. SNELL: Ion antagonism in bacteria as related to antimetabolites. Ann. New York Acad. Sci. **52**, 1249—1259 (1950). — MARCOZZI, V.: Sulla tossicità del magnesio e sul suo antagonismo col calce nelle piante. Nuovo Giorn. bot. ital. **44**, 503—551 (1937). — MCELROY, W. D., and A. NASON: Mechanism of action of micronutrient elements in enzyme systems. Annual Rev. Plant Physiol. **5**, 1—30 (1954). — MEVIUS, W.: Kalzium-Ion und Wurzelwachstum. Jb. wiss. Bot. **66**, 183—253 (1927). — Weitere Beiträge zum Problem des Wurzelwachstums. Jb. wiss. Bot. **69**, 119—190 (1928). — MICHAELIS, L.: The effects of ions in colloidal systems. Baltimore 1925. — MILOVIDOV, P. F.: Physik und Chemie des Zellkerns. 1. u. 2. Teil. Protoplasma-Monogr. **20** (1949 u. 1954). — MISSBACH, G.: Versuche zur Prüfung der Plasmaviskosität. Protoplasma **3**, 327—344 (1928).

NEEB, O.: *Hydrodictyon* als Objekt einer vergleichenden Untersuchung physiologischer Größen. Flora (Jena) **139**, 39—95 (1952). — NETTER, H.: Die Beeinflussung der Alkalisalzaufnahme lebender Pflanzenzellen durch mehrwertige Kationen. Arch. ges. Physiol. **198**, 225—251 (1923). — NIGHTINGALE, G. T., L. G. SCHERMERHORN and W. R. ROBBINS: Some effects of potassium deficiency on the histological structure and nitrogenous and carbohydrate constituents of plants. New Jersey Agric. Exper. Sta. Bull. **499**, 1—36 (1930). — NORTHEN, H. T., and R. T. NORTHEN: Effects of cations and anions on protoplasmic elasticity. Plant Physiol. **14**, 539—547 (1939).

OSTERHOUT, W. J. V.: On the importance of physiologically balanced solutions for plants. I. Marine plants. Bot. Gaz. **42**, 127—134 (1906). — On the importance of physiologically balanced solutions for plants. II. Fresh water and terrestrial plants. Bot. Gaz. **44**, 259—272 (1907). — Die Schutzwirkung des Natriums für die Pflanze. Jb. wiss. Bot. **46**, 121—136 (1909). — Direct and indirect determinations of permeability. J. Gen. Physiol. **4**, 275—283 (1922). — OVERTON, E,: Beiträge zur allgemeinen Muskel- und Nervenphysiologie. III. Mitt. Studien über die Wirkung der Alkali- und Erdalkalisalze auf Skelettmuskeln und Nerven. Arch. ges. Physiol. **105**, 176—290 (1905).

PANTANELLI, E.: Zur Kenntnis der Turgorregulation bei Schimmelpilzen. Jb. wiss. Bot. **40**, 303—367 (1904). — PAULI, W., u. E. VALKÓ: Kolloidchemie der Eiweißkörper. Dresden u. Leipzig: Theodor Steinkopff 1933. — PFEIFFER, H.: Kleine Beiträge zur Bestimmung des IEP von Protoplasten. III. Die Zeit bis zur konvexen Plasmolyse und die Refraktion von planparallelen Zellen bei wechselnder C_H des Mediums. Protoplasma **14**, 83—90 (1932a). — Kleine Beiträge zur Bestimmung des IEP von Protoplasten. IV. Strömungsgeschwindigkeit und Zentrifugalverlagerung des rotierenden Plasmas in *Nitella*-Zellen nach Vorbehandlung mit Puffergemischen. Protoplasma **14**, 90—96 (1932b). — PIRSCHLE, K.: Zur physiologischen Wirkung homologer Ionenreihen. Jb. wiss. Bot. **72**, 335—368 (1930). — Untersuchungen über die physiologischen Wirkungen homologer Ionenreihen. II. Jb. wiss. Bot. **76**, 1—92 (1932). — Vergleichende Untersuchungen über die physiologische Wirkung der Elemente nach Wachstumsversuchen mit *Aspergillus niger*. Planta (Berl.) **23**, 177—224 (1935a). — Weitere Untersuchungen über die vergleichsweise Wirkung der Elemente nach Wachstumsversuchen mit *Aspergillus niger* (Stimulation und Toxizität). Planta (Berl.) **24**, 649—710 (1935b). — PIRSON, A.: Ernährungs- und stoffwechselphysiologische Untersuchungen an *Fontinalis* und *Chlorella*. Z. Bot. **31**, 193—267 (1937). — Über die Wirkung der Alkaliionen auf Wachstum und Stoffwechsel von *Chlorella*. Planta (Berl.) **29**, 231—261 (1939). — Functional aspects in mineral nutrition of green plants. Annual Rev. Plant Physiol. **6**, 71—114 (1955). — PIRSON, A., and L. BERGMANN: Manganese requirement and carbon source in *Chlorella*. Nature (Lond.) **176**, 209 (1955). — PIRSON, A., u. E. GÖLLNER: Zellphysiologische Untersuchungen an der *Lemna*-Wurzel bei verminderter Nitrat- und Phosphatversorgung. Z. Bot. **41**, 147—176 (1953). — PIRSON, A., u. F. SEIDEL: Zell- und stoffwechselphysiologische Untersuchungen an der Wurzel von *Lemna minor* L. unter besonderer Berücksichtigung von K- und Ca-Mangel. Planta (Berl.) **38**, 431—473 (1950). — PIRSON, A., C. TICHY u. G. WILHELMI: Stoffwechsel und Mineralsalzernährung einzelliger Grünalgen. I. Mitt. Vergleichende Untersuchungen an Mangelkulturen von *Ankistrodesmus*. Planta (Berl.) **40**, 199—253 (1952). – PIRSON, A., u. G. WILHELMI: Photosynthese-Gaswechsel und Mineralsalzernährung. Z. Naturforsch. **5b**, 211—218 (1950). — POHL, R.: Ein Beitrag zur Analyse des Streckungswachstums der Pflanzen. Planta (Berl.) **36**, 230—261 (1948). — PRÁT, S.: Plasmolyse und Permeabilität. Biochem. Z. **128**, 557—567 (1922). — Plasmolyse des Cyanophycées. Bull. internat. Acad. Tchèque des Sci., Cl. math. nat. et med. **13**, 96—97 (1923). — Wasserstoffionenkonzentration und Plasmolyse. Kolloid-Z. **40**, 248—251 (1926). — PRINGSHEIM, E. G.: Über Plasmolyse durch Schwermetallsalze. Beih. bot. Zbl. **41** (I), 1—14 (1925). — PRUD'HOMME VAN REINE, W. J. V.: Versuche über die Konsistenz des Protoplasmas. Rec. Trav. bot. néerl. **32**, 467—515 (1935).

REZNIKOFF, P.: Micrurgical studies in cell physiology. V. The antagonism of cations in their actions on the protoplasm of *Amoeba dubia*. J. Gen. Physiol. **11**, 221—237 (1928). — REZNIKOFF, P., and R. CHAMBERS: Micrurgical studies in cell physiology. III. The action of CO_2 and some salts of Na, Ca, and K on the protoplasm of *Amoeba dubia*. J. Gen. Physiol. **10**, 731—738 (1927). — RICHARDS, F. J.: Physiological studies in plant nutrition. III. Further studies of the effect of potash deficiency on the rate of respiration in leaves of barley. Ann. of Bot. **46**, 367—388 (1932). — RICHARDS, F. J., and SHIH SHENG-HAN: Physiological studies in plant nutrition. X. Water content of barley leaves as determined by the interaction of potassium with certain other nutrient elements. Part I, II. Ann. of Bot. N. S. **4**, 165—175, 403—425 (1940). — RIETSEMA, J.: On the influence of p_H on the growth of *Avena* coleoptile sections. Proc. Kon. Ned. Akad. v. Wetensch. C **52**, 1039—1050 (1949). — RUBINSTEIN, D. L.: Das Problem des physiologischen Ionenantagonismus. Protoplasma **4**, 259—314 (1928). — RUGE, U.: Kritische zell- und entwicklungsphysiologische Untersuchungen an den Blattzähnen von *Elodea densa*. Flora (Jena) **134**, 311—376 (1940). — RUHLAND, W., u. C. HOFFMANN: Die Permeabilität von *Beggiatoa mirabilis*. Ein Beitrag zur Ultrafiltertheorie des Plasmas. Planta (Berl.) **1**, 1—83 (1926). — RUNNSTRÖM, J.: The mechanism of fertilization in Metazoa. Adv. Enzymol. **9**, 241—327 (1949). — RUNNSTRÖM, J., and G. KRISZAT: On the effect of adenosine triphosphoric acid and of Ca on the cytoplasm of the egg of the sea urchin, *Psammechinus miliaris*. Exper. Cell Res. **1**, 284—303 (1950).

SAKAMURA, T.: Über die Selbstvergiftung der *Spirogyra* in destilliertem Wasser. Bot. Mag. (Tokyo) **36**, 133—153 (1922). — Beiträge zur Protoplasmaforschung an *Spirogyra*. J. Fac. Sci., Hokkaido Imp. Univ., Ser. V **2**, 287—316 (1933). — Zur Analyse der Salzwirkung auf die pflanzlichen Protoplasten. J. Fac. Sci., Hokkaido Imp. Univ., Ser. V **3**, 101—119 (1934). — SAKAMURA, T., u. H. KANAMORI: Über die Wirkung der Essigsäure, des Ammoniaks und ihrer Salze auf das Protoplasma der Wurzelhaare. J. Fac. Sci., Hokkaido Imp. Univ., Ser. V **4**, 65—98 (1935). — SAKAMURA, T., u. T. S. LOO: Über die Beeinflussung des Pflanzenplasmas durch die H-Ionen in verschiedenen Konzentrationen. Bot. Mag. (Tokyo) **39**, 61—76 (1925). — SCARTH, G. W.: Adhesion of protoplasm to cell wall and agents which cause it. Trans. Roy. Soc. Canada, Sect. V **17**, 137—143 (1923). — Colloidal changes associated with protoplasmic contraction. Quart. J. Exper. Physiol. **14**, 99—113 (1924a). — The action of cations on the contraction and viscosity of protoplasm in *Spirogyra*. Quart. J. Exper. Physiol. **14**, 115—122 (1924b). — The toxic action of distilled water and its antagonism by cations. Trans. Roy. Soc. Canada, Sect. V **18**, 97—104 (1924c). — SCHARRER, K.: Biochemie der Spurenelemente, 3. Aufl. Berlin u. Hamburg: P. Parey 1955. — SCHMALFUSS, K.: Das Kalium. Eine Studie zum Kationenproblem im Stoffwechsel und bei der Ernährung der Pflanze. Naturwissenschaft und Landwirtschaft, H. 19. Freising u. München: P. Datterer & Cie. 1936. — SCHMIDT, H., K. DIWALD u. O. STOCKER: Plasmatische Untersuchungen an dürreempfindlichen und dürreresistenten Sorten landwirtschaftlicher Kulturpflanzen. Planta (Berl.) **31**, 559—596 (1940). — SCHULTE, E.: Untersuchungen über das Auftreten von anomaler (negativer) Osmose im Pflanzenkörper. Protoplasma **29**, 60—98 (1938). — SEIFRIZ, W.: The effects of acids and alcalies on structural properties of protoplasm. Physics **6**, 159—161 (1935). — Protoplasm. New York u. London: McGraw Hill Book Comp. 1936. — Toxicity and chemical properties of ions. Science (Lancaster, Pa.) **110**, 193—196 (1949). — SOROKIN, H., and A. L. SOMMER: Changes in the cells and tissues of root tips induced by the absence of calcium. Amer. J. Bot. **16**, 23—28 (1929). — Effects of calcium deficiency upon the roots of *Pisum sativum*. Amer. J. Bot. **27**, 308—318 (1940). — SPEK, J.: Der Einfluß der Salze auf die Plasmakolloide von *Actinosphaerium Eichhorni*. Acta zool. (Stockh.) **2**, 153—200 (1921). — Studien an zerschnittenen Zellen. Protoplasma **4**, 321—357 (1928). — Das p_H in der lebenden Zelle. Erg. Enzymforsch. **6**, 1—22 (1937). — Das p_H in der lebenden Zelle. Kolloid-Z. **85**, 162—170 (1938). — SPEK, J., u. R. CHAMBERS: Das Problem der Reaktion des Protoplasmas. (Neue experimentelle Studien, ausgeführt an Amöben.) Protoplasma **20**, 376—406 (1934). — STÅLFELT, M. G.: The influence of light upon the viscosity of protoplasm. Ark. Bot (Stockh.) A **33**, Nr 4, 1—17 (1946). — Soil substances affecting the viscosity of protoplasm. Sv. bot. Tidskr. **42**, 17—33 (1948). — The relation between the fluidity of the protoplasm and the insertion and function of the leaves. Physiol. Plantarum (Copenh.) **7**, 354—374 (1954). — STAPP, C., u. H. ZYCHA: Morphologische Untersuchungen an *Bacillus mycoides*; ein Beitrag zur Frage des Pleomorphismus der Bakterien. Arch. Mikrobiol. **2**, 493—536 (1931). — STRUGGER, S.: Untersuchungen über den Einfluß der Wasserstoffionen auf das Protoplasma der Wurzelhaare von *Hordeum vulgare*. II. Sitzgsber. Akad. Wiss. Wien, Math.-naturwiss. Kl., Abt. I **137**, 143—169 (1928). — Untersuchungen an isolierten Kernen der Internodialzellen von *Chara fragilis*. Planta (Berl.) **8**, 717—741 (1929). — Beitrag zur Kolloidchemie des pflanzlichen Ruhekerns. Protoplasma **10**, 363—377 (1930). — Zur Analyse der Vitalfärbung pflanzlicher Zellen mit Erythrosin. Ber. dtsch. bot. Ges. **49**, 453—476 (1931). — Über Plasmolyse mit Kaliumrhodanid. Ber. dtsch. bot. Ges. **50**, 24—31 (1932). — Beiträge zur Physiologie des Wachstums. I. Jb. wiss. Bot. **79**, 406—471 (1934). — SZÜCS, J.: Über einige charakteristische Wirkungen des Aluminiumions auf das Protoplasma. Jb. wiss. Bot. **52**, 269—332 (1913).

TAKAMINE, N.: On the plasmolysis form in *Allium cepa* with special reference to the influence of potassium ion upon it. Cytologia (Tokyo) **10**, 302—323 (1940). — Studies on the plasmolysis form. Cytologia (Tokyo) **11**, 537—590 (1941). — TERRY, R. L.: The surface precipitation reaction in the ovarian frog egg. Protoplasma **40**, 206—221 (1950). — TEUNISSN, P. H.: Ionenreihen in Kolloidchemie und Biologie. Kolloid-Z. **85**, 158—161 (1938). — THIMANN, K. V., and C. L. SCHNEIDER: The role of salts, hydrogen-ion concentration and agar in the response of the *Avena* coleoptile to auxins. Amer. J. Bot. **25**, 270—280 (1938). — TIMMEL, H.: Zentrifugierungsversuche über die Wirkung chemischer Agentien, insbesondere des Kaliums auf die Viskosität des Protoplasmas. Protoplasma **3**, 197—212 (1928). — TRÉNEL, M., u. F. ALTEN: Die physiologische Bedeutung der mineralischen Bodenazidität. Worauf beruht die toxische Wirkung des Aluminiums? Angew. Chem. **47**, 813—820 (1934). —

TROENDLE, A.: Sur la perméabilité du protoplasme vivant. Arch. Sci. physiques et natur., IV. s. **45**, 38—54, 117—132 (1918). — Die Aufnahme von Salzen in die Pflanzenzelle. Denkschr. schweiz. naturforsch. Ges., Abh. 1, **58**, 1—59 (1922).

UMRATH, K.: Die Bildung von Plasmalemma (Plasmahaut) bei *Nitella mucronata*. Protoplasma **16**, 173—188 (1932).

VIRGIN, H. J.: The effect of light on the protoplasmic viscosity. Physiol. Plantarum (Copenh.) **4**, 255—357 (1951). — VOSS, E.: Die Wirkung überoptimaler Konzentrationen einiger Neutralsalze auf kokkoide Bakterien. Arch. Mikrobiol. **18**, 101—132 (1952).

WEBER, F.: Plasmolyseform und Protoplasmaviskosität. Österr. bot. Z. **73**, 261—266 (1924a). — Krampfplasmolyse bei *Spirogyra*. Arch. ges. Physiol. **206**, 629—634 (1924b). — Plasmolysezeitmethode. Protoplasma **5**, 622—624 (1928). — Plasmolyse und "surface precipitation reaction". Protoplasma **15**, 522—531 (1932). — Aluminiumsalz-Wirkung und Plasmolyse-Permeabilität. Protoplasma **17**, 471—475 (1933). — Plasmalemma-Zerstörung und Tonoplasten-Bildung. Protoplasma **21**, 424—426 (1934). — WEIS, A.: Beiträge zur Kenntnis der Plasmahaut. Planta (Berl.) **1**, 145—186 (1926). — WEIXL-HOFFMANN, H.: Beiträge zur Kenntnis der Salzpermeabilität. Protoplasma **11**, 210—277 (1930).—WIECHMANN, E.: Zur Theorie der Magnesiumnarkose. Arch. ges. Physiol. **182**, 74—103 (1920). — WÖSTMANN, E.: Der Einfluß von Kalium und Calcium auf den Wasserhaushalt der Pflanzen. Jb. wiss. Bot. **90**, 335—381 (1942). — WUHRMANN, K.: Der Einfluß von Neutralsalzen auf das Streckungswachstum der *Avena*-Koleoptile. Protoplasma **29**, 361—393 (1938).

YAMAHA, G., u. T. ISHII: Über die Ionenwirkung auf die Chromosomen der Pollenmutterzellen von *Tradescantia reflexa*. Cytologia (Tokyo) **3**, 333—336 (1932). — YAMAHA, G., u. N. SUITA: Über die Wirkung verschiedener Pufferlösungen auf *Spirogyra*-Zellen. Cytologia (Tokyo) **10**, 371—381 (1940).

Elektrophysiologische Phänomene.

Von

Karl Umrath.

Mit 11 Abbildungen.

1. Einleitung. ϵ- und ζ-Potential.

Elektrische Spannungen können nicht innerhalb eines homogenen Mediums auftreten, sondern nur an Phasengrenzen. Hier können Ionen so in der Grenzschicht eingelagert sein, daß sie eine elektrische Doppelschicht bilden, die eine elektrische Spannung zwischen den beiden Phasen oder mit anderen Worten einen Sprung des elektrischen Potentials an der Phasengrenze bedingt. Stellen wir uns vor, die eine Phase sei in der anderen dispergiert, wie etwa einzelne Zellen in einem sie umgebenden Medium, so können die Ionen der elektrischen Doppelschicht in der Grenzschicht zwischen den beiden Phasen festliegen. Es besteht dann eine Spannung zwischen beiden Phasen, die man mit Elektroden, welche man je in eine Phase einbringt, messen kann, aber die Phasen haben keine elektrischen Ladungen, so daß ein durch das System geleiteter elektrischer Strom die Teile der dispersen Phase nicht verschiebt.

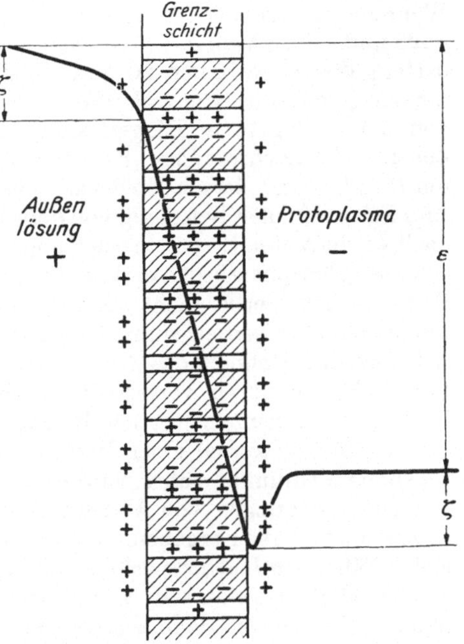

Abb. 1. Schema einer Grenzschicht mit negativ geladenen festen Teilen und positiv geladenen Wasserfäden.

Meist ist die Verteilung der Ionen eine kompliziertere, wobei ein Teil von ihnen in der Grenzschicht festliegt, während sich andere in den unmittelbar an die Grenzschicht anstoßenden, aber gegen sie beweglichen Flüssigkeitsschichten befinden. Abb. 1 gibt ein schematisches Bild eines so bedingten Spannungsverlaufes. Mit Elektroden im Inneren der beiden Medien kann man die Gesamtspannung zwischen ihnen messen, das mit ε bezeichnete, sog. thermodynamische Potential. Die mit ζ bezeichneten sog. elektrokinetischen Potentiale sind Teile des thermodynamischen Potentials. Das ζ-Potential an der Außenseite zwischen einem Teilchen der dispersen Phase und dem Dispersionsmittel oder zwischen einer Zelle und der Umgebungsflüssigkeit ist maßgebend für die elektrophoretische Wanderung des Teilchens oder der Zelle bei Durchleitung eines elektrischen Stromes. Das ζ-Potential beruht ja auf einer elektrischen Doppelschicht, deren einer Teil in der Grenzschicht beider Phasen festliegt, während sich der andere

Teil bis in den Bereich einer flüssigen Phase erstreckt, der gegenüber der Grenzschicht beweglich ist. Da die Grenzschicht so eine Ladung gegenüber der Flüssigkeit erhält, resultiert eine Wanderung der von der Grenzschicht umgebenen Teilchen im elektrischen Strom. Es ist nach dem Gesagten begreiflich, daß adsorbierbare Ionen von Einfluß auf das ζ-Potential sind, weil sie sich in der Grenzschicht verankern, während ihre Gegenionen in der Flüssigkeit beweglich bleiben.

Man sieht, daß ε- und ζ-Potential weitgehend unabhängig voneinander sein können. Leider gibt es keine gut vergleichbaren Bestimmungen von beiden an demselben Objekt. Es haben aber z. B. WINSLOW, FALK und CAULFIELD gefunden, daß durch Hitze getötete Bakterien dieselbe kataphoretische Wanderungsgeschwindigkeit haben wie lebende, während das mit Elektroden zwischen Protoplasma und Außenmedium gemessene ε-Potential beim Abtöten der Zellen, soviel man weiß, verlorengeht oder doch stark reduziert wird. WINSLOW und UPTON haben außer Bakterien auch eine Grünalge, *Chlorella*, und *Saccharomyces apiculatus* untersucht. Alle diese Zellen wandern zur Anode, sind also negativ geladen. Auch alle Messungen mit Elektroden haben eine negative elektrische Spannung des Protoplasmas gegenüber dem Außenmedium ergeben. ζ- und ε-Potential sind hier gleichgerichtet, aber wahrscheinlich dem Betrag nach recht verschieden. Während, wie Tabelle 1, S. 751 zeigt, das ε-Potential von Pflanzenzellen zwischen —25 und —170 Millivolt (mV) liegt und bei Algenzellen in Süßwasser oder destilliertem Wasser zwischen 60 und 170 mV beträgt, fanden WINSLOW und UPTON für das ζ-Potential von *Chlorella* in destilliertem Wasser bei dem günstigsten p_H von 6,5 —23 mV, in Ringerlösung höchstens —13 mV, und bei den anderen Zellen meist noch weniger. RUBINSTEIN und USPENSKAJA haben an plasmolysierten *Helodea*- und *Allium*-Epidermiszellen, meist noch innerhalb der Zellmembran, zum Teil auch nach Befreiung von derselben, eine Bewegung zur Anode beobachtet, nach wiederholten Stromeinschaltungen oder nach längerer Plasmolyse bei einem gewissen Phosphatzusatz und neutraler Reaktion aber Bewegungen zur Kathode. Wenn es sich hierbei nicht nur um eine Mitnahme der Zellen durch die elektroosmotische Wasserbewegung handelte, was unwahrscheinlich ist, so würde es sich um eine Umkehr des ζ-Potentials handeln. Vom ε-Potential ist zwar eine Verminderung durch verschiedene schädigende Eingriffe bekannt, aber keine Umkehr von nennenswertem Betrag.

Da das ζ-Potential einer Partikel von deren Aufladung gegenüber dem umgebenden Medium herrührt, wird es beim isoelektrischen Punkt der Partikel Null. Die Untersuchungen von WINSLOW und UPTON und die von RUBINSTEIN und USPENSKAJA haben ergeben, daß die Wanderung von Zellen bei p_H 2 und 3 in vielen Fällen zur Kathode erfolgt, und daß in den anderen Fällen die Wanderungsgeschwindigkeit schon sehr vermindert ist. Die äußeren Schichten der Pflanzenzellen verhalten sich demnach wie ein Ampholyt mit einem isoelektrischen Punkt bei p_H 2—3 oder darunter. Bei höherem p_H-Wert, also weniger saurer oder alkalischer Reaktion, gibt der Ampholyt H^+-Ionen ab, die sich mit OH^--Ionen der Lösung zu H_2O vereinigen können, und ist selbst negativ geladen. Bei niedrigerem p_H-Wert gibt er umgekehrt OH^--Ionen ab und ist positiv geladen. Daß sich bei manchen Zellen, auch bei sehr tiefem p_H, keine Wanderung zur Kathode beobachten läßt, hat sein Gegenstück darin, daß über p_H 12 die Wanderungsgeschwindigkeit zur Anode vielfach abnimmt, was besonders bei Bakterien von WINSLOW, FALK und CAUFIELD und von WINSLOW und UPTON beobachtet wurde. Es scheint bei so extremen p_H-Werten die Adsorption der in großer Menge vorhandenen und gut adsorbierbaren H^+- bzw. OH^--Ionen eine Rolle zu spielen.

Von Inhaltskörpern der Zelle kennt man nur das Vorzeichen des elektrokinetischen Potentials. Nach der Darstellung in der Monographie von MILOVIDOV

(1. Teil S. 270ff., Literatur 2. Teil) bleibt der Zellkern beim Durchleiten eines Stromes durch die Zelle entweder in Ruhe, oder er bewegt sich zur Anode; im letzteren Fall ist er gegenüber dem Cytoplasma negativ geladen. Der Nucleolus und das Caryotin bzw. die Chromosomen im Kern sind immer negativ geladen.

YAMAHA und ISHII haben Zellbestandteile in 0,1 n KCl-Lösung austreten lassen, die mit HCl verschieden stark angesäuert war, und haben ihre elektrophoretische Wanderung beobachtet. Sie fanden immer im schwach sauren Bereich bei p_H-Werten über dem isoelektrischen Punkt Wanderung zur Anode, also negative Ladung, und unter dem isoelektrischen Punkt Wanderung zur Kathode. Bei verschiedenen Pflanzen fanden sie für Zellkerne isoelektrische Punkte zwischen 3,1 und 4,2, für Leucoplasten zwischen 4,0 und 4,4, für Chloroplasten zwischen 2,8 und 3,2 und für Chromoplasten zwischen 2,2 und 3,1. BUSCH fand für die Chloroplasten von *Selaginella* eine negative Ladung. TOBIAS und SOLOMON haben innerhalb von *Helodea*-Zellen eine Verlagerung der Chloroplasten zur Anode gesehen, während die Protoplasmaströmung weiterging. Auch das Protoplasma war teilweise nach der Seite der Anode verlagert, die Vacuole nach der Kathode. Alle diese Erscheinungen waren reversibel. TOBIAS und SOLOMON schließen mit Recht auf eine negative Ladung der Chloroplasten. HEILBRUNN und DAUGHERTY haben an strömenden Chloroplasten von *Helodea* während der Photosynthese bei elektrischer Durchströmung die Bewegung zur Anode beschleunigt und die zur Kathode verzögert gefunden, bei eingeschränkter oder unterbundener Photosynthese umgekehrt. Der Schluß von HEILBRUNN und DAUGHERTY auf verschiedene Ladung der Chloroplasten scheint mir unsicher, da auch die Geschwindigkeit der Protoplasmaströmung, die ja die Chromatophoren mitnimmt, verändert sein konnte. Das ζ-Potential zwischen der ruhenden Grenzschicht oder Außenschicht des Protoplasmas und dem strömenden Innenplasma kann eine elektroosmotische Flüssigkeits-(Protoplasma-)bewegung bedingen, wenn eine elektrische Spannung in der Längsrichtung der Zelle wirkt. In der lebenden Zelle von *Helodea* scheint mir eine stets negative Ladung der Chloroplasten durch die Untersuchung von TOBIAS und SOLOMON sehr wahrscheinlich; der Vorwurf von HEILBRUNN (S. 101), daß es sich bei diesen Autoren um geschädigte Zellen mit sistierter Protoplasmaströmung handelte, scheint mir auf einer ungenauen Lektüre der Arbeit zu beruhen.

Auch die von SSAWOSTIN beschriebenen Einwirkungen eines Magnetfeldes auf die Protoplasmaströmung von *Vallisneria* und *Helodea* können als Hinweis auf eine elektromagnetische und elektrische Beeinflußbarkeit der Plasmaströmung angesehen werden.

Es sei noch erwähnt, daß man mit zwei gleichen Elektroden eine elektrische Spannung messen kann, wenn man die eine eng mit einem pflanzlichen Gewebe oder einer Zelle in Berührung bringt und die andere in die umgebende Flüssigkeit taucht [YAMAHA, GICKLHORN und UMRATH, UMRATH (2, 3)]. Die Spannung ist allerdings, wie UMRATH (2,3) gezeigt hat, von der Art der Elektroden (Glascapillaren mit 2% Agar und 0,1 n KCl oder Platin oder Gold) stark abhängig. Wenn die Elektroden an verschiedene Gewebe eines Schnittes durch einen Pflanzenteil gebracht werden, so sind die gemessenen Spannungen viel weniger von der Art der Elektroden abhängig, und sie zeigen eine Beziehung zur Färbbarkeit der Gewebe, indem die elektrisch positiven Gewebe mit basischen, die negativen mit sauren Farbstoffen färbbar sind. Die Untersuchungen von GICKLHORN und UMRATH und von UMRATH (2, 3) sind auch in Hinblick auf die elektrostatische Theorie der Färbung von KELLER ausgeführt worden. Leider gibt es keine Untersuchung über die Beziehung der so gemessenen Potentiale zu den ζ-Potentialen, doch dürften ζ-Potentiale oder ihre äußeren Anteile an ihnen beteiligt sein.

2. Die elektrische Spannung zwischen Protoplasma und Außenmedium.

Über die elektrische Spannung des Protoplasmas gegenüber dem Außenmedium sind wir an einigen Zellen durch Messungen mit Mikroelektroden, von denen die eine in die Zelle eingestochen war, während sich die andere im Außenmedium befand, orientiert.

Zunächst muß man sich bei solchen Messungen fragen, ob an den Elektrodenspitzen Spannungen zu erwarten sind. Meist verwendet man als Elektroden mit KCl-Lösung gefüllte Glascapillaren. Die Ableitung zu einem Metalldraht und damit zum Elektrometer erfolgt bei beiden Elektroden in gleicher Weise, etwa durch einen mit Chlorsilber überzogenen Silberdraht. Alle hier vorhandenen Spannungen sind in beiden Elektroden gleich und heben sich daher gegenseitig auf. An den Elektrodenspitzen hat man zunächst mit Diffusionspotentialen gerechnet und eben deshalb die Füllung mit KCl-Lösung gewählt; die K^+- und die Cl^--Ionen haben nahezu gleiche Beweglichkeiten, daher strebt keines wesentlich voranzueilen, und so entsteht kein nennenswertes Diffusionspotential. Wenn die Konzentration des KCl und damit seine Leitfähigkeit genügend groß ist, so drückt es durch diese auch Diffusionspotentiale herab, die durch andere Salze entstehen.

Die tatsächlichen Messungen an Pflanzenzellen haben gezeigt, daß an den Spitzen der Elektroden, die in das Protoplasma eingestochen sind, Membranen entstehen, die für Kationen mehr oder weniger leicht durchlässig sind, für Anionen aber nicht oder doch sehr viel weniger. Der Vorteil einer Elektrodenfüllung mit 0,1 n KCl ist nun vor allem der, daß, solange sich die Elektrodenspitze im Protoplasma befindet, an ihr keine wesentliche Spannung entsteht, auch wenn sich eine Membran bildet, da das Protoplasma Kationen und vor allem Kalium in ähnlicher Konzentration enthält. UMRATH (6) hat gezeigt, daß mit 0,1 n KCl gefüllte Elektroden, die infolge sehr verschieden langen Gebrauches sehr verschieden dichte Membranen an den Spitzen haben und in Wasser Spannungen von 45 bis 70 mV gegeneinander zeigen, im Protoplasma einer *Nitella*-Zelle meist Spannungen von 0, höchstens aber bis 5 mV gegeneinander haben.

Der Einstich der Elektrodenspitze durch die Zellwand ist meist sehr deutlich zu beobachten; dabei tritt nun die Frage auf, ob sich die Elektrodenspitze dann im Protoplasma oder im Zellsaft befindet. Bei gewöhnlichen Pflanzenzellen, abgesehen von den sehr großen Zellen mit besonders dünnem Protoplasmabelag, wie sie etwa bei *Valonia* vorkommen, kann man folgende Argumente dafür anführen, daß sich die Spitze einer eingestochenen Elektrode im Protoplasma befindet: 1. LINSBAUER hat bei seinen Untersuchungen der Protoplasmaströmung an *Chara*-Zellen gefunden, daß die wirksame Kraft ihren Sitz zwischen der äußeren, ruhenden und der strömenden Plasmaschicht hat, so daß schon ein kurzdauernder leichter Druck auf eine Zelle und erst recht eine Verletzung der äußeren Schicht durch den Einstich einer Elektrode zu einer Störung der Strömung und zu einer Anhäufung von Protoplasma an der betroffenen Stelle führt; dieses Protoplasma umgibt die Elektrodenspitze. 2. UMRATH (6) hat an *Nitella*-Internodialzellen Elektroden von oben her eingestochen und unter der Elektrodenspitze mehrfach Partikel hindurchschwimmen gesehen, obzwar die Elektrode ein gewisses Hindernis für die Strömung bildete. 3. Um die Spitze einer in eine Pflanzenzelle eingestochenen Elektrode bildet sich nach Minuten bis Tagen eine Membran[1], welche

[1] Diese Membranbildung wurde von WALKER mikroskopisch beobachtet. Die mir erst bei der Korrektur dieses Artikels bekannt gewordene Arbeit von WALKER ist im nächsten Abschnitt, über das elektrische Potential des Zellsaftes, näher besprochen.

die Elektrodenspitze abkapselt und elektrisch vom Protoplasma isoliert. Dabei nimmt die mit der Elektrode meßbare Spannung und noch mehr ihre Veränderung bei Reizen, d. h. der Aktionsstrom, ab, und der Widerstand an der Elektrode wird sehr groß [UMRATH (7, 10)]. Wenn eine derartige Abkapselung der Elektrodenspitze im Zellsaft und nicht im Protoplasma erfolgen würde, so müßte die Elektrode zunächst, wenigstens vorübergehend, mit dem Protoplasma Kontakt gewinnen; nichts deutet auf einen in diesem Fall zu erwartenden Wechsel in der gemessenen Spannung. 4. Die Abkapselung der eingestochenen Elektrodenspitze wurde ebenso von DIANNELIDIS und UMRATH (2) an Plasmodien von *Physarum polycephalum* beobachtet; diese haben gar keine Zellsafträume, wie sie den meisten Pflanzenzellen mit Cellulosemembranen zukommen.

Tabelle 1. *Spannung des Protoplasmas gegenüber dem Außenmedium bei einigen Pflanzenzellen.*

Pflanze und Zellart	Spannung in mV	Autor
Zellen mit fester Cellulosemembran in Süßwasser:		
Helodea densa, Blattepidermiszelle	—104	UMRATH (10)
Utricularia vulgaris, Zelle der inneren Blasenwand	— 59	DIANNELIDIS und UMRATH (1) und DIANNELIDIS
Utricularia vulgaris, Zelle der äußeren Blasenwand	—164	
Nitella mucronata, Internodialzelle, 20° C	—145	UMRATH (9)
Nitella opaca, Internodialzelle, 20° C	—105	UMRATH (17)
Spirogyra Reinhardii	—103	UMRATH (10)
Vaucheria sessilis und V. uncinata	— 87	UMRATH (10)
Zellen mit Cellulosemembran in Seewasser:		
Valonia macrophysa	—35	UMRATH (13)
Zellen ohne feste Membran:		
Tulipa, Pollenschläuche	—25	GICKLHORN und UMRATH UMRATH (10)
Physarum polycephalum, Plasmodium, ruhend	—37	DIANNELIDIS und UMRATH (2)
Physarum polycephalum, Plasmodium, kriechend auf Agar	—26	
Physarum polycephalum, Plasmodium, kriechend auf Glas	—69	

In Tabelle 1 habe ich alle mir verläßlich erscheinenden Messungen der elektrischen Spannung des Protoplasmas von Pflanzenzellen mit Mikroelektroden zusammengestellt. Die Messungen von BROOKS und GELFAN, die zu den ersten gehören, habe ich nicht in die Tabelle aufgenommen, weil die Autoren, abweichend von allen anderen, das Potential des *Nitella*-Protoplasmas, wahrscheinlich irrtümlich, als positiv angeben. Auch die Messungen von KÜMMEL an *Chara coronata* habe ich nicht aufgenommen. Der von ihr gefundene Mittelwert der Spannung, —140 mV, entspricht ganz dem von *Nitella mucronata* (Tabelle 1), KÜMMEL glaubt aber aus dem Zellsaft abgeleitet zu haben. Sie schreibt: „Die Spitze der Elektrode drang in den Zellsaft und wurde gewöhnlich am Grunde von einem Plasmaballen umkleidet, der manchmal Bewegung zeigte". Ich halte es für wahrscheinlich, daß der Plasmaballen mit der Hauptmenge des strömenden Plasmas in Verbindung stand, und daß er immer vorhanden war. Schließlich habe ich die früheren Messungen an Myxomycetenplasmodien nicht berücksichtigt, da erst DIANNELIDIS und UMRATH (2) erkannten, daß die Abkapselung der eingestochenen Elektrode und andere Umstände viel zu kleine Spannungen und auch solche von falschem Vorzeichen vortäuschen können.

Die nur sehr wenige Pflanzen betreffenden Angaben der Tabelle 1 lassen sich noch durch Schätzung der Spannung zwischen Protoplasma und Außenmedium nach Aktionsstromaufnahmen ergänzen. Wie wir noch sehen werden, bestehen die Aktionsströme bei Pflanzen in einem zeitweisen, völligen oder teilweisen Rückgang der Spannung zwischen Protoplasma und Außenmedium. An tierischen Nerven und Muskeln ist allerdings auch ein Überschießen, eine Potentialumkehr während des Aktionsstroms bekannt. Der Aktionsstrom in einem Blattstiel oder in einer Blattspreite rührt nur von den erregungsleitenden Zellen her, und die inaktiven Zellen und die intercelluläre Flüssigkeit bilden einen Nebenschluß, der die ableitbare Spannung verkleinert. Das Ausmaß der Aktionsströme darf also wohl als eine untere Grenze für die Spannung zwischen dem Protoplasma der erregungsleitenden Zellen und der intercellulären Flüssigkeit angesehen werden. Der Aktionsstrom ist bei *Mimosa pudica* 60—100 mV [UMRATH (4, 11)], bei *Biophytum sensitivum* 40—60 mV [UMRATH (4, 5)], bei *Drosera rotundifolia* 60 bis 120 mV [UMRATH (5)] und bei *Dionaea muscipula* etwa 60 mV [UMRATH (11)].

Man kann nach Tabelle 1 und nach den eben genannten Aktionsstromwerten sagen, daß bei Pflanzenzellen mit festen Cellulosewänden, sowohl von Süßwasserpflanzen als auch von Landpflanzen, das Protoplasma 60—170 mV negativ gegenüber dem umgebenden Medium ist. Bei Meerespflanzen ist nur von *Valonia macrophysa* die Spannung des Protoplasmas gegenüber dem Seewasser bekannt, sie ist —35 mV.

Wir werden noch sehen, daß bei Süßwasserpflanzen der Betrag des Protoplasmapotentials mit zunehmendem Salzgehalt im Außenmedium abnimmt. Gegenüber Meerwasser als Außenmedium hätte ihr Protoplasma wahrscheinlich keine negative Spannung mehr. Auch gegenüber 0,1 n KCl als Außenlösung hat das Protoplasma der Süßwasserpflanzen keine negative Spannung mehr. Deswegen ist es bemerkenswert, daß YAMAHA an Pollenmutterzellen von *Lilium speciosum*, deren Antherenschleim die Leitfähigkeit einer 0,02 n KCl-Lösung hat, bei einer Außenlösung von 0,1 n KCl immer negative Spannungen des Protoplasmas von wenigen bis zu 23 mV gemessen hat. Die Spannungen waren in der Synizesis der Prophase 18—23 mV, in der Telophase I 17—19 mV und in der Telophase II 6—22 mV, sonst waren sie in der Pro- und in der Metaphase höchstens 12 mV, in der Tetrade 4—7 mV. Allerdings konnte YAMAHA die Spannungswerte nur unmittelbar nach dem Einstich der Elektrode ablesen, da das Protoplasma bald koagulierte. Bemerkenswerterweise zeigten die Pollenmutterzellen beim Anlegen einer Elektrode von außen positive Spannungen, die auch in den Telophasen besonders hoch waren, etwa 15 mV, so daß ein von der Kernphase abhängiger, in der Telophase besonders großer Potentialsprung an der Zellgrenzfläche bestehen muß.

Wie die in Tabelle 1 wiedergegebenen Messungen an Pollenschläuchen der Tulpe und an Plasmodien von *Physarum* zeigen, dürfte das Protoplasma von Pflanzenzellen ohne feste Cellulosemembran, wenn sich diese in einem feuchten Medium befinden, nur eine geringe negative Spannung gegenüber dem Außenmedium haben, wahrscheinlich in Zusammenhang mit einem geringen Salzgehalt des Protoplasmas. Vielleicht haben die auf Glas kriechenden Plasmodien mit höherem negativem Potential in Berührung mit der trockeneren Luft einen geringeren Wasser- und einen höheren Salzgehalt.

Wenn sich eine Pflanzenzelle ganz in einem flüssigen Medium befindet, so ist die Spannung zwischen Protoplasma und Außenmedium an allen Stellen der Zelle nahezu gleich, weil der elektrische Widerstand im Zellinneren und auch der im Außenmedium gering ist und der Widerstand an der Zellgrenzfläche im Vergleich dazu ungeheuer hoch ist. Jeder Vorgang, der tendieren würde, die Spannung an der Zellgrenzfläche lokal zu verändern, würde in den flüssigen Medien nur geringe Ionenverschiebungen und an der Zellgrenzfläche eine veränderte Polarisation bedingen, wodurch sich eine neue, aber wieder überall nahezu gleiche Spannung zwischen Protoplasma und Außenmedium einstellen würde. Dies bestätigen Befunde von UMRATH (8, dort besonders Fig. 2 und 3) an *Nitella*-Internodialzellen mit zwei in eine Zelle eingestochenen Elektroden und einer Elektrode im Außenmedium. Anders verhält sich eine Zelle in feuchter Luft; hier ist auch der Widerstand im Außenmedium sehr groß, und örtliche Verschiedenheiten an der Zellgrenzfläche können zu außen abgreifbaren Spannungen führen.

LUND hat diesbezüglich sorgfältige Versuche an der Fadenalge *Pithophora* angestellt. Er hat ein Verhalten gefunden, wie es Abb. 2 nach einer seiner Figuren darstellt. Mit der Spitze des Zellfadens verglichen, zeigt die erste Zelle in ihrem apikalen Teil zunehmend positive Spannungen, die im basalen Teil dieser Zelle und im weiteren Zellfaden gegen die Basis zu zunächst wieder abnehmen und dann in negative übergehen. Diese Spannungsabnahme gegen die Basis des Fadens ist aber nicht gleichmäßig, sondern sie ist von Wiederanstiegen unterbrochen, die sich an den Stellen befinden, an denen die Zellen aneinanderstoßen und in der apikalen Zelle auch dort, wo neue Zellwände erst angelegt werden. Abb. 2 zeigt diese Verhältnisse gut. Die Spannungsunterschiede zwischen verschiedenen Teilen der apikalen Zelle sind in Fig. 1 von LUND fast 20 mV. Sie sind daher wahrscheinlich wesentlich kleiner als die Spannung zwischen Protoplasma und Außenmedium, aber sie kommen doch einem erheblichen Bruchteil von ihr gleich.

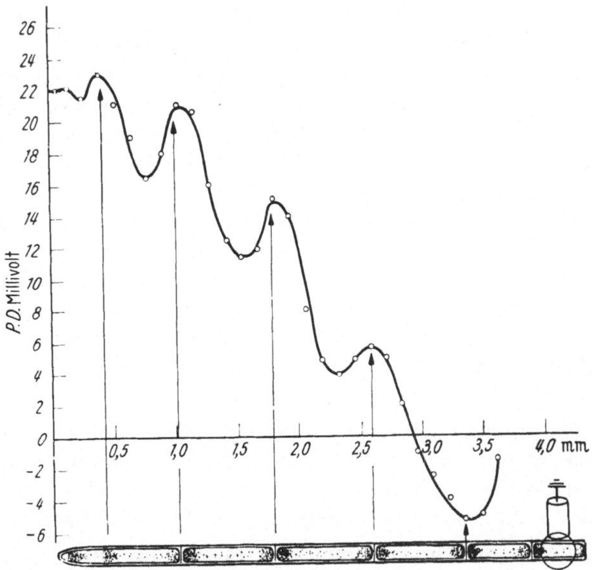

Abb. 2. Spannung an der Außenfläche eines Fadens von *Pithophora* in feuchter Luft nach LUND. [Umgezeichnet aus LUND S. 9.]

An den Myxomycetenplasmodien von *Didymium* haben WATANABE, KODATI und KINOSHITA (1, 2) die Front positiv gegenüber den hinteren Teilen gefunden. Auch wenn die Autoren die Absicht hatten, mit in das Protoplasma eingestochenen Elektroden zu messen (mir sind die japanisch geschriebenen Arbeiten nur aus Referaten bekannt), so dürften, nach den Befunden von DIANNELIDIS und UMRATH (2) an Physarum zu urteilen, ihre Elektroden durch Membranen vom Protoplasma abgekapselt gewesen sein, da die Autoren auch nur sehr geringe Spannungen zwischen Protoplasma und Außenmedium gemessen haben. Immerhin dürfte der Vorderteil eines kriechenden Plasmodiums von *Didymium*, vielleicht außen, um wenige Millivolt elektrisch positiv gegenüber dem hinteren Teil sein.

Bei der Frage nach dem Zustandekommen der Spannung zwischen Protoplasma und Außenmedium kann ich die physikalischen Grundlagen nur kurz streifen. Man hat mit selektiv ionenpermeablen Membranen gerechnet, die gewisse Kationen hindurchlassen, Anionen aber nicht oder doch weniger leicht. BEUTNER vor allem hat mit öligen Grenzschichten gerechnet, in denen manche, besonders organische Ionen einigermaßen gut, andere viel schlechter löslich sind. Wenn eine ölige Schicht zwei wäßrige Lösungen trennt und die Konzentration der im Öl gut löslichen Ionenart nicht von deren Konzentration in den wäßrigen Lösungen abhängt, dann kann sich an einer solchen Ölschicht dieselbe Spannung einstellen, wie sonst an einer selektiv nur für eine Ionenart permeablen Membran. In beiden Fällen erhält man bei einem Elektrolyten mit einwertigen Ionen in der Konzentration c_1 und c_2 auf beiden Seiten der Membran und bei der absoluten Temperatur T die Spannung an der Membran in Millivolt als $0{,}1983 \cdot T \cdot \log c_1/c_2$ oder für 18^0 C und ein Konzentrationsverhältnis von 1:10 57,7 mV.

Durch Versuche mit durch Radioaktivität gekennzeichneten Ionen weiß man, daß die Zellgrenzfläche für alle Ionen bis zu einem gewissen Grad durchlässig ist. Dies bildet eine Schwierigkeit für die Annahme einer selektiv nur für gewisse Ionen durchlässigen Membran. Ich halte es aber nicht für richtig, deswegen mit Diffusionspotentialen zu rechnen, denn aus einer Zelle in Süßwasser diffundiert eben normalerweise kein Salz heraus und von einer Zelle in Meerwasser wird normalerweise nicht Kalium gegen Natrium ausgetauscht, obzwar die Zelle im Inneren viel mehr Kalium und viel weniger Natrium enthält als das Meerwasser. Im einzelnen noch unbekannte, energieverbrauchende Vorgänge erhalten den stationären Ungleichgewichtszustand der Zelle und verhindern dabei das Hinausdiffundieren von Salzen aus ihr, so daß meiner Auffassung nach Diffusionspotentiale gar nicht entstehen.

Schließlich können elektrisch polare Molekeln, die orientiert in die Zellgrenzfläche eingelagert sind, einen Beitrag zu der Spannung zwischen Protoplasma und Außenmedium liefern. Wir werden noch sehen, daß Gründe dafür bestehen, diese Komponente neben einer anderen als wirksam anzusehen.

Wie erwähnt, bilden alle bisher daraufhin untersuchten Pflanzenzellen an den Spitzen und oft um die Spitzen eingestochener Elektroden Membranen mit selektiver Kationenpermeabilität. UMRATH (7) hat an *Nitella mucronata* diese Membranbildung an den Elektroden näher untersucht. Er hat in eine Internodialzelle außer einer mit 0,1 n KCl gefüllten Elektrode noch eine zweite eingestochen, die meist neben 0,1 n KCl noch 0,5 n eines zweiten Kaliumsalzes enthielt. Es zeigte sich, daß die Membranbildung sehr stark durch Oxalat und auch deutlich durch Carbonat gefördert wird. Nach allem scheint die Entstehung eines Calciumniederschlages für die Bildung einer selektiv kationenpermeablen Membran wesentlich zu sein, so wie auch für die TRAUBEschen Zellen die Bildung eines kolloiden Niederschlages wesentlich ist [Literatur bei UMRATH (7)]. UMRATH (6, 7) hat an Elektroden mit dichten Membranen in verdünnten KCl-Lösungen Spannungen gemessen, wie sie nach der oben angegebenen Formel für eine für K^+ durchlässige und für Cl^- undurchlässige Membran zu erwarten sind. Auch konnte UMRATH (7) die Membran an der Elektrodenspitze durch Eintauchen in verdünnte Salzsäure umladen, so daß sie selektiv anionenpermeabel wurde, wie das von anderen natürlichen und künstlichen Membranen schon bekannt war. Man kann sich demnach, grob schematisch, eine solche Membran so vorstellen, wie es in Abb. 1 dargestellt ist, mit negativ geladenen festen Teilen und dazwischen positiv geladenen Wasserfäden. Die positiv geladenen Kationen können sich in den Wasserfäden bewegen, während die negativ geladenen Anionen durch die negative Ladung der festen Teile abgestoßen werden. Säure lädt die Membran um, indem die festen Teile durch Adsorption der H^+-Ionen positiv geladen werden. Da sich die an Elektroden gebildeten Membranen wie selektiv kationenpermeable verhalten, liegt es nahe, auch für die Zellgrenzfläche wenigstens eine Komponente mit ähnlicher Eigenschaft anzunehmen.

An der Zellgrenzfläche selbst wurde die Einwirkung verschiedener Salze, der Salzeffekt, und die Einwirkung eines Salzes in verschiedener Konzentration, der Konzentrationseffekt, untersucht. Ich nenne BEUTNER, der in Zusammenhang mit seiner schon erwähnten Theorie besonders Messungen an stark cutinisierten Pflanzenteilen ausgeführt hat, weiter KÜMMEL. STUDENER hat an *Nitella*-Zellen die Spannung mit einer in die Zelle eingestochenen Elektrode und mit einer im Außenmedium gemessen und ihre Versuche mit $NaNO_3$ und mit KNO_3 mit genügend feiner Abstufung der Konzentrationen über einen großen Konzentrationsbereich ausgeführt. Bei $NaNO_3$ trat in dem ganzen Bereich von 0,0004—0,1 Mol ein gleichmäßiger Konzentrationseffekt auf, der 55% von dem theoretischen für eine kationendurchlässige und anionenundurchlässige Membran ausmacht. Bei weiterer Steigerung der Konzentration war der Effekt größer, und bei 0,25 Mol $NaNO_3$, einer Konzentration, die knapp über der grenzplasmolytischen liegt, war die Spannung an der Zellgrenzfläche sehr viel stärker verringert, als es einem gleichbleibenden Konzentrationseffekt entsprochen hätte. Es ist anzunehmen, daß $NaNO_3$ in so hoher Konzentration die Zellgrenzfläche tiefgreifend verändert, zumal wir gleich sehen werden, daß KNO_3 schon in geringerer Konzentration eine solche Wirkung hat. Daß nicht etwa eine beginnende Plasmolyse für den Effekt verantwortlich ist, geht daraus hervor, daß Plasmolyse mit Rohrzucker die Spannung an der Zellgrenzfläche wenigstens durch mehrere Minuten nicht wesentlich verändert. Bei KNO_3 in geringer Konzentration waren die Spannungen an der Zellgrenzfläche ähnlich wie bei $NaNO_3$; zwischen 0,0004 und 0,002 Mol KNO_3 war der Konzentrationseffekt 62% des theoretisch an einer selektiv kationenpermeablen Membran zu erwartenden. Aber schon bei 0,002 Mol KNO_3 hat STUDENER in vielen Fällen beobachtet, daß die anfänglich nur wenig herabgesetzte Spannung, meist im Anschluß an einige Aktionsströme, ziemlich unvermittelt auf einen deutlich niedrigeren Wert absank. Manche Zellen zeigten sofort eine entsprechend niedrigere Spannung und bei höherer KNO_3-Konzentration war das die Regel. Eine solche Wirkung von Kaliumsalzen bei *Nitella* ist schon von HILL und OSTERHOUT (1, 2) beobachtet und als Kaliumeffekt bezeichnet worden. BLINKS (3) hat bei *Chara coronata* gefunden, daß 0,01 molare Lösungen von NaCl und von KCl zunächst denselben Effekt ergeben, daß aber bei KCl nach dem Auftreten eines Aktionsstromes die elektrische Spannung an der Zellgrenzfläche wesentlich stärker herabgesetzt wird als bei NaCl, womit auch eine wesentliche Abnahme des Widerstandes an der Zellgrenzfläche verbunden ist. STUDENER hat bei höheren KNO_3-Konzentrationen, zwischen 0,01 und 0,05 Mol, ein Konzentrationspotential von 88% des theoretisch an einer selektiv kationenpermeablen Membran zu erwartenden gefunden. Aus all dem ist zu ersehen, daß Kaliumsalze schon in geringerer Konzentration die Zellgrenzschicht weitgehend verändern. Der Salzeinfluß ist also ein sehr komplexer, und wenn man ohne Kenntnis dieser Verhältnisse ein Konzentrationspotential in dem Übergangsgebiet bestimmen würde, so könnte es sich für einen geringen Konzentrationsunterschied sehr hoch ergeben. STUDENER hat auch die Konzentrationspotentiale von 4 Kaliumsalzen mit verschiedenen einwertigen Anionen untersucht und keine sicher feststellbaren Unterschiede gefunden. An der nur kationenpermeablen Membran ist eine solche Einflußlosigkeit der Anionen zu erwarten. So zeigt die Zellgrenzfläche manche Ähnlichkeiten mit einer kationenpermeablen Membran, wenn auch Abweichungen und Besonderheiten auftreten.

Die Argumente dafür, daß elektrisch polare, orientiert in die Zellgrenzfläche eingelagerte Molekeln einen Beitrag zur Spannung zwischen Protoplasma und

Außenmedium liefern, kann ich aus Raummangel nur kurz andeuten. UMRATH (14, 18) hat an *Nitella*-Internodialzellen die Spannung an der Zellgrenzfläche und ihre Veränderungen durch polarisierende Gleichstromstöße wechselnder Richtung mit einer in die Zelle eingestochenen Elektrode und mit einer im Außenmedium gemessen. Der polarisierende Strom wurde auf zweierlei Art zugeleitet, entweder mit einer in die Zelle eingestochenen Elektrode und mit einer im Außenmedium oder mit zwei Elektroden im Außenmedium. Wenn die Polarisation nur durch Ionenstauung an der Zellgrenzfläche erfolgt, so liegen Verhältnisse wie bei der Auf- und Entladung eines Kondensators vor. Die Zeit, die eine solche Kondensatorauf- oder -entladung braucht, ist proportional dem Produkt aus der Kapazität des Kondensators und einer Größe, die von den Widerständen abhängt. Bei Zuleitung des polarisierenden Stromes mit einer in die Zelle eingestochenen Elektrode ist die Kapazität die der ganzen Zellgrenzschichte und die Widerstände sind sehr groß; sie setzen sich zusammen aus dem Widerstand der Zellgrenzschicht, der den Kondensator überbrückt, und dem noch größeren Zuleitungswiderstand der eingestochenen Elektrode. Bei Zuleitung des polarisierenden Stromes durch zwei Elektroden im Außenmedium muß man sich die beiden Zellhälften als zwei hintereinander geschaltete Kondensatoren vorstellen. Jeder Kondensator hat jetzt nur die halbe Kapazität, und durch das Hintereinanderschalten von zwei gleichen Kondensatoren werden die Auf- und die Entladezeit nochmals auf die Hälfte verkürzt. Vor allem sind aber der Widerstand im Zellinneren, der die beiden Kondensatoren verbindet, und der Widerstand im Außenmedium, der sie überbrückt, beide klein. Die Formeln und die zahlenmäßige Auswertung müssen bei UMRATH (14) nachgesehen werden. Es ergab sich, daß theoretisch die Zeit für das Entstehen oder für das Verschwinden der Polarisation bei Stromzuleitung mit einer eingestochenen Elektrode ungefähr 100mal so lang sein sollte, wie bei Stromzuleitung mit beiden Elektroden im Außenmedium. Tatsächlich verhielten sich die Zeiten aber statt wie 100:1 nur wie 5:1. Daraus kann man auf eine Polarisationskomponente schließen, deren zeitliches Entstehen und Vergehen nicht wie eine Kondensatoraufladung von den Widerständen abhängig ist.

Die genauere Analyse des Entstehens und Vergehens der Polarisation hat auch gezeigt, daß ein solcher einzelner zeitlicher Verlauf nicht durch eine e-Potenz darstellbar ist, wie es bei der Auf- oder Entladung eines Kondensators sein müßte, sondern nur durch wenigstens zwei e-Potenzen.

Eine Polarisationskomponente, die sich nicht kondensatorartig verhält, kann man sich durch elektrisch polare, orientiert in die Zellgrenzschicht eingelagerte Molekeln bedingt vorstellen. So wie sich gelöste Molekeln im Lösungsmittel bewegen können, woraus die Diffusion des gelösten Körpers resultiert, so können sich die grenzflächenaktiven Molekeln in der Grenzfläche bewegen und dabei Konzentrationsunterschiede ausgleichen. Dieser Ausgleich von Konzentrationsunterschieden in der Grenzfläche verläuft aber viel rascher als der durch Diffusion in wäßriger Lösung. Ich stelle mir vor, daß die grenzflächenaktiven Molekeln zum Teil in der Zellgrenzschicht orientiert eingelagert sind und zum Teil im Protoplasma gelöst sind. Da sich die elektrisch polaren Molekeln so in die Zellgrenzschicht einlagern, daß ihr positiver Teil zum Außenmedium, ihr negativer zum Protoplasma gerichtet ist, wird ein elektrischer Strom, der durch die Zelle geleitet wird, an der Anode solche Molekeln aus dem Protoplasma orientiert in die Zellgrenzfläche einlagern, an der Kathode sie aus der Zellgrenzfläche heraus in das Protoplasma bringen. Die so entstehenden Konzentrationsunterschiede der orientiert eingelagerten Molekeln werden Ausgleichsbewegungen in der Zellgrenzfläche bedingen.

Die Auffassung von den elektrisch polaren Molekeln in der Zellgrenzfläche findet eine Stütze in Befunden über die elektrische Erregung. NERNST hat eine Theorie entwickelt, nach der der elektrische Strom Salzkonzentrationsänderungen an der Zellgrenzfläche bedingt, auf deren Ausgleich die Diffusion im Zellinneren hinwirkt. UMRATH (1) hat gezeigt, daß, wenn die Zellen ihrer ganzen Länge nach für diesen Ausgleichsvorgang in Frage kommen, es sich um einen Vorgang handeln muß, der sehr viel rascher abläuft als die Diffusion in wäßriger Lösung. Dafür, daß sich der Ausgleichsvorgang über die ganze Zelle erstreckt, hat er verschiedene Beobachtungen angeführt, von denen nur die auf Pflanzen bezüglichen hier erwähnt seien. Reizt man eine Zelle durch Gleichstrom von langer Stromflußzeit, so kann man eine eben noch wirksame Schwellenspannung finden. Verkürzt man die Reizzeit, so muß man, um denselben Effekt zu erhalten, die Spannung erhöhen. Die Reizzeit, bei der man die Schwellenspannung verdoppeln muß, nennt man die Chronaxie. UMRATH (1) hat die Chronaxie für verschiedene Gewebe einer Pflanze, *Mimosa pudica* und *Berberis*, bestimmt und gefunden, daß sie um so größer ist, je länger die erregbaren Zellen des betreffenden Gewebes sind. Die Chronaxien sind den Quadraten der Zellängen proportional, wie es die Theorie verlangt, wenn es sich um einen nach den Diffusionsgesetzen ablaufenden, in allen Zellen ähnlichen Vorgang handelt. Bei *Mimosa pudica* verhalten sich die Chronaxien der verschiedenen Gewebe zueinander wie 1:55, die Produkte aus ihnen und den Quadraten der Zellängen nur wie 1:2. Für verschieden lange Internodialzellen von *Nitella* fand UMRATH (8) entsprechende Chronaxieunterschiede. Schließlich hat UMRATH (15) Internodialzellen von *Nitella* gebogen oder geknickt, so daß sie sich dem elektrischen Strom gegenüber wie zwei kürzere nebeneinanderliegende Zellen verhielten. Er fand dabei stark verkürzte Chronaxien, wie es die Theorie voraussagt, und nach dem Geradestrecken wieder so lange wie ursprünglich an der geraden Zelle.

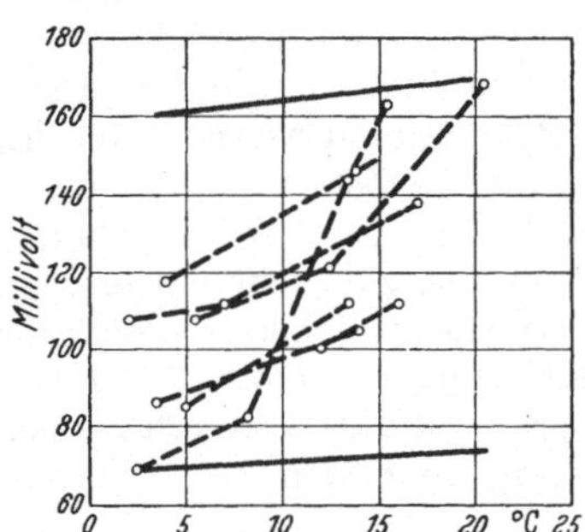

Abb. 3. Abhängigkeit des Plasmalemmapotentials von *Nitella mucronata* von der Temperatur gestrichelt. Ausgezogene Linien proportional der absoluten Temperatur. [Aus UMRATH (9).]

Aus diesen Gründen hat UMRATH (4, 8, 15) in der Theorie der elektrischen Erregung die Diffusion im Protoplasma durch die viel raschere Bewegung grenzflächenaktiver Molekeln in der Zellgrenzschicht ersetzt.

Die Abhängigkeit der Spannung zwischen Protoplasma und Außenmedium von der Temperatur bei Internodialzellen von *Nitella mucronata* gibt Abb. 3 nach einer Untersuchung von UMRATH (9) wieder. Jede gestrichelte Linie stellt einen Versuch mit einer Zelle dar, bei dem die Temperaturänderungen langsam vorgenommen wurden. Die voll ausgezogenen Linien oben und unten in der Abbildung geben Spannungsänderungen proportional der absoluten Temperatur wieder, wie sie an einer kationenpermeablen Membran zu erwarten wären. Man sieht, daß die experimentellen Kurven stärker geneigt sind. Rasche Temperatursteigerungen um wenige Grade bewirken oft einen Rückgang der Spannung und rhythmische Erregungsvorgänge mit Aktionsströmen. Danach scheint der die elektrische Spannung bedingende Mechanismus in der Zellgrenzschicht viel labiler zu sein, als man es bei einer kationenpermeablen Membran erwarten würde.

Es sei hier nochmals daran erinnert, daß LUND bei der Alge *Pithophora* an Zellen in Luft gesetzmäßige Spannungsunterschiede zwischen verschiedenen Stellen der Außenfläche feststellte. Man wird diese eher mit verschiedenen Stoffwechselvorgängen an den verschiedenen Stellen in Zusammenhang bringen wollen, als mit verschiedenen Salzkonzentrationen. LUND selbst denkt an einen Zusammenhang der Spannungen mit Oxydations-Reduktionssystemen. WATANABE, KODATI und KINOSITA (3) fanden, daß die von ihnen am Frontteil von Myxomycetenplasmodien gemessenen Spannungen durch atmungshemmende Stoffe verringert werden.

Schließlich sei noch erwähnt, daß die Zellen im allgemeinen im Inneren Kalium in höherer Konzentration enthalten, als es im Außenmedium vorhanden ist. Die Zellen verfügen also offenbar über einen Mechanismus, der unter Energieaufwand Kalium akkummuliert. Es ist möglich, daß Kalium in gebundener Form durch einen lipoiden Teil der Zellgrenzschicht in die Zelle transportiert wird, während ein für Wasser durchlässiger Teil der Zellgrenzfläche als kationenpermeable Membran wirkt. Wenn aber diese beiderlei Membranteile nicht ein sehr grobes Mosaik bilden, so werden Kaliumionen, die durch den kationenpermeablen Teil der Grenzschicht hindurchtreten, durch den lipoiden Teil der Grenzschicht wieder aufgenommen werden. Ich halte es daher auch deshalb für wahrscheinlich, weil es für die Zelle energiesparend wäre, daß in der Zellgrenzfläche elektrisch polare Molekeln orientiert eingelagert sind, die eine Spannung bedingen, welche ähnlich groß und vielleicht mitunter größer ist als die, welche die Kaliumsalze des Protoplasmas an der kationenpermeablen Membran bedingen würden. So würden die Kaliumionen, trotz der Kationenpermeabilität der Zellgrenzfläche, weitgehend daran gehindert, durch sie hindurchzutreten und an ihr einen äußeren positiven Belag zu bilden.

3. Das elektrische Potential des Zellsaftes.

An den großen Zellen der Meeresalgen *Valonia* und *Halicystis* mit nur sehr dünner Protoplasmaschicht wurden von der Schule von OSTERHOUT Spannungen zwischen dem Zellsaft und dem Seewasser außerhalb gemessen. BLINKS (2) fand den Zellsaft von *Halicystis Osterhoutii* —68 mV und den von *Halicystis* ovalis —80 mV gegenüber dem umgebenden Meerwasser. Später hat BLINKS (4) gezeigt, daß sich bei *Halicystis ovalis* diese Spannung kaum ändert, wenn man mit einer Durchströmungsvorrichtung den Zellsaft durch künstlichen Zellsaft oder durch Seewasser ersetzt oder wenn man außen das Seewasser durch Zellsaft ersetzt. Auch bei *Halicystis Osterhoutii* wird durch diese Eingriffe die Spannung nur auf etwa —45 mV herabgesetzt, wenn der p_H in der Vacuole unter 6 bleibt. Seewasser von einem p_H über 6,5 in der Vacuole führt zu einer Potentialumkehr und macht die Vacuole etwa 30 mV positiv gegenüber dem äußeren Seewasser. Bei *Valonia ventricosa* fand BLINKS (1) den Zellsaft +15—20 mV, bei *Valonia macrophysa* fand ihn DAMON +8 mV gegenüber dem Seewasser außerhalb der Zelle und +62 mV, wenn außen das Seewasser durch Zellsaft ersetzt war. UMRATH (13) hat an *Valonia macrophysa* das Protoplasma zu —20 bis —50 mV, den Zellsaft, wie schon DAMON, zu +8 mV gegenüber dem Seewasser gemessen.

Die Versuche von BLINKS (4) und von DAMON zeigen, daß das Protoplasma unsymmetrisch ist, d. h. daß der Tonoplast, seine Grenzschicht gegen den Zellsaft, sich wesentlich anders verhält als das Plasmalemma, seine Grenzschicht gegen das Außenmedium.

Während wir von *Valonia*- und *Halicystis*-Zellen wenig über die Spannung des Protoplasmas wissen, weil es schwer ist aus ihrem Protoplasma abzuleiten, wissen wir von *Nitella*-Zellen wenig über die Spannung des Zellsaftes. UMRATH [(7), Fig. 2] hat an *Nitella mucronata* mit einer mit 0,1 n KCl + 0,1 n KOH gefüllten Elektrode dieselben Aktionsströme aber oft nur geringere Spannungen des Zellinneren gegenüber dem Außenmedium gemessen als mit einer in dieselbe

Zelle eingestochenen, mit 0,1 n KCl gefüllten Elektrode. Er schließt daraus, daß aus der Elektrode herausdiffundierendes KOH die Plasmaansammlung um die Elektrodenspitze beseitigt und daß diese Elektrode dann das Zellsaftpotential mißt. Danach wäre bei *Nitella mucronata* der Zellsaft elektrisch positiv gegenüber dem Protoplasma und der Aktionsstrom würde nur am Plasmalemma, nicht am Tonoplasten ablaufen, während UMRATH (13) bei *Valonia macrophysa* Aktionsströme am Plasmalemma und am Tonoplasten beobachtete. WALKER hat in junge, noch durchsichtige *Nitella*-Internodialzellen, wahrscheinlich von *Nitella gloeostachys*, Elektroden eingestochen und mikroskopisch bei starker Vergrößerung beobachtet. Wurde die Elektrode nur 20—30 μ tief eingestochen, so reichte ihre Spitze zunächst meist in die Vacuole, sie wurde aber, sobald die Protoplasmaströmung wieder begann, vom Protoplasma umflossen. Wenn die Elektrodenspitze 50—100 μ tief eingestochen wurde, so dauerte es 1 Std und mehr, bis sie vom Protoplasma umflossen wurde, und in dieser Zeit konnte die Spannung des Zellsaftes gemessen werden. Im Protoplasma und nur in diesem beobachtete WALKER auch die Bildung einer Membran um die Elektrodenspitze, ausgehend von der Einstichstelle. Er fand für die Spannung der Vacuole —160 ± 4 mV, für die des Protoplasmas —158 ± 5 mV, also zwischen beiden keinen sicheren Spannungsunterschied. Nach all dem scheint die Spannung des Zellsaftes bei verschiedenen *Nitella*-Arten ähnlich verschieden zu sein wie bei verschiedenen *Valonia*- und *Halicystis*-Arten.

OSTERHOUT (1) hat gezeigt, daß man die Elektrolyte im Zellsaft einer *Nitella*-Internodialzelle verschieben kann, wenn man die Zelle an einer Stelle mit Wasser, an einer anderen mit Rohrzuckerlösung in Berührung bringt. Es entsteht ein Wasserstrom vom Wasser zur Zuckerlösung, der feste und gelöste Inhaltsstoffe des Zellsaftes mitführt und an der Stelle, an der die Zelle mit der Zuckerlösung in Berührung steht, im Zellsaft anreichert. An so behandelten Zellen hat OSTERHOUT (2) Spannungen gemessen und unabhängig von ihm NISHIZAKI. Die Resultate der beiden Autoren sind verschieden. Eine ausführliche Diskussion würde hier zu viel Raum in Anspruch nehmen; es geht aber aus beiden Arbeiten hervor, daß die Elektrolytverschiebung im Zellsaft und vielleicht die Wasserströmung durch das Protoplasma die Spannung, wahrscheinlich am Tonoplasten und am Plasmalemma, verändern.

4. Der elektrische Widerstand des Plasmalemmas.

Schon die allgemein bekannte Tatsache, daß der Durchtritt vieler Stoffe, insbesondere auch der der Ionen, durch die Zellgrenzschicht viel langsamer erfolgt als es der freien Diffusion entsprechen würde, deutet auf einen hohen elektrischen Widerstand des Plasmalemmas hin. Die Widerstandsmessungen sind meist mit Wechselstrom ausgeführt worden, an pflanzlichen Einzelzellen von CURTIS und COLE an *Nitella* und von IWAMURA an *Physarum polycephalum*. Der Widerstand der Zelle ist kein reiner OHMscher Widerstand, bei der Messung in der WHEATSTONEschen Brücke muß man ihn durch einen Widerstand und eine Kapazität kompensieren. Dabei erhält man bei verschiedenen Frequenzen verschiedene Werte für diese Ersatzgrößen. Es sind für die Zelle verschiedene Ersatzschemen aufgestellt worden, auf die man die experimentell gefundenen Werte umrechnen kann. Das einzige, in dem die eingeführten Konstanten von der Frequenz des angewandten Wechselstromes unabhängig sind, ist ein von UMRATH (15) theoretisch abgeleitetes, das dieselben beiden Polarisationskomponenten verwendet, die wir schon S. 756 bei Besprechung der Polarisation an *Nitella* durch Gleichstromstöße wechselnder Richtung kennengelernt haben.

Über den Wechselstromwiderstand von Geweben höherer Pflanzen liegt eine Untersuchung von Paech vor. Auch er mußte in der Wheatstoneschen Brückenschaltung mit Widerstand und Kapazität kompensieren, die beide frequenzabhängig waren. Bei Einwirkung von Narkoticis fand er zunächst eine Widerstandszunahme, der bei höheren Konzentrationen ein rascher irreversibler Abfall folgte. Nach Ausfüllen der Intercellularen mit Flüssigkeit trat nicht, wie man hätte erwarten können, stets eine Erniedrigung, sondern oft eine starke Erhöhung des Widerstandes auf. Das Infiltrieren muß also den Durchtritt von Ionen durch das Plasmalemma hemmen.

5. Der Aktionsstrom.

Unter den Pflanzen ist der Aktionsstrom am besten bei *Nitella mucronata* bekannt. Die verschiedensten Reize, wie Verletzungen und andere mechanische Reize, verschiedene Chemikalien, genügend rasche Temperatursenkungen und

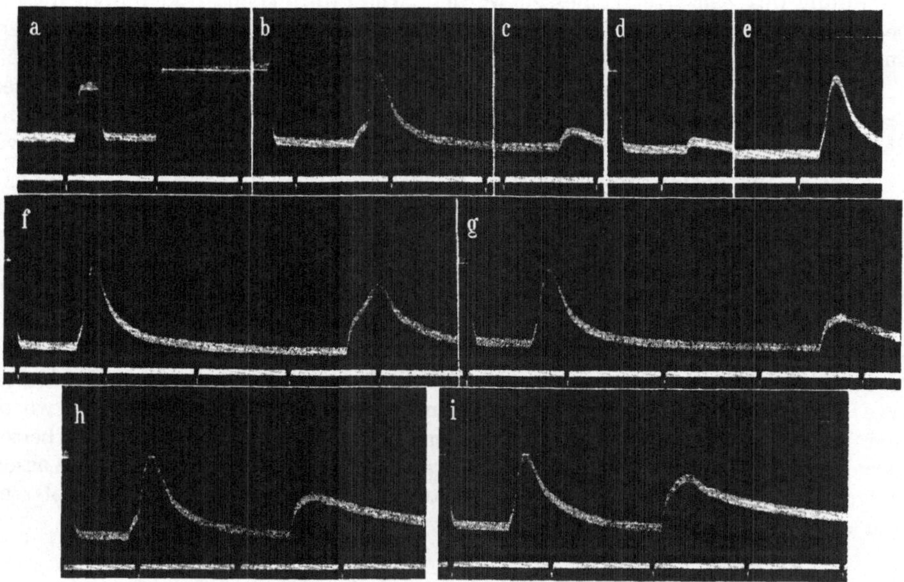

Abb. 4a—i. *Nitella mucronata*, Internodialzelle. Aktionsströme und lokale Potentiale ausgelöst durch Öffnungsinduktionsströme. a Spannung zwischen Protoplasma und Außenmedium = Plasmalemmapotential, Eichkurve 0,1 V, Nullstellung des Elektrometers. b Nullstellung des Elektrometers, Plasmalemmapotential, Öffnungsinduktionsschlag Reizstärke 1,0: lokales Potential und Aktionsstrom. c 2min nach b, Plasmalemmapotential, Reizstärke 1,2: lokales Potential. d 10 min nach c, Nullstellung des Elektrometers, Plasmalemmapotential, Reizstärke 1,0: lokales Potential. e 1 min nach d, Plasmalemmapotential, Reizstärke 1,5: Aktionsstrom. f—i Nullstellung des Elektrometers, Plasmalemmapotential, dann zwei Reize: f Reizstärke 1,5: lokales Potential und Aktionsstrom, nach 30 sec Reizstärke 2,4: lokales Potential und Aktionsstrom; g Reizstärke 1,5: lokales Potential und Aktionsstrom nach 30 sec Reizstärke 1,9: nur lokales Potential: h Reizstärke 1,2: lokales Potential und Aktionsstrom, nach 16 sec Reizstärke 2,4: lokales Potential; i Reize wie in h, Zelle etwas besser erregbar, nach zweitem Reiz lokales Potential und Aktionsstrom. Zeitmarken 10 sec. Original.

elektrische Reize bedingen einen vorübergehenden Rückgang der Spannung zwischen Protoplasma und Außenmedium. Wenn die Reize stark sind, erfolgt der Rückgang der Spannung bei *Nitella mucronata* fast auf Null, und sein Ausmaß ist von der Reizstärke unabhängig. Wenn das Letztere der Fall ist, spricht man vom Aktionsstrom oder Aktionspotential. Solche Aktionsströme zeigt Abb. 4b, e und f bis i nach dem ersten Reiz, wobei die Reizstärke in b 1, in h und i 1,2, in e, f und g 1,5 betrug. Reize, die etwas zu schwach sind, um einen Aktionsstrom auszulösen, also schwächere Reize oder ähnliche bei etwas herabgesetzter Erregbarkeit der Zelle, lösen sog. lokale Potentiale aus, die kleiner als Aktionsströme und in ihrem Ausmaß von der Reizstärke abhängig sind. Die Bezeichnung „lokales Potential“ stammt aus der Tierphysiologie von Fällen, in denen es sich nur über wenige Millimeter mit abnehmender Intensität ausbreitet, während der Aktionsstrom über weite Strecken mit unveränderter Intensität fortgeleitet wird. Bei einer *Nitella*-Zelle in Wasser ist der Widerstand des Plasmalemmas enorm groß gegenüber dem Widerstand im Zellinneren und dem Widerstand des Wassers außen, daher ist die Spannung an der Zellgrenzschicht überall nahezu gleich, und deswegen erstreckt sich auch das „lokale Potential“ gleichmäßig über die ganze Zelle.

Die schwächsten elektrischen Reize bedingen nur eine Polarisation am Plasmalemma. Es wird also je nach der Stromrichtung das Plasmalemmapotential (= die Spannung an der Zellgrenzschicht) vergrößert oder verkleinert, wie es die Abb. 8—11 für Gleichstromstöße wechselnder Richtung zeigen. Die als Reize auch in Abb. 4 angewandten Öffnungsinduktionsschläge sind von so kurzer Dauer, daß ihre Polarisationswirkungen bei einer Registrierung mit dem LINDEMANN-Elektrometer, die in den Versuchen der Abb. 4 und 7—11 angewandt wurde, nicht in Erscheinung treten, weil sie zu gering und kurzdauernd sind.

Die lokalen Potentiale wurden zuerst von UMRATH (8) beschrieben und damals als „Latenzvorgang“ bezeichnet, weil sie in der Latenzzeit zwischen Reiz und Aktionsstrom zu beobachten sind. Das lokale Potential führt bei jeder Richtung des erregenden Stromes zu einer Abnahme des Plasmalemmapotentials und ist daher offenbar eine aktive Äußerung der Zelle. Es liegt ihm wahrscheinlich ein ähnlicher Vorgang zugrunde wie dem Aktionsstrom, nur verläuft der Vorgang beim Aktionsstrom offenbar explosionsartig bis zu einem Maximum, während er beim lokalen Potential nur bis zu einem von der Reizstärke abhängigen Wert ansteigt. Das lokale Potential ist bei herabgesetzter Erregbarkeit, an der ermüdeten Zelle oder im relativen Refraktärstadium herabgesetzt. Dies zeigt Abb. 4d, kleines lokales Potential, im Vergleich zu 4b, größeres lokales Potential mit Aktionsstrom bei gleicher Reizstärke. Das Ausmaß des lokalen Potentials nimmt mit der Reizstärke mehr als proportional zu. In Abb. 4 sieht man lokale Potentiale bei Reizstärke 1 in d, bei 1,2 in c, bei 1,9 beim zweiten Reiz in g und bei Reizstärke 2,4 beim zweiten Reiz in h.

Der Aktionsstrom setzt sich auf das lokale Potential auf, wenn dieses einen gewissen Wert erreicht, der allerdings bei herabgesetzter Erregbarkeit, besonders im relativen Refraktärstadium, größer sein muß als normalerweise. Abb. 4f und i zeigen, daß sich der Aktionsstrom im relativen Refraktärstadium, 30 bzw. 16 sec nach dem ersten Aktionsstrom, erst auf ein sehr großes lokales Potential aufsetzt. Bei gleicher Erregbarkeit setzt sich der Aktionsstrom immer dann auf das lokale Potential auf, wenn dieses ein gewisses Ausmaß erreicht hat, also bei einem stärkeren Reiz früher, Abb. 4e Reizstärke 1,5, als bei einem schwächeren, Abb. 4b Reizstärke 1. Im relativen Refraktärstadium ist der Aktionsstrom von geringerem Ausmaß als normalerweise und stärker herabgesetzt als das lokale Potential. Man vgl. Abb. 4f und i; der zweite Reiz hat in beiden Fällen Reizstärke 2,4, der Reizabstand ist in f 30 sec, in i 16 sec. Der in beiden Fällen gegenüber der nicht refraktären Zelle verkleinerte Aktionsstrom ist in i gegenüber f viel stärker verkleinert als das lokale Potential.

Die Dauer des Aktionsstroms nimmt mit steigender Temperatur ab. UMRATH (9) fand bei *Nitella mucronata* im Bereich von 1—32° C bei einer Temperatursenkung von 10° eine Verlängerung der Anstiegszeit auf das 2,3fache.

Bei *Vaucheria sessilis* und *Vaucheria uncinata* beschreibt UMRATH (10) Aktionsströme, die $^1/_4$—$^5/_6$ des Plasmalemmapotentials rückgängig machen und etwas steiler ansteigen, aber oft langsamer zurückgehen als bei *Nitella mucronata*.

Bei *Spirogyra Reinhardii* beobachtete UMRATH (10) Aktionsströme, die $^1/_4$ bis $^2/_3$ des Plasmalemmapotentials rückgängig machten, ähnlich wie bei *Vaucheria* etwas steiler anstiegen als bei *Nitella mucronata* und ähnlich wie bei letzterer in der Kälte, wenn sie durch Wasser von 0^0 ausgelöst wurden, langsamer verliefen. Bei schwachen Reizen war die Gesamtdauer des Aktionsstroms etwa 7 sec. Bei etwas stärkeren oder wiederholten Reizen war der Rückgang der Aktionsströme etwas verlangsamt. Bei noch stärkeren oder öfter wiederholten Reizen konnte eine Chloroplastenkontraktion eintreten; wenn das der Fall war, ging der Aktionsstrom in den ersten Sekunden gar nicht oder nur minimal zurück und ein deutlicher Rückgang erfolgte erst nach mehreren Minuten. Bei der Chloroplasten kontraktion dürfte die Zelle daher schon geschädigt sein, wenn auch oft noch reversibel.

Bei höheren Pflanzen mit gut ausgebildeter Erregungsleitung, bei einigen Sensitiven und einigen Insectivoren, sind die Aktionsströme von ähnlichem Ausmaß und von ähnlichem Verlauf wie bei den eben besprochenen Algenzellen,

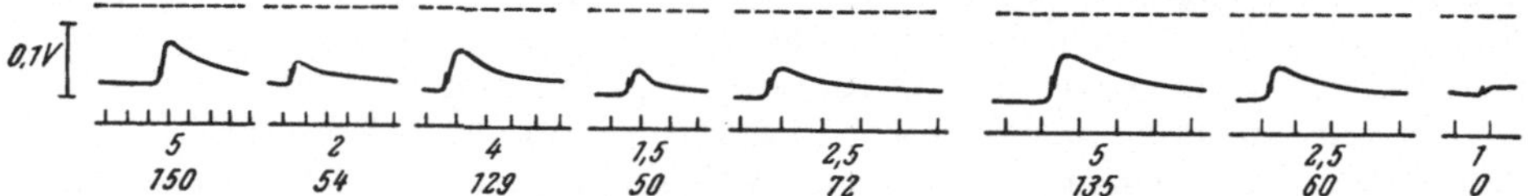

Abb. 5. Junge, 6 mm lange Internodialzelle von *Nitella opaca*. Oben gestrichelt, Nullstellung des Elektrometers, darunter Aktionsströme, darunter Zeitmarken 10 sec, darunter die Reizstärke, zuunterst Zeit der Strömungsstillstände in Sekunden. [Aus UMRATH (17).]

obzwar bei diesen Pflanzen die Aktionsströme außen von den erregungsleitenden Zellen abgeleitet werden und die nicht an der Leitung beteiligten Zellen einen Nebenschluß bilden müssen, der die gemessene Spannung herabsetzt. Das Ausmaß der Aktionsströme einiger dieser Pflanzen ist schon auf S. 752 angeführt.

Es scheint, daß bei höheren Pflanzen mit ihren differenzierten Zellen durchaus nicht alle Zellen Aktionsströme von so großem Ausmaß haben. UMRATH (10) hat an Epidermiszellen der Blätter von *Helodea densa* bei elektrischen Reizen sehr geringe Potentialrückgänge beobachtet, die er wegen ihres geringen Ausmaßes damals nicht für Aktionsströme hielt. Erst als UMRATH (17) bei *Nitella opaca* kleine Aktionsströme beobachtete, kam er zu der Ansicht, daß es sich auch bei den noch kleineren Potentialrückgängen bei *Helodea densa* um Aktionsströme handelte. Auch an der Oberepidermis der Zwiebelschuppen von *Allium cepa* hat UMRATH (12) bei Ableitung von außen sehr kleine Aktionsströme von 1—6 mV beobachtet. Für die Zellen der äußeren und der inneren Schicht der Fangblasen von *Utricularia vulgaris* ergab eine Untersuchung von DIANNELIDIS und UMRATH (1), daß ihre Aktionsströme etwa 70% und vielleicht mehr des Plasmalemmapotentials rückgängig machen.

An Internodialzellen von *Nitella opaca* hat UMRATH (17) auffallende Beobachtungen gemacht. Der Aktionsstrom besteht aus einer rasch ablaufenden Anfangszacke und aus einem anschließenden, viel länger dauernden Teil. Abb. 11 zeigt die Anfangszacke deutlich abgesetzt, die Abb. 5 und 10 zeigen sie weniger deutlich. Bei sehr jungen Internodialzellen von *Nitella opaca* nimmt das Ausmaß des Erregungsvorgangs mit der Reizstärke stufenweise zu, wobei 2—3 Stufen beobachtet wurden. Abb. 5 zeigt eine Reihe solcher Aktionsströme; unter jedem ist die Reizstärke angegeben und darunter die Zeit des Strömungsstillstandes in Sekunden. Bei älteren Internodialzellen sind die Erregungsvorgänge mitunter von der Reizstärke unabhängig und die Aktionsströme haben dann ungefähr das Ausmaß wie bei jungen Zellen nach starken Reizen, ähnlich wie in Abb. 10. Bei alten Internodialzellen macht der Aktionsstrom oft weniger als die Hälfte des Plasmalemmapotentials rückgängig, wie das Abb. 11 zeigt.

Bei dem Myxomyceten *Physarum polycephalum* haben DIANNELIDIS und UMRATH (2) mittels einer in das Protoplasma eingestochenen Elektrode Aktionsströme registriert, die ein Ausmaß von etwa 21 mV hatten und damit 50—70% des Plasmalemmapotentials rückgängig machten. Die Aktionsströme hatten eine Anstiegszeit von 0,6 sec, wie das bei Pflanzen nicht ungewöhnlich ist, aber eine im Vergleich dazu ungeheuer lange Gesamtdauer von im Mittel 36 sec. Schwache Reize bedingten lokale Potentiale von geringerem, von der Reizstärke abhängigem Ausmaß und von etwa 5 sec Dauer. Bei Verstärkung der Reize gingen die lokalen Potentiale kontinuierlich in die größeren und von der Reizstärke unabhängigen Aktionsströme über. Nie setzten sich Aktionsströme auf lokale Potentiale auf, d. h., lokale Potentiale, die deutlich kleiner sind als Aktionsströme, lösen noch keine Aktionsströme aus. TASAKI und KAMIYA beobachteten bei Ableitung von außen und bei auch sonst anderer Methodik bei *Physarum polycephalum* nur lokale Potentiale. Sie stellten als Folge der Reize auch einen Rückgang des Widerstandes fest.

An den Myxomycetenplasmodien von *Didymium* haben WATANABE, KODATI und KINOSITA (1) Schwankungen des elektrischen Potentials beobachtet, die den rhythmischen Änderungen in der Richtung der Protoplasmaströmung vorangingen, an denen von *Physarum polycephalum* haben KAMIYA und ABE elektrische Veränderungen beschrieben, die den Änderungen der Strömungsrichtung folgen und auch auftreten, wenn ein äußerer Druckunterschied das Strömen des Plasmas verhindert. Abb. 6 zeigt zwei solche Beobachtungen von KAMIYA und ABE. Es handelt sich also bei beiden Myxomyceten um Schwankungen der elektrischen Spannung im Rhythmus der Strömung, nur mit verschiedenen Phasenverschiebungen bei den beiden Arten. Obzwar die Veränderungen der elektrischen Spannung und der Strömung, auch nach Ansicht von KAMIYA und ABE, nicht im Verhältnis unmittelbarer Ursache und Wirkung stehen, scheint doch ein innerer Zusammenhang zu bestehen. Es ist bekannt, daß bei den Characeen mit den Aktionsströmen Strömungsstillstände und mit den lokalen Potentialen Strömungsverlangsamungen verbunden sind, wie es UMRATH (8) für *Nitella mucronata* beschrieben hat. Auch an *Tradescantia*-Staubfadenhaaren führen Reize zu auffälligen Strömungsstillständen. An *Physarum polycephalum* haben DIANNELIDIS und UMRATH (2) beobachtet, daß elektrische Reize von der Stärke, wie sie Aktionsströme auslösen, auch die Protoplasmaströmung beeinflussen, indem entweder die Strömungsrichtung ohne Stillstand wechselt oder ein Stillstand von Bruchteilen einer Sekunde bis zu 45 sec eintritt, dem eine Strömung in der alten Richtung oder in der Gegenrichtung folgt. Es liegt daher nahe, die von KAMIYA und ABE an *Physarum* und von WATANABE, KODATI und KINOSITA an *Didymium* beobachteten rhythmischen elektrischen Erscheinungen als spontane lokale Potentiale aufzufassen, die mit den rhythmischen Strömungswechseln verbunden sind.

Die mit dem Erregungsvorgang verbundene Widerstandsabnahme wurde bei *Physarum* schon erwähnt; viel eingehender ist sie bei *Nitella*-Zellen untersucht. COLE und CURTIS haben sorgfältige Messungen mit Wechselströmen durchgeführt. Da sich diese Wechselstrommessungen aber ohne Zuhilfenahme der Mathematik kaum nutzbringend besprechen lassen, will ich hier vor allem auf die anschaulicheren Befunde von UMRATH (14, 18) mit Gleichstromstößen wechselnder Richtung eingehen. Bei diesen wurde die Polarisationsspannung am Plasmalemma, die ja seinen Widerstand bedingt, direkt registriert.

In Abb. 7 ist ein Versuch von UMRATH (14) wiedergegeben, bei dem die Polarisationsspannung einer *Nitella*-Internodialzelle mit einem Oscillographen registriert ist, der eine starke zeitliche Auflösung ermöglicht, aber, weil seine

Verstärkerröhren durch Kondensatoren gekoppelt sind, eine eingeschaltete konstante Spannung statt mit einem konstanten, dauernd bestehenbleibenden Ausschlag, mit einem Ausschlag wiedergibt, der langsam zurückgeht, und zwar ungefähr in 5 sec auf $^1/_3$. In den drei übereinander dargestellten Kurven erfolgte

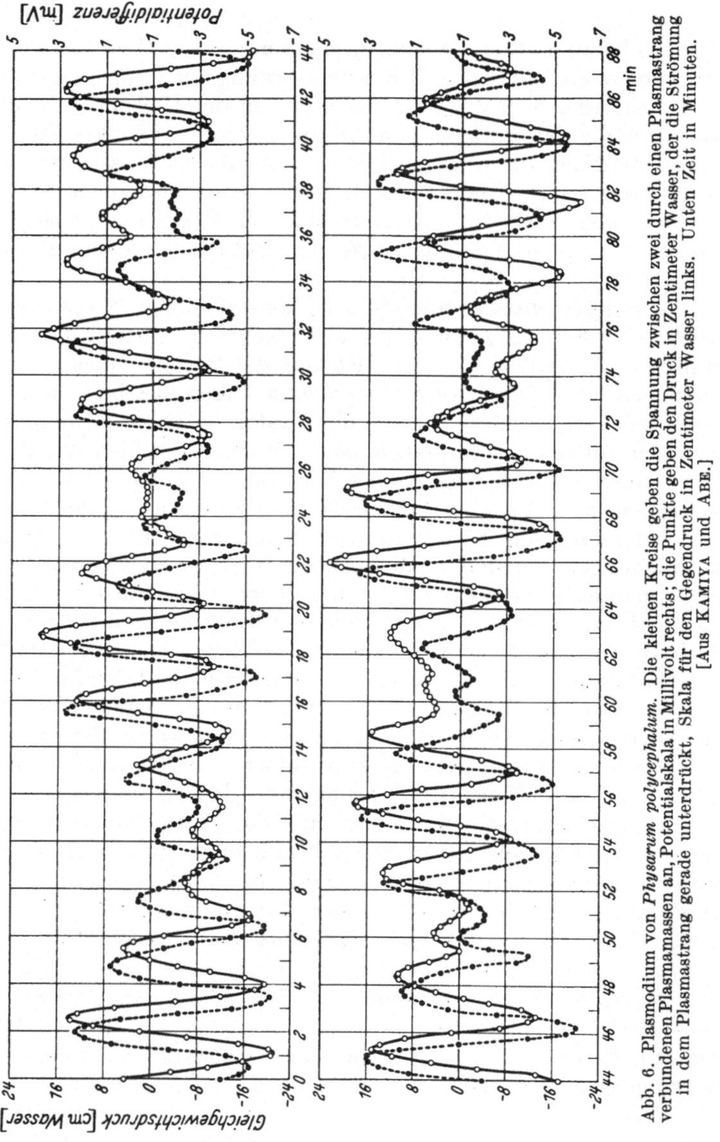

Abb. 6. Plasmodium von *Physarum polycephalum*. Die kleinen Kreise geben die Spannung zwischen zwei durch einen Plasmastrang verbundenen Plasmamassen an, Potentialskala in Millivolt rechts; die Punkte geben den Druck in Zentimeter Wasser, der die Strömung in dem Plasmastrang gerade unterdrückt, Skala für den Gegendruck in Zentimeter Wasser links. Unten Zeit in Minuten. [Aus KAMIYA und ABE.]

die Zuleitung des polarisierenden Stromes in der oberen und in der unteren durch zwei Elektroden im Außenmedium, in der mittleren durch eine in die Zelle eingestochene Elektrode und durch eine im Außenmedium. Die linken Teile der Kurven, vor dem Reiz, habe ich schon im 2. Abschnitt, S. 756, besprochen. Für die durch Ionenstauung am Plasmalemma bedingte kondensatorartige Komponente der Polarisation sind bei Zuleitung des polarisierenden Stromes durch eine in das Protoplasma eingestochene Elektrode, wegen der großen Widerstände, die Zeiten, in denen die Polarisation entsteht und vergeht, sehr lang; man sieht in der Abb. 7 die starke Rundung der Ausschläge an der mittleren Kurve. Bei Zuleitung des polarisierenden Stromes durch zwei Elektroden im Außenmedium muß, wegen der geringeren Widerstände, die Polarisationskomponente durch Ionenstauung sich sehr viel rascher einstellen. Die Komponente, die in der oberen und in der unteren Kurve das Bild beherrscht, weil sie jetzt die langsamere ist, ist die durch elektrisch polare, in der Grenzschicht bewegliche Molekeln bedingte, deren zeitliche Verhältnisse von den Widerständen unabhängig sind. Alle Kurven sind so registriert, daß ein Ausschlag nach oben einer Herabsetzung des Plasmalemmapotentials entspricht. Etwa 1 sec vor dem Reiz ist 0,1 V eingeschaltet, um die Kurve zu senken, damit beim Reiz der Aktionsstrom, der einen Ausschlag nach oben bedingt, die Kurve nicht aus dem empfindlichen Bereich des Oscillographen herausbringt. Der Rückgang nach dem Aktionsstrom erfolgt nicht bis auf das tiefe Niveau vor dem Reiz, weil die Anordnung einen langsam entstehenden Ausschlag nicht wiedergibt, so wie sie einen konstanten nicht dauernd beibehält.

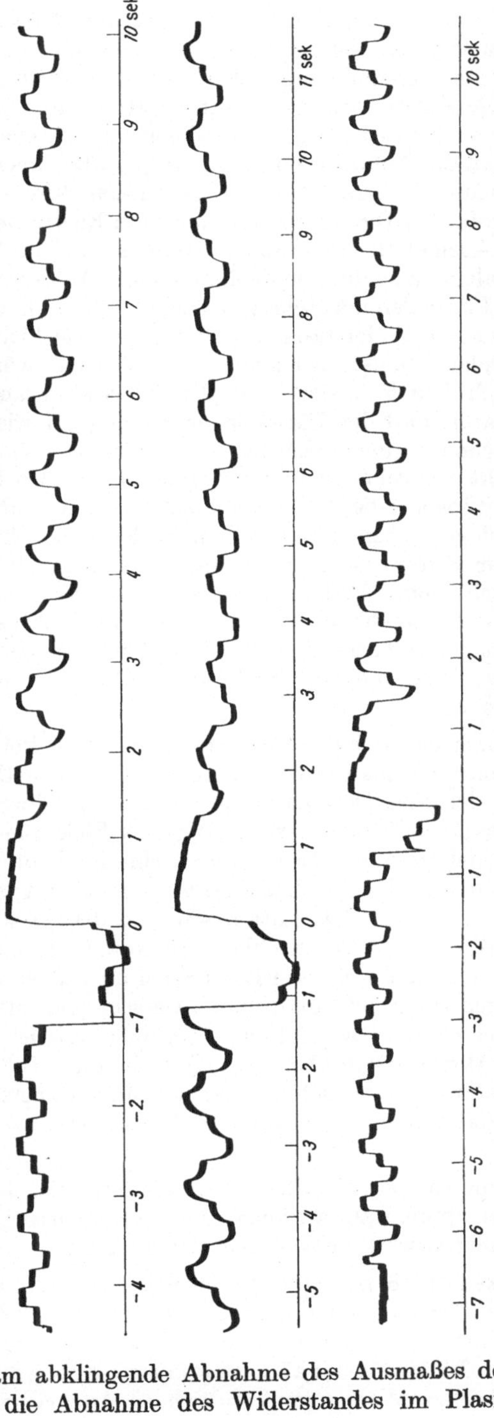

Was hier vor allem interessiert, ist die Veränderung der Polarisation während und nach dem Aktionsstrom. Bei Stromzuleitung durch eine in die Zelle eingestochene Elektrode, bei der mittleren Kurve, sieht man eine zunächst sehr starke und nur sehr langsam abklingende Abnahme des Ausmaßes der Polarisationsausschläge, die auf die Abnahme des Widerstandes im Plasmalemma

Abb. 7. Die durch Gleichstromstöße alternierender Richtung am Plasmalemma von *Nitella mucronata* erzeugte Polarisationsspannung. Oben: Durchströmung von außen; Mitte: Durchströmung von innen; die Stromstöße, die das Plasmalemmapotential durch Polarisation vergrößern und in der Abbildung die Ausschläge nach unten bedingen, haben die doppelte Spannung wie die umgekehrter Richtung. Unten: Durchströmung von außen, bei doppelt so starker Spannung wie oben. Überall etwa 1 sec vor dem Reiz Einschaltung einer Eichspannung von 0,1 V, damit der folgende Aktionsstrom die Kurve nicht zu hoch hebt, dann ein durch einen Öffnungsinduktionsschlag ausgelöster Aktionsstrom, dessen langsamer Rückgang nur zum Teil ersichtlich ist und zum Teil durch die Kondensatorkoppelung im Meßgerät ausgeglichen wird. Zeitmarken sec, Reizmoment mit 0 bezeichnet. [Aus UMRATH (14).]

zurückzuführen ist. Dieselbe Widerstandsabnahme bedingt auch das jetzt rasche Entstehen und Vergehen der Polarisation. Bei Stromzuleitung mit zwei Elektroden im Außenmedium, bei den Kurven oben und unten, ist während des Aktionsstroms die Abnahme im Ausmaß der Polarisation auch sehr groß, aber nicht ganz so groß, wie bei Stromzuleitung mit einer eingestochenen Elektrode und einer im Außenmedium. Nach COLE und CURTIS nimmt der Widerstand einer *Nitella*-Zelle während des Aktionsstroms auf $^1/_{200}$ ab. Wenn das bei *Nitella mucronata* auch so ist, so muß bei Zuleitung mit einer eingestochenen Elektrode das Ausmaß der Polarisation etwa auf $^1/_{200}$ abnehmen, denn der Widerstand der eingestochenen Elektrode ist auch gegen den Widerstand des ruhenden Plasmalemmas groß. Bei Durchströmung mit zwei Elektroden im Außenmedium kommt es für das Ausmaß der Polarisation vor allem auf das Verhältnis des Widerstandes am Plasmalemma zu dem im Zellinneren an. An der ruhenden Zelle ist der Widerstand des Plasmalemmas sehr viel größer. Während des Aktionsstrommaximums ist offenbar der Widerstand des Plasmalemmas kleiner, aber nicht um so viel, daß die Polarisationsausschläge ganz unsichtbar würden. In der Zeit bis etwa $1^1/_2$ sec nach dem Reiz sieht man die Polarisationsausschläge größer werden, als Zeichen dafür, daß der Plasmalemmawiderstand wieder zunimmt. Er ist jetzt offenbar schon wieder wesentlich größer als der Widerstand im Zellinnern, deswegen wirkt sich seine weitere Zunahme bis auf den Ruhewert bei dieser Art der Stromzuleitung auf die Polarisation nicht mehr aus, obzwar sie bei Stromzuleitung mit einer eingestochenen Elektrode noch deutlich in Erscheinung tritt. Auffallend ist, daß bei Stromzuleitung mit zwei Elektroden im Außenmedium die Polarisation nach dem Aktionsstrom ein größeres Ausmaß hat und langsamer entsteht und vergeht als vorher. Dies ist wohl so zu deuten, daß die Beweglichkeit von elektrisch polaren Molekeln in der Grenzschicht geringer geworden ist, oder daß zusätzlich solche Molekeln mit geringerer Beweglichkeit aufgetreten sind.

In Abb. 8 und 9 sind Plasmalemmapotential, Polarisation und Aktionsstrom bei langsamerer Registrierung mit dem LINDEMANN-Elektrometer wiedergegeben. Die Form der Polarisationszacken ist hier allerdings durch die geringere Einstellungsgeschwindigkeit des LINDEMANN-Elektrometers bedingt. In Abb. 8 wurde der polarisierende Strom durch eine ins Protoplasma eingestochene Elektrode und durch eine im Außenmedium zugeleitet; man erkennt auch hier die starke Abnahme der Polarisation während des Aktionsstroms und ihre nur allmähliche Wiederzunahme nachher. In Abb. 9 erfolgte die Zuleitung des polarisierenden Stromes durch zwei Elektroden im Außenmedium und in Abb. 9a, d, e und h waren die Zellen in Aquarienwasser; man sieht, daß der Rückgang der Polarisation während des Aktionsstroms hier geringer ist und daß die Polarisation nach dem Aktionsstrom über lange Zeit verstärkt ist. Es scheint, daß die Polarisation verstärkt bleibt, solange der von außen zugeleitete polarisierende Strom fließt und daß sich der ursprüngliche Zustand erst nach Stromabschaltung wieder herstellt.

Ich bespreche jetzt Einflüsse veränderten Salzgehaltes im Außenmedium auf Plasmalemmapotential, Aktionsstrom und Polarisierbarkeit, die in diesen drei Fällen beachtenswert verschieden sein können.

Wie UMRATH (18) gezeigt hat, wird in destilliertem Wasser gegenüber Aquarienwasser das Plasmalemmapotential um 22% vergrößert, der Aktionsstrom um 6% verkleinert und die Polarisierbarkeit bei Stromzuleitung mit einer eingestochenen Elektrode um 27% vergrößert. Abb. 8b zeigt ein typisches Beispiel von destilliertem Wasser im Vergleich zu Aquarienwasser in Abb. 8a und c.

Bei Zusatz von NaCl oder KCl werden Plasmalemmapotential, Aktionsstrom und Polarisierbarkeit bei Stromzuleitung mit einer eingestochenen Elektrode ver-

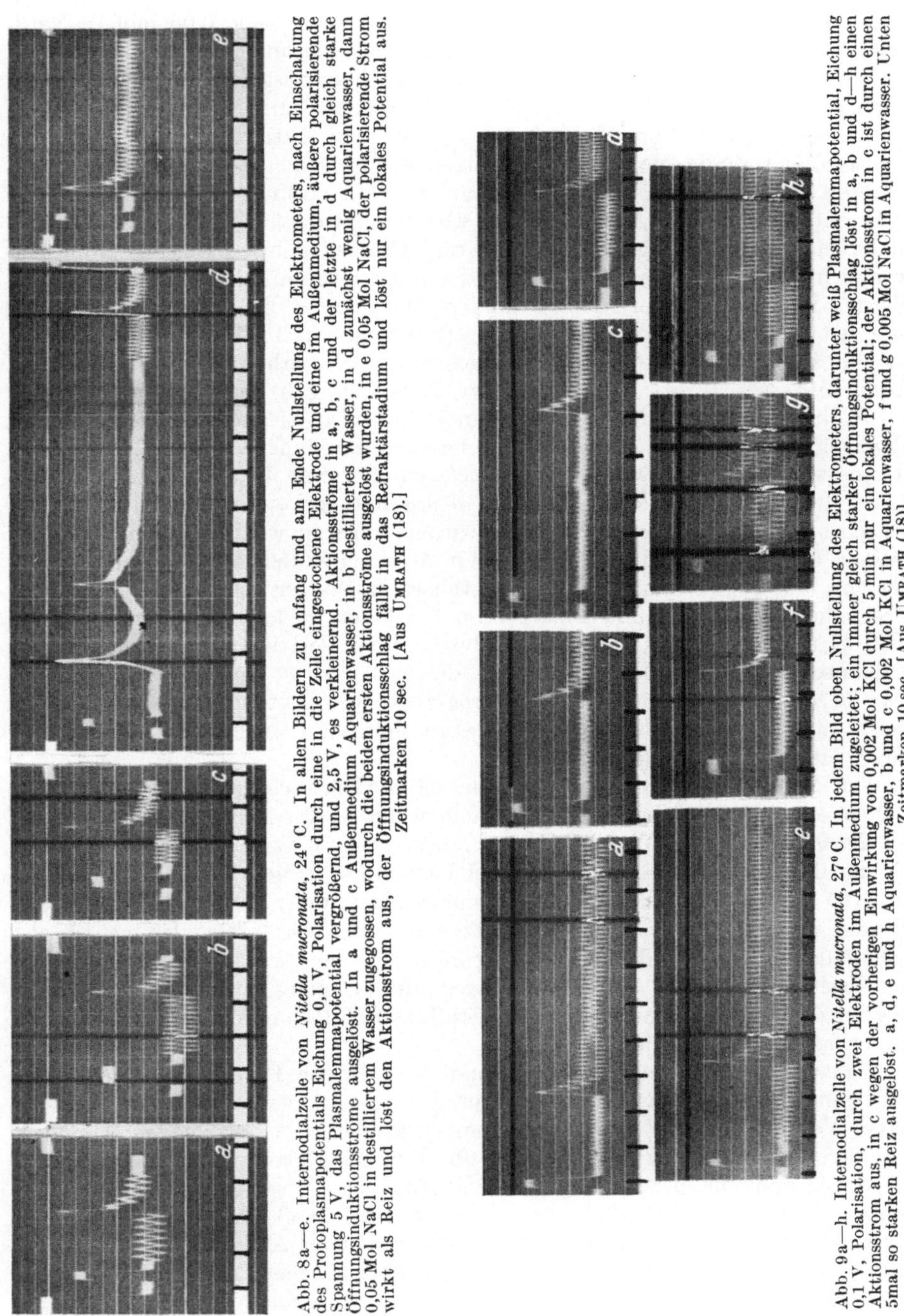

Abb. 8a—e. Internodialzelle von *Nitella mucronata*, 24° C. In allen Bildern zu Anfang und am Ende Nullstellung des Elektrometers, nach Einschaltung des Protoplasmapotentials Eichung 0,1 V, Polarisation durch eine in die Zelle eingestochene Elektrode und eine im Außenmedium, äußere polarisierende Spannung 5 V, das Plasmalemmapotential vergrößernd, und 2,5 V, es verkleinernd. Aktionsströme in a, b, c und der letzte in d durch gleich starke Öffnungsinduktionsströme ausgelöst. In a und c Außenmedium Aquarienwasser, in b destilliertes Wasser, in d zunächst wenig Aquarienwasser, dann 0,05 Mol NaCl in destilliertem Wasser zugegossen, wodurch die beiden ersten Aktionsströme ausgelöst wurden, in e 0,05 Mol NaCl, der polarisierende Strom wirkt als Reiz und löst den Aktionsstrom aus, der Öffnungsinduktionsschlag fällt in das Refraktärstadium und löst nur ein lokales Potential aus. Zeitmarken 10 sec. [Aus UMRATH (18).]

Abb. 9a—h. Internodialzelle von *Nitella mucronata*, 27° C. In jedem Bild oben Nullstellung des Elektrometers, darunter weiß Plasmalemmapotential, Eichung 0,1 V, Polarisation, durch zwei Elektroden im Außenmedium zugeleitet; ein immer gleich starker Öffnungsinduktionsschlag löst in a, b und d—h einen Aktionsstrom aus, in c wegen der vorherigen Einwirkung von 0,002 Mol KCl durch 5 min nur ein lokales Potential; der Aktionsstrom in c ist durch einen 5mal so starken Reiz ausgelöst. a, d, e und h Aquarienwasser, b und c 0,002 Mol KCl in Aquarienwasser, f und g 0,005 Mol NaCl in Aquarienwasser. Unten Zeitmarken 10 sec. [Aus UMRATH (18)].

ringert, bei NaCl alle in etwa gleichem Maß; KCl wirkt stärker als NaCl und verringert Aktionsstrom und Polarisierbarkeit besonders stark. Beide können bei der ersten Einwirkung zu spontanen Aktionsströmen führen, aber NaCl in höherem Maß, und es führt noch sehr lange zu einer Steigerung der Erregbarkeit, während KCl sie verringert. In Abb. 8d sind die beiden ersten Aktionsströme dadurch ausgelöst worden, daß zu der in wenig Wasser liegenden Zelle 0,05 molare NaCl-Lösung zugegeben wurde; dann wurde die Polarisation eingeschaltet und ein Aktionsstrom durch einen Öffnungsinduktionsschlag ausgelöst. In Abb. 8e war die Zelle noch in der NaCl-Lösung; ihre Erregbarkeit war so gesteigert, daß schon der polarisierende Strom als Reiz wirkte und den Aktionsstrom auslöste.

Bei Zuleitung des polarisierenden Stromes durch zwei Elektroden im Außenmedium beobachtete UMRATH (18) nur in NaCl-Lösung keine Zunahme der Polarisation nach dem Aktionsstrom. Abb. 9 zeigt diese Verhältnisse. Abb. 9b und c zeigen die Wirkung von 0,002 molarer KCl-Lösung im Vergleich zu Aquarienwasser in a und d. Man erkennt, daß der Aktionsstrom stärker herabgesetzt ist als das Plasmalemmapotential, besonders bei längerer Einwirkung der Lösung in c, wo auch die Erregbarkeit herabgesetzt ist. In NaCl-Lösung, Abb. 9f und g, ist die Polarisation nach dem Aktionsstrom nicht erhöht, im Gegensatz zu Aquarienwasser, a, d, e und h, und zu KCl-Lösung b und c.

Während man aus den ersten Befunden, auch an tierischen Objekten, wohl allgemein geschlossen hat, daß die Widerstandsabnahme während des Aktionsstroms sehr erheblich ist, hat in der letzten Zeit eine Untersuchung von UMRATH (18) gezeigt, daß sie bei *Nitella opaca* nur gering ist. Abb. 10 zeigt bei *Nitella opaca* eine während des Aktionsstroms nur sehr wenig verringerte Polarisation. In Abb. 10a und b war die Zelle in Aquarienwasser, in c, d und e in destilliertem Wasser; man erkennt bei destilliertem Wasser die Vergrößerung von Plasmalemmapotential und Polarisation bei kaum verändertem Aktionsstrom. Abb. 11 zeigt einen Fall, in dem die Abnahme der Polarisation im aufsteigenden gedehnten Aktionsstromteil 40% war, die größte sonst beobachtete Abnahme war 70%. Während der steilen Anfangszacke des Aktionsstroms von *Nitella opaca* ist die Polarisierbarkeit schwer abzuschätzen, UMRATH (18) hält sie aber nicht für stark vermindert.

Um Mißverständnissen vorzubeugen, sei ausdrücklich erwähnt, daß auch in Fällen sehr starker Widerstandsabnahme während des Aktionsstroms, wie bei *Nitella mucronata*, der Widerstand des Plasmalemmas zwar kleiner werden kann als der des ganzen Zellinhaltes, der bei Längsdurchströmung der Zelle wirksam ist, daß der Widerstand des Plasmalemmas aber immer sehr viel größer bleibt als der einer gleich dicken Schicht Protoplasma. Das Plasmalemma ist also jedenfalls auch während der Erregung noch eine sehr wirksame Barriere gegen den Verlust von Stoffen, und wenn während der Erregung gewisse Stoffe vermehrt aus der Zelle austreten, so bleibt dieser Stoffaustritt doch immer sehr viel geringer, als er bei freier Diffusion wäre.

UMRATH (18) hat an *Nitella*-Internodialzellen auch Polarisationsspannungen angewandt, die das Plasmalemmapotential sehr stark verändert haben und hat dabei Veränderungen des Aktionsstroms gefunden, die verschieden waren, je nachdem, ob der polarisierende Strom durch eine ins Protoplasma eingestochene Elektrode und eine im Außenmedium zugeleitet wurde, wobei das Plasmalemmapotential an der ganzen Zellgrenzschicht in nahezu gleicher Weise verändert wurde, oder ob der Strom durch zwei Elektroden im Außenmedium zugeleitet wurde, wobei das Plasmalemmapotential am kathodischen Zellende herabgesetzt und am anodischen erhöht wurde. Wird das Plasmalemmapotential an der ganzen Zelle herabgesetzt, so führt der Aktionsstrom doch nie zu einer Potentialumkehr, sondern höchstens zu einer Verminderung des Potentials bis auf Null. Wird schon durch die Polarisation das Potential umgekehrt, so läßt sich kein

Aktionsstrom mehr auslösen. Wird das Plasmalemmapotential an der ganzen Zelle erhöht, so kann auch der Aktionsstrom vergrößert werden. Bei starker Erhöhung des Potentials bleibt die Vergrößerung des Aktionsstroms hinter ihr

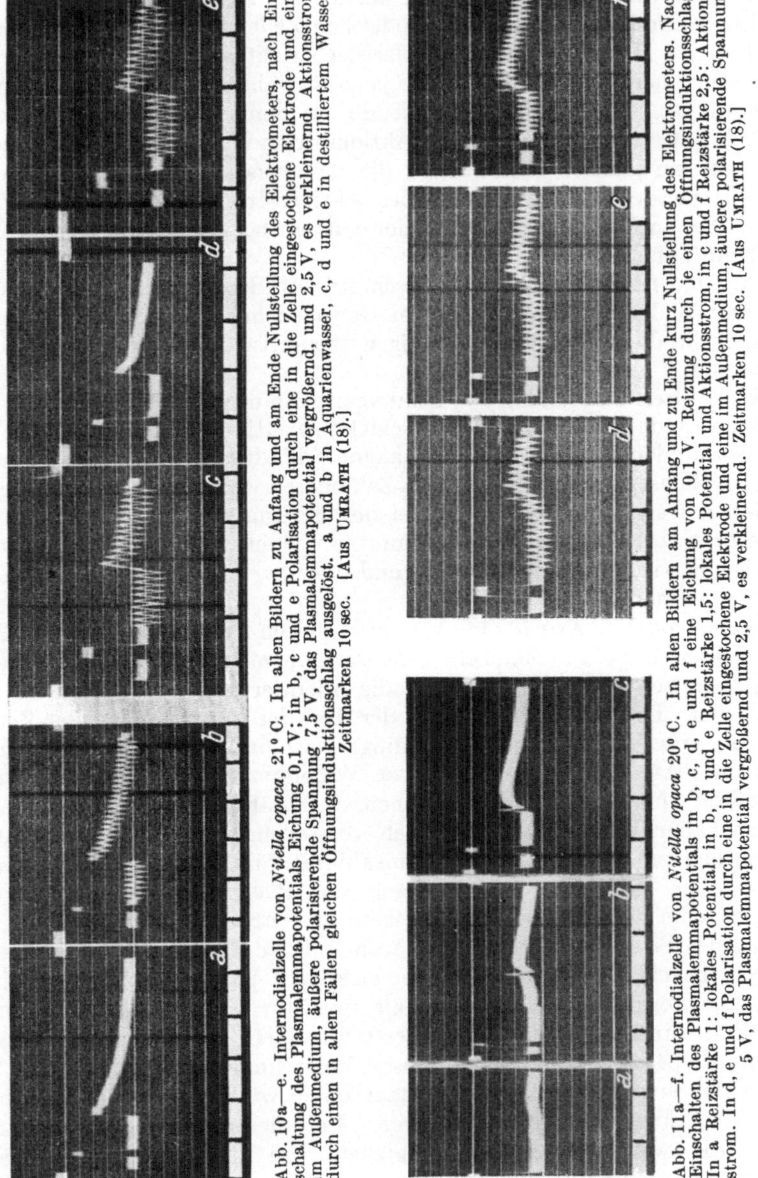

Abb. 10a—e. Internodialzelle von *Nitella opaca*, 21° C. In allen Bildern zu Anfang und am Ende Nullstellung des Elektrometers, nach Einschaltung des Plasmalemmapotentials Eichung 0,1 V; in b, c und e Polarisation durch eine in die Zelle eingestochene Elektrode und eine im Außenmedium, äußere polarisierende Spannung 7,5 V, das Plasmalemmapotential vergrößernd, und 2,5 V, es verkleinernd. Aktionsstrom durch einen in allen Fällen gleichen Öffnungsinduktionsschlag ausgelöst. a und b in Aquarienwasser, c, d und e in destilliertem Wasser. Zeitmarken 10 sec. [Aus UMRATH (18).]

Abb. 11a—f. Internodialzelle von *Nitella opaca* 20° C. In allen Bildern am Anfang und zu Ende kurz Nullstellung des Elektrometers. Nach Einschalten des Plasmalemmapotentials in b, c, d, e und f eine Eichung von 0,1 V. Reizung durch je einen Öffnungsinduktionsschlag. In a Reizstärke 1: lokales Potential, in b, d und e Reizstärke 1,5: lokales Potential und Aktionsstrom, in c und f Reizstärke 2,5: Aktionsstrom. In d, e und f Polarisation durch eine in die Zelle eingestochene Elektrode und eine im Außenmedium, äußere polarisierende Spannung 5 V, das Plasmalemmapotential vergrößernd und 2,5 V, es verkleinernd. Zeitmarken 10 sec. [Aus UMRATH (18).]

zurück. Bei *Nitella mucronata* kann es vorkommen, daß ein Aktionsstrom, der schon ohne Polarisation nur einen Teil des Plasmalemmapotentials rückgängig macht, bei einer Vergrößerung des Plasmalemmapotentials nicht vergrößert wird. An Zellen, deren Aktionsströme durch Narkose verkleinert sind, kann man das regelmäßig beobachten.

Ist, bei Stromzuleitung von außen, die polarisatorische Veränderung des Plasmalemmapotentials an beiden Enden einer *Nitella*-Internodialzelle entgegengesetzt, so führt, bei starker Polarisation, der Aktionsstrom am kathodischen Zellende zu einer Potentialumkehr, ja es kann noch zu einem Aktionsstrom kommen, wenn schon die Polarisation allein zu einer gewissen Potentialumkehr führt. Der Aktionsstrom ist am kathodischen Zellende von längerer Dauer und während seines Maximums ist die Polarisierbarkeit auch bei *Nitella mucronata* nur sehr wenig herabgesetzt, bei der sie ja sonst während des Aktionsstroms stark herabgesetzt ist. Am anodischen Zellende, an dem das Plasmalemmapotential durch Polarisation erhöht ist, ist der Aktionsstrom bei geringer Polarisation vergrößert und bei starker Polarisation, die das Plasmalemmapotential mehr als verdoppelt, verkleinert. Die Dauer des Aktionsstroms ist an diesem Zellende verkürzt. Bei noch stärkerer Polarisation tritt an beiden Zellenden kein Aktionsstrom mehr auf.

Ich komme noch im nächsten Abschnitt auf Beobachtungen über Aktionsströme und Widerstandsänderungen in Geweben höherer Pflanzen zu sprechen und beschließe dieses Kapitel mit einigen theoretischen Bemerkungen über den Aktionsstrom.

Nach der alten Vorstellung von Bernstein kommt das Plasmalemmapotential dadurch zustande, daß die Zellgrenzschicht selektiv kationenpermeabel ist, so daß die Kaliumionen durch das Plasmalemma hindurchdiffundieren können und an dessen Außenseite durch die aus dem Zellinneren wirkenden elektrischen Kräfte der Anionen festgehalten werden. Bei der Erregung nimmt die Permeabilität der Membran zu, ihr Widerstand nimmt ab, auch Anionen können die Zelle ähnlich leicht wie Kationen verlassen und daher verschwindet die Spannung an der Zellgrenzschicht.

Nach der von Umrath (8, 14, 15) entwickelten Vorstellung sind im Plasmalemma elektrisch polare Molekeln orientiert eingelagert, die beim Erregungsvorgang chemisch reagieren, wodurch die orientierten, elektrisch polaren Teile verschwinden. Dadurch verschwindet der Teil der Spannung an der Zellgrenzschicht, der durch diese Molekeln bedingt war, und durch die Lockerung des Gefüges im Plasmalemma nimmt dessen Widerstand ab und es kann zu einer Permeabilitätszunahme für Ionen kommen, die zusätzliche Spannungsänderungen zur Folge haben kann. Man kann sich vorstellen, daß es dabei bei manchen Pflanzenzellen zu einer so hohen Permeabilität für Anionen kommt, daß deswegen auf der Höhe des Aktionsstroms das Plasmalemmapotential so stark zurückgeht oder verschwindet. Man kann sich aber auch vorstellen, daß der Rückgang des Plasmalemmapotentials während der Erregung nahezu ausschließlich durch die chemische Reaktion der elektrisch polaren Molekeln bedingt ist und daß die Permeabilität des Plasmalemmas für Kaliumionen und seine Impermeabilität für Anionen zur Wiederherstellung der Spannung am Plasmalemma führt. Erst später würden nach dieser Vorstellung die Kaliumionen wieder durch elektrisch polare Molekeln ersetzt; dabei würde im relativen Refraktärstadium die Erregbarkeit wiederkehren. Bei dieser Vorstellung erklärt sich zwanglos die weitgehende Unabhängigkeit von Plasmalemmapotential und Aktionsstrom in ihren Ausmaßen. Der während des Aktionsstroms bestehenbleibende Teil des Plasmalemmapotentials ist bei *Nitella opaca* meist beträchtlich, bei *Nitella mucronata* ist es im wesentlichen der in destilliertem Wasser zusätzlich entstehende Teil. Dieser Teil des Plasmalemmapotentials kann entweder durch andersartige elektrisch polare Molekeln, die während des Erregungsvorgangs nicht chemisch reagieren, oder durch die selektive Kationenpermeabilität des Plasmalemmas bedingt sein.

6. Elektrische Spannungen an Geweben und deren Veränderungen bei der Erregung und durch elektrische Ströme.

Wir haben aus der Untersuchung von LUND an der Fadenalge *Pithophora* schon gesehen, daß an Zellfäden in Luft verschiedene Stellen der Außenseite einer Zelle erhebliche Spannungen gegeneinander haben können, die auf Verschiedenheiten des Plasmalemmas an den betreffenden Stellen beruhen müssen. Auf diese Art können auch Spannungsunterschiede zwischen verschiedenen Zellen des Fadens bedingt sein. LUND hat die apikalen Zellen elektropositiv gegenüber den basalwärts gelegenen gefunden, so daß die wachsende Spitzenzelle am stärksten positiv war, aber auch seitlich auswachsende Zellenden waren gegenüber ihrer Umgebung elektrisch positiv. Wie auch Abb. 2 zeigt, sind die Zellgrenzen oft Stellen von Potentialmaxima, also Orte ohne Spannungsabfall, und der basalwärts gerichtete Spannungsabfall im Faden ist im wesentlichen durch die Spannungsabfälle in den apikalen Teilen der Zellen bedingt.

UMRATH (16) ist durch eine Untersuchung an *Nitella*-Sprossen, deren eingehende Besprechung hier zu weit führen würde, zu der Auffassung gekommen, daß die Plasmalemmapotentiale und die Aktionsströme der Zellen gegenüber Wasser als Außenmedium größer sind als gegen die Nachbarzellen, und daß, wenn zwischen zwei benachbarten Zellen der Widerstand des Außenmediums groß genug ist, der Unterschied ihrer Plasmalemmapotentiale gegen das Außenmedium größer ist als der ihrer Plasmalemmapotentiale gegeneinander. Da das natürliche Außenmedium gegenüber dem Protoplasma elektrisch positiv ist, sind demnach in einem Gewebeverband Stellen mit höheren Plasmalemmapotentialen der Zellen elektrisch positiv gegenüber Stellen mit niedrigeren Plasmalemmapotentialen der Zellen. Nach dieser Auffassung von UMRATH ist zu erwarten, daß Spannungen zwischen zwei Stellen eines Gewebes zurückgehen, wenn das ganze Gewebe erregt wird, denn an der durch höheres Plasmalemmapotential positiven Stelle geht die Spannung durch den hier größeren Aktionsstrom stärker zurück.

Bei *Chara vulgaris* hat BROWN an Sprossen, die aus dem Wasser in wassergesättigte Luft ragten, Spannungen an den drei obersten Internodien gemessen. Die obersten, stark wachsenden Teile waren gegen weiter basale elektrisch positiv. Am apikalen Teil eines Internodiums angebrachte mechanische Reize negativierten diesen Teil und verringerten damit die Spannung oder kehrten sie um. Partielle Beleuchtung positivierte die belichtete Stelle, Beleuchtung der ganzen Pflanze verstärkte die apikale Positivität. BROWN hat gezeigt, daß das Licht über die Assimilation wirkt, indem der Effekt mit der Beleuchtungsintensität zuinmmt, im roten und blauen Licht vorhanden ist, im grünen aber nicht und bei Fehlen von Sauerstoff mit der Kohlendioxydkonzentration zunimmt. Offenbar wird zur Aufrechterhaltung der elektrischen Spannungen Energie gebraucht.

An Zwiebelwurzeln hat MARSH (2) gefunden, daß mechanische Reizung die gereizte Stelle elektrisch negativ macht und, bei Ableitung von zwei Stellen, zwischen denen die gereizte Stelle liegt, die apikale Ableitungsstelle negativiert. Es handelt sich also um einen Aktionsstrom an der gereizten Stelle, der sich nicht ausbreitet, und dieser Aktionsstrom ist gegen den apikalen Nachbarteil von größerem Ausmaß als gegen den basalen, so daß der apikale negativiert wird. Später haben AMLONG und BÜNNING an Wurzeln von *Helianthus annuus* bei elektrischer Reizung Aktionsströme und Widerstandsabnahmen nachgewiesen.

Schwache elektrische Ströme erzeugen an Zwiebelwurzeln entgegengerichtete, der Stromstärke proportionale Polarisationsspannungen, wie MARSH (1) und später HOYT (1) gezeigt haben. HOYT (2) hat auch die Wirkung stärkerer Ströme untersucht und hat dabei periodische Schwankungen der Polarisationsspannung und damit auch der Stromstärke gefunden. Wahrscheinlich handelt es sich um Aktionsströme und die sie begleitenden Widerstandsänderungen.

An *Pisum*-Sprossen und an *Avena*-Coleoptilen hat CLARK gefunden, daß Reizung mit Wechselstrom die apikale Zone negativiert und den Widerstand vermindert. Bei Anwendung schwachen Gleichstroms fand er entgegengerichtete Polarisationsspannungen.

An Sprossen und Blattstielen von *Phaseolus multiflorus* fand REHM (2) bei schwachen Strömen entgegengerichtete Polarisationsspannungen, bei stärkeren Strömen, über einer gewissen Schwelle, eine apikale Negativierung, welche die anfänglich bestehende apikale Negativität vergrößerte.

7. Der elektrische Faktor der Wasserbewegung und sonstiger Stoffbewegungen.

BERNSTEIN führt in seiner Elektrobiologie aus, daß das Membranpotential (= Plasmalemmapotential) der Zellen einen Beitrag in ihrer wasserbindenden Kraft liefert. Wenn, wie es in Abb. 1 dargestellt ist, die Membransubstanz negativ geladen ist, dann sind die Wasserfäden in der Membran positiv geladen, so daß eine von außen angelegte Spannung und ebenso das thermodynamische (ε) Potential an der Membran eine Wasserbewegung von der positiven zur negativen Seite der Membran bewirkt. Auf diese elektrisch bedingte Wasserbewegung sind Ionen von Einfluß, insbesondere wird sie durch alkalische Reaktion gefördert und durch saure gehemmt, bis es zu einer Umladung der Membran durch Säure und damit zu einer umgekehrten Wasserbewegung von der negativen zur positiven Seite der Membran kommt. Bei Pflanzenzellen mit festen Zellmembranen führt ein elektrisch bedingter Wassereinstrom zu einer Erhöhung des Turgors, bis der Turgor den rein osmotischen und den elektrischen Kräften das Gleichgewicht hält.

Untersuchungen, die zur Annahme einer elektrisch bedingten „anomalen Komponente des osmotischen Potentials" geführt haben, wurden an verschiedenen Pflanzenzellen von BRAUNER und HASMAN und von STUDENER ausgeführt. Für die Auffassung, daß es sich wirklich um eine elektroosmotische Komponente des Turgors handelt, scheint mir vor allem der Befund von STUDENER zu sprechen, daß gerade sehr geringe Salzzusätze zu einer Glucoselösung, die das Plasmalemmapotential schon beträchtlich vermindern, regelmäßig einen größeren Effekt haben, als es nach den rein osmotischen Verhältnissen zu erwarten wäre. Bei größeren Salzzusätzen scheint nach den Befunden von STUDENER, trotz der zunächst noch etwas größeren Wirkung auf das Plasmalemmapotential, der elektroosmotische Effekt vielfach durch andere Wirkungen der Salze verdeckt.

DIANNELIDIS fand Nektarien mit dünnflüssigem Nektar elektrisch negativ, solche mit dickflüssigem Nektar elektrisch positiv gegenüber dem benachbarten Gewebe. Er hat auch gezeigt, daß diese Spannungsunterschiede nicht Folge des verschieden konzentrierten Nektars sind. Die Wasserdrüsen von *Lathraea* hat schon NOLD und später auch DIANNELIDIS entsprechend ihrem wäßrigen Sekret elektrisch negativ gefunden. Ungereizte Drüsenköpfchen von *Drosera* fand DIANNELIDIS gegenüber der Blattunterseite elektrisch negativ. Nach Aufbringen von Proteinpulver nahm diese negative Spannung zunächst zu, beim physiologischen Abtrocknen nach einigen Tagen sank sie auf sehr niedrige Werte ab. An den Blasen von *Utricularia vulgaris*, die Wasser von innen nach außen transportieren, hat schon NOLD das Blaseninnere positiv gefunden, und DIAN-

NELIDIS hat mit verbesserter Methodik im Blaseninneren im Mittel +108 mV gemessen. DIANNELIDIS hat auch gefunden, daß 0,1 molare Kaliumnitratlösung, welche die elektrische Spannung zwischen Außenmedium und Blaseninnerem aufhebt, auch den Wassertransport stark reduziert oder aufhebt, während 0,1 molare Glucoselösung den Wassertransport bestehen läßt.

Die Befunde von DIANNELIDIS scheinen mir sehr für einen Zusammenhang der elektrischen Spannungen mit dem Wassertransport zu sprechen. DIANNELIDIS selbst diskutiert hierfür zwei Möglichkeiten. Es kann eine elektroosmotisch bedingte Komponente der Saugkraft der Zellen wirksam sein und es kann, bei einem Wassertransport in den Cellulosemembranen außerhalb der Zellen, die Spannung in den Zellmembranen als treibende Kraft wirken. Im ersteren Fall hätte man sich bei der *Utricularia*-Blase den Transport durch das Zellinnere folgendermaßen vorzustellen: Dort, wo die äußere und die innere Zellschicht der Blasenwand aneinanderstoßen, sind die Plasmalemmapotentiale gering und nicht sehr verschieden voneinander, das Plasmalemmapotential der Zellen der inneren Blasenwand gegen das Wasser im Blaseninneren ist aber etwa 110 mV größer als das Plasmalemmapotential der Zellen der äußeren Blasenwand gegen das Wasser außerhalb. Es besteht daher im ganzen eine elektroosmotische Kraft, die Wasser aus dem Blaseninneren nach außen befördert. In ähnlicher Weise mögen auch sonst in der Pflanze elektrische Spannungen am Wassertransport beteiligt sein.

Der Befund von DIANNELIDIS, daß Nektarien mit dickflüssigem Nektar elektrisch positiv gegenüber dem benachbarten Gewebe sind, um so mehr, je dickflüssiger ihr Nektar ist, kann entweder durch eine der positiven elektrischen Spannung des Nektariums parallelgehende Rückresorption von Wasser aus einem zunächst weniger konzentrierten Sekret gedeutet werden oder durch eine mit der positiven Spannung in Zusammenhang stehende Zuckerausscheidung, wie sie KELLER annimmt. Es ist durchaus möglich, daß auch sonstige Stoffbewegungen durch elektrische Spannungen gefördert oder verursacht werden.

8. Elektrische Spannungen und Wachstum.

Nach den Arbeiten von LUND, BROWN, LUND und KENYON, RAMSHORN (1, 2), THOMAS und nach den in dem Buch von LUND enthaltenen Arbeiten von LUND und seinen Mitarbeitern, insbesondere von WILKES und LUND, besteht bei oberirdischen und unterirdischen Pflanzenteilen im allgemeinen ein Zusammenhang zwischen elektrischer Potentialverteilung und Wachstum derart, daß die elektrisch positiven Teile am stärksten wachsen. Eine Ausnahme bildet der Befund von REHM (1), daß bei *Phaseolus multiflorus* nach Dekapitation eine austreibende Achselknospe gegenüber einer gehemmten elektrisch negativ ist und daß eine solche Potentialdifferenz in lateraler Richtung schon an der Basis des Internodiums besteht, das apikal die Knospen trägt. In letzter Zeit haben PILET und MEYLAN die Wurzeln von *Lens culinaris* untersucht. Sie fanden eine elektrisch positive Zone mit dem Potentialmaximum 2 mm von der Spitze, sie umfaßt das Gebiet des Teilungswachstums und größter Längenzunahme und enthält viel reduzierende Zucker und wenig Stärke. Die anschließende Zone des Streckungswachstums fanden sie gegenüber der Wurzelspitze elektrisch negativ, sie enthielt viel Stärke und wenig reduzierende Zucker. Die gegen die Wurzelbasis anschließende Zone fanden sie wieder elektrisch positiv, sie zeigte im Perizykel Anlagen von Nebenwurzeln und war wieder reich an reduzierenden Zuckern und arm an Stärke. Den höchsten Gehalt an Wuchsstoffen haben die beiden ersten Zonen, die größte laterale Wuchsstoffverschiebung findet in der Spitzenzone statt.

Es wurde mitunter angenommen, daß elektrische Kräfte für Wuchsstoffverschiebungen in der Pflanze verantwortlich sind. Befunde von CLARK schließen einen solchen Zusammenhang für den Längstransport des Wuchsstoffes aus. CLARK hat gezeigt, daß eine wäßrige Lösung von Natriumglykocholat 1:100000 den polaren Transport von Heteroauxin in der *Avena*-Coleoptile hemmt, ohne die elektrische Polarität, die Atmung, die Semipermeabilität, das Streckungswachstum oder die Protoplasmaströmung zu beeinflussen. Er hat weiter gezeigt, daß der laterale und der longitudinale Wuchsstofftransport durch verschiedene Mechanismen bedingt sind, da die Einflüsse von Licht, Schwerkraft, elektrischen Einwirkungen, Äthylen und Natriumglykocholat in beiden Fällen verschieden sind.

KOCH hat gefunden, daß Wuchsstoff in Agar zum positiven Pol wandert, wie das auch sonst Säureionen tun. An Wurzeln von *Pisum sativum* und an Hypokotylen der Keimlinge von *Helianthus annuus* und von *Lupinus* fand KOCH Krümmungen zum negativen Pol, die auf Wuchsstoffverschiebungen zum positiven schließen lassen. An *Avena*-Coleoptilen unter Wasser fand er aber Krümmungen zum positiven Pol. Er glaubt, daß wegen der isolierenden Cuticula durch Influenz an der dem negativen Pol zugekehrten Seite der Coleoptile ein positiver Pol entstehe, zu dem der Wuchsstoff wandere. Diese Vorstellung halte ich für falsch. Das von außen angelegte elektrische Feld kann durch Influenz Ladungen erzeugen, die das elektrische Feld im Inneren der Coleoptile höchstens aufheben, aber niemals umkehren können.

In einer Reihe von Arbeiten haben SCHRANK (1—5), SCHRANK und BACKUS und WEBSTER und SCHRANK den Einfluß elektrischer Ströme auf die Wachstumskrümmungen von *Avena*-Coleoptilen untersucht. Eine transversale elektrische Durchströmung bedingt eine Krümmung zum positiven Pol, also eine Wuchsstoffverschiebung zum negativen. WEBSTER und SCHRANK haben gezeigt, daß die elektrisch induzierten Wachstumskrümmungen auch an dekapitierten Coleoptilen auftreten, wenn sie mit Heteroauxin aus einem Agarwürfel versorgt werden. Dabei kann die Heteroauxinversorgung auch erst bis zu 4 Std nach der elektrischen Durchströmung erfolgen. Für die Wuchsstoffverschiebung ist also nicht der von außen zugeleitete elektrische Strom selbst notwendig, sondern eine durch ihn erzeugte, in der Pflanze lange bestehenbleibende Polarität. Die in dieser Zeit nach der Stromabschaltung noch vorhandenen elektrischen Spannungen in der Querrichtung konnten WEBSTER und SCHRANK auch an wuchsstofffreien, nicht wachsenden Colepotilen nachweisen. In einer mir erst bei der Korrektur dieses Artikels bekannt gewordenen Arbeit zeigen WIEGAND und SCHRANK, daß eine elektrische Längsdurchströmung einer Coleoptile eine elektrische Polarität in der Querrichtung bedingt, bei der die durchströmt gewesene Seite elektrisch negativ wird. Die elektrische Querpolarität ist an der Anode des sie erzeugenden Längsstromes größer als an seiner Kathode. Die wuchsstoffbedingten Krümmungen sind bei geringen Wuchsstoffmengen zum negativen Pol der Querpolarität gerichtet, was nach früheren Befunden von SCHRANK auch für andere tropistische Krümmungen gilt. Die elektrische Querpolarität verschwindet 50—70 min nach der elektrischen Längsdurchströmung. An wuchsstofffreien Coleoptilen werden maximale Krümmungen durch Wuchsstoffzufuhr nach 90 min erzeugt und Wuchsstoffzufuhr nach 140 min ist noch wirksam. WIEGAND und SCHRANK halten es für unsicher, ob die elektrische Querpolarität eine mittelbare Ursache der Krümmung ist.

UMRATH und WEBER haben gefunden, daß höhere Konzentrationen von Heteroauxin oder Colchicin, die an Wurzeln zu Keulenbildung führen und das

Längenwachstum sistieren, auch die für wachsende Wurzeln charakteristische Spannungsverteilung in der Längsrichtung aufheben.

Es bewirken also möglicherweise elektrische Spannungen, wenn auch wahrscheinlich indirekt, Wuchsstoffverschiebungen, andererseits verändert ein abnormer Wuchsstoffgehalt die Spannungen in der Pflanze.

9. Viscositätserhöhungen durch elektrische Ströme.

An Stärkescheidezellen von *Phaseolus multiflorus* haben Bersa und Weber bei elektrischer Gleichstromdurchströmung eine reversible Viscositätszunahme mit der Zentrifugierungsmethode festgestellt.

Tobias und Solomon haben an Blattzellen von *Helodea canadensis* beobachtet, daß elektrischer Gleichstrom die Brownsche Molekularbewegung der cytoplasmatischen Partikel am anodischen Ende der Zelle sistiert. Nach Stromöffnung erscheint die Molekularbewegung wieder, anfangend am kathodischen Ende des betroffenen Gebietes. Wenn der Strom stark genug war und genügend lange einwirkte, kam es in der ganzen Zelle zu einer Sistierung der Brownschen Bewegung.

Die Brownsche Molekularbewegung ist eine theoretisch gut bekannte Erscheinung und wenn sie durch verschiedene Reize reversibel sistiert wird, ist der Schluß auf eine reversible Viscositätserhöhung gerechtfertigt. Auch Stillstände der Protoplasmaströmung sind mitunter als Zeichen von Viscositätserhöhungen aufgefaßt worden. Doch ist ein solcher Schluß ungerechtfertigt, solange über die treibende Kraft der Protoplasmaströmung und über ihre Veränderung während des Strömungsstillstandes nichts bekannt ist.

Umrath (8) hat an Zellen von *Nitella mucronata* im Indifferenzstreifen zwischen den entgegengesetzt strömenden Plasmamassen ein Teilchen in Brownscher Molekularbewegung alle halben Minuten zeichnerisch festgehalten und danach die Viscosität nach der von Perrin angegebenen Methode des mittleren Verschiebungsquadrates berechnet. Es wurde durch einige Minuten vor dem Reiz und in den ersten 6 halben Minuten nach dem Reiz, der einen Strömungsstillstand auslöste, gemessen. Messungen an stärker beweglichen Teilchen ergaben für das Protoplasma etwa die 8fache Viscosität des Wassers, ähnlich wie sie Pekarek für das hyaline Protoplasma von *Chara*-Rhizoiden gefunden hatte, und keine sicher nachweisbare Viscositätsänderung während des Stillstandes, insbesondere keine Viscositätszunahme in den ersten 2 min nach dem Reiz. Aus Messungen an weniger beweglichen Teilchen würde sich eine 20mal so hohe Viscosität, etwa die 170fache des Wassers, ergeben. Entweder lagen diese Teilchen in einer äußeren Protoplasmaschicht mit einer sehr hohen Viscosität, oder sie klebten mit einem Teil ihrer Oberfläche am Plasmalemma oder an Chlorophyllkörnern und pendelten nur um eine Mittellage oder rollten etwas. Ihre Bewegungen waren in den ersten $1^1/_2$ min nach dem Reiz vielleicht geringfügig herabgesetzt, in den zweiten $1^1/_2$ min erhöht. Wenn diese Erhöhung nicht nur zufällig und somit scheinbar war, könnte man sie vielleicht mit den zu dieser Zeit auftretenden Glitschbewegungen und dem Wiederbeginn der Strömung in Zusammenhang bringen.

An Staubfadenhaaren von *Tradescantia virginica* hat Umrath (8) nach elektrischen Reizen vorübergehende Strömungsstillstände beobachtet, während denen die Molekularbewegung nicht merklich verändert war. Es ist anzunehmen, daß diese Reize Erregungsvorgänge auslösten und daß zur Sistierung der Molekularbewegung und damit zur Viscositätszunahme stärkere Reize notwendig waren.

Bei ihren Untersuchungen an *Physarum polycephalum* sind DIANNELIDIS und UMRATH (2) ebenfalls zu der Auffassung gekommen, daß elektrische Reize, die eben ausreichen, um Erregungsvorgänge mit Aktionsströmen und Veränderungen in der Protoplasmaströmung auszulösen, noch keine Viscositätszunahmen bedingen, sondern erst stärkere Reize, deren wiederholte Anwendung zu sichtbaren Schädigungen führt.

So scheinen bei durch Schwellenreize ausgelösten Erregungsvorgängen keine Viscositätszunahmen nachweisbar zu sein, wohl aber bei stärkeren oder lang dauernden elektrischen Reizen, deren Folgen zunächst noch reversibel sind.

Literatur.

ARMLONG, H. U., u. E. BÜNNING: Über elektromotorische Kräfte an elektrisch gereizten Wurzeln. Ber. dtsch. bot. Ges. **52**, 445—457 (1934).

BERNSTEIN, J.: Elektrobiologie. Die Wissenschaft, Heft **44**. Braunschweig: F. Vieweg & Sohn 1912. — BERSA, E., u. F. WEBER: Reversible Viskositätserhöhung des Cytoplasmas unter der Einwirkung des elektrischen Stromes. Ber. dtsch. bot. Ges. **40**, 254—258 (1922). — BEUTNER, R.: Die Entstehung elektrischer Ströme in lebenden Geweben und ihre künstliche Nachahmung durch synthetische organische Substanzen. Stuttgart: Ferdinand Enke 1920. — BLINKS, L. R.: (1) Carnegie Institution of Washington Year Book No 28, p. 277, 1929. — (2) Protoplasmic potentials in Halicystis. II. The effects of potassium on two species with different saps. J. Gen. Physiol. **16**, 147—156 (1932). — (3) A secondary bio-electric effect of potassium. Proc. Soc. Exper. Biol. a. Med. **30**, 756—757 (1933). — (4) Protoplasmic potentials in Halicystis. IV. Vacuolar perfusion with artifical sap and sea water. J. Gen. Physiol. **18**, 409—420 (1935). — BRAUNER, L., u. M. HASMAN: Untersuchungen über die anomale Komponente des osmotischen Potentials lebender Pflanzenzellen. Rev. Fac. Sci. Univ. Istanbul, Sér. B **11**, 1—37 (1946). — BROOKS, S. C., and S. GELFAN: Bioelectric potentials in Nitella. Protoplasma **5**, 86—96 (1928). — BROWN, S. O.: Relation between light and the electric polarity of Chara. Plant Physiol. **13**, 713—736 (1938). — BUSCH, G.: Über die photoperiodische Formänderung der Chloroplasten von Selaginella serpens. Biol. Zbl. **72**, 598—629 (1953).

CLARK, W. G.: Electrical polarity and auxin transport. Plant Physiol. **13**, 529—552 (1938). — COLE, K. S., and H. J. CURTIS: Electric impedance of Nitella during activity. J. Gen. Physiol. **22**, 37—64 (1938). — CURTIS, H. J., and K. S. COLE: Transverse electric impedance of Nitella. J. Gen. Physiol. **21**, 189—201 (1937).

DAMON, E. B.: Dissimilarity of inner and outer protoplasmic surface in Valonia. III. J. Gen. Physiol. **15**, 525—535 (1932). — DIANNELIDIS, TH.: Beitrag zur Elektrophysiologie pflanzlicher Drüsen. Phyton **1**, 7—23 (1948). — DIANNELIDIS, TH., u. K. UMRATH: (1) Aktionsströme der Blasen von Utricularia vulgaris. Protoplasma **42**, 58—62 (1953). — (2) Über das elektrische Potential und über den Erregungsvorgang bei dem Myxomyceten Physarum polycephalum. Protoplasma **42**, 312—323 (1953).

GICKLHORN, J., u. K. UMRATH: Messung elektrischer Potentiale pflanzlicher Gewebe und einzelner Zellen. Protoplasma **4**, 228—258 (1928).

HEILBRUNN, L. V.: An Outline of General Physiology, 3. edit. Philadelphia u. London: W. B. Saunders Company 1952. — HEILBRUNN, L. V., and K. DAUGHERTY: The electric charge of protoplasmic colloids. Physiologic. Zool. **12**, 1—12 (1939). — HILL, S. E., and W. J. V. OSTERHOUT: (1) Calculation of bioelectric potentials. II. The concentration potential of KCl in Nitella. J. Gen. Physiol. **21**, 541—556 (1938). — (2) Delayed potassium effect in Nitella. J. Gen. Physiol. **22**, 107—113 (1938). — HOYT, R. C.: (1) Counter-EMFs in onion root. J. Cellul. Comp. Physiol. **29**, 109—130 (1947). — (2) Potential oscillations in the onion root resulting from current flow. J. Cellul. Comp. Physiol. **29**, 131—147 (1947).

IWAMURA, T.: Electric impedance of the plasmodium of a slime mold, Physarum polycephalum. Cytologia (Tokyo) **17**, 322—328 (1952).

KAMIYA, N., and SH. ABE: Bioelectric phenomena in the myxomycete plasmodium and their relation to protoplasmic flow. J. Colloid Sci. **5**, 149—163 (1950). — KELLER, R.: Die Elektrizität in der Zelle, 3. Aufl. Mährisch-Ostrau: Julius Kittls Nachf. 1932. — KOCH, K.: Untersuchungen über den Quer- und Längstransport des Wuchsstoffes in Pflanzenorganen. Planta (Berl.) **22**, 190—220 (1934). — KÜMMEL, K.: Elektrische Potentialdifferenzen an Pflanzen. Planta (Berl.) **9**, 564—630 (1929).

LINSBAUER, K.: Untersuchungen über Plasma und Plasmaströmung an Chara-Zellen. I. Beobachtungen an mechanisch und operativ beeinflußten Zellen. Protoplasma **5**, 563 bis 621 (1929). — LUND, E. J.: Bioelectric fields and growth. Austin: University of Texas Press 1947. — LUND, E. J., and W. A. KENYON: Relation between continous bioelectric currents and cell respiration. I. Electric correlation potentials in growing root tips. J. of Exper. Zool. **48**, 333—357 (1927).

MARSH, G.: (1) The effect of applied electric currents on inherent cellular E. M. F. and its possible significance in cell correlation. Protoplasma **11**, 447—474 (1930). — (2) The effect of mechanical stimulation on the inherent E.M.F. of polar tissues. Protoplasma **11**, 497—520 (1930). — MILOVIDOV, P. F.: Physik und Chemie des Zellkernes. 1. u. 2. Teil. Protoplasma-Monogr. **20** (1949); **21** (1954). Berlin-Nikolassee: Gebrüder Borntráger.

NERNST, W.: Zur Theorie der elektrischen Reizung. Nachr. kgl. Ges. Wiss. Göttingen, Math.-physik. Kl. **1899**, 104—108. — NISHIZAKI, Y.: Bioelectric phenomena accompanying osmosis in a single plant cell. Cytologia (Tokyo) **20**, 32—40 (1955). — NOLD, R. H.: Die Funktion der Blase von Utricularia vulgaris. BBC-Nachr. A **52**, 415—448 (1934).

OSTERHUOT, W. J. V.: (1) Movements of water in cells of Nitella. J. Gen. Physiol. **32**, 553—557 (1949). — (2) Changes in resting potential due to a shift of electrolytes in the cell produced by non-electrolytes. J. Gen. Physiol. **37**, 423—432 (1954).

PAECH, K.: Beiträge zur Bestimmung der elektrischen Leitfähigkeit in lebenden pflanzlichen Geweben. Planta (Berl.) **31**, 265—294 (1940). — PILET, P.-E., et S. MEYLAN: Polarité électrique, auxines et physiologie des racines du Lens culinaris Medicus. Bull. Soc. bot. Suisse **63**, 430—466 (1953).

RAMSHORN, K.: (1) Experimentelle Beiträge zur elektrophysiologischen Wachstumstheorie. Planta (Berl.) **22**, 736—766 (1934). — (2) Wachstums- und elektrische Potentialdifferenzen bei Avena-Coleoptilen. Planta (Berl.) **27**, 219—223 (1937). — REHM, W. S.: (1) Bud regeneration and electric polarities in Phaseolus multiflorus. Plant Physiol. **13**, 81—101 (1938). — (2) Electric response of Phaseolus multiflorus to electrical courrents. Plant Physiol. **14**, 359—363 (1939). — RUBINSTEIN, D. L., and V. USPENSKAJA: Über den isoelektrischen Punkt der pflanzlichen Plasmahaut. Protoplasma **21**, 191—225 (1934).

SCHRANK, A. R.: (1) Electrical and curvature responses of the Avena coleoptile to transversely applied direct current. S. 217—231 in Lund: Bioelectric fields and growth, Austin: University of Texas Press 1947. — (2) Electrical and curvature responses of the Avena coleoptile to transversely applied direct current. Plant Physiol. **23**, 188—200 (1948). — (3) Experimental control of phototropic bending in the Avena coleoptile by application of direct current. J. Cellul. a. Comp. Physiol. **32**, 143—159 (1948). — (4) Influence of longitudinally applied direct current on the electrical polarity and curvature of the Avena coleoptile. J. Cellul. a. Comp. Physiol. **33**, 1—16 (1949). — (5) Control of phototropic bending of the Avena coleoptile by longitudinally applied direct current. J. Cellul. a. Comp. Physiol. **35**, 353—369 (1950). — SCHRANK, A. R., and G. E. BACKUS: The relationship of auxin to electrically induced growth responses in the Avena coleoptile. J. Cellul. a. Comp. Physiol. **38**, 361—376 (1951). — SSAWOSTIN, P. W.: Magnetophysiologische Untersuchungen. I. Die Rotationsbewegungen des Plasmas in einem konstanten magnetischen Kraftfelde. Planta (Berl.) **11**, 683—726 (1930). — STUDENER, O.: Über die elektroosmotische Komponente des Turgors und über chemische und Konzentrationspotentiale pflanzlicher Zellen. Planta (Berl.) **35**, 427—444 (1947).

TASAKI, I., u. N. KAMIYA: Electrical response of a slime mold to mechanical and electrical stimuli. Protoplasma **39**, 333—343 (1950). — THOMAS, J. B.: Electric control of polarity in plants. Trav. bot. néerl. **36**, 373—437 (1939). — TOBIAS, J. M., and S. SOLOMON: Electrically induced polar changes in viscosity in the hyaline protoplasm of Elodea with observations on streaming and plastid charge. J. Cellul. a. Comp. Physiol. **35**, 1—9 (1950).

UMRATH, K.: (1) Zur Theorie der elektrischen Erregung. Biol. generalis (Wien) **1**, 396 bis 481 (1925). — (2) Elektrische Potentiale pflanzlicher Gewebe. Protoplasma **4**, 540—546 (1928). — (3) Zellwandpotentiale lebender und toter Helodea-Blätter. Protoplasma **5**, 444 bis 446 (1928). — (4) Über die Erregungsleitung bei sensitiven Pflanzen mit Bemerkungen zur Theorie der Erregungsleitung und der elektrischen Erregbarkeit im allgemeinen. Planta (Berl.) **5**, 274—324 (1928). — (5) Über die Erregungsleitung bei höheren Pflanzen. Planta (Berl.) **7**, 174—207 (1929). — (6) Potentialmessungen an Nitella mucronata mit besonderer Berücksichtigung der Erregungserscheinungen. Protoplasma **9**, 576—597 (1930). — (7) Die Bildung von Plasmalemma (Plasmahaut) bei Nitella mucronata. Protophasma **16**, 173—188 (1932). — (8) Der Erregungsvorgang bei Nitella mucronata. Protoplasma **17**, 258—300 (1932). — (9) Der Einfluß der Temperatur auf das elektrische Potential, den Aktionsstrom und die Protoplasmaströmung bei Nitella mucronata. Protoplasma **21**, 329—334 (1934). — (10) Über den Erregungsvorgang bei Spirogyra und Vaucheria und über Potentialmessungen an Pflanzenzellen. Protoplasma **22**, 193—202 (1934). — (11) Der Erregungsvorgang bei höheren Pflanzen. Erg. Biol. **14**, 1—142 (1937). — (12) Über den Erregungsvorgang und sonstige reizbedingte Veränderungen in der Oberepidermis der Zwiebelschuppen von Allium cepa. Protoplasma 28, 345—351 (1937). — (13) Über elektrische Potentiale und Aktionsströme von Valonia macrophysa. Protoplasma 31, 184—193 (1938). — (14) Über die Art der elektrischen Polarisierbarkeit und der elektrischen Erregbarkeit bei Nitella. Protoplasma 34, 469—483 (1940). — (15) Zur Theorie der Polarisation an lebenden Zellen. Protoplasma 36, 584—606 (1942). — (16) Über die Natur der elektrischen Potentialdifferenzen an pflanzlichen Geweben. Protoplasma 37, 398—403 (1943). — (17) Über Aktionsstrom und Stillstand der Protoplasmaströmung bei Nitella opaca. Protoplasma 42, 77—82 (1953). — (18) Über die elektrische Polarisierbarkeit von Nitella mucronata und Nitella opaca. Protoplasma 43, 237—252 (1954). — UMRATH, K., u. F. WEBER: Elektrische Potentiale an durch Colchicin oder Heteroauxin hervorgerufenen Keulenwurzeln. Protoplasma 37, 522—526 (1943).

WALKER, N. A.: Microelectrode experiments on Nitella. Austral. J. Biol. Sci. 8, 476—489 (1955). — WATANABE, A., M. KODATI u. S. KINOSITA: (1) Über die Beziehung zwischen der Protoplasmaströmung und den elektrischen Potentialveränderungen bei Myxomyceten. Botanic. Mag. 51, 337—348, japanisch u. dtsch. Zus.fassung 348—349 (1937). Zit. nach Ber. wiss. Biol. 44, 132. — (2) Über die Potentialdifferenzen zwischen den Myxomyceten-Plasmodien und den Außenmedien. Botanic. Mag. 53, 410—414 (1939) [Japanisch]. Zit. nach einem Autoreferat in Ber. wiss. Biol. 53, 418. — (3) Über die Einflüsse des Wuchsstoffs und anderer Substanzen auf die Potentialdifferenzen zwischen dem Plasmodium und dem Außenmedium. Botanic. Mag. 53, 460—465 (1939) [Japanisch]. Zit. nach einem Autoreferat in Ber. wiss. Biol. 53, 418. — WEBSTER jr., W. W., and A. R. SCHRANK: Electrical induction of lateral transport of 3-indoleacetic acid in the Avena coleoptile. Arch. of Biochem. a. Biophysics 47, 107—118 (1953). — WIEGAND, O. F., and A. R. SCHRANK: Curvature responses of electrically stimulated Avena coleoptiles to 3-indol-acetic acid. Arch. of Biochem. a. Biophysics 56, 459—468 (1955). — WINSLOW, C.-E. A., J. S. FALK and M. F. CAULFIELD: Electrophoresis of bacteria as influenced by hydrogen ion concentration and the presence of sodium and calcium salts. J. Gen. Physiol. 6, 177—200 (1923). — WINSLOW, C.-E. A., and M. F. UPTON: The electrophoretic migration of various types of vegetable cells. J. Bacter. 11, 367—392 (1926).

YAMAHA, G.: Potentialmessungen an den sich teilenden Pollenmutterzellen von Lilium speciosum. Proc. Imp. Acad. Tokyo 14, 83—85 (1938). — YAMAHA, G., u. T. ISHII: Über die Wasserstoffionenkonzentration und die isoelektrische Reaktion der pflanzlichen Protoplasten, insbesondere des Zellkerns und der Plastiden. Protoplasma 19, 194—212 (1933).

Narkose und Narkotica.

Unter besonderer Berücksichtigung ihrer Einwirkung auf den aktiven Stofftransport und die Protoplasmapermeabilität.

Nach einem hinterlassenen Manuskriptfragment von

K. Paech †.

Bearbeitet von

Veijo Wartiovaara und **Runar Collander.**

Allgemeines.

Auf den ersten Blick mag es vielleicht befremdlich erscheinen, bei Pflanzen von Narkose oder Anaesthesie zu sprechen, besonders dann, wenn man als wesentlichen narkotischen Effekt die Wirkung auf das Nervensystem der höheren Tiere im Auge hat. Die Analyse dieser Wirkungen führte jedoch dazu, die Narkose auch im tierischen Körper als ein zellphysiologisches Phänomen zu erkennen, und damit wurde es nahegelegt, analoge Erscheinungen auch an Pflanzenzellen zu vermuten. Während man aber beim Tierkörper von der eingetretenen Wirkung her bestimmte chemische Substanzen als Narkotica klassifizierte, hat man bei pflanzlichen Organen meist umgekehrt die nach Applikation von Stoffen, die als Narkotica bekannt waren, auftretenden Veränderungen im plasmatischen Zustand und im Stoffwechsel als charakteristisch für die Narkose bei Pflanzen bezeichnet. Es liegt somit an der Hand, daß der Begriff der pflanzlichen Narkose bei weitem nicht so scharf umschrieben wie derjenige der tierischen Narkose ist. Damit hängt zusammen, daß die Narkoseforschung an pflanzlichen Objekten nicht so zielbewußt oder auf so breiter Basis betrieben worden ist wie die auch praktisch eminent bedeutungsvolle Erforschung der narkotischen Erscheinungen am Tier und Menschen.

Unter Narkoticis versteht man in erster Linie chemisch indifferente organische Anelektrolyte, die reversibel hemmend auf die Funktionen der Zelle im ganzen einwirken. Doch ist schon frühzeitig auch ein anorganisches Gas, nämlich Stickoxydul (Lachgas), als ein typisches Narkoticum erkannt worden, und in neuerer Zeit hat es sich gezeigt, daß selbst Stickstoff und die Edelgase deutliche narkotische Wirkungen entfalten können, vorausgesetzt nur, daß ihre thermodynamische Aktivität einen genügend hohen Betrag erreicht (vgl. Tabelle 1 bei MULLINS 1954). Im allgemeinen läßt sich bei steigenden Dosen des Narkoticums eine Erregungs-, eine Lähmungs- und eine Deletärphase unterscheiden, wobei der Lähmungszustand als Narkose im engeren Sinne zu gelten hat. Es dürfen jedoch nicht alle Hemmungen des Lebensgeschehens als Narkose angesehen werden, und es ist nicht gebräuchlich, die durch HCN oder andere Enzymhemmstoffe hervorgebrachten spezifischen Effekte, auch wenn sie reversibel sind, als narkotisch zu bezeichnen (vgl. dazu HÖBER 1945, S. 355). Auch manche Alkaloide, wie z. B. Morphin, können zwar an einigen Organismen narkoseähnliche Zustände erzeugen, werden aber trotzdem nicht als eigentliche Narkotica bezeichnet, teils weil sie als schwache Basen chemisch nicht indifferent sind, teils aber auch weil ihre Einwirkung auf verschiedene Organismen eine sehr differente ist, wogegen die eigentlichen Narkotica ein auffallend gleichmäßiges Verhalten gegenüber den verschiedenartigsten Organismen aufweisen. In *quantitativer* Hinsicht sind allerdings gewisse charakteristische Unterschiede zwischen der Narkotisierbarkeit verschiedener Organismengruppen vorhanden, indem, wie OVERTON (1901) nachwies, Säugetiere, Vögel, Amphibien, Insekten und Entomostraken zwar alle ungefähr dieselbe Narkoticumkonzentration zu ihrer Narkose bedürfen, wogegen bei den verschiedenen Gruppen der Würmer eine mindestens doppelt so hohe Konzentration und bei den Protozoen und Pflanzen endlich eine etwa 6mal so hohe Narkoticumkonzentration zur Narkose nötig ist. Zum Teil mögen diese Unterschiede allerdings auch davon herrühren, daß zur Kennzeichnung der Narkose innerhalb der verschiedenen Organismengruppen Kriterien benutzt worden sind, die untereinander nicht streng vergleichbar sind.

Die chemische Indifferenz der eigentlichen Narkotica bringt es mit sich, daß sie im allgemeinen nicht mit den Zellkomponenten in Reaktion treten können. In der Regel werden sie somit die Zellen unverändert wieder verlassen. (Eine auffällige Ausnahme liegt unter anderem darin, daß der Äthylalkohol bekanntlich im menschlichen Organismus zu einem beträchtlichen Teil in den Atmungsstoffwechsel einbezogen wird. Narkotisch wirkende Ester können wiederum im Organismus hydrolysiert werden.)

Die viel debattierte Frage nach dem Wirkungsmechanismus der Narkotica kann hier nur verhältnismäßig kurz gestreift werden, da sie bisher fast nur an tierischen Objekten studiert worden ist. Es sei aber auf die zusammenfassenden Darstellungen von WINTERSTEIN (1926), HÖBER (1945), BERGSTERMANN (1951), HEILBRUNN (1952), MCGOWAN (1954), MULLINS (1954) und RUMMEL (1954) hingewiesen sowie auf die Aufsätze in dem Sammelwerk „Mécanisme de la narcose" (1951).

Die Reihe der Narkotica umfaßt so durchaus verschiedenartige Stoffe wie einerseits z. B. Kohlenwasserstoffe, Halogenkohlenwasserstoffe, Äther, Ester, Alkohole, Ketone, Aldehyde und Sulfone sowie auf der anderen Seite Stickstoff, Stickoxydul und Edelgase. Schon diese auffallende Verschiedenheit der Narkotica in chemischer Hinsicht schließt von vornherein jede Narkosetheorie aus, die auf die rein chemischen Eigenschaften der Narkotica basiert ist. Zudem zeigt ja der Umstand, daß auch Edelgase narkotisch wirken können, mit besonderer Evidenz, daß die narkotische Wirksamkeit keine chemische Aktivität voraussetzt.

Unter solchen Umständen bleibt nichts anderes übrig als nachzuprüfen, ob die chemisch so überaus bunte Reihe der Narkotica vielleicht durch irgendeine gemeinsame physikalische Eigenschaft ausgezeichnet sei. In der Tat fanden H. H. MEYER (1899) und OVERTON (1901) unabhängig voneinander, daß eine solche auffallende Eigenschaft allen Narkoticis gemeinsam ist, nämlich ihre ausgesprochene Löslichkeit in lipoiden Lösungsmitteln oder, genauer ausgedrückt, ihr verhältnismäßig großer Verteilungskoeffizient in Systemen von dem Typus Lipoid/Wasser. Zudem stellten sie fest, daß auch in quantitativer Hinsicht eine ausgesprochen positive Korrelation zwischen der Größe der relativen Lipoidlöslichkeit und der narkotischen Wirksamkeit verschiedener Verbindungen besteht. Damit war die Lipoidtheorie der Narkose begründet.

Selbstverständlich darf die Lipoidtheorie der Narkose nicht mit der etwa gleichzeitig von OVERTON aufgestellten Lipoidtheorie der Zellpermeabilität verwechselt oder identifiziert werden, denn die Phänomene der Narkose und die der Permeation sind so grundsätzlich verschieden, daß es *a priori* gut denkbar erscheint, daß die eine dieser Theorien falsch wäre, auch wenn die andere zu Recht bestände. Andererseits steht fest, daß alle eigentlichen Narkotica ein sehr großes Permeationsvermögen gegenüber allen Protoplasten besitzen. Variationen hinsichtlich der Narkotisierbarkeit können also nie auf wechselnde Permeabilität für die Narkotica zurückgeführt werden.

Die zuerst von MEYER und OVERTON beobachtete Korrelation zwischen Lipoidlöslichkeit und narkotischer Wirksamkeit hat sich auch in späteren diesbezüglichen Untersuchungen vollauf bestätigt. So z. B. kamen K. H. MEYER und HEMMI (1935) zu dem Schluß, „daß Narkose eintritt, wenn ein beliebiger, chemisch indifferenter Stoff in einer bestimmten molaren Konzentration in die Zellipoide eingedrungen ist. Diese Konzentration ist von der Tier- und Zellart abhängig, vom Narcotikum jedoch unabhängig“ (vgl. auch K. H. MEYER 1937). Ein Vorbehalt muß allerdings an diese Feststellung geknüpft werden, nämlich der, daß die Lösungsmitteleigenschaften verschiedener Lipoide und lipoidähnlicher Verbindungen selbstverständlich von Stoff zu Stoff etwas verschieden sind, und daß man bis jetzt nicht wissen kann, welche Lipoide gerade zu diesem Vergleich herangezogen werden müßten. Außerdem sind die Lösungsmitteleigenschaften der Lipoide noch recht unvollständig erforscht (COLLANDER 1947, 1949).

Statt von der (relativen) Lipoidlöslichkeit der Narkotica könnte man ebensogut von ihrer Hydrophobie reden, d. h. von ihrer Abneigung zur Bildung von Wasserstoffbrücken oder anderen zwischenmolekularen Bindungen, durch die sie an hydrophile Moleküle gekettet werden könnten. (Bei den Edelgasen erreicht diese Abneigung ihr höchstes denkbares Maß.) Welche Formulierung besser am Platze ist, hängt von dem Zusammenhang ab: bisweilen erscheint die positive Formulierung („Lipoidlöslichkeit“, „Lipophilie“), bisweilen die negative („Hydrophobie“, „Fehlen der zwischenmolekularen Kräfte“) zutreffender.

Auch die Oberflächen- bzw. Grenzflächenaktivität der Stoffe variiert in einem gewissen Ausmaß ihrer Lipoidlöslichkeit und zugleich ihrer narkotischen Wirksamkeit parallel, was dazu geführt hat, daß die Narkose von einigen Autoren als eine Auswirkung der Grenzflächenaktivität gedeutet worden ist. Tatsächlich ist jedoch, wie besonders MEYER und HEMMI (1935) scharf betont haben, der Parallelismus zwischen Lipophilie und narkotischer Wirksamkeit einerseits und Grenzflächenaktivität andererseits bei weitem nicht ausnahmslos. So z. B. gibt es Stoffe genug (Seifen usw.), die trotz stärkster Grenzflächenaktivität nicht oder kaum lipoidlöslich sind und auch nicht narkotisch wirken. Andererseits sind Kohlenwasserstoffe und Halogenkohlenwasserstoffe zwar stark lipophil (hydrophob) und wirken dementsprechend narkotisch, sind aber trotzdem grenzflächeninaktiv. Dies ist in der Tat leicht begreiflich. Nur bei Anwesenheit einer hydrophilen und einer lipophilen Gruppe in ein und demselben Molekül hat das betreffende Molekül die Möglichkeit und zugleich das Bestreben, sich so in eine Grenzschicht einzufügen, daß der hydrophile Teil des Moleküls der wäßrigen Phase und der lipophile der Teil lipoiden Phase zugekehrt ist. Nur solche „amphiphile“ Moleküle sind demnach ausgesprochen grenzflächenaktiv. Narkotische Wirksamkeit brauchen sie aber erfahrungsgemäß nicht zu besitzen. Diese Wirksamkeit variiert vielmehr, jedenfalls in großen Zügen, der Lipophilie symbat, der Hydrophilie dagegen antibat. Es ist somit nicht gut möglich, die Narkose auf die Grenzflächenaktivität der Narkotica zurückzuführen.

Auch wenn wir etwa die Korrelation — oder sogar die direkte Proportionalität — zwischen der Lipophilie und der narkotischen Wirksamkeit als endgültig festgestellt betrachten, folgt daraus aber noch nicht, daß das Problem von dem Mechanismus der narkotischen Wirkung als gelöst anzusehen sei. Eine solche Korrelation oder Proportionalität würde nämlich nur besagen, daß die Lipophilie eine notwendige und zugleich ausreichende (?) Voraussetung für die narkotische Wirksamkeit darstellt, eine Aufklärung des tatsächlichen Mechanismus der Narkose wäre damit aber nicht gegeben. Denn wenn auch experimentelle Ergebnisse von der angedeuteten Art dafür sprechen, daß das Substrat des narkotischen Effekts irgendein hydrophober Zellbestandteil sein dürfte, so ist hiermit weder gesagt, um welchen Zellbestandteil (Plasmahaut ?, Mitochondrien ?, oder vielleicht nur etwa eine hydrophobe Wirkungsgruppe eines bestimmten Enzyms ?) es sich dabei handelt, noch in welcher Weise dieser Bestandteil von den Narkoticis beeinflußt wird.

In der Tat gehen, wie wir sehen werden, die Ansichten hinsichtlich des eigentlichen Mechanismus der Narkose noch heute weit auseinander.

Eine besonders früher häufig verfochtene Auffassung ist die, daß die Narkose durch eine Herabsetzung der normalen Zellpermeabilität zustande kommt. Diese Vorstellung wurzelt vor allem in der Erwägung, daß die Erregung, so wie sie uns etwa bei den Nerven und Muskeln in typischer Form entgegentritt, mit einer plötzlichen Permeabilitätserhöhung einhergeht, und daß die Narkose der Erregung entgegengesetzt sein müßte. Hiergegen kann jedoch eingewendet werden, daß eigentlich nicht die Erregung selbst, sondern eher die Erregbarkeit der Narkose entgegengesetzt ist. Außerdem werden wir weiterhin sehen, daß die Narkose wenigstens bei pflanzlichen Zellen durchaus nicht immer mit einer Permeabilitätsverminderung verknüpft ist: auch narkotische Permeabilitätserhöhungen sind beobachtet worden. Damit scheint erwiesen zu sein, daß die „Permeabilitätstheorie der Narkose“ jedenfalls für pflanzliche Zellen kaum zutrifft — wenigstens nicht in ihrer ursprünglichen Form.

Eine andere oft diskutierte Hypothese der narkotischen Wirkung ist die, daß sie auf eine Hemmung der Zellatmung zurückzuführen sei. In der Tat ist die Narkose wohl immer mit einer Herabsetzung der Atmung verknüpft. Inwieweit aber die Atmungshemmung Narkose hervorruft und inwieweit umgekehrt Narkose eine Atmungshemmung verursacht, das scheint immer noch recht unklar zu sein.

Schließlich sind kolloidale Zustandsänderungen im Protoplasma häufig für die Narkose verantwortlich gemacht worden. So z. B. nimmt Heilbrunn (1952) an, daß die Wirkung der Narkotica sich in erster Linie auf die Lipoproteide der Protoplasten bezieht. Die dabei freigemachten Ca-Ionen sollen dann bestimmte Zustandsänderungen der Plasmakolloide hervorrufen. Wie Heilbrunn zeigt, können in der Tat zahlreiche zellphysiologische Befunde mit solchen Vorstellungen in Einklang gebracht werden.

Immerhin hat man den Eindruck, daß über den Mechanismus der Narkose das letzte Wort noch nicht gesprochen worden ist. Auch fragt es sich, ob wirklich allen unter dem Begriff der Narkose zusammengefaßten reversiblen Leistungshemmungen derselbe Mechanismus zugrunde liegt. Das erscheint besonders zweifelhaft, wenn man sich einerseits vergegenwärtigt, wie außerordentlich spezialisiert die Zellen des tierischen Nervensystems sind, an denen die Phänomene der typischen Narkose in erster Linie studiert worden sind, während man andererseits bedenkt, wie verschiedenartig die unter dem Begriff der pflanzlichen Narkose zusammengefaßten Erscheinungen sind. Ist wirklich etwa die durch Narkotica bewirkte Hemmung der Photosynthese, die Verlangsamung der Plasmaströmung oder die Herabsetzung der Leuchtintensität eines Leuchtbacteriums mit der tierischen Narkose identisch oder auch nur direkt vergleichbar ? Einstweilen scheint eine sichere Beantwortung dieser Frage kaum möglich.

Unter solchen Umständen dürfte es am zweckmäßigsten sein, in diesem Handbuch die narkotische Beeinflussung der verschiedenen Lebensprozesse (Photosynthese, Plasmaströmung, Lichtproduktion usw.) in Zusammenhang mit den betreffenden Lebensprozessen selbst zu behandeln. Nur der Einfluß der Narkotica auf den aktiven Stofftransport und auf die Protoplasmapermeabilität soll bereits an dieser Stelle besprochen werden.

Einfluß der Narkotica auf den aktiven Stofftransport.

Bereits PFEFFER (1886) hat mit der theoretischen Möglichkeit gerechnet, daß lebende Protoplasten die Fähigkeit besäßen, gelöste Stoffe aktiv aufzunehmen, und daß diese Aktivität durch Narkotica gehemmt werden könnte. Während ihm aber die nichtosmotische oder aktive Stoffaufnahme nur als eine hypothetische Denkmöglichkeit vorschwebte, hat dann erst OVERTON (1895, 1896) dieselbe zu einer Realität erhoben, und zwar unter ausdrücklicher Betonung der grundlegenden Bedeutung einer solchen „adenoiden Tätigkeit" für das Leben der Organismen. Unter den Merkmalen, die benutzt werden können, um die aktiven Transportprozesse von den rein osmotischen Vorgängen zu unterscheiden, erwähnt OVERTON, daß nur die letztgenannten „ungefähr gleich schnell" in narkotisierten wie in nichtnarkotisierten Zellen stattfinden. Auf welche konkreten Beobachtungen diese Aussage sich stützt, gibt er allerdings nicht an. Auch später ist diese Frage, wie wir sehen werden, recht wenig untersucht worden.

Als einer der ersten hat TRÖNDLE (1920) die Richtigkeit der in Rede stehenden Anschauung zu beweisen versucht. Auf plasmolytischem Wege fand er, daß die Aufnahme von Salzen seitens Blatt- und Wurzelzellen durch Narkotica (1% Chloralhydrat, 3% Äther) stark gehemmt oder sogar ganz verhindert wurde, wogegen das rein permeatorisch erfolgende Eindringen von Alkaloiden in *Spirogyra*-Zellen (nach der Niederschlagsmethode untersucht) nicht merkbar durch Narkotica beeinflußt wurde. Allerdings sind gegen die von TRÖNDLE benutzte plasmolytische Technik ernste und, wie es scheint, nicht unbegründete Einwände erhoben worden (ZYCHA 1928).

Zuverlässiger in technischer Hinsicht scheinen die von HOAGLAND, HIBBARD und DAVIS (1926) ausgeführten Versuche über den Einfluß von Chloroform, Äther und Kaliumcyanid auf die Bromidaufnahme der *Nitella*-Zellen. Die aufgenommenen Br-Mengen wurden nämlich in dem aus den Zellen isolierten Zellsaft chemisch-analytisch bestimmt. Nach den mitgeteilten Analysendaten zu schließen, wurde die Br-Aufnahme durch die geprüften Gifte etwa um die Hälfte verlangsamt. Doch scheint es sich hier nur um vereinzelte Versuche orientierenden Charakters zu handeln. Außerdem mag es, jedenfalls auf den ersten Blick, scheinen, als ob die Versuchsergebnisse von BÄRLUND (1938) mit denjenigen von HOAGLAND und Mitarbeitern in einem gewissen Widerspruch ständen. BÄRLUND fand nämlich, daß die Aufnahme von Lithiumionen seitens *Chara*-Zellen innerhalb der Fehlergrenzen mit derselben Geschwindigkeit stattfand, einerlei ob die Lösung mit 1,5—2,0 Vol.-% Äthyläther versetzt war oder nicht. (Unter dem Einfluß der betreffenden Ätherkonzentrationen war die Plasmarotation merklich verlangsamt, aber nicht aufgehoben.) Immerhin ist zu beachten, daß die Li-Konzentration bei diesen Versuchen nie auch nur dieselbe Konzentration im Zellsaft wie in der umgebenden Lösung erreichte. Es mag also sein, daß auch von den in ätherfreien Lösungen gehaltenen Kontrollzellen Li-Ionen aus irgendeinem Grunde nicht aktiv aufgenommen wurden.

Die Angaben von LEPESCHKIN (1911) und LULLIES (1925), nach denen die „Salzpermeabilität" der Epidermisprotoplasten von *Rhoeo* durch Narkotica herabgesetzt werden sollen, beziehen sich vielleicht eher auf die aktive Salzaufnahme. Allerdings besteht auch die Möglichkeit, daß die von diesen Autoren beobachtete Verlangsamung der Deplasmolyse nicht auf einer Hemmung der Salzaufnahme, sondern auf einer Erhöhung der Exosmose beruht. Für Untersuchungen dieser Art ist eben die plasmolytische Methode wenig geeignet.

Äther (2 Vol.-%) und Chloralhydrat (0,5—1,0%) hemmen stark die Aufnahme von Sulfosäurefarbstoffen seitens Blütenblattzellen von *Hyacinthus* (COLLANDER

1921). Ebenso fand HERTZ (1922), daß sich die Aufnahme solcher Farbstoffe seitens des Protozoons *Opalina* „wegnarkotisieren" läßt.

Ein Aufhören des Blutens dekapitierter *Sanchezia*-Pflanzen bei der Behandlung des Wurzelsystems mit Äther- oder Chloroformdämpfen ist von HEYL (1933) beobachtet worden. Ebenso fanden SPEIDEL (1939) und VAN ANDEL (1953) an ihren Versuchsobjekten eine kräftige Hemmung der Blutung durch Phenylurethan. Doch ist diese Substanz wohl eher ein Atmungsgift als ein eigentliches Narkoticum. Wenn RUHLAND (1915) bei niedrigen Narkoticumkonzentrationen Verringerung, bei höheren dagegen Verstärkung der Sekretion der salzabsondernden Hautdrüsen der *Plumbaginaceen* fand, liegt es nahe anzunehmen, daß die zuerst genannte Wirkung der eigentlichen Narkose entspricht, während die zuletzt genannte mehr toxischer Natur ist. Weniger durchsichtig ist dagegen der Befund von HEIMANN (1950) an intakten Pflanzen von *Kalanchoë*, daß Chloroform in „sehr kleinen Mengen" eine Zunahme der Guttation bewirkt.

In diesem Zusammenhang sei ein einfacher Versuch erwähnt, der seit etwa 30 Jahren alljährlich mit gutem Erfolg im pflanzenphysiologischen Praktikum der Universität Helsinki ausgeführt wird. Die Guttation in feuchtem Sand wachsender *Avena*-Keimlinge wird eine Weile verfolgt, indem man die Zeit mißt, die nach dem Abwischen der Koleoptilspitze mit einem Stückchen Fließpapier bis zum Erscheinen des folgenden Tröpfchens verstreicht. Wenn eine annähernd konstante Guttationsgeschwindigkeit erreicht ist, wird der Sand mit einer Lösung von 3 Vol.-% Äther in Leitungswasser gründlich gewässert. Die Guttation hört fast sofort auf. Wird hierauf die ätherhaltige Lösung durch reines Leitungswasser verdrängt, so erreicht die Guttation innerhalb weniger Minuten ihre ursprüngliche Geschwindigkeit.

Es leuchtet ein, daß Ergebnisse dieser Art die Hemmung eines aktiven Stofftransportes bedeuten, einerlei ob man die Guttation bzw. das Bluten als das Ergebnis einer aktiven Wassersekretion oder einer Sekretion von gelösten Substanzen, die das Wasser sozusagen mit sich schleppen (VAN ANDEL 1953), betrachtet.

Schließlich sei erwähnt, daß nach OUDMAN (1936) die Aufnahme und/oder der Transport des außerordentlich schwer permeierenden Asparagins seitens der Tentakeln von *Drosera* durch Ätherdämpfe beträchtlich verlangsamt wird. Bemerkenswerterweise fand er aber, daß der Transport auch des verhältnismäßig leicht permeierenden Coffeins deutlich, obwohl schwächer, verlangsamt wurde.

Zusammenfassend kann festgestellt werden, daß die große Mehrzahl der oben referierten Untersuchungen mehr oder weniger klar erkennen läßt, daß aktive Stofftransportprozesse durch subletale Narkoticumgaben gehemmt werden. Allerdings ist das zur Verfügung stehende diesbezügliche Beobachtungsmaterial immer noch relativ knapp.

Einfluß der Narkotica auf die Protoplasmapermeabilität.

Die in der Literatur enthaltenen Angaben betreffs des Einflusses der Narkotica auf die Protoplasmapermeabilität gehen sehr auseinander: nach zahlreichen Angaben sollen Narkotica die Permeabilität herabsetzen, nach etwa ebensovielen bewirken sie umgekehrt eine Erhöhung der Permeabilität, während schließlich in einigen Fällen kein ausgesprochener Einfluß auf die Permeabilität festgestellt werden konnte. An sich braucht diese Unstimmigkeit nicht zu bedeuten, daß die eine oder andere Ansicht fehlerhaft sei. Im Gegensatz zur Hemmung des aktiven Stofftransportes durch Narkose, die gewissermaßen selbstverständlich erscheint, ist nämlich der Einfluß der Narkose auf die eigentliche Permeabilität nicht *a priori* voraussehbar. Wir haben durchaus keinen zwingenden Grund zur Annahme, daß die Permeabilität von den eigentlichen Lebensvorgängen direkt abhängig sein müßte. Vielmehr hängt sie eigentlich nur von der Struktur und chemischen Zusammensetzung des Protoplasten und vor allem seiner Grenzschichten ab. Diese materiellen Eigenschaften des Protoplasmas können aber voraussichtlich durch die Narkose, je nach den Umständen, in verschiedene Richtungen verändert werden. Zu- und Abnahme der Permeabilität sind also etwa gleich gut denkbar. Auch sind Veränderungen solcher Art vorstellbar, daß die Permeabilität für einige Stoffe zunimmt, während die für andere gleichzeitig abnimmt. Außerdem ist natürlich die Konzentration des Narkoticums in diesem Zusammenhang wichtig: niedrige und hohe Konzentrationen können die Permeabilität sehr verschiedenartig beeinflussen. (Daß ausgesprochen toxische Konzentrationen die normale „Semipermeabilität" aufheben, ist allgemein bekannt. Von solchen Wirkungen sehen wir aber im folgenden soweit wie möglich ab.)

Die Erlangung einer klaren Übersicht über das in Rede stehende Gebiet wird in hohem Grade durch das fast vollständige Fehlen umfassender, systematisch durchgeführter Untersuchungen erschwert. Vielmehr haben die diesbezüglichen Beobachtungen größtenteils den Charakter von mehr oder weniger beiläufigen Stichproben. Sogar die besten Untersuchungen auf diesem Gebiet sind begrenzt entweder hinsichtlich der Zahl der untersuchten Objekte, der benutzten Narkotica oder der geprüften Permeantia. Daß zahlreiche Untersuchungen außerdem in technischer Hinsicht durchaus nicht einwandfrei sind, erschwert gleichfalls in hohem Grade das Ausarbeiten eines klaren Bildes auf Grund der jetzt vorliegenden Literatur. Schließlich sei hier noch darauf hingewiesen, daß die Resultate oder mindestens ihre Deutung in einigen Fällen vielleicht auch durch gewisse Vorurteile ungünstig beeinflußt worden sind. Einige Forscher scheinen nämlich davon ausgegangen zu sein, daß die Permeabilität mit der allgemeinen Zellaktivität parallel variieren müsse und daß somit die Narkose eine Permeabilitätsverminderung hervorrufen müsse.

Eine Gruppe der ersten Untersuchungen auf diesem Gebiet bezieht sich auf die Permeation der damals mit besonderer Vorliebe studierten basischen Farbstoffe. LEPESCHKIN (1911, 1932) ging von der theoretischen Erwartung aus, „daß sich die osmotischen Eigenschaften der Plasmamembran auch unter einem gelinden Einfluß anästhesierender Stoffe verändern müssen, und zwar in der Weise, daß alle Stoffe, die sich in diesen Anaesthetica ... schlecht, in Wasser aber gut lösen, während der Narkose langsamer in das Zellinnere permeieren würden und umgekehrt". Wie leicht ersichtlich, ist dieser Ausgangspunkt jedoch falsch gewählt. Denn wenn irgendein Permeans sich z. B. im Verhältnis 1:100 zwischen die Plasmahautlipoide und Wasser verteilt, während seine Verteilung zwischen dem Narkoticum und Wasser 1:10 beträgt, dann liegt selbstverständlich kein Grund vor, zu erwarten, daß sein Durchtritt durch die Plasmahaut infolge der Beladung der Membranlipoide mit Narkoticum abnehmen müßte. Und umgekehrt: wenn ein anderes Permeans sich im Verhältnis 100:1 zwischen Plasmahautlipoid und Wasser, aber im Verhältnis 10:1 zwischen Narkoticum und Wasser verteilt, dann ist nicht zu erwarten, daß ihm das Narkoticum den Durchtritt durch die Plasmamembran erleichtert. Trotz der Unrichtigkeit seines Ausgangspunktes fand LEPESCHKIN seine Erwartungen bestätigt. Nach ihm wird nämlich die Aufnahme des ätherlöslichen Bismarckbrauns durch Äther befördert. wogegen die Aufnahme der in Äther unlöslichen Farbstoffe Methylenblau und Methylgrün durch Äther gehemmt wird. Als Versuchsobjekt bei diesen Experimenten diente *Spirogyra*. Ganz entsprechend fand SEGEL (1915) an anderen Objekten (Wurzelhaare von *Trianea*, Wurzeln von *Lemna*, Blätter von *Elodea*), daß Äther die Aufnahme des Methylenblaus verlangsamt, wogegen die des ätherlöslichen Neutralrots befördert wird. Dagegen konnte RUHLAND (1912) an

Spirogyra keinen deutlichen Einfluß der Narkotica auf die Aufnahme dieser Farbstoffe feststellen. Vermutlich tut man gut, alle diese Ergebnisse mit einer gewissen Zurückhaltung zu beurteilen. Das Protoplasma bietet nämlich den in Rede stehenden basischen Farbstoffen einen überaus kleinen Permeationswiderstand. Infolgedessen ist es durchaus nicht leicht, kleine Änderungen dieses Widerstandes sicher festzustellen, da auch andere Faktoren, wie z. B. die Durchlässigkeit der Zellwand oder die Durchmischung der Außenlösung die Ergebnisse beeinflussen können. Auch die Versuche von HOMÈS (1934) über den Einfluß von Chloral, Äthylalkohol, Äthyläther und Chloroform auf die Aufnahme von Methylenblau seitens verschiedener Objekte sind nicht sehr überzeugend.

Den basischen Farbstoffen schließen sich die Alkaloide an. TRÖNDLE (1920) fand, daß ihr Eindringen in *Spirogyra*-Zellen durch Zusatz von 1% Chloralhydrat nicht merkbar beeinflußt wird. Unter Benutzung derselben Methode — Beobachtung der Zeit, innerhalb welcher ein sichtbarer Niederschlag im Zellsaft entsteht — und desselben Versuchsobjektes gelangte dagegen DEYSSON (1952) zu dem Ergebnis, daß Chloroform, Äther und Benzol in geeigneten Konzentrationen eine starke Beschleunigung des Eindringens von Colchicin, Coffein und Antipyrin verursachen. Dieses an *Spirogyra* gewonnene Ergebnis wurde von DEYSSON auch an anderen Objekten mittels andersartiger (kaum sehr zuverlässiger) Methodik bestätigt.

Eine Stoffgruppe mit ganz andersartigen Eigenschaften repräsentieren die Salze bzw. die Ionen. Auch bei ihnen stehen sehr große Schwierigkeiten der Bestimmung ihrer Permeation entgegen. Die bedeutendsten Schwierigkeiten sind bedingt 1. durch die Kleinheit der Salzpermeabilität der Zellen, 2. durch das Überwiegen nichtosmotischer Transportkräfte sowie 3. durch die elektrischen Ladungen der Ionen, die es bedingen, daß eine Ionenart nie allein permeieren kann. Alle diese Umstände erschweren selbstverständlich auch die Ermittlung des Einflusses der Narkotica auf die Salz- und Ionenpermeabilität. Die angeblich diesbezüglichen Versuche von LEPESCHKIN (1911) und LULLIES (1925) wurden bereits in anderem Zusammenhang erwähnt und als vieldeutig erkannt. Dagegen beziehen sich die *Laminaria*-Versuche von OSTERHOUT (1916) vermutlich eben auf die Ionenpermeabilität, jedenfalls nicht auf den aktiven Salztransport. Diese Versuche wurden nämlich so ausgeführt, daß aus dem Thallus der *Laminaria* Scheiben ausgestanzt wurden, die nach der Art von Geldrollen zwischen Elektroden angebracht wurden, wonach der elektrische Widerstand des ganzen so zusammengesetzten Zylinders gemessen wurde. Da der Widerstand bei Zusatz von Narkoticis reversibel anstieg, deutet dies also auf eine Herabsetzung der Ionenpermeabilität. Mit ähnlicher Methodik hat später PAECH (1940) an anderen Objekten (Früchten, Spargelsprossen) recht ähnliche Ergebnisse erzielt. Nur schade, daß man weder in den Versuchen von OSTERHOUT noch in denen von PAECH mit Sicherheit wissen kann, zu einem wie großen Teil der Strom in diesen Versuchen zwischen den Protoplasten, anstatt durch die Protoplasten hindurchfloß. Mittels ganz andersartiger, an sich sehr zuverlässiger Methodik hat BÄRLUND (1938) Versuche über den Einfluß des Äthyläthers auf das Eindringen von Li-Ionen in *Chara*-Zellen untersucht. Wie bereits erwähnt, konnte dabei kein deutlicher Einfluß der Narkotica beobachtet werden. Vollauf befriedigend sind aber auch diese Versuche nicht, weil aus ihnen nicht genügend klar hervorgeht, in welchem Umfang die stattgefundene Li-Aufnahme als aktiv oder passiv aufzufassen ist. Außerdem ist zu berücksichtigen, daß die in diesen Versuchen benutzten Ätherkonzentrationen recht niedrig waren, nämlich nur 1,5—2,0 Vol.-%. Es ist gut denkbar, daß höhere Konzentrationen einen deutlicheren Einfluß ausgeübt hätten.

Relativ am besten, obwohl noch bei weitem nicht genügend untersucht ist der Einfluß der Narkotica auf die Permeabilität für *Anelektrolyte*. Der aktive Stofftransport kommt bei derartigen Versuchen selten als störender Faktor ernstlich in Betracht. Da außerdem der normale Permeationsmechanismus der Anelektrolyte bereits anfängt, prinzipiell verhältnismäßig gut klargelegt zu sein, ist auch die theoretische Ausdeutung der diesbezüglichen Narkoseversuche etwas einfacher als im Falle der Permeation anderer Stoffe.

Ein Mangel liegt allerdings darin, daß bei derartigen Studien meistens nur wenige Narkotica, wenige Permeantia und wenige Zellobjekte geprüft worden sind. So z. B. sind diejenigen Narkotica (Isobutylurethan, Phenylharnstoff und höhere Alkohole), die von LULLIES (1925) in ihrer Wirkung auf die Anelektrolytpermeabilität der *Rhoeo*-Zellen studiert worden sind, nicht auch an *Characeen*-Zellen geprüft worden, wobei es möglich gewesen wäre, gewisse denkbare Fehlerquellen der *Rhoeo*-Versuche zu vermeiden. Vorläufig bleibt deshalb unaufgeklärt, ob die an dem zuletzt genannten Objekt gewonnenen Versuchsergebnisse womöglich durch eine partielle Exosmose der Zellsaftbestandteile beeinflußt worden sind bei Narkoticumkonzentrationen, die zwar noch keine Exosmose des Anthocyans hervorriefen, aber immerhin nahe an der toxischen Grenze lagen. Die beobachtete Erhöhung der plasmolytischen Grenzkonzentration in Gegenwart von Ca^{++}-Ionen, deren Schutzwirkung gegenüber gefährdeten Protoplasten ja bekannt ist, scheint einem derartigen Verdacht einen gewissen Vorschub zu leisten. Besonders auffällig (und wohl auch verdächtig) erscheint die bei Anwendung von Isobutylurethan von LULLIES erhaltene, etwa um das 9fache herabgesetzte Permeabilität für Glycerin, denn etwas derartiges ist an anderen Zellen und mit anderen Narkoticis nie erreicht worden. Merkwürdig erscheint auch die Angabe, daß sich die Glycerinaufnahme zwar durch Heptyl- und Propylalkohol, nicht dagegen durch Isoamyl- oder Isobutylalkohol hemmen läßt. Das Eindringen des Äthylenglykols wurde angeblich nur durch Isobutylurethan und Phenylharnstoff gehemmt, während das Eindringen des Harnstoffs durch die geprüften Narkotica überhaupt nicht zu hemmen war: oft wurde sogar eine Beschleunigung beobachtet.

Während somit die Ergebnisse von LULLIES recht uneinheitlich ausfielen, gelangten dagegen HÖFLER und WEBER (1926) in ihren offenbar mit großer Umsicht ausgeführten Versuchen zu sehr klaren, übersichtlichen Ergebnissen. Eine Begrenzung liegt allerdings darin, daß sie weder andere Narkotica als Äthyläther noch andere Permeantia als Harnstoff geprüft haben. Statt dessen wurden ihre Versuche auf mehrere Zellobjekte ausgedehnt und die denkbaren Fehlerquellen der von ihnen benutzten plasmometrischen Methode sehr genau beachtet. Es zeigte sich, daß Ätherkonzentrationen zwischen $1^1/_2$ und $2^1/_2$% an allen Objekten die Harnstoffaufnahme um das 1,5—2,5fache beschleunigten. Diese Permeabilitätserhöhung entsprach der narkotisch lähmenden Phase der Ätherwirkung, wie sich aus der gleichzeitigen, reversiblen Unterdrückung der sonst obligaten systrophischen Inhaltsverlagerungen in den plasmolysierten Protoplasten beweisen ließ. Subnarkotische Ätherkonzentrationen (1%) veränderten dagegen die Harnstoffpermeabilität nicht merklich. Die vollständige Reversibilität der studierten Permeabilitätserhöhung zeigte deutlich, daß es sich nicht etwa um eine pathologische, prämortale Erscheinung handelte. Durch diese Untersuchung wurde zum erstenmal erwiesen, daß die bis dahin sehr verbreitete Auffassung, daß Narkose die Permeabilität vermindere, jedenfalls keine allgemeine Gültigkeit besitzt.

Die Untersuchung von HÖFLER und WEBER wird in willkommener Weise durch eine solche von BÄRLUND (1938) ergänzt. Mit einem ganz anderen Objekt

(Chara) und mit gänzlich andersartiger Methodik (quantitative mikrochemische Analyse des Zellsaftes) ausgeführt, führte sie dennoch zu prinzipiell ganz übereinstimmenden Ergebnissen. Von Bedeutung ist aber auch, daß die Untersuchung BÄRLUNDs vier recht verschiedenartige Permeantia betraf. Dabei zeigte es sich, daß sie alle vier in ihrer Permeation durch 2,5% Äther beträchtlich beschleunigt werden, aber, wie es scheint, in deutlich verschiedenem Grade. Die Beschleunigung war nämlich für Hexamethylentetramin 3,8-, für Äthylenglykol gleichfalls 3,8-, für Harnstoff 1,8- und für Trimethylcitrat 1,6fach. (Der Wert für Trimethylcitrat mag allerdings wegen spezieller technischer Schwierigkeiten verhältnismäßig ungenau ausgefallen sein.) Zur Erklärung seiner Versuchsergebnisse weist BÄRLUND darauf hin, daß die unbekannten Lipoide der Plasmahaut seiner Zellen in großen Zügen dieselben Lösungsmitteleigenschaften wie Olivenöl besitzen. Wenn diese Lipoide Äther aufnehmen, wird sich ihr Lösungsvermögen entsprechend ändern und zwar so, daß Permeantia, die sich in Äther leichter als in Olivenöl lösen (dies gilt für alle von ihm untersuchten Stoffe), jetzt reichlicher aufgenommen werden und somit leichter permeieren können.

Hier mögen einige noch nicht veröffentlichte Versuchsergebnisse von WARTIOVAARA und Mitarbeiter kurz erwähnt werden. Es wurde der Einfluß von 2 Vol-% Äther auf die Permeation von verschiedenen Glykolen unter Benutzung von *Nitella*-Zellen als Versuchsobjekt untersucht. Der Äther vergrößerte die Permeabilität etwa um das 3,0—3,6fache, ohne daß dabei irgendwelche klaren Beziehungen zur Lipoidlöslichkeit, Molekülgröße oder zum Permeationsvermögen der betreffenden Glykole zu beobachten waren.

Von ganz anderer Art ist nach den Versuchen von RUHLAND und HEILMANN (1951) der Einfluß der einwertigen aliphatischen Alkohole C_1—C_9 auf die Permeabilität von *Beggiatoa mirabilis*. Die Alkohole bewirken nämlich alle eine deutliche Herabsetzung der Permeabilität. Interessanterweise zeigte es sich dabei, daß diese Hemmung sich mit wachsender Länge der Alkoholmoleküle stufenweise auf immer kleiner-molekulare Permeantia erstreckt. RUHLAND und HEILMANN deuten dies so, daß die Alkohole an den Porenwänden der Plasmaaußenfläche einen monomolekularen, die Porenwände ganz oder zum Teil auskleidenden, kondensierten Film bilden. Dadurch wird der den permeierenden Stoffen zur Verfügung stehende Diffusionsraum um so mehr verengt, je länger die von der Porenwandung abstehenden Paraffinketten sind. Die hydrophile oder hydrophobe Natur der Permeantia soll dabei keine Rolle spielen, was recht eigenartig erscheint, da man eher erwartet hätte, daß hydrophobe (lipophile) Stoffe *ceteris paribus* leichter der hydrophoben Wandschicht entlang beweglich wären als hydrophile Stoffe.

Die Ergebnisse von RUHLAND und HEILMANN sind in ihrer Art beinahe einzig dastehend (vgl. allerdings noch ANSELMINO und HOENIG 1930). Inwieweit es in der Zukunft möglich sein wird, analoge Resultate auch an anderen Objekten zu erzielen, läßt sich natürlich nicht mit Sicherheit voraussagen. Am wahrscheinlichsten dünkt es immerhin, daß sie irgendwie mit den ganz eigenartigen Permeabilitätseigenschaften von *Beggiatoa* zusammenhängen.

Einen entfernten Vergleich mit den Ergebnissen von RUHLAND und HEILMANN ermöglicht vielleicht eine Untersuchung von SAUBERT (1937) über den Einfluß der normalen, einwertigen Alkohole C_1—C_4 auf die Wasserpermeabilität der *Chara*-Zellen. Nur Äthylalkohol rief in einem bestimmten Konzentrationsbereich eine geringe Erhöhung der Wasserpermeabilität hervor, sonst wurden lauter Permeabilitätserniedrigungen beobachtet. Im Falle des Methylalkohols betrug die Herabsetzung bisweilen sogar mehr als 50%. Allerdings ist zu berücksichtigen, daß sämtliche Permeabilitätsänderungen auf osmotischem Wege festgestellt wurden, und daß die quantitative Bestimmung der Wasserpermeabilität mittels solcher Methoden immer eine heikle Sache ist.

Die Resultate von SAUBERT, wonach die Alkohole die Permeabilität herabsetzen, stimmt wie ersichtlich wenigstens in großen Zügen mit den Ergebnissen von RUHLAND und HEILMANN überein, steht dagegen (wenigstens scheinbar) in Widerspruch zu den von BÄRLUND an demselben Objekt, aber mit verschiedenen permeierenden Stoffen, erzielten Ergebnissen. Im Hinblick hierauf muß man bedauern, daß der Einfluß der Alkohole auf die Permeabilität der *Chara*-Zellen für andere Nichtelektrolyte noch nicht untersucht worden ist. Es besteht selbstverständlich die Möglichkeit, daß sich die ausgesprochen polar-nichtpolar gebauten, stark oberflächenaktiven Alkohole anders verhalten als Äthyläther.

Eine narkotische Erhöhung der Permeabilität für Wasser und gelöste Stoffe wird von BÜNNING (1933) angegeben. Weitere Angaben über den Einfluß der Narkotica auf die Zellpermeabilität finden sich in den Arbeiten von GOMPEL (1925), HARVEY (1911), v. KUTHY (1940), LEPESCHKIN (1906), MEDES und McCLENDON (1920) und WEIS (1926). Da sich aber diese Angaben jedenfalls hauptsächlich auf mehr oder weniger toxische Konzentrationen beziehen, erübrigt sich hier ein Eingehen auf sie.

Zur Theorie der Beeinflussung der Protoplasmapermeabilität durch Narkotica.

Zur Erklärung der Einwirkung der Narkotica auf die Permeabilität sind hauptsächlich drei Vorstellungen herangezogen worden, die von BÄRLUND (1938), von RUHLAND und HEILMANN (1951) sowie von SAUBERT (1937) vertreten werden. Ihnen allen ist die Annahme gemeinsam, daß die Narkotica die Plasmagrenzschichten direkt beeinflussen, also nicht etwa auf dem Umwege eines irgendwie abgeänderten Verlaufs der Lebensprozesse der Protoplasten. Es handelt sich also eher um Einwirkungen der Narkotica als um solche der Narkose.

Sehr einfach ist die von BÄRLUND vorgeschlagene Erklärung, die allein die Änderung der lösenden Eigenschaften der Plasmahautlipoide beim Hinzutreten des Narkoticums berücksichtigt. Denselben Ausgangspunkt hatte, wie wir sahen, bereits LEPESCHKIN (1911) gewählt, obwohl ihm dabei ein Gedankenfehler unterlaufen war. Die Erklärung BÄRLUNDS ist zwar an sich einleuchtend, stützt sich aber bis jetzt auf ein sehr knappes Beobachtungsmaterial und kann somit durchaus nicht als bewiesen angesehen werden.

Etwa ebenso einfach und einleuchtend ist die Hypothese von RUHLAND und HEILMANN, sofern man von ihren Prämissen ausgeht. Vorläufig weiß man aber, wie bereits hervorgehoben, nichts davon, ob und inwieweit diese Erklärung auf andere Objekte als *Beggiatoa* anwendbar ist.

Die Erklärungen von BÄRLUND einerseits und von RUHLAND und HEILMANN andererseits schließen einander keineswegs aus. Es leuchtet nämlich schon *a priori* ein, daß Narkotica sowohl Änderungen des Lösungsvermögens der Plasmahautlipoide wie auch Änderungen in der Struktur der Plasmahäute bewirken können und daß die Permeabilität auf diesen beiden Wegen beeinflußt werden kann. Einen Versuch beide zu berücksichtigen, repräsentiert die Erklärung von SAUBERT, die sich teils auf die Theorie von BUNGENBERG DE JONG stützt, wonach die Plasmahäute die Natur von Coacervaten besitzen, und teils auf seine eigenen Beobachtungen über den Einfluß der Alkohole auf Coacervate. Je nach der

Konzentration und Beschaffenheit des Alkohols kann dabei die Struktur des Coacervats entweder verdichtet oder aufgelockert werden, so daß also sowohl Abnahme wie Zunahme der Permeabilität möglich ist.

Die Knappheit des jetzt vorliegenden Erfahrungsmaterials verbietet jede endgültige Stellungnahme zu den oben angedeuteten Versuchen, den Einfluß der Narkotica auf die Protoplasmapermeabilität kausal zu erklären. Vielmehr schließen wir uns ganz der Ansicht von DAVSON an, wenn er seine Darstellung dieses Problems (unter besonderer Berücksichtigung tierischer Zellen) in folgende Deklaration ausmüden läßt: "In concluding, we may state that the experimental work on the influence of narcotic substances on permeability is still too limited to warrant any generalisations either regarding the mechanism of this action or regarding the significance of these changes in respect to narcosis itself; the use of exact methods ... applied with care to different cells under strictly physiological conditions should lead to a rapid advance in this branch of permeability" (DAVSON in DAVSON und DANIELLI 1952, S. 243).

Literatur.

ANDEL, O. M. VAN: The influence of salts on the exudation of tomato plants. Acta bot. néerl. **2**, 445—521 (1953). — ANSELMINO, K. J., u. E. HOENIG: Weitere Untersuchungen über Permeabilität und Narkose. Pflügers Arch. **225**, 56—68 (1930).

BÄRLUND, H.: Einfluß des Äthyläthers auf die Permeabilität der *Chara*-Zellen. Protoplasma (Berl.) **30**, 70—78 (1938). — BERGSTERMANN, H.: 50 Jahre Lipoidtheorie der Narkose von OVERTON und MEYER. Naturwiss. **38**, 128—132 (1951). — BÜNNING, E.: Refraktärstadium, Ermüdung und Narkose bei der Seismonastie. Planta (Berl.) **21**, 324—352 (1933).

COLLANDER, R.: Über die Permeabilität pflanzlicher Protoplasten für Sulfosäurefarbstoffe. Jb. wiss. Bot. **60**, 354—410 (1921). — On "lipoid solubility". Acta Physiol. Scand. **13**, 363—381 (1947). — Die Verteilung organischer Verbindungen zwischen Äther und Wasser. Acta chem. scand. (Copenh.) **3**, 717—747 (1949).

DAVSON, H., and J. F. DANIELLI: The permeability of natural membranes, 2. Aufl. Cambridge: University Press 1952. — DEYSSON, G.: Recherches sur la perméabilité des cellules végétales. Rev. Cytol. et Biol. végét. **13**, 153—315 (1952).

GOMPEL, M.: Sur la pénétrabilité des acides dans les cellules de *Ulva lactuca*. Ann. de Physiol. Physiochim. biol. **1**, 166—177 (1925).

HARVEY, E. N.: Studies on the permeability of cells. J. of Exper. Zool. **10**, 507—556 (1911). — HEILBRUNN, L. V.: An outline of general physiology, 3. Aufl. Philadelphia: W. B. Saunders Company 1952. — HEIMANN, M.: Einfluß periodischer Beleuchtung auf die Guttationsrhythmik. Planta (Berl.) **38**, 137—195 (1950). — HERTZ, W.: Die Vitalfärbung von *Opalina ranarum* mit Säurefarbstoffen und ihre Beeinflussung durch Narkotikum. Pflügers Arch. **196**, 444—457 (1922). — HEYL, J. G.: Der Einfluß von Außenfaktoren auf das Bluten der Pflanzen. Planta (Berl.) **20**, 294—353 (1933). — HOAGLAND, D. R., P. L. HIBBARD and A. R. DAVIS: The influence of light, temperature, and other conditions on the ability of *Nitella* cells to concentrate halogens in the cell sap. J. Gen. Physiol. **10**, 121—146 (1926). — HÖBER, R.: Physical chemistry of cells and tissues. London: Churchill 1945. — HOMÈS, M.-V.: Recherches nouvelles sur les relations entre les qualités physiques et chimiques du milieu et la perméabilité cellulaire chez les végétaux. Mém. Acad. roy. Belg., in 8., Sér. II **13** (1934).

KUTHY, A. v.: Beiträge zum Permeabilitätsproblem. Biochem. Z. **306**, 136—142 (1940).

LEPESCHKIN, W. W.: Zur Kenntnis des Mechanismus der aktiven Wasserausscheidung der Pflanzen. Beih. bot. Zbl. **19**, 409—452 (1906). — Über die Einwirkung anästhesierender Stoffe auf die osmotischen Eigenschaften der Plasmamembran. Ber. dtsch. bot. Ges. **29**, 349—355 (1911). — The influence of narcotics, mechanical agents and light upon permeability of protoplasm. Amer. J. Bot. **19**, 568—580 (1932). — LULLIES, H.: Über die Beeinflussung der Permeabilität von Pflanzenzellen durch Narkotica. Pflügers Arch. **207**, 8—23 (1925).

McGOWAN, J. C.: The physical toxicity of chemicals. J. Appl. Chem. **4**, 41—47 (1954). — *Mécanism de la narcose.* CNRS, Paris 1951. — MEDES, G., and J. F. McCLENDON: Effects of anaesthetics on various cell activities. J. of Biol. Chem. **42**, 541—568 (1920). — MEYER, H. H.: Zur Theorie der Alkoholnarkose. Arch. exper. Path. u. Pharmakol. **42**, 109—118 (1899). — MEYER, K. H.: Contributions to the theory of narcosis. Trans. Faraday Soc. **33**, 1062—1068 (1937). — MEYER, K. H., u. H. HEMMI: Beiträge zur Theorie der Narkose.

Biochem. Z. 277, 39—71 (1935). — MULLINS, L. J.: Some physical mechanisms in narcosis. Chem. Reviews 54, 289—323 (1954).

OSTERHOUT, W. J. V.: The decrease of permeability produced by anesthetics. Bot. Gaz. 61, 148—158 (1916). — OUDMAN, J.: Über Aufnahme und Transport N-haltiger Verbindungen durch die Blätter von *Drosera capensis* L. Rec. Trav. bot. néerl. 33, 351—433 (1936). — OVERTON, E.: Über die osmotischen Eigenschaften der lebenden Pflanzen- und Tierzelle. Vjschr. naturforsch. Ges. Zürich 40, 159—201 (1895). — Über die osmotischen Eigenschaften der Zelle in ihrer Bedeutung für die Toxikologie und Pharmakologie. Vjschr. naturforsch. Ges. Zürich 41, 383—406 (1896). — Studien über die Narkose. Jena: Gustav Fischer 1901.

PAECH, K.: Veränderungen des Plasmas während des Alterns pflanzlicher Zellen, zugleich ein Beitrag zur Kenntnis der Narkose von Pflanzen. Planta (Berl.) 31, 295—348 (1940). — PFEFFER, W.: Über Aufnahme von Anilinfarben in lebende Zellen. Unters. bot. Inst. Tübingen 2, 179—332 (1886).

RUHLAND, W.: Untersuchungen über die Hautdrüsen der Plumbaginaceen. Jb. wiss. Bot. 55, 408—498 (1915). — RUHLAND, W., u. U. HEILMANN: Über die Permeabilität von *Beggiatoa mirabilis* für Anelektrolyte bei Narkose mit den homologen Alkoholen C_1—C_9. Planta (Berl.) 39, 91—120 (1950). — RUMMEL, W.: Theorie der Narkose. Die Narkose, herausgeg. von H. KILLIAN u. H. WEESE. Sutttgart: Georg Thieme 1954.

SAUBERT, G. G. P.: The influence of alcohols on the protoplasmic membrane and colloid models. Rec. Trav. bot. néerl. 34, 709—797 (1937). — SEGEL: Über die Ursache der selektiven Permeabilität des Protoplasmas [Russisch]. Zit. nach LEPESCHKIN 1923. — SPEIDEL, B.: Untersuchungen zur Physiologie des Blutens bei höheren Pflanzen. Planta (Berl.) 30, 67—112 (1939).

TRÖNDLE, A.: Neue Untersuchungen über die Aufnahme von Stoffen in die Zelle. Biochem. Z. 112, 259—285 (1920).

WEIS, A.: Zur Mechanik der Wasserausscheidung aus lebenden Pflanzenzellen. Planta (Berl.) 2, 241—248 (1926). — WINTERSTEIN, H.: Die Narkose, 2. Aufl. Berlin: Springer 1926.

ZYCHA, H.: Über den Einfluß des Lichtes auf die Permeabilität von Blattzellen. Jb. wiss. Bot. 68, 499—548 (1928).

Effects of toxic compounds: stimulation, inhibition, injury, and death.

By

H. B. Currier.

With 7 figures.

I. Introduction.

The effects of chemical substances on plant cells have been studied extensively, and for several reasons. In the realm of fundamental research their use has permitted systematic study of the various cellular functional systems in one way or another; examples are selective respiratory enzyme inhibitors, synthetic growth regulators, and narcotics. In addition, various extremely toxic substances serve as cytological fixing agents, which must possess the capacity of stopping living processes in the shortest possible time in order to preserve original structures. Poisons and the underlying toxicological principles find application in the field of crop protection, where especially rapid advances have been made in the last 20 years. Pesticides have been developed that are lethal to insects but harmless to plants; others that are selective towards fungi. There is today a rapidly expanding group of herbicides, with which it is possible to eliminate plants that grow where they are not wanted. It is encouraging that in a technological field such as weed control many workers are carrying on fundamental research in the physiological (or pathological) effects of poisons on plant cells and tissues. Results of theoretical interest in many instances find immediate application.

That there have been so many studies of toxic effects is understandable when one considers the almost infinite number of organic molecules that can be prepared, the large number of plant species, and the facility with which the chemicals can be applied. Thus there exists in the literature an immense bibliography of the effect of "X" on "Plant species". The biologist finds it difficult to keep pace with the introduction of new compounds by chemical laboratories, due to the careful and time-consuming testing that must be carried out.

Due to the breadth of the subject, certain limitations have been imposed to fit the purposes of this volume. The intention was to deal predominantly with toxicity phenomena in higher plants, with some preference to herbicidal chemicals. An attempt was also made to emphasize those effects observable in the individual cell, relegating secondary importance to responses of tissues, organs, or entire plants. However, to eliminate studies not involving the use or observation of single cells would present an incomplete picture. Responses that are of cellular origin, *e.g.*, photosynthesis and respiration, usually measured on tissues or organs, must of necessity be interpreted in terms of cells.

Distinction between a poisonous and non-poisonous substance must be somewhat arbitrary, since almost any substance if present in excess, or for too long a time, will produce injurious effects. Contrariwise, many well known poisons can be harmless and even stimulating in very slight amounts. With due regard to the difficulties of definition, in the following discussion a *poison* will mean any substance that interferes with the normal functioning of the cell. The meaning of *toxicity* will have to be correspondingly broad, not restricted to the development of irreversible injury, but to all modifications produced by chemicals, transient or persistent, that can be interpreted as injurious.

II. Plant poisons—general occurrence and use.

Naturally occurring poisons are found in soils and waters, in the atmosphere, and in plants, where they may be normal constituents, or where they have been introduced by insects or other organisms. Toxic elements in soil such as boron, selenium, sodium, and chlorine, while interesting from an ecological viewpoint in the distribution of species, constitute problems for agriculture.

Reports of naturally occurring growth inhibitors in all parts of plants are increasing (Bonner 1950a, Grümmer 1953). In some instances definite structures have been determined: trans-cinnamic acid from roots of guayule (Bonner and Galston 1944), 3-acetyl-6-methoxybenzaldehyde from leaves of *Encelia farinosa* (Gray and Bonner 1947), scopoletin (Goodwin and Taves 1950), chelidonic acid (Leopold and Klein 1952), and xanthinin from *Xanthium pennsylvanicum* (Geissman et al. 1954), are examples.

Inhibition of seed germination by naturally occurring substances, both for the same and other species, is a widespread phenomenon (Evanari 1949, Moewus et al. 1951). Coumarin is a well known example, a compound that is finding use as a growth inhibitor in studies of

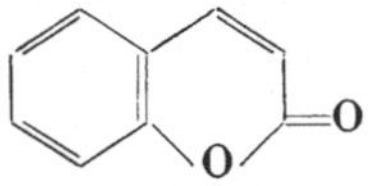

Coumarin
(1,2-benzopyrone).

growth regulation. Unknown toxic materials are present in expressed plant saps and macerates. It seems to be true that such substances are less toxic to the plant producing them than to other species.

There are doubtless numerous substances, as yet undiscovered, that constitute biochemical devices important in the ecological interrelationships of higher plants, and as growth regulators for the individual plant.

Atmospheric poisons that injure or kill plants are usually of industrial origin. Most attention previously has been given to sulfur dioxide (Haselhoff et al. 1932, Thomas and Hill 1937; and various publications of the Boyce Thompson Institute in the USA, cf. Zimmerman 1952). Fluorine compounds, including HF, silicon tetrafluoride or fluosilicic acid, have also caused considerable damage to vegetation (Haselhoff et al. 1932, Romell 1941). Toxic constituents of illuminating gas have received attention, including ethylene, hydrogen cyanide, and carbon monoxide (Crocker 1948).

In a general review of gas damage to plants, Thomas (1951) groups troublesome substances as to toxicity on the basis of fumigation tests. CO, HCN, H_2S, and certain olefins, are least toxic with a threshold concentration in the air of about 50 ppm; HCl, NH_3, and HNO_3 are intermediate, with a threshold of about 10 ppm; Cl_2 and SO_2 are of greater toxicity at 1 ppm; and F_2 and I_2 are extremely toxic at 0.1 ppm or less. Ethylene differs from other olefins in possessing hormonal properties, and is active in high dilution. Mercury vapor is also very toxic.

Inhibition of plant growth by eminations from oils, varnishes and woods has been described by Weintraub and Price (1947). Dormant stored potatoes are believed to evolve substances such as ethylene which suppress their own sprouting (Burton 1952). At the present time, there is an active interest in the extremely toxic constituents of "smog" in the atmosphere of our large cities, apparently a product of civilization. These substances, found to be ozonides by Haagen-Smit et al. (1952), produce characteristic necrotic plant responses. Poisonous contaminants of waters and soils, introduced in by-product disposal, are important industrial problems.

In the laboratory, poisons are useful experimental tools in studies of a wide array of plant functions. Inhibitors, more or less specific, are known to influence mitosis, cell elongation, differentiation, water relations, absorption of nutrients, respiration, and photosynthesis. Cytological fixing agents can be chosen to best preserve the cellular structures to be studied. In the localization of enzymes in the cell, it is important to consider the effect of each constituent in a fixing fluid on the enzymes in question (DANIELLI 1953).

Herbicides. Herbicidal action has become an important field of study for plant physiologists, morphologists, biochemists, and others. The problems are such as to require the collaboration of many specialists. Integrated research programs proceeding at laboratory, greenhouse, and field levels are necessary. Cellular physiology, with focus on the individual cell, is an important segment of the work.

The interesting development of herbicides and their use in weed control is detailed in textbooks (ROBBINS, CRAFTS, and RAYNOR 1952), in monographs on growth-regulating substances (SKOOG 1951, AUDUS 1953, TUKEY 1954), and in various reviews (RADEMACHER 1940, NORMAN et al. 1950, BLACKMAN et al. 1951, CRAFTS 1953b). Table 1 lists representative herbicides in current use, grouped according to manner of application and principal action. It is obvious that in this array of chemicals there is a wide variation in structure and in mode of action. Contact poisons kill only those cells treated; translocated compounds move from place of application to site of toxic action through living parenchyma or phloem tissue. By selective action is meant that a chemical is considerably more toxic to certain species than to others.

Table 1. *Representative types of herbicides.*

Foliar applied, contact action

- Inorganic—arsenicals, chlorates, borates, potassium cyanate (KOCN), cyanamide [$Ca(CN)_2$], kainite ($MgSO_4 \cdot KCl$).
- Substituted phenols—dinitrophenols, pentachlorophenol
- Petroleum oils
- Phenylmercuric acetate (PMAS)
- 3,6-endoxohexahydrophthalate (Endothal)
- Ammonium sulfamate

Foliar applied, translocated

- Auxin type—2,4-dichlorophenoxyacetic acid (2,4-D), 2,4,5-trichlorophenoxyacetic acid (2,4,5-T), 2-methyl, 4-chlorophenoxyacetic acid (MCPA).
- Other growth regulators—maleic hydrazide, 2,2′-dichloropropionic acid.

Root absorbed

- Selective action—trichloroacetates, o-isopropyl-N-phenylcarbamate (IPC), isopropyl-N-(3-chlorophenyl) carbamate (CIPC), 3-(p-chlorophenyl)-1,1-dimethylurea (CMU), N-1-naphthylphthalamic acid (N–1).
- Non-selective—carbon disulfide (CS_2), methyl bromide (CH_3Br), dichloropropane—ene mixture, chloropicrin.
  Borates, chlorates, arsenicals.

The first herbicides used in the field for control of crop weeds were inorganic salts and acids, the list including sulfuric acid, copper sulfate, ammonium sulfate, potassium salts, iron sulfate, and sodium arsenite.

Selectivity was first observed in the use of copper sulfate to control *Brassica* and other broad leaved weeds in grain fields. Kainite and calcium cyanamide were somewhat later applied in Europe in dry powder form both to control weeds and to add deficient K and N respectively to the soil.

Between the years 1915 and 1925 other chemicals were added to the list of useful herbicides: acid arsenicals, carbon disulfide, and sodium chlorate, the latter two applied as soil sterilants.

Another era in chemical weed control began with the introduction of substituted phenols as selective sprays. 3,5-dinitro-ortho-cresol in various salt forms, introduced by BONNET in France (cf. RADEMACHER 1940) some fifty years ago, became the leading selective compound for the control of broad leaved weeds in grain fields; 2,4-dinitrophenol was also used. The success of these substances stimulated a search for other selective chemicals, and 2,4-dinitro-ortho-secondary-butylphenol, or DNSBP (CRAFTS 1945), the amyl derivative, and their salts, became prominent as selectives and fortifying agents in petroleum oils.

OH, O_2N, CH_2—CH, CH_3, CH_3, NO_2

DNSBP

The modern scene began with the discovery that certain compounds having auxin-like properties possessed selective herbicidal activity. In 1945 SLADE, TEMPLEMAN, and SEXTON reported that dilute solutions of naphthaleneacetic acid (NAA) and 2-methyl, 4-chlorophenoxyacetic acid (MCPA) selectively removed seedlings of *Sinapis arvensis* growing in oats *(Avena)*. In the USA ZIMMERMAN and HITCHCOCK (1942) demonstrated the auxin-like properties of 2,4-dichlorophenoxyacetic acid (2,4-D), and in 1944 HAMNER and TUKEY reported similar properties in 2,4,5-trichlorophenoxyacetic acid (2,4,5-T). The principal so-called hormone herbicides in use today are these three substances: 2,4-D, 2,4,5-T, and MCPA.

NH—COOCH, CH_3, CH_3

IPC

O—CH_2—COOH, Cl, Cl

2,4-D

O—CH_2—COOH, Cl, Cl, Cl

2,4,5-T

O—CH_2—COOH, CH_3, Cl

MCPA

Various fractions of petroleum oil find use as both general and selective herbicides (CRAFTS and REIBER 1948). In general, the toxicity increases with the aromatic content. The toxicity of carbamates was reported by FRIESEN (1929), and herbicidal selectivity by TEMPLEMAN and SEXTON in 1945. Root absorbed, these substances are quite toxic to grasses under certain conditions (IVENS and BLACKMAN 1949, SHAW and SWANSON 1953). Trichloroacetic acid and 2,2-dichloropropionic acid, as the sodium salts, are effective grass killers, both when foliar sprayed and when root absorbed. Interesting because of their simple molecular structure, the latter compound especially has the properties of translocatibility and of producing formative effects. Maleic hydrazide, announced by SCHOENE and HOFFMAN (1949), has proved to possess many useful properties. It is used as a temporary growth inhibitor and as a grass killer. Phenylmercuric acetate, with both fungicidal and herbicidal properties, is employed for controlling crabgrass in lawns.

N-1-naphthylphthalamic acid is successful as a selective herbicide in curcurbit crops. Disodium 3,6-endoxohexahydrophthalate is an unusual herbicide because of a slow contact action. It is also an effective defoliant. CMU (BUCHA and TODD 1951) is employed as a

C(=O)OH, CONH—

N-1

NH—CO—N, CH_3, CH_3, Cl

CMU

preemergence herbicide and organic soil sterilant, and as a foliar spray. Two additional chemicals that show promise are 2,3,6-trichlorobenzoic acid, and 3-amino-1,2,4-triazole, and there are many others under investigation.

III. Some underlying principles of toxic action.

a) Properties of the toxicant molecule.

Molecular structure, and the properties related to structure, determine to what extent a substance will penetrate to the site of action, and, once there, to what extent it will react with cellular constituents. The relation of structure to biological activity is broadly discussed in SEXTON'S (1953) monograph. The effectiveness of plant growth regulators is precisely related to the number, position and kind of ring substituents and length of side chain (cf. reviews by THIMANN 1951, VELDSTRA 1953, MUIR and HANSCH 1955).

In hydrocarbon series, decrease in polarity is accompanied by a higher lipid-water distribution ratio and increased toxicity up to a point. So that polarity is first of all important in penetration and accumulation of the toxicant. CRAFTS (1948) considered that the leaf is adapted to absorption of relatively apolar molecules, whereas the root absorbs polar substances more readily. This tendency may be reflected in the relative resistance of root tissues to apolar solvents (CURRIER and PEOPLES 1954). When polar and apolar groups are sufficiently separated, the molecule is surface active, a property that can influence toxicity in one way or another. VELDSTRA (1949) stresses the importance of a certain polar-apolar balance in structure-activity relationships of growth regulators.

The p_H of the solution is important. It has been shown that many poisons penetrate cells more rapidly in undissociated form than as ions (cf. HÖBER 1945, ALBERT 1951). With relation to growth substance interaction, VELDSTRA (1953) notes exceptions to this generalization, and also points out that it is necessary to consider the effect of p_H on the protein and protein complexes of the cell in addition to external effects. Acidic molecules are generally more toxic in acid solution by virtue of the fact that a larger proportion of molecules is in unionized form; conversely, certain weak bases are more toxic in alkaline solution (SIMON and BEEVERS 1952). The activation of 3,5-dinitro-o-cresylate (DNOC) solutions

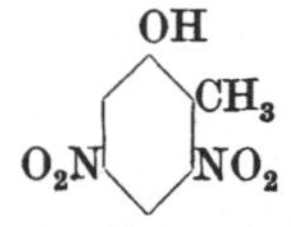

OH
CH_3
O_2N NO_2
DNOC

by ammonium sulfate was explained by CRAFTS and REIBER, (1945) as a lowering of p_H, which resulted in suppressed ionization and increased penetration. The importance of p_H to herbicidal effectiveness has been demonstrated by SIMON and his associates (1949, 1952).

b) Influence of the external molecular environment.

It is not sufficient to consider the nature of the toxicant molecules and simply their number external to the cell to predict or to explain their toxic behavior. The effect of other molecules, those of the solvent and of additional substances present in the solution, must be taken into account. The external molecular environment, as it can be called, through various intermolecular attractions and repulsions, including effects on dissociation, controls in a large measure the *activity* of the toxicant, and here is meant both the thermodynamic activity and the biological activity.

Effect of the solvent. It was suggested by FERGUSON (1939) that the effectiveness of toxic substances can best be compared by the value of their chemical potential in the cellular phase constituting the site of toxic action, rather than by their concentration. The activity externally is theoretically equal to that within the cell. The activity function of G. N. LEWIS (cf. LEWIS and RANDALL 1923) is generally used as an expression of chemical potential. This hypothesis, now referred to as the *Ferguson Principle*, has been supported by a number of different investigators (cf. ALBERT 1951). It holds that indifferent narcotic-like substances, when present at the same activity in a solution, exhibit equal toxicities. Whereas usually there is no relation to concentration, relative saturation values for substances poorly soluble in water lie within 0.1 and 1.0. In contrast, substances acting more or less specifically exhibit activities between 0.001 and 0.1. Interactions between solvent and solute determine solubility and activity of the solute and in this way affect the distribution between cell and external solution. That the nature of the solvent exerts a great influence on the effectiveness of the toxicant can be easily demonstrated. Benzene vapor dissolved in air is lethal to plants at about 0.003 M/liter; dissolved in water, the minimum lethal concentration is in the order of 0.008 M/liter. When paraffin oil is the diluent the threshold concentration is much higher. Molecular structure of indifferent substances is important in toxicity for the most part in the way that it determines physical properties, upon which movement and accumulation in the cell depend. CURRIER and PEOPLES (1954) investigated the phytotoxicity of different low-boiling hydrocarbons by treating several kinds of plants and tissues with vapors diluted with air, and with aqueous solutions. Considering toxicity as the reciprocal of minimum lethal concentration, the order of decreasing toxicity was benzene, cyclohexene, cyclohexane, hexene-1 when applied in vapor form, but this order exactly reversed when used in aqueous solution. The reason is given by solubilities in air (vapor pressure) and in water (Table 2). Thus, molecule for molecule, as supplied in the external solution, change of solvents effected a reversal of toxicity. This emphasizes the distinction between *apparent* and *intrinsic* toxicity. Intrinsic toxicity can be masked by a variety of factors—low permeability of the cells or tissues to the substance, volatility, and effect of diluents or additives.

Table 2. *Water solubility and vapor pressure of hydrocarbons at 25° C.*

	Moles/liter in water	Vapor pressure mm. Hg
Benzene	0.0237	94.5
Cyclohexene . . .	0.0034	87.0
Cyclohexane . . .	0.0013	98.8
Hexene-1	0.00092	183.0

Within homologous series there is an inverse correlation of toxicity with water solubility; otherwise there is no clear relation.

Assuming the site of hydrocarbon accumulation in the cell to be the lipoid phases, and the lipid to have properties similar to those of vegetable oils (cottonseed, corn), the theoretical minimum lethal concentration in lipid was shown to be approximately equal for five different hydrocarbons, 1.9–2.5 M/liter in the case of *Hordeum* leaves. Calculations were made by use of concentration values in the external solution and constants representing the distribution of toxicant between the phases oil/air and oil/water (Table 3). Although solubilities and apparent toxicities vary considerably there is no great difference in the intrinsic toxicity of the five hydrocarbons. MEYER and HEMMI (1935) made similar calculations for a series of animal narcotics and arrived at rather constant values.

Effect of additives. Substances present in a solution other than solvent and toxicant are termed *additives*, which in various ways can influence the toxicity. In many instances, *e.g.* when used with translocated poisons, the additive itself should not be toxic.

Table 3. *Minimum lethal concentrations, distribution ratios, and calculated concentrations of hydrocarbons in cell lipid of barley leaves.*

	Minimum lethal concentrations		Distribution ratios at 25° C		Calc'd concentrations in cell lipid Hydrocarbon in	
	in air mM/L	in water mM/L	oil[1]/air	oil/water	air M/L	Aq. Soln. M/L
Benzene	2.6	8.3	960	206	2.5	1.9
Cyclohexene . .	2.7	1.4	911	1270	2.5	1.9
Cyclohexane . .	3.9	0.91	560	2210	2.2	2.4
Hexene-1 . . .	7.6	0.73	283	3040	2.2	2.5
n-Hexane . . .	6.3	not leth.	305	3590	1.9	>2.5

[1] cottonseed.

Substances that lower interfacial tension at water-lipid interfaces, *surfactants*, in many instances increase herbicidal effectiveness. They may increase the area of contact between spray and plant surface, and probably accelerate penetration across the cuticle and into the cell.

Filming agents such as petroleum oils maintain close contact between the toxicant and plant surface. *Cosolvents* permit more toxicant to dissolve in the solution; examples are alcohols and glycols. *Hygroscopic agents* prevent drying of the solution on the plant, and *deposit builders* provide a thicker film or reservoir of chemical.

Activators increase the effectiveness of the toxicant in various ways. Already in 1920 it was reported by MILLER that the toxicity of a phenol solution towards bacteria could be increased by adding NaCl and other neutral salts, the result of an increased chemical potential of phenol. Boron as an activator is reported to increase the absorption and translocation of 2,4-D (MITCHELL et al. 1953).

In combinations of poisons a greater than additive toxicity is termed *synergism*; less than additive effectiveness is *antagonism*.

c) Morphological relationships.

Foliar absorbed chemicals move either through the cuticle and across the outer epidermal wall to reach protoplasts of epidermal cells, or first through the stomata and into mesophyll protoplasts. There is disagreement as to how important the stomata are as channels of entry. BOYNTON (1954) has summarized the factors involved in the absorption of foliar applied nutrients, concluding that both cuticle and stomata are important. Many herbicides, of low polarity, would be expected to dissolve in the cuticle to a greater extent than mineral salts. Water in bulk does not easily enter the substomatal chamber, even with surfactants present. A study of enzymatically-detached sheets of foliar cuticle led ORGELL (1954) to conclude that passage of chemical substances across the cuticle of some species can be rapid due to the many cracks, imperfections, insect punctures, etc. The way in which surfactants can affect the amount of herbicides and other substances absorbed by cuticle was shown to be related to polarity and charge of the respective molecules, and to p_H.

SCHUMACHER (1942) and LAMBERTZ (1954) have demonstrated the occurrence of plasmodesmata in the outer epidermal walls of *Primula* and other plants, and this too is interesting in the problem of foliar absorption.

Stomata as paths of entry to gaseous toxicants make possible rapid absorption by leaves. Petroleum oils and hydrocarbon solvents in liquid form move readily into leaves when the stomata are open, much more slowly when closed.

Some resistance to movement of chemicals is offered by the cell wall. Since, typically, plant cell walls are composed of hydrophilic polysaccharides, highly water-insoluble substances must be prevented from penetrating. This is probably a factor contributing to the non-toxicity of higher saturated aliphatic hydrocarbons.

The presence of plasmodesmata makes the situation less simple. They may facilitate absorption of gaseous herbicides from the intercellular spaces (VAN OVERBEEK and BLONDEAU 1954). There is so much of a mystery about these protoplasmic connections, as to their exact location, their structural and functional nature, that a decision as to any specific role in absorption is impossible. The response of plants to gaseous hexane and hexene, highly water-insoluble substances, is very rapid, suggesting that it is not simply a diffusion across a hydrated cellulose layer, even though the distance to be traveled is very short. Aside from the plasmodesmata, the exact intimate relationship between the plasma membrane and the cell wall, and how far into the wall structure the protoplast extends, are problems needing clarification.

With regard to toxicants in aqueous solution, the microcapillaries within the walls seem to be large enough to permit the passage of all substances in the molecular state. The rapidity with which the flourescent dye trisodium 3-oxy-5,8,10-pyrenetrisulfonate moves through the cell walls in what STRUGGER (see 1949) has termed the extrafascicular transpiration stream is convincing in view of the large molecular dimensions of the dye.

d) Stimulation and inhibition and their interpretation as toxic phenomena.

Growth stimulation in herbicidal field practice has been reported many times. It may be a matter of concentration of chemical, a level below the lethal range. However, interpretation is not always simple, since explanations other than a direct stimulating effect must be considered. For example, fertilizing action of N-containing herbicides such as cyanamide or sulfamates, or removal of competing microorganisms, may be the answer.

From dosage-response curves, it is clear that many poisons, even those that are extremely toxic, produce in dilute concentration a growth stimulation. It seems that this may be a quite general principle. It is clearly demonstrated by auxins (WENT and THIMANN 1937, SÖDING 1952), and by other substances of unrelated structure, *e.g.*, undecanoic acid (BURSTRÖM 1950), and hydrocarbons (MEITES 1944). Furthermore, in the response-time curve, for a particular lethal concentration, there may be an initial stimulation followed by inhibition. Not conforming to this behaviour is maleic hydrazide, which, in the *Avena* test was found by LEOPOLD and KLEIN (1952) to have only an inhibiting action throughout the range from 10^{-10} to 10^{-3} M/liter. Benzene hexachloride (HOPKINS 1952), also exerted only irreversible effects at various concentrations.

How does a toxicant effect a general stimulation of cellular activity? A great deal has been said about a cellular master reaction, which, when completely identified and understood, would explain the similarity of response invoked by all types of stimuli — chemical, mechanical, electrical, heat and so on.

Auxins are believed to cause dissociation of cellular proteins, on the basis of a decreased protoplasmic viscosity and other evidence (NORTHEN 1942), and, as a result, increased swelling, respiration, and polysaccharide synthesis. In the dissociation, reactive surfaces, *e.g.*, sulfhydryl groups, may be freed and participate in metabolic processes.

According to a general theory of toxicity proposed by SEIFRIZ (1951), changes in molecular cohesion may be a common factor involved in stimulation and inhibition. A stimulant such as caffeine reduces molecular cohesion in protoplasm and a depressant, *e.g.*, chloroform, increases it.

HEILBRUNN (1952), considering the similarities in the irritable process as effected by all of the various stimuli, believes that the fundamental process in stimulation is a loss of calcium from the protoplasm. This theory may explain the decreased viscosity, increased permeability, and other responses evoked by stimulants.

DANIELLI (1950) comments on the concept of a common mechanism, concluding that different stimuli may act on different cellular systems and still produce similar responses. By his reasoning, the cell is so constructed that the different reactions would be "funnelled" in such a way that the response is characteristic of the particular design of cell.

Permeability changes might be involved, but it is often difficult to say whether they are direct or subsequent effects. Increases in water permeability due to auxins in stimulating concentrations are known (POHL 1948, v. GUTTENBERG and BEYTHIEN 1951). Permeability toward sugar (v. GUTTENBERG and MEINL 1952) and toward urea and glycerine (MASUDA 1954) also is increased. It is reasonable that an increased cellular interchange of substances under some conditions can result in stimulation. This may happen, for example, in connection with the change-over from a condition of rest to the active state. Auxin induced water uptake (REINDERS 1942) has been suggested as a possible central process regulating growth in which ATP plays an active part (BONNER and BANDURSKI 1952). The difficulty of interpretation, and the lack of decisive evidence for an active water transfer (BRAUNER and HASMAN 1952, BURSTRÖM 1953a, b, LEVITT 1953) make it more likely that water uptake is a necessary consequence of growth.

ALLEN and SKOOG (1952) suggested that stimulation of seedling growth of *Raphanus sativa* and *Triticum vulgare* by N-phenyl succinimide might involve changes in permeability or interfere with auxin metabolism in some way. The suggestion has been made that non-specific stimulation by chemicals involves the formation of auxins from a precursor (*e.g.*, RAADTS and SÖDING 1948).

e) Selective toxicity.

Of more than theoretical interest, the fact that some poisons affect certain species and not others — or better, *more* than others — is of importance in all branches of biological control. It is the basis for the formulation of effective insecticides and fungicides that do not injure the host plant, and for herbicides that kill weeds and leave the crop unharmed. The many variables to be considered in the problems of selectivity and the relatively few combinations of these variables that have been tested and reported, hold great promise for better selective control of weeds and other pests than we now have. At the same time, our knowledge of basic cellular processes will increase. In addition to combinations of species, the variables remaining include: structure of the toxicant molecule, dosage, composition of solution, time of day, climatic conditions, soil type, and age and stage of growth of plants.

Most of the information as to selectivity among plant species derives from herbicidal field and greenhouse studies. Under those conditions it is a matter of response of the whole plant, and usually whether the plant survives or dies.

Distinction may be made between morphological and biochemical selectivity. In the former, the structure of the plant is such as to prevent sufficient chemical from arriving at the point of action to cause injury. It may be a matter of shape and position of leaves, or character of leaf surface—whether rough and hairy, or smooth. Thickness of cuticle and stomatal distribution may be factors. Buds may be relatively protected or exposed, and there may be differences in the ability to regenerate damaged organs. Distribution of root systems is a factor in connection with soil-applied chemicals. AUDUS (1953) has compiled a useful list of higher plants with designation as to relative resistance toward the auxin herbicides.

Biochemical selectivity differs in that the cause resides in the protoplasm itself. An outstanding example is the well known resistance of grasses *(Gramineae)* to 2,4-D and similar compounds, in contrast to the general susceptibility of broad-leaved species. Here it is not a matter of penetration; resistant species can absorb and translocate an amount of chemical equal to that taken up by the susceptible species. Rather the protoplasm of the two groups of plants differs in some fundamental way. Discussion of this problem — selectivity at the cell level — is continued in a later section.

IV. Methods in toxicity studies.

Usually more than a single type of test is necessary to obtain a precise measure of phytotoxicity of chemical substances. Procedures for herbicides are discussed by BLACKMAN et al. (1951), BLACKMAN (1952), SHAW and SWANSON (1952).

Using entire plants, chemicals may be sprayed on the foliage, or applied to soil or water. The criterion of injury and death may simply be gross appearance — wilting and obvious necrosis. Measure may be made of the change in size, form, or chemical composition, of the whole plant or its parts. In localized treatments, as for example, the single drop technique, accurately known amounts of toxicant can be applied. Various criteria of response include suppression of leaf growth (BROWN and WEINTRAUB 1950), and the bean curvature test (DAY 1952). Radioactive tracers have proved useful (CRAFTS 1953a). Water plants such as *Elodea* (HERBST 1937) and *Lemna* (BLACKMAN and ROBERTSON-CUNINGHAME 1953) are advantageous. For *Lemna*, reproduction, or lack of it, is a simple and satisfactory measure of response.

At the tissue level, discs of leaves or of storage parenchyma are used in studies of absorption, permeability, respiration and other metabolic activities. Leakage of salts (STILES 1927), or of anthocyanin (VELDSTRA 1948), can be measured, the amount proportional to injury. Auxin action of growth regulators, as reflected in changes in cell elongation, is measured by the well known *Avena* coleoptile curvature and straight growth tests, and the pea section test. Tissue culture procedures permit precise control of experimental conditions (GAUTHERET 1945, SKOOG 1950, WHITE 1954). Simple seed germination tests permit measurement of root length.

At the cell level, one may choose microorganisms, detached thin organs such as *Elodea* leaves, flower petals, roots, hairs, and various sections or strips of tissue. Growth may be measured by cell counting techniques such as that used by HOPKINS (1952), by measuring extension growth of cells (WILSKE and BURSTRÖM 1950), or as a further example, by the size of fungal colony (SIMON and BLACKMAN 1949).

Considering toxic responses in the individual cells themselves, reliable criteria of vitality are needed to distinguish the living, the injured and the dead condition. The various tests that have been employed include plasmolysis, vital staining, and microscopically visible changes in form and structure.

Plasmolysis is generally dependable. Cells giving this response are living, those which do not usually are dead. Injury can be detected by the development of anomalous plasmolytic forms (cf. KÜSTER 1951).

Neutral red accumulation in plant vacuoles has long been known to occur only in living cells. The dye has a rather low toxicity and the results are easily observed. However, whereas positive accumulation is a definite indication of the living condition, lack of it is not proof that the cell is dead. The reasons why some living, apparently normal cells do not accumulate the dye are not understood, although p_H relationships doubtless are involved (Zirkle 1937).

Utilization of acridine orange as a vital dye (Strugger 1949) has many applications in connection with fluorescent microscopy. Under prescribed conditions, living protoplasm appears green, dead protoplasm red. It is a very rapid test.

A third important vital stain is tetrazolium salt. Among several derivatives, 2,3,5-triphenyltetrazolium chloride (TTC) is the most generally useful (Kuhn

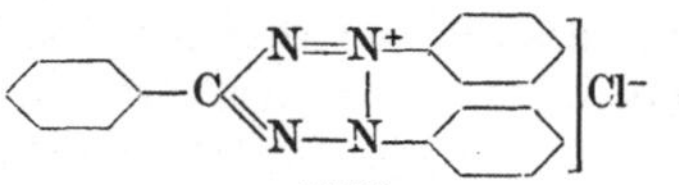

TTC

and Jerchel 1941). TTC is somewhat unique among dyes in that it is colorless in the oxidized state. Intracellular enzymes reduce it to a water-insoluble red formazan that appears as granules, or dissolved in oil droplets, and in later stages as large amorphous and crystalline deposits. The reaction is useful as a criterion of vitality (cf. F. E. Smith 1951, Parker 1953). With *Elodea* leaves as test object, it differentiates between dead and living cells, and allows an estimation of injury. Used in concentrations as low as 0.025%, the toxicity is slight. The method can be made quantitative by extracting the formazan with methanol and determining the amount colorimetrically (van der Zweep 1954). Comparison of the TTC and plasmolytic methods as criteria of vitality in *Elodea* leaves indicate good agreement (Currier and van der Zweep 1955), except where cells were first treated with maleic hydrazide, which inhibited the TTC reaction. Ziegler (1953) earlier reported that iodoacetate inhibits TTC reduction without affecting plasmolyzability. The reaction is partly inhibited by plasmolysis. Thus there are conditions under which the TTC method is unreliable. Another potential disadvantage is that sometimes penetration can be limiting. Formazan production in yeast is accelerated by half saturating the solution with benzene (Currier and Day 1954), probably due to a more rapid penetration of the dye.

In any case, the method to be used must fit the need and the nature of the test objects. Experience with cells in various conditions enables the worker to recognize dead cells by their characteristic appearance — granular, discolored and shrunken protoplast. Although the presence of cytoplasmic streaming is positive proof of life, its absence cannot be used generally to detect dead cells.

V. Mechanisms of toxic action in plants.

At this point some of the various means by which plants are injured or killed by chemicals may be considered. Surprisingly few mechanisms are known with any degree of completeness or certainty. The problems are complex. One of the difficulties is that a substance can act in more than one way. 2,4-D influences respiration, carbohydrate utilization, mitosis, and differentiation, to mention a few of the known effects. Maleic hydrazide is an antimitotic, antiauxin, and a general protoplasmic poison in higher concentrations (Åberg 1953). Gavaudan and Brébion (1951) studied the effects of benzene, chloroform, and acenaphthene in leaves of *Triticum vulgare* on the inhibition of several processes. They found that growth, initiation of karyokinesis, process of karyokinesis, chlorophyll synthesis, and photosynthesis, were not equally sensitive, and suggested that the inhibitions were caused by different mechanisms.

Secondary effects may be more important than the primary reaction. The problem in relation to herbicidal mechanisms was well expressed in 1950 by NORMAN et al.: "In no case has a complete and detailed description of the causative sequence of events been proposed. Information is available as to a number of physiological, biochemical, and morphological derangements which may attend the action of herbicides but, for the most part, it is not clear whether these are to be regarded as causes of death or as incidental responses."

The simplest and best understood toxic action is the osmotic effect. Injury is due both to dehydration, with probable inhibition of enzymes, and, when the external concentration is sufficiently great, to plasmolysis. In the plasmolytic concentration, the intimate relationship between the cell wall and protoplast is disturbed, the protoplasm is subjected to mechanical stress, plasmodesmata are ruptured, and injury results.

General protoplasmic poisons may directly affect proteins, resulting in denaturation and coagulation. Strong acids and bases upset sensitive p_H relationships of proteins, heavy metals cause precipitation, oxidizing and reducing agents react with —SH and other labile groups.

Strong alkalis such as NaOH have a destructive effect on protoplasm, as evidenced by a progressive increase in fluidity, swelling, and vacuolation (SCHINDLER 1938a). Strong acids, as HCl, are sufficiently slow in penetrating that they can act as plasmolyzing agents, but the protoplast is soon killed. Acids differ from alkalis by their fixing and hardening action on protoplasm (SCHINDLER 1938b).

Cells removed some distance from the tissue treated with contact poisons may suffer from starvation, desiccation, a disturbance in hormone balance, or translocated secondary toxic products. Certain chemicals are known to cause rather rapidly a breakdown of young sieve tubes. In particular, maleic hydrazide and 2,4-D have this effect. Such interference in phloem translocation results in systemic injury, which must be considered in addition to the specific cellular effects in the overall picture of toxic action in higher plants. Cell proliferation can also interfere with translocation (EAMES 1950). Growth substances effect closure of stomata (BRADBURY and ENNIS 1952), disturbing normal water relations and gaseous interchange. Root absorption of water is decreased by enzyme inhibitors such as azide (ROSENE 1947) and HCN. BROUWER (1954) suggests on the basis of the HCN effect that the resistance to the water current is located somewhere in the protoplasm. Inhibitors also reduce root pressure in decapitated plants (cf. KRAMER and CURRIER 1950).

Increase in permeability may be considered a mechanism of action, through a direct action on the plasma membrane, injury resulting from a leakage of cell constituents. Certain substances may act to decrease permeability, thus interfering with normal cellular interchange. For example, uranium blocks the uptake of fermentable sugars by yeast, with little influence on other parts of the metabolism (ROTHSTEIN 1954).

Through varying degrees of specific enzyme inhibition, compounds are known that inhibit mitosis, cell enlargement, respiration, photosynthesis, salt accumulation, synthesis of chlorophyll, nitrogen assimilation, and other metabolic sequences.

Anti-auxins produce responses in the plant that result from inhibition of auxin action with subsequent derangement of growth processes. Other compounds produce morphological abnormalities — cell wall overgrowths, giant cells, formative effects in organs, including abnormal growth stimulations that result in galls and tumors.

VI. Visible responses of cells to poisons.

a) Effect on protoplasmic streaming.

One of the most obvious effects of chemical substances on plant cells is the influence on rate of protoplasmic streaming. Those who have watched this phenomenon can understand why it has fascinated many students of cell physiology. EWART in his monograph appearing in 1903 summarized the information on toxic relationships available at the time. More recent reviews include those of SEIFRIZ (1943), HEILBRUNN (1952), VIRGIN (1953) and BOGEN (1954).

Commonly employed test materials are leaves of *Elodea* and *Vallisneria*, *Nitella* cells, *Tradescantia* staminal hairs, root hairs, slime mold plasmodia, and that much studied object, the *Avena* coleoptile. Streaming is considered to be a general phenomenon, characteristic of normal uninjured cells (SEIFRIZ 1943), and not solely a pathological condition.

Some substances only inhibit the rate of streaming or have no effect. Measurement of flow in the *Avena* coleoptile indicates that dinitrophenol (DNP) behaves in this way (OLSON and DU BUY 1940). In the case of a slime mold *(Physarum)*, DNP, HCN, and NaN_3 were so designated (ALLEN and PRICE 1950).

Many substances accelerate streaming at low concentrations, inhibit at higher concentrations. Included are narcotic agents and growth regulators such as IAA and 2,4-D. Caffeine, conforming to its presumed role as a stimulant, accelerated flow in a desmid (WARIS 1951). Ethylene chlorohydrin and thiourea increased the rate in *Elodea* and *Nitella* (MARCY 1937). Indole-3-acetic acid (IAA) in concentrations of from 0.005 to 1 mg/liter, increased streaming in the *Avena* coleoptile (THIMANN and SWEENEY 1937), and in *Avena* root hairs with an optimum of 10^{-3} to 10^{-5} mg/liter (SWEENEY 1944). These reports are in line with the reported effect of auxin in decreasing viscosity (NORTHEN 1942).

As a shock effect streaming is slowed or ceases, sometimes temporarily, when there is an abrupt change in the environment, *e. g.*, exposure to poisons. Such a shock stoppage or retardation may be later followed by acceleration, and, as the concentration is increased, by final stoppage. When, however, the cells are exposed gradually to increasing concentrations, of CO_2 and ethanol for example, they apparently adapt, and the initial depression in rate does not occur.

Stimulative responses must also include the initiation of streaming in quiescent cells. To this was given the name "Chemodinese" by FITTING (1925), who studied the phenomenon in leaves of *Vallisneria*, a plant whose cells normally do not stream unless injured. He (1933) found leaf extracts of this plant and very small amounts of l-histidine (1936) to be effective stimulants. A similar stimulation in *Elodea* leaf cells by widely different herbicidal compounds has been reported (CURRIER 1949).

In general there seems to be no clear relation between rate of respiration and streaming velocity. In some instances limitation of oxygen supply has produced no immediate effect. *Elodea* leaves will remain streaming for days mounted on slides in paraffin or silicone oils. Depression of respiration by KCN concentrations as high as 0.001 M/liter has little concurrent effect on streaming in *Physarum* (ALLEN and PRICE 1950). Dinitrophenol increases respiratory rate at the same time that streaming is inhibited. On the other hand, in the *Avena* coleoptile the effect of indole-acetic acid on the streaming rate is paralleled by changes in the rate of respiration and growth (SWEENEY and THIMANN 1942); and GORENFLOT (1947) found parallel decreases of respiration and streaming in *Elodea* due to the influence of sodium chlorate.

Several odd types of movement in *Elodea* leaves are initiated by chemicals. Two of these have been termed "spiral" and "belt" rotation, induced by alcohol or strontium chloride (SEIFRIZ 1922), and are observed in plasmolyzed cells (KÜSTER 1910). Endothal (disodium 3,6-endoxohexahydrophthalate) also has

$$\begin{array}{c} \text{H} \\ | \\ \text{C} \\ \text{H}_2\text{C} \diagup \quad | \quad \diagdown \text{C} \diagup^{\text{H}}_{\diagdown \text{COONa}} \\ | \quad \text{O} \quad | \\ \text{H}_2\text{C} \diagdown \quad | \quad \diagup \text{C} \diagup^{\text{COONa}}_{\diagdown \text{H}} \\ \text{C} \\ | \\ \text{H} \end{array}$$

Endothal

this type of action, where streaming continues for considerable time in plasmolyzed protoplasts in a quite anomalous fashion (CURRIER and VAN DER ZWEEP (1955). Fig. 1 shows some of these effects.

Interpretation of streaming behavior as a toxic response depends upon the test object to some extent. In cells that normally always stream, *e.g.*, *Tradescantia* staminal hairs, depression of rate or stoppage is considered evidence for at

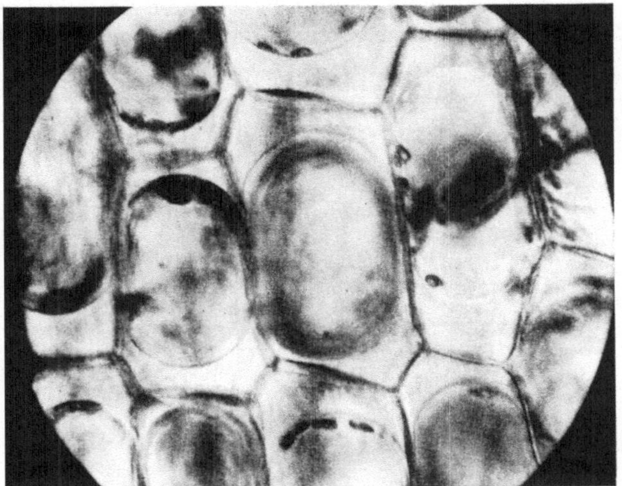

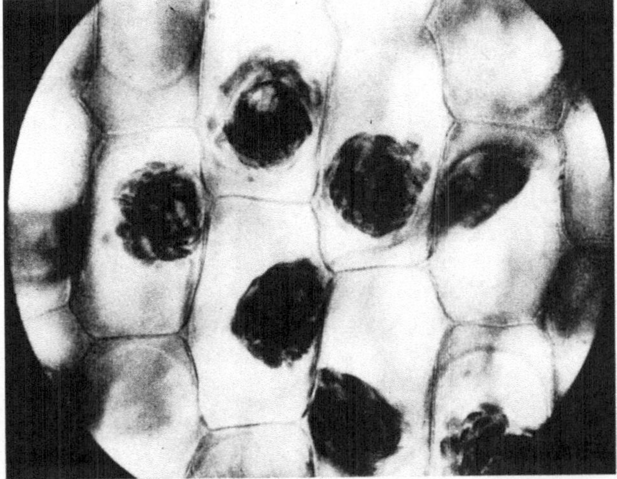

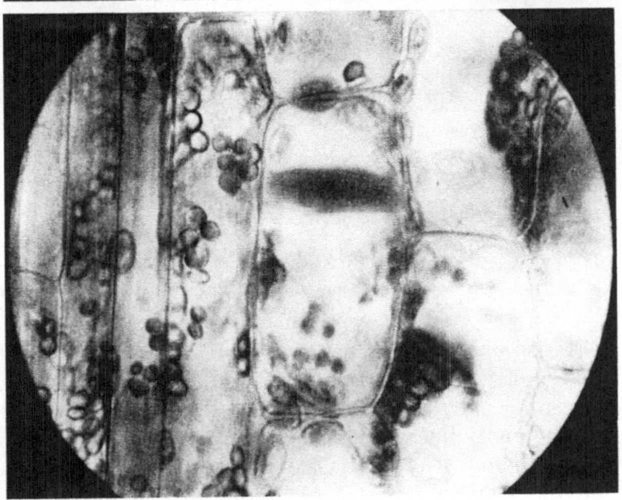

Fig. 1. The stimulative effect of endothal in inducing anomalous streaming behavior in *Elodea canadensis* leaf cells. *Above:* rapid cyclosis of plasmolyzed cells is indicated by the blurred periphery of protoplasts. *Middle:* balling of chloroplasts and a portion of cytoplasm, showing rotation in plasmolyzed cells. *Below:* "belt" streaming indicated by blurred path across the width of the cell in center.

least transient injury. For others, activity may indicate stimulation or incipient injury; often streaming is more rapid in *Elodea* cells adjacent to a wound or dead

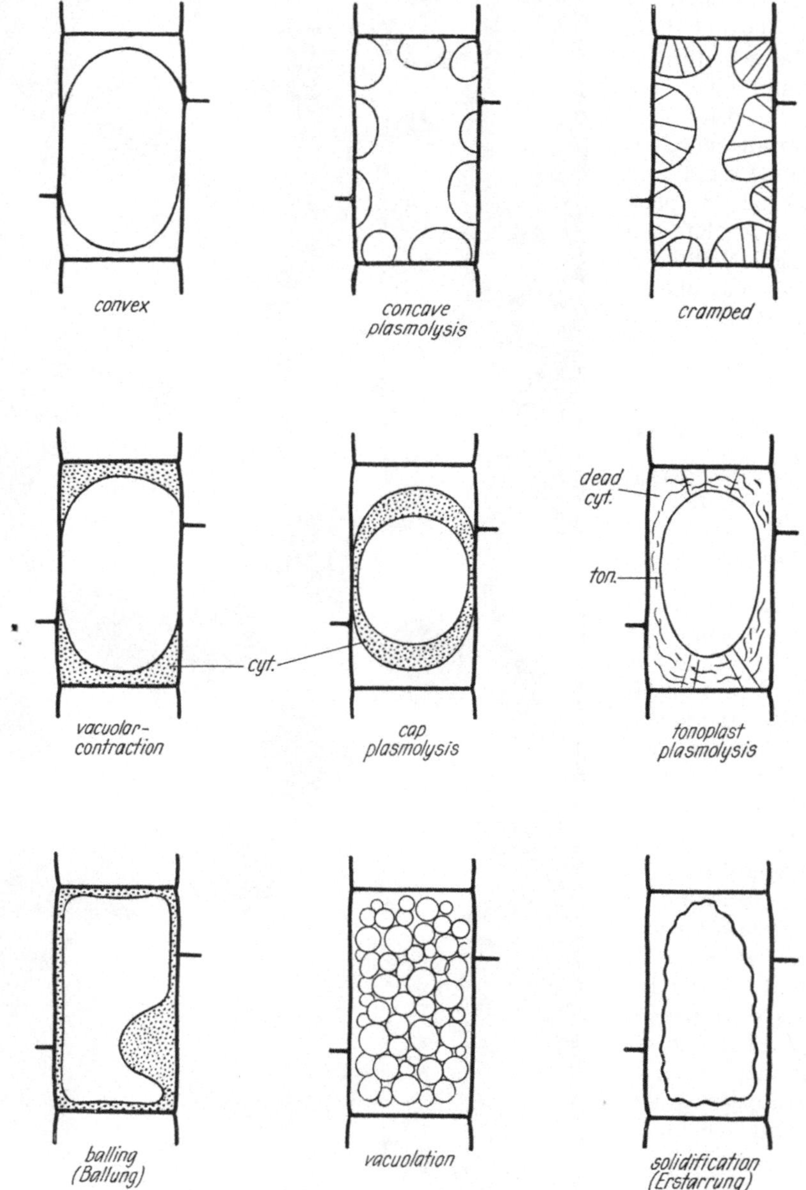

Fig. 2. Types of plasmolytic form (top) and diagrammatic representation of certain other form and structure changes in plant cell protoplasts.

cell. For example DANGEARD et al. (1947) found that colchicine at 1:1000 increased streaming in this plant, while GAVAUDAN and BRÉBION (1951) noted that this same concentration tended to inhibit photosynthesis.

The many factors influencing streaming, in addition to the effect of a particular poison being studied, make interpretation sometimes difficult. Plant species, age and stage of growth, technique of removal of plant part and possible wound stimulation, suddenness of exposure to the poison, length of exposure, and light (which in itself is stimulating) are some of the factors involved. According to VIRGIN (1953) it is generally accepted that low viscosity and rapid flow go hand in hand. The effect of poisons in changing streaming velocity are probably often direct influences on viscosity.

b) Form and structure changes.

The form and visible changes in the structure of the protoplast can provide information as to the condition of the cell. Fig. 2 shows some of the effects. Plasmolytic techniques are useful; as a general rule, convex plasmolysis is typical of the normal uninjured condition, while a concave or cramped pattern indicates injury. The time element is important; the convex form often develops secondarily as the protoplast pulls more and more away from the wall. Age of cells must also be considered, as older cells may exhibit convex shapes even though injured. Calcium salts typically produce changes in the surface which lead to concave plasmolytic forms; sodium and potassium salts promote convex forms. Involved in these form changes is a modification in the adhesion between the cell wall and protoplast surface, perhaps best considered as a "stickiness". Narcotics in dilute concentration can contribute to a more convex shape. Strong alkalis, though destructive, promote low viscosity and convex plasmolysis.

With increasing injury, abnormal forms appear; there is a stiffening of the ectoplasm, and the outline of the protoplast is no longer smooth and glistening. Attempts to deplasmolyze cells under these conditions often results in rupture and escape of vacuolar contents.

Stimulative plasmolysis (Reizplasmolyse) is a general reaction to many kinds of stimulation. It is often observed in intact cells of the wound border of a section. Since it procedes in hypotonic solution its mechanism is fundamentally different from that of normal plasmolysis. There seems to be little molecular specificity with respect to chemical stimulators. Weakly acidic media (KEMMER 1928, p. 20), are known to produce the reaction. Certain plant juices contain unknown substances that are very active in producing this type of plasmolysis.

c) Displacement of cytoplasm.

Systrophic balling is a term applied to the accumulation of cytoplasm in one part of the protoplast, leaving a thin layer in the other parts. 2% ether is known to produce the effect (HÖFLER and WEBER 1926, p. 697). *Formation of intravacuolar cytoplasmic strands* results from chemical stimulation, *e.g.* from chloroform, alcohol and acids (KÜSTER 1951). *Aggregation*, first studied by F. DARWIN in 1876 in *Drosera rotundifolia* tentacle epidermal cells, is a reduction of the vacuolar volume by the intrusion of the cytoplasmic strands and lamellae, thus resulting in the production of many small vacuoles dispersed in the increased volume of cytoplasm. Substances known to produce this effect include urea, H_3PO_4, phosphates, ethanol and ether.

d) Structural cytoplasmic changes.

Swelling of cytoplasm with concomitant reduction in vacuolar volume, *vacuolar contraction*, is a general response to a variety of stimuli (KÜSTER 1951, LEPESCHKIN 1937). It is commonly believed (cf. BOGEN 1951), to be of abnormal occurrence and indicative of injury. The process is similar to stimulative plasmolysis in that it develops in hypotonic solution. Among effective chemical substances one of the most active is neutral red. So common is vacuolar contraction in tissues so stained that it might easily be mistaken for plasmolysis. Phloem parenchyma cells of *Vitis vinifera* respond well (Currier et al. 1955). Other inducing but less effective chemicals are glycerine (Fitting 1920), organic solvents (Loeven 1950), and several herbicidal substances (Currier 1949).

In certain cells stimulation with dilute H_2SO_4 induces a rapid vacuolar contraction, reversible with addition of alkali. This is characteristic of boraginaceous petal cells, clearly shown in *Cerinthe major* (Kenda and Weber 1952). The phenomenon is believed to result from the syneresis of a pectic component of the vacuole.

Hypertonic solutions of Li, Na, and K neutral salts cause swelling of the cytoplasm at the ends of the cell to produce *cap plasmolysis* (Höfler 1928). It is due to the entrance of the salt and the resulting hydration. Ca salts antagonize the action. The effect is reversible in the first stages. Its relation to vacuolar contraction is clear when one plasmolyzes a vacuolar-contracted cell. Neutral red in hypertonic sugar solution will produce a temporary cap plasmolysis (Bogen 1951).

Plasmoptysis or bursting results from a rapid increase of turgor. It may be a consequence of an abrupt increase in external hypotonicity; or, the cause may be a rapid penetration of a substance such as ethanol. Acids, through an increase in colloidal swelling, are effective in the bursting of root hairs (Strugger 1926).

Viscosity changes. Growth substances, *e.g.*, IAA, in stimulating amounts decrease protoplasmic viscosity (Northen 1942). However, the effects are reversible, and the time element appears to be important (Stålfelt 1950), making it difficult to relate concentration of the agent and the direction of viscosity change. Colchicine also apparently has a marked influence on viscosity (Northen 1950, Stålfelt 1949). Fat solvents in dilute concentration lower the consistency, and in higher amounts have the opposite effect (Weber 1922, Heilbrunn 1952).

When the ectoplast, through injury is no longer differentially permeable, the tonoplast, under some conditions, may still contract in response to plasmolyzing solutions. This is known as *tonoplast plasmolysis.* The result is a marked disorganization of the cytoplasm. The inner (upper) epidermis of *Allium cepa* bulb scale is a satisfactory object for demonstrating the phenomenon. If the inner cytoplasm does not swell, and is pulled into grotesque shapes by the contracting tonoplast, the terms "Plasmochise" or "Scheinplasmolyse" are used. KCNS in 1 M concentration is an effective agent (Strugger 1949). At first perfect plasmolysis forms, then the cytoplasm begins to swell and in 10–30 minutes it fills the "extraplasmatische Raum" between the cell wall and contracted protoplast. Only the tonoplast remains living. The effect is not specific and is caused by many different substances.

Separation of phases. Considered as a dehydration of protoplasmic colloidal constituents, *vacuolation* is indicative of injury and can be induced, among other agents, by chemical substances (Küster 1929, Lepeschkin 1937). Alkali hydroxides, *e.g.* KOH, $Ca(OH)_2$, are said to be the prime vacuolators (Küster 1929, p. 145); other active compounds include alkaloids, ammonia, and $(NH_4)_2CO_3$. The action can proceed to a point where the cell contents appear foamy ("schaumige Degeneration"). Vacuolation often follows swelling.

Lipophanerosis. Certain chemical agents cause the collection of normally dispersed protoplasmic lipid into visible droplets, accompanying the necrotic process. This is held to be the mode of action of benzene and similar substances on plant cells (Gavaudan and Brébion 1951). In onion epidermis, and to a lesser extent in other objects, ammonium 4,6-dinitro-2-secondary-butylphenol appears to produce a cytoplasmic degeneration of this type (CURRIER 1949).

Granulation. Normal uninjured cytoplasm is transparent and glistening, with dispersed particulates relatively small and difficult to detect. Injured cells commonly show a condition in which the cytoplasm appears less transparent and the granules larger in size (degeneration granules). Increasing as the cell is irreversibly injured (premortal state), the maximum granulation appears in death of the cell.

A variety of poisons are reported to induce a *solidification* of cytoplasm (Erstarrung), a rigidity of injured but still living protoplasts. The list includes aluminium salts, acids, halogens and especially iodine, saponin, alkaloids, and narcotics (KÜSTER 1929).

It should be emphasized that some of the form and structure changes described above lack agent specificity, and for this reason they may not by themselves be particularly helpful in studies of action mechanisms. They can at least, however, provide subsidiary information. And on the other hand, there are some quite specific effects, and direct observation of cells deserves more of a place in toxicity investigations than it is sometimes given.

e) Plasma membrane effects.

Although the structures of the two principal cellular membranes — the ectoplast (plasmalemma) and the vacuole-limiting tonoplast — are imperfectly known, it is likely that these surfaces are the sites of many kinds of toxic action. Through influences on hydration, sodium and potassium salts increase, and calcium and magnesium salts decrease permeability. Possibly chelating substances withdraw calcium from the membrane to produce injury.

Certain narcotic-like chemicals, if present at too high a concentration, or for too long, cause a disruption of the ectoplast, a phenomenon termed *cytolysis*. This is believed to be the mechanism of acute phytotoxicity of hydrocarbons and petroleum oils (CURRIER 1951, DALLYN and SWEET 1951, VAN OVERBEEK and BLONDEAU 1954). Death of cells under these conditions is due to an irreversible loss of differential permeability of the membrane. Other, chronic (slowly developing, systemic) effects of these substances are otherwise explained (MINSHALL and HELSON 1949, GAVAUDAN and BRÉBION 1951, WEDDING et al. 1952).

Acute toxicity of hydrocarbons is characterized by a "threshold effect". At low concentrations there is little plant response. As the concentration is increased, within a very narrow range the toxicity increases strikingly (CRAFTS and REIBER 1948, CURRIER 1951, IVENS 1952). VAN DER ZWEEP (1954) plotted curves, concentration against amount of killing of *Elodea* for several hydrocarbons, and obtained identical slopes (Fig. 3). This he interpreted as meaning that the mechanism of action is the same in each case, following the reasoning of RICH and HORSEFALL (1952).

The exact manner in which the cytolytic response occurs is not completely known. It is generally agreed that lipids of the membrane must play an important role in the process. VAN OVERBEEK and BLONDEAU (1954) employ the concept of solubilization (MCBAIN 1950) to explain petroleum oil and hydrocarbon injury. The hydrocarbon molecules are believed to become "bound" within the aliphatic "tails" of the amphipatic lipid molecules that are probable constituents of the plasma membrane. By displacing the fatty molecules they "open up" the membrane and increase its permeability.

Amphipatic, or surface active, molecules (surfactants), would be expected to accumulate at the plasma membrane, and to exert a kind of narcotic action. The large number of surfactants available can be classified first according to charge: a) cationic, b) anionic, and c) nonionic. Generally, members of the cationic group, including quaternary ammonium compounds, are markedly toxic, and find use as disinfectants; the anionics are of intermediate toxicity. Nonionics, as a class, are relatively non-toxic, although they may affect permeability relationships.

According to MIRIMANOFF (1953), cationic surfactants used in solution with a toxicant usually exert a simple additive effect; anionics may have a synergistic action, whereas nonionics usually are antagonistic. Satisfactory explanations for the various types of behavior cannot be given. Synergism may be due to an increase in permeability, through an "opening up" effect on the protoplast surface; antagonism to a blocking effect, to a chemical reaction between toxicant and additive, or to solubilization. Cationics are believed to have the property of severing the loose type of union between apo- and co-enzyme groups (KUHN and BIELIG 1940). In addition to charge, spatial relationships of polar and non-polar groups in the molecule are important in the kind of action exerted. But it is properly emphasized (MIRIMANOFF 1949, ENGLER 1950) that there is no clear relation between surface activity (as measured in vitro) and toxicity.

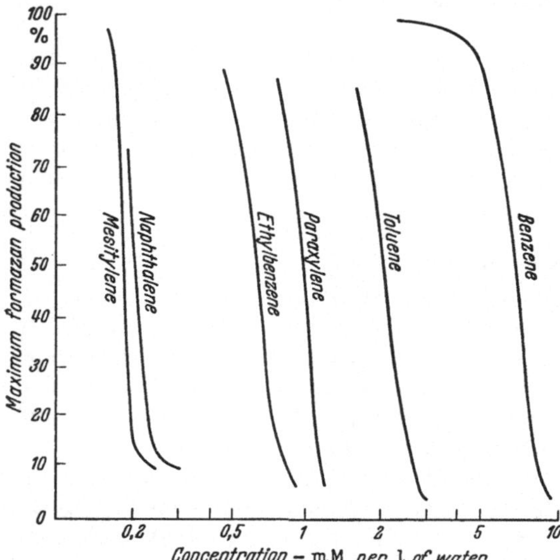

Fig. 3. The relative toxicity of several hydrocarbons in aqueous solution to *Elodea canadensis* leaf cells, using as criterion of injury the amount of TTC reduction subsequent to treatment. (From VAN DER ZWEEP 1954.)

In the use of auxin herbicides it is important that the additive have little or no toxicity, otherwise injury to absorbing and translocating cells will interfere with systemic distribution of the herbicide.

In root absorption tests a few surfactants were found to increase absorption of maleic hydrazide by *Hordeum vulgare*, without exerting any affects by themselves. Due to the lack of stomata and a cuticle, the effect must have been on the protoplasm.

There is increasing evidence that toxic substances may act on enzymic entities in the protoplast surface (ROTHSTEIN 1954). The concept that growth substances may exert their effect by combining with enzymatic surface groups is discussed by VELDSTRA (1953) and by OCHS (1954).

The ectoplast and tonoplast may differ in their response to poisons. OSTERHOUT (1945) found that in *Nitella*, 0.01 M $HgCl_2$ killed the tonoplast before the ectoplast, but when 0.01 M KCl was included in the solution the sequence was reversed. Pyramidon has a differential toxic effect on the inner and outer membranes (GÄUMANN et al. 1954).

Other toxic solutes used as plasmolyzing solutes, for example heavy metal salts (PRINGSHEIM 1925), and KCNS (STRUGGER 1949), act in a way that is indicative of the nature of the ectoplast.

In a study of sorption of chemicals by isolated leaf cuticle, ORGELL (1954) demonstrated some possible interactions between surfactants and various other substances. The important considerations are charge, polarity, and p_H. In Fig. 4, a possible, and, it is emphasized quite speculative, interaction between a cationic surfactant, and safranin, a basic dye, is diagrammatically shown. The figure indicates in A that the presence of the surfactant has a marked effect on the sorption of safranin; in B that at a low p_H the apolar ends of the surfactant are sorbed by the hydrophobic uncharged cuticle, exposing positive charges which repel the dye molecules. At a high p_H (C) the charges are altered such that sorption of the dye is greatly increased. Similar but more complicated relationships might be expected at the plasma membrane-cell wall interface.

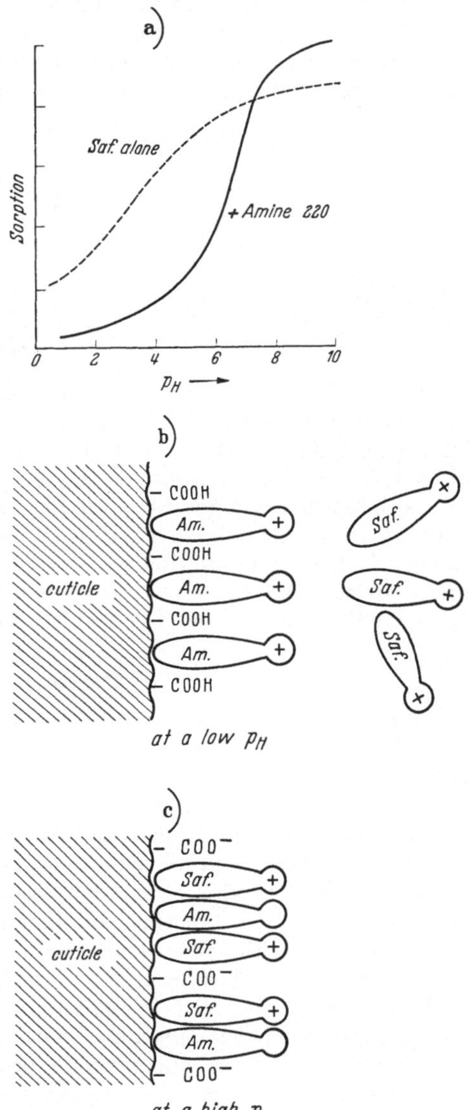

Fig. 4 a—c. Sorption of safranin by isolated leaf cuticle in solutions of varying p_H. Measurements were made both with and without Amine 220, a cationic surface active agent (Carbide and Carbon Corp., USA.). Explanation in text. (After ORGELL 1954.)

f) Nuclear effects.

An especially large number of chemical substances have been used as tools in experimental cytology and genetics. A general treatment of the earlier literature describing effects of toxicants on the mitotic process is available in POLITZER's (1934) monograph.

Attention has long been on colchicine as a specific agent preventing the formation of the mitotic spindle. Since it does not prevent division of the chromosomes, polyploidy results. This is the so-called "c-mitosis" (LEVAN 1938). Certain hydrocarbon solvents also have this action (FERGUSON et al. 1950). Among herbicides, IPC (isopropyl-N-phenyl carbamate) is known to be a mitotic poison, producing marked cytological aberrations in some species (LEFEVER 1939, DEYSSON 1945, ENNIS 1948, IVENS and BLACKMAN 1949). The phenoxyacetic acids, MCPA (DOXEY and RHODES 1949) and 2,4-D (UNRAU and LARTER 1952) are antimitotic.

Chromosome breakage can be induced by a variety of compounds. The effect was observed in a number of species after treatment with ethylurethane, $AlCl_3$, heavy metal salts, alkaloids, nucleic acids and their constituents, seed extracts, and putrescin (OEHLKERS 1953). The fragmentation of chromosomes by maleic hydrazide observed by DARLINGTON and MCLEISH (1951) may explain at least a part of the growth inhibiting action of MH. It is of interest that MH-treated roots continue to elongate even though the meristematic cells show no sign of mitosis.

Stickiness, or adhesion of chromosomes, and inhibition of spindle-formation are produced by many diverse compounds. ÖSTERGREN (1944) demonstrated that effectiveness is generally proportional to the oil/water distribution ratio. Certain substances such as urethane that are more active than would be expected on the basis of oil/water partition are considered to have greater antimitotic specificity.

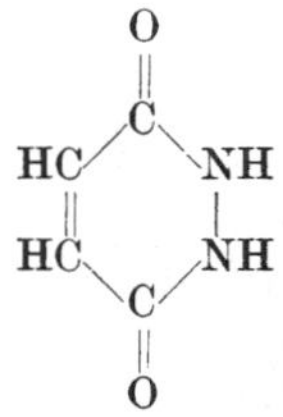

```
        O
        ‖
        C
      /   \
   HC       NH
   ‖        |
   HC       NH
      \   /
        C
        ‖
        O
```

MH

Fungicides, *e.g.*, tetrachloro-p-benzoquinone (YARKAR 1952) and insecticides, *e.g.*, γ-hexachlorocyclohexane (DOXEY and RHODES 1951) cause mitotic disturbances in higher plants. WADA (1952) is studying the effect of various chemicals on mitosis in *Tradescantia* cells *in vivo*.

As the evidence increases, it would seem that antimitotic action is a widespread reaction of plant cells to toxic chemicals. Varying degrees of specificity are shown. In addition to inhibition of cell division, to abnormal growth and development, there must be other effects that interfere with functional relationships between the nucleus and the rest of the cell.

g) Responses of plastids.

Many of the form and structure changes due to chemical agents described for cytoplasm are known in plastids as well (KÜSTER 1937, LEPESCHKIN 1937, KÜSTER 1951, p. 381–390). Swelling and shrinking, vacuolation, lipophanerosis, hardening and breaking apart, and "cellulosic degeneration", are common effects. Characteristic for plastids are agglutination (clumping), fusion or melting together of grana, and pigment changes. Agglutination of chloroplasts is an unspecific effect produced by such assorted compounds as neutral red, chloroform, sulfurous acid, chlorine, ammonia, and phenol. The cause is ascribed to a "stickiness" in the plastid membrane. Acids affect the form of plastids, and aluminum chloride has been found to induce breaking apart in *Spirogyra*.

LÄRZ (1942) observed vacuolation of chloroplasts of *Helodea* and several other plants following treatment with various alkaloids — nicotine, cocaine, atropine, coniine. The phenomenon occurred only in illuminated leaves, and could be prevented by phenylurethane, low temperature, vital staining and plasmolysis. Failure of sugar to move out across the plastid membrane was the suggested cause. An observed plastid osmotic pressure equivalent to 1.2 M of sucrose supports this explanation. LÄRZ believed that alkaloids behave in their sorption characteristics like basic dyes, possibly exerting a condensing action on the membrane.

Further investigations along those lines were carried out by SCHMIDT (1951a) on *Helodea canadensis* and *Helodea densa*. In addition to the alkaloids employed by LÄRZ, ephedrin, arecolin, physostigmin, pervitin, scopolamin, and larocain were used. All acted to produce a reversible vacuolation. Photosynthesis apparently was not completely inhibited at the beginning of the process. SCHMIDT agrees that the effect is probably localized on the plastid membrane, but goes further to suggest that enzymes mediating monosaccharide to oligo- and polysaccharide transformations in the chloroplasts may also be affected. He also stresses the physical properties of the various molecules — solubility, nature of polar and non-polar end groups, as related to this effect.

In another investigation, SCHMIDT (1951b) observed striations (Streifen) in *Helodea* chloroplasts following treatment with dilute solutions of Rhodamin B, methyl green, and crystal violet, attributed to a coarsening of the submicroscopic dimensions of the lamellae. Several pyrene derivatives and quaternary phosphonium compounds induced an irreversible cuplike form of plastid to develop (Näpfchenbildung), indicating a greater swelling capacity on one side than another.

Reduction of chloroplasts, *e.g.*, by chromate salts, is of frequent occurence (cf. STRUGGER 1949). Rubidium chloride at 0.01 M/liter stimulated division of plastids in *Helodea* (LÄRZ 1942). Chlorophyll content as affected by poisons is discussed in a later section.

h) Effects on the cell wall.

Responses of cell walls to chemicals can be grouped as to 1) dissolution processes, 2) thickening, 3) changes in composition, and 4) growth in surface.

1. Dissolution. The middle lamella, of calcium pectate composition, can be solubilized by calcium "competing" substances. Oxalates act in this way by the formation of insoluble $Ca(C_2O_4)$. Pectinases, offered under various trade names, are active in enzymatic maceration of plant tissues (CHAYEN 1952, ORGELL 1954). Other sequestering agents are citrate and derivatives of ethylenediamine tetraacetic acid.

The study of the mechanism of abscission (detachment of leaves, branches, flower parts, seeds, fruit, from the plant) has been given impetus in the last decade by the advent of mechanized agriculture. It has become highly desirable to induce separation of plant parts prior to or during harvesting, for example, defoliation of cotton *(Gossypium)*. Progress in the investigation of physiological relationships is summarized by ADDICOTT and LYNCH (1955).

Chemicals known to induce or accelerate the abscission process include: 1. *unsaturated hydrocarbons:* ethylene, acetylene; 2. *antiauxins:* triiodobenzoic acid (TIBA), trans-cinnamic acid, several maleimides; 3. *enzyme inhibitors:* iodoacetate, phenylmercuric salts, ethyl xanthate; 4. *a variety of substances used as defoliants:* amino triazole, calcium cyanamide, sodium acid cyanamide, calcium cyanate, butyndiol, copper salts, sodium dichromate, sodium chlorate-borate mixture, magnesium chlorate, sodium monochloroacetate, disodium 3,6-endoxohexahydrophthalate (Endothal), sodium ethyl xanthate; 5. *substances used as fruit thinners:* dinitro-ortho-cresol (DNOC), naphthaleneacetic acid (NAA), maleic hydrazide (MH), isopropyl-N-phenyl carbamate (IPC) and 2,4-dichlorophenoxyacetic acid (2,4-D).

These are compounds known to be particularly active. Hundreds of other chemicals tested by various laboratories show only slight activity or none at all (WEINTRAUB et al. 1952). The diversity of structure of many chemicals that have defoliant action suggests that they act indirectly in a common manner. This common effect doubtless involves auxin. Just how an indirect contact toxic action on the leaf blade operates in the process is not understood. Perhaps auxin is produced and is translocated to the abscission region. Workers have stressed such variables as nutritional status, age, etc., as being important in the complex. Generally, a chemical producing a mild injury followed by a slow dehydration of the blade provides a maximum amount of defoliation (JOHNSON 1955).

Of the greatest cytological interest in connection with abscission is the dissolution of the middle lamella, the primary wall, or both, depending upon species. There is a continuing search for naturally occurring pectinases, cellulases, and perhaps other enzyme systems involved in this hydrolytic activity.

Ethylene is the most active abscission-inducing substance known, and a natural product of some plants. A balance between ethylene and auxin may be of importance in the regulation of abscission (HALL 1952) but failure to induce defoliation of cotton in the field with ethylene is interpreted by ADDICOTT and LYNCH (1955) as favoring the auxin gradient theory, in which a relatively lower concentration of auxin on the distal side of the abscission zone than on the proximal side is correlated with accelerated abscission. A gradient in the reverse direction, which can be produced by applying auxin to the leaf blade, retards or prevents separation.

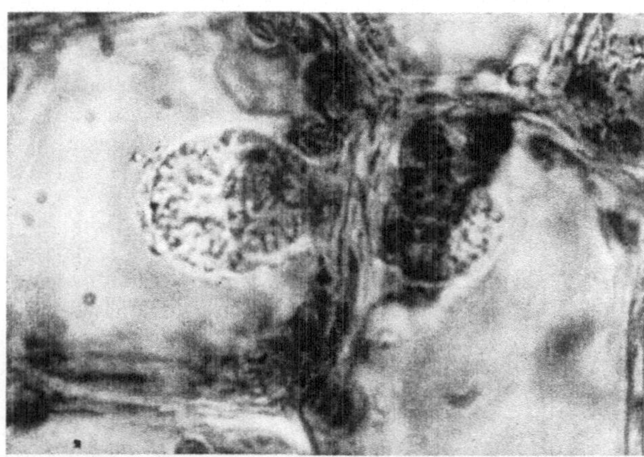

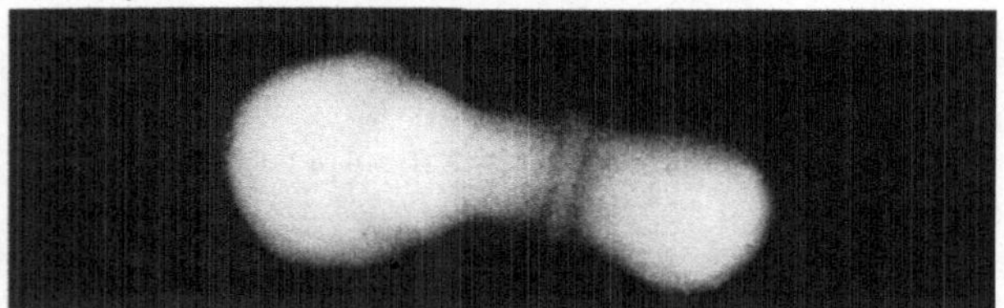

Fig. 5. Cell wall structures ("pegs") produced in *Elodea densa* leaf cells treated with 0.2% borax solution for 6 days. Transferred to dilute aniline blue in alkaline solution there is no staining in bright field, but in UV the structures fluoresce yellow.

In addition to cell wall dissolution, some other cellular changes are known to be induced by defoliants. These include pigment changes, hydrolysis of carbohydrate and N-compounds, increased respiration, a decrease in auxin content, and production of ethylene.

2. Thickening. The term "zellulosige Degeneration" is used by KÜSTER (1951) to describe certain thickening processes, in which various types of incrustations and deposits are laid down on the wall. Peg-like outgrowths in *Elodea* leaf cells, considered to be pectic in nature, result from stimulation especially by boron but also by other compounds such as chromates (MACKE 1939, LÄRZ 1942). Such structures when stained with aniline blue fluoresce yellow in UV (Fig. 5), a response that appears to be fairly specific for callose (CURRIER and STRUGGER 1956).

Deposits of callose on the sieve areas of sieve tubes constitute wall accretions the structure and function of which are not completely understood (ESAU 1953). Eosin promotes callusing of sieve plates (SCHUMACHER 1930); maleic hydrazide does not (CURRIER et al. 1951).

3. Textural changes. Plants suffering from 2,4-D injury display an increased brittleness in stems. This is considered a reflection of a textural change in the cell wall, but the nature of the change is unknown (AUDUS 1953, p. 235).

4. Abnormal growth. Auxins in low concentration stimulate surface extension of growing cell walls, and inhibit in higher concentration. In addition, certain auxins such as 2.4-D cause formative effects, deranged growth of various kinds, involving change in size and shape of cells. Maleic hydrazide and 2,2'-dichloropropionic acid, as non-auxins, also produce such effects. An interesting example is the action of benzimidazole (GALSTON et al. 1953) in producing barrel-shaped cells.

VII. Effect of poisons on certain metabolic functions.

a) Influence on respiratory activity.

Selective inhibitors of respiration have been classified by JAMES (1953) on the basis of cellular reactions believed to be involved (Table 4). These are substances that are known to react with some degree of specificity with particular groupings in enzyme molecules. Dosage and conditions are important considerations in obtaining maximum specificity.

Table 4. *List of selective inhibitors of respiration.* (From JAMES 1953.)

Thiol reagents.—Alkylating agents (iodoacetic acid, iodoacetamide, methyl and ethyl iodoacetate, methyl bromide); trivalent arsenicals (sodium arsenite, lewisite, mapharside); mercaptide formers (p-chlormercuribenzoate, phenylmercuric nitrate); sodium maleate; mild oxidising agents.

Carbonyl reagents.—Hydroxylamine, semicarbazide, hydrogen cyanide.

Reagent forming a fluorophosphate.—Sodium fluoride.

Metal reagents.—Aromatic sulphonic acids (1-naphthol-2-sulphonic acid, suramin); reagents forming complexes with transition metals (hydrogen cyanide, hydrogen or sodium sulphide, sodium azide, sodium pyrophosphate, 8-hydroxyquinoline); other reagents for transition metals (BAL or 2,3-dimercaptopropanol, carbon monoxide); reagents chelating iron (αα'-dipyridyl, *o*-phenanthroline); reagents chelating copper (dieca or sodium diethyldithiocarbamate, potassium ethylxanthate, salicylaldoxime).

Phenols forming inactive complexes with copper proteins.—p-Nitrophenol, resorcinol.

Structural inhibitors.—Pyruvic acid, oxaloacetic acid, acetaldehyde, glyceraldehyde, malonic acid.

Reagent for oxaloacetic acid.—Sodium fluoroacetate.

Reagents affecting phosphate transfers.—Sodium arsenate, sodium azide, 2,4-dinitrophenol, phloridzin.

Narcotics.—Urethane, octyl alcohol, malachite green.

At the same time that they suppress growth, toxic substances can inhibit, or they can stimulate the respiratory rate. Dinitrophenols, at appropriate concentrations, markedly increase respiratory activity, and this is accompanied by inhibition of growth in the *Avena* coleoptile (BONNER 1949). It has been generally accepted that in small amounts DNP uncouples phosphorylation processes, thus interrupting the flow of energy into processes of synthesis and growth (cf. BONNER 1950b). Stronger concentrations lead to inhibition of flavoprotein and decrease of respiration, and finally a denaturation of protein and irreversible injury (SIMON 1953). The dinitro herbicides such as DNSBP (dinitro-sec-butylphenol) have effects on respiration similar to DNP (SWANSON et al. 1953). CIPC (isopropyl N-3-chlorophenylcarbamate) depressed the growth of cotton *(Gossypium)* seedling roots a constant amount throughout a wide range of concentration. Coupled with this was a progressive decrease in respiratory rate (SWANSON et al 1953).

Respiratory responses to 2,4-D depend to some extent upon the concentration. In low amounts there is a reversible stimulation, while at intermediate concentrations a permanent increase continues until the death of the plant. Higher concentrations lead to a continuing depression of CO_2 evolution, sometimes preceded by a brief period of stimulation (NORMAN et al. 1950).

The well known respiratory inhibitors at certain concentrations will stimulate or depress the growth rate just as growth regulators do (F. G. SMITH 1951). Instances are known: a) where the influences on growth and respiration are parallel (*e.g.* HCN), and b) where respiration is inhibited more than growth (iodoacetate). Arsenate inhibits auxin-stimulated growth in the *Avena* coleoptile without influencing the basal respiratory rate (BONNER 1950b, p. 458).

Thus there are many difficulties in attempting to ascribe to a direct effect on respiration, the stimulating or inhibitory action of a chemical.

Metabolic inhibitors depress the growth rate, and the effects are especially pronounced in the presence of optimal amounts of auxin (IAA). This is interpreted by some workers as an interference with an active water transfer promoted by auxin, *i.e.*, water moved and held by energy-requiring nonosmotic forces. The idea, then, is that there is an active component in both solute and water uptake and retention by the cell. There is considerable evidence that salt accumulation is decreased by respiratory inhibitors (cf. ROBERTSON 1951).

In the presence of auxin, uptake of water by potato *(Solanum tuberosum)* tuber tissue is decreased by DNP, azide, arsenite, and fluoroacetate (HACKETT and THIMANN 1950). DNP behaves similarly for *Helianthus tuberosus* storage tissue, according to BONNER et al. (1953), who suggest that auxin might participate in a metabolic sequence in which ATP energizes a water transport system. It is not difficult to visualize such a system, but experimental proof presents difficulties, and alternative explanations such as cell wall growth may not be excluded (BURSTRÖM 1953b).

The plasmolytic method provides a useful experimental approach to the problem. BOGEN and FOLLMANN (1955) found that in the diatoms *Melosira varians* and *Melosira arenaria*, small amounts of DNP within two minutes reduced the Og (osmotic pressure at limiting plasmolysis) as much as 18%. This they interpret as the loss of an amount of water that is normally held by active forces, lost because of disruption of energy supply by the poison. On the basis of vapor pressure measurements (RAST method) of expressed sap, the results could not be explained by an exosmosis of solutes, and the time is short for a growth reaction.

b) Effects on photosynthesis.

Some substances, such as the so-called non-toxic spray oils, can inhibit photosynthesis, possibly by interference with gaseous interchange. Benzene, ether, urethanes, and similar substances depress the rate by a reversible narcotic

$$H_2NCOOC_2H_5$$

Urethane
(Ethyl carbamate)

effect within a narrow concentration range. 2,4-D at 2×10^{-3} M/liter inhibited photosynthesis in *Citrus* leaves and *Chlorella*, the more as the p_H was lowered (WEDDING et al. 1954). Inhibition is also a response of *Phaseolus* and of *Anacharis* to this compound in amounts of 10 to 100 ppm. (FREELAND 1949, 1950). Enzyme inhibitors such as HCN, H_2S, and iodoacetate are sensitive photosynthetic depressants (MEYER and ANDERSON 1952), in addition to their effects on respiration.

c) Inhibition of chlorophyll synthesis.

Certain compounds are known to suppress chlorophyll development without at once being especially toxic in other ways. The existing chlorophyll is not generally affected, but new growth lacks green color. This is one of the effects of certain antibiotics on plants. Sulfanilamide is such a substance, and streptomycin is especially interesting (GRANICK 1951). Other substances include: several tetronic acid derivatives (ALAMERCERY et al. 1951, HAMNER and TUKEY 1951), certain substituted benzoic acids (READY et al. 1952), several carbamates (SHAW and SWANSON 1953), and 3-amino-1,2,4-triazole (SHAW et al. 1953, HAUSER and THOMPSON 1954). The last mentioned, popularly known as amino triazole, is translocated in the plant and shows promise as an herbicide. In addition to this use, such substances can serve in the study of enzymatic processes involved in chlorophyll synthesis. It is noteworthy that molecules of quite different structure are effective.

d) Influence on nitrogen metabolism.

Some rather specific effects on nitrate metabolism are known. *n*-Diamylacetic acid at concentrations of 10^{-6} to 10^{-7} M/liter depressed nitrate assimilation in excised and cultured wheat *(Triticum)* roots; growth inhibition did not occur until 10^{-5} M/liter was reached (BURSTRÖM 1949). In another example, due to structural similarity chlorate is believed to compete for sites normally occupied by nitrate (ÅBERG 1948). It is possible that the ultimate toxic products are chlorite and hypochlorite.

DNP blocks nitrate reduction in the dark in a green alga at physiological p_H values (KESSLER 1955), at the same time that respiration is stimulated.

Acridine orange is reported by BOGEN and KESER (1954) to cause a decrease in protein content of yeast cells and an accumulation of amino acids, and, on the basis of its toxicity, they question the use of this dye as an indicator of vitality.

e) Metabolic and growth patterns modified by auxin herbicides.

The mechanism of action of 2,4-D has been extensively investigated. Measurement of changes in composition of cells and tissues has not been particularly helpful. Initially there is an increase in soluble carbohydrates at the expense

Fig. 6. A series of cotton leaves showing varying degrees of malformation caused by 2,4-D treatment. A normal untreated leaf is at left. At right the structure represents the fusion of two leaves. (From GIFFORD 1953.)

of polysaccharide reserves, but this is probably not the cause of death. N and K are generally reduced. Changes in content of amino acids and proteins, lipids, organic acids, and other constituents have been reported, but interpretation is difficult. The literature is detailed by NORMAN et al. (1950), BLACKMAN et al. (1951), CRAFTS (1953b), and WEINTRAUB (1953).

Nor have attempts to localize the action in individual enzymes been sufficiently revealing to make generalizations possible. In many instances the effect may be stimulating or inhibiting, depending upon the species tested, the part of the plant, length of treatment, method of assay, and other factors.

Herbicidal auxins such as 2,4-D produce formative effects, abnormal patterns of development in roots, stems and leaves. These effects, whatever their nature, have their greatest influence in the youngest cells of the plant, and proportionately less in older tissues. This property is not limited to auxins, being among the responses induced by growth regulators that are not auxins, structurally or functionally, *e.g.*, 2,2'-dichloropropionic acid. Malformations of all kinds appear—swellings, galls, twisted stems, grotesque shapes of leaves. Fig. 6 illustrates some of the assorted leaf patterns produced by 2,4-D in cotton, a species very sensitive to this chemical. The anatomy of the injured leaf is grossly atypical (Fig. 7). Such altered development must be taken into account in the over-all formulation of toxic mechanisms.

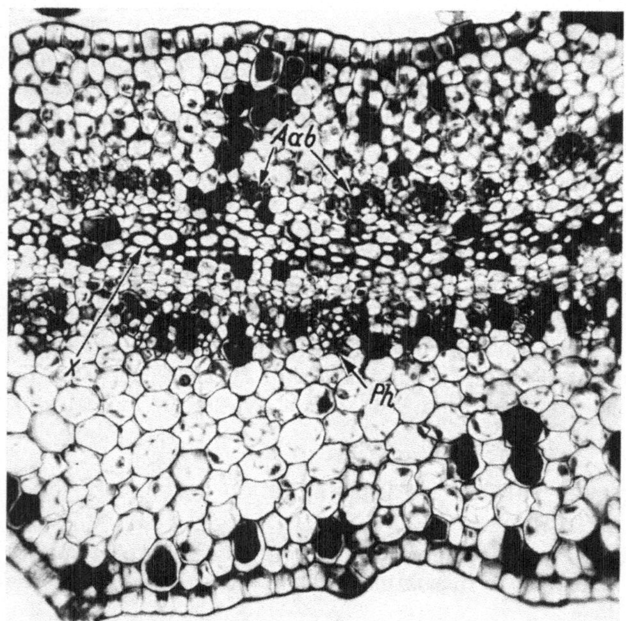

Fig. 7. Transverse section through the lamina of a nearly mature cotton leaf showing formative effects due to 2,4-D treatment. Normal mesophyll is replaced by compactly arranged, relatively undifferentiated tissue, with atypical vascular development. *Aab* adaxial accessory bundle; *Ph* phloem; *x* xylem. × 240. (From GIFFORD 1953.)

WEINTRAUB (1953) attaches a great deal of importance to the formative effect of 2,4-D in connection with its mode of action. Assuming that polarized division of meristematic cells is controlled by endogeneous auxin, the structurally-similar 2,4-D competes for sites or substrates normally occupied by auxin. This results in a disturbance of polarity and a disordered growth pattern, that can be so severe that plants appear to "proliferate themselves to death".

VIII. Cellular bases of herbicidal selectivity.

Study of selective toxicity on the cellular level has received less attention than in relation to entire plants, but it will receive greater emphasis in efforts to discover the mechanisms of biochemical selectivity. Only a beginning has been made by organic chemists to model the molecular structure of the toxicant such that it will react, with lethal consequences, with a characteristic substance present in cells of the unwanted species, this substance lacking in the species to remain unharmed. Or, in line with modern concepts, attention may be directed toward an inhibitor or detoxifier occurring in certain species but not in others.

Progress in these directions will be closely dependent on precise knowledge of cellular constituents — their structure, distribution, and physiological role.

One of the more interesting and distinct selectivities known in weed control is the relative resistance of the members of the *Umbelliferae* to light hydrocarbon oils (CRAFTS and REIBER 1948). It is definitely known that the oils rapidly penetrate the leaves of both resistant and susceptible species, and that morphological exclusion is not the basis of selective behavior. Many weeds in carrots *(Daucus carota)* can now be eliminated with little or no injury to the crop if sprayed at the proper time. The complete explanation of this differential response is not at hand, but it is possible to say that the answers are to be found in the individual cell rather than in some more gross structure or function of organs or tissues.

Individual cells removed from leaf, stem, and root of carrot show the same relative resistance as that possessed by the whole plant (DALLYN and SWEET 1951, VAN OVERBEEK and BLONDEAU 1954, CURRIER and PEOPLES 1954). Carrot and barley *(Hordeum vulgare)* show approximately the same difference in susceptibility when treated with benzene and other low-boiling hydrocarbons where the diluent is air or water (CURRIER and PEOPLES 1954). In fumigation-chamber studies with hydrocarbon vapors, up to the time when foliar odors could be detected in the exhaust, no permanent injury resulted to the plants if they were removed at that point. If not removed, there was leakage of sap into the intercellular spaces, and death of leaf tissue resulted. There is the suggestion, therefore, that selective action is determined at the cell surface, and that there is some unique variation in the nature of the ectoplast of umbelliferous cells. Distinct species differences in erythrocyte "ghost" membranes are evident in the electronmicroscopic photographs of HILLIER and HOFFMAN (1953), and this may also be true for plant cells.

A better known and of course more important selectivity is the relative insensitivity of members of the *Gramineae* to the hormone herbicides, where the mechanism is also protoplasmic. With the idea that some difference in protein structure between grasses and broad leaved plants could be demonstrated, BRIAN and RIDEAL (1952) employed the so-called Langmuir Trough to study the interaction of 2-methyl, 4-chlorophenoxyacetic acid (MCPA) on monolayers of protein, lipoprotein, and lipid material obtained from mono- and dicotyledonous tissue. The results are interesting in that the interaction was greater with dicotyledonous constituents, suggesting that MCPA is more adsorbed in vivo and held inactive by the protoplasm of these plants.

A similar approach was employed by GATENBECK and EHRENBERG (1953) in studying the influence of anesthetics on monomolecular films of cell lipids. Such investigations should aid in determining the exact site of toxic action in the cell. There is, however, always the problem of carrying over results of *in vitro* experiments to the living cell.

The plasmolytic method can be used to study species differences in cellular resistance to chemicals. Sodium carbonate (Na_2CO_3) has proved to be a useful solute for this purpose (HÖFLER 1951). Toxicants can be added to the ordinary plasmolyzing solutions, and some interesting results have been obtained with strong alkalies (CHOLNOKY 1952, 1953) to suggest that a selective action is determined at the cell surface. By adding small amounts of KOH to the KNO_3 plasmolyzing medium, cells of *Calceolaria* were considerably more resistant, on the basis of ability to plasmolyze and deplasmolyze, than cells of certain species of *Senecio.*

Apparently the selectivity shown by plants toward 2,4-D as a hormone type of injury is also displayed by this chemical as a general protoplasmic poison (BIEBL 1953).

An example of different sensitivity exhibited by cells within the same organs is reported by BURSTRÖM (1950). Cells of excised wheat roots in the process of elongation are killed by 10^{-5} M lauric and 10^{-4} undecanoic acids, but meristematic and mature cells are not. The suggested possibility is that the meristematic cells adapt in some manner to the presence of the chemical, with the result that they become relatively insensitive in the elongation phase of growth.

Investigating phenoxyalkylcarboxylic acids, WAIN (1955) finds that as the length of the side chain is increased, the possibilities for selectivity are greater due to the lack, in some species, of enzymes capable of breaking the compound down to the parent phenoxyacetic acid.

Literature.

ÅBERG, B.: On the mechanism of the toxic action of chlorates and some related substances upon young wheat plants. Ann. R. Coll. Sweden **15**, 37–107 (1948). — On the interaction of 2,3,5-triiodobenzoic acid and maleic hydrazide with auxins. Physiol. Plant. **6**, 277–291 (1953). — ADDICOTT, F. T., and R. S. LYNCH: Physiology of abscission. Annual Rev. Plant Physiol. **6**, 211–238 (1955). — ALAMERCERY, J., C. L. HAMNER and M. LATUS: Chlorophyll inhibition and growth regulation by several tetronic acid derivatives. Nature (Lond.) **168**, 85 (1951). — ALBERT, A.: Selective Toxicity. London: Menthuen & Co. 1951. — ALLEN, P. J., and W. H. PRICE: The relation between respiration and protoplasmic flow in the slime mold *Physarum polycephalum*. Amer. J. Bot. **37**, 393–402 (1950). — ALLEN, S. E., and F. SKOOG: Stimulation of seedling growth by seed treatments with N-phenyl-succinimide. Plant Physiol. **27**, 179–183 (1952). — AUDUS, L. J.: Plant growth substances. London: Leonard Hill, Ltd. 1953.

BIEBL, R.: Resistenz pflanzlicher Plasmen gegen 2,4-D. Protoplasma **42**, 193–208 (1953). — BLACKMAN, G. E.: Studies in the principles of phytotoxicity. I. The assessment of relative toxicity. J. of Exper. Bot. **3**, 1–27 (1952). — BLACKMAN, G. E., and R. C. ROBERTSON-CUNINGHAME: The influence of p_H on the phytotoxicity of 2,4-dichlorophenoxyacetic acid to *Lemna minor*. New Phytologist **52**, 71–75 (1953). — BLACKMAN, G. E., W. G. TEMPLEMAN and D. J. HOLLIDAY: Herbicides and selective phytotoxicity. Annual Rev. Plant Physiol. **2**, 199–230 (1951). — BOGEN, H. J.: Über Kappenplasmolyse und Vakuolenkontraktion Planta (Berl.) **39**, 1–35 (1951). — Zellphysiologie und Protoplasmatik. Fortschr. Bot. **15** 212–258 (1954). — BOGEN, H. J., and G. FOLLMANN: Osmotische und nichtosmotische Stoffaufnahme bei Diatomeen. Planta (Berl.) **45**, 125–146 (1955). — BOGEN, H. J., and M. KESER: Eiweiß-Abbau durch Acridinorange bei Hefezellen. Physiol. Plant. **7**, 446–462 (1954). — BONNER, J.: Limiting factors and growth inhibitors in the growth of the *Avena* coleoptile. Amer. J. Bot. **36**, 323–332 (1949). — The role of toxic substances in the interactions of higher plants. Bot. Review **16**, 51–65 (1950a). — Plant Biochemistry. New York: Academic Press 1950b. — BONNER, J., and R. S. BANDURSKI: Studies of the physiology, pharmacology, and biochemistry of the auxins. Annual Rev. Plant Physiol. **3**, 59–86 (1952). — BONNER, J., R. S. BANDURSKI and A. MILLERD: Linkage of respiration to auxin-induced water uptake. Physiol. Plant. **6**, 511—522 (1953). — BONNER, J., and A. GALSTON: Toxic substances from the culture media of guayule which may inhibit growth. Bot. Gaz. **105**, 185–198 (1944). — BOYNTON, D.: Nutrition by foliar application. Annual Rev. Plant Physiol. **5**, 31–54 (1954). — BRADBURY, D., and W. B. ENNIS jr.: Stomatal closure in kidney bean plants treated with ammonium 2,4-dichlorophenoxyacetate. Amer. J. Bot. **39**, 324–328 (1952). — BRAUNER, L., and M. HASMAN: Weitere Untersuchungen über den Wirkungsmechanismus des Heteroauxins bei der Wasseraufnahme von Pflanzenparenchym. Protoplasma **41**, 302–326 (1952). — BRIAN, R. C., and E. K. RIDEAL: On the action of plant growth regulators. Biochim. et Biophysica Acta **9**, 1–18 (1952). — BROUWER, R.: Water absorption by the roots of *Vicia faba* at various transpiration strengths. III. Proc. Kon. Ned. Akad. Wetensch. C **57**, 68–80 (1954). — BROWN, J. W., and R. L. WEINTRAUB: A leaf repression method for evaluating formative activity of plant growth-regulating chemicals. Bot. Gaz. **111**, 448–456 (1950). — BUCHA, H. C., and C. W. TODD: 3-(p-chlorophenyl)-1,1 dimethylurea — a new herbicide. Science (Lancaster, Pa.) **114**, 493–494 (1951). — BURSTRÖM, H.: Studies on growth and metabolism of roots. II. *n*-Diamylacetic acid and assimilation of nitrate. Physiol. Plant. **2**, 332–340 (1949). — Studies on growth and metabolism of roots. III. The adaptation against growth inhibition by aliphatic C_{12} and C_{11} acids. Physiol. Plant. **3**, 175–184 (1950). — Studies on growth and metabolism of roots. IX. Cell elongation and water absorption.

Physiol. Plant 6, 262–276 (1953a). — Growth and water absorption of *Helianthus* tuber tissue. Physiol. Plant. 6, 685–691 (1953b). — BURTON, W. G.: Studies on the dormancy and sprouting of potatoes. New Phytologist 51, 154–162 (1952).

CHAYEN, J.: Pectinase technique for isolating plant cells. Nature (Lond.) 170, 1070–1072 (1952). — CHOLNOKY, B. J.: Beobachtungen über die Wirkung von Kalilauge auf das Protoplasma. Protoplasma 41, 57–68 (1952). — Ein Beitrag zur Kenntnis des Plasmalemmas. Ber. dtsch. bot. Ges. 65, 369–373 (1953). — CRAFTS, A. S.: A new herbicide, 2,4-dinitro secondary butyl phenol. Science (Lancaster, Pa.) 101, 417–418 (1945). — A theory of herbicidal action. Science (Lancaster, Pa.) 108, 85–86 (1948). — Herbicides; their absorption and translocation. J. Agr. a. Food Chem. 1, 51–55 (1953a). — Herbicides. Ann. Rev. Plant Physiol. 4, 253–282 (1953b). — CRAFTS, A. S., and H. G. REIBER: Studies on the activation of herbicides. Hilgardia 16, 487–500 (1945). — Herbicidal uses of oils. Hilgardia 18, 77–156 (1948). — CROCKER, W.: Growth of plants. New York: Reinhold Pulb. Co. 1948. — CURRIER, H. B.: Responses of plant cells to herbicides. Plant Physiol. 24, 601–609 (1949). — Herbicidal properties of benzene and certain methyl derivatives. Hilgardia 20, 383–406 (1951). — CURRIER, H. B., and B. E. DAY: The tetrazolium reaction in yeast. Science (Lancaster, Pa.) 119, 817 (1954). — CURRIER, H. B., B. E. DAY and A. S. CRAFTS: Some effects of maleic hydrazide on plants. Bot. Gaz. 112, 272–280 (1951). — CURRIER, H. B., K. ESAU and V. I. CHEADLE: Plasmolytic studies of phloem. Amer. J. Bot. 42, 68–81 (1955). — CURRIER, H. B., and S. A. PEOPLES: Phytotoxicity of hydrocarbons. Hilgardia 23, 155–173 (1954). — CURRIER, H. B., and S. STRUGGER: Aniline blue and fluorescence microscopy of callose in bulb scales of *Allium cepa* L. Protoplasma 45 552–559 (1956). — CURRIER, H. B., and W. VAN DER ZWEEP: Plasmolysis and the tetrazolium reaction in *Anacharis canadensis*. Protoplasma 45, 125–132 (1955).

DALLYN, S. L., and R. D. SWEET: Theories on the herbicidal action of petroleum hydrocarbons. Amer. Soc. Hort. Sci. Proc. 57, 347–354 (1951). — DANGEARD, P., J. BIRABEN and L. DOUTRE: C. r. Soc. Biol. Paris 141, 1055–1057 (1947). — DANIELLI, J. F.: Cell physiology and pharmacology. New York: Elsevier Publ. Co., Inc. 1950. — Cytochemistry, a new approach. New York: John Wiley & Sons, Inc. 1953. — DARLINGTON, C. D., and J. MCLEISH: Action of maleic hydrazide on the cell. Nature (Lond.) 167, 407–408 (1951). — DAY, B. E.: The absorption and translocation of 2,4-dichlorophenoxyacetic acid by bean plants. Plant Physiol. 27, 143–152 (1952). — DEYSSON, G.: Action simultanée du phenyluréthane et de la colchicine sur les méristèmes radiculaires d'*Allium cepa* L. C. r. Acad. Sci. Paris 220, 367–369 (1945). — DOXEY, D., and A. RHODES: The effect of the plant growth regulator 4-chloro-2-methylphenoxyacetic acid on mitosis in the onion *(Allium cepa)*. Ann. Bot. 13, 105–111 (1949). — The effects of the gamma isomer of benzene hexachloride (Hexachlorocyclohexane) on plant growth and on mitosis. Ann. Bot. 15, 47–52 (1951).

EAMES, A. J.: Destruction of phloem in young bean plants after treatment with 2,4-D. Amer. J. Bot. 37, 840–847 (1950). — ENGLER, V.: Influence des substances tensioactives sur la cellule végétale. Thesis, Université de Genève, Faculté des Sciences 1950. — ENNIS jr., W. B.: Some effects of o-isopropyl N-phenyl carbamate upon cereals. Amer. J. Bot. 35, 15–21 (1948). — ESAU, K.: Plant Anatomy. New York: John Wiley & Sons, Inc. 1953. — EVANARI, M.: Germination inhibitors. Bot. Rev. 15, 153–194 (1949). — EWART, A. J.: On the physics and physiology of protoplasmic streaming in plants. Oxford: Clarendon Press 1903.

FERGUSON, J.: The use of chemical potentials as indicators of toxicity. Proc. Roy. Soc. Lond., Ser. B 127, 387–405 (1939). — FERGUSON, J., S. W. HAWKINS and D. DOXEY: C-mitotic action of some simple gases. Nature (Lond.) 165, 1021–1022 (1950). — FITTING, H.: Untersuchungen über die Aufnahme und über anomale osmotische Koeffizienten von Glyzerin und Harnstoff. Jb. wiss. Bot. 59, 1–170 (1920). — Untersuchungen über die Auslösung von Protoplasmaströmung. Jb. wiss. Bot. 64, 281–388 (1925). — Untersuchungen über den Plasmaströmung auslösenden Reizstoff in den Blattextrakten von *Vallisneria*. Jb. wiss. Bot. 78, 319–398 (1933). — Über Auslösung von Protoplasmaströmung bei *Vallisneria* durch einige Histidin-Verbindungen. Jb. wiss. Bot. 82, 613–624 (1936). — FREELAND, R. O.: Effects of growth substances on photosynthesis. Plant Physiol. 24, 621–628 (1949). — Effects of 2,4-D and other growth substances on photosynthesis and respiration in *Anacharis*. Bot. Gaz. 111, 319–324 (1950). — FRIESEN, G.: Untersuchungen über die Wirkung des Phenylurethans auf Samenkeimung und Entwicklung. Planta (Berl.) 8, 666–679 (1929).

GÄUMANN, E., K. H. RICHLE, A. RIGGENBACH and V. FLÜCK: Über die Wirkung von Pyramidon auf pflanzliche Zellen. Phytopath. Z. 21, 279–310 (1954). — GALSTON, A. W., R. S. BAKER and J. W. KING: Benzimidazole and the geometry of cell growth. Physiol. Plant. 6, 863–872 (1953). — GATENBECK, S., and L. EHRENBERG: The influence of anesthetics on monomolecular films of cell lipids. Ark. Kemi (Stockh.) 5, 333–340 (1953). — GAUTHERET, R.: La culture des tissues. Paris: Gallimard 1945. — GAVAUDAN, P., and G. BRÉBION:

L'échelle des inhibitions fonctionnelles dans la cellule végétale. Exper. Cell Res. **2**, 158–177 (1951). — GEISSMAN, T. C., P. DEUEL, E. K. BONDE and F. A. ADDICOTT: Xanthinin: A plant growth regulating compound from *Xanthium pennsylvanicum*. I. J. Amer. Chem. Soc. **76**, 685–687 (1954). — GIFFORD, E. M.: Effect of 2,4-D upon the development of the cotton leaf. Hilgardia **21**, 605–644 (1953). — GOODWIN, R. H., and C. TAVES: The effect of coumarin derivatives on the growth of *Avena* roots. Amer. J. Bot. **37**, 224–231 (1950). — GORENFLOT, R.: Relation entre les variations de la cyclose, de l'assimilation chlorophyllienne et de respiration de l'*Elodea canadensis* sous l'influence de chlorate de sodium. Rev. Gen. Bot. **54**, 153–185 (1947). — GRANICK, S.: Biosynthesis of chlorophyll and related compounds. Ann. Rev. Plant Physiol. **2**, 115–144 (1951). — GRAY, R., and J. BONNER: An inhibitor of plant growth from the leaves of *Encelia Farinosa*. Amer. J. Bot. **35**, 52–57 (1948). — GRÜMMER, G.: Die gegenseitige Beeinflussung höherer Pflanzen Allelopathie. Biol. Zbl. **72**, 494–518 (1953). — GUTTENBERG, H. v., and A. BEYTHIEN: Über den Einfluß von Wirkstoffen auf die Wasserpermeabilität des Protoplasmas. Planta (Berl.) **40**, 36–69 (1951). — GUTTENBERG, H. v., and G. MEINL: Über den Einfluß von Wirkstoffen auf die Wasserpermeabilität des Protoplasmas. II. Über den Einfluß des p_H-Wertes und der Temperatur auf die durch Heteroauxin bedingten Veränderungen der Wasserpermeabilität. Planta (Berl.) **40**, 431–442 (1952).

HAAGEN-SMIT, A. J., E. F. DARLEY, M. ZEITLIN, H. HULL and W. NOBLE: Investigation of injury to plants from air pollution in the Los Angeles area. Plant Physiol. **27**, 18–24 (1952). — HACKETT, D. P., and K. V. THIMANN: The nature of the auxin-induced water uptake by potato tissue. Amer. J. Bot. **39**, 553–560 (1952). — HALL, W. C.: Eivdence on the auxin-ethylene balance hypothesis of foliar absicission. Bot. Gaz. **113**, 310–322 (1952). — HAMNER, C. L., and H. B. TUKEY: Selective herbicidal action of midsummer and fall applications of 2,4-dichlorophenoxyacetic acid. Bot. Gaz. **106**, 232–245 (1944). — Chlorophyll inhibition and herbicidal action of 3-(alpha-imino-ethyl) 5-methyl tetronic acid. Bot. Gaz. **112**, 525–532 (1951). — HASELHOFF, E., G. BREDEMANN and W. HASELHOFF: Entstehung, Erkennung und Beurteilung von Rauchschäden. Berlin: Gebrüder Bornträger 1932. — HAUSER, E. W., and J. THOMPSON: Effects of 3-amino-1,2,4-triazole and derivatives on nutgrass and Johnson grass. J. Agr. a. Food Chem. **2**, 680–681 (1954). — HEILBRUNN, L. V.: An outline of general physiology. 3rd edition. Philadelphia: W. B. Saunders Company 1952. — HERBST, W.: Das *Helodea*-Blatt als Testobjekt zum Studium der Wirkung von Schädlingsbekämpfungsmitteln. Protoplasma **27**, 455–459 (1937). — HILLIER, J., and J. F. HOFFMAN: On the ultrastructure of the plasma membrane as determined by the electron microscope. J. Cellul. a. Comp. Physiol. **42**, 203–249 (1953). — HÖBER, R.: Physical chemistry of cells and tissues. Philadelphia: The Blakiston Co. 1945. — HÖFLER, K.: Über Kappenplasmolyse. Ber. dtsch. bot. Ges. **46**, 73–82 (1928). — Plasmolyse mit Natriumkarbonat. Protoplasma **40**, 426–460 (1951). — HÖFLER, K., and F. WEBER: Die Wirkung der Äthernarkose auf die Harnstoffpermeabilität von Pflanzenzellen. Jb. wiss. Bot. **65**, 643–737 (1926). — HOPKINS, H. T.: Inhibition of growth by benzene hexachloride isomers and protective effect of glucose as measured by cell counting technique. Plant Physiol. **27**, 526–540 (1952).

IVENS, G. W.: The phytotoxicity of mineral oils and hydrocarbons. Ann. Appl. Biol. **39**, 418–422 (1952). — IVENS, G. W., and G. E. BLACKMAN: The effects of phenylcarbamates on the growth of higher plants. Symposia Soc. f. Exper. Biol. **3**, 266–282 (1949).

JAMES, W. O.: The use of respiratory inhibitors. Annual Rev. Plant Physiol. **4**, 59–90 (1953). — JOHNSON, P. W.: 9th Ann. Cotton Defoliation Conf. Proc. 1955.

KEMMER, E.: Beobachtungen über die Lebensdauer isolierter Epidermen. Arch. exper. Zellforsch. **7** (1928). — KENDA, G., and F. WEBER: Rasche Vakuolen-Kontraktion in *Cerinthe*-Blütenzellen. Protoplasma **41**, 458–466 (1952). — KESSLER, E.: Über die Wirkung von 2,4-Dinitrophenol auf Nitratreduktion und Atmung von Grünalgen. Planta (Berl.) **45**, 94 bis 105 (1955). — KRAMER, P. J., and H. B. CURRIER: Water relations of plant cells and tissues. Annual Rev. Plant Physiol. **1**, 265–284 (1950). — KÜSTER, E.: Über die Verschmelzung nackter Protoplasten. Ber. dtsch. bot. Ges. **27**, 589–598 (1910). — Pathologie der Plastiden. Protoplasma-Monographie 13, II. Berlin: Gebrüder Bornträger 1937. — Pathologie der Pflanzenzelle, Teil I: Pathologie des Protoplasmas. Berlin: Gebrüder Bornträger 1929. — Die Pflanzenzelle. Jena: Gustav Fischer 1951. — KUHN, R., and H. B. BIELIG: Über Invertseifen. I. Die Einwirkung von Invertseifen auf Eiweiß-Stoffe. Ber. dtsch. chem. Ges. **73**, 1080–1091 (1940). — KUHN, R., and D. JERCHEL: Über Invertseifen. VII. Tetrazoliumsalze. Ber. dtsch. chem. Ges. B **74**, 941–949 (1941).

LÄRZ, H.: Beiträge zur Pathologie der Chloroplasten. Flora (Jena) **135**, 319–355 (1942). — LAMBERTZ, P.: Untersuchungen über das Vorkommen von Plasmodesmen in den Epidermisaußenwänden. Planta (Berl.) **44**, 147–190 (1954). — LEFEVER, J.: Similitude des actions cytologiques exercées par le phenylurethane et la colchicine sur des plantules végétales. C. r. Acad. Sci. Paris **208**, 301–304 (1939). — LEOPOLD, A. C., and W. H. KLEIN: Maleic hydrazide as an anti auxin. Physiol. Plant. 5, 91–99 (1952). — LEPESCHKIN, W. W.: Zell-Nekrose und Protoplasma-Tod. Protoplasma Monographien, Bd. 12. Berlin: Gebrüder Bornträger 1937. — LEVAN, A.: The effect of colchicine on root mitosis in *Allium*. Hereditas (Lund) 24, 471–486 (1938). — LEVITT, J.: Further remarks on the thermodynamics of active (non-osmotic) water absorption. Physiol. Plant. 6, 240–252 (1953). — LEWIS, G. N., and M. RANDALL: Thermodynamics and the free energy of chemical substances. New York: McGraw-Hill Book Co., Inc. 1923. — LOEVEN, W. A.: Accumulation of cytoplasm in *Allium* cells as a consequence of exposure to organic substances. I. Proc. Ned. Akad. v. Wetensch. 53, 1599–1609 (1950).

MACKE, W.: Untersuchungen über die Wirkung des Bors auf *Helodea canadensis*. Z. Bot. 34, 241–267 (1939). — MARCY, B.: Effect of ethylene chlorohydrin and thiourea on *Elodea* and *Nitella*. Plant Physiol. 12, 207–212 (1937). — MASUDA, Y.: Über den Einfluß von Auxin auf die Stoffpermeabilität des Protoplasmas. I. Bot. Mag. (Tokyo) 66, 256–261 (1954). — McBAIN, J. W.: Frontiers in colloid chemistry, Chap. 6. New York: Interscience Publ. 1950. — MEITES, M.: Action de l'eau et du benzene sur la cellule végétale. Memoires de la section des Sciences, 3e Serie. Montpellier 1944. — MEYER, B. S., and D. B. ANDERSON: Plant Physiology, 2nd ed. New York: van Nostrand & Co. 1952. — MEYER, K., and H. HEMMI: Beiträge zur Theorie der Narkose. III. Biochem. Z. 277, 39–71 (1935). — MILLER, W. L.: Toxicity and chemical potential. J. Phys. Chem. 24, 562–569 (1920). — MINSHALL, W. H., and V. A. HELSON: The herbicidal action of oils. Proc. Amer. Soc. Hort. Sci. 53, 294–298 (1949). — MIRIMANOFF, A.: Influence des substances tensioactives sur la cellule végétale. Bull. Fed. Internat. Pharm. 238, Annee, no. 3 (1949). — Le comportment de la cellule végétale en présence de toxiques additionnés de substances tensioactives. Protoplasma 42, 250–260 (1953). — MITCHELL, J. W., W. M. DUGGAR jr. and H. G. GAUCH: Increased translocation of plant-growth-modifying substances due to application of boron. Science (Lancaster, Pa.) 118, 354–355 (1953). — MOEWUS, F., L. MOEWUS and E. SCHRADER: Die Wirkung von Cumarin und Parasorbinsäure auf das Austreiben von Kartoffelknollen. Naturforsch. 66, 112–115 (1951). — MUIR, R. M., and C. HANSCH: Growth substances. Annual Rev. Plant Physiol. 6 157–176 (1955).

NORMAN, A. G., C. E. MINARIK and R. L. WEINTRAUB: Herbicides. Annual Rev. Plant Physiol. 1, 141–168 (1950). — NORTHEN, H. T.: Relation of dissociation of cellular proteins by auxins to growth. Bot. Gaz. 103, 668–683 (1942). — Alternations in the structural viscosity of protoplasm by colchicine and their relationship to C-mitosis and C-tumor formation. Amer. J. Bot. 37, 705–711 (1950).

OCHS, G.: Zur Wirkungsweise des Wuchsstoffes auf das Streckenwachstum der Wurzel. Planta (Berl.) 44, 412–436 (1954). — OEHLKERS, F.: Chromosome breaks influenced by chemicals. Heredity (Lond.) 6, Suppl., 95–105 (1953). — ÖSTERGREN, G.: Colchicine, mitosis, chromosome contraction, narcosis and protein chain folding. Hereditas (Lund) 30, 429–467 (1944). — OLSON, R. A., and H. G. DU BUY: Factors influencing the protoplasmic streaming in the oat coleoptile. Amer. J. Bot. 27, 392–400 (1940). — ORGELL, W. H.: Isolation and permeability of isolated cuticle. Ph. D. Thesis, University of California, Davis 1954. — OSTERHOUT, W. J. V.: Differing rates of death at inner and outer surfaces of the protoplasm. III. Effects of mercuric chloride on *Nitella*. J. Gen. Physiol. 28, 343–347 (1945). — OVERBEEK, J. VAN, and R. BLONDEAU: Mode of action of phytotoxic oils. Weeds 3, 55–65 (1954).

PARKER, J.: Criteria of life: some methods of measuring viability. Amer. Scientist 41, 615–618 (1953). — POHL, R.: Ein Beitrag zur Analyse des Streckungswachstums der Pflanzen. Planta (Berl.) 36, 230–261 (1948). — POLITZER, M.: Pathologie der Mitose. Berlin 1934. — PRINGSHEIM, E. G.: Über Plasmolyse durch Schwermetallsalze. Beih. z. Bot. Zbl. 41, 1–14 (1925).

RAADTS, E., and H. SÖDING: Über den Einfluß der Askorbinsäure auf die Auxinaktivierung. Naturwiss. 34, 344–345 (1948). — RADEMACHER, B.: The control of weeds. A symposium on the prevention and eradication of weeds on agricultural land by cultural, chemical and biological means. Bull. Imp. Bur. Past. a. For. Crops (Herb. Publ. Ser.) 27, 1–168 (1940). — READY, D., C. E. MINARIK, D. BRADBURY, H. E. THOMPSON and J. F. OWINGS jr.: Albinism induced by substituted benzoic acids. Plant Physiol. 27, 210–211 (1952). — REINDERS, D. E.: Intake of water by parenchymatic tissue. Rec. Trav. bot. néerl. 39, 1–140 (1942). — RICH, S., and J. G. HORSFALL: The relation between fungitoxicity, permeation, and lipid solubility. Phytopathology 42, 457–460 (1952). — ROBBINS, W. W., A. S. CRAFTS and R. N. RAYNOR: Weed control — a textbook and manual. New York: McGraw-Hill Book Co. 1952. — ROBERTSON, R. N.: Mechanism of absorption and transport of inorganic nutrients in plants. Annual Rev. Plant Physiol. 2, 1–24 (1951). — ROMELL, L. G.: Localized injury to plant organs from hydrogen fluoride and other acid gases. Sv. bot. Tidskr. 35, 271–286 (1941). — ROSENE, H. F.: Reversible azide inhibition of oxygen consumption of onion root tissues. J. Cellul. a. Comp. Physiol. 30, 15–30 (1947). — ROTHSTEIN, A.: The enzymology of the cell surface. Protoplasmatologia II, E, 4, 1–86 (1954).

Schindler, H.: Tötungsart und Absterbebild. I. Der Alkalitod der Pflanzenzelle. Protoplasma 30, 186–205 (1938b). — Tötungsart und Absterbebild. II. Der Säuretod der Pflanzenzelle. Protoplasma 30, 547–569 (1938a). — Schmidt, H.-H.: Studien zur experimentellen Pathologie der Chloroplasten. I. Untersuchungen über die Vakuolenbildungen der Chloroplasten durch Alkaloide und Anaesthetica. Protoplasma 40, 209–238 (1951a). — Studien zur experimentellen Pathologie der Chloroplasten. II. Untersuchungen über die Streifen- uud Näpfchenbildung der Chloroplasten. Protoplasma 40, 507–525 (1951b). — Schoene, D. L., and O. L. Hoffmann: Maleic hydrazide, a unique growth regulant. Science (Lancaster, Pa.) 109, 588–590 (1949). — Schumacher, W.: Untersuchungen über die Lokalisation der Stoffwanderung in den Leitbündeln höherer Pflanzen. Jb. wiss. Bot. 73, 770–823 (1930). — Über plasmodesmenartige Strukturen in Epidermisaußenwänden. Jb. wiss. Bot. 90, 530–545 (1942). — Seifriz, W.: A method for inducing protoplasmic streaming. New Phytologist 21, 107–112 (1922). — Protoplasmic streaming. Bot. Rev. 9, 49–123 (1943). — A molecular interpretation of toxicity. Protoplasma 40, 313–323 (1951). — Sexton, W. A.: Chemical constitution and biological activity, 2nd ed. London: E. a. F. N. Spon, Ltd. 1953. — Shaw, W. C., and C. R. Swanson: Techniques and equipment used in evaluating chemicals for their herbicidal properties. Weeds 1, 352–365 (1952). — The relation of structural configuration to the herbicidal properties and phytotoxicity of several carbamates and other chemicals. Weeds 2, 43–65 (1953). — Shaw, W. C., C. R. Swanson and R. L. Lovvorn: Evaluation of chemicals for their herbicidal properties. Plant Indus. Stat., Beltsville, Maryland, Jan. 1953. — Simon, E. W.: Mechanism of dinitrophenol toxicity. Biol. Revs. 28, 453–479 (1953). — Simon, E. W., and H. Beevers: The effect of p_H on the biological activities of weak acids and bases. I. and II. New Phytologist 51, 163–197 (1952). — Simon, E. W., and G. E. Blackman: The significance of hydrogen ion concentration in the study of toxicity. Symposia Soc. f. Exper. Biol. 3, 253–265 (1949). — Simon, E. W., H. A. Roberts and G. E. Blackman: The p_H factor and the toxicity of 3,5-dinitro-o-cresol, a weak acid. J. of Exper. Bot. 3, 99–109 (1952). — Skoog, F.: Chemical control of growth and organ formation in plant tissues. Ann. Biol. 54, 545–562 (1950). — Plant growth substances. Madison, Wisconsin: Univ. of Wisconsin Press 1951. — Slade, R. E., W. G. Templeman and W. A. Sexton: Plant growth substances as selective weed killers. Nature (Lond.) 155, 497–498 (1945). — Smith, F. E.: Tetrazolium salt. Science (Lancaster, Pa.) 113, 751–754 (1951). — Smith, F. G.: Respiratory changes in relation to toxicity. In Skoog, Plant Growth Substances, p. 111–120. 1951. — Söding, H.: Die Wuchsstofflehre: Ergebnisse und Probleme der Wuchsstoff-Forschung. Stuttgart: George Thieme 1952. — Stålfelt, M. G.: Effects of heteroauxin and colchicine on protoplasmic viscosity. Exper. Cell Res. Suppl. 1, 63–78 (1949). — Stiles, W.: The exosmosis of dissolved substance from storage tissue into water. Protoplasma 2, 577–601 (1927). — Strugger, S.: Untersuchungen über den Einfluß der Wasserstoffionen auf das Protoplasma der Wurzelhaare von *Hordeum vulgare* L. Sitzgsber. Akad. Wiss. Wien, Abt. 1 135, 453–478 (1926). — Praktikum der Zell- und Gewebephysiologie der Pflanze. Berlin: Springer 1949. — Swanson, C. R., W. C. Shaw and J. H. Hughes: Some effects of isopropyl N-(3-chlorophenyl) carbamate and an alkanolamine salt of dinitro ortho secondary butyl phenol on germinating cotton seeds. Weeds 2, 178–189 (1953). — Sweeney, B. M.: The effect of auxin on protoplasmic streaming in root hairs of *Avena*. Amer. J. Bot. 31, 78–80 (1944). — Sweeney, B. M., and K. V. Thimann: The effect of auxins on protoplasmic streaming. III. J. Gen. Physiol. 25, 841–854 (1942).

Templeman, W. G., and W. A. Sexton: Effect of some arylcarbamic esters and related compounds upon cereals and other plant species. Nature (Lond.) 156, 630 (1945). — Thimann, K. V.: The synthetic auxins: Relation between structure and activity. In Skoog: Plant Growth Substances, pp. 21–36. 1951. — Thimann, K. V., and B. M. Sweeney: The effect of auxins on protoplasmic streaming. J. Gen. Physiol. 21, 123–135 (1937). — Thomas, M. D.: Gas damage to plants. Annual Rev. Plant Physiol. 2, 293–322 (1951). — Thomas, M. D., and G. R. Hill: Relation of sulfur dioxide in the atmosphere to photosynthesis and respiration of alfalfa. Plant Physiol. 12, 309–383 (1937). — Tukey, H. B.: Plant regulators in agriculture. New York: Wiley & Sons 1954.

Unrau, J., and E. N. Larter: Cytogenetical responses of cereals to 2,4-D. I. A study of meiosis of plants treated at various stages of growth. Canad. J. Bot. 30, 22–27 (1952).

Veldstra, H.: Actions synergiques das analogues structuraux. Bull. Soc. Chim. Biol. 30, 772–792 (1948). — La balance lipophile hydrophile des composes organiques son importance pour leur action physiologique. Bull. Soc. Chim. Biol. 31, 1–29 (1949). — The relation of chemical structure to biological activity in growth substances. Annual Rev. Plant Physiol. 4, 151–198 (1953). — Virgin, H. I.: Physical properties of protoplasm. Annual Rev. Plant Physiol. 4, 363–382 (1953).

Wada, B.: The effect of chemicals on mitosis studied in *Tradescantia* cells *in vivo*. II. Cytologia 17, 279–286 (1952). — Wain, R. L.: Herbicidal selectivity through specific action of plants on compounds applied. J. Agr. a. Food Chem. 3, 128–130 (1955). — Waris, H.:

Cytophysiological studies on *Merasterias*. III. Physiol. Plant. **4**, 387–409 (1951). — WEBER, F.: Zentrifugierungsversuche mit ätherisierten *Spirogyra*. Biochem. Z. **126**, 21–32 (1921). — WEDDING, R. T., L. C. ERICKSON, and B. L. BRANNAMAN: Effect of 2,4-dichlorophenoxyacetic acid on photosynthesis and respiration. Plant Physiol. **29**, 64–69 (1954). — WEDDING, R. T., L. A. RIEHL and W. A. RHOADES: Effect of petroleum oil spray on photosynthesis and respiration in citrus leaves. Plant Physiol. **27**, 269–278 (1952). — WEINTRAUB, R. L.: 2,4-D; mechanisms of action. J. Agr. a. Food Chem. **1**, 250–254 (1953). — WEINTRAUB, R. L., J. W. BROWN, J. C. NICKERSON and K. N. TAYLOR: Studies on the relation between molecular structure and physiological activity of plant growth regulators. I. Abscission inducing activity. Bot. Gaz. **113**, 348–362 (1952). — WEINTRAUB, R. L., and L. PRICE: Inhibition of plant growth by emanations from oils, varnishes and woods. Smithsonian Misc. Collect. **106**, No 21 (1947). — WENT, F. W., and K. V. THIMANN: Phytohormones. New York: Macmillan & Co. 1937. — WHITE, P. R.: The cultivation of animal and plant cells. New York: The Ronald Press Co. 1954. — WILSKE, C., and H. BURSTRÖM: The growth-inhibiting action of thiophenoxyacetic acids. Physiol. Plant. **3**, 58–67 (1950).

YARKAR, N.: Mitotic disturbances caused by chloranil. Amer. J. Bot. **39**, 540–546 (1952).

ZIEGLER, H.: Über die Reduktion des Tetrazoliumchlorids in der Pflanzenzelle und über den Einfluß des Salzes auf Stoffwechsel und Wachstum. Z. Naturforsch. **8b**, 662–667 (1953). — ZIMMERMAN, P. W.: In U.S. Technical Conference on air pollution (Washington, D. C.). New York: McGraw-Hill Book Co. 1952. — ZIMMERMAN, P. W., and A. E. HITCHCOCK: Substituted phenoxy and benzoic acid growth substances and the relation of structure to physiological activity. Contrib. Boyce Thompson Inst. **12**, 321–343 (1942). — ZIRKLE, C.: The plant vacuole. Bot. Rev. **3**, 1–30 (1937). — ZWEEP, W. VAN DER: Studies in the phytotoxicity of hydrocarbons. Ph. D. Thesis, University of California, Davis 1954.

Resistenz gegen Infektion. Immunität.

Von

Heinz Kern.

Mit 3 Abbildungen.

Bei der Infektion einer Pflanze durch einen Parasiten (z. B. ein Virus, ein Bacterium, einen Pilz oder eine höhere Pflanze) beginnt ein Wechselspiel zwischen Wirt und Parasit, das die beiden zu einer funktionellen Einheit verbindet. Jeder Partner wirkt auf den andern ein, und jeder verändert sich unter der Wirkung des andern. Je nach dem Gewicht der auf beiden Seiten wirksamen Kräfte erkrankt die Wirtspflanze mehr oder weniger heftig.

Vermag die Wirtspflanze einen eindringenden Krankheitserreger erfolgreich abzuwehren, bezeichnen wir sie als *widerstandsfähig* oder *resistent*; kann sie dies nicht, ist sie krankheits*anfällig*. Zwischen den beiden Extremen der höchsten Resistenz (manchmal als Immunität bezeichnet; vgl. S. 831) und der höchsten Anfälligkeit bestehen naturgemäß alle Übergänge.

Auf seiten des Wirtes wird der Resistenz- bzw. Anfälligkeitsgrad durch zwei Gruppen von Faktoren bestimmt:

I. Die bereits präinfektionell vorhandenen *strukturellen und physiologischen Gegebenheiten* in Zellen und Geweben (*passive* Resistenz) und

II. die Fähigkeit der Zellen, auf einen Parasiten oder seine Stoffwechselprodukte in charakteristischer Weise zu *reagieren* (*aktive* Resistenz, Abwehrreaktionen).

Die folgenden Ausführungen beschränken sich auf einige grundlegende zellphysiologische Vorgänge und müssen manche Probleme stark vereinfachen. Für eine eingehendere Erörterung der Beziehungen der Pflanzen zu ihren Krankheitserregern sei auf größere zusammenfassende Darstellungen und die dort zitierte Literatur verwiesen (z. B. FISCHER und GÄUMANN 1929, SAVULESCU 1936, ROEMER, FUCHS und ISENBECK 1938, GÄUMANN 1944, 1951, FERRARIS, CIFERRI und BALDACCI 1948, BUTLER und JONES 1949, LIMASSET und DARPOUX 1951). Die Probleme der Resistenz gegen *Insekten* wurden von PAINTER (1951) umfassend behandelt.

I. Die präinfektionellen Gegebenheiten in Zellen und Geweben.

In pflanzlichen Zellen und Geweben müssen bestimmte Bedingungen erfüllt sein, damit ein Parasit in ihnen leben kann. Vom Pflanzenschutz herkommend, drücken wir uns meist umgekehrt aus: je nach Struktur und chemischer Beschaffenheit setzt eine Zelle oder ein Gewebe dem fremden Organismus verschieden starken *Widerstand* entgegen.

Die in diesem Abschnitt zu behandelnden Resistenzfaktoren sind demnach schon *vor* der Infektion in der Pflanze vorhanden; sie bilden statische und unspezifische Hindernisse (im Gegensatz zur dynamischen, durch den eindringenden Parasiten ausgelösten und mehr oder weniger spezifisch gegen ihn gerichteten

Abwehr durch Zell*reaktionen*). Wir sprechen deshalb von *Ungastlichkeit, passiver Resistenz* oder *Axenie* (GÄUMANN 1951, S. 345) und greifen einige der in Frage kommenden Faktoren heraus.

A. Die Resistenz von Zellwänden und Cuticula.

Die Zellwände erschweren sowohl das *Eindringen* der Parasiten von außen her (Eindringungsresistenz) als auch ihre *Ausbreitung* im Innern der Gewebe (Ausbreitungsresistenz). Gegen außen ist die Pflanze überdies durch die Cuticula geschützt.

1. Die Eindringungsresistenz der Cuticula.

Manche Pilze sind imstande, mit ihren Keimschläuchen die *unverletzte* Körperoberfläche einer Pflanze erfolgreich anzugreifen. Da sie die Cuticula in den wenigsten Fällen enzymatisch auflösen können, müssen sie sie *mechanisch* durchstoßen. Die *Dicke* der Cuticula entscheidet daher an erster Stelle darüber, ob eine Infektion zustande kommt. So können zum Beispiel die Keimschläuche der Basidiosporen von *Puccinia graminis* Pers., des Schwarzrostes des Getreides, in junge Blätter von krankheitsanfälligen *Berberis*-Arten eindringen; erst beim Heranwachsen der Blätter verdickt sich die Cuticula und erschwert die Infektion. Bei resistenten Arten dagegen sind schon die jungen Blätter durch eine dicke Cuticula (und Epidermisaußenwand) gegen einen Befall durch diesen Rostpilz geschützt (MELANDER und CRAIGIE 1927).

2. Die Eindringungs- und Ausbreitungsresistenz der Zellwände.

Der Widerstand der Zellwände wird von manchen Parasiten vorwiegend *mechanisch*, von anderen vorwiegend *enzymatisch* überwunden. Im ersten Fall wächst die Resistenz der Gewebe mit dem Druckwiderstand der Zellwände (z. B. bei der Ausbreitung von *Pythium debaryanum* HESSE in Kartoffelknollen; HAWKINS und HARVEY 1919). Im zweiten Fall müssen die Parasiten die für die Auflösung der Zellwandsubstanzen notwendigen Enzyme (Cellulasen, Ligninasen usw.) bilden können. Enzymatisch wenig leistungsfähige Mikroorganismen vermögen lediglich die hemicellulosischen Mittellamellen anzugreifen (z. B. das *Bacterium carotovorum* (J.) L. et N. in Kartoffelknollen). Andere Erreger bauen wohl die cellulosischen, nicht aber die verholzten Teile der Zellwände ab. Auch manche der holzzerstörenden Pilze greifen vorwiegend die Cellulose und nur in geringerem Maße das Lignin an (Rotfäulen, Destruktionsfäulen; z. B. CARTWRIGHT und FINDLAY 1946, LIESE 1950). Ein erhöhter Ligningehalt kann daher zu einer stärkeren Vermorschungsresistenz des Holzes führen; dies dürfte unter gewissen Voraussetzungen beim Lärchenholz mit zunehmender Meereshöhe der Fall sein (GÄUMANN 1948).

Die Ausbreitungsresistenz der Zellwände wird aber noch durch andere Faktoren mitbestimmt, so durch den *Quellungszustand* der Wandsubstanzen (besonders der Cellulose; GÄUMANN 1930). Die Vermorschbarkeit des Fichten- und Tannenholzes ist am größten, wenn die Bäume während der Wachstumsperiode (wenn sie im Saft stehen) geschlagen werden. In dieser Zeit (Februar bis Juli) ist offenbar unter anderem die Cellulose am stärksten gequollen und damit leichter vermorschbar. Wird das Holz nach der Fällung längere Zeit trocken gelagert, verliert es mehr und mehr Feuchtigkeit, die Cellulose entquillt und wird von den Enzymen der holzzerstörenden Pilze weniger rasch abgebaut. Auf der anderen Seite können ein zu hoher Wassergehalt und die damit verbundene schlechte Durchlüftung des Holzes die Entwicklung mancher Pilze hemmen (MÜNCH 1909).

B. Die Resistenz im Innern der Zellen.

Die Lebensmöglichkeiten eines Parasiten im Wirtsinnern werden weitgehend durch den *Stoffwechsel der Wirtszellen* bestimmt. Veränderungen im Wirtsstoffwechsel beeinflussen den Gehalt der Zellen an Nährstoffen, Wuchsstoffen, Hemmstoffen usw., aber auch ihre physikalisch-chemischen Eigenschaften und nicht zuletzt ihre Reaktionsfähigkeit (S. 831).

In manchen Fällen wissen wir, welche *Einzelfaktoren* dem Parasiten das Leben im Innern des Wirtes leicht oder schwer machen können; sie sollen im folgenden getrennt besprochen werden (wobei jedoch die Resistenz gegen einen bestimmten Erreger wohl in den meisten Fällen durch mehrere Faktoren zusammen bestimmt wird). Bei vielen anderen Beispielen dagegen sind die Reaktionsschritte im Stoffwechsel des Wirtes, die schließlich zu einem bestimmten Grad von Resistenz bzw. Empfänglichkeit führen, noch nicht geklärt; wir können bei ihnen lediglich den Einfluß von Außenbedingungen (der Ernährung, der Temperatur, des Lichtes, einer Vorerkrankung usw.) auf die *Disposition* (d.h. die innerhalb der genotypisch gegebenen Grenzen schwankende Krankheitsbereitschaft) feststellen (Roemer, Fuchs und Isenbeck 1938, S. 84—96, Gäumann 1951, S. 469—555).

1. Der Nährstoffgehalt der Zellen.

Ein Parasit muß in den Wirtsgeweben die zu seiner Ernährung und Vermehrung notwendigen Stoffe finden. Ist der Nährstoffgehalt zu gering, wird seine Entwicklung gehemmt. So geraten Gerstenpflanzen, die nur 2 Std je Tag belichtet werden, in einen Hungerzustand; die Menge der gebildeten Assimilate ist offenbar zu gering, um dem in Frage stehenden Erreger (*Erysiphe graminis hordei* March., Gerstenmehltau) ein normales Wachstum zu ermöglichen, und die Pflanzen erkranken nur schwach. Wird den Blättern solcher Pflanzen Glucose verabreicht, steigt der Befallsgrad (Domsch 1953). — Mit wachsendem Gehalt an verfügbaren Nährstoffen kann sich ein Erreger üppiger entwickeln; so werden Kiefernwurzeln mit steigendem Zuckergehalt für Mycorrhizenpilze anfälliger (Björkman 1942).

Die Beziehungen zwischen Nährstoffgehalt und Krankheitsanfälligkeit treten jedoch nicht immer klar zutage. So ist wohl der Zuckergehalt junger, intensiv wachsender Weizenblätter parallel zu ihrer größeren Anfälligkeit für *Puccinia graminis* Pers. höher als derjenige alter Blätter; doch liegen die Werte bei anfälligen und bei resistenten Sorten ungefähr gleich hoch (Johnson und Johnson 1934); die Resistenzverhältnisse lassen sich also damit höchstens teilweise erklären. In anderen Fällen geht die Resistenz der Wirtsgewebe mit einem *hohen* Zuckergehalt parallel, so bei verschiedenen Pilzen (Yarwood 1934, Caroselli und Feldman 1951), Bakterien (Hughes und Fowler 1953) und Viren (z. B. Bawden und Roberts 1947, Yarwood 1952); verminderte Belichtung der Pflanzen vor der Infektion kann hier die Anfälligkeit erhöhen. Die grundlegenden Mechanismen, vor allem die Zusammenhänge zwischen Wirtsstoffwechsel und *Virus*vermehrung erscheinen jedoch zur Zeit noch wenig klar (vgl. Bawden 1950).

Veränderungen in der *Ernährung der ganzen Pflanze* können sich in mannigfacher Weise auf die Krankheitsresistenz auswirken, doch ist über die einzelnen Glieder der Reaktionskette noch wenig bekannt. Überernährung mit *Stickstoff* setzt die Krankheitsresistenz im allgemeinen herab; dabei dürfte in manchen Fällen vor allem die Reaktionsfähigkeit des Wirtes, in anderen aber auch die Stickstoffernährung des Parasiten wesentlich beeinflußt werden. *Phytophthora infestans* (Mont.) de Bary, der Erreger der Kraut- und Knollenfäule der Kartoffeln, ernährt sich vor allem von den Stickstoffverbindungen der Wirtspflanzen (LEPIK 1939, ALTEN und ORTH 1941); unter den Aminosäuren wirkt jedoch das *Arginin* als Ausnahme schon in geringen Konzentrationen auf den Pilz giftig. Der Gehalt der Kartoffelknollen an den verschiedenen Stickstoffverbindungen läßt sich durch die Düngung (vor allem durch das Verhältnis K:N) beeinflussen; so führt eine steigende Kalizufuhr bei relativ niedrigen Stickstoffgaben zu einer Abnahme des Gesamtstickstoffgehaltes, aber zu einer Zunahme des Arginins; parallel dazu steigt die Krankheitsresistenz. Doch entscheiden diese Verschiebungen sicher nicht allein über das Schicksal der befallenen Pflanzen.

Am Beispiel des *Fusarium lycopersici* Sacc., eines Welkeerregers der Tomatenpflanzen, konnte gezeigt werden, daß nicht nur die Ausbreitungsresistenz der Pflanzen gegenüber dem Parasiten, sondern auch die Resistenz der Wirtszellen gegenüber den *Toxinen* des Parasiten durch schlechte Stickstoffernährung erhöht wird (ZÄHNER 1954); die durch eine bestimmte Toxindosis an Sprossen von Stickstoffmangelpflanzen ausgelösten Nekrosen sind wesentlich schwächer als bei normalen Pflanzen. Die Fusarinsäure (eines der vom *Fusarium lycopersici* gebildeten Welketoxine) greift in verschiedene Zellfunktionen in unterschiedlicher Weise ein: sie hemmt die *Atmung* der normal ernährten und der Stickstoffmangelpflanzen in gleicher Weise, verursacht aber nur bei den normal ernährten Pflanzen eine Erhöhung der *Wasserpermeabilität.*

2. Der Wuchsstoffgehalt der Wirtszellen.

Die pflanzenpathogenen Mikroorganismen benötigen zu ihrem Wachstum bestimmte Wuchsstoffe (Aneurin, Biotin, Mesoinosit, Adermin usw.; FRIES 1938, JANKE 1939, SCHOPFER 1949). Manche unter ihnen sind imstande, die notwendigen Wuchsstoffe selbst zu synthetisieren, sind also wuchsstoff*autotroph*; andere dagegen müssen sie ganz oder teilweise von außen geliefert bekommen, sind also wuchsstoff*heterotroph*. Manche Organismen können nur einzelne Synthesestufen durchführen; so muß *Mucor Ramannianus* Moell. das Thiazol im Substrat vorfinden, kann aber Pyrimidin selbst synthetisieren und aus beiden Komponenten das Aneurinmolekül aufbauen.

Manche Krankheitserreger müssen demnach während ihrer parasitischen Phase bestimmte Wuchsstoffe aus den Wirtszellen aufnehmen. So dürfte der erhöhte Wuchsstoffgehalt der im Saft stehenden Bäume neben anderen Faktoren zur stärkeren Entwicklung und Zerstörungsleistung der holzvermorschenden Pilze beitragen (S. 827). Die Bedeutung des Vorhandenseins oder Fehlens von Wuchsstoffen für die Krankheitsresistenz ist jedoch noch nicht systematisch untersucht worden.

Im Gegensatz zu den eben erwähnten Verbindungen wirken die für höhere Pflanzen wesentlichen Wuchsstoffe auf Mikroorganismen im allgemeinen nicht fördernd, sondern zum Teil hemmend (z. B. Heteroauxin; DÉFAGO 1940) und können offenbar dadurch zum Schutz des Vegetationspunktes gegen Parasitenbefall beitragen (GÄUMANN 1951, S. 372).

3. Die physikalisch-chemischen Bedingungen in der Zelle.

Der *Säuregrad* der Zellen und Gewebe beeinflußt die Lebenstätigkeit der Parasiten und die Wirkung ihrer Enzyme. Dabei sind neben dem p_H an sich auch die Art der vorhandenen Säuren (die teilweise abgebaut und verwertet werden können) und die sonst noch vorhandenen Nährstoffe von Bedeutung.

Pilze sind in bezug auf ihr *Wachstum* für p_H-Schwankungen im allgemeinen wenig empfindlich und gedeihen in einem relativ weiten Bereich ohne große

Unterschiede (z. B. das *Fusarium lycopersici* Sacc. zwischen p_H 2 und 9,5; PRITHAM und ANDERSON 1937). Dagegen dürfte die Bildung, Stabilität und Wirkung der pilzlichen *Toxine* und Enzyme durch das p_H in der Pflanze wesentlich beeinflußt werden.

Viele pflanzenpathogene Bakterien gedeihen im sauren Bereich bis zu einem p_H von etwa 5; ein höherer Säuregrad kann deshalb einen Schutz bewirken (z. B. in Tomaten mit p_H 4,0—4,6; GARDNER und KENDRICK 1921).

Über die Bedeutung des *Redoxpotentials* für die Entwicklung der Parasiten und die Wirkung ihrer Stoffwechselprodukte scheinen im Zusammenhang mit der Resistenzfrage noch keine Untersuchungen vorzuliegen.

4. Osmotische Verhältnisse und Permeabilität.

Damit ein Krankheitserreger im Innern eines pflanzlichen Gewebes leben kann, muß der osmotische Wert seiner Zellen (seine Saugkraft) über demjenigen der Wirtszellen und des Gewebesaftes liegen. Dies dürfte nach den bisherigen Untersuchungen im allgemeinen der Fall sein. In Kartoffelknollen treten nach KLAUS (1943) osmotische Werte von höchstens 11 Atm. auf, während der hier untersuchte Kartoffelparasit [*Alternaria solani* (E. et M.) J. et Gr.] auf einem Nährboden mit 20 Atm. noch normal wächst und erst bei 130 Atm. sein Wachstum ganz einstellt.

Bei manchen Erkrankungen verursachen die Erreger eine *Erhöhung der Permeabilität* des Wirtsplasmas und seiner Grenzschichten (z. B. *Puccinia graminis* Pers. in rostanfälligen Weizensorten; THATCHER 1939, 1942). Auch durch toxische Stoffwechselprodukte von Krankheitserregern (z. B. Welketoxine) und durch Antibiotica kann die Permeabilität erhöht werden (GÄUMANN und Mitarbeiter 1952). In anderen Fällen (z. B. rost*resistenten* Weizensorten; THATCHER 1942) bewirkt der eindringende Erreger (hier wiederum *Puccinia graminis*) eine Permeabilitäts*erniedrigung* der ihm zunächst liegenden Wirtsprotoplasten; dadurch wird offenbar die Stoffaufnahme des Parasiten aus den Wirtszellen zu stark erschwert, und die Infektion abortiert.

5. Toxische Zellinhaltsstoffe.

Die Zellen vieler Pflanzen enthalten Stoffe, die auf pathogene Mikroorganismen giftig wirken. Sie können in einzelnen Fällen nach außen abgeschieden werden und dadurch Krankheitserregern das Eindringen erschweren (z. B. Apfelsäure auf den Blättern von *Cicer arietinum* L.; HAFIZ 1952); in den meisten Fällen wirken sie im Innern der Gewebe als Faktoren der *Ausbreitungsresistenz.* Manche dieser Hemmstoffe lassen sich vorläufig nur grob an Hand ihrer Wirkung, nicht aber auf Grund ihrer chemischen Natur charakterisieren (agglutinierende Stoffe gegen Bakterien in Kartoffelknollen, wachstumshemmende Stoffe gegen Pilze in Preßsäften von resistenten Pflanzensorten, dagegen nicht in solchen von anfälligen Sorten der gleichen Art usw.; s. GÄUMANN 1951, S. 373—374).

In anderen Beispielen ist die chemische Natur der toxischen Zellinhaltsstoffe bekannt; der exakte Nachweis einer Resistenzwirkung stößt jedoch oft auf Schwierigkeiten. So ist freies Allylsenföl für verschiedene parasitische Pilze sehr giftig, das ursprünglich in den Zellen vorkommende Glucosid jedoch nicht (WALKER, MORELL und FOSTER 1937); seine Bedeutung als Resistenzfaktor ist deshalb umstritten. Andere Stoffe wirken wohl auf einzelne Mikroorganismen giftig, können aber von anderen abgebaut und unter Umständen als Nährstoffe verwendet werden.

Die Bedeutung von *Alkaloiden* als Resistenzfaktoren wurde von GREATHOUSE (1938) und von GREATHOUSE und RIGLER (1940) am Beispiel des *Phymatotrichum omnivorum* (Shear) Duggar, des Erregers der Texas-Wurzelfäule, untersucht. Dieser Pilz befällt eine große Zahl von Pflanzenarten mehr oder weniger stark. Die Autoren konnten zeigen, daß zwischen dem Gehalt der Pflanzen an verschiedenen Alkaloiden, der Toxicität dieser Alkaloide für den Parasiten und der Krankheitsresistenz der Pflanzen ein Zusammenhang besteht. Die Wurzeln der resistenten *Mahonia trifoliata* (Mor.) Fed. enthalten Berberidin in wesentlich größerer Menge als für eine Wachstumshemmung des Pilzes in vitro notwendig ist. Das Berberidin findet sich vor allem in den Zellen des Rindenparenchyms eingelagert und bildet dadurch einen Gewebering von erhöhter Ausbreitungsresistenz (GREATHOUSE und WATKINS 1938).

Der vermutete Zusammenhang zwischen dem Gehalt verschiedener Tomatensorten an *Tomatin* (dem Glykosid eines Steroidalkaloids) und ihrer Resistenz gegen *Fusarium lycopersici* Sacc. konnte nicht bestätigt werden, da sowohl die Tomatinkonzentration in den Geweben (vor allem in Wurzeln und Stengelbasis, durch die der Pilz eindringt) wie auch die Toxicität für diesen Parasiten zu gering sind (KERN 1952). Dagegen wurde für Tomatin und das nahe verwandte Demissin eine Giftwirkung auf die Larven des Kartoffelkäfers nachgewiesen (KUHN und GAUHE 1947, KUHN, LÖW und GAUHE 1950). Eine Demissinkonzentration von etwa 0,3% oder darüber im frischen Blatt (die bei *Solanum demissum* vorhanden ist oder in Kartoffelblätter infiltriert werden kann) schützt praktisch gegen Fraß. Ähnlich wirksam ist Tomatin; dagegen ertragen die Tiere auch höhere, in Blätter infiltrierte Konzentrationen an Solanin (einem verwandten Alkaloid aus Kartoffelpflanzen) ohne Schaden.

Bei den gegen *Endothia parasitica* (Murr.) And., den Erreger des Rindenkrebses, widerstandsfähigen Kastanienarten ist der Gehalt an wasserlöslichen *Tanninen* in der Rinde höher als bei den empfänglichen Arten (NIENSTAEDT 1953).

Auch *phenolische* Verbindungen können als Resistenzfaktoren wirken, so Pinosylvin im Kiefernholz und Thujaplicin im Holz von *Thuja plicata* D. Don. gegen einige holzzerstörende Pilze (RENNERFELT 1943, 1948) und Protocatechusäure in den braunen Zwiebelschalen gegen Fäulniserreger (ANGELL, WALKER und LINK 1930).

II. Die Abwehrreaktionen der Zellen.

Eine Zelle kann auf einen eindringenden Parasiten in mannigfaltiger Weise *reagieren*. Ihre Reaktionen sind zum Teil lediglich der Ausdruck einer *Erkrankung*, einer Störung des Stoffwechsels unter der Wirkung des Erregers. Darüber hinaus können sie sich aber — mehr oder weniger direkt und spezifisch — *gegen den eindringenden Parasiten* oder seine Stoffwechselprodukte richten und stellen damit eine *funktionelle Abwehr* der Wirtszellen dar. Diese kann je nach Intensität der Reaktion zu einer Schwächung, Lokalisierung oder Eliminierung des Erregers, also zu mehr oder weniger starker Resistenz führen. Im Gegensatz zur bereits besprochenen passiven Resistenz wird die auf den Abwehr*reaktionen* beruhende aktive Resistenz oft als *Immunität* bezeichnet, doch wird dieser Ausdruck nicht immer eindeutig verwendet.

Bei vielen pflanzlichen Infektionskrankheiten müssen wir auf Grund des Ablaufes der Erkrankung vermuten, *daß* Abwehrreaktionen im Spiele sind. An anderen Beispielen können wir den Verlauf der Abwehrreaktionen in ihrer *Erscheinungsform* verfolgen. Über die ihnen zugrunde liegenden biochemischen

Wirkungsmechanismen und über die cellularpathologischen Anfangsstadien wissen wir jedoch allgemein noch sehr wenig.

Im Anschluß an GÄUMANN (1944, 1951) gliedern wir die Abwehrreaktionen nach drei Gesichtspunkten (Tabelle 1):

1. Nach dem *Objekt*, auf das sie gerichtet sind,
2. nach ihrer *Genese* (ihrer Vorgeschichte) und
3. nach der *Intensität* ihrer Wirkung.

Dabei besprechen wir (stark vereinfacht) die Merkmale der in Tabelle 1 aufgeführten Reaktionsformen; die in der Pflanze tatsächlich ablaufenden Vorgänge folgen natürlich nicht derart starr nur *einem* bestimmten Reaktionstyp.

Tabelle 1. *Übersicht der zu besprechenden Abwehrreaktionen.* (Nach GÄUMANN 1944.)

Objekt der Reaktion	Genese	Wirkungsgrad	Reaktionsform
Der Erreger selbst (antiinfektionelle Abwehrreaktionen)	autonom	normergisch	Plasmatische Abwehrreaktionen (S. 832)
		hyperergisch	Nekrogene Abwehrreaktionen (S. 834)
	induziert	normergisch	Prämunität (S. 834)
Die Stoffwechselprodukte des Erregers (antitoxische Abwehrreaktionen)	autonom	normergisch	Histogene Demarkationen (S. 836)
		hyperergisch	Gummöse Demarkationen (S. 837)
Die eigene Empfindlichkeit (induzierte Toleranz)	induziert		Desensibilisierung (S. 838)

A. Die antiinfektionellen Abwehrreaktionen.

Die antiinfektionellen Abwehrreaktionen richten sich *gegen den Erreger selbst*, und suchen ihn zu schwächen, zu lokalisieren und abzutöten. Wir gliedern sie auf Grund ihrer Genese in zwei Typen.

1. Autonome antiinfektionelle Abwehrreaktionen.

Die Wirtszellen reagieren hier von Natur aus mit ihrer *vollen Leistungsfähigkeit*, und dementsprechend setzen die Abwehrreaktionen schon beim ersten Kontakt mit einem entsprechenden Erreger in voller Stärke ein.

Innerhalb dieser Gruppe scheiden wir weiter in *plasmatische Abwehrreaktionen*, die *normergisch* in einer mittleren bis eher schwachen, von pflanzlichen Zellen zu erwartenden Stärke verlaufen, und in *nekrogene Abwehrreaktionen*, die *hyperergisch*, d. h. in Geschwindigkeit und Stärke unverhältnismäßig übersteigert abrollen.

a) Plasmatische Abwehrreaktionen.

Die plasmatischen Abwehrreaktionen seien an Hand des *Rhizobium leguminosarum* FRANK, das an Hülsenfrüchtlern die Bildung von Wurzelknöllchen auslösen kann, erläutert (z. B. FRED, BALDWIN und McCOY 1932, DIENER 1950, STALDER 1951).

Die Infektion erfolgt durch Wurzelhaare, und die Bakterien dringen als schleimige, fadenförmige Bakterienkolonie (Infektionsfaden) in die Wurzelrinde ein. Aus dem Infektionsfaden gelangen die Bakterien ins Plasma der Wirtszellen und veranlassen dieses zu plasmatischen Abwehrreaktionen, deren Leistungsfähigkeit je nach der Disposition des Wirtes verschieden groß ist.

1. In einer gut ernährten und vor allem ausreichend mit Stickstoff versorgten Pflanze sind die Abwehrreaktionen stark genug, um eine Infektion zu verhindern. Die Bakterien können in den Geweben nicht Fuß fassen, und es werden keine Knöllchen gebildet.

2. Im anderen Extrem, also einer sehr schlecht ernährten Pflanze (Mangel an Stickstoff *und* z. B. an Assimilaten), sind die Abwehrreaktionen zu schwach, um wirksam in Erscheinung zu treten. Die Bakterien überschwemmen sämtliche Wurzelgewebe, treten in die Intercellularräume und Mittellamellen hinaus und führen einen raschen Gewebezerfall herbei, ohne daß Knöllchen gebildet würden.

3. Von größerer theoretischer und praktischer Bedeutung ist der dritte, intermediäre Fall, in dem sich Parasit und Wirt ungefähr die Waage halten, und in dem Wurzelknöllchen gebildet werden. In teilweise unterernährten Pflanzen (Stickstoffmangel) vermögen die Abwehrreaktionen der Wirtszellen die Bakterien zwar nicht zu eliminieren, aber auf die Wurzelrinde zu *lokalisieren*, und es resultiert eine lebensfähige Gemeinschaft von Wirt und Parasit. Die Bakterien reizen die Zellen der Wurzelrinde zur Bildung von Tumoren (Knöllchen), die längere Zeit am Leben bleiben können. Früher oder später geht aber auch diese harmonische Knöllchenphase (während der manche Bakterienstämme atmosphärischen Stickstoff assimilieren können) zu Ende. Die intracellulären Bakterien werden deformiert und schließlich von den Enzymen der Wirtszellen aufgelöst; umgekehrt gelangen Bakterien aus dem Zellplasma heraus in Intercellularräume und Mittellamellen, wo sie den plasmatischen Abwehrreaktionen nicht unterliegen, virulent bleiben und schließlich den Zerfall der Knöllchen herbeiführen.

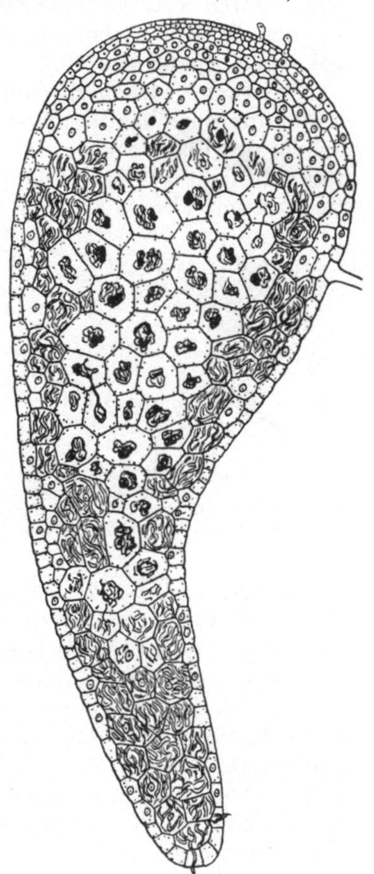

Abb. 1. Orchideenembryo nach 50tägiger Infektion mit *Rhizoctonia mucoroides* Bern. Vergr. 50fach. (Nach BERNARD 1909.)

Zu einem ähnlichen Gleichgewicht kann die Infektion von Orchideenembryonen durch Mycorrhizenpilze führen. Bei einem ausgeglichenen Kräfteverhältnis von Wirt und Parasit wachsen die Pilzhyphen nach der Infektion nicht normal weiter, sondern werden schon in der Randzone des Embryos (Abb. 1) durch die in jeder einzelnen Zelle einsetzenden Abwehrreaktionen deformiert, verknäuelt und damit geschwächt. Der Pilz dringt nur langsam ins Innere des Embryos vor; dabei werden die Hyphen mehr und mehr deformiert und schließlich aufgelöst und verdaut (Abb. 1, Mitte). Der Embryo vermag auf diese Weise den Pilz zu lokalisieren und von der Wachstumszone fernzuhalten (Abb. 1, oben); andererseits gewinnt er bei der Verdauung der Hyphenknäuel Stickstoffverbindungen, Kohlenhydrate oder Wuchsstoffe, die ihm in manchen Fällen überhaupt erst die Entwicklung ermöglichen. — In ähnlicher Weise werden die Hyphen mancher Pilze beim Übergang vom intercellulären zum *intra*cellulären Wachstum durch die Reaktionen des Plasmas zu *Haustorien* deformiert (GÄUMANN 1951, S. 430).

In allen diesen Fällen bleiben chemische Natur, Bildungs- und Wirkungsweise der Agglutinine, Lysine, Pilzhemmstoffe usw. erst noch abzuklären.

b) Nekrogene Abwehrreaktionen.

Wir besprechen hier lediglich das extreme Beispiel des Kartoffelkrebses, ausgelöst durch *Synchytrium endobioticum* (Schilb.) Perc.; bei anderen Krankheiten ist der Ablauf wesentlich komplizierter und mit plasmatischen Reaktionen gekoppelt (GÄUMANN 1951, S. 391—411).

Die verschiedenen Kartoffelsorten können sich gegenüber dem Krebserreger sehr unterschiedlich verhalten. Der erste Schritt ist zwar bei allen gleich: Zoosporen des Pilzes setzen sich auf nichtcutinisierten Stellen von Kartoffelknollen fest, und ihr Protoplast dringt in das Innere einer Epidermiszelle ein. Die Infektion kommt also in allen Fällen zum Haften.

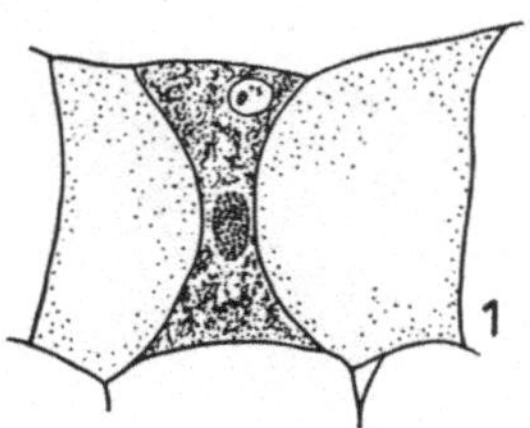

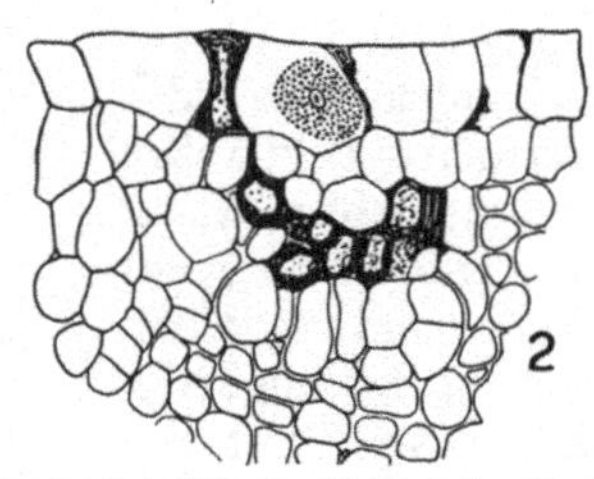

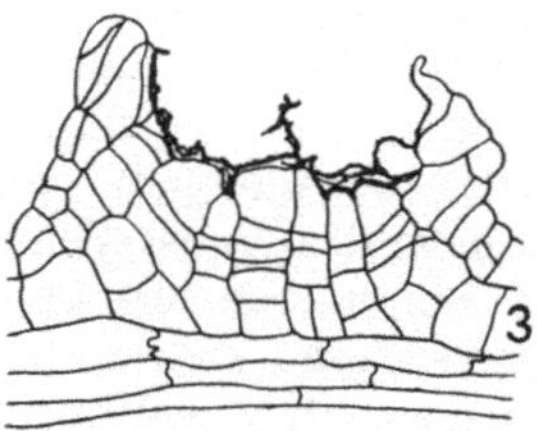

Abb. 2. *1* Hyperergische Reaktion einer krebsinfizierten Epidermiszelle der Kartoffelsorte Ackersegen, einige Stunden nach der Infektion. Der noch gut erhaltene Parasit im Zellinnern ist dunkel punktiert. *2* Isolierung des Infektionsherdes durch eine Zone absterbender Zellen bei der Kartoffelsorte Paulsens Juli. *3* Der Infektionsherd ist abgegrenzt und ausgestoßen worden. Vergr. *1* 825fach, *2* 250fach, *3* 66fach. (*1* Nach KÖHLER 1931; *2* und *3* nach KÖHLER 1928.)

Bei den *krebsanfälligen* Sorten reagiert die befallene Epidermiszelle nicht in auffälliger Weise. Sie erträgt den in ihr lebenden Parasiten, und beide bleiben am Leben. In der Folge kommt es in den umliegenden Zellen zu gesteigertem Wachstum, Zellteilungen und Krebswucherungen; die Knolle wird krank.

Bei den *krebsfesten* Sorten werden schon wenige Stunden oder Tage nach der Infektion starke *hyperergische* Reaktionen sichtbar. Es kann die befallene Epidermiszelle selbst (Abb. 2, *1*) oder eine Gruppe benachbarter, nicht infizierter Zellen (Abb. 2, *2*) in noch nicht genauer untersuchter Weise durch den Erreger entscheidend geschädigt werden und absterben. Bei der Nekrose und Abkapselung des Infektionsherdes geht der Parasit mit zugrunde, weil er sich als obligater Parasit nicht von toter Substanz ernähren kann. Schließlich kann der abgestorbene Infektionsherd abgegrenzt und ausgestoßen werden (Abb. 2, *3*). Die befallene Epidermiszelle (oder der von der parasitischen Wirkung erfaßte Zellkomplex) erträgt den eingedrungenen Parasiten offensichtlich nicht; sie reagiert überempfindlich, *parabiontisch* (im Gegensatz zum eusymbiontischen Wirt-Parasitverhältnis bei den krebsanfälligen Sorten) und geht daran zugrunde; aber die lokale Nekrose führt zur Befallsfreiheit der ganzen Knolle.

2. Induzierte antiinfektionelle Abwehrreaktionen.

Zu den hier zu besprechenden Reaktionen ist die Zelle von Natur aus nur in geringem Maße befähigt. Wohl ist eine prinzipielle Reaktionsbereitschaft vorhanden, doch wird sie erst durch eine *Sensibilisierung* der Wirtszellen beim ersten Kontakt mit einem entsprechenden Erreger *auf die volle Wirkungshöhe gesteigert.* Bei der ersten Infektion *induziert* der Erreger Umstimmungen in den Wirtszellen, die sich bei einer Sekundärinfektion in überraschender Weise auswirken können.

Die Humanmedizin kennt die Sensibilisierung bzw. aktive Immunisierung des Körpers durch den *Erreger* (Überstehen von Kinderkrankheiten, Pockenschutzimpfung) oder durch dessen *Stoffwechselprodukte* (Tetanusschutzimpfung). Ähnliche Probleme stellen sich auch im Leben der Pflanze.

a) Sensibilisierung der Wirtspflanze durch den Erreger (Prämunität).

Bei den Kinderkrankheiten des Menschen verschwindet der Erreger nach der ersten Erkrankung wieder aus dem Körper und hinterläßt im geheilten Organismus eine *erworbene Immunität*, die lange Zeit erhalten bleiben kann. In den Pflanzen ist der Schutz gegen eine zweite Erkrankung in der Regel *an den ersten Erreger gebunden* und hört auf, sobald die Pflanze diesen verliert (infektionsgebundene Immunität, *Prämunität*).

Aus strukturellen Gründen (Fehlen eines generellen Zirkulationssystems, Cellulosewände) sind die Reaktionen in der Pflanze vorwiegend auf die *Einzelzelle* oder auf einen kleinen Gewebeteil beschränkt. Wir haben es deshalb mit einer *lokalen Prämunität* zu tun: der von einer Infektionsstelle ausgehende Schutz erreicht nur die nächste Umgebung (und nicht den ganzen Organismus wie bei der humoralen Prämunität der Humanmedizin).

Bei *Pilz*krankheiten vermag die lokale Prämunität die betreffende Stelle der Wirtspflanze vor einer sekundären Infektion zu schützen. Werden Orchideenkeimlinge mit einem wenig aggressiven Stamm eines *Rhizoctonia*-Wurzelpilzes infiziert, bringen die plasmatischen Abwehrreaktionen den Parasiten schon nach wenigen Zellreihen zum Stehen. Infiziert man den Keimling einige Tage später mit einem stark aggressiven Stamm der gleichen oder einer verwandten Art (der für sich allein den Keimling überwältigen würde), geht die Infektion nicht an (Bernard 1909). Der Schutz ist jedoch begrenzt; wenn der Keimling etwas gewachsen ist, kann in einer gewissen Entfernung von der ersten Infektionsstelle doch eine Sekundärinfektion durch den aggressiven Stamm angehen, und der Keimling geht zugrunde.

Bei *Virosen* vermag die lokale (nicht humorale) Prämunität unter Umständen die *ganze Wirtspflanze* zwar nicht vor einer Sekundärinfektion, aber vor einer Sekundär*erkrankung* zu schützen. Wird eine Tabakpflanze mit einem milden, nahezu symptomlosen Virus aus der X-Gruppe infiziert, verbreitet sich das Virus in der ganzen Pflanze und löst überall lokale Zellreaktionen aus (Salaman 1933). Es kommt zu einer Umstimmung und Leistungssteigerung der Wirtsreaktionen, die in etwa 8 Tagen vollzogen ist. Wird die Pflanze nach dieser Zeit mit einem stark pathogenen Stamm der gleichen Gruppe infiziert, erkrankt sie nicht, sondern ist durch die Vorinfektion geschützt. Der Schutz erstreckt sich nicht auf die Infektion, sondern nur auf die Generalisation und auf die pathogene Wirkung des zweiten Virus; an der Infektionsstelle selbst bleibt dieses erhalten und virulent. Die Schutzwirkung ist ferner gruppenspezifisch, gilt also nur innerhalb der X-Gruppe.

b) Sensibilisierung der Wirtspflanze durch Stoffwechselprodukte des Erregers.

Die Frage, wie weit Pflanzen auf Kulturfiltrate von Krankheitserregern und andere Fremdsubstanzen spezifisch zu reagieren vermögen, ist schon oft untersucht worden, doch sind die Ergebnisse wegen der großen methodischen Schwierigkeiten umstritten (Hess 1949, Gäumann 1951, S. 423). Bei der Infektion von Gerstenpflanzen mit *Helminthosporium sativum* P. K. et B. läßt sich durch vorgängige Behandlung der Keimpflanzen mit Kulturfiltraten dieses Pilzes eine Befallsreduktion erzielen; eine solche tritt aber auch ein, wenn die Körner in konzentrierten Nährlösungen keimen (HESS 1949). Es dürfte sich hier nicht um spezifisch-immunologische, sondern um unspezifische Erscheinungen handeln: jede Störung, welche die Vitalität der Pflanzen herabsetzt, steigert die hyperergische Reaktionsbereitschaft und führt damit zur Befallsfreiheit.

Die Ausscheidung von Stoffwechselprodukten mit antigener Wirkung durch den Parasiten läßt sich an Orchideenknollen zeigen (z. B. GÄUMANN, BRAUN und BAZZIGHER 1950). Bei manchen Orchideen werden die Wurzeln, nicht aber die Knollen von Mycorrhizenpilzen befallen. Läßt man auf Agar den Mycorrhizenpilz gegen ein lebendes Knollenstück wachsen, wird er in einiger Distanz von ihm aufgehalten und es bildet sich ein Hemmhof um die Knolle. Demgegenüber wird eine durch Kälte abgetötete Knolle vom Pilz rasch überwuchert. Der Pilz löst offenbar durch die vor ihm her diffundierenden Stoffwechselprodukte in der Knolle die Bildung von Abwehrstoffen aus, die ihn von einer bestimmten Konzentration an am Weiterwachsen hindern. Die Natur dieser Abwehrstoffe ist noch nicht bekannt.

B. Die antitoxischen Abwehrreaktionen.

Die pathogene Wirkung vieler Krankheitserreger beruht auf ihren *Toxinen* und auf den *nekrogenen Abbauprodukten* der verletzten oder absterbenden Wirtszellen, die sich vom Krankheitsherd aus in die gesunden Gewebe der Pflanze ausbreiten. Die Untersuchungen über die Natur dieser Giftstoffe, die Art und Weise ihrer Ausbreitung, ihre Veränderungen und Bindungen in den Wirtszellen usw. stehen noch sehr in den Anfängen (BLOCH 1953, KERN 1956). Besser als diese biochemischen Vorgänge kennen wir die Reaktionen, mit denen manche Pflanzen als Antwort auf die Giftwirkungen des Parasiten einen Krankheitsherd *abgrenzen*, isolieren und eventuell ausstoßen. Diese antitoxischen Abwehrreaktionen können normergisch (histogene Demarkationen) oder hyperergisch (gummöse Demarkationen) verlaufen.

1. Histogene Demarkationen.

Als Beispiel diene die Schrotschußkrankheit des Steinobstes (NAEF-ROTH 1948). Der Erreger, *Clasterosporium carpophilum* (Lév.) Aderh., dringt in ein Blatt ein und beginnt inter- und intracellulär zu wachsen. Seine Toxine schädigen neben den direkt befallenen auch die benachbarten nicht befallenen Zellen und erzeugen in wenigen Tagen eine braune *Nekrose* von einigen Millimetern Durchmesser. Rund um diese herum (im Abstand von etwa 20 Zellreihen) nimmt hierauf ein mehrere Zellreihen dickes Band auf der ganzen Blattdicke *meristematischen* Charakter an (Abb. 3, *1*). In der Folge bildet sich ein verholztes und verkorktes *Trennungsgewebe*; dieses unterbindet einerseits die Überschwemmung der gesunden Gewebe durch die parasitogenen und nekrogenen Toxine und andererseits die Nährstoffzufuhr aus den gesunden Geweben in den Krankheitsherd. Damit ist die Erkrankung lokalisiert. In manchen Fällen wird der Herd überdies durch Auflösung der Mittellamellen abgetrennt und ausgestoßen (Abb. 3, *2*), und an seiner Stelle bleibt lediglich ein Loch im sonst gesunden Blatt. — Ähnliche Reaktionsfolgen können auch bei der Infektion von Zweigen zu einer Abgrenzung und Ausstoßung des Krankheitsherdes führen.

Zur Auslösung der Nekrose ist die Anwesenheit des Parasiten nicht unbedingt notwendig; sie kann experimentell auch durch die Toxine des Pilzes allein (z. B. Injektion von Mycelextrakt), durch anderweitige organische und anorganische Gifte oder auch durch eine starke Verwundung erfolgen. Der Anstoß zur Demarkationsreaktion ist also nicht unbedingt spezifisch. Die aus dem nekrotischen

Herd diffundierenden, in erster Linie *nekrogenen* Giftstoffe induzieren hierauf das Meristem und führen schließlich zur Abgrenzung und Ausstoßung des Krankheitsherdes.

Die eben geschilderte Demarkationsreaktion richtet sich in erster Linie gegen die parasitogenen und nekrogenen Giftstoffe; sie vermag aber in den meisten Fällen auch das *Clasterosporium* selbst im Infektionsherd zu lokalisieren. Bei anderen Erregern (z. B. *Glomerella rufomaculans* (Berk.) Schr. et Sp., dem Erreger

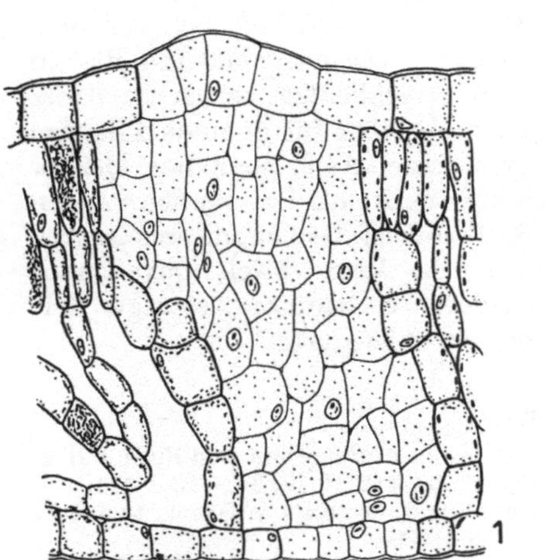

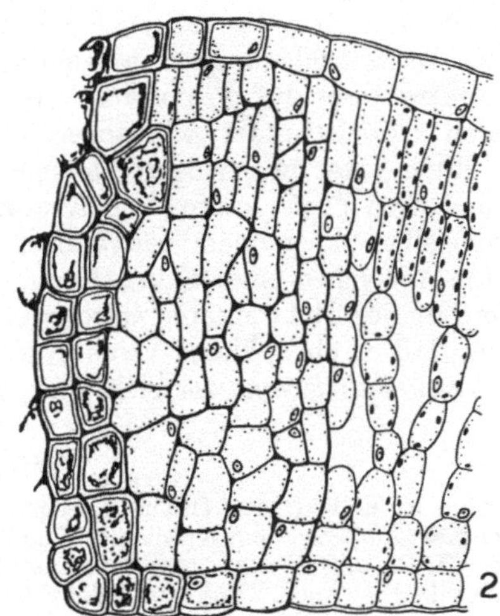

Abb. 3. Abgrenzungsreaktion bei der Schrotschußkrankheit in einem jungen Blatt der Sauerkirsche Englische Morelle. *1* Meristematisches Zellband, 8 Tage nach der Infektion. *2* Wundkork, 15 Tage nach der Infektion; der links vom Trennungsgewebe gelegene Krankheitsherd ist ausgestoßen worden. Vergr. 280fach. (Nach NAEF-ROTH 1948.)

der Bitterfäule an Kirschen; BÖHNI 1949) werden ähnliche antitoxische Abwehrreaktionen ausgelöst; sie verlaufen jedoch gegenüber dem Wachstum des Pilzes zu langsam, und dieser vermag das sich bildende Trennungsgewebe immer wieder zu durchbrechen und dadurch das ganze Blatt zu besiedeln.

2. Gummöse Demarkationen.

Hier antwortet die Pflanze auf die Giftstoffe eines Krankheitserregers nicht mit einem Demarkationsgewebe, sondern mit einer *Gummose*, einer Abkapselung des Krankheitsherdes durch gummi- und gerbstoffartige Verbindungen, Thyllen, Phlobabene usw.

Stereum purpureum Pers., der Erreger des Silberglanzes der Pflaumenbäume, infiziert die Stämme seiner Wirtspflanzen. Sein Mycel besiedelt nur die nächste Umgebung der Infektionsstelle, bildet aber dabei wirksame Toxine. Bei manchen (anfälligen) Sorten gelangen diese ungehindert in die nicht befallenen Gewebe und in die Krone und lösen an den Blättern das typische Krankheitsbild aus. Bei anderen (widerstandsfähigen) Sorten kommt es ebenfalls zur Infektion und zur Bildung eines lokalen Krankheitsherdes. Dieser wird jedoch bald durch heftige Reaktionen des Wirtes mit einer dunklen, gummösen Abwehrzone abgegrenzt und isoliert (BROOKS 1928); der Stoffaustausch und damit die Überschwemmung des Baumes mit Giftstoffen wird dadurch verhindert, und der Baum zeigt trotz der lokalen Infektion keinen Silberglanz.

C. Die induzierte Toleranz.

Eine Virusinfektion kann den pflanzlichen Organismus nicht nur in der Weise umstimmen, daß er z. B. an einer Sekundärinfektion nicht mehr erkrankt; es kann auch im Laufe der Entwicklung einer infizierten Pflanze eine Umstimmung erfolgen, welche eine *Desensibilisierung der neugebildeten Organe* zur Folge hat.

Werden Tabakpflanzen mit dem Virus der Ringfleckigkeit infiziert (PRICE 1932), zeigen sie nach kurzer Zeit auf den Blättern die charakteristischen Symptome als konzentrische blasse Ringe. Mit dem Wachstum der Pflanze breitet sich das Virus auch in allen neugebildeten Blättern aus und läßt sich von dort aus ohne weiteres auf gesunde Pflanzen übertragen. Die neugebildeten Blätter reagieren jedoch immer schwächer auf das Virus; schließlich zeigen sie (trotz des in ihnen zwar in geringerer Menge vorhandenen, aber für gesunde Pflanzen voll virulenten Virus) kaum mehr Krankheitssymptome. Auch wenn diese Blätter mit demselben oder einem verwandten Virus aufs neue infiziert werden, zeigen sie keine Anzeichen einer Erkrankung. Der Stoffwechsel des Organismus ist also im Verlauf der Entwicklung unter der Wirkung des Virus soweit umgestimmt worden, daß er eben dieses Virus nun ohne äußerliche Symptome erträgt. Wie diese *Desensibilisierung* vor sich geht, ist noch nicht geklärt.

Literatur.

ALTEN, F., u. H. ORTH: Untersuchungen über den Aminosäurengehalt und die Anfälligkeit der Kartoffel gegen die Kraut- und Knollenfäule. Phytopath. Z. **13**, 243—271 (1941). — ANGELL, H. R., J. C. WALKER and K. P. LINK: The relation of protocatechuic acid to disease resistance in the onion. Phytopathology **20**, 431—438 (1930).

BAWDEN, F. C.: Plant viruses and virus diseases. Waltham, Mass.: Chronica Botanica Co. 1950. — BAWDEN, F. C., and F. M. ROBERTS: The influence of light intensity on the susceptibility of plants to certain viruses. Ann. Appl. Biol. **34**, 286—296 (1947); **35**, 418 bis 428 (1948). — BERNARD, N.: L'évolution dans la symbiose. Ann. des Sci. natur. Bot. **9**, 1—196 (1909). — BJÖRKMAN, E.: Über die Bedingungen der Mykorrhizabildung bei Kiefer und Fichte. Symbolae bot. Upsalienses **6**, 191 (1942). — BLOCH, R.: Defense reactions of plants to the presence of toxins. Phytopathology **43**, 351—354 (1953). — BÖHNI, E.: Untersuchungen über die Bitterfäule an Kirschen. Phytopath. Z. **15**, 333—375 (1949). — BROOKS, F. T.: Disease resistance in plants. New Phytologist **27**, 85—91 (1928). — BUTLER, E. J., and S. G. JONES: Plant Pathology. London: Macmillan & Co. 1949.

CAROSELLI, N. E., and A. W. FELDMAN: Dutch elm disease in young seedlings. Phytopathology **41**, 46—51 (1951). — CARTWRIGHT, K. ST. G., and W. P. K. FINDLAY: Decay of timber and its prevention. London: H. M. Stat. Office. 1946.

DÉFAGO, G.: Effets de l'aneurine, de ses composants et de l'hétéroauxine sur la croissance de trois parasites du blé. Phytopath. Z. **13**, 293—315 (1940). — DIENER, T.: Über die Bedingungen der Wurzelknöllchenbildung bei *Pisum sativum* L. Phytopath. Z. **16**, 129—170 (1950). — DOMSCH, K. H.: Über den Einfluß photoperiodischer Behandlung auf die Befallsintensität beim Gerstenmehltau. Arch. Mikrobiol. **19**, 287—318 (1953).

FERRARIS, T.: Patologia e terapia vegetale, 5. Aufl., Bd. I/1, bearb. von R. CIFERRI und E. BALDACCI. Milano: Ulrico Hoepli 1948. — FISCHER, E., u. E. GÄUMANN: Biologie der pflanzenbewohnenden parasitischen Pilze. Jena: Gustav Fischer 1929. — FRED, E. B., I. L. BALDWIN and E. McCOY: Root nodule bacteria and leguminous plants. Univ. Wisconsin Stud. Sci. **5**, 343 (1932). — FRIES, N.: Über die Bedeutung von Wuchsstoffen für das Wachstum verschiedener Pilze. Symbolae bot. Upsalienses **3**, 188 (1938).

GÄUMANN, E.: Der Einfluß der Fällungszeit auf die Dauerhaftigkeit des Fichten- und Tannenholzes. Z. d. schweiz. Forstvereins **1930**, Beih. 6, 155; s. a. Angew. Bot. **14**, 387—411 (1932). — Immunreaktionen und Immunität bei Pflanzen. Schweiz. Z. Path. u. Bakter. **7**, 407—441 (1944). — Der Einfluß der Meereshöhe auf die Dauerhaftigkeit des Lärchenholzes. Mitt. schweiz. Anst. forstl. Versuchsw. **25**, 327—393 (1948). — Pflanzliche Infektionslehre, 2. Aufl. Basel: Birkhäuser 1951. — GÄUMANN, E., R. BRAUN u. G. BAZZIGHER: Über induzierte Abwehrreaktionen bei Orchideen. Phytopath. Z. **17**, 36—62 (1950). — GÄUMANN, E., ST. NAEF-ROTH, P. REUSSER u. A. AMMANN: Über den Einfluß einiger Welketoxine und Antibiotica auf die osmotischen Eigenschaften pflanzlicher Zellen. Phytopath. Z. **19**, 160—220 (1952). — GARDNER, M. W., and J. B. KENDRICK: Bacterial spot of tomato. J. Agricult. Res. **21**, 123—152 (1921). — GREATHOUSE, G. A.: Suggested rôle of alkaloids in plants resistant to *Phymatotrichum omnivorum*. Phytopathology **28**, 592—593 (1938). — GREATHOUSE, G. A., and N. E. RIGLER: The chemistry of resistance of plants to *Phymatotrichum* root rot. V. Influence of alcaloids on growth of fungi. Phytopathology **30**, 475—485 (1940). — GREATHOUSE, G. A., and G. M. WATKINS: Berberine as a factor in the resistance of *Mahonia trifoliata* usw. Amer. J. Bot. **25**, 743—748 (1938).

HAFIZ, A.: Basis of resistance in gram to *Mycosphaerella* blight. Phytopathology **42**, 422—424 (1952). — HAWKINS, L. A., and R. HARVEY: Physiological study of the parasitism of *Pythium de Baryanum* on the potato tuber. J. Agric. Res. **18**, 275—297 (1919). — HESS, H.: Ein Beitrag zum Problem der induzierten Abwehrreaktionen im Pflanzenreich. Phytopath. Z. **16**, 41—70 (1949). — HUGHES, L. C., and D. D. FOWLER: Resistance to a plant disease associated with high glucose content of leaf. Nature (Lond.) **172**, 316 (1953).

JANKE, A.: Die Wuchsstoff-Frage in der Mikrobiologie. Zbl. Bakter. II **100**, 409—459 (1939). — JOHNSON, T., and O. JOHNSON: The relationship between sugar content and reaction to stem rust of mature and immature tissues of the wheat plant. Canad. J. Res. **11**, 582—588 (1934).

KERN, H.: Über die Beziehungen zwischen dem Alkaloidgehalt verschiedener Tomatensorten und ihrer Resistenz gegen *Fusarium lycopersici*. Phytopath. Z. **19**, 351—382 (1952). — Problems of incubation in plant diseases. Annual Rev. Microbiol. **10** (1956). — KLAUS, H.: Untersuchungen über *Alternaria solani* usw. Phytopath. Z. **13**, 126—195 (1943). — KÖHLER, E. Fortgeführte Untersuchungen über den Kartoffelkrebs II und III. Arb. biol. Reichsanst. **15**, 135—176, 401—416 (1928). — Über das Verhalten von Synchytrium endobioticum auf anfälligen und widerstandsfähigen Kartoffelsorten. Arb. biol. Reichsanst. **19**, 263—284 (1931). — KUHN, R., u. A. GAUHE: Über die Bedeutung des Demissins für die Resistenz von *Solanum demissum* gegen die Larven des Kartoffelkäfers. Z. Naturforsch. **2b**, 407—409 (1947). — KUHN, R., I. LÖW u. A. GAUHE: Über das Alkaloidglycosid von *Lycopersicum esculentum var. pruniforme* usw. Chem. Ber. **83**, 448—452 (1950).

LEPIK, E.: Über die Rolle der stickstoffhaltigen Bestandteile der Kartoffelknolle bei *Phytophthora*-Fäule. Phytopath. Z. **12**, 292—311 (1939). — LIESE, J.: Zerstörung des Holzes durch Pilze und Bakterien. In MAHLKE-TROSCHELS Handbuch der Holzkonservierung, 3. Aufl., S. 44—111. Herausgeg. von J. LIESE. Berlin-Göttingen-Heidelberg: Springer 1950. — LIMASSET, P., et H. DARPOUX: Principes de pathologie végétale, 2. Aufl. Paris: Dunod 1951.

MELANDER, L. W., and J. H. CRAIGIE: Nature of resistance of *Berberis* spp. to *Puccinia graminis*. Phytopathology **17**, 95—114 (1927). — MÜNCH, E.: Untersuchungen über Immunität und Krankheitsempfänglichkeit der Holzpflanzen. Naturwiss. Z. Land- u. Forstwirtsch. **7**, 54—75, 87—114, 129—160 (1909).

NAEF-ROTH, S.: Untersuchungen über den Erreger der Schrotschußkrankheit des Steinobstes usw. Phytopath. Z. **15**, 1—38 (1948). — NIENSTAEDT, H.: Tannin as a factor in the resistance usw. Phytopathology **43**, 32—38 (1953).

PAINTER, R. H.: Insect resistance in crop plants. New York: Macmillan & Co. 1951. — PRICE, W. C.: Acquired immunity to ring-spot in *Nicotiana*. Contrib. Boyce Thompson Inst. **4**, 359—403 (1932); s. a. Amer. Naturalist **74**, 117—128 (1940). — PRITHAM, G. H., and A. K. ANDERSON: The carbon metabolism of *Fusarium lycopersici* on glucose. J. Agricult. Res. **55**, 937—949 (1937).

RENNERFELT, E.: Die Toxizität der phenolischen Inhaltsstoffe des Kiefernkernholzes gegenüber einigen Fäulnispilzen. Sv. bot. Tidskr. **37**, 83—93 (1943). — Investigations of Thujaplicin, a fungicidal substance in the heartwood of *Thuja plicata* D. Don. Physiol. Plantarum (Copenh.) **1**, 245—254 (1948). — ROEMER, TH., W. H. FUCHS u. K. ISENBECK: Die Züchtung resistenter Rassen der Kulturpflanzen. Berlin: Paul Parey 1938.

SALAMAN, R. N.: Protective inoculation against a plant virus. Nature (Lond.) **131**, 468 (1933); s. a. Philosophic. Trans. Roy. Soc. Lond., Ser. B **229**, 137—217 (1938). — SAVULESCU, TR.: L'immunité aux maladies bactériennes des plantes. Niort: Soulisse-Martin 1936. — SCHOPFER, W. H.: Plants and vitamins. Waltham, Mass.: Chronica Botanica Co. 1949. — STALDER, L.: Über Dispositionsverschiebungen bei der Bildung von Wurzelknöllchen. Phytopath. Z. **18**, 376—403 (1951).

THATCHER, F. S.: Osmotic and permeability relations in the nutrition of fungus parasites. Amer. J. Bot. **26**, 449—458 (1939). — Further studies of osmotic and permeability relations in parasitism. Canad. J. Res. C **20**, 283—311 (1942); s. a. Phytopathology **30**, 24 (1940) und Canad. J. Res. **21**, 151—172 (1943).

WALKER, J. C., S. MORELL and H. H. FOSTER: Toxicity of mustard oils and related sulfur compounds to certain fungi. Amer. J. Bot. **24**, 536—541 (1937).

YARWOOD, C. E.: The comparative behaviour of four clover-leaf parasites on excised leaves. Phytopathology **24**, 797—806 (1934). — Some relations of carbohydrate level of the host to plant virus infections. Amer. J. Bot. **39**, 119—124 (1952).

ZÄHNER, H.: Über den Einfluß der Ernährung auf die Toxinempfindlichkeit von Tomatenpflanzen. Phytopath. Z. **23**, 49—88 (1954).

Vorkommen und Arten des Aktivitätswechsels.

Von

E. Bünning.

Mit 5 Abbildungen.

A. Vorkommen des Aktivitätswechsels.

Ein Wechseln hoher und niedriger Aktivität ist bei Pflanzen schon darum fast selbstverständlich, weil die Pflanzen durchweg wechselnden Außenfaktoren ausgesetzt sind. Mit diesen Außenfaktoren verändert sich sowohl die Höhe als auch die Art der Aktivität. Änderungen in der Höhe der Aktivität werden vor allem beim Messen der Wachstumsintensität deutlich. Änderungen in der Art der Aktivität zeigen sich im Wechseln der dominierenden Entwicklungsprozesse: Perioden der Blattbildung lösen sich mit Zeiten der Blütenbildung, Perioden starker Streckung mit solchen der Zwiebel- oder Knollenbildung usw. ab. Die qualitativen Änderungen der Aktivität sollen uns hier nicht so sehr interessieren (vgl. den Band Entwicklungsphysiologie dieses Handbuchs). Im Vordergrund stehen hier also die quantitativen Veränderungen.

Eine unterschiedlich hohe Aktivität ist beim Verfolgen des Streckungswachstums nicht nur unter dem Einfluß stark schwankender Außenfaktoren feststellbar, sondern auch unter Bedingungen weit geringerer Schwankungen dieser Art, nicht selten sogar bei annähernder Konstanz der Außenfaktoren. Nur wenige Arten von Bäumen und Sträuchern zeigen etwa in den immerfeuchten Tropen ein gleichmäßig bleibendes Wachstum (vgl. VOLKENS 1912, KLEBS 1912, COSTER 1927, DOORENBOS 1953). Unter den nicht in den Tropen heimischen Formen scheint der Prozentsatz solcher Arten ohne ausgeprägten Aktivitätsrhythmus unter experimentell hergestellten, annähernd konstanten Bedingungen noch geringer zu sein (COSTER 1927, DOORENBOS 1953a). Wie weit für die in den immerfeuchten Tropen mit praktisch konstanter, jedenfalls keinem klaren Jahresrhythmus unterworfener Luftfeuchtigkeit und Temperatur andere Außenfaktoren dafür entscheidend sind, daß die meisten dort heimischen oder importierten Arten noch einen Aktivitätsrhythmus zeigen, soll später erörtert werden.

Der Aktivitätswechsel tritt uns hinsichtlich der Laubentfaltung und der Organstreckung meist als Jahresrhythmus entgegen. In den Tropen kann die Länge eines Cyclus aber auch stark von der Jahresdauer abweichen (SIMON 1906, VOLKENS 1912); es können in einem Jahr mehrere Aktivitätscyclen durchlaufen werden, oder auch für einen Cyclus kann mehr als ein Jahr, gelegentlich sogar mehrere Jahre notwendig werden (BÜNNING 1949, STUDHALTER 1955). In den gemäßigten Zonen zeigen manche Pflanzen ebenfalls mehrere Aktivitätsperioden in einer Vegetationszeit (KLEBS 1914). Bekannt ist namentlich das Auftreten einer zweiten Wachstumsperiode während des Sommers (Johannistrieb). Es kann etwa auf das Verhalten von *Fagus* und *Quercus* hingewiesen werden (SPAETH 1912, MAGNUS 1913, DOSTÁL 1927).

Der Wechsel hoher und niedriger Aktivität pflegt bei den Endknospen von Bäumen und Sträuchern, auf die sich die vorhergehenden Sätze beziehen, sehr radikal zu sein. Es darf aber doch nicht angenommen werden, daß während der sog. Ruheperiode überhaupt kein Wachstum mehr stattfindet. Sofern die Außenbedingungen, also namentlich die Temperatur, ein gewisses Minimum nicht unterschreiten, pflegt selbst während der tiefsten Ruhe noch ein gewisses Wachstum zu bestehen. Außerdem sei hier gleich darauf verwiesen, daß während der sog. Ruheperiode oftmals gewisse Differenzierungsvorgänge in den Knospen ablaufen, so daß der Wechsel hoher und niedriger Aktivität teilweise auch mit einem Wechsel von Differenzierungs- und Streckungsphasen gekoppelt ist.

Erwähnt sei hier auch noch, daß im Gegensatz zu den Sproßknospen die Wurzelvegetationspunkte bei einigermaßen günstig bleibenden Außenbedingungen meist keine Ruheperiode zeigen, sondern während des ganzen Jahres gleichmäßig weiterwachsen können (HARRIS 1926).

Besonders auffällig ist auch der Aktivitätswechsel bei Samen. Und hier pflegt die Ruhe ja noch sehr viel tiefer zu sein als bei Knospen. In den ruhenden Samen laufen meist überhaupt keinerlei Entwicklungsvorgänge mehr ab, und der Stoffwechsel kann im Gegensatz zu dem der ruhenden Knospen völlig unterbrochen sein. Nur bei wenigen Arten wachsen die Embryonen bekanntlich ohne Unterbrechung weiter. Man kann hier etwa auf die Embryonen von *Rhizophora* verweisen, die ohne Ruhepause an der Mutterpflanze zu einer Länge von mehreren Dezimetern heranwachsen und nach ihrer Ablösung auch gleich weiterwachsen. Von solchen Ausnahmefällen abgesehen ist der Embryo also zunächst einer Ruheperiode unterworfen. Diese Ruheperiode ist aber nicht etwa nur eine aufgezwungene. Wir werden sehen, daß hier durchaus ähnliche Verhältnisse vorliegen, wie wir sie für die Knospen der Laubbäume fanden. Nur in wenigen Fällen zeigen nämlich die Samen beim Keimversuch unmittelbar eine volle Keimbereitschaft. Meist ist die Keimbereitschaft starken Schwankungen unterworfen.

Ruheperioden können sich ferner in Knollen, Zwiebeln usw. zeigen. Diese Ruheperioden haben ihren Sitz dabei wieder in den betreffenden Vegetationspunkten und sind ganz ähnlicher Natur wie die Ruheperiode der Endknospen von Bäumen usw.

Auch Pollenkörner können vielfach eine Ruheperiode aufweisen, die aber in den meisten Fällen nur wenige Tage beträgt. Meist ist die Ruhe dabei auch nicht so tief wie etwa die der Samen; es läßt sich durchweg noch ein gewisser Stoffwechsel nachweisen. Allerdings kann dieser Stoffwechsel bei einer Trockenlagerung der Pollenkörner sehr niedrig sein oder sogar ganz fehlen, ohne daß die Lebensfähigkeit der Pollenkörner dabei verlorengeht (OKONUKI, BREDEMANN, VISSER).

Eine Ruheperiode finden wir weiterhin bekanntlich bei Achselknospen. Diese Art der Ruhe unterscheidet sich von der der vorhergenannten Fälle dadurch, daß für sie korrelative Einflüsse ausschlaggebend sind, die von den Endknospen bzw. von Blättern ausgeübt werden.

Ebenso wichtig wie Ruheperioden mit einer Dauer von mehreren Monaten sind auch Aktivitätsschwankungen, die sich innerhalb weniger Stunden oder Tage vollziehen. Hier sei über Schwankungen dieser Art nur vermerkt, daß auch bei ihnen innere und äußere Faktoren in komplizierter Weise zusammenwirken.

Allgemein können wir sagen, daß es bei Pflanzen Phänomene des quantitativen und qualitativen Wechselns der Aktivität gibt, bei denen sowohl die Art der regulierenden Faktoren als auch der Zeitabstand zwischen einzelnen Phasen von Fall zu Fall überaus verschieden ist. Wir finden die eigentlich selbstverständliche Beeinflussung der Aktivität durch die verschiedenartigsten Außenfaktoren.

Aber auch dabei stehen wir immer wieder vor Überraschungen. Wir werden etwa sehen, daß auch früher vernachlässigte Faktoren wie die Änderung der Tageslänge einen Einfluß auf den Wechsel von Ruhe und Aktivität haben. Bei den Aktivitätsrhythmen, die ganz oder teilweise noch unter konstanten Außenbedingungen ablaufen, finden wir insofern eine große Mannigfaltigkeit, als es Rhythmen gibt, bei denen gleichartige Phasen einander im Laufe weniger Stunden oder noch schneller ablösen, andere, bei denen diese Periodenlänge mehrere Tage, Wochen oder Monate beträgt. Schließlich werden wir aber auch Fälle kennenlernen, in denen die Periodenlänge mehrere Jahre erreicht.

B. Arten des Aktivitätswechsels.

Zwar gibt es viele Arbeiten, die sich mit der Ruheperiode beschäftigen; aber eine klare Beurteilung lassen die Versuchsergebnisse bisher nicht zu, weil die Bedingungen meist nicht eindeutig genug waren. Komplizierend wirkt schon der angedeutete Umstand, daß sowohl durch äußere Faktoren als auch aus inneren Ursachen Ruheperioden möglich sind. Dieser Unterschied ist schon sehr lange bekannt. JOHANNSEN unterschied von der aus inneren Gründen eintretenden Ruhe die „gezwungene Unwirksamkeit“; ähnlich sprach MOLISCH (1908, 1909) von einer „unfreiwilligen Ruhe“. DOORENBOS (1953) unterscheidet demgemäß von der “dormancy” (dem in der englisch-amerikanischen Literatur üblichen Ausdruck) die “imposed dormancy”. Auch das Begriffspaar “dormancy” und “rest” ist verwendet worden, um diesen Unterschied zum Ausdruck zu bringen. CHANDLER (1925), MEYER und ANDERSON (1952) sowie SAMISH (1954) bezeichnen die aufgezwungene Ruhe als “quiescence”, die freiwillige als “rest”. In beiden Fällen sprechen sie von “dormancy”. Jedoch werden die Ausdrücke “dormancy” und “rest” in der englisch-amerikanischen Literatur durchaus nicht mit einheitlicher Bedeutung angewandt.

DOORENBOS unterscheidet bei der freiwilligen Ruhe weiterhin “summer-dormancy” und “winter-dormancy”. Er verweist damit zweifellos auf einen wesentlichen Unterschied; aber die Bezeichnung selber ist nicht ganz treffend. Unter “summer-dormancy” werden nämlich die Fälle verstanden, in denen die Ruhe durch Faktoren bedingt ist, die innerhalb der betreffenden Pflanze selber liegen, aber doch nicht in der betreffenden Knospe, sondern eben in deren physiologischer Umgebung. Hierher rechnet DOORENBOS z. B. die Ruhe der Achselknospen, die ja bekanntlich korrelativ von den zugehörigen Blättern bzw. von Gipfelknospen abhängt. Unter “winter-dormancy” versteht DOORENBOS die von den in der Knospe selber liegenden Faktoren bedingte Ruhe. Mit allen diesen Bezeichnungsvorschlägen sind schon sehr viele Probleme angedeutet; weiter komplizierend wirkt dabei aber, daß nicht nur die Ruhe und Aktivität selber ihrer Natur nach entweder freiwillig oder unfreiwillig andauern können, sondern auch ihr Beginn ebenso wie ihre Beendigung entweder autonom (= autogen, = endogen) oder aitionom (= aitiogen) eintreten können, da hierfür entweder innere oder äußere Faktoren bzw. (meist) beiderlei Arten von Faktoren gleichzeitig ausschlaggebend sein können. So kann man z. B. die freiwillige Ruhe (“winter-dormancy”, “rest”) als Ausdruck einer von äußeren Faktoren regulierten endogenen Rhythmik ansehen. Äußere Faktoren können die endogene Ruheperiode eintreten lassen, unterbrechen, verlängern, verkürzen usw. Man sieht also, vor welchen Schwierigkeiten man bei der Analyse steht. Und wir verstehen, warum wir von einem physiologischen Verständnis des Wechsels von Ruhe und Aktivität noch weit entfernt sind. Wir können daher auch noch nicht daran denken, die einzelnen Arten der Ruhe und Aktivität für sich zu analysieren. Vor ähnlichen

Schwierigkeiten steht man bei der Beurteilung eines schneller verlaufenden Aktivitätswechsels, etwa angesichts der tagesperiodischen Leistungsschwankungen. Auch hierbei können in einzelnen Fällen äußere Faktoren dominieren oder sogar allein entscheiden. So haben z. B. natürlich Licht, Temperatur und Wasserversorgung einen entscheidenden Einfluß etwa auf Tagesschwankungen der Wachstumsleistungen. Daneben aber besteht in sehr vielen Pflanzen auch bei der Konstanz der Außenfaktoren noch eine Tagesrhythmik der physiologischen Leistungen. Dabei handelt es sich obendrein sehr oft auch um qualitative Veränderungen. Und wieder können hierbei äußere Faktoren stark regulierend eingreifen; es kann sich aber auch hier um eine aitiogen regulierte autogene Rhythmik handeln. Gerade wenn man davon überzeugt ist, daß die Erfindung von Begriffen und Bezeichnungen noch kein Problem löst, sollte man doch diese terminologischen Schwierigkeiten wenigstens kennen; ihre Unkenntnis führt immer wieder zu Mißverständnissen.

Da sich der erste Teil unserer Darstellung auf die oft mit den Jahreszeiten zusammenhängende freiwillige Ruhe ("rest", "winter-dormancy") bezieht, sei über deren Verlauf noch einiges vorweg gesagt. Die Winterruhe findet sich vorwiegend bei Pflanzen der gemäßigten und polaren Regionen. Bei ihnen ist sie aber unterschiedlich tief (vgl. DOORENBOS 1953a). Die Ruhe ist auch in ihren einzelnen Zeitabschnitten nicht gleich tief. Tatsächlich kann man die Ruhe schon im Frühjahr beginnen sehen, wenn man ihren Eintritt danach beurteilt, daß die Entblätterung keine Knospenöffnung mehr bedingt. Dieser Zeitpunkt kann bei einigen Arten im Mai oder Juni, bei anderen aber erst im Herbst gegeben sein (JOST 1891). Die Ruhe wird dann allmählich tiefer, was sich vor allem daraus ergibt, daß es immer schwerer wird, sie durch eines der bekannten Treibverfahren zu unterbrechen. Man unterscheidet etwa eine Vorruhe, Mittelruhe (Hauptruhe) und eine Nachruhe (JOHANNSEN 1906, WEBER 1916). Als diese Unterteilung eingeführt wurde, glaubte man noch, daß die Mittelruhe eine vollständige Ruhe sei, in der jegliches Treibverfahren versagt und der Stoffwechsel ganz ausgeschaltet ist. Jetzt wissen wir, daß hier nur quantitative Unterschiede bestehen. Die tiefste Ruhe wird oft schon zur Zeit des Blattfalls erreicht (vgl. z. B. DOORENBOS 1953b).

Die Winterruhe ist in den ersten Monaten, etwa bis Dezember, eine freiwillige (endogene) Ruhe, womit nicht gesagt sein darf, daß auch ihr Eintritt von äußeren Faktoren unabhängig sei. Später ist diese Ruhe nur noch eine unfreiwillige, d. h. beim Eintreten günstiger Außenbedingungen zeigt sich sofort der Beginn der Wachstumsvorgänge.

Die Ruhe hat ihren Sitz vor allem in den Knospen, aber, wie wir jetzt wissen, auch im Cambium. Man glaubte oft, die Ruhe im Cambium der Stämme sei nur korrelativ von der der Knospen abhängig. Tatsächlich ist ja in vielen Experimenten (vgl. z. B. KÜNNING, KÜNNING und SÖDING) nachgewiesen worden, daß von den sich entfaltenden Knospen korrelative Einflüsse ausgehen, die aktivierend auf die Tätigkeit des Cambiums wirken. Werden die Blätter entfernt, so unterbleibt die Aktivierung des Cambiums. Aber es zeigt sich doch, daß auch im Cambium selber, unabhängig von diesen Korrelationen, ein Aktivitätsrhythmus abläuft (GOUWENTAK 1941); denn die Cambiumaktivität läßt sich durch hormonale Einflüsse erst dann steigern, wenn eine gewisse „innere Bereitschaft" dazu besteht.

Daß die Ruhe in den einzelnen Knospen eines Individuums unabhängig voneinander ist, geht aus der Möglichkeit hervor, eine Knospe durch äußere Faktoren zum Treiben zu veranlassen, während die übrigen ruhen bleiben. Überhaupt darf gesagt werden, daß diese Art der Ruhe von Endknospen überwiegend durch Vorgänge in den Knospen selber gesteuert wird. Deutlich wird das daraus, daß in Pfropfsymbiosen die einzelnen Partner ihre Ruheperiode zu der für sie typischen Zeit unabhängig voneinander durchführen. Und wenn äußere rhythmische Faktoren fortfallen, wie etwa bei den in den Tropen wachsenden Bäumen, so zeigt sich diese Individualität der einzelnen Knospen noch extremer.

Der Winterruhe, die also weitgehend eine Jahresrhythmik der Aktivität darstellt, ist der Aktivitätsrhythmus bei Pflanzen tropischer Regionen in mancher Hinsicht ähnlich. Obwohl es dort keine Ruhe als Anpassung an kalte Perioden gibt, ist doch eine Inaktivität als Anpassung an trockene Monate oft feststellbar. Das braucht sich nicht nur in einem jahreszeitlich gebundenen Laubwechsel solcher Formen zu äußern, sondern kann bei Formen mit mehrjährigen Blättern auch daran erkennbar werden, daß die Stoffwechselleistungen geringer werden. Und auch mehrjährige Pflanzen der gleichmäßig feuchten Tropen zeigen, wie erwähnt, meist einen Aktivitätsrhythmus, also ein periodisches Wechseln von Laubentfaltung, Ruheperioden usw. Bei diesen Aktivitätsrhythmen von Pflanzen gleichmäßig feuchter Tropen ist die Ruheperiode in den meisten Fällen kürzer als die Winterruhe unserer Bäume; sie beträgt (gemessen an der Zeit des Kahlstehens einzelner Äste) oft nur wenige Wochen, in anderen Fällen aber ganz entsprechend der Dauer des „freiwilligen" Abschnitts der Winterruhe unserer Bäume und Sträucher mehrere Monate. Diese oft mit dem Kahlstehen verbundenen Perioden verringerter Aktivität sind wohl physiologisch mit den Aktivitätsrhythmen der winterkalten Regionen verwandt. Das grundlegende Phänomen ist offenbar die Tatsache, daß alle Pflanzen solche Aktivitätsrhythmen mit Cyclenlängen von mehreren Monaten Dauer zeigen und davon ausgehend Anpassungen der Rhythmik an bestimmte Regionen, eben beispielsweise an Regionen mit regelmäßig wiederkehrenden trockenen bzw. kalten Monaten erfolgt sind. Nur bei wenigen Arten höherer Pflanzen kann es praktisch allein von den äußeren Faktoren abhängen, ob die Ruhe eintritt und wie lange sie dauert. So scheint nach den Angaben von VOLKENS die Ruhe von *Albizzia moluccana*, *Artocarpus incisa* und *Morinda citrifolia* beim Wachsen der Pflanzen in gleichmäßig feuchten Tropengebieten unterbleiben zu können. Ähnlich steht es nach den Angaben von CHOUARD bei *Calluna vulgaris*. Und auch einige andere Pflanzen der gemäßigten Zonen zeigen beim Wachsen in den gleichmäßig feuchten Tropen keine Aktivitätsrhythmen mehr (vgl. COSTER 1927). Aber das sind, wie gesagt, seltene Ausnahmen. Häufiger sind solche Ausnahmen bei Farnen; es gibt viele Arten, die in gleichmäßig feuchten Tropen den Eindruck ununterbrochener Aktivität erwecken, aber schon in periodisch trockenen Tropengebieten regelmäßig Ruhepausen, verbunden mit dem Abwerfen von Wedeln, einschieben.

C. Physiologische Eigentümlichkeiten ruhender und aktiver Zellen.

a) Wassergehalt.

Da die Geschwindigkeit biochemischer Reaktionen entscheidend vom Wassergehalt abhängt, hat man sehr früh versucht, das Wesen der Ruhe in einem verminderten Wassergehalt zu sehen. Diese Deutung wird ja auch nahegelegt, wenn man an Zellen mit extrem tiefer Ruhe denkt, also etwa an die der Samen. Tatsächlich ist auch oft nachgewiesen worden, daß die Aktivität durch experimentell hervorgerufene Erniedrigung des Wassergehalts sinkt. Ein Entzug von Wasser durch trockene Luft oder auf anderem Wege führt meistens zu einer Erniedrigung der Aktivität, meßbar etwa an verminderter Wachstumsleistung, verminderter Assimilation oder gehemmtem Wachstum. Bei einigen Pflanzen kann das so weit gehen, daß wir willkürlich zu beliebiger Zeit die Aktivität durch Wasserentzug ganz aufheben oder durch Wasserzufuhr wieder zum Maximum steigern können. Als Beispiel können etwa kleinere Epiphyten, wie namentlich Lebermoose, aber auch viele andere Moose und Farne genannt werden. Im ganzen läßt sich sagen, daß eine willkürliche Inaktivierung und Reaktivierung bei niederen Pflanzen leichter ist als bei höheren. Die höheren Pflanzen haben sich viel stärker an bestimmte Klimarhythmen angepaßt, und ihre Ruhe läßt sich oft nur im Rahmen der natürlich vorkommenden Jahresschwankungen dieses Klimas modifizieren. Jedoch sind auch bei höheren Pflanzen Fälle ziemlich weitgehender experimenteller Änderungen des normalen Aktivitätswechsels beschrieben worden (vgl. DOORENBOS). Dabei handelt es sich zunächst einmal um eine unfreiwillige Ruhe; aber es wäre doch grundsätzlich denkbar, daß auch die freiwillige Ruhe mit einem reduzierten Wassergehalt des Protoplasmas zusammenhängt. Wir müssen diesen Faktor diskutieren, obwohl von vornherein klar ist, daß er nicht allein entscheidet; denn nicht selten sind ja ruhende Organe, etwa manche Knollen, sogar sehr wasserreich.

Ein Mittel der Pflanzenzellen, den Wassergehalt des Protoplasmas zu erniedrigen, besteht in einer Erhöhung der osmotischen Konzentration des Zellsaftes, wodurch ja dem Plasma zwangsläufig Wasser entzogen wird. Tatsächlich findet man sehr häufig in ruhenden Zellen eine erhöhte osmotische Konzentration des Zellsaftes (vgl. die ausführliche Literaturzusammenstellung bei STEINER 1933). Die Erhöhung der osmotischen Werte kann besonders durch Zunahme des Zuckergehalts bedingt sein (PISEK 1950, KNEEN und BLISH 1941, LEVITT 1951). Neben Schwankungen des Zuckergehalts können aber auch solche des Säuregehalts stark mitbeteiligt sein (Abb. 1). Die Schwankungen des Zuckergehalts ihrerseits sind oft durch den Abbau und Wiederaufbau von Stärke verursacht (PISEK 1950). Auf die große Zahl der hierzu durchgeführten Untersuchungen kann nicht eingegangen werden; einige Hinweise müssen genügen. Im Winter pflegt in den Blättern immergrüner Bäume und Sträucher eine Zunahme des osmotischen Wertes und der Zuckerkonzentration feststellbar zu sein, während das Minimum im Sommer erreicht wird (STEINER 1933, 1939, PITTIUS, ULMER). Es können dabei verschiedene Zucker, bei einigen Arten z.B. auch Mannit (ASAI) beteiligt sein. Neuere Untersuchungen liegen vor allem von PISEK vor; er fand bei einigen Arten jahresperiodische Schwankungen der osmotischen Werte, des Stärke- und Zuckergehalts. Obwohl diese Änderungen der osmotischen Werte ein Faktor der Ruhe sein können, vermögen sie diese doch nicht befriedigend zu erklären. Zu beachten ist schon, daß es sehr viele Ausnahmen von der genannten Regel gibt (LEVITT 1941, SCARTH 1944, SIMINOVITCH und BRIGGS 1949). Dabei ist noch besonders zu berücksichtigen, daß es nicht nur einzelne Arten von Pflanzen gibt, bei denen ein Parallelismus des Aktivitätswechsels mit dem Gehalt an osmotisch wirksamen Substanzen fehlt, sondern es kann auch bei ein und derselben Art dieser Parallelismus in einzelnen Jahren vorhanden sein, in anderen fehlen (Abb. 1, vgl. PISEK 1950). Und auch wenn bei einer Art regelmäßig etwa im Winter die höheren osmotischen Werte feststellbar sind, folgt der zeitliche Gang in den Veränderungen der osmotischen Werte doch nicht dem zeitlichen Gang in der Tiefe der Ruhe. Schließlich ist zu beachten, daß die Beobachtungen über einen solchen Parallelismus vielfach gar nicht im Zusammenhang mit dem Grundproblem des Aktivitätswechsels, sondern zur Klärung der Ursachen der Resistenz, namentlich der Resistenz gegen niedere Temperaturen, durchgeführt worden sind. Das ist in einer Hinsicht nicht bedenklich, weil nämlich die Resistenz selber zu einem sehr großen Teil eine Begleiterscheinung der Ruhe ist. Aber dieser Parallelismus zwischen Ruhe und Resistenz ist doch nicht vollständig. Manche Beobachtungen sprechen dafür, daß die gefundenen Schwankungen im Gehalt osmotisch wirksamer Substanzen noch etwas mehr mit dem Aktivitätswechsel als mit der Resistenz zu tun haben. Zum Beispiel läßt sich oftmals schon im Sommer eine deutliche Zunahme der osmotischen Werte feststellen, d. h. zu einer Zeit, in der, wie weiter oben erwähnt wurde, die Ruhe schon beginnen kann, in der aber die Resistenz noch nicht erhöht ist oder sogar erst ihr Minimum erreicht (SIMINOVITCH und BRIGGS 1949).

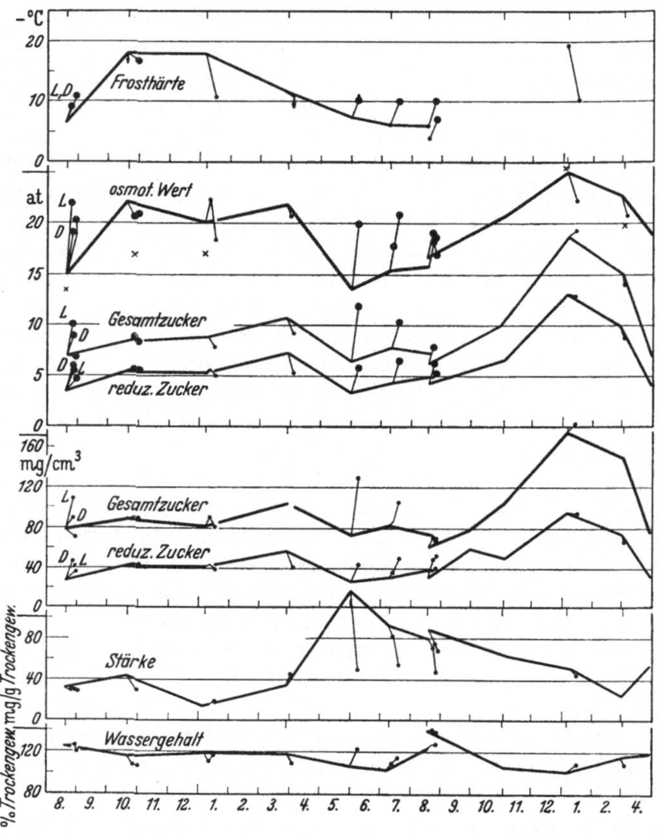

Abb. 1. *Rhododendron ferrugineum*, letztjährige Blätter. Frosthärte, osmotischer Wert, Zucker-, Stärke- und Wassergehalt im Jahrgang 1945—1947 und deren Änderungen in Abhärtungs- (beringte Punkte) und Enthärtungsversuchen (kleine, angehängte Punkte). × Summe der Partialdrucke der Zucker + analysierter Elektrolyte. Mit 16. August 1946 beginnen die Kurven des neuen Jahrganges der Blätter. Abhärtung bei 0 bis —4°. Enthärtung (nur im Winter) bei 16—18°. (Nach PISEK.)

Ebenso sehr wie osmotische Bindungen des Wassers sind auch dessen Bindungen an Kolloide des Protoplasmas beim Studium des Aktivitätswechsels beobachtet worden. Dabei erheben sich ähnliche Schwierigkeiten wie wir sie vorher besprochen haben, zumal es sich auch dabei zu einem sehr großen Teil um Untersuchungen über die Beziehung zwischen der Resistenz und dem Grad der Wasserbindung handelt. Schon aus diesen Gründen kann von einer ausführlichen Erörterung hier abgesehen werden (vgl. CHANDLER 1941, LEVITT 1941, GRANDFIELD 1943, SCARTH 1944, LEVITT 1951).

b) Plasmazustand.

Nicht nur, weil Schwankungen der osmotischen Werte die Aktivitätsänderungen höchstens teilweise erklären können, sondern auch weil die Wasserbindung an Kolloide des Plasmas beteiligt zu sein scheint, werden wir zwangsläufig dazu geführt, kolloidalen Veränderungen eine stärkere Rolle zuzuschreiben. Wenn man solchen kolloidalen Veränderungen vor allem darum eine Rolle zuschreibt, weil der Wasserbindungskraft eine entscheidende Bedeutung zugemessen wird, würde man vor allem an eine Wichtigkeit elektrischer Ladungsänderungen denken. Aber man sollte die Frage nicht von vornherein so sehr einengen, sondern allgemeiner prüfen, welche plasmatischen Veränderungen sich im Zusammenhang mit dem Aktivitätswechsel feststellen lassen; denn es ist doch recht zweifelhaft, ob überhaupt der Grad der Wasserbindung so sehr entscheidend ist und er nicht vielmehr weitgehend nur eine Begleiterscheinung darstellt.

Tatsächlich sind eine ganze Reihe von plasmatischen Veränderungen im Zusammenhang mit dem Aktivitätswechsel gefunden worden. Einen inneren Zusammenhang zwischen den Einzelbeobachtungen können wir bisher nicht feststellen, so daß wir uns mit nebeneinander gereihten Hinweisen begnügen müssen. Und noch einmal wieder muß dabei die Warnung ausgesprochen werden, daß die Beobachtungen zwar auf Erscheinungen verweisen, die mit der Ruhe parallel laufen, es aber bei der Art der durchgeführten Versuche doch völlig zweifelhaft bleibt, ob sie immer zu ihr gehören und einen wesentlichen Bestandteil von ihr darstellen.

Da, wie schon erwähnt, erhöhte Kälteresistenz oft mit der Ruhe parallel läuft, können wir hier zunächst auf die Untersuchungen verweisen, die sich mit den protoplasmatischen Grundlagen der Kälteresistenz beschäftigen. In kälteresistenten Zellen wurde unter anderem eine erhöhte Permeabilität gefunden (LEVITT 1941, SIMINOVITCH und BRIGGS 1949). Viel diskutiert worden sind kolloidale Umwandlungen beim Übergang zur Ruheperiode, die in der Überführung des Plasmas in einen Zustand stärkerer Hydratation bestehen. Das Plasma wird dabei auf Kosten des freien Wassers gelartig. Ein Hinweis auf solche Veränderungen kann darin gesehen werden, daß sich die Plastiden innerhalb des Protoplasten beim Zentrifugieren nicht kälteresistenter Pflanzen in den Zellen leichter umlagern lassen als bei kälteresistenten. Nicht nur hierin, sondern auch in unterschiedlicher Plasmolyseform äußert sich eine Viscositätserhöhung des Protoplasmas der kälteresistenten Pflanzen. Dabei kann es sich also um eine der Ursachen jenes angedeuteten Zustandes verstärkter Wasserbindung handeln (KESSLER und RUHLAND).

Nebenher sei hier erwähnt, daß nicht nur beim Studium länger dauernder Ruheperioden, sondern z. B. auch bei Bakteriensporen ähnliche Beziehungen gefunden worden sind. Diese Sporen brauchen sich von den vegetativen Zellen durchaus nicht durch eine geringere Wassermenge zu unterscheiden, sondern zeichnen sich oft nur durch eine erhöhte Wasserbindungskraft aus. Solche Veränderungen, d. h. ein solcher Übergang in einen mehr gelartigen Zustand, der mit dem Entzug freien Wassers verbunden ist, kann selbstverständlich nicht nur die bei den Untersuchungen in den Vordergrund gestellte Resistenzerhöhung erklären, sondern auch die größere biochemische Trägheit, also die Ruhe.

Auf Unterschiede in der Permeabilität ruhender und aktiver Zellen deuten etwa die Untersuchungen von PIRSON und GÖLLNER. Sie fanden bei *Lemna minor* im Sommer lange Plasmolysezeiten, also offenbar geringe Wasserpermeabilität, außerdem auch eine geringe Harnstoffpermeabilität. Im Winter waren die Permeabilitätswerte höher (auch übrigens wie üblich die osmotischen Werte).

Jedoch ist das beim vergleichenden Studium der Permeabilitätsverhältnisse ruhender und aktiver Gewebe sich ergebende Bild recht unterschiedlich. v. GUTTENBERG und MEINL z. B. fanden bei Kartoffelknollen gerade das Gegenteil, nämlich eine allmähliche Permeabilitätszunahme mit abklingender Ruhe. Danach muß man also sehr skeptisch sein gegenüber der Ansicht, die Höhe der Permeabilität sei ein Maß für Ruhe oder Aktionsbereitschaft.

Es liegen auch etliche mikroskopische Studien über den Unterschied des Protoplasten in Ruhe- und Aktivitätsperioden vor. Nach Untersuchungen russischer Autoren (vgl. HENCKEL 1953) kontrahiert sich der Protoplast beim Beginn der Ruheperiode. Das soll soweit gehen, daß die Plasmodesmen zerreißen. Die gleichen Autoren (HENCKEL, HENCKEL und OKNINA, OKNINA) gaben weiterhin an, daß sich der Protoplast während der Ruhe mit einer sichtbaren Lipoidhülle umgibt, welche einerseits das Austrocknen verhindert, andererseits auch die Permeabilität für Wasser und Sauerstoff vermindert. Diese Veränderungen werden darauf zurückgeführt, daß sich im Herbst hydrophobe Kolloide, besonders Fette und Lipoide anhäufen, die sich auf der Oberfläche des Protoplasten und der Vacuolen sammeln. Bei der Beendigung der Ruhe werden diese Veränderungen wieder rückgängig gemacht, die Lipoidschichten zerstört, die Quellbarkeit des Protoplasmas wieder erhöht und die Plasmodesmenverbindungen neu hergestellt. Über ähnliche Beobachtungen berichtet SATAROVA (1948). Ob etwas Derartiges wirklich möglich ist, kann noch nicht entschieden werden. Namentlich bleibt es wohl noch problematisch, ob ein Zerreißen der Plasmodesmen wirklich als normaler physiologischer Prozeß vorkommt.

Aus diesen Beobachtungen wurde ein umfangreiches Forschungsprogramm entwickelt. Zum Beispiel gab SATAROVA an, daß Äthylenchlorhydrin die oberflächlichen Lipoidschichten der ruhenden Zellen von Kartoffelknollen auflöst, wodurch dann die Permeabilität und das Quellungsvermögen des Protoplasten wieder steigen und damit die Ruhe beendet wird. Samen ohne Ruheperiode zeigten auch nicht die oberflächlichen Lipoidanhäufungen (SITNIKOVA). Die Versuche wurden auf mehrere Objekte mit verschiedenartigen Speichergeweben und Ruheorganen ausgedehnt (SATAROVA, OZOL und LAZARENA). Jedoch ist die Brauchbarkeit der Methoden bestritten worden (LEVITT 1951, PIENIAZEK und WISNIEWSKAJA).

Einen gewissen Anhaltspunkt können auch die Untersuchungen von SIMINOVITCH und BRIGGS sowie von BRIGGS und SIMINOVITCH (1949) geben. In Rindenzellen von *Robinia* ermittelten sie eine Korrelation zwischen der Menge wasserlöslicher Proteine und der Ruhe (bzw. Kälteresistenz). Die Menge wasserunlöslicher Eiweiße bleibt während des ganzen Jahres gleich. In der Phase erhöhter Resistenz ist der Gehalt an wasserlöslichen Eiweißen größer.

TRUCHAČEVA und KEDROVSKIJ untersuchten die jahreszeitlichen Veränderungen der Zellen in der Cambiumzone von *Sambucus racemosa*. Sie fanden neben spindelförmigen Zellen, in denen das Plasma keinen jahreszeitlichen Veränderungen unterworfen ist, vor allem zylindrische Zellen, in denen sehr starke Veränderungen ablaufen. Im Winter besitzen sie ein dichtes Cytoplasma. Im Frühjahr ist eine Vacuolisierung zu beobachten, die Zahl der Nucleoli geht zurück. Schon im Spätsommer gewinnen die Zellen allmählich wieder das Aussehen der winterlichen. Der Stärkegehalt steigt von Mai bis September; im Herbst sinkt er dann wieder ab. Im Winter ist fast gar keine Stärke zu finden.

Bei den bisher erwähnten Beobachtungen lagen insofern immer Schwierigkeiten der Deutung vor, als die festgestellten plasmatischen Veränderungen sowohl Begleiterscheinungen der Ruhe als auch spezifische Äußerungen der Kälteresistenz sein können, beide Phänomene aber ja nicht notwendig parallel laufen. Etwas gemildert wird dieses Bedenken dadurch, daß der Parallelismus von Resistenz und Ruhe wohl wirklich auf gemeinsamen plasmatischen Grundlagen beruht. Die Resistenz ist ebenso wie die Ruhe Ausdruck einer Reaktionsträgheit, die in gleichen plasmatischen Veränderungen begründet liegt. Jedenfalls spricht für diese Auffassung, nach der es also nicht ganz spezifische Veränderungen gibt, die nur mit der Kälteresistenz zu tun haben, auch die

Beobachtung von PISEK und LARCHER, daß auch die jahreszeitlichen Rhythmen der Austrocknungsresistenz mit der Ruhe parallel laufen, d. h. also, auch Austrocknungs- und Kälteresistenz weisen einen gleichartigen jahreszeitlichen Gang

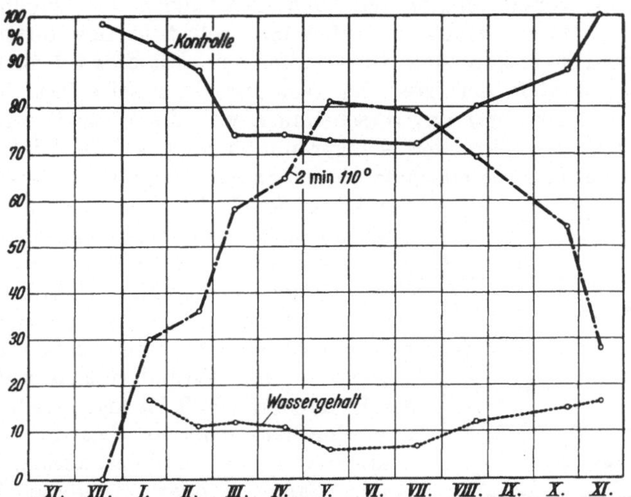

Abb. 2. *Digitalis lutea.* Ordinate: Keimprozent bzw. Wassergehalt. Abszisse: Monate. Dargestellt ist das nach 14 Tage langem Aufenthalt im Keimbett ermittelte Keimprozent bei Kontrollen sowie bei 2 min auf 110° erhitzten Samen, außerdem der Wassergehalt der Samen. Man sieht, daß die Keimfähigkeit und die Hitzeresistenz Schwankungen unterliegen, die in deutlicher Beziehung zu den Änderungen des Wassergehaltes stehen. (Die Samen lagerten bei konstanter Temperatur und Luftfeuchtigkeit.) (Nach BÜNNING und BAUER.)

auf. Weiterhin ist bei jenen Beobachtungen zu berücksichtigen, daß viele der plasmatischen Veränderungen einfach direkte Folgen der äußeren Ursachen sind, die den Eintritt der Ruhe bedingen, ohne notwendigerweise mit dieser selber

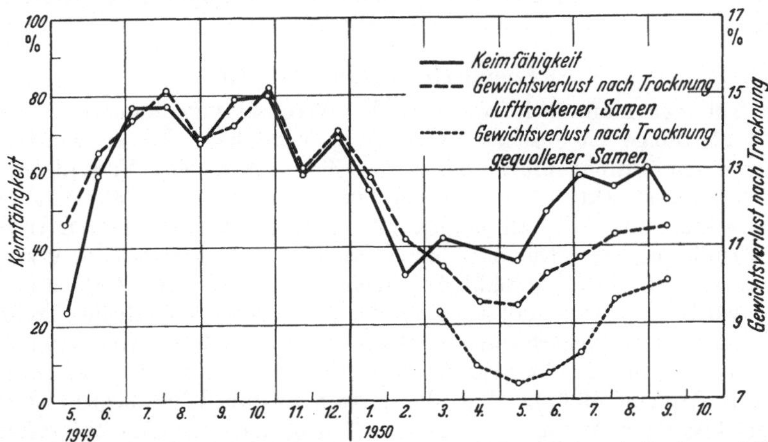

Abb. 3. Zusammenhang zwischen Keimfähigkeit (————), Gewichtsverlust nach Trocknung lufttrockener (........) und Gewichtsverlust nach Trocknung gequollener (— — — —) Teilfrüchte der Futtermalve. (Nach RUGE und LIEDTKE.)

etwas zu tun haben zu müssen. In dieser Hinsicht sind Beobachtungen an ruhenden Samen vielleicht wertvoller, weil hier jene Begleiterscheinungen leichter ausgeschaltet werden können. Mehrfach sind an solchen Samen Schwankungen des Wassergehalts festgestellt worden (CROCKER 1948, RUGE und LIEDTKE 1951). Besonders bemerkenswert ist wohl, daß diese Schwankungen auch dann noch auftreten, wenn die Samen unter konstanten Bedingungen gelagert werden (Abb. 2 und 3) (BÜNNING und BAUER 1952). Aus solchen Beobachtungen folgt, daß die Wasserbindungskraft der Samen merklichen Schwankungen unterworfen ist. Bei Quellungsversuchen konnten auch Änderungen der Quellbarkeit festgestellt werden. Offensichtlich verlaufen also selbst in den lufttrocken aufbewahrten Samen noch kolloidale Umwandlungen. Mit dieser Feststellung darf nicht behauptet werden, der Grad der Wasserbindung sei selbst für die Ruhe entscheidend. Das ist sogar unwahrscheinlich, weil die Ruhe ja auch vom absoluten Wassergehalt mehr oder weniger unabhängig ist, die Tiefe der Ruhe, also die Höhe der Keimbereitschaft eines Samens hängt nicht etwa stark von der Höhe der Luftfeuchtigkeit ab. Aber jene Änderungen der Wasserbindungskraft verweisen doch jedenfalls auf kolloidale Vorgänge, die in einer uns noch nicht näher bekannten Weise für die Ruhe (d. h. bei den Samen für die Keimbereitschaft) entscheidend sein können. Der Parallelismus jener Veränderungen mit der Höhe der Keimbereitschaft ist festgestellt worden. Es wäre durchaus denkbar, daß die hierdurch erkennbaren kolloidalen Veränderungen nicht nur bei den Samen, sondern etwa auch in Knospen einen wesentlichen Teil der Ruhe darstellen. Ähnlich geeignet wie Samen sind zweifellos auch Sporen niederer Pflanzen zum Studium der plasmatischen Grundlagen des Aktivitätswechsels. Daß Sporen eine solche Ruhe haben können, ist lange bekannt (vgl. die ausführlichen Hinweise bei GOTTLIEB). Bei *Sporodinia* sind Permeabilitätserhöhungen gefunden worden, die vielleicht mit der allmählichen Aufhebung der Ruhe zusammenhängen [(Abb. 4) SEILER, vgl. auch WEMMER].

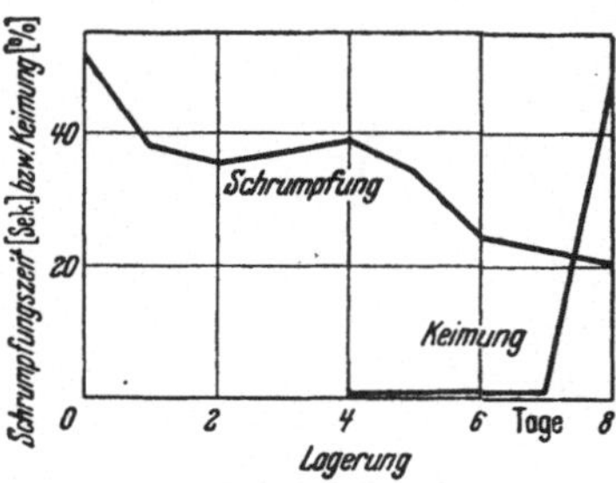

Abb. 4. Nachreifung als Erhöhung der Wasserpermeabilität. Die Schrumpfungsgeschwindigkeit der Zygoten von *Sporodinia* in Glycerin nimmt während des Lagerns zu, d. h. die für diese Schrumpfung notwendige Zeit wird kürzer. Gleichzeitig erhöht sich die Keimbereitschaft, die also offenbar an hohe Wasserpermeabilität gebunden ist. (Nach Versuchen von SEILER.)

c) Chemische Veränderungen.

Sehr früh hat man auch versucht, den Wechsel von hoher und niedriger Aktivität als Folge chemischer Prozesse zu verstehen. Daher wurden viele Untersuchungen durchgeführt mit dem Ziel, chemische Verschiedenheiten in der Ruhe- und Aktivitätsperiode festzustellen. Eine der ältesten Vorstellungen war die, durch die allmähliche Anhäufung von Assimilaten trete eine gewisse Ermüdung ein. Die nach einer längeren Ruhezeit wieder erkennbare Aktivierung wurde durch allmähliche Zuckerbildung erklärt. Zugunsten dieser Deutung wurde angeführt, daß auch bei der Anwendung künstlicher Treibmittel eine Erhöhung des Zuckergehalts auftritt (FISCHER 1891, JOHANNSEN 1896, NIKLEWSKI 1906, SIMON 1906, IRAKLIONOW 1912, HOWARD 1915, GARDNER 1929, DENNY und MILLER 1932,). Eine der letzten ausführlicher begründeten Vorstellungen dieser Art ist die von HOWARD (1915), nach der eine Anhäufung von Produkten der Photosynthese, speziell von Zucker, zu einer allmählichen Hemmung der hydrolysierenden Enzyme führt, wodurch das Wachstum gehemmt und schließlich beendet wird. Durch die während der Ruhe weiterlaufende Atmung werden diese Zucker allmählich verbraucht, so daß die Enzyme wieder aktiv werden können. Es wurde aber schon vor längerer Zeit gefunden, daß der Verlauf der Enzymaktivitäten nicht diesen Forderungen entspricht.

Vor allem muß gegenüber allen diesen Bemühungen, den Aktivitätswechsel aus den elementaren Stoffwechselerscheinungen heraus zu erklären, aber darauf hingewiesen werden, daß der zeitliche Verlauf der Ruhe- und Aktivitätsperioden zum mindesten sehr weitgehend unabhängig ist von der Intensität des Stoffwechsels, also die Ruhe etwa zur gleichen Zeit eintritt, einerlei ob die photosynthetischen Bedingungen im Sommer günstig oder ungünstig waren und die Reaktivierung auch unabhängig davon ist, ob die Temperatur einen stärkeren oder geringeren Verbrauch von Zucker durch Atmung ermöglichte. Allerdings gibt es auch Beobachtungen, die auf eine Beziehung zwischen der Intensität des Stoffwechsels und dem Beginn der Ruhe hinweisen. LAKON fand an *Acer negundo* mit teilweise weißbunten Blättern, daß die Ruhe um so später beginnt, je weniger Chlorophyll die Blätter enthalten. Das legt natürlich den Gedanken einer Erklärung durch das Ausbleiben der Assimilatanhäufung nahe. Zweige mit völlig weißen Blättern zeigten überhaupt keine Ruhe und daher auch keine Erhöhung der Kälteresistenz während des Winters. Auch die Tatsache, daß die Ruheperiode bei Hexenbesen fehlt (SCHELLENBERG), wurde ähnlich erklärt, nämlich durch den Verbrauch der Reservestoffe durch den Parasiten. Jetzt würde man solche Beobachtungen wohl lieber durch Störungen der hormonalen Beziehungen erklären. Daß für den Eintritt der Ruhe andere Faktoren maßgeblich beteiligt sind, wird ja auch klar, wenn wir beobachten, daß die Ruhe nicht etwa zum Zeitpunkt stärkster Anhäufung von Assimilaten ihre größte Tiefe erreicht hat, sondern diese Tiefe dann noch weiterhin zunehmen kann.

Gegen die vor allem von KLEBS vertretene Auffassung, daß der Anhäufung von Reservestoffen eine regulierende Bedeutung zukommt, sprechen auch die neueren Versuche von HENSSEN an *Spirodela polyrrhiza*. Eine Überfüllung der Gewebe mit Reservestoffen ist hier für die Dauersporenbildung nicht ausschlaggebend. Auch eine Beziehung zur Amylaseaktivität konnte nicht gefunden werden.

Hiermit soll nicht behauptet werden, daß solche Schwankungen des Stoffwechsels überhaupt nicht beteiligt seien. Schwankungen des Stoffwechsels und Schwankungen der Fermentaktivität sind schon infolge der erwähnten kolloidalen Veränderungen selbstverständlich. Schwankungen in der Aktivität verschiedener Fermente sind sowohl in ruhenden Knollen und Knospen als auch noch in lufttrocken verwahrten ruhenden Samen mehrfach festgestellt worden.

In Knollen der Kartoffel (*Solanum tuberosum*) wurden Änderungen der Tyrosinaseaktivität gefunden (TODD).

In lufttrocken aufbewahrten Samen sind die Schwankungen der Fermentaktivität meist nur gering (BÜNNING und BAUER). Stärkere Änderungen kann man bei den der Stratifikation unterworfenen Samen beobachten. Unter dieser Stratifikation versteht man bekanntlich die Lagerung der Samen in feuchten Schichten (Sand, Moos usw.) bei niedriger Temperatur. Es ist eine Behandlung, die sich für den schnellen Ablauf der Ruhe bei den Samen vieler Arten als vorteilhaft erwiesen hat. Die Stratifikationsbedingungen entsprechen natürlich weitgehend den in den gemäßigten Zonen in der freien Natur gegebenen. Bei diesen Untersuchungen nun sind starke Erhöhungen der Aktivität mehrerer Fermente [namentlich Katalase (Abb. 5), ferner z. B. Lipase] mit zunehmender Stratifikationsdauer gefunden worden. Aber auch der Gehalt anderer Substanzen unterlag erheblichen Veränderungen. Die Menge titrierbarer Säure, die Zuckermengen usw. können steigen. Die Zuckerzunahme läßt sich deutlich auf eine Stärkehydrolyse zurückführen. Auch verstärkte Atmung, Zunahme der Wasserbindungskapazität, Abbau von Fetten und Eiweißen (damit Anhäufung von Aminosäuren usw.) sind während der Stratifikation gefunden worden. Solche Veränderungen sind im Hinblick auf die genannten Änderungen der Fermentaktivität nicht erstaunlich. Man hat die Vorstellung entworfen, diese Veränderungen, welche ja eine allmähliche Anhäufung eines für den Stoffwechsel leichter zugänglichen Materials bedeuten, seien für die zunehmende Aktivitätssteigerung, d. h. für die Erhöhung der Keimbereitschaft entscheidend. Obendrein wurde die Bedeutung der Bildung osmotisch wirksamen Materials für die erleichterte Samenquellung verantwortlich gemacht (vgl. ECKERSON, JONES, PACK, FLEMION, AFANASIEV und CRESS sowie vor allem die Literaturhinweise bei CROCKER 1948). Bei der Beurteilung all dieser Änderungen muß wieder zur Vorsicht gemahnt werden. Es ist völlig ungewiß, wie weit die festgestellten chemischen Veränderungen wirklich notwendige Bestandteile des Aktivitätswechsels selber sind und wie weit sie nur deren Folgen und Begleiterscheinungen bzw. Vorgänge sind, die auf die besondere Behandlungsart zurückgeführt werden müssen und zwar auch die Keimbereitschaft mit beeinflussen können, aber doch an das eigentliche Wesen des Aktivitätswechsels nicht herankommen. Es sei vor allem darauf hingewiesen, daß der Aktivitätswechsel auch in den ruhenden Samen durchaus nicht nur in einer allmählichen Zunahme der Keimbereitschaft erkennbar wird, sondern vielmehr sehr häufig während der Lagerung auch wieder eine Abnahme festgestellt werden kann. Nicht selten besteht die erste Phase der Samenruhe sogar in einer recht lange anhaltenden Abnahme der Aktivität. Chemische Veränderungen, die in einer Richtung verlaufen, können also nicht das Wesen des Aktivitätswechsels aufhellen.

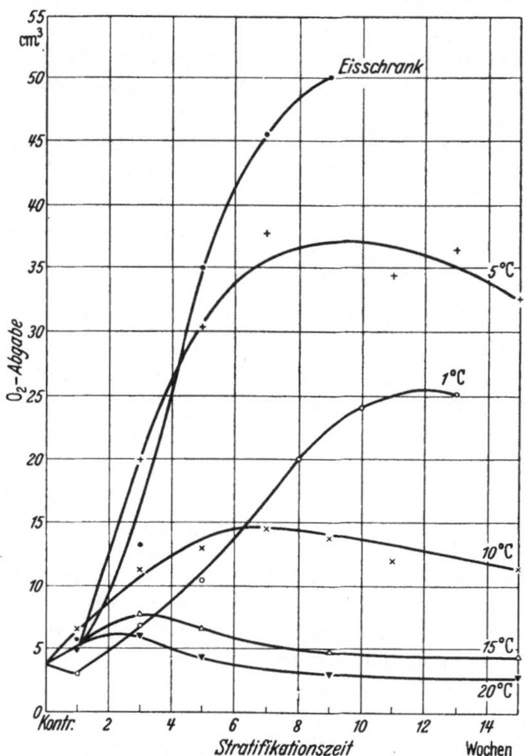

Abb. 5. Katalaseaktivität der Samen von *Sorbus aucuparia* nach verschieden langer (1—15 Wochen) Stratifikation bei verschiedenen Temperaturen. (Aus CROCKER 1948.)

Schließlich sei hier noch auf die Versuche von HENSSEN an Dauersprossen von *Spirodela polyrrhiza* verwiesen. Die Atmung fällt hier während der ersten 14 Tage der Ruheperiode auf den halben Wert und nimmt dann noch weiter ständig ab. Erst während des Austreibens steigt der Sauerstoffverbrauch wieder sprunghaft an. Daraus folgt, daß die während der Ruhe ständig steigende Bereitschaft zum Austreiben nicht von einem Ansteigen des Sauerstoffverbrauches begleitet ist. „Die Intensität der Atmung sagt nichts über die Tiefe der Ruhe aus.“ Die Aktivität der Amylase sinkt bei diesem Objekt zu Beginn der Lagerung. Ende Januar steigt sie wieder an und erreicht schließlich den 8—10fachen Wert des Tiefstandes. Hierbei betsteht eine Parallele mit der Herstellung zur Bereitschaft zum Austreiben. Die Bereitwilligkeit zum Austreiben zeigt sich bei diesem Objekt aber deutlich schon bevor ein Stärkeabbau erkennbar wird. Auch hieraus also folgt wieder, daß die Enzymaktivierung wohl eine Begleiterscheinung, aber nicht die eigentliche Ursache für die Aufhebung der Ruhe ist.

Noch schwieriger ist die Beurteilung der Vorgänge in Knospen. Auch hier sind viele chemische Veränderungen gefunden worden, die den für die Samen genannten teilweise ähnlich sind. Es ist eine allmähliche Fetthydrolyse und eine Anhäufung von Fettsäuren während der Ruhe festgestellt worden (BAHGAT). Auch Veränderungen im Kohlenhydratgehalt, dabei sowohl Abnahme der löslichen wie der unlöslichen Kohlenhydrate, wurden angegeben (McDERMOTT). Bestimmte Polysaccharide können während der Ruhe zunehmen. Die unlöslichen Stickstoffverbindungen können abnehmen, die löslichen zu. SELL und JOHNSTON haben genauere Analysen über die chemischen Veränderungen in Knospen durchgeführt. Diese Autoren arbeiteten mit Knospen, während sonst meist ganze Sprosse genommen werden. Gerade wegen dieser fehlenden Beschränkung auf die Knospen sind natürlich die meisten der älteren Angaben nicht brauchbar; sie stammen aus einer Zeit, in der man noch glaubte, den Aktivitätswechsel im wesentlichen aus den Veränderungen im Gehalt an Nährstoffen in der Gesamtpflanze erklären zu können. SELL und JOHNSTON fanden, daß während der Winterruhe bis kurz vor der Entfaltung der Knospen die Menge reduzierender Zucker abnimmt, dann aber schnell steigt. Die nicht reduzierenden Zucker dagegen nehmen erst zu und dann ab. Der Gehalt an Stärke und anderen Polysacchariden verändert sich zunächst nicht stark, nimmt dann aber während der Knospenstreckung infolge der Mobilisierung von Reservestoffen in Speichergeweben schnell zu. In den stickstoffhaltigen Verbindungen zeigen sich anfangs nur geringe Veränderungen, erst später ist eine Zunahme feststellbar. Der ätherlösliche Lipoidanteil fällt zunächst und steigt während bestimmter Vorgänge der Entfaltung. Der Gehalt an Mineralstoffen ändert sich begreiflicherweise während der Ruhe nur wenig; während der Entfaltungsprozesse erhöht er sich durch Neuzufuhren.

Eine genauere zeitliche Zuordnung der Veränderungen zu den einzelnen Phasen der Ruhe oder doch des Übergangs von der Ruhe zur Aktivität ermöglichen die Untersuchungen von MILLER, GUTHRIE und DENNY. Diese Autoren arbeiteten mit Kartoffelknollen. Allerdings ist zu beachten, daß sie die Ruheperiode experimentell unterbrachen, nämlich durch Einwirkung von Äthylenchlorhydrin, und im wesentlichen die Veränderungen untersuchten, die diese Einwirkung bedingte. Sie fanden eine Steigerung der Kohlendioxydproduktion und eine gleichzeitige Abnahme der Citronensäure, außerdem eine Abnahme der Wasserstoffionenkonzentration und eine Zunahme der Aktivität mehrerer Fermente.

Veränderungen im Eiweißgehalt während der Brechung der Ruheperiode bei Kartoffelknollen untersuchte LEVITT (1954). Da die verschiedenartigsten chemischen Veränderungen untersucht worden sind, ist es kaum zu verwundern, daß unter anderem auch Änderungen im Gehalt an Nucleoproteiden und Nucleinsäuren während des Aktivitätswechsels untersucht wurden (CEL'NIKER). Die Meristeme sind bekanntlich immer sehr reich an Nucleoproteiden. Deren Abnahme während des Wachstums wird mit dem allmählichen Eintritt der Ruhe in Zusammenhang gebracht. Die Reaktivierung hingegen wird aus der allmählichen Wiederzunahme der Nucleoproteide erklärt. Es bedarf kaum einer besonderen Hervorhebung, daß auch hiermit das Wesen der Ruhe nicht erfaßt wird, sondern es sich offensichtlich nur um Begleiterscheinungen handelt.

Das gilt auch für die Untersuchungen von PETROVSKAJA, nach denen Knospen mehrerer Bäume im Winter ein Minimum des Gehalts an Nucleinsäuren (Ribose- und Desoxyribose-Nucleinsäure) enthalten. Vor dem Austreiben besteht ein Maximum an diesen Nucleinsäuren.

Somit müssen wir feststellen, daß alle diese Arbeiten, so wertvoll sie im einzelnen auch sein mögen, uns eigentlich noch nicht viel weiter geführt haben. Auch die Schwankungen in der Atmungsaktivität dürfen nur als Begleiterscheinungen betrachtet werden. Man darf nicht vergessen, daß ja auch ruhende

Samen, die überhaupt nicht mehr notwendig atmen müssen, noch den Aktivitätswechsel zeigen. Immerhin muß hier der Vollständigkeit halber auf diese Atmungsschwankungen doch noch hingewiesen werden. Bei Kartoffeln fand DETTWEILER einen regelmäßigen Atmungsanstieg vor dem Austreiben. Wird durch Fermentinhibitoren die Atmung gehemmt, so unterbleibt auch das Austreiben. Wirksam waren als Hemmstoffe etwa gasförmige Halogenfettsäureester sowie zahlreiche Urethane. Manche solcher Hemmsubstanzen wirken auch dadurch, daß sie schon den Stärkeabbau unterdrücken. Aber es darf nicht übersehen werden, daß das Austreiben nicht identisch ist mit der Beendigung der Ruhe, sondern diese Beendigung nur eine notwendige Voraussetzung für jenes Austreiben ist. Eine andere Voraussetzung ist eben die, daß die Pflanze atmen kann. In das Wesen der Ruhe können wir also so nicht tiefer eindringen. POLLOCK studierte die Atmungsbedingungen in Knospen ausführlich. Bei Beginn der Ruheperiode fand er eine Zunahme des respiratorischen Quotienten, bei Beendigung der Ruhe dessen Wiederabnahme. Solche Veränderungen könnte man nach den auf anderen Gebieten gemachten Erfahrungen mit dem Einfluß bestimmter Substanzen auf den Atmungsvorgang in Zusammenhang bringen, jedoch schreibt POLLOCK der Eintrittsgeschwindigkeit von Sauerstoff eine entscheidende Rolle zu. Auch THORNTON sieht in dem Sauerstoff einen stark regulierenden Faktor nicht nur bei der Ruheperiode von Knospen, sondern auch bei der der Samen. Die Brechung der Ruheperiode, wie sie bekanntlich durch niedrige Temperaturen (vgl. später) leicht erreichbar ist, wird dadurch erklärt, daß von dem geringen Sauerstoff dann größere Mengen für den Abbau der Endprodukte der Anaerobiose verfügbar sind. Auf die Bedeutung der Atmung verweist auch DENNY (1940). Werden *Gladiolus*-Knollen aus der Erde genommen, so beobachtet man eine rasche Zunahme der Atmungsintensität auf das 10fache. Auch dabei ist wieder eine Zunahme des respiratorischen Quotienten feststellbar. Es geht aber klar aus jenen Untersuchungen hervor, daß es sich dabei nicht um eine Begleiterscheinung beendeter Ruhe handelt, sondern nur um eine neben dem Aktivitätswechsel verlaufende Möglichkeit, die Atmung in dem einen oder anderen Sinne zu beeinflussen. Neben den Produkten der Anaerobiose ist auch dem Kohlendioxyd eine regulierende Rolle beim Aktivitätswechsel zugeschrieben worden. Es sei hierzu auf die kritischen Untersuchungen von BURTON (1952a und b) verwiesen, bei denen sich tatsächlich ein Einfluß von CO_2 und anderen Gasen auf den Aktivitätswechsel ergab, obwohl auch diese Arbeitsrichtung keinen eindeutigen Erfolg ermöglicht. Man hat dabei geradezu von einer Autonarkose gesprochen (vgl. BURTON). Neben Kohlendioxyd ist von den eigenen gasförmigen Stoffwechselprodukten der Pflanze vor allem Äthylen wichtig. Starke Wirkungen (bei Kartoffeln) traten aber erst in Gasgemisehen ein, und immer sind die Wirkungen an so hohe Konzentrationen gebunden, daß man solchen Stoffwechselprodukten auch nicht die entscheidende Funktion zuschreiben kann. Wichtiger könnte eine solche Autonarkose durch Kohlendioxyd bei manchen Samen sein, die in noch feuchtem Zustand wasser- und gasundurchlässige Schichten besitzen, die erst durch verschiedenartige Vorgänge beseitigt werden müssen, bevor die Keimbereitschaft hergestellt wird. Auch eine zwischen Samenschale und Embryo befindliche Wasserschicht kann schon genügen, um den Gasaustausch zu hindern, so daß dann ein Austrocknen zur Herstellung der Keimbereitschaft führt. In anderen Fällen sind besondere Gewebeschichten gefunden worden, die erst allmählich verändert, zerrissen oder abgebaut werden müssen, bevor der Stoffwechsel in ausreichendem Maße möglich wird (vgl. die Hinweise bei CROCKER). Jedoch handelt es sich dabei um Sonderfälle, die speziell für die Keimungsphysiologie der Samen interessant sind, aber mit der Ruhe, die sonst auch bei

Samen sehr wohl in Eigenschaften der Embryonen begründet sein kann, eben nur in diesen Sonderfällen zu tun haben.

Fassen wir alle diese Bemühungen zusammen, so können wir sagen, daß sie nicht nur kein einheitliches Bild vermitteln, sondern obendrein ganz klar wird, wie sehr sie zwar Begleiterscheinungen des Aktivitätswechsels sein mögen, aber doch auf keinen Fall diesen erklären können. Höchstens können die festgestellten Faktoren mehr oder weniger stark modifizierend eingreifen. Immer wieder muß darauf hingewiesen werden, daß es ja von den ruhenden Knospen mit mehr oder weniger starkem Stoffwechsel alle Übergänge gibt bis zu den ruhenden Samen ohne nachweisbaren Stoffwechsel. Und es besteht kein Grund zur Annahme, daß das Phänomen des Aktivitätswechsels in den ruhenden Samen wesentlich anderer Natur wäre als das der ruhenden Knospen, zumal diese Phänomene einander so überaus stark bis in Einzelheiten hinein ähneln.

d) Wuchs- und Hemmstoffgehalt.

Die oben genannten Bedenken würden sich naturgemäß auch auf eine mögliche Rolle von Veränderungen im Wuchs- und Hemmstoffgehalt bei der Regulierung des Aktivitätswechsels erstrecken. Trotzdem müssen wir auf diese Untersuchungen hier eingehen, einmal weil solche Veränderungen auch beteiligte Komponenten sein können, andererseits weil bei solchen Substanzen hormonartiger Natur noch mehr als bei den vorher diskutierten chemischen Veränderungen die Möglichkeit besteht, daß sie zwar nicht die primäre Ursache, aber doch ein wesentliches Glied in der Kette der Vorgänge sind, die den Aktivitätswechsel ermöglichen.

Die Beteiligung von Wuchs- und Hemmstoffen an der Ruhe wird natürlich durch mehrere bekannte Erscheinungen nahegelegt. Besonders kann auf die Bedeutung von Hormonen bei der Regulierung der Ruhe von Achselknospen hingewiesen werden. Obwohl dieses Phänomen bisher nicht befriedigend geklärt worden ist, darf doch wenigstens als schon gesichert bezeichnet werden, daß Wuchsstoffe bei den von den Endknospen ausgehenden korrelativen Hemmungen entscheidend beteiligt sind. Diese Korrelation selber soll uns hier nicht näher beschäftigen (vgl. Band XIV). Aber der Hinweis auf sie rechtfertigt doch die Bemühungen, auch für die sonstigen Inaktivierungs- und Aktivierungserscheinungen Wuchs- und Hemmstoffe heranzuziehen.

Zunächst einmal muß auf die Bemühungen hingewiesen werden, den Aktivitätswechsel mit schwankendem Wuchsstoffgehalt in Zusammenhang zu bringen. Und diese Bemühungen begannen mit dem Versuch, den Aktivitätswechsel durch Wuchsstoffzufuhr zu modifizieren. Es sei etwa auf die Untersuchungen von Söding (1937) hingewiesen, nach denen die sich öffnenden Knospen reichlich Wuchsstoffe enthalten, welche auch die Cambiumtätigkeit regulieren. Daß aber auch bei der Cambiumruhe nicht etwa der Wuchsstoffgehalt allein entscheidend ist, haben die Untersuchungen von Gouwentak erwiesen, nach denen Auxinzufuhr allein die Ruheperiode des Cambiums nicht zu brechen vermag. Nun ist zwar nach den weiteren Untersuchungen von Künning und Söding bei diesen Korrelationen wohl ein Gemisch verschiedener Wirkstoffe beteiligt; aber nach den Untersuchungen von Gouwentak sieht es mehr so aus, als seien es überhaupt nicht hormonale Faktoren allein, sondern als wäre auch eine gewisse Bereitschaft des Cambiums notwendig. Auf eine hormonale Regulierung der Ruheperiode deuten weiterhin etwa Untersuchungen von Krasnosselskaja und Richter. Diese Autoren fanden, daß von sich öffnenden Knospen Einflüsse ausgehen, die in der Lage sind, beschleunigend auf die Aktivierung anderer Knospen zu wirken. Einen engen Zusammenhang der Ruheperiode mit dem

Wuchsstoffgehalt hat BOYSEN-JENSEN schon 1935 angenommen. Weiterhin fanden ZIMMERMANN (1936), AVERY, BURKHOLDER und CREIGHTON (1937) einen unterschiedlichen Wuchsstoffgehalt in ruhenden und aktiven Knospen (vgl. weiterhin auch die Angaben von BENNETT und SKOOG 1938). Sehr früh begannen dann auch Versuche über die Möglichkeit einer Beeinflussung der Ruheperiode von Knospen und Samen durch künstliche Wuchsstoffe. Teilweise wurden positive, teilweise negative Resultate erzielt (vgl. AMLONG und NAUNDORF 1938, BORGSTRÖM 1939, NIETHAMMER 1940 sowie die ausführlichen Literaturhinweise bei KRUYT). Aber alle diese Resultate sind eben nicht befriedigend, weil einerseits die Beeinflussungen nicht immer eindeutig sind, andererseits aber auch schon die erwähnten Untersuchungen GOUWENTAKS zur Vorsicht mahnen. Man hat bei den als wirksam vermuteten Hormonen übrigens nicht immer nur an das Auxin gedacht, sondern etwa auch an Glutathion (GUTHRIE 1940), oder an Adenin (GUTHRIE 1941). Eine ganze Reihe anderer Substanzen ist ebenfalls als wirksam gefunden worden, jedoch kann von diesen anderen Substanzen kaum mehr angenommen werden, daß sie normalerweise hormonal die Ruhe regulieren. Wir werden uns daher eher mit ihnen zu beschäftigen haben, wenn wir uns der Regulierung der Ruhe durch äußere Faktoren zuwenden. Dabei muß allerdings betont werden, daß sich eine klare Trennung hier nicht durchführen läßt.

Als man sah, daß Schwankungen im Gehalt an Wuchsstoffen und anderen Hormonen den Aktivitätswechsel in Knospen nicht erklären können, begann man nach Substanzen zu suchen, die diesen Wuchsstoffen entgegenwirken. Man suchte also Hemmstoffe usw. Solche Hemmstoffe sind bekanntlich bei Samen schon sehr lange bekannt und es ist auch festgestellt worden, daß sie tatsächlich auf die Keimbereitschaft einen Einfluß haben. Wieweit sie diese entscheidend bestimmen, soll an dieser Stelle nicht geprüft werden. Eine entscheidende Bedeutung von Hemmstoffen für den Eintritt der Ruhe nimmt z. B. MOLOTKOWSKI an. Er arbeitete mit Hemmstoff aus Samen. Den Eintritt der Ruhe erklärt er wieder aus dem Vorhergehen eines intensiven Stoffwechsels. Die Rolle der Hemmstoffe dagegen wird in der Inaktivierung von Wuchsstoffen gesehen. Zwingend sind diese Schlußfolgerungen nicht. Einen Hinweis auf die Beteiligung von Hemmstoffen bieten auch Beobachtungen von STEWART und CAPLINE. Diese Autoren beurteilen den Hemmstoffgehalt nach der Verminderung der wachstumsfördernden Wirkung von Kokosmilch auf Gewebekulturen von *Daucus carota*. Mit diesem Verfahren fanden sie, daß in Kartoffelknollen, Knospen von *Acer* und Zwiebeln von *Allium cepa* im Zusammenhang mit dem Aktivitätswechsel offenbar Veränderungen des Hemmstoffgehalts ablaufen. Vor allem aber muß auf die sorgfältigen Untersuchungen HEMBERGS verwiesen werden. HEMBERGS Arbeiten beziehen sich zunächst einmal auf die Ruhe der Kartoffelknollen (HEMBERG 1947 und 1952). Der normale Wuchsstoff ist hier wie auch sonst in den meisten Fällen β-Indolylessigsäure. Außerdem findet sich ein neutraler Wuchsstoff, der wohl β-Indolylacetaldehyd ist. Während der Ruheperiode findet man außerdem einen Hemmstoff. Dieser Hemmstoff verschwindet im Lauf der Zeit und zwar nach etwa 10 Wochen ganz. Kurz vor der Keimung zeigt sich außerdem im Periderm eine Zunahme der Menge sauren Wuchsstoffes. Das Abschälen des hemmstoffhaltigen Periderms ermöglicht ein vorzeitiges Austreiben, und zwar sowohl darum, weil der im Periderm enthaltene Wuchsstoff beseitigt wird und andererseits auch darum, weil die geringeren Hemmstoffmengen in den anschließenden Geweben in der trockenen Luft zerstört werden. Die Untersuchungen wurden dann weiterhin auf Endknospen von *Fraxinus excelsior* ausgedehnt (HEMBERG 1949). Die in tiefer Winterruhe (im Oktober) befindlichen

Knospen besitzen sehr viele mit Äther extrahierbare Hemmstoffe (getestet wurde immer an der Unterbindung der β-Indolylessigsäurewirkung auf *Avena*-Koleoptilen). Späterhin nimmt die Menge dieser Hemmstoffe ab. HEMBERG fand weiterhin, daß die Menge an Hemmstoffen durch Einwirken von Äthylenchlorhydrin stark herabgesetzt wird (ebenso war es auch bei Kartoffeln möglich). Das legt natürlich die Vermutung nahe, daß die frühtreibende Wirkung solcher Substanzen sich auch auf eine Beeinflussung des Hemmstoffgehalts zurückführen läßt. Nach diesen Untersuchungen HEMBERGS würde also nicht Schwankungen des Wuchsstoffgehalts eine Rolle zufallen, sondern nur der Hemmstoffgehalt würde durch seine Schwankungen die Ruhe regulieren. Hierzu passen auch Ergebnisse BLOMMAERTS: Papierchromatographisch wurden Veränderungen im Gehalt an Wuchs- und Hemmstoffen nachgewiesen. Es sind dabei mehrere verschiedene dieser Hormone beteiligt.

Schon hingewiesen haben wir darauf, daß auch die Aufhebung der Ruheperiode in Hexenbesen (SCHELLENBERG) auf eine hormonale Regulierung der Ruheperiode hindeutet.

Auch bei Samen hat man versucht, die Ruheperiode mit dem Gehalt an Hemmstoffen in Zusammenhang zu bringen, nachdem sich gezeigt hatte, daß auch hier Wuchsstoffschwankungen nicht ein entscheidender Faktor der Tiefe der Ruhe sein können (vgl. die Hinweise bei KRUYT sowie BARTON 1940 und SHIBUYA). Nur wenige Autoren geben an, daß sie die Ruheperiode einiger Samen mit Wuchsstoffen haben unterbrechen können. Allerdings gibt es Angaben über das allmähliche Auftreten von aktivem Auxin bei allmählich zunehmender Keimbereitschaft, etwa während der Stratifikation (MIROV). Aber da diese Beobachtungen höchstens Teilerklärungen ermöglichten, suchte man bald nach Hemmstoffen, die in Samen tatsächlich ebenso gefunden wurden wie Wuchsstoffe (vgl. z. B. POHL und TEGETHOFF). Es sei etwa auf die Untersuchungen von SMITH und von LUCKWILL verwiesen. Besonders LUCKWILL hat ausführlicher die Änderung des Gehalts an Hemmstoffen während der Stratifikation untersucht. Nun mag es aber wohl richtig sein, daß gerade bei den Samen, die eine Stratifikation benötigen, dem Herausdiffundieren von Hemmstoffen eine große Rolle zukommen kann. Die Untersuchungen LUCKWILLS selber zeigen auch, daß keine klare Beziehung zwischen dem abnehmenden Hemmstoffgehalt und der Herstellung der Keimbereitschaft besteht. Zumindest läßt sich zeigen, daß die Aktivität noch nicht voll erreicht ist, wenn alle Hemmstoffe verschwunden sind. LUCKWILL selber schreibt daher wachstumsfördernden Stoffen eine größere Rolle zu. Auch SMITHS Untersuchungen zeigen, daß keimungshemmende Substanzen allein nicht entscheidend sein können. Man sieht, wie sehr wir zur Zeit unter dem Zwang der Vorstellung stehen, alle solche Vorgänge müßten sich durch das Gegeneinanderwirken von Wuchs- und Hemmstoffen erklären lassen. Aus der Beeinflußbarkeit durch Wuchsstoffe würde natürlich auch noch nicht die entscheidende Rolle dieser Substanzen bei der normalphysiologischen Regulierung des Aktivitätswechsels folgen.

HEMBERG (1955) fand, daß trockene Maiskörner beim Beginn der Lagerungsperiode viel mehr gebundenes Auxin enthalten als am Ende der Ruheperiode. Der Gehalt an freiem Auxin blieb unverändert.

e) Beteiligung physikalischer Faktoren.

Schon angedeutet hatten wir, daß auch physikalische Faktoren einen Einfluß auf den Aktivitätswechsel haben können. Die Durchlässigkeit von Knospenschuppen oder Samenschalen kann verantwortlich sein für die Durchtrittsgeschwindigkeit von Wasser, Sauerstoff, CO_2 usw. Wenn eine Ruhe durch solche Faktoren bedingt ist, würde man natürlich nicht mehr von einer freiwilligen, sondern von einer erzwungenen Ruhe sprechen. Namentlich bei Samen kann solchen Faktoren aber eine sehr große Rolle zufallen (vgl. CROCKER und BARTON 1953). Da eine solche Ruhe aber vom zellphysiologischen Standpunkt nicht so sehr interessant ist, soll sie uns hier nicht ausführlicher beschäftigen. Es sei nur darauf hingewiesen, daß bei manchen Samen die Durchlässigkeit der Samenschalen so gering sein kann, daß keine Keimung erfolgt, bevor diese Samenschale nicht durch äußere Faktoren durchlässig gemacht worden ist. Unter solchen äußeren Faktoren wären etwa zu nennen: Der allmähliche Abbau durch Bakterien und Pilze, die Wirkung von Frost. Wichtig sein kann hierbei sowohl die allmähliche Herstellung der Wasserdurchlässigkeit als auch die allmähliche Herstellung der Sauerstoffdurchlässigkeit. Die Rolle der Sauerstoffkonzentration und der Durchlässigkeit für Sauerstoff ist in mehreren Untersuchungen nachgewiesen worden (vgl. CROCKER und BARTON, SHULL 1911 und 1914, ATWOOD 1914, HARRINGTON 1923, SPÄTH 1932, JOHNSON 1935, THORNTON 1935, STIER 1938).

f) Qualitative Veränderungen während des Aktivitätswechsels.

Zur Beurteilung der Natur des Aktivitätswechsels ist noch ein Hinweis darauf nötig, daß nicht nur quantitative Veränderungen, sondern auch qualitative mit ihm verbunden sein können. Diese qualitativen Veränderungen äußern sich vor allem in den je nach dem Zustand der Pflanze verschiedenartigen Entwicklungsleistungen oder in dem qualitativ unterschiedlichen Ansprechen auf äußere Faktoren.

Sehr ausgeprägt sind diese qualitativen Verschiedenheiten bei den Aktivitätsrhythmen, die etwa dem Tagesgang folgen. Sie äußern sich darin, daß die Pflanze am Tage und in der Nacht unterschiedliche Ansprüche an das Licht und an die Temperatur stellen kann. Auf das Licht kann sie am Tage geradezu entgegengesetzt reagieren wie in der Nacht. Das ist ein wesentlicher Faktor für die Entstehung der photoperiodischen Reaktionen. Hinsichtlich der Temperaturansprüche äußern sich die qualitativen Veränderungen vor allem darin, daß zur Nachtzeit für die pflanzliche Entwicklung meist sehr viel niedrigere Temperaturen optimal sind als am Tage. Offenbar also überwiegen in der Nacht innerhalb der Pflanze andere Vorgänge als am Tage. Das ist die Basis für die thermoperiodischen Erscheinungen. Weiterhin äußern sich die etwa tagesperiodisch schwankenden qualitativen Aktivitätsänderungen auch darin, daß in der Nacht bei sehr vielen Pflanzen viel mehr Zellteilungen stattfinden als am Tage. Vielfach kann man geradezu von einem regelmäßigen Wechsel einer Streckungs- und einer Teilungsphase sprechen.

Aber auch bei dem langsamer, ähnlich etwa einem Jahresrhythmus ablaufenden Aktivitätswechsel zeigen sich qualitative Veränderungen. Ganz besonders kann hier auf die ausgedehnten Untersuchungen von BLAAUW und seinen Mitarbeitern an einer Reihe gärtnerisch wichtiger Pflanzen hingewiesen werden. Nach diesen Untersuchungen kann man bei der pflanzlichen Entwicklung deutlich Perioden der Blattbildung, der Blütenbildung, des Streckungswachstums usw. voneinander trennen. Dabei sind die Zeiten der Blatt- und Blütenbildung oft mit der Phase gekoppelt, die eine scheinbare Ruhe darstellt. Die auf den vorherigen Seiten erörterte Ruheperiode ist also oftmals in Wahrheit gar keine Ruheperiode, sondern eben vielmehr, wenigstens teilweise, eine Periode der Organbildung. Es können dabei auch noch zahlreiche Zellteilungen stattfinden. Auch bei diesem Aktivitätswechsel äußert sich die qualitative Verschiedenheit der Phasen übrigens wieder noch darin, daß zu den einzelnen Zeitabschnitten ganz unterschiedliche Temperaturen optimal sind. Zum Beispiel ist für die

Darwin-Tulpe während der Blütenbildung eine Temperatur zwischen 17 und 20° optimal. In anderen Phasen aber kann das Optimum erheblich unter 10° liegen. Während der Zellstreckungsperioden ändert sich das Temperaturoptimum wiederum kontinuierlich. Es steigt bis über 20°. Diese Bedingungen sind von Pflanze zu Pflanze selbstverständlich verschiedenartig.

Es ist unmöglich, hier im einzelnen auf die zahlreichen Beobachtungen einzugehen, welche zeigen, daß auch diese anderen qualitativen Veränderungen mit dem etwa jahresperiodischen Aktivitätswechsel verbunden sind. Aus der neueren Literatur sei etwa noch die Angabe von GUMPELMAYER erwähnt, daß mit jenem Wechsel auch die Fähigkeit zur Wurzelbildung starken Schwankungen unterworfen ist. LINSER fand eine jahreszeitliche Periodizität von Wachstumskorrelationen zwischen dem Primärblatt und der Koleoptile bei *Avena sativa*. Schließlich sei noch darauf hingewiesen, daß die Koppelung des Wechselns von Ruhe und Aktivität mit qualitativen Veränderungen auch dadurch zum Ausdruck kommen kann, daß der Sproßscheitel jahresperiodische Formänderungen durchlaufen kann (so etwa nach SACHER bei einigen *Pinus*-Arten). Diese Formänderungen zeigen eine enge Korrelation mit dem Wechsel hoher und niedriger Aktivität und naturgemäß nicht nur mit der Lebhaftigkeit, sondern auch mit der Art der Organbildungen.

g) Beziehungen zur Resistenz.

Wir haben mehrfach erwähnt, daß Resistenzänderungen offenbar mit zum Wesen der Ruhe gehören. Das gilt nicht nur für die mehrfach erwähnte Kälteresistenz und die Trockenresistenz, sondern auch die Hitzeresistenz, Strahlen- und Giftresistenz. Es ist ja bekannt, daß ruhende Organe, ganz besonders etwa Samen oder Sporen, wesentlich stärkere Dosierungen aller dieser schädigenden Einflüsse ertragen als aktive Gewebe. Wir haben auch angedeutet, daß hier offenbar ein innerer Zusammenhang besteht: Die Resistenzerhöhung ist ebenso wie die Ruhe Ausdruck einer Verminderung der Labilität. Beide also sind offenbar Folgen gleicher plasmatischer Veränderungen. Diese Beziehungen werden dadurch besonders deutlich, daß wir selbst an trockenen Samen Schwankungen der Resistenz feststellen können, die mit der Tiefe der Samenruhe parallel laufen. Bei der Aktivierung von Knospen oder Samen geht diese Resistenz sehr schnell zurück, und zwar schon (bei Samen) während des Quellungsvorganges. Man hat diese Beziehung bei sehr vielen Untersuchungen festgestellt (vgl. z. B. MITCHELL und BROWN). Auch das Reagieren der Samen auf äußere Reize, etwa auf Lichtreize, kann ja erst dann beginnen, wenn die Samenquellung einsetzt.

D. Zusammenfassung.

Zusammenfassend müssen wir feststellen, daß über das Wesen der Ruhe eigentlich recht wenig bekannt ist. Man hat viele Begleiterscheinungen des Aktivitätswechsels untersucht. Man hat auch die Folgen von Ruhe und Aktivität ermittelt. Ausführlich studiert worden sind die zellphysiologischen Wirkungen von äußeren Faktoren, die Ruhe und Aktivität regulieren, aber eben leider auch zugleich andere Wirkungen in den Zellen hervorrufen. So müssen wir zugeben, daß uns das eigentliche Wesen des Aktivitätswechsels noch nicht bekannt ist. Wir wissen nicht, was eigentlich eine Ruhe- und was eine Aktivitätsperiode auszeichnet. Vor wie großen Schwierigkeiten wir hier stehen, wird am besten aus den Hinweisen über die qualitativen Veränderungen klar. Sehr viel mehr wissen wir über äußere Faktoren, die den Aktivitätswechsel regulieren. Davon wird auf den nächsten Seiten die Rede sein. Am undurchsichtigsten aber sind die Fälle, in denen der Wechsel der Aktivität nicht nur unabhängig von bestimmten äußeren Faktoren bestehen kann, sondern er obendrein unabhängig von solchen Faktoren, d. h. bei einer völligen Konstanz der äußeren Bedingungen weiterläuft. Wahrscheinlich wird ein tieferes Eindringen in die Natur eines solchen Aktivitätswechsels und in das Wesen seiner selbständigen Regulierung nur möglich sein, wenn Objekte untersucht werden, bei denen alle angedeuteten Begleiterscheinungen fehlen. Das heißt, am aussichtsreichsten wäre es wohl, den Aktivitätswechsel in trockenen, unter konstanten Bedingungen verwahrten Samen eingehender zu untersuchen. Mehr negativ als positiv können wir feststellen: Beim Aktivitätswechsel sind Stoffwechselschwankungen, hervorgerufen durch Schwankungen der Fermentaktivitäten, beteiligt. Aber der Wechsel von Ruhe und Aktivität läßt sich nicht aus einem Gegeneinanderwirken der Stoffwechselprozesse erklären. Ebensowenig reicht ein Wechsel von Hormon- und Hemmstoffkonzentrationen zur Erklärung aus. Das Entscheidende und zumindest das Ursprünglichste sind Veränderungen im Protoplasma. Während wir nun zwar Ansatzpunkte zur Erklärung dieser Veränderungen durch äußere Faktoren physikalisch-chemischer Natur haben, können wir bis jetzt in keiner Weise erklären, warum solche Veränderungen der Aktivität auch bei der Konstanz der äußeren Faktoren auftreten. Hierin liegt zweifellos das schwierigste Problem. Gerade in dieser Hinsicht sind Untersuchungen an ruhenden Samen, bei denen das Problem am reinsten, d. h. am wenigsten mit anderen Erscheinungen vermengt ist, vorteilhaft.

Literatur.

AFANASIEV, O., and M. CRESS: Producing seedlings of eastern red cedar (Juniperus virginiana L). Okla. Agric. Exper. Sta. Bull. **256** (1942). — AMLONG, H. U., u. G. NAUNDORF: Über die Bedeutung der Wuchsstoffe für das Frühtreiben. Gartenbauwiss. **12**, 116—120 (1938). — ASAI, T.: Untersuchungen über die Bedeutung des Mannits im Stoffwechsel einiger höherer Pflanzen. I. u. II. Jap. J. of Bot. **6** (1932); 8 (1937). — ATWOOD, W. M.: A physiological study of the germination of Avena fatua. Bot. Gaz. **57**, 386—414 (1914). — AVERY, G. S., P. R. BURKHOLDER and H. B. CREIGHTON: Production and distribution of growth hormone in shoots of Aesculus and Malus, and its probable rôle in stimulating cambial activity. Amer. J. Bot. **24**, 51—58 (1937).

BAHGAT, M.: Some changes occurring in resting tissues of the bartlett pear during the breaking of rest. Doctoral thesis, Univ. California. Berkeley, Calif. 1931. — BARTON, L. V.: Some effects of treatment of seeds with growth substances on dormancy. Contrib. Boyce Thompson Inst. **11**, 229—240 (1940). — Germination of seeds of Juniperus virginiana L. Contrib. Boyce Thompson Inst. **16**, 387—393 (1951). — BENNETT, J. P., and F. SKOOG: Preliminary experiments on the relation of growth-promoting substances to the rest period in fruit trees. Plant Physiol. **13**, 219—225 (1938). — BLOMMAERT, K. J. L.: Growth and inhibiting substances in relation to the rest period of the potato tuber. Nature (Lond.) **174**, 970—972 (1954). — BORGSTRÖM, G.: Theoretical suggestions regarding the ethylene responses of plants and observations on the influence of apple-emanations. Kgl. Fysiogr. Sällsk. Lund Förh. **9**, 1—40 (1939). — BOYSEN-JENSEN, P.: Die Wuchsstofftheorie. Jena: Gustav Fischer 1935. — BREDEMANN, G., K. GARBER, P. HARTECK u. KL. A. SUHR: Die Temperaturabhängigkeit der Lebensdauer von Blütenpollen. Naturwiss. **34**, 279 bis 280 (1947). — BRIGGS, D. R., and D. SIMINOVITCH: The chemistry of the living bark of the black locust tree in relation to frost hardiness. II. Seasonal variations in the electrophoresis patterns of the water soluble proteins of the bark. Arch. of Biochem. **23**, 18—28 (1949). — BÜNNING, E.: Jahres- und tagesperiodische Vorgänge in der Pflanze. Studium gen. **2**, 73—78 (1949). — BÜNNING, E., u. E. W. BAUER: Über die Ursachen endogener Keimfähigkeitsschwankungen in Samen. Z. Bot. **40**, 67—76 (1952). — BURTON, W. G.: Studies on the dormancy and sprouting of potatoes. II. The carbon dioxide content of the potato tuber. New Phytologist **50**, 287—296 (1952). — Studies on the dormancy and sprouting of potatoes. III. The effect upon sprouting of volatile metabolic products other than carbon dioxide. New Phytologist **51**, 154—162 (1952).

CEL'NIKER, JU. L.: Zur Frage nach den physiologischen Ursachen der Periodizität des Wachstums der Bäume. Bot. ž. SSSR. **35**, 445—460 (1950). — CHANDLER, R. C.: Fruit growing. Boston, Mass.: Houghton Mifflin Co. 1925. — Bound water in plant sap and some effect of temperature and nutrition thereon. Plant Physiol. **16**, 785—798 (1941). — COSTER, C.: Zur Anatomie und Physiologie der Zuwachszonen- und Jahresringbildung in den Tropen. Doctors' Thesis Wageningen. Leiden 1927. — CROCKER, W.: Growth of plants. New York: Reinhold Publishing Comp. 1948. — CROCKER, W., and L. V. BARTON: Physiology of seeds. Waltham, Mass.: Published by the Chronica Botanica Comp. 1953.

DENNY, F. E.: Respiration of Gladiolus corms during prolonged dormancy. Contrib. Boyce Thompson Inst. **10**, 453—560 (1939). — DENNY, F. E., and L. P. MILLER: Effect of ethylene chlorhydrin vapors upon dormant lilac tissues. Contrib. Boyce Thompson Inst. **4**, 513—528 (1932). — DETTWEILER, CH.: Zusammenhänge zwischen Austreiben und Atmung bei der Kartoffelknolle. I. Die Hemmung des Austreibens durch Fermentinhibitoren, insbesondere Urethane. Planta (Berl.) **41**, 214—239 (1952). — DOORENBOS, J.: Review of the literature on dormancy in buds of woody plants. Meded. Landbouwhogeschool Wageningen **53**, 1—24 (1953). — Orienterend onderzoek over het forceren van Forsythia en Rhododendron. Meded. Directeur Tuinbouw **16**, 533—543 (1953). — DOSTÁL, R.: Über die Sommerperiodizität bei Quercus und Fagus. Ber. dtsch. bot. Ges. **45**, 436—447 (1927).

ECKERSON, S.: A physiological and chemical study of after-ripening. Bot. Gaz. **55**, 286—299 (1913).

FISCHER, A.: Beiträge zur Physiologie der Holzgewächse. Jb. wiss. Bot. **22**, 73—160 (1891). — FLEMION, F.: Physiological and chemical changes preceding and during the after-ripening of Symphoricarpus racemosus seeds. Contrib. Boyce Thompson Inst. **6**, 91—102 (1934). — Dwarf seedlings, from non-after-ripended embryos of peach, apple, and hawthorn. Contrib. Boyce Thompson Inst. **6**, 205—209 (1934). — Breaking the dormancy of seeds of Crataegus species. Contrib. Boyce Thompson Inst. **9**, 409—423 (1938).

GARDNER, F. E.: Composition and growth initiation of dormant Bartlett pear shoots as influenced by temperature. Plant Physiol. **4**, 405—434 (1929). — GOTTLIEB, D.: The physiology of spore germination in fungi. Bot. Review **16**, 229—257 (1950). — GOUWENTAK, C. A.: Cambial activity as dependent on the presence of growth hormone and the non-resting condition of stems. Proc. Kon. Ned. Akad. Wetensch. **44** (5), 654—663 (1941). — GRANDFIELD, C. O.: Food reserves and their translocations to the crown buds as related to cold and drought resistance in Alfalfa. J. Agricult. Res. **67**, 33—47 (1943). — GUMPELMAYER, E.: Die Bewurzelung von Stecklingen unter dem Einfluß von Heteroauxin im Jahresrhythmus. Phyton **1**, H. 2—4 (1949). — GUTHRIE, J. D.: Rôle of glutathione in the breaking of the rest period of buds by ethylene chlorhydrin. Contrib. Boyce Thompson Inst. **11**, 261—270 (1940). — Sprays that break the rest period of peach buds. Contrib. Boyce Thompson Inst. **12**, 45—47 (1941a). — A preparation from yeast that is active in breaking the rest period of buds. Contrib. Boyce Thompson Inst. **12**, 195—201 (1941b). — GUTTENBERG, H. v., u. G. MEINL: Über die Veränderung der Wasserpermeabilität von Kartoffelknollen während der Lagerzeit und durch Cumarin. Planta (Berl.) **43**, 571—575 (1954).

HARRINGTON, G. T.: Use of alternating temperatures in the germination of seeds. J. Agricult. Res. **23**, 295—332 (1923). — HARRIS, G. H.: Proc. Amer. Soc. Horticult. Sci. **23**, 414—422 (1926). — HEMBERG, T.: Studies of auxins and growth-inhibiting substances in the potato tuber and their significance with regard to its rest-period. Acta Horti Bergiani **14** (5), 133—220 (1947). — Growth inhibiting substances in terminal buds of Fraxinus. Physiol. Plantarum (Copenh.) **2**, 37—44 (1949). — The significance of acid growth-inhibiting substances for the rest-period of the potato tuber. Physiol. Plantarum (Copenh.) **5**, 115—129 (1952). — Studies on the balance between free and bond auxin in germinativy maize. Physiol. Plantarum (Copenh.) 8, 418—432 (1955). — HENCKEL, P. A.: Der Ruhezustand der Pflanzen. Dokl. Akad. Nauk. SSSR. **62**, 409—412 (1948). — Der pflanzliche Ruhezustand als Vorgang der Protoplasmaablösung. Vestn. Akad. Nauk. SSSR. **18**, 16—24 (1948). — The adaptative significance of dormancy in plants. (Presented at 7th Intern. Botan. Congr., Stockholm, July 1950.) Proc. Soc. Internat. Botan. Congr. Uppsala: Almqvist & Wiksells 1953. — HENCKEL, P. A., i. E. Z. OKNINA: Der Ruhezustand der Pflanzen als Ablösungsprozeß des Protoplasten. Trudy Inst. Fiziologii Rastenij im. Timirjazewa **6**, 85—102 (1948). — HENSSEN, A.: Die Dauerorgane von Spirodela polyrrhiza (L.) Schleid. in physiologischer Betrachtung. Flora (Jena) **141**, 523—566 (1954). — HOWARD, W. L.: An experimental study of the rest period in plants. Fifth report., Mon. Agric. Exper. Sta. Res. Bull. **21**, 1—25 (1915).

IRAKLIONOW, P. P.: Über den Einfluß des Warmbads auf die Atmung und Keimung der ruhenden Pflanzen. Jb. wiss. Bot. **51** (4), 515—539 (1912).

JOHANNSEN, W.: Äther- und Chloroformnarkose und deren Nachwirkung. Bot. Zbl. **68** (11), 337—338 (1896). — Das Ätherverfahren beim Frühtreiben, 2. Aufl. Jena: Gustav Fischer 1906. — JOHNSON, L. P. V.: General preliminary studies on the physiology of delayed germination in Avena fatua. Canad. J. Res., Sect. C **13**, 283—300 (1935). — JONES, H. A.: Physiological study of maple seeds. Bot. Gaz. **69**, 128—152 (1920). — JOST, L.: Über Dickenwachstum und Jahresringbildung. Bot. Z. **49**, 485—500 (1891).

KESSLER, W., u. W. RUHLAND: Weitere Untersuchungen über die inneren Ursachen der Kälteresistenz. Planta (Berl.) **28**, 159—204 (1938). — KLEBS, G.: Über die periodischen Erscheinungen tropischer Pflanzen. Biol. Zbl. **32**, 257—285 (1912). — Über das Treiben der einheimischen Bäume, speziell der Buche. Abh. Heidelbg. Akad. Wiss., Math.-naturwiss. Kl., 3. Abh. **1914**. — KNEEN, E., and M. J. BLISH: J. Agricult. Res. **62**, 1—26 (1941). — KRASNOSSELSKAYA, T. A., i. A. A. RICHTER: Transport of break of winter dormancy of buds along branches of woody plants. C. r. Acad. Sci. URSS. **35**, 184—186 (1942). — KRUYT, W.: A study in connection with the problem of hormonization of seeds. Acta bot. neerl. **3** (1), 1—82 (1954). — KÜNNING, H.: Untersuchungen über die Wirkstoffregulation der Kambiumtätigkeit. Planta (Berl.) **38**, 36—64 (1950). — KÜNNING, H., u. H. SÖDING: Über die Wirkstoffregulation der Kambiumtätigkeit. Z. Naturforsch. **4b**, 55 (1949).

LAKON, G.: Über die Festigkeit der Ruhe panachierter Holzgewächse. Ber. dtsch. bot. Ges. **35**, 648—652 (1918). — LEVITT, J.: Frost killing and hardiness of plants. Minneapolis: Burgess Publ. Co. 1941. — Frost, drought and heat resistance. Annual Rev. Plant Physiol. **2**, 245—268 (1951). — Investigations of the cytoplasmic particulates and proteins of potato tubers. III. Protein synthesis during the breaking of the rest period. Physiol. Plantarum (Copenh.) **7**, 597—601 (1954). — LUCKWILL, L. C.: Growth-inhibiting and growth-promoting substances in relation to the dormancy and after-ripening of apple seeds. J. Horticult. Sci. **27**, 53—67 (1952).

MAGNUS, W.: Der physiologische Atavismus unserer Eichen und Buchen. Biol. Zbl. **33**, 309—337 (1913). — MCDERMOTT, J. J.: Changes in the chemical composition of twigs and buds of yellow poplar during the dormant period. Plant Physiol. **16**, 415—418 (1941). — MEYER, B. S., and D. B. ANDERSON: Plant Physiol., 2. Aufl. New York: Chapman a. Hall 1952. — MILLER, C. P., J. D. GUTHRIE and F. E. DENNY: Induced changes in respiration rates and time relation in the change in internal factors. Contrib. Boyce Thompson Inst. **8**, 41—61 (1936). — MIROV, N. T.: A note on germination methods for coniferous species. J. Forestry. **34**, 719—723; (1936) Abstr. Exper. Sta. Rec. **75**, 785 (1936). — MITCHEL, J. W., and J. W. BROWN: Relative sensitivity of dormant and germinating seeds to 2,4-D. Science (Lancaster, Pa.) **106**, 266—267 (1947). — MOLISCH, H.: Über ein einfaches Verfahren, Pflanzen zu treiben. Sitzgsber. Akad. Wiss. Wien, Math.-naturwiss. Kl., Abt. 1 **117**, 87—116 (1908); **118**, 637—690 (1909). — Das Warmbad als Mittel zum Treiben der Pflanzen. Jena: Gustav Fischer 1909. — Warmbad und Pflanzentreiberei. Österr. Garten-Z. **4**, 17—23, 50—59 (1909). — MOLOTKOWSKIJ, G. CH.: Die Bedeutung der Wachstumsinaktivatoren für den Ruhezustand bei Pflanzen. Dokl. Akad. Nauk. SSSR. **68**, 405—408 (1949).

NIETHAMMER, A.: Erwecken ruhender Winterknospen durch Rohwuchsstofflösungen von Pilzen. Gartenbauwiss. **14**, 651—653 (1940). — NIKLEWSKI, B.: Untersuchungen über die Umwandlung einiger stickstofffreier Reservestoffe während der Winterperiode der Bäume. Beih. bot. Zbl., Abt. 1 **19**, 68—117 (1906).

OKNINA, E. Z.: Über die Plasmodesmen ruhender Pflanzenzellen. Dokl. Akad. Nauk. SSSR. **62**, 705—708 (1948). — OKONUKI, K.: Über den Gaswechsel der Pollen. I. Acta Phytochemica (Tokio) **9**, 267—285 (1937). — OZOL, A. M., i A. A. LAZARENA: Der Zustand der Zellen in den Geweben der Triebe während der Winterruhe und Winterresistenz der Juglans-Arten. Dokl. Akad. Nauk. SSSR. **89**, 1111—1114 (1953).

PACK, D. A.: After-ripening and germination of Juniperus seeds. Bot. Gaz. **71**, 32—60 (1921). — PETROVSKAJA, T. P.: Die Veränderung der Nucleinsäuren in ruhenden Knospen. (Russisch.) Dokl. Akad. Nauk. SSSR., N.S. **99**, 475—478 (1954). — PIENIAZEK, I., u. I. WISNIEWSKAJA: The properties of the protoplasm in the tissues of one-year-old fruit-tree shoots in the course of different phenophases. Bull. Acad. Polon. Sci., Cl. II **1954**, 149—152. — PIRSON, A., u. E. GÖLLNER: Beobachtungen zur Entwicklungsphysiologie der Lemna minor L. Flora (Jena) **140**, 485—498 (1953). — PISEK, A.: Frosthärte und Zusammensetzung des Zellsaftes bei Rhododendron ferrugineum, Pinus cembra und Picea excelsa. Protoplasma **39**, 129—146 (1950). — PISEK, A., u. W. LARCHER: Zusammenhang zwischen Austrocknungsresistenz und Frosthärte bei Immergrünen. Protoplasma **44**, 30—46 (1954). — PITTIUS, G.: Über die stofflichen Grundlagen des osmotischen Druckes bei Hedera helix und Ilex Aquifolium. Bot. Archiv **37**, 43—64 (1934). — POHL, R., u. B. TEGETHOFF: Der Hemmstoff des Maisscutellums, ein Wuchsstoffinaktivator. Naturwiss. **36**, 319—320 (1949). — POLLOCK, B. M.: The respiration of acer buds in relation to the inception and termination of the winter rest. Physiol. Plantarum (Copenh.) **6**, 47—64 (1953).

RUGE, U., u. D. LIEDTKE: Zur periodischen Keimbereitschaft einiger Malvenarten. Ber. dtsch. bot. Ges. **64**, 141—150 (1951).

SACHER, J. H.: Structure and seasonal activity of the shoot apices of Pinus lambertiana and Pinus ponderosa. Amer. J. Bot. **41**, 749—759 (1954). — SAMISH, R. M.: Dormancy in woody plants. Plant Physiol. **5**, 183—204 (1954). — SATAROVA, N. A.: Einige Eigentümlichkeiten der Pflanzenzellen während der Sommerruhe. Dokl. Akad. Nauk. SSSR. **62**,

713—716 (1948). — Über einige Eigentümlichkeiten des Protoplasmas in den Zellen der Kartoffelknollen und in Wurzeln der Kautschuk-Pflanze während des Ruhezustandes. Trudy Inst. Fiziologii Rastenij im. Timirjazewa 7, 67—101 (1948). — SCARTH, G. W.: Cell physiological studies of frost resistance. A review. New Phytologist 43, 1—12 (1944). — SCHELLENBERG, H. C.: Zur Kenntnis der Winterruhe in den Zweigen einiger Hexenbesen. Ber. dtsch. bot. Ges. 33, 118—126 (1915). — SEILER, F.: Untersuchungen über Zygotenbildung und Zygotenkeimung bei Sporodinia grandis. Beih. bot. Zbl., Abt. A 54, 235—291 (1936). — SELL, H. M., and F. A. JOHNSTON jr.: Biochemical changes in terminal tung buds during their expansion prior to blossoming. Plant Physiol. 24, 744—752 (1949). — SHIBUYA, T.: The effect of auxin on the germination of seeds. J. Soc. Trop. Agric. 10, 270—273 (1938). — SHULL, C. A.: The oxygen minimum and the germination of Xanthium seeds. Bot. Gaz. 52, 453—477 (1911). — The rôle of oxygen in germination. Bot. Gaz. 57, 64—69 (1914). — SIMINOVITCH, D., and D. R. BRIGGS: The chemistry of the living bark of the black locust tree in relation to frost hardiness. I. Seasonal variation in protein content. Arch. of Biochem. 23, 8—17 (1949). — SIMON, S.: Untersuchungen über das Verhalten einiger Wachstumsfunktionen sowie der Atmungstätigkeit der Laubhölzer während der Ruheperiode. Jb. wiss. Bot. 43, 1—48 (1906). — SITNIKOVA, A. A.: Über den Zustand des Protoplasmas in den Samen der Pflanzen. Trudy Inst. Fiziologii Rastenij im. Timirjazewa 7, 109—114 (1950). — SMITH, B. C.: An investigation of the rest period in the seed of the genus Cotoneaster. Proc. Amer. Soc. Horticult. Sci. 57, 396—400 (1951). — SÖDING, H.: Wuchsstoff und Kambiumtätigkeit der Bäume. Jb. wiss. Bot. 84, 639—670 (1937). — SPÄTH, H. L.: Der Johannistrieb. Ein Beitrag zur Kenntnis der Periodizität und Jahresringbildung sommergrüner Gewächse. Diss. Berlin 1912. — SPAETH, J. N.: Dormancy in seeds of basswood, Tilia americana L. Amer. J. Bot. 19, 835—847 (1932). — STEINER, M.: Zum Chemismus der osmotischen Jahresschwankungen einiger immergrüner Holzgewächse. Jb. wiss. Bot. 78, 564—622 (1933). — Die Zusammensetzung des Zellsaftes bei höheren Pflanzen in ihrer ökologischen Bedeutung. Erg. Biol. 17, 151—254 (1939). — STEWARD, F. C., and S. M. CAPLIN: Investigations on growth and metabolism of plant cells. III. Evidence for growth inhibitors in certain mature tissues. Ann. of Bot., N. S. 16, 477—489 (1952). — STIER, H. L., and H. B. CORDNER: Germination of seeds of the potato as affected by temperature. Proc. Amer. Soc. Horticult. Sci. 34 (1936); 1937, 430—432. — STUDHALTER, R. A.: Tree growth. Bot. Review 21, 1—188 (1955).

THORNTON, N. C.: Factors influencing germination and development of dormancy in cocklebur seeds. Contrib. Boyce Thompson Inst. 7, 477—496 (1935). — Dormancy. In LOOMIS, Growth and differentiation in plants. Ames, Iowa: Iowa State Coll. Press 1953. — TODD, G. W.: Enzyme studies on dormant and active potato tubers. Physiol. Plantarum (Copenh.) 6, 169—186 (1953). — TRUCHAČEVA, K. P., i B. KEDROVSKIJ: Jahreszeitliche Veränderungen der Zellen der Kambialzone des Holunderstammes. Dokl. Akad. Nauk. SSSR., N. S. 78, 1211—1214 (1951).

ULMER, W.: Über den Jahresgang der Frosthärte einiger immergrüner Arten der alpinen Stufe sowie Zirbe und Fichte. Jb. wiss. Bot. 84, 553—592 (1937).

VISSER, T.: Germination and sforage of pollen. Meded. Landbouwhogschool Wageningen 55, 1—68 (1955). — VOLKENS, G.: Laubfall und Lauberneuerung in den Tropen. Berlin: Gebrüder Bornträger 1912.

WEBER, F.: Über eine neues Verfahren, Pflanzen zu treiben: Acetylenmethode. Sitzgsber. Akad. Wiss. Wien, Math.-naturwiss. Kl., Abt. 1 125 (3—4), 189—216 (1916a). — Studien über die Ruheperiode der Holzgewächse. Sitzgsber. Akad. Wiss. Wien, Math.-naturwiss. Kl., Abt. I 125 (5—6), 311—351 (1916b). — Über das Treiben der Buche. Ber. dtsch. bot. Ges. 34, 7—13 (1916c). — WEMMER, L.: Über die Ruheperiode der Zygosporen von Saprolegnia. Arch. Mikrobiol. 21, 217—229 (1954).

ZIMMERMANN, W. A.: Untersuchungen über die räumliche und zeitliche Verteilung des Wuchsstoffes bei Bäumen. Z. Bot. 30, 209—252 (1936).

Regulierende Faktoren des Aktivitätswechsels.

Von

E. Bünning.

Mit 4 Abbildungen.

A. Einleitung.

Der Aktivitätswechsel der Pflanzen kann durch viele äußere Faktoren induziert und reguliert werden. Wenn wir von solchen Regulationen sprechen, so denken wir in erster Linie an einen Aktivitätswechsel, bei dem endogene Komponenten beteiligt sind. Wir meinen also namentlich die Erscheinung der freiwilligen Ruhe oder des freiwilligen Wechsels von qualitativ verschiedenen Aktivitäten. Die Induktion einer nur unfreiwilligen Ruhe durch niedrigere Temperatur, durch Trockenheit usw. soll hier nicht so sehr interessieren[1].

B. Der Temperaturfaktor.

Es ist lange bekannt, daß die Ruheperiode durch den Temperaturwechsel reguliert werden kann. Dieser ist sogar der wichtigste Faktor bei solchen Regulationen. Die Temperatur kann sowohl für den Übergang von hoher zu niedriger Aktivität, als auch für den umgekehrten Prozeß wichtig sein. Dabei sind die kompliziertesten Beziehungen gefunden worden. Sowohl niedrige als auch hohe oder sogar extrem hohe Temperaturen und schließlich außerdem auch noch ein Wechsel von hoher und niedriger Temperatur kann als regulierender Faktor auftreten.

a) Wirkung hoher Temperaturen.

Die Wirkung hoher Temperatur ist lange bekannt. Besonders kann auf die Einführung des Warmbadverfahrens durch MOLISCH (1909a und b) hingewiesen werden. Die im Ruhezustand befindliche Pflanze bzw. einzelne Zweige werden in Wasser von 30—40° getaucht. Erforderlich ist eine Behandlung von mehreren Stunden. Bei dieser Behandlungsart ist wirklich die Temperatur selber ein entscheidender Faktor; es handelt sich nicht etwa einfach um eine Wirkung des Wassers; denn kaltes Wasser bleibt wirkungslos (JESENKO 1912). Aber fraglich ist, ob es sich bei dem Einfluß des Warmbades um einen reinen Temperatureffekt handelt; wir werden darauf zurückkommen. Für *Forsythia* liegen neue Angaben von DOORENBOS (1953b) vor. Die Zweige wurden für mehrere Stunden in Wasser von 30° getaucht und dann bei 15—20° weiter kultiviert. Diese Behandlung erwies sich als schädlich, wenn sie vor dem Eintritt der Winterruhe vorgenommen wurde. Erst kurz vor dem Blattfall, dessen Zeitpunkt hier etwa mit dem der tiefsten Ruhe übereinstimmt, war dieser schädliche Einfluß ganz verschwunden. Die frühtreibende Wirkung machte sich von diesem Zeitpunkt an bemerkbar. Nach etwa 1 Monat, d. h. mit zunehmender Vertiefung der Ruhe, wurde der Effekt geringer.

Auch hohe Lagerungstemperaturen können etwa bei Knollen, Zwiebeln usw. unter bestimmten Bedingungen günstig, d. h. abkürzend auf die Ruhe wirken,

[1] Vgl. Bd. III, S. 696: STOCKER, O., Die Dürreresistenz.

obwohl solche Organe im allgemeinen zur Erreichung dieses Effekts niedrige Temperaturen beanspruchen. Man vergleiche z. B. die Angaben von LOOMIS (1927) und ROSA (1928) für Kartoffeln; von LOOMIS (1934) für *Gladiolus*, von HARTSEMA und LUYTEN (1939) für *Freesia hybrida*. Weitere Hinweise finden sich bei VEGIS (1948). In diesen Fällen handelt es sich jedoch meist um optimale Temperaturen zwischen etwa 20 und 25°, und außerdem muß betont werden, daß diese hohen Temperaturen nur in bestimmten Stadien der Ruhe, in denen meist besondere Differenzierungsvorgänge ablaufen, günstig wirken.

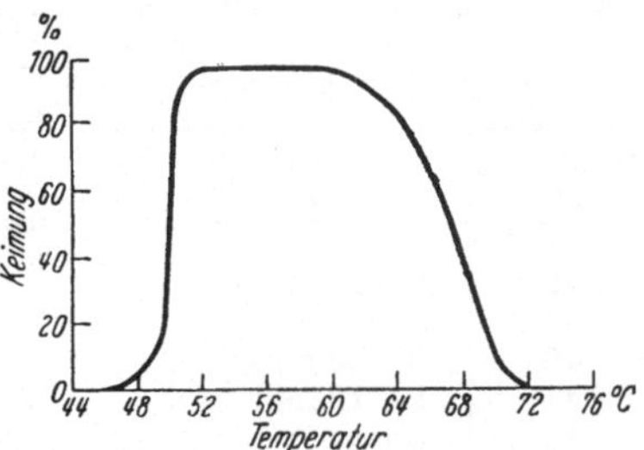

Abb. 1. Beispiel einer Entwicklungsanregung durch extrem hohe Temperatur. Keimung (Ordinate) der Sporen von *Neurospora* nach 20minutiger Erhitzung auf die in der Abszisse genannten Temperaturen. (Nach GODDARD.)

Auch bei Samen und Sporen kann eine vorübergehende Einwirkung hoher Temperatur die Ruheperiode abkürzen. Bei Ascosporen von *Neurospora crassa* beendet eine Erhitzung auf 52—60° die Ruhe (Abb. 1). Die erforderliche Einwirkungsdauer beträgt einige Minuten. Die Aktivierung kann zurückgehen und dann beliebig wiederholt werden (GODDARD). Dafür, daß diese Hitzewirkung auf einer Stimulation der Atmung beruht, spricht, daß ebenso wie durch hohe Temperatur auch durch einige chemische Agentien eine Aktivierung möglich ist und diese dann ebenfalls mit einer Atmungssteigerung verknüpft ist (SUSSMAN). So wirken namentlich bestimmte Furanderivate (EMERSON).

Auch für einige andere Pilze liegen entsprechende Angaben vor (DODGE 1912, GODDARD, GODDARD und SMITH, SHEAR und DODGE 1927, EVANS und CURRAN). Bei Samen müssen wir natürlich die allgemein günstige Wirkung relativ hoher Temperaturen auf den Keimprozeß (das Temperaturoptimum liegt meist höher als bei anderen Entwicklungsvorgängen) unterscheiden von der spezifischen Wirkung auf die Ruheperiode. Eine spezifische Wirkung auf die Ruhe hat z. B. RUGE bei *Malva verticillata* beobachtet (Abb. 2). Die Unterbrechung der Ruheperiode hat hier, wie wohl auch in anderen Fällen, noch einen weiteren Vorteil: Pflanzen, die vor völliger Beendigung der Ruhe keimen, entwickeln sich langsamer; oft bleibt die Entwicklung ganz stecken, und es bilden sich Zwergpflanzen, die erst weiterwachsen, wenn die Ruhe ganz beendet ist. Wir sehen also auch hier schon das später noch etwas ausführlicher zu behandelnde Phänomen, daß die Ruhe der Samen durchaus parallel läuft mit der Ruhe in den Knospen der aus ihnen entstandenen Pflanzen. Es ist das ein wichtiger Hinweis auf die physiologische Verwandtschaft bzw. Identität beider Phänomene.

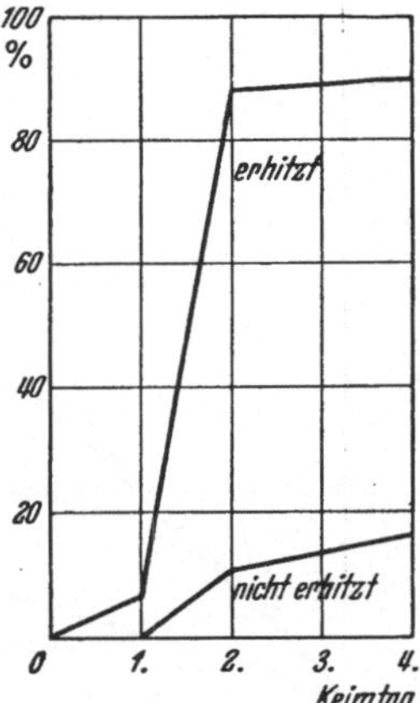

Abb. 2. Keimverlauf bei Samen der Futtermalve (*Malva verticillata*), die 2 Std auf 70° erhitzt waren, im Vergleich zum Keimverlauf nicht erhitzter Samen. (Nach RUGE und KRULL.)

Infolge dieser Erscheinung nun ist bei nicht mit Hitze behandeltem Saatgut die Entwicklung sehr ungleichmäßig. Wird die Ruhe aber durchbrochen, so befinden sich alle Pflanzen im gleichen Aktivitätsstadium, und es kann eine gleichmäßige und schnelle Entwicklung eintreten.

Wir haben hier zunächst nur von der Beendigung der Ruhe durch hohe Temperaturen gesprochen. Bevor wir auf die noch merkwürdigere Induktion einer Ruhe durch hohe Temperaturen eingehen, wollen wir versuchen, jene Ruhebrechung durch Hitze zu analysieren. Für die Einwirkung hoher Temperatur ist eine um so kürzere Zeit erforderlich, je höher die Temperatur ist. So waren für die Winterknospen von *Hydrocharis morsus-ranae* bei der Anwendung des Warmbades im Herbst zur Brechung der Ruhe die in Tabelle 1 angegebenen Badezeiten notwendig, um zu erreichen, daß die Knospen etwa eine Woche später keimten.

Tabelle 1. *Hydrocharis-Winterknospen, Treibwirkung der Temperatur.* (Nach VEGIS.)

Temperatur °C	Erforderliche Einwirkungsdauer
42	60—120 min
45	15—30 min
46	5—20 min
47	5—10 min
48	5 min
50	2 min oder weniger
52	30 sec bis 1 min
55	15 sec oder mehr

Die Temperaturwirkung auf die Beendigung der Ruheperiode zeigt also eine in diesem Ausmaß bei den meisten anderen physiologischen Prozessen nicht bekannte starke Temperaturabhängigkeit. Während bei den meisten Stoffwechselprozessen ein Temperaturkoeffizient zwischen 2 und 3 besteht, ist aus den links wiedergegebenen Zahlen ein Q_{10}-Wert von 58 abzulesen. So hohe Temperaturkoeffizienten sind charakteristisch für Koagulations- und Denaturierungsvorgänge von Eiweißsolen. Demnach kann es sich, auch schon wegen der kurzen Zeit, während der die Temperatureinwirkung erforderlich ist, nicht um eine direkte Beeinflussung des Stoffwechsels handeln, durch die das Warmbad in diesen Fällen zur Beendigung der Ruheperiode führt. Vielmehr dürften kolloidchemische Umwandlungen entscheidend sein, welche zur Folge haben, daß Enzyme freigesetzt werden und dadurch dann sekundär Stoffwechselprozesse beschleunigt werden können. So könnte man sich wohl die Beobachtung von GODDARD und SMITH deuten, nach der bei der erwähnten Hitzeaktivierung der Ascosporen von *Neurospora* eine Aktivierung von Carboxylase im Spiel ist (Abb. 3). Auch bei Samen sind unter dem Einfluß hoher Temperatur (2 Std Heißwasserbad bei 45°) Steigerungen der Katalaseaktivität sowie der Atmung festgestellt worden (LINSKENS).

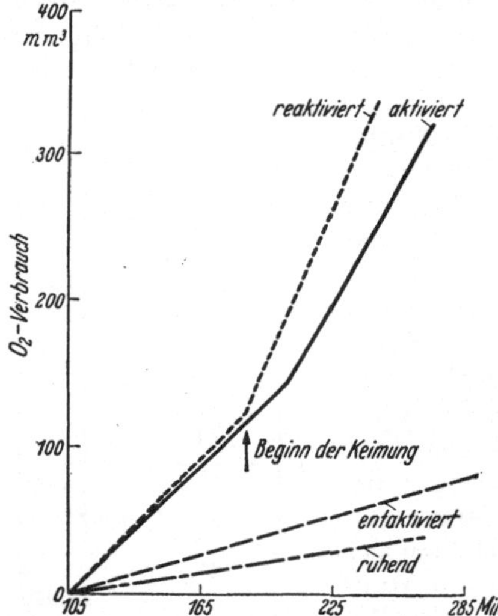

Abb. 3. Erklärung der Temperaturwirkung durch Atmungssteigerung. Atmung der Sporen von *Neurospora* (Kubikmillimeter O_2-Verbrauch), und zwar bei ruhenden Sporen, bei entaktivierten (erneute Rückkehr in den Ruhezustand durch vorübergehende Verhinderung der Atmung), aktivierten (Hitzebehandlung!) und erneut aktivierten (vgl. auch Abb. 1). (Nach GODDARD.)

In den Fällen, in denen eine längere Einwirkungsdauer bei nicht so extrem hoher Temperatur notwendig ist, also bei der Warmbadwirkung an Knospen usw., hat man an eine Förderung der intramolekularen Atmung durch den Sauerstoffmangel bei gleichzeitig hoher Temperatur gedacht. Der Stoffwechsel muß ja durch die hohe Temperatur beschleunigt werden, aber es fehlt der notwendige Sauerstoff zur normalen Atmung. So muß es zur Anhäufung von Produkten der intramolekularen Atmung, wie Acetaldehyd, Brenztraubensäure, Äthylalkohol usw. kommen. Solche Substanzen, die nach Warmbadbehandlung tatsächlich in den Geweben festgestellt werden konnten, können durchaus auch bei direkter Einwirkung frühtreibend wirken (KAKESITA).

Es ist also durchaus zweifelhaft, ob es sich in allen Fällen der Aktivierung durch hohe Temperaturen um ein gleichartiges Auslösungsprinzip handelt. Es scheint einmal Wirkungen zu geben, bei denen Temperaturen um 30—35° am günstigsten sind, nämlich beim Frühtreiben mit Hilfe des Warmbades nach dem MOLISCH-Verfahren. Bei den Turionen von *Hydrocharis* und *Stratiotes* kann aber durch einfaches Warmbad (30⁰) auch bei wochenlanger Einwirkung keine Brechung der Ruheperiode erreicht werden (VEGIS). Man könnte also wohl vermuten, daß es einerseits Wirkungen gibt, bei denen es auf die genannte Schädigung ankommt, also etwa auf die Koagulation oder Denaturierung im Protoplasma, und andererseits Wirkungen, bei denen Stoffwechselbeeinflussungen, namentlich die genannten Produkte der anaeroben Atmung im Spiel sind. Im erstgenannten Fall ist es ganz gleichgültig, ob feuchte oder trockene Hitze wirkt; bei den auf Warmbad ansprechenden Pflanzen aber scheint das Wasser doch auch mitzuwirken. Das Trockenbad ist in diesen Fällen, wie schon MOLISCH fand, wirkungslos (vgl. auch BORESCH 1924). Die vermittelnde Rolle einer begünstigten anaeroben Atmung, die wir schon andeuteten, ergibt sich aus den Angaben von BORESCH über die Möglichkeit, das Warmwasserbad durch Einwirkung von trockener Wärme bei verdünnter Luft zu ersetzen. Weiter fand WEBER (1916), daß auch durch Einwirkung einer Stickstoff-, Kohlendioxyd- oder Wasserstoffatmosphäre die Ruhe unterbrochen werden kann. Umgekehrt kann eine Zufuhr von Sauerstoff während des Warmbades dessen frühtreibende Wirkung verhindern (BORESCH 1928). Das Wasser ist also wohl als sauerstoffarmes Medium wichtig. Auch die erforderlichen Einwirkungsdauern sind hier ja größer als die unter Umständen nur wenige Sekunden notwendige Wirkung bei den Turionen von *Hydrocharis* und *Stratiotes* oder den erwähnten Samen und Sporen. Bei den Turionen, bei Samen und bei Sporen wirkt auch trockene Hitze ebensogut wie feuchte.

Wenn man von der Annahme ausgeht, daß das Wesen der Ruhe immer gleichartig ist, einerlei ob es sich um Samen, Sporen oder Knospen handelt, so wäre also immer ein bestimmter plasmatischer Zustand das entscheidende. Dann müßten alle Wirkungen ruhebrechender Faktoren darauf beruhen, daß dieser spezifische Zustand gewaltsam beseitigt wird. Das mag auf verschiedenem Wege geschehen. Stoffwechseländerungen und Erhöhungen der Fermentaktivitäten wären dann nicht als unmittelbare Ursache der Aktivitätssteigerungen anzusehen, sondern sie wären Ausdruck der gesteigerten Aktivität selber, also eine Folge des allein entscheidenden günstigeren Plasmazustandes. Hierfür spricht auch eine neue Veröffentlichung RUGEs. Bei *Malva* werden durch die keimfördernde Erhitzung auf 70⁰ nicht etwa keimungsfördernde Stoffe gebildet oder aktiviert; auch ein Abbau von Blastokolinen ist nicht im Spiel. Die Bindungskraft der Eiweißmoleküle für Wasser wird als entscheidend angesehen; diese ist, allerdings nur für Wasserdampf, nicht für flüssiges Wasser, bei erhitzten Samen größer. Als primäre Ursache wird eine Denaturierung von Samenproteinen angenommen.

Jetzt müssen wir noch auf die Induktion der Ruhe durch hohe Temperaturen eingehen. Dieses Phänomen hat z. B. VEGIS (1949) bei Winterknospen von *Stratiotes aloides* beobachtet. Diese Knospen befinden sich nach Beendigung der Ruheperiode noch in einer Zwangsruhe, während der sie sich durch einen sehr labilen Zustand auszeichnen. Die Aktivität kann dann durch hohe Temperaturen, die eigentlich nur wenige Stunden zu dauern brauchen, herabgesetzt werden. VEGIS spricht dabei von einer sekundären Ruhe. Bei *Hydrocharis morsus-ranae* begünstigt hohe Temperatur die Entstehung der Ruheknospen (VEGIS 1953). Auch Fälle einer Induktion von Sommerruheperioden durch hohe Temperatur sind beschrieben worden (LAUDE).

Auch für Samen liegen einige Angaben vor, die man ähnlich deuten könnte. DAVIS (1930a) beobachtete, daß Samen von *Ambrosia trifida* nach beendeter Nachreifung bei 30⁰ spärlich keimen. An *Xanthium* machte DAVIS (1930b) ähnliche Beobachtungen. Eine Rolle scheint hierbei zu spielen, daß der Sauerstoffzutritt bei diesen hohen Temperaturen nicht mehr der Stoffwechselsteigerung folgen kann, vielleicht also ein Überwiegen der intramolekularen Atmung eintritt. Auch für Samen von *Taraxacum megalorrhizon* liegen vergleichbare Angaben (Popcov 1935) vor. Weiterhin sei hierzu auf die Ausführungen bei Vegis (1949) verwiesen. Vegis sieht die hohe Temperatur allgemein als einen wichtigen Faktor für die Induktion der Ruhe an.

Kürzlich hat Vegis noch ausführlichere Untersuchungen an *Hydrocharis morsus-ranae* veröffentlicht. Hier ist hohe Temperatur für die Bildung der Ruheknospen entscheidend wichtig. Bei 10° wurden, einerlei wie groß die Tageslänge ist, nie Ruheknospen gebildet, vielmehr wachsen die Pflanzen unter Bildung nichtruhender Ausläuferknospen immer weiter. Die kritische Mindesttemperatur für die Entstehung der Ruheknospen liegt zwischen 10 und 15°. Bei 15, 20 und 25° werden Ruheknospen nur angelegt, wenn die tägliche Beleuchtungsdauer nicht zu lang ist. Vegis betrachtet Licht und Temperatur als antagonistisch wirkende Faktoren. Am einfachsten wird dieses Zusammenwirken von Temperatur und Licht durch Tabelle 2 (nach Vegis) gekennzeichnet.

Tabelle 2. *Hydrocharis morsus-ranae. Bildung von Ruheknospen bei verschiedenen Temperaturen und unterschiedlichen Tageslängen. Angegeben ist, wieviele von je 10 Pflanzen Ruheknospen anlegten.*

Tägliche Beleuchtungsdauer Stunden	10°	15°	20°	25°
0	—	—	—	—
3	—	—	—	—
6	—	10	10	10
9	0	10	10	10
12	0	10	10	10
15	0	10	10	10
18	0	5	10	10
21	0	0	4	10
24	0	0	0	0

Zweifellos handelt es sich bei den genannten Wirkungen hoher Temperatur nicht immer um ein und dasselbe. Es ist ja z. B. zu beachten, daß hohe Temperatur bei Samen etwa den Vorteil eines vorzeitigen Abbaus der Samenschale mit sich bringt, deren Undurchlässigkeit für Wasser und für Sauerstoff bei einigen Arten eine Zwangsruhe bedingt. Ferner ist zu beachten, daß hohe Temperatur in anderen Fällen einfach das Fehlen der direkt stimulierenden niedrigen Temperatur bedeutet, mit deren Einfluß wir uns jetzt näher beschäftigen müssen.

b) Wirkung niedriger Temperaturen.

Die stimulierende Wirkung niedriger Temperaturen auf Samen und besonders auch auf Knospen ist lange bekannt (vgl. Knight 1801). Nach dem gegenwärtigen Stand unserer Kenntnisse darf man wohl vermuten, daß es sich grundsätzlich jedenfalls sehr oft um etwas Gleichartiges handelt wie bei der Vernalisation. Diese Vernalisation selber soll uns hier natürlich nicht interessieren; aber es muß doch zur Ergänzung der Hinweise dieser Seiten auf den betreffenden Abschnitt dieses Handbuches verwiesen werden. Zu den Autoren, die die Brechung der Ruhe durch niedrige Temperaturen schon vor mehreren Jahrzehnten untersuchten, gehörten z. B. Pfeffer (1904) und Howard (1906, 1910, 1915a und b) (vgl. im übrigen die Hinweise bei Doorenbos 1953a). Mit Winterknospen von *Hydrocharis* arbeiteten Wiśniewski (1913) und Matsubara (1931), während die meisten übrigen Angaben sich auf Kartoffelknollen beziehen. Weiterhin liegen dann mehrere Mitteilungen über die Notwendigkeit niedriger Temperaturen für die Beendigung der Ruheperiode bei Nutzpflanzen vor. So hat etwa Gardner (1929) mit Obstbäumen gearbeitet (weitere Literaturhinweise s. bei Vegis 1948). Schon die älteren Angaben zu diesen Fragen lassen erkennen, daß meist Temperaturen zwischen 0 und 5 oder 0 und 10° am stärksten wirken. Diese Tatsachen deuten übrigens darauf hin, daß der gelegentlich unternommene Versuch, ein Durchfrieren der Gewebe als entscheidend anzusehen, in die falsche Richtung weist. Besonders hingewiesen sei hier noch auf die umfangreichen Untersuchungen von BLAAUW und Mitarbeitern; nach diesen Untersuchungen ist niedrige Lagerungstemperatur z. B. günstig für das spätere Streckungswachstum von Narzissen und von *Convallaria majalis* (HARTSEMA und LUYTEN 1933). Weitere Arbeiten aus dem genannten Arbeitskreis beziehen sich auf *Iris*-Knollen (BLAAUW, LUYTEN und HARTSEMA 1939). Weiterhin hat WHITE (1935) mit *Delphinium*, DENNY (1936) mit *Gladiolus*, KIPLINGER (1944) mit *Azalea*, HOWLAND (1945) mit *Daphne cneorum* gearbeitet. Für *Daphne cneorum*

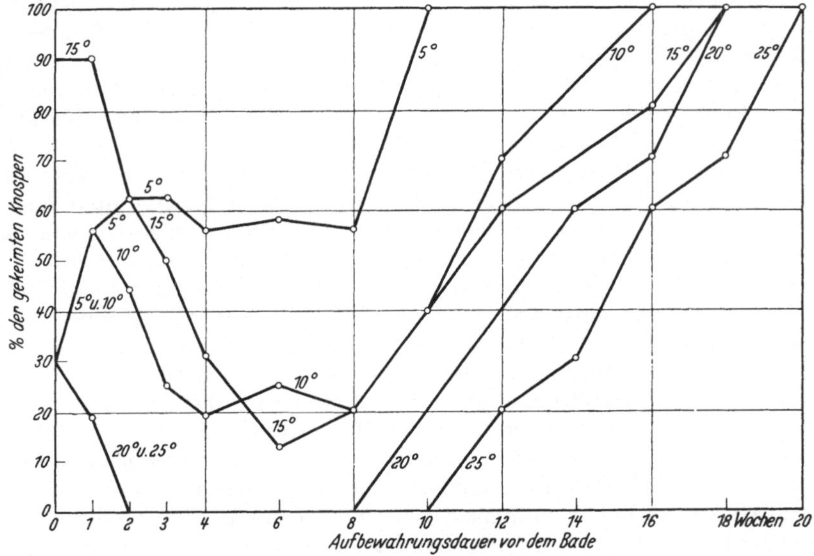

Abb. 4. Keimung von ununterbrochen bei 5°, 10°, 15°, 20° und 25° im Dunkeln verschieden lange aufbewahrten Winterknospen von *Stratiotes aloides*. Zur Einleitung der Keimung wurden die Knospen 1 min in Wasser von 50° gebadet. (Nach VEGIS.)

liegt die optimale Treibtemperatur der Blütenknospen bei 10° oder etwas darüber. *Camellia*-Blütenknospen (BONNER) verhalten sich ähnlich. Die erforderlichen Einwirkungsdauern der niedrigen Temperaturen schwanken zwischen einigen Stunden und mehreren Tagen oder sogar einigen Wochen.

Zu betonen ist, daß die Temperaturoptima zu verschiedenen Abschnitten der Ruhe ganz unterschiedlich sein können. Es kann sich sogar so verhalten, daß in einzelnen Abschnitten hohe, in anderen niedrige Temperatur günstig wirkt.

Wie stark und wichtig der Effekt ist, geht aus der lange bekannten Erscheinung hervor, daß manche Pflanzen unserer Breiten die Knospen kaum öffnen können, oder sogar wegen der fehlenden Entwicklungsbereitschaft der Knospen zugrunde gehen, wenn sie in den Tropen oder in Warmhäusern kultiviert werden (vgl. z. B. SIMON 1906, MOLISCH 1909, WEBER 1916). Es ist unmöglich, hier alle Einzelangaben über die Wirkung der niedrigen Temperaturen zusammenzustellen; man vergleiche hierzu die Literaturangaben bei VEGIS. VEGIS selber arbeitete in neuerer Zeit vor allem mit den Winterknospen von *Stratiotes aloides* (Abb. 4). Seine Versuche ergaben, daß die Knospenruhe um so länger dauert, je höher die Aufbewahrungstemperatur ist. Er hat diese Beziehungen quantitativ im einzelnen ausgearbeitet und diskutiert.

Ähnlich verhalten sich offenbar die Dauersprosse von *Spirodela polyrrhiza* (HENSSEN). Bei 5,5° wird die Ruhe in 2 Monaten aufgehoben, während sie bei 20° fast ein Jahr dauern kann.

Mehrere Angaben liegen auch für Sporen vor. Bei mehreren Pilzen und Algen benötigen Sporen eine vorübergehende Einwirkung niedriger Temperatur, um keimen zu können. Zum Beispiel haben die Sporen von *Oedegonium* normalerweise eine Ruheperiode von 12—14 Monaten; sie können aber schon nach 1 Monat zum Keimen gebracht werden, wenn sie wenige Tage auf einige Grad unter dem Nullpunkt abgekühlt werden (MAINX). Auch für Samen liegen einige Angaben vor. Beispielsweise benötigen Pfirsichsamen für ihre Keimung eine Behandlung mit niedriger Temperatur (LAMMERTS 1943). Weitere Angaben für Samen hat BARTON (1944) gemacht. Gelegentlich ist es so, daß ohne die Temperaturbehandlung wohl die Wurzelbildung erfolgt, aber die Epikotyle nach einem kurzen Wachstum steckenbleiben und erst weiterwachsen können, wenn die Kältebehandlung erfolgt ist. Auch für Samen hat man oft nach einfachen Erklärungen gesucht, etwa gemeint, daß die Zerstörung von Hemmstoffen oder die Mobilisierung von Reservestoffen des Endosperms wichtig sei. Aber dieser Erklärungsversuch, der entsprechend schon früher auch für die Brechung der Ruheperiode bei wasserreichen Knollen usw. vorgebracht worden ist, geht doch wohl am wesentlichen vorbei. Vor allem darf darauf hingewiesen werden, daß nach den Untersuchungen an Kartoffeln die niedrige Temperatur nicht auf die ganzen Reservegewebe einwirken muß, sondern nur auf die Knospen selber. Allerdings können wohl gerade bei Samen die Verhältnisse von Art zu Art recht verschieden sein. Bei *Heracleum sphondylium* zeigt sich während der zur Aktivierung notwendigen Kältebehandlung eine Vergrößerung der Embryonen. Vielleicht wird durch höhere Temperaturen die Mobilisierung der Reservestoffe des Endosperms verhindert. Tiefe Temperaturen hingegen beschleunigen diese Mobilisierung, namentlich den Eiweißabbau. Auch qualitativ unterscheidet sich der Eiweißabbau bei hohen und niedrigen Temperaturen. Bei niedrigen Temperaturen tritt mehr Glycin und Arginin, bei höheren mehr Alanin auf. Glycin und Arginin aber sollen für das Keimlingswachstum vorteilhaft sein (STOKES). Ob diesen Befunden eine allgemeine Bedeutung zukommt, muß offen bleiben.

Ein besonderer Komplex von Erscheinungen ist wohl der, der mit der Herstellung der Keimbereitschaft von Samen durch die sog. Stratifikation, also der Feuchtlagerung, zusammenhängt. Auch diese Stratifikationswirkung hat man früher mit dem begünstigenden Einfluß eines Gefrierens und Wiederauftauens in Zusammenhang gebracht. Späterhin hat sich jedoch gezeigt, daß auch hierbei Temperaturen oberhalb des Nullpunktes (etwa zwischen 0 und 15°) am wirksamsten sind. Das Gefrieren selber mag bei den Samen mancher Wasserpflanzen wichtig sein, wo die sehr undurchlässige Samenschale den Embryonen eine Zwangsruhe auferlegt und daher für die Keimung die vorhergehende Sprengung der Samenschale wichtig ist. Interessanter sind aber natürlich die Fälle, in denen die Ruhe in physiologischen Eigenschaften der ruhenden Embryonen selber begründet liegt. Werden solche Samen mit „freiwillig“ ruhenden Embryonen zur Keimung gezwungen (etwa auch durch chemische Beeinflussungen), so bilden sich zunächst nur Zwergpflanzen. Die Pflanzen können monatelang in diesem Stadium verharren, bis in einer Knospe durch Kältebehandlung die Ruhe gebrochen wird. Unter Umständen kann dieses Fortdauern der Ruhe sogar mehrere Jahre betragen. Der Zwergkeimling braucht also die gleiche Temperaturbehandlung wie der Same. Durchweg scheint dabei das Epikotyl das ruhende Organ zu sein, während die Wurzeln auch ohne die Kältebehandlung weiterwachsen können (FLEMION 1934). Die optimalen Temperaturen liegen verschieden,

z. B. für *Sorbus aucuparia* bei etwa 1°, für *Rhodotypus kerrioides* bei 5°. Nicht selten ist bei diesen Behandlungen eine wechselnde Temperatur günstiger als eine konstante (vgl. die ausführlichen Angaben bei CROCKER 1948).

Daß man als Sitz der Ruhe die Sproßspitzen ansehen muß, geht übrigens auch aus den Beobachtungen von FLEMION und WATERBURY hervor: Werden die Spitzen eines normalen Keimlings auf Sprosse von Zwergpflanzen des genannten Typs gepfropft, so kann die Pflanze jetzt normal wachsen.

Es sind sehr viele Arbeiten über die während der Kältebehandlung, d. h. besonders während der Stratifikation, von Samen eintretenden chemischen Veränderungen veröffentlicht worden. Nur wenige dieser Angaben können hier kurz angedeutet werden (im übrigen sei auf die Darstellung bei CROCKER verwiesen). Es kann während der Stratifikation eine Aciditätszunahme und eine Zunahme der Wasserbindungskraft auftreten. Die Aktivität von Katalase und Peroxydase kann sich erhöhen. Der Fettgehalt kann abnehmen und der Zuckergehalt stattdessen zunehmen. Die Menge der Aminosäuren kann steigen usw. (s. etwa ECKERSON, JONES, PACK, AFANASIEV und CRESS, FLEMION 1933).

Wie verwickelt die Beziehungen bei Samen liegen können, geht auch daraus hervor, daß manche Samen zwei Ruheperioden durchlaufen können, in denen niedere Temperatur wirken muß, während dazwischen ein Zeitabschnitt liegt, in dem hohe Temperatur gefordert wird (vgl. BARTON 1944).

c) Wirkung wechselnder Temperaturen.

Schließlich muß hier noch darauf hingewiesen werden, daß bei manchen Samen eine wechselnde Temperatur am günstigsten wirkt. Eine ausführliche Diskussion ist hier schon darum überflüssig, weil es sich zum großen Teil hierbei gar nicht um eine Regulierung des Aktivitätswechsels handelt, sondern einfach darum, die Keimung zu ermöglichen.

Eine Förderung der Keimung kann nicht schon notwendig als gleichbedeutend mit der Aufhebung der Ruhe betrachtet werden.

Zu der Frage dieses günstigen Einflusses wechselnder Temperaturen bei der Brechung der Ruheperiode von Samen finden sich Angaben bei ROBINSON und bei BÜNNING, CHAUDHRI und ZAIN UL ABIDIN.

C. Lichtwirkungen.

Über eine mögliche Beendigung der Ruheperiode von Knospen durch Licht hat schon JOST (1894) berichtet. Er fand nämlich, daß die Knospen von *Fagus silvatica* in dauernder Dunkelheit ruhen bleiben. Mit den Knospen der gleichen Art gelang es KLEBS (1914) zu zeigen, daß ein Dauerlicht schon geringer Intensität genügt, um die Ruheperiode zu brechen. So ähnlich verhalten sich auch die Knospen mehrerer anderer Pflanzen, z. B. die von *Populus* (VAN DER VEEN 1951a und b). Aber es gibt auch Knospen, bei denen dieses Treibmittel versagt, z. B. schon nach den Angaben von KLEBS (1914, 1917) die von *Quercus*.

Diese älteren Angaben sind vor allem insofern noch unbefriedigend, als einmal die Intensität des Lichtes nicht genauer beachtet wurde, außerdem aber die photoperiodischen Phänomene noch nicht bekannt waren. Nach den jetzt vorliegenden Erfahrungen würden wir sagen, daß viele der früher beschriebenen Einflüsse des Lichtes photoperiodischer Natur sind. Es ist nicht einmal sicher, ob es wirklich einen Gegensatz zwischen solchen Knospen gibt, deren Ruhe durch Licht gebrochen werden kann, und solchen, bei denen das Licht keinen Einfluß hat. Sehr wohl denkbar wäre auch, daß eben nur wie sonst bei photoperiodischen Phänomenen Unterschiede in der optimalen Länge der Tagesdauer bestehen. Hinweise auf eine Beziehung zur Tageslänge finden sich etwa bei KRAMER (1936) und GUSTAFSSON (1938). Letzterer konnte die Ruheperiode von *Pinus* im 16 Std-Tag brechen, wobei dieser Einfluß teilweise einen fehlenden Kälteeinfluß ersetzte.

In der Literatur sind auch immer wieder Angaben veröffentlicht worden über zufällige Beobachtungen hinsichtlich des Einflusses künstlicher Straßenbeleuchtung auf den Aktivitätswechsel der Straßenbäume. Tatsächlich kann man oft sehen, daß unter dem Einfluß dieses Lichtes der Laubfall verzögert wird. Jedoch darf man das nicht ohne weiteres als einen verzögerten Übergang zur Ruheperiode unter dem Einfluß des Lichtes werten. Einerseits bestehen hierbei ja auch Temperatureinflüsse, andererseits darf vor allem nicht übersehen werden, daß der Laubfall nicht identisch ist mit dem Beginn der Ruheperiode, sondern diese nur eine der für ihn (nicht einmal immer) notwendigen Voraussetzungen ist. Auch wenn man nach dem Beginn der Knospenöffnung den Aktivitätswechsel beurteilt, kommt man mindestens nicht zur Annahme eines starken Einflusses des genannten Kunstlichtes (vgl. MATZKE 1936 sowie auch WEBER 1916c).

Die neueren Untersuchungen berücksichtigen dann sorgfältiger die mögliche photoperiodische Natur dieser Lichtwirkung, allerdings wird auch dabei nicht immer unterschieden zwischen dem Einfluß auf den Laubfall und dem Einfluß auf die eigentliche Ruheperiode. DAUBENMIRE (1949) schloß auf Grund der Untersuchung einer größeren Anzahl von Laub- und Nadelhölzern, speziell auf Grund der Registrierung des Dickenwachstums, daß der Holzzuwachs ziemlich unabhängig von der Temperatur an ein bestimmtes Datum, d. h. offenbar an eine bestimmte Tageslänge gebunden ist. Jedenfalls gilt diese Regel für sehr viele Arten, neben denen es offenbar auch andere (nach DAUBENMIRE z. B. *Robinia*) gibt, bei denen der Temperaturanstieg für die Aktivierung ausschlaggebend ist. Aber solche Schlußfolgerungen lassen sich auf Grund von Freilandbeobachtungen mit den viel zu komplizierten Bedingungen doch noch nicht rechtfertigen. OLMSTED fand bei *Acer saccharinum* einen verzögerten Blattfall im 16 Std-Tag und einen beschleunigten im 8 Std-Tag, verglichen mit dem Blattfall im 12 Std-Tag. JESTER und KRAMER beobachteten, daß *Acer rubrum* seine Blätter um so länger behält, je länger die Photoperiode ist. Allerdings ist hierbei wieder zu berücksichtigen, daß der Vorgang des Blattalterns nicht identisch ist mit der Herstellung der Ruhe. Zweifellos hat jedenfalls die Tageslänge einen sehr starken Einfluß auf die Blattalterung. BÜNNING und KONDER fanden, daß bei mehreren Pflanzen die Blattalterung unter Langtagbedingungen schneller erfolgt als unter Kurztagbedingungen, und zwar sowohl bei Langtagpflanzen (etwa *Plantago lanceolata*) wie bei Kurztagpflanzen (*Amaranthus caudatus* und *Perilla ocymoides*). Ausführliche Untersuchungen über den Einfluß der Tageslänge auf die Ruheperiode selber unternahm VAN DER VEEN (1951) bei *Populus*. Die Stecklinge mehrerer *Populus*-Arten wurden im Alter von einigen Jahren unter verschiedenen Tageslängen bei künstlichem Licht kultiviert. Im 9- und 12 Std-Tag hörte das Wachstum nach einigen Wochen auf; es bildeten sich dann Ruheknospen. Dagegen wurde das Wachstum im Langtag und zwar am besten im kontinuierlichen Licht fortgesetzt. Wurden die Pflanzen dann nach der Langtagbehandlung niedriger Temperatur und Kurztagbedingungen ausgesetzt, so behielten sie auffälligerweise ihre Blätter selbst nach mehrmonatiger Fortsetzung des Versuchs. Stecklinge von *Populus robusta* konnten ihr Wachstum unter geeigneten Lichtbedingungen sehr lange Zeit fortsetzen (am besten im kontinuierlichen Licht), ohne daß sie ihre Bätter abwarfen. Allerdings ist bei der Beurteilung auch dieser Ergebnisse eine gewisse Vorsicht geboten. Wir wissen nämlich, daß bei jugendlichen Pflanzen der Aktivitätswechsel viel häufiger und in stärkerem Maße ein passiver ist als bei älteren Pflanzen.

Sodann sind hier die ausführlichen Untersuchungen von WAREING an *Pinus silvestris* und *Fagus silvatica* zu erwähnen. *Pinus silvestris* zeigte im 10 Std-Tag (verglichen mit dem Verhalten im 15 Std-Tag) ein frühes Ende des Sproßwachstums und ebenso eine frühere Beendigung des Längenwachstums der Blätter. Ebenfalls wurde die Cambiumtätigkeit, d. h. das embryonale Wachstum, früher unterbrochen. Der Einfluß auf das Cambium wurde nur beim Übergang zur Ruhe festgestellt, dagegen konnte die Aktivierung des Cambiums im Frühjahr durch Kurztagbedingungen nicht unterbunden werden. WAREING schließt, daß der Übergang zur Winterruhe durch die abnehmende Tageslänge bedingt sei. Die Wiederaufnahme der Cambiumtätigkeit im Frühjahr hingegen wird offenbar durch die Knospenentfaltung bedingt (WAREING 1951). Man kann nicht durch Langtagbedingungen das Cambium direkt aktivieren.

Bei *Fagus silvatica* fand WAREING (1953), daß die Beendigung der Ruheperiode durch Licht eine photoperiodische Reaktion ist. Die Kältebehandlung der Knospen ist hier nicht notwendig. Bei Tageslängen unter 12 Std bleiben die Knospen ruhen, während die Ruheperiode durch Langtagbedingungen unterbrochen wird. Es genügen hierzu wie bei anderen photoperiodischen Reaktionen geringe Beleuchtungsstärken. Auch bei einigen *Rhododendron*-Arten kann die Ruhe durch Langtage gebrochen werden (DOORENBOS 1955). Kontinuierliches Wachstum wird dabei aber nicht erreicht, sondern nur eine Abkürzung der für einen Entwicklungscyclus erforderlichen Zeit auf etwa die Hälfte.

Bei *Hydrocharis morsus-ranae* wird die Bildung von Ruheknospen durch Langtagsbedingungen gehemmt (VEGIS 1955).

Jedenfalls ist also bei einigen Pflanzen eine Regulierung des jahresperiodischen Aktivitätswechsels auch durch die Tageslänge möglich (vgl. hierzu auch noch VEGIS 1953, 1955 und REINDERS-GOUWENTAK und SIPKENS).

Bemerkenswert ist, daß auch die Brechung der Ruheperiode bei Samen durch Licht eine photoperiodische Reaktion sein kann (BLACK und WAREING 1954, 1955, ISIKAWA, BÜNNING, CHAUDHRI und ZAIN UL ABIDIN). Die Samen verschiedener Arten verhalten sich dabei gegenüber den Tageslängen überaus unterschiedlich. Ein weiteres Eingehen auf diesen Effekt erübrigt sich an dieser Stelle, weil die Phänomene besser im Zusammenhang mit den allgemeinen Problemen der Licht- und Dunkelkeimung an anderer Stelle dieses Handbuchs beschrieben werden.

D. Chemische Wirkungen.

Über chemische Einflüsse auf den Aktivitätswechsel haben wir schon bei der Besprechung des Mechanismus physiologisch selbstgesteuerter Aktivitätsrhythmen hingewiesen. Die dort angeschnittenen Teilfragen brauchen hier nicht noch einmal berücksichtigt zu werden.

Eine chemische Beeinflussung des Aktivitätswechsels, und zwar besonders eine Beendigung der Ruheperiode, ist schon vor sehr langer Zeit gefunden worden. Man denke etwa an die Einführung des Frühtreibens durch das Ätherverfahren (JOHANNSEN 1900, 1901, vgl. auch die weiteren Literaturhinweise bei DOORENBOS). Auch einige andere Substanzen sind schon vor längerer Zeit als wirksame Treibmittel gefunden worden, so z. B. Acetylen und in gewissem Maße auch Wasserstoffperoxyd (WEBER 1916).

Auch Cyankali ist früher als ein geeignetes Treibmittel befunden worden (GASSNER und HASSEBRAUK). MATSUBARA konnte Winterknospen von *Hydrocharis* durch Blausäure zum Austreiben veranlassen. Bei *Myriophyllum* hatten RINGEL-SUESSENGUTH entsprechende Erfolge. Neuere Angaben liegen vor von HENSSEN für *Spirodela polyrrhiza*.

Erst in neuerer Zeit aber sind mehr systematische Untersuchungen durchgeführt worden, bei denen überaus wirksame Treibmittel gefunden worden sind. Vor allem kann hier auf die Anwendung von Äthylenchlorhydrin (DENNY 1928) verwiesen werden. Viele dieser Untersuchungen, die im Anschluß an die Entdeckung DENNYS durchgeführt wurden, beschäftigen sich mit der Ruheperiode von Kartoffeln (vgl. etwa MILLER, GUTHRIE). Außer Äthylenchlorhydrin wurde eine ganze Reihe anderer Substanzen, von denen nur Thioharnstoff erwähnt sei, als wirksam gefunden.

Die von einigen Autoren angenommene große Bedeutung des Glutathions für die Regulierung der Ruheperiode könnte man mit der Treibwirkung des Äthylenchlorhydrins in Zusammenhang bringen; denn durch Behandlung mit diesem Treibmittel ließ sich eine Bildung von Glutathion in ruhenden Knollen nachweisen (GUTHRIE 1933). Die Fähigkeit des Glutathions, Ruheperioden zu brechen, ist wiederholt beschrieben worden (GUTHRIE 1940). Nach HEMBERGS (1950) Beobachtungen könnte man vermuten, daß Glutathion das Verschwinden von Hemmstoffen ermöglicht. — Die Wirkung des Glutathions auf Enzyme ist bekannt. Weitere Untersuchungen wurden von den genannten Autoren besonders noch an *Gladiolus*-Rhizomen durchgeführt, außerdem an den Knospen mehrerer Bäume und Sträucher. Es liegen auch zahlreiche Angaben vor über die Art der Wirkung dieser Substanzen. Gefunden wurde eine Aktivierung mehrerer Fermente, eine Zunahme von löslichen Stickstoffverbindungen und eine Aktivierung der Atmung. Alle diese Stoffwechseländerungen hier aufzuzählen, ist unmöglich. Es kann hierzu auf die ausführliche Darstellung bei CROCKER (1948) verwiesen werden.

Ein wichtiger Faktor bei der Regulierung der Aktivität kann auch die Nährstoffversorgung sein. HENSSEN fand, daß die Dauersproßbildung von *Spirodela polyrrhiza* jederzeit durch Nährsalzmangel hervorgerufen werden kann. Ähnliche Angaben hatten schon früher für *Myriophyllum* und *Utricularia* GOEBEL (1891) sowie RINGEL-SUESSENGUTH (1922) gemacht.

Für die chemische Aktivierung von Pilzsporen sind nach EMERSON Verbindungen und ungesättigte Ringsysteme wichtig, z. B. Furfurol und andere Furanabkömmlinge.

E. Wundwirkungen, mechanische Wirkungen.

Zur Vervollständigung sei hier noch darauf hingewiesen, daß auch eine Verwundung in manchen Fällen die Ruheperiode abbricht.

Bemerkenswert ist noch, daß auch eine im bloßen Schütteln bestehende mechanische Reizung nach den Untersuchungen von SCHULZE an Kartoffeln die Ruhe beenden kann. Es kommt dabei anscheinend auf eine Zerstörung der Korkschicht der Knollen an.

Literatur.

AFANASIEV, M., and M. CRESS: Producing seedlings of eastern red cedar. (Juniperus virginiana L.) Okla. Agric. Exper. Sta. Bull. **256** (1942).

BARTON, L. V.: Some seeds showing special dormancy. Contrib. Boyce Thompson Inst. **13**, 259—271 (1944). — BLAAUW, A. H., I. LUYTEN u. A. M. HARTSEMA: Snelle bloei van Hollandsche Irissen. II. Proc. Kon. Akad. Wetensch. Amsterdam **42** (1), 13—22 (1939).—

Black, M., and P. F. Wareing: Photoperiodic control of germination in seed of birch (Betula pubescens, Ehrh.). Nature (Lond.) **174**, 705 (1954). — Growth studies in woody species. VII. Photoperiodic control of germination in Betula pubescens Ehrh. Physiol. Plantarum (Copenh.) **8**, 300—316 (1955). — Bonner, J.: Flower bud initiation and flower opening in the Camellia. Proc. Amer. Soc. Horticult. Sci. **50**, 401—408 (1947).— Boresch, K.: Zur Analyse der frühtreibenden Wirkung des Warmbades. Biochem. Z. **153**, 313—334 (1924). — Zur Biochemie der frühtreibenden Wirkung des Warmbades. III. Biochem. Z. **202**, 180—201 (1928). — Bünning, E., I. I. Chaudhri u. Zain ul Abidin: Die Beziehung von Photo- und Thermoperiodismus bei der Samenkeimung zur endogenen Tagesrhythmik. Ber. dtsch. bot. Ges. **68**, 41—45 (1955). — Bünning, E., u. M. Konder: Tageslänge, Blattwachstum und Blütenbildung. Planta (Berl.) **44**, 9—17 (1954).

Crocker, W.: Growth of plants. New York: Reinhold Publ. Comp. 1948.

Daubenmire, R. F.: Relation of temperature and daylength to the inception of tree growth in spring. Bot. Gaz. **110**, 464—475 (1949). — Davis, W. E.: Primary dormancy, after-ripening, and the development of secondary dormancy in embryos of Ambrosia trifida. Amer. J. Bot. **17**, 58—76 (1930a). — The development of dormancy in seeds of cocklebur (Xanthium). Amer. J. Bot. **17**, 77—87 (1930b). — Denny, F. E.: Storage temperatures for shortening the rest period of gladiolus corms. Contrib. Boyce Thompson Inst. 8, 137—140 (1936). — Dodge, B. O.: Methods of culture and the morphology of the archicarp in certain species of the Ascobolaceae. Bull. Torrey Bot. Club **39**, 139—197 (1912). — Doorenbos, J.: Orienterend onderzoek over het forceren van Forsythia en Rhododendron. Meded. Directeur Tuinbow **16**, 533—543 (1953). — Review of the literature on dormancy in buds of woody plants. Meded. Landbouwhoogeschool Wageningen **53**, 1—24 (1953). — Shortening the breeding cycle of Rhododendron. Euphytica **4**, 141—146 (1955).

Eckerson, S.: A physiological and chemical study of after-ripening. Bot. Gaz. **55**, 286—299 (1913). — Emerson, M. R.: Some physiological characteristics of ascospore activation in Neurospora crassa. Plant Physiol. **29** (5), 418—428 (1954). — Evans, F. R., and H. R. Curran: The accelerating effect of sublethal heat on spore germination in mesophilic areobic bacteria. J. Bacter. **46**, 513—523 (1943).

Flemion, F.: Physiological and chemical studies of after-ripening of Rhodotypos kerrioides seeds. Contrib. Boyce Thompson Inst. **5**, 143—159 (1933). — Dwarf seedlings from non-after-ripened embryos of Rhodotypos kerrioides. Contrib. Boyce Thompson Inst. **5**, 161—165 (1933). — Physiological and chemical changes preceding and during the after-ripening of Symphoricarpos racemosus seeds. Contrib. Boyce Thompson Inst. **6**, 91—102 (1934). — Dwarf seedlings from non-after-ripened embryos of peach, apple, and hawthorn. Contrib. Boyce Thompson Inst. **6**, 205—209 (1934). — Flemion, F., and E. Waterbury: Further studies with dwarf seedlings of non-after-ripened peach seeds. Contrib. Boyce Thompson Inst. **13**, 415—422 (1945).

Gardner, F. E.: Composition and growth initiation of dormant bartlett pear shoots as influenced by temperature. Plant Physiol. **4**, 405—434 (1929). — Gassner, G.: Blausäurebehandlung als Stimulationsmittel im praktischen Pflanzenbau. Angew. Bot. **7**, 74—79 (1925). — Gassner, G., u. K. Hassebrauk: Blausäurebegasungen als Mittel zu schneller Erzielung voller Keimreife. Pflanzenbau **4**, 1—3 (1927). — Goddard, D. R.: The reversible heat activation inducing germination and increased respiration in the ascospores of Neurospora sitophila. J. Gen. Physiol. **19**, 45—60 (1935). — The reversible heat activation of respiration in Neurospora. Cold Spring Harbor Symp. Quant. Biol. **7**, 362—376 (1939). — Goddard, D. R., and P. E. Smith: Respiratory block in the dormant spores of Neurospora tetrasperma. Plant Physiol. **13**, 241—264 (1938). — Goebel, K.: Pflanzenbiologische Schilderungen, Teil II, Kap. IV. 1891. — Gustafson, F. G.: Influence of the length of day on the dormancy of tree seedlings. Plant Physiol. **13**, 655—658 (1938). — Guthrie, J. D.: Change in the glutathione content of potato tubers treated with chemicals that break the rest period. Contrib. Boyce Thompson Inst. **5**, 331—350 (1933). — Rôle of glutathione in the breaking of the rest period of buds by ethylene chlorhydrine. Contrib. Boyce Thompson Inst. **11**, 261—270 (1940). — Sprays that break the rest period of peach buds. Contrib. Boyce Thompson Inst. **12**, 45—47 (1941a). — A preparation from yeast that is active in breaking the rest period of buds. Contrib. Boyce Thompson Inst. **12**, 195—201 (1941b).

Hartsema, A. M., u. I. Luyten: De invloed van lage temperaturen op het snelle strekken en bloeien van Convallaria majalis. I en II. Proc. Kon. Akad. Wetensch. Amsterdam **36**, 120—127, 210—216 (1933). — Snelle bloei van de Narcis (Narcissus Pseudonarcissus var. King Alfred). I en II. Proc. Kon. Akad. Wetensch. Amsterdam **41**, 651—660, 800—809 (1933). — Proeven over het uitloopen van de knollen en het vervroegen van den bloei bij Freesia hybriden. Proc. Kon. Akad. Wetensch. Amsterdam **42**, 438—445 (1939). — Hassebrauk, K.: Über den Einfluß der Blausäure auf die Keimreife der Samen. Angew. Bot. **10**, 407—468 (1928). — Hemberg, T.: The effect of glutathione on the growth-inhibiting substances in resting potato tubers. Physiol. Plantarum (Copenh.) 3, 17—21 (1950). — HENSSEN, A.: Die Dauerorgane von Spirodela polyrrhiza (L.). Schleid. in physiologischer Betrachtung. Flora (Jena) **141**, 523—566 (1954). — HOWARD, W. L.: Untersuchung über die Winterruheperiode der Pflanzen. Diss. Halle-Wittenberg 1906. — An experimental study of the rest period in plants. The winter rest. Mo. Sta. Res. Bull. 1 (1910). — An experimental study of the rest period in plants. Pot grown woody plants. Mo. Sta. Res. Bull. **16** (1915). — An experimental study of the rest period in plants. Fifth report. Mo. Agricult. Exper. Sta. Res. Bull. **21** (1915). — HOWLAND, J. E.: Daphne cneorum — a good novelty pot plant for easter. Florists Exchange **104** (2), 14, 21 (1945).

ISIKAWA, S.: Light-sensitivity against the germination. I. "Photoperiodism of seeds". Bot. Mag. (Tokyo) **67**, 51—56 (1954).

JESENKO, FR.: Über das Austreiben im Sommer entblätterter Bäume und Sträucher. Ber. dtsch. bot. Ges. **30**, 226—232 (1912). — JESTER, J. R., and P. J. KRAMER: The effect of length of day on the height growth of certain forest tree seedlings. J. Forestry **37**, 796 (1939). — JOHANNSEN, W.: Mein Ätherverfahren in der Praxis. Gartenwelt **5** (1900/01). — Das Ätherverfahren beim Frühtreiben mit besonderer Berücksichtigung der Fliedertreiberei. Jena 1900. — JONES, H. A.: Physiological study of maple seeds. Bot. Gaz. **69**, 128—152 (1920). — JOST, L.: Einfluß des Lichtes auf das Knospentreiben der Buche. Ber. dtsch. bot. Ges. **12**, 188—197 (1894).

KAKESITA, K.: Experimental studies on regeneration in Bryophyllum calycinum. Jap. J. of Bot. **5**, 219—252 (1930). — KIPLINGER, D. C.: The effects of cooling and shading the Azalea Coral Bellis on early forcing for Christmas Flowering. Proc. Amer. Soc. Horticult. Sci. **44**, 542—544 (1944). — KLEBS, G.: Über das Treiben der einheimischen Bäume, speziell der Buche. Abh. Heidelbg. Akad. Wiss., Math.-naturwiss. Kl., 3. Abh. **1914**. — Über das Verhältnis von Wachstum und Ruhe bei den Pflanzen. Biol. Zbl. **37**, 373—415 (1917. — KRAMER, P. J.: Effect of variation in length of day on growth and dormancy of trees. Plant Physiol. **11**, 127 (1936).

LAMMERTS, W. E.: Effects of photoperiod and temperature on growth of embryocultured peach seedlings. Amer. J. Bot. **30**, 707—711 (1943). — LINSKENS, H. F.: Untersuchungen über die Änderung des physiologischen Verhaltens von Weizen- und Gerstensamen nach Heißwasser-Bädern. Züchter **20**, 168—187 (1950). — LOOMIS, W. E.: Temperature and other factors affecting the rest period of potato tubers. Plant Physiol. **2**, 287—302 (1927). — Forcing Gladiolus. Proc. Amer. Soc. Horticult. Sci. **30**, 585—588 (1934).

MATSUBARA, M.: Versuche über die Entwicklungserregung der Winterknospen von Hydrocharis morsus ranae. Planta (Berl.) **13**, 695—715 (1931). — MATZKE, E. B.: The effect of street lights in delaying leaf-fall in certain trees. Amer. J. Bot. **23**, 446 (1936). — MILLER, L. P.: Effect of sulphur compounds in breaking the dormancy of potato tubers and in inducing changes in the enzyme activities of the treated tubers. Contrib. Boyce Thompson Inst. **5**, 29—81 (1933). — Effect of various chemicals on the sugar content, respiratory rate, and dormancy of potato tubers. Contrib. Boyce Thompson Inst. **5**, 213—234 (1933). — MOLISCH, H.: Über ein einfaches Verfahren, Pflanzen zu treiben. Sitzgsber. Akad. Wiss. Wien, Math.-naturwiss. Kl., Abt. I **117**, 87—116 (1908); **118**, 637—690 (1909). — Das Warmbad als Mittel zum Treiben der Pflanzen. Jena: Gustav Fischer 1909. — Warmbad und Pflanzentreiberei. Österr. Garten-Z. **4**, 17—23, 50—59 (1909).

OLMSTED, CH. E.: Experiments on photoperiodism, dormancy, and leaf age and abscission in sugar maple. Contributions from the Hull Botanical Laboratory 623. Bot. Gaz. **112**, 365—393 (1951).

PACK, D. A.: After-ripening and germination of Juniperus seeds. Bot. Gaz. **71**, 32—60 (1921). — PFEFFER, W.: Pflanzenphysiologie, 2. Aufl., Bd. II. Leipzig: Wilhelm Engelmann 1904. — POPCOV, A. V.: Note on secondary dormancy of Taraxacum megalorrhizon seeds. C. R. (Doklady) Acad. Sci. USSR. **2**, Nr 8—9 (1935).

REINDERS-GOUWENTAK, C. A., and J. SIPKENS: The influence of photoperiod, dormancy breaking and growth hormone treatment on poplar cuttings. Proc. Kon. Ned. Akad. v. Wetensch., Ser. C **56** (1), 71—80 (1953). — RINGEL-SUESSENGUTH, M.: Über Ruheorgane bei einigen Wasserpflanzen und Lebermoosen. Flora (Jena) **115**, 27—58 (1922). — ROBINSON, R. W.: Seed germination problems in the Umbelliferae. Bot. Review **20** (9), 531—550 (1954). — ROSA, J. T.: Relation of tuber maturity and of storage factors to potato dormancy. Hilgardia **3**, 99—124 (1928). — RUGE, U.: Über die Beeinflussung der Entwicklung der Futtermalve durch Erhitzen des lufttrockenen Saatgutes. Z. Acker- u. Pflanzenbau **95** (1), 73—90 (1952). — Zur Analyse der Saatguterhitzung. von Malvaceen, Beitr. Biol. Pflanz. **31**, 409—417 (1955).

SCHULZE, E.: Mechanische Keimanregung, Schosserbildung und photoperiodisches Verhalten bei Kartoffeln. Z. Acker- u. Pflanzenbau **98**, 385—422 (1954). — SHEAR, C. L., and B. O. DODGE: Life histories and heterothallism of the red bread-mold fungi of the Monilia sitophila group. J. Agricult. Res. **34**, 1019—1041 (1927). — SIMON, S.: Untersuchungen über das Verhalten einiger Wachstumsfunktionen sowie der Atmungstätigkeit der Laubhölzer während der Ruheperiode. Jb. wiss. Bot. **43**, 1—48 (1906). — STOKES, P.: A physiological study of embryo development in Heracleum sphondylium L. III. The effect of temperature on metabolism. Ann. of Bot., N. S. **17**, 157—173 (1953). — The stimulation of growth by low temperature in embryos of Heracleum sphondylium L. J. of Exper. Bot. **4**, 222—234 (1953). — SUSSMAN, A.: The effect of furfural upon the germination and respiration of Ascospores of Neurospora tetrasperma. Amer. J. Bot. **40**, 401—404 (1953).

VEEN, R. VAN DER: Influence of daylength on the dormancy of some species of the genus Populus. Physiol. Plantarum (Copenh.) **4**, 35—40 (1951). — VEGIS, A.: Einfluß tagesperiodisch wechselnder Aufbewahrungstemperatur auf die Streckungsbereitschaft der ruhenden Winterknospen von Stratiotes aloides. Physiol. Plantarum (Copenh.) **1**, 216—235 (1948). — Über den Einfluß der Aufbewahrungstemperatur auf die Dauer der Ruheperiode und die Streckungsbereitschaft der ruhenden Winterknospen von Stratiotes aloides. Symp. bot. Upsalienses **10**, 2 (1948). — Durch hohe Temperaturen bedingter Wiedereintritt des Ruhezustandes bei den Winterknospen. Sv. bot. Tidskr. **43** (2—3), 671—714 (1949). — The significance of temperature and the daily light-dark period in the formation of resting buds. Experientia (Basel) **9**, 462—463 (1953). — Über den Einfluß der Temperatur und der täglichen Licht-Dunkel-Periode auf die Bildung der Ruheknospen. Zugleich ein Beitrag zur Entstehung des Ruhezustandes. Symp. bot. Upsalienses **14**, 1—175 (1955). — VEGIS, A., u. B. VEGIS: Versuche über die frühtreibende Wirkung der Wasserbäder von 35—100° C auf die Turionen von Stratiotes aloides L. Acta Horti Bot. 8, 59—100 (1935).

WAREING, P. E.: Photoperiodic control of leaf growth and cambial activity in Pinus silvestris. Nature (Lond.) **163**, 770—771 (1949). — Growth studies in woody species. III. Further photoperiodic effects in Pinus silvestris. Physiol. Plantarum (Copenh.) **4**, 41—56 (1951). — Growth studies in woody species. V. Photoperiodism in dormant buds of Fagus silvatica L. Physiol. Plantarum (Copenh.) **6**, 692—706 (1953). — WEBER, F.: Über die Abkürzung der Ruheperiode der Holzgewächse durch Verletzung der Knospen, beziehungsweise Injektion derselben mit Wasser (Verletzungsmethode). Sitzgsber. Akad. Wiss. Wien, Math.-naturwiss. Kl., Abt. I **120**, 179—194 (1911). — Studien über die Ruheperiode der Holzgewächse. Sitzgsber. Akad. Wiss. Wien, Math.-naturwiss. Kl., Abt. I **125** (5—6), 311 bis 351 (1916). — WHITE, H. E.: Some studies on the forcing of Delphinium. Proc. Amer. Soc. Horticult. Sci. **33**, 653—654 (1935). — WIŚNIEWSKI, P.: Wplyw niskiej temperatury na przyśpieszenie kielkowania paczkow zimówych zabiścieku (Hydrocharis morsus ranae L.). Kosmos (Lwów) **38**, 1376—1384 (1913).

Nachtrag zu C (Seite 872).

In einer Arbeit von DOWNS und BORTHWICK wird gezeigt, daß im allgemeinen Kurztagbedingungen bei Bäumen die Ruhe induzieren. Aber nur einige Arten können im Langtag in dauernder Aktivität gehalten werden. — Weiterhin erschien eine zusammenfassende Darstellung von WAREING. DOWNS, R. J., and H. H. BORTHWICK: Effects of photoperiod on growth of trees. Bot. Gaz. **117**, 310—326 (1956). — WAREING, P. F.: Photoperiodism in woody plants. Annual Rev. Plant Physiol. **7**, 191—214 (1956).

Über regulierende Faktoren der Samenruhe erschienen noch folgende Zusammenfassungen: KOLLER, D.: The regulation of germination in seeds. Bull. Res. Council Israel D **5**, 85—108 (1955). — TOOLE, E. H., S. B. HENDRICKS, H. A. BORTHWICK and V. K. TOOLE: Physiology of seed germination. Annual Rev. Plant Physiol. **7**, 299—324 (1956).

Endogene Aktivitätsrhythmen.

Von

E. Bünning.

Mit 31 Abbildungen.

A. Einleitung.

Schon in den vorhergehenden Abschnitten wurde wiederholt angedeutet, daß es auch endogene Aktivitätsrhythmen gibt. So haben wir etwa bei der Besprechung der Ruheperiode von Knospen und Samen wiederholt die Beteiligung endogener Faktoren erkannt. Es sei hier noch einmal betont, daß mit dieser Behauptung endogener Faktoren natürlich nicht die Regulierung einer solchen Rhythmik durch äußere Faktoren bestritten wird. Diese ist vielmehr sogar immer entscheidend wichtig. Neben den schon früher angedeuteten jahresperiodischen Aktivitätsrhythmen gibt es auch sehr viel rascher verlaufende. Wenn wir uns bei der Besprechung der den Aktivitätswechsel regulierenden äußeren Faktoren zunächst auf die langsam verlaufende Rhythmik beschränkt haben, so erklärt sich das einfach daraus, daß die Regulierung schneller verlaufender Rhythmen durch äußere Faktoren traditionsgemäß vielmehr im Zusammenhang mit der Bearbeitung entwicklungs- und reizphysiologischer Fragen diskutiert wird. Aber wenn wir hier noch speziell die endogene Regulierung von Aktivitätsrhythmen besprechen, müssen wir versuchen, ein Gesamtbild zu entwerfen. Wir kennen bei Pflanzen endogene Aktivitätsrhythmen, bei denen ein voller Cyclus mehrere Jahre dauert, andere (viel häufigere), bei denen der Cyclus etwa ein Jahr beansprucht, weiterhin Rhythmen mit der Cyclenlänge von einigen Wochen oder Monaten, noch kürzere mit einer Dauer von etwa einem Tag oder gar nur einigen Stunden bzw. Minuten.

B. Rhythmen mit einer Cyclenlänge von mehreren Jahren.

Eine sehr langsam verlaufende Aktivitätsänderung ist der bei vielen Pflanzen nachweisbare Alterungsvorgang. Zwar ist dieser Alterungsvorgang in den meisten Fällen insofern exogener Natur, als für die mit zunehmendem Alter etwa bei Bäumen auftretenden Veränderungen in der Blattgröße, in der Dichte der Aderung sowie die gleichzeitig auftretenden Schwächungen Folgen der erschwerten Nährstoffversorgung der am weitesten von den Wurzeln entfernten Teile sind. Gar nicht selten aber sind hierbei offensichtlich auch endogene Faktoren im strengen Sinne beteiligt. Das trifft unter den Blütenpflanzen in erster Linie für viele Monokotyle zu. Die aus inneren Gründen erfolgende Alterung bedeutet nicht nur eine allmähliche Abnahme der Aktivität, sondern ist oft auch mit qualitativen Änderungen der Leistungen verbunden, namentlich damit, daß durch Erreichung eines bestimmten Alters die Blühbereitschaft hergestellt wird.

Die endogene Natur dieser Aktivitätsänderung kommt darin zum Ausdruck, daß — aber eben nur in solchen durchaus nicht häufigen Fällen — die Zeit bis zur Erreichung der Blühbereitschaft und dem dann oft damit verbundenen

Absterben erblich festgelegt ist und nur in engen Grenzen modifikativ beeinflußt werden kann.

Als besonders extreme Beispiele dürfen wohl einige Pflanzen aus der Verwandtschaft der Agaven gelten, bei denen die Aktivitätsherabsetzung einige Jahrhunderte in Anspruch nehmen kann. So blüht *Fourcroya longaeva* im Alter von etwa 400 Jahren und stirbt dann ab. Manche Agavenarten werden einige Jahre oder einige Jahrzehnte alt und zeigen dann zum Abschluß ebenfalls das Blühen. Dabei ist also die Zeit bis zur Erreichung dieses Stadiums weitgehend erblich determiniert. Zum Beispiel benötigt *Agave sisalana* 5—6 Jahre; in dieser Zeit bildet sie etwa 300 Blätter. Die erbliche Fixierung dieser Zeit ergibt sich nicht nur aus ihrer geringen Variabilität, sondern auch aus dem Auftreten von Mutationen. So wurde bei jener *Agave sisalana* eine Mutation gefunden, die die doppelte Zeit benötigt (DEN DOPP). *Agave cantala* benötigt bis zum Blühen etwa 10 Jahre. Man ist oft geneigt, das Absterben der Pflanzen in solchen Fällen als Folge der Blütenbildung zu betrachten. Der Vegetationspunkt verwandelt sich ja und kann nicht erneut zum blattbildenden Vegetationspunkt werden. Aber das Primäre ist eindeutig das Altern; es führt auch dann zum Absterben der Pflanze, wenn der Vegetationspunkt nicht zum Blütenvegetationspunkt umgewandelt wird. Oft beobachten kann man das auf Pflanzungen von *Agave cantala*. Das Blühen kann hier ausbleiben, aber die Pflanze stirbt dann in einem Alter von 10 Jahren trotzdem ab. Das Altern äußert sich übrigens auch darin, daß die zuletzt gebildeten Blätter relativ klein bleiben (vgl. auch v. DENFFER 1952, BÜNNING 1952). Auch bei der Sagopalme *(Metroxylon rumphii)* äußert sich das Altern schon vor dem Blühen darin, daß kleinere Blätter gebildet werden.

Andere berühmt gewordene Fälle sind die der Bambuseen. Es gibt bekanntlich Bambusarten, die nur sehr selten blühen, so blüht *Arundinaria falcata* alle 28 bis 30 Jahre. *Bambusa arundinacea* alle 32—34 Jahre, *Melocanna bambusoides* alle 45 Jahre. In der freien Natur pflegt ein ganzer Bambusbestand über riesige Flächen hinweg gleichmäßig zu blühen. Dem Blühen folgt dann ein allgemeines Absterben der Rhizome und die Neuentwicklung von Sämlingen, so daß also die Gleichaltrigkeit innerhalb des Bestandes immer gewährleistet ist (vgl. SEIFRIZ und dort zitierte Literatur). Dieser Übergang zur Blütenbildung, der hier eine Begleiterscheinung des Alterns ist, wird ganz offensichtlich durch innere Faktoren reguliert, die auch dann noch im gleichen Zeitpunkt die Blüte entstehen lassen, wenn das Rhizom aus dem Boden entfernt und in einem ganz anderen Gebiet, etwa in einem Gewächshaus, kultiviert wird. Gelegentlich sind übrigens Beobachtungen an Bambuseen gemacht worden, die dafür sprechen, daß dieser lange Cyclus wirklich nur ein Ausschnitt aus einer fortlaufenden Periodizität ist; denn man kann durch sorgfältige Behandlung unter Umständen erreichen, daß die Pflanze das Schwächestadium überwindet und dann ohne abzusterben einen neuen solchen Cyclus beginnt.

Eine Aufzählung weiterer Fälle dieser Art erscheint nicht notwendig. Interessant ist aber, daß es von jenen extremen Fällen mit einmaligen Veränderungen alle Übergänge gibt zu dem Verhalten von Pflanzen, die einen periodischen Verlauf der Aktivität, also mehrere Phasen hoher und niedriger Aktivität im Verlauf ihres Lebens zeigen. Im letzteren Falle gibt es nun auch wieder einzelne Pflanzenarten, bei denen eine Cyclenlänge wie in den genannten Fällen mehrere Jahre benötigt und andererseits die viel häufigeren Fälle, in denen die Länge eines Cyclus der Dauer eines Jahres entspricht. Auch in diesem letzteren Falle können wir wieder solche Typen finden, in denen die Cyclenlänge sich in einem Individuum periodisch wiederholt und solche, bei denen nur ein derartiger Cyclus durchlaufen wird (einjährige und winterannuelle Pflanzen).

Daß eigentlich zwischen den eingangs genannten Pflanzen mit einem individuell kontinuierlichen Aktivitätsabfall, der schließlich zum Blühen und zum Absterben führt, und den mehrjährigen Gewächsen mit nur exogenen Alterserscheinungen kein so tiefgreifender Unterschied besteht, wird klar, wenn wir beachten, daß jene Typen sich von diesen oft genug nur dadurch unterscheiden, daß die neu gebildeten Vegetationspunkte, die eine fortgesetzte Verjüngung mit sich bringen, im letztgenannten Falle am gleichen „Individuum" stehen, im erstgenannten Falle aber sich von ihm als Adventivsprosse lösen. Es ist also im erstgenannten Falle einfach die Trennung, d. h. eine vegetative Vermehrung eingeschoben. So können wir es etwa bei Bananen sehen, die sich nach dem Absterben der Mutterpflanze (hier im Alter von 1—2 Jahren) gleich durch Wurzelschößlinge vermehren. Agaven können sich ähnlich verhalten.

Es gibt also tatsächlich endogene Aktivitätscyclen mit einer Periodenlänge von mehreren oder gar vielen Jahren. Über die physiologischen Grundlagen dieser endogenen Veränderungen wissen wir nichts; wir kennen nicht einmal ihre zellphysiologischen Begleiterscheinungen. Bei der Analyse wird die geringe Modifizierbarkeit zu beachten sein. Einfache Erklärungen durch allmähliche Anhäufung von Assimilaten usw. kommen also nicht in Frage. Wir werden das Problem bei der Besprechung der im wesentlichen wohl hiermit verwandten endogenen Jahresrhythmik nochmals aufgreifen.

Es scheint, daß sich die mehrjährigen endogenen Aktivitätscyclen nicht etwa mit einjährigen überlagern können. Zwar kommt eine Überlagerung mit regelmäßigen Ruheperioden etwa bei den Bambuseen vor, aber dabei scheinen die einjährigen Cyclen doch nur durch aufgezwungene Ruheperioden bedingt zu sein. Sollte sich herausstellen, daß es auch eine Überlagerung mit einjährigen endogenen Rhythmen gibt, so würde das die Problematik noch wesentlich erhöhen.

Andererseits dürfte es sich in den Fällen, in denen die endogen geregelte Jahresrhythmik sich mit mehrjährigen Perioden überlagert, darum handeln, daß hier umgekehrt die überlagerten Perioden mit mehrjährigen Cyclen exogen sind. Der Ausdruck exogen bezieht sich dabei auf die physiologischen Umweltbedingungen der Meristeme, nicht notwendig auf die Umgebung der ganzen Pflanze. Zu denken ist hierbei in erster Linie an die bekannten Fälle des nur im Abstand mehrerer Jahre wiederkehrenden reichen Samenansatzes mehrerer Bäume (z. B. Buche, Tanne oder von tropischen Bäumen mit jährlich einmaligem Laubwechsel, etwa mehrerer Dipterocarpaceen). Bei dieser Regulierung spielt offenbar einerseits der völlige Verbrauch von Reservestoffen in einem „Mastjahr", andererseits aber auch das Walten besonderer auslösender Faktoren, unter denen namentlich trockene Jahre zu nennen sind, eine Rolle (vgl. hierzu auch HUBER 1947). Hingewiesen sei hier auch noch auf die Perioden mit einer Cyclenlänge von etwa 11 Jahren, die bei Jahresringmessungen wiederholt gefunden worden sind. Man hat hierbei an einen Zusammenhang mit dem Auftreten von Sonnenflecken gedacht, so daß auch dieser Rhythmus wohl nur exogener Natur ist.

Jedenfalls also gibt es bis jetzt keinen Anhaltspunkt für die Annahme, daß sich mehrjährige exogene mit einjährigen endogenen Aktivitätsrhythmen überlagern können; nur einer der beiden Rhythmen pflegt jeweils exogen zu sein, der andere ist endogen.

C. Rhythmen mit einer Cyclenlänge von 1 Jahr.

Die Erscheinung einer endogenen Rhythmik mit einer Cyclenlänge von etwa 1 Jahr haben wir schon bei der Besprechung der zellphysiologischen Grundlagen des Aktivitätswechsels kennengelernt. Deutlich demonstriert wird diese Existenz einer endogenen Jahresrhythmik ja einmal an der Tatsache, daß Pflanzen gemäßigter Zonen bei der Übertragung in gleichmäßig feuchte Tropen noch weiterhin den Aktivitätswechsel zeigen und dabei dann die jahreszeitliche Bindung verlorengehen kann. Noch deutlicher wird das Phänomen natürlich an Beobachtungen unter experimentell völlig konstant gehaltenen Bedingungen. Wir haben etwa schon auf das Verhalten lufttrocken aufbewahrter Samen hingewiesen, können aber als weiteres Beispiel etwa auch noch die von HENSSEN mitgeteilte Periodizität der Bildung von Dauerorganen bei *Spirodela polyrrhiza* erwähnen. Wurden die Pflanzen bei 20° im Dauerlicht gehalten und in gewissen Abständen die Fähigkeit zur Dauersproßbildung geprüft, so ergab sich das in der Tabelle 1 dargestellte Bild, welches deutlich die Jahresrhythmik erkennen läßt.

Tabelle 1. *Dauersproßbildung von Spirodela polyrrhiza, die bei konstanten Bedingungen (20°, Dauerlicht) verwahrt wurde. Die Bildung der Dauersprosse wurde nach Übertragung in nährsalzreiche Lösung festgestellt.*

Kultur angesetzt am	Anzahl der je Kolben gebildeten Dauersprosse
November 1950 bis Januar 1951	etwa 100
Februar 1951	70
16. 5. 51	0
25. 5. 51	0
21. 6. 51	0
18. 7. 51	10
4. 8. 51	70
18. 8. 51	70
3. 12. 51	75
18. 12. 51	125
31. 1. 52	70
14. 3. 52	15
31. 3. 52	0
24. 5. 52	0
7. 6. 52	0
10. 7. 52	0

Neuerdings diskutiert VEGIS (1955) noch einmal ausführlich das Problem der endogenen Jahresrhythmik. Bei seinen an anderer Stelle erwähnten Untersuchungen über die Regulierbarkeit der Ruheperiode von *Stratiotes aloides* und *Hydrocharis morsus-ranae* zeigte er, daß sowohl die tägliche Licht-Dunkel-Periode als auch die Temperatur einen starken Einfluß haben. Es ist aber einfach nicht begreiflich, wieso der Verfasser durch das wertvolle Auffinden dieser Faktoren sich für ermächtigt hält, das Mitwirken einer endogenen Rhythmik schlechthin abzustreiten. Selbstverständlich wird ihm jeder zustimmen, wenn er KLEBS zitiert: „Erblich fixiert ist die spezifische Struktur mit allen ihren zahllosen Potenzen; alles, was sich tatsächlich entwickelt, d. h. verwirklicht wird, geschieht unter der notwendigen oder unmittelbaren Einwirkung der Außenwelt.“ Wenn VEGIS sich dagegen wendet, „die Jahresperiodizität des Wachstums und das Eintreten des Ruhezustandes ... als erblich fixierte Eigenschaft“ betrachtet, so kann man ihm sehr weitgehend zustimmen. Kein Forscher vertritt die Auffassung, der jahresrhythmische Aktivitätswechsel sei so, wie er in der freien Natur auftritt, einfach Ausdruck einer erblichen und endogenen Jahresrhythmik. Was an dieser Rhythmik erblich und endogen ist und was von Außenfaktoren reguliert wird, kann jeweils nur durch Versuche geklärt werden. Wenn VEGIS durch seine Versuche zur Klärung der äußeren Faktoren beiträgt, so darf er doch darum nicht einfach die inneren leugnen und über die Beweise für ihr Vorhandensein hinweggehen. Gegen die Versuche SIMONS (1928), nach denen Winterknospen von *Hydrocharis morsus-ranae*, die unter konstanten Außenbedingungen verwahrt wurden, noch einen jahresrhythmischen Aktivitätswechsel zeigten, argumentiert er, hier könne von einer endogenen Jahresrhythmik keine Rede sein, weil ja nach seinen Versuchen die Temperatur und die Tageslänge regulieren können. Mit weiteren Argumenten, die wirklich nicht als solche bezeichnet werden dürften, geht er über die Beobachtungen des Verhaltens von Bäumen in den Tropen hinweg, und die Versuche an Samen diskutiert er gar nicht.

Die Lage ist hier durchaus ähnlich wie bei der Rhythmusforschung in der Zoologie. Durch ein sehr umfangreiches Versuchsmaterial ist nachgewiesen worden, daß auch bei Tieren eine endogene Jahresrhythmik weit verbreitet ist. ASCHOFF (1956) hat einen großen Teil dieses Materials zusammengestellt, und er betonte dabei mehrfach, wie gefährlich und bedenklich es ist, daß man die Tatsache einer Regulierbarkeit dieser Rhythmik durch äußere Faktoren („Zeitgeber") als Argument gegen das Vorliegen einer endogenen Komponente benutzt.

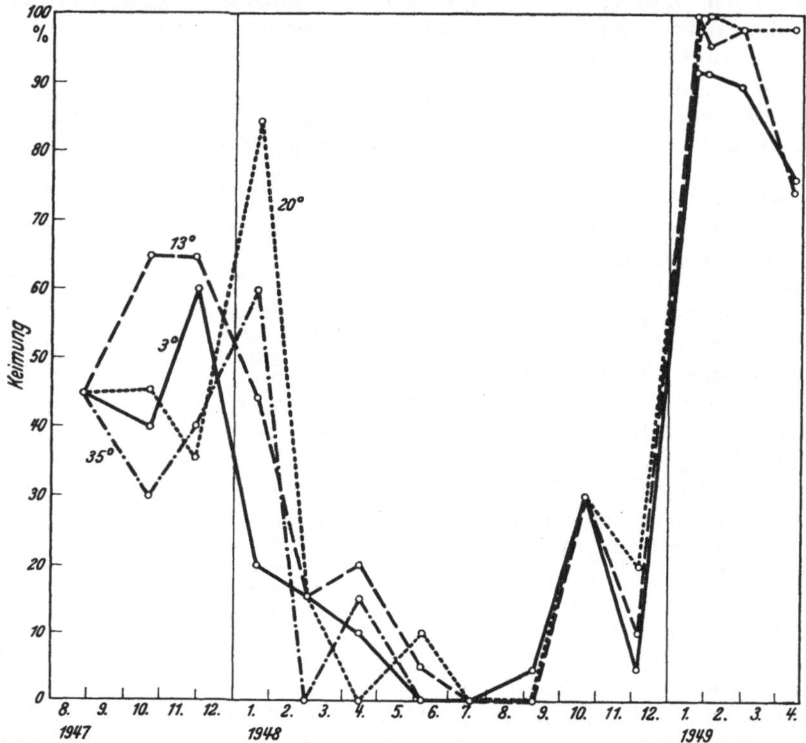

Abb. 1. *Digitalis lutea.* Verlauf der Keimfähigkeit. Angabe des nach 14 Tage langem Aufenthalt im Keimbett festgestellten Keimprozents nach Lagerung der Samen (Ernte 19. August 1947) unter verschiedenen Temperaturbedingungen, die bei den Kurven genannt sind. Die Beobachtung der bei etwa 35° gelagerten Samen konnte nur bis einschließlich Juli 1948 fortgesetzt werden. Obwohl sich jeder Wert nur auf durchschnittlich je etwa 30 Samen bezieht, ist die Ähnlichkeit der Kurven, also die weitgehende Unabhängigkeit in der Lage der Gipfel, namentlich des zweiten (bei dem sich jeder Wert auf 40 Samen bezieht), doch deutlich. Die Abszisse gibt die Jahre und Monate an. (Nach BÜNNING.)

Im übrigen kann gesagt werden, daß die Phänomene der endogenen Jahresrhythmik bei Pflanzen und Tieren sowie die Regulierbarkeit durch äußere Faktoren in beiden Fällen so ähnlich sind, daß man eine gemeinsame physiologische Basis vermuten darf.

Es sei daran erinnert, daß zwar Stoffwechselleistungen beteiligt sind, aber zur Erklärung der Selbststeuerung nicht so einfach, wie es früher gelegentlich geschehen ist, mit einer Rolle der Assimilatanhäufung oder des Wechsels der Mengen von Wuchs- und Hemmstoffen eine Erklärung gefunden werden kann. Diese Änderungen sind vielmehr nicht Ursachen, sondern Begleiterscheinungen des Aktivitätswechsels. Wir haben ja darauf hingewiesen daß der Aktivitätswechsel auch noch in trockenen Samen ohne Stoffwechsel jahresperiodisch verlaufen kann (Abb. 1, vgl. BÜNNING und BAUER, RUGE und LIEDTKE, SIEGEL). Dabei ist noch besonders beachtenswert, daß sich die Geschwindigkeit des

Ablaufs der endogenen Rhythmik als weitgehend von der Temperatur unabhängig erweist. Einerlei ob die Samen bei +45°, bei —22° oder bei einer dazwischenliegenden Temperatur gelagert werden, hat die Länge eines Cyclus immer die gleiche Dauer. Auch Wasserentzug, Lagerung in Sauerstoff, Stickstoff oder Kohlendioxyd und einer Reihe anderer Faktoren können die Geschwindigkeit der Rhythmik nicht merklich beeinflussen (Abb. 2, BÜNNING und MÜSSLE). Man kommt auf Grund solcher Befunde zu dem Ergebnis, daß die Rhythmik im wesentlichen durch physikalische Vorgänge gesteuert wird, deren Natur uns noch unbekannt ist (über endogen-jahresperiodische Keimfähigkeitsschwankungen der Samen vgl. weiterhin EBNER, SCHMIDT, SPERLICH, OKADA, BÜNNING 1949).

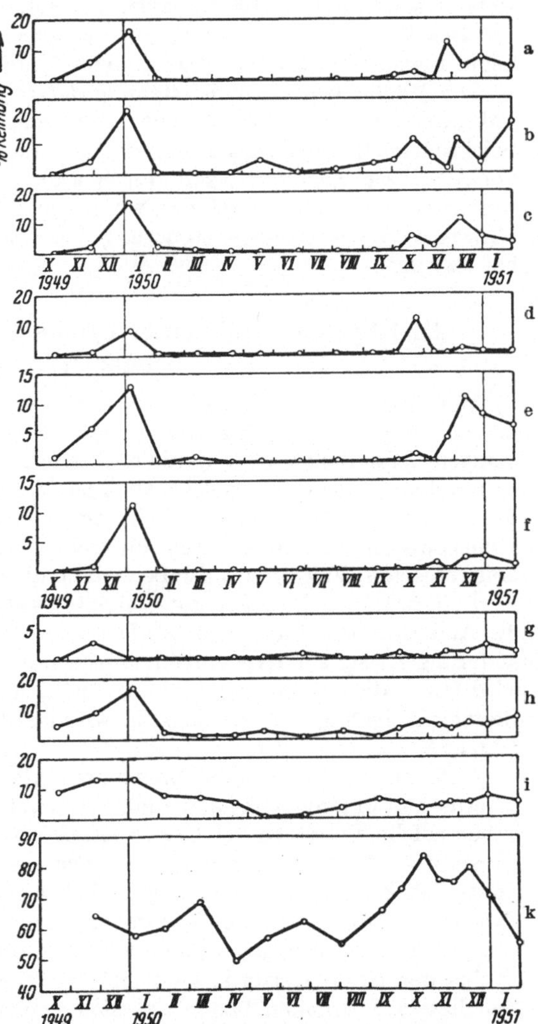

Abb. 2a—k. *Gratiola officinalis.* a Lagerung in Luft; b Lagerung in Sauerstoff; c Lagerung in Kohlendioxyd; d Lagerung in Stickstoff; e vor der Lagerung 1 min erhitzt auf 105°; f 10 min erhitzt; g 40 min erhitzt; h vor der Lagerung mechanisch leicht beschädigt; i stärker beschädigt; k Samen vor jeder Keimprobe angestochen. (Nach BÜNNING und MÜSSLE.)

Wir haben diese Fragen früher schon weitgehend besprochen. An dieser Stelle sei aber noch besonders betont, daß wir die endogene Jahresrhythmik sehr wohl als Resultat von Selektionsvorgängen auffassen dürfen. Immer wieder ist der Gedanke aufgetaucht, man müsse aus der Tatsache des etwa endogen-jahresperiodisch verlaufenden Aktivitätswechsels schließen, die Rhythmik der äußeren Faktoren sei allmählich erblich geworden, es handele sich also um einen Fall der „Vererbung erworbener Eigenschaften". Solche Auffassungen wurden etwa von MUELLER-THURGAU, HOWARD (1915), BORDAGE vertreten (vgl. auch die Diskussion bei KNIEP). Wir wissen aber ja, daß in Gebieten, wo eine solche Selektion nicht erforderlich war, eine viel größere Mannigfaltigkeit endogener Rhythmen besteht. Dazu sei nicht nur auf die schon erwähnten Cyclenlängen von mehreren Jahren verwiesen, sondern auch auf die mehr oder weniger großen Abweichungen von der Dauer eines Jahrescyclus, die in den immerfeuchten Tropen bestehen (vgl. VOLKENS, SIMON, IHERING). Der endogene Charakter der Ruheperiode bei den in solchen Gebieten wachsenden Pflanzen äußert sich auch deutlich in der fehlenden zeitlichen Bindung. Die Angaben hierüber sind zahlreich (vgl. außer den schon zitierten

Autoren KORIBA, RESENDE, DINGLER). Der periodische Charakter der Aktivität ist also offenbar allen Pflanzen gemeinsam. In den immerfeuchten Tropen findet man nur ganz wenige Arten, bei denen er nicht deutlich wird.

Diese Periodizität äußert sich übrigens nicht nur im Laubfall und im Kahlstehen, sondern sehr häufig auch im Holzzuwachs. Natürlich fallen dabei in der Regel die klaren Jahresringe fort; aber der Holzzuwachs hat doch deutlich keine konstante Geschwindigkeit und Qualität.

Die Möglichkeit einer Entstehung der genau an den Jahresrhythmus äußerer Faktoren angepaßten endogenen Rhythmik auf dem Wege von Mutation und Selektion geht auch daraus hervor, daß die Dauer der Ruheperiode ebenso wie die Art ihrer Regulierbarkeit durch äußere Faktoren einer Mutabilität unterworfen ist. Allerdings verfügen wir bis jetzt erst über wenige Angaben, die uns Hinweise auf die erbliche Steuerung dieses Phänomens geben (vgl. die Hinweise bei SAMISH).

D. Rhythmen mit einer Cyclenlänge von einigen Wochen oder Monaten.

Wir finden bei Pflanzen häufig Änderungen in der Quantität und Qualität der Leistung, die sich im Laufe mehrerer Wochen oder einiger Monate bemerkbar machen. Man denke etwa an die im Laufe der Vegetationsperiode sich allmählich ändernden Leistungen. Gelegentlich ist hierfür ein physiologisches „Altern“ der Vegetationspunkte verantwortlich gemacht worden. Jedoch ist man hier leicht Fehlschlüssen ausgesetzt, weil übersehen wird, daß neben den üblicherweise kontrollierten Faktoren (Temperatur und Licht) auch die Tageslänge einen Einfluß hat. So konnten tatsächlich ASHBY und WANGERMANN zeigen, daß die Änderung der Blattform bei *Ipomoea caerulea* nicht eine Folge innerer Veränderungen ist, sondern einfach durch die sich ändernde Tageslänge induziert wird. Aus diesem Grund wird im Frühjahr und im Hochsommer eine andere Blattform bedingt als in dem dazwischenliegenden Zeitabschnitt mit längeren Tagen.

Aber mit diesen Feststellungen wird die Möglichkeit eines physiologischen Alterns aus inneren Gründen nicht ausgeschlossen. WANGERMANN und ASHBY haben solche Vorgänge bei *Lemna minor* nachgewiesen. Wird die Pflanze unter konstanten Außenbedingungen kultiviert, so werden die Tochterglieder kontinuierlich kleiner, bis nach einiger Zeit wieder eine Verjüngung eintritt. Diese Cyclen des Alterns und der Verjüngung dauern mehrere Wochen. Sie sind stark von den Außenbedingungen, besonders von der Temperatur abhängig. Ein voller Cyclus des Alterns erfordert bei 20° 75 Tage, bei 30° ist er wesentlich kürzer. Bei diesem Phänomen handelt es sich zweifellos um etwas anderes als bei der endogenen Jahresrhythmik. Darauf deutet schon die starke Temperaturabhängigkeit hin, die ja in jenem Falle fehlt; außerdem findet sich bei *Lemna* obendrein offenbar auch eine endogene Jahresrhythmik (PIRSON und GÖLLNER).

Es gibt zahlreiche Arten, bei denen der Laubwechsel in Rhythmen verläuft, die eine Periodenlänge von einigen Monaten erkennen lassen. Solche Arten sind in den immerfeuchten Tropen häufig. Besonders bei Kulturpflanzen hat man diese Periodizität genauer beobachtet. Zum Beispiel gibt es beim Tee in einem Jahr etwa „4 flushes“, zwischen denen je eine Ruheperiode liegt (vgl. WIGHT und BARUA).

Wie häufig solche Fälle wie der von ASHBY und WANGERMANN studierte sind, läßt sich nicht entscheiden. Eine besondere Erwähnung verdienen hier aber noch die Beobachtungen über eine endogene Monatsrhythmik. Viele Meeresalgen zeigen bei der Entleerung ihrer Geschlechtszellen eine deutliche Anpassung an den Mondphasenwechsel bzw. an die durch ihn regulierte Periodizität der Springfluten. Zum Beispiel werden bei *Dictyota dichotoma* an den europäischen Küsten die Geschlechtszellen in 14tägigen Abständen, nämlich jeweils bei Einsetzen der Springfluten, entleert. Eine morphologisch gleiche Form dieser Alge an den Küsten von Nord-Karolina erreicht diese Phase nur bei den Vollmondspringfluten. Auch bei *Ulva lactuca* besteht eine 14tägige Periodizität. Die sexuellen und asexuellen Schwärmer treten hier jeweils zur Zeit der Springfluten auf (G. M. SMITH). Diese 14tägige bzw. 28tägige Rhythmik ist wieder weitgehend autonom. Wenn der Zeitpunkt der Springfluten durch Wind vom Normalen abweicht, so halten sich die Pflanzen doch an ihr Schema; offensichtlich handelt es sich wieder um eine durch die Außenrhythmik regulierte endogene Periodizität (HOYT). Die bisher vorliegenden Beobachtungen lassen noch nicht mit Sicherheit den Schluß zu, daß diese endogene Periodizität auch erblich ist. Es bleibt also grundsätzlich noch die Möglichkeit einer Dauermodifikation offen. Über den physiologischen Mechanismus der Selbststeuerung ist bisher gar nichts bekannt.

Nach einer neueren Untersuchung von BROWN, FREELAND und RALPH kann man den Eindruck gewinnen, daß lunare Rhythmen auch bei Landpflanzen vorkommen. Diese Autoren maßen den Sauerstoffverbrauch von Möhren (*Daucus carota*) und Kartoffeln (*Solanum tuberosum*). Sie fanden Rhythmen mit Periodenlängen von 15 bzw. 29 Tagen, die eine enge Beziehung zum Mondphasenwechsel aufwiesen.

E. Rhythmen mit einer Cyclenlänge von etwa 24 Stunden.

a) Einleitung.

Auch bei den tagesperiodischen Vorgängen in der Pflanze müssen wir natürlich unterscheiden zwischen einem exogenen und einem endogenen Aktivitätswechsel. Die Fälle exogenen Aktivitätswechsels interessieren uns hier nicht.

Gerade bei der Diskussion über die endogene Tagesrhythmik sind oft terminologische Mißverständnisse aufgetreten. Daher sei hier noch einmal betont, „nur zur Kennzeichnung, daß sich ein uns entgegentretendes Geschehen bei voller Konstanz der Außenbedingungen abspielt, habe ich ohne eine andere Voraussetzung die Bezeichnung autonom oder autogen benutzt“ (PFEFFER 1907). Die endogene (oder autonome) Tagesrhythmik braucht also nicht einmal erblich zu sein, und wenn sie erblich ist, braucht sie nicht unabhängig von Außenfaktoren aufzutreten. Außenfaktoren können notwendig sein, um sie auszulösen. Außenfaktoren können sie aber vor allem auch regulieren und stark modifizieren. Wenn wir sagen, eine Pflanze besitze eine endogene Rhythmik, so meinen wir nicht mehr, als wenn wir sagen, ein Pendel besitze eine endogene Rhythmik und damit etwa nur meinen, daß, sofern das Pendel angestoßen wird und schwingt, es die Tendenz zu einer Schwingung mit bestimmter Periodenlänge hat, obwohl diese Periodenlänge durch konstante äußere Faktoren oder etwa durch periodisches Anstoßen in gewissen Grenzen modifiziert werden kann.

b) Cyclen mit einer Periodenlänge, die etwas größer ist als 1 Tag.

Bei Blattstielen von *Impatiens sultani*, *Plectranthus fruticosus*, *Coleus blumii* und *Tropaeolum majus* fand TITZ unter Bedingungen konstanter Temperatur und Beleuchtung ein rhythmisches Wachstum (Abb. 3). Jede einzelne Zone des Blattstieles zeigte einen Wechsel von Wachstumsruhe und Wachstumstätigkeit. Während eine Zone sich in Ruhe befindet, wächst die benachbarte und umgekehrt.

Der Zeitabstand zwischen 2 Hochpunkten der Wachstumsgeschwindigkeit war dabei im allgemeinen etwas größer als 1 Tag. Er konnte sogar 2 Tage und mehr erreichen. Ob dieses Phänomen etwas mit der endogenen Tagesrhythmik zu tun hat, läßt sich nicht entscheiden.

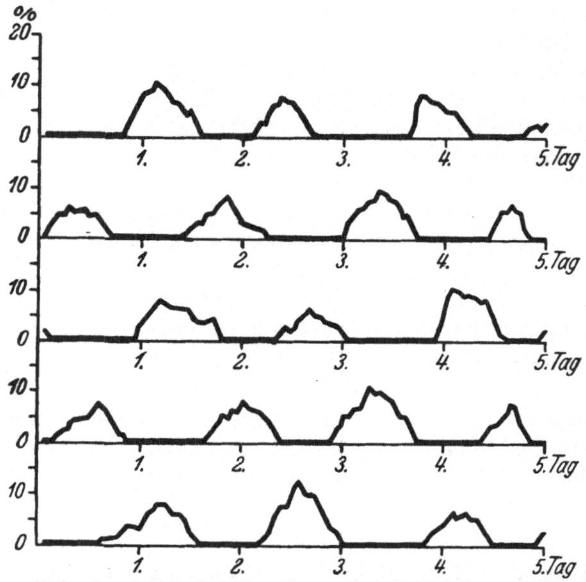

Abb. 3. Zuwachs von fünf nebeneinanderliegenden Zonen eines Blattstieles von *Tropaeolum majus* an fünf aufeinanderfolgenden Tagen mit je 20 Messungen. (Nach TITZ.)

c) Endogene Tagesrhythmik.

Die ersten Beobachtungen über das Vorkommen einer endogenen Tagesrhythmik bei Pflanzen liegen schon sehr weit zurück. Namentlich sind es Beobachtungen über tagesperiodische Blattbewegungen, die schon vor mehr als

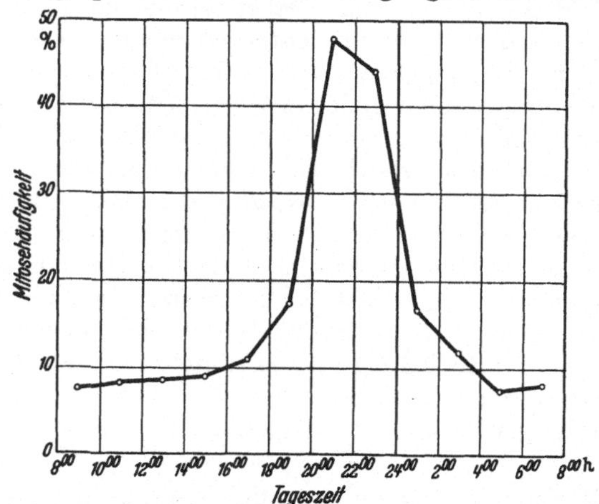

Abb. 4. Tagesperiodische Schwankung der Mitosehäufigkeit im Sproßvegetationspunkt von *Tradescantia zebrina*. Die Ordinate gibt an, welcher Prozentsatz der Kerne sich zu den einzelnen Tageszeiten in Teilung befand. Die Pflanze war im Gewächshaus dem normalen Licht-Dunkel-Wechsel ausgesetzt. (Nach BÜNNING.)

100 Jahren die Ansicht der Beteiligung einer endogenen Komponente aufkommen ließen, ja sogar schon Beobachtungen aus dem 18. Jahrhundert deuten auf diese Möglichkeit hin.

Genauere Angaben wurden in größerer Zahl in der zweiten Hälfte des vorigen Jahrhunderts gemacht. Hingewiesen werden kann etwa auf die Beobachtungen von BARANETZKY (1879) über den rhythmischen Verlauf des Streckungswachstums bei einigen Pflanzen. Sodann seien etwa die Untersuchungen von KARSTEN (1915) und von STÅLFELT (1921) über die Periodizität von Teilungsvorgängen

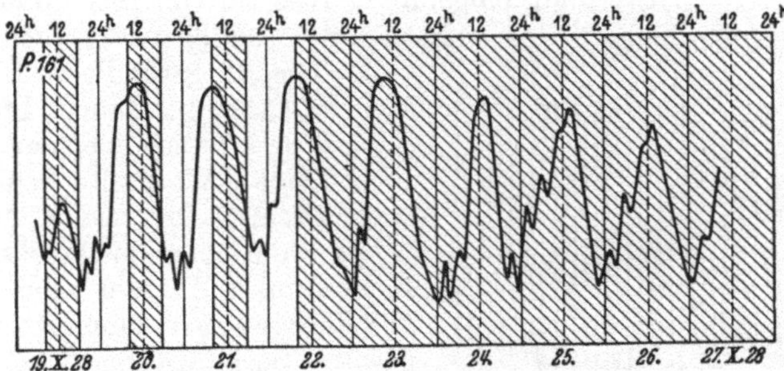

Abb. 5. *Canavalia ensiformis* unter dem Einfluß eines inversen Beleuchtungswechsels, also am Tage verdunkelt, in der Nacht künstlich beleuchtet. Die Blattbewegungen verlaufen so, wie es dem geänderten Beleuchtungswechsel entspricht. (Infolge der Hebelübertragung beim Registrieren entspricht einer Blatthebung eine Kurvensenkung.) Ab 22. 10. wurde konstante Dunkelheit gegeben; die Pflanze setzt die zeitlich verschobene Bewegung fort. (Nach KLEINHOONTE.)

erwähnt. Sehr alt sind auch schon Angaben über die Beteiligung einer endogenen Rhythmik bei Blutungsvorgängen, und schon HOFMEISTER hat hinsichtlich dieses Prozesses die Fortdauer beim Herrschen konstanter Bedingungen hervorgehoben (Abb. 4).

Die ausführlichsten Untersuchungen aber beziehen sich auf die tagesperiodischen Blattbewegungen und zwar sowohl auf Bewegungen von Laubblättern

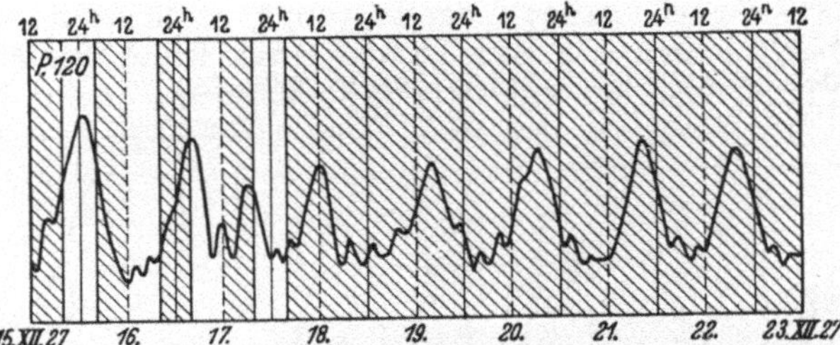

Abb. 6. *Canavalia ensiformis.* Bei 8:8stündigem Licht-Dunkel-Wechsel werden Bewegungen ausgeführt, die mit diesem synchron verlaufen. Jedoch beginnt die Hebung nicht — wie normalerweise — schon in der Dunkelperiode, sondern erst zu Anfang der Lichtperiode. Die Senkungs- (und auch die Hebungs-)maxima werden also (relativ zum Licht-Dunkel-Wechsel) erst spät erreicht. Nachher sieht man in konstanter Dunkelheit das Auftreten der Tagesperiode. (Nach KLEINHOONTE.)

als auch solche von Blütenblättern. Die sorgfältige Bearbeitung dieser Bewegungen hat PFEFFER in den Jahren 1875—1915 eingeleitet, seine Arbeiten zeigen auch sehr deutlich die Beteiligung einer endogenen Komponente. Es ist unmöglich, hier im einzelnen auf alle diese Untersuchungen einzugehen und alle Beobachtungen zu nennen, die für die Beteiligung einer endogenen Tagesrhythmik sprechen. Es sei auch erwähnt, daß schon während der Zeit PFEFFERS und auch hinterher noch gelegentlich damit gerechnet wurde, daß die offenbar endogene Komponente nur durch das Herrschen eines unbekannten Außenfaktors vorgetäuscht wird. Diese Deutung hat sich aber als unrichtig erwiesen.

Hier sei nur ganz kurz gesagt, welche Beobachtungen zur Annahme einer endogenen Rhythmik zwingen, und wieweit diese durch äußere Faktoren reguliert werden kann.

Da die tagesperiodischen Blattbewegungen eine deutliche Beziehung zur Rhythmik äußerer Faktoren, namentlich zum Licht- und Temperaturwechsel, zeigen, müssen wir schließen, daß diese Faktoren normalerweise den tagesperiodischen Gang der Bewegungen regulieren. Besonders klar wird das daraus, daß durch eine Veränderung der zeitlichen Lage von Dunkel- und Hellperioden mehr oder weniger rasch auch eine zeitliche Verschiebung der Bewegungsphasen eintritt (Abb. 5). Außerdem folgen die Bewegungen nicht nur dem normalen, also dem der 24 Std-Rhythmik eingefügten Wechsel von Licht und Dunkelheit, sondern auch einem langsameren, z. B. dem 8:8stündigen und einem schnelleren, z. B. dem 18:8stündigen Licht-Dunkel-Wechsel (Abb. 6). Manche Pflanzen folgen sogar noch sehr schön einem 3:3stündigen Licht-Dunkel-Wechsel. Schließlich aber tritt jedenfalls bei vielen Arten, sowohl bei zu schnellem als auch bei zu langsamem Licht-Dunkel-Wechsel eine Grenze auf. Die Blätter zeigen dann oft zwar noch eine gewisse Beeinflussung durch die Außenrhythmik, jedoch tritt gleichzeitig trotz der Außenrhythmik eine deutliche Tagesperiode zum Vorschein (Abb. 7). Aber auch eine nur geringe Abweichung von der 12:12stündigen Außenrhythmik bedingt schon, daß die Bewegungen nicht mehr ganz normal verlaufen. Die Wendepunkte der Bewegungen treten nämlich so ein, daß sie etwa eine Resultante zwischen dem äußeren Rhythmus und der autonomen 24 Std-Rhythmik darstellen. Es besteht also deutlich innerhalb der Pflanze eine Tendenz zu tagesperiodischen Reaktionen, und eben daraus kann auf das Vorhandensein einer endogenen Tagesrhythmik geschlossen werden, die nur hinsichtlich der Tageszeit der Wendepunkte von den Außenfaktoren reguliert wird. Das trifft für viele Blütenblätter (Abb. 8) ebenso zu wie für Laubblätter.

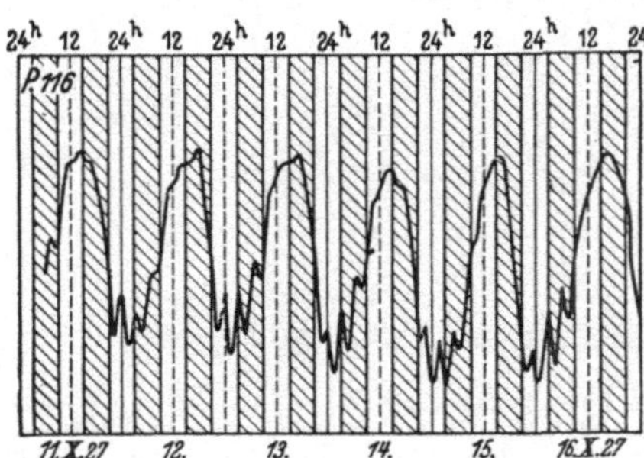

Abb. 7. *Canavalia ensiformis*. Bei 6:6stündigem Licht-Dunkel-Wechsel bleiben die Bewegungen tagesperiodisch; die 24-Std-Autonomie ist also stark ausgeprägt. (Nach KLEINHOONTE.)

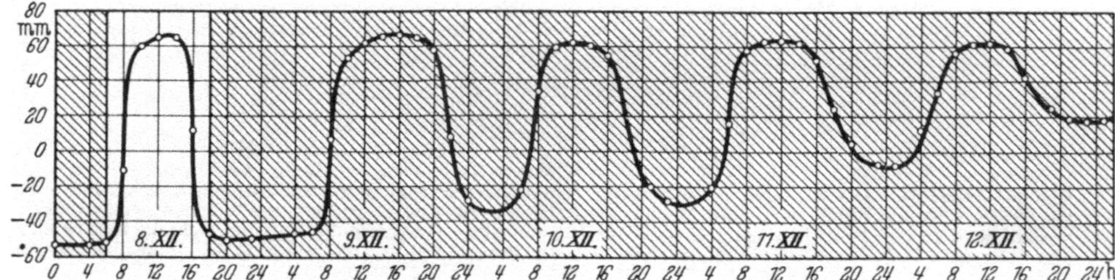

Abb. 8. Öffnungs- und Schließungsbewegungen einer Blüte von *Kalanchoë Bloßfeldiana*, die sich im Dauerdunkel fortsetzen. Die Ordinate gibt die Öffnungsweite an. (Nach BÜNSOW.)

Diese Regulation gelingt außerordentlich leicht, oft schon durch Lichtperioden von etwa 1 min Dauer. Die endogene Natur der Bewegungstendenz kommt auch in den sog. Nachschwingungen zum Ausdruck: Werden die Pflanzen nicht mehr dem Licht-Dunkel-Wechsel ausgesetzt, sondern ihnen kontinuierliche Dunkelheit oder kontinuierliches Licht geboten, so setzt sich die Bewegungsperiodizität fort, und zwar, unabhängig von der Art des vorhergehenden Licht-Dunkel-Wechsels, immer mit einer ungefähr 24 Std-Periodik. Die Bewegungsperiodizität kann sich aber auch dann zeigen, wenn die Pflanzen von der Keimung an bei konstanten Bedingungen gehalten werden. Es ist selbstverständlich, daß in diesem Fall, d. h. bei völligem Fehlen der zeitlichen Regulation, die Wendepunkte bei den einzelnen Individuen zu ganz unterschiedlichen Tageszeiten erreicht werden (vgl. die Literaturhinweise bei BÜNNING 1936, außerdem BÜNSOW, FLÜGEL).

In entsprechender Weise wurde die Beteiligung einer endogenen Tagesrhythmik auch bei mehreren anderen tagesperiodischen Erscheinungen nachgewiesen (Abb. 9).

Erwähnt seien etwa die Wachstumsrhythmen, welche BALL und DYKE bei *Avena*-Koleoptilen beobachteten (Abb. 10). Auch hier ist der Zeitpunkt eines Wechselns zwischen Licht und Dunkelheit für die Regulierung der Rhythmik entscheidend. Untersucht wurde die Auslösung beim Übergang von rotem Dauerlicht zu Dunkelheit. Die Rhythmik setzt sich 2—3 Tage fort. Der erste Gipfel der Wachstumsgeschwindigkeit tritt 16 Std nach dem Anstoß durch die genannte Änderung der Lichtbedingungen auf, der zweite nach 16+24, der dritte nach 16+48 Std. Von diesen Untersuchungen sei noch speziell hervorgehoben, daß sich innerhalb der Grenzen des berücksichtigten Temperaturbereichs überraschenderweise eine Unabhängigkeit der Periodenlänge von der Höhe der Temperatur zeigte.

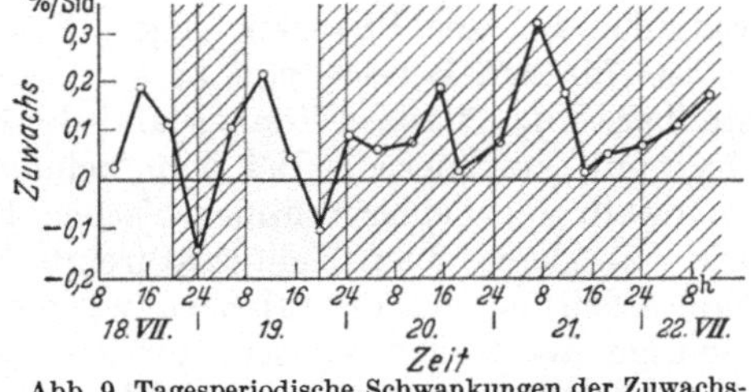

Abb. 9. Tagesperiodische Schwankungen der Zuwachsgeschwindigkeit einer Gewebekultur von *Daucus carota*. Die Kultur war zunächst regelmäßigem Licht-Dunkel-Wechsel ausgesetzt, dann konstanter Dunkelheit. (Nach ENDERLE.)

Bei anderen Objekten hatte sich vorher eine gewisse Temperaturabhängigkeit ergeben, jedoch war diese immer nur relativ gering, d. h. die Abkürzung der Periodenlänge durch zunehmende Temperatur erfolgte immer mit einem Q_{10}-Wert, der wesentlich niedriger war als die für viele stoffwechsel-physiologischen Vorgänge charakteristischen Werte zwischen 2 und 3 (BÜNNING 1931). Zum Beispiel wurde gefunden, daß die Periodenlänge für *Phaseolus multiflorus* (untersucht an den tagesperiodischen Blattbewegungen) bei 15° 29,7, bei 35° 19 Std beträgt (BÜNNING 1931). Bei *Pilobolus* (untersuchter Vorgang: Sporangienträgerbildung) beträgt die Periodenlänge bei 20—25° ungefähr 24 Std, bei 15° 36 Std (UEBELMESSER). Für *Oedogonium* (untersuchter Vorgang: Rhythmik der Sporenbildung) wurde aber ein negativer Temperaturkoeffizient gefunden, die Periodenlänge nimmt bei einem Temperaturanstieg von 17,5° von 20 auf 25 Std

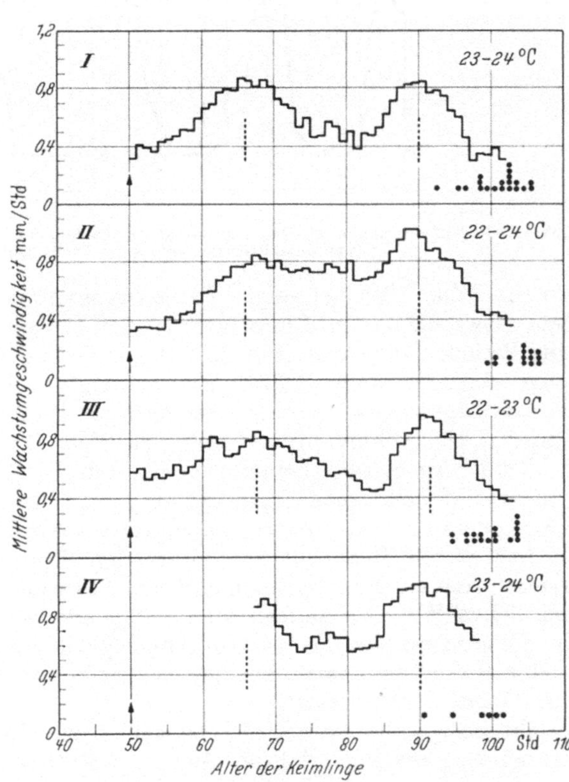

Abb. 10. Endogene Wachstumsrhythmen bei *Avena*-Koleoptilen. Die Registrierung beginnt mit dem durch Pfeile angegebenen Zeitpunkt der Übertragung von rotem Licht in Dunkelheit. Die punktiert eingetragenen senkrechten Linien geben die genaue Länge eines 24 Std-Cyclus an. Abszisse: Alter der Keimlinge in Stunden. Ordinate: Wachstumsgeschwindigkeit in Minuten je Stunde. (Nach BALL und DYKE.)

zu (BÜHNEMANN 1955). Nach BÜNNING und LEINWEBER besteht die Temperaturabhängigkeit (bestimmt für die tagesperiodischen Blattbewegungen von *Phaseolus multiflorus*) nur kurze Zeit nach der Übertragung in die betreffende Temperatur. Befinden sich die Pflanzen längere Zeit unter dem Einfluß dieser Temperatur, so ist (für den Bereich von 15—25°) keine Temperaturabhängigkeit mehr feststellbar. Die Pflanzen können also den Temperaturfehler der endogenen Tagesrhythmik korrigieren. Bei dem benutzten Material betrug die Periodenlänge innerhalb der Fehlergrenzen zwischen 15 und 25° immer 28 Std (LEINWEBER). Die Fähigkeit, einen Temperatureinfluß zu kompensieren, kommt auch darin zum Ausdruck, daß die Pflanzen, wenn sie von mittleren Temperaturen in extrem niedrige übertragen werden, zunächst eine Verlängerung der Perioden zeigen, diese Verlängerung sich aber allmählich ausgleicht und sogar verübergehend eine Verkürz[illegible]zesses

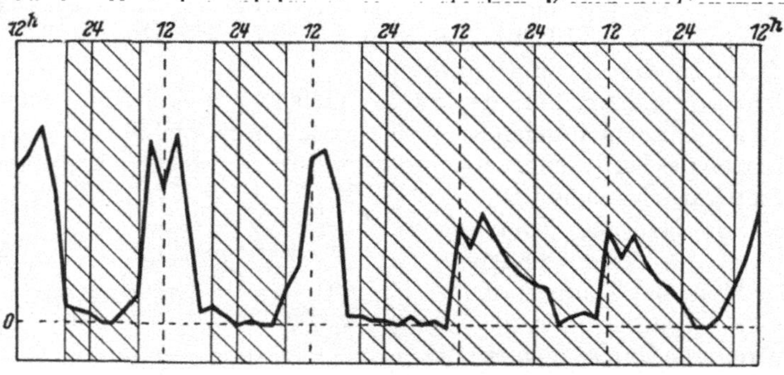

Abb. 11. Verlauf der phototaktischen Sensibilität von *Euglena gracilis* bei 12:12stündigem Tag-Nacht-Wechsel und anschließender konstanter Dunkelheit. (Nach POHL.)

eintritt. Zum Beispiel zeigte *Phaseolus multiflorus* nach der Übertragung aus dem Gewächshaus in Temperaturen von 15° am ersten Tage der Übertragung eine Periodenlänge von etwa 33 Std, im Verlauf von 3 Tagen fiel die Periodenlänge aber auf etwa 24 Std. Ebenso zeigte sich das Einschalten eines Kompensationsprozesses, wenn in eine mittlere Temperatur für einige Stunden eine Periode der Abkühlung auf sehr niedrige Temperatur eingeschoben wurde. Schließlich ist es noch bemerkenswert, daß die Pflanzen nur bei mittleren Temperaturen die endogene Tagesrhythmik gut ablaufen lassen können. Bei extrem hohen oder extrem niedrigen Temperaturen schwindet die Rhythmik viel schneller.

Aus diesen Beobachtungen über den Temperatureinfluß darf geschlossen werden, daß an der Rhythmik mehrere Prozesse in einer zunächst noch unübersichtlichen Weise zusammenwirken. Bemerkenswert ist es aber, daß bei Tieren die Phänomene der Temperaturabhängigkeit ganz ähnlich sind. Durchweg ist auch hier die Temperaturabhängigkeit der Periodenlänge sehr gering (vgl. BROWN und WEBB, PITTENDRIGH).

Um zu zeigen, wieweit die endogene Tagesrhythmik bei Pflanzen verbreitet ist, sei etwa erwähnt, daß bei Algen die Entleerung von Gametangien offenbar von der endogenen Tagesrhythmik mit reguliert werden kann. Bei *Oedogonium cardiacum* untersuchte BÜHNEMANN die Rhythmik der Sporenbildung. Diese wieder endogen tagesperiodische Rhythmik wird wie in anderen Fällen vom vorhergehenden Licht-Dunkel-Wechsel gesteuert. BÜHNEMANN spricht auf Grund seiner Ergebnisse von einer „endogenen Tagesrhythmik der sporogenen Stimmung". Bei *Euglena* fand POHL hinsichtlich des phototaktischen Verhaltens eine endogene Tagesrhythmik (Abb. 11). Für Pilze liegen Angaben von SCHMIDLE und von UEBELMESSER vor, die nachweisen, daß die Sporangienreifung bei

Pilobolus von der endogenen Tagesrhythmik mitreguliert wird (Abb. 12). Auch Beobachtungen von INGOLD und COX über die Sporenentleerung bei *Daldinia*

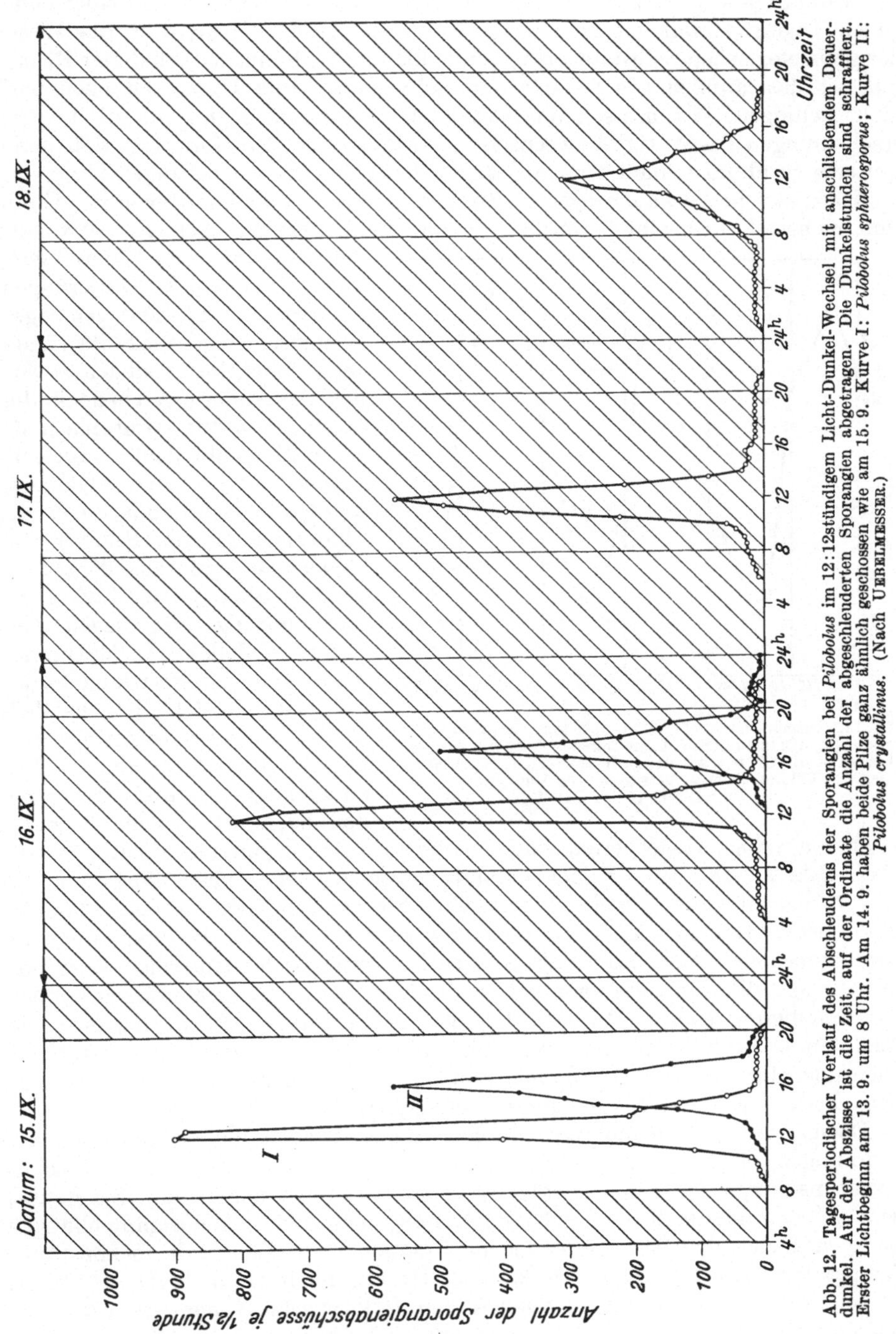

Abb. 12. Tagesperiodischer Verlauf des Abschleuderns der Sporangien bei *Pilobolus* im 12:12stündigem Licht-Dunkel-Wechsel mit anschließendem Dauerdunkel. Auf der Abszisse ist die Zeit, auf der Ordinate die Anzahl der abgeschleuderten Sporangien abgetragen. Die Dunkelstunden sind schraffiert. Erster Lichtbeginn am 13. 9. um 8 Uhr. Am 14. 9. haben beide Pilze ganz ähnlich geschossen wie am 15. 9. Kurve I: *Pilobolus sphaerosporus*; Kurve II: *Pilobolus crystallinus*. (Nach UEBELMESSER.)

demonstrieren die Mitwirkung einer endogenen Tagesrhythmik bei solchen Vorgängen. Dem 6:6stündigen Licht-Dunkel-Wechsel folgte die Rhythmik auch, jedoch dürfte es sich hierbei um das Auftreten von „Oberschwingungen“ handeln, wie sie auch bei Blattbewegungen regelmäßig vorkommen können.

Wesentlich schwieriger als die Feststellung dieses Vorhandenseins einer endogenen Tagesrhythmik ist es nun, deren physiologische Natur, d. h. das Wesen der physiologischen Selbststeuerung, aufzuklären. Auch hinsichtlich dieser Rhythmik ist das, ähnlich wie bei der Jahresrhythmik, trotz Untersuchungen noch nicht gelungen. Wir müssen uns also auch hier damit begnügen, die Begleiterscheinungen des Aktivitätswechsels aufzuzählen in der Hoffnung, daß deren weiteres Studium einmal zur Aufklärung jener Regulation beitragen kann.

Plasmatische Veränderungen im Zusammenhang mit der endogenen Rhythmik sind mehrfach beobachtet worden. Schon die Untersuchungen von STÅLFELT deuten auf eine Beteiligung von Permeabilitätsänderungen. Späterhin sind endogen-tagesperiodische Permeabilitätsänderungen mehrfach beobachtet worden. Auf Änderungen in der kolloidalen Struktur protoplasmatischer Elemente deuten auch die Beobachtungen über endogen-tagesperiodische Änderungen der Chloroplastenform (BUSCH). Bei *Selaginella serpens* sind die Chloroplasten während der Lichtphase flächenförmig und liegen dem Grund der Zelle an. In der Dunkelphase sind sie etwa kugelförmig und liegen in der Mitte der Zelle an der äußeren Epidermiswand. Diese Formänderungen der Plastiden hängen wahrscheinlich mit den ebenfalls regelmäßig nachweisbaren endogen-tagesperiodischen Aciditätsschwankungen zusammen (BÜNNING 1942, BUSCH, Abb. 13). Solche Veränderungen der Chloroplastenform sind besonders beachtenswert, weil sie mit einer Schwankung der in den Plastiden lokalisierten Enzyme in Zusammenhang gebracht werden könnten, denn die Formänderungen sind natürlich Ausdruck physikalisch-chemischer Zustandsänderungen, die ihrerseits auch wieder die Enzymaktivität beeinflussen können. Die Beteiligung einer endogenen Komponente bei diesen Formänderungen der Plastiden kommt darin zum Ausdruck, daß sie beginnen, bevor die betreffende Änderung der äußeren Bedingungen einsetzt; außerdem aber setzt sich auch diese Periodizität unter konstanten Bedingungen fort.

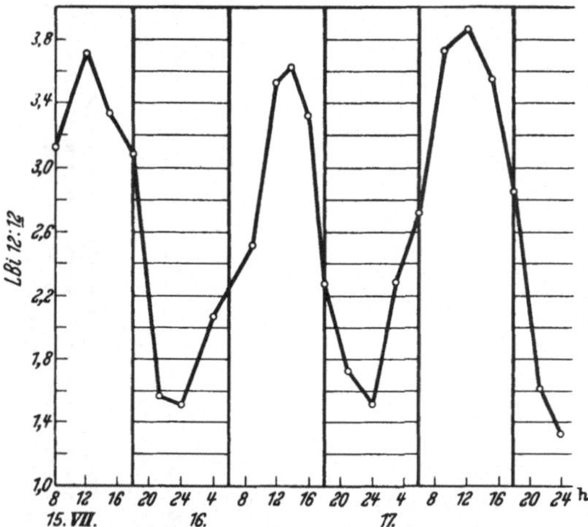

Abb. 13. Formänderung der Chloroplasten von *Selaginella serpens* im 12:12stündigen Licht-Dunkel-Wechsel. Auf der Abszisse ist die Uhrzeit, auf der Ordinate der Längen-Breiten-Index der Chloroplasten aufgetragen. Die Dunkelzeiten sind quer schraffiert. (Nach BUSCH.)

Ebenso liegen auch mehrere Untersuchungen über endogen-tagesperiodische Änderungen der Intensität und Qualität des Stoffwechsels vor. Auf solche deuten natürlich schon die erwähnten Aciditätsschwankungen hin. Sie können etwa mit den mehrfach beschriebenen Atmungsschwankungen zusammenhängen, die auch wieder deutlich eine endogene Komponente zeigen, d. h. in konstanter Temperatur sowie im Dauerlicht oder in Dauerdunkelheit fortlaufen (Abb. 14). Es sei etwa auf die Untersuchungen von SCHMITZ an *Kalanchoë* verwiesen. Jedoch ist es fraglich, ob die Atmungsrhythmik ohne weiteres mit den anderen Phänomenen der endogenen Tagesrhythmik in einen ursächlichen Zusammenhang gebracht werden darf. Die Abhängigkeit vom vorhergehenden Licht-

Dunkel-Wechsel ist nämlich nach PRINZ ZUR LIPPE nicht ebenso wie bei jenen anderen Rhythmen, etwa der Bewegungsrhythmik.

Außer der Atmung schwankt auch die assimilatorische Fähigkeit. Das kommt vor allem darin zum Ausdruck, daß die Photosynthese bei gleichbleibender Lichtintensität nicht gleich hoch bleibt. Man hat das früher zunächst durch die Annahme von Ermüdungserscheinungen infolge der Anhäufung von Assimilaten

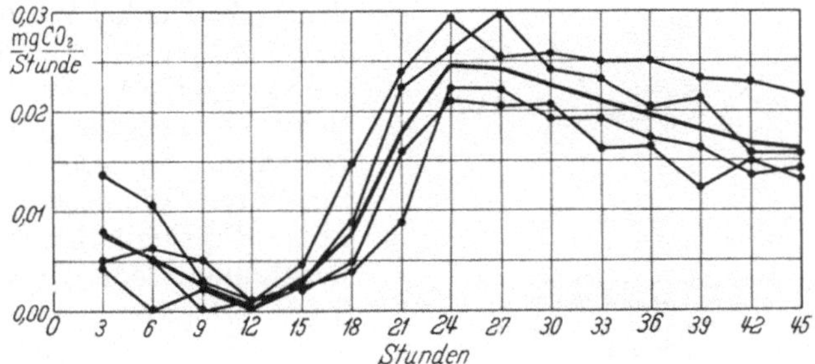

Abb. 14. Verlauf der CO_2-Abgabe bei *Kalanchoe Bloßfeldiana* im Dunkeln nach einem Beleuchtungsrhythmus von 9 Std Licht und 21 Std Dunkelheit. (Nach SCHMITZ.)

erklärt. Späterhin zeigte sich aber, daß diese Schwankungen einen zeitlich ähnlichen Verlauf haben, einerlei ob die Photosynthese relativ hoch oder relativ niedrig ist. Außerdem ergab sich, daß die Pflanze bei der Darbietung künstlicher Lichtperioden je nach ihrem Zustand, d. h. je nachdem in welcher Phase sich die endogene Rhythmik befindet, mit einer unterschiedlich starken Photosynthese antwortet. Dabei verhält es sich offenbar so, daß hohe respiratorische

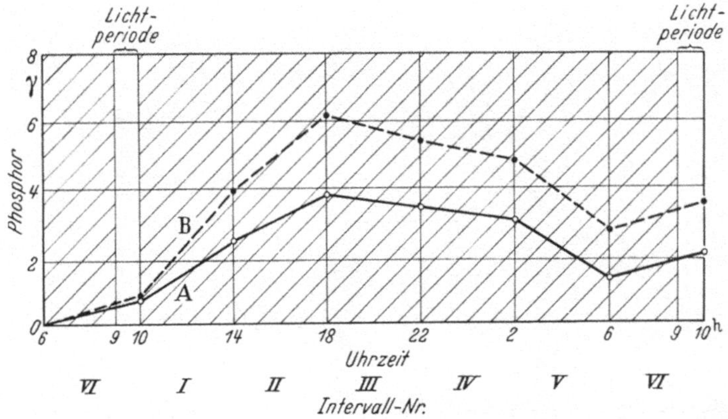

Abb. 15. Mittelwerte der Phosphataseaktivität im 1:23 Std-Rhythmus bei *Phaseolus multiflorus* unter zwei etwas voneinander verschiedenen Bedingungen (Dunkelzeit schraffiert). (Nach EHRENBERG.)

Fähigkeiten mit hohen assimilatorischen alternieren. Es gibt also eine Phase mit etwa 12 Std Dauer, in der die Atmung begünstigt ist, und eine zweite sich ihr anschließende von etwa ebensolanger Dauer, in der die Photosynthese begünstigt ist.

Es ist fast selbstverständlich, daß diesen Schwankungen der unterschiedlichen Stoffwechselleistungen auch Schwankungen der Fermentaktivitäten entsprechen. Tatsächlich sind solche wiederholt gefunden worden. Da es ohnehin bisher nicht gelingt, alle diese Begleiterscheinungen der endogenen Rhythmik miteinander in einen inneren Zusammenhang zu bringen, sei hier nur auf das tatsächliche Ablaufen solcher Schwankungen hingewiesen (DE GROOT, BÜNNING 1942, SISSAKIAN und KOBIAKOVA, EHRENBERG, Abb. 15—17). DE GROOT fand einen periodischen Stärkeabbau, der mit Schwankungen der Amylaseaktivität in Zusammenhang gebracht werden kann (vgl. auch Bünning 1942). Sissakian und Kobiakova beobachteten tagesperiodische Änderungen der Invertaseaktivität und der Oxydoreduktionskapazität. Harder fand tagesperiodische Änderungen der Amylase- und Phosphataseaktivität. Diese Änderungen wurden von Ehrenberg näher untersucht. Die Regulierung der Fermentaktivität durch den Licht-Dunkel-Wechsel ist ähnlich wie etwa die Regulierung der tagesperiodischen Bewegungen. Schwankungen der Amylaseaktivität wurden weiter von Venter an mehreren Pflanzen untersucht. Diese Rhythmik kann sich unter konstanten Bedingungen mehrere Tage fortsetzen.

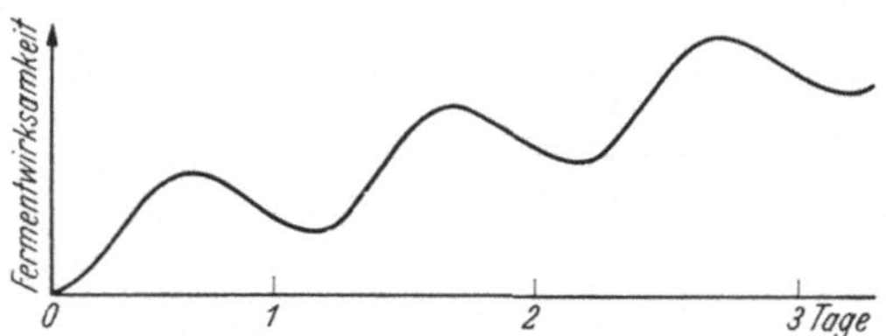

Abb. 16. Schematische Darstellung des Verlaufes der Fermenttätigkeit an drei aufeinanderfolgenden Tagen in den Primärblättern von *Phaseolus multiflorus*. (Nach Ehrenberg.)

Auch die Tatsache, daß die Schichtenbildung beim Stärkekorn autonom-periodisch, und zwar wenigstens annähernd tagesperiodisch verläuft, deutet auf endogene Fermentschwankungen hin. Das Weiterlaufen dieser Schichtenbildung unter konstanten Bedingungen haben Roberts und Proctor sowie Mes und Menge gezeigt. Hess konnte zudem nachweisen, daß dabei innerhalb von 24 Std eine Doppelschicht oder etwas mehr gebildet wird, also offenbar wieder eine enge Beziehung zur endogenen Tagesrhythmik besteht.

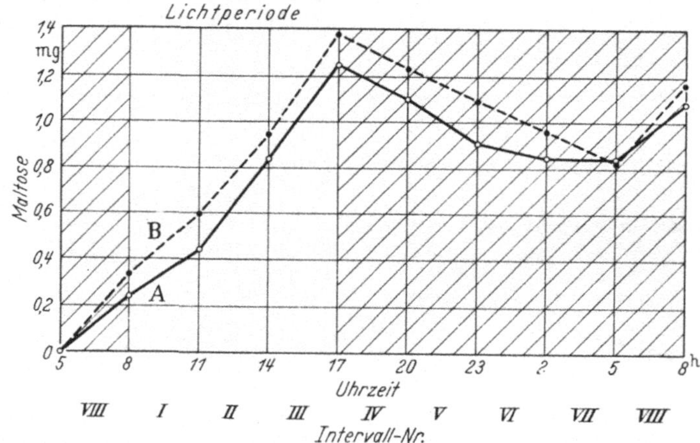

Abb. 17. Mittelwerte der Amylaseaktivität (gemessen an der gebildeten Menge Maltose) in den Primärblättern von *Phaseolus multiflorus* im 9:15 Std Licht-Dunkel-Rhythmus (Dunkelzeit schraffiert) unter zwei etwas voneinander verschiedenen Bedingungen. (Nach Ehrenberg.)

Auf tagesperiodische Stoffwechselschwankungen deuten auch die Beobachtungen über periodische Wärmeproduktion und Geruchsproduktion bei einigen Blüten hin, die sich übrigens meist auch durch den Licht-Dunkel-Wechsel regulieren lassen.

Einige Hinweise zur Analyse des Wesens der endogenen Tagesrhythmik können auch Beobachtungen über die Blutungs- und Guttationstätigkeit geben. Aber eine befriedigende Erklärung ist auch für diese Rhythmik bisher nicht gefunden worden (Abb. 18). Speidel fand, daß diese Rhythmik von der Zeit des Köpfens reguliert wird. Das ist ja nicht weiter verwunderlich, und auch in anderen Fällen sieht es so aus, als könne die Rhythmik (etwa die der Bewegungen) durch starke mechanische Beeinflussungen reguliert werden. Auch bei dieser Rhythmik zeigte sich übrigens wieder die Temperaturabhängigkeit. Durch Atmungsgifte wird die Rhythmik unterdrückt. Allem Anschein nach liegt dem Vorgang eine Periodizität im enzymatischen Abbau der Reservestoffe zugrunde, was ja mit den vorhergehenden Ausführungen harmoniert.

Für die Beteiligung der Atmung an der Blutung sprechen auch die Messungen von GROSSENBACHER über die Schwankungen des Blutungsdruckes. Die Blutungsperiodizität läßt sich auch durch den Licht-Dunkel-Wechsel regulieren (HEIMANN, ENGEL und FRIEDERICHSEN). Zu erwähnen bleibt hier allerdings noch, daß die Guttation auch eine 6stündige Rhythmik zeigen kann, oder auch unregelmäßige Oscillationen mit einem Abstand der Maxima von 4—8 Std auftreten können; und bemerkenswert ist dabei schließlich noch, daß Guttation und Blutung bei gleichem Beleuchtungswechsel einen ungleichen Verlauf zeigen können (HEIMANN). Die Beteiligung einer endogenen und exogenen Rhythmik an den einzelnen periodischen Vorgängen ist offenbar unterschiedlich stark. Auch die Periodizität der Wasseraufnahme ist hier noch bemerkenswert.

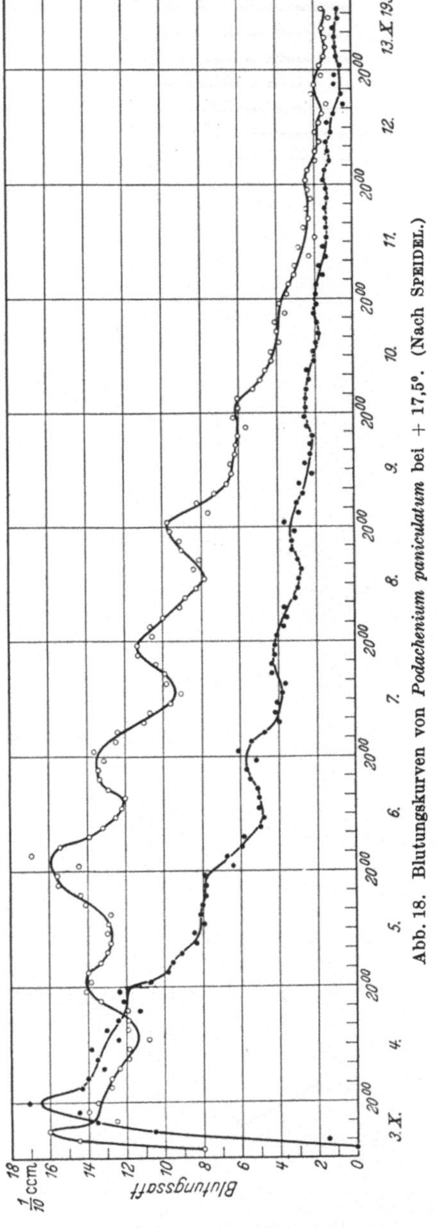

Abb. 18. Blutungskurven von *Podachenium paniculatum* bei + 17,5°. (Nach SPEIDEL.)

Auf periodische Änderungen der Stoffwechselleistungen im Zusammenhang mit der endogenen Rhythmik deuten weiterhin die Untersuchungen von RAU an Keimlingen verschiedener Blütenpflanzen, bei denen sich zeigte, daß Lichtperioden von kurzer Dauer die Substanzproduktion und die Stickstoffaufnahme unterschiedlich beeinflussen, je nachdem zu welchem Zeitpunkt dieses Licht geboten wird. Die Untersuchungen beziehen sich auf *Perilla*, *Amaranthus*, *Lactuca*, *Brassica*, *Plantago* und *Hyoscyamus*. Gearbeitet wurde mit Cyclen von 24 oder 48 Std Länge, wobei die Hauptlichtperiode 6 bzw. 12 Std betrug. Geprüft wurde, wie stark eine in die 18 bzw. 36 Std lange Dunkelperiode eingeschobene 3stündige Zusatzlichtperiode Substanzproduktion und N-Gehalt beeinflußt (Abb. 19). Das Zusatzlicht hatte einen maximalen Effekt, wenn es 6 bzw. 6+24 Std nach Beginn der Hauptlichtperiode geboten wurde. Es zeigen sich also 2 Maxima des Lichteinflusses, die etwa 24 Std auseinanderliegen. Die geringsten Wirkungen hatte Zusatzlicht, das 18 oder 33—42 Std nach dem Beginn der

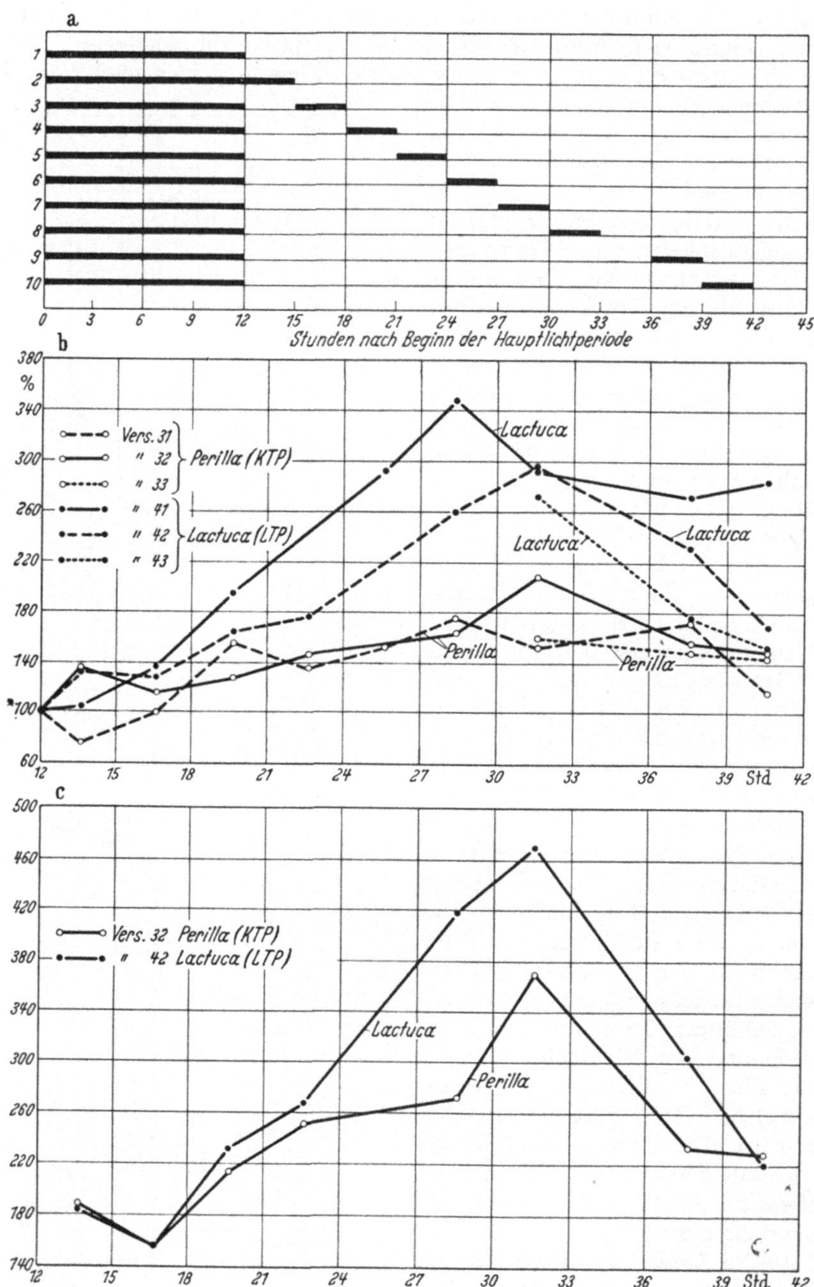

Abb. 19a—c. Unterschiedliche Wirkung eines zu verschiedenen Zeiten nach einer Haupt-Licht-Periode gebotenen Zusatzlichtes auf die Substanzproduktion und die Stickstoffaufnahme bei *Lactuca* und *Perilla*. Oben: Schema der Versuchsanordnung bei 48stündiger Cyclenlänge. Abszisse: Zeit nach Beginn der Haupt-Licht-Periode. Schwarze Striche: Zeit, in der Licht geboten wurde. — Mitte: Substanzproduktion bei Versuchen mit 48stündiger Cyclenlänge. Abszisse: Zeit nach Beginn der Haupt-Licht-Periode. Der Punkt für die betreffende Versuchsgruppe liegt jeweils in der Mitte der Zeit, in der sie Zusatzlicht erhielt. Ordinate: Substanzproduktion (Differenz des Trockengewichts zu Versuchsanfang und -ende) in Prozenten der Gruppe 1. — Unten: Stickstoffaufnahme bei Versuchen mit 48stündiger Cyclenlänge. Abszisse: Wie Abbildung oben; Ordinate: Stickstoffaufnahme während des Versuchs (Differenz des Gesamtstickstoffgehalts zu Versuchsanfang und -ende) in Prozenten der Gruppe 1. Der absolute Wert der Gruppe 1 für die einzelnen Versuche ist aus Tabelle 2 zu ersehen. (Nach Rau.)

Hauptlichtperiode geboten wurde; also auch diese endogen eintretenden Phasen liegen etwa 24 Std auseinander. Ebenso kann auch auf die ähnlich durchgeführten Untersuchungen von CLAUSS verwiesen werden, nach denen die Bildung der Chloroplastenfarbstoffe unter dem Einfluß solcher Beleuchtungen ebenfalls davon abhängig ist, in welchem Zustand der endogenen Rhythmik das Licht geboten wird (Abb. 20). Sehr bemerkenswert sind weiterhin neue Untersuchungen von CLAUSS und RAU über die unterschiedliche Wirkung des Lichtes auf die Chlorophyllbildung in den einzelnen Phasen der endogenen Tagesrhythmik. *Hyoscyamus niger* erhielt Licht-Dunkel-Cyclen von 72 Std Dauer. Jeder Cyclus bestand aus 10 Std Licht und 62 Std Dunkelheit. Innerhalb dieser Cyclen wurden zusätzliche Lichtperioden von 2 Std Dauer geboten. Die zeitliche Lage dieser Zusatzlichtperioden innerhalb der langen Dunkelphasen war von Versuchsreihe zu Versuchsreihe verschieden. Der Einfluß dieser Zusatzlichtperioden auf die Chlorophyllbildung war nun auch von Versuchsreihe zu Versuchsreihe ganz unterschiedlich. Phasen, in denen das Zusatzlicht die Chlorophyllbildung fördert, alternieren mit anderen, in denen es weniger fördert oder sogar hemmt. Dabei liegen Phasen gleicher Kapazität etwa 24 Std auseinander.

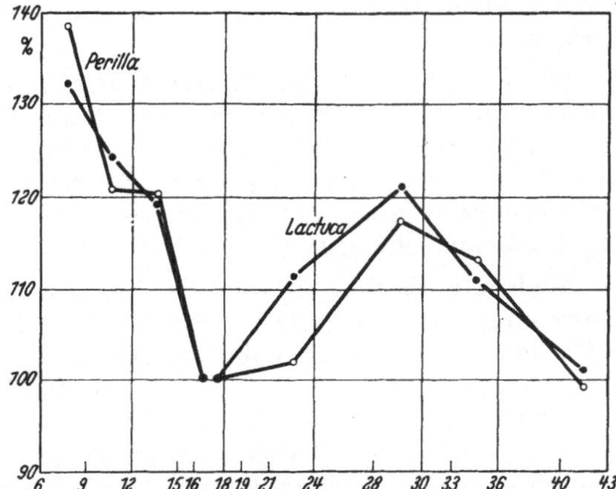

Abb. 20. Unterschiedliche Wirkung eines zu verschiedenen Zeiten nach einer Haupt-Licht-Periode gebotenen kürzeren Zusatzlichtes auf die Chlorophyllbildung bei Keimpflanzen von *Lactuca* und *Perilla*. Abszisse: Stunden der Darbietung des Zusatzlichtes nach der Haupt-Licht-Periode. (Nach CLAUSS.)

Es fehlt nicht an Versuchen, den Mechanismus der physiologischen Selbstregulation der endogenen Tagesrhythmik energetisch zu erklären. Es ist verfrüht, hier Hypothesen aufzuwerfen. Sehr wesentlich ist aber doch wohl, daß eine Phase der Energiespeicherung mit einer Phase des Energieumsatzes regelmäßig, vielleicht in allen lebenden Zellen alterniert und dabei eben ungefähr der Tagesrhythmus eingehalten wird. Das scheint ein ganz grundsätzliches Prinzip in der Arbeitsweise lebender Zellen zu sein.

Berührt wird diese Frage auch von TAMIYA und Mitarbeitern auf Grund von Untersuchungen an *Chlorella*. Die Algen waren einem 12:12 Std-Licht-Dunkel-Rhythmus ausgesetzt. Die Ergebnisse lassen einen Antagonismus der Zelleistungen in der genannten Art erkennen. Auch hier kommt dieser Antagonismus übrigens ähnlich wie bei den besprochenen Formänderungen der Plastiden in wechselnden Zellformen zum Ausdruck. TAMIYA und Mitarbeiter fassen diesen Antagonismus als einen solchen von Photosynthese und formativem Metabolismus auf. Sie bringen in ihrer Darstellung zum Ausdruck, daß sowohl die Phase photosynthetischer Aktivität, als auch die der formativen Aktivität eine bestimmte Zeit andauern muß, bis sie in die entsprechende andere Phase übergeht. An die Phase des formativen Metabolismus ist beispielsweise auch die Bildung der Autosporen gebunden. Das ist ein Schema, welches sich recht gut den beschriebenen Tatsachen über die endogene Tagesrhythmik einfügt. Es paßt sowohl zu den erwähnten Rhythmen der Fermentaktivität und des Alternierens von Assimilation und Dissimilation, als auch zu den beschriebenen Rhythmen der Sporulation. Ferner sei hier an den Tagesrhythmus der Mitose erinnert, der einen solchen

Zusammenhang mit einem Wechsel von Phasen der Energiespeicherung und des Energieumsatzes vermuten läßt.

Erwähnung verdient auch die Untersuchung von GALSTON und DALBERG. Diese Autoren vermuten, daß die endogene Tagesrhythmik mit einer adaptiven Bildung von IES-Oxydase durch allmähliche Auxinanhäufung besteht, so daß nach der entsprechend starken Bildung dieses Ferments Auxininaktivierung und damit eine Phase geringerer Auxinmenge eintritt. Tagesperiodische Schwankungen der Auxinkonzentration sind tatsächlich mit der endogenen Tagesrhythmik verknüpft (vgl. z. B. BECKER). Bei *Kalanchoë blossfeldiana* zeigt sich ein Maximum der Wuchsstoffabgabe aus Laub- und Blütenblättern etwa 3—6 Std nach Beginn der Lichtperiode, in Dauerdunkelheit, d. h. beim Fortfall des vorhergehenden Licht-Dunkel-Wechsels, 6+24, 6+48 Std usw. nach Beginn der letzten Lichtperiode.

Die Idee, daß der Mechanismus der endogenen Tagesrhythmik mit Auxinschwankungen zusammenhängt, hat BALL und DYKE (1956) veranlaßt, den Einfluß von Wuchsstoffen auf die endogen tagesperiodischen Wachstumsrhythmen von *Avena*-Koleoptilen zu prüfen. Es war aber kein Einfluß nachweisbar, der die Vermutung einer entscheidenden ursächlichen Beziehung zwischen Auxinrhythmik und endogener Tagesrhythmik hätte rechtfertigen können.

Noch interessanter ist wohl der Versuch BÜHNEMANNs, die Rhythmik durch Anwendung verschiedener Stoffwechselgifte zu stören. BÜHNEMANN (1955a) hat bei seinen Versuchen über die Rhythmik der Sporenentleerung von *Oedogonium* verschiedene Gifte wie NaCN, Arsenat, 2,4-Dinitrophenol, NaF u. a. benutzt. In allen diesen Fällen blieb aber die Sporenentleerung ein cyclischer Vorgang, solange sie selber überhaupt noch erkennbar war. Irgendeine Verschiebung der Phasen oder eine Verlängerung der Perioden war nicht feststellbar, selbst wenn die Sporenbildung schon sehr stark reduziert worden war. In diesen Versuchen ist ein Punkt noch von ganz besonderem Interesse. Wenn die Gifte längere Zeit eingewirkt haben und ihre Konzentration sich dann allmählich verringert, beginnt der Prozeß der Sporenbildung erneut periodisch. Und dabei ist interessanterweise auch keine Verschiebung der Phasen feststellbar. Die Maxima und Minima treten zu der gleichen Zeit ein wie bei den Kontrollen. Das endodiurnale System läuft also weiter, während von seinen Folgen nichts mehr zu erkennen ist.

Diese Beobachtung könnte zu der Vermutung führen, daß physikalische Vorgänge im Mechanismus der Rhythmik wichtiger sind als chemische. In einer neueren Untersuchung (BÜNNING 1956b) wurde daraufhin der Einfluß verschiedener äußerer Faktoren geprüft. Bemerkenswert ist zunächst, daß das endodiurnale System relativ hitzeresistent ist. Auch in Zellen, die bis nahe zum Absterben erhitzt worden waren, arbeitet es (bei *Phaseolus multiflorus*) unverändert weiter. Eine Störung im Gang der endogenen Rhythmik konnte durch einige Gifte, besonders durch Colchicin, Urethan und vor allem Phenylurethan hervorgerufen werden. Nach der Einwirkung solcher Stoffe zeigen sich Unregelmäßigkeiten im Gang der endogenen Tagesrhythmik und auch eine merkliche Verlängerung der Perioden bis zu etwa 36 Std. Durch viele andere äußere Faktoren, auch z. B. durch extreme Acidität, wird die endogene Rhythmik erst merklich modifiziert, wenn auch starke Schäden an den Pflanzen erkennbar sind. Es ist natürlich sehr auffällig, daß die endogene Rhythmik nach diesen Kenntnissen nur durch Faktoren beeinflußt werden kann, die sonst als Störer der Mitose bekannt sind. Das ist verwunderlich, weil in den betreffenden Fällen Mitoseprozesse überhaupt nicht mehr ablaufen, z. B. die Bewegungen der

Primärblätter ja an Organe gebunden sind, bei denen selbst das Streckungswachstum schon fast beendet ist.

Will man die verschiedenen Befunde zu einem Bild vereinigen, so wird man wohl sagen dürfen, daß offensichtlich innerhalb des Plasmas periodische Zustandsänderungen ablaufen können. Diese periodischen Zustandsänderungen sind für die verschiedenartigen Phänomene der endogenen Tagesrhythmik (Schwankungen von Enzymaktivitäten, Turgor, Wachstum usw.) verantwortlich zu machen. Die gleichen cyclischen Strukturänderungen würden auch für den Mitoserhythmus mitverantwortlich sein. Auch die Mitose ist ja bekanntlich ein Vorgang, der weitgehend tagesperiodisch gebunden sein kann. Fragen wir nach dem Sitz dieser cyclischen Veränderungen, so wird man in erster Linie an die Proteine denken müssen, die normalerweise die Teilungsspindel aufbauen. Aus cytologischen Untersuchungen ist ja die Schlußfolgerung abgeleitet worden, daß der Spindelbildung während der Mitose strukturelle Änderungen in der Form und Anordnung dieser Proteine zugrunde liegen. Wir würden nun zu der Schlußfolgerung gezwungen sein, daß diese periodischen Änderungen in eben diesen Proteinen noch ablaufen, wenn eine ihrer Folgen, nämlich die Mitose, gar nicht mehr möglich ist. Und auch für diese Schlußfolgerung sprechen Tatsachen. Man kann nämlich nachweisen, daß in Zellkernen nicht mehr teilungsfähiger Zellen noch tagesperiodische Strukturänderungen ablaufen. Sie äußern sich vor allem in Volumenschwankungen der Kerne (Bünning unveröffentlicht). Solche Schwankungen sind bekanntlich auch mit der Mitose verbunden (Volumenzunahme in der Prophase). Jetzt sehen wir eben, daß die ihnen zugrunde liegenden strukturellen Veränderungen zwar eine notwendige, nicht aber eine hinreichende Bedingung für die Mitose sind, und daß sie noch cyclisch weiterlaufen, wenn die Mitose gar nicht mehr möglich ist. Das also etwa ist der cytologische Ort, an dem wir die endogene Tagesrhythmik lokalisieren müssen, und wo die weitere Untersuchung dieses Phänomens anzugreifen hat.

Zum Verständnis des Wesens der physiologischen Selbstregulation, die in einer endogenen Tagesrhythmik zum Ausdruck kommt, muß noch erwähnt werden, daß sich das gleiche Phänomen auch bei Tieren zeigt, wo es in neuerer Zeit gründlich untersucht wird. Dabei hat sich eine ganz analoge Regulierbarkeit durch den Licht-Dunkel-Wechsel ergeben wie bei Pflanzen (vgl. z. B. Tribukait, Aschoff). Aber nicht nur die Regulierung durch äußere Faktoren, d. h. besonders durch den Licht-Dunkel-Wechsel, sowie die Art der Fortsetzung unter konstanten Außenbedingungen hat die endogene Tagesrhythmik bei Pflanzen und Tieren gemeinsam. Eine Übereinstimmung scheint auch darin zu bestehen, daß in beiden Fällen die in Dauerlicht oder Dauerdunkelheit feststellbare Abhängigkeit der Periodenlänge von der Temperatur nur sehr gering ist (Brown, Fingerman und Mitarbeiter). Bei Tieren und Pflanzen hat die endogene Tagesrhythmik übrigens auch noch gemeinsam, daß beim Fortfall der regulierenden Außenbedingungen Abweichungen der Cyclenlänge von der Länge des Tages auftreten. Diese Cyclenlängen können dann nämlich zwischen etwa 22 und 26 Std betragen.

Schon die mitgeteilten Tatsachen zeigen, daß mit der endogenen Tagesrhythmik nicht nur quantitative Veränderungen in den Leistungen der Zellen auftreten, sondern auch qualitative. Auf solche qualitative Veränderungen deutet schon die Tatsache hin, daß in der einen Phase die Atmungsvorgänge, in der anderen die Assimilationsvorgänge begünstigt sind.

Noch deutlicher aber wird diese qualitative Verschiedenheit, wenn wir beachten, daß die Pflanzen in den beiden Phasen der endogenen Rhythmik auf Licht geradezu entgegengesetzt reagieren können. In der einen Phase kann Licht

Entwicklungsvorgänge (etwa die Blütenbildung) begünstigen, während es die gleichen Vorgänge bei einer Darbietung in der anderen Phase gerade hemmt. Das ist die entscheidende Grundlage für die photoperiodischen Reaktionen. Jedoch kann darauf an dieser Stelle nicht näher eingegangen werden. Ebenso reagiert die Pflanze in den beiden Phasen der endogenen Rhythmik auch auf die Temperatur ganz unterschiedlich. Zum Beispiel äußert sich das darin, daß das Temperaturoptimum tagesperiodische Schwankungen zeigt und extreme Temperaturen einen unterschiedlich starken Einfluß haben, je nachdem, zu welchem Zeitpunkt sie geboten werden (Schwemmle, Abb. 21). Damit erscheint die endogene Tagesrhythmik auch als eine wichtige Grundlage für die thermoperiodischen Reaktionen.

Mit der endogenen Tagesrhythmik eng verwandt sind wohl die Gezeitenrhythmen, die bei mehreren marinen Tieren und Pflanzen auftreten. Bei ihnen

Abb. 21. *Brassica nigra.* Nach der täglichen 4stündigen Lichtperiode erhielten die Versuchsgruppen 1—5 zu verschiedenen Zeiten der täglichen Dunkelphase 4stündige Perioden niederer Temperatur, und zwar Gruppe 1 4—8 Std (unmittelbar an die Lichtperiode anschließend), Gruppe 2 8—12 Std, Gruppe 3 12—16 Std, Gruppe 4 16—20 Std und Gruppe 5 20—24 Std (unmittelbar vor der Lichtperiode) nach Lichtbeginn. Kontrolle (V) ohne nächtliche Kälteperiode. (Versuchsdauer 7 Tage; von jeder Versuchsgruppe wurde willkürlich ein Topf ausgewählt.) (Nach Schwemmle.)

beträgt die Periodenlänge etwa 12 Std. Endogene Gezeitenrhythmen wurden z. B. von Bracher für *Euglena limosa*, von Fauré-Fremiet für *Chromulina* und für die Diatomee *Hantzschia* beschrieben. Eine solche etwa 12stündige Rhythmik kann man ohne weiteres als Erfolg einer Aufspaltung der 24 Std-Rhythmik verstehen. Auch bei beliebigen anderen Objekten kann man diese Aufspaltung durch bestimmte äußere Faktoren erreichen. Beispielsweise kann man auf den tagesperiodischen Bewegungen durch bestimmte Lichtreize zusätzliche „Oberschwingungen" induzieren, die sich mit dieser Rhythmik überlagern und dann immer wieder zur gleichen Zeit auftreten. Wie eng die Gezeitenrhythmik mit der endogenen Tagesrhythmik verknüpft ist, ergibt sich z. B. aus den Untersuchungen von Brown, Fingerman und Mitarbeitern an der Krabbe *Uca pugnax:* Jede Verschiebung des Gezeitenrhythmus durch besondere Lichtbedingungen bedeutet zugleich auch eine Verschiebung der endogenen Tagesrhythmik. Diese Beziehung zwischen den beiden Rhythmen wird für *Euglena* noch besonders dadurch nahegelegt, daß einerseits bei den marinen Formen taktische Stimmungsänderungen bestehen, die das Auf- und Absteigen mit den Gezeiten ermöglichen, andererseits bei anderen Euglenen auch eine endogene Tagesrhythmik des phototaktischen Verhaltens gefunden wurde (Pohl).

F. Etwa tagesperiodische Rhythmen mit induzierten Cyclenlängen.

Gesondert behandelt werden müssen hier die Rhythmen, die Pirson und Mitarbeiter bei Algen, namentlich bei *Hydrodictyon*, fanden (Pirson, Schön, und Döring, Pirson und Döring, Neeb, Abb. 22—26). Gefunden wurde eine Periodizität des Wachstums, der Photosynthese und der Dunkelatmung. Nach einem 12:12stündigen Licht-Dunkel-Wechsel setzen sich solche periodischen Schwankungen auch im Dauerlicht oder im Dauerdunkel fort. Die Phänomene,

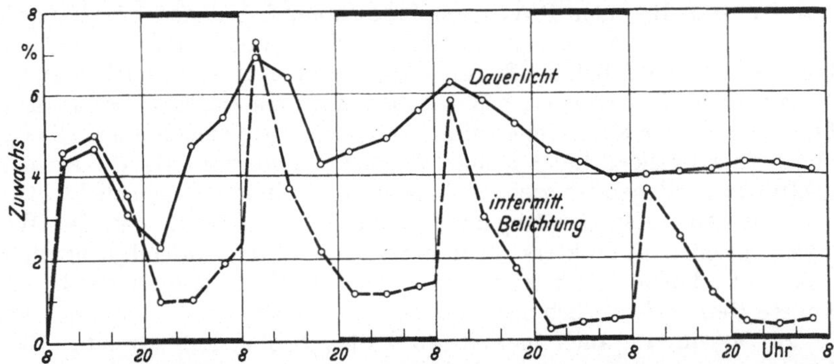

Abb. 22. *Hydrodictyon*. Wachstumsverlauf bei intermittierender (gestrichelte Kurve) und bei kontinuierlicher Belichtung (ausgezogene Kurve). Die Kurvenpunkte stellen die mittlere prozentuale Verlängerung von je fünf kleinen Netzen (Zellänge bei Versuchsbeginn etwa 100 μ) in 4 Std dar. Beleuchtungsstärke 250 Lux. Dunkelzeiten durch dicke Striche gekennzeichnet. (Nach NEEB.)

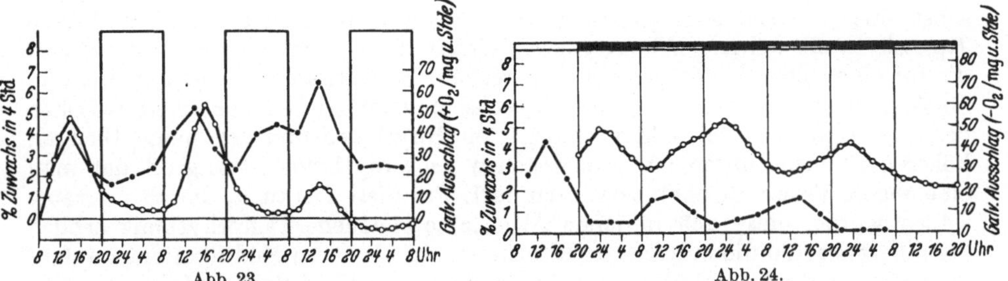

Abb. 23. Wachstumsleistung (Zuwachsprozente in 4 Std, ausgezogene Kurve) und Photosyntheserate (Galvanometerausschlag je Stunde und Milligramm Trockengewicht, gestrichelte Kurve) von *Hydrodictyon* im Dauerlicht nach vorherigem Licht-Dunkel-Wechsel von 12:12 Std. Beleuchtungsstärke 1000 Lux. Temperatur 20°. Die geschlossenen Felder zeigen die Lage der ehemaligen Dunkelzeiten. (Nach PIRSON, SCHÖN und DÖRING.)

Abb. 24. Wachstumsleistung (ausgezogene Kurve) und respiratorischer O_2-Verbrauch (gestrichelte Kurve) von *Hydrodictyon* bei Übergang vom 12:12 Std-Wechsel zu Dauerverdunklung. Die Lage der ehemaligen Lichtzeiten ist durch Markierung über den Kurven gekennzeichnet. Links im Diagramm die letzte Lichtzeit. (Nach PIRSON, SCHÖN und DÖRING.)

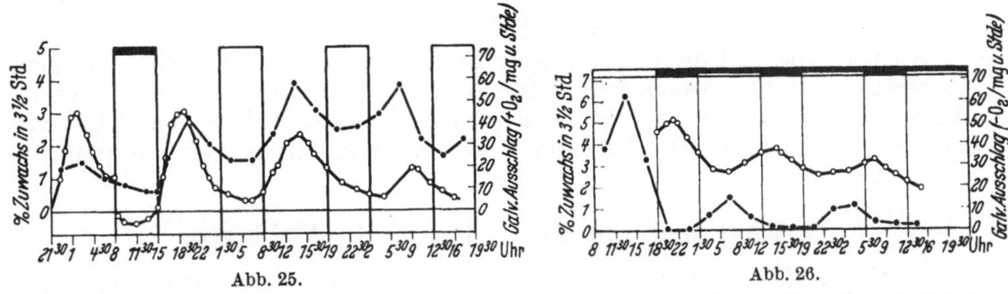

Abb. 25. Wachstumsleistung (Zuwachsprozente in $3^1/_2$ Std, ausgezogen) und Photosynthese (gestrichelt) beim Übergang vom $10^1/_2$:7 Std-Wechsel zu Dauerlicht. Links im Diagramm der letzte Licht-Dunkel-Wechsel. In der Dunkelzeit ist (bei unveränderter Empfindlichkeit) die Atmung eingetragen. (Nach PIRSON, SCHÖN und DÖRING.)

Abb. 26. Wachstumsleistung (ausgezogen) und respiratorischer O_2-Verbrauch (gestrichelt) beim Übergang vom $10^1/_2$:7 Std-Wechsel zu Dauerverdunklung. Links die letzte Lichtzeit. (Nach PIRSON, SCHÖN und DÖRING.)

auch die Regulierbarkeit der Rhythmik durch äußere Faktoren, entsprechen weitgehend dem, was auch für die endogene Tagesrhythmik festgestellt worden ist. Eine besondere Situation scheint bei diesen Algen aber insofern gegeben zu sein, als die Cyclenlänge der bei konstanten Bedingungen auftretenden Periodizität von der Cyclenlänge des vorhergehenden Licht-Dunkel-Wechsels abhängt. Man kann also z. B. auch Rhythmen induzieren, bei denen die Länge eines Cyclus nicht 24, sondern beispielsweise 12 oder 18 Std beträgt. Definitionsgemäß müssen wir natürlich die Schwankungen, die nach einer solchen Induktion unter konstanten Bedingungen beobachtbar sind, als endogen bezeichnen. Es wäre hier eben nur die Cyclenlänge dieser endogenen Rhythmen durch die Cyclenlänge der vorhergehenden Licht-Dunkel-Rhythmik determiniert. Die Anwendung des Ausdrucks „endogen" bezieht sich ja in allen diesen Fällen nicht etwa auf ein spontanes Auftreten der Rhythmik, sondern nur darauf, daß der Wechsel des physiologischen Charakters auch unter konstanten Außenbedingungen erfolgen kann. Daß dazu eine bestimmte Art der Vorbehandlung notwendig ist, daß etwa äußere Reize erforderlich sein können, um die Rhythmik überhaupt deutlich werden zu lassen, ist auch in anderen Fällen beobachtet worden. Die Verschiedenheit der von PIRSON und Mitarbeitern untersuchten Rhythmen von der endogenen Tagesrhythmik besteht also nur darin, daß auch die Cyclenlänge der endogenen Schwankungen von der Art der Vorbehandlung abhängig sein kann. Aber auch das ist wohl keine besonders große Abweichung von dem, was sonst über die endogene Tagesrhythmik bekannt ist. Auch für die endogene Tagesrhythmik höherer Pflanzen liegen einige ältere Angaben von PFEFFER vor, die für eine solche Modifizierbarkeit sprechen. Es sollte also doch geprüft werden, ob sie nicht verbreiteter ist. Die stoffwechselphysiologische Ähnlichkeit zwischen den von PIRSON untersuchten induzierten Rhythmen und den namentlich an höheren Pflanzen untersuchten endogenen Tagesrhythmen ist so groß, daß man eine engere Verwandtschaft vermuten darf. Für diese Deutung, die also voraussetzen würde, daß auch in jenen Algen eine endogene Tagesrhythmik erblich verankert ist und die induzierten Rhythmen nur als deren Modifikationen aufzufassen sind, spricht, daß die Algen für ihr normales Wachstum und ihre normale Entwicklung auf Licht-Dunkel-Rhythmen angewiesen sind, die sich etwa dem 24 Std-Cyclus einfügen. Namentlich im Dauerlicht treten starke Entwicklungsstörungen auf (NEEB). Dieses Phänomen, das der Beziehung zwischen endogener Tagesrhythmik und Photoperiodismus bei höheren Pflanzen entspricht, ist jedenfalls am ungezwungensten durch die Annahme zu erklären, daß jene Algen ebenfalls eine endogene Rhythmik besitzen, die aber, leichter als bei anderen Pflanzen, auch hinsichtlich der Cyclenlänge modifiziert werden kann.

In einer neueren Untersuchung weist SCHÖN darauf hin, daß kein einfacher biochemischer Mechanismus vorstellbar ist, der diese Rhythmen ermöglichen könnte. Bemerkenswert ist vor allem, daß auch durch Zufuhr von CO_2 oder Glucose der Gang der Rhythmik nicht wesentlich modifiziert werden kann.

G. Rhythmen mit noch kürzeren Cyclenlängen.

Nur der Vollständigkeit halber sei hier auch auf das Vorkommen von Aktivitätsrhythmen verwiesen, bei denen eine Cyclenlänge weniger als 1 Tag, oft sogar weniger als 1 Std beträgt. Es handelt sich dabei um Phänomene, die zum großen Teil an anderen Stellen dieses Handbuches (vgl. besonders Band XVII) in einem anderen Zusammenhang zu besprechen sind.

Beim Wachstum findet man oft Schwankungen mit meist nicht sehr regelmäßigen Abständen der Maxima, bei denen dieser Abstand dann einige Stunden oder auch annähernd 1 Tag betragen kann. Derartige Erscheinungen sind sowohl an höheren Pflanzen als auch an Pilzen und bei Bakterienkulturen gefunden worden (Abb. 27). Eine genauere Analyse fehlt. Ebenso sind wiederholt auch Turgorschwankungen beschrieben worden. Zum Beispiel können nach STÅLFELT die Schließzellen der Spaltöffnungen relativ schnell verlaufende Pulsationen aufweisen.

Die Fähigkeit der Pflanze, endogen-periodische Wachstums- und Turgorschwankungen durchzuführen, kommt auch in einer ganzen Reihe von Bewegungserscheinungen zum Ausdruck. Besonders kann hingewiesen werden auf die bekannten autonomen Bewegungen bei *Desmodium gyrans* und mehreren anderen Pflanzen (Abb. 28, vgl. HOSSEUS). Die endogenen kurzperiodischen Pulsationen in Schließzellen von Spaltöffnungen sind hiermit vergleichbar (STÅLFELT 1956). Auch die sog. Cyclonastie und die Circumnutationsbewegungen bei sehr vielen Pflanzen können in diesem Zusammenhang erwähnt werden. Viele Bewegungen beruhen darauf, daß die einzelnen Flanken eines Achsenorgans periodische Schwankungen ihrer Wachstumsaktivität zeigen, dabei aber kein Synchronismus der einzelnen Flanken besteht, diese sich vielmehr in den Phasen stärksten Wachstums ablösen, also eine Welle hoher Wachstumsgeschwindigkeit in einer Richtung um das Achsenorgan herumläuft.

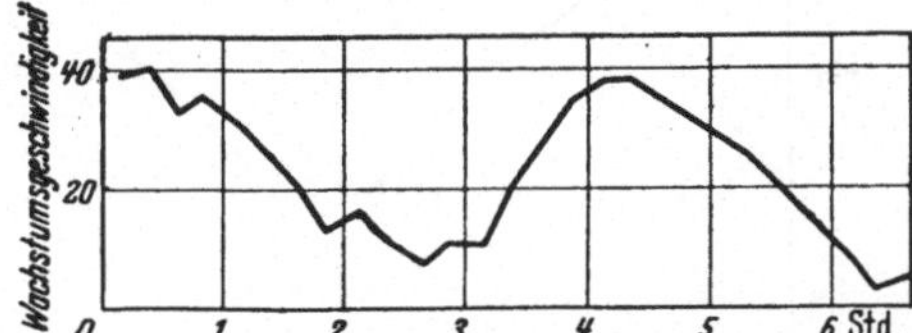

Abb. 27. *Coprinus lagopus*, im Dunkeln gewachsen. Abszisse: Zeit in Stunden. Ordinate: Wachstumsgeschwindigkeit in willkürlichen Einheiten. Endogene kurzperiodische Schwankungen der Wachstumsgeschwindigkeit. (Nach BORRISS.)

Solche autonomen Schwankungen des Wachstums und des Turgors sind natürlich auch immer mit Schwankungen anderer physiologischer Werte verbunden, etwa mit Permeabilitätsschwankungen oder Schwankungen der elektrischen Potentiale.

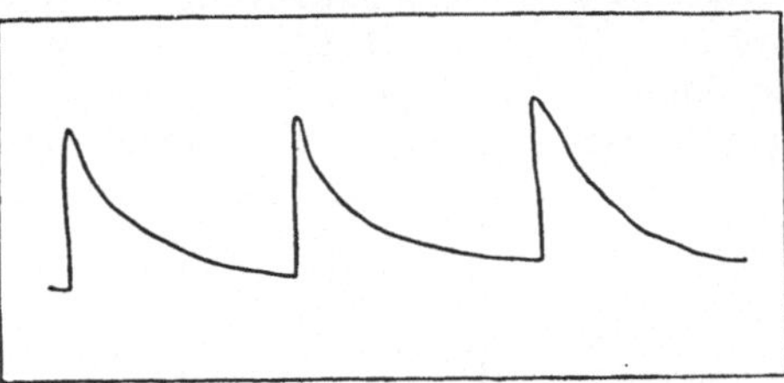
Abb. 28. *Averrhoa bilimbi*. Endonome Blättchenbewegung im Dauerlicht. Jede Bewegung erfordert etwa 3 min. (Nach DAS und PALIT.)

Aus der neueren Literatur können etwa die Untersuchungen ARNALs erwähnt werden, nach denen die Coleoptilen von *Triticum* Nutationsbewegungen mit einer Amplitude von 5—8 mm durchführen, wobei die Länge einer Periode 3 Std beträgt (Abb. 29). ARNAL führt diese Rhythmik auf eine entsprechende Schwankung der Auxinzufuhr zurück, was freilich näherer Begründung bedarf.

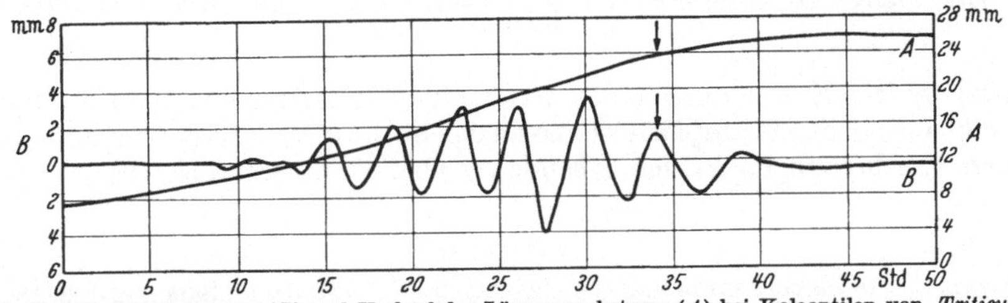

Abb. 29. Nutationsbewegungen (*B*) und Verlauf des Längenwachstums (*A*) bei Koleoptilen von *Triticum*. Die Nutationskurve gibt die Abweichungen der Koleoptilspitze von der Nullage an, der Pfeil den Zeitpunkt des Durchstoßens der Koleoptile vom Primärblatt. (Nach ARNAL.)

Schließlich kann hier noch auf die zahlreichen Arbeiten von TRONCHET und Mitarbeitern (BAILLAUD) verwiesen werden, in denen der Verlauf solcher Nutationsbewegungen ausführlicher beschrieben wird. Erwähnt sei von den Ergebnissen dieses Arbeitskreises nur, daß der Temperaturkoeffizient der Rotationsbewegungen von *Cuscuta odorata* zwischen 10 und 25° etwa 2 beträgt (BAILLAUD).

Auch kurzperiodische Änderungen der Viscosität sind bei Pflanzen beobachtet worden (Abb. 30, STÅLFELT 1946).

Wollten wir diese Phänomene weiterbehandeln, so müßten wir allgemein auf die Erscheinungen rhythmischer Erregbarkeit eingehen, die dem Protoplasma immer zuzukommen scheinen. Hier sei aber nur kurz darauf hingewiesen. Wir

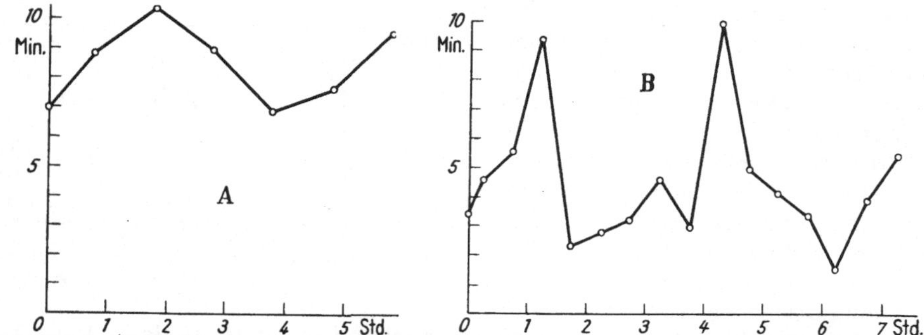

Abb. 30. Periodische Änderungen der Viscosität in Epidermiszellen der Blätter von *Helodea densa*. Abszisse: Zeit in Stunden, Ordinate: relative Werte der Viscosität, gemessen an der zur Chloroplastenverlagerung erforderlichen Zeit des Zentrifugierens. (Nach STÅLFELT.)

finden rhythmische „Erregung" bei der Periodizität der Geißelbewegungen, bei dem von KAMIYA und Mitarbeitern untersuchten rhythmischen Verlauf der Plasmaströmung von *Physarum* und bei der ebenfalls an diesem Objekt gefundenen rhythmischen Torsion eines aufgehängten Plasmastranges (Abb. 31, KAMIYA und

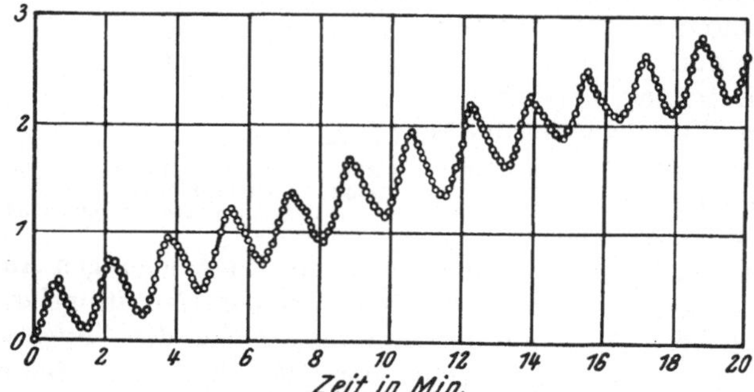

Abb. 31. Rhythmische Schwingungen des unteren Endes eines frei hängenden Plasmodiumstranges von *Physarum*. Abszisse: Zeit in Minuten (Messung alle 5 sec). Ordinate: Torsion (1 Teilstrich = 20°.) (Nach KAMIYA und SEIFRIZ.)

SEIFRIZ). Solche Erscheinungen deuten auf eine allgemein verbreitete rhythmische Kontraktibilität des Protoplasmas, die vielleicht für diese kurzperiodischen rhythmischen Erscheinungen in den lebenden Zellen verantwortlich ist.

Literatur.

ARNAL, C.: Recherches sur la nutation des coléoptiles. Ann. Univ. Saraviensis **2**, 186—203 (1953). — Recherches sur la nutation des coléoptiles. I. Nutation et croissance. Ann. Univ. Saraviensis **2**, 92—105 (1953). — IV. Morphologie du mouvement de nutation. Rev. gén. Bot. **62**, 661—666 (1955). — ASCHOFF, J.: Zeitgeber der tierischen Tagesperiodik. Naturwiss. **41** (3), 49—56 (1954). — Jahresperiodik der Fortpflanzung bei Warmblütern. Studium gen. **8**, 742—776 (1956). — ASHBY, E., and E. WANGERMANN: Studies in the morphogenesis of leaves. IV. Further observations on area, cell size and cell number of leaves of *Ipomoea* in relation to their position in the shoot. New Phytologist **49**, 23—35 (1950). — Studies in the morphogenesis of leaves. V. A note on the origin of differences in cell size among leaves at different levels of insertion on the stem. New Phytologist **49**, 189—192 (1950).

BAILLAUD, L.: Nutation d'une vrille ramfiée de Lagenaria vulgaris. Bull. Soc. Histoire natur. Doubs, **1950**, Nr 54, 101—104. — Remarques sur le mouvement de nutation des tiges volubiles. Ann. Sci. Univ. Besançon Bot. **1951/52**, 29—40. — Action de la température sur la période de nutation des tiges volubiles de Cuscute. C. r. Acad. Sci. Paris **236**, 1986—1988 (1953). — BALL, N. G., and I. J. DYKE: An endogenous 24-hour rhythm in the growth rate of the Avena coleoptile. J. of Exper. Bot. **5**, 421—433 (1954). — The effects of indole-3-acetic acid and 2:4-dichlorophenoxyacetic acid on the growth rate and endogenous rhythm of intact Avena coleoptile. J. of Exper. Bot. **7**, 25—41 (1956). — BARANETZKY, J.: Die tägliche Periodizität im Längenwachstum der Stengel. Mém. Acad. impériale Sci. St.-Pétersbourg, VII. s. **27**, Nr 2 (1879). — BECKER, T.: Wuchsstoff- und Säureschwankungen bei *Kalanchoë Bloßfeldiana* in verschiedenen Licht-Dunkelwechseln. Planta (Berl.) **43**, 1—24 (1953). — BORDAGE, E.: A propos de l'hérédité des caractéres acquis. Bull. Sci. France Belg., VII. s. **54** (1910). — BRACHER, R.: The light relation of Euglena limosa Gard. Part I. The influence of intensity and quality of light on phototaxy. J. Limn. Soc. Bot. **51**, 23—42 (1932). — BROWN, F. A., M. FINGERMAN, M. I. SANDEEN and M. H. WEBB: Persistent diurnal and tidal rhythms of color change in the fiddler crab, Uca pugnax. J. of Exper. Zool. **123**, 29—60 (1953). — BROWN, F. A., R. O. FREELAND and C. L. RALPH: Persistent rhythms of O_2-consumption in potatoes, carrots and the sea weed, *Fucus*. Plant. Physiol **30**, 280—292 (1955). — BROWN, F. A., and H. M. WEBB: Temperature relations of endogenous daily rhythmicity in the fiddler crab, Uca. Physiologic. Zool. **21**, 371—381 (1948). — BÜHNEMANN, F.: Die rhythmische Sporenbildung von Oedogonium cardiacum Wittr. Biol. Zbl. **74**, 1—54 (1955). — Das endodiurnale System der Oedogoniumzelle. II. Biol. Zbl. **74**, 691—705 (1955a). — III. Z. Naturforsch. **10**b, 305—312 (1955b). — IV. Planta (Berl.) **46**, 227—255 (1955c). — BÜNNING, E.: Untersuchungen über die autonomen tagesperiodischen Bewegungen der Primärblätter von *Phaseolus multiflorus*. Jb. wiss. Bot. **75** (3), 439—480 (1931). — Die Entstehung der Variationsbewegungen bei den Pflanzen. Erg. Biol. **13**, 235—347 (1936). — Untersuchungen über den physiologischen Mechanismus der endogenen Tagesrhythmik bei Pflanzen. Z. Bot. **37**, 433—486 (1942). — Zur Physiologie der endogenen Jahresrhythmik in Pflanzen, speziell in Samen. Z. Naturforsch. **4**b, 167—176 (1949). — Über die Ursachen der Blühreife und Blühperiodizität. Z. Bot. **40**, 293—306 (1952). — Endogenous rhythms in plants. Annual Rev. Plant Physiol. **7** (1956a). — Versuche zur Beeinflussung der endogenen Tagesrhythmik durch chemische Faktoren. Z. Bot. **44** (1956b). — BÜNNING, E., u. E. W. BAUER: Über die Ursachen endogener Keimfähigkeitsschwankungen in Samen. Z. Bot. **40**, 67—76 (1952). — BÜNNING, E., u. F. J. LEINWEBER: Die Korrelation des Temperaturfehlers der endogenen Tagesrhythmik. Naturwiss. **43**, 42 (1956). — BÜNNING, E., u. L. MÜSSLE: Der Verlauf der endogenen Jahresrhythmik in Samen unter dem Einfluß verschiedenartiger Außenfaktoren. Z. Naturforsch. **6**b, 108—112 (1951). — BÜNSOW, R.: Endogene Tagesrhythmik und Photoperiodismus bei *Kalanchoë Bloßfeldiana*. Planta (Berl.) **42**, 220—252 (1953a). — Über tages- und jahresrhythmische Änderungen der photoperiodischen Lichtempfindlichkeit bei *Kalanchoë Bloßfeldiana* und ihre Beziehungen zur endogenen Tagesrhythmik. Z. Bot. **41**, 257—276 (1953b). — Über den Einfluß der Lichtmenge auf die endogene Tagesrhythmik bei *Kalanchoë Bloßfeldiana*. Biol. Zbl. **72** (9/10), 465—477 (1953). — BUSCH, G.: Über die photoperiodische Formänderung der Chloroplasten von *Selaginella serpens*. Biol. Zbl. **72**, 598—629 (1953).

CLAUSS, H.: Der Blattfarbstoffgehalt in verschiedenen Licht-Dunkel-Rhythmen gezogener Keimpflanzen. Z. Bot. **42**, 215—245 (1954). — CLAUSS, H., u. W. RAU: Über die Blütenbildung von Hyoscyamus niger und Arabidopsis Thaliana in 72-Stunden-Zyklen. Z. Bot. **44** (1956).

DEN DOPP, J. E. A.: Nicht blühende Sisalpflanzen. Faserforsch. **14**, 9—27 (1939). — DENFFER, D. v.: Die hormonale Beeinflussung pflanzlicher Gestaltungsprozesse. Ber. oberhess. Ges. Naturwiss. Heilk., N. F. **25**, 123—133 (1952). — DINGLER, H.: Versuche über die Periodizität einiger Holzgewächse in den Tropen. Sitzgsber. bayer. Akad. Wiss. München, Math.-physik. Kl. **1911**, 127.

EBNER, H.: Keimungsphysiologie von *Draba verna*, *Thlaspi perfoliatum*, *Holosteum umbellatum* und *Veronica hederifolia*. Österr. bot. Z. **73**, 23—41 (1924). — EHRENBERG, M.: Beziehungen zwischen Fermenttätigkeit und Blattbewegung bei Phaseolus multiflorus unter verschiedenen photoperiodischen Bedingungen. Planta (Berl.) **38**, 244—279 (1950). — ENGEL, H., u. I. FRIEDERICHSEN: Weitere Untersuchungen über periodische Guttation etiolierter Haferkeimlinge. Planta (Berl.) **40**, 529—549 (1952).

FAURÉ-FREMIET, E.: Rhythme de marée d'une Chromulina psammophile. Bull. biol. France et Belg. **84**, 207—214 (1950). — The tidal rhythm of the diatom Hantzschia amphioxys. Biol. Bull. **100**, 173—177 (1951). — FLÜGEL, A.: Die Gesetzmäßigkeiten der endogenen Tagesrhythmik. Planta (Berl.) **37**, 337—375 (1949).

GALSTON, A. W., and L. Y. DALBERG: The adaptive formation and physiological significance of indoleacetic acid oxydase. Amer. J. Bot. **41**, 373—380 (1954). — GROOT jr., G. J. DE:

On the mechanism of periodic movements of variation. Rec. Trav. bot. néerl. 35, 759—833 (1938). — GROSSENBACHER, K. A.: Autonomic cycle of rate of exudation of plants. Amer. J. Bot. 26, 107—109 (1939).

HAGAN, R. M.: Autonomic diurnal cycles in the water relations of nonexuding detopped root systems. Plant Physiol. 24, 441—454 (1949). — HARDER, R.: Über die endogene Tagesrhythmik der Fermentaktivität, Guttation und Blütenbewegung bei *Kalanchoe Bloßfeldiana* und *Phaseolus multiflorus*. Nachr. Akad. Wiss. Göttingen, Math.-physik. Kl. 1949. — HEIMANN, M.: Einfluß periodischer Beleuchtung auf die Guttationsrhythmik. Planta (Berl.) 38, 157—195 (1950). — Abhängigkeit des Blutungsverlaufes von Beleuchtung und Blattzahl. Planta (Berl.) 40, 377—390 (1952). — HENSSEN, A.: Die Dauerorgane von Spirodela polyrrhiza (L.) Schleid. in physiologischer Betrachtung. Flora (Jena) 141, 523—566 (1954). — HESS, C.: Untersuchung über die Rhythmik der Schichtenbildung beim Stärkekorn. Z. Bot. 43, 181—204 (1955). — HOSSEUS, C.: Über die Beeinflussung der autonomen Variationsbewegungen durch einige äußere Faktoren. Diss. Leipzig 1903. — HOWARD, W. L.: An experimental study of the rest period in plants. Pot grown woody plants. Mo. Sta. Res. Bull. 16 (1915). — An experimental study of the rest period in plants. Fifth report. Mo. Agric. Exper. Sta. Res. Bull. 21 (1915). — HOYT, W. D.: The periodic fruiting of *Dictyota* and its relation to the environment. Amer. J. Bot. 14, 592—619 (1927). — HUBER, B.: Physiologische Rhythmen im Baum. Meteorol. Rdsch. 1 (5—6), 144—147 (1947).

IHERING, H. v.: Der periodische Blattwechsel der Bäume im tropischen und subtropischen Südamerika. Englers Jb. Bot. 58, 524—598 (1923). — INGOLD, C. T., and V. J. COX: Periodicity of spore discharge in Daldinia. Ann. of Bot., N. S. 19, 201—209 (1955).

KAMIYA, N.: The rate of the protoplasmatic flow in the myxomycete plasmodium. Cytologia 15, 183—193, 194—204 (1950). — KAMIYA, N., and W. SEIFRIZ: Torsion in a protoplasma thread. Exper. Cell Res. 6, 1—16 (1954). — KARSTEN, G.: Über embryonales Wachstum und seine Tagesperiode. Z. Bot. 7, 1—34 (1915). — KNIEP, H.: Über rhythmische Lebensvorgänge bei den Pflanzen. Verh. physik.-med. Ges. Würzburg, N. F. 44 (2), 107—129 (1915). — Über den rhythmischen Verlauf pflanzlicher Lebensvorgänge. Naturwiss. 3, 462—467, 473—477 (1915). — KORIBA, K.: On the origin and meaning of deciduousness viewed from the seasonal habit of trees in the tropics. I. Seiri seitai (Jap.) 1948. — On the origin and meaning of deciduousness viewed from the seasonal habit of trees in the tropics. II. Seiri seitai (Jap.) 1948.

LAUDE, H. M.: The nature of summer dormancy in perennial grasses. Bot. Gaz. 114, 284—292 (1953). — LEINWEBER, F. J.: Über die Temperaturabhängigkeit der Periodenlänge bei der endogenen Tagesrhythmik von Phaseolus. Z. Bot. 44, 337—364 (1956). — LIPPE, PRINZ ZUR, A.: Über den Einfluß des vorangegangenen Licht-Dunkelwechsels auf die CO_2-Ausscheidung der Primärblätter von Phaseolus multiflorus in anschließender Dunkelheit. Z. Bot. 44, 297—318 (1956).

MENZEL, W.: Über den heutigen Stand der Rhythmenlehre in bezug auf die Medizin. Z. Altersforsch. 6 (1), 26—121 (1952). — MES, M. G., and J. MENGE: Potato shoot and tuber cultures in vitro. Physiol. Plantarum (Copenh.) 7, 637—649 (1954). — MUELLER-THURGAU, H., u. F. SCHNEIDER-ORELLI: Beiträge zur Kenntnis der Lebensvorgänge in ruhenden Pflanzenteilen. I. Flora (Jena), N. F. 1, 309—372 (1910).

NEEB, O.: Hydrodictyon als Objekt einer vergleichenden Untersuchung physiologischer Größen. Flora (Jena) 139, 39—95 (1952).

OKADA, Y.: Studies of Euryale ferox Salisb. V. On some features in the physiology of the seed with special respect to the problem of the germination. Sci. Rep. Tohôku Univ. 5, 41—116 (1930).

PFEFFER, W.: Untersuchungen über die Entstehung der Schlafbewegungen der Blattorgane. Abh. sächs. Ges. Wiss., Math.-physik. Kl. 30, 259—472 (1907). — PIRSON, A., u. H. DÖRING: Induzierte Wachstumsperioden bei Grünalgen. Flora (Jena) 139, 314—328 (1952). — PIRSON, A., u. E. GÖLLNER: Beobachtungen zur Entwicklungsphysiologie der *Lemna minor* L. Flora (Jena) 140, 485—498 (1953). — PIRSON, A., W. J. SCHÖN u. H. DÖRING: Wachstums- und Stoffwechselperiodik bei *Hydrodictyon*. Z. Naturforsch. 9b, 349—353 (1954). — PITTENDRIGH, C. S.: On temperature independence in the clock system controlling emergency time in Drosophila. Proc. Nat. Acad. Sci. U.S.A. 40, 1018—1029 (1954). — POHL, R.: Tagesrhythmus im phototaktischen Verhalten der Euglena gracilis. Z. Naturforsch. 3b, 367—374 (1948).

RAU, W.: Über die Wirkung des zu verschiedenen Tageszeiten gebotenen Lichts auf Substanzproduktion und Stickstoffgehalt von Keimpflanzen. Z. Bot. 42, 305—329 (1954). — RESENDE, F.: Observações sobre ritmo endonómico vegetal em Portugal e suas colónias. I—O ritmo endonómico anual e a floração em Chorisia crispiflora H. B. u. K. Bull. Soc. Portugaise Sci. Naturelles 15, 123—127 (1947). — ROBERTS, E. A., and B. E. PROCTOR: The appearance of starch grains of potato tubers of plants grown under constant light and temperature conditions. Science (Lancaster, Pa.) 119, 509—510 (1954). — RUGE, U., u.

D. LIEDTKE: Zur periodischen Keimbereitschaft einiger Malvenarten. Ber. dtsch. bot. Ges. 64, 141—150 (1951).

SAMISH, R. M.: Dormancy in woody plants. Plant Physiol. 5, 183—204 (1954). — SCHMIDLE, A.: Die Tagesperiodizität der asexuellen Reproduktion von *Pilobolus sphaerosporus*. Arch. Mikrobiol. 16, 80—100 (1951). — SCHMIDT, W.: Die Spiegelung der Jahreszeiten in der Samenaktivität. Forschgn u. Fortschr. 6, 325—326 (1930). — SCHMITZ, J.: Über Beziehungen zwischen Blütenblidung in verschiedenen Licht-Dunkelkombinationen und Atmungsrhythmik bei wechselnden photoperiodischen Bedingungen. Planta (Berl.) 39, 271—308 (1951). — SCHMUCKER, TH.: Physiologische und ökologische Untersuchungen an Blüten tropischer *Nymphaea*-Arten. Planta (Berl.) 16, 376—412 (1932). — SCHÖN, W. J.: Periodische Schwankungen der Photosynthese und Atmung bei Hydrodictyon. Flora (Jena) 142, 347—380 (1955). — SCHWEMMLE, B.: Über die tagesperiodischen Änderungen des Reaktionsvermögens von Keimpflanzen auf niedrige Temperatur. Planta (Berl.) 43, 98—132 (1953). — SEIFRIZ, W.: Gregarious flowering of *Chusquea*. Nature (Lond.) 156, 635—636 (1950). — SIEGEL, S. M.: Effects of exposures of seeds to various physical agents. I. Effects of brief exposures to heat and cold on germination and light sensitivity. Bot. Gaz. 112, 57—70 (1950). — SIMON, S. V.: Zur Keimungsphyisologie der Winterknospen von Hydrocharis morsus-ranae L.; zugleich ein Beitrag zur Frage der Jahresperiodizität. Jb. wiss. Bot. 68, 149—205 (1928). — Studien über die Periodizität der Lebensprozesse der in dauernd feuchten Tropengebieten heimischen Bäume. Jb. wiss. Bot. 54, 71—187 (1914). — SMITH, B. C.: An investigation of the rest period in the seed of the genus Cotoneaster. Proc. Amer. Soc. Horticult. Sci. 57, 396—400 (1951). — SMITH, G. M.: On the reproduction of some pacific coast species of Ulva. Amer. J. Bot. 34, 80—87 (1947). — SPEIDEL, B.: Untersuchungen zur Physiologie des Blutens bei höheren Pflanzen. Planta (Berl.) 30, 67—112 (1939). — SPERLICH, A.: Über den Einfluß des Quellungszeitpunkts von Treibmitteln und des Lichtes auf die Samenkeimung von Alectorolophus hirsutus. Charakterisierung der Samenruhe. Sitzgsber. Akad. Wiss. Wien, Math.-naturwiss. Kl. 1 128, 477—500 (1919). — STÅLEFLT, M. G.: Schwankungen in den Zellteilungsfrequenzen. Sv. bot. Tidskr. 13, 61—70 (1919). — Studien über die Periodizität der Zellteilung und sich daran anschließende Erscheinungen. Diss. Stockholm 1921. — The influence of light upon the viscosity of protoplasm. Ark. Bot. A (Stockh.) 33, Nr 4 (1946). — Die stomatäre Transpiration und die Physiologie der Spaltöffnungen. In Handbuch der Pflanzenphysiologie, Bd. III, S. 351—426. 1956. — SISSAKIAN, N., and A. KOBIAKOVA: The prevailing direction of enzyme action as an index of drought-resistance in cultivated plants. Biochimija 6, 103—111 (1940).

TAMIYA, H., K. SHIBATA, T. SASA, T. IWAMURA and Y. MORIMURA: Effect of diurnally intermittent illumination on the growth and some cellular characteristics of Chlorella. In: Burlew: Algal culture from laboratory to pilot plant. Washington 76—84 (1953). — TITZ, M.: Makroskopische und mikroskopische Wachstumsmessungen an Blattstielen. Bot. Archiv 43, 215—251 (1942). — TRIBUKAIT, B.: Aktivitätsperiodik der Maus im künstlich verkürzten Tag. Naturwiss. 41 (4), 92—93 (1954). — TRONCHET, A., J. TRONCHET et E. CRINQUAND: Nouvelles observation sur le mouvement révolutif des vrilles. Ann. Sci. Univ. Besançon Bot. 1951/52, 1—9. — Sur deux modes de ralentissement dans le mouvement révolutif des vrilles. Bull. Soc. Histoire natur. Doubs 1950, Nr 54, 81—85.

UEBELMESSER, E. R.: Über den endonomen Tagesrhythmus der Sporangienträgerbildung von *Pilobolus*. Arch. Mikrobiol. 20, 1—33 (1954).

VEGIS, A.: Über den Einfluß der Temperatur und der täglichen Licht-Dunkel-Periode auf die Bildung der Ruheknospen. Zugleich ein Beitrag zur Entstehung des Ruhezustandes. Symp. bot. Upsalienses 14 (1), 1—175 (1955). — The significance of temperature and the daily light-dark period in the formation of resting buds. Experientia (Basel) 9, 462—463 (1953). — VENTER, J.: Untersuchungen über tagesperiodische Amylaseaktivitätsschwankungen. Z. Bot. 44, 59—76 (1956). — VOLKENS, G.: Laubfall und Lauberneuerung in den Tropen. Berlin: Gebrüder Bornträger 1912.

WANGERMANN, E., and E. ASHBY: Studies in the morphogenesis of leaves. VII. Part I. Effect of light intensity and temperature on the cycle of ageing and rejuvenation in the vegetative life history of *Lemna minor*. New Phytologist 5, 186—199 (1951). — WIGHT, W., and D. N. BARUA: The nature of dormancy in the tea plant. J. of Exper. Bot. 6, 1—5 (1955).

Altern und Zelltod.

Von

K. Paech† und F. Eberhardt.

Mit 4 Abbildungen.

1. Einleitung.

Viele Beobachtungen über „das Altern" beziehen sich nicht auf die Zellen, sondern auf Organe oder auf die ganze höhere Pflanze. Sicher sind manche Altersveränderungen des vielzelligen Organismus durch analoge Wandlungen seiner Zellen verursacht, aber andere lassen sich wahrscheinlich auf die Bedingungen zurückführen, die der Gewebeverband schafft und die für die Einzelzelle den Wert der Umwelt haben. Der eindeutige Nachweis der möglichen Ursachen, die zum Altern führen und der Erscheinungen, die notwendig mit dem Altern verknüpft sind, wird im Bereich der pflanzlichen Lebewesen dadurch besonders erschwert, daß die höhere Pflanze kein einheitlicher Organismus ist, dessen Glieder alle ungefähr zum gleichen Zeitpunkt angelegt werden, mit ähnlicher Geschwindigkeit sich entwickeln und altern. Am gleichen Individuum bestehen vielmehr Organe in embryonaler Anlage und andere in prämortaler Auflösung harmonisch nebeneinander. Es muß also eine Grenze, die jedoch niemals scharf sein kann, zwischen dem Altern der Zellen und dem des ganzen Organismus gezogen werden. Besonderes Gewicht sollte auf die Erscheinungen gelegt werden, die an Einzellern beobachtet werden. In diesem Beitrag soll, auch soweit es sich um höhere Pflanzen handelt, in erster Linie auf die Wandlungen der *Zelle*, die das Altern begleiten, eingegangen werden, zumal das Altern der ganzen Pflanze und ihrer Organe an einer späteren Stelle im Rahmen der Entwicklungsphysiologie behandelt werden wird (vgl. Band XV). Deshalb werden morphogenetische, entwicklungsphysiologische und genetische Altersprobleme nicht behandelt. Ebenso sollen vergleichende Angaben über die Lebensdauer der Pflanzen in diesem Abschnitt des Handbuches ausgelassen werden (vgl. dazu Molisch, Küster, v. Denffer).

Es versteht sich von selbst, daß dem hier interessierenden physiologischen Alter, d. h. der Veränderung physiologischer Größen, das rein chronologische Alter gegenübergestellt werden kann, das einfach die Dauer der Existenz angibt und physiologisch meist nur als Maßstab für die Geschwindigkeit des physiologischen Alterns eine Rolle spielt. Wenn z. B. Pollenkörner durch Lagerung bei extrem tiefen Temperaturen für unüberschaubare Zeiten bestäubungsfähig bleiben, so sind sie physiologisch unverändert jung, obwohl sie 10000 Jahre alt sein können. Für sie steht die Zeit biologisch gesehen still (Bredemann, Garber, Harteck und Suhr).

Die Lebensdauer der Organe und Gewebe kann durch Abtrennen von der Pflanze oder durch anderweitiges Ausschalten des Einflusses, den die übrigen, besonders die jüngeren Organe ausüben, wesentlich verlängert werden. Schließlich gehen aber alle lebenden Wesen durch einen natürlichen Prozeß zugrunde. Die Geschwindigkeit des Alterns mag also zu verzögern sein, aber ein allen Organismen aufgelegtes Gesetz treibt sie trotz günstiger Umweltbedingungen durch fortschreitende Veränderungen schließlich zum Erlöschen. Wie die begrenzte Lebensfähigkeit der Sporen und Samen zeigt, ist nicht einmal das latente Leben von diesem Gesetz ausgenommen.

Ein besonderes Problem bietet das Altern von Meristemen. Obwohl sie natürlich physiologisch jung bleiben, altert das Organ oder der Organismus, der sie trägt. „Das Altern der embryonalen Zellen ist also durch Faktoren bedingt, die für sie selber Außenfaktoren sind" (BÜNNING 1953). Das chronologische Alter geht an ihnen nicht spurlos vorüber, vor allem dann nicht, wenn sie durch dauernde Erneuerung an einer Pflanze von langer Lebensdauer unter die Umweltbedingungen des alternden Organismus geraten.

Das entscheidende physiologische Altern setzt ein, wenn die Zellen sich zu strecken und zu differenzieren beginnen. „Eine sehr große Bedeutung dürfte bei diesem Aufgeben des Embryonalzustandes bei den mehrzelligen Pflanzen der auf die eine oder andere Weise induzierten Polarität zukommen. Sobald diese nämlich vorhanden ist, führt sie zwangsläufig zu polaren Stofftransporten und damit z. B. auch zwangsläufig zur Fortleitung eines für die Teilung wichtigen Stoffes aus den Zellen" (BÜNNING 1953). Im Hinblick auf die in den Erbanlagen festgelegten Fähigkeiten der Zellen besteht der physiologisch junge Zustand darin, daß nur wenige dieser Potenzen realisiert sind. Je mehr davon sich ausprägen, um so reifer ist die Zelle bzw. das Organ oder der ganze Organismus.

Obwohl das alle Zellen beherrschende, von Umwelteinflüssen unabhängige Altern, das man vielleicht *das immanente Altern* nennen könnte, im allgemeinen eine irreversible Erscheinung ist, gibt es im Bereich der pflanzlichen Organismen echte Rejuvenation, wo ausdifferenzierte Zellen, die sich scheinbar unaufhaltsam auf dem Weg des physiologischen Alterns befinden, in den meristematischen Zustand zurückverwandelt werden und damit gewissermaßen von neuem anfangen dürfen. Bei Einzellern beginnt ein neues Individualleben nach jeder Zellteilung. Bei vacuolisierten Zellen im Gewebeverband reichert sich als erstes Zeichen der Verjüngung Plasma an, worauf auch hier im allgemeinen Zellteilung eintritt. Tatsächlich scheint Plasmawachstum ein wesentliches Merkmal embryonaler, also physiologisch junger Zellen zu sein (BÜNNING 1955). Das physiologische Altern ist offenbar so lange aufgehalten, so lange Plasma gebildet wird oder werden kann. Zellteilung scheint für die Verjüngung nicht unerläßlich zu sein. Bei *Amoeba polypodia* gelang es, durch häufig wiederholte Amputationen eines Teiles der Zelle sie weit über ihr „normales" Alter am Leben zu erhalten (HARTMANN). In ähnlicher Weise kann man durch fortgesetztes Beschneiden die Zelle von *Acetabularia* „unbegrenzt" im Wachstum erhalten (HÄMMERLING). Charakteristisch für ausgewachsene und damit alternde Pflanzenzellen ist ja tatsächlich eine sehr begrenzte Fähigkeit oder Kapazität der Plasmabildung, denn der große Raum der Vacuole bleibt immer ausgespart. Wenn darin Eiweiße angehäuft werden, so handelt es sich um die nach chemischer Zusammensetzung und Funktion von den Plasmaeiweißen ganz verschiedenen Reserveeiweiße.

Der Tod des mehrzelligen Organismus ist unausweichlich. Aber selbst die Einzeller sind nicht unsterblich, denn auch bei ihnen erlischt das Leben des Individuums durch die Teilung. Wenn man den Einzellern Unsterblichkeit zuschreiben wollte, wie das manchmal geschieht, so müßte man sie auch allen anderen Organismen zuerkennen, denn von ihnen bleibt wenigstens ein Teil, nämlich die zu neuen Individuen gewordenen Keimzellen oder vegetative Fortpflanzungsorgane wie Brutknospen, am Leben. So kann man wohl sagen, daß nach unseren jetzigen Kenntnissen das Leben als Ganzes prinzipiell unsterblich ist, während das Individuum, ob Einzeller oder Vielzeller, sich nur einer begrenzten Lebensdauer erfreut (vgl. dazu EHRENBERG). Vom Ende des Lebens beim

Einzeller, das durch die Teilung bestimmt ist, unterscheidet sich der Tod des Vielzellers im wesentlichen nur dadurch, daß hier ein Teil als unbelebtes System, als Leiche, übrigbleibt.

Derartige Restkörper, die schließlich zugrunde gehen, kann man auch schon bei Kolonien höher organisierter Einzeller antreffen. Während z. B. jede Zelle der Flagellatenkolonien von *Gonium* und *Pandorina* vollständig in der Bildung einer Tochterkolonie aufgeht, gehen bei *Volvox* nach Zerfall des Mutterindividuums die nicht in Fortpflanzungszellen umgewandelten Zellen als Leiche zugrunde. Der Übergang von einem Individualtod ohne Restkörper zu einem, bei dem ein Teil als Leiche übrigbleibt, wiederholt sich an verschiedenen Stellen des Systems der niederen Pflanzen. In den geißellosen Grünalgenkolonien *Scenedesmus* und *Pediastrum* vollzieht sich die Teilung noch ohne plasmatischen Restkörper im Gegensatz zu den höher organisierten Chlorophyceen. Auch bei den Desmidiaceen fehlen Reste, die als Leiche bleiben. Bei niederen Pilzen (z. B. *Olpidium* und andere *Myxochytridiales*) geht ebenso der ganze Zellinhalt in Schwärmzellen auf. Andererseits bleiben bei der Fruchtkörperbildung der Myxomyceten plasmatische Anteile als absterbende Restkörper zurück (z. B. Capillitium, Fruchtkörperstiele).

Unter den natürlichen Bedingungen des Pflanzenlebens tritt der Tod einzelner Organe oder des ganzen Individuums unter recht verschiedenen Ursachen ein. *Verhungern*, d. h. mangelnde Zufuhr von Nährstoffen oder Abbau des lebenden Substrates, ist nicht selten. Die ältesten Blätter einjähriger Pflanzen werden von den jüngeren ausgesogen und manche der älteren verhungern dabei. Die untersten Zweige dichtstehender Bäume werden beschattet und sterben ab, lange bevor sie an einem freieren Standort ihr Leben beendet hätten. Der Tod wird hier offensichtlich durch Umweltfaktoren verursacht und hat keine unmittelbare Beziehung zu einem natürlichen Altern der Organe, obwohl es manchmal so scheint, als brächte Hungern manche der für den natürlichen Tod der Organe charakteristischen Erscheinungen hervor, z. B. das Vergilben der Blätter.

Viele Zellen sterben an ihren eigenen Stoffwechselprodukten oder an denjenigen der Nachbarzellen einen Tod durch *Vergiftung*. Geläufige Beispiele bieten die niederen Organismen, die nur bis zu einer gewissen Konzentration ihre eigenen Stoffwechselendprodukte ertragen können: Hefen bis etwa 17% Alkohol, Milchsäurebakterien bis etwa 1,2% Milchsäure usw. In höheren Pflanzen werden Stoffwechselprodukte, die dem lebenstätigen Plasma gefährlich wären, in den Vacuolen abgelagert, die in diesem Zusammenhang als Umwelt angesehen werden müssen, wo auch aggressive Verbindungen gespeichert werden können. Der Tonoplast zeichnet sich durch eine besonders hohe Resistenz gegen Angriffe aus, die das Cytoplasma schädigen. Während z. B. junge Weizenblättchen recht lange ohne Zeichen einer Schädigung in ihren eigenen Zellsaft von mäßiger Acidität eingebettet werden können, gehen Zellen mit stark saurem Vacuolensaft, z. B. *Begonia*-Blätter, zugrunde, wenn ihnen dieser von außen geboten wird. Ähnliches gilt für Zellsäfte, die hohe Konzentrationen von Gerbstoffen, Cumarinen oder Saponinen enthalten. Ein Tod durch Vergiftung kann auch nach längerem Hungern eintreten, wenn aus den Eiweißabbauprodukten Ammoniak freigesetzt und nicht mehr „entgiftet" wird. In diesen Fällen braucht dem Tod natürlich kein Altern vorauszugehen.

Andererseits taucht hier das Problem auf, wieweit Resistenzwandlungen der Tonoplastenmembran im Verlaufe des Alterns mittelbar zu einem Vergiftungstod durch den eigenen Zellsaft führen könnten. Verbindungen zwischen dem Vacuoleninhalt und dem Cytoplasma bestehen ja, wie die Umsetzungen von Stoffen des Vacuoleninhaltes zeigen, und es bedürfte vielleicht nur einer quantitativen

Veränderung der Tonoplasteneigenschaften im Laufe des physiologischen Alterns, um eine Autintoxikation hervorzurufen. Diese Art des indirekten Alterstodes wäre jedoch beschränkt auf die Zellen, in deren Vacuolen Stoffe abgelagert werden, die sich mit dem lebenstätigen Plasma nicht vertragen.

Selbst unter optimalen Umweltbedingungen fallen alle Zellen und Organismen einem „*natürlichen Tod*" anheim, der den Abschluß des immanenten Alterns bildet (vgl. EHRENBERG). Die Zellen können Tage, Wochen, Monate oder viele Jahre alt werden, aber verglichen mit dem hohen Alter mancher Baumarten ist die Lebensdauer der Zellen doch immer relativ kurz. Die ältesten lebenden Zellen hat man im Holzparenchym und in den Markstrahlen von Bäumen sowie in der Rinde und dem Mark bestimmter Kakteen gefunden (vgl. MOLISCH; GOOD, MURRAY und DALE). Unsere Kenntnisse von der Struktur des lebenden Plasmas und ihrer Erhaltung sind heute noch so dürftig, daß es kaum möglich ist, auch nur anzudeuten, worin die Ursachen für den natürlichen Tod tatsächlich liegen. Man hat an „Abnutzungserscheinungen" gedacht, über die aber wohl niemals eine bestimmte, auf das System der lebenden Zelle bezogene Vorstellung entwickelt worden ist. Bei organischen unbelebten Substanzen im kolloidalen Zustand kennt man genügend Alterungserscheinungen, die zu Strukturwandlungen führen. Analoge Phänomene, z. B. eine nachlassende Quellbarkeit, sind häufig am Plasma alternder Zellen beobachtet worden. Es kann also möglich sein, daß einfache physikalisch-chemische Gesetzmäßigkeiten, ohne die für das Leben charakteristische Komplikation, notwendig zur Störung oder Zerstörung der lebenden Struktur führen. Altern und Tod, aufgezwungen durch Gesetze des Unbelebten, würden damit als etwas dem Lebendigen Widerstrebendes erscheinen. Wenn diese Betrachtung auch unbefriedigend erscheint, so kann ihr doch heute noch nichts Beweisendes entgegengehalten werden. Wir dürfen gleichwohl nicht aus dem Auge verlieren, daß sich im Bereich des Plasmas Struktur und Stoffwechsel gegenseitig bedingen. Hier ist das eine ohne das andere nicht mehr bestimmbar. Stoffwechsel und Struktur verhalten sich wechselseitig wie Ursache und Wirkung. Es erscheint deshalb auch theoretisch fast unmöglich, hier noch nach Ursachen und Begleiterscheinungen des Alterns der Zellen zu trennen und nach einer einzigen bestimmten Ursache für das Altern und den Zelltod suchen zu wollen. Wahrscheinlich liegt in dem mit dem Leben gesetzten Urphänomen der Verknüpfung von Plasmastruktur und Stoffwechsel, in der Eigentümlichkeit, daß die lebende Struktur nichts Statisches, sondern eher ein Vorgang ist, auch das Gesetz beschlossen, das die Zellen — über welche verschlungenen Wege auch immer — schließlich zum Tode führt.

Im praktischen Falle ist es häufig nicht ganz einfach, den Zelltod festzustellen bzw. zu entscheiden, ob ein System noch lebendig oder schon tot ist. Es gibt sicher Übergänge, die unter günstigen Bedingungen zum Weiterleben in der Lage wären, unter den obwaltenden aber als tot zu gelten haben. Nach Anaerobiose erscheinen Blätter, nach den üblichen Indizien beurteilt, oft noch als lebend, während sie nach kurzem Aufenthalt in normaler Atmosphäre doch zugrunde gehen (PAECH 1935b). Man kann einen solchen Zustand als prämortal bezeichnen. Eine meßbare „Atmung" kann nur unter bestimmten Bedingungen als zuverlässiges Kriterium des Lebens angesehen werden. Das latente Leben der Samen kann in manchen Fällen wohl ohne nachweisbaren Gasaustausch erhalten sein und andererseits kann sogar ein Brei zerstörter Zellen noch einen der normalen Atmung ähnlichen Gaswechsel fortsetzen. Physikalisch-chemisch wird die Semipermeabilität, also praktisch die Plasmolysierbarkeit und Deplasmolysierbarkeit vacuolisierter Zellen als Kriterium für den lebenden Zustand angesehen. In vielen Fällen ist das zuverlässig, doch kommen auch hier nicht selten Grenzfälle vor, wo die Semipermeabilität, besonders Tonoplastenplasmolyse, erhalten ist, obwohl die Zellen sich in einem prämortalen Zustand befinden. Um eine nicht immer eindeutige Bestimmung des Lebenszustandes scheint es sich auch bei der Anwendung von Fluorescenzfarbstoffen zu handeln (vgl. STRUGGER; BOGEN 1953). Die einer weiteren Anwendung fähige und nach den bisherigen Erfahrungen recht verläßliche Triphenyl-tetrazoliumchlorid-Methode (TTC) gründet sich auf das in allen lebensfähigen Zellen vorhandene System der wasserstoffbeladenen Codehydrasen (vgl. LAKON; KUHN, PFLEIDERER und SCHULZ). Mit fortschreitender Experimentierkunst wird man sicher Methoden kennen lernen, die einen weiteren Komplex der für den lebenden Zustand charakteristischen Merkmale erfassen lassen. Zu den genannten, allgemeiner gerichteten Nachweisen gesellen sich noch einige, die auf spezielle Fälle beschränkt bleiben, so die fortbestehende oder durch Reiz einsetzende Plasmaströmung. Beim Arbeiten mit etiolierten Organen überzeugt die Fähigkeit zum Ergrünen davon, daß die betreffenden Zellen noch voll lebensfähig sind.

Viel schwieriger als unter natürlichen Bedingungen kann zwischen Leben und Tod geschieden werden, wenn experimentelle Eingriffe mit Hemmstoffen, Narkotica oder Giften vorgenommen werden. Dabei können wohl oft die Zweifel, ob es sich noch um lebende oder schon um tote Systeme handelt, nicht ganz behoben werden.

Die cytologischen Veränderungen beim Protoplasmatod hat LEPESCHKIN in einer Monographie eingehend beschrieben.

Nicht so sehr aus sachlich gerechtfertigten Gründen, sondern vor allem wegen der herkömmlichen Verfahren der Untersuchung werden im folgenden getrennt voneinander zunächst die mit dem Altern einhergehenden Wandlungen des Stoffwechsels und danach die physiko-chemischen Begleiterscheinungen des Alterns betrachtet.

II. Altern und Stoffwechsel.

Es ist zu erwarten, daß das Alter der Zellen, Organe oder ganzen Pflanzen einer der inneren Faktoren ist, die den Stoffwechsel stark beeinflussen und doch sind Untersuchungen über diese Abhängigkeit oder über mögliche Wechselwirkungen zwischen Stoffwechsel und Altern der Pflanzen nicht sehr reichlich (vgl. PAECH 1940a).

A. Photosynthese.

Am ehesten hat es noch angeregt, die Beziehung des Zell- und Organalters zu dem entscheidenden synthetischen Vorgang der grünen Pflanzen, der Photosynthese, aufzudecken. Bei wachsenden Organen steht einer eindeutigen Erkennung der altersbedingten Wandlung im photosynthetischen Prozeß die sich meist ebenfalls ändernde Bezugsgröße als Hindernis im Wege. Nach den klassischen Messungen von WILLSTÄTTER und STOLL ergibt sich während der Entfaltung der Blätter von verschiedenen Arten kein einheitlicher Gang der Assimilations*intensität*, ganz gleich, ob Frischgewicht, Trockengewicht oder Blattfläche als Bezugsgröße genommen werden. Eine relativ stetige Veränderung erfahren bei der Laubentfaltung höchstens die sog. *Assimilationszahlen*, d. h. die stündliche Leistung bezogen auf 1 g Chlorophyll (unter sonst optimalen Bedingungen). Obwohl im allgemeinen die Assimilationsleistung je Chlorophylleinheit beim Blattwachstum abnimmt, zeigen die Assimilationszahlen, daß die Produktion der Blattpigmente und die Ausbildung der Funktionstüchtigkeit des plasmatischen Apparates der Photosynthese in den jüngsten Phasen nicht immer mit gleicher Geschwindigkeit voranschreiten. Diese anderen inneren Faktoren der Photosynthese eilen im Frühjahr der Chlorophyllbildung voraus und stehen erst später hinter der reichlichen oder sogar zu reichlichen Bildung der Farbstoffe zurück.

Beim Vergleich verschieden alter Blätter des gleichen Sprosses zeigt sich ebenfalls, daß die Pigmentmenge stark zunimmt und daß die photosynthetische Leistung im allgemeinen unabhängig von der Bezugsgröße ansteigt, aber nicht entfernt im gleichen Maße wie die Chlorophyllmenge, folglich sinkt die Assimilationszahl. Das gilt auch für mehrjährige Blätter. Die Assimilationszahl sinkt jedoch nicht beständig bis zum Abfallen der Blätter. Während des herbstlichen Vergilbens geht zumeist mit der Abnahme des Chlorophyllgehaltes ein Rückgang der Assimilationsleistung einher, so daß die Assimilationszahlen manchmal ihren spätsommerlichen Wert recht lange konstant erhalten. Nicht selten ändert sich die Assimilationsintensität nur unwesentlich während des scharfen Rückganges der Blattpigmente, so daß die Assimilationszahlen während des Vergilbens beträchtlich ansteigen. Aus allen diesen Beobachtungen läßt sich zunächst leider noch keine einheitliche, für alle untersuchten Blätter zutreffende Wandlung der Photosynthese mit fortschreitendem Alter ablesen, obwohl tiefgreifende Änderungen offensichtlich sind.

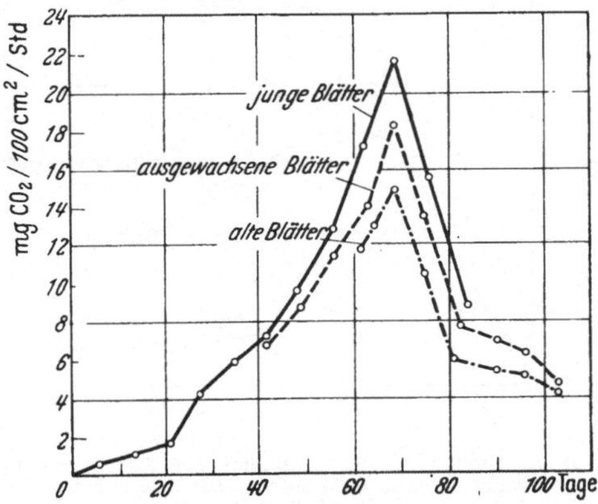

Abb. 1. Einfluß des Alters der Pflanze und des Entwicklungszustandes der Blätter auf die assimilatorische Leistung (Weizenblätter). (Nach SINGH und LAL.)

Unter klarer Scheidung der Überlagerung von Blatt- und Pflanzenalter ist die Assimilationsleistung bei verschiedenen Kulturpflanzen genauer untersucht worden (SINGH und LAL). Die Intensität (mg CO_2 je 100 cm² Blattfläche und Stunde) gleichartiger Blätter (junge, ausgewachsene und alte) steigt mit dem Alter der Pflanzen zunächst an, erreicht bei Weizen und Lein unter den herrschenden Klimabedingungen ungefähr nach 70 Tagen, bei Zuckerrohr nach 200 Tagen ein Maximum und fällt in den späteren Lebensstadien der Pflanze wieder ab. Während dieses „Ganges" liegen die jungen Blätter stets höher als die ausgewachsenen und diese wiederum höher als die noch älteren, wobei der Unterschied zwischen den Gruppen im Zeitpunkt des Assimilationsmaximums am größten und sowohl zu Beginn als auch am Ende der Lebensperiode der ganzen Pflanze fast ganz ausgeglichen ist (vgl. dazu auch DRAUTZ). Die in Abb. 1 aufgezeichneten Verhältnisse können wohl als charakteristisch für die Abhängigkeit der Photosynthese vom Alter bei einjährigen Pflanzen angesehen werden. Das Maximum der Leistung fällt bei Lein und Weizen ungefähr mit der Blüte der Pflanze zusammen. Der anschließende Abfall ist bei diesen beiden Typen sehr steil. Beim Zuckerrohr ist sowohl der Anstieg als auch der Abfall wesentlich flacher. Das sichtbare Vergilben beginnt bei allen untersuchten Pflanzen erst 2—4 Wochen nach Überschreiten des Maximums der Assimilationsintensität.

Auf der Suche nach Faktoren, auf die das beschriebene Verhalten der Assimilationsleistung während des Lebens der Blätter zurückgeführt werden könnte, ließ sich eine lockere direkte Beziehung zwischen photosynthetischer Aktivität der Weizen- und Leinblätter und ihrem Kalium- und Calciumgehalt (bezogen auf Trockengewicht) im Verlauf der Vegetationsperiode nachweisen. Ob das aber als ursächliche Verknüpfung für den steilen Anstieg der Assimilationsleistung in jungen Pflanzen ausreicht, erscheint zweifelhaft. Hingegen ist es wahrscheinlich, daß der spätere scharfe Abfall des Kalium-, Calcium- und Wassergehaltes der Blätter den gleichzeitigen Abfall der Photosynthese hervorruft. Solche Beziehungen dürften allerdings nicht für die Blätter von Holzgewächsen zutreffen, in denen Aschenanalysen ja eine mit dem Alter zunehmende Anreicherung von mineralischen Bestandteilen erkennen lassen.

Coniferen-Blätter, die mehrere Jahre funktionstüchtig bleiben, erscheinen besonders geeignet, den möglichen Einfluß des Alters auf Stoffwechselvorgänge aufzuklären. Da sowohl Frisch- als auch Trockengewicht mit dem Blattalter (bis zu 3 Jahren) zunehmen, wurde der Assimilationsüberschuß bei einer Reihe von *Pinus*-Arten und bei einigen anderen Coniferen auf die Blattzahl bezogen. Dabei weist das Blatt bereits im ersten Sommer, wenn es gerade seine volle Größe erreicht hat, eine maximale Assimilationsleistung (apparente Photosynthese) auf (FREELAND). Diese nimmt dann im Laufe der nächsten beiden Jahre langsam ab, aber auch in dreijährigen Blättern bleibt noch eine beachtlich hohe Assimilationsintensität erhalten.

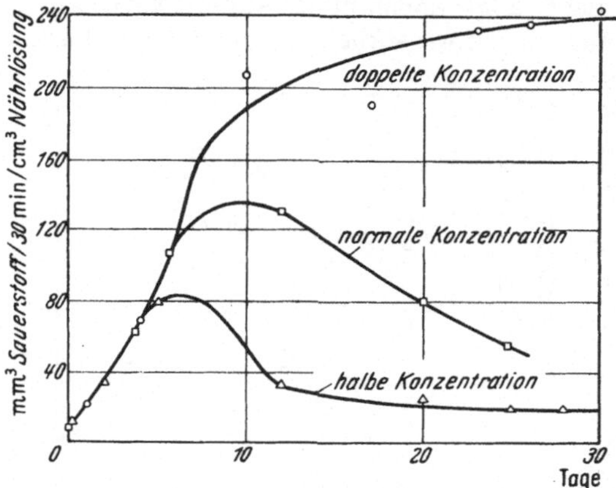

Abb. 2. Einfluß der Konzentration der Nährlösung auf die assimilatorische Leistung von *Chlorella pyrenoidosa*. (Nach VAN HILLE.)

Kulturen von *Chlorella pyrenoidosa* verhalten sich in bezug auf Assimilationszahl und Assimilationsleistung mit dem Alter prinzipiell wie Blätter (HILLE). Die Photosyntheseleistung, sowohl auf die Zellzahl als auch auf das Trockengewicht bezogen, steigt während der ersten Tage der Kultur an und nimmt dann laufend mit dem Alter der Kultur ab (WINOKUR). Hier kann aber durch höhere Konzentration der Nährsalze der Assimilationsabfall bei älteren Kulturen verhindert werden (Abb. 2, HILLE). Die Assimilationszahl jedoch, also die relative Ausnützung des vorhandenen Chlorophylls, sinkt unabhängig von der Konzentration der Nährlösung mit dem Alter der Kultur zunächst rasch, später langsamer bis auf einen längere Zeit konstant bleibenden Wert in älteren Kulturen ab. Dann kann durch Nitratgabe bei gleichzeitiger Anwesenheit genügender Mengen von Kohlenhydraten trotz gleichbleibendem Chlorophyllgehalt die Assimilationsintensität stark gesteigert, die Assimilationszahl also erhöht und somit wenigstens ein scheinbar jugendlicher Zustand hergestellt werden. Da dieser Effekt bei Mangel an Kohlenhydraten ausbleibt, dürfte die Zunahme der Assimilationsleistung wohl auf dem Umwege über eine Eiweißbildung im photosynthetischen Apparat zustande kommen, was jedoch durch Proteinanalysen noch nachzuweisen wäre.

Daß zwischen Eiweißabbau und Vergilben ein sehr enger Zusammenhang besteht, wurde an *Tropaeolum*-Blättern gezeigt (MEYER; MICHAEL). Der auffallend parallele Verlauf der Abnahme des Eiweiß- und Chlorophyllgehaltes reicht bis in die Zeit vor dem sichtbaren Vergilben zurück. Der primäre Vorgang beim Vergilben scheint im Abbau des Chloroplasteneiweißes zu bestehen, der dann den Schwund des Farbstoffes nach sich zieht. Noch einen Schritt weiter führt die Untersuchung isolierter Chloroplasten aus Spinat- und Weizenblättern. Die photochemische Aktivität, d. h. die Intensität der HILL-Reaktion, der Chloroplasten steigt an, solange die Blätter noch wachsen. Danach nimmt die Aktivität der Chloroplasten jedoch ab (CLENDENNING und GORHAM). Da die Leistungsfähigkeit junger Plastiden nicht durch die Cytoplasmafraktion alter Blätter beeinflußt wird, schließen die Autoren, daß die Veränderungen mit dem Blattalter primär in den Plastiden selbst vor sich gehen. Dabei bleibt allerdings noch offen, welche Wandlungen denn im Anfang zu einem Anstieg und später zu einem Nachlassen der photochemischen Aktivität der Plastiden führen.

Die chemische Analyse der Plastiden aus Organen in verschiedenem Entwicklungszustand und Alter bringt Unterschiede zutage, die vielleicht mit den beschriebenen Funktionswandlungen zusammenhängen und für sie verantwortlich sein können (SISSAKIAN). Die Leukoplasteneiweiße von Zuckerrüben enthalten vom August bis Dezember etwa gleiche Anteile von Asparagin- und Glutaminsäure sowie von Alanin und Arginin. (Es wurden im ganzen nur 13 Aminosäuren untersucht.) Besonders stark nahmen während der Lagerung Serin und die Leucine ab. Cystin und Threonin nehmen während der gleichen Zeit bedeutend zu. Der Gesamtgehalt an 8 Aminosäuren verändert sich in den Chloroplasteneiweißen aus Zuckerrübenblättern zwischen August und Oktober nicht wesentlich: Alanin nimmt etwas ab, Threonin und Lysin steigen vorübergehend an und nehmen dann auch ab. Bedeutendere Veränderungen spielen sich in den elektrophoretisch trennbaren Eiweißfraktionen der Chloroplasten ab. Die Wanderungsgeschwindigkeit nimmt teils zu, teils ab, und der relative Anteil von 4 Fraktionen des Chloroplasteneiweißes verändert sich während der Blattalterung von August bis Oktober so, daß in einem mittleren Entwicklungszustand (Anfang September in Moskau) ein Minimum bzw. ein Maximum der betreffenden Fraktion durchlaufen wird.

Bei ungefähr gleichem Gesamtgehalt an Lipoiden in den Chloroplasten während der Blattalterung nehmen die „Freien“ etwas zu, während die „Gebundenen“ stark absinken, so daß das Verhältnis Gebundene : Freie von 0,9 im August auf 0,1 im Oktober fällt. Ein gleichsinniger, noch steilerer Gradient des Lipoidverhältnisses besteht in den aufeinanderfolgenden Blättern einer Zuckerrübenpflanze. Obwohl die Fraktionen „Frei“ und „Gebunden“ zunächst nur durch ihre Löslichkeit (in Äther bzw. in Äthylalkohol) charakterisiert werden können, besteht wahrscheinlich doch ein direkter Zusammenhang mit dem oben erwähnten Rückgang der photochemischen Aktivität der Plastiden in alternden Blättern. Auch der Nucleinsäuregehalt der Plastiden (hauptsächlich Ribonucleinsäuren) fällt mit zunehmendem Alter der Blätter bedeutend ab. Bei Tabakblättern findet jedoch während der Blühphase eine gewisse Angleichung des Nucleinsäuregehaltes der verschieden hoch inserierten Blätter statt, während vorher und nachher die jüngeren durch höheren Nucleinsäuregehalt der Chloroplasten ausgezeichnet sind.

Eine funktionelle Veränderung der Chloroplasten mit dem Alter der Pflanzen geht daraus hervor, daß bei jüngeren Tabakpflanzen (2—6 Monate alt) schon 5 Tage Verdunklung genügen, damit alle Stärke in den Chloroplasten gelöst ist, während bei den ältesten Pflanzen (9—11 Monate alt) selbst nach 30 Tagen Dunkelheit praktisch noch alle Chloroplasten Stärke enthalten (IRMAK). Leider ist dabei nicht angegeben, ob jeweils vergleichbare Blätter, etwa die voll ausgewachsenen, verwendet wurden.

B. Die Atmung im Verlaufe des Alterns.

Obwohl man schon sehr früh ausgeprägte Veränderungen der Atmungsintensität während des Wachstums der Pflanzenorgane erkannt hatte und obwohl die Höhe der Atmung häufig als Maß für die Lebensintensität gewählt und die mit dem Alter angeblich nachlassende Fähigkeit zu Oxydationen als Ursache für alle übrigen Altersveränderungen angesehen wurde, sind systematische Untersuchungen des Atmungsverlaufes, die die eingangs skizzierten komplizierenden Wechselwirkungen zwischen Organen und ganzem Individuum berücksichtigen, doch nur in einem zu geringen Umfang ausgeführt worden, als daß sie ein eindeutiges Bild ergäben. Für alle Aussagen über die Atmungsintensität während des physiologischen Alterns spielt wieder die richtige Bezugsgröße eine entscheidende Rolle. Mit dem Alter ändern sich häufig Wassergehalt und Trockensubstanz derartig, daß die auf Frisch- oder Trockengewicht bezogene Atmungsintensität Schwankungen vortäuscht, die in Wirklichkeit, also bezogen auf die lebende Substanz, gar nicht bestehen oder sogar in entgegengesetzter Richtung verlaufen. Auch hier sollte vielleicht immer die Zellzahl, besser noch ein Maß für die Plasmamenge, etwa der Eiweiß- oder Nucleinsäuregehalt, als zuverlässige Vergleichsgrundlage herangezogen werden.

Wie weit die Bezugsgröße das Bild bestimmen kann, zeigen z. B. die Messungen, nach denen, auf Frischgewicht bezogen, die höchste Atmungsintensität in den äußersten Spitzen (1 mm) von Wurzeln gefunden wird, während auf Eiweiß-N oder Ribose-Nucleinsäuren bezogen die sich streckenden Zellen etwa 4—5 mm von der Spitze entfernt am intensivsten und die Zellen 1—2 mm hinter der Spitze, wo die aktivste Zellteilung vor sich geht, am wenigsten atmen (vgl. GODDARD und MEEUSE).

Unter Beachtung der Überlagerung des Alterns der einzelnen Organe und der ganzen Pflanze kann man sagen, daß im allgemeinen mit Ausnahme der Keimlingsstadien die Atmungsintensität (CO_2-Abgabe) sowohl des Organs als auch des ganzen Individuums erst rasch, später langsamer absinkt (KIDD, WEST und BRIGGS; HOVER und GUSTAFSON; SMIRNOV; RICHARDS). Selbst die Sproßspitze, die ja meristematisch bleibt, zeigt eine mit zunehmendem Alter der Pflanze abfallende Atmungsintensität (vgl. dazu auch SINGH). Demzufolge beginnt jedes später angelegte Blatt mit einer geringeren Atmung als die voraufgehenden und so ergibt sich beim Vergleich ausgewachsener Blätter einer älteren Pflanze, daß die tiefer am Stengel stehenden eine höhere Atmungsintensität haben als die später angelegten (jüngeren!) Geschwister, obwohl im Laufe ihres individuellen Lebens die Atmungsintensität immer mehr nachgelassen hat. Bei Erdbeerpflanzen wird jedoch verneint, daß ein Einfluß des Alters der Pflanze auf die durchschnittliche Atmungsintensität der einzelnen Blätter besteht (ARNEY).

Mit Rücksicht auf die morphologischen Verhältnisse ist die Vorstellung entwickelt worden, daß der Zustand und die Funktion des Vegetationspunktes ganz allgemein von der Versorgung mit Nährstoffen beherrscht wird und daß deshalb durch experimentelle Kürzung der Nährstoffversorgung sogar Jugendstadien wieder hervorgebracht werden können (ALLSOPP).

Charakteristisch für den Atmungsverlauf alternder Blätter ist ein Merkmal, das hier erst jüngst mehr beachtet wurde, obwohl ein ähnliches Phänomen bei reifenden Früchten als "climacteric rise" schon seit langem bekannt war. Sowohl in abgeschnittenen als auch in den an der Pflanze belassenen Blättern steigt kurz vor Beginn der Vergilbung die Atmungsintensität (als O_2-Aufnahme oder als CO_2-Abgabe gemessen) erneut steil an, bis sie nach Überschreiten eines Höhepunktes dann endgültig abfällt (Abb. 3; vgl. JAMES 1953; ARNEY; EBERHARDT 1955). Im Gegensatz zu dem ähnlichen Verlauf der Atmung bei reifenden Früchten, wo der RQ ($= CO_2 : O_2$) wegen Abbau von organischen Säuren oft weit über 1 ansteigt, liegt er bei alternden Blättern, soweit bisher untersucht, nie über der Einheit (EBERHARDT 1955). Er fällt besonders beim Einsetzen des Vergilbens unter 1, was wohl mit dem oxydativen Umsatz von Eiweißabbauprodukten zusammenhängt.

Der klimakterische Atmungsanstieg, der bei Früchten noch ausgeprägter als bei Blättern ist, wurde bisher an Äpfeln, Birnen, Tomaten, Bananen, Vogel-, Holunder- und Erdbeeren u. a. nachgewiesen (vgl. dazu BIALE 1950 und PAECH 1952). Dieses allgemeine Vorkommen ist um so auffallender, als die genannten Früchte morphologisch gar nicht gleichwertig sind. Vielleicht darf man gerade deshalb in diesem vorübergehenden Atmungsanstieg in den späten Phasen der Organentwicklung eine für das Altern wesentliche Erscheinung erblicken. Der Reifungsanstieg der Atmung tritt im allgemeinen sowohl bei Früchten an der Pflanze als auch bei abgetrennt reifenden Früchten auf. Bei *Citrus*-Früchten ist ein Atmungsklimakterium in gewöhnlicher Atmosphäre nicht nachweisbar. Es entwickelt sich aber bei erhöhtem Sauerstoffgehalt (BIALE und YOUNG). Ein einfacher, für alle Früchte gültiger Zusammenhang des Atmungsanstieges mit anderen sinnlich wahrnehmbaren Reifeerscheinungen, etwa der Ausfärbung oder Aromabildung, besteht nicht (vgl. PAECH 1940a; BIALE). Selbst die zunächst vermutete ursächliche Verknüpfung mit dem beim Reifen von Früchten meist gebildeten Äthylen, das den Atmungsanstieg auslösen kann, gilt nicht für alle reifenden Früchte (BIALE, YOUNG und OLMSTEAD). Es hat sich auch keinerlei Beziehung dieses eigenartigen Atmungsverlaufes zu einem leicht analysierbaren inneren Faktor (Zucker-, Säure-, Pektingehalt oder ähnliches) aufdecken lassen. Man mußte deshalb wie bei manchem anderen zunächst nicht erklärbaren physiologischen Phänomen vermuten, daß eine Wandlung im Plasmazustand für den klimakterischen Atmungsanstieg verantwortlich ist.

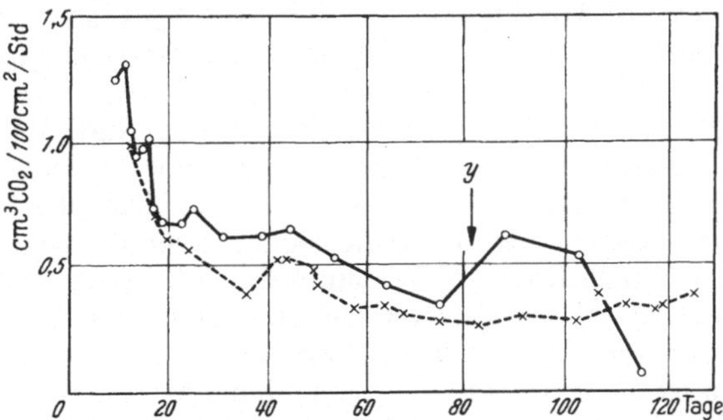

Abb. 3. CO_2-Produktion von Erdbeerblättern, gemessen an der Pflanze. Es sind zwei Kurven aus einer größeren Zahl von Meßreihen ausgewählt. -o-o-o- Blatt, gebildet an einer jungen Pflanze im September, Lebensdauer bis zum darauffolgenden Januar; ×--×--× Blatt an einer alten Pflanze im Juli gebildet, ohne Vergilbung absterbend. Der Zeitpunkt des Vergilbens der September-Blätter ist durch „*y*" angezeichnet. (Aus JAMES, nach ARNEY.)

Präzisere Möglichkeiten diskutieren MILLERD, BONNER und BIALE im Anschluß an ihre Untersuchungen der enzymatischen Veränderungen beim Reifen von Avocado-Früchten. Sie vermuten, daß die Atmungsintensität der Früchte zunächst absinken muß, weil die verfügbaren Acceptoren für die durch die Atmung gebildeten energiereichen Phosphatbindungen abgesättigt werden. Die als Phosphatbindungen anfallende Energie kann also nicht mehr abgenommen werden. Eine gesteigerte Atmung während der Reifung würde dementsprechend bedeuten, daß jetzt Zustände eingetreten sind, in denen Phosphatbindungen freizügiger abgenommen und deshalb durch gesteigerten oxydativen Stoffwechsel nachgebildet werden können. Mit dieser Vermutung stimmt jedenfalls das Verhalten der Früchte gegen 2,4-Dinitrophenol (DNP) überein, das bekanntlich Oxydationen und die daraus resultierenden Phosphorylierungen im Stoffwechsel entkoppelt. Auch bei Äpfeln dürfte das Verhältnis ATP:ADP, das ein Maß für die Sättigung der Phosphorylierungskapazität abgibt, der maßgebende Faktor für die Höhe der Atmungsintensität bei der Reife sein (PEARSON und ROBERTSON).

MILLERD, BONNER und BIALE finden zwar in reifenden Avocado-Früchten eine lösliche Verbindung, die ähnlich wie DNP auf die Atmung wirkt, die also einen oxydativen Umsatz auch ohne gleichzeitige Phosphorylierung zuließe und die gesteigerte Atmung zu einer physiologisch nutzlosen Erscheinung machen würde, womit allerdings die sogleich zu besprechenden Synthesen nicht recht in Einklang zu bringen sind.

Ein zügiger Verbrauch energiereicher Phosphorsäureverbindungen und damit ein Regenerieren verfügbarer Phosphatacceptoren (Adenylsäure oder ADP) findet im allgemeinen bei allerlei Synthesen in der Zelle statt. Und tatsächlich setzen etwa gleichzeitig mit dem Atmungsanstieg eine Reihe von synthetischen Reaktionen ein, deren Produkte meist schon äußerlich an der Ausfärbung und der Aromabildung der Früchte zu erkennen sind. Anthocyane, Carotine, Ascorbinsäure und eine Reihe anderer sekundärer Stoffe entstehen kurz vor, während und nach der Phase des Atmungsanstieges. Diese Synthesen brauchen sicher einen Teil der energiereichen Bindungen auf und geben damit den Weg für die Übernahme neuer Energie aus der Atmung frei. Wenn diese Erklärungsmöglichkeit für den klimakterischen Atmungsanstieg das Richtige trifft, dann bliebe noch zu ergründen, warum die synthetische Phase erst in einem relativ späten Stadium der Fruchtentwicklung einsetzt, obwohl schon lange vorher genügend Mengen an Reservestoffen vorhanden sind.

Besonders bemerkenswert ist, daß an diesen Synthesen in den Früchten auch der *Eiweißumsatz* beteiligt ist. Fast gleichzeitig mit dem klimakterischen Atmungsanstieg in Äpfeln beginnt auch die Synthese von Protein aus aufgespeicherten löslichen N-Verbindungen (Amiden, Aminosäuren, NH_4-Salzen). Wann immer die charakteristische Intensivierung der Atmung einsetzt, ob am Baum, nach der Ernte bei der Lagerung oder künstlich durch Äthylen hervorgerufen, so ist sie stets von einer Zunahme des Eiweißgehaltes der Äpfel begleitet. Wenn der Atmungsanstieg verzögert oder die Höhe des Maximums abgeflacht wird, z. B. durch Anreicherung von CO_2 in der umgebenden Atmosphäre, so resultiert eine entsprechende Verzögerung der Eiweißbildung oder eine im ganzen geringere Zunahme des Proteingehaltes der Früchte (HULME 1949). Das Verhältnis Atmungsintensität zu Eiweißgehalt bleibt in Äpfeln am Baum vom Ende der Zellteilung bis zum Beginn des Atmungsanstieges ungefähr konstant. Man hat diesen R/P-Wert (respiration/protein) in die physiologische Betrachtung der Fruchtreifung eingeführt, denn er ist in gewissen Grenzen unabhängig von den klimatischen Bedingungen und den Bodenverhältnissen, unter denen die Früchte gewachsen sind (GRIFFITHS und Mitarbeiter; HULME 1951, 1954). Bei Äpfeln liegt diese physiologische Konstante für spät reifende, also langsam alternde Sorten bedeutend niedriger als bei frühreifen bzw. schnell alternden Sorten.

Aufschlußreicher als die Sauerstoffaufnahme und die CO_2-Abgabe allein ist manchmal das Verhältnis beider, der RQ ($CO_2:O_2$). Dieser Wert erhebt sich in Früchten meist wesentlich über 1, wenn der Atmungsanstieg beginnt. In Blättern ist ein bedeutend über die Einheit hinausgehender RQ nur während des frühesten Stadiums der Entwicklung gefunden worden (RUHLAND und RAMSHORN). Oft hält sich der RQ schon im sich streckenden Gewebe, auf alle Fälle aber in ausgewachsenen Blättern, sofern sie nicht succulent und dem diurnalen Säurerhythmus unterworfen sind, ziemlich einförmig bei 1,0 (PRINGSHEIM). In vergilbenden Blättern ist er, wie oben bereits erwähnt, im allgemeinen kleiner als 1 (EBERHARDT 1955). Mit Ausnahme der frühesten Stadien der Zellstreckung, in denen die glykolytische Spaltung die oxydative Umsetzung im Stoffwechsel der Blätter und Stengel überwiegt (GODDARD; GODDARD und MEEUSE), läßt sich am RQ keine Wandlung des Atmungssystemes mit dem Alter der Organe ablesen. In Übereinstimmung mit dem Vorherrschen des anaeroben Teiles des Kohlenhydratabbaues in den sich streckenden Zellen steht die Beobachtung, daß junge unentfaltete, noch untergetauchte Blätter von *Nuphar advenum* mehrere Tage anaerob leben können und dabei die CO_2-Abgabe nur wenig einschränken, während ausgewachsene schwimmende Blätter in Anaerobiose bei starkem Abfall der CO_2-Ausscheidung bald geschädigt werden (LAING).

Daß sich wenigstens in den frühen Phasen der Organentwicklung eine Wandlung im Atmungssystem vollzieht, ist besonders durch Anwendung von Atmungshemmstoffen erkannt worden. In Embryonen, jungen Blättern und den Wurzelspitzen junger Keimlinge wird die O_2-Aufnahme bzw. die CO_2-Abgabe zu einem großen Teil, manchmal vollständig durch die Schwermetallgifte CO, HCN und Azid gehemmt (MARSH und GODDARD; DALY und BROWN; JAMES und BOULTER; vgl. auch GODDARD und MEEUSE). Ein großer Teil der Atmung dieser jungen Organe läuft also über Cytochrom als Endoxydase. In den Wurzelspitzen der Gerste wird schon sehr früh das Cytochromsystem durch die Ascorbinsäureoxydase ersetzt, die aber auch noch durch die genannten Gifte hemmbar ist. In den ausgewachsenen Blättern und in älteren Wurzeln von Gerste muß noch eine weitere Veränderung im Atmungsmechanismus vor sich gehen, denn die schwermetallhaltigen Enzyme treten dann noch mehr zurück. Über die Art der an der Atmung ausgewachsener Organe beteiligten Oxydasen ist vorläufig noch nichts Sicheres bekannt. Es ist aber nicht sehr wahrscheinlich, daß die Aufklärung des Atmungssystemes ausgewachsener Organe gleichzeitig auch den Grund und die Notwendigkeit für den Übergang von dem einen zu dem anderen Mechanismus während der Entwicklung der Organe erkennen lassen wird. Auch hier werden wir wohl wieder auf eine vorhergehende Veränderung in der Zusammensetzung oder Struktur des Plasmas als erste Ursache hingewiesen.

Der Übergang von einem enzymatischen System auf ein anderes während der Entwicklung vollzieht sich aber nicht nur bei den Endoxydasen. Auch der Abbau der Hexosen nimmt in jungen und alten Pflanzengeweben einen verschiedenartigen Verlauf. Werden Wurzeln, Stengelinternodien und Blätter verschiedener Pflanzen mit Glucose-1-C^{14} oder Glucose-6-C^{14} gefüttert, so ist im gebildeten Kohlendioxyd das Verhältnis *%C^{14} aus Glucose-6-C^{14} : %C^{14} aus Glucose-1-C^{14}* in jungen Geweben immer größer als in alten Teilen. Daraus geht hervor, daß in jugendlichen Geweben der normale Glykolyseweg vorherrscht, während mit fortschreitendem Alter die Glucose immer stärker auf dem Weg der direkten Oxydation (am 1-C-Atom) veratmet wird. In vielen ausdifferenzierten Organen beträgt der Anteil der direkten Oxydation am Glucoseabbau mindestens 50% (GIBBS und BEEVERS).

Abschließend soll noch erwähnt werden, daß sich keinerlei Anhaltspunkte dafür finden lassen, daß die Atmung in alternden Zellen durch Mangel an Atmungsmaterial erlischt. In vielen Früchten läßt schon der Geschmack erkennen, daß Zucker reichlich vorhanden ist. Auch in den Blättern ist mit dem Alter keineswegs eine Abnahme der Gesamtkohlenhydrate und nicht einmal der alkohollöslichen festzustellen (SMIRNOV). Beim Reifen von Früchten wird zunächst im allgemeinen Stärke gespeichert, die meist schon vor dem Beginn des Atmungsanstieges wieder hydrolysiert wird. Ein Überhandnehmen der hydrolytischen Aktivität macht sich auf verschiedenen Gebieten des Stoffwechsels mit zunehmendem Alter bemerkbar, z. B. bei Eiweißen und Pektinen.

C. Der Eiweißstoffwechsel.

Der Eiweißumsatz bildet insofern einen für die Alterserscheinungen wichtigen Sektor des Stoffwechsels, als es sich dabei um unerläßliche Bausteine des Plasmas handelt. Im Mittelpunkt der Betrachtungen über die Altersabhängigkeit des pflanzlichen Eiweißumsatzes hat meist die immer noch strittige Frage gestanden, ob den Zellen bzw. Organen mit fortschreitendem Alter die Fähigkeit zur Eiweißsynthese verloren geht und dafür eine zunehmende Hydrolyse die lebende Substanz allmählich aufzehrt. Daß dem bei vorsichtiger und kritischer Ausdeutung der Versuchsergebnisse nicht so sein kann, wurde jedoch schon vor längerer Zeit dargelegt (PAECH 1935a). Beim Abblühen, d. h. dem raschen Altern von Blütenblättern, war ein „explosionsartiger Eiweißabbau" als die Ursache für den Turgorverlust, also den Zelltod, angesehen worden (SCHUMACHER 1931). Hier hat sich aber inzwischen herausgestellt, daß der Eiweißschwund eine späte Folge ganz anderer zum Zelltod führender Prozesse ist (SCHUMACHER 1953). Auch hier ist, ähnlich wie das Vergilben der Laubblätter, die Eiweißhydrolyse eine für das Greisenalter charakteristische, prämortale Erscheinung.

Die Hauptschwierigkeit bei der Erforschung des Alterseinflusses auf den Eiweißstoffwechsel liegt in der unsicheren Vorstellung begründet, die man vom Mechanismus des Eiweißaufbaues in pflanzlichen Zellen hat. Weder der ausreichenden Versorgung mit den für die Aminosäuresynthese notwendigen Kohlenstoffgerüsten noch der aus bisher unbekannten Gründen, aber doch offensichtlich begrenzten Speicherfähigkeit der Pflanzenzelle für Eiweißstoffe wurde genügend Aufmerksamkeit geschenkt. Zudem beeinflussen gerade im Stickstoffhaushalt die verschiedenen Organe im Verband der Pflanze einander tiefgreifend, wobei die jüngeren stets dominieren. Ähnliches gilt ja auch für den Wasserhaushalt und für bestimmte mineralische Bestandteile. Nicht nur die wachsende Spitze und die Wurzeln einer krautigen Pflanze wirken durch Wegführung der Spaltprodukte fördernd auf den Abbau in älteren Organen, auch die sich entwickelnden Samen und Früchte sowie die sich füllenden Speicherorgane beeinflussen den Stoffwechsel ausgewachsener Blätter. Nach Abtrennen der Blätter von der Pflanze oder Entfernen der wachsenden Teile wird dieser abbaufördernde und damit indirekt zum Tode treibende Faktor ausgeschaltet (vgl. dazu auch FRANK), wodurch manche scheinbar durch das Alter der Blätter, aber tatsächlich durch die Dominanz jüngerer Organe bedingte „Neigung" des Stoffwechsels ins Gegenteil umgekehrt werden kann. So können alte Gerstenblätter, die im Zustand des Eiweißabbaues sind, nach Abtrennen der jüngeren Sproßteile von der Pflanze und bei verstärkter N-Zufuhr zur Proteinsynthese übergehen (WALKLEY).

Der Eiweißspiegel des Blattes einer krautigen Pflanze (bezogen auf das Frischgewicht!) steigt während der Streckung an. In ausgewachsenen Blättern sinkt er jedoch während der Entwicklung der Blüten alsbald wieder ab, um zur Zeit der Blütenentfaltung ein Minimum zu durchlaufen. Danach nimmt im allgemeinen der Eiweißgehalt wieder zu, zum Teil bis über die im Frühsommer gemessenen Werte. Erst kurz vor Beginn des Vergilbens setzt ein ziemlich rascher Eiweißabbau ein. Die ältesten Blätter nehmen allerdings am Aufstieg nach der Blütezeit meist nicht mehr teil, weil bei ihnen inzwischen das Vergilben schon begonnen hat (vgl. SMIRNOV; MEYER). Bei den Holzgewächsen wird der Eiweißhaushalt der Blätter weniger durch die wachsenden Organe gestört, weil hier die Achsenorgane einen puffernden Reservevorrat darstellen. Die älteren Blätter werden deshalb nicht so stark für den Aufbau der jüngeren Organe in Anspruch genommen (MOTHES).

Obwohl in den Blättern der krautigen Pflanzen die Anforderungen der wachsenden Organe durch einen noch unbekannten Mechanismus den Eiweißumsatz der ausgewachsenen Blätter beherrschen, ist der durch Minima und Maxima charakterisierte Eiweißgehalt während der Lebensdauer des Blattes doch insofern aufschlußreich, als daraus hervorgeht, daß die Blätter auch in einem ziemlich

„hohen Alter“ die Fähigkeit, Eiweiß zu synthetisieren noch erhalten haben, wie die rasche Ergänzung des Eiweißvorrates nach der Auszehrung während der Blütezeit anzeigt. Auch der oben bereits erwähnte Anstieg des Eiweißgehaltes reifender Früchte beweist, daß alternde Organe durchaus zur Proteinsynthese befähigt sind. Die wahren synthetischen Fähigkeiten lassen sich jedoch oft erst nachweisen, wenn in den Zellen Platz für die Aufnahme weiterer Produkte der Synthesen geschaffen worden ist, wie z. B. in ausgewachsenen Blättern während der Blüte. Werden solche Blätter von der Pflanze abgetrennt, so gelingt in Fütterungsversuchen schon während der Blütezeit der Nachweis der Eiweißsynthese (vgl. Tabelle 1 nach PEARSALL und BILLIMORIA). Damit ist übrigens auch die gelegentlich diskutierte Möglichkeit widerlegt, daß der Eiweißabbau während der Blütezeit durch Hormone „gesteuert“ sei.

Von blühenden Pflanzen bilden also mit Ausnahme der allerjüngsten Blättchen des 2. Knotens, die wahrscheinlich während der Versuchsdauer noch kräftig wuchsen, alle Blätter unabhängig von ihrem physiologischen Alter unter den gegebenen Versuchsbedingungen ungefähr gleiche Mengen Eiweiß, nämlich 15 bis 19 mg je 100 g Frischgewicht, was bei den älteren, relativ eiweißärmeren sogar einen höheren Prozentsatz des vorhandenen Gehaltes ausmacht als bei den jüngeren. Vergleichsweise wird bei nicht blühenden Pflanzen, die keine Auszehrung erlitten, sondern sich im Laufe ihrer Entwicklung eher mit Eiweiß gesättigt hatten, durch Fütterung nur in den jüngsten Blättchen eine Synthese erzielt, woraus jedoch nicht auf eine mangelnde Synthesefähigkeit der älteren Blätter geschlossen werden darf, eben weil diese ihre Kapazität für Eiweißstoffe voll ausgenutzt hatten. Wie eine solche Eiweißsättigung in pflanzlichen Zellen zustande kommt, von welchen Faktoren sie bestimmt wird, und ob sie sich mit dem Alter des Organs quantitativ oder qualitativ verändert, das sind noch offene Fragen.

Tabelle 1. *Eiweißbildung der Blätter einer blühenden Coleus-Pflanze bei Zugabe von 3% Glucose + 0,2% Ammoniumnitrat. 15—17° C; 50 Std.*
(Nach PEARSALL und BILLIMORIA 1938.)

Nodus	Eiweiß-N in mg je 100 g Frischgewicht			Zunahme in Prozent
	Versuchsbeginn	Versuchsende	Zunahme	
2. (Spitze)	204	253	49	24,0
3.	179	194	15	8,4
4.	142	161	19	13,4
5.	104	121	17	16,4
6. (Basis)	95	114	19	20,0

Die Zusammensetzung der Gesamteiweiße von *Chlorella vulgaris*-Kulturen verändert sich mit dem Alter der Kultur nur wenig (FOWDEN). Der relative Anteil der meisten der 19 untersuchten Aminosäuren bleibt etwa konstant. Histidin nimmt relativ zu und erreicht nach 35 Tagen einen 1,5mal so hohen Wert wie in 5 Tage alten Kulturen. Eine geringe Zunahme zeigt auch der Anteil von Alanin und Lysin. Nur der Arginingehalt der Eiweiße nimmt mit dem Alter der Algenkulturen merklich ab. Damit kann allerdings noch nichts darüber ausgesagt werden, ob nicht die einzelnen Fraktionen der Eiweiße sich im Laufe des Lebens der Zellen stärker verändern und nur die Summe aller Aminosäuren eine scheinbare Konstanz vortäuscht (vgl. SISSAKIAN). Weitere qualitative Änderungen würde man bei den Nucleoproteiden erwarten, aber außer den erwähnten Verhältnissen in Plastiden (s. S. 915) liegen unseres Wissens keine Untersuchungen darüber vor.

Bei verschiedenen Blütenpflanzen wurde im letzten Teil der Vegetationsperiode ein Schwund an Gesamtstickstoff festgestellt (FRANK). — Über den Eiweißumsatz in reifenden Früchten s. S. 918.

D. Andere Stoffwechselgebiete.

Der Mineralstoffhaushalt verdient eine besondere Erwähnung, weil in ihm eine „Mitursache des Todes der Blätter" gesehen wurde (Molisch 1929). Der Aschengehalt nimmt im allgemeinen entsprechend der Transpirationsintensität zu und erreicht in vielen krautigen und laubabwerfenden Pflanzen unserer Breiten am Ende der Vegetationsperiode Werte von 10—25% des Trockengewichtes. Obwohl vor allem Calcium, das meist einen großen Prozentsatz der Asche ausmacht, in die Vacuole ausgeschieden wird und deshalb den lebenden Teil der Zelle nicht belastet, „muß die auffallende Inkrustierung mit Aschensubstanzen Störungen in der Funktion des Blattes hervorrufen" (Molisch). Immergrüne Blätter, d. h. solche mit einer mehrjährigen Lebensdauer, haben einer geringeren Transpirationsintensität wegen tatsächlich einen wesentlich geringeren Aschengehalt, und die jahrzehntealten Blätter von *Welwitschia* sterben an den Spitzen ab und wachsen an der Basis weiter, so daß auch bei ihnen eine stärkere Anreicherung mit Mineralstoffen vermieden wird. Durch Versuche konnte Molisch weiter unterstreichen, daß die Ansammlung von mineralischen Bestandteilen das Vergilben und den Laubabfall wenigstens beschleunigt, denn eingetopfte Exemplare von Birken und Hainbuchen, die im feuchten Raume gehalten wurden, erhielten ihre Blätter viel länger am Leben als bei ungehemmter Transpiration. Mag tatsächlich bei Blättern die Menge der angesammelten Mineralstoffe mitbestimmend für den Tod der Blätter sein, so kann doch nicht allgemein die fortschreitende Anreicherung mit Aschenbestandteilen als eine Ursache oder auch nur eine enge Begleiterscheinung des Alterns angesehen werden, denn sie bleibt z. B. bei Früchten aus, die abgetrennt von der Mutterpflanze ein natürliches Altern durchlaufen, ohne daß der Mineralstoffgehalt sich verändert.

Qualitative Unterschiede in der Mineralstoffaufnahme junger und älterer Pflanzen sind nachgewiesen worden (Rippel). Daraus geht nicht immer das Überhandnehmen der Calciumaufnahme im Alter, sondern eher manchmal ein Vorauseilen der Kaliumaufnahme in der Jugend hervor. Auf alle Fälle scheinen die für die Salzaufnahme verantwortlichen Mechanismen sich mit dem Alter der Pflanze bzw. der Zellen zu wandeln.

Sekundäre Pflanzenstoffe. Obwohl die intensivste Bildung von allerlei sekundären Pflanzenstoffen häufig im Zusammenhang mit der Streckung der Zellen und Organe oder im unmittelbaren Anschluß daran stattfindet, läßt sich doch in vielen Fällen eine fortschreitende Ansammlung bis in die letzten Lebensstadien der Organe und Pflanzen nachweisen (vgl. Paech 1950). Zwar wurde manchmal die Ansammlung von „Stoffwechselschlacken" — und als solche werden viele sekundäre Stoffe betrachtet — als frühe Ursache für weitere Altersveränderungen und den Zelltod verantwortlich gemacht. Die Einbeziehung sekundärer Pflanzenstoffe in Betrachtungen über den Vorgang des Alterns dürfte aber kaum aus diesen Gründen von Nutzen sein, weil Genese und stoffwechselphysiologische Bedeutung der meisten sekundären Stoffe noch viel zu wenig geklärt sind und dort, wo sie bekannt sind, nicht die ihnen beim Altern zugedachte Rolle spielen. Es könnte aber sein, daß die Synthese sekundärer Stoffe und ihr Verhalten in alternden Organen Hinweise auf grundlegende Wandlungen im Stoffwechsel geben. Unter diesem Gesichtspunkt verdienen in erster Linie die bei der Laubfärbung im Herbst und beim Reifen der Früchte einsetzenden Synthesen von Pigmenten untersucht zu werden. Daß die mit der gesteigerten Atmung einhergehende Anlieferung von Bausteinen und Energie eine wesentliche Voraussetzung für solche Synthesen in den späten Stadien des Lebens von Blättern und Früchten ist, bedarf heute fast keiner besonderen Erwähnung. Durch Zuckerfütterung unter Bedingungen, unter denen der Zucker umgesetzt wird, kann die Anthocyanbildung auch unabhängig von dem fortgeschrittenen Alter der Blätter in Gang gebracht werden; und doch gibt es in der „Neigung“ der Blätter zur Anthocyansynthese Unterschiede, die entweder vom Altersstadium oder aber von den Umweltbedingungen abhängen (EBERHARDT 1954). Im Bereich des Carotinoidstoffwechsels sind sowohl in vergilbenden Blättern als auch in Blütenblättern tiefgreifende Wandlungen mit dem Alter festgestellt worden (SEYBOLD). Ein besonderes Beispiel aus dem Stoffwechsel der Blütencarotinoide mag andeuten, daß möglicherweise die Zusammensetzung der Organe, in diesem Falle der Blütenblätter, durch das fortschreitende Alter der Pflanze verändert wird. *Viola*-Blüten, die im Frühsommer sich entwickeln, enthalten in der Hauptsache Violaxanthin, ein Carotinoid-Epoxyd und wenig von dem furanoiden Auroxanthin, während die im Spätsommer und Herbst entfalteten Blüten zunehmende Mengen des furanoiden Derivates bilden, das als ein umgewandeltes Violaxanthin angesehen werden kann (KARRER und Mitarbeiter). Diese verschiedene Zusammensetzung der Blütenfarbstoffe ist um so bemerkenswerter, als es sich nicht um alternde Blüten, sondern immer um neue Blüten handelt, die an der älter gewordenen Mutterpflanze entstehen. Aber auch hier könnte die fortgeschrittene Jahreszeit die wirkliche Ursache für den veränderten Zustand des Systems der Carotinoidbildung in den Blüten sein.

Rückblickend läßt sich sagen, daß nach den bisher vorliegenden, nicht immer sehr glücklich angestellten Beobachtungen und Versuchen mancherlei Wandlungen des Stoffwechsels im Laufe des Alterns der Zellen, Organe und ganzen Pflanzen nachweisbar sind. Daraus ergibt sich aber noch nicht, daß diese Stoffwechseländerungen mehr als Begleiterscheinungen eines Altersprozesses sind, der seinerseits durch andere Ursachen angestoßen sein muß. Keiner der Stoffwechselvorgänge verläuft von Anfang an so, daß er notwendig zu einer allmählichen Senkung der Lebensfähigkeit oder Lebensintensität der Zellen führen müßte.

Vielleicht sind auch die üblichen stoffwechselphysiologischen Untersuchungen noch zu grob und auf zu komplexe Systeme gerichtet. Die Aufteilung des Zellinhaltes in plasmatische und enzymatische Fraktionen läßt möglicherweise einmal Wandlungen erkennen, die zu einer Schwächung der Zellen führen müssen. Der Unterschied des Plasmas embryonaler und ausgewachsener Zellen wird sich wenigstens in gewissen Zellfraktionen ausprägen. Erste Beobachtungen in dieser Hinsicht sind an der Urease in *Citrullus*-Kotyledonen gemacht worden. Das Enzym erfüllt in den Kotyledonen, in denen es während der Zellstreckung reichlich gebildet wird, keine Funktion. Es deutet wohl „nur“ eine wachstumsgebundene Differenzierung des Plasmas an (WILLIAMS und SHARMA). Die Enzyme sind ja nicht gesonderte, ins Plasma eingelagerte Teilchen, sondern sie sind selbst Plasma, das vielleicht als eine „multikatalytische Einheit“ (SEVAG) angesehen werden darf. Das Auftreten so großer Mengen Urease in den Kotyledonen ist also nur ein Zeichen für einen bestimmten Plasmazustand, der erreicht wird, wenn das Streckungswachstum aufhört.

Wenn man nicht an eine hormonale Auslösung des Altersvorganges denken will, die wohl am sinnvollsten im Zusammenhang mit der Entwicklungsphysiologie diskutiert wird, so verspricht heute noch die Suche nach physikalisch-chemisch faßbaren Wandlungen des Plasmas einen aufschlußreicheren Einblick in die tiefer reichenden, vielleicht sogar ursprünglichen Erscheinungen des Alterns der Zellen.

III. Zellphysiologische Altersveränderungen.

Von den physikalisch-chemischen Veränderungen, die sich im Verlauf des Alterns abspielen, werden im folgenden Abschnitt die Größen besprochen, die von der Zellphysiologie messend erfaßt werden können. Damit sind dem Problemkreis von vornherein Grenzen gesetzt. In die Bearbeitung sind auch Bereiche eingeschlossen, von denen man keine wesentliche Förderung unserer Kenntnis von den zellphysiologischen Grundlagen des Alterns erwarten kann. So wird den Veränderungen der nichtplasmatischen Größen gegenüber denjenigen der protoplasmatischen Faktoren für das Verständnis des Alterns der Zelle die geringere Bedeutung zuzumessen sein. Dennoch müssen die an Preßsäften ermittelten Größen wie osmotischer Wert, p_H-Wert und Redoxpotential in die Darstellung der zellphysiologischen Altersveränderungen einbezogen werden, wenn wir ihnen vielleicht auch nur symptomatische Bedeutung zuerkennen werden. Andererseits wissen wir ja auch bei den protoplasmatischen Größen (Permeabilität, Viscosität, Hydratation usw.) nie mit Sicherheit, ob eine beobachtete Altersveränderung mehr ursächliche Bedeutung besitzt oder mehr Begleiterscheinung ist.

Eine zusammenfassende Darstellung der protoplasmatischen Veränderungen beim Altern von Pflanzenzellen ist von FISCHER (1950) gegeben worden.

A. Vacuole.

Osmotischer Wert. Die Untersuchungen der osmotischen Werte alter und junger Zellen haben bisher je nach Objekt und Methode sehr uneinheitliche Resultate ergeben. Das kann neben anderen Ursachen seinen Grund darin

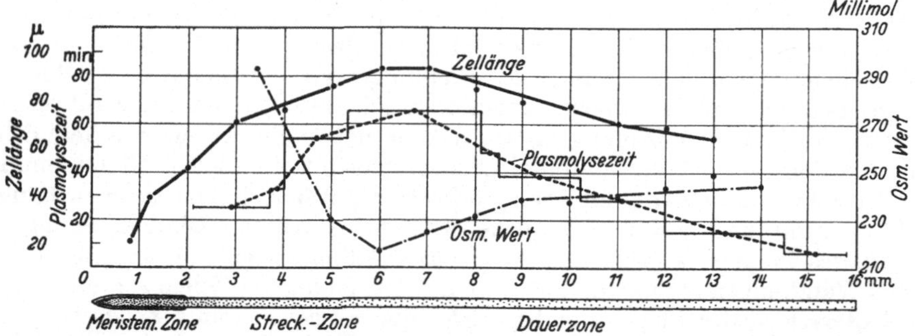

Abb. 4. Vergleich der Gradienten von Plasmolysezeit und osmotischem Wert, sowie Zellänge in der Wurzel von *Lemna minor*. (Alle Werte sind an derselben Wurzel gewonnen.) Die Meßpunkte der osmotischen Werte geben die Mittelwerte von je 10 Zellen wieder. Die Unterschiede zwischen den verbundenen Meßpunkten sind statistisch gesichert. (Aus PIRSON und SEIDEL.)

haben, daß sich die beobachteten Veränderungen des osmotischen Wertes weitgehend unabhängig vom Alter vollziehen. Oder anders ausgedrückt: Der Altersfaktor beeinflußt den osmotischen Wert unter Umständen weniger stark als andere Faktoren, die mittelbar oder unmittelbar auf die Zellsaftkonzentration einwirken können. Darauf deutet auch die kritische Bemerkung FISCHERS (1950) hin, der hervorhebt, wie stark die Unterschiede der Zellsaftkonzentrationen in verschiedenen Teilen der Pflanze von Außenfaktoren abhängen (z. B. von der Wasserversorgung, von Belichtung und Beschattung u. a.). Damit wird von vornherein die Notwendigkeit unterstrichen, bei der Untersuchung von Altersgradienten des osmotischen Wertes unter kontrollierten Außenbedingungen zu arbeiten.

PIRSON und SEIDEL haben derartige Versuche unter definierten Außenbedingungen an der Wurzel von *Lemna minor* durchgeführt und dabei mit Glucose als Plasmolyticum unter anderem gefunden, daß der Gradient des osmotischen Wertes am Ende der Streckungszone, d. h. dort, wo die Zellen ihre größte Länge erreicht haben, ein Minimum durchläuft (Abb. 4). Dieser Gang des osmotischen

Gradienten bleibt auch in einem Kulturmedium mit 0,5% Glucosezusatz bei insgesamt höheren Werten gegenüber zuckerfreiem Medium erhalten. Das Minimum des osmotischen Wertes fällt bei der Wurzel von *Lemna* mit einem Minimum der Permeabilität für Harnstoff zusammen.

Bezüglich einzelner Resultate aus verschiedenen Untersuchungen kann hier im wesentlichen auf die Zusammenstellung von FISCHER (1950) verwiesen werden. Man findet ansteigende wie abfallende Gradienten des osmotischen Wertes. Oft fehlen auch mit dem Alter gerichtete Veränderungen. Bei den kryoskopischen Bestimmungen überwiegen die Angaben für ein Ansteigen des osmotischen Wertes mit dem Alter (SMIRNOV an Tabakblättern, GAIL und CONE an Blättern von *Pinus ponderosa*). In Blütenblättern *(Iris, Gladiolus, Paeonia)* sinkt der osmotische Wert während des Abblühens, so z. B. während der Postfloration bei *Paeonia* von etwa 0,9 auf 0,55 mol Glucose (BANCHER). In der Mehrzahl der Fälle wurde in den Laubblättern höherer Landpflanzen eine Abnahme des osmotischen Wertes mit dem Alter gefunden, während bei *Helodea* und bei Moosen die Angaben über höhere Werte in den basalen Blättern überwiegen.

Tabelle 2. *Vergleich der Redoxpotentiale im Gewebebrei der Beeren von Vitis vinifera in verschiedenen Stadien der Reifung.* (Nach RUMMENI 1954.)

	Redoxpotential Ec mV	Acidität[1] p_H
Unreif. . . .	+84	3,65
Halbreif I . .	+19	3,83
Halbreif II .	—19	3,95
Reif	—35	4,05

p_H-Wert. Veränderungen der Wasserstoffionenkonzentration mit dem Alter sind vor allem vielfach für Blütenblätter angenommen worden, deren Anthocyanfärbung im Verlauf der Blühzeit einen Farbumschlag aufweist. In vielen Fällen führt die Verfärbung von röteren zu blaueren Farbtönen. Danach sollte ein Ansteigen des p_H-Wertes mit fortschreitendem Alter der Blütenblätter erwartet werden. Nun sind aber durchaus nicht allein die Aciditätsverhältnisse für die Farbnuance von Anthocyanen verantwortlich (vgl. PAECH 1950, HAYASHI 1954), demzufolge können zelleigene Anthocyane auch nicht ohne weiteres als Indicatoren für den p_H-Wert des Zellsaftes verwendet werden (DRAWERT 1955). Andererseits ist aber für eine Reihe von Objekten ein Sinken der Acidität mit dem Alter direkt nachgewiesen worden (GAIL und CONE an Blättern von *Pinus ponderosa*, WOOD an Blättern von *Atriplex nummularium*, RUMMENI in Früchten von *Vitis vinifera*, vgl. Tabelle 2). Es sind aber auch p_H-Gradienten mit einem Maximum der Acidität in einer mittleren Altersphase gefunden worden (vgl. PAECH 1940b). Alle diese Beobachtungen können für das Problem der Altersveränderungen in lebenden Pflanzenzellen nur von symptomatischem Wert sein, da sie sich lediglich auf die Acidität des Preßsaftes beziehen. Der p_H-Wert des Zellsaftes ist zudem auch sehr stark von Außenfaktoren abhängig. So fand z. B. GUSTAFSON an Blättern von *Bryophyllum calycinum*, daß bei sonnigem Wetter die jüngsten, bei bedecktem Himmel die ältesten Blätter die geringste Acidität besaßen.

Das Ansteigen des p_H-Wertes bei der Fruchtreifung ist stoffwechselphysiologisch verständlich, da der respiratorische Quotient in dieser Phase über 1 ansteigt und damit auf die Veratmung von Oxy- und Ketosäuren hinweist (vgl. S. 916).

Redoxpotential. Obwohl zum Oxydoreduktionspotential auch Plasmabestandteile beitragen, wird diese Größe hier unter den Vacuolenfaktoren besprochen,

[1] Trotz der parallelen Veränderung der p_H-Werte nimmt die Autorin an, daß die Unterschiede in den gemessenen Redoxpotentialen keineswegs nur auf den — relativ geringfügigen — Unterschieden der Acidität beruhen.

weil die Werte an Preßsäften und Gewebebreien gewonnen wurden. Über Veränderungen des Redoxpotentials pflanzlicher Zellen im Verlauf der ontogenetischen Entwicklung liegen verhältnismäßig wenige Angaben vor, deren Bedeutung für Rückschlüsse auf plasmatische Altersveränderungen meist nur gering ist. Ältere Literatur über dieses Gebiet ist bereits bei PAECH (1940b) in weiterem Zusammenhang kritisch besprochen worden. Die dort zunächst mit gewissen Vorbehalten ausgedeuteten Versuche von KRASSINSKY an Erbsenpreßsaft, von ZILVA, KIDD und WEST an reifenden Äpfeln und von HAMDALLAH an Blättern weisen auf ein wachsendes Reduktionspotential mit steigendem Alter hin. In neuerer Zeit sind ähnliche Resultate von RUMMENI an Gewebebrei von reifenden Weinbeeren gewonnen worden (vgl. Tabelle 2).

Obwohl sich daraus eine gerichtete Veränderung dieses Faktors mit dem Zellalter erkennen läßt, wird eine Deutung der Ergebnisse im Rahmen der übrigen Altersveränderungen vorerst noch nicht möglich sein, ganz abgesehen von der Frage, wieweit sich die an Gewebebreien gewonnenen Resultate auf das intakte Gewebe übertragen lassen.

B. Plasma.

In vielen physiologischen Diskussionen über das Altersproblem stehen kolloidchemische Veränderungen des Protoplasmas im Vordergrund der Betrachtung. Deshalb ist es verständlich, daß die Altersforschung gerade von der Untersuchung protoplasmatischer Faktoren eine weitergehende Klärung ihrer Fragen erhofft. Besonders dem physikalisch-chemischen Vorgang des Alterns von kolloidalen Systemen wird für die biologischen Altersveränderungen Bedeutung beigemessen (vgl. BETHE). Kolloidale Lösungen verändern ihre Eigenschaften im Laufe der Zeit so, daß allmählich eine Abnahme ihrer Ladung und damit eine Verringerung der Hydratation erfolgt. Die Kolloidteilchen können dadurch zu größeren Komplexen zusammentreten, ohne daß es deshalb zu einer Ausflockung kommt. Das System ist aber weniger stabil geworden. Vielleicht kann für diese Fragen das Studium von synthetischen Modellsubstanzen, wie sie heute z. B. als Blutplasmaersatz verwendet werden, weitere Aufschlüsse bringen.

So vielversprechend zunächst die Untersuchung plasmatischer Eigenschaften zu sein scheint, so verwirrend und teilweise widersprechend sind die bisher auf diesem Gebiet erzielten Resultate. Oft wird ein Vergleich der an verschiedenen Objekten gewonnenen Erfahrungen durch die unzureichende Kennzeichnung des Alterszustandes der betreffenden Pflanzen oder Organe erschwert. Darüber hinaus können plasmatische Eigenschaften auch von Faktoren des Zellsaftes beeinflußt werden, die selbst wiederum sehr stark von Außenbedingungen abhängen (z. B. Zellsaftacidität und Harnstoffpermeabilität, BOGEN 1938, DRAWERT 1948).

Permeabilität. Im Verlauf der Entwicklung ändert sich in vielen Fällen die Permeabilität des Protoplasmas für bestimmte Diosmotica. Auch für die Permeabilitätsveränderungen läßt sich jedoch keine allgemeine Regel aufstellen, durch die man den Altersablauf im Hinblick auf diesen plasmatischen Faktor charakterisieren könnte. Wesentlich ist die Erkenntnis, daß die mit dem Altern verbundenen Permeabilitätsveränderungen für verschiedene Diosmotica nicht im gleichen Sinne ablaufen müssen (MARKLUND, zitiert bei FISCHER 1950; RUGE 1943). So kann ein und dasselbe Gewebe im Laufe seiner Entwicklung vom Harnstofftyp zum Glycerintyp übergehen. In den Epidermiszellen der Blattrippen von *Taraxacum pectinatiforme* ist die Permeabilität für Harnstoff und Methylharnstoff bei jungen Blättern größer als bei ausgewachsenen, während Glycerin in das Plasma erwachsener Blätter schneller eindringt (Tabelle 3).

Tabelle 3. *Deplasmolysezeiten (Min.) für Epidermiszellen der Blattmittelrippe von Taraxacum pectinatiforme.* (Nach MARKLUND)

	Junges, nicht ausgewachsenes Blatt	Altes, ausgewachsenes Blatt
Glycerin	$43^1/_3$	$17^1/_3$
Harnstoff. . . .	4	$37^2/_3$
Methylharnstoff .	$1^1/_2$	7
Malonamid . . .	145	330

Auch in Epidermiszellen von *Rhoeo discolor* wurde der Übergang vom Harnstoff- zum Glycerintyp im Laufe der Ausdifferenzierung beobachtet (RUGE 1943). Hier ist die Permeabilität für Erythrit und Glycerin in den Zellen der Blattspitze größer als in Zellen der Basis des gleichen Blattes, während umgekehrt die Permeabilität für Amide (Harnstoff, Sulfoharnstoff, Methylharnstoff) in den Spitzenzellen gegenüber Basiszellen erniedrigt ist.

Daneben findet man den Fall, daß zwar das Permeabilitätsmaximum für zwei verschiedene Diosmotica im gleichen Entwicklungszustand durchlaufen wird, dabei jedoch das Permeabilitätsverhältnis der beiden Substanzen zueinander wechselt (MARKLUND an *Helodea densa*, zitiert bei RUGE 1943; vgl. Tabelle 4).

Tabelle 4. *Permeabilität von Zellen (beiderseits der Mittelrippe am Blattgrund) verschieden alter Helodea-Blätter des gleichen Sproßes für Harnstoff (H) und Methylharnstoff (M).* (Nach MARKLUND)

Beschreibung des Entwicklungszustandes	Blattlänge mm	Deplasmolysezeit in Min.		Verhältnis der Permeabilität von H:M
		H	M	
I Aus der Streckungszone	22—25	240	43	H < M
II Unmittelbar auf die Streckungszone folgend	25—27	4	9	H > M
III Ausgewachsene Blätter	32—33	20	16	H < M

Bei *Ranunculus repens* wurde in den subepidermalen Zellen von normalen (ausgewachsenen) und alten Blattscheiden gleiche Harnstoffpermeabilität gefunden, dagegen war die Glycerinpermeabilität in alten Zellen herabgesetzt (HOFMEISTER). Diesen Verschiebungen des Permeabilitätsverhältnisses mit dem Alter für verschiedene Diosmotica stehen die neuen Befunde von ZICKEL gegenüber, die in ungeteilten „alten" und neugeteilten „jungen" Zellen in der Stengelepidermis von *Bryophyllum daigremontianum* eine gleichbleibende Reihenfolge der Permeationskonstanten von fünf verschiedenen Diosmotica fand. Die Reihenfolge Malonamid < Harnstoff < Glycerin < Acetamid < Glykol wurde auch durch Zusatz von Alkaliionen nicht verändert. Wo zwischen alten und jungen Zellen gesicherte Differenzen für die Permeabilität der reinen Diosmotica auftreten, ist die Stoffaufnahme in den jungen Zellen (d. h. nach vorausgegangener Teilung) niedriger als in ungeteilten Zellen. Wenn die Diosmotica mit Ionen (Li, K, Cs als Nitrate) kombiniert werden, zeigen sich weitere Unterschiede zwischen alten und jungen Zellen. Richtung und Ausmaß der Veränderung der Permeationskonstanten hängen dann außer vom Alter und von der Art des Diosmoticums auch vom Kation und bei neugeteilten Zellen von der Teilungsrichtung (quer oder längs) ab. Eine Verminderung der Harnstoffpermeabilität in jungen, eben durch Teilung entstandenen Zellen wurde auch an *Spirogyra*-Fäden beobachtet (WEBER).

REUTER untersuchte die Harnstoffpermeabilität in *Soja*-Keimblättern und fand in Übereinstimmung mit ihren früheren Befunden an *Sedum praealtum* im grünen Blatt eine hohe Durchlässigkeit für Harnstoff, im vergilbenden aber eine stark herabgesetzte Permeabilität. Die Untersuchung verschiedener Entwicklungsstadien des grünen *Soja*-Keimblattes ließ zunächst eine Zunahme der Harnstoffpermeabilität erkennen. Mit dem Beginn der Vergilbung tritt dann eine starke Verminderung der Durchlässigkeit ein.

Daß die von der Entwicklung abhängigen Permeabilitätsveränderungen nicht durch eine einfache Form eines steigenden oder fallenden Gradienten beschrieben werden können, geht besonders klar aus solchen Untersuchungen hervor, bei denen eine feinere Abstufung der gemessenen Alterszustände angestrebt wurde.

So konnten PIRSON und SEIDEL an der *Lemna*-Wurzel zeigen, daß am Ende der Streckungszone ein Minimum der Harnstoffpermeabilität durchlaufen wird, das auf einen relativ scharf umgrenzten Bereich von kaum mehr als 1—2 mm Wurzellänge beschränkt ist. In dem unmittelbar anschließenden Teil der Dauerzone steigt die Permeabilität bis zu einem Maximum steil an, dem oft nach der Wurzelbasis zu wieder ein Permeabilitätsabfall folgt. In Wurzeln, die bei Kalium- oder Calciummangel kultiviert wurden, verschiebt sich das Permeabilitätsminimum spitzenwärts unter gleichzeitiger Verflachung des Gradienten.

Die Durchlässigkeit der Wurzelhaare für Wasser scheint mit steigendem Alter abzunehmen. ROSENE und ROSENE u. WALTHALL haben die Wasseraufnahme verschieden alter Wurzelhaare an der gleichen Wurzel verglichen und bei *Raphanus sativus* und *Triticum aestivum* gefunden, daß die Geschwindigkeit des Wassereintritts in die Zellen bei älteren Wurzelhaaren wesentlich geringer ist.

Viscosität und Hydratation. Die im Laufe der Entwicklung auftretenden Veränderungen von Viscosität und Hydratation werden hier gemeinsam besprochen. Schon aus methodischen Gründen steht die Messung der Viscosität im Vordergrund; auf den Hydratationsgrad wird meistens nur geschlossen. Im allgemeinen gehen jedoch Änderungen der Viscosität mit solchen der Hydratation parallel, so daß es berechtigt ist, aus Veränderungen der Zähigkeit des Cytoplasmas auf entsprechende Verschiebungen der Hydratation zu schließen (vgl. PAECH 1940b, RUGE 1940).

Während sich bei den bisher behandelten zellphysiologischen Faktoren häufig widersprechende Versuchsresultate gegenüberstanden, zeichnet sich doch wenigstens für die Cytoplasmaviscosität in alternden Zellen ein einigermaßen einheitliches Bild ab. Man ist zwar auch hier bei Anwendung plasmolytischer Methoden zu Einschränkungen hinsichtlich der Aussagen über die Viscosität gezwungen (BORRISS, RUGE 1940), aber es bestehen doch hinreichende Gründe für die Annahme einer Viscositätserniedrigung in alten Zellen (FISCHER 1948, 1949). RUGE fand allerdings im Zusammenhang mit der Analyse des Streckungswachstums an den Blattzähnen von *Helodea densa*, daß Plasmolyseform, Plasmolysezeit und Verlagerungsfähigkeit von Plasmaeinschlüssen durch Zentrifugieren übereinstimmend auf ein Ansteigen des Viscositäts- und Hydratationsgrades von der Mitte der Streckungszone im Blatt bis zu den ältesten Zellen an der Blattspitze hinweisen. Hier muß jedoch berücksichtigt werden, daß keine im eigentlichen Sinne alten Blätter, sondern Blättchen der Sproßknospe untersucht wurden, in denen relativ ältere mit relativ jüngeren Zellen verglichen wurden. Die Befunde von RUGE müssen deshalb nicht im Widerspruch zu den Ergebnissen FISCHERs stehen, der in den Zellen alter, bereits vergilbender Blätter eine herabgesetzte Viscosität beobachtete. Mit Hilfe der Plasmolyseformmethode (unter Berücksichtigung methodisch-kritischer Gesichtspunkte) wurde eine Viscositätserniedrigung übereinstimmend sowohl in alten Blättern von Submersen als auch in solchen von zahlreichen Schwimm- und Landpflanzen gefunden. Wo Unterschiede der Plasmolyseform feststellbar sind, haben die Zellen der jüngeren Blätter die mehr konkaven Formen. Ausnahmen von dieser Regel können auftreten (z. B. *Ranunculus ficaria*, FISCHER 1949). Auch BORRISS fand bei dem Vergleich von Zellen aus verschiedenen Zonen ausgewachsener *Helodea*-Blätter, daß artspezifische Unterschiede in der Veränderung von Plasmolysezeit und Plasmolyseform bestehen können: In ausgewachsenen Blattfeldzellen von *Helodea densa* erreicht die Viscosität ein Minimum, während in den entsprechenden Zellen von *Helodea crispa* nach anfänglichem Absinken der Plasmolysezeit in der

Streckungsphase noch ein starker Anstieg der Viscosität in ausgewachsenen Blattfeldzellen auftritt. Dementsprechend zeigen ausgewachsene Zellen von *Helodea densa* Konvex-, solche von *Helodea crispa* Konkavplasmolyse. BORRISS schließt aus diesem unterschiedlichen Verhalten der beiden Arten, „daß es sich bei den beobachteten Phänomenen (Plasmolysezeitveränderungen) möglicherweise nur um sekundäre, jedenfalls nicht für den Gesamtablauf der Entwicklung entscheidende Faktoren handelt". Die Untersuchung von BORRISS beschäftigt sich ähnlich wie die Arbeit von RUGE nicht mit alten Blättern im eigentlichen Sinne.

In der *Lemna*-Wurzel fallen die Minima des osmotischen Wertes und der Harnstoffpermeabilität mit einem Maximum der Cytoplasmaviscosität zusammen: Am Ende der Streckungszone ist die Plasmolysezeit am größten (PIRSON und SEIDEL, vgl. Abb. 4). Mit fortschreitendem Alter der Zellen, d. h. zur Wurzelbasis hin, nimmt die Viscosität wieder ab. Eine Veränderung der Plasmolysezeitgradienten durch K- oder Ca-Mangelkultur ist nicht nachweisbar, solange das Wurzelwachstum nicht gehemmt ist. Wenn aber der K- bzw. Ca-Mangel zu einer Verlangsamung des Wachstums führt, so verschiebt sich der plasmatische Gradient in Richtung der Wurzelspitze (vgl. auch Permeabilität, S. 928). Die Ionenwirkung ist jedoch unspezifisch. Entsprechende Gradientenverschiebungen können auch durch andere Faktoren ausgelöst werden, die das Wurzelwachstum verlangsamen. Dem Verhältnis K/Ca wird hier kaum eine generelle Bedeutung für die Plasmaviscosität zugemessen werden können. Der in Mangelkulturen mit der Wachstumshemmung verbundene Ablauf des protoplasmatischen Alterns wird durch das K/Ca-Verhältnis offenbar in keiner Weise beeinflußt (PIRSON und SEIDEL). Dagegen zeigen die Analysen von FISCHER (1949) eine stetige Abnahme des Verhältnisses K/Ca mit fortschreitendem Alter und sinkender Cytoplasmaviscosität. In solchen Fällen kann die Ionenwirkung im Sinne einer Hydratationsbeeinflussung gedeutet werden. Es kommen jedoch auch Viscositätsveränderungen bei konstantem Mineralstoffgehalt vor (FISCHER 1949).

Untersuchungen von PAECH (1940b) über die Veränderungen des elektrischen Widerstandes lebender Pflanzengewebe mit dem Alter lassen unter der Annahme lipoidumhüllter Eiweißmicellen auf eine mit dem Alter abnehmende Hydratationsfähigkeit des Protoplasmas schließen. Durch Chloroformnarkose wird der Widerstand in den Geweben erhöht. Diese Widerstandserhöhung kann als eine vermehrte Hydratisierung angesehen werden. In reifenden Früchten (Äpfeln, Tomaten), in Rhabarberblattstielen, *Sempervivum*-Blättern und Spargelstangen nimmt die Widerstandserhöhung unter Chloroformeinwirkung mit fortschreitendem Alter ab. Daraus geht hervor, daß die lipoidumhüllten Eiweißmicellen jüngerer Zellen die Fähigkeit zur Wasseranlagerung in starkem Maße besitzen, während diese Fähigkeit bei alternden Zellen allmählich abnimmt und bei den alten oft ganz verlorengeht. Der Zustand des Plasmas junger Zellen scheint dadurch charakterisiert zu sein, daß es leichter Wasser anlagern und wieder abgeben kann. Diese Elastizität im physikalisch-chemischen Verhalten läßt mit dem Alter stetig nach, wobei das Absinken der Quellfähigkeit schon nachweisbar ist, wenn die Organe noch in Entfaltung stehen. Die Abnahme der Widerstandserhöhung bei Narkose und ihre Deutung lassen sich sinngemäß in die mit plasmolytischen Methoden gewonnenen Erfahrungen über eine Viscositätserniedrigung in alten Zellen einordnen (vgl. FISCHER 1948, 1949).

Resistenz. Altersveränderungen der Resistenz sind vorwiegend in bezug auf Hitze- und Giftresistenz untersucht worden. Die oft beobachtete Erscheinung, daß ältere Blätter vor den jüngeren Frost- oder Trockenschäden erleiden, kann die mit dem Alter nachlassende Viscosität und Hydratation mit den Resistenzerscheinungen in einen sinnvollen Zusammenhang bringen (vgl. dazu FISCHER 1948, 1949). Bei *Rhoeo discolor* sind junge Blätter hitzeresistenter als ältere. Auch innerhalb des *Rhoeo*-Blattes besteht ein entsprechendes Resistenzgefälle von der Basis zur Spitze (BOGEN 1948). Andererseits fand REUTER (1937) bei *Sedum praealtum*, daß die Epidermiszellen grüner Blätter bereits bei 54° C, diejenigen gelber Blätter erst bei 64° C absterben. Gegen einen geradlinigen Abfall des Resistenzgradienten vom jungen zum alten Gewebe sprechen Versuche von SCHEIBMAIR (1938) und Beobachtungen von GICKLHORN (1936). An Zellen von *Plagiochila asplenioides* konnte SCHEIBMAIR eine zunächst mit fortschreitendem Alter abnehmende, dann aber mit der Vergilbung wieder zunehmende Hitzeresistenz beobachten. Diesen Verhältnissen entspricht die Mitteilung GICKLHORNS, daß nach einem Nachtfrost bei *Helianthus anuus* die jüngsten und die schon vergilbenden Blätter überlebten, während die Blätter der mittleren Stengelregion erfroren. Hier können jedoch auch nichtplasmatische Faktoren wie osmotischer Wert, p_H-Wert usw. mitwirken oder zumindest entsprechende Veränderungen erfahren (vgl. KESSLER 1935). Bei manchen Moosen gibt es Altersgradienten der Strahlenempfindlichkeit (BIEBL 1954). Sie sind besonders ausgeprägt bei *Mnium serratum* und bei *Aplozia riparia*. Beide Moose zeigen in jüngeren Blättchen eine höhere Resistenz gegen direktes Sonnenlicht. Bei den meisten Lebermoosen sind die basalen Blattabschnitte resistenter.

Eine Reihe von Untersuchungen läßt eine größere Resistenz junger Zellen gegen toxisch wirkende Chemikalien erkennen (vgl. die Zusammenfassung von FISCHER 1950). Es werden aber auch Beobachtungen mit entgegengesetztem Resultat beschrieben. So fanden z. B. BIEBL und ROSSI-PILLHOFER in ihrer Untersuchung über die Veränderlichkeit der chemischen Resistenz von Epidermen zahlreicher Blütenpflanzen im Laufe der Entwicklung, daß in den meisten Fällen eine Resistenzsteigerung von den jüngsten bis zu den vollentwickelten Stadien gegenüber Mangan- und Zinksulfat auftritt. Manchmal nimmt die Resistenz in alten, überwinterten Blättern wieder ab. Allgemeingültige Aussagen hinsichtlich der Altersveränderungen der chemischen Resistenz lassen sich anscheinend nicht machen, da sich sowohl verschiedene Gewebe innerhalb des gleichen Organes als auch gleiche Gewebe gegenüber verschiedenen Giften unterschiedlich verhalten können. Derartige spezifische Altersveränderungen der Resistenzeigenschaften zeigen Versuche von BIEBL (1947, 1949) und REUTER (1949), von denen einige Resultate in Tabelle 5 zusammengestellt sind.

Von den bisher besprochenen Formen der Resistenz gegenüber chemischen oder physikalischen Faktoren unterscheidet sich die Resistenz gegenüber pflanz-

Tabelle 5. *Veränderungen der chemischen Resistenz mit dem Alter in Abhängigkeit von der Art des Gewebes und von der Art des verwendeten Giftes.* (Nach BIEBL 1947, 1949 und REUTER 1949.)

Autor	Objekt	Gewebe bzw Organ	Behandlung	Resistenz junger (J) und älterer (A) Gewebe
REUTER	*Soja*-Keimblatt	Epidermis Mesophyll	2% Äther	$J < A$ $J > A$
BIEBL (1947)	*Mnium rostratum*	Blatt	H_3BO_3 $ZnSO_4$	$J > A$ $J < A$
BIEBL (1949)	*Calypogeia neesiana*	Blatt	H_3BO_3 (3%) $ZnSO_4$ (3%)	$J > A$ $J < A$

lichen Infektionskrankheiten durch die Mannigfaltigkeit der für die Resistenz in Frage kommenden Ursachen und durch die Art des Kriteriums für den resistenten Zustand. Während bei der Feststellung von Hitze-, Kälte- oder Giftresistenz im allgemeinen das plasmolytische Verhalten bestimmter Gewebe oder Organe und damit als Kriterium der Verlust der Semipermeabilität des Protoplasten beim Zelltod herangezogen wird, ist bei der Beurteilung der Resistenz gegen Infektionen die Ausbildung des spezifischen Krankheitsbildes an der Pflanze maßgebend. Dabei braucht die Resistenz nicht ursächlich mit plasmatischen Faktoren verknüpft zu sein. Insofern gehört dieser Bereich der Resistenzerscheinungen nur zum Teil in den hier besprochenen Fragenkreis. Da aber andererseits gerade die Verschiebung der Krankheitsbereitschaft während der Entwicklung für viele Wirtspflanzen so charakteristisch ist, daß man auch bei Pflanzen Jugend- und Alterskrankheiten unterscheiden kann, müssen diese Phänomene im Hinblick auf das Altersproblem hier kurz behandelt werden. Eine ausführliche Betrachtung darüber findet sich bei GÄUMANN (1951).

Die ontogenetischen Veränderungen der *Befallsresistenz* brauchen hier nicht berücksichtigt zu werden, da sie die mit dem Bau und der Entwicklung der Pflanze zusammenhängende Befallswahrscheinlichkeit betreffen und kein zellphysiologisches Problem darstellen. Altersveränderungen der *Eindringungsresistenz* beruhen in den meisten Fällen auf der unterschiedlichen Cutinisierung und Verkorkung der Oberflächenzellen in der Jugend und im Alter. Hierher gehören deshalb vor allem viele Keimlingskrankheiten, deren Erreger nur in junge oder meristematische Gewebe einzudringen vermögen (z. B. Wurzelbrand bei Cruciferenkeimlingen, Schneeschimmel bei Keimpflanzen des Getreides). Eine entsprechende Verschiebung der Eindringungsresistenz von Jugendanfälligkeit zu Altersresistenz findet man bei Früchten. So wächst z. B. der Perforationswiderstand der Cuticula von Tomaten mit dem Alter und hemmt das Eindringen von *Macrosporium tomato*, ohne daß dabei etwa die Ausbreitungsresistenz gleichsinnige Veränderungen erfährt. Auch bei Weinbeeren besteht im Alter eine Eindringungsresistenz gegenüber *Plasmopara viticola*.

Hinsichtlich der Ausbreitung der Erreger im Gewebe verhalten sich die Früchte gerade umgekehrt wie es eben für die Eindringungsresistenz geschildert wurde: Die *Ausbreitungsresistenz* führt bei Früchten meistens von der Jugendresistenz zur Altersanfälligkeit. Vor allem während der Phase des Reifens sinkt die Ausbreitungsresistenz bedeutend. Beim Holz der Bäume nimmt die Ausbreitungsresistenz den entgegengesetzten Verlauf: Mit zunehmender Celluloseverfilzung, Einlagerung von Gerbstoffen und Phlobaphenen wächst auch die Widerstandskraft gegen Fäulniserreger. Hier ist also die Ausbreitung des Erregers im wesentlichen eine Funktion der Zellwandbeschaffenheit. Kernholz ist resistenter gegen zerstörende Mikroorganismen als Splintholz.

Während den bisher genannten Formen der Resistenzverschiebung gegen Infektionskrankheiten mit dem Alter keine unmittelbare Wirkung einer plasmatischen Veränderung zugrunde gelegt werden kann, so gibt es doch auch Fälle im Bereich der Phytopathologie, bei denen mit der ontogenetischen Entwicklung der Wirtspflanze auch deren Abwehrbereitschaft im Sinne einer protoplasmatischen Resistenz Veränderungen unterliegt. Bei Zuckerrüben tritt ein Befall mit *Cercospora beticola* erst in höherem Blattalter auf. Auch Kartoffeln besitzen eine ausgeprägte Altersanfälligkeit gegen *Phytophthora infestans*. Gerade für den letztgenannten Fall und für die Helminthosporiose des Mohns konnte GRÜMMER zeigen, daß die Infektion mit der Alterschlorose der Blätter einsetzt, d. h. zu einer Zeit, in der der Eiweißstoffwechsel eine grundlegende Umstellung erfährt und die Blattproteine abgebaut werden. Normalgrüne, nichtchlorotische Blätter werden nicht befallen; hingegen wirkt sowohl die normale Alterschlorose als auch die experimentell an jungen Pflanzen erzeugte Chlorose prädisponierend. Daß Veränderungen des Protoplasmas an derartigen Resistenzverschiebungen beteiligt sein können, zeigen auch die Beobachtungen von BERCKS, der mit der serologischen Methode bei Kartoffeln eine Altersresistenz gegen das X-Virus feststellen konnte. Es wird vermutet, daß auch hier der im Alter veränderte Eiweißstoffwechsel an der Verschiebung der Abwehrbereitschaft beteiligt ist. Wenn diese Vermutung zutrifft, so würden die beiden Beispiele an der Kartoffel zeigen, daß die Altersveränderungen der Blattproteine je nach Art des Erregers im einen Fall zu einer erhöhten Anfälligkeit, im anderen zu einer Altersresistenz führen können.

IV. Ausblick.

Aus den voraufgegangenen Betrachtungen über die mit dem Alter ablaufenden Wandlungen des Stoffwechsels und über die physiko-chemischen Begleiterscheinungen des Alterns erwächst nun die Aufgabe, die gemessenen Veränderungen von Stoffwechsel und plasmatischer Struktur sinnvoll zu verbinden und die Fülle der zellphysiologischen und stoffwechselphysiologischen Einzelbeobachtungen zu koordinieren. Einer Verknüpfung der beiden Sphären zu einem einheitlichen Bild steht jedoch die Verschiedenartigkeit der untersuchten Pflanzen entgegen. Besonders erschwert wird eine synthetische Betrachtung dadurch, daß sich die verschiedenen Objekte hinsichtlich ihres Alterszustandes nur schwer vergleichen lassen. Wenn in einer Untersuchung unter „alternd" *die* Zellen verstanden werden, die das Streckungswachstum gerade abgeschlossen haben, so sind solche „alternden" Zellen natürlich jung im Vergleich zu alten Zellen eines Blattes, das in die Phase des Vergilbens eintritt. Aber selbst zell- und stoffwechselphysiologische Daten, die am gleichen Objekt gewonnen wurden, lassen noch keine kausale Verknüpfung zu. So geht aus den Befunden von PIRSON und SEIDEL an der *Lemna*-Wurzel eine weitgehende Selbständigkeit der einzelnen physiologischen Faktoren hervor. Kalium- und Calciummangel beeinflussen die Viscositäts- und Permeabilitätsgradienten in der *Lemna*-Wurzel im Sinne schnelleren Alterns. Die für die Atmungsintensität maßgebenden Strukturen werden aber durch den Mineralstoffmangel unabhängig vom übrigen plasmatischen Verhalten der Wurzelzellen verändert. Während mit dem Alter Plasmolysezeit, Deplasmolysezeit und Sauerstoffverbrauch sinken, werden unter Mangelbedingungen nur die beiden ersten Faktoren verkürzt, die Atmungsintensität jedoch erhöht:

	Kaliummangel	Calciummangel	Alternde Zellen
Plasmolysezeit in Glucose	verkürzt	verkürzt	verkürzt
Deplasmolysezeit in Harnstoff	verkürzt	verkürzt	verkürzt
Osmotischer Wert	erniedrigt	erhöht	ansteigend
O_2-Verbrauch	erhöht	erhöht	erniedrigt

An diesem Beispiel wird deutlich, wie schwierig es ist, die beobachteten Veränderungen einzelner Faktoren zu einem einheitlichen Bild des Altersprozesses zusammenzufügen. Man könnte einen Grund für diese Schwierigkeiten darin sehen, daß die stoffwechselphysiologischen Veränderungen sehr häufig mit Strukturen verbunden sind, deren zellphysiologischer Zustandswechsel von der Protoplasmatik bisher nicht erfaßt werden konnte. So wäre es z. B. für das Verständnis der mit dem Alter einhergehenden Veränderungen der Atmung wesentlich, die protoplasmatischen Zustände der Mitochondrien in verschiedenen Altersstufen zu kennen, da ja die Mitochondrien diejenigen Zellpartikel sind, die mit den Fermentsystemen für Oxydationen und Phosphorylierungen ausgestattet sind. Nachdem sich unsere Kenntnisse über die Lokalisation von Enzymen innerhalb der Zelle in neuerer Zeit wesentlich erweitert haben, wird man mit Hilfe der in diesem Zusammenhang entwickelten Methoden auch dem Problem des Alterns neue Seiten abgewinnen können. Die Zerlegung des Protoplasmas in einzelne Komponenten — Mitochondrien, Chloroplasten usw. — bietet heute keine grundsätzlichen Schwierigkeiten mehr, so daß damit auch die Möglichkeit gegeben ist, die Veränderung einzelner Strukturelemente des Protoplasmas im Verlauf des Alterns zu untersuchen. Die bisher vorliegenden Beobachtungen betreffen vor allem die Mitochondrien, auf deren Bedeutung für das Altersproblem auch FISCHER (1950) hinweist, wenn er schreibt: „Der Hebel, der bei der Auslösung des Alterns angesetzt wird, mag verschiedener Art sein, wenn nur der Ansatzpunkt richtig getroffen wird. Dieser Punkt wird sich vielleicht einmal in Zellstrukturen finden, die der Energielieferung dienen.“ Die Mitochondrien sind die Träger der Enzyme, mit deren Hilfe die bei der Atmung entwickelten Energien in Form von physiologisch nutzbaren Phosphatbindungen festgelegt werden. In einigen Fällen wurde nun gefunden, daß sich Mitochondrien aus alten und jungen Geweben hinsichtlich ihrer Fähigkeit zur oxydativen Phosphorylierung unterscheiden. So verhalten sich die von SLATER und LEWIS aus dem Thoraxmuskel von *Calliphora (Diptera)* isolierten Zellpartikel gegenüber Eingriffen, durch die die Oxydationen von den Phosphorylierungen entkoppelt werden, je nach dem Alter der Fliegen verschieden. Eine Entkoppelung durch 2,4-Dinitrophenol (DNP) führt nur bei Mitochondrien aus jungen Fliegen zu einer Steigerung der Oxydation des Atmungssubstrates (α-Ketoglutarsäure), während der Sauerstoffverbrauch bei Präparaten aus älteren Tieren kaum oder gar nicht durch DNP beeinflußt wird. Auf botanischem Gebiet liegen Beobachtungen über Mitochondrien aus Avocado-Früchten verschiedenen Reifegrades vor (MILLERD, BONNER und BIALE). Mitochondrien, die aus Früchten im Zustand kurz vor der Ausreifung (präklimakterisches Minimum) gewonnen wurden, zeigen eine höhere Atmungsintensität als solche aus Früchten im Reifungsmaximum. Auch die Reaktion auf DNP ist bei den Mitochondrien der beiden Reifestadien verschieden. DNP (10^{-4} und 10^{-5} mol) erhöht den O_2-Verbrauch von Gewebeschnitten aus präklimakterischen Früchten zum Teil um 100%, während bei Geweben im Atmungsklimakterium mit DNP keine Atmungssteigerung erzielt werden kann.

In den genannten Fällen sind an den isolierten Protoplasmapartikeln nur Altersveränderungen von stoffwechselphysiologischen Faktoren gemessen worden. Gerade in der Untersuchung solcher isolierter Zellbestandteile werden sich aber Stoffwechselforschung und Protoplasmatik in der zukünftigen Bearbeitung des Altersproblems treffen können. Welche physikalisch-chemischen Vorgänge den Veränderungen der Eigenschaften der Plasmateile zugrunde liegen, läßt sich bei dem heutigen Stand unserer Kenntnisse noch nicht beurteilen. Vielleicht spielt dabei die Racemisierung optisch-aktiver biologischer Substanzen eine Rolle. W. KUHN mißt der nach physikalisch-chemischen Überlegungen zwangsläufig ablaufenden allmählichen Verminderung des optischen Reinheitsgrades der Zellsubstanzen eine mögliche Bedeutung für das Problem des Alterns bei. Für eine derartige Abnahme des optischen Reinheitsgrades, die Störungen im Ablauf biochemischer Reaktionen zur Folge haben müßte, liegen jedoch noch keine experimentellen Beweise vor.

Literatur.

ALLSOPP, A.: Juvenile stages of plants and the nutritional status of the shoot apex. Nature (Lond.) **173**, 1032—1034 (1954). — ARNEY, S. E.: The respiration of strawberry leaves attached to the plant. New Phytologist **46**, 68—96 (1947).

BANCHER, E.: Die Paeonia-Blüte und ihr zellphysiologisches Verhalten während des Abblühens. Protoplasma **42**, 482—489 (1953). — BERCKS, R.: Weitere Untersuchungen zur Frage der Altersresistenz der Kartoffelpflanzen gegen das X-Virus. Phytopath. Z. **18**, 249—269 (1952). — BETHE, A.: Allgemeine Physiologie. Berlin: Springer 1952. — BIALE, J. B.: Postharvest physiology and biochemistry of fruits. Annual Rev. Plant Physiol. **1**, 183—206 (1950). — BIALE, J. B., and R. E. YOUNG: Critical oxygen concentrations for the respiration of lemons. Amer. J. Bot. **34**, 301—309 (1947). — BIALE, J. B., R. E. YOUNG and A. J. OLMSTEAD: Fruit respiration and ethylene production. Plant Physiol. **29**, 168—174 (1954). — BIEBL, R.: Über die gegensätzliche Wirkung der Spurenelemente Zink und Bor auf die Blattzellen von Mnium rostratum. Österr. bot. Z. **94**, 61—73 (1947). — Vergleichende chemische Resistenzstudien an pflanzlichen Plasmen. Protoplasma **39**, 1—13 (1949). — Lichtgenuß und Strahlenempfindlichkeit einiger Schattenmoose. Österr. bot. Z. **101**, 502—538 (1954). — BIEBL, R., u. W. ROSSI-PILLHOFER: Die Änderung der chemischen Resistenz pflanzlicher Plasmen mit dem Entwicklungszustand. Protoplasma **44**, 113—135 (1954). — BOGEN, H. J.: Untersuchungen zu den „spezifischen Permeabilitätsreihen" HÖFLERS. Planta (Berl.) **28**, 535—581 (1938). — Untersuchungen über den Hitzetod und die Hitzeresistenz pflanzlicher Protoplasten. Planta (Berl.) **36**, 298—340 (1948). — Anfärbung, Schädigung und Abtötung von Hefezellen durch Acridinorange. Arch. Mikrobiol. **18**, 170—197 (1953). — BORRISS, H.: Plasmolyseform und Streckungswachstum. Jb. wiss. Bot. **86**, 784—831 (1938). — BREDEMANN, G., K. GARBER, P. HARTECK u. K. A. SUHR: Die Temperaturabhängigkeit der Lebensdauer von Blütenpollen. Naturwiss. **34**, 279—280 (1947). — BÜNNING, E.: Entwicklungs- und Bewegungsphysiologie der Pflanze, 3. Aufl. Berlin-Göttingen-Heidelberg: Springer 1953. — Grundvorgänge bei der pflanzlichen Formbildung. Naturwiss. Rdsch. 8, 213—221 (1955).

CLENDENNING, K. A., and P. R. GORHAM: Photochemical activity of isolated chloroplasts in relation to their source and previous history. Canad. J. Res., Sect. C **28**, 114—139 (1950).

DALY, J. M., and A. H. BROWN: The in vivo demonstration of cytochrome oxidase in leaves of higher plants. Arch. of Biochem. **52**, 380—387 (1954). — DENFFER, D. v.: Physiologisches Altern. Z. Bot. **41**, 277—287 (1953). — DRAUTZ, R.: Über die Wirkung äußerer und innerer Faktoren bei der Kohlensäureassimilation. Jb. wiss. Bot. **82**, 171—232 (1936). — DRAWERT, H.: Zur Frage der Stoffaufnahme durch die lebende pflanzliche Zelle. IV. Der Einfluß der Wasserstoffionenkonzentration des Zellsaftes und des Außenmediums auf die Harnstoffaufnahme. Planta (Berl.) **35**, 579—600 (1948). — Über die Eignung zelleigener Anthocyane zur p_H-Wert-Bestimmung des Zellsaftes. Protoplasma **44**, 370—373 (1955).

EBERHARDT, F.: Über die Beziehungen zwischen Atmung und Anthocyansynthese. Planta (Berl.) **43**, 253—287 (1954). — Der Atmungsverlauf alternder Blätter und reifender Früchte. Planta (Berl.) **45**, 57—67 (1955). — EHRENBERG, R.: Das Problem des Alterns. Naturwiss. **41**, 296—300 (1954).

FISCHER, H.: Plasmolyseform und Mineralsalzgehalt in alternden Blättern. I. Planta (Berl.) **35**, 513—527 (1948). — Plasmolyseform und Mineralsalzgehalt in alternden Blättern. II. Planta (Berl.) **37**, 244—292 (1949). — Über protoplasmatische Veränderungen beim Altern von Pflanzenzellen. Protoplasma **39**, 661—676 (1950). — FOWDEN, L.: The effect of age on the bulk protein composition of Chlorella vulgaris. Biochemic. J. **52**, 310—314 (1952). — FRANK, H.: Über den Stickstoffverlust bei alternden Pflanzen. Planta (Berl.) **44**, 319—340 (1954). — FREELAND, R. O.: Effect of age of leaves upon the rate of photosynthesis in some conifers. Plant Physiol. **27**, 685—690 (1952).

GÄUMANN, E.: Pflanzliche Infektionslehre. Basel: Birkhäuser 1951. — GAIL, F. W., and W. H. CONE: Osmotic pressure and p_H measurements on cell sap of Pinus ponderosa. Bot. Gaz. 88, 437—441 (1929). — GIBBS, M., and H. BEEVERS: Glucose dissimilation in the higher plant. Effect of age of tissue. Plant Physiol. **30**, 343—347 (1955). — GICKLHORN, J.: Gradienten des Erfrierens von Laubblättern. Protoplasma **26**, 90—96 (1936). — GODDARD, D. R.: Metabolism in relation to growth. Growth **12**, 17—45 (1948). — GODDARD, D. R., and B. J. D. MEEUSE: Respiration of higher plants. Annual Rev. Plant Physiol. **1**, 207—232 (1950). — GOOD, H. M., P. M. MURRAY and H. M. DALE: Studies on heartwood formation and staining in sugar maple, Acer saccharum MARSH. Canad. J. Bot. **33**, 31—41 (1955). — GRIFFITHS, D. G., N. A. POTTER and A. C. HULME: Data for the study of the metabolism of apples during growth and storage. J. Horticult. Sci. **25**, 266—288 (1950). — GRÜMMER, G.: Die Beziehungen zwischen dem Eiweißstoffwechsel von Kulturpflanzen und ihrer Anfälligkeit gegen parasitische Pilze. Phytopath. Z. **24**, 1—42 (1955). — GUSTAFSON,

F. G.: Total acidity compared with actual acidity of plant juices. Amer. J. Bot. **11**, 365—369 (1924).

HÄMMERLING, J.: Persönliche Mitteilung. — HARTMANN, M.: Allgemeine Biologie, 4. Aufl. Stuttgart: Gustav Fischer 1953. — HAYASHI, K.: Überblick der Arbeiten über Anthocyane in Japan unter besonderer Berücksichtigung der Pflanzenfarben. Pharmazie **9**, 584—588 (1954). — HILLE, J. C. VAN: The quantitative relation between rate of photosynthesis and chlorophyll content in Chlorella pyrenoidosa. Rec. Trav. bot. néerl. **35**, 680—757 (1938). — HOFMEISTER, L.: Verschiedene Permeabilitätsreihen bei einer und derselben Zellsorte von Ranunculus repens. Jb. wiss. Bot. **86**, 401—419 (1938). — HOVER, J. M., and F. G. GUSTAFSON: Rate of respiration as related to age. J. Gen. Physiol. **10**, 33—39 (1926). — HULME, A. C.: Abstr. 1. Internat. Congr. of Biochem., p. 493, Cambridge 1949. — The relation between the rate of respiration of an apple fruit and its content of protein. I. The value of this relation immediately after picking. J. Horticult. Sci. **26**, 118—124 (1951). — Studies in the nitrogen metabolism of apple fruit. The climacteric rise in respiration in relation to changes in the equilibrium between protein synthesis and breakdown. J. of Exper. Bot. **5**, 159—172 (1954).

IRMAK, L. R.: The relation between hydrolysis of chloroplast starch of tobacco leaves and the age of the plant. Rev. Fac. Sci. Univ. Istanbul, Sér. B **20**, 117—120 (1955).

JAMES, W. O.: Plant respiration. Oxford: Clarendon Press 1953. — JAMES, W. O., and D. BOULTER: Further studies of the terminal oxidases in the embryos and young roots of barley. New Phytologist **54**, 1—12 (1955).

KARRER, P., E. JUCKER, J. RUTSCHMANN u. K. STEINLIN: Zur Kenntnis der Carotinoid-epoxyde. Natürliches Vorkommen von Xanthophyll-epoxyd und α-Carotin-epoxyd. Helvet. chim. Acta **28**, 1146—1156 (1945). — KESSLER, W.: Über die inneren Ursachen der Kälteresistenz der Pflanzen. Planta (Berl.) **24**, 312—352 (1935). — KIDD, F., C. WEST and G. E. BRIGGS: A quantitative analysis of the growth of Helianthus annuus. I. The respiration of the plant and of its parts throughout the life cycle. Proc. Roy. Soc. Lond., Ser. B **92**, 368—384 (1921). — KÜSTER, E.: Botanische Betrachtungen über Alter und Tod. Abh. theor. Biol. **10**, 1—44 (1921). — KUHN, R., G. PFLEIDERER u. W. SCHULZ: Über die Reduktion von TTC durch Maissamen. Liebigs Ann. **578**, 159 —170 (1952). — KUHN, W.: Mögliche Beziehungen der optischen Aktivität zum Problem des Alterns. Experientia (Basel) **11**, 429—436 (1955).

LAING, H. E.: Respiration of the leaves of Nuphar advenum and Typha latifolia. Amer. J. Bot. **27**, 583—586 (1940). — LAKON, G.: Neuere Beiträge zur topographischen Tetrazoliummethode. Ber. dtsch. bot. Ges. **67**, 146—157 (1954). — LEPESCHKIN, W.: Zell-Nekrobiose und Protoplasma-Tod. Protoplasma-Monographien, Bd. 12. Berlin: Gebrüder Bornträger 1937.

MARKLUND, G.: Vergleichende Permeabilitätsstudien an pflanzlichen Protoplasten. Acta bot. fenn. 18, 1—110 (1936). — MARSH, P. M., and D. R. GODDARD: Respiration and fermentation in the carrot Daucus carota. I. Respiration. Amer. J. Bot. **26**, 724—728 (1939). — MEYER, A.: Eiweißstoffwechsel und Vergilben der Laubblätter von Tropaeolum majus. Flora (Jena) **111**, 85—127 (1918). — MICHAEL, G.: Über die Beziehungen zwischen Chlorophyll- und Eiweißabbau im vergilbenden Laubblatt von Tropaeolum. Z. Bot. **29**, 385—425 (1935). — MILLERD, A., J. BONNER and J. B. BIALE: The climacteric rise in fruit respiration as controlled by phosphorylative coupling. Plant Physiol. **28**, 521—531 (1953). — MOLISCH, H.: Die Lebensdauer der Pflanze. Jena: Gustav Fischer 1929. — MOTHES, K.: Zur Kenntnis des N-Stoffwechsels höherer Pflanzen. 3. Beitr. (Unter besonderer Berücksichtigung des Blattalters und des Wasserhaushaltes). Planta (Berl.) **12**, 686—731 (1931).

PAECH, K.: Über die Regulation des Eiweißumsatzes und über den Zustand der proteolytischen Fermente in den Pflanzen. Planta (Berl.) **24**, 78—129 (1935a). — Experimentelle Studien über die Anaerobiose höherer Pflanzen. Planta (Berl.) **24**, 529—551 (1935b). — Ursache und Verlauf des Alterns bei Pflanzen. I. Z. Altersforsch. **2**, 183—205 (1940a). — Veränderungen des Plasmas während des Alterns pflanzlicher Zellen, zugleich ein Beitrag zur Kenntnis der Narkose von Pflanzen. Planta (Berl.) **31**, 295—348 (1940b). — Biochemie und Physiologie der sekundären Pflanzenstoffe. Berlin-Göttingen-Heidelberg: Springer 1950. — Biologische Grundlagen der Frischhaltung pflanzlicher Lebensmittel. In: Handbuch der Kältetechnik, Bd. 9, S. 223—310. 1952. — PEARSALL, W. H., and M. C. BILLIMORIA: Effects of age and of season upon protein synthesis in detached leaves. Ann. of Bot., N. S. **2**, 317—333 (1938). — PEARSON, J. A., and R. N. ROBERTSON: The physiology of growth in apple fruits. IV. The control of respiration rate and synthesis. Austral. J. Biol. Sci. **7**, 1—14 (1954). — PIRSON, A., u. F. SEIDEL: Zell- und stoffwechselphysiologische Untersuchungen an der Wurzel von Lemna minor L. unter besonderer Berücksichtigung von K- und Ca-Mangel. Planta (Berl.) **38**, 431—473 (1950). — PRINGSHEIM, E. G.: Untersuchungen über den Respirationsquotienten verschiedener Pflanzenteile. Jb. wiss. Bot. **81**, 579—608 (1935).

REUTER, L.: Zur protoplasmatischen Anatomie des Keimblattes von Soja hispida. Österr. bot. Z. **95**, 373—421 (1949). — RICHARDS, F. J.: On the use of simultaneous observations on successive leaves for the study of physiological change in relation to leaf age. Ann. of Bot. **48**, 497—504 (1934). — RIPPEL, A.: Über den Zusammenhang zwischen dem Aufnahmeverlauf der Bodennährstoffe bei den höheren Pflanzen und der Beweglichkeit dieser Stoffe in der Pflanze. (Untersuchungen über physiologisches Gleichgewicht bei Pflanzen. III.) Biochem. Z. **187**, 272—282 (1927). — ROSENE, H. F.: Ageing and the influx of water into radish root-hair cells. J. Gen. Physiol. **34**, 65—73 (1950). — ROSENE, H. F., and A. M. J. WALTHALL: Comparison of the velocities of water influx into young and old root hairs of wheat seedlings. Physiol. Plantarum (Copenh.) **7**, 190—194 (1954). — RUGE, U.: Kritische zell- und entwicklungsphysiologische Untersuchungen an den Blattzähnen von Helodea densa. Flora (Jena) **134**, 311—376 (1940). — Permeabilitätsstudien an jungen und ausdifferenzierten Zellen des Rhoeo-Blattes. Planta (Berl.) **33**, 589—599 (1943). — RUHLAND, W., u. K. RAMSHORN: Aerobe Gärung in aktiven pflanzlichen Meristemen. Planta (Berl.) **28**, 471—514 (1938). — RUMMENI, G.: Redoxpotentiale und Aciditätspotentiale verschiedener pflanzlicher Organe und Gewebe. Ber. dtsch. bot. Ges. **67**, 334—340 (1954).

SCHEIBMAIR, G.: Hitzeresistenzstudien an Mooszellen. Protoplasma **29**, 394—424 (1938). — SCHUMACHER, W.: Über Eiweißumsetzungen in Blütenblättern. Jb. wiss. Bot. **75**, 581—608 (1931). — Weitere Beobachtungen über das Welken ephemerer Blüten. Planta (Berl.) **42**, 42—55 (1953). — SEVAG, M. G.: The protein molecule as a multicatalytic entity. Erg. Hyg. **28**, 424—448 (1954). — SEYBOLD, A.: Untersuchungen über den Farbwechsel von Blumenblättern, Früchten und Samenschalen. Sitzgsber. Heidelberg. Akad. Wiss., Math.-naturwiss. Kl., Abh. 2 **1953/54**. — SINGH, B. N.: The correlation between life duration and respiratory phenomena. Proc. Indian Acad. Sci. **2**, 387—402 (1935). — SINGH, B. N., and K. N. LAL: Investigations of the effect of age on assimilation of leaves. Ann. of Bot. **49**, 291 —307 (1935). — SISSAKIAN, N. M.: Biochimie des plastides. In: Essais de Botanique. Editions de l'Acad. Sci. URSS., Bd. I, S. 224—255. Moskou-Léningrad 1954. — SLATER, E. C., and S. E. LEWIS: Stimulation of respiration by 2,4-dinitrophenol. Biochemic. J. **58**, 337—345 (1954). — SMIRNOV, A. I.: Über die biochemischen Eigentümlichkeiten des Alterns der Laubblätter. Planta (Berl.) **6**, 687—766 (1928). — STRUGGER, S.: Untersuchungen über die vitale Fluorochromierung der Hefezelle. Flora (Jena) **137**, 73—94 (1944).

WALKLEY, J.: Protein synthesis in mature and senescent leaves of barley. New Phytologist **39**, 362—369 (1940). — WEBER, F.: Harnstoffpermeabilität ungleich alter Spirogyra-Zellen. Protoplasma **12**, 129 (1931). — WILLIAMS, W. C., and P. CH. SHARMA: The function of urease in Citrullus seeds. J. of Exper. Bot. **5**, 136—157 (1954). — WILLSTÄTTER, R., u. A. STOLL: Untersuchungen über die Assimilation der Kohlensäure. Berlin: Springer 1918.— WINOKUR, M.: Ageing effects in Chlorella cultures. Amer. J. Bot. **36**, 287—291 (1949). — WOOD, J. G.: The nitrogen metabolism of the leaves of Atriplex nummularium. Austral. J. Exper. Biol. a. Med. Sci. **11**, 237—252 (1933).

ZICKEL, I.: Untersuchungen über zellphysiologische Reaktionen verschieden alter Zellen. Z. Bot. **43**, 73—78 (1955).

Namenverzeichnis. — Author Index.

Die *kursiv* gesetzten Seitenzahlen beziehen sich auf die Literatur.

Page numbers in *italics* refer to the bibliography.

Sachverzeichnis.

(Deutsch-Englisch.)

Ä, Ö, Ü sind wie Ae, Oe, Ue eingereiht.

Bei gleicher Schreibweise in beiden Sprachen sind die Stichworte jeweils einfach aufgeführt.

Subject Index.

(English-German.)

Ä, Ö, Ü are taken as Ae, Oe, Ue.

Where English and German spelling of a word is identical the italiced (German) entry is omitted.